PHOTOSYNTHESIS: FROM LIGHT TO BIOSPHERE

Photosynthesis: from Light to Biosphere

Volume I

Proceedings of the Xth International Photosynthesis Congress,
Montpellier, France, 20–25 August 1995

edited by

PAUL MATHIS
Section de Bioénergétique,
CEA-Saclay, France

KLUWER ACADEMIC PUBLISHERS
DORDRECHT / BOSTON / LONDON

Library of Congress Cataloging-in-Publication Data

International Congress on Photosynthesis (10th : 1995 : Montpellier, France)
Photosynthesis : from light to biosphere : proceedings of the Xth International Photosynthesis Congress, Montpellier, France, 20-25 August 1995 / edited by Paul Mathis.
p. cm.
Includes indexes.
ISBN-13:978-94-010-6562-7
1. Photosynthesis--Congresses. I. Mathis, P. II. Title.
QK882.I56 1995
581.1'3342--dc20 95-47402

ISBN-13:978-94-010-6562-7 e-ISBN-13:978-94-009-0173-5
DOI: 10.1007/978-94-009-0173-5

Published by Kluwer Academic Publishers,
P.O. Box 17, 3300 AA Dordrecht, The Netherlands.

Kluwer Academic Publishers incorporates
the publishing programmes of
D. Reidel, Martinus Nijhoff, Dr W. Junk and MTP Press.

Sold and distributed in the U.S.A. and Canada
by Kluwer Academic Publishers,
101 Philip Drive, Norwell, MA 02061, U.S.A.

In all other countries, sold and distributed
by Kluwer Academic Publishers Group,
P.O. Box 322, 3300 AH Dordrecht, The Netherlands.

Printed on acid-free paper

GENERAL CONTENTS

PREFACE

The advances in scientific knowledge reveal that all major biological processes are much more complex than was originally thought. Among these processes, photosynthesis occupies a unique position, not only because of its basic role in providing energy and organic materials for all living organisms, but also because it integrates a large variety of events which range from the shortest in photophysics up to the longest in ecophysiology. It should also be remembered that photosynthesis has contributed to the build-up of the present atmosphere and still contributes in a key manner to the regulation of global changes.

Not surprisingly, photosynthesis is a field where multidisciplinarity is a general necessity. The most advanced laboratories in photophysics, where time is counted in femtoseconds, also construct mutants or collaborate closely with molecular geneticists. Those who are interested in the regulation of photosynthesis under field conditions, or who use chlorophyll fluorescence for remote sensing, are sent back to understand primary events of energy transfer. Other scientists try to learn how the many proteins that participate in photosynthesis are routed to their location by post-translational events and how they assemble in large complexes : they have to collaborate with bioenergeticists and membrane biochemists.

The 10th International Photosynthesis Congress, which was held from 20 to 25 August 1995 in Le Corum, Montpellier, France, brought together over 1400 researchers from many countries and from all scientific horizons. It was thus an excellent occasion for all these scientists to receive (and communicate) the information which is of such utmost importance in this broad, multidisciplinary field. This publication brings together the vast majority of the contributions which were presented at the congress as plenary conferences, symposium lectures or posters. Chapters have been reorganized a little bit, but no censorship has been imposed on the scientific content of articles, so as to maintain the freshness of the congress. The Montpellier congress has been by far the largest in the series of International Photosynthesis Congresses and this publication thus constitutes a unique document of the research that is done in all over the world, on all aspects of photosynthesis.

The broad and up-to-date scientific content of this book results from the efforts of the International Photosynthesis Committee, the Congress Programme Committee and the Organizing Committee. I thank their members very greatly and specially Myroslawa Miginiac-Maslow for the expertise and the dedication she placed in taking care of the abstracts and of the manuscripts. I also thank Mrs F. Delorme, Mrs B. Fournal and the Chairman agency for setting up the complete edited manuscript. Finally, it is quite appropriate to thank again all the sponsors for their contribution toward the success of the congress.

29 September, 1995 Paul MATHIS

International Photosynthesis Committee

Bertil ANDERSSON (Sweden)
Richard J. COGDELL (United Kingdom)
Arvi FREIBERG (Estonia)
Miguel GUERRERO (Spain)
Arnold J. HOFF (The Netherlands)
Yorinao INOUE (Japan)
Wolfgang JUNGE (Germany), Secretary
Aaron KAPLAN (Israel)
Vyatcheslas V. KLIMOV (Russia)
A.W.D. LARKUM (Australia)
Paul MATHIS (France), Chairman
B. Andrea MELANDRI (Italy)
Prasanna S. MOHANTY (India)
Norio MURATA (Japan)
Donald R. ORT (United States)
Kenneth SAUER (United States)
André VERMEGLIO (France)

Organizing Committee

Paul MATHIS (Chairman)
André VERMEGLIO (Vice Chairman)
Yaroslav de KOUCHKOVSKY (Secretary)
Jean-François MOROT-GAUDRY
Marie-Louise CHAMPIGNY
Anne JOLIOT
Miroslava MIGINIAC
Patrick SETA

Programme Committee

Pierre JOLIOT (Chairman)
Gabriel CORNIC
Pierre GADAL
Jean-Jacques GIRERD
Paul MATHIS
Nicole TANDEAU de MARSAC

National Advisory Committee

G. AUSSENAC
M. CABOCHE
G. CORNIC
Y. de KOUCHKOVSKY
P. DIZENGREMEL
R. DOUCE
J-C. DUVAL
A-L. ETIENNE
P. GADAL
J-P. GAUDILLERE
J.J. GIRERD
C. GRIGNON
C. GUDIN
P. JOLIOT
F. JOSET
H. JUPIN
Y. LEMOINE
M. LUTZ
R. MACHE
P. MATHIS
J-C. MEUNIER
J-F. MOROT-GAUDRY
I. MOYA
G. PELTIER
C. PLANCHON
J. RICARD
P. SETA
N. TANDEAU de MARSAC
D. THOMAS
A. VERMEGLIO
P. VIGNAIS

Patronage

Le Ministre de l'Enseignement Supérieur et de la Recherche
P. CHARTIER, Directeur scientifique de l'ADEME
A. CONESA, Président d'Agropolis
R. DAUTRAY, Haut-Commissaire à l'Energie Atomique
J.P. DECOR, Directeur Scientifique, Rhône-Poulenc Agrochimie
R. DOUCE, Professeur à l'Institut Universitaire de France
P.G. de GENNES, Professeur au Collège de France, Prix Nobel de Physique
J.M. LEHN, Professeur au Collège de France, Prix Nobel de Chimie
G. PAILLOTIN, Président de l'INRA
A. RUELLAN, Directeur du Programme Environnement, CNRS
P. TAMBOURIN, Directeur des Sciences de la Vie, CNRS
Société Française de Biochimie et de Biologie Moléculaire
Société Française de Biophysique
Société Française de Microbiologie
Société Française de Photobiologie
Société Française de Physiologie Végétale

Sponsors

Administrations, Collectivities, Scientific organisations and Agencies

ADEME
CEA (Commissariat à l'Energie Atomique)
CNRS (Centre National de la Recherche Scientifique)
Commission of the European Communauties - DG XII
Conseil Général de l'Hérault
Conseil Régional Languedoc-Roussillon
European Science Foundation
IFREMER
INRA (Institut National de la Recherche Agronomique)
MESR, Programme ACCESS
Montpellier District
UNESCO
Université de Montpellier II

Scientific societies

European Society of Photobiology
IUBMB (International Union of Biochemistry and Molecular Biology)
Société Française de Biophysique
Société Française de Photobiologie
Société Française de Physiologie Végétale

Companies

Bambouseraie de Prafrance
Beckman Instruments France
Boehringer Mannheim
Bruker Spectrospin
Continuum
Du Pont de Nemours
EDF
Goemar
Henkel Corporation
Hoffmann la Roche
Laboratoires Yves Rocher
LVMH (Louis Vuitton-Moët Hennessy)
Pharmacia Biotech
Rhône-Poulenc Agrochimie

Personal contributions

Aghion J.
Eaton-Rye J.
Govindjee
Heber U.
Koyama Y.
Lotina-Hennsen B.
Mimuro M.
Myers J.
Murata N.
Seibert M.
Shioi Y.
Sonoike.K.
Telfer A.

CONTENTS TO VOLUME I

"ABOUT IT AND ABOUT"-CONCERNING SOME EARLY PROBLEMS IN PHOTOSYNTHESIS RESEARCH

D.A.Walker, Robert Hill Institute Field Laboratory
6, Biddlestone Village, Morpeth, Northumberland
NE65 7DT, U.K.

To reflect on the extent to which our understanding of photosynthesis has grown in the last forty years would scarcely require more than a comparison of the present programme with that of the first international congress of photosynthesis. Such is the pace of change that, to someone like myself, already formally retired for seven years, much of what I see is alien if not entirely incomprehensible. So what I shall attempt here, is to give a flavour of times past which will seem as remote to many attending this congress as the political correctness of McCarthy in the United States or that of Lysenko in the former Soviet Union. Since I must inevitably write from a purely personal standpoint the first part of my title is taken from:-

"Myself when young did eagerly frequent
Doctor and Saint and heard great argument
About it and about
But evermore
Came out by that same door as in I went" ....Omar Khyam

My Professor was Meirion Thomas of Crassulacean Acid Metabolism fame. Indeed it was he that introduced that term into the scentific literature but, in the 1950s, neither he nor those in his laboratory regarded 'CAM' as an aspect of photosynthesis. That, if the word from far off California could be believed, might involve sugar phosphates but at Newcastle we were simply preoccupied with the nightly accumulation of malic acid in the leaves of Kalanchoe crenataa. Hence Thomas's suggestion that, for my Ph.D., I might try to identify the enzyme responsible for this dark carboxylation. Rubisco, of course, was still in the future and Axlerod and Bandurski had yet to characterise PEP carboxylase from wheat germ but Ochoa's malic enzyme could actually be made to fix CO_2 and seemed a possibility. Picture then a young man, innocent of all knowledge of enzymology, embarking on this task in a department equally without experience of enzymes; a department without so much as a spectrophotometer, a refrigerated centrifuge, or a cold room. Not that the latter was much of a handicap because, in the winter of 1952, fuel shortages ensured that the laboratory temperature scarcely ever rose above 12° Celsius. Attempts to raise that a degree or two by lighting every available Bunsen burner were usually defeated by the arrival of the professor. Who knows what else he wore but there was a certainly a raincoat on top of an overcoat and beneath that a tweed suit, a pullover and a waistcoat. So garbed he would ride briskly into the University on his bicycle, stride into the lab, and, with a cry of "stuffy in here", throw open all the windows. Lest anyone should think that this sort of thing was peculiar to England at that time I should hasten to add that when I had the good fortune to spend a year with Harry Beevers at Purdue University (Indiana) in 1953/54 things were not so different. There the principal problem in the 'Peirce Conservatory and Small Animal House' was not winter cold but summer heat and for three months the temperature in the Warburg bath had to be brought down each day to 35° Celsius, prior to measurements. Neither were we unconscious, of the cockroaches in the conservatory or the aroma from the small animal house in the cellar beneath our floor. Each day, en route to the nearest refrigerated centrifuge and anxious that my mitochondrial preparation did not warm too much, I had to hurry across an adjacent railway line to the main building. This to the amusement of the janitor who soon came to greet me with cries of "slow down Dave, running ain't going to get you no place."

Ochoa's malic enzyme was NADP (or TPN, as it was then called) specific. No doubt there were graduate students, even then, who were able to avail themselves of the services of the Sigma Chemical Company but, in Newcastle, that was thought to be as sinful as travelling by taxi. So one morning I found myself on my way back, by bus, from the local abattoir, clutching a metal vasculum (polystyrene was unknown) containing livers, from freshly slaughtered pigs, packed in dry ice. Washing facilities were not offered to callow university students so I was obliged to ignore my blood-stained hands and act as though a similarly blood-stained container, encrusted with ice and trailing streams of vapour, like something from a horror movie, was a commonplace sight. It says much for the travelling British public of that time that no one gave it, or me, as much as a second glance. Back at the lab there were galvanised buckets of boiling water and a meat grinder. Bucket biochemistry really was bucket biochemistry and eventually yielded a few milligrams of putative TPN. Chemical reduction gave an appropriate peak at 340 nm (measured in someone's lunch hour in the Physics Department) but how to tell how much of this was NADP and not NAD? So it was back to the abattoir and to the extraction of a known, NADP-specific, isocitric dehydrogenase from pig hearts. Imagine my dismay when this enzyme failed to reduce my painfully extracted TPN. Clutching at straws, I prevailed on a colleague to chromatograph the isocitrate that I had bought from a company long suspected of digging all of its products from the same mine. Sure enough, whatever it was it was not isocitrate. So there were more buckets and lots of ***Bryophyllum*** leaves. Finally, after weeks of graft, there was the immense satisfaction of putting together each laboriously won enzyme, coenzyme and substrate and watching the needle on the spectrophotometer announce that my NADP was indeed NADP. I started to feel like a full fledged biochemist. Of course none of this advanced the cause of science one iota but it was a start and my eventual conclusions that dark carboxylation in CAM was catalysed by PEP carboxylase(1) and decarboxylation by malic enzyme (2), seem to have stood the test of time (3,4).

Experiences of this sort, which may not cause the earth to move but do lots for a young investigators self-esteem, tend to be rather infrequent. The one which made me understand how Archimedes must have felt when he ran naked into the street shouting "Eureka" came much later. By then, I had succeeded in isolating intact chloroplasts (3) and, with my colleagues Chris Bucke, Bill Cockburn and Carl Baldry, learned how to persuade them to fix $^{14}CO_2$ as rapidly as the parent tissue (4-7) but we had not looked for O_2 and there remained the awful possibility that ascorbate, present in our reaction mixtures, might be the hydrogen donor rather than water. Taking advantage of a long friendship with Robin Hill, who influenced my science so much, I therefore travelled once a week from Imperial College, London, to the Biochemistry Department in Cambridge carrying half a batch of pea leaves for chloroplast isolation. Back in London, my colleagues isolated chloroplasts from the other half. Theirs fixed CO_2 at good rates; mine declined to evolve O_2. Robin assured me that this was the case by virtue of looking at my reaction mixtures with a hand spectroscope. In his hands this venerable instrument yielded a wealth of information. I preferred the O_2 electrode (8), to which he introduced me, but the answer was as unequivocal, There was no O_2.We never did establish why pea leaves undergo such rapid post-harvest changes that isolation of active chloroplasts (from them) rapidly becomes impossible but redemption was at hand with the arrival of a vibrating glass O_2 electrode at Imperial College and a shift to spinach bought at Covent Garden or grown in the Chelsea Physic Garden on our behalf. At last I saw CO_2-dependent O_2 evolution for the first time. Like $^{14}CO_2$ fixation it did not start immediately at maximum rates but only after a lag of one or two minutes. This (9) was the first 'Eureka' moment, another came hot on its heels. With the Benson-Calvin Cycle in mind it seemed obvious to repeat the assay with added phosphoglycerate. This time O_2 evolution started immediately on illumination. There was much rejoicing. Pursuing this observation it soon became clear that inorganic phosphate prolonged the induction period whereas phosphoglycerate and triose phosphates shortened it and reversed the Pi inhibition. Obligate exchange, across the chloroplast envelope, of Pi with sugar phosphates was proposed (10) and quickly verified by Heldt and Rapley (11). The 'Pi-translocator' had acknowledged its existence (4,5,12).

Finally, since this congress is again in France and mindful of the historic demonstration by Louis Pasteur, in front of a sceptical academy, of the manner in which left and right handed crystals bend polarised light, it seems appropriate to mention another experiment which can be done with very simple apparatus. If you wish, you can emulate Pasteur by inviting a leaf to tell you which is a solution of D-mannose and which is L-mannose. This is how you do it. It is very easy to induce oscillations in young Barley leaves. All manner of photosynthetic events then oscillate but the easiest to detect and record is chlorophyll *a* fluorescence. (You do not need any sophisticated apparatus from Hansatech or Walz for this purpose. A photodiode, screened by an appropriate optical filter, and linked to a pen-recorder or computer will suffice). Orthophosphate, fed through the petiole, will rapidly suppress these induced oscillations (13). D-mannose will enhance them or reverse their suppression by Pi with equal rapidity. D-mannose does this by sequestering Pi as it is phosphorylated, in the cytosol, by hexokinase. Hexokinase is stereo-specific and fails to respond to L-mannose. I like to believe that this would have made Pasteur smile approvingly. Precisely how the orthophosphate status of the cytosol comes to have such marked effects on photosynthetic oscillations is still uncertain although several theories have been proposed(14). This continues to pose questions of fundamental importance. After all, as Agu Laisk (15) has said, "to understand oscillations is to understand photosynthesis".

REFERENCES

1. Walker, D.A (1956) Nature 178: 593-594
2. Walker, D.A (1957) Biochem J 67: 73-79
3. Walker, D.A (1964) Biochem J 92: 22c-23c

4. Edwards GE, Walker, D.A (1983) C_3C_4, Mechanisms, and Cellular and Environmental Regulation of Photosynthesis, pp 1-542 Blackwell Scientific Publications Ltd Oxford.
5. Walker, D.A (1993) Energy,Plants and Man. pp 1-277 Packard Publishing Ltd., Chichester
6. Heber, U.and Walker, D.A. (1979) Trends in Biochem Sci 4: 252-256
7. Walker, D.A, Cerovic Z.G, Robinson S.P (1986) Methods in Enzymology. Vol 148 pp 145-157
8. Walker, D.A (1987) The use of the oxygen electrode and fluorescence probes in simple measurements of photosynthesis.pp 1-145. Hansatech Instruments Ltd, Pentney, Norfolk PE32 1JL
9. Walker, D.A. and Hill, R.(1967) Biochim Biophys Acta 131: 330-338
10. Walker, D.A. and Crofts A.R (1970) Ann Rev Biochem 39: 389-428
11. Heldt, H.W. and Rapley, L.(1970) FEBS Lett 10 143-148
12. Heldt, H.W.(1976) in 'The Intact Chloroplast' (Barber,J.,ed) pp.215-234. Elsevier Amsterdam.
13. Walker, D.A. and Sivak, M.N.(1986) Trends Biochem. Sci 11: 176-179
14. Walker, D.A. (1992) Photosynthesis Research 34: 387-395
15. Laisk A, Walker, D.A. (1989) Proc. Roy Soc.B. 237: 417-444

LES ENJEUX DE L'AGRICULTURE

G. Paillotin, INRA
147 rue de l'Université, 75338 Paris cedex 07, France

1. Introduction

Au cours des années soixante–dix, une partie de la communauté scientifique de la photosynthèse s'est activement intéressée aux applications énergétiques des processus photosynthétiques (1). Cet intérêt à été provoqué par le choc pétrolier et par le souci d'exploiter l'énergie solaire. L'idée était extrêmement tentante car, en France par exemple, le flux d'énergie solaire reçue par les 18 millions d'hectares de terres arables correspond à 280 tonnes équivalent–pétrole par habitant, pour un besoin de 4 exprimé dans la même unité. Toutefois le rendement de la photosynthèse au champ reste faible, de l'ordre de 0,25%, et dans ces conditions il faudrait utiliser 1 million d'hectares de cultures énergétiques pour couvrir 5% des besoins d'énergie français pour les seuls transports. Comparée aux énergies fossiles concentrées, mais certes non renouvelables, l'énergie solaire souffre du handicap économique de sa faible densité par unité de surface ce qui la rend peu rentable actuellement sauf dans le cas particulier de la gestion des jachères.

Cependant les plantes ne sont pas simplement des convertisseurs d'énergie, ce sont également de remarquables confectionneurs de molécules complexes dont le débouché naturel est l'alimentation de l'homme ou des animaux. Ainsi la principale exploitation de la photosynthèse reste le fait de l'agriculture et de la filière de l'alimentation humaine. La production agricole est pour des raisons que nous venons d'évoquer nécessairement répartie sur un vaste territoire et cette contrainte d'origine énergétique fait jouer un rôle essentiel à l'agriculture : celui de gérer notre espace et notre environnement.

2. Nourrir les hommes.

Le monde comporte aujourd'hui 6 300 millions de bouches à nourrir et les besoins minimum de l'homme sont estimés à 2 200 calories jour. Ces besoins sont aujourd'hui assurés, puisque l'agriculture mondiale produit assez pour que les habitants de la planète disposent de 2 500 calories par jour. Cette donnée moyenne cache cependant d'importantes inégalités : si les habitants des pays industrialisés – à l'exclusion des plus démunis – consomment plus de 3 000 calories/jour, cinquante pays sont toujours en dessous du seuil critique de 2 200 calories/habitant/jour et 800 millions de personnes souffrent de la faim dans le monde (2). L'enjeu de l'alimentation reste d'actualité. Ainsi dans vingt cinq ans la population du monde devrait atteindre 8 500 millions d'habitants dont plus de 80% dans les actuels pays en développement. Compte tenu des capacités techniques de ces pays et des contraintes posées par la préservation de l'environnement, il est douteux qu'on puisse aisément satisfaire les besoins alimentaires de la population du globe.

Examinons maintenant ce que représente pour un pays industrialisé comme la France le volume de la consommation alimentaire :

– Les ménages consacrent 18% de leur budget à l'alimentation ce qui représente 752 milliards de francs en 1992 (3), soit 36 francs/habitant/jour. C'est un poste extrêmement important des dépenses.

– A ces données économiques correspondent également des données physiques très significatives. Ainsi les français consomment-ils près de 40 millions de tonnes de produits alimentaires par an.

– L'industrie agro-alimentaire est donc une industrie de masse et "grand public". Ainsi en France 10% des actifs travaillent dans cette filière qui mobilise beaucoup d'emplois.

Au delà de ces données quantitatives, le consommateur des pays industrialisés recherche sans cesse une amélioration constante de la qualité de son alimentation et ceci sous bien des aspects : la qualité hygiénique, la valeur santé, le service incorporé, la qualité gustative et des représentations hédonistes des produits de la nature.

3. Produire

La production agricole mondiale a connu de grands progrès dans les quarante dernières années et ceci dans tous les pays du monde. Ainsi en France de 1954 à 1992 le volume de la production agricole a été multiplié par 2,5 alors que la population agricole était divisée par 4. A titre d'exemple, le rendement de la production de blé tendre est passé dans la même période de 22 à 66 quintaux/ha. On a appelé "révolution verte" ce remarquable essor de l'agriculture fondé sur les progrès du machinisme, de la génétique, de la fertilisation et des protections des cultures. A noter que le progrès génétique considérable a permis de faire fonctionner de façon optimale la photosynthèse mais ne l'a nullement améliorée dans ses mécanisme de base.

La "révolution verte" s'est cependant essentiellement exprimée là où les conditions pédo-climatiques étaient propices à l'agriculture or bien des pays sont loin de bénéficier d'avantages en la matière.

Un premier paramètre est l'étendue des terres cultivables. La moyenne dans le monde est de 27 ha par habitant, mais si quelques pays disposent de plus de 80 ha/habitant (USA, Argentine, Canada, Australie), d'autres ont moins de 10 ha/habitant (Indonésie, Chine, Belgique, Pays-Bas, Egypte, Japon). Ainsi les pratiques agricoles passent-elles selon les cas de l'extensif à l'intensif. C'est également cette caractéristique physique qui explique pour une grande part les différences de politiques agricoles observées à travers le monde.

Les ressources en eau constituent un deuxième point important dans la production agricole. Là encore les inégalités sont très nettes entre les différents pays. Si le Canada, l'Argentine et l'Australie ne consomment annuellement que quelques pour cent de leurs réserves annuelles, l'Egypte et l'Israël en consomment plus de 90%.

On doit constater que plusieurs pays ont une très forte pression démographique, une faible superficie en terres arables et de maigres ressources en eau. Dans ces conditions extrêmement limitantes il faut rechercher les techniques les plus productives qui préservent cependant des ressources naturelles si limitées. C'est un des enjeux de ce qu'on appelle le développement durable.

Même si la situation est moins dramatique dans les pays industrialisés à agriculture intensive - comme la majorité des pays de l'Union Européenne, la préservation des ressources naturelles (eaux, sols, diversité génétique) devient un souci très fort qui influence grandement les techniques agricoles.

4. Maîtriser l'espace.

L'activité agricole est nécessairement répartie sur le territoire et joue un rôle primordial dans la régulation des relations villes-campagne. Là encore chaque pays a sa caractéristique propre. Les Pays-Bas, la Belgique, le Royaume-Unis ont moins de 10% de leur population en zone rurale, alors que l'Indonésie, la Chine, l'Inde comptent plus de 70% de population rurale. Il reste - même si les agriculteurs ne sont plus majoritaires en zones rurales dans de nombreux pays comme la France - que c'est l'agriculture qui donne vie, rythme et sens à l'essentiel de notre territoire. L'organisation de l'espace en de multiples paysages, l'expression perceptible du rythme des saisons sont des facteurs d'équilibre pour nos sociétés. En témoigne le succès auprès des consommateurs des produits typés, caractérisés par un terroir.

5. Conclusion

Les enjeux de l'agriculture sont majeurs et variés. Rappelons–les :

- Assurer l'alimentation de tous les habitants de la planète ce qui n'est pas acquis dans le futur proche.
- Développer la qualité en articulant mieux la production agricole, les industries de transformation et la distribution.
- Assurer la compatibilité entre développement économique et préservation des ressources naturelles.
- Donner vie et sens au territoire.

Mais ceci étant dit, l'activité scientifique peut–elle jouer un rôle dans la solution de ces problèmes ? A l'évidence oui, lorsqu'on songe à la génétique, à la sélection de plantes plus économes en intrants, de meilleure qualité, résistantes aux ravageurs, oui également aux technologies de conservation et de transformation et on pourrait multiplier les exemples... Mais qu'en est–il de la photosynthèse ?

Il faut noter qu'il s'agit d'un processus extrêmement optimisé dont l'amélioration ne semble pas envisageable et qui n'a en tout cas pas été à la base de la révolution verte. Pourtant une piste mériteraient d'être explorée : c'est celle du rôle des processus photosynthétiques dans la régulation des grands métabolismes de la plante et dans la physiologie de la croissance. L'agriculteur doit dans l'avenir maîtriser autant qu'il est possible ses risques de production en quantité et qualité et sans nul doute la photosynthèse joue t'elle un rôle en la matière. C'est donc à la périphérie de la discipline – mais est–ce bien étonnant ? – que se trouvent sans aucun doute les pistes les plus prometteuses pour ceux qui s'intéressent aux retombées pratiques du progrès de leurs connaissances.

1 Hall, D.O. (1976) in Solar Energy a UK Assessment (The Royal Institution, ed.) pp 265–294, UK Section of International Solar Energy Society, London UK.
2 Rapport mondial sur le développement humain (1994), Programme des Nations Unies pour le Développement, Economica, Paris.
3 Agreste (1994), Ministère de l'Agriculture et de la Pêche, Paris.

1. Antenna systems : structure and function

PHYCOBILIPROTEINS: STUDIES OF BILIN ATTACHMENT

A.N. Glazer, C.D. Fairchild, L.J. Jung, and C.F. Chan, Department of Molecular and Cell Biology, 229 Stanley Hall, University of California, Berkeley, CA 94720-3206, U.S.A.

1. Introduction

The phycobiliproteins are components of photosynthetic light-harvesting complexes of cyanobacteria, red algae, and the cryptomonads (1, 2). Those of cyanobacteria and red algae are very similar to each other and differ from those of the cryptomonads in several fundamental respects (3). The cryptomonad phycobiliproteins will not be discussed here. The building block of the phycobiliproteins is an αβ heterodimer and the α and β subunits of different phycobiliproteins are highly homologous (4). Purified phycobiliproteins are isolated as stable trimers, $(\alpha\beta)_3$, or hexamers, $(\alpha\beta)_6$, occasionally (as for B- and R-phycoerythrin) as $(\alpha\beta)_6\gamma$ together with a "linker" polypeptide (γ) (1, 2). Crystal structure analyses of allophycocyanin (5), C-phycocyanin (6, 7), phycoerythrocyanin (8), and b- and B-phycoerythrin (9,10), show the tertiary and quaternary structure of these proteins to be remarkably similar.

Unlike other proteins involved in photosynthesis, phycobiliproteins bear covalently attached chromophores. The linear tetrapyrrole (bilin) chromophores of phycobiliproteins are attached through thioether bonds from C-3' (and in some instances, from C-18' as well) of the bilin to specific cysteinyl residues of the apophycobiliprotein (1). Four different isomeric bilins are found in cyanobacterial and red algal phycobiliproteins. The structures of these polypeptide-linked bilins are shown in Fig. 1. In C-phycocyanin, the stereochemistry of thioether-linked PCB at C-3' at α-84, β-82 is reported to be *R* and at β-155 to be *S* (6).

Different classes of phycobiliproteins are distinguished by the number of bilin chromophores they carry. Allophycocyanins have one bilin on each subunit, phycocyanins and phycoerythrocyanins have one on the α and two on the β subunit, and phycoerythrins have two or three bilins on the α and three on the β subunit. In some instances, the phycoerythrin-associated linker polypeptides (γ subunits) also carry bilins. The number of distinct phycobiliprotein bilin attachment sites ranges from 8 in organisms with a simple phycobiliprotein composition, (e.g., *Synechococcus* spp. PCC6301 or PCC7002) and PCB as the only bilin, to as many as 22 with three different bilins (PCB, PEB, PUB) in open ocean cyanobacteria such as *Synechococcus* sp. WH8103 or WH8020 (11, 12). The attachment of particular bilins at cysteine residues within homologous sequences in different phycobiliproteins within a single cell requires recognition of the bilin and of the cognate attachment site in the apophycobiliprotein, as well as the formation of bond(s) with the appropriate stereochemistry between the bilin

P. Mathis (ed.), Photosynthesis: from Light to Biosphere, Vol. I, 3–9.

and the apoprotein. The research described here aims at an understanding of the parameters that govern bilin attachment to phycobiliproteins.

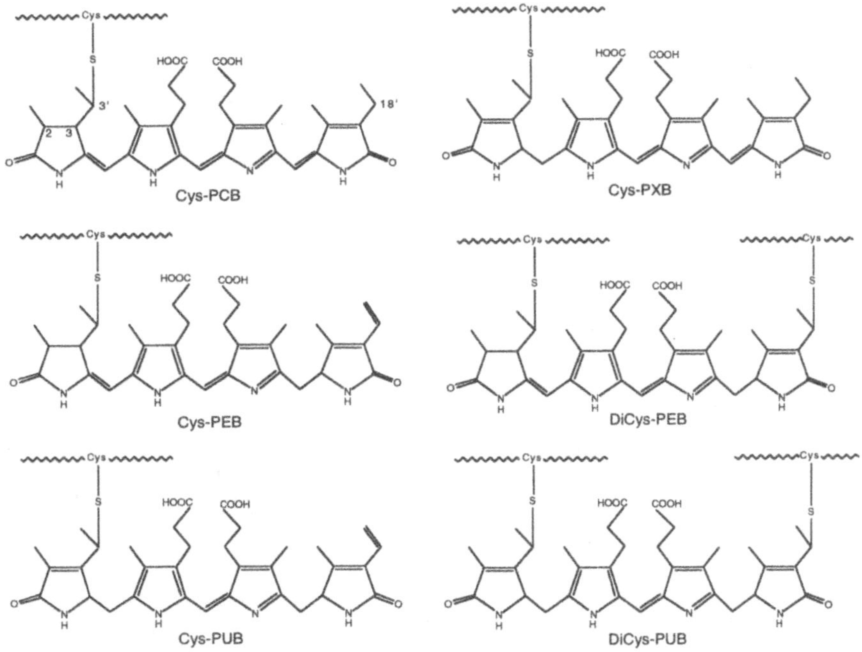

Figure 1. Structure of polypeptide-bound phycocyanobilin (PCB), phycobiliviolin (cryptoviolin; PXB), phycoerythrobilin (PEB), and phycourobilin (PUB).

2. *In vitro* addition of bilins to apophycobiliproteins

The pathway of PEB and PCB biosynthesis has been elucidated in only one organism, the thermoacidophilic red alga *Cyanidium caldarium*. Here, biliverdin formed from heme is converted to PEB, and then PEB is isomerized to PCB (13). Peptide-linked PCB and PEB can be cleaved from phycobiliproteins under reflux in methanol and purified. This treatment results in the elimination of a phycobiliprotein cysteine residue from C-3' of the bilin (see Fig. 1) to yield an ethylidene substituent at C-3 (14, 15). The availability of PCB and PEB with the ethylidene substituent at C-3 and of recombinant apophycobiliprotein subunits expressed in *E. coli* allowed exploration of non-enzymatic bilin addition to apophycobiliproteins.

Native C-phycocyanin carries PCBs at α-84, β-82, and β-155. *In vitro*, PCB and PEB form a complex with apophycocyanin on mixing and then react to form adducts at α-84 and β-82, but not at β-155 (16). The major product of non-enzymatic PCB addition is mesobiliverdin, and that of PEB addition is 15,16-dihydrobiliverdin (Fig. 2). Thus, in contrast to the adduct formed *in vivo*, the *in vitro* reaction in each case leads predominantly to more oxidized adducts with an extra double bond between C-2 and C-3 of ring A (17, 18).

Native C-phycoerythrin α subunit carries PEB at α-82 and α-139. *In vitro* reaction of PEB with recombinant apo-α subunit led primarily to 15,16-dihydrobiliverdin (α-82) or urobilin (α-139) adducts and small amounts of the natural PEB adducts at both α-82 and α-139. PCB reacted primarily at α-82 to form PCB and mesobiliverdin adducts. Both PCB and PEB also formed relatively small amounts of adducts with α-Cys-59, which is not a bilin attachment residue in natural phycoerythrin (19; Fig. 2).

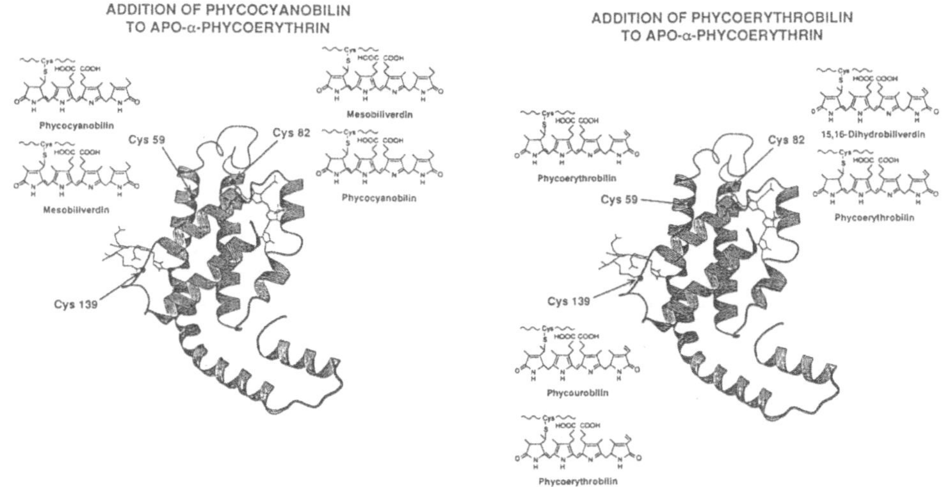

Figure 2 . Products of *in vitro* addition of PCB (left diagram) or PEB (right diagram) to the apo-α subunit of *Calothrix* sp. PCC7601 C-phycoerythrin. For each site, the structure of the major adduct is indicated above that of the lesser adduct.

The results of the *in vitro* addition studies indicate that apophycobiliproteins have non-covalent binding sites for bilins. Rapid *in vitro* covalent addition of bilins takes place, with incomplete site-selectivity and lack of discrimination between PCB and PEB. A mixture of bilin adducts is obtained with the natural PCB and PEB adducts as minor components. The types and proportions of the bilin adducts are different at the different addition sites. Consequently, although the protein environment of a particular cysteine residue does influence the chemical fate of the attached bilin *in vitro* the results of these studies strongly support the view that all bilin addition to apophycobiliproteins *in vivo* is enzyme-catalyzed.

3. Phycobiliprotein α subunit bilin lyases

3.1 Phycocyanin α subunit PCB lyase

The identification of the phycocyanin α subunit PCB lyase and phycoerythrocyanin α subunit PXB lyase resulted from studies of the functions of open reading frames in the C-phycocyanin and phycoerythrocyanin operons. The phycocyanin operon, sequenced in several cyanobacteria, includes two open reading frames, *cpcE* and *cpcF*. Insertional inactivation of either *cpcE* or *cpcF* in *Synechococcus* sp. PCC 7002 resulted in the same distinctive phenotype. These mutants produced a much lower amount of phycocyanin than the wild-type, and the phycocyanin that was produced lacked the PCB chromophore at α-84 (20). However, the α-84 bilin attachment site was competent to react with PCB *in vitro* (21). A minor fraction (~20%) of the phycocyanin from the mutants carried unnatural bilin adducts at α-84 and the predominant adduct was mesobiliverdin. The two phycocyanin β subunit PCBs and the PCBs of allophycocyanin were unaffected in the mutants (21).

Direct biochemical studies of the catalytic activities of recombinant *Synechococcus* sp. PCC7002 CpcE and CpcF decisively confirmed that these polypeptides together function as a specific phycocyanin α subunit PCB lyase (22, 23). CpcE+CpcF (referred to henceforth as CpcEF), but not CpcE or CpcF alone, catalyzed the rapid addition of PCB to the *Synechococcus* sp. PCC7002 apophycocyanin α subunit to form an adduct whose structure corresponded to that of the PCB adduct formed *in vivo*. These experiments show that PCB is indeed the precursor to the polypeptide-bound bilin. Gel permeation chromatography experiments showed that CpcE and CpcF form an enzymatically active 1:1 complex (23).

Surprisingly, CpcEF catalyzed the transfer of bilin from either holo-C-phycocyanin or holophycocyanin α subunit to the *Anabaena* sp. PCC7120 phycocyanin α subunit (22). This is an unanticipated and provocative finding since the fluorescence properties of the donor proteins indicate that the bilin is held in a rigid extended conformation. It is noteworthy that CpcEF causes a reduction in the phycocyanin α subunit fluorescence and strongly affects its absorption spectrum but has no effect on phycocyanin β subunit.

Smith et al. (24) have shown that even at 25°C and pH 7.0, non-covalently bound heme prosthetic groups dissociate slowly from native proteins such as horseradish peroxidase, cytochrome c peroxidase, chloroperoxidase, and leghemoglobin, and can be captured by binding to excess apomyoglobin. These observations indicate that the heme binding pockets in these proteins undergo considerable "breathing" motions. However, spontaneous or catalyzed loss or transfer of covalently attached tetrapyrroles from native proteins had not been reported prior to the work described above.

The CpcEF bilin addition activity exhibits simple Michaelis-Menten kinetics with respect to the apophycocyanin α subunit and to PCB. CpcEF also catalyzes the addition of PEB to the apophycocyanin α subunit. CpcEF shows a preference for PCB relative to PEB both in binding affinity and in the rate of catalysis that appears sufficient to account for the selective attachment of PCB to the apophycocyanin α subunit *in vivo* (23).

3.2 Phycoerythrocyanin α subunit PXB lyase

The rod substructures of the *Anabaena* sp. PCC7120 phycobilisome contain C-phycocyanin and phycoerythrocyanin. Even at low light intensities, phycoerythrocyanin represents no more than 5% of the phycobilisome protein. The β subunits of both proteins carry PCBs at β-82 and β-155, however, C-phycocyanin has PCB at α-84 whereas the phycoerythrocyanin α subunit has PXB at this position (25). The *Anabaena* sp. PCC 7120 operon is made up of five genes. *PecB* and *pecA* encode the β and α subunits of phycoerythrocyanin, *pecC* encodes a linker polypeptide associated with phycoerythrocyanin, and *pecE* and *pecF* which show a high degree of homology to *cpcE* and *cpcF* discussed above (26, 27). Insertional mutants in *pecE* and *pecF*, and an interposon mutant in which a portion of both *pecE* and *pecF* was deleted, were constructed (28). Holo-phycoerythrocyanin was missing from the phycobilisomes of all three types of mutants and the level of the phycoerythrocyanin linker polypeptide was reduced relative to the wild-type. All other phycobilisome components were present at wild-type levels. Holophycoerythrocyanin β subunit was present in all of these phycobilisomes at ~30% of the wild-type level. In contrast, the phycoerythrocyanin α subunit was barely detectable in the *pecE* and *pecF* mutants, but was present in the *pecEF* deletion mutant as a PCB-adduct in a 1:1 ratio with the phycoerythrocyanin β subunit. These results are consistent with the conclusion that *pecE* and *pecF* encode a phycoerythrocyanin α subunit PXB lyase, and indicate that when both polypeptides comprising this lyase are absent, some other bilin lyase is capable of adding PCB to the apophycoerythrocyanin α subunit.

3.3 Characterization of an intergenic suppressor of the defect in bilin attachment in a *Synechococcus* sp. PCC7002 *cpcE* mutant

Pseudo-revertants of *Synechococcus* sp. PCC7002 *cpcE* and *cpcF* mutant strains were observed to arise at high frequency. Analysis of the phycocyanin from a *cpcE* pseudo-revertant, which produced a near wild-type level of phycocyanin with the α subunit carrying PCB, revealed a single amino acid substitution, α-Tyr^{129} --> Cys (21). α-Tyr^{129}, which is conserved in all phycocyanins sequenced to date, forms part of the α-84 bilin binding site and lies within 5Å of the bilin attachment residue α-Cys^{84} (Fig. 3). The genotype of the pseudo-revertant was confirmed by constructing a mutated *cpcA* gene (encoding the phycocyanin α-subunit) with this substitution by site-directed mutagenesis. This mutant gene was then transformed, along with *cpcB*, into a *cpcBAC* deletion strain containing an insertionally inactivated *cpcE*. This strain produced high levels of phycocyanin where most of the α-subunit carried PCB at α-Cys^{84} (21). These observations are consistent with the conclusion that recognition of particular bilin attachments sites by bilin lyases depends at least in part on features of the three-dimensional structure of the apophycobiliprotein. The most likely interpretation of the results seen with the *cpcE* pseudorevertant is that the α-Tyr^{129} --> Cys substitution allows a PCB lyase specific for another phycobiliprotein PCB attachment site to catalyze PCB addition at apophycocyanin α-Cys^{84}.

3.4 Candidate genes for other phycobiliprotein bilin lyases

Certain open ocean cyanobacteria make phycocyanins, designated R-phycocyanins II, which have a PEB at α-84 (29). In the marine *Synechococcus* sp. WH8020, the genes *rpcE* and *rpcF*, are clustered with the genes encoding the α and β subunits of R-phycocyanin II (27, 30). *RpcE has* a 40% average amino acid sequence identity to *cpcE* gene products, and *rpcF* has a 37% average amino acid sequence identity, and a 28% identity to the *Anabaena* sp. PCC7120 PecF protein (27). These homologies suggest strongly that *rpcE* and *rpcF* encode a R-phycocyanin II-α subunit PEB lyase.

The genes encoding the major phycobiliprotein rod components of the phycobilisome of *Synechococcus* sp. WH8020 are clustered in the genome of this organism. In addition to the *rpcE* and *rpcF* genes, four other open reading frames in this gene cluster (*mpeU*, *mpeV*, *cpeY*, *cpeZ*) show suggestive similarities to the known phycobiliprotein lyase genes (30).

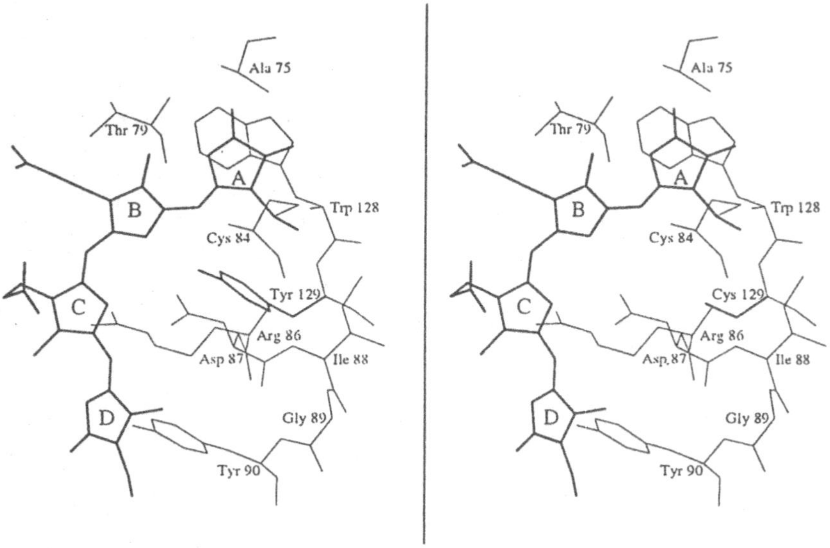

Figure 3. A view of the protein environment of the α-Cys^{84} PCB attachment site and its immediate protein environment in the wild-type and the α-Tyr^{129} --> Cys mutant. This figure is based on the crystal structure of *Mastigocladus laminosus* C-phycocyanin (7).

4. Concluding remarks

The studies of bilin attachment are far from complete. However, some general observations are in order. The linear amino acid sequence around a cysteine residue that serves as a bilin attachment site is insufficient to specify which bilin is to be attached *in vivo* or *in vitro.* Recognition of three-dimensional determinants plays a role. Bilin binding sites exist in apophycobiliproteins. It is likely that all *in vivo* bilin addition to phycobiliproteins is catalyzed by polypeptide site- and bilin-specific lyases. Phycobiliprotein bilin lyases show high selectivity, rather than absolute specificity, for both the bilin and the polypeptide substrate.

Acknowledgments

This research was supported in part by NIH Grant GM28994 and by the Lucille P. Markey Charitable Trust.

References

1 Glazer, A.N. (1994) in Molecular Processes in Photosynthesis (Barber, J., ed.) Advances in Molecular and Cell Biology (Bittar, E.E., ed.) pp. 119-149, JAI Press Inc., Greenwich, Connecticut
2 Sidler, W. (1994) in The Molecular Biology of Cyanobacteria (Bryant, D.A., ed.), pp. 139-216, Kluwer Academic Publishes, Dordrecht, The Netherlands

3 Wilbanks, S.M., Wedemayer, G.J. and Glazer, A.N. (1989) J. Biol. Chem. 264, 17860-17867
4 Apt, K.E., Collier, J.L. and Grossman, A.R. (1995) J. Mol. Biol. 248, 79-96
5 Brejc, K., Ficner, R., Huber, R. and Steinbacher, S. (1995) J. Mol. Biol. 249, 424-440
6 Schirmer, T., Bode, W., Huber, R., Sidler, W. and Zuber, H.. (1986) J. Mol. Biol. 184, 257-277
7 Schirmer, T., Bode, W. and Huber, R. (1987) J. Mol. Biol. 196, 677-695
8 Duerring, M., Huber, R., Bode, W., Ruembeli, R. and Zuber, H. (1990) J. Mol. Biol. 211, 633-644
9 Ficner, R., Lobeck, K., Schmidt, G. and Huber, R. (1992) J. Mol. Biol. 228, 935-950
10 Ficner, R and Huber, R. (1993) Eur. J. Biochem. 218, 103-106
11 Ong, L.J. and Glazer, A.N. (1991) J. Biol. Chem. 266, 9515-9527
12 Swanson, R.V., Ong, L.J., Wilbanks, S.M. and Glazer, A.N. (1991) J. Biol. Chem. 266, 9528-9534
13 Beale, S.I. (1993) Chem. Rev. 93, 785-802
14 Cole, W.J., Chapman, D.J. and Siegelman, W.H. (1967) J. Am. Chem. Soc. 89, 3643-3645
15 Chapman, D.J., Cole, W.J. and Siegelman, W.H. (1967) J. Am. Chem. Soc. 89, 5976-5977
16 Arciero, D.M., Bryant, D.A. and Glazer, A.N. (1988) J. Biol. Chem. 263, 18343-18349
17 Arciero, D.M., Dallas, J.L. and Glazer, A.N. (1988) J. Biol. Chem. 263, 18350-18357
18 Arciero, D.M., Dallas, J.L. and Glazer, A.N. (1988) J. Biol. Chem. 263, 18358-18363
19 Fairchild, C.D. and Glazer, A.N. (1994) J. Biol. Chem. 269, 28988-28996
20 Zhou, J., Gasparich, G.E., Stirewalt, V.L., de Lorimier, R. and Bryant, D.A. (1992) J. Biol. Chem. 267, 16138-16145
21 Swanson, R.V., Zhou, J., Leary, J.A., Williams, T., de Lorimier, R., Bryant, D.A. and Glazer, A.N. (1992) J. Biol. Chem. 16146-16154
22 Fairchild, C.D., Zhao, J., Zhou, J., Colson, S.E., Bryant, D.A. and Glazer, A.N. (1992) Proc. Natl. Acad. Sci. USA 89, 7017-7021
23 Fairchild, C.D. and Glazer, A.N. (1994) J. Biol. Chem. 269, 8686-8694
24 Smith, M.L., Paul, J., Ohlsson, P.I., Hjortsberg, K. and Paul, K.G. (1991) Proc. Natl. Acad. Sci. USA 88, 882-886
25 Bishop, J.E., Rapoport, H., Klotz, A.V., Chan, C.F., Glazer, A.N., Füglistaller, P. and Zuber, H. (1987) J. Am. Chem. Soc. 109, 875-881
26 Swanson, R.V., de Lorimier, R. and Glazer, A.N. (1992) J. Bacteriol. 174, 2640-2647
27 de Lorimier, R., Wilbanks, S.M. and Glazer, A.N. (1993) Plant Mol. Biol. 21, 225-237
28 Jung, L.J., Chan, C.F. and Glazer, A.N. (1995) J. Biol. Chem. 270, 12877-12884
29 Ong, L.J. and Glazer, A.N. (1987) J. Biol. Chem. 262, 6323-6327
30 Wilbanks, S.M. and Glazer, A.N. (1993) J. Biol. Chem. 268, 1226-1235

PROTEIN COMPONENTS OF CHLOROSOMES FROM *CHLOROBIUM TEPIDUM* AND INTERPOSON MUTAGENESIS OF *csmA* AND *csmC* FROM *CHLOROBIUM VIBRIOFORME* 8327D

Soohee Chung[1], Christiane U. Jakobs[1], John G. Ormerod[2], and Donald A. Bryant[1], [1]Dept. of Biochemistry and Molecular Biology, The Pennsylvania State University, University Park, PA 16802 USA; [2]Biology Department, Division of Molecular and Cell Biology, University of Oslo, Box 1050, Blindern, N-0315, Norway

I. Introduction

Chlorosomes are antenna complexes that harbor the light-harvesting pigments, Bchl *c, d*, or *e*, in green bacteria. They appear as oblong bodies attached to the inner surface of the cytoplasmic membrane when viewed by electron microscopy (1). Despite the seemingly close functional relationship between chlorosomes of the two families of green bacteria, Chloroflexaceae and Chlorobiaceae, there appears to be only limited similarities in either the protein composition or amino acid sequences among the protein components. The structure and polypeptide composition of chlorosomes have been most thoroughly investigated in the green gliding bacterium, *Chloroflexus aurantiacus*, for which only 4 polypeptide components of 18, 11, 5.8, and 3.7 kDa were identified (2). The amino acid sequences of these components have been identified by their gene sequences (3, 4). Chlorosomes of the green sulfur bacterium, *Chlorobium tepidum*, were shown by SDS-PAGE analyses and N-terminal amino acid sequence determination to contain 10 polypeptides, ranging from 6.2 to 27 kDa (5). The amino acid sequence of only a single component, the CsmA protein of 6.2 kDa, appears similar to a component of the *C. aurantiacus* chlorosomes.

As part of an ongoing effort to study the structure and function of the protein components of chlorosomes of *Cb. tepidum*, we are cloning the genes for all of the polypeptide components identified by SDS-PAGE, denoted CsmA through CsmJ (5). Using synthetic oligonucleotides as hybridization probes, genes *csmA*, *csmB*, *csmC*, *csmD*, and *csmE* have been cloned, sequenced and transcriptionally characterized (5; S. Chung and D. A. Bryant, in preparation). Genes *csmH* and *csmI* have recently been cloned and sequence analyses of these genes are in progress. To obtain immunological tools for investigating the structure of chlorosomes, five of the protein components were overproduced in *E. coli* by using the IPTG-inducible pET system of Studier et al. (6). To examine the role(s) of the chlorosome polypeptide components, interposon mutagenesis of *Cb. tepidum* and *Cb. vibrioforme* chlorosome genes is being performed using the natural transformation capability of these *Chlorobium* sp. strains. A mutant of *Cb. vibrioforme* specifically lacking the CsmC polypeptide has been constructed by this procedure.

2. Materials and Methods

Cb. tepidum was kindly provided by Dr. Michael Madigan and was grown at 48°C as described (7). Chlorosomes and total chromosomal DNA were prepared, and colony

P. Mathis (ed.), Photosynthesis: from Light to Biosphere, Vol. I, 11–16.

hybridization and Southern hybridization were performed, as previously described (5). Based upon the known amino-terminal sequences of chlorosome polypeptides (5, 8, 9), the degenerate oligonucleotides indicated in Table 1 were synthesized and used as hybridization probes in isolating each of the corresponding genes.

Table 1. Synthetic oligonucleotide probes used in colony hybridization experiments

Genes	Oligonucleotides
csmA	5' GA(A/G)GTTATGGTTGA(A/G)GGTCACTGGGA(A/G)ACTGTTGG
csmB	5' ATGAA(T/C)GG(T/C)AC(T/C)AA(T/C)AT(T/C)GA(T/C)GT
csmC	identified as upstream open reading frame of *csmA*
csmD	5 GCTGA(C/T)GA(A/G)GA(A/G)AA(A/G)AT(C/T)GA(C/T)ACTATGAA
csmE	identified as upstream open reading frame of *csmD*
csmH	5'GA(A/G)GA(A/G)ACCAACATGCC(N)GC(T/C)GC(T/C)GA(A/G)GC
csmI	5' AT(T/C)AT(T/C)AA(T/C)GA(T/C)AA(A/G)AC(N)GC

DNA sequencing was performed on base-denatured, double-stranded templates by following the manufacturer's instructions for using Sequenase™ version 2.0 and ^{35}S-α-thio-dATP. Total RNA isolation, Northern-blot hybridization, and primer extension procedures were described previously (10). SDS-PAGE was performed by the method of Schägger and von Jagow (11) with 6 M urea in the resolving gel. Immunoblotting was performed with Immobilon™ (Millipore) membranes using the methods described (12). Proteins were overproduced in *E. coli* as described (6). Proteins were purified by ion-exchange and/or gel-exclusion column chromatography as indicated in Table 2.

Table 2. Purification of overproduced chlorosome proteins.

Protein	Chromatography
CsmA	CM-Sepharose CL-6B and Sephacryl S-100
CsmB	His•Bind Ni^{2+} and ultrafiltration
CsmC	DEAE-Sepharose CL-6B and Sephacryl S-100
CsmD	His•Bind Ni^{2+} and Sephacryl S-100
CsmE	CM-Sepharose CL-6B and Sephacryl S-100

3. Results and Discussion

Chlorosomes from *Cb. tepidum* have been shown to contain 10 polypeptides by SDS-PAGE analysis and N-terminal amino acid sequencing. In addition to the 9 polypeptides originally reported (5), one additional polypeptide component, denoted CsmJ, was found to comigrate with the 23-kDa CsmF polypeptide. From the sequence information obtained, degenerate oligonucleotides (see Table 1) were synthesized and used as hybridization probes in cloning the corresponding genes. Based upon Southern-blot hybridization experiments, size-fractionated partial genomic libraries were constructed for each gene of interest, and these were screened with radioactively labeled probes in colony hybridizations. A 1.2-kbp *Hin*dIII fragment of *Cb. tepidum* and a 1.6-kbp *Hin*dIII fragment of *Cb. vibrioforme* were cloned for *csmA* and subsequent sequence analysis revealed that the *csmC* gene was located upstream of *csmA*. Northern-blot and primer extension analyses with *Cb. tepidum* total RNA indicated that *csmA* and *csmC* are cotranscribed on a dicistronic message and form an operon having a probable σ^{70}-type promoter. For the monocistronic *csmB* gene, a 1.5-kbp *Sal*I-*Stu*I fragment from *Cb. tepidum* and a 0.9-kbp *Hinc*II-*Eco*RI from *Cb. vibrioforme* were cloned with substantial amounts of flanking sequences. A 1.7-kbp *Hind*III fragment from *Cb. tepidum* was cloned for *csmD* and sequence analysis indicated that *csmE* was the upstream open reading frame. Northern-blot analyses and primer extension experiments with total RNA from *Cb. tepidum* showed that *csmD* and *csmE* are cotranscribed on a polycistronic transcript that probably undergoes extensive processing to create stable transcripts containing *csmED*, *csmE*, and *csmD*. Fragments, 1.3-kbp *Cla*I and a 0.8-kbp *Hind*III, respectively, encoding the *csmI* and *csmH* genes were cloned from *Cb. tepidum* and sequence analyses of these genes are now in progress. Physical maps of several of the genes are shown below.

Figure 1. Physical maps of genes encoding some chlorosome components

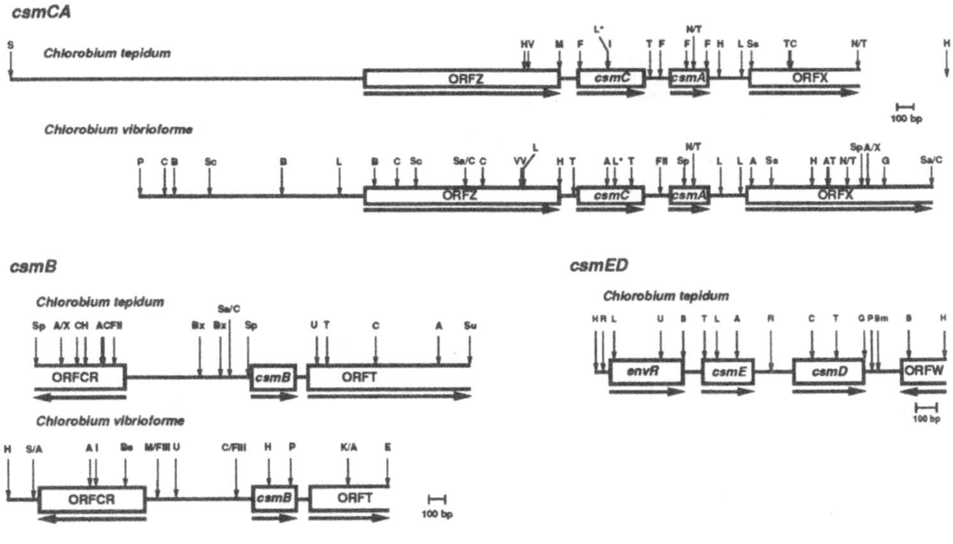

An interesting feature about CsmE is that its amino acid sequence is 49% identical to that of CsmA. Furthermore, the two serine residues around the putative C-terminal processing site (see 5) are conserved, and 13 residues out of 20 amino acids in the C-terminal extension are identical. This suggests that, as in the case of CsmA, the mature CsmE protein might arise from C-terminal processing between the two serines positioned at residues 62 and 63, and the mature form of the protein would have only 62 amino acids.

To obtain immunological tools for investigations of the structure of native chlorosomes, CsmA, CsmB, CsmC, CsmD, and CsmE were overproduced in *E. coli* using the pET system of Studier et al. (6) (Fig. 2A-E, below) to raise antibodies. CsmA (Fig. 2A), CsmC (Fig. 2C), and the mature form of CsmE ($CsmE_m$) (Fig. 2E) were obtained by using pET11 expression vector. CsmB (Fig. 2B) and CsmD (Fig. 2D) were overproduced as fusion proteins by using pET16b that translationally fuses 10 histidines and 11 other residues at the N-termini of the proteins. As indicated in lane 4 of each panel below, each of the proteins could be overproduced by 16 hour-induction with 0.5 mM IPTG (compare to control cells, lanes 2, and uninduced cells, lanes 3) and each was recovered in inclusion bodies (lanes 5). These inclusion bodies were solubilized with 6-7 M urea and loaded onto ion exchange (cation: CsmA and $CsmE_m$; anion: CsmC) or affinity (His•bind Ni^{+2}: CsmB and CsmD) columns. Gel filtration column chromatography was employed for further purification (lanes 6). The N-termini in CsmB and CsmD were removed by factor Xa protease digestion, except for one histidine residue (lane 6a). These proteins were use to raise rabbit antisera which are now being employed in protease mapping and crosslinking studies.

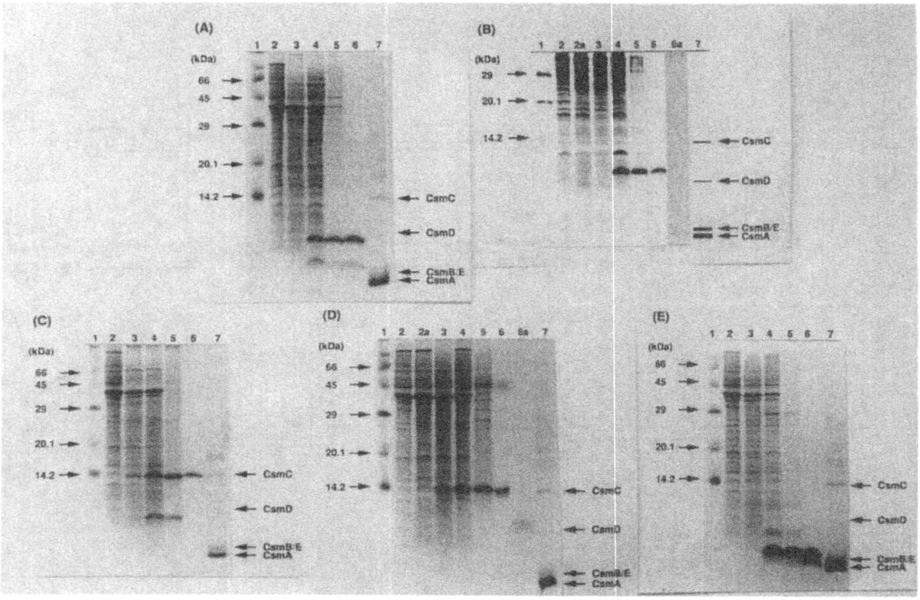

To investigate possible functional roles of CsmA and CsmC by genetic methods, interposon mutagenesis of *csmA* and *csmC* was performed using either the Ω cassette (conferring streptomycin and spectinomycin resistance) or *aphII* gene (conferring kanamycin resistance). *Cb. vibrioforme* or *Cb. tepidum* cells were transformed with each of the mutant gene plasmids constructed from the *csmA* or *csmC* genes from *Cb. vibrioforme*, and transformants were selected at 15 μg ml^{-1} for streptomycin resistance or 40 μg ml^{-1} for kanamycin resistance.

Homologous recombination of the mutated *csmA* genes in *Cb. vibrioforme* 336 (Ω insertion in the same orientation as *csmA*) and *Cb. vibrioforme* 337 (Ω insertion in opposite orientation) was confirmed by Southern-blot hybridization with a gene-specific hybridization probe for *csmA*, but the results indicated that the wild-type and mutant alleles had not segregated. *Hind*III digestion of total chromosomal DNA from the mutant strains showed that the 1.6-kbp wild-type fragment is detected together with fragments generated by the Ω cassette insertion (results not shown). Selection at very high concentrations of streptomycin (600 μg ml^{-1}) with *Cb. vibrioforme* 337 was not successful in forcing segregation. Transformation of *Cb. tepidum* also resulted in merodiploids containing both wild-type and mutant *csmA* genes for both orientations of the Ω cassette (results not shown).

Southern-blot hybridization experiments with total chromosomal DNAs isolated from *csmC* mutant strains of *Cb. vibrioforme* 633 (Ω insertion in the same orientation as *csmC*) and *Cb. vibrioforme* 634 (Ω insertion in the opposite orientation) (Fig. 3A, below) showed that the mutational constructs had integrated correctly into the genome and segregated from the wild-type gene (Fig. 3B); in the double digest with *Nco*I and *Sal*I, a probe specific for *csmC* detected one radioactive fragment for both of the mutant DNAs, each of which is 2.0 kbp larger in size than the wild-type, due to the Ω cassette insertion (compare lane 1, wild-type and lanes 2 & 3, mutants); for *Cb. vibrioforme* 626, where the *csmC* gene is mutagenized with an *aphII* cartridge, *Hind*III digestion resulted in a hybridizing fragment that was 1.3 kbp larger as expected (compare lanes 4, wild-type, and 5, mutant). In Fig. 3C, Southern-blot hybridization experiments with probes specific for the Ω cassette (lanes

Figure 3. Interposon mutagenesis of *csmC* gene of *Chlorobium vibrioforme*

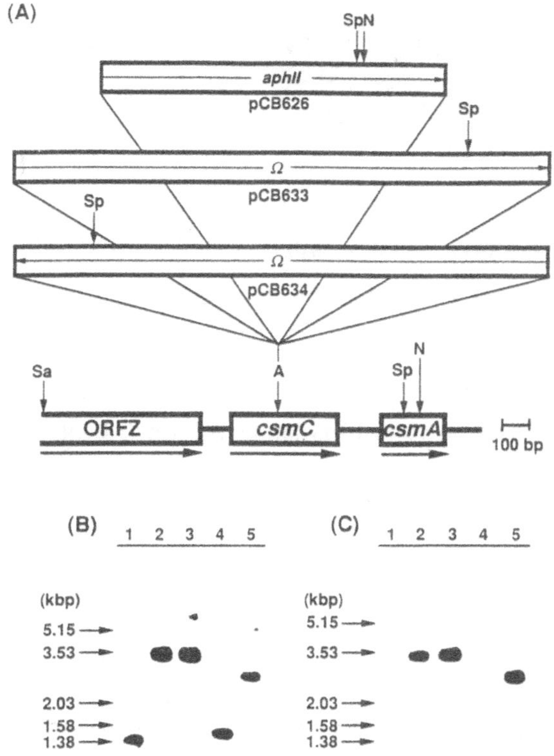

1, 2, and 3) or *aphII* gene (lanes 4 and 5) confirmed that the antibiotic cartridges were integrated into the genome of *Cb. vibrioforme*, indicating that the antibiotic resistances of the mutants are not a result of spontaneous resistance mutations.

To confirm that *csmC* mutants are missing the CsmC protein, total proteins from whole cell extracts of *Cb. vibrioforme* 626 and *Cb. vibrioforme* 633 were separated on 16% polyacrylamide SDS-tricine-urea gel and immunoblotted with antibodies specific for CsmC. As indicated below, the 14.3-kDa CsmC component is missing in mutant strains *Cb. vibrioforme* 633 and 626 (Fig. 4, lanes 2 and 3, respectively). To see if the absence of the CsmC component has any effect on the growth rate of the mutants, doubling times were monitored by the turbidity measurements at 650 nm. Each mutant had a doubling time approximately twice that of the wild-type strain *Cb. vibrioforme* 8327D wild-type, 1.75 hr; *Cb. vibrioforme* 633, 3.5 hr; *Cb. vibrioforme* 626, 3.25 hr). Having established that this general approach can be employed, mutagenesis of genes encoding other chlorosome components is now underway. Biochemical characterization of chlorosomes in such mutants, as well as biophysical analyses of light capture and transfer, will be pursued.

Figure 4. Immunoblot analysis of *Cb. vibrioforme* strains with antibodies to CsmC

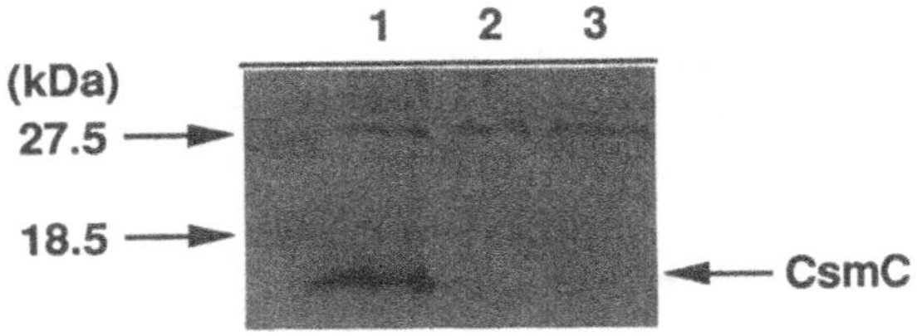

4. References

1. Staehelin, L. A., Golecki, J. R. and Drews, G. (1980) Biochim. Biophys. Acta 589, 30-45
2. Feick, R. G. and Fuller, R. C. (1984) Biochemistry 23, 3693-3700
3. Theroux, S. J., Redlinger, T. E., Fuller, R. C. and Robinson, S. J. (1990) J. Bacteriol. 172, 4497-4504
4. Neidermeier, G., Shiozawa, J. A., Lottspeich, F. and Feick, R. G. (1994) FEBS Lett. 342, 61-65
5. Chung, S., Frank, G., Zuber, H. and Bryant, D. A. (1994) Photosynthesis Res. 41, 261-275
6. Studier, F. W., Rosenberg, A. H., Dunn, J. J. and Dubendorff, J. W. (1990) Meth. Enzymol. 185, 60-88
7. Wahlund. T. M., Woese, C. R., Castenholz, R. W., and Madigan, M. T. (1991) Arch. Microbiol. 152, 81-90
8. Wagner-Huber, R., Brunisholz, R., Frank, G., and Zuber, H. (1988) FEBS Lett. 239, 8-12
9. Wagner-Huber, R., Fischer, U., Brunisholz, R., Rumbeli, M., Frank, G. and Zuber, H. (1990) Z. Naturforsch. 45c, 818-822
10. Dubbs J. M. and Bryant, D. A. (1992) Mol. Microbiol. 5, 3073-3085
11. Schägger, H. and von Jagow, G. (1987) Analyt. Biochem. 166, 368-379
12. Schluchter, W. M. and Bryant, D. A. (1992) Biochemistry 31, 3092-3102

DYNAMICS OF A CYLINDRICAL EXCITON IN CHLOROSOMES OF GREEN BACTERIA

Z.G. Fetisova, V.I. Novoderezhkin, A.S. Taisova, R.E. Uzbekov, A.M. Freiberg (*), K.E. Timpmann (*)
A.N.Belozersky Institute of Phys.-Chem.Biology, Moscow State University, 119899 Moscow, Russia; (*) Institute of Physics, Estonian Acad. Sci., 202400 Tartu, Estonia

1. Introduction

In the past decade, the chlorosome antenna structure and functioning in different green bacteria have been extensively studied (for review see (1)). A major question about the mechanism of excitation energy transfer (EET) within the chlorosome, however, is still open. This problem is most acute in view of the oligomeric organization of the chlorosome pigments. Here, we present the theory of EET within the chlorosome. The theory has been developed for the exciton model of aggregation of BChl *c* (*d* or *e*, depending on species) (see the other paper of Novoderezhkin and Fetisova in this volume). This model is compatible with the key spectral features found *in vivo*, i.e. the exciton level structure, revealed by spectral hole- burning experiments (2), and polarization of all the levels parallel to the long axis of the chlorosome (3). The analytical expression for the time constant of chlorosomal BChl fluorescence decay as a function of the chlorosome size was obtained. Finally,using picosecond fluorescence spectroscopy, we have demonstrated that this theoretical dependence agrees well with the experimental dependence, measured for intact cells of different cultures of the green bacterium *Chloroflexus aurantiacus* with different chlorosome size which is determined by electron microscopic examination of the cells. The data suggest a possible mechanism of EET within the chlorosome, namely, the Forster type transfer of a cylindrical exciton delocalized over tubular BChl *c* aggregate structures.

2. Procedure

All experiments were performed on intact cells of the filamentous nonsulfur thermophilic green bacterium *Chloroflexus aurantiacus* strain Ok-70-fl (collection of Leiden University, Netherlands) used in their own growth medium under strictly anaerobic conditions. Cells were cultivated at 55°C under anaerobic conditions in the light. Cells with chlorosomal antennae of different size (i.e. with different ratios of BChl *c*/BChl *a*, see Fig.1) were adjusted by changing the growth rate (4)

P. Mathis (ed.), Photosynthesis: from Light to Biosphere, Vol. I, 17–22.

and by inhibiting the formation of BChl *c* by gabaculine (0.7, 1.1, 1.6, 2.2 μM concentrations were used) according to Oelze (5).

Electron microscopy studies were performed on Hitachi-11 and Hitachi-12 (Japan) at 75 kV. For electron microscopic examination, the cells were fixed for 30 min in the culture medium at $55^{o}C$ by adding 25% glutaraldehyde to a final concentration of 1% and then at room temperature for 60 min. The samples were postfixed with 1% OsO_4 for 90 min, embedded in Epon-812 and ultrathin-sectioned by standard methods (6). Micrographs of the ultrathin sections were used for morphometric measurements (magnification 50,000x10). Histograms of chlorosomes heights were obtained for cells grown under different conditions. The number of layers of rod elements in a chlorosome was calculated using the results reported in (6).

Spectrally-resolved fluorescence measurements for intact cells of *Chloroflexus aurantiacus* were made with a picosecond spectrochronograph described in (3). Briefly,the picosecond pulse source was a mode-locked CW rhodamine 590 dye laser (pulse duration, 3 ps), synchronously pumped at 76 MHz by a Nd-Yag laser (Ontares, "Coherent"). Emission, viewed at an angle of 90° to the exciting beam (in the reflection mode), was filtered by two single-grating monochromators (Lomo MDR-2), combined in subtractive dispersion mount to avoid extra pulse broadening (bandwidth 4 nm), and registered with a home-made synchroscan streak camera. The time resolution of the measuring system was 10 ps. Data were recorded and processed using a PC IBM computer supported by "B&M Spectronik" OSA 500 optical multichannel analyzer with SIT vidicon.

3. Theory of Energy Transfer between Bacteriochlorophyll Aggregates in a Chlorosomal Antenna

3.1 *Energy Transfer between Tubular Aggregates*

Let us consider the cylinder formed by L linear rows each containing N BChl *c* molecules (Fig.2); n is the number of a molecule in a row; l is the number of a row. Exciton states $|k,\kappa\rangle$ are characterized by two numbers: k and κ, where k takes N integer values 1,2,..N; κ takes L integer values $0,\pm1,\pm2,\ldots$. There are L Davydov components, each having the exciton structure similar to that of a linear chain. The matrix elements of energy transfer from the $|k,\kappa\rangle$ state to trap m or to the $|k',\kappa'\rangle$ state corresponding to another aggregate take the following form ($M_{nl,m}$ and $M_{nl,ml'}$ are interaction energies between nl and m or ml' sites):

$$W^{m}_{k\kappa} = \frac{2}{L\ (1+N)} \left| \sum_{n,l} M_{nl,m} \exp\,(i\Phi l\kappa)\,\sin(k\vartheta n) \right|^2 \quad ; \ \vartheta = \pi/(N+1) \ ; \ \Phi = 2\pi/L$$

$$W_{k\kappa}^{k'\kappa'} = \frac{4}{L^2(1+N)^2} \left| \sum_{n,l} \sum_{m,l'} M_{nl,ml'} \exp(i\Phi l\kappa + i\Phi l'\kappa') \sin(k\vartheta n) \sin(k'\vartheta m) \right|^2 \qquad (1)$$

The energy transfer between tubular aggregates is determined by the interaction between nearest chains (two or three chains from each tubular aggregate). The contribution from each interacting pair of chains (the l-th chain from one aggregate and the l'-th chain from another aggregate) is propotional to $\exp(i\Phi l\kappa + i\Phi l'\kappa')$. The most effective transfer occurs between the $\kappa=0$ and $\kappa'=0$ components. In contrast, the rate of energy transfer between the lowest components $\kappa=L/2$ and $\kappa'=L/2$ being proportional to $\sum(-1)^{l+l'}$ is near zero. This fact should result in a decrease in energy transfer rate at low temperatures.

Notice that the energy transfer rates depend on the overlapping integrals, i.e. Frank-Condon overlapping factors between phonon states accociated with the $|k,\kappa\rangle$, $|k',\kappa'\rangle$ and $|m\rangle$ levels averaged over the initial state. It is very important that the matrix elements for the $|k,\kappa\rangle \rightarrow |k',\kappa'\rangle$ and $|k,\kappa\rangle \rightarrow |m\rangle$ transfer can be nonzero even in the case of dipole forbidden $|k,\kappa\rangle$ states. For example, any $|k,\kappa\rangle$ state can be dipole forbidden due to the spatial symmetry of its wavefunction, but this symmetry will have no effect on the matrix element if molecule m interacts only with a small number of molecules from aggregate. That is why there is no correspondence between the rate constants of energy transfer and the overlapping of the fluorescence and absorption spectra of a donor and acceptor as in the case of the monomer --> monomer transfer.

3.2 *Energy Transfer within a Chlorosomal Antenna*

Consider the model of a chlorosomal antenna containing p layers of rod elements packed in a hexagonal structure (6). Each rod element consists of tubular aggregates of BChl chains; each aggregate from the first layer (nearest to the baseplate) interacts with q BChl *a* (BChl790) molecules. The rate constant of energy transfer from an aggregate of the p-th layer to each of the two nearest aggregates of the (p+1)-th layer is K_A^A; the rate of energy transfer from the BChl *c* tubular aggregate to the BChl790 molecules is K_A^a; the rate of back transfer is K_a^A. We suppose that the antenna is infinite in x and y directions in order to consider the one-dimensional energy migration (i.e. migration to the baseplate in z direction). The rates of the total losses in BChl *c* and BChl *a* chlorosomal antennae are K_c and K_a, respectively. Quantum yields of the losses in the BChl *c* antenna, containing p layers of rods, and in the BChl790 antenna, $\Phi_c(p)$ and $\Phi_a(p)$, are:

$$\Phi_a(1)= \frac{t^+}{(1+ t^+ + t^-)} \qquad ; \quad p = 1$$

$$\Phi_a(p)= \frac{(t^+/p)\, B_p}{(1+ t^+ + t^- + t(1+t^-))\, D_{p-1} + t^2(1+t^-)\, D_{p-2}} \qquad ; \quad p > 1$$

$$D_0=1 \ ; \ D_1= -(1+t) \ ; \ B_1=1$$

$$D_p= -(1+2t)\, D_{p-1} - t^2 D_{p-2} \ ; \quad B_p = D_{p-1} - t\, B_{p-1} \qquad ; \quad p > 1$$

$$t=2K_A^A/K_c; \quad t^+=qK_A^a/K_c; \quad t^-=K_a^A/K_a; \quad \Phi_c = 1- \Phi_a \qquad (2)$$

The time constant of the fluorescence decay in the BChl *c* antenna is

$$\tau(p)= (1- \Phi_a(p))\, \tau_c; \qquad \tau_c=1/K_c \qquad (3)$$

The cells of green bacteria containing chlorosomes of different size can be characterized by the discrete distribution function F(p), where F(p) is the relative number of chlorosomes with p layers. In this case the time constant of the fluorescence decay in the BChl *c* antenna is

$$\tau= \sum_p \tau(p)\, F(p) \qquad (4)$$

If the $\tau(p)$ dependence is linear, then

$$\tau=\tau(\langle p\rangle) \ ; \qquad \langle p\rangle= \sum_p p\, F(p) \qquad (5)$$

4. Results and Discussion

In our experiments several *Chloroflexus aurantiacus* cultures with different BChl *c* contents were used (Fig.1). The corresponding distribution functions, F(p), have been determined by electron micriscopic examination of cells from these cultures (6,7). In line with the results obtained in (7), we have shown that cells with a higher BChl *c* content (Fig.1) exhibit a larger amount of thicker chlorosomes (compare F(p) distributions). For the same cultures the picosecond BChl *c* fluorescence kinetics detected at emission wavelength 750 nm have been measured in intact cells at room temperature. The kinetics observed are strongly biphasic (in Fig.3 only fast components are presented). The experimental data are summarized in the following table:

No.1: $F(p=1, 2, 3, 4)= 0.01, 0.38, 0.58, 0.03$; $\langle p\rangle=2.63$ $\tau=39$ ps
No.2: $F(p=1, 2, 3, 4)= 0.40, 0.60, 0.00, 0.00$; $\langle p\rangle=1.60$ $\tau=26$ ps
No.3: $F(p=1, 2, 3, 4)= 0.77, 0.23, 0.00, 0.00$; $\langle p\rangle=1.23$ $\tau=17$ ps

Theoretical dependences $\tau(p)$ for $t=t^+=50$, $\tau_c=650$ ps, $t^-=0$ (curve 1) and $t=t^+=20$, $\tau_c=275$ ps, $t^-=0$ (curve 2) are shown in Fig.3. The efficiency of energy transfer from the BChl *c* antenna to the baseplate (and the rate constant of fluorescence decay in this antenna) decreases with increasing the chlorosome size. The $\tau(p)$ dependence is nearly linear; that is why the F(p) distribution can be characterized only by the first moment, $\langle p\rangle$. Indeed, the experimental $\tau(\langle p\rangle)$ points are very close to the calculated $\tau(p)$ curves (see Fig.3).

If we use the exact distribution, F(p), the time constants calculated according to Eq.(4) are $\tau=$ 43.69 ps; 23.51 ps and 17.19 ps (for $t=t^+=20$, $\tau_c=275$ ps). They agree well with the experimental values $\tau=$ 39 ps, 26 ps and 17 ps. For our set of parameters ($t=t^+=20$, $\tau_c=275$ ps) we obtain $(K_A^A)^{-1}=$ 27.4 ps; the hopping time between the antenna aggregates $(K_A^A)^{-1}/6=$ 4.4 ps (for the middle layers); $(K_A^A)^{-1}/4=$ 6.6 ps (for the upper layer); $(K_A^A)^{-1}/2=$ 13.2 ps (if there is only one layer); the time of energy transfer to the BChl790 $(qK_A^a)^{-1}=$ 13.7 ps.

These hopping times correspond to the characteristic times for spectral diffusion in an inhomogeneously broadened antenna. A dynamic spectral shift with $\tau=7$ ps was observed in (1). The amplitude of this shift was equal to 4 nm or 80 cm^{-1} and closely corresponded to the inhomogeneous width revealed by spectral hole-burning studies (90-100 cm^{-1} (2)).

Thus, the data presented here are consistent with a possible mechanism of EET within the chlorosome which implies (i) the formation of a cylindrical exciton, i.e. delocalization of excitation over a tubular aggregate of linear single or double exciton-coupled BChl *c* chains with interchain distances of ~2 nm and BChl *c* density, approximating that *in vivo*, and (ii) the Forster type transfer of such a cylindrical exciton between nearest tubular BChl *c* aggregate structures as well as to BChl *a* of the baseplate.

5. Addendum

We are grateful to A.H.M. de Wit for providing us with the *Chloroflexus aurantiacus* (strain Ok-70-fl). The present research was made possible in part by grant No. M6R300 from the International Science Foundation, grant No. 94-04-12707 from the Russian Foundation for Basic Researches and the grant from the Russian Universities Foundation.

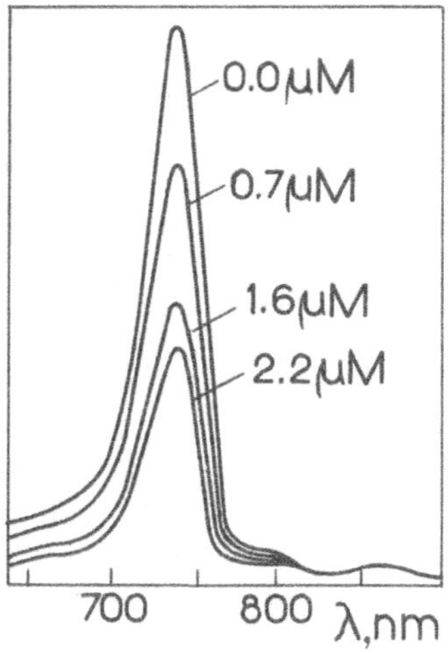

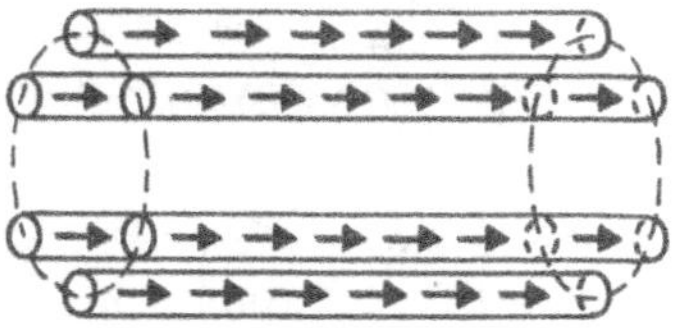

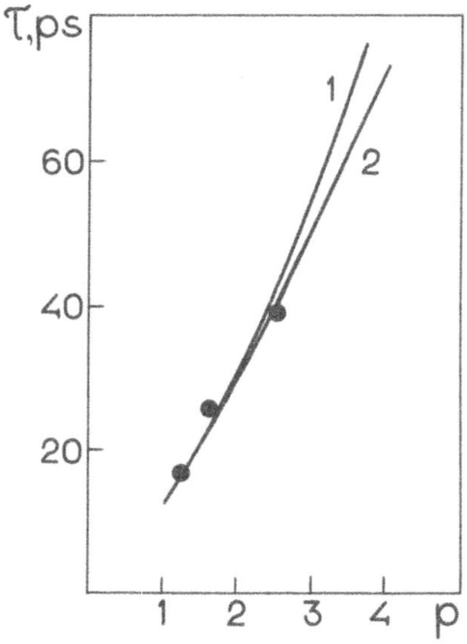

Figure 1. Absorption spectra for *C.aurantiacus* cultures with different BChl *c* contents. Concentrations of gabaculine are shown.

Figure 2. Tubular aggregate model; L=6, N=6; arrows show BChl *c* transition dipole moments.

Figure 3. Calculated dependences τ(p) (solid curves; parameters are listed in the text) and experimental points τ(<p>).

References

1 Lin, S.,Van Amerongen, H., Struve, V.S. (1991) Biochim.Biophys.Acta 1060, 13-24
2 Fetisova, Z.G., Mauring, K., Taisova, A.S. (1994) Photosynth.Res. 41, 205-210
3 Fetisova, Z.G., Freiberg, A.M., Timpmann, K.E. (1988) Nature 334, no. 6183, 633-634
4 Oelze, J. and Fuller, R.C. (1987) Arch.Microbiol. 148, 132-136
5 Oelze, J. (1992) J. Bacteriol. 174, 5021-5026
6 Staehelin, L.A., Golecki, J.R., Fuller, R.C. and Drews, G. (1978) Arch.Microbiol. 119, 269-277
7 Golecki, J.R. and Oelze, J. (1987) Arch.Microbiol. 148, 236-241

ELECTRONIC EXCITATION TRANSFER IN BACTERIAL LIGHT HARVESTING COMPLEXES

S.E Bradforth, R. Jimenez, S.N. Dikshit and G.R. Fleming, Department of Chemistry and The James Franck Institute, The University of Chicago, USA

1. **Introduction**

The increasing availability of structural information has given considerable impetus to spectroscopic studies of the basic light harvesting components of the purple bacterial photosynthetic apparatus. In particular, the beautiful high resolution structure of the peripheral light harvesting protein, LH-2 [1], combined with much lower resolution data for LH-1 [2], suggests these antennas form highly symmetric ring structures and that energy is both stored and transported laterally through the membrane by extremely rapid energy migration within the rings. In this picture, the reaction center protein sits within the large LH-1 ring and the smaller LH-2 rings are in contact with the periphery of the LH-1 ring [3]. The 2.5Å resolution structure of LH-2 from *Rps acidophila* contains a ring of nine B800 bacteriochlorophylls and a ring of 18 B850 bacteriochlorophylls in which each $\alpha\beta$ pair of B850's are more closely spaced (3.4 - 3.5 Å closest distance) than the nearest members of adjacent "pairs" (3.8 - 3.9 Å closest distance). Clearly strong exchange interactions [4] will exist at these separations. In LH-1 it is suspected that the intra- and inter-dimer separations are more distinct than in LH-2 [2], and thus a picture based on comparatively weakly interacting dimers may be reasonable in this case [5].

Returning to LH-2, the structure also reveals the vital light harvesting and protective roles of the carotenoid molecules. A single carotenoid molecule is present for each B800 and B850 "pair." The rhodopin-glucoside molecule is in van der Waals contact with the edge of the B800 ring and then snakes through the structure to lie across the face of the a B850 member of each $\alpha\beta$ pair. The carotenoid molecule serves both to quench triplet chlorophyll molecules and to enhance light harvesting in the blue-green region of the spectrum.

Understanding the functioning of LH-1 and LH-2 requires knowledge of the interactions between the chromophores and between the chromophores and the protein. In general, both types of interactions may be distributed, giving rise to both diagonal and off-diagonal disorder in the electronic states. The strength and timescale of coupling between electronic and nuclear degrees of freedom and between chromophore-localized nuclear motions and those of the protein must also be characterized. To this end, we have carried out two types of measurements: femtosecond fluorescence up-conversion measurements on LH-1 and LH-2, and transient grating and three-pulse photon echo measurements on LH-2. The latter measurements are described elsewhere [6].

P. Mathis (ed.), Photosynthesis: from Light to Biosphere, Vol. I, 23–28.

2. **Experimental**

Fluorescence up-conversion measurements were based on a Ti:sapphire oscillator (Coherent MIRA) and an elliptical reflector up-conversion apparatus [7]. The pulse duration is ca. 110 fs. For fluorescence anisotropy measurements, great care was taken to avoid contamination by up-converted pump light at short times.

3. **The Core Light Harvesting Complex (LH-1)**

Two types of LH-1 complex from *Rhodobacter sphaeroides (Rb. sphaeroides)* were studied -- detergent-isolated complexes and chromophores of the LH-1 only mutant M2192. Data on the isolated complexes and on M2192 excited at 860 nm is given in Ref. [5]. Figure 1 shows isotropic fluorescence data for M2192 excited at 870 nm and detected at 955 nm. Oscillations are clearly apparent in the fluorescence -- a fit indicates the dominant frequency is 105 cm^{-1}, which has a damping time of 500 fs and phase of 1.1 radians. There also appears to be a second oscillatory component with a frequency of ~20 cm^{-1}.

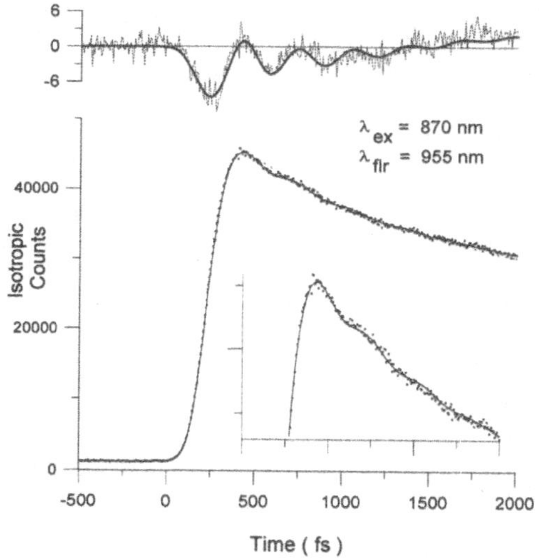

Figure 1. Time-resolved isotropic fluorescence from M2192 membranes excited at 870 nm and detected at 955 nm. The inset shows the oscillations in more detail. The residuals show the presence of a 105 cm^{-1} frequency.

Figure 2 shows parallel and perpendicular signals along with the raw anisotropy *(R(t)* [8]) constructed from these two signals. Particularly striking is that no oscillations are apparent in the anisotropy decay. Also shown is the raw anisotropy calculated from the convoluted fits to the parallel and perpendicular signals. The raw anisotropy differs slightly from the true anisotropy *r(t)* extracted from the data [8]. The final value of the anisotropy $(r(\infty) = 0.06)$ is consistent with earlier picosecond and steady state measurements [9]. The anisotropy decay can be well fit as a sum of two exponential components: $r(t) = r_1 \exp(-t/\tau_1) + r_2 \exp(-t/\tau_2) + r(\infty)$. The major component ($r_1 = 0.26$) has $\tau_1 = 110$ fs; the lesser component ($r_2 = 0.08$) has $\tau_2 = 440$ fs. Note that $r_1 + r_2 + r(\infty) = 0.4$.

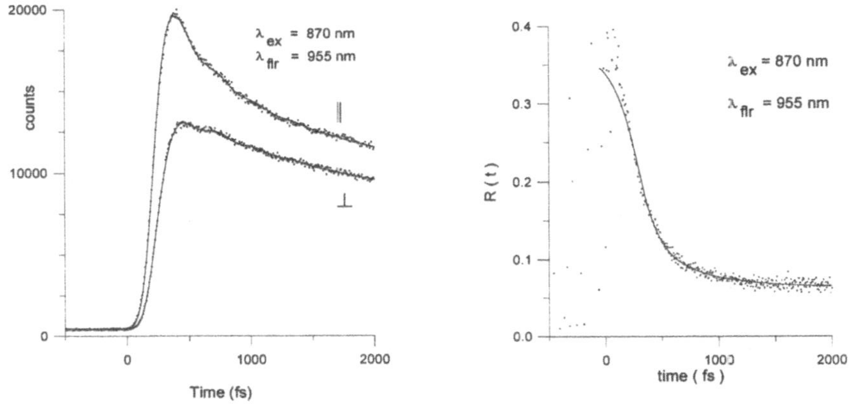

Figure 2. Left: Parallel- and perpendicularly-polarized fluorescence from M2192 membranes excited at 870 nm and detected at 955 nm. **Right:** Experimental raw anisotropy function *R(t)* (dots) and the same function derived from fitted convoluted parallel and perpendicular curves (solid line).

In order to interpret the data in Figures 1 and 2, a detailed model is necessary. No high resolution structure is available for LH-1; the 8.5Å projection map of *Rs. Rubrum* obtained by Karrasch et al. [2] has 16-fold symmetry with 32 pigments per ring, however, various groups have reported numbers ranging from 12 to 24 pigments for different organisms and preparations [10]. Elsewhere, we discuss several models for the LH-1 structure [5]. Here we focus on the Karrasch et al. [2] structure. The B875 chromophores are located on a circle of radius 48Å and are assumed to consist of 16 dimers. The inter-dimer distance is estimated to be 18.6Å. Förster transfer rates are calculated between dimers and disorder is incorporated by selecting each site energy from a Gaussian distribution by a Monte Carlo scheme. Each depolarization decay is calculated by averaging over 5,000 Monte Carlo iterations. Figure 3 shows the resulting calculated anisotropy for a purely homogeneous line shape (250 cm^{-1} fixed for each calculation) and two values of the inhomogeneous width (250 cm^{-1} and 125 cm^{-1}). The 250 cm^{-1} inhomogeneity curve is quite similar to the experimental data. The homogeneous hopping time in this model is 80 fs.

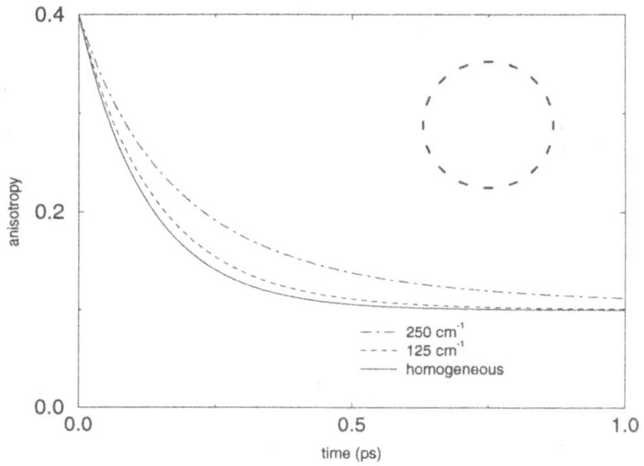

Figure 3. Simulation of anisotropy decay in an inhomogeneous ring consisting of 16 bacteriochlorophyll dimers. The homogeneous anisotropy decays as a single exponential with $\tau = 126$ fs. 125 cm^{-1} inhomogeneity generates a biexponential decay ($\tau_1 = 128$ fs [77%], $\tau_2 = 215$ fs [23%]); 250 cm^{-1} inhomogeneity increases the amplitude of the slow component in addition to lengthening it ($\tau_1 = 150$ fs [71%], $\tau_2 = 500$ fs [29%]). The latter parameters are similar to the experimental data in Figure 2. The entire absorption/emission spectrum is excited/monitored in this calculation.

Returning to the isotropic fluorescence, the 105 cm^{-1} frequency is in good agreement with the dominant frequency observed by Chachisvilis et al. [11] in pump-probe studies. Clearly, our results show the beats are excited state in origin and we suggest that the observation of such a similar frequency to that observed in the reaction center special pair suggests that B875 in LH-1 is dimeric. The damping time for the oscillation is significantly longer than the anisotropy decay. For the model described above and for several related models [5], this implies that the vibrational coherence survives several hops between dimers. Calculations based on Redfield theory suggest that this is indeed possible [12].

Of course, the interpretation given above may be oversimplified. Electronic coupling may be large enough, such that extensive delocalization and motional narrowing take place even at room temperature. Electronic structure calculations and nonlinear spectroscopy applied to B875 may help to clarify which level of description is most appropriate.

4. The Peripheral Antenna Complex (LH-2)

Similar fluorescence studies were made of isolated LH-2 complexes from *Rb. sphaeroides* wild-type in LDAO detergent. Excitation of B850 was at 850 nm and detection at 950 nm. Oscillations are not evident in the magic angle fluorescence detected at room temperature. This result is consistent with the pump-probe data of Chachisvilis et al. [11]. Figure 4 shows the anisotropy decay resulting from B850→B850 transfer and the associated parallel and perpendicular curves.

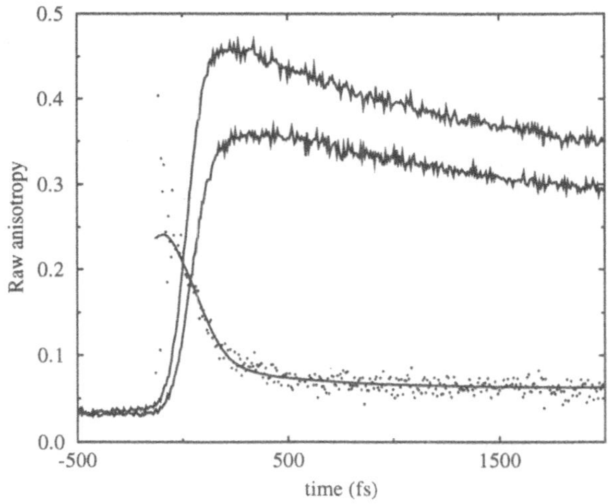

Figure 4. Parallel and perpendicularly-polarized emission from detergent-isolated LH-2 (LDAO) and anisotropy decay. Excitation 850 nm (bandwidth 11 nm), detection 950 nm. The solid line is the reconstructed raw anisotropy from a fit (including convolution) to the parallel and perpendicular data. The fit parameters are given in the text.

The anisotropy decay is about twice as fast as in LH-1 and fits to a sum of two exponential components and an asymptotic value of $r(\infty) = 0.05$. The exponentials are 50 fs ($r_1 = 0.21$) and 517 fs ($r_2 = 0.04$). The r(0) value is 0.31, rather than the expected value of 0.4; this probably reflects limited time resolution. The decreased time scale of energy transfer reflects the smaller size (25Å radius vs. 48Å for LH-1) of the LH-2 B850 ring and the large angle between neighboring pigments. In view of the very close spacing of the B850 molecules (see Introduction), it seems unlikely that a description based solely on a Förster mechanism will prove adequate. We have not therefore attempted to model the data in a similar way to the LH-1 data.

The final values of the anisotropy are lower than 0.1 (expected for polarization randomization in a plane in a homogeneous system) for both LH-1 and LH-2. This is consistent with a small inhomogeneous distribution combined with the effect of spectral selection in the detection step. Simulations show a systematic drop in $r(\infty)$, as the detection window is moved towards the red edge of the steady state fluorescence spectrum. In an inhomogeneous antenna, equilibration produces a nonuniform final distribution; detection removed from the excitation wavelength, on average, excludes the initially-excited molecule and consequently lowers the final anisotropy. Experimentally we observe, in both LH-1 and LH-2, an equivalent effect when the excitation wavelength is moved from red to blue, and the detection wavelength remains fixed, the $r(\infty)$ value decreases [5].

Acknowledgment

This work was supported by a grant from the National Science Foundation.

References

1. McDermott, G., Prince, S.M., Freer, A.A., Hawthornthwaite-Lawless, A.M., Papiz, M.Z., Cogdell, R.J. and Isaacs, N.W. (1995) Nature 374, 517-521.
2. Karrasch, S., Bullough, P.A. and Ghosh, R. (1995) EMBO J. 14, 631-638.
3. Kuhlbrandt, W. (1995) Nature 374, 497.
4. Dexter, D.L. (1953) J. Chem. Phys. 21, 834-850.
5. Bradforth, S.E., Jimenez, R., van Mourik, F., van Grondelle, R. and Fleming, G.R. (1995) J. Phys. Chem (submitted).
6. Joo, T., Jia, Y. D. Jonas, and Fleming, G.R. (1995) J. Phys. Chem. (submitted).
7. Xie, X. Du, M., Mets, L. and Fleming. G.R. (1992) Proc. Intl. Soc. Optical. Eng. (SPIE, Bellingham, WA), p.1640.
8. Cross, A.J. and Fleming, G.R. (1984) Biophys. J. 46, 45-56.
9. Bergstrom, H. Westerhuis, W.H.J., Sundström, V., van Grondelle, R., Niederman, R.A. and Gillbro, T. (1988) FEBS Lett. 233, 12-16; Kramer, H.J.M., Pennoyer, J.D., van Grondelle, R., Westerhuis, W.H.J., Niederman, R.A. and Amesz, J. (1984) Biochim. Biophys. Acta 767, 335-344; van Dorssen, R.J., Hunter, C.N., van Grondelle, R., Korenhof, A.H. and Amesz, J. (1988) Biochim. Biophys. Acta 932, 179-188.
10. Meckenstock, R.U., Krusche, K. Staehlin, L.A., Cyrklaff, M. and Zuber, H. (1994) Biol. Chem. Hoppe-Seyler 375, 429-438; Boonstra A.F., Visschers, R.W., Calkoen, F., van Grondelle, R., van Bruggen, E.F.J. and Boekema, E.J. (1993) Biochim. Biophys. Acta 1142, 181-188.
11. Chachisvilis, M. Pullerits, T., Jones, M.R., Hunter, C.N. and Sundström, V. (1994) Chem. Phys. Lett. 224, 345-351.
12. Jean, J.M. and Fleming, G.R. (1995) J. Chem. Phys. (in press).

AMINO ACID SEQUENCES OF THE LIGHT-HARVESTING PROTEINS OF THE DINOFLAGELLATE *AMPHIDINIUM CARTERAE*

R.G.Hiller, P.M.Wrench and F.P.Sharples. School of Biological Sciences, Macquarie University, NSW, Australia 2109

Introduction

Although the photosynthetic pigments of dinoflagellates have counterparts from most other eukaryotic algal groups, those dinoflagellates which contain peridinin as their principle carotenoid appear to share a common evolutionary origin. Their light-harvesting apparatus has peridinin as the main pigment absorbing in the region from 470nm to 550nm. The peridinin is bound to two distinct proteins; these are the water soluble peridinin-chlorophyll *a* protein (PCP) [1-7] and the intrinsic chlorophyll *a-c*-caroteno-protein (Cac) [8-10]. Until recently, sequence data were not available for either of these proteins. It is now clear, that while PCP has no obvious affinities [11,12] with any other protein, the Cac protein [8,10] is a member of the universal family of pigment-binding proteins having three transmembrane helices, typified by the Cab proteins of higher plants and the Fucoxanthin-chlorophyll-proteins (Fcps) of diatoms [13]. A striking feature of the nuclear-encoded light-harvesting protein genes is that they occur in families and most if not all are expressed. We show here that PCP is no exception but that at least one PCP gene product, present in a low amount, is only 31% identical to that of the more abundant forms which are highly conserved across genera. Variation in amino acid sequence was also detected in Cac proteins.This arises, at least in part, from a different mechanism. cDNA sequencing suggests the Cac proteins are synthesised as a polyprotein which is cleaved at specific arginine residues into different mature proteins [10].

Materials and methods

Amphidinium carterae CS21 was grown to late logphase in Provasoli's enriched seawater at 18°C. Isolation of DNA and RNA, DNA sequencing and conditions for PCR was as previously described [10]. PCP was purified by $(NH_4)_2SO_4$ precitation (70%-90%) and cation exchange chromatography using 5 mM acetate buffer pH 5.1 and a gradient of 0-0.3M NaCl. Fractions [14] eluting between 0.05M and 0.15M NaCl and having an A 478nm/A 280nm ratio > 4.5 were designated 'mainform' PCP (MFPCP) and the fraction eluting at 0.25M NaCl was designated 'High Salt PCP' (HSPCP). Protein sequencing was by automated Edman degradation .

P. Mathis (ed.), Photosynthesis: from Light to Biosphere, Vol. I, 29–34.

Results and discussion

The evidence for the Cac proteins being putative polyproteins was obtained by a) RT-PCR and b) Northern blotting. Degenerate primers, designed on the basis of protein sequence data, generated an amplified DNA of much greater length than expected. This amplified DNA encoded a complete mature Cac protein sequence together with a C-terminal and an N-terminal part of two other Cac proteins. In a further RT-PCR experiment an oligonucleotide encoding amino acids adjacent to the C-terminus was used together with a poly-T primer to amplify DNA which encoded the C-terminus of a mature protein and a complete Cac protein. Successive Cac protein sequences contained no spacer polypeptide but, as deduced from N-terminal protein sequencing, the polyprotein is cleaved at C-terminal arginine residues. Northern blotting showed that Cac mRNA is 6.1kb with some minor hybridisation to species of 2.4kb and 1.5kb whereas the expected size was ~1kb if the 173 residue Cac protein was imported with a normal N-terminal chloroplast-directing leader. A replicate northern blot was probed with PCP and a single band was detected at 1.7kb vs an expected size of ~1.5kb. The sequences of Cac proteins obtained so far are shown in Fig 1.

```
   1  S                              Q                 50        S
a  XFENERGVQD PVGFFDPLGF TADGSVENFK RLAQTEIKHG RVAMLATMGY ITPEITGKLP
b  a--s-l---a -t-------l ss----d--- -rras----- ---------- m-----a-f-
c  a----l--aa -t--w----l ak---mka-- -rras----- -i-------- ----l---f-
d                 ------ -ke-n-d--- -rra------ -i-------- ----l#frfd

                                                      I
                                           100        S       A
a  GYLSPSTGVK YDDIPNGLGA ISKVPAAGWG QMIAYAAFSE LSQDQSAGTP AAEGDFGFKV
b  ----y-qss- fa-v----a- m----vl--a -va--g-vcd v--------- g-a-------
c  ------m-l- -e-------- ------v--a -il---fyc- q-----e-sa gea-------
d  ----r-a--- -a------a- f----fv--v --v--cllv- i-----q-a- -q#-------

               G   A
               A  PI     P      150                   170
a  LTSSDPAELE KKLSAELANG RLAMMAIIGM FFQDGLTGSA WGDWASYTAS PL
b  i--e-eet-k r--ns----- ---------l --------g- y----n--d- --r
c  ---k-eeg-- r--ns-i--- ---------- ---------- y----nf--- --r*
d  --------k- ---r------ ---------- ---------- -----n--e- --r
```

Fig. 1 Amino acid sequences of mature Cac proteins. a, protein-derived sequence with major variants in upper case above; b and c, complete cDNA-derived sequences; d, partial cDNA-derived sequence. (- identity, # a gap, * stop codon).

The DNA-derived sequences are ~70% identical to that obtained by protein sequencing and to each other. Their identity to other Cab proteins is 20%-30% (excluding gaps) with most being at the lower end of this range. A higher identity (39%) is seen with Fcps with very strong identity at the N-terminus. Despite the low overall identity with Cab proteins, it is stronger in regions which could be part of the first and third transmembrane helices. If we assume that the histidine of the sequence ...**KHGR**... binds a conserved chlorophyll *a* molecule in the middle of the first helix then, as shown in Fig 2, there are conserved binding sites for five chlorophyll *a* molecules [15]. Cac and Fcp proteins contain 70-80 less residues than do the Cab proteins. This difference is accounted for in the shortening of the regions N-terminal to the first transmembrane helix and between helices two and three. By analogy with PCP, peridinins would be inserted into the protein and extend to the periphery, some in excitonic pairs as shown by the CD spectra [8,17].

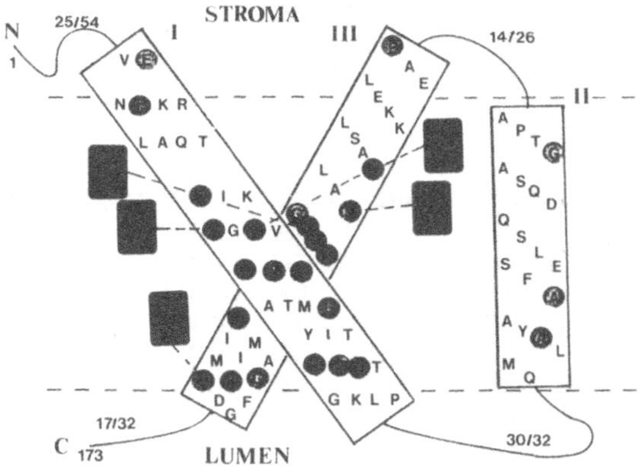

Fig. 2 Sequence of the Cac protein superimposed on the three transmembrane model of LHCP [15]. Circles show identical residues and rectangles chlorophyll *a*. Numbers show residues outside the helices - Cac / LHCP

Under our ion-exchange conditions a considerable portion of the PCP washes through and is composed of PCP forms of lower pI. The fraction taken for protein sequencing and crystallisation (MFPCP) [14] also contains more than one isoform with that of pI 7.6 predominating. All these forms have similar absorbance and fluorescence spectra, an apoprotein of 32kDa and almost identical N-terminal sequences. A small amount of pigmented material is removed from the column at much higher salt concentrations (HSPCP). This has an apoprotein of 34kDa and antibodies to MFPCP react only very weakly with HSPCP. MFPCP and HSPCP can be further distinguished on the basis of fluorescence emission, CD and pigment composition (Figs 3 and 4).

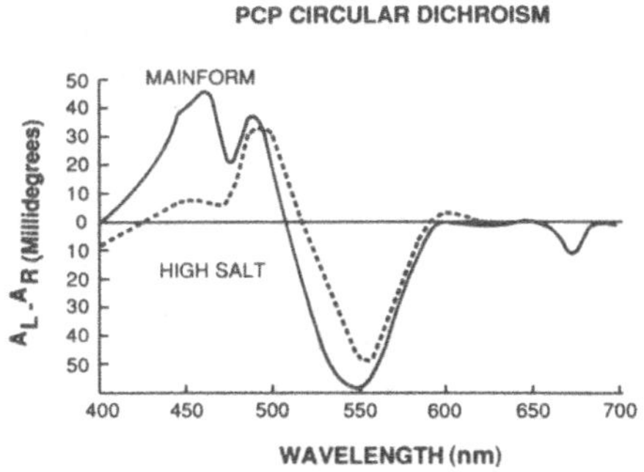

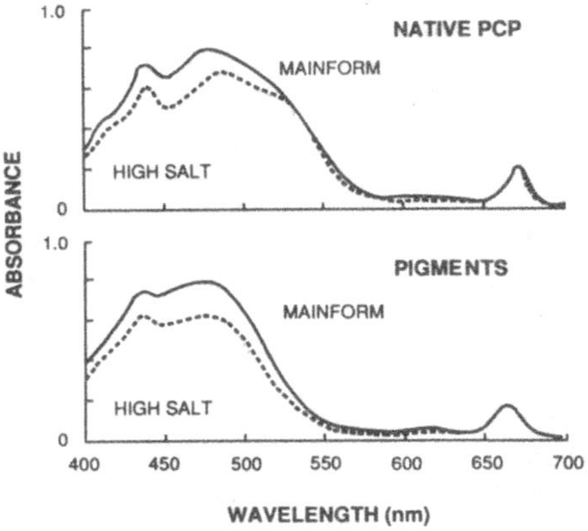

Fig. 3 Circular dichroism spectra of MFPCP and HSPCP
Fig. 4 Absorbance spectra of native PCPs and their extracted pigments.

When the pigments are removed from MFPCP and HSPCP and compared on the basis of equal chlorophyll *a* absorbance, there is is a deficiency in peridinin in HSPCP. Fig 4 indicates that HSPCP has only 6 peridinins instead of 8. This difference might be expected to be reflected in a reduction in peridinin peak height in the CD spectra but the main feature of Fig 4 is the shift of isobestic point from 479nm to 492nm in HSPCP. There are also differences in the spectra in the Soret region. The primary structure of both forms of PCP was determined by a combination of direct protein sequencing and genomic PCR (Fig 5).

```
  1 DEIGDAAKKLGDASYAFAKEVDWNN....GIFLQAPGKLQPLEALKAIDK 46
  1 *G*A**S**FS**T*PI*EKF**GGSSAVAKYIADASASN*RQ*AL*VE* 50
 47 MIVMGAAADPKLLKAAAEAHHKAIG.SISGPNGVTSRADWDNVNAALGRV 95
 51 LLET*LTM****VR**VA**S**LDTAV*N*KL*A*KE*FAA**E**A*M 100
 96 IASVPENMVMDVYDSVSKITDPKVPAYMKSLVNGADAEKAYEGFLAFKDV 145
101 ***ADKQKFAALRTAFPESRELQSSLFAGN..**YE*****DS*K*LTSA 148
146 VKKSQVTSAAGPATV........PSGDKIGVAAQQLSEASYPFLKEIDWL 187
149 *RDASING*NA*VIAEAARSERYVPDGPV*R**KKF***T**IMEKLN*V 198
188 .SDVYMKPLPGVSAQQS...LKAIDKMIVMGAQADGNLLKAAAEAHHKAI 233
199 K*PEIS*Y*ATA*SKDPKMMAPG***TLEVALTMNQ**INN*VY**VR** 248
234 .GSIDATG.VTSVADYAAVNAALGRVIASVPKGTVMDVYNAMAGVTDTSI 281
249 K*ALNTP*F*AERD*F*R**L**AKM*G*ADPAKFKALLT*FP*NA*..L 296
282 PLNMFSKVNPLDANAAAKAFYTFKD.VVQAAQR 313
297 QMAL*A.A**EQ*K**YET*VALTSA** 323
```

Fig. 5 Alignment of MFPCP (**bold** type) and HSPCP sequences. *, identity; . a space, putative chlorophyll binding sites (**H**) are underlined.

The overall identity is only 31% but both proteins possess the same 2-domain structure. Each domain contains a putative chlorophyll binding site, a histidine doublet in MFPCP and a single histidine in HSPCP. The difference in molecular mass seen on SDS-PAGE is partly accounted for by an insertion of 8 aminoacids in the region joining the 2 domains.This insert contains 2 arginine residues, one or both of which are accessible to a specific protease, since on prolonged storage HSPCP is cleaved into two polypeptides whose sequences correspond to the the N-terminii of the 2 domains. All forms of MFPCP were found to have similar N-terminal sequences and we obtained 2 additional complete sequences by RT- and genomic PCR. These are shown in Fig 6 where they are aligned with other complete and partial sequences. The identity between all MFPCPs is very high, between the *Amphidinium* sequences and that of *Symbiodinium* (and the N-terminus of *Alexandrium*) it is ~85% (apoproteins 32kDa) and between the C-terminal domain of *Amphidinium* and *Heterocapsa* (apoprotein 16kDa) it is 59%. Even in the latter case it is much greater than the identity between MFPCP and HSPCP of *Amphidinium.* Our sequence data provide clear evidence that much of the variation in pI of MFPCPs is genomic in origin and we have further shown by genomic PCR that at least 2 MFPCP genes are arranged tandemly, separated by 490bp. There is some disagreement as to the mol ratio of peridinin to chlorophyll *a.* For *Amphidinium* [1] it was found to be 9:2 whereas for *Heterocapsa* it was 4:1 [4] (8:2 for the dimer [6]). We have been unable to detect any difference between these two organisms with respect to their pigment composition and based on the X-ray-derived structure are convinced that the 8:2 ratio is correct. It has been proposed that *Alexandrium* [16] has a unique PCP with 12 peridinin:2 chlorophyll *a*. This requires re-evaluation as the sequence, at least at the N-terminus, is almost identical to that of *Amphidinium* and *Symbiodinium*, the apoprotein is 32kDa and the native absorbance spectrum has an identical A478nm : A670nm ratio (4.1:1) to that of *Amphidinium* MFPCP and several other dinoflagellates [6].

```
   1                                                                ↓60
a  MVRSGKKAVV LAAVAFCATS VVQKSHGFVP SPLRQRAAAA GAAAASAATM FAPAAFADEI
b  ********** **T******* ****TC**** ********** *-****V*** **********
c  ***GAR**IA VG-**VAVAC GL**HLN*** G*RH----** PV***A*SM* M*********
d  .......... .......... .......... .......... .......... .......***
e  ...........................................................***
   61                                                             120
a  GDAAKKLGDA SYAFAKEVDW NNGIFLQAPG KLQPLEALKA IDKMIVMGAA ADPKLLKAAA
b  *****T**** ********** ********** *****A**** ********** ***Q******
c  ********** **S******* ********** *F******** *****E**** *******D**
d  ********** ********** ********** ********** ********** **********
e  ********** *********X ********** *F***X***X *-*HX*
   121                                                            180
a  EAHHKAIGSV SGPNGVTSRA DWDNVNAALG RVIASVPENM VMDVYDSVSK ITDPKVPAYM
b  *******TT* **A******* ********** ***SA***K* *****N**A* **********
c  *********I ********** ***A****I* **V****KAK **A*****TA ****G*****
d  *********I ********** ********** ********** ********** **********
  181                                                  240
a  KSLVSGADAE KAYEGFLAFK DVVKKSQVTS AAGPATVPSG DKIGVAAQQL SEASYPFLKE
b  N***N***** ***A****** *****N**** ********** *T******K* *D*******Q
c  ****N*P*** ***Q***E** ***E*N**AT *SA**V**** ****E**KA* *D*****I*D
d  ********** ********** ********** ********** ********** **********
f  .......... ..........    .......   ........→ ***AD**KK* **********
   241                                                            300
a  IDWLSDVYMK PLPGVSAQQS LKAIDKMIVM GAQADGNALK AAAEAHHKAI GSIDATGVTS
b  ********L* ********** *****R**** ********** ********** *****K****
c  ******I*L* ****KT*PET ********** **KM***L** ********** **********
d  ********** ********** ********** *******L** ********** **********
f  ***T*****A* -*XTANPFDV ********** **AM*SA*** *GL******* *****K***T
   301                                                            360
a  AADYAAVNAA LGRVIASVPK STVMDVYNAM AGA-TDTSIP LNMFSKVNPL DANAAAKAFY
b  *****E**** I***V***** *******K** *SV-***G** ********** **********
c  ****E***** I**LV***** T*******S* **V-V*S*V* N*L******* **V****G**
d  V********* ********** G********* **V-****** ********** **********
f  L***E***** I*HMV**AGM *KT******L **FNLGKDVG PY*M****AA **S**YS*LL
   361
a  TFKDVVQAAQ R
b  ******L*** *
c  ******E*S* *
d  ********** *
```

Fig. 6 Alignment of amino acid sequences of MFPCPs. a,b,d, *Amphidinium;* c *Symbiodinium* [12]; e, N-terminus of *Alexandrium* [16]; f, *Heterocapsa,* **bold** type N-terminal sequence [Pisano and Hiller unpublished], ordinary type [11]. This *Heterocapsa* sequence has been left incomplete as ref [11] has a reading frame error close to the C-terminus. Arrows show the start of the mature proteins. Chlorophyll-binding Histidine residues are boxed.

A major unsolved problem is the location of PCP in relation to the remainder of the photosynthetic apparatus. It has been suggested [12] on the basis of the leader sequence that it may be lumenally located although there is little indication that the thylakoid lumen is distended by lumenally located proteins as it is in cryptophytes .
We speculate in view of its low amount, red shifted fluorescence and its unique primary sequence that HSPCP may function as a linker between MFPCP and the intrinsic light-harvesting components and are actively pursuing its location.

References

1) Haxo,F.T., Kycia,J.H., Somers,G.F., Bennet,A and Siegelman,H.W. (1976) Plant Physiol. 57, 297-303.
3) Chang,S.S. and Trench,R.K. (1984) Proc.Roy.Soc.Lond.B 222, 259-271.
4) Prezelin, B.B. (1987) In Botanical monographs, vol.21 The Biology of Dinoflagellates (ed. Taylor, F.J.R.), pp 174-223 Blackwell, Oxford.
5) Govind,N.S., Roman,S.J.,Inglesias-Prieto,R.,Trench,R.K., Triplett,E.L. and Prezelin,B.B. (1990) Proc. Roy. Soc. Lond.B. 240, 187-195.
6) Inglesias-Prieto,R.,Govind,N.S. and Trench,R.K. (1991) Proc. Roy. Soc. Lond.B. 46, 275-283.
7) Jovine,R.V.M.,Triplett,E.L., Nelson,N.B.and Prezelin,B.B. (1992) Plant Cell Physiol. 33, 733-741.
8) Hiller,R.G., Wrench,P.M., Gooley,A.A., Shoebridge,G.and Breton,J. (1993) Photochem. Photobiol.57, 125-131.
9) Inglesias-Prieto,R., Govind,N.S. and Trench,R.K. (1993) Phil.Trans.R.Soc.Lond.B 340, 381-392.
10) Hiller,R.G., Wrench,P.M and Sharples, F.P. (1995) FEBS Lett.363, 175-178.
11) Triplettt,E.L., Jovine,R.V.M., Govind,N.S., Roman,S.J. Chang,S.S. and Prezelin,B.B. (1993) Mol. Mar. Biol. Biotech. 2, 246-254.
12) Norris,B.J. and Miller,D.J. (1994) Plant Mol. Biol. 24, 673-677.
13) Green,B.R. and Pichersky,E. (1994) Photosynthesis Res. 39, 149-162.
14) Steck,K., Wacker,T., Welte.W., Sharples,F.P. and Hiller, R.G. (1990) FEBS Lett. 268, 48-50.
15) Kuhlbrandt,W., Wang,D.N. and Fujioshi,Y. (1994) Nature 367, 614-621.
16) Ogata,T., Kodama,M., Nomura,S., Kobayashi,M., Nozawa,T., Katoh,T. and Mimuro,M. (1994) FEBS Lett. 356, 367-371.
17) Koka, P. and Song, P-S. (1977) Biochim. Biophys. Acta 495, 220-231.

Ultrafast spectroscopy of the light-harvesting complex and the isolated reaction center from photosystem II

Alfred R. Holzwarth
Max-Planck-Institut für Strahlenchemie, D-45470 Mülheim a.d. Ruhr, Germany

Photosystem (PS) II is the supramolecular multiprotein pigment complex that performs water splitting in organisms active in oxygenic photosynthesis. Its peripheral antenna complex, the so-called light-harvesting complex (LHC) II, binds about 50% of all chlorophyll (Chl) in the biosphere. LHC II is responsible for a variety of fundamental structural, functional and regulational tasks in the thylakoid membrane. Its prime function is the absorption of solar energy and the efficient and rapid transfer to the PS II core and reaction center (RC). However, that is by far not the only task. It is well-known now, although poorly understood, that LHC II also plays a major function in the overall organization of the thylakoid membrane of higher plants and is responsible to a large part for the lateral segregation of PS II and PS I which is essential for the functioning of efficient photo-synthesis. Furthermore there exists overwhelming evidence that LHC II also is involved in the physiological adaptation processes related to the regulation of the energy distribution and usage in and between PS II and PS I. Thus state transitions, involving the phosphorylation of LHC II, and so-called q_E-quenching, related to the xanthophyll cycle pigments present in the LHC complex, also involve the intimate participation of LHC II. Thus understanding the structure and function of LHC II on a molecular level is at the heart of understanding the early reactions in higher plant photosynthesis as well as their regulation. Substantial progress has been achieved recently in the elucidation of the structure of LHC which has been determined to a resolution of about 3.4 Å [1]. This in principle should provide a basis to study and interpret for the first time in

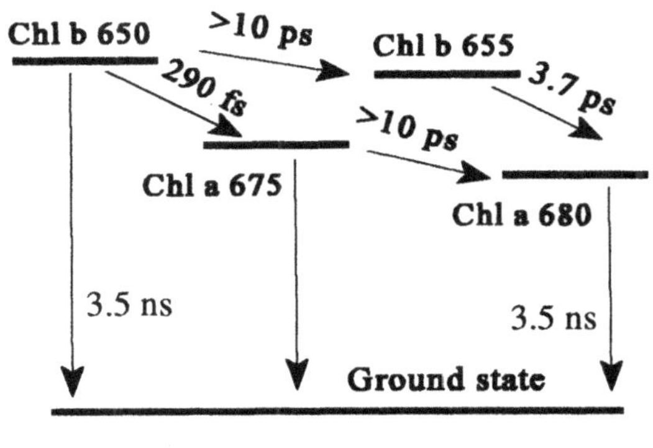

Fig. 1: Minimal kinetic model for the energy transfer steps in the trimeric LHC II complex. The Chl xxx denote spectrally and kinetically distinguishable Chl a and b pools.

P. Mathis (ed.), Photosynthesis: from Light to Biosphere, Vol. I, 35–40.

detail the structure-function relationships in the most important antenna complex from higher plant.

Likewise, since its first isolation substantial progress has been made also in understanding the structure-function relationship in the RC complex of PS II, the so-called D1-D2-cy-b559 complex [2]. Unfortunately so far no high resolution structure has become available of this complex, but a great deal of insight has been obtained by analogy to the highly resolved structure of the purple bacterial RC complex[3]. It is now commonly agreed that the D1-D2 complex in general contains at least two additional Chls as compared to the pigment content of the purple bacterial RC. Thus the Chl/2 Pheo ratio is ≥6. According to current hypothesis, which is quite reasonably supported by a large variety of data, there exists an RC core which is similar to the purple bacterial RC in terms of pigment content and relative arrangement of cofactors, although the exact details of relative orientation and distances remain to be established. This core fairly tightly binds at its periphery at least two additional Chl a. Besides the pressing problem of a detailed structure determination, at present the most important questions with regard to the PS II RC relate to problems like the detailed spectral assignment of the various cofactors in the unstructured absorption, the kinetics and energetics of primary charge separation and the problem whether the isolated D1-D2-cyt-b559 complex indeed possesses the essential properties of the intact complex.

In the following an overview on very recent data on both the LHC II and the D1-D2 complexes is given which aims at the elucidation of some of the questions and problems mentioned above.

Experimental

LHC II has been isolated from spinach leaves as described [4] and has been studied as trimeric complexes. The D1-D2-cyt-b559 complexes have been isolated according to ref. [5]. All measurements were carried out at 4 °C for the D1-D2 complexes and at r.t. for the LHC II complex. Annihilation is found to occur at very low excitation intensity in both the LHC II complex and in the D1-D2 complex. The kinetics shown was obtained at sufficiently low pulse intensity to essentially avoid annihilation effects.

LHC II

In order to study the function of LHC II in the so-called q_E-quenching, the fluorescence kinetics and spectra of isolated LHC II have been measured under various conditions [6]. The fluorescence kinetics at r.t. of LHC II isolated from dark-adapted spinach leaves, when excited into the Chl b absorption band at 650 nm, is fairly simple (data not shown) [6]. The essential components are a ≈4 ps component whose spectrum shows the typical shape of an energy transfer component along with a long-lived ≈3.5 ns component whose emission peaks at 685 nm. The fast component arises from the energy transfer from (part of) the Chl b chromophores to (part of) the Chl a. Neither the kinetics nor the spectra change in LHC II complexes isolated from light-adapted leaves which contains a high amount of the xanthophyll cyle pigment zeaxanthin. Thus the mere presence of zeaxanthin does not quench excited states in LHC II. Substantially shorter lifetimes (the longest lifetime was only 1.4 ns) and more complex kinetics were obtained in aggregated LHC II. The fastest component still remained an energy transfer component of about 3-4 ps. In order to study in more detail the energy transfer function of LHC II we performed a femtosecond transient absorption study upon specific excitation into the Chl b absorption band in the range of 645 and 650 nm. Quantitative analysis of the annihilation curve and kinetics (not shown) is in agreement with the notion that the excitation energy is rapidly ($t \leq 20$ ps) distributed over the entire trimeric complex. Excitation at 650 nm and at sufficiently low intensity avoiding annihilation effects revealed two energy transfer components of 290 fs and 3.7 ps (see Hucke et al., these Proceedings). The latter lifetime is in good agreement with the fluorescence energy transfer component. Analysis leads to the minimal kinetic model shown in Fig. 1. Two spectral pools each of Chl b and Chl a can be distinguished which transfer energy with these two fast lifetimes. The equilibration among the two Chl a pools is not directly observed as an additional kientic component in the data but based on kinetic modelling it is estimated to occur on a time scale of about 10 ps. It is important to note that changing the excitation wavelength in the 640 to 655 nm range has some significant influence on the observed fast lifetimes (not shown). This observation indicates that the Chl b pool is highly spectrally and kinetically heterogeneous and that specific Chl b pigments transfer energy to spectrally distinct specific Chl a pigments before equilibration within the Chl b pool occurs, in agreement with the minimal kinetic model in Fig. 1. In the LHC II structure so far only 12 Chls have been located, notably three less than are supposed to be present in the complex based on biochemical data. This discrepancy, along with the so far unidentified assignment of the resolved Chls to Chl a and Chl b, at present demands extreme caution in any interpretation of the kinetics in terms of the structural data. It follows that the real kinetics of energy transfer will be much more complex than revealed so far which is not surprising given the large number of chromophores and the virtually absent symmetry in the monomeric complex. Unfortunately without additional knowledge on the structural location and spectral positions of the various Chl b and Chl a pigments any assignment of kinetic components to specific energy transfer steps between Chls would be highly speculative and still does not appear to be possible at present.

In view of the absence of any effect of zeaxanthin on the excited state decay in the isolated LHC II complex, any studies aiming to elucidate the postulated role of LHC II (and/or the minor LHC complexes) in the q_E-quenching mechanism have to resort to the use of intact leaves or thylakoids treated appropriately in order to establish the various quenching states. We have chosen to study the quenching effects by time-resolved fluorescence on thylakoids after appropriate treatment of intact chloroplasts. The main aim was to differentiate between the effects of zeaxanthin formation on the one hand and a trans-thylakoid ΔpH gradient on the other hand on the rate constants of the molecular processes located in the antenna (LHC II) and/or in the RC [7](manuscript in preparation). The fluorescence kinetics was measured under the different conditions of control, ΔpH alone, ΔpH+zeaxanthin, and zeaxanthin alone. The kinetics of the PS II Chl fluorescence under these conditions was analyzed in terms of the exciton/radical pair equilibrium model [8]. It follows that in agreement with isolated LHC II zeaxanthin formation in the antenna in the absence of a ΔpH gradient has no effect on the energy quenching. Only in the presence of zeaxanthin and a ΔpH is the rate constant for antenna decay (quenching) increased from its original value of 0.3 ns^{-1} by a factor of ≈2.5. Quite notably a ΔpH alone has a significant effect on the rate of decay of the primary radical pair in the RC (rate constant k_2 in the kinetic model [8]). This effect substantially influences the PS II fluorescence yield and thus the fluorescence induction when the RC is closed but it has only a minor effect with open RCs. The data clearly show that q_E-quenching is associated with zeaxanthin formation only in the presence of ΔpH and that the quenching is indeed located in the antenna and not in the RC. So far it is not known whether and in which way the presence of a ΔpH modifies the zeaxanthin structure, location, or photophysics within the LHC complex such that this efficient quenching is switched on.

D1-D2-cyt-b559

In contrast to bacterial RCs, the absorption spectrum of the D1-D2-cyt *b*-559 complex is dominated by a strong spectral congestion of the Q_y-bands of all pigments, in combination with a substantial inhomogeneous broadening of pigment absorption bands which is of the same order as the spectral separation of the individual pigments. These two factors seriously hamper all optical spectroscopic studies and provide severe difficulties for the elucidation of the energy level ordering in PS II RCs. In a recent study Konermann et al. presented a detailed analysis and interpretation of the D1-D2 absorption spectrum, based on calculated line shapes of individual pigments, taking into account electron-phonon coupling and inhomogeneous broade-ning [9]. This analysis indicates that besides the pigments of the RC core, there exist two or three additional Chls bound to the periphery of the RC core. The additional Chls will be called here "blue", "green" and "red", referring to their spectral positions on the blue, center, or red spectral wing of the P680 absorption band. This notion is well supported by picosecond fluroescence studies on the D1-D2 complex where these additional Chls were resolved by their different kinetics and spectra. The additional Chls transfer their energy relatively slowly with time constants in the range of 7 ("blue" Chl) to 30 ps into the core[10] whereas energy transfer within the core occurs on the subpicosecond time scale[11,12].

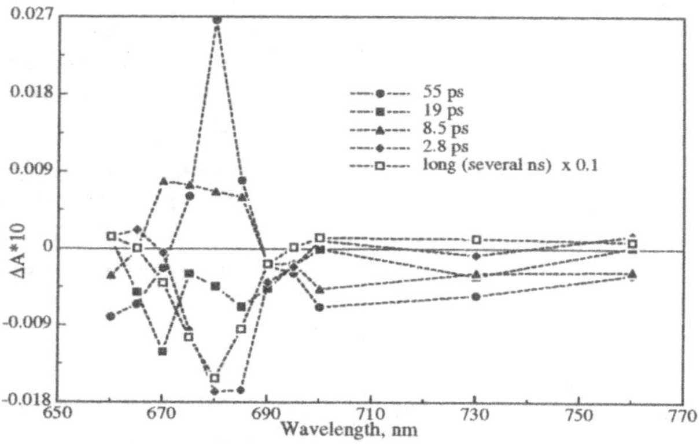

Fig. 2: DADS and lifetimes of the D1-D2 kinetics measured by femtosecond transient absorption and analyzed by global analysis. λ_{exc}=680 nm.

The kinetics of charge separation in the D1-D2 complex has been discussed highly controversially over the past years. Thus lifetimes of 3 ps [11] or 21 ps [13] have been suggested for the *apparent* time constant of charge separation, i.e that time constant with which the major part of radical pair formation is associated. In our view this discrepancy has been caused in the past to a large extent by the lack of explicit kinetic models suitable to rationalize the experimentally observed fluorescence and/or transient absorption kinetics.

We present here some results from femtosecond transient absorption spectroscopy which are suitable to elucidate this problem [14]. Fig. 2 shows the decay-associated difference spectra (DADS) measured under conditions where exciton annihilation can be ignored. Global analysis of decays measured in the range of 650 to 760 nm (Fig. 2) and 535-555 nm (not shown) over a time range of ≥200 ps revealed a complex kinetics with five resolved lifetimes. In view of the complex structure of the D1-D2 RC it is not possible to directly interpret these lifetimes in terms of specific processes in the RC. Rather a kinetic model to rationalize the data is required. The results of such kinetic modelling are shown in Fig. 3. In these data the kinetics of the additional Chls is not fully resolved so far. In fluorescence kinetics three different additional Chls have been found whereas from the transient absorption data only two pools have been resolved so far, i.e. the "blue" Chl and presumably a mixture only of the "green" and "red" Chls. Thus the kinetic scheme shown in Fig. 3 presents only a minimal kinetic model which needs to be refined in the future. The "blue" Chl shows maximal bleaching near 670 nm whereas the difference spectrum of the "green/red" Chls peaks near 680 nm. Within the model the *apparent* lifetime for charge separation is 3 ps. In this model charge separation occurs with an *apparent* rate constant of ≈150-170 ns^{-1}. Note that this is not the *intrinsic* rate constant of charge separation which is defined as the rate constant for electron transfer if the excitation is actually located on P680. The connection between k_{app} and k_{int} is given by the relationship

$$k_{int} = k_{app} / p$$

where p denotes the probability that the excitation is actually located on the RC donor (P680). The value of p can be estimated from the spectral analysis to about 0.25-0.4 and using the kinetic data the rate constant k_{int} is found to be in the range of 250-500 ns^{-1} corresponding to an *intrinsic* charge separation lifetime of 2-5 ps.

Recent detailed kinetic analysis of picosecond fluorescence data analogous to the analysis of the transient absorption data shown (Fig. 3) also confirmed that the *apparent* time constant for charge separation is indeed 3 ps near room temperature[10] in agreement with our previous suggestion[11]. From the kinetic models on the fluorescence data an apparent rate constant for charge separtion in the range of 130-170 ns^{-1} has been obtained. Thus for the first time both transient absorption and fluorescence kinetics give a consistent picture of the primary processes in the D1-D2 complex. An apparent lifetime

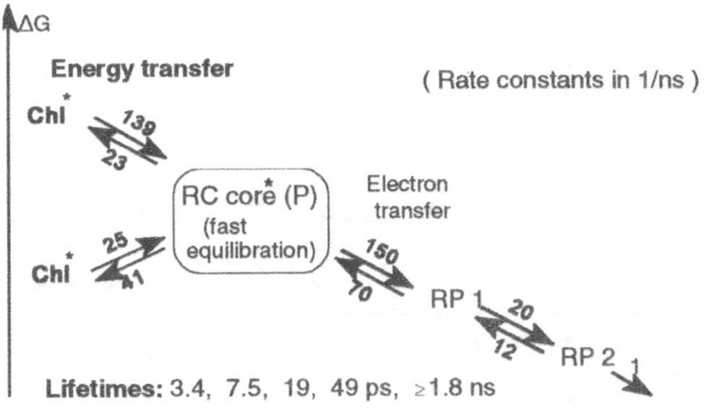

Fig. 3: Minimal kinetic model for the primary processes in the D1-D2 RC complex based on femtosecond transient absorption data. Slow energy transfer processes from teh external Chls are depicted on the left-hand side. On the right-hand side the radical pair processes aare shown.

for charge separation of 21 ps as proposed by Klug and coworkers [13] can be clearly excluded. The latter lifetime would translate into an estimated (Note that no detailed corresponding model has been published) apparent rate of definitely less than 50 ns^{-1} as compared to about 150 ns^{-1} in our case. One may ask the question whether a difference in a factor of 3-4 in the apparent rate constant would indeed have any functional significance or whether this dispute is simply an esoteric argument between groups performing ultrafast spectroscopy. For properly judging this question it is interesting to calculate the quantum yield for charge separation in an intact PS II particle. Such calculation can be carried out using the different rate constants of charge separation and applying them to the exciton/radical pair equilibrium model [15]. Such a calculation reveals that an apparent rate constant of ≈ 150 ns^{-1} allows for a quantum yield of charge separtion in the intact system of ≥ 0.85 whereas a rate constant of ≤ 50 ns^{-1} would allow for a quantum yield of only $\approx 65\%$, i.e. much lower than the actually measured yield. This shows that i) the fast charge separation in the isolated complex is necessary and sufficient to ensure the proper kinetics in the intact complex, and ii) that furthermore the isolated complex has a charge separation rate that is close to the intact system which indicates the essentially native state of the isolated RC. The data reveal furthermore a complex radical pair kinetics which involves several kinetically distinguishable radical pair states. It is not clear at present whether these correspond to different redox states or to different conformational states of the complex.

Acknowledgments
Financial support from the Deutsche Forschungsgemeinschaft (SFB 189 Heinrich-Heine-Universität Düsseldorf and Max-Planck-Institut für Strahlenchemie) and by the European Union (Human Capital and Mobility Grant) is acknowledged. I thank all coworkers and colleagues, most of them cited in the text, who have contributed to this work by their enthusiasm, discussions and cooperation.

References

(1) Kühlbrandt, W., Wang, D.N., Fujiyoshi, Y. *Nature* 1994, 367,614.
(2) Nanba, O., Satoh, K. *Proc.Natl.Acad.Sci.USA* 1987, 84,109.
(3) Michel, H., Deisenhofer, J. *Biochemistry* 1988, 27,1.
(4) Ruban, A.V., Horton, P. *Biochim.Biophys.Acta* 1992, 1102,30.
(5) van Leeuwen, P.J., Nieveen, M.C., van de Meent, E.J., Dekker, J.P., van Gorkom, H.J. *Photosynth.Res.* 1991, 28,149.
(6) Mullineaux, C.W., Gatzen, G., Horton, P., Holzwarth, A.R. *Photosynth. Res.* 1995, submitted,
(7) Wagner, B.C., PhD Thesis Max-Planck-Institut für Strahlenchemie 1995.
(8) Roelofs, T.A., Lee, C.-H., Holzwarth, A.R. *Biophys.J.* 1992, 61,1147.
(9) Konermann, L., Holzwarth, A.R. *Biochemistry* 1995, submitted.
(10) Gatzen, G., Müller, M.G., Griebenow, K., Holzwarth, A.R. *J Phys Chem* 1995 submitted.
(11) Holzwarth, A.R., Müller, M.G., Gatzen, G., Hucke, M., Griebenow, K. *J.Luminesc.* 1994, 60&61,497.
(12) Durrant, J.R., Hastings, G., Hong, Q., Barber, J., Porter, G., Klug, D.R. *Chem.Phys.Lett.* 1992, 188,54.
(13) Hastings, G., Durrant, J.R., Barber, J., Porter, G., Klug, D.R. *Biochemistry* 1992, 31,7638.
(14) Müller, M.G., Hucke, M., Holzwarth, A.R. *J Phys Chem* 1995 submitted.
(15) Schatz, G.H., Brock, H., Holzwarth, A.R. *Biophys.J.* 1988, 54,397.

MOLECULAR NETWORKS AND FUNNELING PROCESS OF ENERGY TRANSFER IN GREEN PHOTOSYNTHETIC BACTERIA

M. Mimuro[1], K. Matsuura[2], K. Shimada[2], Y. Nishimura[3], I. Yamazaki[3], M. Kobayashi[4], Z.-Y. Wang[4] and T. Nozawa[4]
[1]National Institute for Basic Biology, Okazaki 444, [2]Department of Biology, Faculty of Science, Tokyo Metropolitan University, Hachioji 192-03, [3]Department of Chemical Engineering, Faculty of Engineering, Hokkaido University, Sapporo 060, and [4]Department of Biological Engineering, Faculty of Engineering, Tohoku University, Sendai 980-77, Japan.

1. Introduction

The photosynthetic pigment system of green photosynthetic bacteria consists of chlorosomes outside cytoplasmic membranes and BChl *a* pigment-protein complexes in the membranes. The major antenna pigment in chlorosomes is BChl *c* (*d* or *e* depending on species). A fundamental feature of chlorosomes is self assembly of the pigment molecules for functional structure; there have been many arguments on the involvement of proteins in the molecular organization of pigments. Current understanding, however, is that BChl *c* self-aggregates are functional constituents in the rod structure, and polypeptides located in the envelope are not essential for the aggregate structure (Fig. 1A) [1]. In BChl *c* aggregates, heterogeneity in spectral components has been found, that is, spectra are inhomogeneously broadened (Fig. 1B). Energy transfer time between individual pigments is measured to be in a range of 100 fs [2], and the time between the components is also very short (in a range of 10 ps or less) [3]. Ultrafast spectroscopy clearly revealed the coherent nature in BChl *c* aggregates where a very strong molecular interaction is expected [2].

Among the current issues on chlorosomes, followings are especially noteworthy.

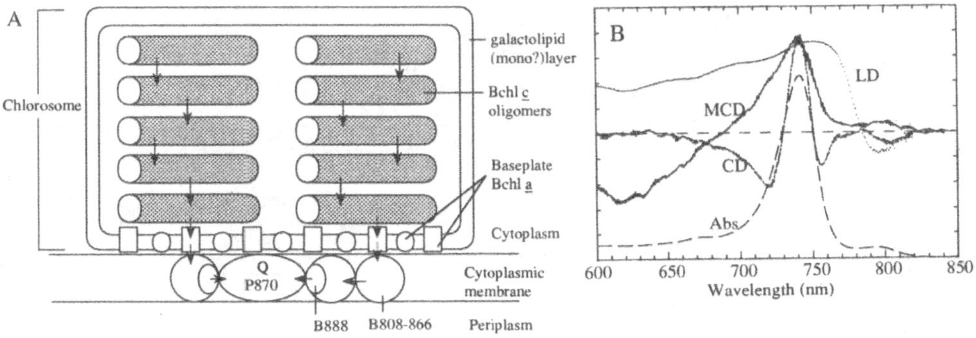

Fig. 1. Schematic structure of antenna system in *Chloroflexus aurantiacus* (A) and the spectral properties of chlorosomes at 25°C (B). For LD, ΔA/A was expressed.

P. Mathis (ed.), Photosynthesis: from Light to Biosphere, Vol. I, 41–46.

(1) Morphology of chlorosomes is mainly determined by rod structure in which self aggregates of BChl *c* molecules are the main factor for the overall structure. (2) Molecular species of BChl *c* is inhomogeneous in terms of side hydrocarbon chains, esterifying alcohols and the stereochemistry at the position of 3^1, and molecular arrangements and optical properties seem to reflect such molecular structures. (3) Localization and orientation of pigments in rods and/or in chlorosomes straightforwardly affect the energy transfer processes and kinetics. Structural aspects of chlorosomes, thus are directly related the functional aspects.

We summarize these issues focusing on our own contribution in this article. A revised model of Bchl *c* arrangement in a rod is presented. Since there are considerable differences between chlorosomes of Chloroflexaceae and Chlorobiaceae in pigment species, overall morphology, intermediary antenna system between chlorosomes and membrane-bound antenna system, we limit our discussion of Chloroflexaceae (*Chloroflexus aurantiacus*, OK-70).

2. Experimental evidences

2. 1. Overall energy transfer in intact cells and comparison with isolated chlorosomes.

We estimated the overall energy transfer time from BChl *c* to RC in intact cells by the time-resolved fluorescence spectra at -196°C. On excitation of BChl *c* at 720 nm, the rise times for the baseplate, B866 and B888 were estimated to be 16 ps, 28 ps and 50 ps, respectively. Changes in the polarization spectrum at 50°C completed within 25 ps after the excitation. This value was faster than those observed in isolated chlorosomes [4], suggesting a possibility that molecular structure in chlorosomes is modified by isolation, although many optical properties are similar to those in intact cells.

2. 2. Hexanol-induced changes in optical properties and energy transfer processes

We analyzed the effect of hexanol on optical properties and energy transfer processes. It has been shown that hexanol induces reversible changes in the aggregation of BChl *c* from higher aggregates peaking at 740 nm to a monomer-like form with a peak at 670 nm. It is supposed that hexanol ligates to Mg atoms of BChl *c*, which inevitably disturb direct molecular interaction of BChl *c* chromophores in the assembly. In the course of hexanol treatment, there are some typical changes in optical properties, thus we can use this treatment to monitor changes in the construction of the higher order architecture of BChl *c* in chlorosomes.

In the case of intact cells, a low concentration of hexanol (0.35 %) induced the enhancement of the 825 nm emission and the decrease in the 810 nm emission at -196°C, whereas the main emission from BChl c was still located at 752 nm. Under these conditions, the transfer efficiency to B866 was decreased, however, the transfer kinetics to B866 was kept almost the same as in control. These suggest that the low-concentration hexanol induced a decrease in the transfer flux to B866.

The hexanol-induced red-shift of the emission around 820 nm was also observed in isolated chlorosomes at -196°C (Fig. 2A). In the absence of hexanol, there exist two fluorescence components in the baseplate at 810 and 825 nm. Excitation spectra indicated that the 825 nm component came from BChl *a*, and 810 nm component was contributed from both of BChl *c* and BChl *a*. By addition of hexanol, the 810 nm component disappeared, suggesting that the origin is related to BChl *c*. This assignment was also supported by CD and MCD spectra based on difference in the sensitivity of pigments to hexanol, and by difference absorption spectrum (Fig. 2B). Thus, this fluorescence band might come from the hybrid of BChl *a* and BChl *c*, even though the actual structure is not yet known.

Struve's group [2] proposed such kind of a complex based on a coherent nature of excited state dynamics. This means that two independent analyses reached to the same conclusion. This complex functions as the energy sink in chlorosomes, thus the possible architecture is a strong interaction between the baseplate BChl *a* complex and a special form of BChl *c* locating in the vicinity of the baseplate.

2. 3. MCD and CD spectroscopy

We introduced a new method, MCD (magnetic circular dichroism), to analyze the electronic state of BChl *c* aggregates [5] (Fig. 1B). In the case of BChl *c*, MCD signal arises from mixing of two excited states induced by the magnetic field (Faraday B term). This is also the case for dimer or higher aggregates. It was shown that intensity of MCD signal depends on the aggregation state [6]. On formation of higher aggregates, an MCD signal intensity per unit absorbance decreased; this phenomenon was confirmed for BChl *d*, RC, LH1 and LH2 isolated from purple bacteria [6]. When we compared the MCD intensity of C744 (a major form of BChl *c* in chlorosomes) with that of BChl *c* in organic solvents, we found that C744 was equivalent to higher aggregates, larger than tetramer. The theoretical background for decrease in the MCD intensity by aggregation is not yet clear. In the CD spectrum, we found two components, each of which show the conserved type spectra with opposite signs in chirality (Fig. 1B) [7]. Recently, Muhlheim group proposed [8] that this difference was assigned to difference in stereochemistry at 3^1 position, that is, R- and S-isomers. A long wavelength form is mainly constructed by the R-form and the short wavelength form, by S-form. Thus, in the rod, when R- and S-isomers independently form the aggregates, difference in chirality of C727 and C747 can be explained [8].

2. 4. Deconvolution of BChl *c* bands into components

We detected several component bands in the spectra. In order to interpret the spectral data consistently, we applied deconvolution procedures to absorption, CD, LD, and MCD spectra. On BChl *c*, we found two major BChl *c* forms (C727 and C744) [7], one minor BChl *c* form (C770) and a special form of BChl *c/a* at 793 nm. The BChl *c* bands in chlorosomes are inhomogeneously broadened. Ultrafast spectroscopy [2] does not necessarily support the presence of plural major forms, however those were consistently found in the four kinds of spectra. Among two major BChl *c* forms, C744

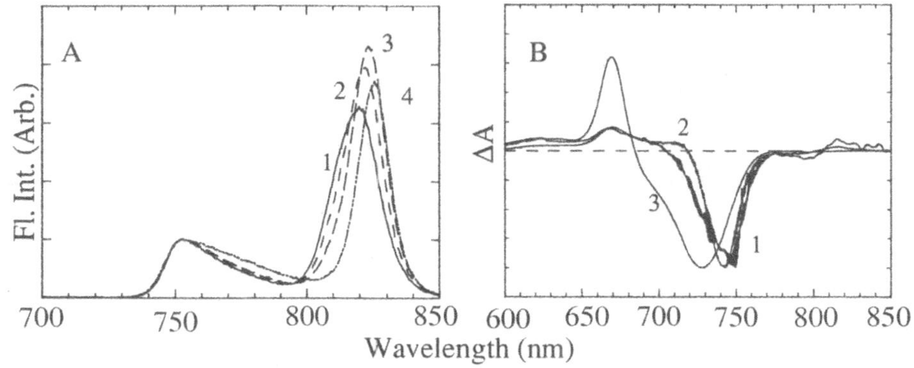

Fig. 2. Fluorescence spectra at -196°C (A) and absorption difference spectra at 25°C (B) of chlorosomes of *C. aurantiacus* induced by hexanol treatment. Numbers indicate the order of dissociation.

can be simulated with two bands at 739 and 752 nm, but it was not the case for C727. The 752 nm band might correspond to the a lower exciton band detected by the hole-burning experiments at 4K [9]. In the short wavelength region of the absorption maximum, we could identify a few bands at 709, 685 and 669 nm, each of which corresponding to tetramer, dimer and monomer of BChl *c*, based on the results obtained on BChl *d* [6]. At -196°C, emissions from monomer and dimer were detected in the cells, but only a little from the tetramer, suggesting its involvement to the energy transfer sequence. Dimer and tetramer were also detected as transient species in the recovery process from hexanol treatment [10]. On BChl *a*, we found the band at 813 nm besides the BChl *c/a* band at 793 nm in the absorption spectrum.

2. 5. Involvement of carotenoids in energy transfer to BChl *c*

Carotenoids are constituent of chlorosomes, and there are two kinds of carotenoid, β- and γ-carotene, in *C. aurantiacus*. Its stoichiometry is estimated to be 1 carotenoid molecule per 4 or 5 BChl *c* [11]. Based on the fluorescence excitation spectrum, γ-carotene is assigned to a more efficient antenna pigment than β-carotene because of a red-shifted peak in the excitation spectrum; the overall transfer efficiency from carotenoids to BChl *c* was estimated to be about 60 %. The transition moment of the S_2 state of carotenoids is nearly parallel to the long axis of rod structure. The energy transfer mechanism from carotenoids to BChl *c* is postulated to be the electron exchange interaction [3], because transition to the S_1 state of carotenoids is forbidden, thus no dipole moment is expected. We have examined theoretically this transfer process by solving the transfer matrix elements [12], and concluded that the coulomb interaction including multipole is favorable for the transfer mechanism. The effect of asymmetry of the conjugated π-electron system was indicated to be much more favorable for the coulomb interaction. For further investigation, analysis of excited state dynamics are important, as we carried out on β-carotene [13].

3. Discussion

3. 1. Energy flow and structural necessity

Over all energy flow in chlorosomes is vectorial to the baseplate and finally to RC in the membranes. Rod structure in the chlorosomes is supposed to be homogeneous in composition, thus the gradient of energy levels to the sink (baseplate) is not necessarily expected. However, time-resolved fluorescence spectra indicates the down hill energy flow among several components to the sink, thus the molecular structure in the rod is expected to satisfy the structural necessity for the energy transfer.

A basic motif for the molecular structure of BChl *c* aggregates is followings, Mg···OH···O=C<. Mg atom is penta-coordinated, and oxygen atoms in the hydroxyl group at 3^1 position ligate to Mg atoms of counterpart. The keto group at 13^1 position forms hydrogen bond with the hydroxyl group at 3^1 position. For elongation of the chain structure, the hydrogen bonds are repeated.

The essential structure of BChl *c* is a linear array along the long axis of rods, thus the energy transfer along this axis is expected to be fast. On the other hand, the energy flow along the circumference of rods is necessary for vectorial flow to the sink. There are two alternative explanations to satisfy the molecular structure for the energy flow to both directions. One is the tilting of the main array (helix structure of strand aggregates) and the other, interaction between linear arrays along the circumference.

LD spectrum indicates that the transition moment of the main BChl *c* absorption band tilts in about 15 to 20 degrees to the long axis of rod structure [4, 7]. If this angle is kept throughout the rods, we can expect two turns of BChl *c* helix consisting of 50 to 60 molecules in one rod. Therefore, at least every 25 or 30 step of energy transfer, the energy can be transferred to an adjacent rod. Since the hopping time of individual transfer is estimated to be 100 fs or less [2], the expected transfer time to an adjacent rod is not long.

On the other hand, even if the main arrays are located parallel to the long axis of rod, the energy flow along circumference is possible when the chlorine face of the double strand is partly overlapped with that of the adjacent strands (Fig. 3). This model was obtained on the basis of the previous two models (Nozawa model [14] and Mimuro-Matsuura model [4, 7]). The main structure is anti-parallel arrangement of two BChl *c* strands, and the interaction along the long axis is the main force for stabilization, as proposed previously. Difference from the previous models is the arrangement of linear arrays, i.e., the unit arrays form ring structure with tilted arrangement, like as in the LH1 [15] and LH2 [16] of purple bacteria. This model is constructed in consideration of the shift of NMR signal due to ring current, space filling and steric hindrance of side chains. This structure enables the two-dimensional energy flow to the long-axis of and circumference of rod structure. Since a strong interaction is not necessarily expected for the direction of circumference, the transfer time might be a range of ps.

This model, however, cannot explain all the properties for rod structure; the presence of plural components and the termination of unit structure [17] are not included in this model.

3. 2. Antenna structure - Conceptual consideration

Three dimensional structures of LH1 [15] and LH2 [16], recently published, show that in the B870 and B850, unit dimer structure will build up the higher order structure, like as a storage ring, consisting of 16 and 9 dimer units, respectively. Two kinds of interaction exist; one within a dimer and the other, inter-dimer. MCD spectra of LH1 and LH2 indicate that the interaction within a dimer is the essential feature in these complexes [6]. As a whole, the storage ring will form one exciton state, as shown by the specific wide absorption band like as B850 and B870. We call this structure "the lateral coupling" of unit building block.

In the case of BChl *c* aggregates, on the other hand, a linear array of double strand is the main building block of the rod structure. As shown in Fig. 2B, in the course of hexanol-induced monomerization, several intermediates were resolved by the

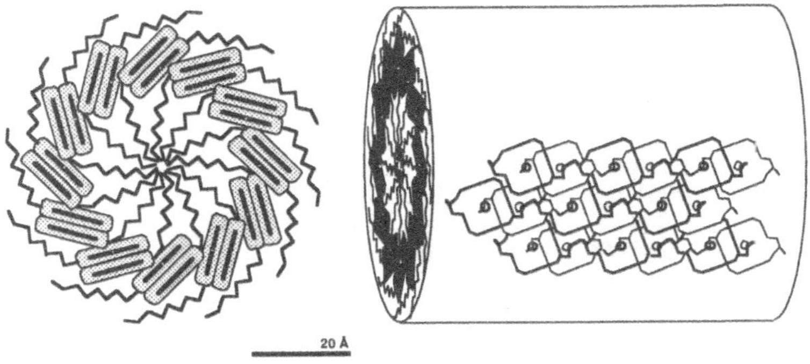

Fig. 3. A tentative model for the molecular structure of major BChl *c* aggregates in the rods of *C. aurantiacus*.

difference and deconvolution of spectra. In the wavelength region shorter than 700 nm, we found the monomer band at 669 nm and the dimer band at 685 nm. This dimer band was shifted to the monomer at the last stage of spectral shift, indicating the dimer is the basic unit for a higher order structure. Probably anti-parallel double strands of BChl *c* aggregates will be dissociated to the dimer units which are stabilized by a pair of ligation of hydroxyl oxygen's to Mg atoms. In this sense, "the dimer unit hypothesis" is applicable to chlorosomes, and we call this type of interaction "the longitudinal coupling" of unit building blocks. This is a conceptual similarity in the principle for the higher order structure of antenna system in photosynthetic bacteria, even though the direction of transition moment is not necessarily determined by the direction of the coupling.

4. General concept in antenna system and future prospect on chlorosome study

BChl *c* aggregates in chlorosomes are unique in that polypeptides are not involved in the stabilization of functional molecular structure, thus the investigations from this view point have progressed. However, as we discussed above, "the dimer unit hypothesis" as building blocks of antenna system and the tilted ring array may become general concepts among chlorosomes and LH complexes of purple bacteria. This contrasts the principle of architecture of antenna system in homo-dimer RC complexes or PS I.

In this article, we do not mention the following important topics; coherent nature of BChl *c* aggregate, general principle for the construction of BChl *c* aggregates in green bacteria including green sulfur bacteria. Furthermore, a strong interaction between BChl *c* molecules and the nature of exited states is not clear. Those will be the targets for the next three years.

Acknowledgments: This work was supported in part by the grant-in-aid from the Ministry of Education, Science, Sport and Culture, Japan, Ciba-Geigy Science Foundation (Japan), and Shimadzu Science Foundation.

References

1 Mimuro, M. et al. (1992) in Research in Photosynthesis (Murata, N., ed.) Vol. 1, pp. 17-24, Kluwer Academic Publishers, Dortrecht, The Netherlands.
2 Savikhin, S. et al. (1994) J. Phys. Chem. 98, 10322-10334.
3 van Grondelle et at. (1994) Biochim. Biophys. Acta 1187, 1-65.
4 Mimuro, M. et al. (1994) Photosynthe. Res. 41, 181-191.
5 Nozawa, T. et al. (1995) Spectrochim. Acta 51A, 125-134.
6 Kobayashi, M. et al. In this Proceedings
7 Matsuura, K. et al. (1993) Photochem. Photobiol. 57, 92-97.
8 Balaban, T.S. et al (1995) J. Mol Struc. 349, 183-186.
9 Fetisova, Z. et al. (1994) Photosynthe. Res. 41, 205-210.
10 Cox, R.P. et al. In this Proceedings
11 Tsuji, H. et al. In this Proceedings
12 Nagae, H. et al. (1993) J. Chem. Phys. 98, 8012-8023.
13 Kandori, H. et al. (1994) J. Am. Chem. Soc. 116, 2671-2672.
14 Nozawa, T. et al. (1994) Photosynthe. Res. 41, 211-223.
15 Karrasch, S. et al. (1995) EMBO J. 14, 631-638.
16 McDermott, G. et al. (1995) Nature 374, 517-521.
17 Chiefari, J. et al. (1995) J. Phys. Chem. 99, 1357-1365.

ASSEMBLY OF TRIMERIC LIGHT-HARVESTING CHLOROPHYLL A/B COMPLEX IN VITRO

S. Hobe, A. Kuttkat, R. Förster, and H. Paulsen
Botanisches Institut III der Universität,
Menzinger Straße 67, D-80638 München, Deutschland

1. Introduction

The major light-harvesting complex of photosystem II (LHCII) in the thylakoid of green plants is most likely organized in trimers. Dreyfuss and Thornber (1) showed that during the greening process, monomeric LHCII is formed first which is then assembled into trimers. LHCII monomers, reconstituted from the overexpressed apoprotein (LHCP) and pigments, trimerize in vitro if lipids, specifically dipalmitoyl phosphatidylglycerol (PG), are present (2). Conversely, isolated LHCII trimers dissociate into monomers when treated with phospholipase which hydrolyzes phospholipids, indicating that phospholipids bound to LHCII stabilize its trimeric form. In this paper we show that trimer formation is stimulated by adding PG to the reconstitution mix for LHCII monomers at a relatively low stoichiometry of a few molecules of lipid per LHCP.

Recently, we identified a trimerization motif, comprising amino acids 16 to 21 of LHCP, which seems to be essential for trimer formation: N-terminal deletions into this segment or amino acid exchanges within the trimerization motif abolish the capability of the mutant protein to form trimeric LHCII in vitro (3). We show here (a) that mutations in the trimerization motif also affect the assembly of LHCP into trimeric LHCII upon insertion into isolated thylakoids, and (b) that another amino acid, W222 near the C terminus of LHCP, also seems to be important for the oligomerization of LHCII.

2. Procedure

The reconstitution of LHCII monomers and trimers in vitro was carried out as described in (2, 3). In short, LHCP or its derivatives, overexpressed in Escherichia coli, were reconstituted with pigments in detergent/lipid solution, separated from detergents, excess pigments and lipids on a DEAE anion exchange column and by subsequent dialysis. The complexes were then concentrated by ultrafiltration, solubilized in dodecylmaltoside and assayed for trimer formation by ultracentrifugation in a sucrose gradient.
For the insertion of LHCP into isolated thylakoids refer to (4) in this volume.

P. Mathis (ed.), Photosynthesis: from Light to Biosphere, Vol. I, 47–52.

3. Results and discussion

3.1 The trimerization motif affects LHCII trimer formation both in detergent solution and in the thylakoid.

The sequence WYxxxR in position 16 to 21 of LHCP is involved in the oligomerization of monomeric LHCII (3): Full-length LHCP and a mutant (Fig. 1) in which 15 N-terminal amino acids are deleted (lane a in Fig. 2) do form trimeric LHCP whereas the deletion of two further N-terminal amino acids (lane b) and the exchange of WY16,17

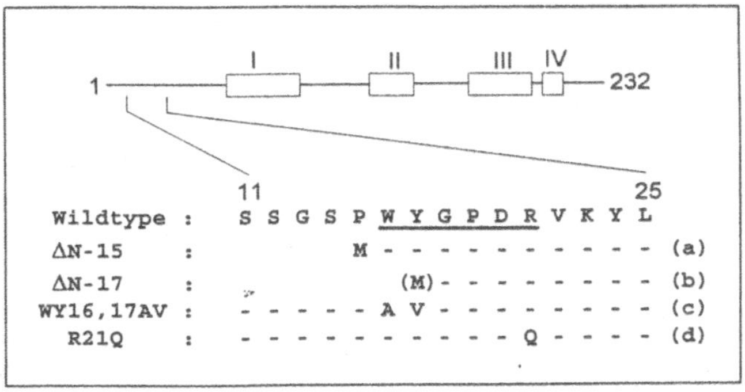

Figure 1. LHCP derivatives used. The boxes I, II, III, and IV correspond to the alpha-helical domains B, C, A, and D in Kühlbrandt et al. (5). Amino acids are numbered with respect to mature LHCP. Dashes represent same amino acids as in the wild type. The mutants are referred to as (a), (b), (c), and (d) in Figures 2 and 3. In mutant (b), the leading M has not been veryfied by protein sequencing. The trimerization motif is underlined.

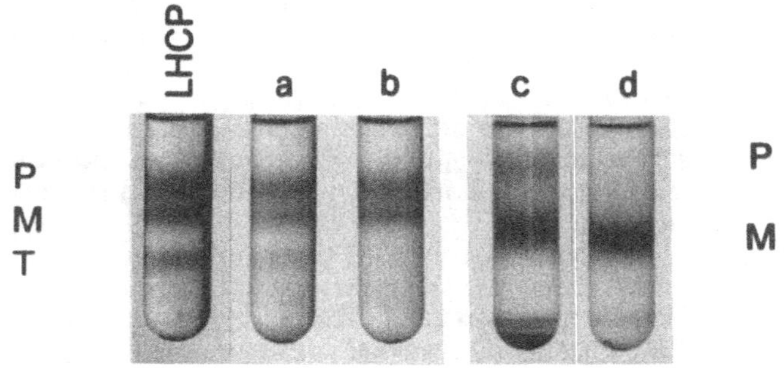

Figure 2. Deletions or mutations in the trimerization motif abolish LHCII trimer formation in detergent solution. LHCP and its mutant derivatives a, b, c, and d (see Fig. 1) were reconstituted and assayed for trimer formation by sucrose gradient centrifugation. T, trimers; M, monomers; P, unbound pigment.

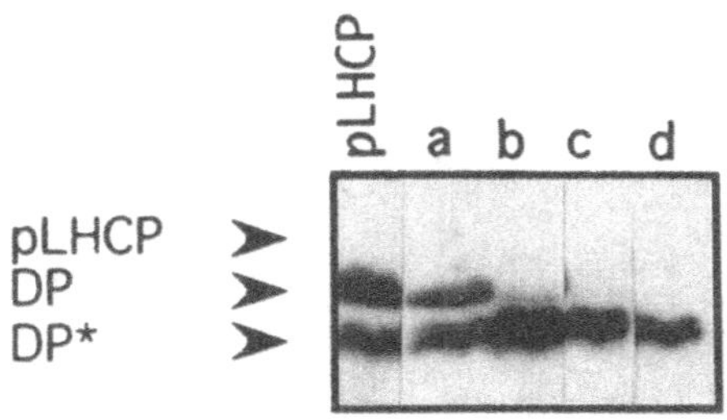

Figure 3. Deletions or mutations in the trimerization motif abolish LHCII trimer formation in the thylakoid. Radioactively labelled pLHCP and the LHCP derivatives a, b, c, and d (Fig. 1) were inserted in isolated pea thylakoids, the thylakoids were treated with thermolysin, electrophoresed on a fully denaturing SDS polyacrylamide gel and autoradiographed (4). The arrows marked pLHCP, DP, and DP* give the position of pLHCP, LHCP-DP, and LHCP-DP*, respectively.

with AV or of R21 with Q (lanes c and d, respectively), completely abolish trimer formation.

Assembly of LHCP and its derivatives into trimeric LHCII can also be observed in another in-vitro assay, namely upon insertion of the protein into isolated thylakoid membranes. The oligomerization of LHCII can be detected by testing the protease sensitivity of the apoprotein: monomeric LHCII gives rise to a 20 kDa digestion product (LHCP-DP*) whereas in trimers about 24 kDa of LHCP (LHCP-DP) are protease-resistent (4). Fig. 3 shows that the mutants b,c and d (Fig. 1) which do not trimerize upon their reconstitution into LHCII monomers in vitro (Fig. 2), do not give rise to LHCP-DP and thus, do not trimerize upon insertion in thylakoids either. We conclude that the influence of structural alterations in LHCP on the oligomerization of LHCII is largely independent from the environment of the complexes whether they are in detergent solution or embedded in the thylakoid membrane.

Interestingly, replacing another positively charged amino acid two positions downstream of R21, K23, with Q does not affect LHCII trimer formation, neither in detergent solution (3) nor during the assembly of LHCII in isolated thylakoids (data not shown). Thus, the formation of stable LHCII trimers seems to require a positive charge in position 21 rather than a net positive charge in this protein segment, indicating that the trimerization motif is involved in specific trimer-stabilizing interactions.

3.2 An aromatic residue in position 222 of LHCP is essential for LHCII trimer formation.

The trimerization motif is not the only domain in LHCP which affects the oligomerization of LHCII. When W222 in pLHCP is exchanged with A, G, or H, the protein is digested to form LHCP-DP* after its insertion into isolated thylakoids (Fig. 4), indicating that only monomeric LHCII is formed. However, when W222 is exchanged with F, trimeric LHCII is formed judging from the re-appearance of LHCP-DP. Preliminary results indicate that the mutations show the same effect on LHCII trimerization in detergent solution (not shown). We conclude that an aromatic amino acid is essential in the 11th position from the C terminus. We have reported earlier that W222 is important for the stability of reconstituted LHCII monomers: in a C-terminal deletion series, complex formation disappears as soon as W222 is deleted although full-length LHCP derivatives with point mutations in position 222 do form stable monomeric complexes with pigments (6). This indicates that W222 and the C-terminal 10 amino acids of LHCP have stabilizing effects in LHCII that are additive or mutually replaceable: only when both

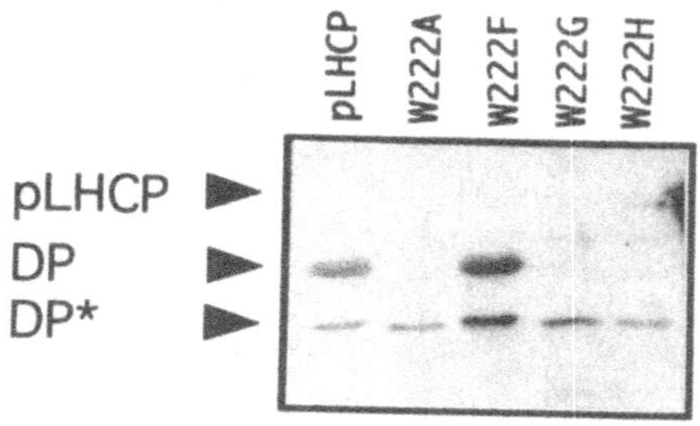

Figure 4. Amino acid exchanges in position 222 of LHCP abolish trimerization. Radioactively labelled pLHCP and its derivatives with W222 exchanged as indicated were inserted in isolated pea thylakoids, the thylakoids were treated with thermolysin, electrophoresed on a fully denaturing SDS polyacrylamide gel and autoradiographed (4). Abbreviations as in Fig. 3.

elements are gone, the protein fails to form stable complexes with pigments. This effect is probably caused by a long-range structural alteration of the protein since the C-terminal 11 amino acids of LHCP are not thought to be involved in direct pigment-protein interactions (5). Similarly, the exchange of W222 may exert its effect on LHCII trimer formation by altering the overall protein structure such that trimerization is disfavored, and not directly affect a contact site between LHCII monomers. Hydrophobic residues like W and F in the loop regions of LHCP are oriented towards the hydrophobic interior of the complex (5) and possibly serve to maintain the loops close to the membrane surface (7). These amino acids in position 222 may position the C terminus in a similar way and thus, stabilize an overall protein conformation that favors trimerization. On the other hand, the atomic model of LHCII trimers in (7) does not exclude that the C terminus of the LHCII apoprotein is involved in direct monomer-monomer interactions.

3.3 Trimerization of LHCII in vitro is stimulated by near-stoichiometric amounts of phosphatidyl glycerol.

Trimeric LHCII dissociates into monomers not only when treated with phospholipase as mentioned above but also when N-terminal residues are cleaved off by chymotrypsin (8). Interestingly, during this treatment the complexes lose their otherwise tightly bound PG, suggesting that the PG that stabilizes LHCII trimers is binding specifically to the N terminal domain of LHCP. In support of this hypothesis, Fig. 5 shows that PG in near-stoichiometric amounts, added during the reconstitution of LHCII monomers, significantly stimulates LHCII trimer formation. The lipid dependence of trimer formation is particularly pronounced between 1 and 5 mol PG added per mol protein, and then levels off into what resembles a saturation curve. The low but significant trimerization found without PG added is probably due to lipids still present in the protein or pigment preparations used for the reconstitution.

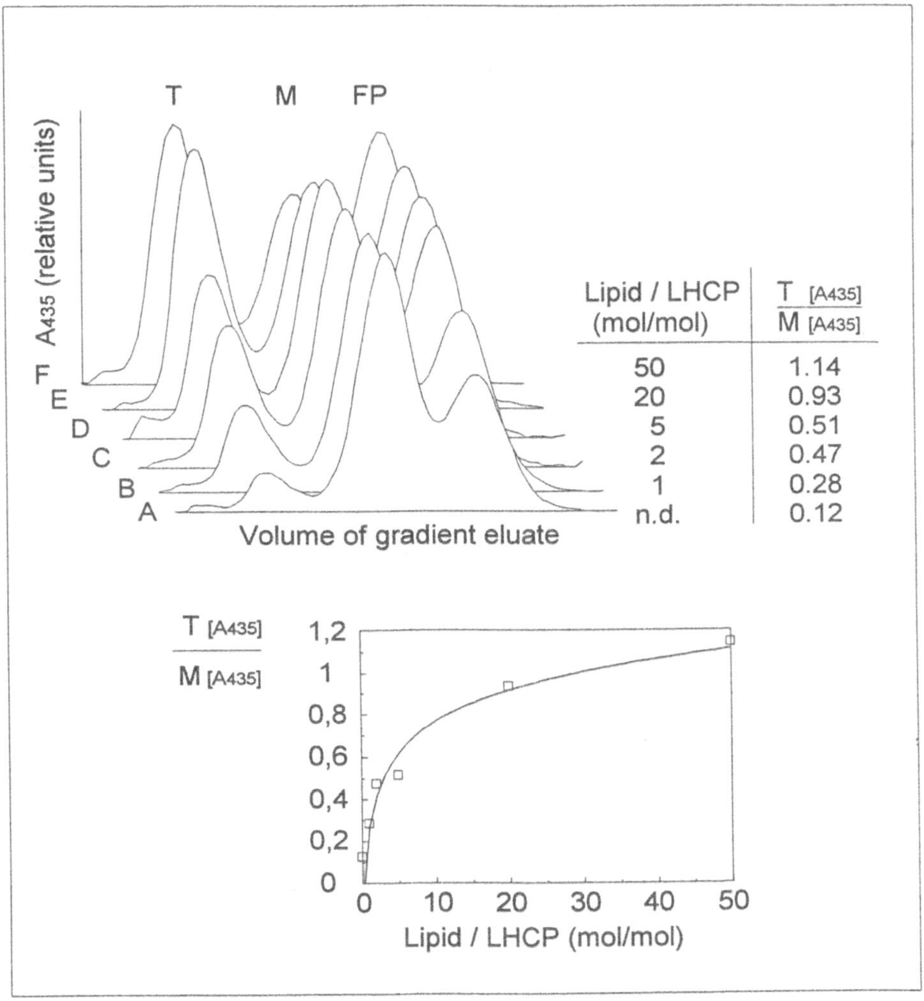

Figure 5. "Titration" of LHCII trimer formation with PG. Bacterially expressed LHCP was reconstituted with pigments in the presence of PG at the lipid:protein ratios given. The resulting LHCII monomers were trimerized and analyzed by sucrose gradient centrifugation (3). Panels A to F give the gradient elution profiles of the sucrose gradients obtained by measuring absorption at 435 nm in a flow-through cuvette with the area under the peaks labelled T, M, and FP being proportional to the chlorophyll (Chl) content in the bands of LHCII trimers, monomers, and unbound pigments, respectively. In the lower panel, the ratio of chlorophyll amounts in the trimer and monomer bands are plotted agains the lipid:protein ratios used in the reconstitutions A to F.

The evaluation of the data in Fig. 5 in terms of binding constant and stoichiometry between PG and LHCII is difficult since it is unknown how much PG stays bound to the LHCP-pigment complexes after the complexes have been washed detergent-free on an ion exchange column (see Procedures). The lipid:LHCP ratios given are the ones present in the reconstitution mix at the beginning of the experiment; these may actually be an over-estimate regarding the situation at the end of the experiment when trimeric LHCII is formed. Therefore we are presently repeating these experiments in collaboration with Dr. A. Trémolières, using radioactively labelled lipids which can be unequivocally quantitated in isolated LHCII trimers and monomers. However, even regarding this ambiguity, the data in Fig. 5 lend strong support to the notion that PG stabilizes trimeric LHCII by stoichiometric binding into (a) specific binding site(s). It is quite possible that the trimerization motiv near the N terminus of LHCP represents such a lipid binding site.

References

1 Dreyfuss, B.W. and Thornber, J.P. (1994) Plant Physiol. 106, 829-839
2 Hobe, S., Prytulla, S., Kühlbrandt, W. and Paulsen, H. (1994) EMBO J. 13, 3423-3429
3 Hobe, S., Förster, R., Klingler, J. and Paulsen, H. (1995) Biochemistry, in press
4 Kuttkat, A., Edhofer, I., Eichacker, L. and Paulsen, H. (1995) in Proceedings of the Xth International Photosynthesis Congress (Mathis, P., ed.) this volume, Kluwer Academic Publishers, Dordrecht, The Netherlands
5 Kühlbrandt, W., Wang, D.A. and Fujiyoshi, Y. (1994) Nature 367, 614-621
6 Paulsen, H. and Kuttkat, A. (1993) Photochem. Photobiol. 57, 139-142
7 Kühlbrandt, W. (1994) Curr. Opin. Struct. Biol. 4, 519-528
8 Nußberger, S., Dörr, K., Wang, D.N. and Kühlbrandt, W. (1993) J. Mol. Biol. 234, 347-356

ON THE NATURE OF THE F695 AND F685 EMISSION OF PHOTOSYSTEM II

Jan P. Dekker, Aukje Hassoldt, Åsa Pettersson, Henny van Roon, Marie-Louise Groot and Rienk van Grondelle
Department of Physics and Astronomy, Vrije Universiteit, De Boelelaan 1081, 1081 HV Amsterdam, The Netherlands

1. Introduction

The native reaction center core antenna complex of photosystem (PS) II contains as main components the membrane-bound core antenna proteins CP47 and CP43, the D1/D2/ Cyt.*b*-559 reaction center complex, a 33 kDa water-soluble extrinsic protein involved in oxygen evolution as well as a number of low-molecular mass proteins. Such complexes are generally characterized by 77K emission bands peaking at 685 and 695 nm (F685 and F695, respectively) [1]. F695 probably arises from the minor 690 nm pigment of CP47 with orientation perpendicular to the plane of the membrane, whereas F685 arises from several ~683 nm absorbing pigments of CP47 and/or CP43.

In this report, we investigate the temperature dependence and quantum yield of the emission of PS II core complexes and of isolated CP43 and CP47 complexes. We conclude that in intact PS II at 4K a large part of the excitation energy is lost in the core antenna and that the 77K F695 emission arises only from those complexes containing the red-most part of the inhomogeneous distribution of the 690 nm CP47 chlorophylls.

2. Materials and Methods

PS II core complexes were obtained from spinach PS II membranes as in [2], while CP47 was purified in the presence of dodecyl maltoside (DM) as in [3]. For the purification of CP43, core preparations as in [2] were solubilized with dodecyl maltoside and separated from the CP47-RC complex and the 28 kDa protein by anion exchange chromatography as in [4], but in which $MgSO_4$ was omitted from the buffers. CP43 was found in the fraction that did not bind to the column. This fraction was dialyzed overnight against 50 mM Tris-HCl (pH 7.8) + 0.003% DM, after which it was applied to a Q-Sepharose column (Pharmacia) equilibrated with 50 mM Tris-HCl (pH 7.8) + 0.03% DM. After the eluate (which mainly contained free Chl) became colorless, CP43 was eluted with the same buffer + 50 mM $MgSO_4$. Its purity was checked by FPLC gel filtration chromatography (see C. Eijckelhoff et al., these Proceedings) and in case of impurities the fraction was recycled through the anion-exchange column as described above.

Low-temperature absorption and fluorescence experiments and fluorescence quantum yield determinations were performed in 65% glycerol glasses with or without 0.03% DM as in [5].

P. Mathis (ed.), Photosynthesis: from Light to Biosphere, Vol. I, 53–56.

3. Results

3.1 CP43

The 4K absorption spectrum of CP43 reveals a broad peak at 669 nm, a much narrower peak at 682 nm and shoulders near 660 and 678 nm (Fig. 1A) and clearly resembles the spectrum of CP43 from *Synechococcus* sp. isolated in SDS [6]. The 682 nm peak is better resolved than in the spectrum reported in [7]. At higher temperatures, the 682 nm peak is broadened to a diffuse shoulder on the low-energy side (Fig. 1A). The 4K emission spectrum peaks at 683 nm (Fig. 1B) and has an extremely narrow bandwidth (fwhm = 5 nm). Upon increasing the temperature the spectrum broadens on the high-energy side to give a well-distinguishable ~670 nm contribution at 260K (Fig. 1B), but the peak wavelength remained at 683 nm at all temperatures (Fig. 2B). Below 80K virtually no 670 nm emission is observed, which suggests that there is almost no uncoupled Chl *a*. The temperature dependence of the overall fluorescence quantum yield resembles that observed before for isolated CP47 [3].

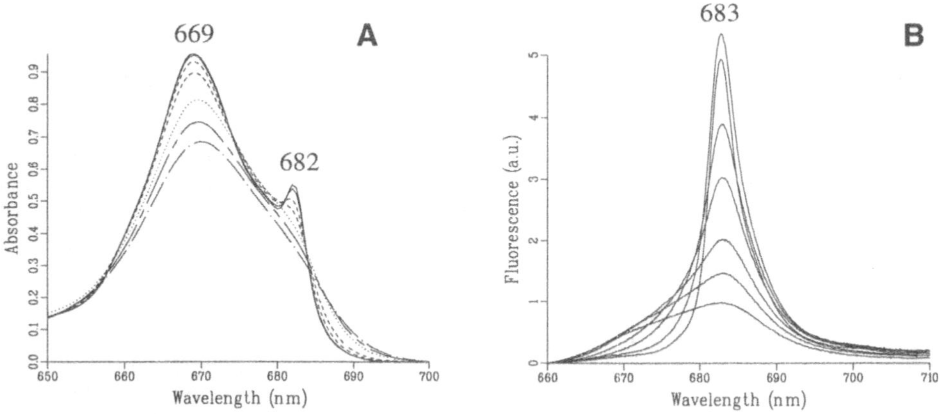

Figure 1. A) Absorption spectra of CP43 at T = 4K, 40K, 77K, 120K, 200K, 250K and 293K. B) Emission spectra of CP43 upon excitation at 605 nm at T = 4K, 40K, 80K, 110K, 160K, 200K and 260K.

3.2 CP47

In [3] we reported OD and emission spectra of CP47 at various temperatures. The 4K emission spectrum peaked between 4K and 80K at 690 nm and is much broader (fwhm = 12.5 nm at 4K) than that of CP43, while above 80K the spectrum gradually shifted to 683 nm (Fig. 2B). The 4K OD spectrum was very similar to the spectrum reported by Chang et al. [8], who also presented evidence that the minor component peaking at 690 nm is the lowest energy component responsible for the steady-state emission at very low temperatures. In [9], however, the 4K OD spectrum of CP47 shows a different fine structure than the spectra in [3,8], while the 4K emission spectrum appeared to be broadened (fwhm ~ 16 nm) and red-shifted to 693 nm.

We reproduced similar OD and fluorescence spectra as in [9] by leaving out the detergent in the glycerol buffer (not shown), due to which we suggest that the different spectroscopic properties in [9] are caused by aggregation of CP47, and not by differences in the isolation procedure or in the conformation of the protein. These results suggest that isolated CP47 in the monomeric form fluoresces at 690 nm between 80K and 4K.

3.3 PS II core

Fig. 2A shows the emission spectra as a function of temperature of PS II core particles prepared according to the method of Ghanotakis and Yocum [2]. Upon decreasing the temperature from 290K to about 90K the emission maximum gradually shifts from 683 to 685 nm (Fig. 2B). At 77K there are two peaks with about equal intensities at 686 and 694 nm. The intensity of the latter peak increases dramatically upon further decreasing the temperature to 4K, while the peak wavelength blue-shifts to 690 nm to give a spectrum with the same peak wavelength and width (fwhm = 12.5 nm) as that of isolated CP47 [3]. Preliminary estimations of the fluorescence quantum yield suggested values of about 1% at room temperature, which gradually increased to about 2% at 77K and 9% at 4K. The latter value is not much smaller than of isolated CP47 (~12% - Ref. 3), from which we conclude that at 4K a considerable part of the excitation energy is lost in the core antenna.

4. Discussion

Upon lowering the temperature from room temperature to about 80K the fluorescence maximum of PS II core complexes shifts from 683 nm to about 686 nm. The shift is in between those observed for isolated CP47 and isolated CP43 (Fig. 2B), which suggests that it arises from an increasing contribution of the 690 nm pigment of CP47 in a well-

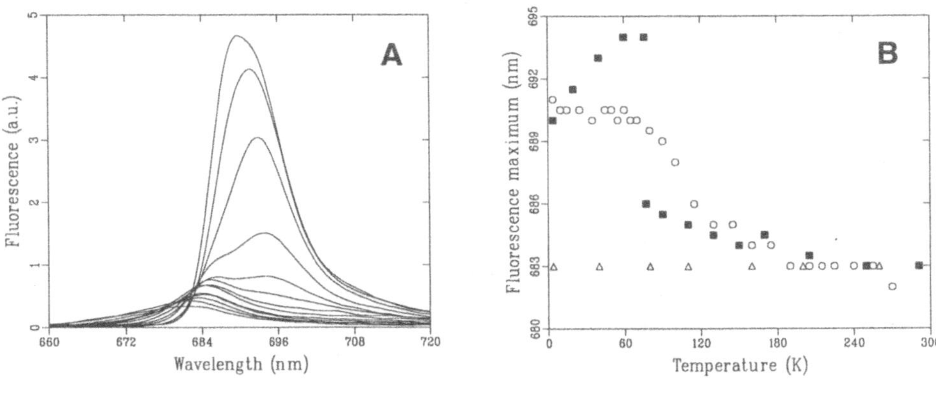

Figure 2. A) Emission spectra of PS II core complexes upon excitation at 590 nm at (from top to bottom) T = 4K, 20K, 40K, 60K, 77K, 90K, 110K, 130K, 150K, 170K, 205K, 250K and 291K. B) Position of the maxima of the emission spectra of PS II core (closed squares), CP47 (open circles) and CP43 (open triangles) preparations as a function of temperature.

equilibrated antenna. Thus, in this temperature range the results can well be explained in terms of an equilibrium between the radical pair and the singlet excited states of basically all core antenna pigments.

Below about 80K the F695 emission is observed in the PS II core complexes, which gradually blue-shifts to 690 nm upon decreasing the temperature to 4K. The blue-shift is accompanied by a strong increase of the fluorescence quantum yield and results at 4K in a spectrum that is virtually indistinguishable from that of isolated CP47. We therefore suggest that almost all 4K emission in the PS II core arises from CP47.

The longer peak wavelength of F695 at intermediate temperatures is explained by the inhomogeneous distribution of the 690 nm pigment of CP47. In those complexes in which this pigment absorbs at a relatively long wavelength the excitation will be trapped more easily on this pigment because more thermal energy is required for the de-trapping to P680. We therefore suggest that only the red-most part of the inhomogeneous distribution of the 690 nm CP47 chlorophyll causes the F695 emission at 77K.

The findings that at 4K the emission spectra of CP47 and the PS II core are virtually the same and that the 4K fluorescence quantum yield of the PS II core is not much lower than of isolated CP47 suggest that at this temperature most of the excitations that hit the (complete distribution of the) 690 nm pigment of CP47 are trapped and lost for charge separation. This, in turn, suggests that the quantum yield of charge separation of the PS II core should be rather low at 4K, which disagrees with recent results from transient absorption changes [10]. Further research is required to clarify this issue.

Acknowledgements

This research was supported by the Netherlands Foundations for Chemical Research (SON) and Life Sciences (SLW) and by the EC contract CT940619.

References

1 Van Grondelle, R., Dekker, J.P., Gillbro, T. and Sundström, V. (1994) Biochim. Biophys. Acta 1187, 1-65
2 Ghanotakis, D.F., Demetriou, D.M. and Yocum, C.F. (1987) Biochim. Biophys. Acta 891, 15-21
3 Groot, M.-L., Peterman, E.J.G., Van Stokkum, I.H.M., Dekker, J.P. and Van Grondelle, R. (1995) Biophys. J. 68, 281-290
4 Dekker, J.P., Bowlby, N.R. and Yocum, C.F. (1989) FEBS Lett. 254, 150-154
5 Groot, M.-L., Peterman, E.J.G., Van Kan, P.J.M., Van Stokkum, I.H.M., Dekker, J.P. and Van Grondelle, R. (1994) Biophys. J. 67, 318-330
6 Breton, J. and Katoh, S. (1987) Biochim. Biophys. Acta 892, 99-107
7 Alfonso, M., Montoya, G., Cases, R., Rodriguez, R. and Picorel, R. (1994) Biochemistry 33, 10494-10500
8 Chang, H.-C., Jankowiak, R., Yocum, C.F., Picorel, R., Alfonso, M., Seibert, M. and Small, G.J. (1994) J. Phys. Chem. 98, 7717-7724
9 Van Dorssen, R.J., Breton, J., Plijter, J.J., Satoh, K., Van Gorkom, H.J. and Amesz, J. (1987) Biochim. Biophys. Acta 893, 267-274
10 Van Mieghem, F., Brettel, K., Hillman, B., Kamlowski, A., Rutherford, A.W. and Schlodder, E. (1995) Biochemistry 34, 4798-4813

SUB-PICOSECOND TRANSIENT ABSORPTION DIFFERENCE SPECTROSCOPY ON LIGHT-HARVESTING COMPLEX II AT 77K

Foske J. Kleima, H. Matthieu Visser, Rienk van Grondelle and Herbert van Amerongen; Dept of Physics and Astronomy and Institute for Molecular Biological Sciences, Vrije Universiteit, De Boelelaan 1081, 1081 HV Amsterdam, The Netherlands

1. Introduction

Light-harvesting complex II (LHCII) is the most abundant pigment-protein complex in higher plants. It functions as an antenna to enhance the efficiency of photosynthesis. The main task of this trimeric protein is to absorb sunlight and channel the excitation energy towards the reaction centre of mainly Photosystem II. Its crystal structure reveals 12 chlorophylls (Chls) and 2 carotenoids and they were assigned to 7 Chl *a*, 5 Chl *b* and 2 lutein molecules [1]. In the atomic model of LHCII all Chl *b* molecules are positioned very close to Chl *a* molecules and one would expect all Chl *b* to Chl *a* energy transfer to take place within hundreds of femtoseconds [1]. In order to refine the atomic model and to get a better understanding of the transfer processes it is important to determine the relative contribution of the picosecond component(s).

Over the last couple of years numerous experiments have been performed on the ultrafast energy transfer in trimeric LHCII [2-9]. Most of these studies concentrated on the energy transfer from Chl *b* to Chl *a*. It was revealed that transfer processes of at most a few hundred of femtoseconds coexist with processes occurring on a time scale of several picoseconds. Only few studies of energy transfer after Chl *a* excitation have been performed and the energy transfer was studied with polarization measurements. No data are available on the spectral evolution after selective excitation of Chl *a*. In this report we present studies of the energy-transfer dynamics in trimeric LHCII at 77K after excitation in the Q_y region of Chl *b* (648.6 nm) and Chl *a* (661.1, 671.6 and 682.3 nm).

2. Materials & Methods

The laser system used for this study consisted of a synchronously pumped dual-jet dye laser, generating ~300 fs (FWHM autocorrelation) pulses around 650 nm. The pulses were amplified to 500 μJ at 30 Hz in a 3-stage dye amplifier. Pump and probe light were generated by focusing part of this beam in a sapphire plate, yielding 'white' light continuum. For the pump pulses, interference filters (1 mm thickness, Ferroperm, Vedbaek, Denmark) with a bandwidth of 5 nm (FWHM) were used to select part of the white light continuum. In a home-built 1-stage dye cuvette this light was amplified to pulses of a few μJ. Group-velocity dispersion (GVD) compensation was obtained with the use of a pair of SF-10 prisms. After GVD compensation, the probe beam was split in a probe and a reference beam; both were focused in the sample and imaged via a 1/4 m spectrograph on the two diode arrays of a double-diode array detector. The intensity of

P. Mathis (ed.), Photosynthesis: from Light to Biosphere, Vol. I, 57–64.

the pump beam was varied by neutral-density filters, the time difference between pump and probe was set by a delay-line. The instrument response width (FWHM) was obtained from the 10-90% rise time of the induced birefringence in CS_2, and was found to be 0.28-0.32 ps. For each measurement 6 scans were made from short to longer delay times. In every scan ~300 shots were taken at each time position. It was checked that the signal remained constant from one scan to the next.

Trimeric LHCII was isolated as described in [5]. Samples were diluted in a buffer containing 20 mM Hepes (pH 7.5), 0.15% *n*-dodecyl-β,D-maltoside and 70% (v/v) glycerol to an OD of 0.7 cm^{-1} at 675 nm at room temperature, and put in acrylic cuvettes with a path length of 0.15 cm. Measurements were performed at 77 K using an Oxford Instruments liquid nitrogen cryostat.

3. Results and Discussion

In fig. 1 the spectral evolution within the first picosecond is shown after selective excitation of Chl *b* at 648.6 nm. At time zero (t = 0) the excitation has just started

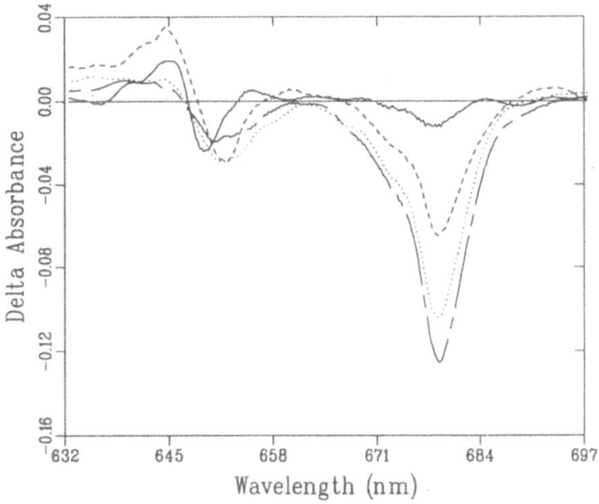

Figure 1. Transient absorption spectra of LHCII (77K) at 0 (solid), 0.25 (dashed), 0.50 (dotted), 0.75 ps (solid dash) after excitation at 648.6 nm. The width of the autocorrelation function is 0.3 ps. The maximum of the autocorrelation function is located between 0 and 0.25 ps.

(the instrument response was 0.33 ps) and more bleaching/stimulated emission (B/SE) is observed in the Chl *b* region than in the Chl *a* region. Already within the duration of the pulse (t = 0.25 ps) the B/SE in the Chl *a* region starts to dominate. Using the fact that the molar extinction coefficient of Chl *b* is approximately 25% smaller than that of Chl *a*, it can be calculated that at t = 0.50 ps approximately 74% of the excitations are located on Chl *a* and at t = 0.75 ps this number has increased to 83%. From this we can conclude that about 17% of the Chl *b* to Chl *a* transfer occurs on picosecond timescale. In the spectra at t = 0 and t = 0.25 ps the strongly bipolar feature in the Chl *b* region is intriguing and it suggests the presence of a set of strongly excitonically coupled Chl *b* molecules. In trimeric LHCII excitonic coupling in this Chl *b* region was indeed observed with circular dichroism spectroscopy but turned out to be absent for monomers [11]. Future transient absorption measurements on monomeric LHCII may therefore shed more light on this issue. In figure 2 the transfer process from Chl *b* to Chl *a* is shown at later times and it is clear that only a small part of the transfer process takes place on picosecond time scale.

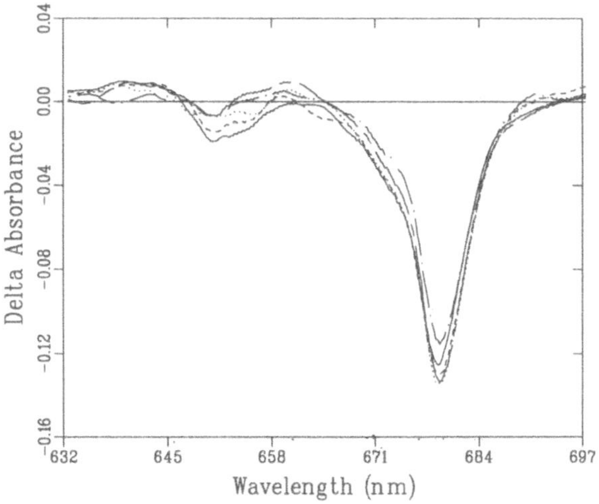

Figure 2. Transient absorption spectra of LHCII (77K) at 0.75 (solid), 1.2 (dashed), 2.4 (dotted), 4.4 (solid dash) and 8.4 ps (solid dot). after excitation at 648.6 nm.

The decrease of the intensity of the Chl *a* signal on this time scale is due to singlet-singlet annihilation which occurred in all our experiments (not shown) on a time scale of several tens of picoseconds, in agreement with the results in [2].
The spectral evolution after excitation at 661.1 nm is shown in fig.3. Already at the onset of the excitation pulse (t = 0 ps) the bleaching/SE band near 680 nm is as large as the B/SE near the excitation wavelength. At t = 0.50 ps (just after the pulse) the major bleaching is located near 680 nm. Note the shoulder around 670 nm reflecting the fact that part of the ultrafast excitation transfer occurs through the 670 nm pool of Chl *a*. Subsequently, a small part of the excitations is being transferred on a picosecond time scale. The spectrum after 20.3 ps is considerably less intense than the spectrum after 5.3 ps due to singlet-singlet annihilation.
Similar experiments as described above were performed with excitation at 671.6 and 682.3 nm. After excitation at 671.6 nm rapid excitation transfer took place and after 1 ps essentially no changes in the shape of the spectrum did occur (not shown). Subsequently, singlet-singlet annihilation took place. Excitation at 682.3 nm led to a B/SE band around 680 nm (FWHM 5 nm). This band broadened to 8 nm on the same time scale as singlet-singlet annihilation took place (10-20 ps).
In conclusion, excitation at the mentioned wavelengths at 77 K leads to very fast spectral equilibration. Most of this spectral equilibration takes place within 1 ps, irrespective of the excitation wavelength. Only excitation at 648.6 and 661.1 nm gives rise to significant contributions with a transfer time longer than 1 ps. The small amplitude of the picosecond Chl *b* to Chl *a* transfer component suggests that it is due to at most one Chl *b* molecule per monomer.

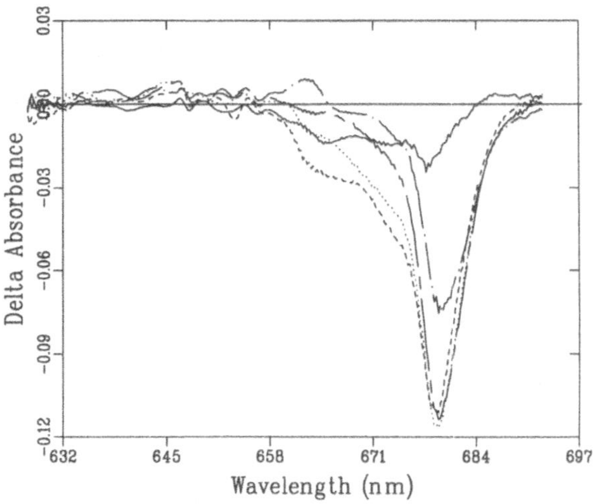

Figure 3. Transient absorption spectra of LHCII (77K) at 0 (solid), 0.50 (dashed), 1.3 (dotted), 5.3 (solid dash) and 20.3 ps (solid dot). after excitation at 661.1 nm.

4. Acknowledgements

The authors thank Dr. J.P. Dekker and Mr. E.J.G. Peterman for useful comments. This research was supported by the Netherlands Organization for Scientific Research (NWO) via the Dutch Foundations for Physical Research (FOM) and for Life Sciences (SLW).

References

1 Kühlbrandt, W., Wang, D.N. and Fujiyoshi Y. (1994) Nature (London) 367, 614-621
2 Bittner, T., Irrgang, K.-D., Renger, G. and Wasielewski, M.R. (1994) J. Phys. Chem. 98, 11821-11826
3 Bittner, T., Wiederrecht, G.P., Irrgang, K.-D., Renger, G. and Wasielewski, M.R. (1995) Chem. Phys. 194, 311-322
4 Kwa, S.L.S., van Amerongen, H., Lin, S., Dekker, J.P., van Grondelle, R. and Struve, W.S. (1992) Biochim. Biophys. Acta 1102, 202-212
5 Savikhin, S., van Amerongen, H., Kwa, S.L.S., van Grondelle, R. and Struve, W.S. (1994) Biophysical J. 66, 1597-1603
6 Eads, D.D., Castner, E.W., Alberte, R.S., Mets, L. and Fleming, G.R. (1989) J. Phys. Chem. 93, 8271-8275
7 Du, M., Xie, X., Mets, L. and Fleming, G.R. (1994) J. Phys. Chem. 98, 4736-4741
8 Mullineaux, C.W., Pascal, A.A., Horton, P and Holzwarth, A.R. (1993) Biochim. Biophys. Acta 1141, 23-28
9 Pålsson, L. O., Spangfort, M.D., Gulbinas, V. and Gillbro, T. (1994) FEBS Lett. 339, 134-138
10 Peterman, E.J.G., Dukker, F.M., van Grondelle, R. and van Amerongen, H. (1995) Biophysical J., accepted
11 Nussberger, S., Dekker, J.P., Kühlbrandt, W., van Bolhuis, B.M., van Grondelle, R. and van Amerongen, H. (1994) Biochemistry 33, 14775-14783

SYNTHETIC METALLOCHLORIN AGGREGATES AS MODELS FOR SUPRAMOLECULAR ANTENNA COMPLEXES IN CHLOROSOMES OF GREEN PHOTOSYNTHETIC BACTERIA

H. Tamiaki,[1] M. Amakawa,[1] A. R. Holzwarth[2] & K. Schaffner[2]
[1]*Department of Bioscience and Biotechnology, Faculty of Science and Engineering, Ritsumeikan University, Kusatsu, Shiga 525-77, Japan*
[2]*Max-Planck-Institut für Strahlenchemie, D-45413 Mülheim an der Ruhr, Germany*

Introduction

Bacteriochlorophyll(BChl)s-*c* and *d* (see Fig. 1) are main light-harvesting pigments in extramembranous antenna complexes (=chlorosomes) of green photosynthetic bacteria [1]. As the model for BChls-*c* and *d*, zinc and magnesium chlorins **1** and **2** with a hydroxymethyl group at the 3-position were prepared (Fig. 1) [2,3]. The synthetic molecules self-aggregated in 1% (*v*/*v*) tetrahydrofuran (THF) and hexane to form oligomers absorbing at longer wavelengths than the monomeric form. Visible (vis) and circular dichroism (CD) spectra of several zinc chlorins showed that the requirements for the aggregation are central zinc, 3^1-hydroxyl, and 13^1-keto carbonyl groups in the molecule. Some assignments of the vibration bands measured by resonance Raman spectroscopy [4] indicated that the oligomers have two characteristic structural features: i) 5-coordinated metallochlorin macrocycle with an axial ligand, and ii) special hydrogen-bonds, C=O···H(X)O···M (= Zn and Mg). A comparison of the spectra of the *in-vivo* and *in-vitro* aggregates suggests that the supramolecular structure of the chlorosomal BChls-*c*, *d* aggregates is similar to those of the artificial oligomers in non-polar organic solvents.

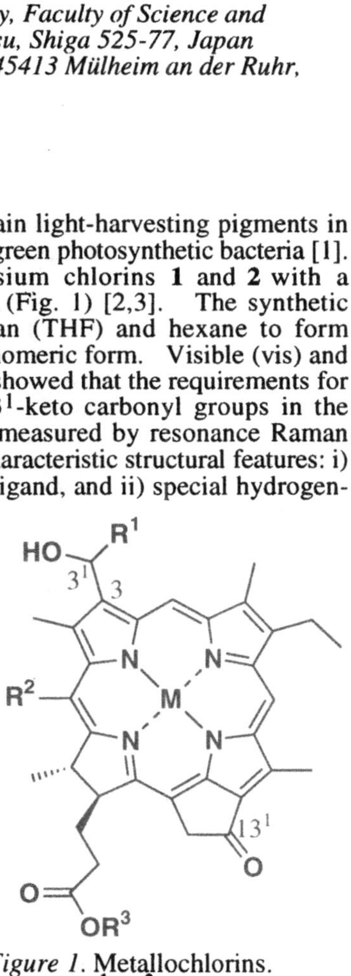

Figure 1. Metallochlorins.
M=Mg, R^1=R^2=Me : BChl-*c*
M=Mg, R^1=Me, R^2=H : BChl-*d*
M=Zn, R^1=R^2=H, R^3=Me : 1
M=Mg, R^1=R^2=H, R^3=Me : 2

Results and discussion

The zinc chlorin model molecule **1** readily aggregates to form oligomers in solid states as well as in non-polar organic solvents [3]. When a concentrated THF solution was highly diluted with hexane, a thin film was formed on the surface of the cuvette. The vis and CD spectra of this film are almost identical with those of oligomers in non-polar solution (Fig. 2), showing that the oligomeric species is exclusive in the solid phase. Infrared (IR) spectra of the aggregate film prepared by removal of the solution and subsequent drying *in vacuo*, were measured on a thin gold film by microscopic reflection absorption spectroscopy (Fig. 3). They exhibit several bands in

P. Mathis (ed.), Photosynthesis: from Light to Biosphere, Vol. I, 61-64

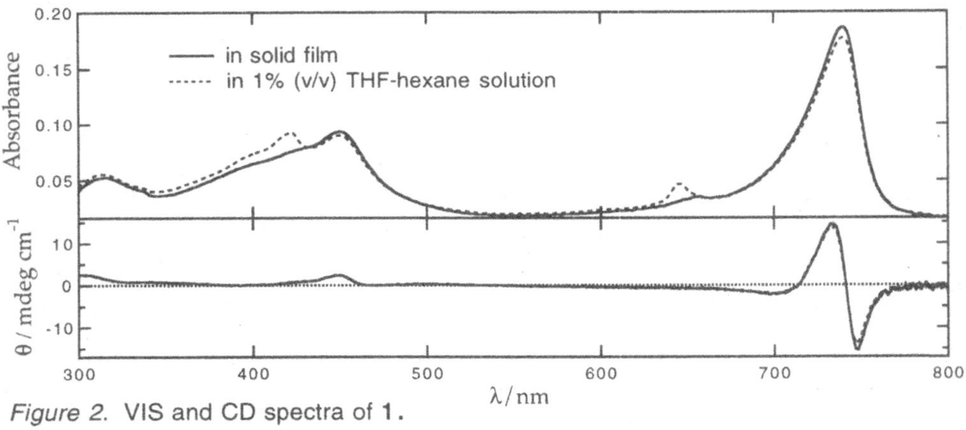

Figure 2. VIS and CD spectra of **1**.

the marker region of 1500–1630 cm^{-1} characteristic of the C–C and C–N stretching vibrations of the chlorin skeleton [5]. These bands are comparable to those in a homogeneous THF solution, where **1** is monomeric and 5-coordinated with an axial solvent molecule. Therefore, zinc in the film should be 5-coordinated with an axial ligand, a pyramidal ZnN_4L arrangement.

Judged by literature data [6], the ester C=O stretching vibrations about 1735–1740 cm^{-1} in the film and in THF are free from any intramolecular interaction. The corresponding band of the 13^1-ketone appears at 1655-cm^{-1} band, a pronounced down shift from 1691 cm^{-1} in neat THF. This must be due to a particularly strong bonding of the ketone to either of the two available sites, *viz.* coordination to central zinc or hydrogen bonding to the 3^1-hydroxyl. In either case, the frequency is lower than literature values (>1660 cm^{-1}) known for similar coordinations [7]. An unequivocal assignment was possible with the help of the synthetic compound **3** which combines two zinc chlorin chromophores each possessing a keto group but lacking hydroxyl. In the methanol-locked bis chromophore complex [8] of **3** in 1% (v/v) MeOH and CH_2Cl_2 (Fig. 4), characterized by the C=O···H(Me)O···Zn bonding arrangement, the C=O stretching vibration is 1655 cm^{-1}. Consequently, the same band frequency of **1** in the solid film is assigned to a ketone analogously hydrogen-bonded to a hydroxyl coordinated to zinc.

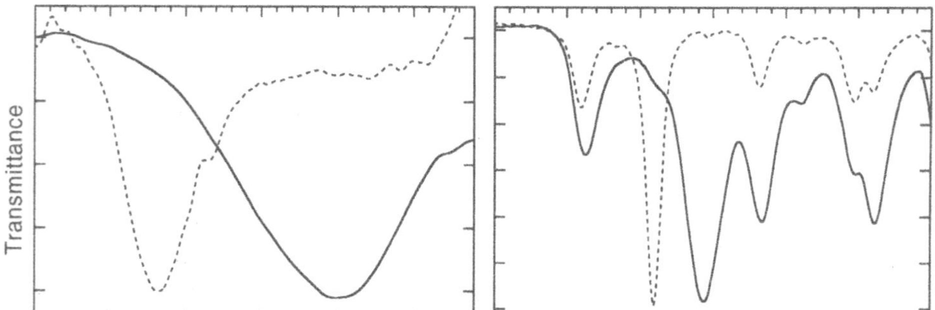

Figure 3. IR spectra of **1** in solid film (——) and in THF solution (-------).

The O–H stretching band of **1** at around 3200 cm^{-1} in the film is down-shifted by about 200-cm^{-1} from the O–H stretching in normal hydrogen-bonding of **1** to a THF oxygen (Fig. 3). This is in accord with an especially strong hydrogen bonding in the film, as in C=O···H(X)O···Zn where the hydrogen bonding-donating character of the hydroxyl hydrogen is increased by the coordination to zinc and therefore the hydroxy group is hydrogen-bonded to the C-13^1 ketone much more strongly than in a usual hydrogen bond.

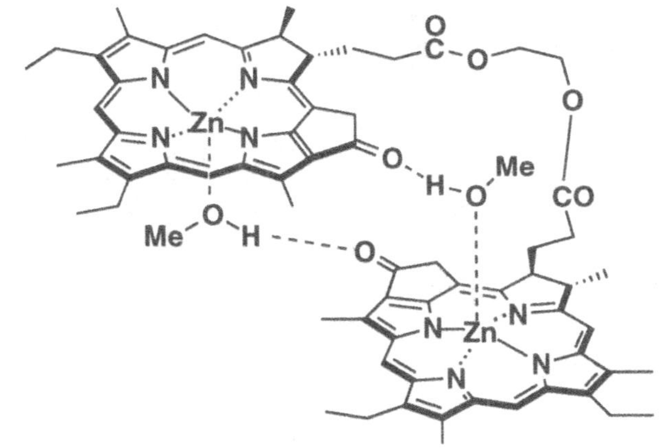

Figure 4. Methanol-locked complex of bis zinc chlorin **3**.

The above results lead to the supramolecular structure of zinc-chlorin aggregates schematically drawn as in Fig. 5.

Like the zinc chlorin **1**, the solid film of the magnesium analog **2** contained exclusively oligomers, as judged by the vis and CD absorption spectra [3]. The IR bands of this film could be assigned as follows, based on comparison with literature data for other magnesium chlorins [9-11]: O–H and C=O stretching vibrations at 3130 and 1650 cm^{-1}, respectively to C=O···H(X)O···Mg, and another carbonyl band at 1733 cm^{-1} to the methyl ester which exhibits no interaction with other functional groups.

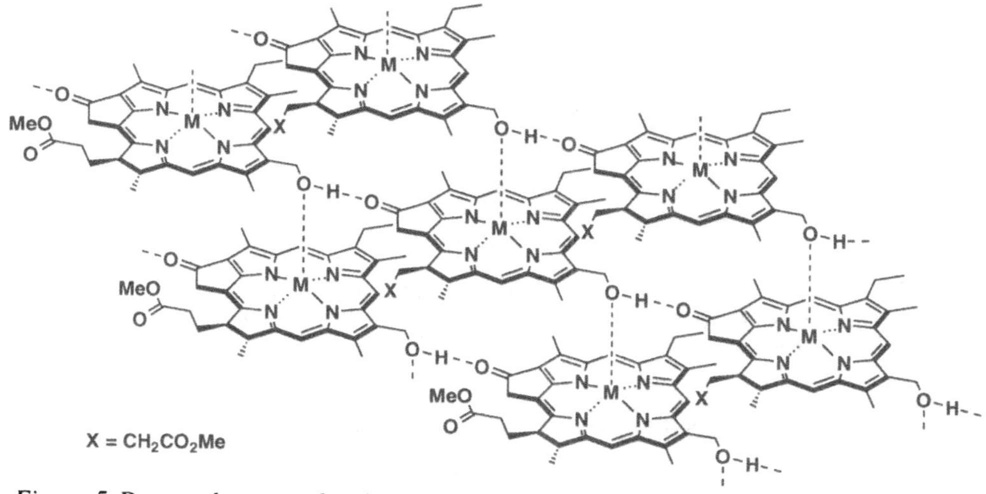

Figure 5. Proposed supramolecular strucuture of zinc and magnesium chlorin aggregates (**1 : M=Zn** and **2 : M=Mg**).

Figure 6. Dependency of metallochlorin IR bands in the solid film on Ct-N distance on the corresponding metalloporphyrins.

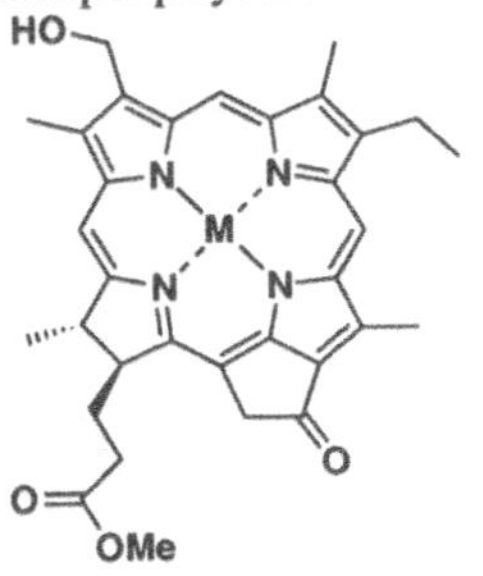

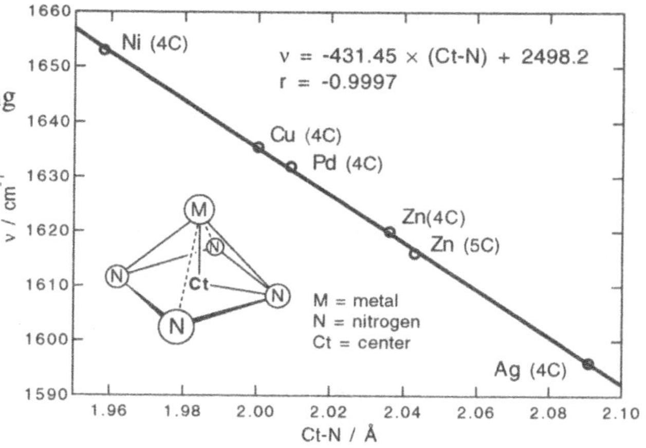

On the other hand, IR spectra of several metallochlorins were measured in the solid film. The highest frequency ν of C–C and C–N vibration bands are linearly correlated with the Ct-N distance between the center of the chlorin macrocycle and the nitrogen atom of the pyrrole determined by crystallographic data of the corresponding metalloporphyrin (Fig. 6), which has already been reported for metallopheophytins-*a* by Fujiwara and Tasumi [5]. Oligomeric Mg-**2** has the band at 1611 cm^{-1} and the Ct-N is estimated to be 2.056 Å from the equation. This value is identical to 2.055 (6) Å in $H_2OMgTPP$ and magnesium in the film is 5-coordinated with an axial ligand.

In summary, the vis, CD and IR data show that the supramolecular arrangements of the artificial magnesium and zinc chlorin aggregates are quite similar (Fig. 5). In natural chlorosomes, BChls-*c* and -*d* self-aggregate to form a similar two-dimensional arrangement and rolling the sheet leads the rod elements of main light-harvesting component in chlorosome found by electron microscopy.

This work was partially supported by Kumagai Foundation for Science & Technology.

References

1. Tamiaki, H. *Coord. Chem. Rev.* in press.
2. Tamiaki, H., Holzwarth, A. R., & Schaffner, K. (1992) *J. Photochem. Photobiol. B: Biol.* **15**, 355–360.
3. Tamiaki, H., Amakawa, M., Shimono, Y. Tanikaga, R., Holzwarth, A. R., & Schaffner, K. in preparation.
4. Hildebrandt, P., Tamiaki, H., Holzwarth, A. R., & Schaffner, K. (1994) *J. Phys. Chem.* **98**, 2192–2197.
5. Fujiwara, M. & Tasumi, M. (1986) *J. Phys. Chem.* **90**, 5646–5650.
6. Ballschmiter, K. & Katz, J. J. (1969) *J. Am. Chem. Soc.* **91**, 2661–2677.
7. Lutz, M. & Mäntele, W. (1991) in *Chlorophylls* (Hugo, S., ed.), pp.855–902, CRC, Boca Raton, FL.
8. Boxer, S. G. & Closs, G. L. (1976) *J. Am. Chem. Soc.* **98**, 5406–5408.
9. Bystrova, M. I., Mal'gosheva, I. N., & Krasnovskii, A. A. (1979) *Mol. Biol.* **13**, 440–451.
10. Katz, J. J., Bowman, M. K., Michalski, J., & Worcester, D. L. (1991) in *Chlorophylls* (Hugo, S., ed.), pp.211–235, CRC, Boca Raton, FL.
11. Chiefari, J., Griebenow, K., Fages. F., Griebenow, N., Balaban, T. S., Holzwarth, A. R., & Schaffner, K. (1995) *J. Phys. Chem.* **99**, 1357–1365.

AN INVESTIGATION OF THE LHC FAMILY OF GENES AND PROTEINS IN THE HOMOSPOROUS FERN *Ceratopteris richardii.*

O. Oberschmidt [1], S. Kim[2], J.M. McGrath[2], M.M. Jansco[2], R. Bassi[3,4], B. Piechulla[1] & E. Pichersky[2]
[1]Institute of Plant Biochemistry, University of Gottingen, 37073 Gottingen Germany, [2]Biology Department, University of Michigan, Ann Arbor MI 48109 USA, [3]Biology Department, University of Padova, Padova Italy, [4]Present address: Faculty of Science, University of Verona, Verona Italy

1. Introduction

At least ten different types of chlorophyll *a/b*-binding (CAB) polypeptides per plant have been found in angiosperms and gymnosperms (1). The expression characteristics of the *Lhc* genes, which encode the CAB proteins, have been examined in great detail in tomato and other plants (2). Their transcription in angiosperms follows a distinct pattern; maximal transcription occurs at midday, whereas no transcription occurs during the night. Control of this pattern is attributed to a circadian 'clock' (3).

In contrast to the extensive investigation of the *Lhc* gene family in seed plants, the corresponding proteins of non-seed plants, such as the ferns and fern allies, have been little characterized. Several *Lhcb* genes (encoding LHCII proteins) have been previously isolated from a homosporous fern, *Polystichum munitum*, but most of these were defective and unlikely to be expressed (4). The single potentially functional gene appeared to encode an LHCII CAB polypeptide, but its mode of expression was not investigated. Here we report the isolation of several *Lhcb* cDNA clones from another homosporous fern species, *Ceratopteris richardii.* We show that there are several distinct LHCII and LHCI CAB proteins in this fern, and that the *Lhcb* genes examined in this study are under the control of a circadian clock. We have also found that these genes are expressed in both gametophytic and sporophytic tissues.

2. Results and Discussion

2.1 Isolation and characterization of LHCI and LHCII: Unstacked *C. richardii* thylakoids (chl *a/b* ratio=2.767) solubilized with 0.06% (w/v) dodecylmaltoside and fractionated by flat bed IEF yielded twelve green fractions. Analysis of each of these fractions on SDS-PAGE revealed that in many cases, some of the proteins were present in more than one IEF fraction. For example, IEF fractions 1 through 3 contain LHCII proteins in the 28-30 kDa range in different relative amounts. To resolve the IEF fractions into smaller, non-denatured chlorophyll-binding complexes, each green band (or a small group of closely spaced bands) were subjected to a sucrose gradient ultracentrifugation. Each sucrose gradient fractionation yielded several green bands, which were analyzed with respect to their absorption spectra and cross-reactivity of their polypeptides (in Western blots) with antibodies against specific CAB and PS proteins. These results are summarized in Table I. The heaviest sucrose gradient band in fractions 4-10 was identified as a PSI-LHCI complex, which had a chl absorption maximum at 679.2 nm. The PSI reaction center was isolated in IEF fraction 4, whereas the PSII reaction center was found in IEF fractions 5-10.

P. Mathis (ed.), Photosynthesis: from Light to Biosphere, Vol. I, 65–68.

An additional 26 kDa polypeptide was present only in fraction 3 and it was tentatively identified as CP24. Its chl *a* absorption maximum is 673.4 nm, which is red-shifted when compared with maize LHCII (674.8 nm). IEF fractions 11 and 12, which appear enriched in polypeptides of 28 and 29 kDa, have low chl *b* content and their chl *a* absorption maximum is at 676 nm; these characteristics are indicative of the presence of the CP26 complex (5).

Protein sequence determination of several electroeluted proteins from LHCI and LHCII was attempted. All were found to be N-terminally blocked, except for one protein in LHCII. The position of this N-terminal sequence, about a dozen residues shorter than in LHCII Type I and II CAB proteins (Fig. 1), strongly suggests that it is derived from an LHCII Type III CAB protein, encoded by an *Lhcb*3 gene. In angiosperms too this is the only LHCII CAB protein that is not N-terminally blocked (6).

Table 1

IEF green band No.	Sucrose gradient band No.	Chl a/b ratio	Chl a Red Max (nm)	Immunolgoical Identification
1	1	1.15	673.80	LHCII
	2	1.15	673.70	LHCII
2	1	1.15	673.50	LHCII
	2	1.37	673.00	LHCII
3	1	1.11	673.10	LHCII
	2	1.22	673.60	LHCII + CP24(?)
	3	1.51	672.90	LHCII + CP24(?)
4	1	9.53	679.00	PSI-LHCI
	2	>20	677.30	PSI RC
	3	3.17	676.30	?
	4	3.32	673.10	LHCII
5 - 7	1	8.48	679.30	PSI-LHCI
	2	>20	675.10	PSII RC
	3	>10	673.90	?
	4	4.41	675.50	?
	5	4.76	673.90	?
8 - 10	1	8.52	679.10	PSI-LHCI
	2	>20	674.80	PSII RC
	3	>10	674.20	PSII RC
	4	4.15	675.20	?
11 - 12	1	3.61	675.90	?
	2	>15	676.10	CP29/CP26

2.2 Isolation and characterization of Lhcb cDNA clones: A cDNA library (7) made from mRNA isolated from light-germinated green gametophytes (*C. richardii* gametophytes require light for germination) was screened with *Lhcb* gene probes from tomato and *Polystichum munitum*. Two full-length and one partial cDNA clones (*Lhcb**1, *Lhcb**2, *Lhcb**3) were obtained and characterized. The amino acid sequences of the proteins encoded by *Lhcb**1 and *Lhcb**2 (Fig. 1) indicate that they are very similar to the previously characterized *P. munitum Lhcb* gene (92-97% amino acid identity) and also to LHCB1 and LHCB2 sequences from angiosperms (89%-91% amino acid identity with the tomato *Lhcb*1*7 gene product). As was shown earlier (4), it is not possible to distinguish LHCB1 and LHCB2 sequences in ferns since the gene duplication that gave rise to the two types may have occurred in the angiosperm lineage after it diverged from the fern lineage. The incomplete cDNA clone of Lhcb*3 encodes a protein that is more divergent from any of these fern and angiosperm sequences (76-80% amino acid sequence identity), but the shortness of the sequence precludes a definite assessment of its character. Southern blots of *C. richardii* genomic DNA with cDNA clone Lhcb*1 (Fig. 1 in ref. [8]) indicate that the number of *Lhcb* genes in this species is >5, and possibly >10, consistent with the data showing multiple LHCII proteins (Table 1). The molecular data show that *C. richardii* has at least one *Lhcb*3-like gene encoding an LHCII Type III CAB protein, and several *Lhcb* genes encoding CAB polypeptides of intermediate LHCII Type I - Type II nature.

```
                                        ↓
1. ..TS.MA...STFA.KAVKLSPSSSEIT--..NG..T.RKTAT.AKPAS...P......VK.......ES..
2. ..TS..A.T.TFS-....KPVN-...............-....SK.--........S...............
3. MAAATASLSSASVIGQQLN----ELSRKVGAGEARVQM-MAPKKSA--PSGSIWYGPDRPLYLGPFSGSPPS
4. ..........SVFS....KPAAS...............RTSS..P.--........................
5. ------------------------------------------------------------------------
6.                                                    DY.......K...
7.                                                  SNDL......VK...

1. ..................................C...........F....R..............S...S
2. ..S...............................S................................S....
3. YLTGEFPGDYGWDTAGLSADPETFAKNRELEVIHARWAMLGALGCVTPELLAKNGVKFGEAVWFKAGAQIFA
4. ..............................................I.D.......................
5. ------------------------------------------------...................S....

1. ..........................V..................V..........................
2. ................................................I.......................
3. EGGLDYLGNPSLVHAQSILAIWACQVILMGAVEGYRVAGGPLGEVEDPLYPGGSFDPLGLADDPEAFAELKV
4. ............I.......................I...................................
5...........N...............V..........G......GL..I....A..........D.......

1. ..I........................L...A......VN.....F..N.V...
2. ......................................A.........N.....
3. KELKNGRLAMFSMFGFFVQAIVTGKGPIENLSDHLADPTVNNAWAYATQFTPGK
4. ................................................N.V...
5. ..I.........................K..........E........N.....
```

Fig. 1: Comparison of sequences of protein encoded by *Lhcb* genes. 1) tomato *Lhcb*1*7, 2) *P. munitum Lhcb**1 (formerly "F3"), 3) *C. richardii Lhcb**1, 4) *C. richardii Lhcb**2, 5) *C. richardii Lhcb**3, 6) N-terminal sequence of a *C. richardii* LHCII protein, 7) N-terminal sequence of tomato LHCB3*1. Dots indicate identity with sequence shown in full. Dashes indicate missing sequence or introduced gaps.

2.3 *Expression of Lhcb genes in C. richardii*: Northern blots of mRNA extracted from green gametophytes show substantial expression of *Lhcb* genes (data not shown). Northern blots from sporophytic leaves (Fig. 2) show that steady-state levels of total *Lhcb*

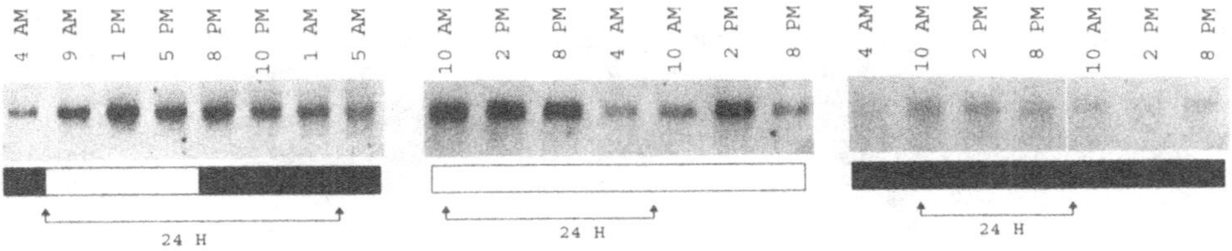

Fig. 2: Northern blots of mRNA extracted from plants grown under different light regimes. mRNAs were extracted at times indicated.

mRNA of plants grown under 12 h light/12 h dark period peak in early afternoon and then decline (see also ref. 9). Some mRNA is seen throughout the dark period, which may indicate that *Lhcb* mRNA in *C. richardii* is rather stable in the dark. In plants subjected to continuous illumination or darkness, the levels of *Lhcb* mRNA continue to oscillate (Fig. 2), indicating that the expression of the *Lhcb* genes in *C. richardii* is under a circadian

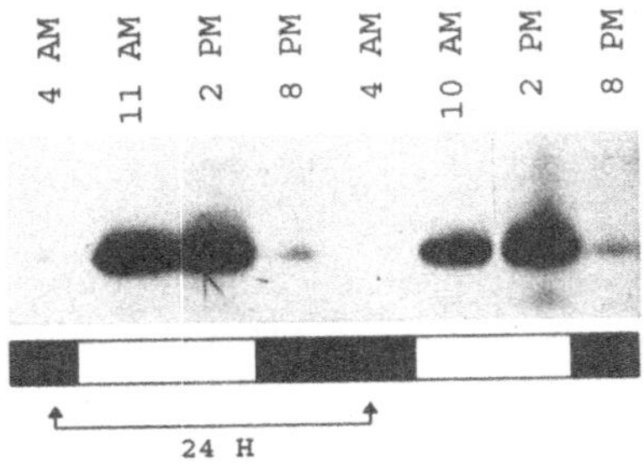

Fig. 3: Primer extension experiment on mRNAs extracted from plants growing under light/dark regime, using a primer specific for *Lhcb**1.

control. Primer extension experiments using gene-specific oligonucleotides show that the amplitudes of oscillation of some *Lhcb* mRNAs is more pronounced, *i.e.*, there is almost no mRNA in the dark period (*e.g.*, *Lhcb**1, Fig. 3), which might indicate that such mRNAs are least stable in the dark (2). The primer extension experiments with both *Lhcb**1 and *Lhcb**2 also showed that both genes are expressed in gametophytic and sporophytic tissues, with Lhcb*2 being more abundant in both tissues (data not shown).

3. REFERENCES

1 Jansson S (1994) Biochim Biophys Acta 1184:1-19.

2 Kellmann J, M Wiesse M, Pichersky E, Piechulla B (1993) Mol Gen Genet 237:439-448

3 Piechulla (1993) Plant Mol Biol 22:533-542.

4 Pichersky E, Soltis P, Soltis D (1990) Proc Natl Acad Sci 87: 195-199.

5 Bassi R, and Simpson D. (1987) In Progress in Photosynthesis Research (edited by J. Biggins), Vol. II, pp 81-88. Martinus Nijhof, Dordrecht.

6 Schwartz E, Stasys R, Aebersold R, McGrath JM, Green BR and Pichersky E (1991) Plant Mol Biol 17:923-925.

7 McGrath JM, Hickok LG, Pichersky E (1994) Plant Syst Evol 189: 193-202.

8 McGrath JM, Hickok LG, Pichersky E (1994) Plant Syst Evol 189: 203-210.

9 Oberschmidt O, Hucking C, Piechulla B (1995) Plant Mol Biol 27: 147-153.

B800 SUBSTITUTION OF THE B800-850 LIGHT HARVESTING COMPLEX OF *RHODOPSEUDOMONAS ACIDOPHILA 10050*

Gelhaye E.*, McDermott G.*, Bandilla M.#, Ucker B.#, Scheer H.# and Cogdell R. J.*

* : Institut of Biomedical and Life Sciences, University of Glasgow, G128QQ,U.K.
# : Botanishes Institut der Universitat Munchen, 8000 Munchen, Deutschland

1. Introduction

The light harvesting apparatus of purple photosynthetic bacteria is comprised of two types of discrete pigment-protein complexes (LH1 and LH2) which contain non covalently bound bacteriochlorophyll (Bchl) and carotenoid molecules. The crystal structure of *Rps. acidophila* 10050 LH2 complex has been recently described (1). The Bchla molecules in this complex are divided into two spectral forms which absorb at 800 nm and 850nm. The B850 molecules are strongly exciton coupled forming a closely interacting ring inside the protein. The B800 molecules are largely monomeric and located on the cytoplasmic side of the membrane (1). The aim of this project was to develop a method to exchange the B800 Bchl in the B800-850 antenna complex from *Rps acidophila* 10050, so that the functionnal effects of changing the type of Bchl could be investigated.

2. Method

The B800-850 complex of *Rps acidophila* 10050 was purified as previously described following solubilization with LDAO (2). At this stage, the purified complex is stored in Tris-HCl 20 mM, LDAO 0.1%, pH 8.0 buffer. The basis of the removal of the B800 absorbing Bchla's is to incubate the complex at pH 5.3. At this pH the complex is unstable in LDAO. However, if the complex is exchanged in to the novel detergent Triton BG the complex is not denatured by incubation at this acid pH. In this case incubation at pH 5.3 using Tris-acetate buffer (final concentration 100 mM) results in slow removal of the 800 nm Bchla's. Reconstitution experiments could be performed by increasing the pH from 5.3 to 8.0 and adding LDAO (final concentration 0.1%). LDAO was found to be required for the reconstitution. Reconstitution could be achieved with either added native Bchla or modified Bchl's. Where modified Bchl's were to be reconstituted the complex was purified by sucrose gradient centrifugation to remove any liberated native Bchla's prior to the reconstitution.

3. Removal of B800 and Reconstitution with Bchla

Absorption spectra

The major effect of the pH decrease is the disappearence of the Bchl monomer Qy band near 800nm (figure 1). Correlated with this decrease, two small broad peaks grow in at approx. 680 nm and 770 nm corresponding to oxidized free Bchl and free Bchl

P. Mathis (ed.), Photosynthesis: from Light to Biosphere, Vol. I, 69–72.

molecules respectively (3). Incubation at pH 5.3 also causes the following smaller spectroscopic changes : (i) a red-shift (2nm) and decrease in intensity (appro. 20%) of the Bchl dimer absorption band at 857 nm, (ii) a red-shift (3nm) and decrease in intensity (appro. 40%) of the Bchl 590nm absorption band (Qx), (iii) a blue-shift (3nm) and decrease in intensity (10-15%) of the carotenoid absorption bands.
The pH increase (from 5.3 to 8.0) allows a large recovery of the 800 nm absorbing band, 70-80 % of the native complex ratio between 857 nm and 800 nm band intensities (figure 1). The other small spectroscopic changes induced by the B800 removal are reversed by the pH increase. The absorption spectra of reconstituted B800-850 and native B800-850 complexes are also quite similar.

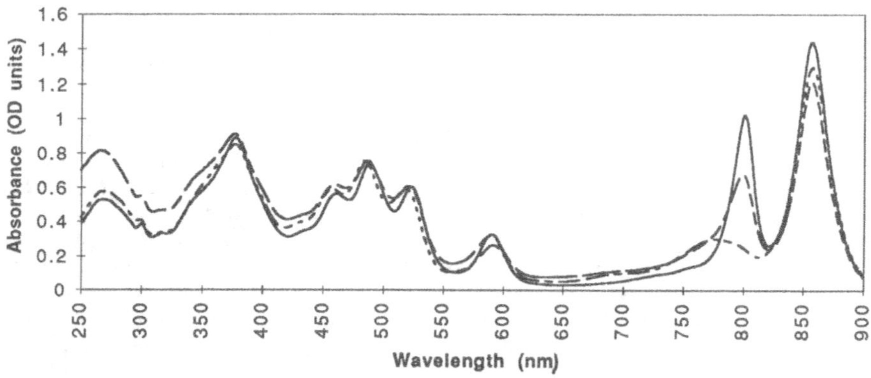

Figure 1 : Absorption spectra of B800-850 (–), B850 (pH decrease from 8.0 to 5.3) (- -) and recB800-850 (pH increase from 5.3 to 8.0) (···).

Fluorescence excitation spectra

Even after the B800 removal due to the pH decrease, the carotenoid molecules are able to transfer energy to 857nm absorbing Bchla's. Fluorescence excitation spectra (emission at 860 nm) performed before and after B800 removal showed that the carotenoid bands have slightly been blue shifted (appro. 3 nm) and that the efficiency of energy transfer from carotenoids to the 857 nm absorbing Bchla's is reduced by appro 55 % in the absence of B800.
The fluorescence excitation spectrum of the reconstituted B800-850 complex again clearly resembles that of the original native complex. The efficiency of energy transfer between carotenoids and 857nm absorbing Bchla's was found to be dependant of the extent of B800 recovery. These results show that the treatment to remove 800 nm absorbing Bchla's and to reconstitute the B800-850 complex doesn't alter significantly the efficiency of energy transfer between carotenoids and Bchla's molecules.

Circular dichroism spectra

Circular dichroism analysis of the light harvesting complex has been performed before and after the B800 removal (figure 2). The pH decrease induces some dramatic changes on the complex CD spectrum. In particular, the monomeric Bchla's signal at 800 nm (800 nm absorbing Bchla's) has nearly disappeared and the dimeric Bchla's signals (857 nm absorbing Bchla's) have been blue-shifted (appro 10 nm) by the pH decrease. The carotenoid bands have been blue-shifted by appro. 3nm after B800 removal. Other modifications due to the applied treatment was observed in the low wavelengths (between 440 and 250 nm), with in particular a dramatic change in the Soret signal. After the pH increase, the CD spectrum changes, induced by the B800 removal, are reversed in the reconstituted complex CD spectrum. In fact, except for the ratio between the monomeric and dimeric Bchla's signals, no significant difference was observed between CD spectra of both reconstituted and native complexes (figure 2). These results show that the treatment to remove the 800 nm absorbing Bchla's and to reconstitute the B800-850 complex doesn't significantly alter the structure of the complex.

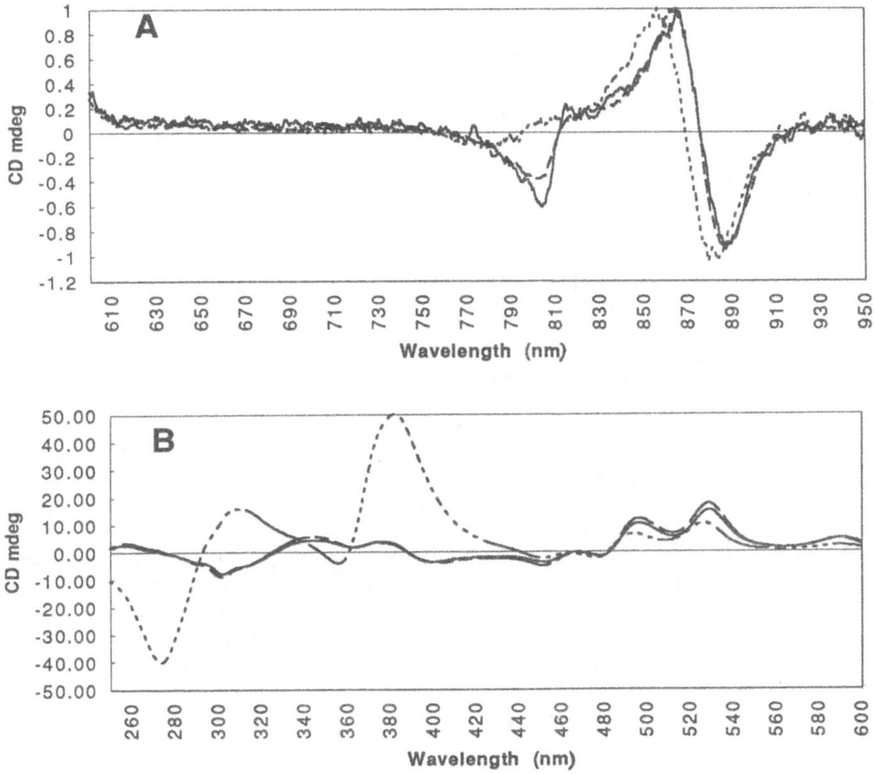

Figure 2 : Circular dichroism spectra of B800-850 (–), B850 (pH decrease from 8.0 to 5.3) (- -) and recB800-850 (pH increase from 5.3 to 8.0) (···). **A**: between 950 and 600 nm, **B**: between 600 and 250 nm.

4. Reconstitution with Acetyl-chla

After B800 removal, the complex was purified in order to remove the liberated BChla's. Reconstitution experiments were performed adding Acetyl-chla, which was obtained as previously described (4). This modified Bchl was chosen because its binding form absorbs at about 685 nm in the *Rps acidophila* 10050 LH2 complex allowing a clear distinction from any 800nm absorbing Bchla (figure 3). Energy transfer has been detected between carotenoids and Acetyl-chla's. A CD spectrum of this B685-850 complex showed the presence of bound modified pigments. These both results confirmed that some Acetyl-chla molecules are really integrated into the complex. In these experimental conditions, we didn't detected any 800 absorbing Bchla's.

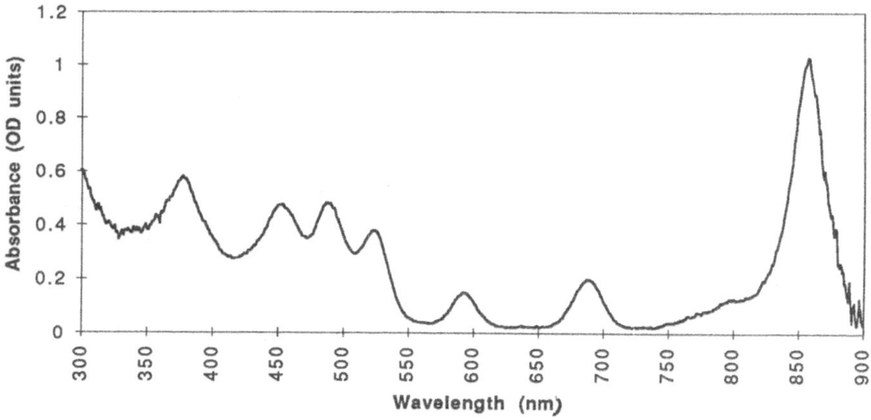

Figure 3 : Absorption spectra of Acetyl-chla reconstituted complex.

5. Conclusion

Here, we report a method for removing the monomeric Bchla's (absorbing at 800nm) from the light harvesting complex LH2 of *Rps. acidophila* 10050. The decrease of the pH allowing the B800 removal induces some changes in the complex structure. However, these phenomena are reversible. In fact, after reconstitution, we don't observe any alteration of the energy transfert efficiency between carotenoids and Bchla's. Furthermore, we didn't detected any structural modifications after reconstitution using circular dichroism. Studies of the effects of these modifications on the function of the antenna complex are in progress in our laboratories.

Acknowledgments

This work was suppported by grants from the BBSRC, E. U. and the British Council.

References

1 McDermott, G., Prince, S. M., Freer, A. A., Hawthornethwaite-Lawless, A. M., Papiz, M. Z., Cogdell, R. J. and Issacs, N. W. (1995) Nature : 374, 517-521.

2 Cogdell, R. J., Durant, I., Valentine, J., Lindsay, J.G. and Schmidt, K. (1983) Biochim. Biophys. Acta. : 722, 427-455.

3 Chadwick, B. W., Zhang, C., Cogdell, R. J. and Frank, A. (1987) Biochim. Biophys. Acta. : 893, 444-451.

4 Smith, J. R. L. and Calvin, M. (1966) J. Am. Chem. Soc. : 88, 4500.

REGULATION OF THE EXCITON DENSITY IN PHOTOSYNTHETIC ANTENNA SYSTEM: "CRUISE CONTROL"

Wiesław I. Gruszecki[a], Alaka Srivastava[b], Magdalena Matuła[a], Zbigniew Krupa[c], Reto J. Strasser[b]
a) Department of Biophysics and c) Department of Plant Physiology, Maria Curie-Skłodowska University, PL-20031 Lublin, Poland; b) Laboratory of Bioenergetics, University of Geneva, CH-1254 Jussy-Geneva, Switzerland

1.Introduction

Efficient operation of the photosynthetic apparatus of plants under unstable illumination requires activity of several adaptation mechanisms at different organisation levels: plant and leaf movement, chloroplast movement, antenna complex migration within thylakoid membrane. Apart from those physiological processes, there exists regulation at the molecular level of photosynthetic antenna system operating via decreasing the singlet exciton density demonstrated by the chlorophyll fluorescence quenching [1]. Excess of energy directed to the photosynthetic apparatus is supposed to be dissipated as heat. In this report it is directly demonstrated that the increase of heat emission from photosynthetic apparatus is correlated with:

1) saturation of photosynthesis
2) singlet excitations quenching

2.Materials and Methods

Winter rye (*Secale cereale* L., cv. Pastar) plants were grown for 10 days in a growth chamber under controlled experimental conditions. Heat emission was measured with a Unipan 605M DSC calorimeter and modulated fluorescence was measured with a PAM101 chlorophyll fluorometer (Waltz) controlled by computer as described previously [2]. Photosynthetic oxygen evolution from chloroplast suspension was measured with a Hansatech oxygen electrode DW1. Chloroplasts were isolated from rye leaves and suspended in 50 mM Hepes-KOH buffer (pH 7.6) containing 0.1 M sorbitol, 5 mM $MgCl_2$, 5 mM NaCl, 2 mM ferricyanide, 2.5 mM NH_4Cl. White actinic light was provided with a halogen lamp equipped with the broad-band filter with transmittance between 350 and 800 nm.

3.Results and Discussion

Independently of light intensity directed to the photosynthetic apparatus of a leaf the steady state chlorophyll fluorescence level expressing singlet exciton density remains constant and close to its initial value [3]. Such a fact is a manifestation of operation of a physiological regulation to maintain constant low level of singlet excitations. Since the rate constant of the singlet energy trapping from antenna system by reaction centre (~ 3 ps^{-1},

P. Mathis (ed.), Photosynthesis: from Light to Biosphere, Vol. I, 73–76.

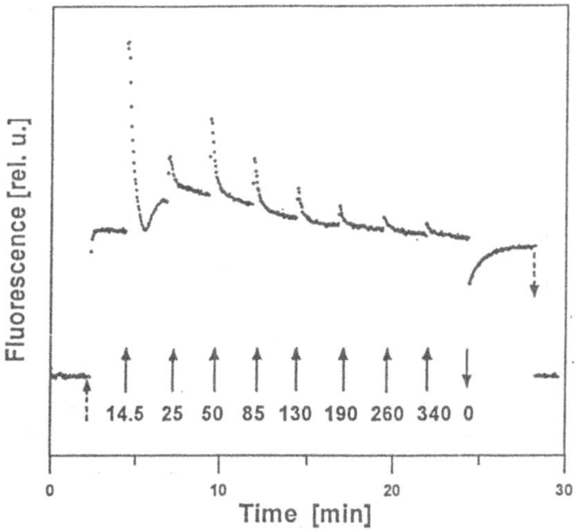

Figure 1. Time course of modulated fluorescence of an intact rye leaf illuminated with very weak modulated red light and white actinic light. Arrows indicate the moment at which modulated light was turned on (↑), turned off (↓) and at which the actinic light intensity was increased to the level indicated below each arrow (↑, in $\mu Em^{-2}s^{-1}$) or turned off (↓).

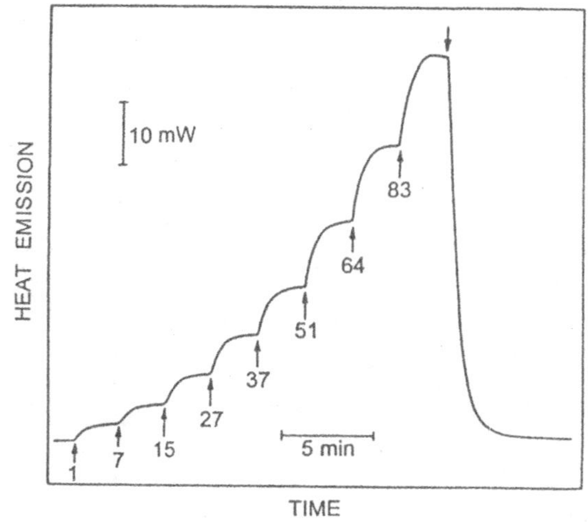

Figure 2. Time course of heat emission from an intact rye leaf illuminated with white light the same as in Fig. 1. Arrows indicate the moment at which light power directed to the leaf was increased to the level indicated below each arrow (↑, in mW).

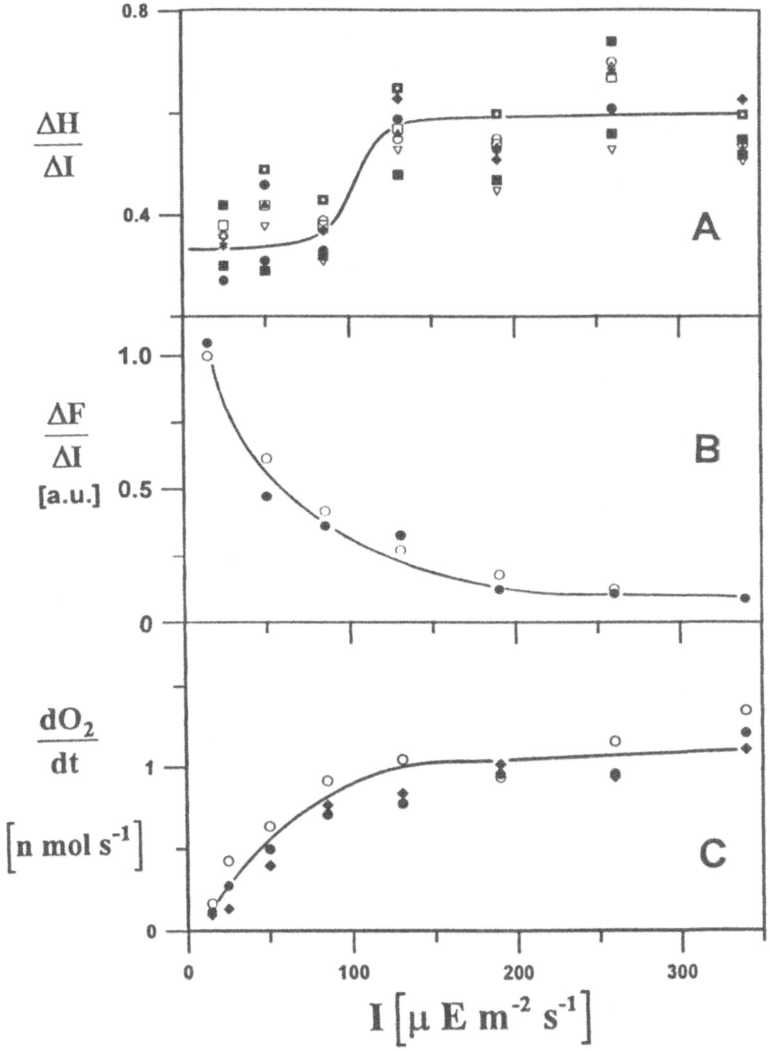

Figure 3. (A) Fraction of light energy directed to a rye leaf emitted as heat ΔH/ΔI; (B) Fraction of light quanta directed to a rye leaf emitted as fluorescence ΔF/ΔI; (C) The rate of photosynthetic oxygen evolution from rye chloroplast dO_2/dt. All three parameters are presented versus light intensity I. Different points denote different experiments. The calculations presented in panels A and B are based on the experiments similar to those presented in Fig. 2 and Fig. 1, respectively.

[4]) is very high relative to other chlorophyll singlet exciton deexcitation processes including the radiative one (~ 1 ns^{-1}, [4]), chlorophyll fluorescence may be recognised as a measure of the density of singlet excitations unutilizable by photosynthesis. This means that excitons which are not utilised by photosynthesis are quenched. Such a quenching is a manifestation of operation of a regulatory mechanism. Fig. 1 presents excitation quenching in rye leaves exposed to the increasing light intensity and Fig. 2 presents heat emission from a rye leaf under the same experimental conditions. A fraction of energy directed to the leaf and emitted as heat ($\Delta H/\Delta I$) and a fraction of light quanta directed to the leaf and emitted as fluorescence ($\Delta F/\Delta I$) calculated on the basis of experiments the same as those presented in Fig. 1 and Fig. 2 are depicted in Fig. 3 versus the current total light intensity. Panel C of Fig. 3 presents also the rate of photosynthetic oxygen evolution from chloroplasts isolated from the same plants under the same light conditions. As it may be seen, the increase of energy dissipation as heat is related to the saturation of photosynthesis. At that time the singlet exciton density monitored by the $\Delta F/\Delta I$ decreases drastically and remains constant and very low. The question arises about the nature of such a regulation (quenching) and about the triggering mechanism to activate this regulation. Since the light-dependent fluorescence quenching operates also in isolated antenna complexes [5,6] it seems that it is an immanent attribute of these molecular assemblies. Rapid energy dissipation via fluorescence and nonradiative singlet deexcitation makes the intersystem crossing leading to chlorophyll triplet state formation a very probable intermediate state in signal transduction leading to activation of regulatory singlet excitation quenching [7]. Several physical mechanisms may be considered as responsible for such a quenching: the singlet-triplet annihilation between excited chlorophyll molecules, triplet-triplet or singlet-singlet excitation transfer from chlorophyll to carotenoids. The latter mechanism was postulated to link the excitation quenching in antenna system accompanying saturation of photosynthesis with accumulation of zeaxanthin due to the operation of the xanthophyll cycle [8,9]. On the other hand, the same effect of the strong light-related excitation quenching would be explained in terms of structural effect to antenna proteins of violaxanthin which concentration in the photosynthetic apparatus is under the control of the xanthophyll cycle [6,10].

References

1. Krause, G.H. and Weis, E. (1991) Ann. Rev. Plant Physiol. Plant Mol. Biol. 42, 313-349
2. Gruszecki, W.I., Wójtowicz, K., Krupa, Z. and Strzałka, K. (1994) J. Photochem. Photobiol. B: Biol. 22, 23-27
3. Havaux, M., Strasser, R.J. and Greppin, H. (1991) Environ. Exper. Bot. 31, 23-32
4. Holzwarth, A.R. (1991) in Chlorophylls (Scheer H., ed.) pp.1125-1151, CRC Press, Boca Raton, USA
5. Jennings, R.C., Garlashi, F.M. and Zucchelli, G. (1991) Photosynth. Res. 27, 57-64
6. Gruszecki, W.I., Kernen, P., Krupa, Z. and Strasser, R.J. (1994) Biochim. Biophys. Acta 1188, 235-242
7. Gruszecki, W.I., Matuła, M., Daniluk, A. and Krupa, Z. (1995) J. Photochem. Photobiol. B:Biol., in press
8. Frank, H.A., Cua, A., Chynwat, V., Young, A., Gosztola, D. and Wasielewski, P.M.R. (1994) Photosynth. Res. 41, 389-395
9. Owens, T.G. (1994) in Photoinhibition of Photosynthesis (Baker, N.R. and Bowyer, J.R., esd.) pp.95-107, Bios Scientific Publishers, Oxford, UK
10. Ruban, A.V., Young, A. and Horton, P. (1994) Biochim. Biophys. Acta 1186, 123-127

LOW INTENSITY FEMTOSECOND ABSORPTION SPECTROSCOPY ON THE LIGHT-HARVESTING CHL A/B COMPLEX LHC II

M.Hucke, M.G.Müller, A.R.Holzwarth
Max-Planck-Institut für Strahlenchemie, D-45470 Mülheim, Germany and
C.Mullineaux, P.Horton, Robert-Hill-Institute,
Department of Molecular Biology and Biotechnology, University of Sheffield, UK

1. Introduction

The main antenna complex present in higher plants and green algae is the Chl a/b containing light-harvesting complex LHC-II. The monomeric species consists of a single polypeptide of MW 26 KDa carrying at least 7 Chl a and 5 Chl b chromophores. The LHC-II complex occurs in vitro preferentially as a trimer. The structure of this complex has recently been determined by electron crystallography to 3.5 Å resolution [3]. These data indicate that the Chls are arranged more or less perpendicular to the plane of the membrane and roughly in two layers parallel to the membrane plane. The Chl-Chl distances vary in the range from about 7 to 11 Å. The chemical identity and exact orientation of the Chls within the complex has not been resolved yet. However, there is good hope that a more detailed structure can be obtained in the near future. A high resolution structure is one of the necessary but still insufficient prerequisites for a detailed interpretation of the very complex energy transfer kinetics.

2. Materials and Methods

LHC-II has been isolated in aggregated form by solubilization of thylakoids with Triton-X-100 and addition of Mg^{2+}. The aggregates were then disaggregated by dialysis and change of the detergent (0.7% digitonin and 0.7% n-octyl-β-D-glucopyranosid). The sample was resuspended in 100 mM sorbitol 5 mM tricine (pH 7.0). The total content of Chl was 129 µM. The Chl a / Chl b ratio was 1.5 as determined by the method of Porra et al. [2]. The aggregation state of the samples was checked by low temperature fluorescence spectra before and after the measurements and was found to be trimeric. All measurements were carried out at 4 °C.

Femtosecond absorption measurements have been carried out using a pump-probe apparatus based on a Ti-Sapphire Oscillator-Amplifier system. Excitation pulses were selected from an optical parametric generator (OPG) yielding pulses of about 80 fs FWHM. The detection pulses were taken from a white light continuum. The samples were contained in a rotating cuvette of 1 mm pathlength. Absorbance of the samples was adjusted to about

P. Mathis (ed.), Photosynthesis: from Light to Biosphere, Vol. I, 77–80.

0.4 at the excitation wavelength of 650 nm which excites preferentially Chl b. A pulse intensity of about 2 x 10^{13} photons /pulse/ cm^2 was used for excitation in the low intensity experiments. The polarization between pump and probe pulses was adjusted to 54.7° in order to exclude contributions by anisotropy decays to the observed kinetics. Absolute ΔA absorption difference spectra were measured. The excitation intensities at the sample were monitored carefully in order to base the annihilation analysis on correct excitation intensity values.

The data were analyzed by the global lifetime method. The results of the analysis are judged on the basis of residual plots and reduced χ^2-values. The assumed model function was a sum of exponentials. Lifetimes τ_i and their decay-associated difference spectra (DADS), which are the plots of the ΔA_i amplitudes for the decay components τ_i as a function of detection wavelength λ_d [4] were obtained. A DADS generally represents a superposition of the difference absorption spectra of the different intermediates detected.

3. Results and Discussion

3.1 *Excitation intensity dependence*

Due to the relative high number of pigments in the trimeric complex (> 36 Chl's) it is important to characterize the threshold for singlet-singlet annihilation processes which modify the kinetics. We measured the intensity dependence of the bleaching signal at 680 nm for long delay times (300 ps) between the exciting and the detecting laser pulse. After such delay times the ultrafast equilibration processes leave the pigment ensemble in an equilibrated state and also all annihilation events are finished. The absorbance change of the equilibrated state, normalized to the excitation intensity, gives a "quantum yield" for excitation survival and is a good measure of the overall annihilation in the system.

The variation of the excitation intensity over two orders of magnitude results in a change in "quantum yield" of a factor of 4-5 as can be seen from Fig.1.

We applied the annihilation theory of Paillotin et al. ("puddle model") [1] and got a good fit to the data. This fit is also consistent with simulations based on the absorption parameters of the sample, the exciting laser pulse intensity, and the assumption of a unity yield of annihilation for multiple excitation of a trimeric complex.

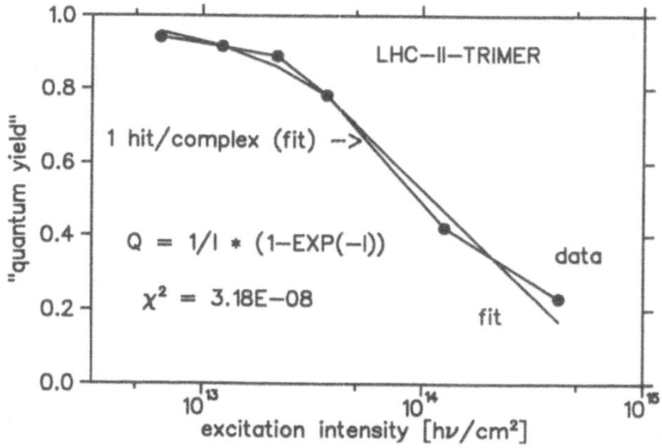

Fig.1: Annihilation curve for 680 nm absorbance change (points) and theoretical annihilation curve (full line).

3.2 *Low intensity transient absorption kinetics*

Excitation at 650 nm with low intensity 2 x 10^{13} photons / pulse /cm^2 yields two fast lifetimes of 290 fs and 3.7 ps (Fig. 2). Interestingly these lifetimes are significantly shorter than observed in earlier experiments with higher excitation intensities (not shown here). This points to a complex kinetic scheme and multiple annihilation components.

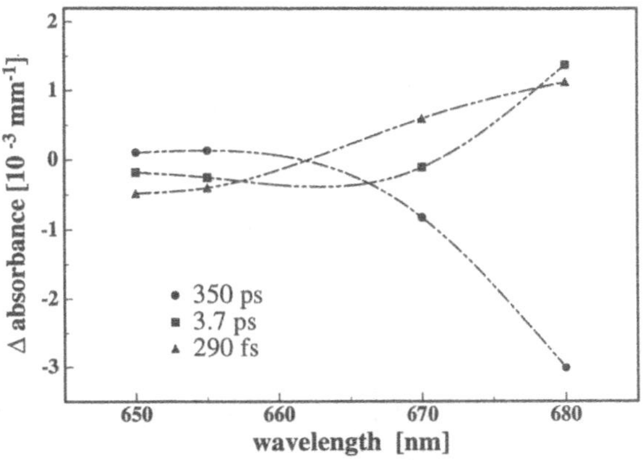

Fig.2: DADS from global analysis for LHC-II trimer; low intensity excitation at 650 nm.

The DADS of the ultrafast component τ_1 starts with negative amplitudes (absorption bleaching) at 650 nm. Above 660 nm the DADS turns positive which indicates a rise term in the transient absorption signal of the longer-lived components. The second lifetime τ_2 also starts with an absorption bleaching at 650 nm. This DADS switches to positive values above 670 nm. The positive amplitude should not be interpreted as an absorption increase but as a delayed rise in the bleaching of the long-lived components. The component τ_3 shows an absorption increase below 660 nm and a negative amplitude (bleaching) at 670 and 680 nm thus contributing the main signal in the Chl a absorption band. The kinetics is significantly dependent on the excitation wavelength (not shown). This is again a hint towards the fact that the real kinetics is much more complex than only tri-exponential.

3.3 *Minimal kinetic model for transient absorption data*

A basic model should include two dominant energy migration paths in order to describe the data in Fig. 2. The spectral positions suggests the assignment of the primarily excited chromophore pools to Chl b species, whereas the excitation energy finally is distributed over at least two different Chl a pools. The approximate maximal bleaching wavelengths of thesepools are 650 and 655 nm for Chl b and 675 and 680 nm for Chl a, respectively. A corresponding kinetic model scheme representing the different Chl pools and the dominant energy transfer paths is given in Fig.3. The additional expected equilibration between the Chl a pools was not resolved in our measurements for unknown reason, presumably because several of the picosecond lifetimes are too close to each other.

Kinetic modelling reveals that this equilibration should occur on the time scale of about 10 ps. The significant contribution of two fast components (τ_1 and τ_2) with zero-crossing amplitudes is in general agreement with recent two-wavelength data of Bittner et al. [6]

although in particular our fastest component is by a factor of about two longer than in the latter work. Our fast lifetime component is in better agreement with the fluorescence up-conversion results of the group of Fleming [5]. The differences in lifetimes probably have two main reasons: First, the experiments of Bittner et al. have been carried out at significantly higher excitation intensity, although the authors claimed to have carried out their measurements in the annihilation free region. There is some discrepancy with the annihilation curve shown in Fig. 1, however. The second and equally important reason for different lifetimes arises due to the fact that a tri-exponential model is certainly not describing the real kinetics. In this system of >12 Chls per monomer spectral and kinetic heterogeneity is very pronounced and the real kinetics of the system is expected to be extremely complex. The measured three-exponential kinetics is thus only an approximation in each case to the real situation. This is also reflected by the pronounced excitation wavelength dependence of the kinetics (not shown).

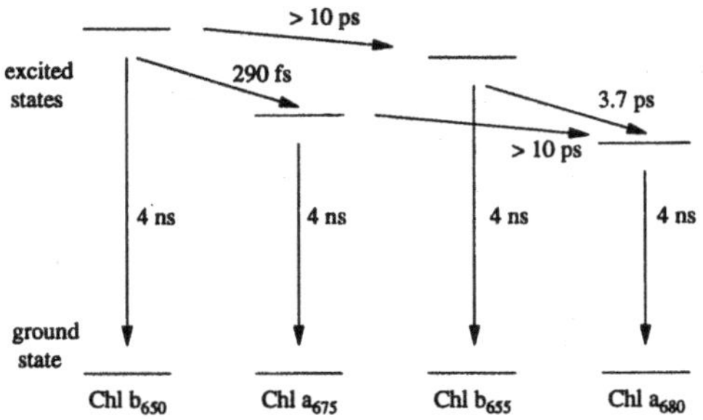

Fig.3: Minimal kinetic model for LHC-II-trimer absorption. The Chl xxx denote the approximate absorption difference maxima of the Chl a and b spectral pools that can be resolved.

Due to the very short chromophore-chromophore distances strong excitonic coupling between neighbouring Chls has to be expected. The absorption transition from the lowest excitonic state to higher excitonic states could explain the pronounced blue shifted excited state absorption observed in the experiments.

Acknowledgments
Financial support by the EC, Human Capital and Mobility Program, is acknowledged.

References

1 Paillotin, G. and Swenberg, C.E. (1979) Ciba Foundation Symposium 61, pp. 201-215, Excerpta Medica, Amsterdam.
2 Porra, R.J., Thompson, W.A. and Kriedemann, P.E. (1989) Biochim. Biophys. Acta 975, 384-394.
3 Kühlbrandt, W. and Wang, D.N. (1991) Nature 350, 130-134.
4 Hucke, M., Schweitzer, G., Holzwarth, A.R., Sidler, W. and Zuber, H. (1993) Photochem. Photobiol. 57, 76-80.
5 Du, M., Xie, X., Mets, L. and Fleming, G.R. (1994) J. Phys. Chem. 98, 4736-4741.
6 Bittner, T., Irrgang, K.-D., Renger, G. and Wasielewski, M.R. (1994) J. Phys. Chem. 98, 11821-11826.

8.5 Å PROJECTION MAP OF THE LIGHT-HARVESTING COMPLEX I FROM *RHODOSPIRILLUM RUBRUM* REVEALS A RING COMPOSED OF 16 SUBUNITS

Simone Karrasch, Per A. Bullough, Robin Ghosh, MRC Laboratory of Molecular Biology, Hills Road, Cambridge CB2 2QH, U.K.

1. Introduction

The light-harvesting complexes of photosynthetic purple sulphur and non-sulphur bacteria are responsible for the highly efficient collection and transfer of light energy to the photosynthetic reaction centres. This results in an initial separation of charge in the reaction centre (RC) and ultimately conversion of the light energy into a chemically useful form. The bacteria generally possess one or more types of membrane-bound light-harvesting complexes, B875 (LHC I), B800-850 (LHC II) and B820 (LHC III), where the numbers refer to the absorption maximum. The bacterial light-harvesting complexes consist of two types of polypeptides, α and β, each comprised of approximately 50 amino acids. *Rhodospirillum rubrum* contains only a single light-harvesting complex with α- and β-polypeptides containing 52 and 54 amino acid residues, respectively. Both polypeptides are predicted to possess a single transmembrane α-helix, which for the α-subunit has been shown by photolabelling studies to be positioned between residues 11 and 29. Two histidine residues (αHis29 and βHis37) which are conserved between species have been implicated in the binding of BChl a. Each αβ-dimer contains two molecules of BChl a and a single carotenoid molecule, spirilloxanthin.

2. Procedures

2.1 *Reconstitution*

Purified LHC I (1) was reconstituted into two-dimensional crystals as described (2), except that sonicated vesicles of DOPC (dioleoyl-9,10- phosphatidylcholin) were added to the β-octylglucoside-solubilised B880 complex in a ratio of 1 mol DOPC/mol αβ-dimer immediately prior to microdialysis.

2.2 *Electron Microscopy and Image Processing*

For electron cryo-microscopy the membranes were adsorbed onto carbon-coated grids that had been glow-discharged in amylamine vapour, blotted at 4°C and high humidity and plunged into liquid ethane at -180°C (3). Grids were transferred to a Gatan 626 coldstage and examined on a Philips CM12 microscope. Low dose images were recorded at 120 kV from nominally untilted crystals. The magnification was 35,000 x and exposure time was 1.0 sec with a beam of about 2.5 μm diameter at the specimen. This resulted in an electron dose of 5-10 electrons / $Å^2$. Images were selected for processing by optical diffraction showing good isotropic resolution. Selected areas were digitised as

P. Mathis (ed.), Photosynthesis: from Light to Biosphere, Vol. I, 81–84.

2000 x 2000 areas in steps of 10 μm using a modified Joyce Loebl Mk 4 densitometer. Image processing and merging of image amplitudes and phases followed procedures previously described (4, 5). The crystals were very small (of the order of 0.3 - 0.5 μm on an edge) and were therefore unsuitable for recording electron diffraction intensities. Therefore we opted to use estimates of the Fourier amplitudes from the image amplitudes (6). However, these amplitudes were significantly modified by the fall off in the phase contrast transfer function (CTF), astigmatism and image or specimen drift so that it was not possible to determine the plane group by examination of the image amplitudes alone. We therefore made an internal comparison of the phases in each image using the program ALLSPACE (7). The program SCALIMAMP (7) was used to correct for the effects of the CTF and secondly to apply a resolution dependent scale factor separately to the amplitudes from each image by comparison of the amplitudes to a reference data set (6). Projection maps were calculated using the corrected and averaged image amplitudes and phases, weighted by their figures of merit as described (8) with phases set to 0° or 180° for centrosymmetric reflections. The ring-like arrangement of density revealed by the projection map was subjected to a Fourier analysis of its rotational power spectrum (9). A rotationally filtered projection map was synthesised using the clear sixteenfold rotational component (53% of the total azimuthal power).

3. Results and Discussion

Fig.1 shows a projection map calculated from a single image of a *p22121* crystal with no symmetry imposed and using unmodified amplitudes with phases corrected for the phase contrast transfer function only. Fig.2 shows the equivalent projection calculated from averaged amplitudes and phases from eight images with symmetry applied and amplitudes sharpened by a resolution dependent scale factor; this was equivalent to applying a negative temperature factor of the order of 500 $Å^2$ to each image. The average phase error is less than random to a resolution of 8.5 Å. In Fig.2 we interpret the rings of positive density to represent protein. One asymmetric unit of the projection contains one ring which from Fig. 4 can be seen to have a clear handedness and the projection in Fig.2 shows both 'top' and 'bottom' views of the ring. Fig.3 shows the result of the rotational power spectrum analysis of the density in Fig.2. Strong sixteenfold and weak thirtytwofold components can be seen above the noise level thus indicating a sixteenfold non-crystallographic rotational symmetry in the projection density. The sixteenfold rotationally filtered map is shown in Fig.4. On average, the ring has an outer diameter of 116 Å and an inner diamter of 68 Å but within individual crystals the rings are elliptical rather than exactly circular. The difference between the two axes varies between 4 and 6%, indicating that the rings possess a degree of flexibility which may help to accomodate the reaction centre within. Each of the sixteen subunits contains three distinct domains. We interpret the inner peak to represent one of the polypeptides. The outer peak probably represents the projected α-helical density from the second polypeptide chain. This helix may be more tilted than the inner helix thus accounting for the smaller projected density. Between the inner and outer peak lies a third peak with the smallest density of all three domains. It is likely that some of this central density represents the pigment molecules bound between the inner and outer helix. Part of this central density may also arise from the tilting of the putative outer helix. A full analysis in three dimensions will resolve this issue. The interpretation of our data has been confirmed by a recent publication (10) where the three-dimensional structure of another light-harvesting complex (LHC II from *Rhodopseudomonas acidophila*) has been solved. X-ray analyses of three-dimensional crystals revealed a ring-like structure with 9 subunits. Each of these subunits contains an arrangement of the two polypeptide chains and the pigments similar to the arrangement we proposed for the LHC I, i.e. the inner helix perpendicular to the membrane plane, the outer helix tilted, and the bacteriochlorophylls between the two polypeptide chains.

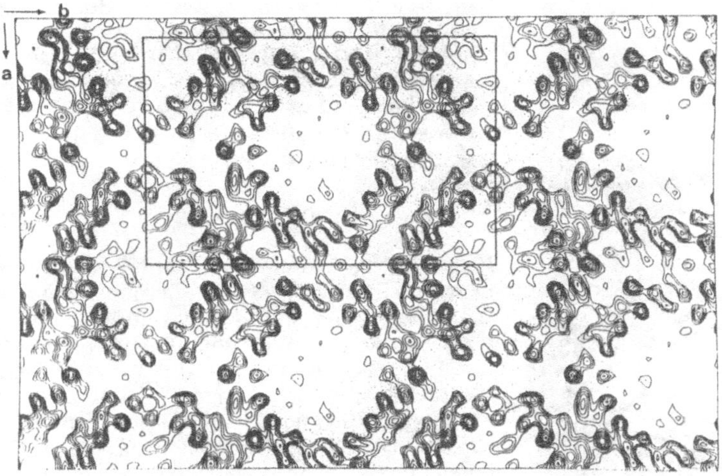

Figure 1 A projection map with *p1* symmetry calculated to 8.0 Å resolution using the Fourier components for the best image with phases and amplitudes modified only for the changes in the phase contrast transfer function. One unit cell (a = 131.0 Å, b = 194.0 Å) is outlined with the a axis vertical and the b axis horizontal. Contours are in steps of 0.5 x r.m.s (root mean square) density with solid lines indicating density above the mean.

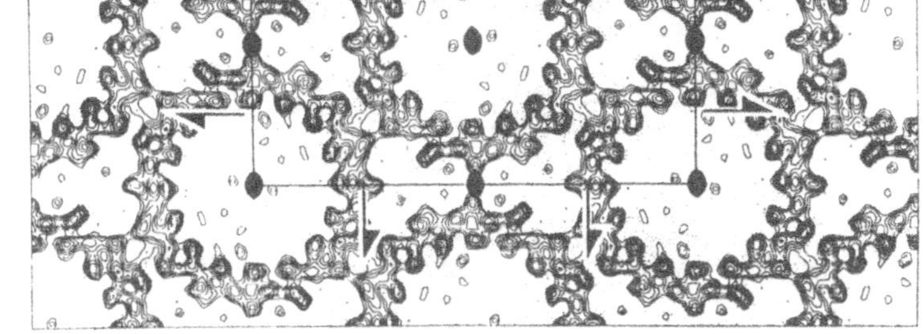

Figure 2 A projection map with $p22_12_1$ plane group symmetry calculated from amplitudes and phases averaged from eight images. One unit cell (a = 128.1 Å, b = 194.2 Å) is outlined with the a axis vertical and the b axis horizontal. The symmetry axes are indicated. Contours are in steps of 0.3 x r.m.s density with solid lines indicating density above the mean.

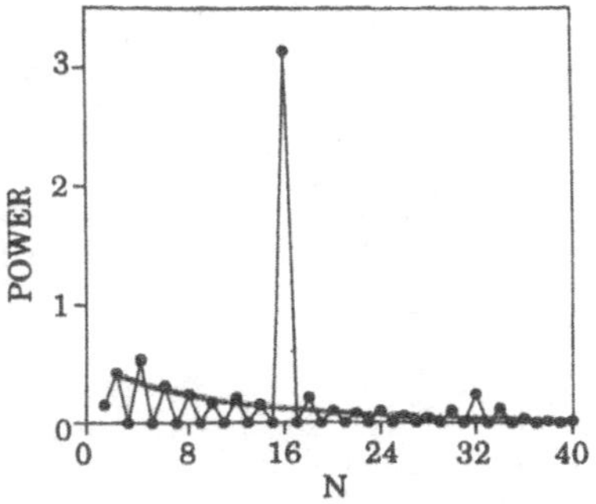

Figure 3 The rotational power spectrum for one ring of density in Fig. 2. The density was boxed off with a circular box of radius equal to the outer radius of the ring. This boxed density was floated into a large square box and the rotational power spectrum analysed with the program RFILTIM (9). Odd orders contribute no power to the noise because of the imposition of twofold crystallographic symmetry. The true noise level is therefore indicated by the even orders only.

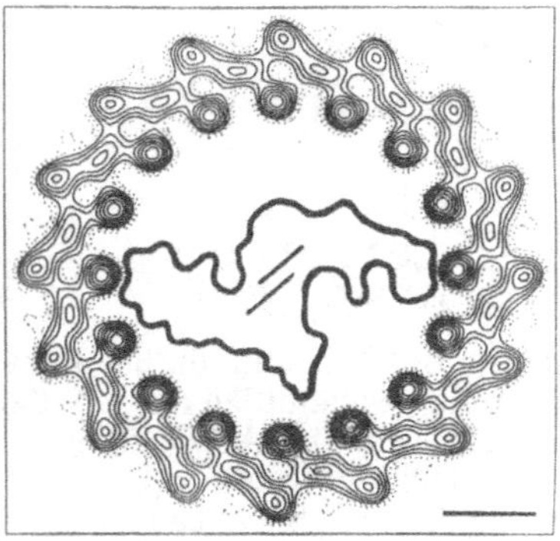

Figure 4 The sixteenfold rotationally filtered image extracted from Fig. 2. Contours have been arbitrarily calculated. The inner and outer peaks probably represent the α- and β-subunits and the density in the middle, at least partially, represents the bacteriochlorophyll pigments. The α- and β-subunits are thought to contain a substantial proportion of their structure as transmembrane helices consistent with the appearance of their density. The helices are separated by 15-20 Å and so cannot be in direct contact with one another. It is possible that the carotenoids fit into the space between the helices. An approximate van der Waals projection of the LM-dimer and the single transmembrane α-helix of the H-subunit from the reaction centre of *Rps. viridis* (11) viewed from above the membrane has been placed arbitrarily within the LHC ring to illustrate the relative dimensions. The porphyrin rings of the BChl b special pair are also shown. Scale bar represents 20 Å.

References

1 Ghosh, R., Hauser, H., and Bachofen, R. (1988). Biochemistry 27, 1004-1014.
2 Ghosh, R., Hoenger, A., Hardmeyer, A., Mihailescu, D., Bachofen, R., Engel, A., and Rosenbusch, J. P. (1993). J. Mol. Biol. 231, 501-504.
3 Subramaniam, S., Gerstein, M., Oesterhelt, D., and Henderson, R. (1993). EMBO J. 12, 1-8.
4 Henderson, R., Baldwin, J. M., Downing, K. H., Lepault, J., and Zemlin, F. (1986).Ultramicroscopy 19, 147-178.
5 Havelka, W. A., Henderson, R., Heyman, J. A. W., and Oesterhelt, D. (1993). J. Mol. Biol. 234, 837-846.
6 Schertler, G. F. X., Villa, C., and Henderson, R. (1993). Nature 362, 770-772.
7 Valpuesta, J. M., Carrascosa, J. L., and Henderson, R. (1994). J. Mol. Biol. 240, 281-287.
8 Bullough, P. A. and Tulloch, P. A. (1991).J. Mol. Biol. 215, 161-173.
9 Crowther, R. A. and Amos, L. A. (1971).J. Mol. Biol. 60, 123-130.
10 McDermott, G., Prince, S. M., Freer, A. A., Hawthornthwaite-Lawless, A. M., Papiz, M. Z., Cogdell, R. J., and Isaacs, N. W. (1995). Nature 374, 512-521.
11 Deisenhofer, J., Epp, O., Miki, K., Huber, R., and Michel, H. (1985) Nature 318, 618-624.

SENSITIVITY OF THE PHOSPHORYLATION OF DIFFERENT PHOSPHOPROTEINS TO Q_o AND Q_i SITE INHIBITORS OF THE CYTOCHROME B/F COMPLEX IN CHLOROPLAST THYLAKOIDS

E. Liker[1], L. Cheng[2], G. Garab[1] and J. F. Allen[2]
[1]Institute of Plant Biology, Biological Research Center, Hungarian Academy of Sciences, Szeged, P.O.Box 521, H-6701, Hungary and
[2]Department of Plant Cell Biology, University of Lund, Box 7007 S-220 07 Lund, Sweden

1. Introduction

Thylakoid membranes exhibit protein kinase activity which is activated by photochemical or chemical reduction of the plastoquinone pool (1,2). The major endogenous substrate for this activity is the light-harvesting chlorophyll a/b binding protein of photosystem II (LHCII), which is phosphorylated by the LHCII-kinase. Besides LHCII, several other thylakoid proteins are phosphorylated: D1/D2, CP43, the psbH gene product (9 kDa phosphoprotein) and several unidentified phosphoproteins.

The site of redox control of the activity(ies) of kinase(s) can be localized to the cytochrome b_6/f complex. Direct evidence that redox activation of the LHCII kinase requires participation of the cyt b_6/f complex was obtained from studies of a cyt b_6/f-deficient mutants of *Chlamydomonas* (3), maize (4) and *Lemna* (5). However, the mechanism of the activation of the LHCII kinase via the cyt b_6/f complex is unknown.

In thylakoid membranes the cyt b_6/f complex functions by transferring electrons from plastoquinol to plastocyanin, and protons are translocated across the membrane (6). There are two quinone binding sites in the cyt b_6/f complex, the Q_o and Q_i site, which are the sites of proton output and input, respectively. There are several reports on the effect of different inhibitors of the cyt b_6/f complex on the phosphorylation of thylakoid membranes (4,8). In most works the investigations of the inhibitory effects were confined to LHCII and the 9 kDa polypeptide.

We carried out a systematic investigation of the effect of Q_o site inhibitors (DBMIB, stigmatellin, bromanil) and Q_i site inhibitors (HQNO, MOA-stilbene) on the phosphorylation of the seven most intensely labelled phosphoproteins, LHCII, D1/D2, and the 56, 43, 20, 18 and 9 kDa proteins. The results show that all the Q_o site inhibitors have inhibitory effect on the thylakoid protein phosphorylation, whereas all the Q_i site inhibitors exhibit no inhibitory effect and some of them even slightly stimulate the phosphorylation. Our results indicate that the Q_o site or cyt b_6 or any component in the complex which locates between these two sites in the electrontransport chain is responsible for the redox control of the thylakoid protein phosphorylation.

P. Mathis (ed.), Photosynthesis: from Light to Biosphere, Vol. I, 85–89.

2. Experimental Procedures

Pea thylakoid membranes were isolated according to Harrison et al. (9).

Protein kinase reactions were initiated by addition of ^{32}P ATP to a final concentration of 0.2 mM, specific activity of 100 μCi/μM ATP in a reaction medium containing 50 mM Hepes, 5 mM $MgCl_2$, 5 mM NaCl, 5 mM NH_4Cl, 0.1 M sorbitol and the inhibitors dissolved in 95 % ethanol (final ethanol concentration did not exceed 5 %). The protein kinase was activated by light. Phosphorylation reactions were terminated with ice cold 80 % acetone after 5 min. SDS-PAGE was performed on 12-25 % gradient gels according to Laemmli (10). Quantitation of the ^{32}P-labelled bands of SDS-gels were performed by phosphorimaging with a Fuji BAS2000 Bio-Imaging Analyzer. Background labelling was substracted for each lane.

3. Results and Discussion

In accordance with earlier reports (cf. 2) we found that DBMIB at 3μM almost fully inhibited the phosphorylation of LHCII, whereas the sensitivity of the 9 kDa phosphoprotein was significantly lower.

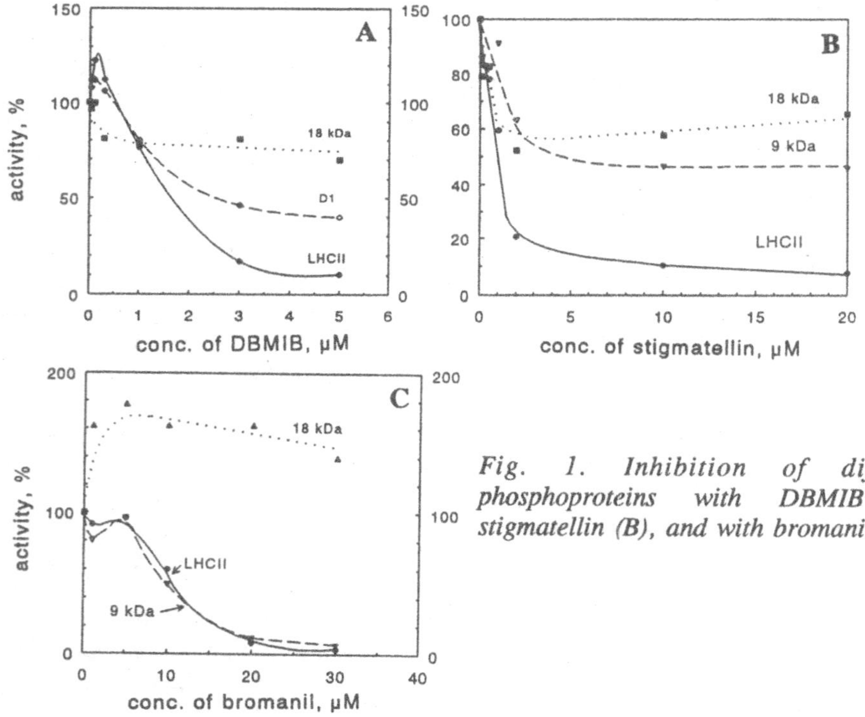

Fig. 1. Inhibition of different phosphoproteins with DBMIB (A), stigmatellin (B), and with bromanil (C).

Our data show that the inhibition of the 56 kDa protein by 3μM DBMIB closely resembles that of LHCII, whereas D1/D2, the 43 and 20 kDa proteins exhibit lower sensitivity. The 18 kDa protein appears to be largely insensitive to DBMIB. (The phosphorylation of this polypeptide was also insensitive to DCMU (data not shown).) Typical titration curves of different phosphoproteins, representatives of these three groups are shown in Fig. 1A.

Phosphorylation of LHCII and the 56 kDa protein could very effectively be inhibited with stigmatellin, which is a potent antibiotic inhibitor of the cyt b_6/f complex at the Q_o site (11) (Fig. 1B). With both DBMIB and stigmatellin the I_{50} values for inhibition of phosphorylation were considerably higher than those for inhibition of the cyt b_6/f complex where I_{50} values were around 50 nM (11,12). Similar shifts of the inhibition of the LHCII kinase was observed earlier with DCMU (cf. 2).

The phosphorylation of the second group of proteins (D1/D2, and the 43, 20 and 9 kDa proteins), which was less sensitive to inhibition by DBMIB than that of LHCII and the 56 kDa protein, was also less sensitive to inhibition with stigmatellin and the 18 kDa protein was the most insensitive polypeptide (data not shown).

Bromanil, which is a halogenated 1,4-benzoquinone and an efficient inhibitor at the Q_o site (12), inhibited the phosphorylation of all phosphoproteins except the 18 kDa protein, the phosphorylation of which was significantly stimulated (Fig. 1C). It must be noted, however, that at the concentration of half inhibitory effect, due to a covalent binding of bromanil to the thylakoids, the electron transport is inhibited irreversibly (13). This may explain why, with this inhibitor, no distinction could be made between the first and second group of phosphoproteins.

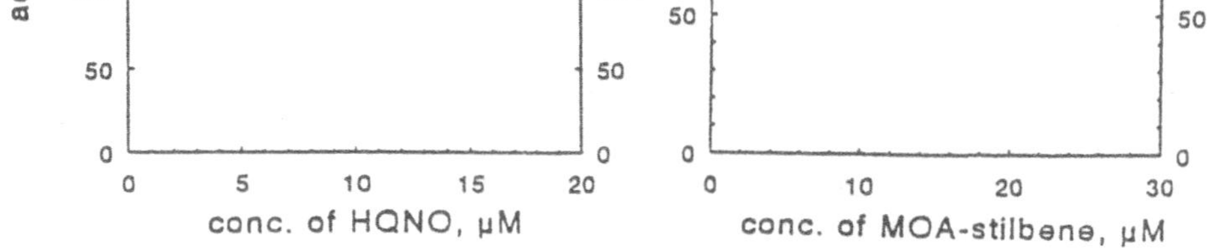

Fig. 2. Effect of HQNO (A) and MOA-stibene (B) on phosphorylation of different phosphoproteins.

HQNO, which is a quinone analogue inhibiting at the Q_i site (14), enhanced significantly the phosphorylation of LHCII and other phosphoproteins of the thylakoid membrane (Fig. 2A). HQNO has been shown to raise the midpoint potentials of the cyt b563 molecules from -150 and - 30 mV to -34 and +57 mV, respectively (15), and to significantly enhance the photoreduction of cyt b563 (13). Earlier, HQNO was reported not to affect phosphorylation (4). On the other hand NQNO, which inhibits at the same site (12), has been shown to enhance phosphorylation of LHCII by 40% (8).

MOA-stilbene, which is a recently introduced Q_i-site inhibitor (6), proved to be ineffective in the phosphorylation of thylakoid membrane proteins (Fig. 2B). In contrast with HQNO, this inhibitor does not alter the redox potentials of cyt b563, and only slightly enhances the reduction of these molecules (6).

In summary, our analysis of the effect of different inhibitors of the cyt b_6/f complex on the seven most intensely phosphorylated polypeptides, shows that the phosphorylation of the proteins, with the exception of the 18 kDa protein, can efficiently be inhibited by Q_o site inhibitors, whereas Q_i site inhibitors were either ineffective or enhance the phosphorylation. Thus, it is concluded that the redox control of the phosphorylation of the thylakoid membranes is found at the Q_o site itself, cyt b_6 or any other component between Q_o and Q_i sites of the cyt b_6/f complex.

Acknowledgements The authors acknowledge the financial support of Swedish Natural Science Research Council, OTKA III/2999 of Hungarian Research Fund and ESF Biophysics of Photosynthesis travel grant to E.L. We thank P.R. Rich for the gift of MOA-stilbene.

References

1 Allen, J.F. (1992) Biochim. Biophys. Acta 1098, 275-335
2 Bennett, J. (1991) Annu. Rev. Plant Mol. Biol. 42, 281-311
3 Lemaire, C., Girard-Bascou, J. and Wollman, F-A. (1987) in Progress in Photosynthesis Research (Biggins, J., ed.), Vol. IV., pp. 655-658, Martinus Dordrecht
4 Bennett, J., Shaw, E.K. and Michel, H. (1988) Eur. J. Biochem. 171, 95-100
5 Gal, A., Shahak, Y., Schuster, G. and Ohad, I. (1987) FEBS Lett. 221, 205-210
6 Rich, P.R., Magdwick, S.A., Brown, S., Jagow, G. and Brandt, U. (1992) Photosynthesis Res. 34, 465-477
8 Gal, A., Schuster, G. Frid, D., Canaani, O., Schweiger, H-G. and Ohad, I. (1988) J. Biol. Chem. 263, 7785-7791
9 Harrison, M.A. and Allen, J.F. (1991) Biochim. Biophys. Acta 1058, 289-296
10 Laemmli, U.K. (1970) Nature 227, 680-685
11 Oettmeier, W., Godde, D., Kunze, B. and Höfle, G. (1985) Biochim. Biophys. Acta 807, 216-219

12 Jones, R.W. and Whitmarsh, J (1988) Biochim. Biophys. Acta 933, 258-268
13 Oettmeier, W., Masson, K. and Dostatni, R. (1987) Biochim. Biophys. Acta 890, 260-269
14 Selak, M.A. and Whitmarsh, J. (1982) FEBS Lett. 150, 286-292
15 Clark, R.D. and Hind, G. (1983) Proç. Natl. Acad. Sci. 80, 6249-6253

ENERGY TRANSFER IN THE LH 2 COMPLEX OF *RB. SPHAEROIDES* AND *RPS. ACIDOPHILA* STUDIED BY SUB PICOSECOND ABSORPTION SPECTROSCOPY

R. Monshouwer[1], I. Ortiz de Zarate[2], F. van Mourik[1], R. Picorel[2], R.J. Cogdell[3] and R. van Grondelle[1]
1 : Dept. of Physics and Astronomy, Vrije Universiteit, De Boelelaan 1081, 1081 HV Amsterdam, The Netherlands
2 : Dept. of plant nutrition, Estación Experimental de Aula Dei (CSIC), Apartado 202, 50080- Zaragoza, Spain.
3 : Department of Botany, University of Glasgow, Glasgow, UK

INTRODUCTION:

The photosynthetic apparatus of the purple bacteria *Rhodobacter sphaeroides* and *Rhodopseudomonas.acidophila* contain two different kinds of antenna complexes: the core antenna LH1 that surrounds the reaction centre (RC) and the peripheral antenna LH2. Light is absorbed by these complexes and the excited state energy is transferred to the RC. The LH2 or B800-850 complex is spectrally characterised by a relatively narrow band absorbing around 800 nm and a broader band at 850 nm. The precise spectral and functional characterisation of LH2 complexes has gained renewed interest since this year the three-dimensional structure of the LH2 complex of *Rps. acidophila* has been elucidated [1]. Accurate determination of the spectral properties combined with precise information about the distances between the pigments, their relative orientations, and the timescale of the transfer processes can provide an important test of the existing models for energy transfer in photosynthetic antenna complexes. We present here low temperature time resolved measurements on isolated LH-2 complexes of *Rb. sphaeroides* and *Rps. acidophila*. We find that the transfer from the B800 complex to the B850 complex takes place with 1.2 for *Rb.sphaeroides* and 1.26 ps for *Rps. acidophila*. Additionally we find that the B800-B850 complex of *Rb. sphaeroides* is very sensitive to photodamage.

MATERIALS AND METHODS:

Both types of isolated complexes were diluted in a potassium buffer (pH 7.8) containing glycerol (80% w/v) and 0.1% LDAO. Pulses were generated using a Mira / REGA combination (Coherent). This Ti-sapphire oscillator / amplifier combination typically produces 4 μJ pulses with a repetition rate of 250 kHz. Both beams were attenuated and the polarisation of the pump beam was rotated using a wave plate and passed through a sheet polariser under magic angle before being recombined with the probe beam in the sample. The combination of lens and beam size yields a beam diameter of about 50 μm in the focus. In a typical measurement about 50 μW was used to excite the sample and in total an average intensity of about 80 μW was focused on the sample. This results in an excitation density of $8\ 10^{13}$ photons/cm^2, and a total average power of 4 W/cm^2. The impulse response of the system was measured using a 1 mm $LiIO_4$ crystal at the place of the sample and was typically 250 fs FWHM. The kinetics were fitted using a home made non linear least square fitting program.

P. Mathis (ed.), Photosynthesis: from Light to Biosphere, Vol. I, 91–94.

RESULTS and DISCUSSION:

Figure 1 shows two typical one colour pump probe measurements in the B800 band of *Rb. sphaeroides.* In general the signal obtained in a transient absorption measurement can consist of several contributions. The actual bleaching, stimulated emission, and the excited state absorption of the product. In the measurements shown in fig 1 the main positive signal is bleaching/stimulated emission due to the creation of the B800* state. This state subsequently decays until a constant remaining induced absorption is reached. This remaining signal has been ascribed to excited state absorption of the B850* state that is populated due to transfer from the B800 to the B850 band.

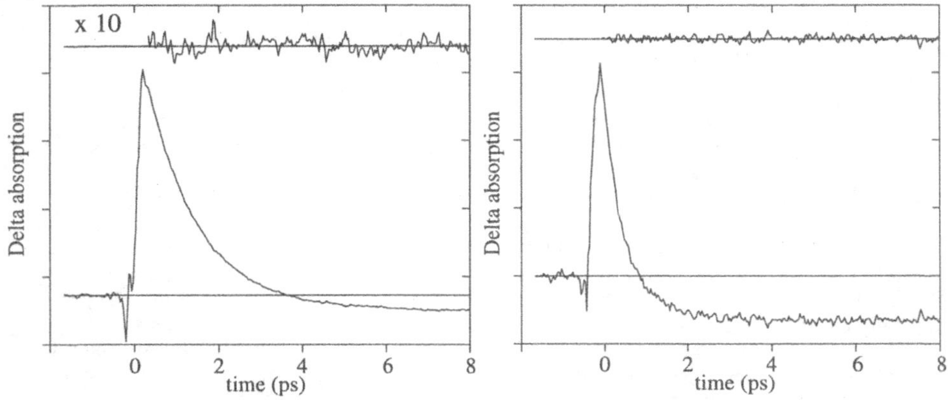

Fig 1 One colour pump probe measurements on the LH-2 complex of Rb. sphaeroides. A) measured at 804 nm. The data could be fitted with a 1.14 ps component (93%) and one of 2.7 ps (7%). B) at 796 nm. This trace was fitted with decay times of 0.4 ps (84%) and 0.9 ps (16%).

By comparing fig 1,A and B, it is evident that the decay of the B800* state is strongly dependent on the wavelength within the B800 band. In the red part of the band the decay is almost mono-exponential. The small long component is variable, and probably an artifact due to some sort of photodamage (see below). Below 800 nm the kinetics become faster and multi-exponential.This wavelength dependence is illustrated in figure 2, where mono-exponential fitting results are plotted versus the laser wavelength. Such a strong wavelength dependence of the decay of the B800* state was already reported by de Caro et al.[2]. They concluded that at low temperature the B800 pigment pool can be subdivided in two pools: one pool around 801 nm can only transfer to the B850 band whereas the other pigments can also transfer excitations to the redmost pool. Our measurements are qualitatively in agreement with these holeburning results. It should be noted however that in these measurements a very small spectral region is examined, whereas in timeresolved measurements short and consequently spectrally broad pulses are used for excitation and detection. This probably causes the multi-exponential decay since a spectrally broad excitation pulse will excite both the redmost pool as well as the pigments that can transfer within the B800 band. We tested this by fitting the measurements in the blue part of the band with one fixed 1.14 ps component and one free component. The resulting fits were satisfactory and the fast component had a value of 0.45 +/- 0.1 ps with a rising relative amplitude towards the blue

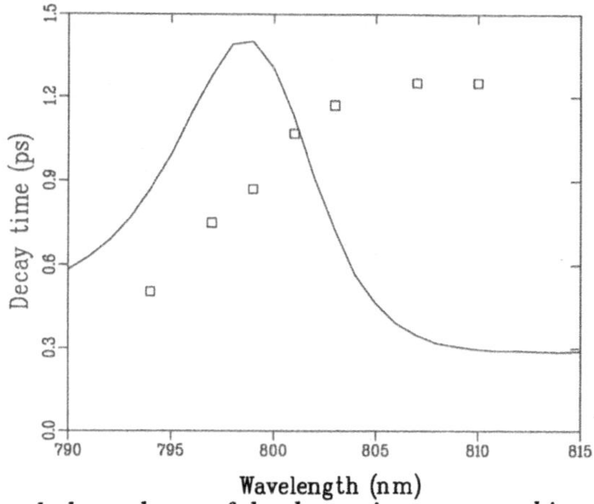

Fig 2 Wavelength dependence of the decay time measured in a one colour pump probe measurement. Even though at some wavelengths the decay was multi-exponential the single exponential result is given for comparison. The solid line is the OD spectrum of the B800 band of Rb. sphaeroides at 77K.

In figure 3 a two colour measurement on the B800 band is shown. In this measurement the blue part of the B800 band was excited and the signal was measured in the red edge of the B800 band. To fit this signal, a rise time of the bleaching/stimulated emission was needed that is quantitatively in good agreement with the fast component needed to fit the one colour measurements in the blue part of the band.

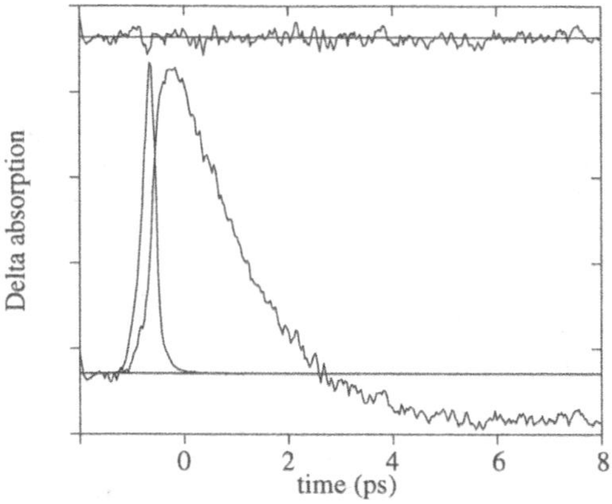

Fig 3 Two colour pump probe measurement on the LH-2 complex of Rb. sphaeroides. The temperature was 77 K. Shown is the measured signal, the cross correlation of pump and probe pulse and the difference with a bi-exponential fit (top). Excitation was at 791 nm, detection at 810 nm. The signal was best fitted with a rise time of 0.44 ps and a fall time of 1.26 ps.

We thus conclude that the transfer time from the B800 to the B850 pool is 1.2 ps at 77 K and that relaxation within the inhomogeneous pigment pool takes place with about 0.45 ps. We note that these results differ from previous measurements by others [3,4]. Furthermore we found that the decay time of the B800* state was strongly dependent on the intensity of the laser light. At intensities exceeding 10 W/cm^2 the decay time became much slower. This effect was permanent, so lowering the power did not restore the initial kinetics.This suggests that some form of photodamage is caused by these light intensities.

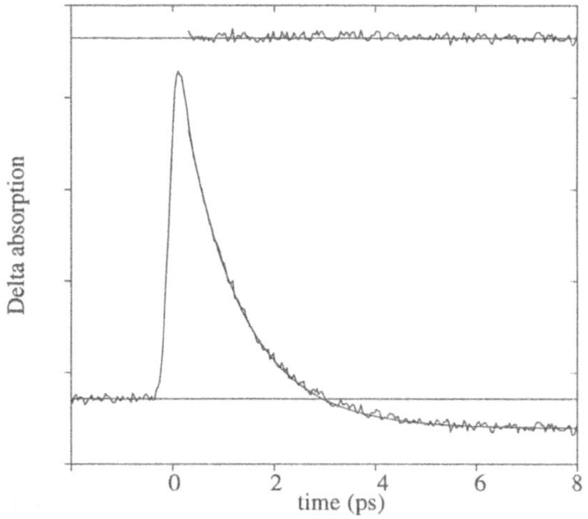

Fig 4 A One colour pump probe measurement on the LH-2 complex of Rps. acidophila at 77K. The wavelength was 803 nm. The data could be fitted with a mono exponential decay of 1.26 ps

Rps. acidophila:

In figure 4 a one colour pump probe measurements of the LH-2 complex of *Rps. acidophila* is shown. Although a complete wavelength dependence should be done to accurately determine the B800->B850 transfer time for this species we think that this decay represents the B800 to B850 transfer time in *Rps. acidophila* at 77 K.The mono-exponentiality seems to suggest that this measurement was done in the "flat" part of the wavelength dependence. Consequentely 1.26 ps is the true B800 to B850 transfer time. This value is very similar to the transfer time in *Rb. spaeroides*

REFERENCES:

1 Mcdermott, G., Prince, S.M., Freer, A.A., Hawthornthwaite-Lawless, A.M., Papiz, M.Z., Cogdell, R.J. and Isaacs, N.W. (1994), Nature 374, 517-521

2 De Caro, C., Visschers, R.W., van Grondelle, R. and Volker, S. J. Phys. Chem. 98, 10584-10590

3 Hess, S., Feldchtein, F., Babin, A., Nurgaleev, I., Pullerits, T., Sergeev, A. and V. Sundström (1993) Chem. Phys. Lett. 216,247-257

4 Visscher, K.J., Gulbinas, V., Cogdell, R.J., van Grondelle, R. and V. Sundström (1993) in Springer Series in Chemical Physics, Vol. 55 (Martin, J.L., Migus, A., Mourou, G.A. and Zewail, A.H. ed.) pp.559-561, Springer-Verlag, Berlin

TRIMERIC LHC II IN PLANTS LACKING 16:1t IN PHOSPHATIDYL-GLYCEROL

E. Selstam[1] and M. Krol[2],
1 Dept of Plant Physiology, University of Umeå S-901 87 Umeå, Sweden;
2 Dept of Plant Science, University of Western Ontario, London, Canada N6A 5B7

Introduction

In the tylakoid membrane of higher plants the Light Harvesting Complex of photosystem two, LHC-II is organised in trimers [1, 2]. In vitro a trimerization of LHC-II occurs spontaneously in the presence of thylakoid lipids [3], and phosphatidyl glycerol (PG) esterified with 16:1 trans acid gives the most rapid reconstitution of LHC-II oligomers [4]. There is also a good correlation between the relative amount of 16:1t in PG and the presence of the oligomeric form of LHC-II in the thylakoid [5].
In the present study, the presence the trimeric form of LHC-II was investigated in a group of plants, that naturally lack 16:1t in the PG.

Materials and methods

Tropical orchids, *Calanthe vestia, Habenaria, Paphopedílum, Lycáste* and wheat (*Triticum aestivum* L.) were grown in a greenhouse and local orchids, *Platantéra bifólia, Goodyéra repens* and *Calypso bulbósa* were collected in the woods near Umeå in June and August. Green leaves were used for lipid extraction and isolation of thylakoid membranes. The composition of acyl groups esterified to PG was analysed by GLC [6] and chlorophyll protein complexes were separated on green gels [7].

Result and Discussion

Acyl groups of PG

The relative composition of acyl groups of PG from seven species of orchids and wheat were investigated, Table 1. Six of the orchids lack 16:1t in PG and in *Platantera bifolia* the amount of 16:1t was much lower than in wheat. The amount of 16:1t will depend on the presence and activity of a specific desaturase that desaturate PG 16:0/18:1 to PG 16:1t/18:1 [8]. Thus in most plants the relative amount of 16:0 + 16:1t in PG is rather constant and approximately 50%. Most investigated orchids had, in comparison to wheat, a relatively low content of 18:3 and instead more of 18:1 and 18:2.

P. Mathis (ed.), Photosynthesis: from Light to Biosphere, Vol. I, 95–98.

TABLE I.
The relative composition of acyl groups esterified to PG isolated from leaves of wheat and some orchids. t = traces

Plant	Acyl group (mole% of total acids)							
	14:0	16:0	16:1t	16:2	18:0	18:1	18:2	18:3
Triticum	-	17	33	-	1	1	4	44
Platanthéra	t	37	10	1	1	21	15	15
Goodyéra	1	47	-	1	5	29	12	5
Calypso	t	46	-	t	7	27	4	4
Calanthe	t	60	-	t	2	14	16	6
Habenaria	-	67	-	1	4	9	9	9
Paphopedílum	-	56	-	3	6	23	6	6
Lycáste	-	21	-	3	7	5	6	57

Chlorophyll protein complexes
Separation of chlorophyll protein complexes on a "green gel" showed that trimers of LHC-II was present after mild solubilisation of thylakoids isolated from *Goodyéra, Platanthéra, Calanthe* as well as in *Triticum* (Figure 1). Most of the LHC-II complexes were in the trimeric form.Thus stable trimers of LHC-II could be isolated from thylakoids independent of the presence of 16:1t in PG. However the stability of the LHC-II complexes were correlated to the acyl group composition of PG. LHC-II trimers from *Platanthéra* (16:1t) were more resistant to SDS solubilisaton thapn LHC-II trimers of *Goodyéra* (16:0) (Figure 2). Trimers of LHC-II has also been isolated from a mutant of *Arabidopsis thalina* lacking 16:1t, and these mutant trimers were much more sensitive to NaCl during solubilization than wild type 16:1t-trimers [9].

A B C D

CP1
LHC-II trim
LHC-II mono
FP

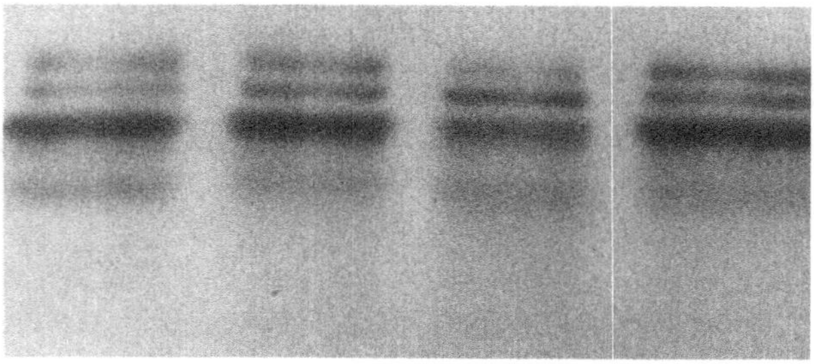

Figure 1. Electrophoretic separation of chlorophyll protein complexes from A, *Goodyéra*; B, *Platantéra*; C, wheat; D, *Calanthe*.

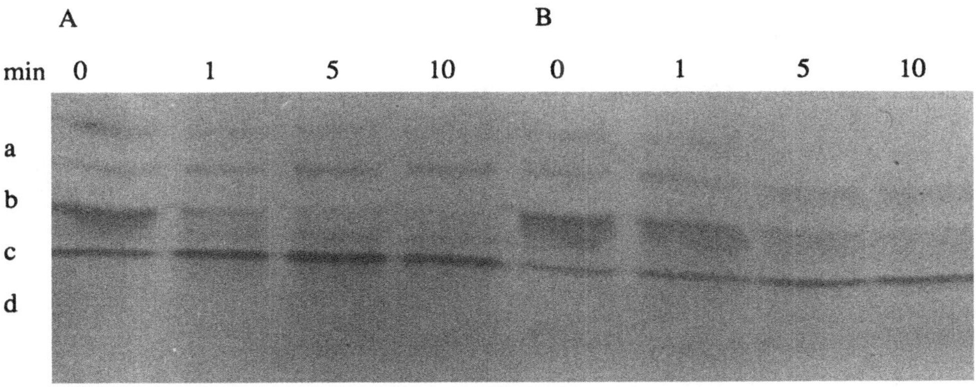

Figure 2. Electrophoretic separation of chlorophyll protein complexes from A, *Goodyéra* and B, *Platantéra*. The thylakiods were solubilised 0, 1, 5, 10 minutes with SDS, before addition of deoxycolate [7]. a, CPI; b, LHC-II trimers; c, LHC-II monomers; d, free pigment.

Spectral properties of LHC-II trimers

Low temperature fluorescence emission spectra was measured of the LHC-II trimers isolated by gelelectrophoreses. The fluorescense spectra were almost the same in all plants. The fluorecense maxima showed a small variation from 687nm in wheat(16:1t), 692nm in *Platanthéra*(16:1t) and 691nm *Goodyéra*(16:0) and *Calante*(16:0).

Conclusions

These results show that presence of PG-16:1t is not necessary for the formation of stable LHC-II trimers. PG-16:1t makes the trimers more resistent to SDS solubilisation, this might indicate that PG-16:1t is a better lipid than PG-16:0 in the LHC-II trimer.

References

1 Kühlbrandt, W. and Wang, D.N. (1991) Nature 350, 130-134
2 Dreyfuss, B.W. and Thornber, J.P. (1994) Plant Physiol. 106, 829-839
3 Hobe, S., Prytulla, S., Kühlbrandt, W. and Paulsen, H. (1994) EMBO 13, 3423-3429
4 Rémy, R., Trémolières, A. and Ambard-Bretteville, F. (1984) Photobiochem. Photobiophys. 7, 267-276
5 Huner, N.P.A., Krol, M., Williams, J.P., Maissan, E., Low, P., Roberts, D. and Thompson J.E. (1987) Plant Physiol. 84, 12-18

6 Wingsle, G., Mattson, A., Ekblad, A., Hällgren, J.-E. and Selstam, E. (1992) Plant Science 82, 167-178
7 Huner, N.P.A., Campbell, D., Krol, M. Hayden, D.B., Myscich, E.M., Basalyga, S. and Williams, J.P. (1992) Plant Physiol. 99, 830-836
8 Browse, J. and Somerville, C. (1991) Annu. Rev. Plant Physiol. Plant Mol. Biol. 42, 467-506
9 McCourt, P. Browse, J., Watson, J., Arntzen, C.J. and Somerville, C. (1985) Plant Physi 78, 853-858

SPECTRAL HOLE BURNING AS A TEST FOR EXCITON MODELS OF CHLOROSOMAL ANTENNA

Vladimir I. Novoderezhkin and Zoya G. Fetisova
A.N.Belozersky Institute of Physico-Chemical Biology
of Moscow State University, Moscow 119899, Russia

1. Introduction

The chlorosomal antenna of green photosynthetic bacteria contains several thousands of bacteriochlorophyll (BChl) *c,d* or *e* molecules in an aggregated state (1). The strong orientational ordering of the BChl transition dipoles *in situ* suggested that an elementary BChl aggregate has the form of a linear chain (2). Subsequently, plausible models of assembling of BChl chains were proposed (1,3,4). According to one group of models the chlorosomal antenna consists of noninteracting BChl chains (linear single chain, linear antiparallel double chain, or zig-zag double chain) (for review see (1)). Another group of models assumes tubular arrangement of tight- packed single (3) or double (4) BChl chains due to self-aggregation of pigments into an entire macrocycle network.

Information about the BChl aggregation is available due to recent spectral hole burning studies performed on three species of intact cells from both families of green bacteria, *Chlorobiaceae* and *Chloroflexaceae* (5-7). The experiments have shown the following features of hole burning spectra (HBS) which are basically similar for all species investigated:

(*i*) The total width of a hole-burning spectrum is always equal to that of a preburnt spectrum, exceeds the inhomogeneous width (90-100 cm^{-1}) and does not depend on temperature. This means that the large homogeneous width of the absorption spectrum (350 cm^{-1} for *Cf.aurantiacus*, see Fig.1) is determined by exciton levels rather than by phonon wings, i.e. exciton-phonon coupling is weak.

(*ii*) A hole-burning spectrum consists of a resonant zero-phonon hole (ZPH) corresponding to the lowest exciton level (1,600-2,000 cm^{-1} red-shifted with respect to the monomer absorption maximum) and of a broad non-resonant hole corresponding to the higher levels (broadened due to relaxation). The integral intensity of ZPH is considerably less than that of the broad component. Thus, the higher exciton level, blue-shifted by 200-400 cm^{-1} (depending on species) with respect to the

P. Mathis (ed.), Photosynthesis: from Light to Biosphere, Vol. I, 99–102.

lowest one, has the most part of the total oscillator strength (Fig.1).

In this paper, we demonstrate that neither the models of noninteracting BChl chains nor tubular models with tight-packed BChl chains can explain the *in vivo* exciton level structure. Only the models of exciton-coupled BChl chains with a low packing density, approximating that *in vivo*, can provide a reasonable explanation for the exciton level structure revealed by hole-burning spectroscopy.

2.Results

2.1 Noninteracting BChl chains

In the case of noninteracting single chains the lowest exciton state of BChl aggregate has most of the total oscillator strength. This is a common property of single chain aggregates with $M<0$, where M is the matrix element of resonant interactions between nearest neighbouring molecules. The higher levels give only relatively weak lines. Such an exciton level structure is in contradiction with that revealed by experimental hole-burning data (5-7). Thus, the single chain models must be ruled out.

In a double chain each exciton state splits into two states (Davydov splitting) with the energies depending on the interaction energies $M'(n-m)$ between molecules n and m from different chains. For BChl aggregates $M'>0$ for an antiparallel configuration and $M'<0$ for a zig-zag chain (for example, for the zig-zag model $M'(0)= M'(1)=-450$ cm^{-1} (1)). In both cases the oscillator strength of the lowest Davydov component is much greater than that of a higher component. So, the absorption spectrum comprises intense line corresponding to the lowest exciton level of the lowest Davydov component and weak shorter wavelength peaks, this fact being inconsistent with the hole-burning experiments. Thus, the double chain models must be also ruled out.

2.2 Tight-packed BChl chains models (tubular models)

In the case of tubular models BChl aggregate has a form of the cylinder formed by L parallel linear chains each containing N BChl molecules. Exciton states are characterized by two numbers: k and κ, where k takes N integer values 1,2,..N; κ takes L integer values $0,\pm1,\pm2,...$ There are L Davydov components, each having the exciton structure similar to that of a linear chain. If the BChl transition dipoles are parallel to the long axis of a tubular aggregate, only the $\kappa=0$ component of the absorption spectrum is dipole allowed. We suppose that $M<0$, $M'<0$, and the absolute values of M and M' are comparable (3). In this case the dipole allowed $\kappa=0$ component is the lowest among other L components. So, the absorption spectrum is the same as for the single or double chain models. The same insoluble problems arise in the case of the tubular aggregate of L double chains (see (4)). Thus, the tubular models of tight-packed BChl chains (3,4) cannot explain the spectroscopic data.

2.3 Alternative models with low density of BChl chains packing

The models of noninteracting BChl chains (1) and the models with tight-packed BChl chains (3,4) correspond to the two limiting cases of BChl chains aggregation. Consider the intermediate case of tubular aggregate with low density of BChl chains packing. To explain the spectral features revealed by hole-burning experiments (5-7) in the framework of a tubular model, we must assume that M is about -750 cm^{-1} and M' is about 50-100 cm^{-1}. In this case the gap between the dipole allowed $\kappa=0$ level and the lowest $\kappa=L/2$ level (weakly nonforbidden due to site inhomogeneity) is 4M'=200-400 cm^{-1}. The 50-100 cm^{-1} interaction energy corresponds to the interchain distance of 2 nm or more. This strongly suggests that the spatial organization of L chains into a tubular structure cannot be determined by self-aggregation of pigments into an entire macrocycle network as it was postulated in (3,4).

According to the data on supramolecular organization of chlorosomes (8) the BChl *c*, *d*, or *e* pigments are thought to be associated with six subunits organized hexagonally in rods. Assuming that there is one linear (single or double) BChl chain per each subunit we obtain L=6 (Fig.2). The low-temperature absorption spectrum for tubular model with L=6; N=8; M=-750 cm^{-1}; M'=75 cm^{-1}; phonon frequency Ω=50 cm^{-1} (we assume a weak exciton-phonon coupling); and homogeneous line width $\Gamma/2$= 3 cm^{-1} is shown in Fig.3A. We neglected line broadening due to relaxation in order to obtain well resolved lines. Relative intensities of Davydov components $\kappa=0$, $\kappa=\pm1$, $\kappa=\pm2$ and $\kappa=3$ are 100, 27, 3 and 1.6. The $\kappa=\pm1$, $\kappa=\pm2$ and $\kappa=3$ components are supposed to be weakly nonforbidden due to site inhomogeneity; the corresponding intensities are proportional to the square of their displacement from the dipole allowed $\kappa=0$ state. The k=1 level of the $\kappa=0$, $\kappa=\pm1$, $\kappa=\pm2$ and $\kappa=3$ components corresponds to the -1,250 cm^{-1}, -1,325 cm^{-1}, -1,475 cm^{-1} and -1,550 cm^{-1} lines (Fig.3A). The k=3 level of the $\kappa=0$ and $\kappa=\pm1$ components corresponds to the -550 cm^{-1} and -625 cm^{-1} lines (Fig.3A).

The -1,550 cm^{-1} line corresponds to the lowest 752 nm component of HBS and the 754 nm peak of CD; the -1,250 cm^{-1} and -550 cm^{-1} lines correspond to the 738 nm and 713 nm components of CD; the most intense -1,250 cm^{-1} line also corresponds to the 742 nm maximum of HBS (Fig.1).

To calculate HBS we introduced the 150 cm^{-1} homogeneous line width (FWHM) for all the exciton levels except of the lowest (k=1,κ=3) level and the 100 cm^{-1} inhomogeneous width (FWHM). The hole profiles are shown in Fig.3B. Arrows show the burning frequencies. Sharp ZPH can be burned only when the burning frequency corresponds to the long-wave side of the spectrum. Calculated HBS have the same features as experimental ones. In particular, the energy gap between ZPH and a broad non-resonant hole, their relative intensities, and the large width of non-resonant hole can be explained by our model.

This work was supported in part by the Russian Foundation for Basic Research (Grant No.94-04-12707) and International Science Foundation (Grant No.M6R000).

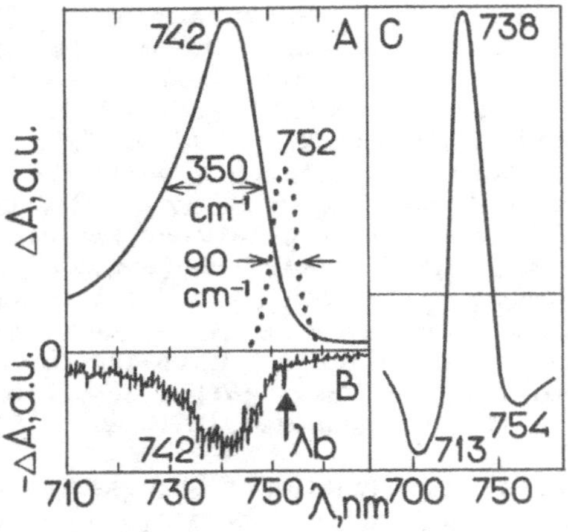

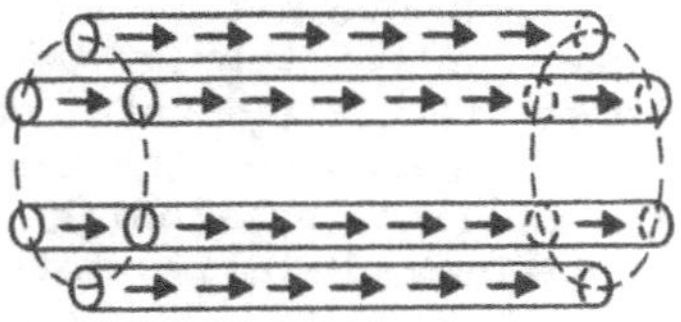

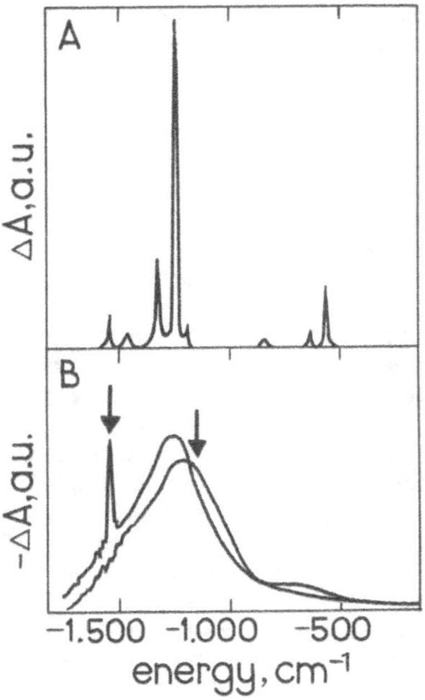

Figure 1. A: Absorption spectrum for *C. aurantiacus*. B: Difference hole burning spectrum (1.8K). C: CD spectrum (295K).

Figure 2. Tubular model (L=6); Arrows show transition dipoles.

Figure 3. A: Homogeneous absorption spectrum for the tubular model with L=6, N=8 and 3 cm^{-1} line width. B: Hole burning spectrum for 150 cm^{-1} homogeneous width and 100 cm^{-1} inhomogeneous width.

References

1. Alden,R.G., Lin,S.H. and Blankenship,R.E. (1992) J. Luminesc. 51, 51-66
2. Fetisova,Z.G., Freiberg,A.M. and Timpmann,K.E. (1988) Nature 334, 633-634
3. Holzwarth,A.R. and Schaffner,K. (1994) Photosynth. Res. 41, 225-233
4. Nozawa,T., Ohtomo,K., Suzuki,M., Nakagawa,H., Shikama,Y., Konami,H. and Wang,Z. (1994) Photosynth.Res. 41, 211-223
5. Fetisova,Z.G.and Mauring,K.(1992) FEBS Lett. 307, 371-374
6. Fetisova,Z.G.and Mauring,K.(1993) FEBS Lett. 323, 159-162
7. Fetisova,Z.G., Mauring,K. and Taisova,A.S. (1994) Photosynth. Res. 41, 205-210
8. Staehelin,L.A., Golecki,J.R. and Drews G. (1980) Biochim. Biophys. Acta 589, 30-45

A HOLE-BURNING STUDY OF Fe-LIMITED AND Fe-REPLETE CELLS OF *DUNALIELLA TERTIOLECTA*

[1]T. Polívka,[1]J.Pšenčík,[1]P.Kroh,[1]D.Engst,[2]O. Prášil, [3]P.G. Falkowski,
[1]J. Hála
[1]National Research Centre of Photosynthesis and Global Climate Change, Department of Chemical Physics, Charles University, Ke Karlovu 3, 121 16 Prague 2, [2]Institute of Microbiology, Czech Academy of Sciences, 379 81 Třeboň, Czech Republic; [3]Oceanographic and Atmospheric Sciences Division, Brookhaven National Laboratory, Upton, New York 11973, U.S.A

1. Introduction

Iron is the most abundant transition metal in the Earth's crust. Since the evolution of photosynthesis in the anaerobic Archean ocean, iron has been used both as a prosthetic group and as an electron carrier. In the modern oxygenic environment, Fe (III) is virtually insoluble in seawater, and consequently, large regions of the world oceans appear to be limited by iron. The limitation is physiologically manifested by large decreases in the quantum yield of photochemistry in PSII and increases in the F_o yield of chlorophyll fluorescence that corresponds with large increase in the 1.2 ns lifetime in the F_o state. These effects suggest that Fe limitation alters the efficiency of excitation energy transfer (EET) from the antenna to PSII reaction centers [1,10].

The method of persistent spectral hole-burning (PSHB) can be used for determination of very fast excited state lifetimes of chlorophylls in pigment-protein complexes of photosynthetic systems. PSHB also provides further information about pigment-protein complexes by characterization of electron-phonon coupling (Huang-Rhys factor S) and the mean frequency of protein phonons (Ω) [2, 3]. The holewidths (δ_{HB}) are related to the total relaxation time that is at low temperatures (4.2 K) determined predominantly by the excited state lifetime T_1. The PSHB has been measured in fluorescence of the antenna complexes CP43 and CP47 of *Synechococcus sp.* [4], CP47 of wild type and mutant of *Synechocystis sp.* PCC 6803 [5] and PSHB in absorption spectra of LHCII and CP47 isolated from higher plants were published in [6,7]. The aim of this paper is to compare the EET in antennae of iron-limited and iron-replete whole cells of *Dunaliella tertiolecta*.

2. Procedure

The spectroscopic characterization of iron-induced changes in the light harvesting processes was performed in whole cells of marine unicellular chlorophyte *Dunaliella tertiolecta*. Growth conditions and preparation of samples are described in [1]. Cells were stored in the dark in liquid nitrogen before measurements. The relative abundance of some Photosystem II proteins was determined by protein analysis [10] and is summarized in following table.

P. Mathis (ed.), Photosynthesis: from Light to Biosphere, Vol. I, 103–106.

	D1	CP43	CP47	LHCII	Cyt b_{559}
Fe-replete cells	1	1	1	1	1
Fe-limited cells	0.35	0.42	0.36	1.08	9.3

Low temperature fluorescence was detected after excitation in the Soret absorption band by the 456 nm or 488 nm lines of a *cw* Carl Zeiss Jena ILA 190 Ar^+ laser. The persistent holes were burnt into fluorescence spectra by a *cw* dye laser Spectra Physics 375B with bandwidth less than 0.3 cm^{-1}. The typical burning power ranged from 1-100 $mW \cdot cm^{-2}$. Special low temperature cuvettes were filled with sample and immersed directly in liquid helium in a Leybold Heraeus BBK 100 optical cryostat. A detailed description of experimental set-up can be found in [8].

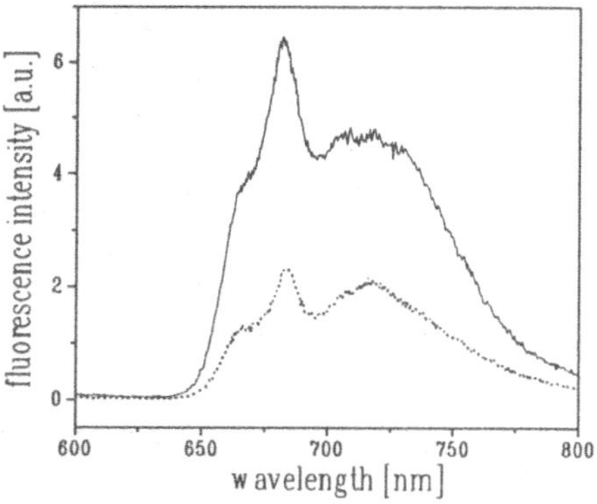

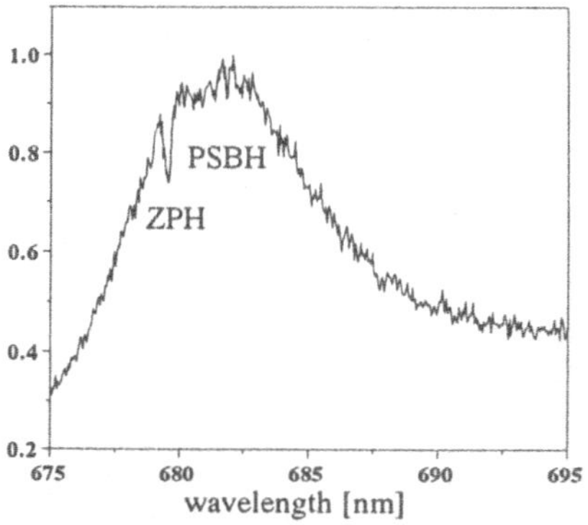

Figure 1. (left) The low temperature (77K) fluorescence spectra of both Fe-limited (solid line) and Fe-replete (dotted line) whole cells under excitation at 488 nm.

Figure 2. (right) Persistent resonant zero phonon hole (ZPH) and phonon side-band hole (PSBH) of the Fe-limited cells (λ_{HB} = 679.5 nm).

3. Results and Discussion

The low temperature (77K) fluorescence spectra of both Fe-limited and Fe-replete cells are shown in *Figure 1*. The fluorescence yield of Fe-limited cells is nearly 3-fold higher due to a loss of PSII reaction centers relative to LHCII antenna. Both spectra show an intense PSII fluorescence band with the maximum at 683.5 nm (Fe-replete) and 682 nm (Fe-limited), respectively. A blue shift of the fluorescence band of Fe-limited cells is caused by excessive LHCII antenna proteins. The fluorescence band near 666.5 nm probably belongs to free chlorophyll molecules. PSI fluorescence is situated in the spectral region above 720 nm [9], but the origin of high fluorescence between 700 - 720 nm is still unclear. Three separated bands in the fluorescence spectrum of Fe-limited cells can be distinguished in spectral region above 700 nm (maxima 707 nm, 716.5 nm, 730 nm). Our investigation was focused on the fluorescence band originating from PSII. Detailed spectra of this region at liquid helium temperature (4.2K) are shown in *Figure 4*. The shape of absorption bands at 4.2 K were the same for both Fe-limited and Fe-replete cells. This supports the idea that most of changes caused by Fe-depletion are changes in EET.

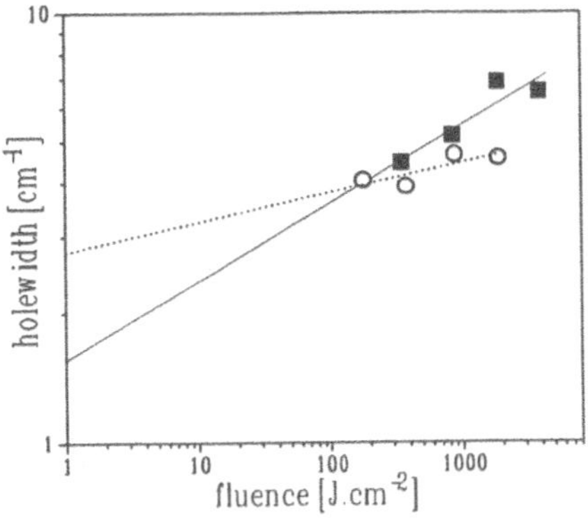

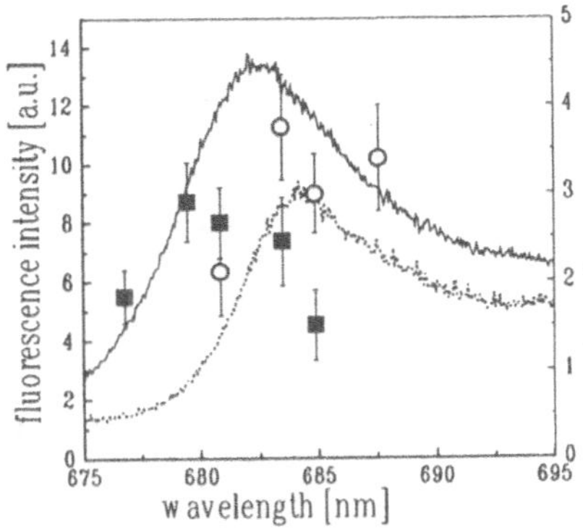

Figure 3. (left) Holewidth dependencies $\delta_{HB} = \delta_{HB}$ (P. t) in logarithmic plot obtained for the Fe-limited (solid line,■) and Fe-replete (dotted line, O) cells at $\lambda_{HB} = 685$ nm.
Figure 4. (right) Low temperature (4.2 K) fluorescence spectra of Fe-limited (solid line) and Fe-replete (dotted line) cells and a wavelength dependence of the holewidth for Fe-limited (■) and Fe-replete (O) cells.

A typical PSHB fluorescence spectrum of Fe-limited cells is shown in *Figure 2*. The spectrum consists of the resonant zero phonon hole (ZPH) and the red shifted phonon side-band hole (PSBH). The related Huang-Rhys factor S and the half of the Stokes shift D are the same ($S \approx 0.9$, $D = 24 - 27$ cm^{-1}) for both Fe-limited and Fe-replete cells. The related mean phonon energy of the protein is $\Omega \approx 30$ cm^{-1}. This indicates that no dramatic changes in the protein structure of PSII antenna complexes occur under Fe-limited conditions. To eliminate the saturation effects we measured the holewidths δ_{HB} as a function of the burning fluency P·t (P - burning power, t - burning time) and extrapolated the obtained holewidths to zero burning power. These dependencies are presented in *Figure 3* for both Fe-limited and Fe-replete samples at the burning wavelength 685 nm. Furthermore, the wavelength dependence of the ZPH was determined in the spectral range 675-688 nm (*Figure 4*). The related lifetimes of excited states (T_1) at particular wavelengths were calculated from the ZPH widths. A 2-fold slower EET was found in Fe-limited cells above 683 nm: $T_1 \approx 8$ ps for Fe-limited cells, $T_1 \approx 4$ ps for Fe-replete cells. This slowing can be explained by the missing core proteins in some PSII. While the amount of LHCII is approximately the same for both samples, the amount of core antenna protein CP43 and CP47 is markedly reduced (see table). The path of energy transfer from the outer antenna to the core antenna is interrupted in some of PSIIs which leads to the increase of fluorescence intensity and a decrease of EET rate between antenna protein complexes.

Higher burning powers (above 150 mW·cm^{-2}) resulted in very broad holes at 683 nm in all hole-burning spectra (*Figure 5*). The width of the holes was 180 cm^{-1}. This is in good agreement with the holewidths observed in absorption spectra [7,8]. These broad holes correspond to very fast excitonic relaxation of pigments coupled within particular proteins. This result indicates that the lack of iron has no effect on ultrafast energy transfer between chlorophyll molecules within particular proteins.

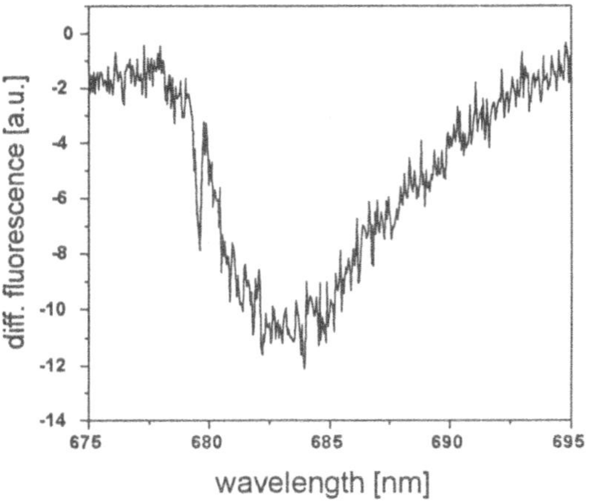

Figure 5. The difference PSHB spectrum of Fe-limited cells obtained for 6 min burning exposure (180 mW cm^{-2}) at λ_{HB} = 679.5 nm.

References

1 Greene, M.R., Geider, J.G., Kolber, Z., Falkowski P.G. (1992) Plant. Physiol.,100, 565-575

2 Jankowiak, R. and Small, G.J. (1993) in The Photosynthetic Reaction Center (Deisenhofer J. and Norris J.R.,eds.) Vol. 2, pp. 133-177, Academic Press, Inc., San Diego, USA

3 Vácha, M., Pšenčík, J., Adamec, F., Ambrož, M., Dian, J., Boček, J., Komenda, J. and Hála, J. (1992), in Research in Photosynthesis (Murata N., ed.), Vol 1, pp. 335, Kluwer Academic Publishers, Dodrecht

4 Hála, J., Vácha, M., Dian, J., Ambrož, M., Adamec, F., Prášil, O., Komenda, J., Nedbal, L., Vácha, F., Mareš, J. (1991) J. Lumin.,48&49, 295-298

5 Polívka, T., Engst, D., Dian, J., Kroh, P., Pšenčík, J., Vácha, M., Nedbal, L., Vermaas, W.F.J., Hála, J. (1994) Spectral Hole-Burning and Related Spectroscopies: Science and Applications, Vol. 15 OSA Technical Digest Series, pp. 318-321, Optical Society of America, Washington DC

6 Chang H.C., Jankowiak R., Yocum C.F., Picorel R., Alfonso M., Seibert M. and Small G.J. (1994) J. Phys. Chem., 98, 7717-7724

7 Reddy N.R.S., van Amerongen H., Kwa S.L.S., van Grondelle R. and Small G.J. (1994) J. Phys. Chem., 98, 4729-4735

8 Pšenčík, J., Vácha, M., Adamec, F., Ambrož, M., Dian, J., Boček, J., Hála, J. (1992) Spectral Hole Burning and Luminescence Line Narrowing, Third topical meeting, Ascona, Switzerland, Technical Digest Series , pp. 115-118, Opt. Soc. Am., Vol. 223, Washington DC

9 van Grondelle, R., Dekker J.P., Gillbro, T., Sundstrom, V. (1994) Biochim. Biophys. Acta 1187, 1-65

10 Vassiliev, I.R., Kolber, Z., Wyman, K.D., Falkowski, P.G., Mauzerall, D., Shukla, V.K. (1995) Plant Physiology, in press.

EXCITON DYNAMICS AND DELOCALIZATION LENGTH IN B850 MOLECULES OF LH2 OF *RHODOBACTER SPHAEROIDES*.

Tõnu Pullerits, Mirianas Chachisvilis and Villy Sundström
Department of Chemical Physics, Lund University, Box 124, 22100 Lund, Sweden

1. Introduction

The antenna system of photosynthetic purple bacteria consists of two different light-harvesting complexes – the core antenna LH1 surrounding the reaction center, and the peripheral antenna LH2. The LH2 antenna has two distinc spectral bands – B800 and B850. Recently the crystal structure of LH2 from *Rhodopseudomonas (Rps.) acidophila* was determined to a resolution of 2.5 Å (1) revealing a ring of nine αβ-polypeptide pairs, each containing two Bchl *a* molecules of B850 and one Bchl *a* of B800. This is certain to stimulate strong interest in LH2 complexes and hopefully will eventually lead to a deeper understanding of the fundamentals of primary photosynthesis.

Intermolecular excitation transfer in photosynthesis is generally described by the Förster dipole-dipole resonance mechanism. Measurements of the transfer between spectrally different antenna forms (2) and energy migration among similar pigment molecules studied in experiments with relatively limited time resolution (3) have been well described by this incoherent hopping theory.

On the other hand also the exciton concept has been applied for the photosynthetic antenna systems (4, 5). In that case excitation is delocalized over a number of pigment molecules and the dynamics occurs through the relaxation between different exciton states. However, often it is not so straightforward to unambiguously distinguish in experiment the two qualitatively different kinetic processes, incoherent hopping and exciton relaxation. The actual dynamics can be a combination of these limiting cases. For example, smaller sections of the full system might behave as a small exciton whereas the dynamics on a larger scale may correspond to the hopping-like transfer of this small exciton. An argument against the pure incoherent Förster mechanism is the recent observation of coherent nuclear motions in the antenna complexes of photosynthetic bacteria (6). The vibrational coherence is preserved for approximately the same time as the estimated single step transfer time in these systems. It means that

P. Mathis (ed.), Photosynthesis: from Light to Biosphere, Vol. I, 107–110.

one of the main assumptions of the Förster theory that the vibronic relaxation occurs much faster than the excitation transfer is not fulfilled. Furthermore, the structural data (1) show a densely packed Bchl *a* system, where exciton interactions are quite strong.

At first we present the anisotropy decay studies of the photosynthetic purple bacterium *Rhodobacter (Rb.) sphaeroides.* With 40 fs time resolution we have measured the isotropic and anisotropic pump-probe signal decays in B850 band. Simulations of the initial dynamics based on an incoherent hopping approach strongly suggest that the initial dynamics is to a considerable extent due to exciton relaxation. We have estimated the exciton delocalization length by comparing the experimental absorption difference spectra with relevant calculations based on the structural and spectroscopic data.

2. Results and Discussion

2.1 *Polarized studies*

In Fig. 1. the anisotropic and isotropic decays at room temperature at 864 nm are presented. Both decays have a fast subpicosecond component of 130 fs and 70 fs for anisotropic and isotropic decay, respectively. Qualitatively it is understandable that for a ring-like structure of spectrally inhomogeneous Bchl *a* molecules it takes a few jumps before anisotropy drops considerably because the transition dipole moments of the nearest B850 molecules are nearly parallel. At the same time the spectral equilibration inside of the inhomogeneous distribution function (IDF) takes place faster. However, direct numerical simulations show that for the B850 ring of 18 Bchl *a* molecules at room temperature in case of the excitation into the red side of the absorption band the fast component of the anisotropy decay is three times longer than the corresponding component of the isotropic decay. We have found similar contradictions also for other temperatures and wavelengths. This result strongly suggests that the entity which acts as an elementary excitation is not a single excited Bchl *a* molecule. The really hot question to be answered is: how big is this elementary excitation. Is it a dimer, or is the full ring acting as an exciton, or is the reality something in between?

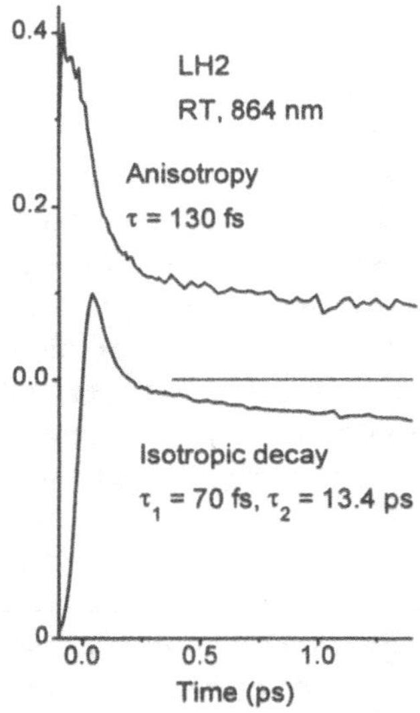

Fig. 1. Anisotropic and isotropic decays in B850.

2.2 *Exciton delocalization length in B850*

The nonlinear optical properties of aggregates are sensitive to the delocalization length of the electronically excited states. For example in a linear aggregate of two-level molecules one can calculate the exciton delocalization length from the approximate expression $\Delta\omega \approx 3\pi^2 V/(N+1)^2$, where $\Delta\omega$ is the difference between pump-pulse induced bleach (ground state to one-exciton transition) and pump-pulse induced absorption (one-exciton to two-exciton transition) maxima, V is the dipole-dipole interaction between nearest neighbors and N is the exciton delocalization length (7). However, monomeric Bchl *a* has a strong excited state absorption in the region of the Q_y transition and therefore B850 is much more realistically described as an aggregate of multilevel (not two-level) molecules where one- to two-exciton transition has a significant intramolecular component where two photons double excite a single molecule. It has been shown recently that this may influence quite significantly the calculated induced absorption spectra of the aggregates (8).

Unfortunately, the strength and energy of the monomeric induced absorption for Bchl *a* is not a well-defined quantity. From available experimental studies of Bchl *a* in solutions (9-11) we have extracted out that the strength of the monomeric induced absorption is about half of the strength of the Q_y transition, its dipole moment is nearly parallel to the Q_y transition dipole moment and located 100 cm^{-1} to the blue compared with the Q_y transition. The calculated absorption difference spectra are presented in Fig. 2. For the sake of clarity we have used a very narrow homogeneous width of the transitions. We have calculated the interactions between Bchl *a* molecules by using published structural information (1). According to our estimations the nearest neighbor interaction in an αβ-polypeptide is ~ 450 cm^{-1}, for adjacent polypeptides ~ 320 cm^{-1}. We have assumed that the monomeric transition is due to the dispersive protein shift at 800 nm. One can clearly see the dependence of the bleaching (positive) and induced absorption (negative) on the exciton delocalization length. In Fig. 3. we have made a comparison of the calculated spectra for the thermalized population with the experimental absorption difference spectrum recorded 2 ps after the excitation with a femtosecond laser pulse into the B800 band. The calculated spectra are shifted in the way that they have the bleaching maximum at 856 nm in accordance with the measurements. The best correspondence between calculated and measured spectra are

Fig. 2. Calculated absorption difference spectra for a ring of B850 molecules with different exciton delocalization lengths.

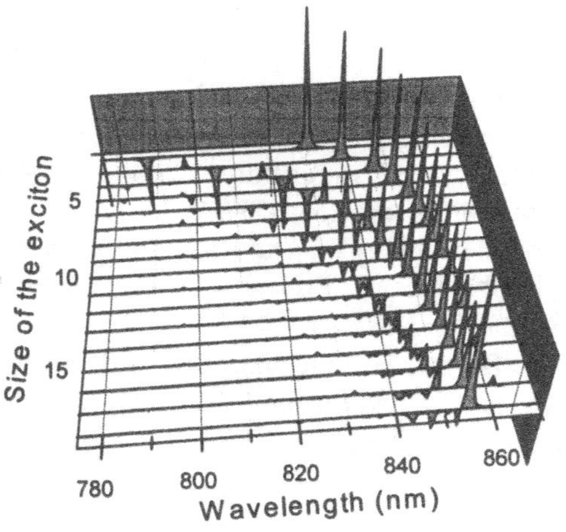

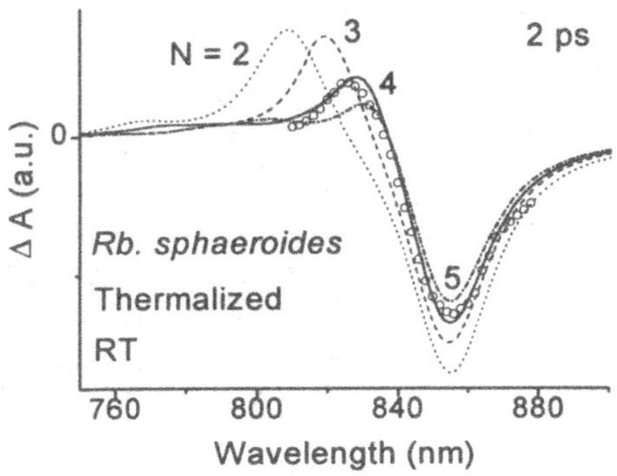

Fig. 3. Experimental (circles) and calculated (lines) absorption difference spectra of B850.

for N = 4. However, it has to be pointed out that this result is sensitive to the assumed parameters for the monomeric induced absorption. For example, if we assume that the strength of the monomeric induced absorption is equal to the strength of Q_y transition, then we would get a reasonable agreement for N equal to two or three. Therefore, for a more definite value of the delocalization length very accurate information about the excited state absorption of Bchl *a* monomer is needed.

At this stage we can conclude that our results do not support the idea that the exciton in B850 is delocalized over the full ring (at least 2 ps after the excitation at B800). Available information suggests that the exciton delocalization length in B850 is 4. However, because of the lack of accurate enough data of the monomeric excited state absorption of Bchl *a* we formulate our final result as N = 4±2.

References

1 McDermott, G., Prince, S.M., Freer, A.A., Hawthornthwaite-Lawless, A.M., Papiz, M.Z., Cogdell, R.J. and Isaacs, N.W. (1995) Nature 374, 517-521

2 Hess, S., Visscher, K.J., Pullerits, T., Sundström, V., Fowler, G.J.S. and Hunter, C.N. (1994) Biochemistry 33, 8300-8305

3 Pullerits, T., Vissher, K.V., Hess, S., Sundström, V., Frieberg, A., Timpmann, K. and Van Grondelle, R. (1994) Biophys. J. 66, 236-248

4 Pullerits, T., Chachisvilis, M., Jones, M.R., Hunter, C.N. and Sundström, V. (1994) Chem. Phys. Lett. 224, 355-365

5 Novoderezhkin, V.I. and Razjivin, A.P. (1994) Photosynth. Res. 43, 9-15

6 Chachisvilis, M., Pullerits, T., Jones, M.R., Hunter, C.N. and Sundström, V. (1994) Chem. Phys. Lett. 224, 345-351

7 Van Burgel, M., Wiersma, D.A. and Duppen, K. (1995) J. Chem. Phys. 102, 20-33

8 Knoester, J. and Spano, F.C. (1995) Phys. Rev. Lett. 74, 2780-2783

9 Becker, M., Nagarajan, V. and Parson, W.W. (1991) J. Am. Chem. Soc. 113, 6840-6848

10 Savikhin, S. and Struve, W.S. (1994) Biophys. J. 67, 2002-2007

11 Hess, S., Åkesson, E., Cogdell, R., Pullerits, T. and Sundström, V. (1995) Biophys.J. (accepted)

LHC AGGREGATION IN INTACT LEAVES OF TOBACCO PLANTS STRESSED BY CO_2 STARVATION.

Pavel Šiffel & František Vácha
Institute of Plant Molecular Biology, Academy of Sciences of the Czech Republic, Branišovská 31, 370 05 České Budějovice, Czech Republic

1. Introduction

Aggregation of the light-harvesting complex II (LHC II) was proposed to regulate dissipation of the excitation energy to heat [1, 2]. Most evidences supporting this hypothesis have been obtained on LHC II aggregates prepared from detergent solubilized chlorophyll-protein complex [1, 2, 3]. In this contribution, the formation of the LHC II aggregates *in vivo* and their relation to non-photochemical quenching (q_{NP}) and zeaxanthin cycle is presented. LHC II aggregates were detected by its characteristic emission band near 700 nm measured in low temperature fluorescence spectra.

2. Procedure

Plant cultivation and treatment:
Plantlets of tobacco (*Nicotiana tabacum* L. cv. White Burley) were cultivated in sterile conditions on agar medium [4] in the glass vessels (covered by Al foils) in controlled chamber under 20 °C, 12 hours photoperiod and irradiance of 50 μmol PAR $m^{-2} s^{-1}$ [see ref. 5]. Then the CO_2 absorber based on the sodium hydride (Natrocalcid, Chemopetrol Spolana Neratovice, Czech Republic) was added and the Al-covers were replaced with hermetically sealed glass stoppers. CO_2 concentration in the vessels with CO_2 absorber was well below CO_2 compensation point (25 - 30 ppm, [6]).

In vivo chlorophyll fluorescence
The ratio of variable (F_V) to maximal (F_M) fluorescence and q_{NP} were determined on leaf disks using a PAM fluorometer.

Fluorescence and absorption spectra
Fluorescence emission and excitation spectra were measured at 77 K using diluted leaf homogenates [7], otherwise indicated in the text. The intensity of the 699 nm emission band was accessed by deconvolution of the spectra into Gaussian components peaking at 681 nm (LHC II), 685 and 695 nm (PS II), 699 nm (LHC II aggregates), 729 and

P. Mathis (ed.), Photosynthesis: from Light to Biosphere, Vol. I, 111–114.

740 nm (PS I) and 712 and 738 nm (vibrational bands). The spectra used for the deconvolution have been normalized on Chl content in leaf homogenate and on the fluorescence intensity of rhodamine B in the sample. Spectra were taken with a Fluorolog spectrofluorometer (SPEX, U.S.A.) using the spectral bandwidth of 2 nm.

Absorption spectra of tobacco were measured with whole leaves frozen on a plexiglass holder in liquid nitrogen. Spectra were recorded with a PU 88OO spectrophotometer (*Phillips-Pye Unicam*, U.K.) using a bandwidth of 2 nm.

Isolation of LHC II

The LHC II was isolated using mild LiDS polyacrylamide gel electrophoresis according to Anderson [8].

Zeaxanthin

Amount of zeaxanthin was determined by C-18 reverse phase HPLC using mixture of acetonitrile, methanol and water. Capacity of the zeaxanthin cycle was accessed as the total amount of zeaxanthin induced by irradiation of leaves with intensive white light (1000 μmol PAR $m^{-2} s^{-1}$) for 20 min under nitrogen atmosphere.

3. Results and Discussion

CO_2 depletion resulted in a gradual degradation of both photosystems, indicated by suppression of emission bands at 685, 695 and 735 nm and by the shift of the maximum at 685 nm to 681 nm (Fig. 1). The new band at 697-700 nm (further referred to as F_{699})

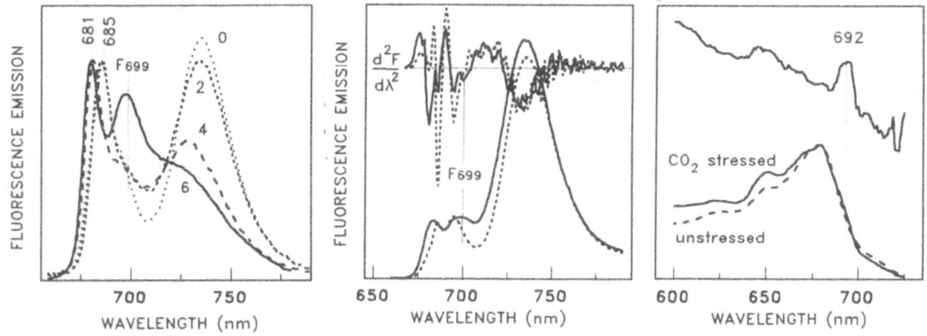

Figure 1. (left) Changes of the fluorescence emission spectrum of tobacco plants upon CO_2 starvation. Days of CO_2 starvation are indicated at the spectra. Spectra are normalized at the short-wavelength maximum.

Figure 2. (centre) Formation of F_{699} upon CO_2 starvation measured on whole leaves after 2 days of CO_2 starvation. Spectra are normalized at 685 nm.

Figure 3. (rigth) Formation of the 691-693 nm absorption band in CO_2 starved plants measured on whole leaves. Spectra are normalized at the maximum.

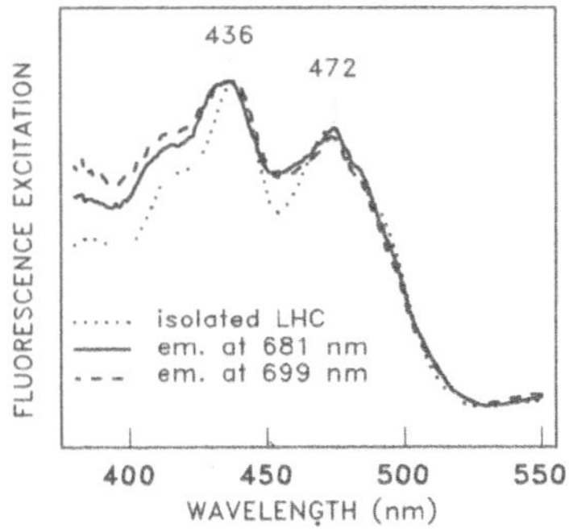

Figure 4. Excitation spectra of tobacco plants after 6 days of CO_2 starvation. Emission wavelength at 681 nm - broken line, 699 nm - full line. Excitation spectrum of the detergent-solubilized LHC - dotted line. Spectra are normalized at 436 nm.

appeared in the emission spectra of tobacco plants after four or five days of CO_2 starvation. F_{699} was more pronounced in the spectra of whole leaves in comparison to homogenates since the re-absorption reduced the intensity of the most intensive 685 nm band (Fig. 2). With whole leaves, the LHC aggregates were detected even in plants stressed for two days. In absorption spectra, a new shoulder with the maximum at 691-693 nm was formed (Fig.4). Both the F_{681} and F_{699} showed identical excitation spectra (Fig. 5) which were close to the spectrum of LHC II isolated by electrophoresis. This shows that both emission bands found in stressed tobacco plants were excited by light absorbed in LHC II, and supports the notion that F_{699} originates from the LHC II aggregates. These results manifest the formation of LHC aggregates *in vivo* upon CO_2 starvation.

The LHC aggregation has been shown to be stimulated by the zeaxanthin [1]. Both the LHC aggregation and zeaxanthin content have been related to the q_{NP} [1, 2, 3, 9]. Therefore, we have compared the kinetic of q_{NP} with the kinetics of the intensity of the F_{699}, of the amount of zeaxanthin and of the capacity of zeaxanthin cycle during CO_2 starvation. The amplitude of the 699 nm Gaussian component increased during the

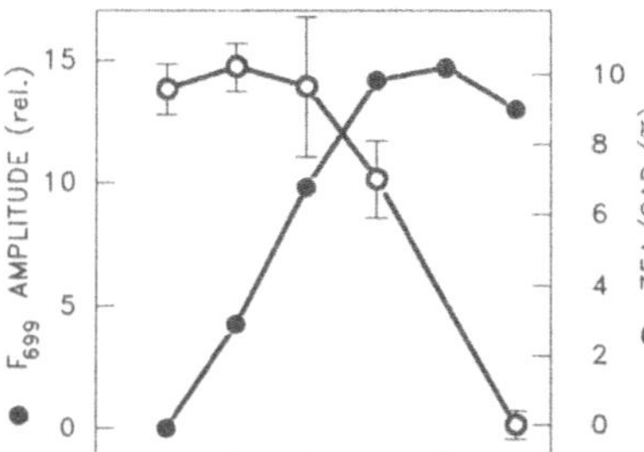

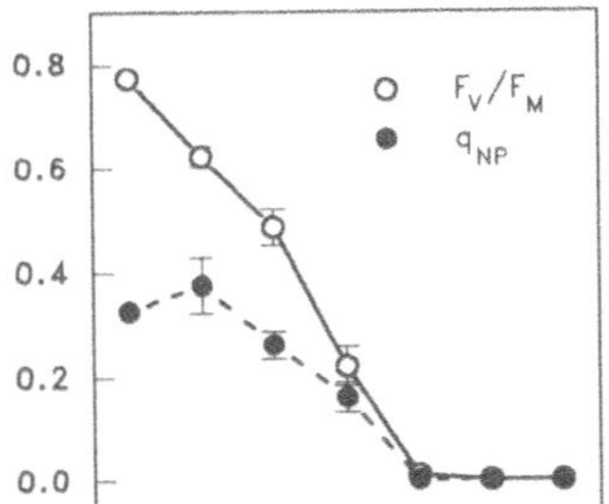

Figure 5. (left) The kinetics of the amplitude of F_{699} compared with the capacity of zeaxanthin cycle (amount of zeaxanthin relative to the total amount of all carotenoids) during CO_2 starvation.

Figure 6. (right) Kinetics of F_V/F_M and q_{NP} during CO_2 starvation.

stress (Fig. 6), however, q_{NP} and the capacity of zeaxanthin cycle declined (Figs. 6 and 7). No zeaxanthin was found throughout the stress unless the leaves were irradiated by bright light. Also the reversible photoinduced changes of emission near 700 nm, which had been presented to support an involvement of the LHC II aggregation in the regulation of excitation energy dissipation [3], did not correspond to the q_{NP} development under CO_2 starvation: the stressed plants, showing the depression in q_{NP}, exhibited more pronounced decrease near 700 nm then the non-stressed ones (Fig. 7).

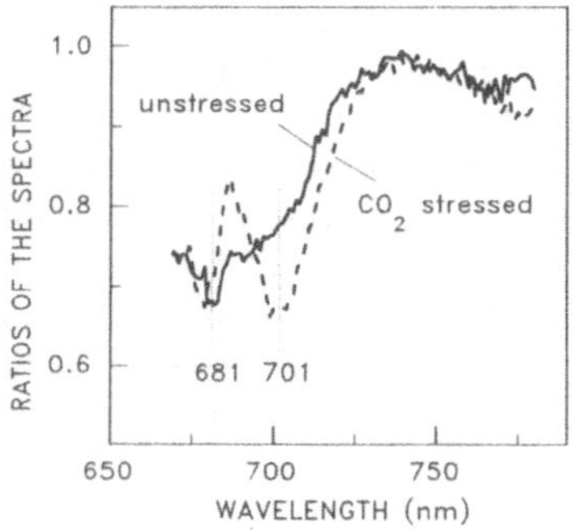

Figure 7. Reversible light-induced changes in the emission spectrum of whole leaves of unstressed and stressed plants. Ratios of the spectra of irradiated (red actinic light produced by PAM fluorimeter: 150 μmol PAR m^{-2} s^{-1} for 7 min) to dark adapted (10 min) leaf segments are depicted. Stressed leaves with F_V/F_M of approximately 0.3 were taken for measurement.

The contradictory responses of the intensity of the F_{699} amplitude and the extent of F_{699} photoinduced changes on the one hand, and q_{NP} and capacity of zeaxanthin cycle on the other hand indicate, that the LHC II aggregates formed under CO_2 starvation can not be related to the regulation of excitation energy dissipation by the q_{NP}.

References

1 Horton, P., Ruban, A.V., Rees, D., Pascal, A.A., Noctor, G. and Young, A. (1991) FEBS Lett. 242, 1-4
2 Horton, P. and Ruban, A.V. (1992) Photosynth. Res. 34, 375-385
3 Ruban, A.V., Rees, D., Noctor, G.D., Young, A. and Horton, P. (1991) Biochim. Biophys. Acta 1059, 355-360
4 Murashige, T. and Skoog, F. (1962) Physiol. Plant. 15, 473-497
5 Pospíšilová, J., Solárová, J. and Čatský, J. (1992) Photosynthetica 26, 3-18
6 Šantrůček, J., Šiffel, P., Polišenská, M. and Pešková, H. (1991) Photosynthetica 25, 203-209
7 Šiffel, P., Kutík, M. and Lebedev, N.N. (1991) Photosynthetica 25, 395-407
8 Anderson, J.M. (1980) Biochim. Biophys. Acta 591, 113-126
9 Deming-Adams, B. (1990) Biochim. Biophys. Acta 1020, 1-24

ROLE OF THYLAKOID LIPIDS IN THE ASSEMBLY AND STRUCTURAL FLEXIBILITY OF CHIRALLY ORGANIZED MACROAGGREGATES OF THE LIGHT-HARVESTING CHLOROPHYLL A/B PROTEINS

I. Simidjiev[1], V. Barzda[1], I. Horvàth[2] and G. Garab[1]
Institute of Plant Biology[1] and Biochemistry[2], Biological Research Center, Hungarian Academy of Sciences, Szeged, P. O. Box 521, H-6701, Hungary

1. Introduction

The interactions between lipids and proteins within biological membranes are fundamentally important for the understanding the structure and function of many membrane proteins. One of only five membrane proteins with a known structure which has been vigorously investigated is the major chlorophyll a/b-containing light-harvesting complex of photosystem II (LHCII). Its polypeptide contains 232 amino acids which binds noncovalently a minimum of 12 chlorophylls (1), two xantophylls (2) as well as two different lipids, PG (phosphatidylglycerol) and DGDG (digalactosyldiacylglycerol) (3). The stable form of LHCII is a trimer. The protein isolated from thylakoid membranes with mild, non-ionic detergent, always carries five different thylakoid lipids. Two of them, DGDG and PG, have special roles. PG participates directly in trimer formation and subunit-subunit interactions. DGDG is essential for crystal formation and thus for the structural integrity of the complexes (3). Beside its primary role, harvesting the light and transferring the excitation energy to the reaction centers, LHCII plays an important role in mediating the stacking of membranes in (4). In addition, LHCII participates in different regulatory processes that control photophysical processes in the antenna (5,6). This type of regulation requires a flexibility in the molecular organization of the antenna system.

LHCII has been shown to play an important role in the assembly of chirally organized macrodomains of PSII particles and the spatial separation of the two photosystems (7). Macroaggregation of PSII particles is likely to be responsible for the connectivity of PSII reaction centers (8). Purified LHCII also readily forms large macroaggregates with long range chiral order (9).

Chloroplast thylakoid membranes have been shown to be capable of undergoing light-induced reversible changes in their chiral macrostructures, detected as light-induced reversible changes in the circular dichroism (CD) bands with psi-type features (10). These changes are likely to play a role in the regulation of the dissipative pathways in the membranes (Istokovics et al., this Proceedings). Further, it has been shown that the structural flexibility of the thylakoid membranes is due to the inherent

P. Mathis (ed.), Photosynthesis: from Light to Biosphere, Vol. I, 115–118.

flexibility of the chiral macrodomains. It has also been reported that not only thylakoid membranes but chirally organized macroaggregates of purified LHCII also possess the ability of undergoing light-induced reversible structural changes (Barzda et al., this Proceedings).

The light-induced reversible structural changes in LHCII macroaggregates showed large biological variations, which depended mainly on the conditions of the solubilization of thylakoid membranes. In this work, we analyzed the differences in the lipid content of different LHCII preparations and the lipid and fatty acid composition of the samples with respect to their ability to form lamellar macroaggregates and to undergo light-induced reversible changes in their macrostructure.

2. Experimental Procedures

LHCII was isolated from two-week old pea leaves (*Pisum sativum* L.) according to (11) with minor modifications. As the solubilization step with Triton X-100 was critical in obtaining LHCII macroaggregates exhibiting light-induced reversible changes in their macrostructure, solubilization was performed with different detergent concentrations between 0.6 and 0.8 % (v/v). As revealed from Western blots for D1 protein, the samples contained no contamination of PSII reaction centers.

Thylakoid lipids were isolated following the procedure of (12). Lipids were separated on TLC plates (Kieselgel 60, Merck). All organic solvents were saturated with N_2 and contained butylated hydroxytoluene (1 μg/ml). Fatty acids were methylated and analyzed on GLC.

CD spectra were recorded in a CD6 dichrograph (Jobin-Yvon), the Chl concentration was 20 μg/ml. The dichrograph was equipped with a side-illumination attachment; the photomultiplier was protected with crossing filters against light from the beam of a 650 W tungsten lamp, which also passed through a heat filter of 10 cm water. Before the measurements the samples were kept for 20 min in the dark.

3. Results and Discussion

Shown on Fig 1, three independent LHCII preparations, which exhibit differences in the amplitude and rate of ΔCD. SDS-PAGE, using the method of silver staining, revealed the presence of 27 and 25 kDa polypeptides in about the same proportions in the three preparations (data not shown). Chlorophyll a/b ratio was also essentially the same, about 1.1, in the three preparations. Thus, the differences in ΔCD could not be accounted for by differences in the polypeptide composition or in the chlorophyll a/b ratio of the samples.

On the other hand, good correlation was found between the amplitude of ΔCD (Fig. 1a) and the lipid concentration (Fig. 1b) of the samples. The differences can be found not only in the total lipid amount, but also in the amount of different lipid species (Fig. 1c). Further, we demonstrate that the highly unsaturated linolenic (18:3) and linoleic (18:2) fatty acids correlate best with the amplitude of ΔCD (Fig. 1d). We also found, that there is a critical amount of total lipid concentration below which ΔCD could not be observed (data not shown). This suggests that lipids play an essential role in the structural flexibility of LHCII aggregates.

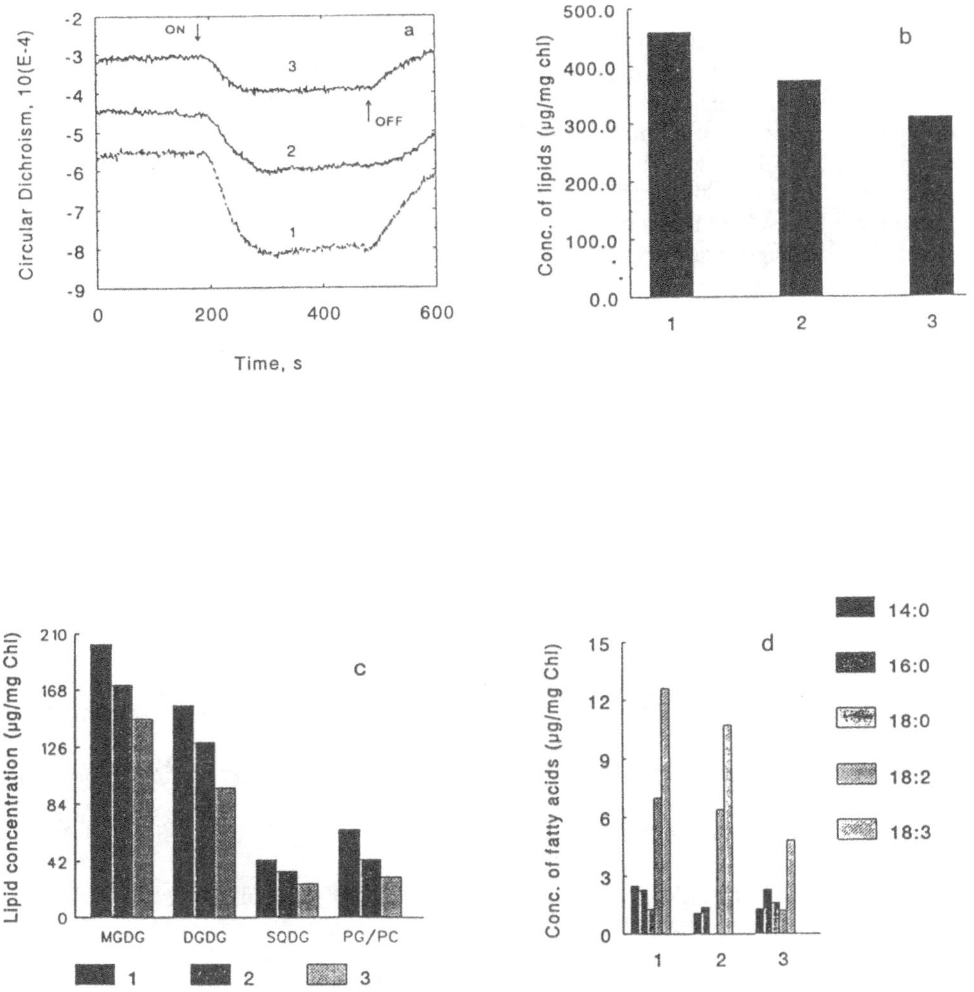

Figure 1. Correlation between ΔCD and lipid content and composition of different LHCII preparations: ΔCD at 495 nm (a), total lipid concentration (b), distribution of different lipid species (c), and fatty acid composition (d) in three different preparations. Concentration of linolenic acid (18:3) is ten times higher than plotted in (d).

The higher initial rate and larger amplitudes of ΔCD are explained on the basis that lipids play an essential role in the structural integrity of LHCII and facilitate the formation of 2-D and 3-D crystals (13). Indeed, our ultrastructural investigations showed that in all cases when ΔCD was observed, LHCII was found to form stacked lamellar structures. As it is shown elsewhere (Barzda et al., this Proceedings) ΔCD can be eliminated by disintegrating the macroaggregates into smaller particles (100-200 nm in diameter). In line with these findings, we observed that ΔCD of LHCII depends on the content of thylakoid lipids in the macroaggregates, and in samples deficient in lipids no lamellar-type aggregation of the complexes occurred, and these preparations did not exhibit ΔCD.

Further support to the conclusion that lipids play a crucial role in the formation and structural flexibility of chirally organized macrodomains of LHCII is provided by experiments with the addition of thylakoid lipids to LHCII preparations. As it will be reported elsewhere, we found that the addition of different lipids facilitated the formation of LHCII macroaggregates of different structures, which also differed substantially in their ability to undergo light-induced structural rearrangements.

Acknowledgments

This work was supported by a grant from the Hungarian Research Fund, OTKA III/2999.

References

1. Kühlbrandt, W., Wang, D.G., Fujiyoshi, Y. (1994) Nature 367, 614-621.
2. Bassi, R., Pineau, B., Dainese, P., Marquardt, J. (1993) Eur. J. Biochem. 212, 297-303.
3. Nussberger, S., Dörr, K., Wang, D.G., Kühlbrandt, W. (1993) J. Mol. Biol. 234, 347-356.
4. Butler, P.J.G., Kühlbrandt, W. (1988) Proc. Natl. Acad. Sci. USA 85, 3797-3801.
5. Allen, J.F. (1995) Physiol. Plant. 93, 196-205.
6. Horton, P., Ruban, A.V., Walters, R.G. (1994) Plant Physiol. 106, 415-420.
7. Garab, G. (1992) in Research in Photosynthesis (Murata, N., ed.) Vol. I. pp. 171-178, Kluwer Academic Publishers, Dordrecht, Boston, London.
8. Lavergne, J., Trissl, H.W., (1995) Biophys. J. 68, 2474-2492.
9. Barzda, V., Mustárdy, L., Garab, G. (1994) Biochemistry 33, 10837-10841.
10. Garab, G., Leegood, R.C., Walker, D.A., Sutherland, J.C., Hind, J. (1988) Biochemistry 27, 2430-2434.
11. Krupa, Z., Huner, N.P.A., Williams, J.P., Maissan, E., James, D.R. (1987) Plant Physiol. 84, 19-24.
12. Sprague, S.G., Staehelin, L.A. (1987) Methods Enzymol. 148, 319-327.
13. Kühlbrandt, W. (1994) Current Opinion in Structural Biology 4, 519-528.

LOCALIZED EXCITATIONS AND EXCITONS IN PHOTOSYNTHESIS. LH2 – A TESTING GROUND OF ENERGY TRANSFER.

S. Hess*, M. Chachisvilis*, T. Pullerits*, M.R. Jones+, G.J.C. Fowler+, C.N. Hunter+ and V. Sundström*
*Department of Chemical Physics, Lund University, 22100 Lund, Sweden
+Krebs Institute for Biomolecular Research and Robert Hill Institute for Photosynthesis, University of Sheffield, Sheffield S10 2UH, U.K.

1. Introduction

Light-harvesting antenna pigments of photosynthetic organisms greatly enhance the spectral coverage and efficiency of light collection, as compared to what would be possible if the photochemical reaction center (RC) were the only light-absorbing entity. For efficient light energy conversion the absorbed light has to be transported through the array of pigments to the RC. Knowledge about how this is realized is now rapidly accumulating. Protein crystallography and ultrafast spectroscopy combined with site directed mutagenesis appear to be particularly useful techniques in this work. The three-dimensional structure provided by crystallography sets the scene for events that may be fine-tuned by mutagenesis and examined by ultrafast spectroscopy on an atomic time scale, to yield a fundamental understanding of the structure-function relationship of these pigment systems.

The recently published structures of LHCII (1) of green plants and LH2 (2) of purple bacteria have shown that pigment densities may be very high in light-harvesting systems, leading to quite strong dipole-dipole interactions between the involved chromophores. This makes it necessary to carefully consider both the incoherent Förster and coherent exciton mechanisms when interpreting experimental results of energy transfer dynamics in these and similar systems.

In this work we summarize our recent studies of excitation dynamics and properties in LH2 complex and LH2 $\rightarrow$ LH1 transfer in purple bacteria.

2. Results and discussion

2.1 *B800 dynamics*

Fig. 1 shows the one-color transient absorption kinetics of B800 at 790, 800 and 810 nm at 77 K, measured with ~ 80 fs pulses. The B800 excited state is seen to decay in a non-

P. Mathis (ed.), Photosynthesis: from Light to Biosphere, Vol. I, 119–122.

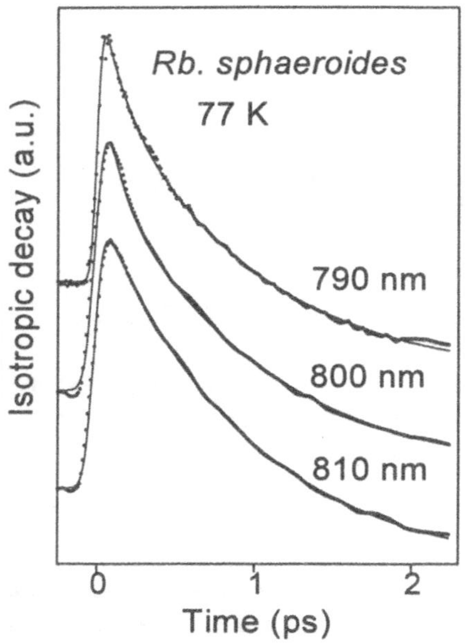

Fig. 1. Isotropic pump-probe kinetics of the B800 band at different wavelengths.

Fig. 2. Anisotropy decay of the B800 band at different wavelengths.

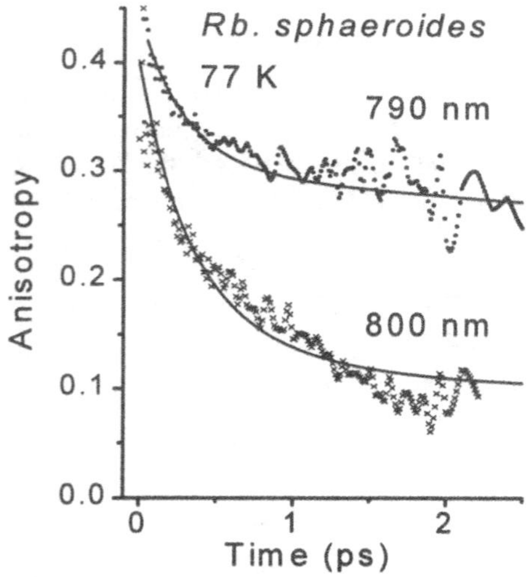

exponential fashion, characterized by two time constants, 0.35 ± 0.05 ps and 1.8 ± 0.2 ps. The slower time constant is related to the B800 → B850 transfer whereas the 0.35 ps time is assigned to the energy transfer within the spectrally inhomogeneous B800 band. A measurement of the absorption anisotropy within the B800 band (Fig. 2) yields a similar decay time of 0.3 ps as well as a slower decay (1.5 – 5 ps) of lower amplitude. In order to interpret these results we have performed computer simulations of the excitation transfer described as incoherent hopping in a system of spectrally inhomogeneous antenna complexes, using a Master equation approach (3). The simulations show that the measured B800 dynamics is well described as energy transfer with a characteristic average nearest neighbor pairwise transfer time of 0.35 ps among ~ 10 Bchl molecules in a circular arrangement (4), in good agreement with the LH2 structure (2). Interestingly enough, a random arrangement of the B800 molecules cannot describe our result.

2.2 *B800→B850*

At room temperature the B800 to B850 energy transfer was measured by exciting B800 with 80 fs pulses at 805 nm and probing the appearance of B850 excitations with ~ 100 fs pulses in the wavelength range 830-875 nm, derived from an optical parametric oscillator (OPO) pumped by the 805 nm Ti:Sa laser pulses (5). As is seen from the kinetics of Fig. 3, measured in the red and blue wings of the B850 spectrum, the build-up time of the B850 excited state is 0.7 ps. At wavelengths close to the isosbestic point of the slow B850 decay,

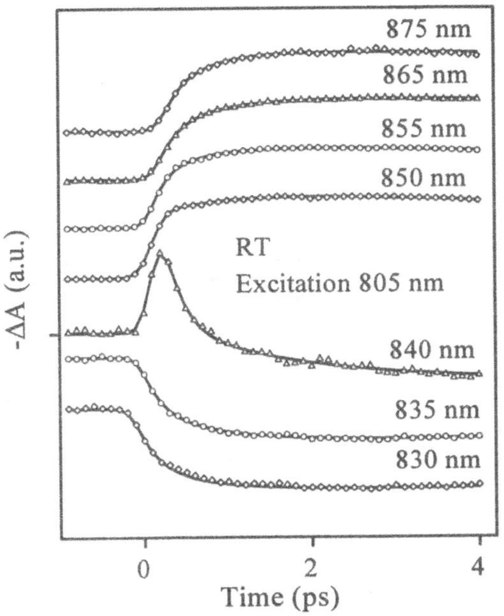

Transient absorption kinetics of LH2 excited into the B800 band and probed at different wavelengths within the B850 band.

the rise is apparently faster. This is a result of very fast energy equilibration within B850, and to it associated spectral changes (5). At 77 K the B800→B850 transfer occurs with a time of 1.8 ± 0.2 ps in the wild-type B850 complex. This energy transfer was also measured in a series of blue-shifted B850 mutants (B839, B826), and was found to be well described by an incoherent energy transfer model (6). A simple comparison with the crystal structure of LH2 suggests that the measured rate of B800 → B850 transfer corresponds mainly to energy transfer from one B800 to two B850 Bchl *a* molecules. However, a detailed modeling of this process has to await the precise coordinates of all pigment molecules. In addition it will be necessary to examine how the B800 → B850 transfer depends on the nature of the B850 excited state.

2.3 *B850 (7)*

The estimates based on the published crystal structure of LH2 (2) show that the dipole-dipole interaction in B850 band is quite strong (7). This puts at sharp focus the character of the excited states and the mechanism of excitation transfer in B850. The polarized room temperature measurements in the red wing of the B850 band show that the anisotropy decay is 1.4 – 1.8 times the isotropic lifetime. A simulation of the incoherent excitation transfer within the circle of 18 inhomogeneously broadened B850 Bchl *a* molecules suggests that in case of Förster transfer the anisotropy decay time would be three times longer than the isotropic decay time. This result strongly indicates that exciton state relaxation, at least partially, is responsible for the observed femtosecond anisotropy dynamics. By comparing the measured and calculated absorption difference spectra of B850 we have concluded that this exciton is delocalized over $N = 4 \pm 2$ Bchl *a* molecules (7).

2.4 *LH2→LH1 transfer (5)*

There is very little information available on the rate of energy transfer between the outer (LH2) and inner (LH1) antenna systems. Measurements with picosecond time resolution

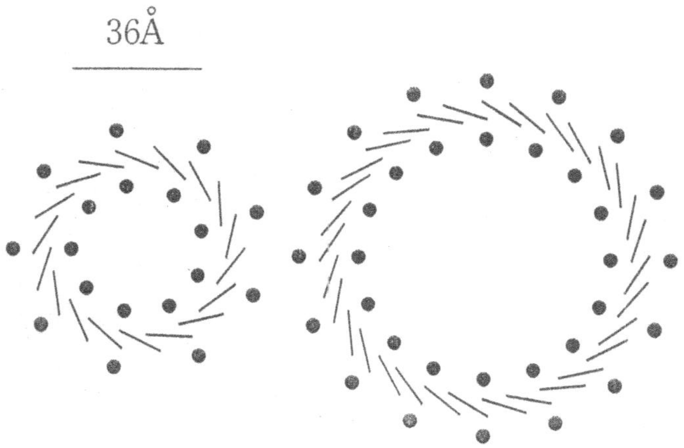

Fig. 4. Model for the association of LH2 and LH1 circles.

(8) and indirect information obtained from the overall antenna energy equilibration time, suggest that this transfer step occurs on the few ps time scale. In a femtosecond two-color measurement (5) we have now obtained direct information on the LH2 → LH1 transfer step. The build-up of B875 excitations in *Rb. sphaeroides* at room temperature was found to occur with a time constant of 3.3 ps, following excitation of B800. Since B800 → B850 transfer and energy equilibration within B850 and B875 all are much faster (≤ 0.7 ps), this time represents the B850 → N875 transfer (5). With the help of these data a model for the association of LH2 and LH1 circles was constructed (Fig. 4), which suggests that the distance between the B850 and B875 molecules involved in LH2 → LH1 energy transfer is 27-32 Å, consistent with the structural data of LH2 (2) and LH1 (9).

References

1 Kühlbrandt, W., Wang, D.N. and Fujiyoshi, Y. (1994) Nature 367, 614-621

2 McDermott, G., Prince, S.M., Freer, A.A., Hawthornthwaite-Lawless, A.M., Papiz, M.Z., Cogdell, R.J. and Isaacs, N.W. (1995) Nature 374, 517-521

3 Pullerits, T. and Freiberg, A. (1992) Biophys. J. 63, 879-896

4 Hess, S., Åkesson, E., Cogdell, R., Pullerits, T. and Sundström, V. (1995) Biophys.J. (submitted)

5 Hess, S., Chachisvilis, M., Timpmann, K., Jones, M.R., Hunter, C.N. and Sunström, V. (1995) Proc. Natl. Acad. Sci. USA, (accepted)

6 Hess, S., Visscher, K.J., Pullerits, T., Sundström, V., Fowler, G.J.S. and Hunter, C.N. (1994) Biochemistry 33, 8300-8305

7 Pullerits, T., Chachisvilis and Sundström, V. (1995) Contribution in this book.

8 Zhang, F.G., Van Grondelle, R. and Sundström, V. (1992) Biophys. J. 61, 911-920

9 Karrasch, S., Bullongh, P.A. and Ghosh, R. (1995) EMBO J. 14, 631-638

CAROTENOID COMPOSITION OF *DIGITALIS PURPUREA* IN RELATION TO NON-PHOTOCHEMICAL QUENCHING.

Jeanette I. Webster[1], Andrew J. Young[2] and Peter Horton[1].
[1]Robert Hill Institute, Sheffield University, Sheffield S10 2TN, UK; [2]School of Biological and Earth Sciences, John Moores University, Liverpool, UK.

1. Introduction

Excess light energy not used for photosynthesis can cause inhibition of the plant photosystems by the production of destructive species, such as oxidised P_{680} and oxygen free radicals which lead to pigment and protein damage [1,2]. Plants have a mechanism of dissipating the excess energy safely. This mechanism of non-photochemical quenching (qN) [3,4] can be subdivided into a number of components the largest of which is the ΔpH-dependent non-photochemical quenching termed qE [4,5].There are various mechanisms proposed for qE which differ in the site of quenching i.e. in the reaction centre [6] or the light-harvesting antennae of photosystem II (PSII) [7-9], and in the identity of the quencher. The various models for qE in the antennae all include a dependence on the transthylakoid ΔpH, a role for the xanthophyll cycle, and a structural change in the light-harvesting complex due to protonation .

In addition to qE there is a slowly relaxing component of qN referred to as qI; the origin of qI is not understood but it includes a contribution from a sustained photoprotective quenching resembling qE in that it is sensitive to nigericin [10]. This portion of qI has been correlated with both the presence of zeaxanthin and the structural change of the LHCII. It has been suggested by Gilmore and Björkman that this sustained qN arises because of a stable ΔpH in darkness [11]

This paper describes a detailed study of the fluorescence characteristics and carotenoid composition of *Digitalis purpurea* grown under different light intensities; this plant species has been shown previously to exhibit a high level of reversible qN [12].

2. Materials and Methods

Digitalis purpurea plants were grown under a range of light intensities from 80 to 600 μmol m^{-2} s^{-1} provided either by 400W metal halide discharge lamps (Osram) (MH) or PLL fluorescent tubes (Philips) supplemented with tungsten filament bulbs (FT). Plants were used approximately 6 to 8 weeks after germinated seeds had been potted out into larger pots.

P. Mathis (ed.), Photosynthesis: from Light to Biosphere, Vol. I, 123–126.

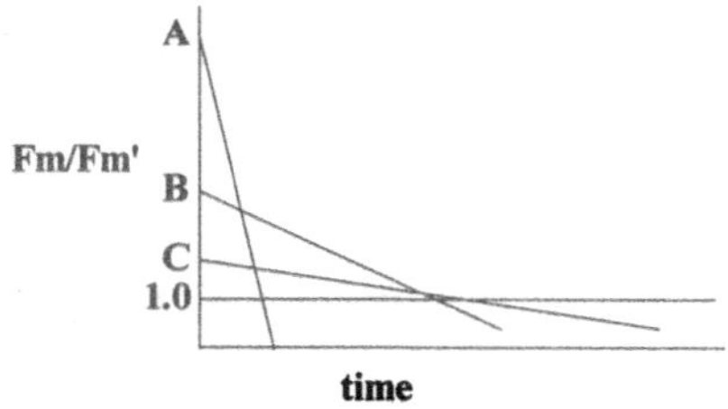

Figure 1. Analysis of the dark relaxation of quenching of chlorophyll fluorescence. The components of the quenching were calculated from the intercepts of the y axis at A, B and C. $qN = A-1$, $qN_s = C-1$, $qN_f = A-B$ and $qN_m = B-C$.

Plants were dark-adapted overnight and chlorophyll fluorescence of 3.14 cm^3 leaf discs were measured in a Hansatech leaf disc electrode using a Walz PAM 101 chlorophyll fluorometer. The data obtained was analysed by the following method: Fm/Fm' was plotted against the time of the recovery period (fig 1). Three straight lines were fitted to the data points to resolve the three components of qN. The fast relaxing component of qN (qN_f) is known to be dependent on the ΔpH and is termed qE; the medium relaxing component of qN (qN_m) which has a relaxation time of 17 minutes is possibly similar to the slow component of qE as discussed above [10,11]. The third component qN_s probably arises from damage to PSII reaction centre. The total non-photochemical quenching, qN, was calculated by the equation: qN = (Fm/Fm')-1, in which Fm' corresponds to the last saturating pulse applied during the light-treatment period. This point was also used as the 0 time point in plotting the above graph.

The light-treatment used for the analysis of chlorophyll fluorescence was repeated and the leaf discs were frozen for carotenoid analysis. Carotenoid analysis was carried out by reversed phase HPLC as described by Johnson et al [13].

3. Results and Discussion

The effect of light intensity on qN, qN_f and qN_s are shown in fig 2. qN was saturated at an actinic light intensity of 1100 μmol m^{-2} s^{-1} for all the growth conditions shown. qN_f and qN_s both show similar saturation curves as qN. Plants grown under 400 μmol m^{-2} s^{-1} have a much larger qN_f at the saturation point than the plants grown at 150 and 600 but the plants grown at 600 μmol m^{-2} s^{-1} have a large qN_s.

Fig 3 shows the maximum extent of qN in plants grown under a range of light intensities; total qN increased continuously up to 500 and thereafter declines slightly at 600.

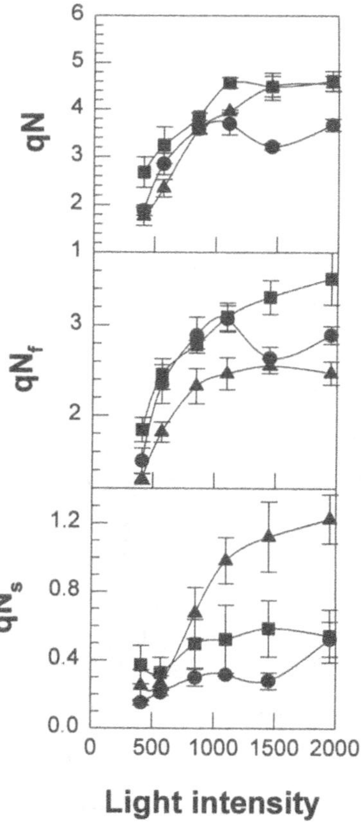

Figure 2. Components of quenching of chlorophyll fluorescence for Digitalis purpurea *grown at 150 (●), 400 (■), and 500 μmol m^{-2} s^{-1} (▲).*

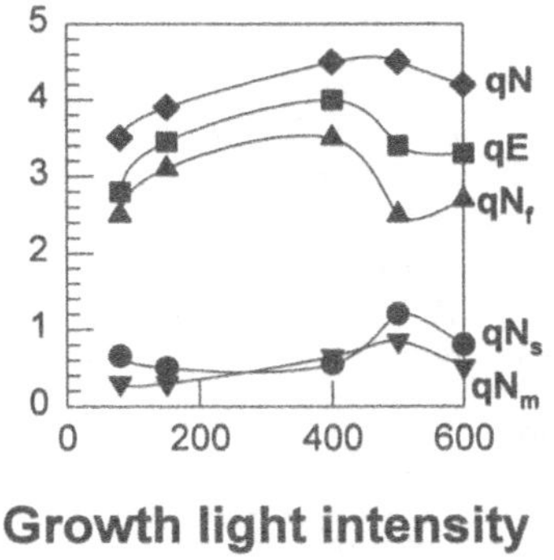

Figure 3. The components of quenching of chlorophyll fluorescence in Digitalis purpurea *with growth light intensity in μmol m^{-2} s^{-1}.*

Unexpectedly, qN_f whilst increasing from 80 to 400 showed a marked decline at 500 and 600. Whereas qN_m and qN_s increased to 500 μmol m^{-2} s^{-1} and then decreased slightly to 600 μmol m^{-2} s^{-1} (fig 3). As growth light intensity increases beyond 400 then qN_f decreases and qN_m continues to increase. Since it has previously been suggested that this medium relaxing component of qN is associated with both the xanthophyll cycle [9] and a structural change in the LHCII [10] then it can be assumed that this form of qN is involved in photoprotection and is similar to qE. Hence qN_f and qN_m can be added together to give the total qE (labelled qE in fig 3). This total qE again follows the patterns seen with qN_f in that qE increases with growth light intensity up to a maximum of 400 μmol m^{-2} s^{-1} and then decreases slightly with further increases in growth light intensity.

The de-epoxidation state (DES) of the xanthophyll cycle pool shows similar saturation curves to those of qN and qN_f (fig 4). The maximum DES also increases with growth light intensity up to a maximum of 500 μmol m^{-2} s^{-1} and then shows a slight decrease with a further increase in growth light intensity (fig 5).

The xanthophyll cycle pool size mimics the characteristics of qE in that the pool size increases to a maximum at a growth light intensity of 400 μmol m^{-2} s^{-1} and then decreases with a further increase in growth light intensity (fig 5). There have been no previous data reports that have examined plants growth over a range of light intensities. Demmig-Adams and co-workers have previously shown that the size of the xanthophyll cycle pool increases with growth environment and hence with qE. This previous study only compared plants growing in the extremes of shade, at 12 μmol m^{-2} s^{-1} which has a pool size of 9-15 % of the total carotenoid with a plant growing in full sunlight, at 2000 μmol m^{-2} s^{-1} which has a pool size of 25-26 % of the total carotenoid [14].

4. Conclusion

The data shows that the responses of the xanthophyll cycle and qN to growth light intensity are complex. It is shown that the xanthophyll cycle pool size shows a peak declining at both lower and higher light intensities. qN_f shows a very similar response.

The data suggests that at least two components of qN are involved in the adaptation of plants to growth light intensity: (1) a fast relaxing component of qN (qN_f) which is involved at all light intensities and, (2) a medium relaxing component of qN (qN_m) which appears to be induced when maximum qN has been reached. It is not known whether these

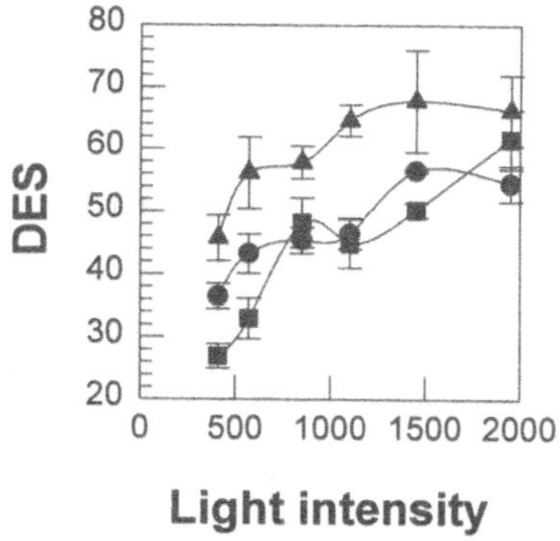

Figure 4. Changes in DES for Digitalis purpurea *grown at 80 (●), 150 (■), and 600 μmol m^{-2} s^{-1} (▲).*

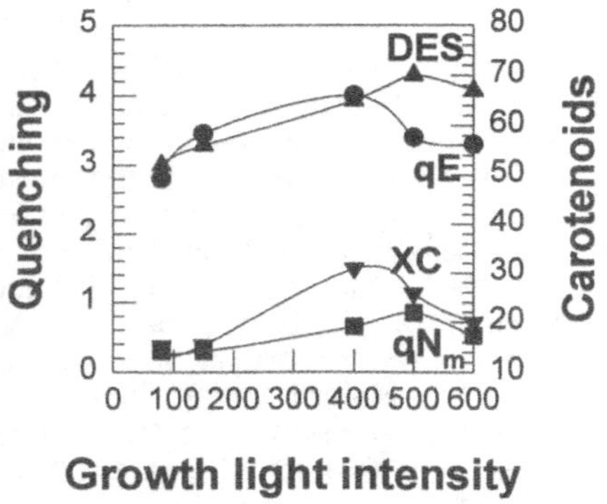

Figure 5. Changes in DES (■), Xanthophyll cycle pool size (▲) and qE (●) in Digitalis purpurea *with growth light intensity in μmol m^{-2} s^{-1}.*

components have the same mechanism or not. Since the DES is correlated to qN_m and the xanthophyll cycle pool size is correlated to qE then it can be assumed that these are perhaps both involved in photoprotection but maybe to different extents. High levels of sustained qN are associated with very high levels of de-epoxidation which may be necessary for the stable alteration in the LHCII. It has been suggested that this medium component of qN, which has been found to be sensitive to nigericin is associated with both the xanthophyll cycle and the structure of the LHCII [10]. Previously Andersson has shown that as growth light intensity increases then the 27kDa polypeptide in the outer peripheral pool of the LHCII is lost [15]. Polypeptide analysis of *Digitalis purpurea* suggests that the 27kDa polypeptide is lost with increasing growth light intensity (data not shown). It is possible that this loss of the 27kDa polypeptide is involved in these alterations in qN.

References

1. Osmond, C.B. (1981) Biochim Biophys Acta **639**, 77-98.
2. Kyle, D. J., Ohad, I. and Arntzen, C. J. (1984) Proc Natl Acad Sci USA **81**, 4070-4072.
3. Schreiber, U., Schliwa, U. and Bilger, W. (1986) Photosynth Res **10**, 51-61.
4. Horton, P. and Bowyer, J. R. (1990) In. Methods in Plant Biochemistry vol 4 (Harwood, J. L., and Bowyer, J. R., eds.) pp.259-296, Academic Press, London.
5. Krause, G. H. and Weis, E. (1991) Annu Rev Plant Physiol Plant Mol Biol **42**, 313-349.
6. Krieger, A., Moya, I. and Weis, E. (1992) Biochim Biophys Acta **1102**, 167-176.
7. Horton, P., Ruban, A. V., Rees, D., Pascal, A. A., Noctor, G. and Young, A. J. (1991) FEBS lett **292**, 1-4.
8. Gilmore, A. M. and Yamamoto, H. Y. (1992) Proc Natl Acad Sci USA **89**, 1899-1903.
9. Demmig-Adams, B. (1990) Biochim Biophys Acta **1020**, 1-24.
10. Ruban, A. V. and Horton, P. (1995) Plant Physiol **108**. 721-726.
11. Gilmore, A. M. and Björkman, O. (1994) Planta **192**, 537-544.
12. Johnson, G. N., Young, A. J., Scholes, J. D. and Horton, P. (1993) Plant Cell Environ **16**, 673-679.
13. Johnson, G. N., Scholes, J. D., Horton, P. and Young, A. J. (1993) Plant Cell Environ **16**, 681-686.
14. Demmig-Adams, B. and Adams III, W. W. (1992) Plant Cell Environ **15**, 411-419.
15. Mäenpää, P. and Andersson, B. (1989) Z Naturforsch **44c**, 403-406.

STRUCTURE AND MAGNESIUM BINDING OF PEPTIDE FRAGMENTS OF LHCII IN ITS PHOSPHORYLATED AND UNPHOSPHORYLATED FORMS STUDIED BY MULTINUCLEAR NMR.

Dalibor Stys[1], Torbjörn Drakenberg[2], Michael D. Spangfort[3], Sture Forsén[2] and John F. Allen[1].
[1]Plant Cell Biology, Box 7007, Lund University, S-220 07 Lund, Sweden,
[2]Physical Chemistry 2, and [3]Biochemistry, Lund University, Box 124, S-221 00 Lund, Sweden

1. Introduction

It is believed that upon phosphorylation, some LHCII complexes move towards unappressed regions of the thylakoid and act as light-harvesting antennae for photosystem I [1,2,3]. A structure at 3.4 Å resolution has been described for the major part of non-phosphorylated LHCII [4] but gives no information regarding the N-terminal domain that contains the phosphorylation site. Attempts to isolate pure phosphorylated LHCII have so far been unsuccessful.

Appression of thylakoid membranes, also called "stacking", is induced by the presence of cations. For a long time there has been a dispute about the involvement of specific and non-specific interactions in this process [1,2,3,5,6]. Barber [1] suggested the dominance of unspecific electrostatic interactions in both stacking and lateral segregation of proteins, Mullet and Arntzen [8] considered dominance of protein-protein interactions involving LHCII in stacking and Allen [3] proposed dominance of protein-protein interactions in both stacking and lateral segregation. If unspecific, electrostatic shielding of electrostatic repulsion between lamellae in the presence of salts induces appression, then the salt concentration may be several orders of magnitude lower in the interlamellar space than in the rest of the solution [7]. This prediction is in clear disagreement with experimental evidence from thylakoids, where the membranes, although sometimes highly appressed, are more enriched in cations than it would be needed for compensation of the surface charge. The electrostatic equilibrium concentration may be restored by EDTA treatment, by which tight complexes may be disrupted. An alternative hypothesis could thus be proposed by which the interlamellar interactions in stacking are dominated by specific binding of ions rather than by purely electrostatic effects.

It is known that LHCII has a strong tendency to aggregate[5]. The three-dimensional structure of the stromal part of LHCII [4,8] is relatively flat and contains a number of carboxylic acid groups. It may also contain a phosphatidylglycerol (PG) headpiece which is known to be tightly bound to LHCII [9]. Several molecules of digalactosyldiacylglycerol (DGDG) are necessary for proper crystallisation of LHCII [8]. Carboxylic acids, phosphate groups and saccharides are known to bind cations.

A number of laboratories have produced data on the relative effectiveness of mono- and di-valen cations in cation-dependent processes in thylakoid membranes and in lipid vesicles. There is an apparent similarity of the effectiveness of Mg^{2+} relative to monovalent cations for stacking of thylakoids [10], for aggregation of LHCII [5], and for association of PG vesicles [11]. Such similarity may be just coincidental and the whole phenomena needs proper analysis.

P. Mathis (ed.), Photosynthesis: from Light to Biosphere, Vol. I, 127–130.

In systems containing a number of non-equivalent binding sites, the occupancy of binding site is determined by the local activity of ions [11]. The local activity is determined by electrostatics, by the affinity of neighbouring specific binding sites and by the position of individual binding sites on the surface [11]. However, in equilibrium conditions, the activity of cations should be the same all over the whole sample. In the case of thylakoids, a strong ion binding site is introduced when a phosphate group is attached to a membrane protein. We may well assume non-equlibrium conditionsin thylakoid memebranes, as it is known from fractionation experiments that it takes at least 30 min for the membrane to equilibrate for new conditions. Thus, a timescale for the equilibration exceeds that of phosphorylation-induced chages and local non-equilibrium conditions may be expected.

In the few known examples of magnesium binding to phosphate groups on protein [12,13] the binding constants are always of the order of few hundreds to thousands M^{-1}, largely exceeding the binding constant of the phosphate group of lipids [11,14]. Here we describe results obtained with small synthetic peptide fragments of the LHCII phosphorylation site. In any case the incorporation of the phosphate group into a protein a the membrane surface will significantly affect the binding of ions at weaker sites.

2.Results

2.1 Comparison of NMR spectra of the phosphorylated and non-phosphorylated LHCII fragments

Major differences between the spectra of the non-phosphorylated and phosphorylated peptides are clearly observed at all pH values (Figure 1). We will therefore concentrate th discussion to the spectra at pH 5.3, where the assignment of resonances was least complicated by spectral overlap and by exchange of NH protons.

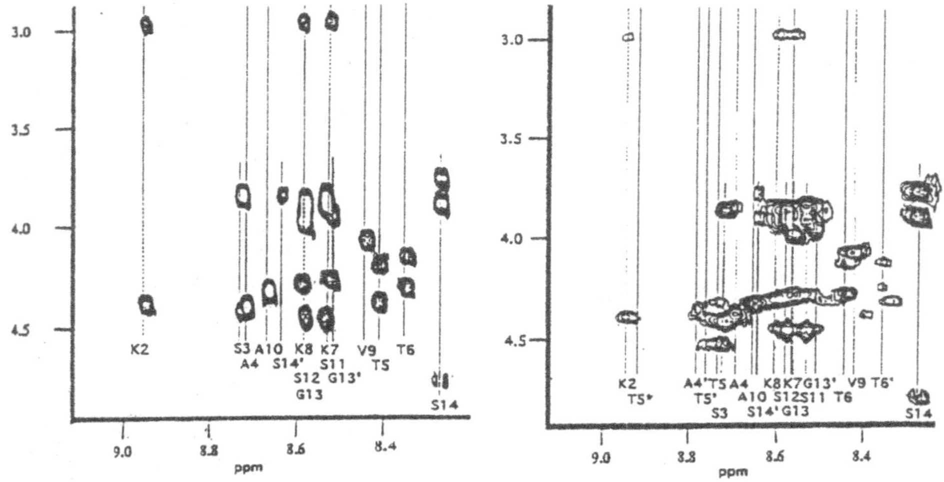

Figure 1. Segments of ^{1}H NMR TOCSY spectrum of unphosphorylated LHCII fragment and of phosphorylated LHCII fragment at pH 5.3

Both the non-phosphorylated and phosphorylated forms of the peptide RKSAT(PO_3)TKKVASSGSP show non-random-coil chemical shifts for most of the protons (Figure 1). Major differences in chemical shifts between the two forms of the peptide are observed for protons of Thr 5 and 6, and these can be attributed to changes in covalent structure and bond properties. In addition, changes in chemical shifts are greater in the region Ser 3 - Lys 8 than in the C-terminal part of the molecule. This indicates that changes in structure (or stabilisation of one prevalent structure) occur in the N-terminal part of the molecule upon phosphorylation. Further strong evidence for the existence of a different preferred structure in each form of the peptide is the observed doubling of the resonances of Ser 14 and Gly 13 in the non-phosphorylated form and in those of Ala 4, Thr(PO_3) 5 and Thr 6 in addition to Ser 14 and Gly 13 in the phosphorylated form. This doubling is most likely caused by cis/trans isomerism around the peptide bond between Ser 14 and Pro 15. The doubling of resonances for residues 4 to 6 shows that there exists at least one more populated conformation in which the C-terminus is sufficiently close to residues 4 to 6 to affect their chemical shifts. In addition, several non-sequential mainchain-to-sidechain NOE crosspeaks are seen in the spectrum of the phosphorylated peptide whereas in the non-phosphorylated peptide NOESY crosspeaks are observed only between the sidechains of Lys 7 or Lys 8 and Ser 14 NH. There is, however, a clear indication of a preferred tertiary structure in the non-phosphorylated peptide between residues 7 and 15, and of extension of such a structure right up to the N-terminus of the phosphorylated peptide.

The pH-dependent changes in the spectrum were consistent with pKa's of 4.5 an 5.9 for the single and double protonations, respectively, of the phosphate group, consisten with previous data on the protonation of the phosphate group of various phosphoproteins given in [15].

2.2 Magnesium binding to peptides

The binding constant of Mg^{2+} ions to both phosphorylated and unphosphorylated peptide was determined from the concentration dependence of line broadening of the ^{25}Mg NMR signal at 25°C (Fig.2).

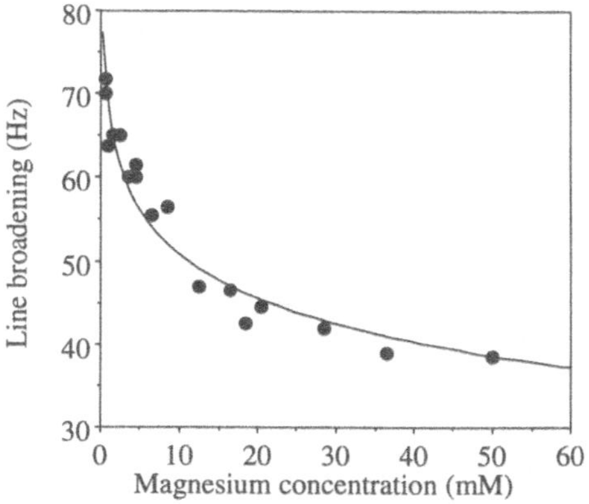

Figure 2. Dependency of line broadening of $^{25}Mg^{2+}$ signal in presence of 2.1 mM phosphorylated LHCII fragment on concentration of Mg^{2+}.

No magnesium binding site is located on the unphosphorylated peptide. In contras the phosphorylated peptide binds magnesium. The calculation of the binding constant of Mg^{2+}-phosphopeptide complex gives a value of 110 M^{-1}. The binding constant is low in comparison to known binding constants of phosphate group which are between 250 and 25000 M^{-1}in casein and casein-derived peptides [12] where the magnesium binding is to cluster of serine phosphates. The binding constants to phosphate groups on lipids and sugars are, however, of the order of 5-20 M^{-1}.

Proton NMR spectroscopy was performed in 8 mM solution of the phosphorylated peptide in concentrations of $MgCl_2$ that were increased stepwise up to 150 mM. Small differences were found for the chemical shifts of the NH protons. The differences in chemical shift were uniform in direction, varying between 0.04 and 0.01. In the work of Wahlgren et al. [13] changes in chemical shifts were found also for the α protons surrounding the metal-binding phosphoserines. These changes [13] were one order of magnitude larger than those found in our work, indicating that the magnesium-induced change in the structure is very small, if any.

3. Conclusions

The proton NMR spectra of the phosphorylated as well as the unphosphorylated peptide gave indications of the existence of a preferred structure in solution. Marked differences between the proton NMR spectra of the phosphorylated and unphosphorylatec fragments were observed. These differences may be interpreted either as a change in preferred structure or as a stabilisation of one of the preferred conformations of the N-terminal part of the peptide. The structure of the peptide furthermore varied with changes in the protonation of the phosphate group.

The phosphate group on proteins is known to be one of the strongest magnesium binding groups in biochemical systems[14]. The hypothesis [16] that stacking of thylakoi membranes is determined to a large extent by binding of magnesium to lipids seems to have some experimental support [5,9,11]. The binding constant of 110 M^{-1} for magnesium to the phosphopeptide is almost one order of magnitude higher than the value 12 M^{-1} found in the literature for PG [11]. In phosphorylated LHCII will the phosphate group at the N-terminus favourably compete with the sites on lipids unless the phosphate group is buried in the structure..

Acknowledgements:
This work was funded by research grants awarded by Swedish Natural Research Council (NFR) and by Royal Physiographical Society in Lund. D.S. received a FEBS short term fellowship in the early stage of this work and the postdoctoral fellowship from NFR.

4. References:

1 Barber, J. (1983). Photobiochem .Photobiophys. 5, 181-190.
2 Allen, J.F. (1992). Trends Biochem. Sci. 17, 12-17.
3. Allen, J.F. (1992) Biochim. Biophys. Acta 1098, 275-335.
4 Kühlbrandt W., Wang D.N. and Fujioshi Y. (1994). Nature, 367, 614-621.
5 Mullet J.E. and Arntzen C. (1980) Biochim. Biophys. Acta 589, 100-117
6 Anderson J.M. and Andersson B. (1982) Trends. Biochem. Sci. 7, 288-292
7 Dubois M., Zemb T, Belloni L., Delville A., Levitz P and Setton R. (1992) J. Chem. Phys. 96, 2278-2286
8 Kühlbrandt W. (1994) Curr. Opin. Struct. Biol. 4, 519-528
9 Nußberger S., Dörr S., Wang D.N. and Kühlbrandt W. (1993) J. Mol. Biol. 234, 347-56
10 Wollman F.A. and Diner B. (1980) Arch. Biochem. Biophys.
11 Cevc G. Biochim. Biophys. Acta (1990) 1031-3, 311-382
12 Wahlgren N.M., Leonil J., Dejmek P. and Drakenberg T. (1993) Biochim. Biophys. Acta 1202, 121-128.
13 Wahlgren Wahlgren N.M., Dejmek P., Drakenberg T., J. (1990) Dairy Res. 57, 355-364.
14 Cowan J.A. (1995) The Biological Chemistry of Magnesium. VCH publishers, New York.
15 Ho, C., Magnusson, J.A., Willson, J.B., Magnusson, N.F., Curland, R.G. (1969). Biochemistry, 8, 2074-2082.
16 Menikh, A. and Fragata, M. (1993), Eur. Biophys. J., 22, 249-258.

PHOSPHORYLATED PHOTOSYSTEM I ANTENNA PROTEINS IN BARLEY

J. Knoetzel, D.U. Meyer & L.H. Grimme
Institute of Cell Biology, Biochemistry and Biotechnology,
University of Bremen, Leobener Str./NW II, 28334 Bremen,
Germany

1. Introduction

The photosystem I (PSI) of barley has a light-harvesting antenna consisting of four different chlorophyll *a/b*-binding proteins of 21, 22, 23 and 25 kDa encoded for by the genes *Lhca4*, *Lhca1*, *Lhca2* and *Lhca3*, respectively (1). They are synthesized as precursor proteins in the cytoplasma, posttranslationally imported into the chloroplast, cleaved to the mature size and inserted into the thylakoid membrane.

One of the most frequent forms of posttanslational modification of proteins is their phosphorylation. Protein phosphorylation at serine and threonine residues are involved in regulation of metabolism and intracellular signal transduction. About 20 years ago, it has been shown that phosphoproteins with molecular masses from 7-70 kDa are involved in photosynthesis (2). Chloroplast phosphoproteins have been reported in thylakoid membranes, the stroma, the ribosomes, the inner and outer envelope membranes, and in the space between the envelopes. In thylakoid membranes, the polypeptides LHCB1 and 2 of LHCII, the PSII proteins D1 and D2, CP 43 and the 9-kDa *psbH* gene product are modified by phosphorylation on very different sequences (3). Reaction centre or light-harvesting proteins of PSI are not known to be phosphorylated.

Here, we provide evidence that the *Lhca* proteins of barley photosystem I are phosphorylated at threonyl and seryl residues. Proteolysis of LHCA2 shows that serine phosphorylation occurs on the amino terminal surface exposed segment and both serine and threonine residues are probably phosphorylated in the stroma-exposed amino acid chain connecting the membrane-spanning α-helices 2 and 3.

2. Material and Methods

To isolate the *Lhca* proteins, PSI-200 complexes were isolated from chloroplasts of seven-day-old barley seedlings (4). The polypeptides were separated on 16-21% gradient gels.The bands containing the *Lhca* proteins with apparent molecular masses of 21, 22, 22.5, 23 and 25 kDa were cut out, electroeluted in 50 mmol/l Tris-190 mmol/l glycin, 0.1% (w/v) SDS for 5-16h with the Model 422 electro-eluter (Biorad, Munich, FRG).

P. Mathis (ed.), Photosynthesis: from Light to Biosphere, Vol. I, 131–134.

Before cleavage, ultrafiltration of the eluted proteins was done in Centricon microconcentrators (Amicon, Witten, FRG) with endoproteinase Lys-C digestion buffer (25 mmol/l Tris-HCL, pH 8.5, 1 mmol/l EDTA), and the proteins finally concentrated to about 0.2 μg/μl.

The *Lhca2* protein was cleaved with endoproteinase Lys-C (EC 3.4.21.50, Boehringer, Mannheim, FRG) for 3-8h at 36°C, the cleavage was stopped by dilution with electrophoresis sample buffer and the fragments were separated on 16.5% separating gels with 10% spacer gel (5). After electrophoresis, the polypeptides were blotted onto Immobilon (Millipore, Bedford, MA, USA) for 3h at 50V in the Mini Trans-Blot cell (Biorad). Immunoblot assays were done according to (6). Phosphorylated threonine and serine were detected with monoclonal anti-phosphothreonine and anti-phosphoserine antibodies (Sigma, Deisenhofen, FRG).

3. Results and Discussion

The four proteins LHCA4, 1, 2 and 3 that have apparent molecular masses of 21, 22, 23 and 25 kDa serve as antenna of barley PSI. PSI-200 preparations of barley contain an additional antenna protein with an apparent molecular mass of 22.5 kDa (1) that could be identified as *Lhca1* protein after N-terminal sequencing. Using monospecific antibodies against LHCA1, 3 and 4 (7) and LHCA2 (6) in immunoblot assays with pea PSI-200, additional LHCA1 and 2 protein bands could be detected (not shown).

To investigate whether posttranslational phosphorylation could be responsible for the different electrophoretic mobilities of the respective proteins we used monoclonal antibodies against phosphorylated threonine and serine in immunoblot assays with PSI-200 proteins of barley (Fig. 1A). Anti-phosphothreonine predominantly reacts with LHCA1 and 4 although LHCA2 and 3 were also detectable. Anti-phosphoserine reacts predominantly with LHCA2, but all other *Lhca* proteins react with the antibody. This results were confirmed using the isolated LHCA1-4 in immunoblot assays (Fig. 1B). Each of the proteins reacted both with anti-phosphothreonine and anti-phosphoserine indicating the existence of multiple phosphorylation sites.

To determine whether multiple sites on a *Lhca* protein are being phosphorylated we used endoproteinase Lys-C to digest LHCA2 from barley with 211 amino acids and a calculated molecular mass of 23161 Da. We observed the digestion of the resulting peptide products by electrophoretic separation (Fig. 2). Digestion yielded main fragments that migrate with apparent molecular weights of 16.1, 9.4, 7.6, 3.4, 3.1 and 2.7 kDa. Different incubation times from 3-8h indicate no further digestion (not shown). The fragments were electroblotted onto PVDF membranes which were used in immunoblot assays (Fig. 2). Anti-phosphothreonine reacted with fragments of 16.1 and 9.4 kDa, anti-phosphoserine labelled two fragments with apparent molecular masses of 16.1 and 7.6 kDa.

An endoproteinase Lys-C fragment derived from the sequence of barley LHCA2 with a calculated molecular mass of 7643 Da corresponds to the 70 amino

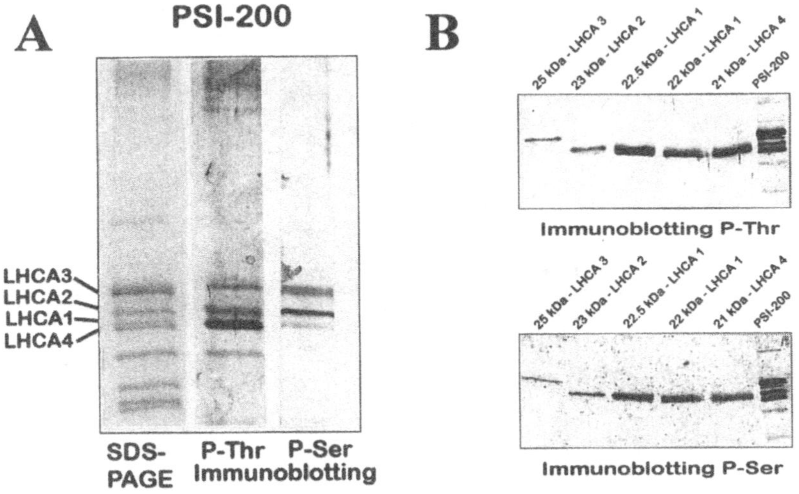

Figure 1. Immunoblot analysis. A. PSI-200 complexes were isolated from 7d-old barley seedlings, polypeptides separated on 16-21% gradient SDS-PAGE and electroblotted onto PVDF membrane. B. The PSI antenna proteins LHCA1-4 with apparent molecular masses of 21, 22, 22.5, 23 and 25 kDa were electroeluted from gel slices after SDS-PAGE of PSI-200 polypeptides, subsequently electrophoresed and transferred to PVDF membrane. Protein blots were incubated with anti-phosphothreonine (P-Thr) or anti-phosphoserine (P-Ser). All *Lhca* proteins react with both antibodies indicating multiple phosphorylation sites on each protein.

acids of the amino terminus of the protein with three phosphorylatable serine residues. This fragment is recognized by anti-phosphoserine. A partially cleaved fragment with a calculated molecular mass of about 15.6 kDa from amino acid 71-211 (C-terminus) or, alternatively, a fragment with a calculated mass of about 17 kDa consisting of the amino acids 1-155 could correspond to the second fragment that reacts with anti-phosphoserine. Both include the stroma exposed loop between the membrane spanning helices 2 and 3 with one seryl residue. However, the 17 kDa fragment contains the N-terminal serine residues and thus our experiments allow not to conclude whether seryl residues in the connector between helix 2 and 3 could be phosphorylated.

All possible fragments that could correspond to the anti-phosphothreonine labelled peptide with the apparent molecular weight of 9.4 kDa (a 8.9-kDa fragment with the amino acids 130-211 or a 9.4-kDa fragment (amino acids 71-155) or a 10.4-kDa fragment (amino acids 116-211)) contain the stroma-exposed amino acid chain connecting the helices 2 and 3 with three or four threonyl residues. Thus, the absence of a reaction of anti-phosphothreonine with the 7.6-kDa amino terminal peptide together with the labelling of the 16.1-kDa fragment could mean that only the threonines between helices 2 and 3 are phosphorylated whereas N-terminal threonines remain unphosphorylated in LHCA2 of barley.

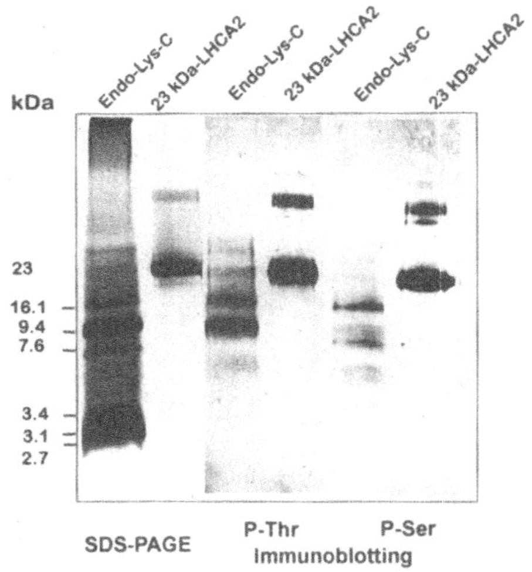

Figure 2. Proteolysis of the 23-kDa *Lhca2* protein from barley with endoproteinase Lys-C (Endo-Lys-C). The digestion (3h at 36°C) yielded fragments migrating at apparent molecular masses of 16.1, 9.4, 7.6, 3.4, 3.1 and 2.7 kDa. Anti-phosphothreonine (P-Thr) reacted with the 16.1 and 9.4-kDa fragments, anti-phosphoserine (P-Ser) with the 16.1 and 7.6-kDa fragments.

Determination of the exact number and placement of the phosphoseryl and phosphothreonyl residues will require further study. Additional study will also be needed to determine the role of phosphorylation and what subset of phosphorylation sites is necessary for its function(s). Preliminary results show that LHCA2 is already phosphorylated when detectable in thylakoid membranes. That could mean that phosphorylation of *Lhca* proteins takes place during or after import into chloroplasts as an obligatory step.

Financial support from the Deutsche Forschungsgemeinschaft (Kn 276/3-2) is greatfully acknowledged.

References

1 Knoetzel, J., Svendsen, J. and Simpson, D.J. (1992) Eur. J. Biochem. 206, 209-215
2 Bennett, J. (1977) Nature 269, 344-346
3 Allen, J.F. (1992) Biochim. Biophys. Acta 1098, 275-335
4 Bassi, R. and Simpson, D. (1987) Eur. J. Biochem. 163, 221-230
5 Schägger, H. and von Jagow, G. (1987) Anal. Biochem. 166, 368-379
6 Høyer-Hansen, G., Hønberg, L.S. and Simpson, D.J. (1985) Carlsberg Res. Commun. 50, 23-35
7 Król, M., Spangfort, M.D., Huner, N.P.A., Öquist, G., Gustafsson, P. and Jansson, S. (1995) Plant Physiol. 107, 873-883

CHLORINA-MUTANTS OF BARLEY LACKING SPECIFIC LHCA-PROTEINS

B. Boßmann[1], S. Jansson[2] & J. Knoetzel[1]
[1]Institute of Cell Biology, Biochemistry and Biotechnology, University of Bremen, Leobener Straße/NW II, 28334 Bremen, Germany; [2]Dept. of Plant Physiology, Umeå University, 90187 Umeå, Sweden

1. Introduction

At least ten different chlorophyll (Chl) *a/b* binding proteins serve as collectors of light energy in plant photosynthesis. Six different genes, *Lhcb*1-6 code for the light-harvesting proteins of photosystem II (PSII) while four genes *Lhca*1-4 encode the antenna proteins of photosystem I (PSI) (1).

LHCA1 and LHCA4 are thought to form a dimer in the pigment-protein complex LHCI-730 which is characterised by a strong 77K fluorescence emission peak at 730 nm. The isolated pigment-protein complex LHCI-680 fluoresces at 680 nm, consists of the two polypeptides LHCA2 and LHCA3 (2) and can be further fractionated into LHCI-680A and LHCI-680B containing LHCA3 and 2, respectively, suggesting that these two proteins are not intimately associated. The role of this four proteins in the two PSI antenna complexes LHCI-730 and LHCI-680 in energy transfer, photoprotection or proton funnelling is unresolved.

One molecular tool to study their functions are mutants lacking individual *Lhca* proteins. Using specific antibodies that have been raised against the light-harvesting chlorophyll *a/b*-binding proteins LHCA1-4 of PSI, we identified three *chlorina* mutants lacking specific *Lhca* proteins. *Chlorina-f2*^{101} (clo-*f2*^{101}) had an incomplete LHCI-680 complex lacking LHCA2. *Chlorina-*^{126} and *chlorina-*^{134} were missing LHCA1 and LHCA4, which means that the LHCI-730 complex was removed.

2. Material and Methods

Wild-type barley (*Hordeum vulgare* L.) and thirteen chlorophyll *b*-less mutants from the Carlsberg mutant collection were grown at 22°C and a light intensity of 75 µmol photons·m^{-2}·s^{-1} in a 14 h light/10 h dark cycle for 7-8 days except *chlorina-*^{111} and *chlorina-*^{130} which were grown at 15°C and a higher light intensity of 250 µmol photons·m^{-2}·s^{-1}. For the isolation of thylakoid membranes, intact chloroplasts were isolated according to (3), subsequently broken and the membrane proteins applied to denaturing SDS-PAGE on 16-21% gradient gels (4).

P. Mathis (ed.), Photosynthesis: from Light to Biosphere, Vol. I, 135–138.

Proteins were transferred to a PVDF-membrane (Immobilon, Millipore, Bedford, MA, USA) in a Mini-Trans-Blot-Transfer Cell (Biorad, Munich, FRG) at 20 V for 12 h. Immunoblot assays were done with monospecific antibodies against the proteins LHCA1, 3 and 4 (5) and LHCA2 (6).

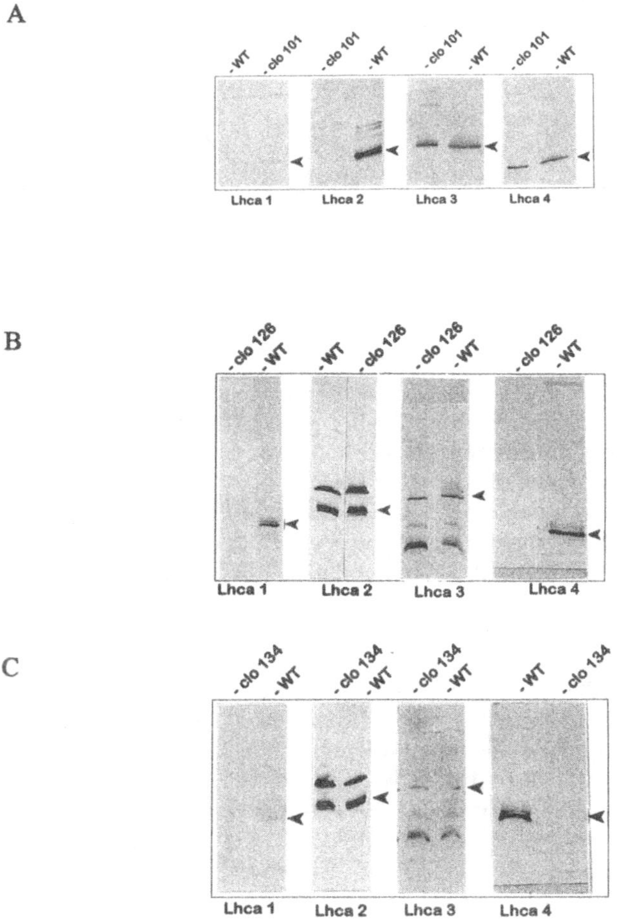

Figure 1. Immunoblot analysis of the thylakoid membranes from clo-[101] (A), clo-[126] (B) and clo-[134] (C) with the *Lhca*1-4 antibodies. Thylakoid membranes were separated on 16-21% gradient SDS-PAGE, transferred onto PVDF membrane and protein-blots were incubated with anti-LHCA1-4. Antibodies reacted against antenna proteins LHCA1-4 with apparent molecular masses of 22, 23, 25 and 21 kDa, respectively. *Chlorina-f2*[101] lacks LHCA2 (LHCI-680B), clo-[126] and clo-[134] lack the two proteins of LHCI-730, LHCA1 and 4.

Table 1. Relative amounts of *Lhca* proteins in *chlorina* mutants of barley.

A. Mutants allelic to *chlorina-f2*				
	LHCA1	**LHCA2**	**LHCA3**	**LHCA4**
clo-*f2*	xxx	xxx	xxx	xxx
clo-*f2*[101]	xxx	o	xxx	xxx
clo-*f2*[107]	xxx	xx	xxx	xxx
clo-*f2*[109]	xxx	xxx	xxx	xxx
clo-*f2*[122]	xxx	xxx	xxx	xxx
clo-*f2*[123]	xxx	xxx	xxx	xxx
B. Mutants with increased 77K fluorescence at 720 nm				
	LHCA1	**LHCA2**	**LHCA3**	**LHCA4**
clo-[111]	xxx	xxx	xxx	xx
clo-[117]	xxx	xxx	xxx	xx
clo-[119]	xxx	xx	xxx	xx
clo-[125]	xx	xx	xx	x
clo-[126]	o	xx	o	xx
clo-[130]	xxx	xxx	xxx	xx
clo-[134]	o	xx	o	xx

xxx wild-type level; xx reduced level as compared to wild-type; x trace amounts; o absent

3. Results and Discussion

Barley photosystem I has four different chlorophyll *a/b*-binding proteins of 21, 22, 23 and 25 kDa. The PSI core is surrounded by a monolayer of eight LHC polypeptides (7). Since the four *Lhca* proteins seem to be presented in approximately equal amounts, this implies that there are probably two copies of each *Lhca* protein bound per reaction-centre of PSI (8). The four proteins are found in two supramolecular complexes LHCI-730 and LHCI-680. Our investigations with the conditional barley mutant *chlorina*-[104] which lacks LHCA2, the LHCI-680B protein, under restrictive growth conditions of 15°C and high light intensity (250 µmol photons·m^{-2}·s^{-1}) led to the conclusion that excitation energy is transferred from LHCI-730 via LHCI-680 to the reaction centre of PSI (9). After this results we decided to find barley mutants in which other *Lhca* proteins are genetically removed. 31 new *chlorina* mutants were initially characterised by 77K fluorescence (10) and grouped according to emission characteristics. We selected mutants allelic to *chlorina-f2* and mutants expected to lack *Lhca* proteins (increased yield of 77K fluorescence at 720 nm arising from Chl *a*-antenna close to P700) (10).

The *Lhca*1 antibody reacted against a thylakoid membrane protein with an apparent molecular mass of 22 kDa and anti-LHCA4 recognized an 21 kDa protein. The monoclonal anti-LHCA2 specifically detected a 23 kDa and the anti-LHCA3 reacted against a barley protein with apparent molecular masses of 25 kDa consistent with (2,5). Most of the *chlorina* mutants allelic to clo-*f2* showed wild-type levels or slightly reduced amounts of *Lhca* proteins compared to WT. The same was true for clo-^{111}, clo-^{117} and clo-^{119} (Tab. 1). Significantly reduced levels of *Lhca* proteins were found in four *chlorina* mutants (Tab. 1). *Chlorina-f2*^{101} had LHCA1, LHCA3 and LHCA4 on a wild type level but lacked LHCA2 (Fig. 1A). This means that clo-*f2*^{101} has an incomplete LHCI-680 complex, similar to the conditional barley mutant *chlorina*-^{104} which lacks LHCA2 under restrictive growth conditions (9). Immunoblot analysis revealed that clo-^{126} and clo-^{134} had reduced levels of the light-harvesting proteins LHCA2 and LHCA3 (LHCI-680) compared to WT, but lack the dimeric LHCI-730 complex containing LHCA1 and LHCA4 (Fig. 1B and 1C). Clo-^{125} has reduced levels of LHCA1, 2 and 3, and only traces of LHCA4 were detectable.

The three *chlorina* mutants clo-*f2*^{101}, clo-^{126} and clo-^{134} will enable us to get more insight into the roles of the individual *Lhca* proteins in excitation energy transfer, light-harvesting and photoprotection. A first interesting aspect is that *chlorina*-^{126} despite lacking LHCI-730 is exhibiting 77K fluorescence at 730 nm. This is supporting the hypothesis that both LHCI subcomplexes originally emit fluorescence at long wavelength.

Financial support from the Deutsche Forschungsgemeinschaft (Kn 276/3-2) is greatfully acknowledged.

References

1 Jansson, S. (1994) Biochim. Biophys. Acta 1184, 1-19

2 Knoetzel, J., Svendsen, I. and Simpson D.J. (1992) Eur. J. Biochem. 206, 209-215

3 Nielsen, V.S., Mant, A., Knoetzel, J., Møller, B.L. and Robinson, C. (1994) J. Biol. Chem. 269, 3762-3766

4 Fling, S.P. and Gregerson, D.S. (1986) Anal. Biochem. 155, 83-88

5 Król, M., Spangfort, M.D., Huner, N.P.A., Öquist, G., Gustafsson, P. and Jansson, S. (1995) Plant Physiol. 107, 873-883

6 Høyer-Hansen, G., Hønberg, L.S. and Simpson, D.J. (1985) Carlsberg Res. Commun. 50, 23-35

7 Boekema, E.J., Wynn, R.M. and Malkin, R. (1990) Biochim. Biophys. Acta 1017, 49-56

8 Bruce, B.D. and Malkin, R. (1988) J. Biol. Chem. 263, 7302-7308

9 Knoetzel, J. and Simpson, D.J. (1991) Planta 185, 111-123

10 Simpson, D.J., Machold, O., Høyer-Hansen, G. and von Wettstein, D. (1985) Carlsberg Res. Commun. 50, 223-238

THE THREE-DIMENSIONAL STRUCTURE OF ALLOPHYCOCYANIN

Katjusa Brejc, Ralf Ficner[¶], Stefan Steinbacher and Robert Huber
Max-Planck-Institut für Biochemie, Martinsried, Germany
[¶]Present address: EMBL, Heidelberg, Germany

1. Introduction

Phycobilisomes are extramembranous light-harvesting protein-chromophore complexes found in the cyanobacteria and the eukaryotic rhodophyta (1) Phycobilisomes absorb light in the range of 450 nm to 650 nm, and transfer the excitation energy to the reaction centres with an overall quantum efficiency of over 90% (2). Structurally, they consist of two distinct parts: a core and peripheral rods radiating from the core (3,4). Each part is build up of disc-like structures, composed of phycobiliproteins and linker polypeptides. The core consists predominantly of allophycocyanin discs (APC, A_{max}=650 nm). In the peripheral rods, phycocyanin discs are proximal to the core (PC, A_{max}=620 nm), while phycoerythrin discs (PE, A_{max}=560 nm) form the tip of the rods. Each phycobiliprotein is composed of two different subunits, α and β, forming a stable ($\alpha\beta$) monomer. Each subunit carries one or more pigment molecules, the phycobilins, which are open-chain tetrapyrroles covalently bound to cysteine residues.

APC is the minor component of the phycobiliprotein fraction in phycobilisomes. The two subunits, α and β, consist of 160 and 161 amino acid residues, respectively (5). Each subunit has only one pigment, phycocyanobilin (PCB). APC has been isolated and crystallised from the cyanobacterium *Spirulina platensis* (6), and the three-dimensional structure has been solved recently by multiple isomorphous replacement techniques (6).

2. Molecular Structure

The two APC subunits show the same folding pattern (Figure la.), with some minor changes in the spatial arrangement of the secondary structure elements (Figure1b.).

The three-dimensional structure of APC confirms the general principle of phycobiliprotein architecture (7,8,9). The major differences arise from deletions and insertions involved in the protein-pigment interactions (Figure 2.). In the α-subunit of APC, a deletion of two residues in the BE loop changes the loop conformation and results in protein-pigment interactions which are not present in C-PC from *Fremyella diplosiphon* (PCFR) and in b-PE from *Porphyridium cruentum* (PEPC) (Figure 2a.). The GH lopp differs in the α-subunits of all three phycobiliproteins. Compared to PCFR, in APC there is an insertion of two residues; the GH loop in PEPC is enlarged and contains

P. Mathis (ed.), Photosynthesis: from Light to Biosphere, Vol. I, 139–142.

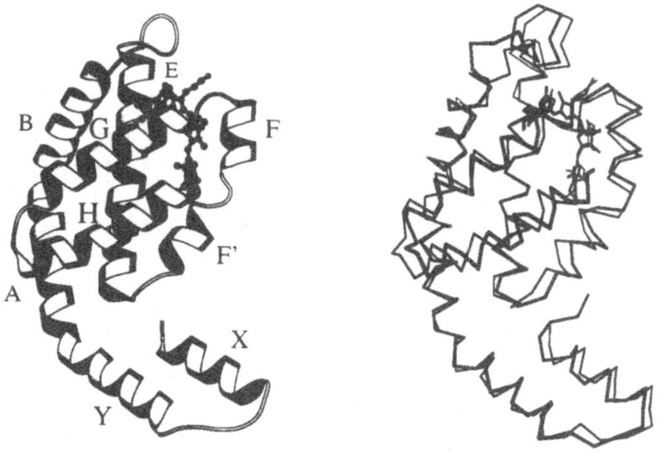

Figure 1. (a) Ribbon representation of the APC α-subunit with chromophore. Helical segments are marked X, Y, A, B, E, F', F, G and H. (b) Superposition of the C^α backbone of the α (thin line) and β-subunit (thick line) of APC.

the attachment site for the additional chromophore (Figure 2a.). The BE loop is highly conserved in the β-subunits of the three phycobiliproteins. The residues in the BE loop participate in interactions with two chromophores: β84 chromophore and the 3α84 chromophore from the neighbouring monomer in the trimer(Figure 4.). Additionally, it contributes to monomer-monomer contacts. The major difference between the three β-subunits is located in the GH loop (Figure 2b.). APC is the only phycobiliprotein which does not contain an additional chromophore binding site. Compared to PCFR the GH loop in APC is ten residues shorter. The GH loop in PEPC has five additional residues, compared to the PCFR structure, which are directed towards the double-linked chromophore (Figure 2b.).

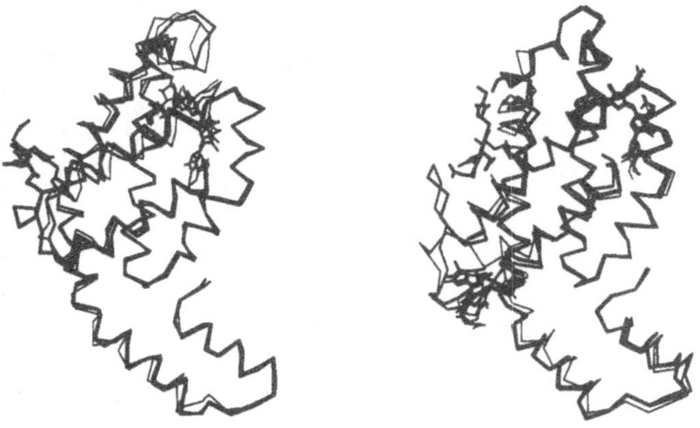

Figure 2. Superposition of the C^α backbone of APC (very thin line), PCFR (thin line) and PEPC (thick line): (a) α-subunit, (b) β-subunit.

3. Structure and environment of the chromophores

APC carries two PCB chromophores at positions α84 and β84. Both sites are topologically equivalent to the pigment binding sites in PCFR and PEPC. The stereochemistry and configuration of the chromophores in APC are identical with the PCB pigments in PCFR. In all three phycobiliprotein structures conserved "canonical" residues keep the chromophore in an extended conformation: Argα86, Aspα87 and Thrβ77 for the α84 chromophore and Argβ86 and Aspβ87 for the β84 chromophore.

4. Spectral differences between APC and PC

The absorption spectra of the monomers and trimers of PC are almost identical (near 620 nm), i.e. the oligomerisation does not influence the spectral properties of either chromophore. In APC, the monomer spectrum is very similar to that of the PC monomer; in contrast, the spectrum of the trimer is significantly different. The trimeric form of APC has a sharp maximum near 650 nm. Exciton interactions between the neighbouring chromophores (10) or a changed chromophore conformation due to monomer-monomer interaction (11) have been proposed as the cause for the red shift in APC trimers.

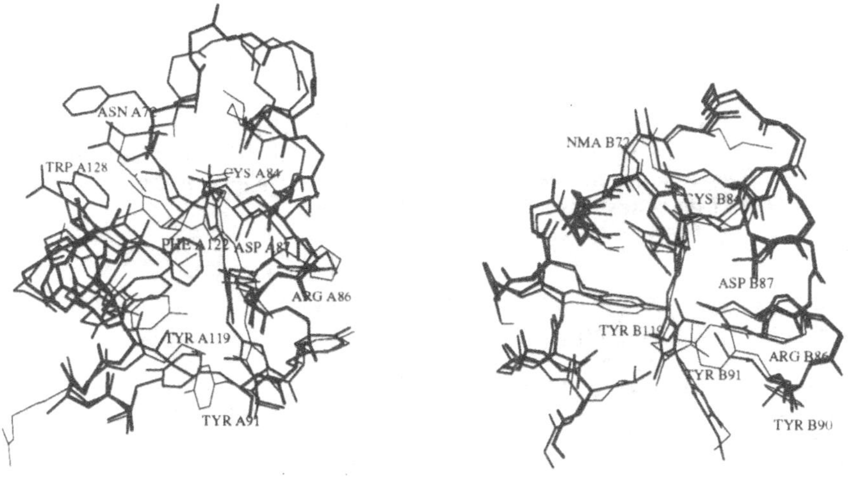

Figure 3. Superposition of the C^{α} backbone of APC (thin line) and PCFR (thick line). (a) The α84 chromophores and their interaction with neighbouring residues in the monomer (note that the α84 pigment is buried in the trimeric form). (b) The β84 chromophores and their interaction with the neighbouring residues in the monomer.

Structure-based comparison of the APC and PCFR chromophore binding regions shows a significant conformational change in the α84 chromophore due to the different protein environment of the α and the neighbouring β-subunit (Figure 3a.). The only conserved residues which surround and interact with the α84 chromophore are Argα86, Aspα87, Tyrα90 and Tyrα91. The amino acid substitutions Trpα128Ala and Tyrα129Ile reduce hydrophobic interactions in APC compared to PCFR. The smaller deviation from planarity of the α84 chromophore in APC probably contributes to the shift of the trimer absorption maximum to a longer wavelength. As shown in Figure 3b. the differences between the chromophore environment in APC and PCFR structures are small in the β-subunit. The only non conserved amino acid residues are β90 and β91. In APC, one of the tyrosines Tyr (β90) has an almost parallel orientation to the D ring and may interact with the aromatic system of the β84 chromophore.

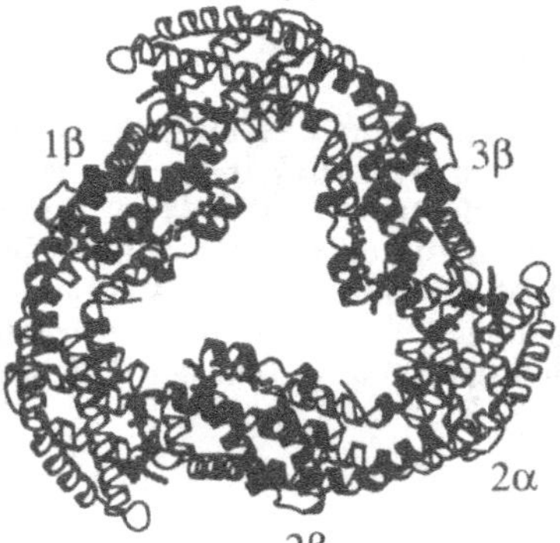

Figure 5. APC trimer projected down the crystallographic three-fold axis. Ribbon representation of the α-subunit (light grey), the β-subunit (dark grey) and chromophores (black sticks). The α84 chromophore is situated at the periphery of the trimer, while the β84 chromophore points towards the central channel.

References

1 Gantt, E. and Conti, S.F. (1966) J. Cell Biol. 29, 423-434
2 Sauer, K. (1975) in Bioenergenetic of Phtosynthesis (Govindjee, ed.) pp.115, Academic Press, San Francisco, U.S.A.
3 Mörschel, E., Koller, K.-P. and Wehrmeyer, W. (1980) Arch. Microbiol. 125, 43-51
4 Bryant, D.A., Guiglielmi, G., Tandeau de Marsac, N., Castets, A. and Cohen-Bazire, G. (1979) Arch. Microbiol. 123, 113-127
5 Sidler, W., Gysi, J., Isker, E. and Zuber, H. (1981) Hoppe-Seyler's Z. Physiol. Chem. 362, 611-628
6 Brejc, K., Ficner, R., Huber, R. and Steinbacher, S. (1995) J. Mol. Biol. 249, 424-440
7 Schirmer, T., Bode, W. and Huber, R. (1987) J. Mol. Biol. 196, 677-695
8 Duerring, M., Huber, R., Bode, W., Ruembeli, R. and Zuber, H. (1990) J. Mol. Biol. 611, 633-644
9 Ficner, R., Lobeck, K., Schmidt, G. and Huber, R. (1992) J. Mol. Biol. 228, 935-950
10 MacColl, R., Csatorday, K., Berns, D.S. and Traeger, E. (1981) Arch. Biochem. Biophys. 208, 42-48
11 Murakami, A., Mimuro, M., Ohki, K. and Fujita, Y. (1981) J. Biochem. 89, 79-86

LOW TEMPERATURE ABSORPTION, FLUORESCENCE AND HOLE BURNING SPECTROSCOPY OF ALGAE AFFECTED BY UV-B STRESS.

D. Engst, P. Kroh, T. Polívka, J. Pšenčík, *L. Nedbal, *O. Prášil and J.Hála
Photosynthesis and Global Climate Change Research Centre; Department of Chemical Physics, Faculty of Mathematics and Physics, Charles University, Ke Karlovu 3, 121 16 Prague 2, Czech Republic, *Institute of Microbiology, Czech Academy of Sciences, 379 81 Třeboň, Czech Republic

1. Introduction

The antropogenic depletion of stratospheric ozone layer results in increasing levels of UV-B radiation (280-320 nm) affecting both the terrestrial and aquatic ecosystems [1]. The elevated UV-B radiation affects various fundamental functions of the plant cell and induces multiple protection mechanisms. The fluorescence emitted from chlorophyll molecules of photosynthetic apparatus reflects both damaging and protective mechanisms. The suppression of photosynthetic activity is accompanied by decline of variable and, frequently, of maximal fluorescence. Constant fluorescence mostly rises during a short term exposure.

The UV-B induced changes in fluorescence emission justify the hypothesis that, in parallel to inhibition of electron transfer activity, also the exiton transfer in photosynthetic antennae is affected by the exposure. The anticipated modification of the exciton transfer parameters are investigated here using the low temperature absorption, fluorescence, and hole burning spectroscopy.

2. Procedure

Culture of the chlorococal alga *Scenedesmus quadricauda* [2] was grown synchronously in 14/10 hours light/dark cycle, bubbled with air enriched by 2% CO_2 at 25°C and exposed to photosynthetically active radiation of 250 $\mu E.m^{-2}.s^{-1}$ provided by tungsten filament bulbs [3]. The algae were harvested in the first half of the light period, resuspended at the chlorophyll concentration about 50 µg/ml and exposed to UV-B. Samples of the algae were taken during the exposure, centrifuged and resuspended in cultivation medium mixed with glycerine. Then, the samples were placed in the cuvettes for low temperature spectroscopy and stored in liquid nitrogen.

The UV-B radiation was produced using the 200 W high pressure mercury lamp screened by UV-B filter (Carl Zeiss, Jena 1602) and cellulose acetate film. The spectrum of the radiation incident on the sample surface is shown on *Figure 1.* . The total UV-B power ($\approx$ 0,3 W/m^{-2}) applied on the samples was determined using SED 240 detector (International Light). The samples of algae were collected after 10, 20 and 40 minutes of UV-B exposure.

P. Mathis (ed.), Photosynthesis: from Light to Biosphere, Vol. I, 143–146.

The variable fluorescence of the algae (Walz fluorimeter) declined during the UV-B exposure approximately in parallel to the decline of photosynthetic activity measured by oxygen evolution. Typical variable fluorescence measured before, during and after UV-B irradiation is shown on *Figure 2.*.

Low temperature (77 and 4,2 K) absorption and fluorescence spectra were measured together with persistent spectral hole-burning (PSHB) on the home built set-up described in [4]. The resolution of the PSHB both in fluorescence and in absorption spectra were limited by used double grating monochomator ≈ 0.3 cm^{-1}.

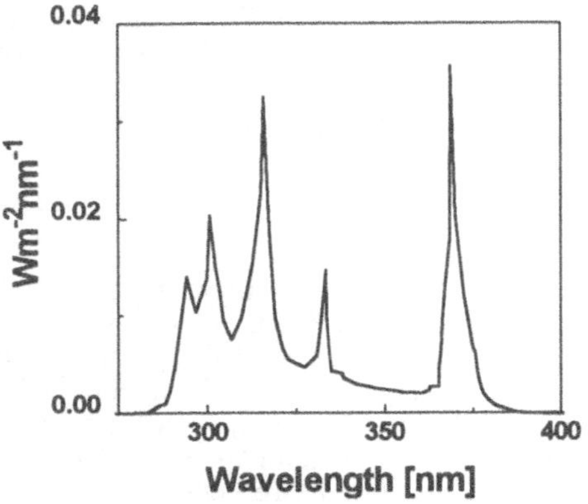

Figure 1. Spectrum of applied UV-B irradiation

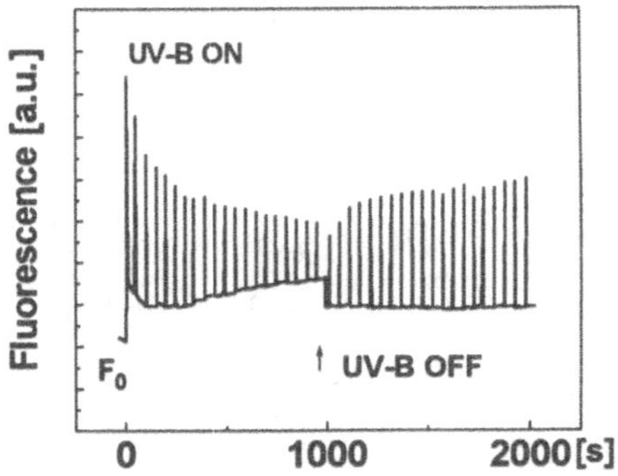

Figure 2. Variable fluorescence of Scenedesmus measured during and after UV-B irradiation.

3. Results and Discussion

The applied maximal doses of the UV-B radiation causes ≈ 50% decrease of photosynthetic activity as determined from the variable fluorescence and oxygen evolution rates. *Figure 3.* shows the resonance region of low temperature (4.2 K) absorption and fluorescence spectra of UV-B treated and control algae. All the spectra shown in *Figure 3* were normalised to the fluorescence signal of PS I (near to 7150 Å).

Following UV-B effects and trends can be found in the low temperature absorption and fluorescence spectra. The applied UV-B exposition was not enough intense to produce any significant changes (≤ - 4%) in the red part of absorption spectra corresponding to a bleaching or a destruction of photosynthetic pigments. On the other hand the applied UV-B exposition density was sufficiently high to produce an increasing of relative fluorescence intensity caused by global decrease of excited energy transfer (EET) efficiency in antennae of PS II: in LHC II (≈ +30% near 6800 Å), in CP 43 and in CP 47 (≈ +16% near 6880 Å). An decreasing of relative fluorescence of RC (≈ -47% near 6950 Å) reflects an interruption of EET between CP 43, 47 and RC of PS II.

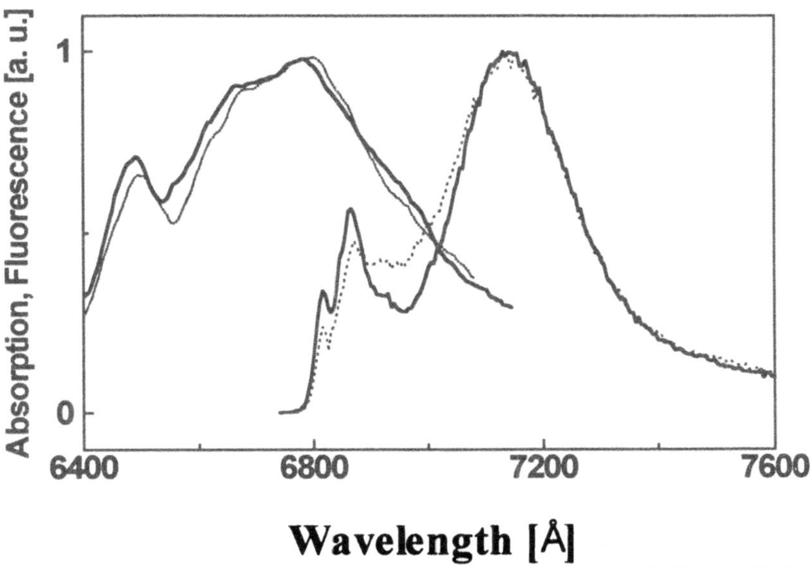

Figure 3. Resonance region of low temperature absorption (left) and fluorescence (right) spectra of UV-B treated (full curves) and control (dotted curves) Scenedesmus.

Typical PSHB in absorption spectra are shown on *Figure 4-a)*. The upper and lower spectrum belongs to the control (untreated) and UV-B irradiated algae, respectively. Both spectra consist of narrow zero phonon holes (ZPH) at the wavelengths of the burning laser and red shifted satellite phonon side band holes (PSBH). Linear and logarithmic dependencies of the ZPH widths $\delta_{HB} = \delta_{HB}$ (P.t) on the burning exposition (P.t) are shown in *Figure 4-b,c)*. To remove a power broadening, the holewidths $\delta_{HB}(0)$ extrapolated to zero burning exposition were determined. All the PSHB data of untreated green algae fit well with the PSHB data obtained on related photosynthetic systems obtained from higher plants [5, 6] (1,5 $cm^{-1} < \delta_{HB}(0) < 2$ cm^{-1}). The corresponding EET rate constant, which represents EET between particular pigment-proteins, was $6 < (\mathbf{k_{EET}})^{-1} < 9$ ps. The Huang-Rhys factor **S** characterising electron-phonon coupling (representing pigment-protein interaction) was determined from the ratio of areas of ZPH and PSBH. The **S** value was approximately 1.5 times higher in UV-B stressed algae ($0.8 < \mathbf{S} < 1.2$). On the other hand the distance between ZPH and PSBH representing the mean frequency of protein vibrations $\mathbf{\Omega}$ was ~ 24 cm^{-1} in all samples.

Broad-band holes were found in the fluorescence spectra besides ZPH and PSBH after prolonged hole burning. In these experiments higher laser power density was applied ($\geq$ 100 $mWcm^{-2}$) during the burning period ($\geq$ 10 minutes) (*Figure 4-d*). These data fit well with the results observed in absorption of untreated systems from higher plants [7]. The hole-widths were 170 cm^{-1}, due to very fast (< 100 fs) exciton migration between antenna pigments in particular antenna pigment-proteins. They were not significantly affected by UV-B irradiation.

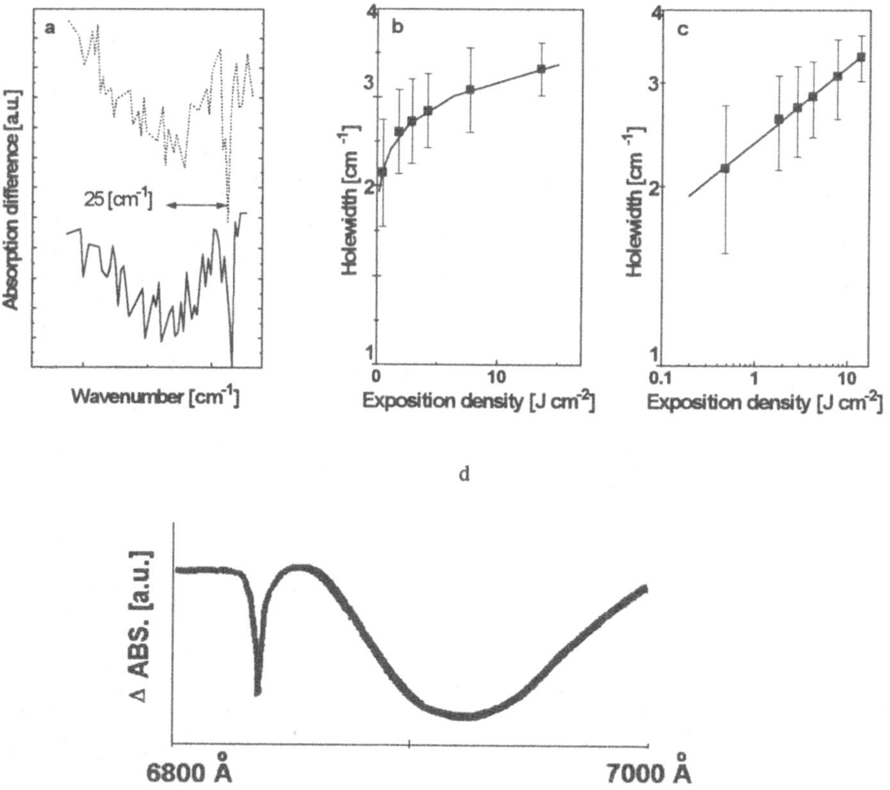

Figure 4-a) Zero phonon holes (ZPH) and phonon side band holes (PSBH) burned in the absorption spectra of Scenedesmus; UV-B irradiated (top), control (bottom); -b) linear and -c) logarithmic dependencies of the ZPH widths $\delta_{HB} = \delta_{HB}(P.t)$ *on the burning exposition (P.t) -d) fluorescence spectrum with a broad-band hole*

References

1 Tevini, M, UV-B Effects on Terrestrial Plants and Aquatic Organisms, Progress in Botany, Vol. 55, Springer-Verlag Berlin Heidelberg 1994
2 Komárek J., Fott B.: Das Phytoplankton des Süsswassers 7, *Chlorophyceae, Chlorococcales*. In: Hüber-Pestalozzi G. ed. 1044 pp. (1983)
3 Knoppová, J.,. Masojídek, J., Pokorný, J., (1992), Photosynthetica 28(4) ,541-547
4 Kroh, P., Polívka, T., Engst, D., Pšenčík, J.,: Czech. J. Phys. A *(accepted in 1995)*
5 Vacha, M., Psencik, J., Adamec, F., Ambroz, M., Dian, J., Nedbal, L., Hala, J. (1994) J. Lumin., 60&61, 523-526
6 Reddy N.R.S., van Amerongen H., Kwa S.L.S., van Grondelle R. and Small G.J. (1994) J. Phys. Chem., 98, 4729-4735
7 Chang H.C., Jankowiak R., Yocum C.F., Picorel R., Alfonso M., Seibert M. and Small G.J. (1994) J. Phys. Chem., 98, 7717-7724

CONTRIBUTION OF PLASTOQUINONE QUENCHING TO SATURATION PULSE-INDUCED RISE OF CHLOROPHYLL FLUORESCENCE IN LEAVES

D. M. Kramer*[1], G. DiMarco† and F. Loreto†
*Department of Plant Biology, University of Illinois at Urbana-Champaign, Urbana, Illinois 61801 U.S.A.
†CNR-Istituto di Biochimica ed Ecofisiologia Vegetali, Roma, Italy.

The saturation pulse-induced rise in fluorescence has been attributed to blockage of photosystem II (PSII) photochemistry, i.e. the elimination of photochemical quenching by PSII, (or q_P). The efficiency of photosynthesis can, in principle, be calculated from a true measure of q_P because the yield of fluorescence reflects the competition for excitons in the pigment bed among photochemistry, fluorescence and all other dissipative processes or quenchers (11). Blocking just the photochemical de-excitation pathway will result in an increase in fluorescence related to the rate of the photochemical trapping. We have found that, under certain conditions, a substantial fraction of the saturation pulse fluorescence rise is caused by the relief of non-photochemical quenching attributable to oxidized PQ (q_{PQ}), a quenching process previously observed in thylakoids (e.g. 2). The effect of this non-photochemical quenching will contribute to the breakdown of the relationship between yields of linear electron transport measured by fluorescence techniques and CO_2 fixation (3). We introduce a technique with which it might be possible to estimate the extent of q_P.

MATERIALS AND METHODS

Plants material was collected on or near the University of Illinois farms. Atrazine (5 μM in distilled water) was vacuum infiltrated into leaves after gentle abrasion with 400 grit carborundum. Controls were similarly abraded, but were infiltrated with distilled water. Infiltration with water had variable effects, decreasing maximal fluorescence in some species. Only leaves that showed minimal effects of water infiltration were used.

The fluorimeter used in these studies was similar to that developed by Kramer et al. (4) for observation of electron transfer reactions on the microsecond time scale, and will be described in more detail elsewhere. The measuring beam was comprised of 3 μs pulses of orange-red light from four light-emitting diodes (LEDs, HLMP-8104, 637 nm peak emission) filtered through a 650 nm 10 nm bandpass interference filter (Oriel) to block infrared light. Fluorescence excited by the LED pulses was detected by a PIN photodiode (Hamamatsu 1773) protected by two layers of Kodak Wratten 89B filter covered by one layer of red glass filter (Schott RG695). Steady-state actinic illumination was provided by white light (from 10 to 2000 μmole $m^{-2}s^{-1}$) from a heat-filtered quartz halogen lamp (Schott KL1500) introduced through a fiber optic bundle. The steady-state could be punctuated by saturation pulses (800 ms of more than 10000 μmole $m^{-2}s^{-1}$ white light) from a heat-filtered quartz-halogen lamp (PLS2, Hansatech, Norfolk, UK) modified to allow direct timing control by the computer. Alternatively, a flash sufficiently intense and short (5 μs at half intensity) to provide a single-turnover of >95% of PSII centers was provided by a xenon lamp (FX201, EG&G, Salem,

P. Mathis (ed.), Photosynthesis: from Light to Biosphere, Vol. I, 147–150.

Ma.) filtered through a blue color glass filter (Corning 4-96). Probe pulses from the LEDs to measure fluorescence yield were given at one second intervals during the steady-state actinic illumination, in the microsecond range, starting at 45 μs, after xenon flash excitation, and every 100 ms during the saturation pulses. Illumination and detection systems were merged at 45° angles to the leaf surface by a mechanical mounting.

Fig. 1 is a diagram of the protocols used for illumination and fluorescence measurements. The solid curve represents the fluorescence yield measured by the LED probe pulses to give a time course similar to that provided by the PAM pulse-saturation technique. The base fluorescence yield, F_0, was measured after a single measuring pulse given to a leaf dark-adapted for 20 min. During saturation pulses, the fluorescence yield increased, giving F_M and F'_M, the maximal fluorescence yields in dark-adapted and light-exposed materials, respectively. When the saturating single-turnover xenon flashes were applied, (Fig. 1, inset), F_{MXD} was estimated by extrapolating from the complete decay curve of LED pulses after fitting the data to two exponential decay components. The steady-state fluorescence yield measured under constant light is termed F_S.

1.

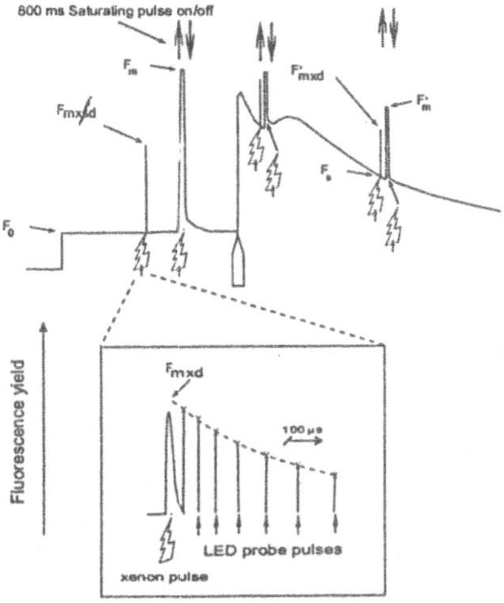

RESULTS AND DISCUSSION.

Figure 2 shows variable fluorescence values obtained from detached soybean leaves by after various techniques to block PSII. The value of $(F_{MXD}-F_0)/F_0$ of about 1.89 was obtained by fitting the decay curve after a single xenon flash in a water-infiltrated leaf to three exponential components and extrapolating to zero time. When 5 μM atrazine was included in the infiltrate, one flash produced a maximal Fv/F0 about 5% higher than $(F_{MXD}-F_0)/F_0$. The difference is probably due to an overlap of the slow rise phase of fluorescence (see in the atrazine trace) on the fluorescence decay in the control leaf (5). Upon a second actinic flash, 300 ms after the first, there was an increase of about 7% in F_V/F_0. There was no difference between the 2nd and subsequent flashes. As has been previously noted (5) the difference between the first and second flashes is not due to a subsaturating flash, since reducing the flash intensity by 50% had no effect on the fluorescence extents (not shown). Upon application of an 800 ms saturation pulse (above 10,000 μmole photons m^{-1} s^{-1}), to a water-infiltrated leaf, $(F-F_0)/F_0$ attained a value of about 3.5. When atrazine was introduced, saturation pulse-induced $(F-F_0)/F_0$ was lowered by about 37% to 2.2, very close to the value after the second xenon flash in the presence of atrazine.

We suggest that the difference between maximal fluorescence values attained in the presence and absence of atrazine is due to direct quenching by oxidized plastoquinone (q_{PQ}) (2). With the saturation pulses, photons arrive at the PSII centers too slowly to effectively compete with the electron transfer from Q_A to Q_B. In order for all Q_A to be reduced and the

2.

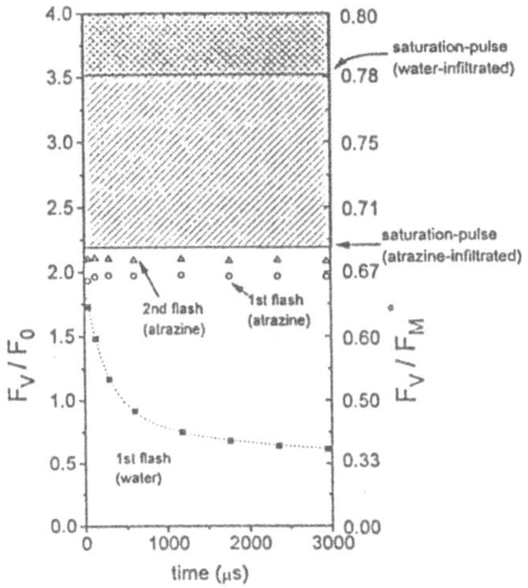

PSII centers to be closed, the PQ pool must first be completely reduced. This is possible because the turnover of the cytochrome b_6f complex and/or the reactions at the acceptor side of PSII become rate limiting during the pulse. Thus, it is likely that all measurements of maximal fluorescence with the pulse-saturation technique are made in the presence of a reduced PQ pool. However, when PSII is blocked by atrazine, continued turnover of PSI will insure complete oxidation of PQ. The same explanation can be used for the difference between the maximal fluorescence values attained with saturation pulses and single-turnover pulses. With the pulse-probe technique, excitation pulses of < 5 µs are used to transiently place PSII centers in a closed state (with Q_A reduced), allowing the measurement of fluorescence yield when all PSII photochemistry is blocked, but before any changes in the PQ pool redox state have occurred.

It has been previously demonstrated in thylakoids that PQ quenches about 20% of the maximal fluorescence (2). In a survey of different plants, including species of *Amaranthus*, *Polygonium*, *Helianthus*, *Ipomea* and *Convolvulus*, we measured a differences in $(F_M-F_0)/F_0$ between water- and atrazine-infiltrate leaves in the range of 25 to 38%. These variations can be explained by the presence of a variable amounts of non-photochemical quenchers. The fluorescence yield is inversely proportional to the sum of the rate constants of various deactivation pathways and therefore, the relative modulatory effect of a change in a rate constant will depend upon the fluorescence yield (2, 6). Under conditions where the fluorescence yield is high, a weak quencher will have a relatively large effect. Thus, the magnitude of q_{PQ} will depend not only upon the concentration of PQ but also upon the total allotment of other quenchers.

Since quenching by the PQ pool is not due to photochemistry, it should not be included in estimates of photochemical efficiency. We therefore recalculated PSII photochemical yield, based on our estimates of maximal fluorescence values in the absence of reduced PQ. On the right-hand scale of Fig. 2, the Fv/Fm, commonly used to estimate PSII photochemical yield (see above) is calculated considering the corresponding fluorescence values as F_M. The 37% difference between $(F-F_0)/F_0$ in leaves infiltrated with water and atrazine translates into a calculated decrease in PSII efficiency of about 10%. If we use F_{MXD} as an estimate of F_M, we calculate a PSII yield of 0.66 (16% lower than that calculated using the saturation pulses). Because of the overlap of the slow fluorescence rise phase, and the inability to excite all PSII centers with each xenon flash, F_{MXD} should be considered an underestimate of the maximal fluorescence with the PQ pool oxidized. Keeping this in mind, we can use it as a qualitative probe of the effects of q_{PQ} on fluorescence measurements.

In order to estimate the possible effects of q_{PQ} on fluorescence-derived estimates of PSII photochemical efficiency, we measured, in the same leaves under a range of constant illumination intensities, both F'_M-F_S/F'_M, using saturation pulses, and $F'_{MXD}-F_S/F'_{MXD}$, using single-turnover pulses. Measurements made with xenon flashes and with saturating pulses were comparable when the photochemical efficiency was low (i.e. at high light intensity), but deviated at high efficiency . The most likely explanation is that, at low efficiency, the PQ pool was substantially reduced and the effect of q_{PQ} was low. This is also consistent with the expectation that, when PSII efficiency is high, its electron acceptor, PQ, should be predominantly oxidized. Also consistent with this interpretation are the kinetics of fluorescence decay after the xenon flashes (see corresponding letters in Fig. 3a and 3b) which slow dramatically as the light intensity was increased. We interpret this as indicating a lack of acceptor PQ at high light.

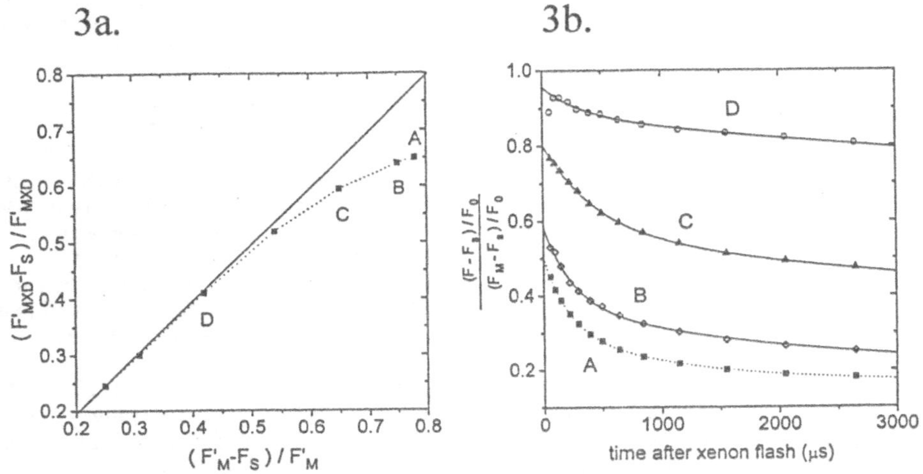

We conclude that q_{PQ} is likely to impact on PSII yield measurements under conditions of high PSII efficiencies (i.e. at low light with high Calvin cycle activities), resulting in an overestimate of photochemical efficiency. Because the measurements using xenon pulses are likely to be underestimates of the true F_M (see above), it is not possible at present to give a quantitative estimate of the extent of this effect. However, the predicted effect should be measurable, at the low light levels where it should manifest itself as a non-linearity between F'_M-F_S/F'_M and CO_2 uptake or O_2 evolution. However, respiratory CO_2 release or O_2 uptake make large contributions to photosynthesis measurements at low light levels and uncertainties in estimating the amount of respiration to correct for possibly may have obscured the effect (3).

Acknowledgments: The work was supported by CNR-RAISA fellowship and a U.S. Department of Energy Integrative photosynthesis research training grant DOE 92-ER-20095. We thank R. Larimor for identifying the plant material.

1. Genty, B., Briantais, J.M. & Baker, N.R. (1989) Biochim. Biophys. Acta 990: 87-92.
2. Vernotte, C., Etienne, A.L. & Briantais, J.M. (1979) Biochim. Biophys. Acta 545, 519-527.
3. Oberhuber W., Edwards G.E. (1993) Plant Physiol. 101: 507-512.
4. Kramer D.M., Robinson H.H. & Crofts AR (1990) Photosynth. Res. 26: 181-193.
5. Robinson, H.H. and Crofts, (1987) In Biggin, J. ed., Progress in Photosynthesis Research. pp. 429-433
6. Lavorel, J. & Etienne, A.L. (1977) in Primary Processes of Photosynthesis (Barber, J., ed.) pp. 203-268, Elsevier, Amsterdam.

Room temperature investigation of the LH2 antenna of *Rb. sphaeroides* and *Rps. acidophila* by non-linear polarization spectroscopy in the frequency domain (NLPF)

F. Nowak[1], B. Voigt[1], D. Leupold[1], M. Bandilla[2], H. Scheer[2], R. J. Cogdell[3]

[1]Max-Born-Institut für Nichtlineare Optik und Kurzzeitspektroskopie, D-12474 Berlin, Germany,
[2]Botanisches Institut der LMU, D-80638 München, Germany
[3]Division of Biochemistry and Molecular Biology, University of Glasgow, Glasgow G12 8Q Q, UK

Introduction

The peripheral antenna LH2 of *Rhodobacter (Rb.) sphaeroides* has two NIR absorption bands, according to their maxima this antenna is also called B800-850. There has been continuous interest in its spectroscopic properties for a long time. Nevertheless results to relevant questions are still controversial. One of these is the type of B850 band broadening. On the one hand, by means of hole burning spectroscopy at 4K, the absorption band has been found to be homogeneously broadened[1], on the other hand an inhomogeneous[2] or heterogeneous broadening with two forms[3] (by means of ps-resolved absorption and fluorescence at 77K) as well as with at least three subbands by NLPF at room temperature (RT)[4] has been observed .
In the following we report on further LH2 investigations of chromatophores and isolated B800-850 antennas of *Rb. sphaeroides* at RT with an improved NLPF spectrometer. They confirm and further specify the B850 substructure and also give first NLPF hints concerning the interaction B800 - B850. In addition to that, B800-850 antennas of *Rps. acidophila,* whose crystal structure has been described recently[5], have been investigated.

Material and Methods

The preparation of the antenna complexes of *Rps. acidophila* strain 10050 and of *Rb. sphaeroides* 2.4.1 as well as that of the chromatophores of the latter have been described elsewhere[6,7].
The principle of the method as well as the NLPF apparatus have been described earlier[7]. However, essential improvements have been performed since which are specified in detail elsewhere[8,9]:

- a change of the direction in which the pump beam passes the sample cell. Whereas up to now pump- and probe beam were focused in opposite directions

P. Mathis (ed.), Photosynthesis: from Light to Biosphere, Vol. I, 151–154.

into the sample, we now use an arrangement in which both beams are passing the cell at a right angle between them.

- a selected 5mm flow-through dye-laser cuvette (HELMA) which is of low birefringence is used as sample cell. Transmission in the absorption maximum of the samples was about 5%.

These changes resulted in a sensitivity one to two orders of magnitude higher than in experiments described before. NLPF spectra could be obtained down to photon flux densities of $10^{23}cm^{-2}\ s^{-1}$ in the pump pulse. The measuring range was extended from less then 800nm to more than 860nm by the use of the laser dye Styryl 9 (LAMBDA Physik, Göttingen) dissolved in propylencarbonate.Experimental data were fitted on the basis of theoretical line shape functions described in [4,10].

Results and Discussion

NLPF spectra of B800-850 of *Rb. sphaeroides* have been recorded at different probe wavelengths covering the complete range of measuring between 800nm and 860nm. In each case we have measured highly resolved spectra around the resonance of pump and probe wavelength as well as broad-band spectra covering the complete tuning range of the pump laser. Spectra have been detected with varying intensities of the pump laser, but here we will concentrate on the discussion of spectra obtained for the lowest intensities corresponding to an average photon flux density of $10^{23}cm^{-2}s^{-1}$.

The first hint on spectral inhomogeneity of the 850nm band can be seen directly in the spectra, because changing of the probe wavelength results in a shift of the maxima of the broad band NLPF spectra of several nanometers. NLPF spectra probed at several wavelengths in the 850nm band (for an example see fig. 1 below) can be fitted sufficiently with a common set of parameters by using three homogeneously broadened subbands in this spectral region. This result confirms and qualifies earlier measurements[4]. The positions and the homogeneous widths of these bands are given in tab. 1.

Regarding the highly resolved NLPF spectrum around $\lambda_{Probe} = \lambda_{Pump} = 860.395$nm (fig. 1 top), we find strong indication of another subband outside the range of measuring. This fact follows from the strong asymmetry shown in the narrowbanded spectrum: Whereas the overall full width at half maximum points towards an "averaged" energy relaxation time of about 100 to 200ps, this asymmetry indicates that there must be a comparatively long-living component located at about 870nm. Assuming a fourth substructure at 870nm with $T_2 = 40$fs, the narrowbanded spectra can now be fitted with the known parameters for the three bands from tab. 1. The results are also given in tab. 1.

λ / nm	Γ_{hom} / cm^{-1}	T_2 / fs	T_1 / ps
829 ± 1	110 ± 15	50 ± 7	200 ± 20
845 ± 1	220 ± 40	24 ± 4	190 ± 20
855 ± 1	120 ± 14	43 ± 5	160 ± 20
(870)	(130)	(40)	1150 ± 200

tab. 1: position of subbands and corresponding T_1 - and T_2 - times of B800-850 of *Rb. sphaeroides*

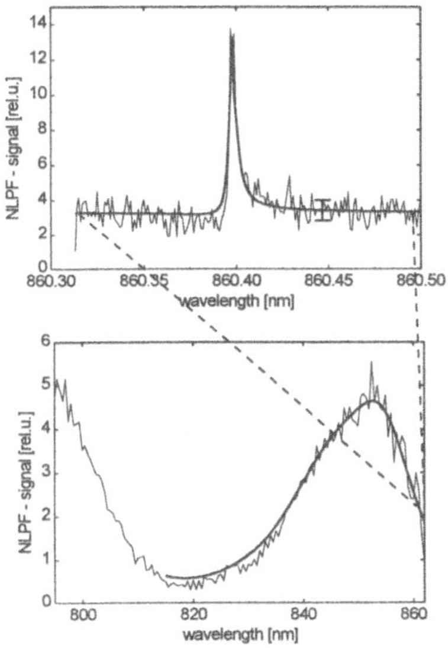

fig. 1: NLPF spectra of *Rb. sphaeroides*
$\lambda_{Probe} = 860.395$nm
photon flux density of pump beam:
10^{23} $cm^{-2}s^{-1}$

In fig. 2 there is shown an example for a NLPF-spectrum measured at a pump intensity comparable to those discussed above. In contrast to them the probe wavelength has now been in the 800nm band. The greater noise shown in this spectrum is the result of the much shorter life time of this band in the ps range leading to a lower amplitude of the NLPF-signal. The most striking feature is the fact that the 850nm band can be seen clearly in the spectrum, which can neither be explained by the spectral overlap of the band at 800nm with the most prominent subbands of B850 indiviually nor by up-hill energy transfer.

Similar measurements have been performed with B800-850 complexes of *Rps. acidophila*. A spectrum probed at 850nm is shown in fig. 3. By "shifting" the corresponding spectrum of *Rb. sphaeroides* about 5nm to the red, the measurements for both species can be brought to a good agreement at least for the 850nm band. Therefore it should be possible to fit the spectra for *Rps. acidophila* with a similar set of parameters as it was possible for *Rb. sphaeroides*. This indicates that from the view of NLPF there are no fundamental differences in the light-harvesting complexes of both purple bacteria. Therefore the general assumption of a great similarity in the arrangement of the bacteriochlorophylls in the LH2 antenna of both species is supported. Assuming a ring with two different BChl binding sites of 18 BChl a molecules responsible for the 850nm band[5] for *Rb. sphaeroides* and regarding theoretical calculations [11] concerning excitonic states of circular aggregates, at least two branches with two electronic transitions in each can be expected. They do not necessarily correspond to the four subband transitions shown in tab. 1: first, our results for λ_{Probe} in the 800nm band can be interpreted as a coupling of at least the three most prominent subband transitions (828, 845, 855nm) and second, recent fs pump-probe experiments[12] suggest at least one vibronic component

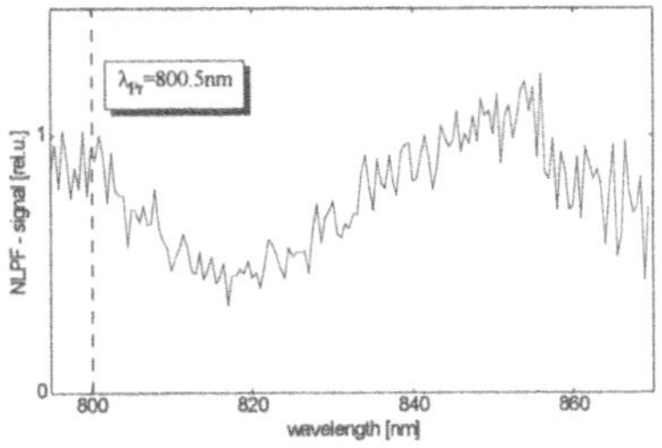

fig. 2: NLPF-spectrum of *Rb. sphaeroides* λ_{Probe} = 800.5nm

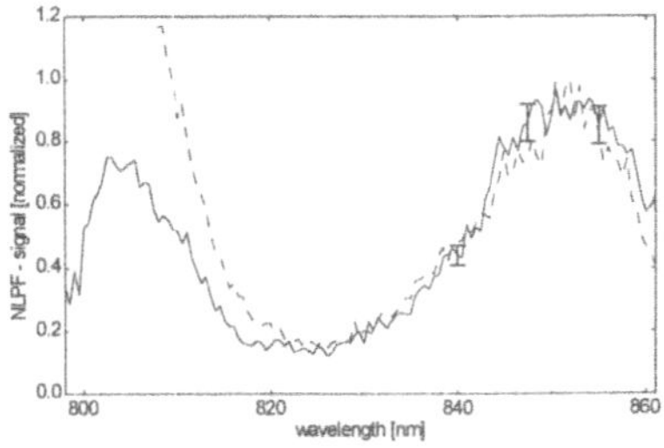

fig. 3.: NLPF spectra of *Rps.acidophila*, λ_{Probe} = 850nm (dotted) and of *Rb. sphaeroides*, λ_{Probe} = 840 nm (solid); the latter shifted 5nm to the red

(110cm^{-1}) besides pure electronic transitions. Therefore it is not possible for the moment to assign our subbands to certain transitions. Further relevant information is included in the dependence of NLPF spectra on intensity as well as in planned NLPF experiments at cryogenic temperatures.

Typical spectra we obtained for chromatophores of *Rb. sphaeroides* (not shown) can be fitted with a similar set of parameters as the spectra for the B800-850 complexes in the case of λ_{Probe} = 840nm, the spectrum probed at 857nm looks completely different: its maximum is shifted towards the red, and its FWHM is much smaller. The highly resolved spectra also show T_1 times about one order of magnitude shorter than those obtained for the complexes. Additionally, we observed a much weaker intensity dependence of the spectra than we found for B800-850. This is valid for the broad- as well as for the narrow band spectra.

Acknowledgements

This work was supported by the Deutsche Forschungsgemeinschaft, Bonn (AZ Le729/1-1, 1-2 and Sche 140/13-1, 13-2)

[1]Jankowiak, R., Hayes, J. M., Small, G. J. (1993) Chem. Rev. 93, 1471 - 1502

[2]van Grondelle, R., Dekker, J. P., Gillbro, T., Sundström, V. (1994) Biochim. Biophys. Acta 1187, 1 - 65

[3]Sundström, V., van Grondelle, R. in Laser Spectroscopy of Biomolecules (Korppi-Tommola, J. E. I., ed.) pp. 146 - 155, SPIE Vol. 1921

[4]Leupold, D., Voigt, B., Nowak, F., Ehlert, J., Hirsch, J. (1994) Lietuvos fizikos zurnalas 34, 339 - 343

[5]McDermott, G., Prince, S. M., Freer, A. A., Hawthornthwaite-Lawless, Papiz, M. Z., Cogdell, R. J., Isaacs, N. W. (1995) Nature 374, 517 - 521

[6]Cogdell, R. J., Durant, I., Valentine, J., Lindsay, J. G., Schmidt, K (1983) Biochim. Biophys. Acta 722, 427-455

[7]Leupold, D., Voigt, B., Pfeiffer, M., Bandilla, M., Scheer, H. (1993) Photochem. Photobiol. 57, 24 - 28

[8]Voigt, B. et al. in prep.

[9]Lokstein, H.,Leupold, D.,Voigt, B., Nowak,F., Ehlert,J.,Hoffmann, P.,Garab, G.(1995) Biophys.J. in print

[10]Mory, S., Neef, E. (1991) Exp. Techn. Phys 39, 385 - 388

[11]Novoderezhkin, V. I., Razjivin, A. P. (1995) Biophys. J. 68, 1089 - 1100

[12]Chachisvilis, M. T., Pullerits, M. R., Jones, C., Hunter, N., Sundström, V. (1994) Chem. Phys. Lett. 224, 345 - 351

COMPARISON OF BACTERIOCHLOROPHYLL BINDING REQUIREMENTS IN THE CORE LIGHT-HARVESTING COMPLEXES OF *RHODOSPIRILLUM RUBRUM* AND *RHODOBACTER SPHAEROIDES*

P.S. Parkes-Loach[1], C.M. Davis[1], C. Cook[1], K.A. Meadows[1], M. Bandilla[2], H. Scheer[2] & P.A. Loach[1]. [1]Department of Biochemistry, Molecular Biology & Cell Biology, Northwestern University, 2153 Sheridan Rd., Evanston, IL 60208, USA; [2]Botanisches Institute, Universität München, Menzinger Str. 67, D8000 München 19, Germany.

1. *INTRODUCTION*

The ability to reconstitute the core light-harvesting complex (LH1) of many photosynthetic bacteria from their individual polypeptide, bacteriochlorophyll (BChl) and carotenoid components, allows for the systematic evaluation of binding elements that stabilize the complex (1-3). From studies in which analogs of BChl were examined with the LH1 α- and β-polypeptides isolated from *Rhodospirillum rubrum*, several structural requirements were observed (4): the Mg atom is required, presumably for coordinating a ligand provided by the protein, the acetyl group at position C3 (see Fig. 1 for the structure of BChl and the numbering system used for its atoms) is important, presumably for participating in hydrogen bonding interactions with groups on the protein, the carbomethoxy group at position $C13^2$ is important either for hydrogen bonding or because of its influence on the binding of the $C13^1$ carbonyl group, and the bacteriochlorin oxidation state of the macrocycle is required, suggesting a highly restrictive overall geometry for fitting the binding site.

The examination of polypeptide interactions with BChl analogs has been extended to include the LH1 α- and β-polypeptides of *Rhodobacter sphaeroides*. This bacterium was chosen because it has only about 30% amino acid sequence homology in its LH1 and reaction center (RC) polypeptides compared to those of *R. rubrum*. Also, it has a secondary LH complex, LH2, whereas *R. rubrum* does not, and the spectral properties (absorption and circular dichroism (CD) spectra) of LH1 are somewhat different from those of *R. rubrum*.

2. *METHODS & MATERIALS*

The preparations of the polypeptides used in these studies and all reconstitution assay procedures were as previously described (2,4). For most experiments, a standard set of concentration conditions were chosen to assay reconstitution with BChl analogs. These consisted of 3 μM each of the α- and β-polypeptides and a concentration of BChl or analog between 1 and 3 μM. All BChl analogs used were obtained as reported before (4) except for: [3-viny]-BChl a_p and [13^2-OH]-BChl a_p were prepared as described in (5); BChl e_7 trimethylester and 13^1a-oxa-[13^2-oxo]-BChl a_m were prepared as described in (6) and metal substituted BPh derivatives were prepared as described in (7).

P. Mathis (ed.), Photosynthesis: from Light to Biosphere, Vol. I, 155–158.

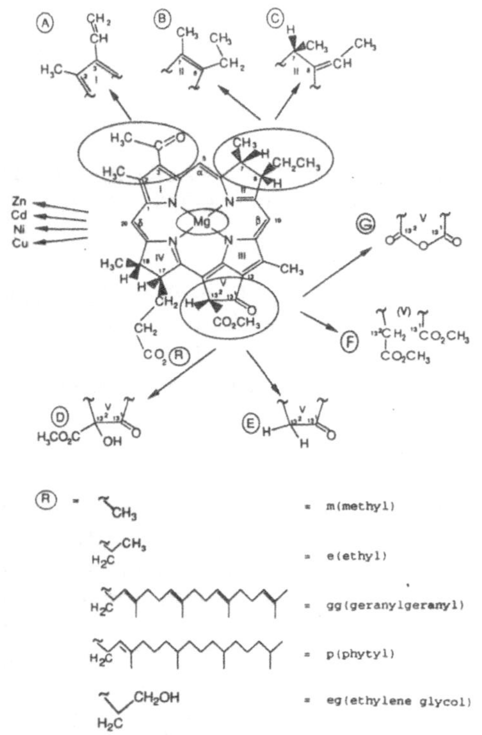

Abbreviation	Macrocycle modification(s)	Central metal	Esterifying alcohol
BChl a_p	unmodified	Mg	p
BChl a_{gg}	unmodified	Mg	gg
BChl a_e	unmodified	Mg	e
BPh a_p	unmodified	none	p
Cd(II)BPh a_p	unmodified	Cd	p
Zn(II)BPh a_p	unmodified	Zn	p
Ni(II)BPh a_p	unmodified	Ni	p
Cu(II)BPh a_p	unmodified	Cu	p
BChl b_p	C	Mg	p
pyro-BChl a_p	E	Mg	p
pyro-BChl a_{eg}	E	Mg	eg
CHl a_p	A + B	Mg	p
Chl b_p	A + B[1] ([1]where 7-CH_3 is replaced by CHO)	Mg	p
BChl e_7	F	Mg	tme[2]
[3-vinyl]-BChl a_p	A	Mg	p
[13_2-OH]-BChl a_p	D	Mg	p
[3-acetyl]-Chl a_p	B	Mg	p
13^1a-Oxa-[13^2-Oxo]-BChl a_m	G	Mg	m

[2]tme = trimethyl ester

Figure 1. Structures of BChl *a* and analogs.

3. RESULTS & DISCUSSION

3.1 Analogs that support subunit- and LH1-type complex formation

Of the analogs tested (Fig. 1), only molecules with changes in the esterifying alcohol, the side chain at position 8 and certain metal derivatives formed subunit-type complexes. Three BChl analogs that had different esterifying alcohols at position $C17^4$ (phytyl, geranylgeranyl, ethyl) formed native-like subunit and LH1 complexes with either the α- and β-polypeptides of *R. rubrum* or *Rb. sphaeroides*. With both sets of polypeptides, the analog with the shortest esterifying alcohol (ethanol) formed a less stable subunit-type complex as evidenced by the significant amount of absorbance near 780 nm, indicating free BChl, but all three derivatives formed stable LH1-type complexes. With BChl *b*, the red shifts observed in forming subunit-type and LH1-type complexes with the α- and β-polypeptides of *Rb. sphaeroides* (to 853 nm and 901 nm, respectively) were the same as those with the α- and β-polypeptides of *R. rubrum* and represent the same magnitude of shifts in energy as observed with BChl *a* (4).

Four metal derivatives of BPh (Zn, Cu, Ni and Cd) were prepared and tested in reconstitution with the α- and β-polypeptides of both *R. rubrum* and *Rb. sphaeroides*. In keeping with expectations, Cd(II)BPh and Zn(II)BPh readily formed subunit-type complexes with the α- and β-polypeptides of either *R. rubrum* or *Rb. sphaeroides*(Fig.2). Ni(II)BPh also interacted with both sets of polypeptides to form a subunit-type complex,

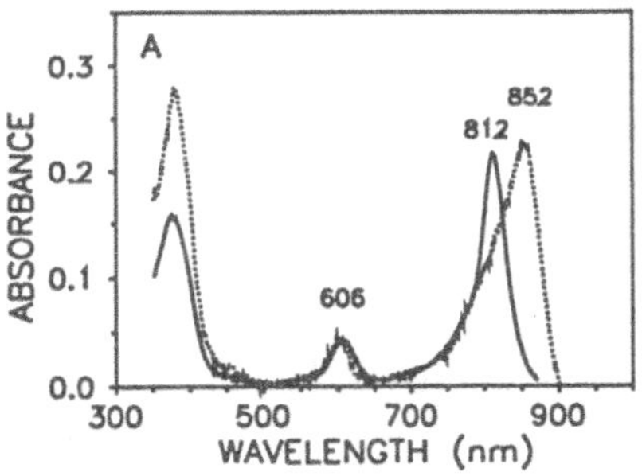

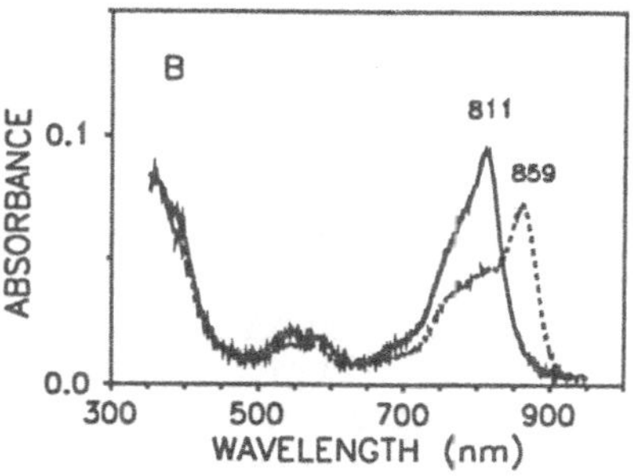

Figure 2. Reconstitution of Cd(II)BPh (A) and Ni(II)BPh (B) to form subunit-type (solid) and LH1-type (dashed) complexes.

but was not as stable as those formed with the Cd(II)BPh, Zn(II)BPh or native BChl *a* (Fig. 2). The LH1-type complexes for Cd(II)BPh, Zn(II)BPh and Ni(II)BPh complexes were readily formed but were less completely formed (Fig. 2) than in the case of BChl and were not as kinetically stable.

3.2 Analogs that did not support subunit- or LH1-type complex formation

Analogs that failed to show evidence for formation of subunit-type and LH1-type complexes under our assay conditions included compounds such as [3-acetyl]-Chl a_p, indicating the importance of the bacteriochlorin oxidation state, and pyro BChl a_{eg}, BChl e_7 trimethyl ester and 13^1a-oxa-[13^2-oxo]-BChl a_m, indicating the importance of ring V. Extensive self-aggregation was observed with three of the BChl analogs. Especially interesting was the derivative 13^1a-oxa-[13^2-oxo]-BChl a_m which displayed an extensively red-shifted Q_y band (1008 nm; Fig. 3), the furthest red band reported for protein-free, porphyrin-based aggregates. It is comparable to that observed for LH1 in the BChl-*b*-containing *Rps. viridis* (1015 nm).

3.3 Analogs supporting subunit-type but not LH1-type complex formation

When the subunit complex is induced to associate to form LH1, specific interactions between the subunits occur, perhaps involving the bound BChl (3). It therefore seems reasonable that a BChl analog could be found that would form a stable subunit-type complex but would not form an LH1-type complex with the α- and β-polypeptides present. The [3-vinyl]-BChl *a* analog does show such behavior with the

Figure 3. 13^1a-oxa-[13^2-oxo]-BChl a_m in 50 mM phosphate buffer, pH 7.5 containing 0.64% OG (solid) and in methanol containing KOH (dashed).

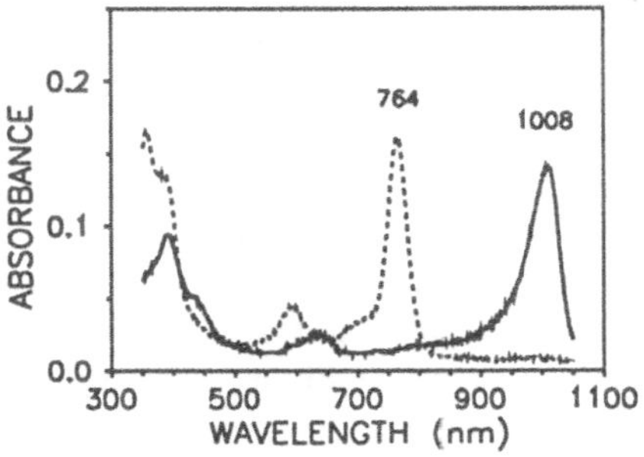

α- and β-polypeptides of *R. rubrum*. The Q_y absorption band shifted from 739 nm to 762 nm and the CD spectrum was very similar, albeit blue-shifted, to those of the subunit complex with BChl *a* but no further red-shift was observed under LH1-forming conditions. When tested with the α- and β-polypeptides of *Rb. sphaeroides*, there was evidence for only slight interaction under special conditions to form a subunit-type complex.

3.4 Differences with polypeptides of R. rubrum *and* Rb. sphaeroides

As discussed above, [3-vinyl]-BChl *a* interacts differently with polypeptides of *R. rubrum* compared with those of *Rb. sphaeroides*. A second example where such differentiation was observed was with the derivative, [13^2-OH]-BChl *a*. Reconstitution with this derivative and the α- and β-polypeptides of *Rb. sphaeroides* resulted in smaller red shifts than with native BChl *a*, but both subunit-type and LH1-type complexes were formed. On the other hand, [13^2-OH]-BChl *a* does not form stable subunit-type or LH1-type complexes with the α- and β-polypeptides of *R. rubrum* (4).

4. CONCLUSIONS

Comparable results were obtained with most of the BChl analogs and the polypeptides of *R. rubrum* or *Rb. sphaeroides*, indicating the conservation of BChl binding sites through evolution. In general, high specificity is exhibited as few changes can be made to the BChl structure without preventing complex formation. Except for one analog, [3-vinyl]-BChl *a*, the specific structural requirements for BChl to form a subunit complex were the same as those required for forming LH1. This fact makes it likely that the interactions between BChl and the protein in the subunit complex are retained in LH1. As a result of association of these subunits, at least one site of additional interaction occurs between BChl and the protein in LH1 compared with the subunit complex. This interaction appears to involve a hydrogen bond to the carbonyl group at position $C3^1$. It will be of considerable interest to evaluate the amino acids of the LH1 polypeptide in terms of their possible involvement in hydrogen bonding to BChl.

5. REFERENCES

(1) Parkes-Loach PS, Sprinkle JR & Loach PA (1988) Biochemistry 27, 2718-2727.

(2) Loach PA, Parkes-Loach PS, Davis CM & Heller BA (1994) Photosyn Res, 40, 235-246.

(3) Loach PA and Parkes-Loach PS (1995) in Advances in Photosynthsis: Anoxygenic Photosynthetic Bacteria, (Blankenship RE, Madigan MT and Bauer CE, eds) Kluwer Academic Publishers, in press.

(4) Parkes-Loach PS, Michalski TJ, Bass WJ, Smith U & Loach PA (1990) Biochemistry 29, 2951-2960.

(5) Struck A, Cmiel E, Katheder I & Scheer H (1990) FEBS Lett 268, 180-184.

(6) Struck A, Beese D, Cmiel E, Fischer M, Müller A, Schäfer W & Scheer H (1990) in Reaction Centers of Photosynthetic Bacteria (Michel-Beyerle, M.E., ed.), pp 313-326, Springer, Berlin.

(7) Hartwich G (1995) Ph.D. Dissertation, Technische Universität München.

MINIMAL STRUCTURAL REQUIREMENTS FOR FORMATION OF THE SUBUNIT COMPLEX OF LH1 OF PHOTOSYNTHETIC BACTERIA

K.A. Meadows, J.W. Kehoe & P.A. Loach. Department of Biochemistry, Molecular Biology & Cell Biology, Northwestern University, 2153 Sheridan Rd., Evanston, IL 60208, USA

1. INTRODUCTION

Reconstitution methodology has allowed for the systematic probing of structure-function relationships in LH1 using a variety of analogs of BChl and the α- and β-polypeptides (1-4). As a result of these studies, a core region of each polypeptide has been identified that seems to contain all the necessary structural features for formation of the subunit complex (Fig. 1). The core region for LH1 formation is somewhat larger.

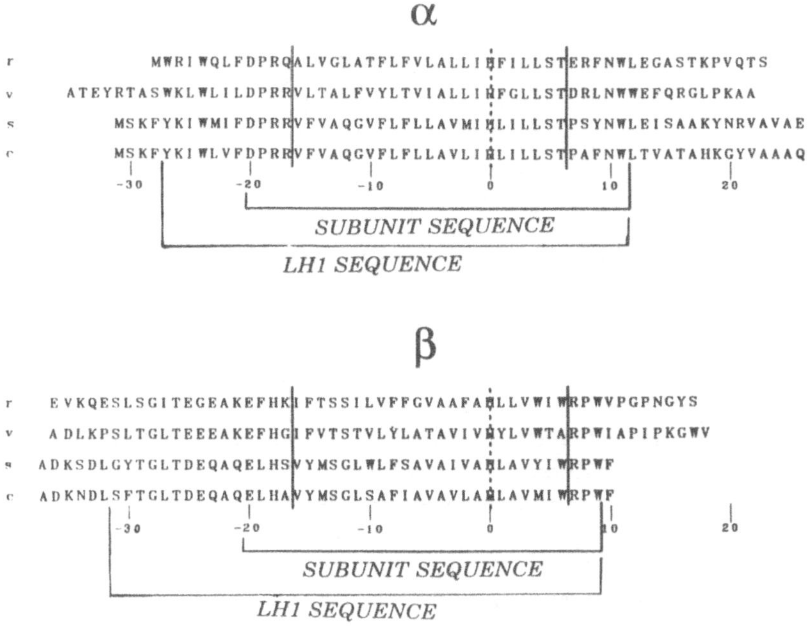

Figure 1. Amino acid sequences of the α- and β-polypeptides of LH1 of the bacteria *Rs. rubrum* (r), *Rps. viridis* (v), *Rb. sphaeroides* (s) and *Rb. capsulatus* (c).

P. Mathis (ed.), Photosynthesis: from Light to Biosphere, Vol. I, 159–162.

We have found the strategy of *gradual reductionism* to be an effective approach to identifying structural requirements. Of considerable help in evaluating requirements for subunit formation, was the observation that a subunit-type complex could be reconstituted using only the β-polypeptide of either *R. rubrum*, *Rb. sphaeroides* or *Rb. capsulatus* (1,4). These subunit-type complexes were incapable of associating to form LH1-type complexes unless an α-polypeptide was included in the system.

2. RESULTS & DISCUSSION

For determination of the minimal structural requirements for formation of the subunit complex, a comparison of the properties of the native β-polypeptides of *R. rubrum* and *Rb. sphaeroides* with shortened polypeptide analogs is instructive. Results with two significantly shortened β-polypeptides are shown in Fig. 2. Both the absorption spectra and CD spectra (not shown) of the systems with shortened polypeptides are indistinguishable from those of the native systems. Because the stability of the subunit complex seems to be the net result of contributions by several binding elements (5), it is important to determine the K_{Assoc} for each native system and to compare it to those obtained with the modified systems. As may be seen from Table 1, K_{Assoc} for the shortened polypeptides ($\beta_{EFHK..to\ C\text{-}term}$ of *R. rubrum* and $\beta_{LHSV..to\ C\text{-}term}$ of *Rb. sphaeroides*) are only about four-fold smaller (less than 1 kcal mol^{-1}) than that of the native polypeptides, indicating that the N-terminus removed in each case is not very important for formation of the subunit complex. In both cases, it is, however, important for stabilization of LH1 (3).

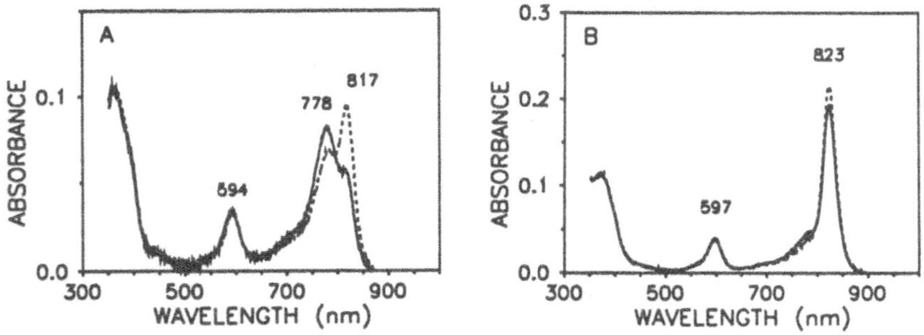

Figure 2. Absorption spectra of the reconstituted subunit-type complex at 0.90% (solid) and 0.75% OG (dashed) with the truncated β-polypeptides of *R. rubrum* (EFHK..to C-terminus) and *Rb. sphaeroides* (LHS..to C-terminus) and BChl.

If four additional amino acids are removed from the N-terminus of the truncated β-polypeptide (EFHK..to C-term) of *R. rubrum*, a substantial decrease (about 100-fold) is seen in the stability of the subunit complex (see truncated $\beta_{IFTS..to\ C\text{-}term}$ in Table 1). Although apparently not an absolute requirement, we assume that it is the strictly conserved His at position -18 which is involved as one of the important binding elements in stabilizing the subunit complex.

A great deal of evidence exists to indicate the importance of His_0 of each

TABLE 1

Association Constants of Reconstituted Subunit Complexes

System	Polypeptide (μM at 0.75% OG)	K_{Assoc} (M^3) X 10^{-16} 0.90 % OG	K_{Assoc} (M^3) X 10^{-16} 0.75% OG
Rs. rubrum	native β (3.7)	1.2	19
	truncated $\beta_{EFHK..\,to\,C\text{-}term}$ (4.4)	0.3	5
	truncated $\beta_{IFTS..\,to\,C\text{-}term}$ (12.0)	0.02	0.13
Rb. sphaeroides	native β (1.8) + native α (1.8)	19	≥300
	native β (6.1)	12	175
	truncated $\beta_{LHSV..C\text{-}term}$ (1.8) + native α (1.8)	5	47
Chemically synth	$31mer_{ELHS..C\text{-}term}$ (3.3) + native α (3.0)	7	53
Rb. capsulatus	native β (6.1)	-	11

polypeptide of LH1 in providing a coordinating ligand to BChl. On the basis of model studies, the minimum stabilization energy contributed by His interaction with BChl is likely to be between 5.0 and 6.0 kcal mol^{-1} (5). Since there are two BChl molecules bound in each subunit complex (assuming $\alpha_1\beta_1$•2BChl), this stabilization energy is approximately 11 ± 1 kcal mol^{-1} and accounts for about half of the total binding energy stabilizing the subunit complex (5).

Several amino acid side chains near the C-terminal end of the β-polypeptide have also been implicated in stabilizing the subunit complex. From comparison of the activities of the β-polypeptide of several bacteria, Trp_{+6} was suggested to be involved in stabilization of the subunit complex (4). Based on the comparison of K_{Assoc} for subunit formation with the β-polypeptide of *Rb. sphaeroides* with that for formation with the β-polypeptide of *Rb. capsulatus* (Table 1), Tyr_{+4} of the *Rb. sphaeroides* polypeptide (which is a Met in the β-polypeptide of *Rb. capsulatus*) has also been implicated in stabilizing the subunit complex (4). With these considerations along with the locations of highly conserved amino acids in the region expected to be near the BChl binding sites, the amino acids at positions +4, +6, +7, and +9 of the β-polypeptide are prime candidates for providing stabilizing interactions. Using a combination of methods including carboxypeptidase treatment, site-directed mutagenesis and chemical synthesis, preliminary data indicate that all four groups are important (unpublished results).

In what promises to be a very powerful approach, we have successfully chemically synthesized a polypeptide containing 31 amino acids with a sequence identical to that of the native β-polypeptide of *Rb. sphaeroides* from Glu_{-20} to the end of the C-terminus (K.A. Meadows, unpublished). Its activity and K_{Assoc} were very similar to those of the enzymatically shortened polypeptide (Fig. 2 and Table 1). Thus it is now possible to synthesize polypeptide analogs which differ by as little as a single atom. This should be immensely helpful both in evaluating the contribution to binding by different structural elements and in preparing specifically-labelled polypeptides to enable NMR, RR and FTIR analyses.

3. CONCLUSIONS: Minimal Requirements for Forming the Subunit Complex

On the basis of knowledge about LH1, its subunit complex and the recent crystal structure of LH2 (6), the following structural components for the subunit complex of LH1 are proposed.

(1) An α-helical polypeptide of 29 amino acids in length is sufficient to fully support formation of a subunit complex. The α-helical structural assignment for the complete length of the polypeptide is based on CD and analogy to the LH2 crystal structure (6). The length of 29 amino acids is based on results with shortened polypeptides such as those cited above and the fact that neither Glu_{-20} nor the C-terminal Phe_{+10} of the native β-polypeptide is required for full activity (3). Thus, a polypeptide of 29 amino acids contains all the structural requirements for forming the subunit complex with a native K_{Assoc}.

(2) Out of the 29 amino acids, a sequence of 23 amino acids with apolar side chains (from position -16 to position +6) seems to be required for spanning the membrane. A His must occur in this apolar segment at position 0 for providing the coordinating ligand to BChl. Furthermore, an amino acid with a very short side chain (Ala or Gly) is required at positions -4 and -7 to accomodate the BChl aromatic ring.

(3) BChl a_p, BChl a_{gg} or BChl *b* is required, although a shorter esterifying alcohol is acceptable, as is another central metal such as Cd, Ni or Zn (C. Cook, unpublished).

(4) A His at postition -18 is required to help stabilize (by approximately 3.0 kcal mol^{-1}) the subunit complex (3).

(5) Two to four sites of cross-hydrogen bonding near the C-terminus of the α- and β-polypeptides are proposed to be essential for stabilizing the subunit complex. The existence of such bonding, only one of which is observed in the published crystal structure of LH2, would explain how a subunit complex can be isolated in stable form from LH1 complexes and, also how the β-only system can form the same type of subunit complex. Many of the above minimal requirements have been accommodated in proposed models (5), two versions of which have features strikingly similar to one of the two dimeric pairs of BChl that exist in the structure of LH2 (6).

4. REFERENCES

(1) Parkes-Loach P.S., Sprinkle J.R. and Loach P.A.(1988) *Biochemistry* 27: 2718-2727.

(2) Parkes-Loach,, P.S., Michalski, T.J., Bass, W.J., Smith, U. and Loach, P.A. (1990) *Biochemistry*, 29: 2951-2960.

(3) Meadows K.A., Iida K., Tsuda K., Recchia P.A., Heller B.A., Antonio B., Nango M. and Loach P. (1995) *Biochemistry*, 34: 1559-1574.

(4) Loach P.A., Parkes-Loach P.S., Davis C.M. and Heller B.A. (1994) *Photosyn Res*: 40: 235-246.

(5) Loach P.A. and Parkes-Loach P.S. (1995) in Advances in Photosynthsis: Anoxygenic Photosynthetic Bacteria (Blankenship R.E., Madigan M.T. and Bauer C.E., eds.) Kluwer Academic Publishers, in press.

(6) McDermott G, Prince SM, Freer AA, Hawthornthwaite-Lawless AM, Papiz MZ, Cogdell RJ, and Isaacs NW (1995) *Nature* 374: 517-521.

HYDROPHOBIC MODIFICATION OF ANTENNA CHLOROPHYLL IN CHLOROBIUM DURING GROWTH WITH ACETYLENE

Niels-Ulrik Frigaard[1] and John Ormerod[2]

[1]Institute of Biochemistry, Odense University, Campusvej 55, DK-5230 Odense M, Denmark and [2]Biology Dept., Oslo University, Blindern, N-0316 Oslo, Norway.

1.Introduction

Anaesthetic gases such as nitrous oxide or acetylene inhibit specifically the synthesis of the bulk antenna chlorophyll in green bacteria (1). In this work we have used acetylene to probe the structure of the antenna in Chlorobium vibrioforme. The pronounced ability of the green sulfur bacteria to grow at very low light intensities is due largely to the efficiency of this antenna, which is contained in the chlorosomes (Fig. 1).

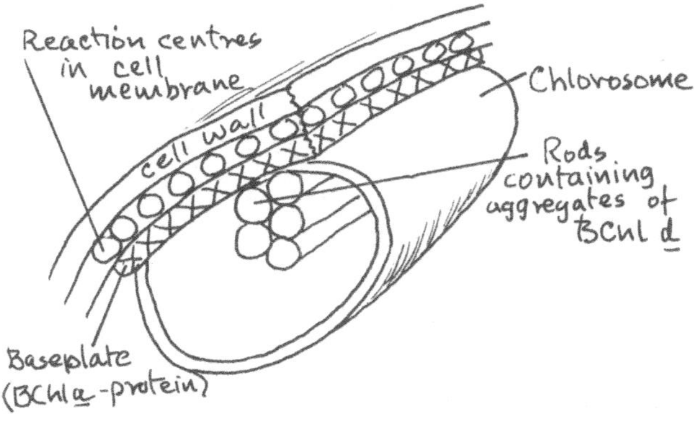

Fig. 1. Sketch showing structural details of chlorosome in cell.

Chlorosomes have a more complicated structure than the membrane antennae found in other photosynthetic organisms. Chlorosomes contain rod shaped structures, believed to consist of aggregates of bacteriochlorophyll (BChl) d (2). These aggregates are presumably stabilised by one or more of the chlorosome polypeptides, of which there are nine, ranging in size from 6 to 27 kD (3). Light regulates the size and BChl d content of the chlorosomes

P. Mathis (ed.), Photosynthesis: from Light to Biosphere, Vol. I, 163–166.

(4). This regulation is accompanied by changes in the chemical structure of the BChl d molecules synthesized: at low light ntensity the BChl d is more hydrophobic than at high (5,6). The hydrophobicity depends on the length of the hydrocarbon substitutions on the Chl d molecule (7).The substituent on C8 can be ethyl, propyl, isobutyl or neopentyl, that on C12, methyl or ethyl and that on C20, H or methyl. The significance of the differences in hydrophobicity is not at present understood.

2. Methods

The organism used was *Chlorobium vibrioforme* strain 8327. Growth rates were measured in tube cultures containing, where indicated, 1 ul ethrel /ml (1). Acetylene treatment was carried out as described in (1) and HPLC of extracts of the pelleted cells according to (8).

3. Results and Discussion

Cells grown in the presence of 2-3 mM acetylene have less than 5 % of the BChl d content of control cells (Fig. 2) while the synthesis of BChl a (mostly 810 nm) is unaffected.

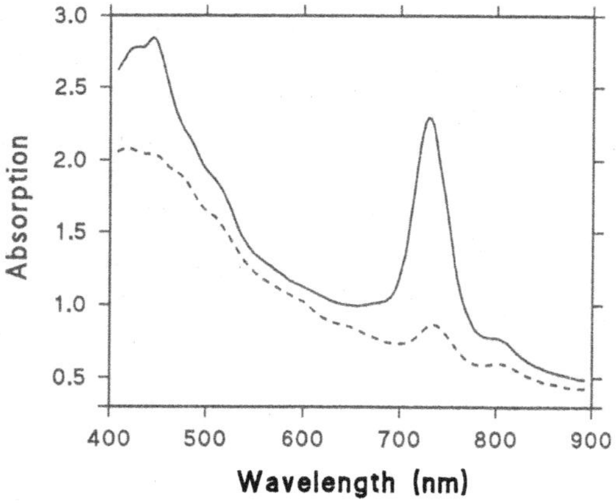

Fig. 2. Absorption spectra of cultures grown without (full line) and with (dashed line) acetylene.

Analysis by HPLC shows that growth in the presence of acetylene causes a pronounced increase in hydrophobicity of the BChl d compared with the control cells (Fig 3).

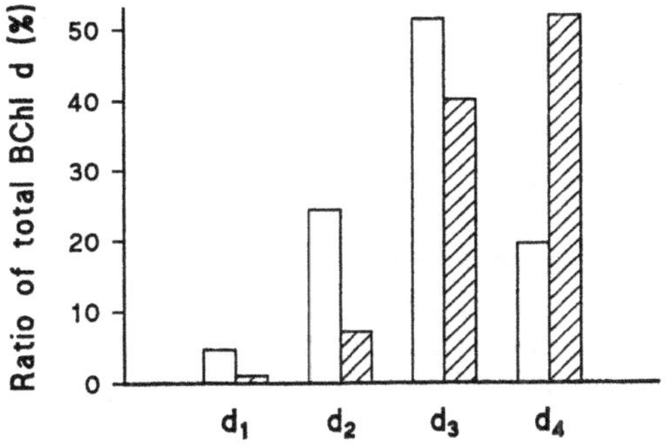

Fig. 3. Histogram showing homologue composition of bacteriochlorophyll d. Open columns, control culture; shaded columns, culture grown with acetylene. d1 - d4, indicates homologues of increasing hydrophobicity.

In the absence of acetylene, such a degree of hydrophobicity would only be achieved by slow growth at very low light intensity (6). In these experiments, the light intensity for growth of the cells was between 2 and 3 klux- this is above the saturation value.

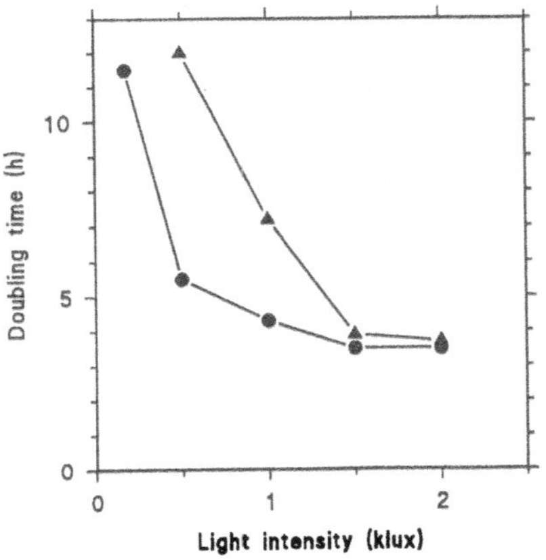

Fig. 4. Cell doubling time (h) as a function of light intensity in the absence (●) and presence (▲) of ethrel.

The effect of light intensity on growth rate is shown in Fig. 4. For practical reasons, ethylene (ethrel) was used in this experiment in amounts giving a similar effect to that of the acetylene used in the earlier experiments. Surprisingly, light saturation is achieved at about the same intensity for both control and treated cultures, in spite of the large difference in antenna size. The results show that increase in hydrophobicity occurs during growth in the presence of acetylene. Earlier work (5, 6) has shown that growth at low light intensity has the same effect. The question is then, are these two factors- low light intensity and acetylene, influencing hydrophobicity by the same mechanism? This seems unlikely, considering that the cells were grown at light intensities above the saturation value, and at this intensity, the acetylene culture grew just as fast as the control. Presumably, neither of the cultures was energy limited at this intensity. Another way of looking at the problem is from the point of view of the way the chlorosome antenna is built up. There is evidence for inhomogeneity of the BChl within the chlorosome (9), with the long-wavelength absorbing molecules closest to the baseplate. In agreement with this we observed a red shift of a few nm in the BChl d spectrum and fluorescence maximum in acetylene grown cells. In the situation where BChl d synthesis is inhibited by acetylene, the small amount available would form a thin layer next to the baseplate. There may be a structural requirement for hydrophobicity associated with the formation of this layer.

The high growth rate of the acetylene cells may be due to a very efficient energy transfer mechanism from the chlorosome, and fluorescence measurements were carried out to test this idea. The results were inconclusive.

References

(1) Ormerod, J.G. Nesbakken,T. and Beale, S.I. (1990) J. Bacteriol. 12, 1352-1360

(2) Holzwarth, A. F., Gribenow, K. and Schaffner, K. (1990) Z. Naturforsch. 45c, 203-206.

(3) Chung, S., Frank, G., Zuber, H. and Byant, D. A. (1994) Photosyn. Res. 41, 261-275.

(4) Broch-Due, M., Ormerod, J. G. and Strand Fjerdingen, B. (1978) Arch. Microbiol. 116, 269-274.

(5) Broch-Due, M and Ormerod, J. G (1978) FEMS Microbiol. Letts. 3, 305-308.

(6) Huster, M. S. and Smith, K. M. (1990) Biochemistry 29, 4348-4355.

(7) Smith, K.M. (1994) Photosyn. Res. 41, 23-26.

(8) Frigaard, N-U., Larsen, K. L. and Cox, R. P (1995) Submitted to FEMS Micobiol. Ecol.

(9) Otte, S. C. M., van der Heiden, J. C., Pfennig, N. and Amesz, J. (1991) Photosyn. Res. 28, 77-87

THE SIZE OF THE CORE ANTENNA IN PURPLE BACTERIA

Christof Francke and Jan Amesz,
Department of Biophysics, Huygens Laboratory, University of Leiden,
P.O. Box 9504, 2300 RA Leiden, The Netherlands

1. Introduction

Purple bacteria have a relatively simple antenna system consisting of a core antenna protein complex (LHI) and, in many species, one or more peripheral antenna complexes (LHII). For understanding the mechanism of photosynthesis in purple bacteria and for constructing structural and functional models of the antenna, it is important to know the size of the antenna in relation to the number of reaction centres. This applies in particular to the core antenna. Although it is generally assumed that its size is constant, independent of species and culture conditions, the actual numbers found in the literature are rather variable and range from about 25 to 40 BChls/ RC [1-5]

We decided to reinvestigate this problem by determining the BChl/BPhe ratio in a number of BChl *a* and BChl *b* containing species that have no peripheral antenna. Since only the reaction centre contains BPhe, the BPhe content is a measure for the number of reaction centres. The results support the model according to which the core antenna consists of 12 so-called α-β heterodimers per reaction centre [6]. The in vivo extinction coefficients of BChl *a* and *b* were also determined.

2. Materials and Methods

Various purple bacteria (listed in Table 1) were grown at different light intensities. The bacteria were selected on basis of literature data which indicated the presence of only a core antenna in these species and this was confirmed by the absorption spectra. The cells were harvested and the pigments were extracted quantitatively, using a 50-100 fold excess of an ice-cold 1:1 (v:v) mixture of methanol and acetone. The suspension was subjected to a mild sonication for 1 min at 4 °C. The extracts were filtered over a 0.45 µm teflon membrane filter, dried at reduced pressure with a rotary evaporator and subsequently dissolved in eluent.

Pigments were separated at 4°C on a normal phase silica HPLC column with a mixture of n-hexane, 2-propanol, methanol (100/2/2 v/v/v) as eluent. Pigment elution was monitored by means of a diode array detector.

3. Results

Representative chromatograms of pigment extracts are given in Fig. 1. They show two major peaks, due to BChl *a* and to BPhe *a*, as was confirmed by the corresponding

P. Mathis (ed.), Photosynthesis: from Light to Biosphere, Vol. I, 167–170.

absorption spectra in eluent. BChl (but not BPhe) in *Rhodospirillum rubrum* is known to be esterified with geranyl-geraniol [7]. This was reflected in the chromatogram (Fig. 1A), which showed a peak of BChl a_{GG} together with a peak of BPhe a_P. The virtual absence of BPhe a_{GG} (less than 0.4 % of the total amount of BChl) in the elution pattern of *R. rubrum* demonstrates that essentially no conversion of BChl to Bphe occurred during the extraction procedure. This shows that the chromatograms gave a reliable picture of the BChl to BPhe ratio in the bacteria tested, and thus of the BChl/ RC stoichiometry.

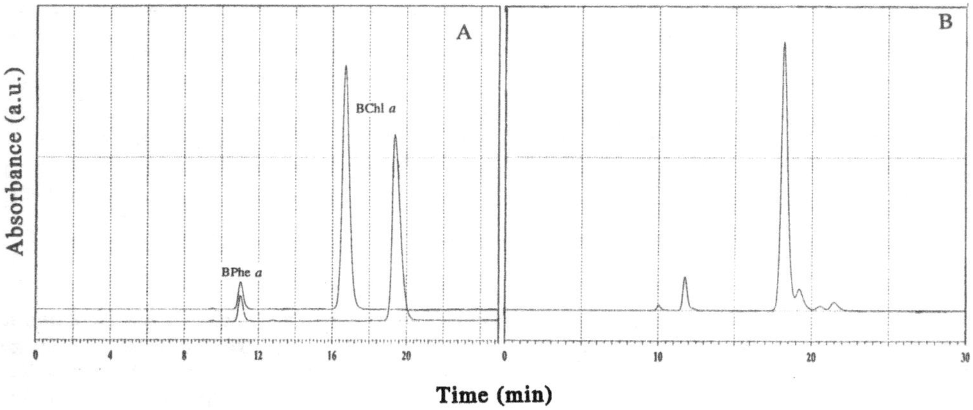

Figure. 1. Elution profiles of the pigments extracted from (A) *Rb. sphaeroides* R26 (top) and *R. rubrum* G9 (bottom) and (B) *Rps. marina* monitored at 754 nm.

The nature of the BChls found in the various bacteria is given in the fourth column of Table 1. *R. sodomense* and *Rhodopseudomonas marina* (Fig. 1B) contained small amounts of BChl a_{GG} (8-10 % and 3.3 % of the total amount of BChl, respectively), together with minor amounts of two other unidentified BChls *a* (0.5 % and 1.9 % in *R. sodomense*, 1.4 % and 4.3 % in *Rps. marina*). It is known that the last step in the biosynthesis of BChl a_P is the conversion of BChl a_{GG} by the subsequent hydrogenation of three double bonds of geranyl-geraniol [8]. Considering the elution times, it is conceivable that the unidentified BChls are esterified with the intermediate hydrogenation products, dihydro and tetrahydro geranyl-geraniol (Δ2,10,14-phytatrienol and probably Δ2,14-phytadienol) [9]. Small amounts of BChl a_{GG} have also been observed in *R. photometricum* [10] and *Chromatium okenii* [11]. BChl a_{GG} was also present in the other species, albeit in amounts smaller than 1%.

A small amount of pigment eluting somewhat faster than BPhe a_P and b_P, with an identical absorption spectrum, was found in all bacteria studied. The amount of this pigment was usually 2 - 10 % of the total amount of BPhe. No other peaks were observed in the chromatograms.

From the ratio of the peak areas of BChl and BPhe in the chromatograms and the relative extinction coefficients the ratio in which these pigments are present in the bacteria, and thus the BChl/ RC stoichiometries, could be calculated. The relative extinction coefficients of BChl and BPhe in eluent were determined by extraction of the corresponding reaction centres. The stoichiometries varied between 27 and 31. They did not vary with the light intensity, as can be seen in Table 1.

We also determined the in vivo extinction coefficients of BChl at the Q_Y maxima.

These were determined by dissolving the dried extracts in diethylether and comparing the absorbance of the extract with the absorbance of the sample before pigment extraction. Extinction coefficients of 91 mM^{-1} cm^{-1} for BChl *a* [12] and of 106 mM^{-1} cm^{-1} for BChl *b* [13] in diethyl-ether were used for the calculation. A (small) correction was made for the presence of BPhe.

Table 1. Pigment-reaction centre stoichiometries and in vivo extinction coefficients for various species of purple bacteria. The second column lists the light intensities during growth. Column four gives the number of BChls per reaction centre. The last column gives the in vivo extinction coefficients at the Q_y maximum of the antenna absorption (in parentheses), with a standard error of about 10%.

Species	Light (klux)	Main type of BChl	BChls /RC	in vivo ε ($mM^{-1}cm^{-1}$)
Rb. sphaeroides R26	0.8	a_P	31.5 ± 3.2	130 (866 nm)
Rps. marina	0.8	a_P	29.3 ± 2.9	147 (883 nm)
	3.5		28.9 ± 2.9	
Rps. viridis	0.8	b_P	27.9 ± 3.6	130 (1019 nm)
	3.5		26.9 ± 2.8	
Rps. sulfoviridis	0.8	b_P	26.7 ± 4.0	166 (1014 nm)
R. rubrum G9	0.8	a_{GG}	31.2 ± 4.5	142 (874 nm)
R. rubrum S1	0.8	a_{GG}	29.9 ± 3.0	140 (880 nm)
	3.5		29.4 ± 3.0	
R. sodomense	4	a_P	28.6 ± 4.0	140 (875 nm)
T. pfennigii	0.8	b_P	29.0 ± 3.0	148 (1024 nm)

4. Discussion

Our results clearly support the notion that the size of the core antenna is the same in all species of purple bacteria; the results obtained with 8 species and strains all fall in the range of 27 - 31 BChls/ RC, corresponding to a size of the core antenna of 23 - 27 BChls. These numbers are in excellent agreement with those obtained by means of protein analysis of *R. rubrum* S1 and G9 and *Rps. marina* [4,5] and by means of annihilation measurements with *Rps. viridis* [3]. Our experiments had to be confined to species and strains that lack a peripheral antenna, but data obtained by Ueda et al. [5] with wild-type *Rhodobacter sphaeroides* and *Chromatium vinosum* indicate the same stoichiometry. The same applies to the results obtained by photochemical bleaching of isolated core-reaction centre complexes [2], although the numbers are more variable than ours.

These results support a model according to which 24 core antenna proteins are arranged with hexagonal symmetry in 12 so-called α-β heterodimers around the reaction centre [6]. This model, which is based on electron microscopy and electron diffraction of membranes of *Rps. viridis* and other BChl *b* containing purple bacteria [14], membranes of *Rps. marina* [15] and isolated core-reaction centre complexes of *R. molischianum* [16], thus may apply to all purple bacteria. A stoichiometry of 16 α-β units per reaction centre was recently proposed by Karrash et al. [17]. This number, however, is based on electron microscopy of two-dimensional crystals of isolated LHI complexes (without reaction centre), and thus may not be representative for the arrangement in the membrane.

The in vivo extinction coefficients determined by us for BChl *a* containing species are in good agreement with those reported earlier for *R. rubrum* and *Rb. sphaeroides* R26 [18]. The coefficient for *Rps. viridis* is lower than found by Trissl et al. [19]; however, using their extinction coefficient instead of the one reported by Oelze [13] the number obtained (159 mM^{-1} cm^{-1}) was comparable.

Acknowledgements

The authors wish to thank A.H.M. de Wit for culturing the bacteria and Ms. S.J. Jansen for preparing the reaction centres. This work was supported by the Netherlands Organisation for Scientific Research (NWO) via the Netherlands Foundation for Chemical Research (SON) and by the European Community (contract no. SCI* CT92-0796). Thanks are due to Dr M.T. Madigan for a culture of *R. sodomense*.

References

1 Aagaard, J. and Sistrom, W.R. (1972) Photochem. Photobiol. 15, 209-225
2 Dawkins, D.J., Ferguson, L.A. and Cogdell, R. (1988) in Photosynthetic Light Harvesting Systems (Scheer, H. and Schneider, S., eds.) pp. 115-127, Walter de Gruyter & Co., Berlin New York
3 Deinum, G. (1991) Doctoral thesis, University of Leiden, The Netherlands
4 Meckenstock, R.U., Brunisholz, R.A. and Zuber, H. (1992) FEBS Lett. 311, 127-135
5 Ueda, T., Morimoto, Y., Sato, M., Kakuno, T., Yamashita, J. and Horio, T. (1985) J. Biochem. Tokyo 98, 1487-1498
6 Zuber, H. and Brunisholz, B.A. (1991) in Chlorophylls (Scheer, H., ed.) pp. 627-703, CRC Press, Boca Raton, FL
7 Walter, E., Schreiber, J., Zass, E. and Eschenmoser, A. (1979) Helv. Chim. Acta 62, 899-920
8 Bollivar, D.W., Wang, S., Allen, J.P. and Bauer, C.E. (1994) Biochemistry 33, 12763-12768
9 Shioi, Y., Fukae, R. and Sasa, T. (1983) Biochim. Biophys. Acta 722, 72-79
10 Künzler, A. and Pfennig, N. (1973) Arch. Mikrobiol. 91, 83-86
11 Gloe, A. and Pfennig, N. (1974) Arch. Microbiol. 96, 93-101
12 Smith, J.H.C. and Benitez, A. (1955) in Modern Methods of Plant Analysis (Peach, K. and Tracey, M.V., eds.) Vol. 4, pp. 142-196, Springer Verlag, Berlin
13 Oelze, J. (1985) Methods in Microbiology 18, 257-284
14 Engelhardt, H., Baumeister, W. and Owen-Saxton, W. (1983) Arch. Microbiol. 135, 169-175
15 Meckenstock, R.U., Krusche, K., Brunisholz, R.A. and Zuber, H. (1992) FEBS Lett. 311, 135-138
16 Boonstra, A.F., Germeroth, L. and Boekema, E.J. (1994) Biochim. Biophys. Acta 1184, 227-234
17 Karrash, S., Bullough, P.A. and Ghosh, R. (1995) EMBO J. 14, 631-638
18 Clayton, R.K. (1963) in Bacterial Photosynthesis (Gest, H., San Pietro, A. and Vernon, L.P., eds.) pp. 495-500, The Antioch Press, Yellow Springs, OH
19 Trissl, H.W., Breton, J., Deprez, J., Dobek, A. and Leibl, W. (1990) Biochim. Biophys. Acta 1015, 322-333

THE MAJOR LIGHT-HARVESTING ANTENNA OF *PROCHLOROCOCCUS MARINUS* IS SIMILAR TO CP43', A CHL BINDING PROTEIN INDUCED BY IRON LIMITATION IN CYANOBACTERIA

Julie La Roche[1], Frederic Partensky[2] & Paul Falkowski[1]
[1]Dept. of Applied Science, Brookhaven National Laboratory, Upton, NY USA 11973, [2]Station Biologique, CNRS UPR 4601 et Université Paris 6, BP 74, 29682 Roscoff cx, France

1. Introduction

Prochlorococcus marinus, a minute unicellular marine photosynthetic organism is the most recently discovered Prochlorophyte (1). Prochlorophytes, also comprised of *Prochloron didemni* and *Prochlorothrix hollandica* , differ from cyanobacteria by the absence of phycobilisomes and the presence of chl a/b protein complexes as major light harvesting systems. Phylogenetic studies (2, 3) suggested that the three known Prochlorophytes have evolved from three different cyanobacterial lineages, *Prochlorococcus* being most closely related to oceanic cyanobacteria. *P. marinus* is extremely abundant throughout the euphotic zone of most of the intertropical region of the world's ocean (4) as well as in the Mediterranean Sea (5). Striking phenotypic differences in ratios of divinyl (DV)-chla/chlb and growth irradiance optima have been observed between clones of *Prochlorococcus* (6) suggesting that photoadaptation to a specific ecological niche may be common in this species. We have begun a study of the light harvesting complexes in the Sarg- and Med-types of *P. marinus* isolated from the Sargasso Sea and surface Mediterranean waters, respectively. Here we show that the light-harvesting chl a/b binding protein of *Prochlorococcus* is homologous to CP43', the product of the iron-regulated *isiA* gene in cyanobacteria.

2. Material and methods

P. marinus strain Sarg and *Prochlorococcus* sp. clone ccmp1378 (Med-type) were obtained from Dr. S.W. Chisholm and from the CCMP culture collection (Bigelow lab for ocean science, West Boothbay Harbor, ME), respectively. Culture conditions were as described (7). PSII reaction center densities were determined from single turnover oxygen flash yields (8). PSI reaction centers were measured by absorption difference spectroscopy in isolated thylakoid membranes in the presence of excess

P. Mathis (ed.), Photosynthesis: from Light to Biosphere, Vol. I, 171–174.

ascorbate and methyl viologen (9). Whole cell absorption spectra were obtained with an opal glass technique (10). PSI action spectra were obtained *in vivo* in the presence of 10 μM DCMU. For this measurement, the oxidized *minus* reduced absorption difference spectra were continuously measured at 697 nm with a 720 nm reference beam in a dual wavelength mode; the change in the amplitude of the absorption difference was recorded from 400 to 600 nm using a Jablon grating monochrometer and a Oriel high pressure 500 W xenon source. The action spectra for PSII was inferred from the 77 K fluorescence excitation keyed to the 685 nm emission band. Thylakoids from *Prochlorococcus* cells were prepared as described (11), solubilized in octyl-glucoside (OG) and loaded on a 3-40% linear sucrose gradient containing 3 mM OG, then centrifuged at 200K X g for 36 hours. Three bands were harvested through the tube wall using a 22 gauge needle. The protein samples were precipitated, solubilized and separated by SDS-PAGE. Proteins were transferred to immobilon for microsequencing. The *isiA*-like gene was initially amplified from Med-type DNA by PCR using degenerate primers designed from the N-terminal sequence of the 35 kDa protein from the chl b enriched fraction and from a region conserved in both *psbC* and *isiA* genes. The product was used as a probe to screen a plasmid library enriched in a 3.5 kb Hind III fragment from Med genomic DNA and the positive clone was sequenced.

3. Results and discussion

The light harvesting chl binding proteins of Prochlorophytes are approximately 35 kDa in size and contain Chl b, zeaxanthin and carotene as accessory pigments. Among the three known Prochlorophytes, the photosynthetic apparatus of *Prochlorothrix* has been most extensively studied leading to some debate as to whether or not the light harvesting antenna of this organism can transfer energy to both PSI and PSII or is primarily associated with PSII (12,13). In contrast little is known about the PSI and PSII of *Prochlorococcus*. We found that the ratio of PSII:PSI averaged ca 1.4+/- 0.2 in the Sarg strain grown at low light (8 $\mu E\ m^{-1}s^{-1}$). This ratio is in contrast to that of cyanobacteria, where PSII:PSI ratios are typically less than 1 (14). The action spectra for PSI show clear sensitization in wavebands corresponding the absorption by DV-Chl b as well as those corresponding to DV-Chl a and the PSI action spectra is similar to that of PSII (Fig 1). These results strongly suggest that the main LHC serves both photosystems with comparable efficiency.

N-terminal sequencing of an abundant 35 kDa protein from the DV-chl b enriched fraction indicated a strong similarity with the *isiA* gene product, a cyanobacterial gene which is itself homologous to the *psbC* gene encoding CP43. The nucleotide sequence of a 3 kb fragment containing the *isiA* gene of the Med strain as well as southern blots (Results not shown) suggest that only one gene is present. Similar preliminary results have been obtained for the Sarg strain. The *isiA* gene of *Prochlorococcus* has approximately 55% identity with two other cyanobacterial *isiA* genes and alignments with other *isiA* and *psbC* genes (Fig 2) suggest that the *Prochlorococcus* protein has retained the six postulated transmembrane spans of CP43

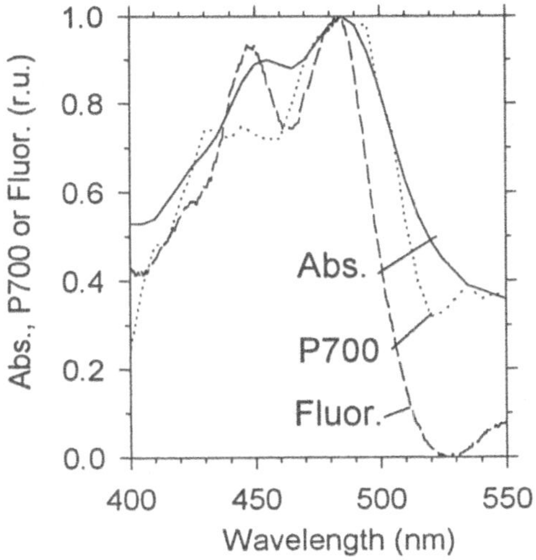

Fig. 1. Normalized action spectrum of P700 (P700) and corresponding absorption spectrum (Abs.) of Prochlorococcus cells grown at low light. A 77K fluorescence excitation spectrum of thylakoid membranes (Fluor.) is superimposed.

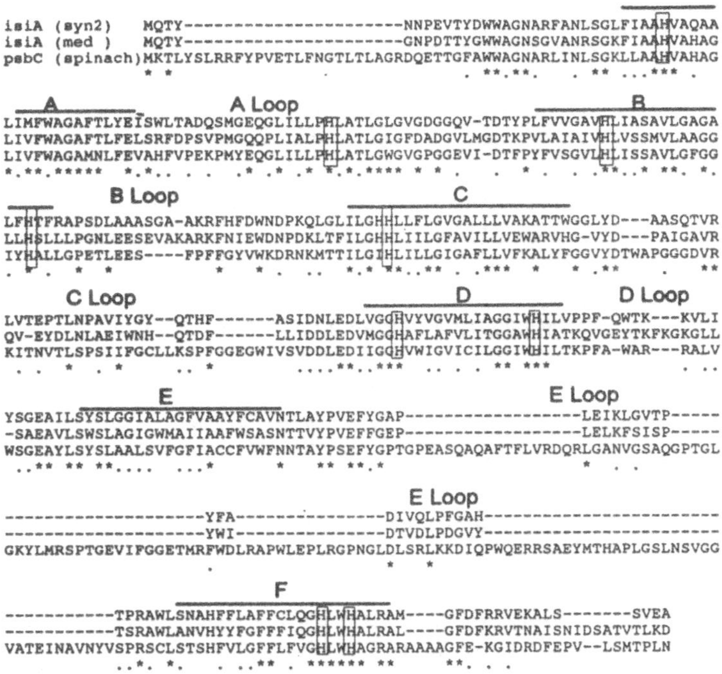

Fig. 2. Amino acid alignment of isiA genes products (Synechococcus strain R2, PIR accession number A30189, and this study) with CP43 (spinach PIR accession number F2SP44). A-F represent the postulated transmembrane spans, based on hydropathy plots (16).

but lacks a large portion of the E-loop of CP43 (15). These observations suggest that the light harvesting protein of *Prochlorococcus* as well as CP43' are likely to have evolved from a deletion of the E-loop of CP43, and that the differences in the E-loop may play a role in the differentiation of the function between CP43 and CP43'. In cyanobacteria the *isiA* gene is part of the *isiAB* operon where the *isiB* gene codes for flavodoxin and both genes are heavily expressed under iron-limitation. Cyanobacteria drastically reduce the abundance of PSI, CP43 and phycobilisomes when under iron stress. The role of CP43' has not yet been established but it has been suggested that it may replace CP43 under iron-limited growth conditions. This hypothesis was not supported by experiments with cyanobacterial *psbC* mutants deficient in oxygen evolution and others have suggested that CP43' could have a function in light harvesting in iron-limited conditions (16). Our study demonstrates that in *Prochlorococcus*, the major light harvesting protein is encoded by a constitutively expressed *isiA*-like gene. Our findings lend support for a light harvesting role for CP43' in cyanobacteria as well (17). Ecological and biogeographical studies of *Prochlorococcus* have produced a clear picture of the ecological niche of this species. It is dominant in oligotrophic oceanic waters that contain very little dissolved iron and in which cyanobacteria do not generally thrive (18). We propose that evolutionary pressures in an oceanic environment in which iron progressively became scarce as a result of oxygenic photosynthesis, led to the difference in light-harvesting systems and to a separation of niches between cyanobacteria and *Prochlorococcus*.

4. References

1. Chisholm S.W. et al. (1988) Nature (Lond.) 334:340-343.
2. Urbach E., Robertson D, Chisholm SW (1992) Nature (Lond.) 355:267-269.
3. Palenik BP, Haselkorn (1992) Nature (Lond.) 355:265-267.
4. Olson RJ et al (1990) Deep Sea Res. 37:1033-1051.
5. Vaulot D et al. (1990) Limnol Oceanogr 35:1156-1164.
6. Partensky F et al. (1993) Plant Physiol. 101:285-296.
8. Chisholm et al. (1992) Arch. Microbiol. 157:297-300.
9. Sukenik A et al (1987) Nature (Lond.) 327:704-707.
10. Falkowski PJ, Owens TG (1980) Plant Physiol. 66: 592-595.
11. Dubinsky Z., Falkowski PG, Wyman K (1986) Plant Cell Physiol. 27:1335-1349.
12. Post et al (1992) Biochim Biophys Acta 1100:75-82.
13. Bullerjahn GS, Post AF (1993) Critical Rev Microbiol 19:43-59.
14. van der Staay GWM, Staehelin LA (1994) J Biol Chem 269:24834-24844.
15. Fujita Y et al. (1994) In The molecular Biology of Cyanobacteria (Bryant D.A. ed.) pp. 677-692, Kluwer Academic Publishers, Dordrecht, The Netherlands
16. Straus NA (1994) In The molecular Biology of Cyanobacteria (Bryant D.A. ed.) pp. 731-750, Kluwer Academic Publishers, Dordrecht, The Netherlands
17. Burnap RL, Troyan T, Sherman LA (1993) Plant Physiol 103:893-902.
19. Campbell L, Vaulot D (1993) Deep-Sea Res. 40: 2043-2060.

The Chl *a/b* Antenna from Prochlorophytes is Related to the Iron Stress -induced Chl *a* Antenna (IsiA) from Cyanobacteria

G.W.M. van der Staay[1], A. Ducret[2], R. Aebersold[2], R. Li[3], S.S. Golden[3], R.G. Hiller[4], P.M. Wrench[4], A.W.D. Larkum[5] and B.R. Green[1]

[1]Botany Dept. , Univ. of British Columbia; Vancouver, B.C., Canada; [2]Molecular Biotechnology Dept., Univ. of Washington, Seattle, WA, USA; [3]Biology Dept. Texas A&M Univ., College Station, TX, USA; [4] Biological Sciences, Macquarie Univ., North Ryde, Australia; [5] Biological Sciences, Univ. of Sydney, Sydney, Australia

Introduction

The presence of Chl *b* in the prochlorophytes has led to speculations about the relationship between prochlorophytes and the Chl *b* containing chloroplasts of green algae and higher plants (1,2). However, morphological characteristics not related to the photosynthetic apparatus and phylogenetic analyses placed the prochlorophytes with the cyanobacteria rather than with the chloroplasts (1-3). The only possible link left between prochlorophytes and chloroplasts is the Chl *a/b* antenna. In *Prochloron* and *Prochlorothrix*, this antenna is composed of polypeptides with apparent molecular masses above 30 kDa (1,2), in contrast to the Chl *a/b* (CAB) antenna polypeptides of chloroplasts which have molecular masses below 30 kDa (4). The *Prochlorothrix* antenna has four polypeptides of 32 to 38 kDa (5) with a Chl *a/b* ratio of 4 (2). In *Prochloron* and *Prochlorococcus*, the antennae contain chemically different Chl molecules: the *Prochloron* antenna binds a Chl c-like pigment as well as Chl *a* and Chl *b* in a ratio of about 2 (6); the *Prochlorococcus* antenna binds divinyl Chls *a* and *b* in a 1 to 1 ratio (7).

We have cloned and sequenced genes for the Chl a/b antenna in *Prochlorothrix* and shown they match tryptic peptide sequences from both *Prochlorothrix* and *Prochloron*.

Procedures

The Chl *a/b* antenna from *Prochlorothrix* was isolated on a non-denaturing gel (5). The antenna band migrating close to the major PS I band was used since it was devoid of co-migrating PS II. The excised band was run on a denaturing Tris-Tricine gel (8) and stained with copper (9). The antenna proteins were eluted, run on a denaturing Tris-glycine gel (5) and processed for the sequencing of N-termini and tryptic fragments as described (10). All other methods were according to standard protocols.

Results and Discussion

Two antenna polypeptides of *Prochlorothrix*, with apparent molecular masses of 32 kDa (P32) and 38 kDa (P38), were purified (Fig. 1), and subjected to sequencing of the N-terminus and of tryptic fragments. Two of the five peptide sequences determined for the 32 kDa protein appeared to match the iron stress-induced Chl *a* antenna protein of cyanobacteria, CP43' (11). This protein, a product of the isiA gene

P. Mathis (ed.), Photosynthesis: from Light to Biosphere, Vol. I, 175–178.

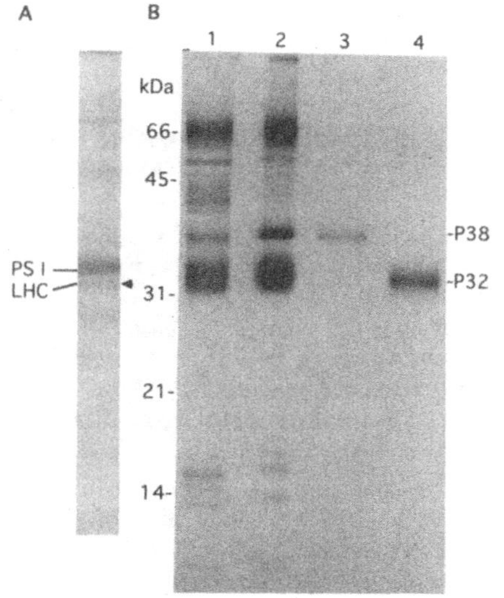

Figure 1
Isolation of the 32 and the 38 kDa antenna polypeptides
A)Separation of Chl-protein complexes by non-denaturing electrophoresis . The Chl a/b antenna (LHC) band (arrow) migrating close to PS I was selected because traces of PS II polypeptides were present in other antenna bands.
B) Denaturing gel demonstrating the purification of the proteins. Lane 1) whole thylakoids, 2) band from non-denaturing gel, 3) 38 kDa antenna protein (P38), 4) 32 kDa protein (P32).

(12), is closely related to the PS II core Chl a antenna protein, CP43. A DNA fragment of *Prochlorothrix*, which was originally isolated because it was recognized by a CP43 probe, was found to contain two related genes and part of a third gene (Figure 2).

The protein sequence of the first gene (P32) matches the sequences of the tryptic peptides from the 32 kDa protein (Fig. 2). As judged from the protein sequencing data, N-terminal processing removes the methionine. As found for many hydrophobic proteins, the calculated molecular mass (38.1 kDa) is higher than the mass deduced from electrophoretic migration.

No tryptic fragments matched the protein encoded by the second gene. Since this protein has a calculated molecular mass (37.8 kDa) comparable to P32, it can not be excluded that it co-migrated with P32 but escaped detection because it was present in a lower amount. At least two minor antenna components which migrate slightly slower than the major P32 band and are not always well separated from P32 have been described (5). We propose the name P33 for this polypeptide.

The third protein shows a match with one sequence determined for P38. Therefore, we assume that this gene encodes P38. The N-terminus of the mature protein was blocked.

As shown in Figure 2, all three genes show a high sequence homology to CP43' from cyanobacteria. No homology to any sequences of members of the Lhc gene family (4) was found. The sequences of two tryptic peptides from the *Prochloron* antenna protein are also homologous to CP43' (Fig. 2). Furthermore, an antenna protein of *Prochlorococcus* is closely related to CP43' too (La Roche et al., these Proceedings). This shows that the Chl a/b antenna of prochlorophytes and CP43' from cyanobacteria had a common ancestor. The results support the idea that prochlorophytes are not a unified group (3). Since an antenna with

Figure 2: Alignment of protein sequences.
The protein sequence of the iron-stress induced protein from Synechocystis PCC 6803 (13) and the protein sequences of the three antenna proteins from Prochlorothrix derived from the DNA sequence are shown. Gaps were inserted to produce maximal alignment. Identical amino acids are shaded. The sequences of the tryptic fragments of the 32 and the 38 kDa protein and a 32 kDa antenna protein from Prochloron are shown in boxes.

```
PCC6803      MQTYGNDTVQYE-------------WWAGNARFADQSGLFIAAHVAQAALTAFWAGAFTLF
P32          MATTATPEYG---------------WWAGNSRFALQSGKWLSAHIAQYALITFWAGGITLF
P33          MQTYGSRNVDYG-------------WWAGNSRFADRTGLWLAAHVAQASFIVLWAGAITLF
P38          MEECSCDNRFRRGNNEPAGFSLDEQWWAGNIRLVDLSGQLLGAHIAHAGLIAFWAGSITVL
P32 protein  ATTATPEYG
Prochloron   MQAYGNPDEYGGD-------------AGNSRLA

PCC6803      EISRFDPTQAMGDQGLILLPHLATLGWGVGDGGQIVDTYPYFVIGRIHLIASAVLGAGALF
P32          ELARYNPDVSMGEQGLILIPHLATLGWGIGSGGQVVDTYPYFVIGVIHLVASAVFGAGALY
P33          EVARYTPSIPMGEQGLILIPHLAALGIGVGAGGVVVNTFPYTAIGAVHLISSFVFAFGAIF
P38          EVARYVPDVPFYEQGLGLLPHLATLGFGIGPDGTVVDTYPYFVIGILHLVTSAVLGAGGLF

PCC6803      HTLRAPADLSTLKGQGK-KFHFTWENPQQLGIILGHHLLFLGAGALLLAGKAMYWGGLYDA
P32          HALKGPEDLSQSDFEFAKNFHFEWDDAAKLGNILGHHLLTLGYAALLFVIWLR-FHGVYDS
P33          HVRFGPESIGEGSDS---KFAFSWDDPKQLGLILGHHLLLLGLQAIFFVGWIR-FHGVYDS
P38          HTFKGPAILAEGGALAP-KFHYDWGDTKQLSLILGHHLLLLGI......
P32 protein      GPEDLXQXDFEFA NFPFEWDDAAQA
P38 protein      GPAILAEGGALAP
Prochloron                          KFDFDDGDGKKLGFITGM

PCC6803      TTQTVRLVSQPTLDPLVIYGYQTHFAS-------ISSLEDLVGGHIFVGFLLIGGGIWHILV
P32          TIGEVRVVTNPGATILSVLFEYGWFTPDHNPYFVNNLEDLASGHAYIAVVLLAGGFWHINQ
P33          TIGDVRIVSNPGATILSVIFEYGWFTPEHNPFFVNNLEDLASGHAFIGVVLLSGGVWHITQ

PCC6803      PPLGWAKKVL--LFSGEAILSYSLGGIALAGFVAAYFCAVNTLAYPPEFYGPPLAIKLGIF
P32          APFPWAQRLLASLFSPEGLLSASLAGLSMAGFAAAYFSAVNTLAYPVEFFGPPLELKFSVA
P33          EPFAWAQRLLASLFSAEGLLSSALAGLSMLGFAAAYFSAVNTLAYPIEFFGPPLELQFNIM
P32 protein                                                            FSVA

PCC6803      PYFADTVELPMHAHTSRAWLANAHFFLAFFFLQGHLWHALRALGFDFKRVNKRLIPCKPSH
P32          PYFVDTIDLPNGAHTARAWLCNVHFFLAFFVLQGHLWHALRALGFDFKRIPQALGSLSGEA
P33          PYFVDSVDLPAGVYSARAWLCNVHFYLAFFVLQGHLWHAIRAIGFDFKKVSNAFATLSN--
P32 protein  PYFVDTIDLPNGA                          ALGFDFK
```

the potential to bind Chl *b* (or maybe the inability to discriminate between different Chl molecules) is already present in cyanobacteria, only the acquisition of Chl *b* (or related Chl molecules) is required. In those organisms where the protein is being made constitutively, the phycobilisomes could have been lost without affecting light-harvesting capacity. In the cyanobacteria examined so far, only one copy of the isiA gene has been found. In *Prochlorothrix*, the isiA gene has evolved into at least three different genes. The products of this gene appear to be combined in one antenna complex (5). So far, no evidence for a 'specialization' as reported for the members of the CAB-gene family of plants has been found. However, the isiA genes are part of a gene family in the broader sense since the isiA gene is closely related to CP43, a Chl *a*-binding antenna complex. This means that the "CP43 gene family" has members that bind a variety of different chlorophyll species in varying ratios in different organisms. The only connection with the Chl *a/b* and Chl *a/c* antenna family of eukaryotes is that the latter family also have some members that are stress-induced, the ELIPS (14,15).

References

1) Post, A.F & Bullerjahn, G.S. (1994) FEMS Microbiol. Rev.13, 393-414
2) Matthijs, H.C.P., van der Staay, G.W.M. & Mur, L.R. (1994) in The Molecular Biology of Cyanobacteria (Bryant, D.A., ed.) pp. 49-64, Kluwer Academic Publishers, Dordrecht, The Netherlands
3) Bryant, D.A. (1992) Current Biology 2, 240-242
4) Jansson S (1994) Biochim. Biophys. Acta 1184: 1-19
5) van der Staay, G.W.M & Staehelin, L.A. (1994) J. Biol. Chem. 269, 24834-24844
6) Larkum, A.W.D., Scaramuzzi, C., Cox, G.C., Hiller, R.G., Turner, A.G. (1994) Proc. Natl. Acad. Sci. USA 91, 679-683
7) Chisholm, S.W. Frankel, S.L., Goericke, R., Olson, P.J., Palenik, B., Waterbury, J.B., West-Johnsrud, L. & Zettler, E.R. (1992) Arch. Microbiol. 157, 297-300
8) Schägger, H. & von Jagow, G. (1987) Anal. Biochem. 166, 368-379
9)Lee, C., Levin, A. & Branton, D. (1987) Anal. Biochem. 166, 308-312
10) Schwartz, E., Shen, D., Aebersold, R., McGrath, J.M., Pichersky, E. & Green, B.R. (1991) FEBS Lett. 280, 229-234
11)Burnap, R.L, Troyan, T. & Sherman, L.A. (1993) Plant Phys. 103, 893-902
12) Laudenbach, D.E. & Straus, N.A. (1988) J. Bact. 170, 5018-5026
13) Ferreira, F. & Straus, N.A. (1993) GenBank, accession. no. L26530
14) Green, B.R. & Pichersky, E. (1994) Photosynth. Res. 39, 149-162
15) Green, B.R. & Kühlbrandt, W. (1995) Photosynth. Res. 44, 139-148

A THERMAL BROADENING ANALYSIS OF THE LIGHT HARVESTING CHLOROPHYLL a/b COMPLEX II ABSORPTION SPECTRUM IN TERMS OF SUB-BANDS .

Giuseppe Zucchelli, Flavio M. Garlaschi, Laura Finzi and Robert C. Jennings
Centro CNR Biologia Cellulare e Molecolare delle Piante
Dipartimento di Biologia, Università di Milano, via Celoria 26,
20133 Milano, Italy

1. Introduction

The major light harvesting antenna complex of higher plants is the chlorophyll a/b protein complex II (LHCII). It binds around 50% of the total chl and is mainly associated with Photosystem II. Crystallography indicates a trimeric basic unit with ≈12 chls, in each monomer subunits, at a nearest neighbour distance of ≈9-14Å (1). The absorption (A) spectrum of LHCII is characterised by considerable spectral congestion with at least 5-7 separate transitions suggested in the literature between 660nm and 684nm plus contributions due to chl*b* around 650nm (2,3,4). The presence of a low frequency phonon mode, as suggested by hole-burning (4), responsable for the homogeneous band broadening, and the inhomogeneous band broadening, due to statistical fluctuations and sample inhomogeneity, implies that the different transitions constituting the A spectrum are represented by bell-shaped curves, even at low temperature. An important problem is to characterise these sub-bands. To date the hole burning thechnique has yielded such information only for a single band on the low energy side of the LHCII A spectrum. At higher energy this information cannot be obtained presumably due to the rapid energy transfer inside chromophore array. To throw more light on the LHCII properties, A spectra have been analysed, in the temperature (T) range 70-300K, by numerical sub-band decomposition. This approach yields information on the reorganisation energy and the inhomogeneous broadening of each sub-bands.

2. Materials and Methods

LHCII has been prepared as previously described (2). The absorption spectra have been measured in the temperature range 70-300K using a temperature controlled nitrogen flow cryostat (MMR Tech. USA) and an OMAIII (EG&G) with an intensified diode array mounted on a spectrograph (150 groove mm^{-1}) as detector system and an halogen lamp as light source. Each spectrum is the sum of $2x10^5$ sequential acquisitions to obtain an high signal to noise ratio. Decomposition of the A spectra have been performed as previously described (2) in terms of a linear combination of "asymmetric gaussian" bands.

P. Mathis (ed.), Photosynthesis: from Light to Biosphere, Vol. I, 179–182.

3. Results and Discussion

The LHCII A spectra and the relative sub-bands are shown in fig.1 for the two extreme temperatures analysed. This description differs from initial attempts (5) by the presence of two more sub-bands. In particular, two sub-bands in the 655-660nm wavelength interval, instead of one, describes better the "structure" there observed at low T. An additional band is now present in the spectral region around 670nm. More than one band around this wavelength is suggested in the literature using different techniques as A, CD (see e.g. 3) and also by deconvolution of the 71K absorption spectrum using as "response function" a

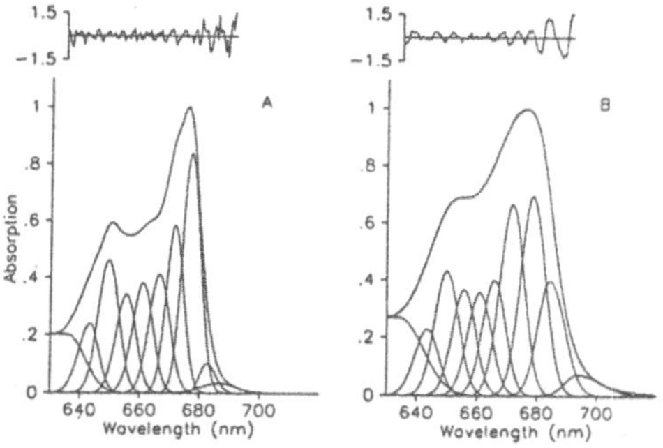

Figure 1. The LHCII absorption spectra measured at 71K (A) and 296K (B). The relative sub-band decompositions are also shown with the calculated absorption and the residuals plot.. The percentage difference between measured and calculated spectra are <0.2% along the spectra. χ^2 =0.1 for both spectra.

gaussian of 140-170 cm^{-1} FWHM (data not shown). The increased number of sub-bands increases the numerical difficulties due to the greater cross correlations between parameters. A number of fits have been performed moving around the minimum in the parameters hyperspace of each spectrum obtained by the fitting procedure so that the percentage differences between measured and calculated spectra remain below 0.2%. In this way we have estimated mean parameters used in the temperature analysis (table 2).
Analysing the FWHM of each sub-band versus T (fig.2) yields information about the optical reorganisation energy $S\nu_m$, due to electronic coupling (S) to a mean low frequency phonon mode (ν_m) (6), using $FWHM^2 = 7.7S\nu_mT + FWHM^2_{inh}$. The inhomogeneous contribution to the absorption spectra can also be obtained. This last term is interpreted as due to: (a) statistical fluctuations in the chl binding site, which leads to a statistical distribution of the absorption energy, and (b) possible sample inhomogeneity. Only the sub-bands in the 661-678nm wavelength range have been analysed as the contribution of the higher vibrational sub-bands of the red-most bands becames important for the shorter wavelength sub-bands. The values of the product of the coupling factor S and the mean phonon fequency ν_m ,obtained by linear fits (fig. 2), are shown in table 3 with the $FWHM_{inh}$. The $S\nu_m$ values are in the range 9-15cm^{-1} that indicates, using a ν_m~20cm^{-1}, as suggested by hole burning experiments (7), a coupling in the range 0.45-0.75. The inhomogeneous contribution is in the range 117-170cm^{-1} and increases towards shorter

Table 2. Decomposition sub-band parameters between 70-296K.

	71K	100K	130K	170K	210K	250K	270K	296K
λ_{max} (nm)	656.0	656.0	656.0	656.0	656.0	656.0	656.0	656.0
FWHM (cm^{-1})	182.2±0.4	186.7±0.2	194.4±0.2	197.4±0.6	201.4±0.3	208.3±0.7	216.0±4.0	211.7±2.1
area (%)	13.3	13.5	13.1	12.8±0.1	12.3	12.1±0.1	12.8±1.0	11.7±0.2
λ_{max} (nm)	661.5	661.5	661.5	661.5	661.5	661.5	661.5	661.5
FWHM (cm^{-1})	180.3±1.8	181.6±1.0	192.4±1.0	198.9±2.0	205.1±1.0	211.7±2.0	215.2±3.5	218.0±2.3
area (%)	15.0±1.0	14.1±1.0	13.1±1.0	13.0±0.2	12.5±0.1	12.3±0.2	11.1±1.0	11.95±0.2
λ_{max} (nm)	667.0	666.5	666.5	666.5	666.5	666.5	666.5	666.5
FWHM (cm^{-1})	177.1±6.0	175.8±1.8	181.6±1.5	183.6±2.8	202.9±5.4	205.3±6.2	210.1±7.1	210.8±6.2
area (%)	16.1±1.0	15.4±1.0	14.4±1.0	13.2±1.0	14.4±1.0	13.8±1.0	13.6±1.0	13.3±1.0
λ_{max} (nm)	672.0	672.0	672.0	672.0	672.0	672.0	672.0	672.0
FWHM (cm^{-1})	152.8±1.9	159.1±1.1	167.7±2.8	177.5±3.2	196.1±5.6	201.1±3.1	205.6±2.3	206.8±4.4
area (%)	20.8±1.7	21.2±.6	21.8±.7	20.6±1.0	20.1±1.7	20.1±1.9	20.4±1.4	19.8±2.5
λ_{max} (nm)	677.5	677.5	678.0	678.0	678.5	678.5	678.5	678.5
FWHM (cm^{-1})	152.3±4.9	162.5±4.9	165.4±2.8	182.1±3.5	197.5±2.8	210.9±1.0	211.9±2.9	213.9±4
area (%)	29.6±1.6	28.3±1.2	27.9±1.0	28.1±1.0	25.8±1.0	25.2±1.0	25.1±1.2	24.2±1.0
λ_{max} (nm)	683.0	683.0	683.0	683.0	683.0	684.0	684.5	684.5
FWHM (cm^{-1})	142.2±7.1	164.9±3.2	178.4±1.3	195.5±3.0	212.8±5.5	225.6±7.0	228.0±3.6	227.0±3.3
area (%)	4.2±1.4	6.1±.9	8.1±.5	10.1±.5	12.6±.6	13.4±1.0	14.4±.8	15.8±1.5

Table 3. Optical reorganisation energy and the inhomogeneous bandwidth.

λ_{max}	661.5	667	672	678
$S\nu_m$ (cm^{-1})	9.6±0.4	8.8±1.0	12.4±0.6	15.2±0.8
$FWHM_{inh}$ (cm^{-1})	163±2	155±4	126±2	117±5

wavelength bands. This may well be due to a greater contribution of the vibrational sub-bands of the red-most Q_y bands at these wavelengths. The $S\nu_m$ values obtained are in reasonable accordance with the limited hole burning data for LHCII. Similar values for $S\nu_m$ have also been obtained for the accessory pigments sub-bands of the

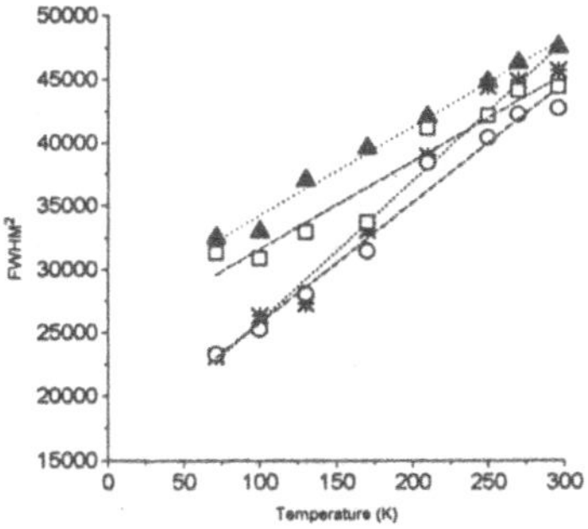

Figure 2. Temperature dependence of the FWHM^2 of chla sub-bands. ▲= 661nm (·····); □=667nm (-·-·-), O=672nm; (-·-·); ✱=678nm (- - -).

D1/D2/cytb559 complex associated to chl*b* (8). These data, with the hole burning data (7), seems to indicate similar properties for the chl*b* in an host protein matrix. These physical parameters determine a FWHM of accessory/antenna chl*a* at room temperature which is much smaller than that measured in solvents as well as smaller Stokes shifts. (≈1-2 nm) (9).

Another interesting point, already described (2), is the extreme "temperature sensitivity" of the contribution to the LHCII A of the 684nm sub-bands (table 2 and fig.3). Its contribution decreases smoothly with T in a reversible process (data not shown). A

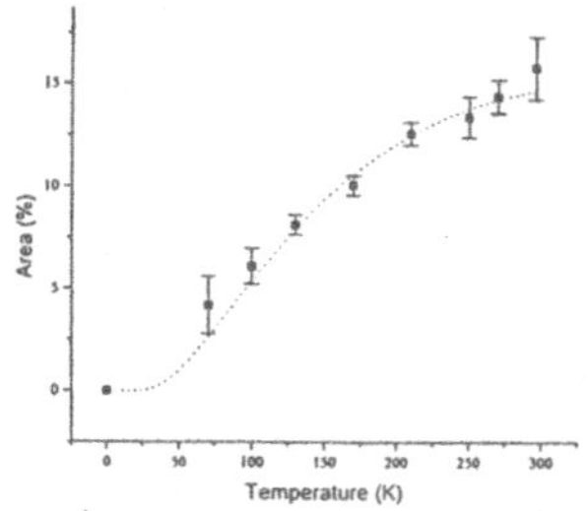

Figure 3. The percentage contribution to the Q_y absorption of the 684nm sub-band with respect to temperature. Also shown is the best fit using equation 1

suggestion to explain this behaviour is, in the context of a two level system in harmonic approximation, to hypothise the presence of a normal mode with a frequency that determines energy states populated at all the temperatures analysed and energy spaced by multiple integers of this value. The A contribution can then be calculated using the Bose-Einstein statistics. The 684nm is, in this context, the first transition at energy lower than the 0-0 transition (1-0 transition). Its contribution to the A can be described by

$$A \propto S\langle n\rangle \sum_{i=0}^{\infty} \frac{S^{2i}(1+\langle n\rangle)^i \langle n\rangle^i}{(1+i)!\,i!}; \qquad \langle n\rangle \equiv \left(e^{\frac{h\nu_m}{k_B T}} - 1\right)^{-1} \tag{1}$$

where $\langle n\rangle$ is the thermal mean occupation number. This expression has been used to fit the area data of the 684nm sub-bands versus temperature (fig.3). The parameter errors are high, however values of ν_m~120 cm^{-1} for the mode frequency and a coupling of S~0.4 is obtained. This energy value is very similar to the energy spacing between sub-bands obtained by numerical decomposition (table 2).

References

1 Kühlbrandt, W., Wang, D.N. and Fujiyoshi, Y. (1994) Nature 367, 614-621
2 Zucchelli, G., Dainese, P., Jennings, R.C., Breton, J. Garlaschi, F.M. and Bassi,R. (1994) Biochemistry 33, 8982-8990
3 Nussberger, S.N., Dekker, J.P., Kühlbrandt, W., van Bolhuis, B.M., van Grondelle, R. and van Amerongen, H. (1994) Biochemistry 33,14775-14783
4 Reddy, N.R.S., van Amerongen, H., Kwa, S.L.S., van Grondelle, R. and Small, G.J. (1994) J.Phys. Chem. 98, 4729-4735
5 Zucchelli, G., Jennings, R.C. and Garlaschi, F.M. (1990) J. Photochem. Photobiol. B: Biol. 6, 381-394
6 Hayes, J.M., Gillie, J.K., Tang, D. and Small, G.J. (1988) Biochim. Biophys. Acta 932, 287-305
7 Jankowiak, R., Hayes J.M. and Small, G.J. Chem. Rev. (1993) 93, 1471-1502
8 Cattaneo R., Finzi L., Zucchelli G., Garlaschi, F.M. and R.C. Jennings (1995) Proc. Xth international Congress of Photosynthesis
9 Zucchelli, G., Jennings, R.C. and Garlaschi F.M. (1992) Biochim. Biophys. Acta 1099, 163-169

THERMAL EQUILIBRATION OF EXCITED STATES IN ANTENNA OF PSI-200

Roberta Croce[1], Flavio Garlaschi[1], Giuseppe Zucchelli[1], Roberto Bassi[2] and Robert C. Jennings[1].
1. Centro CNR. Dip. Biologia. Univ. Milano. Via Celoria 26. ITALIA; 2.Lab. di Fotosintesi. Facoltà di Scienze MM.FF.NN. Univ. Verona. Strada Le Grazie. ITALIA.

1.Introduction

In previous studies on PSI-200 the room temperature (RT) steady state fluorescence (F) spectra are characterised by a dominant band near 680nm and a less pronounced emission at longer wavelengths, near 720nm (e.g. 1,7). As PSI displays rather strong absorption (A) at wavelengths above 700nm this emission spectrum seems to correspond to a non thermodynamically equilibrated state, as has been previously suggested (2). In this study we have undertaken a careful thermodynamic analysis of the A/F relationship in PSI-200 prepared with the mild detergent octylglucopyranoside (OG). It is demonstrated that at T>200K emission is thermally equilibrated with about 85% of the intensity associated with at least three red absorbing forms. Stokes shifts are near 5nm.

2. Materials and methods

The PSI-200 complex was prepared by solubilisation of thylakoid membranes of maize using 2% OG. Purification was by sucrose gradient ultracentrifugation in the presence of 0.8% OG. A preliminary polypeptide characterisation, by two-dimensional gel electrophoresis and immunoblotting with antibodies against the PSI core , psaC, psaD, psaE and psaH shows this preparation to be biochemically similar to others previously reported (3,4). The chl a/b ratio was 11-12 as determined in (5). A & F spectra were measured between 75 and 280K in the same instrument (EG & G OMA Mod. 1460) essentially as previous described (6). F spectra are corrected for the instrumental response.

3. Results and Discussion

Initially we investigated the effects of detergent on PSI-200 spectra. In Fig.1 it can he seen that in the absence of the detergent the major emission band is broad and has a maximum around 722nm. Near 684nm a distinct shoulder is apparent. Increasing the detergent concentration had the effect of increasing the 680nm emission. Previously

P. Mathis (ed.), Photosynthesis: from Light to Biosphere, Vol. I, 183-186

published PSI-200 spectra , using Triton X-100 solubilisation, are similar to spectra in Fig. 1 measured in the presence of detergent. This detergent induced increase in the short wavelength emission may be understood in terms of some uncoupling of the outer antenna complex LHCI-680. We therefore suggest that such commonly used detergents as Triton X-100 lead to some antenna uncoupling, which does not occur with OG.

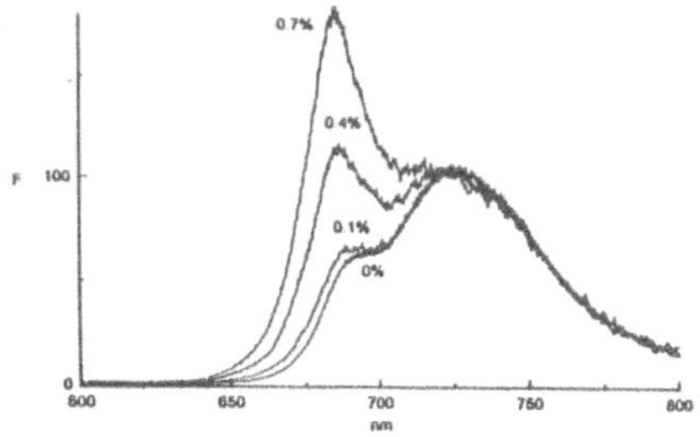

Fig.1 Fluorescence spectra of PSI-200 in 5mM Tricine pH 7.8, 50mM sorbitol at different OG concentrations.

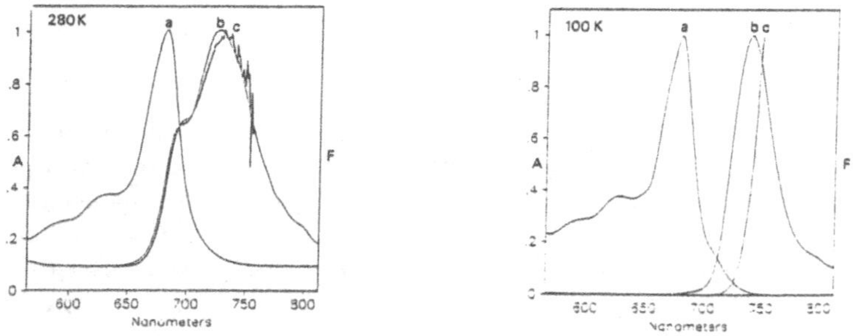

Fig.2 Comparison between calculated (c) and measured (b) F spectra for PSI-200 at 280K (A) and 100K (B). Emission spectra were calculated from the measured A spectra (a) as described in the text.

In Fig. 2 the question of thermal equilibration is addressed To this end the measured emission spectra are compared with F spectra calculated from the measured A spectra by the Stepanov equation for samples without detergent.

$$F(\nu)/A(\nu) = C\,\nu^3 \exp(-h\nu/kT) \qquad \text{equation 1}$$

This expression connects the F and A spectra for a system in which excited states are thermally equilibrated. This means that equilibration occurs within the excited singlets of each pigment as well as between the different energy levels of the spectral forms. It is evident (Fig. 2A) that the calculated and measured spectra (RT) are in close agreement, thus indicating a substantial excited state equilibration prior to F emission. The slight blue shift of the measured spectrum with respect to that calculated is probably due to preequilibration decay components (7). We estimate that the F intensity of the main, red-most emission band, is around 0.85. Thus it is concluded that in a thermally equilibrated PSI around 85% of excited states are associated with the red-most chl forms near RT. A similar analysis has been performed over the temperature range 100-280K. At

temperatures below 200K thermal equilibration is lost (Fig. 2B). This can be explained in terms of a marked temperature effect on transfer rates between antenna pigment sites and/or pools (4) while the overall trapping rate by reaction centres is little changed (8).
In Fig. 3 the emission spectra between 75-280K are presented. The expected thermodynamic red shifting on lowering the temperature is apparent with almost complete elimination of the 684nm emission. Of particular interest is the presence of three clear emission structures in the red-most spectral region, which are located at approximately 720nm, 730nm and 740nm. These structures are most apparent in the higher temperature range and are also present in samples incubated with detergent (OG 0.7%; data not shown). Previous studies have suggested the presence of two red emission bands, located at about 720nm and 735nm (e.g.9, 4).

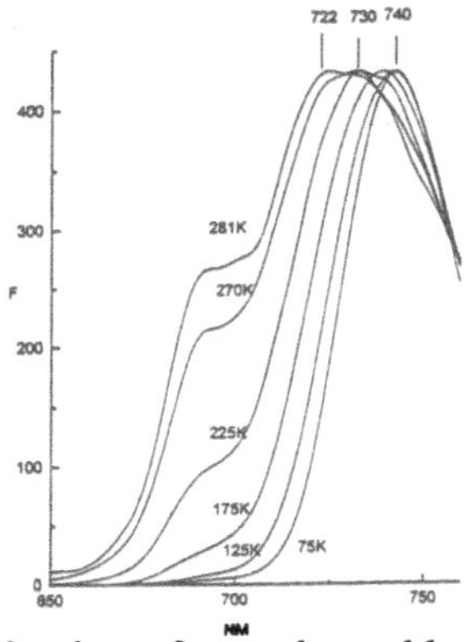

Fig. 3 Fluorescence spectra measured at different temperatures. Vertical lines indicate the approximate positions of F structures.

We have attempted to determinate the origin bands of the three above described red emission structures. To this end emission spectra measured between 75 and 280K (fig.3) were analysed by a linear combination of gaussian subbands decomposition technique (10). In this analysis (Fig.4) the spectral interval of interest is well described by three gaussian subbands with λmax values at 720nm, 730nm, 742nm. The wavelength position of these subbands is constant over the entire temperature range studied. We do not attempt to interpret the significance of the very long wavelength subband (λmax 755nm) as this, though necessary for the gaussian description, lies in a rather structureless spectroscopic region which is also subject to rather large measurement errors, and will in addition contain a substantial contribution from vibrational bands.

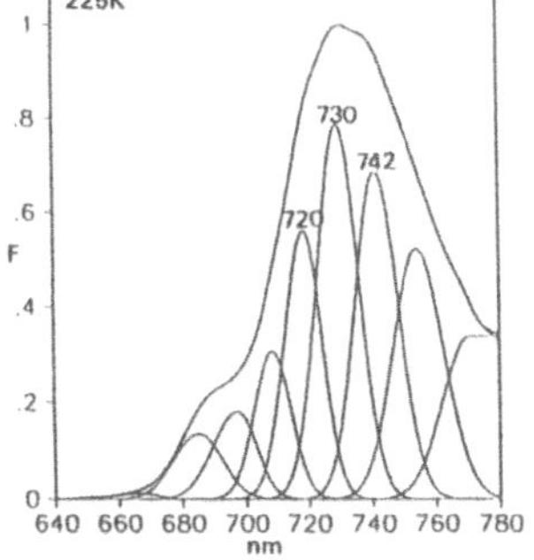

Fig.4 Gaussian subband decomposition of F spectrum measured at 225K. The sum of gaussians is superimposed on the measured spectrum.

On the assumption that the three gaussian subbands (λmax 720nm, 730nm, 742nm) represent with reasonable accuracy fluorescence origin bands we have calculated for RT spectra the associated A structures by means of the Stepanov relationship (equation 1). Results of these calculations are shown in fig 5 where it can be seen that A bands are located at 714nm, 725nm and 738nm. Thus Stokes shifts are approximately 4-6nm, values which are much greater than for non-red shifted chl forms, which are in the 1-2nm range (10,11). These thermodynamically consistent Stokes shifts, are considerably less than those often suggested for the red forms of PSI (eg. 9,12).

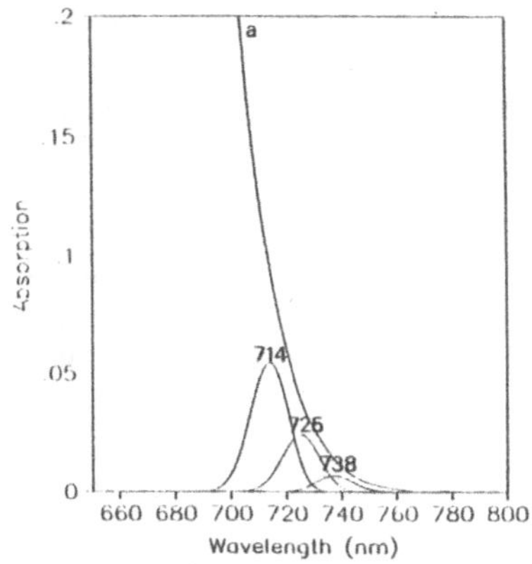

Fig. 5 Absorption bands calculated from the three red emission bands (720nm, 730nm and 742nm) using equation 1 and assuming thermal equilibration in excited states.The curve (a) is the red tail of the absorption spectrum (RT).

4. References

1. Mimuro, M., Yamazaki, I., Tamai, N., Yamazaki, T. and Fujita, Y. (1986) Progress in Photosynth Res. (Biggins, J. Ed.) 1, 91-94.
2. Trinkunas, G. and Holzwarth, A.R. (1994) Biophys. J. 66, 415-429.
3. Bassi, R. and Simpson, D. (1987) Eur. J. Biochem. 163, 221-230.
4. Palsson, L-O., Tjus, S.E., Andersson, B. and Gillbro, T. (1995) BBA, 1230, 1-9.
5. Porra, R.J., Thompson, W.A.,and Kriedemann, P.E. (1989) BBA, 975, 384-394.
6. Garlaschi, F.M., Zucchelli, G., Giavazzi, P. and Jennings, R.C. (1994) Photosynth. Res. 41, 465-473.
7. Turconi, S., Weber, N., Schweitzer, G., Strotmann, H. and Holzwarth, A.R.(1994) Biochim. Biophys. Acta 1187, 324-334.
8.Jia, Y., Jean, J.M., Werst, M.M., Chan, C.K. and Fleming, G.R. (1992) Biophys. J. 63, 259-273.
9. Wittmershaus, B.P. (1986) Progress in Photosynth Res. (Biggins J. Ed.) 1, 74-82.
10.Zucchelli, G., Jennings, R.C. and Garlaschi, F.M. (1992) Biochim. Biophys.Acta 1099, 163-169.
11. Peterman, E.J.G., Dekker, J.P., van Grondelle, R., van Amerongen, H. and Nussberger, S. (1994), Proc. Int. Workshop on Pigment-Protein Complexes. Preila (Lithuania). Lith J Physics, 34, 301-305
12 Woolf, V.M., Wittmershaus, B.P., Vermaas, W.F.J. and Tran, T.D. (1994) Photosynth. Res. 40, 21-34.

THERMAL BROADENING OF GAUSSIAN SUBBANDS OF THE REACTION CENTRE COMPLEX OF PHOTOSYSTEM II.

Cattaneo, R., Finzi, L., Zucchelli,G., Garlaschi, F., Jennings, R.C.
Centro CNR, Dipartimento di Biologia, Universita' di Milano,via Celoria, Milano, Italy.

1. Introduction.

In recent years there has been much interest in understanding the detailed spectroscopic properties of the D1/D2/cytb559 complex, which binds the reaction centre (RC) of photosystem II (PSII; eg 7,1). Apart from its intrinsic interest this is essential for describing the dynamics of energy flow in RCs of PSII. This communication analyses absorption (A) spectra of the RC complex measured between 80K-300K, in terms of a linear combination of gaussian bands, in an approach which takes into account all available, relevant, spectroscopic data.

2. Materials and Methods.

The D1/D2/cytb559 complex was prepared according to Chapman et al. (2) as described previously (3). The pigment stoichiometry was determined to be approximately 6chl/2pheophytin (11) for this preparation. A spectra were measured between 80K-300K with the previously described experimental setup (10) and analysed using the same gaussian subband decomposition programme as before (10).The appropriateness of the gaussian model to describe the homogeneously and inhomogeneously broadened zero-zero transition of chls for the temperature range used here has been previously discussed (5).

3. Results and Discussion.

The present analysis is based on the gaussian subband decomposition technique used together with the following experimental observations.

1. A pheophytin band near 680nm (8,9,3) which has an FWHM near 2K of about 4 nm and less than 12nm at room temperature (RT), as indicated by hole burning (8), ADMR (9) and A (3) spectroscopies.

P. Mathis (ed.), Photosynthesis: from Light to Biosphere, Vol. I, 187–190.

2. An A structure near 674nm at RT, as indicated by light induced bleaching A difference spectra (4).
3.P680 which absorbs near 680nm, as indicated by A (4) and hole burning spectroscopies (8).
4. The following excited state optical reorganisation energies (Sν; where S is the electron phonon coupling and ν is the average phonon frequency), as indicated by hole burning measurements:

a. For P680, Sν $\cong$40cm^{-1} (8).

b. For the 680nm pheophytin, Sν $\cong$15-20cm^{-1} (8).

c. For accessory (antenna) chl, Sν $\cong$10-20cm^{-1}(10).

5. The site inhomogeneously broadened bandwidth ($FWHM_{inh}$) for all pigment sites is around 100-150cm^{-1}, as indicated by ADMR (9) and hole burning (8) measurements of chl/protein complexes.

Thermal broadening studies on A bands yield information on Sν and $FWHM_{inh}$ (points 4 and 5 above) . In the hypothesis of linear electron phonon coupling to a single average phonon frequency (11) the half bandwidth of an homogeneously broadened A band ($FWHM_{hom}$) is given by:

$$FWHM^2{}_{hom} = 8 \cdot \ln 2 \cdot S\nu^2 \mathrm{Coth}\frac{h\nu\, c}{2kT} \quad (1)$$

This expression may be approximated by equation 2 for the temperature range 70K-300K, as ν has a value near 20cm^{-1} (2,8,11):

$$FWHM^2{}_{hom} \cong 16 \cdot \ln 2 \cdot \frac{k}{hc} \cdot S\nu\, T \quad (2)$$

In addition to homogeneous broadening, A bands are subject to inhomogeneous broadening due to statistical fluctuations in the protein matrix and also due to some particle heterogeneity. The total band width (FWHM) is therefore given by:

$$FWHM^2 = FWHM^2_{hom} + FWHM^2_{inh} \quad (3)$$

Thus analysis of $FWHM^2$ versus T yields Sν and $FWHM_{inh}$.

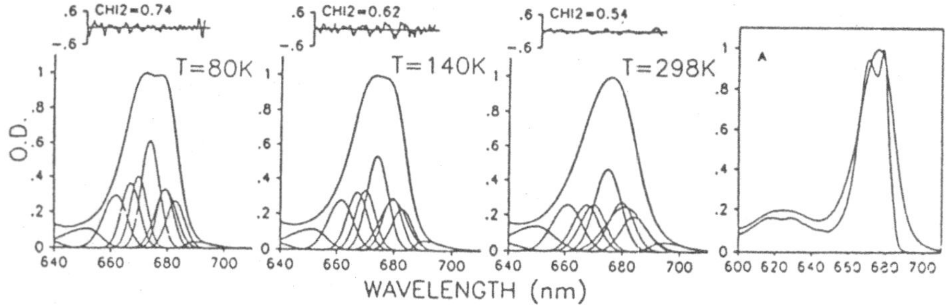

Figure 1. The sub-band decomposition description of D1/D2/cytb559 absorption spectra measured at 80K, 140K and RT.A: comparison between the absorption spectra measured at RT and 10K.

We have previously published an A subband analysis of the RC complex in terms of 5 bands with RT A_{max} values at 660nm, 669nm, 675nm, 680nm 681nm. The 680nm and 681nm subbands represented P680 and pheophytin. This description satisfied points 1-3 above, but when analysed as a function of T did not satisfy criteria 4 and 5. In order to obtain fits which satisfied all five criteria it was necessary to include two additional gaussians. This seven gaussian subband fit is shown in figure 1 for several of the temperatures analised and the subband parameters are given in Table 1. The shortest wavelength gaussian in figure 1 is considered to represent vibrational transitions and is not taken into consideration here.

In Figure 2 the thermal broadening analysis is represented with the Sν and $FWHM_{inh}$ parameters presented in Table 2.

Table 1. Gaussian parameters for the sub-band decomposition of D1/D2/cytb559 absorption spectra measured between 80-300K.

bands	80K	100K	140K	180K	220K	260K	RT
I λ_{max} (nm)	662.5	662.2	662.0	662.0	661.9	661.8	661.5
FWHM (nm)	9.7	9.9	10.1	10.3	10.6	11.0	11.9
AREA (%)	14.3	14.0	4.0	13.8	13.1	13.4	13.9
II λ_{max} (nm)	667.5	667.5	667.6	667.7	667.7	667.6	668.0
FWHM (nm)	7.5	7.9	8.3	9.0	9.3	9.8	10.4
AREA (%)	13.8	13.9	13.5	13.2	12.8	12.7	12.1
III λ_{max} (nm)	670.5	670.5	670.5	670.5	670.6	670.6	670.5
FWHM (nm)	6.9	7.3	7.9	8.6	9.2	9.5	10.2
AREA (%)	13.8	13.8	13.3	13.1	12.9	12.2	12.0
IV λ_{max} (nm)	674.6	674.6	674.6	674.6	674.7	674.8	675.3
FWHM (nm)	7.4	7.8	8.5	9.1	9.7	10.2	10.6
AREA (nm)	22.6	22.6	22.9	22.8	23.0	23.2	22.4
V λ_{max} (nm)	680.0	680.2	680.2	680.4	680.5	680.8	681.3
FWHM (nm)	8.8	9.6	10.9	12.1	13.3	14.2	15.5
AREA (%)	14.5	14.6	15.3	16.0	16.5	17.0	17.3
VI λ_{max} (nm)	679.7	679.7	679.9	680.0	680.1	680.1	680.3
FWHM (nm)	6.7	7.1	7.7	8.4	9.0	9.5	10.2
AREA (%)	11.2	11.3	11.3	11.5	11.9	11.6	12.7
VII λ_{max} (nm)	683.4	683.3	683.2	683.0	682.9	683.3	684.3
FWHM (nm)	7.5	7.9	8.4	9.0	9.6	10.3	11.0
AREA (%)	9.8	9.9	9.7	9.6	9.8	9.8	9.6

Figure 2 The square of the FWHM as a function of temperature for the sub-band description of the D1/D2/cytb559 absorption spectra. A: P680 (□), Pheo (•) and sub-band 684nm (*).B: sub-band 674nm (*), 670nm (•) and 667nm (□).

The following points of interest emerge from this description:
1. For a P680 with Sv 40cm^{-1}, and pheophytin (A_{max} 680nm) with Sv near 15cm^{-1} a major subband near 684nm is required. This band can be eliminated from the fit only by assuming what seem to be unreasonably high Sv values for P680 (Sv 80cm^{-1}).
2. The 684nm subband has an Sv value similar to that of all other accessory pigments.
3. The P680 subband has an intensity which is close to that of a chl monomer. We emphasise that in the presence of the 684nm pigment band plus the pheophytin with A_{max} near 680nm, P680 *cannot* have an intensity which significantly exceeds that of a monomer.

Table 2. Optical reorganisation energy (Sv) and $FWHM_{inh}$ for gaussian sub-bands.

	P680	Pheo	684 nm	674 nm	670 nm	667 nm
Sv (cm^{-1})	43	16	16	16	16	15
$FWHM_{inh}$ (cm^{-1})	95	104	122	130	117	146

Acknowledgements:
We wish to thank Dr. A. R. Holzwarth and Mr. M. Reus of the Max-Planck-Institut fur Strahlenchemie, Mulheim a.d. Ruhr, for kindly measuring the absorption spectrum at 10K.

References:
1.Seibert, M. (1993) in The Photosynthetic Reaction Center (Deisenhofer, J., & Norris, J., Eds.) vol.1, pp 317- Academic Press, New York.
2.Chang, H. C., Jankowiak, R., Reddy, N. R. S., Yocum, C. F., Picorel, R., Seibert, M., & Small,G. J. (1994) J. Phys. Chem. 98, 7725-7735.
3.Chapman, D. J., Gounaris, K., & Barber, J. (1988) Biochim. Biophys. Acta 933, 423-431.
4.Garlaschi, F. M., Zucchelli, G., Giavazzi, P., & Jennings, R. C. (1994) Photosynthesis Res. 41, 465-473.
5.Zucchelli, G., Dainese, P., Jennings, R. C., Breton, J., Garlaschi, F. M., & Bassi, R. (1994) Biochemistry 33, 8982-8990.
6.Zucchelli, G., Jennings, R. C., & Garlaschi, F. M. (1990) J. Photochem. Photobiol. B: Biol. 6,381-394.
7.Jennings, R. C., Garlaschi, F. M., Finzi, L., & Zucchelli, G. (1994) in Proceedings of International Workshop on Pigment Protein Complexes: Structure and Spectral Properties. Special Issue of Lith. J. Physics 34, 293-300.
8.Tang, D., Jankowiak, R., Seibert, M., Yocum, C. F., & Small, G. J. (1990) J. Phys. Chem. 94, 6519-6522.
9.van der Vos R., van Leeuwen P. J., Braun, P., & Hoff, A. J. (1992) Biochim. Biophys. Acta 1140, 184-198.
10.Reddy, N. R. S., van Amerongen, H., Kwa, S. L. S., van Grondelle, R., & Small, G. J. (1994) J. Phys. Chem. 98, 4729-4735.
11.Hayes, J. M., Gillie, J. K., Tang, D., & Small, G. T. (1988) Biochim. Biophys. Acta 932, 287-305.

PURIFICATION AND CRYSTALLIZATION OF THE LIGHT-HARVESTING COMPLEX OF MANTONIELLA SQUAMATA (PRASINOPHYCEAE)

Welte, C., Nickel, R., Wild, A.
Institute of General Botany of the Johannes Gutenberg University,
D-55099 Mainz, Germany

1.Introduction

One of the most intriguing questions about photosynthesis is still the process of the absorption of solar energy by antennae systems. A great diversity of light-harvesting systems has evolved enabling a variety of photosynthetically active organisms to populate almost every living space on our planet. Could these differences manifest themselves just in a variability of pigment composition or did totally different stategies develop to harvest the light? Unfortunately there is still not enough known about pigment-protein interactions and the spatial orientation of the pigments within the light-harvesting complex to answer this question. Better knowledge of structural details would help to elucidate the molecular mechanisms underlying the light-harvesting process. Structural information can be obtained in electron- or x-ray-diffraction studies. Both techniques, though, require crystals of the antenna complex.

We established a purification procedure for the light-harvesting complex (LHC) of Mantoniella squamata which led to successful three-dimensional crystallization. In contrast to higher plants, the green flagellate Mantoniella squamata has only one type of light-harvesting complex that delivers the radiant energy to both photosystems without any preference [1]. Three different chlorophyll species (a,b and c) exist within the antenna complex [2]. Thus, Mantoniella squamata is regarded a primitive representative of the chlorophyll a/b light-harvesting line [3].

As the three-dimensional structure of the LHC II of higher plants has already been resolved at 3.4 Å [4], a comparison of the three-dimensional structure of both complexes might be interesting and provide insights into the evolution of light-harvesting systems.

2.Procedure

A detailed description of the isolation and purification procedure of the LHC of Mantoniella squamata will be given in a forthcoming publication [5]. The cells were mechanically disrupted and the LHC solubilized by detergents. Separation from other proteins and purification was achieved by sucrose-density centrifugation, ion-exchange chromatography and gel filtration.

P. Mathis (ed.), Photosynthesis: from Light to Biosphere, Vol. I, 191–194.

3.Result and Discussion

High amounts of the LHC of Mantoniella squamata with a purity apparently sufficient for crystallization were obtained (Fig. 1).

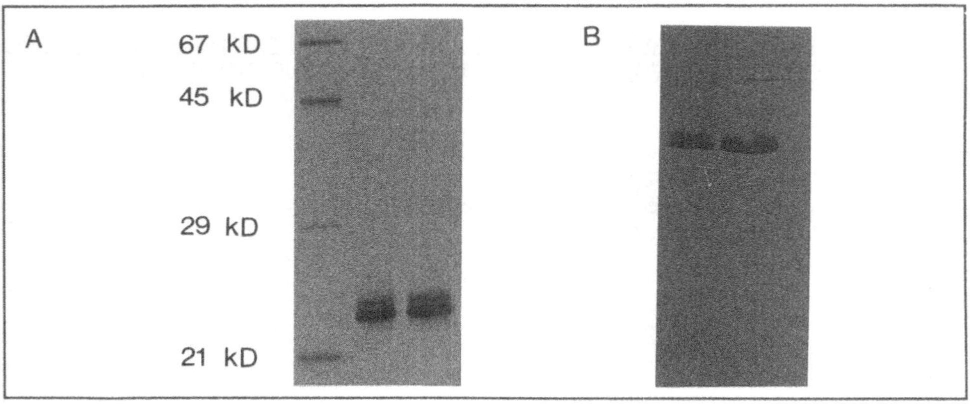

Fig. 1
Polypeptide composition of the purified LHC. **A**:Under denaturing conditions; Lane 1 contains protein standards as indicated, Lane 2 and 3 purified LHC. The LHC of Mantoniella squamata consists of two different polypeptides (23.5 and 25 kDa). **B**:Native conditions.

Purity alone is not the only prerequisite for successful crystallization. The purification procedure must produce native and homogeneous preparations. In this particular case homogeneity can not only be defined by the missing of contaminating proteins, in addition the purified pigment-protein complex should be free of lipids and exist in a single dispersion state.

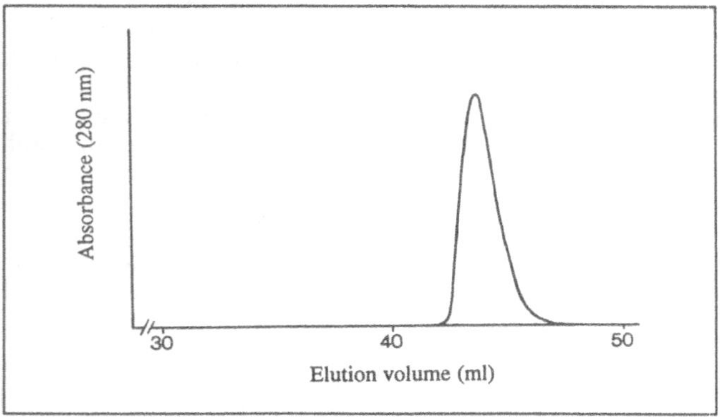

FIG. 2
Gel filtration chromatography
The LHC eluted in one peak from a gel filtration column (Sephacryl S-300, Pharmacia) in 0.5 mM dodecyl maltoside.

The chromatographical procedures removed most of the lipids present in the pigment-protein preparation. There was, however, evidence that some residual lipids remained associated with the LHC of Mantoniella squamata. It has been shown that several lipids are essential to the stability of the trimer in higher plant LHC [7] and therefore for crystallization. If we assume that the LHC of Mantoniella squamata also exists as a trimer [5], lipids might perform a similar function here. The purified pigment-protein complex exists in one oligomeric state as judged by gel filtration (Fig. 2). By calibrating the gel-filtration column, the molecular mass of the complex in detergent solution could be determined to be 170 kDa. Assuming 20-25 % of the apparent molecular weight is due to the incorporated detergent [8] the calculated molecular mass of approximately 120-135 kDa fits well the trimeric model. Further evidence supporting a trimeric organization comes from comparison of the mobility in native gel electrophoresis of the LHC of Mantoniella squamata and the LHC-trimer of higher plants. A trimeric organization for the LHC of Mantoniella squamata would also explain the results of freeze-fracture studies [9], where particles of the LHC of Mantoniella squamata are of a size equivalent to the LHC II of higher plants.

Besides gel filtration and gel electrophoresis, negative stain-electron microscopy is useful in the analysis of the aggregation state of isolated membrane proteins [6]. Our purification procedure resulted in mixed protein-detergent micelles most of them small and homogeneous [5].

It was necessary to confirm that the pigment-protein complex retained the functions of light absorbtion and energy transfer during our extensive purification procedure. Our spectroscopic measurements (Fig. 3) revealed the outstanding stability of the LHC of Mantoniella squamata, a great advantage for crystallization experiments.

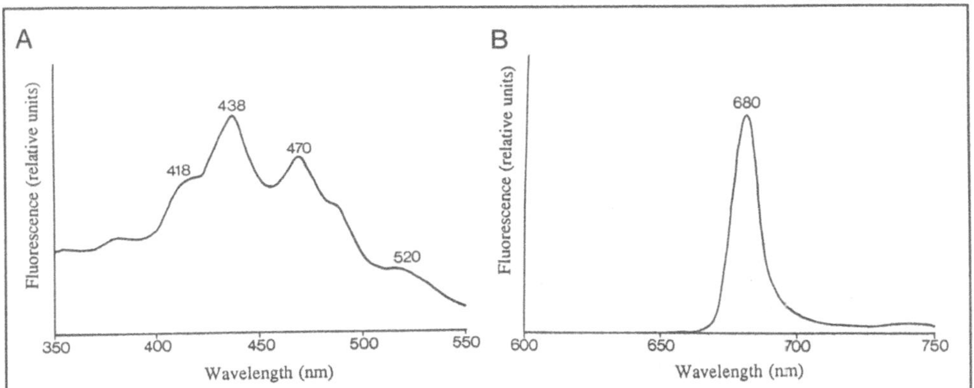

Fig. 3
Spectroscopic characterization of the purified LHC.
A:Fluorescence excitation spectra at 77 K. Emission wavelength 680 nm. **B**:Fluorescence emission spectra at 77 K. Excitation wavelength 475 nm.

The preparation of Mantoniella squamata LHC provided the ideal material to start crystallization experiments. These were accomplished by vapor diffusion using the hanging-drop technique. Crystals formed at a chlorophyll concentration of about 4 mg/ml. The starting concentration of precipitant in the crystallization solution was 2-5 % polyethylene glycole 4000 (w/v) and 300-500 mM sodium chloride.

Crystallization occured at pH 6.5 by vapor diffusion against 20 % polyethylene glycole 4000 (w/v) and 1 M sodium chloride. Besides 1 % (w/v) octyl glucoside, small amphiphilic molecules seemed to be necessary for the crystallization of the LHC. We had success with either heptane triol (Fig. 4/A) or benzamidine (Fig. 4/B). In contrast to higher plant LHC [10], the small amphipiles did not denature the complex. Crystallization was allowed to proceed in the dark at a temperature of 20°C. Within 2-3 weeks crystals appeared in the drop.

The final successful crystallization of the LHC of Mantoniella squamata seems to be the best indication for the suitability of the established purification procedure.

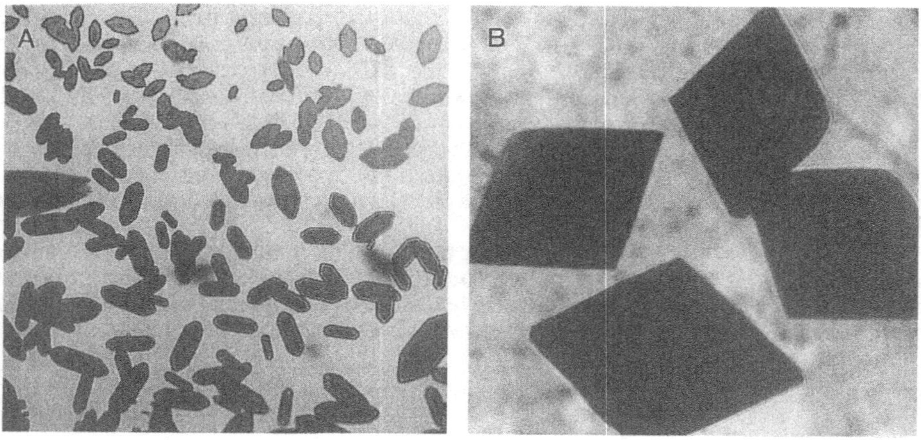

Fig. 4

Crystals of the LHC of Mantoniella squamata. **A**:Hexagonal crystals of approximately 10-50 μm, grown in the presence of 0.5 % (w/v) heptane triol. **B**:Rhombic crystals of 0.1- 0.2 mm in size appeared with the addition of 3 % (w/v) benzamidine.

References

1 Schmitt, A., Herold, A., Welte, C., Wild, A. and Wilhelm, C. (1993) Photochem.Photobiol. 57 (1),132-138

2 Wilhelm, C. (1987) Biochem.Biophys.Acta 892, 23-29

3 Schmitt, A., Frank, G., Staudenmann, W., Zuber, H. and Wilhelm, C. (1994) Photosynth.Res. 40, 269-277

4 Kühlbrandt, W., Wang, D.N. and Fujiyoshi, Y. (1994) Nature 367, 614-621

5 Welte, C., Nickel, R. and Wild, A. Biochem.Biophys.Acta (Bioenergetics), in press

6 Boekema, E.J. (1991) Micron Microsc. Acta 22 (4), 361-369

7 Nußberger, S., Dörr, K., Wang, D.N. and Kühlbrandt, W. (1993) J.Mol.Biol. 234, 347-356

8 Rhiel, E., Lange, W., Mörschel, E. (1993) Biochem.Biophys.Acta 1143, 163-172

9 Dekker, J.P., Betts, S.D., Yocum, C.F., Boekema, E.J. (1990) Biochem. 29, 3220-3225

10 Kühlbrandt, W. (1991) in Crystallization of Membrane Proteins (Michel, H., ed.) pp.155-166, CRC Press, Boston,The United States

ORGANISATION OF PHOTOSYSTEM I ANTENNA

ÉVA SÁRVÁRI, GABRIELLA MALATINSZKY and PÉTER NYITRAI
Department of Plant Physiology, Eötvös University
H-1445 Budapest, P. O. Box 330, Hungary

1. Introduction

Photosystem I (PSI) in higher plants is composed of a core complex (CC) anchoring the electron transport cofactors together with about 100 chlorophyll-a (Chl-a) molecules serving as inner antenna, and of a light-harvesting complex (LHCI) which binds the other half of Chls in PSI including Chl b (1,2). 5-6 polypeptides of 11-25 kDa is reported to build LHCI, most of which (20-25 kDa and 17 kDa corresponding to Lhca1-4 and PsaF, respectively) have been shown to bind Chl (3-5). Lhca1-4 gene products are organised into two subcomplexes designated LHCI-680 and LHCI-730, respectively (6,7). Under certain circumstances LHCII was also found to bind to PSI particles (7-14). In spite of the great progress in the isolation techniques (15,16) and in the molecular biology of the pigment-binding polypeptides (17), the available information on the organisation of the different Chl-proteins in the holocomplex are controversial. In the present work four PSI particles of quite different antenna composition were isolated, their structure was analysed and discussed in connection with the organisation of PSI antenna.

2. Procedure

Maize (*Zea mays L.*) seedlings grown in the dark for 4 days and then illuminated with white light (100 μmol photons m^{-2} s^{-1}) for 72 h were used in the experiments. Mesophyll (M) and bundle sheath (BS) chloroplasts were isolated by a differential grinding method (18). Thylakoids were prepared as in Ref. 19. Solubilisation of membranes with dodecyl sucrose : nonyl glucoside : lithium dodecyl sulfate = 4.5:4.5:1, non-denaturing Deriphat PAGE, and denaturing SDS PAGE were carried out as described in Ref. 20. Fluorescence spectra of excised green bands were measured with an MPF-44B Perkin-Elmer spectrofluorometer at 77K.

3. Result and Discussion

Using mainly glucosidic surfactants for the solubilisation of thylakoids, and a mild green gel method of high resolution in the region of larger complexes (20), four different PSI particles were isolated from maize M and BS thylakoids (Fig. 1: bands 1-4). BS preparations were highly deficient in PSII-related complexes: the amounts of bands 5, 6 corresponding to different-sized CCII as well as the CCII content of band 3 decreased significantly as did bands 10 and 11 containing CP43 and CP29, respectively (Fig. 1A).

P. Mathis (ed.), Photosynthesis: from Light to Biosphere, Vol. I, 195–198.

This made possible to isolate PSI complexes from BS thylakoids in a purer form and in a greater amount (added up to 55% of the Chl applied to the gel in contrast to the 39% found in M thylakoids). LHCII-like bands (bands 7-9) were present in the BS preparations, too. However, the main LHCII band of M thylakoids (band 9) was the faintest one among them (Fig. 1A).

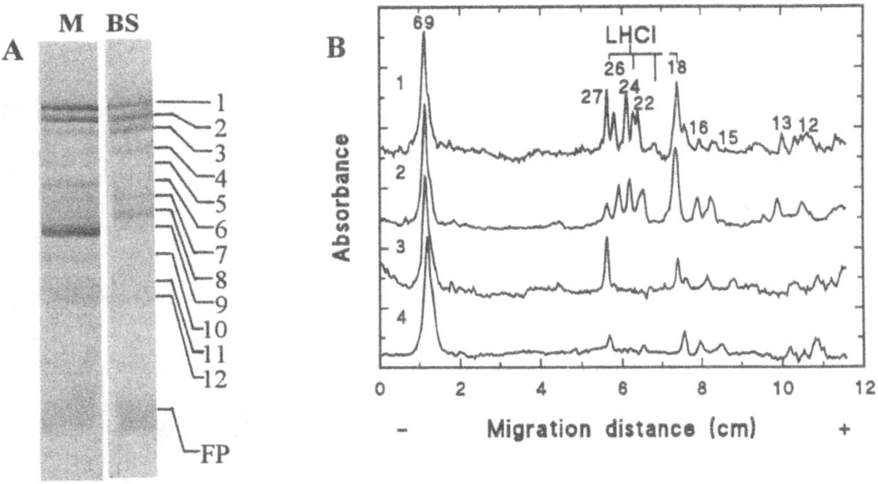

Figure 1. (A) Chl-protein bands (numbered) isolated from M and BS thylakoids solubilised with a detergent/Chl ratio of 10, and resolved in a 7-12% gradient gel. Free pigment (FP) band contained mainly carotenoids, displaced Chl was in the range of 3-6%. (B) Densitograms of stained polypeptide patterns of bands 1-4 isolated from BS thylakoids. Numbers shown on densitogram 1 mean molecular masses in kDa.

The composition and properties of PSI particles isolated from M and BS thylakoids were about the same. The following complexes (Fig. 1B) were isolated (in the order of increasing mobility): PSI particles containing CCI and LHCI and a complex built of LHCII-like polypeptides of 27 kDa, and having a Chl a/b ratio of 5.5 (band 1), PSI composed of CCI and LHCI (band 2), or CCI and the LHCII-like antenna (band 3) with Chl a/b ratios around 7.5 and 14, respectively, and CCI containing only Chl a (band 4). The removal of the LHCI complex displaced the 22-26 kDa polypeptides and strongly diminished the amount of a 18 kDa polypeptide, which sometimes was separated into two bands and the lower one disappeared under similar conditions, and shifted the long wavelength fluorescence emission maximum from around 733 nm (bands 1, 2) to around 725 nm (bands 3, 4) in agreement with the results obtained in Refs. 1,21. The 27 kDa polypeptide found in PSI particles reacted with a polyclonal α-LHCII (not shown). Higher detergent/Chl ratios decreased the amount of the largest PSI complex (band 1) in favour of the smaller ones (Fig. 2) and the LHCII oligomer band, however, without changing their composition (only their fluorescence emission properties referred to a somewhat disconnected energy transfer in the complex). This means that LHCII disappeared together with the other parts of the complex, which excludes the possibility that some larger LHCII oligomers just run together with the PSI particles. Otherwise such species were found neither in LHCII nor in PSII (BBY) preparations in our system, and PSI bands originating from BS thylakoids having much less LHCII and being more sensitive to solubilisation contained the same amount of

LHCII. The LHCI-containing complex (band 2) was the most stable among the particles proving a stronger binding of LHCI than LHCII in the complex.

The LHCII efficiently transferred the absorbed light energy to the long wavelength Chl forms of PSI at 77K, and the trimeric complex was highly fluorescent after isolation from the PSI complexes (Fig. 3). In case of band 3, however, the efficiency of this energy transfer decreased (not shown).

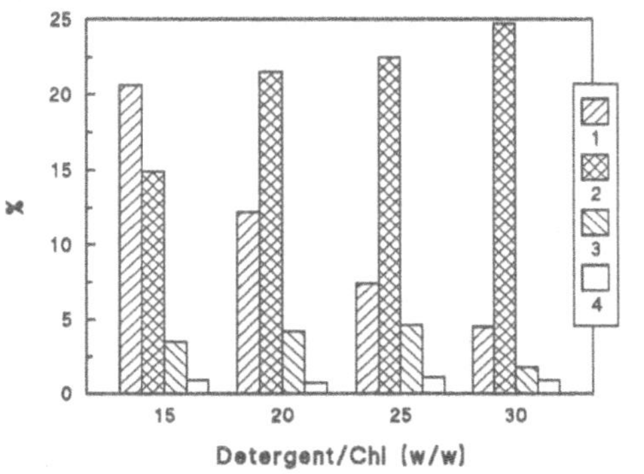

Figure 2. Changes of the relative amounts of PSI particles as a function of detergent/Chl ratio.

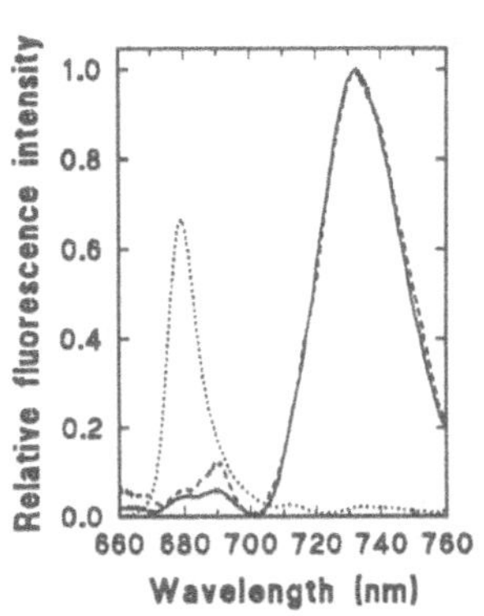

Figure 3. 77K fluorescence emission of the largest PSI particles excited at 440 (---) or 470 (—) nm, and that of the disconnected LHCII complex (...) measured in identical set-up.

LHCII connected to PSI in different amounts was published to be responsible for the heterogeneous antenna size of PSI in grana margins and stroma thylakoids (13). It also appeared in PSI particles isolated with lower detergent/Chl ratios (7). Moreover, the association of LHCII to PSI was found to be stronger under phosphorylating (10) or rather under non-phosphorylating (12) conditions. Either only 27 kDa polypeptides (10) or 27 and 25 kDa species (12,13) appeared in PSI preparations. The heterogeneity of PSI particles in thylakoids as the source of our different PSI complexes cannot be totally ruled out, but the facts that no PSI particles containing higher and lower amounts of LHCII were isolated, that the same populations were found in M and BS thylakoids of maize, as well as the behaviour of PSI bands under stronger solubilising conditions (see Fig. 2) do not seem to favour this explanation. The LHCII found in our PSI particles, though reacted with α-LHCII, was somewhat different from the bulk LHCII of M thylakoids. It contained only LHCII polypeptide(s) of higher molecular mass in agreement with (10) that is found to build up the mobile LHCII complexes (10,22), and the maximum/650 nm ratio in its fluorescence excitation spectrum was higher than that of the bulk LHCII. However, thylakoids were isolated under dim laboratory light so they could not be in a phosphorylated state. Moreover, the isolation was carried out in the absence of $MgCl_2$, which is stated to stabilize binding of LHCII to PSI (9). Therefore, our results show that a special LHCII population constitutes the part of the PSI antenna.

There are several recommendations concerning the organisation of the PSI antenna based on the composition of isolated PSI particles (7,11,23,24) as well as on electron microscopy and computer image analysis (25). Models suggest an energy transfer pathway of LHCII → LHCI-680 → LHCI-730 → CCI together with either a similar spatial arrangement of complexes (7,11,24) or a parallel binding of antenna complexes to the CCI (17,25). Preparations of CCI binding either only LHCI-680 (23) or only LHCI-730 (7) were published. Our data show the presence of CCI binding exclusively LHCII (band 3). These results can be explained assuming that all the antenna complexes are bound directly to CCI and may be bound to each other as well, and different associations of the complexes are stable under different solubilisation conditions. However, the fact that the energy transfer from LHCII to the CCI is somewhat disconnected in band 3 reveals that the presence of LHCI complexes facilitates the transfer of absorbed light energy from LHCII to CCI in PSI complexes, in agreement with recent models (7,11).

Research grants from OTKA (T-5503) and EEC (CIPA CT930202) are greatfully acknowledged.

References

1 Mullet, J.E., Burke, J.J. and Arntzen, C.J. (1980) Plant Physiol. 65, 814-822
2 Haworth, P., Watson, J.I. and Arntzen, C. J. (1983) Biochim. Biophys. Acta 724, 151-158
3 Vainstein, A., Peterson, C.C. and Thornber, J.P. (1989) J. Biol. Chem. 264, 4058-63
4 Anandan, S., Vainstein, A. and Thornber, J.P. (1989) FEBS Lett. 256, 150-154
5 Ikeuchi, M., Hirano, A. and Inoue, Y. (1991) Plant Cell Physiol. 32, 103-112
6 Lam, E., Ortiz, W. and Malkin, R. (1984) FEBS Lett. 168, 10-14
7 Bassi, R. and Simpson, D. (1987) Eur. J. Biochem. 163, 221-230
8 Remy, R. and Ambard-Bretteville, F. (1984) FEBS Lett. 170, 174-180
9 Williams, R.S., Allen, J.F., Brain, A.P.R., Ellis, R.J. (1987) FEBS Lett. 225, 59-66
10 Bassi, R., Giacometti, G.M. and Simpson, D.J. (1988) Biochim. Biophys. Acta 935, 152-165
11 Knoetzel, J., Svendsen, I. and Simpson, D.J. (1992) Eur. J. Biochem. 206, 209-215
12 Harrison, M.A. and Allen, J.F. (1992) Eur. J. Biochem. 204, 1107-1114
13 Andreasson, E. and Albertsson, P-Å. (1993) Biochim. Biophys. Acta 1141, 175-182
14 Dreyfuss, B.W. and Thornber, J.P. (1994) Plant Physiol. 106, 841-848
15 Thornber, J.P., Morishige, D.T., Anandan, S. and Peter, G.F. (1991) in Chlorophylls (Scheer, H., ed.) pp. 549-585, CRC Press, Boca Raton, USA
16 Bassi, R., Rigoni, F., Giacometti, G.M. (1990) Photochem. Photobiol. 52, 1187-1206
17 Jansson, S. (1994) Biochim. Biophys. Acta 1184, 1-19
18 Woo, K.C., Anderson, J.M., Boardman, N.K., Downton, W.J.S., Osmond, C.B. and Thorne, S.W. (1970) Proc. Natl. Acad. Sci. USA 67, 18-25
19 Fuad, N., Day, D.A., Ryrie, I.J. and Thorne, S.W. (1983) Photobiochem. Photobiophys. 5, 255-262
20 Sárvári, É. and Nyitrai, P. (1994) Electrophoresis 15, 1068-1071
21 Preiss, S., Peter, G.F., Anandan, S. and Thornber, J.P. (1993) Photochem. Photobiol. 57, 152-157
22 Peter, G.F. and Thornber, J.P. (1991) J. Biol. Chem. 266, 16745-16754
23 Bassi, R., Machold, O., Simpson, D. (1985) Carlsberg Res. Commun. 50, 145-162
24 Thornber, J.P., Peter, G.F., Morishige, D.T., Gómez, S., Anandan, S., Welty, B.A., Lee, A., Kerfeld, C., Takeuchi, T. and Preiss, S. (1993) Biochem. Soc. Transact. 21, 15-18
25 Boekema, E.J., Wynn, R.M. and Malkin, R. (1990) Biochim. Biophys. Acta 1017, 49-56

cDNA DEDUCED AMINO ACID SEQUENCES OF MAIZE CP24 AND CP26, THE TWO MAJOR ZEAXANTHIN-BINDING PROTEINS OF PHOTOSYSTEM II.

P. Dainese[1], E. Bergantino[1], S. Sechi[2], R. Bassi[3] and E. Pichersky[4].

[1]Dipartimento di Biologia, Universitá di Padova; [2]National Institute of Health, Bethesda; [3]Facoltà di Scienze MM.FF.NN, Universitá di Verona; [4]Natural Science Building, Ann Arbor, University of Michigan.

INTRODUCTION

The antenna of photosystem II (PSII) is organised in an elaborate set of chl *a/b* (CAB) proteins, including the major light harvesting complex LHCII and three other proteins - CP29, CP26 and CP24 - called the minor antenna complexes since they bind only the 15% of the total chlorophyll content of PSII (1,2,3). They also bind most of the xanthophyll violaxanthin, which under high light intensity is converted to zeaxanthin (4,5). Thus they are the site of the zeaxanthin-dependent energy dissipation which protects the photosynthetic apparatus from photoinhibition (6). This conclusion is also strengthened by the binding of dicyclohexylcarbodiimide (DCCD), inhibitor of the nonphotochemical quenching, to CP26 and CP29 (7).
The heterogeneity of the PSII CAB protein was shown both on the basis of their organisation and function within the PSII unit and by molecular genetic studies. For example, three types of genes (*Lhcb1*, *Lhcb2*, *Lhcb3*) encoding LHCII CAB proteins have been characterised (8,9,10). However, in respect to the minor antennae of PSII, in many species the correspondence of a given *Lhc* gene to his product is still unclear. In particular, some ambiguities exist about the exact identities of the genes coding for the polypeptides of CP29 and CP26, in part because of the different operational definitions of these complexes employed by various researchers and also due to the difficulty in separating the proteins of these complexes (11,12,13). CP24 has also scarcely been investigated in different species. Here we report the isolation of CP24 and CP26, the two major zeaxanthin containing proteins, in their native pigment-binding form from maize mesophyll chloroplasts. The purified proteins of these complexes were partially sequenced. CP24 and CP26 cDNAs were also isolated and sequenced. This work clearly establishes the identities of the CP24 and CP26 proteins and their genes in *Zea mays*.

MATERIALS AND METHODS

Isolation of the CP24 and CP26 complexes. PSII-enriched membranes from *Zea mays* seedlings were prepared as already described (3). The membranes were solubilised in 0.4% dodecylmaltoside and subjected to sucrose gradient ultracentrifugation (3): CP24 and CP26 were then isolated by isoelectrofocusing from sucrose bands 4 and 2 respectively (4).
Isolation of CP24 and CP26 apoproteins. The pigment-protein complexes were subjected to two subsequent preparative SDS-PAGE, using the Tris-sulphate system with urea (1) for the first run and the Tris-Tricine system (4) for the second run, after excision of the Coomassie-stained bands.
Immunoblotting and HPLC analysis were described elsewhere (4).

P. Mathis (ed.), Photosynthesis: from Light to Biosphere, Vol. I, 199–202.

Purification of proteolytic peptides and amino acid sequencing. After electroelution, the purified apoproteins were digested either with endoproteinase Lys C or with trypsin. The digestion products were separated by a Tris-Tricine SDS-PAGE and blotted onto PVDF filters using the CAPS buffer (14). The filters were stained with Red Ponceau and the proteolytic fragments excised from them. Samples were sequenced with an Applied Biosystems Model 475A sequencer equipped with an "on line" model 120A PTH amino acid analyser. In all our PAGE experiments, the acrylamide and urea solutions were deionised with mixed amberlite resin and 1 mM dimethylamino-ethanthiol-hydrochloride was included in the running buffer, to prevent N-terminal blocking.

Cloning of *Lhcb5* and *Lhcb6* cDNAs. We screened a Lambda Zap II cDNA library (a generous gift of Dr. Alice Barkan) prepared from mRNAs isolated from maize plants grown in the dark for 7 days and then subjected to 24 h of light. The screening was performed by using radiolabeled fragments of tomato *Lhcb5* and *Lhcb6* (*Cab9* and *Cab10B* respectively in 11 and 15). Probe labelling and library screening were performed as previously described (15). DNA was sequenced by the enzymatic method (16).

RESULTS AND DISCUSSION

Isolation and partial sequencing of the antenna apoproteins.

We isolated CP26 and CP24 from a PSII-enriched membrane preparation by a combination of non-denaturing fractionation methods: sucrose gradient ultracentrifugation and isoelectrofocusing in the presence of the non-ionic detergent dodecylmaltoside. The identities of the complexes were verified from their absorption spectra, HPLC analysis, SDS-PAGE and immunoblot. HPLC analysis of the purified pigment proteins indicated that CP24 lacked neoxanthin and both CP24 and CP26 were highly enriched in violaxanthin, consistent with previous reports (4,5). Since several polypeptides belonging to the CAB proteins of PSII have similar MW, we isolated the apoproteins from each individual complex by two subsequent preparative SDS-PAGE, as described in Materials and Methods, to avoid the presence of minor contaminations. The identification of the Coomassie stained bands was always verified by immunological assay with specific antibodies. The proteins, electroeluted from the gel, were then subjected to proteolysis. For CP26 four peptides were sequenced and they yielded sequences of 19, 17, 23 and 23 residues (Fig. 1), corresponding in total to one third of the whole apoprotein. On the other hand, the analysis of CP24 was not successful, since both the isolated uncut protein and the proteolytic fragments contained multiple N-terminals.

fragment 1	H_2N-	KPAQKPKPAAVSXXXPDIK
fragment 2	H_2N-	WYGPDRXIYLPDELLDD
fragment 3	H_2N-	FGANXGPEAVWFKTGALLEDKNT
fragment 4	H_2N-	VYPGGAFDPLGLADDPDTAAELK

Figure 1: partial amino acid sequence of the proteolytic fragments obtained from the purified CP26 apoprotein.

Characterisation of *Lhcb5* and *Lhcb6* clones.

Two complete cDNA clones encoding maize *Lhcb5* were isolated and were found to contain 957 and 1073 nucleotides respectively. Both sequences contain an open reading frame of 283 codons. The proteins encoded are highly similar, but there are appreciable differences at the 5' and 3' flanking regions. In the coding region, 23 nucleotide substitutions out of a total of 27 are silent (data not shown). The two deduced CP26 apoproteins (Fig. 2) differ only at four positions, one of which is localised in the transit peptide. The protein sequences encoded by these two maize *Lhcb5* genes are distinct from the peptide sequences of the LHCB5 protein obtained by direct sequencing (Fig. 1), with a total difference of 11 or 12 residues out of the total of 82 residues sequenced from the polypeptide fragments.

A full-length *Lhcb6* cDNA clone contained an open reading frame of 247 codons. It was identified as maize CP24 apoprotein by its high sequence identity with the deduced sequences of the previously cloned *Lhcb6* genes of tomato (15) and spinach (17), and with the determined primary sequence of barley (18) (Fig. 3).

```
                                         ↓
1     MAALAPSKIL GTKLNFAGSS RYATAAPTAG AQKIVSLFSK KPAQKPKPSA VSSSSPDISD
1     MAALAPSKIL GTQLNFAGSS RYATAAPTAG AQKIVSLFSK KPAQKPKPAA VSSSSPDISD

61    ELAKWYGPDR RIYLPDGLLD RSEVPEYLTG EVPGDYGYDP FGLGKKPEDF AKYQAYELIH
61    ELAKWYGPDR RIYLPDGLLD RSEVPEYLTG EVPGDYGYDP FGLGKKPEDF AKYQAYELIH

121   ARWAMLGAAG AVIPEACNKF GANCGPEAVW FKTGALLLDG NTLSYFGNSI PINLVVAVIA
121   ARWAMLGAAG AVIPEACNKF GANCGPEAVW FKTGALLLDG NTLNYFGNSI PINLVVAVIA

181   EVVLVGGAEY YRIINGLDLE DKLHPGGPFD PLGLASDPDQ AAILKVKEIK NGRLAMFSMF
181   EVVLVGGAEY YRITNGLDLE DKLHPGGPFD PLGLASDPDQ AAILKVKEIK NGRLAMFSMF

241   AFFIQAYVTG EGPVENLAKH LSDPFGNNLL TVISGAAERT PSL
241   AFFIQAYVTG EGPVENLAKH LSDPFGNNLL TVISGAAERT PSL
```

Figure 2: Amino acid sequences of the precursor CP26 proteins of maize, as deduced from the nucleotide sequences of the *Lhcb5.1* and *Lhcb5.2* cDNAs. Vertical arrow, putative cleavage site of the preprotein; bold characters, amino acid differences between the two gene products; double underline, regions of the protein determined by direct protein sequencing; horizontal arrows, hydrophobic regions corresponding to putative membrane spanning domains.

Multiple genes for the CP26 protein in maize.

The two maize clones that we have characterised fit well with the sequences (a total of 82 aa) obtained from partial sequencing of the purified CP26 protein. Whereas in tomato, barley and scot pine (19) only one copy of *Lhcb5* gene was found, maize contains at least three genes: we have sequenced two different cDNA clones and at least one additional gene must exist, since the amino acidic sequences obtained did not exactly match those deduced from the two cDNAs (see Figs. 1 and 2). Of the four peptides, 1 and 2 cover almost the entire N-terminal region preceding the first of the two regions of strong similarity found in all *Lhc* genes (9,19). Fragment 3 is located in between the second and third transmembrane helices. The fourth one is situated past the second transmembrane helix and partially covers the second region of high similarity immediately before the third helix. The last fragment appears to be the most dissimilar to the proteins encoded by the two characterised cDNA clones, with 6 differences in the amino acid composition. However, the analysis of the structural features of a hypothetical CP26 protein with the sequence of this peptide shows no appreciable differences; neither the hydrophobicity plot nor the beta-turn probability profile are significantly different from that of other CP26 proteins.

The maize CP24.

Lhcb6 was the first gene of a PSII minor antenna complex to be isolated (15,17). Moreover, the position of the cleavage site of the transit peptide in the CP24 precusor is well established, since a N-terminal sequence of the mature protein is available both from barley (18) and spinach (20). In tomato two tandemly linked genes were isolated, and they encode almost identical proteins (95% identity in the mature peptide). The presence of additional CAB

```
                                         ↓
1     MARAHFRCRA QDPFLGARRA LANAAAKPGP RRALVVAAAA AAKKSWIPAI KSDAEIVNPP

61    WLDGSLPGDF GFDPLGLGKD PAFLKWYREA DVIHGRWAMA AVLGIFVGQA WSGIPWFEAG

121   ADPTRIAPFS FGSLLGTQLL LMGWVESKRW VDFFNPDSQA VEWATPWSRT AENFANFTGE

181   QGYPGGKFFD PLALAGTSRD GVYIPDVDKL ERLKLAEIKH ARIAMLAMLA FYFEAGQGKT

241   RLGALGL
```

Figure 3: Amino acid sequence of the precursor CP24 protein deduced from the nucleotide sequence of the *Lhcb6* cDNA. The arrow points to the N-terminal residue of the mature protein, as inferred by homology with the barley (18) and spinach (20) protein; other symbols as in legend to Fig. 2.

polypeptides is so far an unresolved issue (1,17,21). In maize, at least two polypeptides can be observed, but they do not separate well in the Tris-sulphate gel system (1). The major differences between the maize CP24 protein and the spinach and tomato ones is found in the transit peptide. The maize transit peptide is shorter (37 aa vs. 50 and 47/45 respectively) and shows some similarity only with barley. The sequence of the mature form of the protein is also more similar to barley than to the dicot species, with the main differences found in the N-terminal region: the tomato and the spinach sequences show a 83% similarity in the first 23 aa, while this value falls to 57% and 60% when maize is compared to spinach and tomato. Those differences could be explained on the basis of the evolutionary distance between monocots and dicots, and the generally high level of sequence divergence found at the N-termini of all CAB proteins.

REFERENCES

1. Bassi R, Hoyer-Hansen G, Barbato R, Giacometti GM, Simpson DJ (1987), J. Biol. Chem. 262: 13333-13341.
2. Bassi R, Rigoni F, Giacometti GM (1990), Photochem. and Photobiol. 52: 1187-1206.
3. Bassi R, Dainese P (1992), Eur. J. Biochem. 204: 317-326.
4. Bassi R, Pineau B, Dainese P, Marquardt J (1993), Eur. J. Biochem. 212: 297-303.
5. Ruban AV, Young AJ, Pascal AA, Horton P (1994), Plant Physiol. 104: 227-234.
6. Demmig-Adams B (1990), Biochim. Biophys. Acta 1020: 1-24.
7. Walters RG, Ruban AV, Horton P (1994), Eur. J. Biochem. 226: 1063-1069.
8. Morishige DT, Thornber JP (1990), in Current Research in Photosynthesis (Baltscheffsky M ed), vol II pp. 261-264, Kluwer Academic Publishers, Dordrecht, The Netherlands.
9. Green BR, Pichersky E, Kloppstech K (1991), Trends in Biochem. Sci. 16: 181-186.
10. Schwartz E, Stasys R, Aebersold R, McGath JM, Green BR, Pichersky E (1991), Plant Mol. Biol. 17: 923-925.
11. Pichersky E, Subramaniam R, White MJ, Reid J, Aebersold R, Green BR (1991), Mol. Gen. Genet. 227: 277-284.
12. Morishige DT, Thornber JP (1992), Plant Physiol. 98: 238-245.
13. Sorensen AB, Lauridsen BF, Gausing K (1992), Plant Physiol. 98: 1538-1540.
14. Matsudaira P (1987), J. Biol. Chem. 262: 10035-10038.
15. Schwartz E, Pichersky E (1990), Plant Mol. Biol. 15: 157-160.
16. Sanger F, Nicklen S, Coulsen A (1977), Proc. Natl. Acad. Sci. USA 74: 5463-5467.
17. Spangfort M, Larsson U, Ljunberg U, Ryberg M, Andersson B (1990), in Current Research in Photosynthesis (Baltscheffsky M ed), vol II pp. 253-256, Kluwer Academic Publishers, Dordrecht, The Netherlands.
18. Morishige DT, Anandan S, Jaing JT, Thornber JP (1990), FEBS Lett. 2: 239-242.
19. Jansson S (1994), Biochim. Biophys. Acta 1184: 1-19.
20. Murata N, Kajiura H, Fujimura Y, Miyao M, Murata T, Watanabe A, Shimozaki K (1987), in Progress in Photosynthesis Research (Biggins J ed), vol I pp. 701-704, Martinus Nijhoff Publishers, Dordrecht, The Netherlands.
21. Dunahay TG, Staehelin LA (1986), Plant Physiol. 80: 429-434.

The GenBank accession numbers for the cDNA sequences reported in this article are U23188, U23189 and U23190. This work was supported by the Italian Ministry of Agriculture and Forestry Grant 4.7240.90.

A SPECTROSCOPIC INVESTIGATION OF AGGREGATION OF CHLOROPHYLL *a* IN SOLUTION

D.M. Gazdaru[1], L.C. Petcu[2], E.M. Franken[3], T.J. Aartsma[3]
[1]Faculty of Physics, Bucharest University, P.O.Box MG-11, 76900 Bucharest-Magurele, Romania; [2]Faculty of Medicine, Ovidius University, 8700 Constanta, Romania; [3]Department of Biophysics, Huygens Laboratory, University of Leiden, P.O.Box 9504, 2300 RA Leiden, The Netherlands.

1. Introduction

Many absorption spectroscopic studies have been carried out on chlorophyll *a* (chl *a*) aggregates to elucidate the difference in optical absorption between chlorophyll *a in vivo* and in solution /1-5/.
In particular, the dimeric chlorophyll *a* species having the absorption peaks around 680 and 700 nm, have received attention due to their importance as models of P680 and P700, the reaction centres of PS I and PS II, respectively /6-9/.
It is generally recognized that chlorophyll *a* at low temperatures forms a mixture of monomers, dimers and higher aggregates, depending on the nature of the solvents and on chlorophyll *a* concentration /4, 7, 10/. The photo-excited triplet state of dimeric chlorophyll *a* has been investigated at about 4.2 K by the fluorescence-detected magnetic resonance spectroscopy (FDMR) and at 77 K by the flash photolysis /11, 4/.
The present work is a spectroscopic study on the chlorophyll *a* aggregates. It presents the absorption spectra of the chlorophyll *a* in dry 3-methylpentane (3-MP) over the temperature range 300-77 K and the time-resolved transient (T-S) absorption spectrum of the dimeric chlorophyll *a* in 3-MP rigid matrix at 77 K.

2. Procedure

Chlorophyll *a* extracted from spinach was chromatographed on the powdered sugar column, according to the method of Strain and Svec /12/. The absorption maxima as well as the band intensity ratios were in accordance with the literature data /13/.
The purified chlorophyll *a* was dissolved in n-hexane and stored in a freezer at -20° C.
Before the preparation of the samples, the three co-distillations with acetone were made and the resultant film of chlorophyll *a* was heated at 60° C for 30 min. at 10^{-2} mm Hg pressure to remove the residual solvent. This procedure reduces the water content of the chlorophyll *a*. Dry 3-methylpentane was used to obtain a 10^{-4} M chlorophyll *a* solution. All manipulations were carried out in a controlled atmosphere, N_2-purged glove dry-box.
The absorption spectrum at room temperature was recorded on a Shimatsu spectrophotometer.
The absorbance difference spectrum was obtained by using a conventional flash photolysis apparatus (single beam device), the light source being a tungsten xenon lamp. A cryogenic device was used for obtaining the spectra at low temperature.

P. Mathis (ed.), Photosynthesis: from Light to Biosphere, Vol. I, 203–206.

3. Results and Discussion

3.1 *Results*

The absorption spectrum of 10^{-4} M chlorophyll *a* in dry 3-MP presents, at room temperature, an asymmetric band in red region, specific to the chlorophyll *a* anhydrous dimer. The peak is situated at 662 nm and the shoulder is at 678 nm (*Fig. 1*).

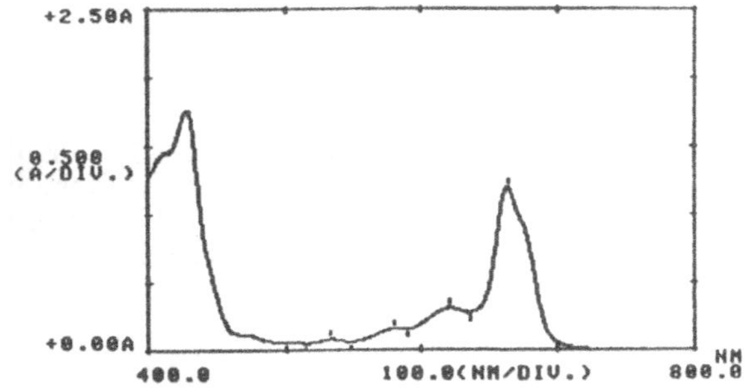

Figure 1. Absorption spectrum of 10^{-4} M Chl *a* in dry 3-MP at room temperature.

The absorption spectra of the sample over the temperature range 235-77K (*Fig.2*) indicate that the chlorophyll *a* monomer transforms itself into the A-710 species. The Gaussian deconvolution of the red band of the absorption spectrum obtained at 77 K (*Fig.3*) shows the presence of the species : A-671 nm, A-689 nm and A-710 nm.

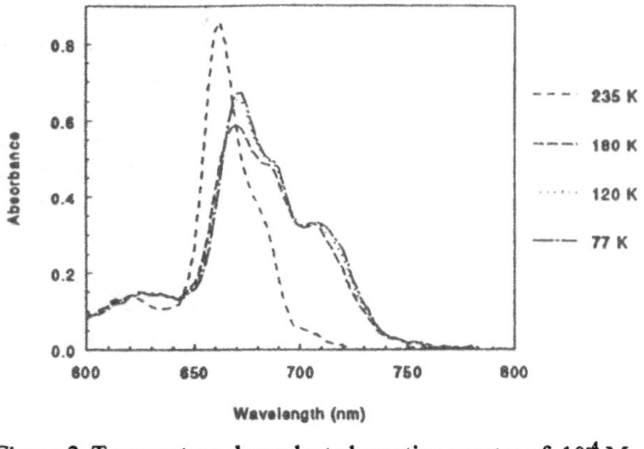

Figure 2. Temperature-dependent absorption spectra of 10^{-4} M Chl *a* in dry 3-MP.

The absorbance difference spectrum (*Fig. 4*) measured at 250μs after a flash has a positive absorption peak at 676nm and a negative peak at 712 nm accompanied by a minor satellite at 682 nm. The kinetics of the absorbance change at 710nm is not expressed by a single exponential function of time (*Fig. 5*). The lifetime of the fast transient species was determined to be 1.36ms over the wavelength range of 620- 840 nm.

3.2 *Discussion*

The main species existing in the sample at room temperature (A-678) is the chlorophyll *a* anhydrous dimer, according to the literature reports /14, 15/. A small quantity of the chlorophyll *a* monomer also exists.

The three chlorophyll *a* species presented in the absorption spectra at low temperature were interpreted as : monomer, anhydrous dimer /14, 15/ and hydrated dimer [(chla $H_2O)_2$ /9/ or (chla $2H_2O)_2$ /8/]. Analysing the temperature dependence of the absorption spectra, we conclude that the A-662 nm absorbing species at 235 K is most likely a precursor of the A-710 nm absorbing species at lower temperature.

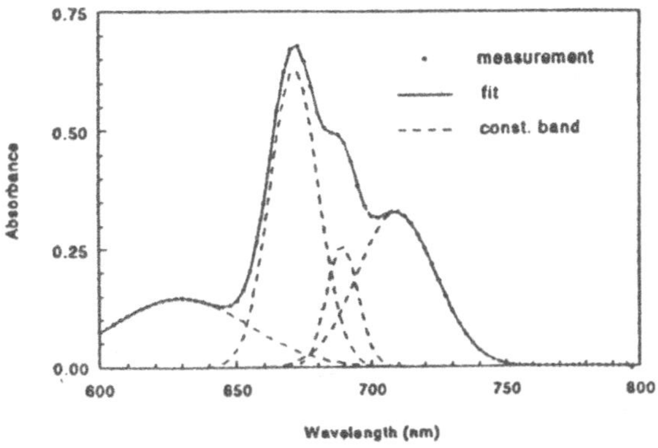

Figure 3. Absorption spectrum of 10^{-4} M Chl *a* in dry 3-MP at 77 K.

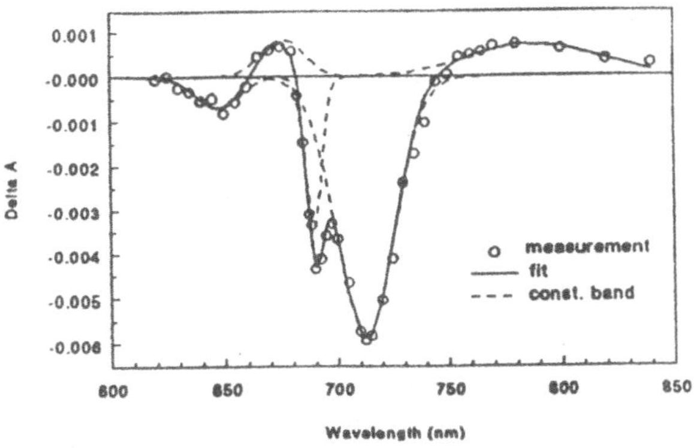

Figure 4. Absorbance difference spectrum of 10^{-4}M Chl *a* in dry 3-MP at 77 K, measured at 250 μs after flash.

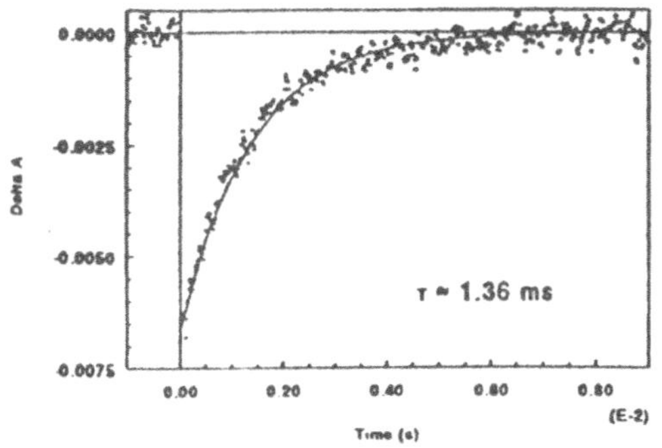

Figure 5. Kinetics of absorbance change of 10^{-4} M Chl *a* in dry 3-MP at 710 nm (77K).

The spectral shifts observed for the three species as function of temperature may be due to some configuration changes of theirs in glassy matrices by cooling. The appearance of the chlorophyll *a* hydrated dimer shows that the water was not completely removed by co-distillation through the vacuum line.

The double band in the red spectral region (682, 700 nm) of the transient difference spectrum was tentatively attributed to exciton splitting of the energy levels of two chlorophyll *a* molecules in a dimer, according to Döring et al. /16/. They obtained for the redox reaction of chlorophyll *a* in chloroplasts a spectrum closely resembling to the one we obtained.

It also could be possible that the band together with its minor satellite be attributed to the energy levels of the two kinds of the dimer. In this case the difference spectrum is a "sum" of the spectra of two transient species which have approximately the same lifetime.

4. Acknowledgement

This work was performed in the Department of Biophysics of Huygens Laboratory, University of Leiden, within the Tempus Project. D.M. Gazdaru and L.C. Petcu wish to express their gratitude to Professor Jan Amesz for his collaboration in the framework of the above mentioned project.

References

1 Brody, S.S. and Broyde, S.S. (1968) Biophys. J. 8, 1511-1533
2 Cotton, T.M., Loach, P.A., Katz, J.J. and Ballshmitter, K. (1978) Photochem. Photobiol. 27, 735-749
3 Balmy, C., Broyde, S.S. and Hoa, G.H.B. (1969) Photochem. Photobiol. 9, 445-454
4 Hoshino, M., Ikehara, K., Imamura, M. and HamaY. (1983) Bull. Chem. Soc. Jpn. 56, 208-211
5 Agostiano, A., Cosma, P. and Della Monica, M. (1991) J. Photochem. Photobiol. A: Chem., 58, 201-213
6 Sauer, K., Lindsay Smith, J.R. and Schultz, A.J. (1965) J. Am. Chem. Soc. 88, 2681
7 Shipman, L.L., Cotton, T.M., Norris, J.R. and Katz, J.J. (1976) J. Am. Chem. Soc. 98, 8222
8 Fong, F.K., Koester, V.J. and Galloway, L. (1977) J.Am. Chem. Soc. 99, 2372
9 Boxer, S.G. and Closs, G.L. (1976) J. Am. Chem. Soc. 98, 5406
10 Hoshino, M., Ikehara, K., Imamura, M., Saki, H. and Hama, Y. (1981) Photochem. Photobiol. 34, 75
11 Kooyman, R.P.H., Schaafsma, T.J. and Kleibeuker, J.F. (1977) Photochem. Photobiol. 26, 235
12 Strain, H.H. and Svec, W.A. (1966) in The Chlorophylls (Vernon L.P. and Seely G.R., eds.) pp.21-26, Academic Press, New York, USA
13 Svec, W.A. (1978) in The Porphyrins, Vol. V (Dolphin D., ed.) pp.342-399, Academic Press, New York, USA
14 Katz, J.J., Schipman, L., Cotton, T.M. and Janson, T.R. (1978) in The Porphyrins, Vol. V (Dolphin D., ed.) pp.401-458, Academic Press, New York, USA
15 Fong, F.K. and Koester, V.R. (1976) Biochim. Biophys. Acta 423, 52-64
16 Döring, G., Bailey, J.L., Kreutz, W., Weikard, J. and Witt, H.T. (1968) Naturwissenschaften 55, 219-220

TWO KINDS OF CHLOROPHYLL POLYMERS

S.Y.Yang
Shanghai Institute of Plant Physiology, Chinese Acad.Sciences
300 fenglin Road, Shanghai , 200032, China

Introduction

In photosynthetic systems the excitation energy is absorbed by so called antenna pigments and subsequently transported through a series of energy transfer steps to some special supramolecular structures, the reaction centres, where it is used for photochemical charge separation. Chlorophyll a takes a principal role in these energy transformation processes, and exists in different states in antenna complexes and reaction centres.

Though there is indirect evidence that the difference in aggregate state of chlorophyll or in the type of linkage with the micro envionment had different characteristics, and participated in differentprocesses, it is desirable to observe the behaviors of different aggregates in vitro, where the composition of the aggregates is simple and well defined.

In vitro the chlorophyll a-dioxane aggregates was different from chlorophyll a-water aggregates. in the absorption spectrum and photochemical properties etc,. In comparion of them it might provide a clue to illustrate the property and function of chlorophyll a *in vivo*.

Materials and Methods

Chlorophyll a: Chlorophyll a was prepared from an acetone extract offresh spinach leaves using a microcrystalline cellulose chromatographiccolumn.[1]
Chlorophyll a-water aggregates: Chlorophyll a was dissolved in warm, water-saturated hexane. The solution was kept at 5 ℃ over night or longer, to form chlorophyll a-water aggregates. The suspension then centrifuged,the precipitate was re-suspended in hexane.
Chlorophylla-dioxane aggregates: Chlorophyll a was disolved in warm 2% dioxane solution in hexane. After standing for 4 hours, the suspension was centrifuged. The precipitate was re-suspended in hexane.

P. Mathis (ed.), Photosynthesis: from Light to Biosphere, Vol. I, 207–210.

Photoelectric potential of chlorophyll a membrane: The change of thephotoelectric potential of chlorophyll a membrane is measured bya pHS-3 pH meter in mV range, and recorded by Dahua XWT-246 chartrecorder. the light source is a tungsten lamp,15V and 75W. Heat is filtered by water. Light intensity is about 1.6 mmole quanta m^{-2} s^{-1}.

Results and Discussion

The electric properties of chlorophyll a aggregates:

Two platinum electrodes were dipped in the chlorophyll a-dioxane aggregate suspension. When a potentiol difference of 400-800V was applied between two electrodes,the chlorophyll a-dioxane aggregates migrated toward the anode and deposited there to form a chlorophyll a-dioxane aggregate membrane,but when the electrodes were dipped in the chlorophyll a-water aggregate suspention, the chlorophyll a-water aggregates migrated toward the cathods and deposited there to form a membrane of chlorophyll a-water aggregates .The migrating speed increased considerably when the suspension was illuminated during electroplating.

This result indicated that chlorophyll a-dioxane aggregates carried negative charge, while the chlorophyll a-water aggregates carried positive charge. Under illumination , the total charge of chlorophyll a aggregates was increased, resulting in the speeding up of migration.

2, The photoelectric potential of the membrane of chlorophyll a aggregates

A platinum electrode electroplated with chlorophyll aaggregates and another one as reference electode were dipped in a pH 6.8 phosphate buffer.and linked to a pH meter to measure the electric potential. The electropotential of the membrane of chlorophyll a-dioxane agregates increased rapidly at the beginning of illumination,then reached a steady state. When light was turned off,the electropotential of the membrane returned slowly to the original value.The change could be repeated by turning on and off the light. While the electropotenial of the membrane of chlorophyll a-water aggregates decreased rapidly at the beginning of illumination, and reached steady state more quickly than that of the chlorophyll a-dioxane, the photoelectric potential of chlorophyll a-dioxane aggregates changed in the opposite diraction with that of the chlorophyll a-water aggregate, apparently due to the difference in their photoelectic properties.

When electroplating, the chlorophyll a aggregates deposited in same orientation on the surface of electrode. So under illumination the extent of the change in the photoelectric potential of the membrance was larger than that of the membrane with irregular structure.(For example, the membranewas prepared by planted an acetone solution of chlorophyll a on the electrode surface directly. The chlorophyll a aggregates could obsorb light energy and be transformed into an energized state. So under illumination the photoelectric potential change of the chlorophyll a aggregate membrane did not reach maximum immediately, and decayed slowly after illumination. The pattern of change and the decay rate of the photoelectric potential of the membrane were similar to those of delayed light emission of chlorophyll aggregates.

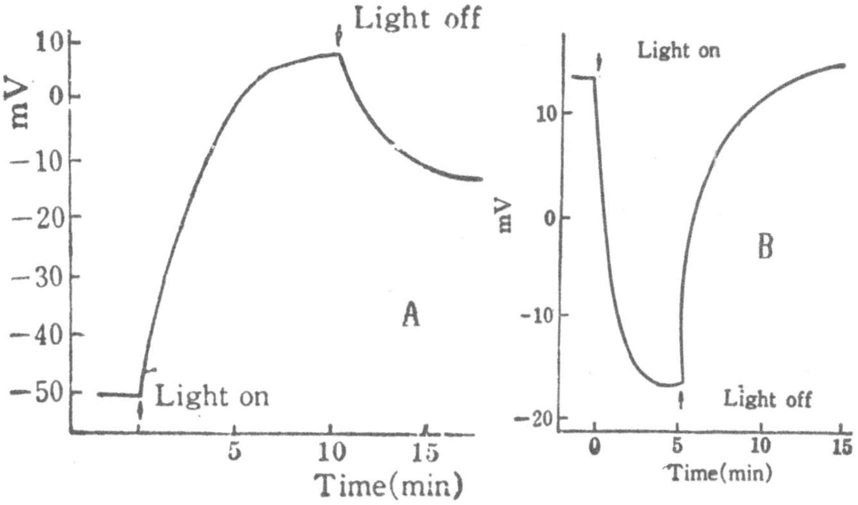

FIG.1 The electropotential change of chlorophyll a aggregates under illumination.
a: chlorophyll a-dioxane aggregates ,b: chlorophyll a-water aggregates. The light intensity is 1.6 mmolequanta $m^{-2}s^{-1}$.

3. Effects of $K_3Fe(CN)_6$ and $K_4Fe(CN)_6$ on photoelectropotential of chlorophyll a aggregates membrane .

Table 1 showed that the effects of the $K_3Fe(CN)_6$ and $K_4Fe(CN)_6$ on the photoelectric potential change of chlorophyll a-dioxane aggregates is difference to that of chlorophyll a-water aggregates under illumination.

$K_3Fe(CN)_6$ increased the positive photoelectric potential of chlorophyll a-dioxane aggregates membrane significantly, while the $K_4Fe(CN)_6$ decreased it. $K_4Fe(CN)_6$ increased the negative photoelectropotential of chlorophyll a-water aggregates membrane siginificantly, while $K_3Fe(CN)_6$ decreased it.This shows that the chlorophyll a-dioxane aggregates is different from shows that the chlorophyll a-water aggregates is different from the chlorophyll a-water aggregates in redox properties.The former is sensitive to an oxidant and the latter to a reductant.

Table 1. The effects of $K_3Fe(CN)_6$ and $K_4Fe(CN)_6$ on the photoelectric potential of chlorophyll a aggregate membrane.

	Photoelectric potential mV	
	CHl a-H_2O aggregates	Chl a-dioxane aggregates
CK	-12	+26
$K_3Fe(CN)_6$	-5	+61
$K_4Fe(CN)_6$	-38	+14

The solution containted 0.05 mol/L phosphate. The final concentration of $K_3Fe(CN)_6$ and $K_4Fe(CN)_6$ was 1mmol/L. The light intensity was 1.6 mmole quanta m^{-2} s^{-1}.

In vivo, if there were two kinds of chlorophyll aggregates difference in the characteritics between which correspond to that between the chlorophyll a-dioxane aggregates and chlorophyll a-water aggregates, It might explan why the chlorophyll in different state has different functions in the energy transformation processes.

From these results it can be inferred that the functions of the chlorophyll aggregates in photosynthetic systems under illumination are the following :

1.To absorb the light energy ,transfer light energy between chlorophyll molecules and to the reation centre.

2. To store the light energy and turn itself to an energized state,then transfer the energy between photosynthetic pigments and to reaction centres[2].

3. To establish an electric potential difference, produce reduced sitesand oxidized sites, thus results in a redox reaction take place.

This work was supported by grants from National Natural Science Foundation of China.

Refernce

1. S.Y.Yang Lumin Display Devices 6 149-153 (1985)
2. S.Y.Yang in Methods in Emzymology 214 202-207.ed. L.Packer. Academic press.

ALTERED BACTERIOCHLOROPHYLL AND CAROTENOID BINDING IN COMBINATORIAL LH2 MUTANTS OF *RHODOBACTER CAPSULATUS*

Qinghui Hu[1], Simon Delagrave[2], Douglas C. Youvan[2] and Robert A. Niederman[1]
[1]Department of Molecular Biology and Biochemistry, Rutgers University, Piscataway, NJ, 08855-1059, USA; [2]Palo Alto Institute for Molecular Medicine, Mountain View, CA, 94943, USA.

1. Introduction

Site-directed mutagenesis, in which one or at most two amino acids are changed at a time, has the potential limitation of sampling insufficient regions of a total protein sequence to allow isolation of modified proteins with more substantially informative phenotypes. Moreover, modifications based upon molecular modeling, in which protein structure is predicted from amino acid sequences, have inherent limitations in defining functionally important residues. Even when an atomic resolution structure is available, crucial positions may not be obvious, as indicated by the failure to symmetrize electron flow in modified bacterial reaction centers.

In contrast, combinatorial mutagenesis, especially when constraints can be placed on mutant screening sizes by using phylogenetic data to limit amino acid target sets, allows for rapid isolation of mutants with desired phenotypes from a large library of expressed proteins and the assessment of larger areas of protein sequence space [1]. This circumvents the need for structure-function predictions in conventional site-directed mutagenesis and maximally exploits protein diversity, useful in creating improved binding sites or catalytic capabilities, as in the engineering of new antibodies for expression in phage display libraries [2]. Combinatorial mutagenesis may also result in expression of proteins in which mutations that result in an unstable structure can be compensated for by residue changes at other sites, and may give rise to phenotypes from multiple mutations differing from the sum of the characteristics of the individual amino acid substitutions.

The LH2 (B800-850) peripheral antenna complex of *Rhodobacter capsulatus*, in conjunction with automated colony screening by digital imaging spectroscopy, has served as a useful model protein for development of combinatorial mutagenesis strategies [3-6]. This integral membrane protein is composed of a heterodimer of α- and β-polypeptides consisting of 60 and 49 amino acid residues, respectively [7, 8], with a typical conserved tripartite structure [9], as well as a γ-subunit [10]. The stability and low-temperature optical properties of the LH2 complex formed by several combinatorial mutants with residue changes in the C-terminal half of the β-polypeptide are described in this report. Many of their phenotypic characteristics can be explained by the atomic-resolution structure determined recently for the LH2 complex of *Rhodopseudomonas acidophila* by X-ray crystallography [11], thereby illustrating the utility of combinatorial mutagenesis approaches for revealing crucial structure-function relationships.

P. Mathis (ed.), Photosynthesis: from Light to Biosphere, Vol. I, 211–214.

2. Experimental Procedures

The strains used in this study and the strategies employed in their isolation are presented in Table 1.

Table 1

Combinatorial *R. capsulatus* mutants used in this study

Strain[1]	Phenotype	Sequence[2]	Ref.
Wild type	LH2+	RVFGAMALVAHILSAIATPWLG	[4]
S2	pLH1[3]car-[4]	RVFLAMGLVAHILAAIATPWLG	[4]
S4	pLH1	RVFVAMGVVAHILAGIATPWLG	[4]
S5	pLH1 car-	RVFLAMAAVAHILAGIATPWLG	[4]
P13	pLH2[5]	RVFGAMAVVAHILAFIASPWLG	[4]
REM1.8	pLH2	RVFGAMALVAHILSAILIWPVL	[5]
REM1.9	pLH2	RVFGAMALVAHILSAIALWPLV	[5]
1.4.1	pLH1	ILFLWVALVAHILSAIATPWLG	[6]
1.4.2	pLH1 car-	RVWAAGALVAHILSAIATPWLG	[6]
1.4.4	pLH1	YQWLLWALVAHILSAIATPWLG	[6]
1.4.8	pLH1	LMYAWLALVAHILSAIATPWLG	[6]

[1]Strains S2, S4, S5 and P13 obtained by using phylogenetic constraints to simultaneously mutagenize LH2-β polypeptide in six residues modeled to lie on the same face of transmembrane α-helix as His liganding B850 BChl [4]; strains REM1.8 and REM1.9 with mutations in six C-terminal residues generated with optimization algorithms using information obtained by previous rounds of mutagenesis [5]; mutants 1.4.1, 1.4.2, 1.4.4 and 1.4.8 obtained by random mutagenesis of residues 28-33 (RVFGAM)[6]. Combinatorial libraries were expressed in *R. capsulatus* U71 (LH2 chromosomal deletion background with expression of LH1 and RC inactivated by point mutation); several of original pLH1 strains also expressed in LH1⁻ LH2⁻ chromosomal deletion background to assure that spectral properties did not arise from complementation of LH2-α by LH1-β.

[2]C-terminal half (residues 28-49) of LH2-β polypeptide; altered residues underlined.

[3]Pseudo-LH1 antenna lacking B800 with red shifted B850 band.

[4]Carotenoid depleted.

[5]Pseudo-LH2 antenna with blue shifted B850 band.

Procedures for growing strains semiaerobically in the dark in the presence of tetracycline (2.5 μg/ml) and isolation of chromatophores by isopycnic sedimentation on sucrose density gradients, are described in [12]. Samples were prepared and spectroscopy was performed as described previously [13], using a Johnson Research Foundation DBS-3 dual-wavelength spectrophotometer modified for fluorometry and equipped with an Oxford DN1704 variable temperature liquid nitrogen cryostat; fluorescence polarization measurements were made with Polaroid HR filters. Levels of bacteriochlorophyll *a* (BChl) and the carotenoid spheroidenone were estimated as in [14].

3. Results and Discussion

When the LH2-β apoprotein was mutated at six residues located on the same face of the membrane-spanning α-helix as the His 38 residue, which provides the fifth ligand for B850 BChl, a number of mutants were isolated that formed a B800-less pseudo LH1 (pLH1) complex [4] with ~30-nm red shifts in their B850 band at 77 K (Table 2). Since the carotenoid content was also greatly reduced in two of these strains (S2 and S5), their B850 red shifts can be explained by polyene depletion, since a pLH1 complex lacking B800 BChl can be obtained by extraction of carotenoids from membrane-bound LH2 [16]. Because pLH1 is normally unstable both *in vitro* [16] and *in vivo* [17], combinatorial mutagenesis apparently resulted in the introduction of compensa-tory residue changes in these strains, permitting assembly of carotenoid-depleted pLH1 complexes. The B850 band of the carotenoid-depleted preparations in [16] was dissociateded by the solubilization

Table 2

Pigment content and 77 K near-IR absorption maxima of mutant chromatophores

Strain	Near-IR maxima (nm)	BChl/carotenoid (mol/mol)	ε_{mol} (300 K) (redmost λ)
U43 (pU2922)[1]	889	0.80	135
U15g (pU2922)[2]	801, 858	2.4	110
S2	886	>4.8	155
S4	889	2.6	157
S5	886	>5.7	174
P13	802, 858[3]	1.2	167
REM1.8	802, 846	1.2	231
REM1.9	802, 844	1.4	199
1.4.1	884	2.7	198
1.4.2	876	3.5	182
1.4.4	874	1.6	213
1.4.8	868	2.0	169

[1]Control strain ($RC^+LH1^+LH2^-$), containing LH1 (B880) core complex as sole antenna, constructed by complementation of $RC^-LH1^-LH2^-$ strain U43 with plasmid pU2922 [15].

[2]Pseudo-wild-type control strain ($RC^+LH1^+LH2^+$) constructed by complementation of RC^-LH1^- strain U15g with plasmid pU2922 [15].

[3]A low level of a residual LH1-like complex absorbing at ~890 nm, apparently arising from the strain U71 background, was formed in the pLH2 containing strains and could be removed by extraction of the pLH2 complex with 1% (wt/vol) LDAO and purification by sucrose gradient centrifugation. This was also the apparent basis for the increased band width and anomalously red shifted fluorescence emission spectra obtained with the pLH1 containing strains (Stokes shifts of 32-42 nm vs. 18 nm in LH1 control strain and of 15 to 19 nm for the pLH2 complex purified from strains REM1.8 and REM1.9), but the pLH1 complexes were not sufficiently stable to permit isolation by these procedures (see text).

with lauryldimethyamine *N*-oxide (LDAO) used to isolate the unaltered complex; an assessment of the detergent stability of the strain S2, S5 and 1.4.2 carotenoid-less pLH1 complexes by this procedure, as a further test of the stability-conferring properties of possible compensatory residue changes, showed that their BChl-protein associations were also labile in the presence of LDAO. Thus, gentler procedures will be necessary to obtain purified preparations of the pLH1 complex for further spectro-scopic analyses of the red shift of their B850 band, since the presence of contaminating LH1-like material in the U71 background apparently interferes with measurements on isolated chromatophores. Alternatively, it will be useful to express the pLH1-containing strains in a more appropriate deletion background.

Random mutagenesis of residues 28-33 yielded strains 1.4.1, 1.4.2 1.4.4 and 1.4.8 which also had pLH1 phenotypes [6]. While carotenoid depletion in strain 1.4.2 may account for the absence of a B800 band, it is possible that the ~10- to 20-nm B850 red shifts (Table 2) in this mutant and in strains 1.4.4 and 1.4.8, are due in part to introduction of new hydrogen bonds between B850 BChl and the tryptophan or tyrosine residues replacing βPhe30, a residue that is in contact with these BChl molecules in the LH2 crystal structure [11]. The B800-less phenotypes of the carotenoid-containing strains 1.4.1, 1.4.4 and 1.4.8 can be explained by exchanges in the crucial βArg28, which forms a strong hydrogen bond with the 2-acetyl carbonyl of B800 BChl in the crystalline complex [11]. These possibilities are currently being tested by resonance Raman spectroscopy.

An optimized strategy for mutagenesis of the six C-terminal residues of the LH2-β apoprotein yielded mutants REM1.8 and 1.9, in which the penultimate and completely conserved PW residue pair was inverted to WP [5], resulting in a 12- to 14-nm blue shift of the B850 band. The atomic resolution structure [11] shows that this tryptophan residue promotes α- and β-apoprotein associations; their disruption may result in an altered arrangement of the exciton coupled ring of B850 BChls, a possibility supported by a preliminary fluorescence polarization analysis of the pLH2 complexes isolated from these strains.

References

1 Arkin, A.P. and Youvan, D.C. (1992) Bio/Technology 10, 297-300
2 Smith, G.P. (1985) Science 228, 1315-1317
3 Youvan, D.C. (1994) Nature 369, 79-80
4 Goldman, E.R. and Youvan, D.C. (1992) Bio/Technology 10, 1557-1561
5 Delagrave, S., Goldman, E.R. and Youvan, D.C. (1993) Prot. Engn. 6, 327-331
6 Delagrave, S., Goldman, E.R. and Youvan, D.C. (1995) Prot. Engn. 8, 237-242
7 Youvan, D.C. and Ismail, S. (1985) Proc. Natl. Acad. Sci. USA 82, 63-67
8 Drews, G. (1985) Microbiol. Rev. 49, 59-70
9 Zuber, H. (1990) in Molecular Biology of Membrane-Bound Complexes in Phototrophic Bacteria (Drews, G. and E. A. Dawes, E.A., eds.) pp. 161-180, Plenum Press, New York
10 Shiozawa, J.A., Welte, W., Hodapp, N. and Drews, G. (1982) Arch. Biochem. Biophys. 213, 473--485
11 McDermott, G., Prince, S.M., Freer, A.A., Hawthornthwaite-Lawless, A.M., Papiz, M.Z., Cogdell, R.J. and Isaacs, N.W. (1995) Nature 374, 517-521
12 Olivera, L.M., Westerhuis, W.H.J. and Niederman, R.A. (1994) Biochim. Biophys. Acta 1185, 318-326
13 Westerhuis, W.H.J., Farchaus, J.W. and Niederman, R.A. (1993) Photochem. Photobiol. 58, 460-463
14 Olivera, L.M. and Niederman, R.A. (1993) Biochemistry 32, 858-866
15 Bylina, E.J., Robles, S. and Youvan, D.C. (1988) Israel J. Chem. 28, 73-78
16 Zurdo, J. Fernandez-Cabrera, C. and Ramirez, J.M. (1993) Biochem. J. 290, 531-537.
17 Lang, H.P. and Hunter, C.N. (1994) Biochem. J. 298, 197-205.

SPECTRAL PROPERTIES OF MONOMERIC BACTERIOCHLOROPHYLL *C* IN NEMATIC LIQUID CRYSTAL.

A.Dudkowiak[1], C.Francke[2], A.Planner[1], R.Cegielski[1], I.Hanyż[1], J.Amesz[2] & D.Frąckowiak[1]
[1]Institute of Physics, Poznań University of Technology, Piotrowo 3, 60-965 Poznań, Poland, [2]Department of Biophysics, Huygens Laboratory, University of Leiden, P.O.Box 9504, 2300 RA Leiden, The Netherlands

1. Introduction

Bacteriochlorophyll *c* (BChl *c*) is the main light-absorbing pigment in various species of green sulfur bacteria [1,2]. It is oriented to a high degree in chlorosomes [3-7]. The spatial arrangement of the pigment molecules, their mutual interactions and their interactions with the chlorosomal proteins, have not yet been fully elucidated [7-9].
It appears to be preferable to study a simpler ordered model simulating some aspects of the properties of the native system. We therefore tried to mimic the natural situation, where the BChl *c* are oriented in a regular manner [3-5], by introducing BChl *c* in a nematic liquid crystal (LC), because of its fluid and oriented structure, has proven to be a good model system for studying the anisotropy parameters of pigment molecules [10-12] and allowed to investigate the interaction between close-lying, uniaxially oriented, and uniformly distributed pigment molecules [10,12].

2. Procedure

3^1R-8,12 diethyl farnesyl BChl *c* (R[E,E] BChl c_F [13]) and bacteriopheophytin (BPhe) *c* were obtained from *Prosthecochloris aestuarii*. The pigments were isolated and introduced into a nematic LC as previously described [14]. Samples prepared with "dry" chloroform will be referred to as "dry" samples; those prepared from chloroform saturated with water as "wet" samples. The following LC solvents were used: p-methoxy-benzylideno-p'-butylaniline (MBBA) / p-ethoxybenzolideno-p'-butylaniline (EBBA) (3/2, v:v) and 4-(trans-4'-n-hexylcyclohexyl)-isothiocyanato-benzene (6CHBT). The pigmented LC was situated in a cell with windows covered by an orienting (SiO_x) layer. Macroscopically, such a cell can be treated to first approximation as uniaxially oriented with the axis parallel to the plane of the windows.
Absorption and fluorescence spectra were measured using a single beam spectrophotometer [3,15]. Four polarized components of fluorescence were measured for the oriented cells: VVV, VHV, VHH, VVH, (H - horizontal; V - vertical). The first and the last letter refer to the direction of the electric vector of the exciting and fluorescence light beam, respectively, and the middle letter refers to the position of the orientation axis of the LC matrix. Delayed luminescence (DL) was measured with an apparatus described elsewhere [16]. All DL spectra were taken with an additional delay of 200 ns with respect to the laser peak, in order to eliminate the prompt fluorescence contributions.

P. Mathis (ed.), Photosynthesis: from Light to Biosphere, Vol. I, 215–218.

3. Results and Discusion

The absorption spectra of [E,E] BChl c_F in the nematic LC matrix are shown in Fig.1A. The absorption spectra of BChl *c* in nematic solvent were slightly red-shifted with respect to those in chloroform due to interaction of the pigment with the LC [17]. From the position, the shape and the FWHM of the Q_y absorption band we conclude that BChl *c* in the LC is in the monomeric form.

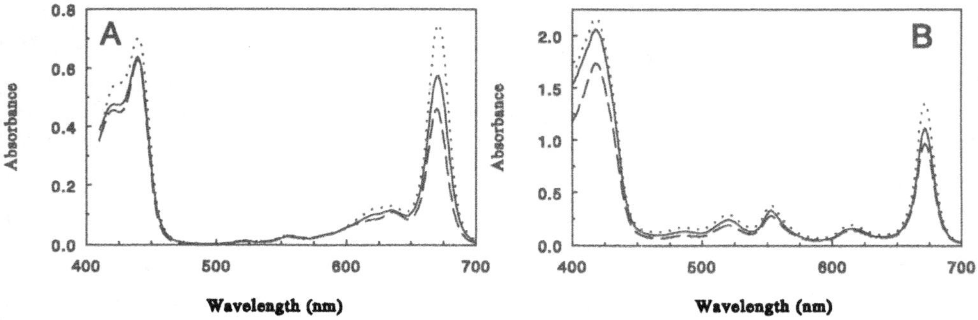

Figure 1. Polarized absorption spectra of BChl *c* (A) and BPhe *c* (B) in "dry" LC (MBBA/EBBA and 6CHBT, respectively, non-polarized component - solid line, parallel one - dotted line, perpendicular one - dashed line).

Figure 1 shows that BChl *c* and BPhe *c* in a liquid crystals introduced to these solvents from dry chloroform are uniaxially ordered to a high degree. The degree of orientation (S) and the average angles (Θ) between the direction of various absorption transition moments (TMs) of pigment and the LC orientation axis, were calculated from an analysis of the spectra of Fig.1. For the BChl *c* samples the orientation parameter S is higher in the red band region than in the Soret one. The angle between LC orientation axis and the Q_y TMs calculated for the BChl *c* equals 46° (for the "wet"), 48° (for the "dry") and for the "dry" BPhe *c* sample about 49°. Whereas the average angle for the TMs in the Soret range is about the same and differs in case of BChl *c* from the Q_y TMs by about 5 degrees. Thus, we may conclude that the y and x polarized TMs of BChl *c* are probably inclined at a similar angle with respect to the LC axis direction as well as hydration and the type of the matrix changes the average inclination angle only slightly. Figure 2 shows the polarized fluorescence of BChl *c* in MBBA/EBBA. The main maximum of the fluorescence band is located at 678 nm. It is shifted to somewhat longer wavelength with respect to its position in chloroform [14]. The fluorescence was polarized to a high degree for the "dry" sample, the intensity of the emission in the perpendicular component was independent of the sample rotation (VVH=VHH) (Fig.2A). This demonstrates the strong influence of polarized photoselection on the emission anisotropy. As usual for a uniaxially oriented sample [18], four coefficients of emission anisotropy were calculated. The anisotropy coefficients showed that the presence of water or the lack of Mg atom perturbs the arrangement of fluorescent molecules to a higher degree and has a stronger influence on the TMs distribution function around the preferential orientation axis than on the average angle of their inclination. This distribution function cannot be obtained from absorption anisotropy, but a change in it can be indicated by different fluorescence anisotropy results.

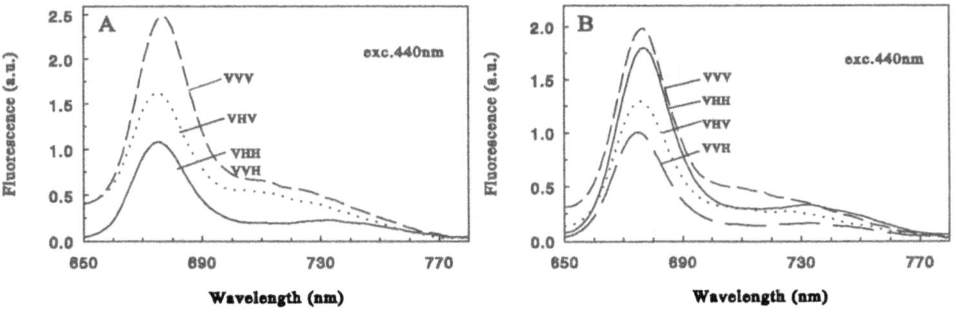

Figure 2. Polarized fluorescence spectra of BChl *c* in LC (MBBA/EBBA): (A) "dry" sample, (B) "wet" sample. Polarized components notation given in text.

Figure 3 shows the DL spectra of [E,E] BChl c_F in chloroform and in the LC. These spectra show that the DL is predominantly emitted by monomeric species. The position of the main DL maximum is similar to that of the fluorescence spectrum (Fig.2).

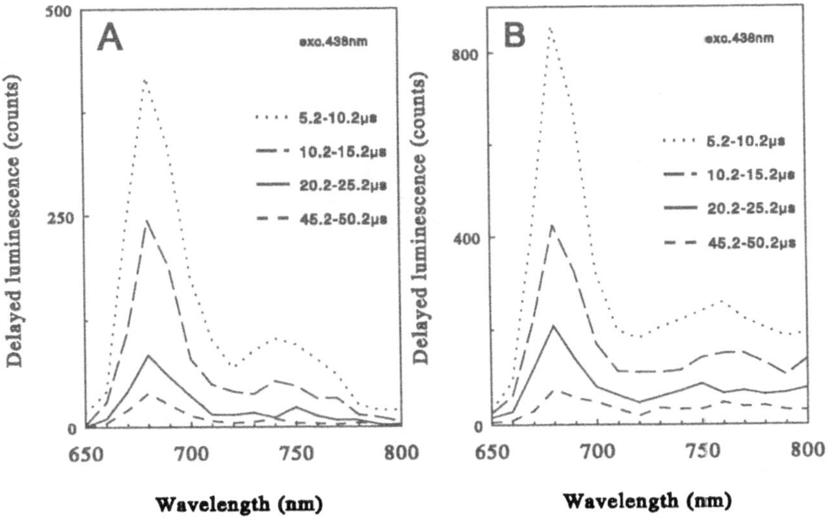

Figure 3. Delayed luminescence spectra of BChl *c* in chloroform (A) and in LC-MBBA/EBBA (B). Time windows are given in figure. Decay time τ: (A) 17$\pm$2 μs, (B) 20$\pm$2 μs.

The position of the bands are independent of the solvents (Fig.3), the excitation wavelengths and the light polarization (results not shown). Because the DL of BChl *c* and BPhe *c* in chloroform and in a LC matrix exhibit similar properties of those previously observed for Chl *a* [19], the mechanism of the DL generation seems to be the same. This means that the DL is generated by the ionization and delayed recombination of pigment molecules [20]. Also because of similarity of this DL to that observed for

BChl *c* by Krasnovsky *et al.* [21] we cannot exclude that the observed DL could be at least partially generated by the thermal activation from the triplet to the excited singlet state. It is interesting that the DL decay times τ have similar value for all samples measured in non-polarized light whereas τ of BChl *c* in LC depends on the orientation of emitting molecule with respect to LC matrix.

4. Acknowledgements

Supported by the Netherlands Foundation for Chemical Research (SON), financed by the Netherlands Organization for Scientific Research (NWO), by Poznań University of Technology grant 62-112/2 and by an ESF short-term travel grant to A.D.

References

1 Holt, S.C., Conti, S.F. and Fuller, R.C. (1966) J.Bacteriol. 91, 311-323
2 Hoff, A.J. and Amesz, J. (1991) in Chlorophylls (Scheer H., ed.) pp.723-738, CRC Press, Boca Raton
3 Otte, S.C.M., van Heiden, J.C., Pfennig, N. and Amesz, J. (1991) Photosynth. Res. 28, 77-87
4 van Dorssen, R.J., Vasmel, H. and Amesz, J. (1986) Photosynth. Res. 9, 33-46
5 van Dorssen, R.J., Gerola, P.D., Olson, J.M. and Amesz, J. (1986) Biochim. Biophys. Acta 848, 77-82
6 Blankenship, R.E., Brune, D.C. and Wittmershaus, B.P. (1988) in Light-Energy Transconduction in Photosynthesis. Higher Plants and Bacterial Models (Stevens S.E. Jr. and Bryant D.A.) pp.32-46, American Society of Plant Physiology, Rockville, MD
7 Olson, J.M. (1990) Biochim. Biophys. Acta 594, 33-51
8 Wechsler, T., Suter, F., Fuller, R.C. and Zuber, H. (1985) FEBS Lett. 181, 173-178
9 Holzwarth, A.R. and Schaffner, K. (1994) Photosynth. Res. 41, 225-233
10 Frąckowiak, D., Bauman, D. and Stilman, M.J. (1982) Biochim. Biophys. Acta 681, 273-285
11 Frąckowiak, D., Hotchandani, S. and Leblanc, R.M. (1983) Photobiochem. Photobiophys. 6, 339-350
12 Frąckowiak, D., Bauman, D., Manikowski, H. and Martynski, T. (1977) Biophys. Chem. 6, 369-377
13 Smith, K.M. (1994) Photosynth. Res. 41, 23-26
14 Dudkowiak, A., Francke, C., Amesz, J., Planner, A., Hanyż, I. and Frąckowiak, D. (1995) Spectrochim. Acta (submitted)
15 Francke, C., Otte, S.C.M., van Heiden, C.J. and Amesz, J. (1994) Biochim. Biophys. Acta 1186, 75-80
16 Planner, A. and Frąckowiak, D. (1991) Photochem. Photobiol. 54, 445-449
17 Frąckowiak, D., Szurkowski, J., Szych, B., Hotchandani, S. and Leblanc, R.M. (1986) Photobiochem. Photobiophys. 12, 9-19
18 Frąckowiak, D., Dudkowiak, A., Zelent, B. and Leblanc, R.M. (1991) J. Fluorescence 1, 225-234
19 Frąckowiak, D., Planner, A. and Goc, J. (1993) Photochem. Photobiol. 58, 737-744
20 Frąckowiak, D., Zelent, B., Malak, H., Cegielski, R., Goc, J., Niedbalska, M. and Ptak, A. (1995) Biophys. Chem. 54, 95-107
21 Krasnovsky, A.A. Jr., Cheng, P., Blankenship, R.E., Moore, T.A. and Gust, D. (1993) Photochem. Photobiol. 57, 324-330

OLIGOMERIC STATES OF THE LHI AND LHII COMPLEXES FROM *RUBRIVIVAX GELATINOSUS* IN DETERGENT SOLUTIONS

V. Jirsakova, J.L. Ranck and F. Reiss-Husson
Centre de Génétique Moléculaire, C.N.R.S. F91198 Gif-sur-Yvette, France

1. Introduction

The membranes of several purple bacteria contain two types of antenna, LHI (B870) and LHII (B800-850) which mediate light energy collection. Both are multimeric aggregates of (αβ) polypeptide units and pigments. Recent X-ray diffraction studies (1) show that LHII from *Rps. acidophila* consists of cylindrical rings built by 9 α and 9 β with their associated pigments ; at a lower resolution, electron crystallography of LHI from *R. rubrum* also indicated the presence of rings but comprising 16 αβ in this case (2). A related topic is the state of association of these complexes when isolated in detergent solution, which a priori could differ from that in the crystalline state or in the membrane. Thus *Rps. marina* LHI in octylglucoside was proposed to be a (αβ) dodecamer (3), and *Rps. molischianum* LHII in various detergents an octamer (4). In this work we have investigated the degree of association of isolated *Rv. gelatinosus* LHI and LHII using analytical centrifugation and electron microscopy.

2. Procedure

LHI was purified from *Rv. gelatinosus* S1 (strain isolated by R. Uffen) as described (5). LHII was solubilised from chromatophores by LDAO and purified by ion-exchange in presence of this detergent. Samples were concentrated by ultrafiltration and dialyzed extensively against the buffer used for the centrifugation experiments : for LHI, 10 mM Tris-HCl, pH 8.0, 1 mM EDTA, 2 mg/ml decanoylsucrose ($C_{10}S$, Calbiochem); for LHII, same buffer except for 1 mg/ml LDAO (Fluka) instead of $C_{10}S$.
Lipids bound to the purified antennae were identified by thin layer chromatography and quantified by P and fatty acid analysis of total lipid extracts by gas phase chromatography (in collaboration with A. Trémolières).
The densities of buffers and of the antenna samples were measured at 5°C with a DMA06 densitometer (Anton Paar, Graz, Austria) for determining buoyant factors.
Buoyant molecular weights were measured by sedimentation equilibrium at 5°C in a Beckman Optima XL-A ultracentrifuge. The concentration of antenna as a function of radial distance was determined by scanning at 376 nm, i.e. in the Soret band.
Samples were examined by electron microscopy after negative staining with uranyl acetate, using a Philips CM12 electron microscope operating at 80KV.

P. Mathis (ed.), Photosynthesis: from Light to Biosphere, Vol. I, 219–222.

3. Results and Discussion

3.1 Molecular weight of the LHII antenna.

Biochemical analyses of the purified LHII indicated that its elementary units consisted of (αβ/3Bchl/1.3car). A large heterogeneity in carotenoid composition was observed, and several phospholipids were still present, the major ones being phosphatidyl-glycerol and -ethanolamine, amounting to 3-4 molecules per elementary unit (Jirsakova, unpublished). The equilibrium sedimentation runs yielded linear plots of the logarithm of LHII concentration c against the square of the radial displacement r indicating that a single sedimenting particle was monitored. The experimental parameter is :

$$M_p(1-\phi'\rho) = (2RT / \omega^2)(d\ln c / dr^2) \quad (1)$$

where M_p is the LHII molecular weight (comprising pigments), ρ is the buffer density (g/cm^3), ω the angular velocity and $(1-\phi'\rho)$ the buoyant density factor defined by eq. (2):

$$(1-\phi'\rho) = (1 - \nu_p\rho) + \delta_{det}(1 - \nu_{det}\rho) + \delta_{lip}(1 - \nu_{lip}\rho) \quad (2)$$

where ν_p is the partial specific volume of the antenna, δ_{det} and δ_{lip} the amounts (w/w) of detergent and lipid bound to the antenna, respectively, and ν_{det} and ν_{lip} their partial specific volumes.

Thus M_p can be determined if the buoyant factor is known. We determined this factor by measuring the density ρ_{sol} of LHII solutions (previously equilibrated with the sedimentation buffer) as a function of antenna concentration (6) ; a plot of ρ_{sol} versus c is used for getting the slope at infinite dilution, which is equal to the buoyant factor (Fig. 1).

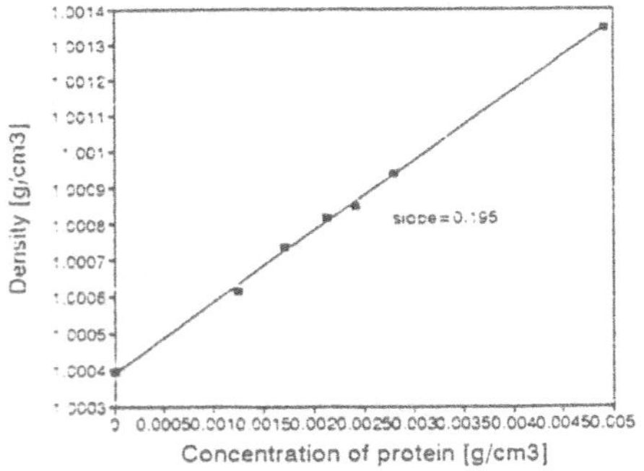

Fig. 1 Determination of the buoyant density factor of LHII

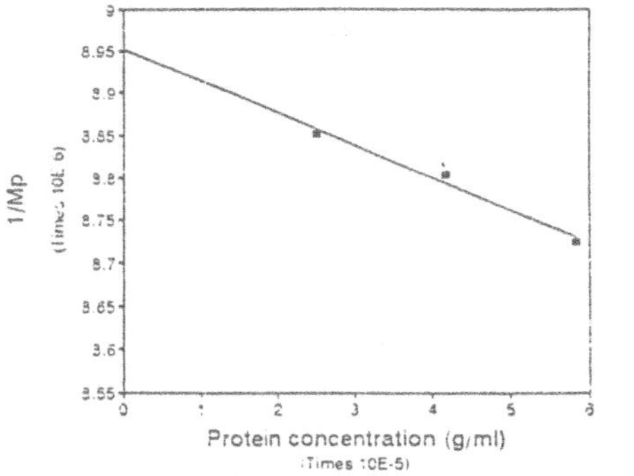

Fig. 2 Extrapolation at infinite dilution of M_p^{-1} values for LHII

Table I	LHII	LHI
ρ (g/cm^3)	1.00039	1.00525
ν_p (5°C, g/cm^3)	0.777	0.795
ν_{det} (5°C, g/cm^3)	1.114	0.769
$(1 - \phi'\rho)$	0.195	0.71
M_p	**111700 ± 3200 Da**	**180650 ± 8600 Da**
Oligomeric degree	**7.0 ± 0.2**	**12.6 ± 0.6**
δ_{det} (g/g prot.)	0.38	2.23

We performed equilibrium sedimentations at different antenna concentrations, and the dependance of M_p on the concentration was taken into account (4) by extrapolating at infinite dilution (Fig. 2) according to eq. (3) :

$$(M_{p,c})^{-1} = (M_{p,c=0})^{-1} + Bc \quad (3)$$

In Table I we present the parameters used in these calculations and the results. The molecular weight determined for the antenna corresponds to an heptamer of elementary (αβ/3Bchl/1.3car) units. Using the experimental value of the buoyant density factor the amount of bound LDAO could be computed by using eq. (2) (Table I) ; 182 LDAO molecules were bound per heptamer, leading to 156 KDa for the whole complex including detergent and lipids .

3.2 Molecular weight of the LHI complex

Biochemical analyses of purified LHI show that it was composed of (αβ/2Bchl/2hydroxyspheroidene) units (5) to which 2-3 phosphatidylethanolamine molecules were still bound (Jirsakova, unpublished).

The buoyant factor of LHI samples in $C_{10}S$ solutions (Table I) was measured by densitometry as described above. Sedimentation velocity experiments indicated that several components were present in these solutions (not shown); in equilibrium sedimentation runs, plots of $\ln(c)$ versus r^2 were not linear, and were fitted with at least 2 lines of different slopes. We took into account only the initial slope of two of these plots for obtaining the minimum molecular weight of LHI antenna (Table I), which indicated that the smallest oligomer is a dodecamer. The amount of $C_{10}S$ bound to the antenna could also be estimated by using eq.(2); a large amount of bound detergent was found (Table I) leading to about 600 KDa for the detergent-dodecamer complex, in agreement with previous gel filtration experiments (5).

3.3 Electron microscope images of isolated LHI and LHII

Negatively stained images representative of LHI and LHII are shown in Fig. 3.

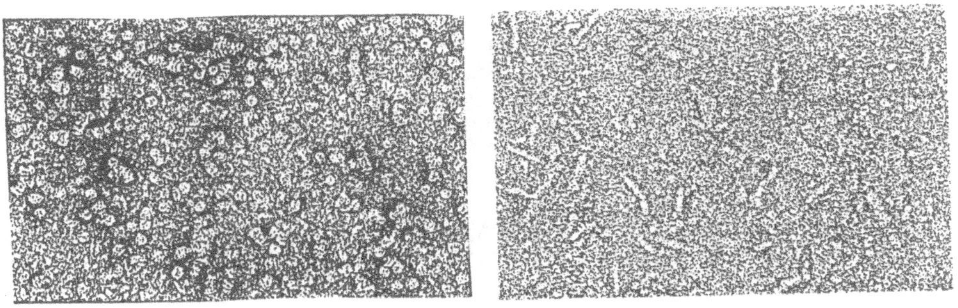

Fig. 3. Negatively stained images of antenna complexes. Left : LHI in decanoyl-sucrose; magnif. x 105 000 (1 cm = 952 Å). Right : LHII in LDAO; magnification x 135 000 (1 cm = 740 Å)

LHI samples contained ring-like particles, with a central pit of 25Å width and an outer diameter of about 180 Å when viewed from the top; their thickness could be measured in the stacks seen from the side, which seem to be formed by units 35-40 Å thick. The isolated ring-like particles probably corresponded to dodecamer/detergent complexes.

Indeed taking into account the molecular weight determined for the LHI/detergent complex, the known amount of bound detergent and the values of the partial specific volumes, the volume of the sedimenting particle could be calculated. This value (about 680×10^3 Å^3) corresponds to a ring-like particle with a central hole of 25Å and an outer diameter of 160Å, if it were supposed to be 35Å thick.

Images of LHII samples were quite different. A few globular particles were viewed from the top ; most of them, viewed from the side, formed rod-like aggregates with an outer diameter of 75 Å and variable length. The resolution was not sufficient for determining the substructure of the stacks. Because of the same reason, the particles when viewed from the top were too small for being precisely distinguished from the background. We suppose that they represent the heptamer/detergent complexes detected in the more diluted samples used for centrifugation. In support of this hypothesis, we can calculate a particle volume of 136 560 Å^3 considering the molecular weight determined for the LHII heptamer, the amount of bound LDAO and the partial specific volumes of both ; supposing a particle thickness of 35 Å this volume corresponds to a ring of 75 Å outer diameter with a central hole of 25 Å. The presence of numerous stacked particles is typical for negatively stained membrane proteins samples (7) ; it could also be explained by the much higher concentrations used for electron microscopy than for centrifugation experiments.

In conclusion, the presence of dodecamers as the smallest aggregates of LHI in decanoyl-sucrose solutions, and of heptamers for LHII in LDAO solutions, which is indicated by the centrifugation experiments, is compatible with the electron microscope images of these solutions. The degree of oligomery might depend on the detergent ; indeed we observed some differences in the electron microscope images of LHI samples when the detergent was dodecyl-maltoside (Jirsakova and Ranck, unpublished). While the existence of dodecamers was often proposed for LHI in several bacterial species (in vivo and in vitro), other degrees were more recently observed for *R. rubrum* LHI (2) and *Rps. molischianum* LHII (4). *Rv. gelatinosus* LHII heptamers also differ from the *Rps. acidophila* LHII trimers which are the building blocks of the ring-like nonameric crystalline assembly (1). Species-linked differences are possible, especially for *Rv. gelatinosus* LHII because of peculiar features already noted in the primary sequence of its α polypeptide (i.e. a long hydrophobic C-terminal extension) (8). Oligomery, which is a general trend for these antennae, probably also depends on physical parameters such as the state of the sample (crystalline or in solution), temperature, pH, and so on.

References

1 Mc Dermott, B., Prince, S.M., Freer, A.A., Hawthornthwaitelawless, A.M., Papiz, M.Z., Cogdell, R.J. and Isaacs, N.W. (1995) Nature 374, 517-521

2 Karrasch, S., Bullough, P. A. and Ghosh, R. (1995) EMBO J. 14, 631-638

3 Meckenstock, R. U., Brunisholz, R. A. and Zuber, H. (1992) FEBS Letters 311, 128-134

4 Kleinekofort, W., Germeroth, L., van den Broek, J. A., Schubert, D. and Michel, H. (1992) Biochim. Biophys. Acta 1140, 102-104

5 Jirsakova, V. and Reiss-Husson, F. (1993) Biochim. Biophys. Acta 1183, 301-308

6 Butler, P.J.G. and Kühlbrandt, W. (1988) Proc. Natl. Acad. Sci. USA 85, 3797-3801

7 Boonstra, A.F., Visschers, R.W., Calkoen, F., Van Grondelle, R., Van Bruggen, E.F.J. and Boekema, E.J. (1993) Biochim. Biophys. Acta 1142, 181-188

8 Brunisholz, R.A., Suter, F. and Zuber, H. (1994) Eur. J. Biochem. 222, 667-675

LIGHT-HARVESTING COMPLEXES FROM A BROWM ALGA. BIOCHEMICAL AND MOLECULAR STUDY.

L. Caron, D. Douady, B. Rousseau, M. Quinet-Szely and C. Berkaloff.
URA CNRS 1810, Ecole Normale Supérieure, 46 rue d'ULM, 75230 PARIS.

A great part of oceanic primary production is performed by chromophytic algae belonging to several classes. Due to specific pigments, chlorophyll c and special xanthophylls, these organisms are well fitted for harvesting the blue and green photons which are the most abundant under-sea.

The light-harvesting protein complexes (LHC) of Chromophytes are composed of polypeptides with molecular masses ranging from 17 to 22 kDa, which is significantly smaller than their Chlorophyte counterparts. These polypeptides are in vivo assembled in macromolecular complexes, with molecular masses ranging from 120 to 700 kDa: the molecular architecture of these native complexes is still poorly understood. The polypeptide subunits are encoded by a nuclear multigene family, the study of these genes is still at its beginning. In some species but not in all, these genes are clustered on the genome. The total number of genes is not yet established.

Chloroplasts of Chromophytes are known to be surrounded by four membranes, the two peripheral being of ER origin. So, the model of protein targeting to these chloroplasts has to take into account this ultrastructural feature.

In this work, we have solubilized by n-octyl-β,D-glucopyranoside (OGP) the native LHC from a brown alga and separated by DEAE ion-exchange chromatography or isoelectrofocusing fractions with distinct pigment composition. We have cloned and sequenced a LHC gene from the macrophyte brown alga *Laminaria saccharina* (L.) Lamouroux and compared it with the recently published cDNA sequences from *Macrocystis pyrifera*, the only brown alga LHC sequence known at this time, and with the other LHC ones from Chromophytes and Chlorophytes. We also discuss the characteristics of the transit sequence

1. LHC heterogeneity: separation of several discrete subunits

Native LHC fraction was solubilized from chloroplasts with digitonin and isolated on sucrose gradient (1). This fraction was treated with O.6% (OGP) (w/v) and the subunits separated in two ways:

1.1 DEAE separation
Two LHC fractions were separated by DEAE exchange column chromatography with a 0 to 400mM NaCl gradient. The elution buffer was 10mM sodium phosphate pH7.5, O.6% OGP. Both were enriched in fucoxanthin and Chl c relative to native LHC. Only one of

P. Mathis (ed.), Photosynthesis: from Light to Biosphere, Vol. I, 223–226.

them, fraction 2, contained violaxanthin (Table1). The polypeptides of the two fractions have molecular masses of 20 kDa.

The fluorescence excitation spectra indicated a good connexion between harvesting pigments (Chl c and fucoxanthin) and Chl a (Figure 1,A).

1.2 IEF separation

Before loading on isoelectric focusing gel, LHC fraction was dialysed against 1% glycine and centrifuged. The pellet was resuspended in an adequat volume of 1% glycine buffer + O.6 %OGP to obtain 150 mg Chla/ml. IEF was performed on a 4% ultrodex gel containing O.6% OGP, pH range 3-7, using a Multiphor II apparatus.

Three fractions were separated and when compared to native LHC, fraction 1 is enriched in fuco- and violaxanthin, fraction 2 in fucoxanthin and Chl c, fraction 3, the most abundant, was slightly depleted in fucoxanthin and Chlc, it contains no violaxanthin (Table1). They were composed of polypeptides in the 20 kDa range. A weak difference was observed on denaturing gel between the molecular masses of peptidic components of the fractions. This has to be confirmed by other analysis methods (data not shown).

The fluorescence excitation spectra indicated that the connexion between light-harvesting pigments and Chl a was preserved in fraction 3 (Figure 1,B).

Pigments	native LHC	DE AE		IEF		
		F 1	F 2	F 1	F 2	F 3
Chl a + Chide	100,0	100,0	100,0	100,0	100,0	100,0
Chl c	26,3	33,2	30,8	0,0	76,2	21,2
Fucoxanthin	95,8	120,0	114,7	123,7	138,2	89,5
Violaxanthin	8,5	0,0	3,5	20,8	1,5	0,0
ß-carotene	1,3	1,7	0,5	2,0	2,0	0,0

Table 1: Pigment compositions of the fractions obtained by DEAE and IEF. Comparison with native LHC. F1, F2, F3: Fractions 1, 2, 3 respectively.

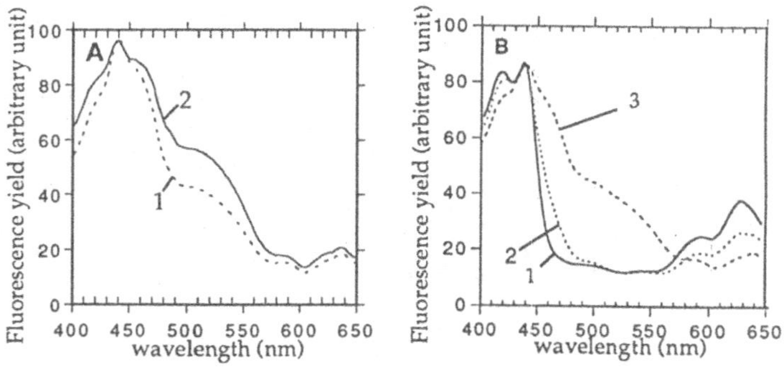

Figure 1: Fluorescence excitation spectra at room temperature of the fractions obtained by DEAE and IEF. A: DEAE, B: IEF. 1, 2, 3: fractions 1, 2, 3 respectively.

The two methods separated fractions with different pigment composition and in some of them the connexions between light-harvesting pigments and Chl a were preserved. Experiments are in progress to correlate the results obtained by the two methods.

2. Cloning and sequencing of a gene coding for a LHC protein.

Nucleic acids were extracted from sporophytes of *L. saccharina* according to Apt *et al.* (2) The polysaccharides were precipited with 25% ethanol before purification of nuclear DNA on CsCl gradient.

The genomic DNA was digested with EcoRI and a genomic bank was constructed in EMBL4. To screen the bank, a PCR probe was obtained from two oligonucleotides, the sequences of which have been deduced from LHC partial protein sequences (3). Several clones positively reacted with the probe. One of them, containing a 17kb insert, has been subcloned in Bluescript SK+. Finally, double-stranded sequencing of a 1.8kb subclone was performed.

The coding DNA sequence corresponding to a LHC protein is 672 bp long and the deduced amino acid sequence, presented here, corresponds to the 46 amino acids of the transit peptide plus 178 amino acids of the mature protein (Figure2).

```
MLTLIPLVLAFLSXASAALTLCFCFCSAFNGAALTTPVKASSAMKM*
------MKS:VMAV:C::APG-:RGP::::::::::SA::C:::::*

SFESEIVXQAPLGFWDPLXLLADADQERFERLRYVEVKHGRIRMLA
::::::GA::::::::::G::::::::::::::::::::::::A::

IAGHLTQQNTRLPGMLSNSANLSFADMPNGVAALSKIPPGGLAQIF
:::::::::::::::::::::::::::::::::::::::A::::::

GFIGFLELAVMKNVEGSFPGDFTNGGNPFASSWDNMSEETQASKRA
A::::::::::::::::::::::L:::::GA::A::::::::::::

IELNNGRAAQMGILALMVHEELNNQPYIINNLVGASYTFN
::::::::::::::::::::::::K::V::D:::::::::
```

***:presumed N-terminus for the mature proteins.**

Figure 2: Amino acid sequence alignements. Homology between *Laminaria saccharina* and *Macrocystis pyrifera* LHC (2).

Several LHC sequences deduced from cDNA have been published (2) for an other Laminariale: *Macrocystis pyrifera*. There is a high homology between *M. pyrifera* and *L saccharina* LHC mature proteins: 87 to 92% of the amino acids are identical according to the appropriate *M. pyrifera* LHC sequences. An alignment between the *L. saccharina* sequence and one of those of *M. pyrifera* is shown. The dots indicate identical aminoacids. Concerning the transit peptide, its COOH-part is very homologous but the NH2-part presents very few homology with their *M. pyrifera* counterparts.

3. Comparison with the LHC proteins of Chla/b plants.

All LHC proteins of photosynthetic organisms with Chl a contain 3 hydrophobic domains. It has been suggested that they are transmembranar helices (4)

The structure of the light-harvesting Chla/b protein complex of green plants has been determined by electron crystallography of 2-D crystals (5) and has confirmed the model of three membrane-spanning α-helices (MSH). Two of them (A and B) are held together by ion pairs formed by charged residues that also serve as chlorophyll ligands. These ion pairs would provide strong attractive forces between the two helices and are likely to stabilize the protein in the membrane. The third MSH, which links the two others MSHs, also binds chlorophylls. Two molecules of lutein, which is the major xanthophyll of the Chla/b LHCs, have been found by monomer.

Brown alga LHC proteins show similarity with the green plant ones. Two hydrophobic domains can be aligned with the two helices A and B, they contain Arg (R) and Glu (E) residues at such positions which could form two ion pairs. The third hydrophobic domain can't be aligned with any helice of the Chla/b LHCs. According to the program of secondary structure predictions of Garnier et al 1987, this region only partially fits a transmembrane helice; however it has to span the membrane to maintain the correct orientation of the two A and B helices. This region also exhibits the most extensive variability between the LHC of Chlc-containing algae.

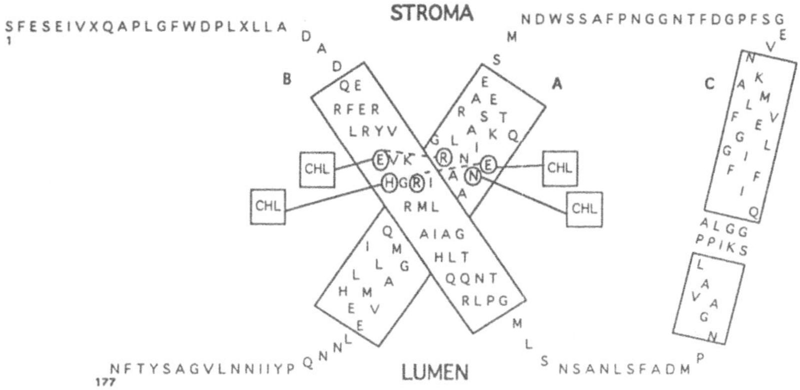

Figure 3: Hypothetical structure of *L. saccharina* LHC according to the model proposed by Kühlbrandt *et al.*(5) for green plant LHC.

Where are bound the pigments?

4 Chla are likely to be electrostatically associated with the ion pair residues as proposed for Chla/b LHCs. Concerning the Chlc molecules, as Chla they have a tetrapyrrol ring; they probably are linked by the ring to the protein, however they have no phytyl chain, but a negatively charged carbonyl group. We did not recognize other chl sites in the sequence. Are molecules of Chlc situated on the MSH C?

The major xanthophyll is fucoxanthin in brown algae and the isolated complexes contain about the same amount of fucoxanthin and Chla, i.e. a much higher amount of xanthophylls than in green plant LHC monomer. Are they located between the helices, inside the complexes or at the periphery? Furthermore, As green plant ones, the brown alga LHC contain violaxanthin, involved in the « xanthophyll cycle », and thus in photoprotection processus. Its location also remains to determine.

References

1 Caron *et al.* (1985) *Plant Cell Physiol.* 26,131-139.
2 Apt *et al.* (1995) 246,455-464
3 Douady *et al.* (1994) *Biochemistry* 33, 3165-3170.
4 Green et al (1992) in research in Photosynthesis (Murata, N. ed.)Vol I, pp195-202, Kluwer Academic Publishers, Dordecht, the Netherlands.
5 Kühlbrandt *et al.*(1994) *Nature*.367, 614-621.

DECREASE OF THE PS I ANTENNA COMPLEX WITH INCREASING GROWTH IRRADIANCE IN *PORPHYRIDIUM CRUENTUM*

Shi Tan, Gregory R. Wolfe*, Francis X. Cunningham, Jr and Elisabeth Gantt
Dept. Plant Biology and MAES, University of Maryland, College Park, MD 20742; * Dept. Botany, Arizona State University, Tempe, AZ 85287, USA

1. Introduction

In contrast to higher plants and green algae, most of the Chl in cells of the red alga *Porphyridium cruentum* is attributable to PS I (1,2). We previously found that the Chl/P700 ratio in *P. cruentum* decreases by about 20% with increasing PFD (6 to 280 μmol photons$\cdot$ $m^{-2}\cdot$ s^{-1}), but the stoichiometry of the photosystems remains relatively constant over this entire range of growth irradiance (2). With a fractionation procedure developed for *P. cruentum* it is possible to recover >90% of the total Chl in Chl-binding fractions upon detergent solubilization of thylakoid membranes (3). Most of the Chl is energetically coupled to PS I holocomplexes (52% of total Chl), and the majority of Chl present in a second fraction (16% of total Chl) is probably also assignable to PS I. With this method for isolation of PS I holocomplexes it is now possible to quantitatively assess the size and composition of PS I in cells of *P. cruentum* acclimated to growth under different PFD of white light.

2. Materials and Methods

Porphyridium cruentum (ATCC 50161) was grown at 18°C in 5% CO_2 in air (2), and acclimated for many generations in continuous light at 10 (low light, LL), 50 (medium light, ML), 150 (high light, HL), and 280 μmol photons$\cdot$ $m^{-2}\cdot$ s^{-1} (very high light, VHL). Cells were harvested in exponential growth at 3 x 10^6 cells $\cdot$ mL^{-1}. For thylakoid fractionation and PS I isolation the method of Wolfe et al. (3) was followed. Thylakoids isolated in 0.5 M sucrose, 2.5 mM Na-EDTA, and 10 mM $NaPO_4$ buffer (pH 7.0), 4°C, were concentrated on a sucrose gradient, and rinsed in 1 M NaBr in 0.3 M sucrose, 20 mM NaCl, 5 mM $MgCl_2$, 50 mM Tris-HCl (pH 7.8). After rinsing in 0.4 M sucrose, 50 mM MES (pH 6.0), 10 mM NaCl they were solubilized (30 min) in dodecyl-β-D-maltoside (β-DM) with a final concentration of 0.66% (w/v) yielding a detergent:Chl ratio of 37:1 (w/w).

PS I was recovered in the lower (LGB) of two Chl-containing bands from a linear gradient of 15 to 30% sucrose in 50 mM MES (pH 6.0), 10 mM NaCl, 10% glycerol (v/v) and 0.04% β-DM by centrifugation in a Beckman 60 Ti rotor for 16 h at 252,000 x g, followed by two separations on Q Sepharose Fast Flow (Pharmacia) anion exchange columns. PS I-mediated electron transport rates were measured as O_2-uptake by the

P. Mathis (ed.), Photosynthesis: from Light to Biosphere, Vol. I, 227–230.

Mehler reaction using ascorbate-reduced DCPIP (5 mM ascorbate and 0.1 mM DCPIP) as an electron donor and methyl viologen as an electron acceptor (50 μM) (4).

A polyclonal antiserum was prepared in rabbits to LHC I from purified holocomplex and after SDS-PAGE. Polypeptides separated on polyacrylamide gels were transferred to Immobilon-P membranes and immunoprobed with LHC I antiserum of *P. cruentum*, and LHC II antisera of *Euglena gracilis* and *Chlamydomonas reinhardtii* (gift from Dr. J. K. Hoober's laboratory). Chl and P700 determinations were as in (3).

3. Results and Discussion

In cells acclimated for many generations over their entire light growth range the reaction center stoichiometry (Q_A/P700) remains relatively constant (2) (Table 1). However, Chl/P700 decreases with increasing growth light intensity. PS I activity, measured as O_2-uptake by the Mehler reaction, was 26% greater in thylakoids of cells acclimated at 280 μmol photons· m^{-2}· s^{-1} (VHL) than in cells acclimated at 10 μmol photons· m^{-2}· s^{-1} (LL), indicating a change in the light absorbance capacity of PS I. In isolated PS I holocomplexes the PS I activity also was 22% greater in VHL cells than in LL cells.

Table 1. About 25% decrease of Chl in cells and PS I holocomplexes of LL→VHL.

Cell content:		
	Chl/P700 (mol/mol)	Q_A/P700
LL	181 ± 8	0.54 ± 0.06
ML	172 ± 13	0.45 ± 0.11
HL	152 ± 7	0.43 ± 0.02
VHL	148 ± 1	0.53 ± 0.01
PSI Holocomplex:		
	Chl/P700 (mol/mol)	% of VHL
LL	165 ± 3.5	125
ML	156 ± 23.1	118
HL	148 ± 18.0	112
VHL	132 ± 9.4	100

Examination of PS I holocomplexes on SDS-PAGE from LL→VHL showed an overall decline in the 19.5 to 23.5 kDa region, with a notable decrease of three polypeptides (19.5, 21.0 and 22 kDa) in VHL-complexes relative to LL-complexes (Fig. 1). These polypeptides belong to a novel LHC I complex that binds Chl *a* and zeaxanthin and is functionally coupled to the RC 1 core (5). At least six different polypeptides are reactive with anti-LHC I. We suggest that the decrease in PS I Chl antenna size observed with increasing irradiance is attributable to changes occurring in the LHC I-antenna complex, although it is not known if all bind Chl *a* and zeaxanthin and in what amounts.

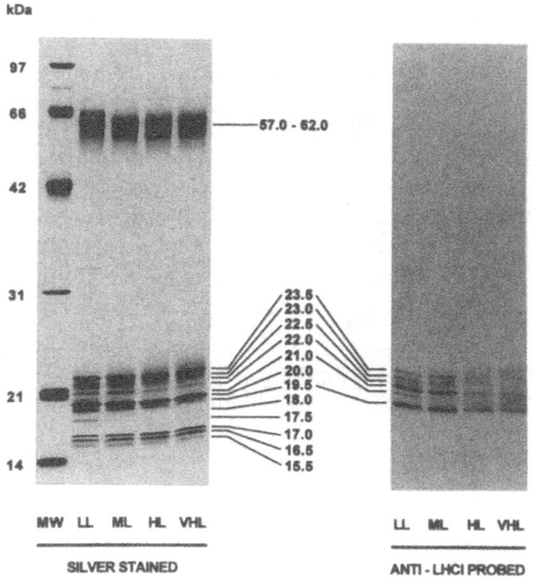

FIG. 1 In PS I holocomplexes (at equal Chl) LHC polypeptides decrease from LL→ VHL, notably 19.5, 21 and 22 kDa. Six polypeptides on right react with anti-LHC I.

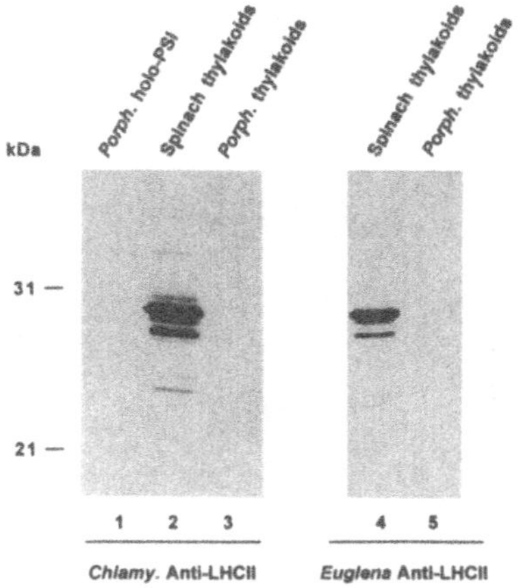

FIG. 2 Lack of LHC II in *P. cruentum* is indicated by immunoprobing with antisera to LHC II apoproteins from *Chlamydomonas reinhardtii* and *Euglena gracilis*. Load per lane was 1.5 μg Chl. 12% SDS-PAGE.

Confirmation for a Chl-binding antenna complex associated with PS II core complexes in *P. cruentum* has not been found. LHC II-type polypeptides were not observed in functionally active PS II preparations (3), nor did we detect polypeptides that showed immunocross-reactivity with LHC II specific antisera (made to *Chlamydomonas* and *Euglena* LHC II) (Fig. 2).

Evidence that the variable polypeptide components of the PS I holocomplex actually bind Chl and carotenoid pigments is indirect but convincing. First, these polypeptides, associated with the holocomplex, are immunologically related to Chl *a/b* and Chl *a/c* binding LHC polypeptides of plants and algae (5). Second, the removal of these polypeptides from the PS I holocomplex, using a procedure that does not yield free Chl pigment, is accompanied by removal of much of the Chl and most of the zeaxanthin of the PS I holocomplex (5). Third, considerable sequence similarity with *Lhca* genes of higher plants (Jansson 1994) is evident near the N-terminus of the first *P. cruentum* cloned gene sequence (**P K N L D G S M A G D V G F D P L G F**; residues identical to consensus *Lhca* sequence are underlined) and from partial amino acid sequence determinations of five LHC I polypeptides (unpublished data in collaboration with R. Aebersold and A. Ducret).

Conclusions

Major changes in Chl/P700 in thylakoids grown under different light intensities are attributable to decreases in PS I.

PS I holocomplexes of low light grown cells contain 25% more Chl than in high light grown cells.

Evidence for LHC II is lacking in *P. cruentum*.

Preliminary evidence from gene sequence comparison indicates that LHC I polypeptides of Chl*a/b*-containing plants and red algae are related.

References

1 Ley, A.C. and Butler W.L. (1980) Plant Physiol 65, 714-722

2 Cunningham, F.X. Jr., Dennenberg R.J., Mustardy L., Jursinic P.A. and Gantt E. (1989) Plant Physiol. 91, 1179-1187

3 Wolfe, G.R., Cunningham F.X. Jr., Grabowski, B. and Gantt, E. (1994) Biochim. Biophys. Acta 1188, 357-366

4 Allen, J.F. and Holmes, N.G. (1986) in Photosynthetic Energy Transduction: A Practical Approach (Hipkins M.F. and Baker N.R. eds.) pp.103-142, IRL Press, Oxford

5 Wolfe, G.R., Cunningham F.X. Jr., Durnford D., Green, B.R. and Gantt, E. (1994) Nature 367, 566-568

6 Jansson, S. (1994) Biochim. Biophys. Acta 1184, 1-19

THEORETICAL ANALYSIS OF FEMTOSECOND COHERENT NUCLEAR MOTION IN PHOTOSYNTHETIC REACTION CENTERS

E. Gaizauskas and L. Valkunas, Institute of Physics,
Gostauto 12, 2600 Vilnius, and Vilnius University,
Sauletekio al 10, 2054 Vilnius, Lithuania

1. Introduction

Photoinitiated primary stages of the electron transfer in the photosynthetic reaction center (RC) are completed within a few ps time domain, and the femtosecond time-resolved spectroscopy seems to be very promising in investigating these ultra-fast events of charge separation [1,2]. However, in addition to the high resolution, it reveals new phenomena which are due to the coherence created by light-matter interaction that is absent in the conventional picosecond spectroscopy results. Under these conditions, when the dephasing time T_2 (either electronic or vibrational) occurs comparable to or even longer than the pulse duration used in pump-probe measurements, the coherent light-matter interaction greatly modifies the material response with the pulse duration scale and has to be taken into consideration even for relatively week fields [2-8]. The fact that the primary electron donor is a dimer can also create the additional resonances, for instance, a two-photon resonance, and should be taken into account.

2. Equation of motion

In the case of resonance excitation conditions, the electronic degrees of freedom are usually well approximated as a two-level system characterized by energies ε_1 and ε_2, and, consequently, by the transition frequency ω_{21} ($\hbar\omega_{21} = \varepsilon_2 - \varepsilon_1$). In time-resolved pump-probe spectroscopy the system under consideration is subjected by the total field of both pulses, therefore, the applied electric field has to be written as a superposition of the two quasimonochromatic plane-wave input pulses:

$$E(t) = \frac{1}{2}[\Im_L(t)\exp(ik_L r - i\omega_L t) + \Im_P(t-\tau)\exp(ik_P r - i\omega_P t)] + c.c., \quad (1)$$

where $\Im_{L,P}(t)$ is the (complex) slowly varying electric field envelope function, indexes L,P define pump and probe pulses, respectively, and the definition c.c. means complex-conjugated

P. Mathis (ed.), Photosynthesis: from Light to Biosphere, Vol. I, 231–234.

terms. The electric fields are all assumed to be linearly polarized in the same direction. In what follows we will confine ourselves to the so-called time-resolved difference absorption technique, where the absorption changes of the second pulse induced by the first one are monitored as a function of τ.

Kinetics of these absorption changes is determined by the time evolution of the light-matter interacting system and can be obtained in principle by solving the corresponding 2x2 matrix equation for the density operator ρ of the two-level molecules:

$$i\hbar\frac{\partial\rho}{\partial t} = -[\rho, H_0 + H_1] + decay\ term, \tag{2}$$

where:

$$H_0 = \frac{1}{2}\begin{bmatrix}\varepsilon_2 & 0\\ 0 & \varepsilon_1\end{bmatrix}, \quad H_1 = -E(t)\begin{bmatrix}0 & \mu\\ \mu^* & 0\end{bmatrix}, \tag{3}$$

with μ being the dipole transition matrix element and the *decay term* determines the excitation relaxation which is due to the interaction of the two-level system with its surrounding and other degrees of freedom which are not taken into account in the dynamical processes under consideration. Together with electronic degrees of freedom, molecules are also characterized by the intramolecular vibrations the active modes of which can give the evident effect on the changes of the optical density corresponding to the electronic transition in femtosecond time-resolved spectroscopy [3,4]. In our description it means that the electronic energies ε_i in general are potential surfaces $\varepsilon_i(Q)$, where Q is the generalized vibrational co-ordinate. Thus, the transition frequency also possesses the same kind of dependence, i.e. $\omega_{21} = \omega_{21}(Q)$. We consider the field frequencies ω_L and ω_P to be close to the resonance of the two-level system, e.g. $\Delta_{L,P}(Q) << [\varepsilon_2(Q) - \varepsilon_1(Q)]$ (where $\Delta_{L,P}(Q) = \omega_{21}(Q) - \omega_{L,P}$) and, thus, we use the usual rotating frame approximation. On these assumptions we write the well-known optical Bloch equations for the slowly varying part of density matrix elements (where the dominant time dependent $\exp(-i\omega_L t)$ was excluded), as following:

$$\begin{aligned}\frac{\partial\rho_{12}}{\partial t} &= i\Delta_L(Q)\rho_{12} + i[\Lambda_L^* \exp(-ik_L r) + \Lambda_P^* \exp(-ik_P r - i\Delta t)]n_{12} - \frac{\rho_{12}}{T_2},\\ \frac{\partial n_{12}}{\partial t} &= 2i[\Lambda_L \exp(ik_L r) + \Lambda_P \exp(ik_P r)]\rho_{12} - c.c.,\end{aligned} \tag{4}$$

where $\Lambda_{L,P} = \mu\Im_{L,P}(t)/\hbar$ are Rabi frequencies for pump and probe fields, $n_{12}=\rho_{11}-\rho_{22}$, $\Delta=\omega_L-\omega_P$. In order to take into account the relaxation of the polarization, the phenomenological term (resulting from the *decay term*) of the form $(-\rho_{12}/T_2)$ was included into the right hand side of the appropriate equation. The details of the theoretical analysis are described in [6].

The vibrations generated during the excitation pulse action also undergo the time evolution. We assume that potential surfaces in the ground and excited electronic states are parabolic and there are no changes of the vibration frequencies in the electronically excited state. For the sake of simplicity a single vibration mode will be considered. Thus, during the excitation two oscillators are activated:

one of them corresponds to the vibrational movement of the "hole" in the ground state while another describes the vibrations in the electronically excited state [8]. Both oscillations are treated classically, the phenomenological parameter of the damping is also taken into account. The initial conditions of both oscillations are as follows:

$$Q_1(0)=\frac{S\Omega+\Delta_L^0}{\sqrt{2S\Omega}},\qquad Q_2(0)=\frac{S\Omega-\Delta_L^0}{\sqrt{2S\Omega}},\tag{5}$$

where $Q_{1,2}$ determine the vibrational co-ordinates in the ground and excited states, respectively, Ω is the vibration frequency, S is the Huang-Rhys factor which determines the shift of the potential surfaces, $\Delta_L^0=\omega_{21}^0-\omega_L$ and ω_{21}^0 determines the minimal distance between both surfaces.

3. Results and Discussion

The induced absorption changes evaluated according to eq. 4 are shown in Fig. 1. Parameters used for these calculations are as follows: $\Delta_L^0=0$, $\Omega=30ps^{-1}$, $\sqrt{2S}=1$, $\eta=10ps^{-1}$, $T_2=50fs$ and the pulse duration of excitation and probing is 60 fs. Kinetics of the selective probe for different values of Δ_P^0 are depicted in Fig. 1a, while the integrated kinetics is presented in Fig. 1b.

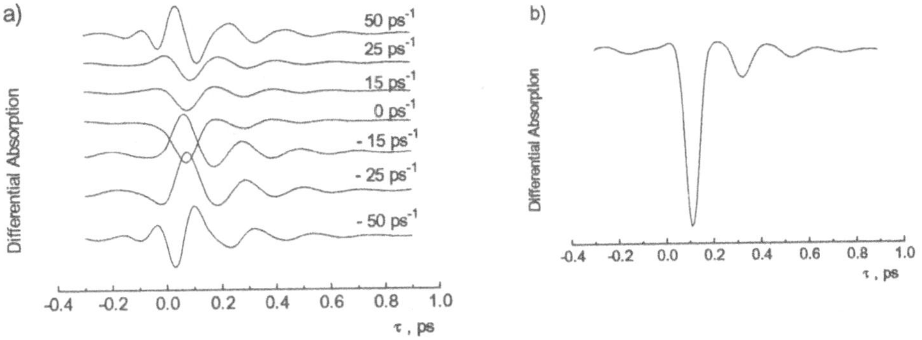

Fig. 1. Differential absorption kinetics. a) - selective absorption kinetics at different detuning of the probing pulse frequency Δ_P^0 (the values are depicted above the corresponding curves), b) - the integrated difference absorption kinetics.

According to eq. 5, the excitation conditions $\Delta_L^0=0$ correspond to the excitation of the vibrations in the electronically ground and excited states at the initial conditions $Q_{1,2}(0)=\frac{1}{2}\sqrt{2S}$, i.e. in the middle of the maximally possible values. The effect of the gradual changes in the probing frequency evidently is seen by changing the detuning Δ_P^0 from a level of far above the maximally possible excitations (-50 ps^{-1}) to the level of the far below (50 ps^{-1}). All of them demonstrate the oscillating behaviour after the excitation pulse action which are evidently due to the nuclear motion. Nevertheless, few typical features should be pointed out. First, oscillation with nuclear frequency (30 ps^{-1}) can be more clearly distinguished in the integrated signal while the selective signal possesses the oscillations on the sub-harmonic frequencies (see Fig. 2). This effect is due to the oscillating movement in both (excited and ground) electronic states. The vibration dephasing rate η is responsible for the decay of these oscillations. Second, the phase of oscillations above and below of the minimum of the transition frequency are opposite. Third, the effect of the detuning of the probing pulse frequency from the optical transition of the two-level system [7] is also seen in the selective pump-probe conditions. It gives the effect at the negative delay times of the probing pulse and can be well explained in the frequency domain.

Thus, both oscillating features can be easily distinguished from each other according to their difference in the scale of the delay time. Electronic dephasing time (T_2) determines the scale of the damping of those oscillations. At $\Delta_P^0=0$, no oscillations are observed and the signal corresponds to the coherent spike which occurs due to the interaction of both pulses. It is also noteworthy, that the integrated kinetics does not indicate any oscillations at the negative delay times and the whole kinetics is also mainly determined by the coherent interaction between both pulses.

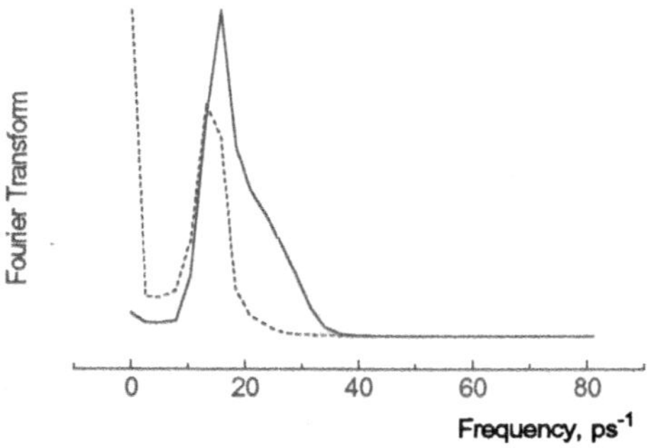

Fig. 2. Fourier transform of the integrated (solid line) and selected at Δ_P^0= -50 ps^{-1} (dashed line) signals.

Optical transitions to the higher electronic state of the dimer during the excitation create additional oscillating features [9].

4. Acknowledgement

This research was made possible in part by Grant No. LH6100 from the Joint Fund Program of Lithuanian Government and International Science Foundation.

5. References

1. J. Breton, J.-L. Martin, G.R. Fleming and J.-C. Lambry (1988) *Biochemistry* 27, 8276-8284.
2. M. Du, S.J. Rosenthal, X. Xie, T.J. Di Magno, T.J. Schmidt, D.K. Hanson, D.K. Schriffer, J.R. Noris and G.R. Fleming (1992) *Proc. Nat. Acad. Sci. USA* 89, 8517-8521.
3. M.H. Vos, F. Rappoport, J.-C. Lambry, J. Breton and J.-L. Martin (1993) *Nature* 363, 320-325.
4. M.H. Vos, M.R. Jones, C.N. Hunter, J. Breton and J.-L. Martin (1994) *Proc. Nat. Acad. Sci. USA* 91, 12701-12705.
5. R.J. Stanley and S.G. Boxer (1995) *J. Phys. Chem.* 99, 859-863.
6. E. Gaizauskas and L. Valkunas (1994) *Opt. Comm.* 109, 75-80.
7. M. Chachisvilis, H. Fidder and V. Sundström (1995) *Chem. Phys. Lett.* 234, 141-150.
8. D.M. Jonas, S.E. Bradforth, S.A. Passino and G.R. Fleming (1995) *J. Chem. Phys.* 99, 2594-2608.
9. E. Gaizauskas and L. Valkunas (1994) *Lith. J. Phys.* 34, 384-388.

MOLECULAR ASSEMBLIES OF LIGHT-HARVESTING POLYPEPTIDES / PORPHYRIN DERIVATIVES AS AN ARTIFICIAL MODEL FOR THE LIGHT-HARVESTING POLYPEPTIDE COMPLEX OF PHOTOSYNTHETIC BACTERIA

Kouji Iida,[§] Pamela S. Parkes-Loach,[¶] Paul A. Loach,[¶] and Mamoru Nango[†]
[§]Nagoya Municipal Industrial Research Institute, Rokuban-3-chome, 4-41, Atsuta-ku, Nagoya 456 JAPAN.
[¶]Department of Biochemistry, Molecular Biology, and Cell Biology, Northwestern University, Evanston IL 60208-3500. USA.
[†]Department of Applied Chemistry, Nagoya Institute of Technology, Gokiso-cho, Showa-ku, Nagoya 466 JAPAN.

1. Introduction

In purple photosynthetic bacteria such as *R. rubrum*, photon is collected and transferred to the reaction center (RC) by light-harvesting (LH) protein/bacteriochlorophyll (BChl*a*) complex (1-2). It is known that BChl*a* has 777 nm absorbance in acetone but the LH complex of *R. rubrum* has 870 nm absorbance of BChl*a*, implying that BChl*a*s are oriented and aggregated in the LH polypeptides. The primary structures of the LH polypeptides, α– and β– polypeptides consist of central hydrophobic core buried in hydrophobic domain of the membrane and hydrophilic segments on both N and C-terminals (Scheme 1)(3). BChl*a* is likely to bind to the histidine residue on the hydrophobic core of the LH polypeptides and to interact with amino acids such as tryptophan through hydrogen bond (4) and hydrophobic interaction in the domain of the LH polypeptides so that the orientation and aggregation of BChl*a* are regulated and the Qy band of BChl*a* is red-shifted. In this paper, molecular assembly of the α– and β– polypeptides of the LH proteins from *R. rubrum* with mesoporphyrin derivatives, M_1MPMME in octylglycopyranoside (OG) was examined to provide insight into the effect of porphyrin structure on formation of the LH complex of the photosynthetic bacteria.

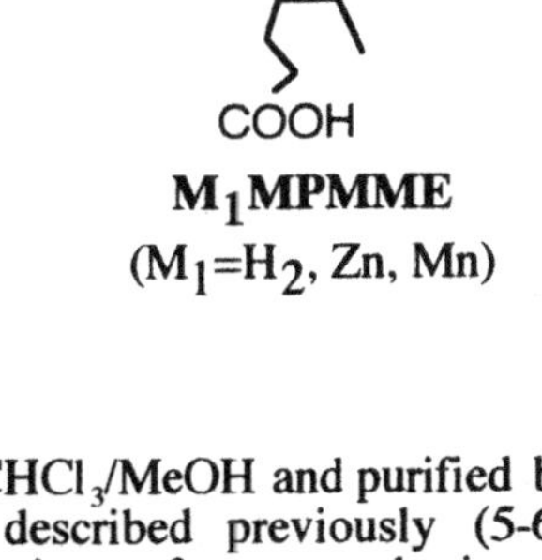

M_1MPMME
(M_1=H_2, Zn, Mn)

2. Procedure

LH polypeptides from *R. rubrum* were extracted by $CHCl_3$/MeOH and purified by Sephadex LH-60 gel chromatography and HPLC as described previously (5-6). M_1MPMME (M_1= H_2, Zn, and Mn) was prepared by reduction of protoporphyrin and following acid-hydrolysis. Zinc and manganese complexes were also prepared as described previously. ^{1}H NMR and Fab mass spectra support unambiguously the assigned structure. The reconstitution of the LH complex by M_1MPMME was carried out as follows. Each

P. Mathis (ed.), Photosynthesis: from Light to Biosphere, Vol. I, 235–238.

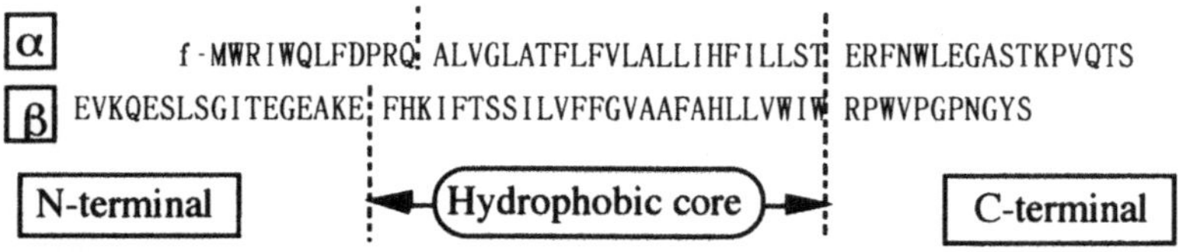

R. rubrum LH polypeptides

Scheme 1

5 nmol of α and β polypeptide was dissolved in 4.5% OG (50 mM Mops buffer, pH 7.2) and diluted to 0.9% OG by buffer and then 10 nmol of M_1MPMME in acetone was added. The solution was further diluted to 0.8% (*subunit condition* as an analogy of LH-BChl*a in OG solution*) and was chilled overnight (*LH1 condition* as an analogy of LH-BChl*a in OG solution*).

3. Results and Discussion

Figure 1 illustrates UV/Vis. spectra of ZnMPMME in the absence and presence of α– and β– polypeptides in OG solution. As is apparent from Figure 1(a), the Soret band is shifted from 408 nm at 0.9% OG solution to 414 nm at the subunit condition and

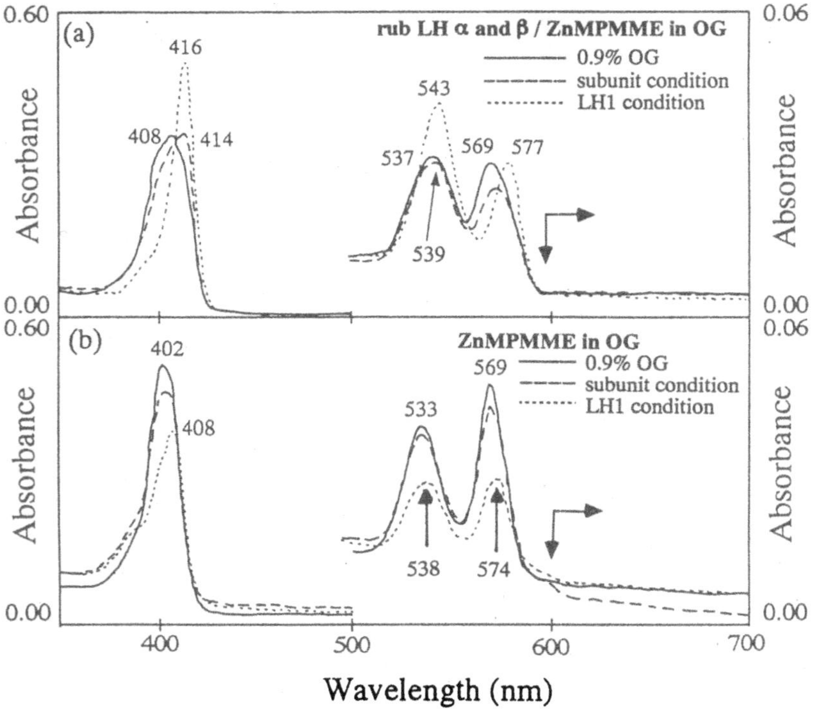

Figure 1

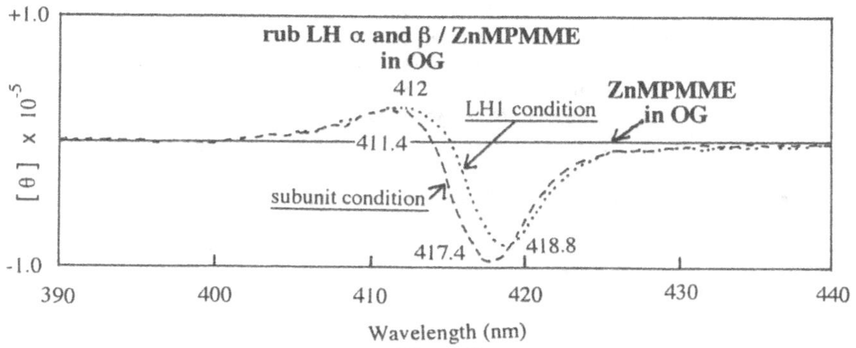

Figure 2

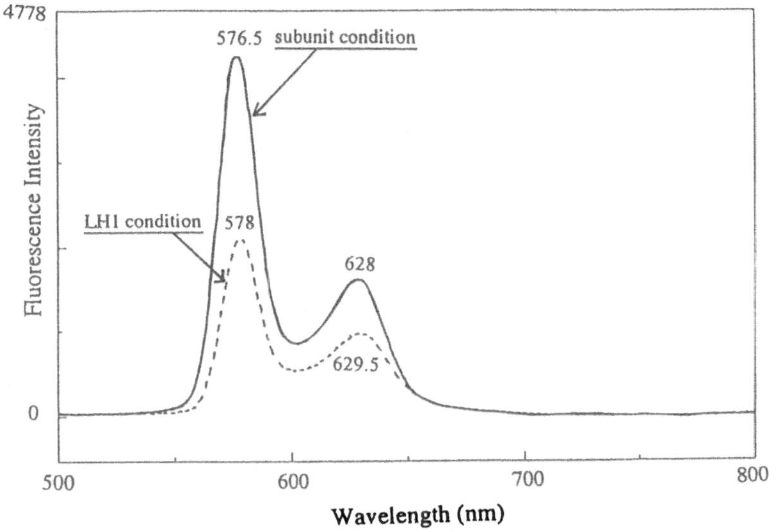

Figure 3

further to 416 nm at the LH1 condition, in which the shape of the Soret band was sharpen especially at the LH1 condition. Furthermore, the Q band is also red-shifted at the LH1 condition. However, the Soret bands of H_2MPMME and MnMPMME in the presence of the LH polypeptides showed no red-shift (data not shown). As is apparent from Figure 1(b), both the Soret band and Q band of ZnMPMME in the absence of the LH polypeptides are red-shifted, in which the shift is not large in comparison to that in the presence of the LH polypeptides. These results imply that ZnMPMME is likely to bind with α– and β–polypeptides as is analogous to those observed in LH polypeptides / BChl*a* complex.

Figure 2 illustrates CD spectra of ZnMPMME in the absence and presence of α– and β–polypeptides in OG solution . As is apparent from Figure 2, ZnMPMME shows Cotton effect at the corresponding Soret band in the presence of LH polypeptides (both the *subunit and LH1 condition*), indicating that ZnMPMME is oriented due to the presence of α and β polypeptides. However, the CD spectrum of ZnMPMME in the absence of the LH polypeptides shows no Cotton effect.

Figure 3 illustrates fluorescence spectra of ZnMPMME in the presence of the LH polypeptides when excited at the Soret band. As is apparent from Figure 3, the fluorescence spectrum of ZnMPMME at the LH1 condition is quenched by 50% of that at the subunit condition, implying that ZnMPMME is more aggregated at the LH1 condition than that at the subunit condition. Thus, it is likely that ZnMPMME easily coordinates with the LH polypeptides at the LH1 condition.

ZnMPMME also complexed with α– or β–polypeptide alone, consistent with the results obtained using the mixture of α– and β–peptides with ZnMPMME, in which the Soret band was red-shifted and the CD spectra showed the Cotton effect at the corresponding Soret band (data not shown). The CD spectra in the presence of α-polypeptide alone is larger by 6 folds in comparison to those obtained using β–polypeptide alone and with the mixture of α– and β– polypeptides, implying that ZnMPMME is rather oriented in the presences of α-polypeptide alone than in the presence of β–polypeptide alone and of the mixture of α– and β–polypeptides (7).

In conclusion, these spectroscopic properties of ZnMPMME in the presence of the LH polypeptides were analogous to those obtained using the LH polypeptides and BChl*a* in OG solution and also with the LH complex of photosynthetic bacteria. Thus, the zinc mesoporphyrin derivative is useful to study on an artificial model for formation of the LH complex of photosynthetic bacteria.

References

1 Grondelle, R. van, Dekker, J. P., Gillbro, T., and Sundstrome, V. (1994) Biochim. Biophys. Acta 1187, 1-65

2 Hawthornthwaite, A. M. and Cogdell, R. J. (1993) in The Clorophylls (Scheer, H., ed.) pp. 493-528, CRC, Boca Raton, FL, USA

3 Zuber, H. and Brunisholtz, R. A., (1993) in The Clorophylls (Scheer, H., ed.) pp. 627-703, CRC, Boca Raton, FL, USA

4 Olsen, J. D., Sockalingum, G. D., Robert, B., and Hunter, C. N. (1994) Proc. Natl. Acad. Sci. USA 91, 7124-7128

5 Miller, J. F., Hinchigeri, S. B., Parkes-Loach, P. S., Callahan, P. M., Sprinkle, J. R., Riccobono, J. R., and Loach, P. A., (1987) Biochemistry, 26, 5055-5062

6 Parkes-Loach, P. S., Sprinkle, J. R., and Loach, P. A., (1988) Biochemistry, 27, 2718-2727

7 Meadows, K. A., Iida, K., Tsuda, K., Recchia, P. A., Heller, B. A., Antonio, B., Nango, M., and Loach, P. A., (1995) Biochemistry, 34, 1559-1574

Two-dimensional crystallization and structure determination of LHII from *Rhv. sulfidophilum* at 7Å resolution

H. Savage, M. Cyrklaff, G. Montoya, W. Kühlbrandt & I. Sinning
EMBL, Meyerhofstr. 1, D-69117 Heidelberg, Germany

1. Introduction

The absorption of a photon by the light-harvesting (LH) antenna complexes is the initial event of bacterial photosynthesis. LH complexes are integral membrane protein pigment complexes which transfer the excitation energy rapidly and efficiently to the reaction center (RC). At the RC a charge separation takes place and electrons are transferred across the membrane. In the membrane of most photosynthetic purple bacteria two different LH complexes are found at increasing distances from the RC, referred to as B880 (LHI) and B800-850 (LHII) complexes, according to their absorbance maxima (1). The structure and function of LH complexes have been studied in great detail using biochemical and spectroscopic methods, but only recently the three-dimensional structure of LHII from *Rps. acidophila* strain 10050 has been determined using X-ray crystallography (2). After many years of expecting LHII to be a hexamer (3), the X-ray structure revealed a very simple and elegant arrangement of nine αβ heterodimers in a doughnut shaped complex. A similar arrangement was described for LHI from *Rps. rubrum* based on a 8.5 Å projection map from 2D crystals (4). LHI is a doughnut formed by 16 αβ heterodimers - just big enough to accommodate the RC. An arrangement of LHII different from a hexamer or nonamer has so far only been described for *R. molischianum*, for which an octamer has been postulated based on native gels and analytical ultracentrifugation (5).
Recently we reported a new isolation protocol for LHII from the purple bacterium *Rhodovulum (Rhv.) sulfidophilum* (previously called *Rhodobacter sulfidophilus*) . An initial structure analysis at low resolution from 2D crystals using electron microscopy was part of the same report (6). We have now collected higher resolution data to 7Å and the results are described here.

2. Procedure

Cells of *Rhv. sulfidophilum* have been grown and the LHII complex was purified as described earlier. A procedure for 2D crystallization was developed based on that used for 3D crystallization (6). The LHII complex was reconstituted with dimyristoyl phosphatidylcholine (DMPC) at different ratios. The detergent concentration was lowered by microdialysis in glass capillaries as described by Kühlbrandt (7). A ratio of 70:1 of lipid to protein was found to be optimal in producing well ordered crystalline arrays. Protein concentration was 1 mg/ml. The samples for cryo-electron

P. Mathis (ed.), Photosynthesis: from Light to Biosphere, Vol. I, 239–242.

microscopy were embedded in glucose and in vitreous water as described in (8). Electron micrographs were recorded in a JEOL 2000EX electron microscope. A more detailed description will be given elsewhere. The images were processed using the MRC image processing package (9). The CCP4 suite of crystallographic programs (10) was used to calculate Fourier projection maps. One of the ring shaped complexes was extracted from a high resolution map and a Fourier analysis of its power spectrum undertaken. Using the extracted map segment, a rotationally filtered projection map was synthesised with the nine-fold rotational component.

3. Results and Discussion

2D crystals of the tubular and vesicular type were obtained (6). The tubular crystals were 0.5 to 0.6 μm wide and a few μm long (Figure 1a). Flattened vesicles measured about 1-2 μm in diameter. So far, two different symmetries were observed. In the previous study a tetragonal unit cell was assigned with parameters a = b = 157Å (6). Due to the low resolution it was not possible to resolve the oligomerization state of the LHII complex, which was within the range of heptamer to nonamer. From the projection map calculated using $P42_12$ symmetry an octamer seemed most likely (Figure 1b). In the current study, using cryoelectron microscopy we found another crystal type which diffracted to higher resolution. Indexing revealed an orthorhombic unit cell, plane group $P2_12_12$ with cell parameters of a= 160Å and b = 140Å. The individual LHII complexes are organised as ring like structures with an outer diameter of approximately 76 Å, similar to our previous data from the tetragonal crystals. A projection map is shown in figure 2a. Figure 2b shows the projection map for one particle after rotational filtering.

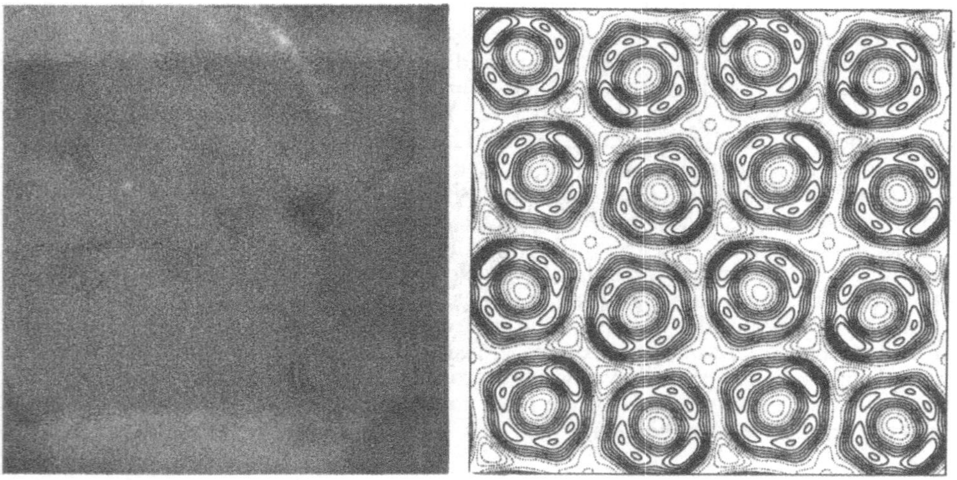

a **b**

Figure 1: a) LHII 2D tubular crystal of the reconstituted protein in DMPC, negatively stained with uranyl acetate; b) projection map of negatively stained LHII complex at 18 Å resolution, two images have been merged and $P42_12$ symmetry applied (6).

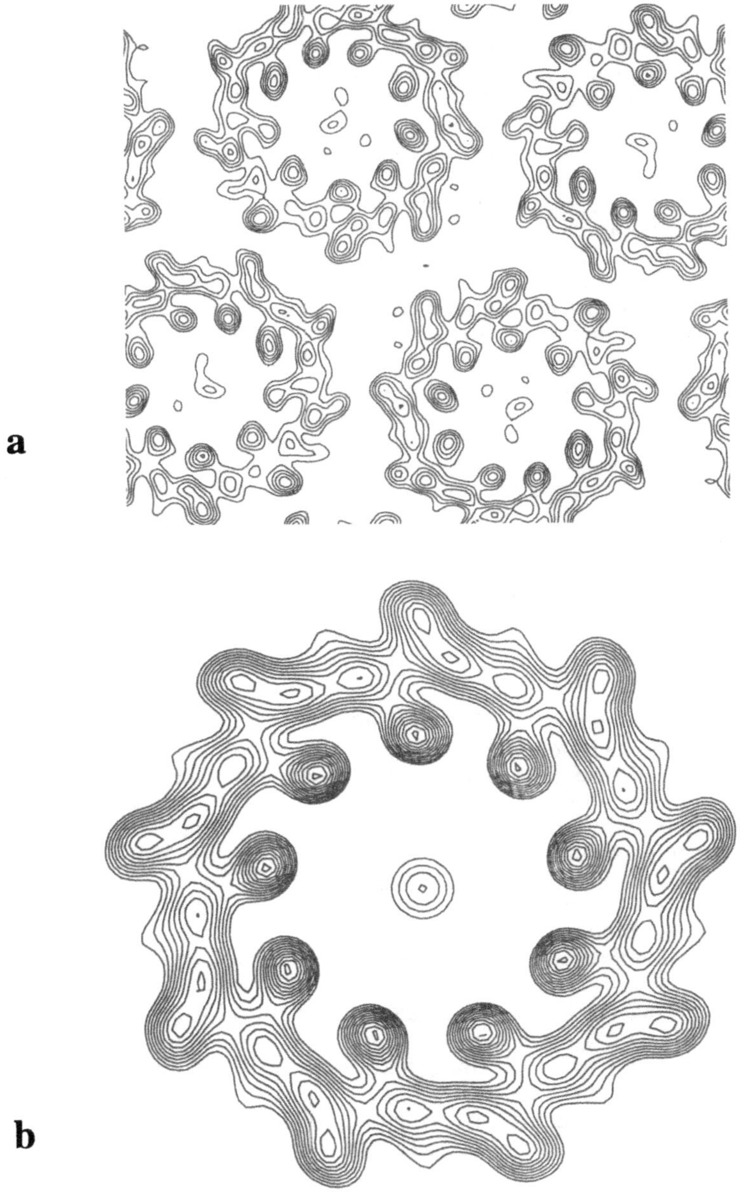

Figure 2: a) projection map calculated from four images at 7Å resolution with $P2_12_12$ symmetry applied; b) rotationally filtered projection map for one particle excised from a. The outer diameter is about 76Å.

A nine fold rotational symmetry within each ring is clearly observable in Figure 2a and b. Around the central hole, two rings with nine peaks in each, representing the α-helices of the α and β subunits are present. This organization and also the dimensions of the ring are very similar to the ones described for the X-ray structure (2). Furthermore, the inner helices are perpendicular to the membrane plane, whereas the outer ones appear to be tilted. The density between the two helices (Figure 2b) has been attributed to eighteen bacteriochlorophyll molecules of B850, which are known to be oriented with the tetrapyrole macrocycles perpendicular to the membrane plane close the periplasmic surface of the membrane. There is no obvious density for the B800 molecules, being closer to the cytoplasmic surface of the membrane and the tetrapyroles oriented parallel to the membrane plane. However, comparing the projection map of Figure 2b with the corresponding map of LHI from *Rsp. rubrum* (4), it is clear that in LHI the electron density is much more carved out around the helices. We attribute this to the absence of B800 in LHI. Therefore, the additional density between the outer helices in the projection maps shown in Figure 2a,b was interpreted as the contribution from the B800 tetrapyroles.
The ring-like arrangement of the αβ heterodimers in LHII is an elegant and simple construction to allow delocalization of the excitation energy and its very efficient transfer to LHI and the RC (11). We have to wait for more structures in order to see whether different oligomerization states are present in different bacteria and how big the variability might be. So far no data are available on the assembly state of LHII in the native membrane. However, based on the X-ray structure, which reveals details on how the pigments are interacting with the protein (2) it seems unlikely that this complex could change its oligomerization state due to detergent treatment.

References

1. Zuber, H. and Brunisholz, R.A. (1993) in The Chlorophylls (H. Scheer, ed.) pp. 627-704, CRC Boca Raton.
2. McDermott, G., Prince, S.M., Freer, A.A., Hawthornthwaite-Lawless, A.M., Papiz, M.Z., Cogdell, R.J. and Isaacs, N.W. (1995) Nature 374, 517-525.
3. Papiz, M.Z., Hawthornthwaite, A.M., Cogdell, R.J., Woolley, K.J., Wightman, P.A., Ferguson, L.A. and Lindsay, J.G. (1989) J. Mol. Biol. 209, 833-835.
4. Karrasch, S., Bullough, P.A. and Ghosh, R. (1995) EMBO J. 14, 631-638.
5. Kleinekofort, W., Germeroth, L., van den Broek, J.A., Schubert, D. and Michel, H. (1992) Biochim. Biophys. Acta 1140, 102-104.
6. Montoya, G., Cyrklaff, M. and Sinning. I. (1995) J. Mol. Biol. 250, 1-10.
7. Kuehlbrandt, W., (1992) Quart. Rev. Biophys. 25, 1-49.
8. Adrian, M., Dubochet, J., Lepault, J. and McDowell, A.W. (1994) Nature 308, 32-36.
9. Henderson, R., Baldwin, J., Downing, K.H., Lepault, J. and Zemlin, F., (1986).Ultramicroscopy 19, 147-178.
10. Collaborative Computational Project, No.4 (1994) Acta Cryst. D50, 760-766.
11. Kühlbrandt, W. (1995) Structure 3, 521-525.

EQUILIBRATION OF EXCITONS AND DOUBLE EXCITED LIGHT HARVESTING COMPLEX OF PHOTOSYSTEM II

Thorsten Schrötter, Thomas. Renger, Joachim Voigt
Humboldt-Universität zu Berlin, Institut für Physik, Invalidenstr. 110,
D-10115 Berlin, Germany

1. Introduction

The most abundant pigment-protein complex in higher plants is the light harvesting complex II (LHC II) associated with the photosystem II (PS II). This chlorophyll (Chl) a/b binding protein complex consists mainly of 25-27 kDa proteins organized into trimers. Their main function is the generation of electronically excited states referred to as excitons by absorbing solar energy and to transfer it to the place of charge separation, the so-called reaction center [for review 1, 2, 3].
In the past the incoherent Förster transfer (weak coupling limit) was preferred to describe the energy transfer within and between pigment protein complexes. The best structure data of LHC II [4] demonstrate smallest distances of 8 Å between the Chl's, which give rise to strong excitonic interaction. For this reason the discussion of measured ultrafast transfer times [5] in terms of the hopping like Förster mechanism might be wrong because of the delocalisation of excitons.
Investigations with the pump probe technique under quasi-stationary conditions were performed to get information about the spectral distribution of excitons in the LHC II. The nonlinear character of a pump probe experiment contains the possibility of excited state absorption (ESA). Therefore it was necessary in the interpretation of the data to include transitions from single to double excited LHC II monomer. Notice, because of the delocalization of the exciton over the monomer we have to define the terms ground state absorption, stimulated emission and ESA in relation to the monomer and not to the single Chl molecule. The hamiltonian of 12 coupled singlet transitions of chlorophyll molecules was diagonalized using dipole-dipole approximation to describe the interaction of the pigments.

2. Measurements

The LHC II proteins have been separated after β-dodecyl maltoside solubilization (detergent/Chl ratio 10/1 [wt/wt]) of salt-washed (NaCl and $CaCl_2$) PS II membrane fragments by sucrose density gradient centrifugation in the presence of 0.05% β-DM [6].
The present experiments were performed by using two independently tuneable dye lasers to generate the pump and the probe beams respectively, with the following specifications (pulse duration 10 ns; spectral width smaller than 0.01 nm; parallel polarisation between the beams) The absorption changes ($\Delta OD = OD_0 - OD$) were detected by the probe beam (density max. $1 \cdot 10^{15}$ photons$\cdot$cm$^{-2}\cdot$pulse^{-1}) from 640 to 690 nm. The pump beam intensity amounted to $6 \cdot 10^{18}$ photons$\cdot$cm$^{-2}\cdot$pulse^{-1} (for more details see Ref. 7).
The pump probe spectra of LHC II at room temperature are shown in Fig 1. Independent from the pump beam wavelength two maxima appear at about 650 nm and 680 nm. The first one corresponds well with the absorption maximum in the Chl b region. However the second maximum in the Chl a region is red-shifted by about 5 nm.
The independence of the absorption changes from pump beam wavelength suggests the assumption of

P. Mathis (ed.), Photosynthesis: from Light to Biosphere, Vol. I, 243–246.

thermally distributed excited states. The absorption changes in the range from 660 nm to 690 nm can be fitted assuming Boltzmann-distributed excited states [7,8]. However the absorption changes around 650 nm were at first unclear. To understand them a more sophisticated treatment is needed which takes account for ESA.

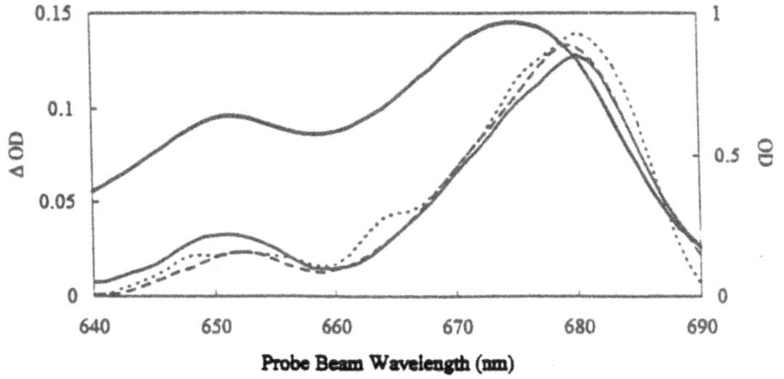

Fig. 1 Absorption changes (ΔOD) of LHC II as a function of the probe beam wavelength at different pump beam wavelengths (—650nm; ··· 660nm; - - - 680nm). Linear absorption spectrum for comparison (thick solid line). Excitation intensity: $6{\cdot}10^{18}$ photons·cm^{-2}·pulse^{-1} (adopted with modification from Ref. 7)

3. Calculations of excitonic interaction

Applications of the exciton theory to a dimer or larger systems of coupled pigments are known (for a see review Ref. 2 and citations therein).

The hamiltonian $H = \sum_{i}^{12} H_0^i + \sum_{i,j;i\neq j}^{12} V^{i,j}$ describes a LHC II monomer binding 12 Chl's, where $V^{i,j}$ is the Coulomb interaction in dipole-dipole approximation. $H_0^i = H_{Chl} - \Delta E_i$ is the Hamiltonian of the uncoupled pigment H_{Chl} modified by pigment-protein interaction ΔE_i (binding energy). Additional to the ground state $\psi^{(0)} = \prod_{i}^{12} \varphi_i$ and the 12 single excited states $\psi_i^{(1)} = \sum_{j}^{12} c_j^i \varphi_j^* \prod_{k=1,k\neq j}^{12} \varphi_k$, the 66 double excited states $\psi_i^{(2)} = \sum_{j=1}^{12} \sum_{k=j+1}^{12} c_{jk}^i \varphi_j^* \varphi_k^* \prod_{l=1,l\neq j,k}^{12} \varphi_l$ will be considered, where φ_i and φ_i^* are the wavefunctions of the ground (S_0) and first excited singlet (S_1) state of Chl i, respectively. The Hamiltonian was diagonalized to determine the new energy levels and the dipole strength of all possible transitions.

The dipole-dipole interaction was estimated using the structure date of LHC II by Kühlbrandt et al. [4]. Although Kühlbrandt et al. didn't resolve the structural differences between Chl a and Chl b, they were able to give reasons to assign the single Chl molecules to Chl a and Chl b. We followed the assignment of Kühlbrandt et al. The reached resolution either did not allow to see the orientation of the molecular dipole moments. Therefore for the direction of the dipole moment two limiting cases were introduced, „in line“ all dipole moments lie in the membrane plane or „sandwich“ all moments stand perpendicular in the membrane. The original dipole strengths of Chl a and Chl b were amounted to 15.5 D^2 and 12.5 D^2 for dipoles in "in line" configuration and to 16.85 D^2 and 11.57 D^2 in the "sandwich" case, respectively. To create an absorption spectrum the homogeneous and inhomogeneous broadening was

taken into consideration by using a gaussian line shape function (FWHM 6 nm at 1.1K). To get a good agreement with the experimental spectrum the binding energy had to be varied as a fit parameter to simulate different binding sites of Chl at the protein.
Figure 2 shows the fitted absorption spectrum of LHC II at 1.1 K [9] for both dipole directions. The excitonic interaction produces a splitting of energy levels, well known from the degenerated dimer and a redistribution of the dipole strength. In the case of LHC II monomers the splitting is relatively small (a few meV), assuming the assignment of Chl a and b by Kühlbrandt et al. So for a good fit it was necessary to spread the molecular energy levels over the whole absorption spectrum by varying the binding energy of the single Chl molecules at the protein. However the remaining effect of excitonic interaction let arise a strong redistribution of the dipole strength depending on the dipole direction. (For the exact data see Ref. 10.) In good agreement with hole burning results [11] the lowest energy level was determined to be at 680 nm with half of the dipole strength of the main transition at 676 nm.

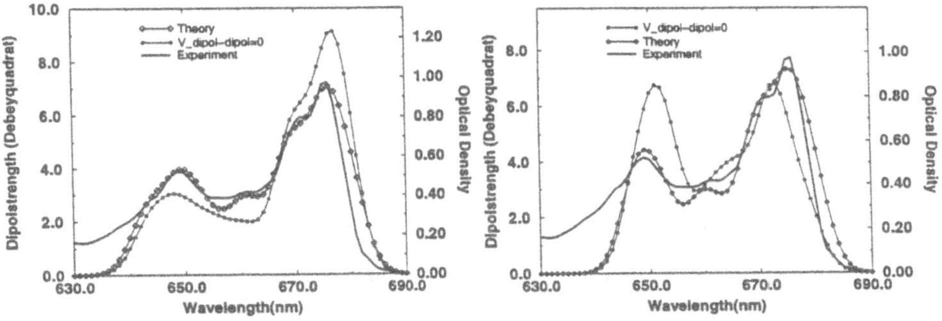

Fig. 2 Fitted absorption spectrum of LHC II at 1.1 K. for different dipole arrangements left: "sandwich", right: "in line"

4. Discussion

In our pump probe experiments the following effects contribute to the absorption changes of the probe beam:

- ground state depletion
- stimulated emission
- excited state absorption as a transition from a single to double excited monomer

The calculated pump probe spectra for both limiting cases of dipole orientation is shown on the left side of Figure 3. Because of the independence of measured pump probe spectra from the pump wavelength we assumed Boltzmann distributed single excited monomers. The spectrum of stimulated emission results by shifting the transitions from ground to single excited states (Stokes shift 5 nm). At room temperature a gaussian line shape function (FWHM 12 nm) was used to take account for the line broadening mechanism. Hereby the inhomogeneous broadening of the excitonic interactions stronger effects the linewidth of the transition between one and two exciton states than the transition from the ground state of the monomer. We had to assume a 1.4 times smaller line width for the letter. For these fits the ground states were depleted by 65%.
With the assumption made above we got a qualitatively good agreement of measured and calculated pump probe spectra. The measured pump probe spectrum doesn't fall in between the two calculated spectra for the cases of dipole orientation in the whole wavelength region. The two maxima near 650 nm and 680 nm can be reproduced. The course of the pump probe spectrum originates from the redistribution of dipole strength within the transitions to single and to double excited LHC II monomers by the excitonic interactions (see fig. 3 (right). The differences between measured and calculated spectra

seem to result from the limited completeness of the structure data and/or the validity of the other assumptions made for instance using a gaussian line shape function fitting 1.1 K absorption spectrum. We can conclude that the transitions to double excited LHC II monomers have to be taken into consideration to describe the pump probe spectrum. The excitonic interaction markedly influences the relation between the absorption cross section of ground state and ESA.

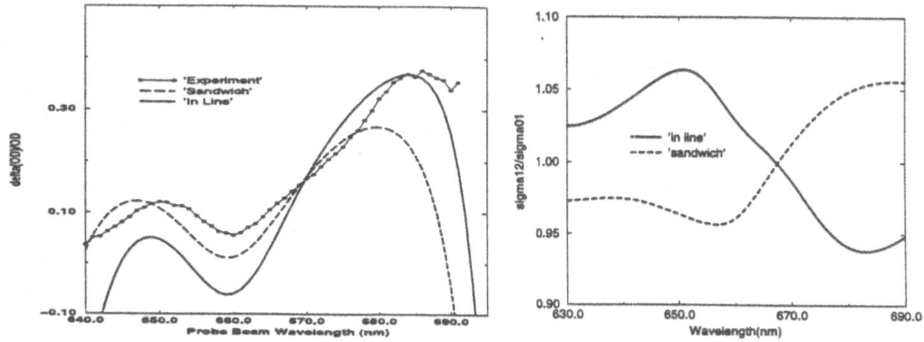

Fig. 3 left: Fitted relative absorption changes ΔOD/OD of LHC II as a function of probe beam wavelength for different dipole arrangements (dashed line: "sandwich", full line: "in line", line with circles measured), right: Quotient between absorption cross section for the transitions from single to double and the ones from ground to single excited LHC II monomer as a function of wavelength

Acknowledgement

We thank G. Renger and K.D. Irrgang for helpful discussion and for providing us with the used sample material. The financial support by the Deutsche Forschungsgemeinschaft (Sonderforschungsbereich 312) is gratefully acknowledged.

References

1. Renger, G. (1992) in Topics in Photosynthesis (Barber, J., ed.) Vol. 11, pp. 45-99 Elsevier Amsterdam
2. van Grondelle, R., Dekker, J.P., Gillbro, T., Sundstrom, V. (1994) Biochim. Biophys. Acta 1187, 1-65
3. Jansson, S. (1994) Biochim. Biophys. Acta 1184, 1-19
4. Kühlbrandt, W., Wang, D.N., Fujiyoshi, Y. (1994) Nature 367, 614-621
5. Bittner, T, Irrgang, K.D., Renger, G., Wasilewski, M.R. (1994) J. Phys.chem. 98, 11821-11826
6. Irrgang, K.D., Boekema, E.J., Vater, J., Renger, G. (1988) Eur. J. Biochem. 178, 209-217
7. Kehrberg, G., Voigt, J., Schrötter, T., Renger, G. (1995) Biochim. Biophys. Acta in press
8. Schrötter, T., Voigt, J., Kehrberg, G. (1994) Photosynthetica 30(4), 603-612
9. Voigt, J., Maciy, K., Schrötter, T. (1994) Photosynthetica 30(4), 567-574
10. Renger, T., Voigt, J., May, V. in preparation
11. Reddy, N.R.S., van Amerongen, H., Kwa, S.L.S., van Grondelle, R., Small, G.J. (1993) J. Phys. Chem. 98, 4729-4735

THE EFFECT OF pH ON LHCII
Montpellier, France, 20-25 August 1995

A. A. Pascal[1,*], A. V. Ruban[1], A. J. Young[2] & P. Horton[1]
[1]Robert Hill Institute, Dept of Mol. Biol. & Biotechnol., University of Sheffield, Sheffield S10 2TN, U.K. [2]School of Biol. & Earth Sci., Liverpool John Moores University, Byrom Street, Liverpool L3 3AF, U.K. *current address: SBPM / DBCM, Bât. 532, CE-Saclay, 91191 Gif-sur-Yvette, CEDEX, France.

1. Introduction

Aggregation of the chlorophyll-*a/b* light-harvesting complex of photosystem II (LHCII) results in dramatic changes in spectral properties. These include a reduction in fluorescence intensity of about 90% [1] and shifts in absorption and fluorescence peaks [2], and reflect specific changes in pigment conformations [3]. Similar organisational changes *in vivo* (on a smaller scale) may be responsible for the down-regulation of light-harvesting efficiency in strong light conditions ("high-energy-state" quenching of chlorophyll fluorescence, q_E; [4]). We report observations relating the aggregation state of LHCII *in vitro* with the pH of the medium.

2. Procedure

LHCII was isolated from *Spinacia oleracea* (L.) var. *subito* by a sucrose density gradient of triton-solubilised thylakoids [5] (based on [6]). Violaxanthin-containing samples (isolated from non-light-treated leaves) are referred to as "dark" LHCII while those exhibiting 50-70% violaxanthin conversion to zeaxanthin (from light-treated leaves; [7,8]) are termed "light" LHCII. Pigment compositions were determined by reversed-phase HPLC [9]. The extent of violaxanthin deepoxidation is expressed as the deepoxidation state, DES [5]. 77 K fluorescence and 298 K absorption spectra were obtained as described previously [2,5] on an EG&G OMA-Vision fluorescence spectrophotometer fitted with a home-built cryostat, and an SLM DW2000 UV/Vis scanning spectrophotometer, respectively. Chlorophyll fluorescence yields were measured using a Walz PAM 101 chlorophyll fluorimeter in a Hansatech DW2/2 aqueous-phase oxygen electrode chamber maintained at 293 K. Samples were removed from the chamber following fluorescence stabilisation for absorption measurement or were immediately frozen in liquid nitrogen for 77 K fluorescence measurements. Simultaneous pH readings were taken with a Corning semi-micro combination electrode inserted into the DW2/2 chamber. All experiments were performed in 5 mM citrate, 5 mM MES, 5 mM tricine, at pH 7.80 except where indicated otherwise.

P. Mathis (ed.), Photosynthesis: from Light to Biosphere, Vol. I, 247–250.

3. Results and Discussion

LHCII in the aggregated state has a low fluorescence yield and two 77 K emission peaks, at 680 & 700 nm (***Figure 1***). Solubilisation in non-ionic detergent yields complexes in the trimeric state, characterised by a high fluorescence intensity and the disappearance of the peak at 700 nm. However, if solubilisation is performed at pH 4.50 (***Figure 1***) the fluorescence yield is only 50% that at pH 7.80 and the *F*700/*F*680 ratio at 77 K is higher (0.18 compared to 0.15, calculated from integrated peak areas), indicating that the lower fluorescence level is probably a result of less efficient disaggregation. This was confirmed by green gels of LHCII solubilised at 0.35 mgchl/ml (not shown). Note also that the effect appears to be greater in complexes containing zeaxanthin ("light").

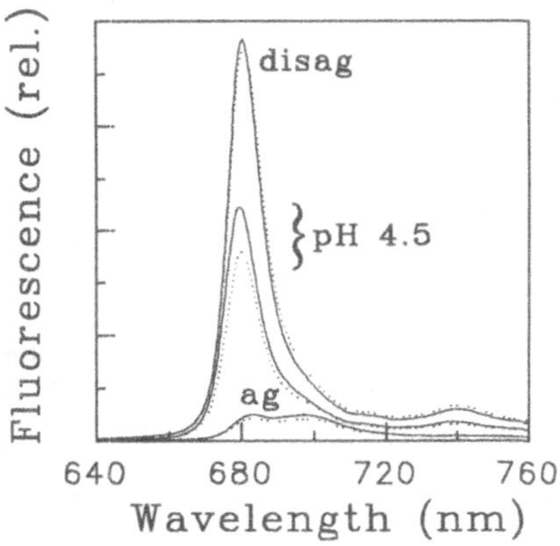

Figure 1 : *77 K fluorescence emission spectra of LHCII at 10 μgchl/ml at pH 7.80 (ag) and upon solubilisation with 0.3% OG + 0.2% digitonin at pH 7.80 (disag) and 4.50. Solid lines - "dark", dotted lines- "light".*

LHCII partially solubilised in 0.04% Triton X-100 exhibits fluorescence quenching upon acidification to pH 4.0-4.5, the fluorescence intensity being reduced by up to 25% (***Figure 2***). No further quenching was observed below pH 4.0 (not shown), and the effect was 20-60% reversible upon addition of KOH to raise the pH back to 7.8. Quenching was only observed with *partially*-solubilised LHCII; addition of a higher triton concentration or other detergents resulted in a higher initial fluorescence yield, but one which could not be quenched by lowering the pH (not shown). pH-quenching was inhibited by both antimycin A and dibucaine, with I_{50}'s of 0.3 & 400-600 μM respectively (not shown). DCCD increased the relative reversibility of quenching with KOH without inhibiting quenching itself.

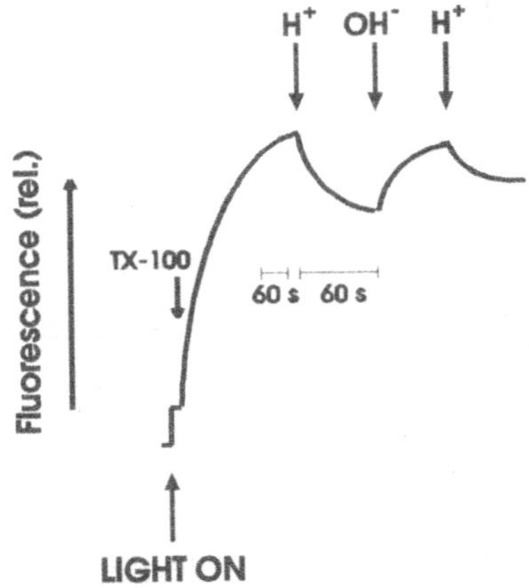

Figure 2 : *Low-pH-induced quenching of chlorophyll fluorescence in "dark" LHCII. After equilibration of the fluorescence level (5-8 min), small volumes of HCl (KOH) were added to lower (raise) the pH to 4.5 (7.8).*

Absorption and 77 K fluorescence spectra were obtained for samples taken directly from the sample chamber at various points (***Figure 3***). ***Figure 3a*** shows absorption spectra at pH 7.70 before (ag) and after (disag#1) addition of detergent and upon acidification to pH 4.31 (disag#2). As evidenced by the difference spectrum the effect of lowering the pH on the absorbance is similar to that associated with aggregation, with positive

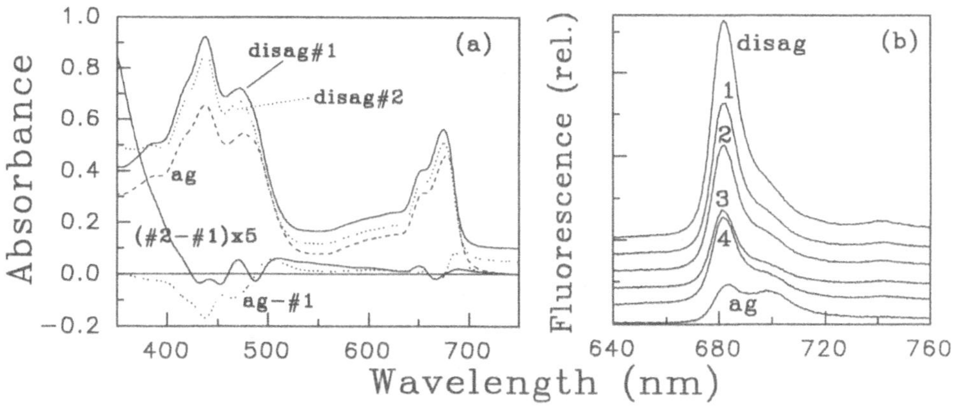

Figure 3 : *Changes in absorption **(a)** and 77 K fluorescence emission **(b)** spectra associated with pH-quenching in LHCII. **(a)** Absorbance at pH 7.70 (disag#1) and after acidification to pH 4.31 (disag#2). **(b)** Emission at pH 7.70 (disag) and upon successive acidification to pH 6.84 (1), 6.03 (2), 4.91 (3) & 4.07 (4). (ag - spectra of sample prior to solubilisation.)*

bands at 505-650, 650-660 & 685-690 nm [2,5]. Note also the appearance of a large scattering signal from the low pH sample, which is not associated with aggregation *per se* (compare with aggregated spectrum) or with acidification of the detergent/buffer solution alone (not shown). However this signal is negligible above 550 nm and so it is reasonable to assume that no artefacts have been introduced to give rise to the peaks in the red region at least.

77 K fluorescence emission spectra of the sample in the aggregated (ag) and disaggregated (disag) states at pH 7.64, and upon successive acidification down to pH 4.07 (1-5), are given in ***Figure 3b***. The effect of lowering the pH is clearly to increase the relative contribution of *F*700 to the spectra (as well as reducing the fluorescence intensity).

A pH titration of fluorescence quenching was performed between pH 7.8 & 4.5 for two sets of samples ("dark" &"light" LHCII isolated together in each case; ***Figure 4***). The first set of samples (***Figure 4a***) showed an increase in DES due to prior light treatment of the leaves of 0.5 to 67.5 %, while for the second set (***Figure 4b***) this was 0.0 to 47.0 %. In both cases the apparent pK_a was higher for the "light" sample, and in the first case (pK_a shift from 4.96, "dark", to 5.37, "light") is comparable to the shift seen in pH-quenching of thylakoids [7] and for q_E in chloroplasts [8]. Points corresponding to the absorption and fluorescence spectra (***Figure 3***) are indicated by inverted triangles (A used for absorption spectrum disag#2; 1-4 used for 77 K fluorescence).

Clearly the aggregation state of isolated LHCII, and therefore its spectral properties, are highly flexible with respect to the pH of the surrounding medium. In particular the fluorescence of pre-solubilised LHCII is quenched by lowering the pH, the pK_a being in the range 4.5 to 6.0 depending on the sample used. This phenomenon has similar properties to q_E, the mechanism of energy dissipation in strong light conditions, in terms

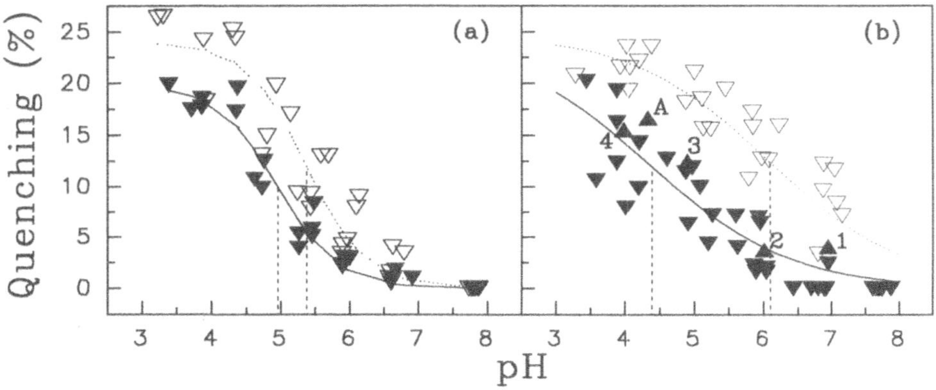

Figure 4 : *Titration of pH quenching of chlorophyll fluorescence in LHCII solubilised with 0.04% Triton X-100 at 10 μgchl/ml.* ***(a)*** *&* ***(b)*** *represent different sets of samples (filled symbols - "dark", open symbols - "light"). Samples removed for absorption & fluorescence spectra (***Figure 3***) are indicated by inverted triangles. Lines represent fitting of data to the equation* $y=c/(1+10^{(x-m)})$ *(c, m constants); dashed lines are drawn for* pK_a *values (= m),* ***(a)*** *4.96 & 5.37 and* ***(b)*** *4.38 & 6.10.*

of spectral characteristics, the effects of inhibitors, and the potentiating effect of zeaxanthin. These data therefore support the hypothesis that organisational changes within the photosystem II antenna are responsible for protective energy dissipation [4].

References

1 Mullet, J.E. and Arntzen, C.J. (1980) Biochim. Biophys. Acta 589, 100-117
2 Ruban, A.V. and Horton, P. (1992) Biochim. Biophys. Acta 1102, 30-38
3 Ruban, A.V., Horton, P. and Robert, B. (1995) Biochem. 34, 2333-2337
4 Horton, P., Ruban, A.V., Rees, D., Noctor, G.D., Pascal, A.A. and Young, A.J. (1991) FEBS Letts 292, 1-4
5 Pascal, A.A. (1995) Ph.D. Thesis, University of Sheffield, U.K.
6 Burke, J.J., Ditto, C.L. and Arntzen, C.J. (1978) Arch. Biochem. Biophys. 187, 252-263
7 Noctor, G., Rees, D., Young, A; and Horton, P. (1991) Biochim. Biophys. Acta 1057, 320-330
8 Rees, D., Noctor, G., Ruban, A.V., Crofts, J., Young, A. and Horton, P. (1992) Photosynth. Res. 31, 11-19
9 Barry, P., Young, A.J. and Britton, G. (1990) J. Exp. Bot. 41, 123-129

THE PIGMENT-PROTEIN INTERACTIONS OF SOME UNUSUAL LIGHT-HARVESTING ANTENNAE : A RAMAN STUDY.

A .Gall [1], V. Yurkov [2], R.J.Cogdell [3], A. Vermeglio [2] and B. Robert [1].

[1]DBCM/CEA and URA 1290/CNRS,Centre d'Etudes de Saclay, F-91191 Gif sur Yvette Cedex, France. [2]SBC/DPVE/CEA, Centre de Cadarache, F-13108 St. Paul lez Durance, France. [3]Division of Biochemistry and Molecular Biology, University of Glasgow, Glasgow G12 8QQ, Scotland.

1. Introduction.

In the bacteriochlorophyll*a*-containing photosynthetic bacteria light energy is collected by large arrays of light-harvesting (LH) pigment-protein complexes who's structural properties allow the captured energy to pass to the photochemical reaction centres (RC) where the transduction into chemical energy occurs (1). In wild-type purple bacteria the RC is surrounded by a ring of *core*, or LHI, antenna. Some species also synthesize a second type of light-harvesting complex known as the peripheral, or LHII, antenna. Depending on their environmental conditions a subset of these peripheral antenna-producing species can produce two distinct species of LHII thereby increasing their overall efficiency in capturing incident radiation (2).

The minimal unit of the LH-antenna contains two transmembrane-spanning α- and β-polypeptides (although a third, the γ-polypeptide, may be present), BChls and carotenoids (2). In the LHI antenna two BChls are present which give rise to a single absorption maximum between 870 nm and 890 nm. In the LHII antenna the α/β heterodimer contains three BChls. Two of these are involved in the 850 nm absorption band while the third has been assigned to the 800 nm absorption band. In the recently solved crystal structure of the LHII antenna from *Rhodopsudomonas acidophila* (1) has shown that it is comprised of a nonameric ring of α/β LHII minimal units arranged in two concentric rings (3). The pigment-protein H-bonding in this LHII complex, as well as other LH antenna, have been suggested to play a rôle in fine tunning the NIR absorption maxima (4,5,6) thus allowing heterogenous energy tansfer to occur towards the RC. This H-bond hypothesis is further examined here to see if the rôle of these BChla macrocycle H-bonds are universal in LH antenna. In this study we make a preliminary report, at the chromophore level, on the spectroscopically, and biochemically, unusual Bacteriochloroyhyll*a*-containing LH complexes named below. *Erythromicrobium ramosum* E5 and *Roseococcus thiosulfatophilus* RB3 are two BChl*a*-containing species which spectroscopically and biochemically unusual light-harvesting antenna (2). The LHII antenna of *E. ramosum* is a B798-832 complex. *R. thiosulfatophilus* RB3 has a RC-LHI with a NIR absorption maximum of 856 nm.

P. Mathis (ed.), Photosynthesis: from Light to Biosphere, Vol. I, 251–254.

The *Chromatium purpuratum* BN5500 B830-complex and the two *Rhodopseudomonas palustris* 2.6.1 LHII (B800-850) complexes are described elswhere(6).

2. Procedure

The B830-complex from *C. purpuratum* BN5500 and the B800-850 complexes from *Rps. palustris* 2.6.1. were grown, and purified as described elsewhere (4,7). Complexes from *E. ramosum* E5 and *R. thiosulfatophilus* RB3 were prepared as previously described (2). Room-temperature FT-RR and low-temperature (10K) Resonance Raman spectra were obtained by as described elsewhere (8,9).

3. Results and Discussion.

Figure 1 illustrates the room-temperterature absorption spectra of the 830nm-absorbing LHII antennae from *C. purpuratum* and *E. ramosum* (*Figure 1A*) and the RC-LHI *core* conjugates from *R. thiosulfatophilus* and *E. ramosum* (*Figure 1B*). The FT-RR spectra in *Figure 2* indicate differences between the proteins in this study and previous investigations (9,10,11). The RC-B871 complex from *E. ramosum* (*Figure 2A*) shows a

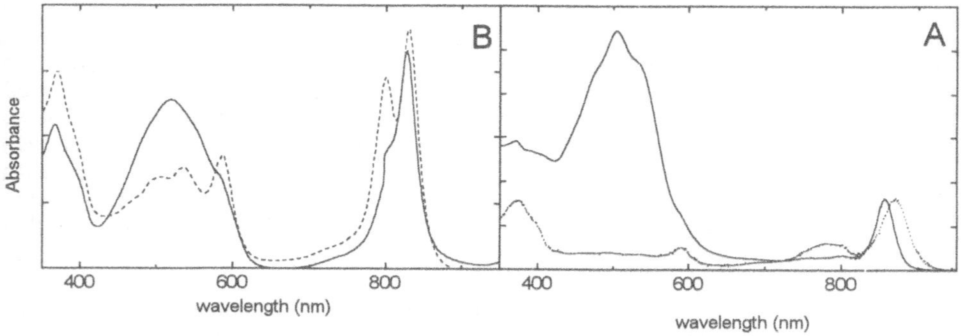

Figure 1. Room-temperature absorption spectra: (A) Absorption spectra of the B830 complexes from *C. purpuratum* (solid line) and *E. ramosum* (dotted line). (B) Absorption spectra of the "core" complexes from *R. thiosulfatophilus* (solid line) and *E. ramosum* (dotted line).

typical LHI Raman spectrum (10,11). However, the "blue shifted" RC-B856 complex from *R. thiosulfatophilus* (*Figure 2B*) has Raman shifts at 1654 and 1669 cm $^{-1}$ suggesting that their may be free 2-acetyl carbonyl groups in this complex. In *Rps. acidophila* the B800-820 complex (*Figure 2C*) has previously been shown to have free 2-acetyl carbonyl groups since the 1671 cm $^{-1}$ band (in the B800-850 complex) is down shifted to 1665 cm $^{-1}$, with an associated reduction at 1633 cm $^{-1}$ (10). In this report similar differences occur in between the *C. purpuratum* B830-complex (*Figure 2D*) and *Rps. acidophila* B800-850. This suggests that free 2-acetyl carbonyl groups are present in the *C. purpuratum* B830-complex. The B798-832 complex from *E. ramosum* (*Figure 2E*) is, however, more complex. It has Raman bands at 1649, 1669 and at

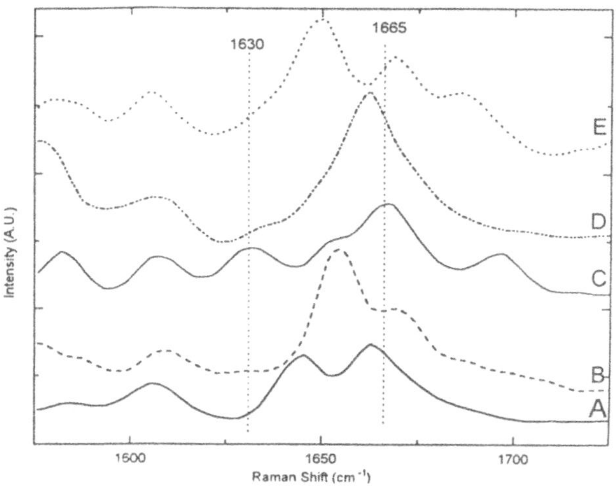

Figure 2. Room-temperature FT-RR spectra of the carbonyl region from: (A) *E. ramosum* RC-B871, (B) *R. thiosulfatophilus* RC-B856, (C) *Rps. acidophila* B800-820, (D) *C. purpuratum* B830-complex, (E) *E. ramosum* B798-832. *Rps. acidophila* spectra courtesy of Dr. J.N.Sturgis, DBCM/CEA,C.E. Saclay, F-91191 Gif sur Yvette, France.

1687 cm^{-1} but does tenatively also suggest a reduction in 2-acetyl H-bonding. In the high 850 nm-absorbing and low 850 nm-absorbing LHII antennae from *Rps. palustris* the major differences in the absorption spectra occur in the NIR region. Relative to the high 850 nm-absorbing complex, the low 850 nm-absorbing LHII shows a change in the 800 nm-absorbing region and a large decrease in the 850 nm-absorbing band. Such large differences in the NIR absorption spectra would allow us to postulate that large changes would be observed in FT-RR measurements in the 1620-1710 cm^{-1} carbonyl region. However, this is belived not to be the case. The only difference is at 1629 cm^{-1} (*Figure 3A*). Using classical RR spectroscopy the *Rps. palustris* complexes were excited with the 363.8 nm line from an Argon laser (Innova 100, Coherent) which coincides with the absorption maximum of the soret band of Bchl*a*. The RR spectra clearly show differences between the two LHII forms (Figure 3B). In the low 850 nm-absorbing LHII there is an increase *ca.* 1660 cm^{-1} associated with a decrease *ca.* 1630 cm^{-1}. This may associated with the 2-acetyl carbonyl groups on the 800 nm absorbing BChls but seems unlikely. This study has shown further evidence that the degree of BChl*a* macrocycle H-bonding contributes to the NIR absorption maxium in bacterial light-harvesting antennae. In particular, in LHII complexes the loss of 2-acetyl H-bonds is associated with a "blue shift" in the absorption maximum from *ca.* 850 nm to *ca.* 820-830 nm.

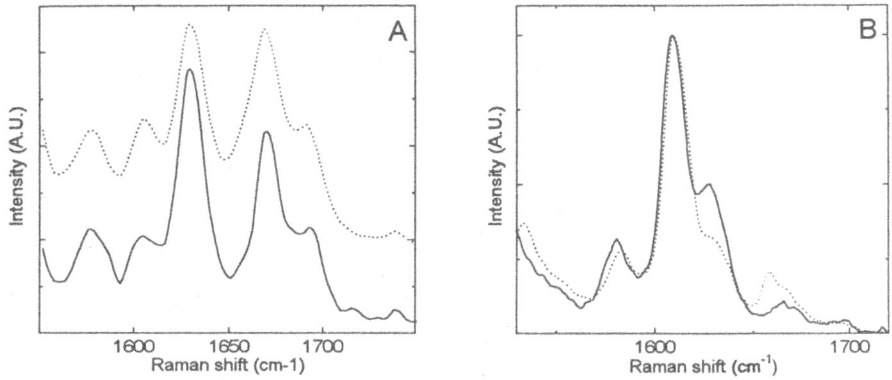

Figure 3. Raman analysis of the high 850nm-absorbing B800-850 complexes (solid lines) and the low 850nm-absorbing B800-850 complexes (dotted lines) from *Rps. palustris* LHII antenna. (A) Room-temperature FT-RR spectra of the 1620-1710 cm^{-1} carbonyl region excited at 1064nm. (B) Low temperature (10K) RR spectra of the carbonyl region excited at 363.8 nm.

This study has been unable to elucidate the molecular interactions in *Rps. palustris* LHII at the level of the chromophores, however, this complex mechanism is under further investigation. Primary sequence analysis is presently being undertaken for all the unknown sequences and once complete will aid a better understanding of the BChl-protein interactions in these complexes

References.

1 McDermott, G., Prince, S.M., Freer, A.A., Hawthornthwaite-Lawless, A.M., Papiz, M.Z., Cogdell, R.J., and Isaacs, N.W. (1995) Nature. 374, 517-521

2 Yurkov, V., Gad'on, N., and Drews, G. (1993) Arch. Mirobiol. 160, 372-376

4 Cogdell, J.C., Hawthornthwaite A. M., Evans, M. B., Ferguson, L. A., Kerfeld, C., Thornber, J. P., van Mourik, F., and van Grondelle, R. (1990) Biochim. Biophys. Acta. 1019, 239-244

6 Zuber, H., and Brunisholz, R.A. (1986) in Chlorophylls. (Scheer, H., ed.) CRC Press. London.

7 Cogdell, R.J., Durant, I., Valentine, J., Lindsay, J.G., and Schmidt, K (1983) Biochim. Biopys. Acta. 722, 427-45.

8 Mattioli, T.A.., Hoffmann, A., Robert, B., Schrader, B., and Lutz, M (1991) Biochemistry. 30, 4648-4654

9 Robert, B., and Lutz, M. (1985) Biochim. Biopys. Acta. 807, 10-23

10 Sturgis, J.N., Jirsakova, V., Reiss-Husson, R., Cogdell, R.J., and Robert, B. (1995) Biochem. 34, 517-523.

11 Robert, B., Vermeglio, A., Steiner, R., Scheer, H., and Lutz, M. (1988) in Photosynthetic Light-Harvesting systems (Scheer, H., and Schneider,S., eds.) pp355-363, de Gruyter and Co., Berlin, Germany.

A kinetic study of the reassociation of the antenna complex of *Rhodospirillum rubrum* in detergent solution.

James Sturgis and Bruno Robert.
SBPM/DBCM CEA and URA 1290 CNRS, Centre d'Etudes de Saclay, 91191 Gif sur Yvette Cedex, FRANCE.

1. Introduction

The core light-harvesting antennae complexes function to collect light-energy and rapidly funnel this energy to the reaction centre (1). These complexes are composed of aggregates of a minimal structural unit containing two bacteriochlorophyll molecules and one carotenoid molecule and two small polypeptides (α and β) each containing about 50 amino acids (2). This minimal unit aggregates to form an apparently cyclic complex containing 16 subunits that probably encloses the reaction centre (3). This cyclic structure is somewhat larger than the 9 subunit cyclic aggregate that has been described for the peripheral antenna complex of *Rhodopseudomonas acidophilla* (4).

In the presence of n-octyl-β-D-glucopyranoside (βOG) it is possible to dissociate the core antennae derived from a number of bacteria into a subunit form (5, 6, 7, 8). This subunit form with an absorption maximum near 820 nm has been variously reported to contain one or two minimal heterodimeric units (9, 10, 11, 12, 13, 14). Upon treatment with higher concentrations of βOG the B820 subunit form further dissociates into a monomeric units composed of polypeptide bound bacteriochlorophyll (14), that absorb near 777 nm. This stepwise dissociation process is completely reversible, and in particular the interconversion between the B777 and B820 spectral forms appears to be an equilibrium in the micellar phase (14).

We have previously investigated this equilibrium under steady state conditions to determine the reaction order and the energetics of the association reaction (14). These studies showed a reaction order of four and so strongly suggested a structure for the B820 subunit containing four bacteriochlorophyll molecules in disagreement with much spectral data. In this paper we report an initial investigation of the kinetics of the interconversion between the B820 and B777 spectral forms, this investigation suggests that there are two B820 forms one of which has a dimeric structure, and the second of which may well be tetrameric.

2. Procedure

The core antenna, B873, from *R. rubrum* was prepared, from the carotenoid deficient strain G9^{+}, as described previously (14). B820 subunits were prepared by titration of

P. Mathis (ed.), Photosynthesis: from Light to Biosphere, Vol. I, 255–258.

isolated purified B873 with βOG, obtained from either Sigma or Calbiochem. Throughout all antenna preparations the material was protected from light.

Stopped flow measurements were performed on a stopped flow absorption apparatus interfaced to a microcomputer with a Biologic data acquisition system. Data were transferred to a Sun SPARC station 2 for kinetic analysis using programmes written for the purpose.

3. Results and Discussion

In figure 1 is shown the kinetics of the absorption changes that follow mixing of an equilibrium mixture of the B777 and B820 antenna forms with buffer containing βOG

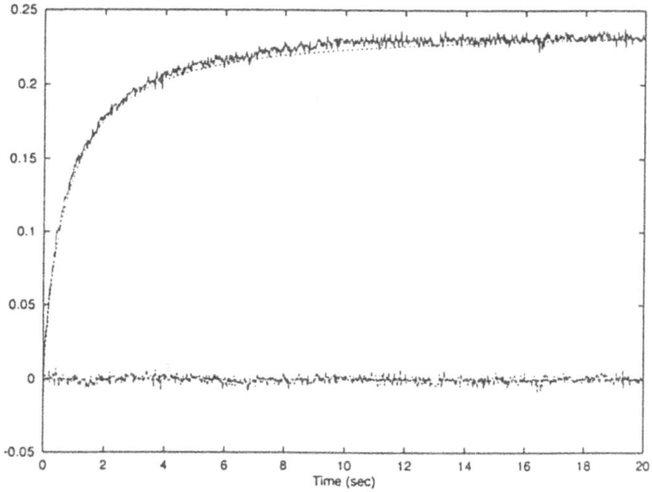

Figure 1. Change in absorption at 823 nm measured following mixing of a mixture of B777 and B820 in 30 *mM* βOG with 15 *mM* βOG, at 20°C. Also shown are a fitted second-order curve, and the residual errors.

at less than the critical micelle concentration, which in these experiments was about 23 *mM*. The rapid absorption rise that occurs during the first few seconds is clearly non-exponential. Attempts to fit this curve with an integrated second order rate equation are reasonably successful (as shown) while attempts to fit the observed kinetic with either an integrated fourth order rate equation, or a sum of exponentials were less successful.

An investigation of the wavelength dependence of the amplitude of the second order reaction curve fitted to the reassociation kinetic is shown in figure 2. This illustrates clearly that there are no detectable spectral intermediates, during the association, and that the entire spectral change can be reasonably described by a second order reaction in which two B777 subunits come together to give a B820 heterodimeric subunit. This is entirely in agreement with the spectroscopic evidence (9, 10, 11).

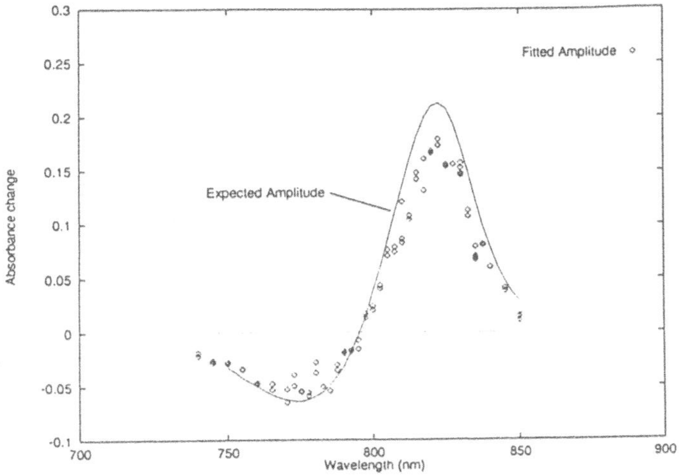

Figure 2. Comparison of the amplitudes of reassociation kinetics measured at different wavelengths with that expected from equilibrium measurements. The kinetic measurements were made using the same conditions as in figure 1.

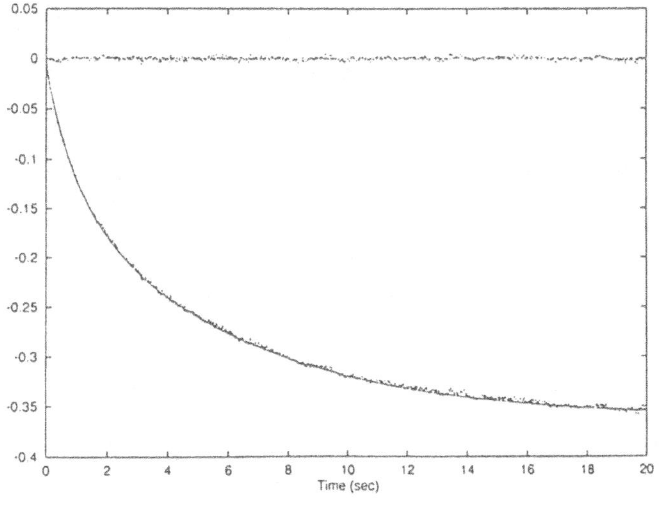

Figure 3. Dissociation kinetic measured at 823 nm measured following mixing of a mixture of B777 and B820 in 30 *mM* βOG with 30 *mM* βOG, at 20°C. Also shown are a fitted double exponential curve, and the residual errors.

Figure 3 shows the absorption kinetic associated with the dissociation of the B820 subunit form when it is mixed with βOG solutions at above the critical micellar concentration. This kinetic is complex, and has been fitted with a double exponential decay containing two components with rate constants of about 1.2 sec^{-1} and 0.2 sec^{-1}. This fit gives a satisfactory result except on a fast timescale where there is some evidence for an additional component that is believed to be a detergent mixing artefact. This double exponential kinetic appears to be indicate a second step in the association and dissociation of B820 that is not visible in the absorption changes, but nevertheless influences the dissociation kinetics. It is possible, though not shown here, that this second spectroscopically invisible step is due to a homodimerisation of the heterodimeric B820, thus giving a tetrameric complex.

It is interesting to note, that the time scale of these association reactions, several seconds, are very similar to those of the kinetics that have been previously observed in the refolding of bacteriorhodopsin (15). It is thus tempting to speculate that this time-scale is common to different systems where the docking of transmembrane α-helices in a detergent or lipid environment occurs.

J.S. was supported by a fellowship from the EEC (Human Capital and Mobility programme).

References

1 van Grondelle, R: (1985) *Biochim. Biophys. Acta* **811** 147-195.
2 Cogdell, R.J., Lindsay, J. G., Valentine, J., and Durant, I. (1982) *FEBS Lett.* **150** 151-154.
3 Karasch S., Bullough, P.A., and Ghosh, R. (1995) *EMBO J.* **14** 631-638.
4 Mc Dermott G., Prince, S.M., Freer, A.A., Hawthornthwaite-Lawless, A.M., Papiz, M.Z., Cogdell, R.J., and Isaacs, N.W. (1995) *Nature* **374** 517-521.
5 Loach, P.A., Parkes, P.S., Miller, J.F., Hinchigeri, S., and Callahan, P.M. (1985) In *Molecular Biology of the Photosynthetic Apparatus* (Antzen, C., Bogorad, L., Bonitz, S. and Steinback , K. eds.) pp197-209 Cold Spring Harbour Press, Cold Spring Harbour, NY.
6 Chang, M.C., Meyer, L., and Loach, P.A. (1990) *Photochem Photobiol* **52** 873-881.
7 Heller B.A. and Loach, P.A. (1990) *Photochem Photobiol* **51** 621-627.
8 Meckenstock R.U., Brunisholz R.A. and Zuber , H. (1992) *FEBS Lett* **311** 128-134.
9 Visschers, R.W., Chang, M.C., van Mourik, F., Parkes-Loach, P.S., Heller, B.A., Loach, P.A., and van Grondelle, R.. (1991) *Biochemistry* **30** 2951-2960
10 Visschers, R.W., Nunn, R., Calkoen, F., van Mourik, F., Hunter, C.N., Rice, D.W., and van Grondelle, R. (1992) *Biochim. Biophys. Acta* **1100** 259-266.
11 van Mourik, F., van der Oord, J.R., Visscher, K.J., Parkes-Loach, P.S., Loach, P.A., Visschers, R.W. and van Grondelle, R. (1991) *Biochim. Biophys. Acta* **1059** 111-119.
12 Miller, J.F., Hinchigeri, S.B., Parkes-Loach, P.S., Callahan, P.M., Sprinkle, J.R., Riccobono, J.R. and Loach, P.A. (1987) *Biochemistry* **26** 5055-5062.
13 Ghosh, R., Hauser, H., and Bachofen, R. (1988) *Biochemistry* **27** 1004-1014.
14 Sturgis, J.N. and Robert, B. (1994) *J. Mol. Biol.* **238**, 445-454.
15 Booth, P.J., Flitsch, S.L., Stern, L.J., Greeenhalgh, D.A., Kim, P.S., and Khorana, H.G. (1995), *Struct. Biol.* **2** 139-143.

INFLUENCE OF METAL EXCHANGE ON ABSORPTION SPECTRA OF THE LIGHT HARVESTING COMPLEX OF *RHODOSPIRILLUM RUBRUM*

A. Näveke[*,§], H. Scheer[*], B. Robert[§] & J. Sturgis[§]
[*] Botanisches Institut, Universität München, Germany
[§] SBPM/ DBCM and URA CNRS 1290, CE Saclay, Gif-sur-Yvette, France

1. Introduction

We are investigating the assembly of the light-harvesting antennae of the photosynthetic purple bacterium *Rhodospirillum rubrum (Rsp. rub.)* as a model system in order to elucidate the nature of the interactions that stabilize the structure of membrane proteins. The antennae of photosynthetic bacteria gather sunlight and funnel the energy to the neighbouring reaction center, where it is converted into chemical potential energy. The antennae complexes of *Rsp. rub.* are composed of aggregates of two types of small transmembrane polypeptides together with associated bacteriochlorophyll a (BChla) pigments. According to a two-step model, the folding and association of membrane proteins involves first formation of lipid bilayer spanning α-helical segments. These subsequently interact and give rise to the functional tertiary and quaternary structures[1]. The antenna complex B873 of *Rsp. rubrum*, when treated with n-Octyl-D-glucopyranoside (βOG), forms a smaller complex B820 itself composed of B777 subunits. Upon further addition of detergent, the B777 subunit becomes the only spectral form, the process reverses when the detergent is removed[2,3]. Sturgis and Robert[4] have calculated that the enthalpy decreases by 175 kJ/mole when B820 is formed. B777 and B820 complexes containing modified pigments help to determine the contribution of the bound chromophores to the total energy difference and therefore help to understand the role of the BChla molecules in the stabilization of the antenna complex. We have studied the effect of exchanging the native bound pigments with Zinc-Bacteriophaeophytin a (Zn-BPhea).

2. Procedure

Starting material for the synthesis of Zn-BPhea was BChla, extracted with methanol from *Rsp. rub.* carotenoidless strain G9+[5]. Purification was done on DEAE-Cellulose (Whatman) in chloroform[6]. The magnesium ion was exchanged with zinc by transmetalation via the labile cadmium complex[7]. Zn-BPhea was purified on a reversed phase column (LiChroprep RP-18, Merck) in acetone-HEPES (25 mM, pH 7.85) buffer (80:20 v/v)[8]. The purity of the material was checked by visible/near infrared absorption spectroscopy and high performance liquid thin layer chromotography (HPTLC, RP-8, Merck).
Rsp. rub. B873 was purified from washed, reaction center depleted, chromatophores[9] of the carotenoidless strain G9+. Antennae were extracted with 2% dodecyl-maltoside

P. Mathis (ed.), Photosynthesis: from Light to Biosphere, Vol. I, 259–262.

(BIOMOL, Hamburg, Germany) before ion exchange chromatography (IEC) on a Resource Q® column (Pharmacia) and elution with a 0 to 1M NaCl gradient in 10 mM Tris-HCl buffer (pH 8.0) containing 0.05% dodecyl-maltoside. The purified B873 was dialysed against 10 mM Tris-HCl (pH 8.0). B820 and B777 subunits were prepared by titration of purified B873 with βOG (BIOMOL).
Zn-BPhea was dissolved in acetone containing 10% βOG (w/v) to an optical density (OD) of 20 - measured in the Q_y - and dried under Argon. To this pigment-detergent mixture a solution was added that contained B820 and B777 in a ratio of about 1:1. The total concentration of antenna complex corresponded to 7.5 μM BChla, and a fivefold excess of Zn-BPhea (OD ratio) was used. This mixture was incubated for 1h at 30°C (more recent experiments have shown that incubation for 5 min at room temperature yields the same results), put on ice for 15 min and dialysed against 10 mM Tris-HCl (pH 8.0). Antennae complexes and free pigment were separated by IEC as described above. However, in place of Resource Q®, DEAE-Cellulose was used. Pigments were extracted with acetone, transferred into diethyl ether, dried under argon and redissolved in diethyl ether. Absorption spectra were measured using a Cary 5 spectrophotometer.

3. Results and Discussion

The first antennae complexes containing modified chromophores were formed by reconstitution from purified polypeptides and pigments[10]. For the first time, we here report the exchange of BChla with a modified pigment on the level of an antennae subunit that is an integral protein-pigment complex. We detected a remarkable shift to the blue for the Q_y transition for the entirely reconstituted antennae form containing Zn-BPhea (Figure 1).

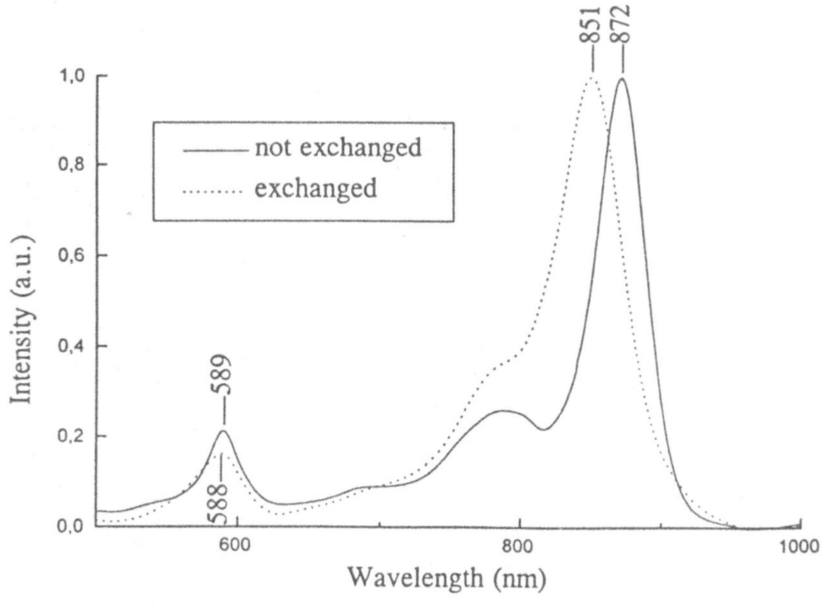

Figure 1: The Q_x and Q_y region of the absorption spectra of purified B873 antenna (—) and of the entirely reconstituted antenna after the exchange of BChla with Zn-BPhea (···)

Qualitativley, this fits well with the shift seen when magnesium is replaced by zinc in the free pigment (Figure 2). The absorption spectra of the extracted pigments demonstrate that the exchange was quantitative since we could not detect any BChla in the complex that contained Zn-BPhea (Figure 2).

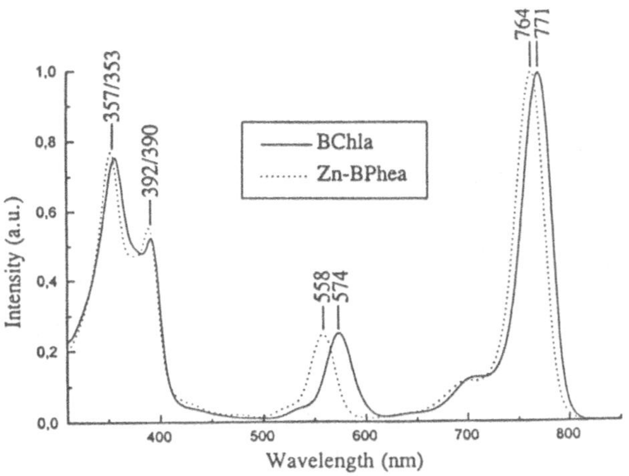

Figure 2: Absorption spectra of BChla extracted from the purified B873 antenna (—) and of Zn-BPhea (···). The spectrum of Zn-BPhea used for the exchange and the spectrum of the pigment extracted from the reconstituted antenna after the exchange of BChla with Zn-BPhea were identical

Two subunit forms with clearly blue shifted Q_y absorption maxima reversibly dissociated from the purified antannae complex that contained Zn-BPhea when it was treated with βOG (Figure 3). This behaviour is well known for B873 [2,3].

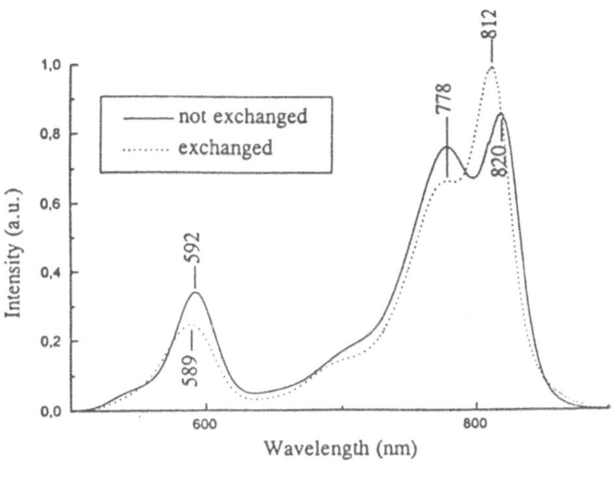

Figure 3: The Q_x and Q_y region of the absorption spectra of the subunits generated with βOG from purified B873 antenna (—) and from the entirely reconstituted antenna after the exchange of BChla with Zn-BPhea (···)

The equlibrium between these two complexes varies with temperature, which is also known for B820 and B777[4]. This dependence supports our interpretation that Zn-BPhea has replaced the native chromophore in the binding pocket. This will be confirmed by Resonance Raman spectroscopy in the immediate future. Nevertheless, the polypeptides were present in approximately stoichometric amounts, estimated by the UV absorption band and SDS gel electrophoresis.
While isolated polypeptides can form structures that are not accessible to pigment binding and reconstitution, the exchange of BChla with modified pigments on the level of B820/B777 should allow us to obtain antennae complexes in high yields. Furthermore, an exchange system for B820/B777 will give acces to the relative binding constants for various bound pigments, quantities which can not be determined in the reconstitution system. The temperature dependence of the equilibrium between the two subunit forms B820 and B777 in βOG solution will allow to calculate the thermodanymic quantities of the association process such as enthalpy and entropy.
It will be of great interest to form antennae complexes with pigments systematically altered at positions probably involved in the protein-pigment contact, but only one position at one time. Their vibrational spectra could again improve the insight in the structure of bacterial light-harvesting systems which has been largely exetended, recently[11,12]. The exchange of BChla with pigments containing other central ions than magnesium in the antennae complex will yield spectral probes which are much less fluorescent than the native complexes. This will greatly improve the possibilities of Resonance Raman spectroscopy. Excitation in the Q_y transition, for example, should be possible for Ni- and Co-BPhea.

4. References

1. Popot, J.L. & and Engelmann, D. (1990) Biochemistry 29, 4031-4037
2. Ghosh, R., Hauser, H. & Bachofen, R. (1988) Biochemistry 27, 1004-1014
3. Parkes-Loach, P.S., Sprinkle, J.R. & Loach, P.A. (1988) Biochemistry 27, 2718-2727
4. Sturgis, J. & Robert, B. (1994) J. Mol. Biol. 238, 445-454
5. Struck, A. (1991) Thesis, Universität München, Germany
6. Omata, T. & Murata, N, (1983) Plant Cell Physiol. 24, 1093-1100
7. Scheer, H. & Hartwich, G. (1995) in Anoxygenic Photosynthetic Bacteria (Blankenship, R.E., Madigan, M.T. & Bauer, C.E., eds.), Kluwer Academic Press, Dordrecht, Netherlands, in press
8. Helfrich, M. (1995), Thesis, Universität München, Germany
9. Cogdell, R.J., Lindsay, J.G., Valentine, J. & Durant, I. (1982) FEBS Lett. 150, 151-154
10. Chang, M.C., Callahan, P.M., Parkes-Loach, P.S., Cotton, T.M. & Loach, P.A. (1990) Biochemistry 29, 421-429
11. Karrash, S., Bullough, P.A. & Ghosh, R. (1995) EMBO J. 14, 631-638
12. McDermott, G., Prince, S.M., Freer, A.A., Hawthornthwaite-Lawless, A.M., Papiz, M.Z., Cogdell, R.J. & Isaacs, N.W. (1995) Nature 374, 517-521

A. Näveke was supported by a doctoral fellowship from the DAAD (Deutscher Akademischer Austauschdienst). J. Sturgis was supported was supported by a post doctoral fellowship from the Commission of the European Union (HCM program).

SPECTRAL MOMENTS FACILITATE THE ANALYSIS OF CIRCULAR DICHROISM SPECTRA
Application to Chlorosomes from *Chloroflexus aurantiacus*

Oscar J.G. Somsen, Rienk van Grondelle and Herbert van Amerongen, Dept. of Physics and Astronomy and Institute for Molecular Biological Sciences, Vrije Universiteit, De Boelelaan 1081, 1081 HV Amsterdam, The Netherlands

1. Introduction

The spectral properties of pigment-protein aggregates are affected by two types of interaction. Firstly, interaction between the pigments leads to mixing of their excited states, and consequently to excitonic splitting of absorption (OD) spectra. Circular dichroism (CD, the difference in absorption of left- and right circular light) spectra are especially sensitive to this type of interaction.
Secondly, interaction with the protein matrix is responsible for so-called site-shifts of the absorption bands as well as for the widths of these bands. Inhomogeneous broadening is caused by a variation in site-shifts of individual pigments, while further broadening results from coupling of protein vibrations to electronic states in the pigments (here referred to as homogeneous broadening).
The analysis of broadened excitonic spectra provides valuable information about the structure of pigment-protein complexes. Unfortunately, the effects of spectral broadening are difficult to simulate and the site-shifts are often unknown. We demonstrate, that these problems are largely circumvented by using spectral moments.
Strong excitonic interactions are observed in chlorosomes, large light-harvesting complexes of green bacteria. They are composed of small amounts of protein and Bchl *a*, and up to 20000 Bchl *c*, *d* or *e* pigments, organized in protein-less cylindrical aggregates of 5-10 nm diameter (see e.g. [1] and references therein). The Q_y transition dipole moments are oriented at an angle of 15° with the cylinder axis [2, 3].
Even for freshly prepared chlorosomes, the CD spectra vary strongly [3]. However, most CD spectra of *Chloroflexus aurantiacus* chlorosomes can be reproduced by a linear combination of two S-shaped basis spectra, named type I and II [3]. The zero crossing of the type I spectrum coincides with the OD maximum while its magnitude is strongly variable. The type II spectrum is blue shifted by approximately 10 nm and of more or less constant magnitude. The variability of the CD spectra, is not due to differences in preparation [3], but reflects the presence of two or more structural organizations. The absence of corresponding changes in the OD and LD (linear dichroism) spectra indicates that the differences between these structures are small.
Many models for chlorosomes consider a linear aggregate as the main structural element (see e.g. [4]) but these models cannot explain the strong CD signals. As will be shown below, cylindrical models as e.g. in [1,5] give rise to strong excitonic CD and can in fact explain the observed variability.

P. Mathis (ed.), Photosynthesis: from Light to Biosphere, Vol. I, 263–266.

2. Theory

Excitonic interaction in a complex of N pigments, is defined by the transition values V_{nm} of the electronic interaction between pigments n and m, and the site-energies W_n (only singly excited states $|\Phi_n>$ are included). They constitute the interaction matrix:

$$H_{nm} = \langle\Phi_n|H|\Phi_m\rangle = \delta_{nm}W_n + V_{nm} \tag{1}$$

and the excitonic spectrum is obtained from its eigenvectors:

$$Hx_J = E_J x_J \tag{2}$$

The eigenvalues E_J are equal to the energy levels of the excitonic transitions of the complex, and the coefficients x_{Jn} determine their dipole- and rotational strengths:

$$D_J = \sum_{n,m=1}^{N} x_{Jn}x_{Jm}(\mu_n \cdot \mu_m)$$
$$R_J = \frac{D_J^L - D_J^R}{4} = -\frac{\pi}{2\lambda}\sum_{n,m=1}^{N} x_{Jn}x_{Jm}(r_{nm} \cdot \mu_n \times \mu_m) \tag{3}$$

where μ_n is the transition dipole moment of each pigment, $r_{nm} = r_m - r_n$ the relative position between two pigments, and λ the wavelength of the light [6].
The result is a stick spectrum of narrow excitonic transitions. The finite width of the absorption bands often merges these transitions into a single, structured spectrum. Unfortunately, simulation of such spectra (if possible at all) requires extensive numerical calculations. The site energies are a further complicating factor. They are difficult to calculate, and may have a considerable effect on the shape of the excitonic spectra.
These problems are largely circumvented by using spectral moments. Several moments of OD and CD spectra are discussed elsewhere [6]. Here we consider the first moment of the rotational strengths:

$$R^{[1]} = \sum_{J=1}^{N} R_J E_J = -\frac{\pi}{2\lambda}\sum_{n,m=1}^{N} H_{nm}(r_{nm} \cdot \mu_n \times \mu_m) \tag{4}$$

For conservative ($R^{[0]} = \Sigma R_J = 0$) CD spectra, $R^{[1]}$ is an indicator of the magnitude. As may be seen from eq.4, it is obtained by summation over all interactions and diagonalization of the interaction matrix is not required. In addition, spectral broadening does not alter the first moment of the excitonic CD spectrum [6], and it is independent of site-shifts, even though such shifts may radically influence the shape of the spectrum itself.
The value of $R^{[1]}$ can directly be extracted from experimental spectra if the width of the bands is larger than the splitting between them. In that case, a typical conservative CD spectrum has an S shape, similar to that of the first derivative (OD') of the OD spectrum. Its amplitude is determined by $R^{[1]}$, which can be calculated from three experimentally accessible parameters: the peak-peak amplitude (CD_{pp}) of the CD, the amplitude of the absorption band (OD_{max}), and its full width at halve maximum (Δ_{FWHM}). By assuming Gaussian shaped absorption bands, it can be shown [6] that:

$$\frac{R^{[1]}}{D^{[0]}} = \frac{\text{magn(CD)}}{4\varsigma^2 \text{magn(OD')}} = 0.058\frac{\Delta_{\text{FWHM}}\text{CD}_{\text{pp}}}{\text{OD}_{\text{max}}} \tag{5}$$

The sign of $R^{[1]}$ is determined by the shape of the CD spectrum. It is expressed relative to the total dipole strength $D^{[0]} = \Sigma D_J$. The factor ς^2=1.5 is needed because broadening of the CD and OD spectra differ [6]. The right hand side of eq.5 is experimentally accessible while the left hand side is directly related to the structural model.

3. Results and Discussion

The chlorosome CD can now be quantized. With eq.5, one estimates from the experimental spectra [3] that the magnitude of the type I basis spectrum varies: $0 < R^{[1]}/D^{[0]} < 0.07$ cm^{-1}, while the type II spectrum remains constant at $R^{[1]}/D^{[0]} = -0.02$ cm^{-1}. We attempt to explain this variability by small structural changes in a cylindrical aggregate, sufficiently long to ignore edge effects. Only Q_y transitions are considered. The cylinder is defined by the positions and dipole orientations of its pigments:

$$r_n = \begin{bmatrix} r\cos(x_n/r) \\ r\sin(x_n/r) \\ d_n \end{bmatrix} \qquad \hat{\mu}_n = \begin{bmatrix} \sin\theta\cos(x_n/r+\phi) \\ \sin\theta\sin(x_n/r+\phi) \\ \cos\theta \end{bmatrix} \tag{6}$$

The displacements x_n and d_n form a regular lattice on the rolled out surface of the cylinder around the origin $x_0 = d_0 = 0$. The angles θ and ϕ define the orientation of the transition dipole moments. The allowed transitions cluster in two regions near the energies:

$$E_0 = \sum_n V(n) \qquad \text{and} \qquad E_1 = \sum_n V(n)\cos(x_n/r) \tag{7}$$

where $V(n) = V_{0n}$ is the strength of the interaction between pigments *0* and *n*. Most of the dipole strength is concentrated near the transition at E_0 which is oriented parallel to the symmetry axis. Its red shift confirms that $V(n) < 0$ for the most prominent interactions. The CD is a sum of two S shaped spectra. One is formed by two opposite rotational strengths near E_0 and E_1 (type I). The other is centered around E_1, which is somewhat less red shifted (type II). If $x_n \ll r$ for the relevant interactions, the first moment of the rotational strengths is [6]:

$$\frac{R^{[1]}}{D^{[0]}} \approx \frac{\pi}{2}\sum_n V(n)\sin\theta^2\frac{x_n}{r\lambda}(\cot\theta\sin\phi x_n - d_n) \tag{8}$$

The two terms correspond the magnitudes of the type I and II spectrum, respectively. A change of ϕ has little effect on the structure but can explain the variability of the type I amplitude. Changes in d_n, x_n or θ are less likely, since these would affect the OD and LD spectra, and also the type II CD spectrum.

A rough estimate of the required changes can be made by assuming that eq.8 is dominated by one nearest-neighbor interaction. Since pigments with either $x_n = 0$ or $d_n = 0$ do not contribute to the type II spectrum, we consider the interaction to a diagonal neighbor at $d_n = x_n = 0.5$ nm, and use θ=15° and r = 2.5 nm. With these parameters, the magnitude of the second term ($R^{[1]}/D^{[0]} = 0.02$ cm^{-1}) requires an interaction strength $V(n) = -700$ cm^{-1},

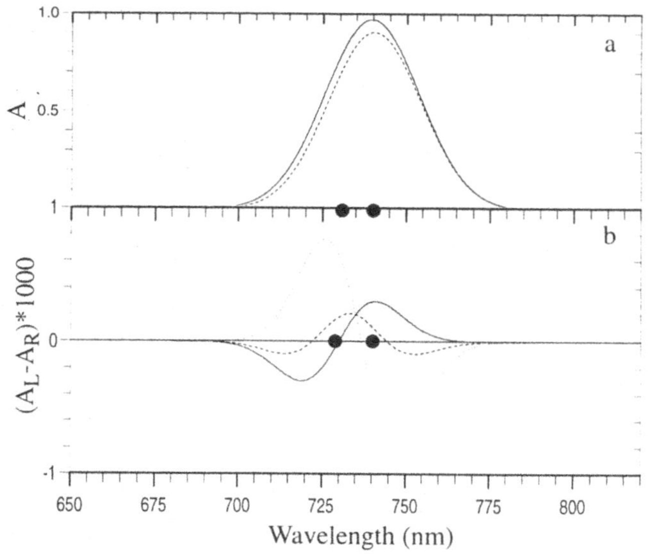

Figure 1. Reconstructed chlorosome spectra, based on the analysis in the text. a) OD (solid) and LD (dashed) spectrum. b) CD of a cylindrical aggregate with ϕ = 0° (solid), ϕ = 16° (dashed) and ϕ = 90° (dotted). Positions of E_0 and E_1 are marked with a dot on the axis; θ = 15°. V = -700 cm^{-1}, $d_n = x_n$ = 0.5 nm, r = 2.5 nm

comparable to the value expected from a dipole-dipole calculation. By varying ϕ between 0° and 90°, the first term in $R^{[1]}/D^{[0]}$ varies between 0 and 0.07 cm^{-1}, equal to the experimentally observed variation. Fig. 1, shows simulated OD, LD and CD spectra for several values of ϕ. The variation is similar to that in the experimental spectra [3]. One should note that d_n and x_n increase for non-nearest-neighbor interactions. It is clear that small variations in the cylindrical organization can explain the experimentally observed variations in the CD spectra of chlorosomes.

Acknowledgments

This work was financially supported by the Netherlands Organization for Scientific Research (NWO). The authors thank prof. Walter S. Struve for valuable discussions.

References

1 Holzwarth, A.R. and Schaffner, K. (1994) Photosynth. Res. 41, 225-233
2 Van Amerongen, H., van Haeringen, B., van Gurp, M. and van Grondelle, R. (1991) Biophysical J. 59, 992-1001
3 Griebenow, K., Holzwarth, A.R., van Mourik, F. and van Grondelle, R. (1991) Biochim. Biophys. Acta 1058, 194-202
4 Alden, R.G., Lin, S.H., Blankenship, R.E. (1992) J. Luminescence 51, 51-66
5 Nozawa, T., Ohtomo, K., Suzuki, M., Nakagawa, H., Shikama, Y., Konami, H. and Wang, Z.-Y. (1994) Photosynth. Res. 41, 211-223
6 Somsen, O.J.G. (1995) thesis Vrije Universiteit, Amsterdam, The Netherlands

PH EFFECT ON THE SUPRAMOLECULAR ASSEMBLIES OF BACTERIOCHLOROPHYLL C AGGREGATES IN AQUEOUS ORGANIC MEDIA

K. Uehara,[1] T. Ishii,[1] and M. Mimuro[2]
[1]Department of Applied Biosience, Research Institute of Advanced Science and Technology, Osaka Prefecture University, 1-2 Gakuencho, Sakai 593, Japan; [2]National Institute of Basic Biology, Myodaiji, Okazaki 444, Japan

1. Introduction

Chlorosomes contain about 10,000-16,000 molecules of bacteriochlorophyll c (BChl c) in self-aggregated forms absorbing at 740-750 nm and rodlike BChl c aggregates which finally form bandle-structure in chlorosomes have been proposed to act as a functional form in green photosynthetic bacteria.[1] However, molecular structure and mechanism for aggregation are still open to question. We have found that BChl c homologs form artificial aggregates in an aqueous medium containing lipid, monogalactosyl digliceride, and show optical properties similar to those of BChl c in intact chlorosome.[2] In order to clarify the role of the media on the organization in aggregates, we investigated the effects of pH and salt concentration on the formation of artificial aggregates of BChl c in aqueous organic media.

2. Materials and methods

[E,E]BChl c_F and [P,E]BChl c_F (Fig. 1) were isolated from Chlorobium limicola strain 6230 and purified according to the method described by Olson and Pedersen[3]. These BChl c's were dissolved in tetrahydrofuran (THF), acetone, methanol, and dimethyl sulfoxide (DMSO). For the preparation of aggregates, a phosphate buffer solution was added dropwise into the organic solution of BChl c under magnetic stirring. Absorption and CD spectra were recorded to identify the formation of aggregates. All operations were performed under dime light. 6 v/v% organic solvent showed the largest pH effect, irrespective of organic solvent used.

3. Results and Discussion

Figure 2 shows a typical example of absorption spectra of [P,E]BChl c_F at various pH conditions. The location of the absorption maximum of the Qy band of a BChl c aggregate depends on the pH of the aqueous organic solution. A large red-shift of the absorption maximum and a broadened band width with a light-scattering were observed under the low pH condition (pH 4.49), suggesting formation of higher aggregates. Figure 3 shows the pH effect for a few organic aqueous solvent systems on the aggregation of [P,E]BChl c_F. The absorption maxima were red-shifted in the order of methanol, acetone and THF, and in the respective solvent systems, red-shift of the maxima was observed with decrease in pH. Pheophytinization at acidic condition (pH 4.49) was remarkably supressed for the

P. Mathis (ed.), Photosynthesis: from Light to Biosphere, Vol. I, 267–270.

present BChl c aggregation: almost no observable bacteriopheophytin c was detected even 1 hr later after mixing. Figure 4 shows the concentration effect of phosphate buffer adjusted at pH 4.49 on the aggregation of [P,E]BChl c_F for various 6 v/v% organic aqueous media. With increase of buffer concentration, the absorption maximum of Qy band shifted to longer wavelength.

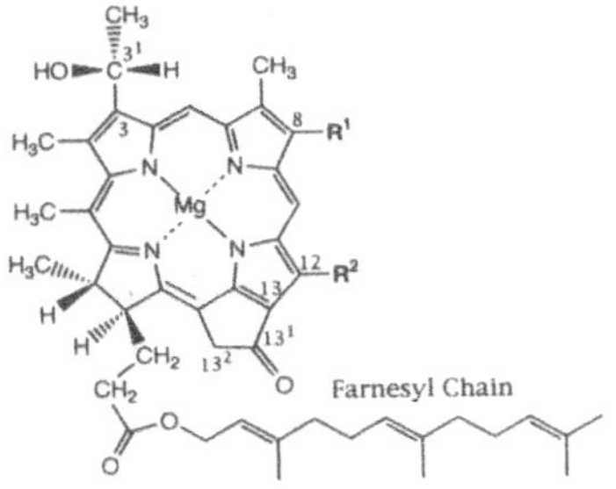

BChlc	R^1	R^2
[E, E]BChl c $_F$	ethyl	ethyl
[P, E]BChl c $_F$	propyl	ethyl

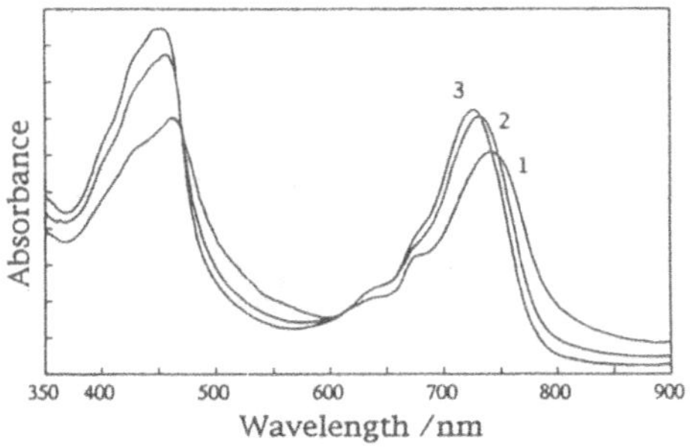

Fig. 1 Chemical structure of BChl c homologs.

Fig. 2 Absorption spectra of [P,E] BChl c $_F$ in 6% (vol/vol) Acetone aqueous solution. pH4.49(1), pH7.17(2), pH9.18(3)

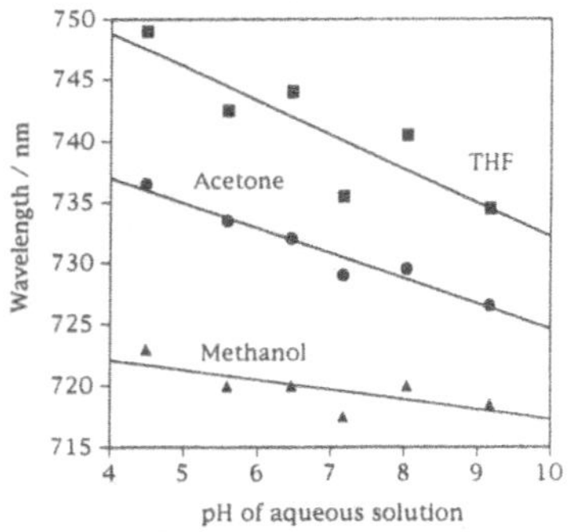

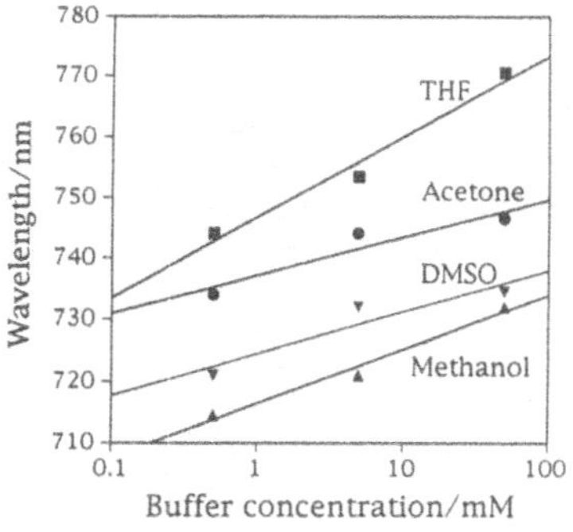

Fig. 3 pH effect on the aggregation of [P,E] BChl c_F in 6%(vol/vol) organic aqueous solution for 10mM phosphate buffer.

Fig. 4 Buffer concentration effect on the aggregation of [P,E] BChl c_F in 6% (vol/vol) organic aqueous solution for pH4.49 phosphate buffer.

[E,E]BChl c_F showed less aggregation ability than [P,E]BChl c_F as we pointed out in previous papers[2,4], however, similar pH effects were observed for each 6 v/v organic aqueous media (data not shown). Figure 5 a and b show CD spectra of each aggregates of [P,E]BChl c_F and [E,E] BChl c_F formed in 6 v/v% THF aqueous media, respectively. They are quite different from each other as we reported. [2] [E,E]BChl c_F showed a symmetrical excitonic pattern and a strong pH dependency in intensity: ellipticity increased significantly with increasing pH value. It was suggested that as the aggregation of [E,E]BChl c_F proceeds to further steps in the acidic region (pH 4.49), the CD signal diminishes markedly, as similarly observed in chlorosome.[5] Figure 6 shows salt effects of NaCl and KCl for [P,E]BChl c_F and [E,E]BChl c_F aggregation in 6 v/v% acetone aqueous solution in the presence of 1 mM Tris buffer (pH 7.2). The wavelength of Qy band red-shifted with increasing salt concentration added. NaCl was more effective than KCl for both [E,E]BChl c_F and [P,E]BChl c_F. Buffer concentration effect shown in Fig. 4 might be superimposed by the salt effect. These effects might be induced by salting out the solvent (water) molecules surrounding the BChl c aggregate to enhance the hydrophobic interaction.

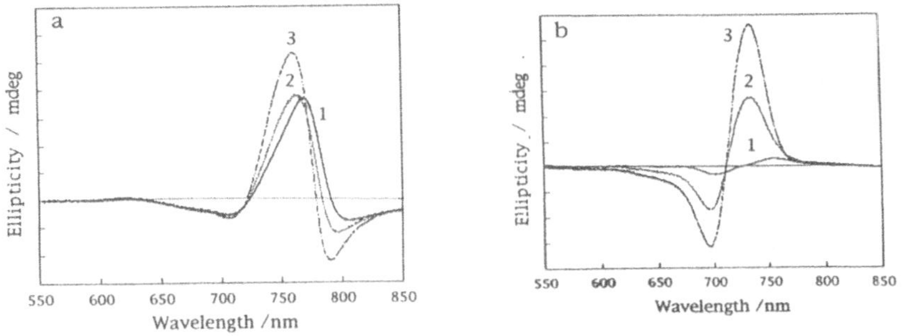

Fig. 5 CD spectra of [P,E] BChl c_F (a) and [E,E]BChl c_F (b) in 6% (vol/vol) THF aqueous solution. pH4.49(1), pH7.17(2), pH9.18(3).

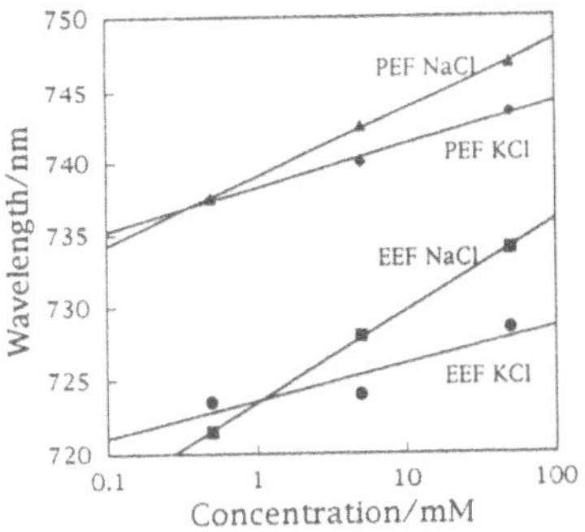

Fig. 6 Salt concentration effect on the aggregation of [P,E] BChl c_F (PEF) and [E,E] BChl c_F (EEF) in 6% (vol/vol) acetone aqueous solution for 1mM Tris buffer (pH7.2).

As the bonding pattern, keto carbonyl group hydrogen-bonded to hydroxy group which is simultaneously coordinated to the magnesium atom, Mg...O-H...O=C< , is proposed for both small BChl c aggregate (Type I) and large-chlorosome-like BChl c (Type II) .[6] We pointed out the possibility for involvement of the ester group in some organic solvents.[7]

We interpret our observation like followings. The pH and the salt effects observed in the present study suggest the role of hydrophobic interaction for the elongation of pre-existing unit building blocks. Intermediates for rod-like structure, that is, aggregates larger than tetramer denoted as [A] in Fig. 7, contains terminal pigments whose Mg atom is ligated to water, organic solvents and hydroxyl groups. These suppress a further elongation of aggregates through Mg...OH...O=C< networks. Proton and salt remove such chemical species from Mg atoms, which gives rise to a formation of larger aggregates, so-called Type II aggregates.

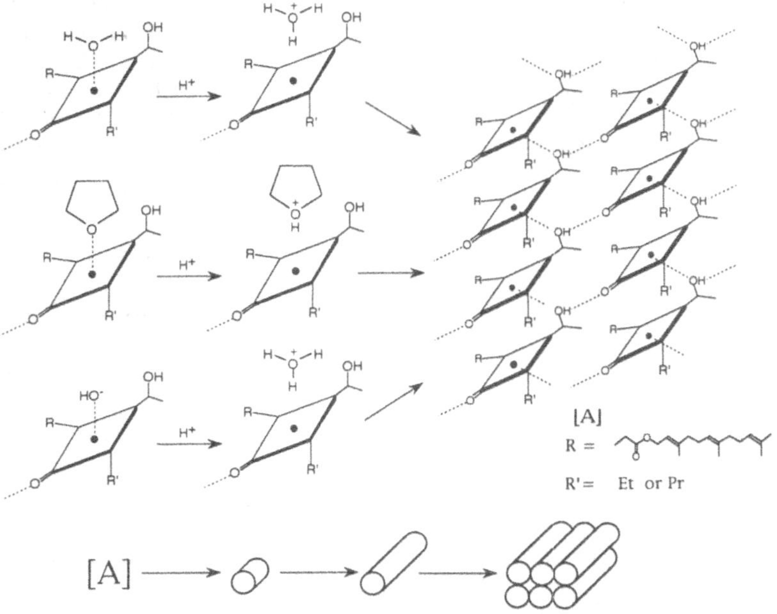

Fig. 7 A possible mechanism of the pH effect on the aggregation of BChl c in an organic aqueous solution. Closed circle in rhombus means Mg atom of BChl c. [A] is an intermediate of type II aggregates.

References

1 Olson, J. M. (1980) Biochim. Biophys. Acta 594, 33-51

2 Uehara, K., Mimuro M., Y. Ozaki, and Olson, J. M. (1994) Photosynth. Res. 41, 235-243

3 Olson, J. M. and Pedersen, J. P. (1990) Photosynth. Res. 25, 25-37

4 Uehara, K., Ozaki,Y., Okada, K., and Olson, J. M. (1991) Chem. Lett., 909-912

5 Matsuura, K., Hirota, H., Shimada, K., and Mimuro, M. (1993) Photochem. Photobiol., 57, 92-97

6 Chiefari, J., Griebenow, K., Griebenow, N., Balaban, T. S., Holzwarth, A. R., and Schafner, K. (1995) J. Phys. Chem., 99, 1357-1365

7 Sato, H., Uehara, K., Ishii, T., and Ozaki, Y. (1995) Biochemistry, 34, 7854-7860

IN VITRO RECONSTITUTION AND PIGMENT BINDING PROPERTIES OF RECOMBINANT CP29 AND CP24.

E. Giuffra, D. Cugini, A. Pagano, R. Croce, D. Sandonà and R. Bassi.
Università di Verona-Facoltà di Scienze MM.FF.NN.. Strada Le Grazie-37134 Verona - Italia.

Introduction

The "minor proteins" of higher plants photosystem II antenna complex (CP29, CP26, CP24) bind about 15% of the total chlorophyll of PSII. For this reason, their function as light harvesters appears unlikely. An increasing body of evidence suggests that the minor proteins are involved in the regulatory mechanisms that prevent overexcitation and photoinhibition of photosystem II, such as the NPQ (Non-Photochemical Quenching) (1) and the CP29 specific reversible phosphorylation (2). The structural basis of the regulative function of the minor chlorophyll proteins is basically unknown due to the difficulties in their biochemical isolation in sufficient amounts and in their native form. In order to overcome these problems we have attempted the reconstitution of the CP29 and CP24 proteins from the apoprotein overexpressed in *E.coli.* We report on the reconstitution with pigments "*in vitro*" of recombinant apoproteins overexpressed in *E. coli*, on their pigment binding properties and on the effect of a sigle aminoacid mutation.

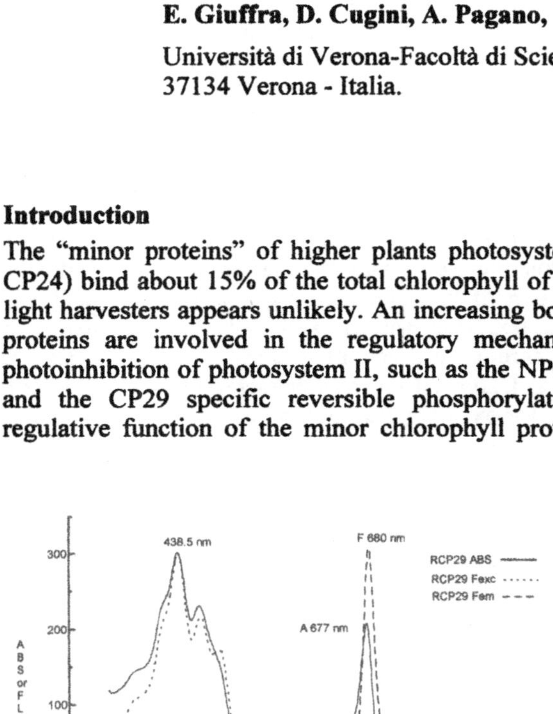

Fig. 1: Absorption and fluorescence spectra of reconstituted recombinant CP29.

2. Results and discussion

2.1 Construction of CP29 (*Lhcb4*) and CP24 (*Lhcb6*)-expression plasmids. In order to overexpress plant CP29 and CP24 in *E. coli*, the maize *Lhcb4* cDNA (Bergantino et al., in

P. Mathis (ed.), Photosynthesis: from Light to Biosphere, Vol. I, 271–274.

prep.) and *Lhcb6* (Dainese et al., these proceedings) were mutagenized by PCR to be put "in frame" in an expression vector of the pDS series.

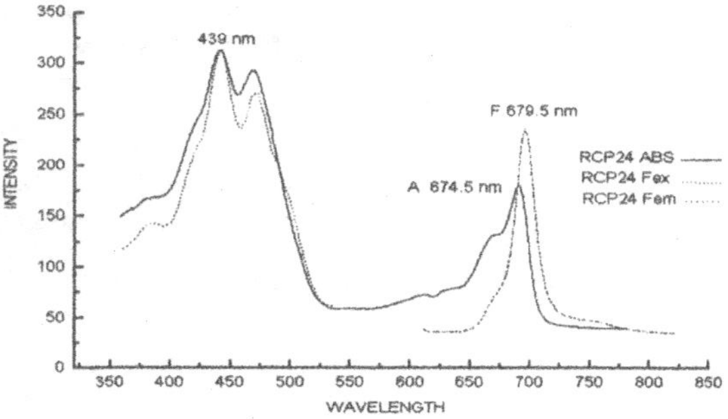

Fig. 2: Absorption spectrum of reconstituted CP24

2.2 Reconstitution of CP29 and CP24 complexes: In the basic reconstitution experiments (3,4) a chl a to chl b ratio of 3.0 and 1.9, similar to the ratio in the native complexes (5), were used. Unbound pigments were separated from the reconstituted complex by sucrose gradient ultracentrifugation in the presence of DM. The comparison of the absorption and fluorescence spectra of the reconstituted protein with the native one, purified from thylakoid membranes, suggested that part of the pigments might still be unspecifically bound to a correctly reconstituted complex. This was checked by further purification of the preparation from sucrose gradient through a preparative Deriphat-PAGE green gel. CP29 and CP24, after elution from the gel, showed spectral characteristics very similar to the native proteins (Fig. 1 e 2): rCP24, for example did not bind neoxanthin as expected. However, while CP24 showed an absorption spectrum almost identical to native CP24, in the case of CP29, recombinant minus native difference absorption spectra showed a positive chl b contribution peaking at 653 nm suggesting it binds chl b in excess. A more complete analysis of rCP29 pigment binding properties was then performed in order to clarify the reasons for the excess chl b bound to the complex. The stoichiometry between chlorophyll and protein was then determined. Native CP29 bound 6 chl a and 2 chl b moles per mole of CP29 apoprotein according to previous determinations (5) while recombinant CP29 bound an additional chl b molecule consistently with its spectral properties (6 chl a and 3 chl b). The removal of the exceeding chl b molecule from the recombinant CP29 was attempted by extensive washing of the complex after binding to a ion exchange column. This procedure was, however, ineffective and the complex, eluted from the column, had essentially unchanged characteristics thus suggesting that the binding of the third chl b was not more labile than that of the pigments bound to the native protein. This suggests the additional pigment is incorporated during folding in a precise site rather than unspecifically bound. 2.3. Pigment requirement for CP29 reconstitution. In order to identify the pigments indispensable for complex formation, we performed the reconstitution experiment with isolated pigments rather than with total pigment extracts from thylakoids. As in the case of LHCII (3,4), only lutein was indispensable for the formation of a stable complex. In this case, no green bands of reconstituted monomer were present in the sucrose gradient. The reconstitution products obtained in the other conditions were also challenged for their stability in the harsher conditions of SDS-PAGE confirming that the complexes reconstituted with the omission of neoxanthin, violaxantin and β-carotene had similar stability.

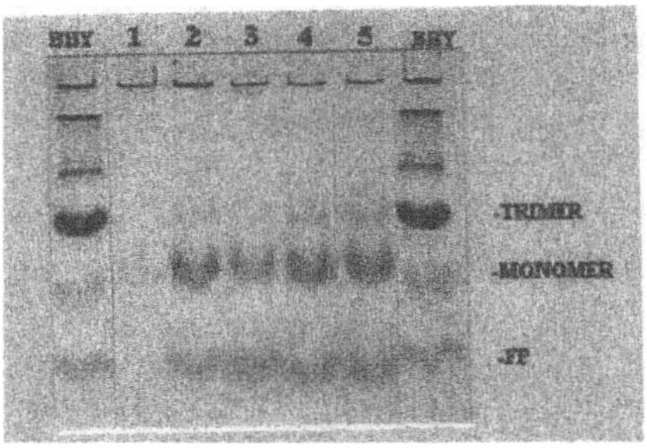

Fig. 3: Deriphat-PAGE (green) gel of recombinant CP29 reconstituted with various pigment mixtures. 1), - Lutein; 2),- violaxanthin; 3), -neoxanthin; 4), - β-carotene; 5), all pigments. FP, free pigments. 2 μg chl were loaded in each slot.

<u>Dependence of r CP29 composition on supplyed chla/b ratio</u>. CP29 was then reconstituted with pigment mixtures containing different chl a to chl b ratios. Different reconstitution efficiencies were observed: complexes obtained with mixtures with chl a only or with traces of chl b (chl a/b ratio=100) only occasionally survived Deriphat-PAGE but when intermediate chl a/b ratios were used (i.e. chl a/b=8) a stable complex was obtained although with low yields. Biochemical data and absorption spectra show that three stable complexes with different chl b content were obtained. The mixtures with chl a/b ratios lower than 2, between 2 and 6 and between 6 and 8 yielded reconstituted complexes with, respectively, chl a/b ratios of 1.25, 2.0 and 2.75 and pigment/protein

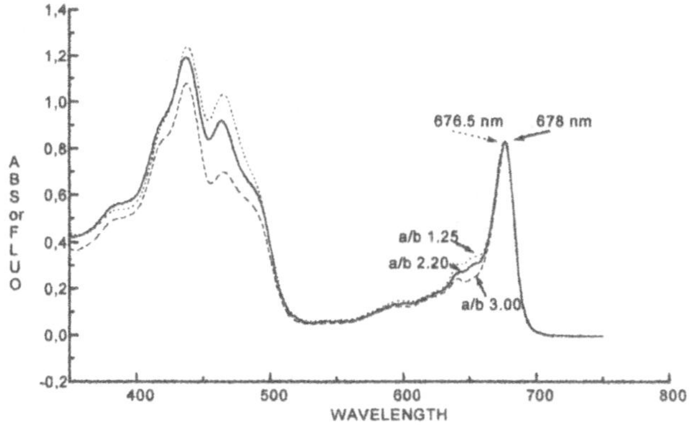

Fig. 4: Absorption spectra of rCP29 reconstituted in the presence of different chl a/b ratios

stoichiometry of 9, 9 and 8 respectively. Of the two chl b contributions peaking respectively at 641 and 653 nm only the amplitude of the second is affected. The major chl a peak was largely unaffected, only the complex with the highest chl b content (chl a/b=1.25) had a peak value shifted from 677 to 676 nm. Fluorescence excitation spectra indicates that in all the three isoforms, chl b can efficiently transfer excitation energy to chl a. These data contrast with the previous results on LHCII (3,4) showing that the pigment binding during folding is indipendent from the reconstitution mixture. Nevertheless, it should be noted that the purification procedure used was quite different respect to previous work with LHCII (3,4). The biochemical data (chl a/b and pigment/protein ratios) and absorption spectra are consistent with the view that a rCP29 almost dentical to the nCP29 (2 chl b + 6 chl a) is obtained with chl a/b ratio in the mixture between 6 and 10 while increasing chl b supply yields the occupation of a chl b site previously empty (chl a/b ratio= 2.0; 3chl b + 6 chl a) and then the substitution of a chl a with a chl b molecule (chl a/b ratio= 1.25; 4 chl b + 5 chl a).

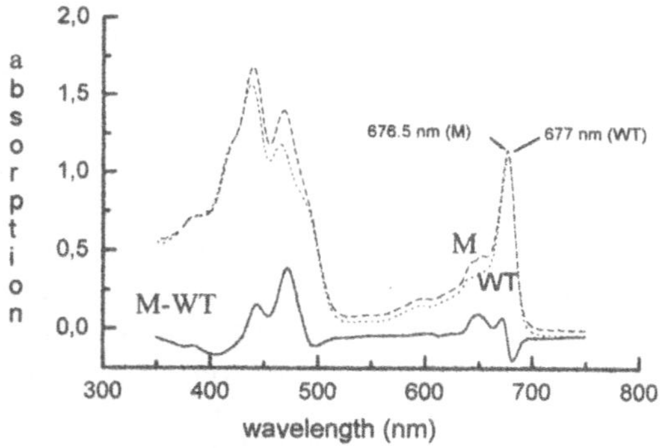

Fig. 5: Absorption spectra and difference spectrum of rCP29 and of its E166Q mutant

The present data confirm that CP29 binds a lower number of chl molecules per polypeptide than LHCII, which bind 7 chl a and 5 chl b (6). It is therefore likely that one chl a and three chl b binding sites, not occupied, but at least in part conserved in CP29, may be filled depending on the relative chromophore concentrations during folding. The presence of such potential chl binding sites was assayed by constructing a mutant in a chl binding residue: the mutation E166Q of CP29 restores, in fact, the glutamine residue which binds chl b6 in LHCII (E131) (6) . The spectra in Fig.5 and the chl a/b ratios of 1.5 (M) vs 2 (WT), show that the mutation restores the chl b6 binding site of LHCII. This result supports the idea of an overall similar structure organisation of chl a/b proteins while suggests that the single mutation approach may be useful in the attribution of chl absorption forms to porfirin molecules located in the crystal structure.

References

1) Bassi,R., Pineau, B., Dainese, P. & Marquardt J., (1993). Eur. J. Biochem., 212: 297-303.

2) Bergantino,E., Dainese, P., Cerovic, Z., Sechi, S. and Bassi R., 1995. J. Biol. Chem., 270: 8474-8481.

3) Paulsen, H, Rumler,U. and Rudiger,W. (1990) Planta 181, 204-211.

4) Plumley, F.G. and Schmidt ,G.W. (1987) Proc. Natl. Acad. Sci. U.S.A.84, 146-150.

5) Dainese, P. and Bassi, R. (1991) J. Biol. Chem. 66, 8136-8142.

6) Kuhlbrandt, W. Wang, D.N. and Fusijoshi, (1994) Nature 367, 614-621.

CRYSTALLISATION OF THE LIGHT HARVESTING COMPLEX LH2 (B800-850) FROM *RUBRIVIVAX GELATINOSUS*.

J.B. Arellano[1], S. M. Prince[2], G. McDermott[2], E. Halloren[1] & R. J. Cogdell[1].
[1]Davidson Building, Division of Biochemistry and Molecular Biology, Institute of Biomedical and Life Sciences and [2]Joseph Black Building, Chemistry Department. University of Glasgow, Glasgow G12 8QQ UK.

Introduction

The photosynthetic light harvesting antenna of purple bacteria is a highly organized arrangement of pigment-proteins. Recently the structure of the B800-850 light harvesting antenna (LH2) from *Rps. acidophila* strain 10050 has been determined by X-ray crystallography (McDermott *et al.*, 1995). The crystal structure shows that the active assembly consists of two concentric cylinders of helical protein subunits which enclose the pigments molecules. Eighteen bacteriochlorophyll *a* molecules sandwiched between the helices form a continuous overlapping ring and a futher nine are positioned between the outer helices with the bacteriochlorin rings perpendicular to the transmembrane helix axis.

The homology of α and β polypeptides is very high in the LH2 of purple bacteria (Zuber and Brunisholz, 1991). So the arrangement of bacteriochlorophyll and carotenoid molecules in the LH2 antenna of purple bacteria family should be very similar. However, *Rubrivivax gelatinosus* has an unusual LH2 complex, because the α apoprotein has an extremely long C-terminal portion and there is evidence to indicate that this particular α polypeptide spans the membrane twice (Zuber and Brunisholz, 1991) and secondly, in the β polypeptide the second conserved His near the N-terminus is substituted by Gln. Recently, McDermott *et al.*, 1995 have proposed the N-formyl-methionine of the a polypeptide binds the B800 bacteriochorophyll in *Rps. acidophila*. However, the sequenced of α polypeptide of *Rubrivivax gelatinosus* does not show a similar N-formyl-methionine (Brunisholz *et al.*, 1994).

In this paper we present the strategy to crystallise the LH2 complex of *Rubrivivax gelatinosus* pointing out the results which improve the size and shape of these crystals.

Material and Methods

Cells from *Rubrivivax gelatinosus* 1.5.1 and 1.4.9. were grown photosynthetically at 30° C on Pfennings medium (Pfenning, 1969). The cells were harvested by centrifugation at 3,500 rpm for 20 minutes at 4°C using a cool

P. Mathis (ed.), Photosynthesis: from Light to Biosphere, Vol. I, 275–278.

spin centrifuge, resuspended in 20 mM Tris-HCl pH 8.0 and disrupted by passage through a French press at 8000-10000 psi (Cogdell *et al.*, 1983). Then, the broken cells were adjusted to an absorbance of 25 cm^{-1} at 850 nm, solubilised with 2% Lauryldimethylamine N-oxide (LDAO) for 3 h at 4° C and then centrifuged at 10,000g for 10 minutes at 4°C to remove any unsolubilised material. The supernatant was then layered onto a sucrose step gradient. The gradient consisted of 5 ml steps of 0.8, 0.6, 0.4 and 0.2 M sucrose prepared in 20 mM Tris-HCl pH 8.0 containing 0.1% (v/v) LDAO. The gradients were then centrifuged at 180000 g for 16 h at 4°C. Two pigmented bands were obtained (Dawkins *et al.*, 1988). The upper band contained the LH2 complex and the lower band the RC-LH1 complexes. The LH2 complex was further purified by DE 52 anion exchange chromatography. The LH2 fractions with a 850nm/270nm ratio higher than 2.4 were collected and concentrated with 30M amicon centricon. Then, LH2 fractions of 2 mls were loaded onto the Superdex 200 FPLC column equilibrated with 20 mM Tris-HCl pH 8.0 and 0.1% LDAO and finally 'cleaned up' by molecular sieve chromatography. Those fractions with a 850 nm /270 nm higher than 3.0 were pooled and concentrated in a 30M amicon centricon. At this stage the LDAO could be exchange for β-octylglucoside (BOG).

Crystallisation trials were carried out using the sitting-drop method (Cogdell *et al.*, 1990).

Results

The ratio of the absorbance at 850 nm (due to bacteriochlorophyll a) to that at 270 nm (due to aromatic amino acids) has been used as a parameter to determine the purity of the LH2 complex (Cogdell *et al.*, 1990). A ratio higher than 3.0 is optimal for the crystallization assays.

Our crystallisation trials were set up at a pH 9.0-9.5, conditions where the complex was stable for several weeks and at a temperature between 4 and 18°C. Two different amphiphiles were tested 1,2,3-heptanetriol and Benzamidine hydrochloride. However, the later one was the only one which gave us satisfactory results. Using 2.5% Benzamidine and 0.1 (w/v) % LDAO we tested a range of precipitants. Crystals were obtained using as precipitants a mixture of NaCl and K_2HPO_4 in the sample well and $(NH_4)_2SO_4$ in the reservoir well. 350 mM NaCl was used constantly in all our assays and a range of $(NH_4)_2SO_4$ and K_2HPO_4 concentrations were assayed. This study has revealed how to control two important factors in crystallisation: Nucleation and size of crystals. Firstly $(NH_4)_2SO_4$ pH 9.33 in a range of 2.4 - 1.6 M in the reservoir well showed a effect in the number of crystals in the sample well. The number of nuclei decreased when $(NH_4)_2SO_4$ concentration decreased. Secondly, K_2HPO_4 was used in the sample well in a range of concentration between 0.9 and 1.4. In those trials where K_2HPO_4 was used in lower concentrations the crystal had a flat and featureless shape. However as increasing the K_2HPO_4 concentration the crystal showed a three dimensional structural shape. The combination of both salt effects gave us crystals with a right size and shape to study by X-ray diffraction. Our conclusions from these studies are shown in the Table 1 where we have summarised the effectiveness of amphiphiles, detergents and precipitants we have tried for crystal formation.

Crystals of the B800-850 light harvesting antenna were exposed to a monochromatic synchrotron X-ray beam of wavelength 0.87 Å at CCLRC Daresbury Laboratory's station 9.6. A number of crystals were screened. The crystal samples were found to diffract constantly to a resolution of 6.0-8.0 Å. Some anisotropy of diffraction resolution was observed and crystalline disorder (such as splitting or smearing of diffraction intensities) was present. The crystal habit appears different from that found in *Rhodopseudomonas acidophila*, although at this resolution and with the crystalline disorder encountered characterisation of the crystals of *Rubrivivax gelatinosus* is not possible. The diffraction intensities appear to be compatible with the monoclinic system with perhaps a face centred lattice.

Amphiphile and detergent	Precipitant (in)	Precipitant (out)	Time (weeks)	Type and number of crystal
2.5% Benzamidine 0.1% LDAO	350mM NaCl 0.9-1.1 M K_2HPO_4 pH 9.20	2.4-3.0 M $(NH_4)_2SO_4$ pH 9.33	1-2	Many tiny and thin needles
2.5% Heptanetriol 0.1% LDAO	350mM NaCl 0.9-1.1 M K_2HPO_4 pH 9.20	2.3 M $(NH_4)_2SO_4$ pH 9.33	-	Amorphus precipitant
2.5% Benzamidine 0.1% LDAO	350mM NaCl 1.0 M K_2HPO_4 pH 10.20	2.3 M $(NH_4)_2SO_4$ pH 9.33	-	Phase separation
3.5% Benzamidine 0.1% LDAO	350mM NaCl 20 mM $MgCl_2$	2.3 M $(NH_4)_2SO_4$ pH 9.33	-	Amorphus precipitant
2.5% Benzamidine 1% BOG	350mM NaCl 1.0 M K_2HPO_4 pH 9.20	2.3 M $(NH_4)_2SO_4$ pH 9.33	2-3	Many small and irregular cubes
2.5% Benzamidine 0.1% LDAO	350mM NaCl 1.2-1.4 M K_2HPO_4 pH 9.20	2.2-2.4 M $(NH_4)_2SO_4$ pH 9.33	2-4	Several thick needles (12 Å)
2.5% Benzamidine 0.1% LDAO	350mM NaCl 1.2-1.4 M K_2HPO_4 pH 9.20	1.6-2.2 M $(NH_4)_2SO_4$ pH 9.33	4-6	A few thick needles (6-8 Å)

Table 1. A summary of the results of trying to crystallize the B800-850 complex from *Rubrivivax gelatinosus* in the presence of different detergents, amphiphiles and precipitants.

Conclusion

We have obtained a few crystals which have been examined by X-ray crystallography. So far the assayed LH2 crystals from *Rubrivivax gelatinosus* have had a resolution of 6-8 Å using in the sample well 0.1(w/v) % LDAO, 2.5 (w/v) % Benzamidine-hydrochloride, 350 mM NaCl, 1.3 M K_2HPO_4 and LH2 complex with a final absorbance of 80 cm^{-1} at 850 nm and 1.6-2.2 M $(NH_4)_2SO_4$ pH 9.33 in the reservoir well. Since now crystal size and shape can be controlled we expect in future crystallisation trials have more resolved diffraction pattern.

Acknowledgement

This work was supported by a grant from the EC and the BBSRC.

References

Brunisholz, R. A., Suter, F. and Zuber, H. (1994). Eur. J. Biochem. 222: 667-675.

Cogdell, R. J., Durant, I., Valentine, J., Lindsay, J. G. and Schmidt, K. (1983). Biochim. Biophys. Acta, 722: 427-435.

Cogdell, R. J., Woolley, K. J., Ferguson, L. A. and Dawkins, D. J. (1990). In: Michel, H. (ed) Crystallization of Membranes Proteins, pp. 125-136. CRC Press, Boca Raton.

Dawkins, D. J., Ferguson, L. A. and Cogdell, R. J. (1988). In: Scheer, H. and Schneider, S. (eds) Photosynthetic Light-Harvesting System, pp. 115-127. Walter de Gruyter and Co., Berlin.

Pfenning, N. (1969). J. Bacteriol. 99: 597-602.

McDermott, G., Prince, S. M., Freer, A. A., Hawthornthwaite-Lawless, A. M., Papiz, M. Z., Cogdell, R. J. and Isaacs, N. W. (1995). Nature, 374: 517-521.

Zuber, H. and Brunisholz, R. A. (1991). In: Scheer, H. (ed) Chlorophylls, pp. 627-703. CRC Press, Boca Raton.

FEMTOSECOND ENERGY TRANSFER KINETICS IN INTACT CHLOROSOMES AND BCHL *c* AGGREGATES FROM GREEN BACTERIA

Sergei Savikhin[1], Paula I. van Noort[2], Yinwen Zhu[2], Robert E. Blankenship[2], and Walter S. Struve[1]

[1]Ames Laboratory-USDOE and Department of Chemistry, Iowa State University, Ames, IA 50011 and [2]Department of Chemistry and Biochemistry and Center for the Study of Early Events in Photosynthesis, Arizona State University, Tempe, AZ 85287-1604, USA

1. Introduction

Chlorosomes in green sulfur and filamentous bacteria are structurally unique among known light-harvesting assemblies. Unlike other antennae (where proteins determine the pigment organization), the $\sim 10^4$ bacteriochlorophyll (BChl) *c/d/e* pigments in a chlorosome are assembled into aggregates with minimal protein content [1-3], enclosed in an ellipsoidal lipid monolayer appressed to the inside of the cytoplasmic membrane. Excitations in the 740-750 nm BChl *c/d/e* antenna pass through a 790-800 nm BChl *a* antenna prior to trapping at reaction centers [4]. Much spectroscopic evidence for the lack of protein involvement in the BChl *c/d/e* organization has emerged from comparative studies of intact chlorosomes and their reconstituted BChl - lipid aggregates [5-7]. Extraction of chlorosomes from *Chloroflexus aurantiacus* or *Chlorobium limicola* yields a protein-free phase containing BChl *c* and BChl *a* monomers, carotenoids, and lipids. Dilution with aqueous buffer causes self-assembly of BChl *c* pigments into aggregates, encased in lipid micelles 80-100 nm in diameter. We have compared the femtosecond kinetics of intact chlorosomes and BChl *c* - lipid aggregates from *Cf. aurantiacus* and *Chlorobium tepidum*, as well as [E,E] BChl c_F - dodecylmaltoside aggregates.

2. Materials and Methods

Chlorosome samples from *Cf. aurantiacus* and *Cb. tepidum* were prepared according to the method of Gerola and Olson [8]; the aggregates were reconstituted according to Miller et al. [9]. [E,E] BChl c_F aggregates were prepared by diluting pure [E,E] BChl c_F (extracted from *Cb. tepidum* chlorosomes) in a pH 7.4 buffer containing 0.15 mM dodecylmaltoside and 10 mM phosphate. The BChl *c* steady-state Q_y absorption band maxima occur at ~740 nm and 746 nm for intact chlorosomes and aggregates from *Cf.*

P. Mathis (ed.), Photosynthesis: from Light to Biosphere, Vol. I, 279–282.

aurantiacus; at ~ 746 nm and 742 nm for intact chlorosomes and aggregates from *Cb. tepidum*; and at ~720 nm for [E,E] BChl c_F - maltoside aggregates. The Ti:sapphire laser and detector system have been described previously [10].

3. Results

3.1 Anisotropies

One-color absorption difference anisotropies *r(t)* are shown in Fig. 1 for each of the intact chlorosomes and reconstituted aggregates. The wavelength in each case is near the BChl *c* absorption band maximum, where the signal is dominated by photobleaching and stimulated emission (PB/SE). Intact *Cf. aurantiacus* chlorosomes show little anisotropy decay ($r(0) = 0.4$, $r(\infty) = 0.36$), indicating considerable long-range as well as short-range collinearity in their BChl *c* transition moments. Reconstituted aggregates from the same species show appreciably less long-range order ($r(\infty) = 0.21$). A similar distinction arises between intact chlorosomes and aggregates from *Cb. tepidum* ($r(\infty) = 0.25$ and 0.18,

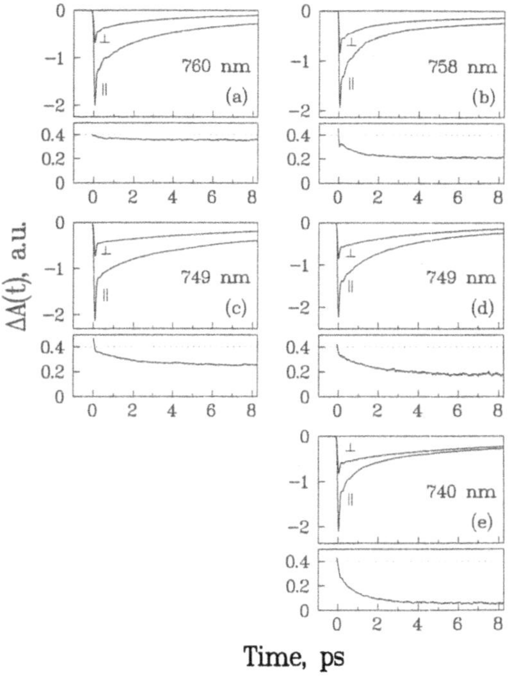

Figure 1. One-color anisotropic profiles $\Delta A_{\parallel}(t)$, $\Delta A_{\perp}(t)$ and anisotropic decays *r(t)* at the indicated wavelengths: (a) *Cf. aurantiacus* chlorosomes; (b) BChl *c* - lipid aggregates from *Cf. aurantiacus*; (c) *Cb. tepidum* chlorosomes; (d) BChl *c* - lipid aggregates from *Cb. tepidum*; (e) [E,E] BChl c_F - maltoside aggregates.

respectively). Essentially random long-range ordering appears in the [E,E] BChl c_F - maltoside aggregates, where $r(\infty) = 0.06$. Hence, the encapsulating lipid appears to influence the long-range pigment organization in the aggregates.

3.2 Isotropic decays

The one-color isotropic PB/SE decays at wavelengths to the red of the BChl *c* steady-state absorption band maximum are usually multiexponential, with major lifetime components 1-2 ps and 5-15 ps, for both the intact chlorosomes and the reconstituted aggregates. An additional 100 fs - 1 ps component appears in [E,E] BChl c_F - maltoside aggregates, which do not show 1-2 ps components. Since 5-15 ps PB/SE decay components are often observed in the reconstituted aggregates, their attribution to BChl $c \rightarrow$ BChl *a* energy transfers in intact chlorosomes appears doubtful. They are likely due to internal BChl *c* energy transfer kinetics, as has recently been proposed [10].

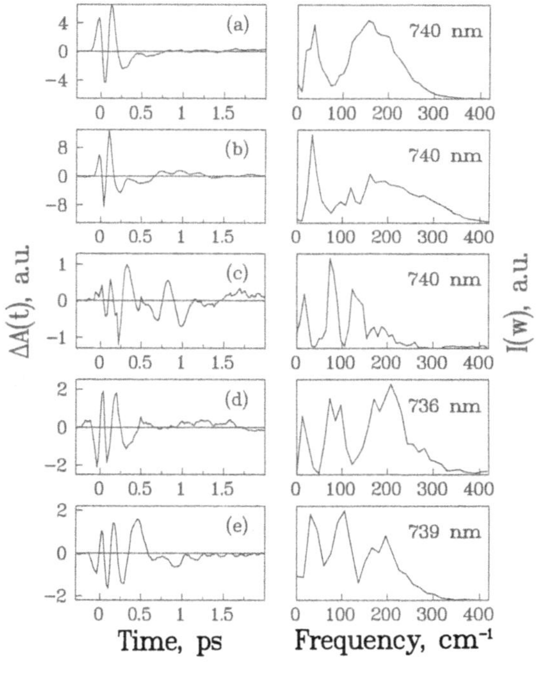

Figure 2. Oscillating parts of BChl *c* isotropic profiles and Fourier transform spectra at the indicated wavelengths: (a) *Cf. aurantiacus* chlorosomes; (b) aggregates from *Cf. aurantiacus*; (c) *Cb. tepidum* chlorosomes; (d) aggregates from *Cb. tepidum*; (e) [E,E] BChl c_F - maltoside aggregates.

3.3 Coherent oscillations

The isotropic decays at very early times show coherent oscillations, which are typically damped within 1-2 ps at room temperature. The oscillations are shown in Fig. 2, with their Fourier transform spectra. The Fourier spectra of the two native chlorosomes are easily distinguished. Moreover, the Fourier spectrum of BChl *c* - lipid aggregates from *Cb. tepidum* differs slightly from that of the intact chlorosomes. Recent Q_y-excitation resonance Raman spectra of individual BChl *c* pigments from *Cb. tepidum* (D. Bocian, private communication) suggest that the relevant low-frequency modes in the coherent nuclear motion are strongly influenced by the nature of the alkyl substituents on the BChl macrocycle. The differences between the Fourier spectra of intact *Cf. aurantiacus* and *Cb. tepidum* chlorosomes may then stem from contrasting alkyl substituents in the BChl 8,12 positions (predominantly [E,M] and a mixture of [E,E] and [P,E], respectively). The differences between the Fourier spectra of intact chlorosomes and aggregates from *Cb. tepidum* may be due to contrasting alkyl group conformations (e.g. ordered in chlorosomes, versus random in the aggregates). Alkyl group conformations strongly influence their vibrational frequencies in the model compound octaethylporphyrin, which crystallizes in several different space groups distinguished by the ethyl group conformations relative to the macrocycle plane [11].

References

1 Krasnovsky, A. A. and Bystrova, M. I. (1980) BioSystems 12, 181-194

2 Oelze, J. and Golecki, J. R. (1995) in Anoxygenic Photosynthetic Bacteria (Blankenship, R. E., Madigan, M. T., and Bauer, C. E., eds.) pp. 259-278, Kluwer Academic Publishers, Dordrecht

3 Blankenship, R. E., Olson, J. M., and Miller, M. (1995) in Anoxygenic Photosynthetic Bacteria (Blankenship, R. E., Madigan, M. T., and Bauer, C. E., eds.) pp. 399-435. Kluwer Academic Publishers, Dordrecht

4 Betti, J. A., Blankenship, R. E., Nagarajan, L. V., Dickinson, L. C., and Fuller, R. C. (1982) Biochim. Biophys. Acta 680, 194-201

5 Hirota, M., Moriyama, T., Shimada, K., Miller, M., Olson, J. M., and Matsuura, K. (1992) Biochim. Biophys. Acta 1099, 271-274

6 Lutz, M. and van Brakel, G. (1988) in Green Photosynthetic Bacteria (Olson, J. M., Ormerod, J. G., Amesz, J., Stackebrandt, E., and Trüper, H. G., eds.) pp. 23-34, Plenum Press, New York

7 Hildebrandt, T., Tamiaki, H., Holzwarth, A., and Schaffner, K. (1994) J. Phys. Chem. 98, 2192-2197

8 Gerola, P. D. and Olson, J. M. (1986) Biochim. Biophys. Acta 848, 69-76

9 Miller, M., Gillbro, T., and Olson, J. M. (1993) Photochem. Photobiol. 57, 98-102

10 Savikhin, S., Zhu, Y., Lin, S., Blankenship, R. E., and Struve, W. S. (1994) J. Phys. Chem. 98, 10322-10334

11 Li, X.-Y., Czernuszewicz, R. S., Kincaid, J. R., Stein, P., and Spiro, T. G. (1990) J. Phys. Chem. 94, 47-61

ON THE DEPENDENCE OF LHC II FLUORESCENCE LIFETIMES ON TEMPERATURE

Matthias Seydack, Harald Redlin and Joachim Voigt
Institute of Physics, Humboldt-University Berlin, Invalidenstr. 110,
D-10115 Berlin, Germany

1. Introduction

In the last time a remarkable progress could be achieved in the investigation of photosynthetic antenna systems. This includes the structure /1,2/ as well as the mechanisms of their function /3/. To gain a deeper insight into these mechanisms, i. e. the details of the energy transfer within the antennae and the transfer to the reaction centre, the competing processes, like radiative and nonradiative decay, have to be studied. In this respect kinetic information, as obtained from fluorescence or transient absorption is very helpful.
Recently, great effort has been made to extend kinetic measurements towards shorter times. Using ps- and sub-ps laser-spectroscopy, these experiments are usually performed at one fixed temperature only. In contrast to earlier investigations, nowadays measurements are mostly carried out at 80 K or below. The use of low temperatures is vital in eliminating the influence of vibrations, in the chlorophyll as well as in the protein. This is particularly important for pigment - protein complexes from higher plants in order to obtain more distinctly structured spectral features.
On the other hand, the dependence of the spectral and kinetic properties of an antenna system on the temperature can be expected to offer more information about the interaction between electronic /excitonic properties and vibrations. According to the authors' knowledge no such data have been published for photosynthetic antennae regarding the decay kinetics yet. In this poster first results shall be presented from an investigation of fluorescence kinetics of the light harvesting complex from photosystem II in the temperature range from 16 K to 295 K.

2. Materials and Methods

LHC II trimers were isolated from spinach according to the method of Berthold as modified by Irrgang et al. /4/. The sample consisted of trimeric pigment - protein complexes solubilized in the detergent β-DM. The fluorescence spectrum at 77 K showed no obvious sign of aggregation. Neither a peak nor a shoulder was visible at 695 nm. Only a decomposition into gaussian bands required a minor band at this wavelength.
Figure 1 shows the experimental arrangement used to measure the fluorescence decay. To adjust the temperature of the sample a refrigerator cryostat operating in the temperature range from 16 to 295 K was used. The cooling rate amounted to about 5...10 K/min.

P. Mathis (ed.), Photosynthesis: from Light to Biosphere, Vol. I, 283–286.

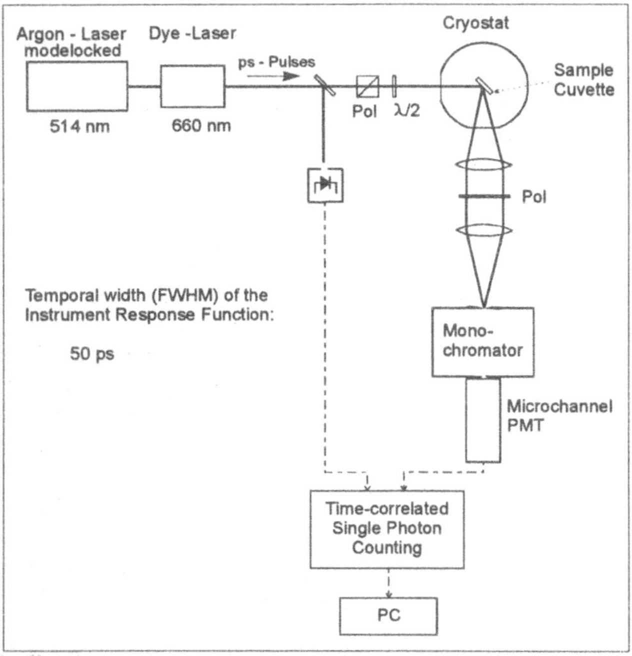

Figure 1. Experimental arrangement for the measurement of the fluorescence decay

The sample was situated in a quartz cuvette with 2 mm pathlength. For the measurement at each temperature a fresh sample was used.

3. Results

The fluorescence decay was measured at nine particular temperatures in the range from 16 K to 295 K. Typical time courses of the fluorescence intensity are shown in figure 2

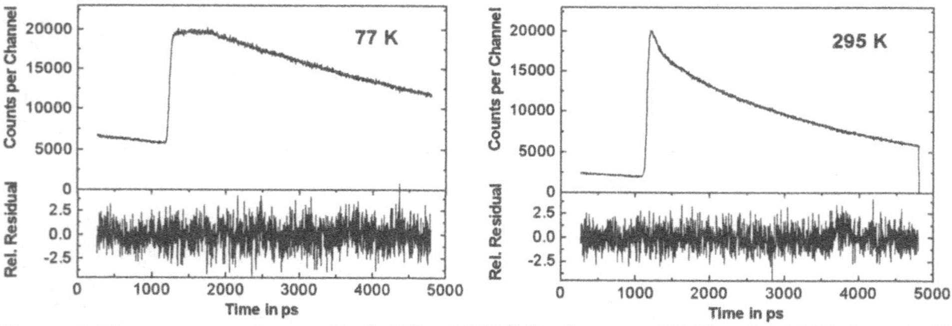

Figure 2. Fluorescence decay of solubilized LHC II trimers at 77 K and 295 K. For the fit the following parameters are used: 77 K: $\tau_1 = 6.75$ ns, $a_1 = 0.71$; $\tau_2 = 3.2$ ns, $a_2 = 0.24$; $\tau_3 = 90$ ps, $a_3 = 0.045$; $\tau_4 = 8$ ps, $a_4 = -0.8$. 295 K: $\tau_1 = 4.0$ ns, $a_1 = 0.49$; $\tau_2 = 1.1$ ns, $a_2 = 0.18$; $\tau_3 = 0.16$ ns, $a_3 = 0.33$; $\tau_4 = 8$ ps, $a_4 = 0.3$.

for the temperatures 77 K and 295 K. For all temperatures the experimental data could be fitted with a sum of four exponentials. The residuals demonstrate the good quality of the fits.

Two of the components (1, 2) have a lifetime in the ns region corresponding to the known components with $\tau = 4$ ns and $\tau = 2$ ns at room temperature /5/. The lifetime of the other two components is in the ps region. In figure 3 and 4 the temperature dependence of the lifetime parameters is depicted. Within the investigated temperature range each lifetime remains in its order of magnitude. Hence it is fully justified to assign the components measured at different temperatures to the same physical processes in the sequence of their lifetimes.

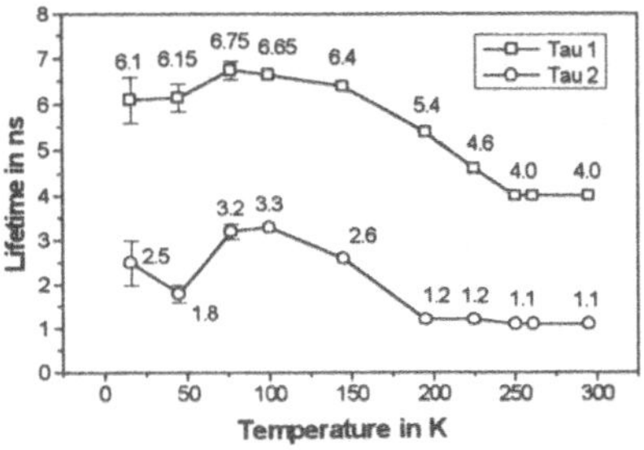

Figure 3. Temperature dependence of lifetimes of the ns-components in the fluorescence decay of LHC II

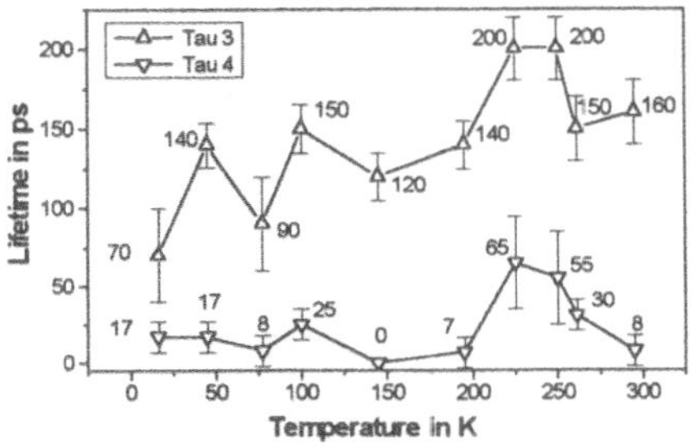

Figure 4. Temperature dependence of lifetimes of the ps-components in the fluorescence decay of LHC II

The lifetimes of the two ns components show a temperature dependence of the same kind with a slight increase between 16 K and 77 K and a marked decrease from 150 K to 200 K or 250 K respectively. The lifetime of the third component exhibits in contrast a continuous increase from about 70 ps to 180 ps distributed over the entire temperature range. Errors in the lifetime determination being greater than for the components 1 and 2 prevent to discover the temperature course more exactly.

The shortest component could be determined only with considerable uncertainty. Its contribution to the fluorescence yield is small so that errors in the longer components have a large effect. Further the lifetime is in the order of the experimental time resolution. Influences of instrument response errors can not be excluded completely. The amplitude of component 4 has a negative sign at some temperatures and a positive one at the others without any obvious rule. To sum it up, such a short component is always necessary to obtain a good fit but the determination of its parameters is too faulty to allow further statements.

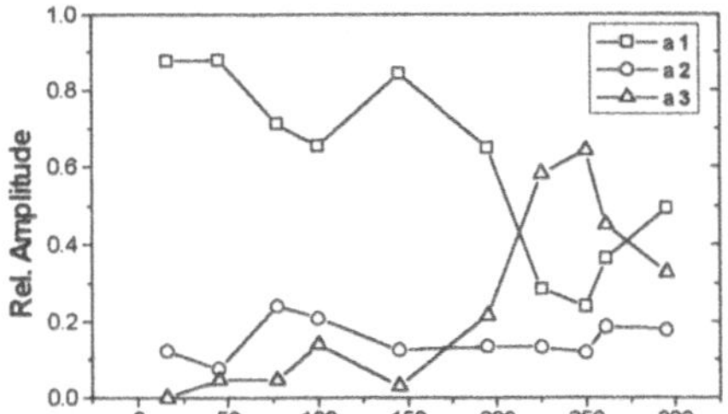

Figure 5. Temperature dependence of amplitudes of the components in the fluorescence decay of LHC II

In figure 5 the temperature dependence of the amplitude parameters is depicted for the components 1 to 3. The components 1 and 3 exhibit a mutually mirror-inverted course. That suggests that there might be a dependence between them. Except for the temperature range from 200 K to 260 K component 1 has the greatest amplitude and consequently the greatest share on the fluorescence yield. At temperatures below 200 K the dominance is very pronounced. The amplitude of component 3 shows no significant temperature dependence.

For a first interpretation of the obtained data a simple two-level scheme shall be used. It is obvious that this is only sufficient to model one component. Here it will be applied to component 1 representing the main decay process of Chl a. Two decay channels are included, a radiative and a nonradiative one. The radiative lifetime is mainly determined by the properties of Chl a and therefor assumed to be independent on temperature. To describe the thermal properties of the nonradiative channel several empirical approaches were tried. The combination of an activation term with a polynomial prefactor gave the best agreement with the experimental data (Figure 6).

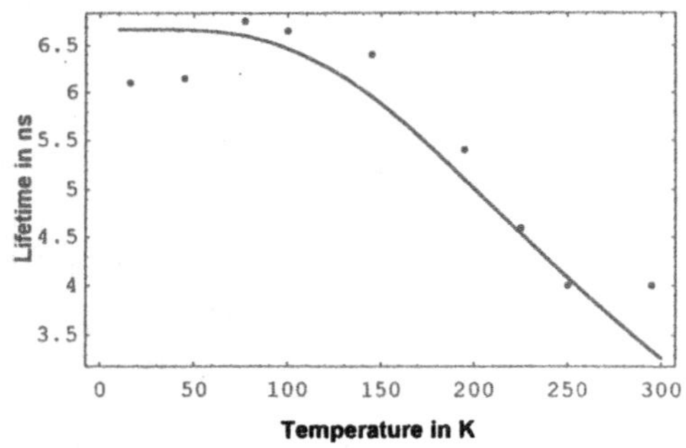

Figure 6. Temperature dependence of the fluorescence lifetime in a two-level scheme with experimental data points from component 1

$$\tau_F = \frac{1}{k_1 + k_2}, \qquad k_1 = 1.5 \cdot 10^8\, s^{-1}$$

$$k_2 = 2.7 \cdot 10^8 s^{-1} \frac{T^2}{80000K^2} \exp\left(-\frac{200K}{T}\right)$$

The same model can in principle be applied to component 2 assuming that both components result from pools of pigments that are not connected with each other but have a similar physical nature. Possibly component 2 represents a small fraction of aggregated LHC II particles though they are not visible in the fluorescence spectrum. Such a supposition is supported by investigations of intentionally aggregated LHC II (not shown here) performed under the same conditions.

Further investigations are necessary to enable the construction of models describing the temperature dependence of all lifetimes and amplitudes.

This work was supported by the Deutsche Forschungsgemeinschaft (SFB 312).

5. References

1 Kühlbrandt, WE., Wang, D. N., Fujiyoshi, Y. (1994) Nature 367, 614-621
2. McDermott, G. et al. (1995) Nature 374, 517-521
3. van Grondelle, R., Dekker, J. P., Gillbro, T., Sundstrom, V. (1994) BBA 1187, 1-65
4. Irrgang, K.-D., Boekema, E. J., Vater, J., Renger, G. Eur. J. (1988) Biochem. 178, 209-217
5. Liu, B., Napiwotzki, A., Eckert, H.-J., Eichler, H.-J., Renger, G. (1993) BBA 1142, 129-138

NONLINEAR POLARIZATION SPECTROSCOPY IN THE FREQUENCY DOMAIN: SPECTRAL SUBSTRUCTURE AND ULTRAFAST EXCITON DYNAMICS IN HIGHER PLANT LIGHT-HARVESTING COMPLEX II

H. Lokstein[1], D. Leupold[2], B. Voigt[2], F. Nowak[2], J. Ehlert[2], P. Hoffmann[1]
1) Institut für Biologie, Humboldt-Universität zu Berlin, Unter den Linden 6, D-10099 Berlin, Germany;
2) Max-Born-Institut für Nichtlineare Optik und Kurzzeitspektroskopie, PF 1107, Rudower Chaussee 6, D-12474 Berlin, Germany

1. Introduction

In higher plants efficient light capture and regulated excitation energy transfer (EET) to the reaction centers is assured by chlorophyll (Chl) a/b containing light-harvesting complexes (LHCs) [1,2]. Most abundant and investigated is LHC II, the major photosystem II associated antenna. Although it is known to 3.4 Å resolution [2] reflection of the structure in the LHC II optical spectra is not understood, yet [1]. In particular, spectral substructures ("Chl-forms") in the Chl a/b Q_y absorption region, their origin as well as functional implications are a matter of debate. Nonlinear Polarization Spectroscopy in the Frequency Domain (NLPF) - a novel in photosynthesis research laser-spectroscopic technique (yielding information on ps/fs-exciton dynamics as well as on spectral sub-structure and band-broadening) - provides a promising approach to the above issues [3-5]. A major advantage of NLPF is its feasibility for the investigation of biological samples at room temperature (RT). Here we report on NLPF experiments with isolated (monomeric and trimeric) LHC II. Although trimeric LHCII most probably represents the native state, attention is focussed mainly on LHC II monomers, since trimers are an extremely complex system (with at least 36 Chls a+b) complicating unambiguous interpretation of the data. The results demonstrate the wealth of information in the NLPF profiles as well as contribute to the better understanding of the highly complex LHC II optical spectra and exciton dynamics.

2. Material and Methods

LHC II was prepared from pea (*Pisum sativum* L.) leaves according to ref. [6]. Trimeric LHC II was obtained by incubation at 0.35 mg Chl ml^{-1} in Tricine buffer (50 mM, pH 7.8) containing 0.5% octylglycoside (OG)/digitonin [7]; dissociation into monomers was induced by RT-incubation at 2 mg Chl ml^{-1} for 48 to 120 h in 1.8% OG [8]. The aggregation state was assessed by circular dichroism (CD) and 77K-fluorescence emission (spectra not shown). The principle of NLPF has been described before [3,9] and is briefly accounted for in Fig. 1. The NLPF set-up was essentially the same as described [3,4] *cf.* also [5]. Modifications (allowing to record spectra at a one order of magnitude lower pump photon flux density, PFD) have been introduced since: i) samples were flown through a selected (for low birefringence) 5-mm-cell; ii) pump and probe beams were directed orthogonally to one another through the cell, iii) DCM/propylene carbonate (tunable from 620 to 680 nm) was used as laser dye for both, pump and probe beam. Pump PFD was varied from ~ $6 \cdot 10^{23}$ up to $6 \cdot 10^{25}$ photons cm^{-2} s^{-1}. NLPF spectra were corrected for probe laser fluctuations as well as for the pump laser power spectrum.

P. Mathis (ed.), Photosynthesis: from Light to Biosphere, Vol. I, 287–290.

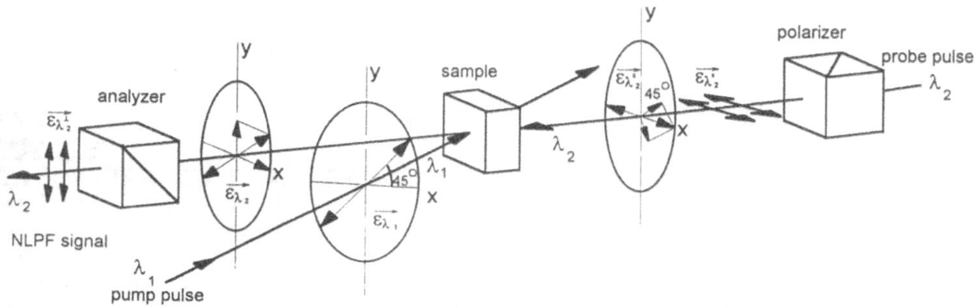

***Figure 1*: The principle of NLPF spectroscopy**. (Redrawn from [4]). The sample is probed by a highly monochromatic, linearly polarized beam (λ_2; fixed in the absorption band to be analyzed). The beam is blocked behind the sample by a perpendicularly oriented polarizer (analyzer). A pump field (λ_1; linarly polarized at 45° to the λ_2 polarization plane, tuned over the investigation range) induces dichroism and birefringence in the sample, thus generating a polarization component orthogonal to the original λ_2 polarization. This component (NLPF signal) is detected behind the analyzer as a function of $\Delta\lambda_{1,2}$.

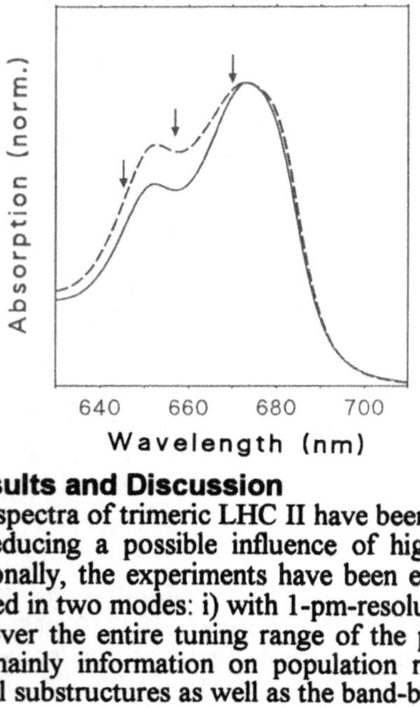

***Figure 2*: Q_y-Absorption spectra** (normalized) of trimeric (solid line) and monomeric LHC II (dashed line). Arrows indicate NLPF probe wavelengths (λ_2 = 645, 657 and 670 nm, from left to the right).

3. Results and Discussion

NLPF spectra of trimeric LHC II have been obtained at lower pump PFDs than previously thus reducing a possible influence of high-PFD phenomena (*vide infra*, *cf.* also [4]). Additionally, the experiments have been extended on monomeric LHC II. Spectra were recorded in two modes: i) with 1-pm-resolution in the immediate vicinity of λ_2; in 0.5-nm-steps over the entire tuning range of the pump laser. Whereas the high-resolution scans yield mainly information on population relaxation (T_1), broad-range scans can reveal spectral substructures as well as the band-broadening mode(s).

3.1. Spectral substructure in the Q_y-region

At first sight broad features distant from λ_2 (*arrows*) are apparent (Fig. 3). The shift of their maxima upon variation of λ_2 indicates heterogeneity (i.e. the existence of several sub-bands). These subspecies are essentially homogeneously broadened [4]. A further hint, that inhomogeneous broadening might be negligible at RT comes from the observation that the (narrow) peak is not shifted from λ_2 at low pump PFD (Fig. 4b; a shift indicating spectral cross-relaxation requires inhomogeneity as a precondition). Thus spectra were analyzed in terms of a heterogeneous-homogeneous model, *cf.* [3,4]. In principle, the description of the NLPF profiles with the proposed Chl-forms, e.g. [8], is possible; however, a reasonable fit of the λ_2 = 645-nm spectrum is obtained with only two of them (649, 671 nm, Fig. 3a). Whereas a subband around 680 nm is required to fit the LHC II-trimer spectra at all λ_2, (data not shown) for monomers it is needed only at λ_2 = 670 nm. One may speculate, that this reflects coupling due to vicinity/similar orientation of the 680-nm species transition dipole with shorter-wavelength forms of adjacent monomers in a trimer or a commonly shared excitonic transition. However, a final statement concerning the subbands is not yet possible. The limited tuning range of the dye lasers precluded definite investigation beyond 680 nm. Low temperature NLPF experiments, currently to be set up in our lab, may be helpful as well as extraction of information from the PFD-dependency of the NLPF spectra (*vide infra*).

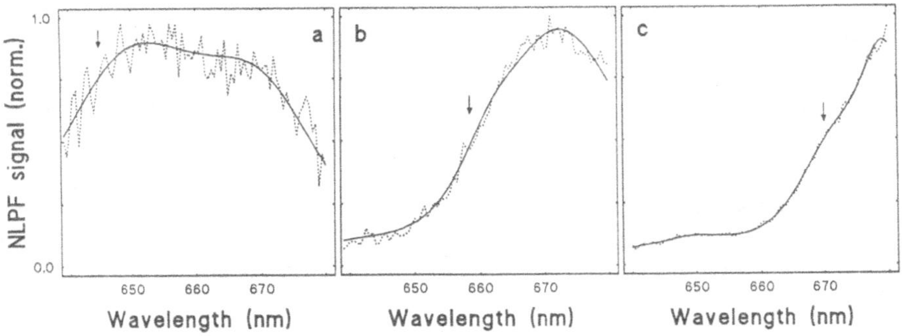

***Figure 3*: NLPF spectra of monomeric LHC II.** Spectra were recorded at the lowest pump PFD ($6 \cdot 10^{23}$ photons cm^{-2} s^{-1}) for λ_2 = 645 (a), 657 (b) and 670 nm (c). Solid lines correspond to the fits (see text).

3.2. Dephasing

Theoretical estimates of the order of phase relaxation times (T_2) in photosynthetic antenna systems diverge considerably [10,11]. Experimental data (in particular for RT-conditions) are scarce, *cf.*, e.g. [12]. From the width of the subbands (taking the above heterogeneous-homogeneous model) one can estimate T_2. Regardless of the number of forms, T_2 fall in the range of some 10 to 60 fs, thus corroborating previous findings with LHC II-trimers and aggregates [4]. Notably, this is in the same order as recently inferred for Chl b to Chl a EET rates in LHC II [13]. The knowledge of T_2 is of importance for mechanistic interpretations and modelling of EET processes [1]. Furthermore, the question of the participation of coherent phenomena in facilitating ultrafast antenna EET has been raised [14].

3.3. Pump-beam intensity dependence of the NLPF profiles

NLPF profiles are markedly altered upon increasing the pump-beam PFD (Fig. 4). The development of a second, blue-shifted maximum in the broad-range spectra with increasing PFD may reflect excited-state absorption (Fig. 4a). Broadening of the peak in the vicinity of λ_2 (indicating T_1 shortening, Fig. 4b) may be interpreted either as probing the usually shorter life-times of the higher excited states and/or indication of processes like exciton-exciton annihiliation. The latter is corroborated by the observation that aggregates

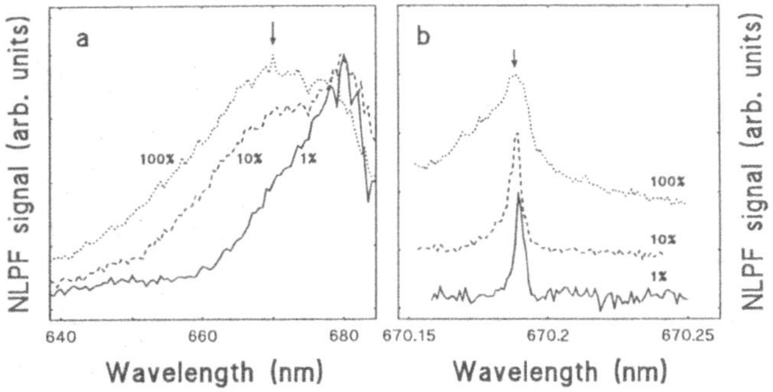

***Figure 4*: Pump intensity dependence of the NLPF signal.** (a) Overall NLPF profile changes in LHC II monomers at $\lambda_2 = 670$ nm. (b) Bandwidth changes in the immediate vicinity of λ_2. Numbers correspond to the percentage of the maximum pump PFD ($6 \cdot 10^{25}$ photons cm^{-2} s^{-1})

of LHC II proved more susceptible for T_1 shortening (annihilation is known to depend critically on domain size) [4]. Since our evaluation software does not yet contain respective expressions, we defer at he moment any quantitative statements.

Acknowledgements

We are grateful to Dr. G. Garab (BRC, Szeged) for providing CD-spectroscopy facilities. This work was supported by the Deutsche Forschungsgemeinschaft (grants Ho 1757/2-1 and Le 729/2-2).

References

[1] Van Grondelle, R., Dekker, J.P., Gillbro, T., and Sundstöm, V. (1994) *Biochim. Biophys. Acta* **1187**, 1-65.
[2] Kühlbrandt, W., Wang, D.N. and Fujiyoshi (1994) *Nature* **367**, 614-621
[3] Leupold, D., Voigt, B., Pfeiffer, M., Bandilla, M. and Scheer, H. (1993) *Photochem. Photobiol.* **57**, 24-28
[4] Lokstein, H., Leupold, D., Voigt, B., Nowak, F., Ehlert, J., Hoffmann, P. and Garab, G. (1995) *Biophys. J.*, in press
[5] Nowak, F., Voigt, B., Leupold, D., Bandilla, M., Scheer, H., and Cogdell, R., *these proceedings*
[6] Krupa, Z., Huner, N.P.A., Williams, J.P., Maissan, E. and James, D.R. (1987) *Plant Physiol.* **84**, 57-61
[7] Mullineaux, C.W., Pascal, A.A., Horton, P. and Holzwarth, A.R. (1993) *Biochim. Biophys. Acta* **1141**, 23-28
[8] Nußberger, S., Dekker, J.P., Kühlbrandt, W., van Bolhuis, B.M., van Grondelle, R. and van Amerongen, H. (1994) *Biochemistry* **33**,14775-14783
[9] Song, J.J., Lee, J.H. and Levenson, M.D. (1978) *Phys. Rev.* **A 17**, 1439-1447.
[10] Kenkre, V.M., and Knox, R.S. (1976) *J. Lumin.* **12/13**, 193-197.
[11] Nedbal, L., and Szöcs (1986) *J. Theor. Biol.* **120**, 411-418.
[12] Bittner, T., Voigt, J., Kehrberg, G., Eckert, H.J., and Renger, G. (1991) *Photosynth. Res.* **28**, 131-139.
[13] Du,M., Xie, X., Mets, L., and Fleming, G.R. (1994) *J. Phys. Chem.* **98**, 4736-4741.
[14] Chachisvilis, M., Pullerits, T., Jones, M.R., Hunter, C.N., and Sundström, V. (1994) *Chem. Phys. Lett.* **224**, 345-351.

NONPHOTOCHEMICAL QUENCHING OF CHLOROPHYLL FLUORESCENCE IN LEAVES: INFLUENCE OF PHOTOSYSTEM II ANTENNA SIZE AND VIOLAXANTHIN DE-EPOXIDATION

Heiko Härtel & Heiko Lokstein
Institut für Biologie/Lehrstuhl für Pflanzenphysiologie, Humboldt-Universität zu Berlin, Unter den Linden 6, D-10099 Berlin, Germany

1. Introduction

Higher plants are able to dissipate excess absorbed light energy non-radiatively as heat. This energy dissipation pathway can be quantified as nonphotochemical quenching of chlorophyll (Chl) fluorescence (NPQ). The main proportion of NPQ, the thylakoid ΔpH-dependent one, relaxes rapidly (within a few minutes) upon darkening of leaves. The mechanism of this rapidly reversible NPQ (often also referred to as qE) is a matter of controversy. At least two alternative models exist to explain its origin. It may arise: i) from charge recombination within the photosystem (PS) II reaction centre (RC) [1] or, ii) in the light-harvesting pigment/protein complexes of PSII (LHCII) [2-4]. The finding, that the extent of NPQ formation is often correlated with the light-induced activity of the xanthophyll cycle, i.e. an enzymatic conversion of violaxanthin (V) via antheraxanthin (A) into zeaxanthin (Z), has led to the assumption that Z is essentially involved in this process [2]. In addition, it has been proposed that all Z-independent rapidly reversible NPQ is explained by A accumulation [5]. However, the build-up of near maximum levels of qE has been reported in the complete absence of Z [6, 7]. This has led to the conclusion that Z is not obligatorily necessary in this process; it only enhances ΔpH-dependent NPQ which is in turn brought about solely by aggregation of LHCII that may be mediated by protonation of negative charges of amino acid residues [3]. A promising approach to gain new information about NPQ is the use of plants with a modified LHCII complement [4, 8, 9]. Here we report on measurements of NPQ development and the involvement of V de-epoxidation in plants comprising a step-wise reduced antenna system.

2. Material and Methods

Plants of barley wild type (WT) and the Chl b-less mutant *chlorina* 3613 were cultivated either under continuous light (CL) or intermittent light (IML) as described [9]. Modulated Chl fluorescence was assessed and pigments were quantified by HPLC analysis as in [4]. Applying the Stern-Volmer formalism, NPQ of the maximum fluorescence yield, F_m, (SV_N) and its rapidly (within 5 min of darkness) reversible component (ΔSV_N) were calculated as in Härtel and Lokstein [9] where SV_N is defined as $(F_m/F_m')-1$ and F_m' reflects F_m in the respective energized state. Leaves were dark-adapted for 60 min prior to light exposure. Treatment with dithiothreitol (DTT) was performed essentially as described [9]. The de-epoxidation state of the xanthophyll-cycle pigments (DPS) was calculated as DPS = (0.5A+Z)/(V+A+Z). The dependencies of DPS and NPQ on the photon flux density (PFD) were monitored in the range from 0-1300 µE $m^{-2}s^{-1}$. To verify the antenna composition in our material, Western blot analyses were done similar as in [8].

P. Mathis (ed.), Photosynthesis: from Light to Biosphere, Vol. I, 291–294.

3. Results and Discussion

The plants used in this study comprise a step-wise reduced PSII antenna size (Fig. 1). Based on the model of antenna organization [10] one can differentiate between: i) the peripheral LHCII antenna, consisting of mixed trimers of the *Lhcb1* and *Lhcb2* gene products (present only in the WT), ii) the inner LHCII antenna, comprising the polypeptides Lhcb3-Lhcb6 (largely abundant in the CL-grown mutant) and, iii) the core antenna, CP43 and CP47 (the only antenna present in the IML-grown mutant).

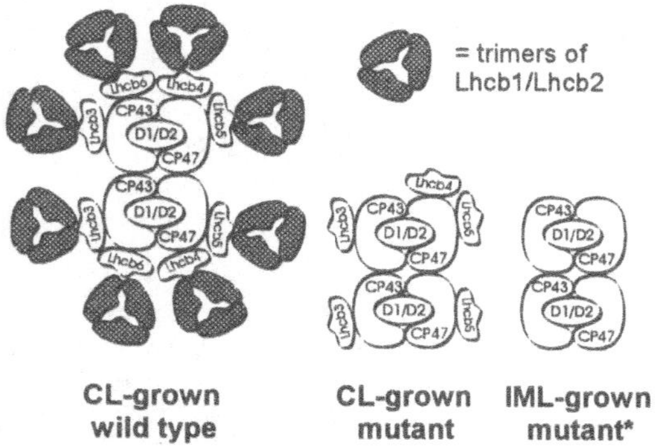

Figure 1. Model of the PSII light-harvesting antennae in mature WT barley leaves and in leaves of the Chl b-less mutant *chlorina* 3613 grown either under CL or IML-conditions. The model based on our Western blot analysis was adapted according to the proposed scheme of Jansson [10] for the higher plant light-harvesting antenna organization. (*traces of Lhcb5; D1 and D2 represent the RC proteins)

The differences in SV_N between the plants compared reveal that the capability to develop NPQ is linked to the presence of certain antenna polypeptides (Fig. 2A). It is evident: the more the PSII antenna size is reduced, the lower is the capacity to form NPQ. Notably, as apparent from the data obtained with the IML-grown mutant a significant proportion of overall NPQ (about 40% after 15 min of light exposure) emanates from the RC/core antenna complex.

The significance of xanthophyll-cycle operation for NPQ formation becomes obvious by treatment of leaves with the V de-epoxidase inhibitor DTT (Fig. 2B). SV_N values largely approach to the same level that was obtained with IML-grown mutant leaves in the absence of DTT, clearly indicating the need of xanthophyll-cycle activity for the

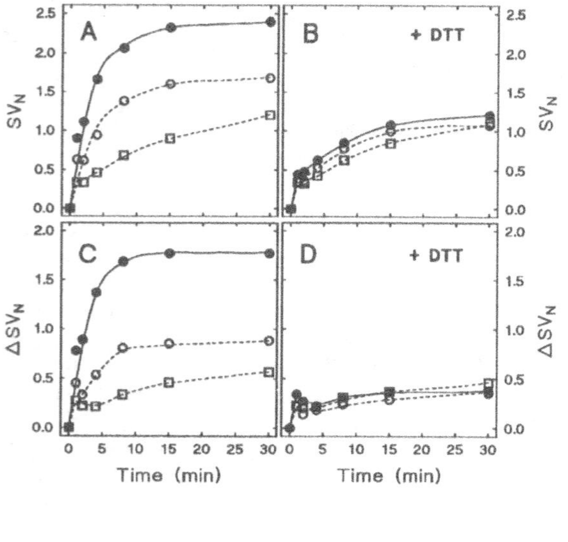

Figure 2. Kinetics of the development of overall NPQ (SV_N) and its rapidly (within 5 min of darkening) reversible component (ΔSV_N) in control (A, C) and DTT-treated (B, D) leaves in dependence on the PSII antenna size. All leaves were exposed to photosynthesis-saturating light of 1300 µE $m^{-2}s^{-1}$. Symbols represent: CL-grown WT (closed circles), CL-grown mutant (open circles) and IML-grown mutant (open squares). Note the different scales of the y-axis for SV_N and ΔSV_N.

development of that proportion of NPQ which is linked to the presence of the inner and peripheral LHCII antenna. The finding that NPQ arising from the RC/core antenna complex is independent from the activity of the xanthophyll cycle is not surprising since the core antenna complex does not bind xanthophyll-cycle pigments [11]. In parallel to overall NPQ its rapidly reversible component was found to be reduced in a similar manner both in control (Fig. 2C) and DTT-treated leaves (Fig. 2D). Again, two fractions of ΔSV_N are apparent: a xanthophyll cycle independent and a xanthophyll cycle dependent one, the latter is dominating in the WT. A quantitative inspection of the data reveals that the increase in the SV_N levels with increasing PSII antenna size is nearly exclusively of the qE type, i. e. rapidly reversible and hence, ΔpH-dependent.

To further address the possible role of V de-epoxidation in NPQ development, the DTT-sensitive component of NPQ was plotted against the corresponding DPS (Fig. 3A). Notably, with onset of light exposure leaves contained no Z and only traces of A. In the CL-grown WT, SV_N first rises steeply to a value of 1.15 (about 90% of the final level); followed by a further slight increase. The strong initial rise in SV_N is accompanied by a rather low V de-epoxidation (attaining a DPS of 0.2) whereas the later - only slight - increase occurred along with strong V de-epoxidation. Obviously, NPQ is more sensitive to A+Z at lower than at higher xanthophyll concentrations. In CL and IML-grown mutants the initial rise in SV_N was either strongly reduced or even absent. This observation could point to an essential contribution of the peripheral LHCII antenna to NPQ build-up. Accordingly, the smaller rise of SV_N in the CL-grown mutant could reflect a minor contribution of the inner LHCII antenna to NPQ development, since in IML-grown mutant leaves the lack of the inner and peripheral LHCII corresponds to the almost complete absence of the DTT-sensitive NPQ proportion. Interestingly, the slope of the later minor increase of SV_N in the WT equals that obtained in the antenna-depleted plants (dashed lines in Fig. 3A), suggesting a similar (relatively low) sensitivity of this part of NPQ formation to A and Z accumulation.

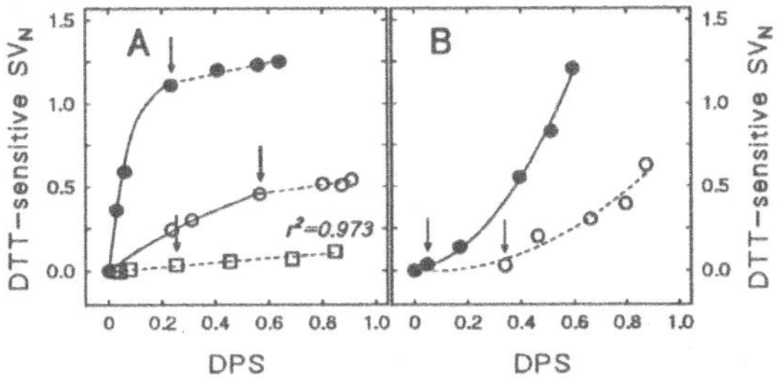

Figure 3. Relationship between DTT-sensitive SV_N and the DPS obtained either during induction of NPQ (A) or in the steady state by varying PFD (B) in leaves in dependence on the PSII antenna size. DTT-sensitive SV_N was calculated from the difference of SV_N in the absence and in the presence of DTT. Arrows mark the values obtained after 4 min of light exposure (A) or at 170 μE $m^{-2}s^{-1}$ (B). The V+A+Z pool sizes prior to light exposure were 64 (WT), 79 (CL-grown mutant) and 649 mmol (mol Chl a)$^{-1}$ (IML-grown mutant). For symbols cf. Fig. 2.

Concerning the role of the xanthophyll cycle in NPQ development the situation is more complicated owing to the fact that the presence of antenna polypeptides influences both the availability of V for and the rate of its de-epoxidation (Fig. 3A). Moreover, an elevated ratio of V+A+Z to Chl is observed in antenna-depleted plants (see legend of Fig. 3). To make this complex issue more transparent, the arrows in Fig. 3 mark the DPS values obtained either after the same duration of light exposure (in A) or the same PFD (in B).

Plotting the PFD-dependent DTT-sensitive SV_N increase vs. DPS a curved relationship is obtained for both CL-grown WT and mutant leaves (Fig. 3B). In general, NPQ appears to be more sensitive to A+Z formation in response to a PFD increase than during the induction period at high constant PFD (cf. Figs. 3A and B). Moreover, the steeper rise in

the PFD dependency obtained for the WT suggests a closer coupling of DPS and NPQ than in the mutant (Fig. 3B). In the latter a significant fraction of A and Z (about 35%) is present without affecting NPQ. Most probably, it is this fraction which is also superimposed on the DPS increase in the mutant during NPQ development (Fig. 3A). Although difficult to show unequivocally, based on both kinetics of V de-epoxidation and LHCII polypeptide and pigment composition it appears to be a fraction of 'free' (not bound to LHCII proteins) and hence, more rapidly photoconvertible V.
If qE is determined by the product of [A+Z] and lumen acidity [5], the intial steep rise in NPQ at low DPS in the WT during induction should mainly result from increasing [H^+] accumulation, presumably in localized membrane domains, which themselves may be situated within the LHCII [7, 12, 13]. However, if qE formation is only the result of H^+-driven conformational changes of LHCII as suggested [3, 6, 7], one would expect its maximum build-up even in the presence of DTT, i.e. without xanthophyll-cycle operation. This is obviously not the case. The fact, that near maximum qE build-up is associated with the accumulation of only low amounts of A and Z can account for the contradictory reports concerning the role of the xanthophyll cycle in this process (cf. [2, 5, 9] vs. [3, 6, 7]). In the later period of induction, V de-epoxidation appears to lag behind NPQ development; a strong A+Z accumulation results in a very low further NPQ increase in the WT (Fig. 3A). It might be the result of LHCII-protein embedded V becoming accessible for the de-epoxidase in response to the protonation. This idea would also explain the linearization of the relationship between NPQ and DPS in the steady state (cf. Fig. 3B).
Recent work has indicated the importance of the minor LHCII-polypeptides in the formation of qE. They are enriched in xanthophyll-cycle pigments [11] and seem to be involved in H^+-channeling [12]. However, it appears that solely their presence is not sufficient to establish maximum qE. In this respect, it has yet to be shown whether the lower capacity for qE build-up in the *chlorina* mutant is due to the reduced level of minor polypeptides and/or the peripheral LHCII. Regardless of this open question, two major conclusions emerge from our study: i) structural modifications within the LHCII establish the basis of the major proportion of rapidly reversible NPQ *in vivo*, and ii) xanthophyll-cycle operation is an obligatory requirement therein.

4. Acknowledgements

Dr. Bernhard Grimm (IKP Gatersleben) is thanked for help with the Western blot analyses. This study was supported by the Deutsche Forschungsgemeinschaft (Ho 1757/1).

References

1 Krieger A, Weis E (1993) Photosynth Res 37, 117-130
2 Demmig-Adams B (1990) Biochim Biophys Acta 1020, 1-24
3 Horton P, Ruban AV, Walters RG (1994) Plant Physiol 106, 415-420
4 Lokstein H, Härtel H, Hoffmann P, Renger G (1993) J Photochem Photobiol B: Biol 19, 217-225
5 Gilmore AM, Yamamoto HY (1993) Photosynth Res 35, 67-78
6 Noctor G, Rees D, Young A, Horton P (1991) Biochim Biophys Acta 1057, 320-330
7 Noctor G, Ruban AV, Horton P (1993) Biochim Biophys Acta 1183, 339-344
8 Jahns P, Krause GH (1994) Planta 192, 176-182
9 Härtel H, Lokstein H (1995) Biochim Biophys Acta 1228, 91-94
10 Jansson S (1994) Biochim Biophys Acta 1184, 1-19
11 Bassi R, Pineau B, Dainese P, Marquardt J (1993) Eur J Biochem 212, 297-303
12 Walters RG, Ruban AV, Horton P (1994) Eur J Biochem 226, 1063-1069
13 Renganathan M, Dilley RA (1994) J Bioenerg Biomembr 26, 117-125

QUENCHING OF CHLOROPHYLL FLUORESCENCE IN THE MINOR CHLOROPHYLL A/B BINDING PROTEINS OF PHOTOSYSTEM II

A.V. Ruban[1], A.J. Young[2] and P. Horton[1],
[1]Robert Hill Institute, Department of Molecular Biology and Biotechnology, University of Sheffield, Western Bank, Sheffield S10 2TN, U.K; [2]School of Biological and Earth Sciences, Liverpool John Moores University, Byrom Street, Liverpool, U.K.

1. Introduction

When plants are exposed to excess light, a protective dissipative process is induced that removes the excess of absorbed energy, the non-photochemical quenching of chlorophyll fluorescence (qE). Two factors synergistically promote qE - the synthesis of zeaxanthin (zea) by de-epoxidation of violaxanthin (viol) [1] and the ΔpH [2]. It has been proposed that zea is required directly as a quencher of chl fluorescence [1] or alternatively that the xanthophyll cycle is a regulator of quenching [3]. A variety of data indicate that quenching occurs in LHCII [reviewed in 3] and because the major complex, LHCIIb, can be manipulated *in vitro* to give rise to fluorescence quenching, it was proposed that qE occurs by the same mechanism [3]. Recently it was shown that dilution of detergent solubilised LHCIIb into a low detergent medium gave rise to quenching that resembled qE in some detailed aspects [4]: low pH and the presence of dibucaine enhanced quenching whereas antimycin A caused inhibition, exactly similar to their effects on qE [5]. Most importantly, addition of viol was found to *inhibit* quenching whilst zea was without effect, strongly suggesting the indirect role of these carotenoids in qE. Recent work has suggested that two minor components of LHCII, LHCIIa and LHCIIc (CP29 and CP26 respectively) play an important role in qE: they are enriched in viol and zea, binding one molecule per polypeptide compared to less than 1 per trimer for LHCIIb [6]; inhibition of qE in isolated chloroplasts by the carboxy modifying agent DCCD is associated with covalent binding to these complexes [7]. Protonation of these carboxy amino acid residues could result in a structural change in these proteins, causing alteration in pigment associations [8], quenching resulting from either formation of a chl-chl dimer or by bringing chl into close proximity with a quenching carotenoid. There have been no studies of the behaviour of the minor LHCII components *in vitro*; in order to further test the applicability of the principles of the LHCII model for qE it is important to investigate whether they show a similar flexibility in photophysical properties as LHCIIb; similarly it needs to be established whether those conditions known to be important for qE have similar effects on the minor complexes.

2. Materials and Methods

LHCIIa, LHCIIb and LHCIIc were isolated from spinach as described previously [6]. LHCII solubilised in 200 μM n-dodecyl ß-maltoside (DM) was diluted into 6 μM DM with constant stirring at a chl concentration of 2 μg/ml to induce quenching of chl fluorescence [4], which was quantified as (Fm-F)/F, where Fm is the level recorded for a sample diluted into 200 μM DM and F is the quenched fluorescence. After 45 s HCl was

P. Mathis (ed.), Photosynthesis: from Light to Biosphere, Vol. I, 295–298.

added to lower the pH to 5.0. Purified viol and zea dissolved in ethanol were added to a concentration of 10 µg/ml and 20 µg/ml respectively.

3. Results

Dilution of LHCIIa, LHCIIb and LHCIIc into 6 µM DM results in a spontaneous quenching which slows down after approx. 10 sec (Fig. 1). Acidification to pH 5.0 then causes an immediate and strong acceleration of quenching for all complexes, the total final quenching was 5.2, 2.7 and 7.8 for LHCIIa, LHCIIb and LHCIIc respectively Quenching at low pH in LHCIIb was only weakly and irreproducibly inhibited by DCCD, but, in contrast, there was substantial and consistent reversal of quenching in LHCIIa and LHCIIc following addition of DCCD (Fig. 1). Other reagents known to strongly affect qE in thylakoids [5] were tested: dibucaine, which accelerates the formation of qE, stimulated quenching in all three complexes when present before acidification, but the final level at pH 5.0 was unaltered; antimycin A, a strong inhibitor of qE, caused inhibition of both the spontaneous and low pH induced quenching for LHCIIb but was without affect on LHCIIa and LHCIIc. It has previously been shown that addition of viol inhibits both the spontaneous and pH-dependent quenching in LHCIIb, whereas zea had no effect [14]. More dramatic effects of viol and zea were observed with the minor complexes (Fig 2). As for LHCIIb, viol inhibited the spontaneous quenching and also reduced the extent of quenching at low pH in LHCIIa. However, in the case of LHCIIa, addition of zea caused a strong decrease in fluorescence intensity and only a small amount of additional quenching was observed upon acidification; the

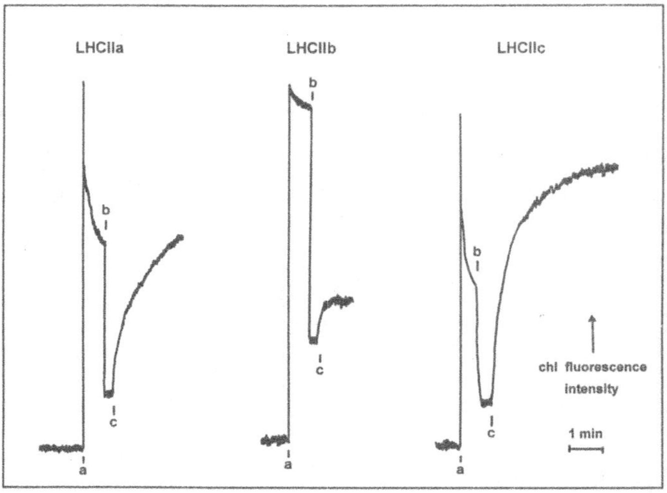

Figure 1. Fluorescence quenching in LHCIIa, LHCIIb and LHCIIc upon dilution into 6 µM DM (a), acidification (b) and addition of DCCD (c).

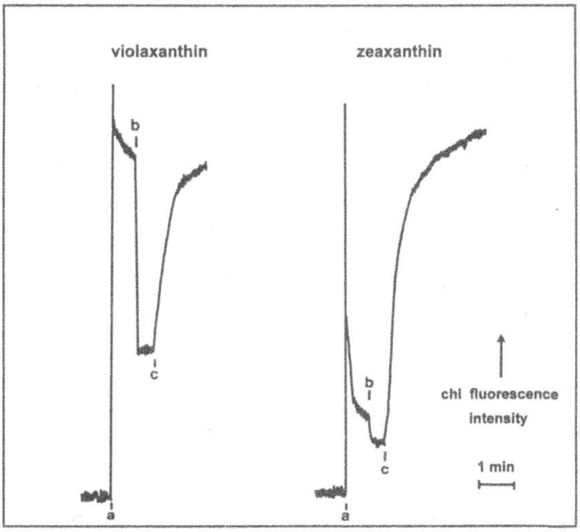

Figure 2 Effects of viol and zea on fluorescence quenching in LHCIIa. The experiment was carried out as for Fig. 1 with xanthophylls added prior to dilution.

final extent of quenching was increased in the presence of zea by about 20%. All of the quenching observed in the presence of zea was reversed by DCCD. Results similar to those observed for LHCIIa were also found with LHCIIc (not shown).

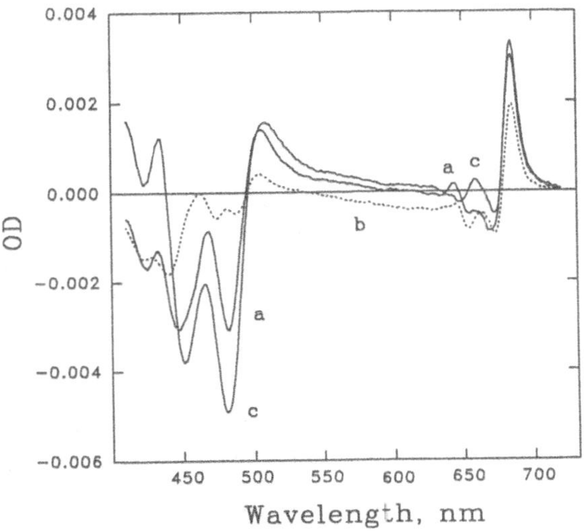

Figure 3 Absorption difference spectrum for quenched-minus-unquenched states of LHCIIa (a), LHCIIb (b) and LHCIIc (c). The unquenched sample was dissolved in 200 μM DM and the quenched in 6 μM at pH 5.0

Quenching in LHCIIb brought about by alteration in detergent concentration or low pH brings about changes in the absorption spectrum [4,9]; absorption difference spectra show similar features for all the LHCII components: negative binds at 440-450 nm and 480-490 nm arising from chl and xanthophyll, a positive band near 510 nm and changes in the Q_y band of chl a and b (Fig. 3). It is significant that the extent of the absorbance changes are larger for LHCIIa and c compared to LHCIIb, consistent with the larger quenching observed in these complexes. Particularly interesting is that the changes in xanthophyll absorption are much stronger for the minor complexes than for LHCIIb; whereas the extent of the change in the chl a band near 685 nm is approximately in proportion to the differing extents of quenching in the three complexes, there is a 6-10 fold increase in the absorption change at 485 nm, which may be caused by their enrichment in xanthophyll cycle carotenoids [6].Quenching of LHCIIb *in vitro* has also been associated with formation of long wavelength emitters, particularly a 700 nm band at 77 K [9]. Emission spectra at 77 K also show a peak near 700 nm in the quenched state for the minor LHCII (Fig. 4). The intensity of 700 nm emission relative to that near 680 nm is a characteristic index of aggregation-dependent quenching, and is in the order c>a>b, correlating with the extent of quenching at room temperature.

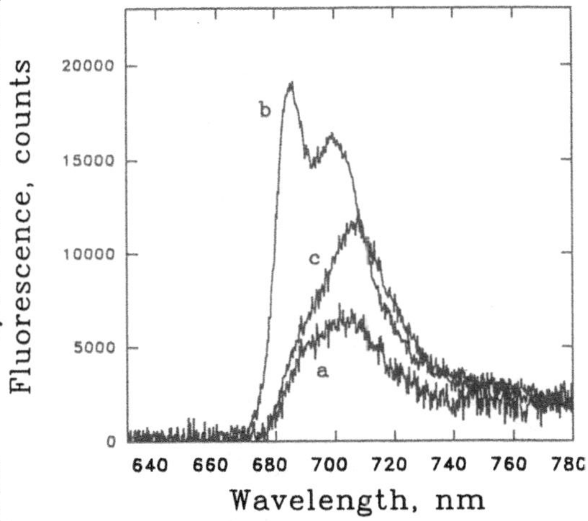

Figure 4 Fluorescence spectra at 77 K of LHCIIa (a), LHCIIb (b) and LHCIIc (c) in the fully quenched state obtained following acidification to pH 5.0.

4. Discussion

The fluorescence yield of LHCIIa, LHCIIb and LHCIIc can be manipulated by lowering of the detergent concentration and altering the pH. The similarity of the absorption and fluorescence spectral changes indicate that the nature of the change in LHCII structure underlying the increase in quenching is similar for all the LHCII components. These changes are associated with increased protein-protein interaction, which have been shown in the case of LHCIIb to be the formation of macromolecular aggregates, and are reflected in dramatic changes in spectral properties of bound chl and xanthophyll [3,8,9]. The features of the quenching in isolated LHCII resemble in many respects qE in thylakoids. The pH requirement for quenching in LHCII (pK 6.2, not shown) is similar to that observed for pH-dependent quenching in thylakoids (pK 5.7) and to indirect estimates of the lumen pH *in vivo*. Conditions which control qE (DCCD, antimycin, dibucaine) have the similar effects on quenching, and in particular the quenching in all LHCII components is controlled by the xanthophyll cycle carotenoids, namely an inhibition and stimulation of quenching by viol and zea respectively. A significant feature of the data presented here is the DCCD-sensitivity, the stronger quenching and the greater control by the viol:zea ratio for the minor complexes. These features provide further evidence for an important role of these complexes in the control of energy quenching *in vivo* and the strong stimulation of quenching by zea without pH lowering also suggests that they play a key role in the long-lived quenching state referred to as qI. There is considerable debate concerning the possible mechanism(s) of zea-mediated qE [10]; the S_1 energy level of zea means that it could directly quench chl fluorescence by singlet-singlet interaction [11], but although the stimulation of LHCII quenching by added zea is consistent with this mechanism, it is hard to explain the inhibitory effect of viol in these terms. Moreover, zea had a similar effect to dibucaine and low pH and quenching in all cases was reversed by DCCD. Viol and zea respectively suppress or stimulate a quenching process within LHCII for which there is an obligatory requirement for protonation (as for qE *in vivo*) as predicted by the LHCII model for qE [1,4,10]. Clearly, the role of the xanthophyll cycle carotenoids in the regulation of quenching is complex and may include both 'direct' de-excitation of excited state chl (production of zea) and indirect effects through control of protein structure (removal of viol). Only a systematic study of a range of carotenoids with different structure will be able to resolve the relative importance of these two effects (see Phillip *et al*, these proceedings).

References

1 Demmig-Adams (1990) Biochim. Biophys. Acta 1020, 1-24.
2 Briantais, J.-M., Vernotte, C., Picaud, M. and Krause, G.H. (1979) Biochim. Biophys. Acta 548, 128-138.
3 Horton, P. and Ruban, A.V. (1994) in: Photoinhibition of Photosynthesis - from molecular mechanisms to the field (Baker, N.R. and Bowyer, J.R. eds). pp 111-128, Bios Scientific Publishers, Oxford
4 Horton, P., Ruban, A.V., Rees, D., Pascal, A.A., Noctor, G. and Young, A.J. (1991) FEBS Lett. 292, 1-4.
5 Ruban, A.V., Horton, P. and Young, A.J. (1994) Biochim. Biophys. Acta 1186, 123-127.
6 Ruban, A.V., Young, A.J., Pascal, A.A. and Horton, P. (1994) Plant Physiol. 104, 227-234.
7 Walters, R.G., Ruban, A.V. and Horton, P. (1994) Eur. J. Biochem. 226, 1063-1069.
8 Ruban, A.V. and Horton, P. (1992) Biochim. Biophys. Acta 1102, 30-38.
9 Ruban, A.V., Horton, P. and Robert, B. (1995) Biochemistry 34, 2333-2337
10 Horton. P., Ruban, A.V. and Walters, R.G. (1994) Plant Physiol. 106, 415-420.
11 Frank, H.A., Cua, A., Chynwat, V., Young, A.J., Goztola, D. and Wasielewski, M.R. (1994) Photosynth. Res. 41, 389-395.

DCCD BINDS TO LUMEN-EXPOSED GLUTAMATE RESIDUES IN LHCIIc
Implications for the mechanism of photoprotective energy dissipation

Robin G. Walters and Peter Horton
Robert Hill Institute, Sheffield University, Sheffield S10 2TN, UK

1. Introduction

The role of the light-harvesting complexes of photosystem II (LHCII) in protecting against the deleterious effects of excess illumination is now well-established. The thermal dissipation of energy (qE) by LHCII, manifested as quenching of chlorophyll fluorescence, is known to depend on the acidification of the thylakoid lumen which results from light-induced proton translocation; it is inferred that a reorganisation of the light-harvesting apparatus is prompted by the protonation of specific aminoacid residues or pigments [1].
The reagent dicyclohexylcarbodiimide (DCCD) has been widely used to identify active sites in proton-translocating complexes; it covalently modifies carboxyl residues within hydrophobic domains. Treatment of thylakoids with DCCD inhibits energy dissipation, while also inhibiting proton translocation from the site of water-splitting to the lumen [2-3]. We have recently shown that the only PSII proteins bound by DCCD are two minor LHCII complexes - LHCIIa (also referred to as CP29) and LHCIIc (CP26) - and that the major complexes (LHCIIb) are not modified [4]. One possibility is that DCCD-reactive sites are directly involved in energy dissipation, perhaps acting to sense the pH of the lumen. This paper describes the identification of two DCCD binding sites on LHCIIc, and discusses the significance of the results in terms of current models for qE.

2. Materials and Methods

2.1 *Isolation and treatment of LHCIIc*
LHCII complexes were isolated by isoelectric focusing as previously described [4]. [^{14}C]-DCCD-labelled LHCIIc apoprotein was prepared as previously described [4], by treating LHCIIc (35 μg.ml^{-1}) with 50 μM [^{14}C]-DCCD (RT, 15 min), following which DCCD and detergent were removed by extraction with methanol-chloroform. Digestion of labelled protein with TPCK-treated trypsin (10 μg.ml^{-1}, 37°C overnight) was carried out in 1% octylglucoside, 0.1% SDS, 0.06% dodecylmaltoside, 50 mM Hepes (pH 7.5); protein was then precipitated with methanol-chloroform. For chemical cleavage with formic acid or cyanogen bromide, protein was dissolved in 70% (v/v) formic acid, and incubated either overnight at 37°C, or at room temperature for 4 hours after the addition of a few crystals of CNBr; cleaved protein was freeze-dried, washed in distilled water, and again freeze-dried.

P. Mathis (ed.), Photosynthesis: from Light to Biosphere, Vol. I, 299–302.

2.2 *Identification and sequencing of ^{14}C-labelled LHCIIc fragments*
Protein fragments were separated by Tris-tricine electrophoresis [5], and electroblotted onto PVDF membrane. After Coomassie staining and autoradiography, labelled protein bands were excised and sequenced using an Applied Biosystems Sequenator model 473A. Where appropriate, the eluates for each sequencing cycle were collected and freeze dried; after dissolving in water and adding scintillant, the eluted label was counted for a period sufficient to register a minimum of 2000 counts.

3. Results and Discussion

3.1 *Microsequencing of [^{14}C]-DCCD-labelled peptide fragments*
Cleavage of [^{14}C]-DCCD-labelled LHCIIc apoprotein with trypsin, formic acid or cyanogen bromide has allowed the location of the site(s) at which DCCD binds to be more accurately determined. The molecular weights of these fragments were approximated from their migration during Tris-tricine electrophoresis (for the smallest fragments, such estimates are subject to significant error), and their start-points were determined by comparison of their N-terminal amino-acid sequences with previously-characterised LHCIIc sequences. Using this approach, a large number of DCCD-labelled fragments were identified (see Figure 1); no single region of the protein was common to all of them, thus it was clear that there are at least two DCCD-binding sites on LHCIIc. Their locations were identified as regions of overlap, common to several DCCD-binding fragments.

3.2 *Determination of labelled residues*
The strong homology between light-harvesting proteins implies that the main features of the LHCIIb structure [6] are also present in LHCIIc. The two DCCD-binding fragments are therefore located towards the lumenal ends of the two tilted transmembrane helices, B and A, the latter fragment also including amphiphilic helix D. Inspection of the LHCIIc consensus sequence revealed that each of these two regions contain more than one acidic aminoacid, each of which were considered candidate DCCD-binding residues (DCCD binds acidic residues in hydrophobic domains). In order to further investigate

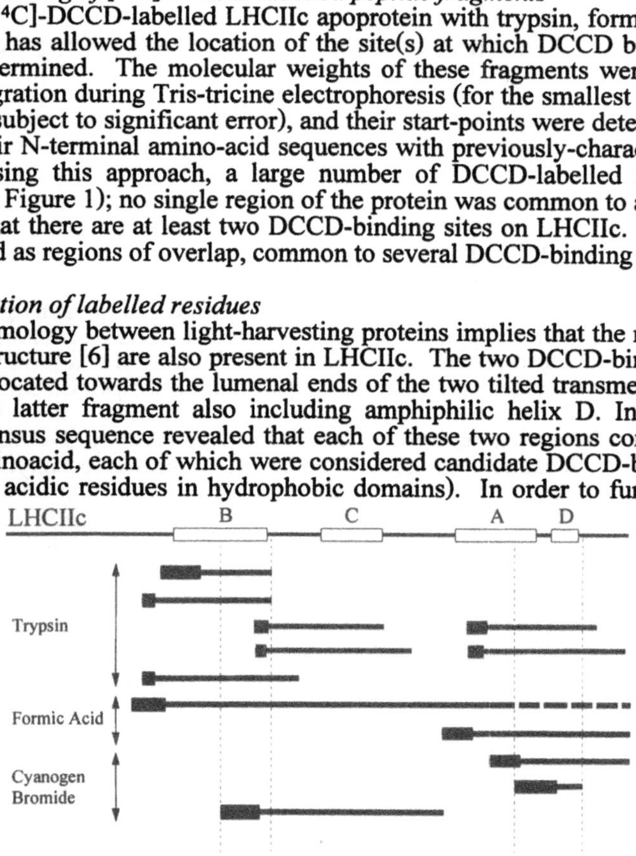

Figure 1. Identification of [^{14}C]-DCCD-labelled fragments of LHCIIc apoprotein. *Labelled fragments of LHCIIc (solid bars) were generated by cleavage with trypsin, formic acid or CNBr. Their molecular weights were estimated via their electrophoretic mobility, and their locations determined by comparison of N-terminal aminoacid sequence data (thick bars) with the consensus sequence for LHCIIc apoprotein. In two cases, aminoacid sequence data were ambiguous due to the presence of two fragments (grey). The presumed locations of helices A-D within the apoprotein are as shown.*

the identity of the binding sites, eluate was collected during sequence analysis of appropriate LHCIIc fragments and the released ^{14}C label determined.

No convenient sites for commonly-used proteolytic cleavage agents were identified on the N-terminal side of several of the candidate residues. The most useful fragments were generated by cleavage with CNBr (see Figure 1), but even such cleavage sites were up to 22 aminoacid residues distant from possible DCCD-binding residues. Figure 2 shows the results for label release during sequence analysis of these two CNBr-generated fragments; there was a measurable level of label released during every sequencing cycle - we believe that not only were there unavoidable losses due to protein degradation during sequencing (despite repetitive yields of up to 95%), but the ^{14}C-labelled N-acyl-urea residue formed by DCCD-binding was itself susceptible to degradation.

Figure 2 shows that ^{14}C release is significantly above background only for those residues corresponding to 218-219 and 110-111 in tomato [7]; these positions are the same at which there are peaks in glutamate yield, indicating that it is the residues corresponding to Glu_{218} and Glu_{110} in tomato which are bound by DCCD. Conversely, Glu_{98} (Binding Site 2) is clearly not DCCD-binding. We interpret the lag in the release of Glu and ^{14}C at these positions as being the cumulative effect of low levels of non-cleavage during each sequencing cycle. Glu peaks observed at residues 208 and 214 during analysis of Binding Site 1 are due to degradation of the glutamine residues at these positions.

3.3 *The role of LHCIIc in energy dissipation*

Comparison of the aminoacid sequences of LHCIIb and LHCIIc shows that the DCCD-binding residues identified above are both potentially exposed to the lumen; Site 1 is on the amphiphilic D helix, and Site 2 is a few residues beyond the C-terminal end of the B helix. It has been suggested that the residues bound by DCCD are involved in pigment co-ordination [8]; however, it appears that this is not the case, since the homologous residues in LHCIIb are not involved in chlorophyll or carotenoid binding [6].

We have previously shown that DCCD inhibits the formation of photoprotective energy dissipation (qE) in thylakoids and binds only to LHCIIa and LHCIIc [2, 4]. The lumen pH is the primary signal responsible for inducing qE, and these two Glu residues are in ideal positions to detect the lumen pH; therefore, one strong possibility is that one or

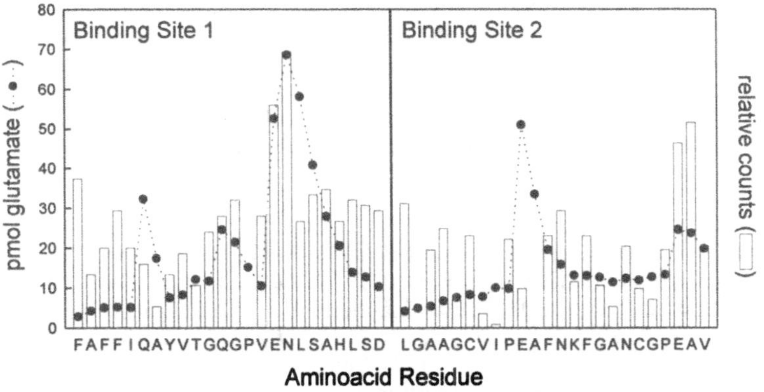

Figure 2. Label associated with fractions collected during sequence analysis of [^{14}C]-DCCD-labelled LHCIIc fragments. *Eluate for each cycle during protein sequencing was collected, freeze-dried, and ^{14}C content was determined by scintillation counting. Glutamate yield for each sequencing cycle are also shown. Aminoacid sequences correspond to LHCIIc residues 203-226 (Site 1) and 89-112 (Site 2) in tomato [7].*

	Binding Site 2			Binding Site 1		
		*			*	
LCHIIa	123	LsVEWL.... ....TGVTWq	134	229	AATGKGpLNN WATHLSDPLH	248
LHCIIb 28 kDa	81	VFPELLaRNG .VKFGEAVWF	99	200	IVTGKGPlEN LADHlADPVn	219
LHCIIb 27 kDa	76	vFPEiLsKNG .VkFGEAVWF	94	195	IVtGKGPIeN LxDHvaDPVA	214
LHCIIb 25 kDa	69	itPEVLqKWV rVdFKEPVWF	88	190	IVTGKgPLEn LlDHLDnpVA	209
LHCIIc	101	IIPEAfNKFG AnCGPEAVWF	120	216	YVTGeGPvEN LaAHLSDPFG	235
LHCIId	70	FVGQAW.... ....sGIPWF	81	200	GQ.GKTPLGA LGL	211

Figure 3. Sequence comparison of LHCII proteins near DCCD-binding sites. *Consensus sequences compiled from published data for angiosperm LHCII proteins were aligned in the region of LHCIIc DCCD-binding sites (*). The residues homologous to the binding sites and another highly conserved region (see text) are shaded. Invariant residues are shown in upper case. Residues are numbered according to [9].*

both of these DCCD-binding sites is directly involved in the formation of qE. Alternatively, the DCCD-binding sites on LHCIIa may be the ones important for qE, while those on LHCIIc have some other function. A more detailed comparison of LHCII sequences (Figure 3) shows that neither LHCIIa nor LHCIId have the two Glu residues bound by DCCD in LHCIIc; Site 1 is on a loop which is absent in these two minor complexes, while conserved asparagine and glycine residues are found in place of the Glu at Site 2. LHCIIa must therefore bind DCCD at wholly different sites; work is in progress to identify the residues involved.

Treatment of thylakoids with DCCD has been shown to inhibit the release into the lumen of protons derived from water-splitting [3]. Since DCCD is known to block proton channels by binding to active acyl aminoacids, it is possible that these DCCD-binding glutamates form part of the proton release pathway rather than being directly involved in qE. Comparison of LHCII consensus sequences also shows the presence of a conserved region in LHCIIc which is altered or absent in other LHCII apoproteins (see Figure 3); included in this region is a lysine residue which the LHCIIb structure suggests might be able to interact with both the DCCD-reactive glutamates. We suggest that these three residues together may form a proton channel involved in the release of protons from water-splitting; it is unclear whether this putative channel plays a role in qE.

References

1. Horton, P., Ruban, A.V. and Walters, R.G. (1994) Plant Physiol. 106, 415-420
2. Ruban, A.V., Walters, R.G. and Horton, P. (1992) FEBS Lett. 309, 175-179
3. Jahns, P. and Junge, W. (1990) Eur. J. Biochem. 193, 731-736
4. Walters, R.G., Ruban, A.V. and Horton, P. (1994) Eur. J. Biochem. 226, 1063-1069
5. Schägger, H. and von Jagow, G. (1987) Anal. Biochem. 166, 368-379
6. Kühlbrandt, W., Wang, D.N. and Fujiyoshi, Y. (1994) Nature 367, 614-621
7. Pichersky, E., Subramaniam, R., White, M.J., Reid, J., Aebersold, R. and Green, B.R. (1991) Mol. Gen. Genet. 227, 277-284
8. Crofts, A.R. and Yerkes, C.T. (1994) FEBS Lett. 352, 265-270
9. Jansson, S. (1994) Biochim. Biophys. Acta 1184, 1-19

A RHODOPSIN AS PHOTORECEPTOR IN CHROMATIC ADAPTATION OF THE CYANOBACTERIUM *CALOTHRIX SP.*

Jeroen H. Geerdink[a], Andrea Haker[a], Hans C.P. Matthijs[a,b], Wouter D. Hoff[b], Klaas J. Hellingwerf[b] & Luuc R. Mur[a]

Department of Microbiology, ARISE[a] & BioCentrum Amsterdam[b], University of Amsterdam, Nieuwe Achtergracht 127, 1018 WS Amsterdam, the Netherlands

1. Introduction:

In a variety of circumstances, cyanobacteria have to be able to sense their environment, to take notion of nutrient concentrations, light conditions, etc.

Few actual receptors involved in signal transduction have been identified, examples are the *NTRC* (Nitrate) and *pho* (Phosphate) systems (1,2). The primary receptor in light colour sensing, used in cyanobacteria with Complementary Chromatic Adaptation (CCA) is still at question. Different suggestions have been made, like phytochrome or phycoerythrin (3,4). Recently, retinal has been proposed as a possible photoreceptor for CCA (5,6). Here we present data supporting this role for retinal in light colour sensing. We have obtained data from 2 similar, but slightly different cyanobacteria: *Fremyella diplosiphon* and *Calothrix sp.* (PCC 7601).

The physiology and molecular mechanisms involved in CCA have been well documented (1,7). Nicotine was reported to inhibit CCA by blocking retinal synthesis (5,6). Retinal was shown to be present in *Fremyella diplosiphon* and *Calothrix sp.* cells. In this report we focus on direct retinal binding to cell extracts of *Calothrix*.

2. Materials and Methods:

Cell extracts were prepared by breaking the cells by passage through a french pressure cell twice at 20,000 psi. Unbroken cells were removed by centrifugation.

Retinal identification was performed according to (8) using cell extract. Cells were harvested from a 20 l vessel (OD_{750} 0.25-0.5).

In the retinal binding studies, two cuvettes are filled with the same cell extract or -fraction, and a control differential spectrum from 350 to 750 nm is taken (Aminco DW2000). While the cuvettes are still in the apparatus, 10 μl retinal solution (in ethanol) is added to the measuring-cuvette, and simultaneously 10 μl ethanol to the reference cuvette. Immediately a number of scans are taken.

P. Mathis (ed.), Photosynthesis: from Light to Biosphere, Vol. I, 303–306.

The control is subtracted from a subsequent scan. The time between retinal addition and the first spectrum acquired is 50 sec at the 540 nm point.
Membrane preparations were obtained by centrifugation of cell extracts for 90 min. at 120,000 g. Membranes were further purified by sucrose gradient fractionation (40-60%). The green fraction was collected and sucrose was removed by dialysis. The purified membranes were solubilized by treatment with detergent. Out of a range of detergents and -concentrations tested, Zwittergent 12 at a concentration of 10:1 detergent/Chl a (w/w) proved to efficiently dissolve membranes without interfering with the retinal binding. Solubilized membranes were separated on a sucrose gradient (7-25%). Ten fractions of equal volume were pumped out.

3. Results:

3.1 *Retinal presence:*
Retinal could be identified in cells of *Calothrix* (fig.1) and also in *Fremyella* (5,6). The actual amount of retinal found (between 100 and 3000 molecules per cell) could be regarded as an indication for a receptor function.

3.2 *Retinal binding:*
Absorbance changes as a result of retinal binding were monitored with membrane solutions of two cyanobacterial species: *Calothrix sp.* and *Oscillatoria agardhii*. A characteristic 540 nm peak appears only in *Calothrix* (fig.2). The retinal binding experiment was also performed with cells from an orange- and a green light grown culture, and during 10 days of adaptation of these cultures to the opposite light colour.

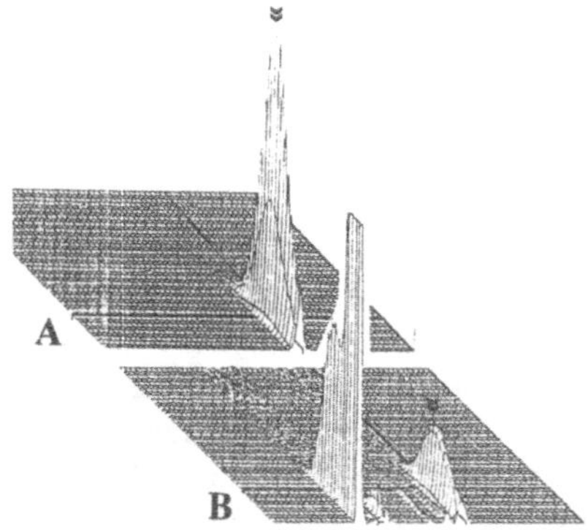

Fig.1: HPLC 3-D-plots of pure retinal (A) and extract of *Calothrix* cells (B).
A: Abs.max.: 382 nm; ret.time: 9,3 min
B: Abs.max.: 382 nm; ret.time: 9,1 min

Adaptation from green to orange light yields gain of a 540 nm peak, and a decrease of a 660 nm peak. Reversed adaptation renders the opposite effect (Fig.3).

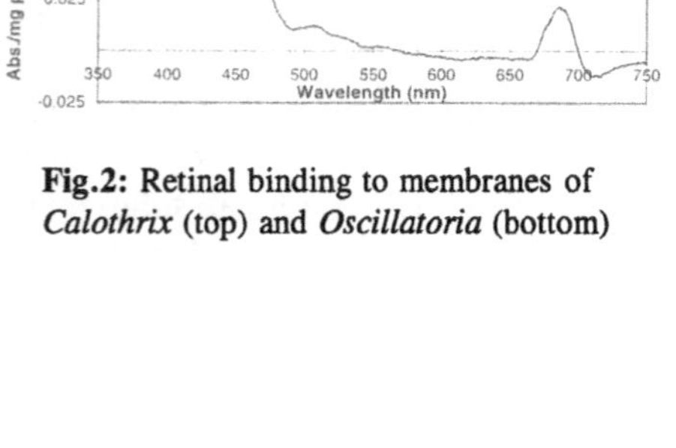

Fig.2: Retinal binding to membranes of *Calothrix* (top) and *Oscillatoria* (bottom)

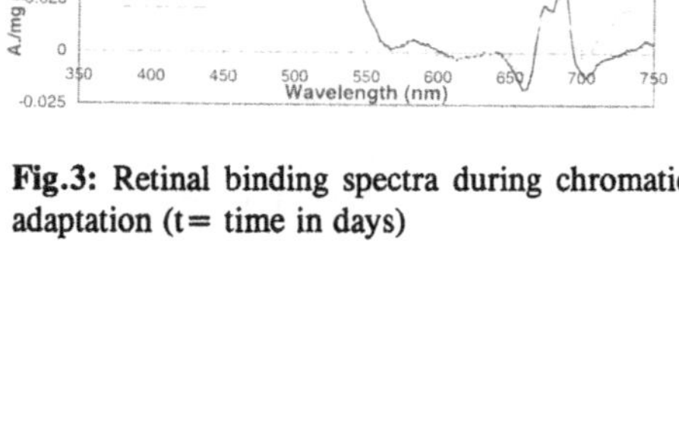

Fig.3: Retinal binding spectra during chromatic adaptation (t= time in days)

The development of the specific peaks appeared to be very quick (fig.4): within the first spectrum taken after addition (= 50 sec. at 540 nm). Although peak quantification was not performed, varying extract concentration may point to a linear dependence of the 540 nm signal to protein concentration. Varying the retinal concentration may indicate a non-linear relation (fig.5, not all data shown). (Note: the actual absolute absorption difference is very small (range -0.005 to 0.015 AU) with respect to the absorbance of the membranes proper (range 0.5 - 1 AU)).

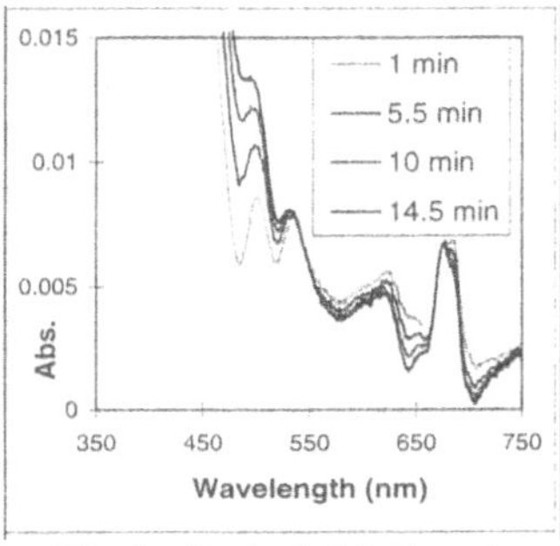

Fig.4: Retinal binding in time (cells grown in orange light, 10 μM retinal added at t=0)

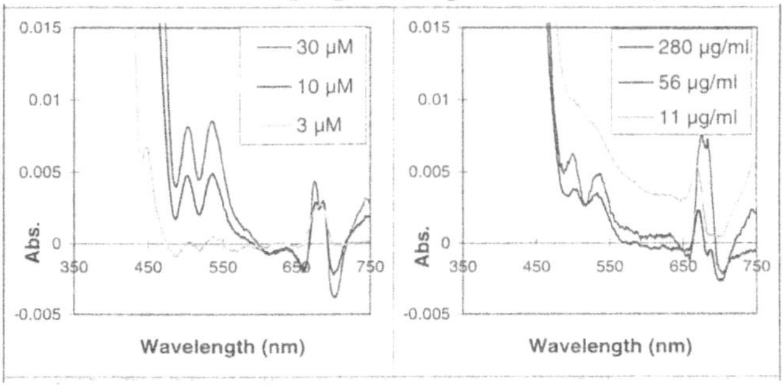

Fig.5: Dependence of retinal binding of retinal (left) and protein (right) concentration (orange light grown *Calothrix* cells)

3.3 *Purification of retinal binding protein:*

The first step in the purification was the identification of the putative retinal binding protein as a membrane protein (fig.6). After a pre-purification using a sucrose gradient and subsequent detergent treatment, the solubilised membranes were fractionated on a second sucrose gradient (7-25%). The retinal binding protein (i.e. the 540 nm peak) could be localised in a specific fraction (10-12% sucrose) (fig.6).

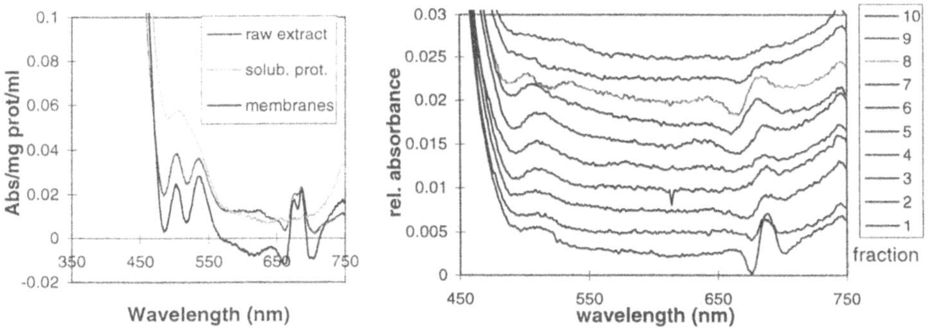

Fig.6: Retinal binding in cell fractions (left) and in membrane (sucrose) fractions of orange light grown *Calothrix* (total protein about 30 μg/ml in all sucrose fractions)

4. Discussion:

Retinal identification in *Calothrix sp.* indicated the possible involvement of a rhodopsin in light-colour signalling. Such a role was earlier suggested by reconstitution of CCA in nicotine inhibited cells of *Fremyella*, through addition of retinal (5,6). Retinal addition to broken cell preparations yielded typical spectral changes in different cyanobacteria. However, a specific peak with its maximum at 540 nm was only seen in *Calothrix sp.*. Furthermore, this 540 nm complex becomes visible only in the orange light adapted cultures, and not in the green. When adapting cultures to the other light colour, this pattern adapts too. The formation of the retinal complex happens in less than one minute. Subsequent absorption changes seem to be less specific. The amount (peak area) may appear linear with the membrane concentration, but not with the retinal amount added. The maximum for the green light response in the action spectrum for complementary chromatic adaptation (5) is 540 nm, just as our product peak. Peaks matching the orange light response maximum (660 nm) were also seen, but these seem unspecific, appearing in more than one organism, adaptation state and fraction. Also this wavelength area gives some trouble in measurements because of small chlorophyll absorbance changes between cuvettes. However, both the 540 and 660 nm peak areas respond to light colour changes and in opposite directions. The retinal binding protein could be localised in the membranes, and after detergent treatment (Zwittergent 12) and sucrose gradient purification in a specific fraction (10-12%).

In conclusion: A retinal-protein complex absorbing light in a region around 540 nm is probably involved in light colour sensing and subsequent triggering of complementary chromatic adaptation. This protein appears to be membrane bound. It has been partially purified by detergent solubilisation and sucrose gradient fractionation. Ongoing experiments include ^{3}H-retinal labelling and subsequent SDS-PAGE and autoradiography. Several major questions concerning the putative rhodopsin, including the non-characteristic small half band-width (about 30 nm) of the absorbance peak in the 540 nm spectral region are also topic of ongoing research.

5. References:

1) Tandeau de Marsac, N., Houmard, J. (1993) FEMS Microbiol. Rev. **104**, 119-90

2) Scanlan, D.J., Mann, N.H. and Carr, N.G. (1993) Mol. Microbiol. **10**: 181-191

3) Diakoff, S., Scheibe, J. (1972) Action Spectra for Chromatic Adaptation in *Tolypothrix tenuis* Plant Physiol. **51**, 382-385

4) Björn, L.O., Björn, G.S. (1980) Photochem. Photobiol. **32**, 849-552

5) Hoff, W.D., Matthijs, H.C.P., Schubert, H., Crielaard, W., Hellingwerf, K.J. (1995) Biophys. Chem. **56**, 193-199

6) Hellingwerf, K.J., Crielaard, W., Hoff, W., Matthijs, H.C.P., Mur, L.R., van Rotterdam, B.J. (1994) Antonie van Leeuwenhoek **65**, 331-347

7) Grossman, A.R., Schaefer, M.R., Chiang, G.G., Collier, J.L. (1993) Microbiol. Rev. **57**, 725-49

8) Gualtieri, P., Pelosi, P., Passarelli, V., Barsanti, L. (1992) Biochim. Biophys. Acta **1117**, 55-59

Andrea Haker (permanent adress: Universität Rostock, FB Biologie, Freiligrathstr. 7/8 18057 Rostock, Germany.) was supported by a COMETT scolarship.

ENERGY TRANSFER IN THE PHOTOSYNTHETIC ANTENNA SYSTEM OF THE PURPLE NON-SULFUR BACTERIUM *RHODOPSEUDOMONAS CRYPTOLACTIS*.

H. Kramer [1], G. Deinum [1], A.T.Gardiner [2], R.J. Cogdell [2], C.Francke [1], T.J. Aartsma [1], J.Amesz[1].
[1] Department of Biophysics, University of Leiden, P.O. Box 9504, 2300 RA Leiden, the Netherlands and [2] Department of Botany, University of Glasgow, G12 8QQ, Glasgow, U.K.

1. Introduction

Rhodopseudomonas cryptolactis [1] is able to synthesize more than one spectral type of peripheral antenna complex. At low light intensity it synthesizes B800-820, at high intensity B800-850 and a mixture of both at intermediate light intensity (Fig. 1). This offered us the opportunity to see whether the conclusions of our study on the energy transfer properties of *Rps. acidophila* [2] were generally applicable.

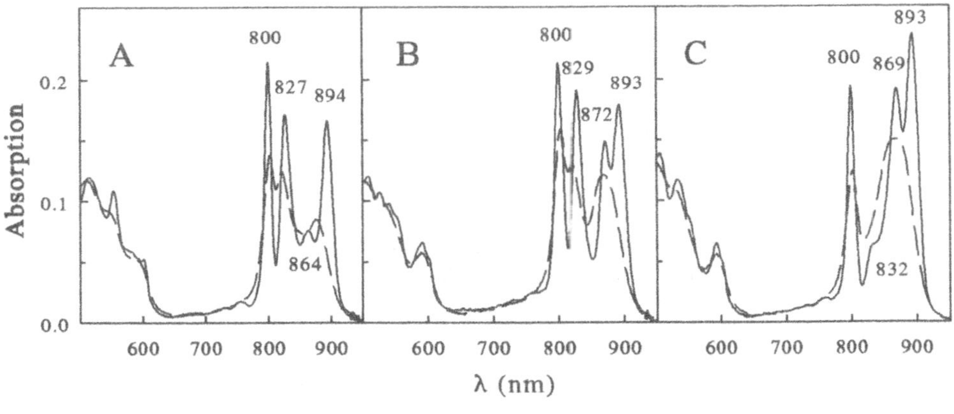

Figure 1. Absorption spectra of membranes of Rps. cryptolactis grown at low (A), intermediate (B) and high (C) light intensity, at 6 K (solid line) and 298 K.

2. Materials and methods

Membrane fragments of *Rhodopseudomonas cryptolactis* were suspended in a buffer containing 20 mM Tris (pH=8.0) and 30 % glycerol (v/v). The fluorescence emission and annihilation measurements were performed under anaerobic conditions. For the transient absorption measurements ascorbate (40 mM) and PMS (20 μM) was added in order to keep the primary donor in the reduced state. For the fluorescence measurements

P. Mathis (ed.), Photosynthesis: from Light to Biosphere, Vol. I, 307–310.

the sample was excited either by a frequency doubled Nd-YAG laser with a maximum pulse energy of about $5mJ/cm^2$ at 532 nm and a pulse duration of approximately 25 ps or by a low intensity xenon flash with filters providing a band width of 40 nm, centred at 520 nm. Transient absorption measurements were performed with a synchronously pumped dye laser in a one colour pump-probe configuration. The angle between the polarisation of pump and probe beams was set at the magic angle of 54.7^0 by means of a Glan-Thomson polarizer. The instrumental time response was 4-6 ps.

2. Results

Variable fluorescence measurements (Fig. 2), revealed no significant increase of the fluorescence yield in the region below 830 nm indicating that back transfer of energy to the B800-B820 complex is severely restricted. The increase of fluorescence from B800-850 was somewhat lower than from B880 which can be the case if energy equilibration between the two complexes is not extremely fast. This confirms the notion that only the B880 complex is in direct contact to the reaction centre complex.

From emission spectra at high intensity we concluded that the extent of annihilation in B880 exceeds that in the peripheral antennae. This is due to a rapid energy transfer from peripheral to core antenna.

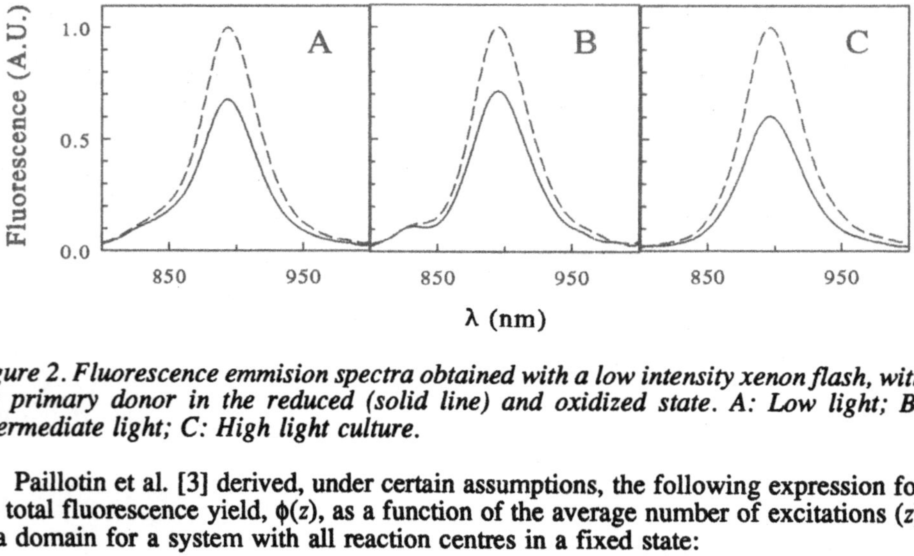

Figure 2. Fluorescence emmision spectra obtained with a low intensity xenon flash, with the primary donor in the reduced (solid line) and oxidized state. A: Low light; B: Intermediate light; C: High light culture.

Paillotin et al. [3] derived, under certain assumptions, the following expression for the total fluorescence yield, $\phi(z)$, as a function of the average number of excitations (z) in a domain for a system with all reaction centres in a fixed state:

$$\phi(z) = \phi_o \left(1 + \sum_{k=1}^{\infty} \frac{1}{k+1} \frac{(-z)^k}{(r+1)\ldots(r+k)} \right) \qquad (1)$$

r denotes the ratio of the mono-excitation decay rate and the bi-excitation decay rate. ϕ_0 describes the fluorescence yield at low excitation energies. By comparing the calculated curves with measurements of the fluorescence yield versus the pulse intensity, we obtain information about the efficiency of annihilation (r) and about the number of BChls between which energy transfer can occur (domain size). The energy transfer efficiencies needed for analyzing the data were obtained from fluorescence excitation measurements. We used extinction coefficients at the Q_y maxima of 140 $mM^{-1}cm^{-1}$ for the core BChls and 184 $mM^{-1}cm^{-1}$ for the peripheral BChls.

The annihilation curves recorded at 930 nm (Fig. 3) could be fitted with r=1.3, r=2.5 and r=5 for the low light, intermediate light and high light culture, respectively. However, most of the excitations will reach B880 by energy transfer from the peripheral complexes. This fraction is difficult to determine exactly, but as judged from the absorption spectra it will be more than 60% for all three cultures. This means that annihilation processes in B800-820 and B800-850, albeit less than in B880, cannot be neglected. We therefore determined the annihilation curves for fluorescence emitted by the peripheral antennae and applied a correction to the annihilation curves at 930 nm.

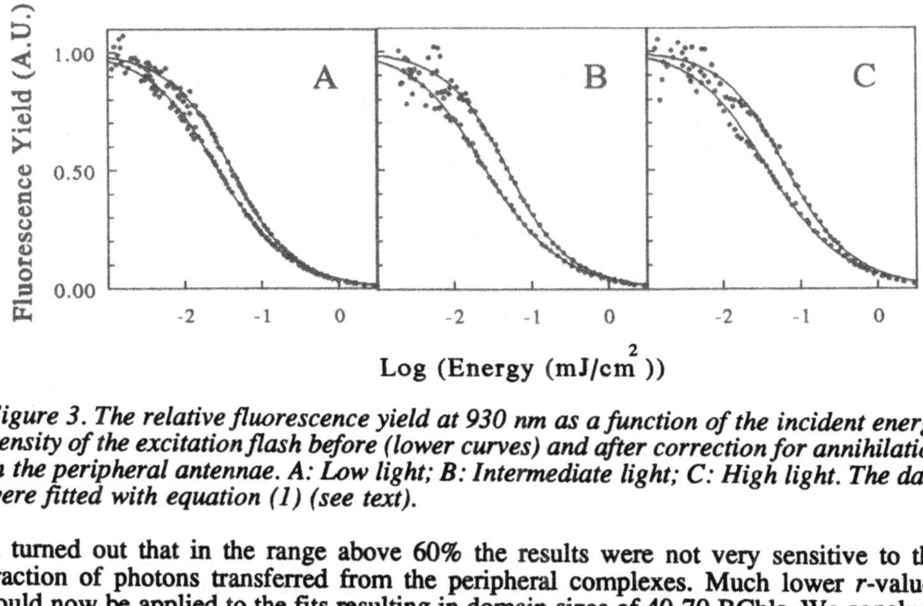

Figure 3. The relative fluorescence yield at 930 nm as a function of the incident energy density of the excitation flash before (lower curves) and after correction for annihilation in the peripheral antennae. A: Low light; B: Intermediate light; C: High light. The data were fitted with equation (1) (see text).

It turned out that in the range above 60% the results were not very sensitive to the fraction of photons transferred from the peripheral complexes. Much lower r-values could now be applied to the fits resulting in domain sizes of 40-70 BChls. We conclude that the B880 antenna complex consists of relatively small units of interconnected BChls in all three cultures. The size of these units does not appear to be much larger than the number of BChls associated with one reaction centre.

One colour transient absorption measurements with the reaction centres in the oxidized state are shown in Fig. 4A-D. The decays of the initial bleaching are ascribed to energy transfer to B880 and varied from 12 to 15 ps. The increase of the absorbance is ascribed to excited state absorption of the core antenna and the subsequent decay to energy transfer to closed reaction centres (about 200 ps). A minor long lived kinetic component was observed at all wavelengths. Fig. 4E-F shows the transient absorption signals obtained at 900 nm for reaction centres in the oxidized and in the reduced state, showing time constants of 170 and 47 ps for trapping by closed and open reaction centres, respectively.

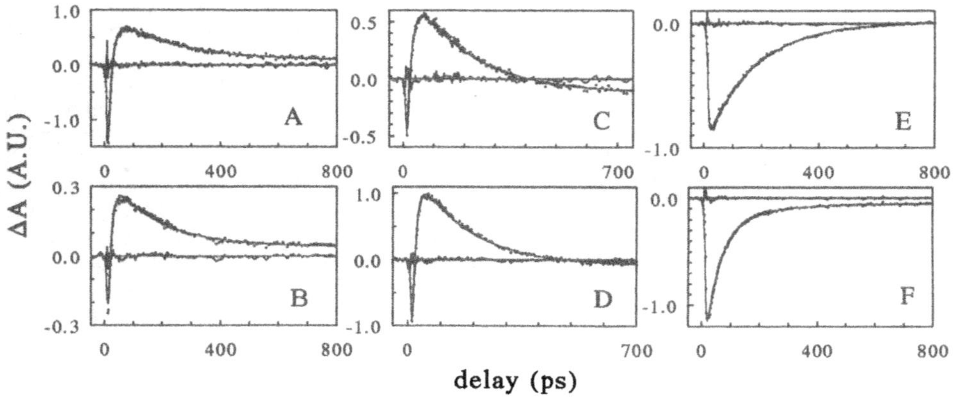

Figure 4. One colour time resolved transient absorption measurements with the primary donor in the oxidized state, measured at 820 nm (A: low light, B: intermediate light culture), at 864 nm (C: intermediate light; D: high light) and at 900 nm (E: intermediate light). Panel. F shows the decay at 900 nm with open reaction centres.

3. Conclusions

Like in *Rps. acidophila*, the excitation energy can be transferred from B880 to B800-850, but not to B800-820. In contrast, however, to *Rps. acidophila*, the domains for energy transfer for B880 appear to be about the same for the three cultures studied and to contain one or two core-reaction centre complexes only. The model proposed for *Rps. acidophila*, where the B880 complexes are energetically connected by large arrays of B800-850 complexes [2] therefore does not appear to apply to *Rps. cryptolactis*. It has to be replaced by one where each B880 unit is associated with its own complement of B800-850, without an extensive interconnecting network of B800-850 complexes.

Our time resolved absorption measurements revealed trapping rates of $(50\ ps)^{-1}$ and $(170\text{-}200\ ps)^{-1}$ for open and closed reaction centres, respectively. The time constant of energy transfer from B800-850 to B880 was 12-15 ps. The time constant of energy transfer from B800-820 to other complexes is almost the same (15 and 12 ps) in the low and in the intermediate light culture. This indicates that energy transfer to B880 occurs at the same rate as to B800-850, in spite of the different overlap integrals.

References

1. Stadtwald-Demchick, R., Turner, F.R. and Gest, H. (1990) FEMS Microbiol. Lett. 71, 117-122
2. Deinum, G., Otte, S.C.M., Gardiner, A.T. Aartsma, T.J., Cogdell, R.J. and Amesz, J. (1991) Biochim. Biophys. Acta 1060, 125-131
3. Paillotin, G., Swenberg, C.E., Breton, J. and Geacintov, N.E. (1979) Biophys. J. 25, 512-533

PROTONATION OR CHEMICAL MODIFICATION OF A His RESIDUE ELICITS COMPLETE BLUE SHIFT OF LH2 800-nm BACTERIOCHLOROPHYLL

J. M. Ramírez and C. Fernández-Cabrera. Centro de Investigaciones Biológicas, Velázquez 144, 28006 Madrid, Spain

1. Introduction

The light-harvesting protein 2 (LH2) of Rhodobacteriaceae is an intrinsic pigment-protein complex of the photosynthetic membrane that contains two short hydrophobic polypeptides (α and β), bacteriochlorophyll (Bchl), and a carotenoid in the molar ratio of 1:1:3:1.5 (1) (or 1:1:3:1?, see 2, 3). The crystal structure of *Rhodopseudomonas acidophila* LH2, that has been recently reported (3), shows that both polypeptides are mainly α-helical and have transmembrane orientations. Two spectrally distinct Bchl populations, with Q_y ($S_0 \rightarrow S_1$) transitions near 800 and 850 nm respectively, correspond to two fractions of the pigment in different environments (3). While the macrocycles of the 850-nm Bchls are within the hydrophobic core of the membrane and lie perpendicular to its plane, those of the 800-nm Bchls are also intramembrane located, but parallel to the membrane, closer to its cytoplasmic surface and in a relatively polar environment. The carbonyl O of the N-terminal formyl of the α polypeptide seems to be bound to the central Mg of 800-nm Bchl as an axial ligand. This pigment is also H-bonded by its 2-acetyl carbonyl to an Arg residue of the β subunit. A water molecule and a βHis residue are also present in the binding site of the pigment.

It will be shown in the present report that the protein unfolding reagent, urea, exposes to the solvent the binding site of 800-nm Bchl in *Rhodobacter capsulatus* membrane-linked LH2, since the location and shape of the 800-nm Bchl band becomes pH-dependent in the presence of high urea concentrations (above 5 M). In contrast, the Q_y band at 850 nm remains unchanged. The p*K* of the urea-dependent acid-base transition suggests that the protonation of βHis20 causes the loss of the native red shift of the band, a conclusion consistent with the presumable proximity of such a residue to 800-nm Bchl (3).

2. Procedure

The strain BY1424 of *R. capsulatus*, a nonphototrophic mutant that lacks LH1 and the photochemical reaction center (4), was grown in the dark and in the absence of air as previously described (5). Intracytoplasmic membrane vesicles isolated from this strain, which contain only LH2 as the major Bchl-protein constituent, were used throughout this work as preparations of membrane-linked LH2. Solubilized LH2 preparations were obtained by extracting membrane vesicles isolated from strain BY1424 with dodecyldi-

P. Mathis (ed.), Photosynthesis: from Light to Biosphere, Vol. I, 311–314.

methylamine-*N*-oxide (LDAO), using a procedure which will be described elsewhere.

Urea (8 M) stock solutions in mixtures of 50 mM potassium phosphate plus 50 mM Tris-HCl, adjusted to several pHs, were stored frozen in glass tubes until use. The assay mixtures contained intracytoplasmic membrane vesicles or solubilized LH2 (A^{850} ranging from 6 to 12 cm^{-1}) and buffered urea at the appropriate concentration and pH. The mixtures were incubated at 22 °C. Their absorption spectra were obtained at the same temperature using cells of 1- or 0.5-mm optical path in the secondary sample compartment of a double beam spectrophotometer (Shimadzu, UV-3000), equipped with a S1 photomultiplier.

3. Results and Discussion

When preparations of LH2-containing membrane vesicles were incubated in urea above 5 M at neutral or acidic pHs, a blue shift of the 800-nm Bchl band to about 770 nm was observed (Fig.1). In contrast, the Q_y band of the major Bchl fraction, located near 850 nm, remained largely unaltered. As shown also in Fig. 1, the extent of the band shift was reduced at basic pHs, indicating that it was related to the protonation of a group of p*K* close to 7. Besides, the pH-dependent band shift did not occur in the absence of urea and happened only to a limited extent when urea concentration was kept below 4 M (Fig. 2). These observations suggest that the protonable group that was associated to the band shift only equilibrated with the solvent when the 800-nm Bchl region of membrane-linked LH2 was unfolded by urea. The spectral alterations caused by acidic urea were reversible, since the native 800-nm band was almost completely restored when urea concentration was lowered by dilution and the pH was raised above 8.0; if urea was diluted out at more acidic pHs, the reversion was slower and incomplete (not shown).

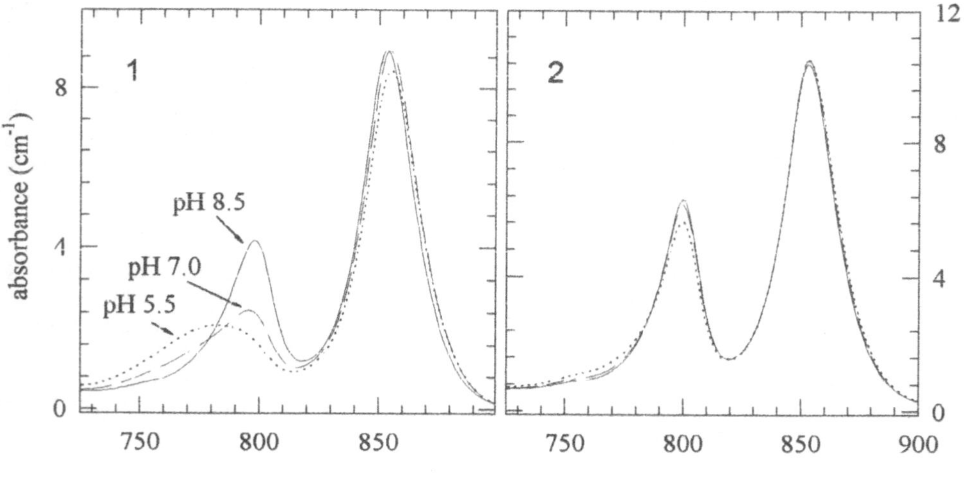

Figure 1. Effect of pH on the near infrared absorption spectra of LH2-containing membrane vesicles after 50 min in the presence of 6.4 M urea.

Figure 2. Near infrared spectra of membrane vesicles suspended in 1.2 M urea for 50 min, at the same pHs of Fig. 1.

Concentrated urea had a different effect on detergent-solubilized LH2. As shown in Fig. 3, incubation in 7 M urea at neutral pH resulted in the slow attenuation of both the 800- and the 850-nm bands and in the appearance of a broad band near 770 nm. The two bleachings and the colouring at 770 nm were correlated, suggesting that the observed changes mainly reflected the simultaneous shifts to 770 nm of both the 800- and the 850-nm transitions (Fig. 4). And since only about 20 % of those spectroscopic changes could be reversed by tenfold urea dilution (not shown), they seem likely to mainly arise from LH2 dismantling. The regression lines of Fig. 4 do not go through the origin, indicating early changes at 770 and 800 nm that were not accompanied by a bleaching at 850 nm. At mildly acidic pHs, the irreversible changes were faster than at pH 7, and at pH 5.5 both native transitions quickly shifted to 770 nm, even in the absence of urea (not shown). Thus, we were unable to elicit a selective and reversible blue shift of the 800-nm component of solubilized LH2 similar to that observed in membrane-linked LH2. Such differences between the micellar and the membrane-bound forms of LH2 indicate that the detergent fails to provide a protein environment similar to that provided by the membrane. The higher stability of the protein region located within the hydrophobic membrane core (11) -where 850-nm Bchl is bound (3)- is the likely reason why the membrane-linked protein did not dismantle when the more polar region which binds 800-nm Bchl was unfolded by urea (Fig. 1). Apparently, the detergent was unable to stabilize the 850-nm Bchl region to a comparable extent.

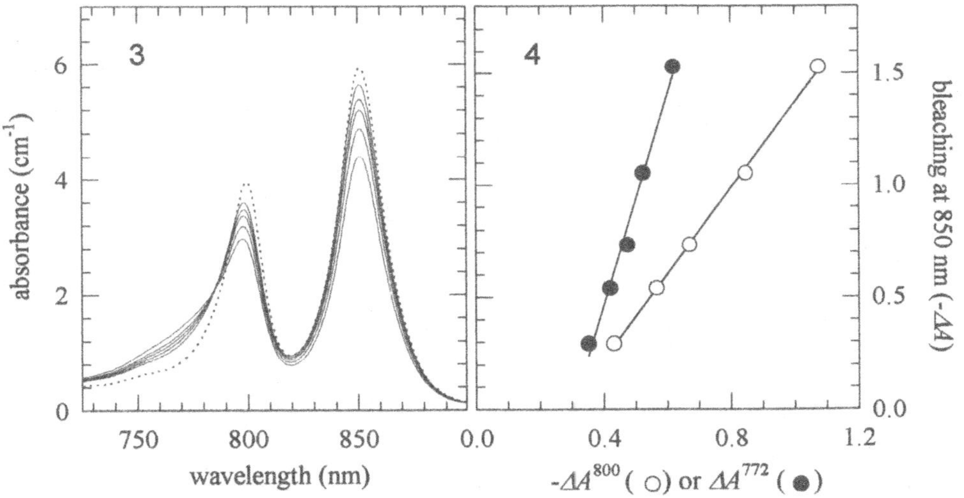

Figure 3. Near infrared absorption spectra of solubilized LH2 in the presence of 7 M urea at pH 7. Consecutive spectra (solid lines) were scanned just, 10, 20, 30 and 40 min after urea addition. The spectrum of LH2 in the absence of urea is also shown (dots).

Figure 4. Correlation among the changes of intensity of the Q_y bands (data of Fig. 3).

If it is assumed that the structure of membrane-linked *R. capsulatus* LH2 is similar to the reported crystal structure of the *R. acidophila* protein (3), only βHis20, among the residues which are located in the 800-nm Bchl region of the protein, may have a p*K* close to 7, since the terminal α and β Met residues are blocked and there are no Cys residues in either subunit (6). In accordance with the involvement of a His residue in the urea-, pH-dependent blue shift of the 800-nm band, it was observed that the shift was enhanced and made irreversible at basic pHs by the His modifying reagent, diethylpyrocarbonate (J.M. Ramírez, unpublished). In *R. acidophila* LH2 the corresponding βHis residue, at position 12, is not bonded to 800-nm Bchl, although it is close to the putative formyl group that coordinates Bchl Mg and to the buried water molecule that is adjacent to the pigment (3). It seems therefore that, although the βHis20 residue of *R. capsulatus* is not likely to be directly involved in pigment binding, its protonation or chemical modification disturbs the native environment of 800-nm Bchl. It should be mentioned here that replacement of the homologous βHis residue by Ser in a *R. sphaeroides* mutant led to a LH2 protein that lacked the 800-nm Bchl component (7).

While the Q_y transition of monomeric Bchl *a* in solution is a broad band centered around 770 nm, the protein-bound pigment exhibits Q_y transitions which are narrower and red shifted by several tens of nm. It has been proposed that H-bonds between the coupled carbonyls of Bchl on one side and donor groups of the polypeptide residues on the other contribute to the native red-shifts of protein-bound Bchls (8-10). Current experimental evidence is in agreement with this proposal for the 800-nm Bchl component of LH2. Thus, while in the *R. acidophila* protein the 2-acetyl carbonyl of the pigment is H bonded to a conserved βArg residue (8), a *R. sphaeroides* mutant in which the homologous Arg was changed into Glu contained an altered '800-nm' Bchl with a Q_y band very similar to that of the free pigment, i.e., the transition had lost the native red shift and resembled therefore the blue-shifted band that results when the membrane-linked *R. capsulatus* protein is treated with urea at mildly acidic pHs (Fig.1). These similarities suggest that protonation (or chemical modification) of *R. capsulatus* βHis20 in the presence of urea interferes with the putative H bond between the conserved βArg28 residue and the 2-acetyl carbonyl of the pigment.

References

1 Evans, M.B., Cogdell, R.J., and Britton, G. (1988) Biochim. Biophys. Acta 935, 292-298
2 Shiozawa, J.A., Welte, W., Hodapp, N., and Drews, G. (1982) Arch. Biochem. Biophys. 213, 473-485
3 McDermott, G., Prince, S.M., Freer, A.A., Hawthornthwaite-Lawless, A.M., Papiz, M.Z., Cogdell, R.J., and Isaacs, N.W. (1995) Nature 374, 517-521
4 Scolnik, P.A., Walker, M.A., and Marrs, B.L. (1980) J. Biol. Chem. 255, 2427-2432
5 Zurdo, J., Fernández-Cabrera, C., and Ramírez, J.M. (1993) Biochem. J. 290, 531-537
6 Brunisholz, R.A. and Zuber, H (1992) J. Photochem. Photobiol. B: Biol. 15,113-140
7 Visschers, R.W., Crielaard, W., Fowler, G.J.S., Hunter, C.N., and Van Grondelle, R. (1994) Biochim. Biophys. Acta 1183, 483-490
8 Callahan, P.M. and Cotton, T.M. (1987) J. Am. Chem. Soc. 109, 7001-7007
9 Fowler, G.J.S., Sockalingum, G.D., Robert, B., and Hunter, C.N. (1994) Biochem. J. 299, 695-700
10 Sturgis, J.N., Jirsakova, V., Reiss-Husson, F., Cogdell, R.J., and Robert, B. (1995) Biochemistry 34, 517-523
11 Haltia, T. and Freire, E. (1995) Biochim. Biophys. Acta 1228, 1-27

LIGHT INDUCED REVERSIBLE STRUCTURAL CHANGES IN CHIRALLY ORGANIZED MACROAGGREGATES OF PURIFIED CHLOROPHYLL A/B PROTEIN COMPLEXES

V. Barzda[1], I. Simidjiev[1], A. Vianelli[2], R. Jennings[2] and G. Garab[1]
[1]Institute of Plant Biology, Biological Research Center, Hungarian Academy of Sciences, Szeged, P.O. Box 521, H-6701 Hungary, and
[2]Dipartimento di Biologia, Universita di Milano, Via Celoria 26, 20133 Milano, Italy

1. Introduction

In chloroplasts, the large majority of the pigment molecules belong to the antenna system, which supply the reaction centers with excitation energy. The efficiency of the energy migration toward the reaction centers and the quantum yield of charge separation are largely determined by the molecular architecture of the pigment system. The structure of the antenna and the reaction centers appears to be optimized for an operation with a nearly 100% quantum yield. On the other hand, it has been demonstrated that chloroplasts possess protective mechanism(s) by the aid of which the membranes are capable to dissipate substantial amounts of the excess light energy absorbed by photosystem II (PSII) [1].

As shown by fluorescence induction, after brief exposures of membranes to excess radiation the system continues to dissipate the excitation energy even in low light, and regains the efficient utilization of the absorbed quanta only gradually in about a minute. This strongly suggests that structural changes are involved in the tuning of photophysical pathways in the membrane. However, the nature and the physical mechanism of the putative structural changes associated with the regulation of utilization and dissipation of the excitation energy in the pigment system have not been elucidated.

By measuring light-induced CD changes (ΔCD) in thylakoid membranes, earlier we have shown that granal thylakoid membranes are capable to undergo light-induced reversible changes in their macrodomains possessing long-range chiral order [2]. Our recent investigations (Istokovics et al., this Proceedings) revealed that these reversible structural changes are associated with reversible changes in the fluorescence yield. In other terms, the light-induced changes in the chiral macroorganization appear to bring about alterations in the photophysical pathways (an increased dissipation in excess light). Further, it was shown that the changes could be attributed to the inherent structural flexibility of the chiral macrodomains. On the other hand, the light-harvesting Chl a/b protein of PSII (LHCII) has been shown to play a crucial role in the organization of macrodomains in the membrane. Purified LHCII complexes also readily form chiral macroaggregates [3]. These facts and the characteristics of light-induced

P. Mathis (ed.), Photosynthesis: from Light to Biosphere, Vol. I, 315–318.

reversible changes in the membrane imply that LHCII macroaggregates are directly responsible for the light-induced structural rearrangements in the membranes.

Here we show that macroaggregates of purified LHCII with long-range chiral order are indeed capable of undergoing light-induced, reversible structural changes, which can be detected as changes in their main circular dichroism bands (ΔCD). These changes are associated with reversible changes in the fluorescence yield. Detailed characterization of these reversible structural changes revealed many similarities those in thylakoid membranes. It is proposed that long-range energetic coupling inside the macroaggregates is responsible for the occurrence of the changes in the chiral macroorganization and the photophysical pathways in the antenna system.

2. Experimental Procedures

LHCII was isolated from two-week-old pea (*Pisum sativum L.*) leaves according to the method of [4] with minor modifications. For obtaining LHCII macroaggregates exhibiting ΔCD, the concentrations of Triton X-100 was varied between 0.6 and 0.8% (v/v), the optimum depended on the age and growth conditions of the leaves. The Chl a/b ratio varied between 1.0 and 1.15. In these preparations SDS-PAGE revealed the presence of the 25 and 27 kDa apoproteins. Minor complexes could also be detected by silver staining, but were present in very small amounts. As judged from Western blots for the D1 protein, the samples contained no contamination from PSII reaction centers.

CD spectra were recorded in a Jobin Yvon CD6 dichrograph. The dichrograph was equipped with a side-illumination attachment; the photomultiplier was protected with a Corning CS 4-96 bandpass glass filter against actinic red light (a Corning 2-64 cut off filter) from the beam of a 650 W tungsten lamp. The actinic beam also passed through a heat filter of 10 cm water. For measurements above 620 nm the two filters were exchanged.

Fluorescence induction kinetic measurements on thylakoid membranes and LHCII-aggregates were performed in a pulse-amplitude modulation Chl-fluorimeter (Walz, PAM 103), or in a fluorimeter described in [5]. The Chl concentration of LHCII was adjusted to 5 μg/ml and 20 μg/ml for fluorescence and CD measurements, respectively.

3. Results and Discussion

Upon illumination, macroaggregates of LHCII exhibit significant changes in the psi-type CD bands (Fig. 1). (Psi, polymer or salt-induced; cf. [6].) Ultrastructural investigations revealed that in all cases when ΔCD was observed, LHCII was found to be assembled into stacked lamellar aggregates (data not shown, cf. [7]). On the other hand, ΔCD could easily be eliminated by disintegrating the aggregates into smaller particles of 100-200 nm in diameter (Fig. 1). (The state of aggregation of the sample was judged by negative staining electron microscopy and by measuring the fluorescence emission spectra [8].)

We found that ΔCD was not confined to a particular set of dipoles but the changes could be detected in all CD bands with psi-type features. On the other hand, it was shown that ΔCD did not affect the excitonic CD bands. Thus, it can be concluded that the observed changes reflect overall structural rearrangement rather than changes in the individual trimers. In perfect agreement with this notion, the light-minus-dark difference spectrum was dominated by anomalously shaped, i.e. non-conservative, broad bands, which are typical for psi-type aggregates (cf. [6]) and cannot be accounted for an origin of short-range interactions (data not shown).

Temperature-dependence of ΔCD of LHCII macroaggregates (not shown), which was similar to that observed in thylakoid membranes (Istokovics et al., this Proceedings) provides further support to the conclusion that ΔCD originates from structural changes affecting the chiral macroorganization of the complexes.

As shown in Fig. 2, the initial rate of the light-induced changes of CD is linearly proportional with the intensity of the exciting light, without any indication for saturation. These data clearly show that the changes are driven by light. It must be noted, however, that at the light intensities used in our experiments and with the lifetime of the excited state of the Chl molecules, the concentration of the excitations in the macroaggregate at any instant is extremely low compared to the concentration of Chl molecules. This, and the disappearance of ΔCD in small aggregates, strongly suggest that the effect of light must concern a collective property of the macroaggregates (cf.[6], see also Barzda, V., Garab, G. Gulbinas, V. and Valkunas, L., in this Proceedings.)

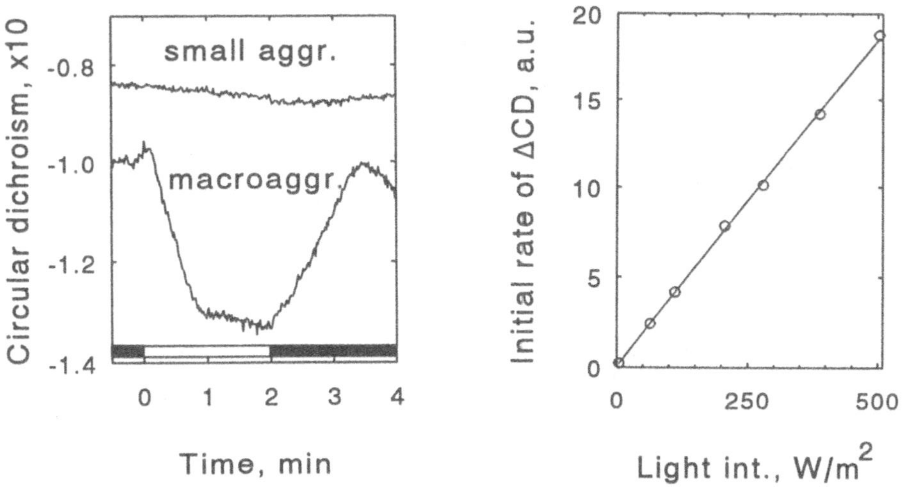

Figure 1. Light induced CD changes at 495 nm in macroaggregates and small aggregates of LHCII. Illumination was with an actinic light of 500 W/m^2 (open bar).

Figure 2. Dependence of the initial rate of the light induced CD changes on the intensity of actinic light. CD changes were monitored at 495 nm.

Light-induced reversible changes in fluorescence yield in LHCII was observed by Jennings and coworkers, who suggested that they play a role in dissipating excess excitation energy [5]. For correlating our ΔCD measurements with these light-induced alterations in the photophysical pathways, we studied fluorescence transients under essentially the same conditions as in ΔCD measurements. Although, the reversible fluorescence changes were not completely eliminated upon disintegration of LHCII macroaggregates, the relative contribution of the reversible changes diminished considerably. This shows that the reversible structural changes (ΔCD) are associated with reversible alterations in the photophysical pathways (ΔF). (Detailed comparative analysis of ΔCD and ΔF are in progress and will be reported elsewhere (V. Barzda, G. Zuchelli, L. Finzi, G. Garab and R. Jennings, in preparation).)

Concerning the nature of the changes, in analogy with the hypothesized proximity changes between the stacked thylakoid membranes, we propose that the reversible changes are due to alterations in the stacking of adjacent sheets formed by LHCII.

The physical mechanism of this phenomenon remains to be explored. We propose that long-range energetic coupling inside the macroaggregates with chiral order plays an important role. Excitation of LHCII lamellae may induce changes in the close vicinity of the excited molecules, e.g. via local heat produced during the rapid relaxation processes. This, via energetic coupling may extend to a larger area and thus affect the stacking interactions. It is a close assumption that such changes in the stacking would relax relatively slowly, in accordance with the relatively slow restoration of the CD signal in the dark following the illumination.

Acknowledgements.

This work was supported by an OTKA III/2999 grant (Hungary). V.B. is grateful to ESF, Programme "Biophysics of Photosynthesis" for travel grant.

References

1 Horton, P., Ruban, A.V. and Walters, R.G. (1994) Plant Physiol. 106, 415-420.
2 Garab, G., Leegood, R.C., Walker, D.A., Sutherland, J.C. and Hind, G. (1988) Biochemistry 27, 2430-2434.
3 Garab, G. (1992) in Research in Photosynthesis (Murata, N., ed.), pp. 171-178, Kluwer Academic Publishers, Dordrecht, Boston, London.
4 Krupa, Z., Huner, N.P.A., Williams, J.P., Maissan, E. and James, D.R. (1987) Plant Physiol. 84, 19-24.
5 Jennings, R.C., Garlaschi, F.M. and Zucchelli, G. (1991) Photosynth. Res. 27, 57-64.
6 Keller, D. and Bustamante, C. (1986) J. Chem. Phys. 84, 2972-2979.
7 McDonnel, A. and Staehelin, L.A. (1980) J. Cell Biol. 84, 40-56.
8 Ruban, A.V. and Horton, P. (1992) Biochim. Biophys. Acta 1102, 30-38.

LONG DISTANCE MIGRATION OF THE EXCITATION ENERGY AND FLUORESCENCE QUENCHING MECHANISMS IN CHIRAL MACROAGGREGATES OF LHCII

V. Barzda[1,2], G. Garab[1], V. Gulbinas[2] and L. Valkunas[2]
[1]Institute of Plant Biology, Biological Research Center, Hungarian Academy of Sciences, Szeged, P. O. Box 521, H-6701, Hungary, and
[2]Institute of Physics, A. Gostauto 12, Vilnius 2600, Lithuania

1. Introduction

The role of the light harvesting antenna in photosynthesis is to absorb the solar energy and transfer the excitation energy to the reaction center where charge separation takes place. Under 'normal conditions', virtually no quantum losses occur in the antenna system. On the other hand, considerable evidence has been accumulated for the existence of different regulatory mechanisms which control the dissipation of excess excitation energy in the antenna [1]. The energy dependent non-photochemical quenching, q_E, which is responsible for dissipating a substantial part of the excess radiation, has been proposed to be correlated with the macroaggregation of the light harvesting chlorophyll a/b complex of PSII (LHCII) in the membranes [2].

In thylakoid membranes, LHCII has been shown to mediate the formation of chirally organized macrodomains, three-dimensional macroaggregates with sizes commensurate with the wavelength of the measuring light [3]. Purified LHCII also easily forms chiral macroaggregates and liquid crystalline sheets [4]. As shown by fluorescence measurements, macroaggregation of LHCII creates new channels of energy dissipation [2], the physical mechanism of which remains to be identified.

We carried out a comparative study of the absorbance, circular dichroism (CD), and 77 K and room temperature fluorescence spectra, and picosecond absorbance kinetics under singlet-singlet annihilation conditions in different aggregation states of purified LHCII. Our data provide experimental evidence that chiral macroaggregates of LHCII constitute a structural basis for long-distance migration of the excitation energy. Further, we found that the variations of fluorescence yield are correlated with (i) the size of macroaggregates, and (ii) the intensity of excitonic interactions between Chl molecules in trimers and monomers of LHCII.

2. Experimental Procedures

LHCII was prepared from pea leaves according to [5]. Aggregation and disaggregation of LHCII was controlled by the addition of $MgCl_2$, and Triton-X 100 or n-Dodecyl β-D-maltoside (DM), respectively. Size of macroaggregates was determined in light and electron microscopes.

P. Mathis (ed.), Photosynthesis: from Light to Biosphere, Vol. I, 319–322.

CD spectra were measured in a Jobin-Yvon CD6 dichrograph, and expressed in absorbance units. Fluorescence measurements were performed with a setup described earlier [6]. Determination of exciton-exciton annihilation dynamics was carried out by means of pump-probe absorption spectroscopy, which is described elsewhere [7].

3. Results and Discussion

A comparison of the absorbance spectra of small aggregates (d ≈ 100-200 nm) and large aggregates (d ≈ 4-6 μm) revealed no marked differences (data not shown). Thus, macroaggregation appears not to bring about new excitonic interactions between pigment molecules. These interactions are contained in the trimers of LHCII [8].

In contrast with the absorbance, formation of macroaggregates resulted in dramatic changes in the CD, the appearance of psi-type CD bands, the amplitude of which has been shown to depend on the size of macroaggregates [9].

Bleaching kinetics due to singlet-singlet annihilation in LHCII show that both the macroaggregates (A) and the small aggregates of LHCII trimers (B) display nonlinear kinetics (Fig. 1), i.e. dependence of decay on the intensity of the exciting pulse but non-linearity is more pronounced in the macroaggregates.

Analysis of the traces were carried out by means of the following equation:

$$\frac{dn}{dt} = I(t) - kn - \gamma(t)n^2, \qquad (1)$$

where n is the concentration of excitations of Chl *a* in domains, k is the excitation decay rate due to the natural relaxation and trapping by various quenching centers, γ is the relaxation rate due to the singlet-singlet annihilation, I(t) is the rate of formation of excited molecules (S_1) upon single excitation pulse [7].

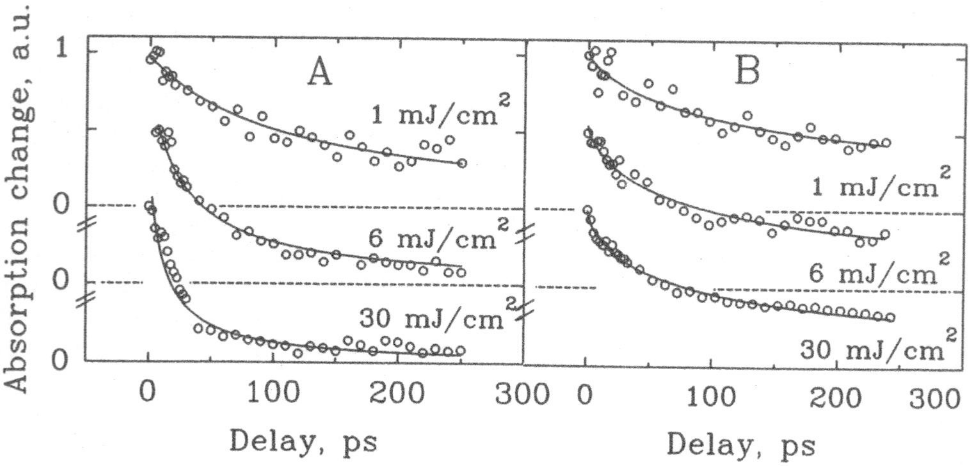

Figure 1. Absorbance transients due to singlet-singlet annihilation in large (A) and small (B) aggregates of LHCII. Open circles, data points; lines, fitted curves (eq. (1)).

Our analysis shows that small aggregates of LHCII behave like energetically disordered structures where the excitations migrate through percolation type cluster [10] of the pigment molecules creating the local minima for excitations. In contrast, the hyperbolic decay observed in the macroaggregates is typical for non-linear relaxation processes in large 3-dimensional structures where the non-linear rate parameter γ becomes time-independent. These data provide the first direct experimental evidence that LHCII macroaggregates constitute a structural basis for long-range migration of the excitation energy or its delocalization over large psi-type domains (cf. [11]).

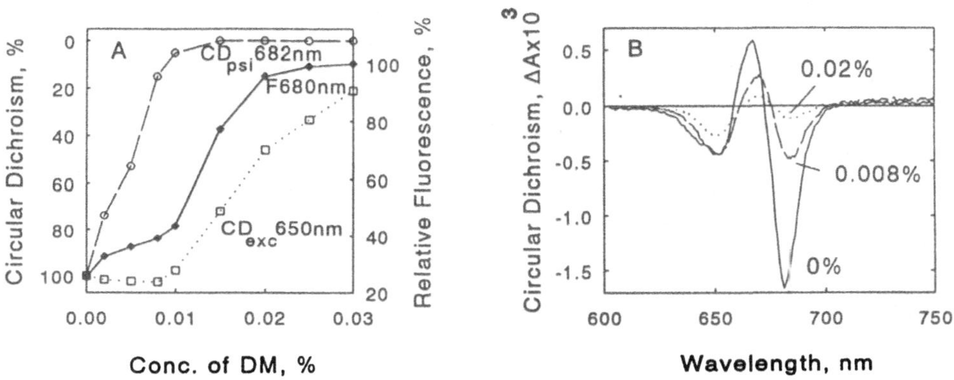

Figure 2. (A) Effect of gradual disintegration of LHCII macroaggregates on the intensity of the room temperature fluorescence emission (—) and on the amplitudes of psi-type (---) and excitonic CD bands (···). (B) CD spectra of LHCII at three different concentrations of n-Dodecyl β-D-maltoside.

Fig. 2A shows that upon gradual disintegration of LHCII macroaggregates the fluorescence yield increases substantially (cf.[2,7]). These changes are accompanied with characteristic changes in the CD signals of different physical origins.

In the concentration range between 0 and 0.01% DM, the 30-50% increase in the fluorescence yield is due to the disaggregation of the macroaggregates (d ~ 4-5 μm) into smaller aggregates (d ~ 100-200 nm). This is associated with the gradual disappearance of the psi-type CD bands (Fig. 2B, cf. also [7]). It is important to point out that under these conditions the excitonic CD bands are unaffected, indicating that the structure of the trimers remain intact. Further disintegration of LHCII, above 0.01% DM, brings much larger increase in the fluorescence yield. However, these gradual changes in the fluorescence are associated with a gradual degradation of the excitonic CD signal. This indicates that disintegration by the detergent brings about significant alterations in the molecular organization of complexes (cf. also [12]).

Based on the above findings we propose that two quenching mechanisms must be considered for explaining the variations of the fluorescence yield of LHCII.

1. The dramatic increase in the fluorescence yield of non-aggregated LHCII

(trimers and monomers) is explained by the gradual degradation of the excitonic interactions. An excitonic band can serve as a quencher if the optically allowed state of the excitonic band is situated not on the lowest energy level of the band. Conversely, upon disrupting the excitonic interactions in the complexes the quenching ceases, and the fluorescence yield increases.

2. If the size of aggregates remains in the range between about 5 μm and 100 nm, the decrease (~50%) in the fluorescence yield upon macroaggregation can be explained with a model involving 'impurity'-like quenching centers. In small aggregates these centers are disconnected from the bulk pigments, whereas in macroaggregates, because of the long-range migration of the excitation energy, they can quench the fluorescence. Minor components, e.g. the 29 kDa pigment-protein complex (cf. [13]) or complexes with high xanthophyll content [14], can act as 'impurities'. It seems plausible that light-induced reversible changes in chiral macroorganization in thylakoid membranes and LHCII macroaggregates (see Istokovics et al., and Barzda et al. in this Proceedings) involve changes in the size of aggregates, or in the size of the domains in which the excitation energy can diffuse. As shown by fluorescence measurements, these structural changes are associated with changes in the dissipative pathways in the antenna system.

Acknowledgements. This work was supported by an OTKA III/2999 grant (Hungary).

References

1 Horton, P., Ruban, A.V. and Walters, R.G. (1994) Plant Physiol. 106, 415-420.
2 Horton, P., Ruban, A.V., Rees, D., Pascal, A.A., Noctor, G. and Young, A.J. (1991) FEBS Lett. 292 (1,2), 1-4.
3 Garab, G. (1992) in Research in Photosynthesis I (Murata, N., ed.), pp. 171-178, Kluwer Academic Publishers, Dordrecht, Boston, London.
4 Kühlbrandt, W. (1994) Current Opinion Struct. Biol. 4, 519-528.
5 Krupa, Z., Huner, N.P.A., Williams, J.P., Maissan, E. and James, D.R. (1987) Plant Physiol. 84, 19-24.
6 Garab, G., Sundqvist, C., Mustárdy, L. and Faludi-Dániel, A. (1980) Photochem. Photobiol. 31, 491-494.
7 Barzda, V., Garab, G., Gulbinas, V. and Valkunas, L. (1995) submitted to Biochim. Biophys. Acta.
8 Nussberger, S., Dekker, J.P., Kühlbrandt, W., Van Bolhuis, B.M, Van Grondelle, R. and Van Amerongen, H. (1994) Biochemistry 33, 14775-14783.
9 Barzda, V., Mustárdy, L. and Garab, G. (1994) Biochemistry 33, 10837-10841.
10 Klymko, P.W. and Kopelman, R. (1983) J. Chem. Phys. 87, 4565-4567.
11 Keller, D. and Bustamante, C. (1986) J. Chem. Phys. 84, 2972-2979.
12 Gülen, D., Knox, R.S. and Breton, J. (1986) Photosynthesis Res. 9, 13-20.
13 Bassi, R., Rigoni, F. and Giacometti, G.M. (1990) Photochem. Photobiol. 52, 1187-1206.
14 Gilmore, A.M., Hazlett, T.L. and Govindjee, (1995) Proc. Natl. Acad. Sci. USA 92, 2273-2277.

ASSEMBLY OF LIGHT-HARVESTING CHLOROPHYLL-A/B PROTEIN IN ISOLATED THYLAKOID AND PROTHYLAKOID MEMBRANES

A. Kuttkat, I. Edhofer, L. Eichacker, and H. Paulsen
Botanisches Institut III der Universität,
Menzinger Straße 67, D-80638 München, Deutschland

1. Introduction

During the greening process, light-harvesting chlorophyll a/b protein (LHCP) is assembled with chlorophyll (Chl) a, Chl b and xanthophylls in (or near) the thylakoid membrane to yield the major light-harvesting complex (LHCII) of photosystem II. The role of pigments in this assembly process and in the co-regulation of LHCP and chlorophyll accumulation is not yet understood. Pigments bound to LHCP are thought to stabilize the protein whereas non-assembled LHCP is proteolytically degraded in the thylakoid. On the other hand, the binding of pigments has been shown to induce the refolding of LHCP in vitro (1), suggesting a more active role of pigments during the folding of LHCP in the thylakoid and maybe even during the translocation of the protein into the membrane (2).

LHCP integrates into isolated thylakoid membranes in vitro in the presence of a "stromal factor" and Mg-ATP or GTP (3-5). LHCP does not integrate into prothylakoid membranes isolated from etioplasts, even when stroma from mature or greening plastids is present (4). This may either be due to the lack of chlorophyll in the etioplast membranes, or to the early developmental stage of these membranes, or both. In this paper we show that LHCP insertion into etioplast membranes is stimulated by chlorophyll synthesis in these membranes in the dark from exogenously added substrates.

2. Procedure

Radiolabelled pLHCP and its derivatives were prepared by over-expressing pea cab gene constructs in Escherichia coli (1) in a growth medium containing (^{35}S)methionine. The protein was solubilized in 8 M urea immediately prior to the insertion assays. Greening chloroplasts were isolated from 7 to 9 day-old pea seedlings (Pisum sativum var. Golf) grown under a 16 h light-8 h dark regime (6) and separated into washed-thylakoid and stroma fractions (7). Barley etioplasts were isolated from barley seedlings grown in darkness for 4 days.

Integration of LHCP and its derivatives into thylakoids and protease digestion assays were essentially following the procedure of Yuan et al. (8). Protein electrophoresis on fully denaturing SDS-polyacrylamide gels and on partially denaturing gels was as described earlier (1) and according to Allen and Staehelin (9), respectively. For the

P. Mathis (ed.), Photosynthesis: from Light to Biosphere, Vol. I, 323–326.

insertion of pLHCP into barley thylakoids or prothylakoids with simultaneous synthesis of Zn-pheophytin (Zn-Phe), membranes isolated from 8.5x10^7 plastids were additionally incubated in the dark with 0.25 nmol Zn-pheophorbide (Zn-Pheide) a, 0.2 nmol Zn-Pheide b (10), and digeranyl pyrophosphate (GGPP) (11).

3. Results and discussion

3.1 Monomeric and trimeric LHCII in the thylakoid have different protease sensitivities. Upon insertion into the thylakoid membrane, LHCP is largely protected against proteases except for an N-terminal fragment of about 1 kDa. The protease-resistant 24-kDa LHCP fragment (LHCP-DP) is taken as evidence for successful insertion into the membrane (7, 12, 13). Another digestion product of about 20 kDa (LHCP-DP*) has been observed by us and others (14) upon protease treatment of membrane-inserted LHCP. Both fragments appear and increase simultaneously; their ratio is independent of the duration of or the temperature during protease treatment (A. Kuttkat and H. Paulsen, unpublished), indicating that these different digestion products reflect different states of LHCP after its insertion. When native LHCII trimers are dissociated into monomers by heat treatment of the thylakoid, the 24 kDa digestion product gradually disappears whereas the shorter digestion product increases (Fig. 1) suggesting that LHCP-DP and -DP* result from protease digestion of LHCII trimers and monomers, respectively. Consistently, when radioactively labelled pLHCP is inserted into isolated thylakoids, heat-

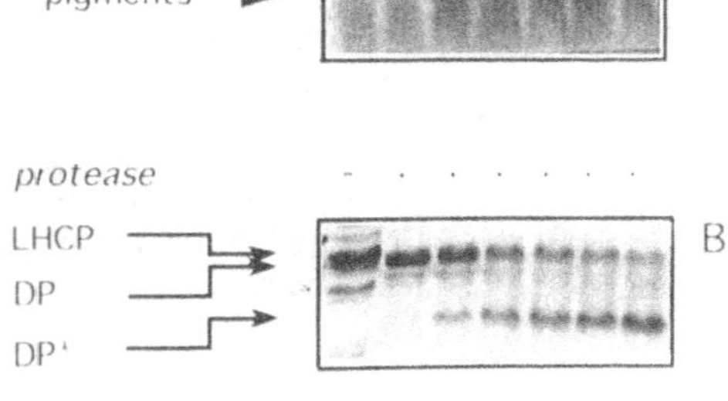

Figure 1. Dissociation of LHCII into monomers gives rise to a shorter protease digestion product (LHCP-DP).* **(A)** Isolated pea thylakoids were incubated at 60 °C for the time periods indicated, solubilized, and run on a partially denaturing gel. The bands representing LHCII trimers, monomers, and unbound pigment are marked in the margin. **(B)** Aliquots of the thylakoid membranes were treated with 0.1 ug/ml thermolysin for 30 min at room temperature, run on a denaturing SDS polyacrylamide gel and stained with Coomassie. The arrow marked LHCP indicates the position of undigested LHCP.

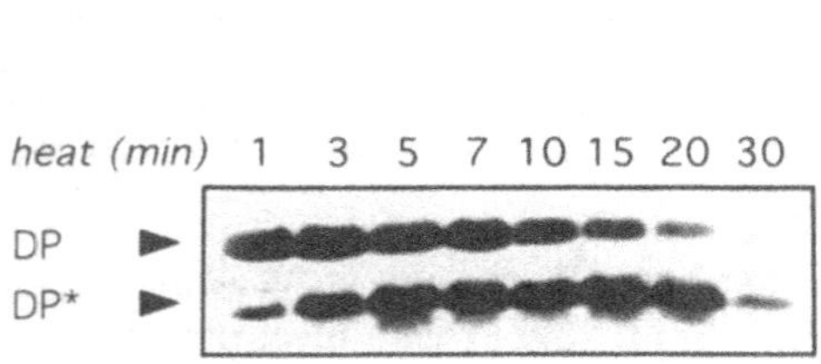

Figure 2. Heat treatment of thylakoids after the integration of radiolabelled pLHCP increases the amount of the shorter digestion product LHCP-DP.* Upon insertion of pLHCP into isolated thylakoids, these were incubated at 60 °C for the times indicated and then protease-treated and electrophoresed as described in Fig. 1B. The arrows marked DP and DP* give the positions of LHCP-DP and LHCP-DP*, respectively.

treated for various times, and then protease-digested, the same progressive shift from predominantly LHCP-DP to predominantly LHCP-DP* is observed (Fig. 2). Insertion of LHCP yields the same results (not shown); pLHCP was chosen in these experiments and the ones described below because it is more easily distinguishable from LHCP-DP than LHCP.

The correlation between trimeric and monomeric LHCII and the two different protease-resistant LHCP fragments is further corroborated by thylakoid insertion experiments with LHCP mutants that, after their reconstitution with pigments, do not trimerize: these mutants give rise only to the protease digestion product LHCP-DP* but not -DP (15). Isolated or reconstituted LHCII, in its trimeric or monomeric form, in detergent solution is also protease-digested into LHCP-DP and -DP*, respectively, indicating that protease protection of LHCP is due to the pigments bound rather than to its localization in the membrane. This leads to the conclusion that LHCP inserted into isolated thylakoids is complexed with pigments and can even assemble into trimeric LHCII.

3.2 Newly synthesized chlorophylls enable prothylakoids from etioplasts to insert LHCP. Incubation of pLHCP with membranes from barley etioplasts does not give rise to protease protection of the protein, even when stroma from mature pea chloroplasts is present in order to provide the stromal factor (Fig. 3, lane 1). However, when the membranes are supplemented with Zn-pheophorbides a and b (Zn-Pheide a,b) as well as digeranyl pyrophosphate (GGPP) so that they can synthesize the Chl-a,b analogs Zn-pheophytin (Zn-Phe) a,b, an LHCP fragment becomes protease-resistant (Fig. 3, lane 4) which corresponds to LHCP-DP* (Figs 1 and 2). The addition of only Zn-Pheide a,b or only GGPP (Fig. 3, lanes 2 and 3, respectively) does not yield any protein-resistant LHCP band, indicating that in fact the synthesis of the chlorophyll analogs is necessary. After illuminating the etiolated barley for 16 or 48 h, pLHCP inserted into the greening membranes gives rise to both LHCP-DP and -DP* regardless of whether Zn-Pheides and/or GGPP have been added, showing that (a) enough endogenous chlorophylls a and b have been formed to render inserted LHCP protease-resistant, and (b) the developmental state of the plastids supports the formation of trimeric LHCII as indicated by the appearance of LHCP-DP.

Zn-Pheides have been used in these experiments since they are more stable than chlorophyllides under the experimental conditions (10). The Zn analogs of chlorophylls are almost as efficient in forming stable LHCP-pigment complexes as their natural Mg

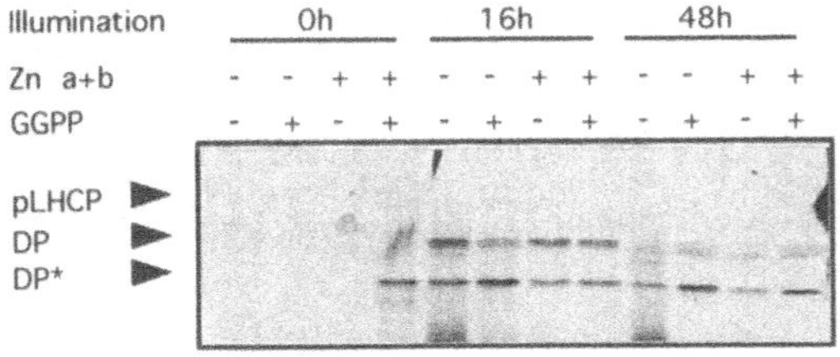

Figure 3. Chlorophyll synthesis stimulates the insertion of LHCP into prothylakoids from etioplasts. Membranes were isolated from barley etioplasts or plastids from barley illuminated for the times given. The membranes were supplemented with Zn-Pheide a, Zn-Pheide b, and GGPP where indicated. The membranes were protease-treated and electrophoresed as in Fig. 1B. Zn a+b: Zn-Pheide a,b; other abbreviations as in Figs 1 and 2.

analogs (16). Zn-Pheide and GGPP-dependent formation of protease-resistant LHCP fragments has also been observed in preliminary experiments with a homologous insertion system by using pea etioplast membranes.

We cannot decide at this point whether chlorophylls (or their analogs) are necessary for the integration of LHCP into the thylakoid or the stabilization of the inserted protein towards protease, or both. However, our data clearly show that the only component that etioplasts are lacking in order to be capable of stably inserting LHCP is chlorophylls, and no other possibly developmentally regulated factor. However, the capability of etiolated membranes to form trimeric LHCII as indicated by the appearance of LHCP-DP is not restored by Zn-Phe synthesis. This means that either a minimum amount of monomeric LHCII has to be accumulated before the onset of oligomerization and that this threshold level is reached only upon more massive light-dependent chlorophyll synthesis, or that trimerization is dependent on some other developmentally regulated factor(s) as has been suggested by Dreyfuss and Thornber (17). We are presently testing these possibilities by varying the amounts of chlorophyll precursors supplemented and by adding other components like lipids that might act as such factors.

References

1 Paulsen, H., Rümler, U. and Rüdiger, W. (1990) Planta 181, 204-211
2 Paulsen, H. (1995) Photochem. Photobiol. 62, in press
3 Cline, K. (1986) J. Biol. Chem. 261, 14804-14810
4 Chitnis, P.R., Nechushtai, R. and Thornber, J.P. (1987) Plant Mol. Biol. 10, 3-11
5 Li, X.X., Henry, R., Yuan, J.G., Cline, K. and Hoffman, N.E. (1995) Proc. Natl. Acad. Sci. USA 92, 3789-3793
6 Schindler, C., Hracky, R. and Soll, J. (1987) Z. Naturforsch. Sect. C Biosci. 42, 103-108
7 Yuan, J.G., Cline, K. and Theg, S.M. (1991) Plant Physiol. 95, 1259- 1264
8 Yuan, J.G., Henry, R. and Cline, K. (1993) Proc Natl Acad Sci USA 90, 8552-8556
9 Allen, K.D. and Staehelin, L.A. (1991) Anal. Biochemistry 194, 214-222
10 Helfrich, M. and Rüdiger, W. (1992) Z. Naturforsch. Sect. C. Biosci. 47, 231-238
11 Lindsten, A., Welch, C.J., Schoch, S., Ryberg, M., Rüdiger, W. and Sundqvist, C. (1990) Physiol. Plant. 80, 270-285
12 Auchincloss, A.H., Alexander, A. and Kohorn, B.D. (1992) J. Biol. Chem. 267, 10439-10446
13 Huang, L.Q., Adam, Z. and Hoffman, N.E. (1992) Plant Physiol. 99, 247- 255
14 Reed, J.E., Cline, K., Stephens, L.C., Bacot, K.O. and Viitanen, P.V. (1990) Eur. J. Biochem. 194, 33-42
15 Hobe, S., Kuttkat, A., Förster, R. and Paulsen, H. (1995) in Proceedings of the Xth International Photosynthesis Congress (Mathis, P., ed.) this volume, Kluwer Academic Publishers, Dordrecht, The Netherlands
16 Paulsen, H., Hobe, S. and Eisen, C. (1991) in Regulation of Chloroplast Biogenesis (Argyroudi-Akoyunoglou, J.H., ed.) pp.343-348, Plenum Publishers, New York
17 Dreyfuss, B.W. and Thornber, J.P. (1994) Plant Physiol. 106, 829- 839

GENETIC MANIPULATION OF *LHCB4*, A GENE ENCODING ONE OF THE MINOR LIGHT-HARVESTING COMPLEXES, IN *ARABIDOPSIS THALIANA*.

Carol A. McAuley[1], Tristan A. Dyer[2] and Peter Horton[1]
[1]Robert Hill Institute, Sheffield University, Sheffield S10 2TN, UK
[2]John Innes Centre, Colney Lane, Norwich, NR4 7UH, UK

1. Introduction

The light-harvesting antenna of Photosystem II (LHCII) is comprised of four pigment-protein complexes which vary in their pigment and polypeptide composition [1]. The polypeptides are encoded by the *Lhcb* genes, which are members of the *Lhc* gene family. *Lhcb1*, *Lhcb2* and *Lhcb3* encode the three polypeptides of the bulk light-harvesting complex, LHCIIb; and *Lhcb4*, *Lhcb5* and *Lhcb6* encode the polypeptides of the minor-light harvesting complexes LHCIIa, LHCIIc and LHCIId, respectively [2].

The light-harvesting complexes are believed to be arranged around the reaction centre core in such a manner that the minor LHCIIs form an inner antenna, and the bulk LHCIIb forms an outer antenna. In this way, the minor LHCIIs may act together as a connector complex, passing excitation energy from the bulk LHCIIb to the reaction centre proteins [2].

In addition to its function in light absorption, a role for LHCII in the protection of the photosynthetic apparatus against excess light has been established [3,4]. The dissipation of surplus excitation energy as heat, measured as the non-photochemical quenching of chlorophyll fluorescence, qN, is thought to occur within the light-harvesting antenna [5]. The major component of qN under conditions of excess light is qE, which is dependent on the pH gradient across the thylakoid membrane (ΔpH). Although the precise mechanism for qE is uncertain, two models have been suggested. The first, proposed by Demmig-Adams and co-workers is that the reversible de-epoxidation of violaxanthin via the xanthophyll cycle forms zeaxanthin, which acts directly as a quencher molecule by accepting excess excitation energy from chlorophyll [4,6]. However, it would appear that the formation of zeaxanthin is not a prerequisite for the induction of qE, but instead activates its formation [7]. The second model, proposed by Horton and co-workers suggests that qE results from a change in the conformational state of the LHCII antenna within the thylakoid membrane, possibly involving LHCII aggregation, which may be brought about synergistically by both protonation of certain LHCII polypeptides and by zeaxanthin formation [3,5,8]. The changes in protein conformation brought about by pH-dependent changes may increase chlorophyll-chlorophyll and chlorophyll-xanthophyll interactions which could promote quenching.

P. Mathis (ed.), Photosynthesis: from Light to Biosphere, Vol. I, 327–330.

Both protonation and zeaxanthin formation are, therefore, key events in the formation of qE. Recent studies have shown that two of the minor light-harvesting complexes, LHCIIa and LHCIIc, are bound by DCCD, a qE inhibitor [9]. DCCD binds to carboxy-amino acid residues, suggesting that amino acid residues on these two proteins may be involved in the induction of qE, specifically in detecting the lumen pH. These two protein complexes are also enriched in xanthophyll cycle pigments, with between 50% and 80% of the violaxanthin in LHCII being associated with the minor complexes [10,11]. The presence of both xanthophyll cycle components and DCCD-binding sites implies that LHCIIa and LHCIIc play a vital role in the formation of qE.

To investigate the role of LHCIIa, an novel approach is being attempted. By introducing an antisense copy of the *Lhcb4* gene into *Arabidopsis thaliana* plants, it may be possible to reduce the amount of *Lhcb4* polypeptide produced. If LHCIIa is as closely involved in excitation transfer as has been suggested, both with regards to quenching and to light capture, its absence could alter the capacity of the plant for both processes. Conversely, by creating plants with overexpressed *Lhcb4* genes, it may be possible to obtain plants with an increase in levels of the LHCIIa complex.

2. Materials and Methods

2.1 *Plasmid construction*

Three binary vectors were created using the vectors described in Gleave (1992) [12]. pG27 is a control plasmid containing the β-glucuronidase (GUS) reporter gene. pAL427 contains a copy of the *Lhcb4* gene in an antisense orientation with respect to the promoter and pSL427 contains a copy of the *Lhcb4* gene in a sense orientation. The constructs were cloned by first introducing the appropriate gene in the correct orientation into the primary cloning vector pART7 [12], using different restriction enzymes as seen in Figure 1. Digestion using *Not*1 removes the expression cartridge containing the gene cloned between the CaMV 35S promoter and the octopine synthase (ocs) terminator sequences, which can then be cloned into the binary vector pART27.

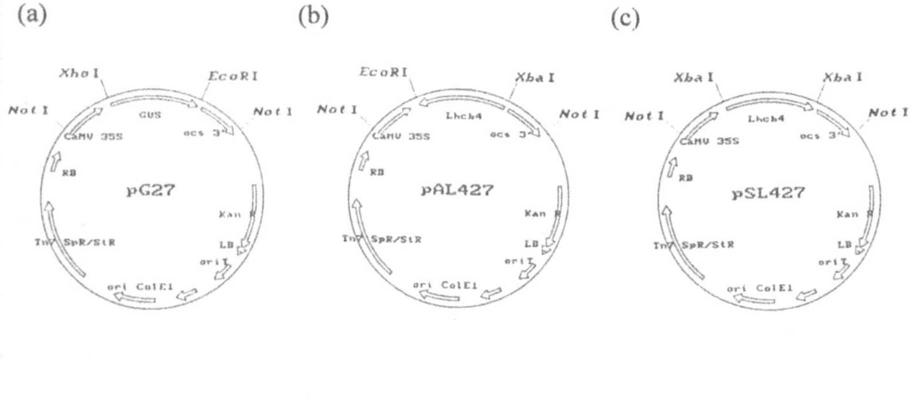

Figure 1. Plasmid maps showing (a) pG27 and the position of the GUS gene, (b) pAL427 and the position of the antisense Lhcb4 gene, and (c) pSL427 and the position of the sense Lhcb4 gene.

2.2 *Introduction of the binary vectors into Agrobacterium tumefaciens*
The binary vectors had to be introduced into *Agrobacterium* before they could be used to transform *Arabidopsis*. The *Agrobacterium* strain used was C58C1 pGV2260 [13]. This strain is resistant to the antibiotics rifampicin and carbenicillin, but is sensitive to kanamycin. The binary vectors were introduced into C58C1 using a standard triparental mating protocol and the *Agrobacterium* colonies containing the plasmids were identified by selecting for resistance to rifampicin, which is contained on the *Agrobacterium* DNA, and spectinomycin, which is contained on the binary plasmid.

2.3 *Transformation of Arabidopsis thaliana*
The transformation protocol used was a modified protocol based on that described by Valvekens et al [14], as used at the John Innes Centre, Norwich, UK. The *Arabidopsis* seedlings were initially grown in liquid induction media (IM), which allows better root growth and easier access to the root tissue compared with plants grown on solid induction media. The root tissue was removed and cut into small segments, which were co-cultivated with the *Agrobacterium* strain containing the binary vector. The root explants were maintained on shoot inducing media (SIM) until green calli appeared and began to regenerate into shoots. At this stage, the shoots were transferred to germination media (GM), where they developed buds and set seed.

3. Results and Discussion

3.1 *Transformation of Arabidopsis thaliana.*
Transformation using pG27 has yielded transformed plants, establishing that the protocols and plasmids used are effective in introducing DNA into the *Arabidopsis* plants. Seeds from nine independent transformants have been obtained and grown on media supplemented with kanamycin ($50\mu g ml^{-1}$). The resulting progeny segregate with an approximately 3:1 ratio of kanamycin resistant:sensitive plants. Statistical analyses of these results suggest that the DNA introduced into the plant genome has been inserted at a single locus.

3.2 *Future work*
Transformations of *Arabidopsis* plants with pAL427 and pSL427 are currently in progress. Once these plants have been obtained, it will be possible to characterise differences between transformed and untransformed plants. Using parameters such as chlorophyll a:b ratio, PSII content, LHCII content and composition *etc.* it will be possible to determine whether or not any changes in the level of LHCIIa complex within the plant have an effect on both the quantum yield of photosynthesis and the capacity for qE formation, as described above.

The light-harvesting antennae may also have a function in the adaptation of the photosynthetic apparatus to changes in environmental light regime as described in Walters and Horton (1994) [15]. This study showed that light intensity controlled the levels of LHCII complexes associated with the PSII reaction centre, in that plants grown under low light ($100\mu Em^{-2} s^{-1}$) had increased amounts of LHCII compared with plants grown under higher light regimes ($400\mu Em^{-2}$). Far-red enriched light also affected the photosynthetic apparatus by altering the ratio of PSI:PSII. As LHCIIa is found closely associated with the PSII reaction centre proteins, it can be suggested that this protein complex may have a function in the acclimation response. It may be interesting to see how, if at all, levels of LHCIIa change with respect to altered light regime, i.e. whether

LHCIIa behaves as a LHC protein and is affected by light intensity, or whether it behaves as a reaction centre protein and is affected by spectral quality.

3.3 *Concluding remarks*

The pigment-protein complex LHCIIa would appear to be involved in a number of regulatory responses within the photosynthetic apparatus. This study offers a novel approach to attempt to clarify the function of LHCIIa within the PSII antenna complex.

References

1. Peter, G.F. and Thornber, J.P. (1991) J. Biol. Chem. 266, 16745-16754
2. Jansson, S. (1994) Biochim. Biophys. Acta 1184, 1-19
3. Horton, P. and Ruban, A.V. (1992) Photosyn. Res. 34, 375-385
4. Demmig, B., Winter, K., Kruger, A. and Czygan, F.-C. (1987) Plant Physiol. 84, 218-224
5. Horton, P., Ruban, A.V. and Walters, R.G. (1994) Plant Physiol. 106, 415-420
6. Demmig-Adams, B. (1990) Biochim. Biophys. Acta 1020, 1-24
7. Noctor, G., Rees, D., Pascal, A.A. and Horton, P. (1991) Biochim. Biophys. Acta 1059, 355-360
8. Horton, P., Ruban, A.V., Rees, D., Pascal, A.A., Noctor, G. and Young, A.J. (1991) FEBS Lett. 299, 1-4
9. Walters, R.G., Ruban, A.V. and Horton, P. (1994) Eur. J. Biochem. 226, 1063-1069
10. Ruban, A.V., Young, A.J., Pascal, A.A. and Horton, P. (1994) Plant Physiol. 104, 227-234
11. Bassi, R., Pineau,B., Dainese, P. and Marquandt, J. (1993) Eur. J. Biochem. 212,297-303
12. Gleave, A. (1992) Plant Mol. Biol. 20, 1203-1207
13. Deblaere, R., Bytebier, B., De Greve, H., Deboeck, F., Schell, J., Van Montagu, M. and Leemans J. (1985) Nucleic Acids Res. 13, 4777-4788
14. Valvekens, D., Van Montagu, M. and Van Lijsebettens, M. (1988) Proc. Natl. Acad. Sci. USA 85, 5536-5540
15. Walters, R.G. and Horton, P. (1994) Planta 195, 248-256

FEATURES AND FUNCTIONS OF CHLOROPHYLL A/B-BINDING PROTEINS IN LIVERWORT (*MARCHANTIA POLYMORPHA*)

R. Kilian and C. Schäfer

Lehrstuhl Pflanzenphysiologie, Universität Bayreuth, Universitätsstr. 30, 95440 Bayreuth, Germany

1. Introduction

Much is known about chlorophyll a/b binding (cab) proteins of higher plants and algae (1, 2). Whereas in different higher plant species considerable similarities are found for these proteins (amino acid sequence, structure etc.) the different groups of algae appear to exceed a larger variety. Only a few investigations deal with pigment protein complexes of Bryophytes (3, 4) and studies which may profit from the improved techniques in pigment protein isolation are still missing. Bryophytes, which represent a transition state from lower to higher plants, could probably provide important cues on the evolution of the cab proteins (1). The present study reveals for the first time distinct differences in molecular weight and isoelectric point of cab proteins between a Bryophyte (*Marchantia polymorpha*) and a higher plant (*Chenopodium rubrum*).

2. Procedures

The experiments were performed with photoautotrophic culture cells of *C. rubrum* and *M. polymorpha* (5). The isolation of thylakoid membranes and PSII membranes was done in the presence of protease inhibitors as described in (5). PSII membranes (BBY) were prepared according to (6) and washed with NaBr (2 M) before handling. Fully denaturing SDS PAGE was performed as described in (7) in the presence of 6 M urea. 2-D Electrophoresis was done as follows: High salt washed PSII membranes were solubilized in 2% (w/v) dodecyl-ß-D-maltoside (DM) (DM:Chl ratio 12.5 in *M. polymorpha* and 10.0 in *C. rubrum*). The isoelectric focusing (IEF) gels (in glas tubes, 1 mm diameter 100 mm length) contained 4 % (w/v) acrylamide:bisacrylamide (37.5:1), 20 % (w/v) gycerol, 1 % (w/v) Servalyte pH 3-7 and 1 % (w/v) Servalyte pH 6-8. After prefocusing IEF was performed at 400 V for 8 h (Tube Cell, Modell 175 from Biorad, München, Germany). The whole IEF gels were denaturated, placed on a 1 mm slab gel and submitted to a fully denaturing electrophoresis. Flatbed IEF was done according to (8), but the matrix contained 1.4 % (w/v) Servalyte pH 3-7, 0.6 % (w/v) Servalyte pH 5-8 and additionally 20 % (w/v) glycerol.

P. Mathis (ed.), Photosynthesis: from Light to Biosphere, Vol. I, 331–334.

Immunodetection: After fully denaturing PAGE the proteins were transferred to nitrocellulose by semidry blotting (7) and blots were developed as described in (9) using antibodies against LHCIb (10), CP29 and CP26. Chlorophyll determinations were done according to (11). UV/VIS absorption spectra were recorded on a UVICON 941 spectrophotometer (Kontron, Zürich, Switzerland).

3. Results and Discussion

3.1 *1-D Electrophoresis and immunodetection*: The apparent molecular weight (MW) of the PSII reaction center proteins (D1, D2, CP43, CP47 and OEE) was known from earlier studies (12 and unpublished results). For identification of PSII cab proteins the enrichment of certain protein bands in the PSII membranes compared to the thylakoids (data not shown) and cross-reactivity with antibodies against different cab proteins (Figure 1) were used as criteria for the identification of PSII cab proteins.

Figure 1: SDS-PAGE and immunoreaction pattern of *M. polymorpha* high salt washed PSII membranes (left) and pure LHCII (right). For each antibody the pattern of immunoreactions is drawn schematically (center). The band thickness indicates the strength of immunoreaction and horizontal lines assign the immunosignals to the respective protein bands on the gel. Molecular weight markers are given on the left. [1]) Pure LHCII was isolated by flatbed IEF, [2, 3]) polyclonal antibodies were prepared against the respective proteins in *C. rubrum* ([2]) and *Zea mays* ([3]).

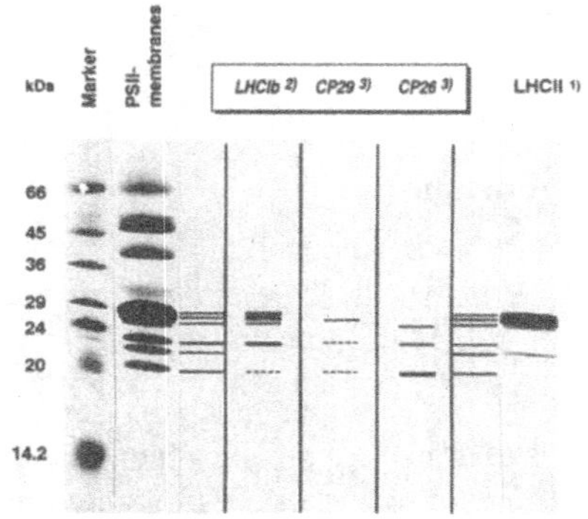

Immunodetection gave the following results: *i*) Antibodies against LHCIb of *C. rubrum* showed a strong cross-reactivity with two apoproteins of LHCII (28.0 kDa, 27.5 kDa) and a 23.5 kDa protein. *ii*) Antibodies against CP29 (*Zea mays*) reacted with a 27.0 kDa protein, and (weakly) with a 23.5 kDa and a 20.0 kDa protein. *iii*) Antibodies against CP26 (*Zea mays*) reacted with a 26.0 kDa protein, a 23.5 kDa protein and very strongly with a 20.0 kDa protein. *iv*) Apoproteins of cab proteins with molecular weights above 28.0 kDa could not be detected. *v*) None of the above antibodies reacted with the 22.5 kDa apoprotein of LHCII.

We conclude from these data that proteins with the apparent molecular weights of 20.0, 23.5 and 27.0 kDa may represent the minor cab proteins in *M. polymorpha*, while LHCII consists of 28.0, 26.5, 26.0 and 22.5 kDa apoproteins.

3.2 *2-D Electrophoresis* (Figure 2): Tube gel IEF indicated that the 20.0 and the 23.5 kDa proteins of *M. polymorpha* were spread over a rather broad pH range. The maximum density of both proteins was found around pH 6.0, and additionally at pH 6.5 for the 20.0 kDa protein. Thus these proteins have a considerably higher pI than the respective proteins of *C. rubrum* (around pH 4.5 for CP24 and CP26, Table 1). A more precise determination of the pI of the proteins of *C. rubrum* will need a more extended pH gradient (8).

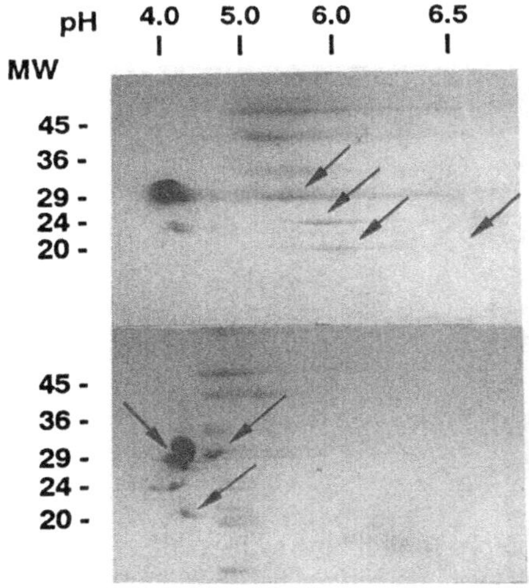

Figure 2: 2-D Electrophoresis of PSII membrane proteins. Pigment protein complexes were separated in the first dimension by non-denaturing IEF (pH gradient as indicated) and in the second dimension by fully denaturing PAGE. Coomassie stained gels are shown for *M. polymorpha* (upper part) and *C. rubrum* (lower part). MW is given in kDa. Arrows indicate the location of CP24, CP26 and CP29 in *C. rubrum* and the corresponding proteins in *M. polymorpha* (20.0, 23.5 and 27.0 kDa).

cab protein	*M. polymorpha*		*C. rubrum*	
	apparent MW	pI	apparent MW	pI
LHCII	28	4.2	28	4.4
	26.5	4.2	27.5	4.25
	26	4.3	27	4.4
	22.5	4.3	24	4.0, 4.3
CP29	27	5.4-5.5	29	4.8
CP26	23.5	5.9-6.1	27	4.4
CP24	20	6.0-6.1, 6.5	20.5	4.5

Table 1: Molecular weight (kDa) and isoelectric points of cab proteins from *C. rubrum* and *M. polymorpha*.

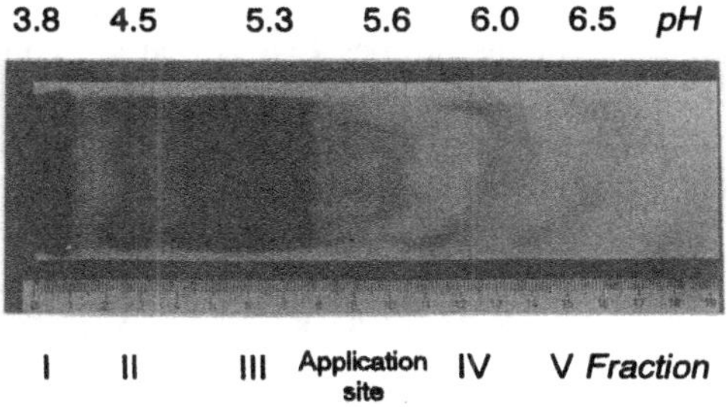

Figure 3: Flatbed IEF of high-salt washed PSII membranes from *M. polymorpha*. Fractions I-IV were eluted, concentrated, washed with elution buffer (25 mM Hepes, pH 7.6, 0.06% DM) and used for electrophoresis and the determination of absorption spectra.

3.3. *Flatbed IEF for isolation of the complexes and further analysis.*
Flatbed IEF (8) was optimized for the isolation of the putative minor cab proteins from *M. polymorpha* (Figure 3). Based on the protein composition (Figure 4, left) five different green fractions (I-V) were distinguished: pure LHCII (I), PSII core proteins together with different portions of cab proteins (II, III), a mixture of a 23.5 kDa and a 20.0 kDa protein (IV), a 20.0 kDa protein plus changing amounts of lower molecular weight proteins (mainly 19.0 kDa, V). The amount of these latter proteins increased considerably during further handling (data not shown) and therefore it appears that they represent degradation products. The absorption spectrum of fraction I showed to be a typical LHCII spectrum (maxima at 437, 472, 652 and 673 nm, Chl a/b-ratio 1.24 (1, 13, 14)). Absorption spectra of fractions IV and V resembled those published for CP26 and CP24, respectively (13). These data show that besides LHCII also the putative minor cab proteins with 23.5 kDa and 20.0 kDa could be isolated by IEF.

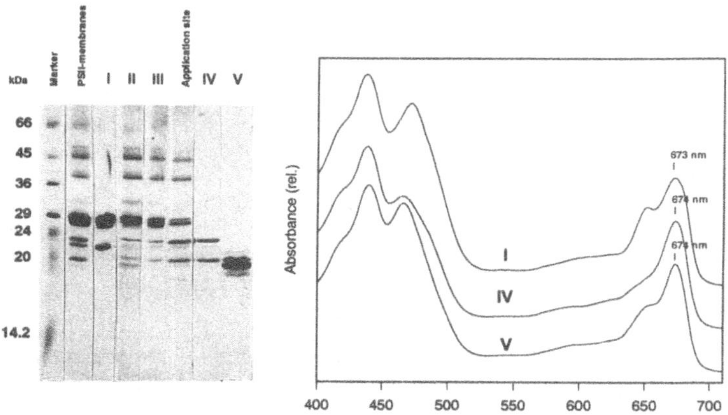

Figure 4: Characterization of IEF fractions by fully denaturing SDS PAGE (I-V) and absorption spectroscopy (I, IV, V). Long-wavelength maxima are indicated. Chl a/b ratios were 1.24 (fraction I), 1.98 (fraction IV) and 1.19 (fraction V).

In the present work we have shown that *M. polymorpha* contains at least three different PSII cab proteins besides LHCII that are equivalent to the minor cab proteins of higher plants (CP24, CP26 and CP29). These minor cab proteins in *M. polymorpha* can be distinguished from those in higher plants by their lower apparent molecular weight and their less acidic isoelectric point.

4. Addendum

We want to thank Prof. R. Bassi (Verona, Italy) and Dr. Volkmar Schmid for the provision of antibodies and for many valuable discussions, Helga Simper for excellent technical assistance and the Deutsche Forschungsgemeinschaft for financial support (Scha 468/2).

References

1 Jansson, S. (1994) Biochim. Biophys. Acta 1184, 1-19
2 Paulsen, H. (1995) Photochem. Photobiol. 62, in press
3 Aro, E.-M.(1982) Plant Sci. Lett. 24, 335-345
4 Long, Z., Wang, S.-Y., Nelson, N. (1989) Gene 76, 299-312
5 Schäfer, C., Schmid, V., Roos, M. (1994) J. Photochem. Photobiol. B 22, 67-75
6 Dunahay, T.G., Staehelin, L.A., Seibert, M., Ogilvie, P.D., Berg, S.P. (1984) Biochim. Biophys. Acta 764, 179-193
7 Schäfer, C., Vogg, G., Schmid, V. (1993) Planta 189, 433-439
8 Dainese, P., Hoyer-Hansen, D., Bassi, R. (1990) Photochem. Photobiol. 51, 693-703
9 Schmid, V., Schäfer, C. (1994) Planta 192, 473-479
10 Schmid, V (1995) Verlag Shaker, Aachen, ISBN 3-8265-0450-X
11 Porra, R.J., Thompson,W.A., Kriedemann, PE.(1989) Biochim. Biophys. Acta 975, 384-394
12 Schmid, V., Peter, S., Schäfer, C. (1995) Photosynth. Res., in press
13 Thornber, J.P., Peter, G.F., Morishige, D.T., Gomez, S. , Anandan, S., Welty, B.A., Lee, A., Kerfeld, C., Takeuchi, T., Preiss, S. (1993) Biochem. Soc. Trans. 21, 15-18
14 Dunahay, T.G. and Staehelin, L.A.(1986) Plant Physiol. 80, 429-434.

STRUCTURAL INFORMATION ON LIGHT-HARVESTING COMPLEX II AS OBTAINED FROM EXCITON CALCULATIONS AND POLARIZED SPECTROSCOPY.

Demet Gülen[a], Rienk van Grondelle[b] and Herbert van Amerongen[b]; [a]Dept. of Physics, Middle East Technical University, Ankara, 06531, Turkey; [b]Dept of Physics and Astronomy and Institute for Molecular Biological Sciences, Vrije Universiteit, De Boelelaan 1081, 1081 HV Amsterdam, The Netherlands

1. Introduction

A large of the structure of light-harvesting complex II (LHCII), the major antenna of green plants, has recently been modeled to a resolution of 3.4 Å [1]. The model reveals 2 carotenoids (Cars) and 12 chlorophylls (Chls) per monomeric subunit. The Cars were assigned to lutein because of stoichiometry and symmetry. A preliminary assignment of the identities of all Chls has been proposed [1]: the 7 Chls that are in close contact with the two central Cars present in each monomer are Chl *a*; the remaining 5 Chls are Chl *b*. However, it should be stressed that this assignment is preliminary. Essential information that is lacking at the current resolution (i.e. the position of the phytyl chains, 5th ring) only allowed the Chls to be modeled as tetrapyrrole rings. No distinction could be made between the x- and y-axis which are oriented in the molecular plane of the Chls.
The atomic model is of general importance for addressing any question that relates to the determination of the structure-function relationship. Important aspects of the structure-function relationship one would like to understand are: the character of the excited states, the mechanism(s) of energy transfer, and the mechanism(s) leading to efficient quenching of the Chl triplets. Since the publication of the previous LHCII model [2] in which many structural features of the C_3-symmetric Chl *a/b* protein were apparent, there have been extensive spectroscopic studies and a wealth of information on the excited states has been accumulated, especially for the Q_y region (640-680 nm) [3-8]. It is an enormous challenge to correlate the structure of LHCII to its spectroscopic properties which in turn are intimately related to the function.
Based on the densely packed pigment content, the existence of excitonic interactions among the Chls can be expected and this has been substantiated by a number of spectroscopic studies [4-8]. However, two current sources of uncertainty of the structural model -the identities of the Chls and the directions of the x- and y axes, along which the most important transition dipole moments are oriented- generate a large number of possible molecular conformations. In principle, a considerable amount of additional information is available through the key features of the steady-state spectra. In this respect, the sensitivity of polarized spectroscopy (circular dichroism CD and linear dichroism LD) on the molecular conformations is particularly valuable. The potential of accurate LD measurements is illustrated in [9] where the orientations of the Q_y transition dipole moments of LHII from *Rb. sphaeroides* were determined and which recently turned out to be in good agreement with the orientations in the crystal structure of LHII from *Rps. acidophila* [10]. In the present study exciton calculations were performed in

P. Mathis (ed.), Photosynthesis: from Light to Biosphere, Vol. I, 335–338.

order to investigate whether the assignment of the Chl identities [1] can explain at least the global features of the CD and LD spectra. Since for each Chl there are two possible orientations for the x- (and y) axis, this leaves us with 2^{12} possible combinations for the orientations of the most important transition dipole moments which are approximately parallel to either the x- or the y-axis in Chl. For each of the 2^{12} possible conformations the CD and LD spectra were calculated and essential aspects of the spectroscopic characteristics were compared to the experimentally determined ones. It turns out that these calculations give important spectroscopic and structural information and a large reduction in the number of possible conformations.

2. Materials & Methods

An exciton formalism that takes into account all important low-lying x- and y-polarized red and Soret transitions of the Chls was numerically implemented. Excitonic states obtained on the basis of dipole-dipole interactions in the point dipole approximation and the subsequent introduction of the spectral width effects through symmetric Gaussians have been used. Several detailed representations of the exciton formalism used in the simulations may be found in numerous review articles [11].
Simulations are carried out for the monomeric LHCII model [1] which contains 7 Chl *a* and 5 Chl *b* molecules. There are 792 possibilities in which 7 out of 12 Chls are Chl *a;* for our calculations we used the conformation from [1] where all Chl *a* molecules are lying next to the central Cars. We have taken the Q_y direction along one of the two diagonals in the tetrapyrrole ring in the LHCII model. However, uncertainties exist about the exact orientations of the Chls and the exact orientations of the transition dipoles with respect to the pigments [7, 12].
In the simulations several reasonable sets of parameters for the unperturbed site energies, the magnitudes of the transition dipole moments and the widths of the Gaussian envelopes are adapted by following the values accepted in the literature [8, 13,14]. Only the globally striking characteristics of the LD and CD spectra of the monomeric LHCII in the Q_y region are used to evaluate the simulation results and comparison between theory and experiment is done on the basis of relatively mild restrictions in view of the uncertainties mentioned above.
$A_{\|}$ and $A_{\perp}$ are defined as the absorption of light with polarization parallel and perpendicular to the trimeric plane and A is the isotropic absorption. The reduced LD (LD_r) is then defined according to $LD_r = (A_{\perp} - A_{\|}) / 3A$ and it can theoretically vary between -0.5 and 1.0 [9]. It was determined in [3] that the values of the reduced dichroism above 660 nm (called LD_a because this is dominated by Chl *a*) and between 640 and 660 nm (here called LD_b) are close to -0.3 and 0.0, respectively. Given the uncertainties in the dipole directions of individual pigments and in the exact contributions of the absorption of Chl *a* and Chl *b* in the indicated regions, it was somewhat arbitrarily decided that only those conformations are acceptable for which
$-0.5 < LD_a < -0.15$ and $-0.25 < LD_b < 0.25$.
The CD in the Chl *a* region of the spectrum is strongly biphasic (with a crossing point around 670 nm) being positive on the short wavelength side (peaking around 665 nm) and negative on the long wavelength side (peaking around 675 nm) [4, 5]. Calculated spectra are regarded to be unacceptable if either these signs are reversed or the CD signal is dominantly monophasic. The CD in the Chl *b* region has not been used as a selection criterion yet.

3. Results and Discussion

The strongest reduction of possible conformations is obtained by considering LD_a. Using the given criterion, an approximately 20-fold reduction is obtained in the number of possible conformations. The restrictions on LD_b and CD each lead to an additional 2-fold reduction. Eventually, a combination of the three restrictions lead to about 50 conformations which survived out of the 4096 (2^{12}) starting conformations.
Following the elimination on the LD selection, the Chl *a* region conformations of the surviving cases are found not to be scattered over the whole conformational space available, but rather belong to certain subclasses (around 10 out of 2^7) in which at least 5 of the 7 Chl *a* molecules adopt one of the two possible'y' directions. Moreover, the CD signals in the Chl *a* region belonging to each Chl *a* conformational subclass display common characteristics irrespective of the different Chl *b* orientations. Each of the restrictions used has unequivocally forced one specific Chl molecule to a definite orientation.

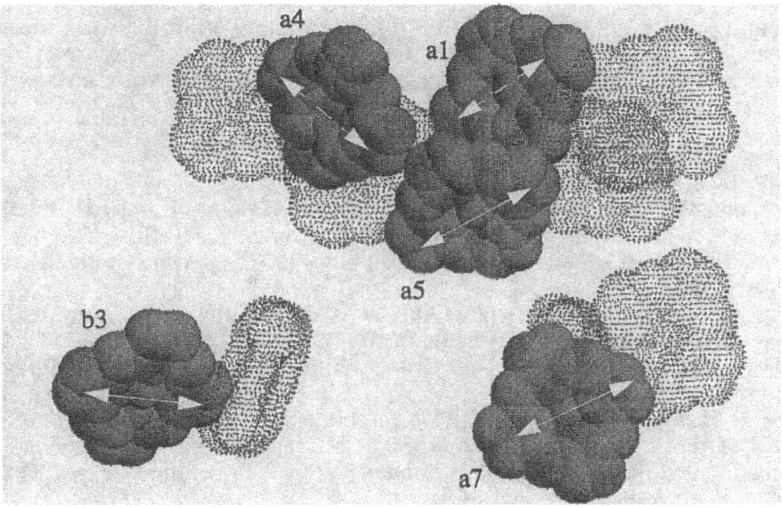

Figure 1. Calculated orientations for the y-axis in 5 Chls is indicated by double-headed arrows. Dashed arrows are slightly less certain. The C_3-symmetry axis is oriented vertically. See text for further details.

The orientations of Chl *a*5, Chl *a*7 and Chl *b*3 turn out to be uniquely defined in all allowed conformations. Apart from that the orientations of Chl *a*1 and Chl *a*4 do not vary except in a very few cases. These results are illustrated in figure 1, where these 5 Chls are represented as tetrapyrrole rings. Based on the present calculations no conclusions could be drawn about the orientations of the other (dotted) chlorophylls.
LD spectra for many allowed conformations have in common that the long wavelength half of the Chl *a* spectrum has a polarization which is lying more in the plane of the trimer than the remaining part of the spectrum in the Q_y region, in agreement with experimental results [3]. Moreover, almost all possible conformations lead to a weak red-most Chl *a* band which was experimentally observed for trimers [6, 7] and

monomers [7].
It should be stressed that the above calculations were performed according to the assignment of the Chl identities in [1] which is only one of the 792 possible cases, assuming 7 of the 12 Chls to be Chl *a*. As was demonstrated the proposed arrangement leads to calculated CD and LD spectra which are in global agreement with experimental spectra only for a limited number of Chl orientations. At this moment we are addressing the question whether also information can be obtained concerning the nature of the Chls. With the results obtained, strategic choices can be made, for mutating specific Chl-binding amino acids in order to reveal the orientations and identities of the corresponding chlorophylls. In the future, the selection method will be further refined by using more details of polarized spectra; furthermore essential features from triplet minus singlet absorption difference spectra and (sub)picosecond excitation transfer data will be included.

4. Acknowledgements

D. G. acknowledges a grant from EMBO for performing part of the work mentioned in this paper. The authors thank Dr. J.P. Dekker and Mr. E.J.G. Peterman for useful discussions. Dr. W. Kühlbrandt is acknowledged for providing us with the coordinates of the tetrapyrrole rings.

References

1 Kühlbrandt, W., Wang, D.N. and Fujiyoshi Y. (1994) Nature (London) 367, 614-621
2 Kühlbrandt, W. and Wang, D. N. (1991) Nature (London) 350, 130-134
3 Van Amerongen, H., Kwa, S.L.S., van Bolhuis, B.M. and van Grondelle, R (1994) Biophys. J. 67, 837-847
4 Nussberger, S., Dekker J.P., Kühlbrandt, W., van Bolhuis, B.M,, van Grondelle, R. and van Amerongen, H. (1994) Biochemistry 33, 14775-14783
5 Hemelrijk, P.W., Kwa, S.L.S., van Grondelle, R. and Dekker, J.P. (1992) Biochim. Biophys. Acta 1098, 159-166
6. Reddy, N. R. S., van Amerongen, H., Kwa, S. L. S., van Grondelle, R. and Small, G. J. (1994) J. Phys. Chem. 98, 4729-4735
7 Peterman, E.J.G., Dekker, J.P., van Grondelle, R., van Amerongen, H.and Nussberger, S. (1994) Lith. J. Phys. 34, 301-305
8 Krawczyk, S., Krupa, Z. and Maksymiec, W. (1993) Biochim. Biophys. Acta 1143, 273-281
9 Van Amerongen, H., van Haeringen, B., van Gurp, M. and van Grondelle, R. (1991) Biophys. J. 59, 992-1001
10 McDermott, G., Prince, S.M., Freer, A.A., Hawthornthwaite-Lawless, A.M., Papiz, M.Z., Cogdell, R.J. and Isaacs, N.W. (1995) Nature (London) 374, 517-521.
11 Pearlstein, R. M. (1991) in Chlorophylls (Scheer H., ed.), pp. 1047-1078, CRC Press, Boca Raton, USA
12 Moog, R.S., Kuki, A., Fayer, M.D. and Boxer, S.G. (1984) Biochemistry 23, 1564-1571
13 Gülen, D. and Knox, R.S. (1984) Photobiochem. Photobiophys. 7, 277-284
14 Eads, D.D., Castner, E.W., Alberte, R.S., Mets, L. and Fleming, G.R. (1989) J. Phys. Chem. 93, 8271-8275

THE PSII-S POLYPEPTIDE - AN ATYPICAL CAB PROTEIN

Christiane Funk[1,2], Gernot Renger[1], Iwona Adamska[3], Bertil Andersson[2] and Wolfgang P. Schröder[2]
[1]Max-Volmer Institute, TU-Berlin, Straße des 17. Juni 135, D-10623 Berlin, Germany, [2]Arrhenius Laboratories, Department of Biochemistry, Stockholm University, S-10691 Stockholm, Sweden, [3]Institute for Botany, University of Hannover, Herrenhäuser Straße 2, D-30419 Hannover, Germany

1 Introduction

A PSII subunit of 22 kDa (1), encoded by the nuclear *psb*S gene (2,3), exhibits unique properties but has so far escaped unambiguous assignment of its functional role. This hydrophobic protein was isolated via different methods and reported to contain neither pigments nor metallic cofactors (1-3). It was therefore surprising, when the total sequence determination showed that the PSII-S protein is related to the Chl a/b binding (CAB) proteins (4-6). Biochemical data for this possibility were obtained by mild SDS-PAGE analysis (7) and recently by preparative isoelectric focusing (8). Here we report the isolation of the Chl a/b binding PSII-S protein and its biophysical properties. Further results are given that in contrast to other Chl-binding proteins the PSII-S protein is stable without bound pigments.

2. Procedure

Spinach was grown in water culture for 8 weeks with 12 hours alternating dark and light (350 $\mu E/m^2s$). Etiolated spinach plants were grown in darkness for 8 days (9). Isolation of the PSII-S protein from green and etiolated spinach plants was performed via preparative isoelectric focusing (8) after solubilisation of membrane fragments (9) with OGP. Protein analysis was performed by SDS-PAGE combined with immuno blotting. Chlorophyll content and protein concentration were determined by standard procedures. Pigment analysis was either performed by HPLC (10) or by thin layer chromatography (11). Steady state fluorescence was measured at 77K and fluorescence decay curves were monitored by time-correlated single photon counting technique (12).

3. Results and Discussion

The PSII-S protein was isolated in a native state by using a procedure based upon non-ionic detergents (OGP) and preparative isoelectric focusing (8). A major green band focusing at pH 6.1 was immunologically identified as highly purified PSII-S protein. Determination of the chlorophyll- and protein-content of the isolated PSII-S protein after acetone extraction revealed a Chl a/b ratio of 6:1 and a Chl/protein ratio of 5±1. A more detailed analysis of the pigment composition the purified PSII-S protein was obtained by reversed phase HPLC (10). The results reveal that per 100 mol Chl a the PSII-S protein binds 17 mol Chl b, 6 mol neoxanthin, 5 mol violaxanthin, 12 mol lutein and 1 mol β-carotene. The analysis also indicated the presence of small amounts of antheraxanthin and zeaxanthin.
The absorption spectrum of the purified protein at room temperature exhibits a peak at

P. Mathis (ed.), Photosynthesis: from Light to Biosphere, Vol. I, 339–342.

671 nm (Fig. 1A), typical for Chl a species. Moreover, a second peak/shoulder in the region between 640 to 650 nm is lacking, consistent with the low amount of Chl b. The fluorescence emission spectrum measured at 77K using an excitation wavelength of 435 nm is characterised by a maximum at 672 nm and a small vibronic 1->0 satellite near 730 nm. These features are typical for Chl a, but the overall band maximum is blue-shifted compared with other isolated CAB proteins (Fig. 1B).

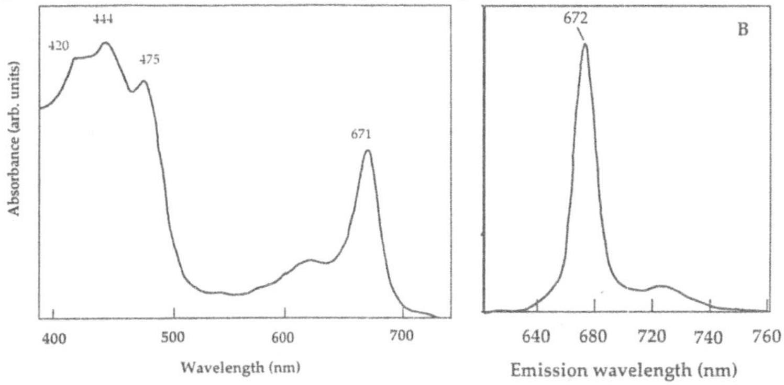

Fig. 1: Room-temperature absorption spectrum (A) and 77K-fluorescence emission spectrum (B) of the purified PSII-S protein; emission spectrum obtained upon excitation at 435 nm.

Time-resolved fluorescence measurements of the isolated PSII-S protein provide further information on the properties of chromophores bound to the protein matrix (Fig.2). A numerical analysis of the data yields triexponential kinetics (lower part of the figure). The fit procedure leads to the following life times and normalised amplitudes (given in brackets): τ_1 = 1.8 ns (a_1 = 0.09), τ_2 = 4.4 ns (a_2 = 0.60), τ_3 = 6.1 ns (a_3 = 0.31). The first two kinetics closely resemble those observed in solubilised non-aggregated LHCII (12). However, in marked contrast to LHCII a third kinetics with a life time of 6.1 ns is observed in the isolated PSII-S protein (Fig. 2). Similar values have been observed for free chlorophyll in liposomes (12). It is therefore tempting to assign the 6.1 ns component to chlorophyll molecules which are comparatively loosely associated with the protein. These time resolved fluorescence data suggest that about 70% of the chlorophyll is bound to the isolated PSII-S protein in the same or similar way as in a native CAB protein. The remaining 30% of the chlorophyll molecules are inferred to be still associated with the protein but in a disordered manner. A weak excitonic coupling of all pigments of the PSII-S protein was also seen in CD measurements (not shown).
Two important conclusions can be gathered from the experimental findings presented so far: i) the PSII-S protein is pigment binding, and ii) the mode of pigment binding differs in some respect from that of typical CAB proteins. In order to analyse in more detail the latter phenomenon and its possible relation to the function of the PSII-S protein, experiments were performed in etiolated and greening spinach seedlings. An inspection of the data obtained on light induced accumulation of different proteins readily shows that the PSII-S protein exhibits peculiar properties: in marked contrast to all other Chl-binding proteins and the ELIPs, the PSII-S protein attains relatively high levels in etiolated seedlings and is already present to a significant extent (> 50%) in the dark, i.e. under conditions of complete absence of chlorophyll (Fig. 3). These properties could be due to either a different overall folding pattern of the protein (A) or the binding of chlorophyll precursors and/or carotenoids (B). The former explanation appears to be plausible because the PSII-S protein is predicted to form four transmembrane helices

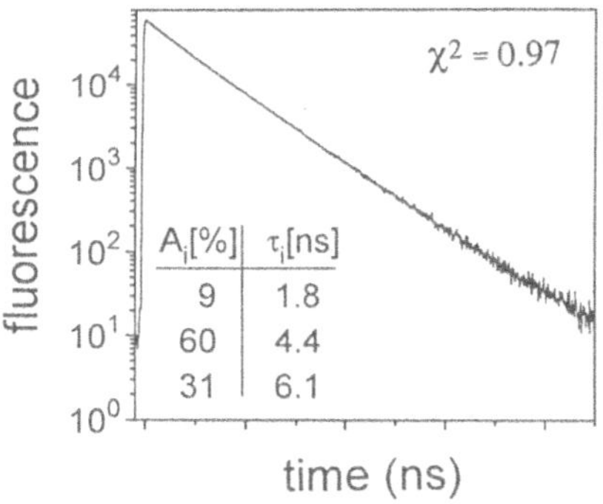

Fig. 2:
Fluorescence lifetime measurement of the purified PSII-S protein at 4°C. Fit quality is given by c^2_{Red}-values.

compared to the three of other CAB proteins. This structural feature permits the folding as a classical four helix bundle with helix-helix interactions giving rise to additional stability. To check for the alternative possibility (B) pigment analyses were performed by thin layer chromatography after acetone precipitation of the PSII-S protein isolated by isoelectric focusing from plants at different stages of greening (etiolated, after 4 or 12 hours of greening, mature species) (Fig. 4). It shows that the PSII-S fraction isolated from etiolated plants contained neither chlorophyll precursors nor carotenoids. Accordingly, a stabilisation by pigments other than Chl (explanation B) is not very likely. Furthermore, the pigment composition of the PSII-S protein isolated from plants after 4 or 12 hours of greening is virtually the same as that of the sample from mature plants. This finding together with the data of Fig. 3 shows that after 4 hours of greening the PSII-S protein is the only known Chl a/b-binding protein present in spinach plants.

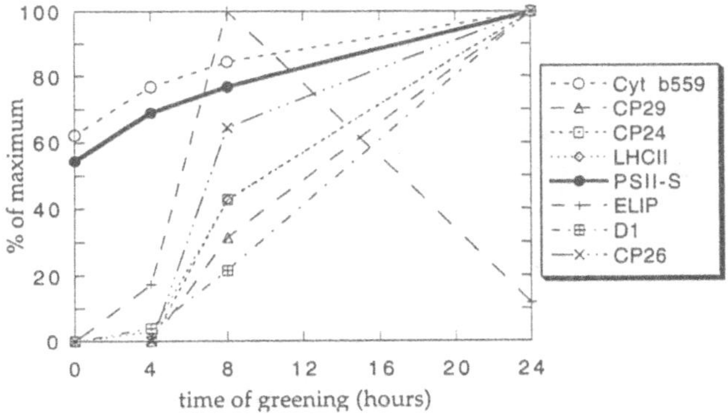

Fig. 3:
Accumulation of the PSII-S protein in plastid membranes during greening of etiolated seedlings in comparison to other PSII subunits.

The combined biochemical and biophysical data of this study suggest that the PSII-S protein is not an conventional light-harvesting protein, but rather might play a regulatory role in biosynthesis and assembly of the other chlorophyll-binding proteins of PSII. It is conceivable that during early stages of development, the PSII-S protein as the only pigment-binding protein could serve as a transient chlorophyll storage protein as a kind of ligand chaperone. In this way newly synthesised chlorophyll molecules would not be free species in the membrane, but kept in association with β-carotene thereby avoiding photo-damage via formation of chlorophyll triplets and singlet oxygen. Moreover, the chlorophylls would be present in a ligation competent state in the PSII-S protein prior to the subsequent accumulation of D1, D2, CP47, CP43 and the various Chl a/b binding

proteins. In mature thylakoid membranes the PSII-S protein could be involved in transient chlorophyll binding during high-light induced turn-over of PSII reaction centres and acclimative degradation of LHCII (13). It is also known that photosynthetic reaction centres can acquire chlorophyll molecules from the antenna proteins (14) and the PSII-S could possibly mediate such a process.
Furthermore, the carotenoids binding to the PSII-S protein and its presence in different mutants (unpublished) that lack all conventional CAB proteins (15), suggest the possibility that this protein could also play a role in violaxanthin/zeaxanthin interconversion and non-photochemical quenching.

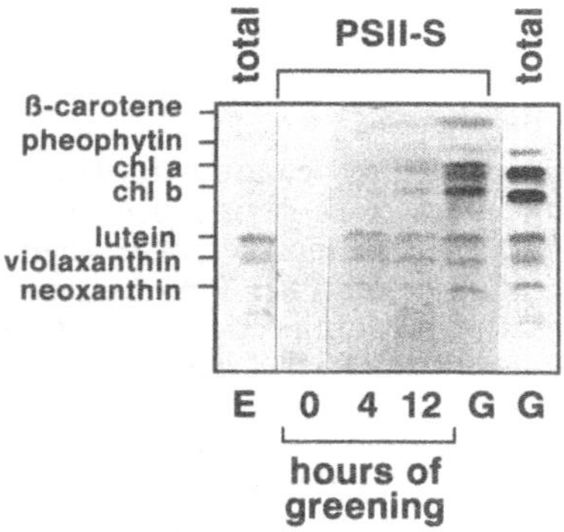

Fig. 4: Pigment composition of the PSII-S protein isolated from etiolated spinach seedlings after different times of illumination in comparisson to total pigment composition in etiolated (E) and green (G) plants.

4. References

1. Ljungberg, U., Åkerlund, H.-E. & Andersson, B. (1986) *Eur. J. Biochem. 158*, 477-482.
2. Wedel, N., Klein, R., Ljungberg, U., Andersson, B. & Herrmann, R. G. (1992) *FEBS Lett. 314*, 61-66.
3. Kim, S., Sandusky, P., Bowlby, N. R., Aebersold, R., Green, B. R., Vlahakis, S., Yocum, C. F. & Pichersky, E. (1992) *FEBS Lett. 314*, 67-71.
4. Bowlby, N. R. & Yocum, C. F. (1993) *Biochim. Biophys. Acta 1144*, 271-277.
5. Mishra, R. K. & Ghanotakis, D. F. (1993) *Photosynth. Res. 36*, 11-16.
6. Green, B. R. & Pichersky, E. (1994) *Photosynth. Res. 39*, 149-162.
7. Funk, C., Schröder, W. P., Green, B. R., Renger, G. & Andersson, B. (1994) *FEBS Lett. 342*, 261-266.
8. Funk, C., Schröder, W. P., Napiwotzki, A., Tjus, S.E., Renger, G. & Andersson, B. (1995) *Biochemistry*, in press.
9. Adamska, I., Kloppstech, K. & Ohad, I. (1993) *J. Biol. Chem. 268*, 5438-5444.
10. Thayer, S. S. & Björkman, O. (1990) *Photosynth. Res. 23*, 331-343.
11. Boardman, N. K. & Thorne, S. W. (1971) *Biochim. Biophys. Acta 253*, 222-231.
12. Liu, B., Napiwotzki, A., Eckert, H.-J., Eichler, H. J. & Renger, G. (1993) *Biochim. Biophys. Acta 1142*, 129-138.
13. Lindahl, M., Yang, D.-H., Andersson, B. (1995) *Eur. J. Biochem. 231*, 503-509.
14. Argyroudi-Akoyunoglou, J.H., Akoyunoglou, A., Kalosakas, K. & Akoyunoglou, G. (1982) *Plant Physiol. 70*, 1242-1248.
15. Palomares, R., Herrmann, R. G. and Oelmüller, R. (1993) *Eur. J. Biochem.* **217**, 345-352

Acknowledgements
The authors would like to thank the Swedish Natural Science Research Council, the Swedish Forestry and Agricultural Research Council and the Deutsche Forschungsgesellschaft for financial support. We also thank Prof. B.R. Green for taking an active interest in the project and for useful comments.

Sub-picosecond energy equilibration via energy transfer in the light harvesting antenna of LH-1 only mutants of *Rb. sphaeroides*, at RT and 4K; experiments and modelling

H. M. Visser, O. J.G. Somsen, F. van Mourik and R. van Grondelle.
Department of Physics and Astronomy, Vrije Universiteit Biological Sciences, De Boelelaan 1081, 1081 HV Amsterdam, the Netherlands.

1. Introduction

The first event in photosynthesis is absorption of the incident photon by specialised pigment-protein complexes and results in an excited state that first is rapidly transferred between pigments and then captured by a reaction centre in the form of a charge-separated state (1). This contribution describes the dynamics of the energy transfer within the core, light-harvesting-1 (LH-1) complex of the purple bacterium *Rhodobacter (Rb.) sphaeroides*. The samples were excited at 600 nm in the Q_x region and the transient difference spectra were recorded using a spectrophotometer described in (2). Instrument response time as observed from the 10-90% rise-time of the induced birefringence in CS_2 was 350 fs. The optical density (OD) at the peak absorption was 1.0. The experiments were performed on mutants containing LH-1 as their only pigment-protein complex, so that all observed dynamics can be ascribed to processes within LH-1. Similar experiments on membranes of *Rs. rubrum* (2) at room temperature (RT) revealed a sub-picosecond 12 nm redshift. In isolated B820 antenna subunits, no such shift was observed. Therefore, the red-shift was explained as the establishment *via energy transfer* of a Boltzman equilibrium over the states within the inhomogeneous distribution function (IDF). The *rate* at which the equilibration takes place is a measure for the rate of energy transfer and from the *size* of the shift the width of the IDF can be deduced.

2. Results and Discussion

The experiments on LH-1 of *Rb. sphaeroides* at RT could be described using the same theoretical framework. A similar value for the width of the IDF, 400 cm^{-1} (FWHM) was found, the optimal transfer time between two neighbouring sites, τ_0, was found to be 280 fs in this antenna, somewhat slower than for *Rs rubrum*. Whereas the extent of equilibration at RT is straightforward to predict (2), this is not the case at low temperatures. Then only down-hill energy transfer occurs and excitations may get trapped in a local energy minimum. Using the R^{-6} dependence of the Förster energy transfer and assuming a 100 fs optimal transfer time between neighbouring antenna sites and a square lattice, it will take a locally trapped excitation 4^6*0.1 = 410 ps to 'jump' four lattice distances. Thus, within a few hundred picoseconds after excitation, the energetically lowest site of a cluster or 'domain' of 64 can be reached. Assuming a Gaussian IDF, one calculates that this energy is lower than the average energy by an amount corresponding to the width of the IDF. Here we take a width of 200 cm^{-1} (FWHM) for the IDF at 4 K. Similar values, 240-

P. Mathis (ed.), Photosynthesis: from Light to Biosphere, Vol. I, 343–346.

260 cm^{-1} were used to model excitation energy-selective fluorescence spectra on B820 (3). Fig. 1 shows the result of the experiment. Compared to RT, the 4 K Δ OD spectra are about twice as narrow, like the steady-state absorption spectra. Again, a red-shift is observed. In fig. 2, the position of the isosbestic point as a function of delay time is plotted. Most of the shift takes place in the first 20 ps. Between 20 and 160 ps, the isosbestic point shifts from 137 to 151 cm^{-1} from its first detectable position.

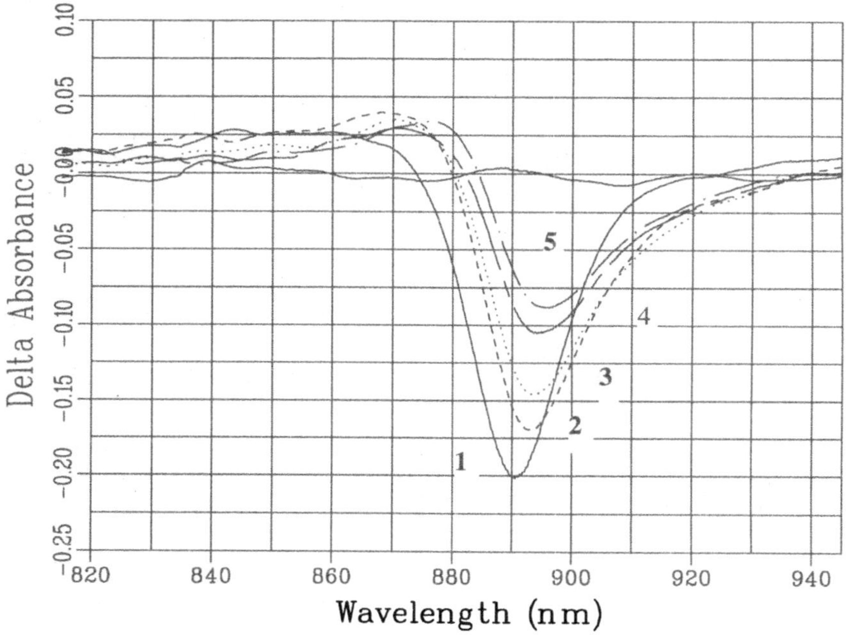

Fig. 1. *Transient difference spectra recorded for membranes of Rb.sphaeroides M2192 at 4K upon Q_x excitation at 600 nm. Relative delay times: 1: 0 fs, 2: 0.85 ps, 3: 1.7ps, 4: 5.5 ps, 5: 13.8 ps.*

The line in fig. 2 is the result of a simulation. The energy transfer rate between subunits at different energy was calculated using a Förster overlap integral between the homogeneous fluorescence and absorption spectrum as proposed for B820 in (3). The width of the IDF was taken 200 cm^{-1}, the transfer time between optimally overlapping subunits was 400 fs. As a lattice we used the optimally packed 12-ring structure shown in fig. 3. where we assume that each subunit corresponds with a dimer of two BChl *a* 's. We note that a different arrangement was proposed recently (4). To investigate the influence of the lattice structure on the dynamics of the equilibration, we compared simulations on the three different lattices shown in fig. 3. In all simulations, periodic boundary conditions were imposed to mimic large systems. The result of this comparison is shown in fig. 4. These simulations were performed for RT, using a 400 cm^{-1} width of the IDF. The simulated equilibration for the hexagonal and the square lattices yields very similar results, in the case of the 12-ring a slower phase is present, probably due to the lower connectivity and density of this structure.

In conclusion, we can model the observed sub-picosecond red-shifting of the transient Δ OD spectra using a model employing a reasonable set of parameters, both at RT and at

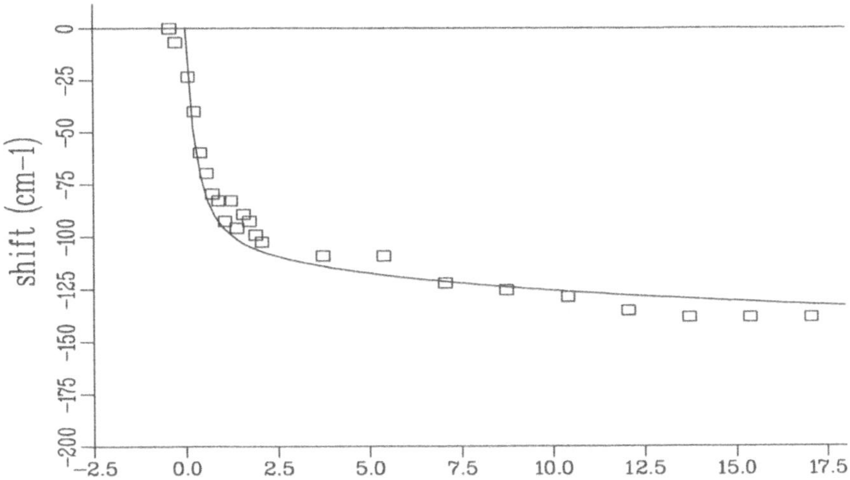

Fig. 2. *Position of the isosbestic point in the transient difference as a function of time delay (in ps) between excitation and probe pulse. Line: simulation.*

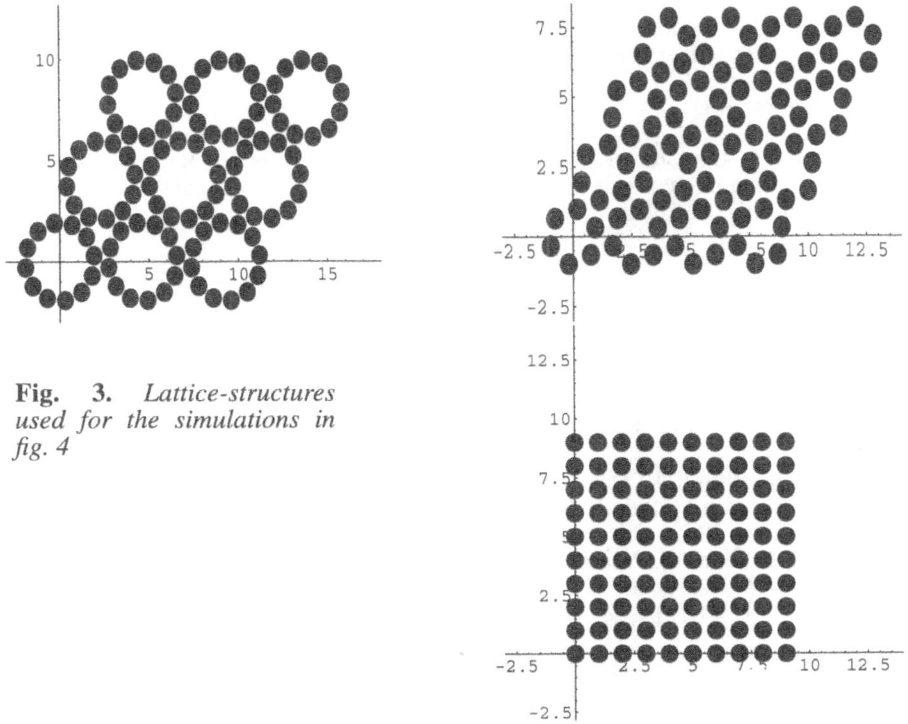

Fig. 3. *Lattice-structures used for the simulations in fig. 4*

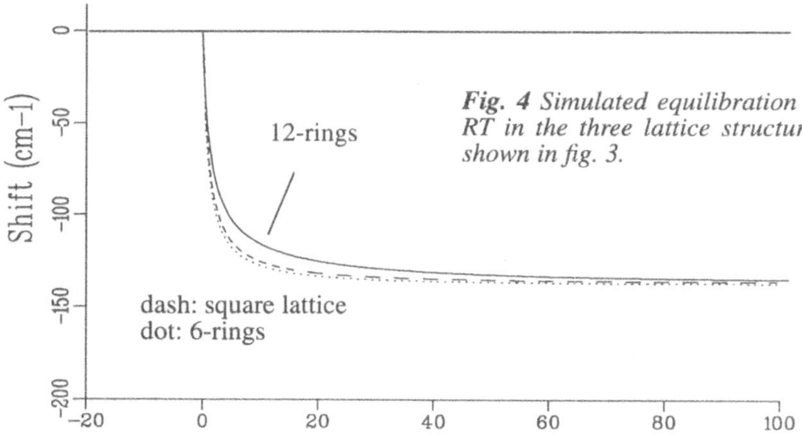

Fig. 4 Simulated equilibration at RT in the three lattice structures shown in fig. 3.

4K. The value found for the transfer time between neighbouring subunits is in agreement with earlier estimations from singlet-singlet annihilation studies (5), as well as recent time-resolved depolarization studies (6). Including inhomogeneity in the description of the excited state dynamics of the LH-1 antenna leads to a natural explanation of the observed 'shrinking' of the domains upon lowering sample temperature from RT to 4K (7). It is of interest that a similar theoretical framework has been used to describe sub-picosecond energy transfer in conjugated polymers and triplet energy transfer in glassy benzophenone at low temperature (8).

3. Acknowledgement.

This research was supported by the Netherlands Organization for Scientific Research (NWO) via the Dutch Foundation for Life Sciences (SLW) and the EC via contracts CT920796 and CT930278.

4. References

1 Van Grondelle, R.; Dekker, J.P.; Gillbro, T. and Sundström, V. **1994**. *Biochim. Biophys. Acta,* 1187, 1-65.
2 Visser H.M., Somsen O.J.G., Van Mourik F., Van Stokkum I.H.M. and Van Grondelle R., *Biophys. J,* in press.
3 Pullerits T., Monshouwer R., Van Mourik F. and Van Grondelle R. **1995**. *Chem. Phys.* 194, 395-407.
4 Karash, S., Bullough P.A. and Ghosh R. **1995**. *EMBO. J.* 14, , 631-638.
5 Bakker, J.C.G., Van Grondelle R., and Den Hollander W.Th.F.. **1983**. *Biochim. Biophys. Acta* 725: 508-518
6 Bradforth, S. Jimenez R., Van Mourik F., Van Grondelle R and Fleming G.R. *J. Phys. Chem,* submitted for publication.
7 Deinum G., T. J. Aartsma, R. van Grondelle and J. Amesz. **1989**. *Biochim. Biophys. Acta* 976:63-69.
8 Kersting R., Lemmer U., Mahrt R.F., Leo K., Kurz H., Bässler H. and Gobel E.O. **1994**. *Phys Rev. Lett.* 70, 3820-3823.

COMPARISON OF THE STACKING OF CHLOROPHYLLS IN CHLOROSOMES VERSUS AGGREGATES OF BACTERIOCHLOROPHYLL *C* AND CHLOROPHYLL *A* USING 2-D MAS NMR SPECTROSCOPY

Gert-Jan Boender[1], Teodor Silviu Balaban[3], Alfred R. Holzwarth[3], Kurt Schaffner[3], Jan Raap[1], Stephan Prytulla[2], Hartmut Oschkinat[2], & Huub J.M. de Groot[1]
[1]Leiden Instute of Chemistry, Gorlaeus Laboratoria, Leiden University, P.O. Box 9502, 2300 RA Leiden, The Netherlands; [2]EMBL P.O. Box 10 22 09, D-69117 Heidelberg, Germany; [3]Max Planck-Institut für Strahlenchemie, P.O. Box 10 13 65, D-45413 Mülheim an der Ruhr, Germany

1. Introduction

Probably structurally the most simple system for photosynthetic light-harvesting is the chlorosome, which is thought to contain extended bacteriochlorophyll *c* (BChl *c*) aggregates that are self assembled by intermolecular forces [1]. The mechanism of aggregation and the precise structure of the BChl *c* assemblies are currently a matter of debate. Since chlorophyll assemblies are micro-crystalline solids that are neither accessible to diffraction, nor to solution NMR techniques , we are now developing a solid-state NMR strategy for structure elucidation of amorphous multispin labelled solids using 1-D CP/MAS and 2-D CP/MAS correlation spectroscopy [2]. In the present note we summarise recent advancements towards a detailed structural characterisation of uniformly ^{13}C enriched chlorosomes of *Chlorobium tepidum* [3]. In addition, a comparison with two model systems, self-assembled dry [U-^{13}C]BChl *c* and [U-^{13}C]chlorophyll *a* (Chl *a*) aggregated with H_2O is provided. With absorption mode 2D MAS NMR correlation spectroscopy intermolecular dipolar correlations can be obtained, which provide spatial structural information to atomic resolution [2]. Using the novel structural refinement techniques, we find convincing evidence that the chlorophyll stacking in chlorosomes, dry Bchl *c* and Chl *a*/H_2O aggregates is highly similar.

2. Procedures

Uniformly ^{13}C enriched Chl *a* was isolated from a cell culture of *Chenopodium rubrum* (13.7 mg, >95% ^{13}C). The chlorophyll *a* /water aggregates were obtained by exposing the Chl *a* molecules to H_2O according to the procedures described in [4] except that the hydration of the chlorophyll oligomers was performed in n-heptane.
Uniformly ^{13}C enriched intact chlorosomes were isolated from a cell culture of *Chlorobium tepidum,* which was grown anaerobically at 47 °C under low-light conditions. After the isolation of the chlorosomes the [U-^{13}C]BChl *c* were isolated from the same culture. The aggregating was done in hexane [3].

3. Results and discussion

Fig. 1 shows the 2-D MAS NMR correlation spectra of the [U-^{13}C] chlorosomes, BChl *c* aggregates and Chl *a*/H_2O aggregates. The lines in fig. 1 illustrate the nearest neighbour

P. Mathis (ed.), Photosynthesis: from Light to Biosphere, Vol. I, 347–350.

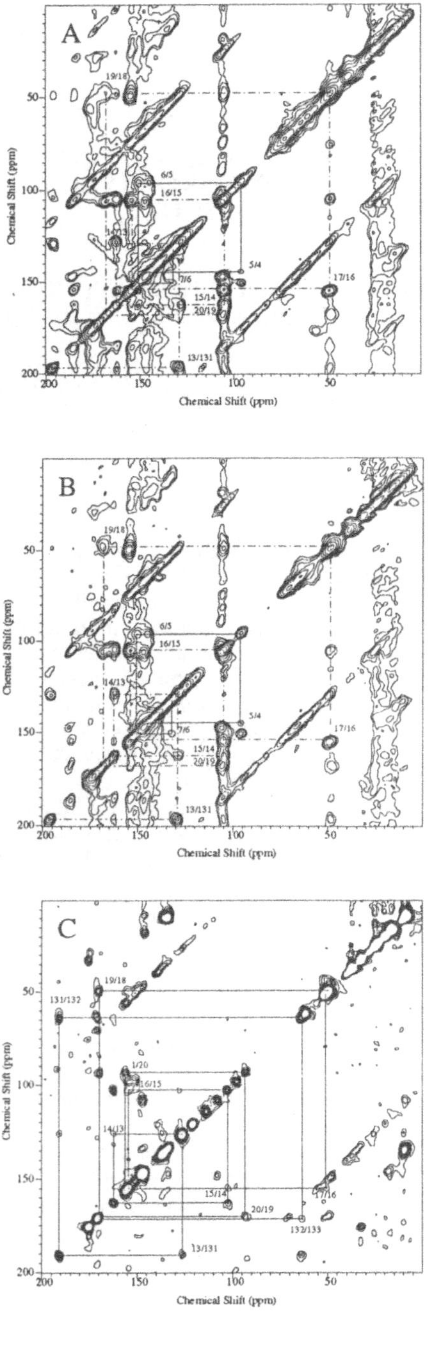

correlations along the ^{13}C framework of the molecule, leading to unambiguous assignments of NMR signals [2,3].

It appears that the chemical shifts of the BChl *c* in the chlorosomes and in the self aggregated form are identical within the experimental error. This strongly suggests that BChl *c* is organised in the same way in the chlorosomes and in the aggregates indicating that self-organisation of the BChl *c* molecules is the main mechanism supporting the structure of the chlorosomes. However, some resonances are subject to an aggregation shift $\Delta\sigma \geq 3$ ppm with respect to the solution value (cf. table 1). In fig. 2 the pattern of these aggregation shifts is visualised. The data suggest that there are globally two regions of intermolecular interaction in the aggregated forms.

In addition to the chemical shift information, intermolecular correlations which correspond to a short (<5 Å) intermolecular distances where obtained from the spectra. For instance, we observed in the spectra of the chlorosomes and the BChl *c* aggregates intermolecular correlations between the 13^1 and the 19 position, and the 14 and the 3^1 position of the BChl *c* molecule. In the spectra of the Chl *a*/H_2O

Fig. 1 Contour plot of a ^{13}C-^{13}C correlation spectrum of [U-^{13}C]chlorosomes collected at T ~ 278 K (A) and of [U-^{13}C]BChl c aggregates collected at room temperature (B), with a spinning speed $\omega_r/2\pi = 8000 \pm 3$ Hz and with mixing time $\tau_m = 1$ ms. In (C) the data for [U-^{13}C]Chl a/H_2O aggregates collected at room temperature with $\tau_m = 0.8$ ms, and $\omega_r/2\pi = 10000 \pm 5$ Hz are shown. The dotted and solid lines indicate two sequences of nearest neighbour correlations.

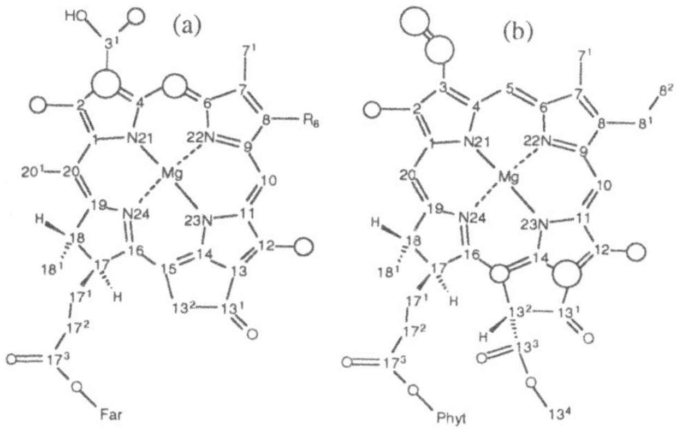

Table 1 Aggregation shifts ($\Delta\sigma \geq 3$ ppm) for BChl c and Chl a .

	BChl *c*	Chl *a*/H_2O
no	$\Delta\sigma$	$\Delta\sigma$
2^1	3,0 (1)	3,0 (0,5)
3^2	3,0 (0,5)	5,5 (0,5)
3^1		5,5 (0,5)
3	5,5 (1)	
5	4,5 (0,5)	
13		5,5 (0,5)
15		3,5 (0,5)
12^1	3,5 (1)	3,5 (0,5)

Fig. 2. Visualisation of the aggregation shifts of BChl c (a) and Chl a (b) with spheres which radius is proportional to the shift. The IUPAC numbering is used. The Chl a molecule has a phytyl chain and BChl c a farnesyl at position 17.

aggregates intermolecular correlations between the 13^1 and the 20, the 14 and the 2, and the 12 and the 4 are observed. This implies stacking according to fig. 3. The shifts induced by the ring-current of the π-system of the macro aromatic cycle of a neighbouring molecule are about +3 ppm for 2^1 and 12^1 and +5.5 ppm for a carbon above the Mg if the distance between two molecules is 3.5 Å [5]. This in agreement with the first Hunter-Sanders rule that the π-systems of two neighbouring chlorophylls are parallel, with an interplanar separation of 3.4-3.6 Å [6]. It implies that the distance between 19 and 13^1 and the distance between the 3^1 and 14 in the BChl *c* stacking is ca. 4 Å, while the distance between the 13^1 and the 20, the 14 and the 2 and the 12 and the 4 in the Chl *a* stacking is ca. 3.5 Å. Hence the NMR provides convincing evidence that the stacking is planar, and highly similar in BChl *c* and Chl *a*/H_2O.
Small-angle neutron scattering studies indicated that the shape of the aggregates form extended tubular micelles [7]. It was argued that this shape could be generated by rotating the ring ca. 16° along an axis perpendicular to the plane of the molecule [8]. The accuracy of our experiment is not yet sufficient to estimate the tilt angle and further refinement will be necessary. For this reason the chlorophylls in fig. 3 are not yet rotated with respect to each other. It is thought that aggregation of BChl *c* involves the interaction of the 3^1-OH oxygen with the Mg of the other molecule, and that the OH hydrogen should be hydrogen bonded to the 13^1-keto group of a third molecule [8]. The 3^1-OH of the BChl *c* is situated consequently above the Mg and the 3^2 above the edge of the neighbouring molecule. The expected ring current is ~ 3 ppm, which correlates well with the observed aggregation shift. The H_2O molecule in the Chl *a*/H_2O aggregate is expected to fulfil a similar functional rôle as the 3^1-OH of the BChl *c* in the aggregates [9]. In this case the 3^1 and 3^2 should be located above the centre of a neighbouring molecule, which could very well produced a ring current explaining the aggregation shift of ~ 5,5 ppm. Summarising, we conclude that the information about the aggregate structures obtained thus far with NMR yields the schematic arrangements of the chlorophylls shown in fig. 3.

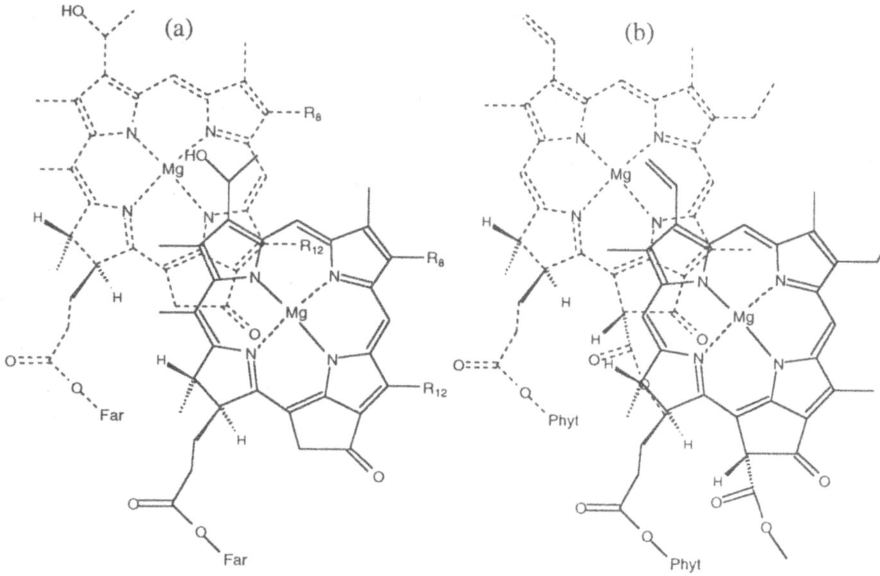

Fig. 3. Schematic representation of the proposed stacking of BChl c molecules in the chlorosomes and the BChl c aggregates (a) and for the Chl a molecules in the Chl a/water aggregates (b).

References

1 Holzwarth, A.R., Griebenow, K. and Schaffner K. (1992) J. Photochem. Photobiol. A: Chem. 65, 57-61
2 Boender, G.J., Raap, J., Prytulla, S., Oschinat, H. and de Groot, H.J.M. (1995) Chem. Phys. Lett 237, 502-508
3 Balaban, T.S., Boender, G.J., Holzwarth, A.R., de Groot, H.J.M., and Schaffner K. submitted
4 Katz, J.J., Bowman, M.K., Michalski, T.J. and Worcester, D.L. in Chlorophylls (Scheer, H. ed.) p.211-236, CRC press, Boca Raton
5 Giessner-Prettre, C. and Pullman B (1971) J. Theor. Biol. 31, 287-294
6 Hunter, C.A. and Sanders, J.K.M.(1990) J. Am. Chem. Soc. 112, 5525-5534
7 Worcester, D.L., Michalski, T.J., and Katz, J.J. (1986) Proc. Natl. Acad. Sci. USA 83, 3791-3795
8 Holzwarth, A.R. and Schaffner, K. (1994) Photosynthesis Research. 41, 225-233
9 Chow, H.-C., Serlin, R. and Strouse, C.E. (1975) J. Am. Chem. Soc. 97, 7230-7242

Correlation of non-conservative CD signals, absorption spectra and dimer structure. Application to the B820-subunit.

M.H.C. Koolhaas, G. van der Zwan, F. van Mourik, R. van Grondelle.
Department of Physical Chemistry
Vrije Universteit
Amsterdam.

The B820-antenna subunit is an ideal model system. It is formed by two Bchl*a* molecules. The eigenfunctions of the dimer are calculated using excitonic interaction between the product monomer eigenfunctions. The relative contributions of the monomer transitions to the dimer spectrum depends on the dimer transition, interaction energies and the monomer energy levels.

The absorption spectrum can be calculated using a two-two-level-system. The CD spectrum of the B820 subunit is nonconservative in the range of 700 to 850 nm. The smallest system that leads to a nonconservative CD spectrum is a two-three-level-system. The two-three-level-system associated with two possible transitions from the ground state leads to correct CD and absorption spectra in the range from 550 nm to 850 nm. The involved monomer energy levels are: E_0=0 Joules, E_{Q_y}=2.48 10^{-19} Joules (i.e. 800 nm) and E_{Q_x}=3.30 10^{-19} Joules (i.e. 600 nm). The values of the monomer transition vectors are $|\vec{\mu}_{Q_y}|$=6.3 Debye and $|\vec{\mu}_{Q_x}|$=1.87 Debye [1]. The vectors are along the long and the short molecular axes respectively.

The Q_y part of the spectra, i.e. from 700 to 850 nm, is of main interest. The spectrum in the region from 550 to 850 nm will be calculated under the condition that the Q_y part is nonconservative. Calculations on the line intensities in the absorption spectrum yields the angle between $\vec{\mu}_{1Qy}$ and $\vec{\mu}_{2Qy}$. The excitonic interaction energy is directly calculated from the line positions. The CD spectrum is nonconservative if $\vec{\mu}_{1Qy}$, $\vec{\mu}_{2Qy}$ and $\vec{R}_{12}$ are in a single plane. For this configuration the contributions from the Q_y vectors to the spectrum are negligible. The CD spectrum is dominated by the Q_x transition vectors: $\vec{\mu}_{1Qx}$ and $\vec{\mu}_{2Qx}$. The contributions of these latter transitions are of higher order than from Q_y and therefore nonconservative CD spectra are low in intensity.

For every orientation of $\vec{R}_{12}$ there is only one orientation of monomer 2 that is consistent with the experimental line positions in the range 700-850 nm and the polarization of the absorbed light. All possible combinations of rotation and translation of monomer 2 are shown in the first figure.

P. Mathis (ed.), Photosynthesis: from Light to Biosphere, Vol. I, 351–354.

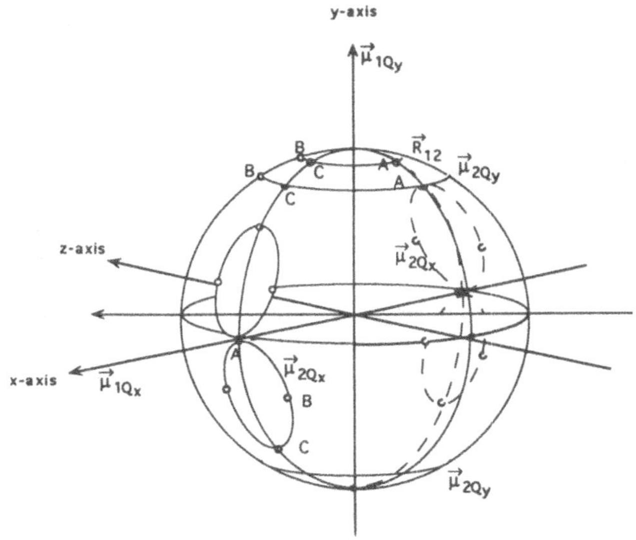

Figure 1: All possible orientations and translations of monomer 2.

There are exceptions though. At positions A and C the CD are exceptions since intensities are zero. At C all vectors are in one plane so the intensity is zero. At A there is no mixing of the Q_x monomer wavefunctions into dimer wavefunctions belonging to the two lowest excited states.

The intensities of the lines in the CD spectrum as a function of the orientation and translation of monomer 2 from C to A (see figure 1) are depicted below.

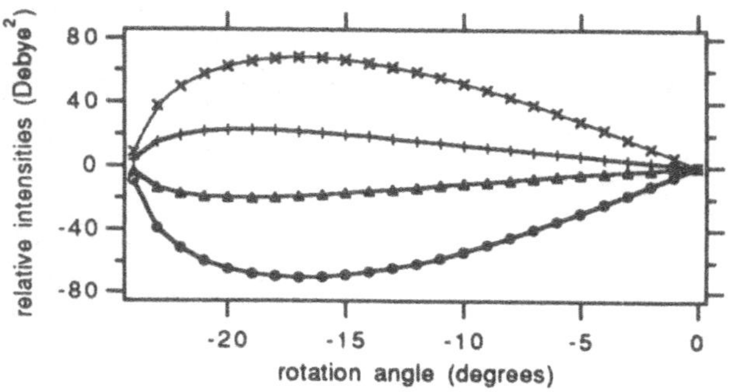

Figure 2: Rotation along the upper circle (see figure 1).
Line positions of the triangles are at 820 nm, the plus-signs are at 783 nm, the circles are at 600 nm and the crosses are at 599 nm.

The intensities change as a function of the orientation of $\vec{R}_{12}$, since the relative orientations of the Q_x vectors in respect to $\vec{R}_{12}$ changes.
The restriction that $\vec{R}_{12}$, $\vec{\mu}_{1Q_y}$ and $\vec{\mu}_{2Q_y}$ must be in a common plane is illustrated in the next figure. Molecule 2 is slighly rotated around the axis perpendicular to the molecular plane of monomer 1. At 0.5° monomer 2 is at orientation B. A change of 2 degrees in the orientation of monomer 2 leads to a conservative spectrum.

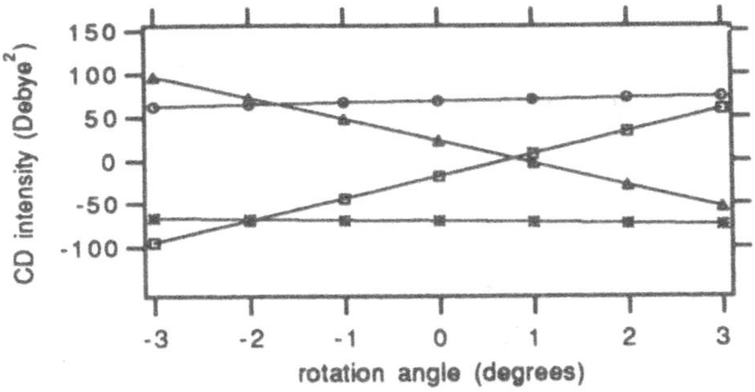

Figure 3: Rotation perpendicular to the circle.
Line positions of the squares are at 820 nm, the triangles are at 783 nm, the stars are at 599 nm and the circles are at 600 nm.

The line shapes of absorption and CD spectra must have the same physical origin. Mechanisms of inhomogeneous broadening can be variation in $|\vec{R}_{12}|$, fluctuations in the orientations and diagonal disorder. The structure of the dimer must be rather rigid, since the intensities of the non-conservative CD spectrum change drastically with structural modifications. This conclusion is supported by calculations using a Gaussian distribution in $|\vec{R}_{12}|$: The calculated lineshapes are very dissimilar to the experimental ones. The lineshapes in the absorption and the CD spectra calculated using the diagonal disorder model are in agreement with the experimental lineshapes [2]. Using this model the peak ratio of the lines in the region 700-850 nm is similar to the spectra obtained by Chang [3] when the angle between the Q_y transition vectors is 180-24 degrees.

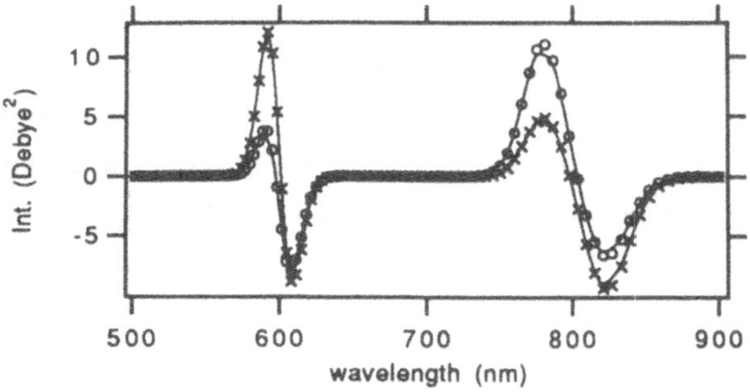

Figure 4: The CD spectrum at B is denoted with circles and B_T with stars.

The peak ratio between the lines in the 550-700 nm region and the corresponding ones 700-850 nm area must determine the direction of $\vec{R}_{12}$ and so the orientaton of monomer two.

CONCLUSIONS

The combination of a non-conservative CD spectrum and the absorption spectrum leads to the following conclusions. For every $\vec{R}_{12}$ the orientation of the momomer 2 is determined in the molecular frame of monomer 1: $\vec{R}_{12}$, $\vec{\mu}_{1Q_y}$ and $\vec{\mu}_{2Q_y}$ are in one plane.

The intensities of non-conservative CD signals are very sensitive to changes in the orientations of the monomers in the dimer. The non-conservativity originates from higher order (perturbations) transitions. So non-conservative CD signals are low in intensity.

The diagonal disorder model fits the line shapes nicely both in the CD spectrum and in the absorption spectrum. A consequence of the model is that the monomers are orientated anti-parallel relative to each other. The structure of the dimer is essentially rigid.

References

[1] Visschers, R.W., Chang, M.C., Mourik, F. van, Parkes-Loach, P.S, Heller, B.A., Loach, P.A., and Grondelle, R. van. (1991) Biochemistry 30,5734-5742.

[2] Koolhaas, M.H.C., Mourik, F. van, Zwan, G. van der, and Grondelle, R. van (1994) J. of Luminescence 60 and 61 515-519.

[3] Chang, M.C,Callahan, P.M., Parkes-Loach, P.S, Cotton, T.M., and Loach, P.A (1990) Biochemistry 29, 421-429.

ENERGY TRANSFER AND EXCITON COUPLING IN ISOLATED B800-850 COMPLEXES OF *CHROMATIUM TEPIDUM*.

John T.M. Kennis, Alexandre M. Streltsov, Thijs J. Aartsma, Tsunenori Nozawa* and Jan Amesz.

Department of Biophysics, Huygens Laboratory, University of Leiden, P.O. Box 9504, 2300 RA Leiden, The Netherlands; *Department of Biochemistry and Engineering, Tohoku University, Sendai 980, Japan.

1. Introduction

In purple bacteria, the so-called core antenna, also referred to as LH1, is associated with the reaction centre and usually absorbs in the same wavelength region. In many species of purple bacteria there is also a peripheral antenna, called LH2, located further away from the reaction centre and absorbing at shorter wavelength.

Recently, the crystal structure of the B800-850 (LH2) complex of *Rhodopseudomonas acidophila* has been determined by means of X-ray crystallography [1]. The complex appeared to consist of a ring of nine α and β subunits in a nine-fold circular symmetry, which bind eighteen BChl 850 molecules, nine BChl 800 molecules and nine carotenoids. The nine BChl 800 molecules form a ring on the cytoplasmic side, and they lie 21 Å apart. On the periplasmic side, eighteen BChl 850 molecules form a ring, and the centre-to-centre distance between the molecules is about 9 Å. Femtosecond transient absorption studies on isolated LH2 complexes of *Rhodobacter sphaeroides* have indicated that the energy transfer from BChl 800 to BChl 850 takes place in less than one ps [2,3].

In this paper we describe the results of a spectrally resolved transient absorption study with a time resolution of 200 fs on isolated B800-850 complexes of *Chromatium tepidum*. Energy transfer from BChl 800 to BChl 850 was found to occur in 0.8 ps. From the relative magnitudes of the absorbance difference signals of excited BChl 800 and BChl 850 we conclude that the spectral properties of BChl 850 are largely determined by strong exciton interactions between BChl molecules in a symmetric arrangement.

2. Materials and methods

B800-850 complexes of *C. tepidum* were isolated as described earlier [4]. Absorbance difference spectroscopy with a time resolution of 200 fs was performed by means of a home-built amplified femtosecond laser system with optical multichannel analyzer (OMA) detection. All measurements were performed at room temperature.

P. Mathis (ed.), Photosynthesis: from Light to Biosphere, Vol. I, 355–358.

3. Results

Figure 1 shows the time-resolved spectra upon excitation at 800 nm with pump and probe beam polarised at the magic angle at increasing delays between -0.33 and 3 ps. At a delay of 0.13 ps, a bleaching near 800 nm was observed, which is due to formation of excited BChl 800. The bleaching is due to a decrease of absorption, caused by ground state depletion and to stimulated emission from the excited level to the ground state. At increasing delays, the amplitude of the bleaching near 800 nm progressively decreased, whereas a bleaching appeared around 860 nm, indicating energy transfer from BChl 800 to BChl 850. This bleaching was accompanied by an absorbance increase at wavelengths shorter than 845 nm. At a delay of 3 ps, energy transfer from BChl 800 to BChl 850 was completed, and the absorbance difference spectrum was now entirely due to excited BChl 850. The bleaching, with a maximum at the long-wavelength side of the absorption band, can be attributed to ground state depletion of BChl 850 combined with stimulated emission. The absorbance increase at wavelengths shorter than 845 nm can be assigned to excited state absorption by BChl 850. It is obvious that the final bleaching near 860 nm at a delay of 3 ps was considerably larger than the initial bleaching at 0.13 ps near 800 nm.

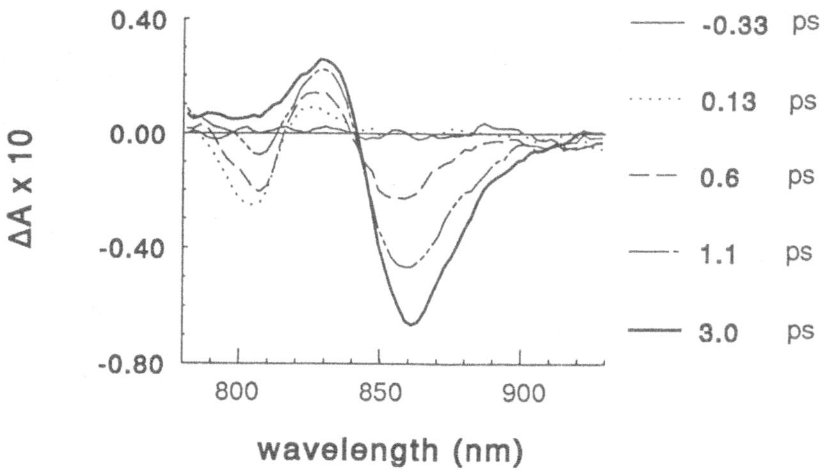

Figure 1: Absorbance difference spectra upon excitation at 800 nm at the delays indicated.

At increasing delays, the maximum of the bleaching around 860 nm shifted to the red by 5 nm with a time constant of about 1 ps. There are several mechanisms that may underlie this red shift, such as vibrational relaxation, exciton scattering to lower lying exciton levels, or energy transfer between aggregated B800-850 complexes.

Figure 2 gives the kinetics at 802 nm and 860 nm. At 802 nm they show a bleaching which decayed with a single time constant of 0.8 ps into an absorbance increase. The kinetics at 860 nm showed a rise time of 0.8 ps. After about 3 ps, the signal decreased somewhat, which is presumably a consequence of singlet-singlet annihilation. These results indicate that energy transfer from BChl 800 to BChl 850 occurs with a time constant of 0.8 ps, which is essentially the same value as that derived from the decay of excited BChl 800 in *Rb. sphaeroides* [2,3]. This suggests that the distances between BChl 800 and BChl 850 in LH2 complexes from *Rb. sphaeroides* and *C. tepidum* are not much different.

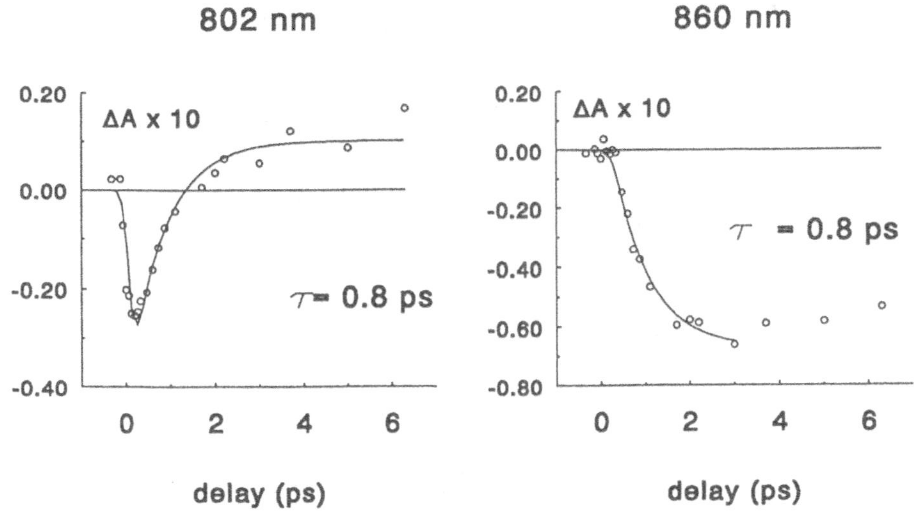

Figure 2: Kinetics of absorbance changes at 802 nm and 860 nm upon excitation at 800 nm. At 802 nm, the signal decays with a single exponent of 0.8 ps. At 860 nm, the signal is fitted with a rise time of 0.8 ps.

4. Discussion

Our results unambiguously indicate that energy transfer from BChl 800 to BChl 850 in isolated B800-850 complexes of *C. tepidum* occurs with a time constant of 0.8 ps. This is reflected both by the decay of excited BChl 800 and by the rise of excited BChl 850. It may be noted that in previous time-resolved studies of B800-850 complexes from *Rb. sphaeroides*, either only the decay of excited BChl 800 was monitored during one-colour measurements [3], or the rise time of excited BChl 850 did not match the decay of excited BChl 800 [2].

The most peculiar aspect of the data presented here is the relative magnitude of the absorbance difference signal measured for excited BChl 850 as compared to that of excited BChl 800. In Figure 1, the second spectrum gives the time-resolved absorbance difference at a delay of 0.13 ps, and it is the maximum bleaching that is seen for excited BChl 800. The fifth spectrum is taken at a delay of 3 ps, when all excitations have been transferred to BChl 850. Obviously, the absorbance difference signal induced by excitation of BChl 850 is much larger than that by excitation of BChl 800. More precisely, the integrated bleaching of BChl 850 is five times larger than that of BChl 800.

In the B800-850 complex from *Rps. acidophila* the BChl 800 molecules are 21 Å apart, and to a first approximation one may regard them as monomers. BChl 800 in *C. tepidum* probably occurs as dimers with two exciton components of equal amplitude [4]. Excitation of BChl 800 then also produces a monomeric bleaching. It thus appears that the bleaching observed upon excitation of BChl 850 is five times larger than for monomeric BChl. A similar conclusion was obtained by comparing the amplitude of the LH1 bleaching to that caused by photooxidation of the primary donor [5,6].

Our results may qualitatively be understood as follows [7]. Let us assume that the arrangement of BChl 850 molecules in the B800-850 complex of *C. tepidum* is similar to that in the corresponding complex of *Rps. acidophila.* The crystal structure of the *Rps. acidophila* LH 2 complex indicates a highly symmetric ring structure, and the distances between BChl 850 molecules lead to strong dipolar intermolecular couplings of about 500 cm^{-1}. We thus are dealing with a circular aggregate of strongly interacting molecules. Strict selection rules for optical transitions apply for such aggregates, and the oscillator strength of the constituent monomers is concentrated in a few optically allowed transitions [7-9]. Excitation of the aggregate then leads to ground state depletion of a transition carrying the oscillator strength of several monomers.

From the singly excited exciton level, extensive absorption can take place to the doubly excited exciton level [7], which accounts for the observed absorbance increase at wavelengths shorter than 845 nm.

In conclusion, our results indicate that the spectral properties of BChl 850 in the B800-850 complex of *C. tepidum* are largely determined by strong dipolar interactions between BChl 850 molecules in a circular symmetric arrangement, which lead to concentration of oscillator strength in a few optically allowed transitions. These conclusions are based on the assumption that the structure of the B800-850 complex of *C. tepidum* is similar to that of *Rps. acidophila,* and is the same in crystalline form and in solution.

6. Acknowledgements

The investigations were supported by the Life Science Foundation (SLW), which was subsidised by the Netherlands Organisation for Scientific Research (NWO).

7. References

[1] MacDermott, G., Prince, S.M., Freer, A.A., Hawthornthwaite-Lawless, A.M., Papiz, M.Z., Cogdell, R.J., Isaacs; N.W. (1995) Nature 374, 517-521.
[2] Shreve, A.P., Trautman, J.K., Frank, H.A., Owens, T.G., Albrecht, A.C. (1991) Biochim. Biophys. Acta 1058, 280-288.
[3] Hess, S., Feldchtein, F., Babin, A., Nurgaleev, I.; Pullerits, T., Sundström, V. (1993) Chem. Phys. Lett. 216, 247-257
[4] Nozawa, T., Fukada, T., Hatano, M., Madigan, M.T. (1986) Biochim. Biophys. Acta 852, 191-197.
[5] Kennis, J.T.M, Aartsma, T.J., Amesz, J. (1994) Biochim. Biophys. Acta 1188, 278-286.
[6] Xiao, W., Lin, S., Taguchi, A.K.W., Woodbury, N.W. (1994) Biochem. 33, 8313-8322.
[7] Novoderezhkin, V.I., Razjivin, A.P. (1993) FEBS Lett. 330, 5-7.
[8] Reddy, N.R.S., Picorel, R., Small, G.J. (1992) J. Phys. Chem. 96, 6458-6464.
[9] Pearlstein R.M. and Zuber, H. (1985) in Antennas and Reaction Centers of Photosynthetic Bacteria (Michel-Beyerle M.E. ed.) pp 53-61; Springer Verlag, Berlin.

THERMOLUMINESCENCE STUDIES ON THE MECHANISM OF PHOTON PROTECTION

C. Hagen[1], A. A. Pascal[2], P. Horton[2] & Y. Inoue[3]
[1]Institute of Botany, Friedrich Schiller University Jena, Am Planetarium 1, D-07743 Jena, Germany; [2]Robert Hill Institute, Dept. of Mol. Biol. & Biotechnol., University of Sheffield, P.O. Box 594, Sheffield S10 2UH, UK; [3]Solar Energy Research Group, RIKEN Institute, Hirosawa 2-1, Wako, Saitama 351-01, Japan

1. Introduction

Hypothetically, the molecular mechanism for the main part of non-photochemical chlorophyll (Chl) fluorescence quenching (q_N), i.e. the high-energy-state Chl fluorescence quenching (q_E) [1], (1) is based on protonation-induced aggregation of the light-harvesting pigment-protein complex of photosystem II (PS II, LHC II), (2) violaxanthin (Vio) cycle carotenoids modulate this structural change, and (3) LHC II aggregation allows one or more quenching Chl-Chl interactions [2, 3].

The increase of the Chl-Chl interaction strength, perhaps by dimerization of closely associated Chl molecules of LHC II, results in specific changes in absorption and fluorescence properties [3]. In order to investigate structural changes in LHC II, we used low-temperature thermoluminescence (TL). Two basic characteristics of the broad-banded TL component peaking at 110 K (Z-band) [4] have made it promising for studies on the molecular mechanism of q_E. (1) It seems to arise mainly from recombination of charged Chl species of the peripheral antenna system [5, 6]. (2) The emission intensity has been shown to depend on the interaction strength of the Chl molecules involved [5].

In a previous part of this study, we have shown that Z-band emission is influenced by the capacity of excitation energy dissipation in the pigment bed [7]. In this paper, we summarize results on the influence of structural changes of the antenna system on Z-band emission.

2. Procedure

Thylakoid membranes and LHC II were prepared as described in the previous part of this study [7]. Thylakoid membranes were uncoupled by 2 μM nigericin, and de-epoxidation was performed as described before [7]. LHC II aggregates were resuspended by gentle sonication at pH 7.6 or pH 5.2. Up to 12 mM dodecyl maltoside (DM) was present to control the aggregation state. In samples containing antimycin A, 20 μM were added before lowering the pH. The final Chl concentration was 300 μg Chl ml^{-1} for measurements of Z-band and 77 K fluorescence emission spectra. Pigment analysis was as described previously [7, 8].

Thermoluminescence glow curves and Z band emission spectra were obtained as described previously [7, 9]. Thermoluminescence spectra represent accumulated data from 12 (thylakoids) or 10 (LHC II) measurements. Steady state fluorescence emission

P. Mathis (ed.), Photosynthesis: from Light to Biosphere, Vol. I, 359–362.

spectra of thylakoid and LHC II preparations were detected at 77 K (excitation: 460 to 480 nm; emission slit: 2 nm; optical pathlength: 0.4 mm; Model RF-540, Shimadzu). Fluorescence intensities were normalized to the emission of fluorescein as an internal standard [10].

3. Results and Discussion

At least three competitively acting processes might be considered to influence simultaneously the Z-band emission yield during LHC II aggregation, and during induction of q_E in thylakoid membranes. (1) Enhancement of fluorescence quenching should diminish similarly Z-band TL by radiationless deactivation of singlet-excited Chl. (2) The heat release involved in fluorescence quenching processes can detrap Z-band emitting charges [7]. (3) Aggregation-induced reinforcement of Chl-Chl interactions increases the charging yield [5]. Figure 1 depicts the relationship between fluorescence and Z-band emission characteristics in dependence on the aggregation state of LHC II. The strong fluorescence decrease upon aggregation (Fig. 1B) was not paralleled by the reduction of Z-band emission (Fig. 1A). The simplest explanation of this discrepancy is that LHC II aggregation causes an increase in the efficiency of Z-band charging, together with a decrease in Z-band intensity as a result of energy dissipation. The resultant effect would be a smaller decrease in Z-band emission than in fluorescence intensity. At longer wavelengths, the enhancing effect of aggregation on Z-band charging was obviously

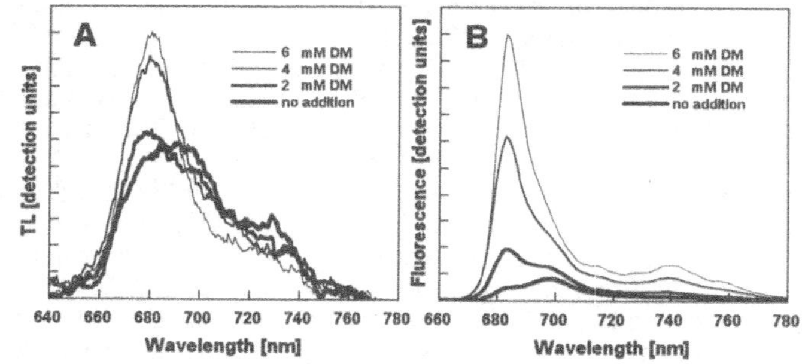

Figure 1: Aggregation-dependent changes in Z-band emission (A) and 77 K fluorescence (B) spectra of LHC II isolated from dark-adapted leaves (pH 7.6).

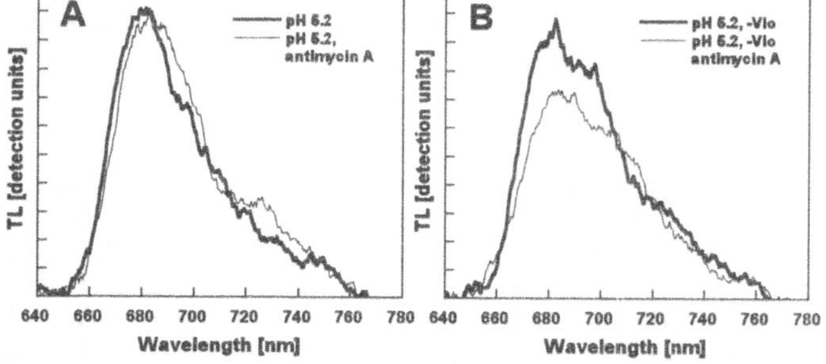

Figure 2: Influence of antimycin A on acidification-induced changes in Z-band emission spectra of microaggregated LHC II (3 mM DM). The PS II antenna complexes were isolated from dark-adapted (A) or light-activated (B) leaves.

dominant since stronger Z-band emission above 695 nm was detected at low detergent concentrations (Fig. 1A). Similar results were obtained in Vio-depleted material (not depicted), except for stronger aggregation found at a low concentration of DM.

The protonation-induced fluorescence quenching in microaggregates of LHC II formed at low detergent concentrations is a more appropriate model *in vitro* for q_E-related processes *in vivo* [3, 11]. Even more useful is the action of antimycin A as an inhibitor of protonation-induced fluorescence quenching and aggregation in LHC II [11]. It provides a functional link to the q_E mechanism since it has been shown to inhibit q_E in chloroplasts without decreasing the transthylakoid proton gradient, irrespective of the level of Vio de-epoxidation [12]. We observed the detergent-like activity of antimycin A in Vio-depleted LHC II only. In contrast to a decrease of 77 K fluorescence in the absence of antimycin A (20.4 % quenching at 698 nm), aggregation induced enhancement of Z-band emission between 660 and 710 nm in Vio-depleted LHC II microaggregates (Fig. 2B). Samples with an high Vio content showed reduced fluorescence quenching (6 %), and minor differences between the Z-band emission spectra of both variants (Fig. 2A).

Uncoupled thylakoid membranes were used to compare aggregation of isolated LHC II with protonation-induced conformational changes in the membrane-bound pigment bed. In order to change specifically the relative content of q_N-related xanthophylls, ascorbate-induced de-epoxidation of Vio was performed in the dark at pH 5.2 to activate the de-epoxidase [7, 13]. For samples retaining the dark-adapted (Vio-enriched) state, de-epoxidation was prevented by 5 mM dithiothreitol (DTT). The strongest 77 K fluorescence quenching upon acidification was found in Vio-depleted thylakoid membranes (20.1 % at 694 nm). The inhibition of de-epoxidation reduced this value to 11 %, while antimycin A reduced it to 2 %.

Protonation-induced and antimycin A-sensitive enhancement of Z-band emission at wavelengths typical for LHC II aggregation (690 to 700 nm) was stronger in Vio-depleted than in Vio-enriched thylakoid membranes (Fig. 3). The xanthophyll-independent increase at wavelengths below 680 nm might be discussed in parallel with the different peaks in fluorescence quenching spectra of q_E in dark-adapted (680 nm) or light-activated material (700 nm) [14], thus indicating a basic pH-dependent LHC II aggregation [3]. The Z-band increase upon acidification at wavelengths characteristic for PS I-emission might originate from parallel aggregation of LHC I (Fig. 3A). Antimycin A inhibited completely the spectral changes in the PS II waveband, but only slightly at wavelengths above 730 nm

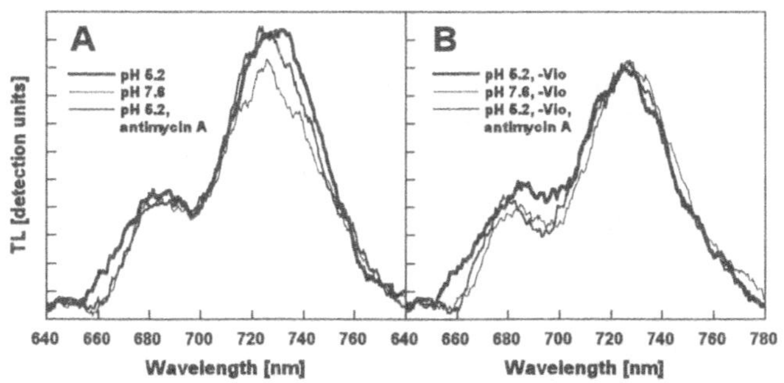

***Figure 3*: Influence of acidification, antimycin A and Vio de-epoxidation on spectral characteristics of Z-band emission in uncoupled thylakoid membranes. If noted, 20 μM antimycin A were added before lowering of the pH. (A): Control samples. De-epoxidation was inhibited by 5 mM DTT. (B): Vio-depleted samples.**

(Fig. 3). This slight inhibition can be explained with changes in LHC II, too, since a small increase during LHC II aggregation was observed in this waveband (Fig. 1A). Thylakoid membranes after Vio de-epoxidation did not exhibit a pH-dependent enhancement of Z-band emission above 705 nm (Fig. 3B). That might be due to xanthophyll-dependent dissipation processes in PS I causing local heating and Z-band discharge in LHC I. It was shown previously that Z-band emission in the PS I waveband is more sensitive to detrapping than at PS II wavelengths [7].

In LHC II microaggregates, pH-dependent and antimycin A-sensitive formation of larger aggregates was accompanied by an increase in Z-band emission (Fig. 2B), although fluorescence was diminished. A similar pH-dependent and antimycin A-sensitive Z-band increase at LHC II-related wavelengths was detected in thylakoid membranes (Fig. 3B). These observations are consistent with a stronger enhancing than quenching influence of aggregation-amplified Chl-Chl interactions on Z-band TL, especially at an already high level of those interactions as compared to trimeric LHC II. It suggests that q_E in the membrane-bound PS II antenna is accompanied by reinforced interactions between Chl molecules. Additionally, our results support the view that changes in the Vio cycle carotenoids have a direct structural impact on the aggregation level of LHC II, thus modulating photon protection by control of the LHC II organization in the energized thylakoid membrane [3, 15].

References

1 Briantais, J.-M., Vernotte, C., Picaud, M. and Krause, G.H. (1979) Biochim. Biophys. Acta 548, 128-138

2 Horton, P., Ruban, A.V., Rees, D., Pascal, A.A., Noctor, G. and Young, A.J. (1991) FEBS Lett. 292, 1-4

3 Horton, P. and Ruban, A.V. (1994) in Photoinhibition of Photosynthesis - from Molecular Mechanisms to the Field (Baker, N.R. and Bowyer, J.R., eds.), pp. 111-128, Bios Scientific Publishers, Oxford

4 Vass, I. and Inoue, Y. (1992) in The Photosystems: Structure, Function and Molecular Biology (Barber, J., ed.), pp. 259-294, Elsevier, Amsterdam

5 Sonoike, K., Koike, H., Enami, I. and Inoue, Y. (1991) Biochim. Biophys. Acta 1058, 121-130

6 Noguchi, T., Inoue, Y. and Sonoike, K. (1993) Biochim. Biophys. Acta 1141, 18-22

7 Hagen, C., Pascal, A.A., Horton, P. and Inoue, Y. (1994) Photochem. Photobiol., in press

8 Gilmore, A.M. and Yamamoto, H.Y. (1991) J. Chromatogr. 543, 137-145

9 Sonoike, K. and Inoue, Y. (1992) J. Lumines. 51, 129-137

10 Ruban, A.V., Rees, D., Noctor, G.D., Young, A. and Horton, P. (1991) Biochim. Biophys. Acta 1059, 355-360

11 Ruban, A.V., Young, A. and Horton, P. (1994) Biochim. Biophys. Acta 1186, 123-127

12 Gilmore, A.M. and Yamamoto, H.Y. (1992) Proc. Natl. Acad. Sci. USA 89, 1899-1903

13 Pfündel, E. and Dilley, R.A. (1993) Plant Physiol. 101, 65-71

14 Ruban, A.V., Young, A.J. and Horton, P. (1993) Plant Physiol. 102, 741-750

15 Osmond, C.B. (1994) in Photoinhibition of Photosynthesis - from Molecular Mechanisms to the Field (Baker, N.R. and Bowyer, J.R., eds.), pp. 1-24, Bios Scientific Publishers, Oxford

EXCITED STATE DYNAMICS IN PHOTOSYNTHETIC ANTENNA COMPLEXES STUDIED WITH ACCUMULATED PHOTON ECHOES.

R.J.W. Louwe and T.J. Aartsma
Department of Biophysics, University of Leiden
P.O. Box 9504, 2300 RA Leiden, The Netherlands

1.Introduction

Generally, energy transfer is described either by a random walk proces (incoherent) or as wavelike (coherent) energy transfer, depending on the homogeneous linewidth compared to the strength of the dipolar interaction between neighbouring pigments. The coherence of the electronic excited states characterized by the optical dephasing time T_2 can be measured directly at low temperatures by the accumulated photon echo (APE) technique. APE measurements are complementary to those of population relaxation time (T_1) in transient absorption measurements. For details of the APE technique we refer to [1, 2]. In this study we present results of APE measurements on the BChl *a* pigment-protein complexes from green sulphur bacteria (FMO) and on the main light harvesting complex of PSII of green plants (LHCII).
For a description of the structure and the spectroscopic features of FMO and LHCII we refer to the literature and references therein [3–6].

2.Materials and Methods

APE experiments were performed at 1.4 K in a home built helium bath-cryostat, at other temperatures in an Oxford flow cryostat. The pulses of the dye laser were almost transform limited (pulse duration 2ps FWHM). The energies of the 'pump' and 'probe' beam were 500 μW and 50 μW, respectively and were focussed into the sample by a lense with a focal length of 20 cm.
The samples were prepared as described in [4] for the FMO-complexes and in [6] for the LHCII particles. The optical density was adjusted to 1 in a 2 mm cuvet at the maximum of the Q_Y absorption band at cryogenic temperatures. Glycerol (66% v/v) was added to the sample to prevent cracking upon cooling.

3.Results

3.1 Measurements on the FMO-complex

The experiments on the FMO-complex show strong dependence of the dephasing times with the wavelength of excitation in the Q_Y band, ranging from sub-picosecond decays at 790 nm to decays of several hundred picoseconds at 830 nm.

P. Mathis (ed.), Photosynthesis: from Light to Biosphere, Vol. I, 363–366.

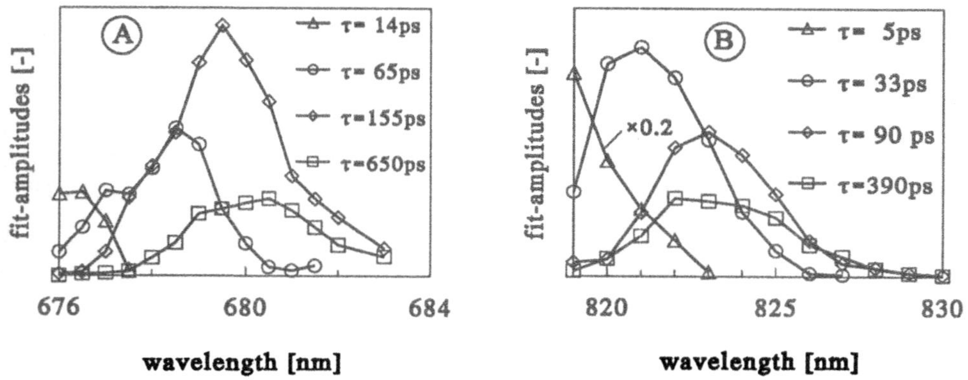

Figure 1: Decay associated spectra of the various time constants in the APE decays of LHCII (a) and FMO (b).

The best fit required at least two exponentials at the longest wavelengths and an increasing number of time constants at shorter wavelengths. Using a global analysis procedure we could fit all the echo traces in the wavelength region of 819 nm to 830 nm with a fixed set of time constants: 390, 90, 33 and 5 ps ($T_2/2$). A plot of the amplitudes versus excitation wavelength (Decay Associated Spectrum, DAS, fig. 1a) shows two bands centered at approximately 824 nm and one at 821 nm, respectively. A fourth band, associated with the fastest time constant is only partially resolved, due to the presence of faster components at shorter wavelengths exceeding the present time resolution.

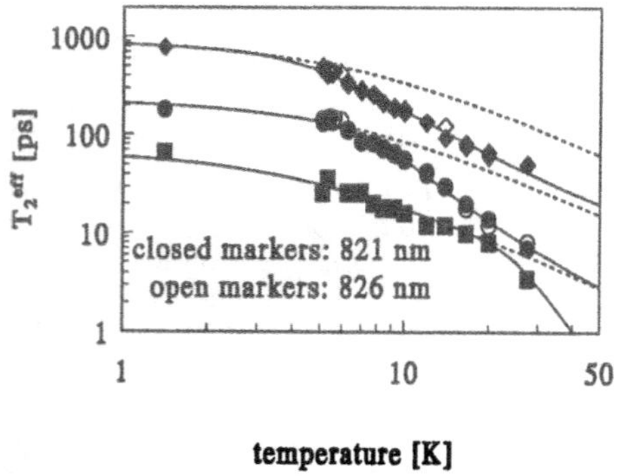

Figure 2: Temperature dependence of the total dephasing time of FMO measured at 821 nm and 826 nm. Note the correspondence between time constants at 821 and 826 nm.

Table 1: Parameters used in the fit on the temperature dependence of the total dephasing time of FMO(eq. 1).

T_1	a	b	ΔE
[ps]	[$ns^{-1}K^{-\alpha}$]	[ns^{-1}]	[cm^{-1}]
455	0.09	2.10	13
114	0.38	345	28
68	2.10	62832	125

The temperature dependence of the photon echo decays was measured between 1.4 K and 30 K at 821 nm and at 826 nm. Fig. 2 shows the total dephasing time T_2

of the lowest exciton states of the FMO-complex obtained from the fits of the echo decays at 821 and 826 nm, as a function of temperature between 1.4 K and 30 K.

The dephasing rates are strongly temperature dependent and increase rapidly with increasing temperature.

3.2 Measurements on LHCII

Measurements on LHCII at 1.4 K show similar photon echo decay patterns as FMO, ranging from (sub-) picosecond decays at 670 nm to decays of several hundred of picoseconds at 683 nm. At wavelengths below 665 nm we did not observe any photon echo decay, probably due to absence of an effective coupling between the electronic states at these wavelengths and the bottleneck (triplet-) state. Global analysis in the wavelength range 676 nm to 683 nm reveals four time constants of 650 ps, 155 ps, 65 ps and 14 ps, and a DAS which shows three bands centered at 681 nm, 679.5 nm , 678.5 nm and a partially resolved fourth band, respectively (fig. 1b).

4. Discussion and conclusions

The dipolar interaction between the pigment molecules in FMO as well as in LHCII exceed the homogeneous linewidth as determined by the APE measurements by several orders of magnitude. Therefore, at least at low temperatures, energy transfer is more appropiately described in terms of coherent energy transfer and dephasing occurs via interaction of the chromophores with the protein. The positions of the gaussian bands in the DAS qualitively match those found in exciton calculations [7] and the positions of the zero phonon holes in holeburning [8]. The time constants found in the photon echo decays are therefore interpreted as representing the dephasing times of specific excitonic levels in the Q_Y manifold, the gaussian bands in the DAS of fig. 1 corresponding to the lowest (inhomogeneously broadened) transitions. At the highest energies within the manifold more downward relaxation paths are available due to transitions within the manifold and the (phonon assisted) relaxation will be faster than the relaxation of lower levels. None of the photon echo decays contained an ultrafast decay component characteristic for a phonon side band as observed in photon echo decays of the primary donor of reaction centers, indicating that the electron-phonon coupling in FMO and LHCII is weak.

For FMO the best fit of the temperature dependence of the total dephasing times is obtained by a combination of a power law [9], combined with an exponential activation term [10]:

$$\frac{1}{T_2} = \frac{1}{2T_1} + a\, T^{\alpha} + b\, \frac{e^{(-\Delta E/kT)}}{1 - e^{(-\Delta E/kT)}} \tag{1}$$

Here we take for α the value of 1.3, according to observations in other disordered systems [11–14]. The term with $T^{1.3}$ decribes the dephasing due to coupling of a chromophore to a distribution of two-level systems which is characteristic of an amorphous host [9]. At higher temperatures an exponential activation is required to describe the temperature dependence. Here we find activation energies, which more or less match the energy differences between the lowest electronic states in the manifold and therefore conclude that the corresponding proces involves thermally activated population of higher electronic states within the exciton manifold. Note that thermally activated processes have been observed in other systems with specific

electronic structure [15, 16]. The other parameters of the fit are summarized in Table 1. Extrapolation of the temperature dependence of the dephasing times to 0 K yields the depopulation rates of 455 ps, 114 ps and 68 ps for the lowest transitions, which are somewhat faster compared to time resolved fluorescence data [17], indicating that some spectral diffusion may take place on the timescale of our experiment.
The fits of photon echo decays of LHCII and FMO are very similar and we conclude therefore that also in LHCII dephasing at low temperatures is governed by relaxation within the Q_Y manifold through interactions of the pigments with the protein. The DAS of the fits on the photon echo decays of LHCII shows four bands, whereby the lowest transition is located approximately at 681 nm, in agreement with the position of the fluorescence origin [18]. In persistent holeburning studies on FMO and LHCII much shorter dephasing times were reported than we observe [19, 8]. Although the reason for this discrepancy can not be unambiguously identified, one origin might be the different timescales of the experiments (6 orders of magnitude), implying a substantially different contribution of spectral diffusion in both experiments.

6. Acknowledgements

We thank Christof Francke and Marta Germano for help with preparation of the samples. The investigations were supported by the Life Science Foundation (SLW) which is subsidized by the Netherlands Organization for Scientific Research (NWO).

References

1 Hesselink,W. H. and Wiersma,D. A. (1979) Phys. Rev. Lett. **43**, 1991–1994.
2 Louwe,R. J. W. and Aartsma,T. J. (1994) J. Luminescence **58**, 154–157.
3 Tronrud,D. E., Schmidt,M. F. and Matthews,B. W. (1986) J. Mol. Biol. **188**, 443–454.
4 Miller,M., Cox,R. P. and Olson,J. M. (1994) Photosynth. Res. **41**, 97–103.
5 Kuhlbrandt,W., Wang,D. N. and Fujiyoshi,Y. (1994) Nature **367**, 614–621.
6 Hemelrijk,P. W., Kwa,S. L. S., van Grondelle,R. and Dekker,J. P. (1992) Biochim. Biophys. Acta **1098**, 159–166.
7 Lu,X. and Pearlstein,R. M. (1993) Photochem. Photobiol. **57**, 86-91.
8 Johnson,S. G. and Small,G. J. (1991) J. Phys. Chem. **95**, 471–479.
9 Völker,S. (1989) in: Relaxation processes in molecular excited states, Fünfschilling,J ed., Kluwer Academic Publishers Dordrecht, p. 113–242.
10 Jackson,B. and Silbey,R. (1983) Chem. Phys. Lett. **99**, 331–334.
11 Thijssen,H. P. H., Dicker,A. I. M. and Völker,S. (1982) Chem. Phys. Lett. **92**, 7–12.
12 Fidder,H., de Boer,S. and Wiersma,D. A. (1989) Chem. Phys. **139**, 317-326.
13 Hegarty,J., Broer,M. M., Golding,B., Simpson,J. R. and MacChesney,J. B. (1983) Phys. Rev. Lett. **51**, 2033–2035.
14 Fear,B. L., Stout,R. P., Hayes,J. M. and Small,G. J. (1983) J. Chem Phys. **78**, 7013–7015.
15 Aartsma,T. J. and Wiersma,D. A. (1980) Chem. Phys. Lett. **54**, 415–419.
16 Morsink,J. B. W., Hesselink,W. H. and Wiersma,D. A. (1982) Chem. Phys. **71**, 289–294.
17 Zhou,W., LoBrutto,R., Lin,S. and Blankenship,R. E. (1994) Photosynth. Res. **41**, 89–96.
18 Kwa,S. L. S., Groeneveld,F. G., Dekker,J. P., van Grondelle,R., van Amerongen,H., Lin,S. and Struve,W. S. (1992) Biochim. Biophys. Acta **1101**, 143–146.
19 Reddy,N. R. S., van Amerongen,H., Kwa,S. L. S., van Grondelle,R. and Small,G. J. (1994) J. Phys. Chem. **98**, 4729-4735.

MORPHOLOGY AND SPECTROSCOPY OF CHLOROSOMES BY ALCOHOL TREATMENTS

Z.-Y. Wang [1], G. Marx [1], Y. Imamura [1], M. Mimuro [2] and T. Nozawa [1]
[1] Department of Biochemistry and Engineering, Faculty of Engineering, Tohoku University, Sendai 980-77(Japan) and [2] National Institute for Basic Biology, Myodaiji, Okazaki, Aichi 444(Japan)

1. Introduction

Chlorosomes are known as major light-harvesting apparatus in green photosynthetic bacteria and are attached to the cytoplasmic side of the inner cell membrane. The light energy absorbed by chlorosomes is transferred to membrane antenna pigments, via the so-called baseplate, and finally to the reaction centre located in cytoplasmic membrane. The main light absorbing pigment in chlorosomes has been determined to be bacteriochlorophyll *c* (BChl *c*, or BChl *d*, *e* in some species) which has in vivo absorption maximum at about 740 nm in near infrared region, corresponding to a red-shift of 70 ~ 80 nm from that of its monomer form as observed in polar organic solvents. Some small amounts of other pigments, such as carotenoid and BChl *a*, have also been found in the chlorosomes as minor components.

In the present study, we report results on the morphologic changes of chlorosomes isolated from *Chlorobium tepidum* by alcohol treatments along with the corresponding spectral properties. We used negative staining electron microscopy to obtain the local information on the shape and size of chlorosomes at different stages of alcohol treatment. At the same time, dynamic light scattering(DLS) was applied to gain statistical information on the distribution of hydrodynamic diameter of particles, a measure of the size for chlorosomes. For spectroscopic measurements, magnetic circular dichroism (MCD) was used in addition to absorption and circular dichroism (CD). The intensities and signs of MCD are known to be very sensitive to the changes in electronic state of chromophores and these are practically useful for detecting very weak electron transitions (the chlorin Q_x band) as the transitions will be significantly enhanced by mixing with closely related states of x- or y-polarization in the case of BChl *c*.

2. Experimental

2.1 *Chlorosome isolation*

Cells of *Chlorobium tepidum* were grown as described by Wahlund et al.[1]. Chlorosomes were isolated by the method of Gerola and Olson [2] using 2 M NaSCN, 10 mM potassium phosphate (pH7.0), and purified by ultracentrifugation on a continuous sucrose density gradient. Fresh chlorosome samples were used throughout the experiments.

P. Mathis (ed.), Photosynthesis: from Light to Biosphere, Vol. I, 367–370.

2. 2 *Alcohol treatments*

Three alcohols (1-hexanol, 1-butanol and phenol) were used in the treatment of chlorosomes. Buffer saturated with each alcohol was prepared by adding certain amount of alcohol to 20 mM potassium phosphate solutions in a separating funnel. After shaking, the solutions were allowed to stand for at least 30 min before separation of the alcohol saturated buffers. Alcohol concentrations in the saturated solutions were estimated to be 59 mM for 1-hexanol [3], 910 mM for 1-butanol [4] and 900 mM for phenol [5], respectively. The alcohol treated chlorosomes were diluted in fresh buffer to a final alcohol concentration equal to the half saturated concentration.

2.3 *Spectroscopic measurements*

Absorption spectra from 250 nm to 900 nm were measured using a Beckman DU-640 spectrophotometer with a 1-cm quartz cuvette. Both CD and MCD spectra were recorded on a Jasco J-720w spectropolarimeter with a cuvette of 1 cm pathlength. For MCD measurement, an electromagnet was used to produce an external magnetic field of 1.5 T. MCD spectra were obtained by subtracting the negative signals from the positive ones and correspond to those obtained under a magnetic field of 3.0 T. All spectroscopic measurements were conducted at room temperature.

2.4 *Morphologic measurements*

Transmission electron microscopy was performed on a Hitachi H-800 operating at 150 kV. The samples were negatively stained with 4 % sodium phosphowolframate (pH 7.2) on collodion-coated 400 mesh copper grids with carbon reinforcement. Distribution of hydrodynamic diameters of chlorosomes was measured on a Otsuka Electronics DLS-700 dynamic light scattering spectrophotometer equipped with a 5 mW He-Ne laser (632.8 nm wavelength). A photomultiplier was set at scattering angle of 90° and the scattered light was analysed by an autocorrelator with a sampling time of 4 μsec and correlation channel of 1024. Computer analysis of the time decay of the autocorrelation function using the methods of histogram and cumulant provides a characterisation of the distribution of correlation times associated with the polydispersity of the chlorosome particles.

3. Results and Discussion

3.1 *Spectroscopy*

Figure 1 shows the absorption spectra of chlorosomes at different stages of 1-hexanol treatment. Chlorosomes in phosphate buffer shows an absorption peak at 743 nm in the near infrared region (A), a characteristic of aggregated form of BChl *c* molecules. After treated with 1-hexanol saturated buffer (59 mM), the 743 nm peak disappeared, instead a 671 nm peak with a similar height but only half the bandwidth has been observed (B), corresponding to the monomeric form of BChl *c* in polar organic solvents. The reaction was almost completed within 15 min and the most drastic changes occurred over the first 5 min. When the hexanol treated chlorosomes were diluted into a solution with 29.5 mM(half the saturated concentration) of 1-hexanol,

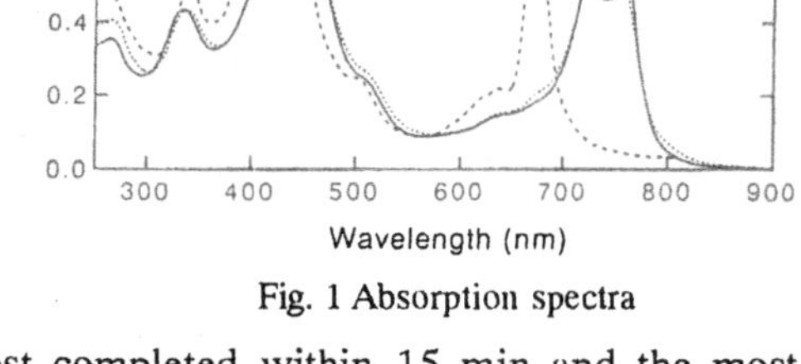

Fig. 1 Absorption spectra

a peak closely resembling that in the original spectrum of chlorosomes was observed at 741 nm (C). These results are in good agreement with those previously reported by Brune et al.[6] and Matsuura et al.[3]. Similar results have been observed for the chlorosomes treated with 1-butanol and phenyl alcohols. A new feature we have found is that the degree of saturation, not the concentration of alcohol, is the key factor for the complete conversion. This result can be understood by assuming that only the unhydrated alcohol molecules are effective in altering the aggregated structure of BChl *c* in chlorosome, while below the saturated concentration all alcohols are considered to be hydrated by water molecules and be unable to coordinate to the Mg atom of BChl *c*.

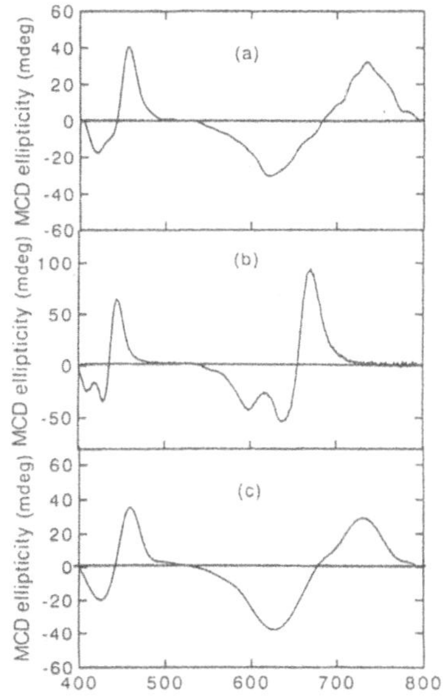

Fig. 2 MCD spectra

Figure 2 shows the corresponding MCD spectra of chlorosomes under the same conditions as described in Fig. 1. For the native chlorosomes(Fig. 2a), there is a broad positive peak around 740 nm assigned as the Q_y band of aggregated BChl *c*, and another broad maximum of opposite sign around 622 nm assigned as Q_x or a mix of Q_x and $Q_{y1\text{-}0}$ of BChl *c*. It is noted that the Q_y band appears to be composed of several components. When treated with 59 mM 1-hexanol, all MCD bands became much narrower, having the same shape as monomeric BChl *c* (Fig. 2b). The 740 band shifted to 672 nm and the 622 nm band splitted into two bands: Q_x at 599 nm and $Q_{y1\text{-}0}$ at 638 nm. The $Q_{y1\text{-}0}$ band was stronger than that of Q_x. By the following dilution up to 29.5 mM 1-hexanol, the MCD spectrum showed a similar but more uniform shape with almost the same intensity compared with that of the native chlorosomes. A broader band shape in 29.5 mM hexanol than that in 59 mM indicates that plural components in the native chlorosomes were also recovered by dilution of hexanol concentration. The MCD results obtained are consistent with those from CD, but show more detailed features in the spectrum, especially for the Q_x and $Q_{y1\text{-}0}$ bands that are usually very weak in the absorption and CD spectra.

3.2 *Morphology*

Electron micrographs shows that the untreated chlorosomes have elongated bodies of uniform shape and size with rounded ends and typical dimension of 150 nm × 60 nm, similar to those observed by Staehelin et al.[7]. After treatment with 1-hexanol saturated buffer, remarkable changes have been found in size, shape and distribution of the chlorosomes. The originally rice-grain-shaped chlorosome particles became stick-like form with sharp ends and increased more than two times in length. A considerable variation was observed in the size and shape of the hexanol treated chlorosomes. Dilution of the hexanol treated chlorosomes to a solution of 29.5 mM 1-hexanol brought about further changes in the size and shape which were also different from those of untreated chlorosomes. The chlorosomes were more rounded than the original with a mix of larger and smaller particles varying in shape. The overall size was of about the same order of that for native chlorosomes. Apparently, significant changes in the morphology of chlorosomes have been caused by the hexanol treatment at each step and these changes seem to be irreversible.

Figure 3 shows the size distribution histograms of the chlorosomes at different stages of 1-hexanol treatment. The fresh chlorosomes have a narrow distribution centred at 105 nm in the hydrodynamic diameter (Fig. 3a). This result is in good agreement with the observed uniformity for the same sample by electron microscopy. By treatment with the 1-hexanol saturated buffer, the distribution shifted toward greater hydrodynamic diameters, peaked at about 240 nm, and became much broad (Fig. 3b). After the following dilution to 29.5 mM 1-hexanol buffer, the distribution was further broadened with the average value back to about 180 nm. The behaviour observed from the changes in hydrodynamic diameter during this process is well consistent with the trend revealed by electron microscopy that fresh chlorosome particles with uniform shape increased more than two times in length by treatment with 59 mM 1-hexanol and then changed back to an intermediate average size in the following treatment with 29.5 mM 1-hexanol.

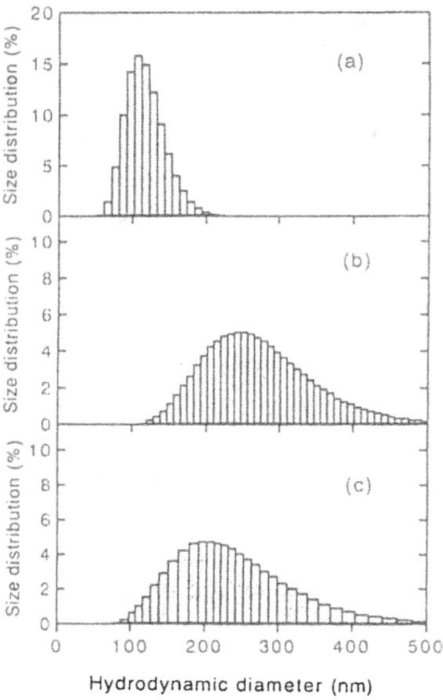

Fig. 3 Distribution of hydrodynamic diameter.

Both EM and DLS results have shown significant changes in the size, shape and distribution of chlorosomes during the process of alcohol treatment, and these changes are illustrated in Scheme I. If we use the average hydrodynamic diameter d=105 nm determined by DLS for the native chlorosomes and assume that the width (short axis) is 80 nm from the micrograph, then the length (long axis) is estimated to be 160 nm based on Perrin's formula. This is in good agreement with the value that we can measure from the micrograph. Similarly, an average length of about 700 nm was estimated for the extremely prolate chlorosomes treated with 59 mM 1-hexanol. Since the chlorosome particles obtained by dilution to 29.5 mM 1-hexanol can be approximated to have a nearly rounded shape, the average hydrodynamic diameter of 183 nm represents the actual dimension.

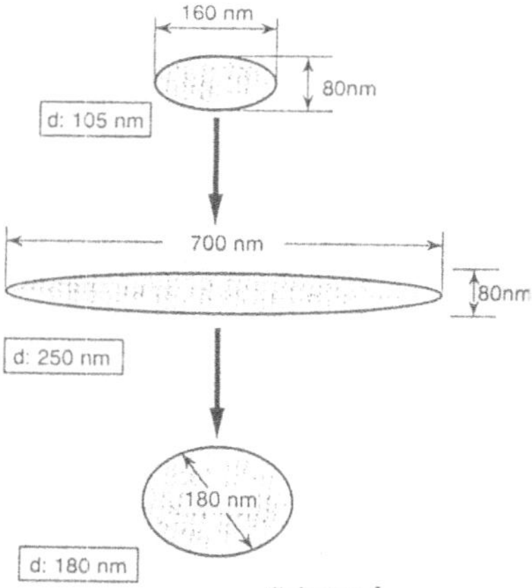

Scheme I

References

1 Wahlund, T. M., Woese, C. R., Castenholz, R. W. and Madigan, M. T. (1991) Arch. Microbiol. 156, 81-90.
2 Gerola, P. D. and Olson, J. M. (1986) Biochim. Biophys. Acta 848, 69-76.
3 Matsuura, K. and Olson, J. M. (1990) Biochim. Biophys. Acta 1019, 233-238.
4 Booth, H. S. and Everson, H. E. (1948) Ind. Eng. Chem. 40, 1491-1493.
5 The Merck Index(1968), 8th Ed.(Stecher, P. G., Windholz, M. and Leahy, D. S., ed.) pp 810, Merck & Co., Inc., Rahway, N. J., USA.
6 Brune, D. C., Nozawa, T. and Blankenship, R. E. (1986) Biochem. 26, 8644-8652.
7 Staehelin, L. A., Golecki, J. R. and Drews, G. (1980) Biochim. Biophys. Acta 589, 30-45.

Temperature dependent fluorescence lifetimes and quantum yield of the B820 subunit of LHI antenna complex of purple bacterium *Rhodospirillum Rubrum*

Vesa Helenius, René Monshouwer and Rienk van Grondelle
Department of Physics and Astronomy, Free University of Amsterdam,
De Boelelaan 1081, 1081 HV Amsterdam

1. Introduction

The light harvesting antenna of purple bacterium *Rs. Rubrum* consists of a core antenna (LHI) only. The core antenna complex (absorbing at 875 nm, B875) can be isolated from the membrane by using detergents. The B820 subunit can be isolated from the core complex by using detergent n-Octyl-ß-D-glucopyranoside (OG)[1,2]. The B820 consists of two short polypeptides α,ß and two molecules of BChl *a* (αß-BChl a_2). Each of the polypeptides bind a BChl *a* by a histidine residue. The structure is believed to be closely related to the basic building block of LHII antenna complex that was recently crystallized by McDermott et al. [3].

In this paper we report the temperature dependence of the B820 fluorescence quantum yield and lifetimes. The results are discussed by using the energy gap law of non-radiative decay processes [4,5].

2. Experimental

The stock solution of B820 particles were prepared as described elsewhere [1]. For optical measurements the stock solution was diluted in 60% (wt/vol) glycerol, 50 mM phosphate buffer (pH 7.8) and 1.15% wt/vol OG. For fluorescence measurements an OD^{820} of 0.1/cm was used. Fluorescence spectra were measured by a Chromex 500IS spectrograph and a Chromspec CCD detection (1.3 nm bandwidth). Fluorescence was excited by a Ti:Sapphire laser by directly exciting the 820 absorption band. Excitation intensity was about 2 mW/cm^2. Time resolved fluorescence was measured by single photon counting. Part from a femtosecond white light continuum was selected by a narrow band interference filter (at 812 nm) for excitation. The detection wavelength was selected a by long-pass filter at 830 and an interference filter at 840. The white light continuum was pro-

P. Mathis (ed.), Photosynthesis: from Light to Biosphere, Vol. I, 371–374.

duced by a Coherent regenerative Ti:Sapphire amplifier operating at 250 kHz. An instrumental response of 90 ps was obtained.

3. Results and discussion

The fluorescence spectrum of B820 peaks at 826 nm (4K) when excited at 819 nm. The Stokes shift is then 107 cm^{-1}. At higher temperatures the fluorescence maximum gradually shifts 3 nm to the blue. The fluorescence spectrum shows no vibrational structure even at 4K when excited at 819 nm. Excitation on the red edge of the absorption band (> 840 nm) reveals some vibronic structure (not shown). The fluorescence intensity shows a strong temperature dependence between 150K and 300K. At low temperatures the intensity is nearly constant. Around 200K the fluorescence intensity starts to decrease rapidly. Fig 1. shows the fluorescence spectra of B820 from 77K to 298K. The spectra are not corrected for changes in the transmission.

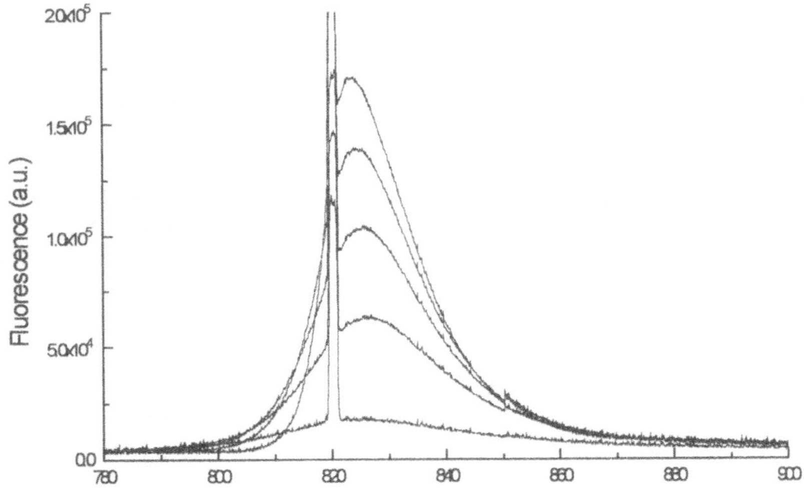

Fig. 1. Fluorescence spectra of B820 at 77K (upper curve), 150K, 210K, 250K and 298K.

The changes in the fluorescence quantum yield as a function of temperature are explained by assuming that the rate of nonradiative decay is temperature dependent. We use the "strong coupling limit" of the energy gap law [4] to fit the temperature dependence of the quantum yield. An activation energy of 1450 cm^{-1} and average frequency of 150 cm^{-1} for the promoting mode are found for the non-radiative decay rate constant.

The fluorescence yield of B820 (fig. 2) is fitted with the expression:

$$k_{nr}(T) = \frac{C}{(k_B T_{eff})^{1/2}} \exp\left(\frac{-E_A}{k_B T_{eff}}\right).$$

where T_{eff} is the effective temperature: $k_B T_{eff} = \frac{1}{2}\hbar\omega_m \coth(\beta\hbar\omega_m/2)$.

The k_{nr} is the sum of the nonradiative decay rates, C is a constant, $\beta=1/k_BT$ and ω_m is the mean vibrational frequency of the active modes of the non-radiative decay. E_A is the activation energy for the non-radiative decay.

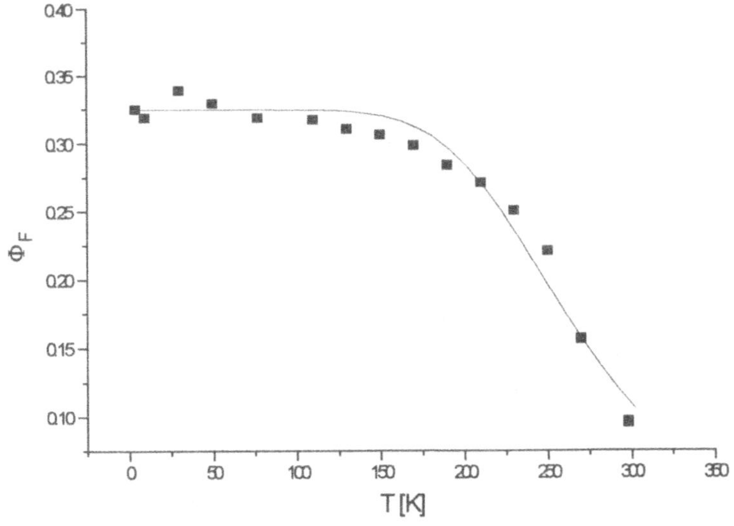

Fig 2. Fluorescence quantum yield Φ_F of B820 as a function of temperature. A laser dye IR140 in ethanol was used as a reference. Φ_F of 100% is assumed for IR140.

Fluorescence lifetimes of B820 follow the same pattern as the quantum yield. At low temperatures the decay is single exponential (2.3 ns at 4.2K). As temperature is raised the lifetime gradually becomes shorter (2.12 ns at 200K). Between 200K and 298K the lifetime drops to 0.55 ns.

In the strong coupling limit the displacement between the ground state potential and the excited state potential should be large, at least for the modes involved in the internal conversion. The rather small Stokes-shift of B820 implies that the coupling is not very strong.

The weak coupling limit for the non-radiative decay predicts a weak temperature dependence that is mostly determined by the high-frequency vibrations (CH stretching modes)[7]. What is found is an activated process around 200K and a strong dependency on the temperature (fig. 2). The above equation is derived from the assumption that the vibrational frequencies don't change in the excited state. A more general treatment [7] taking into account the frequency change leads to a same type of expression where the transition probability is determined by the activation energy E_A and the mean vibrational frequency ω_m. We propose that the difference in the ground state and excited state vibrational frequencies may explain why the strong coupling expression fits the non-radiative decay rate of B820, even though the coupling strength is only intermediate.

4. Conclusion

Fluorescence quantum yield and lifetimes of the B820 subunit of core antenna of Rs. Rubrum are reported at temperature range from 4.2K to 300K. The temperature dependency of the fluorescence quantum yield of B820 is strong. The non-radiative decay process is activated around 200K. The fluorescence quantum yield is nearly constant from 4K (0.32) to about 200 K. Between 200K and room temperature the quantum yield drops to 0.10. The fluorescence lifetimes follow the same temperature dependence as the quantum yield. The lifetime we observe is shorter than the one observed by Chang et al. (0.55 ns compared to 0.72 ns). The difference may be due to a difference in detergent concentration and the glycerol buffer we used. The temperature dependence is modeled by using the energy gap law in the strong coupling limit. An activation energy of 1450 cm^{-1} and average frequency of 150 cm^{-1} for the promoting mode are found for nonradiative decay rate constant. The strong coupling limit expression of the energy gap law fits the quantum yield results, although the electron-phonon coupling in B820 is only intermediate.

References

1 Miller, J.F., Hinchigeri, S.B., Parkes-Loach, P.S., Callahan, P.M., Sprinkle, J.R., Riccobono, J.R. and Loach, P.A. (1987) Biochem. 26, 5055-5062

2 Visschers, R.W., Chang, M.C., van Mourik, F., Parkes-Loach, P.S., Heller, B.A., Loach, P.A. and van Grondelle, R. (1991) Biochemistry 30, 2951-2960

3 McDermott, G., Prince, S.M., Freer, A.A., Hawthornthwaite-Lawless, A.M., Papiz, M.Z., Cogdell, R.J. and Isaacs, N.W. (1995) Nature 374, 517-521

4 Englman, R. and Jortner (1970) J. Mol. Phys. 18, 145-164

5 Groot, M.-L., Peterman, E.J.G., van Stokkum, I.H.M., Dekker, J.P. and van Grondelle, R. (1995) Biophys. J. 68, 281-290

6 Chang, M.C., Callahan, P.C., Parkes-Loach, P.S., Cotton T.M. and Loach, P.A. (1990) Biochemistry 29, 421-429

7 Freed, K.F. and Jortner, J. (1970) J. Chem. Phys. 52, 6272-6290

SINGLET AND TRIPLET STATE EMISSION STUDIES ON ZINC TETRA PHENYL PORPHYRIN ENCAPSULATED LIPOSOMES

Rupali Chaudhuri* and K.K.Rohatgi Mukherjee, Jadavpur University, Calcutta 700 032, India *Present address: Biophysics Division, S.I.N.P. 37 Belgachia Road, Calcutta 700 037, India

Introduction

In photosynthesis and various photobiological processes, photoinduced electron-transfer is the basic phenomenon which initiates various photochemical events[1,2].In such photosensitised processes the reaction occurs against the thermodynamic potential gradient, the energy of the photon absorbed by the photosensitiser being used to promote the uphill chemical events[1-3] .For high efficiency therefore, such photoinduced redox reactions can best be stimulated by carrying out the reaction in heterogeneous media in which the oxidising and reducing equivalents are prevented from recombination. The key concept is organisation for which liposomes or bilayer lipid membranes are used[2-6].

The overall efficiency of transmembrane charge seperation depends on various factors. In our earlier studies on chlorophyll system extracted from spinach leaves, the fluorescence quenching technique[8] was useful in assesing the location of the photosensitiser within the liposomes. Here we present photophysical studies using Zinc tetraphenyl porphyrin as the photosensitiser in liposome assemblies. The two lipids egg phosphatidyl choline (EPC) and di-palmitoylphosphatidyl choline (DPPC) were used for liposome preparation. The EPC liposomes were further modified using cholesterol.

Materials and methods

Zinc tetraphenyl porphyrin (ZnTPP), the electron donor ethylene diamine tetra acetic acid (EDTA), the electron acceptor 1,5 anthraquinone disulphonate (AQDS) and the lipids EPC, DPPC and cholesterol were purchased from Sigma and used as such.Liposomes were prepared in the usual way[9] . ZnTPP was added during film formation and EDTA solution during sonication. EDTA was removed from the outer pool of the liposome by passing the suspension through a Sephadex G 25 coloumn. The final composition in the liposomal systems were ZnTPP:Lipid ~ 1:280 with about 8 -10 ZnTPP molecules per vesicle. For flash excitation studies the second harmonics (532nm) of the Nd-YAG laser was used and all solutions thoroughly degassed.

Results and discussion

Absorption studies: Considerable changes in the absorption spectrum of ZnTPP are observed when incorporated in EPC liposomes. In DPPC liposomes, the spectrum is sharp and similar to that in benzene(Fig 1). except the red shift of the Soret and other bands. Table 1 gives the absorption maxima (λ^{abs}_{max}) of each component under various conditions.

P. Mathis (ed.), Photosynthesis: from Light to Biosphere, Vol. I, 375–378.

Fluorescence quenching studies in EPC liposomes: For fluorescence quenching studies of ZnTPP in EPC liposomes the samples were irradiated at the Soret band and emission studied at 655nm. Since preliminary observation indicated that no quenching was observed by the addition of the oxidative quencher in the absence of EDTA in the inner pool of liposomes, all systems contained 0.5M EDTA in the inner pool of all liposomal systems. For each system (at varying cholesterol content) the quenching data were plotted as a function of quencher concentration [Q], according to the Stern Volmer (SV) equation,

$$F/F_0 = 1 + K_{SV}[Q] \qquad \text{.......................................(1)}$$

where F and F_0 are the fluorescence intensities in the absence and presence of the quencher at the emission maxima and K_{sv} is the quenching constant.

With the oxidative quencher AQDS , the SV plots for quenching of ZnTPP fluorescence in EPC liposomes (Fig 2) show that except for the unmodified liposome, the plots are linear with unit intercept. Nonlinear SV-plots are known to occur in heterogeneous systems where all the fluorescent molecules are not located within the sphere of action of the quencher molecules. For such cases the Lehrer[10] modification of the SV equation is used. Considering that only a fraction f_a is available for quenching the SV equation will be valid only for the available fraction, so that,

$$F_0 / F_0 - F = 1/f_a + 1/f_a K_{sv}[Q] \qquad \text{.......................................(2)}$$

where F_0, F and K_{sv} have there usual meanings. From the plot of F_0 / (F_0- F) vs

Table 1—Absorption spectral data for various components studied in suitable solvents or in liposomal suspensions

Components	Absorption peak (nm)
Egg-PC (EPC)	275
Cholesterol	270
Methyl viologen	258
ZnTPP in benzene	422, (510), 550, 589
ZnTPP+ EPC	432, (520), 560, 600
ZnTPP+ EPC + 8% Cholesterol	430, (520), 560, 600
ZnTPP + DPPC	427, 560, 600

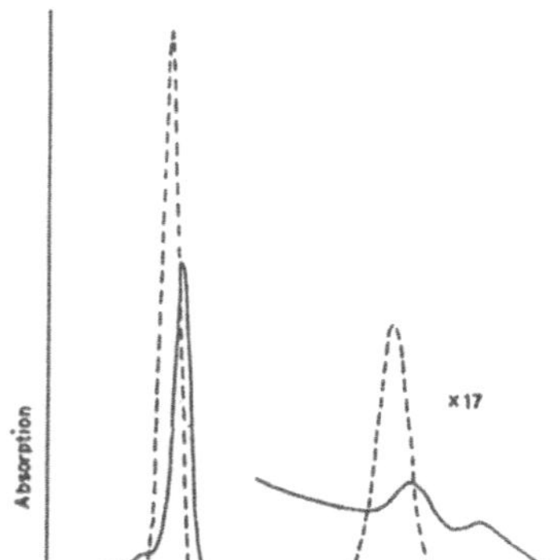

Fig. 1—Absorption spectra of ZnTPP in benzene (---) and in liposome (–)

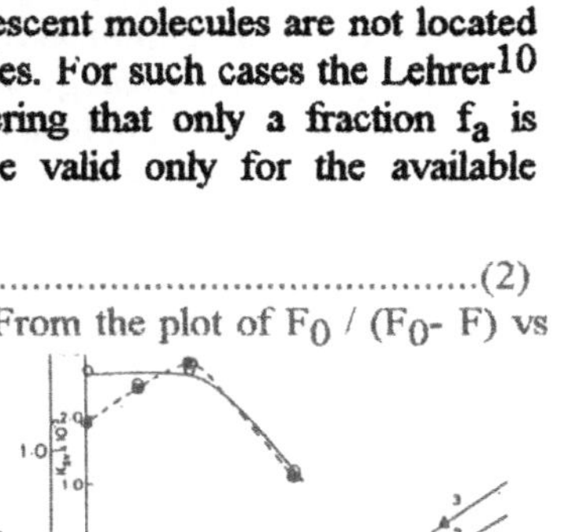

Fig. 2—Stern-Volmer plots for the quenching of fluorescence of ZnTPP by 1,5-AQDS²⁻ in EPC liposomes modulated by addition of cholesterol [(1) 0%, (2) 7.5%, (3) 15%, (4) 30%. Inset: Calculated values of K_{sv} for the four cases plotted as a function of % cholesterol (---) from unmodified S-V equation, (–) from modified S-V equation as presented in Fig. 3]

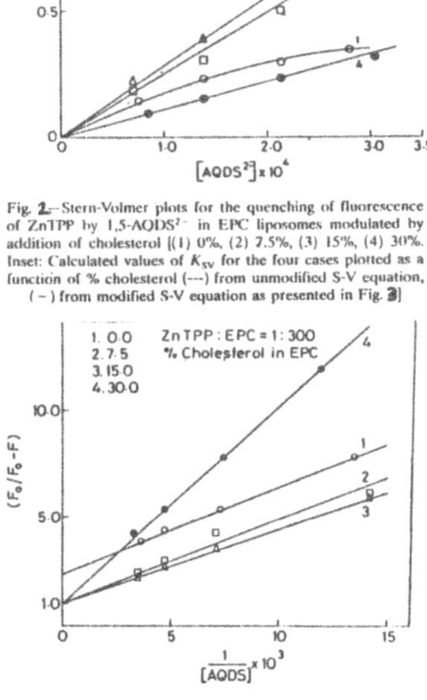

$[Q^{-1}]$(Fig 3) the values of f_a and K_{SV} thus obtained are presented in Table 2. Here it is observed that in the unmodified liposomes nearly 60% of the fluorescence is not available for quenching. The remaining 40% is quenched with the same efficiency as in systems 2 and 3, $K_{SV} = 2.7 \times 10^3\ M^{-1}$. Since the systems 2 and 3

Table 2—Stern-Volmer (SV) quenching constants (K_{SV}) calculated from Stern-Volmer plots, direct and modified, for quenching of fluorescence of ZnTPP in EPC liposomes with varying cholesterol content and 0.5*M* EDTA in the inner pool [The quencher 1,5-$AQDS^{2-}$ added in the outer pool in phosphate buffer, *p*H = 7, T = 301K; excitation at the Soret band and fluorescence at 605 nm and 655 nm]

Liposome No.	Cholesterol (mol %)	Soret band λ_{abs} (nm)	$K_{SV}^{(1)a}$ $\times 10^{-3} (M^{-1})$	$K_{SV}^{(2)b}$ $\times 10^{-3} (M^{-1})$	f_a^d
1	0	427	1.9^c	2.7	0.4
2	7.5	426	2.4	2.5	1.0
3	15.0	425	2.8	2.7	1.0
4	30.0	423	1.1	1.2	0.8

[a] $K_{SV}^{(1)}$ = S-V quenching constant from the slope of direct S-V equation $(F_0/F) - 1$ vs [Q] plot.
[b] $K_{SV}^{(2)}$ = S-V quenching constant from modified S-V equation $F_0/(F_0 - F)$ vs [Q] plot with intercept $1/f_a$.
[c] Calculated from the initial slope of the nonlinear unmodified SV plot.
[d] f_a = fraction available for quenching.

have the same K_{SV} as in system 1 it can be concluded that the photosensitizer is within the same sphere of action of the quencher. At 30% cholesterol concentration, a sharp decrease in quenching is observed (Fig 2,inset); the fraction available for quenching is only 80% . The drastic lowering of the K_{SV} values are in sharp contrast to those of the chlorophyll-complex incorporated liposomes[7], where curved SV-plots were obtained for highly modulated liposomes.

Fluorescence quenching in DPPC liposomes: The ZnTPP incorporated DPPC liposomes with 0.5M EDTA (Lip A) and 0.25M EDTA (Lip B) in the inner pool and varying [AQDS] in the outer pool,show linear SV plots on irradiation with 427nm. The values of K_{SV} thus obtained are:$1.46 \times 10^3 M^{-1}$ for Lip A and $0.27 \times 10^3\ M^{-1}$ for Lip B (Fig 4)Consequently K_{SV} is not a linear function of [EDTA]in the inner pool. The value of K_{SV}/[EDTA] shows that doubling of [EDTA] ~ triples the quenching efficiency thus confirming the role of EDTA in the overall quenching of the liposome encapsulated ZnTPP by AQDS.

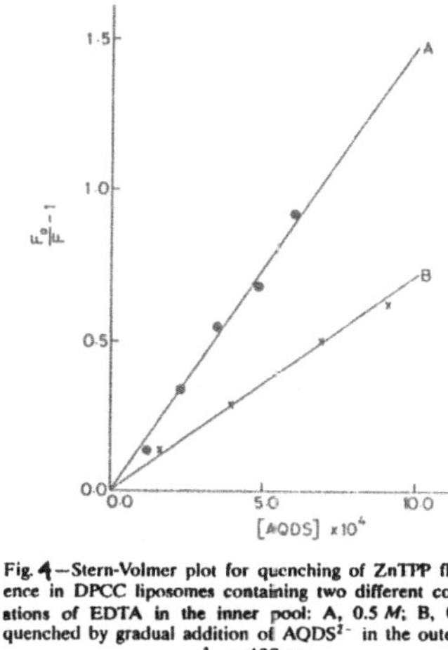

Fig. 4—Stern-Volmer plot for quenching of ZnTPP fluorescence in DPCC liposomes containing two different concentrations of EDTA in the inner pool: A, 0.5 *M*; B, 0.25 *M*; quenched by gradual addition of $AQDS^{2-}$ in the outer pool, $\lambda_{exc} = 427$ nm

Flash kinetic studies in DPPC liposomes: The singlet state quenching efficiency of photosensitisers has been shown to be not very effective for the generation of charge seperated species[11].The triplet state dynamics of transmembrane electron-transfer of ZnTPP encapsulated in DPPC liposomes were evaluated with the help of flash kinetic studies for Lip A, Lip B and Lip C with the oxidative quencher AQDS in the outer pool of the liposome A. The difference spectrum of Lip A,shows bleaching between 400-570nm (Fig 5) with a broad negetive peak at 465nm due to ground state absorption. The positive peak at 465nm appears for the triplet absorption of ZnTPP and so the decay at 465nm is studied. It is noted that:

1) Lip A generates greater concentration of transients as compared to Lip B.

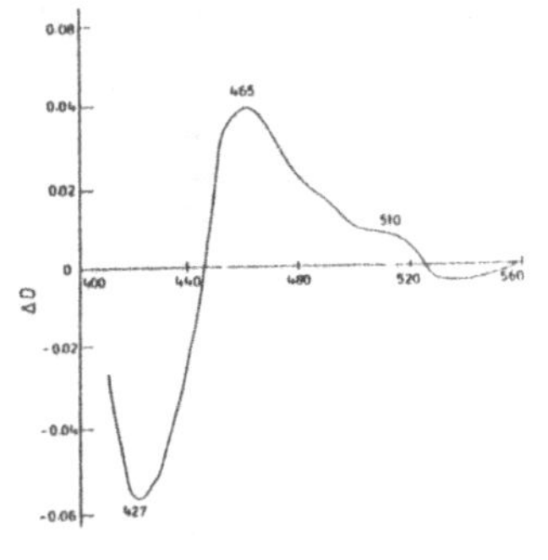

Fig. 5 -- Triplet-triplet end-of-the pulse absorption spectra of ZnTPP in DPCC liposome, Lip A. Bleaching observed between 400 and 450 nm with a negative peak at 427 nm. A positive peak develops at 465 nm with shoulder at 510 nm, assigned to T-T absorption

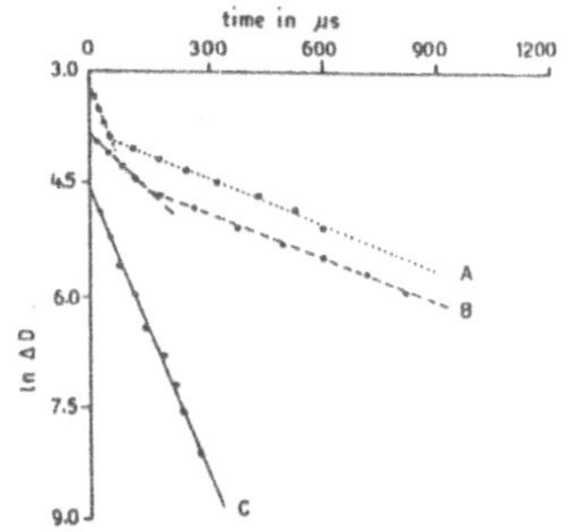

Fig. 6 —Kinetic treatment of triplet decay of ZnTPP-DPPC (1:276) vesicle suspensions at 465 nm [(...o...) Lip A, 0.5*M* EDTA in the inner aqueous phase; (—o—) Lip B, 0.25*M* EDTA in the inner aqueous phase; (– o –) Lip C, $4 \times 10^{-4}M$ $AQDS^{2-}$ added to the outer pool of the vesicles]

2) in both cases the decay is biphasic; an initial fast decay followed by a slow one and 3) on addition of AQDS, the trace decays completely and merges with the base line (Fig 6).

Conclusion A reductive quenching initiated by electron-transfer from EDTA to the photosensitiser ZnTPP is the likely mechanism for quenching of ^{1}ZnTPP emission in liposomes. Addition of cholesterol to EPC liposomes affects the availability of ZnTPP which helps to shuttle the electron across the bilayer[12]. Flash photolysis studies confirms the role of EDTA concentration in generating $ZnTPP^{-}$.; the decay is biphasic. On addition of the electron acceptor AQDS, the triplet decays completely.

Acknowledgement: One of us (RC) wishes to thank Dr.S.C. Bera, A.Karmakar of J.U., Cal 32, and Dr. P.K.Sengupta and J.GuhaRoy of. S.I.N.P. Cal 37. and the CSIR Delhi, India for the grant no. 9/489(16)/91-EMR-I.

References

1. Ghosh,J.C.and Gupta,S.B.(1934)J.Ind.Chem.Soc.11 pp65
2. .Seely, G.R (1978)Photochem.Photobiol. 27 pp639
3. Gratzel,M (1982) Pure and Appl.Chem. 54 pp2367
4. Calvin,M.(1979) Energy Research 3 pp73
5. Fendler,J.H.(1983)J.Chem. Edu. 60 pp872;(1985) J.Phys. Chem. 89 pp 2730
6. Ford,W.F. and Tollin,G. (1983). Photobiol. Photochem. 88 pp441
7. Hurley,J.K. and Tollin,G.(1982) Solar Energy 22 pp187
8: Chatterjee. M.: Bera. S.C.Chatterjee. A.: Nandy. P and Rohatgi Mukherjee. K.K. (1993) J.Photochem. Photobiol.72 pp209
9. Bartzri.S. and Korn.E.D.(1973)Biochem.Biophys.Acta. 29 pp1015
10. Lehrer, S.(1971) Biochemistry, 10 pp325
11. .Rentzepis. P.M: Huppert,D. andTollin. G.(1976) Biochim. Biophysics. Acta. 440 pp356
12. Ilani. A ;Woodle. M and Mauzerall. D.(1989) Photochem Photobiol. 49 pp 673

The Light-Harvesting Chl a/b/Mg-DVP-Complex of *Mantoniella squamata*: Supramolecular Structure and Sequence Analysis

Erhard Rhiel, Anke Randolf, Erhard Mörschel
FB Biologie-Botanik, Philipps-Universität, D-35032 Marburg, FRG

1. Introduction

M. squamata, a member of the Prasinophyceae, an ancient group belonging to the Chlorophyta, has a LHC, which simultaneously contains the chls a, b and Mg-2,4-Divinylpheoporphyrin a_5-monomethylether (Mg-DVP) [1]. A LHC harboring three different chl species is a unique feature and up to now only detected in *M. squamata* and *Micromonas pusilla*. From their pigment composition one would suggest a biochemical and structural relationship between the chl a/b LHCs of higher plants and the chl a/c complexes of the chromophytes. Therefore, we characterized this LHC and sequenced the genes, in order to obtain information on structure and evolution of the chl-binding proteins.

2. Results and Discussion

Polypeptide composition: The chl a/b/Mg-DVP LHC of *M. squamata* is resolved in polypeptides of Mrs 22,000. They are supposed to be the products of related individual genes, because they differ slightly in their amino acid compositions. The Mr of the LH polypeptides from *M. squamata* is significantly smaller than that of the chl a/b polypeptides of higher plants which are in the range of 25,000 to 30,000 [2]. However, the Mr of the chl a/b/Mg-DVP LHC polypeptides is similar to the chl a/c polypeptides of the LHCs of chromophytes and cryptomonads, which are all in the range of 17,000 to 22,000 [3].

The LH polypeptides are arranged in a native complex with an apparent molecular weight of 80,000. This Mr including an equivalent of about 14 chl molecules of Mr 900 fits directly to a trimeric structure of the LHC aggregates of *M. squamata*.

Freeze fracture analysis of thylakoids: The PF-thylakoid faces contained mainly 7.5 nm particles, which are correlated with free LHC-complexes [4]. The size of 7.5 nm fits well to trimeric LHC of higher plants. Normally, the fracture plane cleaves the thylakoids along the inner hydrophobic space between the two membrane leaflets. This cleavage pattern is evident in thylakoids of *M. squamata*. However, as an additional unusual feature, the fracture planes expose the exoplasmic surface (ES) together with the protoplasmic fracture faces (PF), which are shown in Fig.1 together with the stratification pattern of appressed thylakoids [5].

The exoplasmic membrane surface (ES) is covered by protein complexes, purported LHC-particles, which are arranged in regular, most likely hexagonal, para-crystalline arrays with a lattice repeat distance of about 9 nm. The "trigons" visible in these arrays are merly

P. Mathis (ed.), Photosynthesis: from Light to Biosphere, Vol. I, 379–382.

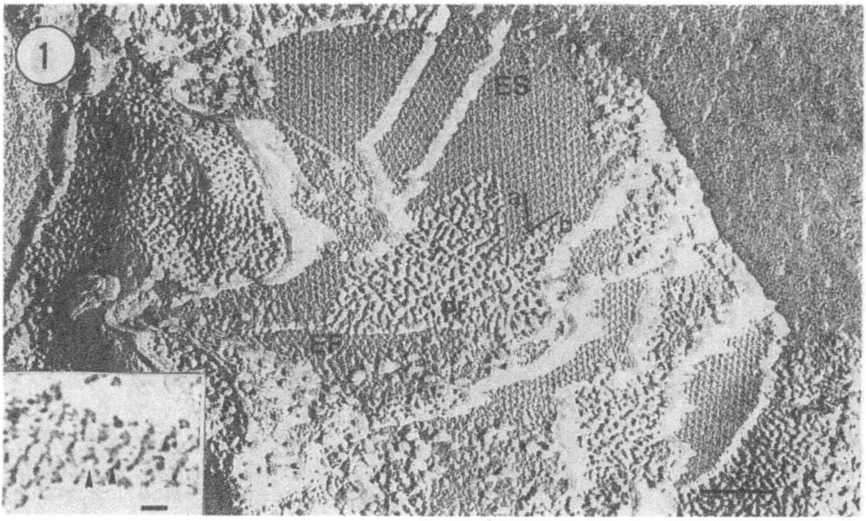

Fig. 1. Freeze-fractured thylakoids of *M. squamata*. The PF and ES particles are correlated with LHC-complexes. Inset: ES-face-latice. Bar: 100 nm; Bar of inset: 10 nm.

parts of the lattice and thus are likely the contribution of more than one particle. The apparent discrepancy between the lattice repeat of about 9 nm and the 7.5 nm size of the particles, is explained by the intercalation of lipid molecules between the particles of the lattice [4].

Reconstitution of the chl a/b/Mg-DVP LHC in liposomes: The chl a/b/Mg-DVP LHC was reconstituted into soybean lecithin liposomes to analyze its supramolecular structure. The freeze fracture preparations show a random distribution of the chl a/b/Mg-DVP LHC on both fracture faces of the proteoliposomes as well as unregular and regular aggregation. The size of the incorporated complexes was about 7.5 nm. The regular, well arranged arrays show a dense packing of the protein complexes in a purported hexagonal pattern with lattice repeats of about 10.0 nm (a) and 8.6 nm (b) and comparable values of 8.5 to 8.8 nm (a) and 7.9 to 8.1 nm (b) determined by optical diffraction. The structures of the crystalline lattice are similar to those observed in thylakoids of *M. squamata* (Fig. 1, inset) and for the chlorophyll a/b LHC. From profile views, the height of the complexes could be calculated to be 6 to 7 nm [4].

Molecular analysis: RNA species were isolated, total RNA, poly(A)+ and poly(A)- RNA translated *in vitro* and the products were electrophoresed and fluorographed . Indirect immunprecipitation with protein A sepharose resulted in a highly specific precipitation reaction for both total RNA and poly(A)+ RNA . A single protein species with a Mr of about 27,000 and a weaker signal of about 50,000 (dimer) were detected. The Mr 27,000 polypeptide was about 5,000 larger in molecular weight compared to the native subunits. The difference in molecular weight was attributed to an amino terminal transit peptide which is cleaved from the precursor during import into the chloroplast. No precipitation signal could be observed when poly(A)- RNA was used. Thus, the polypeptides of the chl a/b/c LHC are encoded by nuclear DNA.

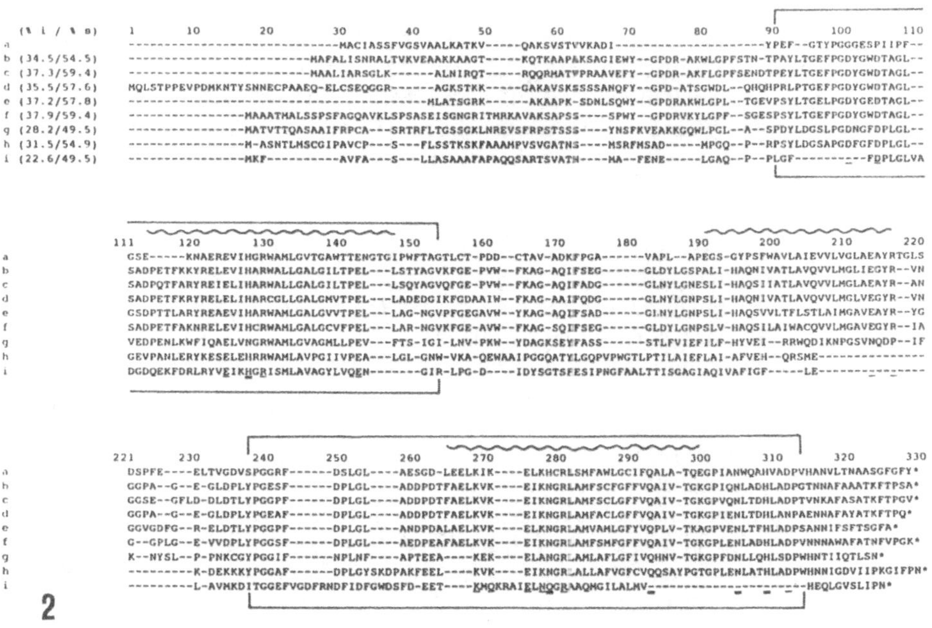

Fig. 2 Amino acid sequence of LHC1*1 of *M.squamata* (a) in comparison to the LHCII of *C. moewusii* (b), *D. tertiolecta* (c), *D. salina* (d), *E. gracilis* (e), *L. esculentum* (f) and LHCI-IV (g), LHCI-I-I (h) both from *L. esculentum* and FCP-3 from *P. tricornutum*. Regions of high homology are boxed; wavy lines show membrane spanning helices.

In order to identify the genes encoding subunits of the LHC of *M. squamata*, we chose the approach of cDNA synthesis and antibody screening. We obtained a set of four different full-sized clones encoding precursor molecules of the chl a/b/Mg-DVP LHC subunits of *M. squamata* [6] Synthesized fusionproteins of Mr's 27,000 - 31,000 were localized in inclusion bodies as analyzed by immunogold-labeling of ultrathin sections. The four clones are 97% identical within the proposed gene-coding region, starting from the Met and ending at the TAA-stop codon, as only seven amino acids are exchanged in the clones (Accession numbers: X69524, X69525, X 73539, X73540). The nucleotide sequences of the four clones differ significantly 5'-upstream from the assumed start codons Met and 3'-downstream from the proposed TAA-stop codons. This is an indication, that the LHC of M. sqamata is encoded by at least four different genes, most likely representing a multigene family. The nuclear genes corresponding to the plasmid cDNA clones are preliminary named Lhc1*Msq1-4 according to current nomenclature.

The amino acid sequence of the chl a/b/Mg-DVP LHC precursor molecule of *M. squamata* LHC1*1, deduced from the nucleotide sequences of the plasmid pER-21 is shown in Fig.2. The initiator codon Met initiates an open reading frames of 231 amino acids.The precursor protein would have a Mr of about 24,200, which is in agreement with the size of Mr 27,000 determined for the primary translation products.

The LHC-polypeptides of *M. squamata* are related in sequence and tripartite structure to those of green algae and higher plants and share at the most 34 - 38% homology with them. Surprisingly, the lowest homology of only 22% is found between the LHC of *M. squamata* and the fucoxanthin-chl a/c LHC of the diatom *Phaeodactylum tricornutum* [8]. Thus, *M. squamata* is placed in the phylogenetic tree in line with the green algae.
As a major feature, the whole N-terminal part, of about 30 amino acids, and consequently the reversibly phosphorylated threonine residue is absent from the LHC polypeptides. Therefore, the LHC is unable to alternate photosystem preferences and directs its energy to the nearest neighbour. We suppose that the acquisition of the regulatory phosphorylation site may have occured very early in the development of the Chlorophyceae. The lack of this domain may also diminish the attractive forces of the stacked thylakoids, because small gaps are observed between adjacent thylakoids [4].
A transmembrane helix, comparable to helix I of LHCII is not predicted by hydropathy plots, however proposed by sequence similarities comprising amino acids 53 - 83. The luminal half of this helix contains a stretch of polar and charged amino acids which may supply binding sites forMg-DVP, which lacks the phytol tail and might need a somewhat different environment than chl a or b.
The LHC precursor contains an N-terminal transit sequence which is removed during import. The cleavage site follows alanine [7]; the N-terminal aspartic acid of the mature protein is blocked by acetylation. The first quarter of the import peptide corresponds more to that of PSI-3 than to comparable import peptides of higher plant LHCs. No conserved sequence domains were found.
The possible existence of introns within the genes on the genome was investigated by PCR. Single PCR products of the expected size were amplified from insert DNAs. Also, single products of identical size were synthesized when genomic DNA was used as template. Thus, most likely no introns exist within the two gene coding regions of LHC1*1 and LHC1*2.

Addendum: This study was generously supported by the Deutsche Forschungsgemeinschaft. We thank Dr. W. Kühlbrandt for the optical diffraction measurements.

References

1 Fawley, M.W., Steward, K.D. and Mattox, K.R. (1986) J. Mol. Evol. 23, 168-176
2 Green, R.B., Pichersky, E. and Kloppstech, K. (1991) Trends Biochem. Sci. 16, 181-186
3 Rhiel, E., Mörschel, E. and Wehrmeyer, W. (1987) Eur. J. Cell Biol. 43, 82-92
4 Krämer, P., Wilhelm, C., Wild, A., Mörschel, E. and Rhiel, E. (1988) Protoplasma 147, 170-177
5 Rhiel, E., Lange, W. and Mörschel, E. (1993) Biochim. Biophys. Acta 1143, 163-172
6 Rhiel, E. and Mörschel, E. (1993) Mol. Gen. Genet. 240, 403-413
7 Schmitt, A., Frank, G., James, P., Staudenmann, W., Zuber, H. and Wilhelm, C. (1994) Photosynthesis Research 40, 269-277

POSSIBLE OCCURRENCE AND CHARACTERISTICS OF TWO TYPES OF PHYCOBILISOMES DURING ONTOGENESIS OF RED ALGA *PORPHYRA YEZOENSIS*

Ding-Ji SHI[1], Guo-Hong PENG[2], Jian-Xin CHEN[3], Hong-Jie WANG[4], Ping JIN[1], Li-Jie FENG[1], Hong ZHANG[3], Xiu-Geng FEI[2], Cheng-Kui TSENG[2] & Pei-Song TANG[1]

[1]Dept. of Photosynthesis, Institute of Botany, Academia Sinica, Beijing 100044; [2]EMBL, Institute of Oceanology, Academia Sinica, Qingdao 266071; [3]KLDSSM, Institute of Chemistry, Academia Sinica, Beijing 100080, [4]Capital Normal University, Beijing 100037, P.R. China.

INTRODUCTION

The conversion of solar energy by green plants (photosynthetic cells) is through multiways, particularly by algae[1], This multiformity of light conversion does not only mean converting light energy into reduced biochemicals but also absorbing and transferring light energy [2,3]. Recent research has proved that light-harvesting pigment systems (LHP) are various, such as chlorophyll a/b binding protein in chlorophyta, flucoxanthin-chlorophyll a/c complex in diatoms, dinoflagellates, chrysophytes and brown algae, and phycobilisomes in cyanophyta and rhodophyta.

Among these antenna pigment systems, phycobilisomes may be the best investigated one. There have been a number of works on their morphology, biochemistry and molecular biology and responsibility to environmental factors [4,5,6]. However, little is known of the relationship between phycobilisomes and ontogenesis. Usually, the type of phycobilisomes varies with the species.

The life cycle of *Porphyra yezoensis* consists of two phases [7]: conchocelis (sporophyte) and the thallus (gametophyte) which all carry out photosynthesis. Comparing photosynthetic characteristics of these two phases [8] two types of phycobilisomes were unexpectedly isolated. Here we reported the isolation of them and their features.

MATERIALS AND METHODS

1. Culture conditions

Free-living conchocelis of *Porphyra yezoensis* (in early-phase, and late-phase) and thalli were grown at 18℃ in PES medium[9] under white light at 60 $\mu M^{-2}S^{-1}$(late phase), 20 $\mu M^{-2}S^{-1}$ (early phase), and for 10h per day. Thalli were cultured at 10℃. All materials came from the same cell strains.

2. Isolation and preparation of phycobilisome

Phycobilisomes of *P. yezoensis* were isolated according to Gantt et al. [10] and Nies et al. [11]with some modifications.

3.Absorption spectra

The Absorption spectra of phycobilisomes were measured at room temperature by using a UV-3000 type double-wavelength dual-beam spectrophotometer, the range of wavelength was 360-720nm[12].

4. Fluorescence emission spectra

77K fluorescence emission spectra[12] were measured by using a Hitachi Model MPF-44 fluorescence spectrophotometer. .

P. Mathis (ed.), Photosynthesis: from Light to Biosphere, Vol. I, 383–386.

5. Picosecond fluorescence decay kinetics

The fluorescence decays were measured by a time-correlated single-photon counting method. The excitation was done at 300nm with a pulse width<1ps (FWHM) and a repetition rate of 4MHz. Emission wavelength was selected with a homemode subtractine double grating monochromator and detected using a Hamamatsu R2809U01 microchannel plate photomultiplier tube. The instrument response time was ca.50ps. Decays were detected at 570nm, 650nm, 680nm, respectively and accumulatedto 2000 counts in the peak channel. The data were analysed individually and globally by iteratine re-convolation based on the Marquardt fiting algorithum, assuming multiexponential decay kinetics. All measurements were carried out at room temperature (20+1℃) in quartz cell, optical path-length 1cm, active modelock YAG, synchron.

6. Electron Microscopy

Phycobilisome preparations were negatively stained at pH6.8 with 2% phosphotungstic acid [9], or stained with 1% lead. acetate. Grids were examined with Hitachi H-600. electron microscopy.

RESULTS

1.The isolation of phycobilisomes

The last procedure in the method was sucrose gradient centrifugation, two types of phycobilisomes were isolated (Fig.1A). The phycobilisomes from early-phase conchocelis (CPBS) appeared only one violet band and with smaller molecular weight. The phycobilisomes from thalli (TPBS) were contained in only one red band with larger molecular weight. Interestedly, late-phase conchocelis possessed both CPBS and TPBS. Fig.1B showed CPBS located in the region of 0.70M sucrose and TPBS in the region of 0.85M sucrose.

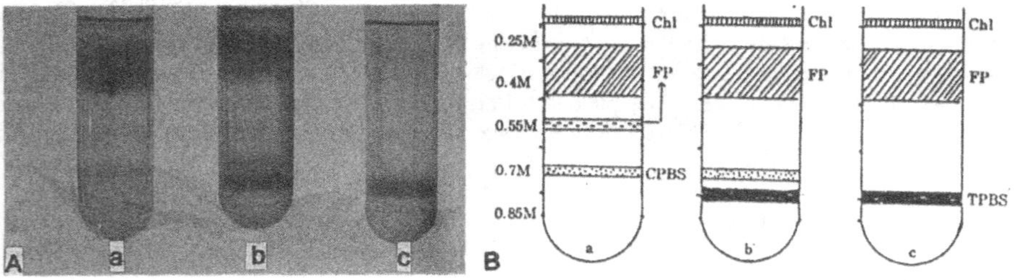

Fig.1 Sucrose density gradient sedimentation profile from different development stages of. *P yezoensis*. A. Photograph of gradient centrifugation, B. Model of sedimentation profile. a: early-phase conchocelis; b: late-phaseconchocelis; c.thallus; Chl:Clorophyll; FP:free phycobiliprotein; PBS:phycobilisome.

2. Spectra analysis of the phycobilisomes.

2.1. Absorption spectra in room temperature(Fig.2).

From the spectra, both CPBS and TPBS had 5 absorption peaks: A498, A545 and A568 from phycoerythrin; A618 from phycocyanin; and A653 from allophycocyanin. As the same as the measurement with intact cells, CPBS contained more phycocyanin and allophycocyanin and TPBS had more phycoerythrin.

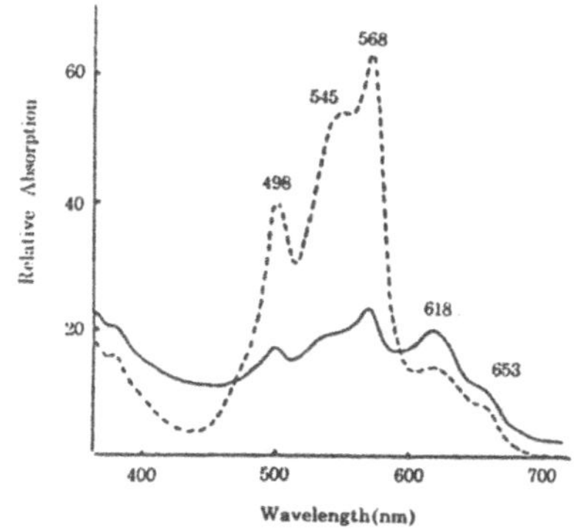

Fig.2 Absorption spectra of phycobilisomes from the conchcelis in early-phase(---), and thallus(—) of *P.yezoensis* at room temperature.

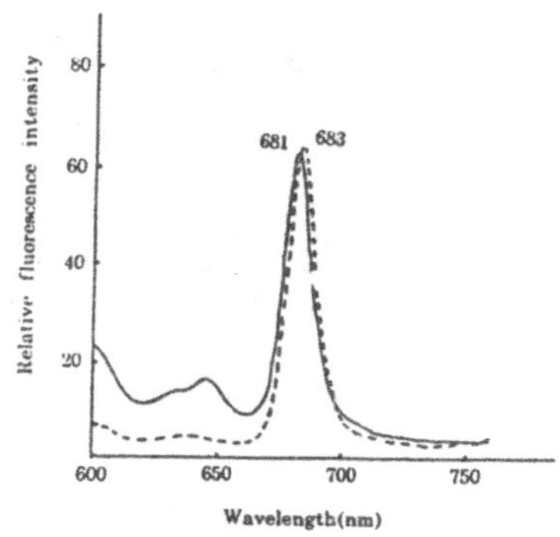

Fig.3 Fluorescence emmission spectra of phycobilisomes from theconchocelis in early-phase (---),and thallus(—) of *P.yezoensis* at 77K,excitation at 580nm

2.2. Low-temperature fluorescence emission spectra (Fig.3)

Fluorescence emission from CPBS occured 681nm and the emission from TPBS at 683nm. These single peaks proved the intactness of phycobilisomes.

From 77K fluorescence emission spectra, the ratio of the heights of the PSI fluorescence peak (F720) with 540nm and 436nm excitation provided a rough indicator of the relative efficiency of energy transfer from phycobilisomes to PSI. Table 1 showed the comparison of the energy transfer between two types of phycobilisomes and two photosystems in early-phase conchocelis and the thallus. At state2 (pretreated with dark) CPBS transfered energy to PSI less than TPBS. And at state 1 (pretreated with blue. green and red light) CPBS transfered more energy to PSI. than TPBS. This implied CPBS may be closed to PSII and TPBS may prefer to PSI.

Table 1. Excitation energy transfer from of phycobilisomes to PSI inearly-phase conchocelis and the thallus of *P.yezoensis*.

Materials	Pretreatments with light			
	Dark	Blue	Green	Red
	(Ratio between F730 ex.540nm/F730 ex.436nm)			
early-phase	0.923	1.422	1.370	2.025
thallus	1.290	1.600	0.978	1.292

2.3. Picosecond fluorescence decay kinetics

The results of the fluorescence decay Global analysis showed that the short-lived component with the lifetimes 55ps in CPBS had only one negative amplitude (-71.7) at 680nm and the components with 36ps in TPBS had two negative ones (-66.5 at 650nm and-23.2at 680nm). This meaned for energy transfer only from 650nm to 680nm in CPBS, and in TPBS there were two transfers: From 570nm to 650nm and from 650nm to 680nm. The former was more than the latter.

Another difference between CPBS and TPBS was that the transfer from 650nm to 680nm was only one way related to 55ps in CPBS and two ways in TPBS: one from 36ps and another from 97ps.

2.4. TEM of negative sraining CPBS and TPBS

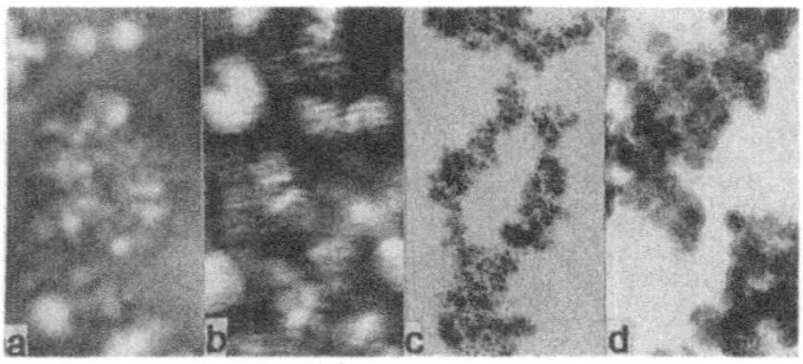

Fig. 4 TEM of negatively stained CPBS (a, c) and TPBS (b, d) of *P. yezoensis.*

Fig. 4 showed the shapes of purified CPBS and TPBS negative staining with phosphotungstic acid (Fig. 4 a, b) or with lead acetate (Fig. 4 c,d) were different. The CPBS looked like bundles of rod-shaped with smaller particles, and the TPBS appeared ellipsoidal shape. This means the CPBS was similar to the phycobilisomes from cyanophyta and the TPBS closed to that from rhodophyta.

DISCUSSION

Since the phycobilisomes were discovered by Gantt & Conti[14] in 1965, the ultramolecular proteins have been confirmed in a number of species of cyanophyta and rodophyta[3,5]. Up to now, three types of phycobilisomes can be distinguished by their ultrastructural shapes: (1) bundles of rod-shaped elements from extraordinary cyanobacterium *Gloeobacter violaceus* only; (2) hemidiscoidal commonly distributed in other cyanobacteria; (3) hemiellipsoidal wide occured in rhodophyta. The morphology of phycobilisomes varies with the source organism. These types seem to be related to algal phylogeny.

In this work, the two types of phycobilisomes (CPBS and TPBS) were isolated by routine method. CPBS was only from both early-or late-phases conchocelis and TPBS occured in both thalli and late-phase conchocelis. Fluorescence emission spectra proved all of them were intact. TEM showed CPBS and TPBS possessed different shapes. These data evidenced the types seem to be also related to ontogeny. The late-phase conchocelis contained two types of phycobilsomes provided that one cell encluded not only one type of phycobilisomes. Algarra et al.[15] reported phycobilisome heterogeneity in *P.unbilicalis* and they concluded the existence of two kinds of phycobilisomes as intermediary steps of phycobilisomes organization in algae exposed to rapid changes in environmental factors. In our experiment, the alga was grown in culture and adapted stable conditions. Therefore, two types of phycobilisomes occured may mainly be intrinsic.

As the shape of CPBS was similar to the phycobilisomes from cyanophyta and TPBS appeared typical phycobilisomes of rhodophyta. The change of types of phycobilisomes during ontogenesis in fact reflected the phylogensis of algae. It may be reasonable to prospect that two or more types of phycobilisomes should be found in other species of rhodophyta and cyanophyta.

REFERENCES

1. Tang, P.-S. 1979. Trends in Biology (1):1-8 (in Chinese).
2. Tseng, C.-K. & Zhou,B.-C. 1983. Anthology on Evolution. pp.34-43, Scientific Press, Beijing.
3. Glazer, A. N. 1989. J. Biol. Chem., 264:1-4.
4. Grossman, A. R. et al;. 1993. Microbiol. Rev. 57: 725-749.
5. Reuter, W. & Muler, C. 1993. J. Photochem. Photobiol. B: Biol., 21: 3-27.
6. Bryant, D. A. 1994. The Molecular Biology of Cyanobacteria, Kluwer Academic Pub.
7. Tseng, C.-K., & Zhang, D.-R. 1954. Acta Bot. Sin. (3) : 287-302.
8. Peng, G.-H. et al;. 1995. Acta Bot. Sin. (in press).
9. Provasoli, L., 1968.Culture and Collections of Algae. pp. 63-75, Jap.Sociey of Plant Physiologists.
10. Gantt, E. et al. 1979. Plant Physiol., 63: 615-620.
11. Nies, M. et al. 1980. Planta, 150:330-337.
12. Shi, D. -J. et al. 1983. Botanical Research. No.1, 207-215.
13. Mullineaux, C. W. 1994. Biochim. Biophys. Acta. 1184: 71-77
14. Gantt, E. & Conti, S. F. 1965. J. Cell Biol. 26, 365.
15. Algarra, P, et al. 1990. Plant Physiol., 92:570-576.

2. Reaction centers : purple bacteria and PSII

ULTRAFAST INFRARED AND VISIBLE SPECTROSCOPY OF BACTERIAL REACTION CENTERS

W. Zinth[1], T. Arlt[1], H. Penzkofer[1], P. Hamm[1], M. Bibikova[2], B. Dohse[2], D. Oesterhelt[2], M. Meyer[3], H. Scheer[3]
[1] Institut für Medizinische Optik der Universität München
Barbarastr.16, 80797 München, Germany
[2] Max-Planck-Institut für Biochemie
Am Klopferspitz, 82152 Martinsried, Germany
[3] Botanisches Institut der Universität München
Menzinger Str. 67, 80638 München, Germany

1. Introduction

Since ten years femtosecond experiments have been used to study the primary electron transfer (ET) of photosynthetic reaction centers (RC) [1-5]. During that period, important information has been obtained on the dynamics of the primary ET. At present the interest on photosynthetic ET is focused on several still interesting topics:
(i) Basic theory for ET in photosynthetic RC: adiabatic versus nonadiabatic theory [6,7].
(ii) Primary electron donor: Why uses nature an excitonically coupled pair of bacteriochlorophyll molecules?
(iii) The electron transfer itself:
- Asymmetry of ET
- Intermediates used in ET
- Optimisation of photosynthesis

In this paper we address these questions: At first we present data from mutated RCs where a change in the energetics of the special pair P leads to a speed-up of the ET step away from P^*. These result will demonstrate convincingly the role of the accessory bacteriochlorophyll B_A as the primary electron acceptor. In the second part, results from experiments performed in the mid infrared region are presented. They reveal an unexpected 200 fs transient evolving in the excited electronic state of the special pair P. These data could be used to obtain a better understanding of the dimeric nature of P^* and of the asymmetry of electron transfer.

2. Observation of $P^+B_A^-$ in Spectral Regions with Reduced Interference from Other Intermediates

Within the nonadiabatic description of primary ET a system of rate equations governs the evaluation of the population of the intermediate states and exponential kinetics of the absorbance changes result [4]. In the stepwise ET model the electron is transferred with the time constant τ_1 from P^* to B_A, with τ_2 to the secondary acceptor, the bacteriopheophytin H_A and with τ_Q to the quinone Q_A.

P. Mathis (ed.), Photosynthesis: from Light to Biosphere, Vol. I, 389–394.

$$(1) \qquad P^* \xrightarrow{\tau_1} P^+B_A^- \xrightarrow{\tau_2} P^+H_A^- \xrightarrow{\tau_Q} P^+Q_A^-$$

The population of an intermediate and its visibility are strongly reduced, when the intermediate is faster depopulated than it is populated. In this situation, $\tau_1 > \tau_2$, a change in absorption cross-section $\Delta\sigma_{B/H}$ between the intermediates $P^+B_A^-$ and $P^+H_A^-$ causes a weak amplitude a_2 of the absorbance change occurring with the time constant τ_2 of:

$$(2) \qquad a_2 \propto \frac{\tau_2}{\tau_2 - \tau_1} \cdot \Delta\sigma_{B/H}$$

The reduction of visibility is striking for the radical pair state $P^+B_A^-$ in wild type RC. E. g. in RC of *Rhodopseudomonas (Rps.) viridis* where $P^+B_A^-$ is populated from P^* with 3.5 ps (monoexponential fit of the P^* decay) and depopulated with 0.65 ps [5] the amplitude a_2 is small. The visibility of state $P^+B_A^-$ (i.e. the amplitude a_2) will be improved by a decrease of the initial ET time constant τ_1. E. g. for a reduction of τ_1 from 3.5 ps to 1 ps, the amplitude a_2 will be increased by a factor of ≈ 8. In this case a_2 should become clearly visible.

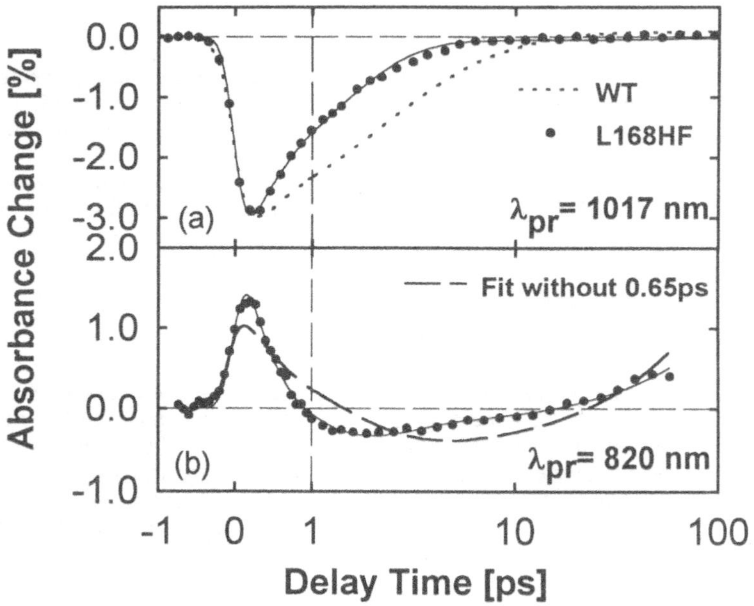

Figure 1. Time resolved absorption differences $\Delta A(t_D)$ for mutated RC of *Rps. viridis* L168HF. Probing wavelengths are set to positions where the excited electronic state P^* causes strong stimulated emission (1017 nm, Fig. 1a) and where B_A absorbs strongly (820 nm, Fig. 1b). The broken curves represent the results from wild type RC (Fig. 1a) and from a fit without the 0.65 ps kinetic component (Fig. 1b).

Different possibilities allow to rise the ratio $\tau_2/(\tau_2 - \tau_1)$ by modifying the RC:
(i) Changing τ_2 by introducing a charge at or around H_A. (However this change may also influence the energetics of $P^+B_A^-$ leading to unexpected results).
(ii) Modifying the chromophore at position B_A or modifying its surrounding.(This may alter all parameters in Eq. (2)).
(iii) Modifications at the special pair P, far away from B_A^-. (This change should allow to obtain clear results since it is expected to change only τ_1).

Suitable modifications at P can be achieved by introducing or removing hydrogen bridges between amino acid residues and the special pair. This procedure changes the redox potential P/P^+ and the free energy of P^+. Recently those mutations were presented for RCs from *Rb. sphaeroides* [8,9] and also realised for RC of *Rps. viridis.* [10].

The experiments are performed with a femtosecond laser/amplifier system based on a CPM-dye laser described in detail in Ref. [11]. Fig. 1 presents the results from experiments on the mutant L 168HF [10] where the replacement of histidine L168 by a phenylalanine leads to the removal of one hydrogen bond. As a consequence, the energy of the cation radical P^+ is decreased relative to wild type RC by $\approx$650 cm^{-1} (value deduced from the change in P/P^+ redox potential). The experiments testing the stimulated emission of P^*, e. g. at the probing wavelength 1017 nm (Fig. 1a) clearly demonstrate an accelerated decay with a time constant of $\tau_1 = 1.1$ ps (instead of 3.5 ps found in wild type RC shown as a broken curve in Fig. 1a). This strong speed-up of τ_1 increases the visibility of the $\tau_2 = 0.65$ ps process (see Eq. 2). Very striking becomes the 0.65 ps component in the Q_y absorption band of B_A: At 820 nm the 0.65 ps kinetic is by far the dominating contribution to the absorption change. Here the strong absorption decrease related with the 0.65 ps kinetic is due to a bleach of the $Q_y(B_A)$ absorption band upon formation of $P^+B_A^-$. A series of experiments on other mutates RC gives similar results [11]. They strongly confirm the presence of $P^+B_A^-$ in primary ET and with that the stepwise reaction model of Eq. 1.

3. Femtosecond Spectroscopy in the Mid Infrared - Ultrafast Processes in the Excited Electronic State of the Special Pair

A set of the transient absorbance data of RC from *Rb. sphaeroides* covering the spectral range from 1000-1800 cm^{-1} and a temporal range of 0-1 ns was recorded [12]. (Spectral resolution, 4-7 cm^{-1}, temporal resolution, 100 fs). The experimental set-up used to measure these data has been described recently in detail [12,13].

At most wavelengths in the investigated spectral range absorption increases immediately after optical excitation. However, a detailed analysis shows that the initial signal rise is slower than expected from the temporal response function of the experimental set-up [12,14]. This delay is well recognised around 1460 cm^{-1} (see Fig. 2) and 1225 cm^{-1}. A numerical analysis shows that the instantaneous absorption increase is weak. The main part of the absorption rises delayed with a time constant of 200±100 fs. It is interesting to note, that a similar kinetic component (time constant 250 fs) has been found recently in dynamic hole burning experiments performed in the P absorption band in the near IR [15].

In the following we want to focus on the spectral and temporal properties in the low-frequency spectral regions <1600 $cm^{-1}$where the C=C, C-C, C-N stretching and C-H bending modes of the chlorophyll tetrapyrolle ring system determine the transient absorbance changes. A striking feature is found in this part of the spectrum: Fig. 3 (lower part) shows

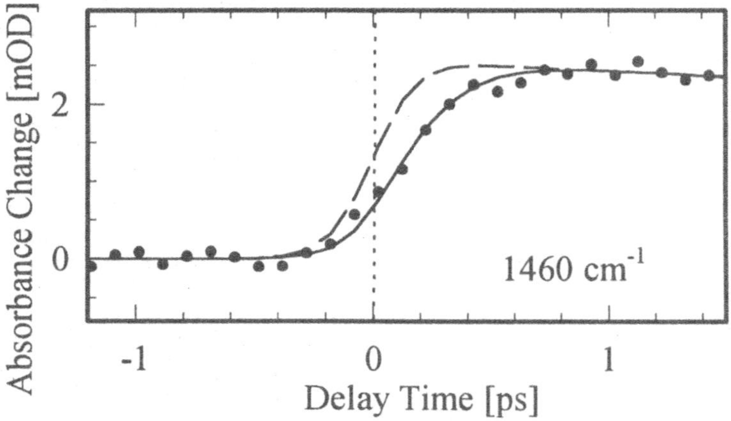

Figure 2. Transient absorption change measured in the IR at 1460 cm^{-1} at the peak of an intense IR-absorption band. The absorption rise (solid line and points) is delayed with respect to instrumental response function (dashed line). The rise of the absorption can be modelled by a exponential function with a time constant of 200±100 fs.

a pronounced positive absorption background in connection with strong and broad positive bands in P* (recorded at $t_D = 1$ ps, central trace of Fig. 3) and P^+ (bottom trace) at slightly different peak positions. As indicated above, these positive bands do not appear instantaneously after electronic excitation but within a delay time of 200 fs (see Fig. 2). From the complete set of transient data one could deduce the spectrum of the absorbance change present immediately after optical excitation. We call the related initial state P^{**}. Its spectrum is plotted on the top of Fig. 3. As expected from the time dependent detection (e. g. from Fig. 2), the absorption increase in this state is still weak. Absorbance strongly increases only subsequently with the 200 fs transition to P^*. Qualitatively, these observations can be explained as follows: As a consequence of excitonic coupling between two chromophores to a dimer (the special pair P), low lying electronic transitions appear when the RCs populate P* or P^+ [14,16]. Due to the low excitation frequency of these electronic transitions (1000-3000 cm^{-1}), coupling between electronic and distinct vibronic transitions is expected, leading to a transfer of oscillator strength from the electronic transition to the (in general) weak vibronic transitions. This coupling increases the absorption cross section in the vibrational part of the spectrum of P* and P^+. Within this model, some observations can be explained qualitatively: (i) the positive P*/P- and P^+/P-difference spectra, (ii) the fact that the absorbance changes of the special pair dominate the difference spectra.

However the observation that the positive bands are not present immediately after electronic excitation but rise within an ultrafast 200 fs relaxation process has still to be explained: The amplitude of the 200 fs kinetic component is too large to be understood as a simple vibrational relaxation process. Therefore the electronic structure of the special pair (and with that the electronic-vibronic-coupling mechanism) must change considerably during that relaxation. It is known, that the excitonic coupling of the bacteriochlorophyll

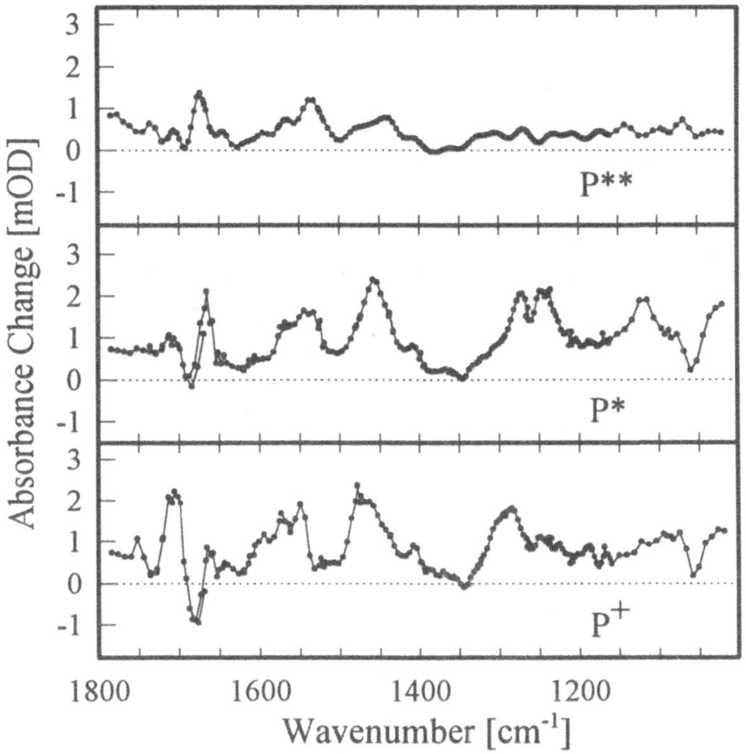

Figure 3. Time resolved difference spectra taken in the mid infrared. The spectrum at the top was calculated from all absorbance data. It represents the state P^{**} immediately populated by optical absorption. The spectra of states P^* and P^+ were recorded at a delay time of 1 ps (P^*) and 10 ps (P^+) when the RCs are predominantly in the states P^* and $P^+H_A^-$. (The weak contributions from H_A are neglected).

molecules in the special pair depends critically on the arrangement of both molecules. In addition quantum mechanical arguments indicate that in a symmetric special pair P, the coupling between electronic and vibronic transition should be weak since no intramolecular charge separation within P is possible. Under these conditions a strong 200 fs kinetic component can be explained as follows: Immediately after electronic excitation, the RCs populate an almost symmetric initial excited state with uniform charge distribution (small charge transfer character). Later on - in parallel with the 200 fs component - a motion along a low frequency mode (for example a motion of both bacteriochlorophyll molecules or parts of them against each other) occurs destroying the symmetry of the dimer and leading to an increased charge separation in the dimer. As a consequence, this reaction may facilitate ET along the active branch and may present one reason for the "unidirectionality" of the electron transfer.

References

1. Martin, J.L., Breton, J., Hoff, A.J., Migus, A., and Antonetti, A.,(1986) Proc. Natl. Acad. Sci. USA 83, 957.
2. Breton, J., Martin, J.L., Migus, A., Antonetti, A., Orzag, A., (1986) Proc. Natl. Acad. Sci. USA 83, 5121.
3. Holzapfel, W., Finkele, U., Kaiser, W., Oesterhelt, D., Scheer, H., Stilz, H.U., and Zinth, W., (1989) Chem. Phys. Letters 160, 1.
4. Holzapfel, W., Finkele, U., Kaiser, W., Oesterhelt, D., Scheer, H., Stilz, H.U., and Zinth, W., (1990). Proc. Natl. Acad. Sci. USA 87, 5168.
5. Dressler, K., Umlauf, E., Schmidt, S. Hamm, P., Zinth, W., Buchanan, S., and Michel, H., (1991) Chem. Phys. Letters 183, 270.
6. Parson, W.W., Warshel, A., (1993) in: The photosynthetic reaction center *Vol. II*, ed. Deisenhofer, J., and Norris, J.R., (1993) (Academic Press, San Diego) p. 23.
7. Bixon, M., Jortner, J., and Michel-Beyerle, M.E., (1991) Biochim. Biophys. Acta 1056, 301.
8. Murchison, H.A., Alden, R.G., Allen, J.P., Peloquin, J.M., Taguchi, A.K.W., Woodbury, N.W., Williams, J.C.; (1993) Biochemistry 32, 3498.
9. Williams, J.C., Alden, R.G., Murchison, H.A., Peloquin, J.M., Woodbury, N.W., Allen, J.P., (1992) Biochemistry 31, 11029.
10. Bibikova, M., Arlt, Th., Zinth, W., and Oesterhelt, D., Same proceedings.
11. Arlt, T., Schmidt, S., Kaiser, W., Lauterwasser, C., Meyer, M., Scheer, H., and Zinth, W., (1993) Proc. Natl. Acad. Sci. USA 90, 11757.
12. Hamm, P., Zurek, M., Mäntele, W., Meyer, M., Scheer, H., and Zinth, W., (1995) Proc. Natl. Acad. Sci. USA 92, 1826.
13. Hamm, P., Wiemann, S., Zurek, M., and Zinth, W., (1994) Opt. Lett. 19, 1642.
14. Hamm, P., and Zinth, W., (1995, in press).
15. Streltsov, A.M., Yakovlev, A.G., Shkuropatov, A.Ya., Shuvalov, V.A., (1995) FEBS Letters 357, 239.
16. Walker, G. C., Maiti, S., Cowen, B. R., Moser, C. C., Dutton, P. L., and Hochstrasser, R. M., (1994) J. Phys. Chem. 98, 5778.

PROTEIN AND BACTERIOPHEOPHYTIN RESPONSE TO Q_A REDUCTION IN PHOTOSYNTHETIC BACTERIAL REACTION CENTERS FROM *RB. SPHAEROIDES* AND *RP. VIRIDIS* INVESTIGATED BY $^1H/^2H$ EXCHANGE AND LIGHT-INDUCED FTIR DIFFERENCE SPECTROSCOPY

Jacques Breton and Eliane Nabedryk

SBE/DBCM, CEA-Saclay, 91191 Gif-sur-Yvette, France

1. Introduction

In the photosynthetic bacterial reaction center (RC), light induces the rapid transfer of an electron from a dimer of bacteriochlorophylls to a molecule of bacteriopheophytin (H_A) and then to a primary quinone acceptor (Q_A). Although the crystal structure of the RC suggests that the interactions between the protein and the cofactors are important both for the geometrical organization of the electron transfer pathway and for fine tuning the energy levels of the cofactors, the precise role of the protein in assisting the electron transport and in preventing wasteful back reactions is still poorly understood. For long range electron transfer reactions to proceed, the configuration should have the same energy in the reactants and products. The reactants undergoing changes of charge, the solvent contributes to the reorganization energy. In the case of the RC, the necessary balance of energy is provided by the protein nuclear rearrangements. However, the specific contributions of the different parts of the protein to this reorganization energy have not yet been identified. In addition, the relaxation effects could be either localized in the close vicinity of the cofactors or distributed at more distant sites depending on the electrostatic of the system. Furthermore, the protein movements (structural rearrangements) following electron transfer will proceed on various time scales.

Among the techniques that could reveal localized nuclear rearrangements of the RC protein upon electron transfer reactions, light-induced FTIR difference spectroscopy affords the sensitivity to investigate changes of the individual atomic bonds which are altered by the charge separation process. This technique, which provides information on all the bonds of both the cofactors and the protein that are affected by the photoreaction, is in itself rather unspecific. However, there are a few favorable situations where selectivity can be achieved as, e.g., for the photoreduction of the primary (Q_A) or secondary (Q_B) quinone acceptors in *Rb. sphaeroides* and *Rp. viridis*. In this case, the Q_A^-/Q_A and Q_B^-/Q_B FTIR spectra of quinone-depleted RCs reconstituted with isotopically labeled ubiquinones have allowed the quinone modes perturbed by the photoreaction to be discriminated from those of the RC protein for both Q_A (Breton et al., 1994a–c; Brudler et al., 1994) and Q_B photoreduction (Breton et al., 1995a,b). The challenge then remains to assign the non-quinonic vibrations observed in these spectra to specific parts of the protein.

In the present communication, we describe preliminary $^1H/^2H$ exchange experiments aimed at the investigation of possible contributions from the protein backbone and side chains as well as from the other cofactors in the Q_A^-/Q_A FTIR spectra.

P. Mathis (ed.), Photosynthesis: from Light to Biosphere, Vol. I, 395–400.

2. Results

The Q_A^-/Q_A light-minus-dark FTIR difference spectra of native *Rb. sphaeroides* RCs incubated in ^1H_2O and ^2H_2O are shown in Figures 1a and 1b, respectively. In these spectra, the negative bands pertain to vibrations from the Q_A state and the positive ones belong to those of the Q_A^- state. After the Q_A site has been depleted of its native ubiquinone and reconstituted with vitamin K_1 (a menaquinone with a phytyl chain), the corresponding spectra are presented in Figures 2a and 2b. The Q_A^-/Q_A spectra of *Rp. viridis* RCs in ^1H_2O and ^2H_2O are shown in Figures 3a and 3b, respectively. While the spectra in ^1H_2O are almost indistinguishable from those previously reported for the corresponding RCs (Breton et al., 1991, 1992, 1994a; Nabedryk et al., 1991), those in ^2H_2O show a number of significant differences. These differences reflect a specific change upon deuteration of the frequency and/or intensity of some of the modes of the protein, of the quinone, and of the other cofactors that are affected by Q_A photoreduction. They are best visualized by calculating double-difference spectra (^1H_2O-minus-^2H_2O) between the two individual Q_A^-/Q_A spectra by minimizing the residual signals. Although these double-difference spectra (Figures 1c, 2c, and 3c) are not unique solutions and may therefore appear arbitrary, most of the bands in these spectra, and notably all those that will be discussed below, are quite insensitive to large variations (typically ± 20%) in the coefficient used for the subtraction. The two double-difference spectra for *Rb. sphaeroides* RCs (Figures 1c and 2c) have many bands in common. With the exception of the 1400- to 1500-cm^{-1} range where the semiquinone C$\cdots$C and C$\cdots$O vibrations are found, the slight differences observed between these two spectra appear mostly related to uncertainties due to the subtraction procedure described above and to small variations in the extent of deuteration of the samples. Analysis of a larger data set will be needed to ascertain the possible significance of these slight differences. Larger differences can be noticed when comparing these two double-difference spectra for *Rb. sphaeroides* to that obtained for *Rp. viridis* RCs (Figure 3c), notably between 1640 and 1700 cm^{-1}.

3. Discussion

3.1 *Effect of ^2H_2O on the Q_A Vibrations*:

The vibrations of the neutral Q_A in *Rb. sphaeroides* RCs have been previously determined by isotope labeling of the quinones (Breton et al., 1994a,c, 1995b; Brudler et al., 1994). Bands at 1660, 1628, and 1601 cm^{-1} have been assigned to the C_1=O (free from interaction with the protein), C=C, and C_4=O vibrations of ubiquinone in the RC, respectively. The negative bands at 1628 and 1601 cm^{-1} are present both in ^1H_2O (Figure 1a) and ^2H_2O (Figure 1b) and do not shift upon $^1H/^2H$ exchange (Figure 1c). These observations demonstrate that the strong hydrogen bond between the C_4=O carbonyl of Q_A and the protein (most probably at the imidazole ring of His M219) is left unperturbed by the deuteration owing either to an identical strength of the hydrogen or deuterium bond or to a lack of exchange at this site. For vitamin K_1 in the Q_A site of *Rb. sphaeroides* (Breton et al., 1992, 1994a), the two carbonyl groups absorb at 1651 and 1640 cm^{-1} and the main C=C mode gives rise to the band at 1585 cm^{-1} clearly seen in Figures 2a and 2b while for the native menaquinone in *Rp. viridis* RCs only the C=O band at 1636 cm^{-1} and the C=C band at 1585 cm^{-1} (Figure 3a,b) have yet been assigned (Breton et al., 1992, 1994a). Similarly to the case of ubiquinone in *Rb. sphaeroides*, none of these menaquinone vibrations appear significantly affected by the deuteration (Figures 2c and 3c).

The semiquinone bands appear much more sensitive to deuteration than those of the neutral species. In the case of ubiquinone in the Q_A site of *Rb. sphaeroides*, the low

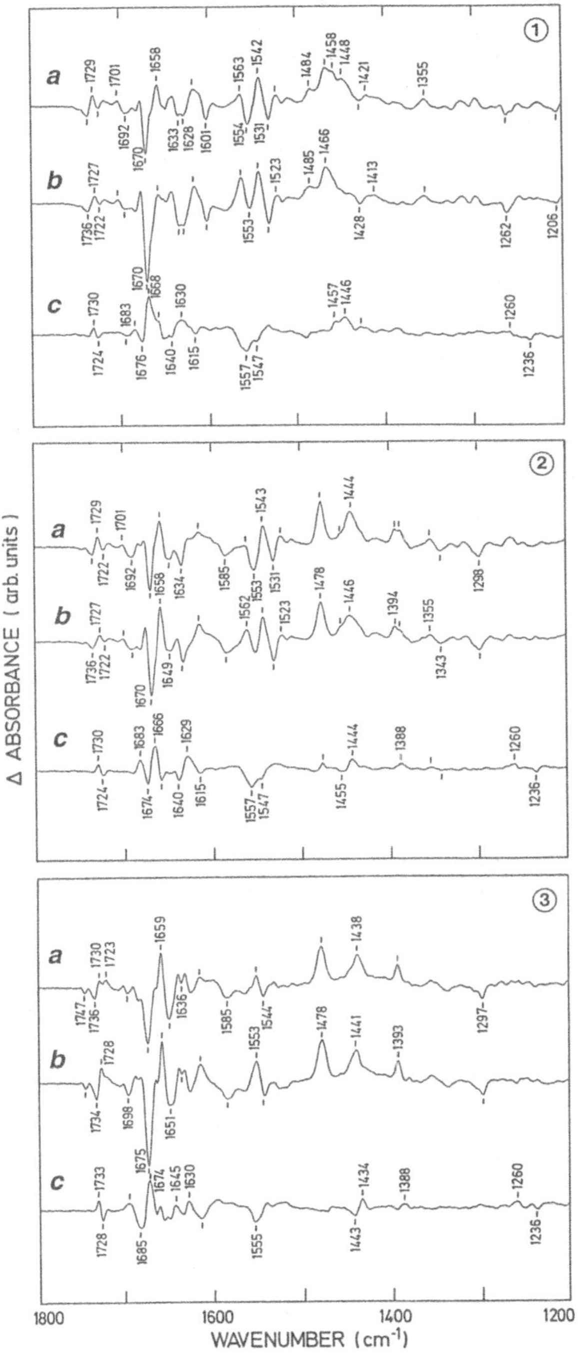

Q_A^-/Q_A light-minus-dark FTIR difference spectra at 5°C of (1) native *Rb. sphaeroides* RCs, (2) Q_A-depleted *Rb. sphaeroides* RCs reconstituted with vitamin K_1, and (3) native *Rp. viridis*. The RCs were incubated in ^1H_2O (a) or ^2H_2O (b). Experimental conditions are described in Breton et al., 1994a. The double-difference spectra (^1H_2O-minus-^2H_2O) are shown in 1c, 2c, and 3c, respectively. Resolution is 4 cm^{-1}.

frequency side of the 1466-cm^{-1} anion band (Figure 1a) decreases considerably in amplitude upon exchange (Figure 1b). Using selectively ^{13}C-labeled ubiquinones, several tentative assignments of the semiquinone C$\underline{\cdots}$C and C$\underline{\cdots}$O vibrations have been proposed for Q_A^- (Breton et al., 1994c; Brudler et al., 1994). The bands at 1484, 1466, and 1448 cm^{-1} have been tentatively assigned either to C$\underline{\cdots}$C, $C_1\underline{\cdots}$O, and $C_4\underline{\cdots}$O modes, respectively (Breton et al., 1994c) or to $C_1\underline{\cdots}$O, $C_4\underline{\cdots}$O, and C$\underline{\cdots}$C vibrations, respectively (Brudler et al., 1994). The strong decrease in amplitude of the 1448-cm^{-1} band upon $^1H/^2H$ exchange (Figure 1) appears much more compatible with the assignment of this band to a hydrogen-bonded $C_4\underline{\cdots}$O mode than to a C$\underline{\cdots}$C mode. In the case of vitamin K_1 in *Rb. sphaeroides* and of the native menaquinone of *Rp. viridis*, a triplet of bands characterizes the semiquinone state (Figures 2a,b and 3a,b). By using ^{18}O-labeled vitamin K_1 to reconstitute the Q_A site of *Rb. sphaeroides* (Breton et al., 1994a), the bands at the lower and upper wavenumbers have been assigned primarily to C$\underline{\cdots}$C modes while the C$\underline{\cdots}$O modes have been proposed to contribute mostly to the middle frequency band at 1444 cm^{-1} in *Rb. sphaeroides* (Figure 2a) and at 1438 cm^{-1} in *Rp. viridis* (Figure 3a). Upon $^1H/^2H$ exchange (Figures 2b and 3b), it is the central band that is the most affected by the deuteration, giving rise to the differential signals at 1455/1444 cm^{-1} in *Rb. sphaeroides* (Figure 2c) and at 1443/1434 cm^{-1} in *Rp. viridis* (Figure 3c). The dominant effect of the exchange on this middle band agrees with our previous assignment of this band to C$\underline{\cdots}$O modes. This observation also lends further support to our assignment of the 1448-cm^{-1} band of ubiquinone in the Q_A site of *Rb. sphaeroides* to the $C_4\underline{\cdots}$O mode. In addition, these spectra (Figures 2 and 3) suggest a slightly stronger hydrogen bonding interaction ($\approx$6 cm^{-1}) of the C$\underline{\cdots}$O modes of the menaquinone in *Rp. viridis* than in *Rb. sphaeroides*.

3.2 *Effect of ^2H_2O on the Non-quinone Vibrations*:

Among the non-quinone contributions, three distinct spectral regions will be discussed:

- **Around 1550 cm^{-1}**, typical of the amide II absorption (60% N-H bending; 40% C-N stretching), a large negative signal at $\approx$1556 cm^{-1} appears in all the double-difference spectra (Figures 1c, 2c, and 3c). This band thus most probably contains contributions from amide II vibrations. The absence of a correlated positive band, corresponding to the downshifted amide II' mode at $\approx$1450 cm^{-1}, can be explained by a small extinction coefficient of this band as well as by its overlap with the C$\underline{\cdots}$O modes of Q_A^-. The W-shaped signal, with negative lobes at 1554 and 1531 cm^{-1} and a positive one at 1542 cm^{-1} (Figures 1a,b and 2a,b), does not shift in ^2H_2O. On the other hand, it shifts by 14 cm^{-1} upon $^{15}N/^{14}N$ substitution and by 12 cm^{-1} upon uniform ^{13}C labeling of the RC protein (J.B., E.N, and A. Verméglio, unpublished). Such shifts are consistent with an assignment to amide II modes and indeed correspond to those observed for the amide II band in the absorption spectrum of the RCs. This W-shaped signal is thus assigned to the perturbation upon Q_A reduction of non-exchangeable N-H groups from the backbone. A similar $^{13}C/^{12}C$ shift is also observed for the differential signal at 1553/1544 cm^{-1} in *Rp. viridis*. The different shape of the FTIR signals in this frequency range for *Rb. sphaeroides* and *Rp. viridis* indicates the participation either of non-conserved residues in the Q_A^-/Q_A spectra of RCs of the two species or of conserved residues in different conformations and/or environments.
- **In the 1700-to 1630-cm^{-1} frequency range**, typical of the absorption region of amide I modes as well as of a number of side chains, the overlap of many signals in the double-difference spectra makes it difficult to propose reliable assignments. The increase in amplitude upon deuteration with no frequency shift of the negative band of largest amplitude in the Q_A^-/Q_A spectra, at 1670 cm^{-1} for *Rb. sphaeroides* (Figures 1a and 2a) and at 1675 cm^{-1} for *Rp. viridis* (Figure 3a), appears due to the downshift of a negative component present in ^1H_2O at $\approx$1675 cm^{-1} in *Rb. sphaeroides* and at

1685 cm^{-1} in *Rp. viridis*. The downshift of these modes upon $^1H/^2H$ exchange would explain the differential signals centered at ≈1671 cm^{-1} in *Rb. sphaeroides* (Figures 1c and 2c) and at 1680 cm^{-1} in *Rp. viridis* (Figure 3c). Although quite large, these shifts are still compatible with an assignment to amide I C=O modes. The small positive band at 1683 cm^{-1} in the double-difference spectra for *Rb. sphaeroides* and at 1698 cm^{-1} for *Rp. viridis* seems due to the downshift of a positive band present at these frequencies in the Q_A^-/Q_A spectra in ^1H_2O. These bands could possibly shift to ≈1675 cm^{-1} in *Rb. sphaeroides* or ≈1685 cm^{-1} in *Rp. viridis* although larger downshifts, such as those experienced upon deuteration of Arg, Gln, or Asn side chains, are also fully compatible. More extensive work with ^{15}N-labeled RCs is in progress to help solve such ambiguities.

- **Above 1700 cm^{-1}**, the main contributors to the FTIR difference spectra are the C=O groups from ester carbonyls of the pigments and from the protonated carboxylic residues Glu and Asp. The effect of $^1H/^2H$ exchange on these spectra should be revealing inasmuch as it could affect the carboxylic residues but should leave the ester groups unperturbed. Complex differential signals are seen in this spectral region, with notably a main component centered at 1732 ± 1 cm^{-1} in the Q_A^-/Q_A spectra of *Rb. sphaeroides* and *Rp. viridis* (Figures 1a, 2a, and 3a). The precise peak frequency of the positive and negative lobes of these differential signals varies for the spectra recorded in ^1H_2O and ^2H_2O, showing the overlap of these differential signals with at least one other ^2H_2O-sensitive component. Such component is revealed upon $^1H/^2H$ exchange as a differential signal centered at 1727 cm^{-1} in *Rb. sphaeroides* and at ≈1730 cm^{-1} in *Rp. viridis* (Figures 1c, 2c, and 3c), indicating the participation of a protonated carboxylic residue.

It is known (Verméglio & Clayton, 1977) that Q_A photoreduction leads to characteristic electrochromic shifts of the electronic transitions of the bacteriopheophytin electron acceptor (H_A). It is thus of interest to compare the Q_A^-/Q_A and the H_A^-/H_A FTIR difference spectra for the same species. Negative bands at ≈1746 cm^{-1} and at 1732 cm^{-1}, that are common to the H_A^-/H_A FTIR difference spectra of both species (Nabedryk et al., 1995), have been tentatively assigned to the bleaching of ester C=O groups of the bacteriopheophytin. Furthermore, a weakly ^2H_2O-sensitive signal overlapping with these bands was proposed to originate from the C=O of the COOH group of Glu L104 hydrogen-bonded to the 9-keto carbonyl of H_A. It is thus tempting to speculate that both Glu L104 and the ester C=O groups of H_A contribute to the Q_A^-/Q_A signals between 1720 and 1750 cm^{-1} with only the vibrations of the former being sensitive to $^1H/^2H$ exchange. In this scheme, the differential Q_A^-/Q_A signal centered at 1727 cm^{-1} in *Rb. sphaeroides* and at ≈1730 cm^{-1} in *Rp. viridis* observed upon $^1H/^2H$ exchange (Figures 1c, 2c, and 3c) would correspond to the electrostatic response of Glu L104, while the differential signal of larger amplitude and centered at 1732 ± 1 cm^{-1} in the Q_A^-/Q_A spectra of both species in ^1H_2O and ^2H_2O would represent the downshift of a ester C=O group of H_A in response to the electrostatic field of Q_A^-. The most likely candidate for the differential signal centered at 1732 cm^{-1} is the 10a ester C=O which is proposed to be hydrogen-bonded to Trp L100 in both *Rb. sphaeroides* and *Rp. viridis* structures. On the other hand, the very small negative 1746-cm^{-1} signal in the Q_A^-/Q_A spectra of both species could correspond to a minor electrostatic perturbation of a free 7c ester C=O that would be responsible for the bleaching observed also at ≈1746 cm^{-1} in the H_A^-/H_A FTIR difference spectra of both species. Taking this scenario one step further, it becomes conceivable that the 9-keto group of H_A, which has been proposed to absorb at 1677 cm^{-1} in *Rb. sphaeroides* and at 1681 cm^{-1} in *Rp. viridis* (Nabedryk et al., 1995), could also contribute to the large negative bands at 1670 and 1675 cm^{-1} in the Q_A^-/Q_A spectra of *Rb. sphaeroides* and *Rp. viridis*, respectively.

Independent support for this scheme involving an electrostatic perturbation of the vibrations of H_A in the Q_A^-/Q_A spectra can be found upon observing that the large differential signal centered at 1722 cm^{-1} in the Q_A^-/Q_A spectra of PS II (Berthomieu et al., 1990, 1992), and that is essentially insensitive to $^1H/^2H$ exchange, corresponds precisely to the frequency of a ^2H_2O–insensitive band that bleaches in the H_A^-/H_A FTIR difference spectra of D_1D_2 PS II RCs (Nabedryk et al., 1990). Thus the ≈10–cm^{-1} shift between the frequency of the 10–a ester C=O of H_A in the RCs of purple bacteria and of PS II is closely correlated in both their Q_A^-/Q_A and H_A^-/H_A spectra. Furthermore, the bleaching of bands at 1739 and 1677 cm^{-1} in the H_A^-/H_A FTIR difference spectra of PS II has corresponding differential bands in the Q_A^-/Q_A spectra (Berthomieu et al., 1990, 1992). Additional support for these assignments is being seeked for by investigating the FTIR spectra of mutants at the Glu L104 site in *Rb. sphaeroides* (collaboration with J. Williams and J. Allen) and *Rp. viridis* (collaboration with M. Bibikova and D. Oesterhelt) as well as at the Gln D_1–130 site of *Synechocystis* (collaboration with P. Nixon).

In conclusion, the present FTIR study of the effect of $^1H/^2H$ exchange on Q_A photoreduction provides the first evidence that an ionized cofactor can perturb the vibrational properties of polar groups of distant (>10 Å) amino acid residues and cofactors through long range electrostatic interactions.

References

Berthomieu, C., Nabedryk, E., Mäntele, W., & Breton, J. (1990) *FEBS Lett.* 269, 363–367.

Berthomieu, C., Nabedryk, E., Breton, J., & Boussac, A. (1992) in *Research in Photosynthesis* (Murata, N., Ed.) Vol. II, pp. 53–56, Kluwer Academic Publishers, The Netherlands.

Breton, J., Thibodeau, D. L., Berthomieu, C., Mäntele, W., Verméglio, A., & Nabedryk, E. (1991) *FEBS Lett.* 278, 257–260.

Breton, J., Burie, J.–R., Berthomieu, C., Thibodeau, D. L., Andrianambinintsoa, S., Dejonghe, D., Berger, G., & Nabedryk, E. (1992) in *The Photosynthetic Bacterial Reaction Center II: Structure, Spectroscopy, and Dynamics* (Breton, J., & Verméglio, A., Eds.) pp. 155–162, Plenum Press, New York.

Breton, J., Burie, J.–R., Berthomieu, C., Berger, G., & Nabedryk, E. (1994a) *Biochemistry* 33, 4953–4965.

Breton, J., Burie, J.–R., Boullais, C., Berger, G., & Nabedryk, E. (1994b) *Biochemistry* 33, 12405–12415.

Breton, J., Boullais, C., Burie, J.–R., Nabedryk, E., & Mioskowski, C. (1994c) *Biochemistry* 33, 14378–14386.

Breton, J., Boullais, C., Berger, G., Mioskowski, C. & Nabedryk, E. (1995a) *Biochemistry* 34, in press.

Breton, J., Nabedryk, E., Mioskowski, C., & Boullais, C. (1995b) in Proceedings Feldafing–III Workshop on *Reaction Centers of Photosynthetic Bacteria* (Michel-Beyerle, M.–E., Ed.) Springer, Berlin, in press.

Brudler, R., de Groot, H. J. M., van Liemt, W. B. S., Steggerda, W. F., Esmeijer, R., Gast, P., Hoff, A. J., Lugtenburg, J., & Gerwert, K. (1994) *EMBO J.* 13, 5523–5530.

Nabedryk, E., Andrianambinintsoa, S., Berger, G., Leonhard, M., Mäntele, W., & Breton, J. (1990) *Biochim. Biophys. Acta* 1016, 49–54.

Nabedryk, E., Berthomieu, C., Verméglio, A., & Breton, J. (1991) *FEBS Lett.* 293, 53–58.

Nabedryk, E., Andrianambinintsoa, S., Dejonghe, D., & Breton, J. (1995) *Chem. Phys.* 194, 371–378.

Verméglio, A. & Clayton, R. K. (1977) *Biochim. Biophys. Acta* 461, 159–165.

ASYMMETRIC PRIMARY QUINONE HYDROGEN BONDING IN *RHODOBACTER SPHAEROIDES* REACTION CENTERS

Huub J.M. de Groot
Leiden Institute of Chemistry P.O. Box 9502,
2300 RA Leiden, The Netherlands

1. Introduction

The photosynthetic reaction center of *Rb. sphaeroides* contains two identical ubiquinone-10 (UQ_{10}) cofactors [1-3]. The primary quinone (Q_A) is tightly and permanently bound to the RC and operates as a one-electron redox switch in the process of photosynthetic photochemical energy conversion, due to specific asymmetric interactions with the RC protein complex involving the two quinone carbonyl functionalities. There is converging evidence from both magnetic resonance and infrared spectroscopy for a hydrogen bond at the 4-C=O, when the primary acceptor is in the semiquinone state $Q_A^{\cdot-}$ during the photosynthetic photochemistry process. This was deduced already at an early stage from ^{1}H hyperfine interactions with the electronic spin determined with ENDOR [4]. Using isotopically labeled UQ_{10} or ubiquinone-3 (UQ_3) it was confirmed by FTIR measurements [5,6], while Q-band EPR of the same UQ_{10} labeled RC samples revealed that the semiquinone spin density on 4-C is larger than on 1-C. This implies that a larger fraction of the additional (negative) charge density associated with the extra electron is on 4-O compared with 1-O, in agreement with the picture of hydrogen bonding at the 4-C=O [7]. According to ESEEM measurements on Zn-reconstituted RC after illumination at room temperature and fast freezing, the hydrogen bonding partner is His M219, since an interaction of the electronic spin with a ^{14}N nuclear quadrupole moment was observed, characteristic for an imidazole nitrogen [8].

To arrive at a complete understanding of the fundamental molecular mechanisms behind primary quinone function, detailed knowledge of the neutral and non-magnetic quinone ground state Q_A to atomic resolution is also a prerequisite. Unfortunately, in the three-dimensional structure of the RC the description of the environment of Q_A is incomplete, leaving space for different amino acids interacting with the bound quinone. The first X-ray structures suggest the possibility of specific hydrogen-bond interactions at both carbonyls, 1-C=O with the Ala M260 backbone nitrogen and 4-C=O with the side chain of Thr M222 [1,2]. According to the most recent X-ray data of another crystal morphology, diffracting to the highest resolution of 0.265 nm, the 1-C=O is at the shortest distance of 0.28 nm from the peptide nitrogen of Ala M260, within H-bonding range. The $N^{\partial(1)}$-H functionality of the imidazole ring of His M219, involved in H-bonding in the semiquinone state in the Zn reconstuted RC [8], is at a somewhat longer distance of 0.32 nm from the 4-C=O and could form a hydrogen bond with this carbonyl already in the quinone ground state [3].

P. Mathis (ed.), Photosynthesis: from Light to Biosphere, Vol. I, 401–406.

At the pseudo-atomic level, model compound binding studies revealed asymmetric protein-cofactor interactions in the ground state, as only one keto group and a ring are required for binding [9-11]. In contrast, on the protein side, site-directed mutatenesis studies have shown that the replacement of the indole side chain of Trp M252 by a phenyl (Phe) or phenol (Tyr) moiety is not sufficient to maintain the quinone affinity of the binding pocket [12]. In addition, a net binding free energy of –3.6 kcal/mol was measured [11].

Detailed magnetic resonance and infrared studies with genuine atomic selectivity of essentially unmodified RC preparations are indispensable for an improved characterization of the quinone ground state. Such investigations were impossible until specifically isotope enriched ubiquinones were prepared *via* organic chemistry total-synthesis [13]. Using isotope labels has the obvious advantage that the RC are essentially unperturbed, compared with, for instance, investigations involving mutants or chemical modification. The quinone ground-state is diamagnetic and therefore inaccessible to electron paramagnetic resonance techniques. For large membrane proteins and complexes like the RC, the novel high-resolution solid state magic angle spinning (MAS) NMR technique, in conjunction with selective or multispin isotope enrichment, is the method of choice for magnetic resonance studies to atomic resolution. Over the past few years, this NMR technology was developed into a routinely applicable method for research in bacterial photosyntesis [14], and very recently the first comprehensive study of the primary quinone ground state was performed [15]. Selective isotope enrichment is also essential for infrared studies, to assign unambiguously dominant internal vibrations to the normal modes corresponding with the IR absorption bands [5,6].

With spectroscopic data to atomic resolution of the quinone ground state now available, new questions emerge regarding the specific binding interactions of importance for the molecular mechanisms involved in one-electron redox switching. According to the MAS results obtained thus far, the Q_A binding pocket is apolar and somewhat heterogeneous on the scale of the NMR, and compared to other details of the RC complex. This is in agreement with the X-ray structural data. The MAS signal of the 1-^{13}C=O is a well-resolved narrow line, even at ambient temperature in hydrated precipitated samples. In contrast, the 4-^{13}C=O could only be detected at temperatures below $T \sim 255$ K. This asymmetry should be related to destructive interference of a dynamic process with the CP/MAS in the vicinity of the 4-C=O [15]. The FTIR reveals a dramatic ground state shift of 60 cm^{-1} for a mode dominated by the 4-C=O internal vibration [5,6], while the mode of the coupled C=O/C=C vibrations is at a higher frequency than the 4-C=O specific mode [5]. On the other hand, the binding shift of the 1-C=O is small [5,6]. As yet, these observations have not been mimicked satisfactorily simultaneously in models, neither in solution nor in the solid state. In particular, both NMR and FTIR responses of quinones are remarkably ***insensitive*** to hydrogen bonding capability and polarity of solvents [15,16]. Although some hydrogen bonding NMR shifts of up to a few ppm are in principle expected, these could be compensated by shifts from other sources. Hence the NMR response of the Q_A can be reconciled with hydrogen bonding at one or both carbonyls. In the recent literature the dramatic 4-C=O IR shift is correlated with strong hydrogen bonding effects [5,6,17], but recent assignments of Q_B IR absorptions using UQ_{10} labeling revealed only small shifts, which also had to be interpreted in terms of hydrogen bonding [18].

The purpose of the present paper is to discuss in some detail the effect of hydrogen bonding on the MAS NMR response. Since pure asymmetric hydrogen bonding can not be modeled in the reality of e.g. a solvent, hydrogen-bond type interactions with benzoquinone were modeled in the virtual reality of the computer, at the molecular mechanics and semi-empirical Hartree-Fock level, in an attempt to investigate the effect of an asymmetric hydrogen bond on a quinone keto functionality. The results of these computations are related to the spectroscopic data already available in the literature.

2. Procedure

Molecular mechanics calculations to approximate hydrogen bonding configurations were performed using the Hyperchem package on a 486 personal computer. The internal MM2 force field of the programme was used. RHF calculations were done with the same system or with the MOPAC 93 package using a Silicon Graphics Indy personal workstation.

3. Result and Discussion

The first MAS NMR data provide evidence for dynamic quinone-protein interactions affecting the 4-C=O, in parallel a neat solid state type 1-C=O response, and no evidence for an effect on the net atomic charges. The isotropic shifts and the anisotropies were determined with site-specific ^{13}C labeling at the 1, 2, 3 and 4-^{13}C positions. The shifts of both the 1 and 4 carbonyls upon binding of the ubiquinone to the protein are negligible, while the chemical shift anisotropies are only slightly different [15]. This is in strong contrast with the 60 cm^{-1} shift of the IR absorption dominated by the 4-C=O internal mode, as both techniques are sensitive to the local electronic configurations.
The interpretation of NMR data in terms of structural details like hydrogen bonding goes in two steps: first the relation between the NMR response and the variation of the electronic properties of the molecule has to be evaluated, and then the relation between the electronic structure and the molecular structure should be assessed.
Regarding the first step, the NMR and FTIR observations can be reconciled most readily when the paramagnetic contribution to the chemical shielding $\sigma_p = \sigma_p^{xx} + \sigma_p^{yy} + \sigma_p^{zz}$, which takes into account the hybridization effects, is expressed in terms of a simple LCAO molecular orbital scheme [19],

$$\sigma_p^{zz} = -\frac{e^2\hbar^2}{2m^2c^2\Delta E}\langle r^{-3}\rangle\left(Q_{AA}^{zz} + \sum_{B\neq A} Q_{AB}^{zz}\right), \tag{1}$$

with σ_p^{xx} and σ_p^{yy} obtained by cyclic permutation. ΔE is the average excitation energy of the excited electronic states in the absence of a magnetic field. The summation runs over neighbours B of A, the atom with the nucleus of interest and σ_p is proportional to the average value of r^{-3} for the 2p orbitals. The Q_{AA}^{zz} is expressed in terms of the generalized charge densities $P_{i_A j_A}$,

$$Q_{AA}^{zz} = 2 - 2\left(P_{x_A x_A} - 1\right)\left(P_{y_A y_A} - 1\right) + 2P_{x_A y_A}^2, \tag{2}$$

while the nearest neighbour bond effect Q_{AB}^{zz} is expressed in terms of the generalized bond orders $P_{i_A j_B}$,

$$Q_{AB}^{zz} = -2P_{x_A x_B}P_{y_A y_B} + 2P_{x_A y_B}P_{y_A x_B}. \tag{3}$$

According to these expressions, bond order changes may thus be expected to affect the principal components of the chemical shift tensor through the terms that depend on $P_{i_A j_B}$, but these variations are at present beyond the sensitivity limit of the CP/MAS technique [15].
Although the tensor elements implicitly contribute to the isotropic shift, it is generally assumed, based on a wide range of experimental observations, that the isotropic shift

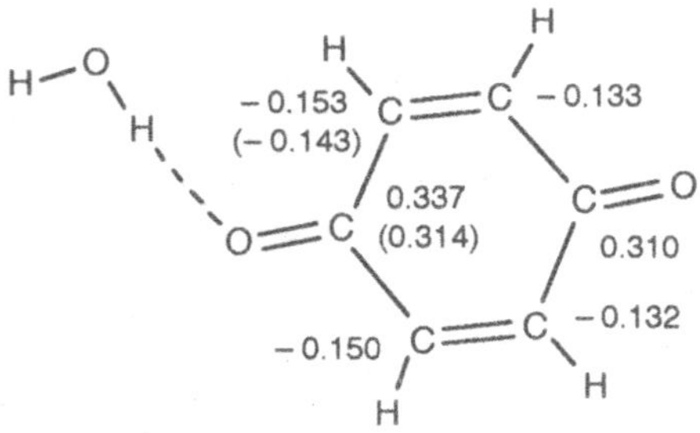

Figure 1. Asymmetric hydrogen bonding between H_2O and benzoquinone, and fractional charge densities. Values in parenthesis are for benzoquinone.

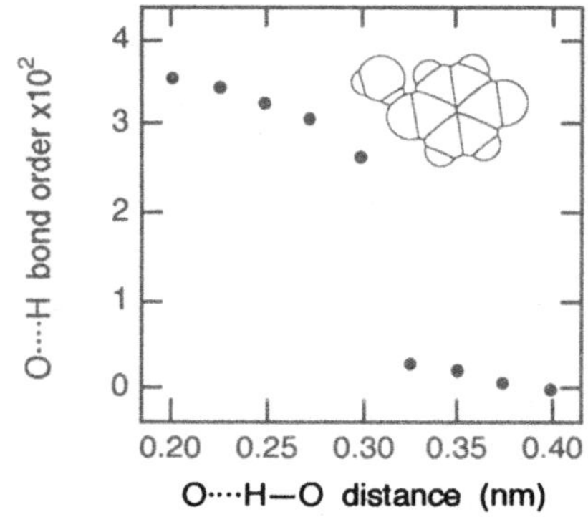

Figure 2. Variation of O····H bond order with O····H—O distance. The inset illustrates the penetration of orbitals for the configuration of fig. 1.

varies approximately linearly with the net charge density associated with the atom of interest, while the FTIR measures primarily the bond order. In this respect the two spectroscopic techniques are complementary and it was concluded that there is at present no conflict between the NMR shift anisotropy analysis and the quite dramatic 60 cm^{-1} shift of the C=O stretch mode observed with infrared spectroscopy [15].

The question then arises: could a hydrogen bond induce a substantial bond order change while leaving the charge density essentially unaffected? In an attempt to provide some insight into this problem, some very simple hydrogen bonding configurations were simulated in the computer. For instance, fig. 1 shows a H_2O molecule in approximate H-bond position with respect to a benzoquinone keto functionality. From the molecular modelling and subsequent Hartree-Fock internal energy minimization an O····H—O distance of 0.273 nm is inferred, which is characteristic for a normal H-bond. The 'formation' of the H-bond corresponds with an energy-difference of –3.1 kcal/mole, which is encouraging. In the calculations, the asymmetric H-bonding provokes a difference of 20 cm^{-1} between the two normal modes dominated by the 1-C=O and 4-C=O internal vibrations with respect to the symmetric benzoquinone *in vacuo*, and the mode representing the coupled C=O/C=C vibrations is at a lower frequency than the 4-C=O specific mode.

Using a standard relation for the variation of the chemical shift with charge of 150 ppm per positive electronic equivalent [20], a downfield hydrogen bonding shift of ~ 3.5 ppm can be estimated. This is not crucial on the scale of the solid state NMR and could be compensated by other mechanisms. On the other hand, such a shift is also not observed for quinones in solution and the charge difference may simply be an artifact of the computational procedures, since it is definitely associated with the formation of the H-bond.

A hydrogen bond only exists if the distance between the two partners is longer than the corresponding covalent distance, but smaller than the sum of the van der Waals radii of the respective ions and the atomic orbitals penetrate into each other. This is illustrated in fig. 2, which shows the calculated O····H bond order versus O····H—O distance. At distances less than ~ 0.31 nm, corresponding to the sum of the van der Waals radii, there is significant O····H bond order, while for increasing distance it rapidly vanishes.

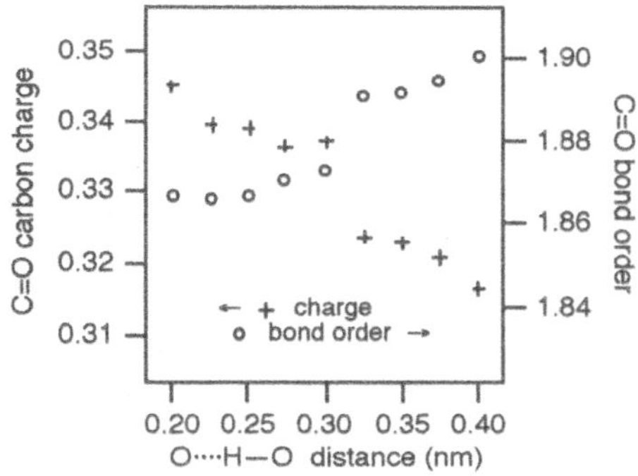

Figure 3. Fractional charge and C=O bond order versus O····H—O distance

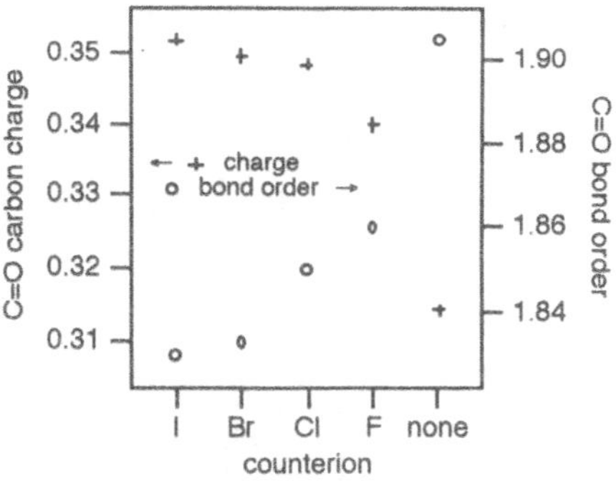

Figure 4. Fractional charge and C=O bond order for increasing counterion size

The actual charge transfer in the formation of a H-bond is small, but the proximity of the counterion could possibly give rise to significant mutual polarization of both participants in the H-bonding event. To estimate the mutual polarization effects for the configuration of fig. 1, the water molecule was moved along the O····H—O bond vector and the effect on the charge and the bond order was calculated. The results are depicted in fig. 3. At this semi-empirical level the effects on the charge and the bond order are apparently correlated. However, in the H-bond range, below ~ 0.3 nm, both quantities are quite insensitive to the variation of the O····H—O distance. This is in line with the experimental observations for quinones in different solvents.

Another important aspect of H-bonding, in particular for a protein, is the effective size of the counterion. In fig. 4 the effect of a series of monovalent negative counterions of increasing size in a hydrogen bonding geometry on the local electron configuration of the keto group is shown. Interestingly, the charge variations are comparable to those listed in fig. 3, and are small for the larger diffuse counterions. The bond order variations, on the other hand, are quite substantial, also for the softer counterions. Going from no counterion at all to a very diffuse I^-, would correspond to an estimated downfield NMR shift of ~ 6 ppm, which could in principle still be compensated by some other mechanism in a protein environment. On the other hand, a bond order change of 0.7 could give rise to a substantial IR shift, ~ 40 cm^{-1} [21].

The present calculations suggest that the NMR and FTIR observations are compatible. In this sense the NMR now fully supports the interpretation in terms of a strong hydrogen bond between the 4-C=O and His M219 promoted by the FTIR spectroscopy. On the other hand, this interpretation is not yet unique. A reasonable alternative may still be a H-bond at the 1-C=O, which is at present not necessarily excluded by the IR, and even to some extent supported by the NMR, in conjunction with a strong interaction at the 4-C=O, different from hydrogen bonding and involving dynamic character, which may be related to the "Kleinfeld effect" [22]. This would be in line with the recent X-ray and mutagenesis results [3,12], and would imply a small structural change involving the quinone upon reduction. Additional NMR experiments aiming at an indirect detection of proton resonances will be crucial for gaining further insight into the specific Q_A-protein interactions, while a full vibrational analysis, adressing such details as the mode inversion, with the coupled C=O/C=C vibrations at a higher frequency than the 4-C=O specific mode, will also be required. Finally, only part of the quinone entity was studied using isotopes, while, for instance, also the methoxy groups may be of importance for the Q_A function.

References

1 Allen, J.P., Feher, G., Yeates, T.O., Komiya, H. and Rees, D.C., (1988) Proc. Natl. Acad. Sci. USA 85, 8487-8491
2 Chang, C.-H., El-Kabbani, O., Tiede, D., Norris, J.R. and Schiffer, M. (1991) Biochemistry 30, 5352-5360
3 Ermler, U., Fritzsch, G., Buchanan, S. and Michel, H. (1994) Structure 2, 925-936
4 Feher, G., Isaacson, R.A., Okamura, M.Y. and Lubitz, W. (1985) in Antennas and Reaction Centers of Photosynthetic Bacteria (Michel-Beyerle M.E. ed.) pp.177-189, Springer, Berlin, Germany
5 Brudler, R., de Groot, H.J.M., van Liemt, W.B.S., Steggerda, W.F., Esmeijer, R., Gast, P., Hoff, A.J. Lugtenburg, J. and Gerwert, K. (1994) EMBO J. 13, 5523-5530
6 Breton, J., Boullais, C., Burie, J-R., Nabedryk, E. and Mioskowski, C. (1994) Biochemistry 33, 14378-14386
7 van de Brink, J., Spoyalov, A., Gast, van Liemt, W., Raap, J., Lugtenburg, J. and Hoff, A. (1994) FEBS Lett. 353, 273-276
8 Bosch, M., Gast, P., Hoff, A., Spoyalov, A. and Tsvetkov, Y. (1995) Chem. Phys. Lett. 239, 306-310
9 Gunner, M.R., Braun, B.S., Bruce, J.M. and Dutton, P.L. (1985) in Antennas and Reaction Centers of Photosynthetic Bacteria (Michel-Beyerle M.E. ed.) pp.298-305, Springer, Berlin, Germany
10 Warncke, K., and Dutton, P.L. (1993) Biochemistry 32, 4769-4779
11 Warncke, K., and Dutton, P.L. (1993) Proc. Natl. Acad. Sci. USA 90, 2920-2924
12 Stiltz, H.U., Finkele, U., Holtzapfel, Lauterwasser, C., Zinth, W. and Oesterhelt, D. (1994) Eur. J. Biochem. 223, 233-239
13 van Liemt, W.B.S., Steggerda, W.F., Esmeijer, R. and Lugtenburg, J. (1994) Rec. Trav. Chim. Pays-Bas 113, 153-161
14 de Groot, H.J.M., in Advances in Photosynthesis, Biophysical Techniques in Photosynthesis (Amesz, J. and Hoff, A.J. ed.) in press
15 van Liemt, W.B.S., Boender, G.J., Gast, P., Hoff, A.J., Lugtenburg, J. and de Groot, H.J.M. (1995) Biochemistry, in press
16 Bauscher, M. and Mäntele, W. (1992) J. Phys. Chem. 96, 11101-11108
17 Brudler, R., de Groot, H.J.M., van Liemt, W.B.S, Gast, P., Hoff, A.J., Lugtenburg, J. and Gerwert, K. in Proceedings of the Feldafing Meeting on Photosynthesis, in press
18 Brudler, R., de Groot, H.J.M., van Liemt, W.B.S., Steggerda, W.F., Esmeijer, R. Gast, P., Hoff, A.J., Lugtenburg, J., and Gerwert, K. (1995) FEBS Lett., in press.
19 Pople, J.A. (1962) J. Chem. Phys. 37, 53-59
20 Spiesecke, H. and Schneider, W.G. (1961) Tetrahedron Lett. 468-472
21 Josien, M.-L., Fuson, N., Lebas, J.-M. and Gregory, T.M. (1953) J. Chem. Phys. 21, 331-340
22 Kleinfeld, D., Okamura, M.Y., and Feher, G. (1984) Biochemistry 23, 5780-5786.

THE ROLE OF THE PROTEIN IN STABILIZING THE CHARGE SEPARATED STATES IN BACTERIAL PHOTOSYNTHESIS

M.R. Gunner and Barry Honig#
Department of Physics, City College of New York
138th St. and Convent Ave., New York, NY 10031 U.S.A.
#Department of Biochemistry and Molecular Biophysics
Columbia University, 630 West 168th St. New York, NY 10032 U.S.A.

1. Introduction

The early stages in photosynthesis in all organisms starts with absorption of a photon by a molecule that functions as the primary donor (1). The photon's energy promotes an electron in the donor to the excited singlet state. The excited donor then transfers an electron to an intermediate acceptor 10 or more Ångstroms away to form a charge separated state. This long-range electron transfer occurs rapidly, in a few picoseconds. It occurs with high quantum yield so that close to 100% of the RCs that absorb a photon form the charge separated state. In addition, in many RCs this electron transfer occurs without slowing down as the temperature is lowered to cryogenic temperatures. Thus, photosynthetic RCs can produce a stable state in which a positive and negative ion is produced rapidly and at low temperature. The role of the protein in this process can now be explored using the x-ray crystal structures of the RCs from the purple non-sulphur bacteria from Rps. viridis (2) and Rb. sphaeroides (3, 4).

When the structures for the bacterial RCs were elucidated the protein could be seen to provide a scaffolding for the cofactors that function as electron donor and acceptor. In addition, the protein and surrounding membrane protect the cofactors from water. A rigid arrangement of the cofactors in a low dielectric environment will facilitate rapid electron transfer with a small reaction ΔG (5, 6). The question arises whether the protein plays only these passive roles so that efficient, charge separation could occur if the cofactors were fixed in any medium with a low dielectric constant.

The formation of any charge separated or ion pair state occurs in opposition to the attractive force between the positive and negative charge. There are three methods to stabilize the ion pair: (1) The intrinsic electron affinity of the electron donor and acceptor can be large and favorable; (2) The reaction can take place in a polarizable medium which interacts with the two ions to solvate and stabilize them; (3) The electron transfer reaction can take place in a static field oriented with the proper polarity along the direction of electron transfer. Estimates of the cost of charge separation with the cofactors isolated from the protein

P. Mathis (ed.), Photosynthesis: from Light to Biosphere, Vol. I, 407–411.

suggest that the charge separated states would be higher in energy than the donor excited singlet state in the absence of the protein. Calculations of the electrostatic potentials throughout the Rps. viridis RC show that there is a substantial field that favors the charge separated state (7). In addition, asymmetry in the field explains why electron transfer occurs along the L branch of the protein.

2. Procedure

Calculations were carried out the Rps. viridis RC structure 1PRC from the Brookhaven Data Bank.(8, 9) All crystallographic water molecules were treated as part of the protein. The influence of the membrane was estimated by including a cube 72x72x33 Å of low dielectric material surrounding the protein. Calculations reported for the isolated cofactors include all six cofactors and the 4 histidines that are axial ligands to the bacterio-chlorophylls. The phytyl tails of the bacteriochlorophylls and bacteriopheophytins are truncated at the ether link between the atoms O2A and C1. O2A is included in the calculations.

The program DelPhi solves Poisson's equation for the potential in and around a protein given a distribution of charges and of regions of water with a high dielectric constant and protein with a low dielectric constant (10, 11). The RC structure is combined with atomic radii to define the shape of the region that is removed from water. The charge distribution is defined by CHARMM charges placed on each atom center. The potential was calculated at each cofactor (see (12) for a more complete account of the method). All ionizable residues except for Glu L 104 were assumed to be ionized. The charges on the neutral and ionized cofactors were taken from (13). DelPhi also provides the reaction field or solvation energy of the isolated cofactors, which is the free energy of transfering the isolated cofactors from vacuum to a medium with a higher dielectric constant.

3. Results and discussion

Figure 1 outlines the estimated free energies of the different charge separated states given the cofactor positions in Rps. viridis RCs. The first panel shows the energy levels if charge separation were carried out in vacuum. P* is placed 1.4 eV above the ground state as it is in the protein. Using this value in vacuum has no consequences for the rest of the analysis. In isolation from the protein the difference between the L and M branch cofactors is negligible so they are not differentiated in the figure. The difference in free energy between P* and P^+H^- were taken from the ZINDO calculations of Thompson and Zerner (14). This places the charge separated state 1.03 eV above P*. Thus, charge separation would not be expected to occur in vacuum.

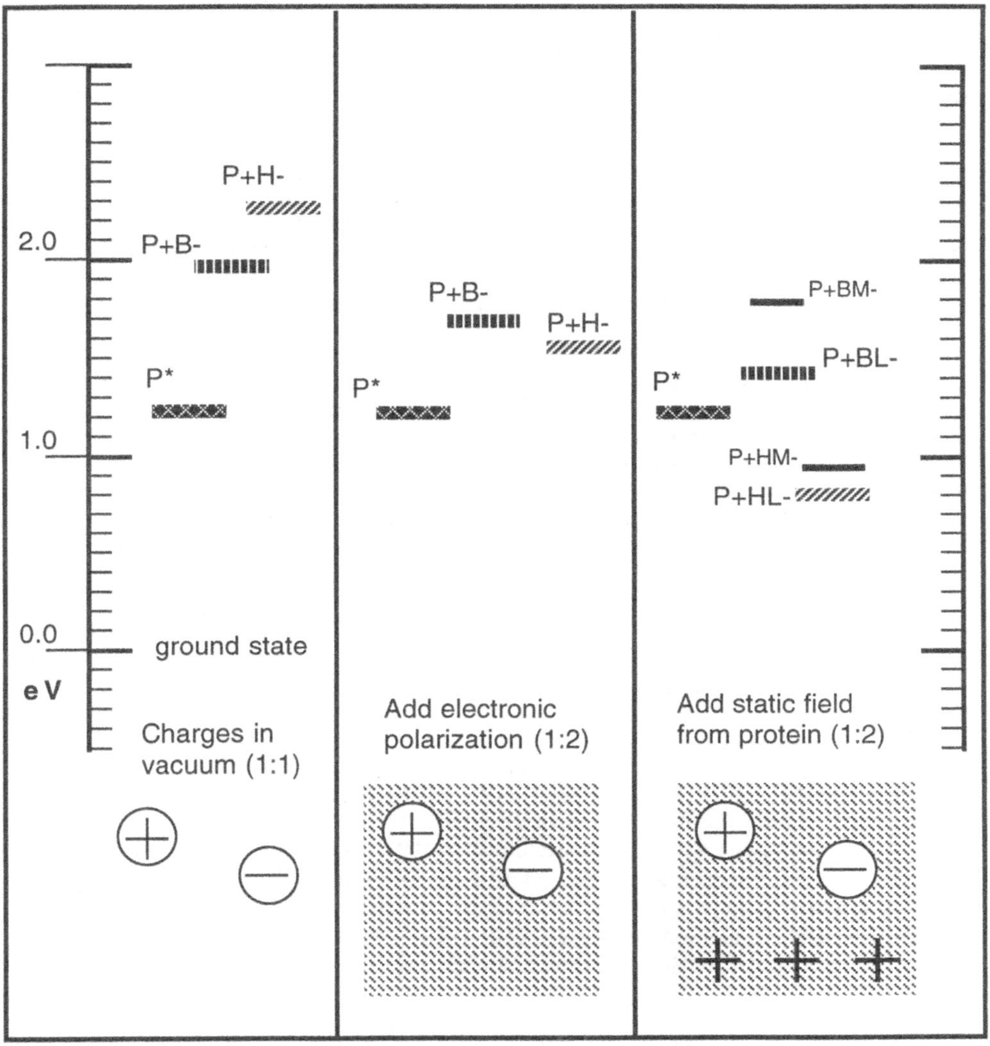

Figure 1: The relative free energy of the different charge separated states of the Rps. viridis RC cofactors: in vacuum; in a uniform medium with a dielectric constant of 2; and in the protein.

The free energy level of P^+B^- is placed relative to P^+H^- using the following arguments: Given the attraction to P+ it should take 0.5 V to move the electron from B which is 11 Å from P to H where the distance between cation and anion is 18Å. However, H (a BPh) is easier to reduce than B (a BChl). In solution BPh is reduced at a potential 0.25 V more positive than BChl (15). This difference can be explained with an electrostatic analysis where the difference in midpoints is due to the BChl having an axial ligand while BPh does not. The extra ligand makes BChl harder to reduce by shielding the cofactor from solvent and so diminishing the solvation energy and by placing a negative partial charge on the atom liganding the BChl. Moving the BChl into vacuum removes any difference in solvation energy between the BChl and BPh, but the impact of the charges on the ligand becomes greater. The result is that the difference in redox potential is quite insensitive to the environment. P^+B^- is therefore placed 0.25 V lower in energy than P^+H^- in vacuum.

The second panel shows the effect of moving the isolated cofactors into a medium with a dielectric constant of 2. This dielectric constant represents the reorganization of the electrons in the medium in response to the newly formed charge distribution. The effective dielectric constant either at very low temperature or for very short time will approach 2 as a lower limit. The reaction field or solvation energy of P* is assumed to be the same as that of the ground state, so its position relative to the ground state is unchanged. The reaction field energy of P^+B^- and P^+H^- are calculated to be -0.35 and -0.68 V respectively. The larger value for P^+H^- is expected since the reaction field energy increases as the charges in an ion pair are further separated. This term therefore partially balances the unfavorable Coulomb cost of separating charge. In addition B has an axial His ligand which H lacks which diminishes B's reaction field energy. Thus, if the RC cofactors were placed in a uniform medium with a small dielectric constant formation of P^+H^- would be unfavorable. Addition of the reaction field energy places P^+H^- below P^+B^-.

The third panel shows the effect of the potential found in the protein on the free energy of the charge separated states. Since the potential varies throughout the protein the states with electron transfer along the L and M branches are now distinct. A dielectric constant of 2 is used for the protein and for a low dielectric slab that models the membrane. P is at a more negative potential than either H. This stabilizes the P^+H^- states. The static field favors $P^+H_L^-$ by 0.78 eV. Formation of $P^+H_L^-$ is now thermodynamically allowed. The calculated ΔG between P* and $P^+H_L^-$ is -0.42 eV. This is larger than the experimental value for this state as it is initially formed (16). However, there are estimates of the free energy for this electron transfer monitored at later time that are closer to this calculated result (1).

The static field in the protein stabilizes $P^+B_L^-$ relative to P* by -0.31 eV. While this is insufficient to make reduction of B_L- favorable, the state is calculated to be only 0.15 eV above P*. This small energy gap would place $P^+B_L^-$ in a position to facilitate electron transfer by a super exchange mechanism.

Comparison of the state energies for reduction of the L and M branch cofactors shows that the static field is asymmetrical. The difference is more pronounced at the monomer BChls than the BPhs. Thus, the potential at B_M is actually 0.05 V more negative than P, which destabilizes electron transfer. Thus, the discrimination between the L and M branches appears to be made at the Bs. The potential at H_M is 0.12 eV more negative than at H_L. Thus, reduction of either H is favorable, with the reduction of the active BPh favored.

The analyses of the electrostatic potentials throughout the protein show that the protein plays an essential role in directing electron transfer down the L branch and stabilizing the initial charge separated state. The potential is primarily from ionized residues throughout the protein. More that 40 residues are involved in providing the static field that stabilizes the charge separated state and ensures the proper direction of electron transfer. These sites are distributed throughout the protein (7).

5. Acknowledgements

This work was supported by the NIH grant GM-48726 (MRG) and NSF grant MCB-0304127 (BH). We would like to thank R. Friesner, A. Warshel, and W.W. Parson for illuminating discussions.

References

1. Gunner, M. R. (1991) Current Topics in Bioenergetics 16, 319-367
2. Michel, H., Epp, O. and Deisenhofer, J. (1986) The EMBO Journal 5, 2445-2451
3. Yeates, T. O., Komiya, H., Chirino, A., Rees, D. C., Allen, J. P. and Feher, G. (1988) Proc. Natl. Acad. Sci. USA 85, 7993-7997
4. Ermler, U., Fritzsch, G., Buchanan, S. K. and Michel, H. (1994) Structure 2, 925-936
5. Closs, G. L. and Miller, J. M. (1988) Science 240, 440-447
6. Dutton, P. L. and Moser, C. C. (1994) Proc. Natl. Acad. Sci. 91, 10247-10250
7. Gunner, M. R., Nicholls, A. and Honig, B. (submitted) J. Phys. Chem.
8. Bernstein, F. C., Koetzle, T. F., Williams, G. J. B., Meyer, E. F., Brice, M. D., Rodgers, J. R., Kennard, O., Shimanouchi, T. F. and Tasumi, M. (1977) J. Mol. Biol. 112, 535-542
9. Deisenhofer, J. and Michel, H. (1989) The EMBO Journal 8, 2149-2170
10. Sharp, K. A. and Honig, B. (1990) Annu. Rev. Biophys. Biophys. Chem. 19, 301-332
11. Honig, B. and Nicholls, A. (1995) Science 268, 1144-1149
12. Gunner, M. R. and Honig, B. (1991) Proc. Natl. Acad. Sci. USA 88, 9151-9155
13. Parson, W. W., Chu, Z.-T. and Warshel, A. (1990) Biochim. Biophys. Acta 1017, 251-272
14. Thompson, M. A. and Zerner, M. C. (1991) J. Am. Chem. Soc. 113, 8210-8215
15. Cotton, T. M. and Van Duyne, R. P. (1979) J. Am. Chem. Soc. 101, 7605-7615
16. Woodbury, N. W., Parson, W. W., Gunner, M. R., Prince, R. C. and Dutton, P. L. (1986) Biochim. Biophys. Acta. 851, 6-22

MAGNETIC RESONANCE STUDIES OF BACTERIAL REACTION CENTERS: EFFECTS OF HYDROGEN BONDS ON THE ELECTRONIC STRUCTURE OF $P^{+\cdot}$ AND $I^{-\cdot}$

W. Lubitz, F. Müh, J. Rautter, F. Lendzian, J. P. Allen[†], and J. C. Williams[†]

Max-Volmer-Institut, TU Berlin, Str. des 17. Juni 135, D-10623 Berlin, Germany; [†] Department of Chemistry and Biochemistry, Center for the Study of Early Events in Photosynthesis, Arizona State University, Tempe, Arizona 85287-1604, USA

1. Introduction

The initial charge separation step in bacterial reaction centers (RCs) leads to the formation of the radical pair state P^+I^- [1], in which P is the primary donor, a bacteriochlorophyll (BChl) dimer, and I is the intermediate acceptor Φ_A, a bacteriopheophytin (BPh). To understand this process from first principles it is necessary to know both the spatial [2,3] and the electronic structure of the donor and acceptor molecules. The electronic structure can be obtained from EPR studies of the radical ions formed in the ET process. Since P^+I^- has a lifetime of only 200 ps and thus cannot be investigated directly by transient EPR techniques, these cofactors must be stabilized separately in their paramagnetic states $P^{+\cdot}$ and $I^{-\cdot}$. The distribution of the unpaired electron in the frontier orbitals can then be obtained via a determination of the hyperfine couplings (hfcs) using cw ENDOR and TRIPLE resonance [4,5].
An important factor that influences the electronic structure of $P^{+\cdot}$ and $I^{-\cdot}$, and hence their function in the ET process, is the pigment-protein-interaction. This can be altered by site-directed mutagenesis, allowing an exchange of specific amino acid residues in the vicinity of the cofactors. In this communication we report on the effects of H-bonds on the electronic structure of $P^{+\cdot}$ and $I^{-\cdot}$ in RCs from *Rhodobacter (Rb.) sphaeroides*.

2. Experimental Procedures

The construction of mutant RCs has been described in detail [6]. RCs were isolated by standard procedures [7] using the detergent LDAO (N-dodecyl-N,N-dimethylamine-N-oxid). Detergent exchange was performed by washing RCs bound to a DEAE-column with 0.2% (w/v) SB12 (N-dodecyl-N,N-dimethyl-3-ammonio-1-propanesulfonate) or 0.05% (w/v) Triton X-100, respectively, in 10 mM Tris/HCl pH 8.0, 1 mM EDTA and eluting with 300 mM NaCl followed by dialysis. $P^{+\cdot}$ was generated by illumination of RCs with actinic light [4]. $I^{-\cdot}$ was photoaccumulated at ambient temperatures and freeze-trapped as described [8]. The ENDOR and Special TRIPLE spectra were recorded using a Bruker ESP 300E spectrometer equipped with home-built ENDOR/TRIPLE accessories as previously described [4].

P. Mathis (ed.), Photosynthesis: from Light to Biosphere, Vol. I, 413–418.

3. Results and Discussion

3.1 The primary donor radical cation $P^{+\cdot}$

In recent years the electron spin density distribution of $P^{+\cdot}$ in *Rb. sphaeroides* has been studied in great detail using ENDOR and Special TRIPLE resonance spectroscopy [9]. These investigations showed that the unpaired electron is asymmetrically distributed over the two halves of the dimer, P_L and P_M (Figure 1). Furthermore it was shown that the spin density ratio of $\rho_L/\rho_M \approx 2$ is highly conserved in RCs of several Bchl *a* containing purple bacteria [4]. Studies of site-directed mutants of *Rb. sphaeroides* revealed that H-bonds formed to the 2-acetyl or 9-keto groups (see Figure 4 for numbering) of the dimer lead to dramatic shifts of the spin density between the two halves [5]. The fraction of spin density located on P_L is illustrated in Figure 1(A) for eleven mutants with different H-bond patterns; the residues which have been changed by site-directed mutagenesis [6] are shown in Figure 1(B).

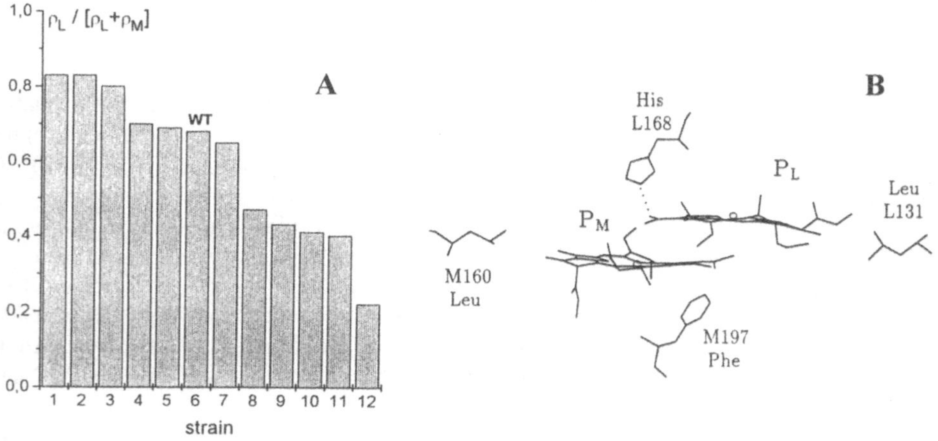

Figure 1: Fraction of spin density on P_L from ENDOR/TRIPLE [5] for different strains of *Rb. sphaeroides* (A) 1: LH(M160); 2: LH(M160) + FH(M197); 3: LH(M160) + LH(L131) + FH(M197); 4: LH(M160) + HF(L168); 5: LH(M160) + LH(L131); 6: wild type (WT); 7: FH(M197); 8: LH(L131); 9: HF(L168); 10: HF(L168) + FH(M197); 11: LH(L131) + FH(M197); 12: LH(L131) + HF(L168). The strains differ in the H-bonds between P and surrounding amino acid residues in the positions shown in (B), from [3].

The observed shifts of spin density can be understood in the framework of a simple MO model (Figure 2), in which the frontier orbitals of P are linear combinations of the respective orbitals of P_L and P_M. H-bonds influence the energies of the individual MOs leading to changes in their energy difference ($\Delta\alpha$) and their contributions to the dimer orbitals. This is reflected in the spin density distribution of $P^{+\cdot}$. These changes can be traced back to a direct charge effect of the H-bond donor. In addition, a rotation of the acetyl group - caused by a hydrogen bond to this group - has an impact on $\Delta\alpha$. The latter effect is nicely demonstrated by the removal of the only H-bond to P in the wild type (shown in Figure 1B) existing between histidine L168 and the 2-acetyl group of P_L. In RCs of the mutant HF(L168) only 43% of the spin density of $P^{+\cdot}$ is located on P_L as compared with 68% in the wild type (Figure 1A, strains 9 and 6). By introducing H-bonds to the 9-keto groups of either BChl of P the spin density can be shifted back and forth between the dimer halves (see e.g. Figure 1A, strains 4 and 12 or 1 and 8). The related change of the Coulomb energy for such a H-bond is approximately 80 meV [5].

This set of mutants presents an ideal system for studying the relationship between the electronic structure of $P^{+\cdot}$ and the observed ET-rates in the RC. Since the spin density distribution also reflects the distribution of the positive charge in the HOMO, reactions leading to the reduction of $P^{+\cdot}$ may be affected. We found that the rates for the ET from cytochrome c_2^{2+} to $P^{+\cdot}$ in these mutants [10] show deviations from the exponential dependence on ΔG^0 that correlate with the fraction of spin density on P_L. This result indicates the existence of a favorable ET pathway between the heme of cytochrome c_2 and the L-half of the dimer [11].

The charge separation and recombination processes in the RC could also be affected by protein conformational changes accompanying the ET. We recently observed the existence of two spectroscopically distinct forms of the primary donor in its neutral and oxidized state [4,12]. The second state of $P^{+\cdot}$ was found to have a spin density distribution of approximately 5:1 in favor of P_L [4]. Since such a state may play a role in stabilizing the charge separation, we have further investigated its formation by varying a number of experimental parameters. We found that the properties of the detergent used for RC solubilization have an impact on the spectroscopic properties of $P^{+\cdot}$. For example, the use of SB12 - a detergent with a larger zwitterionic head group than LDAO - causes a considerable shift of the spin density towards P_L (Table 1) as compared with LDAO (Figure 3). Interestingly, the change in the spin density ratio correlates with a shift of the long wavelength P/P* optical absorption from 865 to 853 nm and a significant decrease in oscillator strength of the characteristic $P^{+\cdot}$ band at 1250 nm.

Table 1: Spectroscopic properties of RCs from *Rb. sphaeroides* R26.1 in LDAO or SB12.

	LDAO	SB12
$A(5a_L)^a$	5.61	6.80
$A(1a_L)^a$	3.95	4.92
$A(5a_M)^a$	3.41	1.53
$A(1a_M)^a$	1.48	0.60
$A(\beta_L)^b$	9.50	13.30
	8.46	11.60
ρ_L/ρ_M	2.0	5.5
λ_P/nm	865	853
k_{AP}/s^{-1}	10.3	8.0

[a] hfcs (MHz) of methyl protons (± 30 kHz); [b] hfcs (MHz) of β_L protons in positions 3 and 4 (± 50 kHz); [c] ratio of spin densities of $P^{+\cdot}$ on P_L and P_M; [d] absorption maximum of lower energy P band; [e] $P^+Q_A^-$ charge recombination rate, [f] E. Schlodder, unpublished.

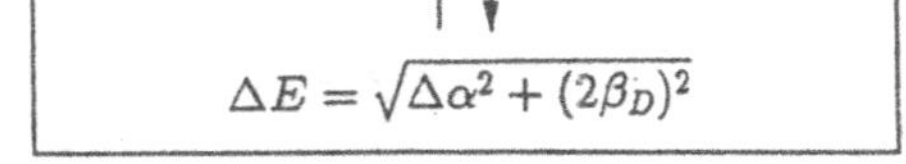

Figure 2: Energy level scheme showing the HOMOs of the monomeric dimer halves P_L and P_M having different Coulomb energies ($\Delta\alpha = \alpha_L - \alpha_M$) and the two supermolecular orbitals; β_D is the resonance integral between P_L and P_M.

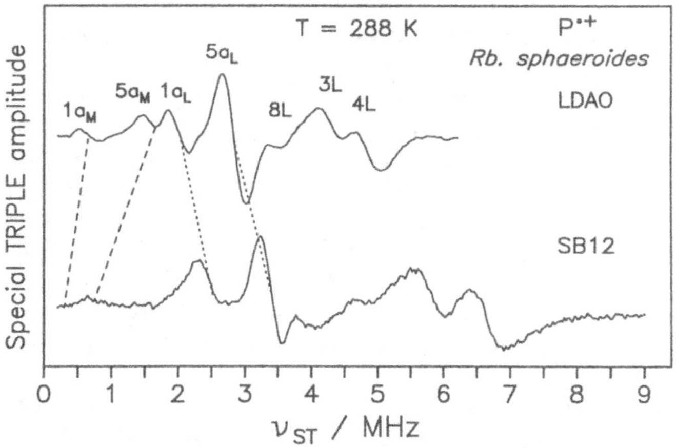

Figure 3: ^{1}H Special TRIPLE spectra of $P^{+\cdot}$ in RCs from *Rb. sphaeroides* R26.1 solubilized in 0.1% LDAO (top) and 0.2% SB12 (difference spectrum showing the dominant species ≥ 80%, bottom). The assignment of the signals to molecular positions (see Figure 4) in P_L and P_M is given; for hfcs see Table 1. The change of line positions in the two RC states are indicated by dashed (P_M) and dotted lines (P_L).

On the basis of our MO model (Figure 2) the effect can be understood by assuming that the detergent causes a protein structural change that affects the coupling of the dimer halves. Model calculations [13,14] predict both a blue shift of the P/P* transition and a decrease in intensity of the $P^{+\cdot}$ band upon increasing the distance between P_L and P_M. Thereby the orbital interaction described by the resonance integral β_D is reduced. The resulting supermolecular orbitals are then closer in energy to the individual MOs (Figure 2). Such an effect will significantly enlarge the fraction of spin density on P_L as is confirmed by the experiment (Figure 3). Assuming that $\Delta\alpha = 55$ meV remains constant we estimate $\beta_D = 55$ meV (SB12) whereas $\beta_D = 160$ meV in LDAO [4]. Note that this is different from the H-bond effects, which alter $\Delta\alpha$. Detergent effects as described here have also been observed in RCs of other purple bacteria, e.g. *Rb. capsulatus* [4,12].
An interesting result is that the $P^+Q_A^-$ charge recombinatiom rate k_{AP} is diminished by about 25% in SB12 as compared with LDAO (Table 1). The change in rate seems to be correlated with the electron (hole) distribution in $P^{+\cdot}$. It can be speculated that such a stabilization of light-induced charge separation is important for the efficiency of the ET process. To better understand the assumed structural changes we are presently investigating the effects of various detergents on the RC of *Rb. sphaeroides*.

3.2 The intermediate acceptor radical anion $I^{-\cdot}$

The electron spin density distribution of photoaccumulated $I^{-\cdot}$ in RCs from *Rb. sphaeroides* has already been investigated by ENDOR [8,15]. The state was identified as $\Phi_A^{-\cdot}$, although deviations of the hfcs from those of BPh $a^{-\cdot}$ *in vitro* were observed. The differences could be attributed to the formation of a H-bond between glutamic acid (Glu) L104 and the 9-keto group of Φ_A. The existence of such a H-bond had been verified by H_2O/D_2O-exchange experiments [8]. Replacement of Glu L104 by leucine in RCs from *Rb.*

capsulatus showed changes in the optical spectra, but the direction of charge separation remained unaffected and the changes of the ET rates were quite small [16].

Table 2: ^{1}H hfcs A (MHz) of the 1a and 5a methyl protons of I^{-} in RCs from *Rb. sphaeroides* wild type and the mutant EL(L104).

	1a			5a		
	$A_{\perp}{}^{b}$	$A_{\parallel}{}^{b}$	$A_{iso}{}^{a}$	$A_{\perp}{}^{b}$	$A_{\parallel}{}^{b}$	$A_{iso}{}^{a}$
wild type	7.6	9.3	8.2	8.8	11.0	9.5
EL(L104)	6.3	7.9	6.8	9.5	11.4	10.1

[a] $A_{iso} = (2A_{\perp} + A_{\parallel})/3$; [b] values were obtained by fits to axially symmetric tensors. Slight deviations from axial symmetry (± 0.2 MHz) improved the fits of the experimental spectra (see Figure 4).

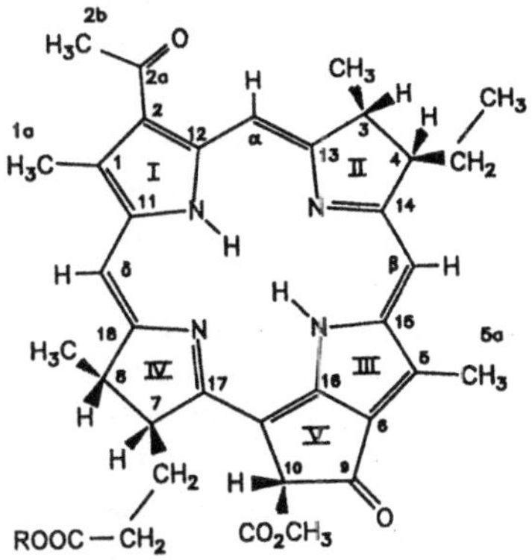

Figure 4: Molecular structure of BPh *a*, same numbering for BChl *a* (R = phytyl).

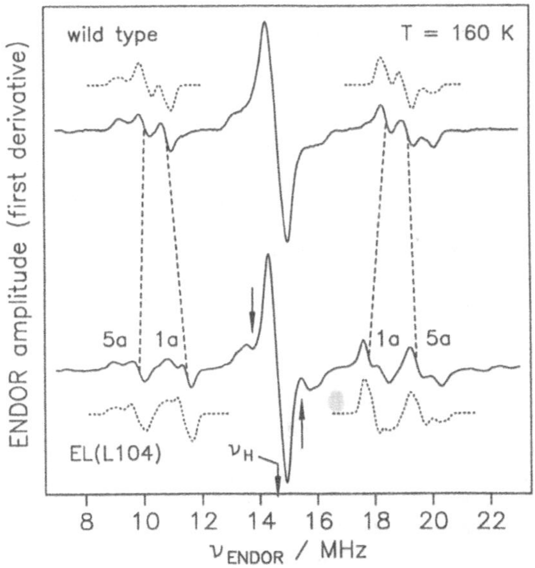

Figure 5: ^{1}H ENDOR spectra of freeze-trapped I^{-} in frozen RCs from *Rb. sphaeroides* wild type (top) and the mutant EL(L104) (bottom). Simulations of the lines belonging to the methyl proton hfc tensors (pos. 1a and 5a) are shown as dotted lines (see Table 2). The shifts of these hfcs between wild type and mutant are indicated by dashed lines.

We constructed an analogous mutant of *Rb. sphaeroides* and confirmed the effect on the optical absorption spectra (not shown): At 20 K the BPh-Q_x-region shows two peaks (534 and 546 nm) in the wild type, but only one (536 nm) in the mutant EL(L104). Thus the

hydrogen bond shifts the Q_x absorption of Φ_A by 10 nm to the red. The absorption maximum of the trapped I^{-} radical anion in the visible region at 278 K is shifted from 631 nm in the mutant to 662 nm in the wild type (not shown), which is in good agreement with transient optical spectra [16,17]. The optical I^{-}-spectra show no indications for the reduction of a BChl, it is therefore assumed that I^{-} is identical with Φ_A^{-} in both cases. The ENDOR spectra (Figure 5) show that the spin density distribution of Φ_A^{-} is significantly altered by the hydrogen bond as judged from the hfcs of the 1a and 5a methyl groups, for which the principal hf tensor components were deduced (Table 2, Figure 5). The data clearly indicate a considerable change of the LUMO of Φ_A^{-} upon removal of the H-bond. The spectrum of the mutant (Figure 5) shows holes in the matrix range indicated by arrows. The missing line pair corresponds to a component of the hfc tensor that is within experimental error identical to that of the bridging proton of Glu L104 as determined by H_2O/D_2O-exchange experiments [8]. This corroborates the existence of the hydrogen bond in wild type RCs and its absence in the mutant. The unexpected magnitude of the spin density shift in Φ_A^{-} indicates additional structural changes upon loss of the H-bond.
We are currently investigating further mutants with changes in the vicinity of both BPhs and the accessory BChls in order to better understand the electronic structure of the intermediate acceptor and its role in RC function.

The authors thank I. Geisenheimer (TU Berlin) for assistance with the RC preparations and E. Schlodder (TU Berlin) for providing kinetic data. This work was supported by DFG (Sfb 312, TP A4), NSF (MCB-9404925) to J.P.A, and NIH (GM45902) to J.C.W.

References

1 Kirmaier, C. and Holten, D. (1993) in The Photosynthetic Reaction Center II (Deisenhofer, J. and Norris, J.R., eds.) pp. 49-70, Academic Press, San Diego; Zinth, W. and Kaiser, W., ibid., pp. 71-88
2 Chirino, A.J., Lous, E.J., Huber, M., Allen, J.P., Schenck, C.C., Paddock, M.L., George, F., and Rees, D.C. (1994) Biochemistry 33, 4584-4593
3 Ermler, U., Fritzsch, G., Buchanan, S.K., and Michel, H. (1994) Structure 2, 925-936
4 Rautter, J., Lendzian, F., Lubitz, W., Wang, S., and Allen, J.P. (1994) Biochemistry 33, 12077-12084
5 Rautter, J., Lendzian, F., Schulz, C., Fetsch, A., Kuhn, M., Lin, X., Williams, J.C., Allen, J.P., and Lubitz, W. (1995) Biochemistry 34, 8130-8143
6 Williams, J.C., Alden, R.G., Murchison, H.A., Peloquin, J.M., Woodbury, N.W., and Allen, J.P. (1992) Biochemistry 31, 11029-11037
7 Feher, G. and Okamura, M.Y (1978) in The Photosynthetic Bacteria (Clayton, R.K. and Sistrom, W.R., eds.) pp.349-386, Plenum Press, New York
8 Feher, G., Isaacson, R.A., Okamura, M.Y., and Lubitz, W. (1988) in The Photosynthetic Bacterial Reaction Center (Breton, J. and Vermeglio, A., eds.) pp.229-235, Plenum Press, New York
9 Lendzian, F., Huber, M., Isaacson, R.A., Endeward, B., Bönigk, B., Möbius, K., Lubitz, W., and Feher, G. (1993) Biochim. Biophys. Acta 1183, 139-160
10 Lin, X., Williams, J.C., Allen, J.P., and Mathis, P. (1994) Biochemistry 33, 13517-13523
11 Rautter, J., Lendzian, F., Lin, X., Williams, J.C., Allen, J.P., and Lubitz, W. (1995) Proc. Feldafing III Workshop on Reaction Centers of Photosynthetic Bacteria, Structure and Dynamics, Springer, Berlin, in press
12 Wang, S., Lin, S., Lin, X., Woodbury, N.W., and Allen, J.P. (1994) Photosyn. Res. 42, 203-215
13 Warshel, A. and Parson, W.W. (1987) J. Am. Chem. Soc. 109, 6143-6152
14 Reimers, J.R. and Hush, N.S. (1995) J. Am. Chem. Soc. 117, 1302-1308
15 Feher, G., Isaacson, R.A., Okamura, M.Y., and Lubitz, W. (1987) Biophys. J. (Abstr.) 51, 377a
16 Bylina, E.J., Kirmaier, C., McDowell, L., Holten, D., and Youvan, D.C. (1988) Nature 336, 182-184
17 Kirmaier, C., Laporte, L., Schenck, C.C., and Holten, D. (1995) J. Phys. Chem. 99, 8903-8909 and 8910-8917

FROM PHOTONS TO PROTONS IN THE PHOTOCYCLE OF BACTERIAL REACTION CENTER

P. Maróti, Sz. Osváth, Cs. Tápai, D. K. Hanson* and P. Sebban**
JATE University Szeged, Institute of Biophysics
Egyetem utca 2. Szeged, Hungary H-6722
*Center Mech. Biol. and Biotechnology, Argonne National Laboratory, Argonne, USA
**Centre de Génétique Moleculaire, CNRS, Gif/Yvette, France

Introduction

The membrane bound RC protein plays central role in energetics of photosynthetic bacteria by converting different forms of free energy: light energy into phosphate potential via redox energy and proton electrochemical gradient [1]. The absorption of photons initiates photochemical electron transfer from the excited dimer (P^*) through a chain of electron acceptor molecules, ultimately reducing the primary and secondary quinones Q_A and Q_B, respectively. The electron transfer is coupled to proton binding by RC to solvate (stabilize) semiquinone anions (Q_A^- and Q_B^-) and later to reduce Q_B to hydroquinol, Q_BH_2 [2]. The doubly reduced quinone dissociates from the RC and is reoxidized by the cytochrome bc_1 complex. The vectorial transport of protons from the cytoplasmic to the periplasmic side of the membrane builds up proton electrochemical gradient which drives ATP synthesis.

Proton transport is a key process in utilization of light energy. Some proton transport proteins, e.g. bacteriorhodopsin pump protons directly through the dielectric core of the biological membrane. On the other hand, photosynthetic organisms use indirect way by decreasing the energetic barrier of the membrane for the protons: electrons will be transported through the membrane via redox chain and form subsequently electrically neutral species with the H^+-ions. Thus the protons can pass the membrane with significantly less penalty. The photosynthetic electron transport can be viewed as accessory step that prepare the RC for protonation. Taking this point of view, it is surprising that much less effort has been placed on the proton transport than on electron transfer reactions in RCs.

In this work special emphasis will be put on energetics and kinetics of proton binding to RC as primary step in bacterial proton transport. The results of site-directed mutagenesis and structure-based electrostatic calculations will help us to understand the electrostatic role of the protein in proton binding and the possible pathways of H^+-ions to Q_B. The pH-dependence of the rate of the photocycle will be used to demonstrate how the limitations of quinone/quinol exchange and protonation become the bottleneck of the photocycle in the neutral and alkaline pH-ranges under intense continuous illumination.

Methods

The details of construction of mutants and measurements of electron and proton transfer characteristics were performed as described in cited references.

P. Mathis (ed.), Photosynthesis: from Light to Biosphere, Vol. I, 419–424.

The program DELPHI, a generous gift from Dr. Barry Honig, Columbia University, New York was used to solve the Poisson-Boltzmann equation for the electrostatic potential in and around the RC. The coordinates were taken from the Brookhaven Protein Data Bank (structure 4RCR, date 09-SEP-91, resolution 2.8 Å).

Electrostatic profile of the RC

The electrostatic field of the RC was calculated using classical electrostatics as in [3,4]. Figure 1 demonstrates the electrostatic contour levels of 2.kT/e ($\approx$50 mV) in a two-dimensional slice through the main cofactors of P, Q_A and Q_B at low and high pH values. Negative potential is represented by dashed lines and positive potential by solid lines. The surface of the protein is indicated by a heavy solid line. The electrostatic field drives the H^+-ions from the bulk phase to the interior of the protein upon light excitation.

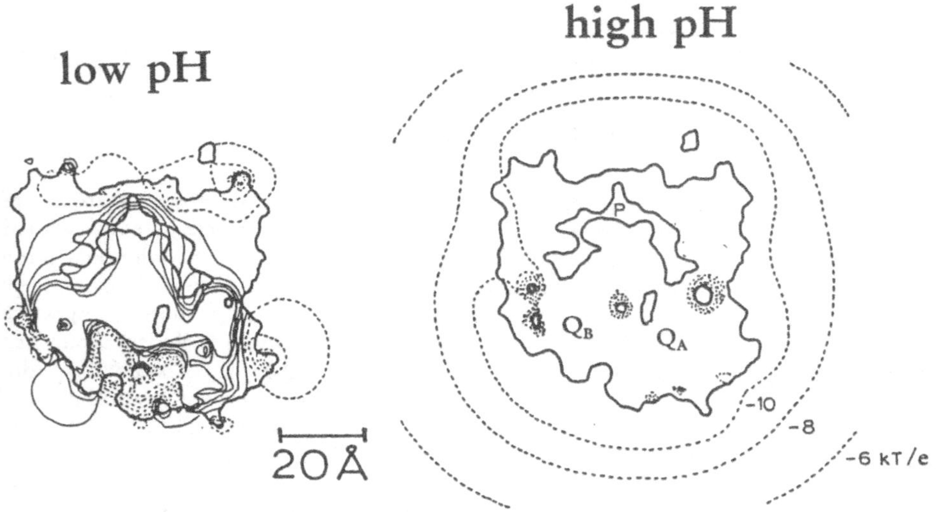

Figure 1 Electrostatic profile of RC at low (Glu and Asp are negatively and His, Lys and Arg positively charged) and high (Glu, Asp, Cys and Tyr are negatively and Arg positively charged) pH values.

The clusters of acidic residues around Q_B binding pocket produce increasing negative surface potential by alkalization of the solution and in some places, the focusing of the electric field in the Q_B pocket can be observed. These effects will facilitate the uptake H^+-ions by the RC in the alkaline pH range where the concentration of H^+-ions and thus the rate of bimolecular collision with the protein dramatically decreases.

Proton uptake due to solvation of semiquinone anions

The light-generated anion radical species, Q_A^- and Q_B^-, are effectively solvated by the protein. The anionic charges induce pK_a shifts in ionizable side-chain groups of various amino acid residues (see above for calculation). The differential compensation of the two semiquinone charges, which includes proton binding by the protein, provides much of the driving force for transfer of the first electron from Q_A^- to Q_B. Studies of site-specific mutants have identified Glu the charged forms of the quinones and also appear to interact electrostatically with each other: the ionization of Asp L213 suppresses the ionization of Glu L212, which has an unusually high $pK_a \approx 9.6$. Electron and proton transfer in the quinone acceptor complex could be partially accounted for by a simple electrostatic model of interacting Glu L212 and Asp L213, and Q_B^- [12,13]. Protonation of the two amino acids under acidic (Asp L213) and alkaline (Glu L212) conditions determine the pH dependence of the one-electron equilibrium constant of electron transfer from Q_A^- to Q_B [2,4,14,15].

Recent studies on flash-induced proton binding stoichiometry of RCs with substituted quinones at the primary quinone binding site indicated the possible contribution of Tyr H40 in the stabilization of the primary semiquinone [16]. Glu L212 has been identified by time resolved and steady state infrared spectroscopy and significant protonation of Glu L212 has been shown at pH region well below its proposed pK_a value (e.g. ≈ 0.5 H^+/RC at pH 7) [17]. It was concluded that the pH behaviour of Glu L212 could not be explained by the titration of a single isolated residue, but the titration must be unusually broad due the strong interaction with other charged groups in the Q_B site [3,4].

These observations call for re-examination of measurements on pH dependence of electron and proton transfer in RCs of wild type and mutants and revision of the interpretation in terms of the key residues. The responsibility of Asp L213 is also unclear probably due to the location of this residue as part of the acidic charge-cluster involving Asp L210 and Arg L217, amongst others, providing the potential for strong interactions between these groups [6]. The charge cluster can be studied by mutation of protonatable residues for non-protonatable groups. The effect of substituting lysine for Glu L212 is exceptionally interesting as the expected large effects of a net positive charge were not seen and the quinone function was not greatly impaired [18]. Lys L212 may form charge pair with a residue other than Asp L213.

The results from IR measurements require a better understanding of the electrochromic shift of bacteriopheophytins that have been used for tracking interquinone electron transfer. The measured absorption change contains components both from electron transfer and from ionic (proton) relaxation around the quinones [19-21]. Separation of these events is of great importance. One possible way is the observation of dimer fluorescence. The yield of fluorescence of the dimer in the RC is highly dependent on the redox state of the RC: it is high in the closed ($PQ_A^-Q_B$) and low in the open ($PQ_AQ_B^-$) state. The yield of fluorescence is much more sensitive to the position of the electron in the quinone complex than to the ionic configuration in the surroundings, thus the kinetics of the electron transfer will control the yield of dimer fluorescence [see 22].

The study of kinetics of proton binding accompanying semiquinone formation has revealed that much of the H^+-ion uptake appeared to be directed at residues that were not immediately accessible, and the kinetics of proton binding were not simply limited by diffusion of proton donor species [23,24]. In D_2O, little or no solvent isotope effect was observed at pH/D=7, but the ratio of rate constants of proton binding (k_H/k_D) increased to about 3 at higher pH. The kinetic behaviour implied rate-limiting intraprotein proton transfer to specific buried residues in the protein matrix.

Proton delivery pathways to Q_B

After the second light-driven electron transfer Q_BH_2 is formed, and net proton uptake from the aqueous phase must occur. The proton delivery pathway to the quinone head group involves several ionizable amino acid residues in and around the Q_B pocket. Analysis of site-directed mutants indicated the activity of two terminal pathways for proton delivery to Q_B.

One of them involves Ser L223 through Asp L213, whereas the other involves Glu L212 via possibly Asp L213 [2]. Recent Monte Carlo calculations [25] and high resolution X-ray data [26] reveal water molecules located around the Q_B site; their involvement in proton delivery is also likely. Proton uptake measurements of RC films with variable water content would certainly reveal the importance of bound and mobile water molecule in proton transfer.

Not only internal water molecules but subtle structural changes can contribute to proton delivery. Voids which can also be created by mutations, can be filled by water molecules facilitating the transport of H^+-ions in the interior of the protein. Their effect (together with the possible modification of the H-bond network of internal water molecules and protonatable amino acid residues) was studied by double mutants and revertants. It was shown that the compensation mutations (second site revertants) can occur far away from Q_B [11,27] indicating the central role of long-range electrostatic interaction in function of the protein [15,28].

Limitation of proton uptake under continuous illumination

The vast majority of kinetic and energetic data have been obtained from flash kinetic measurements. From physiological point of view, it could be interesting to see how these results will predict the electron and proton transfer characteristics observed under continuous

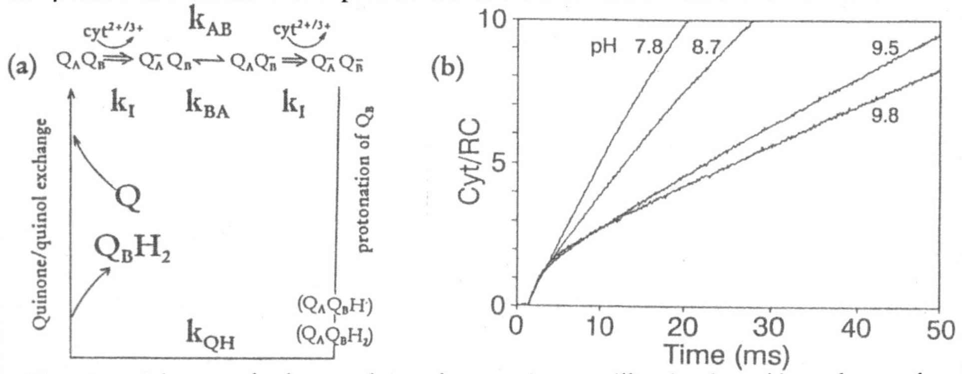

Figure 2 Scheme of photocycle under continuous illumination (a) and cytochrome photooxidation at different pH values ($k_I = 2.10^3$ s^{-1}).

illumination.

The photocycle under stacionary conditions can be modelled by two light and dark reactions alternatively connected in series (Fig. 2a). The rates of the two light reactions (k_I) are equal, k_{AB} and k_{BA} represent the rates of forward and back reactions in semiquinone equilibration, respectively, and k_{QH} denotes the apparent rate of protonation of Q_B and quinone/quinol exchange which depends on the nature of the quinone and detergent. The photocycle can be tracked either by observation of cytochrome oxidation or proton binding. The initial rate is k_I which decreases to a stacionary value of

$$k_{obs} = \frac{k_I}{1 + \frac{k_I}{2} \cdot \left(\frac{1}{k_{QH}} + \frac{1}{k_{AB}}\right) + \frac{k_{BA}}{2 \cdot k_{AB}}} \qquad (1)$$

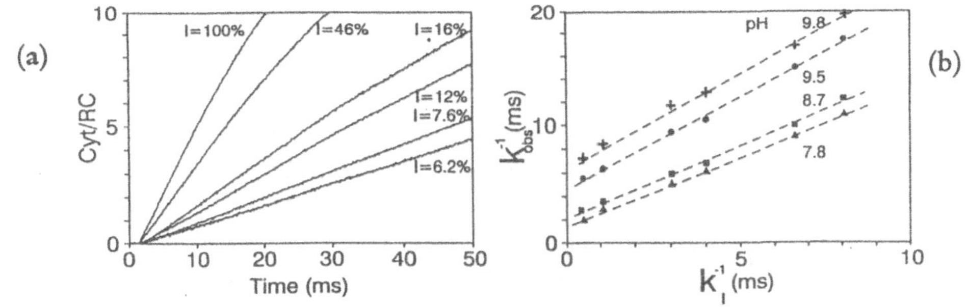

Figure 3 Cytochrome photooxidation at different light intensities (a) and double reciprocal plot to determine k_{QH} according to Eq. (1) (b).

after a transient period much longer than the slowest step in the photocycle (Fig. 2b). If k_{obs} is measured under different light intensities (Fig. 3a), k_{QH} can be derived from interception of a double reciprocal plot (k_{obs}^{-1} vs. k_I^{-1}) (Fig. 3b). According to Fig. 4, k_{QH} is limited by the rate of quinone/quinol exchange below pH 7 and by the rate of protonation in the alkaline pH range.

Inspection of the X-ray structure, the isoprenoid tail of the ubiquinone paves the way for the head group at dissociation from the RC. An aliphatic residue on the L subunit, Leu L232, is positioned close to the tail of Q_B in such a way that it may regulate the diffusion of the quinone/quinol into and out of the RC. The equivalent amino acid on the Q_A site is Trp M266, the indole ring of which interacts with the tail of the primary quinone and could prohibit the release of Q_A under physiological conditions. Measurement of quinone/quinol exchange rate of site-directed mutants at Leu L232 would bring us closer to understanding the export of reducing equivalents from RC at molecular (atomic) level.

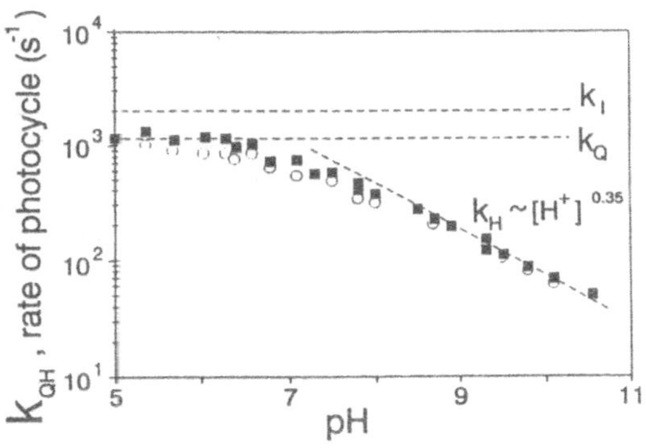

Figure 4 PH-dependence of the rate of photocycle.

CONCLUSIONS

The detailed knowledge of the atomic coordinates of the bacterial RCs has permitted a close scrutiny of structure/function relationships not only of the quinones but of the protein itself with its internal water structure. Protonatable groups were identified as intrinsic part of the redox reactions, providing charge compensation (important for semiquinone formation) and forming channels for the movement of H^+-ions to Q_B^{2-}. The nature and position of these groups give rise to electrostatic profile that determines the kinetics and energetics of proton transport. Fine tuning or dramatic variations of proton delivery pathways can adapt the photocycle to changes in bulk phase pH values, buffering capacities and primary structure (mutations) of the reaction center.

ACKNOWLEDGEMENTS: This work was supported by the Hungarian Science Foundation (OTKA 17362/95), Balaton project (NP-715/95-F), Foundation for Hungarian Education and Research (FEFA III 1034) and International Human Frontier Science Program (RG-329/95 M).

REFERENCES

1 Nicholls, D.G. and Ferguson, S.J. (1992) Bioenergetics 2, Acad. Press, H. Brace Jovanovich, Publishers
2 Okamura, M.Y. and Feher, G. (1992) Annu. Rev. Biochem. 61, 861-896
3 Gunner, M. and Honig, B. (1992) in The Photosynthetic Bacterial Reaction Center (Breton J. and Vermeglio A., eds.) pp. 403-410, Plenum Press, New York
4 Beroza, P., Fredkin, D., Okamura, M.Y. and Feher, G. (1995) Biophys. J. 68, 2233-2250
5 Paddock, M.L., Rongey, S.H., Feher, G. and Okamura, M.Y. (1989) P. Natl. Acad. Sci. 86, 6602-6606
6 Paddock, M.L., Rongey, S.H., McPherson, P.H., Juth, A., Feher, G. and Okamura, M.Y. (1994) Biochemistry 33, 734-745
7 Takahashi, E. and Wraight, C.A. (1990) Biochim. Biophys. Acta 1020, 107-111
8 Takahashi, E. and Wraight, C.A. (1991) FEBS Letters 283, 140-144
9 Takahashi, E. and Wraight, C.A. (1992) Biochemistry 31, 855-866
10 McPherson, P.H., Schönfeld, M., Paddock, M.L., Okamura, M.Y. and Feher, G. (1994) Biochemistry 33, 1181-1193
11 Maróti, P., Hanson, D.K., Baciou, L., Schiffer, M. and Sebban, P. (1994) PNAS 91(12), 5617-5621
12 Shinkarev, V.P., Takahashi, E. and Wraight, C.A. (1992) in The Photosynthetic Bacterial Reaction Center II. Structure, Spectroscopy and Dynamics (Breton J. and Vermeglio A., eds.) pp. 375-387, Plenum Press, New York
13 Shinkarev, V.P. and Wraight, C.A. (1993) in The Photosynthetic Reaction Center (Deisenhofer J. and Norris J., eds.) Vol 1, pp. 193-255, Academic Press, New York
14 Baciou, L., Bylina, E.J. and Sebban, P. (1993) Biophys. J. 65, 652-660
15 Sebban, P., Maróti, P. and Hanson, D.K. (1995) Biochimie (in press)
16 Kálmán, L. and Maróti, P. (1994) Biochemistry 33, 9237-9244
17 Hienerwalder, R., Grzybek, S., Fogel, C., Kreutz, W., Okamura, M.Y., Paddock, M., Breton, J., Nabedryk, E. and Mantele, W. (1995) Biochemistry 34, 2832-2843
18 Takahashi, E., Shinkarev, V.P., Maróti, P. and Wraight, C.A. (1992) in: "Structure, Function and Dynamics of the Bacterial Reaction Center" (eds. Breton J and Vermeglio A) NATO workshop, Cadarache pp 63-64.
19 Brzezinski, P., Okamura, M.Y. and Feher, G. (1992) in The Photosynthetic Bacterial Reaction Center II. (Breton J. and Vermeglio A., eds.) pp. 321-330, Plenum Press, New York
20 Tiede, D.M. and Hanson, D.K. (1992) in The Photosynthetic Bacterial Reaction Center II. Structure, Spectroscopy and Dynamics (Breton J. and Vermeglio A., eds.) pp. 342-350, Plenum Press, New York
21 Kálmán, L., Turzó, K. and Maróti, P. (1993) Photosynthetica 28(2), 185-194
22 Osváth, Sz., Laczkó, G., Sebban, P. and Maróti, P. (this Proceedings)
23 Maróti, P. and Wraight, C.A. (1989) Biophys. J. 55, 428a
24 Takahashi, E., Maróti, P. and Wraight, C.A. (1992) in Electron and Proton Transfer in Chemistry and Biology (Diemann E. ed.) pp. 219-236 Elsevier Publ. Amsterdam
25 Beroza, P., Fredkin, D., Okamura, M.Y. and Feher, G. (1992) in The Photosynthetic Bacterial Reaction Center II. (Breton J. and Vermeglio A., eds.) pp. 363-374, Plenum Press, New York
26 Ermler, U., Michel, H. and Schiffer, M. (1994) J. Bioenergetics Biomembranes 26, 5-15
27 Hanson, D.K., Baciou, L., Tiede, D.M., Nance, S.L., Schiffer, M. and Sebban, P. (1992) Biochim. Biophys. Acta 1102, 260-265
28 Sebban P., Maróti, P., Schiffer, M. and Hanson, D.K. (1995) Biochemistry 34, 8390-8397

A PROPOSAL FOR THE STRUCTURE OF THE P_{680} PIGMENTS IN PHOTOSYSTEM II.

Bengt Svensson, Paul J.M. van Kan and Stenbjörn Styring, Department of Biochemistry, Arrhenius Laboratories for Natural Sciences, Stockholm University, S-10691 Stockholm, Sweden

1. Introduction

The membrane proteins D1 and D2 constitute the photosystem II (PSII) reaction centre and form a heterodimer that binds the redox components necessary for the primary electron transfer reactions. The three dimensional structure of the PSII reaction centre has not yet been determined. However, there exists a considerable structural and functional homology with the photosynthetic reaction centres from purple bacteria [1,2] for which the crystallographic structure has been determined [3,4]. A similar orientation and distance between the two pheophytins, the Fe^{2+} ion and some of the four to six chlorophylls, that is generally bound in PSII reaction centre preparations can therefore be expected [5]. The functional analogy is almost complete for the acceptor side [1]. However, there exists several striking differences in both function, orientation, and binding of the chlorophylls constituting P_{680} and the accessory chlorophylls. P_{680} has a much higher midpoint potential than the primary donors P_{870} in *Rb. sphaeroides* and P_{960} in *Rps. viridis*. PSII is therefore able to extract electrons from water.

In purple bacteria two bacteriochlorophylls constitute a dimer with considerable excitonic interaction, which acts as the primary electron donor. The exciton coupling between pigments in the entire PSII core complex is significantly lower than in its bacterial counterpart [6,7]. Still the primary donor in PSII may be constituted of two symmetrically placed chlorophyll molecules, with parallel conformation. This is supported by the fact that the histidines ligating the central Mg^{2+} ion of the bacteriochlorophylls in the purple bacteria are conserved as D1-His198 and D2-His198 in PSII.

The accessory pigments of the PSII reaction centre are not bound as their counterparts in purple bacteria. The histidines coordinating the accessory chlorophylls in purple bacteria are not conserved to PSII [2,8]. It is possible though that accessory pigments in PSII are bound close to P_{680} in the cleft between the D1 and D2 proteins, but they would have other polar groups as ligands. Additionally, accessory chlorophylls in PSII have been proposed to be ligated by D1-His118 and D2-His118 on the B helices [2]. Chlorophylls bound at these positions would be oriented perpendicular to the membrane (Svensson et al. unpublished) and such an orientation is experimentally supported by the finding that at least some of the accessory pigments in PSII have an almost perpendicular angle with the membrane [9].

P. Mathis (ed.), Photosynthesis: from Light to Biosphere, Vol. I, 425–430.

In this communication, we propose a structure of the primary donor, P_{680}. To achieve this, we have utilised information from exciton interaction theory [10] in combination with molecular mechanics modelling to fit available spectroscopic data [11-14]. The basis of this study is the structural model for PSII [15] (Svensson *et al.*, in preparation). The PSII model includes the most central parts of the D1 and D2 proteins and seven cofactors (four chlorophylls, two pheophytins and a Fe^{2+} ion). The model is experimentally well supported for the regions of the redox active pheophytin, the acceptor side Fe^{2+} ion and Tyr_D. Supporting experimental results are also available for the surroundings of Tyr_Z. It is therefore challenging to model the structure of the central chlorophylls within this protein structure.

In the theoretical model, that is used as starting point in this work, all included chromophores are found at the contact surfaces between the D1 and D2 proteins. The pheophytins are bound in the centre of the hydrophobic span of the proteins, probably in an almost identical fashion as in the purple bacteria. The non-heme Fe^{2+} is bound on the C2 symmetry axis between the helices D and E on both the D1 and D2 protein. The chlorophylls constituting P_{680} are symmetrically bound as a dimer and ligated to D1-His198 and D2-His198 on the D helices. In our model two accessory chlorophylls are positioned in the cleft between the D1 and D2 proteins corresponding approximately to the position of the accessory pigments in *Rps. viridis.*

2. Methods

Optimisation of the pigment positions in the PSII model [15] (Svensson et al., in preparation) were obtained by making small translational and rotational movements of the included six porphyrin chromophores, in combination with excitonic interaction calculations. The starting position of the chromophores was as in the *Rps. viridis* structure. Subsequently, small changes in the conformation of the pigments were introduced. With each change a limited CHARMM force field minimisation [16] was performed. Side chains within a radius of 5Å from any moved pigment atom and the flexible ring substituents on the pigment molecules were allowed to adopt for the new conformation. The total interaction energies were analysed after minimisations by 50 steps of steepest descent and 200 steps of adopted-basis Newton-Raphson. The limitation in movement was determined by the increasing Van der Waals energy, when two atoms were forced to close to each other. In this way the degrees of freedom of the possible chromophore displacements within the protein were explored, giving the limitation that the backbone structure may not be distorted.

The pair wise excitonic interactions between the chromophores were calculated on all pigment structures obtained in the previous step using the point-dipole approximation [10]. In this approximation the excitonic interaction energy V_{12} in wavenumbers (cm^{-1}) between a pair of chromophores is represented by the formula:

$$V_{12} = \frac{1}{4\pi\varepsilon_r\varepsilon_0} \cdot \frac{\bar{\mu}_1 \cdot \bar{\mu}_2 - 3(\bar{\mu}_1 \cdot \bar{r}_{12})(\bar{r}_{12} \cdot \bar{\mu}_2)}{|\bar{r}_{12}|^3}$$

The $\bar{\mu}_1$ and $\bar{\mu}_2$ are the transition dipole moments centred on the chromophore monomers. The transition dipole strength μ^2 for chlorophyll *a* was assumed to be 23 D^2. The dipole strength for pheophytin *a* is assumed to be 13.8 D^2 i.e. 0.6 times that of chlorophyll *a*. The transition dipole moments $\bar{\mu}$, used in the formula, were taken as the square roots of 23 and 13.8 D^2. $\bar{r}_{12}$ is the distance vector between the centre points of the chromophores. The ε_r and ε_0 are the relative and absolute dielectric constants, respectively. The first part of the expression has the value of 5.04 [10].

The spectral positions of the absorption bands were determined from the diagonal element of an interaction matrix of all pair wise chromophore interactions. The exciton coupling contribution from two additional chlorophylls ligated by D1-His118 and D2-His118 on the B helices was also computed.

3. Results

We have optimised the position of the chromophores and their protein environments, with the focus on the pigments constituting the primary donor P_{680}. Energy minimisations and potential energy calculations were performed while translating and rotating the pigment molecules. The pigment conformations, that were possible to adopt within the protein model, were evaluated with exciton interaction calculations to determine which structure would best agree with interaction energies determined from difference spectra for P_{680} (Table 1).

Table 1. **Spectroscopic data on P_{680}**

λ (nm) $P_{antisym}$	λ (nm) P_{sym}	Interact. V (cm^{-1})	Rel. int.	Ref.
683.5/679.5	668-669	120-170	16:1	[11]
679	669	110	24:1	[12]
684.5	-	-	-	[13]
681-684	667	150-190	30:1	[14]

The initial attempt included a pigment conformation homologous to that in *Rps. viridis* with a chlorophyll *a* dimer. In this conformation (Figure 1), the distance between the central points of the porphyrin rings of the two chlorophylls is 7.3 Å and the angle between the Q_Y transitions is 141°. The interaction energy was calculated to 399 cm^{-1} which is far to high in comparison with spectral data for P_{680} (Table 2).

Therefore, we started the optimisation procedure by moving the chlorophylls stepwise further apart in order to reduce the excitonic coupling. Our criterion for a solution was that the P_{680} pigments could be accommodated in the modelled protein structure. The best solution we found involved a 1.5Å symmetric translation of the two chlorophylls in the dimer along an axis parallel to the membrane plane and the chlorophyll ring planes in combination with a rotation of 5.5° outwards around the normals to the chlorophyll ring plane. The centre to centre distance in this structure is 10.1Å and the angle between the Q_Y transitions is 152° giving a exciton interaction energy of 185 cm^{-1} for the two chlorophylls of the dimer (Table 2).

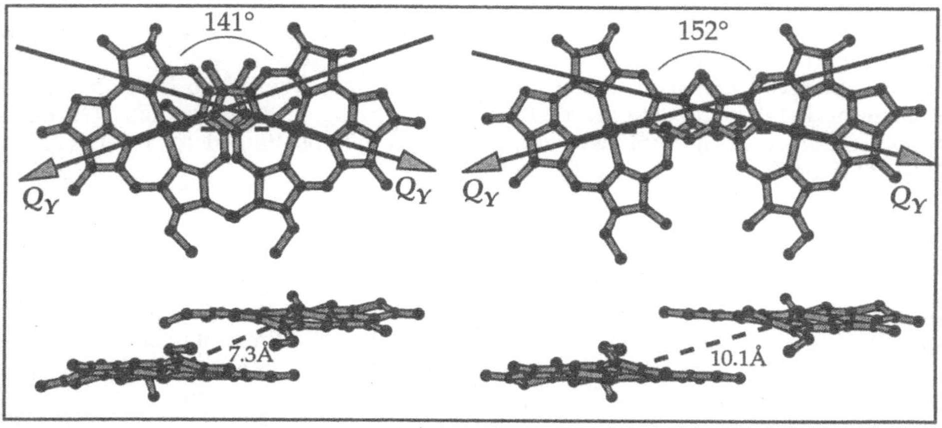

Figure 1. The chlorophylls in the *Rps. viridis* conformation and in the new modelled PSII conformation. The Q_Y transition axis are shown.

Table 2. **Exciton calculation results**

Chl *a´*s. as in *Rps. viridis*			**Chl *a´*s. adopted to PSII**	
λ (nm)	Intensity		λ (nm)	Intensity
690	1.781	$P_{antisym}$	680	1.886
654	0.219	P_{sym}	663	0.114
Interact. V	Rel. intensity		Interact. V	Rel. intensity
399 cm^{-1}	8:1		185 cm^{-1}	17:1
7.3 Å		Distance	10.1 Å	
141°		Angle	152°	

This result shows that, it is possible to accommodate two chlorophylls in a weakly coupled pair in the structural model of PSII. Of the two chlorophylls in the dimer one is bound to the D1 protein, where D1-His198 provides a ligand to the central Mg^{2+} ion. The other is associated with the D2 protein and ligated by D2-His198. Very interestingly, after moving the two chlorophylls apart, we observed that other specific protein - chromophore interactions appeared in the structural model. Polar groups on the D1 and D2 proteins came close enough to the chlorophyll molecules to actually form hydrogen bonds. The hydrogen bonds are proposed to be formed between D1-Thr286 and the O1D oxygen atom (PDB nomenclature) of the 13^2 ester group in $Chl_{D1,P680}$ (Figure 2a) and between D2-Ser283 and the corresponding oxygen on $Chl_{D2,P680}$ in the dimer (Figure 2b). The hydrogen bonds are strong with a bond distance of respectively 2.7Å and 2.6Å.

Another protein - pigment interaction of importance is a close Van der Waals contact between D2-Trp192 and the D2 associated chlorophyll of the dimer. This tryptophan is stacked against ring D of the chlorophyll and shows an interaction, possibly a weak hydrogen bond to the O2D oxygen of the 13^2 ester group of $Chl_{D2,P680}$. It is important to note that at least two, probably three, hydrogen bonds between the chlorophylls in P_{680} and the protein were proposed from FTIR work [17].

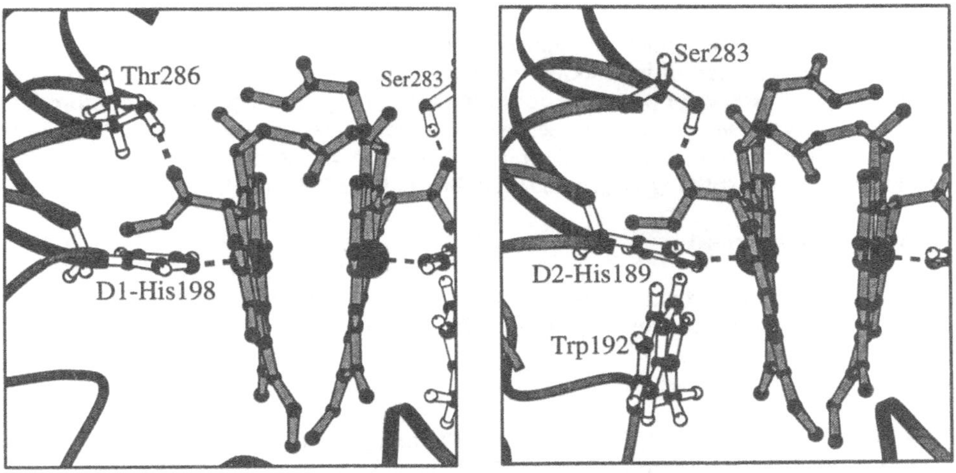

Figure 2. Structural environment of the dimer chlorophylls showing the potential hydrogen bonds between D1-Thr286 and $Chl_{D1,P680}$ and between D2-Ser283 and $Chl_{D2,P680}$.

Our proposal for the structure of P_{680}, which also is supported by Durrant *et al.*, 1995, can be called a separation model. Several suggestions exists for the structure of the primary donor in PSII. In some of these, one chlorophyll is rotated or deleted. We have tried to accomodate many different chlorophyll conformations that fit the spectroscopic data to the structural model. However, configurations with a chlorophyll rotated so that the ring plane no longer is close to perpendicular to the membrane plane or when one clorophyll is deleted, cannot be accommodated without major perturbations of the structural model. We argue that such proposals for the structure of P_{680} are unlikely since the model is proven to be in very good agreement with experimental results for other regions of the D1/D2 heterodimer, including donor side regions.

Our interpretation based on some of the available spectra for P_{680}, is that both chlorophylls with considerable exciton interaction and chlorophylls with very small interaction with P_{680} are present in PSII reaction centre preparations. We have simulated this using a model with six chlorophylls per two pheophytins bound to the PSII reaction centre. In this model are two accessory chlorophylls bound close to the P_{680} dimer and two are bound more distant. In our calculations, we have found that the closely associated chlorophylls contribute by approximately 75 cm^{-1} to the excitonic coupling in the PSII reaction centre. However, the more distant pigments contribute by approximately 10 cm^{-1}, which is almost negligible in the context of the dominant coupling between the two chlorophylls of the dimer. Considering this, we have only included the four most central chlorophylls in our modelling approach presented here.

4. Conclusions

In the structural model of the PSII reaction centre, we have accommodated two chlorophyll *a* molecules in a weakly coupled dimer. The centre to centre distance of the chlorophylls in the pair in PSII is with a separation of 10.1 Å (Table 2, Figure 1) larger than the distance of 7.3 Å in the *Rps. viridis* structure. The two chlorophylls are parallel and the distance between their ringplanes is 3.4Å. The larger separation produces a more evenly distributed exciton interaction between *all* the pigments in PSII. This has been described as a multimer of pigments constituting P_{680} [7]. In our preferred conformation we have observed two, possibly three, hydrogen bonds between the chlorophylls in the dimer and residues on the D1 and D2 proteins. The residues interacting most strongly with the chlorophylls constituting P_{680} are D1-Thr286, D2-Ser283 and D2-Trp192.

5. Acknowledgement

This work was supported by the Swedish Natural Science Research Council. P.v.K acknowledges post. doc. fellowships from NWO and EMBO.

6. References

1 Rutherford, A.W. (1987) in Progress in Photosynthesis Research (Biggins, J. Ed.) Vol. I, pp. 277-283, Martinus Nijhoff Publishers, Dordrecht, The Netherlands.
2 Michel, H. and Deisenhofer, J. (1988) Biochemistry 27, 1-7.
3 Deisenhofer, J., Epp, O., Miki, K., Huber, R. and Michel, H. (1985) Nature 318, 618-624.
4 Allen, J.P., Feher, G., Yeates, T.O., Komiya, H. and Rees, D.C. (1987) Proc. Natl. Acad. Sci. U.S.A. 84, 6162 6166.
5 Satoh, K. (1993) in The Photosynthetic Reaction Center (Deisenhofer, J. & Norris, J.R. Eds.) Vol. I, pp. 289-318, Academic Press, San Diego, CA, USA
6 Van Gorkom, H.J. and Schelvis, J.P.M. (1993) Photosyn. Res. 38, 297-301.
7 Durrant, J.R., Klug, D.R., Kwa, S.L.S., van Grondelle, R., Porter, G. and Dekker, J.P. (1995) Proc. Natl. Acad. Sci. 92, 4798-4802.
8 Kuhn, M., Fromme, P. and Krabben, L. (1994) Trends Biochem. Sci. 19, 401-402.
9 Kwa, S.L.S., Newell, W.R., van Grondelle, R. and Dekker, J.P. (1992) Biochim. Biophys. Acta 1099, 193-202.
10 Pearlstein, R.M. (1991) In Chlorophylls (Scheer, H., Ed.) pp. 1047-1078, CRC Press, Boca Raton, FL, USA.
11 Van Kan, P.J.M., Otte, S.C.M., Kleinherenbrink, F.A.M., Nieveen, M.C., Aartsma, T.J. and Van Gorkom, H.J. (1990) Biochem. Biophys Acta 1020, 146-152.
12 Braun, P., Greenberg, B.M. and Scherz, A. (1990) Biochemistry 29, 10376-10387.
13 Carbonera, D. Giacometti, G. and Agostini, G. (1994) FEBS Lett. 343, 200-204.
14 Kwa, S.L.S, Eijkelhoff, C., van Grondelle, R. and Dekker, J.P. (1994) J. Phys. Chem. 98, 7702-7711.
15 Svensson, B., Etchebest, C., Tuffery, P., Smith, J. and Styring, S. (1992) in Research in Photosynthesis (N. Murata Ed.) Vol. II, pp 147-150, Kluwer Academic Publishers, Dordrecht, The Netherlands.
16 Brooks, B.R., Bruccoleri, R.E., Olafson, B.D., States, D.J., Swaminathan, S. and Karplus, M. (1983) J. Comp. Chem. 4, 187-217.
17 Noguchi, T., Inoue, Y. and Satoh, Ki. (1993) Biochemistry 32, 7186-7195.

CHARACTERIZATION OF PHOTOSYNTHETIC SUPRAMOLECULAR ASSEMBLIES USING SMALL ANGLE NEUTRON SCATTERING.

D.M. Tiede[1], P. Marone[1], A.M. Wagner[1] and P. Thiyagarajan[2]
[1]Chemistry Division D-200 and [2]Intense Pulsed Neutron Source
Argonne National Laboratory, Argonne, IL 60439 USA

1. Introduction

We are using small angle neutron scattering (SANS) to resolve structural features of supramolecular assemblies of photosynthetic proteins in liquid and frozen solutions. SANS resolves the size, shape, and structural homogeneity of macromolecular assemblies in samples identical to those used for spectroscopic assays of photosynthetic function. Likely molecular structures of the supramolecular assemblies can be identified by comparing experimental scattering data with scattering profiles calculated for model supramolecular assemblies built from crystal structures of the individual proteins. SANS studies of the *Rhodobacter sphaeroides* reaction center, RC, presented here, show that the detergent solubilized RC exists in a variety of monomeric and aggregation states. The distribution between monomer and aggregate was found to depend strongly upon detergent, temperature and nature of additives, such as ethylene glycol used for low temperature spectroscopy and polyethylene glycol used for crystallization. Likely aggregate structures are being identified by fitting the experimental scattering profiles with those calculated for model aggregates built-up using the RC crystal structure. This work establishes the foundation for using SANS to identify intermediates in the RC crystallization pathways, and for determining likely structures of complexes formed between the RC and its physiological reaction partners, cytochrome c, and the LHI antenna complex.

2. Procedures

Deuterated RCs were isolated from *Rb. sphaeroides* R-26 cells grown on >98% ^{2}H-enriched medium, using procedures described previously (1). RCs were placed in either 0.06% (w/v) lauryldimethylamine-N-oxide, LDAO, or 0.8% n-octyl-b-D-glucoside, OG, by washing a RC loaded DEAE Sephadex column with either buffer, and eluting with the same buffer containing 0.28 M NaCl. The $^2H_2O/^1H_2O$ ratios were adjusted by dialysis to be 5% and 17% for the LDAO and OG samples respectively. These ratios were chosen to "contrast match" the neutron scattering length densities of the detergents to that of the solvent, thus effectively eliminating the detergent contribution to SANS. Deuterated RCs were used to maximize their scattering amplitude. SANS data were collected at the Intense Pulsed Neutron Source of Argonne National Laboratory, using the small-angle

P. Mathis (ed.), Photosynthesis: from Light to Biosphere, Vol. I, 431–436.
Kluwer Academic Publishers.

diffractometer. Scattering profiles were calculated from crystal structures using a previously described algorithm (2).

3. Results and Discussion

3.1 Effect of Detergents

SANS data show that RCs in LDAO are monodispersed and unaffected by ionic strength. This can be illustrated by standard Guinier plots (3), in which the logarithm of the measured scattered neutron intensity, $\ln[I(q)]$, is plotted versus q^2, where $q=4\pi\sin(\theta/\lambda)$, θ is the half scattering angle, and λ is the neutron wavelength. The square root of the slope will be proportional to the radius of gyration, Rg, for homogenous, non-interacting particles of any type in the region where $qRg<1$, and it will be proportional to the z-averaged Rg for polydisperse systems. Figure 1 shows Guinier plots for LDAO solubilized RCs in low (10 mM tris) and high (10 mM tris, 1 M NaCl) ionic strength. Both plots are well fit by single slopes corresponding to Rg values of 31±2 Å and 29±1 Å respectively. The consistency of the slopes shows that the RC particle size is uniform and essentially independent of ionic strength.

Significantly different behavior is seen for RCs solubilized in OG solutions, as illustrated by the Guinier plots in figure 2, shown for samples in both 10 mM Tris and 10 mM Tris, 1 M NaCl. Both show increases in slopes at low q-values, reflecting the presence of RC aggregates and size dispersity in these samples. In addition, the slopes of Guinier plots for samples at low ionic strength are steeper than those at high ionic strength. The fitted regions in the Guinier plots in figure 2 correspond to Rg values of 40±1 Å and 32±1 Å for the low and high ionic strength samples respectively. These results suggest that the RC particle size distribution in OG solutions is shifted by high ionic strength to smaller sizes, and approach those seen in LDAO solutions.

Likely structures for RC particles in detergent solubilized solutions can be identified by fitting experimental SANS data with scattering profiles calculated from the RC crystal coordinates. Figure 3 shows the comparison between the scattering profile calculated for RC monomers based upon the crystal structure (4) and the experimental data measured for LDAO solubilized RCs at high ionic strength. The close agreement between both curves shows i) that LDAO solubilized RCs are predominately monomeric with a structure consistent with the crystal coordinates, and ii) that the contrast matching technique has successfully removed scattering from the detergent, allowing the scattering curves to be analyzed in terms of the deuterated RC components exclusively. This latter fact establishes the validity of using SANS measurements on deuterated photosynthetic proteins under detergent contrast matching conditions to study structures of supramolecular assemblies without interference from detergent components.

Preliminary modeling corroborates the finding that OG solubilized RCs are a mixture of monomeric and aggregated forms. The distribution shifts toward a monomer at high ionic strength. The aggregates appear to be approximately linear, and span the size range from dimers to at least octamers.

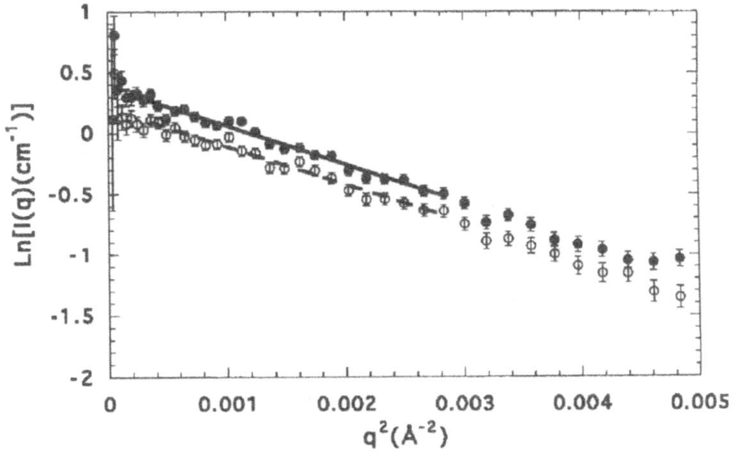

Figure 1 Guinier plots for RCs in 0.06% (w/v) LDAO, 10 mM Tris pH 7.8, ●, and the same buffer with 1 M NaCl,○.

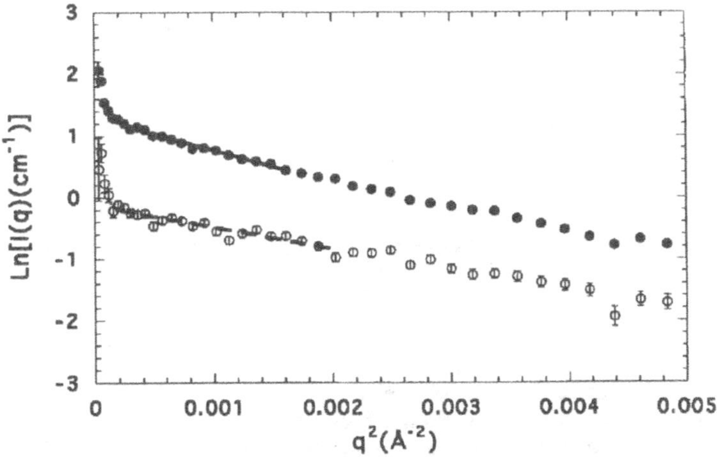

Figure 2 Guinier plots for RCs in 0.8% (w/v) OG, 10 mM Tris pH 7.8, ●, and the same buffer with 1 M NaCl,○.

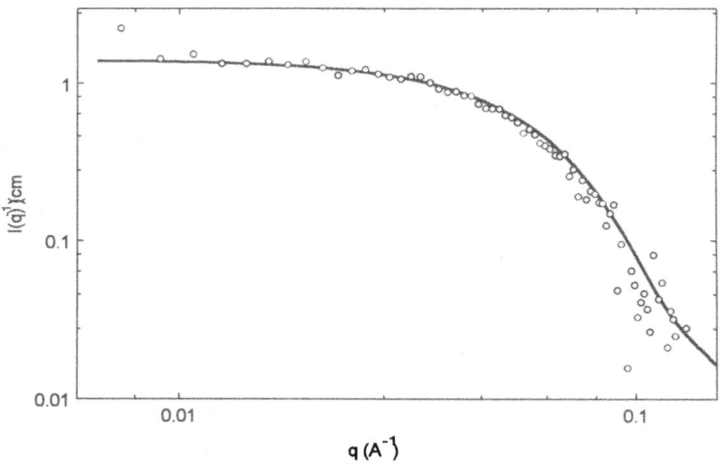

Figure 3 Experimental SANS profile for RCs in 0.06% LDAO, ●, and calculated profile from RC crystal coordinates, --.

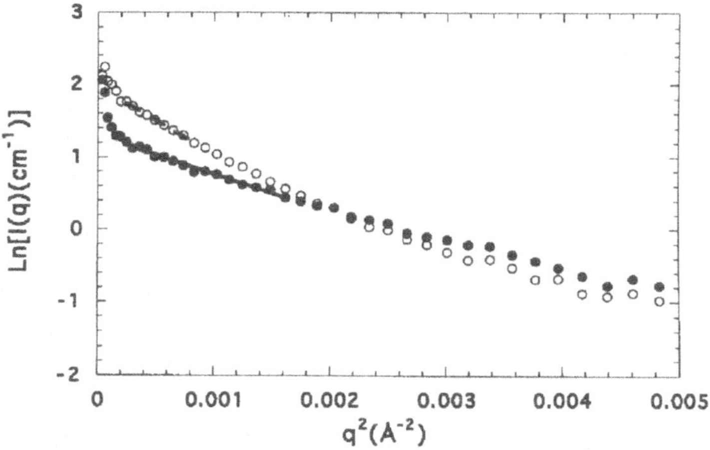

Figure 4 Guinier plots for RC in 0.8% OG, 10 mM Tris, pH 7.8, 22°C, ●, and 6°C, ○.

3.2 Effect of Temperature

Temperature in the range 22°C to 6°C was not found to affect the SANS of LDAO solubilized RCs, reflecting the independence of RC particle size on temperature in this range. In contrast, significant temperature-dependent changes were seen for SANS profiles of OG solubilized RCs. Figure 4 shows Guinier plots of SANS data for an OG solubilized RC sample measured at 22°C and 6°C. The data collected at 6°C shows increased scattering intensity at low q values. The slopes for equivalent portions of Guinier plots yield Rg values of 53±1 Å and 39±1 Å for the 6°C and 22°C data respectively. These results show that an appreciable shift toward larger RC particle sizes occurs upon cooling OG solubilized RCs that was not seen in LDAO solutions.

For both LDAO and OG solubilized RCs extensive three-dimensional aggregation occurs upon solution freezing. A scattering profile for LDAO solubilized RCs at -250°C is shown in figure 5. The significant increase in scattering at low q values reflects the increase in particle size. Aggregation is expected to occur due to phase separation during freezing of aqueous samples. However, the scattering profiles for frozen solutions also show a dramatic decrease in scattered intensity in the q-range 0.01 to 0.1 $Å^{-1}$. Scattered intensity in this region arises from the RC shape. The SANS data for frozen solutions show that all information about RC shape is lost. This result can only occur in the present experiments if the interface is lost between the RC and the contrast matched water and detergent phases. Such conditions imply the existence of aggregation in which there are extensive, direct RC-RC contacts. Recovery from this state upon thawing is found to take one to several hours.

3.3 Effect of Co-Solutes

Solutions of 50% ethylene glycol or 50% glycerol are typically used for spectroscopic characterization of RCs at low temperature. For both LDAO and OG solubilized RCs, these co-solutes were found to cause an aggregation of RCs at 22°C that could not be altered by changes in ionic strength nor increases in detergent concentration. Figure 5 shows the scattering profile for LDAO solubilized RCs in 50% ethylene glycol at -250°C. Only slight changes in scattering profiles were seen upon cooling. Preliminary modeling suggests that under the conditions used for low temperature spectroscopy RCs exist predominately as approximately linear aggregates.

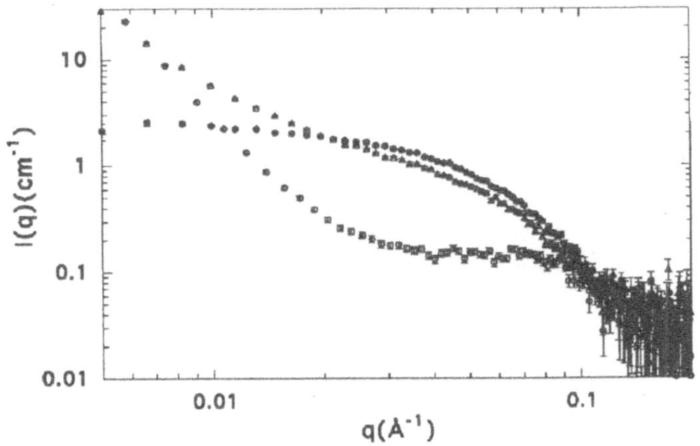

Figure 5 SANS for RCs in 0.06% LDAO at 22°C, •, -250°C, ○, and in 50% ethylene glycol -250°C, Δ.

Polyethylene glycol, PEG, which is used for the crystallization of RCs, was also found to cause RC aggregation at concentrations below the crystallization/precipitation threshold. The similarity in the induction of RC aggregation by PEG, ethylene glycol and glycerol suggests that the effect is not reagent specific, but is driven by a generalized osmotic effect. The extent of RC aggregation was found to increase dramatically as the PEG concentration approaches the crystallization threshold of 13.6% (w/v) in 0.28 M NaCl. The results indicate that < 20% of RCs are likely to be in the monomeric form in crystallization mixtures. Experiments that remove RC aggregates from crystallization mixtures suggest that the disaggregation of RCs is likely to be a critical, rate-determining step in RC crystallization, and that this is the event that needs to be optimized for successful crystallization.

4. Conclusions

The SANS data presented here characterize the aggregation state of the bacterial RC under conditions typically used in assays of photosynthetic function. The results show that in aqueous buffers, RCs solubilized by LDAO are more monodispersed than with OG, and that OG solubilized RCs also show a surprising temperature dependent aggregation. The use of glass forming aqueous mixtures, solution freezing, and PEG additions used for crystallization caused prominent aggregation of RCs in both detergents. This work establishes protocols for the investigation of the structures of supramolecular formed between the RC and it physiological reaction partners.

5. Acknowledgments

Support for D.M.T., A.M.W., P.T. and the IPNS by DOE Office of Basic Energy Sciences, under Contract W-31-109-ENG-38 and for P.A.M. by NASA Microgravity Biotechnology Program M951-ES-3-004-251. D.G. Wozniak is acknowledged for excellent technical help.

References

1 Wraight, C.A. (1979) Biochim. Biophys. Acta 548, 309-327
2 Bradley, K.F., Chen, S-H., Bellissent-Funel, M-C. and Crespi, H.L. (1994) Biophys. Chem. 53, 37-44
3 Guinier, A. and Fournet, G. (1955) in Small Angle Scattering, Wiley, New York
4 Chang, C-H., El-Kabbani, O., Tiede, D.M., Norris, J.R. and Schiffer, M. (1991) Biochemistry 30, 5352-5360

CURRENT QUESTIONS ON OPTICAL PROPERTIES AND ENERGY TRANSFER IN THE REACTION CENTER OF PHOTOSYSTEM II

Hans J. van Gorkom, Department of Biophysics, Huygens Laboratory, P.O.Box 9504, NL-2300 RA Leiden, The Netherlands

1. Introduction

Since the previous Photosynthesis Congress the study of isolated Photosystem II reaction centers (PS II RCs) has been persued with vigor and we have had two ESF workshops dedicated to the subject (Muehlheim, August 1993, and Wye, April 1995). Nevertheless some of the most basic questions about the structure and function of the PS II RC remain unanswered or controversial. Here I will try to summarize, from my point of view, current issues with respect to pigment arrangement and the kinetics of energy tansfer and charge separation in the isolated RC.

We all believe that the PS II RC is structurally similar to the purple bacterial reaction center (PB RC) and base our models on that. This applies especially to the electron acceptor side, where the arrangement and interactions of the (active) pheophytin, Q_A, non-heme iron, and Q_B are basically the same [1]. The notion that the core of the PS II RC must therefore be a heterodimer of the D1 and D2 proteins [2] similar to the L/M pair in the PB RC, and its subsequent isolation [3,4] have greatly enhanced our confidence in the use of the known PB RC structure [5] as a starting point to model PS II. There is even a tendency to be not only more interested in but also more sceptical about results that do not fit the comparison to purple bacteria. Yet there are striking differences.

2. Pigment content

The finding that the isolated PS II RC contains 6 chlorophyll (Chl) molecules [6-8], instead of the 4 expected on the basis of the bacterial analogy, is not generally accepted. The number of 6 is based on the notion that PS II contains two pheophytin (Pheo) molecules [9]. The only published verification of that notion in isolated RCs seems to be the ratio of 2 Pheo : 80 kDa protein determined by Montoya et al. [10], but it fits the bacterial analogy and is not contested. Assuming the presence of 2 Pheo, the 6 Chls follow from the Chl/Pheo ratio determined by HPLC [6-8, 11] and, independently, also from its 5.9-fold decrease in isolating the RC from core particles which had 35±3 Chl/Q_A and 35±5 Chl/P680 (and 38±2 Chl/2 Pheo by HPLC) [8].

The fact that the earliest preparations were thought to have 4-5 Chls was probably due to inadequate determination procedures and not to the long exposure to Triton X-100, as has sometimes been assumed, because it is now clear that extensive Triton X-100 treatment may remove one of the two ß-carotenes but does not decrease the Chl/Pheo ratio [11-13]. Also the method to determine the Chl/Pheo ratio by comparison of the absorption

P. Mathis (ed.), Photosynthesis: from Light to Biosphere, Vol. I, 437–442.

spectra of an acetone extract before and after complete pheophytinization [10] appears to be unreliable [11]. This is unfortunate, because its use has led to the idea that PS II RCs with 4 Chls could be isolated [10, 14-16]. The possibility to isolate such a preparation could not be confirmed by other groups. Unlike the preparation of Vacha et al. [17], who seem to have succeeded in removing one of the Chls absorbing near 670 nm, the putative "4 Chl preparations" have spectral properties surprisingly similar to those of 6 Chl preparations used by other groups (especially to the preparation of van Leeuwen et al. [8], which also has one carotene per two Pheo [18]). Eijckelhoff and Dekker [11] conclude that the "4 Chl" preparations contain 6 and the "5 Chl" and "6 Chl" in the same publications must be assumed to contain more than 6, presumably due to incomplete removal of CP47 [16].

In view of the available evidence the "native" Chl content of the D1D2 complex is most likely 6 and numbers below 5 or above 7 can probably be excluded. To evaluate the pigment content of our preparations a simple and reliable recipe to determine the ratio of Chl:Pheo:Car in a preparation would be extremely useful at this stage. A simple formula based on the room temperature absorption spectrum of an 80 % acetone extract, as proposed by Eijckelhoff and Dekker [11], would be ideal. Until we have a rigorously calibrated formula it is recommendable to keep such an absorption spectrum, from 400 to 750 nm with suitably amplified 520-560 nm region, as a routine documention for every preparation used.

3. Slow energy transfer

The issue of Chl content is interesting in view of slow (tens of ps) energy transfer processes observed in the RC, which indicate much larger distances between pigments than in the PB RC. A conservative model for the pigment arrangement in the PS II RC would place 4 of the 6 Chls at positions corresponding to those of the BChls in the PB RC and the two additional Chls situated sufficiently far away from the central cluster of 4 Chls to explain the slow energy transfer [e.g. 19, 20]. Likely binding sites for those Chls would be the conserved H118 on both D1 and D2, and the distance of those residues to the central pigments could indeed be large enough to explain the slow energy transfer to P680 and slow oxidation by $P680^+$ [21, 22, 23]. In this model, the 2 extra Chls would be expected to have little interaction with the other pigments and most likely exhibit a well defined monomeric absorption spectrum with a Q_Y absorption maximum near 670 nm, in agreement with [21, 24].

In this picture the remaining pigments, absorbing in the 680 nm region, would be arranged like those in the PB RC, with some modification to decrease the special pair interaction. There is a possibility of extensive exciton delocalization [20] and anyway the equilibration of the excited state between these pigments should be very fast (perhaps 0.1 ps [25] or 0.5 ps [19], but see below). Evidence for long-wavelength, slowly transferring pigments or even irreversible excitation traps obtained by measurements at low temperature is not inconsistent with this picture.

However, also at room temperature long wavelength excitation may lead to energy transfer limited charge separation. Charge separation times of 21 ps and 3 ps have been proposed and the discrepancy appears to be hard to resolve [19, 21, 26, 27]. Suppose both are right, then either the preparations used are different or heterogeneous, or the 21 ps process is rate-limited by energy transfer. At intermediate temperatures (77 and 95 K) both components have now been observed in two independent studies using different preparations (Germano et al., these proceedings; 28). Presently the most likely interpretation may be that the 21 ps component is selectively excited at wavelengths above 690 nm and reflects energy transfer limited charge separation. If this is not due to incomplete removal of core antenna pigments [29] it would imply that the long wavelength pigments in the PS II RC do not form a single, rapidly equilibrating cluster as expected from the bacterial analogy. This consideration leads me to question the validity of the presumed 100 fs equilibration of the excited state over the majority of the RC pigments.

4. Fast energy transfer

At short wavelengths (around 665 nm) the absorbance may be dominated by the slowly transferring 'extra' Chls at large distance from P680, as concluded in [21] and perhaps supported by the small accumulated photon echo signals in this region (P. Schellenberg et al., these Proceedings). Yet large sub-ps components have been observed also on short wavelength excitation [25, 27, 28]. The absence in this case of a pronounced 1-3 ps component suggests that the fs changes are not due to equilibration with P680, even though the sub-ps absorbance recovery is accompanied by a bleaching on the long wavelength side and the reverse happens on long wavelength excitation. Comparing the results of different groups [25, 19, 28, M. Germano et al., these Proceedings], there appears to be a correlation of the fs decay time with the duration of the pump pulse, and of the spectral width of the initial bleaching with that of the spectral overlap between pump pulse and absorption spectrum, while positive changes are seen to the high and low energy side of that. Moreover, the polarization dependence of the ~0.1 ps component in my view points to a coherence effect rather than energy transfer induced depolarization (although such an effect may occur in 0.4-0.9 ps [19, 27]). Fig.1 illustrates the kinetics on excitation with a 55 fs pump pulse as measured by M. Germano et al., these Proceedings. The transient bleaching during the pulse is large at magic angle (solid line) at nearly vanishes at perpendicular orientation (dashed line) between pump and probe.

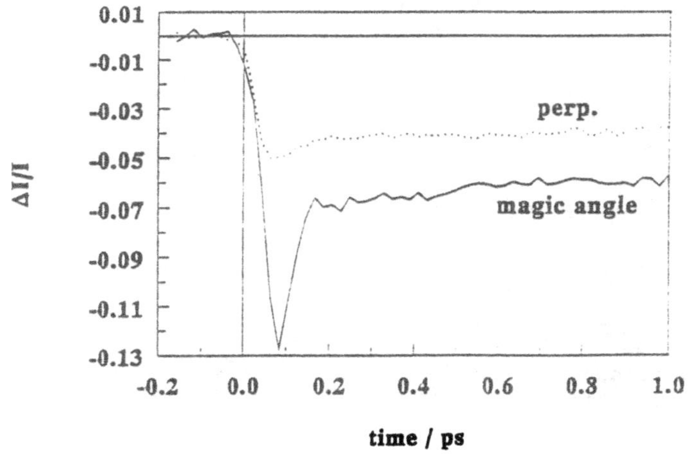

Fig.1 Absorption transients at 682 nm with a wide-band 695 nm pump pulse of 55 fs at room temperature, as measured by M. Germano et al., these Proceedings.

The femtosecond changes clearly depend on the absorbance of the PS II RC. but to what extent they reflect energy transfer between its pigments remains to be shown.

5. Pigment arrangement and exciton delocalization

There is still considerable room for speculation in this field. We do not know that the central pigment group contains all four Chls homologous to the bacterial arrangement; there could be three, or even two. One might speculate about alternative binding sites, like the conserved His 62 in helix I of the D2 protein, or somewhere outside as a 'linker' Chl between the RC and core antenna proteins [16, 30]. And although there probably are two Pheos there seems to be no compelling evidence that the 'inactive' Pheo is at the position and orientation expected.

We do know that the active Pheo is probably where the corresponding BPheo is in the PB RC, that for efficient charge separation a pigment must be present at a position close to that of the adjacent accessory BChl in the PB RC, that this pigment might well be the one where the triplet state of P680 is localized at low temperature [31], and that 'P680' comprises at least two interacting pigments.

The photoaccumulation of reduced Pheo at 10 K by Breton [32] has shown several points relevant to this discussion. My interpretation of Breton's data is illustrated in Fig.2.

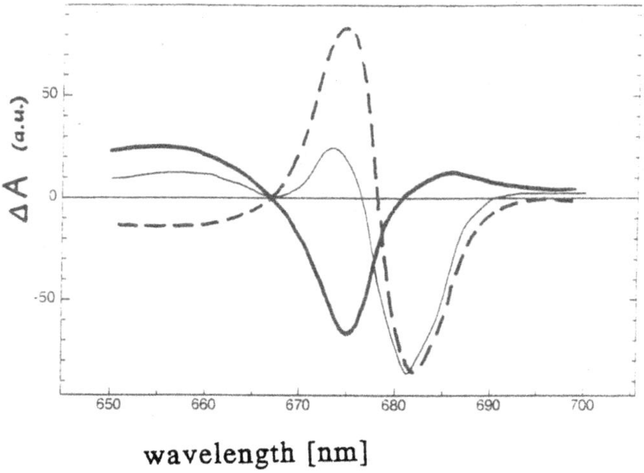

Fig. 2 Absorption (thin line) and Linear Dichroism (heavy line) difference spectra induced by phototrapping of Pheo⁻ at 10 K as measured by J. Breton [32], rescaled to illustrate that their difference (dashed line) may be attributed to a blue-shift of adjacent Chls.

First of all, the associated changes in Linear Dichroism (heavy line) were fully consistent with the expected absorption difference spectrum of reduction of a Pheo molecule to its π radical anion [33], oriented as the BPheo in PB RCs, without complications by interactions with other pigments. Playing with multimer models involving 'intramolecular' charge separation [20] might become very useful when LD and CD spectra can be predicted. Secondly, the measured (isotropic) absorption difference spectrum (thin line) was complicated by the expected blue-shift of nearby Chls (dashed line), just as the the phototrapping of $BPheo^-$ in the PB RC is accompanied by a blue-shift of the adjacent accessory BChl and to a lesser extent of the special pair [34-36]. The spectral shape of the absorption difference spectrum on the long wavelength side is the same as that observed on $P680^+Pheo^-$ or $^3P680$ formation and includes not only the ~680 nm band but also the

684 nm shoulder [37].

As pointed out before [38], the important conclusion here is that the blue-shift of these absorption bands attributed to P680 Chls does not show up in the LD difference spectrum. This indicates that the involved transition moments are oriented at the magic angle with the normal to the membrane, in agreement with [39], and supports the proposal [40] that P680 is made up of two Chls with approximately parallel Q_Y transition moments oriented at the magic angle, one placed like the bacterial B_A and the other at the position of the bacterial special pair (the orientation of P_B would be consistent).

The fact that the 684 nm shoulder appears prominently in the difference spectra not only of ^{3}P680 [37,39] but also of P680$^+$Pheo$^-$ and of Pheo$^-$ [37, 38] does not support its assignment to a distant 'linker' Chl [16, 30]. To the extent that its apparent removal by treatments used to isolate the presumed "4 Chl preparations" cannot be explained by removal of antenna proteins (see above) its decrease more likely results from decreased exciton interactions than from loss of a Chl. The assignment of the 684 nm shoulder to spectral heterogeneity of P680 [38], with the 684 nm band perhaps being the 'native' form [41,42], still holds.

If one of the 'special pair' molecules is missing or some other important structural difference decreases the coupling between the 'active' and 'inactive' branches of the RC, the Pheo and Chl of the 'inactive' branch could form a separate long wavelength cluster, distinct from P680. A stronger coupling of the 'inactive' Pheo in this cluster might explain its appearance in ADMR [39] and hole-burning [43] measurements and perhaps also prevent electron transfer from this cluster to P680$^+$.

If the 'active' Pheo has its own, well defined Q_Y absorption band around 676 nm and its reduction causes a large blue-shift of the adjacent Chl, oxidation of that Chl may be expected to cause a large red-shift of the Pheo band. In my view this is the cause of the pronounced 675 nm bleaching in P680$^+$$Q_A^-$ difference spectra at low temperature [42, 44], which might be verified by measuring the associated LD difference spectrum. The absence of Pheo red-shift at room temperature suggests that in that case the positive charge is not localized on the Chl next to Pheo but on that (or those) near the position of the bacterial special pair. The position of the positive charge in P680$^+$ may depend on the local electrical field and could even be S-state dependent.

References

1. Mathis, P. and Rutherford, A.W. (1987) in: New Comprehensive Biochemistry 15: Photosynthesis (Amesz, J., ed.) pp. 63-96. Elsevier, Amsterdam.
2. Trebst, A. (1986) Z. Naturforsch. 41c, 240-245.
3. Nanba, O. and Satoh, K. (1987) Proc. Natl. Acad. Sci. USA 84, 109-112.
4. Tang, X.-S., Fushimi, K. and Satoh, K. (1990) FEBS Lett. 273, 257-260.
5. Deisenhofer, J., Epp, O., Miki, K., Huber, R. and Michel, H. (1985) Nature (London) 318, 618-624.
6. Kobayashi, M., Maeda, H., Watanabe, T., Nakane, H. and Satoh, K. (1990) FEBS Lett.260,138-140.
7. Gounaris, K., Chapman, D.J., Booth, P., Crystall, P., Giorgi, L.B., Klug, D.R., Porter, G. and Barber, J. (1990) FEBS Lett. 265, 88-92.
8. Van Leeuwen, P.J., Nieveen, M.C.,, van de Meent, E.J., Dekker, J.P. and van Gorkom, H.J. (1991) Photosynth. Res. 28, 149-153.
9. Kobayashi, M., Watanabe, T., Nakazato, M., Ikegami, I., Hiyama, T., Matsunaga,T. and Murata, N. (1988) Biochim. Biophys. Acta 936, 81-89.
10. Montoya, G., Yruela, I. and Picorel, R. (1991) FEBS Lett. 283, 255-258.
11. Eijckelhoff, C. and Dekker, J.P. (1995) Biochim. Biophys. Acta 1231, 21-28.
12. Tang, D., Jankowiak, R., Seibert, M. and Small, G.J. (1991) Photosynth. Res. 27, 19-29.
13. De Las Rivas, J., Telfer, A. and Barber, J. (1993) Biochim. Biophys. Acta 1142, 155-164.

14. Yruela, I., van Kan, P.J.M., Müller, M.G. and Holzwarth, A.R. (1994) FEBS Lett. 339, 25-30.
15. Aured, M., Moliner, E., Alfonso, M., Yruela, I., Toon, S.P., Seibert, M. and Picorel, R. (1994) Biochim. Biophys. Acta 1187, 187-190.
16. Chang, H.-C., Jankowiak, R., Reddy, N.R.S., Yocum, C.F., Picorel, R., Seibert, M. and Small, G.J. (1994) J. Phys. Chem. 98, 7725-7735.
17. Vacha, F., Joseph, D.M., Durrant, J.R., Telfer, A., Klug, D.R., Porter, G. and Barber, J. (1994) Proc. Natl. Acad. Sci. USA 92, 2927-2933.
18. Van Leeuwen, P.J. (1993) Doctoral Thesis, Leiden University.
19. Holzwarth, A.R. Müller, M.G., Gatzen, G., Hucke, M. and Griebenow, K. (1994) J. Lumin. 60, 479-502.
20. Durrant, J.R., Klug, D.R., Kwa, S.L.S., van Grondelle, R., Porter, G. and Dekker, J.P. (1995) Proc. Natl. Acad. Sci. USA 92, 4798-4802.
21. Schelvis, J.P.M., van Noort, P.I., Aartsma, T.J. and van Gorkom, H.J. (1994) Biochim. Biophys. Acta 1184, 242-250.
22. Van Gorkom, H.J. and Schelvis, J.P.M. (1993) Photosynth. Res. 38, 297-301.
23. Koulougliotis, D., Innes, J.B. and Brudvig, G.W. (1994) Biochemistry 33, 11814-11822.
24. Vacha, F., Joseph, D.M., Durrant, J.R., Telfer, A., Klug, D.R., Porter, G. and Barber, J. (1994) Proc. Natl. Acad. Sci. USA 92, 2929-2933.
25. Durrant, J.R., Hastings, G., Joseph, D.M., Barber, J., Porter, G. and Klug, D.R. (1992) Proc. Natl. Acad. Sci. USA 89, 11632-11636.
26. Wiederrecht, G.P., Seibert, M., Govindjee and Wasielewski, M.R. (1994) Proc. Natl. Acad. Sci. USA 91, 8999-9003.
27. Klug, D.R., Rech, T., Joseph, D.M., Barber, J., Durrant, J.R. and Porter, G. (1994) Chem. Phys. 194, 433-442.
28. Visser, H.M., Groot, M.-L., van Mourik, F, van Stokkum, I.H.M., Dekker, J.P. and van Grondelle, R. (1995) J. Phys. Chem., in press.
29. Schelvis, J.P.M., Germano, M., Aartsma, T.J. and van Gorkom, H.J. (1995) Biochim. Biophys. Acta 1230, 165-169.
30. Chang, H.-C., Small, G.J. and Jankowiak, R. (1995) Chem. Phys. 194, 323-333.
31. Van Mieghem, F.J.E., Satoh, K. and Rutherford, A.W. (1991) Biochim. Biophys. Acta 1058, 379-385.
32. Breton, J. (1990) in: Perspectives in Photosynthesis (Jortner, J. and Pullman, B., eds.) pp. 23-38, Kluwer Academic Publishers, Dordrecht.
33. Fujita, I., Davis, M.S. and Fajer, J. (1978) J. Am. Chem. Soc. 100, 6280-6282.
34. Tiede, D.M., Prince, R.C. and Dutton, P.L. (1976) Biochim. Biophys. Acta 449, 447-467.
35. Okamura, M.Y., Isaacson, R.A. and Feher, G. (1979) Biochim. Biophys. Acta 546, 394-417.
36. Florin, S. and Tiede, D.M. (1987) in: Progress in Photosynthesis Research (Biggins, J., ed.) Vol.1, pp.205-208, Martinus Nijhoff Publishers, Dordrecht.
37. Van Kan, P.J.M., Otte, S.C.M., Kleinherenbrink, F.A.M., Nieveen, M.C., Aartsma,T.J. and van Gorkom, H.J. (1990) Biochim. Biophys. Acta 1020, 146-152.
38. Otte, S.C.M., van der Vos, R. and van Gorkom, H.J. (J. Photochem. Photobiol. B: Biol. 15, 5-14.
39. Van der Vos, R., van Leeuwen, P.J., Braun, P. and Hoff, A.J. (1992) Biochim. Biophys. Acta 1140, 184-198.
40. Van Gorkom, H.J. and Schelvis, J.P.M. (1993) Photosynth. Res. 38, 297-301.
41. Kwa, S.L.S., Eijckelhoff, C., van Grondell, R. and Dekker, J.P. (1994) J. Phys. Chem. 98, 7702-7711.
42. Hillmann, B., Brettel, K., van Mieghem, F, Kamlowski, A., Rutherford, A.W. and Schlodder, E (1995) Biochemistry 34, 4814-4827.
43. Tang, D., Jankowiak, R., Seibert, M., Yocum, C.F. and Small, G.J. (1990) J. Phys. Chem. 94, 6519-6522.
44. Hillmann, B. and Schlodder, E. (1995) Biochim. Biophys. Acta 1231, 76-88.

NEW CONCEPT OF ENERGY TRAPPING IN PURPLE BACTERIA. CHARGE TRANSFER-POLARON MODEL.

A.Yu. Borisov, A.N.Belozersky Institute of Physico-Chemical Biology, Moscow State University, Building "A", Moscow 119899, Russia.

SUMMARY.

A new concept of the reaction center (RC) is developed. The concept accounts for a number of fakts still waiting for an explanation: a) why the primary electron donor is a pair of BChls? b) recent data /1/ on a small portion of excitations which escapes from RC special pair to core BChls; c) why the second P_2-P800-BPH-Q brunch of RC is not active? d) why the fluorescence spectra of RCs are enormously red-shifted relative to their absorption spectra?

INTRODUCTION.

For the first time the idea that the role of the primary RC trap is played by the charge-transfer state accompanied by polarization of water moleculеs was advocated and theoretically grounded in the work /2/. At that time however all experimental data available fit well the trapping-limited PSU model and subpicosecond trapping proposed in /2/ looked too ambiguous. However, experiments performed on the temperature dependence of primary RC dynamics /3/, on hole-burning /4-6/ and Stark effect /7/ suggested rather similar conclusions. Recent works (/8/ and others) in which the Stokes shift increase was observed in the 10^{-13}s. time domain also seem to correlate with our idea.

THEORETICAL BACKGROUNDS.

Ample experimental and theoretical evidence has proved that dimers of dye molecules are usually asymmetric and do not exhibit the C2 symmetry. That is why being excited, they often produce excimer or charge-transfer (CT) states which means that a good portion of an excited electron is transferred to one dimer molecule leaving an equivalent positive charge on the other. Also bear in mind that CT-states are created in negligible time period, $< 10^{-14}$s. due to a small activation barrier for an excited π-electron between two closely spaced dye molecules.

$$D\tilde{}D + h\nu \longrightarrow [^{+q}D\tilde{}D^{-q}]^* = CTS^*$$

D~D are interacting molecules; $h\nu$ is light quantum; q^+, q^- are charge

P. Mathis (ed.), Photosynthesis: from Light to Biosphere, Vol. I, 443–446.

densities in dimer molecules excited into the CT-state.

The appearance of electric dipole in a dimer in the CT-state necessarily induces a response in its close environment - polarization of mobile charged particles mostly protons. Physicists call such a state polaron. In polar media it is known to consume an appreciable portion of energy (ΔW) up to 0.1eV /9/. We believe that this phenomenon is bound to occur in reaction centers and that it provides the primary process of an excitation trapping which precedes the P^F-state.

Now what kind of effects should we expect from such a model?

a) Despite a hydrophobic nature of the membrane interior it could hardly be free from water molecules due to their high concentration outside of the membrane. According to /10/ bound water accounts for about 20% of RC weight. Even if water molecule has two bonds via oxygen and hydrogen, there will be some mobility left owing to the shift of the second hydrogen atom. Besides, proton in M208 tyrosin of the RC was recently proved to be mobile and shift /11/. Quantum-mechanical calculations have yielded the time span for so called "free electron oscillations" (without energy barriers) to approximate $(1\text{-}3)\cdot 10^{-17}$s. /12/. A proton should have this time considerably longer due to its greater inertia mass. Besides its polarization under actual conditions is associated with overcoming energy barriers which entails two consequences:

a) proton polarization absorbs a portion of electronic energy from CTs of RC which induce this process. <u>This portion was estimated for a water molecule about 5-7 Å apart from P_2 to vary in the range of 35 - 100meV depending on the mutual position of a water molecule and RC /2/</u>. Thus the energy portion ~0.1eV assimilated by water molecules embedded within ~ 10Å from the P870 must be subtracted from the initial CT-state when it converts into the charge-transfer-polaron state (CTPS), equilibrated with a polaron in its environment.

b) the activation barrier for proton polarization slows down the process according to the Boltzman law. We expect these barriers to be of 1-2 Kcal/M. Thus we expect this slowing-down to be about 10-100-fold. Together with the above estimation of the proton "free oscillation time" it brings us into the time domain of hundreds-tens of femtoseconds for the formation of polaron CTP-state in P_2.

<u>**The hypothesis suggested.**</u>

Electron expell from excited $P865^*$ to its primary accepter is not the primary process of exciton trapping. A super short CHARGE-TRANSFER-state must exist in bacterial RC which precedes the formation of the CTP-state with a polaron in the vicinity of the P865. This state is responsible for the primary excitation trapping at least in purple bacteria.

Our prefícted model may be advanced as follows,

$$P_2^* \underset{K}{\overset{K_{pol}}{\rightleftharpoons}} \text{CTPS} \underset{K_{rev}}{\overset{K_d \tilde{=} 3\cdot 10^{11}s^{-1}}{\rightleftharpoons}} P_2^+X^- \qquad (1)$$

the reaction with the K_d/K_{rev} constants is the same as in the currently accepted model. CTPS are known to be manifested by red shifted, widened absorption, to have low fluorescence yield, to quench excited states in closely-spaced monomers. All these qualities are present in bacterial P_2.

Now consider advantages of the CTP-state predicted.

1. The CTPS-model makes clear why RCs need a pair of BChls acting as an excitation trap. The CTPS cannot be created in a single molecule.

2. Due to dimer assymmetry in CTP-state the excited electron is transferred not to either of two but to a definite molecule. In the RC dimers such an electron-accepting BChl molecule appears to be closer to one P800-BPh-Qa brunch, the fact that apparently explains its predominance.

3. Our hypothesis establishes the reasons for the enlarged red shift of the fluorescence spectra of the RC special pairs relative to their absorption bands. In the context of our model, absorption in the long-wave band corresponds to the creation of the excited CTPS in RC under "intact" environmental conditions. Then in a time span of tens-hundreds of femtosecond this state drops by some tens of millielectronvolt and CTPS thus created emits fluorescence from a lowered position!

4. The energy differences between the absorption peaks of RC and those of its core antenna BChls are as great as 25 and 70 meV respectively for Rs.rubrum and Rps.viridis cells. This is very unfortunate from the point of view of Foerster's theory because hampers the useful process of excitation trapping by RC and accelerates the reversed process of excitation detrapping from RC back to antenna BChl. However here we have to reject the formal application of the Foerster theory. Recall that this theory was developed only for pair-wise excitation transfer between molecules /13/. Foerster did not consider dimers as participants in this process and we must take this fact into account. As it was stated above, in the course of the P_2^* --> CTPS transition the electronic energy must decrease by about 0.05-0.1eV in < 1ps. If so, the reversed exciton transfer from P_2^* to its closest core BChl must be considerably slowed down because the newly formed CTP-state becomes even lower relative to the core BChl singlet excited states.Therefore we have to conclude that this reversed excitation transfer proceeds within a considerably longer time >> 10 picoseconds! It should be a typical Foerster-type process with interchromophore distances ~30 Å in accordance with the data in /14/ and suggestions in the works /15,16/. That is why the excitation lifetimes measured for purple bacteria with different mutual positions of core and RC absorption peaks (~ 60-80ps, /15/) reveal no essential difference as it could be expected from the formal application of the Foerster theory. We have examined the PSU model shown in the figure below. It represents a RC and may be described by a simplified system of five (or seven-eifht if BChls of the core antenna are included) equations. These systems were solved by app lication of the PC-program /17/ for the following conditions: $K_{2,3}^{-1}$ was varied within < 1ps; $K_{1,2} = 3 \cdot 10^{12} s^{-1}$ /18/; $K_{1,2} = 100\ K_{2,1}$; $K_{3,2} = 0.1\ K_{2,3}$ (the data obtained were not much sencitive to this ratio and to the value of the BChl deactivation constants which were equal to $10^9\ s^{-1}$ for all molecules). The values for the other rate constants were borrowed from the literature data (see reviews /15,19/). In more detail this modeling is accounted in the work /20/.

The modeling in /20/ has revealed that the $P865^*$ conversion into the $P865^+X^-$ state is preceded by a rapid rise of the CTPS in a time ~ 0.1-0.5ps. In general model with a core complement this rise-time depends on both the energy migration rates between the core BChl and and that of the CTPS conversion into the $P865^+X^-$ state. In these conditions the CTPS signal reaches its maximum at about 1.0ps and decays along with the rise of $P865^+X^-$ signal. Our modeling of RC + core BChl has shown that the risetime for the $P865^*$-->$P865^+X^-$ approximates to 3.0ps /20/ and that P^+X^- formation kinetic is in good agreement with numerous literature data available.

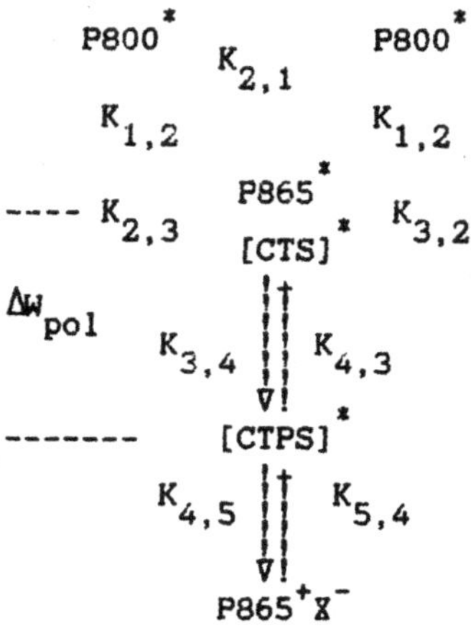

REFERENCES.

1. Abdourakhmanov,I.A. Danielius,R.V.and Razjivin,A.P. (1989) FEBS Letters, 245,47-50.
2. Fok, M.V. and Borisov,A.Y.(1981) Studia Biophysica, 84, N2,115-124.
3. Parson,W.W.and Shuvalov,V.A.(1981) Proc.Nat.Acad.Sci.USA,78,957-960.
4. Meech,S., Hoff,A.J.and Wiersma,D.A.(1985) Chem.Phys.Lett.121,287-292.
5. Boxer,S.G., Lockhart,D.J.and Middendorf,T.R. (1986) Chem.Phys.Lett. 123, N6,476-482.
6. Shuvalov,V.A. Klevanik,A.V. Ganago,A,O. Shkuropatov,A.Y.and Gubanov V.S. (1988) FEBS Letters, 237,57-60.
7. Lockhart,D.J. and Boxer,S,G. (1987) Biochemistry,26,664-668.
8. Vos,M. Jones,M.R. Hunter,C.N. Breton,J. Lambry,G.C.and Martin,J.L. (1994) Biochem. 33,6750-6757.
9. Birks,J.R. (1975) in Reports Progr. Phys., V.38, pp.903-974.
10. Nox,P.P. Kononenko,A.A.and Rubin,A.B. (1979)Bioorgan.Chem. 5,879-896.
11. Schenk,G.C, Parson,W.W., Holten,D., Kirmaer,C. (1992) Biochim.Biophys Acta.1187,1171-1174.
12. Landau,L.D.and Lifshitz,I.M. (1963) in Theoretical Physic, Nauka edition (USSR) V.3,pp 117-118.Moscow.
13. Foerster,T. (1948) Ann.Phys.2,55-75.
14. Stark,W., Kuhlbrandt,W., Wildhaber,J., Wehrli,E.and Muhlethaler K. (1984) EMBO J.,3,777-783.
15. Grondelle,van R., Dekker,J.P., Gilbro,T.and Sundstrem,W.(1994) Biochim.Biophys.Acta,1187,1-65.
16. Timpmann,K., Fu Geng Zhang, Freiberg,A.and Sundstrem,V. (1993) Biochim.Biophys.Acta, 1183, 85-193.
17. Borisov, A.Y. and Vinogradov, A.A. (1992) Biol.Membr. 6(4),569-586.
18. Akhmanov,S.A., Borisov,A.Y.,Kozlovsky,V.S., Rotomskis,R.V., Piskarskas,A.S. and Razjivin,A.P. (1982) Biophysica (Soviet) 23, 912-913.
19. Hoff,A.J. (1993) Molecular Physics, 78,N 4,799-819.
17. Borisov A.Y. (1995) Biochem.and Molec.Biol.Internat. 35,N4,833-840.

CHLOROPHYLLa/ORGANIC SEMICONDUCTOR JUNCTION - POSSIBLE MODEL FOR PHOTOINDUCED CHARGE SEPARATION IN PHOTOSYNTHETIC REACTION CENTERS

Laura Tugulea and Stefan Antohe
Faculty of Physics, Bucharest University, P.O.Box MG - 11,
Bucharest - Magurele, 76900 Romania

1. Introduction

The primary photophysical and photochemical principles underlying natural photosynthesis are amenable to be studied by various model systems. The design and study of photosynthesis mimics can lead also to the development of new bioengineering strategies in solar energy conversion technology. The photosynthetic reaction centers are using the light energy to photogeneration of charges with very high quantum yield and to quickly separate the negative and positive charges by a series of fast and efficient electron transfer steps. The RC can be considered as a photovoltaic device of molecular dimension [1] that uses light energy to generate charge-separated states. Several photovoltaic devices fabricated with chlorophyll are reported in literature[2-4] and even a semiconductor model of plant photosynthesis has been proposed [5]. In this paper we deal with another type of chlorophyll *a* based photovoltaic cell, which contains a p-n heterojunction between chlorophyll *a* (Chla), as p-type semiconductor and 5,10,15,20 - tetra(4-pyrydil)21H,23H - porphyne (TPyP) as n-type semiconductor. The results of the study of electrical and photovoltaic properties of ITO/Chla/TPyP/Al structures are presented and discussed. The cell behaviour under illumination, suggests that the singlet excitons are the precursors of the charge carriers and that the space-region is at Chla layer/ organic layer interface.

2. Procedure

Chla was prepared from fresh spinach leaves as described in [4]. TPyP (Aldrich - Chemie Company) was used without other purification. The investigated cells were prepared by the following procedure. The ITO coated glass, providing the transparent conducting substrate, was chemicaly cleaned to remove any residual grease. On these substrates a thin layer of Chla of various thickness was electrodeposited from a suspension/solution of Chla in n-pentan. The film thickness was controlled by varying the molar concentration of Chla solution and/or by varying the electrodeposition time. The film thickness was estimated from the absorbance spectra and from interferometric microscopy measurements. On top of Chla layer was deposited a TPyP layer, about 100 nm thick, by conventional vacuum evaporation. The organic layer was vacuum sublimated at 10^{-3} Pa from a quartz container heated to 280°C and covered with a glass wool plug. Finally, a lightly transparent Al layer was evaporated on the top of TPyP

P. Mathis (ed.), Photosynthesis: from Light to Biosphere, Vol. I, 447–450.

layer. The active area, defined by the Al electrode area, was 0.3 cm^2. Schottky-type cells ITO/Chla/Al and ITO/TPyP/Al were also prepared. Films of a common component in these cells, i.e Chla, TPyP, Al, were prepared in such a way as to have the same thickness and morphology. The current - voltage characteristics under dc bias were measured using a stabilised power supply in connection with a helipot potentiometer, giving a resolution of 20 mV. Device current and voltage output measurements were made with a Philips microameter and a Keithley 614 electrometer, respectively. All measurements were performed in atmospheric air and at room temperature. For the illumination of the cells, an 1000W halogen lamp was used. A monochromator (in the spectral range 300 - 800 nm) was used to obtain the spectral responses of the cells. The optical absorption spectra of the component layers were obtained on a double beam UV - VIS spectrophotometer Lambda2S Perkin Elmer.

3. Results and Discussion

3.1 Dark current density - voltage (J-U) characteristics

Several ITO/Chla/TPyP/Al cells have been studied. The thickness of TPyP layer was about 100 nm and the thickness of Chla layer was in the range 150-220 nm. The forward bias direction corresponded to a positive voltage on the ITO electrode with respect to Al electrode. All the (J-U) characteristics are nonlinear and asymetric, presenting a small rectification ratio (10-20), at the voltage of 3V. Since Chla (considered as a p-type semiconductor) and TPyP (considered as a n-type semiconductor) make ohmic contacts with ITO and Al electrodes respectively, we infer that the rectification shown in the bipolar J-U characteristics is due to the energy barrier at the Chla/TPyP interface. This barrier is also responsible for the photovoltaic properties of the cells and the obtained open circuit photovoltage of about 500 mV is consistent with the difference in work function of Chla and TPyP, forming the barrier.

The current transport mechanism in the forward - biased direction through the ITO/Chla/TPyP/Al cell was investigated by plotting lnJ versus U in the range (0-3V). The applied voltage range was limited to ± 3V to prevent the breakdown of the cells. A typical (J-U) characteristic is presented in a logarithmic plot - Figure 1. Two distinct regions can be well delimited: at low voltages, the slope of the logJ versus logU plot is approximately equal to unity, while at higher voltages (U>1V) the slope is approximately equal to three. For applied voltages greater than 1V, the slope of logJ versus logU plot is 2.89 (Figure 1) and this superquadratic power dependence suggests that J may be a space-charge-limited current in the presence of trap states. Because there is a double injection, another phenomenon like charge recombination becomes very important. In this case, the role of space charge is less important and sometimes can be neglected. Due to the recombination, the space charge is reduced

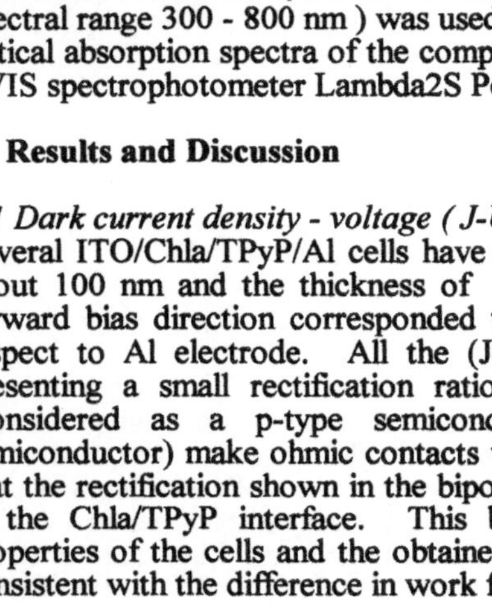

Figure 1. Current voltage characteristic of ITO/Chla(<200nm)/TPyP(100nm) cell in the dark, at rootemperature

and the current is limited by the recombination together with space charge appearence. In the majority of cases in which the structures and the working conditions are similar to ours, the J-U characteristic is given by

$$J = B \cdot U^3/d^3 \qquad (1)$$

where B is a constant and d is the distance between the injecting electrodes. The obtained value (2.89) for the slope of logJ versus logU plot (Figure 1) led us to the conclusion that at high forward voltages (U>1V) the double injection of carriers is dominating and the J-U characteristic is fitted by equation (1).

3.2 Photovoltaic properties

In order to obtain more information concerning the charge generation in the p-n heterojunction cells, action spectra have been obtained by plotting short-circuit currents against the wavelength of excitation. For all action spectra, the photocurrents are devided by the number of incident photons for each wavelength and then normalized to unity. In Figure 2 the action spectra of the cell ITO/Chla (~220nm)/TPyP(100nm)/Al are presented, when illuminated through ITO electrode(a) and through Al electrode (b), respectively. When illumination is made through Al electrode, an agreement between the action spectrum and the absorption spectrum of Chla is observed and an inverse type spectrum is obtained when illuminating through ITO electrode. This behaviour suggests that the singlet excitons are the precursors of the of the charge carriers and that the space-charge region is at the Chla/TPyP interface.

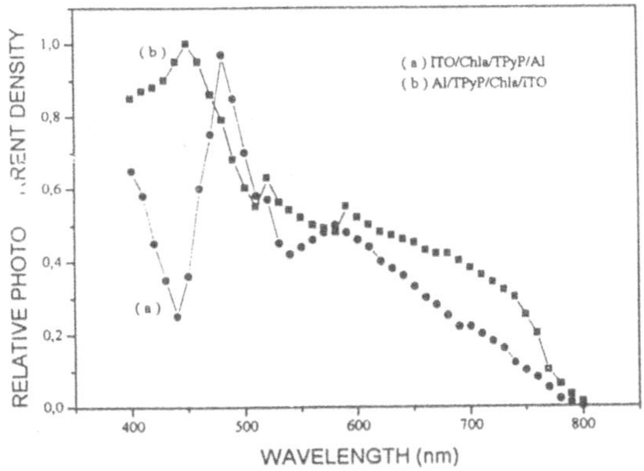

Figure 2. Action spectra of p-n heterojunction cell ITO/Chla(~220nm)/TPyP(100nm)/Al, illuminated through ITO(a) and through Al(b)

Figure 3.B shows the action spectrum of the cell ITO/Chla(<200nm)/TPyP(100nm)/Al (1), together with the action spectra of the Schottky type cells: ITO/Chla(<200nm)/Al (2) and ITO/TPyP(100nm)/Al(3),respectively. These spectra are obtained by illumination trough the ITO electrode. Figure 3.A shows the optical absorption spectrum of Chla/TPyP (a), Chla (b) and TPyP (c) layers, used in the fabrication of sandwich cells. In all p-n hetrojunction cells, the observed polarity of the photovoltage (ITO electrode always positive with respect to Al electrode) is strongly indicative for the following mode of operation: the photocurrent flows from n-type porphyrin to p-type Chla in the interior of the ITO/Chla/TPyP/Al cell. The electric field in the space-charge region at Chla/TPyP interface separates the electron-hole pairs formed in the boundary layer under illumination and drives the electrons into the n-type zone and the holes into the p-type zone. Only those excitons created at the Chla/TPyP interface, or having diffused into it contribute to the photocurrent. The inverse type relationship between the absorption spectrum of Chla layer and the action spectra of the ITO/Chla/TPyP/Al cells could be explained as follows. The excitons created close to ITO electrode by the photons strongly absorbed by Chla, have to travel a long distance to arrive at the Chla/TPyP interface and then the dissociation of excitons is small and consequently the photocurrent is feeble. The photons weakly absorbed in Chla layer, penetrate deeply in the layer, thus creating excitons in a region near to the Chla/TPyP interface where, because of the built-in electric field, the charge separation is efficient and a larger current is measured.

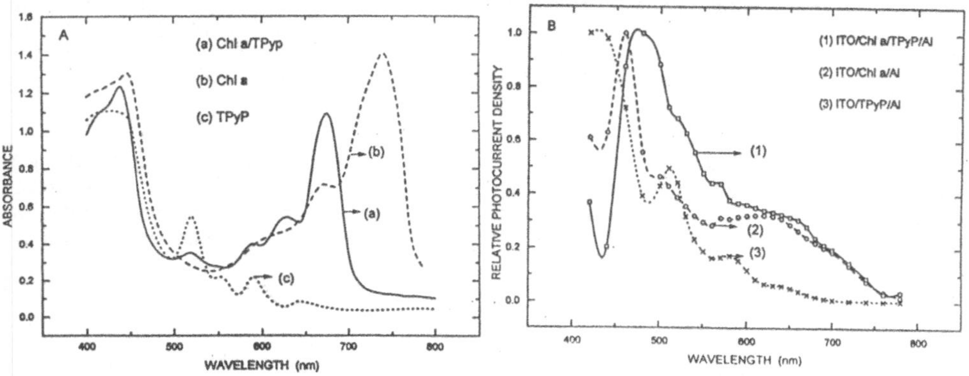

Figure 3. (A) Absorption spectrum of : Chla/TPyP films(a), Chla layer(b), TPyP layer (c) (B) Action spectra of: ITO/Chla/TPyP/Al(1), ITO/Chla/Al(2), ITO/TPyP/Al(3)

The photocurrent generation in the p-n heterojunction cell, as compared with that in the Schottky cells, is strongly enhanced at the wavelengths corresponding to the absorption peaks of the organic and biologic layer. The photocurrent measured at 460 nm is about three times greater in the case of the ITO/Chla/TPyP/Al cell than that of ITO/Chla/Al cell, while the photocurrent of p-n heterojunction cell measured at ~680 nm is about seven times greater than that of ITO/TPyP/Al cell. The appearence of a large photocurrent at wavelengths corresponding to the Soret region of Chla absorption spectrum can be explained by the photogeneration of charges in TPyP layer in addition to the charge generation in Chla layer. The larger photocurrent of p-n heterojunction cell at ~680 nm is due to the charges photogenerated mostly in Chla layer. The absence of the 740 nm peak, present in the red region of the absorption spectrum of Chla layer in the Schotky (one layer) cell- Figure 3A(b) and in the p-n heterojunction cell with thicker Chla layer (Figure 2), is explained by the loss of water in thin Chla layer (<200 nm), during the procedure of TPyP film evaporation on top of it. The evaporation was done in a vacuum of ~ 10^{-5} Torr, at room temperature. The chlorophyll a species present at the interface Chla/TPyP is probably the dimer and the two-layer heterojunction cell could be used as a possible model for charge photogeneration in the PSII reaction center.

References

1 Gust, D. and Moore, Th.A. (1993) in The photosynthetic reaction center, vol.II (Deisenhofer, J. and Norris, J.R., eds.) pp.419, Academic Press

2 Tang, C.W. and Albrecht, A.C.(1975) J.Chem.Phys. **62**, 2139

3 Ségui, J., Hotchandani,S., Baddou, D. and Leblanc, R.M. (1991) J. of Phys. Chem. **95**(22), 8807

4 Tugulea, L. and Antohe, S.(1992) in Research in Photosynthesis, vol.II, pp.845-849 (N.Murata ed.), Kluwer Academic Publishers, Dordrecht, The Netherlands

5 Ti Tien, H. and Chen, J.W. (1992) Solar Energy, **48** (3), 199-204

DIRECT AND INDIRECT EFFECTS OF 2.2'-BIPYRIDYL, A PHOTODYNAMIC HERBICIDE, ON PHOTOSYNTHETIC APPARATUS

A. Mostowska, M. Siedlecka & E. Parys
Dept. of Plant Physiology, Warsaw University, Krakowskie Przedmiescie 26/28, 00-927 Warsaw, Poland.

1. Introduction

Photodynamic herbicides represent a group of compounds which induce accumulation of chlorophyll precursors in the darkness and cause photooxidative damages of plants during subsequent illumination [1, 2]. In previous study on photodynamic herbicides [3, 4, 5] we revealed that δ-aminolevulinic acid or the complexing agents: 1.10-phenathroline, 2.2'-bipyridyl (2.2'-B) and 8-hydroxyquinoline inhibit the development of the photosynthetic apparatus in etiolated pea and cress plants after exposure to light. In the present work we would like to answer the following questions: what is the effect of 2.2'-B on the development of photosynthetic apparatus during pea ontogenesis, what are the sites of damage of photosynthetic apparatus in green pea seedlings, what is the primary, direct target of 2.2'-B action and what are indirect effects ?

2. Procedure

2.1 8-day old, etiolated pea plants *Pisum sativum* L. var. Illowiecki were treated with 1 - 5 mM doses of 2.2'-B and subsequently exposed to white light with intensity of 5 W/m^2 for 6 or 12 hours. Detailed procedure was as described in [4].

2.2 8-day old green seedlings were treated with 5 - 30 mM 2.2'-B and subsequently exposed to white light with intensity of 100 W/m^2 for 9 hours. Detailed procedure was as described in [5].

Material for electron microscopy was taken from plants after each treatments and prepared as described in [4]. Chlorophyll (Chl) and carotenoid (Car) analysis were determined according to [6, 7]. Photosynthesis (Pn) and the respiration were determined as described in [8]. The measurements of fast fluorescence were obtained according to [9].

P. Mathis (ed.), Photosynthesis: from Light to Biosphere, Vol. I, 451–454.

3. Results and Discussion

3.1 2.2'-bipyridyl (2.2'-B) in low doses (1 - 5 mM) inhibits grana formation (*Fig. 1*), Chl and Car accumulation and causes strong dilation of endoplasmic reticulum (ER) cisternae (*Fig. 1*) in etiolated and subsequently illuminated pea seedlings. The reason of reduced greening process are related to chelating properties of 2.2'-B, standard photooxidative mechanism, or/and LHCP accumulation within ER cisternae[3, 4].

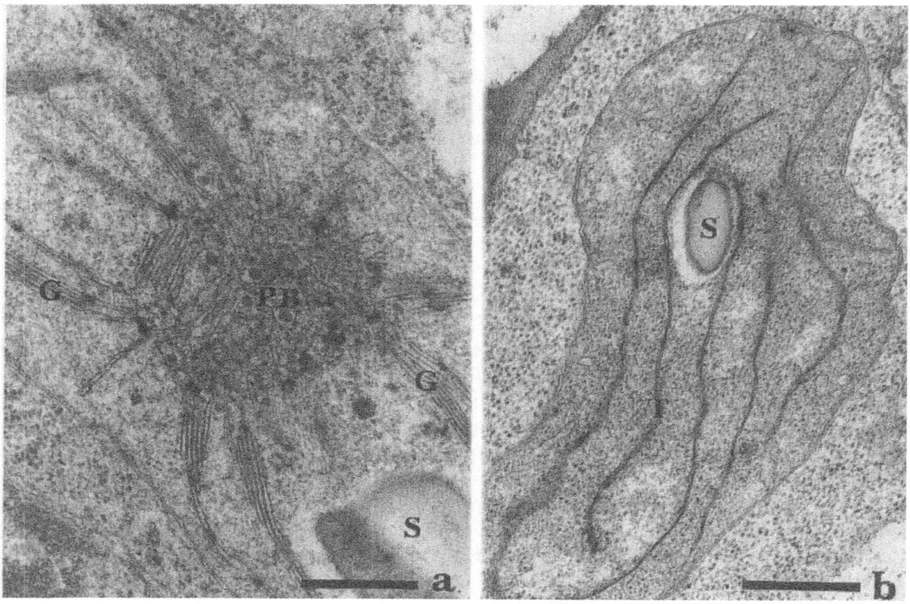

Figure 1 Plastids in mesophyll cells of pea seedling, control (*Fig. 1a*) or treated with 5 mM of 2.2'-B (*Fig. 1b*) grown in darkness and illuminated for 12 h. G granum, PB prolamellar body, S starch grain, T thylakoid. Bars represent 0.5 μm.

3.2 Application of higher (5 - 30 mM) 2.2'-B concentrations on green pea plants revealed that these plants were resistant to 5 mM of 2.2'-B with respect to morphology, pigment level and structure of mesophyll cells, but 10 and 30 mM 2.2'-B with illumination caused lost of turgor and photooxidative damages seen at the ultrastructural (*Fig. 2*) and functional level (*Figs. 3, 4*). 2.2'-B caused two groups of reactions: rapid - directly after exposure to light, affected probably by processes which took place during dark period, monitored by the decrease of Chl fluorescence and of photosynthetic rate (*Fig. 4*) and slow - consisting of destructive changes in mesophyll cell ultrastructure (dilation of thylakoid membranes and ER cisternae, disruption of the chloroplast envelope, *Fig. 2*) and a decrease of Chl (*Fig. 3*) and Car contents. Slow reactions were induced by light and required longer illumination time. We consider function of photosynthetic apparatus, probably of photosystem II (PS II) as a primary target of 2.2'-B action. Other dysfunctions are the result of the PS II damage, however they are also induced by light and based on mechanism of photosensitizing of porphyrins.

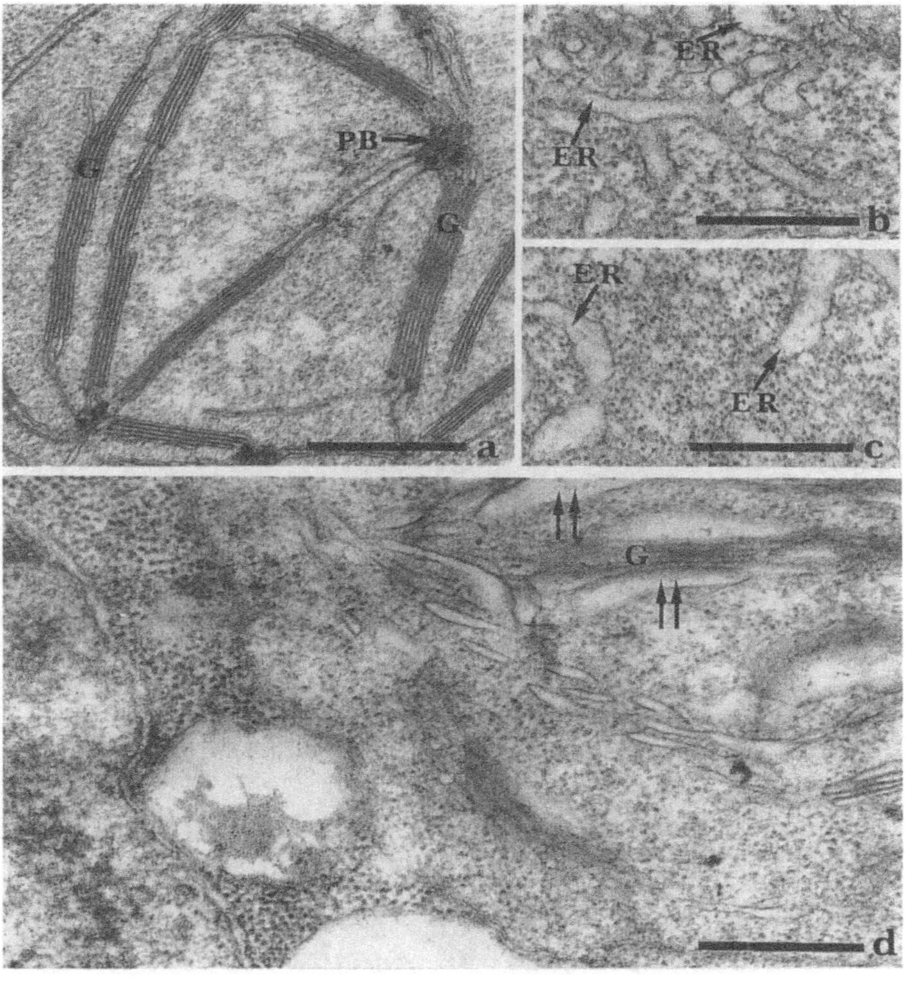

Figure 2 Portions of mesophyll cells of control pea seedlings fixed after a 12- h dark period followed by 9 h of illumination (*Fig. 2a*), or treated with 10 mM 2.2'-B and illuminated for 4 h (*Fig. 2b*), or treated with 30 mM 2.2'-B and illuminated for 6 h (*Figs. 2c,d*). ER endoplasmic reticulum, G granum, PB prolamellar body, double arrow indicates swollen thylakoid. Bars represent 0.5 μm.

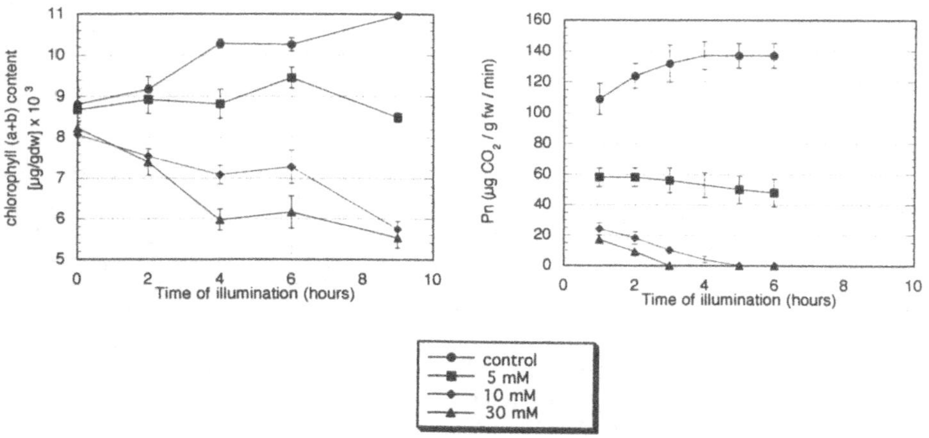

Figure 3 *Figure 4*

Figure 3 The effect of 2.2'-B (5 - 30 mM) and subsequent illumination on chlorophyll content in pea seedlings (mean + SE)

Figure 4 The effect of 2.2-B (5 - 30 mM) and subsequent illumination on net photosynthesis rate (Pn) in pea seedlings (mean + SE)

References:

1. Rebeiz, C.A., Montazer-Zouhoor, A., Hopen, H.J., Wu, S.M. (1984) Enzyme Microbiol. Technol. 6, 390-401
2. Mostowska, A. (1992) Photosynthetica 26, 19-31
3. Kittsteiner, U., Mostowska, A., Rüdiger, W. (1991) Physiol. Plant. 81, 139-147
4. Mostowska, A., Kittsteiner, U., Rüdiger, W. (1991) Protoplasma 161, 23-30
5. Mostowska, A., Fiett, D. and Parys, E. (1991) Env. Exp. Bot. 31, 385-395
6. Lichtenthaler, H. K. and Wellburn, A. R. (1983) Biochem. Soc. Trans. 603, 591-592
7. Davies, B. H. (1976) in Chemistry and biochemistry of plant pigments. Vol. 2. (Goodwin T. W., ed.) pp. 38-165, Academic Press, London, New York, San Francisco .
8. Parys, E. (1990) Plant Physiol. Biochem. 28, 711-717
9. Oquist, S. and Wass, R. (1988) Physiol. Plant. 73, 211-217

NOVEL PHENOLIC INHIBITORS OF ELECTRON TRANSFER IN PHOTOSYSTEM II
Montpellier, France, 20-25 August 1995

[1]Klimov V.V., [1]Zharmukhamedov S.K., [1]Allakhverdiev S.I., [2]Kolobanova L.P., [2]Baskakov Yu.A. [1]Institute of Soil Science and Photosynthesis, Russian Academy of Sciences, Pushchino, Moscow Region, 142292; [2]Institute of Chemical Protection of Plants, Moscow.

INTRODUCTION
It has been shown earlier [1,2] that a known phenolic herbicide, dinoseb, is capable (along with blocking the electron transfer between plastoquinones Q_A and Q_B [3]) of redox interaction with the components of photosystem 2 (PS-2) reaction center (RC). This results in a faster decay of the pair [$P680^+$ $Pheo^-$] evidently due to acceleration of both oxidation of reduced pheophytin ($Pheo^-$) and reduction of oxidized chlorophyll P680 ($P680^+$), the primary electron acceptor and donor of PS-2. The new cyclic electron transfer involving dinoseb can compete with the functional reactions of $Pheo^-$ oxidation and $P680^+$ reduction and contribute to the general inhibition of PS-2 only at relatively high concentrations of this inhibitor [1,2]. Recently, a new group of highly-efficient PS-2 inhibitors, derivatives of perfluoroisopropyldinitrobenzol, has been found [5,7] and here we describe in detail the inhibtory effects of one of them, 4-[methoxy-bis-(three-fluorinemethyl)-methyl]-2,6-dinitrophenyl hydrazone methylketon named as K-15. The data obtained show that redox interaction with components of PS-2 reaction center leading to formation of short cyclic electron flow around PS-2 is the basis for the inhibitory effect of these compounds.

Procedure
Pea chloroplasts and subchloroplast O_2-evolving PS-2 membrane fragments (DT-20) were isola-ted as described earlier [1]. Photoinduced changes (ΔF) of chlorophyll fluorescence yield (F) related to photoreduction of PS-2 electron acceptors, Q_A and Pheo, as well as absorbance changes (ΔA) related to photoreduction of Pheo and photooxidation of P_{680} were measured in a 1-cm cuvette with phosphoroscopic set-up [1]. The measurements were carried out in the medium containing: Tris-HCl buffer (pH 7.8), 35 mM NaCl and 2 mM $MgCl_2$. Complete (>97%) removal of Mn from DT-20 fragments was performed using 1 M Tris-HCl buffer (pH 8.0), and 0.5 M $MgCl_2$ [6]. The rate of photoinduced oxygen evolution was measured with a Clark-electrode. K-15 was synthesized as decribed earlier [4].

Results and Discussion
K-15 at concentration of 0.6 μM, 2 μM and 7 μM inhibits photoinduced oxygen evolution in pea chloroplasts by ~30 %, ~60 % and 90 %, respectively (Fig.1). Similar effects are observed for suppression of ΔF related to photoreduction of Q_A. On the other hand, K-15 did not inhibit photoreduction of $NADP^+$ in chloroplasts in the presence of reduced 2,6-dichlorophenolindo-phenol (DCPIP) as electron donor, when only PS-1 is involved in the photoreaction. From the known PS-2 inhibitors, dinoseb has the chemical structure most similar to that of K-15 type compounds [5]. However, the inhibitory effeciency of K-15 considerably (by a factor of 80-100) exceeds that of dinoseb (Fig. 1). Furthermore, contary to dinoseb, K-15 suppresses both electron transfer and photoinduced ΔF in the same region of concentrations (Fig. 1). Hence, in the case of K-15 type inhibitors, contrary to dinoseb, the inhibition of both electron transfer and ΔF is evidently based on the same mechanism.

Fig.2 shows the results of the comparision of the inhibitory action of K-15, diuron and dinoseb on photoinduced ΔF of PS-2 in DT-20. At concentration of 18 nM (curve 3) K-15 decreases the level of maximal F (F_{max} equal to the sum F_o + ΔF) by 15% as a result of decrease of ΔF. In the presence of 80 nM and 360 nM K-15, the ΔF is diminished by 57 %

P. Mathis (ed.), Photosynthesis: from Light to Biosphere, Vol. I, 455–458.

and 86 %, respectively (curves 4, 5). Contrary to diuron, K-15 practically does not change the F_o-level as well as the rate of dark relaxation of ΔF (reflecting reoxidation of Q_A^-). Besides, its inhibitory effect on ΔF retaines at prolonged (more than 5 min) illumination by actinic light even when K-15 is used at concentration of 0.05-0.1 μM. Repeated (2-3 times) washing of PS-2 preparations poi-soned with K-15 using centrifugation did not lead to removal of the inhibitor from PS-2 (judging from preservation of the inhibitory effects). Effeciency of inhibition of photoinduced

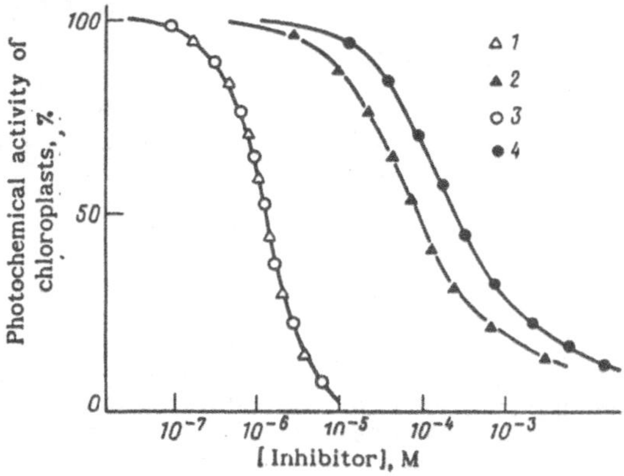

Fig.1. Dependence of photochemical activity of pea chloroplasts measured as O_2 evolution in the presence of ferredoxin (20 mM), ferredoxin-NADP$^+$-reductase (20 mM) and 5 mg/ml NADP$^+$ (1,3) and as photoinduced ΔF of PS-2 related to photoreduction of Q_A (2,4) upon concentration of K-15 (1,2) or dinoseb (3,4). Chlorophyll concentration is 100 μg/ml, 20° C.

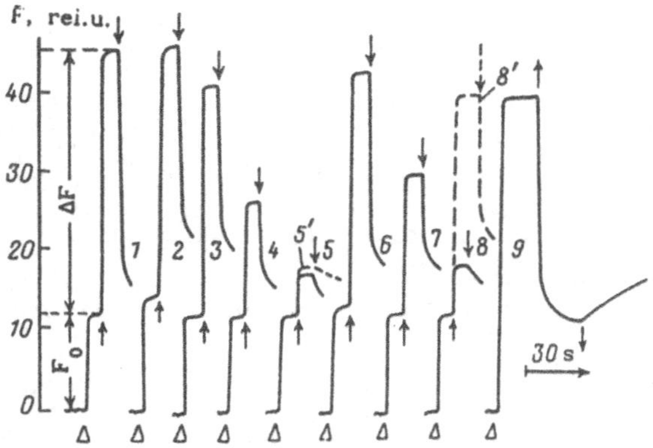

Fig.2. Kinetics of photoinduced ΔF related to photoreduction of Q_A in PS-2 under aerobic conditions in untreated DT-20 membrane fragments (1-7, 9, solid lines) without additions (1) and after the addition of: 0.5 μM diuron (2); K-15 at concentration 20 nM (3), 80 nM (4) and 360 nM (5) in the absence of other additions and after the addition of: 0.8 mg/ml dithionite to "5" (9); dinoseb at concentration of 1 μM (6) and 5 μM (7). Curves 8, DT-20 particles after complete removal of Mn without (solid line) and in the presence (dashed line) of 20 μM $MnCl_2$. Curves 5', dashed line, is same as 5 but after addition of 20 μM $MnCl_2$. Chlorophyll concentration is 10 μg/ml, 20° C.

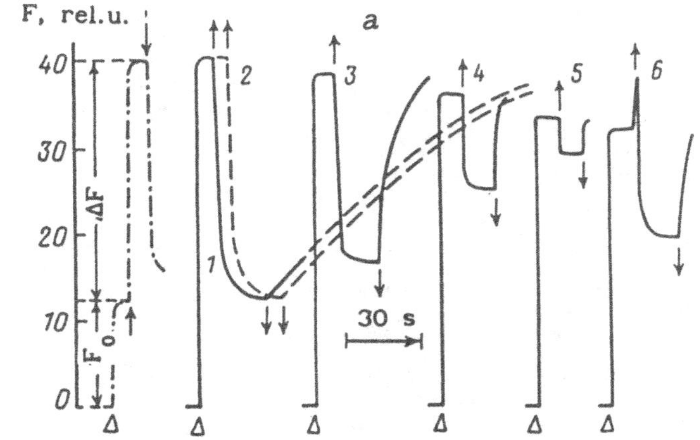

Fig.3. Kinetics of negative photoinduced ΔF related to reversible photoreduction of Pheo in PS-2 membrane fragments (DT-20) under anaerobic conditions (after addition of 10 mM glucose, 50 un/ml glucoseoxidase, 1000 un/ml catalase, 5 mM sodium ascorbate and 1 μM carbonyl cyanide-m-chlorophenylhydrazone) in the absence of other additions (1) and after addition of: 50 μM of diuron (2), K-15 at concentration of 38 nM (3), 65 nM (4) and 360 nM (5) or 10 μM of DCPIP (6). Kinetics of ΔF related to photoreduction of Q_A in the same sample under aerobic conditions is shown by dash-dotted line, left. Chlorophyll concentration is 10 μg/ml, 20° C

ΔF by K-15 is not practically changed upon addition of artificial PS-2 electron donors: - 0.1-20 mM Mn^{2+} (curve 5), 2 mM sodium ascorbate, 1 mM diphenylcarbazide, 1 mM NH_2OH (not shown). Addition of dithionite (0.8 mg/ml) completely eliminates the ΔF inhibition induced by K-15. F reaches level F_{max} and actinic light induces the decrease of F (curve 9) (related to photoreduction of $Pheo^-$ [7]). Dithionite causes dark reduction of K-15 as follows from disappearence of characteristic absorption bands The value of redox potential for reduction of K-15 and its derivatives is equal to -0.4 : -0.5 V [8].

Photoreduction of Pheo in PS-2 can be also observed in the absence of dithionite, under anaerobic conditions [7]. In this case, F reaches the F_{max}-level due to photoreduction of Q_A by the measuring light and illumination by actinic light causes a decrease of F (Fig.3) related to reversible Pheo photoreduction. The addition of K15 at concentration of 38 nM leads to considerable (by a factor of 15-25) acceleration of the dark relaxation of -ΔF reflecting dark oxidation of $Pheo^-$ (Fig.3). At higher concentrations of K-15 (65 nM and 360 nM) a considerable decrease of the value of photoinduced -ΔF (in addition to the decrease of the F_{max}) is observed. Similar effects were observed for ΔA related to photoreduction of Pheo.

Photooxidation of the primary electron donor of PS-2, P_{680} (measured in DT-20 fragments lacking Mn by ΔA at 678 nm in the presence of potassium ferricyanide and silicium molibdate) was inhibited by K-15 at concentration 22 nM and 80 nM by 17 % and 57 %, respectively. Similar suppression of ΔA at 678 nm was observed upon the addition of Mn^{2+} (0.1-3 mM) reflecting reactivation of electron donation to RC of PS-2 [6,7].

Analysis of the data obtained shows that for all PS-2 activities measured in this work a 50% inactivation is observed when the ratio K-15/Rc is close to 1:1.

Inhibition of photoinduced oxygen evolution and $NADP^+$ reduction in chloroplasts up-on the addition of K-15 reveals their capability of interrupting the electron transfer from water to $NADP^+$. The site of this inhibitory action is PS-2 since the inhibitors do not suppress photoreduction of $NADP^+$ from $DCPIPH_2$ in PS-1. This conclusion is confirmed by data on inhibition of photoinduced ΔF of PS-2 in both chloroplasts and PS-2 preparations. Comparison of inhibition of ΔF in PS-2 caused by K-15, diuron and dinoseb (Fig.2) reveals that, contrary to diuron or low concentrations of dinoseb [1], K-15 does not induce the increase of F_o-level indicating the absence of the block in electron transport on the acceptor side of PS-2. On the other hand, the inhibitory action of K-15 is not due to blocking electron flow on the donor side of PS-2 since it suppresses photoinduced ΔA related to photooxidation of P_{680}. In addition, photoinduced ΔF inhibited by K-15 are not reactivated by subsequent addition of Mn^{2+} or other electron donors. The decrease of ΔF and the lowering of F_{max} by K-15 can not be also explained by acceleration of Q_A^--reoxidation. Indeed, K-15 does not accelerate the dark relaxation of ΔF. Besides, the lowering of F_{max} by K-15 under anaerobic conditions is not accompanied by the appearance of positive ΔF related to photoreduction of Q_A. Pronounced acceleration of the dark relaxation of ΔF related to photoreduction of Pheo, and as consequence of that, the decrease of -ΔF (Fig.3), show that in the presence of K-15 the life-time of the RC state with the reduced Pheo is considerably shortened due to fast reoxidation of $Pheo^-$. Then both the decrease of the value of positive ΔF under aerobic conditions and lowering the F_{max}-level under anaerobic conditions can be interpreted as a result of acceleration of the decay of the pair [P_{680}^+ $Pheo^-$] induced by K-15 due to $Pheo^-$ oxidation. Acceleration of the decay of the [P_{680}^+ $Pheo^-$] by K-15 is supported by suppression of the long-lived components of chlorophyll F caused by this compound [2]. The K-15 inhibitory action can not be explained only by taking electron from $Pheo^-$, since we do not observe exhaustion of the inhibitory action of K-15 due to its photoaccumula-tion in a reduced state. On the other hand, acceleration of the Pheo photoreduction after addition of K-15 to PS-2 fragments lacking Mn in the presence of dithionite [7] directly indicates capability of the reduced form of K-15 to donate electron to RC of PS-2. This idea is also supported by suppression of ΔA related to photooxidation of P_{680} by K-15 and its analogues [7]. Besides, these compounds inhibit the photoinduced EPR signal II related to photooxidation of the secondary electron donor,. Z [9]. The data are consistent with the idea that K-15 acts as a mediator for the back electron transfer between $Pheo^-$ and P_{680}^+ (or Z^+) leading to formation of a short circuit of electron transfer [7]. The high binding constant of K-15 to PSII membrane fragments [7], the lack of competition between K-15 and diuron for the binding site [10], inhibiton of PS-2 photoreactions in isolated D1/D2/cyt b_{559} complex lacking Q_A and Q_B [11] as well as redox interaction of K-15 with $Pheo^-$ led to the suggestion that K-15 occupies the Q_A-binding site and serves as a substitute of Q_A (in contrast to diuron or atrazine which replace Q_B) [7, 10].

References

1 Klimov, V.V., Allakhverdiev, S.I., and Zharmukhamedov, S.K. (1989) Plant Physiol., (Russian) 36, 770-777

2 Allakhverdiev, S.I., Zharmukhamedov, S.K., Klimov, V.V., Vasil'ev, S.S., Korvatovsky, B.N. and Pashchenko V.Z. (1989) Biological Membranes (Russian) 6, 1147-1153

3 Trebst, A. and Draber W. (1979) in Advances in Pesticide Science, pp.223-234, Oxford, Pergamon Press

4 Konstantinova, N.V., Kolobanova, L.P., Trofimova, G.I., Livshic, B.R. and Baskakov, Yu.A. (1980) in Chemical means of plant protection (Russian) (Promenkov V.K., ed.) VNIIChSZR and NIITEChIM Press, Moscow.

5 Klimov, V. V., Zharmukhamedov, S. K., Allakhverdiev, S. I., Kolobanova, L. P. and Baska-kov, Yu. A. (1993) Biological Membranes (Russian) 10, 565-570

6 Klimov, V.V., Allakhverdiev, S.I., Shuvalov, V.A. and Krasnivsky, A.A.(1982) FEBS Lett., 148, 307-312

7 Klimov, V.V., Zharmukhamedov, S.K., Allakhverdiev, S.I., Kolobanova, L.P. and Baskakov Y.A. (1992) Biological Membranes (Russian) 9, 565-575

8 Kisselev, B.A., Suponeva, E.P., Zharmukhamedov, S.K. and Klimov, V.V. (1993) Biological Membranes (Russian) 10, 571-580

9 Zharmukhamedov, S.K., Allakhverdiev, S.I., Klimov, V.V. and Proskuryakov, I.I. Biochemistry (Russian) (submitted)

10 Zharmukhamedov, S.K., Klimov, V.V. and Allakhverdiev S.I. (1995) Biochemistry, (Russian) 60, 962-969

11. Klimov V.V., Zharmukhamedov S.K., De Las Rivas J. and Barber J. (1995) Photosynth. Res., (in press)

A new mutation in the *puf* L gene responsible for the terbutryn resistance phenotype in *Rubrivivax gelatinosus.*

S. Ouchane, M. Picaud & C. Astier.
CNRS, Centre de Génétique Moléculaire, Photosynthèse Bactérienne
bât 24, Avenue de la Terrasse 91198 Gif sur Yvette CEDEX FRANCE.

1. Introduction

The primary photochemical reactions of bacterial photosynthesis take place within a membrane-bound pigment-protein complex known as the photochemical reaction center (RC). RCs have been isolated from many species of purple bacteria. The best characterized are those from *Rhodopseudomonas (Rps.) viridis* and *Rb. sphaeroides* which have been crystallized and the structures of which are known at atomic resolution [1, 2, 3]. The RC of purple bacteria is composed of subunits called L, M, and H and in some species, includes a tetrahemic cytochrome c protruding into the periplasmic space. The pigments are burried in LM core polypeptides, allowing transmembrane, unidirectional electron transfer from the dimer of bacteriochlorophylls (P) to the primary quinone acceptor, the quinone (Q_A). The secondary acceptor (Q_B) is bound in a pocket called the Q_B-site localized in the L polypeptide. After accepting two electrons and two protons it dissociates from the RC and enters the quinone pool. Subsequent steps of proton translocation using the bc1 complex create a pH gradient, which drives ATP synthesis. Different classes of herbicides have been known for many years to inhibit photosynthesis at the level of the RC by blocking the electron transfer from Q_A^- to Q_B. These herbicides, which also inhibit electron transfer in photosystem II (PSII) in higher plants, green algae and cyanobacteria, act by competing with Q_B for its binding site. The genes coding for the subunits of RCs are organized in operons. The nucleotide sequence of the *puf* operon has been determined for five species belonging to the α subclass of purple bacteria and one species *Rx. gelatinosus* belonging to the β subclass [4]. In this work, we describe the phenotypic characterization and the identification of the mutation of a mutant of *Rx. gelatinosus*, strain 1 [5] resistant to terbutryn.

2. Materials and Methods

Rx. gelatinosus strain 1 [5] was grown anaerobically under photoheterotrophic conditions. *Echerichia coli* XL1B was used as host for cloning in plasmid. A mutant of *Rx. gelatinosus* was isolated by its ability to grow photosynthetically in the presence of 200 µM terbutryn under 600 lux. Chromatophores from *Rx. gelatinosus* were prepared by differential centrifugation after disruption of cells with a French press. Flash excitation spectroscopy and inhibition of electron transfer were measured as described [6]. Molecular methods were performed according to Sambrook et al [7]. Sequencing of both strands were performed using the dideoxy chain termination method of Sanger et al. [7] with sequenase vers. 2.0 kit, Amersham. Sequences of products of at least 3 independent PCR reactions were determined.

P. Mathis (ed.), Photosynthesis: from Light to Biosphere, Vol. I, 459–462.

3. Results

3-1 Determination of herbicide resistance levels

Sensitivity to terbutryn of isolated photosynthetic membranes and cross resistance to other herbicides were analysed. Electron transfer was measured by the reduction of the electron donor P^+ after one saturating flash, followed by measurement of the variation of absorption at 785nm. Figure 1 shows the inhibition curves in chromatophores prepared from wild type bacteria and from the terbutryn resistant mutant in the presence of various concentrations of terbutryn. I_{50} values which undicated the concentration needed to block half of RCs are presented in table 1.

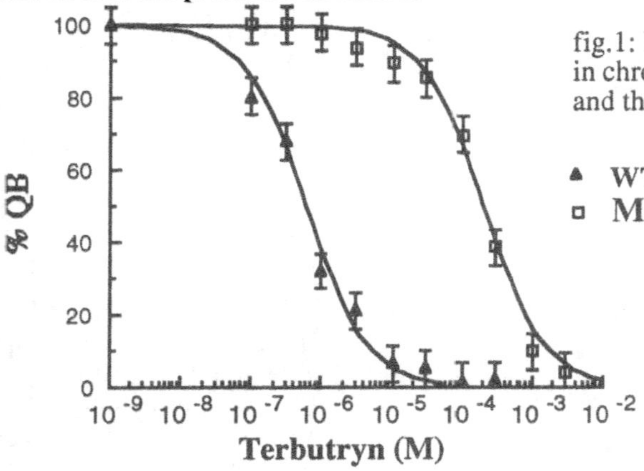

fig.1: Inhibition curves for terbutryn in chromatophores from the wild type and the mutant.

	I50 wild type	I50 mutant	R/S
Terbutryn	6. 10-7	2. 10-4	300
Atrazine	4. 10-5	2.5. 10-4	6
O-phenanthroline	3. 10-6	9. 10-6	3

Table 1: I50 values (in M) from Rx. gelatinosus strain 1 of the wild type and the mutant. R/S is the ratio of the I50 of the mutant to that of the wild type.

For o-phenanthroline, Rx. gelatinosus wild type is much more sensitive than other purple bacteria such as Rb. capsulatus and Rb. sphaeroides , I50=2.10-4 M [6, 7].

3-2. Molecular analysis

3-2-1. Sequence of the *pufL* gene of *Rx. gelatinosus*, strain 1:

The *puf* operon sequence from *Rx. gelatinosus* strain IL144 was published by Nagashima et al. [4]. We used oligonucleotides according to this sequence to clone and sequence the *puf* operon of strain 1. From this analysis the organisation of the *puf* operon from *Rx. gelatinosus* strain 1 was determined, and found to be the same as in the IL144 strain. The *pufL* nucleotide sequence of strain 1 has been submitted to GenBank with accession number U30310. There are 20 nucleotide differences between the two strains. Only one difference gives rise to an amino acid substitution at position L279 (C-terminus): AAC (Asn) in strain 1 to AAG (Lys) in strain IL144.

3-2-2. Mapping of the mutation in the *pufL* gene of the mutant:

As most of the mutations responsible for herbicide resistance phenotypes were mapped to "Q_B pocket" of the polypeptide L, we investigated this region of *pufL* from the mutant. Sequencing of PCR DNA fragment revealed a point mutation in this region of the *pufL* gene. This mutation results in the substitution of Glycine L192 by an Aspartic acid. Furthermore, this mutation created a new *Bst*YI restriction site. Figure 2 presents the results of a Southern blot of genomic DNA from wild type and mutant digested by *Bst*YI and hybridized with a radiolabelled probe overlapping the position at 192.

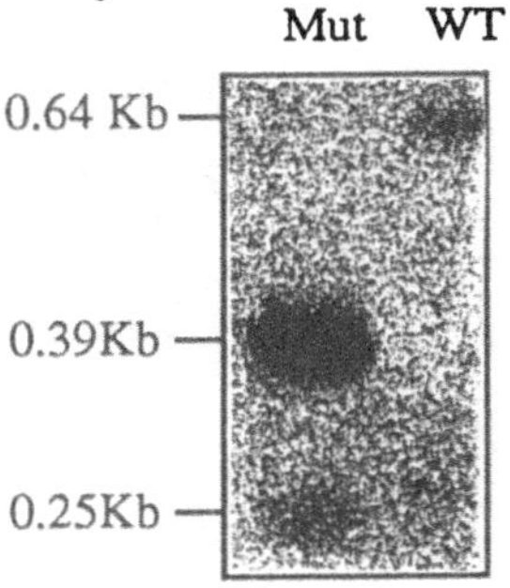

fig.2: Southern blot analysis of genomic DNA from the wild type and the mutant digested with *Bsty*I restriction endonuclease.

In the mutant DNA, the fragment of 0.64 kb detected in the wild type DNA is cut in two fragments of 0.39 kb and 0.25 kb. This result confirms in the mutant genomic DNA the presence of the mutation found in the PCR fragment.

4. Discussion

The new mutation observed in the terbutryn resistant mutant L-G192A, leads to a 300 fold resistance to this herbicide on the acceptor side of RC of *Rx. gelatinosus* and induced slight cross resistances to other herbicides compared to wild type.

The quinone site is formed by residues of the L subunit, located in the transmembrane helices D and E, the connecting helix DE and the loop between DE and E. The substitution in the mutant is located on the D helix of the L subunit. All mutations conferring herbicide resistances described until now in purple bacteria are located in helix DE, between helices DE and E and in helix E [9,10,11,12]. This mutation is the first one located in this helix. This region of L subunit is highly conserved in many purple bacteria and also in the corresponding part of D1 of PSII in organisms performing oxygenic photosynthesis [4] (fig. 4).

					190							
Rs.rubrum	A	L	A	L	H	G	S	L	I	L	S	A
Rb.sphaeroides	A	L	A	L	H	G	A	L	V	L	S	A
Rb.capsulatus	A	L	A	M	H	G	A	L	V	L	S	A
Ro.denitrificans	A	L	A	L	H	G	G	L	I	L	S	A
Rps.viridis	A	L	G	L	H	G	G	L	I	L	S	V
Rx.gelatinosus	A	M	S	M	H	G	<u>G</u>	L	I	L	S	A
D1 of PS II	F	S	A	M	H	G	S	L	V	T	S	S
					215							

fig.4: Amino acid sequence comparison of a part of D helix of polypeptide L from six purple bacteria and of D1 from oxygenic photosynthetis performing organism [4].

Its high conservation suggests that it plays an important role in the electron transfer process. It has been poorly studied, probably because no mutation mapping in this region was found in purple bacteria. Only two mutations mapping there were described in cyanobacteria at positions 211 and 219 of D1 [14, 15] and one in green algae at position 219 [16]. Why does this L-G192D mutation confer terbutryn resistance? Glycine is the smallest residue and is uncharged. It is able to confer flexibility to a polypeptide chain. Aspartic acid is bulky and charged. Glycine 192 might leave enough space to accommodate binding of terbutryn and aspartic acid might hinder it. However, it is surprising that this L192 amino acid plays a role in terbutryn binding if we consider the results obtained with other purple bacteria. In the crystallographic RC structure of *Rps. viridis* [9], residue L192 is located far from the terbutryn binding site, too far to be in direct interaction with the herbicide molecule or QB atoms. Using the refined coordinates of *Rb. sphaeroides* Y RC [3], we found a minimal distance of 10 Å between the QB cycle and Gly L192 Cα Therefore, the binding site for terbutryn (and perhaps also for QB) should be different in *Rx. gelatinosus* compared to these 2 other bacteria. Besides, comparisons of amino acid sequences of the helix D region around position 192 from several purple bacteria (fig. 4) have shown that the most divergent sequence is that of *Rx. gelatinosus* , with 3 important differences at positions 187, 188 and 189. Met, Ser and Met are less hydrophobic and bulkier than Leu, Gly and Leu. In spite of the high conservation between purple bacteria RCs, including *Rx. gelatinosus,* which belongs to the β subclass of purple bacteria , our results on the mutant and the different amino acid sequences of helix D suggest that the structure of this part of the Q_B-site of *Rx. gelatinosus* RC could be different compared to the RCs already known. The role of helix D will be further investigated using other new mutations.

High sensitivity of *Rx. gelatinosus* to *o*-phenanthroline was found ($I_{50}=3\cdot10^{-6}$ M). *Rps. viridis* is the only wild type strain described in litterature which presents a high affinity for this herbicide ($I_{50}=3\cdot10^{-5}$ M) [17], I_{50} values are much higher for *Rb. capsulatus* and *Rb. sphaeroides* ($I_{50}=2\cdot10^{-4}$ M)[6, 7]. An enhanced sensitivity to *o*-phenanthroline has also been observed in a *Rb. capsulatus* mutant bearing the L-T226A mutation ($I_{50}= 5\cdot10^{-6}$ M) [6]. Sensitivity to *o*-phenanthroline of *Rx. gelatinosus* confirmed that the amino acid in position 226 of L subunit is important for the affinity of RC to this herbicide. All strains resistant to *o*-phenanthroline have a threonine in position 226 and sensitive strains a serine or an alanine.

References

[1] Deisenhofer, J., Epp, O., Sinning, I. and Michel, H. (1995) J. Mol. Biol. 246, 429-457.

[2] Ermler, U., Fritzsch, G., Buchanan, S.K. and Michel , H. (1994) Structure 2, 925-936.

[3] Arnoux, B., Gaucher, J.F., Ducruix, A. and Reiss-Husson, F. (1995) Acta Cryst. D51. 368-379.

[4] Nagashima, K.V.P., Ohyama, M.K., Shimada, S.K.(1994) J.Biol.Chem. 269,1-8.

[5] Uffen, R.L. (1976) Proc. Natl. Acad. Sci. USA 73, 3298-3302.

[6] Baciou, L., Bylina, E.J. and Sebban, P. (1993) Biophysical J. 65, 652-660.

[7] Sambrook, J., Fritsch, E.F. and Maniatis, T. Molecular Cloning, A Laboratory Manual Second Edition. Cold Spring Harbor Laboratory Press 1989

[8] Shopes, R.J. and Wraight, C.A. (1985) Biochim. Biophys. Acta 806, 348-356.

[9] Sinning, I. (1992) Trends Biocem. Sci. 17, 150-154.

[10] Sinning, I. and Michel, H. (1986) Z. Naturforsch. 42c, 751-754.

[11] Bylina, E.J., Jovine, R.V.M. and Youvan, D.C. (1989) Bio/Technology 7, 69-74.

[12] Paddock, M.L., Williams, J.C., Rongey, S. H., Abresch, E.C., Feher, G. and Okamura, M. Y. (1988) Photosynth. Res. 17: 75-96.

[13] Svensson, B., Vass, I. and Styring, S. (1991) Z. Naturforsch. 46c, 765-776.

[14] Ajlani, G., Kirilovsky, D., Picaud, M. and Astier, C. (1989) Plant Mol. Biol. 13, 469-479.

[15] Gingrich, J., Buzby, J.S., Stirewalt, V. L. and Bryant, D.A. (1988) Photosynth. Res. 16, 83-99.

[16] Johanningmeier, U., Bodner, U. and Wildner, G.F. (1987) FEBS Lett. 211, 221-224.

[17] Sinning, I., Michel, H., Mathis, P. and Rutherford, A.W. (1989) Biochemistry 28, 5544-5553.

LONG RANGE ELECTROSTATIC EFFECTS IN BACTERIAL REACTION CENTERS

Jean-Dominique DELCROIX[1], Marianne SCHIFFER[2], Deborah K. HANSON[2] and Pierre SEBBAN[1]
[1] Centre de Génétique Moléculaire, CNRS, Gif/Yvette, France
[2] Argonne National Laboratory, Argonne, Il., USA

We have previously reported the characterization of two suppressor strains isolated from the photosynthetic incompetent reaction center double mutant L212Glu/L213Asp -> Ala/Ala from *Rhodobacter (Rb). capsulatus* [1]. These strains have recovered their abilities to grow photosynthetically by mutating an additional residue distant from the active site, the secondary quinone electron acceptor, Q_B. The two additional mutations are M231Arg -> Leu (≈ 15 Å from Q_B) and M43Asn -> Asp (≈ 9 Å from Q_B) in the class 3 and class 4 strains, respectively. We have shown that these "electrostatic" mutations i) restore to some extent the initial free energy gap between the Q_A^- (the first quinone electron acceptor) and Q_B^- states observed in the wild type, increased by more than 60 meV in the double mutant, ii) accelerate the second electron transfer rate about 15 times compared to the double mutant. This is still more than 500 times smaller than in the wild type.

We present here preliminary data concerning two phenotypic revertant strains, isolated from the photosynthetic incompetent quadruple mutant: L212Glu/L213Asp -> Ala/Ala, M246Ala/M247Ala -> Glu/Asp (M246Ala and M247Ala being the equivalent to L212Glu and L213Asp in the Q_A proteic environment). These two strains, RQrev4B and RQrev5B, respectively, have a wild type Q_A pocket but still carry the L212Glu/L213Asp -> Ala/Ala mutations in the Q_B pocket and an additional mutation: L217Arg -> Cys (≈ 10 Å from Q_B) and M5Asn -> Asp (≈ 16 Å from Q_B), respectively.

Figure 1 shows the pH dependence of k_{BP}, the rate constant of $P^+Q_B^-$ charge recombination (P is the primary electron donor of the reaction center) in the reaction centers from the RQrev4B strain, the double mutant and the wild type. The observed decreased rate of about 10 times in the double mutant suggest an increase of about 60 meV of the differential stabilization energy between Q_B^- and Q_A^- compared to the wild type. It is of interest to note that the L217Arg -> Cys mutation restores the initial free energy barrier between these states, at pH 8. The electrostatic effect of this mutation on the local potential detected by Q_B is therefore stronger than that due to the M231Arg -> Leu or M43Asn -> Asp mutations in the class 3 and class 4 suppressor strains, respectively, i.e. about 30 meV [1,2]. The low and high pH dependencies of k_{BP} in the wild type are still observed in the RQrev4B reaction centers, although with a somewhat smaller amplitude than in the wild type. The presence of these variations of k_{BP} in a strain which lacks L212Glu and L213Asp exclude the possibility that these variations exclusively reflect the ionization states of these residues in the wild type.

P. Mathis (ed.), Photosynthesis: from Light to Biosphere, Vol. I, 463–465.

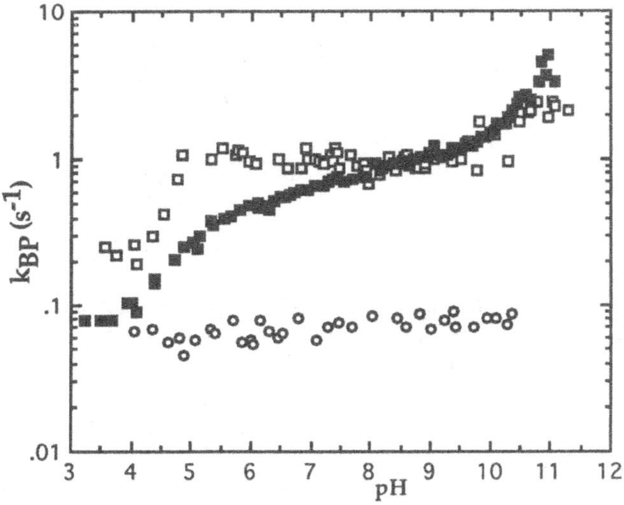

Figure 1: pH dependence of the rate constant of $P^+Q_B^-$ charge recombination in reaction centers from the wild type *Rb. capsulatus* (■), the L212Glu/L213Asp -> Ala/Ala double mutant(o) and the RQrev4B strain (L212Glu/L213Asp -> Ala/Ala + L217Arg -> Cys) (□)

Figure 2 displays the pH dependence of k_{AB}^2, the rate of second electron transfer between Q_A^- and Q_B^-, in the wild type, the double mutant, the class 3 and 4 suppressor strains and the RQrev4B and RQrev5B phenotypic revertant strains. This process is intimately related to the transfer of the first proton to Q_B and therefore its value reflects the efficiency with which protein drives protons from the cytoplasm to Q_B. In the double mutant, at pH 7, this rate is diminished about 5000 times compared to the wild type (≈ 5500 s^{-1}). The photosynthetic capabilities are recovered in the four phenotypic revertant strains by accelerating this process by a factor of only about 6 (RQrev5B) to 25 (class 3 suppressor strain). This process does not seem to be limiting in the function of the photosynthetic apparatus since these strains grow under photosynthetic conditions with generation times similar to that of the wild type. However, the k_{AB}^2 value measured in the RQrev5B strain seem to be a threshold value under which the system cannot function.

All additional mutations restore a more negative environment near Q_B, suggesting an important electrostatic role for the protein. As we have previously mentioned this role would be to keep the pKs of the groups involved in proton conduction to Q_B, high enough for an efficient proton transfer and delivery to Q_B.

Lastly, the similar values of k_{AB}^2 measured in the four revertant strains, with mutations taking place in different part of the protein suggest a common process. This could be to open up channels were water molecules could diffuse, accelerated by the negative potential present near Q_B. It is also of possibility that if protons follow in the wild type a channel of structured water molecules from the cytoplasm to Q_B, as recently suggested [3,4], the electrostatic mutations somewhat replace the electric role of L213Asp in the wild type.

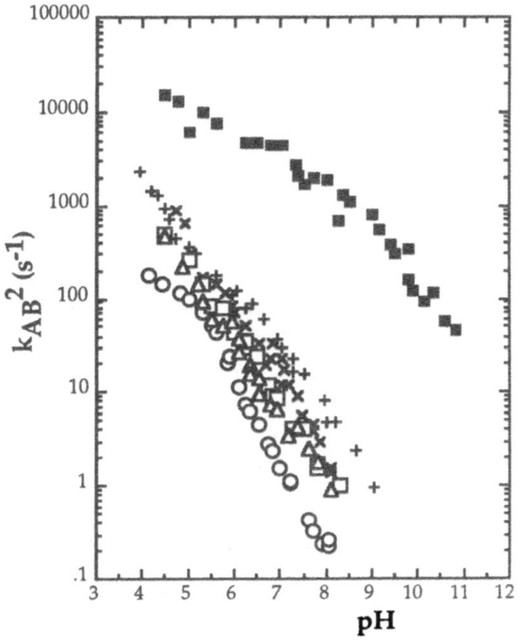

Figure 2: Rate of the second electron transfer from
Q_A^- to Q_B^- in reaction centers from:
(■), the wild type *Rb. capsulatus*
(o), the L212Glu/L213Asp -> Ala/Ala double mutant
(□) the RQrev4B strain (L212Glu/L213Asp-> Ala/Ala + L217Arg -> Cys)
(Δ) the RQrev5B strain (L212Glu/L213Asp-> Ala/Ala + M5Asn -> Asp)
(x) the class 4 rev strain (L212Glu/L213Asp-> Ala/Ala + M43Asn -> Asp)
(+) the class 3 rev strain (L212Glu/L213Asp-> Ala/Ala + M231Arg -> Leu)

Acknowledgements:
This work was supported by North Atlantic Treaty Organization Grant CRG920725 and National Science Foundation - Centre National de la Recherche Scientific Grant CDP900350. P. M. was supported by the French Ministère de la Recherche et des Universités, by the European Community (PECO fellowship) and OTKA (1978-1991). D. K. H. and M. S. were supported by the United States Department of Energy, Office of Health and Environmental Research, under Contract No. W-31-ENG-38, and by Public Health Service Grant GM36598.

References:

1 Maróti P., Hanson D.K., Baciou L., Schiffer M., Sebban P. (1994) *Proc. Natl. Acad. Sci. USA* 91, 5617-5621.

2 Sebban P., Maróti P., Hanson D.K., Schiffer M. (1995) *Biochemistry* 26, 34, 8390-8397.

3 Ermler, U., Fritzsch, G., Buchanan, S.K. and Michel, H. (1994) *Sructure* 2, 10, 925-936.

4 Baciou, L. and Michel, H. (1995) *Biochemistry* 34, 25, 7967-7972.

ELECTROSTATIC INTERACTION BETWEEN L212GLU AND Q_A^- IN BACTERIAL REACTION CENTERS

Jaroslava MIKSOVSKA[1], Peter MARÒTI[2], Marianne SCHIFFER[3]
Deborah K. HANSON[3] and Pierre SEBBAN[1]
[1] Centre de Génétique Moléculaire, C.N.R.S., Gif/Yvette, France
[2] Depart. of Biophysics, University of Szeged, Szeged, Hungary
[3] Argonne National Laboratory, Argonne, Il., USA

The photosynthetic reaction center transforms light excitation energy into chemical free energy. Coupling of electron and proton transfer reactions occurs at the level of secondary quinone acceptor, Q_B. The electrostatic influence of the protein over quite long distances (≈ 15 Å) on these processes has been demonstrated [1]. In particular, the possible influence on the local potential near Q_B was suggested [1-4]. Long distance interactions between the Q_B protein pocket and that surrounding the first quinone acceptor, Q_A (situated at about 18 Å from Q_B), was also very recently suggested [5]. Consistently, it was reported that time dependent protein relaxation following Q_A^-formation on the first flash may involve the ionization state of L212Glu situated at 17 Å from Q_A [6,7].
To further investigate this possibility we have measured the pH dependence of the proton uptake stoichiometry by the Q_A^- state (H^+/Q_A^-) in wild type reaction centers from *Rhodobacter (Rb.) capsulatus* and from a revertant strain isolated from the photosynthetic incompetent double mutant L212Glu/L213Asp -> Ala/Ala, which has retained L212Glu -> Ala as single mutation.
Figure 1 shows the pH dependence of the H^+/Q_A^- curves in reaction centers from both strains. The line represents a fitting of the wild type data assuming the presence of four interacting groups with Q_A^-, and according to equation 1:

$$H^+/Q_A^- = \sum_i \left(\frac{1}{1+10^{(pH-pK_{iQ_A^-})}} - \frac{1}{1+10^{(pH-pK_{iQ_A})}} \right) \qquad (1)$$

where $pK_{iQ_A^-}$ and pK_{iQ_A} denote the pK of a ionizable group, i, when Q_A is reduced and oxidized, respectively.

P. Mathis (ed.), Photosynthesis: from Light to Biosphere, Vol. I, 467–470.

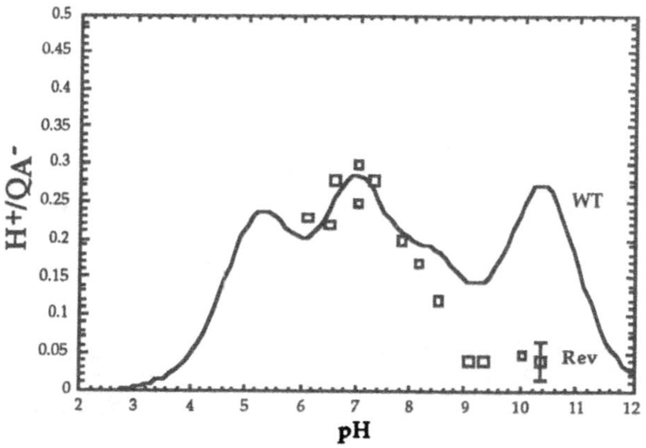

Figure 1: Stoichiometry of proton uptake by the Q_A^- state in reaction centers from the wild type *Rb. capsulatus* and from the L212Glu -> Ala revertant strain. The line is by fitting the wild type data with equation 1.

It can be deduced that, for the wild type, the pK_{QA} and pK_{QA^-} of the group situated in the high pH range are 10.1 ± 0.10 and 10.6 ± 0.25, respectively. If the data of the revertant strain are nearly superimposable to the wild type below pH 8, it is noteworthy that the revertant almost totally fail of taking up protons above pH 9. We have previously shown by analyzing the pH dependence of the Q_A^- <-> Q_B^- equilibrium constant curve that a group interacting with Q_B has a pK of about 10.1 in the dark state of the reaction center. In *Rb. sphaeroides*, this high pH group has previously been suggested to be L212Glu [8]. We show here that it is likely that L212Glu is involved in an interaction with Q_A^- at 17 Å apart.

To further analyze this problem, we have studied the pH dependence of the free energy stabilization of the $P^+Q_A^-$ state, where P is the primary electron donor of the reaction center. This was achieved in the double mutant strain and in the revertant strain in comparison to the wild type. In the reaction centers of the three strains, the native Q_A (ubiquinone$_{10}$) was replaced by a low potential quinone, the 1-amino-5-chloro-anthraquinone. It was previously shown that in such modified reaction centers charge recombination from the $P^+Q_A^-$ state proceeds via a thermal activated state, i. e., a relaxed state of P^+I^-, where I is the intermediate bacteriopheophytin molecule [9-11]. Under these conditions the free energy level of the $P^+Q_A^-$ state is very sensitive to the protonation events involving ionizable groups interacting with Q_A^-, P^+, and /or I^-.

Figure 2 shows the pH dependence of k_{AP}, the rate constant of $P^+Q_A^-$ charge recombination in the reaction centers from the wild type, the revertant strain and the double mutant. Clearly, destabilization of the $P^+Q_A^-$ state occurs in the wild type reaction centers above pH 8, whereas in the double mutant and in the revertant, k_{AP} remains relatively unchanged in the same pH range. This is consistent with the lack of progressive stabilization of the Q_A^- state in the wild type above pH 8.5 as suggested by the proton uptake data.

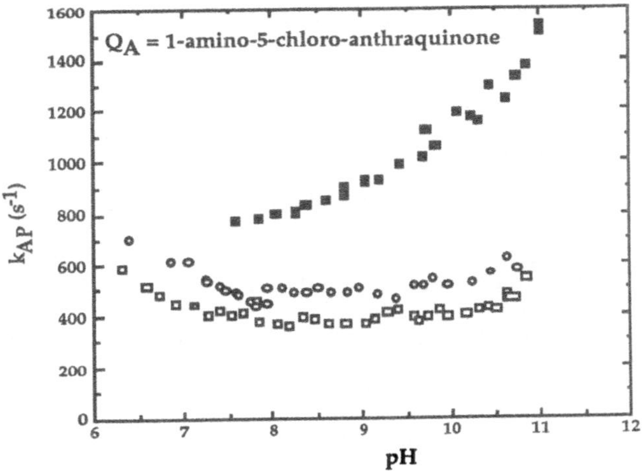

Figure 2: pH dependence of k_{AP}, the rate constant of charge recombination, in reaction centers from the wild type *Rb. capsulatus* (■), the L212Glu/L213Asp -> Ala/Ala double mutant (o) and the L212Glu -> Ala revertant strain(□)

Conclusion

We show here that in two strains in which L212Glu is replaced by an alanine, the reaction centers fail to take up proton upon Q_A^- formation. This is supported by H^+/Q_A^- proton uptake (unpublished data) of the double mutant and the class 4 suppressor strain (L212Glu/L213Asp -> Ala/Ala + M43Asn -> Asp) isolated from the double mutant [2]. The H^+/Q_A^- values measured for those strains are superimposable to those of the revertant in the whole pH range.

This suggest that L212Glu is involved in the Q_A^- proton uptake detected in the wild type above pH 8.5. Interestingly, it is then likely that the redox state of Q_A, influences the pKs of ionizables groups situated in the Q_B vicinity. This may be of importance in either the first electron transfer process to Q_B or in modulating the pKs of the groups involved in proton transfer to Q_B.

Acknowledgments:

This work was supported by North Atlantic Treaty Organization Grant CRG920725 and National Science Foundation - Centre National de la Recherche Scientific Grant CDP900350. P. M. was supported by the French Ministère de la Recherche et des Universités, by the European Community (PECO fellowship) and OTKA (1978-1991).D. K. H. and M. S. were supported by the United States Department of Energy, Office of Health and Environmental Research, under Contract No. W-31-ENG-38, and by Public Health Service Grant GM36598.

References:

1 Hanson, D.K., Tiede, D. M., Nance, S. L., Chang, C.H. and Schiffer, M. (1993) *Proc. Natl. Acad. Sci. USA* 90, 8929-8933.

2 Maróti P., Hanson D.K., Baciou L., Schiffer M., Sebban P. (1994) *Proc. Natl. Acad. Sci. USA* 91, 5617-5621.
3 Sebban P., Maróti P., Hanson D.K., Schiffer M. (1995) *Biochemistry* 26, 34, 8390-8397.
4 Sebban, P., Maròti, P. and Hanson, D.K. (1995) *Biochimie* 77.
5 Baciou, L. and Sebban, P. (1995) *Photochem. Photobiol.* 62, 2, 271-278.
6 Brzezinski, P., Okamura, M.Y. and Feher, G. (1992) in The Photosynthetic Bacterial Reaction Center II (Breton, J. and Vermeglio, A., eds) pp. 321-330, Plenum press, New York, USA.
7 Tiede, D. M. and Hanson, D. K. (1992) in The Photosynthetic Bacterial Reaction Center II (Breton, J. and Vermeglio, A., eds) pp. 341-350, Plenum press, New York, USA.
8 Paddock, M. L., Rongey, S. H., Feher, G., Okamura, M. Y. (1989) *Proc. Natl. Acad. Sci. USA* 86, 6602-6606.
9 Woodbury, N. W., Parson, W. W., Gunner, M., Prince, R. C. and P. L. Dutton (1986) *Biochim. Biophys. Acta* 851, 6-22.
10 Shopes, R. J. and C. A. Wraight (1987) *Biochim. Biophys. Acta* 893, 409-425.
11 Kleinfeld, D. M., Okamura, M. Y. and G. Feher (1985) *Biophys. J.* 48, 849-852.

SITE-SPECIFIC MUTAGENESIS AT HISTIDINE 118 OF THE PHOTOSYSTEM II D1 PROTEIN OF *CHLAMYDOMONAS REINHARDTII*

Ronald S. Hutchison & Richard T. Sayre,
Departments of Plant Biology and Biochemistry, The Ohio State University, 2021 Coffey Rd., Columbus, OH 43210 USA

1. Introduction

Symmetry related histidine residues are located at positions 118 in the B transmembrane spans of the Photosystem II D1 and D2 proteins. These residues have been proposed to coordinate chlorophyll monomers which may be analogous to the accessory chlorophylls of the bacterial photosynthetic reaction center (1). The predicted distance (30Å) from H118 to the Chl special pair is similar to the distance predicted from the accessory Chl to P680, based on energy transfer dynamics. Significantly, the symmetry related histidine residues located in the C-D alpha helix (L and M subunits), which bind the accessory Chl in the bacterial reaction center, are not conserved in PSII. H118 also has been proposed to coordinate Chl_Z, a chlorophyll monomer believed to be involved in cytochrome b_{559} donation to P680+ (2). We report here mutations that convert the histidine at position 118 (conserved among all D1 sequences determined to date) to leucine and arginine. A histidine to leucine mutation at position 117 (118 in *Chlamydomonas*) has previously been made in the D2 protein of *Synechocystis* PCC 6803, resulting in the inability of PSII to evolve oxygen (see 3). Likewise, we find that mutations at H118-D1 result in an inability to evolve oxygen and reduce DCPIP via a light dependent oxidation of PSII artificial donors. In addition, the H118-D1 mutants have a reduced accumulation of the D1 protein in dark- grown cells and an enhanced sensitivity of the D1 protein to turnover in the light (H118L only). Based on Chl fluorescence decay measurements, EPR analysis of the failure to accumulate the Tyr_D radical, and the inability to form the charge separated state giving rise to the A_T thermoluminescence band, it appears as though charge separation is impaired in the H118 mutants. These results suggest that the mutagenesis of H118-D1 disrupts the functional and structural integrity of the PSII reaction center; results that are consistent with its possible participation in coordination of an accessory chlorophyll.

P. Mathis (ed.), Photosynthesis: from Light to Biosphere, Vol. I, 471–474.

2. Methods

The plasmids (pH118LD1 and pH118RD1 which changes H118 to H118L and H118R, respectively) in the *psb*A gene and contains the *aad*A gene, encoding resistance to spectinomycin and streptomycin (4) were shot into *Chlamydomonas* strain CC744 containing a *psb*A deletion. The cells were resuspended in Tris-Acetate-Phosphate (TAP) media after transformation and plated on TAP plates containing 85 ug/ml spectinomycin. Resistant colonies were transferred to 3 ml of TAP containing 25 ug/ml streptomycin and grown for 1 week. 100 ul of the cell suspension was plated on TAP plates containing 200 ug/ml spectinomycin to begin the secondary screen. DNA was isolated from cells using the S&S Elu-Quik DNA purification kit. The isolated DNA was subjected to PCR analysis using the primers complimentary to the *psb*A gene. Using this PCR screening procedure, we identified mutant colonies containing the *psb*A gene by identifying the introduced diagnostic restriction site (Hinf I for H118R, Apl I for H118L). After further selection, Southern analysis of chloroplast DNA, isolated from the mutants and WT, was performed using the intron-less *psb*A gene as the probe. For biochemical analysis, all cells were grown heterotrophically in Tris-Acetate-Phosphate media. Oxygen evolution was carried out according to Roffey *et al.* (1994) on the Hansatech CB1 electrode (5,6). For DCPIP reduction measurements, assays were performed on NH_2OH extracted thylakoids with NH_2OH and DPC used as donors. Manganese determinations were carried out using PS II particles (1 mg Chl) digested with 1 N nitric acid. Elemental analysis was carried out at REAL labs (Wooster, USA). Fluorescence measurements were carried as described previously (6). EPR was performed in collaboration with Dr. Russ Hille on a Bruker ESP-300. The following instrument settings were used for detection of the Tyr_D signal: modulation amplitude, 5 G; microwave power, 5 mW. Dark-adapted PSII particles were first scanned at 150°K, re-warmed to room temperature, illuminated for one minute and scanned again at 150°K. Thermoluminescence studies were carried out according to Kramer et al. (1994).

3. Results and Discussion

	Oxygen Evolution umol O2/mg chl/hr		DPC/DCPIP assay umol DCPIP/mg chl/hr.	[Mn] Mn/250 Chl
	DMBQ	PBQ	DCPIP	
Wild-type	222±30	163±15	35±4	4.2
H1118L	0	0	0	<1.5
H118R	0	0	0	<1.5

Table 1. Biochemical properties of H118R&L

An intron-less *psb*A plasmid linked to the *aad*A gene (encodes spectinomycin and streptomycin resistance) developed by Minagawa and Crofts (1994) was used to introduce the H118R and H118L mutations (4). Southern blots of chloroplast DNA probed with the coding sequence of *psb*A indicated that the mutant strains were homoplasmic for the mutation.

Furthermore, the presence of the 9 kb EcoR I fragment indicated that the introduced *psbA* gene underwent homologous recombination with regions flanking the *psbA* deletion. In order to determine if the mutants assemble a stabile PSII complex, we examined the D1 protein levels by western blot analysis. Thylakoids and PSII particles were isolated from light- and dark- grown cells. The thylakoids of both mutants showed lower D1 levels compared to wild-type (WT) in both light- (10% and 47% of WT for H118L and H118R, respectively) and dark- (37% and 49% for H118L and H118R, respectively) grown samples. PSII particles have wild-type D1 levels, with the exception of the light grown H118L, which had only 4% of the WT levels. Neither H118R or H118L were able to evolve oxygen, assemble an intact tetra-Mn water oxidizing complex (H118L and H118R had <1.5 Mn/250 Chl; WT had 4.2 Mn/250 Chl), or reduce DPIP using artificial donors to PSII.

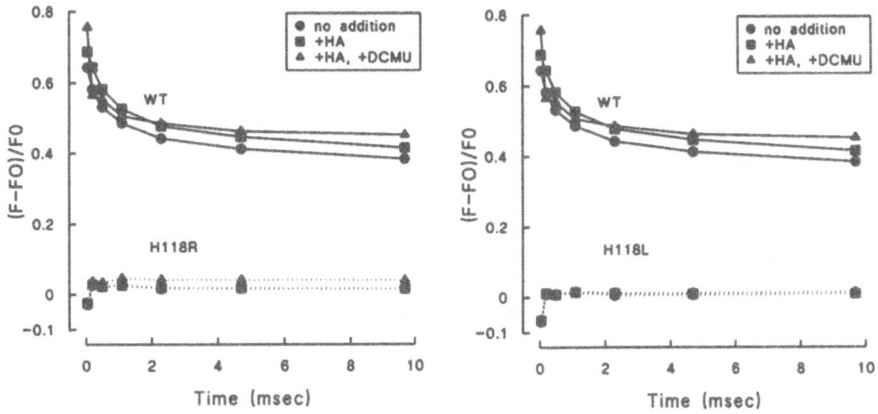

Figure 1. Fluorescence decay kinetics of dark-grown thylakoids

In order to characterize the charge separated state, flash-induced chlorophyll fluorescence decay analysis was performed on dark-grown mutant and wild-type thylakoids (Fig. 1). The variable fluorescence (Fv) rise for H118R was less than 10% of the wild-type while the mutant H118L had no Fv, indicating a loss of functional charge separation. The fluorescence decay of H118R in the absence of DCMU was more rapid than in the presence of DCMU, indicating some electron transfer between Q_A^- to Q_B, although much less than wild-type.

In order to determine the basis for the inability to evolve oxygen, we assayed for the ability to form Tyr_D^o by EPR. Unlike wild-type thylakoids grown in the dark and extracted with NH_2OH, we observed no recognizable Tyr_D^o signal following illumination. Instead, a broad, symmetrical radical was generated in the mutant with low quantum yield relative to the formation of Tyr_D^o in wild-type. This radical is one third the size of the Tyr_D^o signal from WT PSII particles. The inability to form Tyr_D^o may result, however, from photodamage by the high light intensities required to photosaturate the sample. Another indication of the ability to form a charge separated state is the formation of the A_T thermoluminescence band in non oxygen evolving thylakoids. Unlike dark- grown WT thylakoids, neither H118R or H118L dark- grown thylakoids were able to accumulate an A_T thermoluminescence band. These mutants also showed no absorbance change at 515 nm (Δ515), an electrochromic shift indicative of an ability to form a charge separated state. Recently, Durrant et al. (1995) has suggested that P680 may consist of loosely coupled chlorophylls and pheophytins including the Chl monomer (7). Furthermore, the orientation of the P680 triplet state is not consistent with its location on the special pair (assuming it is structurally similar to the bacterial reaction center). The loss of the ability to form a charge separated state in the H118 mutants is not inconsistent with its involvement in the coordination of a Chl which participates in the formation of $P680^+$.

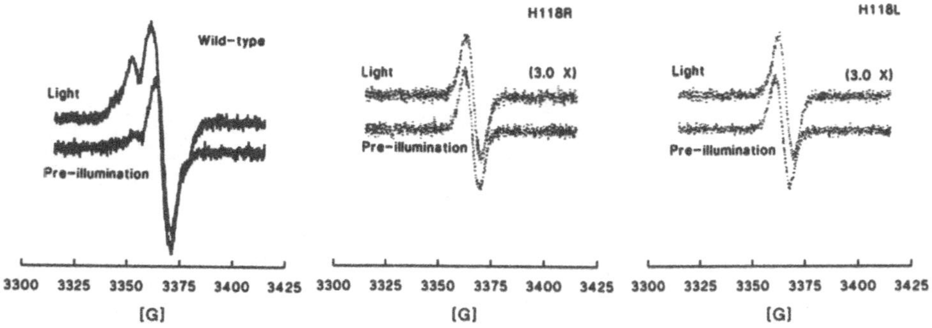

Figure 2. Try_D^o EPR signals from PSII particles obtained from dark-grown cells

References:

1 Schelvis, J.P.M., van Noort, P.I., Aartsma, T.J., and van Gorkum H.J. (1994) Biochem. Biophys. Acta **1184** 242-250

2 Koulougliotis, D., Innes, J.B., and Brudvig G.W. (1994) Biochemisty **33** 11814-11822

3 Pakrasi, H.B. and Vermaas, W.F.J. (1992) in The Photosystems: Structure, Function, and Molecular Biology. 231-258 Topics in Photosynthesis. vol. 11

4 Minagawa, J. and Crofts A. (1994) Photo. res. **42(2)** 121-131

5 Roffey, R.A., Golbeck, J.H., Hille, C.R., and Sayre, R.T. (1991) Proc. Natl. Acad. Sci. **88** 9122-9126

6 Roffey, R.A., Kramer, D.A., Govindjee, and Sayre, R.T. (1994) Biochem. Biophys. Acta **1185** 257-270

7 Durrant, J.R., Klug, D.R., Kwa, S.L.S., van Grondelle, R., Porter, G., and Dekker, J.P. (1995) Proc. Natl. Acad. Sci. **92** 4798-4802

PHOSPHORYLATION OF PSII CORE IN RELATION TO D1 TURNOVER IN HIGHER PLANTS

B. Geiken, A. Cona, M.T. Giardi and A.K. Mattoo*
IBEV-CNR, Via Salaria km 29.3, 00016 Monterotondo Scalo, Italy
**Plant Mol.Biol. Lab, USDA/ARS, Beltsville, MD 20705, USA*

It is well known that reversible protein phosphorylation is used as one of the signals for metabolic regulation in plants. It appears to be fundamental in the transduction of extracellular signals into intracellular biochemical changes [1]. In most cases, both the phosphorylation and dephosphorylation reactions are active at the same time and the relative activities and kinetic constants determine the steady-state level.

The photosynthetic membrane contains several phosphoproteins. The most conspicuous among these are components of the light-harvesting chlorophyll a/b complex, CP43, D1, D2, PsbH and an unknown PSII protein at about 9 kDa. The phosphorylation of chloroplast proteins has been recently reviewed by [2] and [3,4].

The D1 protein is a central component of the PSII reaction center and also the target site of triazine and urea-type herbicides. This protein undergoes at least five post-translational modifications during its life cycle: C-terminal processing, removal of the initiating methionine residue, N-acetylation of the resulting N-terminal threonine residue, covalent palmitoylation, and O-phosphorylation of the N-terminal threonine of the mature protein. Processing of D1 occurs on stroma-exposed membranes while palmitoylation and phosphorylation occur in spatially distinct grana membranes. Light-dependent rapid turnover is a characteristic of D1. A scission in the protein between the amino-acid residues 248-247 yields an N-terminal breakdown product of about 23 kDa which has been suggested as the primary degradation product of D1 *in vivo* [5]. Phosphorylation of D1 results in the formation of a less mobile electrophoretic form [6,7,8]. The role of phosphorylation in D1 metabolism has not yet been deciphered. We have studied the implications of the phosphorylation process on D1 turnover under physiological and extreme environmental conditions. The present results, correlated with previous experiments reported in the literature, indicate that in higher plants phosphorylation has multiple effects on D1 turnover, acting in a regulatory and protective way at the levels of synthesis, assembly and degradation.

Results

Four heterogeneous PSII core populations isolated by isoelectrofocusing of spinach grana particles have now been recognized [9-13]. These four complexes, referred to as **a**, **b**, **c** and **d** in the increasing order of phosphorylation, show either a different level

P. Mathis (ed.), Photosynthesis: from Light to Biosphere, Vol. I, 475–478.

of phosphorylated PSII core polypeptides or a different content of the *psb*H gene product. The most phosphorylated PSII core population **d** has about 40 times more phosphorylated D1 protein than that in the core complex **a.** The complex **d** is unable to transfer electrons from DPC to DCPIP. These differently phosphorylated PSII core populations have different affinities for the binding of photosynthetic herbicides and different sensitivities to photoinhibitory conditions [9-11].

We studied the turnover of D1 as influenced by phosphorylation in pea leaves (*Pisum sativum*) analysing D1 turnover in the differently phosphorylated populations by pulse-chase experiments *in vivo* with ^{32}P-orthophosphate and ^{35}S-methionine. Under physiological conditions after a short-term pulse with ^{35}S-methionine in the light, the population containing non-phosphorylated D1 was the first to incorporate radioactivity, while the populations containing phosphorylated D1 incorporated radioactivity in a longer-term experiment. A complete interconversion among the heterogeneous phosphorylated populations was observed after three subsequent periods of light-dark-light cycles. The pulse-chase experiments and immunoblot analysis showed that the phosphorylated populations of grana membranes accumulated the 23 kDa breakdown fragment of D1. The 23 kDa product was found to be phosphorylated. When protein synthesis was blocked with chloramphenicol, the D1 protein of the least phosphorylated populations was observed to disappear even more rapidly than that in phosphorylated population, in accordance with [14] who reported that the non-phosphorylated D1 is less stable than the phosphorylated form. It is possible that under these conditions the phosphorylated fragments are more stable than the non-phosphorylated ones and are easily detected since they accumulate. It seems that the accumulation of the breakdown product of D1 is even more dramatic when the population containing non-phosphorylated D1 form was less than 10% (under strict light conditions). We have also recently observed that treatment of leaves with inhibitors of phosphorylation affected D1 turnover within the stacked regions of the membrane (Geiken et al., unpublished). However, enhanced PSII protein phosphorylation was correlated to accelerated D1 synthesis in drought-stressed leaves [13]. Degradation of the D1 protein was also enhanced under these conditions and formation of many phosphorylated fragments was observed.

Discussion

Phosphorylation of PSII polypeptides in concert with D1 turnover seems to offer two protective mechanisms for the structural and functional integrity of PSII [5,7,12]. In line with this idea, it has recently been suggested that D1 turnover leads to down-regulation of PSII activity under high light stress [14,15]. The recent experimental results indicate that the phosphorylation process as a whole may have a protective function against deleterious effects of strong irradiance. Phosphorylation of all PSII core proteins results in a destabilization of the core and in a change of the Q-B pocket activity [10] leading to a partial inhibition of the light-saturated rate of oxygen evolution, which may reflect a decreased affinity of plastoquinone for the Q_B site [17].

The role of phosphorylation in D1 turnover is intriguing. Apparently contradictory results exist in the literature, for example: phosphorylation reduces the speed of D1 degradation under strong illumination [14,18]; phosphate deficiency causes an inhibiton of D1 degradation always under strong illumination; an increased turnover of D1 was signaled under diverse stress conditions [19,20,21] correlated with an increase in phosphorylation. All these results are difficult to be reconciled with a single mechanism. However, it is important to note that strong illumination may have another effect than low light [13,viz. these proceedings). Under stress conditions an increased phosphorylation occurs perhaps as a mechanism of protection to avoid the PSII core depletion due to an increased susceptibility to light. A possible explanation, taking into account most of the results, was given by [21]. They suggest that, although phosphorylation may apparently make D1 more stable, it might act indirectly on D1 degradation at the level of a putative D1 protease system. If the D1 protease itself is under a phosphorylation-dephosphorylation control, then different degrees to which it is phosphorylated relative to the D1 protein may be pivotal in determining the stability or instability of D1.

Our recent results indicate that *in vivo* the D1 synthesis occurs through a non-phosphorylated D1 form, whereas its degradation proceeds through the phosphorylated form. The assembly of D1 into the stacked regions may involve phosphorylation. In conclusion, phosphorylation seems to have multiple effects on D1 metabolism in higher plants acting as a regulatory and protective signal.

Acknowledgements
This work was supported by RAISA special CNR grant. Dr Geiken was supported by the programme "Human Capital and Mobility" of the EC.

References

1 Bennett J (1977) *Phosphorylation of chloroplast membrane polpypeptides.* Nature **269**, 344-346
2 Bennett J (1991) *Protein phosphorylation in green plant chloroplasts.* Annu Rev Plant Physiol **42**, 281-234
3 Allen JF (1992) *Protein phosphorylation in regulation of photosynthesis.* Biochim Biophys Acta **1098**, 275-335
4 Allen JF (1993) *Redox control of gene expression and the function of chloroplast genomes-an hypothesis.* Photosynth Res **36**, 95-102
5 Mattoo AK, Marder JB and Edelman M (1989) *Dynamics of the photosystem II reaction center.* Cell **56**, 241-246
6 Callahan FE, Ghirardi ML, Sopory SK, Mehta AM, Edelman M and Mattoo AK (1990) *A novel metabolic form of the 32 KDa-D1 protein in the grana-localized reaction center of photosystem II.* J Biol Chem **265**, 15357-15360
7 Elich TD, Edelman M and Mattoo AK (1992) *Identification, characterzation and resolution of the in vivo phosphorylated form of the D1 photosystem II reaction center protein.* J Biol Chem **267**(5), 3523- 3529

8 Elich TD, Edelman M and Mattoo AK (1993) *Dephosphorylation of photosystem II core proteins is light-regulated in vivo.* EMBO J **12**, 4857-62
9 Giardi MT, Rigoni F, Barbato R (1992) *PSII core phosphorylation heterogeneity: differential herbicide binding and electron transfer regulation in PSII particles from Spinach.* Plant Physiol **100**, 1948-1954
10 Giardi MT (1993a) *Phosphorylation and disassembly of photosystem II core as an early stage of photoinhibition.* Planta **190**,107-113
11 Giardi MT, Komenda J and Masojidek J (1994) *Involvement of PSII phosphorylation in the sensitivity to strong light.* Physiol Plant **92**,181-187
12 Giardi MT, Cona A and Geiken B (1995) *Photosystem II core phosphorylation heterogeneity and the regulation of electron transfer in higher plants.* J Bioelectrochem Bioenerg, in press
13 Giardi MT, Cona A, Kucera T, Geiken B, Masojidek J and Mattoo AK (1995) *Long-term drought stress induces structural and functional reorganization of photosystem II.* Planta, in press
14 Rintamäki E, Kettunen R, Tyystjärvi E and Aro EM (1995) *Light-dependent phosphorylation of D1 reaction centre protein of photosystem II: hypothesis for the functional role in vivo.* Physiol Plant **93**, 191-195
15 Öquist G, Chow WS and Anderson M (1992) *Photoinhibition of photosynthesis represents a mechanism for the long-term regulation of photosystem II.* Planta **186,** 450-460
16 Critchley C and Russell AW (1994) *Photoinhibition of photosynthesis in vivo: the role of protein turnover in photosystem II.* Physiol Plant **92,** 188-196
17 Horton P and Lee P (1984) *Phosphorylation of chloroplast thylakoids decreases the maximum capacity of photosystem II electron transport.* Biochim Biophys Acta **767,** 563-567
18 Anderson JM and Aro EM (1994) *Grana stacking and protection of photosystem II in thylakoid membranes of higher plants under sustained high irradiance.* Photosynth Res **41,** 315-326
19 Godde D and Hefer M (1994) *Photoinhibition and light dependent turnover of D1 reaction-centre polypeptide of photosystem II are enhanced by mineral-stress conditions.* Planta **193,** 290-299
20 Dannehl H, Herbik A and Godde D (1995) *Stress-induced degradation of the photosynthetic apparatus is accompanied by changes in thylakoid protein turnover and phosphorylation.* Physiol Plant **93,** 179-186
21 Bracht E and Trebst A (1994) *Hypothesis on the control of D1 protein turnover by nuclear coded proteins in Chlamydomonas reinhardtii.* Z Naturforsch **49c**, 439-446

PSII CORE PHOSPHORYLATION HETEROGENEITY

T. Kucera, J.-M. Briantais*, M. Hodges* and M.T. Giardi**
* *I.B.P., Bat 630, Université Paris Sud, Orsay Cedex, France;*
** *IBEV-CNR, Via Salaria km 29.3, 00016 Monterotondo Scalo, Italy*

Many types of PSII heterogeneity have been observed at the level of the antenna and photochemistry, both at the donor and acceptor side [1,2].
The heterogeneity in the rate of reduction of the primary stable PSII acceptor (Q_A) in the presence of DCMU was attributed to the existence of two types of PSII centers. The fast sigmoidal phase was related to PSIIα and the subsequent slow phase to PSIIβ. It has been suggested that they correspond to two types of PSII reaction center differing, among other properties, in their antenna size and location within the thylakoid membrane. PSIIα was thought to reside in grana partitions, whereas PSIIβ was located in the stroma-exposed membrane regions [2,3]. However, this interpretation contrasts with some experimental data like the observation that the slow phase of fluorescence induction is eliminated by high concentrations of DCMU indicating the existence of PSII units that vary in their sensitivity to DCMU [4]; while it has been shown that PSII located in stroma-exposed regions and PSII present in grana regions are probably equally sensitive to DCMU. Moreover, fluorescence induction curves of DCMU-poisoned chloroplasts have been resolved by mathematical analysis into three and even four phases; a major, rapid sigmoidal phase followed by two or three slower exponential phases [1,5]. Another type of heterogeneity within PSII is associated with the reducing-side. Through several approaches, it has been demonstrated that the secondary stable electron acceptor, Q_B, is not reduced by some PSII centers, termed non-B-type centers [1].

The photosynthetic membrane contains several phosphoproteins comprising components of the light-harvesting chlorophyll a/b complex and several PSII proteins, among these CP43, D1, D2 and PsbH [6].

Results

Four heterogeneous PSII core populations have been isolated by isoelectrofocusing of spinach grana particles solubilized with non-ionic detergent [7-10]. The four complexes, referred to as **a**, **b**, **c** and **d** in the increasing extent of phosphorylation, showed either a different level of phosphorylation on PSII core polypeptides or a different content of the *psb*H gene product. The most phosphorylated population **d** was about 40 times more phosphorylated in D1 than the core complex **a** and was inactive in electron transfer. These PSII core populations have been shown to have different affinities for the binding of photosynthetic herbicides and different sensitivities to photoinhibitory conditions.

In the present work we analyze the distribution of PSII core phosphorylation hetero-

P. Mathis (ed.), Photosynthesis: from Light to Biosphere, Vol. I, 479–482.

geneity in different plant species. We report a modified PSII core phosphorylation heterogeneity in a chlorophyll deficient mutant of wheat that does not exhibit a part of functional PSII heterogeneity shown by Q_A measurement using 1) chlorophyll fluorescence induction in the presence of DCMU and 2) chlorophyll fluorescence increase upon flash illumination.

A yellow-green mutant of wheat, whose photosynthetic apparatus has previously been studied [11], has also been observed to lack a part of functional PSII heterogeneity since it gives rise to a chlorophyll fluorescence induction curve with DCMU-poisoned leaves. In this study we have thoroughly examined the mutant. Thylakoids and PSII enriched particles were isolated from the wheat WT and mutant with a Triton X-100: Chl ratio of 15. SDS-PAGE analysis of the membranes showed that the PSII polypeptide composition of the mutant is qualitatively the same as that observed in the WT, although the relative stoichiometry of the polypeptides was quite different. Using polyclonal antibodies raised against the PSII components, it was observed that all of the main PSII polypeptides were present in the mutant. Immunoblot analysis with polyclonal antibodies raised against a 64 kDa protein, attributed to the LHCII kinase, indicated that the wheat mutant was considerably depleted compared with the WT. Micrograph analyses showed that the mutant contains appressed and non-appressed regions. The mutant was depleted in thylakoid content per leaf area, but despite the partial depletion of PSI previously observed, it was still possible to separate PSII enriched grana particles and stroma lamellae using the digitonin method.

Phosphorylation of PSII proteins was observed *in vivo*, by incubating the cut petiole of the seedlings in the presence of ^{32}P-orthophosphate in the light. The radioactivity incorporated into the mutant was considerably different compared to the WT. For the WT the phosphorylation pattern was typical of that observed in higher plants: radioactivity was incorporated into CP43, D2, D1, LHCII, 22 kD, 9 kD, 8 kD and 5 kD proteins. The mutant showed an altered phophorylation pattern: there was almost no radioactivity incorporated into the CP43, D2 and 9 kD proteins while a high level of radioactivity was found in the D1, 22 kD, 8kD and 5 kD proteins. Despite the depletion of a 64 kD kinase, the mutant LHCII showed some degree of phosphorylation. The phosphorylation state of the mutant was also tested using monoclonal antibodies against phosphorylated-threonine, confirming the above mentioned radiolabelled profiles.

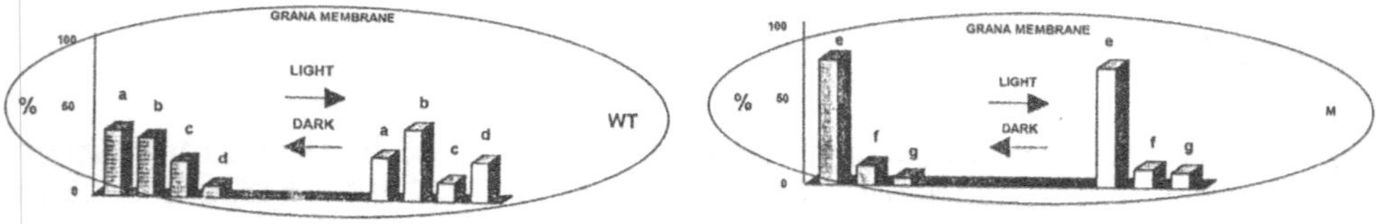

Fig.1 Relative light-dark distribution of the phosphorylated PSII core populations in the wheat mutant and its wild-type.

Applying the isoelectrofocusing method followed by sucrose gradient centrifugation [7-10] it was found that the mutant lacked the usual pattern of four phosphorylated PSII core populations (Fig.1). In the mutant, a preponderant PSII core population **e** (about 80%) and small quantities of core populations **f** and **g** were obtained, instead of two main populations **a** and **b** and small quantities of **c** and **d**. In particular, a PSII core population similar to population **b**, with sucrose gradient position at 0.57M, was absent. The least phosphorylated population **e** of the mutant, compared to the least phosphorylated population of the WT, shifted towards more negative pIs, as expected by the lack of phosphorylation, while the shift was not observed in the minor populations **f** and **g**. The room-temperature fluorescence characteristics of the yellow-green mutant and of the WT were compared (as in [12,13]. Since the thylakoids from wheat seedlings were easily damaged and gave very poor electron transfer and Fv/Fm signals, fluorescence measurements were carried out *in vivo*. After 1h dark adaptation, leaves of the mutant exhibited a significantly higher Fv/Fm ratio than the WT (Table I). This difference can be explained by a lower amount of PSI which preferentially affects Fo. As judged by the so-called O-I rise in fluorescence induced by low light intensities, both the WT and mutant plants contained similar amounts of non-B-type PSII reaction centres (Table I). However, the mutant plants differed from the WT in that the area above the fluorescence induction curve in the presence of DCMU could be fitted by a single component. The mutant did not show the slow β-phase and the fast α-phase was slower than in the WT, as expected due to the decreased antenna size of the mutant.

Fluorescence parameters	**WILD-TYPE**	**MUTANT**
Fv/Fm	**0.79±0.02**	**0.86±0.01**
Fo-i/Fv	**0.038±0.009**	**0.031±0.008**
Complementary area of fluorescence induction in the presence of DCMU		
α	**108ms(0.42)**	**140ms(1.00)**
β	**430ms(0.58)**	-
IDFmax/2 (amplitude)	**0.05(0.65)**	**0.13(1.00)**
	0.23(0.35)	

Table I. Comparison of different chlorophyll fluorescence parameters of WT and mutant wheat leaves. IDF max/2 is the relative flash intensity induced by a micro-second flash leading to half-saturation of variable fluorescence.

As DCMU penetration and translocation into the leaf could differ in the two genotypes and this could result in artifactual information on PSII heterogeneity, Q_A photoreduction in 1h dark-adapted leaves was also analysed (as in [14]). Figure 2 shows the increases of PSII fluorescence upon a single turn-over flash (A) and the corresponding semilog plot analysis (B) in the WT and the wheat mutant leaves. Half increase of fluorescence ($\Delta F = 1/2\ \Delta Fmax$) requires a higher flash intensity in the mutant than in the WT, again explainable by smaller PSII antenna size. Fig.2 shows that two phases of Q_A

photoreduction are observed in the WT leaves but only a single component is present in the mutant, confirming what was previously seen in the case of fluorescence induction in continuous light in DCMU-treated leaves.

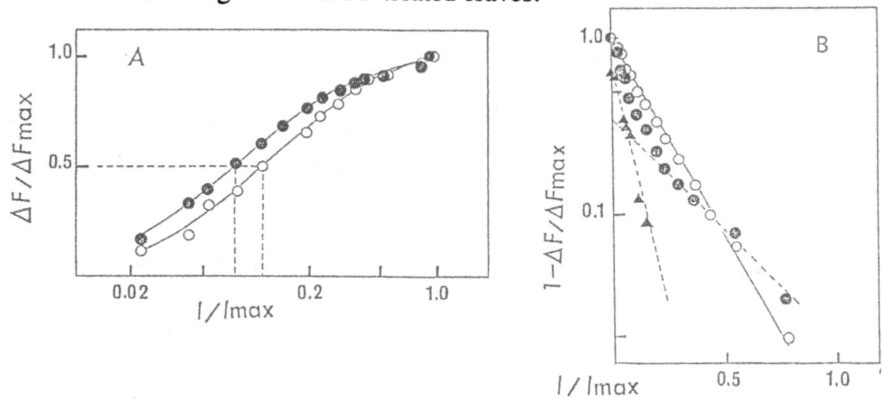

Fig. 2 (A) Relative increase of PSII fluorescence upon a single turnover flash (ΔF/ΔFmax versus I/Imax, I being the intensity of the flash). (B) Semilog plot analysis (complement of ΔF/ΔFmax versus I/Imax). Closed and open symbols represent data obtained respectively with wild-type and mutant wheat leaves.

In conclusion, a correlation between the PSII core phosphorylation and the functional heterogeneity described by PSII variable fluorescence seems to exist. We suggest that some of the observations on the functional PSII heterogeneity could be explained by the existence of differently phosphorylated PSII core populations.

References

1 Black MT, Brearly TH and Horton P (1986) Photosynth Res **8**, 193-207
2 Krause GH and Weis E (1991) In: Ann Rev Physiol Plant Mol Biol, Briggs et al. (eds), Vol **42**, pp 313-349, Elsevier Publishers, The Netherlands
3 Guenther JE and Melis A (1990) Photosynth Res **23**, 195-203
4 Hodges M and Barber J (1986) Biochim Biophys Acta **848**, 239-243
5 Hsu BD and Lee J-Y (1995) Aust J Plant Physiol **22**, 195-200
6 Allen J (1992) Biochim et Biophys Acta **1098**, 275-335
7 Giardi MT, Rigoni F and Barbato R (1992) Plant Physiol **100**, 1948-1954
8 Giardi MT (1993a) Planta **190**,107-113
9 Giardi MT, Komenda J and Masojidek J (1994) Physiol Plant **92**, 181-187
10 Giardi MT, Kucera T, Briantais J-M and Hodges M (1995) Plant Physiology, in press
11 Di Marco G, D'Ambrosio N, Giardi MT, Massacci A and Tricoli D (1989) Photosynth Res **21**, 117-122
12 Hodges M, Boussac A and Briantais JM (1987) Biochim Biophys Acta **894**, 138-145
13 Hodges M, Cornic G, and Briantais JM (1989) Biochim Biophys Acta **974**, 289-293
14 Genty B, Harbisson J, Briantais JM and Baker NR (1990) Photosynth Res **25**, 219-257

DEVELOPMENT OF A PHOTOSYSTEM I-LESS STRAIN OF *SYNECHOCYSTIS* SP. PCC 6803 FOR ANALYSIS OF MUTATIONS IN THE PHOTOSYSTEM II PROTEINS D2 AND CP43.

Svetlana Ermakova-Gerdes[1)], Sergey Shestakov[2)] and Wim Vermaas[1)],
[1)]Department of Botany and Center for the Study of Early Events in Photosynthesis, Arizona State University, Tempe AZ 85287-1601, USA; and [2)]N. Vavilov Institute of General Genetics, Moscow 117809, Russia.

1. Introduction

The cyanobacterium *Synechocystis* sp. PCC 6803 has a number of unique features that make it particularly suitable for genetic modification and structural and functional analysis of photosystem II (PS II) (see [1, 2]). Furthermore, PS I-less strains of *Synechocystis* 6803 have been developed [3, 4] that have a number of additional traits advantageous for detailed characterization of PS II. Use of these strains allows functional characterization *in vivo* of PS II by spectrometry and fluorescence techniques, which is complicated in wild-type *Synechocystis* 6803 by the unfavorable PS I/PS II ratio. Even though several useful PS II preparation procedures have been developed for wild-type *Synechocystis* 6803, they are inapplicable for isolation of oxygen-evolving particles from a number of PS II mutants. In part this is due to a number of vigorous purification steps necessary to separate PS II particles from contamination with PS I. The use of the PS I-less strain allows in many cases to substitute *in vitro* analysis of mutant PS II particles with less intrusive *in vivo* techniques, as well as to improve the quality of isolated PS II particles when *in vitro* characterization is still necessary.

In this communication we describe the development of a recipient PS I-less *Synechocystis* 6803 strain designed for construction, screening and analysis of mutations in D2 and CP43 PS II proteins. We have deleted *psb*C and both copies of the *psb*D genes from the PS I-less strain earlier developed in our lab [4] by replacing the corresponding regions of the chromosome with different drug-resistance cassettes. Use of this deletion strain facilitates introduction of mutated copies of *psb*DI and *psb*C into the *Synechocystis* 6803 genome.

2. Procedure

Strains and growth conditions. PS I-less strains of *Synechocystis* 6803 were cultivated in BG-11 medium, in which $NaNO_3$ was partially substituted with NH_4NO_3 (final concentration of ammonia was 4.5 mM). The medium was supplemented with 15 mM glucose and 10 mM TES/NaOH, pH 8.0. When necessary, BG-11 was supplemented with antibiotics in the following concentrations: chloramphenicol, 10 μg/ml; erythromycin, 15 μg/ml; and spectinomycin, 15 μg/ml. Cultures were grown under dim light (5 $\mu mol\ m^{-2} s^{-1}$) conditions.

P. Mathis (ed.), Photosynthesis: from Light to Biosphere, Vol. I, 483–486.

3. Result and Discussion

3.1. *Construction of a Synechocystis 6803 strain lacking the psaAB, psbDIC and psbDII genes.* *Synechocystis* 6803 contains two copies of *psb*D per genome; *psb*DI overlaps with *psb*C and *psb*DII is located in a different locus of the chromosome. Plasmid pDELDII described in [5] was used for deletion of *psb*DII. Upon transformation of the chloramphenicol-resistant PS I-less strain [4, 6] with pDELDII, 85% of the coding region of *psb*DII (a *Hinc* II/*Sma* I fragment) in the *Synechocystis* 6803 genome was replaced with a spectinomycin-resistance cassette obtained from pHP45Ω [7]. Spectinomycin-resistant transformants were selected and transformed subsequently with the plasmid pdelDICK-Er (Fig. 1). This plasmid was constructed from pDICK (Fig. 1A) described in [5] by substituting the *Nhe* I/*Cla* I DNA fragment that contained *psb*DI, *psb*C and 80% of the kanamycin-resistance marker with a 1.7 kb *Nhe* I /*Cla* I fragment of the erythromycin-resistance cassette from pRL425 [8]. Transformants were propagated under dim light conditions in the presence of erythromycin and of 20 μM atrazine to allow segregation. Several cycles of restreaking were necessary for a complete loss of the *psb*DIC and *psb*DII genes. To make sure that homozygosity had been attained at these loci, PCR was performed using chromosomal DNA of several transformants as templates and appropriate sets of primers to amplify the *psb*DIC operon and the *psb*DII gene. The PS I -less strain that had lost these genes did not produce any PCR-product of the sizes expected from the wild-type *psb*DIC and *psb*DII genes, but gave rise to PCR-fragments of sizes corresponding to these genes replaced by drug-resistance cassettes (Fig. 2).

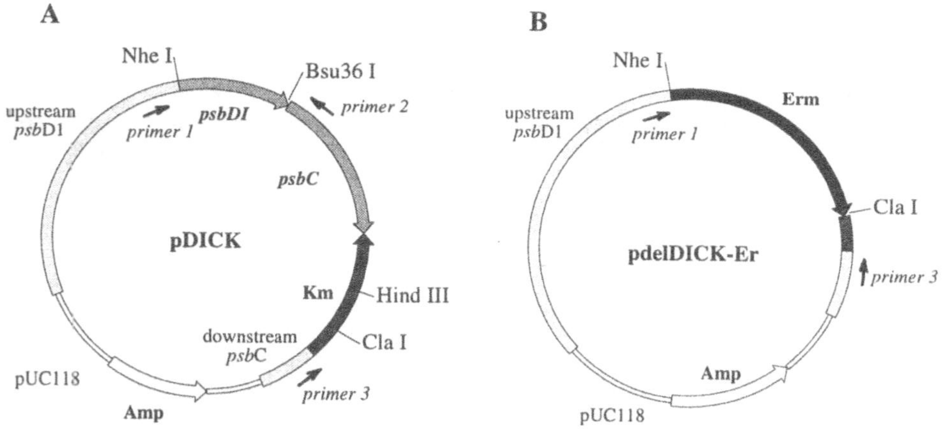

Figure 1. Maps of plasmids pDICK (A) and pdelDICK-Er (B). Positions of the primers used in PCR amplification of the *psb*DIC operon (described in Fig. 2) are indicated.

3.2. *Characterization of the Synechocystis 6803 strain lacking PS I and the D2 and CP43 proteins.* 77 K fluorescence emission spectra obtained upon 435 nm excitation of whole cells of the PS I-less strain and of the PS I-less, *psb*DIC⁻/DII⁻ strain are presented in Fig. 3. As expected, both spectra lack the fluorescence emission band at 725 nm, characteristic of the intact PS I complex. Importantly, the fluorescence peak at 695 nm, which is know to originate from chlorophyll associated with PS II, is absent in the PS I-less, *psb*DIC⁻

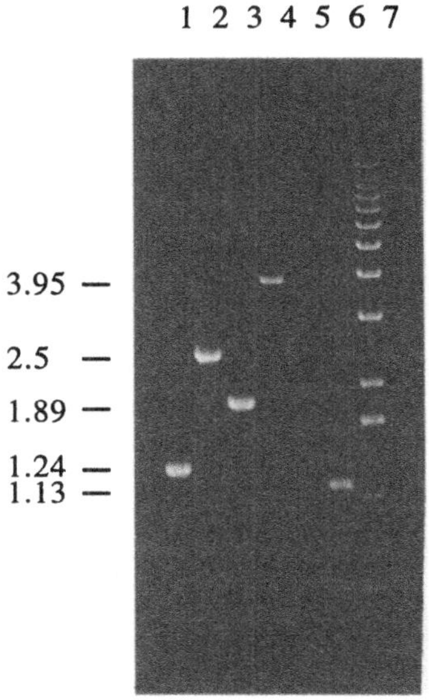

Figure 2. Amplification by PCR of the *psb*DII and *psb*DIC genes using total chromosomal DNA of the PS I-less strain (lanes 1, 4, 6) and of the PS I-less, psbDIC⁻/DII⁻ strain (lanes 2, 3, 5) as templates. Lane 7: 1 kb ladder. Estimated sizes of PCR products in kb are indicated. Lanes 1 and 2: amplification of the *psb*DII region. Lanes 3 and 4: amplification of the entire *psb*DIC locus. One primer located upstream and the other downstream of the *psb*DIC operon (primers 1 and 3 in Fig. 1) were used. Lanes 5 and 6: amplification of *psb*DI. For this experiment primer 2 located in the area deleted in pDICK-Er (Fig. 1) and primer 1 were used. The absence of a PCR product in lane 5, as well as the differences in sizes of PCR products from DNA of the control and mutant strains in lanes 1 - 4 indicate that in the mutant genome all copies of *psb*DIC and psbDII have been substituted with drug-resistance cassettes.

/DII⁻ strain. This shows that functional PS II centers do not assemble in the absence of D2 and CP43. This is in agreement with the observation in PS I-containing strains that deletion of the *psb*DIC and *psb*DII genes leads to complete absence of D1 from thylakoid membranes and to a drastic reduction in the amount of the CP47 protein [9]. A similar phenotype was observed in a PS I-less, *psb*DIC⁻/DII⁻ *Synechocystis* 6803 strain obtained from a PS I-less mutant that was reported to grow only under light-activated heterotrophic growth conditions and not in dim continuous light [10].

An emission band at 685 nm is still present in the spectrum of the PS I-less, *psb*DIC⁻/DII⁻ strain (Fig. 3). The 77 K fluorescence excitation spectra monitoring emission at 685 nm indicates that this peak originates from chlorophyll (data not shown). The total amount of chlorophyll in this strain as determined by methanol extraction of whole cells was found to be $20.3 \pm 4.4\%$ of that of the PS I-less strain, which is about 4.5% of the chlorophyll present in wild-type *Synechocystis* 6803. This is a high chlorophyll content for a strain that lacks PsaA, PsaB, D1, D2, CP43 and presumably has a low amount of CP47 (the PS I-containing *psb*DIC⁻/DII⁻ mutant contained less than 5% of the wild-type CP 47 amount [9]). We are currently working on identifying which protein(s) this chlorophyll may be associated with.

The PS I-less, *psb*DIC⁻/DII⁻ strain can grow well at light intensities as high as 30 μmol m^{-2} s^{-1} while the parental PS I-less strain does not tolerate light brighter than 6 μmol m^{-2} s^{-1}. This supports the concept that functional PS II is the main cause of light sensitivity in PS I-less strains. This property allows to use the constructed strain for direct positive selection by light of mutants impaired in PS II.

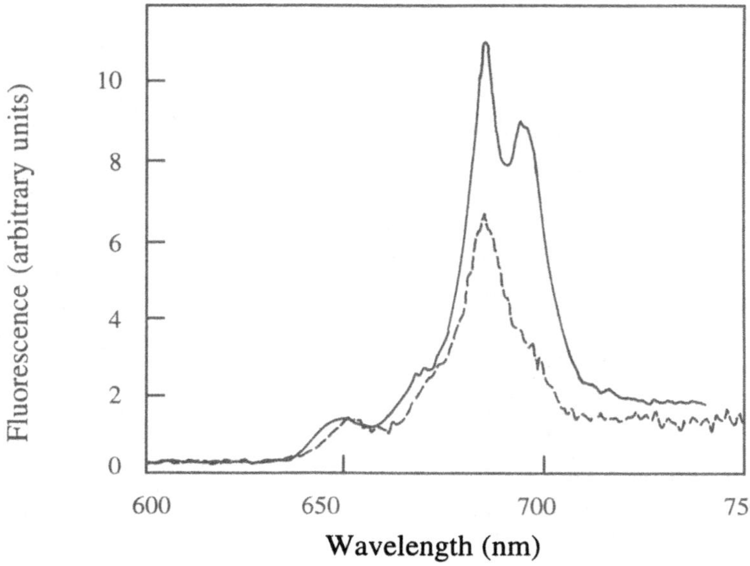

Figure 3. 77 K fluorescence emission spectra of whole cells of the PS I-less (——), and the PS I-less, *psb*DIC$^-$/DII$^-$ (– – –) strains of *Synechocystis* 6803. Excitation at 435 nm. The chlorophyll concentration was 1μg/ml. Spectra were normalized at 650 nm.

4. Acknowledgments

This work was supported by NSF grant MCB-9316857 to W.V., in particular by a supplement for S.E. under the NSF/Russian Academy of Sciences collaborative program. S.S. was supported in part by a grant from the International Science Foundation (MMA000) and the Russian Foundation for Basic Research.

References

1 Shestakov, S.V., Reaston, J. (1987) Oxford Surv. Plant Mol. Cell Biol. 4, 137-166
2 Williams, J.G.K. (1988) Methods Enzymol. 167, 766-778
3 Smart, L.B., Anderson S.L. and McIntosh, L. (1991) EMBO J. 10, 3289-3296
4 Shen, G., Boussiba, S. and Vermaas, W.F.J. (1993) Plant Cell 5, 1853-1863
5 Vermaas, W.F.J., Charité, J. and Eggers, B. (1990) in Current Research in Photosynthesis (Baltscheffsky M., ed.), vol. I, pp.231-238, Kluwer Academic Publishers, Dordrecht, The Netherlands
6 Boussiba, S. and Vermaas, W. (1992) in Research in Photosynthesis (Murata N., ed.), vol.III, pp.433-436, Kluwer Academic Publishers, Dordrecht, The Netherlands
7 Prentki, P. and Kirsch, H.M. (1984) Gene 29, 303-313
8 Elhai, J. and Wolk, P. (1988) Gene 68, 119-138
9 Yu, J. and Vermaas, W.F.J. (1990) Plant Cell 2, 315-322
10 Smart, L.B., Bowlby, N.R., Anderson, S.L., Sithole, I. and McIntosh, L. (1994) Plant Physiol. 104, 349-354

PHOTODEGRADATION OF D1-D2 PSII REACTION CENTER HETERODIMER

I. S. Booij, M. Swegle, S. Dube[1], M. Edelman[2] & A. K. Mattoo
Plant Molecular Biology Lab., USDA/ARS, Beltsville, MD 20705, USA
[1]Center for Agricultural Biotechnology, UMCP, MD 20742, USA
[2]Dept. Plant Genetics, Weizmann Inst. Science, Rehovot, Israel

1. Introduction

Light is intimately associated with the dynamics of photosystem II (PSII) reaction center proteins, D1 and D2. The D1 protein is a major product of the chloroplast protein synthesizing apparatus (1) and has received considerable attention. D1 is synthesized in the light (2) and processed on unstacked stroma lamellae, and then translocated to stacked grana lamellae. During translocation to and/or integration into grana, the D1 protein is covalently modified by palmitic acid (3). D1 palmitoylation is light-dependent. Following its integration into grana PSII complex, D1 as well as other PSII core proteins undergo reversible light-dependent phosphorylation (4). The mature grana-located D1 protein is rapidly degraded in the light compared to other chloroplast proteins (5-7). The half-life of D1 protein varies as a function of light intensity (8). The PSII inhibitor diuron inhibits degradation (5), palmitoylation (9) and phosphorylation (10) of the D1 protein. Since diuron (DCMU) displaces Q_B from its pocket, these results suggest a central role for the Q_B niche in the life cycle of the D1 protein.

The PSII reaction center has also been shown to be the main site of UV_B damage to the photosynthetic apparatus (11). UV_B radiation increases the degradation rate of D1 (8), very likely involving its Q_B site (12). It seems that the D1-D2 heterodimer, and not just the D1 protein, may be the target of UV_B radiation damage (13). In *Spirodela*, D2 is relatively stable in visible light compared to D1, but both these proteins are rapidly degraded in UV_B. The effect of UV_B radiation was found to be more detrimental in the presence of a background of visible light (13). An interesting possibility of UV_B damage to PSII reaction center is that the PSII stability is limited by sensitization of D2 for rapid turnover in the UV_B in the presence of a background of visible light. While D1 can be replenished because of its rapid synthesis in the light (2), the D2 protein synthesis being not as rapid as that of D1, it is not as easily replaced. Confirmation of D1-D2 heterodimer as the target for UV_B damage will provide an *in vivo* measure of plant adaptation or sensitivity to UV_B radiation. This is relevant to the increasing awareness that the depletion of the ozone layer will allow more UV_B penetration to the biosphere with a potential to negatively impact agricultural quality and productivity. It has been predicted that a decline of 1% in stratospheric ozone will result in the reduction of 1% in crop yields (14). To combat potential damage to crops, it is important to identify and monitor the key cellular targets sensitive to UV_B radiation. To ascertain if indeed the D1-D2 heterodimer is the protein target of UV_B damage to crop plants, we studied the D1 and D2 degradation

P. Mathis (ed.), Photosynthesis: from Light to Biosphere, Vol. I, 487–490.

rates in different light conditions using known UV_B-sensitive (CNS) and tolerant (Williams) cultivars of soybean (see Table 1).

2. Materials and methods

The soybean cultivars, CNS (UV_B sensitive) and Williams (UV_B tolerant), were grown under greenhouse conditions using only natural light. For studying D1 and D2 degradation rates, leaves were radiolabeled for 3 h with [^{35}S]methionine at 50 μmol/m^2/s PAR and then chased in 1 mM methionine under UV_B radiation with and without a background of visible light. The rates of D1-D2 degradation were quantified by densitometric scans of X-ray films (8, 10), while the steady-state levels were determined immunologically (15). The level of the 43-kDa PSII protein was taken as an internal control to normalize the data.

3. Results and Discussion

The rates of D1 and D2 degradation in visible light were found to be of the same order in CNS as that in the Williams cultivar (Fig. 1). At low UV_B radiation dose, D1 and D2 were relatively stable (compare Figs. 1 and 2) but when plants were irradiated at a higher UV_B dose, the degradation rate of both D1 and D2 proteins was higher in CNS than in Williams cultivar (Fig. 2). Interestingly, when the lower dose of UV_B was given in the presence of a background of visible light, a greater enhancement of D1 and D2 degradation rates was seen in CNS as well as Williams but the effect was more pronounced in CNS versus Williams (Fig. 1).

These data show that D2 protein, which is normally a stable protein, is degraded under a mixture of UV_B and visible radiation. It is possible that UV_B sensitizes the normally stable Q_A niche in D2 and in the presence of high visible light intensities D2 degradation is synergistically enhanced. These data confirm the studies of Jansen *et al.* (13) with *Spirodela* and extend the generalization that the D1-D2 heterodimer is the protein target of environmentally-relevant UV_B radiation damage in crop plants.

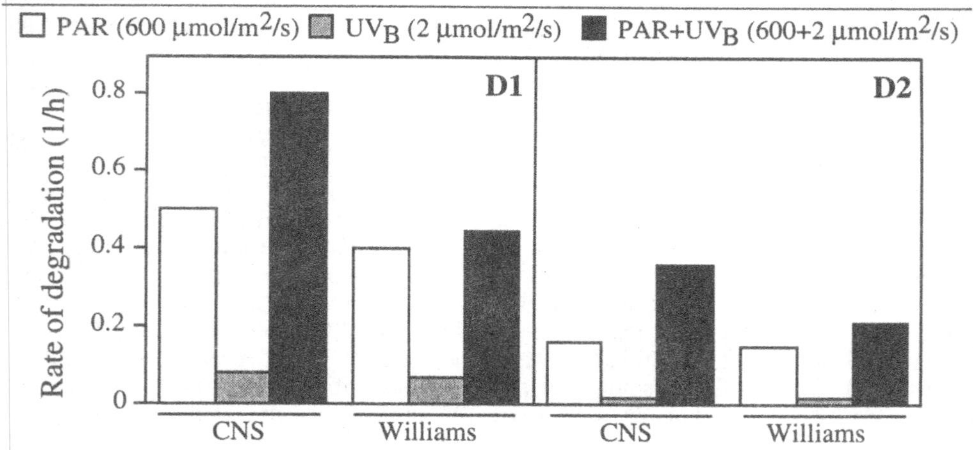

Figure 1: D1 and D2 degradation in CNS and Williams soybean cultivars. Leaf discs from each cultivar were pulse-labeled and chased under the indicated light conditions. Thylakoid proteins were fractionated by urea-PAGE and the rates of D1-D2 degradation quantified.

UV_B-tolerant soybeans (cv. Williams) show a reduced rate of D1-D2 degradation under environmentally-relevant UV_B radiation as compared to that in the UV_B-sensitive cultivar CNS (Fig. 1). These data suggest that genetic traits that make Williams cultivar tolerant to UV_B damage also protect the PSII reaction center.

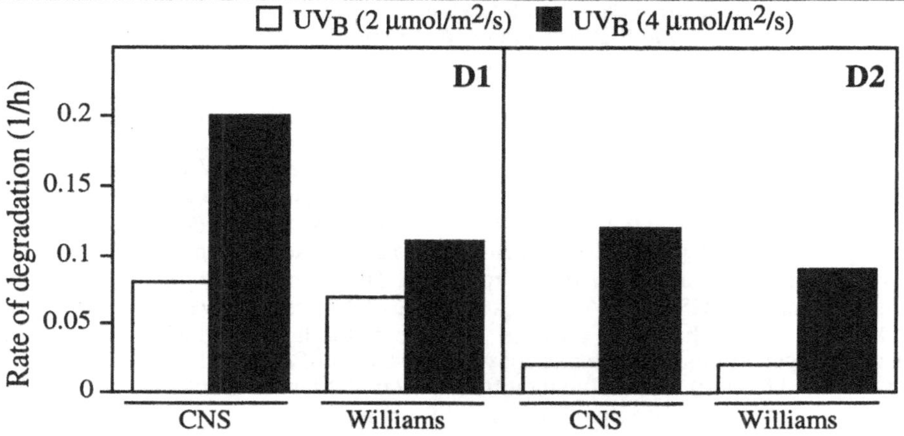

Figure 2: Effect of UV_B radiation on D1 and D2 degradation in CNS and Williams cultivars. Leaf discs from each soybean cultivar were pulse-labeled and chased under the indicated light conditions. Thylakoid proteins were fractionated by urea-PAGE and the rates of D1-D2 degradation quantified.

Table 1: Effect of UV_B radiation on growth and yield of two contrasting soybean cultivars in the greenhouse. Values represent % change from non-treated controls (adapted from 20).

Soybean cultivar	Height	Seed yield	Leaflet expansion	Leaf area	Aerial dry biomass	UV_B absorbing compounds
CNS	- 21	- 50	- 56	- 51	- 51	+ 3
Williams	- 16	+ 25	- 27	- 4	- 3	+ 25

Plants respond to UV_B by avoidance and repair processes that include screening the irradiation via increased production of epidermal flavonoids and reduction in leaf transmittance (16-18), repairing damaged DNA (19), and scavenging damaging oxygen free radical species (16). In the case of the two soybean cultivars studied here, UV_B-tolerant soybeans (cv. Williams) are known to accumulate more UV_B absorbing compounds (for example, flavonoids) than the UV_B-sensitive ones (cv. CNS) when exposed to UV_B radiation (Table 1), suggesting that bioscreening may be a key mechanism for plant protection against UV_B stress.

Other data on soybean cultivars with contrasting responses to UV_B radiation indicate that specific leaf weight, an estimate of leaf thickness (21), was unchanged in sensitive cultivars but increased by 24% in tolerant ones (16) upon exposure to UV_B radiation. Thicker leaves can mitigate UV_B radiation through additional tissue absorbance, thereby prevent a greater proportion of UV_B from reaching sensitive organelles in mesophyll tissues. Differences in UV_B sensitivity in soybean cultivars thus appear to be correlated with a combination of anatomical and biochemical differences, especially leaf thickness and accumulation of UV_B absorbing compounds. The compounds most associated with this function are flavonoids which accumulate in the vacuoles of epidermal cells (16, 22). Interestingly, they do not interfere with photosynthesis because they only weakly absorb visible light, which is consistent with the similar degradation rates observed for the D1 and D2 proteins in visible light for both CNS and Williams cultivars (Fig. 1). The highest flavonoid levels in UV_B-tolerant plants result presumably from *de novo* synthesis involving gene expression. Chalcone synthase and phenylalanine ammonia-lyase are key enzymes in the flavonoid biosynthesis pathway. The chalcone synthase gene is controlled by an UV-inducible promoter (23). A reasonable protection mechanism for plants, therefore, seems to be the accumulation of flavonoids which could bioscreen the interior of the leaf and thereby shield intracellular targets from UV_B radiation. Indeed, *Arabidopsis thaliana* mutant plants unable to synthesize chalcone synthase have reduced levels of leaf flavonoids and are more vulnerable to UV_B damage (24). Since many crop plants appear to be UV_B sensitive, a useful outcome of these experiments would be to find a way to decrease plants' sensitivity to UV_B. The overproduction of flavonoids by overexpression of CHS and PAL genes linked to UV_B-responsive promoter sequences could be one approach for this purpose. Because of the lack of an effective bioassay to measure UV penetration into the mesophyll tissues, quantifying the levels of D1 and D2 in plants exposed to UV_B and visible light for a period of a few hours can form a good test to evaluate tolerance or sensitivity of plant germplasm to UV_B.

References

1. Reisfeld, A. *et al.* (1982) Eur. J. Biochem. 124, 125-129
2. Mattoo, A.K. *et al.* (1989) Cell 56, 241-246
3. Mattoo, A.K. and Edelman, M. (1987) Proc. Natl. Acad. Sci. USA 84, 1497-1501
4. Elich, T.D. *et al.* (1993) EMBO J. 12, 4857-4862
5. Mattoo, A.K. *et al.* (1984) Proc. Natl. Acad. Sci. USA 81, 1380-1384
6. Wettern, M. and Ohad, I. (1984) Israel J. Bot. 33, 253-263
7. Gounaris, K. *et al.* (1987) FEBS Lett. 211, 94-98
8. Greenberg, B.M. *et al.* (1989) Proc. Natl. Acad. Sci. USA 86, 6617-6620
9. Mattoo, A.K. *et al.* (1993) in Post-translational Modifications in Plants (Battey N.H. *et al.* eds.) pp. 65-78, University Pess, Cambridge, UK
10. Elich, T.D. *et al.* (1992) J. Biol. Chem. 267, 3523-3529
11. Melis, A. *et al.* (1992) Biochim. Biophys. Acta 1100, 312-320
12. Jansen, M.A.K. *et al.* (1993) in Photosynthesis Responses to the Environment, Current Topics in Plant Physiology, Vol. 8 (Yamamoto H.Y. and Smith C.M. eds.) pp. 142-149, American Society of Plant Physiologists, Rockville, Maryland
13. Jansen, M.A.K. *et al.* (1995) submitted
14. Coohill, T. (1991) Photochem. Photobiol. 54, 859-870
15. Callahan, F.E. *et al.* (1989) Plant Physiol. 91, 629-635
16. Murali, N.S. *et al.* (1988) Photochem. Photobiol. 48, 653-657
17. Tevini, F.E. *et al.* (1991) Photochem. Photobiol. 53, 329-333
18. Wilson, M.I. and Greenberg, B.M. (1993) Plant Physiol. 102, 671-677
19. Pang, Q. and Hays, J.B. *et al.* (1991) Plant Physiol. 95, 536-543
20. Reed, H.E. *et al.* (1992) Crop Sci.. 32, 1214-1219
21. Mirecki, R.M. and Teramura, A.H. (1984) Plant Physiol. 74, 475-480
22. Caldwell, M.M. *et al.* (1983) Physiol. Plant. 58, 445-450
23. Lipphardt, S. *et al.* (1988) EMBO J. 7, 4027-4033
24. Li, J. *et al.* (1993) Plant Cell 5, 171-179

INFLUENCE OF HYDROGEN BONDS ON THE ENERGY AND ELECTRON TRANSFER IN REACTION CENTERS OF *RHODOBACTER SPHAEROIDES*.

V.Z.Paschenko, S.S.Vasil'ev, V.V.Gorokhov, P.P.Knox, A.B.Rubin
Department of Biology, M.V.Lomonosov State University,
Moscow 119899, Russian Federation

Introduction

The key processes of photosynthesis are the efficient transfer of excitation energy to photochemical reaction centers (RC) and its trapping in the form of a stable separated charges. One of the most suitable objects for such study is the RC complexes of purple bacteria for which the spatial structure has already been established [1,2]. After excitation of a special photoactive bacterochlorophyll dimer P870 electron is transferred from P870* to bacteriopheophytin (Bph) thus forming the primary radical pair [$P680^+BPh^-$]. The primary charge separation is stabilized by a rapid $Bpheo^-$ reoxidation by a special quinone molecule (Q_A). Functional groups P870, Bph and Q_A are incorporated in a protein complex, consisting of polypeptides L and M. High quantum efficiency of electron transfer is provided by a series of intermediate states where forward reactions proceed much faster than backward ones. These reactions are coupled with molecular relaxation processes which may also play a regulatory role by fine tuning of the reaction coordinates. Vibrational relaxations with excitation of accepting protein modes occurs within 10^{-12}-10^{-13} s [3]. A recent analysis on the efficiency of electron transfer processes in the reaction centers of purple bacteria led to the conclusion that relaxation processes, presumably protein dielectric polarisation in close proximity of the redox active groups are of functional relevance [4,5].

Experimental evidence of the changes in electron transfer rates induced by modifications of the protein matrix have been reported previously. It was shown that $H_2O \rightarrow D_2O$ exchange and addition of solvents which affect the protein dynamics by interfering with hydrogen bonds lead to the retardation of electron transfer from Bph^- to Q_A [6,7]. The degree of the effect was found to be correlated with the ability of agent to penetrate into hydrophobic areas of the reaction center and its proton-accepting ability [8]. The observed effects were discussed in the framework of the model in which initial energy level of the primary padical pair [$P870^+BPh^-$] relaxes with time and electron is finally stabilized at BPh.

The present work concerns with the possible role of hydrogen bonds in the [$P870^+BPh^-$] formation.

Procedure

Reaction centres were isolated from *Rhodobacter sphaeroides* as described in [9]. During experiments the samples were kept in 10 mM sodium phosphate buffer (pH 7.0).

P. Mathis (ed.), Photosynthesis: from Light to Biosphere, Vol. I, 491–494.

Concentration of P870 was about $0.5x10^{-3}$ M. To accelerate the cycle of light-induced reactions N,N,N',N' tetrametil-phenilen-diamine (10^{-4} M) and sodium ascorbate (10^{-3} M) were added to the samples. Reaction centers were modified by addition of 40% v/v DMSO. Pump-probe measurements of the absorption changes in the BPh anion-radical band (665 nm) were performed with the laser picosecond spectrometer as described in [6]. Samples were excited by 5 ps (FWHM) laser pulses with wavelength 533 (main frequency doubled output) nm or 590 nm (frequency doubled stimulated Raman scattering output) derived from Nd^{3+}-dopped [KGd(WO_4)] laser system operating at repetition frequency 2 Hz. Experimental data was analysed with the set of kinetic equations describing light induced reactions in the reaction centres.

Results and discussion

Fig. 1. shows the kinetic traces of optical density changes at 665 nm (Fig. 1.) measured with excitation at P870 absorption band. After addition of DMSO kinetic of the Bph^- formation was somewhat slower, compared with the control. To gather from this data information on the rate constant of charge separation a simplified kinetic pattern with the relaxed excited state $[P870^*BPh]_r$ was considered:

$$[P870^*BPh] \xrightarrow{k_r} [P870^*BPh]_r \xrightarrow{k_e} [P870^+BPh^-] \qquad (1)$$

It was implied that back reactions are much slower than forward ones. The value of k_r = 0.2 ps^{-1} estimated in [10] was used and it was supposed also that charge separation occurs mainly from $[P870^*BPh]_r$ state. The electron transfer to the primary quinone was not taken into account because this process is much slower (by 2 orders of magnitude) than the charge separation.

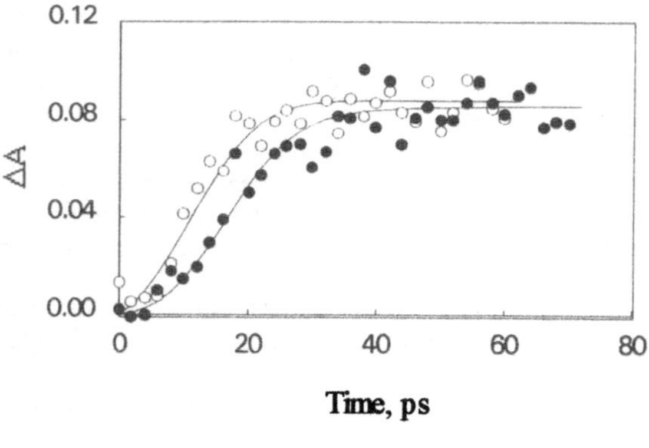

Fig. 1. Absorption changes at 665 nm, λ_{ex}=590 nm (P870 Q_X absorption band), o - control; •- modified.

Analysis of the experimental kinetic for the control samples (Fig. 1) gave the value of τ_e=5.7 ± 0.5 ps. This time in good agreement with our previous work [11] and results, obtained by other groops [12,13]. Analysis of the kinetic curve measured with modified samples revealed that τ_e value in this case is equal to 10.8±0.7 ps what corresponds to almost 2-fold decrease of the rate of charge separation in reaction centers modified by DMSO. Kinetic of this process was found to be mono-exponential what means that all reaction centers in the sample were modified.

The results of absorption measurements of the samples excited in Bph band are presented in Fig. 2. According to [12,13] in this case the main part of the excitation energy is transferrred to P870 with characteristic time of 10 ps inducing subsequent charge separation. The reaction sequence in this case is described by the following scheme:

$$[BPh^{*}P870BPh] \xrightarrow{k_m} [P870^{*}BPh] \xrightarrow{k_r} [P870^{*}BPh]_r \xrightarrow{k_e} [P870^{+}BPh^{-}] \qquad (2)$$

Where k_m is the rate constant of the energy transfer from BPh* to P870. After analysis of the $[P870^{+}BPh^{-}]$ formation kinetic in the framework of model (scheme 2) the following values of the rate constants were calculated: τ_m=10±1.5 ps and τ_e=5.4±0.8 ps. The value t_e is close to that obtained under direct excitation of P870 (see above). The value τ_m =10 ps is in good agreement with the lifetime of the energy migration from Bph* to P870 measured in [12].

In modified samples (closed circles, Fig. 2) lifetimes of the both processes significantly increased: τ_m=15.5±1.8 ps and τ_e=10.4±1.5 ps. Thus, for modified samples the value of t_e is close to that obtained in experiments with direct excitation of P870. The decrease in k_m under these conditions can be due to the retardation of the excitation transfer from BPh* to the photoactive pigment caused by modification of hydrogen bonds and/or protein conformation; it can be explained by a change of the Bph* interaction with its close surrounding. As a consequence shifts of the energy levels of Bph* and P870 after relaxation, as well as changes in overlapping of radiation and absorption spectra of the energy donor (Bph*) and acceptor (P870) may appear.

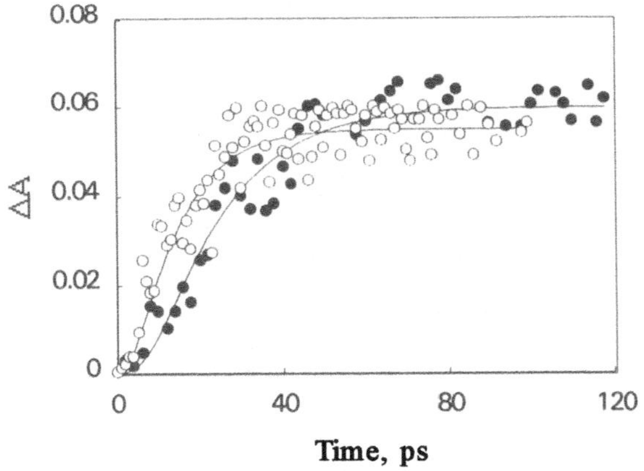

Fig. 2. Absorption changes at 665 nm, λ_{ex} =530 nm (Bph Q_x absorption band) ,o - control; • - modified.

The two-stage electron transfer between P870 and BPh suggested in our work can be influenced particularly by dielectric polarization processes in the surroundings of the photoactive pigments leading to the stabilization of the intermediate state. We suppose that this stabilization, connected with reorganization of the environment, occurs prior to the charge separation. Quite probable participants for the processes of rapid repolarization are molecules of the bound water and small polar groups, i.e. a specific "target" for our treatments. Moreover, the modification of relaxation processes may lead to incomplete overlapping of the $[P870^{*}BPh]_r$ and $[P870^{+}BPh^{-}]$ levels as well. This disturbs the activationless character of the electron transfer from P870 to BPh, what can be expressed by change of the Frank-Condon factor in the expression for the rate of electron transfer derived in [3]. Studies on the molecular structure of the bacterial RC revealed the presence of a series of hydrogen bonds between the components of RC and their protein surroundings and also the possibility of the presence of water molecules near the photoactive BChl dimer [1,2].

Relaxation processes connected with movements of light polar groups in protein surrounding may lead to the rapid (within 10^{-12} s) change of a special pair energy level and consequent change in the efficiency of excitation trapping by RC. We found that the process of charge separation have at least two stages sensitive to the modification of the hydrogen bonds: energy transfer to the photoactive pigment of the RC and formation of the primary radical pair [$P870^{+}BPh^{-}$]. In our previous works [6,7] we observed similar effects on electron transfer from BPh^{-} to Q occuring after $H_2O \rightarrow D_2O$ exchange The efficiency of the energy trapping by the special pair may depend on relaxation processes and be connected with the charge separation between components of the photoactive dimer P870.

Thus, the changes observed in the kinetics of the excitation energy transformation in the photosynthetic RC of bacteria as a result of the modification of hydrogen bonds indicate that the small H-containing molecular groops in the RC structure can be the key participants of these processes. They are capable to accept energy at the steps of energy transfer to the photoactive pigment, primary charge separation and electron transfer to Q. Therefore, the efficiency of excitation trapping can be also controlled by the relaxation processes, this is not a single-stage process and it is connected with charge separation between the components of P870 dimer after its excitation.

This work was supported by the Russian Fund for Fundamental Investigation (Grants Ns 95-04012597a and 95-04-11857a). S.S.V. gratefully acknowledges support from the Alexander von Humboldt Foundation.

M.V. Lomonosov Moscow State University

REFERENCES

1 Deisenhofer J., Michel H.(1989) EMBO J. 8, 2149-2170.
2 Mc Dermont G., Prince S.M., Freer A.A. et al. (1995) Nature 374, 517-521
3. Jortner J. (1980) Biochim. Biophys. Acta 594, 193-230
4. Xu, D & Schulten, K. (1994) Chem. Phys. 182, 91-117
5. Krishtalik,L.I. (1995) Biochim. Biophys. Acta 1228, 58-66
6. Paschenko V.Z., Korvatovsky B.N, Logunov S.L. et al. (1987) FEBS Lett. 214, 28-34 (1987)
7. Rubin A.B., Kononenko A.A., Shaitan K.V. et al. (1994) Biophysics 39, 173-195
8. Knox P.P., Godik V.BPh., Kononenko A.A. et al.(1990) Biofizika 35, 621-623
9. Zakharova N.I.,.Fabian M., Uspenskaya N.Ya. et al.(1981) Biochimiya 46, 1703-1711
10. Treutlein H., Schulten K., Brunger A.T. et al (1992) PNAS USA 80, 75-79
11. Paschenko V.Z., Korvatovsky B.N., Kononenko A.A. et al. (1985) FEBS Lett. 191, 245-248.
12. Moscowitz E., Malley M.M. (1978) Photochem. Photobiol. (1978) 27, 55-59
13. Holten D. Hoganson C., Windsor M.W. et al. (1980) Biochim. Biophys. Acta 592, 461-477

Wavelength dependence of the stimulated emission decay of membrane bound *Rb. sphaeroides* reaction centers and observation of a BChl-a anion involved in electron transfer

Lucas M.P. Beekman[‡], Mike R. Jones[§], Ivo H.M. van Stokkum[‡] & Rienk van Grondelle[‡]
[‡]Dept. of Physics & Astronomy, Free University of Amsterdam, De Boelelaan 1081, 1081 HV Amsterdam, The Netherlands. [§]Department of Molecular Biology and Biotechnology, University of Sheffield, Western Bank, Sheffield S10 2UH, UK

1. Introduction

In reaction centers (RCs) of photosynthetic purple bacteria, upon excitation of the primary electron donor (P → P*) a sequence of electron transfer steps results in a charge separation across the photosynthetic membrane. Despite thorough experimental and theoretical work there is still not a full consensus about the exact pathway of electron transfer. The initial electron transfer step, which is related to the loss of P*, in isolated RCs is approximately 3.5 ps [1], whereas in membrane bound RCs it was found to be 4.5 ps [2,3]. The electron is transferred from P, either to the nearby bacteriochlorophyll (BChl) molecule, B_L, after which the electron is rapidly (0.9 ps) transferred to a bacteriopheophitin (Bphe) molecule [1], or directly to the latter within 3.5 ps.
Furthermore, it was found that both spontaneous and stimulated emission decays of P* could not be accurately approximated by a single exponential, but rather a bi-exponential decay [1,3], or a distribution of decays [4,5]. This non-exponentiality of the electron transfer reaction has been described in terms of a distribution in both the driving force (ΔG), or electronic coupling (V) or a combination of the two [4,5,6], which are important parameters in theoretical descriptions of electron transfer.

In this paper we present data focusing on both multiexponential P* decay as well as on the mechanism of electron transfer. We have performed kinetic measurements at different wavelengths varying from 860 up to 1060 nm. We observe an acceleration of the decay of stimulated emission to the red-side of the emission band and present evidence for the transient appearance of a Bchl anion intermediate at approximately 1020 nm [1].

2. Materials and Methods

The membrane bound bacterial RCs were prepared as described in [3] and pump-probe measurements with an approximate 200 fs time resolution were performed as described in [3,5]. The sample was excited at 802 nm and single wavelength detection was achieved by selecting the probe wavelength from a generated continuum using bandpass filters and a 1/8 m monochromator. RCs were kept in an active state by pre-reduction of the quinone molecules with 200 mM Na-ascorbate. The data was analyzed using a Global analysis program described elsewhere [7].

P. Mathis (ed.), Photosynthesis: from Light to Biosphere, Vol. I, 495–498.

3. Results & Discussion

Stimulated emission

In Fig 1A we have plotted several typical traces which were obtained for the membrane bound RCs, in the wavelength range from 880-950 nm. The signal in this region is dominated by stimulated emission, with two other contributions due to groundstate bleaching of P and an absorption increase due to the formation of P^+. The ground-state bleaching signal follows the shape of the ground state absorption band of P, whereas the stimulated emission band resembles the emission spectrum. The P^+ absorption gives a broad contribution on the red side of the P-ground-state bleaching. The ground-state bleaching is limited by the instruments response, persists after formation of P^+ and is best modeled with an infinite component. P^+ is formed with the loss of the excited state giving it the same kinetics as P*, and is thus indistinguishable.
The data was analyzed using a model with three kinetic components of which one was taken to be infinite on the timescale of the experiment. The value of the two other decay times was estimated as well as their pre-exponential factors at each wavelength. In Fig 1B we have plotted the relative contribution of both decaying components as a function of wavelength. The decay times resulting in the best fits to the kinetics were 4.1 and 16 ps. As can be seen from Fig 1B the relative contribution of the slower component decreases going to the red in the spectrum. This results in an increase of average lifetime from 4.7 ps at 950 nm to 5.4 ps at 890 nm. Thus there is a significant difference in the charge separation properties of spectrally different membrane bound RCs.

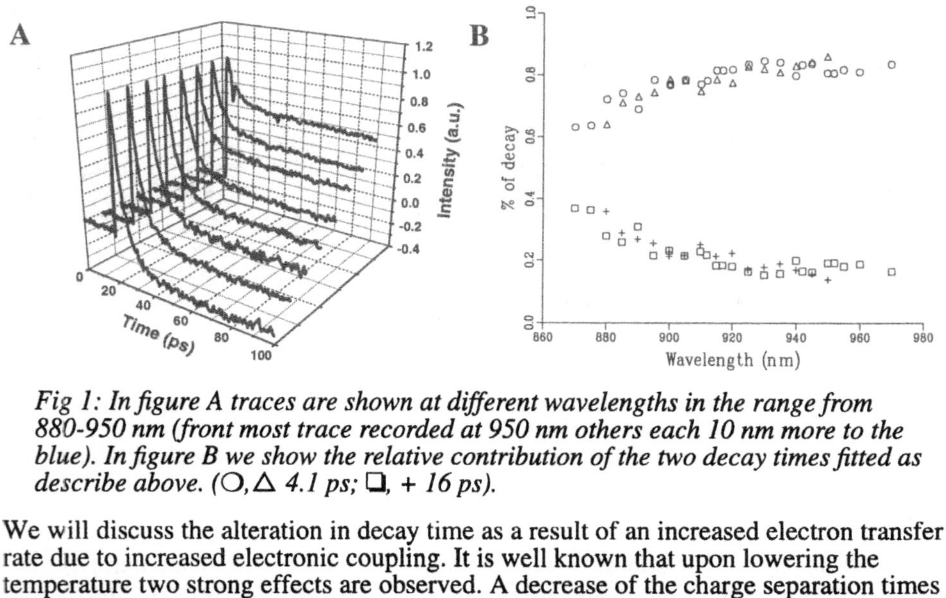

Fig 1: In figure A traces are shown at different wavelengths in the range from 880-950 nm (front most trace recorded at 950 nm others each 10 nm more to the blue). In figure B we show the relative contribution of the two decay times fitted as describe above. (○,△ 4.1 ps; ❑, + 16 ps).

We will discuss the alteration in decay time as a result of an increased electron transfer rate due to increased electronic coupling. It is well known that upon lowering the temperature two strong effects are observed. A decrease of the charge separation times (3.5 ps at RT to 1.2 ps at 4 K) and a spectral red shift of P upon cooling (860 nm at RT and 890 nm at 77-4 K).
Won & Friesner [8] have modelled the temperature dependence of the RC spectrum. For a consistent explanation of the shift of P they had to assume that upon cooling the protein becomes more rigid thereby reducing the mobility of the chromophores. They concluded

that at room temperature the two Bchl molecules that compose the dimer P would be further apart in the excited state (P*) resulting in a decreased of exciton coupling as compared to low temperature. This would explain the redshift of the P band. Furthermore the temperature dependence of the electron transfer in biology can be explained by a contraction of the protein, upon cooling [9].

Contraction of the protein brings the pigments closer together, increasing the dipole-dipole interaction (R^{-3} dependence). In the electron transfer process the dependence on a distance is found in the electronic coupling term. The electronic coupling is described by $V^2 = V_o{}^2 * \exp(-\beta R)$, with V_o being the maximum electronic coupling constant at zero distance and β being the exponential coefficient describing the distance dependence, and R the distance. β has been estimated to be 1.4 $Å^{-1}$ for electron transfer reactions in biology [10].

From the crystal structure we estimate that the center to center distance between the two Bchl molecules in P is about 7.7 Å. At room temperature P absorbs at about 860 nm and from holeburning experiments the inhomogeneous broadening was estimated to be approximately 150 cm^{-1} [11] which corresponds with ~10 nm. If we assume that the spectral variation is mainly due to a variation in distance variation between the two Bchls of P, we find that a 0.4 Å variation in distance would already explain a spectral variation of 10 nm. This means that a fluctuation in compression of the protein with about 5% is enough to explain the spectral effect. Similarly, for the edge to edge distance between P and B_L, which is about 5.4 Å, a variation of 5% which is about 0.3 Å, will result in an almost two fold difference in the charge separation rate. Due to the large homogeneous width of the P absorption band one will always observe a mixture of, red and blue, or more or less densely packed RCs. This will prevent us from seeing the pure "red" or "blue" RCs in kinetic measurements, however the ratio of the two will vary going to edges of the spectrum, as was observed in our experiments

We take this result as an indication that the dispersive kinetics in RCs is at least to some extent due to a distribution in the electronic coupling (δV), due to a variation in the distance between P and B_L.

Analysis of B^- data:

We have measured a similar set of kinetic traces as given in Fig 1A for the region 960-1060 nm although now on a timescale of only 20 ps. This data set was also analyzed using the global analysis program. To obtain good fits we needed at least 4 kinetic components, with decay times of 200 fs, 1.8 ps and 5 ps and an component of infinite lifetime. The spectrum of the infinite component followed the P^+ spectrum and the 5 ps component followed the tail of the stimulated emission, which is still visible to some extent at these wavelengths [1] (see Fig 2 A ❑). The 200 fs component probably describes the excitation energy transfer from B* to P [12]. The 1.8 ps component is peaked around 1020 nm, and has a width of about 40-50 nm. This result is very similar to that presented in [1], where the 1020 nm band was described as arising from a Bchl anion.
The spectrum of the 1.8 ps component was however not entirely satisfactory, since it carries some amplitude around 960 nm. This is probably due to the fact that, in this fit, the decay of stimulated emission was estimated using a single exponential. If we use the two decay constants obtained from the stimulated emission (4.1 and 16 ps) we find a life time for the 1020 nm feature of about 1.5 ps and a satisfactory spectral form for it to arise from a Bchl anion (see Fig 2A).

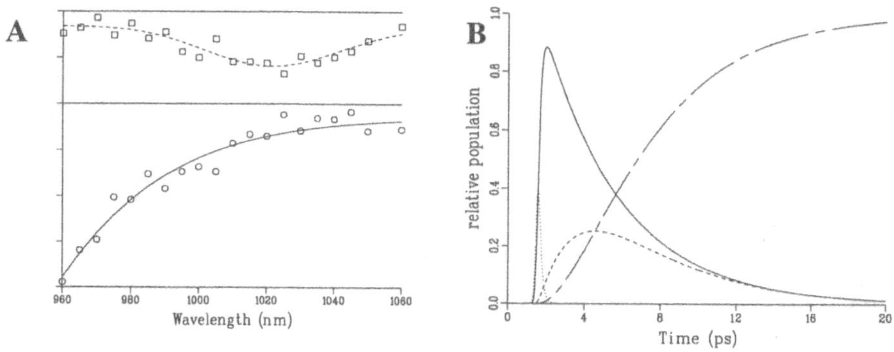

Fig 2: In figure A the DAS are shown calculated from the data in the 960-1060 nm region (○ 4.1 ps, □ 1.5 ps). The first two were fixed to the values found from the 850-970 nm region. In Fig. B the concentration profile is given for a sequential model, note that in this case the stimulated emission decay is modelled with only one component. The model is $B^* \rightarrow P^* \rightarrow P^+B^- \rightarrow P^+I^-$

The analysis described above was performed using a parallel model. However, the electron transfer process can be best described using a sequential model. For the analysis of the data we used a sequential model: $B^* \rightarrow P^* \rightarrow P^+B^- \rightarrow P^+I^-$. To obtain reasonable spectra for the intermediates in the model we had to include the P^+B^- intermediate in the electron transfer process, with a 1.5 ps lifetime. Preliminary measurements in the 780-820 nm region support these findings.

Acknowledgments

LB was supported by the Dutch foundation for Life Sciences. MRJ was supported by the Wellcome Trust and the Biotechnology and Biological Sciences Research Council of the United Kingdom. MRJ is a BBSRC Senior Research Fellow.

References

1 Artl, T., Schmidt, S., Kaiser, W., Lauterwasser, C., Meyer, M., Scheer, H. & Zinth, W. (1993), *PNAS*, vol. 90, 11757.
2 Schmidt,., Artl, T., Hamm, P., Lauterwasser, C., Finkele, U., Drews, G. & Zinth, W. (1993) *BBA*, 1144, 385
3 Beekman et al. *Biochemistry* accepted for publication
4 Jia, Y. DiMagno, T.J., Chan, C-K., Wang, Z. Du, M., Hanson, D.K., Schiffer, M., Norris, J.R., Flemming, G.R. & Popov, M.S. (1993) J. Phys. Chem. 97, 13180
5 Beekman et al. Manuscript in preparation.
6 Small, G.J., Hayes, J.M. & Silbey, R.J. (1992) *J. Phys. Chem.*, 96, 7499-7501
7 Stokkum, I.H.M. van, et al. (1994) *Proc. Kon. Ned. Acad. v. Wetensch.*, 94 (1), 43
8 Won, Y. & Friesner, R.A. (1988), In "The Photosynthetic Reaction Center" Eds. Breton, J. & Vermeglio, A., Plenum Press NewYork, Nato ASI series, pp. 341
9 Hales, B.J. (1975) Biophys. J., vol 16, pp. 471-480
10 Moser, C.C., Keske, J.M., Warncke, K., Farid, R.S. & Dutton, P.L. (1992), *Nature*, vol. 355, 796
11 Raja, N., Reddy, S., Lyle, P.A. & Small, G.J. (1992), *Photosyn. Res.*, 31, 167.
12 Jia Y., Jonas, D.M., Joo, T., Nagasawa, Y., Lang, M.J., & Fleming, G.R. (1995), *J. Phys. Chem.*, 99, 6263

Analysis of the Photosystem II reaction center and its complex with CP47 using gelfiltration chromatography and electron microscopy

Camiel Eijckelhoff[1], Henny van Roon[1], Egbert J. Boekema[2] and Jan P. Dekker[1]
[1]*Department of Physics and Astronomy, Vrije Universiteit, De Boelelaan 1081, 1081 HV Amsterdam, The Netherlands;* [2]*BIOSON Research Institute, Biophysical Chemistry, University of Groningen, Nijenborgh 4, 9747 AG Groningen, the Netherlands*

1. Introduction

The purity of the Photosystem II reaction center (PS II RC) has often been a matter of debate and so has it's pigment stoichiometry. Recently, we presented an analysis of a number of HPLC and spectroscopic pigment quantitation methods [1] and concluded that some of the previous controversies were caused by differences in the pigment quantification methods and that the PS II RC complexes obtained by short and long incubation with Triton X-100 contain about 6 Chl *a* and 1-2 β-Car per 2 Pheo *a* molecules. In this report, we analyse the purity of several PS II RC preparations by combining FPLC gelfiltration and diode array detection, which allowed us to study both spectroscopic differences and differences in size between particles. In this way, not only the purity of the PS II RC and CP47-RC complexes could be addressed, but also the aggregate size (monomer or dimer). The results indicate that PS II RC complexes purified *without* the help of Triton X-100 contain 6.5 Chl *a* and 2.0 β-Car per 2.0 Pheo *a* molecules. By combining sucrose density gradient centrifugation and gelfiltration we were also able to isolate dimeric CP47-RC complexes, and an analysis of these dimers by electron microscopy (EM) and image analysis is presented.

2. Materials and Methods

Gelfiltration experiments were performed using a Superdex 200 HR 10/30 column (Pharmacia). The buffer used as the mobile phase contained 20 mM BisTris, 20 mM NaCl, 10 mM $MgCl_2$ at pH 6.5 (BT-buffer) and 0.03% dodecyl maltoside (DM), and was pumped with an FPLC-pump (Pharmacia) at a flowrate of either 25 ml/h (for analytical purposes) or 40 ml/h (for purification purposes). The experiments were performed at room temperature in darkness. The sample was applied to the column using a Pharmacia injection valve fitted with a 200 μl sample loop. The sample volume amounted therefore 0.8% of the bed volume which is within the recommended sample volume range. Samples were filtrated through a 0.45 μm PVDF filter prior to injection. A complete gelfiltration run takes approximately 40 minutes at a flowrate of 25 ml/h. For detection a diode array detector (Waters990) was used in the wavelength range of 280 - 750 nm with

P. Mathis (ed.), Photosynthesis: from Light to Biosphere, Vol. I, 499–502.

2 nm resolution.
PS II RC complexes were isolated from CP47-RC using a either a short Triton X-100 incubation step as in [1] or a combination of $LiClO_4$ and DM as in [2]. The latter RC's were purified further on a sucrose density gradient composed of sucrose solutions in BT-buffer + 0.03% DM containing respectively 25-20-18-16-13-8% (w/v) sucrose. The centrifuge tubes had a volume of 12 ml and were loaded with 200 to 300 µl sample. They were centrifuged overnight using a swing-out rotor (TST41) at 40,000 rpm. The 16% layer containing the PS II RC was carefully harvested, discarding the part ~ 50 mm above the 18% layer (monomeric CP47-RC complex) to avoid contamination. Pigment analysis was performed as in [1].

CP47-RC complexes were isolated as in [2] and further purified by sucrose density centrifugation as described above. After centrifugation, the particles were found in two bands in the 18 and 20% layers. For analysis by EM, complexes were prepared using the droplet method with 2% uranyl acetate as the negative stain. EM was performed with a JEOL JEM 1200-EX electron microscope using 80 kV at 60.000x magnification. The images were analysed using alignment strategies and multivariate statistical analysis plus classification [3,4].

3. Results

3.1 PS II Reaction Center

In order to assess the composition and purity of isolated PS II RC complexes, we combined the FPLC gelfiltration setup with diode array detection. In this way different particles can be recognized, not only on the basis of their retention times, but also on the basis of their differences in spectroscopic properties. A good way to characterize the different PS II complexes is the ratio of absorption at 416 and 435 nm, which was found to be 0.92, 1.20 and 0.75 for the CP47-RC, RC and CP47 complexes, respectively. The retention times of the pure complexes were found to be 25, 26-27 and 30 minutes, respectively. The results with the PS II RC preparation obtained by a short Triton X-100 treatment of CP47-RC complexes [1] suggest no significant contamination by CP47-RC, CP47 or other pigment protein complexes (not shown). The results presented in Fig.1A indicate that RC complexes isolated using $LiClO_4$ and DM, however, are significantly contaminated with CP47-RC and CP47, because the $A416/A435$ ratio reveals shoulders with lower ratio's on both sides of the main fraction. In Fig.1B it is shown that the additional purification using a sucrose density gradient has removed the CP47-RC and most of the CP47 contamination. An additional preparative gelfiltration step appeared to be sufficient to remove this residual contamination (not shown).

The pigment composition of the most purified fractions was estimated by HPLC chromatography and spectroscopic analysis [1] to be 6.5 Chl *a* and 2.0 β-Car per 2.0 Pheo *a*. The Chl-content is virtually identical to the content found in Triton-treated preparations, but the β-Car content is higher [1], suggesting that Triton extracts primarily β-Car from the complex. The 4K absorption spectrum (not shown) revealed well separated absorption bands at 671 and 679 nm, as well as a clear shoulder at 683 nm, which indicates that this shoulder is not due to a contamination of CP47 (as suggested in [5]).

3.2 CP47-RC complex

When we examined isolated CP47-RC complexes with the gelfiltration technique, we usually found a shoulder with varying amplitude on the lefthand side of the main peak.

The particles in the shoulder show the same *A416/A435* ratio of 0.92, which suggests that they do not represent CP43-containing core complexes. Analysis of the 18 and 20% sucrose layers harvested from the sucrose density gradient, revealed that the two particles were separated to considerable extents, elution times of 23.5-24 and 25 minutes. We therefore conclude that these particles represent monomeric and dimeric forms of the CP47-RC complex, respectively.

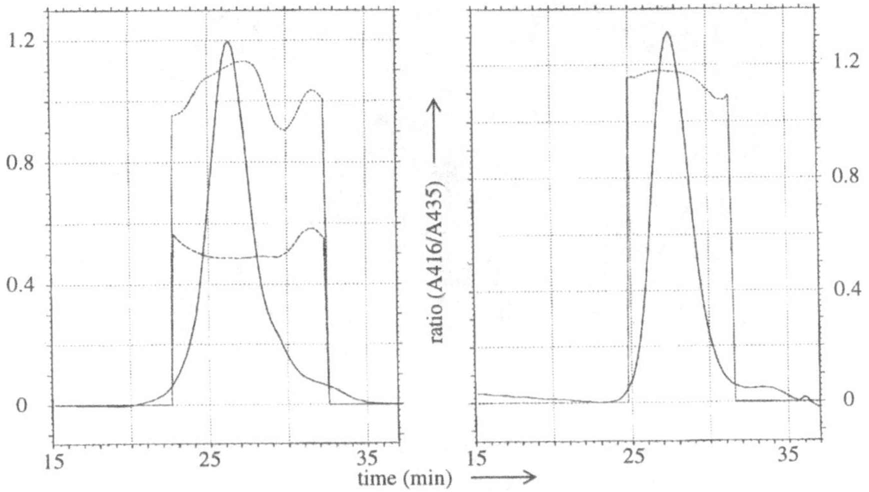

Figure 1: Gelfiltration chromatograms (416 nm) of a PS II preparation isolated with $LiClO_4$ and DM, before (A) and after sucrose density gradient centrifugation (B)

For analysis by EM, we used freshly prepared dimers, because under our experimental conditions the dimers readily disintegrated into monomers. After image analysis procedures, two types of projections (Fig.2A+B) were found, which we attribute to dimeric CP47-RC complexes in two positions on the carbon support film (flip- and flop type). Three masses can be distinguished within these projections, as compared to four in the core particle [4], in agreement with the loss of CP43. The data, however, do not allow an unambiguous attribution of CP43 to a certain mass in the core complex yet, because it cannot be excluded that the dimeric associations in the core- and the CP47-RC complex are different.

4. Discussion

The results presented in this report indicate that it is possible to completely purify PS II RC complexes without the help of Triton X-100 as in [2], but with additional sucrose density centrifugation and gelfiltration steps. The pigment stoichiometry and low-temperature optical properties of these preparations closely resemble those of the PS II RC complexes prepared by a short Triton X-100 treatment [1], except that only without Triton it appeared possible to retain the two β-Car molecules. The shape of the gelfiltration chromatogram suggests that these preparations consist of a homogeneous set of PS II RC particles. Because these particles eluted from the gelfiltration column at considerably later times than the monomeric CP47-RC complex, we conclude that they are monomeric.

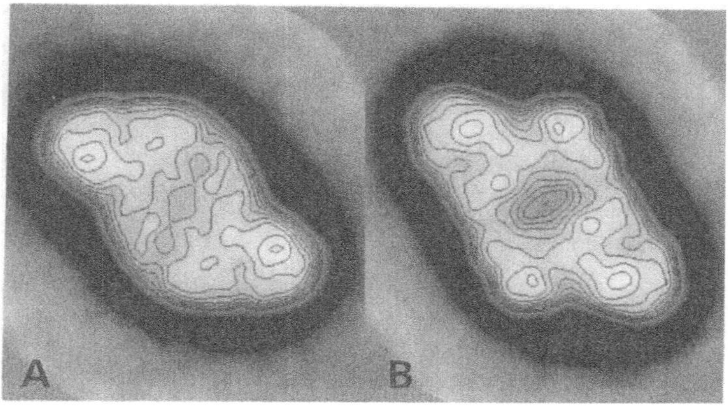

Figure 2: Two types of projections of the dimeric CP47-RC complex: (A) flip-type and (B) flop-type; mirrored after analysis to facilitate comparison

Acknowledgements

This research was supported by the Netherlands Foundation for Chemical Research (SON) and by the EC contract CT940619.

References

[1] Eijckelhoff, C. and Dekker, J.P. (1995) Biochim. Biophys. Acta 1231, 21-28
[2] Dekker, J.P., Bowlby, N.R. and Yocum, C.F. (1989) FEBS Lett. 254, 150-154
[3] Harauz, G., Boekema, E. and Van Heel, M. (1988) Meth. Enzymol. 164, 35-49
[4] Boekema, E.J., Hankamer, B., Bald, D., Kruip, J., Nield, J., Boonstra, A.F., Barber, J. and Rögner, M. (1995) Proc. Natl. Acad. Sci. USA 92, 175-179
[5] Chang, H-C., Jankowiak, R., Reddy, N.R.S., Yocum, C.F., Picorel, R., Seibert, M. and Small, G.J. (1994) J. Phys. Chem. 98, 7725-7735

FEMTOSECOND ABSORBANCE CHANGES IN PS II REACTION CENTRES

M. Germano, M.H. Vos[1], J.-L. Martin[1], T.J. Aartsma and H.J. van Gorkom
Department of Biophysics, Huygens Laboratory, Leiden University, P.O. Box 9504, 2300 RA Leiden, The Netherlands; [1] Laboratoire D'Optique Appliquée, École Polytechnique - E.N.S.T.A., 91120 Palaiseau, France

1. Introduction

At room temperature, the Q_Y transitions of all six Chls and two Pheos of the isolated photosystem II reaction centre form a single broad absorption band around 675 nm. Nevertheless, two pigment pools can be distinguished, with absorption bands centred around 670 and 680 nm, respectively. Selective excitation at 685 nm leads to charge separation in 3 ps, whereas excitation at 665 nm produces charge separation in 25-30 ps, indicating that absorbance on the short wavelength side is mainly due to accessory Chls that transfer their excitation energy only slowly to P680 even at room temperature (1). The long wavelength absorbance is attributed to a central group of pigments homologous to the BChls and BPheos in the purple bacterial RC. At room temperature, energy transfer between these pigments must be faster than 3 ps. The extent of exciton interaction between them is currently a matter of debate.

Here we report absorbance measurements with femtosecond resolution using selective excitation of the long-wavelength pigments, at room temperature and at 95 K.

2. Materials and Methods

PS II RCs with Chl:Pheo:Car=6:2:1 (2) were used at an optical density of 0.5 (at 675 nm, path length 1 mm) after addition of oxygen scavengers as in (1) and (for 95 K measurements) 60% glycerol.

At room temperature, the sample was in a quartz cuvette moving back and forth perpendicular to the laser beams and was replaced after 45000 pulses. Position and shape of the Q_Y band were unchanged after measurements. At 95 K, the sample was in a Plexiglas cuvette in an Oxford Instruments flow cryostat and was stable.

The set-up described in (3), equipped with a CCD camera that was read out between two laser shots, was used to measure transient spectra from 630 to 840 nm with 1 nm resolution. Excitation was provided at 30 Hz by a 55 fs (FWHM) pulse of 10 nm spectral width (FWHM), centred at 685 nm for 95 K and at 695 nm for 293 K

P. Mathis (ed.), Photosynthesis: from Light to Biosphere, Vol. I, 503–506.

measurements, both exciting about 10% of the RCs. Pump and probe polarisations were set at the magic angle. Kinetic traces at 660-695 nm were globally fitted by convolution of the 65 fs (FWHM) cross-correlation of pump and probe with a sum of exponentials, assuming wavelength-independent kinetics.

3. Results and discussion

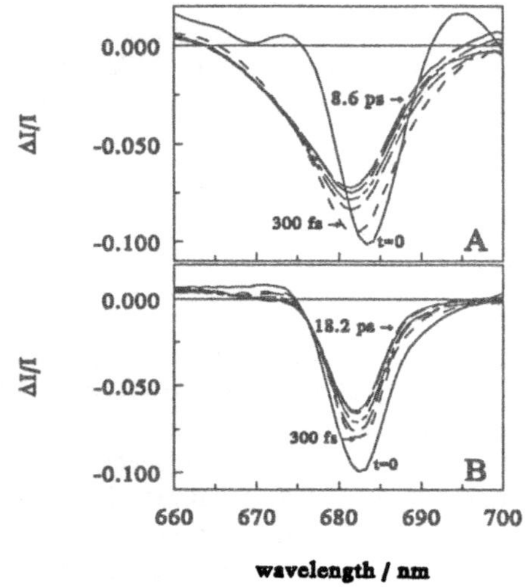

Figure 1. Transient absorption difference spectra at room temperature (A) and T=95 K (B) at delay times between 0 and 20 ps.

Figure 1 shows transient absorption difference spectra in the Q_Y region at 293 K with 695 nm excitation (A) and at 95 K with 685 nm excitation (B), at time delays from 0 to 20 ps. At both temperatures, an initial bleaching around 683 nm with the spectral width of the pump pulse is observed. The additional features at room temperature come mainly from the chirp of the continuum pulse. At both temperatures the difference spectrum changes into that of the radical pair, $P680^{+}Pheo^{-}$. The changes in the ps time domain (figure 2A), where the charge separation takes place, are small. Most of the amplitude of the absorbance changes is associated with a very fast process, of the order of the cross correlation between pump and probe. At room temperature, the presence of the chirp prevented detailed analysis. The

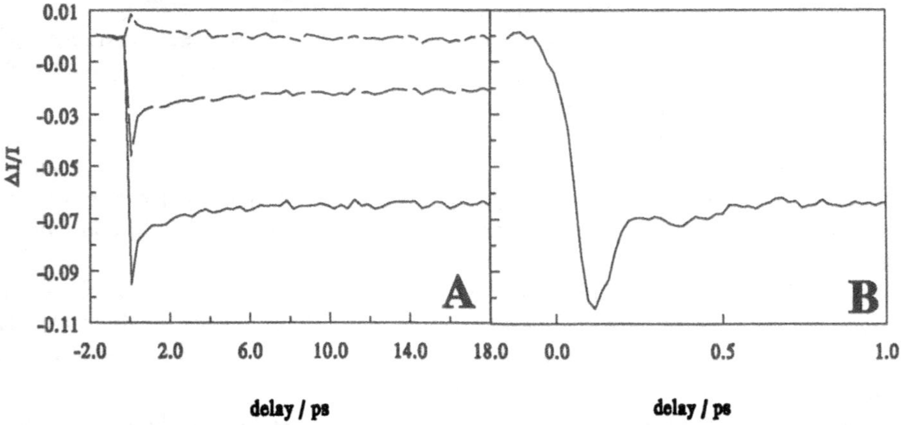

Figure 2. Kinetics in the ps (A) and fs (B) time domain at T=95 K of the absorbance changes induced by excitation with 55 fs pulses of the long wavelength-absorbing pigments.(— - — 670 nm; ——— 682 nm; — — 687 nm).

kinetics on the sub-ps domain at 95 K are illustrated in figure 2B. The initial decay of the flash-induced changes is very fast, much less than 100 fs if fitted by an exponential, and followed by oscillatory features with a period of about 300 fs. Using short pulses, pump-probe coherence and cross phase modulation inevitably contribute to the signal around t=0, but an independent experiment showed that the initial fast transient bleaching of the 682 nm band is not dominated by these effects. The oscillation may reflect vibrational coherence in the excited state, as in bacterial RCs (4), or in the ground state, by impulsive stimulated Raman scattering, but the lack of a phase shift in the spectral dependence of this feature may indicate coherence between near-lying (~80 cm^{-1}) electronic levels as the origin of the oscillations. In view of the presumed density of electronic states in the Q_Y

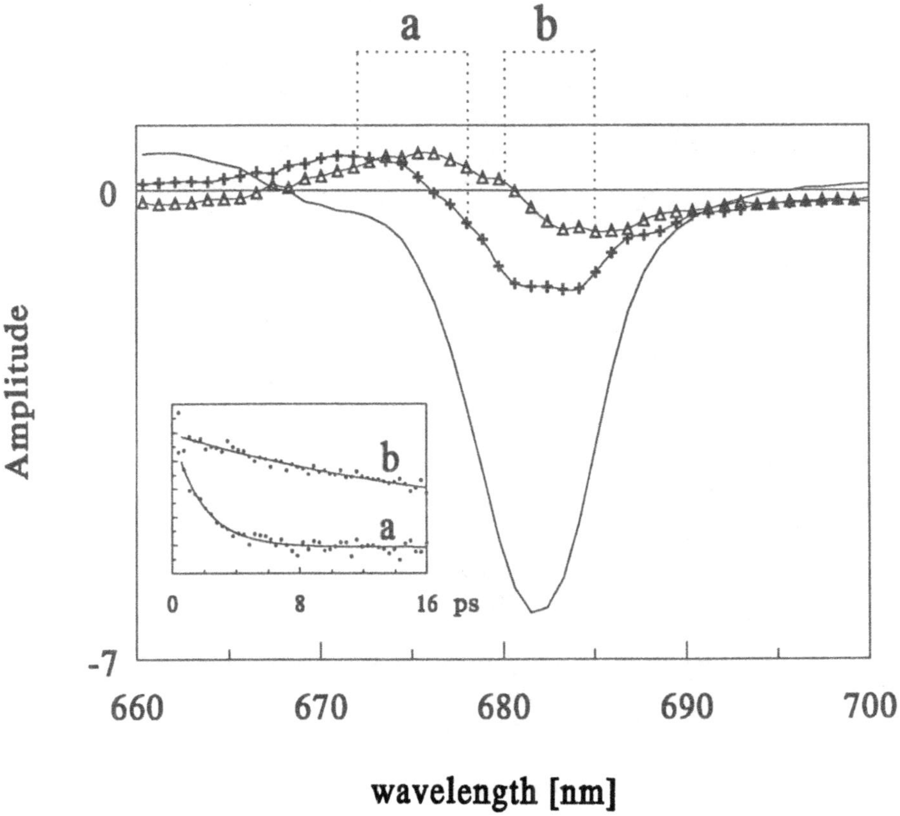

Figure 3. Decay associated spectra of the 1.9 ps component (+) the 20-30 ps component (Δ) and the non-decaying component (solid line). In order to better illustrate the time constants, the kinetics at, respectively, 685 and 680 nm and 678 and 672 nm were subtracted, yielding traces **a** and **b** in the inset. These are approximately monoexponential, with fitted time constants of 1.9 ps (**a**) and ~20 ps (**b**).

band (5), the PS II RC is a likely candidate for such effects to occur. Altogether, it is clear that either vibrational or electronic coherence contributes to the signal, and that care must be taken in interpreting the ultrafast spectral development in terms of excitation equilibration only.

Three time constants and a non-decaying component were required to adequately fit the kinetic traces between 660 and 695 nm. The time constants that gave the best fits for all wavelengths were ~65 fs, 1.9 ps and 20-30 ps. The later value is only approximate, as the time window scanned in the experiment (0-20 ps) was too narrow for a precise determination. The spectrum of the non-decaying component (line in figure 3) agrees well with reported absorption difference spectra of the primary radical pair (6). The decay associated spectra of the 20-30 ps (Δ) and 1.9 ps (+) components are similar in shape to those reported in (7) at room temperature, but the amplitudes, especially of the 1.9 ps component in the 680-685 nm region, are much larger. On the basis of other studies (1,8,9) we can attribute the 1.9 ps component to charge separation upon direct excitation of P680 and the 20-30 ps component to energy transfer-limited charge separation, but the data shown here do not rule out other interpretations.

Similar data were obtained by Visser et al. (Visser,H.M. et al., submitted) at 77 K with an 0.3-0.4 ps pump pulse. Comparison confirms that the femtosecond component results mainly from the fact that the spectral width of the bleaching initially corresponds to that of the pump pulse.

We thank Mr. Christof Francke for measuring the 4K absorption and fluorescence spectra, and Ms. Saskia Janssen for preparing the samples. Financial support was provided by the European Science Foundation and by the Portuguese Joint for Scientific and Technical Research (JNICT).

References

1 Schelvis,J.P.M., van Noort,P.I., Aartsma,T.J. and van Gorkom,H.J. (1994) Biochim. Biophys. Acta 1184, 242-250

2 van Leeuwen,P.J., Nieveen,M.C., van de Meent,E.J., Dekker,J.P. and van Gorkom,H.J. (1991) Photosynth. Res. 28, 149-153

3 Martin,J.-L. and Vos,M.H. (1994), Meth. Enz. 232, 416-430

4 Vos,M.H., Rappaport,F., Lambry,J.-C., Breton,J. and Martin,J.-L. (1993) Nature 363, 320-325

5 Durrant,J.R., Klug,D.R., Kwa,S.L.S., van Grondelle,R., Porter,G. and Dekker,J.P. (1995), Proc. Nat. Acad. Sci. USA, 92, 4798-4802

6 van Kan, P.J.M., Otte,S.C.M., Kleinherenbrink,F.A.M., Nieveen,M.C., Aartsma,T.J. and van Gorkom,H.J. (1990), Biochim. Biophys. Acta, 1020, 146-152

7 Klug,D.R., Rech,T., Joseph,D.M., Barber,J., Durrant,J.R. and Porter,G. (1995) Chem. Phys. 194, 433-442

8 Wasielewski,M.R., Johnson,D.G., Govindjee, Preston,C. and Seibert,M. (1989) Photosynth. Res. 22, 89-99

9 Jankowiak,R., Tang,D., Small,G.J. and Seibert,M. (1989) J. Phys. Chem. 93, 1649-1654

THE ROLE OF LONG-WAVELENGTH ANTENNA PIGMENTS IN ENERGY TRANSFER IN PHOTOSYSTEM II.

J.P.M. Schelvis[1], M. Germano, T.J. Aartsma and H.J. van Gorkom
Department of Biophysics, Leiden University, P.O. Box 9504,
2300 AA Leiden, The Netherlands.

1. Introduction

In isolated Photosystem II (PS II) reaction centers (RCs), preferential excitation of the accessory chlorophylls (Chls) results in a 30 ps energy transfer to the primary electron donor, P680, prior to charge separation [1]. We argued that such a slow energy transfer would lead to a low yield of photosynthesis in more intact PS II if the energy would be equilibrated over the antenna Chls and the accessory Chls in the RC [2]. This would also be in contradiction with Schatz et al. [3], who assumed that in PS II core particles the excitation energy is equilibrated over all the antenna and RC pigments within 10 ps. Instead we proposed [2] that the accessory Chls are bypassed when energy is transferred from the antenna to P680 via a pheophytin (Pheo), which receives the excitation energy from a long-wavelength antenna pigment, e.g. the 690 nm Chl in CP47 [4].

More intact PS II retains the secondary electron acceptor, Q_A, a plastoquinone molecule. The redox state of Q_A is of decisive importance for the fate of the charge separation. When Q_A is neutral, trapping of the excitation takes 30-100 ps [5-7] and the primary charge separation is stabilized by electron transfer to Q_A in 300 ps [8,9]. When Q_A has been prereduced, trapping has been reported to take 200 ps, followed by charge recombination of the radical pair in equilibrium with the excited state in 2 ns [3,5].

In order to investigate the pathway of energy transfer in PS II or to verify its presumed rapid equilibration, we performed picosecond transient-absorption measurements on spinach PS II core particles with Q_A in its reduced state.

2. Material and Methods

PS II core complexes containing about 35 Chl *a*, were prepared according to [10]. The activity of the preparations was checked by measuring the oxygen yield and the photoreduction of Q_A. Total absence of PS I was extensively verified by fluorescence emission and excitation spectra at 6 K. 10 μM DCMU was added to prevent any electron transfer beyond Q_A. 65 μg/ml catalase, 4 mM glucose, and 65 μg/ml glucose-oxidase

[1] current address: Department of Chemistry, Michigan State University, East Lansing, MI 48824, USA

P. Mathis (ed.), Photosynthesis: from Light to Biosphere, Vol. I, 507–510.

was added to the sample to remove oxygen. The samples had an absorbance of 1 at 675 nm and all experiments were performed at 295 K.

The transient absorption measurements were performed using the dye-laser system described in [11] operated in the one-color pump-probe mode at a repetition frequency of 76.6 MHz. The spectral bandwidth was less than 1 nm and the pulse duration was about 7 ps. The samples were placed in a spinning cell (6000 rpm) with a radius of 55 mm. The average excitation density was 10^{13} photons cm^{-2} per pulse, which was low enough to avoid singlet-singlet annihilation. Accumulation of the triplet state of P680 is not expected, taking into account its short lifetime and low yield [7].

3. Results and Discussion

Transient absorption changes of PS II core complexes are shown in Fig. 1. Such changes were measured at wavelengths from 670 to 706 nm and fitted by a non-decaying component and a set of three exponentials with lifetimes of 21±6 ps, 80-200 ps, and about 1.5 ns. The intermediate lifetime decreased systematically from 200 ps near 670 nm to 80 ps around 700 nm and may be composed of two spectrally different components (Fig. 2).

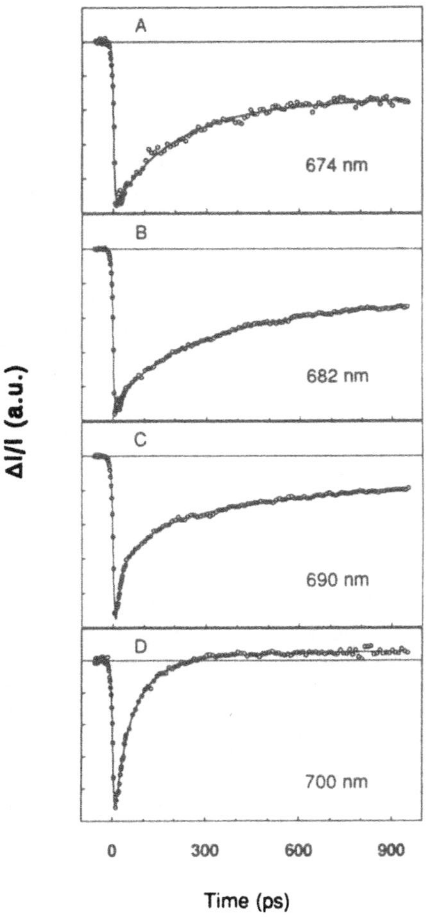

Fig. 1. Transient absorption changes in PS II core particles in the presence of Q_A^- at 295 K induced and detected with pump and probe parallel at 674 (A), 682 (B), 690 (C) and 700 nm (D). The data are plotted as circles and the solid line represents a fit of an offset and three exponentials (see text).

When all traces were fitted with a 21 ps, a variable intermediate lifetime, a 1.5 ns, and a stable component, the amplitudes shown in Fig. 3 were obtained for parallel (A) and perpendicular (B) polarization between the pump and probe beam. The amplitudes are normalized to the amount of absorbed photons. The low absorption of PS II core at wavelengths above 700 nm makes the amplitudes in that region less reliable.

The spectrum of the 1.5 ns component (▲, Fig. 3) resembles that of a corresponding lifetime component observed in dual wavelength measurements [5,8] and it is attributed to recombination of the primary radical pair in equilibrium with the excited state. The spectrum of the end level (Δ, Fig. 3) has a significant (but small) contribution only at shorter wavelengths than 680 nm. Its maximum is around 670 nm, and not 682 nm, therefore, we attribute the end level to a triplet or excited state of a Chl uncoupled from the energy transfer process, and does not involve P680.

The amplitudes of the fast (●) and intermediate (○) components are polarization dependent. Above 700 nm, these two components are responsible for the complete decay of the bleaching and the initial absorbance change is larger than can be explained by ground-state bleaching. Therefore, we attribute part of the change to stimulated emission and these two components to processes involving redistribution or trapping of excitations.

The presence of at least two components and the wavelength dependence of the lifetime of the intermediate component (see Fig. 2) exclude thermal equilibration of the excitation in less than 10 ps, as postulated by Schatz et al. [3]. The 200 ps trapping time in the presence of Q_A^-, as determined by Schatz et al.[5] using 675 nm excitation, is not inconsistent with our data, but we find that the lifetime gradually decreases to about 80 ps beyond 700 nm. This may be due to a variation in the average trapping time with wavelength, because longer-wavelength pigments are positioned closer to the trap reducing the 'first passage time'. The possibility that the 80 ps component might come from a PS I contamination [12] has been ruled out. The 21 ps component may reflect charge separation after direct excitation of a small pool of long-wavelength pigments including P680. In that case,

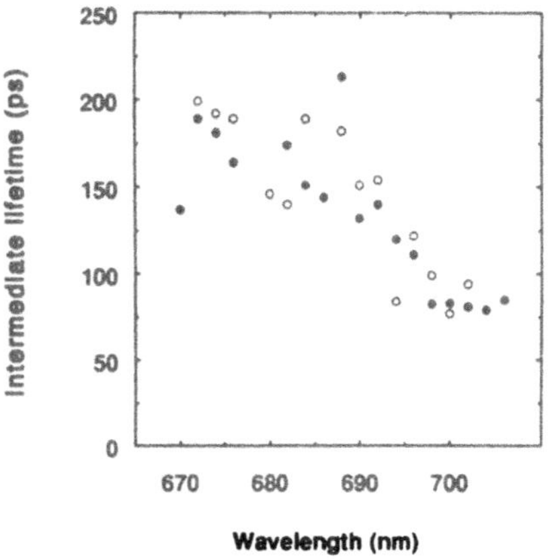

Fig. 2. Wavelength dependence of the lifetime of the intermediate component. Pump an probe beams parallel (●) and perpendicular (○).

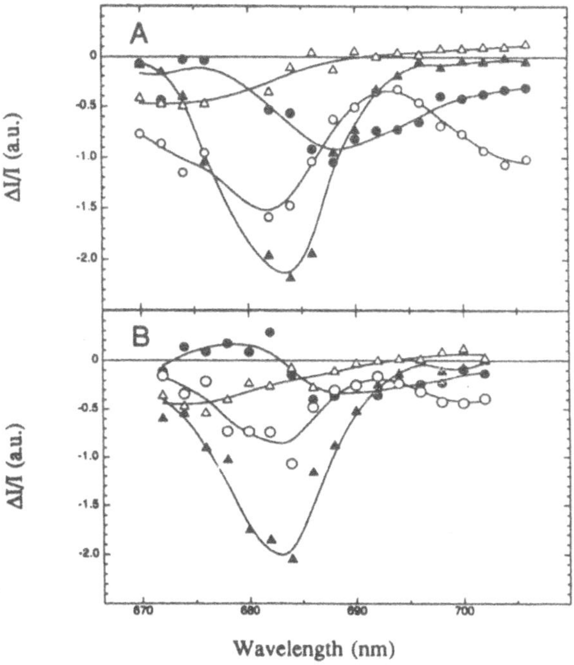

Fig. 3. Contribution of the 21 ps (●), 80 to 200 ps (○), and 1.5 ns (▲) components and the end level (Δ) to the absorption changes induced and probed with pump and probe either parallel (A) or perpendicular (B).

the intermediate component would reflect energy transfer from the antenna to this small pool. The spectrum of the 21 ps component and its pronounced polarization dependence suggests that this pool comprises only a few pigments, absorbing selectively on the long-wavelength side, with a significantly different average orientation with respect to the other antenna pigments. This might support the hypothesis proposed in [2], but two-color pump-probe experiments and measurements with open RCs will be needed to arrive at an unambiguous assignment of the fast and intermediate components.

In RCs (D1-D2-cyt *b* 559 particles) isolated from these core particles according to [10] we do not find a 21 ps component upon long-wavelength excitation, but charge separation in 3 ps [1]. It is tempting to speculate that the 21 ps charge separation time upon 694 nm excitation of isolated RCs reported in [13,14] has the same origin as the 21 ps component reported here, and is due to a small fraction of the centers retaining (part of) the core antenna. The 25 ps trapping time observed by Freiberg et al. [15] in RC-CP47 complexes may support this view.

We thank mr. C. Francke for low-temperature fluorescence-measurements.

4. References

1. Schelvis, J.P.M., Van Noort, P.I., Aartsma, T.J. and Van Gorkom, H.J. (1994) Biochim. Biophys. Acta 1184, 242-250.
2. Van Gorkom, H.J. and Schelvis, J.P.M. (1993) Photosynth. Res. 38, 297-301.
3. Schatz, G.H., Brock, H. and Holzwarth, A.R. (1988) Biophys. J. 54, 397-405.
4. Van Dorssen, R.J., Breton, J., Plijter, J.J., Satoh, K., Van Gorkom, H.J. and Amesz, J. (1987) Biochim. Biophys. Acta 893, 267-274.
5. Schatz, G.H., Brock, H. and Holzwarth, A.R. (1987) Proc. Natl. Acad. Sci. USA 84, 8414-8418.
6. Hodges, M. and Moya, I. (1988) Biochim. Biophys. Acta 935, 41-52.
7. Van Mieghem, F.J.E., Searle, G.F.W., Rutherford, A.W. and Schaafsma, T.J. (1992) Biochim. Biophys. Acta 1100, 198-206.
8. Nuijs, A.M., Van Gorkom, H.J., Plijter, J.J. and Duysens, L.N.M. (1986) Biochim. Biophys Acta 848, 167-175.
9. Eckert, H-J., Wiese, N., Bernarding, J., Eichler, H-J. and Renger, G. (1988) FEBS Lett. 240, 153-158.
10. Van Leeuwen, P.J., Nieveen, M.C., Van de Meent, E.J., Dekker, J.P. and Van Gorkom, H.J. (1991) Photosynth. Res. 28, 149-153.
11. Van Noort, P.I., Gormin, D.A., Aartsma, T.J. and Amesz, J. (1992) Biochim. Biophys. Acta 1140, 15-21.
12. Schelvis, J.P.M., Germano, M., Aartsma, T.J. and Van Gorkom, H.J. (1995) Biochim. Biophys. Acta 1230, 165-169.
13. Hastings, G., Durrant, J.R., Barber, J., Porter, G. and Klug, D.R. (1992) Biochemistry 31, 7638-7647.
14. Durrant, J.R., Hastings, G., Joseph, D.M., Barber, J., Porter, G. and Klug, D.R. (1993) Biochemistry 32, 8259-8267.
15. Freiberg, A., Timpmann, K., Moskalenko, A.A. and Kuznetsova, N.Y. (1994) Biochim. Biophys. Acta 1184, 45-53.

Temperature Dependent Sub-ps Transient Spectroscopy of the Photosystem II Reaction Center

Marie-Louise Groot, Frank van Mourik, H. Matthieu Visser, Jan P. Dekker and Rienk van Grondelle
Faculty of Physics and Astronomy, Vrije Universiteit, De Boelelaan 1081, 1081 HV Amsterdam, The Netherlands

1. Introduction

Time resolved studies of several groups on the reaction center of Photosystem II have revealed the presence of various kinetic and spectral components with lifetimes in the range of sub-ps to ns. While the longer lifetimes are generally ascribed to recombination of the long-lived radical pair $P680^{+}Pheo^{-}$ formed upon charge separation, the origin of the different sub-ps to ps components is controversial and such processes as energy transfer, charge separation and exciton relaxation have been proposed. At RT e.g., charge separation has been reported to occur in 3 ps [1,2,3,4,5], but Klug and co-workers [6 and references therein] have found that radical pair formation occurs primarily (for ~70%) with a 21 ps time constant.
We have performed time-resolved measurements on the RC as a function of temperature, since we expect that the particular temperature dependence of the various components will yield information on their origin.

2. Materials and Methods

PS II RC (*D_1-D_2-cytochrome b-559*) complexes were isolated from spinach by means of a short Triton X-100 treatment of CP47-RC complexes as described earlier [7,8]. The samples were characterized by a ratio of the room temperature absorptions at 416 nm and 435 nm of 1.20 and contained Chl *a*, Pheo *a* and β-Car in a ratio of 6.4 / 2.0 / 1.6 [8]. The samples were diluted in a buffer containing 20 mM BisTris (pH 6.5), 20 mM NaCl, 0.03% *n*-dodecyl-β,D-maltoside and 80% (v/v) glycerol to an optical density of ~1.0 at 675 nm in a 1.5 mm pathlength. The laser system was described elsewhere [9], the instrument response of the system was improved to be 280 fs. The repetition rate of the system is 30 Hz, therefore no accumulation of RCs in the triplet state occurs, excitation energy was 50 nJ/pulse. Excitation was at ~685 nm with a narrow-band (5 nm fwhm) interference filter. Usually 6 scans, consisting of 40 delay positions, were taken for data collection. Effects of divergence of the pump beam were minimised by aperturising just before the focusing lens. Spectra were corrected for the energy of the pump beam at the position of the sample, which was measured as a function of the delay line.

P. Mathis (ed.), Photosynthesis: from Light to Biosphere, Vol. I, 511–514.

3. Results and Discussion

Absorption difference spectra were collected in the Q_y region upon excitation at 685 nm at 77 K, 150 K and 240 K. Fig 1a shows examples of ΔOD spectra at 150 K at 7 delay times, in fig 1b the result of a global analysis of the 150 K dataset using three kinetic components is shown.

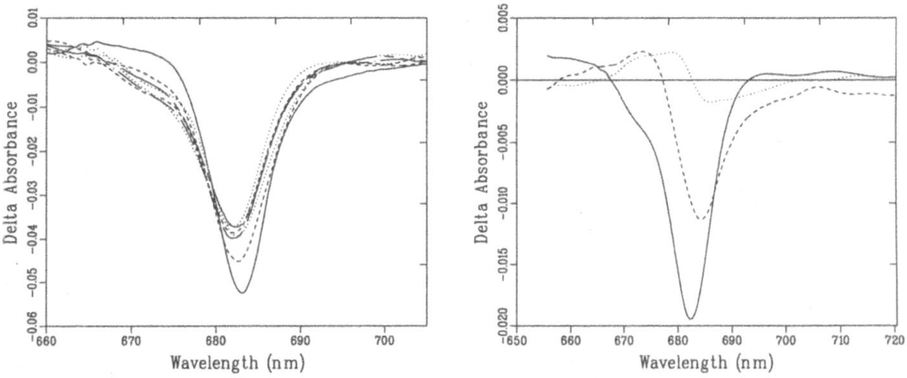

Figure 1 150 K dataset (a) ΔOD spectra measured time-delays of 0.6 ps (solid), 1.25 ps (dash), 4.75 ps (dot), 21 ps (chain-dash), 55 ps (chain-dot), 155 ps (dash) and 1.3 ns (dot).
(b) Decay-associated spectra of the 0.6 ps (dash), 38 ps (dot) and 33 ns (solid) components

The datasets collected at all three temperatures can be fitted with three components, a sub-ps component, a component that accelerates from 40 ps at 77 K to 18 ps at 240 K, and a long-lived ns component.
The sub-ps component, which has a very similar DAS spectrum at all temperatures, accelerates from 0.7 ps at 77 K to 0.4 ps at 240 K. It may be that the 0.4 ps component at 240 K is a mixture of two lifetimes such as the 100 fs and 3 ps lifetimes reported at RT by Klug et al. [6]. Interference of these two components will occur at 240 K rather than at 77 K since at that temperature the equilibrium distribution is different from that initially excited, which results in a significant ΔOD spectrum of the energy transfer component.
The DAS-spectrum of the longest-lived component, which corresponds to the radical pair state $P680^+Pheo^-$, is very similar for all datasets, except that the spectrum is broader at higher temperature and the shoulder at ~672 nm less well resolved.
The blue positive signal in both the fast and the intermediate component is probably due to the ingrowth of the shoulder in the radical pair spectrum. Since, at least at low temperature, an energy transfer process to ~672 nm absorbing states is highly unlikely. Upon 685 nm excitation, both P680 and the pigment(s) that in holeburning experiments [10] at T = 4 K have been identified to trap excitations are excited. We ascribe the 0.7 ps lifetime to the intrinsic charge separation time from P680, in line with other low temperature measurements [11, 12], though we note that a faster equilibration process may be mixed in with this component. This leaves the (set of) pigment(s) degenerate with P680 as the

origin of the intermediate component.
A closer study of the 77 K dataset has revealed that the fit for the intermediate component is not unambiguous, because for the 77 K dataset (but not for the other two), a good fit can also be obtained including a fourth component. This now leads to a fit at 77 K in which a 16 ps lifetime and a long, 1 ns lifetime have very similar DAS-spectra (see fig. 2a and b). Due to the nature of the experiment, the long lifetime can not be determined accurately, good fits were also obtained with e.g. 500 ps. The similarity of the DAS spectra of the second and third component suggests that they arise from a similar process.

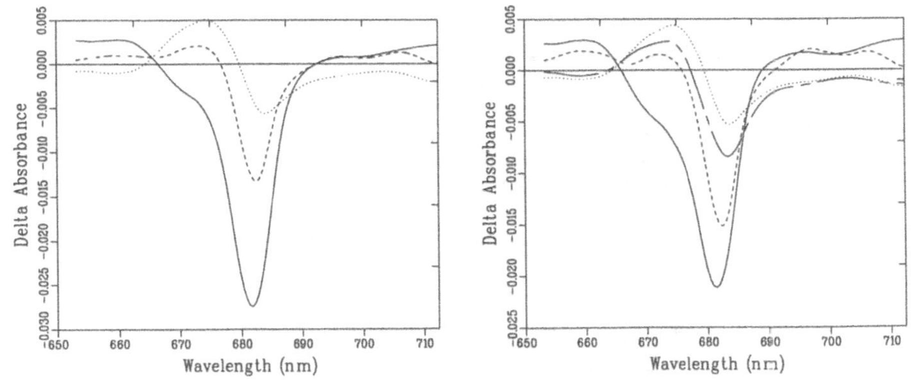

Figure 2 77 K dataset (a) Decay-associated spectra of a 3 component fit: 0.7 ps (dash), 40 ps (dot) and 4 ns (solid)
(b) Decay-associated spectra of a 4 component fit 0.5 ps (dash), 16 ps (dot), 1 ns (chain-dash) and 13 ns (solid)

The separation in a shorter and a longer component with similar spectrum is an indication that these components actually describe a process which has a broad distribution of lifetimes. At 150 K apparently the extreme long lifetimes have disappeared, i.e. the distribution of lifetimes may be narrower, whereas at 240 K this component is represented by a single lifetime of 18 ps.
We propose that the distribution of lifetimes is due to the inhomogeneity of the bands involved in the energy transfer process. At 77 K, kT = 54 cm^{-1}, the inhomogeneous width of the absorption bands involved is ~ 110 cm^{-1}, or 5 nm. Especially if the bands are shifted with respect to each other, this leads to an intrinsically broad distribution of energy transfer times.
Apart from P680 and the pigments degenerate with P680 there are pigments in the RC that absorb at higher energy. The holeburning results [10] and the measurements of [13] indicate that direct energy transfer between the 'trap' state and P680 is very slow. To explain the temperature dependence of the intermediate component it is therefore more likely that energy transfer occurs from the 'trap' level to a level that lies at a few nm shorter wavelength. From that level, the excitation may be quickly transferred and trapped in the radical pair state. This would gives rise to the observed broad distribution and temperature dependence of the rate of radical pair formation. This explanation implies that at all temperatures a significant part of the radical pair formation is energy transfer limited.

Acknowledgements

We thank dr. I.H.M. van Stokkum for his help with the data analysis and H. van Roon for the expert preparation of the PS II reaction centre particles. This research was supported by the "Nederlandse organisatie voor Wetenschappelijk Onderzoek (NWO)" via the "Stichting voor Fundamenteel Onderzoek der Materie (FOM)" and the "Foundation for Life Sciences (SLW)".

References

1 Wasielewski, M. R.; Johnson, D. G.; Seibert, M. and Govindjee *Proc. Natl. Acad. Sci. USA* **1989** 88, 524-5
2 Roelofs, T. A.; Gilbert, M.; Shuvalov, V. A. and Holzwarth, A. R. *Biochim. Biophys. Acta.* **1991**, 1060, 237-244
3 Gatzen, G. K.; Griebenow, M. G.; Müller, and Holzwarth, A. R. *In* Research in Photosynthesis, Vol. II, N. Murata, editor. Kluwer Academic Publishers, Dordrecht, **1992**, 69-72
4 Schelvis, J. P. M.; van Noort, P. I.; Aartsma, T. J.; and van Gorkom, H. J. *Biochim. Biophys. Acta* **1994**, 1184, 242-250
5 Wiederrecht, G. P.; Seibert, M.; Govindjee, Wasielewski, M. R. *Proc. Natl. Acad. Sci. USA*, **1994**, 91, 8999-9003.
6 Klug, D. R.; Rech, Th.; Joseph, D. M.; Barber, J.; Durrant, J. R.; Porter, G. *Chem. Physics,* **1995**, 194, 433-442. and references therein
7 Kwa, S. L. S.; Newell, W. R.; van Grondelle, R.; and Dekker, J. P. *Biochim. Biophys. Acta* **1992**, 1099, 193-202
8 Eijckelhoff, C., and Dekker, J.P. *Biochim. Biophys. Acta* **1995**, 1231, 21-28
9 H. M. Visser, M.-L. Groot
10M.-L. Groot, F. den Hartog, J. P. Dekker, R. van Grondelle, S. Völker, to be published
11Wasielewski, M. R.; Johnson, D. G.; Govindjee, Preston, C. and Seibert, M. *Photosynth. Res.* **1989b**, 22, 89-99
12Jankowiak, R.; Tang, D.; Small, G. J. and Seibert, M. *J. Phys. Chem.* **1989**, 93, 1649-1654
13Groot, M.-L., Peterman, E.J.G, van Kan, P.J.M.,van Stokkum, I.H.M., Dekker, J.P. and van Grondelle, R. (1994) Biophys. J. 67, 318-330

DYNAMICS OF REACTION CENTERS WITH MODIFIED SECONDARY ELECTRON ACCEPTOR

H. Huber, T. Nägele, W. Zinth, J. Wachtveitl
Institut für Medizinische Optik, Universität München, Germany
M. Meyer, H. Scheer
Botanisches Institut, Universität München, Germany

Introduction

The photoinduced electron transfer in the bacterial reaction center (RC) of *Rb. sphaeroides* R26.1 has been studied extensively in the last years. It is now well established, that the electron transfer steps along the active branch predominantly follow a sequential mechanism. After photoexcitation of the bacteriochlorophyll dimer (P) anions of the monomeric bacteriochlorophyll *a* (BChl a) B_A^-, bacteriopheophytin *a* (BPhe a) H_A^- and primary ubiquinone Q_A^- are formed with time constants of 3 ps, 0.9 ps and 200 ps, respectively. Strong evidence for the reaction via B_A came from experiments on pigment modified samples, where the state B_A^- is in equilibrium with an energetically raised state $P^+H_A^-$ and is populated to a much higher extent than in native RCs [1,2]. We report experiments on RCs where 3-acetyl-Phe a and 3-vinyl-BPhe a are introduced at the BPhe a-sites $H_{A/B}$ [3]. These two pigments are especially interesting, since they both represent a structural link between Phe a used as electron acceptor in plant photosystems and the bacterial BPhe a. The bacteriochlorin type pigment 3-vinyl-BPhe a differs from BPhe a only by its substituent at the C3-position (vinyl instead of acetyl), which results in similar absorption characteristics of the two pigments, with only weak shifts of the main bands in the Q_Y- and the Q_X-region of 3-vinyl-BPhe a to higher frequencies. The 3-acetyl-Phe a differs from BPhe a in the hydrogenation state of its ring II macrocycle and as a chlorin type pigment shows spectral features more similar to Phe a. The absorption spectra of the corresponding RCs are shown in Fig. 1.

Time resolved experiments

The transient absorbance changes were recorded in pump-and-probe experiments with a temporal resolution of 150 fs. Light pulses for the measurements were generated by a Ti:Sapphire laser system operating at a wavelength of 870 nm, that has been described elsewhere [3]. Time resolved measurements in the BChl a-anion band (max. around 1020 nm) showed, that wild type (WT) RCs exhibit a pronounced subpicosecond kinetic

P. Mathis (ed.), Photosynthesis: from Light to Biosphere, Vol. I, 515–518.

component, which can be assigned to a signal originating from the reduced B_A^--intermediate [4]. In addition, Phe a RCs exhibit a strong long lasting absorption increase at that wavelength, that is related with a population of $P^+B_A^-$ in equilibrium with $P^+H_A^-$ (Fig. 2, broken line) [2].

The dynamics of 3-acetyl-Phe a RCs is comparable to WT RCs in this spectral range, although the early absorption increase occurs somewhat slower. At long delay times only a very weak relative absorbance decrease occurs (Fig. 2, dashed-dotted line).
Measurements taken at probing wavelengths of 907 nm (P*-decay), 850 nm (ground state bleaching) and 690 nm (maximum of the $H_{A/B}$-band) show no significant difference to WT data in comparable spectral regions [3].

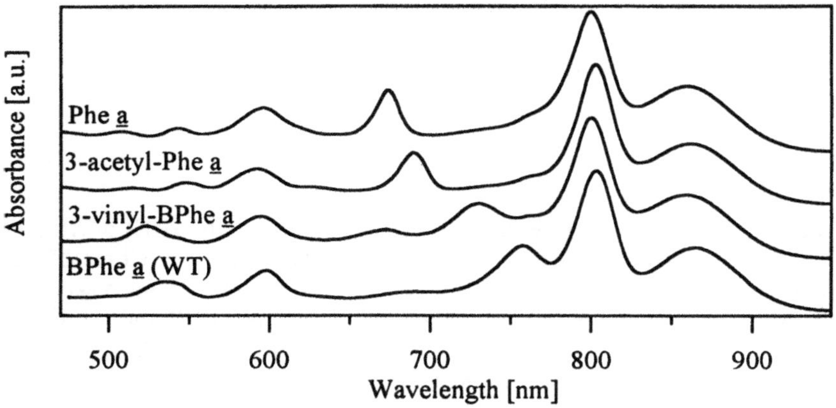

Figure 1. Absorbance spectra of the different modified RCs with the incorporated chromophore at the sites $H_{A/B}$.

We can conclude for RCs containing 3-acetyl-Phe a: The change of the double bond character of ring II - with the 3-acetyl-group being conserved as in BPhe a - mainly affects the absorption properties of the RCs. The dynamic characteristics, however, is very similar to the one found in WT.

The spectral and kinetic properties of 3-vinyl-BPhe a RCs are completely different: The absorption spectrum is more similar to WT (Fig. 1), the changes in kinetics, however, are drastic (Fig. 2). The data obtained at $\lambda_{pr} = 907$ nm display a slower initial decay of the gain signal, which can be fit with a time constant of 5.8 ps [3]. The time resolved measurement in the P-band (at $\lambda_{pr} = 850$ nm) demonstrates a slow decay at long delay times which reflects ground state recovery: In contrast to WT RCs, the P^+-formation is not stably maintained, since about 30 % of P ground state is repopulated within the timescale of the experiment. The time constant used to model this decay is 310 ps. This constant represents an eigenvalue of the rate equation system. It is determined by the microscopic rates of the P^+-decay and the electron transfer from H_A^- to Q_A [3]. A long lived absorption due to the BChl a-anion-radical state could be detected at $\lambda_{pr} = 1017$ nm (Fig. 2, dotted line). After a small initial absorption decrease, caused by the stimulated emission of P*, the signal rises to a pronounced maximum at intermediate delay times. The curve subsequently decreases and a positive absorption difference due to the spectrally broad P^+-band remains at late times. The model curve shown in Fig. 2 is composed of a 1.5 ps and a 5.8 ps component of approximately equal amplitude for the absorption increase and a 310 ps component for the decay of the signal. In analogy to the Phe a containing RCs [1,2], the additional long lived absorption in 3-vinyl-BPhe a RCs on the 10 - 350 ps timescale must be due to a population of the state $P^+B_A^-$, since at a probing wavelength of 1017 nm absorption is only expected from P^+ or B_A^-.

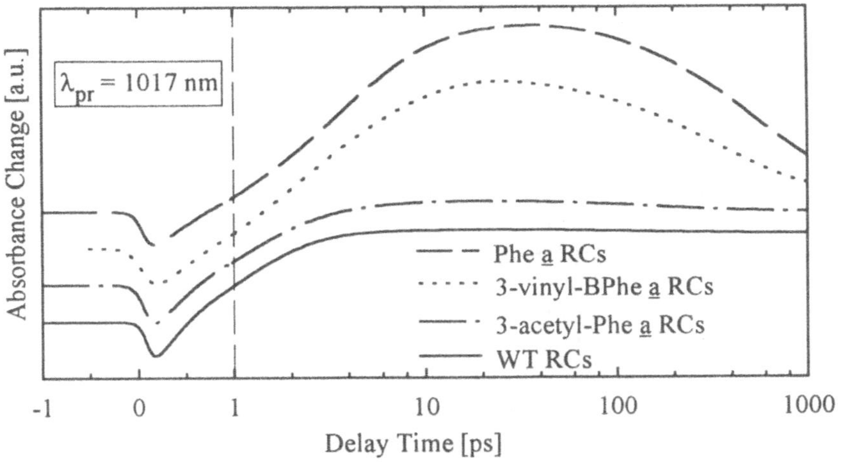

Figure 2. Transient absorption changes of the modified RCs in the $P^+B_A^-$-band. The figure shows best fits to the measured data. RCs that contain pheophytins with a vinyl side-group at the C3-position have a long lasting absorption due to thermal repopulation of the state $P^+B_A^-$ from $P^+H_A^-$.

Discussion

The interpretation of the observations follows non-adiabatic electron transfer theory. The electron transfer model introduces back rates, that are connected to the forward rates via the principle of detailed balance. Additionally, recombination reactions from P* and $P^+B_A^-$ to the groundstate P are considered. This model was described in detail elsewhere [2]. Estimates of the energetics of the intermediate states and the microscopic rates can be obtained by using the observed time constants, the efficiency of $P^+Q_A^-$-formation and the concentration of P* and $P^+B_A^-$ (in equilibrium with $P^+H_A^-$) within this reaction model. The model unfurls that the change in free energy of $P^+H_A^-$ - caused by the exchange of bacteriopheophytin H_A - has drastic consequences on amplitudes and time constants of the observed absorption transients.

The most striking feature of the experiments around 1017 nm is the long lasting absorption exhibited by RCs containing pheophytins with a C3-vinyl group, reflecting a long lived population of the state $P^+B_A^-$. This feature can be reproduced by the model, if the free energy of the state $P^+H_A^-$ is raised to such an extent, that thermal repopulation of $P^+B_A^-$ as well as P* is possible. The calculations deliver an energetic upshift of the state $P^+H_A^-$ in vinyl containing RCs of 1350 to 1400 cm^{-1} ($\approx$170 mV). This reduces the free energy difference between $P^+B_A^-$ and $P^+H_A^-$ to 175 cm^{-1} (20 mV) (in WT: 1550 cm^{-1} (190 mV)) and enables thermal population of $P^+B_A^-$. For free energy differences of 600 cm^{-1} and more the long lasting population decreases beyond the detection limit of our experiments. Therefore this absorption signal can not be detected in 3-acetyl-Phe a containing RCs.

It is interesting to compare the changes in free energy calculated from the reaction model with the changes in redox midpoint potential of the chromophores in solution [3]. The redox values of the vinyl-type chromophores 3-vinyl-BPhe a and Phe a are about 170 mV (1370 cm^{-1}) lower than the one for BPhe a. This difference in redox potential agrees well with the free energy values deduced from the time resolved experiments. The redox potential for 3-acetyl-Phe a is 70 mV (560 cm^{-1}) below the one of BPhe a. Assuming that after incorporation of this pigment in the RC the free energy of $P^+H_A^-$ is shifted up by a similar value compared to WT, the reaction model predicts a very small long lived population density $P^+B_A^-$ of approx. 1 % for RCs containing 3-acetyl-Phe a. This value is at the detection limit of the measurements shown in Fig. 2 (dashed-dotted line) and thus no significant difference in long time kinetics to WT RCs can be observed.

We can therefore conclude that the change in energetics that is caused by the incorporation of the different types of chromophores can be well described by the redox potential change of the pigments. The direct comparison of all four types of RCs demonstrates the role of different functional groups of BPhe a in the ET process. The modifications of ring II strongly influence the absorption properties of the chromophores $H_{A/B}$ and hence should reflect the modified electron density in the molecule. The side-group at the C3-position (acetyl or vinyl) determines the energetics of the intermediate $P^+H_A^-$ via the redox potential. This altered energetics leads to pronounced changes in the reaction kinetics of the isolated RCs.

1 S. Schmidt, T. Arlt, P. Hamm, H. Huber, T. Nägele, J. Wachtveitl, M. Meyer, H. Scheer and W. Zinth (1994) Chem. Phys. Letters 223, 116-120

2 S. Schmidt, T. Arlt, P. Hamm, H. Huber, T. Nägele, J. Wachtveitl, M. Meyer, H. Scheer and W. Zinth (1995) Spectrochim. Acta, in press

3 H. Huber, M. Meyer, T. Nägele, I. Hartl, H. Scheer, W. Zinth and J. Wachtveitl (1995) Chem. Phys., in press

4 T. Arlt, S. Schmidt, W. Kaiser, C. Lauterwasser, M. Meyer, H. Scheer and W. Zinth (1993) Proc. Natl. Sci. USA 90, 11757-11761

Cu^{2+} INHIBITS ELECTRON TRANSPORT THROUGH PSII. AN EPR STUDY.

C. Jegerschöld[1], J. B. Arellano[2], W. P. Schröder[1], M. Barón[3], P. J. M. van Kan[1] and S. Styring[1]. [1]Dept. of Biochem., Stockholm Univ., S-10691 Stockholm, Sweden. [2]Dept. of Biochem., Univ. of Glasgow, Glasgow G12 8QQ, U.K. [3]Dept. of Biochem., Cell and Mol. Biol. of Plants, Est. Experim. del Zaidín, CSIC, Profesor Albareda 1, E-18008 Granada, Spain.

1. Introduction

Cu^{2+} is a potent inhibitor of photosynthetic activity and it has become clear that PS II is more sensitive to Cu^{2+} than PS I. The target for Cu^{2+}-binding within PS II is unclear and reports on Cu^{2+} inhibiting the function on either the donor- or the acceptor side in PS II have been presented (see refs. in ref. 1). Here, we have used EPR-spectroscopy to directly monitor the effects of Cu^{2+} on the function of the redox components in PS II.

2. Materials and Methods

Preparation of PSII membranes and Tris-washing of the PSII membranes to remove the water oxidising complex were performed as in ref. 1 and stored in the following buffer that was also used for all experiments: 50mM Mes-NaOH pH 6.0, 15mM NaCl, 5mM $MgCl_2$, 0.4M Sucrose. The addition of Cu^{2+} to the PSII membranes was done by adding 1000-1500Cu^{2+}/PS II (6-7mM $CuSO_4$) and incubating at room temperature for 10 or 20 min. Then, either of two treatments (at 4°C) followed before resuspending to 2-3mg Chl/mL: Either the sample was washed once in buffer to remove excess Cu^{2+} (C in Table 1; Figs. 1 and 4) or the sample was washed 3 times with EDTA followed by one buffer wash (D in Table 1; Figs. 1 and 4) to test the reversibility of the Cu^{2+} treatment. A and B in Table 1 and Figs. 1 and 4 denote control samples not incubated with Cu^{2+} before a bufferwash (A) or an EDTA wash (B). EPR-samples were prepared as described (1). Chlorophyllconc., atomic absorption, and protein measurements were made as in ref. 1.

3. Results and Discussion

3.1. Cu^{2+} addition affects the protein structure of PS II. Incubation of PSII membranes with Cu^{2+} followed by one buffer wash (Table 1, C) leads to the loss of the 16 kDa extrinsic protein while the 33- and 23 kDa proteins remain bound. Furthermore, 2 Mn-ions were lost (Table 1, C) leading to the irreversible loss of O_2-evolution (not shown). Moreover, $Cytb_{559}$ was fully oxidised and its EPR-signal (g_Z-line; (2)) shifted to g=2.96 indicating that 100% of a low potential form of $Cytb_{559}$ was induced (Table 1, C). Treatments known to bring about changes from high potential reduced $Cytb_{559}$ to low

P. Mathis (ed.), Photosynthesis: from Light to Biosphere, Vol. I, 519–522.

Table 1. Structural effects of copper-treatment of PS II membranes

Samples; treatment		Cu ions PS II[a]	Mn ions PSII[a]	Protein content	Cyt b559 HP/LP[d]
A	PS II control; buffer-washed	0	4	normal	70/30
B	PSII control EDTA-washed	0	4	normal	not. det.
C	PS II + copper; buffer-washed[b]	250	2	90% loss of 16 kDa	0/100
D	PS II + copper; EDTA-washed[c]	3-4	2	100% loss of the 33, 23 and 16 kDa	0/100

a. Metal determination with atomic absorption (based on 220 chlorophylls/PSII centre).
b. Cu^{2+} incubation for 10 min. followed by one buffer wash. See Mat. and Meth.
c. Cu^{2+} incubation for 10 min. followed by three EDTA washes and one bufferwash.
d. HP=High Potential reduced Cytb559. LP=Low Potential oxidised cytb559.

potential oxidised Cytb559 include washing both the 23- and 16 kDa extrinsic proteins away or selectively removing Ca^{2+} from PS II (2 and refs. therein, 3). Our results indicate that the 16 kDa subunit by itself may influence the redox state of Cytb559. Another possibility is that Cu^{2+} induces a loss of Ca^{2+} together with the 2 lost Mn-ions resulting in a change in Cytb559. EDTA washing of a Cu^{2+} sample (Table 1, D) removed also the 23- and 33 kDa proteins. In a control, not treated with Cu^{2+}, all extrinsic proteins and 4 Mn/PSII remain bound after the EDTA washing procedure (Table 1, B).

3.2. The presence of Cu^{2+} in PS II blocks the oxidation of Tyr_Z. The function of the tyrosine electron donors Tyr_D and Tyr_Z were investigated in the presence and absence of Cu^{2+} in Tris-washed PSII membranes. Tyr_D^{ox} gives rise to the Signal II_{slow} (inner traces in Fig. 2 and Fig. 3, inset) which is stable for about an hour in Tris-washed PSII membranes. Incubation with Cu^{2+} had no effect on the amplitude or shape of Signal II_{slow}. Therefore SII_{slow} could be used as an internal spin standard amounting to 1 radical per PSII centre. At room temperature the reduction of Tyr_Z^{ox} occurs in ≈130 ms in Tris-washed PSII membranes (lacking the water oxidising complex). This allows the observation of Tyr_Z^{ox} (denoted Signal II_{fast}) as a doubling of the amplitude of Signal II brought about by continuous illumination into the EPR cavity (Fig. 2, lower spectra, outer trace and Fig. 1, Signal II_{fast}, A). However, in the presence of Cu^{2+}, Tyr_Z could not be oxidised during illumination at room temperature (Fig. 1, Signal II_{fast}, C) indicating an impairment of its function by the presence of Cu^{2+}. Instead a new radical with a simple shape, g= 2.0028 and linewitdh 9.5±0.3 G was observed during the illumination (Fig. 1, Radical RT, C and Fig. 2, upper spectra, outer trace). After EDTA washes (Fig. 1, Signal II_{fast}, D) ≈70% of Tyr_Z^{ox} again was induced indicating: (i) the reversibility of the Cu^{2+} induced block in Tyr_Z oxidation (ii) most of the 2-3 Cu^{2+}/PSII remaining after the EDTA washes do not bind to PSII, at least not close to Tyr_Z. Furthermore, the formation of another light-induced radical, in the presence of Cu^{2+} instead of Tyr_Z^{ox} indicated that the primary charge separation was still operational.

Illumination at 200K allows only one charge separation. We therefore illuminated a Cu^{2+} sample (Fig. 1, 200K, C and Fig. 3) to investigate which other redox-components in PS II that were still functional after one charge separation. The results

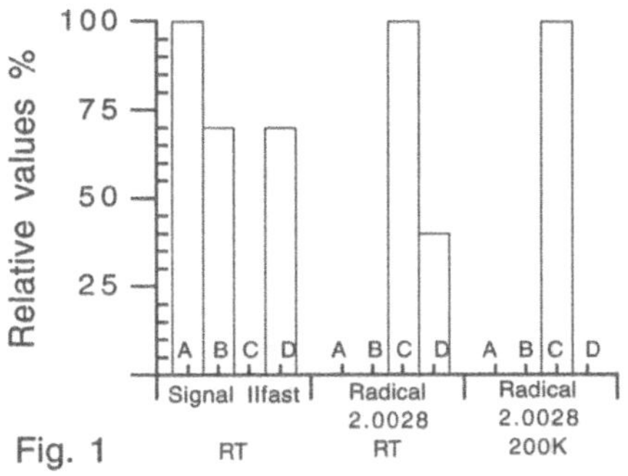

Fig. 1

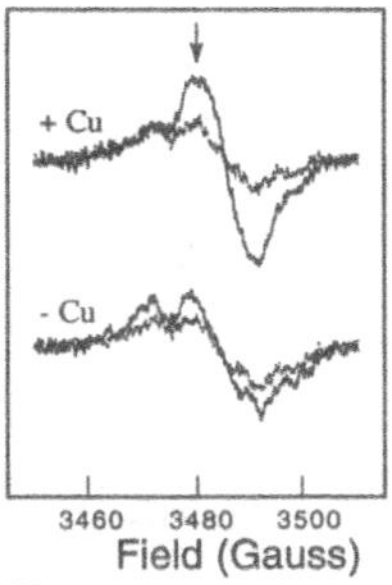

Fig. 2

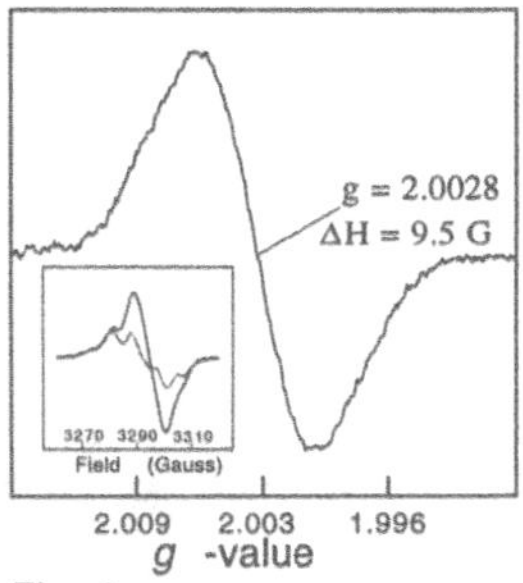

Fig. 3

Figs. 1, 2, 3; Effects of Cu^{2+} on Donor Side radical induction in Tris washed PSII membranes. EPR spectra (standard settings, ref. 1) at room temperature (RT) are shown in Fig.2 or at 77K after ***illumination at 200K in Fig.3. The amount of light induced radicals (Signal II_{fast} and the radical with g=2.0028) is presented in Fig. 1.***

showed that 1.0 radical/PSII with the same characteristics as the radical formed at room temperature was induced (Fig. 3). This shows that the same pathway is operational at room temperature and 200K. In the control samples (Fig. 1, 200K, A and B) no radical at g=2.0028 was induced by 200K illumination. The reversibility of Cu^{2+} binding on the donor side of PSII was further demonstrated by the lack of a radical at 2.0028 in the EDTA washed sample (Fig. 1, 200K, D).

3.3. Removal of added Cu^{2+} reveals that the acceptor side had been irreversibly affected by Cu^{2+}. Two EPR-signals have been characterised from reduced Q_A. The Q_A^--Fe^{2+} signal depends on the presence and correct magnetic coupling of both Q_A^- and the ferrous non-haem iron (Fig. 5, inset, upper trace). However, in the absence of a coupling to the non-haem iron, a featureless spectrum with $g \approx 2.004$ and linewidth 9.5±0.5 G assigned to Q_A^- can be observed (4, 5). In our experiments Tris-washed PSII membranes were either illuminated at 200K (Fig. 4, 200K) or reduced with dithionite (Fig. 4, Dit) to reduce Q_A. In case Cu^{2+} was present, the Q_A^--Fe^{2+} EPR-signal could not be detected due to the large overlapping epr-signal from Cu^{2+} (Fig. 4, 200K, C). However, if Cu^{2+} was removed, by reduction either to Cu^+ (Fig. 4, Dit, C) or by EDTA washes (Fig. 4, 200K and Dit, D; Fig. 5, inset, lower trace), the Q_A^--Fe^{2+} signal could be observed in 30% of the centres.

In addition and very interestingly, a radical with the characteristics of a Q_A^- not coupled to a Fe^{2+} (Fig. 5) was observed in 70% of the centres in the chemically reduced samples (Fig. 4, Dit, C and D). Together, these results show that Q_A could be reduced chemically in all centres but that the Cu^{2+} incubation had modified 70% of the centres irreversibly. This indicates that the non-haem iron had been released or replaced by Cu^{2+} in these centres. The presence of 30% Q_A^- Fe^{2+} in samples before and after the EDTA washes shows that it is the Cu^{2+} treatment and not the EDTA washings that induce the acceptor side modification. Finally, 200K illumination of an EDTA washed Cu^{2+} treated sample (Fig. 4, 200K, D) induced about 25% of the uncoupled Q_A^- radical.

Why then is the free Q_A^- radical not induced on top of the donor side g=2.0028 radical after 200K illumination of a Cu^{2+} containing sample? An interesting explanation could be that, in the presence of paramagnetic Cu^{2+}, free Q_A^- relaxes too fast to be detected. Therefore, only the donor side radical is observed. Only in the absence of

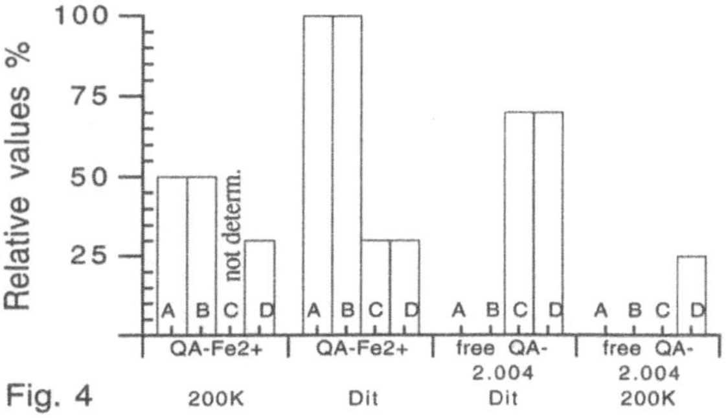

Fig. 4

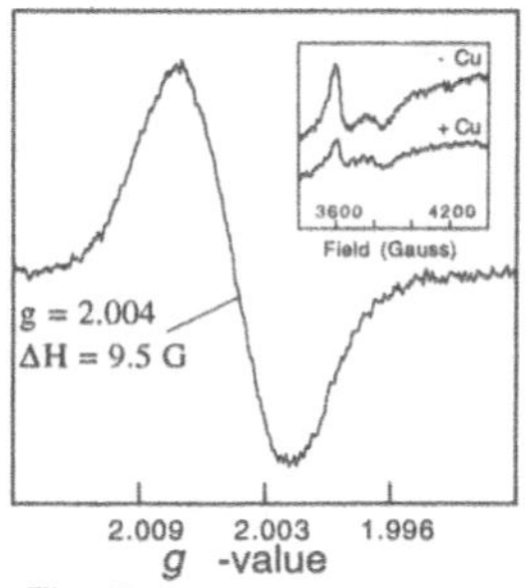

Fig. 5

***Figs. 4, 5:** Effects of Cu^{2+} on Acceptor Side radical induction in Tris washed PSII membranes. EPR spectra (standard settings, ref. 1) from free Q_A^--radical at g=2.004 is shown in Fig. 5 and the Q_A^--Fe^{2+} signal in Fig. 5, inset. (EDTA washed, ± Cu^{2+}, dithionite reduced). The amount of the signals are presented in Fig. 4.*

Cu^{2+} ions (after EDTA wash) or in the presence of diamagnetic Cu^+ (after dithionite reduction) the uncoupled Q_A^- signal can be observed. This hypothesis would also imply that Cu^{2+} binds in the vicinity of Q_A, perhaps in the Fe^{2+} site.

4. Conclusions

Cu^{2+} addition induces a reversible block in Tyr_Z oxidation and an irreversible modification of the acceptorside close to Q_A. The primary charge separation is still operational as seen by the (i) possibility to oxidise another radical (g=2.0028) instead of Tyr_Z (Figs. 1-3) and (ii) the ability to reduce Q_A in all centres (Figs. 4,5). This has been shown qualitatively in this communication. Experiments are under way to investigate the stoichiometric relationships and time dependencies for the Cu^{2+} effects on the function of (i) the donor side and (ii) the acceptor side and (iii) the structure of PSII. This makes it possible to use Cu^{2+} as a probe to the structure and function of Tyr_Z and Q_A in PS II.

Financial support is gratefully acknowledged from; the NFR (S.S.), Sven and Lilly Lawski Foundation (C.J.) EMBO and MWO (P. v K.) and ESF Photosynthesis Programme (J.B.A.).

References

1. Jegerschöld, C., Arellano, J.B., Schröder, W.P., van Kan, P.J.M., Barón, M. and Styring, S. (1995) Biochemistry, in press.
2. Miller, A.-F. and Brudvig, G.W. (1991) Biochim. Biophys. Acta 1056, 1-18
3. McNamara, V. P. and Gounaris, K. (1994) in Proceedings from the Second Robert Hill Symp. on Photosynth. (Barber, J., ed.) pp. 50-51, Imperial College of Science, Technology and Medicine, London
4. Klimov, V.V., Dolan, E., Shaw, E.R. and Ke, B (1980) Proc. Natl. Acad. Sci. USA 77, 7227-7231
5. Sanakis, Y., Petrouleas, V. and Diner, B.A. (1994) Biochemistry 33, 9922-9928

DYNAMIC STUDIES ON THE PRIMARY PHOTOCHEMICAL REACTION IN THE ISOLATED PHOTOSYSTEM II REACTION CENTER BY TIME-RESOLVED ABSORPTION AND FLUORESCENCE SPECTROSCOPY

Jian-Min Hou, Ting-Yun Kuang, Zhen-Bao Yu, Chong-Qin Tang, Kun-Yun Yang, Pei-Song Tang
Laboratory of Photosynthesis, Institute of Botany, Academia Sinica, Beijing 100044, P. R. China

Tong Ye, Yan Cui, Yao-Dong Chen, Shui-Cai Wang, Xun Hou
State Key Laboratory of Transient Optics Technology, Xi'an Institute of Optics and Precision Mechanics, Academia Sinica, Xi'an 710068, P. R. China

1. INTRODUCTION

In spite of the primary reaction of the isolated Photosystem II (PSII) reaction center had been studied using time-resolved spectroscopic and other techniques by a number of groups, the time constants of energy transfer and primary charge separation in the PSII reaction center are still subject to discussion[1-3].

In this paper we investigated the dynamic of the primary photochemical reaction in the PSII reaction center using time-resolved absorption and fluorescence spectroscopy. A 350 ps fluorescence component may be attributed to the energy transfer associated with β–carotene (β–Car) molecule. The possible energy transfer scheme is proposed. The preliminary data showed the time constant of primary charge separation is about 3-ps.

2. MATERIALS AND METHODS

PSII particles was prepared according the reference [4]. PSII core complex and PSII reaction center D1/D2/Cyt b-559 complex were isolated mainly as described by Nanba and Satoh [5], and collected in the elution from the column respectively.

The subnanosecond fluorescence measurement was obtained by the home-build setup [6]. A mode-locking Argon ion lasers with 200 mW, 82 MHz, 200 ps pulse at 488 nm or 514 nm is used as the excitation beam. The fluorescence decays were detected by the fast photomultiplier and analyzed by the boxcar averager in the range of 650 nm to 740 nm with a resolution of 5 nm. The final time resolution was about 100 ps.

The picosecond pump-probe absorption difference and the picosecond fluorescence measurement were also obtained by the home-build setup [7]. For picosecond absorption difference and picosecond fluorescence measurement the 80 fs, 82 MHz femtosecond Ti:saphire laser pulse which is pumped with Argon ion laser is frequency-doubled, which produced doubled light used as excitation beam and remain fundamental light pulses as probe beam. The wavelengths of the pump and probe pulse were 720 nm to 840 nm and 360 to 420 nm respectively. The absorption decays were analyzed by optical multichannel analyzer (OMA). The fluorescence decays were detected by the frequency up-turn conversion and the photon counting technique. The time resolutions of both system were about 470 fs.

P. Mathis (ed.), Photosynthesis: from Light to Biosphere, Vol. I, 523–526.

Lifetimes were calculated by the global analysis after deconvolution and the Marquardt fitting algorithm assuming multiexponential kinetics. The quality of the fits was assessed using a reduced X^2 criterion and plots of weighed residuals.

3. RESULTS

3.1 The subnanosecond fluorescence measurement

The peak of the long-wavelength absorption band of the isolated PSII reaction center shifted by less than 1 nm before and after the experiment measurement, which suggest that the sample activity loss is less than 10 %. Table 1 showed that the analysis result at 488 nm excitation wavelength of the isolated PSII reaction center complex. The chi square X^2 is 1.01. Four components were obtained, in which the 350 ps lifetime is main fluorescence component with relative amplitude of 74%. When the excitation wavelength was 514 nm, it also obtained four components. However the short component changed into 100 ps as shown in Table 2 with the chi square X^2=1.00. The 100 ps lifetime is main fluorescence component with relative amplitude of 73%.

Table 1. The analytical result of PSII reaction center at excitation of 488 nm

Component	Lifetime (ps)	Relative amplitude (%)
1	350	74
2	1500	0.6
3	6000	24
4	38000	1.4

Table 2. The analytical result of PSII reaction center at excitation of 514 nm

Component	Lifetime (ps)	Relative amplitude (%)
1	100	73
2	1500	5
3	6000	20
4	38000	2

3.2 The picosecond fluorescence and pump-probe absorption difference measurement

Table 3 and Table 4 showed the analytical results of picosecond fluorescence and picosecond pump-probe absorption difference measurement. In the pump-probe measurement we obtained three components for PSII particles and two components for PSII core complex. In the picosecond fluorescence experiment we obtained three components for PSII particles and four components for PSII core complex.

4. DISCUSSION

4.1 The primary charge separation process

The primary charge separation time constant is still a matter to discussion. Some authors considered that this time constant is about 3 ps at room temperature, while others suggested 21 ps time constant based on their experimental results [1-3]. Our observation showed that 3-3.5 ps and

22 ps components for PSII particles, when the PSII light-harvesting complex is removed only 3 ps component was obtained. This preliminary finding supported the 3 ps time constant proposed by Wasielewski and co-workers [3].

Table 3. The analytical result of PSII particles and PSII core complex by pump-probe absorption difference measurement

Component	Lifetime (ps)	Relative amplitude (%)
PSII particles		
1	1.5	2
2	3.0	17
3	22	22
PSII core complex		
1	1.6	0.2
2	3.5	29

Table 4. The analytical result of PSII particles and PSII core complex by picosecond fluorescence method

Component	Lifetime (ps)	Relative amplitude (%)
PSII particles		
1	0.8	2
2	2.7	9
3	23	43
PSII core complex		
1	3.5	6
2	12	8
3	25	26
4	100	20

4.2 The energy transfer process

It had been obtained 100 to 500 ps component with time-resolved spectroscopy [8-10]. However, the origin of this component remains to be clarified. It is considered as slow energy transfer, or charge recombination, or nuclear relaxation, or heterogeneity of samples [9]. Till now there are not direct experiment evidence.

Our result showed that the 100 to 500 ps component is dependent upon the excitation wavelength. When we used 488 nm as excitation pulse 350 ps component is obtained, and 514 nm excited the PSII reaction center 100 ps component is observed. As we know that β–Car absorbed in the range of 400 to 500 nm, while pheophytin a (Pheo) absorbed in the 500 to 550 nm [5,11]. Therefore, 350 ps component is related to β–Car molecule in the PSII reaction center and 100 ps is associated with Pheo.

It had reported that energy transfer time constants in the PSII reaction center between Chla and primary electron donor P680, Pheo and P680 were about 20-30 ps, 40-70 ps respectively [8,12]. In our experiment the time resolution is limited to 100 ps, we observed 100 ps component may be attributed to energy transfer between Pheo/Chla and P680. Barber and co-workers and our group had been observed that in the presence of artificial electron acceptor or donor β–Car molecule is damaged by exposure of strong light conditions [13-14]. This finding suggested that β–Car molecule is most probably near P680. Therefore energy transfer may occurred between β–Car and P680. Here we proposed a possible energy transfer scheme in the PSII reaction center as in Fig. 1.

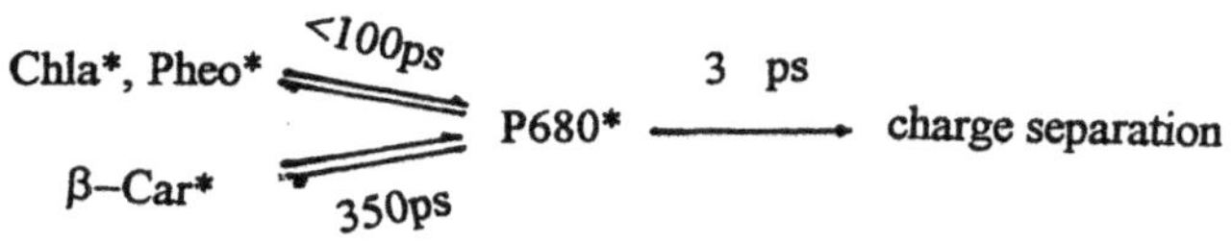

Fig. 1. The possible energy transfer model in the PSII reaction center

5. ACKNOWLEDGMENTS

This work is supported by the Pandeng Project of the China Commission for Science and Technology and the National Natural Science Foundation of China and Academia Sinica.

REFERENCES

1. Van Grondelle, R., Dekker, J. P., Gillbro, T., Sundstrom, V. (1994) Biochim. Biophys. Acta, 1187, 1-65
2. Rech, T., Durant, J. R., Joseph, D. M., Barber, J., Porter, G., Klug, D. R. (1994) Biochemistry, 33, 14768-14774
3. Wiederrecht, G. P., Seibert, M. R., Govindjee, Wasielewski, M. R. (1994) Proc. Natl. Acad. Sci. USA, 91, 8999-9003
4. Kuwabara, T., Murata, N. (1982) Plant Cell Physiol., 23, 533-539, 1982
5. Nanba, O., Satoh, K. (1987) Proc. Natl. Acad. Sci. USA, 84, 109-112
6. Ye, T., Cui, Y., Hou, J.-M., Chen, Y.-D., Yu, Z.-B., Wang, S.-C., Tang, C.-Q., Kuang, T.-Y., Hou, X. (1995) Acta Photonica Sincica, in press
7. Tong Ye (1994) Ph. D. Dissertation, Xi'an Institute of Optics and Precision Mechanics, Academia Sinica
8. Roelofs, T. A., Kwa, S. L. S., Van Grondelle, R., Dekker, J. P., Holzwarth, A. R. (1993) Biochim. Biophys. Acta, 1143, 147-157
9. Seibert, M. (1993) In The Photosynthetic Reaction Center (Deisenhofer, J., Norris, J. R., eds.) pp. 319-356, Academic Press, San Diego, California, USA
10. Crystall, B., Booth, P. J., Klug, D. R., Barber, J., Porter, G. (1989) FEBS Lett., 213, 241-244
11. Satoh, K. (1993) In The Photosynthetic Reaction Center (Deisenhofer, J., Norris, J. R., eds.) Academic Press, San Diego, California, USA
12. Tang, D., Jankowiak, R., Seibert, M., Yocum, C. F., Small, G. J. (1990) J. Phys. Chem., 94, 6519-6522
13. Telfer, A., De las Rivas, J., Baber, J. (1991) Biochim. Biophys. Acta, 1060, 106-114
14. Tang, C.-Q., Kuang, T.-Y., Yu, Z.-B., Peng, D.-C. & Tang, P.-S. (1994) Acta Botanica Sinica 36(4), 257-264

HYDROGEN-BONDING EFFECT ON EPR SPECTRA OF ^{13}C-LABELED UBISEMIQUINONES *IN VITRO*

T.N. Kropacheva[1], J. Raap[2], J. Lugtenburg[2], A.J. Hoff[3]
[1]Dept. of Chemistry, Udmurt State University, Izhevsk, 426037, Russia;
[2]Dept. of Chemistry, Leiden University, P.O.Box 9502, 2300RA Leiden, The Netherlands;
[3]Dept. of Biophysics, Leiden University, P.O.Box 9504, 2300RA Leiden, The Netherlands

1. Introduction

The biological importance of quinones (Q) is related to their electron carrier properties, i.e. the ability to act as one or two-electron acceptors. In the reaction centers of *Rb.Sphaeroides* two ubiquionones serve as both primary and secondary acceptors, Q_a and Q_b respectively, the reduction process of which is schematicly shown below (1):

$$DQ_aQ_b \rightarrow DQ_a^{\cdot-}Q_b \rightarrow DQ_aQ_b^{\cdot-} \rightarrow DQ_a^{\cdot-}Q_b^{\cdot-} \rightarrow DQ_aQ_b^{2-} \rightarrow DQ_aQ_bH^- \rightarrow DQ_aQ_bH_2$$

The differences in Q_a and Q_b behavior (the number of accepted electrons; the midpoint potentials; the proton binding) result from the different protein environment of both ubiquinones. Several types of spectroscopy (optical, EPR/ENDOR, IR) are presently used to explore the native binding sites of quinones in the reaction centers. All these studies are usually accompanied by preliminary investigations of model systems (solvents of different polarity, detergent solutions, films, frozen samples). For EPR studies it is advantageous to use reaction centers reconstituted with ^{13}C-labeled quinones, because ^{13}C-hyperfine coupling constants are known to be very sensitive to the environment (2). In the present work the hydrogen-bonding effect on ^{1}H and ^{13}C isotropic hyperfine couplings of ubisemiquinones was studied in a number of different solvents by using the simultaneous electrochemical-EPR technique. The anion radicals of ubiquinones-0 and 10 were generated inside the microwave cavity of EPR spectrometer in a coaxial electrolysis cell consrtucted according to the literature (3). Compared to a conventional EPR flat cell, this cell provides more accurate potential control, a higher sensitivity and allows to use solvents with high dielectric constants. The cell can be used to study the mechanism of complex electrode reactions that include paramagnetic intermediates.

2. Procedure

The design of the EPR-electrochemical cell was essentially the same as in (3) exept for the following. The working electrode was constructed like a 4 cm-length helix from 1 mm diameter gold wire. It tightly fits along the inner wall of 4 mm i.d. quartz tube. The auxiliary electrode (0.5 mm diameter platinum wire) was positioned inside the helix along

P. Mathis (ed.), Photosynthesis: from Light to Biosphere, Vol. I, 527–530.

its axis. The leads to both electrodes were mounted in a stop-cork of the quartz tube, so that the constraction can be easily disassembled and washed. The tap attached to the stop-cork allowed the cell to be connected to a vacuum line for the degassing procedure. The whole constraction was mounted directly into the center of the microwave cavity of JEOL JES-RE2X spectrometer. For aqueous solutions the standard mini-reference silver-silver chloride electrode was placed in the side-arm of the cell. All measurements were done at room temperature. The EPR spectra were simulated using the JEOL software.

3. Results and Discussion

The environmental effect on semiquiones formation during the ubiquinones reduction cycle was studied by electrochemical generation of radicals in different solvents. The aqueous solutions, which are more characteristic for all *in vivo* processes, were examined first. With a magnetic field adjusted at the central peak of the EPR spectrum the changes in EPR signal intensity were monitored as a function of electrode potential in an alkaline solution of ubiquinone-0 (UQ-0) (Fig.1). The cyclic voltammogram of UQ-0 recorded under the same conditions shows one wave, which can be attributed to the direct two-electron reduction ($Q + 2e \rightarrow Q^{2-}$) or to the situation when two individual waves ($Q + e \rightarrow Q^{\cdot -}$ and $Q^{\cdot -} + e \rightarrow Q^{2-}$) are not resolved. The later is possible if the potentials for both steps differ less than 0.1V. The comparison of EPR signal with the voltammogram indicates that the signal increases when the reduction begins and continues to increase after the cathodic peak. Such behavior can be explained assuming that semiquinone radicals are formed through two reactions: $Q + e \rightarrow Q^{\cdot -}$ and $Q + Q^{2-} \rightarrow 2Q^{\cdot -}$. The second process is essential since the EPR signal still increases even when according to the voltammogram only dianion Q^{2-} occures. The degradation of EPR signal during the direct scan is connected with the further reduction of $Q^{\cdot -}$ as well as with the decrease of Q concentration during the electrolysis. At slow scan rates the signal disappeares comletely, while at fast scan rates the process is diffusionally controlled. When the potential sweep is reversed, a new build-up of radicals is observed which is parallel to the increase of anodic current ($Q^{2-} - e \rightarrow Q^{\cdot -}$, $Q^{\cdot -} - e \rightarrow Q$). Thus, the potential-determined behavior of the EPR signal qualitatively represents the complex combination of electrode and solution reactions that lead to the semiquinone formation.

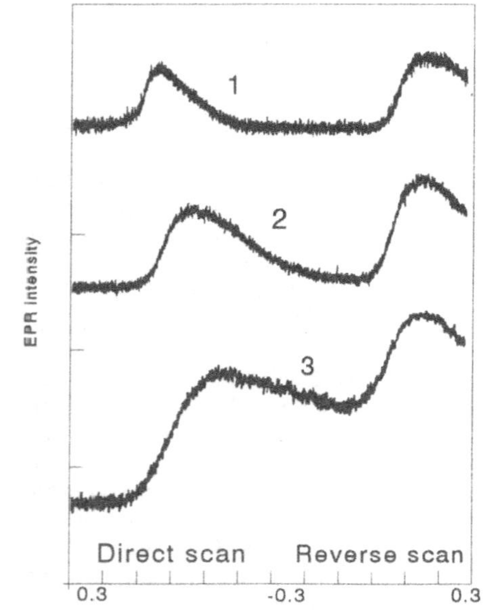

Figure 1. EPR intensity during ubiquinone-0 reduction-oxidation cycle in aqueous solution at pH 9.4. Scan rate: 1-2 mV/s; 2-5 mV/s; 3-10 mV/s.

The EPR specra of ubisemiquinone-0 were not detected in acidic and neutral solutions; only at pH above 8.5-9.0 the $Q^{\cdot -}$radicals are stable enough to be measured. The instability of semiquinone is due to its protonation followed by fast reduction ($QH^{\cdot} + e \rightarrow QH^{-}$) (4).

Besides, when the protonated forms of Q^{2-} exist the disproportionation equilibria
$QH^- + Q \leftrightarrow 2Q^{\cdot -} + H^+$, $QH_2 + Q \leftrightarrow 2Q^{\cdot -} + 2H^+$
are pH dependent, so that for obtaining a sufficient amount of $Q^{\cdot -}$ one needs to use alkaline solutions.
The generation of anion radicals in a number of solvents revealed changes in proton and ^{13}C-splitting constants when the solvent polarity was varied. The changes in splittings induced by a solvent are due to the redistribution of electron and spin densities upon solvatation (5). In all polar aprotic solvents (DMF, DMSO, DME, MeCN), which are capable to only non-specific solvatation of a radical, the measured splittings are almost the same. For the protic solvents (alcohol / water) there are the systematic changes in splitting constants with an increase of the solvent's ability to hydrogen bonding interactions.
The solvent effect becomes clearer when H-splittings are measured in a serious of mixed solvents with different mole fraction of a protic component (Fig.2). For UQ-0 the changes in splittings with the increase of solvent polarity are opposite for two groups of protons. The increase of splittings for the methyl protons shows the increase of spin density on the ajacent (i.e.5-C) carbon. On the contrary, the spin density on 6-C atom gradually decreases upon addition of a protic solvent. By analyzing the set of possible resonance

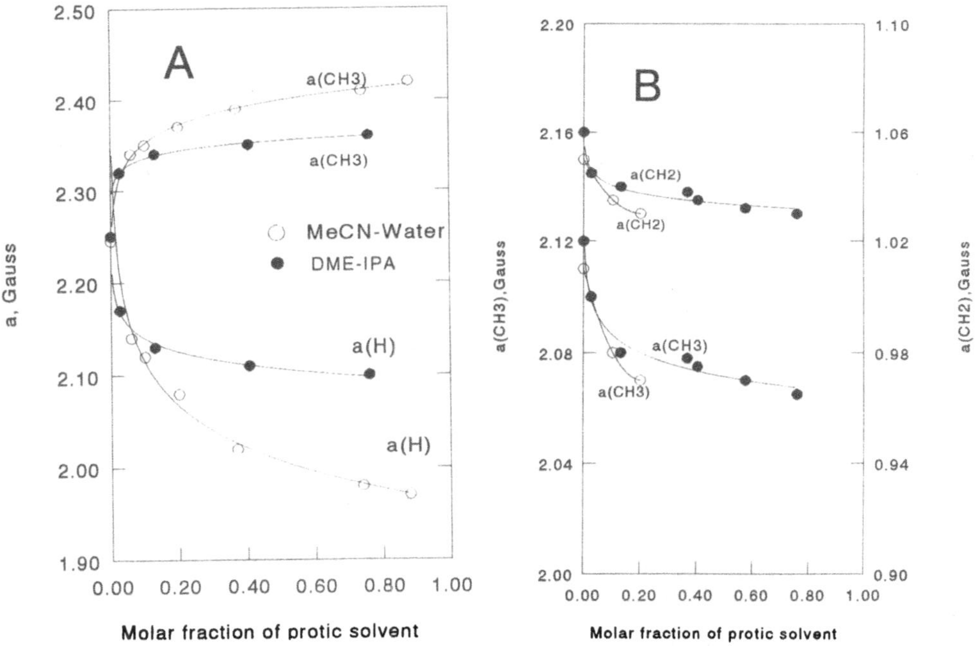

Figure 2. Solvent dependence of isotropic hyperfine H-splitting constants for UQ-0 (A) and UQ-10 (B). Abbreviations: MeCN - acetonitrile; DME - dimethoxyethane; IPA - isopropyl alcohol.

structures it is possible to conclude that spin density increases on the 1,3,5-C atoms compared to 2,4,6-C only when the 1-O atom is more electronegative. Thus, for UQ-0 the carbonyl oxygens are not equvivalent and the first hydrogen bond is formed via 1-O atom.
The preferable solvatation of 1-O atom can explain the less important solvent effect on

EPR spectra of UQ-10 compared to UQ-0 . The intermolecular hydrogen bond formed by the methylene proton of polyisoprene chain with the neighbouring 1-O atom reduces the ability of ubisemuquinone-10 to external hydrogen bonding. In addition, in UQ-10 the 1-O atom is screened by bulky hydrophobic chain from the interactions with the polar molecules. The stronger effect of water compared to isopropanol on the H-splittings for both ubiquinones agrees with the high solvatation power of water, which is the highest of all studied solvents.

The solvent effect on the isotropic ^{13}C-hyperfine splittings was studied for specifically ^{13}C-labeled ubiquinones with the ^{13}C atom at the fourth (carbonyl) position (IUPAC numbering). An increase of alcohol content in mixed solvents caused a much more appreciable change in ^{13}C-splittings compared to proton splittings (Fig.3). The study of the differences in the linewidth of the low and high field ^{13}C-lines shows the change of the ^{13}C-splitting sign from the negative to the positive values upon increasing hydrogen-bonding of ubisemiquinone-10. The high sensitivity of carbonyl ^{13}C-splittings in ubisemiquinones to the external hydrogen-bonding interactions will allow to use ^{13}C-labeled quinones for studying with EPR the binding properties of UQ-10 in photosynthetic reaction centers.

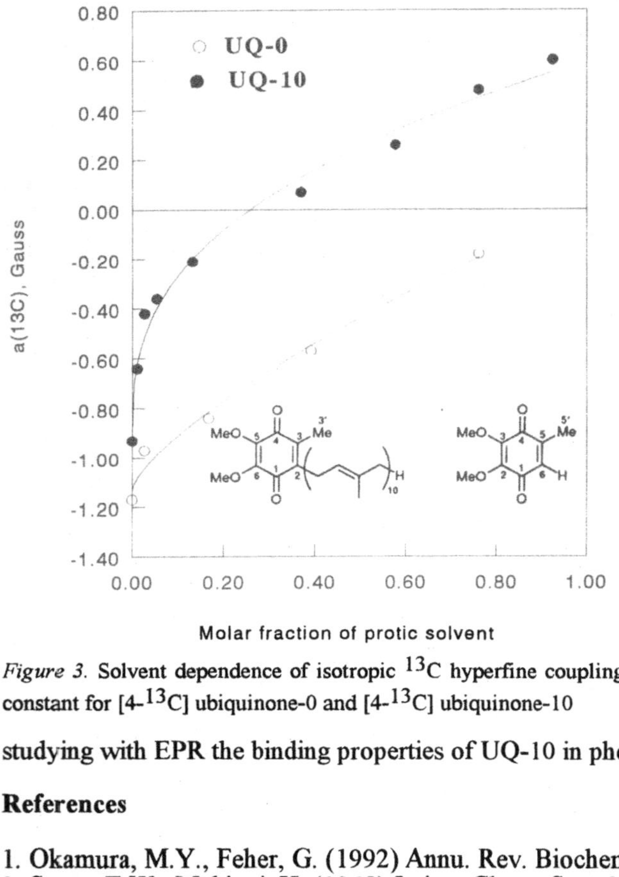

Figure 3. Solvent dependence of isotropic ^{13}C hyperfine coupling constant for [4-^{13}C] ubiquinone-0 and [4-^{13}C] ubiquinone-10

References

1. Okamura, M.Y., Feher, G. (1992) Annu. Rev. Biochem. 61, 861-896
2. Stone, E.W., Maki , A.H. (1965) J. Am. Chem. Soc. 87, 454-458
3. Allendoerfer, R.D., Martinchek, G.A., Bruckenstein, S. (1975) Analytical Chemistry 47, 890-894
4. Chambers, J.Q. (1988) in The Chemistry of Quinonoid Compounds (Patai S. and Rappoport Z., eds.), pp. 719-757, John Willey and Sons Ltd., London
5. Reichardt, C. (1988) in Solvents ant Solvent Effect in Organic Chemistry, VCH, Germany

TOPOLOGY OF CHROMOPHORES AND COFACTORS IN PSII AS DERIVED FROM THE LOCAL ELECTROCHROMIC BANDSHIFTS†

D.A. Cherepanov[1], M. Haumann[2], W. Junge[2] and A. Mulkidjanian[2]

[1]A.N.Frumkin Institute of Electrochemistry, Moscow, 117071, Russia,
[2]Abt. Biophysik, Universität Osnabrück, D-49069 Osnabrück, Germany

1. Introduction

Recent structural models of PSII from plants [1,2] are shaped along the known structure of bacterial RC [3]. The water oxidizing system of PSII, however, seems to call for a different structure to cope with the extremely high redox potential of the cofactors. The electrochromic difference spectra induced by a negative charge on Q_A, a positive charge on D1 Tyr-161 (Y_Z) and a positive charge on the Mn cluster were measured with PSII core particles and deconvoluted into Gaussian components. Our results suggest an arrangement of pigments and redox cofactors that differs from the structure of bacterial RC.

2. Materials and Methods

Oxygen evolving PSII core particles from pea seedlings were suspended at 5 μM of chlorophyll in a medium with 5 mM $CaCl_2$ and 5 mM MES, pH 6.5. Deactivation of the Mn-cluster was achieved by TRIS-washing. Samples were dark-adapted for 15 min. DCBQ and hexacyanoferrate(III) were used as electron acceptors. Difference spectra were deconvoluted into Gaussian components by a numeric Powell optimization procedure. The electrochromic shifts were related to the topology by using the data on electrochromism of chlorophylls *in vitro* and in the bacterial RCs as well as the results of INDO/s calculations [4].

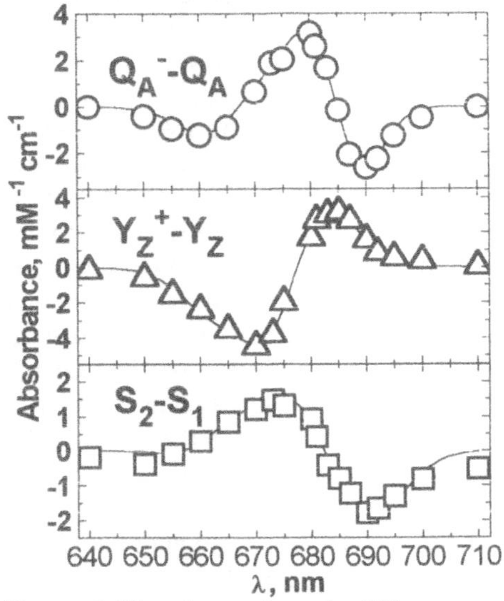

Figure 1. The electrochromic difference spectra measured in core particles of PSII for $Q_A \Rightarrow Q_A^-$ (○), $Y_Z \Rightarrow Y_Z^+$ (△) and $S_1 \Rightarrow S_2$ (□) transitions.

3. Results

The difference spectra of electrochromic absorbance transients $Q_A^- - Q_A$, $Y_Z^+ - Y_Z$ and $S_2 - S_1$ are given on *Figure 1*. Points are

P. Mathis (ed.), Photosynthesis: from Light to Biosphere, Vol. I, 531–534.

experimental and lines are calculated with obtained fit parameters. The following gross features are obvious: 1) the response to Q_A^- formation is composed of a red shift centered at about 670 nm and a blue shift centered at about 685 nm; 2) the response to Y_Z^+ formation is caused by a red shift centered at about 678 nm; 3) the response to the $S_1 \Rightarrow S_2$ transition is mainly caused by a blue shift at about 681 nm. The measured spectra could be singularly deconvolved into four Gaussian components, and from analogy with the electrochromic difference spectra in photosynthetic bacteria [5-7] and known spectral properties of PSII tentatively attributed to accessory chlorophyll *a* (Chl_A), two chlorophylls of the primary donor (P_A and P_B) and pheophytin *a* ($Pheo_A$), respectively. The results of the fit are shown as the solid lines on *Fig. 1* and are included in *Table 1*.

Table 1. The fit parameters of the four-component deconvolution.

Peak position λ_0, nm	Bandwidth w, nm	Q_A^--Q_A $\Delta\lambda_0$, nm	Y_Z^+-Y_Z $\Delta\lambda_0$, nm;	P^+	S2-S1 $\Delta\lambda_0$, nm	Attributed to
667.3	15.7	+0.088	+0.141	-	+0.015	Chl_A
677.1	14.4	0	+0.252	-	+0.045	P_B
680.5	20.9	0	+0.125	1.44%	-0.269	P_A
684.6	11.7	-0.233	+0.006	-	-0.002	$Pheo_A$

4. Discussion

In purple bacteria, the major component of the relevant electrochromic bandshifts caused by the $BQ \Rightarrow BQ_A^-$ transition is a red shift of the respective Q_y peaks of $Bpheo_A$ and a smaller blue shift attributed to the accessory bacteriochlorophyll [5-7]. A comparison of the INDO/s-calculated bandshifts with the experimental data obtained for *Rps. viridis* demonstrated good predictability of the theory [4]. The electrochromic difference spectrum for the reaction $Q_A \Rightarrow Q_A^-$ shown on *Fig. 1* is in line with previously published [8,9]. The appearance of an electron on Q_A caused a blue shift of $Pheo_A$ and a smaller red shift of Chl_A that implies an inverted orientation of $Pheo_A$ ring relative to Q_A as compared with the situation in bacteria. The difference spectrum Y_Z^+-Y_Z is similar to the one obtained in [10]. Its asymmetry is readily understood as an admixture of 1.44% bleaching of *P* at 680.5 nm (*Table 1*) that corresponds to the difference of 110 mV in midpoint potentials of *P* and Y_Z. L-Arg135 which takes the position of Y_Z in the bacterial RC lies almost on the line of the Q_y transitions of both bacteriochlorophylls of the special pair. A positive charge being placed at this position should induce a strong blue shift of the nearest bacteriochlorophyll BP_A and a weaker red shift of BP_B. Thus, the resulting red shift of the spectral components 677.1 and 680.5 nm that we attributed to *P* was opposite to the expectations based on the analogy with the bacterial RC. The difference spectrum S_2-S_1 resembles the ones published in [9,11]. The common feature of these difference spectra is the dominant blue shift of a band at 680 nm in contrast to a red shift in this domain in the Y_Z^+-Y_Z spectrum. This excludes a position of the manganese cluster in a line with Y_Z and *P*.

The putative topology of the chromophores and redox cofactors in PSII that emerges from our data is illustrated on *Figure 2*. Chl_A peaking at 667 nm revealed red bandshifts in response to the formation of Q_A^- and Y_Z^+ that placed this molecule between Q_A and

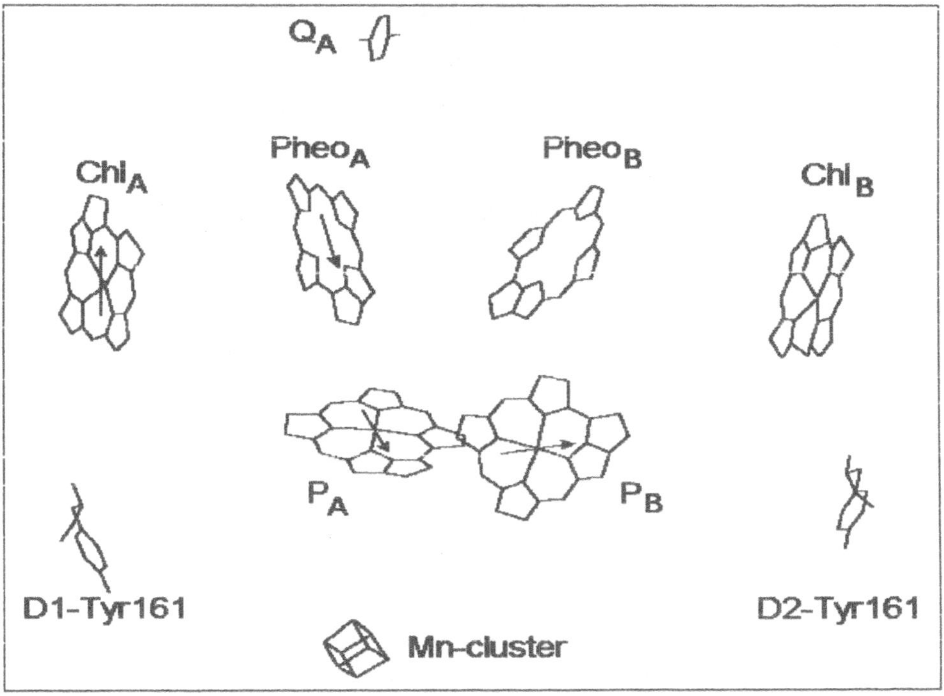

Figure 2. Proposed arrangement of chromophores and redox cofactors in PSII obtained from the electrochromic spectra and the structure of the bacterial RC [12]. Arrows indicate the changes in the dipole moment of chlorin ring during excitation [4].

Y_Z. The current structural models place Chl_A in branch A of PSII between *P* and $Pheo_A$ [1,2]. Because both histidines that coordinate the central magnesium atoms of the accessory chlorophylls in the bacterial RC are absent in the primary structures of D1 and D2 [3] these chlorophylls probably take other positions. Besides, it is difficult to understand how the deleterious oxidation of the accessory chlorophylls *a* with normal oxidation potential of +0.7 V can be avoided if they are so closely disposed to *P* with E_m of more than +1.1 V. The primary sequences of D1 and D2 contain two strictly conserved histidines in the positions 118 somewhat closer to the lumenal ends of helices B that might coordinate chlorophyll molecules whose ring planes are perpendicular to the membrane [3]. The substitution of the corresponding His117 on D2 of *Synecocystis* 6803 by Arg, Asp, Gln, Cys slowed down the growth of the mutants and the substitution by residues that are unable to coordinate a Mg^{2+} ion destroyed the phototrophic growth probably by inhibiting the assembly of PSII (W. Vermaas, personal communication). It is reasonable to assume that the accessory chlorophylls in PSII are anchored by His-118 of D1 and D2.

To explain the very selective response of the spectral components which peaked at 677 and 680.5 nm to a positive charge either on Y_Z or on the manganese cluster we propose that the *y*-axis of P_B is oriented as in the bacterial RC parallel to the membrane and the *y*-

axis of P_A is also essentially parallel to the membrane plane but perpendicular to the Y_Z - P_B line and to the assumed *y*-axis of P_B. The blue bandshift in response to a positive charge in the Mn-cluster then implies that the latter is located in the plane of P_A ring and close to the line extending its *y*-axis.

5. Conclusions

1) The accessory chlorophylls in PSII are placed at different positions than the bacterial counterparts, farther away from *P* and presumably anchored to His-118 on D1 and D2.
2) The chlorin plane of Chl_A on D1 (λ_{max}=667 nm) is roughly perpendicular to the membrane plane, its ring V is pointing to Q_A.
3) Ring V of $Pheo_A$ on D1 (λ_{max}=684.7 nm) is pointing away from Q_A in contrast to its bacterial counterpart and to Chl_A.
4) *P* is composed from two chlorophyll *a* molecules, P_A and P_B, with peaks at 680.5 and 677 nm, respectively. The orientation of the *y*-axis of P_B is similar to its bacterial counterpart, while the y-axis of P_A is orthogonal to the former (in contrast with the situation in the bacterial RC). Both *y*-axes are oriented about parallel to the membrane.
5) The intermediate electron carrier Y_Z (Tyr-161 on the D1 subunit) is located approximately coaxially with the *y*-axis of P_B.
6) The manganese cluster is located out of the line formed by Y_Z and *P*.

†) This study was conducted in the framework of a collaborative project supported by the International Association for the Promotion of Cooperation with Scientists from the Independent States of Former Soviet Union (INTAS-93-2852).

References

1. Svensson, B., Vass, I., Cedergren, E. and Styring, S. (1990) EMBO J. 9, 2051-2059
2. Ruffle, S.V., Donnelly, D., Blundell, T.L., and Nugent, J.H.A. (1992) Photosynth. Res. 34, 287-300
3. Michel, H. and Deisenhofer, J. (1988) Biochemistry 27, 1-7
4. Hanson, L.K., Fajer, J., Thompson, M.A., and Zerner, M.C. (1987) J. Am. Chem. Soc. 109, 4728
5. Shopes, R.J. and Wraight, C.A. (1985) Biochim. Biophys. Acta 806, 348-356
6. Verméglio, A. and Clayton, R.K. (1977) Biochim. Biophys. Acta 461, 159-165
7. Tiede, D.M. and Hanson, K. (1992) in The Photosynthetic Bacterial Reaction Center II (Breton, J. and Verméglio, A., eds.) pp 341-350, Plenum Press, New York
8. Schatz, G.H. and Van Gorkom, H.J. (1985) Biochim. Biophys. Acta 810, 283-294
9. Velthuys, B.R. (1988) Biochim. Biophys. Acta 933, 249-257
10. Dekker, J.P. (1985) Electron Transfer in the Oxygen-Evolving System of Photosynthesis, Ph. D. Thesis, Amsterdam
11. Saygin and Witt, H.T. (1985) FEBS Lett. 187, 224-226
12. Deisenhofer, J., Epp, O., Sinning, I., Michel, H. (1995) J. Mol. Biol. 246, 429-457

L PROTEIN SUPPORTED QUINONE EXCHANGE AT QA SITE IN PSII CORE COMPLEX

Shinichiro Ozawa, Hisashi Hoshida and Yoshinori Toyoshima
Graduate School of Human & Environmental Studies, Kyoto University,
Yoshida-nihonmatsu-cho, Sakyo-ku, Kyoto 606-01, Japan

1. Introduction

QA quinone in the photosynthetic reaction center of purple bacteria was successfully replaced with other quinones. Analysis of the kinetic data obtained with the quinone exchanged reaction centers on the basis of Marcus theory provided a basic concept of determining the rate of electrontransfer between given donor and acceptor pairs in the bacterial reaction centers[1-2]. Contrary to these cases, systematic replacement of QA quinone in PSII complex by exogenous quinones has not yet been performed, because of complexity in the protein composition of the PSII in plants. We previously reported that plastoquinone-9 (PQ-9) reinserted in the PQ-9 depleted PSII core complex does function as QA, when the complex was simultaneously reconstituted with L protein encoded by *psb*L gene[3]. In this report, we demonstrate that the L protein supported QA reconstitution is applicable to replacement of QA quinone in the PSII core complex with several exogenous quinones. It is also shown that the PSII complex which lost its electron transport activity by photoinhibitory treatment under anaerobic condition is readily reactivated by the L protein supported reconstitution of QA quinone, giving an direct evidence for a damage at QA site to occur as an early event upon photoinhibitory illumination.

2. Materials and Methods

2.1 *Quinone exchange of PSII core complex*

PSII core complex which consists of CP47/CP43/D1/D2/Cytb-559/H/I/L/K/n-5.0/n-4.1 proteins (RC) was prepared from fresh spinach leaves as described in [4]. Depletion of PQ-9 from the RC was done by using a mixed detergent system (17.5mM n-octyl-ß-D-glucopyranoside(OG) and 10mM n-octyl-ß-D-thioglucopyranoside(OTG)) as in [5]. In the resulting complex (depl-RC1), PQ-9 was scarcely detected (less than 0.1 PQ-9/reaction center)and CP43, L, K proteins, two nuclear coding proteins (n-5.0, n-4.1) and a partial portion of H protein were lost. The protein components including L protein liberated from RC into the detergent solution was separated from PQ-9 and lipids by column chromatography as in [5] and named as total protein extract (TPE). PQ-9 was isolated from the total extract as in [6]. The other quinones and digalactosyldiglyceride (DGDG) were purchased from Sigma Chemical Company.

PQ-9 and Cytb-559 were analyzed as in [6] and [7], respectively. The concentration of reaction center in each RC preparation was determined on the basis of Cytb-559 assuming

P. Mathis (ed.), Photosynthesis: from Light to Biosphere, Vol. I, 535–538.

Cytb-559/RC=1[8].

Reconstitution of the depl-RC1 with each exogenous quinone was performed as follows. A given quinone and DGDG were colyophilized and solubilized in small volume of a solution(17.5mM OG, 10mM OTG and 5mM Tris (pH 7.5)). To this solution, the depl-RC1 and TPE, respectively were added. The mixture was incubated for 30 min on ice and dialyzed against 25 mM phosphate buffer (pH 7.5) for 18 h on ice, changing the buffer solution four times. At the end of dialysis the solution was diluted 5-fold with the same buffer and finally centrifuged at 280,000 x g for 80 min at 4 ℃. The resulting pellet was collected as quinone exchanged RC1 (Q-RC1) and resuspended in a solution of 0.05 % DM and 50 mM Tris (pH 7.5).

2.2 *Photoinhibition of PSII core complex and reactivation by quinone exchange*

Photoinhibitory treatment of PSII core complex(RC) which is active in the reducing side of reaction center but not in water oxidation was undertaken as follows. The sample solution containing 1.2 μM RC and 5mM Mn^{2+} as an artificial electron donor was put in a shielded glass tube of 2-cm in inner diameter immersed in a water jacket kept at 20℃ by a circulating thermostated water and illuminated with white light of 7,000 $\mu E \cdot m^{-2} \cdot s^{-1}$ provided by a halogen lump equipped with UV cut-filter. In some experiments, DCMU was added to the solution at a concentration of 100 μM. When the photoinhibitory treatment was done under anaerobic condition, the sample solution was subjected to 20 cycles of degas by reducing pressure and injecting nitrogen prior to illumination. After photoinhibitory treatment, the sample solution was immediately chilled to 0℃ in darkness and submitted to the photoinduced electron transport measurements or reconstitution experiments. Disintegration and reconstitution of the photoinhibited RC were performed as described in [5].

2.3 *Measurement of the QA activity in PSII*

The QA activities of the RCs before and after given treatments were assayed by measuring the photoreduction rate of dichlrophenolindophenol (DCPIP) in the presence of diphenylcarbazide (DPC) as an electron donor as in [3].

3. Results and discussion

3.1 *Quinone exchange*

The depl-RC1 was reconstituted with each of eleven exogenous quinones and DGDG(125 μg/1 nmol Cytb-559) together with or without TPE, changing the amount of the quinone. Although L protein had been confirmed to be dominantly responsible for the restoration of the QA activity in the reconstituted RC by our previous works, in this work, TPE in stead of purified L protein was used for supporting the exchange of the quinone in RC. The eleven quinones examined are plastoquinones with two(PQ-2) and nine (PQ-9) units of isoprenoid, ubiquinones(UQ) without isoprenoid(UQ-0), with two(UQ-2), six(UQ-6), seven(UQ-7), nine(UQ-9) and ten(UQ-10) units of isoprenoid, menaquinone-4 (MQ-4), tetramethylbenzoquinone (TMBQ), 2,5-dibromo-3-methyl-6-isopropyl-1,4-benzoquinone (DBMIB). The QA activities determined from the electron transport rate from DPC to DCPIP upon illumination are plotted against the amount of quinone added in the reconstitution medium in Figure 1. PQ-9 restored the QA activity in the resulting particle and the activity increased with increasing amount of PQ-9 until the molar ratio of PQ-9 to reaction center in the reconstitution medium about 2 to 1 and saturated. This behavior was observed both in the reconstituted particles with and without TPE, but the saturation level in the particles without TPE was limited to 13 % of the original RC, while that with TPE went up to 43 %. PQ-2 did not stand for PQ-9 as QA quinone. Significant recovery of the QA activity with PQ-2 was not observed even in the reconstituted particle with TPE. MQ-4, TMBQ and DBMIB did not recover the QA activity in the depl-RC1 at all, independent of the presence and absence of TPE. UQs except for UQ-0 function as QA quinone when reinserted in the depl-RC1. The QA activity of the UQ-RC1 increased with increasing amount of UQ and saturated. The saturation level of the activity in the UQ-RC1 was significantly enhanced by the addition of TPE in the reconstitution medium and slightly increased with increasing number of isoprenoid units of UQ used. The later behavior was not observed in the UQ-RC1 obtained without TPE. Thus, the recovery of QA activity of the Q-RC1 depended on number of isoprenoid unit as well as the type of quinone ring and the difference of quinone much strongly reflected on the QA activity of the reconstituted particles with than without TPE.

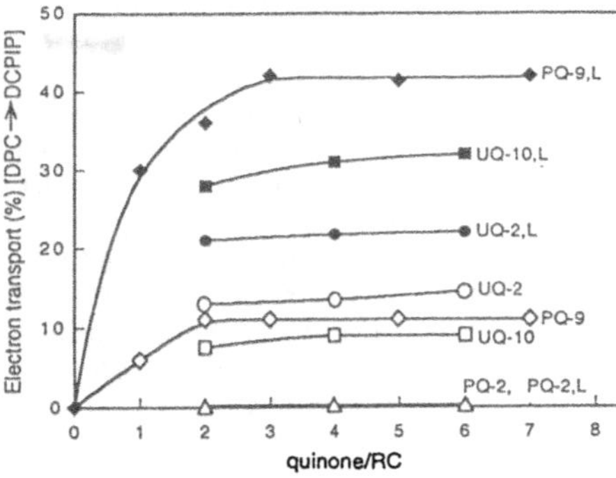

Fig.1 The QA activity in quinone-exchanged RC1 with and without L protein.

Reconstitution of a PSII reaction center complex consisting of D1/D2/Cytb-559/I firstly isolated by Nanba and Satoh (depl-RC2) with PQ-9 [8] and exogenous quinones, such as decylplastoquinone or DBMIB had been reported[9]. In these cases, no protein was used for supporting the reconstitution of QA function. So we examined the effect of TPE on the recovery of the QA activity in this complex with the eleven exogenous quinones. Reconstitution of depl-RC2 with each of the eleven quinones was carried out with DGDG and with or without TPE. Recovery of the QA activity in the resulting particles was observed, when the quinones which restore the QA activity in the depl-RC1 were used, but not with other quinones. However, the saturation level of the restored QA activity did not exceed 13% of the original level and was not enhanced by the addition of TPE at all. TPE exhibited no effect on the recovery of QA activity in depl-RC2, suggesting besides L protein, possible involvement of CP47 and/or H protein in normal functioning of QA in PSII complex.

3.2 Reactivation of photoinhibited PSII by L protein supported quinone exchange

Illumination of RC with high light (7,000 $\mu E \cdot m^{-2} \cdot s^{-1}$) both under aerobic and anaerobic conditions resulted in reduction of electron transport activity of the complex. As shown in Figure 2, after 10 and 30 min illumination, the activity decreased to 30 and 15 % of the original level, respectively. This reduction of the activity was not recovered by 24h dark adaptation. Reduction of the amplitude of 1479 cm^{-1} band in light induced FTIR difference spectrum by photoinhibitory illumination of RC indicated that the inhibition of electron transport occurred between P680 and QA site.

Judging from SDS-PAGE profile, RC proteins, including D1 did not degraded after photoinhibitory treatment under anaerobic condition in spite of the serious reduction of electron transport activity, suggesting that inhibition in electron transport is due to some damage of a redox molecule itself or the local structure surrounding the molecule. If it occurs at the QA site, one may expect that electron transport activity reduced by photoinhibition under anaerobic condition is recovered by the quinone reconstitution at QA site.

PSII core complexes before (RC) and after (RC*) photoinhibitory illumination were disintegrated by treatment with the mixed detergent(OG/OTG) solution used in the quinone exchange experiments to prepare depl-RC, depl-RC*, total extract (TE) and TE* to be used for reconstitution. Reconstitution was carried out for four combinations, depl-RC/TE depl-RC*/TE, depl-RC/TE* and depl-RC*/TE* and the electron transport activity from DPC to DCPIP and light induced FTIR difference spectrum were measured for the resulting particles(R-RC and R-RC*s).

Figure 2 illustrates the electron transport activities of RC* and the R-RC* from depl-RC*/TE*. The electron transport activity reduced in RC* was recovered in all of three type R-RC*s. The light induced FTIR difference spectra also showed the recovery of the QA function in R-RC*s. The electron transport activity of the R-RC* from depl-RC/TE* was 70 % of that of the original RC, irrespective to the exposure period, which is equivalent to the activity of the R-RC from depl-RC1/TE(control), indicatingthat PQ-9 and L protein were not damaged during the photoinhibitory treatment under anaerobic condition. While, the activities of R-RC*s from depl-RC1*/TE and from depl-RC*/TE* decreased with increasing the period exposed to high light, although the activity was recovered to some extent by reconstitution at least within 60 min of photoinhibitory treatment. These results indicate that release of PQ-9 at QA site occurs as an early event of photoinhibition at reducing side, irrespective of degradation of PSII proteins.The recovery of electron transport activity in the R-RC* from depl-RC*/TE was much more significant, when photoinhibitory illumination was done in the presence of DCMU suggesting a protective effect of DCMU against damage of the photochemical reaction center under photoinhibitory illumination.

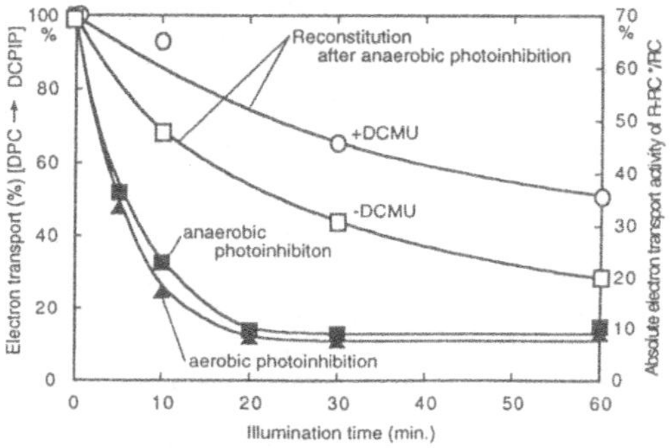

Fig.2 Recovery of electron transport activity in photoinhibited RC by reconstitution.

Reference

[1] Woodbury, N. W., Parson, W. W., Gunner, M. R., Prince, R. C. and Dutton, P. L.(1986) Biochem. Biophys. Acta 851, 6-22.

[2] Moser, C. C., Keske, J. M., Warncke, K., Farid, R. S. and Dutton, P. L. (1992) Nature, 355, 796-802.

[3] Kitamura, K., Ozawa, S., Shiina, T. and Toyoshioma , Y.(1994) FEBS Lett. 354, 113-116.

[4] Dekker, J.P., Bowlby, N.R. and Yocum, C.F. (1989) FEBS Lett. 254,150-154.

[5] Araga,C., Akabori,K., Sasaki, J., Maeda, A., Shiina, T. and Toyoshima,Y. (1993) Biochem. Biophys. Acta 1142, 36-42.

[6] Akabori, K., Tsukamoto, H., Tsukihara, J., Nagatsuka, T., Motokawa, O. and Toyoshima, Y. (1988) Biochem. Biophys. Acta 932, 345-357.

[7] Miyazaki, A., Shiina, T., Toyoshima, Y., Gounaris, K. and Barbaer, J. (1989) Biochem. Biophys. Acta 975, 142-147.

[8] Chapman, D.J., Vass, I. and Barber, J. (1991) Biochim. Biophys. Acta 1057, 391-398.

[9] Mathis,P.,Satoh, K. and Hasson, O. (1989) FEBS Lett. 251, 241-244.

PROTON UPTAKE AND FREE ENERGY CHANGES ASSOCIATED WITH REDUCTION OF Q_B IN ILE L229 ---> MET MUTANT REACTION CENTER OF *RHODOBACTER SPHAEROIDES*

Júlia Tandori, László Nagy, Szabolcs Osváth and Péter Maróti
JATE University Szeged, Institute of Biophysics
Egyetem utca 2. Szeged, Hungary H-6722

1. Introduction

The secondary quinone (Q_B) in the reaction center (RC) of photosynthetic bacteria *Rhodobacter sphaeroides* is the ultimate electron and proton acceptor. The reducing equivalents (electrons and protons) are exported from the RC in pairs in form of doubly reduced quinone, hydroquinol, Q_BH_2, and are then replaced by a quinone from the pool. The atomic structure of the Q_B binding domain (Fig. 1) revealed that the binding pocket is highly hydrophobic and surrounded by a cluster of strongly interacting acidic residues (not shown). The carbonyl oxygens (O-2 and O-5) of the head group of Q_B form hydrogen bonds with N-δ1 atom of His^{L190} and the side chain of Ser^{L223}, respectively.

In this study, the non-protonatable residue, Ile^{L229} in the vicinity of Q_B was mutated to Met. This amino acid makes the Q_B binding stable being in van der Waals contact with it [1,2]. Although not involved directly in the proton transfer to Q_B, the mutation modifies the pK_a value of the key residue Glu^{L212} via subtle structural changes, thus influences the pH-profile of the flash-induced proton binding. The rate of multiple turnover (photocycle) under physiological conditions (continuous illumination), however, is not affected.

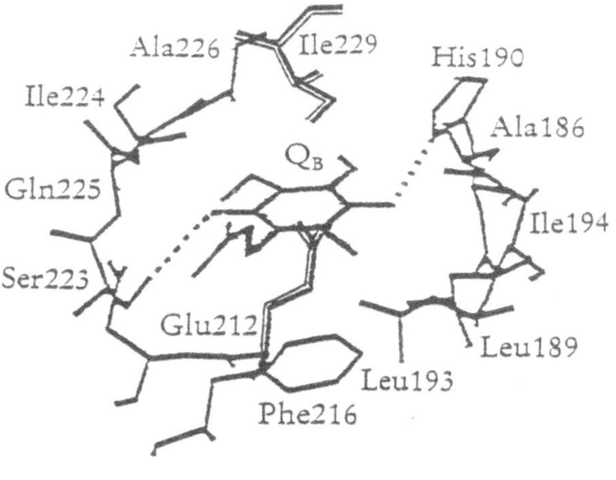

Figure 1. Amino acids of the L subunit forming the Q_B binding pocket. Hydrogen bonds to Q_B (dashed lines) and amino acids involved in this study (double lines) are indicated.

2. Materials and Methods

The spontaneous mutant from the wild-type *Rhodobacter sphaeroides* R-26 was selected for resistance to 200 μM atrazin [2]. Absorption changes were measured by a single-beam spectrophotometer of local design. The excitation was achieved either by a xenon flash lamp

P. Mathis (ed.), Photosynthesis: from Light to Biosphere, Vol. I, 539–542.

(type EG&G FX200, $t_{1/2} \approx 8.5\ \mu s$) passed through a 760 nm cut-off filter or by a laser diode (Laser Diode, Inc. type LCW-100, emission wavelength 810 nm). The analog signal of the photomultiplier (EMI 9558) was converted with an A/D converter card (Advantech PCL818) attached to an IBM-compatible 386 AT computer. Flash-induced proton binding was detected and calibrated by pH-indicator dyes (bromocresol purple (pH 5.5-7.5), o-cresol red (pH 7.0-9.0) and o-cresolphthalein (pH 8.7-10.0)) and pH electrodes as described earlier [3,4].

3. Results and Discussion

The free energy difference between $Q_A^-Q_B$ and $Q_AQ_B^-$ can be derived either from the difference in proton uptake associated to $Q_AQ_B^-$ and $Q_A^-Q_B$ redox states of the RC [4]:

$$\Delta G_{AB} = k_B \cdot T \cdot \ln 10 \cdot \int \left(\frac{H^+}{Q_B^-} - \frac{H^+}{Q_A^-}\right) d(pH) + const \tag{1}$$

or from one-electron equilibrium constant, K_{AB} in the acceptor quinone complex [2,3]:

$$\Delta G_{AB} = -60\,meV \cdot \log K_{AB} \tag{2}$$

where $k_B \cdot T$ is the Boltzmann-term and H^+/Q_B^- and H^+/Q_A^- are the proton uptake by the RC in reduced state of the secondary and primary quinone, respectively. The pH-dependence of the flash-induced proton binding was experimentally measured in the presence of fast electron donor (ferrocene) to the oxidized dimer (P^+) in both redox states of the RC (Fig. 2).

The proton uptakes for both semiquinones show similar pH dependences, being smaller for Q_A^- at low (pH≤6) and high (pH≥9) pH values. In these regions the electrostatic interaction between semiquinones and ionized forms of key titrable residues of Asp[L213] and Glu[L212] controls the binding of protons from the aqueous phase, respectively [5-7]. The integration of the difference in proton uptake is related to the change in free energy ΔG_{AB}. By using Eqs. (1) and (2), the proton uptake and electron equilibria in the acceptor quinones are in good agreement (Fig. 3).

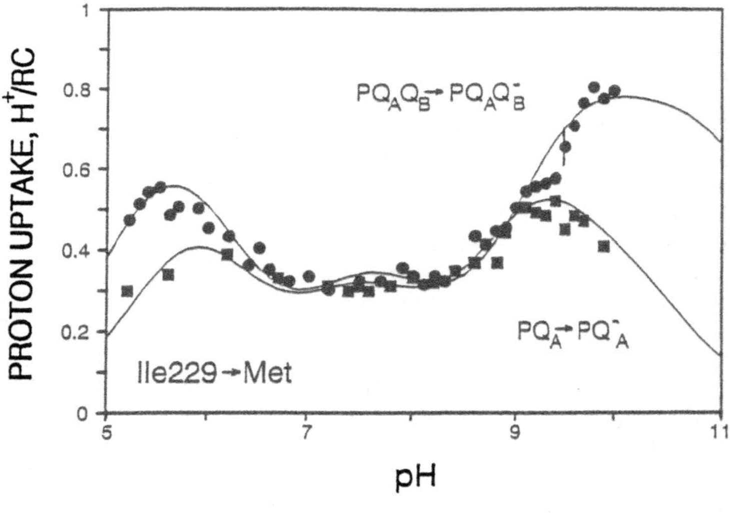

Figure 2. H^+-ion binding by Q_A^- (■) and Q_B^- (●) states of the mutant RC. 1 μM RC, 100 μM ferrocene, 40 μM UQ_{10} and 0.03% Triton X-100.

K_{AB} in Eq. (2) was determined from two types of assays: 1. charge recombination (○,●) when

P remains in oxidized state and 2. multiple (cytochrome and semiquinone) turnover (▵,▴) of the RC when P^+ is always re-reduced. Figure 3. clearly shows that the redox state of P does not influence the free energy gap between $Q_A^-Q_B$ and $Q_AQ_B^-$. The solid lines were drawn according to Eq. (2) using data of proton binding measurements (Fig. 2 for mutant and [4] for native RC). The integration constants were chosen to give agreement with results from electron transfer measurements in the neutral pH region. In the acidic and alkaline pH range, the stabilizing effect of the interaction of the two key residues with the reduced quinones can be clearly seen in both the wild-type and the mutant. The destabilization of the $Q_AQ_B^-$ state relative to $Q_A^-Q_B$ by about 60 meV at all pH values and an upward shift of about 0.5 pH unit in pK_a value of the residue with the highest pK_a can be well demonstrated in the mutant. At pH > 10, the inversion of semiquinone equilibrium can be observed: the state $Q_AQ_B^-$ becomes energetically less favorable than the $Q_A^-Q_B$ state.

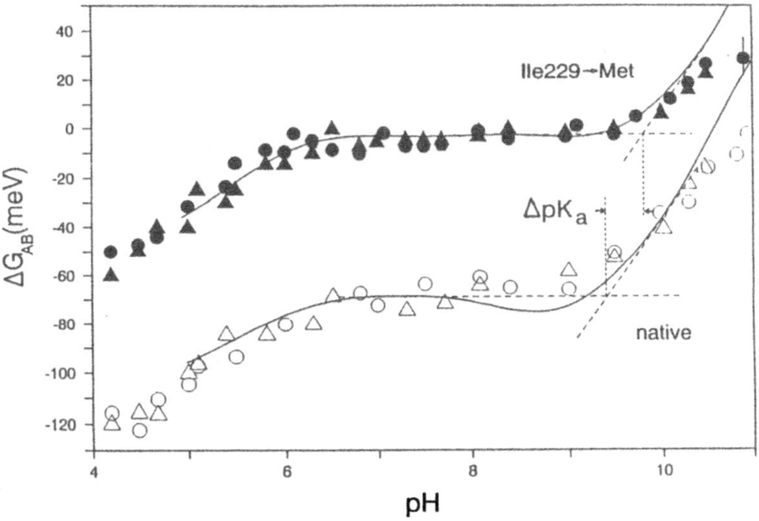

Figure 3. PH dependence of the free energy difference, ΔG_{AB}, between $Q_A^-Q_B$ and $Q_AQ_B^-$ states of native and mutant RCs with UQ_{10} as Q_B.

As the yield of charge stabilization on Q_B decreased, one can expect lower photosynthetic efficiency in the mutant RC. However, the mutant had similar photocycle rate under continuous illumination (I = 1 W·cm^{-2}) as the wild-type (Fig. 4). The photooxidation of cytochrome by RC measures the flow of electrons and protons through the RC photocycle [1]. For each electron transferred to the quinones, 1 cytochrome c molecule is oxidized in the RC. As two cytochrome molecules are oxidized per completely reduced quinone (Q_BH_2), the photocycle rate is one-half of the cytochrome turnover rate. At the onset of excitation, the cytochrome molecules are rapidly oxidized by a rate determined mainly by the intensity of the illumination and K_{AB}. After the turnover of 2-3 cytochromes/RC, the rate is controlled by the rate of the slowest step in the photocycle. At acidic or neutral pH, this step is usually the photochemistry (light reactions), but at higher pH values it can be the quinone/quinol exchange mainly with the protonation as the bottleneck. A reduced rate of cytochrome photooxidation after a transient period of 2-3 cyt/RC in the mutant compared to that of wild-type would be an excellent indication of decreased quinone/quinol exchange rate. The native and mutant RCs showed identical turnover rate (560 cyt/RC·s^{-1}) at pH 7 and only minor differences were observed at higher pH values. The mutation did not have influence on the quinone/quinol exchange rate in the photocycle under conditions of external quinone saturation.

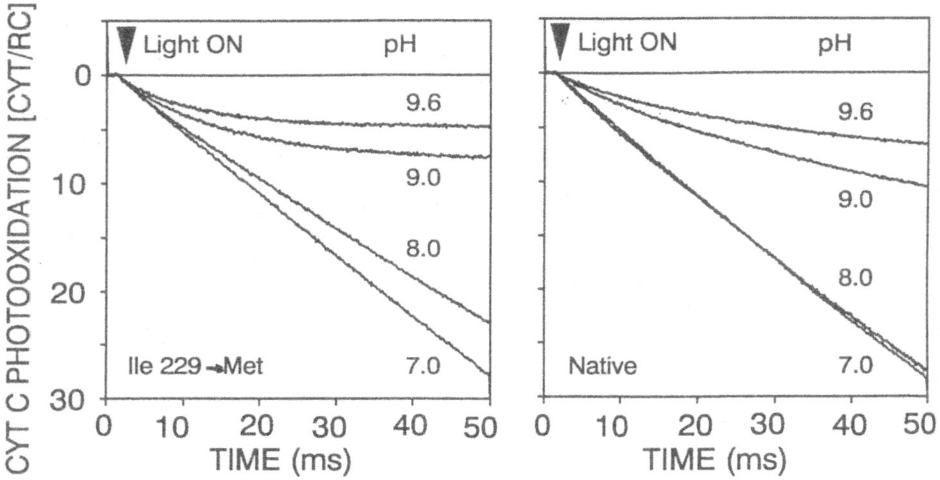

Figure 4. Cytochrome photooxidation in RCs at different pH values. 1 μM RC, 50 μM cyt c, 5 mM TRIS, 0.03 % Triton X-100 and 5 mM UQ_o (saturation at Q_B for both strains [1]).

4. CONCLUSIONS

The Ile^{L229} --> Met mutation affects the electrostatic interaction of Q_B^- with Fe^{2+} and with Glu^{L212} and thus the protonation state of this key amino acid residue. That could possibly occur by subtle displacement of the quinone ring in the binding pocket: moving off the non-heme iron atom and coming closer to Glu^{L212}. As a consequence, the free energy gap between $Q_AQ_B^-$ and $Q_A^-Q_B$ decreases by about 60 meV at all pH values and the pK_a value of Glu^{L212} (the observed protonatable residue with highest pK_a) increases by 0.5 pH units, relative to those in wild-type. The quinone/quinol exchange rate is not affected by the mutation.

ACKNOWLEDGEMENTS: This work was supported by the Hungarian Science Foundation (OTKA 17362/95) and Foundation for Hungarian Education and Research (FEFA III 1034).

References

1 Paddock, M.L., Rongey, S.H., Abresch, E.C., Feher, G. and Okamura, M.Y. (1988) Photosynth. Res. 17, 75-96
2 Tandori, J., Nagy, L., Puskás, Á., Droppa, M., Horváth, G. and Maróti, P. (1995) Photosynth. Res. (in press)
3 Maróti, P. and Wraight, C.A. (1988) Biochim. Biophys. Acta 934, 314-328 and 329-347
4 McPherson, P.H., Okamura, M.Y. and Feher G. (1988) Biochim. Biophys. Acta 934, 348-368
5 Paddock, M.L., Rongey, S.H., Feher, G. and Okamura, M.Y. (1989) Proc. Natl. Acad. Sci. USA 86, 6602-6606
6 Takahashi, E. and Wraight, C.A. (1992) Biochemistry 31, 855-866
7 Maróti, P. (1993) Photosynth. Res. 37, 1-17

TIME-RESOLVED EPR SPECTROSCOPY OF THE TRIPLET STATE OF THE PRIMARY ELECTRON DONOR P680 IN PS II.

Paul J.M. van Kan[1], Ivo H.M. van Stokkum[2] and Stenbjörn Styring[1]
[1]Department of Biochemistry, Arrhenius Laboratories for Natural Sciences, Stockholm University, S-106 91 Stockholm, Sweden.
[2]Faculty of Physics and Astronomy, Free University, De Boelelaan 1081, 1081 HV Amsterdam, The Netherlands.

1. Introduction

In the reaction centre of Photosystem II (PS II), absorption of a photon leads to an efficient charge separation reaction. An electron is moved across the photosynthetic membrane, between the primary electron donor P680 and a pheophytin *a* molecule acting as primary acceptor. In intact centres, the electron is moved further from pheophytin to the secondary acceptor, a quinone molecule Q_A. This secondary step, together with the subsequent reduction of $P680^+$ by Tyr_Z, provides sufficient stabilisation of the charge-separated state to allow oxidation of water and reduction and exchange of the second quinone acceptor Q_B. In photo-inhibited centres [1] or in centres where Q_A is already reduced or absent [2], the charge stabilisation cannot proceed and charge recombination takes place instead. This results in the formation of a triplet (T) state ^{3}P680, which possesses a strong electron spin-polarisation when a magnetic field is applied [3]. This electron spin-polarisation is inherited from the primary radical pair, which is formed in its singlet state (S) but decays to the $T_{|0>}$ state within its lifetime (S-$T_{|0>}$-mixing in a magnetic field) [3].
The distinctive spectrum from ^{3}P680 as observed in PS II membranes in an X-band ESR experiment [2] yielded zero-field splitting parameters for the spin-polarised triplet that were corroborated in ADMR studies, which also provided estimates of the triplet sublevel decay rates [4, 5]. The spin-polarised triplet state from P680 is readily detected in isolated PS II reaction centres. Optical measurements showed that the triplet yield is about 30% at room temperature [6] and more than 80% at liquid helium temperatures [7, 8]. The high yield of ^{3}P680 is a direct consequence of the long lifetime of the radical pair in isolated reaction centres. This lifetime is around 30 ns at 280 K [6], and even longer at cryogenic temperatures: up to 100 ns at 10 K [7].
In these proceedings, we present time-resolved X-band CW-EPR and FT-EPR measurements of the flash-induced spin-polarised triplet signal of P680, from 4 K up to 42 K. These kinetic measurements allow us to discuss both transverse and longitudinal spin-relaxation within the P680 triplet manifold. The results obtained strongly support

P. Mathis (ed.), Photosynthesis: from Light to Biosphere, Vol. I, 543–546.

the notion that the triplet state is located on a Chl *a* monomer at low temperature, but that this Chl *a* molecule is part of a dimer or even cluster of chlorin residues constituting the primary electron donor P680.

2. Experimental procedure

Sample preparation. Reaction centres of PSII (D1-D2-cyt b_{559} protein complexes) were used with an absorption maximum of the preparation at 675.5 nm. The absorbance at this wavelength was 14 cm^{-1}. EPR tubes were filled with 200 µl of the suspension, in dim green light, under an argon atmosphere. The tubes were frozen in the dark to form opaque, polycristalline samples and were stored in liquid nitrogen.
EPR measurements. CW-X-band EPR was performed at 9.24 GHz with a Bruker ESP 300 spectrometer equipped with a waveform digitiser. The sample frequency was 250 kHz. Data acquisition was performed with the ESP 300 program of the spectrometer. EPR signals were collected in 16 ms time windows. These time frames were repeated at about 1 Hz and the signals were added. The time resolution of the experiments was determined by the output filter of the detector channel on the EPR spectrometer, which was set to an RC time of 80 µs. FT-EPR measurements at 9.68 GHz were performed on a Bruker ESP 380 instrument equipped with a BLG resonator with optical access.
A Quanta Ray GCR-3 Nd:YAG laser was synchronously triggered by the spectrometer during the data acquisition period. Actinic flashes at 532 nm were of 10 ns duration with an energy density of 20-25 mJ·cm^{-2} at the entrance of the grid in the microwave cavity.
Temperature control and calibration. An Oxford He flow cryostat and temperature controller were used for measurements between 4 K and 42 K. The He flow was kept high for the measurements at 4 K to prevent heating of the sample due to the flashes.

3. Results and Discussion

CW-EPR

The flash-induced CW-EPR signal of ^{3}P680 at 4 K is shown in Figure 1. The signal with is six 'turning points' is qualitatively comparable with previous results, obtained with continuous illumination. [2]. The (powder) spectrum can be simulated very well with zero field splitting parameters |D| and |E| much like those for monomeric ^{3}Chl *a* [4, 5]. The anisotropy of the *g*-tensor can be neglected at these magnetic fields.

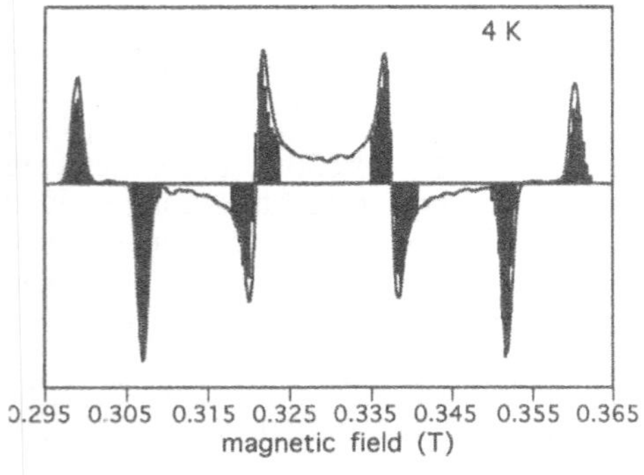

Figure 1. EPR spectrum of ^{3}P680 in isolated PS II RCs at 4 K obtained by plotting the amplitudes of the flash-induced signals *vs.* magnetic field The data are represented as a filled histogram. Microwave power: 4 µW, mod. ampl.: 1.0 mT, sum of 128 flashes. Solid line: Simulation of the powder spectrum of P680 in the $T_{|0>}$ state. Zero-field parameters used for the simulation were |D| = 287 *10^{-4} cm^{-1} and |E| = 44 * 10^{-4} cm^{-1}. The *g*-value used was 2.0027. Gaussian spin packets were given a FWHM linewidth of 1.5 mT.

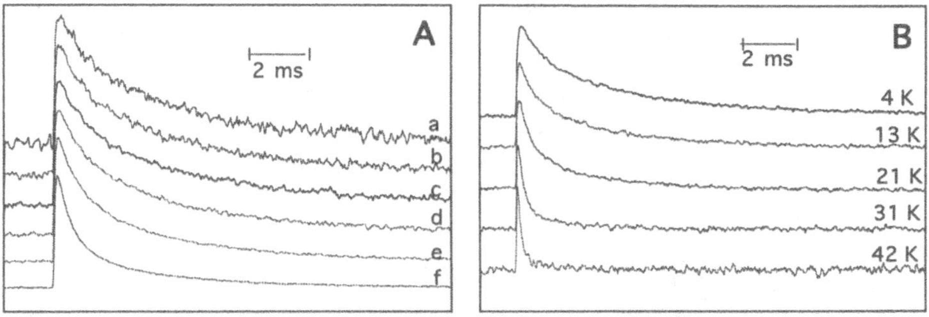

Figure 2. (**A**) Kinetics at 4 K of the flash-induced EPR signal of ^{3}P680 at 0.2990 T (Z turning point) as a function of microwave power. (a) 130 nW; (b) 250 nW; (c) 1 μW; (d) 4 μW; (e) 16 μW; (f) 63 μW. (**B**) Temperature dependence of the decay kinetics of the flash-induced EPR signal at 0.2990 T.

The simulation assuming full $T_{|0>}$ polarisation shows a good correspondence with the measured amplitudes. This indicates that electron-spin polarisation is not dependent on orientation.

Relaxation of the triplet state is most easily studied at the field extremes of the spectrum, where only triplets with their dipolar z-axis parallel to B_0 resonate. Kinetic traces measured at the low-field extreme of the spectrum are shown in Figure 2. The leftmost frame, Figure 2A, shows the decay of the triplet signal as a function of microwave power. As expected, the decay is faster for relatively high microwave powers (trace e-f). Very interestingly, though, the decay times measured do not depend on microwave power, once one applies less than 4 μW, (trace a-d). Even at a level of 0.13 μW (a), the decay of the EPR signal is non-monoexponential. This indicates that the |z> level of ^{3}P680 has more than one decay pathway. If the initial electron spin polarisation is indeed high, this means that apart from a decay directly to the ground state by intersystem crossing, there is an additional pathway *via* the other triplet sublevels. Since the intersystem crossing rates of the three triplet sublevels are generally different [3-5], such a mechanism would indeed yield more complicated kinetics instead of a single-exponential decay. This phenomenon distinguishes ^{3}P680 from a monomeric ^{3}Chl *a* molecule in solution. A possible explanation might be, that the triplet state is localised on one Chl *a* moiety, but that this pigment interacts with the protein or, more challenging, with other pigments that constitute P680.

Experimentally, this hypothesis was tested by measuring the decay from the |z> level at various temperatures (Figure 2B). The decay rate is observed to increase with increasing temperature. Moreover, this rate becomes independent of the orientation of the triplet (equal for all three sublevels) above 40 K (not shown). Comparison with optical measurements of the decay rate of ^{3}P680 [8], indicates that intersystem crossing rates show very little temperature dependence from 4 to 40 K. Hence, it is the interaction between the sublevels of P680 that is increasing with increasing temperature. At 40 K, transitions between the sublevels of P680 occur at a rate faster than all intersystem crossing rates, leading to a rapid collapse of the spin polarised EPR signal.

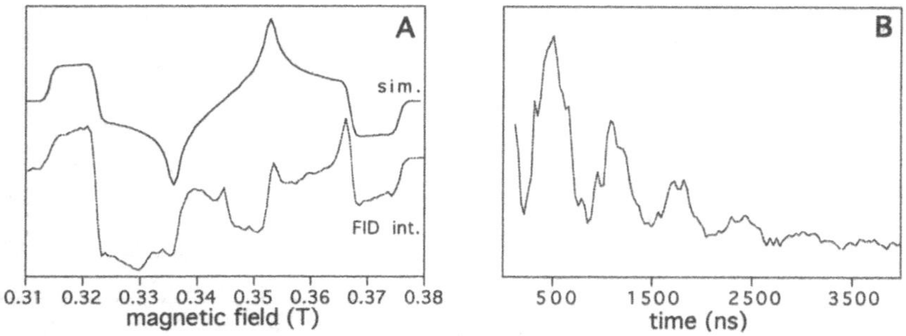

Figure 3. **(A)** FID detected EPR spectrum of ^{3}P680 at 15 K compared with the integral of the simulation of Fig. 1. **(B)** Hahn echo decay at 0.3143 T (Z turning point) measured also at 15 K.

FT-EPR In Figure 3, we show some preliminary results obtained by application of strong microwave pulses, 5 μs after a laser flash. The spectrum in Figure 3A is obtained by integrating the free induction decay of a single π/2 pulse. One may readily observe the (integrated) spectrum in the low and high field regions. The centre of the spectrum is distorted, which might have a number of causes, still under investigation. Interestingly, it is possible to observe the FID with a spectrometer deadtime of 100 ns, at least for some orientations. Two pulse electron-spin echo experiments (Hahn echo) for the |z> level show a transverse magnetisation decay of about 900 ns at 15 K. This value is considerably longer than previous estimates [5]. The Hahn echo amplitude is strongly modulated with a period of about 300 ns which would suggest a strong (hyperfine) interaction of about 3 MHz. At this stage, more investigations are needed of the relaxation behaviour of the electron spin on the primary donor P680, which might elucidate some of the structural properties that underly its function as a high-potential redox component in Photosystem II.

4. Acknowledgements

We gratefully acknowledge the support of the Swedish Reserch Council (NFR), the Netherlands Organization for Scientific Research (NWO) and the Knut and Alice Wallenberg Foundation.

References

1 Vass, I. and Styring,, S. (1993) Biochemistry 32, 3334-3341.
2 Van Mieghem, F.J.E., Nitschke, W., Mathis, P. and Rutherford, A.W. (1989) Biochim. Biophys. Acta 977, 207-214.
3 Budil, D.E. and Thurnauer, M.C., (1991) Biochim. Biophys. Acta 1057, 1-41.
4 Den Blanken, H.J., Hoff, A.J., Jongenelis, A.P.J.M. and Diner, B.A. (1983) FEBS Lett. 157, 21-27.
5 Angerhofer, A., Bernlocher, D. and Robert, B. (1993) Z. Phys Chem. 182, 167-180.
6 Danielius, R.V., Satoh, Ki., Van Kan, P.J.M., Plijter, J.J., Nuijs, A.M. and Van Gorkom, H.J. (1987) FEBS Lett. 213, 241-244.
7 Van Kan, P.J.M., Otte, S.C.M., Kleinherenbrink, F.A.M., Nieveen, M.C., Aartsma T.J. and Van Gorkom, H.J. (1990) Biochim. Biophys. Acta 1020, 146-152.
8 Groot, M.-L., Peterman, E.J.G., Van Kan, P.J.M., Van Stokkum, I.H.M., Dekker, J.P. and Van Grondelle, R. (1994) Biophys.J. 67, 318-330.

ELECTRON TRANSFER FROM Q_A^- TO P^+ : EFFECTS OF $\Delta G°$ AND TEMPERATURE ON RATE AND REORGANIZATION ENERGY

J.M. Ortega[1,3], X. Lin[2], J. C. Williams[2], J.P. Allen[2] and P. Mathis[1]

1 - DBCM-SBE, CEA-Saclay, 91191 Gif-sur-Yvette Cedex, France ; 2 - Dept. of Chem. and Biochem. (Center for the Study of Early Events in Photosynthesis), ASU, Tempe AZ 85287-1604, USA ; 3 - permanent address ; Inst. Bioquimica Vegetal y Fotosintesis, Univ. de Sevilla - CSIC, 41080 Sevilla, Spain

1. Introduction

Photosynthetic reaction centers offer one of the most elaborate cases for biological electron transfer. Such processes are characterized by a rate (k) which can be expressed according to classical electron transfer theory (Marcus) (1,2):

$$k = f(d)\,(4\pi\lambda k_B T)^{-1/2} \exp[-(\Delta G° + \lambda)^2/4\lambda k_B T] \qquad \text{Eq. 1}$$

where d is the distance between donor and acceptor, λ is the reorganization energy, k_B is the Boltzmann constant, T is the temperature and $-\Delta G°$ is the free energy variation (driving force, $-\Delta G° = E_A - E_D$, difference of redox potential of Acceptor and Donor).

We measured the rate of electron transfer from Q_A^- to P^+ (charge recombination) in isolated reaction centers of the purple bacterium Rhodobacter sphaeroides and varied two parameters of eq. 1 :

- Temperature T, by studying the reaction between 293 and 10 K ;
- Driving force $-\Delta G°$, by using mutants in which the potential of the P/P$^+$ couple is varied from + 410 mV to + 765 mV (+ 505 mV for the wild-type), leading to variations of $-\Delta G°$ from 425 meV to 780 meV (3).

Electron transfer from Q_A^- to P^+ in wild type reaction centers has been studied in many laboratories, for several species of purple bacteria. In Rb. sphaeroides, the reaction has been shown to be direct and activationless. In addition, it has been shown that its rate accelerates when the temperature is lowered : the $t_{1/2}$ is approximately 60 ms at room temperature and 20 ms at cryogenic temperatures. This surprising property, which is found in many purple bacteria (4-9), has not yet been satisfactorily explained. One reason often advocated is a contraction of the protein which would reduce intermolecular distances (10,11). Our results lead us to conclude that, in Rb. sphaeroides, the reorganization energy for the $P^+Q_A^- \rightarrow PQ_A$ reaction decreases by $\approx$ 250 meV as the temperature decreases from 293 K to 10 K. This decrease explains the acceleration of the reaction when the temperature is lowered.

2. Material and methods

♦ Reaction Centers. All strains were prepared by complementation of a pufLM deletion strain of Rb. sphaeroides. They correspond to various situations for hydrogen bonds established between the primary donor P and the L or M polypeptide, and the midpoint potential of P/P$^+$ varies accordingly from + 410 to + 765 mV (3).

For spectroscopic measurements, isolated reaction centers were in 15 mM Tris-HCl pH 8.0, 0.025 % LDAO and 1 mM EDTA. Glycerol was added to a final proportion of 60 % per volume. 20 μM terbutryn was added in order to prevent electron transfer from Q_A^- to Q_B. Final concentration of reaction centers was around 1 μM.

♦ Measurement of electron transfer kinetics.

The cuvette with reaction centers was placed in a thermostated compartment (10 - 293 K). Charge separation was induced by a 8-ns laser pulse, and the time course of P^+ concentration was monitored by recording the light absorption at 1250 nm where P^+ has an absorption band. After the initial charge separation, the state

P. Mathis (ed.), Photosynthesis: from Light to Biosphere, Vol. I, 547–550.

$P^+Q_A^-$ is formed in less than 1 ns. Since there was no good electron donor to P^+ (no reduced cytochrome added) and since Q_B was displaced by terbutryn, the state $P^+Q_A^-$ decayed slowly by charge recombination.

3. Results

Following flash excitation, the absorption at 1250 nm increases due to the formation of P^+. It then decays because of electron return from Q_A^- to P^+. This reaction has already been studied in detail in wild type reaction centers and we have confirmed previous observations (e.g. refs. 4-9):

- the $t_{1/2}$ is about 60 ms at room temperature ;
- the reaction accelerates by a factor of approximately 4 when the temperature is decreased.
- fitting the decay with one single exponential is roughly satisfactory, but two exponential components are needed for obtaining a good fit. The rates from the two-component fit differ by a factor of 3-5.

In mutants, the results display the same general properties as in the wild-type. Differences are observable (see below) in :

- the rate of the reaction at high temperature ;
- the extent of variation in the rate between 293 K and 10 K ;
- the relative proportion of the two phases which have to be used for a better fit of the kinetics.

We present here a few key results obtained at 293 K and 10 K utilizing a one-exponential analysis of the kinetics. For the 8 strains studied, the rates of electron transfer from Q_A^- to P^+, at 10 K and 293 K, have been plotted versus the $-\Delta G°$ of the reaction (Fig. 1).

4. Discussion

The series of mutants used in this work has already been well characterized, in particular for the P/P^+ midpoint potential (3), for the effect of $-\Delta G°$ on the $P^+Q_A^-$ back reaction at room temperature (3), and on electron donation from cytochrome c_2 to P^+ (12).

The plot of Fig. 1 allows a first confrontation with the Marcus theory. For the present discussion, we assume that the strains differ only in the P/P^+ potential (i.e. in the $-\Delta G°$ of the reaction), and that mutations have no effect on the P-Q_A distance and on the reorganisation energy λ. The results in Fig. 1 show that at room temperature, the rate continues to increase as $-\Delta G°$ increases and does not reach a maximum, while at low temperature, the rate reaches a maximum when $-\Delta G°$ is approximately 550 meV and does not continue to increase. Equation 1 predicts that the maximum rate is obtained when $-\Delta G° = \lambda$, and so Fig. 1 immediately indicates that λ is smaller by several hundred meV at 10 K than at 293 K.

The Marcus theory, however, assumes that the vibrational modes coupled to electron transfer are populated at the temperature of measurement. This assumption is not valid at 10 K if medium or higher energy modes are involved. The data have thus been analyzed according to equation 2, a more general expression that uses both medium and high frequency modes, in which quantum effects are considered (13,14).

$$k = f(d)\,(\varpi_M)^{-1} \sum_{i=0}^{\infty} \{[e^{-S'} S'^{i}/i\,!]\,[e^{-S(2n+1)}]\,[(n+1)/n]^{p/2}\, I_p\,[2S\,(n(n+1))^{1/2}]\} \qquad \text{Eq. 2}$$

where $S = \lambda_M / h\nu_M$, $S' = \lambda_L / h\nu_L$ (subscript M :medium frequency mode ; subscript L : high frequency mode) (total reorganization energy: $\lambda = \lambda_M + \lambda_L$) ; $n = 1/[\exp(h\nu_M/k_BT) - 1]$ which is the population of the medium frequency mode ; $p = -(\Delta G° + ih\nu_L/h\nu_M)$; and I_p is a modified Bessel function of order p.

A good fit is then obtained with the following parameters :

- at 293 K, λ = 795 meV with strong coupling to a medium frequency mode (130 cm^{-1}) and weak coupling to a high frequency mode (1600 cm^{-1}).

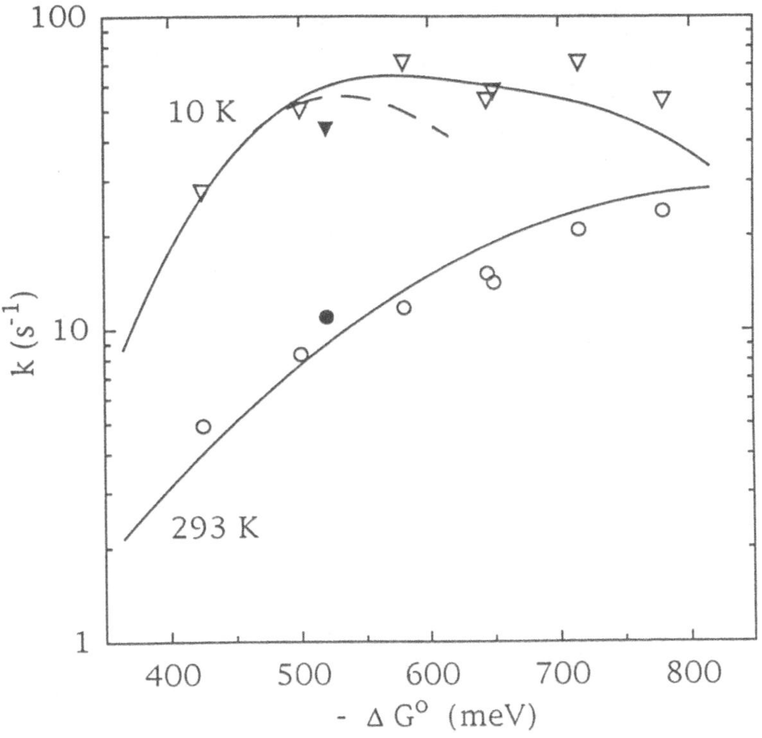

Fig. 1 - The rates of electron transfer from Q_A^- to P^+ at 10 K and at 293 K, using a one-exponential analysis of the kinetics, are plotted versus -ΔG° for the eight strains of Rb. sphaeroides. The rates for wild-type reaction centers are indicated by filled symbols. The lines correspond to a fit of the data using eq. 2. At 10 K, the continuous line corresponds to a fit with three vibrational modes (a 600 cm^{-1} mode whose contribution is significant only for large values of -ΔG°) and the dashed line is a fit with only two vibrational modes.

- at 10 K when $-\Delta G° < \lambda$, $\lambda = 540$ meV with the same medium and high frequency modes used for the 293 K fit.

- at 10 K when $-\Delta G° > \lambda$, an additional contribution was required, so a fit of all the 10 K data involved the sum of two rates. One rate is defined by the process that dominates at room temperature and in the non-inverted region at low temperature ($\omega_M = 130$ cm^{-1}, $\omega_L = 1600$ cm^{-1}). The other rate dominates in the inverted region at low temperature and is strongly coupled to both a medium frequency mode (130 cm^{-1}) and a high frequency mode (600 cm^{-1}).

- at 293 K and 10 K, since the additional process required to fit the 10 K rates only contributes significantly in the inverted region, the 293 K rates can also be described as a sum of the same two processes. The 600 cm^{-1} mode contributes an additional approximately 200 meV to the total reorganization energy at both low and high temperatures.

Both the room temperature and low temperature rates can be fit by adjusting only one parameter, λ, and the decrease in λ accounts for the increase in rate at low temperature. The present results thus lead to a consistent interpretation of the well-known phenomenon that the reaction accelerates when the temperature decreases. The decrease of λ at low temperature has been proposed by Dutton and Mosser (15).

Two kinds of approaches have been used previously to study the effect of ΔG° on the rate of the same reaction :

- chemical removal of ubiquinone Q_A and replacement by quinones that have different redox potentials (16,17) ;
- orientation of reaction centers in bilayers or films, and modification of ΔG° by application of an electric field (18-20)

A comparison of room temperature results shows a reasonable agreement between various studies. Concerning vibrations coupled to electron transfer, our conclusions are similar to those reached by Gunner and Dutton (16) who fit their low temperature data using a strong coupling with 120 cm^{-1} vibrations and a weak coupling with 1,600 cm^{-1} vibrations. The present study is however probably the first one where ΔG° and temperature can be varied in a large range, under conditions where reaction centers are intact. A decrease of reorganization energy when temperature decreases offers a good interpretation to the data and is physically meaningful.

Acknowledgements
This work was supported by the European Science Foundation, Programme on Biophysics of Photosynthesis (J.M.O.) and by grant MCB-9404925 from the NSF.

References

1. Marcus,R.A. and Sutin,N. (1985) Biochim. Biophys. Acta 811, 265-322
2. Moser,C.C., Keske,J.M., Warncke,K., Farid,R.S. and Dutton,P.L. (1992) Nature 355, 796-802
3. Lin,X., Murchison,H.A., Nagarajan,V., Parson,W.W., Allen,J.P. and Williams,J.C. (1994) Proc. Natl. Acad. Sci. USA 91, 10265-10269
4. Hsi,E.S.P. and Bolton,J.R. (1974) Biochim. Biophys. Acta 347, 126-133
5. Loach,P., Kung,M. and Hales,B.J. (1975) Ann. N. Y. Acad. Sci. 224, 297-319
6. Romijn,J.C. and Amesz,J. (1976) Biochim. Biophys. Acta 423, 164-173
7. Clayton,R.K. (1978) Biochim. Biophys. Acta 504, 255-264
8. Parot,P., Thiery,J. and Vermeglio,A. (1987) Biochim. Biophys. Acta 893, 534-543
9. Sebban,P. (1988) Biochim. Biophys. Acta 936, 124-132
10. Hales,B.J. (1976) Biophys. J. 16, 471-480
11. Feher,G., Okamura,M.Y. and Kleinfeld,D. (1987) in Protein structure : Molecular and electronic reactivity (Austin,R., Buhks,E., Chance,B., De Vault,D., Dutton,P.L., Frauenfelder, H. and Goldanskii,V.I. Eds), 399-421, Springer Verlag, New York
12. Lin,X., Williams,J.C., Allen,J.P. and Mathis,P. (1994) Biochemistry 33, 13517-13523
13. Jortner,J. (1976) J. Chem. Phys. 64, 4860-4867
14. Bixon,M., Jortner,J. and Michel-Beyerle,M.E. (1995) Chem. Phys. in press
15. Dutton,P.L. and Mosser,C.C. (1994) Proc. Natl. Acad. Sci. USA 91, 10247-10250
16. Gunner,M.R. and Dutton,P.L. (1989) J. Am. Chem. Soc. 111, 3400-3412
17. Gunner,M.R., Robertson,D.E. and Dutton,P.L. (1986) J. Phys. Chem. 90, 3743-3795
18. Popovic,Z.D., Kovacs,G.J., Vincett,P.S., Alegria,G. and Dutton,P.L. (1986) Chem. Phys. 110, 227-237
19. Feher,G., Arno,T.R. and Okamura,M.Y. (1988) in The Photosynthetic Bacterial Reaction Center, (Breton,J. & Verméglio A. Eds)pp. 271-287, Plenum, New York
20. Franzen,S. and Boxer,S.G. (1993) J. Phys. Chem. 97, 6304-6318

RECONSTITUTION OF THE PLASTOQUINONE POOL OF INSIDE OUT VESICLES, PSII MEMBRANE FRAGMENTS AND CORE COMPLEXES FROM HIGHER PLANTS

J. Kurreck, A. G. Seeliger, F. Reifarth, M. Karge, G. Renger
Technical University Berlin, Straße des 17. Juni 135, 10623 Berlin, Germany

1. Introduction

The acceptor pool capacity of inside-out vesicles (ISO-vesicles) and PSII membrane fragments is limited to 2-3 plastoquinones, compared with 7 plastoquinone molecules in thylakoids. Therefore measurements of the characteristic quarternary oscillation patterns of the oxygen yield without exogenous electron acceptors with a conventional Joliot-type electrode are restricted to one or at most two periods. More drastic changes are observed in PSII core complexes, where also the Q_B-site is severely modified [1]. This study describes attempts to reconstitute the plastoquinone pool of ISO-vesicles and PSII membrane fragments and to improve the function of the Q_B-site of PSII core complexes.

2. Materials and Methods

ISO-vesicles were prepared from market spinach as described in [2] with slight modifications. PSII membrane fragments were isolated according to the procedure of [3] as outlined in [4]. Preparations of PSII core complexes were obtained as in [5]. Plastoquinone-9 (PQ-9) was extracted from spinach chloroplasts by high-performance liquid chromatography (HPLC) using a reversed phase nucleosil $100C_{18}$ column ([6], with modifications).

In reconstitution experiments quinones were added at a final ratio of 15 quinone molecules per reaction centre in ISO-vesicles and PSII membrane fragments and 30 quinone molecules per reaction centre in PSII core complexes. After 30 min incubation on ice the samples were sonicated for 15 min in an ice-cooled sonication bath. The sonication step is essential for successful reconstitution. All control samples were prepared under the same conditions but without quinone addition.

Flash-induced oxygen oscillation patterns were measured with a modified Joliot-type electrode [7] as described in [8]. The probabilities of misses, double-hits and the S_i populations before flash excitation were determined by a least-squares fit method within the framework of an extended Kok model ([9] and References therein). Flash-induced changes of the fluorescence quantum yield were monitored with a home-built equipment [10]. Measurements of the fluorescence-induction were performed as in [11].

3. Results and Discussion

3.1 Oscillation patterns of oxygen yield in control and PQ-9 treated samples.

Figure 1 shows oscillation patterns of the oxygen yield as a function of the flash number in dark-adapted control and PQ-9 reconstituted PSII membrane fragments (A), ISO-vesicles (B) and core particles (C). The control samples exhibit typical features known

P. Mathis (ed.), Photosynthesis: from Light to Biosphere, Vol. I, 551–554.

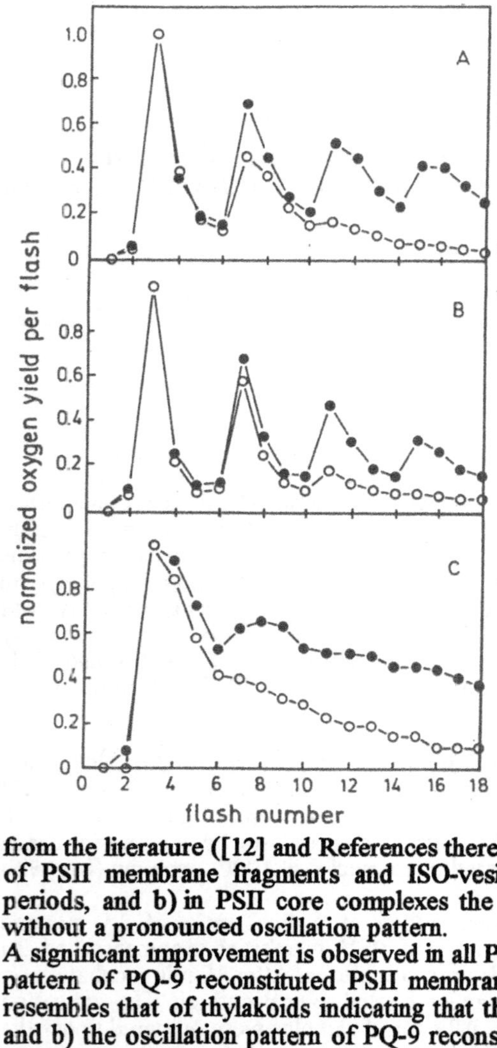

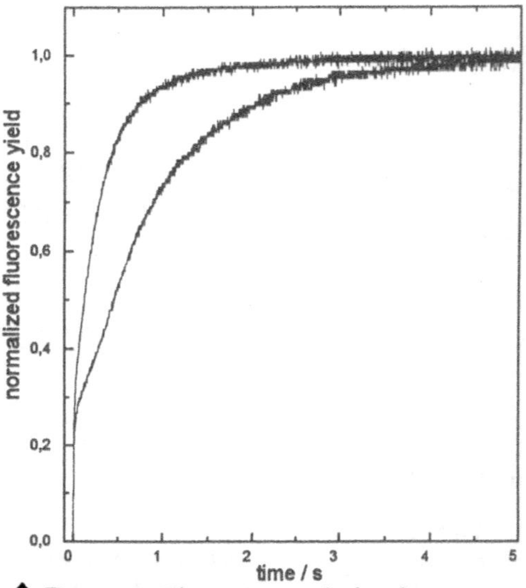

↑ *Figure 2.* Fluorescence induction curves of control (upper trace) and PQ-9 reconstituted (lower trace) PSII membrane fragments.

← *Figure 1.* Oxygen yield as a function of the flash number in dark-adapted PSII membrane fragments (A), ISO-vesicles (B) and PSII core complexes (C). O control, ● PQ-9 reconstituted samples

from the literature ([12] and References therein): a) the normalized oxygen yield per flash of PSII membrane fragments and ISO-vesicles fades after at most two quarternary periods, and b) in PSII core complexes the oxygen yield declines after the third flash without a pronounced oscillation pattern.

A significant improvement is observed in all PQ-9 reconstituted samples: a) the oscillation pattern of PQ-9 reconstituted PSII membrane fragments and ISO-vesicles qualitatively resembles that of thylakoids indicating that the effective pool size is markedly increased, and b) the oscillation pattern of PQ-9 reconstituted core complexes is more pronounced but still markedly different from that of thylakoids. An evaluation within the extended Kok-model according to [11] reveals that this feature is mainly due to a drastic increase of the probability of misses ($\alpha \approx 0.22$) and the higher apparent population of S_0 and S_{-1}. The large α-value suggests that the severely modified Q_B site in PSII core complexes [1] cannot be restored by the reconstitution procedure. This idea is supported by measurements of flash induced transient changes of the fluorescence yield. [vide infra].

3.2 Fluorescence induction curves of control and PQ-9 reconstituted PSII membrane fragments.

Fluorescence induction curves were measured in order to check by an independent method the increase of the effective pool size in PQ-9 reconstituted PSII membrane fragments. The results depicted in Figure 2 indicate that the area over the normalized induction curves increases significantly in the reconstituted sample. A quantitative evaluation leads to a ratio of 2.6 for the areas in PQ-9 reconstituted and control samples, respectively.

3.3 Flash induced changes of the normalized fluorescence yield in control and PQ-9 treated samples.

The kinetics of Q_A^- reoxidation were monitored via the decay of laser flash induced changes of the fluorescence yield. Figure 3 shows typical traces in dark adapted samples after excitation with the first two actinic flashes (dark time between flashes 1 s) of a train. In PSII membrane fragments and ISO-vesicles the relaxation is dominated by two kinetics with τ_1 of the order of 1 ms and τ_2 = 10 - 20 ms. The faster kinetics describe the electron transfer from Q_A^- to $Q_B(Q_B^-)$, the slower are ascribed to a diffusion limited Q_A^- reoxidation by endogenous PQ-9 ([13] and References therein). Reconstitution with PQ-9 has only minor effects. The extent of these two kinetics after the nth flash is a measure of endogenous PQ-9 that is available for Q_A^--reoxidation. Accordingly, the normalized sum of these amplitudes taken over a flash train of more than 20 flashes should provide a measure of the endogenous pool size. The evaluation of the experimental results (data not shown) reveals that the endogenous pool comprises 2-3 PQ-9 molecules in the control and 6-7 molecules in reconstituted samples. The ratio of these values with a maximum of 3.0 is in perfect agreement with that obtained from the fluorescence induction curves (vide supra).

In PSII core complexes, characterized by slow Q_A^- reoxidation, PQ-9 reconstitution leads to a significant acceleration of the decay kinetics only in the first flash.

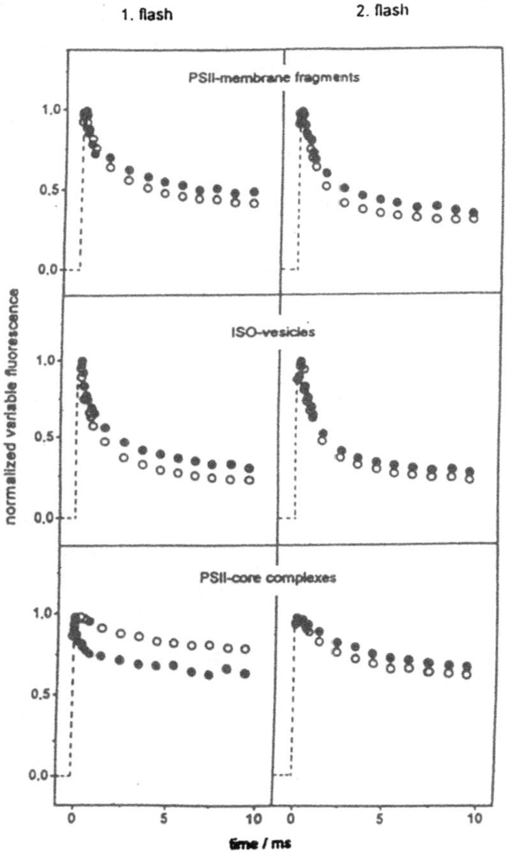

Figure 3. Normalized laser flash induced change of the relative fluorescence quantum yield as a function of time in dark adapted PSII membrane fragments, ISO-vesicles and PSII core complexes. O control, ● PQ-9 recontituted samples.

3.4 Reconstitution with other quinones.

In order to analyze the possibility of specific effects due to the chemical nature of the quinone molecules the same reconstitution procedure was performed with phenyl-p-benzoquinone (Ph-p-BQ), 2,3 dimethyl-5-decyl-p-benzoquinone (DMDBQ) and ubiquinone-45 (UQ-9). The oscillation patterns of the oxygen yield measured in these samples are shown in Figure 4. The results reveal that none of these quinones is as efficient as PQ-9. A closer inspection of the data indicates that DMDBQ with the same head group as PQ-9 leads to a partial reconstitution while

UQ-9 with the same hydrophobic tail as PQ-9 and the lipophilic Ph-p-BQ are inefficient under these conditions. Fluorescence measurements (data not shown) lead to the same order of efficiency. These results provide clear evidence for the existence of structural determinants that affect the reconstitution of an endogenous quinone pool. Therefore, it seems reasonable to assume that PQ-9 interacts with specific binding sites (most likely located in proteins) rather than being incorporated as mobile species into the lipid phase of the membrane. Apart from mechanistic questions that will be analyzed in forthcoming studies, the PQ-9 reconstituted samples are of practical relevance because they permit studies of the oscillation pattern of the water oxidase over several periods in the absence of exogenous electron acceptors. Experiments of this type are under progress.

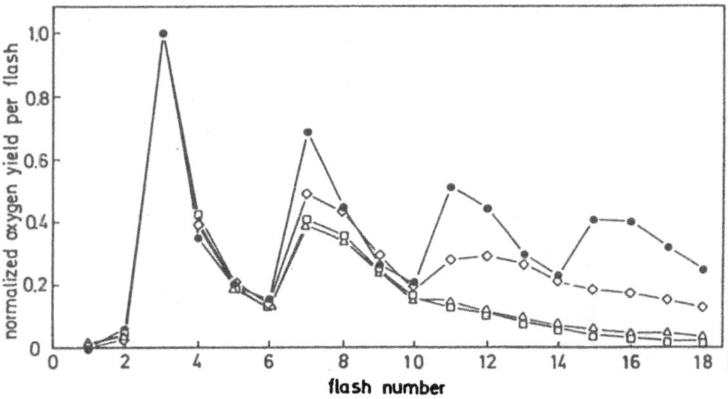

Figure 4. Oxygen yield per flash as a function of the flash number in dark adapted PSII membrane fragments reconstituted with PQ-9 (●), DMDBQ (◊), UQ-9 (Δ) and Ph-p-BQ(□).

Acknowledgement
The financial support by Deutsche Forschungsgemeinschaft (Re 354/11-3) and Fonds der Chemischen Industrie to G. R. is gratefully acknowledged.

References

1 Gleiter, H. M.; Haag, E.; Inoue, Y. and Renger, G. (1993) Photosynth. Res. 35, 113-126
2 Akerlund, H. E. and Anderson, B. (1983) Biochim. Biophys. Acta 725, 34-40
3 Berthold, D. A.; Babcock G. T. and Yocum, C. A. (1981) FEBS Lett. 134, 231-234
4 Völker, M.; Ono, T.; Inoue, Y. and Renger, G. (1985) Biochem. Biophys. Acta 806, 25-34.
5 Haag, E.; Irrgang, K. D.; Boekema, E. J. and Renger, G. (1990) Eur. J. Biochem. 189, 47-53
6 MacMillan, F.; Lendzian. F.; Renger, G. and Lubitz, W. (1995) Biochemistry 34, 8144-8156
7 Joliot, P. (1972) Methods Enzymol. 24, 123-134
8 Gleiter, H. M. (1993) Ph.D. Thesis, Technical University, Berlin, Germany
9 Messinger, J.; Wacker, U. and Renger, G. (1991) Biochemistry 30, 7852-7862
10 Gleiter, H. M.; Ohad, N.; Hirschberg, J.; Fromme, R.; Renger, G.; Koike, H. and Inoue, Y. (1990) Z. Naturforsch. 45c, 252-258
11 Renger,G. and Kayed, A. (1987) Biochim. Biophys. Acta 894, 261-269
12 Kurreck, J.; Seeliger, A. G.; Reifarth, F.; Karge, M. and Renger, G. (1995), submitted
13 Renger, G.; Eckert, H.-J.; Bergmann, A.; Bernarding, J.; Liu, B.; Napiwotzki, A.; Reifarth, F. and Eichler, H. J. (1995) Aust. J. Plant Physiol. 22, 167-181

ELECTROCHROMIC BAND-SHIFT OF CAROTENOID IN *RHODOPSEUDOMONAS VIRIDIS*

Masayuki Hara[1], Christopher C. Moser[2] and P. Leslie Dutton[2]
[1]NIBH, AIST, MITI, 1-1 Higashi, Tsukuba, Ibaraki 305, JAPAN
[2]Johnson Research Foundation of Molecular Biophysics, University of Pennsylvania, B501 Richards Building, 37th & Hamilton Walk, Philadelphia PA 19104, U.S.A.

1. Introduction

It have been reported that carotenoid molecules in chromatophore membrane of purple photosynthetic bacteria such as *Rhodobacter Sphaeroides* and *Rhodobacter Capsulatus* shows electrochromic band-shift (Stark shift) of absorption spectra when potential is applied across the membrane whereas no band-shift was reported in *Rhodopseudomonas viridis* (1 - 3) It has been thought that we can not apply the potential across the lipid bilayer by the salt-induced potential methods (1) because the chromatophore prepared from *Rps. viridis* are not sealed vesicles but flat sheets (4). We show that carotenoid in *Rps viridis* actually shows a band-shift by applied potential by application of potential to the electro-deposited film of chromatophore in this paper. This is the first report about the electrochromic band-shift of carotenoid in *Rps. viridis*.

2. Materials and Methods

Chromatophore in 10 mM Tris-HCl (pH8.0) solution prepared as described previously (5). Chromatophore suspension was dialysed against distilled water for overnight to decrease the conductivity, sonicated for 5 s to prevent aggregation and used for electro-deposition (ED) with the absorbance of 100 at 1020 nm.

Electro-deposition procedure was carried out as previously described with slight modifications (6). Chromatophore membrane was electrophoretically deposited on the area (11 x 19 mm) of an indium-tin oxide (ITO) plate by applying square wave potential of 20V for 25 ms repeatedly with the interval of 75 ms between an ITO plate and a cupper plate as an anode and an cathode, respectively with a U-shaped rubber gasket of 1mm thickness. The deposition was done for 10 s. The chromatophore-deposited ITO plate (thichness of the chromatophore layer was about 2.0 um) was dried slowly, coated with polyethylene (MW = 700) with a thickness of 500 nm as an micro insulater layer and then coated with platinum with a thickness of 10 nm as a counter electrode in vacuum condition as shown in Fig. 1(A).

Time-resolved voltage-induced absorption changes were monitored with photomultiplier tubes in a single beam configuration in argon as previously described (6). Data was averaged 500 times in most case of those measurement.

3. Results

Absorption spectrum of the chromatophore was not severely changed as shown in Fig. 1(B). First and second derivative spectra were shown in Fig. 1(C). The dried E.D. film showed light-induced spectrum derived from Stark shift of carotenoid whereas chromatophore suspension of this species in aqueous buffer did not (data not shown).

P. Mathis (ed.), Photosynthesis: from Light to Biosphere, Vol. I, 555–558.

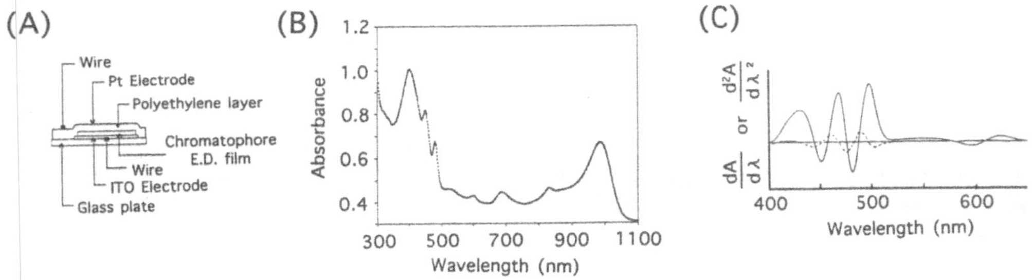

Figure 1. Schematic representation (A), absolute absorption spectrum (B), first (solid line) and second (dashed line) derivative spectra (C) of the E.D. film.

Figure 2(B) shows typical data traces of potential-induced absorbance change of the sample. Absorbance at 495 nm increased immediately after we applied the potential. However, decrease in absorbance at 482.5 nm is composed of fast and slow phase. We considered the fast phase as the Stark shift . We can not conclude about the slow phase now. It may be derived from potential-induced electron transfer reactions in RC (6) as previously reported, or charging/discharging phenomina of the E.D. film as a capacitor.

Figure 2(C) shows potential-induced spectra of the sample. The spectrum caused by 95 V in Fig. 2(C) has the positive peaks at 420-435 nm, 465 nm, 495 nm and negative peaks at 450 nm, 480 nm and 605 nm. The spectrum resembles the first derivative of absorption spectrum in Fig. 1(C). It shows that the potential-indued spectrum in the range of 420-500 nm mostly derives from electrrochromic band-shift (Stark shift) of carotenoid molecule in chromatophore.

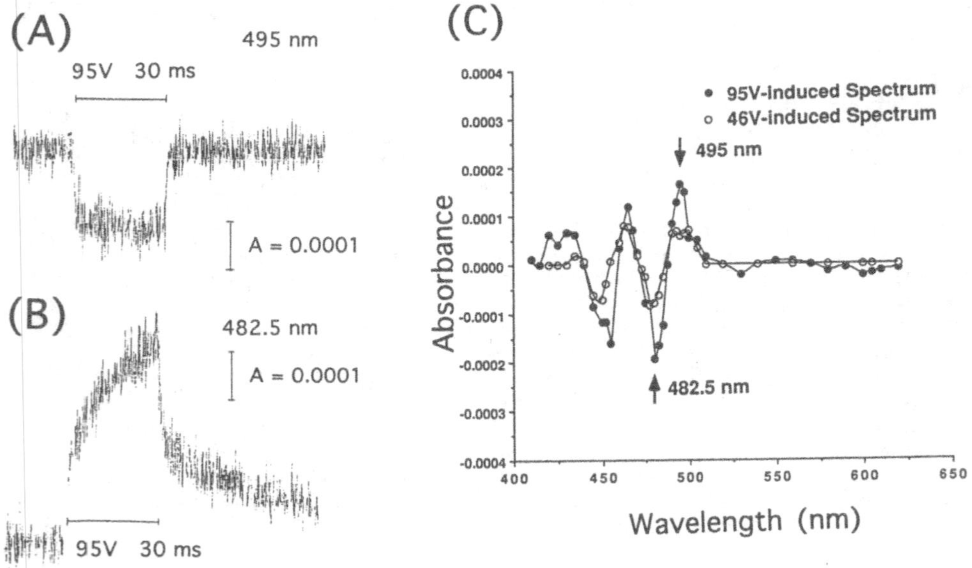

Figure 2. Example of potential-induced absorbance change at 495 nm (A), that at 482.5 nm (B) and potential induced spectra of the E.D. film (C). Potential of 95V (closed circle) or 45 V (open circle) was applied to the E.D. film.

We investigated the potential dependency of the Stark shift at 495 nm as shown in Fig. 3(A). The fitted curve has a symmetrical shape within experimental errors. The absorbance change was almost linear to square of potential in both of minus and plus region as shown in Fig. 3(B) and 3(C), respectively. The fitted curve of the potential dependency of the band-shift measured at 482.5 nm was also symmetrical as shown in Fig. 3(D). The absorbance change was almost linear to square of potential as shown in Fig. 3(E) and 3(F).

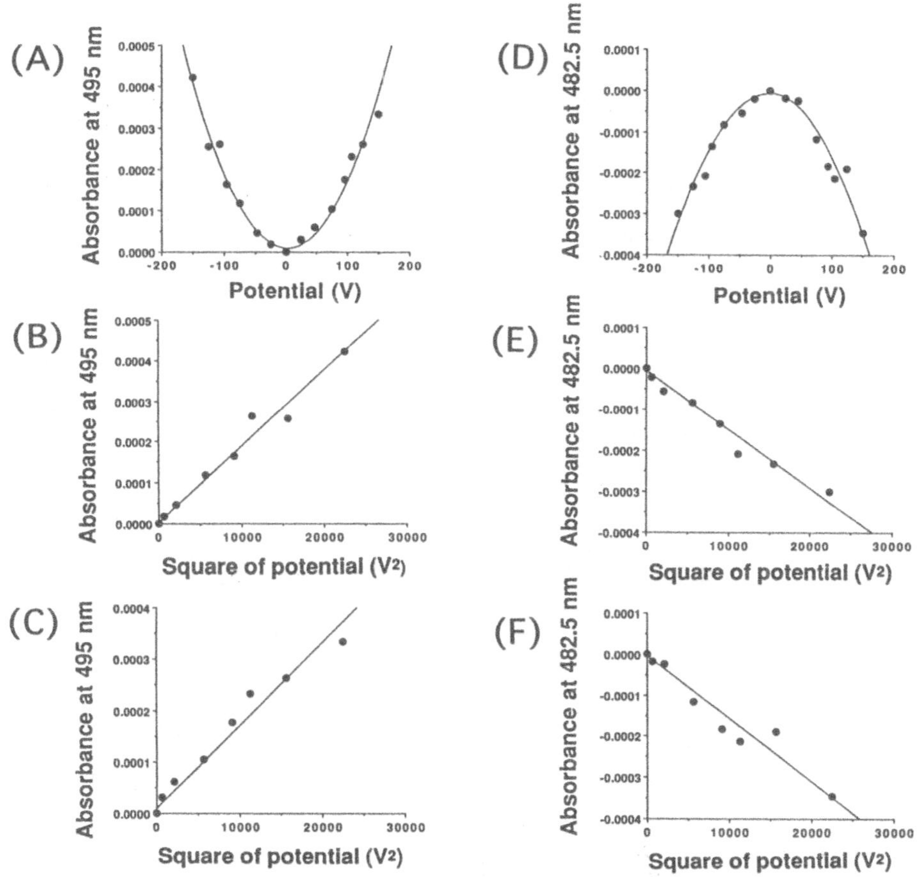

Figure 3. Potential dependency of the Stark shift of carotenoid at 495 nm (A) and that at 482.5 nm (D). Absorbance change in (A) in plus and minus region were replotted versis square of applied potential (V^2) in (B) and (D), respectively. Absorbance change in (D) in plus and minus region were replotted in the same manner in (E) and (F), respectively.

4. Discussion

The change in extinction coefficient of a sterically fixed molecule induced by electric field E is given as follows (1).

$$\Delta\varepsilon = \frac{\lambda^2}{hc}\Delta\mu \mathrm{E}\cdot\cos\theta\cdot\frac{\partial\varepsilon}{\partial\lambda} + \frac{1}{2}\frac{\lambda^2}{hc}\cdot\Delta\alpha \mathrm{E}^2\cdot\cos^2\phi\frac{\partial\varepsilon}{\partial\lambda} + \frac{1}{2}\frac{\lambda^4}{h^2c^2}\cdot\Delta\mu^2\cdot\mathrm{E}^2\cdot\cos^2\theta\frac{\partial^2\varepsilon}{\partial\lambda^2} + \cdots\cdots$$

$\Delta\mu = \mu$ excited - μ ground : the difference of the dipole moment, E : electric field, ε : extinction coefficient,
$\Delta\alpha = \alpha$ excited - α ground : the difference in polarizabilities of the ground and excited state λ : wavelength,
θ : the angle between the electric field and the change in the permanent dipole, h : Plank's constant,
ϕ : the angle between the electric field and the change in polarizabiliry, c : the velociry of light

First term of the formula contributes dominantly in case of salt-induced potential application reported with chromatophore of other bacterial species (1) because potential was applied asymmetrically to each chromatophore membrane.

In our experiment, the chromatophore membranes were flattened and stacked in the dried E.D. film. If half population of chromatophore membranes in the E.D. film is in upside down orientation, first term in the equation will be cancelled to zero. Therefore the spectral change mostly depends on the seond and third term. The amplitude of the Stark shift will be linear to the square of F. This is accord with our experimental results shown in Fig. 3 and 4. If the second term mostly contributes to the spectral change, the Stark spectrum will be first derivative of the absolute absorption spectrum. When the third term mostly contributes to it, the Stark spectrum will be the second derivative of the absorption spectrum. Potential-induced spectra in Fig. 2(C) shows that both terrms contributed ot the Stark spectrum in our experiment. Contribution of the second term is larger than that of the third term as described in the results section.

We have Applied 95V on the sample plate with the thickness of 2.5 um including the chromatophore E.D. film and a polyethylene layer in the measurement in Fig. 2. The amplitude of the potential was 38 mV/nm if dielectric constant of polyethylene and that of the E.D. film were the same. Potential in the range of 0 - 200 mV was applied on chromatophore membrane of other bacterial species in most case of salt-induced potential experiment reported by now. Amplitude of the potential is 40 mV/nm if we make the assumption that 200 mV of salt-induced potential was applied lipid bilayer with the thickness of 5 nm. This value is comparable to our experimantal data. The Stark shift of carotenoid should be observed also by the salt-induced potential method when we were to apply 40 mV/nm to the chromatophpre membrane. From this comparison, we conclede that difficulty in detecting the Stark shift in suspension of chromatophore from*Rps. viridis* is mostly derived from topology (not being sealed) of each chromatophore membrane although carotenoid molecule is capable of showing the Stark shift.

5. References

(1) C.A. Wraight, R.J. Cogdell and B. Chance (1978) in The Photosynthetic Bacteria (R.K. Clayton and W.R. Sistrom eds.), p. 471-511, Plenum Press, New York.
(2) N.G. Holmes, C.N. Hunter, R.A. Niederman and A.R. Crofts (1980) FEBS Letter 115, 43-48.
(3) G.D. Webster, R.J. Cogdell and G. Lindsay (1980) Biochim. Biophys. Acta 591, 321-330.
(4) J.S. Jacob and K.R. Miller (1983) Arch. Biochem. Biophys. 223, 282-290.
(5) G. Alegria and P.L. Dutton (1991) Biochim. Biophys. Acta 1057, 239-257.
(6) G. Alegria, C.C. Moser and P.L. Dutton (1992) in The Photosynthetic Bacterial Reaction Center II (J. Breton and A. Vermeglio eds.), p. 313-319, Plenum Press, New York.

THE LOW PH FORM OF REDOX-ACTIVE COMPONENTS IN THE DONOR SIDE OF PHOTOSYSTEM II STUDIED BY PULSED EPR AND ENDOR

Hiroyuki MINO[1], Andrei V. ASTASHKIN[1,2], Asako KAWAMORI[1] , Taka-aki, Ono[2] and Yorinao INOUE[2]

[1]*Faculty of Science, Kwansei Gakuin University, Nishinomiya 662, Japan*
[2]*Institute of physical and Chemical Research (RIKEN) Wako 351-01,Japan*

1. Introduction

Tyrosines play an important role as electron carriers in some biological systems such as PS II and ribonucleotide reductase. The reaction center of PS II contains two redox-active tyrosine residues, Y_Z and Y_D, the tyrosines 161 of the D1 and D2 polypeptides, respectively. The essential function of Y_Z is to mediate the electron transfer from the Mn cluster in OEC to the oxidized primary electron donor P680. On the other hand, Y_D functions only as an auxiliary electron donor or acceptor, depending on its oxidation state. The oxidized Y_D gives an EPR signal (Sig. II_s), that shows a remarkable stability at room temperature. On the contrary, the lifetime of Y_Z^+ in oxygen-evolving PS II is very short and its EPR signal (Sig. II_{vf}) is difficult to observe. The elimination of the OEC by washing with a Tris buffer or the inhibiting of OEC such as Ca-depleted PS II result in the increase of the lifetime of Y_Z^+, so that its EPR signal (Sig. II_f) is observed more easily.

2. Materials and Methods

The oxygen evolving PS II membranes (300-500 μM O_2/mg Chl per hour) were prepared from market spinach by the method of Kuwabara and Murata [1]. The membranes were suspended in a 0.8 M Tris buffer (pH 8.5), stirred for 30 min at 0°C under the room light and centrifuged at 35000 × *g* for 20 min. To prepare the sample with various pH, the pellet was washed in the buffers of 0.4 M Sucrose /20 mM NaCl with addition of 20 mM Mes/NaOH (pH 5.5, 6.0, 6.5 at 0°C) or 20 mM Mops/NaOH (pH 7.0 and pH 7.5 at 0°C), or 20 mM Hepes/NaOH (pH 8.0 at 0°C).

For Ca-depletion, the oxygen evolving PSII membranes were prepared from market spinach by the method of Berthold, Babcock and Yocum [2] with modifications described in [3]. The membranes at a Chl concentration of 0.5 mg/ml were suspended in a buffer 400 mM sucrose, 20 mM NaCl, 10 mM citrate/NaOH (pH 3.0) at 0°C for 5 min. Then, 10% vol. of 400 mM sucrose, 20 mM NaCl and 500 mM Mops/NaOH (pH 7.5) were added to adjust the final pH at about 6.5. The treated membranes were washed and resuspended in a final buffer of 400 mM sucrose, 20 mM NaCl and 20 mM Mes/NaOH at pH 6.5.

To generate and trap Y_Z^+, the samples were illuminated by a 500 W tungsten-Br_2 lamp for 10 sec at 253K for tris treated PS II and at 246K for Ca-depleted PS II through a 10 cm-thick water filter and frozen in liquid N_2 within one second after illumination.

P. Mathis (ed.), Photosynthesis: from Light to Biosphere, Vol. I, 559–562.

3. Results and Discussion

3.1 *Tris-treated PS II*

Fig. 1 shows the X-band CW EPR spectra of the trapped Y_Z^+ at several pH values. They were obtained by substraction of the spectra of Y_D^+ recorded after the dark incubation of the sample for 30 min at 0°C from the spectra in the illuminated sample. The shape of the Y_Z^+ EPR spectrum at pH above 6.5 is similar to that obtained in previous works [4]. Below pH 6.5, however, the Y_Z^+ spectrum is found to be narrower and a slight decrease of the g-value is evident. On the other hand, the spectrum of Y_D^+ shows no change in the observed pH range.

Traces 1,2 and 3 in Fig. 2 show the pulsed ENDOR spectra of the Y_D^+ at pH 7.5, the Y_Z^+ at pH 7.5 and the Y_Z^+ at pH 5.5, respectively. The Y_Z^+ spectrum is obtained by subtraction of Y_D^+. Those ENDOR peaks are labelled A to G. The component A can be assigned to the perpendicular orientation of the hyperfine tensor of the protons at the ring carbons C_3 and C_5. Other two lines may be assigned to the perpendicular (B) and parallel (C) orientations of the hyperfine tensor axes of one of the β-methylene protons with respect to the direction of the static magnetic field B_o. From the frequencies of B and C, the isotropic hyperfine constant $a_\beta \approx 28.8$ MHz can be determined for the β-methylene proton.

The shape of the Y_Z^+ ENDOR spectrum at pH 7.5 is very similar to that of Y_D^+. The prominent peak D of the ring proton line, same as for Y_D^+, is assigned to the hyperfine splitting of 19.8 MHz. A major difference consists in the shift of the β-methylene proton line in the Y_Z^+ spectrum to the higher frequencies E. From the position of this line, the hyperfine constant $a_\beta \approx 34$ MHz can be estimated. On the other hand, the spectrum of Y_Z^+ at pH 5.5 (trace 3 in Fig. 2) is remarkably different. It does not exhibit a well-separated

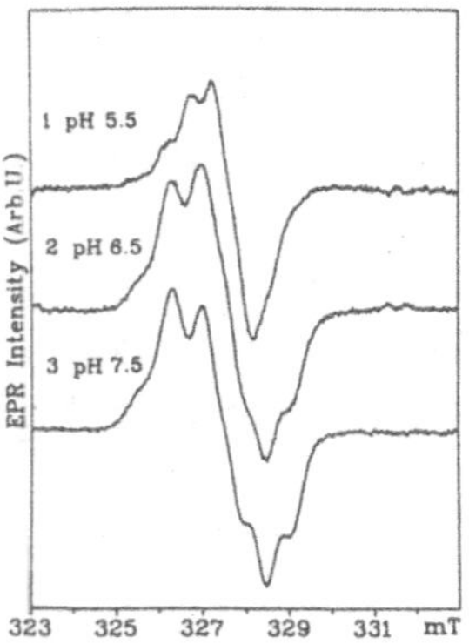

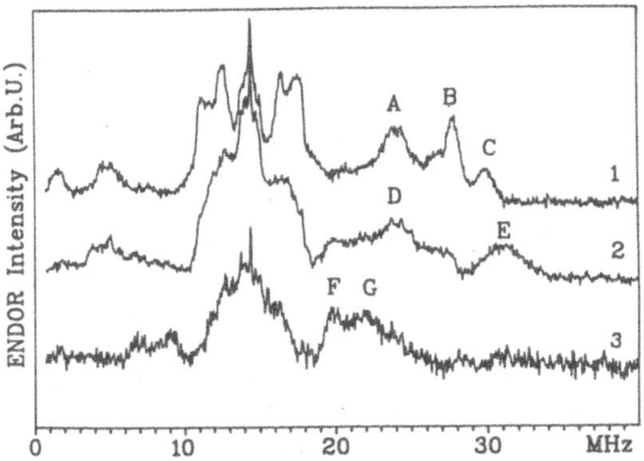

Fig.1. (left) pH-dependence of the X-band EPR spectra of $Y_Z^{+\cdot}$ EPR conditions: m.w. frequency, 9.21 GHz; m.w. power, 0.05 mW; modulation frequency, 100 kHz; modulation amplitude, 0.4 mT; temperature, 80K.

Fig.2. (right) Davies pulsed ENDOR spectra of Tris-treated PSII. Trace 1, Y_D^+ signal at pH7.5; Trace 2, Y_Z^+ signal at pH7.5; Trace 3, Y_Z^+ signal at pH5.5. The measurement temperature was 30K and the m.w. pulses of 96 ns, 48 ns and 96 ns duration were used. The separation between the first (preparation) and the second m.w. pulse was 8 μs. The separation between the second and third m.w. pulses was 200 ns. The duration of the rf pulse applied between the first and second m.w. pulses was 6 μs.

On the other hand, the spectrum of Y_Z^+ at pH 5.5 (trace 3 in Fig. 2) is remarkably different. It does not exhibit a well-separated β-methylene proton line, but shows a broad response situated symmetrically with respect to the proton matrix line, with main maxima, F and G, corresponding to the hyperfine splittings of 11 and 15.5 MHz, respectively.

For interpretation of these two ENDOR peaks with hyperfine splittings of 11 and 15.5 MHz, we have found that four protons are included to give the spectral structure and overall width observed for Y_Z^+. We assume two of these four protons to be the 3,5-ring protons. The other two protons are supposed to belong to the β-methylene group. The hfs constants of the 2,6-ring protons are expected to have much smaller hyperfine constants and to contribute to the matrix region of the ENDOR spectra.

Using McConnell relations for the isotropic hfi constant of a ring proton and β-methylene proton of an aromatic organic radical, we can estimate the spin density on the ring carbons C_3 and C_5 to be $\rho_{3,5} \sim 0.22$ and the dihedral angle of β-methylene protons to be 57.2° and 62.8°, or both 60° from the p_z-orbital of the ring electron. Y_D^+ and the high-pH form of Y_Z^+ have similar spin density distributions and orientations of the β-methylene group with respect to the ring plane. In the low-pH form of Y_Z^+ the spin densities on the carbons $C_{1,3,5}$ are slightly reduced and the orientation of the β-methylene group is noticeably changed. These small structural alterations, however, result in remarkably different ENDOR spectra of Y_Z^+ at high and low pH . The alterations of the Y_Z^+ molecular structure and spin density distribution with decreasing pH are, probably, caused by the changes in the close environment of this amino residue induced by the excess H^+ present in the media at low pH. It is interesting that the characteristic pH value of 6.0 to 6.5 for the transition between the high- and low-pH forms of Y_Z^+ spectra correlates with the pK_a value for the loss of a proton from the histidine imino nitrogen in PS II . It is possible therefore that the pH effect on Y_Z^+ is related to the modification of the histidine amino acid residue in its closest vicinity.

3.2 Ca-depleted PS II

Two low and high pH forms of Y_Z^+ were found in Tris treated PS II. Naturally in oxygen evolving PS II the two forms of Y_Z^+ would be expected to exist. However, the observation of Y_Z^+ has been difficult because of fast reduction by Oxygen Evolving Complex (OEC). Recently trapped Y_Z^+ can be observed more easily in Ca-depleted PS II, which includes OEC, by illumination at 246 K [5]. The trapped Y_Z^+ converted to the abnormal S_2 state during dark incubation at 273 K. The trapped Y_Z^+ at pH 6.5 showed a similar EPR line shape as the low pH form in tris treated PS II.

No attempt has not yet succeeded in trapping Y_Z^+ beyond the S_2 state Among several discussions about the origin of the S_3 signal, the proposal, that Y_Z^+ is coupled with the Mn-cluster, seems to be plausible [6]. We have determined the spin number of the S_3 signal as following.

The field-sweep ESE spectra measured at T=6K in the S_2 and S_3 states of Ca-depleted PS II are shown in Fig. 3 by traces 1 and 2, respectively. 50μM DCMU was added to the stable S_2 state of the Ca-depleted sample to prevent further S-state advancement. At the magnetic field (B_o=354 mT), the maximum higher field peak of the S_3 signal, we have studied the dependence of the S_3 ESE signal amplitude on the intensity of the m.w. field B_1. B_1 dependence of Y_D^+ signal was observed at the field position of 346 mT for comparison. Fig.4 showed the B_1-dependence for S_3 ESE amplitude, which was obtained by subtraction of background signals. It is seen from Fig. 4 that the B_1-dependence for S_3 is considerably different from that for Y_D^+. As for both signals the g-factors are close to 2, this difference is explained by the values of the transition operator matrix elements $<S_x>$ inducedby two-pulse ESE sequence with similar amplitudes and the lengths of t_p and $2t_p$. The ESE signal intensity for resonant spins is proportional to $\sin^3(2g\beta B_1<S_x>t_p)$.

Thus the S_3 spectrum shape and the signal amplitude dependence on the intensity of the microwave field were interpreted in terms of a dipole- and exchange-coupled radical pair with $S \approx 1$ and the radical distance separated 5.7 Å.

Gilchrist et al. [6] suggested the dipolar coupled spin may be the Mn-cluster with a distance 4.5Å from Y_Z^+. On the other hand, Kodera et al. estimated that the distance between Y_Z^+ and the Mn-cluster at 15-20Å based on the T_1 measurement of the trapped Y_Z^+ [5]. To resolve the difficulty to assign the origin of the S_3, more investigations should be performed.

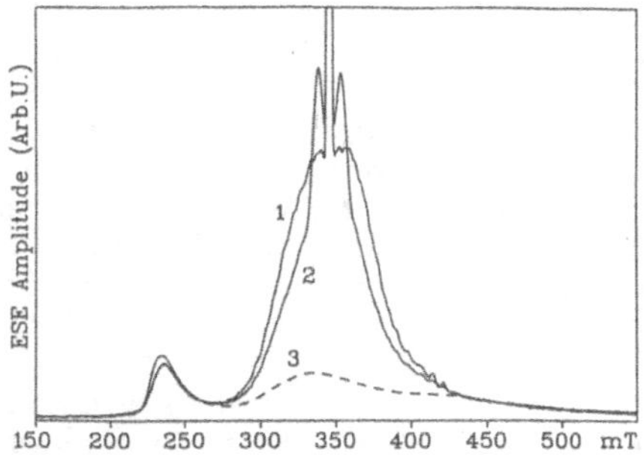

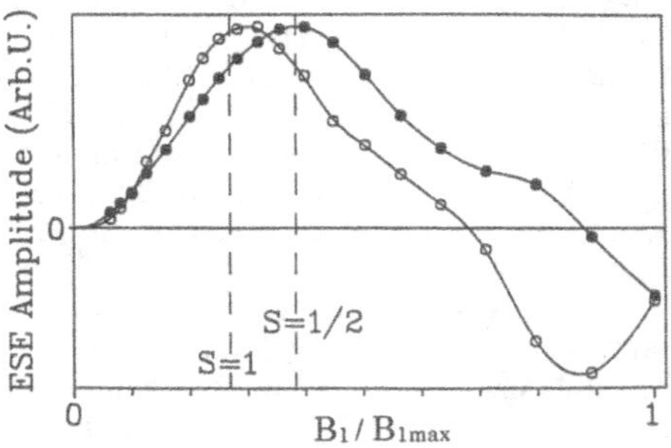

Fig.3. (left) Primary field sweep ESE observed at τ=200 ns. Trace 1, Ca-depleted PS II in the S_2 state. Trace 2, the same sample after illumination for one 1 min at 0°C with 100μM DCMU. Dashed trace 3, the Cyt b559 spectrum. The measurement conditions: T=6K, pulse repetition rate 1 kHz, pH 6.5.

Fig.4. (right) B_1-dependences of the primary ESE signal amplitude in the Ca-depleted PS II sample illuminated for 1 min at 0°C. Open circles show the B_1 dependence of S_3 signal given by the difference between the measurements at B_o=354 mT and at at B_o=364 mT (back ground signal). Filled circles, measured at B_o=346 mT (the signal due to Y_D^+). The solid lines are the spline-interpolations. B_{1max} is the maximum B_1 fed to the resonator in the current experiment. Vertical dashed lines labelled "S=1" and "S=1/2" are the maximum positions in the B_1-dependences expected for the corresponding spin values.

Reference

[1]Kuwabara, T. & Murata, N.(1982) Plant Cell Physiol. 23, 533-539.
[2]Berthold, D. A., Babcock, G. T. and Yocum, C. F. (1981) FEBS Lett. 134, 231-234.
[3]Ono, T. and Inoue, Y. (1989) Biochim. Biophys. Acta 973, 443-449.
[4]Mino, H. & Kawamori, A.(1994) Biochim. Biophys. Acta.1185, 213-220.
[5]Kodera, Y.,Hara, H.,Astashkin, A V.,Kawamori,A. and Ono,T. (1995) Biochim. Biophys. Acta. in press
[6]Gilchrist,M.L., Ball, J. A., Randall, D.A. and Britt, R.D. (1995) Proc. Natl. Acad. Sci. USA. in press.

ESEEM AND PULSED ENDOR STUDIES OF PRIMARY ACCEPTOR QUINONE IN PHOTOSYSTEM II.

Asako KAWAMORI[1], Hiroyuki MINO[1], Andrei V.ASTASHKIN[2], Shigeki KUROIWA[3] and Kozo AKABORI[3]

[1]*Faculty of Science, Kwansei Gakuin University, Nishinomiya 662, Japan;* [2]*Institute of Physical and Chemical Research (RIKEN), Wako 351-01, Japan,* [3]*Faculty of Integrated Arts and Sciences, Higashi-Hiroshima 724, Japan.*

1. Introduction

In PS II, the primary electron acceptor plastoquinone, Q_A, is situated in vicinity of a non-heme Fe^{2+} ion. The magnetic interaction of Q_A^- with Fe^{2+} leads to a broad EPR spectrum that gives only a restricted information about the quinone-iron complex structure. When Fe^{2+} is removed from PS II, substituted by Zn^{2+} or converted into the diamagnetic state by the NaCN treatment, a 0.9 mT wide EPR spectrum of Q_A^- is observed [1,2]. The ENDOR spectroscopy was applied to the Fe-depleted PS II to determine the hfi of Q_A^- molecular protons and the protons forming the hydrogen bonds with Q_A oxygen atoms [3]. The hfi of two hydrogen-bond protons were found to be different, probably, due to the steric hindrances imposed by the isoprenyl chain R (see Fig.1).

Recently we have used the ESEEM spectroscopy to probe the binding site of Q_A^- in Zn-substituted PS II RC [4]. It was established that Q_A is coordinated by the >NH group in the imidazole ring of histidine (probably, histidine 215 of the D2 protein) and by the peptide group of another (probably, alanine 261 of the D2 protein) amino acid residue. In this work we extend these studies to Fe-depleted and NaCN-treated PS II RC preparations.

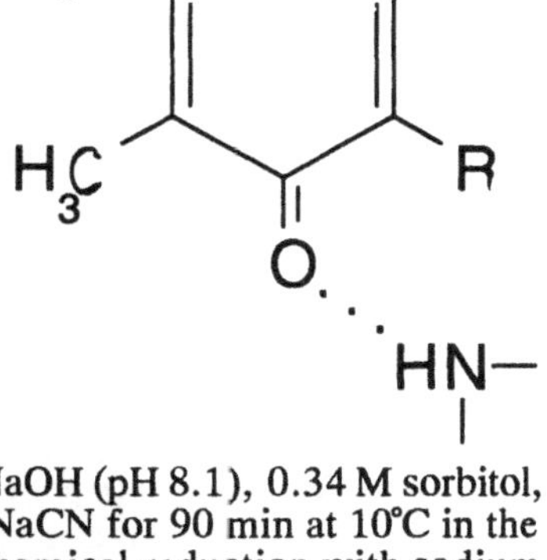

Fig.1. The molecular structure of the plastoquinone-9 (Q_A). The hydrogen-bonds to two >NH groups are shown. R is $-(CH_2CHC(CH_3)CH_2)_9H$.

2. Materials and Methods

The Zn-substituted (Zn-s) PS II RC complexes were prepared and characterized by the method of flash-induced transient absorption as described in [4], with Fe-depleted (Fe-d) PS II RC complexes being the intermediate stage of the preparation. NaCN treatment (CN-t) was done by incubating the PS II RC in a solution of 100 mM Tricine NaOH (pH 8.1), 0.34 M sorbitol, 10 mM NaCl, 5 mM $MgCl_2$, 10 mM $NaHCO_3$, 340 mM NaCN for 90 min at 10°C in the darkness. The anion radicals Q_A^- were generated by a chemical reduction with sodium dithionite (see [2,4]).

P. Mathis (ed.), Photosynthesis: from Light to Biosphere, Vol. I, 563–566.

The experiments have been performed on a pulsed EPR spectrometer ESP380 (Bruker) using a stimulated ESE sequence of 3×16 ns m.w. pulses. In the Davies ENDOR measurements, the m.w. pulses were 96, 48 and 96ns long. The intervals between the 1st and 2nd, 2nd and 3rd m.w. pulses were 8 μs and 200ns, respectively. The duration of the r.f. pulse applied between the 1st and 2nd m.w. pulses was 6 μs. The measurement temperature was 30K.

3. Results and Discussions

3.1 ESEEM spectra of Q_A^-

The EPR spectra of Q_A^- recorded in this work are single lines with a width of ≈0.9 mT. The ESEEMs were measured at static magnetic field B_o=346 mT corresponding to the centers of Q_A^- EPR spectra. Spectra 1,2 and 3 in Fig.2 show the sums of cosine Fourier transforms of six normalized positively-phased (see [4]) stimulated ESEEMs measured at the intervals τ=120,160,200,240,280 and 320 ns between the 1st and 2nd m.w. pulses in Zn-s, Fe-d and CN-t PS II RC, respectively. The narrow lines at the frequencies of 0.53, 0.88, 1.46, 2.02 and 2.88 MHz in spectrum 1 are explained in [4] as pure nuclear quadrupole resonance (NQR) transitions of two ^{14}N nuclei (denoted N_I and N_{II}) with the frequencies:

$$\nu_o = 2k\eta, \qquad \nu_- = k(3-\eta), \qquad \nu_+ = k(3+\eta) \tag{1}$$

where k is a quadrupole coupling constant and η is an asymmetry parameter of the nqi tensor. The broad lines at 4.3 and 5 MHz are the Δm_I=2 transitions with the frequencies:

$$\nu_{\Delta m=2} = 2[(\nu_N \pm A/2)^2 + k^2(3+\eta^2)]^{1/2} \tag{2}$$

where ν_N is a Zeeman frequency of ^{14}N and A is a secular component of the hfi. Using Eqs.(1,2), the nqi (k≈0.39 MHz, η≈0.69 for N_I and k≈0.82 MHz, η≈0.52 for N_{II}) and hfi (A≈1.9 MHz for N_I, N_{II}) parameters were estimated. N_I was assigned to the amino ^{14}N in the imidazole ring of histidine and N_{II} - to a peptide ^{14}N. The spin density is delocalized on these ^{14}N through the hydrogen-bonds between the >NH groups and the oxygen atoms of Q_A^-.

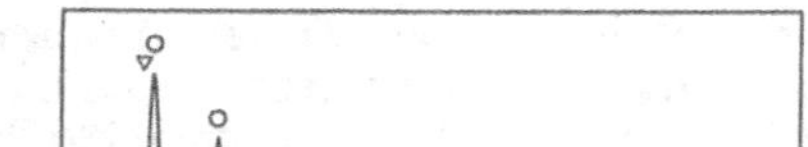

Fig.2. Traces 1,2 and 3, see text. Trace 4, cosine FT of the normalized stimulated ESEEM at τ=120ns in CN-t PS II RC. Open circles and triangles mark the NQR transitions of the histidine and peptide ^{14}N, respectively. The closed circle and trisangle mark the Δm_I=2 transitions of the histidine and peptide ^{14}N, respectively.

In the ESEEM spectrum of CN-t PS II RC (Trace 3 in Fig.2) the imidazole ^{14}N lines are absent. On the other hand, the intensity of the peptide ^{14}N lines is about two times higher than in Zn-s sample. It is assumed [2] that at high pH a CN^- group substitutes one histidine residue at the equatorial binding site of Fe^{2+}, which results in the conversion of the paramagnetic iron into the diamagnetic state. It appears that the Fe – CN binding leads to the change of the position of the histidine imidazole ring bringing its >NH group outside the distance range where it can be bound to Q_A. Q_A then coordinates to the second peptide group that might belong to the same histidine or to another amino acid residue.

From spectra 3,4 we find $k \approx 0.87$ MHz, $\eta \approx 0.42$, $A \approx 2.17$ MHz for one peptide ^{14}N (N_A) and $k \approx 0.82$ MHz, $\eta \approx 0.45$, $A \approx 2.05$ MHz for another one (N_B). The correspondence between N_A, N_B in CN-t and N_{II} in Zn-s PS II RC is not obvious. However, the nqi parameters of N_{II} are closer to those of N_B implying that N_B and N_{II} might refer to the same ^{14}N.

The NQR lines of the histidine imidazole ^{14}N (N_I) and the peptide ^{14}N (N_{II}) are seen in the spectrum of Fe-d PS II RC (Trace 2 in Fig.2). The N_I contribution to spectrum 2 is only ≈30% of that in Zn-s PS II RC (Trace 1) indicating the loss of a Q_A binding to the >N_IH group in ≈70% of the PS II RC particles. However, as seen from the intensity of the peptide ^{14}N lines, no compensating binding to another peptide >NH group occurs.

Thus, the ESEEM data show the structure of the Q_A binding site in Zn-s PS II RC to be similar to that in bacterial RC [5]. The Fe-depletion and NaCN-treatment mostly modify the histidine side of Q_A binding site that exhibits more loose binding or is more strongly affected by the treatments. The Fe-d sample is heterogeneous on the microscopic level, with only ≈30% of the PS II particles preserving the coordination of Q_A to histidine. This percentage, however, might depend on the sample preparation. As the Fe-depletion is an intermediate stage in preparation of the ZnS sample, we conclude that the rearrangements of the Q_A binding structure are reversible. This is, probably, true also for the CN-t PS II RC that shows a reversible EPR signal after the removal of cyanide [2].

3.2 ENDOR spectra of Q_A^-

Traces 1,3 and 5 in Fig.3 are the Davies ENDOR spectra of Q_A^- in Zn-s CN-t and Fe-d PS II RC, respectively. They are symmetrized with respect to the proton Zeeman frequency to improve the S/N ratio. The attempt to separate the contribution of the hydrogen-bond protons to these spectra by the H-D substitution was only partially successful. There were no noticeable effects in the D_2O-washed Zn-s PS II RC sample. In CN-t sample the pair of lines at 13.2 and 15.9 MHz, and in the Fe-d sample the broad peaks at 11.9 and 17.2 MHz were suppressed.

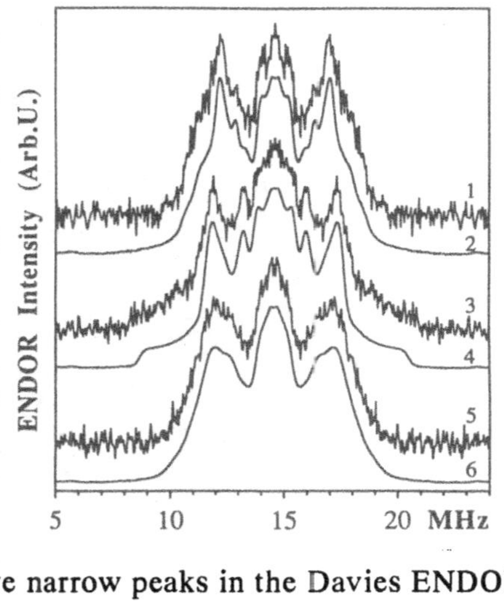

Fig.3. Traces 1,3 and 5, Q_A^- Davies ENDOR spectra in Zn-s, CN-t and Fe-d PS II RC, respectively. Traces 2,4 and 6, the simulations of the spectra 1,3 and 5, respectively, with the parameters given in the text.

The analysis of the ^1H ESEEM spectra similar to that in [4] gives the anisotropic hfi value of ≈-5 MHz for one of the hydrogen-bond protons. Thus, one can assign the outer peaks in the spectra in Fig.3 to the hydrogen-bond proton. The numerical simulations allow one to estimate the dihedral angles of two β-methylene protons of the isoprenyl chain of Q_A to be close to 60° and 300° (or 120° and 240°) in all studied samples. They show also that these protons do not give narrow peaks in the Davies ENDOR spectra. Therefore, we ascribe the narrow features with splittings greater than 2 MHz to the hydrogen-bound protons.

The anisotropic hfi values $T_\perp$ for two hydrogen-bond protons obtained by the simulations of the ENDOR spectra are:

-3.5 and -4.4 MHz (Zn-s), -2.8 and -5.9 MHz (CN-t), -3.8 and -5 MHz (Fe-d) (3)

Traces 2,4 and 6 in Fig.3 are the simulations with these parameters for Zn-s, CN-t and Fe-d PS II, respectively. The spectrum of (plastoquinone-9)$^-$ / C_3D_7OD was used in these calculations to approximate the contributions of the methyl and α-protons. The individual linewidth for the hydrogen-bond protons used in the simulation for the Fe-d PS II ($\approx$0.7 MHz) is significantly greater than that for Zn-s and CN-t PS II ($\approx$0.3 MHz). This agrees with the heterogeneity of the Fe-d sample observed in the ESEEM measurements. It is probable that the quinones that have lost their binding to histidine bind to another partner. As no ^{14}N ESEEM is observed for this partner, it might be an OH group of some amino acid residue.

References

1 Klimov, V.V., Dolan, E., Shaw, E.R. and Ke, B.(1980) Proc.Natl.Acad.Sci.USA **77**, 7227

2. Sanakis, Y., Petrouleas, V. and Diner, B.A. (1994) Biochemistry **33**, 9922

3. MacMillan, F., Lendzian, F., Renger, G. and Lubitz, W. (1995) Biochemistry **34**, 8144

4. Astashkin, A.V., Kawamori, A., Kodera, Y., Kuroiwa, S. and Akabori, K. (1995) J. Chem. Phys. **102**, 5583.

5. Deisenhofer, J. and Michel, J. (1989) EMBO J. **8**, 2149

PHOTOSYSTEM II ELECTRON FLOW REQUIRES BOUND BICARBONATE

Oscar J. de Vos[1], Jack J.S. van Rensen[1] and Govindjee[2]
[1]Department of Plant Physiology, Agricultural University, 6703 BD Wageningen, The Netherlands; [2]Department of Plant Biology, University of Illinois, Urbana, IL 61801, USA

1. Introduction

Carbon dioxide is required for photosynthesis as it is fixed by ribulose 1,5-bisphosphate carboxylase to produce carbohydrates. In addition, CO_2 (or HCO_3^-) is also involved in Photosystem II (PSII) electron transport, both on its electron donor (see e.g. 1,2), as well as its electron acceptor (see e.g. 3-5) side. This phenomenon is known as the bicarbonate effect. Experiments in several laboratories (6-8) have provided new information on this important phenomenon. Incubation of plant thylakoids and cyanobacteria in media with anions (formate, acetate, nitrate, etc.) inhibits electron flow. This inhibition is reversed fully and uniquely by the addition of bicarbonate (see e.g., 9, 10). Bicarbonate has been shown to be liganded to the non-heme iron in the Q_A-Fe-Q_B complex of PSII (11, 12).

Jursinic and Stemler (13) discussed two hypotheses for the bicarbonate effect: (1) the "bound-bicarbonate" hypothesis: here, a high-affinity binding of bicarbonate is required for high rates of electron flow; and (2) the "inhibitory-anion" hypothesis, also called the "empty-site" hypothesis. Here, for high rates of electron flow, the binding site may be entirely free of any inhibitory anion, and bicarbonate functions only to displace the inhibitory anions. These two explanations differ in the residency state of the binding site for bicarbonate when normal (rapid) electron flow occurs. Jursinic and Stemler (13) failed to demonstrate the presence of CO_2 in CO_2-depleted thylakoid suspensions in low (10 mM) NaCl medium that had high rates of electron flow. Thus, they concluded that the inhibitory-anion or "empty-site" hypothesis is correct, implying that bicarbonate is not a requirement for PSII reactions at this site. On the other hand, Cao et al. (14) using a kinetic model, were unable to distinguish between what they had called the essential and non-essential activator models, respectively, equivalent to the "bound bicarbonate" and "empty-site" hypothesis.

In this paper, we demonstrate CO_2 release under the experimental conditions of Jursinic and Stemler (13) and, thus, support the "bound bicarbonate" and not the "empty-site" hypothesis.

2. Materials and Methods

Broken chloroplasts (thylakoids) were prepared from 10 to 15 day old pea plants (*Pisum sativum* L. var. Rondo) and 7 to 10 day old maize plants (*Zea mays* L.) grown in climate-controlled rooms. Thylakoids were isolated as described earlier (15), suspended in a medium containing 0.4 M sorbitol, 20 mM tricine-NaOH (pH 7.8), 10 mM NaCl, 5 mM $MgCl_2$, 2 mM Na-ascorbate and 0.2% (w/v) bovine serum albumin, and stored at -80°C. Thylakoids were thawed on ice before use.

Maximum rates of O_2 evolution in continuous saturating light was measured as described earlier (15). Thylakoids were pretreated exactly as described by Jursinic and Stemler (13). As was already indicated by Jursinic and Stemler, it is difficult to measure small changes in CO_2-concentration in the gas phase above a thylakoid

P. Mathis (ed.), Photosynthesis: from Light to Biosphere, Vol. I, 567–570.

suspension using a closed LICOR infra-red gas analyzer system. They had to take special precautions (such as covering the entire instrument with plastic) to avoid, e.g., leakage. To avoid this problem, we have chosen to use an open infra-red gas analyzing system, and to inject samples taken from the gas phase above the thylakoid suspension. Carbon dioxide concentrations in the gas phase above the samples were measured with a type 225-MK2 infra-red gas analyzing system (IRGA; Analytical Development Co. Ltd., Hertsfordshire, England). This IRGA system is based on a differential measurement method. In both the sample and the reference cell the reference gas (pure N_2) was passed. The difference in the concentration of CO_2 between the sample and the reference cell caused a difference in absorption of IR radiation. This exerted a difference in pressure between the measuring units (which are separated by a membrane) of both cells. This pressure difference was measured. Calibration of the IRGA was carried out with a calibration gas which contained 400 ppm CO_2 in N_2 gas. There was a perfect linearity in the response of our system. The flow rate of reference N_2 gas through both the analyzing and the reference cells was 350 $ml.min^{-1}$. Because of the positive pressure in our system, we encountered no problem of leakage of CO_2 into the analyzer.

Thylakoids were incubated at 20°C in the dark for 30 min in a reaction medium containing 50 mM sodium phosphate and 10 mM NaCl at pH 6.4 at a concentration of 50 μg $Chl.ml^{-1}$, whereafter they were collected by centrifugation at 1000 x g for 8 min and resuspended in 3.8 ml reaction medium at a concentration of 4.2 mg $Chl.ml^{-1}$. The thylakoid suspension was transferred into a septum vial with a magnetic stir bar. To facilitate equilibration between CO_2 and HCO_3^-, 100 μl of carbonic anhydrase solution (20 $\mu g.ml^{-1}$, pH 6.4) was added to a final concentration of 0.5 $\mu g.ml^{-1}$. To examine whether the high affinity binding sites of the pretreated thylakoids were depleted of bicarbonate or not, 100 μl of CO_2-depleted 4 M sodium formate solution at pH 6.4 was injected to a final concentration of 100 mM. Both the carbonic anhydrase and the formate solution were prepared in CO_2-depleted reaction medium at pH 6.4 and stored in septum vials. The gas volume above the thylakoid suspension in the septum vial was 4.8 ml. The thylakoid suspension was incubated at 20°C for 15 min in room light and was continuously stirred during the incubation period. A sample of 0.3 ml was taken from the gas phase above the thylakoid suspension and its CO_2 content was analyzed by injecting the sample into the IRGA via a rubber septum into the gas stream which then entered the analyzing sample cell. We also measured controls in which, instead of thylakoid suspension, reaction medium was incubated.

3. Results and Discussion

Stemler (16) and Jursinic and Stemler (13) concluded that PSII electron flow can proceed normally when the bicarbonate binding site at PSII is empty of all monovalent anions, including bicarbonate. Govindjee et al. (17) had challenged Stemler's conclusion at pH 6 (16), based on the fact that there was much more CO_2 release at pH 6.5 than at pH 6. The conclusion of Jursinic and Stemler (13) was, however, based on the following result, done at pH 6.4.

In maize thylakoids, incubated in the dark for 30 min in a CO_2-free medium at pH 6.5 containing 10 mM NaCl, the rate of O_2 evolution without the addition of bicarbonate was found to be only 5% lower than when bicarbonate was added two min before the start of the measurement. This must have been an almost uninhibited condition, in which the bicarbonate binding site has bicarbonate bound according to the bound-bicarbonate hypothesis, and is empty, according to the inhibitory-anion hypothesis. A choice in favour of the empty-site hypothesis was made when Jursinic

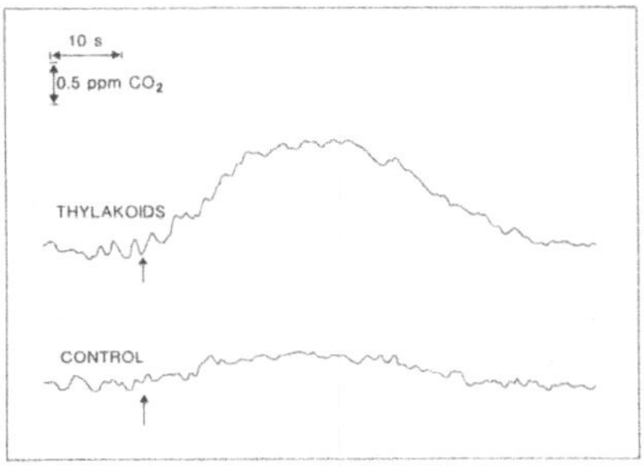

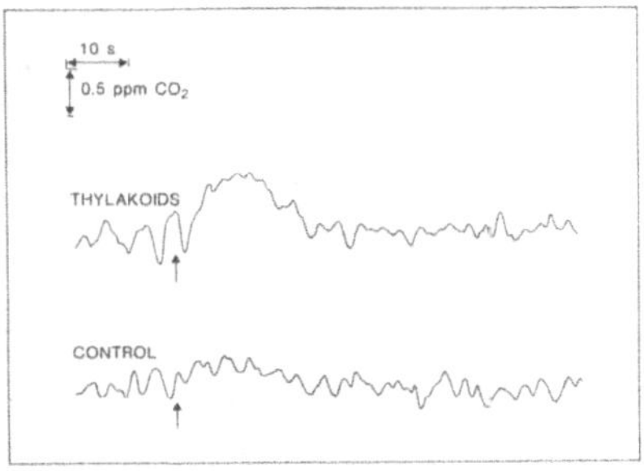

Figure 1. (left) CO_2 release by pea thylakoids after injection of 100 mM formate. At the time of the upward arrow, 0.3 ml gas sample was injected in the IRGA, taken from the gas phase above a thylakoid suspension which was incubated with 100 mM formate during 15 min. Control is from an experiment in which the same medium was used without thylakoids. One representative result out of four experiments.
Figure 2. (right) CO_2 release by maize thylakoids after injection of 100 mM formate. Same protocol as in Fig. 1. One representative result out of four experiments.

TABLE I. The rate of oxygen evolution by maize and pea thylakoids after incubation in the dark at 20 °C for 30 min in a CO_2-free medium containing 50 mM sodium phosphate and 10 mM NaCl at pH 6.4 (n=5). Then, after collection by centrifugation and resuspension of the thylakoids in fresh CO_2-depleted reaction medium to 50 μg $Chl.ml^{-1}$, 10 mM bicarbonate was added and, after a further 2 min dark incubation, the electron acceptor system (5 mM NH_4Cl, 40 μM DCBQ and 0.4 mM FeCy) was added and oxygen evolution was measured (A. "+ 10 mM $NaCO_3$") The B. "no addition" sample was treated in the same way, but no bicarbonate was added. A third sample (C. "+ 100 mM $NaHCO_2$") was again treated in the same way, but CO_2-free formate was added. Oxygen evolution in μmol O_2.mg $Chl^{-1}.h^{-1.}$ The horizontal line D. "CO_2 release" indicates the number of CO_2 released per total PSII reaction centers upon the addition of 100 mM formate to sample B.

	maize thylakoids	pea thylakoids
A. + 10 mM $NaHCO_3$	198 ± 13	177 ± 11
B. no addition	178 ± 14	153 ± 14
Percent inhibition	10 ± 2	13 ± 3
C. + 100 mM $NaHCO_2$	61 ± 6	37 ± 7
Percent inhibition	69 ± 3	79 ± 3
D. CO_2 release	0.54 ± 0.04	1.04 ± 0.10

and Stemler (13) did not observe any CO_2 release after injection of 100 mM formate to thylakoids, as used above. Because there occurred an unidentified, what they called, respiratory process in their spinach and pea thylakoids, they did not use spinach and pea.

In this work we have repeated this critical experiment. First, we could not find significant differences between CO_2 and O_2 exchanges in peas and maize, used in this research, in darkness as reported by Jursinic and Stemler (13). We observed a small inhibition (10 - 13%) after incubation of pea thylakoids in the dark for 30 min in a medium containing 10 mM Cl^- (Table I). This is a somewhat larger inhibition than the 5% found by Jursinic and Stemler (13). Since the inhibition is partly reversed by the addition of bicarbonate, this result indicates that this treatment removes bicarbonate to a certain extent from its binding site at PSII. The critical result, however, was that we found a substantial release of CO_2 by the addition of formate to NaCl-treated thylakoids of not only peas (Fig. 1), but also of maize (Fig. 2). This means that, although the thylakoids were incubated in CO_2-free medium for 30 min, they still have bicarbonate bound at their binding site which may be removed by the addition of formate. On the basis of the results presented above, we have calculated the number of CO_2 released by the addition of 100mM formate to be in the range of 0.6 to 1 (Table I).

In conclusion: in both maize and pea thylakoid suspensions depleted of CO_2 without any inhibitor, but with 10 mM NaCl in the medium, high rates of electron flow are associated with the presence of 0.6 to 1 CO_2 per photosystem II reaction center. These data support the essential activator model for bicarbonate, the details of which remain to be elucidated.

We thank Dr. L.W.A. van Hove for technical advice on the IRGA measurements. Govindjee thanks an ongoing discussion with Alan Stemler and a fellowship from the Wageningen Agricultural University.

1 El-Shintinawy, F. and Govindjee (1990) Photosynth. Res. 24, 189-200.
2 Klimov, V.V., Allakhverdiev, S.I., Feyziev, Ya.M. and Baranov, S.V. (1995) FEBS Lett. 363, 251-255.
3 Govindjee and Van Rensen, J.J.S. (1978) Biochim. Biophys. Acta 505, 183-213.
4 Govindjee and Van Rensen, J.J.S. (1993) in The Photosynthetic Reaction Center (Deisenhofer, J. and Norris, J.R., eds.), Vol. 1, pp. 357-389, Academic Press Inc., San Diego.
5 Diner, B.A., Petrouleas, V. and Wendoloski, J.J. (1991) Physiol. Plant. 81, 423-436.
6 Cinco, R.M., MacInnis, J.M. and Greenbaum, E. (1993) Photosynth. Res. 38, 27-33.
7 Petrouleas, V., Deligiannakis, Y. and Diner, B.A. (1994) Biochim. Biophys.Acta 1188, 271-277.
8 Demeter, S., Janda, T., Kovacs, L., Mende, D. and Wiesner, W. (1995) Biochim. Biophys.acta 1229, 166-174.
9 Good, N.E. (1963) Plant Physiol. 38, 298-304.
10 Jursinic, P.A. and Stemler, A. (1988) Photosynthesis Res. 15, 41-56.
11 Diner, B.A. and Petrouleas, V. (1990) Biochim. Biophys.acta 1015, 141-149.
12 Hallahan, B.J., Ruffle, S.V., Bowden, S.J. and Nugent, J.H.A. (1991) Biochim. Biophys. Acta 1059, 181-188.
13 Jursinic, P.A. and Stemler, A. (1992) Biochim. Biophys. Acta 1098, 359-367.
14 Cao, J., Ohad, N., Hirschberg, J., Xiong, J. and Govindjee (1992) Photosynth. Res. 34, 397-408.
15 Snel, J.F.H. and Van Rensen, J.J.S. (1984) Plant Physiol. 75, 146-150.
16 Stemler, A. (1989) Plant Physiol. 91, 287-290.
17 Govindjee, Weger, H.G., Turpin, D.H., Van Rensen, J.J.S., De Vos, O.J. and Snel, J.F.H. (1991) Naturwissenschaften 78, 168-170.

THE COUPLING OF LIGHT-INDUCED ELECTRON TRANSFER AND PROTON UPTAKE: REFINED CRYSTAL STRUCTURES OF MODIFIED REACTION CENTRES FROM *RHODOPSEUDOMONAS VIRIDIS*.

C. Roy D. Lancaster and Hartmut Michel, Max-Planck-Institut für Biophysik, Heinrich-Hoffmann-Str. 7, D-60528 Frankfurt am Main, Germany

1. Introduction

In photosynthetic reaction centres (RCs), light-induced electron transfer is coupled to proton uptake from the cytoplasm at the binding site of the secondary quinone, Q_B. In the original structure of the RC from *Rhodopseudomonas (Rp.) viridis* (PDB entry 1PRC, [1-2]), the Q_B site was poorly defined because it was only 30% occupied in the standard RC crystals. Recently, quinone reconstitution experiments have yielded crystals with full quinone occupancy of the Q_B site. Subsequent X-ray analysis and refinement has led to a well-defined Q_B-site model, which shall be discussed below. Also, the structure of a Q_B-depleted RC will be presented. Furthermore, the complex of the RC with the antibiotic stigmatellin [3], a chromone-type electron transfer inhibitor, will be introduced. Apart from inhibiting quinone reduction, both in photosystem II and in photosynthetic RCs, stigmatellin is equally active at inhibiting quinol oxidation, for instance in the photosynthetic bc_1-complex, in the cytochrome b_6f-complex of chloroplasts, and in the mitochondrial bc_1 -complex.

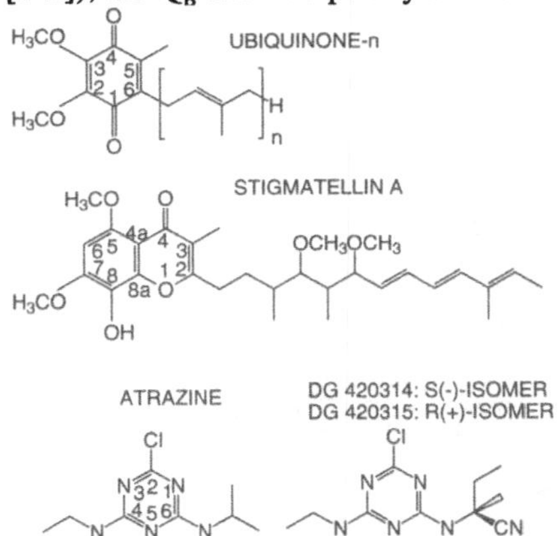

Figure 1. Chemical structures of Q_B and Q_B-site inhibitors. The native Q_B in the RC of *Rp. viridis* is ubiquinone-9. See text for details

The Q_B site is also a well-established site of triazine action. Although the triazine binding site has been localised by X-ray crystallographic analysis of the RC-terbutryn complex at a resolution of 2.9 Å [4], description of the exact nature of triazine binding and its effect on the structure of the protein has had to await the refinement of high resolution structures. With the data collected recently on RC complexes with atrazine and two chiral atrazine derivatives (cf. Fig 1), such a detailed description is now possible.

The data collected to resolve the Q_B-site related issues introduced above are of sufficient quality to improve the original model in parts other than the Q_B site. Possibly biologically relevant modifications include the asymmetry of binding of the primary quinone Q_A, the structure of the carotenoid [5], and the observation of more than a dozen additional tightly-bound water molecules between the cytoplasmic surface and the Q_B site (these will be presented elsewhere).

P. Mathis (ed.), Photosynthesis: from Light to Biosphere, Vol. I, 571–574.

2. Methods

Reaction centres were purified and crystallised essentially as described earlier [6], with Q_B removal achieved by a one hour illumination (900-1000 nm) of the protein in the presence of 20 mM ascorbate between column runs. RCs were reconstituted with 50 μM ubiquinone-2, 170 μM atrazine, 200 μM chiral atrazine derivative, and 400 μM stigmatellin, respectively. Diffraction data were collected using monochromatic synchrotron radiation at wavelengths of 0.92-1.0 Å at the EMBL X11 and MPG/GBF-BW6 beamlines at HASYLAB/DESY Hamburg. Details of data processing and crystallographic refinement will be given elsewhere.

3. Results

The crystallographic R-factors ($R = \Sigma||F_o|-|F_c||/\Sigma|F_o|$) for the structures are currently between 17.6% and 19.2% [5]. Of particular interest is the (DG-420314) complex, in that it is defined by over 100,000 unique reflections to a high resolution limit of 2.3 Å. This is the best-defined reaction centre structure to date and the R-factor currently stands at 18.4% for all data between 10.0 and 2.30 Å.

3.1 Quinone Reconstitution

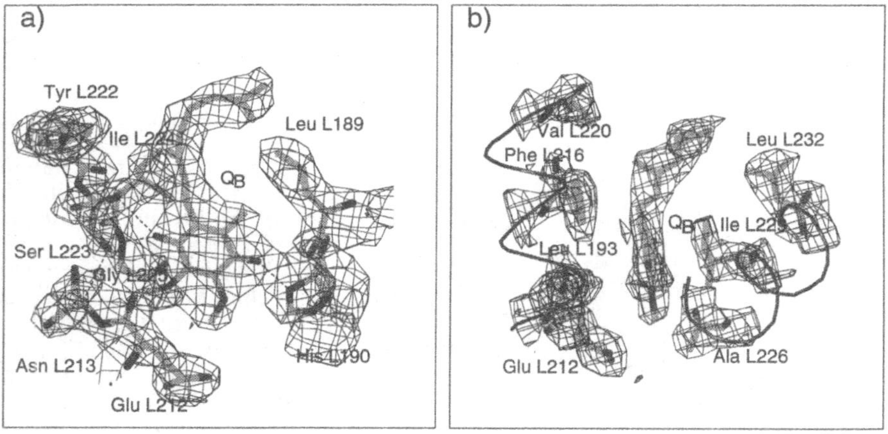

Figure 2. a) Electron density (2|Fo|-|Fc|, αc) at the Q_B-binding site. Contouring is at a level of 1.3 σ above the mean density of the map. Hydrogen bonds are indicated as thin black dashes. **b)** Side view of panel a. The phenylalanine ring is oriented approximately parallel to the quinone ring system. This figure was generated with O [7].

The electron density at the Q_B-binding site in the Q2-reconstituted complex is displayed in Figure 2. All atoms are well defined in terms of electron density. Compared to the previous model (1PRC), the hydrogen bonds assigned between the proximal carbonyl Q_B oxygen (O4) and His L190 Nδ and between the distal carbonyl oxygen (O1) and the peptide backbone NH of Gly L225 are maintained. However, the distal side of the molecule has moved by approximately 0.7 Å. As a consequence, the hydrogen bond from

Ser L223 Oγ is more favourably donated to the side chain Oδ of Asn L213 [5] and not to the distal carbonyl oxygen of Q_B, as was suggested earlier [1]. Other discrepancies to the previous model include different side chain orientations for Glu L212 and Asn L213 [5] and the reduction of the interplanar angle between the aromatic ring of Phe L216 and the quinone ring of Q_B from 12.9° to 5.6° [8].

3.2 Quinone Depletion

In the Q_B-depleted preparation, six water molecules and a detergent replace the quinone and form hydrogen bonds with the respective protein residues. Differences between the quinone-reconstituted and the quinone-depleted structure consist of a small (~0.3 Å) movement of the main chain of the L220-L226 loop. The most marked side chain changes are found for Glu L212, which forms a hydrogen bond to one of the new waters, and Asn L213 (approx. 0.8 Å), causing a slight contraction of the pocket. The orientations of Glu L212 and Asn L213 in the Q_B-depleted structure are very similar to the original 1PRC model, indicating that the structure at the Q_B site in the original model is dominated by the 70% Q_B-depleted RCs in the crystal.

3.3 Stigmatellin Binding

The $2|F_o|-|F_c|$ electron density at a contour level of 1.3 σ (to be presented elsewhere) of the stigmatellin complex allows the unequivocal placement of all atoms relevant for the binding of the compound. In analogy to quinone binding, the chromone ring of stigmatellin lies in the same plane as does the quinone ring, with the phenyl ring of Phe L216 oriented roughly parallel to it. Compared to quinone binding, the stigmatellin carbonyl oxygen O4 substitutes for the proximal carbonyl O4 in accepting a hydrogen bond from the His L190 Nδ. The hydroxyl O8 replaces the distal carbonyl as an acceptor in a bifurcated hydrogen bond from the backbone amides of Ile L224 and Gly L225. Additionally, a hydrogen bond is donated by the hydroxyl O8 to the Oγ of Ser L223. Also, the imidazole Hδ of His L190 lies exactly in the plane formed by the stigmatellin O4, its proximal methoxy oxygen, and His L190 Nδ and is located equidistantly from each of these two stigmatellin oxygens, thus establishing a three-centre hydrogen bond.

3.4 Triazine Binding

Refinement of the RC complex with DG-420314 has yielded a highly reliable model of triazine inhibitor binding. At a contour level of 1.5 σ above the mean $2|F_o|-|F_c|$ electron density (not shown), unequivocal assignment of the substituents at the chiral centre of the inhibitor is possible, as is the case for all atoms relevant for triazine binding. In addition, novel principles of triazine class inhibitor binding have emerged. These include, two tightly-bound water molecules, one which is involved in accepting a hydrogen bond from His L190 Nδ and in donating a hydrogen bond to the N3 of the triazine ring, the other completes the web with hydrogen bonds to the first water molecule and to Glu L212 Oε1. A subtle change (~0.3 Å) in the conformation of the L220-L226 loop is observed, which brings the backbone carbonyl oxygen of Tyr L222 within hydrogen bonding distance of the triazine cyanobutylamino NH. In the original terbutryn complex [4], only the two hydrogen bonds from the backbone amide of Ile L224 to the N5 of the triazine ring and from the aminoethyl NH to Ser L223 Oγ had been observed. This means that the triazine molecule is bound to the protein directly by three hydrogen bonds on the distal side, and only indirectly via water molecules (and four hydrogen bonds) on the proximal side.

4. Conclusions

In all of the structures introduced here, the side chain Oγ of Ser L223 appears to be involved in the **donation** of a hydrogen bond to the side chain Oδ of Asn L213. The existence of this hydrogen bond in the wild type structure is further corroborated by the change in side chain orientation observed for Asn L213 in the X-ray structure of the triazine-resistant mutant T1, where one of the mutations is Ser L223→Ala [9]. Ser L223 may act as an **acceptor** of a hydrogen bond in those cases where the ligand is able to donate a hydrogen bond to this position. This is the case for stigmatellin and also for triazine inhibitors, but not for Q_B in its oxidised, i.e. quinone form. With the absence of a hydrogen bond, the presence of the Ser L223 Oγ so close to the distal carbonyl oxygen is electrostatically unfavourable for quinone binding. This finding explains the observed 9-fold increase in quinone affinity for the aforementioned Ser L223-less *Rp. viridis* mutant T1 [10]. While not favourable for quinone binding, the close association of the two oxygen atoms is compatible with the importance of Ser L223 in stabilising the reduced quinone and for efficient proton donation to Q_B as demonstrated for the *Rp. viridis* RC [11] and for the *Rb. sphaeroides* RC [12]. Since the first proton is most probably taken up by the distal carbonyl oxygen, such a stabilisation of a monoprotonated doubly reduced Q_BH^- intermediate could be similar to the hydrogen bonding pattern observed for the stigmatellin complex.

All of these six structures are important for the modelling of the Q_B site in the D1 protein of the chloroplast photosystem II [13-14].

5 Acknowledgements

Ubiquinone-2 was kindly provided by Dr. Thomas Link, Universität Frankfurt. DG-420314 and DG-420315 were supplied by Dr. Schäfer, Degussa AG, Hanau.

6 References

[1] Deisenhofer, J. and Michel, H. (1989) *EMBO J. 8*, 2149-2169.

[2] Deisenhofer, J., Epp, O., Sinning, I. and Michel, H. (1995) *J. Mol. Biol. 246*, 429-457.

[3] Kunze, B., Kemmer, T., Höfle, G. and Reichenbach, H. (1984) *J. Antibiotics 37*, 454-461.

[4] Michel, H., Epp, O., & Deisenhofer, J. (1986) *EMBO J. 5*, 2445-2451.

[5] Lancaster, C.R.D and Michel. H. (1995) In: Michel-Beyerle, M.E. (ed.) *Reaction Centers of Photosynthetic Bacteria*. Springer Verlag, Berlin., in the press.

[6] Michel, H. (1982) *J. Mol. Biol. 158*, 567-572.

[7] Jones, T.A,. Zou, J.Y., Cowan, S.W. and Kjeldgaard, M. (1991) *Acta Cryst A47*, 110-119.

[8] Lancaster, C.R.D., Ermler, U. and Michel, H. (1995) In: Blankenship, R.E., Madigan, M.T., and Bauer, C.E. (eds.), *Anoxygenic Photosynthetic Bacteria*. Kluwer Academic Publishers, The Netherlands, pp. 503-526

[9] Sinning, I., Koepke, J. and Michel, H. (1990) In: Michel-Beyerle, M.E.(ed.) Reaction Centers of Photosynthetic Bacteria.. Springer Verlag, Berlin., pp. 199-208.

[10] Sinning, I., Michel, H., Mathis, P. and Rutherford, A.W. (1989). *Biochemistry 28*, 5544-5553.

[11] Leibl, W., Sinning, I., Ewald, G., Michel, H. and Breton, J. (1993). *Biochemistry 32*, 1958-1964.

[12] Paddock, M.L., McPherson, P.H., Feher, G. and Okamura, M.Y. (1990) *Proc Natl Acad Sci USA 87*, 6803-6807.

[13] Egner, U., Hoyer, G.-A. and Saenger, W. (1993) *Biochim. Biophys. Acta 1142*, 106-114.

[14] Sobolev, V. and Edelman, M. (1995) *Proteins 21*, 214-225.

CHARACTERIZATION OF A SITE-DIRECTED MUTANT (D1-ARGININE 269-GLYCINE) OF *Chlamydomonas reinhardtii*

Jin Xiong, Ronald Hutchison*, Richard Sayre*, and Govindjee
Dept. of Plant Biology, Univ. of Illinois, Urbana, IL 61801; *Depts. of Biochemistry and Plant Biology, Ohio State Univ., Columbus, OH 43210 USA

Introduction

Numerous studies have demonstrated that bicarbonate has a strong positive influence on the electron and proton transfers on the acceptor side (see review 1) of photosystem II (PSII) as well as on the electron transfer on its donor side (2, 3). Depletion of bicarbonate or its displacement with other small anionic molecules (*e.g.* formate) substantially slows the kinetics of electron transfer from Q_A^- to Q_B. Although the identity of the bicarbonate binding sites remains unknown, one of its sites is expected to be near the non-heme Fe based on EPR studies of bicarbonate-depleted membranes. In fact, bicarbonate may provide a fifth ligand for the non-heme iron. Since bicarbonate is anionic, it is expected that the binding would be electrostatic in nature and therefore positively charged amino acid residues are likely to participate in bicarbonate binding. One likely candidate for the bicarbonate binding site is arginine 269 (R269) of the D1 protein. Based on the homology studies with the bacterial photosynthetic reaction center (1), this residue is expected to be located on the stromal side of the putative transmembrane helix E and in the close vicinity of the non-heme iron binding site. In this report, we describe the phenotype of the R269G mutant, in which arginine 269 has been converted to a non-conservative glycine. This mutant is characterized by its inability to evolve oxygen, slowed Q_A^- to Q_B electron transfer, a reduction of the formate enhanced Q_A^-Fe EPR signal, and structural instability of the PSII complex. These results suggest that the mutation somewhat destabilizes the PSII complex resulting in reduced numbers of PSII centers capable of charge separation and may indirectly affect the bicarbonate binding.

Methods

Site-directed mutations were generated in the chloroplast encoded *psbA* gene of *Chlamydomonas reinhardtii* by conventional Kunkel's method. Co-transformation of the mutagenized plasmid was done with a plasmid containing the 16S rRNA gene conferring spectinomycin resistance, according to the procedures of Roffey et al. (4). A diagnostic SalI restriction site was engineered into the *psbA* gene fragment containing the R269G mutation to identify transformants. Mutants were judged as homoplasmic for the R269G mutation on the basis of Southern blot analysis of chloroplast DNA and sequencing of the mutagenized region. For biochemical and biophysical studies all cells were grown in a heterotrophic Tris-Acetate-Phosphate (TAP) medium in total darkness. Thylakoid membranes and PSII particles were isolated according to (5). Electron transfer, Mn quantification, and fluorescence measurements were done as described earlier (4 and 5).

P. Mathis (ed.), Photosynthesis: from Light to Biosphere, Vol. I, 575–578.

Herbicide binding studies were conducted using ^{14}C Terbutryn according to established procedures (6). A long-term formate treatment of the intact cells was done at pH 6.5 as described by El-Shintinawy et al. (7). Low temperature (77 K) chlorophyll fluorescence emission spectra were measured using a Perkin Elmer LS-5 fluorescence spectrophotometer with a red-sensitive photomultiplier and with a chlorophyll concentration of 30 µg/ml in the sample. The excitation wavelength was set at 435 nm and the monochromator bandwidths 10 nm for excitation and 3 nm for emission. The spectra were corrected for the wavelength dependence of the photomultiplier sensitivity and normalized at the PSI emission band at 715-nm (F715). EPR studies were performed on a Bruker ESP-300. The following instrument settings were used for detection of the Q_A^-Fe EPR signal: modulation amplitude, 28G; temperature, 5 K; and microwave power, 32 mW. PSII particles were illuminated at 1,000 µE/m^2/s in the presence of 100 mM formate for 10 min at 77 K at a chlorophyll concentration of 5 mg/ml. For Tyr_D EPR measurements, dark-adapted PSII particles were first scanned at 150 K, warmed to room temperature, illuminated for 1 min and scanned at 150 K.

Results and Discussion

A description of the isolation of the homoplasmic R269G mutant is presented elsewhere by the authors (R. Hutchison et al., in preparation). Since R269 may be located near the stromal end of the fifth transmembrane span of the D1 protein, its conversion to a glycine residue may disrupt the alpha-helical structure, prematurely terminating the transmembrane span and destabilizing the PSII complex. In order to determine whether the mutation destabilized the complex, we measured D1 protein levels and quantified the low temperature fluorescence emission intensity in cells, thylakoids and PSII particles. Western blot analysis indicated that the levels of D1 protein in dark grown whole cells, thylakoids and PSII particles were 113, 83 and 88% of the wild type levels, respectively. Low temperature fluorescence emission spectra of the wild type and the mutant were measured in an attempt to characterize the PSII assembly (Fig. 1). Our deconvolution analysis (with the help of Vladimir Shinkarev) of the spectra revealed that there is a reduction in both 685-nm (F685, CP43) and 695-nm (F695, CP47) bands for the mutant cells and the thylakoids with the F695 reduced to a greater extent. The ratio of F695/F685 of the mutant samples is reduced by 20-30% compared with the wild type. This indicates a differential reduction in these PSII antenna complexes *provided* the mutation had not caused changes in excitation energy transfers among these complexes and the PSII reaction center. Since there were no changes made in the CP43 and CP47 genes, the proposed reduction, especially of CP47, could well be related to the changes in the D1/D2 complexes to which the antenna complexes are associated. As was shown for several *Synecocystis* 6803 mutants, the reduction in CP47 correlates well with the reduction of the PSII core proteins (8). Therefore a lowered F695 band may indicate a lowered PSII assembly in this mutant.

In order to determine whether the mutation affected rates of electron transfer, we measured rates of oxygen evolution in the presence of various quinone acceptors. Under no conditions were we able to detect oxygen evolution. Furthermore, the inability to evolve oxygen was associated with a reduced Mn content in PSII particles. Wild type PSII particles typically have 4.2 Mn/250 Chl while R269G PSII particles had 1.0 Mn/250 Chl. In addition, the mutant was unable to effectively (5% of the wild type rate) oxidize artificial donors (hydroxylamine, diphenylcarbazide) to PSII using DPIP as an electron acceptor. In order to further characterize the ability to form a charge separated state and to characterize the bicarbonate effect on the electron transfer of Q_A^- to the plastoquinone pool, we measured the flash induced chlorophyll fluorescence (PSII) rise and decay in dark-grown whole cells in the presence and absence of formate and/or bicarbonate (pH 6.5). The decay kinetics of the first flash is shown in Fig. 2, in which the ratio of Fv/Fo for R269G was approximately 44% of that of the wild type, indicative of a loss of functional centers (to be discussed by J. Xiong et al., in preparation). In addition, the rate of fluorescence decay was substantially slower than for wild type, indicating that electron transfer from Q_A^- to Q_B was slowed. Addition of formate (25 mM) to the wild type cells reduced not only the rate of fluorescence decay consistent with a reduction in the electron transfer rate from Q_A^- to Q_B, but the level of the Fv/Fo suggesting an additional effect on the donor side of PSII. The formate effects were partially reversed by bicarbonate (2.5 mM). Similarly, the addition of formate to R269G produced a small but detectable reduction in the rate of fluorescence decay which was partially reversed by addition of bicarbonate. Herbicide binding experiments (6), using ^{14}C-Terbutryn, showed that the mutant had dramatically reduced Terbutryn affinity relative to the wild type. This suggests a large change in the Q_B pocket.

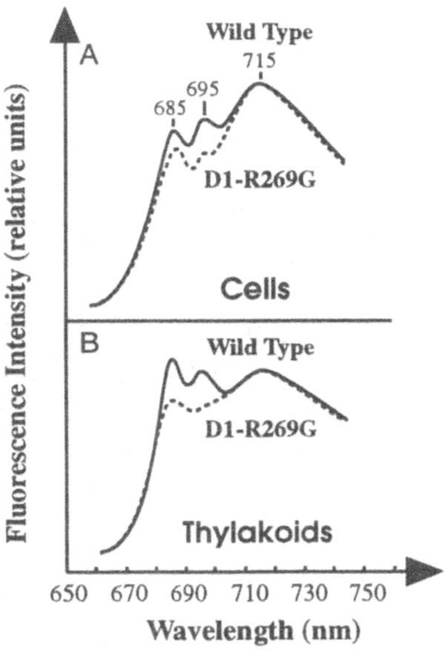

Fig. 1: 77 K chlorophyll fluorescence emission spectra.

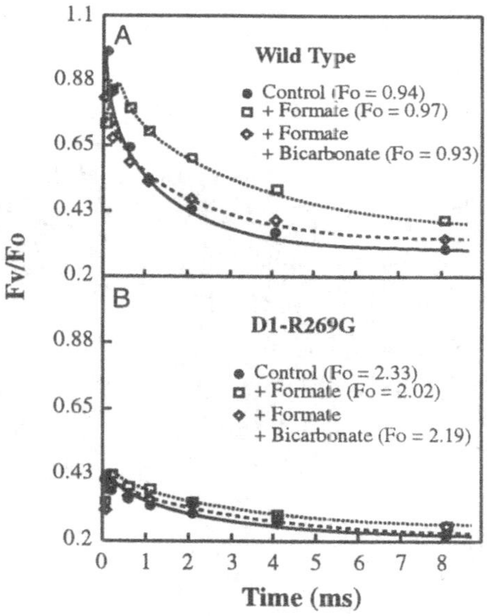

Fig. 2: Fluorescence decay kinetics (first flash) ± formate and bicarbonate.

Since western blot analyses indicated that there may be some reduction in the number of functional PSII complexes in R269G thylakoids and PSII particles, further studies were undertaken to quantify the number of functional PSII centers. Low temperature fluorescence measurements may be taken to support this conclusion if we suggest that the predicted reduction in CP43 and CP47 reflect reduction in PSII reaction centers. Since Tyr_D is dark stable and accumulates in the light (assuming no rate limitation in oxidation of Tyr_D), an estimation of the number of functional PSII complexes can be made by quantification of the Tyr_D. EPR signal in non-oxygen evolving PSII particles. There was a 68% reduction in the level of the Tyr_D EPR signal in R269G compared to the wild type (data not shown).

Since R269G does not evolve oxygen and since the formate effect on the fluorescence decay was difficult to quantify in R269G, we measured the effect of formate on the g = 1.83 Q_A^-Fe EPR signal. In the presence of formate, the g = 1.83 signal is substantially enhanced in PSII particles. Furthermore, the effects of structural perturbations on the bicarbonate binding site can be ascertained on the basis of the g value and line-width of the Q_A^-Fe EPR signal (9). As shown in Fig. 3, the addition of formate to R269G generated approximately 1/3 the Q_A^-Fe EPR signal produced in wild type. Furthermore, the signal had the same g value and line width as WT.

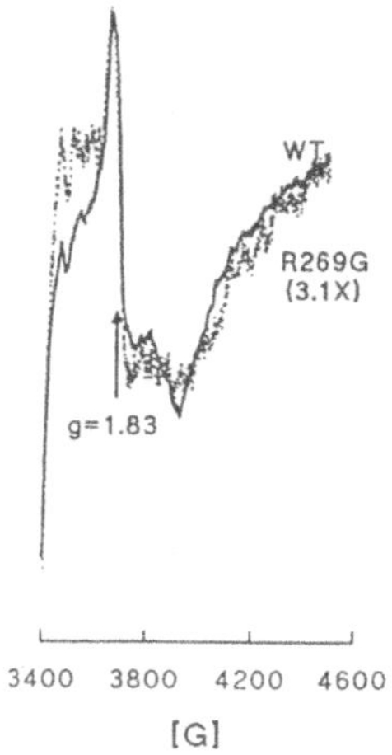

Fig. 3: Formate-enhanced Q_A^-Fe EPR signal.

The results of this study suggest that the D1 residue R269 is critically important in maintaining the structural integrity of the PSII complex and in facilitating electron transfer between Q_A^- and Q_B. The introduction of a helix breaking residue destabilizes the complex. Based on quantification of the Tyr_D, Q_A^-Fe EPR signals, and measurements of the variable chlorophyll fluorescence level, it is estimated that the steady state level of functional R269G PSII complexes is about 1/3 of that of the wild type. It is unlikely, however, that R269 is directly involved in bicarbonate binding since formate induces a structurally unperturbed Q_A^-Fe EPR signal, although at reduced levels. In addition, the residue, which is suggested to be located in the interface of D1 and D2 proteins, is considered to play a delicate but important structural role for maintaining the proper D1-D2 conformation (unpublished homology model of J. Xiong, S. Subramaniam and Govindjee).

References:

1. Govindjee and Van Rensen, J.J.S. (1993) In *The Photosynthetic Reaction Center*, Deisenhofer, J. and Norris, J. eds., Vol. I, pp. 357-389, Academic Press, Inc., San Diego.
2. El-Shintinawy, F. and Govindjee. (1990) Photosynth. Res. 24:189-200.
3. Klimov, V.V., Allakhverdiev, S.I., Feyziev, Y.M., and Baranov, S.V. (1995) FEBS Lett. 363:251-255.
4. Roffey, R.A., Golbeck, J.H., Hille, C.R., and Sayre, R.T. (1991) Proc. Natl. Acad. Sci. USA 88: 9122-9126.
5. Roffey, R.A., Kramer, D., Govindjee and Sayre, R.T. (1994) Biochim. Biophys. Acta 1185:257-270.
6. Vermaas, W., Charité, J., and Shen, G. (1990) Biochemistry 29:5325-5332.
7. El-Shintinawy, F., Xu, C., and Govindjee. (1990) J. Plant Physiol. 136:421-428.
8. Haag, E., Eaton-Rye, J.J., Renger, G., and Vermaas, W.F.J. (1993) Biochemistry 32:4444-4454.
9. Deligiannakis, Y, Petrouleas, V., and Diner, B.A. (1994) Biochim. Biophys. Acta 1188: 260-270.

PHOSPHORYLATION OF PSII POLYPEPTIDES PREVENTS DEGRADATION OF THE D1 PROTEIN

Volker Ebbert and Doris Godde, Lehrstuhl für Biochemie der Pflanzen, Ruhr-Universität, D-44780 Bochum, Germany

1. Introduction

Phosphorylation of thylakoid proteins is known to regulate photosynthetic electron transport by affecting the antenna size of PSII (1). Recently, it has been proposed that phosphorylation of the D1 protein is induced by a photoinhibitory damage to PSII (2) and might regulate its rapid turnover either by preventing (3) or stimulating its degradation (4). To study the regulation of thylakoid protein phosphorylation and its effects on D1 protein turnover, we measured incorporation of $^{32}P_i$ and ^{14}C leucine into thylakoid proteins in intact chloroplasts with high activities of CO_2 fixation. The organello system should posses similar regulatory mechanisms as intact leaves. However, it has the advantage that substrates can easily be applied in defined concentrations.

2. Procedure

Isolation of intact spinach chloroplasts, pulse-modulated chlorohyll fluorescence measurements, light-dependent phosphorylation of thylakoid proteins, D1 protein synthesis, SDS-PAGE, immunoblotting and phosphoimaging were carried out as described in (5). Fractionation of thylakoids into grana, stroma and intermediate membranes was carried out according to (6).

3. Results and Discussion

To test whether phosphorylation of the PSII D1 reaction centre polypeptide depends on a photoinhibitory damage to PSII, the incorporation of $^{32}P_i$ into the D1 protein was determined after a 15 min illumination with increasing light intensities. It was compared with the chlorophyll fluorescence parameter F_v/F_m which indicates the functionality of PSII. As shown in Fig.1 the D1 protein in intact chloroplasts was already maximally phosphorylated at low light intensities of 125 µmoles m^{-2} s^{-1}. With increasing light intensities up to 2500 µmoles m^{-2} s^{-1}, the level of D1 protein phosphorylation decreased about 20%. This decrease was probably caused by an inhibition of the phosphorylation

P. Mathis (ed.), Photosynthesis: from Light to Biosphere, Vol. I, 579–582.

reaction at higher light intensities (5). Illumination with low light intensities of 125 μ moles m^{-2} s^{-1} for 15 min did not result in any loss of PSII function as determined by F_v/F_m (5). An accumulation of inactivated PSII centres started after a 15 min illumination with light intensities around 800 μmoles m^{-2} s^{-1}. This shows clearly that D1 protein phosphorylation is not induced by a photoinhibitory damage to PSII.

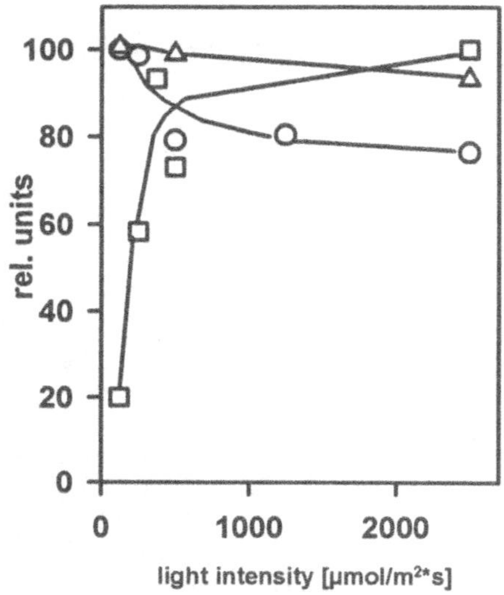

Fig.1. PSII function determined as F_v/F_m, D1 protein synthesis and D1 protein phosphorylation after a 15 min illumination with increasing light intensities

Δ F_v/F_m
○ ^{32}P labelling of the D1 protein
□ ^{14}C labelling of the D1 protein

To get information about D1 protein turnover, D1 protein synthesis, measured as light dependent incorporation of ^{14}C leucine into the D1 protein, was compared to the actual D1 protein content determined by Western-blotting. As shown in Fig.1 D1 protein synthesis reached its maximum at a light intensity of 800 μmoles m^{-2} s^{-1}. No decrease in D1 protein content was observed even when light intensities were increased further (5). Thus, D1 protein degradation also reached its maximum at the same light intensity, where also photoinhibition started to occur. Obviously, accumulation of photoinactivated PSII centres occurred when D1 protein turnover becomes limiting. However, a strong correlation between the degree of D1 protein phosphorylation and D1 protein turnover was not observed.

Recent results have indicated that the phosphorylated D1 protein is no substrate for proteolytic degradation. To test this hypothesis, we have separated the D1 protein into two populations differing in their retardation on a LiDS gel (Fig.2a). By labelling with either $^{32}P_i$ or ^{14}C leucine it could be shown that the upper form was phosphorylated, but not labelled with ^{14}C leucine, whereas the lower form was labelled with ^{14}C leucine but not phosphorylated. This indicates, that in the light an exchange between the phosphorylated and the non-phosphorylated D1 protein was not possible. Only during the first 10 min, when D1 protein phosphorylation was not complete, a slight incorporation of ^{14}C leucine into the upper form of the D1 protein was possible (Fig.2b). However, when the maximum of D1 phosphorylation was reached, no incorporation of ^{14}C leucine into this pool was possible.

To test weather D1 protein degradation occurs via its phosphorylated form, the distribution of ^{14}C labelled D1 protein between its two forms was monitored for a 1,5 h chase period. During this time 30 % of radioactively labelled D1 protein was degraded (Fig. 3b). The content of two different forms of the D1 protein stayed constant during this experiment (Fig.3a). ^{14}C leucine was again only incorporated into the non-phosphorylated form of the D1 protein. It did not migrate to the phosphorylated form during the degradation. Thus, it can be excluded that the D1 protein degradation occurs via the phosphorylated form.

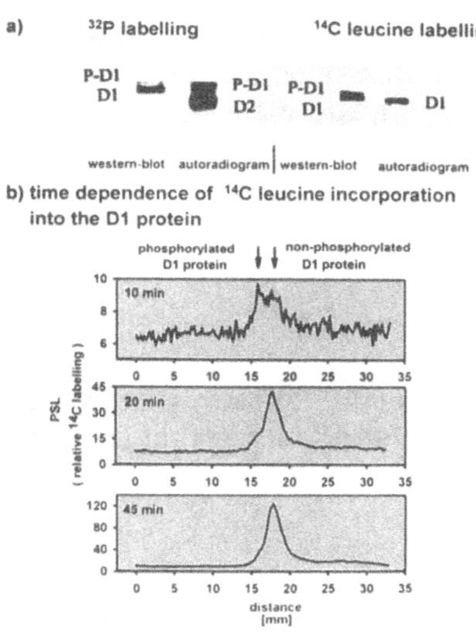

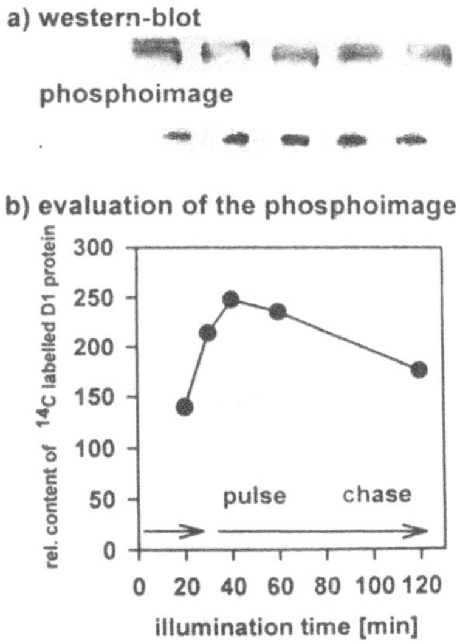

Fig.2 Incorporation of ^{14}C leucine and $^{32}P_i$ into the two forms of the D1 protein

Fig.3 Only the non-phosphorylated D1 protein is degraded in a pulse-chase experiment

Recently evidence has been provided that there is no turnover of phosphate groups esterified of the D1 protein in intact chloroplasts in the light (5). When chloroplast were illuminated for 30 min before $^{32}P_i$ was added, no labelling of PSII polypeptides was observed (Fig.4), whereas a labelling of LHCII polypeptides was still possible. The PSII polypeptides could only labelled in dark adapted chloroplasts which had not been illuminated. Together with the results presented here, this shows that once the D1 protein is phosphorylated, it is stabilised and cannot be degraded. Only in the non-phosphorylated PSII centres, the D1 protein is rapidly degraded and synthesised in the light.

To summarize, phosphorylation of the D1 protein creates a stable pool of PSII centres which is not involved in the PSII repair cycle. Under our conditions, the phosphorylated D1 protein was rapidly degraded and resynthesised. So far, it has been assumed that the degradation of the D1 protein is induced by an activation of PSII leading to a still non-identified damage to the protein (3). According to this hypothesis, the non-phoshorylated PSII centres taking part in D1 protein turnover are continuously damaged and repaired. Such an inactivation was not detected in the phosphorylated PSII centres. This might be

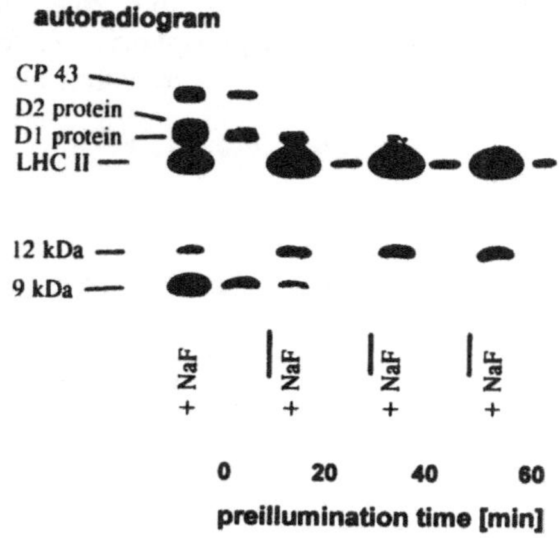

Fig.4. Possibility of thylakoid protein phosphorylation during increasing times of preillumination.

After preillumination $^{32}P_i$ was added. Radioactive labelling of thylakoid proteins was determined after 10 min illumination at the same light intensity.

due to an additional protective effect of phosphorylation on PSII activity as proposed in (7). On the other hand, the assumption might be wrong that degradation of the D1 protein requires a damage to the protein moiety itself.

References

1. Allen, J.F. Biochim. Biophys. Acta 1098, 275-335
2. Aro, E.-M., Virgin, I. and Andersson, B. (1993) Biochim. Biophys. Acta 1143, 113-134
3. Elich, T.D., Edelman, M. and Mattoo, A.K. (1994) J. Biol. Chem. 267, 3523-3529
4. Aro, E.-M., Kettunen, R. and Tyystjärvi, E. (1992) FEBS Lett. 297, 29-33
5. Ebbert, V. and Godde, D. (1994) Biochim. Biophys. Acta 1187, 335-346
6. Leto, K.J., Bell, E., McIntosh, L. (1985) EMBO J. 4, 1645-1653
7. Harrison, M.A. and Allen, J.F. (1991) Biochim. Biophys. Acta 1058, 289-296

COMPLEX FORMATION AND ELECTRON TRANSFER BETWEEN CYTOCHROME C_2 AND REACTION CENTERS OF *RHODOBACTER SPHAEROIDES*: A CHEMICAL CROSS-LINKING STUDY

Friedel Drepper, Pierre Dorlet and Paul Mathis
Section de Bioénergétique / DBCM, CEA Saclay, 91191 Gif-sur-Yvette, France

1. Introduction

Within the cyclic electron flow of the purple bacterium *Rhodobacter sphaeroides* a soluble cytochrome (cyt) c_2 reduces the primary donor, P, stabilizing the light-induced charge separation in the reaction center (RC). This reaction provides a model to study electron transfer between a small soluble protein and a multiprotein complex. Many *intra*protein electron transfer reactions have been interpreted in terms of three main factors determining their rate and direction, the distance between cofactors, the free energy (ΔG^0) and the medium reorganization coupled to the electron transfer (1). A common understanding is rendered more difficult for *inter*protein electron transfers, i.e. when only the formation of a temporary complex positions the cofactors appropriately for electron transfer. In this case the final distance between them as well as their redox potentials determining the ΔG^0 may depend on the docking process.

The reactions between cyt c_2 and isolated RCs from *Rb. sphaeroides* have been actively studied. The kinetics of P^+ reduction measured by flash absorption spectroscopy were explained by a model assuming one well-defined RC-cyt complex (2). More complex kinetics observed in several studies suggested an additional intermediate bound state of the cyt (3,4), the properties of which have not been elucidated. The effect of temperature (2) and altered driving force (5) on the rate of P^+ reduction by bound cyt c_2 have been used to estimate the reorganization energy for this reaction. However, both approaches rely on estimates for several parameters, including the ΔG^0 under working conditions, that have to be known for a quantitative analysis. The preliminary structure of a co-crystallized complex positioned the cyt towards the M side of the periplasmic surface (6) in agreement with one of the structural models of an electrostatically stabilized RC-cyt complex that have been proposed (7). In a previous work, cyt c_2 has been cross-linked to the L subunit of the RC using a 10 Å cross-linker (8). In this work we describe conditions for a zero-length cross-linking of the RC and cyt c_2. The results serve as a basis for an analysis of both the docking and the electron transfer.

2. Experimental Procedures

Protein Purification. Reaction centers were isolated from *Rb. sphaeroides 2.4.1* according to (3). The buffer was exchanged to 10 mM HEPES, 0.025 % LDAO. Cyt c_2 was isolated from a strain of *Rb. sphaeroides* overexpressing cyt c_2 (kindly provided by Dr. P. Richaud, Cadarache) as described in (9) except that the first two columns were omitted.

P. Mathis (ed.), Photosynthesis: from Light to Biosphere, Vol. I, 583–586.

Chemical Cross-linking. The standard reaction mixture for cross-linking contained RCs at 6 μM with 20 μM cyt c_2, 10 mM HEPES, pH 7.5, 0.025% LDAO, 1mM ascorbate, 60 μM 1,4-naphtoquinone. The incubation was started by addition of N-hydroxysulfosuccinimide (NHS, adjusted to pH 7.5) and 1-ethyl-3-(3,3-dimethylamino-propyl)-carbodiimide (EDC) to final concentrations of 12 mM and 6 mM, respectively. After 3 h in the dark at 22°C the reaction was stopped by the addition of ammonium acetate (50 mM). After 10 min the mixture was diluted 10-fold with 10 mM HEPES, pH 7.5, 0.025% LDAO and concentrated using Centricon 100 filters (Amicon). The washing procedure was repeated two times. The filtrate contained the excess soluble cyt c_2. It was further purified and used for control experiments. For oxidizing conditions during the incubation, ascorbate was replaced by 10 mM ferricyanide and the cyt was preoxidized by ferricyanide washing.

Flash Absorption Spectroscopy. Kinetics of flash induced absorbance changes were measured essentially as described in (2). Flash excitation with a 10 ns pulse at 595 nm was provided by a dye laser pumped by a frequency-doubled Nd:YAG laser. Kinetics of electron transfer were monitored at 1283 nm in the absorption band of P^+ with a dwell time of 100 ns. The absorption transients were fitted to a sum of exponentials using a modified Marquardt algorithm installed by Dr. P. Sétif.

Redox Potentiometry. Midpoint potentials of isolated cyt were determined using the method of mixtures (10). In flash titrations the redox potential was measured with a platinum electrode *vs.* a calomel reference electrode in an argon purged cuvette. The ambient potential was changed by addition of ferricyanide and ascorbate. The following redox mediators were used: 50 μM 2,4-naphtoquinone, 100 μM TMPD, 50 μM di-hydroxybenzoquinone, 10 μM as initial concentrations of each ferri- and ferrocyanide.

3. Results and Discussion

Subunit Composition of Covalent Complexes. The cross-linking products were characterized by SDS-gel electrophoresis (Fig. 1). Two major effects on the electrophoretic profile of isolated RCs (lane 1) were observed after incubation of the RCs with EDC and NHS in the presence of cyt c_2 (lane 3). A new doublet of bands appeared at apparent molecular weights of 40 and 42 kDa. These bands are also visible in the heme stain (lane 4) indicating the presence of cytochrome in the cross-linking product. Secondly, the relative intensity of the band at 24.7 kDa decreased by about 85 % compared to the control indicating the involvement of subunit M in the covalent complex. A low intensity band at 32 kDa was obser-

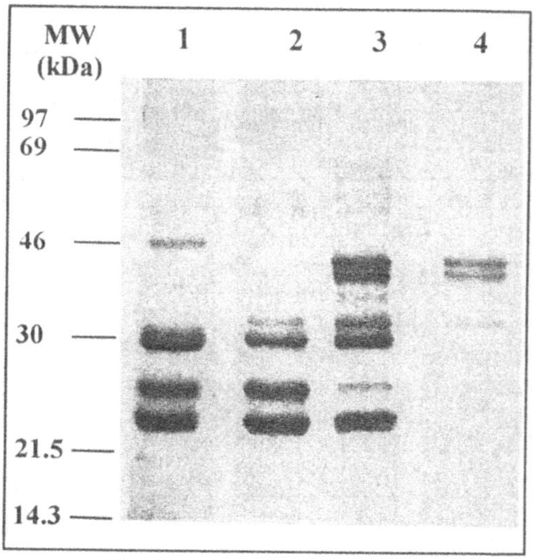

Figure 1: 10 % SDS-polyacrylamide gel electrophoresis of RC (lane 1), RC after incubation with EDC and NHS in the absence (lane 2) and in the presence of cyt c_2 (lanes 3,4). Proteins were stained by Coomassie blue (lane 1-3) or heme staining with 3,3',5,5'-tetramethylbenzidine (lane 4).

ved after incubation both in the absence and in the presence of cyt. The identity of this latter band was not further analyzed.

Kinetics of Electron Transfer. The function of the RC-cyt complexes was investigated measuring the flash-induced kinetics of P^+ reduction. We observed two major kinetic phases (Fig. 2a) with halftimes of 700 ns (± 100 ns) and 60 μs (± 10 μs) under standard conditions, both accounting for about 40 % of total photooxidized P. About 20 % of total P^+ decayed in a few ms and was attributed to RCs undergoing back-reaction from acceptor side quinones or being reduced by ascorbate. Fig. 2b shows the P^+ reduction at 0.5 M NaCl that is almost not altered compared to that without added NaCl (2a). By contrast, the reaction between RCs and soluble cyt c_2 under these conditions typically shows no detectable contribution of the 700 ns phase and a bimolecular reaction slowed down by almost two orders of magnitude. In the presence of 60 % glycerol the halftimes of the two kinetic phases of P^+ reduction are slightly decreased to 400 ns and 40 μs whereas the amplitudes are the same as in the absence of glycerol (Fig. 2c). Since both phases are even slightly accelerated at increased viscosity, their rate can not be limited by a diffusional reorientation but should be attributed to an electron transfer within a covalently stabilized complex.

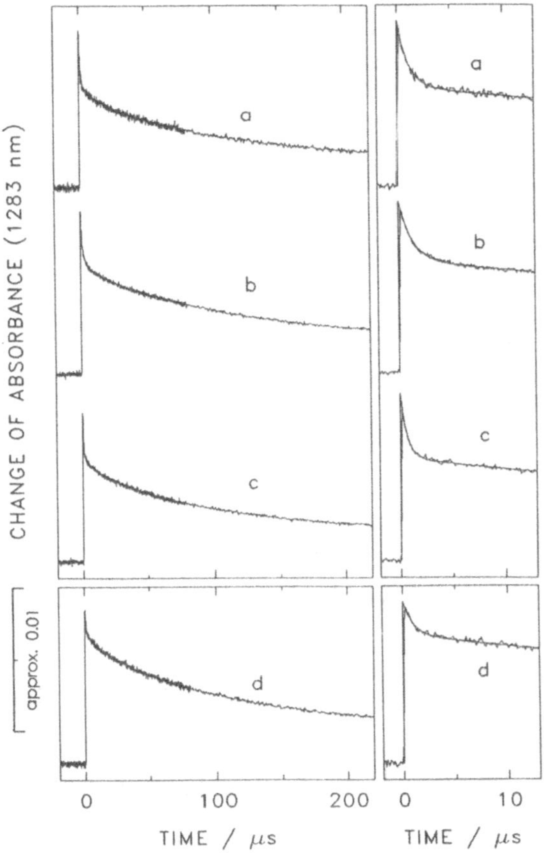

Effect of oxidation states. The effect of the ambient redox potential during cross-linking was studied by varying the concentrations of ferro- and ferricyanide. Fig. 2d shows P^+ reduction kinetics resulting when the reaction mixture had been poised to a redox potential above 560 mV. The relative amplitude of the 60μs phase increased to 55% on the expense of the fast phase (20%). By contrast, if ferro- instead of ferricyanide was present during incubations almost equal amplitudes for the 700ns and 60μs phase were observed (not shown).

Figure 2: Flash-induced absorbance changes of RCs with covalently bound cyt c_2. *Trace a*: 2 μM RCs in 10 mM HEPES, 0.025% LDAO, 100 μM TMPD, 60 μM naphtoquinone and 1mM ascorbate; *trace b*: plus 0.5 M NaCl; *trace c*: same buffer as in *a* plus 60 % glycerol. On the right side the fast absorption changes are shown on an expanded time scale. P^+ reduction kinetics were fitted to a sum of two exponentials plus a constant (solid lines). *Trace d* shows the effect of oxidizing conditions (E_h >560 mV) during cross-linking (see Procedures).

Electrochemistry. Redox titrations of the flash-induced absorbance changes at pH 7.5 gave midpoint potentials of 360 and 370 mV for the intermediate and the fast kinetic phase, respectively, being 20 to 30 mV higher than the E_m of cyt c_2 in solution under the same conditions (Table 1). This indicates that binding to the RC has a small effect on the redox properties of the heme. From the total amplitude of photooxidation in a flash as a function of the ambient potential we determined an E_m for the primary donor in the cross-linked complex of 520 mV compared to 500 mV in the free RC (Table 1).

redox couple	cyt^{ox}/cyt^{red} free		cyt^{ox}/cyt^{red} bound[b]		P^+/P [c]	
	(control)	(EDC/NHS)[a]	(700ns phase)	(50μs phase)	(free RC)	(RC-cyt^{ox})
E_m (mV)	340	340	370	360	500	520

Table 1: Midpoint potentials of cyt and P in solution and in the complex. Estimated errors are ±10 mV. Conditions: 10 mM HEPES, pH 7.5, 25°C.
[a] Recovered from incubation with cross-linker,
[b] from amplitudes of kinetic phases of P^+ reduction,
[c] from total amplitudes of flash-induced P^+.

4. Conclusions

Zero-length cross-linking stabilized a 1:1 covalent complex between subunit M of the RC and cyt c_2. Kinetics of P^+ reduction show i) a fast phase with a halftime of 700 ns similar to that observed for electron transfer in the noncovalent complex and ii) a slow phase ($t_{1/2}$ = 50 μs) that is attributed to a cyt bound less favorably for electron transfer. It may represent one of the intermediate bound states previously observed in experiments with soluble cyt. The yield of the 700 ns phase is decreased when both cyt and P are oxidized during the cross-linking. A slight increase of the midpoint potential of bound cyt by 20 to 30 mV compared to the E_m of free cyt c_2 may be explained by a decreased solvent exposure of the heme within the RC-cyt complex.

Acknowledgements
We thank Sandra Andrianambinitsoa for the assistance with protein isolation and Pierre Richaud for providing us with cells of *Rb. sphaeroides* overexpressing cyt c_2 and for his advices concerning its isolation protocol. FD was supported by a fellowship from the European Molecular Biology Organization.

References

1 Moser, C.C., Keske, J.M., Warncke, K., Farid, R.S. and Dutton, P.L. (1992) Nature 355, 796-802
2 Venturoli, G., Mallardi, A. and Mathis, P. (1993) Biochemistry 32, 13245-13253
3 Wachtveitl, J., Farchaus, J.W., Mathis, P. and Oesterhelt, D. (1993) Biochemistry 32, 10894-10904
4 Tiede, D.M. Vashishta, A.-C. and Gunner, M.R. (1993) Biochemistry 32, 4515-4531
5 Lin, X., Williams, J.C., Allen, J.P. and Mathis, P. (1994) Biochemistry 33, 13517-13523
6 Adir, M., Okamura, M.Y. and Feher, G. (1994) Biophysical J. 66, A127
7 Tiede, D.M. and Chang, C.H. (1988) Isr. J. Chem. 28, 183-191
8 Rosen, D., Okamura, M.Y., Abresch, E.C., Valkirs, G.E. and Feher, G. (1983) Biochemistry 22, 335-341
9 Meyer, T.E. and Cusanovitch, M.A. (1985) Biochim. Biophys. Acta 807, 308-319
10 Pettigrew, G.W., Meyer, T.E., Bartsch, R.G. and Kamen, M.D. (1975) Biochim. Biophys. Acta 430, 197-208

SELECTIVE EXTRACTION OF PIGMENTS FROM THE PS II REACTION CENTER

Tatsuya Tomo[1,2], Masami Kobayashi[3], Kimiyuki Satoh[1,2]
([1]Natl. Inst. Basic Biol., Okazaki, Japan, [2]Okayama Univ., Okayama, Japan, [3] Tsukuba University, Tsukuba, Japan)

Introduction

The D1 and D2 proteins provide the binding sites for pigments, such as chlorophyll (chl) a, pheophytin (pheo) a and β–carotene (car), and furnish machinery for the primary charge separation in photosystem II (PS II)(1) (2). The D1 and D2 proteins are significantly homologous in the amino acid sequence to the L and M subunits of purple bacterial reaction center. On the other hand, the pigment stoichiometry of the photochemical reaction center of purple bacteria is established to be 4 Bchl, 1 carotenoid and 2 Bpheo. However, the chromophore stoichiometry of PS II reaction center is the subject of debate which is partly due to strong Qy overlapping absorption spectra of pigments. Typical preparations of the PS II reaction center so far analyzed can be classified into two types, in terms of pigment stoichiometry. In one type, the chl a to pheo a ratio is two, consistent with its purple bacterial counterpart. The other type of preparation, which seems to be more native than the other, contains two additional chl a and one additional β–car. In the present study, the selective extraction of pigments was attempted by using organic solvent (ether), in order to understand the functional of organization of pigments in the photosystem. By the ether treatment, we have succeeded in preparing the PS II reaction center which contains 0,1 or 2 β–car per reaction center.

Materials and Methods

1. Preparation of the PS II reaction center

The PS II reaction center was isolated from spinach grana thylakoids by the method previously described(3). The detergent in the buffer solution was replaced by 0.15 % sucrose-monocaprate.

2. Ether treatment

The ether treatment of PS II reaction center was carried out in a way essentially similar to that used for PS I (4).

3 Quantitation of pigments, metals and redox components.

Quantitation of pigments in the extracts were carried out according to Kobayashi et al. using HPLC(5). Metal content was estimated by ICP (inductively coupled plasma) atomic emission using a SPS2100A spectrometer(SEIKO Ind. Corp.). Cyt b559 contents were estimated spectroscopically using difference millimollar absorption coefficients of 15 cm^{-1}(6).

4. Photochemical activity measurement

P. Mathis (ed.), Photosynthesis: from Light to Biosphere, Vol. I, 587–590.

The photo-accumulation of reduced pheo a was measured by the absorbance change induced by continuous illumination of saturating light from a xenon lamp (500 W) at 4 °C with a Hitachi-557 spectrophotometer equipped with a cross-illumination system. The reaction mixture contained methyl viologen(2μM) and Na-dithionite(2 mg /ml).

Result

Pigment content in PS II reaction center varied significantly by ether treatment. When the lyophilized powders of PS II reaction center was treated twice with ether solution of different water saturations, i. e., 0, 50 and 70 %, there were a remarkable decrease in the β–car absorption region, by the increase in water saturation in ether solution. The Qy region of chl a and pheo a absorption also decreased by the treatment, but the extent was relatively small (less than 18 % of the control) and this accompanied with the blue shift (2-3 nm) of the absorption maximum. These features in the absorption spectra are shown in Fig. 1

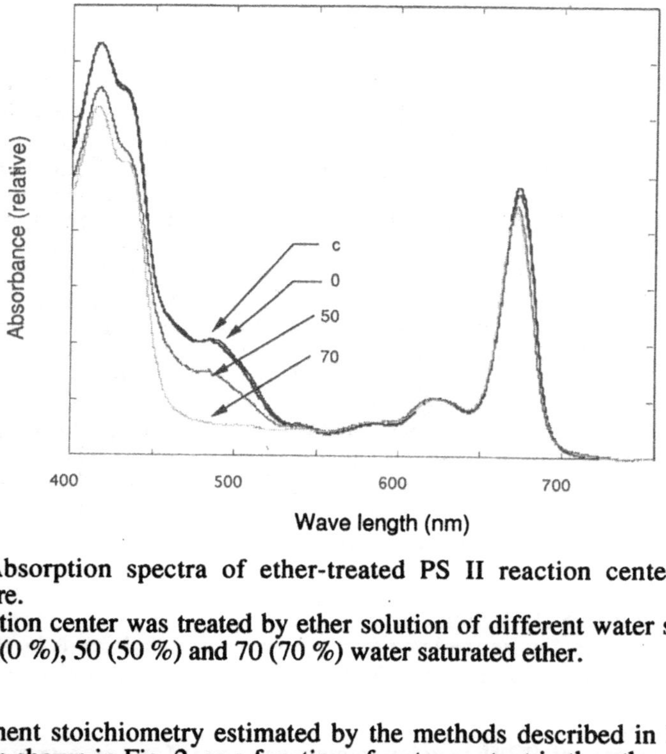

Fig. 1. Absorption spectra of ether-treated PS II reaction center at ambient temperature.
PS II reaction center was treated by ether solution of different water saturations: c, control; 0 (0 %), 50 (50 %) and 70 (70 %) water saturated ether.

The pigment stoichiometry estimated by the methods described in Materials and Methods is shown in Fig. 2, as a function of water content in the ether solution. The starting material contained 6.23 ± 0.26 chl a and 2.18 ± 0.24 β–car per 2 pheo a molecules. The free chl a of this reaction center was less than 2 %(7). The dry ether treatment did not bring about any change in the pigment composition. However, by the treatment with 20-50 % water saturated ether solution, the amount of β–car decreased to 1.08 ± 0.15 per two pheo a. The further increase in the water content in the ether solution (more than 60 % saturation) resulted in the complete loss of β–car.

A remarkable fact is that the PS II reaction center with no β–car associated still retains the photochemical activity of pheo⁻ photoaccumulation. However, the activity estimated by the level of pheo⁻ decreased to some extents corresponding to the amount of β–car, i. e., the reaction center containing one β–car kept its activity of about 60 % of the control, and that with no β–car, the activity decreased to about 35 %. Thus, by manipulating water content in the ether solution, we can prepare the PS II reaction center with 0, 1 or 2 β–car.

The chl a content, on the other hand, was more resistant to the ether-treatment. A loss of chl a equivalent to one chl a per two pheo a was attained by the treatment with 30 % water saturated ether solution. However, no further decrease was observed by increasing the water content. The partial loss of chl a from the reaction center was also confirmed by an ICP atomic emission spectroscopy: the Mg (chl a) to Fe (cytb559) ratio was estimated to be 4.87 ± 0.09 for the preparation treated with 30 -70 % water saturated ether solution. On the other hand, the content of pheo a, as well as of the cyt b559, did not change by the ether treatment. Thus the extraction of pigments from the PS II reaction center is selective in the ether treatment.

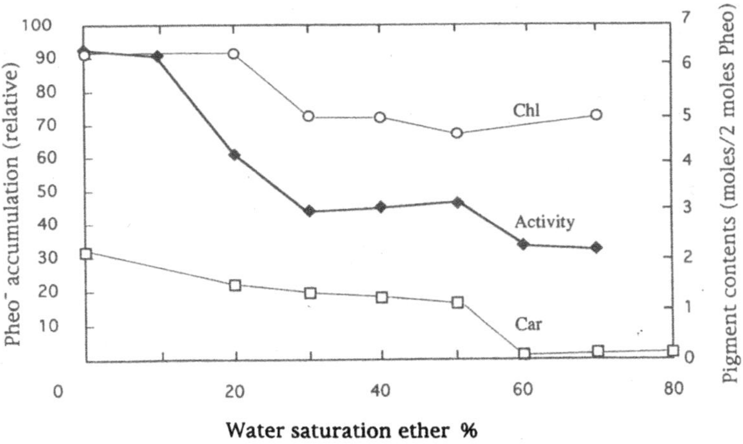

Fig. 2. The pigment stoichiometry and the photochemical activity of pheophytin reduction in the PS II reaction center treated with ether solution of different water saturations.

Discussion

The pigment contents in the PS II reaction center have been a matter of discussion for recent years. For carotenoid, some preparations contained 2 β–car per 2 pheo a (per reaction center), however, several reports showed that one out of the two was easily extracted by detergent treatment(8)(9)(10). In the present study, we used ether solution of different water saturations for the extraction. The water content in the ether solution was shown to be critical for the extraction. Thus by manipulation the water content, we have succeeded in preparing the PS II reaction center with 0 or 1 β–car, in addition to the preparation with 2 β–car. An interesting fact is that the reaction center with no β–car still has relatively high activity of pheo$^-$ photoaccumulation. However, this preparation was highly sensitive to the irradiation with strong light. Further analysis of the characteristics of these preparations are now in progress.

Acknowledgment
We thank Mr. A. Murakami(NIBB, Japan) for measuring the ICP spectroscopy. We thank Drs. S. Itoh, M. Mimuro and M. Iwaki(NIBB, Japan) for valuable discussions and Dr. M. Mimuro for critical reading of the manuscript.

Reference

1. Nanba, O. and Satoh, K. (1987) *Proc. Natl. Acad. Sci. USA* **84,** 109-112
2. Tang, X., Fushimi, K. and Satoh, K. (1990) *FEBS Lett.* **273,** 257-260
3. Tomo, T. and Satoh, K. (1994) *FEBS Letters* **351,** 27-30
4. Ikegami, I., Itoh, S. and Iwaki, M. (1995) *Plant Cell Physiology* **36,** 857-864
5. Kobayashi, M., Maeda, H., Watanabe, T., nakane, H. and Satoh, K. (1990) *FEBS letter* **260,** 138-140
6. Cramer, W. A., Theg, S. M. and Widger, W. R. (1986) *Photosynth. Res.* **10,** 393-403
7. Mimuro, M., Tomo, T., Nishimura, Y., Yamazaki, I. and Satoh, K. (1995) *Biochimica Biophysica Acta in press*
8. Rivas, J. D. L., Telfer, A. and Barber, J. (1993) *Biochimica et Biophysica Acta* **1142,** 155-164
9. Montoya, G., Yruela, I. and Picorel, R. (1991) *FEBS letter* **283,** 255-258
10. Yruela, I., Kan, P. J. M. v., Muller, M. G. and Holzwarth, A. R. (1994) *FEBS letter* **339,** 25-30

TIME RESOLVED INFRARED AND STEADY-STATE FOURIER TRANSFORM INFRARED SPECTROSCOPY OF NATIVE AND MUTANT REACTION CENTERS OF *RHODOBACTER SPAEROIDES*

C. Fogel[1], S. Grzybek[1], R. Hienerwadel[3], M. Y. Okamura[4], M. L. Paddock[4], J. Breton[3], E. Nabedryk[3] and W. Mäntele[1]
[1]Institut für Physikalische Chemie, Universität Erlangen, Egerlandstr. 3, 91058 Erlangen, Germany; [3]SBE/DBCM, CEA Saclay, 91191 Gif-sur-Yvette Cédex, France; [4]Department of Physics, University of California San Diego, La Jolla, California 92093-0319, USA.

1. Introduction

In photosynthetic bacterial reaction center (RC), reduction of the secondary acceptor quinone Q_B is accompanied by proton transfer of nearby amino acid side chains. These reactions can be visualized by infrared (IR) difference spectroscopy. In *Rhodobacter sphaeroides* RC, proton uptake by the residue Glu L212 in the Q_B pocket upon $Q_A^-Q_B \rightarrow Q_AQ_B^-$ electron transfer was recently demonstrated by time resolved IR and steady-state Fourier transform infrared (FTIR) spectroscopy [1-3]. Since studies of site-directed mutants have shown the influence of other protonatable residues, we have analysed the steady-state Q_B^-/Q_B FTIR difference spectra and transient $Q_A^-Q_B \rightarrow Q_AQ_B^-$ IR signals of several mutants in the Q_B site as a function of pH. In the 1680-1760 cm^{-1} region, the variation of the intensity of several modes with pH and their absence or presence in mutants lead us to propose assignments for the groups involved in the cluster of ionizable residues around Q_B.

2. Materials and Methods

Sample Preparation: Native and mutant RCs from *Rb. sphaeroides* were prepared as previously described [4-6]. For IR sample preparation, RCs were transferred to 100 mM buffer: Dimethylpiperazine (pH 4.5), Cacodylic acid (pH 6), TRIS (pH 8) and CAPS (pH 10.5) containing 100 mM KCl and 10 mM UQ_{10} in DMSO. For steady-state experiments 0.1 mM $K_4[Fe(CN)_6]^{2+}$ and 2 mM cytochrome *c* were added. With the preadjusted pH, RCs were concentrated to approximately 0.5-1 mM using ultrafiltration cells with a 30 kDa exclusion limit. A sample of approximately 10 µl of the concentrated RC was transferred to the demountable IR microcell and further reduced in water content by a gentle stream of nitrogen to a final volume of 1-2 µl.
Spectroscopy: Time-resolved IR signals (µs to s) were recorded with the kinetic IR photometer using tunable IR diode lasers described in [7, 8]. Only laser monomodes in the range from approximately 1700 to 1780 cm^{-1}, spaced every 4-5 cm^{-1}, were used. Signals were averaged over 100 flashes. RC samples were excited by a pulsed dye laser flash at 595 nm (15 ns, ca. 5 mJ). To guarantee precisely the same IR laser mode structure for different samples, up to ten IR cuvettes were mounted in a computer-controlled and thermostated sample wheel. The samples can thus be measured sequentially, allowing a long dark adaptation time for those not exposed to the IR beam. Steady-state FTIR spectra were recorded on a modified Bruker IFS 25 spectrophotometer. To obtain photochemically induced Q_B^-/Q_B difference spectra, 40 flashes were

P. Mathis (ed.), Photosynthesis: from Light to Biosphere, Vol. I, 591–594.

averaged with reduction of the oxidized Primary Donor (P^+) by cytochrome *c* and electrochemical rereduction of cytochrome *c* between flashes as described in [9]. Time resolved, as well as steady-state experiments, were performed at 4 °C.

3. Results and Discussion

Figure 1 shows photochemically obtained steady-state Q_B^-/Q_B FTIR difference spectra for native RC and several mutants in the spectral region between 1760 cm^{-1} and 1670 cm^{-1}. The FTIR frequencies listed here are 4 cm^{-1} lower than the values reported in [1-3]. This can be explained by differences in data processing.

Spectra of native RC samples with pH values of 4.5, 8.0 and 11.0 are plotted in figure 1a. With respect to pH 8 the large positive peak at 1724 cm^{-1} increases at pH 4.5 by approximately 25 % and at pH 11 by approximately 45 %. In addition, a shift of this peak to 1721 cm^{-1} is seen at pH 11, which may be the consequence of a decrease of a small negative peak at 1736 cm^{-1} with high pH. The small absorbance band at 1704 cm^{-1} does not change significantly at pH 8 and 11 but grows and shifts to 1700 cm^{-1} at pH 4.5. A shift of 4 cm^{-1} to higher wavenumbers is observed for the large negative peak at 1680 cm^{-1} for increasing pH. A strong pH dependency is also observed for the 1692 cm^{-1} peak, which approximately grows eightfold from pH 4.5 to 11.

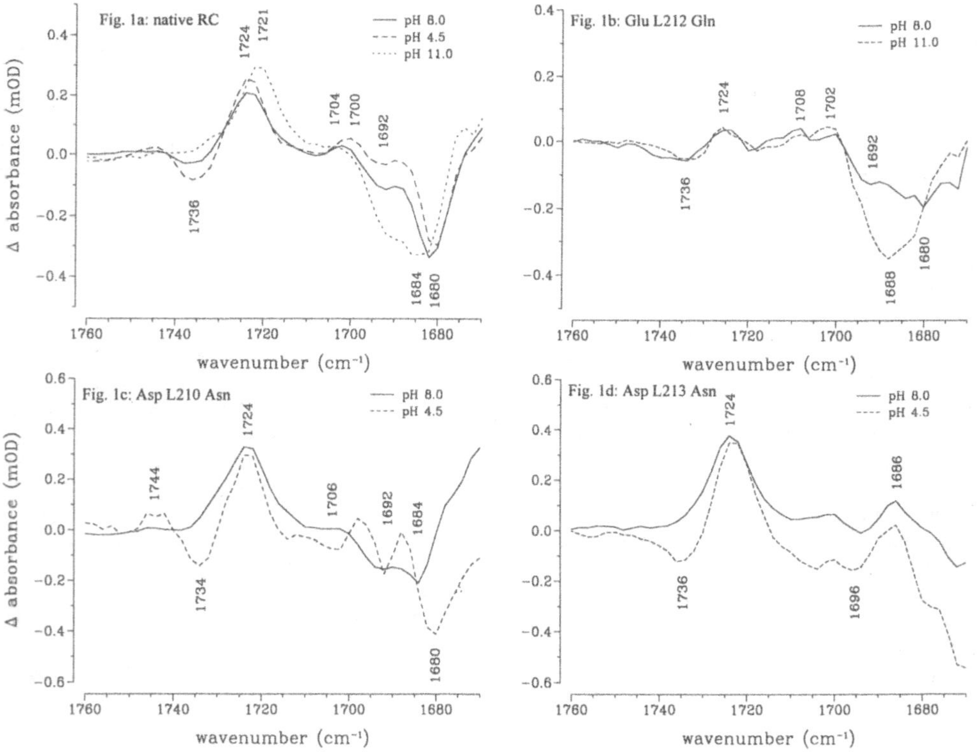

Figure 1 a-d: Q_B^-/Q_B difference spectra, photochemically induced. All spectra are normalized on the 1476 cm^{-1} difference band (data not shown).

Figure 1b shows two spectra of the Glu L212→Gln mutant at pH 8.0 and 11.0. The large positive band at 1724 cm^{-1} is missing for both. No pH dependence can be found in the spectral region above 1700 cm^{-1}. Below 1700 cm^{-1} the pH dependence of Glu L212→Gln RCs is approximately the same as for native RCs: At pH 8, the negative 1692 cm^{-1} band exhibits the same amplitude as the native sample, whereas the 1680 cm^{-1} band shows only 50 % of its native amplitude. Both modes increase by a factor of 2-3 at pH 11.
Figure 1c shows spectra of Asp L210→Asn samples at pH 4.5 and 8.0. Both exhibit the same positive absorbance band at 1724 cm^{-1}, however, the low pH spectrum has a smaller half-width than the one at pH 8. Two negative bands can be seen at 1692 cm^{-1} and 1684 cm^{-1} in the spectrum at pH 8. While the first one seems to be pH independent, the second one is strongly growing and shifts to 1680 cm^{-1} for low pH.
Spectra of Asp L213→Asn at pH 4.5 and 8.0 are shown in figure 1d. Both spectra show the positive band at 1724 cm^{-1} and a small negative peak at 1696 cm^{-1}. A further negative band can be seen in the low pH spectrum at 1736 cm^{-1}. In contrast to the spectra of the native samples, the spectra of Asp L213→Asn show a small positive absorbance at 1686 cm^{-1}.

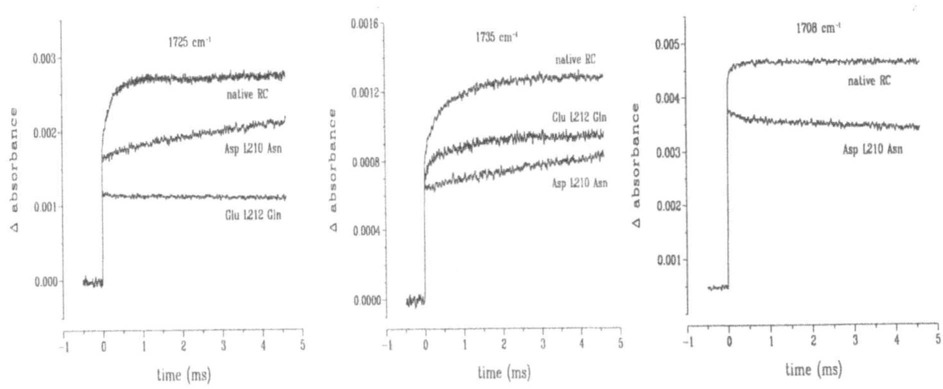

Figure 2 a-c: Time resolved IR signals: Native RC at pH 7, mutant RCs at pH 8.

Figure 2 shows time resolved IR signals of native and two mutant RCs measured at 1725, 1735 and 1708 cm^{-1}. A rapid absorbance increase after the flash is observed in all kinetics corresponding to the reaction $PQ_AQ_B \rightarrow P^+Q_A^-Q_B$, which cannot be resolved with our instrument resolution (approximately 500 ns). The initial non resolved fast rise was cut off for reasons of display.
In figure 2a at 1725 cm^{-1} the immediate rise is followed by a slow biphasic transient with 200 μs and 1 ms half-times, in native RCs attributed to the reaction $D^+Q_A^-Q_B \rightarrow D^+Q_AQ_B^-$ The mutant Asp L210→Asn also shows a biphasic transient, but the amplitude of the fast phase is small and the millisecond phase slows down to 4 ms half-time. The signal reaches its full amplitude after 20 ms (data not shown). In the Glu L212→Gln mutant RC this transient is absent. Instead, a small negative signal with a considerably slower phase can be observed.
At 1735 cm^{-1} (figure 2b) the native RC sample shows a similar biphasic kinetic signal, but with the millisecond phase around 30 % smaller. For Asp L210→Asn the 1735 cm^{-1} kinetic signal is observed to be five times slower than that of native RC and again, the

full amplitude is reached after 20 ms. Glu L212→Gln shows a biphasic transient with a small fast phase in the 100 μs range and a slow phase in the ms range.
In figure 2c the transient signals of native RC and the Asp L210→Asn RC mutant at 1708 cm^{-1} are plotted together. While the native RC sample as at the other wavenumbers shows the biphasic reaction only with smaller amplitudes, the positive kinetic amplitude in Asp L210→Asn is missing. Instead a small negative signal can be observed.

In FTIR difference spectra of the $Q_B \rightarrow Q_B^-$ transition, the absorbance increase at 1724 cm^{-1} appears to be responsible for the kinetic signal at that wavenumber [2]. Similar to this argument the small peak at 1704 cm^{-1} matches the kinetic signal at 1708 cm^{-1}. The fact that this signal is missing in the FTIR spectrum as well as in the time resolved IR spectrum of the Asp L210→Asn mutant RC may indicate that Asp L210 is related to this signal. We note however, the discrepancy between the amplitude of the 1706 cm^{-1} signal in the DN L210 FTIR spectrum reported in [3] and the one shown in figure 1c.
The absorbance increase at 1724 cm^{-1} in the pH 11 sample of native RC supports the assignment of that band to the protonation of Glu L212 connected with the electron transfer from Q_A to Q_B [1-3] and fits into the model that Glu L212 with a pK_a of 9.3 [10] is mainly deprotonated at pH 11 and can thus be easier protonated by the photoinduced charge separation.
Furthermore the FTIR spectra demonstrate a strong pH dependence of the 1680-1692 cm^{-1} band. An increase of this negative band upon increasing pH matches an increase of the 1724 cm^{-1} signal. This seems compatible with a model, where the 1680-1692 cm^{-1} band belongs to a proton donor for Glu L212 in native RC. Replacement of the acceptor Glu L212 by a non protonatable residue at pH 8 reduces the donor signal. At pH 11 another group may be ionized and thus be able to accept that proton instead of Glu L212. This idea is supported by the fact that the time resolved IR signals at 1680 to 1690 cm^{-1} have the same kinetic parameters as the 1724 cm^{-1} band (data not shown).
Further kinetic studies between 1670 and 1780 cm^{-1} are on their way to fully analyse the proton transfer mechanisms upon reduction of Q_B.

4. References

1 Hienerwadel, R., Nabedryk, E., Paddock, M.L., Rongey, S., Okamura, M.Y., Mäntele, W. & Breton, J. (1992) Research in Photosynthesis (Murata, M., Ed.) **1**, 437-440, Kluwer Akademic Publishers. Dordrecht, The Netherlands.

2 Hienerwadel, R., Grzybek, S., Fogel, C., Kreutz, W. Okamura, M.Y., Paddock, M.L., Breton, J., Nabedryk, E. & Mäntele, W. (1995) Biochemistry **34**, 2832-43.

3 Nabedryk, E., Breton, J., Hienerwadel, R., Fogel, C., Mäntele, W., Paddock, M.L., Okamura, M.Y. (1995) These Proceedings.

4 Paddock, M.L., McPherson, P.H., Feher, G. & Okamura, M.Y. (1990) Proc. Natl. Acad. Sci. U.S.A. **87**, 6803-6807.

5 Paddock, M.L., Rongey, S.H., Feher, G. & Okamura, M.Y. (1989) Proc. Natl. Acad. Sci. U.S.A. **86**, 6602-6606.

6 Paddock, M.L., Rongey, S.H., McPherson, P.H., Juth, A., Feher, G. & Okamura, M.Y. (1994) Biochemistry **33**, 734-745.

7 Mäntele, W., Hienerwadel, R., Lenz, F., Riedel, W.J., Grisar, R. & Tacke, M. (1990) Spectrosc. Int. **2**, 29-35.

8 Hienerwadel, R., Thibodeau, D., Lenz, F., Nabedryk, E., Breton, J., Kreutz, W. & Mäntele, W. (1992) Biochemistry **31**, 5799-5808.

9 Bauscher, M., Leonhard, M., Moss, D.A. & Mäntele, W. (1993) Biochim. Biophys. Acta **1183**, 59-71.

10 Okamura, M. Y. & Feher, G. (1992) Annu. Re. Biochem. **61**, 861-896.

TIME-DEPENDENT RADICAL PAIR RELAXATION IN CHROMATOPHORES OF AN ANTENNA-FREE MUTANT FROM *RHODOBACTER CAPSULATUS*

Marc G. Müller, Dieter Dorra, Alfred R. Holzwarth, Max-Planck-Institut für Strahlenchemie, Mülheim/Ruhr, Germany
Nasser Gad'on, Gerhart Drews, Institut für Biologie II, Mikrobiologie, Albert-Ludwigs-Universität, Freiburg, Germany

1. Introduction

Picosecond radical pair kinetics in the antenna-free mutant of *Rb. capsulatus* was studied by time-resolved fluorescence spectroscopy of the excited primary electron donor P*. The data indicate a complex radical pair relaxation, proceeding up to at least 1 ns at room temperature. The time-dependent ΔG values are calculated for open and closed reaction centers and the differences are discussed in terms of kinetic models.

The highly multiexponential decay of the excited primary electron donor P* in reaction centers (RCs) of purple bacteria [1] gave rise to various speculations about the mechanism and possible time-dependent thermodynamics of the early electron transfer process. The most extreme models concern either a dynamic solvation effect with

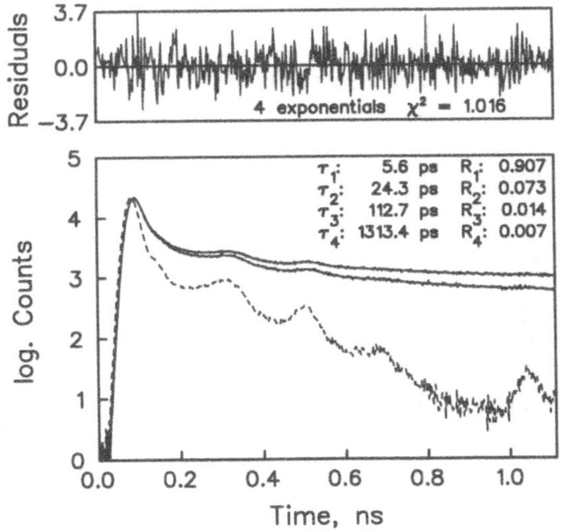

Fig. 1. Bottom: Fluorescence decays from chromatophores of an antenna-free mutant from Rb. capsulatus with reaction centers open (lower curve) and closed (upper curve). The dashed line indicates the system response function. Inset: Lifetimes and relative amplitudes. The values shown were obtained by a four-component single-decay lifetime analysis for the case with open RCs. Top: Plot of the Residuals for the fit.

P. Mathis (ed.), Photosynthesis: from Light to Biosphere, Vol. I, 595–598.

conformational cooling [2] or a static distribution in the free energy gap due to a conformational distribution leading to different rates for charge separation and recombination [3]. Here we concentrate on the analysis of the picosecond fluorescence decay of the excited special pair in terms of a time-dependent free energy ΔG(t). Special attention is given to the redox state of the secondary electron acceptor Q_A, i.e. oxidized (open RCs) or reduced either photochemically or chemically (closed RCs). The direct influence of the redox state on the primary kinetics is demonstrated in Fig. 1.

2. Materials and Methods

Chromatophores were prepared from the antenna-free *Rb.-capsulatus* mutant strain U43 (pTX AB5) [4]. The measurements were carried out by time-correlated single-photon counting up to 30000 counts in the peak channel with a full width at half maximum of the system temporal response function of about 30 ps [1]. During the measurements the sample was pumped through a flow cuvette and the reservoir was kept anaerobically on ice. The sample was excited by 1-2-ps pulses from a Ti:S laser with an energy of ≈4 nJ at a wavelength of 800 nm which corresponds to the maximum absorption of the accessory bacteriochlorophylls. A near-infrared detection system was used [1]. Typically 5 decays were recorded between 870 and 920 nm for each case.

Data analysis was performed by global lifetime as well as global target analysis directly on the original data. For the calculation of the time-dependent free energy ΔG we invoked a quasi-continuous sequential multi-state kinetic model whose rate constants and consequently the ΔG_i value for each state *i* were determined by the target analysis. Hence the total ΔG(t) relative to P* was obtained by summing up all individual ΔG_i values relative to P* weighted by the corresponding time-dependent population density of each state *i*.

3. Results and Discussion

The fluorescence decays of open and closed RCs reveal remarkable differences in the kinetics (c.f. Fig. 1) which show up also in the results of the target analysis (Fig. 2 and 3). For the target analysis we restricted ourselves to sequential schemes containing one fluorescing excited state P* and 4 non-fluorescing radical pair states (Fig. 3). No direct protein relaxation between 'non-nearest' radical pair states was assumed. Thus these models follow approximately the model of conformational cooling. (Note that target analysis always can only proof a model for its consistency with the data.) In both cases of open and closed RCs the first step, which is the charge separation to RP_1, is the dominating fastest process with reasonable high rate constants of approx. 200 and 100 ns^{-1}, respectively. However, the corresponding ΔG values in both cases are very small indicating that additional relaxation processes have to follow in this model. For open RCs the following relaxation processes slow down (Fig. 2*A*). Simultaneously the ΔG gaps per step become smaller up to a certain point where it increases again by 50-60 meV (200 cm^{-1} = 25 meV ≈ 17°C). Qualitatively the same behavior holds also for the case of closed RCs (Fig. 2*B*), but quantitatively the jump in the last step is more pronounced from +8 meV to -72 meV. Whereas the description of the data within this

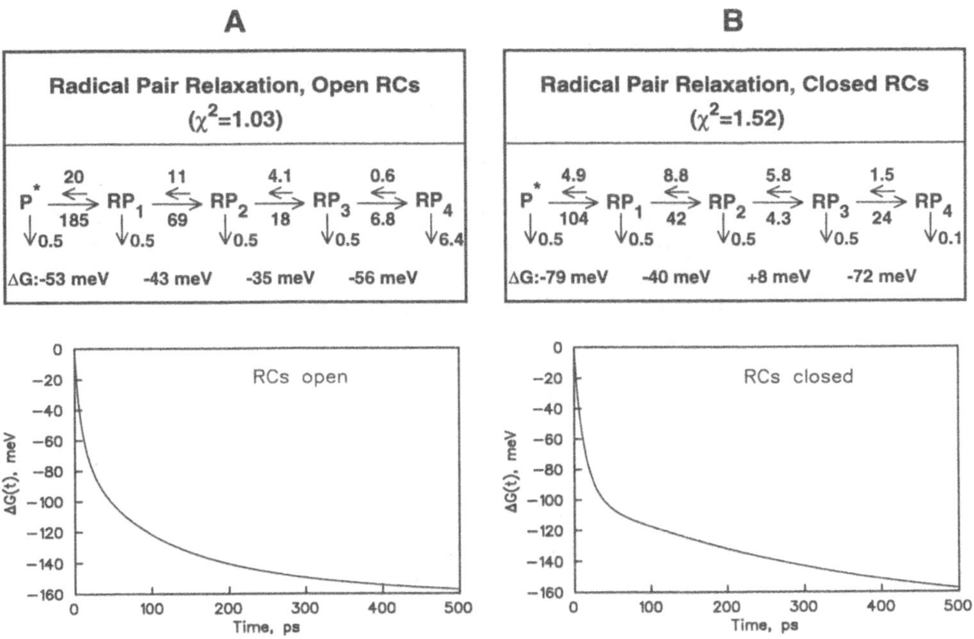

Fig. 2. Quasi-continuous multi-state kinetic models for charge separation and radical pair relaxation in chromatophores with open RCs (A) and closed RCs (B). Rate constants are given in units of ns^{-1} and were obtained by target analysis. These correspond to lifetimes of 5, 14, 48, 126, 233 ps for open RCs and 9, 21, 32, 286, 6532 ps for closed RCs. The $\Delta G(t)$ functions were calculated as described in section 2. Note that all radical pair states (RP) are non-radiative states (c.f. Fig. 3).

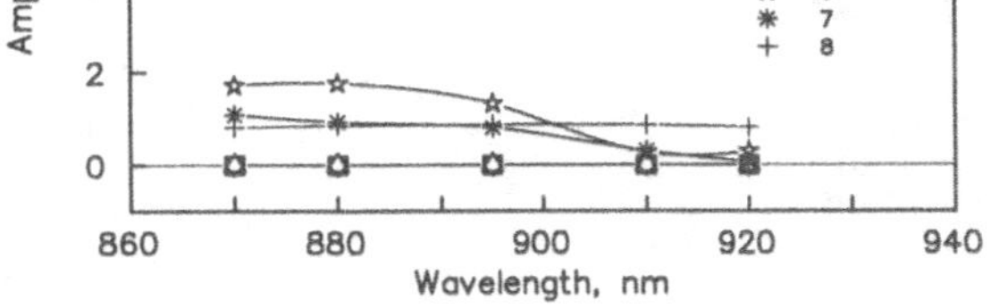

Fig. 3. Species-associated spectra as determined by global target analysis for chromatophores with open RCs. Note that within the model there is only one spectrum for the fluorescence of the excited primary donor P. Spectra 6-8 are strongly blue-shifted and account for minor amounts of non-coupled BChl fluorescence in the analysis.*

model is perfect for open RCs, it has to be noted that the target analysis does not yield as satisfactory results for closed RCs (the χ^2 value is 1.52 instead of 1.03). Increasing the complexity of the model may improve the fit. The time-dependent plots of ΔG show remarkable differences between open and closed RCs. Whereas on the time scale of up to 100 ps the RP looses energy more rapidly, the protein relaxation seems to be delayed for closed RCs in the time range between 100 and 400 ps (Fig. 2).

Our results are in agreement with a dynamic conformational cooling model of the radical pair. However at present we cannot rule out the possibility of an additional static disorder contributing to the decay kinetics which may be likely. A purely heterogeneous system based on a substantial static disorder would also give formally good results in the target analysis. It is quite difficult at present to distinguish formally between these models. We should like to note, however, that a model of static heterogeneity requires a drastic distribution in the forward rate of electron transfer. Given the relatively small free energy gaps for the fast reactions, and assuming that the majority of the RCs has a conformation and electron transfer parameters that puts them near the flat top part of the Marcus curve, such a drastic variation in the forward rates is not expected. At present we prefer the quasi-continuous sequential conformational cooling model over a static distribution of ΔG, although the real situation might well involve some admixture of the latter to a so far unknown content.

Acknowledgment

This work was supported by the Deutsche Forschungsgemeinschaft (Sonderforschungsbereich 189, Heinrich-Heine-Universität, Düsseldorf).

References

[1] Müller, M.G., Griebenow, K. and Holzwarth, A.R. (1991). *Biochim. Biophys. Acta* 1098: 1-12.

[2] Peloquin, J.M., Williams, J.C., Lin, X., Alden, R.G., Taguchi, A.K.W., Allen, J.P. and Woodbury, N.W. (1994). *Biochemistry* 33, 8089-8100.

[3] Bixon, M., Jortner, J. and Michel-Beyerle, M.E. (1995). *Chem. Phys.*, in print.

[4] Dörge, B., Klug, G., Gad'on, N., Cohen, S.N. and Drews, G. (1990). *Biochemistry* 29, 7754-7758.

THE WATER CHAINS AROUND Q_A AND Q_B AND OTHER STRUCTURAL ASPECTS OF THE REACTION CENTRE FROM *RB. SPHAEROIDES*

G. Fritzsch, U. Ermler and H. Michel
Max-Planck-Institut für Biophysik, Heinrich-Hoffmann-Str. 7,
60528 Frankfurt a.M., Germany

1. Introduction

By means of trigonal crystals the three-dimensional structure of the reaction centre (RC) from *Rhodobacter (Rb.) sphaeroides* has been reinvestigated (1,2). The crystals diffract to 2.6 Å. and are suitable to determine the positions of fixed water molecules in the RC. In contrast to earlier studies on orthogonal RC crystals (3,4,5), many supposed differences in cofactor binding between the two bacterial RCs from *Rb. sphaeroides* and *Rhodopseudomonas (Rps.) viridis* (3,4,5) could not be confirmed.

2. Procedure

The crystals of the RCs have been prepared according to (1,2). The protein was solubilized with 0.5% N-dodecyl-N,N-dimethylamine-N-oxide (LDAO). By anion-exchange chromatography with DEAE-sepharose the RCs have been purified up to an OD_{280}/OD_{850}-ratio of 1.2 to 1.3. Crystals were grown by the vapour diffusion method with sitting drops. About 100 µM (10.6 mg ml^{-1}) protein with 1.0 M potassium phosphate, pH 7.0 ± 0.5, 3% heptane-1,2,3-triol, 1 to 10% dioxane and 0.1% LDAO were equilibrated against 1.5 M potassium phosphate. (The heptanetriol was purchased from Oxyl. With the heptanetriol available from Fluka, a concentration of 1.5 to 2.0%

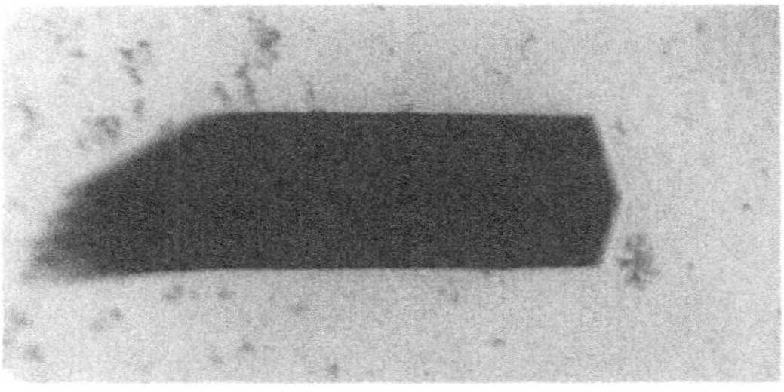

Figure 1. Trigonal Crystal of the RC from *Rb. sphaeroides*. Length: ca. 1 mm, thickness: 0.4 mm.

P. Mathis (ed.), Photosynthesis: from Light to Biosphere, Vol. I, 599–602.

is more favorable for the growth of trigonal crystals.) Crystals appear after one or several weeks (Fig. 1). They are up to 3 mm long and sometimes more than 1 mm thick. They belong to the trigonal space group $P3_121$ with unit cell dimensions of 141.4 Å, 141.4 Å, 187.2 Å.

3. Result and Discussion

For the determination of the RC structure 56 141 unique reflections have been used. The crystallographic R-factor is 18.6%. Only few differences between the structures of the RCs from *Rb. sphaeroides* and *Rps. viridis* are found (1). Thus, the ring V of the D_A bacteriochlorophyll is bent towards helix C in the RC of *Rb. sphaeroides* whereas it is bent towards helix D in the RC of *Rps. viridis*. The distances between the centres of mass of the cofactors along the active branch are about 0.2 Å larger in the *Rb. sphaeroides* than in the *Rps. viridis* RCs.

An inspection of the pattern of 165 crystallographically detected water molecules shows different "chains" of fixed waters with hydrogen bonding distances between each other. A remarkable water chain with a length of 23 Å extends from the quinone Q_B through the H-subunit to the cytoplasmic side of the RC from *Rb. sphaeroides* (Fig. 2, Table).

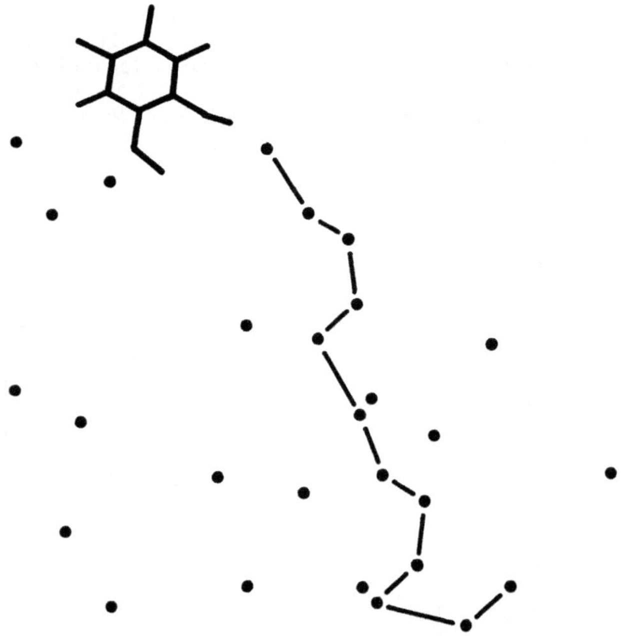

Figure 2. The water chain near Q_B. The head group of the quinone is shown above, the cytoplasmic side is below. Fixed water molecules are represented as black spheres. Distances of 2.8 to 3.4 Å between water molecules are connected by straight lines.

The chain consists of 12 water molecules separated by distances between 2.8 to 3.4 Å. Most probably, this water chain facilitates the proton transfer from the cytoplasmic side to the Q_B molecule. The water molecule w65 is 3.0 Å away from the proximal ether oxygen of Q_B and can therefore interact with the quinone. At the cytoplasmic end of the chain, the water molecule w124 is accessible from the solvent. Interruption of the Q_B water chain reduces the rates of the proton uptake and of the second electron transfer to Q_B (6).

Another water chain exists in a zone 5 to 12 Å away from the Q_A molecule (Fig. 3, Table). 10 water molecules with hydrogen bonding distances are located at the interface of the H- and M-subunit with a minimal separation of 5.2 Å to the Q_A molecule (distance from water molecule w98 to the distal ether oxygen of Q_A). Thus, at variance with the Q_B water chain, there is no hydrogen bonding interaction between Q_A and the chain-forming water molecules. w17 and w18 represent the end-points of the chain. They are in contact with the outside of protein. It is not known whether this structure of fixed water molecules near Q_A plays a physiological role. It might be involved in the uptake by the protein of the fractional proton upon reduction. The Q_A and Q_B water chains are not symmetry-related.

Q_B		
3.0		
w65		
3.3		**bulk phase**
w66		w18
3.0		3.1
w31	**Q_A** - 5.2 -	w98
2.8		3.0
w56		w14
3.0		2.9
w55		w16
3.3		2.8
w135		w40
2.8		3.0
w11		w94
2.7		3.0
w127		w41
2.9		2.8
w93		w99
3.2		2.8
w137		w42
3.4		3.2
w136		w17
3.4		**bulk phase**
w124		
bulk phase		

Table. Schemes of the two water chains near the quinones Q_B and Q_A. The distances between the water molecules are given in Å. There is no hydrogen bonding interaction between the water chain near Q_A and the Q_A molecule itself.

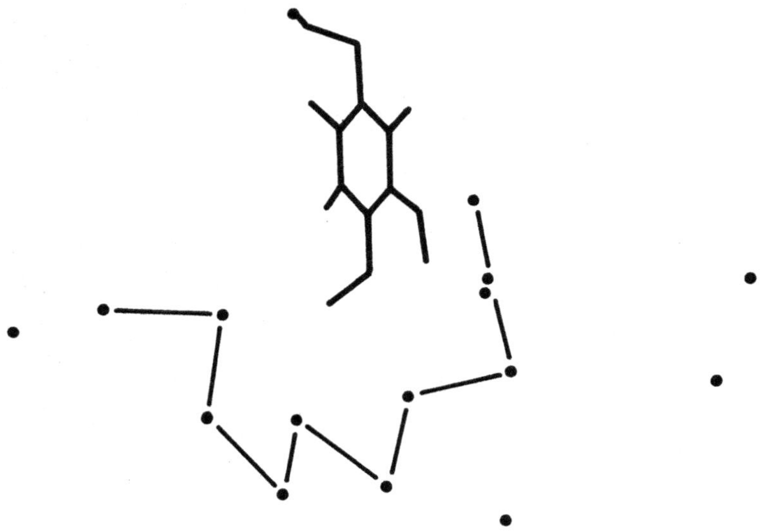

Figure 3. The water chain near Q_A. The head group of the quinone is shown above. Fixed water molecules are represented as black spheres. Distances of 2.8 to 3.2 Å between water molecules are connected by straight lines.

In addition to the amino acid residues, cofactors and fixed water molecules, one phosphate anion and 11 LDAO molecules have been detected crystallographically. Most of the LDAO molecules touch the RC at the hydrophobic zone of the protein surface that is in contact with the lipids of the membrane in vivo.

Acknowledgements: We thank the staff of the EMBL outstation at DESY (Hamburg) for the help in use of their facilities.

References

1 Ermler, U., Fritzsch, G., Buchanan, S.K. and Michel (1994) Structure 2, 925-936
2 Buchanan, S.K., Fritzsch, G., Ermler, U. and Michel, H. (1993) J. Mol. Biol. 230, 1311-1314
3 Yeates, T.O., Komiya H., Chirino, A., Rees, D.C., Allen, J.P. and Feher, G. (1988) Proc. Natl. Acad. Sci. USA 85, 7993-7997
4 El-Kabbani, O., Chang, C.H., Tiede, D., Norris, J. and Schiffer, M. (1991) Biochemistry 30, 5352-5360
5 Allen, J.P., Feher, G., Yeates, T.O., Komiya, H. and Rees, D.C. (1988) Proc. Natl. Acad. Sci. USA 85, 8487-8491
6 Baciou, L. and Michel, H. (1995) Biochemistry 34, 7967-7972

Temperature Dependence of the Recombination Fluorescence from PS II Core Complexes

B. Hillmann[1], I. Moya[2] and E. Schlodder[1]
[1]Max-Volmer-Institut für Biophys. Chemie, TU-Berlin, Str. des 17. Juni 135, 10623 Berlin, [2]Lab. pour l'Utilisation du Rayonnement Electromagnétique, Bât 209-D, Centre Universitaire Paris-Sud, 91405 Orsay Cedex

Introduction

In Photosystem II, that is responsible for the oxidation of water, the trapping of excitation energy occurs by electron transfer from the excited singlet state of the primary donor, P680*, to an intermediate electron acceptor, a pheophytin *a* molecule (Ph). The electron is subsequently transferred to a plastoquinone (Q_A) in about 300 ps to stabilise the light induced charge separation. When the electron transfer to Q_A is blocked, charge recombination during the lifetime of $P680^+Ph^-$ can regenerate P680*, which equilibrates with the antenna pigments, and gives rise to long-lived recombination fluorescence. Measurements of the recombination fluorescence as a function of temperature offer the possibility to determine the thermodynamic parameters ($\Delta G°$, $\Delta H°$, $\Delta S°$) of the primary charge separation step. In this work we resolved long-lived ($\tau \geq 10$ ns) fluorescence components from PS II core complexes with singly reduced Q_A. Their temperature dependence indicates that the standard free energy gap between the states $P680^+Ph^-$ and P680*Ph decreases nearly linearly when the temperature is lowered.

Materials and Methods

O_2- evolving PS II core complexes containing about one photoreducible Q_A per 40 Chls were prepared from the thermophilic cyanobacterium *Synechococcus elongatus* and further purified as described previously [1] in order to remove completely the phycobiliproteins from the preparation.

For fluorescence measurements at low temperature, the samples were diluted 2:1 (v/v) with glycerol. Q_A was singly reduced by addition of approx. 4 mM $Na_2S_2O_4$ in the dark. For measurements under conditions of oxidised Q_A, the sample was in the presence of 2 mM ferricyanide and 200 µM 2,5-dichloro-p-benzoquinone.

The fluorescence decay was measured using time-correlated single photon counting as described previously [2] except for the following modifications. The samples were excited by 640 nm pulses from a pulsed laser diode. The photon flux incident on the sample was $< 10^{14}$ photons per s per cm^2, corresponding to less than 1 excitation per 4 s. Fluorescence at 685 nm was selected with a monochromator (Jobin & Yvon M25, bandwidth 5.5 nm) and detected by a microchannel plate photomultiplier (Hamamatsu). Detection was at a right angle to excitation. The cuvette (path length 0.5 mm) was positioned in a variable temperature cryostat (5 K - 300 K) at 45^0 to the exciting beam. The full width at half maximum of the instrument response function measured with scattered light from the

P. Mathis (ed.), Photosynthesis: from Light to Biosphere, Vol. I, 603–606.

sample at the excitation wavelength was about 100 ps. Different time scales ranging from 4 ps up to 160 ps per channel have been used. 2048 channels were collected. Most measurements were performed on the 320 ns time scale in order to determine precisely the parameters of long-lived components. The decay kinetics were analysed by a sum of exponential decay components convoluted with the instrument response function. On-line deconvolution was done by the Fluomarqdt software developed at LURE (Orsay).

Results and Discussion

At room temperature, the time courses of fluorescence from PS II core complexes with Q_A in the oxidised and the reduced state have been analysed. Typical results of fitting the fluorescence decays measured on different time scales to a sum of five exponentials are presented in Table I.

redox state of Q_A	$\tau 1$	$\tau 2$	$\tau 3$	$\tau 4$	$\tau 5$
oxidised	50 ps 80 %	400 ps 10 %	1.5 ns 9.5 %	5.5 ns 0.5 %	–
singly reduced (Q_A^-)	100 ps 40 %	1.5 ns 33 %	4 ns 25.4%	12 ns 1.5 %	36 ns 0.1 %

Table I: Lifetimes (τ_i) and relative amplitudes (A_i) of fluorescence decay components at 293 K. The parameters may vary by up to 50 % using different time scales or different samples.

In the presence of oxidised Q_A, the fluorescence kinetics are dominated by two subnanosecond decay components. Upon reduction of Q_A, the fluorescence yield increases by a factor of about 7 mainly due to the appearance of ns decay components with lifetimes of about 1.5 and 4 ns. These results are in agreement with previously measured fluorescence decay curves for similar preparations [2-5]. However, we could additionally resolve long-lived components ($\tau \geq 10$ ns) with small amplitudes, when Q_A was singly reduced by the addition of 4 mM dithionite (see Fig. 1). With the exception of Ref. 5, these phases have not been observed before, which might be due to the shorter time-scales used in the other studies. Fluorescence decays measured with PS II complexes in the absence of electron acceptors at higher excitation intensities exhibit long-lived components with nearly identical lifetimes (not shown). Under these conditions Q_A is kept reduced to a large extent by the excitation light.

The nanosecond decay components in the presence of Q_A^- can be attributed to radical pair recombination fluorescence. This attribution is supported by similar time constants found for the decay of flash induced absorbance changes associated with $P680^+Ph^-$ [1,6]. The multiexponential decay may be explained in accordance with earlier work by the assumption that a stabilisation of the radical pair occurs during the lifetime due to the relaxation in the free energy of the radical pair [6,7]. The assignment of the slowest components ($\tau \geq$ 10 ns) to recombination fluorescence seems unambiguous, since none of the pigments has such a long excited state lifetime.

In samples with singly reduced Q_A, the recombination fluorescence has been studied as function of temperature between 20 and 300 K. Fig. 1 shows the fluorescence decays at three temperatures. The remaining amplitudes after 50 ns have been expanded by a factor of 200 illustrating that the amount of long-lived fluorescence is higher and that the decay kinetics are slower at 154 K than at 260 K. The longer lifetimes of recombination fluorescence at 154 K are in accordance with results of absorbance difference spectroscopy [8]

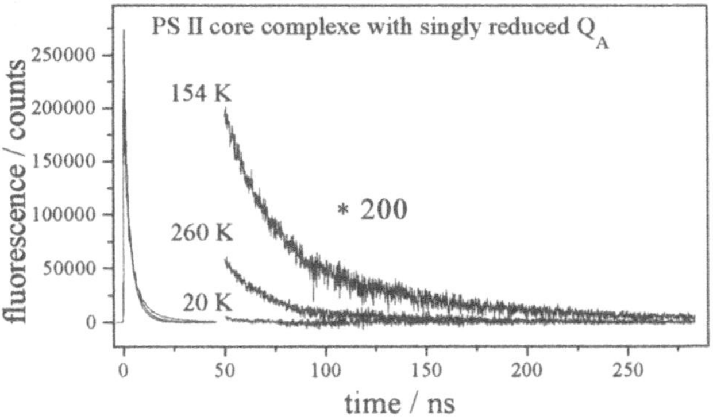

Fig. 1: Fluorescence decays from PS II complexes with reduced Q_A at three temperatures.

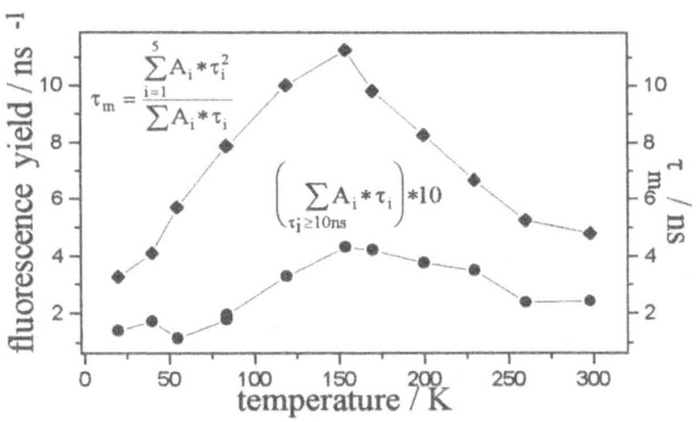

Fig. 2: Yield of long-lived fluorescence (•) and average lifetime τ_m (♦) as function of temperature.

indicating a slower decay of P^+Ph^- at low temperature. Fig.2 shows the average lifetimes and relative yields of long-lived components ($\tau \geq 10$ ns) as a function of temperature. The maximum is observed at about 150 K. At this temperature the relative yield of components with lifetimes of ≥ 10 ns correspond to about 30 %. The temperature dependence of the long-lived fluorescence components is surprising, since the thermal repopulation of P* from P^+Ph^- giving rise to recombination fluorescence should drastically decrease when the temperature is lowered. It strongly suggests that the free energy gap between P* and P^+Ph^- decreases upon lowering the temperature. In order to obtain more quantitative information, $\Delta G°$ has been calculated assuming that the equilibrium between the excited singlet state of the antenna, A*, and P^+Ph^- is established rapidly compared to the lifetime of the radical pair. Then the equilibrium constant for the reaction A* $\Leftrightarrow$ P^+Ph^- is equal to the ratio of the initial amplitude of the fluorescence to the amplitude of the recombination fluorescence [7]. In order to avoid uncertainties in the origin of the 1-6 ns components, only decay phases with lifetimes of $\tau_i \geq 10$ ns were taken into account as recombination fluorescence. The observation of small components with $\tau = 1.5$ and 5.5 ns in the presence of oxidised Q_A at 293 K (see Tab. 1) may indicate contributions from uncoupled antenna chlorophylls. Additionally, a 5 ns component with an emission maximum at 691 nm arises when the temperature is lowered from 70 to 20 K. This component may be assigned to antenna Chls, which become thermally uncoupled at low temperature because they absorb at slightly longer wavelengths than P. The amplitudes for the recombination fluorescence ($\tau_i \geq 10$ ns) were corrected for the yield of long-lived primary radical pair formation measured by time resolved absorption difference spectroscopy [8]. In order to determine the initial amplitude of the fluorescence the fastest component was fixed to 100 ps in agreement with the free running fits on shorter time scales. Fig. 3 (♦) shows the resulting

ΔG° in PS II core complexes with singly reduced Q_A as function of the temperature. For the sake of simplicity, we used ΔG°(A* ⇔ P*) ≈ kT∗ln N (N = number of antenna pigments) neglecting differences in the 0-0 transition energies between the excited and ground states of P and A. ΔG°(A* ⇔ P*), given by the solid line in Fig. 3, was obtained with N = 40. The circles (•) represent ΔG°(P* ⇔ P^+Ph^-) = ΔG°(A* ⇔ P^+Ph^-) - ΔG°(A* ⇔ P*). The ΔG°(P* ⇔ P^+Ph^-) calculated in this work for room temperature (≈ -170 meV) corresponds to the ΔG° between P* and a relaxed state of P^+Ph^-. For comparison, ΔG° between P* and the initially formed state of P^+Ph^- has been determined to be about -100 meV [3,6]. The variation of ΔG° with temperature is only slightly effected by the corrections and approximations described above. Similar temperature dependencies of ΔG° (P* ⇔ P^+Ph^-) have been reported for bacterial reaction centres [9] and D1/D2/cyt *b*-559 complexes [10]. The temperature dependence of ΔG° can be explained within the sequential relaxation model [7-10] in which the charge separated state is stabilised after formation by protein conformational changes. When the temperature is lowered, these conformational motions are gradually frozen out. As a consequence the relaxation in free energy of the radical pair state becomes less effective and thermally repopulation can easier occur. Below 150 K, kT becomes smaller relative to the free energy gap between P* and P^+Ph^- which results in the decrease of the long-lived components.

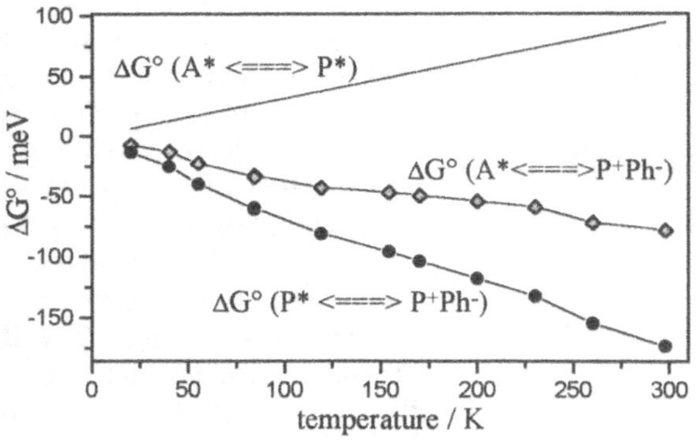

Fig.3: Temperature dependence of the free energy gap between the excited antenna (A*), the excited donor (P*) and the primary radical pair P^+Ph^-.

Acknowledgement We thank A. Krieger for participation in the initial state of the project. This work was supported by grants from LURE, project BF 017, and the DFG (SFB312)

References

1 Van Mieghem, F., Brettel, K., Hillmann, B., Kamlowski, A., Rutherford, A. W. and Schlodder, E., (1995) Biochemistry 34, 4798-4813
2 Hodges, M. and Moya, I. (1988) Biochim. Biophys. Acta 935, 41-52
3 Schatz, G. H., Brock, H. and Holzwarth, A. R., (1987), Proc. Nat. Acad. Sci. USA 84, 8414-8418
4 Van Mieghem, F. J. E., Searle, G. F. W., Rutherford, A. W. and Schaafsma, T. J. (1992) Biochim. Biophys. Acta 1100, 198-206
5 Liu, B, Napiwotzki, A., Eckert, H.-J., Eichler, H. J. and Renger, G. (1993) BBA 1142, 129-138
6 Schlodder, E. and Brettel, K. (1988) Biochim. Biophys. Acta 933, 22-34
7 Woodbury, N. W. T. and Parson, W. W. (1984) Biochim. Biophys. Acta 767, 345-361
8 Schlodder, E. and Brettel, K. (1990) in Current Research in Photosynthesis (Baltscheffsky, M., Ed.) Vol. I, pp 447-450, Kluwer, Dordrecht
9 Peloquin, J. M., Williams, J. C., Lin, X., Alden, R. G., Taguchi, A. K. W., Allen, J. P. and, Woodbury N. W. (1994) Biochemistry 33, 8089-810
10 Booth, P. J., Crystall, B., Barber, J., Porter, G. and Klug, D. R. (1991) Biochemistry 30, 7573-7586

PHOTOSELECTIVE EXCITATION OF P680 ?

Stephen Merry, Shigeichi Kumazaki*, James Durrant, Yasuhiro Tachibana, James Barber, George Porter and David Klug
Molecular Dynamics and Photosynthesis Research Groups, Centre for Photomolecular Sciences, Department of Biochemistry, Imperial College, London, SW7 2AZ
* current address- Institute for Molecular Science, Myodaiji, Okazaki

1) Introduction

A key issue in elucidating the primary photochemistry of photosynthetic reaction centres is the discrimination between energy and electron transfer processes. In studies of the primary charge separation in bacterial reaction centres, contributions to the experimental data from energy transfer processes can be readily avoided by selective excitation of the primary electron donor, P870. In PS II, however, selective excitation of P680 is greatly complicated by the extensive spectral overlap between this species and the other reaction chlorins. Moreover, this high degree of spectral and therefore energetic degeneracy allows extensive femtosecond equilibration of excitation energy between the reaction centre chlorins prior to electron transfer[1]. Charge separation in PS II therefore proceeds from an equilibrated pool of chlorin excited states, even when P680 has been directly excited. Indeed, we have previously suggested that dipole/dipole coupling between these 'core' pigments may result in extensive coherent delocalisation of the excited states - the P680 multimer model[2].

Between one and three reaction centre chlorophylls are bound on the periphery of the reaction centre complex, and transfer energy to the 'core' chlorins relatively slowly (10- 50 ps) [3]. For the study of the primary electron transfer kinetics, it is important to select excitation conditions which minimise contributions to data associated with slow energy transfer from these peripheral chlorins. We have previously shown that such slow energy transfer kinetics are greatly reduced when excitation pulse centred at 694 nm are used rather than pulses centred at 665 nm. However, recent studies have also reported the presence of a state absorbing at 680 nm - 684 nm [4,5,6] which, at least at low temperatures, also transfers energy relatively slowly to P680.

In this paper we use femtosecond absorption anisotropy measurements to demonstrate that energy transfer between two states absorbing near 680 nm occurs in

P. Mathis (ed.), Photosynthesis: from Light to Biosphere, Vol. I, 607–610.

500fs. This provides further evidence that at room temperature, our 694 nm pulse selectively excites RC 'core' pigments in fast energy transfer equilibrium with P680.

2) Materials and Methods

The transient absorption spectrometer used has been described elsewhere [7]. The Photosystem II-Reaction Centres (PS II-RC) samples were isolated from pea seedlings as before [7]. Data were collected using either parallel or perpendicular polarised pump and probe beams, and globally analysed as previously. Isotropic data and anisotropy decays were generated by using the following relationships:-

$$\Delta OD_{isotropic} = \frac{(\Delta OD_{para} + 2\Delta OD_{perp})}{3}$$

$$\text{anisotropy}, r(t) = \frac{(\Delta OD_{par} - \Delta OD_{perp})}{(\Delta OD_{par} + 2\Delta OD_{perp})}$$

Using these procedures, initial anisotropy values of 0.4 were obtained for the laser dye DMOTCI at all wavelengths between 660 nm and 690 nm.

3) Results

The transient absorption kinetics observed at probe wavelength of 683 nm are shown in figure 1, following excitation with 694 nm pulses. Kinetic global analysis of such data collected at probe wavelengths between 660 nm and 690 nm resolved two exponential components with lifetimes of approx. 100 and 500 fs, in addition to a component which did not decay on this 0 - 1.5 ps timescale. The isotropic data exhibits a recovery of the initial bleach with a 100 fs time constant, which we have previously shown results from energy transfer ~680 nm absorbing states to ~670 nm absorbing states. In contrast the anisotropy decays more slowly, with approximately a 500 fs time constant.

Figure 2 shows the absorption difference spectra at 250 fs and 1 ps, corresponding approximately to spectra before and after the 500 fs decay observed in the anisotropy data. These spectra are dominated by the bleach / stimulated emission of the Q_y absorption bands. These data clearly demonstrate the relatively small change in the isotropic transient spectrum associated with this component. Moreover, it is clear this component is not associated with appearance of bleach / stimulated emission at other wavelengths, in contrast for example to the 100 fs which results in the appearance of bleach / stimulated emission between 670 nm - 675 nm [1]. The small changes in the isotropic data associated with the 500 fs component, when combined with the large change in the anisotropy signal (0.4 to 0.2) indicate this component must be assigned to a high degree of population transfer which occurs with this time constant. The large change in anisotropy indicates that the transitions between which these transfers occur must have a large angular separation.

FIGURE 1: Transient absorption spectra of the PSII-RC obtained with 694 nm excitation, probing at 683 nm

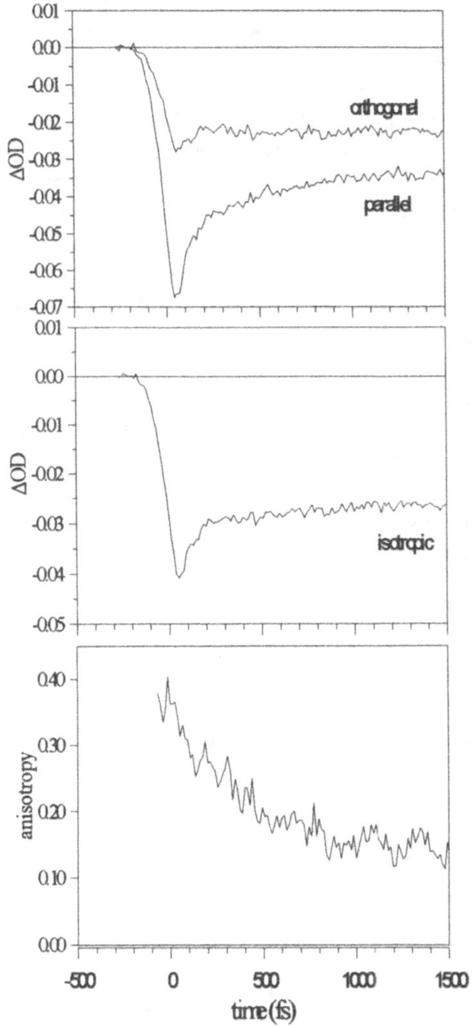

FIGURE 2: Transient spectra of the PSII-RC obtained at 250 fs and 1 ps for isotropic configuration

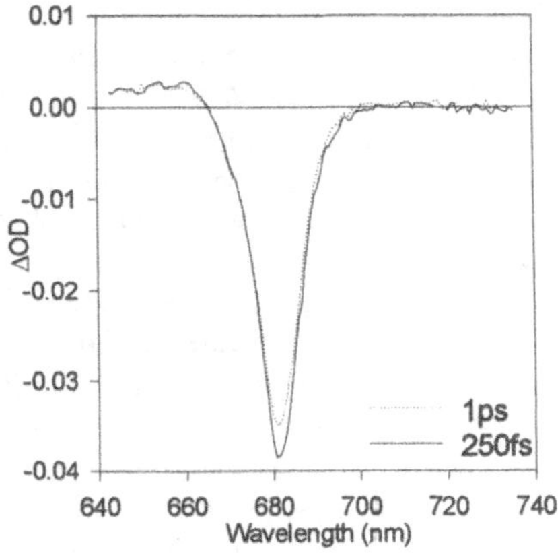

In the isotropic data, the small 500 fs component indicates that the population transfer is between two states of approximately similar oscillator strengths, with a maximum ratio of 1:2. The 500 fs component is not accompanied by any shift in peak of the transient spectra in figure 2. The population transfer is therefore between two pools of pigments absorbing near 680 nm

4) Discussion

We have observed that there are two sub-picosecond energy transfer steps occurring in the `core` of the Photosystem II reaction centre. The fastest step, the 100 fs component, results in the equilibration of energy between states with absorption maxima at 680-683 nm and those at 670-675 nm. The second step, the 500 fs component, results from the equilibrium between two states both with maxima in the range 680-683 nm. The two equilibrated states have a large angular separation between their dipole moments. Numerical simulations indicate the oscillator strengths of these states can not differ by more than a factor of two. These states may either correspond to different pigments, or within the multimer model, to different exciton states of a delocalised system[2].

It is likely that our observation of two ~ 680 nm absorbing states in PS II reaction centres is related to previous low temperature observations of 680-684 nm absorbing state in addition to P680 by other groups[4,5,6]. At low temperatures, this state slowly transfers energy to P680. We have, however, demonstrated that at room temperature these states equilibrate in 500 fs. This timescale is much faster than that observed for radical pair formation in isolated PS II-RC's. This therefore provides further evidence that 694 nm excitation of PS II-RC's results primarily in excitation of reaction core pigments, whose excited states equilibrate prior to charge separation, and avoids contributions from slow energy transfer kinetics associated with peripheral chlorophylls.

5) References

1) Durrant, J.R., Hastings, G., Joseph, D.M., Barber, J., Porter, G., Klug, D.R. (1992) Proc. Nat. Acad. Sci. USA, 11632-11636

2) Durrant, J.R., Klug, D.R., Kwa S.L.S., van Grondelle, R., Porter, G., Dekker, J.P. (1995) Proc. Nat. Acad. Sci. USA, 4798-4802

3) Rech, T., Durrant, J.R., Joseph, D.M, Barber, J., Porter. G., Klug, D.R. (1994) Biochemistry,14768-14774

4) Groot, M-L., Petermann, E.J.G., van Kan, P.J.M., van Stokkum, I.H.M., Dekker, J.P., van Grondelle, R. (1994) Biophysics. J.,67, 318-330

5) Tang, D., Jankowiak, R., Seibert, M., Yocum, C.F., Small, G.J. (1990) J. Phys. Chem.,94,6519 -6522

6) Otte, S.C.M., van der Vos, R., van Gorkom, H.J. (1992) J. Photochem. Photobiol. B: Biol, 15, 5-14

7) Klug, D.R., Rech, T., Joseph, D.M., Barber, J., Durrant, J.R., Porter, G. (1995) Chemical Physics, 194, 433-442

THE INFLUENCE OF ENERGY LEVEL DISORDER ON THE CHARGE SEPARATION / TRAPPING KINETICS IN PHOTOSYSTEM TWO.

James R. Durrant, George Porter, James Barber and David R. Klug.
Molecular Dynamics and Photosynthesis Research Groups, Centre for Photomolecular Sciences, Department of Biochemistry, Imperial College, London SW7 2AU, U.K..

1. Introduction

Experimental studies of the primary charge separation in the photosystem II reaction centre (PS II-RC) have resolved complex kinetics [1, Klug et al. these proceedings]. While the majority of charge separation occurs with a 21 ps time constant, up to a third of the charge separation proceeds much faster with a 3 ps time constant. In addition, a 100 ps component observed in both time resolved absorption and emission studies may reflect an additional slower phase of the charge separation kinetics. Similarly complex kinetics have recently also been reported by Holzwarth and co-workers (unpublished data). In this paper we consider possible physical origins for these complex kinetics.

At room temperature, PS II-RC chlorins may be loosely grouped in two types, core pigments and peripheral chlorophylls. Excitation energy equilibration between the core pigments occurs on a sub-picosecond time scale [2], while energy transfer from between peripheral chlorophylls to the core takes 10 - 30 ps [3-6]. It has been suggested that the 3 ps charge separation component results from direct excitation of core pigments, while the 21 ps results from excitation of 'peripheral' chlorophylls and is rate limited by a slow energy transfer step [5,6]. However, we have previously argued that our 694 nm excitation conditions minimise excitation of peripheral chlorophylls [2-4]. Moreover, we have recently shown (Merry et al, these proceedings) that the 694 nm pulses excite two ~ 680 nm transitions which equilibrate in 500 fs and must therefore both be attributed to RC core pigments. The relatively large amplitude of the 21 ps charge separation component observed under these conditions is therefore only consistent with it originating from direct excitation of RC core pigments.

We now consider two alternative models for the complex charge separation kinetics. The complex kinetics could originate from multiple steps in the formation of the final radical pair state $P680^{+}Ph^{-}$. Such steps could include both electron transfer steps and energetic

P. Mathis (ed.), Photosynthesis: from Light to Biosphere, Vol. I, 611–614.

relaxation processes. This multiple step model will be discussed in detail elsewhere. Alternatively, the complex kinetics could result from static energy level disorder of the radical pair, resulting in a distribution of electron transfer rate constants. Such a model has been proposed to account for similarly multiexponential kinetics observed for primary charge separation in the reaction of purple bacteria [7]. In this paper, we use numerical simulations to determine whether this static disorder model can indeed account for the complex kinetics observed for PS II.

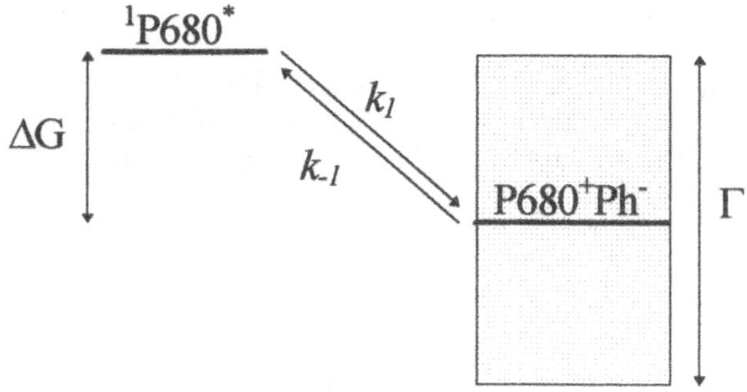

Figure 1, illustrating static energy level disorder (inhomogeneous broadening) of the free energy gap for primary electron transfer in PS II.

2. Theoretical Methods

The dependence of the rate of electron transfer, k_1, upon the free energy gap between RC^* and $P680^+Ph^-$ is given by:

$$k_1 = k_0 \exp\left(\frac{-(\Delta G - \lambda)^2}{4\lambda k_B T}\right)$$

where λ is the reorganisational energy of the reaction. The rate for reverse electron transfer k_{-1} is given by:

$$k_{-1} = k_1 \exp\left(-\Delta G / k_B T\right)$$

and the observed rate of equilibration between RC^* and $P680^+Ph^-$, corresponding to the observed rate of charge separation, is therefore given by $k_{ob} = k_1 + k_{-1}$.

Numerical simulations of the influence of energy level disorder were obtained by ensemble averaging k_{ob} over 100 reaction centres, where for each reaction centre the free energy gap ΔG was selected at random from a Gaussian distribution with mean $\langle \Delta G \rangle$ and width (FWHM) Γ.

3. Results and Discussion

Figure 2 compares log /linear plots of the experimentally observed charge separation kinetics, determined from time resolved absorption and emission data [1, Klug et al., these proceedings] with that of a numerical simulation.

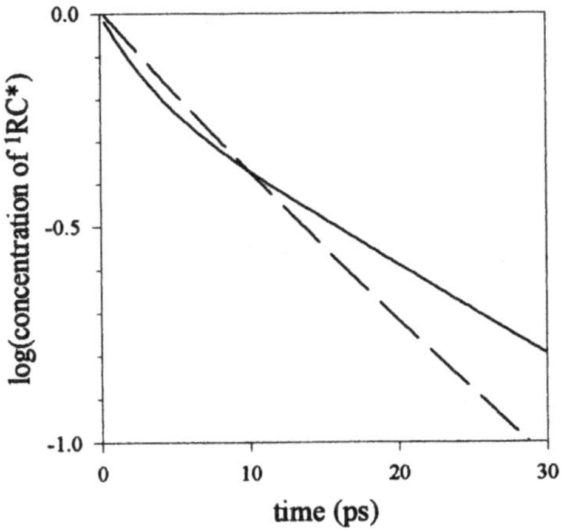

Figure 2. Log/linear plot of the experimentally observed charge separation kinetics (solid line) versus simulated kinetics (dashed line). Numerical simulation used $\langle \Delta G \rangle$ = 0.06 eV, λ = 0.2 eV and Γ= 0.1 eV.

The numerical simulations required knowledge of the parameters $\langle \Delta G \rangle$, Γ and λ. Previous experimental studies have indicated $\langle \Delta G \rangle \sim$ 0.06 - 0.11 eV and $\Gamma \sim$ 0.06 - 0.1 eV (FWHM) [8,9]. Previous reports of ultrafast charge separation in PS II even at liquid helium temperatures indicate λ is of similar magnitude to $\langle \Delta G \rangle$. Typical values for λ for this type of electron transfer are ~0.2 eV. Simulations were conducted for a range of values of each of these parameters: $0 < \langle \Delta G \rangle < 150$ meV, $0 < \Gamma < 100$ meV and $50 < \lambda < 250$ meV. However, qualitatively similar results were obtained for all values of $\langle \Delta G \rangle$, Γ and λ within these ranges, and a typical result is shown in figure 2 (dashed line). The value of k_0 used did not affect the linearity of the simulated curve, and therefore its value was treated as a free fitting parameter.

The experimental trapping kinetics shown in figure 2 are clearly non-linear on this log/linear plot, reflecting the multiexponential kinetics discussed above. In contrast, the simulated kinetics are almost linear, and can therefore be reasonably fit by a single exponential component. Similarly linear simulated kinetics were obtained for *all* simulation conducted for the parameter ranges given above. A even greater discrepancy was found between the experimental and simulated data when a 100 ps charge separation component was included and the simulation conducted over a longer timescale. It can thus be concluded that unless static energy level disorder is an order of magnitude greater than that reported experimentally, this disorder can not account for the observed multiexponential charge separation kinetics.

The amplitude of the energy level disorder used in these simulations is relatively large. For example for the simulation shown in figure 2, $\langle \Delta G \rangle$ was less than Γ, and therefore a significant proportion of the charge separation was endothermic. Despite this large distribution of free energy gaps, the predicted effect on the charge separation kinetics is negligible. This contrasts to the charge recombination kinetics, where the distribution of equilibrium constants between RC^* and $P680^+Ph^-$ would be expected to result in highly multiexponential kinetics, as experimentally observed [7, 9].

We thus conclude that energy level disorder can not account for the multiexponential charge separation kinetics experimentally observed in PS II. Elsewhere we consider in detail alternative multistep models for these complex kinetics.

The financial support of the BBSRC, EC, RITE and the Royal Society is gratefully acknowledged.

4) References

1) Kumazaki, S., Joseph, D.M., Crystall, B., Durrant, J.R., Barber, J., Porter, G., Yoshihara, K. and Klug, D.R. (1995) submitted to Biochemistry.
2) Durrant, J. R., Hastings, Joseph, D. M., Barber, J., Porter, G., and Klug, D. R. (1992) *Proc. Natl. Acad. Sci. USA 89*, 11632-11636.
3) Rech, T., Durrant, J. R., Joseph, D. M., Barber, J., Porter, G., and Klug, D. R (1994) *Biochemistry,* **33,** 14768-14774.
4) Vacha, F., Joseph, D.M., Durrant, J.R., Telfer, A., Klug, D.R., Porter, G., and Barber, J. (1995) *Proc. Natl. Acad. Sci. USA*, **92,** 2929-2933..
5) Holzwarth, A.R., Müller, M.G., Gatzen, G., Hucke, M. and Griebnow, K. (1994*) J. Luminescence 60,* 497-502.
6) Schelvis, J. P. M., van Noort, P. I., Aartsma, T. J. and van Gorkom, H. J. (1994) *Biochim. Biophys. Acta 1184*, 242-250.
7) Ogrodnik, A., Keupp, W., Volk, M., Aumeier, G. and Michel-Beyerle, M.E. (1994) J. Chem. Phys. 98, 3432-3439.
8) Groot, M-L., Peterman, E.J.G., van Kan, P.J.M., van Stokkum, I.H.M., Dekker, J.P. and Rienk van Grondelle (1994) Biophys. J. 67, 318-330.
9) Booth, P.J., Crystall, B., Ahmad, I., Barber, J., Porter, G. and Klug, D.R. (1991) Biochem. 30, 7573-7586.

COMPARISON OF PS II PRIMARY PHOTOCHEMISTRY IN HIGHER PLANT, *SYNECHOCYSTIS* AND *SYNECHOCYSTIS* MUTANTS.

Linda B. Giorgi, Peter J. Nixon, Steve A.P. Merry, D. Melissa Joseph, James R. Durrant, Javier De Las Rivas*, James Barber, George Porter and David R. Klug.

Molecular Dynamics and Photosynthesis Research Groups, Centre for Photomolecular Sciences, Department of Biochemistry, Imperial College, London, SW7 2AY, U.K. *Department of Biochemistry, Faculty of Science, University of the Basque Country, PO Box 644, Bilbao, Spain.

1. Introduction

Site-directed mutagenesis has been found to be a useful tool in identifying residues of the protein matrix that affect primary photochemistry of purple bacterial reaction centres. Site-directed mutagenesis of PS II has mostly employed the cyanobacterium *Synechocystis* PCC 6803 (1), although it is now also possible to make mutations in *Chlamydomonas* (2). Comparison of the primary structures of the *Synechocystis* D1 and D2 polypeptides with those of the higher plant, *Pisum sativum*, (3) indicates several significant differences in primary structure between the two organisms. Of particular relevance to this paper, is the difference in the residue at position D1-130. In wild type *Synechocystis*, the D1-130 residue is a glutamine, while in *Pisum sativum* it is a glutamate; this residue has been suggested to lie close to the putative binding site of the pheophytin electron acceptor (4).

Charge separation in Photosystem II (PS II) reaction centres results in the oxidation of P680, the primary electron donor of PS II, and the reduction of the pheophytin electron acceptor (Ph), leading to the radical pair state $P680^{+}Ph^{-}$. In this paper we compare the quantum yield of radical pair formation, as estimated from nanosecond absorption data, in higher plant (*Pisum Sativum*) PS II and in the PS II reaction centre preparation isolated from wild type *Synechocystis* and from two different D1-130 mutants of *Synechocystis*, the D1-130 glutamate mutant and the D1-130 leucine mutant.

2. Materials and Methods

The procedure used for the isolation of a PS II reaction centre preparation from *Synechocystis* and its mutants is described in (5). These particles contain approximately 7-8 chlorophyll-a, 2 pheophytin-a, 1 cytochrome b-559 and 1 β-carotene (5). Higher plant PS II reaction centre particles were isolated from *Pisum sativum* as described previously (6).

P. Mathis (ed.), Photosynthesis: from Light to Biosphere, Vol. I, 615–618.

Nanosecond transient absorption measurements, at 820 nm, were carried out as described previously (7), with modifications as described in (5). The excitation pulses, at 630 nm or 680 nm, were non-saturating. The samples were prepared as described in (5) and they all had the same optical density (OD) at the excitation wavelength being used. The data analysed were an average of 100 flashes. The absorption decays were fit to three exponentials; one lifetime was fixed (at 4.7 μs) to represent the electronically filtered decay of triplet P680 in the PS II reaction centres (see ref. 7), while all other parameters were free running. During the course of these experiments, the activity of the samples was reduced by less than 10 % (5).

3. Results and Discussion

Fig. 1 compares the nanosecond absorption kinetics, monitored at 820 nm, of the higher plant PS II reaction centre and of the wild type and mutant *Synechocystis* PS II reaction centre preparations. Measurements at 820 nm monitor the presence of $P680^{+}Ph^{-}$ and

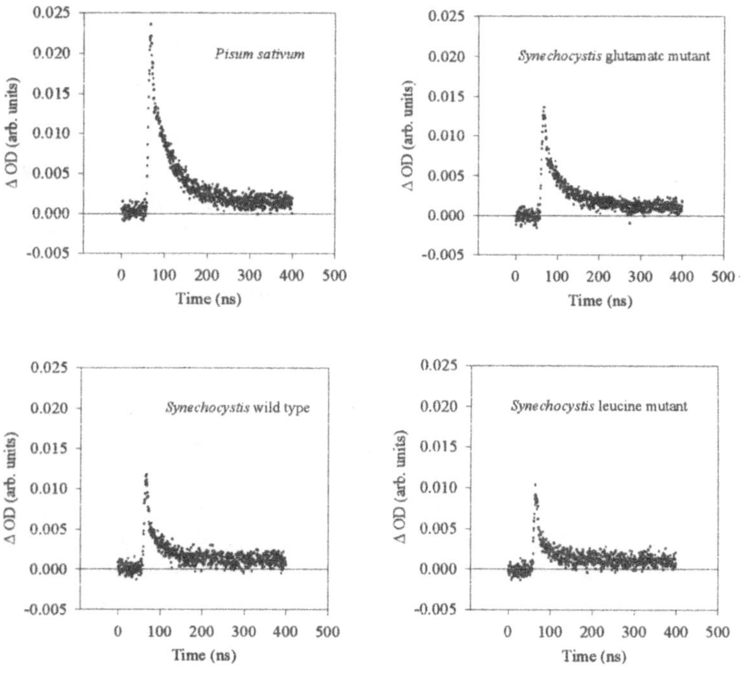

Figure 1. Comparison of the nanosecond decay kinetics, monitored at 820 nm and following excitation at 680 nm, for PS II reaction centres isolated from higher plant (*Pisum sativum*), wild type *Synechocystis* and the D1-130 glutamate and leucine *Synechocystis* mutants.

triplet P680; contributions from unbound chlorophyll singlet excited states are relatively small (see below).

Analysis of the decay kinetics in Fig. 1 resulted in the following lifetimes, with absolute amplitudes (x 10^{-2}), being given in parentheses. Higher plant (*Pisum sativum*): 48 ns (3.26) and 4.8 ns (3.43); WT *Synechocystis* : 36 ns (1.07) and 4.8 ns (2.84); D1-130 glutamate: 45 ns (1.70) and 4.9 ns (2.57); D1-130 leucine: 44 ns (0.55) and 6.4 ns (2.39). (The amplitude of the 4.7μs component, due to an electronic filter, was of the order of 0.23 x10^{-2} for all samples). The nanosecond components represent charge recombination of the primary radical pair (7), and therefore, their amplitudes can be taken to be a measure of the quantum yield of radical pair formation (ϕ_{RP}) in the sample. Since the OD's of the samples at the excitation wavelength used were identical and the excitation pulses were non-saturating, the amplitude of the nanosecond components in the different samples are proportional to the quantum yield of radical pair formation.

The ϕ_{RP} in the different samples was estimated by comparing the total amplitude of the nanosecond decay kinetics in Fig. 1. However, since the fast nanosecond component may also contain a contribution from chlorin singlet states formed in active reaction centres, the ϕ_{RP} was also estimated by comparing the relative amplitudes of the long-lived nanosecond component, which is solely attributed to charge recombination of the primary radical pair (7). In both cases, the amplitudes are normalised to 100 % for ϕ_{RP} in the isolated higher plant PS II reaction centres. The results of these analyses are given in Table I. The same results were obtained for either 630 nm or 680 nm excitation. Using 630 nm excitation, all chlorophylls present, whether they be uncoupled or bound to active

	Higher plant	Wild Type	Glutamate	Leucine
Quantum yield of radical pair formation estimated from:				
(i) total amplitude of nanosecond decay kinetics	100 %	60 %	65 %	46 %
(ii) relative amplitude of long-lived nanosecond component	100 %	55 %	81 %	36 %

Table I

Estimated quantum yield of radical pair formation in PS II reaction centres isolated from the higher plant, *Pisum sativum*, and from wild type and mutant *Synechocystis*.

reaction centres, will be excited; using 680 nm excitation, only chlorophyll bound to active reaction centres, will be excited preferentially. Since the nanosecond absorption data is identical for either excitation wavelength, it can be concluded that very little uncoupled chlorophyll is present in these samples. This has been confirmed, at least for the wild type *Synechocystis* PSII preparation, by time-resolved fluorescence measurements (data not shown, but see ref. 5).

From Table I, therefore, it appears that, independent of the way in which the quantum yield of radical pair formation is estimated from the nanosecond data, the relative quantum yield of radical pairs formed in isolated PS II reaction centres, at 1 ns, is in the order: higher plant > glutamate mutant > wild type > leucine mutant. This order is, in fact, apparent by inspection of the nanosecond data presented in Fig. 1. Furthermore, a similar observation has also been made by simulation of the spectra of the non-decaying component resolved in the ultrafast data obtained for these different samples and estimating the contribution of radical pair states to the spectra (5). This analysis give the rather surprising conclusion that PS II isolated from the glutamate mutant has a higher quantum yield of radical pair formation than that isolated from wild type *Synechocystis*. It is not clear whether this difference is present *in vivo* or is a consequence of the isolation procedure. It is interesting to note that Oren-Shamir et al. (8) observe that the quantum yield for the photoaccumulation of Ph^-, in their wild type *Synechocystis* PS II reaction centre preparation, is about half of that activity observed in PS II reaction centres of higher plants.

In conclusion, nanosecond transient absorption measurements indicate that the quantum yield of $P680^+Ph^-$ formation is affected by the identity of the D1-130 residue. We find that, in isolated PS II reaction centres, $\phi_{RP\ \text{higher plant}} > \phi_{RP\ \text{glutamate mutant}} > \phi_{RP\ \text{wild type}} > \phi_{RP\ \text{leucine mutant}}$. What is most remarkable is that the quantum yield of radical pair formation in PS II isolated from wild type *Synechocystis* can be increased by mutating the D1-130 residue so that it is the same as in higher plants.

The financial support of the BBSRC, EC, RITE and the Royal Society is gratefully acknowledged.

References

1) Pakrasi, H.B. and Vermaas, W.F.J. (1992) in The Photosystems: Structure, Function and Molecular Biology (Barber, J. ed.), pp. 231-257, Elsevier Science Publishers.
2) Roffey, R.A., Golbeck, J.H., Hille, C.R. and Sayre, R.T. (1991) Proc. Natl. Acad. Sci. USA 88, 9122-9126.
3) Svensson, B., Vass, I. and Styring, S. (1991) Z. Naturforsch. 46C, 765-776.
4) Michel, H. and Deisenhofer, J. (1988) Biochemistry 27, 1-7.
5) Giorgi, L.B., Nixon, P.J., Merry, S.A.P., Joseph, D.M., Durrant, J.R., De Las Rivas, J. Barber, J., Porter, G. and Klug, D.R. (1995) J. Biol. Chem., submitted.
6) Chapman, D.J., Gounaris, K. and Barber, J. (1991) in Methods in Plant Biochemistry (Rogers, L., ed.) pp. 171-193, Academic Press, London.
7) Booth, P.J., Crystall, B., Ahmad, I., Barber, J., Porter, G. and Klug, D.R. (1991) Biochemistry 30, 7573-7586.
8) Oren-Shamir, M., Sai, P.S.M., Edelman, M. and Scherz, A. (1995) Biochem. 34, 5523-5526.

L PROTEIN, OVEREXPRESSED IN *E.COLI*, RESTORES NORMAL FUNCTIONING OF Q_A IN THE ISOLATED D1/D2/CP47/Cytb-559/I/(H) PHOTOSYSTEM II CORE COMPLEX

Tohru Kobayashi, Shinichiro Ozawa, Takashi Shiina and Yoshinori Toyoshima
Graduate School of Human and Environmental Studies, Kyoto University
Yoshida-nihonmatu-cho, Sakyo-ku, Kyoto 606-01, Japan

1. Introduction

L protein encoded by *psb*L gene in plastid DNA of the higher plants is a component of the PS II reaction center complex. This protein consists of 37 amino acid residues and has a single putative transmembrane α-helix in its carboxyl-terminal region [1,2]. Recently, we purified L protein from spinach PSII particles to examine its role on the photoinduced electron transport through the primary quinone acceptor (Q_A) in PS II. Plastoquinone-9 (PQ-9) reinserted into PQ-9-depleted PSII reaction center core complex consisting of D1/D2/CP47/Cytb-559/I/(H) normally functioned as Q_A, when L protein was simultaneously reconstituted in the complex, suggesting the L protein to be involved in the regulation of the electron transfer at Q_A site in the PSII.

Determination of the amino acid residues in the L protein crucial for the Q_A function is the first step to understand the actual role of the L protein on the electron transfer in the reducing side of PSII. It would be realized, if different types of mutant of L protein were available, since a method of functional reconstitution of the PSII complex with the L protein had been established[3,4]. Therefore, we attempted to construct a system for overproduction of the L protein in *Escherichia coli* by use of an expression vector pMAL-c2[5]. Here we describe the construction of a system for overproduction, purification and assay of the L protein .

2. Materials and methods

2.1 *Overexpression of L protein in E.coli*

A 127 bp DNA fragment containing tobacco *psb*L gene[6] was ligated to XmnI and Bam HI sites of the pMAL-c2 vector to yield a plasmid pMAL-psbL.

E. coli JM109 transformed with plasmid pMAL-psbL was cultured in 100 ml of LB medium containing ampicillin at 37 °C and synthesis of the recombinant protein was induced by isopropyl-β-D-thiogalactopyranoside (IPTG). Cells were harvested, suspended in solution A(20 mM Tris-HCl (pH7.5), 0.2 M NaCl, 1 mM EDTA, 10mM 2-mercaptoethanol, 1 mM NaN_3), sonicated on ice for 1.5 min in total with 30 times interval of 7 sec each and submitted to a centrifugation at 18,000 x g for 20 min at 4 °C. The pellet was solubilized in solution A additionally containing 17.5 mM n-octyl-β-D-glucopyranoside(OG) and 10 mM n-octyl-β-D-thioglucopyranoside (OTG) and treated with an amylose resin (New England Biolabs) column to isolate the recombinant MBP-L fusion protein.

P. Mathis (ed.), Photosynthesis: from Light to Biosphere, Vol. I, 619–622.

The vector pMAL-c2 was designed so that the expressed protein can be cloven to give the desired protein by factor Xa[8]. In this work, however, we used trypsin instead of factor Xa for cleavage of the fusion protein, because we had found that the L protein isolated from spinach PSII will not be digested by trypsin. Since the carboxy-terminus of the linker sequence connecting MBP and the L protein is Arg, the L protein was expected to be cloven from the fusion protein by trypsin digestion. Trypsin digestion of the fusion protein was performed as follows. To the solution containing the fusion protein, trypsin was added to give a final concentration of 0.55 μM and incubated at 37 °C for 24 hr . After addition of trypsin inhibitor (0.64 μM), the solution was incubated for additional 1 hr, dialyzed against a buffered solution of 5mM Tris-HCl, (pH 7.5) for 3 hr at 4 °C and loaded on a DEAE-Sephacel (Pharmacia Biotec.) column equilibrated with solution B (17.5 mM OG, 10 mM OTG, 5 mM Tris-HCl (pH 7.5)). After extensive washing with solution B, the column was subjected to elution with 10 mM NaCl. The fractions which show a single band at 5.0 kDa on SDS-PAGE were collected and finally passed through an S-Sepharose (Pharmacia Biotec.) column to remove trace amount of trypsin from the solution.

2.2 *Assay of Overexpressed L protein*

Plastoquinone-9 (PQ-9) depleted PSII reaction center core complex consisting of D1/D2/CP47/Cytb-559/I/(H) (depleted RC), was isolated from spinach PSII particles by treating the particles with solution containing mixed detergents as described in ref. [3, 9]. Judging from the protein compositions revealed by SDS-PAGE, a part of the depleted RCs in the preparation retained H protein encoded by *psb*H. The crude extract from the PSII particles by the treatment with the mixed detergent solution were collected, from which PQ-9, total lipids and L protein were isolated. Digalactocyldiglyceride(DGDG) was purchased from Serdary Research Laboratories INC and used without further purification .

Reconstitution of the depleted RC with PQ-9 together with the lipids and L proteins from spinach and *E. coli* was carried out changing the amount of each L protein, PQ-9 and the lipids as indicated in ref.[3,4]. To separate out the free PQ-9 and L protein and other proteins such as Cytb-559 dissociated from the RC during the depletion and reconstitution procedure from the assembled complexes, some of reconstituted RC preparations were subjected to a sucrose gradient centrifugation at 40,000 x g for 10 hr in the presence of 0.05% DM at 4 °C and several fractions after centrifugation were collected for the assay of the Q_A activity.

The Q_A activity of the RCs was assayed by measuring photoreduction rate of 2,6-dichlorophenolindophenol (DCPIP) in the presence of diphenylcarbazide (DPC) as an electron donor. Photoreduction rate of DCPIP was determined spectrophotometrically by following absorbance change at 605 nm upon illumination with continuous red light provided by a tungsten lamp of 1 kW through a pair of glass filters (Toshiba IRA-25 and R-65) and a 10 cm water layer as described in ref.[3].

3. Results and discussion

3.1 *L protein was overproduced* in *E. coli*

Our initial attempts to produce the L protein in *E. coli* failed mainly due to instability of the expression product in the bacteria. We overcame this problem with plasmid pMAL-psbL which was designed to produce the L protein as a fusion protein with maltose-binding protein (MBP) in the bacteria under the control of tac promoter. Upon addition of IPTG, one product which migrated at the position of 53kDa on SDS-PAGE was induced in *E. coli* JM 109 transformed with pMAL-psbL and accumulated as the most abundant protein in the cells (10% of the total cell protein content). The 53kDa product collected from the sonicated cells by a centrifugation at 18,000 x g was solubilized by using a mixed detergent solution of 17.5 mM OG and 10 mM OTG and purified by an amylose resin column.

Chromatographic purification of L protein from the fragments produced upon trypsin digestion was performed on a DEAE-Sephacel column. As expected from our previous experience[3], a protein of 5 kDa was eluted with 10 mM NaCl. The fractions which showed a single band at 5 kDa on the silver stained gel (Fraction 2) were pooled. Although any other significant band was not detected on the gel, very weak but detectable trypsin activity remained in this fraction. Trypsin activity was completely removed from the solution by passing the fraction through an S-Sepharose column and the resulting preparation was submitted to the amino-terminal amino acid analysis. The sequence of 18 amino acids from amino-terminal of the 5 kDa protein was TQSNPNEQNVELNRTSLY which agreed with that of the L protein of various plants including spinach reported [9]. Although the carboxy-terminus of the obtained 5 kDa protein was not determined, it migrated on the SDS-PAGE at the exactly same position as the L protein from spinach PSII did, indicating the obtained protein to be equivalent to the spinach L protein in molecular mass, consequently in primary structure.

3.2 *The E.coli L protein recovered the Q_A function in the PSII complex*

The E.coli L protein thus obtained was submitted to the reconstitution experiments to test the activity of the L protein for supporting the recovery of the Q_A function in the PSII core complex. We previously revealed that restoration of the Q_A activity in the PQ-9 depleted PSII reaction center core complex upon reconstitution with PQ-9 was greatly enhanced, when it was carried out together with the spinach L protein. So the effectiveness of the L protein from *E. coli* in restoration of the Q_A activity in the reconstituted PSII complex was compared with the case of the L protein isolated from spinach PSII. The reconstitution of the PQ-9 depleted RC with PQ-9, DGDG and the L proteins from the two different origins were carried out changing the amounts of the L proteins and PQ-9 in the reconstitution medium and the Q_A activities of the original,depleted and reconstituted RCs were measured. In the reconstitution experiments, the amount of DGDG added in the reconstitution medium was fixed to be 120 μg/nmol of Cytb-559 at which the maximum recovery of the Q_A activity had been observed in the previous work. The results are shown in the figure.

The recovery of the Q_A activity was detectable in the RC reconstituted with PQ-9 and DGDG alone, but it did not exceed 11 % of the original level. When the reconstitution was carried out together with the L protein isolated from spinach PSII, the apparent Q_A activity in the resulting RC increased with increasing the amounts of the L protein added in the reconstitution medium and became saturated at about

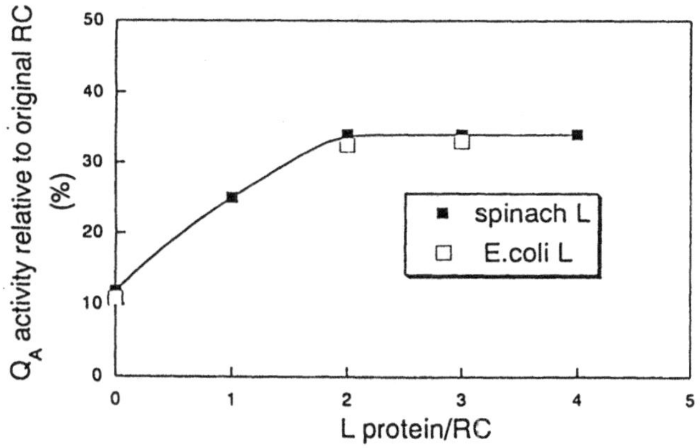

Q_A activity in the reconstituted RCs with spinach and E.coli L protein
The abscissa indicates the amount of L protein added to the reconstitution medium relative to RC.

35 % of the original PSII. The E. coli L protein was also effective and any significant difference was not recognized between the spinach and E. coli L proteins in the effectiveness to recover the Q_A activity in the resulting complexes. In our previous work, the saturation of the apparent Q_A activity at 35 % was suggested to be partially due to the presence of a large amount of the RC complexes which do not assemble with the L protein even after reconstitution. The apparent Q_A activity in the RC reconstituted with the L protein was demonstrated to increase up to 50 % upon partial removal of the unassembled complexes by means of a sucrose gradient centrifugation[3].

The spinach L protein was isolated from the detergent extracts of PSII particles by two step anion exchange column chromatography in the presence of the detergents. The fraction collected as the L protein exhibited only a single band at 5 kDa on the silver stained SDS-PAGE suggesting that it was not contaminated with any other protein. However, the possibility that the unknown components other than the L protein, such as special types of lipids and carbohydrates incidentally eluted in the L protein fraction acted as the actual effective factor to the Q_A activity had not been ruled out. The present works provided the direct evidence for the involvement of the L protein in the Q_A function and a novel approach to understand the mechanism by which the L protein regulates the electron transport in the reducing side of PSII. We are now overproducing several types of mutant L proteins with deletion or point-mutation to be submitted to the reconstitution experiments.

Finally it should be noted that the isolation of the recombinant fusion protein from the membrane fraction of *E. coli* , its digestion with trypsin and purification of the L protein from the total fragments after trypsin digestion of the fusion protein were succeeded in the presence of a mixed detergent system consisting of 17.5 mM OG and 10 mM OTG, but not in the absence of the detergents. Since factor Xa was confirmed to be active in this detergent system, the present expression system might be used to the overproduction of the other membrane proteins. We are now examining the applicability of this system to produce several proteins consisting the PSII complex.

References

[1] Cushman, J.C., Christopher, D.A., Little, M.C., Hallick, R.B. and Price, C.A. (1988) Curr. Genet. 13, 173-180.
[2] Webber, A.N., Hird, S.M, Packman, L.C., Dyer, T.A. and Gray, J.C. (1989) Plant Mol. Biol. 12, 141-151.
[3] Kitamura, K., Ozawa, S., Shiina, T. and Toyoshima, Y. (1994) FEBS Lett. 354, 113-116.
[4] Araga, C., Akabori, K., Sasaki, J., Maeda, A., Shiina, T. and Toyoshima, Y. (1993) Biochim. Biophys. Acta 1142, 36-42.
[5] Maina, C.V., Riggs, P.D., Grandea III, A.G., Slatko, B.E., Moran, L.S., Tagliamonte, J.A., McReynolds, L.A. and Guan, C. (1988) Gene.74, 365-373.
[6] Shinozaki, K., Ohme, M., Tanaka, M., Wakasugi, T., Hayashida, N., Matsubayashi, T., Zaita, N., Chunwongse, J., Obokata, J., Yamaguchi-Shinozaki, K., Ohto, C., Torazawa, K., Meng, B.Y., Sugita, M., Deno, H., Kamogashira, T., Yamada, K., Kusuda, J., Takaiwa, F., Kato, A., Tohdoh, N., Shimada, H. and Sugiura, M. (1986) EMBO J. 5, 2043-2049.
[7] Nagatsuka, T., Fukuhara, S., Akabori, K. and Toyoshima, Y. (1991) Biochim. Biophys. Acta, 1057, 223-231.
[8] Nagai, K. and Thogersen, H.C.(1987) Methods Enzymol. 153,461-481.
[9] Ikeuchi, M.,Takio, K. and Inoue, Y. (1989) FEBS Lett. 242, 263-269.

MECHANISM OF ELECTRON FLOW THROUGH THE Q_B SITE IN PHOTOSYSTEM II. 2.ANALYSIS OF REACTION MECHANISM AT THE Q_B AND PQ SITE IN PHOTOSYSTEM II CORE COMPLEXES

Koike, H., Yamashita, M., Kashino, Y. and Satoh, K.
Dept. Life Sci., Fac. Sci., Himeji Inst. Tech., Harima Science Garden City, Hyogo, 678-12, Japan

1. Introduction

We recently measured oxygen-evolution activities supported by various benzoquinone (BQ) derivatives and estimated binding affinities to elucidate the reaction mechanism and the structure of Q_B-binding pocket in oxygen-evolving photosystem II (PS II) particles prepared from thermophilic cyanobacteria [1,2]. We found that externally added BQ derivatives accept electrons at two different sites: 1) Q_B site; BQ is bound to the Q_B site in place of PQ and accepts two electrons at the site and is replaced by another oxidized BQ molecule. 2) PQ site; BQ is reduced *via* plastoquinol (PQH_2) which has been reduced at the Q_B site.
Similar experiments using PS II core complexes from higher plants had been quite difficult because the Q_B site in those complexes had been shown to be modified as indicated by the change of dependency of oxygen evolution on electron acceptor [3]. Recently, however, we succeeded in isolation of PS II core complexes from spinach which still retain most of the Q_B intact [4]. Furthermore, the binding affinities of the Q_B site in these complexes for various quinones has been found to be almost the same as those in their starting material, PS II membrane fragments [5].
In the present work, we show that, in the PS II core complexes prepared from spinach, BQs are reduced at the two sites as in the case of cyanobacterial PS II particles.

2. Materials and Methods

PS II membrane fragments were prepared from spinach as described in [6] with slight modifications. PS II core complexes were prepared by treating the PS II membrane fragments with heptylthioglucoside (HTG) as shown in [7] with slight modifications [4]; PS II membrane fragments were treated with 1.4 - 2.0 % HTG in a buffer containing 20 mM $NaHCO_3$ to keep PQ bound at the Q_B site. The HTG particles were finally precipitated by centrifugation (35,000 *g*, 15 min) and suspended in a medium containing 0.4 M sucrose/10 mM NaCl/5 mM $CaCl_2$/40 mM 2-*N*-morpholinoethane sulfonic acid (Mes)-NaOH (pH 6.0) and kept frozen at -85 ℃ until use. The PS II complexes retain three extrinsic proteins and intact Q_B.
Oxygen evolution was measured at 30 ℃ using a Clark-type oxygen electrode

P. Mathis (ed.), Photosynthesis: from Light to Biosphere, Vol. I, 623–626.

(Rank-Brothers co. ltd.) with saturating yellow light. The reaction mixture was the same as that used to suspend HTG particles except various amounts of BQs were added [2].

3. Results

PS II extracts electrons and releases H^+ from H_2O using light energy and transfers electrons to PQ. As pointed out by Satoh et al. [8], the PS II particle is considered as water:plastoquinone oxidoreductase. The externally added BQs which act as electron acceptors are thus considered as substrates for the oxidoreductase. If this is the case, oxygen-evolving activities measured as a function of concentrations of BQs must obey Michaelis-Menten type kinetics [1,2].

Fig. 1A shows a double reciprocal plots of rates of oxygen evolution as a function of 2,3-dimethyl-*p*-benzoquinone (2,3-DMBQ) concentration (Lineweaver-Burk plots). The plot yielded a straight line in the absence of 3-(3,4-dichlorophenyl)-1,1-dimethylurea (DCMU). Intersections with the ordinate and abscissa indicate V_{max} and K_m for the reaction, respectively. In the presence of 80 nM DCMU, at which concentration oxygen evolution activity is only partially inhibited, the plot yielded another straight line which shared the same V_{max} value. A set of the plots indicates that DCMU acts as a competitive inhibitor for 2,3-DMBQ. Since DCMU is known to bind to the Q_B site and replaces PQ resulting in the inhibition of electron transport [9], the results suggest that 2,3-DMBQ binds to the Q_B site and accepts electrons from $Q_A^{\cdot}$.

When tetramethyl-*p*-benzoquinone (duroquinone, DQ) is used as an electron acceptor, the double reciprocal plots in the absence and presence (80 nM) of DCMU also yielded two straight lines. However, the mode of inhibition was different from that with 2,3-DMBQ. Intersection with the abscissa shared the same value (K_m), but that with ordinate was different (Fig. 1B). The plot showed a typical non-competitive inhibition. This indicates that the site where DQ accepts electrons is different from the DCMU- binding site. This result can be explained if we assume that DQ does not have a binding affinity to the Q_B site in higher plants as shown in PS II particles from cyanobacterium [2]. In this case, it is considered that DQ is reduced *via* PQ which was retained in the Q_B site and DCMU acts as an inhibitor which interrupts electron transfer before the site of DQ reduction.

Double reciprocal plots of other quinone derivatives such as trimethyl-*p*-benzoquinone, mono- or dichloro-*p*-benzoquinone showed complex features. Fig. 1C shows one of the examples in which 2,6-dichloro-*p*-benzoquinone (2,6-DCBQ) is employed as an electron acceptor. The plots were composed of two straight lines both in the absence and presence of 80 nM DCMU. When the concentration of 2,6-DCBQ was increased, the activity deviated from a straight line giving rise to a decrease in oxygen-evolution rate. The phenomenon is explained if 2,6-DCBQ accepts electrons both at Q_B site and *via* PQH_2 which had been reduced at the Q_B site (denoted as PQ site). At low concentrations of 2,6-DCBQ, electron transfer *via* PQH_2 is dominant, while as increasing the concentration of 2,6-DCBQ, the electron transfer reaction at the Q_B site becomes dominant. If the electron transfer rate to 2,6-DCBQ at the Q_B site is lower than that *via* PQH_2, the rate should decrease at higher concentrations because electron flow *via* PQH_2 must be inhibited by the binding of 2,6-DCBQ to the Q_B site.

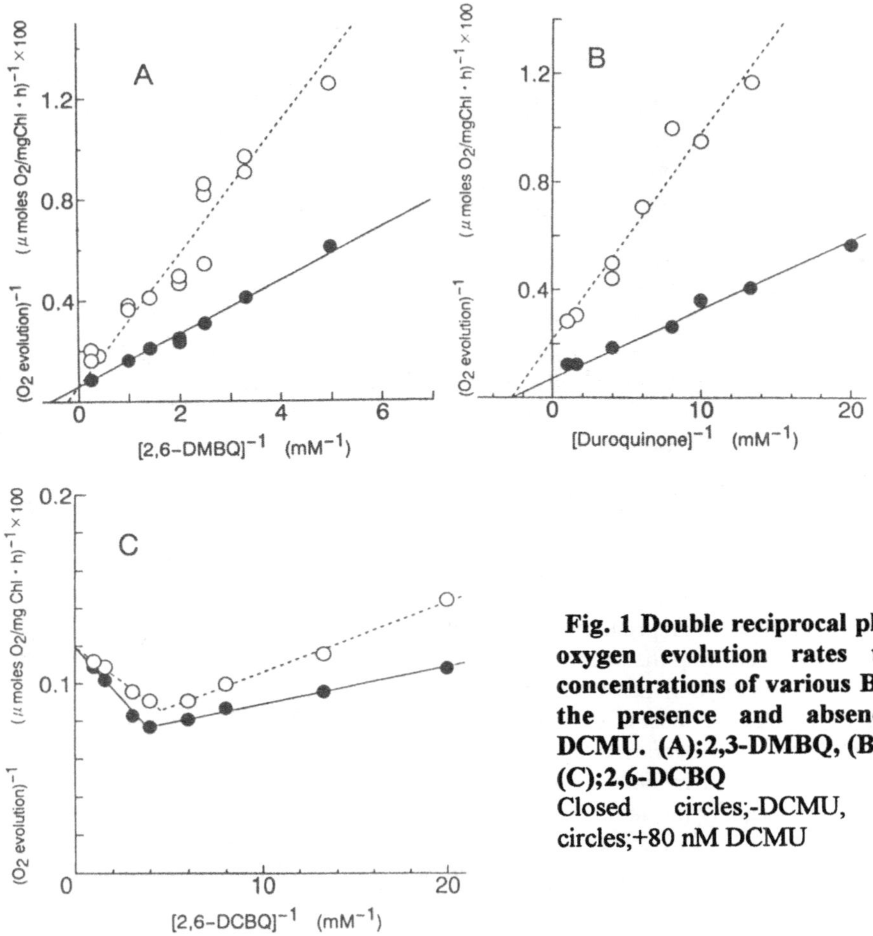

Fig. 1 Double reciprocal plots of oxygen evolution rates *versus* concentrations of various BQs in the presence and absence of DCMU. (A);2,3-DMBQ, (B);DQ, (C);2,6-DCBQ
Closed circles;-DCMU, open circles;+80 nM DCMU

The deviation does not always result in a decrease in the absolute oxygen-evolution activity. When monochloro-*p*-benzoquinone (MCBQ) is used as an acceptor, the total activity was increased at higher concentrations. This is also explained by the assumption that MCBQ could be reduced at two sites (the Q_B site and PQ site), while V_{max} at the Q_B site is higher than that *via* PQH_2. It was indicated that most of the BQs are reduced at two sites in the reducing side of PS II.

4. Discussion

Benzoquinone derivatives tested in the present work is classified into three groups; 1) BQs belonging to the first group such as 2,3-DMBQ or 2,6-DMBQ interact only at the Q_B site. DCMU acts as a competitive inhibitor in this case. 2) The second group contains only DQ which is reduced only at the PQ site. DCMU in the presence of DQ acts as a non-competitive inhibitor. 3) Most of the BQs tested are classified into the third group, and are assumed to be reduced through the both sites.

The following reaction scheme can be drawn for concentration dependence of BQs in the third group. At low concentrations of BQs, the reduction of BQs proceeds *via* both the PQ and the Q_B site. When concentrations of BQs are increased, reduction of BQs at Q_B site becomes prominent and if the quinone has high affinity to the site, inhibition of electron flow through the endogenous plastoquinone becomes clear. At the infinite concentration, all the Q_B site is occupied by BQ so that the rate can be expressed by V_{max} at Q_B site. Using computer simulation and dependency of oxygen evolution on quinone concentrations, we have recently introduced a method to calculate the rate of the reduction of endogenous plastoquinone [2]. The obtained value was 2,800 μ mol O_2/(mg Chl · h) which corresponded to 1,000 μ mol O_2/(mg Chl · h) in PS II membrane fragments (Kashino et al. in preparation).

The values for K_m and V_{max} at PQ site were not obtained for 2,3- and 2,6-DMBQ, since the double reciprocal plot yielded a single straight line and DCMU acted as a competitive inhibitor. This might be due to a fairly low affinity of BQs at the PQ site to measure by oxygen electrode. When electron transport rate was measured by monitoring the reduction of 2,3-DMBQ spectrophotometrically (Hill reaction), double reciprocal plots were composed of two straight lines (data not shown).

2,3-DMBQ is, on the other hand, considered as a PQ derivative in which it has no isoprene side chain in a quinone ring. The activity of electron transport of other PQ derivatives in which the number of isoprene unit was changed from 1 to 3 (PQ_1 - PQ_3) could also be measured by monitoring the reduction of quinones spectrophotometrically (to be published elsewhere). The PQ derivatives tested have affinities both at Q_B and PQ site.

We have indicated that most of BQs are reduced at two sites in PS II particles prepared from a cyanobacterium, *Synechococcus vulcanus* [2]. Taking into account the facts that both particles do not have membraneous structure, the affinities of BQs to the Q_B site of the particles is principally the same between *S. vulcanus* and spinach if the site is not modified by the detergent used to isolate the particles. We could deduce from these facts that the configuration of Q_B site is essentially the same between cyanobacteria and higher plants.

References

1. Satoh, K., Koike, H., Ichimura, T. and Katoh, S. (1992), *Biochim. Biphys. Acta,* 1102:45-52
2. Satoh, K., Oh-hashi, M., Kashino, Y. and Koike, H. (1995) *Plant Cell Physio*l. 36:597-605
3. Ikeuchi, M. and Inoue, Y. (1986) *Arch. Biochem. Biophys.* 247:97-107
4. Kashino, Y., Koike, H. and Satoh, K. in *Research in Photosynthesis* (Murata, N. ed.) vol.II,pp.163-166, Kluwer Academic Publishers
5. Kashino, Y., Yamashita, M., Koike H. and Satoh, K. (1995) in *Proc. Xth Internatl. Congress Photosyn.* in press
6. Berthold, D.A., Babcock, G.T. and Yocum, C.F. (1981) *FEBS Lett.* 134:231-234
7. Enami, I., Kamino, K., Shen, J.-R., Satoh, K. and Katoh, S. (1989) *Biochim. Biphys. Acta* 977:33-39
8. Satoh, K., Ohno, T. and Katoh, S. (1985) *FEBS Lett.* 180:326-330
9. Babcock, G.T., in *Photosynthesis* (1987) (J. Amesz ed.) pp.125-158, Elsevier, Amsterdam

ISOLATION OF PHOTOSYSTEM II CORE COMPLEXES FROM A CYANOBACTERIUM WITHOUT USING COLUMN CHROMATOGRAPHY

Hirokazu Egashira, Yasuhiro Kashino, Hiroyuki Koike & Kazuhiko Satoh
Department of Life Science, Faculty of Science, Himeji Institute of Technology, Harima Science Garden City, Hyogo 678-12, Japan.

1. Introduction

Isolation of oxygen-evolving photosystem II (PS II) core complexes had been performed by many workers from higher plants and cyanobacteria [1~3]. Recently, however, it was shown that the plastoquinone molecules at the Q_B sites of these particles had been removed [4] and, irrespective of their high oxygen-evolving activities, these sites had already been modified as shown by the change in the dependency of the oxygen evolution on electron acceptors [3]. This modification seems to be induced during treatments of the complexes with high concentrations of detergents or during detergent treatments at the purification step using an ion exchange column. These processes had been thought to be necessary because PS II core complexes in higher plants are bound tightly to large amounts of light-harvesting chlorophyll *a/b* protein complexes and PS II core complexes in cyanobacteria are associated with huge phycobiliprotein complexes, phycobilisomes. Here we will report a method to isolate and purify PS II core complexes from the thermophilic cyanobacterium, *Synechococcus vulcanus*, without using column chromatography.

2. Procedure

2.1 Cell culture

The thermophilic cyanobacterium, *S. vulcanus*, was grown at 55 ℃ for 2 days as described in [5].

2.2 Isolation of thylakoid membranes

The collected cells were treated with lysozyme and thylakoid membranes were prepared as reported by Koike and Inoue [6]. Thylakoid membranes were stored at - 80 ℃ before

P. Mathis (ed.), Photosynthesis: from Light to Biosphere, Vol. I, 627–630.

use.

2.3 Isolation of crude PS II core complexes

All the isolation procedures were carried out at 4 ℃ or on ice if otherwise noted. The thylakoid membranes were suspended in the solution A (50 mM Mes-NaOH (pH 6.7), 10 mM NaCl, and 5.0 mM $MgCl_2$) added with 0.20 M sucrose and then centrifuged at 35,000*g* for 15 min to remove unbound phycobiliproteins (PB). The pellet was resuspended in solution A with 0.20 M sucrose to give chlorophyll *a* (chl *a*) concentration of 2 mg/ml. For further removal of PB, the washed thylakoids were treated with 0.60 % (w/v) Triton X-100 for 10 min at 20 ℃ in the dark. After centrifugation (35,000*g* 15 min), the pellet was washed 3 times using solution A with 0.20 M sucrose to remove the remaining Triton X-100. The triton-treated thylakoid membranes (2 mg chl.*a*/ml) were further treated with 0.90 % (w/v) n-heptyl- β-D-thioglucoside (HTG) and were incubated for 10 min at 20 ℃ in the dark. Unsolubilized thylakoids and PS I particles were precipitated by centrifugation (35,000*g* 15 min), and PS II core complexes were collected from the supernatant by ultra centrifugation (300,000*g* 1.5 hr). These preparations were mostly free from PS I particles but were contaminated with PB and ATPase complexes (see Fig. 1).

2.4 Other procedures

Polypeptide compositions of various preparations were analyzed by sodium dodecylsulfate-polyacrylamide gel electrophoresis (SDS-PAGE) according to the method of Laemmli [7] as modified by Ikeuchi and Inoue [8]. Polypeptides on the gel were visualized by staining proteins with coomassie brilliant blue R-250. Oxygen-evolving activities were measured with a Clark-type oxygen electrode at 30 ℃ in solution A added with 1.0 M sucrose [9].

3. Result and Discussion

3.1 Purification of PS II core complexes

The crude PS II core complexes were layered onto a linear sucrose gradient (0.40 ~ 1.2 M sucrose in solution A with 0.90 % (w/v) HTG), and after ultra centrifugation (300,000*g* 2 hr), a dark green band appeared at sucrose concentrations of 0.75 M. This step reduced the amount of ATPase and some minor contaminating proteins (see Fig. 1, lane 4). Fractionation using ammonium sulfate was found to be very efficient in removing the PB (Fig. 1, lane 5). After replacing the detergent, HTG, with 0.80 % (w/v) n-octyl- β -D-glucoside (OGP), most of the PB was precipitated by addition of 30 % saturation of ammonium sulfate, and PS II core complexes were collected as a fraction of 50 ~ 60 % saturation of ammonium sulfate. The remaining ammonium sulfate was washed out by ultra centrifugation (300,000*g* 1 hr). After the ultra centrifugation, the pellet was resuspended in solution A contained 0.90 %(w/v) HTG and 0.20 M sucrose. The α, β subunit bands of ATPase were removed almost completely by the second sucrose gradient centrifugation (Fig. 1, last lane). The ammonium sulfate fraction was layered onto a linear sucrose gradient (0.40 ~ 1.2 M sucrose in solution A with 0.90 %(w/v) HTG) and the dark green band was collected at approx. 0.75 M sucrose after ultra centrifugation (300,000*g* 2 hr). The PS II core complexes were considerably purified from the view point of polypeptide compositions. Note that an unknown band at around 10 kDa increased its density with the increase in the purity of the PS II core complexes. There also remains some other unknown bands at around 15 kDa and 45 kDa.

3.2 Oxygen-evolving activity

Table.1 shows oxygen-evolving activities of the PS II preparations at each purification step. In this study, we used 2,5-dichlorobenzoquinone (2,5-DCBQ) or duroquinone (DQ) as an electron acceptor. Recently, Satoh et al. [10] showed that DQ has almost no affinity to the Q_B site in PS II particles. In other words, oxygen-evolving activities can not be observed if there is no Q_B left in the particles when DQ is used as an electron acceptor. Although the activity reduced to about a half of the crude preparation, the purified core complexes still showed high oxygen-evolving activity with 2,5-DCBQ and even with DQ as electron acceptors. The oxygen evolution was severely inhibited by 3-(3,4-dichlorophenyl)-1,1-dimetylurea (DCMU) and the ratio of the activities with 2,5-DCBQ and DQ as electron acceptors remained almost constant after isolation of PS II core complexes, showing that the Q_B site was not largely modified [3]. Flash-induced absorbance changes due to redox changes of Q_A suggested at least 50 % of the Q_B site of the purified PS II core complexes is intact and maintained plastoquinone at this site (data not shown).

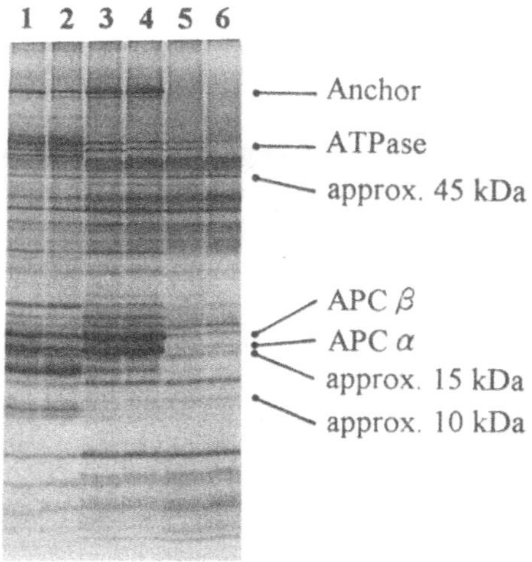

Fig. 1 Polypeptide patterns of PS II preparations at each purification step.
Lane 1, thylakoids; lane 2, triton-treated thyakoids; lane 3, HTG extracts; lane 4, a sucrose density gradient fraction; lane 5, ammonium sulfate fraction; lane 6, a final preparation. The gel was stained with CBB.

APC: Allophycocyanin

Table 1. Oxygen-evolving activities of PS II preparations at each purification step.

Preparation	2, 5-DCBQ -DCMU	2, 5-DCBQ +DCMU	DQ
Thylakoids	990	50	220
Triton-washed Thylakoids	990	0	220
HTG Extracts	2450	20	1030
A Sucrose Gradient Fraction	2940	50	1140
An Ammonium Sulfate Fraction	1030	20	430
A Final Preparation	1020	20	400

(μ moles O_2 / Chl *a* • hr)

4. Conclusions

PS II core complexes were purified from the thermophilic cyanobacterium, *S. vulcanus*, without using column chromatography. The purified PS II core complexes had high oxygen-evolving activity and retain Q_B-plastquinone at about 50 % of the Q_B site. The polypeptide composition of the core complexes showed that some unknown polypeptides are associated with the complexes.

References

1. Satoh, K., Ohno, T. and Katoh, S. (1985) FEBS Lett. 180, pp. 326-330.
2. Ikeuchi, M., Yuasa, M. and Inoue, Y. (1985) FEBS Lett. 185, pp. 316-322.
3. Tang, X.-S. and Satoh, K. (1985) FEBS Lett. 179, pp. 60-64.
4. Kashino, Y., Koike, H. and Satoh, K. (1993) in; Research in Photosynthesis, ed. by N. Murata, vol. 2, pp. 163-166, Kluwer Academic Publishers.
5. Yamaoka, T., Satoh, K. and Katoh, S. (1978) Plant Cell Physiol. 19, pp. 943-954.
6. Koike, H. and Inoue, Y. (1983) in; The Oxygen Evolving System of Photosynthesis (Y. Inoue, A.R. Crofts, N. Govindjee Murata, G. Renger and K. Satoh, eds.) pp. 257-263. Academic Press, Tokyo.
7. Laemmli, U.K. (1970) Nature. 227, pp. 680-685.
8. Ikeuchi, M. and Inoue, Y. (1988) Plant Cell Physiol. 29, pp. 1233-1239.
9. Katoh, S., Satoh, K., Yamagishi, A. and Yamaoka, T. (1975) Plant Cell Physiol.
10. Satoh,K., Oh-hashi, M., Kashino, Y. and Koike, H. (1995) Plant Cell Physiol. 36(4), pp. 597-605.

PROPERTIES OF n-HEPTYL-β-D-THIOGLUGOSIDE-EXTRACTED PHOTOSYSTEM II CORE COMPLEXES FROM SPINACH.

Yasuhiro Kashino, Mari Yamashita, Hiroyuki Koike and Kazuhiko Satoh
Department of Life Science, Faculty of Science, Himeji Institute of Technology, Harima Science Garden City, Hyogo 678-12, Japan

1. Introduction

It is the primary necessity to obtain a photosystem II (PS II) preparation as intact as possible to investigate its precise structure or function. PS II enriched membrane fragments prepared from higher plants using Triton X-100 (so called BBY) [1] are very good preparations in that sense. But, they maintain large amounts of light-harvesting chlorophyll *a/b* protein (LHCII). Enami et al [2] reported a smart method to remove most of the LHCII from BBYs using n-heptyl-β-D-thioglucoside (HTG). However, we found that the PS II core complexes were damaged at the Q_B site. We have, therefore, introduced a modified method to isolate PS II core complexes (HTG-PS II particles) which still retain most of the Q_B site intact. The point is to add sodium bicarbonate to the isolation and dialysis mediums [3]. Here, we will show the binding affinity of the Q_B site for 2,6-dichlorobenzoquinone (DCBQ) in the PS II core complexes is almost the same as that in BBY, and show that our PS II preparation is useful for the study of PS II.

2. Materials and Methods

HTG-PS II particles were prepared as described in [3] from spinach grown at a local farm. In brief, BBYs (3 mg Chl/ml) were treated with 1.7% (w/v) HTG (Dojindo, Kumamoto, Japan) in a medium containing 20 mM $NaHCO_3$, 20 mM Hepes-NaOH (pH7.0), 0.4 M NaCl, 5 mM $CaCl_2$ and 1 M sucrose for 10 min on ice in the dark. After addition of 2 volumes of solution containing 20 mM $NaHCO_3$, 20 mM Hepes-NaOH (pH7.0), 0.4 M NaCl, 5 mM $CaCl_2$ and 0.5 M sucrose, the mixture was kept on ice in the dark for 5 min and then centrifuged at 35,000xg for 15 min. The resultant supernatant was dialyzed against a solution containing 20 mM $NaHCO_3$, 40 mM Mes, 10 mM NaCl and 0.4 M sucrose for 3 h (The dialysis solution becomes pH6.0 in this condition without adjustment. The solution was changed every one hour.), and subsequently centrifuged at 35, 000xg for 15 min.

Polypeptide profiles were analyzed by SDS-polyacrylamide gel electrophoresis as reported by Ikeuchi et al. using gels containing 16-22% acrylamide and 7.5 M urea [4].

Absorbance changes of Q_A at 415 nm induced by a series of one hundred flashes with 10 μs of half-height width were detected with a Photal flash spectrophotometer (Otsuka electronics, Osaka, Japan). The flash was fired at 1 Hz and each trace was stored and averaged. The HTG-PS II particles were diluted to 3.43 μg Chl/ml in a reaction medium containing 0.4 M sucrose, 10 mM NaCl, 5 mM $CaCl_2$ and 40 mM Mes-NaOH (pH6.0) and the reaction was measured at 25°C.

P. Mathis (ed.), Photosynthesis: from Light to Biosphere, Vol. I, 631–634.

Flash-induced fluorescence intensity changes were monitored by a Pulse-Amplitude-Modulator (PAM) manufactured by Heinz Waltz (Effeltrich, Germany). The data were transferred to a Panasonic oscilloscope VP570A and 64 traces were averaged. The reaction mixture and temperature were the same as above except the concentration of HTG-PS II was increased to 9.3 μg Chl/ml. The concentration of BBY corresponded to 21 μg Chl/ml.

3. *Results and Discussion*

Figure 1 shows the polypeptide profiles of BBY and HTG-PS II particles. LHCII is the major component in BBY (lane 1). In contrast, HTG-PS II exhibits simple subunit composition and contains almost no LHCII reflecting high chlorophyll *a* to *b* ratio ordinarily over 12, although putative CP29 is present (lane 3). As indicated in [4], the presence of sodium bicarbonate during the isolation procedure was useful to keep the Q_B site intact. However, polypeptide profiles have not been changed by the presence of sodium bicarbonate. This means that only local modifications occurred in the absence of bicarbonate. The only notable difference is that the band of D1 and D2 heterodimer is one of dominant bands in lane 2 (indicated by arrow head). Lane 4 shows the polypeptide pattern of PS II complexes precipitated before dialysis. Peripheral polypeptides such as 24 and 17 kDa proteins are removed because of a high salt concentration, which means that the decrease of the salt concentration during dialysis restores those polypeptides to their original position. On the other hand, 33 kDa protein, which is also a peripheral protein, remains to be attached to the PS II complexes. This is why the HTG-PS II particles give high rates of oxygen evolution, implying intactness of the donor side of PS II.

To investigate the nature of the acceptor side in HTG-PS II, binding affinities of the Q_B site and the turnover rate of plastoquinone at this site were determined by means of the measurements of flash-induced absorbance changes of Q_A and fluorescence decay curves.

The 3-(3,4-dichlorophenyl)-1,1-dimethylurea (DCMU) inhibits electron flow at the Q_B site in competition with Q_B. One can obtain competition coefficient of analogous electron acceptors such as benzoquinone against DCMU by measuring the shift of I_{50} value (half inhibition concentration) of DCMU for the electron flow from Q_A^- to Q_B [5]. The fast decay of absorbance at 415 nm after flash-induced rapid increase was attributed to oxidation of Q_A^- by Q_B. Actually, this fast decay component is completely suppressed by the addition of 10 μM DCMU (data not

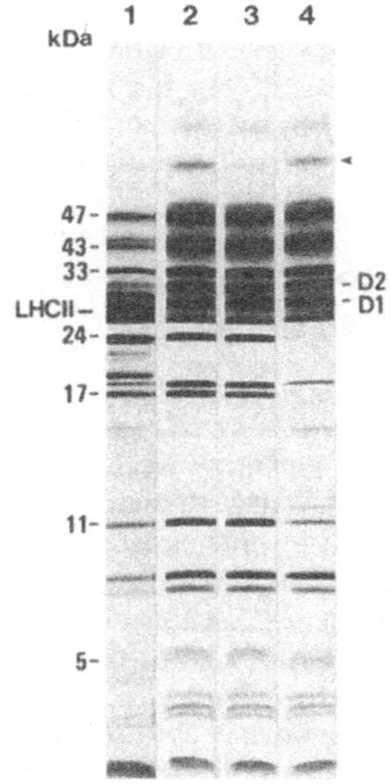

Figure 1. Polypeptide profiles of PS II particles prepared using HTG. Lane 1, BBY; lanes 2 and 3, HTG-PS II prepared without and with $NaHCO_3$, respectively; lane 4, HTG-PS II particles before dialysis. Gel contained 16-22% acrylamide and 7.5 M urea. 6 μg Chl were loaded to each lane.

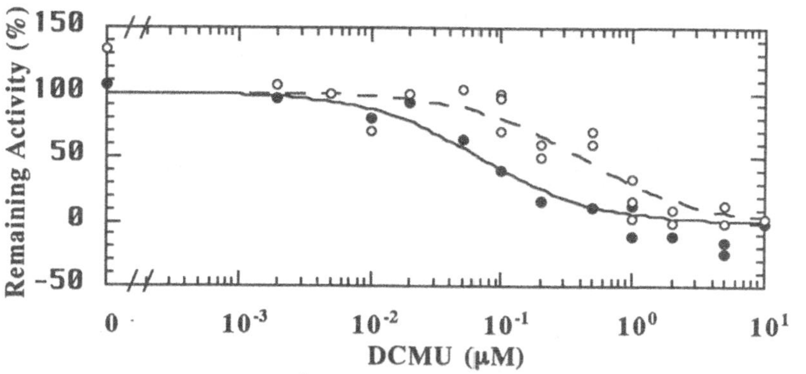

Figure 2. Inhibition curves of Q_A^- oxidation by DCMU measured by decrease of the fast absorbance decay component at 415 nm at 25 °C. The reaction mixture contained 0.4 M sucrose, 10 mM NaCl, 5 mM $CaCl_2$, 40 mM Mes-NaOH (pH6.0) and HTG-PS II of 3.43 μg Chl/ml. ●; 0.1 mM ferricyanide, ○; 0.5 mM 2,6-DCBQ.

shown, but see Figure 2). The degree of inhibition depends on the concentration of DCMU, and the inhibition curve shifted to higher concentration of DCMU compared to ferricyanide. Ferricyanide is supposed not to bind to the Q_B site and do not compete with DCMU [6]. The increase of I_{50} corresponds to the degree of competition between DCMU and benzoquinone at the Q_B site. So, *Kb*, which corresponds to *Km*, can be calculated by the ratio of I'_{50} in the presence of additional quinones to I_{50} in the presence of ferricyanide using the following formula [5].

$$Kb = \frac{[BQ]}{I'_{50}/I_{50} - 1}$$

Here, [BQ] represents the concentration of benzoquinone derivatives used. The *Kb* values for 2,5- and 2,6-DCBQ in HTG-PS II were calculated to be 0.2 mM and 0.1 mM, respectively.

It is very difficult to determine the *Kb* values in BBY by this method because the absorbance changes are quite small on a chlorophyll basis due to a large antenna size. Therefore, we tried to measure the electron flow from Q_A^- to Q_B by fluorescence decay curves using a PAM system.

The variable fluorescence from chlorophyll *a* reflects redox changes of Q_A. The flash-induced fluorescence increase was followed by a fast decay with a half time of hundreds microseconds as is the case of absorption change of Q_A (data not shown). This fluorescence decay component is also inhibited by DCMU indicating that it surely reflects electron flow from Q_A to Q_B (Figure 3). I'_{50} in the presence of benzoquinone is higher than I_{50} in the presence of ferricyanide. Using these I'_{50} and I_{50} values, *K'b* was also calculated by the formula shown before. It was necessary to use higher chlorophyll concentrations because of a technical reason. The value of *K'b* did not depend on the PS II concentration used (data not shown) and the obtained values were 0.4 mM for 2,5-DCBQ and 0.2 mM for 2,6-DCBQ in the case of BBY.

Using *Kb* values of 2,6-DCBQ, being 0.2 mM for BBY and 0.1 mM for HTG-PS II, the rate of the electron flow through the QB site was calculated by computer simulation. The simulation program was the same as used in [5] but was slightly modified. The rate was 1,000 μmoles O_2/mg chl•h for BBY and 2,800 μmoles O_2/mg chl•h for HTG-PS II [7]. These values are shown on a chlorophyll basis. So, these are comparable to each other because the amount of chlorophyll per reaction center of BBY is about 3 times higher than that of HTG-PS II [2]. This indicate that the damage of HTG-PS II particle caused during the preparation process is almost negligible and that the removal of membrane and LHCII does not influence the nature of the QB site.

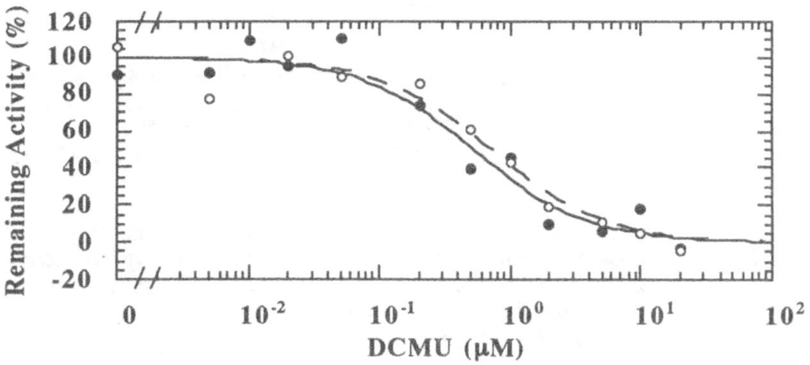

Figure 3. Inhibition curves of fuluorescence fast decay component by DCMU. The reaction mixture and temperature were the same as Figure 2. The concentration of BBY was 21μg Chl/ml. ●; 0.1 mM ferricyanide, ○; 50μM 2,6-DCBQ.

References

1. Berthold, D. A., Babcock, G.T. and Yocum, C. F. (1081) *FEBS Lett. 134*: 231-234.
2. Enami, I., Kamino, K., Shen, J. -R, Satoh, K. and Katoh, S. (1989) *Biochim. Biophys. Acta 977*: 33-39.
3. Kashino, Y., Koike, H. and Satoh, K. (1992) *Research in Photosynthesis* (edited by Murata, N., Kluwer, Dordrecht), vol II; 163-166.
4. Ikeuchi, M and Inoue, Y (1988) *Plant and Cell Physiology 29*: 1233-1239
5. Satoh, K., Oh-hashi, M., Kashino, Y. and Koike, H. (1995) *Plant and Cell Physiology 36*: 597-605.
6. Satoh, K., Koike, H., Ichimura, T. and Katoh, S. (1992) *Biochim. Biophys. Acta 1102*: 45-52.
7. Koike, H., Yoneyama, X, Yamashita, M., Kashino, Y. and Satoh, K. (1996) *in this book.*

INVESTIGATION OF THE BINDING DOMAIN IN THE *RHODOBACTER SPHAEROIDES* CYT C_2:REACTION CENTER COMPLEX BY SITE-DIRECTED MUTAGENESIS OF ASP-M184 AND ASP-L155 TO LYS.

Rongey, S. H., Feher, G. and Okamura, M. Y.
University of California, San Diego, Physics Dept.
9500 Gilman Drive, La Jolla, Ca. 92093

INTRODUCTION

In *Rb. sphaeroides* the water soluble cytochrome (cyt) c_2 shuttles electrons to the membrane bound Reaction Center (RC). Electron transfer from cyt c_2 to the RC is biphasic: there is a first order phase ($\tau \cong 1\mu s$) due to cyt c_2 bound to the RC and a second order phase ($k_0=10^9\ M^{-1}s^{-1}$) which is limited by the collision rate between the cyt c_2 and the RC (1). Cyt c_2 and the RC are attracted to one another at long range by electrostatic interactions; both the collision rate and binding affinity for cyt c_2 and the RC decrease with increasing ionic strength (see *e.g.* 2, 3). Cyt c_2 binding occurs on the periplasmic surface of the RC (4). Chemical modification of Lys residues on the surface of cyt c_2 demonstrated that residues surrounding the heme crevice are important for the association of cyt c_2 with the RC (5). Thus, the Lys residues surrounding the cyt c_2 heme crevice are thought to interact electrostatically with Asp and Glu residues on the periplasmic surface of the RC(6-8).

To identify residues on the periplasmic surface of the RC that govern the attraction and binding of cyt c_2, Asp residues were changed to Lys residues using site-directed mutagenesis. The Asp → Lys changes were made at two positions: L155 and M184. The resulting mutants are: DK(L155) (Asp-L155 → Lys) and DK(M184) (Asp-M184 → Lys). These residues were targeted because computer docking models predict both residues to interact with Lys residues on cyt c_2 (6,7), and because they are related to each other by the pseudo 2-fold symmetry axis of the RC (see Fig. 1). The effects of these mutations on the RC-cyt c_2 interaction and electron transfer are discussed in this work. Specifically, regions important for attraction and binding of the two proteins are determined, and compared to different docking models.

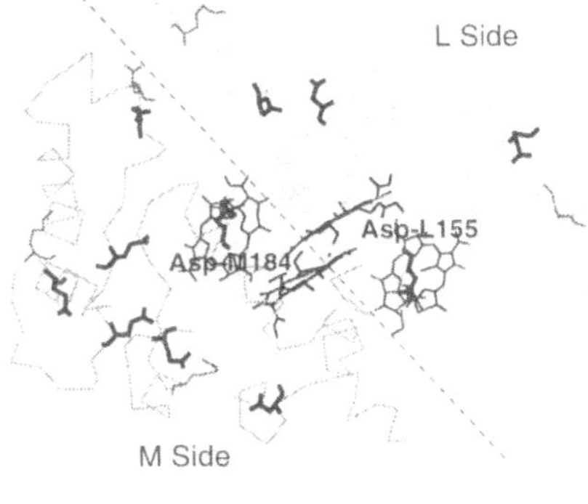

Figure 1 *View along the membrane normal facing the periplasmic surface of the RC. The dotted line separates the L and M subunits. Asp and Glu residues are shown with thick lines; Lys residues are shown with thin lines.*

P. Mathis (ed.), Photosynthesis: from Light to Biosphere, Vol. I, 635–638.

MATERIALS AND METHODS

Sample preparation The construction of the site-directed mutants was performed as described (9,10). The mutations were incorporated into a M13 vector containing portions of the *puf* operon coding for the L or M subunits of the RC. The mutated DNA sequences were introduced into a pRK vector, reconstructing the *puf* operon (9, 10) and then transferred into the *Rb. sphaeroides* deletion strain ΔLM1. RCs were isolated from semi-aerobically grown cells as described (11). Cyt c_2 was isolated from *Rb. sphaeroides* as described (12) and was ≥ 98% reduced. Protein purity was characterized by optical absorbance ratios: $A_{280} / A_{802} < 1.30$ for RCs and $A_{280} / A_{417} \leq 0.25$ for cyt c_2. Q_0 (2,3-dimethoxy-5-methyl benzoquinone) was obtained from Aldrich and is >99% pure. Q_0H_2 was formed by reducing Q_0 with hydrogen gas in the presence of platinum black (Aldrich).

Electron transfer rate measurements The kinetics of the absorption changes accompanying electron transfer were recorded on a modified Cary 14 spectrophotometer (Varian). All measurements were performed at 23 °C in 10mM Hepes, 0.04% β-D-maltoside and 0.1mM EDTA at pH 7.5. Prior to the measurement 0.1mM Q_0 and 0.1mM Q_0H_2 were added as a redox buffer, ensuring that cyt c_2 was reduced before each laser flash. The reaction of cyt c_2 with the RC was monitored optically at either 595nm or 865nm. From the kinetics of the optical absorption changes the rate of electron transfer from reduced cyt c_2 to the photooxidized donor was determined. Photo-excitation of the RCs was accomplished with either a ruby laser (694nm) or a chemical dye laser (590nm, rhodamine-6G dye). The fast kinetics (μs) were measured at 595nm with the ruby laser. The slower kinetics and the dissociation constant (K_D) of the RC:cyt c_2 complex was determined using either laser and monitoring at 595nm or 865nm.

The reaction of cyt c_2 with the RC was modeled using a single binding site, with a dissociation constant K_D. In the dark, an equilibrium is reached between the free proteins and the bound complex. Upon excitation by a short light pulse, D is oxidized to D^+. The equilibrium fraction of RCs with a bound cyt c_2 will have D^+ rapidly reduced to D, oxidizing cyt $c_2^{2+} \rightarrow$ cyt c_2^{3+} in the process. The RCs without cyt c_2 bound will react with free cyt c_2, by a second-order diffusion process. The fraction of RCs with cyt c_2 bound is determined from the amplitude of the fast phase. The concentration of free cyt c_2 ($[C]_{free}$) is calculated by subtracting the amount of cyt c_2 that reacted in the fast phase (the RC and cyt c_2 react in a 1:1 stoichometry) from the total amount of cyt c_2 present. The second-order rate constant (k_0) of the slower reaction was determined by plotting the observed rate versus $[C]_{free}$; k_0 is equal to the slope of this plot. The dissociation constant (K_D) is equal to the slope of the plot of $[C]_{free}$ versus the ratio of bound to free RCs.

RESULTS AND DISCUSSION

Characterization of mutant RCs The fast kinetics (μs) were measured by monitoring the reduction of $D^+ \rightarrow D$ following flash induced charge separation of the RCs with the ruby laser. The time constants for the fast phase are the same, within experimental error, for native and mutant RCs (Table 1). Although the rate of the fast phase of the reaction is independent of cyt c_2 concentration, the proportion of the fast phase increases with increasing cyt c_2 concentration. This behavior is consistent with electron transfer from a bound state. The extent of cyt c_2 binding to the RC is equal to the fraction of RCs that reacted with a fast phase; this method assumes that all of the bound complexes react with fast kinetics (i.e. the actual amount of bound cyt c_2 would be greater if some of the cyt c_2 is bound in a configuration in which electron transfer cannot proceed at a fast rate). Changing Asp-M184 → Lys weakens binding 600 fold, while changing Asp-L155 → Lys weakens binding only ~10 fold (Table 1).

The fraction of RCs without cyt c_2 bound prior to the laser flash react with free cyt c_2. The rate k_0 of this reaction depends on the cyt c_2 concentration according to a second-order process. The Asp-M184 → Lys change reduces k_0 over 10 fold while Asp-L155 → Lys reduces k_0 only 20% (Table 1).

Parameter	Strain		
	Native	DK(L155)	DK(M184)
τ_e, (μs)	0.8 ± 0.1	0.9 ± 0.1	0.8 ± 0.2
k_0, ($M^{-1}s^{-1}$)	9 x 10^8	7 x 10^8	7 x 10^7
K_D, (M)	5 x 10^{-7}	6 x 10^{-6}	3 x 10^{-4}

Table 1 *Parameters describing the interaction of cyt c_2 with native and mutant RCs. τ_e: time constant for the fast electron transfer phase. k_0: second-order rate constant. K_D: dissociation constant for the RC - cyt c_2 complex.*

Mapping the cyt c_2 binding site on the RC The results discussed above show that changing a single Asp → Lys on the surface of the RC can greatly affect the interaction of cyt c_2 with the RC. In particular, the position of the altered residue is a determining factor in how the RC - cyt c_2 complex is affected. Both mutants transfer an electron to the RC from the bound cyt c_2:RC complex, with rates equal to the native rate. These results suggest that cyt c_2 binds to the mutant RCs in essentially the same position and orientation as it does in native RCs.

The second-order rate (k_0), which is limited by the collision frequency of cyt c_2 and the RC, provides a measure of the attraction of cyt c_2 to the RC. Since this process is governed by diffusion in an electrostatic field, a reduction of k_0 indicates an unfavorable change in the electrostatic environment surrounding the region of the RC that is approached by cyt c_2. The results show that Asp-M184, which has a larger effect, is closer to the cyt c_2 binding region than Asp-L155.

The stability of the cyt c_2:RC complex is indicated by K_D; the larger its value the easier the complex dissociates. Comparison of the changes in K_D for the mutant RCs identifies the relative importance that particular residues have on binding of cyt c_2 to the RC. This information, along with a knowledge of the three dimensional structure of the RC, provides a map of regions of the RC surface that are important for cyt c_2 binding.

The mutations reported here are symmetrically located about the pseudo 2-fold symmetry axis of the RC and are relatively far from one another on the RC surface (Fig. 1). Changing Asp-M184 → Lys weakens cyt c_2 binding much more than the change at L155 does (see Table 1), showing that the bound cyt c_2 is closer to M184 than L155.

Comparison of different docking models The three models for binding of cyt c_2 to the RC (6, 7, 13) allow us to make predictions of the changed interactions in the two mutants. For the Allen model (6), cyt c_2 was positioned directly over the RC bacterichlorophyll dimer and then rotated to maximize the number of charge pair interactions. The Tiede model (7) used electrostatic calculations to find an energy minimum for the cyt c_2:RC complex; cyt c_2 is shifted toward the M side of the RC and has many charge pair interactions. For these two models the interactions between the RC and cyt c_2 are classified into charge pairs (or salt-bridges) for residues close to one another or as separated charge pairs (or possible salt-bridges) for residues nearby (>5Å) but too far to form salt-bridges. By assuming that the interactions of salt-bridges are greater than the interactions for separated charge pairs or possible salt-bridges, we have predicted the relative effects expected for changing the L155 or M184 residues (see Table 2). The Adir model (13) is based on the structure of the cyt c_2:RC co-crystal; cyt c_2 is shifted toward the M side of the RC even more than it is in the Tiede model. From this model, changing Asp-M184, which is near the bound cyt c_2, should have large effects while changing Asp-L155 (≈ 20Å from the cyt c_2) should have small effects. The interactions for the Asp-L155 and Asp-M184 residues for each model and the predicted importance of these residues are summarized in Table 2. Only the structural model based on the co-crystal (13) is consistent with the experimental results (*i.e.* M184 > L155).

Model	Asp-L155	Asp-M184	Predicted Effect
Allen *et al.*	salt-bridge	possible salt bridge	L155 > M184
Tiede *et al.*	separated charge pair	separated charge pair	L155 ≅ M184
Adir *et al.*	≈ 20Å away	4 Å from BCC methyl group of cyt c_2 heme	M184 > L155

Table 2 *The interactions of the Asp-L155 and Asp-M184 residues with lysines on cyt c_2 and the relative importance of their mutations on the cyt c_2 - RC interaction predicted by the different docking models.*

CONCLUSIONS

The replacement of Asp at either residue L155 or M184 affects the interaction of cyt c_2 with the RC. The change at M184 reduces the RC's affinity much more than the change at L155 does. These results show that cyt c_2 interacts with the M side of the RC's periplasmic surface more than it does with the L side (see Fig. 1). Although both the Adir and Tiede models have cyt c_2 interacting with the M side of the RC, our results agree better with the model based on the structure of the cyt c_2:RC co-crystal (13).

ACKNOWLEDGEMENTS

We thank Andrea Juth for technical assistance. Work supported by NSF (MCB93-04782) and NIH (GM 13191).

REFERENCES

(1) Overfield, R. E., Wraight, C. A. and Devevault, D. (1979) *FEBS Lett* **105**, 137-142.

(2) Moser, C. C. and Dutton, P. L. (1988) *Biochemistry* **27**, 2450-2461.

(3) Wachtveitl, J. Farchaus, J W., Mathis, P. and Osterhelt, D. (1993) *Biochemistry* **32**, 10894-10904.

(4) Prince, R. C., Baccarini-Melandri, A., Hauska, G. A., Melandri, B. A. and Crofts,A. R. (1975) *Biochim. Biophys. ACTA* **387**, 212-227.

(5) Long, J. E., Durham, B., Okamura, M. and Millett, F. (1989) *Biochemistry* **28**, 6970-6974.

(6) Allen, J., P., G. Feher, Yeates, T. O., Komiya, H., and D. C. Rees. (1987) *Proc. Natl. Acad. Sci. USA* **84**, 6162-6166.

(7) Tiede, D. M. and Chang, C. (1988) *Israel J. Chem.* **28**, 183-191.

(8) Mathis P. (1994) *Biochim. Biophys. ACTA* **1187**, 177-180.

(9) Paddock, M. L., Rongey, S. H., Feher, G. and Okamura, M. Y. (1989) *Proc. Natl.*

(10) Rongey, S. H. (1994) Ph.D. Thesis, University of California, San Diego.
Acad. Sci. USA **86**, 6602-6606.

(11) Paddock, M. L., Rongey, S. H., Abresch, E. C., Feher, G. and Okamura, M. Y. (1988) *Photosynthesis Research* **17**, 75-96.

(12) Bartsch, R. G. (1978) in *The Photosynthetic Bacteria* (Clayton, R. K. and Sistrom, W. R., Eds.) Chapter 13, pp. 249-279 (Plenum Press, New York).

(13) Adir, N., Okamura, M. Y. and Feher, G. (1994) *Biophys. J. (Abstracts)* **66**, A127; Adir, N., Axelrod, H. L., Beroza, P., Isaacson, R. A., Rongey, S. H., Okamura, M. Y. and Feher, G. (1995) submitted to *Biochemistry*.

SUPPRESSOR MUTATION IN THE H-SUBUNIT PARTIALLY RESTORES PROTON TRANSFER IN ASP-L213 → ASN MUTANT REACTION CENTERS FROM *RHODOBACTER SPHAEROIDES*.

Senft, M.E., Paddock, M.L., Rongey, S.H., Feher, G. and Okamura, M.Y.;
University of California, San Diego, Physics Dept., 9500 Gilman Drive, La Jolla, Ca. 92093

INTRODUCTION

Light-induced electron transfer in the photosynthetic bacterial reaction center (RC) results in the reduction and protonation of the bound quinone molecule, Q_B. This occurs through two sequential electron transfer steps, the second coupled to protonation of the quinone molecule:

$$Q_A^- Q_B^- + H^+ \xrightarrow{k_{AB}^{(2)}} Q_A Q_B H^-$$

The RC is composed of three subunits (L, M and H) and nine cofactors, including the primary and secondary quinones Q_A and Q_B. Several amino acid residues of the L subunit located near Q_B have been shown to be important for the proton-coupled electron transfer ($k_{AB}^{(2)}$), including Ser-L223 (1) and Asp-L213 (2-5). Replacement of these two residues with ones incapable of efficient proton transfer resulted in a drastic reduction of $k_{AB}^{(2)}$ by factors of 10^2 and 10^4, respectively, and the inability of the bacteria to grow under photosynthetic conditions. Spontaneous revertants have been isolated and characterized from several of these lethal mutations that contain suppressor mutations in either the L or M subunit (6-9). Several of these are at sites located at a relatively large distance (~ 15 Å) from the original mutation and near several H subunit residues (10) and bound water molecules (11-12). No suppressor mutations associated with the H subunit had been previously reported. We have developed and used a strategy to isolate, identify and characterize H suppressor mutations. Here, we present preliminary data for an H suppressor mutant of the DN(L213) (Asp-L213 → Asn) mutation.

MATERIALS AND METHODS

The construction of the site-directed replacement of Asp-L213 with Asn [DN(L213)] was described in (5). The *puf* operon, carrying the genes coding for the L and M subunits, was reconstructed in the pRK404 plasmid and transferred into an *Rb. sphaeroides* deletion strain ΔLM1, which contained the gene coding for the H subunit (*puh* A). The pFF1 plasmid had a chloramphenicol marker; it is incompatible with the pRK404 plasmid which had a tetracycline marker. DNA manipulations were performed as described in (5) or as recommended by the manufacturer's kit (Amersham). Photosynthetic growth was tested as described in (13). RCs were isolated from semi-aerobically grown cells and purified to an optical absorbance ratio of $A_{280}/A_{802} < 1.30$ (14). Measurements were performed at 23 °C in 10mM Hepes pH 7.5, 0.04% β-D-maltoside and 0.1mM EDTA except where noted. Charge recombination kinetics were determined by monitoring 865nm in RCs with only Q_A (k_{AD}) or with active Q_B (~5-10 Q/RC) (k_{BD}) following a saturating laser flash (see *e.g.* ref. 5). Kinetics for the forward

P. Mathis (ed.), Photosynthesis: from Light to Biosphere, Vol. I, 639–642.

electron transfer rates following a saturating laser flash were measured at 750nm and 450nm for the first ($k_{AB}^{(1)}$) and the second electron transfer ($k_{AB}^{(2)}$), respectively.

RESULTS AND DISCUSSION

Revertant isolation. The complemented deletion strain carrying the Asp-L213 → Asn mutation was found to be photosynthetically deficient. However a small number (~ 10^{-5} of the viable cells spread onto the plate) of colonies did grow under photosynthetic restriction (5). These colonies are photosynthetic revertants. Twelve revertants were screened for potential H subunit suppressor mutations (see next section).

Identification of an H suppressor revertant. The suppressor mutation was determined, based on the characteristics of the revertant DNA and the isolated RCs, to be either on the *puf* operon (i.e. the L or M subunit gene) carried by the pRK404 plasmid or on the *puh* A operon (i.e. the H subunit gene) on the *Rb. sphaeroides* chromosome. We first screened the twelve revertants for suppressor mutations located on the *puf* operon. One pRK404 plasmid bearing the *puf* operon was removed from each revertant, transferred into the *Rb. sphaeroides* deletion strain ΔLM1 possessing the wild type *puh* A operon (the H subunit gene), and tested for growth under photosynthetic conditions. Growth indicated that viable RCs were created, implying that the suppressor mutation was on the *puf* operon (i.e. on the L or M subunit gene). Seven of the twelve revertants were found to have a suppressor mutation on the *puf* operon by this method and did not require further screening. It is likely that the remaining five revertants contained a suppressor mutation on the *Rb. sphaeroides* chromosome. However, the pRK404 is a multiple copy plasmid, and, therefore, these revertants may have possessed pRK404 populations heterogeneous with respect to the presence of a suppressor mutation. For these cases, since only one plasmid from each revertant was screened, it is possible that plasmids without a suppressor mutation were tested.

To eliminate this possibility, we screened the five revertants for the presence of the suppressor mutation on the *Rb. sphaeroides* chromosome. The pRK404 was removed from each revertant by inserting (and selecting for) the incompatible plasmid, pFF1, which cannot stably coexist in the same cell with the pRK404 plasmid. This procedure created revertant deletion strains that contained the revertant *Rb. sphaeroides* chromosome, and hence *puh* A, but lacked the *puf* operon. To three of the five revertant deletion strains, the pRK404 was reintroduced and photosynthetic viability tested. Each of these complemented revertant deletion strains grew photosynthetically, indicating that the suppressor mutation was on the chromosome as suggested by the first screening.

We further corroborated the location of the suppressor mutation to be on the H subunit (i.e. on *puh* A, the H subunit gene) by isolating and characterizing the RCs of one of the complemented revertant deletion strains (DN(L213)/H rev.). The characteristics of the DN(L213)/H rev. RCs differ from those of either the native or the DN(L213) RCs (Table 1). Since the L and M subunits contain only the DN(L213) replacement, the suppressor mutation must be on the H subunit. The characteristics of the DN(L213) revertant and the DN(L213)/H rev. RCs were essentially the same, indicating that the revertant contained no other mutations in the L or M subunit genes.

Table 1 Kinetic rate constants for native and mutant RCs at pH 7.5 (± 10% precision).

RC type	Rate (s^{-1})			
	k_{AD}	k_{BD}	$k_{AB}^{(1)}$	$k_{AB}^{(2)}$
native	8.4	0.67	6,500	1,300
DN(L213)	8.7	0.04	500	0.2
DN(L213)/H rev.	7.9	1.7	1,900	70

Characteristics of the DN(L213) H revertant. The isolated RCs were characterized with respect to their charge recombination kinetics and electron transfer rates to the secondary quinone Q_B (Table 1). The charge recombination rate constant k_{AD} was the same for native, DN(L213) and DN(L213)/H rev. RCs indicating that no major structural changes are present near the Q_A site due to the mutations. The recombination rate k_{BD} was increased in the DN(L213)/H revertant RCs indicating altered electrostatic interactions that are unfavorable to Q_B^- formation, but may be favorable to internal protonation of some species involved in proton transfer.

The forward electron transfer rate $k_{AB}^{(1)}$ differs in the three RC types. It is smallest in the DN(L213) RCs and appears to be partially restored by the H suppressor mutation. The proton-coupled electron transfer $k_{AB}^{(2)}$ was greatly reduced in the DN(L213) RCs compared to native RCs (2-5). For the DN(L213)/H rev. RCs, $k_{AB}^{(2)}$ was greater than that measured in DN(L213) RCs but smaller than in native RCs (Figure 1). The pH dependence of $k_{AB}^{(2)}$ is similar for the DN(L213)/H rev. and native RCs - decreasing with increasing pH from pH 5 to pH 10. Thus the suppressor mutation does partially compensate for the blockage of proton transfer created in the DN(L213) mutant (Asp-L213 → Asn).

The characteristics of the DN(L213)/H revertant discussed here are similar to those of the Arg-M233 → Cys (or Leu) suppressor mutation(s) previously reported (8,15). The revertant RCs display increased k_{BD}, indicating altered electrostatics near Q_B^-, and partial restoration of $k_{AB}^{(2)}$.

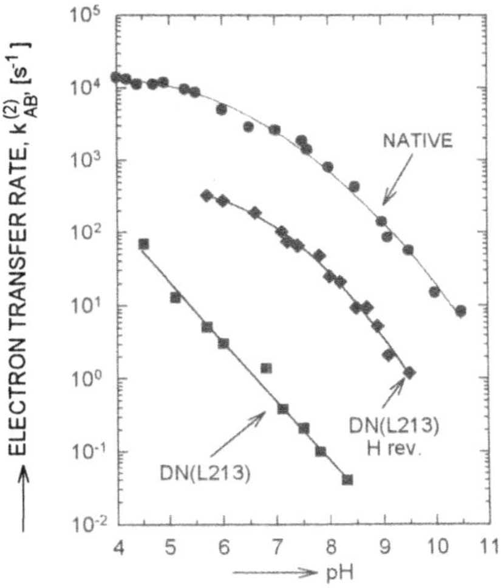

Figure 1. Measured $k_{AB}^{(2)}$ for native, DN(L213) and DN(L213)/H rev. as a function of pH. Conditions: 0.04% maltoside, 0.1mM EDTA, 2.5mM each of Hepes, Caps, Ches, Pipes, tricine and citrate.

Although the position of the H suppressor mutation has not yet been identified, we know that being on the H-subunit it must be distant from Q_B and Asp-L213, similar to several reported suppressor mutations of the Asp-L213 → Asn (Ala) replacements (8,9,15). Those suppressor mutations introduce a greater negative potential near Q_B by replacing either a positive residue with a neutral one (*e.g.* Arg → Cys, Leu) or a neutral residue with a negative one (*e.g.* Asn → Asp). Given the similarities between those revertants and the H revertant reported here, it is likely that the DN(L213)/H suppressor mutation involves a similar type of replacement. Possible candidates include Arg-H117, Arg-H177 or Lys-H130 replaced by a neutral species (*e.g.* Cys, Leu, Ser, Gly, Gln) or Gln-H174 replaced by a potentially negatively charged Glu residue. Many of these H residues are located near bound water molecules (11,12) which may also play an important role in the mode of action of the reversion.

From these and other results (16,17) it appears that many H subunit residues are important for RC function through electrostatic influences. These interactions may be a major reason for the differences in properties between Q_A and Q_B and appear to influence the proton-coupled elecron transfer $k_{AB}^{(2)}$. Even though the H residues are distant from Q_B and have smaller effects than the nearby L subunit residues, the cumulative effect of the numerous charged H subunit residues may be significant in fine tuning the properties and functions of Q_B.

CONCLUSIONS

In this work we describe a procedure to identify H suppressor mutations that compensate for detrimental mutations on the L or M subunit genes. The characteristics of one of the H suppressor mutations are given. The revertant displayed increased proton-coupled electron transfer $k_{AB}^{(2)}$ (10^2-fold at pH 7.5) compared to the original DN(L213) lesion, however $k_{AB}^{(2)}$ remained reduced (~10-fold at pH 7.5) compared to native RCs. The revertant also displayed an increased charge recombination rate k_{BD}, indicating altered electrostatic interactions between the protein and Q_B^-. These results are similar to previously reported suppressor mutations to the DN(L213) lesion, which are located at a relatively large distance (> 10 Å) from Q_B and Asp-L213. Thus, the mode of action of the H suppressor mutation may be similar to the previously reported suppressor mutations and may involve altered electrostatics for proton transfer and/or modifications of internal water structure. We are at present in the process of identifying the DN(L213)/H suppressor mutation.

ACKNOWLEDGEMENTS

We thank Don Helinski and his lab for the gift of the pFF1 plasmid. Work supported by NIH (GM 41637, GM 13191) and NSF (MCB89-15631).

REFERENCES

(1) Paddock, M.L., Rongey, S.H., Feher, G. & Okamura, M.Y. (1989) *Proc. Natl. Acad. Sci.* USA 86, 6602-6606.

(2) Takahashi, E. & Wraight, C.A. (1990) *Biochim. Biophys. Acta* 1020, 107-111.

(3) Rongey, S.H., Paddock, M.L., McPherson, P.H., Feher, G. & Okamura, M.Y. (1991) *Biophys. J.* 59, 142 (abstr.).

(4) Takahashi, E. & Wraight, C.A. (1992) *Biochemistry* 31, 855-866.

(5) Paddock, M.L., Rongey, S.H., McPherson, P.H., Juth, A., Feher, G. & Okamura, M.Y. (1994) *Biochemistry* 33, 3, 734-745.

(6) Hansen, D.K., Baciou, L., Tiede, D.M., Nance, S.L., Schiffer, M. & Sebban, P. (1992a) *Biochim. Biophys. Acta* 1102, 260-265.

(7) Hansen, D.K., Nance, S.L. & Schiffer, M. (1992b) *Photosynth. Res.* 32, 147-153.

(8) Okamura, M.Y., Paddock, M.L., McPherson, P.H., Rongey, S.H. & Feher, G. (1992) in *Research in Photosynthesis* (Murata, N., Ed.) Vol. 1, pp. 349-356, Kluwer Academic Publishers, Dordrecht, The Netherlands.

(9) Maroti, P., Hansen, D.K., Baciou, L., Schiffer, M. & Sebban, P. (1994) *Proc. Natl. Acad. Sci.* USA 91, 5617-5621.

(10) Allen, J.P., Feher, G., Yeates, T.O., Komiya, H. & Rees, D.C. (1988) *Proc. Natl. Acad. Sci.* USA 85, 8487-8491.

(11) Ermler, U., Fritzsch, G., Buchanan, S.K. & Michel, H. (1994) *Structure* 2, 925-936.

(12) Lancaster, C.R.D., Ermler, U. & Michel, H. (1995) in *Anoxygenic Photosynthetic Bacteria* (Blankenship, R.E., Madigan, M.T. & Bauer, C.E., Eds.) Kluwer Academic Publishers, The Netherlands, in press.

(13) Rongey, S.H., Paddock, M.L., Feher, G. & Okamura, M.Y. (1993) *Proc. Natl. Acad. Sci.* USA 90, 1325-1329.

(14) Paddock, M.L., Rongey, S.H., Abresch, E.C., Feher, G. & Okamura, M.Y. (1988) *Photosynth. Res.* 17, 75-96.

(15) Sebban, P., Maroti, P., Schiffer, M. & Hanson, D.K. (1995) *Biochemistry* 34, 8390-8397.

(16) Takahashi, E. & Wright, C. (1995) *Biophys. J.* 68, 95 (abstr.).

(17) Rongey, S.H., Juth, A.L., Paddock, M.L., Feher, G. & Okamura, M.Y. (1995) *Biophys. J.* 68, 247 (abstr.).

THE ASSOCIATION OF DIFFERENT DETERGENTS WITH THE PHOTOSYNTHETIC REACTION CENTER PROTEIN OF *RHODOBACTER SPHAEROIDES* R26 AND THE EFFECTS ON ITS PHOTOCHEMISTRY.

P.W. Hemelrijk, P. Gast, H.J. van Gorkom and A.J. Hoff. Dept. of Biophysics, Huygens Lab., POBox 9504, 2300 RA Leiden, The Netherlands

1. Introduction

The number of detergent molecules associated with the RC protein is directly related to the size of the detergent belt, and is therefore an important factor for crystallization experiments. This number has been determined in solution by chromatography using ^{14}C-labelled detergents [1] and in crystals by low-resolution neutron diffraction using H_2O/D_2O contrast variation [2-4], but these methods are impractical to use for selecting suitable detergents for crystallization attempts. Recently, we introduced a novel way to measure the detergent/RC ratio [5]. By this method, the detergent is removed by incubating the RCs with detergent-adsorbing polystyrene divinylbenzene beads (Bio-Beads) [6]. Using RCs of *Rps. viridis* we found that detergent-free RCs can be completely redissolved by adding a sufficient amount of detergent. By monitoring the resolubilization as a function of protein concentration, the detergent/RC ratio was determined. It was found that 260 ± 30 LDAO molecules were associated with the RCs of *Rps. viridis*, and that this number was reduced by about 50% when 5% 1,2,3 heptanetriol was added. The method is rapid and simple and should be generally applicable to detergent-solubilized membrane proteins.

In this work we have used the above mentioned method to study the solubilization process of reaction centers of *Rhodobacter (Rb.) sphaeroides* R26 with different types of detergent in the presence of 1,2,3 heptanetriol, for a series of detergents with varying chain length, and as a function of the pH of the medium. We also investigated the effects of detergent on the photochemistry of the RC, and found significant differences in the charge-recombination rate of the RCs when solubilized with different detergents.

2. Materials and methods

N,N-dimethyltridecylamine-*N*-oxide (C_{13}DAO), *N,N*-dimethyldodecylamine-*N*-oxide (C_{12}DAO or LDAO), *N,N*-dimethylundecylamine-*N*-oxide (C_{11}DAO), *N,N*-dimethyldecylamine-*N*-oxide (C_{10}DAO), *N,N*-dimethylnonylamine-*N*-oxide (C_9DAO), *N,N*-dimethyloctylamine-*N*-oxide (C_8DAO), n-Octyl-ß-D-glucopyranoside (ß-OG) and 1,2,3-heptanetriol (HT) were purchased from Sigma Chemicals, Triton X-100 from Merck, polystyrene divinylbenzene Bio-Beads SM-2 from Bio-Rad Laboratories. Horse heart type III cytochrome c (Cyt c) (Sigma) was reduced by sodium dithionite and purified by gel filtration. RCs from *Rb. sphaeroides* R26 were isolated as in Ref. [7]. The LDAO detergent was removed according to the method described earlier [5].The minimal amount of detergent needed to solubilize the RCs was determined as in Ref.[5]. The absorption spectrum of detergent-free RCs was recorded for a suspension of finely dispersed particles. The spectrum was not corrected for scattering and flattening, which did not affect the peak positions of the various bands. Flash-induced absorption difference measurements were carried out with a split-beam apparatus [8]. Saturating excitation flashes were provided using a Xenon lamp producing flashes of 10-20 μs duration with a repetition rate of 0.2 Hz.

3. Results and discussion

3.1 RC solubilization

We have investigated the solubilization by LDAO and a series of detergents differing from LDAO only in the length of the apolar alkyl chain. Table 1 lists the resulting detergent/RC ratios for all detergents used and the minimal solubilization concentration. Rather extreme values are found for the detergent/RC ratio using the two shortest alkyl chain detergents C_9DAO and C_8DAO (730 ± 200 and 2800 ± 1000 respectively). These numbers appear to be unrealistic, since the surface of the protein cannot accommodate such a large number of detergent molecules. It is conceivable that either the detergent is present in "empty" micelles before solubilization or that the RC is solubilized in detergent-structures other than simple micelles. The value found for the minimal solubilization

P. Mathis (ed.), Photosynthesis: from Light to Biosphere, Vol. I, 643–646.

concentration is close to the reported CMC for all detergents (especially evident in the alkyl amine *N*-oxide series of detergents, which shows a 300-fold increase of the CMC between the longest and the shortest alkyl chain) indicating that RC solubilization occurs practically at the CMC.

The effect of HT on the detergent/RC ratio and the CMC was also investigated. The results are shown on the right-side of Table 1. Except for Triton X-100, the detergent/RC ratios are reduced to about 45-55% in the presence of 5% HT. The CMC values of C_{12}DAO and C_{11}DAO are also somewhat affected by the presence of 5% HT, while the CMC values of the other detergents appear not to be changed.

Detergent	CMC [mM]	#Det./RC	$CMC_{5\%HT}$ [mM]	#Det./$RC_{5\%HT}$
ß-OG	20.0 ± 1 (22-23)	370 ± 100	19 ± 1	200 ± 60
Triton X-100	0.53 ± 0.10 (0.2-0.9)	51 ± 5	0.50 ± 0.15	44 ± 15
C_{13}DAO	0.53 ± 0.10 (0.52)	87 ± 15	0.50 ± 0.1	39 ± 10
C_{12}DAO	0.90 ± 0.30 (1.1)	150 ± 20	0.40 ± 0.2	67 ± 20
C_{11}DAO	5.0 ± 0.4 (6.0)	165 ± 50	3.6 ± 0.5	80 ± 40
C_{10}DAO	12 ± 1 (15)	200 ± 100		
C_9DAO	50 ± 3 (50.8)	730 ± 200		
C_8DAO	145 ± 10 (162)	2800 ± 1000		

Table 1. Left side: minimum solubilization concentrations and detergent/RC ratios for eight different detergents for RCs of Rb. sphaeroides R26. The numbers between parentheses represent reported CMC values [9,10]. The CMC, of C_{13}DAO and C_{10}DAO were estimated using Traube's rule [11]. Right-side: CMC values and detergent/RC ratios for five different detergents for RCs of Rb. sphaeroides R26 in the presence of HT.

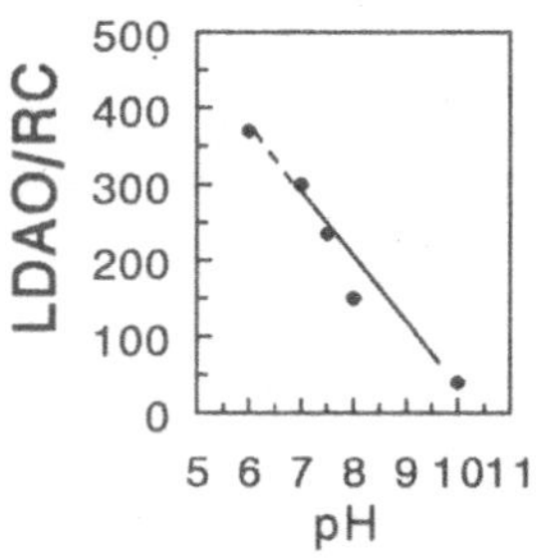

Fig. 1. LDAO/RC ratio of RCs of Rb. sphaeroides R26 as a function of pH.

The solubilization of detergent-free RCs from *Rb. sphaeroides* R26 by LDAO was found to depend on pH (Fig. 1). At pH 6.0 we found no change in the CMC within experimental error, while the LDAO/RC ratio increased to 370. The same effect was observed for normal RCs: they precipitated at pH 6.0 when the detergent concentration was insufficient. At pH 10.0 only 40 LDAO molecules associated with the detergent-free RC. At pH 6.5 the detergent-free RCs could not be completely resolubilized. A similar effect has been found with RCs from *Rps. viridis* [5]. This value is close to the pH at which *Rps. viridis* crystals grow, and may indicate that at this pH the RC protein molecules strongly interact, because of minimum electrostatic repulsion. This pH value may thus correspond to the iso-electric point of the protein.

3.2 Photo-activity

As with RCs from *Rps. viridis* [5], no difference is observed between the absorption spectra of "normal" RCs from *Rb. sphaeroides* R26 and RCs that had been resolubilized after Bio-Bead treatment. After resolubilization of the RCs, the absorption maximum of the primary donor was found at 865 nm for all detergents tested. However, differences were found between detergent-free and resolubilized RCs, as shown in Fig. 2A. The room-temperature absorption band at 865 nm of the primary donor was shifted to about 846 nm when the detergent was removed, while the two other bands in the Q_y region, at 802.5 and 758.5 nm, displayed small band-shifts of 1 nm and 2.5 nm, respectively. Also at 1.2 K the absorption band of the primary donor was blue shifted, from 892 nm for resolubilized RCs to 881 nm for detergent-free RCs (data not shown). Figure 2B shows the shift of the absorption maximum as a function of the amount of LDAO added (open circles) and the concomitant light-scattering (closed circles). The results demonstrate that before the RCs are dissolved (indicated by the drop in the scattering level), the absorption band has already returned to almost its original position. This shows that the detergent binds to the RC at concentrations below those needed for resolubilization. The band-shift did not exhibit an isosbestic point, so it seems to be a continuous shift rather than a transition between two discrete states of the RC. Hellingwerf [6] found a similar but smaller band-shift to 850 nm for the primary donor in a solution from which 66% of the detergent had been removed from the RCs. These results suggest that the band-shift of the primary donor may be used as an indicator of the amount of detergent bound to the RC.

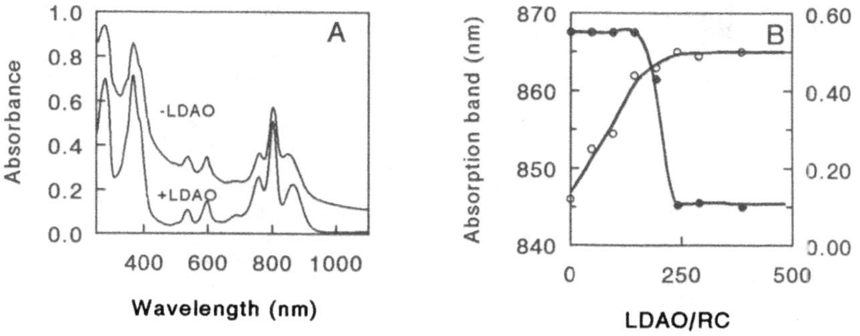

Fig. 2. (A) Absorption spectrum of detergent-free RCs (-LDAO) and with LDAO resolubilized RCs (+LDAO). (B) Shift of the absorption band of the primary donor as a function of the amount of detergent added (open circles) and the concomitant change in the scattering (closed circles).

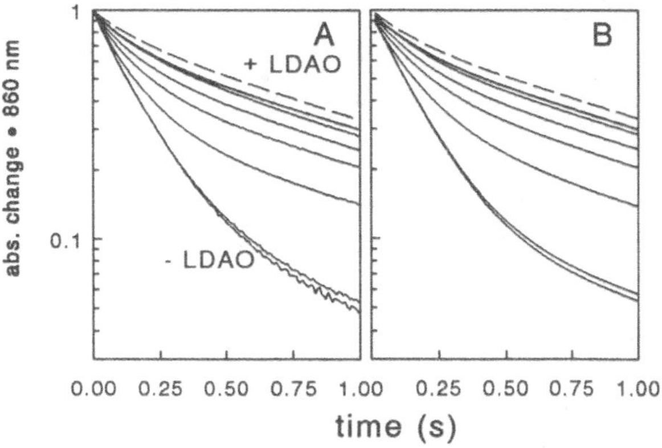

Fig. 3. (A) Absorption changes measured at 860 nm in RCs with increasing amount of LDAO; normal RCs: dotted line. (B) Global analysis of the kinetics using a bi-exponential decay function.

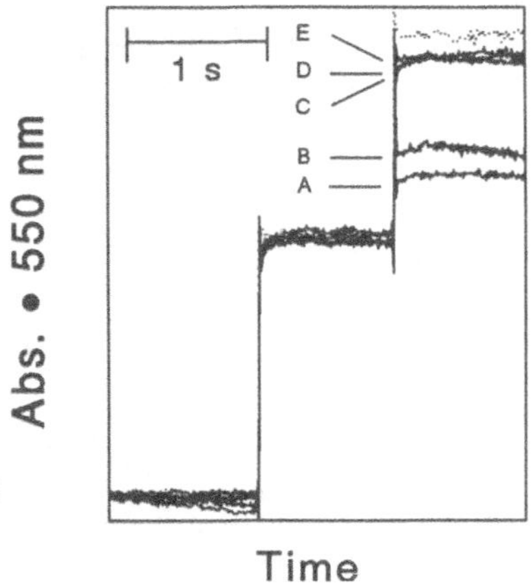

Fig.4. Cytochrome oxidation in a two-flash experiment measured at 550 nm in RCs with increasing LDAO/RC ratio: A, 0; B, 100; C, 200; D, 300; E, 500. Normal RCs: dotted line.

We have probed the photo-activity of detergent-free RCs by measuring the recovery of the Q_y absorbance of the primary donor after its flash-induced bleaching. Figure 3 shows the absorption changes at 865 nm after flash illumination of detergent-free RCs with different amounts of LDAO, and "normal" detergent-solubilized RCs. In normal RCs a slow recovery, with a time constant of 1.2 s is found, which is attributed to charge recombination from the state $P^+Q_B^-$ [11]. For detergent-free

LDAO/RC	Q_B^- (cyt)	Q_B^- (kin)
0	11	24
100	31	34
200	55	69
300	64	70
500	67	70
Normal	76	88

Table 2. Percentage of ET to Q_B found from the kinetics (kin) and from cyt oxidation (cyt) for RCs with variable LDAO/RC ratio.

RCs, a faster and clearly biphasic recovery is observed with time constants of 160 ms and 1.2 ms. When sufficient LDAO is added to the sample to resolubilize the RCs, the relative amplitude of the fast phase decreases and the recovery approaches that in normal RCs. In the presence of 10 mM *o*-phenanthroline mainly the fast time constant is observed, both with and without detergent, suggesting that it is due to $P^+Q_A^-$ recombination [12]. The presence of both phases in detergent-free RCs rules out a rapid equilibrium between Q_A^- and Q_B^- and indicates that in some centers electron transfer to Q_B is blocked. When measuring the kinetics of P^+-decay of detergent-free RCs at wavelengths ranging from 830 nm to 890 nm we did not observe any kinetic differences, indicating that the two types of reaction centers have the same absorption spectra. Fig 3B shows the results of a global analysis of the kinetics using a model in which the slow phase is due to $P_+Q_B^-$ backreaction (1.2 s) and the fast phase to $P^+Q_A^-$ backreaction (160 ms). Further evidence for the blocking of electron transfer to Q_B in detergent-free RCs was obtained by measuring cytochrome oxidation at 550 nm in a two-flash experiment with samples to which cytochrome was added (Fig.4): Table 2 shows the fraction of Q_B^- formed determined from the simulation of the kinetics and those obtained from the cytochrome oxidation on the second flash. Except for samples with low detergent, where the light flash may not be saturating due to the scattering, good agreement is obtained between both experiments.

4. Conclusions

Detergents are capable of association with detergent-free RCs of *Rb. sphaeroides* R26 below the CMC. The RCs, however, start to solubilize only when the detergent concentration in the solution has reached the CMC. The minimal number of detergent molecules that is associated with the RC after solubilization depends on the type of detergent. It increases with decreasing number of C-atoms for *N, N* alkylamine-*N*-oxide detergents, is found to be pH dependend for LDAO and decreases by a factor two in the presence of 5% HT for all detergents tested except Triton X-100. In detergent-free RCs the electron transfer to Q_B is inhibited (80-90%); it can be almost completely restored upon re-solubilization.

5. Acknowledgements

Ms S. Jansen expertly prepared the RCs. PG thanks Rob Louwe for help with the global analysis. This work was sponsered by the Netherlands Foundation of Chemical research (SON) and the Royal Academy of Arts and Sciences (KNAW).

References

1 Møller, J.V. and le Maire, M. (1993) J. Biol. Chem. 268, 18659-18672.
2 Roth, M., Lewitt-Bentley, A., Michel, H., Deisenhofer, J., Huber, R. and Oesterhelt, D. (1989) Nature 340, 659-662.
3 Zinth, W., Kaiser, W. and Michel, H. (1983) Biochim. Biophys. Acta 723, 128-131.
4 Roth, M., Arnoux, B., Ducruix, A. and Reiss-Husson, F. (1991) Biochemistry 30, 9403-
5 Gast, P., Hemelrijk, P.W. and Hoff, A.J. (1994) FEBS Lett. 337, 39-42.
6 Hellingwerf, K. (1987) J. Bioenerg. Biomembr. 19, 203-223.
7 Feher, G. and Okamura, M.Y. (1978) in The Photosynthetic bacteria, (Clayton, R.K. and Sistrom, W.R. Eds.), pp. 349-386, Plenum Press, New York and London.
8 Visser, J.W.M. (1975) Photosynthetic Reactions at Low Temperature, Doctoral Thesis, Leiden.
9 Mukerjee, P. and Mysels, K.J. (1970) in: Critical Micelle Concentrations of Aqueous Surfactant Systems, National Bureau of Standards.
10 Michel, H. (1991) in Crystallization of Membrane Proteins, (Michel, H. Ed.), pp.209-211, CRC Press, Boston.
11 Traube, J. (1891) Ann. Chem. 265, 27-55.
12 Kleinfeld, D., Okamura, M.Y. and Feher, G. (1984) Biochim. Biophys. Acta 766, 126-140.

THE STRUCTURAL ENVIRONMENT OF THE TYROSYL RADICALS IN PHOTOSYSTEM II.

Bengt Svensson and Stenbjörn Styring, Department of Biochemistry, Arrhenius Laboratories for Natural Sciences, Stockholm University, S-10691 Stockholm, Sweden

1. Introduction

The D1 and D2 proteins constitute the PSII reaction centre and bind the primary donor P_{680}, the pheophytins, the quinones, Q_A and Q_B and probably the Mn-cluster. Oxidised P_{680} extracts electrons from water via the donor side components, the Mn-cluster and Tyr_Z. On the donor side there also exists an accessory electron donor to P_{680}, Tyr_D. The redox active tyrosine residues Tyr_Z, on the D1 protein and Tyr_D, on the D2 protein are in their oxidised form neutral radicals.

The three dimensional structure of the PSII reaction centre has not yet been determined. However, the structures are known for the purple bacterial reaction centres [1,2], where the L and M subunits show considerable homology with the D1 and D2 proteins in amino acid sequence, topology and function [3-5]. The similarities has allowed us and others to model the structure of the D1/D2 heterodimer starting from the bacterial reaction centre [6-9; Svensson et al, in preparation]. Despite its limitations, homology based modelling can in fortunate cases be very profitable. Such a case is our work on Tyr_Z and Tyr_D [6,7], where the model was verified by a combination of site-directed mutagenesis and spectroscopic methods [10-12].

A new extended PSII model was computed, including the core of the D1/D2 heterodimer, a chlorophyll dimer, two monomeric chlorophylls, two pheophytins and a Fe^{2+} ion, [8, Svensson et al, in preparation]. In this model we have considered all experimental data, which provide reliable structural information. The model agrees well with experimental results for the regions of the redox active pheophytin [13-15] and the acceptor side Fe^{2+} ion [4]. On the donor side the model is in agreement with the earlier modelling attempts and experimentally supported for the region of Tyr_D [10,11]. Supporting experimental results are also available for the surroundings of Tyr_Z [12].

We have earlier presented detailed structural descriptions of the refined PSII model for the regions of Tyr_D, the redox active pheophytin [8] and the primary donor P_{680} [Svensson et al., these proceedings]. Here, we describe the detailed structure for structural environment of Tyr_Z and for a comparison also the structure around Tyr_D. Based on the structural model, hypothesis about the function and hydrogen bonding interactions are discussed for both Tyr_Z and Tyr_D.

P. Mathis (ed.), Photosynthesis: from Light to Biosphere, Vol. I, 647–650.

2. Methods

The basis for the modelling of PSII is the crystal structure of the reaction centre from *Rps. viridis.* [1]. The procedure can be summarised as follows: Assignment of structurally conserved parts of protein backbone and side chains between PSII and *Rps. viridis*; transfer of atomic coordinates, including PSII cofactors; positioning of non-conserved amino acid side chains using a rotamer library [16]; minimisation of the potential energy with CHARMM [17]; optimisation of side chain and pigment interactions using energy calculations; and evaluation of the structural quality. The modelling procedure is described in more detail in Svensson et al. 1992 [8] and Svensson et al., in preparation.

3. Results and discussion

The redox active tyrosyl residues, Tyr_D and Tyr_Z, have by site directed mutagenesis been identified as Tyr161 on the D2 protein and Tyr161 on the D1 protein, respectively. The tyrosines are located symmetrically by the C2 axis near the lumenal ends of helix C on the D2 and D1 proteins, respectively. They are found on the side of the C helix which faces the special pair. Their protein environments are constituted by residues from the C and D transmembrane helices as well as the lumenal C-D loop and the CD helix.

Tyr_D : Tyr_D is placed in a hydrophobic environment constituted of many aromatic residues [7]. The most dominant interaction in the model is the hydrogen bond to D2-His190. The distance between the Nε1 nitrogen in the histidine and the phenolic oxygen in the tyrosine is 2.8 Å. A hydrogen bond for Tyr_D is supported by ESEEM and ENDOR [18,11]. A close proximity between Tyr_D and most probably a histidine, has been proposed from kinetic EPR data [19]. It is quite possible that D2-His190 also holds a structural role in connecting the C-D loop with the turn where the E helix breaks by hydrogen bonding to the peptide oxygen of D2-Leu294 (Fig. 1).

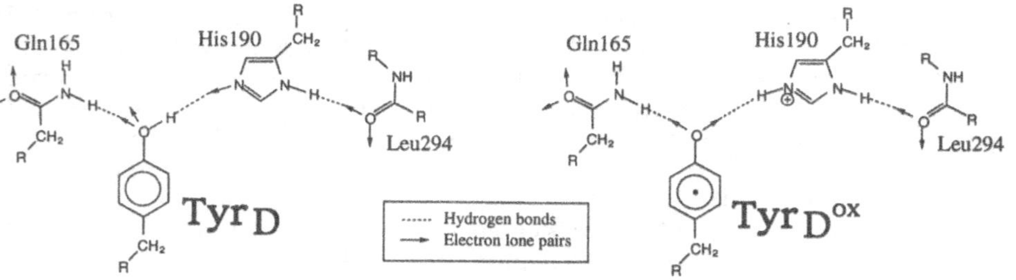

Figure 1. A proposed model for the hydrogen bonding pattern and proton donation upon oxidation of Tyr_D.

Another residue that probably interacts strongly with Tyr_D is D2-Gln165, which with its amide group comes close to the phenolic oxygen of the tyrosine to possibly form another hydrogen bond (Fig. 1). D2-Phe186 and D2-Ala291 are in Van der Waals contact with the aromatic ring of Tyr_D. D2-Phe186 is occupying the space between Tyr_D and the primary donor, approximately midways between these. Important for Tyr_D is also D2-Pro162, the residue next in sequence on the same helix.

The proposed strong interaction between D2-His190 and Tyr_D was verified by EPR spectroscopy of the Tyr_D^{ox} radical in mutants of the cyanobacterium *Synechocystis* 6803 [10,11]. Similar investigations also supported interaction between Tyr_D^{ox} and D2-Gln165 and D2-Pro162 [10,20].

Tyr_Z: Tyr_Z is found in a much more hydrophilic environment than Tyr_D [7]. Towards the lumenal side from Tyr_Z there is a cluster of polar and charged residues: D1-Gln165, D1-Asp170 and D1-Glu189 (Fig. 2). They come close in space and might constitute part of the binding site for the Mn-cluster within a few Å from Tyr_Z. D1-Asp170 has been verified to be a Mn ligand [21]. Also very interestingly is a mechanistic model that recently has been proposed, where Tyr_Z^{ox} act as an hydrogen atom abstractor directly coupled to the water splitting chemistry of the Mn-cluster [22,23, Tommos et al., these proceedings]. Upon oxidation Tyr_Z releases its proton to form a neutral radical. It is possible that the polar residues close to Tyr_Z, together with D1-His190, could participate in transfering this proton to the lumen. The distance between the phenolic oxygen on Tyr_Z and D1-His190 is in the model approximatly 4Å. Thus, there is no hydrogen bond with D1-His190 and only a weak electrostatic interaction is expected. Indications that no well ordered hydrogen bond is present for Tyr_Z^{ox} has been obtained [23,24]. D1-His190 may be involved in a hydrogen bond the peptide oxygen of D1-Leu297, which is symmetrical to the proposed hydrogen bond close to Tyr_D. A mutation of D1-His190 to either phenylalanine or tyrosine in *C. reinhardtii.* results in loss of oxygen evolution. However, the spectra of Tyr_Z^{ox} in the mutant D1-H190F is very similar to the wild type spectrum [12], which suggests that D1-His190 is not in close interaction with Tyr_Z. This result is in agreement with prediction from the structural model. Tyr_Z is in the model in Van der Waals contact with D1-Phe186, D1-Gln165 and D1-Ala294 on the D1 protein. D1-Phe186 is placed between Tyr_Z and P_{680} and is a likely part in the electron transfer pathway (Fig. 2).

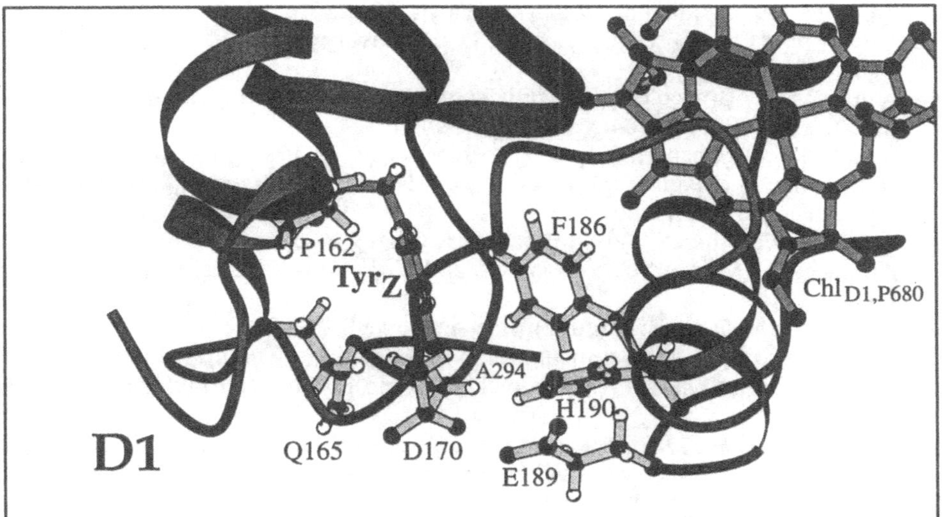

4. References

1 Deisenhofer, J., Epp, O., Miki, K., Huber, R. and Michel, H. (1985) Nature 318, 618-624.
2 Allen, J.P., Feher, G., Yeates, T.O., Komiya, H. and Rees, D.C. (1987) Proc. Natl. Acad. Sci. U.S.A. 84, 6162 6166.
3 Michel, H. and Deisenhofer, J. (1988) Biochemistry 27, 1-7.
4 Rutherford, A.W. (1987) in Progress in Photosynthesis Research Vol. I (Biggins, J. Ed.) pp. 277-283, Martinus Nijhoff Publishers, Dordrecht, The Netherlands.
5 Sayre, R.T., Andersson, B. & Bogorad, L. (1986) Cell 47, 601-608.
6 Svensson, B. Vass, I. Cedergren, E. and Styring, S. (1990) EMBO J. 9, 2051-2059.
7 Svensson, B. Vass, I. and Styring, S. (1991) Z. Naturforsch. 46c, 765-776.
8 Svensson, B., Etchebest, C., Tuffery, P., Smith, J. and Styring, S. (1992) in Research in Photosynthesis (N. Murata Ed.) Vol. II, pp 147-150, Kluwer Academic Publishers, Dordrecht, The Netherlands.
9 Ruffle, S.V., Donnely, D., Blundell, T.L. and Nugent, J.H.A. (1992) Photosyn. Res. 34, 287-300.
10 Tommos, C., Davidsson, L., Svensson, B., Madsen, C., Vermaas, W. and Styring, S. (1993) Biochemistry 32, 5436-5441.
11 Tang, X.-S., Chisholm, D.A., Dismukes, C., Brudvig, G.W. and Diner, B.A. (1993) Biochemistry 32, 13742-13748
12 Roffey, R.A., van Wijk, K.-J., Sayre, R.T. and Styring, S. (1994) J. Biol. Chem. 269, 5115-5121
13 Lubitz, W., Isaacson, R.A., Okamura, M.Y., Abresch, E.C., Plato, M. and Feher, G. (1989) Biochim. Biophys. Acta 977, 227-232
14 Moenne-Loccoz, P., Robert, B. and Lutz, M. (1989) Biochemistry 28, 3641-3645
15 Nabedryk, E., Andrianambinitsoa, S., Berger, G., Leonhard, M., Mäntele, W. and Breton, J. (1990) Biochim. Biophys. Acta 1016, 49-54
16 Tuffery, P., Etchebest, C., Hazout, S. and Lavery, R. (1991) J. Biomol. Struct. Dynam. 8, 1267-1289.
17 Brooks, B.R., Bruccoleri, R.E., Olafson, B.D., States, D.J., Swaminathan, S. and Karplus, M. (1983) J. Comp. Chem. 4, 187-217.
18 Rodrigies, I.D., Chandrashekar, T.K., Babcock, G.T. (1987) in Progress in Photosynthesis (Biggins, J. Ed.) Vol. I., pp. 119-133, Martinus Nijhoff Publishers, Dordrecht, The Netherlands.
19 Vass, I. and Styring, S. (1991) Biochemistry 30, 830-839.
20 Tommos, C., Madsen, C., Styring, S. and Vermaas, W. (1994) Biochemistry 33, 11805-11813.
21 Nixon,P.J. and Diner,B.A. (1992) Biochemistry 31, 942-948
22 Hoganson, C.W., Lydakis-Simantiris, N., Tang, X.-S., Tommos, C., Warncke, K., Babcock, G.T., Diner, B.A., McCracken, J. and Styring, S. (1995) Photosyn. Res. in press.
23 Tommos, C., Tang, X.-S., Warncke, K., Hoganson, C.W., Styring, S., McCracken, J. Diner, B.A. and Babcock, G.T. (Submitted) J. Am. Chem. Soc.
24 Tang, X.-S., Zheng, M., Chisholm, D.A., Dismukes, G.C. and Diner, B.A. (Submitted) Biochemistry

Y_Z as a Hydrogen-Atom Abstractor in the Water-Oxidizing Chemistry of Photosystem II

Cecilia Tommos and Stenbjörn Styring
Department of Biochemistry,
Arrhenius Laboratories for Natural Sciences,
Stockholm University, S-106 91 Stockholm, Sweden

The Photosystem II/Oxygen-Evolving Complex (PSII/OEC) catalyzes the light-driven oxidation of water to dioxygen in oxygenic organisms. Scheme 1 conveys the conventional view of the redox-components and chemistry involved (1). The water-splitting reactions are thermodynamically driven by the ~ 1.3 V generated by the light-induced oxidation of the PSII reaction-center chlorophyll complex P_{680}. The oxidizing equivalents created at P_{680}^+ are transferred, one by one, to a tetra-nuclear Mn-cluster, identified as the catalytic core of the OEC, via the tyrosyl residue Y_Z (tyrosine 161 on the D1 protein). Y_Z re-reduces P_{680}^+ in the sub-µs time range, which prevents wasteful charge-recombination reactions and provides a quantum yield close to one. The $(Mn)_4$-cluster binds substrate water and, by cycling through five redox states, denoted S_0-S_4, accumulates the four oxidizing equivalents required for the water splitting reaction. Oxidation of water occurs upon formation of the S_4 state, dioxygen is released, and the system relaxes back to the S_0 state. The two inorganic cofactors, chloride and calcium, are also essential for the process (2). PSII contains a second redox-active tyrosine, Y_D, positioned at 161 on the D2 protein. The Y_D tyrosine functions as an auxiliary electron donor not directly involved in the water-splitting chemistry of PSII (1).

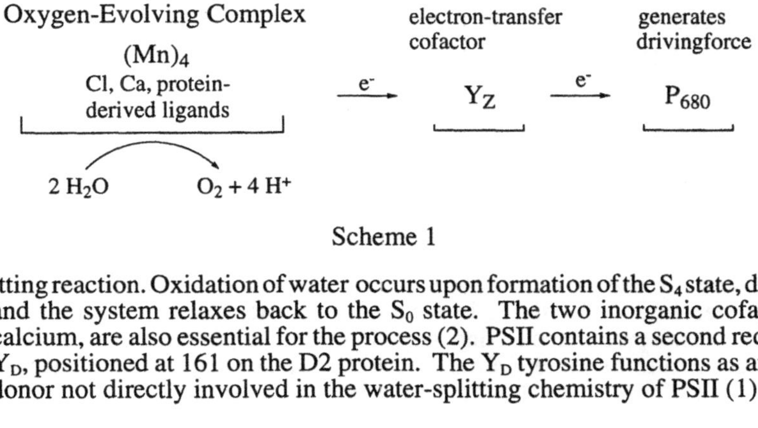

Scheme 1

Properties of the two redox-active tyrosines

On the basis of $^1H_2O\backslash^2H_2O$-exchange data from ENDOR and ESEEM studies, $Y_D^{\cdot}$ was proposed to be a neutral, deprotonated radical with its phenol oxygen involved in a hydrogen-bond (1). The exchange rate was extremely slow, indicating a shielded site for the Y_D radical. From modeling studies, it was suggested that Y_Z and Y_D form hydrogen-bonds to nearby D1-His190 and D2-His190 (D2-His189 in *Synechocystis*), respectively (3). The prediction made for the D2 protein was later experimentally supported by site-directed mutagenesis studies. Upon replacement of D2-His189, both the normal EPR spectrum of

P. Mathis (ed.), Photosynthesis: from Light to Biosphere, Vol. I, 651–654.

$Y_D^{\bullet}$ (4, 5), as well as the ENDOR detected hydrogen-bond (5), were removed. These results support the proposed 'proton-rocking' model for the $Y_D \backslash Y_D^{\bullet}$ redox chemistry where, upon oxidation, the proton migrates to the base (D2-His189) and, upon re-reduction, returns to the tyrosine (1).

In contrast, no pronounced $^1H_2O \backslash ^2H_2O$-exchangeable features could be observed in the $Y_Z^{\bullet}$ ENDOR spectrum (6, 7). These results indicate that $Y_Z^{\bullet}$ is not involved in a hydrogen-bonding interaction equivalent to the well-ordered hydrogen-bond detected for the $Y_D^{\bullet}$ radical. This rationalizes the observation that point-mutations at D1-His190 do not significantly alter the $Y_Z^{\bullet}$ EPR spectrum (8, 9). However, the D1-190 histidine is essential for the oxygen-evolving reactions of PSII (5, 9, 10). Point-mutations at this position impede the electron transfer between Y_Z and P_{680}^+ dramatically. Moreover, Debus *et al.* report oxygen-evolution activety of the D1-H190 mutants only upon changing the histidine to another relatively large, basic residue (10). These observations have been rationalized by postulating the following events (6, 11). The Y_Z tyrosine is hydrogen-bonded to D1-His190 in its reduced form. Upon oxidation, the phenol O-H bond breaks, the proton becomes fully associated with the histidine, and the neutral, deprotonated $Y_Z^{\bullet}$ radical is formed. The histidine deprotonates rapidly, and the proton is transported out of the radical site towards the lumenal bulk medium. The suggestion that the proton resides only transiently on the histidine is in agreement with FTIR data (8). In addition, proton ejection studies on PSII core preparations show a stoichoimetric release pattern upon each S-state transition (12), consistent with the proposed mechanism (6, 11). Finally, a proton-transfer mechanism, initiated by oxidation of Y_Z and mediated by D1-His190, may provide an explanation for the positive trap involved in the A_T thermoluminescence band, but, without invoking a *redox-active* histidine. Upon charge-recombination, the equilibrium shifts, and the ejected proton returns to the tyrosine via the D1-190 histidine residue (for a discussion of the A_T band, see 13 and refs. therein).

In addition to the observed difference in hydrogen-bonding status of the two tyrosyl radicals, the motional dynamics of the redox centers differ. A detailed ESEEM study rendered evidence of low rotational mobility (< 4°) about the C_{para}-C_β bond for $Y_D^{\bullet}$ in *Synechocystis* (14). This is in contrast to the Y_Z radical in the same organism, where both ENDOR and ESEEM line shapes show evidence of motional broadening (6). Quantitative information could be derived from the ESEEM data, which showed a ~ 14° variation in the dihedral angle about the C_{para}-C_β bond of $Y_Z^{\bullet}$ (6). These results are in agreement with observations made by Kawamori and co-workers, who proposed higher mobility at the Y_Z site, as compared to the Y_D site, from temperature-dependent ENDOR studies in spinach (15). Another distinction between Y_Z and Y_D is suggested from modeling studies that predict that the Y_Z site is more hydrophilic than is the protein environment around Y_D (16).

To summarize, for the Y_D tyrosine the phenol oxygen is proposed to be hydrogen-bonded to nearby D2-His189, the site is hydrophobic and shielded from the bulk medium. In addition, there is evidence of low motional disorder in the radical site. In contrast, Y_Z is not involved in a well-ordered hydrogen-bond in its oxidized form and is proposed to slough its phenol proton upon oxidation. Moreover, the protein environment of the Y_Z radical is suggested to be more hydrophilic with a higher degree of mobility in comparison to the Y_D site. These observations appear to be inconsistent with Y_Z functioning as a pure electron-transfer cofactor; in order to lower reorganization during redox change and maximize the rate, one expects that the electron-transfer process will occur in a hydrophobic environment and that changes in structure upon oxidation and reduction will be minimized. The $Y_Z \backslash Y_Z^{\bullet}$ couple shows several properties that clearly deviate from these principles that govern simple and rapid electron transfer. The site is hydrophilic, and the de- and reprotonation of Y_Z upon redox change are expected to involve considerably more nuclear motion as compared to the proposed 'proton-rocking' motion of the $Y_D \backslash Y_D^{\bullet}$ couple.

A mechanistic model for the function of Y_Z in PSII

Recently, a hydrogen-atom abstraction function for Y_Z in the water-splitting chemistry of PSII has been proposed (6, 11). The overall reaction proposed in this model is shown in Scheme 2. In this mechanism, the $(Mn)_4\backslash Y_Z$ unit forms the catalytic center of the Oxygen-Evolving Complex; the $(Mn)_4$-cluster provides binding sites for substrate and delocalizes oxidizing equivalents, and $Y_Z^{\bullet}$ acts by abstracting hydrogen-atoms from substrate aquo- or hydroxy-species, ligated to the Mn-ions in the cluster, on each S-state transition. The proposed hydrogen-abstraction function accounts for the differences observed between Y_Z and Y_D, as described above, and the paradoxical behavior of Y_Z relative to that expected from a pure electron-transfer cofactor. Moreover, studies of other radical-containing metalloenzymes show a growing number of systems where the radical performs a hydrogen-atom abstraction function (see 17 and refs. therein). In general terms, a metal-containing cofactor generates a nearby radical species, by reacting with oxygen or by facilitating the homolytic cleavage of a bond. The radical, in turn, initiates the catalytic cycle by abstracting a hydrogen-atom from substrate. Examples include the prostaglandin H synthase enzyme in which the heme reacts with peroxide to generate a $Fe^{IV}/Y^{\bullet}$ species; the tyrosyl radical initiates the catalytic reaction by abstracting a hydrogen-atom from bound substrate. A second example is provided by ribonucleotide reductase. Upon activation of oxygen at its di-iron active site, a tyrosyl radical is generated, which, directly or indirectly, participates in hydrogen-atom abstraction at the ribonucleotide substrate.

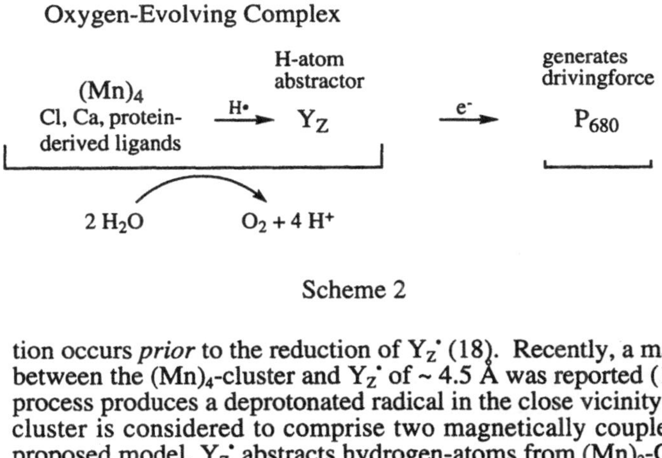

Scheme 2

The complete S-state cycle of the proposed mechanistic model occurs as postulated in the following steps. Upon each turnover, Y_Z is oxidized by P_{680}^+ and its phenol proton is sloughed from the radical site towards the lumen. This is consistent with the observation that a large part of the proton release that accompanies water oxidation occurs *prior* to the reduction of $Y_Z^{\bullet}$ (18). Recently, a magnetic dipole-dipole distance between the $(Mn)_4$-cluster and $Y_Z^{\bullet}$ of ~ 4.5 Å was reported (19). Thus, the proton ejection process produces a deprotonated radical in the close vicinity of the metal center. The Mn-cluster is considered to comprise two magnetically coupled $(Mn)_2$ dimers (20). In the proposed model, $Y_Z^{\bullet}$ abstracts hydrogen-atoms from $(Mn)_2$-OH_2 to produce $(Mn)_2$-OH and further to $(Mn)_2$=O in the S_0-S_1 and S_1-S_2 transitions, respectively. The same procedure occurs in the other dimer, whose substrate binding site was protected by Cl until after S_2 formation (21), in the higher S-state transitions to produce two $(Mn)_2$=O species upon S_4 formation. The two oxo-ligands (each of which carries a formal charge of -2) become oxidized in a concerted fashion, the oxygen-oxygen bond is formed, and dioxygen is released as the system resets to the S_0 state. This model accounts for the report by Messinger *et al.* which shows one fast and one slowly exchanging substrate water molecule in the S_3 state, and for the observation that Y_Z reduction in the S_3-S_4-S_0 transition rate-limits oxygen evolution (23). Moreover, the model in Scheme 2 recasts the distinction between chemical and electrostatic protons (18), and shows how the release of chemically produced protons can occur upon oxidation of Y_Z, rather than upon its reduction.

Acknowledgements
The work and ideas presented in this manuscript is the result of a collaborative effort and we gratefully acknowledge G. T. Babcock, J. McCracken, K. Warncke, C. W. Hoganson and N. Lydakis-Simantiris at Michigan State University, and B. A. Diner and X.-S. Tang at Du Pont de Nemours and Co. C.T. acknowledge support by the Foundation BLANCEFLOR Boncompagni-Luovisi, née Bildt during a research visit 1994 at Michigan State Univesity. S.S. acknowledge support by the Swedish Natural Science Research Council.

References
1 Diner, B. A. and Babcock, G. T. (1995) in Oxygenic Photosynthesis: The Light Reactions (Ort, D., Yocum, C., eds.) Kluwer Academic Publishers, Dordrecht, in press
2 Debus, R. J. (1992) Biochim. Biophys. Acta 1102, 269-352
3 Svensson, B., Vass, I., Cedergren, E. and Styring, S. (1990) EMBO J. 9, 2051-2059
4 Tommos, C., Davidsson, L., Svensson, B., Madsen, C., Vermaas, W. and Styring, S. (1993) Biochemistry 32, 5436-5441
5 Tang, X.-S., Chisholm, D. A., Dismukes, G. C., Brudvig, G. W. and Diner, B. A. (1993) Biochemistry 32, 13742-13748
6 Tommos, C., Tang, X.-S., Warncke, K., Hoganson, C. W., Styring, S., McCracken, J., Diner, B. A. and Babcock, G. T. (1995) J. Am. Chem. Soc., in press
7 Tang, X.-S., Zheng, M., Chisholm, D. A., Dismukes, G. C., Diner, B. A. (submitted) Biochemistry
8 Bernard, M. T., MacDonald, G. M., Nguyen, A. P., Debus, R. J. and Barry, B. A. (1995) J. Biol. Chem. 270, 1589-1594
9 Roffey, R. A., van Wijk, K. J., Sayer, R. T. and Styring, S. (1994) J. Biol. Chem. 269, 5115-5121
10 Chu, H.-A., Nguyen, A. P., Debus, R. J. (1995) Biochemistry 34, 5839-5858
11 Hoganson, C. W., Lydakis-Simantiris, N., Tang, X.-S., Tommos, C., Warncke, K., Babcock, G. T., Diner, B. A., McCracken, J. and Styring, S. (1995) Photosyn. Res., in press
12 Lübbers, K., Haumann, M. and Junge, W. (1993) Biochim. Biophys. Acta 1183, 210-214
13 Kramer, D. M., Roffey, R. A., Govindjee and Sayre, R. T (1994) Biochim. Biophys. Acta 1185, 228-237
14 Warncke, K., Babcock, G. T. and McCracken, J. (1994) J. Am. Chem. Soc. 116, 7332-7340
15 Mino, H. and Kawamori, A. (1994) Biochim. Biophys. Acta 1185, 213-220
16 Svensson, B., Vass, I. and Styring, S. (1991) Z. Naturforsch 46, 765-776
17 Sigel, H. and Sigel, A. eds. (1994) Metal Ions in Biological Systems Vol. 30, Marcel Dekker, New York, USA
18 Lavergne, J. and Junge, W. (1993) Photosyn. Res. 38, 279-296
19 Gilchrist, Jr., M. L., Ball, J. A., Randall, D. W. and Britt, R. D. (1995) Proc. Natl. Acad. Sci. USA, in press
20 Yachandra, V. K., DeRose, V. J., Latimer, M. J., Mukerji, I., Sauer, K. and Klein, M. P. (1993) Science 260, 675-679
21 Rutherford, A. W., Zimmermann, J.-L. and Boussac, A. (1992) in The Photosystems: Structure, Function and Molecular Biology (Barber, J., ed.) pp. 179-229, Elsevier, Amsterdam, The Netherlands
22 Messinger, J., Badger, M. and Wydrzynski, T. (1995) Proc. Natl. Acad. Soc. USA 92, 3209-3213
23 Babcock, G. T., Blankenship, R. E. and Sauer, K. (1976), FEBS Lett. 61, 286-289

EPR CHARACTERISATION OF $Q_A^{\cdot-}$ FROM BIOSYNTHETICALLY AND CHEMICALLY SUBSTITUTED Zn(II) REACTION CENTRES FROM *RHODOBACTER SPHAEROIDES* AND *RHODOPSEUDOMONAS VIRIDIS*.

A.T. Gardiner, F. Lendzian, F. MacMillan, S.G. Zech, R. Bittl, M. Kuhn, W. Lubitz.
Max-Volmer-Institut für Biophysikalische und Physikalische Chemie, Technische Universität Berlin, Straße des 17. Juni 135, 10623 Berlin, Germany.

1. Introduction

In the bacterial reaction centre (RC) magnetic coupling of the semiquinone anion radicals to the nearby high spin Fe(II) ion broadens the EPR line and prevents detection of their intrinsic magnetic parameters. The successful replacement of the high spin Fe(II) (S = 2) by the diamagnetic Zn(II) (S = 0) decouples $Q_A^{\cdot-}$ (and $Q_B^{\cdot-}$) from the iron and allows investigation of the semiquinone anion radicals by stationary and time resolved, spin-polarised EPR (ESP) [1] as well as electron nuclear double resonance (ENDOR) [2]. Two approaches have been developed for the replacement of Fe(II) by Zn(II) in the *Rhodobacter (Rb.) sphaeroides* RC; either (i) chemically, through the action of chaotropic agents [3] or (ii) biosynthetically, by culturing the cells in media containing an elevated concentration of Zn(II) with a correspondingly decreased concentration of Fe(II) [4, 5].

2. Methods

Biosynthetically substituted *Rb. sphaeroides* R26.1 were grown photoheterotrophically in succinate based media containing 20ppm Zn(II) and 0.04ppm Fe(II), *Rhodopseudomonas (Rps.) viridis* was cultured in a slightly higher Fe(II) concentration (0.2ppm) to allow for biosynthesis and function of cytochrome. The *Rb. sphaeroides* RC preparation and procedure for producing chemically exchanged RCs is outlined in [3] except that KSCN instead of LiSCN was used here. *Rps. viridis* RCs in LDAO were prepared as described in [6]. The preparation of *Rb.sphaeroides* mutant HC(M266) ZnRCs in which the His at position 266 on the M-subunit has been changed to Cys was done as described in [7].

EPR experiments have been performed in X- (9GHz) and Q-band (35GHz). Q-band cw-spectra have been recorded on a Bruker ER 200D spectrometer with a ER 050QG bridge, ER 4118CF-O He-cryostat and modified ER 5106QT resonator, controlled by a PC. Time-resolved transient EPR in X- and Q-band have been performed on a ESP 380E spectrometer. In X-band the pulsed mw-bridge ESP-380-1010 and a ER 4118 X-MD-5 probehead were used. In Q-band the same bridge and resonator were used as for the cw-experiments. For light excitation a Spectra Physics GCR 130 Nd-YAG laser (532nm) with a typical pulse width on 10ns and 10mJ was used

P. Mathis (ed.), Photosynthesis: from Light to Biosphere, Vol. I, 655–658.

To determine the amount of uncoupled quinone in the biosynthetically substituted preparations, P^+ was produced by ferricyanide oxidation and $Q_A^{\bullet-}$ by anaerobic dithionite reduction in the dark in high buffer concentration after the RCs had been exchanged into cholate. If 100% $Q_A^{\bullet-}$ is uncoupled from the high spin Fe(II) the integrated signal intensity of $Q_A^{\bullet-}$ should be of the same magnitude as that of P^+. After proper scaling to correct for differences in the ODs of the samples and measuring under identical conditions the integrals of the two signals yielded approximate amounts of uncoupled $Q_A^{\bullet-}$ (assumed to be, but not proven by this method, equivalent to the Zn(II) content) of 6%±3% and 5%±2% for *Rb. sphaeroides* R26.1 and *Rps. viridis* respectively. This agrees well with the results of atomic absorption measurements yielding ≤10% Zn incorporation into the RCs.

3. Results and Discussion

Although the amount of uncoupled $Q_A^{\bullet-}$ is fairly low determination of the $Q_A^{\bullet-}$ g-tensor components from biosynthetically Zn(II) substituted *Rb. sphaeroides* R-26.1 (Q_A is ubiquinone-10) and *Rps. viridis* (Q_A is menaquinone-9) RCs was possible. The obtained Q-band EPR spectra are presented in Figure 1. A similar spectrum of $Q_A^{\bullet-}$ from Fe-free PSII [8] is presented for comparison. The spectra were analysed with a simulation and fit routine for powder EPR spectra from which the obtained *g*-tensor values are collected in Table 1.

Table 1. Comparison of g-tensor principal values of $Q_A^{\bullet-}$ in ZnRCs and related model quinone radical anions in isopropanol.

	$Q_A^{\bullet-}$			Model compounds[a]		
	Rps. viridis ZnRC	*Rb. sphaeroides* ZnRC	PSII Fe-free	$UQ_{10}^{\bullet-}$	vit. $K_1^{\bullet-}$	$PQ_9^{\bullet-}$
g_{xx}	2.0059(1)	2.0065(1)	2.0073(1)	2.00645(5)	2.00579(5)	2.00631(5)
g_{yy}	2.0050(1)	2.0053(1)	2.0054(1)	2.00550(5)	2.00498(5)	2.00521(5)
g_{zz}	2.0021(1)	2.0021(1)	2.0023(1)	2.00221(5)	2.00218(5)	2.00225(5)

[a] Q-band (34GHz) *g*-tensor values of frozen model semiquinone anion radicals in isopropanol. The measured *g*-tensor components are assigned to the molecular axes as follows: $g_{xx} > g_{yy} > g_{zz}$, where x is along the C=O bond direction and z is perpendicular to the quinone plane. Vitamin K_1 is structurally similar to menaquinone. Values in brackets are errors in the last digit.

It has been shown for *Rb. sphaeroides* that the *g*-tensor components of biosynthetically and chemically substituted ZnRCs agree within experimental errors [9]. Comparison of the *g*-tensor principal values of $Q_A^{\bullet-}$ of the different photosystems with those of the respective model quinone anion radical in isopropanol shows good agreement between the respective g_{yy} and g_{zz} values. However, in all three systems the g_{xx} values of $Q_A^{\bullet-}$ are consistently larger than those of the respective model compounds in isopropanol. This has been found earlier for *Rb. sphaeroides* R26.1 RCs [9] and for Fe-free PSII preparations [8] and was attributed to different asymmetric hydrogen bonding and/or π-π interactions with aromatic amino acid residues[8, 9]. In recent FTIR investigations of Q_A in *Rb. sphaeroides* RCs reconstituted with specifically 13C-labelled

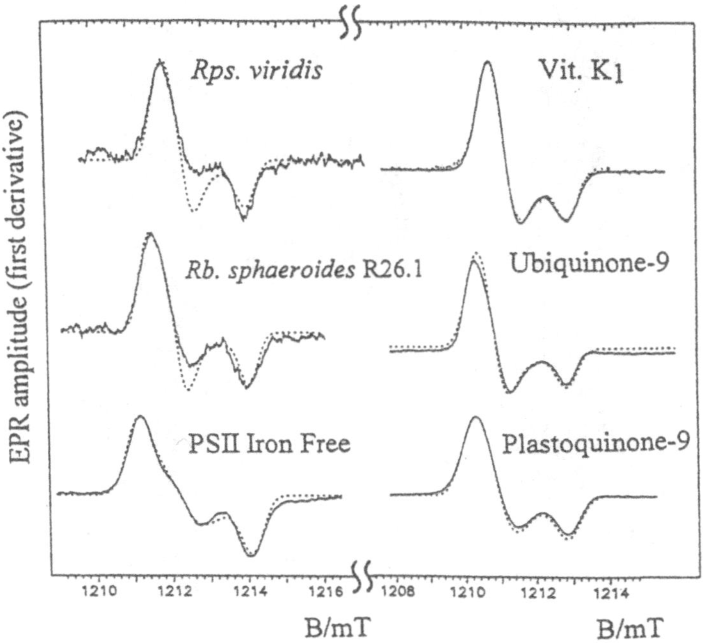

Figure 1. Q-band EPR spectra at T = 80K of chemically reduced $Q_A^{\cdot-}$ in different RCs and the corresponding model quinone radical anions.The dashed lines represent computer fits of the powder EPR spectra from which the *g*-tensor values were obtained.

ubiquinones only one strong H-bond was observed to His(M219) [11, 12]. It is interesting that we observe the same effect for *Rps. viridis*. It has been shown by W-band (95GHz) EPR that in particular the g_{xx}-value of quinone anion radicals is significantly increased in aprotic solvents where hydrogen bonding is prevented [10].

Structural information about the relative orientation of P^+ and $Q_A^{\cdot-}$ in the RCs can be obtained from transient EPR on the spin-polarised secondary radical pair [1]. For native iron containing bacterial RCs only the P^+ part is present in the spectrum. Due to the large Fe(II) content in the biosynthetically Zn-substituted RCs the same is true for these preparations. Figure 2A shows the spin-polarised spectrum for *Rb. sphaeroides* recorded at 100K which agrees well with the published spectrum of native (Fe(II)-containing) RCs recorded at 13K [13]. In chemically Zn(II) substituted RCs from *Rb. sphaeroides* the amount of Fe-decoupled (Zn(II)-substituted) RCs was large enough to record transient EPR spectra of the radical pair P^+ and $Q_A^{\cdot-}$ (Figure 2B). The observed spectrum agrees well with that reported earlier for *Rb. sphaeroides* ZnRCs (95% Zn [1]). Furthermore, ZnRCs of the mutant HC(M266) [7] also show a strong spin-polarised $P^+Q_A^{\cdot-}$ signal (Figure 2C) in agreement with the oservation that this mutant incorporates ≥95% Zn in the RC when grown on enriched Zn(II) media.

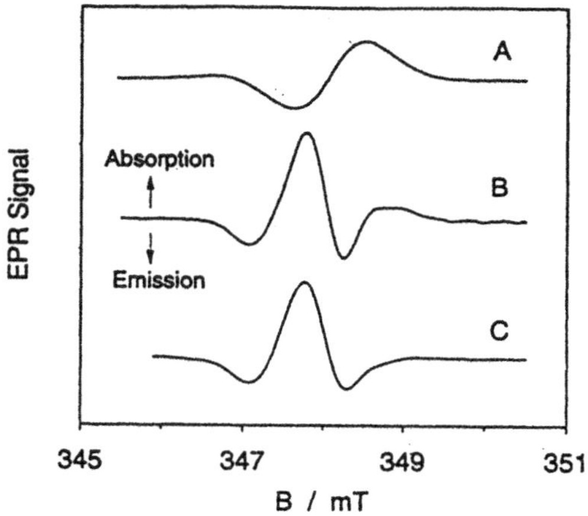

Figure 2. Spin-polarised X-band EPR spectra of *Rb. sphaeroides* RCs:
(A) Biosythetically substituted ZnRCs
(B) Chemically exchanged ZnRCs
(C) HC(M266) ZnRC.
The spectra were extracted from 2D (time/field) data sets. T = 100K.

Work is in progress to establish the conditions necessary for chemical exchange of Fe(II) with Zn(II) in *Rps. viridis*. This would offer the possibility of a detailed structural comparison of the radical pair in both purple bacteria.

We thank J.C. Williams and J. Allen (ASU, Tempe) for providing the *Rb. sphaeroides* HC(M266) mutant. ATG is grateful to the Alexander von Humboldt Stiftung for financial support. This work was supported by Deutsche Forschungsgemeinschaft, Sfb 312 (TP A4)

References

1. Füchsle, G., Bittl, R., van der Est, A., Lubitz, W., Stehlik, D. (1993). Biochim. Biophys. Acta 1142,23-35
2. Feher, G., Isaacson, R.A., Okamura, M.Y. & Lubitz, W. (1985) Springer Series in Chemical Physics, pp 174-189. (Vol. 42), Springer-Verlag, Berlin.
3. Debus, R.J., Feher, G. & Okamura, M.Y. (1986) Biochemistry 25,2276-2287.
4. Buchanan, S.K., Dismukes, G.C. & Prince R.C. (1988) *FEBS Letts.* 229,16-20.
5. Ferris, K.F., (1987) Ph.D Thesis, Princeton University.
6. Thornber, J.P., Seftor, R.E.B. & Cogdell, R.J. (1981) FEBS Letts. 134,235-239
7. Isaacson, R.A., Abresch, E.C., Lendzian, F., Boullais, C., Paddock, M.L., Mioskowski, C., Lubitz, W. & Feher, G. (1995) in Proc. Feldafing III- Workshop on Reaction Centres of Photosynthetic Bacteria Structure and Dynamics, Springer, Berlin, in press.
8. MacMillan, F., Lendzian, F., Renger, G. & Lubitz, W. (1995) Biochemistry 34,8144-8156
9. Isaacson, R.A., Lendzian, F., Abresch, G., Lubitz, W. & Feher, G. (1995) Biophys. J. 69,311-322.
10. Burghaus, O., Plato, M., Möbius, K., MacMillan, F. & Lubitz, W. (1993) J. Phys. Chem. 97,7639-7647.
11. Breton, J., Burie, J.R., Boullais, C. Berger, G. & Nabedryk E (1994) Biochemistry 33,12405-12415.
12. Brudler, R., de Groot, H:J:M., van Liemt, W.B.S., Steggerda, W.F., Esmeijer, R., Gast, P., Hoff, A.J., Lugtenburg, J & Gerwert, K. (1994) EMBO J. 13,5523-5530.
13. Snyder, S.W., Morris, A.L., Bondesson, S.R., Norris, J.R. & Thurnauer, M.C. (1993) J. Am. Chem. Soc. 113,3774-3775.

EPR, ENDOR AND ESEEM INVESTIGATION OF THE ELECTRON ACCEPTOR RADICAL ANION $Q_A^{-\bullet}$ IN PHOTOSYSTEM II (PS II) REACTION CENTRES.

F. MacMillan[#], J. Kurreck, N. Adir[$], F. Lendzian, H. Käss, F. Reifarth, G. Renger and W. Lubitz.

Max-Volmer Institut für Physikalische und Biophysikalische Chemie, Technische Universität, Berlin, D-10623 Berlin, Deutschland, [$] Department of Physics, University of California, San Diego, La Jolla, CA 92093-0319, USA. [#] Present Address DBCM-SBE, CEA-Saclay, F-91191 Gif-sur-Yvette, France

1. INTRODUCTION

In Photosystem II (PS II) a plastoquinone molecule transiently bound to a special pocket (referred to as the Q_B-site) is reduced to hydroquinone via a sequence of two univalent redox steps by the semiquinone of a tightly bound plastoquinone, $Q_A^{-\bullet}$, acting as electron donor. The functional and structural organisation of this process is similar to that of anoxygenic purple bacteria [1]. A non-heme high-spin Fe^{2+} is located between Q_A and Q_B. Its magnetic interaction with the semiquinone radicals $Q_A^{-\bullet}$ and $Q_B^{-\bullet}$ severely restricts their analysis by magnetic resonance methods. In general, two different approaches should be suitable to circumvent this problem; a) removal of this iron centre, and b) transformation of Fe^{2+} from its high spin (S=2) form into the low spin state (S=0). Methods have been developed for both, iron removal [2,3] and spin state transition by cyanide treatment [4]. Furthermore, PS II core complexes were isolated that exhibit an X-band EPR signal typical for $Q_A^{-\bullet}$ without magnetic interactions with Fe^{2+} [5]. The present report describes; comparative Q-band EPR and ^{1}H-ENDOR measurements of $Q_A^{-\bullet}$ in all three sample types and of the isolated PQ-9 radical anion in frozen alcoholic solution and ^{14}N-ESEEM measurements of $Q_A^{-\bullet}$. Implications of the experimental results are discussed in terms of interactions of $Q_A^{-\bullet}$ with the surrounding protein, in particular by hydrogen bonding.

2. MATERIALS AND METHODS

PS II membrane fragments (PS II mf) and core complexes were prepared as outlined in [6,7] and [5] respectively. Iron-depletion and cyanide treatment of PS II mf were done as described in [3] and [4], respectively. For replacement of exchangeable protons the samples were freeze-dried and resuspended in D_2O buffer three times. Control samples were prepared using the same procedure except that resuspension was in H_2O buffer. ENDOR spectra of these control samples show that the lyophilisation steps are virtually without effect (data not shown). The H_2O/D_2O exchange was done after iron removal but must be done before cyanide treatment. All samples were dark adapted at 0°C, to generate $Q_A^{-\bullet}$ $Na_2S_2O_4$ (100 fold molar excess) was added before shock freezing in liquid nitrogen. Plastoquinone-9 (PQ-9) was isolated from spinach and purified with HPLC [3,8]. The anion radical was generated in slightly basic deoxygenated 2-propanol [10]. Q-band EPR spectra were recorded on a computer controlled Bruker ER200D

P. Mathis (ed.), Photosynthesis: from Light to Biosphere, Vol. I, 659–662.

spectrometer using a standard Bruker cavity equipped with an Oxford ESR 935 cryostat. ENDOR experiments were performed using a Bruker ESP300E spectrometer with a self-built, high sensitivity TM_{110} cavity as described [9]. ESEEM experiments were performed on a Bruker ESP380E spectrometer.

3. RESULTS AND DISCUSSION

Fig. 1 shows ^{1}H-ENDOR spectra of $Q_A^{-\bullet}$ in Fe-free (A) and cyanide treated (B) PS II mf and a special PS II core preparation (C) The spectra exhibit line pairs due to the protons of the quinone as well as protons involved in hydrogen bonding between the quinone oxygens and the protein. An analysis of $Q_A^{-\bullet}$ in the Fe-free samples and of the PQ-9 anion radical in 2-propanol was recently given [3]. The $Q_A^{-\bullet}$ spectra of the other two preparations show pronounced differences. From (H_2O-D_2O) ^{1}H-ENDOR difference spectra of $Q_A^{-\bullet}$ in all three PS II preparations (see e.g. Fig. 1D) the hyperfine (hf) tensors of protons in hydrogen bonds were obtained. For comparison similar experiments were performed for PQ-$9^{-\bullet}$ in 2-propanol [3]. The analysis of these spectra reveals shifts both of the proton hf tensors of $Q_A^{-\bullet}$ and of the hydrogen bonds. For example, the hf tensor values of the latter protons indicate weaker hydrogen bonding to $Q_A^{-\bullet}$ in Fe-free than in cyanide treated PS II samples. This change could be explained by a distortion of the Q_A binding site caused by removal of the iron. On the other hand, ligation with cyanide could also lead to structural changes that affect the interaction of $Q_A^{-\bullet}$ with the surrounding protein. The hydrogen bonds of PQ-$9^{-\bullet}$ in 2-propanol are only slightly weaker than those for $Q_A^{-\bullet}$ in cyanide treated PS II mf, but stronger than those observed for $Q_A^{-\bullet}$ in Fe-free PS II mf.

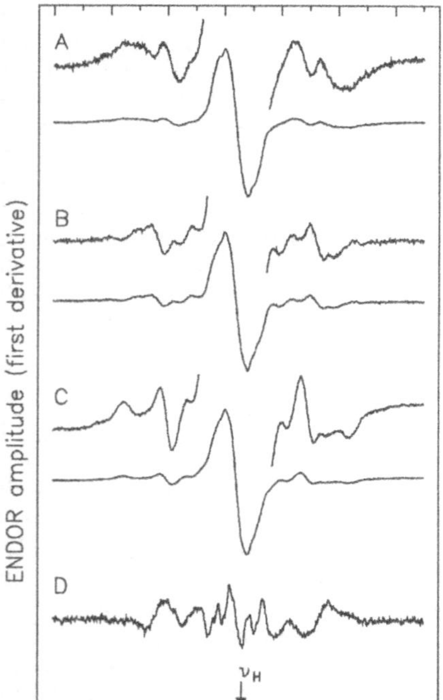

Figure 1:

^{1}H-ENDOR of $Q_A^{-\bullet}$ in A) iron-depleted PS II mf; B) cyanide treated PS II mf; C) untreated core complexes. D) ^{1}H-ENDOR difference spectrum (H_2O-D_2O) of $Q_A^{-\bullet}$ in Fe-free PS II mf. Exptl. conditions: T=20K, mw power = 6 mW, rf power 100 W, 12.5 kHz fm : deviation ±100 kHz, scan time = 120 min;

The structural differences deduced from the ^{1}H hf tensors of $Q_A^{-\bullet}$ in the different samples should also be reflected in their g tensor values. Fig. 2 shows Q-band (34 GHz) EPR spectra of $Q_A^{-\bullet}$ in Fe-free and cyanide treated PS II mf samples and of PQ-9$^{-\bullet}$ in 2-propanol. The g-tensor principal values obtained from fits of the EPR powder spectra (dotted lines, Fig.2) are given in Table 1.

Pronounced differences between the three samples are observed mainly for the g_{xx} component which is the most affected by the hydrogen bond situation [10,11,12]. The high g_{xx} value of $Q_A^{-\bullet}$ in the Fe-free sample is indicative for weak asymmetric H-bonds or H-bonds to only one of the quinone C-O groups. In addition, π-stacking from a nearby aromatic amino acid (Trp 253 D2) could play a role in determining the large g_{xx} value. In the cyanide treated sample g_{xx} is also shifted slightly with respect to PQ-9$^{-\bullet}$ in 2-propanol. The shifts of the g values found for $Q_A^{-\bullet}$ in the different preparations corroborate the different H-bond situations deduced from the ^{1}H-ENDOR spectra. A similar correlation between g-shift and H-bond situation has also been observed for $Q_A^{-\bullet}$ in bacterial reaction centres (RCs) of *Rb. sphaeroides* [12].

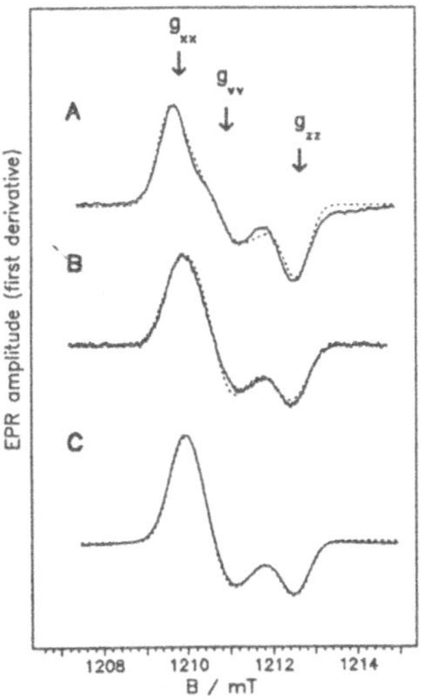

Table 1. G tensor principal values

Radical	g_{XX}	g_{YY}	g_{ZZ}	1/3 Tr g
$Q_A^{-\bullet}$ /A	2.0073	2.0054	2.0023	2.0050
$Q_A^{-\bullet}$ /B	2.0065	2.0052	2.0022	2.0046
PQ-9$^{-\bullet}$ /C	2.0063	2.0052	2.0022	2.0046

A) iron-depleted; B) cyanide treated

Figure 2. Q-band EPR spectra at 34 GHz. A: $Q_A^{-\bullet}$, Fe-free mf, B: $Q_A^{-\bullet}$, cyanide treated mf, C: PQ-9$^{-\bullet}$, 2-propanol. Exptl. condits.; P_{MW}, 2.5 μW; field modulation depth, 0.2 mT, frequency, 12.5 kHz; accumulation time, 10 min; T= 80K.

Recently, ^{14}N hyperfine interactions have been observed in electron spin echo envelope modulation (ESEEM) spectra of $Q_A^{-\bullet}$ both in bacterial RCs [13] and plant PS II preparations [14,15] indicating coupling to a nitrogen atom from the protein surrounding in the vicinity of the semiquinone.

Fig. 3 shows stimulated echo (3 pulse) ESEEM spectra from all three $Q_A^{-\bullet}$ samples studied in this paper. In each of these spectra characteristic lines (*) are observed at 2.9, 2.1 and 0.8 MHz which are assigned to the zero field nuclear quadrupolar resonances, ν_+, ν_- and ν_0 respectively, of one ^{14}N nucleus (for details see refs. 13-15). The effective ^{14}N hf coupling obtained from the double quantum transitions $\nu_{dq} \approx 5.4$ MHz is 2.3 MHz. From the quadrupolar transitions we obtain a coupling e^2Qq/h = 3.4 MHz and an asymmetry parameter η = 0.49 in excellent agreement with refs. [14,15]. These values are typical for ^{14}N in the peptide backbone. In the Fe-free sample and in particular in the core preparation a second ^{14}N quadrupolar tensor is detected which is very similar to those observed in refs. 13 in bacterial RCs and 14 in PS II and is assigned to the N^{δ}-H of a histidine residue.

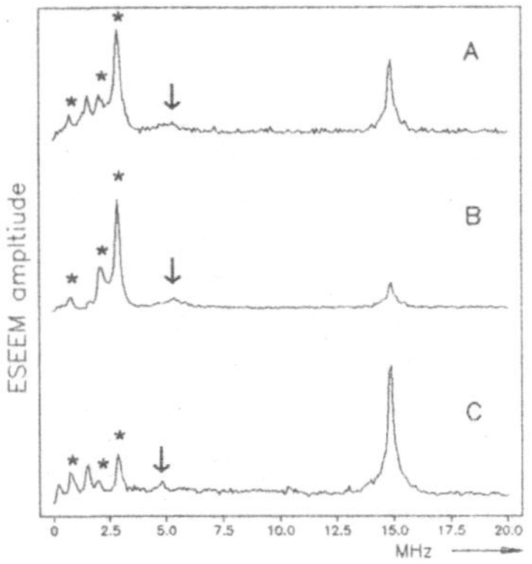

Fig. 3. Stimulated echo ESEEM spectra of $Q_A^{-\bullet}$ in A) iron-depleted mf, B) cyanide treated mf and C) untreated core complex PS II preparations at 20K. The NQR transitions of the peptide backbone ^{14}N are indicated by asterisks, the double quantum transition ν_{dq} is marked by an arrow. Skyline projection of 2D-data sets. Exptl. conditions. : T x τ = 512x128 pts; start values, T_0=240ns, τ_0=160ns; time increments, $\Delta T = \Delta\tau$ = 16ns; $\pi/2$ pulse, 16ns; a four step phase cycle was employed; total accum. time for each spectrum ca. 1.5 h.

This indicates a close contact between $Q_A^{-\bullet}$ and backbone and His nitrogens, probably via hydrogen bonds, which differ in their relative strengths in the investigated three preparations. Since all three samples have been treated in different ways in order to decouple $Q_A^{-\bullet}$ from the magnetic interaction with the high spin Fe^{2+} it is not obvious at present which of these preparations most closely resembles the native system. Work is in progress to investigate this problem in more detail.

Work supported by DFG (Sfb 312, TP A2 and A4) and United States Department of Agriculture grant (USDA 93-37306-9579)

REFERENCES

1. Crofts, A. R. & Wraight, C. A. (1983) Biochim. Biophys. Acta, 726, 149-185.
2. MacMillan, F., Gleiter, H., Renger, G. & Lubitz, W. (1990) in Current Research in Photosynthesis (Baltscheffsky, M., Ed.) Vol. 1, pp 531-534.
3. MacMillan, F., Lendzian, F., Renger, G. & Lubitz, W. (1995) Biochemistry 34, 8144-8156.
4. Sanakis, Y., Petrouleas, V. & Diner, B. A. (1994) Biochemistry 33, 9922-9928.
5. Adir, N., Okamura, M. Y. & Feher, G. (1992) Research in Photosynthesis (Murata, N., Ed.), Vol. II, 195-198.
6. Berthold, D.A., Babcock, G.T. & Yocum, C.F. (1981) FEBS Lett. 134, 231-234.
7. Völker, M., Ono, T., Inoue, Y. & Renger, G. (1985) Biochim. Biophys. Acta, 806, 25-34.
8. MacMillan, F. (1993) Doctoral Thesis, Freie Universität Berlin.
9. Zweygart, W., Thanner, R. & Lubitz, W. (1994) J. Magn. Reson., A109, 172-177.
10. MacMillan, F., Lendzian, F. & Lubitz, W. (1995) Mag. Reson. Chem., in press.
11. Burghaus, O., Plato, M., Rohrer, M., Möbius, K., MacMillan, F. & Lubitz, W. (1993) J. Phys. Chem. 97, 7639-7647.
12. Isaacson, R. A., Lendzian, F., Abresch, E. C., Lubitz, W. & Feher, G. (1995) Biophys. J., 69, 311-322.
13. Bosch, M. K., Gast, P., Hoff, A. J., Spoyalov, A. P. &Tsvetkov, Yu. D. (1995) Chem. Phys. Lett., 239, 306-312.
14. Astashkin, A. V., Kawamori, A., Kodera, Y., Kuroiwa, S. & Akabori, K. (1995) J. Chem. Phys., 102, 5583-5588.
15. Deligiannakis, Y., Boussac, A. & Rutherford, A.W. (1995) Biochemistry, submitted.

FEMTOSECOND SPECTROSCOPY OF PSII REACTION CENTERS: NEW RESULTS

Scott R. Greenfield[a], Michael R. Wasielewski[a], Govindjee[b], and Michael Seibert[c]
[a]Chemistry Division, Argonne National Laboratory, Argonne, Illinois 60439, USA. [b]Department of Plant Biology, University of Illinois, Urbana, Illinois 61801, USA. [c]Basic Sciences Division, National Renewable Energy Laboratory, Golden, Colorado 80401, USA

1. Introduction

The water-oxidation process of photosynthesis is driven by the conversion of light energy into chemical potential in the photosystem II (PSII) reaction center (RC) where rapid energy transfer to the primary electron donor, P680, and charge separation between P680 and the active pheophytin (Pheo) occur. We have reexamined these processes in isolated spinach PSII RCs using femtosecond pump-probe methods at lower pump energy and higher S/N than our previous studies (1,2). Furthermore, we have pumped directly into both the high and low energy sides of the 676-nm absorption band of the RC while probing at 544 nm, where the disappearance of the Pheo ground state can be monitored.

2. Procedure

PSII RCs (6 Chl/2 Pheo) were isolated from market spinach by standard techniques (3). Samples were run at 5 °C under anaerobic conditions using an enzymatic oxygen-scrubbing system (4). Pheo was phototrapped in the presence of freshly made 1.5-mM dithionite. Time-resolved pump-probe kinetic spectroscopy was carried out with 180-fs time resolution using a Ti-sapphire laser-driven optical parametric amplifier (5) with the pump laser polarized at the magic angle relative to the polarized probe beam. The spectral width of the excitation pulses was about 7 nm (FWHM). Pump energies were 50 nJ to 1 μJ, and the spot size was 250 μm. The S/N levels represented by these data are around 50-60, and the analyses reported are averages of fits of individual data sets at 544 nm (i.e., not global analyses).

3. Results and Discussion

Time evolution of transient absorption changes excited by pumping PSII RCs at 683 nm (where P680 absorbs) or 665 nm (where RC antenna pigments absorb) and probing at 544 nm (Fig. 1 inset) elicited a rapid increase in absorption within the time of the pump pulse, and, subsequently, three distinct kinetic components representing decreases in absorption:

P. Mathis (ed.), Photosynthesis: from Light to Biosphere, Vol. I, 663–666.

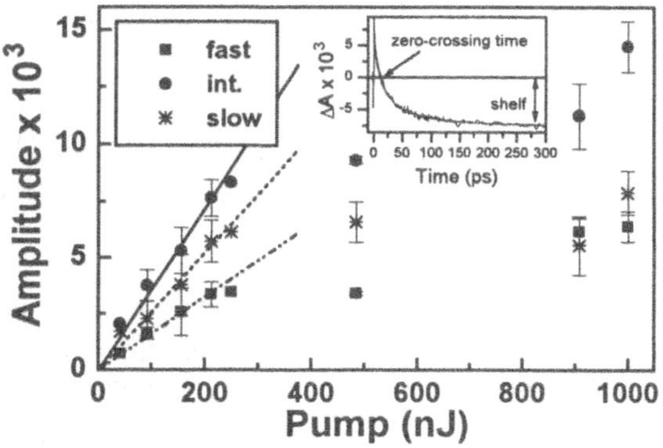

Figure 1. Amplitudes of the three kinetic components vs. pump energy. The inset shows a typical kinetic trace. Pump, 683 nm; probe, 544 nm.

a fast component with τs on the order of 1-3 ps, an intermediate one with τs of 10-25 ps, and a slow one with τs ranging from 50-100 ps. The amplitudes of the three bleach components versus pump energy are linear up to 250 nJ at both pump wavelengths (Fig. 1 shows 683 nm data). Between 250 nJ and 1 μJ the amplitudes continue to increase but at a lesser slope. Zero-crossing times (ZCT, defined in Fig. 1 inset) increase with increasing pump energy from a minimum of about 10 ps at 683 nm excitation and 20 ps at 665 nm excitation. ZCTs provide a convenient way of comparing from laboratory to laboratory the actual pump intensity "seen" by a sample. Whereas the amplitudes of the three kinetic components increase with pump energy, there does not seem to be a consistent change in the relative percentage of the three components over the pump energy range investigated (data not shown). However, the percentage of intermediate time component appears greater, and the lifetimes of all three components increase somewhat when pumping at 665 nm vs. 683 nm.

To further examine this point, we obtained transient absorption spectra of PSII RCs at 500 ps and 1 ps after low energy pump pulses at 665 nm and 683 nm. Figure 2A shows that at 500 ps, transient absorption spectra in the 500-nm to 600-nm range are independent of pump wavelength. This indicates that energy transfer from antenna pigments, absorbing at 665 nm, to P680 is close to 100% efficient. On the other hand, Figure 2B shows that at short time there is a substantial difference in the spectra generated by the two pump wavelengths. The 665 nm minus 683 nm difference spectrum at 1 ps is rather featureless, and the species it represents is probably a Chl pigment excited state that disappears with a τ of about 29 ps. These data provide information about the "donor" state that transfers energy to P680 and also explains why more intermediate component is generated by 665-nm vs. 683-nm pump light on a relative basis, as mentioned above.

In an attempt to determine which of the three kinetic components is associated with primary charge separation, we reduced the RCs by phototrapping $Pheo^-$ in the presence of dithionite. Figure 3 shows that under our conditions, we pre-reduced about half of the active Pheo on average over the time period that the spectra were obtained. The negative "shelf"

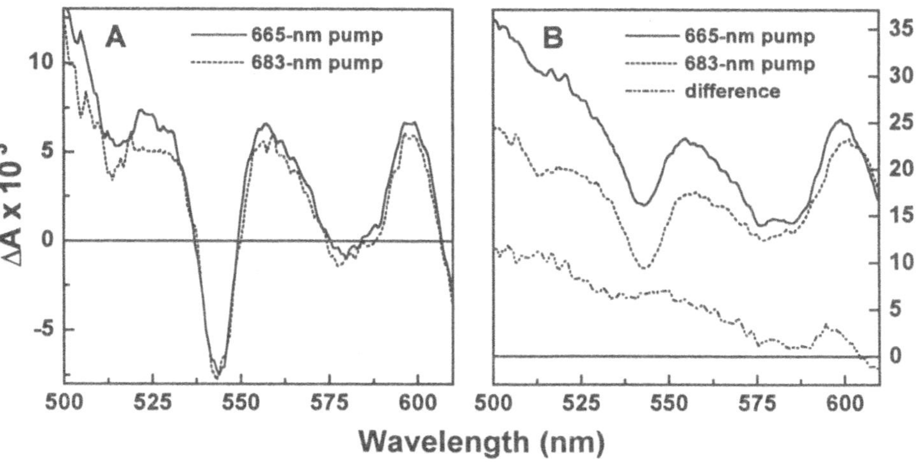

Figure 2. Transient absorption spectra at 500 ps (A) or 1 ps (B) after 665 nm or 683 nm pump pulses. The absorption of PSII RCs are approximately equal at these two wavelengths, and the pump energy of the flashes were also the same (~250 nJ).

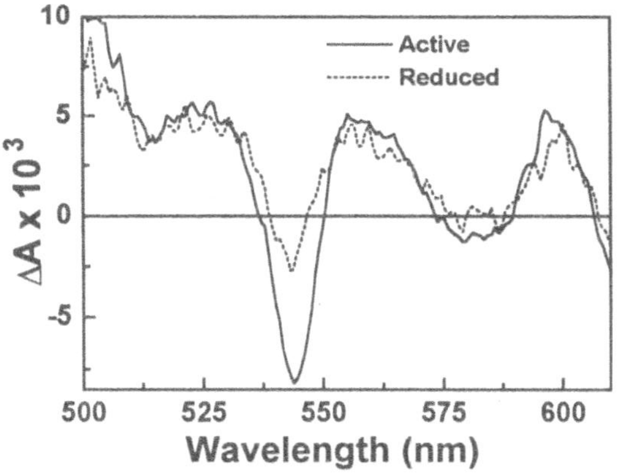

Figure 3. Transient absorption spectra of active and reduced ($Pheo^-$ phototrapped) PSII RCs 500 ps after a 683-nm flash. On average, about half the active Pheo remained in a pre-reduced state over the 10 min it took to obtain the data.

(Fig. 1 inset) value at 500 ps reflects the amount of Pheo reduced by the pump flash. Independent EPR confirmation that our phototrapping technique forms $Pheo^-$ was obtained. Figure 4 shows relative changes in amplitude of the shelf and the three kinetic components after phototrapping $Pheo^-$. Although the data should be considered preliminary at this point, it appears that both the intermediate and slow components lose amplitude when Pheo is pre-reduced. If the samples are allowed to re-oxidize (indicated by a more negative shelf at 500 ps after the pump flash), the amplitudes of both intermediate and slow components increase. This experiment could be complicated by loss of function of some RCs in the sample since experiments (not shown) indicate that the amplitudes of all three kinetic components as well as their lifetimes decrease in "dead" centers.

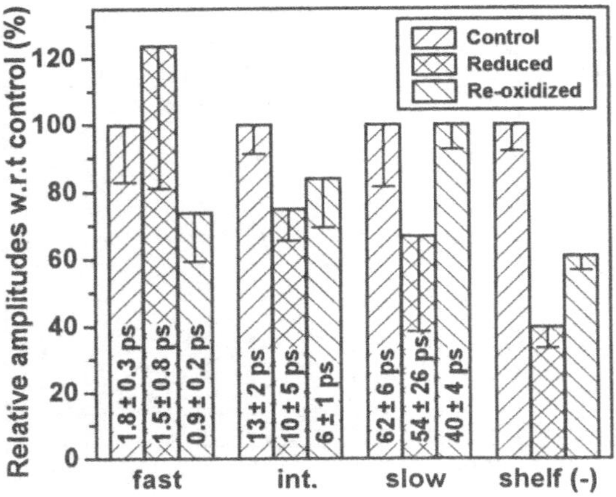

Figure 4. Changes in amplitude of the three kinetic components when Pheo is pre-reduced prior to a pump flash. Control means active samples. Re-oxidized samples were measured ca. 1 hr after phototrapping $Pheo^-$. The numbers on the columns are τs. Half error bars are also shown.

In contrast to our previous report (2), we observe that pre-reduction of Pheo under our current conditions decreases somewhat the amplitude of the intermediate and slow components monitored at 544 nm, but not the fast (3-ps) component. However, since previous studies do not as yet eliminate the possibility that the fast component is associated with charge separation (2,6,7), we suggest that charge separation in isolated PSII RCs is heterogeneous and that the different transient absorption kinetic components monitored at 544 nm may all be convoluted functions of energy transfer and charge separation.

5. Acknowledgements

This work was supported by the Division of Chemical Sciences, U.S. Department of Energy (MS, MW) and by UIUC Urbana (G). SG is a DOE Distinguished Postdoctoral Fellow. The authors thank S. Toon for preparing the RCs used in this study and Drs. L. Utschig and M. Thurnauer for examining reduced samples by EPR.

References

1. Wasielewski, M. R., Johnson, D. G., Seibert, M., and Govindjee (1989) *PNAS 86*, 524-528.
2. Wiederrecht, G., Seibert, M., Govindjee, and Wasielewski, M.R. (1994) *PNAS 91*, 8999-9003.
3. Picorel, R., Chumanov, G., Cotton, T. M., Montoya, G., Toon, S., and Seibert, M. (1994) *J. Phys. Chem. 98*, 6017-6022.
4. McTavish, H., Picorel, R., and Seibert, M. (1989) *Plant Physiol. 89*, 452-456.
5. Greenfield, S. R. and Wasielewski, M. R. (1995) *Opt. Lett. 20*, 1394-1396.
6. Roelofs, T. A., Kwa, S.L.S., van Grondelle, R., Dekker, J.P., and Holzwarth, A.R. (1993) *Biochim. Biophys. Acta 1143*, 147-157.
7. Klug, D. R., Rech, T., Joseph, D.M., Barber, J., Durrant, J.R., and Porter, G. (1995) *Chem. Phys. 194*, 433-442.

FEMTOSECOND STUDIES OF THE PRIMARY CHARGE SEPARATION IN DIFFERENT PREPARATIONS OF PHOTOSYSTEM II.

Brent Donovan, Larry A. Walker II, Charles F. Yocum, and Roseanne J. Sension*, Department of Chemistry, University of Michigan, Ann Arbor, Michigan, 48109, USA.

1. Introduction

Recent femtosecond and picosecond measurements of the primary charge separation event in photosystem II have resulted in a controversy with regard to the time constant for the primary charge separation process. Klug and coworkers have published a series of papers in support of a ca. 20 ps effective time constant for the appearance of the charge-separated state [1-3]. Their conclusions are supported by the experiments of McCauley et al. [4] and Freiberg et al. [5]. On the other hand, several groups have published data interpreted as supporting a ca. 3 ps time constant for the primary charge separation [6-9]. An important aspect of this controversy is that the femtosecond spectroscopic measurements of the only two groups to have performed systematic subpicosecond studies of charge separation in PSII are in disagreement with each other [3,9]. In light of this disagreement we have performed a series of subpicosecond transient absorption measurements on the reaction center of PSII. In these studies, we have concentrated on the effect of sample preparation and laser intensity on the observed transient absorption signals. The most interesting finding is the sensitivity of the observed transient absorption signals to laser intensity.

2. Materials and Methods

2.1 *Sample Preparation*

Subchloroplast PSII membranes (BBY) and oxygen-evolving PSII reaction center complexes (OEC) were prepared from spinach and stored in the dark at -70 °C [10]. Three different D1-D2-Cyt_{b559} isolation protocols were used. The first method of isolation (Prep A) is a variant of the original Nanba and Satoh [11] procedure that has been modified to improve the reproducibility and stability of the resulting D1-D2-Cyt_{b559} complex [12]. This method involves extensive exposure to the detergent Triton X-100 which is then replaced with the milder detergent β-dodecyl-D-maltoside (DM). The second method (Prep B) uses DM to remove the light-harvesting polypeptides from the PSII membranes and extracts the core antenna proteins by addition of a 10% Triton X-100 solution [13]. The Triton X-100 is replaced by 0.03% DM before the D1-D2-Cyt_{b559}

P. Mathis (ed.), Photosynthesis: from Light to Biosphere, Vol. I, 667–670.

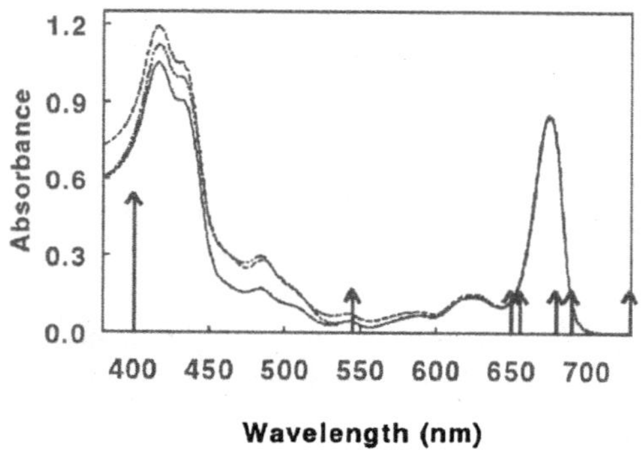

Figure 1. Absorption spectra of the D1-D2-cyt b559 complex prepared according to the three different protocols discussed in the text. Solid line -Prep A, Dashed line -Prep B, Dot-dashed line -Prep C. The arrows indicate the pump and probe wavelengths used in this study.

complex is eluted from the column. The third method (Prep C) avoids the use of Triton X-100 by using a combination of DM and the chaotropic agent $LiClO_4$ [14]. Visible absorption spectra of the three preparations are shown in Fig. 1. The harshest preparation method (Prep A) appears to remove a carotenoid molecule that remains in the other two preparations. Otherwise the spectra are very similar. For all three preparations the Q_y band peaks at 675.5 nm.

2.2 *Spectroscopic Techniques*
A 1 kHz regeneratively amplified Titanium Sapphire laser system was used to provide 150 fs pulses at 800 nm. The second harmonic of half of this beam was used as the excitation pulse (400 nm). The other half of the laser beam is used to generate a white light continuum from which the probe pulses were selected by a series of 10 nm bandpass interference filters. All measurements were made with the pump and probe polarizations set at the magic angle of 54.7° with respect to each other. All experiments were performed under anaerobic conditions produced by the addition of catalase, glucose and glucose oxidase [6]. The sample was kept on ice and flowed through a jacketed, 1mm path length flow cell cooled to 13 °C.

3. Results and Discussion

3.1 *General*
Transient absorption measurements were obtained for a variety of probe wavelengths. Data obtained at 680 nm and 690 nm are shown in Fig. 2. Data obtained at 545 nm, 650 nm, 656 nm and 730 nm are shown in Fig. 3. For all of the scans shown the pump intensity was attenuated until the data obtained at 680 nm were clearly linear with pump intensity. When higher pump intensities were used, various fast components appeared at most probe wavelengths. These components are assigned to singlet-singlet annihilation processes resulting from the multiple excitation of individual reaction center complexes. In the regime where the signals are linear, all of the data can be fit by assuming a model consisting of two components, an instantaneous bleach or absorption with a lifetime > 5 ns and a component having a lifetime of 20 ± 2 ps. The only exception

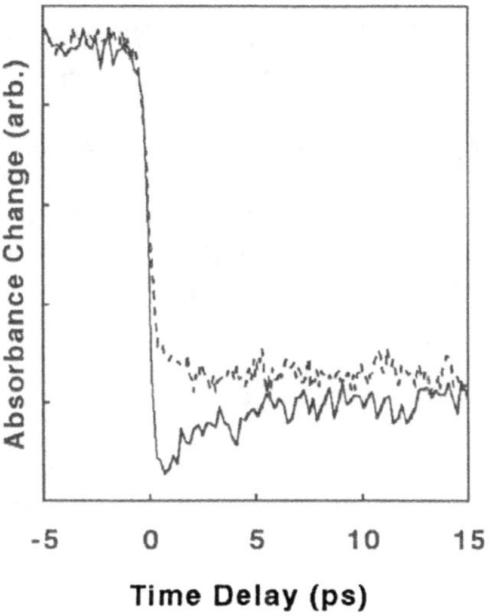

Figure 2. Transient absorption data obtained at 680 nm (dashed line) and 690 nm (solid line) in the Q_y region of the PSII reaction center spectrum. Both of the scans shown were obtained using Prep A.

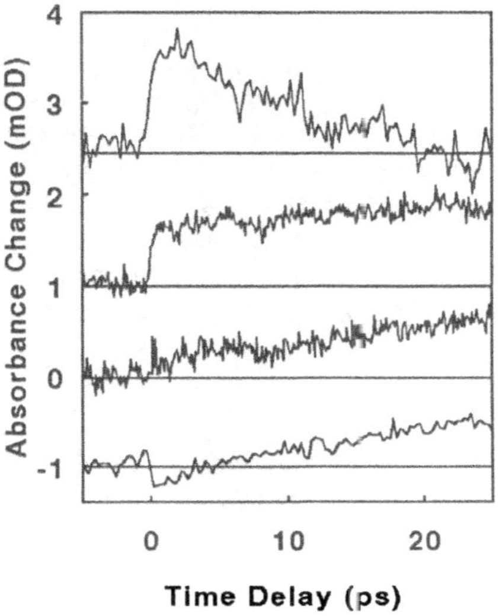

Figure 3. Transient absorption data obtained from top to bottom at 545 nm (Top), 650 nm, 656 nm, and 730 nm. The scans at 545 nm, 650 nm and 656 nm were obtained using Prep B, while the scan at 730 nm was obtained using Prep A.

only exception to this is seen at 690 nm where it was also necessary to include a fast exponential decay of ca. 2 to 3 ps in order to fit the data. The presence of a fast component of small amplitude is not excluded in the remaining data, but the inclusion of such a component does not improve the quality of the fit.

3.2 *Sample Preparation Protocols*

Data were obtained at 650 nm, 680 nm and 690 nm for all three of the sample preparation protocols. Data were obtained at 545 nm for protocols B and C, and at 730 nm for protocols A and C. For all of the samples the results are qualitatively the same. A fast component of ca. 2 to 3 ps is only required to fit the data at 690 nm. At all of the remaining wavelengths the only time-varying component has a lifetime of 20 ± 2 ps. Although the kinetic signals observed are the same for all three of the sample preparation protocols, the samples are not identical. The samples prepared using protocol A are consistently less stable and have to be replaced on average much more often than samples prepared according to the other two protocols.

4. Summary and Conclusions

The results presented here are most consistent with a 20 ± 2 ps time constant for the primary charge separation process in photosystem II reaction center preparations. The bleach of the Pheophytin Q_x band at 545 nm and the decay of stimulated emission and appearance of the ion-pair absorption band at 730 nm both occur on a ca. 20 ps time scale. The Pheophytin anion absorption at 650 and 656 nm also appears on a ca. 20 ps time scale. Any faster contribution to the signals observed at these wavelengths is of a much smaller amplitude. The only possible support for a 3 ps charge separation process is observed at 690 nm. This may be significant, however, as stimulated emission from P680 is expected to make the largest contribution to the observed signal in this spectral region.

5. Acknowledgement

This research was supported by a National Science Foundation grant (MCB-9418390) to R.J.S. C.F.Y. was supported by NSF (MCB-9314173) and USDA-NCGRIP-92-37306-7662.

6. References

1. Hastings, G., Durrant, J.R, Barber, J., Porter, G., and Klug, D.R. (1992) Biochemistry, 31, 7638.
2. Durrant, J.R., Hastings, G., Joseph, D.M., Barber, J., Porter G, and Klug, D.R. (1993) Biochemistry, 32, 8259.
3. Klug, D.R., Rech, T., Joseph, D.M., Barber, J, Durrant, J.R., and Porter, G. (1995) Chem. Phys. 194, 433.
4. Freiberg, A., Timpmann, K., Moskalenko, A.A., and Kuznetsova, N.Y. (1994) Biochim. Biophys. Acta, 1184, 45.
5. McCauley, S.W., Baronavski, A.P., Rice, J.K., Ghirardi, M.L., and Mattoo, M.K. (1992) Chem. Phys. Lett. 198, 437.
6. Schelvis, J.P.M., van Noort, P.I., Aartsma T.J., and van Gorkom, H.J. (1994) Biochim. Biophys. Acta, 1184, 242.
7. Holzwarth, A., Müller, M.G., Gatzen, Hucke G.M., and Griebenow, K. (1994) J. Luminescence, 60&61, 497.
8. Wasielewski, M.R., Johnson, D.G., Seibert, M., and Govindjee, (1989) Proc. Natl. Acad. Sci. USA, 86, 524.
9. Wiederrecht, G.P., Seibert, M., Govindjee, and Wasielewski, M.R. (1994) Proc. Natl. Acad. Sci. USA, 91, 8999.
10. Ghanotakis, D.F., de Paula, J.C., Demetriou, D.M., Bowlby, N.R., Peterson, J., Babcock, G.T., and Yocum, C.F. (1988) Biochim. Biophys. Acta, 974, 44.
11. Nanba O. and Satoh, K. (1987) Proc. Natl. Acad. Sci. USA, 84, 109.
12. Chapman, D.J., Gounaris, K., and Barber, J. (1991) Methods in Plant Biochemistry, 5, 171-193.
13. van Leeuwen, P.J., Nieveen, M.C., van de Meent, E.J., Dekker, J.P., and van Gorkom, H.J. (1991) Photosynthesis Research, 28, 149.
14. Dekker, J.P., Bowlby, N.R., and Yocum, C.F. (1989) FEBS Lett., 254, 150.

THE PRIMARY ELECTRON DONOR OF THE PHOTOSYNTHETIC BACTERIA *CHLOROFLEXUS AURANTIACUS* CHARACTERIZED BY NEAR-INFRARED FOURIER TRANSFORM (FT) RAMAN SPECTROSCOPY.

A. Ivancich[‡], R. Feick[#], A. Ertlmaier[#], and T. A. Mattioli[‡].
[‡] Section de Biophysique des Protéins et des Membranes, DBCM, CEA and URA CNRS 1290. Centre d'Etudes de Saclay, 91191 Gif-sur-Yvette, France.
[#]Max-Planck Institut für Biochemie, D-8033 Martinsried, Germany.

1. Introduction

Although the thermophilic bacterium *Chloroflexus* (*Cf.*) *aurantiacus* belongs taxonomically to the phototrophic green bacteria, its reaction center (RC) resembles that of purple bacteria such as *Rhodobacter* (*Rb.*) *sphaeroides*. The taxonomical classification of *Cf. aurantiacus* is based on the localization of the major antenna pigment (BChl *c*) in structures known as chlorosomes (1-3) which is characteristic for green bacteria. Biochemical and spectroscopical characterization have shown that *Cf. aurantiacus* RC (4-8) is composed of three BChl *a* and three BPhe *a* molecules and only two protein subunits with about 40 % similarity to the respective L and M subunits of *Rb. sphaeroides* and *Rp. viridis* RCs (5). This pigment-protein composition makes it the smallest functional RC isolated thus far. Despite a structure-function relationship similar to that of purple bacteria RC, there are, however, interesting subtle differences in respect to the physico-chemical properties of the primary photochemical reactions (13).

Near-infrared Fourier transform (FT) Raman spectroscopy has proved to be an ideal tool for characterizing bacteria reaction centers, in particular those of *Rb. sphaeroides* (14). The 1064 nm radiation used to excite the RC Raman spectrum selectively enhances the vibrational modes of the primary donor (P) over those of other chromatophores and of the protein. Thus, a pre-resonance spectrum of the neutral P (P°) as well as a resonance spectrum of the oxidized P ($P^{+\bullet}$) can be obtained independently (14, 15). In this work, structural information concerning the protein hydrogen bonds to the primary donor of *Cf. aurantiacus* is deduced from the comparative analysis of the FT Raman spectra of *Rb. sphaeroides* and *Cf. aurantiacus* RCs.

2. Procedure

The isolation and preparation of reaction centers from *C. aurantiacus* and *Rb. sphaeroides* R26 RCs were done as previously described (5, 14). Near-infrared Fourier transform (FT) Raman spectra were recorded using a Bruker IFS 66 interferometer coupled to a Bruker FRA 106 Raman module as described elsewhere (14). The samples were excited with 160 mW of 1064 nm. radiation produced by a diode-pumped Nd:YAG laser; the obtained spectral resolution is at 4 cm^{-1}. Samples were contained in a sapphire cell (14) and measurements were performed at room temperature.

P. Mathis (ed.), Photosynthesis: from Light to Biosphere, Vol. I, 671–674.

RCs were chemically treated by either Na ascorbate (or diathionite) or potassium ferrycianide to poise their primary donor in its reduced (P°) and oxidized ($P^{+\bullet}$) states.

3. Result and Discussion

Near-infrared FT Raman spectra of RCs from *Rb. sphaeroides* and *Cf. aurantiacus* were recorded at room temperature. A comparison of the room temperature FT Raman spectra of both species for the spectral region of 1550-1750 cm^{-1} is shown in Figure 1. These spectra corresponds to RCs of *Cf. aurantiacus* (a) and *Rb. sphaeroides* (b) treated with dithionite and ascorbate, respectively. As it has been mentioned above, the 1064 nm excitation wavelength selectively enhances the preresonance spectrum of the P° in the case of *Rb. sphaeroides* RCs. The reduced primary donor from *Cf. aurantiacus* absorbs at 865 nm and its vibrational modes predominate in the spectrum of Figure 1a.

3.1 The carbonyl region of the FT Raman spectrum for the reduced primary donor.

In this region, the P° spectrum from *Rb. sphaeroides* (Figure 1b) shows four bands which were assigned to the π-conjugated carbonyl groups of the two BChl *a* molecules that constitute the primary donor (15). The 1679 cm^{-1} and the 1691 cm^{-1} frequencies were only consistent with keto carbonyl frequencies, thus they were attributed to both non-hydrogen bonded C_9 keto carbonyls vibrational modes of P_M and P_L, respectively. Consequently, the 1620 cm^{-1} and 1653 cm^{-1} band were attributed to the C_2 acetyl carbonyl vibrational modes with the distinction that the first was hydrogen bonded while the second remained free from interactions. Moreover, the 1620 cm^{-1} band has been shown to correspond to the P_L acetyl carbonyl vibrational mode which is hydrogen bonded to His L168, while the 1653 cm^{-1} band originated from the P_M C_2 carbonyl groups for which there is no known hydrogen bond donor from the polypeptide (see table I in 14).

Similarly, the FT Raman spectrum of the P° from *Cf. aurantiacus* (Figure 1A) showed in the carbonyl region three distinguishable bands at 1632 cm^{-1}, 1648 cm^{-1}, and 1696 cm^{-1} which as in the case of *Rb. sphaeroides* disappeared upon oxidation; thus they were attributed to the neutral state of P (13). Since each BChl *a* molecules has two π-conjugated carbonyls to the porphyrin macrocycle, one C_2 acetyl carbonyl and one C_9 keto carbonyl group, the presence of three carbonyl bands in the spectrum of Figure 1a indicated that more than one BChl *a* molecule constitutes the P° of *Cf. aurantiacus,* as it was the case for *Rb. sphaeroides* (14). This is further evidence for the dimeric character of the primary donor in *Cf. aurantiacus.*

In order to assign the above mentioned bands, it is noteworthy that the C_2 acetyl carbonyl group of BChl *a* may exhibit vibrational frequencies ranging from ca. 1660 cm^{-1} if it is free from interactions, down to 1620 cm^{-1} if engaged in hydrogen bond; the magnitude of the downshift depends on the strength of the H-bond. Similarly, the C_9 keto carbonyl, in general, can be seen to vibrate from ca. 1690 cm^{-1} down to 1660 cm^{-1} (14).

Considering these facts, the 1632 cm^{-1} band can unambiguously be attributed to a C_2 acetyl carbonyl group engaged in a hydrogen bond and the 1696 cm^{-1} band should correspond to a C_9 keto carbonyl group free from protein interactions. As mentioned above, if the primary donor is a dimer there should be another band in the carbonyl region that corresponded to the vibrational modes of the second C_2 acetyl carbonyl. The

1648 cm^{-1} band was the likely candidate for the second C_2 acetyl carbonyl. Finally, as there was no other band to attribute to the second C_9 keto carbonyl group, the band for this group must be also under the intense 1696 cm^{-1} . Both keto carbonyls appeared to vibrate at the same frequency and are both not engaged in hydrogen bonds. The shoulder at 1678 cm^{-1} does not correspond to a carbonyl contribution as in the case of *Rb. sphaeroides* RC but to the BPhe *a* (13).

Thus, the resulting hydrogen bond pattern for the carbonyl groups of the reduced primary electron donor of *Cf. aurantiacus* RC was the following: i) one C_2 acetyl carbonyl (vibrating at 1632 cm^{-1}) was hydrogen bonded while the other was free (1648 cm^{-1}); ii) both C_9 keto carbonyl groups were not engaged in hydrogen bonds (1696 cm^{-1}) and seemed to share similar protein environments. Although this hydrogen bond pattern was similar to that of the P from *Rb. sphaeroides*, the two remarkable points are that the strength of the interaction is different for the H-bonded C_2 acetyl carbonyls in each species; the C_9 keto carbonyl groups in *Cf. aurantiacus* had similar protein enviroments in contrast to that of *Rb. sphaeroides* RC. A comparison of the primary structures of the RCs from both species upon which we based our assignments of the residues responsible for the above mentioned H-bond pattern for *Cf. aurantiacus* was done elsewhere (13).

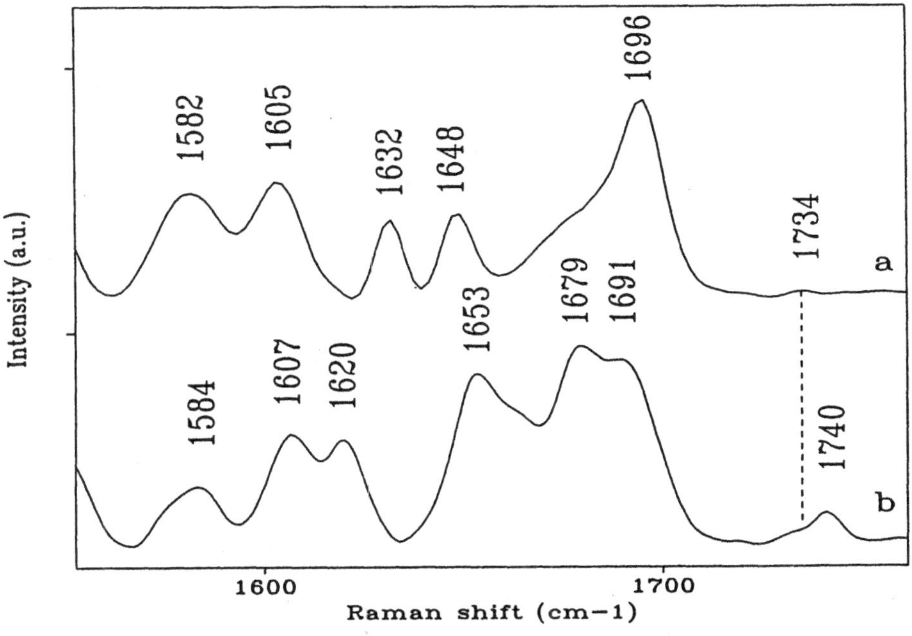

Figure 1. Room temperature P° FT Raman preresonance spectra of *C. aurantiacus* (a) and *Rb. sphaeroides* R26 strain (b) reaction centers. Spectra represent the coaddition of 4000 interferograms.

3.2 Other bands

The 1607 cm^{-1} band in Figure 1b has been unambiguously assigned to the C_aC_m

stretching mode of a BChl species with a single axial ligand (14). Moreover, the crystal structure at atomic resolution of *Rb. sphaeroides* RC shows that a (conserved) histidine residue is the sole axial ligand to each BChl *a* molecule of the primary donor.
Thus, the 1605 cm^{-1} band in the FT Raman spectrum from *Cf. aurantiacus* (Figure 1b) can be attributed to the C_aC_m methine bridge stretching modes of the BChl *a* molecules that constitute the primary donor. The same narrow bandwidth, as in the case of *Rb. sphaeroides* (14), is indicative of a sole axial ligand to each BChl molecule.

It is noteworthy the difference in relative intensities of the 1582-1584 cm^{-1} band in Figure 1a and 1b, respectively. This frequency has been attributed to BPhe *a* contribution from the *in vitro* FT Raman spectrum in the same preresonance conditions (14). The fact that the BPhe *a* molecules contribute to some extent more in the FT Raman spectrum of *Cf. aurantiacus* than in that of *Rb. sphaeroides* is evident in the mid-region FT Raman spectrum as well as in the Resonance Raman spectrum (13). Finally, the 1734 cm^{-1} in both spectra from Figure 1 are likely due to the C_{10} carbometoxy C=O groups (14).

4. Addendum

Abbreviations: BChl, bacteriochlorophyll; BPhe, bacteriopheophytin; *Rb.*, *Rhodobacter*; *Cf.*, *Chloroflexus*; FT, Fourier transform; RC, reaction center.

References

1 Olson, J.M. (1980) *Biochim. Biophys. Acta 594*, 33-51.
2 Feick; G. and Fuller, R. C. (1984) *Biochemistry 23*, 3693-3700.
3 Gibson, J., Ludwig, W., Stackebrandt, E. and Woese, C. R. (1985) *System. Appl. Microbiol. 6*, 152-156.
4 Pierson, B. and Thornber, J. P. (1983) *Proc. Natl. Acad. Sci. U.S.A. 80*, 80-84.
5 Shiozawa, J. A., Lottspeich, F. and Feick, R. (1987) *Eur. J. Biochem. 167*, 595-600.
6 Shiozawa, J. A., Lottspeich, F., Oesterhelt, D. and Feick, R. (1989) *Eur. J. Biochem. 180*, 75-84.
7 Ovchinnikov, Yu. A., Abdulaev, N. G., Zolotarev, A. S., Shmukler, B. E., Zargarov, A. A., Kutuzov, M. A., Telezhinskaya, I. N. and Levina, N. B. (1988) *FEBS Lett. 231*, 237-242.
8 Ovchinnikov, Yu. A., Abdulaev N. G., Shmukler B. E., Zargarov, A. A., Kutuzov, M. A., Telezhinskaya, I. N., Levina N. B. and Zolotarev, A. S. (1988) *FEBS Lett. 232*, 364-368.
9 Allen, J. P., Feher, G., Yeates, T. O., Komiya, H. and Rees, D. C. (1987) *proc. Natl. Acad. Sci. U.S.A. 84*, 5730-5734.
10 Yeates, Y. O., Komiya, H., Chirino, A., Rees, D. C., Allen, J. and Feher, G. (1988) *Proc. Natl. Acad. Sci. U.S.A. 85*, 7993-7997.
11 El-Kabbani, O., Chang, C.-H., Tiede, D., Norris, J. and Schiffer, M. (1991) *Biochemistry 30*, 5361-5369.
12 Ermler, U., Fritzsch, G., Buchanan, S. K. and Michel, H. (1994) *Structure 10*, 925-936.
13 Ivancich, A., Feick, R., Ertlmaier, A. and Mattioli, T.A. (submitted).
14 Mattioli, T. A., Hoffman, A., Robert, B., Schrader, B. and Lutz, M. (1991) *Biochemistry 30*, 4648-4654.
15 Mattioli, T. A., Williams, J. C., Allen, J. P. and Robert, B. (1994) *Biochemistry 33*, 1636-1643.

MAGNETIC CIRCULAR DICHROISM OF PHOTOSYSTEM II REACTION CENTER COMPLEX

Chong-Qin Tang, Kun-Yun Yang, Jian-Min Hou, Ting-Yun Kuang & Pei-Song Tang
Lab. of Photosynthesis, Institute of Botany, Acdemia Sinica, Beijing 100044, P. R. China
Shen Zi-wei, Lin Wei, Zhang xu-fang, Zhao Nan-ming
The state key Lab. of Biomembrane and membrane biotechnology, Tsinghua University, Beijing, 100084

1. Introduction

In the long wavelength Q band of PS II-RC absorption spectrum, serious overlap exists due to pigments of different kinds and different states [1]. However, the absorption spectrum in this area is one of the most sensitive indications of photochemical activity and configuration of pigments [2]. We notice that, during past ten years, magnetic circular dichroism (MCD), as an effective method analyzing the electronic structure of molecules, had been applied to biochemical fields widely [3]. It contributes significantly to research on several proteins with metal prophyrins as cofactors. However, there were few reports about application of MCD method in photosynthesis field. The limited number of MCD reports of chlorophyll and its derivatives were based on the analysis of organic chemistry[4-6]. It is known that most pigments in PSII-RC, such as chlorophyll a(Chl a), pheophytin a (Pheo a) and heme in Cyt b559, are pigments of metal porphyrin type [7], so the research of its MCD will certainly improve our understanding of pigment structure and its binding state on proteins of PSII-RC. This paper first reports the MCD spectrum of PS II-RC and a part of primary assignments of its peak components.

2. Materials and methods

2.1 *Materials*

D1/D2/Cyt b559 complex was isolated from spinach as described by Nanba and Satoh [7].
Chl a and B-Carotene (B-Car) standard samples were dissolved in acetone as 10ug/ml for MCD measurement. Pheo a was prepared by using HCl to eliminate magnesium of Chl a [5].

P. Mathis (ed.), Photosynthesis: from Light to Biosphere, Vol. I, 675–678.

2.2 *MCD spectrum measurement*

The MCD spectrum was measured by Jasco J-500C spectropolarimeter and its MCD accessory. PS II-RC sample was diluted with a buffer of 50 mM Tris-HCl (pH7.2) containing 30mM NaCl, 2mM β-dodecyl maltoside (A675=1.5). The optical path of curette was 0.5 cm. In room temperature, we added a magnetic field of 1.5T parallel to the direction of light and then measured the circular dichroism (CD) spectrum. This CD spectrum was subtracted by the CD spectrum measured without magnetic field. The resultant difference spectrum was the MCD spectrum.

3. Results and discussions

3.1 *MCD spectrum of Chl a*
Figure 1 shows the MCD of Chl a standard sample, which is similar to previously reported results[5,6]. According to previous theoretical calculations and the comparisons with experiment results[4-6], each peaks in figure 1 can be assigned as follows. The positive peak with the greatest magnitude at 662nm in long wavelength region derives obviously from Qy (0-0) transition. This transition is also very clear in absorption spectrum. The negative peak at 628nm

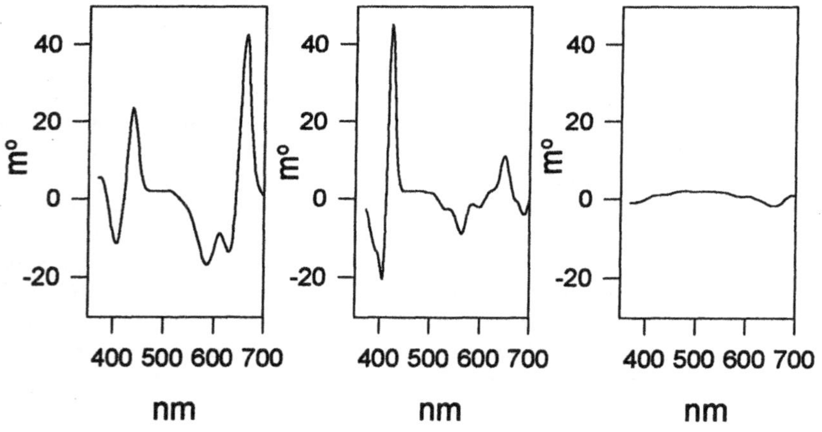

Fig.1 MCD of Chl a Fig.2 MCD of Pheo a Fig.3 MCD of β-Car

with relatively small magnitude comes from Qy (1-0) transition. It still can be found in absorption spectrum. The greatest negative peak at 585 nm is Qx (0-0) transition, and this signal already can not be easily discerned in absorption spectrum. As to Qx (1-0) transition, in our MCD spectrum and previously reported MCD [5] of Chl a which is isotropically dissolved in organic solvent, it does not appear near 540nm as theoretical calculation. However, Frackowiak etc. (1987) [6] dissolved Chl a in liquid crystal to result in an orderly arrangement, Q(1-0) signals appear at 539nm. This phenomenon can be observed in following MCD spectrum of PS II-RC.

3.2 *MCD spectrum of Pheo a*

The MCD spectrum of Pheo a (Fig.2) has an obvious character that its B band signal is remarkable intensive. Its Qx(1-0) transition can be seen at 531nm. As in Chla, the negative peaks at 564nm and 598nm and the positive peak at 650nm can be assigned to Qx(0-0) and Qy(1-0).

3.3 *MCD spectrum of β-Car*

Deduced from its structure, β-Car should contribute weak MCD signals. Actually measured MCD spectrum of β-Car was shown in Fig. 3, with no obvious MCD signal. Without the overlap of β-Car signal, the MCD spectrum of PS II-RC becomes more simple, and it is more easy to discern the signals of other chromophores.

3.4 *MCD spectrum of PSII-RC*

The MCD spectrum of PSII-RC (Fig.4) shows similar character with that of Chla and Pheo a, but some obvious differences still exist. First, compared with the MCD spectrum of Chl a, the relative magnitudes of B band and Q band are changed, probably as the result of the overlapping of Pheo a 's MCD signals. Furthermore, there may be influence from Cyt b559, because according to many reports of MCD spectrum of proteins containing hemes, their MCD signals only appear near 405nm in Chl a absorption B band . The other obvious character of PSII-RC's MCD spectrum is that the strength proportions of each transition component in Q region suffer great change. As to Chl a, the relative strength of the two negative peaks corresponding to Qx(0-0) and Qy(1-0) increases. This change shows the perturbation of the PSII-RC protein microenvioronment to the electronic energy level structure of pigments. Since the number of Chl a in RC is three times as many as that of Pheo a[7], this change mainly comes from Chl a. For the same reason, Qx(1-0) signal at 540nm also mainly comes from Chl a. Although the Chl a in organic solvent does not show Qx(1-0) signal (see Fig.3 and the results of Houssier etc.[5]), Frackowick etc. (1987) [6] observed Q(1-0) of Chl a in orderly liquid crystal solvent. It appeared at 539nm. Beddard and Porter[8] suggested that Chl a in liquid crystal has similar state to the Chl a binding to protein, and could be used as a model for protein-binded Chl a. In our experiment, Chl a in PSII-RC is binded to proteins, so the appearances of Qx (1-0) substantiate the speculation of Beddard and Porter. In addition, we do not exclude the possible contribution of pheo a's Qx(1-0) transition to negative peak at 540nm, for the spectrum of Pheo a undergoes red shifts when binded with proteins, as can be seen from absorption spectrum.

Compared with MCD spectrum of standard Chl a, the longer wavelength side of Qy(0,0) peak of PSII-RC has different shape. It indicates the existence of at least two kinds of Chla. At 680nm, a shoulder peak can be distinguished. It may be the signal of P680. After photoinhibition, to which P680 is excessively sensitive, the signal of 680nm decreased (Fig. 5). This result verified the signal of P680.

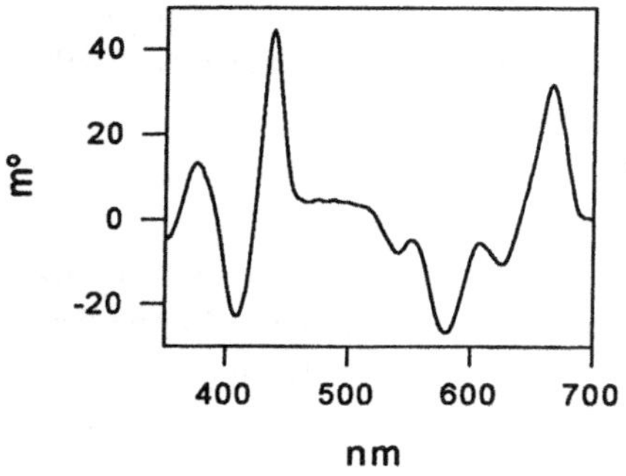

Fig.4 MCD of PSII-RC

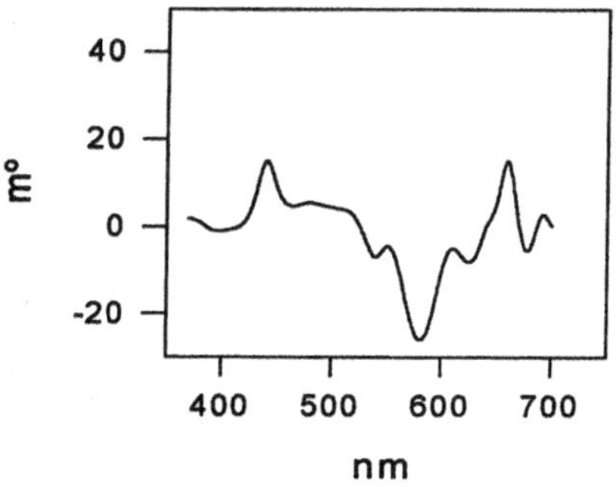

Fig.5 MCD of PSII-RC after photoinhibition

The discernment of P680 component in absorption spectrum needs fourth derivative spectrum, the temperature of liquid nitrogen or computer-aided resolution of spectrum. So MCD spectrum of PSII-RC can supply more information than absorption. In addition, because β-Car has no MCD signals, MCD spectrum of PSII-RC is simplified, which is helpful to further analysis. Therefore, MCD technique is suitable for research on structure and function of PSII-RC.

References

1 Garlaschi, F. M., Zucchelli, G., Giavazzi, P. (1994) Photosynth. Res. 41: 465-473
2 Chapman, D.J., Gounaris, K. and Barber, J. (1989) Photosynthetica 23, 411-426
3 Holmquist, B., Vallee, B. L. (1980) Method in Enzymol. 130, 270-291
4 Briat, B., Schooley, D. A., Records, R. et al. (1967) J. Amer.Chem. Soc. 89: 6170-6177
5 Houssier, C., Sauer, K. (1970) J. Amer.Chem. Soc. 92: 779-791
6 Frackowiak, D., Bauman, D., Manikowski, H. et al. (1987) Biophy. Chem. 28:101-114
7 Nanba, O. and Satoh, K. (1987) Proc. Natl. Acad. Sci. U.S.A. 84, 109-112
8 Beddard, G. S., Porter, G. (1976) 260: 366-378

Light Induced Changes in Photosynthetic Pigment–Protein Complexes Detected by Synchrotron Radiation

Chongci Li, Zhiying Zhang, Yi Sheng
Beijing Agricultural University, Beijing 100094, P.R.China
Wei Wang, Xiangming Gu, Beijing Synchrotron Radiation Facility, P.O.Box 918, Beijing 100039, P.R.China
Tingyun Kuang, Chongqin Tang, Kebin Wang, Fuhong Zhao, Academia Sinica, Institute of Botany, Beijing 100044, P. R. China

1. INTRODUCTION

Absorption spectra or CD (circular dichroism) spectra obtained from ordinary spectrophotometer or spectropolarimeter are limited to wavelength longer than about 190 nm. Because wavelength shorter than 190 nm, the absorption of air increases quickly with the decreasing of wavelength. Furthermore, we have no convenient strong light source with wavelength under 190 nm. But spectra of this shorter wavelength range are very important and useful in measuring and determining conformational changes of Amino acids;proteins; pigment-protein complexes etc.

In a synchrotron,light energy are emitted from and transmitted in high vacumm space, and it is strong enough for some spectral measuring work. We built a device on a radiation beam from < Beijing Electron Positron Collider >, and obtained a light beam with wavelength down to about 165 nm.

2. EQUIPMENT

The device is shown in fig.1 schematically, where a light beam from the synchrotron comes from left, and by mirrors M1 and M2 it is reflected to a concave mirror and focused on the slit S1 of a vacumm monochromator (VM-502). After passing the monochromator, the light beam becomes a monochromatic light beam,and is reflected by mirror M3 to a calcium fluoride window W2.All these mirrors, monochromator and light beam are in high vacumm with vacumm pressure less than 0.0001 Pa. Going out from the window W1,the light beam enters the sample chamber, where presure can be a little less than 1 Pa or as open air pressure.

The light beam from synchrotron is highly linear polarized, so if we put a 1/4 wavelength plate in the optical path

* The Project Supported by NSFC

P. Mathis (ed.), Photosynthesis: from Light to Biosphere, Vol. I, 679–682.

with proper orientation, the light beam will be changed to circular polarized. Instead of a 1/4 wave plate we used a photo-elastic modulator (PEM), which can be used as a 1/4 wave plate of different wavelength when apply different electric voltage on the modulator. This is very convenient when variable wavelength is used in a "Wavelength Scan". Furthermore,when an AC voltage is applied on the modulator, the light beam will be "clockwise circular polarized" and "anti-clockwise circular polarized" alternatively. This is extremly important in CD measuring, and was known as producing "right-revolving" and "left-revolving" circular polarized light alternatively

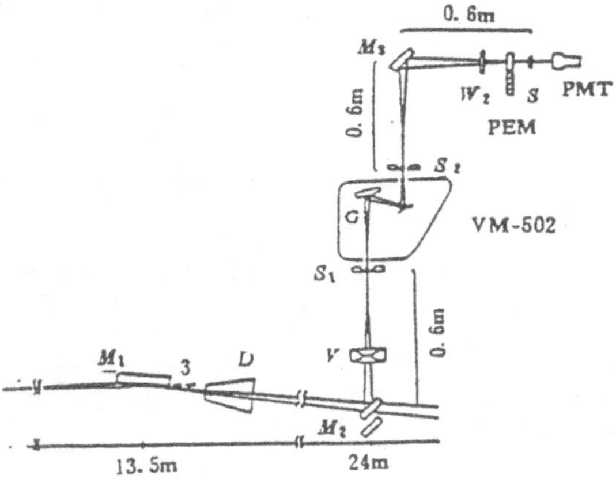

fig. 1

After passing the PEM, the light beam goes through the sample, and then to the photoelectron-multiplier (PMT). Due to the charity of the sample, the absorption of the right-revolving and the left-revolving circular polarized light are different, This is so called CIRCULAR DICHROISM, and thus produce different current in the photoelectron multiplier tube,and this difference can be used to measure circular dichroism.

3.MATERIAL AND METHOD

a, Pigment-protein complex D1/D2/cyt b559. This complex was extracted from PSII reaction center of spinach by a method describsd by Nanba and Satoh [1] and Kuang [2].

b, Pigment-protein complex LHCII2 from cucumber leaves. This complex was prepared as the method used by Xu [3] and Wang [4].

All these complexes were kept in darkness carefully till they were put into the light beam for measuring.

Every complex was measued its CD spectrum first, and this is the "dark spectrum", and then the complex was illuminated by a strong white light of intensity about 150 mW per square cm. for 1 minute, then CD spectrum was measured again, and this spectrum is named "illuminated spectrum".

with proper orientation, the light beam will be changed to circular polarized. Instead of a 1/4 wave plate we used a photo-elastic modulator (PEM), which can be used as a 1/4 wave plate of different wavelength when apply different electric voltage on the modulator. This is very convenient when variable wavelength is used in a "Wavelength Scan". Furthermore,when an AC voltage is applied on the modulator, the light beam will be "clockwise circular polarized" and "anti-clockwise circular polarized" alternatively. This is extremly important in CD measuring, and was known as producing "right-revolving" and "left-revolving" circular polarized light alternatively

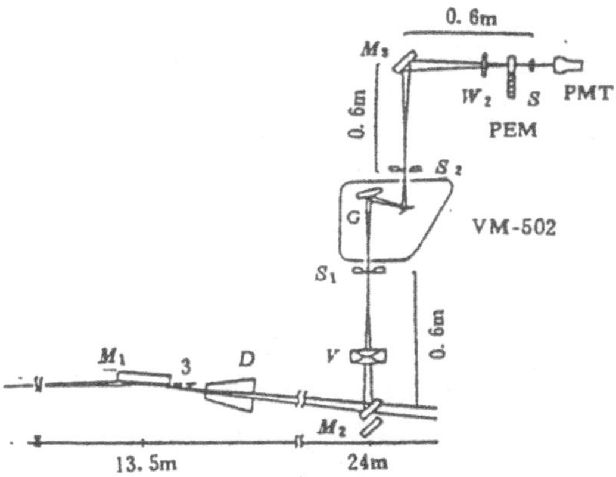

fig. 1

After passing the PEM, the light beam goes through the sample, and then to the photoelectron-multiplier (PMT). Due to the charity of the sample, the absorption of the right-revolving and the left-revolving circular polarized light are different, This is so called CIRCULAR DICHROISM, and thus produce different current in the photoelectron multiplier tube,and this difference can be used to measure circular dichroism.

3.MATERIAL AND METHOD

a, Pigment-protein complex D1/D2/cyt b559. This complex was extracted from PSII reaction center of spinach by a method describsd by Nanba and Satoh [1] and Kuang [2].

b, Pigment-protein complex LHCII2 from cucumber leaves. This complex was prepared as the method used by Xu [3] and Wang [4].

All these complexes were kept in darkness carefully till they were put imto the light beam for measuring.

Every complex was measued its CD spectrum first, and this is the "dark spectrum", and then the complex was illuminated by a strong white light of intensity about 150 mW per square cm. for 1 minute, then CD spectrum was measured again, and this spectrum is named "illuminated spectrum".

Table 1

D1/D2/cyt b559				LHCII2			
DARK		LIGHT		DARK		LIGHT	
WL	CD	WL	CD	WL	CD	WL	CD
199	59.86	199	45.875	199	-39.5	199	22.16
218	-163.0	218	-136.5	218	-148.5	219	-167.5
229	-106.6	230	-95.59	234	-111.5	230	-138.9
240	-28.57	239	-31.81	239	-68.96	240	-49.26

Table 2

D1/D2/cyt.b559				LHCII2			
DARK		LIGHT		DARK		LIGHT	
Hx	18.243	Hx	16.712	Hx	15.43	Hx	13.71
Bs	0.1645	Bs	2.918	Bs	0.362	Bs	6.650
Bt	46.739	Bt	46.199	Bt	49.26	Bt	46.39
Rd	34.854	Rd	34.167	Rd	34.95	Rd	33.24

Reference:

1. Nanba, O. and Satoh, K. (1987) Proc. Natl. Acad. Sci. U.S.A. 84, 109-112
2. Kuang, T. Research in Photosynthesis I.2.267 (Editor: Norio Murata)
3. Xu,C.H. et.al. (1981) Acta Botanica Sinica 23, 40-46
4. Wang,K.B. (1991) Chinese Bulletin of Botany 8(4):56-57
5. Chen,Y.H. and Yang,J.T. (1974) Biochem. 13, 3350-3359
6. Chang,C.T. and Yang,J.T. (1978) Anal.Biochem.13,91
7. Li,C. (1989) Current research in photosynthesis Vol 2 Vol.2, 5.281 (Editor: Baltscheffsky)

ROLE OF THE WATER CHAIN IN THE REACTION CENTER FROM *RB. SPHAEROIDES*

Laura Baciou and Hartmut Michel
Max-Planck-Institut für Biophysik, Heinrich Hoffmann-Str. 7, D-60528 Frankfurt/Main, Germany

Introduction

In the photosynthetic bacteria, a membrane pigment-protein called reaction center (RC) converts light energy into electrochemical energy by performing a sequence of photoinduced electron and proton transfers. The final gate for electrons and protons is a quinone molecule, Q_B, which is buried inside the RC. According to current models, the conduction of protons to Q_B occurs through the protein matrix along a series of side chains of protonatable residues and/or water molecules. A chain of bound water-molecules was recently identified in the photosynthetic reaction center from *Rhodobacter sphaeroides* by X-ray crystallography [1]. The participation of water molecules in the proton transfer to Q_B has recently been proposed [2-5]. To test the role of the water chain in the proton transfer from the cytoplasmic surface of RC to Q_B, we have designed and constructed RCs from *Rb. sphaeroides* in order to interrupt the continuity of the water chain. Selectively, the neutral residue Pro L209 (located in the L subunit about 9Å from Q_B) was changed to the aromatic residue Phe (PL209F), Tyr (PL209Y) or Trp (PL209W). The effects of the mutations on the coupling of the electron and the proton transfer have been studied.

Materials and Methods

The procedure used for the construction of the PL209Y and PL209F mutants and the purification of the modified RCs have been previously described [6]. Similarly, a 37-mer synthetic oligonucleotide containing the mutation (TGG) was used for the construction of the PL209W mutant. Bacterial growth and measurements of electron and proton transfer rates were performed as described in [6]. All electron and proton transfer measurements were carried out at 21°C on a home made apparatus described previously [7].

Results

Growth studies using plates show that all mutants were photosynthetically competent. Generation times of the mutant strains when grown under photosynthetic conditions in liquid culture are slightly longer than in the native strain. In order to

P. Mathis (ed.), Photosynthesis: from Light to Biosphere, Vol. I, 683–686.

understand the effects of the mutation on the cyclic function of the RCs, the formation and disappearance of the semiquinone species after successive saturating flashes were followed at 450nm, at pH 7.5.

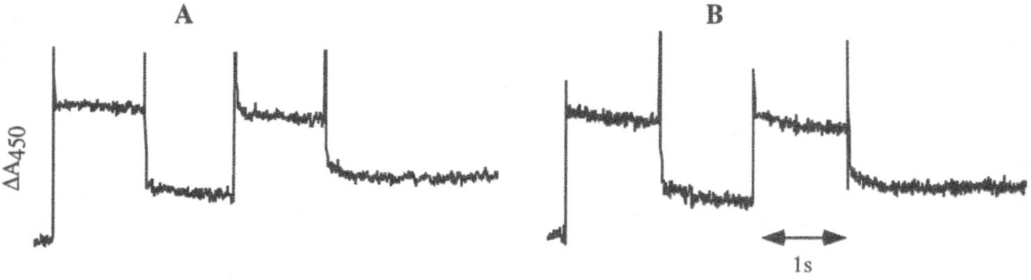

Figure 1: semiquinone oscillation in the PL209Y (**A**) and PL209F (**B**) mutants measured at 450nm in the presence of an exogenous electron donor to P^+ (ferrocene 100 mM) and excess UQ_6 (100 mM) at pH 7.5.

The semiquinone oscillation pattern observed in the PL209Y and PL209F RCs (fig. 1) was essentially identical to that observed in WT. However, in the PL209F mutant, a decrease of the rate of the Q_BH_2 formation was observed.The PL209W mutant displays only slight semiquinone oscillations (data not shown) presumably due to poor Q_B binding (35% Q_B reconstitution at pH 6.8).

Table 1: Electron and proton transfer rates (s^{-1}) and free energy gap between the $Q_A^-Q_B$ and $Q_AQ_B^-$ states (ΔG°, meV) in the RCs from WT and PL209 mutants of *Rb. sphaeroides* (pH 7.5)

	k_{BP}	-ΔG°	$k_{AB}^{(1)}$	$k_{AB}^{(2)}$	k_{H+}
WT	1.0±0.1	55±5	7690±150	1390±50	1110±30
PL209Y	1.3±0.1	51±5	8330±150	455±10	360±10
PL209F	0.5±0.1	73±5	1090±50	90±5	95±5
PL209W	2.3±0.1	25±5	190±10	-	-

Table 1 summarises the rate constants of electron and proton transfer in mutants and WT RCs at pH 7.5. The PL209Y RCs show first electron transfer ($k_{AB}^{(1)}$) and $P^+Q_B^-$ recombination rates (k_{BP}) similar to that observed in WT RCs. This observation suggests that the mutation does not lead to a structural change and does not affect the electrostatic environment of Q_B. In the PL209F mutant, $k_{AB}^{(1)}$ is 7-8 times smaller than WT and $P^+Q_B^-$ is stabilized of about 20 meV. The effect of the mutation in the PL209F mutant on

$k_{AB}^{(1)}$ is surprising compared to the lack of changes in the PL209Y mutant. However, the pH dependence pattern of PL209F RCs previously described [6] is very close to that of the wild type with a similar turning point at pH 8 (attributed to the titration of GluL212 [8, 9]), suggesting that the pK of GluL212 is probably not affected by the mutation. In PL209W RCs, preliminary studies show that the change of ProL209 to Trp leads to faster $P^+Q_B^-$ recombination rate and about 45 times smaller $k_{AB}^{(1)}$ than in WT RCs. The introduction of the large side-chain of Trp has probably induced significant structural changes.

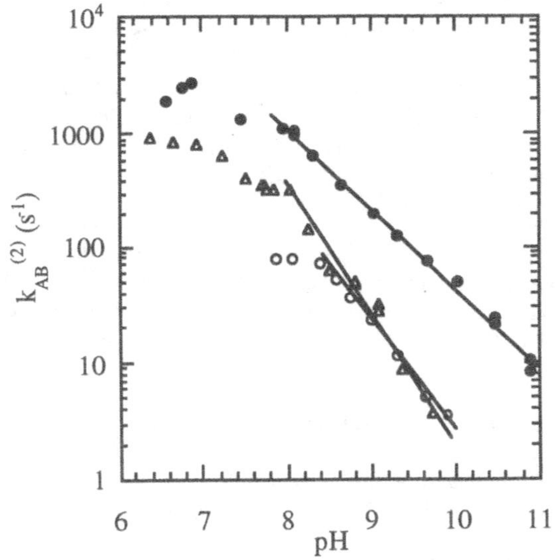

Figure 2: pH dependence of $k_{AB}^{(2)}$, the second electron transfer rate in the RCs from the WT (l), the PL209Y (Δ) and PL209F (o) mutants of *Rb. sphaeroides*. Conditions: 3 μM RCs, 0.1% LDAO, 100mM NaCl, 40mM cyt c, 1mM sodium ascorbate, 450nm.

In L209Y and L209F, the second electron transfer rates ($k_{AB}^{(2)}$) is slowed down over the whole pH range (fig. 2) and displays above pH 8 a slope of ≈ 1 decade/pH unit which is significantly higher than in the WT (0.6 decade/pH unit). Similarly to the $k_{AB}^{(2)}$ values, the rates of proton uptake (k_{H^+}) upon the second flash are decreased in the mutants (Table 1). It is interest to note that k_{H^+}, within experimental errors, are essentially the same as $k_{AB}^{(2)}$. This agrees well with the idea of coupling of the second electron transfer and first proton transfer processes.

Discussion

In this work we have constructed and characterized RCs from *Rb. sphaeroides*, in which Pro L209 (in *Rps. viridis*, the homologous residue is an alanine residue) situated close to the water chain, was changed to Phe, Tyr and Trp. These mutant RCs were designed in order to interrupt the chain of firmly bound water molecules, which was recently identified in the trigonal crystal form of *Rb. sphaeroides* RCs [1]. The substantial slowing down of the second electron transfer and of the proton transfer in the PL209Y and PL209F mutants suggests that the transfer of at least the first proton coupled to the double reduction of Q_B has been affected due to a partial interruption of the pathway used in the WT. The tyrosine is large enough to interrupt the water chain, but small enough to avoid severe clashes with surrounding amino acids. However, in the PL209F mutant, the introduction of a hydrophobic phenylalanine in this region surrounded by a cluster of strongly interacting charged residues might have disturbed the interaction and/or induced a slight structural reorganization of this cluster causing the observed changes of $k_{AB}^{(1)}$. The slope of ≈ 1 decade/pH unit in the PL209Y and PL209F suggests for the mutants a higher non-specific diffusion process for proton transfer using other possible parallel pathways. This hypothesis is consistent with the smaller effect of the mutations on the proton transfer rate than of the mutations of Ser L223, Asp L213 or Glu L212 [10], which may serve as final gates for the protons. In the WT, the chain of firmly bound water molecules may be the most efficient pathway for proton conduction from the cytoplasm to Q_B. Whether the chain of water molecules has indeed been interrupted in the mutants is currently being investigated by X-ray crystallography. We further investigate the role of the water-chain in proton transfer to Q_B by varying the side chain at L209.

References

1. Ermler, U., Fritzsch, G., Buchanan, S. K., & Michel, H. (1994) *Structure 2*, 925-936
2. Beroza, P., Fredkin, D. R., Okamura, M. Y., & Feher, G. (1992) in *The Photosynthetic Bacterial Reaction Center* II (Breton, J., & Verméglio, A., eds.), Plenum Press, New-York, pp. 363-374
3. Hanson, D. K., Tiede, D. M., Nance, S. L., Chang, C. C., & Schiffer, M. (1993) *Proc. Natl. Acad. Sci. USA. 90*, 8929-8933
4. Maroti, P., Hanson, D. K., Baciou, L., Schiffer, M., & Sebban, P. (1994) *Proc. Natl. Acad. Sci. USA. 91*, 5617-5621
5. Okamura, M. Y., Paddock, M. L., McPherson, P. H., Rongey, S., & Feher, G. (1992) in *Research in Photosynthesis*, Vol. I (Murata, N., ed.) pp 349-356, Kluwer Academic Publishers, Dordrecht, The Netherlands
6. Baciou L. and Michel H. (1995) *Biochemistry* 34, 7967-7972
7. Baciou, L., Rivas, E., & Sebban, P. (1990) *Biochemistry 29*, 2966-2976
8. Paddock, M. L., Rongey, S. H., Feher, G., & Okamura, M. Y. (1989) *Proc. Natl. Acad. Sci. USA. 86*, 6602-6606
9. Takahashi, E., & Wraight, C. A. (1992) *Biochemistry 31*, 855-866
10. Okamura, M. Y., & Feher, G. (1992) *Ann. Rev. Biochem. 61*, 861-896

ELECTRON TRANSFER AND TRIPLET STATE FORMATION IN PHEOPHYTIN *A* - MODIFIED REACTION CENTERS FROM *RHODOBACTER SPHAEROIDES* (R-26) AT LOW TEMPERATURES

E.M. Franken[1], A.Ya. Shkuropatov[2], T.J. Aarstma[1], P. Gast[1] and A.J. Hoff[1]
[1]Department of Biophysics, University of Leiden, P.O. Box 9504, 2300 RA Leiden, The Netherlands; [2]Institute of Soil Science and Photosynthesis, R.A.S., Pushchino, Moscow region 142292, Russian Federation

1. Introduction

Recently, it has been shown that bacteriopheophytin *a* (BPheo) at the $H_{L,M}$ binding sites in reaction centers (RCs) from *Rhodobacter (Rb.) sphaeroides* R-26 can be selectively exchanged by plant pheophytin *a* (Pheo) molecules [1]. (Pheo-)modified RCs were found to preserve overall photochemistry leading to formation of the charge-separated state $P^+Q_a^-$, but the quantum yield of this state had diminished to 70-80% at room temperature [2,3]. The long-living population of $P^+B_L^-$ and P^*, simultaneously with $P^+H_L^-$, has been directly observed in modified RCs by Schmidt *et al.* [3]. They proposed a reaction scheme in which the free energy of $P^+B_L^-$ lies approx. 450 cm^{-1} below P^*, and 180 cm^{-1} above $P^+H_L^-$ [3].

Since this scheme suggests a thermally activated population of the $P^+B_L^-$ and P^* states, we studied the effect of temperature on the quantum yield of $P^+Q_a^-$ in modified RCs. We also report on photo-induced charge separation and triplet state formation in these modified RCs, in order to extend the characterisation of this preparation.

2. Materials and methods

RCs from *Rhodobacter sphaeroides* R-26, isolated as in [6], were modified as described by Scheer et al. [1]. A rough estimation, based on the decrease of the BPheo Q_y absorption band at 758 nm shows that approximately 80% of bacteriopheophytin molecules are replaced by Pheo molecules in the modified RCs used in this study.

All measurements were performed in a 10 mM Tris-HCl / 0.1 % LDAO buffer at pH 8.0, containing 66% (v/v) glycerol. For measurements on the triplet state, the primary quinone acceptor of the RCs (Q_a) was pre-reduced by freezing the sample under continuous illumination in the presence of 50 mM sodium ascorbate.

Flash induced absorption difference kinetics were measured with a single-beam-spectrophotometer. Excitation flashes were provided by a Q-switched frequency doubled Nd-YAG laser (15 ns FWHM, 532 nm) or by a tunable BBO-OPO laser (6 ns FWHM).

For room temperature measurements the sample was placed in a 1-cm path length cuvette. Low temperature measurements were performed in an Utreks helium flow cryostat with the sample in a 1-mm cuvette placed under 45° with respect to measurement and excitation beam. All samples had an OD of about 0.2 at 865 nm.

3. Results and discussion

P. Mathis (ed.), Photosynthesis: from Light to Biosphere, Vol. I, 687–690.

3.1 $P^+Q_a^-$ *state formation*

Similar to native RCs, modified RCs are photochemically active down to 10 K. Fig. 1 shows the spectra of absorbance changes induced in modified and native RCs at 10 K by non-saturating flashes. Both spectra were normalized at 895 nm. These spectra are better resolved than those reported at 80 K [5] and 300 K [2], showing, in particular, the negative shoulder at approx. 815 nm, which is a characteristic feature of the low temperature ground state absorption spectra [6]. The spectrum measured for native RCs agrees well with spectra reported earlier.

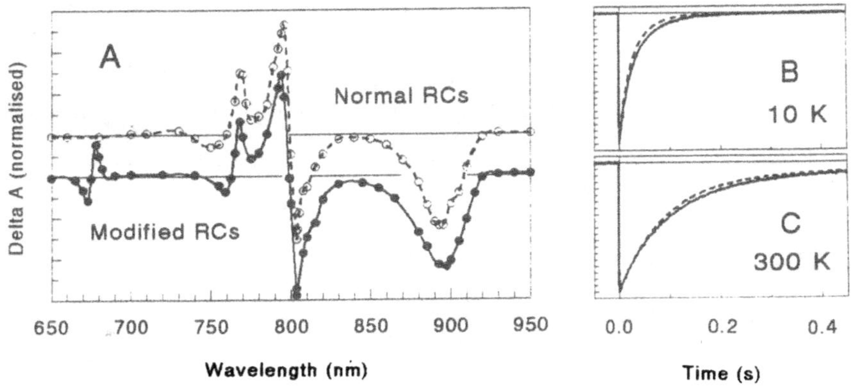

Figure 1. Normalised $P^+Q_a^-$ absorption difference spectra at 10 K (excitation at 895 nm) (A) and decay kinetics measured at 870 nm (excit. at 532 nm) at 10 K (B) and 300 K (C) of modified (solid line) and native (dashed line) RCs of Rb. sphaeroides *R-26.*

The most pronounced differences between the spectra of absorbance changes measured for modified and native RCs are seen, as expected, in the spectral regions of the Q_y transition of BPheo (730-775 nm) and Pheo (660-685 nm). In the latter region the new feature corresponds to a red shift of the Q_y absorption band of Pheo introduced in the RCs [5]. The interpretation of the absorbance changes in the spectral region 760-768 nm is more complicated. Whatever the origin of these absorbance changes, their presence in the $P^+Q_a^-$ spectrum could hint towards the possibility that the residual BPheo molecules in modified RC are mainly localized at the H_L-binding site. In that case, about 40% of modified RCs still contains BPheo at the H_L site.

Fig. 1B and C show the normalized kinetics of the flash-induced absorbance changes that reflect light-induced generation and recombination of the $P^+Q_a^-$ state in native and modified RCs at 10 K and 300 K. The kinetics were monitored at 870 nm (the 'isosbestic point' for the long-wavelength Q_y band of the special pair P). The charge recombination time in native RCs is about 26 ms at 10 K, which is in general agreement with previous reports, remaining constant with rising temperature until about 100 - 150 K, and then increasing gradually to 70 ms for native RCs at 300 K. At all temperatures, the recombination time for modified RCs is slightly longer than for native RCs. The recombination time between Q_a^- and P^+ is thought to be very sensitive to the changes in the distance between P and Q_a. The results presented here indicate that the exchange of BPheo by Pheo does not lead to drastic structural alterations.

3.2 *Triplet state formation*

Recently the triplet state belonging to the primary electron donor (^{3}P) has been detected in modified RCs at low temperatures by MODS and flash-EPR [5] techniques. Here we report on a study of the triplet states in modified RCs by means of optical detection.

Fig. 2A shows the normalised flash-induced triplet-minus-singlet (T-S) absorbance-difference spectra detected at 10 K for modified and native RCs with pre-reduced Q_a. The T-S spectra of both kind of RCs are similar, and agree with spectra previously reported for the ^3P state in native RCs [4]. In Fig. 2B, were the 780-830 nm region (BPheo and accessory BChl Q_y transitions) is enlarged, the distinct differences are visible in the oscillations present in the ^3P spectra around 800 nm. The oscillations for native RCs are comparable to those in earlier reports [4]. In the ^3P spectrum of modified RCs all oscillations are shifted 2 nm to the blue resulting in a much smaller absorption decrease around 796 nm. The shift in oscillations correlates with an approx. 2-nm blue shift of the 802-nm band in the absorption spectrum of modified RCs caused by exchange of BPheo with Pheo (not shown). The observed oscillations seem to be mainly due to shifts and / or intensity changes of the 802 and 812 nm Q_y absorption bands of $B_{L,M}$ upon ^3P formation [4]. The preservation of the oscillations in the T-S spectrum of modified RCs suggests that the interactions of P and the accessory bacteriochlorophyll are not dramatically altered by the modification procedure.

At all temperatures native and modified RCs have about indentical lifetimes of the ^3P state. Between 10 K and 150 K ^3P decays with about 140 µs. At higher temperatures the lifetime decreases with temperature to approx. 40 µs at 250 K, thus being in good agreement with the results earlier reports for native RCs.

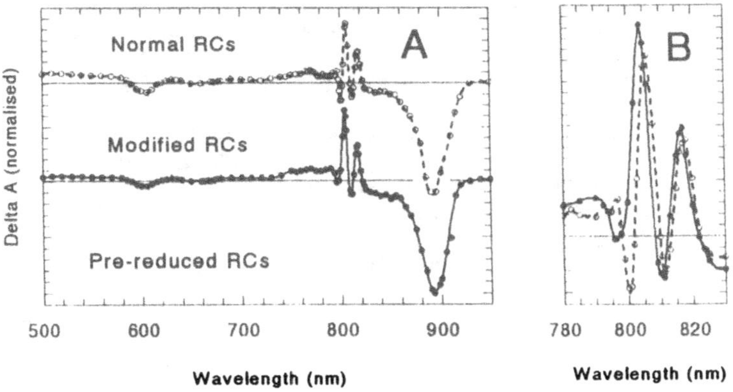

Figure 2. Normalised optically detected T-S spectra for pre-reduced modified (solid line) and native (dashed line) RCs at 10 K. Non-saturating excitation at 895 nm. Spectral region (A) 500 - 950 nm and (B) 780 - 830 nm. Spectral resolution ≤ 1 nm.

3.3 *Temperature dependence of the $P^+Q_a^-$ yield*

Fig. 3 presents the temperature dependence of the relative quantum yield of $P^+Q_a^-$ formation in modified RCs, measured as the ratio of the P band bleachings of modified and native RCs, excited with non-saturating light flashes. The quantum efficiency in native RCs has been reported earlier to be about 100 % at all temperatures.

The room temperature measurements revealed a yield of the $P^+Q_a^-$ state formation in modified RCs of 80 ± 8 %, confirming recently reported values of 79% [2] and 70% [3]. Both measurements at 870 nm and at the P bleaching maximum showed that the yield of $P^+Q_a^-$ in modified RCs is independent of temperature, within the experimental error of ± 10 %.

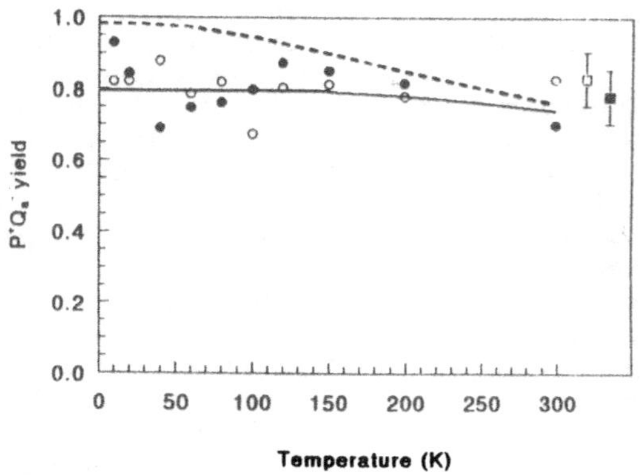

Figure 3. Temperature dependence of the relative yield of $P^+Q_a^-$ state formation for modified RCs. Filled circles: measured at 870 nm; open circles: measured at maximal bleaching of P band. Non-saturate excitation at 870 nm. Squares: separate room temperature measurement (detection and non-saturate excitation at 865 nm). Solid line: fit according to this work; dashed line: fit according to parameters of Schmidt et al. [3].

In order to compare the experimentally measured temperature dependence with that predicted by the scheme proposed by Schmidt *et al.* [3] we calculated the temperature dependence of the $P^+Q_a^-$ yield for the parameter-set presented there, thus with recombination to the ground state only from $P^+B_L^-$. Rate constants associated with energetically downward transfer processes were assumed to be temperature independent, while the temperature dependence of upward processes was described by a factor $e^{-\Delta E/kT}$, where ΔE is the energy difference involved. The calculated $P^+Q_a^-$ yield increased gradually from 80 % at room temperature to over 95 % at 10 K, which clearly deviates from the measured yield (figure 3, dashed line). To improve the fit, we allowed in addition a recombination path via $P^+H_L^-$, and took for the recombination times of both $P^+B_L^-$ and $P^+H_L^-$ the lifetimes measured for the P^+I^- state for pre-reduced RCs (1.5 ns for modified [2] and 12 ns for native RCs). The resulting fit turned out to be quite well (figure 3, solid line). Using these new parameters, the ps measurements at 1020 nm presented in [3] could be fitted equally well as with the original parameter-set.

In conclusion, these results suggest that recombination to the ground state via the $P^+H_L^-$ state must be included in the kinetic model proposed by Schmidt *et al.* [3], in order to account for the temperature dependence of the $P^+Q_a^-$ yield.

Acknowledgements

The work was financed by the Netherlands Foundation for Scientific Research (S.O.N.).

References

1 Scheer, H., Meyer, M. and Katheder, I. (1992) in: The Photosynthetic Bacterial Reaction Center II: Structure, Spectroscopy and Dynamics (Breton, J. and Vermeglio, A. eds.), NATO ASI Series. Ser. A: Life Sciences 237, pp.49-57, Plenum Press, New York
2 Shkuropatov, A.Ya. and Shuvalov, V.A. (1993) FEBS Lett. 322, 168-172
3 Schmidt, S., Arlt, T., Hamm, P., Huber, H., Nagele, T., Wachtveitl, J., Meyer, M., Scheer, H. and Zinth, W. (1994) Chem. Phys. Lett. 223, 116-120
4 Vrieze, J., Thesis (1994), Leiden university
5 Shkuropatov, A.Ya., Proskuryakov, I.I., Shkuropatova, V.A., Zvereva, M.G. and Shuvalov, V.A. (1994) FEBS Lett. 351, 249-252
6 Feher, G. (1971) Photochem. Photobiol. 14, 373-387

Electrostatic influences of Glu^{M236} on the proton conduction pathway of photosynthetic reaction centers from *Rhodobacter sphaeroides*

Eiji Takahashi and Colin A. Wraight
Department of Plant Biology, MC-051
University of Illinois, Urbana, IL 61801-3838 USA

1. Introduction

In the bacterial photosynthetic reaction center (RC), the secondary quinone (Q_B) binding site is buried within the protein. Light-induced electron transfer events in the RC result in double reduction of quinone, accompanied by double protonation, leading to the conversion of a quinone to quinol (Q_BH_2). Protonatable residues near the Q_B binding site are important both for proton delivery in the production of quinol and for stabilizing the singly reduced anionic semiquinone, Q_B^-. These roles have been investigated through structure-function studies employing site-directed mutagenesis (see refs. 1,2).

In *Rb. sphaeroides,* Glu^{L212} and Asp^{L213} comprise an inner shell of residues with particular functional importance for proton transfer to Q_B (3-5). Asp^{L213} is crucial for delivery of the first proton, coupled to transfer of the second electron, while Glu^{L212} is important for delivery of the second proton, following the second electron transfer. These two residues, however, do not represent a unique solution to the problem of transferring protons to the Q_B site. In closely related bacterial reaction centers from *Rps, viridis* and *Rs. rubrum,* the native residue at L213 is a non-protonatable asparagine. In these species, an aspartic acid is found at M44 (or M43, depending on the species), which, in the case of *Rb. sphaeroides* and *capsulatus,* is asparagine. Furthermore, in species containing aspartic acid at L213, proton transfer function is restored when mutation of Asp^{L213} to alanine or asparagine is accompanied by mutation of Asn^{M44} to aspartic acid (6-8). Residues farther away from the Q_B binding site also influence the proton conduction pathway. Suppressor mutations to site-directed mutants with a non-ionizable residue at L213 have been found more than 17 Å from Q_B, e.g., $Arg^{M233}\rightarrow$Leu or Cys (7,8).

This dependence of function on charge distribution suggests the general importance of a negative electrostatic potential in the proton transfer functions of this domain. The recent identification of a continuous chain of water leading from the cytoplasmic surface to the Q_B binding site in *Rb. sphaeroides* RCs elevates the likely importance of water molecules in the proton conduction pathway (9). As part of a general exploration of the importance of internal water molecules in the proton conduction pathway, we have made several mutations at position M236, a glutamate residue which appears to interact with several water molecules in the chain.

2. Materials and Methods

The general mutagenesis procedures have been described previously (10). Mutations were obtained by *in vitro* mutagenesis, using oligonucleotides with altered

P. Mathis (ed.), Photosynthesis: from Light to Biosphere, Vol. I, 691–694.

codons for amino acid residue 236 of the M-subunit gene. Growth conditions for mutant cells, as well as the RC preparation procedure, have been described previously (5). Mutant RCs were characterized by optical spectroscopy performed on a kinetic spectrophotometer of local design. The assay solution was 2.5 mM KCl, 0.02% Triton X-100, 20 μM ubiquinone (Q-10), 1 mM buffer (succinate, citrate, Mes, Pipes, Tris, Ches, or Caps, depending on the pH) and 1 μM reaction center. The kinetics of P^+ dark decay following a flash ($P^+Q_AQ_B^-$ charge recombination) were measured at 430 nm. The kinetics of the first electron transfer from Q_A^- to Q_B were measured at 397 nm. The kinetics of the second electron transfer were monitored at 450 nm with 1-200 μM ferrocene as donor to P^+.

3. Results

3.1 *$P^+Q_AQ_B^-$ charge recombination*

In the presence of functional Q_B, the rate of decay of P^+ (k_{obs}) reflects charge recombination from the $P^+Q_AQ_B^-$ state. This slow back reaction is a simple (inverse) measure of the one-electron equilibrium constant (K_{AB}): $P^+Q_A^-Q_B \leftrightarrow P^+Q_AQ_B^-$:

$$K_{AB} = k_{AP}/k_{obs} - 1$$

where k_{AP} is the rate constant for decay of P^+ in the absence of functional Q_B. The pH dependence of k_{obs} reveals differences in the protonation properties of the $P^+Q_A^-Q_B$ and $P^+Q_AQ_B^-$ states, and is shown for Wt and mutant RCs in Figure 1. These indicate a shift in an apparent pK, associated with the $P^+Q_A^-Q_B$ state, from ~9.5 in the Wt to ~8.0 in the Glu^{M236}→Gln and Leu mutants and ~7.5 in the Gly mutant. No shift was detected in the Asp mutant, although the back reaction rate was slower than the Wt at all measured pH.

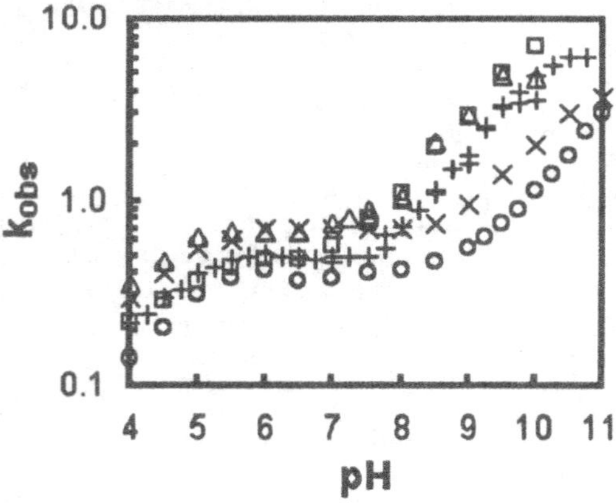

Figure 1. The pH dependence of the $P^+Q_AQ_B^-$ charge recombination, k_{obs} (s^{-1}), in Wt and M236 mutant RCs. Symbols: Wild type (×), Glu→Asp (o), Glu→Gln (Δ), Glu→Gly (□), Glu→Leu (+).

3.2 *Kinetics of the first electron transfer*

The pH dependences of the kinetics of the first electron transfer, from Q_A^- to Q_B ($k_{ET}^{(1)} = k_{AB}^{(1)} + k_{BA}^{(1)} \sim k_{AB}^{(1)}$ at pH < 10), are shown in Figure 2. In mutant RCs the pH dependence of $k_{ET}^{(1)}$ was shifted in a manner similar to the shift in k_{obs}, with Glu^{M236}→Leu and Gln mutants exhibiting pK shifts to ~8.5, and Gly mutant to ~7.5, while the Asp mutant showed a pH dependence very similar to that of the wild type RC.

3.3 *Kinetics of the second electron transfer*

The pH dependences of the kinetics of the second electron transfer, from Q_A^- to Q_B^- (($k_{ET}^{(2)} = k_{AB}^{(2)} + k_{BA}^{(2)} \sim k_{AB}^{(2)}$ at pH < 9), are shown in Figure 3. The Glu^{M236}→Asp

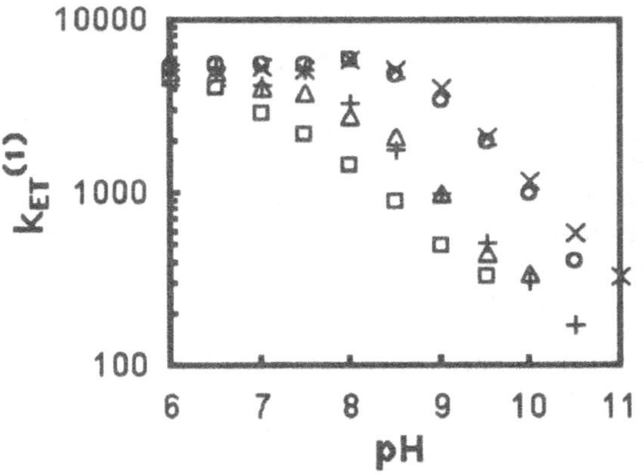

Figure 2. The pH dependence of the rate of the first electron transfer kinetics (s^{-1}) in Wt and mutant RCs, measured at 397 nm. Symbols: As for Figure 1.

mutant behaved very similar to the Wt. The $k_{ET}^{(2)}$ for the other 3 mutants were similar with rates that are slightly slower than the Wt at all measured pH, although the wild type rate is approached in the low pH region, (pH < 6.0).

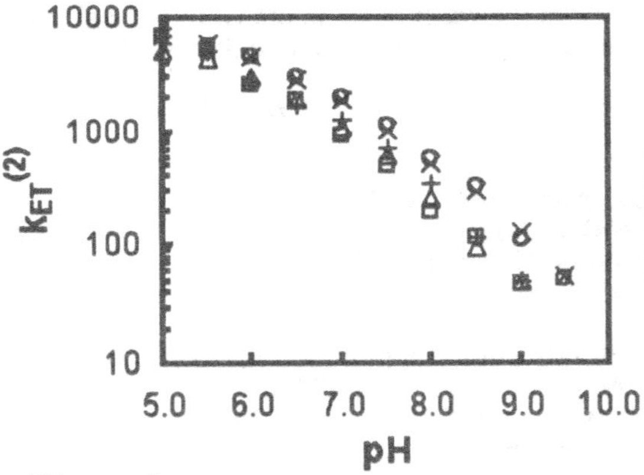

Figure 3. The pH dependence of the rate of the second electron transfer kinetics (s^{-1}) in Wt and mutant RCs, measured at 450 nm. Symbols: As for Figure 1.

4. Discussion

The likely participation of water molecules in proton transfer to Q_B has been widely suggested (1,9,11-13). Internal water molecules can be modified indirectly by mutating residues to bring about rearrangement of the waters, possibly resulting in altered proton transfer behavior. Baciou and Michel (14) attempted to interrupt the water chain by mutating Pro^{L209} to Tyr or Phe. The major effect was on the proton-coupled process of second electron transfer, and the authors suggested that the chain of bound water may represent the most efficient pathway of proton transfer.

In this study we took the approach of mutating a residue, Glu^{M236}, that appears to make contact with several water molecules in the water chain (9). The Glu^{M236}→Gly, Gln and Leu mutants displayed noticeable effects on the first electron transfer kinetics (Figure 2) as well as the back reaction kinetics, k_{obs} (Figure 1), which is related to the one-electron transfer equilibrium between Q_A and Q_B. However, the effects were not dramatic compared to those seen in mutants of Glu^{L212} and Asp^{L213}, two residues close to Q_B (5-6Å from carboxylate to nearest quinone carbonyl), or compared to mutation of Glu^{H173} (13), slightly farther away from the Q_B site (< 9Å).

Although interpretation is equivocal, mutational "neutralization" of Asp^{L213} and Glu^{H173} appears to induce shifts in the functional pK of the $P^+Q_A^-Q_B$ state from ~9.5 to ~5 and ~7, respectively (5,13). Similar "neutralization" of Glu^{M236} (16 Å from Q_B) induces a shift to ~8.0. These effects, which reflect electrostatic influences on the stability of Q_B^-, are consistent with the relative distances of the residues from the Q_B pocket. Alteration of Glu^{M236} to Asp results in pH dependences of $k_{AB}^{(1)}$ that is very close to the native Glu. However, the pH dependence of k_{obs} indicates functional pKs for the $P^+Q_A^-Q_B$ and $P^+Q_AQ_B^-$ states that are both higher than the Wt, resulting in increased values of K_{AB} at all pHs. Whether this - or any other - apparent pK shift represents true pK behavior is uncertain, however, due to complex interactions between ionizable residues and the influence of surface charge on pH titrations via surface pH effects.

Previous work has shown a strong dependence of the Q_B protonation reactions on the electrostatic environment provided by widely distributed residues. However, the very small effect of the M236 mutants on the second electron transfer represents only a minor influence of Glu^{M236}, in spite of its evident effect on the stability of Q_B^- (first electron behavior). This indicates that the influence of this residue is insufficient to change the rate-limiting steps in coupled proton and electron delivery, perhaps because the location of such steps is removed from the vicinity of Q_B itself and therefore under the influence of different parts of the potential field. Also, disruption of the water chain in the Glu^{M236} region may still leave alternate proton conduction pathways operational, with negligible effect on the conduction rate.

This work was supported by NSF grant to C.A.W. (MCB 92-08249).

References

1. Okamura, M.Y. and Feher, G. (1992) Annu. Rev. Biochem. 61, 861-896.
2. Takahashi, E. and Wraight, C.A. (1994) *in* Advances in Molecular and Cell Biology (Barber, J., ed.), pp. 197-251, JAI Press, Greenwich, CT.
3. Paddock, M.L., Rongey, S.H., Feher, G. and Okamura, M.Y. (1989) Proc. Natl. Acad. Sci. USA 86, 6602-6606.
4. Takahashi, E. and Wraight, C.A. (1990) Biochim. Biophys. Acta 1020, 107-111.
5. Takahashi, E. and Wraight, C.A. (1992) Biochemistry 31, 855-866.
6. Hanson, D.K., Baciou, L., Tiede, D.M., Nance, S.L., Schiffer, M. and Sebban, P. (1992) Biochim. Biophys. Acta 1102, 260-265.
7. Hanson, D.K., Tiede, D.M., Nance, S.L., Chang, C-H. and Schiffer, M. (1993) Proc. Natl. Acad. Sci. USA 90, 8929-8933.
8. Okamura, M.Y., Paddock, M.L., McPherson, P.H., Rongey, S.H. and Feher, G. (1992) *in* Research in Photosynthesis (Murata, N., ed) Vol.1, pp. 349-356, Kluwer Academic Publ., Dordrecht, The Netherlands.
9. Ermler, U., Fritzsch, G., Buchanan, S.K. and Michel, H. (1995) Structure 2, 925.
10. Takahashi, E., Maróti, P. and Wraight, C.A. (1989) *in* Current Research in Photosynthesis (Baltscheffsky, M., ed.) pp. 169-172, Kluwer Academic Publ., Dordrecht, The Netherlands.
11. Beroza, P., Fredkin, D.R., Okamura, M.Y. and Feher, G. (1992) *in* The Photosynthetic Reaction Center II (Breton J. and Verméglio A., eds), pp. 363-374, Plenum Press, New York.
12. Maróti, P., Hanson, D.K., Baciou, L., Schiffer, M. and Sebban, P. (1994) Proc. Natl. Acad. Sci. USA 91, 5617-5621.
13. Takahashi, E. and Wraight, C.A. (1995) Proc. Natl. Acad. Sci. USA In press.
14. Baciou, L. and Michel, H. (1995) Biochemistry 34, 6967-7972.

ISOLATION OF OXYGEN EVOLVING PSII PARTICLES AND PSII REACTION CENTRES FROM THE FILAMENTOUS CYANOBACTERIUM *SPIRULINA PLATENSIS*.

Eva Šetlíková [1], Petr Budáč [1] and Danuše Sofrová [2]

[1] Department of Autotrophic Microorganisms, Institute of Microbiology, Academy of Sciences of the Czech Republic, 37981 Třeboň;

[2] Department of Biochemistry, Faculty of Sciences, Charles University, Albertov 2030 , 12 840 , Praha 2; Czech Republic

1. Introduction

One of us participated recently in developing a procedure for isolation and purification of Photosystem 2 reaction center from a thermophilic cyanobacterium *Synechococcus elongatus* [1]. Immobilized metal (Cu^{2+}) affinity chromatography (IMAC) has been used as the terminal step. The result was a functional reaction center consisting of D1, D2 and cyt *b559* and with the molar ratio of pigments similar to PS2 reaction centres isolated from higher plants. This is at variance with an earlier report on a cyanobacterial RC [2].

In this communication we show that (1) *Spirulina* also yields „higher plant like" PS2 reaction centers; (2) the IMAC method of RC2 purification works well also with a filamentous cyanobacterium; (3) there are noticeable differences in the suitability of various detergents for the extraction of PS2 core particles used for RC2 isolation.

The structural identity of cyanobacterial and chloroplast RC2 has been lately confirmed by Oren-Shamir et al. [3]. Vácha et al. [4] showed that the IMAC (Cu^{2+}) method can be used also for purification of chloroplast RC2 starting from BBY particles.

2. Material and methods

The filamentous cyanobacterium *Spirulina platensis* was grown in a chemostat described in [1] at 30°C and with a dilution rate of 0,05 h^{-1}. The irradiance was regulated between 50 and 200 Wm^{-2} so as to give a high PS2 / PS1 ratio (up to 2:1).

Analytical methods have been described in [1]. - For immobilized metal affinity chromatography Chelating Sepharose Fast Flow was packed in a HR 10/10 column (Pharmacia - LKB). The diameter of the column was 10 mm and its height 80 mm. The column and all the solutions were kept at 2 to 4°C.

Buffers used: **A:** 30 mM K_2HPO_4 , 5 mM $MgCl_2$, pH = 7,5; **B:** 30 mM K_2HPO_4, 500 mM mannitol, 5 mM $MgCl_2$, 35 mM NaCl, 1 mM $CaCl_2$ and 1 mM ε-aminocaproic acid, pH = 6,5; **C:** buffer A with added ε-aminocaproic acid, 1 mM; **E:** 100 mM Na_2HPO_4, 50 mM NaCl, 10 mM $MgCl_2$, 20 μM $CaCl_2$, 0,02 % SB12, pH = 6,5; **F:** 30 mM Na_2HPO_4, 5O mM NaCl, 1 mM $MgCl_2$, 2 mM $CaCl_2$, 500 mM mannitol, pH = 6,2.

P. Mathis (ed.), Photosynthesis: from Light to Biosphere, Vol. I, 695–698.

mechanically disintegrated with glass beads by two 1 minute shaking runs in a laboratory made vibratory disintegrator cooled during operation to -2 °C. Glass beads were separated from the homogenate by filtration and cell debris were pelletted by 15 minutes of centrifugation at 6000 g. Two centrifugations in buffer C removed most of the phycobilisomes. The homogenate was then sedimented at 24000 g (30 min, 4 °C).

We have compared the suitability of the following detergents for the solubilization of PS2 complexes from the thylakoids of *Spirulina*: Triton X-100, dodecylmaltoside (DM), lauryldimethylaminooxide (LDAO), sulfobetain 12 and a mixture of DM and octylglucoside (DM+OG). The thylakoids were resuspended in buffer B at a concentration of 1 mg.ml^{-1} of Chl *a* and the detergent was added at a weight ratio relative to Chl *a* r_w. The suspension was incubated for *t* minutes at *T* °C with continuous stirring. The specification of the incubation parameters is given in the following table:

detergent	Triton X-100	SB12	LDAO	DM	DM + OG
r_w	15	3,5	7	5	5+9
T °C	4	25	2	2	4
t min	30	30	30	15	2+3

After incubation the suspension was centrifuged at 150 000 g for 2 hours at 2 °C. The resulting pellet contained thylakoid membrane fragments enriched with PS1. PS2 particles remained in the supernatant contaminated with proteins of various molecular weights and with a small amount of PS1 complexes. To reduce this contamination the supernatants were subjected to a centrifugation in a 20 to 45 % sucrose gradient, at 160 000 g and 2°C for 16 hours. The protein composition of the complexes in various zones is shown in Fig 1. The extract produced by (DM + OG) was not loaded to the sucrose gradient but precipitated by addition of PEG 6000 and centrifugation at 360 000 g for 40 minutes as described in [3]. This procedure yields PS2 particles of a quality comparable to the best results from the gradients and is much faster.The suitability of the detergents for a selective extraction of PSII complexes decreases in the order DM+OG > DM > LDAO > SB 12 > Triton X-100 (Fig. 1). In the same order the contamination increases with proteins of M_r between 45 and 70. These may be products of cross-linking between constituents of the complexes detached during isolation; similar products have been described in [5] as a result of PS2 photoinhibitory degradation. Their removal is an important step in the purification of the PS2RC.

4. Purification of the reaction centre complex

For isolation and purification of the reaction centre complex we have used the procedure described in [1]. Loading of the Chelating Sepharose column with Cu^{2+} and with the solubilized PSII particles (approximately 2 mg of Chl *a*) was the same as in [1] and also the phycobiliproteins were satisfactorily removed with the phosphate buffer E. For efficient strippping of CP43 and CP47 from the RC, however, octylglucoside (1 %) had to be substituted for $LiClO_4$ in buffer F. Fig. 1. F shows the removal of the inner antenna and the protein composition of the resulting RC.

3. Isolation of PS2 core particles

For isolation of PS2 oxygen evolving core particles about 10 g (d.wt.) of *Spirulina* cells were centrifuged, washed in buffer A and resuspended in buffer B. The cells were then

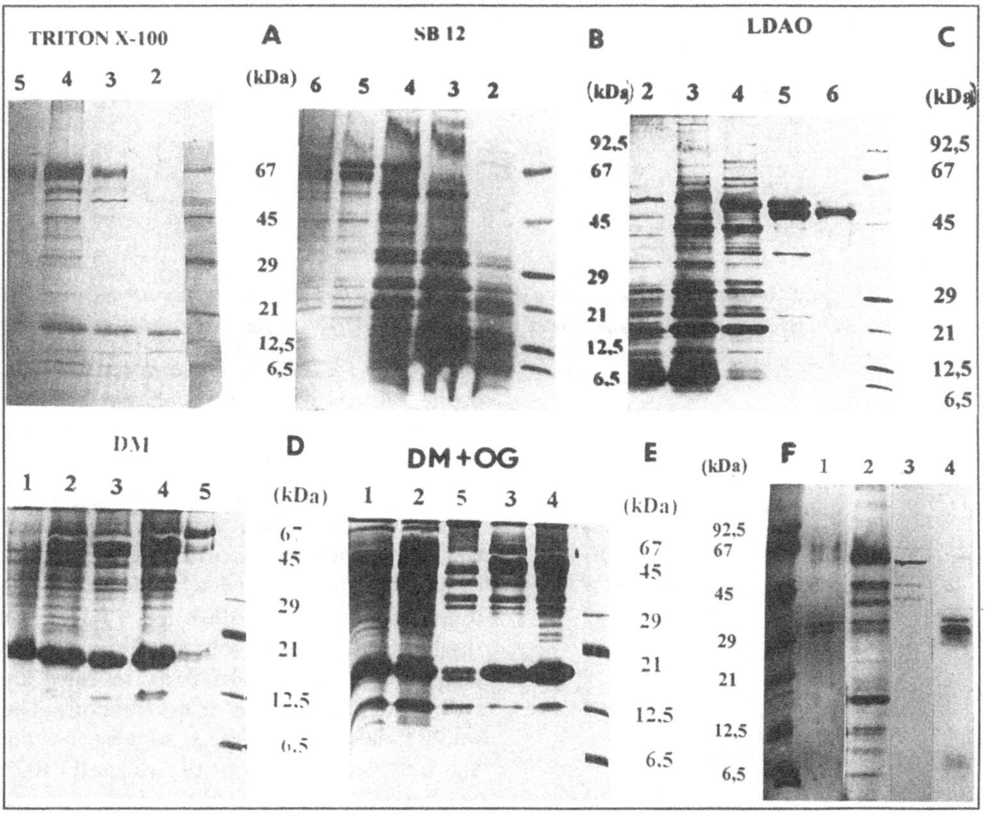

Fig.1.PAGE showing protein composition of various isolated complexes. ***A*** *to* ***D*** *- fractions from the 20 to 45 % discontinuous sucrose gradient. For sharp bands the position is given below as % of sucrose solution in the layer (t-top, b-bottom), other samples are from diffuse zones. Number of the fraction chosen for RC purification in bold type. The type of detergent is given in the figures.* ***A*** *- 3:30; 4:45; no suitable fraction.* ***B*** *- 2:20t; 3: 20b;* ***4****:30t; 6:40b.* ***C*** *- 2:on top of 20; 3:20t;* ***4****:20b; 6:35b.* ***D*** *- 1:on top of 20; 2:20t; 3:20b; 4:30; 5:35.* ***E*** *- fractions from the PEG procedure; 1: thylakoid membranes; 2: supernatant and 3: sediment after (DM+OG) treatment; 4: supernatant and* ***5****: sediment after PEG 6000 precipitation.* ***F*** *- purification of the RC from DM PS2 particles (lane 2) includes washing out the CP43, CP47 and aggregates of higher m.w.(lane 3). Eluate of RC complexes: lane 4. The same eluate for LDAO particles as starting material shows a significant contamination with aggregates.*

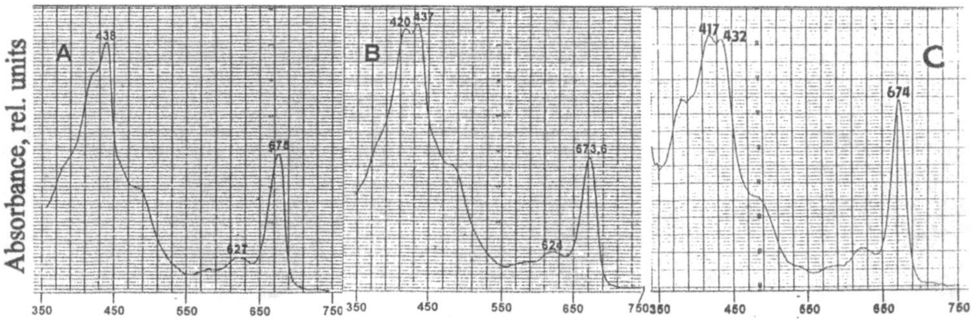

*Fig. 2. Absorption spectra of oxygen evolving PS2 particles extracted with LDAO (**A**) and with DM (**B**) and of reaction centre complexes purified from DM particles (**C**).*

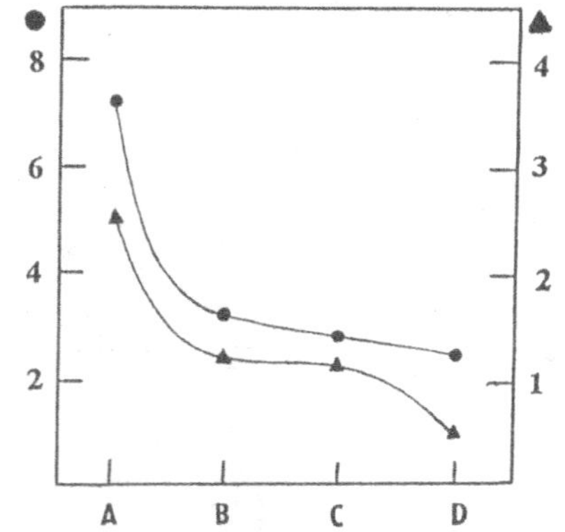

Fig. 3. Molar ratios Chl a/Pheo a (circles) and Car β/Pheo a (triangles) in successive IMAC elution fractions

When the PS2RC of *Synechococcus* was being eluted from the column [1] the consecutive fractions showed characteristic variation in the molar ratio of pigments, starting with 13 Chl *a* and 8 Car β molecules and ending with 5 Chl *a* and 1 Car β molecule per two Pheo *a* molecules. In the middle fractions a stable ratio was found Chl *a* : Car β : Pheo $a \cong 6:2:2$. A similar trend can be observed during the elution of *Spirulina* RC2 as revealed by HPLC pigment analysis of the fractions. The results illustrated in Fig. 3. suggest that the pigment stoichiometry of isolated RC2 complexes is a subtle value which depends on slight changes of the isolation and purification procedure.

4. References

[1] Šetlíková E, Ritter S, Hienerwadel R, Kopecký J, Komenda J, Welte W and Šetlík I (1995) Photosynth Res 43:201-211

[2] Gounaris K, Chapman DJ and Barber J (1989) Biochim Biophys Acta 973:296-301

[3] Oren-Shamir M, Maruthi Sai PS, Edelman M and Scherz A (1995) Biochemistry 34: 5523-5526

[4] Vácha F, Joseph DM, Durrant JR, Telfer A, Klug DR, Porter G and Barber J (1995) Proc Natl Acad Sci USA 92:2929-2933

[5] Mori H, Yamashita Y, Akasaka T and Yamamoto Y (1995) Biochim Biophys Acta 1228: 37-42

DYNAMIC EFFECTS ON SPECTRA OF THE PHOTOSYNTHETIC REACTION CENTER

Bruno Källebring[1] and Sven Larsson[2]
[1]Department of Biochemistry and Biophysics, Göteborg University and Chalmers University of Technology, S-413 90 Göteborg; [2]Department of Physical Chemistry, Chalmers University of Technology, S-412 96 Göteborg,

1. Introduction

Theoretical investigations of the photosynthetic reaction center have been made much easier thanks to the crystallographically solved structures of the bacterial reaction center [1,2]. The reaction center is a very fine tuned system where the first excited state (S1) decays to or interacts with the charge transfer state (CT) [3,4]. So far, no satisfactory theoretical explanation for the very high rate of the primary charge separation has been given. Including the dynamics in the system by means of molecular dynamics (MD) might shed some light on the mechanisms important for this very fast electron transfer. We have performed MD on relevant chromophores in the reaction center and sampled coordinate sets at certain time intervals to include the dynamics of the system. We present some results on calculations with CNDO/S where the input data are obtained by molecular dynamics.

2. Procedure

To make MD on the whole reaction center is a tremendous task and extremely time consuming. Instead we selected those parts of the reaction center which are directly involved in the first step of the primary charge separation. From the crystal structure of *Rps.Sphaeroides* the two bacteriochlorophylls of the primary electron donor (P) were singled out and also the two accessory bacteriochlorophylls. The surrounding protein was chosen in such a way that within a sphere of 15 Ångström from the carbonyl oxygen on the fifth ring of the L part of P all residues totally or partly inside this sphere were included in this subset of the reaction center. For the MD calculations the Biosym Technologies insightII software was used with the MD module Discover. In the MD calculation the backbone of the protein was kept fixed while the side chains and chromophores were allowed to move. After initial energy minimization the MD was run for 1 ps before any coordinate sets were sampled. Then with time intervals of 50 fs coordinate sets were sampled from the system. This was done during a period of 1 ps. From this series of coordinate sets a further reduction of atoms had to be done to be able to do molecular orbital calculations with CNDO/S. From each coordinate set P and the accessory bacteriochlorophyll L was singled out. The Tyr M210 was also included which is laying right between P and the accessory bacteriochlorophyll. With these reduced coordinate sets CNDO/S calculations were done to see how the total energy, the spectrum

P. Mathis (ed.), Photosynthesis: from Light to Biosphere, Vol. I, 699–701.

and charge transfer states changed with time. In the calculation no reaction field was included and no extra charges in order to promote electron transfer from P to the accessory chlorophyll. To get the spectrum the CNDO/S includes CI with singly substituted Slater determinants. The singly substituted Slater determinants do not improve the ground state energy but the energies of the excited states are improved and may be compared to the experimental spectrum.

3. Results and Discussion

The CNDO/S calculations showed well localized molecular orbitals which makes it possible to distinguish between localized excitation states and CT states. Molecular orbital calculations though, tend to over estimate the energy of the CT states. This is due to the difficulty to include the surrounding media effects into the calculations [5,6]. One way to do this is to include a reaction field [7] into the calculations. The reaction field takes into account the instantaneous polarization of the protein. Since the reaction field is a direct response to the transferred charge the effect has to be calculated iteratively and is rather computing intensive. The advantage with including the reaction field and get the CT state at about the same energy level as the S1 state is that the matrix element of the avoided crossing can be calculated directly [8]. The relative value of the coupling can although, be estimated by examining the energy level of the CT state. The larger the coupling is the lower in energy the CT state comes [5,6]. So by monitoring the energy level of the CT state with time we should be able to tell if the dynamics of the system is important for the electronic coupling and the electron transfer rate.

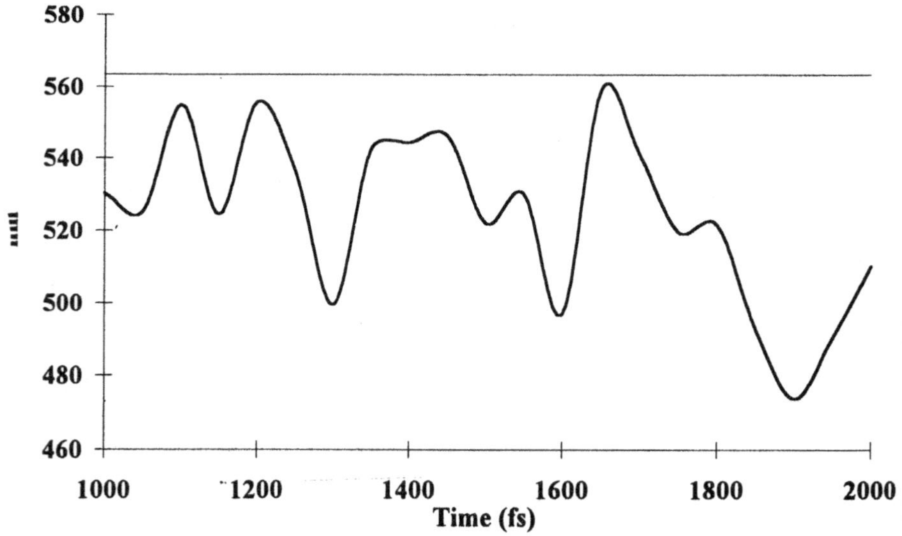

Figure 1. Graph showing the CT state energy fluctuation with time starting after 1 ps. Top line indicating the CT state energy of the energy minimized system.

Our results show that there is some change in coupling with time but the largest coupling is achieved when calculating on the energy minimized system. This is also in agreement with what can be seen experimentally. The lack of activation energy in the primary charge separation suggest a very small temperature dependence [9]. If this simple method can describe the shape of the spectrum will be investigated further. Closer sampling interval and under a longer period of time is required.

4. Acknowledgment

We are grateful to Prof. Harry B. Gray for stimulating discussions. We are also grateful to NFR for financial support.

References

1. Diesenhofer, J., Epp, O., Miki, K., Huber, R. and Michel, H. (1984) *J.Mol.Biol.* **180**, 385
2. Chang, C.H., Tiede, D., Tang, J., Smith, U., Norris, J. and Schiffer, M. (1986) *FEBS Lett.* **205**, 82; Refined structure: Ermler, U., Fritzsch, G., Buchanan, S.K., and Michel, H. (1994) *Structure*(London), **2**, 925
3. Hayes, J.M. and Small, G.J. (1986) *J.Phys.Chem.* **90**, 4928
4. Lockhart, D.J. and Boxer, S.J. (1987) *Biochem.* **26**, 664
5. Petke, J.D. and Maggiora, G.M. (1986) *J.Chem.Phys* **84**, 1640
6. Larsson, S. and Volosov, A. (1987) *J.Chem.Phys.* **87**, 6623
7. Onsager, L. (1936) *J.Am.Chem.Soc.* **58**, 1486
8. Larsson, S., Broo, A., Källebring, B. and Volosov, A. (1988) *Int.J.Quantum Chem., QBS* **15**, 1
9. Holszapfel, W., Finkele, U., Kaiser, W., Oosterhelt, D., Scheer, H., Stilz, H.U. and Zinth, W..(1988) *Chem.Phys.Letters* **146**, 13

SITE-DIRECTED MUTAGENESIS OF THE *psbA* GENE IN THE GREEN ALGA, *CHLAMYDOMONAS REINHARDTII*

Yuichiro Takahashi, Keiichiro Utsumi, Aya Hatano, Yumiko Yamamoto and Kimiyuki Satoh
Department of Biology, Faculty of Science, Okayama University, Tsushima-naka, Okayama 700 Japan.

1. Introduction

The D1 protein is one of the subunits of photosystem II (PSII) reaction center (1), which is encoded by the chloroplast *psbA* gene. This protein is synthesized as a precursor (pD1)(2, 3) and then processed at its carboxy terminus (4-6). However, physiological significance of the presence of carboxyl terminal extension in D1 has not been established. Determination of carboxyl terminal residue of D1 revealed that the precursor protein is processed at the carboxyl side of Ala at position of 344 (5, 6). Comparison of amino acid sequences of D1 deduced from *psbA* genes of different organisms examined so far reveals that the carboxyl terminal sequence of the mature protein is highly conserved (6). On the contrary, the sequence and the length of carboxyl extension is poorly conserved, except for the residue at 345 position of pD1; Ala in most organisms and Ser in *C. reinhardtii* and some cyanobacteria (7).

In the present study, we have constructed a transforming vector for site-directed mutagenesis of the *psbA* gene and mutated Ser345 in pD1 of *C. reinhardtii* to Gly, Cys, Val and Phe to study the effect of the amino acid substitutions at this position on the processing of pD1 *in vivo*.

2. Results and Discussion

The chloroplast *psbA* genes of *C. reinhardtii* are present in the inverted repeats and contain four introns (8). To construct a vector which allows to introduce site-directed mutations to the *psbA* gene (fig. 1), a 5.0 kb EcoRI/XhoI fragment containing exon 5 and 3' flanking region of the *psbA* gene was cloned in the Bluescript vector from *Chlamydomonas* chloroplast BamHI fragment Ba12 according to Rochaix (9). The *aadA* cassette which confers resistance against spectinomycin and streptomycin was excised

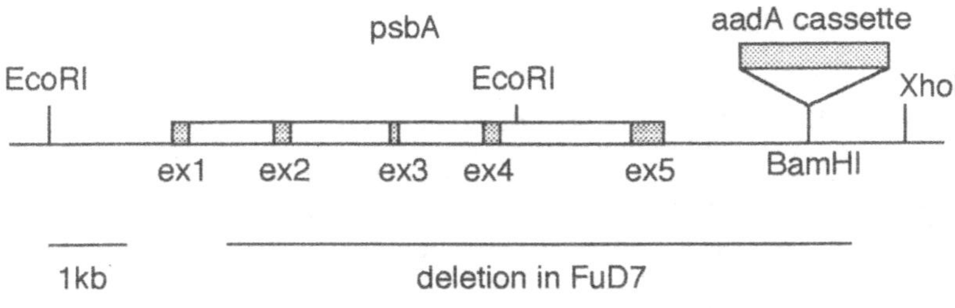

Fig. 1 The vector for site directed mutagenesis of the *psbA* gene. Bar in the bottom indicates a deletion in Fud7 strain.

P. Mathis (ed.), Photosynthesis: from Light to Biosphere, Vol. I, 703–706.

from pUC-atpX-AAD plasmid (10) and was inserted at BamHI restriction site downstream of the *psbA* gene (pEX-50-AAD). A 6.0 kb EcoRI fragment R14 according to Rochaix (9) containing exons 1-4 of the *psbA* gene was subsequently inserted at EcoRI site in the pEX-50-AAD. The resulting plasmid pR12-EX-50-AAD entirely covers a deletion that ranges from a part of the first intron extending to the 3' flanking region of the *psbA* gene in the chloroplast genome of Fud7 strain (fig. 1).

Using this vector, we have introduced point mutations into the *psbA* gene in order to substitute Ser345 to Gly, Cys, Val and Phe. The vector was introduced into the chloroplasts of Fud7 cells and the putative transformants were selected on agar plates of Tris-acetate-phosphate (TAP) medium containing spectinomycin. Southern blotting analysis of total DNA from the transformant cells indicated that the DNA introduced in the chloroplast has been integrated into the chloroplast genome by homologous recombination and the deletion in Fud7 chloroplast genome has been replaced by wild type or mutated *psbA* gene (data not shown).

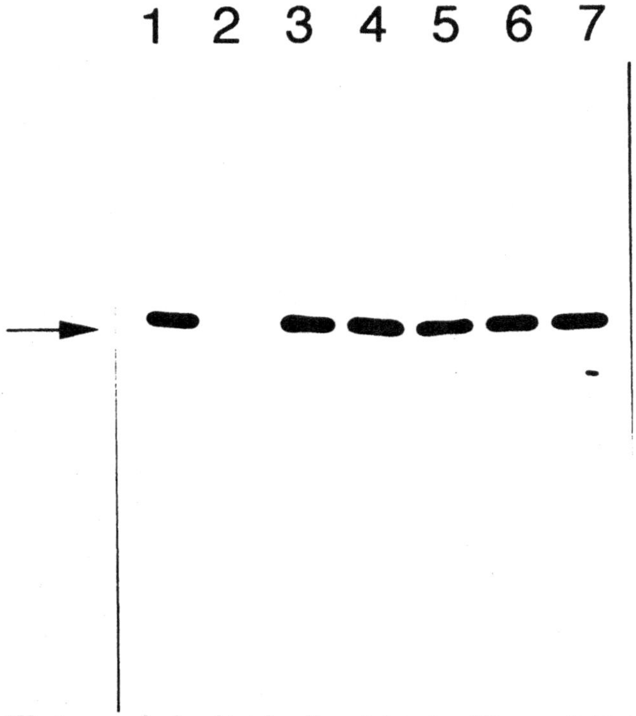

Fig. 2 Western analysis of total cell proteins. 1, wild type, 2; Fud7, 3; control strain (*aadA* cassette at BamHI site), 4; Ser345Cys, 5; Ser345Gly, 6; Ser345Val, 7; Ser345Phe. Total cell proteins from cells of 5ug chl were loaded on each lane. Proteins on the nitrocellulose filter were probed with antibodies against D1 protein.

Fig. 2 shows Western blotting analyses of total cell proteins. The cells were grown to mid-log phase in TAP liquid culture in the dark. Total cell proteins were separated by SDS-polyacrylamide gel electrophoresis, blotted onto nitrocellulose filter and probed with antibodies raised against D1 protein. While the D1 is absent in Fud7 cells as expected, all transformants accumulated D1 at wild type level. It is also noted that migration of the signal in each transformant is identical to that of mature D1 in wild type, suggesting that pD1 in these transformants is processed. Fluorescence induction transients of the dark adapted cells of the transformants showed that they have wild type PSII activities (data not shown). These results suggest that the mutations introduced into the pD1 have no effect on the functional assembly of the PSII complex at least in the dark.

In order to test if cells of the transformants can grow photoautotrophically, they were spotted on agar plate of TAP medium in the dark or of minimal medium (HSM) in the light (1,500 or 50,000 lux) (fig. 3). The four transformants appeared to grow like wild type under both heterotrophic and photoautotrophic conditions. It is expected that if the processing rate of pD1 is reduced by these mutation, the photoautotrophic growth rate of the transformant cells may be affected, since the D1 protein turns over rapidly under illumination. However, all four transformants grew normally even under strong light condition (fig. 3 C). These results strongly suggest that the mutations at the position of 345 did not affect rapid replacement of the damaged D1 protein, which should require efficient processing of pD1.

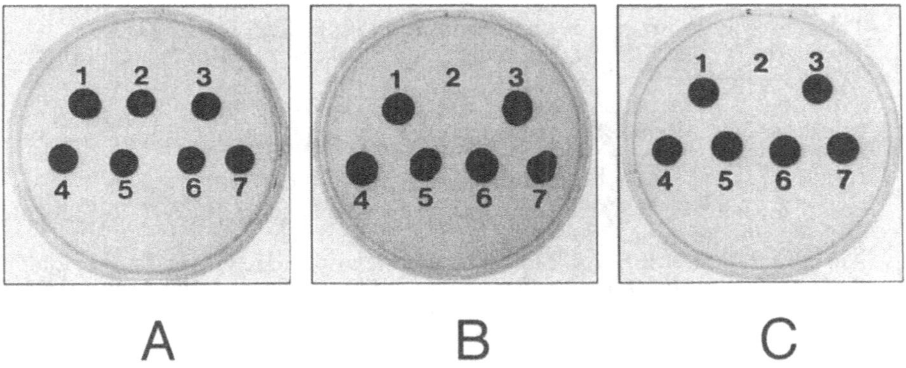

Fig. 3 Spot tests. Cells from 1; wild type, 2; Fud7, 3; control, 4; Ser345Cys, 5; Ser345Gly, 6; Ser345Val, 7; Ser345Phe were spotted on agar plates of either TAP (A) or HSM (B and C) medium and grown in the dark (A) or under illumination of 1,500 lux (B) or 50,000 lux (C).

The present results demonstrated that the processing of pD1 is not affected by replacement of Ser345 to Gly, Cys, Val and Phe, although this residue is relatively well conserved. The enzyme responsible for cleavage of pD1 has been purified from higher plants (11). A gene encoding the protein, *ctpA*, has been cloned in a cyanobacterium, *Synechocystis* sp. PCC6803 (12). It was reported that synthetic carboxyl-terminal oligopeptides are cleaved correctly at the processing site by the partially purified enzyme from spinach (13). The rate of cleavage was largely reduced by the substitution of Ala345 by Gly and Val. However, the results obtained by these *in vitro* analyses are not consistent with those obtained in this study. This discrepancy may arise from organisms used since the residue at position of 345 is Ser in *C. reinhardtii* while Ala in spinach. The possibility remains that processing rates observed by *in vivo* analysis may not represent the actual rate of carboxyl-terminal cleavage of pD1. However, another possibility cannot be ruled out that difference in the reaction condition in both systems including the size and structure of substrate may be the origin of this discrepancy.

3. Acknowledgements

We thank Dr. J.-D. Rochaix for Fud7 strain and *psbA* plasmid and Dr. M. Goldschmidt-Clermont for the *aadA* cassette. This work was supported by the Grant in Aid for Scientific Research on Priority Areas (04273101 and 06259212) and General Research (06404003 and 06640842) from Ministry of Education, Science and Culture, Japan and Human Frontier Science Program.

5. References

1. Nanba, M.and Satoh, K. (1987) Proc. Natl. Acad. Sci. USA 84, 109-112.
2. Minami, E. and Watanabe, A. (1985) Plant Cell Physiol. 26, 839-846.
3. Reisfeld, A., Mattoo, A. K. and Edelman, M. (1982) Eur. J. Biochem. 124, 125-129.
4. Marder, J. B., Goloubinoff, P. and Edelman, M.(1984) J. Biol. Chem. 259, 3900-3908.
5. Takahashi, M., Shiraishi, T. and Asada, K. (1989) FEBS Lett. 240, 6-8.
6. Takahashi, Y., Nakane, H., Kojima, H. and Satoh, K. (1990) Plant Cell Physiol. 31, 273-280.
7. Svensson, B., Vass, I. and Styring, S. (1991) Z. Naturforsch, 46c, 765-776.
8. Erickson, J. M., Rahire, M. and Rochaix, J.-D.(1984) EMBO J. 3, 2753-2762.
9. Rochaix, J.-D. (1878) J. Mol. Biol. 126, 597-617.
10. Goldschmidt-Clermont, M. (1991) Nucleic Acids Res. 19, 4083-4089.
11. Fujita, S., Inagaki, N., Yamamoto, Y., Taguchi, F. and Satoh, K. (1995) Plant Cell Physiol. in press.
12. Anbudurai, P R., Mor, T. S., Ohad, I., Shestakov, G, E., Gadzhiev, A., Lind, L. K. and Pakrasi, H. (1994) Proc. Natl. Acad. Sci. USA, 91, 8082-8086.
13. Taguchi, F., Yamamoto, Y. and Satoh, K. (1995) J. Biol. Chem. 270, 10711-10716.

CONTROL OVER ELECTRON FLOW THROUGH THE *RHODOBACTER SPHAEROIDES* REACTION CENTRE

B.J. van Rotterdam, W. Crielaard, H.V. Westerhoff and K.J. Hellingwerf.
E.C. Slater Institute for Biochemical and Microbiological Research, BioCentrum Amsterdam, University of Amsterdam, Nieuwe Achtergracht 127, 1018 WS Amsterdam, The Netherlands.

1. Introduction

Light-driven free-energy transduction by bacterial reaction centres in a reconstituted liposomal system [1] leads to the build up of a proton-motive-force over the energy transducing membrane, after the addition of appropriate redox-mediators [1]. This liposomal system is being applied in our studies, as a model system for bacterial photosynthesis, to *quantitatively* describe the control of the membrane potential ($\Delta\psi$) over the individual steps in the *Rhodobacter sphaeroides* reaction centre (RC). In order to observe this (putative) back-pressure effect, it is essential to create conditions under which maximal control over light-driven electron transfer lies within the reaction centre and not in the reaction between it and cytochrome *c* (cyt *c*) or ubiquinol-0 (UQ_0H_2), the two redox mediators used in the liposomal system to short-cut cyclic electron transfer in the absence of a b/c_1-complex and/or the reaction between these two mediators themselves [1]. The kinetics and reaction mechanism of the $UQ_0H_2 \leftrightarrow$ cyt *c* reaction, as analysed from light-induced redox changes in a mixture containing purified, solubilized *Rb. sphaeroides* $RCLH_1$-complexes, ubiquinon/ol-0 (UQ_0/H_2) and horse heart cytochrome *c*, are described in the present study. Similar studies have been performed by Rich and Bendall, who analysed the reaction mechanisms between ferric cyt *c* and a series of substituted *p*-benzoquinols [2] and menadiol [3], but ***not*** UQ_0. It was concluded from these studies that the dominant reductant, in the reactions containing benzoquinones, is the deprotonated semiquinone formed by dismutation with the quinone and that the reduction rate by the menadiol is governed by the oxidation of the anionic-quinol. Yamazaki and Ohnishi [4] previously showed that reduction of ferric-cyt *c* by benzoquinol was accelerated by the addition of quinone, which confirms these conclusions.

P. Mathis (ed.), Photosynthesis: from Light to Biosphere, Vol. I, 707–710.

Below we will not only present the kinetic analysis and the mechanism of the reaction between UQ_0H_2 and cyt *c*, but also how these new parameters affect shifting control towards the RC, during the overall light-induced electron flux in the liposomal system. Control coefficients were calculated using our 12 state kinetic model for light induced cyclic electron transfer through the RC (of which a preliminary version is presented in [5]).

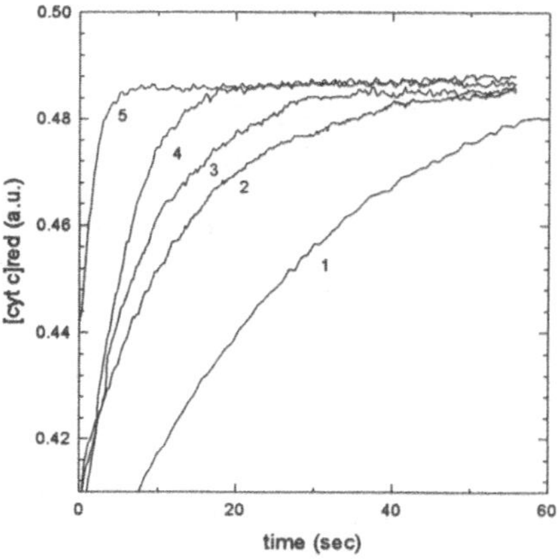

Figure 1. Cyt *c re*-reduction after light induced oxidation by $RCLH_1$. [RCLH1] = 60 nM, $[UQ_0]^{ini}$ = 400 μM, $[\text{cyt } c^{2+}]^{ini}$ = 28.65 μM, $[\text{cyt } c^{3+}]^{ini}$ = 1.35 μM, $[UQ_0H_2]^{ini}$ = 0, 25, 50, 100 and 225 μM for 1, 2, 3, 4 and 5 respectively.

2. Results and Discussion

2.1 *Mechanism of cyt c reduction*

The light-induced oxidation of cyt *c*, in the system described was followed spectroscopically ($OD_{550-540}$). After a steady-state reduction level of cyt *c* was reached, illumination was switched off and the re-reduction of cyt *c* was followed in time until equilibrium was retained. Figure 1 shows that at a fixed cyt *c* concentration (i.e. at a fixed $OD_{550-540}$) the cyt *c* reduction rate increases as a function of the amount of ubiquinol added.

Analysis of these and subsequent experiments revealed that the cyt *c* reduction rate shows a first order dependence on both [cyt c^{3+}] (data not shown) and [ubiquinol] (Fig. 2) and a minus first order dependence on the proton activity (Fig. 2). Hence we propose the following reaction mechanism (free protons are omitted in the following equations):

$$UQ_0H_2 + \text{cyt } c^{3+} \underset{k_{cq}}{\overset{k_{qc}}{\leftrightarrow}} UQ_0H^{\cdot} + \text{cyt } c^{2+} \quad (1)$$

$$UQ_0H^{\cdot} + \text{cyt } c^{3+} \underset{k_{ccq}}{\overset{k_{qqc}}{\leftrightarrow}} UQ_0 + \text{cyt } c^{2+} \quad (2)$$

The average values of the rate constants, which were determined from 16 measurements with initial ubiquinol concentrations varying from 0 to 400 μM, are listed in Table 1.

	determined ($M^{-1}.sec^{-1}$)	sem (n=16)◆	according to[2]★ ($M^{-1}.sec^{-1}$)	ΔE (mV)
k_{qc}	8.17×10^{2}	1.06×10^{2}	2.7×10^{2}	-126
k_{cq}	10.3×10^{4}		3.4×10^{4}	
k_{qqc}	1.2×10^{9}			626
k_{ccq}	4.50×10^{-2}	1.22×10^{-2}		

★Rich and Bendall calculated k_{qc} for UQ_1 and suggested that it should be similar for UQ_0 [2]. ◆ sem: standard error of the mean.

Table I. Measured and calculated values for the forward and backward rate constants and redox potential differences of the reaction between ubiquinol/ubisemiquinone and ferrous/ferric cyt *c* (k_{qc} & k_{cq}) and ubisemiquinone/ubiquinone and ferrous/ferric cyt *c* (k_{qqc} & k_{ccq}) at pH 8.0.

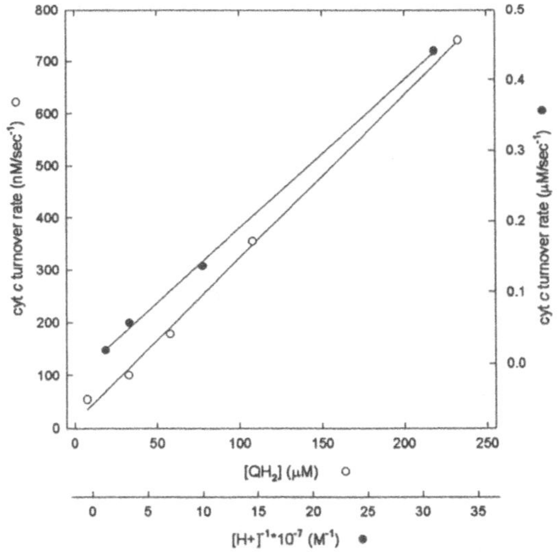

Figure 2. Dependence of the cytochrome *c* reduction rate after light induced oxidation. on : ○ the $[UQ_0H_2]$ concentration, [RCLH1] = 60 nM, $[UQ_0]$ = 400 μM, [cyt c^{2+}] = 28.7 μM, [cyt c^{3+}] = 1.3 μM. and ● the proton activity , $[UQ_0]$ = 150 μM, $[UQ_0H_2]$ = 50 μM, [cyt c^{3+}] = 8.8 μM.

2.2 *Optimising control through the reaction centre*

The new rate-constants of and mechanism for cyt *c* reduction were incorporated in the 12-state kinetic model [5] and the *time-dependent* light-induced redox changes were evaluated using the metabolic modelling program SCAMP [6]. To reveal the conditions under which the control of the RC on cyclic electron flux (C^{J}_{RC}) is maximal, the control of the local rates through the reaction centre were calculated and summed. Scaled flux control coefficients were calculated using Reder analysis [7] by calculating the steady state rate through a specific step (i) and then analyse the response of J(flux) to an introduced first-order multiplier (α) in the rate equation. In equation format: $R^J_\alpha = C^J_i \varepsilon^i_\alpha$. Because ε^i_α=1 the response of J to α equals the control of the rate through step i on J. Figure 3 shows that that the control of the reactions in the reaction centre on the overall electron flux J is maximised when using **1)** low light intensities, **2)** high cyt *c* concentrations and **3)** a total electron concentration of 100 μM ($0.6<C^J_{RC}<0.8$). At low electron concentrations, e.g. al low [UQ_0H_2] (see new mechanism and constants), the control of the RC will never exceed 0.4.

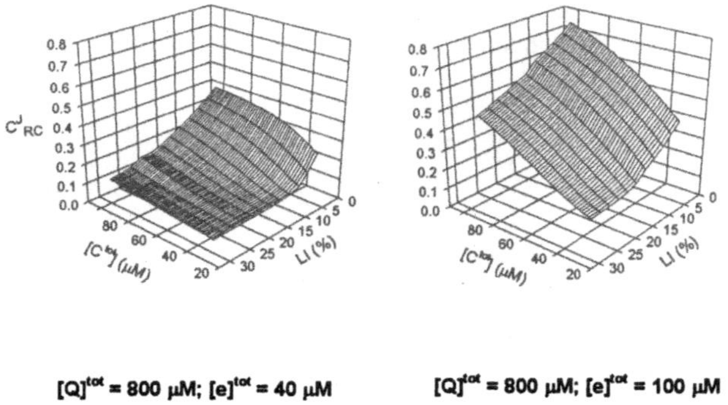

Figure 3. Control of the reactions in the reaction centre on the overall steady state flux as a function of the light intensity, cytochrome *c* and electron concentration.

References

[1] Molenaar, D., Crielaard, W., and Hellingwerf, K.J. (1988) Biochem. 27, 2014-2023.

[2] Rich, P.R. and Bendall, D.S. (1980) Biochim. Biophys. Acta 592, 506-518.

[3] Rich, P.R. and Bendall, D.S. (1979) FEBS Lett.105, 189-194.

[4] Yamazaki, I. and Ohnishi, T. (1966) Biochim. Biophys. Acta 112, 469-481.

[5] Hellingwerf, K.J., Crielaard, W. and Westerhoff, H.V. (1993) in Modern Trends in Biothermokinetics (Schuster, Rigoulet, Ouhabi and Mazat eds.), pp. 45-52, Plenum Press, New York, USA

[6] H. M. Sauro (1993) Comp. App. Biosci. 9, 441-450.

[7] Reder, C. (1988) J. Theor. Biol. 135, 175-201.

STATIC SIDE CHAIN CONFORMATIONAL DISTRIBUTION IN THE Y_Z TYROSINE RADICAL IN MANGANESE-DEPLETED, Y_D^- PHOTOSYSTEM II

Kurt Warncke[a], Xiao-Song Tang[b], Cecilia Tommos[c], Gerald T. Babcock[a], Bruce A. Diner[b], and John McCracken[a], [a]Department of Chemistry, Michigan State University, East Lansing, Michigan 48824, [b]Central Research and Development Department, Experimental Station, E. I. Du Pont de Nemours and Company, Wilmington, Delaware 19880, [c]Department of Biochemistry, Arrhenius Laboratories for Natural Science, Stockholm University, S-10691 Stockholm, Sweden

1. Introduction

The Y_Z tyrosine radical, $Y_Z^•$, functions nominally in photosystem II as an obligatory intermediate in the transfer of electrons generated by oxidation of water at the tetranuclear manganese center to the photooxidized primary donor chlorophyll , $P680^{+•}$ [1,2]. Recently, $Y_Z^•$ has been trapped by freezing photosystem II core preparations from the blue-green alga *Synechocystis* PCC 6803 depleted of the manganese center to cryogenic temperatures during illumination [3,4]. The spectrally-overlapping $Y_D^•$ species was eliminated by mutation of Y_D to redox-inert phenylalanine (D2-Tyr160Phe mutant). These approaches have allowed determination of the $Y_Z^•$ structure by specific ^2H-labelling and electron paramagnetic resonance (EPR) spectroscopic techniques [3,4] to a high resolution comparable with that of the $Y_D^•$ structure [5,6].

A dramatic difference between the trapped $Y_Z^•$ and $Y_D^•$ in manganese-depleted photosystem II is revealed by electron spin echo envelope modulation (ESEEM) spectroscopy: The ESEEM lineshapes arising from hyperfine coupling between the two specifically-incorporated β-methylene ^2H nuclei and unpaired π-spin density at ring carbon atom C_1, $\rho_{\pi C1}$, in $Y_Z^•$ are broadened to an extent beyond that expected for the dipolar hyperfine anisotropy arising from the random orientations of photosystem II complexes in the sample [4]. Broadening of the β-methylene ^2H lineshapes in the tyrosine model radical trapped in a low temperature aqueous glass was demonstrated previously to be caused by a "frozen-in" static distribution in the dihedral angle, θ, about the C_1-C_β bond [7]. This is because the β-methylene ^2H isotropic coupling exhibits a $\cos^2\theta$-dependence [8]. The broadening in $Y_Z^•$ was treated in a preliminary analysis by assuming a uniform distribution in θ [4]. Equilibrium thermodynamic parameters that characterize the motion of the phenol ring prior to establishment of the static distribution, including the form of the constraining potential, can be obtained from analysis of the C_1-C_β rotamer population distribution [9]. Here, we apply these methods to resolve the static rotamer distribution and local librational potential about the C_1-C_β bond in $Y_Z^•$.

2. Materials and Methods

The strains of *Synechocystis* PCC 6803 used, the construction of the mutants, growth of cells, purification of photosystem II complexes and trapping of $Y_Z^•$ have been described [3,4]. Envelope modulation was collected by using the three-pulse, 90°-τ-90°-T-90°, microwave pulse sequence. The application of the microwave pulse-swapping/envelope division technique employed here to the ^2H hyperfine coupling in organic radicals has been described [7,10]. The ESEEM spectra were obtained by Fourier transformation of the summation envelope modulation. Computer simulation

P. Mathis (ed.), Photosynthesis: from Light to Biosphere, Vol. I, 711–714.

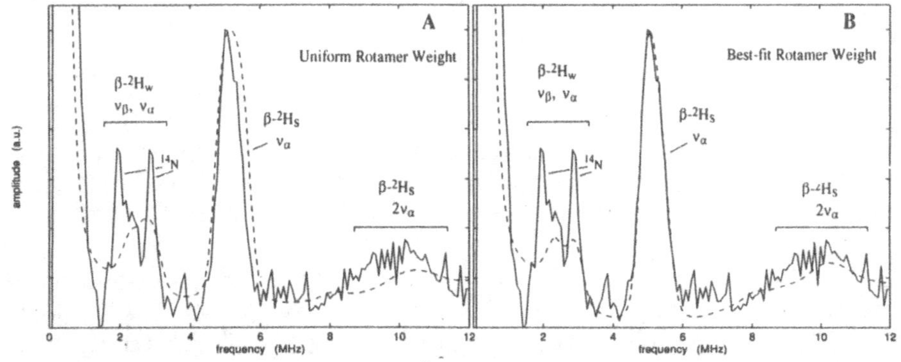

Figure 1. ESEEM spectra of the β-methylene ^2H-methylene hyperfine interaction in $Y_Z^{\bullet}$ (solid line) and spectrum simulated (dotted line) by assuming a distribution in C_1-C_β conformation. *A.* Uniform rotamer weighting. *B.* Arbitrary rotamer weighting. Conditions: microwave frequency, 11.208 GHz; magnetic field, 0.4004 T; $\tau+T_0$,140 ns; $\nu_N(^2H)$=2.6 MHz. abscissa scale: 0-12 MHz.

of the summation ESEEM, and assumptions regarding the symmetries of the β-H hyperfine tensors and their dependence on θ, have been elaborated [7].

3. Results and Discussion

3.1 *Summation ESEEM Spectra of the β-methylene ^2H Coupling in the Y_Z radical*

Figure 1 shows envelope-divided ESEEM spectra of β-methylene ^2H-labelled/per-protonated $Y_Z^{\bullet}$. The spectra were obtained by Fourier transformation of the summation of $^2H/^1H$ quotient envelope modulation collected at ten τ values from 176 to 1056 ns [4]. The spectra show two dominant features centered at ~0.8 and 5.0 MHz at 0.4004 T, that are assigned to the ν_β and ν_α frequencies of one set of strongly-coupled β-methylene ^2H nuclei (β-^2H_s). Weak $\Delta m_I=\pm2$ features for the α-electron spin manifold, $2\nu_\alpha$, are also observed. Narrow features in the spectra at 0.8, 2.0 and 2.8 MHz arise from the nuclear quadrupole coupling of ^{14}N nuclei in contaminant chlorophyll radicals in the samples. The broad, featureless spectral intensity at the base of these characteristically sharp ^{14}N features in the region around the free ^2H Larmor frequency (ν_N) is assigned to a second, weakly coupled set of β-methylene-^2H nuclei (β-^2H_w).

3.2 *ESEEM Simulations incorporating a distribution in C_1-C_β conformation*

The simulated ^2H ESEEM in Figure 1 incorporates a static distribution in θ about the C_1-C_β bond. A range of C_1-C_β rotamer orientations corresponding to $|\theta_R|$=12°-24° is necessary to account for the full extent of the β-^2H_s ν_α feature. A value of $\rho_{\pi C1}$=0.37 ±0.03 provided the best match with experiment. Figure 1A shows the simulation corresponding to an equal spectral contribution from each rotamer. The uniform rotamer weighting does not give an exact match with the experimental lineshape. Figure 1B shows that the simulated ESEEM obtained with an arbitrary weighting of the different rotamer contributions generates excellent agreement with experiment. Thus a distribution in θ causes the broadened β-methylene ^2H ESEEM lineshapes in $Y_Z^{\bullet}$.

The disorder in the β-methylene ^2H ESEEM lineshapes could, in principle, be caused by a distribution in $\rho_{\pi C1}$. However, the extent of the distribution necessary to simulate the observed ESEEM [11], $0.32<\rho_{\pi C1}<0.42$, is incommensurate with the relatively constant values of $\rho_{\pi C1}$ observed for model and protein-bound tyrosine radicals, including those that do and do not form phenol-O···H hydrogen bonds [4]. In addition, the ρ_π values at the C_2, C_3, C_5 and C_6 positions in

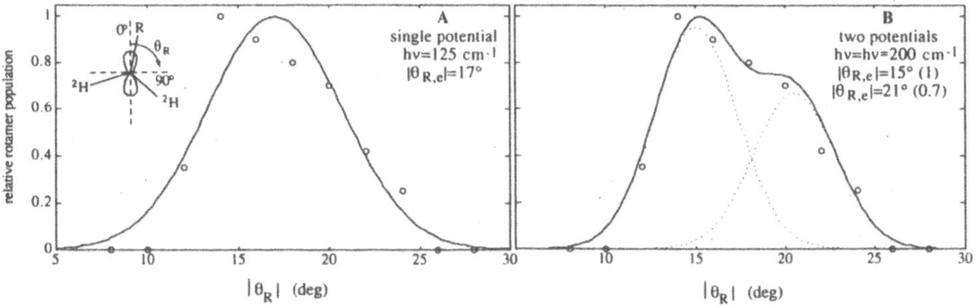

Figure 2. Relative rotamer distribution for $Y_Z^{\bullet}$ (circles) and population distribution functions for a single harmonic oscillator (*A*) and two overlapping harmonic oscillators (*B*). Relative contributions are in parentheses, T=200 K.

$Y_Z^{\bullet}$ are the same within 0.01 as those in model and other protein-bound tyrosine radicals [4]. Therefore, we propose that the broadening of the β-methylene ^2H spectral features is not caused by a distribution in $\rho_{\pi C1}$. This conclusion rules out a significant influence of disorder in phenol-O···H hydrogen bonding on $\rho_{\pi C1}$ value.

3.3 *Rotamer Distribution in the Y_Z radical*

Figure 2 shows the observed $Y_Z^{\bullet}$ relative rotamer distribution given by the arbitrary weighting for the mutual simulation of ESEEM spectra collected at 0.3282 and 0.4004 T. There is a small preference for conformations at one boundary of the rotation. The distribution is static from 4.2 K, where the ESEEM experiments are performed, to at least 108 K, because the β-^1H ν_α feature in transient-electron nuclear double resonance (ENDOR) spectra of the same samples [4,12] obtained at 108 K displays a spectral extent corresponding to the β-^2H features in the ESEEM spectra. The temperature at which C_1-C_β torsional motion "freezes-in" is uncertain. We estimate that motion stops at ~200-250 K, which corresponds to temperatures over which $Y_Z^{\bullet}/Y_Z$-mediated electron transfers effectively cease [1].

Figure 2A shows a fit of the observed rotamer distribution to the angular distribution function computed for a single harmonic potential. The asymmetry in the experimental rotamer distribution cannot be accomodated by one harmonic potential. Thus, the ring is not rotating freely in a cavity in the protein, within a potential governed by interactions with the adjacent peptide backbone atoms [9]. This result also indicates that the ring does not protrude directly into an aqueous phase. The failure of the single harmonic potential model suggests that the motion prior to rigidification, and hence the distribution in θ, is controlled predominantly by non-bonded contacts of the ring with a surrounding polypeptide matrix. Thus, there is static disorder in the protein structure surrounding the Y_Z site. This indicates that ring motion at elevated temperatures is "gated" by the protein matrix.

The distribution can be well-fitted by assuming two or more overlapping harmonic potentials which differ in equilibrium angles and zero point energies, and hence relative populations. Figure 2B shows a representative, but not unique, fit using two harmonic potentials for which equivalent vibrational frequencies are assumed to constrain the number of variable parameters. Additional overlapping potentials yield better fits. Therefore, if the librational potential for each photosystem II complex in the sample ensemble is assumed to be harmonic, there are ≥2 minima in the protein-ring conformation.

3.4 *$Y_Z^{\bullet}$ side chain mobility and water oxidation catalysis*

The photosystem II core preparations examined in this work are depleted of the manganese center [3]. Thus, the static distribution in $Y_Z^{\bullet}$ C_1-C_β conformations, as well as disordered hydrogen bond interactions [3,4] and enhanced matrix ENDOR temperature dependence relative to $Y_D^{\bullet}$ [13], might reflect structural instability in the protein introduced by removal of the nearby (separation distance ~4.5 Å [14]) manganese center. Structural influences would be local to the Y_Z site on the D1 polypeptide, since treatments that perturb or remove the manganese center cause negligible (<1°) changes in the θ values of $Y_D^{\bullet}$ [5] and the C_1-C_β conformation in $Y_D^{\bullet}$ in manganese-depleted photosystem II is discrete [6]. If true, this result shows that the conformational and hydrogen bonding state, and hence the energy of $Y_Z^{\bullet}$, is responsive to the structure of the manganese site. Thus, ligand field changes accompanying redox and associated substrate-bonding changes of the manganese center may be coupled directly to $Y_Z^{\bullet}/Y_Z$ electron transfer energetics.

Broadened β-methylene H lineshapes of extent comparable to those observed by ^{2}H ESEEM have, however, also been observed in calcium-depleted photosystem II [14], where the manganese center is maintained. Further, as shown above, the phenol ring is not undergoing simple free libration. These findings intimate that mobility of $Y_Z^{\bullet}$ may be integral to photosystem II turnover. Indeed, the disorder in $Y_Z^{\bullet}$ and the proximity of the manganese center have lead to proposals of a direct water hydrogen or water proton abstractor function of $Y_Z^{\bullet}$ in the water-oxidation mechanism [4,12,14].

Acknowledgements

This work was supported by grants from NIH (G.T.B., J. M.), USDA (G.T.B.) and NRI/CGP/USDA (B.A.D.) & support from Foundation Blanceflor Boncompagni-Ludovisi (C.T.).

References

1 Babcock, G. T., Barry, B. A., Debus, R. J., Hoganson, C. W., Atamian, M., McIntosh, C., Sithole, I. and Yocum, C. F. (1989) Biochemistry 28, 9557-9565

2 Hoganson, C. W. and Babcock, G. T. (1994) in Metal Ions in Biological Systems, Volume 30 (Sigel, H. and Sigel, A., eds.) pp. 77-107, Marcell Dekker, New York

3 Tang, X.-S, Zheng, M., Chisholm, D. A., Dismukes, G. C. and Diner, B. A. (1995) Biochemistry, submitted

4 Tommos, C.; Tang, X.-S.; Warncke, K., Hoganson, C. W.; Styring, S.; McCracken, J.; Diner, B. A.; Babcock, G. T. (1995) J. Am. Chem. Soc., in press

5 Rigby, S.E.J., Nugent, J.H. and O'Malley, P.J. (1994) Biochemistry 33, 1734-1742

6 Warncke, K., Babcock, G. T. and McCracken, J. (1994) J. Am. Chem. Soc. 116, 7332-7340

7 Warncke, K. and McCracken, J. (1995) J. Chem. Phys., in press

8 Heller, C. and McConnell, H. M. (1960) J. Chem. Phys. 32, 1535-44

9 Warncke, K., Babcock, G. T. and McCracken, J. (1995) J. Phys. Chem., submitted

10 Warncke, K and McCracken, J. (1994) J. Chem. Phys. 101, 1832-1841

11 Warncke, K., Tommos, C., Tang, X.-S., Diner, B. A., Babcock, G. T., and McCracken, J. (1995) Biochemistry, submitted

12 Hoganson, C. W., Lydakis-Simantiris, N., Tang, X.-S., Tommos, C., Warncke, K., Babcock, G. T., Diner, B. A., McCracken, J. and Styring, S. (1995) Photosyn. Res., in press

13 Mino, H. and Kawamori, A. (1994) Biochim. Biophys. Acta 1185, 213-220

14 Gilchrist, M. L., Ball, J. A., Randall, D. W., Britt, R. D. (1995) Proc. Natl. Acad. Sci., in press

CHARACTERIZATION OF A RHODOBACTER CAPSULATUS REACTION CENTER MUTANT WITH THE Q_B SITE MADE MORE LIKE Q_A SITE

Jiali. Li, W.J. Coleman*, D.C. Youvan#, M.R.Gunner
Physics Dept. City college of New York, CUNY, 138th St. and Convent Ave., NY, NY 10031. *Dupont CR&D, Experimental Station, P.O.Box 80173, Wilmington, DE,19880. #Palo Alto Institute, Mountain View, CA 94043

1.Introduction

The isolated reaction center complex (RC) from the purple photosynthetic bacteria Rb. capsulatus has three protein subunits, L, M, and H, and four bacteriochlorophylls (BChl), two bacteriopheophytins (BPh), and two quinones, Q_A and Q_B. When RCs are activated by light, a special pair of BChls (P) is excited and an electron is transferred by way of BPh to Q_A forming the semiquinone (Q_A^-) producing the state $P^+Q_A^-Q_B$. Q_A^- then reduces the quinone bound at the Q_B site to form $P^+Q_AQ_B^-$. The two quinones are both UQ_{10}. However, they are functionally different presumably because of differences in their binding sites. Q_A, tightly bound, in a relatively hydrophobic environment, is a one electron acceptor. Q_B, is loosely bound, in a more polar region, and functions as a sequential two electron acceptor.

The amino acid residues near the Q_B site are mainly from the L subunit of the protein surrounded by residues from the M and H subunits. Coleman et. al. produced a genetically engineered Rb. capsulatus RC where the 42 residues, M220 to M261, from the Q_A binding site replace the 35 residues, L193 to L227, in the Q_B binding site (1). This produces a protein where many of the residues near Q_B have been replaced by residues from the Q_A site. The original mutant was photosynthetically inactive. A6D1 is a photosynthetically active revertent (1) which retains the Q_A insert intact, but has the additional changes Met^{M144}--> Ile and Ala^{M145} -->Ser. This mutant RC provides a very good system to study how changing the Q_B site to be more like the Q_A site effects the photochemical cycle, especially the Q_B affinity and the Q_B dependent electron transfer rates.

2. Procedure

2.1 *Sample preparation*

Generation of the A6D1 mutant of Rb. capsulatus is described in reference (1). The wild type and mutant bacteria were grown in the dark. Reaction centers were prepared by a modification of the protocol in (2). Purified wild type RCs had 1.1-1.2 quinones as indicated by 10 to 20% slow back recovery kinetics, but the A6D1 RC preparations had no Q_B activity. UQ_4 was obtained from Sigma and was solubilized in ethanol. Horse heart cytochrome c was obtained from Sigma and was reduced by Sodium Borohydride (Sigma). Ferrocene (Aldrich) was diluted in ethanol.

P. Mathis (ed.), Photosynthesis: from Light to Biosphere, Vol. I, 715–718.

2.2 *Spectroscopy*
RCs were characterized by optical spectroscopy performed on a flash spectrometer (University of Pennsylvania Biomedical Instrumentation Group). The activating light was a 10us xenon flash.

3. Result and Discussion

3.1 *The Rate of Charge Recombination from Q_A^- and Q_B^-*
When quinone occupies only the Q_A site, charge recombination from $P^+Q_A^-$ to reform PQ_A occurs as a nearly single exponential process with a rate (k_{AP}) of 8 s^{-1} in both wild type and mutant RC (data not shown). Thus, the large scale modification of the Q_B site has little or no impact on the function of Q_A.

Charge recombination (Figure 1a) from $P^+Q_B^-$ to reform the ground state (k_{BP}) was determined in the presence of excess UQ4 (100-200 UQ4/RC). k_{BP} for the mutant is significantly slower ($\leq$0.05 s^{-1}) than for the wild type (0.85 s^{-1}). If charge recombination from $P^+Q_B^-$ occurs via $P^+Q_A^-$ then the difference between k_{AP} and k_{BP} can be used to derive the equilibrium constant between Q_A and Q_B (3). Given $K_{AB} = k_{AP}/k_{BP} - 1$, K_{AB} is $\approx$10 for the wild type and 200 for the mutant RCs. Thus, the Q_B^- state is stabilized by 80 meV in the mutant RC relative to the wild type protein. The removal of Glu L 212, and Asp L210 and L213 near Q_B may stabilize the negative charge on Q_B in the mutant. A slow rate of charge recombination from $P^+Q_B^-$ is also found in L212/L213 double mutants with Rb. capsulatus and Rb. sphaeriodes (4, 5).

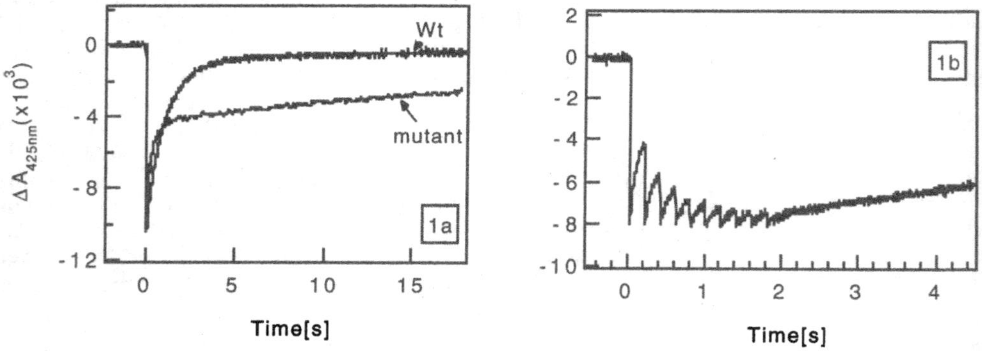

Figure1. Charge separation and recombination monitored by the absorbance change at 425 nm as a function of time. pH=7.3, 0.04% LDAO, 0.5μM RCs, 100 μM UQ_4. The rate constants (1a): Wild type, k_{AP}=7.6 s^{-1}, k_{BP}=0.85 s^{-1}; mutant, k_{AP}=7.4 s^{-1}, k_{BP}=0.027 s^{-1}. (1b). mutant RCs repetitively excited at an interval of 0.2 s..

3.2 *QB reconstitution*
When UQ4 is added to RCs with only Q_A, charge recombination is seen to be biphasic with rates k_{AP} and k_{BP}. The reconstitution of the Q_B site can be determined by the fraction of the RCs that exhibit the rate k_{BP}. In addition, the ratio of cytochrome c oxidation on the first and second flashes can discriminate between RCs with and without Q_B. Thus, if Q_B is not

reduced after the first flash, no additional cytochrome c oxidation occurs on latter excitation (6).

Figure 2 shows the UQ_4 titration curves with wild type and mutant RCs. Using the fraction of slow back reaction as a monitor, the K_d value for the wild type is 4.4±0.5 μM and 11.9±0.6 μM for the mutant. When the RCs are saturated with UQ, 90% of the wild type and 50% of the mutant RCs show slow recombination kinetics. With the cytochome c assay, with flashes spaced 0.5 s apart, the K_d's are 7.8±0.9 μM and 11±0.6 μM with an asymptote of 98% and 90% respectively for wild type and mutant RCs . Thus, despite the large scale rearrangement of the Q_B site, the affinity of the binding site for UQ is weakened by less than a factor of three. However, the yield of $P^+Q_B^-$ is significantly smaller in the mutant RCs than in the wild type protein.

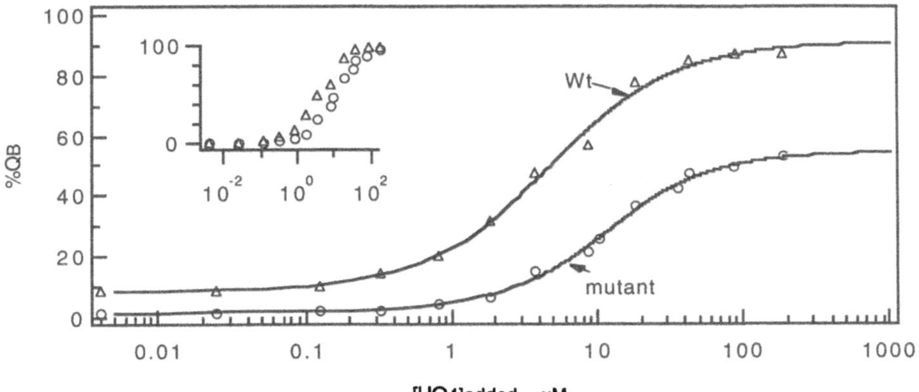

Figure 2. UQ_4 titration to the Q_B site of RCs from wild type (o) and mutant (Δ) RCs. Conditions were the same as Figure 1. RCs with Q_B is obtained from the fraction of the charge recombination that occured at k_{BP}.

3.3 *The Rate of Electron Transfer From Q_A to Q_B.*

The small yield of reaction centers in the $P^+Q_B^-$ state following a single flash appears to be due to a rate of electron transfer from Q_A^- to Q_B that is comparable to the rate of electron transfer from Q_A^- to P^+. Figure 1b shows the result of repetitive excitation under conditions where the RCs are saturated with UQ4. After the first flash approximately 50% of the charge recombination is at the faster rate (k_{AP}) and 50% at k_{BP}. On each additional flash half of the population that showed k_{AP} now shows the slower recombination rate characteristic of RCs in the state $P^+Q_B^-$. Thus, the low yield of $P^+Q_B^-$ after the first flash does not indicate that half the RCs have an incompetent Q_B site. Rather the quantum yield for formation of $P^+Q_B^-$ is 50% for all of the mutant RCs. If k_{AP} is 8 s^{-1}, a quantum yield of 50% suggests the rate of electron transfer k_{AB} is 8 s^{-1}.

Since cytochrome c oxidation is not seen if Q_A is still reduced, the ratio of cytochrome c oxidized on the first and second flash can be used to assay the electron transfer rate from Q_A to Q_B. Varying the time interval between the first and second flashes, the rate of electron transfer from Q_A to Q_B was determined to be 14 s^{-1}. This is consistent with the estimate of the rate obtained from the quantum yield of Q_B^-.

3.4. *Why is k_{AB} so slow?*
The mutant RCs described here represent a large scale rearrangement of the Q_B site. The affinity of Q_B is surprisingly relatively unchanged. The $P^+Q_B^-$ state in the mutant appears to be 80 meV more stable than in the wild type protein, a result also found in a number of mutants that change only selected residues in the Q_B site (7). Thus, the mutant and the other L212/I213 double mutants show comparable results. However, the largest change in the mutant is to slow the electron transfer from Q_A to Q_B. In the wild type this reaction occurs in a multiphase process with rates from 10^3 to 10^4 (4, 8). In contrast, in this mutant the reaction occurs at 14 s^{-1}, 100 to 1000 times slower.

It is not clear why the electron transfer rate is so much slower in the mutant than the wild type RCs. The reaction has a larger, favorable ΔG than in the wild type protein. The activation energy of the reaction is approximately 400 meV, comparable to the value found in wild type RCs (3). An increased electron transfer distance between Q_A and Q_B because the quinone is bound at an alternate site could contribute to the slower rate (9, 10). In addition, the large change in rate points to some required change in protein conformation or proton transfer that is more difficult to accomplish in the mutant RCs.

4. Acknowledgments
This work was supported by NIH grant GM-RO1-48726.

References

1. Coleman, W.J. & Youvan, D.C. (1993) Nature 366, 517-518.
2. Prince, R.C. & Youvan, D.C. (1987) Biochim. Biophys. Acta 890, 286-291.
3. Mancino, L.J., Dean, D.P. & Blankenship, R.E. (1984) Biochim. Biophys. Acta 764, 46-54.
4. Baciou, L., Bylina, E.J. & Sebban, P. (1993) Biophysical J. 65, 652-660.
5. Takahashi, E. & Wraight, C.A. (1992) Biochemistry 31, 855-866.
6. Giangiacomo, K.M. & Dutton, P.L. (1989) Proc. Natl. Acad. Sci. USA 86, 2658-2662.
7. Takahashi, E. & Wraight, C.A. (1994) Advances in Molecular and Cell Biology 10, 197-251.
8. Okamura, M.Y. & Feher, G. (1992) Annu. Rev. Biochem. 61, 861-896.
9. Moser, C.C., Keske, J.M., Warncke, K., Farid, R. & Dutton, P.L. (1992) Nature 355, 796-802.
10. Deisenhofer, J. & Michel, H. (1989) The EMBO Journal 8, 2149-2170.

QUINONE REPLACEMENT ALLOWS DIRECT OPTICAL MEASUREMENT OF A PREVIOUSLY UNREPORTED PHASE OF ELECTRON TRANSFER FROM Q_A^- TO Q_B^- IN *RHODOBACTER SPHAEROIDES* REACTION CENTERS

Dan Gilroy and M.R. Gunner

Department of Physics Rm J419, City College of New York, 138th St and Convent Ave., New York, NY 10031 (212)650-5557 (voice) (212)650-5503 (fax)

1. Introduction

Reaction centers of purple non-sulfur photosynthetic bacteria (RCs) store the energy of a photon by a series of electron transfers[1]. Following absorption of a photon by a dimer of bacteriochlorophyll (P) the primary quinone (Q_A) is reduced in 150 ps forming $P^+Q_A^-$. Electron transfer from Q_A to the secondary quinone (Q_B) occurs within milliseconds (at rate k_{AB}) yielding $P^+Q_B^-$. This reaction results in sub-stoichiometric proton uptake beginning the coupling of electron transfer and proton uptake important for energy conservation[2-5]. This is the only intra-RC electron transfer that is temperature dependent, pointing to the role of protein or proton motions that may control its rate. However, this reaction is difficult to study since in RCs from *Rhodobacter sphaeroides* the same redox cofactor resides in both the Q_A and Q_B sites. Secondary signals have been used to monitor either changes in the spectrum of the bacteriopheophytins adjacent to the quinones at 750 or 770 nm or small differences in the absorbance of UQ_A^- and UQ_B^- at 397 nm[6-10].

The Q_A^- and Q_B^- can be clearly resolved if quinones with different semiquinone anion spectra can be placed in the Q_A and Q_B sites. Menaquinone has a different semiquinone spectra than ubiquinone[11]. Specific quinone substitution succeeds because the Q_A site has a preference for menaquinone over ubiquinone, while the Q_B site has no measurable affinity for menaquinone, at least under conditions when Q_A is a UQ. The replacement of Q_A by MQ is expected to modify the Q_A^- to Q_B^- electron transfer very little. *Rb. sphaeroides* RCs with MQ in the Q_A site and UQ at the Q_B site have the same quinone as the native RCs from *Rps. viridis*, [12], so this reconstitution mimics a natural RC. In addition, although MQ is 140 meV harder to reduce in solution than UQ, in the Q_A site of *Rb. sphaeriodes* RCs MQ_A and UQ_A appear to have reduction potentials that differ by less than 30 meV [13]. The results found using quinone substitution can be compared with another, recent direct method for monitoring electron tranfer from Q_A^- to Q_B^- using time-resolved IR absorption[14].

These specially prepared RCs, where the Q_A^- to Q_B^- transient is approximately 2.5 times that of the native RCs, clearly show that a substantial portion of the Q_A to Q_B electron transfer has a fast rate, i.e. >30,000/sec, in addition to slower processes at approximately

P. Mathis (ed.), Photosynthesis: from Light to Biosphere, Vol. I, 719–722.

5000/sec and 1200/sec. This fast process has also been seen monitoring Stark shifts in the BPh spectrum [10].

2. Proceedure

Sample preparation. Rhodobacter sphaeroides R-26 reaction centers were isolated in the usual way using LDAO detergent extraction followed by purification using ammonium sulfate and DEAE chromatography[15]. Quinones were removed with orthophenathroline using the method of Okamura [16], with some modifications [17]. This typically yielded RCs with no Q_B and 5-15% Q_A.

Several methods were used to reconstitute the quinone sites. The most effective was the method of Wraight [18], using quinones dissolved in Triton 100X. The quinone sites were reconstituted sequentially with Menaquinone (Vitamin K1, Fluka) at a concentration of 1-2 quinones/RC. Then Q_B site was reconstituted with UQ (UQ10 or UQ4, Sigma or Fluka) with a ratio of 5 quinones/RC. Typical samples contain nearly 100% menaquinone at the Q_A site and approximately 70-80% ubiquinone at the Q_B site. The quinone occupying the Q_A site was determined from the charge recombination kinetics following the addition of orthophenanthroline to remove Q_B. $P^+ MQ_A^-$ and $P^+UQ_A^-$ charge recombination occurs at 15/sec and 10/sec respectively. This shows that exchange between quinone sites is limited if the RCs are kept in low detergent after quinone reconstitution. Reconstituted samples were generally diluted in 10 mM TRIS pH 8.1 buffer. For measurements at different pHs, the following buffers were used: MES (pH 5.2-5.8), PO4 (pH 6.2-7.8), TRIS (pH 8.1-8.6), and CHES(pH 9.2-10.2).

Spectroscopy. Light induced absorption changes were measured using a University of Pennsylvania Biomedical Instrumentation Group flash spectrometer. Flash activation was either with a 10 μs xenon flash or a 5 ns pulsed YAG laser pumping a flourescent dye cell.

3. Results and Discussion

Figure 1 shows the light activated change in RC absorbance in the presence of 200μM ferrocene, which serves as an electron donor to the oxidized chlorophyll dimer. The donation of an electron to P^+ prevents the Q_A^- or Q_B^- to P^+ charge recombination, and effectively traps the electron on the quinone, allowing the observation of a stable semiquinone anion absorbance change [19]. When UQ is added to the Q_B site of RCs with either UQ_A or MQ_A, the difference spectra with ferrocene is now predominately due to Q_B^-. The spectra from UQ_A^- and UQ_B^- are different in the regions around 397 nm and around 450 nm. One previous approach to measuring the rate of electron transfer has been to monitor the kinetics of the absorbance changes at either of these wavelengths [6, 7]. However, comparison of the UQ_A^-, UQ_B^-, and MQ_A^- spectra show that, as reported earlier [12], when MQ is present in the Q_A site, the semiquinone signal is much larger and broader than with UQ as Q_A. The change in absorbance associated with a shift from the MQ_A^- state to the UQ_B^- state is approximately 2.5 times the absorbance change found at 397 nm in RCs with the native quinone in both sites. This is consistent with published semiquinone spectra [11].

Figure 2 shows a flash activated kinetic trace for a sample containing MQ_A and UQ_B. The log plot clearly shows the process is not characterized by a simple, single exponential rate. Rather a minimum of 4, reasonably distinct rates can be found. The slowest represents the

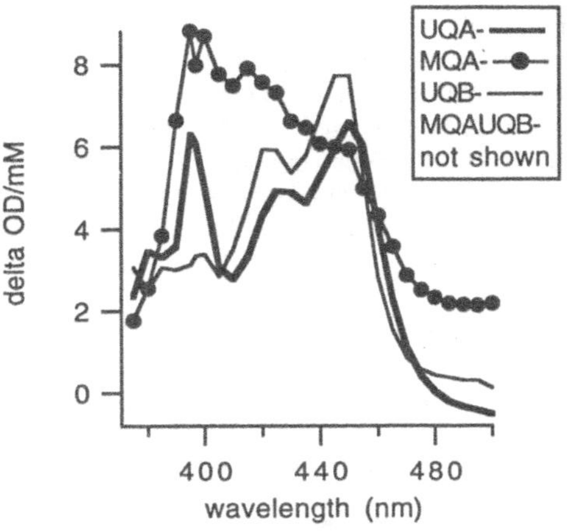

Figure 1.

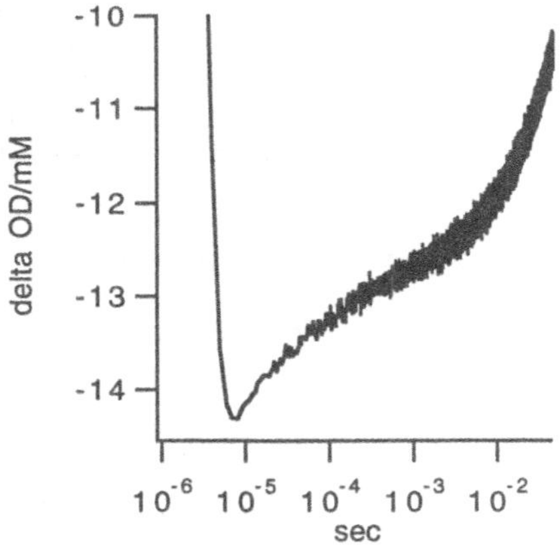

Figure 2.

15/sec decay of $P^+MQ_A^-$ state in RCs without Q_B. The three faster rates, include 5000/sec and 1200/sec components, which have been previously reported and are consistent with other recent work[9], and a new >30,000/sec component. This fast component can be seen to represent a significant fraction of the electron trasfner process. Similar kinetic components are also seen in the smaller UQ_A^- to UQ_B^- transient at 397 nm. No similar transients are seen in RCs without Q_B.

The expected amplitude for the electron transfer can be derived from measurement of the spectra of trapped MQ_A^- or UQ_B^- (see figure 1). The difference between the trapped Q_A^- and Q_B^- spectra at 400 nm is 5.1 OD/mM. The intial amplitudes of the 5000/s and 1200/s is approximately 30% of the expected amplitude. The addition of the >30,000/s component yields a total of amplidues that is within 60% of that expected. This discrepancy may point to additional, faster components of the electron transfer from Q_A^- to Q_B^-. The delay time in the current measurements is 5-10 μs.

The temperature and pH dependence of the rates that are slower than 10,000/s have been characterized between pHs 6 and 8.5 and 0-25 °C. The activation energy is 2.8±1.4 and 9.0±1.2 kcal/mole for the faster and slower process respectively. At 25 °C the amplitude of the faster rate of these two rates is approximately 2 times that of the slower component. However, at 0°C there is more slow than fast component. The decrease in the amplitude of the faster component provides an additional mechanism for slowing the overall electron transfer process as the temperature is lowered. Between pH 6 and 8.6 the rates slow by less than a factor of 2 as the pH as increased. Analysis of data containing multiple exponentials is notoriously difficult, however preliminary characterization of the fastest component suggests that the activation energy of this process is <1 kcal/mole. The amplitude associated with this fast rate remains the major fraction of the total amplitude between 0-25° C and pH 7-10 (data not shown).

The complexity of the electron transfer kinetics from Q_A^- to Q_B^- may reflect heterogeneity of the Q_B environment in an ensemble of RCs. This part of the protein is characterized by an interconnected group of acidic residues [4, 20] Calculations of the proton occupancy of these ionizable residues show that there are a large number of possible protonation states with similar energies. This would imply that RCs exist with a variety of sub-states. Since each state has a different distribution of ionizable residues, the free energy of Q_B^- would be different in each state. In addition, the rate and activation of needed conformational changes could also be state dependent. Evidence for heterogeneity in the Q_B^- site has been seen by time resolved I.R. measurements that show sub-stoichiometric protonation of Glu L 212 following reduction of Q_B[14]. If the multiple kinetic components are due to different ionization states of the protein the inter conversion between these states must be slower than the electron transfer kinetics.

Acknowledgments: We are grateful for the financial support of NIH grant RO1-GM48726.

References

1. Feher, G., J.P. Allen, M.Y. Okamura, and D.C. Rees, 1989, Nature, 339, p. 111-116.
2. Maroti, P. and C.A. Wraight, 1988, Biochim. Biophys. Acta, 934, p. 314-328.
3. McPherson, P.H., M.Y. Okamura, and G. Feher, 1988, Biochim. Biophys. Acta, 934, p. 348-368.
4. Gunner, M.R. and B. Honig, 1993, in *The Photosynthetic Bacterial Reaction Center: Structure, Spectroscopy and Dynamics II,* ed. J. Breton and A. Vermeglio, Plenum, New York, p. 403-410.
5. Brischwein, M., B. Scharf, M. Engelhard, and W. Mantele, 1993, Biochemistry, 32, p. 13710-13717.
6. Vermeglio, A. and R.K. Clayton, 1977, Biochim. Biophys. Acta, 461, p. 159-165.
7. Wraight, C.A., 1979, Biochim. Biophys. Acta, 548, p. 309-327.
8. Kleinfeld, D., M.Y. Okamura, and G. Feher, 1984, Biochim. Biophys. Acta, 766, p. 126-140.
9. Takahashi, E., P. Maroti, and C.A. Wraight, 1992, in *Electron and Proton Transfer in Chemistry and Biology,* ed. A. Muller, Elsevier, p. pgs. 219-236.
10. Tiede, D.M. and D.K. Hanson, 1992, in *The Photosynthetic Reaction Center II,* ed. J. Brenton and A. Vermeglio, Plenum Press, New York, p. 341-350.
11. Swallow, A.J., 1982, in *Function of Quinones in Energy Conserving Systems,* ed. B.L. Trumpower, Academic Press, New York, p. 59-72.
12. Shopes, R.J. and C.A. Wraight, 1985, Biochim. Biophys. Acta, 806, p. 348-356.
13. Woodbury, N.W., W.W. Parson, M.R. Gunner, R.C. Prince, and P.L. Dutton, 1986, Biochim. Biophys. Acta., 851, p. 6-22.
14. Hienerwadel, R., D. Thibodeau, F. Lenz, E. Nabedryk, J. Breton, W. Kreutz, and W. Mantele, 1992, Biochemistry, 31, p. 5799-5808.
15. Clayton, R.K. and R.T. Wang, 1971, Methods Enzymol, 23, p. 696-704.
16. Okamura, M.Y., R.A. Isaacson, and G. Feher, 1975, Proc. Natl. Acad. Sci, USA, 72, p. 3492-3496.
17. Woodbury, N.W.T., W.W. Parson, M.R. Gunner, R.C. Prince, and P.L. Dutton, 986, Biochim. Biophys. Acta, 851, p. 6-22.
18. McComb, J.C., R.R. Stein, and C.A. Wraight, 1990, Biochim. Biophys. Acta, 1015, p. 156-171.
19. Clayton, R.K., *Photosynthesis: Physical Mechanisms and Chemical Patterns.* 1980, Cambridge: Cambridge University Press.
20. Beroza, P., D.R. Fredkin, M.Y. Okamura, and R. Feher, 1995, Biophys. J., 68, p. 2233-250.

PHOTOPROTECTION OF NON-PHOTOACTIVATED PHOTOSYSTEM II

Anja Krieger[1], Sebastian Heimann[1] and Giles Johnson[2]

[1]Julius-von-Sachs Institut für Biowissenschaften, Universität Würzburg, Mittl. Dallenbergweg 64, D-97082, Würzburg, Germany.
[2]Institüt für Botanik, Universität Münster, Schloßgarten 3, D-48161, Münster, Germany. (current address: School of Biological Sciences, University of Manchester, 3.614 Stopford Building, , Oxford Road, Manchester, GB-M13 9PT England) (corresponding author)

Introduction

The first step in the synthesis of Photosystem II seems to be the assembly of the D1-D2-b_{559} reaction centre. Assembly of the manganese complex required for water oxidation and so full activity of PSII, photoactivation, is the last step to occur and requires terminal cleavage of the D1 polypeptide. It also requires light to drive oxidation of manganese so enabling the formation of a stable tetra-manganese complex. The overall quantum yield of photoactivation is very low (for refs see [1]).

Prior to photoactivation, the PSII reaction centre is extremely vulnerable to photoinhibition. In a fully active PSII centre, P_{680}^+, the oxidised primary donor, is reduced on a nano-micro second timescale by the tyrosine residue Z which is in turn reduced by the water splitting complex. In the absence of the water splitting complex, donation from Z is slowed and oxidised Z is stable on a microsecond timescale. Under such conditions, the probability of charge recombination occurring, with the electron on Q_A or Q_B rereducing P/Z via the excited state, P^+Pheo^-, will be increased. This recombination reaction has been implicated in some photoinhibtion processes [2]. Charge recombination via the excited state leads to a high yield of triplet chlorophyll formation. this can, in turn, interact with molecular oxygen to form highly reactive singlet excited oxygen (see Scheme 1).

The Role of Q_A Redox Potential

Measurements of the redox mid-point potential of the charge pair Q_A/Q_A^- in active and inactive PS II have shown that, whilst in active centres a potential of -80 mV is seen, this is shifted to +65 mV in samples where the oxygen evolving complex is inhibited [3,4,5]. This is the case whether O_2 evolution is inhibited by complete destruction of the water

P. Mathis (ed.), Photosynthesis: from Light to Biosphere, Vol. I, 723–726.

splitting complex (by tris or NH_2OH treatment) or by removal of Ca^{2+} (by low pH treatment or salt washing). Further, in samples of PSII where oxygen evolution has never been active, for example in dark-grown or LF1 mutant cells of *Scenedesmus obliquus*, Q_A is found to be in the high potential form [5]. Activation (or reactivation) of oxygen-evolving capacity causes Q_A to shift to the low potential form.

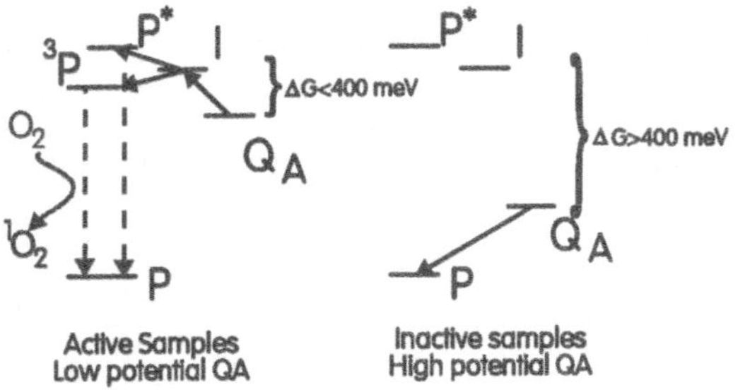

Scheme 1: *Charge recombination reactions in active and inactive Photosystem II. By analogy with observations in bacterial systems, it is postulated that when the ΔG between the states IQ_A^- and I^-Q_A is less than 400 meV then charge recombination between P^+ and Q_A^- proceeds via the excited state, leading to a high probability of triplet formation. When the ΔG is greater than 400 meV direct or indirect charge recombination- to the ground state of P- occurs, avoiding possible triplet formation. In inactive samples the ΔG is estimated to be 475 meV, compared with 330 meV in active systems*

This shift in potential of Q_A might have a role in the prevention of charge recombination reactions in PSII, so protecting the reaction centre from photoinhibition. It has been observed in bacterial reaction centres that the route for charge recombination is determined by the free-energy gap between the states IQ_A^- and I^-Q_A (see [6]). When this gap less than 400 meV, recombination proceeds via formation of reduced pheophytin. When the gap is greater than 400 meV then recombination occurs directly to the ground state of P, avoiding repopulation of the excited state and hence possible triplet formation.

In active PS II, the free energy gap between IQ_A^- and I^-Q_A is estimated to be 330 meV [7]. If the shift in the mid-point potential of Q_A of 145 meV is taken into account, this implies a free energy gap. in inactive centres, of 475 meV. Hence, the shift in the mid-point potential of Q_A may serve to prevent charge recombination and therefor triplet formation in reaction centres prior to photoactivation.

A Possible Role for Cytochrome b_{559}?

Whilst the shift in the Q_A mid-point potential may prevent charge recombination via the excited state and so formation of triplet chlorophyll and oxidative damage to the reaction centre (and the cell in general), inactive PSII is still vulnerable to photoinhibition. The charge pair P_{680}/P_{680}^+ is estimated to have a redox mid-point potential of approximately 1V, enough to oxidise other chlorophyll molecules and amino acid residues. P_{680} is rereduced in centres where the water splitting complex is absent by Tyr Z, which is also highly oxidising. The long lifetime of highly reactive species in centres where water oxidation is inactive, increases the likelihood of damaging oxidative reactions occurring that may be involved in so-called "donor-side" photoinhibition. It has been suggested that cytochrome b_{559} may play a role in limiting such damage [8].

In PSII membranes (Figure 1) and thylakoid membranes (not shown) in which oxygen evolution has been inhibited by hydroxylamine treatment, cytochrome b_{559} is oxidised with a half-time of 50-100 ms. (By comparison, P_{680} rereduction, measured as an absorbance change at 820 nm, has a half-time of 120 µs [9]). P_{680} is reduced in such samples by Tyr Z. So, under conditions where the water splitting complex is absent, cytochrome b_{559} is able to donate electrons, directly or indirectly, to Tyr Z^{+}/P_{680}^{+} This donation may help to limit photoinhibition caused by oxidising reactions of these species.

If cytochrome b_{559} is to provide adequate protection against oxidative photoinhibition, it must be able donate electrons to the reaction centre repeatedly as long as the water splitting complex is inactive. When PSII membranes are illuminated in the absence of an exogenous electron donor, reduction of light-oxidised cytochrome b_{559} proceeds only extremely slowly. By contrast, if illumination is in the presence of a PSII donor, such as Mn^{2+}, the oxidised cytochrome is rereduced much more rapidly (minimum t½= 16 s- see below). The addition of Mn^{2+} after illumination does not bring about cytochrome reduction, implying that electrons are not donated directly from Mn^{2+} to b_{559}. That rereduction of the cytochrome requires multiple turnover of the reaction centre suggests that donation of electrons may occur via the plastoquinone pool, either via the Q_B site or, conceivably, via another quinone binding site. This reduction is significantly less efficient than the oxidation and hence, under steady-state conditions, will limit the donation of electrons to P_{680}/Z by cytochrome b_{559}. This implies that an efficient electron cycle around cytochrome b_{559} could not occur.

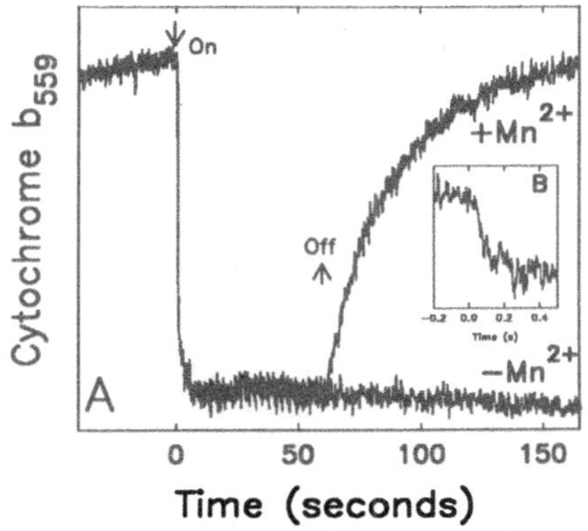

Figure 1: A. Light-induced changes in the redox state of cytochrome b_{559} in NH_2OH washed PSII membranes in the presence (red) or absence (blue) of 100 µM $MnCl_2$. Light (2000 $\mu Em^{-2}s^{-1}$) was switched on and off as indicated and the redox state of b_{559} followed using a Diode-Array Spectrophotometer, as described by Klughammer et al. (Ref. 10) B. Data recorded in the presence of 100 µM $MnCl_2$, light intensity- 10000 $\mu Em^{-2}s^{-1}$

Efficient donation of electrons to P_{680}^{+} by cytochrome b_{559} may, however, be undesirable. Photoactivation requires oxidation of Mn^{2+} by P_{680}^{+}. If repeated efficient donation from b_{559} were to occur, this would compete with Mn^{2+} oxidation, preventing assembly of the water splitting complex. Cytochrome b_{559} might instead simply provide and "emergency route" for preventing excessive oxidative reactions in PSII prior to photoactivation.

Could Cytochrome b_{559} have a Direct Role in Photoactivation?

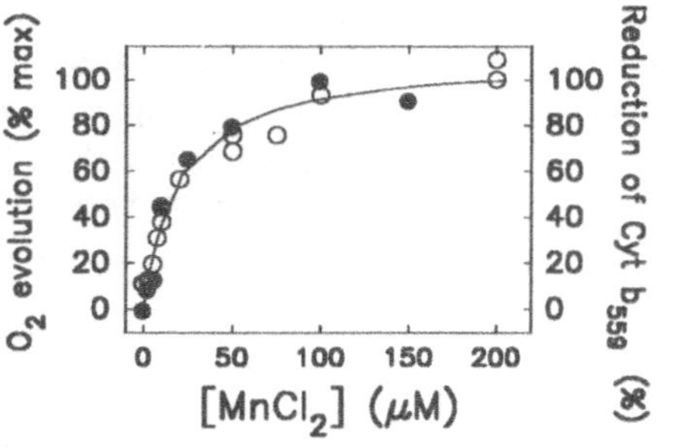

Figure 3: Closed symbols: relationship between rereduction of photo-oxidised cytochrome b_{559} and the concentration of $MnCl_2$ in NH_2OH-treated PSII membranes. Data expressed as the percentage of total photo-oxidised b_{559} reduced 120 seconds after the end of illumination, recorded as in Fig. 2. Open symbols: the relationship between $MnCl_2$ concentration and the degree of reactivation of O_2 evolution in NH_2OH treated PSII membranes. Data expressed relative to the maximum reconstitution observed. All conditions, other than $MnCl_2$ concentration as in Fig. 1B. Black curve represents a Michaelis-Menten curve fitted through both data sets. K_m= 20 μM

The kinetics for the rereduction of cytochrome b_{559} following illumination in the presence of Mn^{2+} are dependant on the concentration of Mn^{2+} used (Figure 2). The Km for Mn^{2+} for this reaction is 20 μM. Photoactivation can be induced *in vitro*, in NH_2OH treated PSII in the presence of Mn^{2+}. Ca^{2+} and DCPIP (e.g.[11]). The Km for Mn^{2+} for this process is also 20 μM. This coincidence implies that the rereduction of cytochrome and photoactivation may in some way be directly connected. Indeed cytochrome b_{559} may play a direct role in photoactivation.

Acknowledgments

We would like to thank Bill Rutherford, Engelbert Weis and Ulrich Schreiber for their help and support

References

1. Debus, R.J. (1992) Biochim Biophys Acta 1102, 269-352
2. Keren, N., Gong, H. and Ohad, I. (1995) J. Biol. Chem. 270, 806-814
3. Krieger, A. and Weis, E. (1992) Photosynthetica.27, 89-98
4. Krieger, A., Rutherford, A.W. and Johnson, G.N. (1995) Biochim Biophys Acta 1229, 193-201
5. Johnson, G.N., Rutherford, A.W. and Krieger, G.N. (1995) Biochim Biophys Acta 1229, 202-207
6. Woodbury, N.W., Parson, W.W., Gunner, M.R., Prince, R.C. and Dutton, P.L. (1986) Biochim Biophys Acta 851, 16-22
7. De Groot, B.G. and van Gorkom, H.J. (1981) Biochim Biophys Acta 635, 445-456
8. Thompson, L.K. and Brudvig G.W. (1988) Biochemistry 6653-6658
9. Haveman, J.and Mathis P., (1976), Biochim. Biophys. Acta 440, 346-355
10. Klughammer, C., Kolbowski, J. and Schreiber U. (1990) Photosynth. Res. 25, 317-327
11. Cheniae, G.M. and Martin, L.F. (1972) Plant Physiol. 50, 87-94

SEVERAL PROPERTIES OF *RHODOCYCLUS TENUIS* REACTION CENTER.

I. Agalidis and F. Reiss-Husson
C.G.M., Bât. 24, C.N.R.S., F- 91198 Gif-sur-Yvette, France.

1. Introduction

Rc. tenuis belongs to the β subclass of Proteobacteria (1). Up to date the properties of its photosynthetic unit have not been described. Like some other photosynthetic bacteria, it contains a tetraheme cyt c tightly associated to the reaction center (RC). In the present paper we report several characteristics of a purified RC preparation.

2. Materials and methods

2.1 Reaction center (RC) isolation
Rc. tenuis (strain ATCC 25093) was grown anaerobically in the light. Chromatophores were incubated with dodecyl-dimethyl amineoxide ($C_{12}DAO$) and a crude RC extract was obtained; further purification of the RC was achieved by ammonium sulfate precipitation followed by ion exchange chromatography in presence of dodecylmaltoside. Best purity index ($A_{280\ nm}/A_{800\ nm}$) was 1.3.

2.2 *Electrophoresis*
SDS polyacrylamide gel electrophoresis on minigels was carried out (2) with 12.5% polyacrylamide in the resolving gel and 4% in the stacking gel. Heme staining was performed according to (3), then the gel was destained and stained further with Coomassie Blue or silver.

2.3 *Extraction and identification of quinones*
Quinones were extracted and freed from polar lipids using a modified Bligh and Dyer method and identified by thin -layer chromatography as already described (4).

2.4 *Spectral measurements*
Absorption spectra were recorded on a Cary 2300 equipped for cross illumination. Flash-induced absorption changes were measured with a home made spectrophotometer (5).

3. Results and Discussion

A typical absorption spectrum of *Rc. tenuis* RC is shown in Fig.1. It displays features similar to those of RCs isolated from purple bacteria which contain Bchl a and a carotenoid.

P. Mathis (ed.), Photosynthesis: from Light to Biosphere, Vol. I, 727–730.

The absorbance of the tetraheme cytochrome c in the Soret band is barely visible ; in fact during the isolation procedure the cyt c is gradually detached and only 20% remains bound to the purified RC (not shown). Except for *Rps. viridis*, in other purple bacteria like *Chloroflexus auriantiacus, Chromatium vinosum* and *Chromatium tepidum* the tetraheme cyt c is liberated from the RC when the latter is isolated from the membrane (6). Photochemical activity is evidenced by the bleaching of the donor band at 870nm (not shown) and the appearance of the near IR band at 1260 nm (attributed to $P870^+$) (see Fig. 1, right). We should point out that the position of this last band is redshifted by 10 nm relative to that observed in *Rb. sphaeroides* (7). It has been shown that spectral characteristics of P and P^+ correlate with the number of H bonds engaged by the carbonyl groups of the dimer (8). One would then suggest that the acetyl carbonyl of P_M is H bonded. This hypothesis is confirmed by Fourier transform infrared Raman resonance (FTIR) spectra of the donor in *Rc. tenuis* (Mattioli and Agalidis, unpublished results).

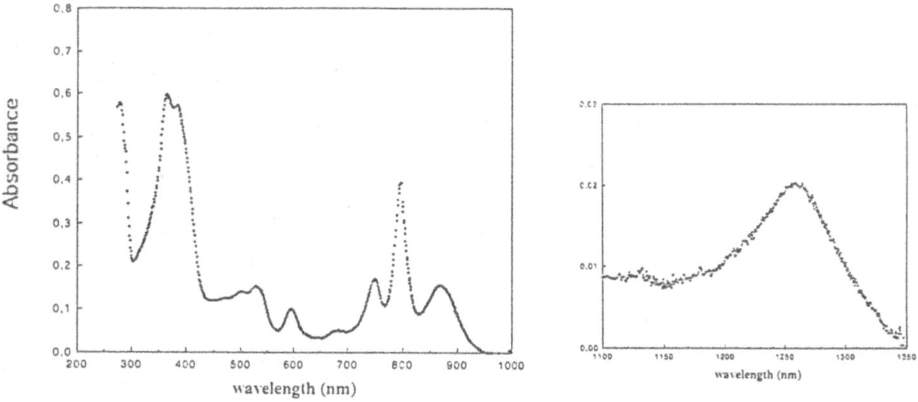

Figure1. Absolute absorption spectrum of purified *Rc. tenuis* reaction centers suspended in 10mM Tris pH 8 containing 0.2 mg/ml $C_{12}M$. Left: : recorded in the dark. Right : spectrum recorded between 1100-1350 nm in presence of continuous saturating light.

The electrophoretic pattern of purified *Rc. tenuis* RC displays three bands with apparent M.W. of 25, 29 and 36 kDa which likely correspond to L, M and H subunits respectively (Fig.2). A weak band at 43 kDa was revealed by heme staining (but it was not stained by Coomassie Blue and very weakly by silver) ; it should correspond to the residual tetraheme cyt c.

Extraction of quinones allowed us to identify Q_A with menaquinone 8 and Q_B with ubiquinone 8. The same quinone composition was determined in *Rubrivivax gelatinosus* RC (9).

The flash induced absorbance change at 430 nm shows the formation of $P^+Q_B^-$ radical pair which decays with a mean time of about 10 sec in the purified RCs whereas in chromatophores this recombination reaction lasts only for 1sec. This slowing down of the rate in the solubilised RC must be due to the interaction of the detergent with the Q_B pocket. Q_B was still present in 85% of this preparation and binary oscillations of Q_B^- demonstrate that the secondary quinone functions as a two electron gate. In the presence of terbutryn or $C_{12}DAO$, Q_B is unbound and after a flash $P^+Q_A^-$ radical pair is formed and relaxed with a mean time of 35ms. This recombination time was unchanged relative to that determined in chromatophores; it is close to the values measured in bacterial RCs which have Q_A as menaquinone and Q_B as ubiquinone (10-12).

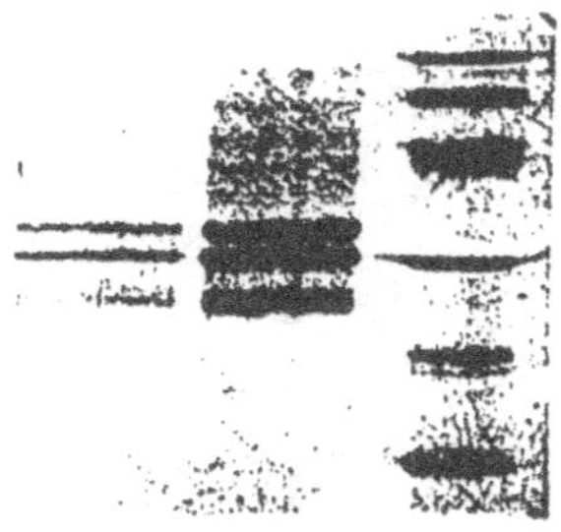

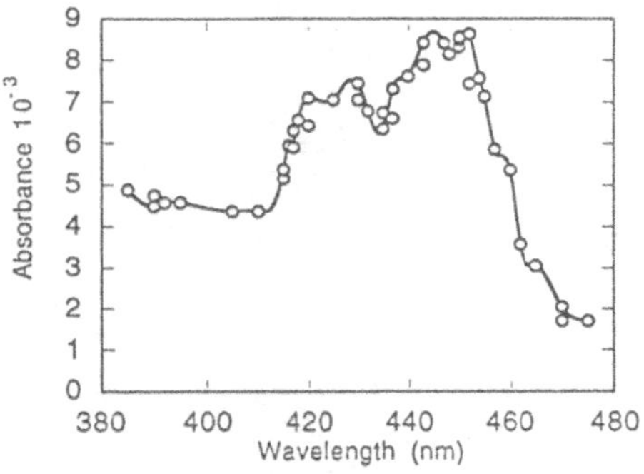

Fig. 2 SDS PAGE electrophoresis of *Rc. tenuis* RC (2 left lanes) and markers, after Coomassie Blue staining.

Fig. 3 Flash-induced difference spectrum of Q_A^-. Signal amplitude was measured 4-6 sec after the flash. 2 μM RCs, in 10 mM Tris buffer pH 8 containing 40 μM terbutryn, 0.2 mg/ml $C_{12}M$ and 2 mM DAD.

In presence of diaminodurol (DAD) (a fast reductant of P^+) and terbutryn, a long life (20min) flash-induced menasemiquinone Q_A^- is formed in *Rc. tenuis* RCs (see Fig.3). Unexpectedly, the absorption profile of this species displays several differences when compared with those of menasemiquinones photoinduced in RCs isolated from other bacteria (11, 13-14). We observe several broad peaks centered at 420, 430 and 450 nm the latter being the highest, whereas in *Rps. viridis* and *R. gelatinosus* the semimenaquinone has a main narrow peak at 412 nm and shoulders at 400, 430 and 450 nm (9, 14). In fact the absorption band of the menasemiquinone in vitro is centered at 395 nm (15). It is assumed that in the RC the additional multiple peaks and/or shoulders between 400-450 nm essentially reflect electrochromic shifts of the nearby pigments induced by the charged quinone (14). Therefore the spectral changes in Q_A^- state of *Rc. tenuis* could result from a different interaction with the pigments and/or aminoacid residues in the Q_A pocket. We should point out that *Rc. tenuis* and *R. gelatinosus* RCs have the same pigment and quinone composition. The marked spectral differences observed when the primary quinone is photoinduced could originate from primary sequence or structural differences in the Q_A binding pocket. Determination of the primary sequences of the RC polypeptides would be required for explaining these observations.

References

1 Willems, A, Gillis, M. and De Ley, J. (1991) Int. J. of Syst. Bact. 41, 65-73
2 Fling, S.P. and Gregerson, D.S. (1986) Anal. Biochem. 155, 83-88

3 Thomas, P.E., Ryan, D. and Levin, W. (1976) Anal. Biochem. 75, 168-176
4 Bligh, E.G. and Dyer, W.J. (1959) Can. J. Biochem. Physiol. 37, 911-917
5 Maroti, P., Hanson, D.K., Baciou, L., Schiffer, M. and Sebban, P. (1994) Proc. Natl. Acad. Sci. U.S.A. 91, 5617-5621
6 Nitschke, W. and Dracheva, S. M. (1995) in Anoxygenic Photosynthetic Bacteria (Blankenship, R. E., Madigan, M.T. and Bauer C. E. eds.) Kluver Academic Publishers in press
7 Wang, S., Lin, S., Lin, X., Woodbury, N.W. and Allen, J.P. (1994) Photosynth. Res. 42, 203-215
8 Wachtveitl, J., Farchaus, R.D., Lutz, M., Robert, B. and Mattioli, T.A. (1993) Biochemistry 32, 12875-12885
9 Agalidis, I., Rivas, E. and Reiss-Husson, F. (1990) Photosynth. Res. 23, 249-255
10 Nozawa, T., Trost, J.T., Fukada, T., Hatano, M., Mc. Manus, J.D. and Blankenship, R.E. (1987) Biochim. Biophys. Acta 894, 468-476.
11 Agalidis, I. and Reiss-Husson, F. (1992) Biochim. Biophys. Acta 1098, 201-208
12 Leguijt, T., Parot, P., Vermeglio, A., Crielaard, W. and Hellingwerf, K.J. (1993) Biochim. Biophys. Acta 1183, 292-300
13 Vasmel, H. and Amesz, J. (1983) Biochim. Biophys. Acta 724, 118-122
14 Shopes, R.J. and Wraight, C.A. (1985) Biochim. Biophys. Acta 806, 348-356
15 Rao, P.S.P. and Hayon, E. (1983) Biochim. Biophys. Acta 191, 516-533

ELECTRON TRANSFER REACTIONS IN CRYOPHILIC ALGAE

Atsuko Kanazawa[1], Antony R. Crofts[1,2] and David M. Kramer[3]
[1]Program in Biophysics and Computational Biology, [2]Dept. of Microbiology and [3]Dept.of Plant Biology. University of Illinois at Urbana-Champaign, 393 Morrill Hall, 505 S. Goodwin Ave., Urbana, IL 61801

1. Introduction

Cryophilic plants (species that grow well at low temperatures) occupy unique and important ecological niches in the Arctic, Antarctic and high mountain regions, growing under harsh environments of low temperature, low water potential, high doses of ultra-violet radiation, constant freezing and thawing cycles and severe nutrient limitations, while playing important roles in the carbon and nitrogen nutrient cycles. The reactions of the photosynthetic electron transfer chain have been found to be highly temperature-dependent, or to be sensitive to cold (chilling) stress in temperate plants. We surmise that these sites will have been subject to selective pressures during the evolutionary adaptation of cryophilic organisms to cold environments, and thus are the most likely to have important differences.

We have obtained ten strains of cryophilic algae isolated from the surface of Antarctic snow, and are making a comparative analysis of the energetics of photosystem II (PSII) in them. In this paper, we will discuss the electron transfer reactions of PSII of *Chlorella antarctica* in comparison with temperate species.

2. Materials and methods

Ten strains of cryophilic algae were obtained gift from Dr. L. Kovacik in the Institute of Botany, Trebon, Czech Republic. *C. antarctica* was grown in Tris-acetate-phosphate (TAP) media, pH = 7.0, stirred constantly under low light (approx. 10-20 $\mu Em^{-2}s^{-1}$) at +4°C until well-grown and dark green. The culture was washed and resuspended in phosphate buffer (15 mM, pH = 7.0), in the presence of 1 nM nonactin and 10 µM benzoquinone (BQ). Thylakoids were isolated according to the procedure in (1), resuspended in 15mM phosphate buffer, pH 7.0, with 10 µM BQ. Fluorescence kinetics were measured by a house-constructed multi-flash fluorimeter with a vertical cuvette attached to a rapid temperature controller to probe fluorescence changes associated with the forward reactions from Qa^- to Qb^-, and the back reactions from $S2Qa^-$ in the presence of 10 µM DCMU. A house-constructed thermoluminescence (TL) photometer was also used to observe the back reactions from $S2Qa^-$ and from $S2Qb^-$. The kinetics of the reaction from Qa to Qb were deconvoluted into three components as in (2).

3. Results and discussion

P. Mathis (ed.), Photosynthesis: from Light to Biosphere, Vol. I, 731–734.

In Fig. 1, we show the kinetics of Qa to Qb electron transfer after a flash of light, monitored by fluorescence, in spinach (a) or *C. antarctica* (b) at a range of temperatures. At +15°C, the Qa to Qb electron transfer is more rapid in spinach than in *C. antarctica*. At 0°C, however, the rates are comparable, and at -15°C, the spinach electron transfer is nearly completely blocked, whereas a significant rate is still detected in *C. antarctica*.

Arrhenius plots of the most rapid first-order rate constants for the Qa to Qb electron transfer, derived from kinetic traces (Fig. 1c), show that the rate of Qa to Qb electron transfer is more temperature-dependent in spinach than in *C. antarctica*. Under some growth conditions, the extent of the slowest component increased dramatically as the temperature was lowered. We are currently investigating this phenomenon.
The plot of spinach data shows a distinct break at approximately +4°C, as seen for other PSII reactions from temperate species (3), but this is not seen in *C. antarctica*. The plot of *C. antarctica* data shows only a weak dependence on temperatures. Activation-less or nearly activation-less electron transfer reactions have been observed previously in PSII and bacterial reaction centers between donor cytochromes and the primary donor chlorophylls, between the primary donor and the pheophytin, and between the pheophytin and Qa (4). Such behavior for the Qa to Qb electron transfer has not been published yet, but the laboratory of Dr. M. Gunner has observed a small phase of apparently activation-less Qa to Qb electron transfer in isolated bacterial reaction centers (personal communication). The activation barrier for Qa to Qb transfer was not changed when *C. antarctica* was grown at +22°C, indicating that adaptation to higher temperature did not affect this reaction.

The rise in fluorescence after flashes was also drastically affected by low temperatures in spinach (see Fig. 1a). The effect in spinach has been previously observed by Joliot (5). It is most likely due to inhibition of P_{680}^{+} re-reduction by a decrease in the equilibrium constant for reduction of P_{680}^{+} by Y_Z and a slowing of the S1 to S2 transition (6). The absence of the effect in *C. antarctica* (Fig. 1b) suggests that the donor side reactions are adapted to low temperatures.

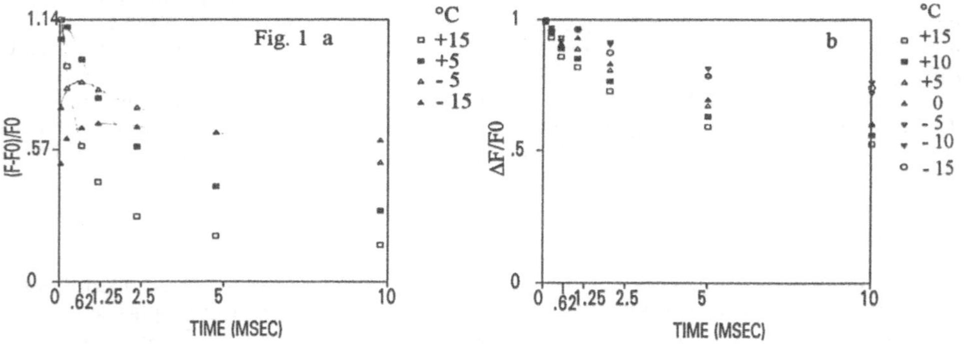

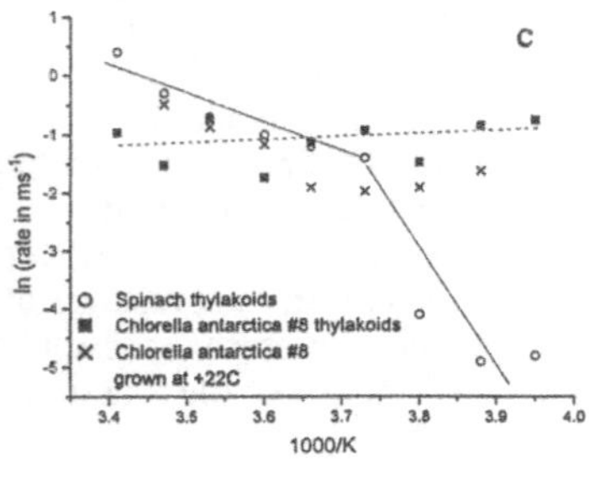

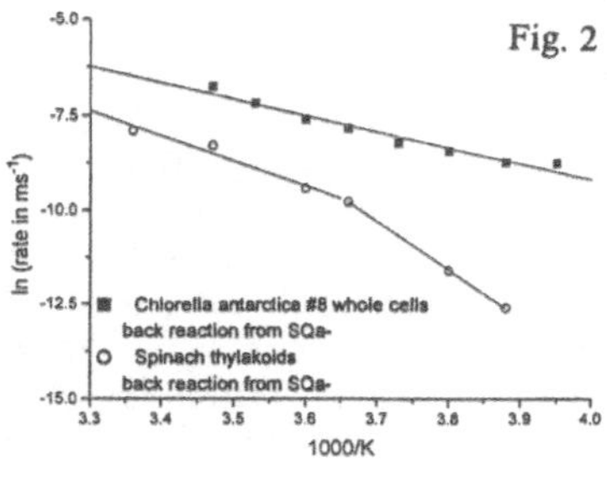

When DCMU is added, it replaces plastoquinone (PQ) from the Qb binding site and blocks the normal oxidation pathway for Qa^-. The recombination reactions occur by a reversal of the forward reactions and thus often used to probe the energetics of electron transfer from the S-states and Qa. The Arrhenius plot shown in Fig. 2 shows that the rates of recombination from $S2Qa^-$ to S1Qa are more rapid in *C. antarctica* than spinach at all temperatures, but show a less steep temperature dependence (a lower activation barrier). Again, a break in the rates occurs near +4°C in spinach, but not in *C. antarctica*.

Another technique to monitor the energy barriers of recombination reaction is thermoluminescence (TL) (see reviews 7,8). As illustrated by Figs. 3a-d, we have found TL to be very useful diagnostic tool for rapidly screening PSII centers for cryophilic low temperature behavior. Samples were flash-illuminated at +10°C and rapidly cooled to -50°C. Each figure shows the intensity of TL as the temperature was raised at 1°C/s, in the presence and absence of DCMU. DCMU displaces PQ from the Qb site, so that traces taken in the presence of DCMU represent recombination from $S2Qa^-$ to S1Qa, while traces taken in the absence of DCMU represent recombination from $S2Qb^-$ to S1Qb. Since $S2Qb^-$ is at a lower energy that $S2Qa^-$, the activation barrier for recombination is higher and hence the TL band is observed at a higher temperature. In spinach thylakoids (Fig. 3d), recombination from $S2Qa^-$ (the "Q" band) and from S2 Qb^- (the "B" band) are observed at +17 and +35°C, similar to previous observations in other temperate species. In *C. antarctica* grown at +4°C, however, these bands (Fig. 3a) appear at +7 and +17°C (10 to 20°C lower than in temperate species). This is consistent with the fluorescence experiments (above) showing a lower activation barrier for recombination. When *C. antarctica* was grown at 22°C for one day, the Q and B TL bands shifted to +18 and +26°C. similar to the bands observed in spinach. When +22°C-grown cells were transferred back to low temperatures, the effect was reversed within 24 hours, showing that it was not due to permanent PSII damage. Thylakoid membranes, lacking soluble components, isolated from cold-grown *C. antarctica* (Fig. 3b) behaved similarly to whole cells (Fig. 3a), implying that aqueous phase components are not responsible for the cryophyte's behavior.

The difference in Q and B band peak temperatures was significantly in *C. antarctica* than in spinach, suggesting a lower equilibrium constant for electron transfer from Qa to Qb. This was confirmed by taking the ratio of the back reactions rates from $S2Qa^-$ and $S2Qb^-$, which is proportional to the equilibrium constant between Qa and Qb (9). Recombination from $S2Qa^-$ was measured by decay of fluorescence in the presence of DCMU (see above) while that from $S2Qb^-$ was measured by observing the decay of the TL B band when different delay times were given after illumination, but before cooling.

Fig. 3

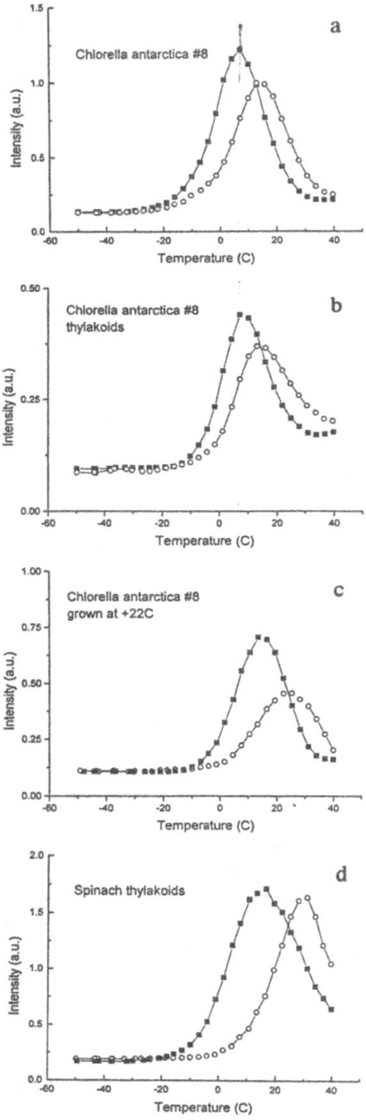

There was nearly a 2-fold lower equilibrium constant for this reaction in *C. antarctica* than in spinach (data not shown).

From the above studies, we can construct comparative models for PSII energetics in a temperate and a cryophilic species. Fig. 4 depicts the activation barriers between various energy states of PSII derived from the above studies and from Kanazawa et al (10). In temperate species, a high activation barrier restricts electron transfer from Qa to Qb at low temperatures, but in *C. antarctica*, this barrier is lowered. When *C. antarctica* is grown at +4°C, the recombination from $S2Qa^-$ has a significantly lower activation barrier than that found in spinach of in *C. antarctica* grown at +22°C. This suggests that either the enthalpy that drives the reduction of Qa by I^-, or the reduction of P_{680}^+ by the S-states, is lower in *C. antarctica* grown at +4°C. We conclude that *C. antarctica* possesses distinct evolutionary traits for photosynthetic growth at low temperatures.

Fig. 4

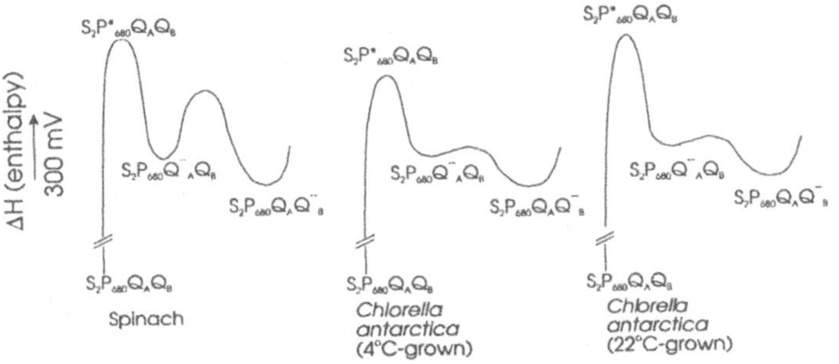

Acknowledgment: This work was supported by Department of Energy grant DOE DEF 602-86 ER 13594.

References

1 Roffey, R.A., Golbeck, J.H., Hille, C.R. and Sayre, R.T. (1991) Proc. Natl. Acad. Sci. U.S.A. 88, 9122-9126.

2 Crofts, A.R., Baroli, I.M., Kramer, D.M. and Taoka, S. (1993) Z. Naturforsch. [C] 48, 259-266.

3 Gleiter, H.M., Haag, E. and Renger, G. (1990) in Current Research in Photosynthesis (Baltschffsky ed.) Vol. 1, pp.479-482, Kluwer Academic Publishers, Dordrecht, The Netherlands.

4 Devault, D. (1980) Quat. Rev. Biophys. 13, 387-564.

5 Joliot, A. (1974) in Proceedings of the Third International Congress on Photosynthesis (Avron M., ed.) pp.315-322, Elsevier Scientific Publishing Company, Amsterdam, The Netherlands.

6 Mauzerall, D. (1972) Proc. Natl. Acad. Sci. U.S.A. 69, 1358-1362.

7 Sane, P.V. and Rutherford, A.W. (1986) in Light Emission by Plants and Bacteria (Govindjee, Amesz J. and Fork D.C. eds.) pp.329-362, Academic Press, New York.

8 Inoue, Y. and Shibata, K. (1982) in Photosynthesis (Govindjee ed.) Vol. 1, pp.507-533, Academic Press, New York.

9 Robinson, H.H. and Crofts, A.R. (1987) in Progress in Photosynthesis Research (Biggins J. ed.) pp.429-432, Martinus Nijhoff Publishers, Dordrecht, The Netherlands.

10 Kanazawa, A., Kramer, D.M. and Crofts, A.R. (1992) in Research in Photosynthesis (Murata N. ed.) Vol. II, pp. 131-134, Kluwer Academic Publishers, Dordrecht, The Netherlands.

ELECTRIC FIELD EFFECTS ON THE STEADY-STATE FLUORESCENCE EMISSION OF *Rb. sphaeroides* CHROMATOPHORES
Montpellier, France, 20-25 August 1995

Jana L. Steiger and Kenneth Sauer
Structural Biology Division, Lawrence Berkeley Laboratory and Department of Chemistry, University of California, Berkeley, CA 94720 (U.S.A.)

1. Introduction

The charge separation process in the reaction center of *Rhodobacter. sphaeroides* creates a transitory dipolar state across the intracytoplasmic membrane in whole cells and in chromatophores which are derived from those membranes. Energies of these charge separated states are expected to change when an external electric field is applied; theoretical calculations have been made to predict the effect of external electric fields on the rate of charge separation in the reaction center (1). External electric fields have been applied as an energetic probe of isolated *Rb. sphaeroides* reaction centers in a variety of spectroscopic experiments (2-4). Analysis of the results is complicated by lack of orientation of the reaction centers to the applied field.

In this paper we utilize differential potassium ion diffusion potentials to generate transient electric fields across chromatophore membranes of *Rb. sphaeroides*, allowing experiments to be performed in a near *in vivo* state and simultaneously taking advantage of the natural unidirectional orientation of chromophores in the membrane. The fluorescence emission of the chromatophores is monitored as a function of applied electric field strength, because fluorescence emission and charge separation are competitive pathways for excitation loss. Similar experiments on spinach thylakoid membranes have demonstrated an electric field effect influencing the charge separation and recombination processes in the PSII reaction center (5, 6). Our results on *Rb. sphaeroides* are consistent with an electric field effect on the reaction center and an effect in the opposite direction on the antenna pigment/protein complexes present in the chromatophore preparation.

2. Procedure

Rhodobacter sphaeroides 2.4.1 was grown anaerobically at moderate light intensity. Chomatophores were prepared from cells harvested in the mid log phase of growth, typically three days after inoculation, after the method of Crielaard *et al.* (8). The chromatophores were washed with buffer containing 800 mM sucrose, 100 mM KCl, 5 mM $MgCl_2$, and 5 mM Hepes, pH 7.5, and resuspended to give $A_{800} = 1$. Valinomycin was added to 1 µM, and the solution was dark-adapted for at least one hour before measurements. Where specified, 10 mM sodium dithionite ($Na_2S_2O_4$, Sigma) was used to reduce the membrane-bound quinone Q_A (7).

Diffusion potentials were applied through a rapid mixing technique which has been described in detail elsewhere (5). In the presence of the potassium-specific ionophore

P. Mathis (ed.), Photosynthesis: from Light to Biosphere, Vol. I, 735–738.

valinomycin, a change in external potassium chloride concentration generates a transient potassium ion diffusion potential across the membrane. Optical measurements were made approximately 350 ms after mixing. The ratio of the flow rate of salt solution to the flow rate of chromatophore solution was 13:1, allowing a minimum external potassium concentration of 7.1 mM. The chromatophore absorbance after mixing was $A_{800} = 0.08$.

Fluorescence measurements were made with a Hellma QS flow cell of path length 1.5 mm in the sample compartment of a SPEX fluorolog 1680 fluorimeter using single-photon counting detection with a near-infrared extended photomultiplier tube (Hamamatsu R3310-02). Excitation was at 514 nm or 800 nm with a xenon lamp (FWHM = 8 nm).

Measurements were made in either the Fo or Fm states. The Fo state, corresponding to open reaction centers, was obtained for dark-adapted samples under flowing conditions where the sample was exposed only briefly to the measuring light (approximately 30 ms). Under these conditions the measuring light did not produce any detectable actinic effect. The Fm state, or closed reaction center state, was obtained by stopping the pump and allowing the sample to remain in the excitation beam. Under these conditions charge separation is minimal. In the Fm state attained by stopping the pumps, an electric field effect was found to persist for several seconds. This is consistent with observations of the duration of electrochromic effects generated by potassium diffusion potentials in *Rb. sphaeroides* chromatophores (8). The difference (Fm-Fo) is known as the variable fluorescence and has a maximum value at 890 nm (7).

3. Results and Discussion

Orientation of the chromatophore membrane preparation was confirmed to be identical to that reported in the literature for chromatophores (9) through the direction of the electrochromic carotenoid bandshift, as well as the direction of proton translocation.

Internally positive diffusion-generated electric fields applied across this oriented system result in a decrease of the fluorescence emission in the Fo state. Fig. 1 shows the fluorescence emission spectra with an internally positive (Fpos) and internally negative (Fneg) applied electric field. Total applied electric field strength is 108 mV. The change in fluorescence intensity is observed over the entire emission band. The spectrum of the electric-field induced change (ΔF, Fneg-Fpos) is also shown in Fig. 1, with a maximum at 890 nm similar to the zero field variable fluorescence spectrum (Fm-Fo). The spectrum of the electric field effect in the Fm state is also similar to the variable fluorescence emission. Because the difference spectrum does not display a first- or second-derivative lineshape of the emission spectrum, it does not represent a traditional Stark effect.

The change in fluorescence at 890 nm in both the Fo and Fm states is linear within the accuracy of the measurements with the logarithm of the external potassium concentration as shown in Fig.2 for both positive and negative applied voltages. Negative applied voltages result in a linear increase of fluorescence yield. The applied electric field strength was calculated from the Nernst equation, in the range -68 mV to +40 mV. The average value of the slope of five experiments was 0.8% per 10 mV for the Fo state, and approximately double in the Fm state; 1.6% per 10 mV.

Addition of sodium dithionite to a concentration of 10 mM reduces the primary quinone Q_A and produces a reversal of the electric field effect. Application of negative internal potentials now leads to a decrease in fluorescence intensity. The effect is linear over the accessible voltage range. The magnitude of the Fo state effect is greater than that seen for untreated chromatophores, about 1.5% per 10 mV (compare with Fig. 2). The electric-field induced effect is also maximal at 890 nm, which is similar to the variable fluorescence under dithionite-treated conditions.

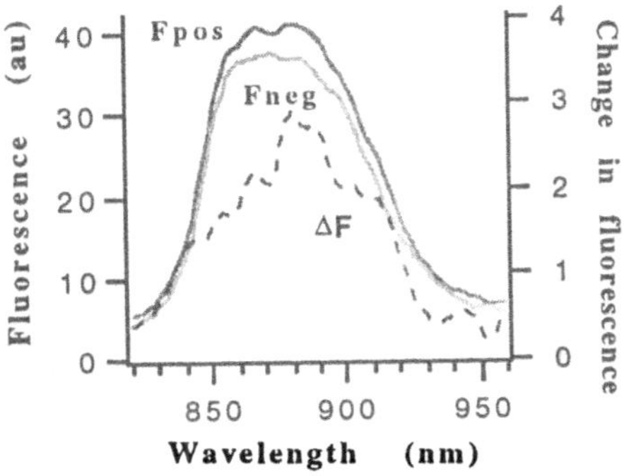

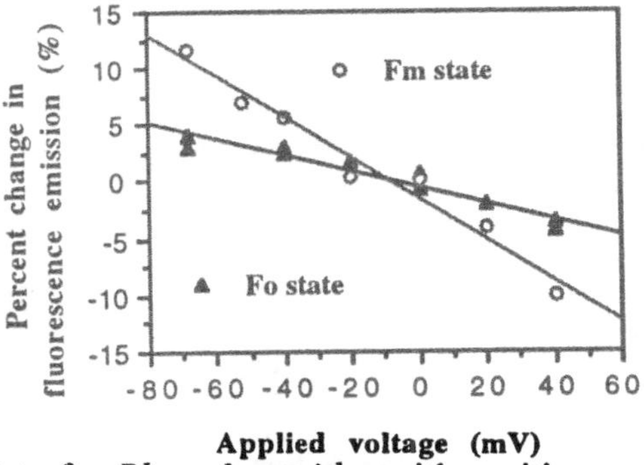

Fig. 1: Fluorescence emission in the Fo state for *Rb. sphaeroides* with positive and negative applied fields. The difference spectrum is shown in dashed line.
Fig. 2: Percent change in fluorescence (ΔF/Fmax) in the Fo and Fm state as a function of applied field strength at 890 nm.

Contrary to our expectation, fluorescence quenching occurs under conditions which are expected to decrease the probability of charge separation, i.e., with an internally positive applied electric field which is oriented to oppose light-induced trans-membrane dipole formation in the reaction center. For spinach thylakoid membranes, salt-induced electric fields oriented to oppose primary charge separation resulted in a fluorescence increase (5). Futhermore, the electric field effect in spinach thylakoids vanished under conditions of trap closure, whereas the observed electric field effect is increased in the Fm state of *Rb. sphaeroides*. It is apparent that the results from spinach thylakoids cannot be extended in a straightforward manner to salt-induced electric field effects on *Rb. sphaeroides* chromatophores.

One possible explanation for the direction of the effect in *Rb. sphaeroides* would arise from charge separation in the inverted Marcus region, in which case raising the free energy of the charge separated state with an internally positive electric field would bring the curve crossing closer to the activationless region and result in an increase in the rate of charge separation. However, this would not explain the enhanced electric field effect in the Fm state, when the probability of primary charge separation is minimal.

Another possibility is an electric field effect on the fluorescence emission from the antenna pigments as well as an effect on charge separation in the reaction center. An electric field effect on the antenna fluorescence of opposite sign and greater magnitude than the electric field effect on the reaction centers is consistent with our results. An electric field effect on the fluorescence of antenna pigment/protein complexes is not unprecedented. Gottfried et al. have observed unusual fluorescence quenching under applied fields for isolated antenna complexes oriented isotropically in PVA matrices (10). The reason proposed for this fluorescence quenching was an electric field-influenced coupling to a state with rapid non-radiative relaxation. If the explanation of an electric-field enhanced coupling to a non-radiative decay state is valid and also applies to the situation in chromatophores, our experiments would imply that the decay path is present under zero-field conditions because negative internal fields result in increased fluorescence emission.

Addition of 10 mM sodium dithionite, which reduces the primary quinone acceptor in the chromatophore preparation, results in a dramatic reversal of the electric field effect onfluorescence emission. Reduction of the primary quinone permits charge separation and enhances charge recombinationin the reaction center (11). The electric field effect in these circumstances may reflect an effect on charge recombination in the reaction center.

However, it should be noted that in photosystem II of spinach thylakoids the electric field effect on fluorescence emission due to charge recombination is relatively small (6).

Another possibility is that the decreased probability of primary charge separation with reduced Q_A indicates that excitation will be more likely to escape the reaction center and be emitted as fluorescence from the antenna. Therefore the observed fluorescence emission under these conditions will consist of excitation which has a higher residence time in the reaction center. For *Rb. sphaeroides* and other purple photosynthetic bacteria there is currently debate whether the excitation transfer is trap-limited (12) as in PSII, or diffusion-limited (13). That the fluorescence change with reduced Q_A corresponds to a direction expected for an electric field induced change on the charge separation in the reaction center indicates that residence time of excitation is a critical factor in determining sensitivity of fluorescence emission to the reaction center state.

In this scheme the normal oxidized quinone state cannot be described as a fully trap-limited system since a large proportion of the fluorescence emission apparently does not reflect processes in the reaction center. However, the observed increase in the electric-field induced fluorescence change under conditions of trap closure (the Fm state) indicates that the fluorescence emission is not entirely insensitive to the state of the reaction center. The best characterization of excitation transfer in this system may be an intermediate description between trap-limited and diffusion-limited.

Acknowledgments

This work was supported by the Director,Office of Energy Research, Office of Basic Energy Sciences, Energy Biosciences Division, of the US Department of Energy under Contract No. DE-AC03-76SF00098

We would like to thank Dr. Holger Dau for helpful discussions.

References

1. Bixon, M. and Jortner, J. (1988) J. Phys. Chem. 92, 7148-7156.
2. Lockhart, D. J., Hammes, S. L., Franzen, S. and Boxer, S. G. (1991) J. Phys. Chem. 95, 2217-2226.
3. Ogrodnik, A., Langenbacher, T., Bieser, G., Siegl, J., Eberl, U., Volk, M. and Michel-Beyerle, M. E. (1992) Chem. Phys. Lett. 198, 653-658.
4. Popovic, Z. D., Kovacs, G. J., Vincett, P. S., Alegria, G. and Dutton, P. L. (1986) Biochim. Biophys. Acta 851, 38-48.
5. Dau, H. and Sauer, K. (1991) Biochim. Biophys. Acta 1098, 49-60.
6. Dau, H. and Sauer, K. (1992) Biochim. Biophys. Acta 1102, 91-106.
7. Jursinic, P. A. (1986) in Light Emission by Plants and Bacteria, (Govindjee, Amesz, J. and Fork, D. C., ed) pp. 291-328, Academic Press, Orlando.
8. Crielaard, W., Hellingwerf, K. J. and Konigs, W. N. (1989) Biochim. Biophys. Acta 973, 205-211.
9. Matsuura, K. and Nishimura, M. (1977) Biochim. Biophys. Acta 483-491.
10. Gottfried, D. S., Stocker, J. W. and Boxer, S. G. (1991) Biochim. Biophys. Acta 1059, 63-75.
11. Schenck, C. G., Blankenship, R. E. and Parson, W. W. (1982) Biochim. Biophys. Acta 680, 44-59.
12. Muller, M. G., Drews, G. and Holzwarth, A. R. (1993) Biochim. Biophys. Acta 1142, 49-58.
13. Beekman, L. M. P., Van Mourik, F., Jones, M. R., Visser, H. M., Hunter, C. N. and Van Grondelle, R. (1994) Biochem. 33, 3143-3147.

DYNAMICS OF ELECTRONIC AND VIBRATIONAL COHERENCES IN PHOTOSYNTHETIC REACTION CENTERS

M. Hayashi[a)], M. Sugawara[a)], Satoru. Suzuki[a),b)] and S. H. Lin[a),b)]
a) Institute of Atomic and Molecular Sciences, Academia Sinica, P.O.Box 23-166, Taipei, 106 Taiwan, R.O.C.
b) Department. of Chemistry and Biochemistry and the Center for the Early Events in Photosynthesis, Arizona State University, Tempe, AZ 85287-1604, USA

1. Introduction

The conversion of solar energy into chemical energy is one of the most important functions of photosynthesis. Recently, femtosecond (fs) time-resolved spectroscopy have revealed that the energy conversion process consists of a series of electron transfer (ET) reactions leading to the separation of charge across the photosynthetic membrane.

In analyzing the experimental data of ET rates, the conventional ET theory is commonly used. Vibrational relaxation (VR) is assumed to be much faster than ET so that the vibrational equilibrium is established before ET takes place. However, the primary ET in RCs occurs in 1 to 4 ps. This ultrafast ET implies that the assumption should be examined. To elucidate the mechanism of the initial ET in RCs, detailed quantum chemical calculations have been performed for the electronic structure of the special pair P, the electronically excited state $P^*$and its cation P^+. Both steady-state and time-resolved spectroscopy have provided these calculations with important information on vibrational modes and charge separated states. In particular, time-resolved IR measurements can observe transient behaviors of specific vibrational modes of molecular groups. In measuring the primary ET rates in photosynthetic RCs, fs transient absorption techniques have typically been used. From the uncertainty principle, we can see that both population and coherence have contributions to the observed fs time-resolved spectra in the fs time regime. Quite recently, the vibrational coherence effects have been observed in several ultra-short time-resolved spectra both of the D_{LL} mutant RCs of *Rb. capsulatus* [1] and of the RCs of *Rb. sphaeroides* (*R-26*) by Vos *et al.*[2].

In this work, we shall present the studies of the dynamics and spectroscopy of photosynthetic RCs. We analyze the experimental fs spectra of *Rb. sphaeroides* (*R-26*) which usually consist of contributions both from population and from coherence. We will theoretically demonstrate how temperature affects fs transient dynamics appearing in the stimulated emission spectra of *Rb. sphaeroides* (*R-26*). We shall also investigate

P. Mathis (ed.), Photosynthesis: from Light to Biosphere, Vol. I, 739–742.

coherent dynamics of an electron transfer system with an arbitrary strength of the electronic coupling between neutral and charge separated states. Solving the generalized master equation (GME) numerically and incorporating the optical processes perturbatively, we will present stimulated emission signals calculated for both weak and strong electronic coupling cases, and investigate how the spectral features depend on the strength of the electronic coupling constant.

2. Theoretical

We consider a model for the RC of photosynthetic bacteria in which an electronically excited state *b* (its vibrational manifold $\{v\}$) couples with a charge separated state *c* (its vibrational manifold $\{j\}$) *via* electronic interaction $H' = J(|b\rangle\langle c| + h.c.)$. To treat the dynamics of the system, we use the stochastic Liouville equation. Here the master equations for the system are given by

$$\frac{d\rho_{vv}^{bb}}{dt} = -\frac{2}{\hbar}\sum_{j} \mathrm{Im}(H'_{cj,bv}\rho_{vj}^{bc}) + \sum_{v'} (\Gamma_{v'v'}^{vv}\rho_{vv}^{bb} - \Gamma_{vv}^{v'v'}\rho_{v'v'}^{bb}) \quad (1)$$

$$\frac{d\rho_{jj}^{cc}}{dt} = -\frac{2}{\hbar}\sum_{v} \mathrm{Im}(H'_{cj,bv}\rho_{vj}^{bc}) + \sum_{j'} (\Gamma_{j'j'}^{jj}\rho_{jj}^{cc} - \Gamma_{jj}^{j'j'}\rho_{j'j'}^{cc}) \quad (2)$$

$$\frac{d\rho_{vj}^{bc}}{dt} = -(i\omega_{vj}^{bc} + \Gamma_{vj}^{vj})\rho_{vj}^{bc} + \frac{i}{\hbar}H'_{cj,bv}(\rho_{vv}^{bb} - \rho_{jj}^{cc}) \quad (3).$$

$$\frac{d\rho_{vv'}^{bb}}{dt} = -(i\omega_{vv'}^{bb} + \Gamma_{vv'}^{vv'})\rho_{vv'}^{bb}. \quad (4)$$

The initial conditions of Eqs. (1-4) are determined by This model allows the system to undergo vibrational relaxation via the vibrational transition rate constants $\Gamma_{v'v'}^{vv}$ and $\Gamma_{vv}^{v'v'}$. Here Γ_{vj}^{vj} and $\Gamma_{vv'}^{vv'}$, respectively, denote the dephasing constant of the vibronic coherence $bv \leftrightarrow cj$ and that of the vibrational coherence $v \leftrightarrow v'$. The initial conditions of Eqs. (1-4) are determined by the pumping optical process [3,4]. We derive an expression for the time-resolved stimulated emission spectra that are associated with the imaginary part of the transient susceptibility [3,4]:

$$\tilde{\chi}''(\omega_{pu}, \omega_{pr}, \tau) = \frac{-\vec{\mu}_{ba} \otimes \vec{\mu}_{ba}}{\hbar} \sum_{vv'} \mathrm{Re}\{\rho_{vv'}^{bb}(\omega_{pu}, \tau) F_{vv'}^{bb}(\omega_{pr})\} \quad (5).$$

where $F_{vv'}^{bb}(\omega_{pr})$ is associated with the probing optical process [3,4].

3. Result and Discussion

Now we shall demonstrate how the transient dynamics of the electronically excited state b affects the time-resolved spectra of the stimulated emission. For this purpose, based on the experimental conditions and absorption spectrum of the RCs (R26) of *Rb. sphaeroides* reported by Vos *et al.*[2]. we employ physical parameters: $\omega_{pu} = 11494.2\ cm^{-1}$, $T_{pu} = 30$ fs, $\gamma_{pu} = 180\ cm^{-1}$, $T_{pr} = 80$ fs, $\gamma_{pu} = 66\ cm^{-1}$, $\omega_{ba} = 10649.2\ cm^{-1}$. Although the RCs involves many vibraional modes, only $\omega = 100\ cm^{-1}$ plays an important role in the optical pumping and probing processes since the coupling constants of the other modes are small in the $P \rightarrow P^*$ transition. This implies that all the other modes are in $\{v = 0\}$ states of the b state after the optical excitation and the transient vibrational dynamics of the b state can be described in terms of $\omega = 100\ cm^{-1}$. We first solve Eqs. (1-5) by applying perturbation theory with respect to the electronic coupling interaction. Figure 1 shows the temperature dependence of the time-resolved stimulated emission spectra at $\omega_{pr} = 10661.0\ cm^{-1}$. The dotted, broken and solid line in Fig. 1 denote the time-resolved profile at 10, 130 and 290 K, respectively. From this figure, we can easily see inverse temperature dependence of the rate at which the intensity of each time-resolved profile decreases. Now we shall investigate how the strength of the electronic coupling affect on stimulated emission signals for a model system. For this purpose, we solve Eqs. (1-4) numerically. Figure 2 shows simulation of the stimulated emission signals as a function of probing frequency and delay time for the weak coupling case ($J = 10\ cm^{-1}$). The spectra are depicted in a contour plot and the intensity of the signals becomes high as the color is dark. In Fig. 2, we find a peak at around 10200 cm^{-1} immediately after the optical excitation by the pump pulse. There are oscillatory shifts of the peak position in the subsequent time region (< 700fs). Those shifts are due to a coherent vibrational motion of the neutral state b. We find that the intensity of the stimulated emission signals decreases as the population of the state b is transferred to the optically forbidden state c through the electronic coupling. Figure 3 shows the calculated stimulated emission signals for the strong case ($J = 40 cm^{-1}$).

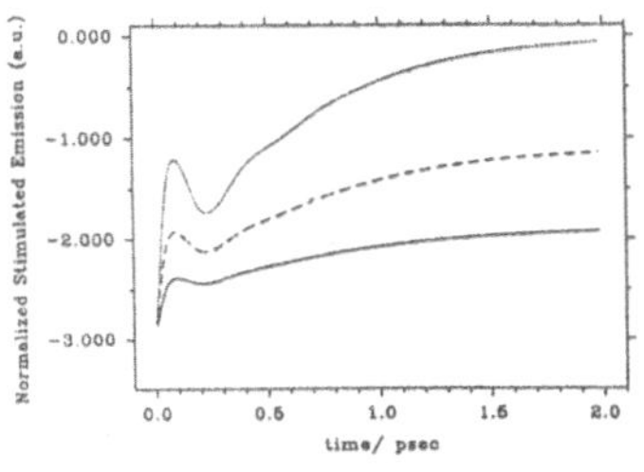

Figure 1 Temperature dependence of the stimulated emission spectra

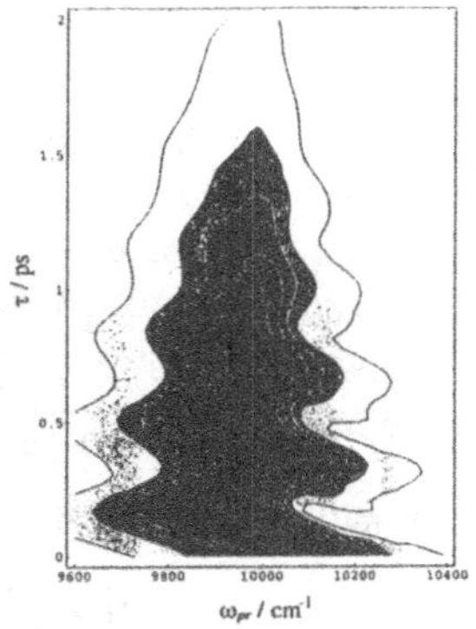

Figure 2 Simulation for the time-resolved stimulated emission spectra for the weak coupling case

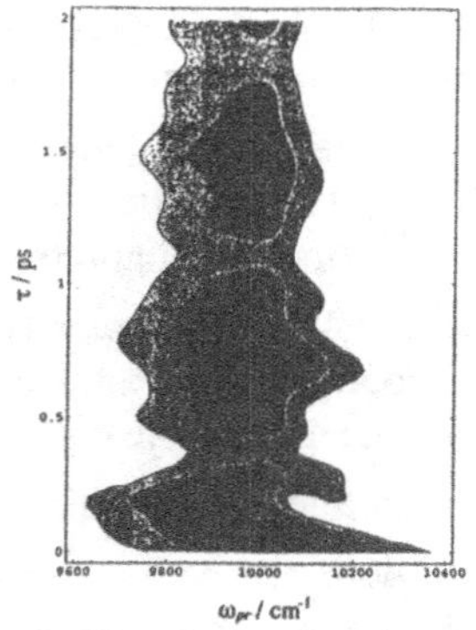

Figure 3 Simulation for the time-resolved stimulated emission spectra for the strong coupling case

In this case, although the slight peak shift is found in an early time (< 300 fs), there is no oscillatory feature in the peak positions in the subsequent time. In other words, there is no significant probing pulse frequency dependence other than in the amplitude of the signals.

In conclusion, on the basis of the density matrix method, we have solved transient dynamics that takes place during the pump-probe measurement. Without assuming thermally equilibrated ET, our calculations still show inverse temperature dependence. We have also shown effects of electronic and vibraional coherent motions on the stimulated emission signals. We have demonstrated that rapid vibrational coherent motions result in significant phase changes in quantum beats as a function of probing frequency. These phase changes, in addition to VR, lead to transient Stokes shifts in the stimulated emission spectra. Electronic coherence, on the other hand, prevents such phase change in quantum beats.

Acknowledgements
This work was supported by NSF and NSC(ROC).

References

1. M. H. Vos, F.Rappaport, J.-C.Lambry, J.Breton and J.-L.Martin, (1993) Nature 363 12701-12705.
2. M.H.Vos, M.R.Jones, C.N.Hunter, J. Breton, and J.-L.Martin, (1994) Proc. Natl. Acad. Sci. USA 91 12701-12705.
3. X.Z.Gu, M.Hayashi, S.Suzuki and S.H.Lin (1995) Biochim. Biophys. Acta 1229 215-224.
4. Satoru Suzuki, H.C.Sung, M.Hayashi and S.H.Lin Physica A (in press).

FTIR STUDY OF THE NON-HEME IRON BINDING SITE IN PHOTOSYSTEM II: BICARBONATE ^{13}C-LABELING AND EXCHANGE BY GLYCOLATE.

Rainer Hienerwadel and Catherine Berthomieu
CEA Saclay, DBCM/SBE Bat 532, 91 191 Gif-sur-Yvette, Cedex.

1. INTRODUCTION

The non-heme iron in photosystem II (PS II) exhibits a pH-dependent midpoint potential ($E_{m,7} \approx 410$ mV, -60 mV/pH) and can be oxidized by ferricyanide on the second time scale. Upon light-induced charge separation, the oxidized iron is reduced by electron transfer from Q_A^- (reviewed in (1)). This allows the investigation of the iron binding site by light-induced FTIR difference spectroscopy resulting in the Fe^{2+}/Fe^{3+}-spectra (2, 3). In this report, Fe^{2+}/Fe^{3+}-spectra obtained with ^{12}C- and ^{13}C- isotope labeled bicarbonate are compared to reveal the coordination sensitive ν_{as} and $\nu_s(CO)$ stretching modes of this putative ligand. Bicarbonate exchange by glycolate modifies the midpoint potential of the iron and its pH dependence. Glycolate binding was proposed to induce a lowering of the pK_a of the protein residue(s), which deprotonate upon iron oxidation in presence of bicarbonate (4, 5). The vibrational modes of the(se) residue(s) can be detected by comparison of Fe^{2+}/Fe^{3+}-spectra in presence of bicarbonate and glycolate.

2. MATERIALS AND METHODS

The IR samples consisted in a pellet of Tris-washed PS II enriched membranes sandwiched between two CaF_2 windows in the presence of ferricyanide as described in (2). The Fe^{2+}/Fe^{3+}-difference spectra were obtained by taking 64 scans (32 scans for measurements at pH 6) before and 2 s after a saturating flash. The measurements were cycled with 80 s between flashes for an average of 24 h at 4 °C. Precise controls of the reaction conditions are described in (3).

3. RESULTS AND DISCUSSION

Fig. 1-A shows the comparison of Fe^{2+}/Fe^{3+}-spectra obtained in presence of ^{12}C- (thick line) or ^{13}C-bicarbonate (thin line). The signals affected by ^{13}C-bicarbonate labeling appear in the "^{12}C-*minus*-^{13}C" difference spectrum of Fig. 1-B. Positive bands correspond either to modes of ^{12}C-bicarbonate in the Fe^{2+}-state or to modes of ^{13}C-bicarbonate in the Fe^{3+}-state. Inversely, the negative bands correspond either to modes of ^{12}C-bicarbonate in the Fe^{3+}-state or to modes of ^{13}C-bicarbonate in the Fe^{2+}-state. Assignments are made by comparison with spectra of ^{12}C- and ^{13}C-

P. Mathis (ed.), Photosynthesis: from Light to Biosphere, Vol. I, 743–746.

bicarbonate in solution (Fig. 2). In solution, bands at 1648 and 1363 cm^{-1} are downshifted by 62 and 22 cm^{-1} upon ^{13}C-labeling and are assigned to the ν_{as} and ν_s(CO)- modes of bicarbonate, respectively. In PS II (Fig. 1), signals at 1530 ±10 and 1339 cm^{-1} are assigned to the ν_{as}- and ν_s- (CO)- modes of bicarbonate in the Fe^{2+}-state. A signal at 1258 cm^{-1} is further assigned to the δ(COH) bending mode (Fig. 1-B). In the Fe^{3+}-state, a signal at 1228 cm^{-1} is clearly identified and assigned to the ν_s(CO)- mode of bicarbonate, while a mode at 1658 ±20 cm^{-1} is tentatively assigned to the corresponding ν_{as}(CO)- mode.

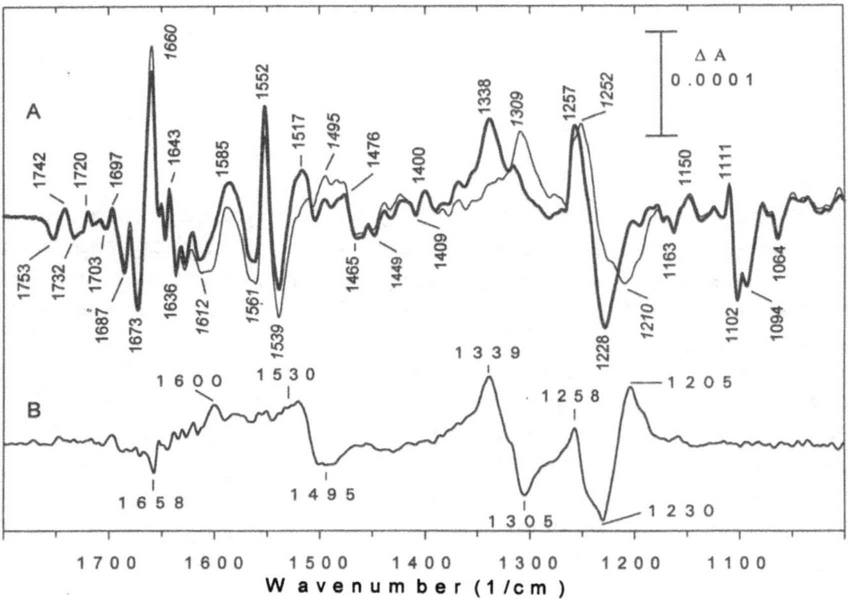

Figure 1: A) Fe^{2+}/Fe^{3+}-spectra spectra obtained in presence of ^{12}C- (thick line) and ^{13}C- bicarbonate (thin line) at pH 8. B) Difference between Fe^{2+}/Fe^{3+}-spectra in presence of ^{12}C- and ^{13}C-bicarbonate.

The low frequency of the ν_{as}(CO) bicarbonate mode at 1530 ±10 cm^{-1} in the Fe^{2+}-state as compared to the vibrations observed in even strongly hydrogen bonded ionic bicarbonate (6) is taken as experimental evidence that bicarbonate is coordinated to the iron. In the Fe^{2+}-state, the frequency difference ν_{as}-*minus*-ν_s (CO) of 192 cm^{-1} is further strongly indicative that bicarbonate is a bidendate ligand of the non-heme iron in the Fe^{2+}-state (7). In the Fe^{3+}-state, the frequency difference ν_{as}-*minus*-ν_s (CO) is strongly enhanced to 430 ±20 cm^{-1} and indicates that bicarbonate is a monodendate ligand of the iron. The contribution of bicarbonate and not carbonate both in the Fe^{2+}- and Fe^{3+}- states is suggested by the effect of $H/^2H$ exchange on the ν_s(CO)- modes and by the presence of a ^{13}C-sensitive signal at 1258 cm^{-1} attributed to a δ(COH)- mode of bicarbonate in the Fe^{2+}-state.

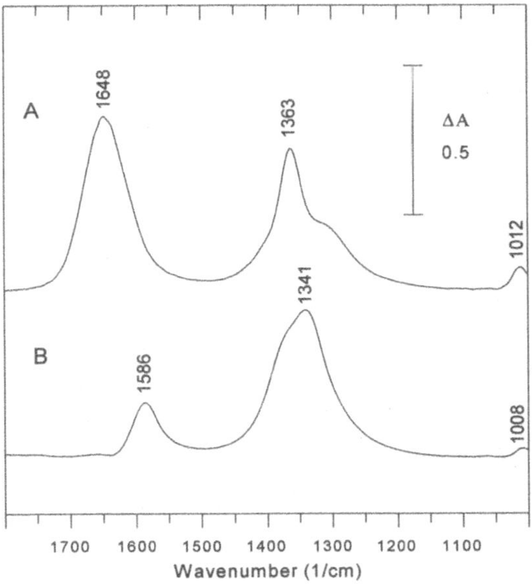

Figure 2: Spectra of ^{12}C- (A) and ^{13}C-bicarbonate (B) in H_2O.

Fig. 3 compares the Fe^{2+}/Fe^{3+}-spectra obtained in presence of bicarbonate (Fig. 3-A) and glycolate (Fig. 3-B) at pH 6. As expected bicarbonate signals at 1338 and 1229 cm^{-1} are not observed in Fig. 3-B. Instead, small and broad positive signals are observed at 1421 cm^{-1} and 1322 cm^{-1} which are tentatively assigned to the ν_{as} and $\nu_s(CO)$-modes of glycolate in the Fe^{2+}-state and which could indicate bidendate ligation. Above 1500 cm^{-1} overlap with strong protein modes make assignments rather difficult. Broad negative signals at ≈1505 or ≈1630 cm^{-1} could account for the $\nu_{as}(CO)$-modes of glycolate in the Fe^{3+}-state. More precise assignments will be possible with ^{13}C-isotope labeled glycolate.

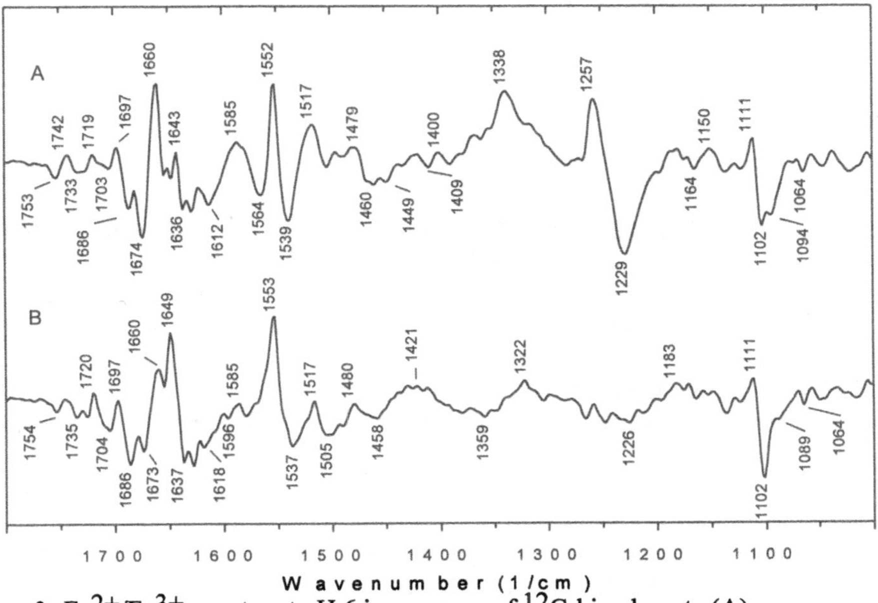

Figure 3: Fe^{2+}/Fe^{3+}-spectra at pH 6 in presence of ^{12}C-bicarbonate (A) and glycolate (B).

The intensity of the ν(CO) vibration of the hydroxyl group of glycolate depends on the type of coordination (8) and should appear in the 1050 to1085 cm^{-1} region. We exclude the possibility that this vibration appears at around 1089 cm^{-1} (Fig. 3-B) and cancels the negative signal at 1094 cm^{-1} in presence of bicarbonate (Fig. 2-A), since this signal is affected in the same way upon bicarbonate exchange by glyoxylate (not shown). The absence of a clear signal of the hydroxyl group could indicate that the coordination of glycolate is not altered upon iron oxidation.

The sharp bands observed in the 1500 to 1750 cm^{-1} region of Fig. 3 are attributed to changes of protein modes upon iron reduction. Changes in the 1650 to 1660 cm^{-1} region indicate different responses of peptide carbonyl(s) upon iron oxidation in presence of bicarbonate and glycolate. Signals in the 1550 to 1600 cm^{-1} and 1700 to 1750 cm^{-1} region could arise from deprotonated and protonated carboxylic acid groups of Glu and Asp side chains. The Fe^{2+}/Fe^{3+}-spectra obtained in presence of bicarbonate (Fig. 3-A) exhibit negative bands at 1094, 1102 cm^{-1} and a small positive one at 1111 cm^{-1} which we propose to arise from histidine ligands of the non-heme iron (2,3). The signals at 1102/1111 cm^{-1} could arise from neutral histidine(s) side chains while the signal at 1094 cm^{-1} is proposed to arise from a histidine ligand that deprotonates upon iron oxidation by comparison with IR data from metal-(4-merhyl)imidazole model complexes (3). This histidine could determine the midpoint potential of the iron in presence of bicarbonate. The absence or strong reduction of this signal in presence of glycolate (Fig. 3-B) is in agreement with the possible lowering of the pK_a of this residue upon glycolate binding.

ACKNOWLEDGMENTS: We thank A. Boussac, J. Breton, Y. Deligiannakis, E. Nabedryk and V. Petrouleas for helpful discussions. R.H. was supported by an EMBO fellowship.

REFERENCES

1 Diner, B.A. Petrouleas, V., and Wendoloski, J.J. (1991) Physiol. Plant. 81, 423-436
2 Hienerwadel, H., Boussac, A., and Berthomieu, C. (1993) in *Fith international Conference of the Spectroscopy of biological Molecules* (Theophanides, J. and Fotopoulos, N., Eds.) pp 317-318, Kluwer Acad. Publ., Dordrecht, The Netherlands.
3 Hienerwadel, R., and Berthomieu, C. *submitted.*
4 Deligiannakis, V., Petrouleas, V., and Diner, B.A. (1994) Biochim. Biophys. Acta 1188, 260-270
5 Petrouleas, V., Deliagiannakis, Y., and Diner, B.A. (1994) Biochim. Biophys. Acta 1188, 271-277
6 Nakamoto, K. Sarma, Y. A., and Ogoshi, H. (1965) J. Chem. Phys. 43, 1177-1181
7 Deacon, G.B. and Phillips, R.J. (1980) Coord. Chem. Rev. 33, 227-250
8 Larsson, R. (1965) Acta Chem. Scand. 19, 783-790

INFRARED MODES OF TYROSINE $_D$ AND TYROSINE $_D$ RADICAL IN PHOTOSYSTEM II: COMPARISON WITH CRESOL.

Rainer Hienerwadel, Alain Boussac*, Jacques Breton and Catherine Berthomieu

*CEA Saclay (*URA CNRS 1290) Section de Bioénergétique, DBCM, 91 191 Gif-sur-Yvette, Cedex, France.*

1. Introduction

Two redox-active tyrosine residues located on the D1 and D2 polypeptides of the photosystem II (PS II) reaction center have been extensively characterized by EPR spectroscopy. Tyr_Z (D1-161), which is involved in the electron transfer pathway between P_{680} and the oxygen-evolving complex, forms a transient radical. The symmetry-related Tyr_D (D-160), which has a still mostly unknown function, easily forms the stable radical $Tyr_D°$ (1,2). This different behaviour of the two tyrosines may be due to local structural asymmetry. FTIR difference spectroscopy can probe the structure and bonding interactions of a cofactor which changes ionization state upon illumination even in large membrane protein complexes such as PS II. Due to the specificity of the FTIR sample conditions, this approach requires stringent controls of the state generated. Furthermore, the interpretation of the IR difference spectra *in vivo* is based on comparison with model compound spectra and isotope or mutagenesis effects.

Previously, FTIR spectra assigned to $Tyr_Z°/Tyr_Z$ and $Tyr_Z°Tyr_D°/Tyr_DTyr_Z$ were reported and the ν(CO) mode of the radical was assigned at 1480-1477 cm^{-1} for $Tyr_Z°$ and at 1473 cm^{-1} for $Tyr_D°$ respectively (3). However, similar spectra were found by the same group in the presence of hydroxylamine that reduces the tyrosine radicals and were assigned to chlorophyll oxidation (4). In all the light-induced experiments reported above, the plastoquinone Q_A is photoreduced concomitant to the tyrosine or chlorophyll oxidation. The spectra in (3,4) are actually similar to Q_A^-/Q_A spectra reported in (5,6). This led us to suspect that the spectra reported in (3,4) are dominated by Q_A^-/Q_A IR signals.

In this paper, we report the first $Tyr_D°/Tyr_D$ FTIR difference spectrum obtained in spinach PS II membranes. IR modes of the neutral and radical Tyr_D are discussed by comparison with radical-*minus*-neutral FTIR spectra obtained by UV-irradiation of cresol (p-methyl-phenol) at 10 K.

2.Materials and Methods

Tris-treated PS II-enriched membranes were resuspended at pH 6.5 in the presence of formate to avoid oxidation of the non-heme iron. The FTIR sample was made of a membrane pellet deposited between two CaF_2 windows (Chl concentration > 15 mg/ml). Ferricyanide (20 mM) was added to the sample as a fast electron acceptor. The FTIR difference spectrum was obtained by measuring two sets of interferograms (80 scans,

P. Mathis (ed.), Photosynthesis: from Light to Biosphere, Vol. I, 747–750.

duration ≈14 s), one before and the other 2 s after the illumination of the sample. The same FTIR difference spectra were obtained with single laser flash or with 1 s illumination with a laser diode at 680 nm, the later illumination giving larger signals. Signals from several light-cycles (every 12 min.) were averaged. The cresol radical was obtained at pH 8 (0.2 M concentration) as described for phenol in (7). The EPR sample consisted of a fraction of the PS II membrane pellet deposited on a mylar sheet.

3. Results and Discussion

EPR spectroscopy was used to monitor the state of the sample before and at various times after illumination. *Figure 1*-A-a shows the typical Tyr° EPR spectrum light-induced under one Hz illumination at 4°C. The light-induced signal decays slowly but completely in ≈ 10 min. in the dark (*Figure 1*-A-b to -f).

The kinetic trace of Tyr° decay recorded at 4°C (*Figure 1*-B) is biphasic with one phase having $t_{1/2} \approx$ 140 ms and the other phase $t_{1/2} \approx$ 70 s. Identical kinetics were observed with a single flash (not shown). The fast disappearing component is assigned to Tyr_Z°, while the slow phase observed in *Figure 1* is assigned to the decay of Tyr_D°. In the FTIR measurement conditions, only the species with the slow decay (Tyr_D°) is measured. *Figure 1* further shows that no detectable Chl^+ is formed under these conditions.

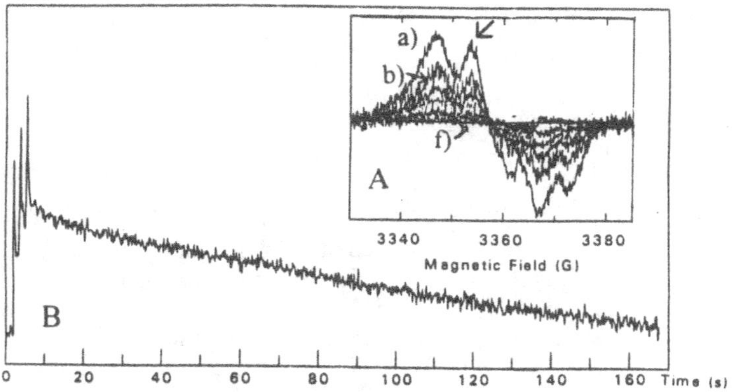

Figure 1. Tyr° EPR signal at 4°C. A- a) recorded under 1 Hz laser flash illumination, b) to f) at increasing times after illumination, time const. 40 ms. B- Kinetic trace recorded at the position of the arrow upon a sequence of three laser flashes performed every 200 s.

The FTIR difference spectrum obtained at pH 6.5 on Tris-washed PS II is shown in *Figure 2-A*. In the inset, the absorption region of ferri- and ferrocyanide shows that ferricyanide is reduced while Tyr_D is photooxidized. This shows that ferricyanide acts as a fast electron acceptor of Q_A^-. The decay of the FTIR signals in the 1800-1000 cm^{-1} range follows the decay of the ferricyanide signal (not shown) indicating that ferrocyanide reduces Tyr_D° on a slow time scale. Moreover, no IR bands which could be assigned to the non-heme iron (8) or Q_A (5,6) electron acceptors are detected in the FTIR difference spectrum. We thus conclude that the spectrum in *Figure 2* displays only the

contributions from Tyr_D and Tyr_D°, together with the changes of neighbouring protein groups induced by the tyrosine oxidation.
Figure 2-B shows the radical-*minus*-neutral spectrum obtained with cresol at pH 8. The modes of neutral cresol (negative signals) are assigned at 1615 and 1603 cm^{-1} $\nu_{8a,b}(CC)$, 1522 and 1513 cm^{-1} $\nu_{19a,b}(CC)$, 1241 $\nu_{7a}(CO)$, 1173 $\delta_{9a}(CH)$ and 1105 cm^{-1} $\delta_{15}(CH)$ (9). Signals at 1437 and 1383 cm^{-1} are largely downshifted upon $H/^2H$ exchange and therefore assigned to cresol δ(OH) modes (not shown, 13).

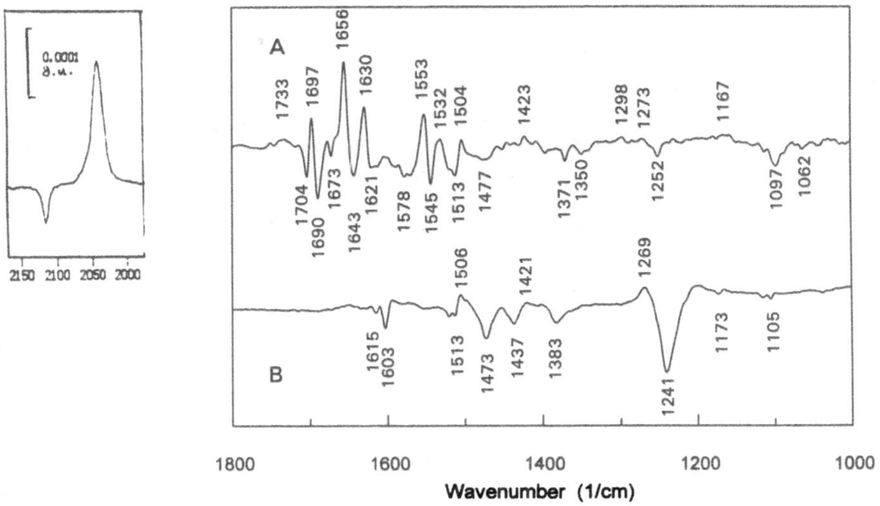

Figure 2. A- Tyr_D°/Tyr_D FTIR difference spectrum recorded at pH 6.5, 4cm^{-1} resolution, 19200 scans. A-inset ν(CN) modes of ferri- (2116 cm^{-1}) and ferrocyanide (2036 cm^{-1}). B- Cresol°/Cresol FTIR difference spectrum obtained by UV irradiation at 10 K, pH 8.

A FTIR study of various radicals of tyrosine, phenol and ethylphenol was recently presented in (7). Cresol is the best model for tyrosine in PS II since the IR mode of neutral cresol (9) are within a few wavenumber of those of tyrosine in solution (10) and since its solubility at pH 8 allows comparison of radical-minus-neutral spectra with those in PS II at comparable pH. Several bands are similar in the Cresol°/Cresol and the Tyr_D°/Tyr_D- spectra. Negative bands at 1513 and 1252 cm^{-1} in the Tyr_D°/Tyr_D spectrum are assigned to the $\nu_{19}(CC)$ and $\nu_7(CO)$ modes of neutral Tyr_D side-chain, respectively, by comparison with the negative bands in Figure 2-B and literature data on tyrosine (10). These frequencies provide an experimental evidence that the neutral Tyr_D is protonated. For tyrosinate, these modes would appear at 1500 and 1268 cm^{-1}, respectively (7,10). At ≈1615-1603 cm^{-1}, the tyrosine signals probably overlap with other protein modes. In addition, signals at 1371 and 1350 cm^{-1} are also tentatively assigned to Tyr_D-side chain.
At pH 8, radical formation is followed by deprotonation of cresol to give the neutral radical Cresol° (11). The IR modes of Cresol° are observed at 1506, 1421 and 1269 cm^{-1}. For radicals of tyrosine, phenol and ethylphenol IR signals at 1556-1550, 1513-1503 and 1290-1288 cm^{-1} were assigned to the ν_8 (CC), $\nu_{7a}(CO)$ and 14 (CC+CH) modes, respectively (7). The IR mode at 1513-1503 cm^{-1} corresponds to a mode observed by RR

spectroscopy at 1517-1502, which was tentatively assigned to a ν(CO) mode of the radicals (12). Study of cresol radicals with ^{2}H- and ^{18}O-labeling were performed to precise these assignments (13). The IR signal of radicals at 1290-1288 cm^{-1} is downshifted when these radical is hydrogen bonded, or formed at pH 8 (1275 cm^{-1} for phenoxyl and 1269 cm^{-1} for Cresol°). At pH 8, it is proposed that an hydrogen bond could exist between the radical and the proton photoreleased at 10 K (7).

For Tyr_D° in PS II, the positive signal observed at 1504 cm^{-1} is assigned to the ν7a (CO) mode. Signals at 1421 and 1275 cm^{-1} in *Figure 2-A* are also tentatively assigned to Tyr_D° side chain modes. As discussed above, the mode at 1275 cm^{-1} could indicate that Tyr_D° is hydrogen bonded in PS II. The assignments proposed for the Tyr_D°/Tyr_D spectrum presented in this paper are in disagreement with those in (3). In particular, in the RR and FTIR study of tyrosine and phenoxy radicals, the ν_{7a} (CO) mode was never observed below 1490 cm^{-1}, even for radicals hydrogen bonded to an imidazole group (7). In (3) overlap with Q_A^-/Q_A signals probably prevents detection of the Tyr_D° or Tyr_Z° modes. Combination of spectroscopy and site directed mutagenesis (14) as well as modelling studies (15) have shown that Tyr_D° is in interaction with D2-His189. Further analysis of the Tyr_D°/Tyr_D spectrum will imply studies in *Synechocystis* 6803 with specifically labeled tyrosines and histidines.

References

1 Debus, R.J. (1992) Biochim. Biophys. Acta 1102, 269-352.
2 Barry, B.A. (1993) Photochem. Photobiol. 57, 179-188.
3 MacDonald, G.M. and Barry, B.A. (1992) Biochemistry 31, 9848-9856; MacDonald, G.M., Bixby, K.A. and Barry, B.A. (1993) Proc. Natl. Acad. Sci. USA 90, 11024-11028; Bernard, M.T., MacDonald, G.M., and Barry, B.A.(1995) J. Biol. Chem. 270, 1589-1594.
4 MacDonald G.M., Steenhuis, J.J. and Barry B.A (1995) J. Biol. Chem. 270, 8420-8428.
5 Berthomieu, C., Nabedryk E., Mäntele W. and Breton J. (1990) FEBS Lett. 269, 363-367; Berthomieu C., Nabedryk, E., Breton, J. and Boussac, A., (1992) in Research in Photosynthesis (Murata, N. Ed.) Vol. II, pp 53-56, Kluwer Acad Pub, Dordrecht.
6 Noguchi, T., Ono, T. and Inoue, Y. (1992) Biochemistry 31, 5953-5956.
7 Berthomieu, C. and Boussac, A. (1995) Biospectroscopy 1, 187-206.
8 Hienerwadel, R., Boussac, A. and Berthomieu, C. (1992) Fifth Int. Conf. on the Spectroscopy of Biological Molecules (Theophanides T.et al. Eds.) p. 317-318, see also Hienerwadel, R. and Berthomieu, C., these proceedings.
9 Takeuci, H., Watanabe, N. and Harada, I. (1988) Spectrochim. Acta 44A, 749-761.
10 Dollinger, G., Eisenstein, L. and Lin, S.-L. (1986) Biochemistry 25, 6524-6533.
11 Dixon, W.T. and Murphy, D. (1976) J. Chem. Soc. Faraday Trans II 72, 1221-1230.
12 Beck, S.M. and Brus, L.E. (1982) J. Chem. Phys. 176, 4700-4704; Tripathi, G.N.R. and Schuler, R.H. (1984) J. Chem. Phys. 81, 113-121; Johnson, C.R., Ludwig, M. and Asher, S.A. (1986) J. Am. Chem. Soc. 108, 905-912; Tripathi, G.N.R. and Schuler, R.H. (1988) J. Phys. Chem. 92, 5129-5133.
13 Berthomieu, C. and Boussac, A. (1995) manuscript in preparation.
14 Tang, X.-S., Chrisholm, D.A., Dismukes, G.C., Brudwig, G.W. and Diner, B.A. (1993) Biochemistry 32, 13742-13748.
15 Svensson B., Vass I., Cedergren, E. and Styring, S. (1990) EMBO J. 7, 2051-2059.

FEMTOSECOND INFRARED SPECTROSCOPY OF PHOTOSYNTHETIC REACTION CENTERS

P. Hamm, M. Zurek, W. Zinth, Institut für Medizinische Optik,
Universität München, Germany

1. Introduction

The reaction center (RC) is the key unit for the primary charge separation and energy storage in photosynthesis. The RC, a large chromophore-protein complex, consists of two chromophore chains with the potential to transfer electrons. Both branches (A and B) start with a dimer of two strongly coupled bacteriochlorophyll (BChl) molecules, the special pair P, followed by a BChl monomer molecule (B_A and B_B), a bacteriopheophytin (BPhe) molecule (H_A and H_B) and finally by a quinone (Q_A and Q_B). However, after electronic excitation of the special pair, an electron is transferred through only one of these electron transfer (ET)-branches (the active A-branch) in a multi step reaction, while the other branch is not involved. This so-called 'uni-directionality' of ET is one of the remaining questions of bacterial photosynthesis.
Two main experimental techniques have been applied to investigate the structure-function relationship of the RC: (i) The structure of the chromophore-protein-complex was determined by X-ray scattering with atomic resolution. (ii) Time resolved spectroscopy in the visible or the near IR was applied to characterize the kinetics of the photoreaction as well as the electronic states of the chromophores. While X-ray structure analysis gives only a static picture of the arrangement of the chromophores as well as of the surrounding protein, time resolved VIS/NIR-spectroscopy yields dynamic information. However, direct information on the protein or the structural arrangement of the pigments is lacking. In order to obtain more information on structural modifications of the chromophores and on the protein during the photoreaction, time resolved vibrational spectroscopy in the mid-IR spectral range was started recently [1-4]. It will be shown that this technique promises new information on the reaction mechanism itself and on the electronic structure of P^* and P^+ [3,5]. In this paper we will present transient absorbance spectra of the RC's from the bacterium *Rb. sphaeroides* (26) measured in a spectral range between 1000-1800 cm^{-1} with a spectral resolution of 4-7 cm^{-1} and a temporal resolution of better than 200 fs.

P. Mathis (ed.), Photosynthesis: from Light to Biosphere, Vol. I, 751–754.

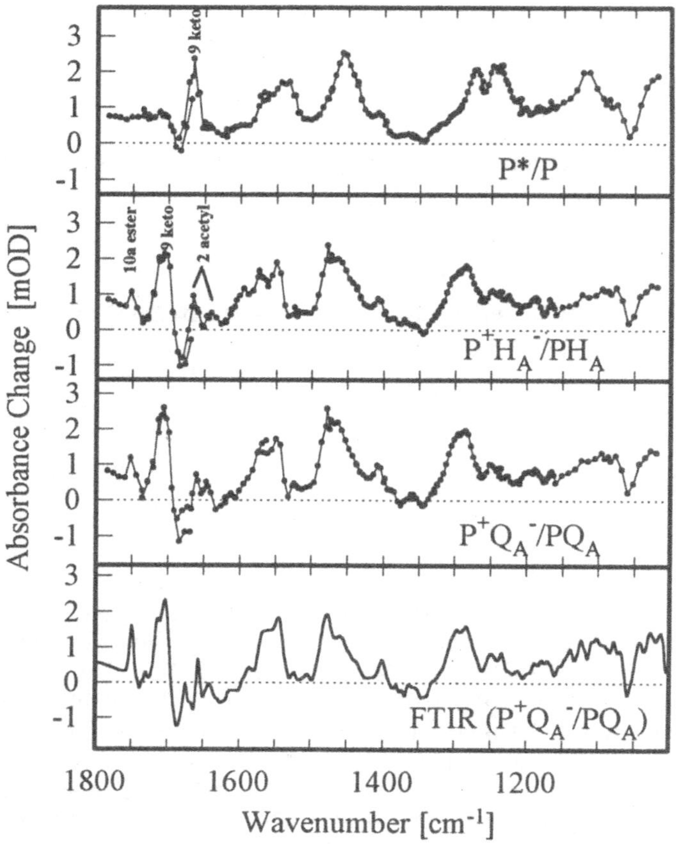

Figure 1. Transient difference spectra of the intermediate states P* (t_D=1 ps), $P^+H_A^-$ (10 ps) and $P^+Q_A^-$ (1000 ps) in comparison with a light induced steady state FTIR spectrum.

Table 1. C=O modes of the special pair P

	ground state P	excited state P*	oxidized state P^+
10a-ester C=O	1736 cm^{-1}	not affected	1752 cm^{-1}
9-keto C=O	1683 cm^{-1}	1666 cm^{-1}	1706 cm^{-1}
2-acetyl C=O (P_L)	1626 cm^{-1}	not affected	1646 cm^{-1}
2-acetyl C=O (P_M)	1654 cm^{-1}	not affected	1666 cm^{-1}

2. Results and Discussion

Throughout the entire investigated spectral range, an initial absorption increase occurs. A detailed analysis shows that in general, the signal rises not instantaneously. This effect is particularly pronounced in certain spectral regions where strong initial positive difference band are found (around 1460 cm^{-1} and 1240 cm^{-1}, see also Fig.1 top). The initial absorption rise can be fit well by assuming that the instantaneous absorption increase is small and that the signal rises with a subsequent time constant of 200±100 fs. This effect is discussed in detail in [3,5].

The subsequent temporal evolution can be well described by two well known kinetic components of ET: The dominant feature shows up the same time constant (3.4 ps) as the decay of the initial state P* while the $P^+H_A^- \rightarrow P^+Q_A^-$-ET step (220 ps) is observed with smaller amplitudes at certain spectral positions (mainly at frequencies >1600 cm^{-1}). The fast ET step from $P^+B_A^- \rightarrow P^+H_A^-$ (in 0.9 ps) which was observed in VIS/NIR measurements is not resolvable in the IR experiments. In addition, within the present signal to noise level, there is no evidence for further kinetic components, which might be assigned to slow protein relaxation due to a response on the changed charge distribution.

In Fig. 1, transient difference spectra at distinct delay time positions related to the intermediate states P* (t_D=1 ps), $P^+H_A^-$ (10 ps) and $P^+Q_A^-$ (1000 ps) are shown in comparison with a light induced steady state FTIR spectrum (which corresponds to the state $P^+Q_A^-$ at a average delay time of several milliseconds). Within the experimental error, the time resolved and the steady state $P^+Q_A^-$-difference spectra are identical indicating that after the $P^+H_A^- \rightarrow P^+Q_A^-$-ET step, no further reactions occur. When comparing the $P^+H_A^-$-difference spectrum with that of the state $P^+Q_A^-$ one also finds remarkable similarities with small differences especially at frequencies >1600 cm^{-1}. On the other hand, pronounced modifications of the transient difference spectra occur between P* and $P^+H_A^-$ (see also Fig. 2). Therefore, it can be concluded that the IR absorbance changes are predominantly caused by modifications of the special pair P (or of the protein surrounding P) and only to a small extent by the other chromophores. This also explains why the $P^+B_A^- \rightarrow P^+H_A^-$ ET-step is not observable in the present experiments.

The discussion of the difference spectra is divided into two spectral regions:

(i) In the spectral regions >1600 cm^{-1} absorption bands of the C=O groups of the chromophores and of the protein (Amide I band) are expected. Here, a comparison with steady state data found in the literature allows to give a detailed assignment of distinct absorption bands (see [1] and references therein). While in the state P* the 9-keto mode is red shifted, one observes a blue shift of the 10a-ester mode, the 9-keto and the 2-acetyl modes of P_L and P_M already in the state $P^+H_A^-$, exactly as known from steady state FTIR and Raman measurements. A band assignment is given in Tab. 1. In order to suppress the dominating contribution of the special pair and to enhance bands of H_A and Q_A, double difference spectra are shown in Fig. 2. Tentative assignments are given in Tab. 2

(ii) In the low frequency range (<1600 cm^{-1}) the C=C, C-C, C-N stretching and C-H bending modes of the BChl's tetrapyrolle ring are expected. Unfortunately, due to the lack of exact model calculations, a detailed assignment of the difference bands to distinct modes is not possible at the moment. However, a striking feature is found especially in

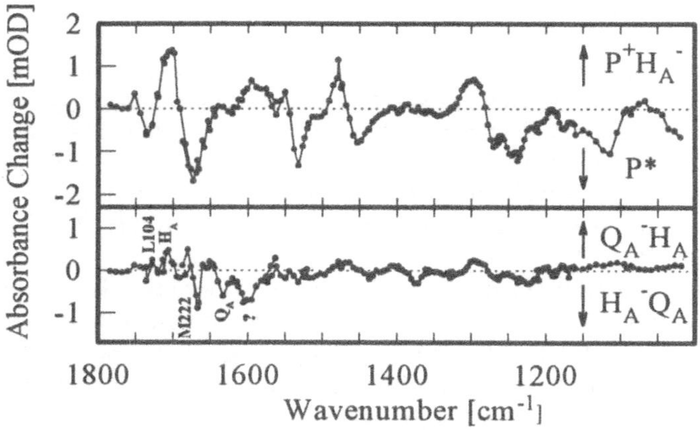

Figure 2. Double difference spectra suppressing the dominating contribution of the special pair P

Table 2. C=O modes of H_A and Q_A

	Sign	Quinone Q_A	BPhe H_A
1605 cm^{-1}	-		?
1633 cm^{-1}	-	Q_A, C=O stretch	
1667 cm^{-1}	-	Thr M222, Amide I	
1706 cm^{-1}	+		H_A, 9-keto C=O
1728 cm^{-1}	+	Glu L104, -COOH	

this part of the spectrum: a pronounced positive absorption background in connection with three strong broad positive bands which are observed in P* and P^+ at slightly different peak positions. In addition, as shown above, these positive bands are connected to the ultrafast 200 fs process mentioned above. A qualitative discussion of these features is given in [3,5].

References:

[1] Hamm, P., Zurek, M., Mäntele, W., Meyer, M., Scheer, H. and W. Zinth (1995) Proc. Natl. Acad. Sci. USA 92, 1826-1830

[2] Maiti, S., Walker, G.C., Cowen, B.R., Pippenger, R., Moser, C.C., Dutton P.L. and Hochstrasser, R.M. (1994) Proc. Natl. Acad. Sci. USA 91, 10360-10364

[3] Hamm, P. and Zinth, W. (1995) J. Phys. Chem., in press

[4] Walker, G.C., Maiti, SA., Cowen, B.R., Moser, C.C., Dutton, P.L. and Hochstrasser, R.M. (1994) J. Phys. Chem. 98, 5778-5783

[5] Zinth et al., see this issue

TRAPPING AND ENERGY TRANSFER IN THE D_1D_2-cyt-b-559 REACTION CENTER COMPLEX AT 1.5 K.

P. Schellenberg, R.J.W. Louwe, H.J. van Gorkom, T.J. Aartsma
Department of Biophysics, University of Leiden
P.O. Box 9504, 2300 RA Leiden, The Netherlands

1.Introduction

Much less is known about the primary processes in PS II reaction centers than about those in bacterial reaction centers. This is partly due to the fact that there is no x–ray structure so far, although there is some structural analogy with bacterial reaction centers. Another difficulty arises from the fact that, unlike bacterial reaction centers, the different chromophores in PS II reaction centers all have similar absorption frequencies of their electronic transitions.

One of the most challenging questions in this respect is the nature of the primary donor P680. Despite the analogies between PSII and bacterial reaction centers, there must exist significant differences as to the structure and electronic properties of the primary donor. One indication is the small shift of the chlorophyll *a* absorption bands, another the extraordinarily high oxidizing power of P680.

Accumulated photon echo methods were employed for measuring the loss of coherence of the optically excited state of the PSII reaction center. At sufficiently low temperature this dephasing is dominated by population relaxation, described by a time constant T_1, which can be compared with results of incoherent transient absorption measurements. Additionally, pure dephasing (T_2^* processes) may contribute, especially at higher temperatures. Photon echo experiments are usually explained in terms of the rotating-frame picture analogous to NMR, except that the magnetization vector is replaced by a pseudo-electric polarization vector. In a 3-pulse photon-echo, the phase information is transformed into the population domain. The accumulated photon echo is a variant of the 3-pulse photon-echo, and can be observed when the excited state decays into a long lived bottleneck state (e.g. a triplet state) causing an absorbance modulation in the frequency domain [1]. This modulation is enhanced by consecutive excitation with a stream of pulse-pairs. The modulation depth depends on the dephasing time T_2 relative to the fringe spacing which is determined by the time delay between the two excitation pulses. The APE-signal efficiency, therefore, depends on the dephasing time T_2.

2. Materials and Methods

The experimental set-up consists of a dye laser synchronously pumped by a frequency doubled mode-locked Nd:YAG-laser. The detection was performed by using a double modulation technique as described before [2], which improves the signal-to-noise

P. Mathis (ed.), Photosynthesis: from Light to Biosphere, Vol. I, 755–758.

ratio. The laser pulse width as determined from the autocorrelation signal was about 1.5 ps for fully mode-locked pulses and about 0.4 ps for stochastic excitation.

PS II reaction centers were isolated from spinach according to the procedure of van Leeuwen et al [3] and diluted 3–fold with glycerol prior to cooling of the sample to 1.5 K in a Helium –bath cryostat.

3. Results and Discussion

A typical APE–decay curve for Photosystem II reaction centers at 1.5 K is shown in Fig. 1. Similar curves were obtained over the wavelength range of 664 to 685 nm. The APE-amplitude is substantially diminished at excitation wavelengths below 675 nm. At all wavelengths the APE-decay can be fitted by a convolution of the instrument reponse function with a sum of three exponentials. The observed time constants for the coherence decay are 1.5 ps, 35 ps and 250 to 400 ps, with the longest time constant slightly increasing to the red wing of the spectrum. The relative amplitudes of the three decay components are strongly wavelength dependent, as shown in Fig. 2.

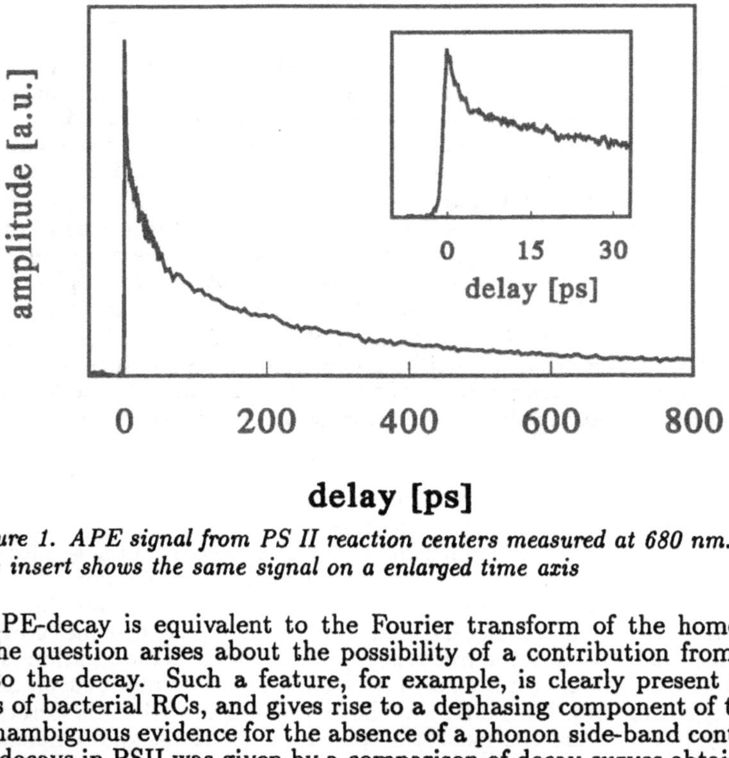

Figure 1. APE signal from PS II reaction centers measured at 680 nm. The insert shows the same signal on a enlarged time axis

Since the APE-decay is equivalent to the Fourier transform of the homogeneous lineshape, the question arises about the possibility of a contribution from phonon side-bands to the decay. Such a feature, for example, is clearly present in APE-decay curves of bacterial RCs, and gives rise to a dephasing component of the order of 100 fs. Unambiguous evidence for the absence of a phonon side-band contribution to the APE-decays in PSII was given by a comparison of decay curves obtained with transform limited pulses of 1.5 ps and with stochastic excitation, giving a time resolution of about 400 fs. The amplitudes of these decays are very similar, which proves that there is no significant component faster than 1.5 ps. We conclude, therefore, that the electron-phonon coupling of the pigments contributing to the APE-decay of

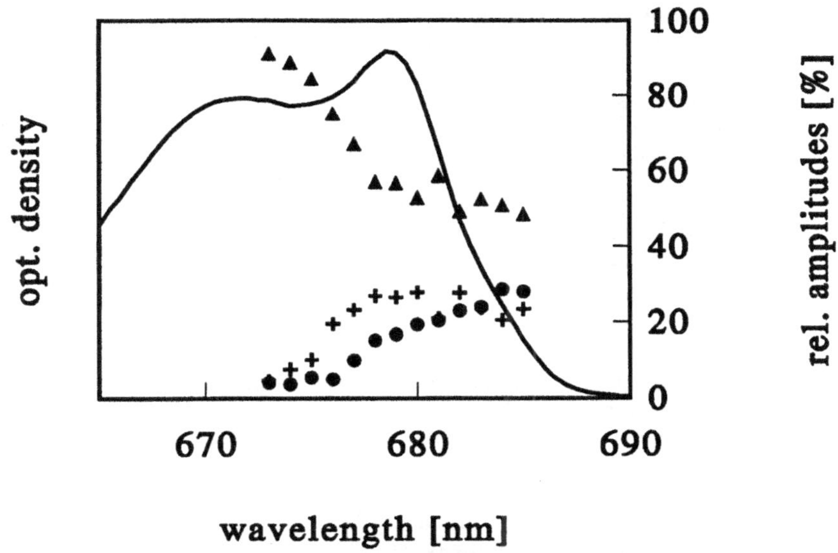

Figure 2. Relative amplitudes of the APE decay rates as a function of probing wavelength. △ : 1.5ps component, + : 35ps component, ○ : 250-400ps component

PSII is weak, at least much weaker than in bacterial RCs. Hence, the observed time constants are attributed to the intrinsic dynamics of electronic excitations in PSII.

A lower Debye–Waller factor of PS II compared to that in bacterial reaction centers is in agreement with Stark–experiments by Lösche et al. [4], who observed a significantly lower charge transfer character of the excited state in PS II compared to the reaction center of *Rb. sphaeroides*. However the lower Debye–Waller–factor is apparently in contradiction to hole burning measurements [5].

The non-exponential APE-decay may arise from a dispersion in electron-transfer rates as found in bacterial reaction centers. However, considering the variation in time constants, this seems unlikely. More likely, the various time constants are related to the congestion of absorption bands in PSII so that, at a specific wavelength, several pigments (or electronic states) can be excited. The distinct step in the relative contribution of the 1.5 ps components to the APE-decay (see Fig. 2) points to participation of two different processes, depending on the wavelength of excitation. This does not exclude the possibility of additionally having a dispersive kinetics for each of the contributing chromophores.

Based on low temperature pump-probe measurements [6], the 1.5 ps component in the long-wavelenth region can be assigned to primary electron transfer following the direct excitation of the pigments involved, namely P680 and the active pheophytin. The 35 ps component may reflect energy–transfer–limited charge separation upon excitation of other long–wavelength pigments isoenergetic with P680. These might be the inactive pheophytin or a 'trap' chlorophyll as postulated by Kwa [7]. Apparently, the contribution of these pigments to the absorbance around 675 nm is only small. The 250-400 ps component could result from trapping of the excited state on such long wavelength pigments in reaction centers where their energy happens to be lower than that of P680* and charge separation cannot occur [8]. In this case, the decay time is likely to be determined by T_2 rather than by the excited state lifetime.

In previous photon echo experiments on PS II, only very long decay times (217 ps and 698 ps) could be observed [9]. These decay components might correspond to the long decay component of our measurements. Why the short decay rates have not been observed in this study is not clear. Reaction centers prepared according to the original procedure of Satoh were used, which is known to be an unstable preparation. Therefore the reaction centers used in their measurements may have been photochemically inactive, and only signals from uncoupled chromophores may have been detected.

4. Conclusions

The phase decay kinetics of PS II follows a rather complicated pattern, which originates from the presence of several different absorbers around 680 nm. Three decay rates are observed due to excitation of different pigments or electronic states. The magnitude of these rates and their assignment are basically consistent with other time-resolved measurements on PSII.

The electron–phonon coupling is significantly lower in PS II compared to bacterial reaction centers. This is a strong indication for a significantly different situation in P680 compared to the special pair in bacterial reaction centers.

5. Acknowledgements

The investigation was supported by the Life Science Foundation (LSF) which is subsidized by the Netherlands Organization for Scientific Research (NWO). P.S. thanks the European Community for a grant from the Human Capital and Mobility Program (contract No. ERBCHBGCT930361)

References

1 Hesselink, W.H. and Wiersma, D.A. (1979) Phys. Rev. A 43, 1991-1994
2 Louwe, R.J.W. and Aartsma, T.J. (1994) J. Lumin. 58, 154-157
3 van Leeuwen, P.J., Nieveen, M.C., van de Meent, E.J., Dekker, J.P. and van Gorkom, H.J. (1991) Photosynth. Res. 28, 149-153
4 Lösche, M., Satoh, K., Feher, G., Okamura, M.Y. (1988) Biophys. J. 53, 270
5 Chang, H.-C., Jankowiak, R., Reddy, N.R.S., Yokum, C.F., Picorel, R., Seibert, M. and Small, G.J. (1994) J. Phys. Chem. 98, 7725-7735
6 Wiederrecht, G.P., Seibert, M., Govindnee and Wasielewski, M.R. (1994) Proc. Natl. Acad. Sci. USA 91, 8999-9003
7 Kwa, S.L.S. (1993) PhD thesis, Free University Amsterdam
8 Groot, M.-L., Petermann, E.J.G., van Kan, P.J.M., van Stokkum, I.H.M., Dekker, J.P. and van Grondelle, R. (1994) Biophys. J. 67,318-330
9 Vink, K.J., De Boer, S., Plijter, J.J., Hoff, A.J.and Wiersma, D.A. (1987) Chem. Phys. Lett 142, 433-438

DETERMINATION OF THE COMPLEX HOMOGENEITY AND PIGMENT STOICHIOMETRY OF ISOLATED PS II REACTION CENTRES

D. Zheleva, F. Vacha, B. Hankamer, A. Telfer and J. Barber
Photosynthesis Research Group, Department of Biochemistry,
Imperial College of Science, Technology and Medicine, London SW7 2AY, UK

1. Introduction

Since the isolation and purification by Nanba and Satoh (1) and Barber *et al.* (2) of the photochemical reaction centre of photosystem II (PS II), our knowledge about the primary energy transfer and charge separation reactions within the reaction centre has increased significantly (3). The D1, D2, cyt b559 and I polypeptides together constitute the most stable form of the isolated complex which is capable of ultrafast electron transfer from a Chl *a* species called P680 to a nearby Pheo *a* molecule. Despite substantial interest in the structure and function of the isolated PS II RC, its pigment composition is still a matter of considerable debate. In their first report Nanba and Satoh (1) evaluated, by reversed-phase HPLC, the Chl *a* : Pheo *a* : β-Car molar ratio in the RC from spinach to be around 5 : 2 : 1. A number of later reports have suggested that the reaction centre of PS II binds 6 Chl *a*, 2 Pheo *a* and 2 β-Car molecules (4-6). It has been reported that the number of Chls bound to the reaction centre complex can be reduced to 4 by extensive washing of the preparation when bound to an ion-exchange column (7), but later reinvestigation by ourselves (unpublished) and by Eijckelhoff and Dekker (6) did not confirm these results. However using copper affinity chromatography Vacha *et al.* (8) prepared PS II RC from *Pisum sativum* with a pigment stoichiometry 5 Chl *a*, 2 Pheo *a* and 1 β-Car. Two main sources of confusion regarding the pigment stoichiometry of the PS II RC may be due to the use of different preparation procedures and pigment quantification methods. In this report we used gel permeation and reversed phase chromatography to analyse the homogeneity and pigment content of PS II RC preparations isolated by either Cu-affinity (8) or ion-exchange (9) chromatography.

2. Material and methods

PS II RC were isolated from *Pisum sativum* according to Chapman *et al.* (9) and Vacha *et al.* (8). The first type (50 μg Chl *a*.ml^{-1}) were solubilized with Triton X-100 (1%) and subjected to Cu-affinity chromatography. HPLC size exclusion analysis was carried out using a Zorbax GF - 450 column 9.4/250mm (Jones Chromatography). The mobile phase consisted of 0.2M Tris-HCl pH 7.2, 0.05% dodecyl maltoside (DM) and was passed through the column at a rate of 0.5 ml min^{-1}. The elution profiles were monitored at 418 nm. For pigment analyses, the pigments were extracted with 80% acetone at 4°C under dim light conditions, vortexed for 30 s and centrifuged for 1 min. The pigments were resolved by passing the supernatant through an ODS-1 Spherisorb column (Anachem) in a 68% MeOH,

P. Mathis (ed.), Photosynthesis: from Light to Biosphere, Vol. I, 759–762.

30% ethyl acetate, 2% water mobile phase at a flow rate of 1 $ml.min^{-1}$ and detected at 453 nm and 663 nm. Chl *a* and β-Car peaks were calibrated using the extinction coefficients 76.79 $mM^{-1}.cm^{-1}$ at 663.6 nm in 80% acetone for Chl *a* (10) and 139 $mM^{-1}.cm^{-1}$ at 452 nm in 100% hexane for β-Car (11). The pheophytin *a* standard, used for calibration was produced by acidifying Chl *a* standard with 2 mM HCl.

3. Results and Discusion

3.1 *Size exclusion HPLC analyses*

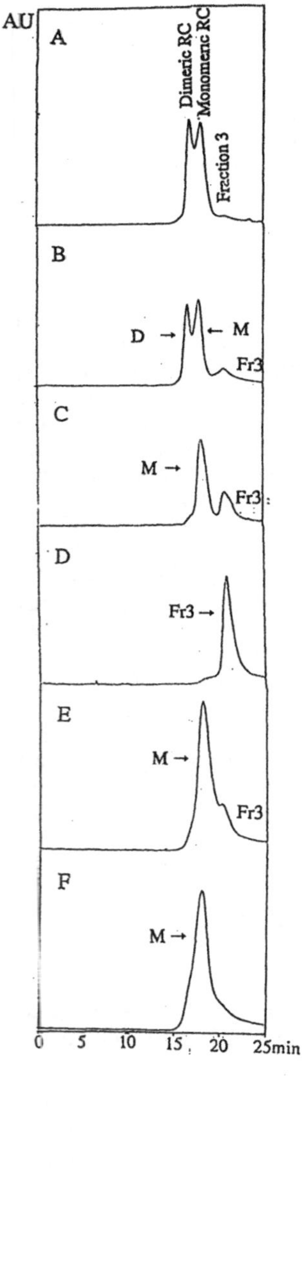

PS II RC isolated from *Pisum sativum* by the method of Chapman *et al.* (9) were resolved into two populations by size exclusion HPLC (Fig. 1A). Using a gel filtration calibration kit (Pharmacia) the molecular mass of these two populations of particles were calculated to be 180 kD and 390 kD. The two fractions were purified and subjected to SDS-PAGE, gel permeation and reversed-phase chromatography. SDS-PAGE protein profiles of the two resolved populations were very similar, containing D1, D2, cyt b559 and D1/D2 heterodimer (data not shown). The subunit composition and molecular mass determinations suggest that the 390 kD and 180 kD peaks correspond to dimeric and monomeric reaction centre fractions, respectively. A similar value for the molecular mass of monomeric RC has been reported by Tang *et al.* (12). Reinjection of the purified dimeric RC fraction onto a size exclusion column resulted in its partial monomerisation and the separation of a low molecular weight population (Fig 1B). This latter fraction (Fraction 3) was purified (Fig. 1D) and subjected to SDS-PAGE. As no protein components were detected in Fraction 3 and on the basis of room temperature absorption spectra of this fraction (data not shown) we concluded that it was composed of solubilized chlorophyll *a* and β-carotene incorporated into dodecyl maltoside micelles having a molecular weight of about 50kD.

The PS II RC prepared by the method of Chapman *et al.* (9) was subjected to Cu affinity chromatography prior to size exclusion HPLC. The elution profile of this RC preparation (Fig 1E) showed it to consist predominantly of a monomeric RC form. Our results suggest that the dimeric reaction centre component dissociates during the Cu affinity chromatography step. This is consistent with the finding that PS II RC isolated by the Cu affinity chromatography procedure of Vacha *et al.* (8) were purely monomeric (Fig 1F).

Figure 1. Size exclusion HPLC elution profile of A) RC isolated by the method of Chapman *et al.* (9), B) Dimeric RC, C) Monomeric RC, D) Fraction 3, E) RC isolated by the method of Chapman *et al.* (9) and purified by Cu affinity chromatography, F) RC isolated by the method of Vacha *et al.*(8). The column was loaded with RC, corresponding to 1.5 μg Chl *a*.

3.2 *Reversed-phase HPLC pigment analysis*

The data obtained from reversed-phase HPLC pigment analyses (Table 1) show that dimeric RCs are associated with a slightly higher Chl *a* : β-Car : Pheo *a* ratio (6.25±0.13 : 2.28±0.31 : 2) than their monomeric counterparts (5.89±0.07 Chl *a* : 1.88±0.27 β-Car: 2 Pheo *a*). The pigment stoichiometry data of the PS II RC isolated according to (9) (6.01±0.07 Chl *a*: 2.13±0.29 β-Car: 2 Pheo *a*) are in good agreement with the fact that this preparation consists of a mixture of monomeric and dimeric RCs. When these RCs were subjected to Cu affinity chromatography we observed that a monomerisation process occured simultaneously with a reduction in the Chl *a* : β-Car : Pheo *a* molecular ratio (5.52±0.16 : 0.82±0.27 : 2). These RC, purified by Cu affinity chromatography contained free pigments (Fraction 3, Fig. 1E) and monomeric RC. We suggest that monomerisation of the dimer results in the release of pigment molecules (see Fig. 1B). The lowest Chl *a* : Pheo *a* ratio (5.11±0.04 : 2) was obtained for PS II RC prepared by the method of Vacha *et al.* (8), which consisted of purely monomeric RC and lacked Fraction 3.

Table 1. Reversed-phase HPLC data for pigment stoichiometries of different PS II RC preparations (the data represent the average of at least 3 different extractions)

RC	Chl a : β-Car : Pheo a (using extinction coefficient dependent HPLC calibration)	Chl a : Pheo a (using "areas" method)
RC prepared by method of Chapman *et al.* (9)	6.01±0.07 : 2.13±0.29 : 2	6.01±0.05 : 2
Dimeric RC	6.25±0.13 : 2.28±0.31 : 2	6.18±0.04 : 2
Monomeric RC	5.89±0.07 : 1.88±0.27 : 2	5.88±0.05 : 2
RC prepared by method of Chapman *et al.* (9) and subjected to Cu affinity chromatography	5.52±0.16 : 0.82±0.27 : 2	5.57±0.21 : 2
RC prepared by method of Vacha *et al.* (8)	5.11±0.04 : 0.83±0.17 : 2	5.17±0.05 : 2

The estimation of the Chl *a* : Pheo *a* ratio is strongly dependent on the extinction coefficients used. We therefore also determined RC Chl *a* : Pheo *a* stoichiometries by calculating the ratio of the peak areas of Chl *a* before and after pheophytinisation with 2mM HCl. This ("areas" method) is independent of extinction coefficients and eliminates related errors. The data obtained using the "areas" method are presented in the third column of Table 1 and show less than 1% difference from the data obtained by the extinction coefficient dependent HPLC calibration (see Material and Methods section).

Our results clearly show that the PS II RC isolated by ion-exchange chromatography are heterogeneous and contain monomeric as well as dimeric forms of the reaction centre complex. Whether or not the PS II RC exists as a dimeric complex *in vivo* or is a product of purification induced aggregation has not yet been conclusively established. However monomerisation of the dimeric RC complexes was observed after they were subjected to gel permeation chromatography (Fig.1B), high light treatment (Fig.2), Triton X-100 or strong DM solubilisation (data not shown). These data are consistent with the hypothesis that the

PS II exists as a dimer *in vivo* and that the monomeric RC are the result of detergent or light induced dimer dissociation.

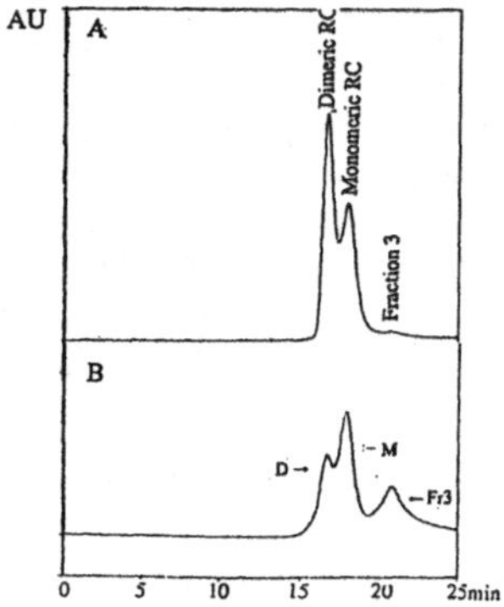

Figure 2 Size exclusion HPLC elution profile of A) RC - control and B) RC, illuminated with 4000 $\mu E,m^{-2}.s^{-1}$ for 10 min at 20°C.

Monomeric RC are shown to be associated with lower Chl *a* : β -Car : Pheo *a* ratios than their dimeric counterparts. Furthermore the pigment stoichiometry of the RC preparation is dependent on the relative monomer-dimer levels. Seasonally related variations of growth conditions of the pea plants from which the PS II RC were isolated, resulted in preparations which differed in their relative monomer : dimer ratios and pigment stoichiometry, respectively. The exact cause of this variation is not as yet fully understood. Factors which could have influenced the monomer-dimer ratio and pigment stoichiometry of the PS II RC include photoinhibition and phosphorylation levels of the thylakoid membranes prior to isolation of the reaction center complex.

References

1. Nanba, O. and Satoh, K. (1987) Proc. Natl. Acad. Sci. USA 84, 109-112.
2. Barber, J., Chapman, D.J. and Telfer, A. (1987) FEBS Lett. 220, 67-73.
3. Van Grondelle, R., Dekker, J.P., Gillbro, T. and Sundstrom, V. (1994) Biochim. Biophys. Acta 1187, 1-65.
4. Kobayashi, M., Maeda, H., Watanabe, T., Nakane, H. and Satoh, K. (1990) FEBS Lett. 260, 138-140.
5. Gounaris, K., Chapman, D.J., Booth, P., Crystall, B., Giorgi, L., Klug, D.R., Porter, G. and Barber, J. (1990) FEBS Lett. 265, 88-92.
6. Eijckelhoff, C. and Dekker, J.P. (1995) Biochim. Biophys. Acta (in press).
7. Yruela, I., Van Kan, P.J.M., Muller, M.G. and Holzwarth, A.R. (1994) FEBS Lett. 339, 25-30.
8. Vacha, F., Joseph, D.M., Durrant, J., Telfer, A., Klug, D.R, Porter, G. and Barber, J. (1995) Proc. Natl. Acad. Sci. USA 92, 2929-2933.
9. Chapman, D., Vass, I. and Barber, J. (1991) In: Methods in Plant Biochemistry, Vol.5 (Rogers, L. ed) pp 171-193 Acad. Press, LTD.
10. Porra, R., Thompson, W. and Kriedemann, P. (1989), Biochim. Biophys. Acta 975, 384-394.
11 Zechmeister, L. and Polgar, A. (1943) J. Am. Chem. Soc. 65, 1522 - 1528.
12 Tang, X-S., Fushimi, K. and Satoh, K. (1990) FEBS Lett. 273, 257-260.

The Semiquinone-Isoprene Conformation of the Primary Electron Acceptor (Q_A) of Higher Plant PSII (Plastosemiquinone) Differs from that in Bacterial Reaction Centers(Ubisemiquinone or Menasemiquinone) by *ca* 90°

Ming Zheng [a], Vasili Petrouleas [b] and G. Charles Dismukes [a]
[a] Department of Chemistry, Princeton University, Princeton, NJ 08544 USA [b] Institute of Material Science, NCSR 'Democritos',15310 Aghia Paraskevi Attikis, Greece

Introduction

The iron-quinone complex in oxygenic PSII reaction centers is similar in many respects to its counterpart in the bacterial reaction center.[1,2] However, in non oxygenic photosynthetic reaction centers, Q_A can be either menaquinone or ubiquinone, whereas in PSII only plastoquinone-9 (PQ-9) is found. Moreover, these two classes of quinones are not capable of functionally replacing one another in electron transport at comparable concentrations.[3-5] Is this because the quinones interact differently with the reaction center environments of their respective proteins? And if so, why did Nature find it necessary to evolve the Q_A site?

We offer answers to some of this question in this work by investigating the structure of the PQ-9 anion radical, both free in solution and bound to the Q_A site of PSII in spinach and the cyanobacterium Synechocystis PCC 6803, using Electron Nuclear Double Resonance (ENDOR) spectroscopy. To eliminate the magnetic influence of the paramagnetic Fe^{2+}, a recently developed method is used which relies on binding of the strong-field ligand cyanide to Fe^{2+}, thereby converting it from its high spin (S = 2) state to the low spin (S = 0) state.[6]

Results

PQ-9 anion radical. **Fig. 1** summarizes the isotropic ^1H hyperfine constants obtained by ENDOR and gives their assignments based on simulation of the experimental EPR lineshape. Our isotropic hyperfine couplings agree with those reported by Pedersen for the anion radical on the basis of EPR,[7] and by Macmillan et al based on their EPR and ENDOR measurements.[8]

EPR spectra of Q_A^- EPR measurements were done using cyanide-treated spinach PSII particles in both H_2O and ^2H_2O buffers and Synechocystis PSII particles in H_2O buffer, as shown in **Fig. 2** a to e. ^2H_2O exchange sharpens the EPR signal in both spinach PSII membranes and core complexes, with evidence for additional resolved hyperfine splittings on both the low field and high field sides in spinach PSII membranes (arrows).

ENDOR spectra of Q_A^-. ^1H ENDOR measurements on spinach PSII membranes and Synechocystis PSII core complexes in H_2O buffer, as shown in **Fig. 3** a and b, respectively, reveal the same number of peaks, although the a-a', b-b' and c-c' transitions are sharper in Synechocystis PSII core complexes than in spinach PSII membranes. Hence, the Q_A^- structure is essentially the same in both organism.

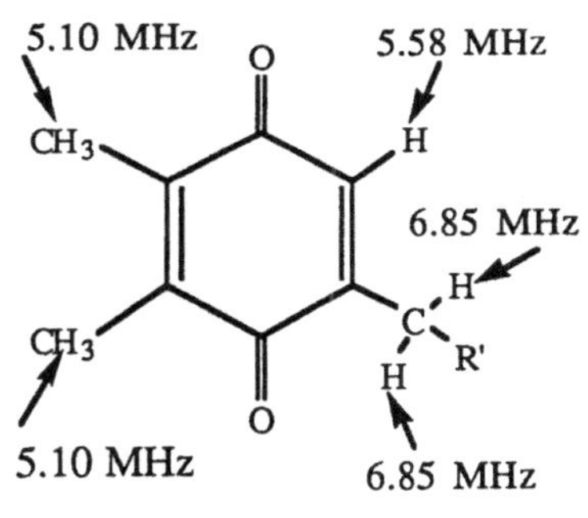

Fig. 1 Hyperfine assignments for PQ-9$^{-\cdot}$

P. Mathis (ed.), Photosynthesis: from Light to Biosphere, Vol. I, 763–766.

Special Triple spectra of Q_A^-. **Fig.4** shows the Special TRIPLE spectra for the high frequency coupling region (h.f.c ≥ 2 MHz) obtained from spinach PSII membranes and core complexes at 135K and 200K, and in H_2O and ^2H_2O buffer. The pair of resolved transitions labelled d-d' and e-e' in PSII membranes seen at 135K merge into a single unresolved, broad peak at 200 K. In PSII core complexes sample, all peaks are broader and the corresponding pair of transitions is not resolved and appear as a single broad peak at temperatures between 135K and 200K.

^2H_2O exchangeable proton(s). The sharpening of the EPR spectrum of Q_A^- upon ^2H_2O exchange indicates that ionizable protons are coupled to Q_A^-. ^1H Special TRIPLE spectra (**Fig. 4f**) reveal that ^2H_2O exchange decreases a broad peak underlying the sharp transitions d-d' and e-e' at 1.3 ~2 MHz (A = 2.6 ~ 4.0 MHz) and another peak at 2.5 ~ 2.7 MHz (A = 5.0 ~ 5.4 MHz).

Table 1. ^1H hyperfine couplings of $Q_A^{-\cdot}$ in PSII measured at 135K.

Transitions	Hyperfine couplings /MHz	Assignment
a - a'	0.38	matrix H
b - b'	0.98	matrix H
c - c'	1.59	matrix H
d - d'	2.73	CH_2 (A_{per})
e - e'	3.30	CH_2 (A_{per})
f - f'	4.00	-H (Apara)
g - g'	5.29	CH_3 (A_{per})
h - h'	5.71	CH_3 (A_{per})
i - i'	6.55	CH_3 (A_{para})
j - j'	7.36	CH_3 (A_{para})
k - k'	8.60	-H (Aper)

Discussion

Solution conformation of PQ-9 anion radical. The measured hyperfine couplings for the PQ-9 anion radical can be used to derive the spin density distribution of the unpaired π electron and the

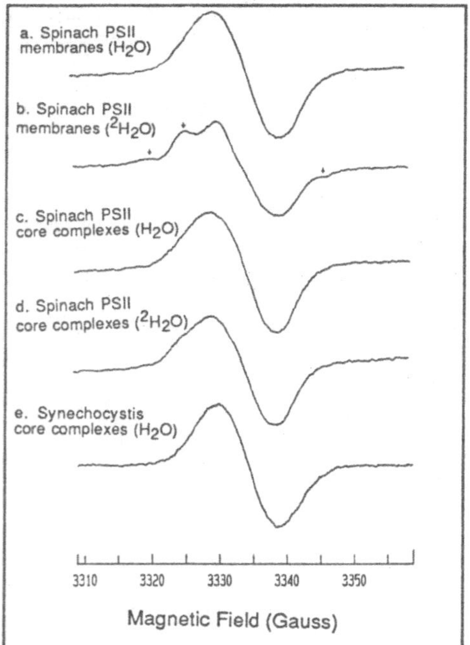

Fig. 2 Q_A^- EPR spectra

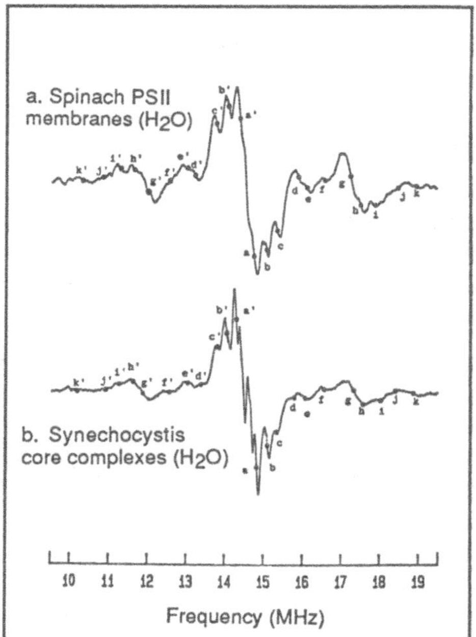

Fig. 3 ^1H ENDOR of Q_A^-

conformation of the β-methylene protons by using the McConnell[9] and the Mclachlan formulae.[10-12] Based on these formulae, it has been determined before that semiquinone anion radicals of Vitamin E, Vitamin K1, ubiquinone-10 have the same conformation,[13] as shown in **Fig. 5a**. For PQ-9 anion radical, our results show that the dihedral angle C6C5CβCγ is determined to be 0° or 180°, placing the isoprenyl ring in the plane of the quinone ring(**Fig. 5c**), and differing by 90° to that for Vits E, K_1 and UQ-10 anion radicals.

Q_A^- ^1H ENDOR assignments. The assignment of ^1H ENDOR resonances observed for Q_A^- is given in Table 1. Transitions g-g', h-h' i-i' and j-j' are characteristic of the methyl rotor hyperfine tensors,[14-17] and we assign them accordingly. The f-f' and k-k' peaks are assigned to the α proton, based on model system studies.[18-20] The d-d' and e-e' peaks are assigned to the perpendicular components of the two β methylene proton hyperfine tensors, based on their temperature dependences and on model compound studies.[21]

Conformation of Q_A^-. Because the d - d' and e - e' transitions are very close to each other in frequency, the two protons see roughly the same spin density. This argues strongly for a symmetric placement of the two β methylene protons about the aromatic ring, a conformation identical to that for PQ-9 anion radical in solution (**Fig. 5d**).

PSII-Q_A^-hydrogen bonding. With insufficient information at this stage to determine the spin density at O-carbonyl, it's not possible to provide an accurate hydrogen bonding distance based upon the observed exchangeable ^1H hyperfine couplings. Nevertheless, these couplings (2.6~4MHz and 5.0~5.4MHz) can be compared with those for the hydrogen bonded protons in 1,4-benzoquinone anion

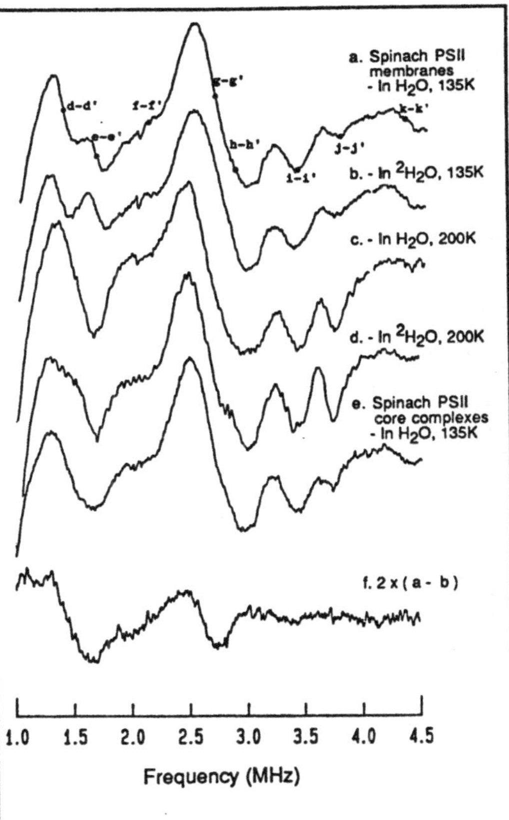

Fig. 4 Special Triple of Q_A^-

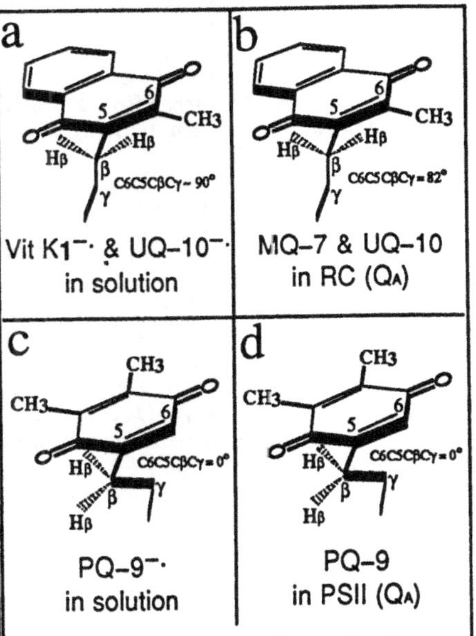

Fig. 5 Isoprene conformation

radical in aqueous solvent (2. 8MHz and 5.9MHz)[19] and to Q_A^- in the *Rb. sphaeroides* RC (4.7MHz, 6.4MHz and 9.0MHz).[17,22] The much larger hyperfine couplings for the H-bonded protons of Q_A^- in bacterial RCs indicates an appreciable larger spin density on the carbonyl oxygen atoms, or alternatively, shorter (i.e. stronger)H-bonds.

Conformation of Q_A: PSII vs. bacterial RC's. Our ENDOR data suggest that the conformation of Q_A^- in PSII is the same as conformation of PQ-9 anion radical free in solution. For the *Rb. sphaeroides* RC's, according to a recent X-ray structure at 2.65 Å resolution,[23] the C6C5CβCγ dihedral angle for UQ-10 at Q_A site is 82°, a number that is very close to what is found for the UQ-10 anion radical in solution (**Fig. 5b**). For the *Rps. viridis* RC, the C6C5CβCγ dihedral angle for MQ-7 at the Q_A site is also 82°. This conformation is again the same as that of Vit K_1 anion radical, a napthoquinone analog of MQ-7 anion radical, in solution (**Fig. 5**). Therefore, the conformation of the PQ-9 isoprenyl chain relative to the semiquinone head in the Q_A site of higher plant PSII differs by *ca* 90° versus both types of structurally characterized bacterial reaction centers. This result offers a structural explanation for the difference in quinone binding affinity observed in PSII and bacterial reaction centers.[3-5]

We propose that the PSII and bacterial RCs proteins differ at the Q_A site in terms of accomodating isoprene conformations differing by 90° in the C6C5CβCγ dihedral angle.

Acknowledgements

Supported by the NIH, GM39932. MZ acknowledges NIH training grant GM08309. NATO travel grants (870733 and OUTR.LG940992) are acknowledged. We also acknowledge Drs. X.-S. Tang and B. A. Diner for providing a sample and commenting on the manuscript.

References

1 Mattoo, A. K.; Marder, J. B.; Edelman, M., *Cell* **1989**, *56*, 241-246.
2 Michel, H.; Deisenhofer, J., *Biochemistry* **1988**, *27*, 1-7.
3 Krogmann, D. W.; Olivero, E.; Duane, W., *J. Bio. Chem.* **1962**, *237*, 3292-3295.
4 Trebst, A.; Eck, H.; Wagner, S. In *Photosynthetic Mechanisms of Green Plants*Natl. Acad. Sci.: Washington, D. C., 1963; pp 174-194.
5 Klimov, V. V.; Dolan, E.; Shaw, E. R.; Ke, B., *Proc. Natl. Acad. Sci. USA* **1980**, *77*, 7227-7231.
6 Sanakis, Y.; Petrouleas, V.; Diner, B. A., *Biochemistry* **1994**, *33*, 9922-9928.
7 Pederson, J. A., *Phytochemistry* **1978**, *17*, 775-778.
8 MacMillan, F.; Lendzian, F.; Renger, G.; Lubitz, W., *Biochemistry* **1995**, *34*, 8144-8156.
9 McConnell, H.; Chesnut, D., *J. Chem. Phys.* **1958**, *28*, 107-117.
10 McLachlan, A. D., *Mol. Phys.* **1958**, *1*, 233-240.
11 Lykos, P. G., *J. Chem. Phys.* **1960**, *32*, 625-626.
12 Heller, C.; McConnell, H. M., *J. Chem. Phys.* **1960**, *32*, 1535-1539.
13 Das, M. R.; Connor, H. D.; Leniart, D. S.; Freed, J. H., *J. Am. Chem. Soc.* **1970**, *92*, 2258-2268.
14 Hyde, J. S.; Rist, G. H.; Eriksson, L. E. G., *J. Phys. Chem.* **1968**, *72*, 4269-4276.
15 O'Malley, P. J.; Babcock, G. T., *J. Chem. Phys.* **1984**, *80*, 3912-3913.
16 Kurreck, H.; Bock, M.; Bretz, N.; Elsner, M.; Kraus, H.; Lubitz, W.; Muller, F.; Geissler, J.; Kroneck, P. M. H., *J. Am. Chem. Soc.* **1984**, *106*, 737-746.
17 Feher, G.; Isaacson, R. A.; Okamura, M. Y.; Lubitz, W. In *Antennas and Reaction Centers of Photosynthetic Bacteria. Structure, Interaction, and Dynamics*; M. E. Michel-Beyerle, Ed.; Springer: 1985; Vol. 42; pp 174-189.
18 Hales, B. J., *J. Am. Chem. Soc.* **1975**, *97*, 5993-5997.
19 O'Malley, P. J.; Babcock, G. T., *J. Am. Chem. Soc.* **1986**, *108*, 3995-4001.
20 Burghaus, O.; Plato, M.; Rohrer, M.; Mobius, K.; MacMillan, F.; Lubitz, W., *J. Phys. Chem.* **1993**, *97*, 7639-7647.
21 Kevan, L.; Kispert, L. D. In *Electron Spin Double Resonance Spectroscopy*John Wiley & Sons, Inc.: 1976; pp 199-202.
22 Lubitz, W.; Abresch, E. C.; Debus, R. J.; Isaacson, R. A.; Okamura, M. Y.; Feher, G., *Biochim. Biophys. Acta* **1985**, *808*, 464-469.
23 Ermler, U.; Fritzsch, G.; Buchanan, S. K.; Michel, H., *Structure* **1994**, *2*, 925-936.

THE EFFECT OF PS II-NON Q_B REDUCING CENTERS ON THE MEASUREMENT OF THE FLUORESCENCE QUENCHING PARAMETERS qP AND qN

Philip D. Laible, Jeffrey L. Mullert, Philip M. Ginsburg and Thomas G.Owens, Section of Plant Biology, Cornell University, Ithaca, NY 14853-5908 USA

1. Introduction

The measurement of chlorophyll (chl) fluorescence is widely used in both basic and applied studies as a probe of photosynthesis, physiology and stress response. Using modern instrumentation, the techniques of chl fluorescence can be equally applied to isolated photosynthetic complexes, membranes, intact plant tissues or whole plants. One of the primary techniques used in the analysis of chl fluorescence is the separation and quantifying of processes that regulate fluorescence yield into those that are related to photochemistry in the PS II reaction center (photochemical quenching or qP) from those that are independent of PS II photochemistry (nonphotochemical quenching or qN).

The calculation of qP and qN is based upon comparisons of fluorescence emission measured with either open or closed PS II reaction centers in dark adapted samples (qN = 0; Fo and Fm, respectively) or in illuminated samples ($0 \leq qN \leq 1$; Fo' and Fm', respectively)

$$qP = (Fm' - F)/(Fm' - Fo') \qquad qN = 1 - (Fm' - Fo')/(Fm - Fo)$$

where F is the fluorescence emission from the illuminated sample (1). An inherent assumption in the calculation of qP and qN is that PS II is homogeneous with respect to the effects of illumination. That is, for any given light intensity, the probability that the PS II trap will be open or closed is the same for all PS II (a similar argument applies to qN). However, it is well established that PS II is heterogeneous with respect to both antenna size (α/β heterogeneity) and its ability to reduce Q_B, the secondary quinone acceptor of PS II (2,3). PS II units that are unable to reduce Q_B do not participate in electron transport or oxygen evolution and have been called inactive or non-Q_B centers (3). The observation that the rate of Q_A^- oxidation in the non-QB centers is about 1.5 s, or about 1000 times slower than for active PS II centers (4), indicates that the non-QB centers will be closed at very low actinic intensities (<10 μmol photons m^{-2} s^{-1}) and remain closed at higher intensities where the majority of active PS II center are open. Thus partitioning of fluorescence quenching into qN and qP will be biased by the fact that some fraction of total PS II centers (the non-QB centers) will be completely closed under virtually all actinic intensities. As a result, measured values of qP and qN may underestimate the actual values in active PS II centers.

In the present study, we report simulations of the effect of PS II non-QB centers on the measurement of qP and qN and present a simple method to estimate the fraction of

P. Mathis (ed.), Photosynthesis: from Light to Biosphere, Vol. I, 767–770.

non-QB centers in intact leaf tissues.

2. Materials and Methods

Simulation of chl fluorescence quenching parameters was accomplished using MathCAD (MathSoft, Inc. Cambridge, MA). Fluorescence parameters were measured on intact leaves (market spinach) as previously described (5). Fluorescence induction kinetics and the effects of exogenous quinone acceptors were measured in thylakoid membranes as in Owens (6). Thylakoid membranes from market spinach were prepared and resuspended in 0.4 M sorbitol, 5 mM MES-KOH (pH 6.5), 20 mM KCl, 2 mM $MgCl_2$, and 1 µM nigericin and were stored on ice for less than one hour before use.

3. Results and Discussion

Our simulations of the effect of non-QB centers on measured fluorescence quenching parameters as a function of the fraction of non-QB centers (NR) are derived from calculation of the effects of qN on the initial (Fo) and variable (Fv) fluorescence components in active and inactive PS II centers. The total initial fluorescence Fo^{total} (=1)

$$Fo^{total} = NR \cdot Fo^{NR} + (1-NR) \cdot Fo^{R} = 1$$

is the sum of the weighted contributions from active (Fo^R) and inactive (Fo^{NR}) centers. The variable fluorescence for active and inactive centers (Fv^R and Fv^{NR}) is determined by the fluorescence ratios V^R and V^{NR}, respectively

$$Fv^{NR} = V^{NR} \cdot Fo^{NR} \qquad Fv^{R} = V^{R} * Fo^{R}$$

and the maximum fluorescence in dark adapted (qN = 0) sample (Fm) is

$$Fm = Fo^{NR} + Fv^{NR} + Fo^{R} + Fv^{R}$$

The effect of steady state actinic illumination is to induce both qP and qN in the sample. However, for intensities >10 µmol photons m^{-2} s^{-1}, all of the non-QB centers will be closed while for most physiological intensities the majority of the active centers will remain open. Direct measurements of qN on inactive PS II centers have not been made, but we assume that qN has identical effects on active and inactive centers because both types of PS II units contain the minor LHC II polypeptides where qN is thought to be localized (7). Thus the parameters Fo', F and Fm' are given by

$$Fo' = (1-qN)(Fo^{NR} + Fo^{R})$$
$$F = (1-qN)(Fo^{NR} + Fo^{R}) + (1-qN)\,Fv^{NR} + (1-qN)(1-qP)\,Fv^{R}$$
$$Fm' = (1-qN)Fm$$

Fixed values of qN (affecting active and inactive centers) and qP (affecting only active centers) were initially assigned, and then the measured values of qP and qN were simulated using the above equations. A representative data set is shown in Figure 1 for assigned values of qP = 0.8, qN = 0.3 and $V^R = V^{NR} = 4$ (the same ratio of Fv/Fo in active and inactive centers). The data show that the measured value of qN is independent of the fraction of non-QB centers present and identical to the assigned value. In contrast, the measured values of qP decrease linearly with the fraction of inactive centers. Because inactive centers have a smaller antenna size than active centers, simulations were also run for values of V^{NR} equal to 3 or 2 (Figure 1, curved lines). The effect of lower Fv/Fo in

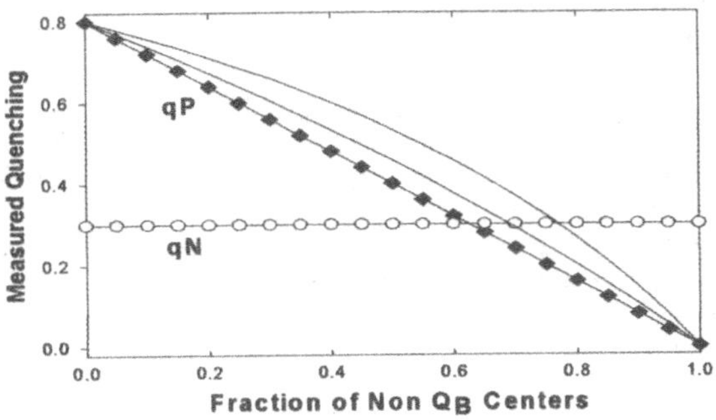

Figure 1. Simulated values for measured quenching parameters qN (open symbols) and qP (filled symbols). Data were calculated for actual values of qP = 0.8 and qN = 0.3 and assuming Fv/Fo = 4 in both active and inactive centers. Additional curved lines represent qP calculated for Fv/Fo values of 3 and 2 in inactive centers.

the inactive centers is small such that the actual value of qP is approximately equal to the measured value of qP divided by (1-NR). Thus measurements of qP can be expected to significantly underestimate the actual values of qP for physiological values of NR (0.1-0.25)(3,8).

To date, most reliable measurements of the fraction of inactive PS II centers have been accomplished in thylakoid membrane preparations (3,8). Under moderate actinic illumination, addition of compounds that accept electrons from the plastoquinone pool keep the active PS II centers open, and the extent Fv is a measure of the amount of inactive PS II. A modification of the technique described by Cao and Govindjee (8) is shown in Figure 2. Actinic illumination (6.6 µmol photon m^{-2} s^{-1}) was sufficient to close all inactive centers but only some of the active centers. Increasing concentrations of dimethyl quinone (DMQ) quench the variable fluorescence from the active centers (capable of reducing the PQ pool) but do not affect the inactive centers. The fraction of the total variable fluorescence (Fm-Fo) that was not quenched by DMQ (Fp-Fo) at saturating quinone is an estimate of the fraction inactive (non-QB) centers. Averages over three independent thylakoid preparations yield a value of 12% ± 4% in good agreement with previous estimates (3,8).

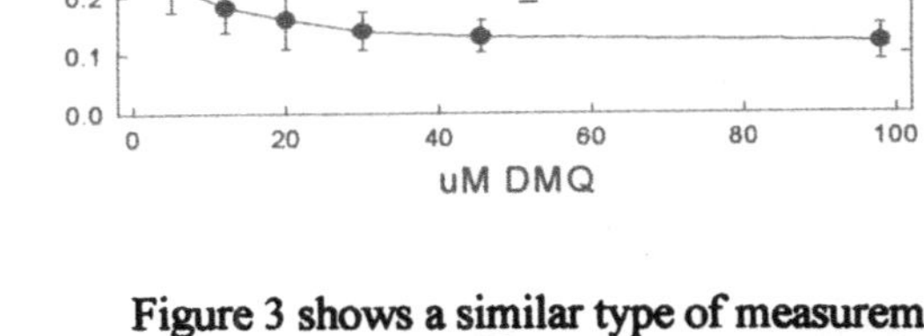

Figure 2. Quenching of variable fluorescence in active PS II centers by DMQ. Spinach thylakoids (30 ug chl/ml) were illuminated for 7 s with 6.6 umol photons m^{-2} s^{-1} in the presence of DMQ and the peak fluorescence was recorded. Data was normalized to maximum variable fluorescence measured in the presence of DCMU and corrected for quenching of Fm by DMQ.

Figure 3 shows a similar type of measurement made on intact spinach leaves using the PAM fluorometer. These measurements were also based on knowledge that low intensity light will rapidly close inactive PS II centers, but at the same intensity the active centers will be kept open by linear electron transport. The figure shows the extent of the variable fluorescence induced by a 7 s actinic "pulse" (Fp-Fo) normalized to the maximum variable fluorescence (Fm-Fo) measured by a 0.3 s saturating pulse (10000 μmol photon m-2 s-1). The intersection between the rapid fluorescence rise at low actinic intensity (closing of inactive centers) and the slower rise at higher intensity (closing of active centers) yield an estimate of 12% for the fraction of inactive centers, in good agreement with the thylakoid measurements in Figure 2. Thus, it is possible to estimate the fraction of inactive PS II centers in intact leaves without resorting to thylakoid preparation. With this information, it is also possible to correct the measured value of qP for the presence on inactive centers using the data in Figure 1.

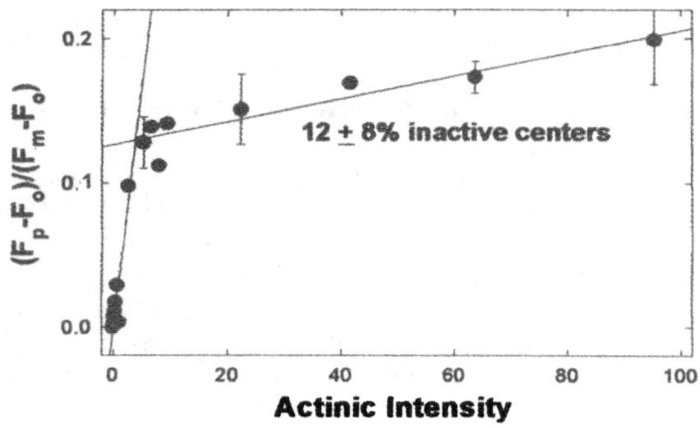

Figure 3. Estimation of the fraction of inactive centers in intact spinach leaves using the PAM fluorometer. The peak fluorescence (Fp) was recorded during a 7 s actinic pulse at the indicated intensity (actinic + 1.6 kHz modulated measuring beam). Data were normalized to the maximum variable fluorescence measured with a 0.3 s saturating pulse.

4. Acknowledgements The authors acknowledge financial support from the US Department of Agriculture and HATCH funds to TGO. JLM was supported by a summer internship from the Cornell Plant Science Center and PDL was supported by an National Institutes of Health training grant.

5. References

(1) Krause, G.H. and Weis, E. (1991) Ann. Rev. Plant Physiol. Plant Mol. Biol. 42:313-349

(2) Melis, A. (1985) Biochim. Biophys. Acta 808:334-342

(3) Grann, T. and Ort, D.R. (1986) Biochim. Biophys. Acta 852:320-330

(4) Chylla, R.A.and Witmarsh, J. (1989) Plant Physiol. 90:765-772

(5) Ting,C.S. and Owens, T.G. (1992) Plant Physiol. 100:367-373

(6) Owens, T.G. (1986) Photochem. Photobiol. 43:535-544

(7) Bassi,R., Pineau, B., Dainese, P. and Marquart, J. (1993) Eur. J. Biochem. 212:297-303

(8) Cao, J. and Govindjee (1990) Biochim Biophys Acta 1015:180-188

ISOLATION OF REACTION CENTRES FROM PLANT PHOTOSYSTEM II AND PHOTOSYNTHETIC PURPLE BACTERIA BY RAPID PERFUSION CHROMATOGRAPHY.

Margrit Roobol-Boza[1], Susana Shochat[2,3], Peter Gast[3] and Bertil Andersson[1,2]
[1]Dept. of Biochemistry, Stockholm University, Sweden, [2]Dept. of Biological Chemistry, Hebrew University of Jerusalem, Israel,[3]Dept. of Biophysics, Leiden University, The Netherlands

1. Introduction

In photosynthesis research pure and functional preparations of reaction centres (RC) have been of major importance. In the case of photosynthetic purple bacteria such preparations have been instrumental for elucidating the principal molecular details of the primary steps in photosynthetic energy conversion. The extensive studies of photosystem II (PSII) have advanced mainly due to preparations of PSII membranes, oxygen evolving core complexes and reaction centre particles (see 1). Photosynthetic reactions centres are comprised of several hydrophobic membrane proteins and therefore require quite elaborate purification methods involving solubization by detergents and subsequent isolation by centrifugation and/or chromatographic techniques.These isolations normally require several hours, or even days, for completion and there is consequently a need for more efficient and gentle separation techniques for such membrane proteins. The most recent development in the separation of soluble proteins came in the early 1990s with the introduction of perfusion chromatography (2). With this technique separation times could be drastically decreased, because the porous matrix optimises the flow through and around its surface. In addition, it provides high resolution and loading capacity. We have recently introduced perfusion chromatography for the very rapid isolation of several membrane proteins (3,4). Here we describe its use for the isolation RC from plants and photosynthetic bacteria.

2. Materials and Methods

2.1 *Isolation of PSII reaction centres*
Spinach were used to isolate PSII membranes according to (5) and finally resuspended in 20 mM Bis-Tris, pH 6.5, 20 mM $MgCl_2$, 5 mM $CaCl_2$, 10 mM $MgSO_4$, 0.4 M sucrose. PSII core complexes were isolated mainly according to (6) but using anion exchange perfusion chromatography (POROS Q-20; PerSeptive Biosystems attached to a FPLC). The core complexes were used as starting material for isolation of PSII RC particles using the same chromatographic material.

P. Mathis (ed.), Photosynthesis: from Light to Biosphere, Vol. I, 771–774.

2.2 *Isolation of bacterial reaction centres*

Rhodobacter sphaeroides (ΔLM1 deletion strain complemented with wild type genes) was grown semi aerobically in the dark (7). Chromatophores were obtained as in (3) and resuspended in 100 mM potassium phosphate, pH 7.5. The RC were extracted with 0.25% lauryl dimethylamine oxide (LDAO) by two incubations for 30 min of chromatophores resuspended to a concentration of A_{865}=30/cm. After the second incubation crude RC were precipitated with 30% ammonium sulphate, resuspended in 10 mM Tris-HCl, pH 8.0, 0.1% LDAO and desalted using Sephadex G-50 before loading the suspension onto a POROS Q-20 column.

2.3 *Analytical methods*

Proteins were analysed by SDS-PAGE and western blotting. Electron transport activity was monitored as silicomolybdate reduction at 500 nm in the presence of $MnCl_2$. The purity of bacterial RC was estimated by the ratio A_{280}/A_{802}. HPLC analysis of pigments was carried out on reverse phase columns.

3. Results

3.1 Isolation of PSII reaction centres

Isolated PSII RC particles consist of the D1 and D2 proteins, cytochrome b_{559} and the *psb*I and *psb*W gene products. The original isolation procedures (8,9), as well as later developments, are all based upon solubization with mild detergents followed by conventional anion exchange chromatography. In this communication we will demonstrate how perfusion chromatography can be used to dramatically reduce the time for isolation of highly pure spinach PSII RC particles.

PSII membranes in the resuspension buffer (2 mg chl/ml) were solubilized in 0.03% (w/v) dodecyl maltoside (DM) by gentle stirring on ice for 15 min and standing for 10 min. Unsolubilzed material was removed by centrifugation at 40 000 x g for 20 min. The supernatant was loaded onto a small column (1 ml) packed with POROS Q anion exchange beads equilibrated with resuspension buffer. Elution at a flow rate of 1ml/min washes LHCII from the column and after 10 min the eluate is practically colourless. The PSII core complexes are eluted from the column with 250 mM $MgSO_4$ after a total elution time of 15 min. Their high purity was judged from their polypeptide composition as well as a high rate of oxygen evolution (950 μmol O_2/mg chl h).

These core complexes (500 μg chl/ml) in 20 mM Bis-Tris, pH 6.0, 0.04% DM and 5 mM $LiClO_4$ were used as starting material for isolation of PS II RC using another perfusion chromatography step. After solubilization for 15 min on ice in an equal volume of 20 mM Bis-Tris, pH 6.0, 2.5% DM, 1% Triton X-100 and 4 M $LiClO_4$. The solution (approximately 2 ml) was then desalted on a Sephadex G-25 column and transferred to the POROS Q column (1 ml) equilibrated with starting buffer. After injection of the sample, and washing with 2 ml of resuspension buffer, the actual elution was initiated by increasing the $LiClO_4$ concentration to 90 mM (Fig.1). After 4 ml of elution the concentration was further increased to 125 mM $LiClO_4$ for 4 ml followed by further 4 ml of 200 mM $LiClO_4$. Finally, the column was eluted with 6 ml of 300 mM $LiClO_4$. To protect the sensitve PSII RC exposure to light was avoided and the temperature was kept at 4 ^{o}C. The first and major peak (Fig. 1) eluted with the void volume and contained undissociated core particles and residual amounts of LHCII. Basically the same kind of material was recovered in the 2nd peak and in the material eluting at the highest $LiClO_4$ concentration. In contrast, the third peak which elutes only 6-7 minutes after starting this chromatographic step contained pure PSII RC

(Fig 1, Table I). They contained only the D1 and D2 proteins, cytochrome b599 as well as the *psbI* and *psbW* gene products with no or residual contamination of CP47 and with a typical 417/435 nm absorption ratio of 1.2. Their elctron transport activity was 600 meqi./mg chl h measured as silicomolybdate reduction in the presence of Mn^{2+} ions.

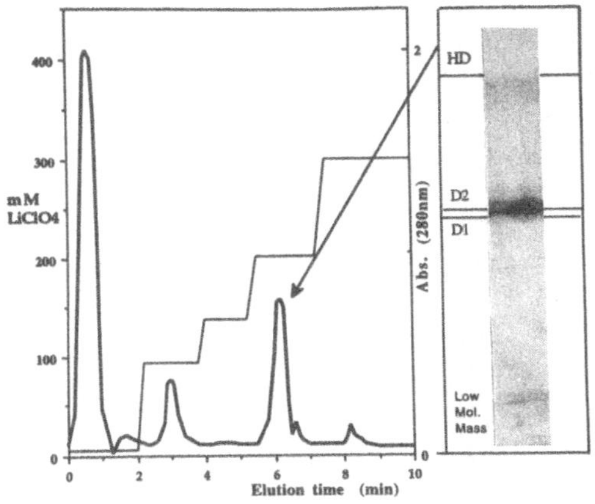

Table 1. Properties of isolated PSII RC

Abs $_{max}$ red region	676 nm
Abs 417nm/435 nm	> 1.2
Chl a per 2 pheo	6
ß-Carotene per 2 pheo	2
Cyt b_{559} per 2 pheo	1-2
Yield	6%

Fig. 1. Perfusion chromatography of solubilized PSII core particles to yield pure reaction centre particles in 6 - 7 min.

3.2 Isolation of bacterial reaction centres

The bacterial RC, composed of the three subunits L, M and H, is embedded in the cytoplasmic membrane and surrounded by the two light harvesting antenna complexes B_{870} and $B_{800\text{-}850}$. Procedures for isolation of RC from purple bacteria include their extraction from chromatophores by solubilization with a suitable detergent (usually LDAO) and isolation by a combination of ammonium sulphate fractionation and traditional ion exchange chromatography (10). A potential problem with these procedures, particularly for cells grown in the dark, is a contamination of light harvesting complexes and consequently several separation steps requiring several hours, are needed for satisfactory purity of RCs.

Perfusion chromatography was applied to the LDAO extract following the ammonium sulphate fractionation. The precipitate was dissolved in 10 mM Tris-HCl pH 8.0, 1 mM EDTA, 0.1% LDAO and loaded onto the POROS Q column (1ml) equilibrated in the same buffer. Free pheophytin and non coloured proteins are washed with the loading low salt buffer. The elution was performed with a 100-300 mM NaCl gradient in the same buffer. A first peak eluting at the beginning of the gradient represents about 20% of the RCs (Fig. 2). A second peak containing the rest of the RCs is eluted at around 200 mM NaCl. The former peak does not represent a clean fraction since its A280/A802 is 2. In contrast the second and major peak contains pure RCs as judged by a low ratio A208/A802 of 1.1-1.3. Moreover, this material contained only the L, M and H subunits as revealed by SDS-PAGE (Fig. 2). Following the gradient, pure light harvesting complexes were eluted as a broad peak at higher NaCl concentrations (400-600 mM). This isolation of the bacterial RCs can also be applied on a preparative scale using a 5 ml POROS Q-50 column.

4. Future perspectives

When chromatographic isolation of a protein mixture can be completed within minutes new experimental possibilities, which do not only relate to purification, become available. Thus, perfusion chromatography should allow isolation of photosynthetic proteins in transient functional states or different stages of biosynthesis or turnover. The possibility to use the method as an analytical tool through "chromatographic screening", should make it possible to systematically search for optimal solubilization parameters for functional and crystallographic analyses of membrane proteins as well as finding conditions for maintained ligand binding during isolation. Finally, perfusion chromatography allows the combination of sequential purifications in a short time which should be beneficial for isolation of proteins present in substochiometric amounts which is the case for many of the auxiliary enzymes of photosynthesis.

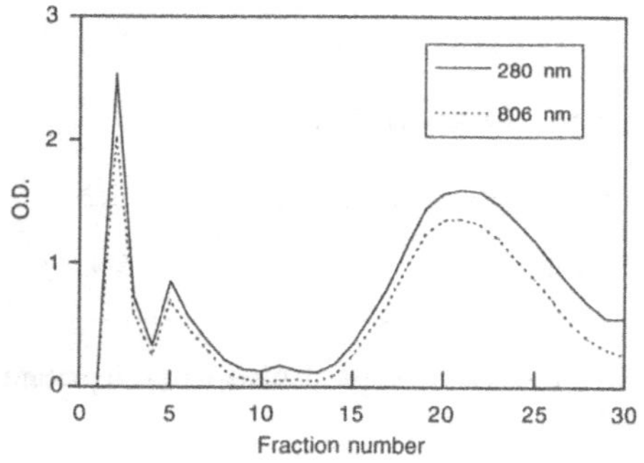

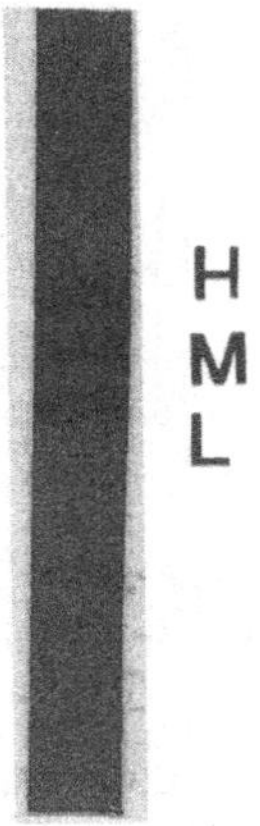

Fig. 2. Perfusion chromatography for isolation of bacterial reaction centres. Elution was performed using a gradient of 100-300 mM NaCl. Polypeptides were analyzed by SDS-PAGE

5. References

1 Seibert, M. (1993) in: The Photosynthetic Reaction Centre (Deisenhofer, J. and Norris, J. R., eds.) vol 1, pp. 319-357, Academic Publishers, San Diego, USA.

2. Regnier, F. E. (1991) Nature 350, 634-635

3. Roobol-Boza, M., Shochat, S., Tjus, S.E., Hagman, Å., Gast, P. and Andersson, B.(1995) Photosynth. Res. in press

4. Tjus, S. E., Roobol-Boza, M., Pålsson, L.-O. and Andersson, B. (1995) Photosynth. Res. in press

5. Ghanotakis, D. F., Babcock, G. T. and Yocum, C. F. (1984) Biochim. Biophys. Acta 765, 388-39.

6. Van Leeuwen, P. J., Nieveen, M. C., van der Meent, E.-J., Dekker, J. and van Gorkum, H. J. (1991) Photosynth. Res. 28, 149-153

7. Paddock,, M. L., Rongey, S. H., Feher, G. and Okamura, M. Y. (1989) Proc. Natl. Acad. Sci. USA 86, 6602-6606.

8. Nanba, O. and Satoh, K. (1987) Proc. Natl. Acad. Sci., U.S.A. 84, 109- 112

9. Barber, J., Chapman, D. J. and Telfer,, A. (1987) FEBS Lett. 220, 67-73

10. Gingras, G. (1978) . in: The Photosynthetic Bacteria (Clayton, R. K. and Sistrom, W. R., eds.) pp 119-132, Plenum Press, New York.

THE BINDING ENVIRONMENT OF THE REDUCED PRIMARY QUINONE ELECTRON ACCEPTOR, Q_A^-, OF PSII.

X.-S. Tang[1], J.M. Peloquin[2], G.A. Lorigan[2], R.D. Britt[2] and B.A. Diner[1], - 1) CR&D, E. I. Du Pont de Nemours & Co., Wilmington, DE, USA; 2) Department of Chemistry, University of California, Davis, CA, USA.

Introduction

A number of recent developments have made it possible to probe in detail the protein environments of the redox cofactors of Photosystem II. These include: the availability of structural information derived from the crystal structure of the reaction center of the purple non-sulfur photosynthetic bacteria, the ability to introduce site-directed mutations into genes encoding the PSII reaction center polypeptide D1 and D2, the development of growth conditions and the isolation of mutants that permit the specific introduction into reaction centers of amino acids bearing stable magnetic isotopes and the use of new magnetic resonance and vibrational spectroscopic techniques to obtain structural information from labeled reaction centers.

The x-ray crystallographic structure of the reaction centers of the purple non-sulfur photosynthetic bacteria indicate that Q_A is hydrogen bonded by an imidazole nitrogen of M-His 217, also a ligand to the non-heme iron, and by the peptide nitrogen of M-Ala258; furthermore, M-Trp250 is also very close to Q_A [1]. It has been speculated that Q_A in Photosystem II is probably coordinated by the structurally homologous PSII residues, D2-His214, D2-Ala260 and D2-Trp253.

We describe in the present paper some of our recent work directed at obtaining structural information on the environments of the primary quinone electron acceptor, Q_A, of Photosystem II using cw and pulsed magnetic resonance techniques, isotopic-labeling and site-directed mutagenesis.

Materials and Methods

The glucose-tolerant strain of the cyanobacterium, *Synechocystis* PCC6803 [2] was used in the current experiments. A histidine-tolerant mutant was obtained from blue-green wild type *Synechocystis* as described previously [3].

Wild-type and mutant cells were grown photoheterotrophically at 30 °C under constant illumination for 3-4 days in 18-liter carboys containing BG-11 [4] plus 5 mM glucose, and bubbled with 5% CO_2 in air. For ^{15}N-global labeling experiments, wild type cells were grown in the presence of ^{15}N-labeled sodium nitrate as the sole nitrogen source. To reverse-label histidine in Photosystem II, histidine-tolerant strain cells were

P. Mathis (ed.), Photosynthesis: from Light to Biosphere, Vol. I, 775–778.

grown with ^{14}N-histidine in the presence of ^{15}N-labeled sodium nitrate as a background nitrogen source [3].

Intact PSII core complexes were isolated from *Synechocystis* cells according to the procedure described by Tang and Diner [5]. They were then passed through a hydroxyapatite column as described by Rögner et al. [6], and desalted by passage through a gel filtration column equilibrated with 50 mM Tricine-NaOH buffer (pH 7.8) containing 0.4 M sucrose, 15 mM NaCl and 5 mM $MgCl_2$. The complex was further incubated overnight in this buffer. Cyanide-treatment of PSII core complexes was carried out according to Sanakis et al. [7] but with some modifications [8].

Low-temperature continuous-wave EPR spectra were recorded at X-band with a Bruker ER200D spectrometer equipped with an upgraded computer software system (ESP-300 model) and an Oxford ESR-910 helium flow cryostat. ESEEM spectra were recorded with a laboratory-built pulsed EPR spectrometer [9].

Results and Discussion

Sanakis et al. [7] have shown that treatment of BBY PSII membranes with CN^- results in a conversion of the non-heme iron from high (S=2) to low spin (S=0), decoupling the semiquinone anion, Q_A^-, from the non-heme iron, and narrowing the Q_A^- EPR signal to a width of 9.5G. We have adapted this procedure to PSII core complexes isolated from the cyanobacterium *Synechocystis* 6803 [8]. Furthermore, we have recently found that treatment of PSII core complexes with high concentrations of phosphate buffer on the hydroxyapatite column [7], followed by incubation of the complexes at pH 7.8, can also decouple Q_A^- from the metal center, and results in a narrowing of the Q_A^- EPR signal (Fig. 1). This narrow Q_A^- EPR signal has been used for ESEEM studies of Q_A^-.

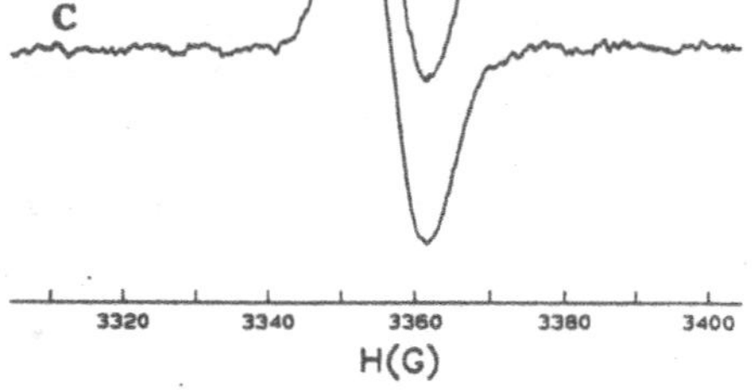

Fig.1 EPR spectra of PSII core complexes in the presence of 50 mM dithionite in the dark. A). intact PSII core complex; B). NaCN-treated core and C). phosphate-treated core. The samples were suspended in 50 mM Tricine-NaOH buffer (pH 7.8) containing 0.4 M sucrose, 15 mM NaCl and 5 mM $MgCl_2$. Temperature: 10 K; Microwave power: 200 nW; Modulation frequency: 100 kHz; Modulation amplitude: 2 G.

Identical Q_A^- ESEEM spectra were obtained from PSII core complexes treated with either cyanide or phosphate as shown in Fig. 2A & B. Four sharp modulation peaks at 0.75, 1.52, 2.06 and 2.84 MHz (tau=355 ns) plus a broad double quantum peak at approx. 5 MHz (tau=213 ns, data not shown) were observed. When the complex was treated at pH 9.2 with NaCN, the 1.52 MHz feature disappeared, but no change was observed for the remaining three peaks (Fig. 2C). This result show that the 1.52MHz component can be titrated in the pH range 7.8-9.2, and that the origin of the 1.52MHz feature is different from that of the other three peaks.

To investigate the origin of the ESEEM components, we grew the *Synechocystis* cells in the presence of $^{14}NO_3^-$ or $^{15}NO_3^-$ as the sole nitrogen source, and purified the PSII core complexes from these cells. Fig. 2D shows the Q_A^- ESEEM spectra of ^{15}N-labeled PSII core complexes, and clearly demonstrates that all four sharp ESEEM features plus the broad double quantum feature observed in the ^{14}N-sample (Fig 2B) arise from nitrogen. A small, sharp feature at 1.42MHz observed in the ^{15}N-sample is from a weakly coupled ^{15}N. Together with results from the pH-dependence experiment (Fig.2C), we conclude that there are at least two different nitrogens associated with Q_A^-.

Extrapolating from the x-ray crystallographic structure of the purple non-sulfur bacterial reaction centers, one would expect that the nitrogens observed here probably come from imidazole nitrogen of histidine, peptide nitrogen of alanine or the ring nitrogen of tryptophan. The features at 0.75, 2.06 and 2.84 MHz are consistent with the zero field nuclear quadrupole couplings of a peptide nitrogen [10] (Table 1), possibly from Ala260. This tentative assignment is consistent with that made by Atashkin et al. [11]. However, to make a definitive assignment, we need more direct experimental evidence.

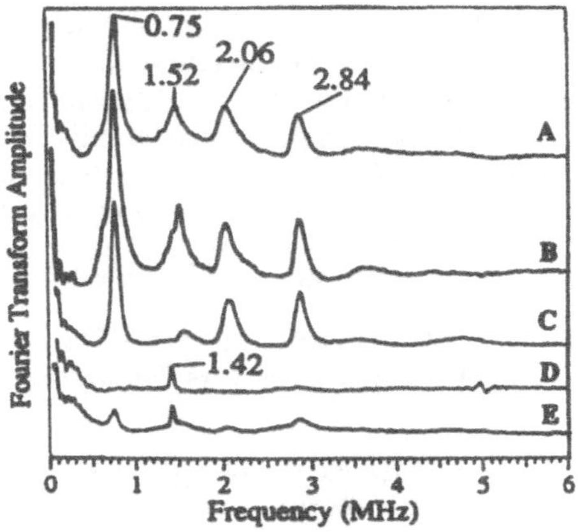

Fig.2 Three pulse Q_A^- ESEEM spectra of *Synechocystis* PSII core complexes. A). cyanide-treated ^{14}N-sample, pH 7.8; B). phosphate-treated ^{14}N-sample, pH 7.8; C). same as A, but pH 9.2; D). phosphate-treated, labeled with $^{15}NO_3^-$, pH 7.8; E). phosphate-treated, labeled with ^{14}N-histidine plus $^{15}NO_3^-$, pH 7.8. Temperature:4.2K; tau: 355ns.

Table 1. Comparison of nitrogen couplings obtained from Q_A^- and the nuclear quadrupolar frequencies of peptide nitrogen at zero field in model compound.

Source	υ_+ [MHz]	υ_- [MHz]	υ_0 [MHz]	Reference
peptide NH glycylglycine glycylalanine glycylalanine	2.59-2.90	1.98-2.20	0.62-0.72	[10]
nitrogen associated with Q_A^-	2.84	2.06	0.75	this work

Using a histidine tolerant strain of *Synechocystis* [3], we have incorporated ^{14}N-histidine into PSII core complexes in which all nitrogens except for these of histidine have been labeled with ^{15}N. Under our experimental condition, 85% of the histidine incorporated into protein is in the ^{14}N form [4]. Fig. 2E shows the three pulse Q_A^- ESEEM spectra of such PSII core complexes. Aside from a small scrambling of the ^{14}N from histidine, all four sharp nitrogen peaks plus the broad double quantum peak observed in the globally labeled ^{14}N-sample are absent upon ^{14}N-histidine/$^{15}NO_3^-$ labeling. This result clearly demonstrates that histidine is not responsible for these nitrogen couplings, and is consistent with the tentative assignment of 0.75, 2.06 and 2.84 MHz couplings to a peptide nitrogen. Our results disagree with a recent suggestion of Atashkin et al. [11], assigning the 1.52 MHz component to an imidazole nitrogen of histidine, based only on homology with the bacterial reaction center.

We have recently constructed several mutants at the D2-Ala260 and D2-Trp253 sites. Further experiments are planned using these mutants and isotopic labeling in order to make definitive assignments for the nitrogen couplings associated with Q_A^-.

Conclusion

1). Treatment of PSII core complexes with high concentrations of phosphate, followed by incubation at pH 7.8, can magnetically decouple Q_A from the non-heme iron, resulting in a narrowing of the Q_A^- EPR signal to a linewidth of 9.5 G.

2). Four sharp ESEEM modulation peaks and a broad double quantum peak have been observed for Q_A^-. $^{14}N/^{15}N$-labeling experiments demonstrate that all of them come from nitrogen. Their pH-dependence shows that these nitrogen couplings come from at least two different nitrogen sources.

3). The imidazole ring nitrogens of histidine are not responsible for any of these nitrogen couplings.

4). The features at 0.75, 2.06 and 2.84 MHz are tentatively assigned to the zero field nuclear quadrupole couplings of a peptide nitrogen.

Acknowledgments

B.A.D. gratefully acknowledges the support of the NRI/CRGP of the U.S.D.A. R.D.B. gratefully acknowledges support from NIH and U.S.D.A..

References:

1). Ermler, U., Fritzsch, G., Buchanan, S.K. and Michel, H. (1994) *Stucture* 2: 925-936.
2). Williams, J.G.K. (1988) *Methods Enzymol.* 167: 766-778.
3). Tang, X.-S., Diner, B.A., Larsen, B.S., Gilchrist, M.L., Lorigan, G.A. and Britt, R.D. (1994) *Proc. Natl. Acad. Sci. USA* 91: 704-708.
4). Rippka, R., Deruelles, J., Waterbury, J.B., Herdman, M. and Stanier, R.Y. (1979) *J. of Gen.* Microbiol. 111: 1-61.
5). Tang, X.-S. and Diner, B.A. (1994) *Biochemistry*, 33: 4594-4603.
6). Rögner, M., Nixon, P.J. and Diner (1990) *J. Biol. Chem.* 265: 6189-6196.
7). Sanakis, Y., Petrouleas, V. and Diner, B.A. (1994) *Biochemistry* 33: 9922-9928.
8). Koulougliotis, D., Tang, X.-S., Diner, B.A. and Brudvig, G.W. (1995) *Biochemistry* 34:2850-2856.
9). Sturgeon, B.E. and Britt, R.D. (1992) *Rev. Sci. Instrum.* 63: 2187-2192.
10). Edmonds, D.T. (1977) *Phys. Rep.* 29C: 233-290.
11). Atashkin, A.V., Kawamori, A., Kodera, Y., Kuroiwa, S. and Akabori, K. (1995) *J. Chem. Phys.* in press.

MUTAGENESIS OF HIS D1-198 AND HIS D2-197 IN *SYNECHOCYSTIS PCC 6803*: EFFECTS ON THE PRIMARY DONOR OF PHOTOSYSTEM II

W.J. Coleman[a], P.J. Nixon[b], W.F.J. Vermaas[c] and B.A. Diner[a]

[a]E.I. DuPont de Nemours & Co., Central Research and Development, Experimental Station, P.O. Box 80173, Wilmington, DE 19880-0173 USA, [b]Dept. of Biochemistry, Imperial College, London, SW7 2AY, UK, [c]Dept. of Botany, Arizona State University, Tempe, AZ 85287-1601 USA

Introduction

Structural models of the PS II reaction center (RC) have been proposed based on an overall sequence similarity between the D1/D2 polypeptides of cyanobacteria, algae and plants and the L/M polypeptides of the purple non-sulfur bacteria [1,2]. Since detailed X-ray crystallographic structures are available for the RCs of the bacteria *Rhodopseudomonas viridis* [3] and *Rhodobacter sphaeroides* [4], it has been suggested that there may be a correspondence between His L173 of the bacterial RC and His D1-198 of the PS II RC, as well as between His M200 and His D2-197. In the bacteria, His L173 and His M200 are the ligands which coordinate the central Mg atoms of each of the bacteriochlorophylls (Bchls) comprising the so-called special pair. These two Bchls together form the primary donor of the bacterial RC [for a review, see ref. 5]. Mutating this His to Leu or Phe in the L-subunit (or its equivalent in the M-subunit) causes major changes in the structure and photochemistry of the bacterial primary donor [5-8]. It therefore might be expected that if the analogous His residues in the D1/D2 polypeptides have a similar function, then mutating these residues should also have a significant effect on the spectroscopic properties of P680, the primary donor of PS II (if P680 is structurally similar to the bacterial special pair). We have made a number of changes in the side chains of His D1-198 and His D2-197 by site-directed mutagenesis, and we report here the preliminary characterization of these mutants in the cyanobacterium *Synechocystis PCC 6803*.

Procedure

The *Synechocystis* psbA3 and psbD1 genes were mutated using the *in vitro* method of Kunkel et al. [9]. Deletion strains and plasmids [10,11] and protocols for mutagenesis and transformation [11,12] and isolation of PS II core complexes [13,14] have been described previously. A Joliot-type flash detection spectrophotometer was used for optical absorption/fluorescence measurements [15,16]. The PS II RC concentration was assayed on a per cell basis relative to WT by measuring the variable fluorescence yield of Chl in the presence of 20 μM DCMU and 20 mM NH_2OH in cells pre-treated with 0.3 mM *p*-benzoquinone and 0.3 mM potassium ferricyanide. For EPR spectra of the P680 triplet, PS II core complexes were resuspended in buffer containing glycerol and 40 mM MOPS (pH 7.0). Samples were chemically reduced with sodium dithionite and benzyl viologen [17] and frozen in quartz EPR tubes in darkness in liquid N_2 until use. Liquid-helium EPR measurements were performed on a Bruker ER200D spectrometer fitted with an Oxford cryostat as in [17].

P. Mathis (ed.), Photosynthesis: from Light to Biosphere, Vol. I, 779–782.

Results and Discussion

Table 1 summarizes the results of photoautotrophic growth tests and measurements of O_2-evolving activity and PS II reaction center concentrations in each of the mutants as a percentage of the WT levels.

Table 1. Comparison of PS growth, O_2-evolution and RC concentration between WT PS II and several mutants.

Mutant	PS Growth	O_2-Evolution	RC Conc.
WT	+	100%	100%
D1-His198Gln	+	100	80-100
Lys	-	100 (photoinhibited)	40
Cys	+	70-100	80-90
Ala	+	50-70	50-60
Asn	+ (poor)	~50	10-20
Glu	-	20-25	30-40
Ser	-	~20	27-33
Thr	-	~10	16-23
Val	-	~10	5-10
Leu	-	0	0
D2-His197Gln	+	40	80
Leu	-	0	0

The results summarized in Table 1 indicate that a number of the mutants are able to grow photosynthetically and evolve oxygen. However, unlike in the purple bacteria, the His to Leu mutation does not permit RCs to be assembled. In order to further examine the effects of these mutations on the PS II primary donor, we measured $P680^+$-P680 difference spectra, the kinetics of donor side charge recombination and rates of $P680^+$ reduction by the secondary donor, Y_Z. The optical data show a displacement of up to 2.5 nm (to the blue) in the $P680^+$-P680 difference spectra compared to WT. The kinetic data are consistent with decreases relative to WT of up to 90 mV in the reduction potential of $P680^+$/P680 (Nixon et al., manuscript in preparation). We also obtained low-temperature EPR spectra of the triplet state in purified PS II core complexes from some of the mutants. Preparations of several D1-H198 mutants yielded triplet spectra whose zero-field splitting (ZFS) parameters are virtually identical to WT PS II (Figure 1). The ZFS parameters are summarized in Table 2.

Based on the results of earlier mutagenesis experiments on His L173 and His M200 (M202) in the bacterial RC [6-8,18], we anticipated that there might be major spectral changes in the PS II primary donor as a result of changing His D1-198 and His D2-197. In the bacterial RC, when large hydrophobic residues such as Leu or Phe replace His L173 or His M200, bacteriopheophytin replaces Bchl on that half of the special pair (although initially no small side chains incapable of ligating the Mg atom were examined). Nevertheless, even though Chl binding is maintained in the PS II mutants that assemble RCs, it was expected that changing the central Chl ligand might have noticeable

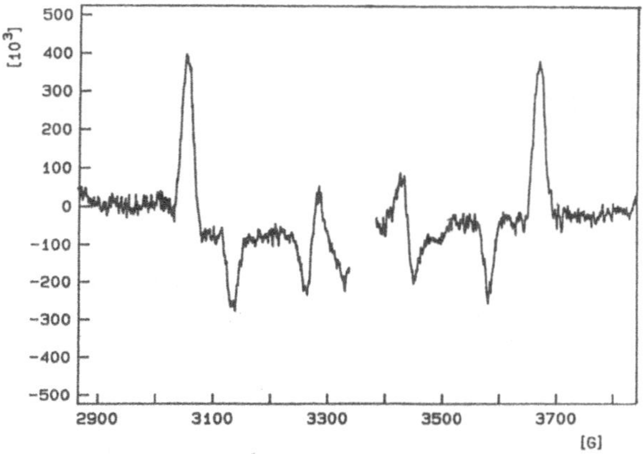

Figure 1. Triplet state EPR spectrum of purified PS II core complexes of the mutant D1-His198Ala. This is a light-minus-dark spectrum. The PS II concentration was approximately 1 mg ml^{-1}. Temperature:4.2 K; Microwave power:63 μW (35dB); Sweep width:1000 G; Frequency:9.416 GHz; Modulation frequency:100 kHz; Modulation amplitude:20 G; Time constant: 82 ms; Number of scans:10.

Table 2. Triplet ZFS parameters for several His D1-198 mutants.

Mutant	$\|D\|$	$\|E\|$
WT	287 x 10^{-4} cm^{-1}	44 x 10^{-4} cm^{-1}
D1-His198Gln	289	44
Cys	286	44
Ala	288	44
Glu	289	44

effects on the ZFS parameters of the triplet state, as has been reported for Chls ligated by various solvents [19]. In PS II, it appears that a residue having a small aliphatic side chain, such as Ala, can be introduced without perturbing the putative Chl binding site or removing the Mg atom from the Chl, possibly because there is enough room for a water molecule to enter and bind to the Mg atom. Larger hydrophobic residues such as Leu or Val, however, do not appear to be tolerated at this position. This differential sensitivity might result from structural differences between PS II and the bacterial RC in the Chl binding pocket. If, for example, there are fewer hydrogen bonds holding the Chls of the putative PS II dimer in place, then the PS II RC might be more sensitive than the bacterial RC to the introduction of large, destabilizing side chains or the loss of Mg coordination. Thus, a Mg-water-protein or Mg-protein bond might be required for stabilizing the assembly of the PS II RC. Most likely, D1-His198 and D2-His197 both coordinate Chls that are associated with a P680 pigment ensemble through exciton coupling [20]. Alternatively, one or the other His may yet coordinate a Chl on which the positive charge of $P680^+$ is localized. In such a case, however, one is forced to admit that the Mg^{2+} ligand has little influence on the properties of the primary donor. There is, however, precedent for this latter possibility in the bacterial RC, where substituting Gly for His L173 and His M200 (M202) in *Rb. sphaeroides* has little effect on the optical and redox properties of the primary donor (S. Boxer, personal communication).

Acknowledgments

B.A.D. gratefully acknowledges the support of the NRI/CRGP of the U.S. Dept. of Agriculture. P.J.N. was supported by the Royal Society and the BBSRC. W.F.J.V. was supported by the D.O.E. and the U.S.D.A.

References

1. Michel, H and Deisenhofer, J (1988) *Biochemistry 27*,1-7.
2. Trebst, A (1986) *Z. Naturforsch. 41*, 240-245.
3. Michel, H, Epp, O and Deisenhofer, J (1986) *EMBO J. 5*, 2445-2451.
4. El-Kabbani, O, Chang, C-H, Tiede, D, Norris, JR and Schiffer, M (1991) *Biochemistry 30*, 5361-5369.
5. Coleman, WJ and Youvan, DC (1990) *Annu. Rev. Biophys. Biophys. Chem. 19*, 333-367.
6. Bylina, EJ and Youvan, DC (1988) *Proc. Natl. Acad. Sci. USA 85*, 7226-7230.
7. Bylina, EJ and Youvan, DC (1990) In *Current Research in Photosynthesis* (Baltscheffsky, M, ed.) Vol. 1, pp. 53-59, Kluwer, Dordrecht.
8. Schenck, CC, Gaul, D, Steffen, M, Boxer, SG, McDowell, L, Kirmaier, C and Holten, D (1990) In *Reaction Centers of Photosynthetic Bacteria: Feldafing II* (Michel-Beyerle, ME, ed.) pp. 229-238, Springer Verlag, Berlin.
9. Kunkel, TA, Roberts, JD and Zakour, RA (1987) *Methods Enzymol. 154*, 367-382.
10. Nixon, PJ, Trost, JT and Diner, BA (1992) *Biochemistry 31*, 10859-10871.
11. Vermaas, W, Charité, J and Eggars, B (1990) In *Current Research in Photosynthesis* (Baltscheffsky, M, ed.) Vol. 1, pp. 231-238, Kluwer, Dordrecht.
12. Nixon, PJ and Diner, BA (1992) *Biochemistry 31*, 942-948.
13. Rögner, M, Nixon, PJ and Diner, BA (1990) *J. Biol. Chem. 265*, 6189-6196.
14. Tang, XS, Zheng, M, Chisholm, DA, Dismukes, GC and Diner, BA (1995), submitted.
15. Metz, JG, Nixon, PJ, Rögner, M, Brudvig, GW and Diner, BA (1989) *Biochemistry 28*, 6960-6969.
16. Joliot, P, Béal, D and Frilley, B (1980) *J. Chem. Phys. 77*, 209-216.
17. van Mieghem, FJE, Satoh, K and Rutherford, AW (1991) *Biochim. Biophys. Acta 1058*, 379-385.
18. Bylina, EJ, Kolaczkowski, SV, Norris, JR and Youvan, DC (1990) *Biochemistry 29*, 6203-6210.
19. Schaafsma, TJ (1982) In *Triplet State ODMR Spectroscopy* (Clarke, RH, ed.) pp. 291-365, Wiley, New York.
20. Durrant, JR, Klug, DR, Kwa, SLS, van Grondelle, R, Porter, G and Dekker, JP (1995) *Proc. Natl. Acad. Sci. USA 92*, 4798-4802.

SOLUBILIZATION AND PURIFICATION OF PHOTOSYNTHETIC UNITS FROM RHODOSPIRILLUM RUBRUM
Montpellier, France, 20-25 August 1995

Janine KESSI and Reinhard BACHOFEN
Institute of Plant Biology, University of Zürich, Zollikerstrasse 107, CH-8008 Zürich, Switzerland.

1.Introduction

In recent years a detailed structural characterization of purified reaction centers (RC) and light harvesting complexes (LH) from several purple non sulfur bacteria has been achieved using X-ray diffraction or analysis of electron microscopy imaging. In this work we present the solubilization and purification of photosynthetic units from *Rhodospirillum rubrum* (LHI-RC-complex) in a stable native form using the short-chain lecithin diheptanoyl-sn-phosphatidylcholine (DHPC). The complex is characterized by its absorption and near-IR CD spectra, SDS-PAGE and bacteriochlorophyll content.

2.Material and methods

The carotinoidless mutant $G9^+$ of *Rhodospirillum rubrum* was grown in Sistrom medium at 30 ^oC, constant pH and increasing light intensity as described in [1]. Chromatophores were prepared in 50 mM Tris.HCl pH 8.0, and the cells were broken using a French Press.
Chromatophores were solubilized with 15 mM DHPC in 10 mM Tris.HCl pH 7.5 containing 150 mM NaCl and 2 mM $MgCl_2$.
The photosynthetic units (LHI-RC complex) were purified using successively DEAE-Sepharose chromatography and ultracentrifugation in a sucrose gradient. The column was eluted with a 50 mM NaCl step gradient in Tris.HCl pH 8.0 containing 2 mM $MgCl_2$ and 2 mM DHPC. Ultracentrifugation was performed in the same buffer containing 150 mM NaCl and the sucrose gradient was 0.1-0.4 M.
Absorption spectra were recorded with an Uvicon 860 spectrophotometer and near-IR CD spectra with a J-500 C spectropolarimeter.
The activity of the reaction center was determined by light-induced photobleaching measured at 600 nm using an Aminco DW-2 spectrophotometer in the dual mode equipped with an actinic light source. The changes in absorbance at 600 nm compared with 640 nm upon illumination were recorded.
The bacteriochlorophyll (BChl) was extracted three times using acetone:MeOH 7:2 and the absorption coefficient of the 870 nm peak was computed using an absorption coefficient at 772 nm in acetone:MeOH 7:2 of 68 $mM^{-1}.cm^{-1}$ [2] to be 146 $mM^{-1}.cm^{-1}$. The LHI-BChl/RC ratio were evaluated analyzing the absorption spectra and using absorption coefficients of 146 $mM^{-1}.cm^{-1}$ and 330 $mM^{-1}.cm^{-1}$ for the 870 nm peak and the 802 nm peak, respectively.

P. Mathis (ed.), Photosynthesis: from Light to Biosphere, Vol. I, 783–786.

3.Results and discussion

Chromatophores of *Rhodospirillum rubrum* were solubilized using the short-chain lecithin DHPC. The solubilization curves are sigmoidal reaching a plateau at a detergent/protein ratio of 2-3 µmol/mg, corresponding to 75-90 % solubilized protein. Proteins solubilized at a DHPC/protein ratio of 3 µmol/mg are stable up to a ratio of 8 µmol/mg (Fig.1). Absorption spectra, near-IR CD spectra and RC-activity of the solubilized membrane show the high preservation of the threedimensional structure of the BChl-containing proteins in presence of DHPC (Fig. 2 and 3). A specific solubilization of the RC-complex attempted by varying the DHPC/protein ratio from 0.4 to 8.0 µmol/mg was unsuccessful as well as attempts to separate the RC itself from the LHI-RC-complex using either gel filtration, ion-exchange chromatography or ultracentrifugation in a sucrose gradient. These results indicate the inability of DHPC to separate the two complexes.

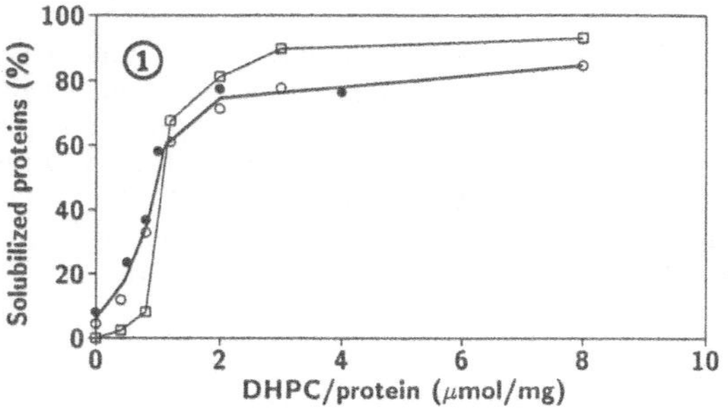

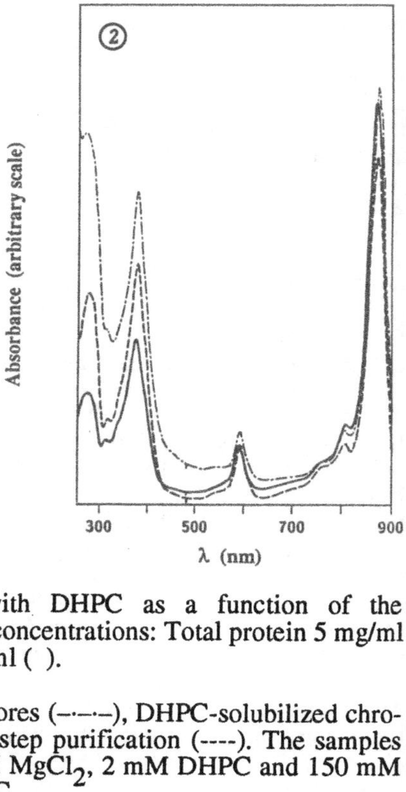

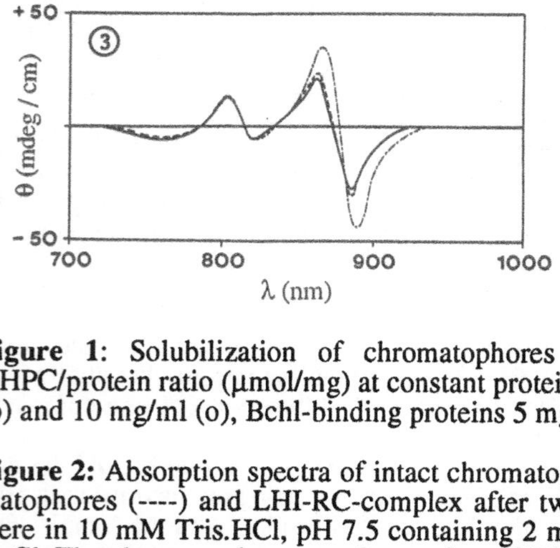

Figure 1: Solubilization of chromatophores with DHPC as a function of the DHPC/protein ratio (µmol/mg) at constant protein concentrations: Total protein 5 mg/ml (o) and 10 mg/ml (o), Bchl-binding proteins 5 mg/ml ( ).

Figure 2: Absorption spectra of intact chromatophores (—·—·—), DHPC-solubilized chromatophores (----) and LHI-RC-complex after two step purification (----). The samples were in 10 mM Tris.HCl, pH 7.5 containing 2 mM $MgCl_2$, 2 mM DHPC and 150 mM NaCl. The chromatophore sample was free of DHPC.

Figure 3: Near-IR CD spectrum of intact chromatophores (—·—·—), DHPC-solubilized chromatophores (----) and LHI-RC-complex after two step purification (----). The samples were as in Figure 2.

A LHI-RC-complex was then purified from chromatophores solubilized at a DHPC/protein ratio of 3 µmol/mg by DEAE-Sepharose chromatography followed by sucrose gradient centrifugation. About 50% of the total of protein including succinate dehydrogenase, NADH dehydrogenase and bc_1-complex were separated from the BChl-containing proteins by the DEAE-Sepharose chromatography. Succinate dehydrogenase and NADH-dehydrogenase were eluted in broad peaks at 150 mM NaCl, the BChl-containing proteins in a sharp peak at 200 mM NaCl and the bc_1-complex in a small peak at 250 mM NaCl. The most important contaminant of the sharp BChl-containing protein peak was the F_0F_1-ATPase, which was separated in a second purification step using ultracentrifugation in a 0.1-0.4 M sucrose gradient (Fig. 4). narrowness and symmetry of the BChl-binding protein peak obtained in the second purification step give evidence for a good homogeneity of the particle size of the purified LHI-RC complex. Its SDS-PAGE analysis is presented in Fig.5b. It shows the three bands L, M and H of the subunits of the RC-complex at molecular masses of about 25 kDa, 28 kDa and 32 kDa as well as the broad band located near the front corresponding to the subunits of the LHI-complex together with a diffuse but rather heavily stained band H' located slightly above the H-subunit of the RC-complex (33-35 kDa) of so far unknown origin. SDS-PAGE analysis of the ATPase peak is presented in Fig. 5c. Apart from the contamination of the BChl-binding proteins it agrees well with SDS-PAGE analysis of the purified ATPase presented in [3].

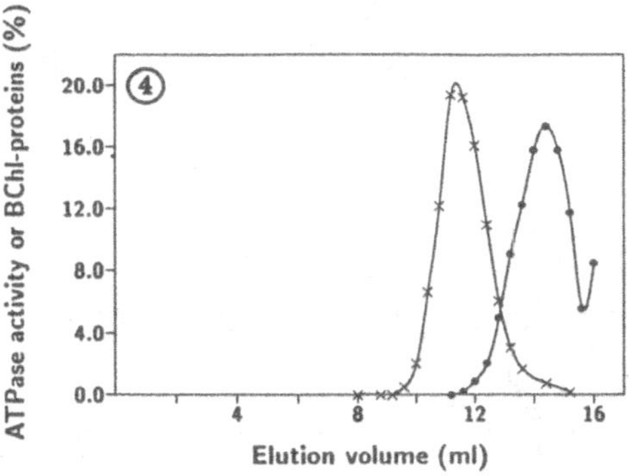

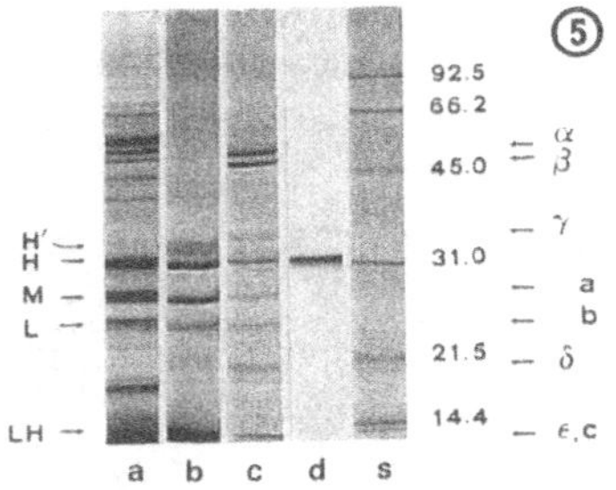

Figure 4: Ultracentrifugation patterns of BChl-binding proteins and F_0F_1-ATPase (60'000 x g, 48 h, 0.1-0.4 M sucrose gradient). (o) F_0F_1-ATPase activity; (x) BChl-binding proteins (absorbance at 870 nm).

Figure 5: SDS-Page of: a) Chromatophores; b) LHI-RC-complex (L, M and H designate the subunits of the reaction center and LH the subunits of the light harvesting complex); c) F_0F_1-ATPase (α, β, γ, ε, δ, a, b and c designate the subunits of the ATPase, molecular weights given in (Slooten and Vandenbranden, 1989); d) heme staining of the DEAE-Sepharose fraction eluting at 250 mM NaCl; s) standard proteins with their molecular weights in kDa.

Absorption and near-IR CD spectra of the purified complex show the good preservation of its threedimensional structure (Fig. 2, 3). Furthermore, the spectra were stable up to 6 days after purification when the samples were kept at 4 ^{0}C in the dark, demonstrating the high stability of the complex in presence of 2 mM DHPC. The RC-activity is similar to that measured in intact chromatophores.

The BChl was quantitatively extracted from samples of BChl-binding proteins and its concentration was computed using an absorption coefficient in acetone:MeOH 7:2, v/v of 68 $mM^{-1}.cm^{-1}$ at 772 nm [2]. This yielded an in vivo absorption coefficient at 870 nm of 146 $mM^{-1}.cm^{-1}$, which is in good agreement with literature values [2, 4]. The absorption coefficient of the 802 nm peak was calculated from the absorption coefficient of the RC at 870 nm [2] using an A_{802}/A_{870} ratio of 2.3 [1, 2, 5], to be 330 $mM^{-1}.cm^{-1}$. Using these absorption coefficients the LHI-BChl/RC-complex ratio was evaluated by analyzing the absorption spectra and resulted in a value of 24.4 ± 0.5 mol LHI-BChl/mol RC for the two step purified LHI-RC-complex. This value is close to the one from *Rps marina* [6]. Assuming a stoichiometry of 2 BChl/αβ polypeptide [6], [7], our data fit well with results from purified LHI-complexes analyzed by electron microscopy from *Rsp. marina* [8] *and R. molischianum* [9], where the presence of 12 αβγ and αβ-complexes organized in a ring surrounding a central core with a diameter of about 50 A is suggested. This structure is also suggested by low resolution projections obtained from EM images of natural membranes from *Rps. viridis* [10] and *E. halochloris* [11], all showing a 6-fold symmetry. A similar structure of 12 αβ pairs is furthermore proposed for the threedimensional crystals of the LHI-complex from *Rb. sphaeroides* [12].

We expect that the isolation of this new, stable, well-defined DHPC-solubilized LHI-RC-complex will be useful for biophysical studies on energy transfer and may lead to complexes that can be crystallized.

4.References

1. Kessi J. (1994) Ph. D. Thesis, University of Zürich
2. Van der Rest M. and Gingras G. (1974) J. Biol. Chem. 249, 6446-6453
3. Slooten L. and Vandenbranden S. (1989) Biochim. Biophys. Acta 975, 148-157
4. Picorel R.,. Bélanger G. and Gingras G. (1983) Biochemistry 22, 2491-2497
5. Snozzi M. and Bachofen R. (1979) Biochim. Biophys. Acta 546, 236-247
6. Meckenstock R.U., Brunisholz R.A. and Zuber H. (1992a) FEBS Lett. 311, 128-134
7. Broglie R.M., Hunter C.N., Delepelaire P., Niederman R. A., Chua N.H. and Clayton R.K. (1980) Proc. Natl. Acad. Sci. 77, 87-91
8. Meckenstock R.U., Krusche K. Brunisholz R.A. and Zuber H. (1992b) FEBS Letters 311, 135-138
9. Boonstra A.F., Germeroth L. and Boekema E.J. (1994) Biochim. Biophys. Acta 1184, 227-234
10. Miller K.R. (1982) Nature 300, 53-55
11. Engelhardt H., Engel A. and Baumeister W. (1986) Proc. Natl. Sci. USA 83, 8972-8976
12. Nunn R.S., Artymiuk P.J., Baker P.J., Rice D.W. and Hunter C.N. (1992) J. Mol. Biol. 228, 1259-1262

THE TRIPLET STATE $^3P_{680}$ IN PHOTOSYSTEM II STUDIED WITH TIME-RESOLVED EPR

A. Kamlowski, A. van der Est and D. Stehlik
Institute of Experimental Physics, Free University Berlin, Arnimallee 14, 14195 Berlin, Germany

1. Introduction

The nature of the primary donor, P_{680}, in the Photosystem II (PSII) reaction center (RC) in not known with certainty. However, similarities between the PS II D_1D_2cytb_{559}-complex and the RC of purple bacteria (bRC) suggest that P_{680} is a chlorophyll-a (Chl) dimer [1]. Although such a structure is often assumed, it is hard to reconcile with various experimental data, most notably with low temperature EPR on the triplet state of the primary donor, $^3P_{680}$. Its' fine structure (zfs) parameters, $|D|$ and $|E|$, resemble those of a Chl monomer [2] and the principal axes show a peculiar orientation to the membrane, i. e. with an angle of 30° between the Chl plane and the membrane similar to the orientation of one of the accessory bacteriochlorophylls in the bRC [3]. In addition, optical studies reveal a weaker excitonic coupling for P_{680} [4] than for the 'special pair' in bRC. In this contribution the temperature dependence of the spin-polarized transient EPR spectra of $^3P_{680}$ in D_1D_2cytb_{559}-complexes will be presented. Drastic changes in the spectra occur with increasing temperature which are interpreted as evidence for delocalization of the triplet excitation over two or more Chl molecules with different orientations.

2. Material & Method

The EPR spectra reported here are from D_1D_2cytb_{559}-comlexes which were prepared from spinach according to [5] and kindly provided by A. R. Holzwarth and co-workers. The preparations consist of approx. six Chl per two pheophytin-a (Pheo).
Time-resolved EPR experiments were performed at X-Band (microwave frequency $\nu_{MW} \sim 9$ GHz) as described in [6]. Light excitation was achieved with a Nd:YAG-pumped MOPO-system (GCR170 and MOPO710 from Spectra Physics) using an excitation wavelenght of $\lambda_{exc} = 640$ nm. Note that the EPR signal is recorded in direct detection mode as a function of both the magnetic field and the time after laser excitation: $S = S(\mathbf{B}_0, t)$. The EPR spectra have been obtained from the full data set by integrating the EPR signal over a chosen time window (see [6]), and are displayed with absorptive (A) and emissive (E) signals up and down, respectively.

3. Results & Discussion

Fig. 1 shows the spin-polarized transient EPR spectrum of D_1D_2cytb_{559}-complexes at 100 K together with a simulation. The spectrum is readily attributed to $^3P_{680}$

P. Mathis (ed.), Photosynthesis: from Light to Biosphere, Vol. I, 787–790.

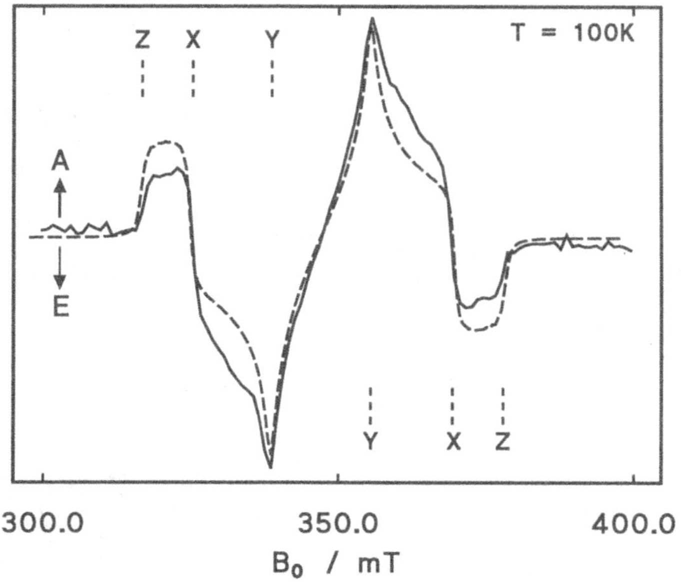

Figure 1. X-Band transient spin-polarized EPR spectra of D_1D_2cytb_{559}-complexes from spinach at T = 100 K (solid line), and simulation of the spectrum (dashed line) with T_0-population of $^3P_{680}$ and the following zfs parameters: $|D| = 0.0290\,cm^{-1}$ and $|E| = 0.0044\,cm^{-1}$. Experimental parameters: $\nu_{MW} = 9.7977$ GHz; 100.0 mT sweep with 101 field points around 349.6 mT; 128 events were averaged at each field point; $\lambda_{exc} = 640$ nm, digital boxcar-window: 0.3...2.0 μs after laser excitation. The signal decays with $\tau \approx 2\,\mu$s (microwave power: 10 mW) at all field orientations.

because the spin-polarization pattern AEEAAE (A - absorption, E - enhanced emission at the canonical orientations X, Y or Z) is characteristic for selective T_0-population which results from S-T_0-mixing within the primary radical pair, $P^{+\cdot}_{680}$Pheo$^{-\cdot}$, followed by charge recombination to $^3P_{680}$. This assignment is supported by the correspondance of the experimental spectrum and the simulation which uses exclusive T_0-population of the triplet state and zfs parameters identical to those of a Chl-monomer triplet state. The rise-time of the EPR signal was found to be of the order of 100...200 ns for T $\leq$ 100 K (not shown) which agrees with optical studies of the magnetic field dependent lifetime of $P^{+\cdot}_{680}$Pheo$^{-\cdot}$ (see e.g. [7]).

Fig. 2 shows the temperature dependence of the spin-polarized EPR spectrum in D_1D_2cytb_{559}-complexes for three selected temperatures. Raising the temperature from 100 K to 160 K and 200 K alters the shape of the transient EPR spectrum completely. The low temperature AEEAAE-pattern still contributes at 160 K, but the amplitudes at the Z-orientations decrease relative to those in the central part of the spectrum.

At 200 K a drastically different spin-polarized EPR spectrum is observed which also appears to contribute to the spectrum at 160 K. At 200 K the spectrum exhibits nearly no intensity where the outer field regions of the low temperature Z-orientation occur. The remaining E/A-pattern is rather broad and without obvious edge features.

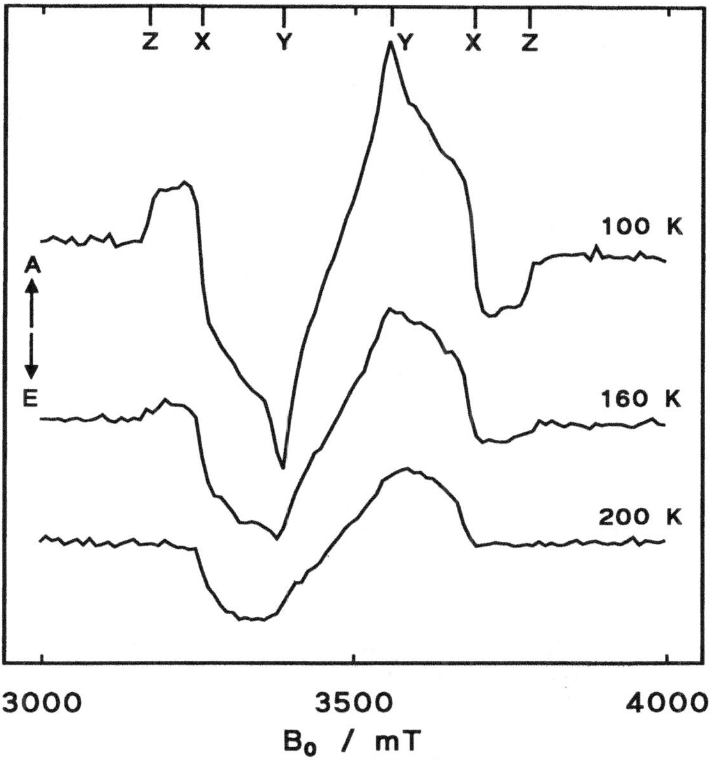

Figure 2. Temperature dependence of the time-resolved EPR spectrum in D_1D_2cytb_{559}-complexes. Experimental parameters as in Fig. 1 except for: $\nu_{MW}(160\text{ K}) = 9.7904$ GHz and $\nu_{MW}(200\text{ K}) = 9.7861$ GHz.

It is worth noting, that in the whole temperatures range between 10 K and 250 K the kind of a spin-polarized EPR spectra shown in Fig. 2 is detectable. This is in contrast to the temperature dependence of the primary donor triplet state, $^3P_{700}$, in Photosystem I (PS I), in which the EPR spectrum is nearly lost in the noise level between 50 K and 200 K [8]. This points to different dynamics being responsible for the temperature induced changes of the spectra of $^3P_{700}$ and $^3P_{680}$.

In oriented PS II membranes with doubly reduced quinone acceptor, Q_AH_2, preliminary studies indicate a corresponding temperature effect. In addition, the orientation of the principal axes of $^3P_{680}$ changes drastically at 200 K with respect to 10 K (not shown).

One way to account for the observed temperature dependence is a linear superposition of two spectra: the "usual" monomeric low temperature spectrum which dominates at $T \leq 30$ K and the high temperature spectrum at $T \geq 200$ K. At temperatures above ~ 30 K the observed spectra can be simulated as weighted sums of both spectra. In principle, this behaviour can be explained with an activation barrier for triplet delocalization in line with a temperature-dependent equilibrium between a 'localized' and 'delocalized' triplet excitation. Correspondingly, the spectrum at low temperature remains assigned to a triplet excitation trapped on a monomeric Chl.

The narrowed spectrum at high temperature, on the other hand, can be interpreted as a triplet excitation which involves fast hopping or delocalization among different molecular sites. In PS I the high temperature EPR spectrum of $^3P_{700}$ can be understood as a result of triplet delocalization between the two Chls of the primary donor P_{700} [8], in which both Chl planes are oriented perpendicular to the membrane. In contrast, simulation of the high temperature spectrum in PS II observed here requires delocalization between sites with differently oriented principal axes systems. What remains intact is the exclusive T_0-population of $^3P_{680}$ consistent with the formation from the primary radical pair, $P_{680}^{+\cdot}Pheo^{-\cdot}$. An arrangement which nearly simulates the 200 K spectrum is constituted by two Chls: one has an orientation reminescent of the accesory BChl in bRC and the other is oriented as one of the special pair BChls. Such an arrangement is similar to that proposed in ref. [9]. This arrangement can account for the narrowing of the spectrum but could not simulate satisfactory the broad E/A pattern. This may be due to further spin dynamics not considered yet.

4. Conclusion

It has been shown that up to 200 K a spin-polarized transient EPR spectrum of $^3P_{680}$, the triplet state of the primary donor in PS II, can be observed in D_1D_2cytb_{559}-complexes. The temperature dependence is due to a superposition of two different spectra. The spectrum at 200 K which is narrowed with respect to 10 K, cannot be accompained with a monomeric character of P_{680}.

5. Acknowledgement

We thank A.R. Holzwarth for providing us with the PS II RC from spinach. This work was supported by the Deutsche Forschungsgemeinschaft (Sfb 312, TP A1).

6. References

[1] Michel, H. and Deisenhofer, J. (1988) Biochemistry 27, 1–7.
[2] Van Mieghem, F.J.E., Satoh, K., Rutherford, A.W. (1991) Biochim. Biophys. Acta 1058, 379–385.
[3] Rutherford, A.W., Patterson, D.R., Mullet, J.E. (1981) Biochim. Biophys. Acta 635, 205–214.
[4] Hillmann, B., Brettel, K., van Mieghem, F.J.E., Kamlowski, A.,Rutherford, A.W., Schlodder, E. (1995) Biochemistry 34, 4814-4827, and refs. therein.
[5] Van Leeuwen, P.J., Nieveen, M.C., van de Meent, E.J., Dekker, J.P., van Gorkom, H.J., (1991) Photosyn. Res. 28 149–153.
[6] Bittl, R., van der Est, A., Kamlowski, A., Lubitz, W., Stehlik. D. (1994) Chem. Phys. Lett. 220, 349–358.
[7] Volk, M., Gilbert, M., Rousseau, G, Richter, M., Ogrodnik, A., Michel-Beyerle, M.-E. (1993) FEBS. Lett 336, 357–362.
[8] Sieckmann, I., Brettel, K., Bock, C., van der Est, A., Stehlik, D. (1993) Biochemistry 32, 4842–4847.
[9] Noguchi, T., Inoue, Y., Satoh, K. (1993) Biochemistry 32, 7186–7195.

INSERTIONAL INACTIVATION OF THE CHLOROPLAST *psbE* GENE IN THE GREEN ALGA *Chlamydomonas reinhardtii.*

F. Morais, J. Barber and P.J. Nixon
Photosynthesis Research Group, Department of Biochemistry
Imperial College of Science, Technology and Medicine
London SW7 2AY, UK

1. Introduction

In thylakoid membranes, photosystem II (PSII) catalyses the photooxidation of water to molecular oxygen and the reduction of plastoquinone to plastoquinol (reviewed in [1]). One of the components of PSII is cytochrome b559, a transmembrane haemprotein located within the PSII reaction centre complex and probably arranged in a heterodimer structure of the α and β subunits (reviewed in [2]). As with all the other three components of the isolated PSII reaction center complex (D1, D2 and PsbI), the α and β subunits of cytochrome b559 are encoded by plastid genes, namely *psbE* and *psbF*, respectively (3). The function of this cytochrome is still uncertain but recent evidence suggests that it probably plays a role in protecting the reaction centre from photoinhibitory damage, by providing PSII with an alternative, possibly cyclic, route of electron flow (reviewed in [4]).

The unicellular green alga, *Chlamydomonas reinhardtii*, is an excellent organism to study the structure/function relationships within the photosynthetic apparatus. This organism can be grown heterotrophically in the presence of acetate as a carbon source and, in addition, its plastome can be transformed using the biolistic technique (5). Recently it was found that in *C. reinhardtii* the *psbE* and *psbF* are not co-transcribed with the downstream *psbL* and *psbJ* genes (operon *psbEFLJ*) as happens in all higher plant and cyanobacterial species so far examined (6, 7). Instead, the *psbE* gene lies approximately 7.3kb away from *psbF* and is transcribed in the opposite direction. This unique organisation allows us to construct *psbE* mutants without affecting the transcription of the other genes, therefore helping us to understand the function of this cytochrome in an eukaryotic chloroplast.

To obtain insights into the role of this cytochrome, we have inactivated the *psbE* gene of *Chlamydomonas reinhardtii* through directed chloroplast gene disruption via biolistic transformation and we have examined some of the properties of the *psbE*-inactivated transformants.

2. Procedure

Chlamydomonas reinhardtii wild-type strain CC-125 (kindly provided by Prof. E. H. Harris of the *Chlamydomonas* Genetics Center, Department of Botany, Duke University, Durham, NC, USA) was grown in Tris-acetate-phosphate (TAP) and high-salt minimal (HSM) media (8). Bombardment of the cells with tungsten particles coated with a plasmid

P. Mathis (ed.), Photosynthesis: from Light to Biosphere, Vol. I, 791–794.

DNA was performed essentially as described in (9). The transformants were maintained on TAP medium containing 100µg/ml spectinomycin and 10µM DCMU.
Plasmid p78, which consists of the 15.1kb *Pst*I-4 fragment from the chloroplast genome of *C. reinhardtii* (CC-125) cloned into pBR322 (kindly provided by Prof. E.H. Harris), and the spectinomycin-resistance cassette (kindly provided by Dr. M. Goldschmidt-Clermont) were used for the generation of plasmid constructs.
Thylakoid membranes from *C. reinhardtii* cells were prepared according to (11), in the presence of a cocktail of protease inhibitors. Samples containing 5µg chlorophyll were subjected to electrophoresis on 12% polyacrylamide gels containing 6M urea. Protein blotting onto nitrocellulose filters, incubation with polyclonal antibodies and colour development to visualise immunodecorated protein bands were performed as described in (12).

3. Results and Discussion

Plasmid pF1 was constructed by restriction digestion of p78 with *EcoR*I followed by religation. This derivative of p78 contains the *psbE* gene with upstream and downstream flanking sequences of 8kb and 2kb, respectively. Insertion of a spectinomycin-resistance cassette (*aadA*) at an *AlwN*I site, 23 codons within the coding region of the *psbE* gene led to the generation of plasmid pF1aad (Figure 1). After bombardment of *C. reinhardtii* wild-type cells with this construct, using a particle gun, five putative transformants were selected for their ability to grow on TAP plates containing spectinomycin.

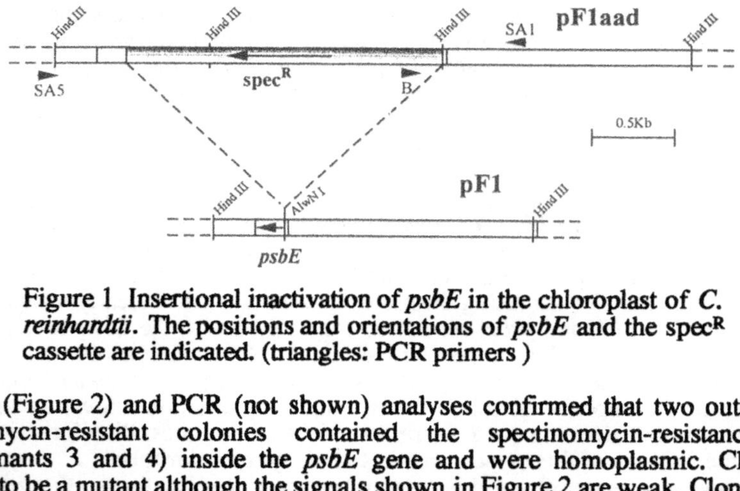

Figure 1 Insertional inactivation of *psbE* in the chloroplast of *C. reinhardtii*. The positions and orientations of *psbE* and the spec^R cassette are indicated. (triangles: PCR primers)

Southern (Figure 2) and PCR (not shown) analyses confirmed that two out of the five spectinomycin-resistant colonies contained the spectinomycin-resistance cassette (transformants 3 and 4) inside the *psbE* gene and were homoplasmic. Clone 2 also appeared to be a mutant although the signals shown in Figure 2 are weak. Clone 1 showed an intact *psbE* gene and did not contain the spec^R cassette. Presumably, a spontaneous mutation gave rise to spec^R in this strain. Clone 5 appeared to be a mutant also. However an additional *aad*A hybridisation band is present in this mutant (data not shown) so it was not analysed further.
Mutants 3 and 4 were not able to grow in minimal medium (data not shown) and light-saturated oxygen evolution assays (data not shown) indicated that they lacked any PSII-mediated electron transport activity (water to benzoquinone). Western immunoblotting analysis (Figure 3) showed the absence of D1, D2, CP47, 33kDa and a reduced amount of CP43 (~5%) in thylakoid membranes of mutants 3 and 4, isolated from either dark or

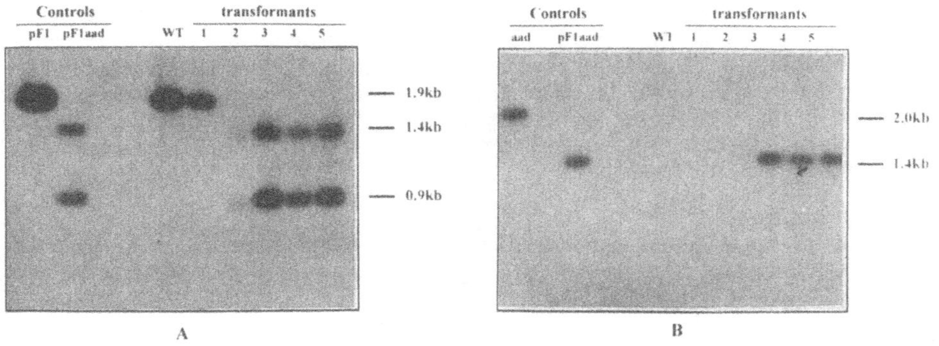

Figure 2 DNA gel blot hybridisation of the total DNA from wild-type (WT) and putative transformants (1-5), following digestion with *Hind*III. Controls were ~1ng of pF1, pF1aad and a 2kb *aadA* cassette: (A) probed with a *psbE*-specific probe and (B) probed with *aad*A-specific probe.

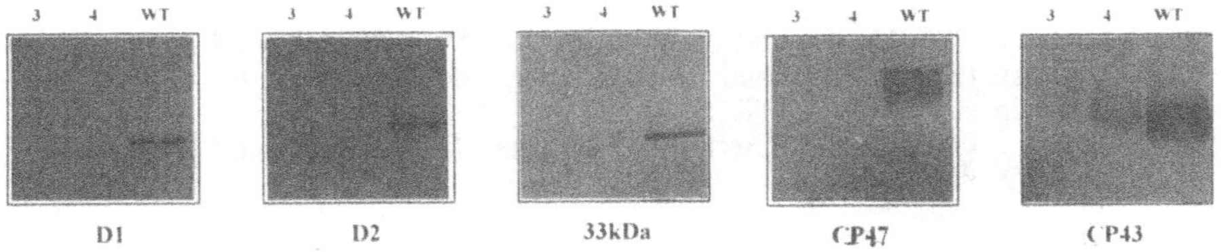

Figure 3 Western blots of thylakoid membrane polypeptides of wild-type (WT) and transformants (3 and 4), immunostained with antibodies raised against D1, D2, 33kDa, CP47 and CP43 PSII proteins.

low light grown cells. The absence of these proteins in dark grown cells excludes the possibility of a rapid light-induced degradation of PSII. All these results indicate that a functional PSII complex is not assembled, which leads to an inability of the cells to grow photoautotrophically.

Genetic modification of cytochrome b559 indicates that it plays a crucial structural rule in the assembly and/or stability of PSII. In *Synechocystis* 6803, deletion of the *psbEFLJ* operon, or of the individual genes *psbE* and *psbF*, has resulted in mutant strains with no PSII activity (reviewed in [4]). In all these cyanobacterial mutants, steady-state levels of the two reaction center polypeptides D1 and D2 were very low, whereas three other PSII proteins, CP47, CP43 and 33kDa were present at near normal amounts (13). However, in the presence of appropriate protease inhibitors, D1 was detected at significant levels (14). In contrast with this result we did not detect D1, CP47 and 33kDa in the presence of protease inhibitors. CP43 was detected at ~5% of Wt levels.

Differences in phenotype between *C. reinhardtii* and cyanobacterial photosynthetic mutants have previously been observed. Disruption of the *psbO* gene encoding the 33kDa *psbO* protein leads to the rapid turn-over of the PSII complex in *C. reinhardtii* (15). However, loss of this protein in *Synechocystis* still allows photoautotrophic growth and oxygen-evolving activity is retained although at a reduced level (16, 17). Similarly, loss of the *psaC* protein of PSI in *C. reinhardtii* prevents stable accumulation of this complex (5) whereas in the cyanobacterium *Anabaena variabilis* the PSI complex is stable, but deficient in photochemical activity (18). More recently, it has been shown that *psbK* mutants in *C. reinhardtii* are unable to accumulate substantial amounts of PSII and to grow photoautotrophically (19), in clear contrast with similar cyanobacterial mutants (20, 21). As suggested by (19), it may not necessarily be possible to elucidate the functional role of a protein in the chloroplast of eukaryotic cells from the phenotypes observed in photosynthetic mutants from cyanobacteria. However, there is no evidence yet for differences in missense mutants.

Acknowledgments

This work was financially supported by BBSRC and the Royal Society. F.M. is a recipient of a JNICT/PRAXIS XXI studentship from Portugal and P.J.N. is a Royal Society University Research Fellow.

References

1 Vermaas, W. (1993) Ann. Rev. Plant. Physiol. Plant Mol. Biol. 49, 457-481
2 Cramer, W.A., Tae, G.-S., Furbacher, P.N. and Böttger, M. (1993) Physiol. Plant. 88, 705-711
3 Herrmann, R.G., Alt, J., Schiller, B., Widger, W.R. and Cramer, W.A. (1984) FEBS Lett. 176, 239-244
4 Whitmarsh, J., Samson, G. and Poulson, M. (1994) in Photoinhibition of photosynthesis (Baker, N.R. and Bowyer, J.R., eds.) pp.75-93, bIOS Scientific Publishers Limited, Oxford, UK.
5 Takahashi, Y., Goldschmidt-Clermont, M., Soen, S.-Y., Franzen, L.G. and Rochaix, J.-D. (1991) EMBO J. 10, 2033-2040
6 Alizadeh, S., Nechushtai, R., Barber, J. and Nixon, P. (1994) Biochim. Biophys. Acta 1188, 439-442
7 Mor, T.S., Ohad, I., Hirschberg, J. and Pakrasi, H.B. (1995) Mol. Gen. Genet. 246, 600-604
8 Harris, E.H. (1989) The Chlamydomonas Sourcebook. Academic Press, Inc. New York
9 Boynton, J.E. and Gillham, N.W. (1993) Methods in Enzymol. 217, 367-382
10 Goldschmidt-Clermont, M. (1991) Nucl. Acids Res. 19, 15, 4083-4089
11 Diner, B.A. and Wollman, F.-A. (1980) Eur. J. Biochem. 110, 521-526
12 Shipton, C.A. and Barber, J. (1991) Proc. Natl. Acad. Sci. USA 88, 6691-6695
13 Pakrasi, H.B., Diner, B.A., Williams, J.K.G. and Arntzen, C.J. (1989) Plant Cell 1, 591-597
14 Shukla, V.K., Stanbekova, G.E., Shestakov, S.V. and Pakrasi, H.B. (1992) Mol. Microbiol. 6, 947-956
15 Mayfield, S.O., Bennoun, P. and Rochaix, J.-D. (1987) EMBO J 6, 313-318
16 Mayes, S.R., Cook, K.M., Self, S.J., Zang, Z. and Barber, J. (1991) Biochim. Biophys. Acta 1060, 1-12
17 Burnap, R.L. and Sherman, L.A. (1991) Biochemistry 30, 30440-30446
18 Mannan, R.M., Whitmarsh, J., Nyman, P. and Pakrasi, H.B. (1991) Proc. Natl. Acad. Sci. USA 88, 10168-10172
19 Takahashi, Y., Matsumoto, H., Goldschmidt-Clermont, M. and Rochaix, J.-D. (1994) Plant Mol. Biol. 24, 779-788
20 Ikeuchi, M., Eggers, B., Shen, G., Webber, A., Yu, J., Hirano, A., Inoue Y. and Vermaas, W. (1991) J. Biol. Chem. 266, 11111-11115
21 Zhang, Z.-H., Mayes, S.R., Vass, I., Nagy, L. and Barber, J. (1993) Photosyn. Res. 38, 369-377

INDUCTION OF FLUORESCENCE IN ISOLATED REACTION CENTERS OF *RHODOBACTER SPHAEROIDES*

Szabolcs Osváth, Gábor Laczkó, Pierre Sebban* and Péter Maróti
JATE University, Szeged, Institute of Biophysics
Egyetem utca 2., Szeged, Hungary H-6722
*Centre de Génétique Moléculaire CNRS, Gif-sur-Yvette, France F-91198

1. Introduction

The conversion of light into electrochemical energy is mediated in bacterial photosynthesis by a protein-pigments complex called the reaction center (RC) [1]. The primary electron donor (P) is a dimer of bacteriochlorophyll and is located on the periplasmic side of the membrane. The two quinones of the acceptor complex on the cytoplasmic side of the membrane in *Rhodobacter (Rb.) sphaeroides* are chemically identical but functionally different [2]. The primary quinone (Q_A) functions only as a one-electron carrier under physiological conditions. It is tightly bound into a relatively hydrophobic part. The secondary quinone (Q_B), however, is situated in a more polar region and functions as a two-electron acceptor.

The energy of absorbed photons in the antenna is transferred to P and initiates charge separation with a quantum yield of near unity [3]. The separated charges are stabilized by subsequent redox states of $P^+Q_A^-$ and $P^+Q_B^-$. If P^+ is re-reduced by an exogenous electron donor (D), the RC will be ready for additional charge separation. The proton-coupled electron transfer leads to cyclic reduction of quinones (photocycle) [4].

Fluorescence of bacteriochlorophyll (BChl) either from the RC [5-7] or from the antenna [8-10] can be used to monitor redox changes in the RC. The relatively large changes in BChl fluorescence yield of photosynthetic bacteria have been used mainly to study energy transfer and trapping (for a recent review, see [11]) but rarely as a probe for electron transfer in the RC.

The dimer fluorescence is centered around 910 nm [12-14] and has much lower quantum yield (4.10^{-4}) than the light-harvesting BChl in chromatophores or whole cells (2.10^{-2} to 1.10^{-1} depending on the strain) [6,13].

Here, the changes in yield of dimer fluorescence upon rectangular shape of illumination ("Kautsky-effect", see [15]) is used to track electron and proton transfer in the photocycle of isolated RC from *Rb. sphaeroides*. The application of laser diode for monochromatic excitation permitted of 1) the significant reduction of scattering light, 2) very high cyclic rate of the quinones under steady-state conditions and 3) the rapid onset of illumination thus the exact determination of initial value of fluorescence. The effectiveness of dimer fluorescence as a supplementary method to absorption change techniques in study of electron and proton transfer reactions in RCs at room temperature is demonstrated .

P. Mathis (ed.), Photosynthesis: from Light to Biosphere, Vol. I, 795–798.

2. Materials and methods

Reaction centers from *Rb. sphaeroides* were prepared as described earlier [16]. Secondary quinone activity was reconstituted by UQ_6 (Sigma). Ferrocene and reduced cytochrome c (cyt c^{2+}) were used as external electron donor to P^+.

Fluorescence excited by a laser diode (Laser Diode, Inc. Type LCW-100, emission wavelength 810 nm, power 500 mW) was detected by a red-sensitive photomultiplier (Hamamatsu Type R3310-03) and digitalized using a Philips PM 3350A digital storage oscilloscope connected to an IBM AT compatible PC [17]. The photomultiplier was protected from the exciting light by a red cut-off filter (850 nm) and a 915 nm interference filter of approximately 15 nm bandwidth. Charge recombination kinetics and cytochrome turnover were monitored by absorption changes at 865 nm and 550 nm, respectively.

3. Results and Discussion

3.1 Minimum model

On illumination of RC from *Rb. sphaeroides* after a period of darkness, the intensity of dimer fluorescence measured at 915 nm passes through a characteristic sequence of changes before settling to its steady-state value. The quantitative interpretation meets with complications of rates and pool-sizes both on acceptor and donor sides of the RC. The minimum model for interpretation of basic phenomena of fluorescence induction should contain at least two cofactors (P and Q_A) with four possible redox states which are arranged according to their relative fluorescence yields in Figure 1. The kinetic traces of fluorescence reflect the competition of the photochemical (light) reaction (rate k_I) with the thermal reactions of recombination (rate k_{PA}) and charge stabilization on both the acceptor and the donor sides of the RC (rates k_A and k_D, respectively). While the donor pool makes one-electron chemistry, the quinone pool is involved in two-electron chemistry [1,2,4]. To assure formal symmetry between the acceptor and donor pool in our model, Q_B was taken as part of the acceptor pool (i.e. it consists of the quinone pool and Q_B). The observed yield of fluorescence is determined by the concentrations of the fluorescence emitting species P and of fluorescence quencher Q_A. In the absence of external donor (D) to the dimer, the yield of fluorescence decays with an observed rate constant of $k_{obs} = k_I + k_{PA}$, because the nonfluorescent ($P^+Q_A^-$) states are accumulating. In the presence of very fast ($k_D >> k_I$, k_A) electron donor to P^+, however, the yield of fluorescence increases with observed rate constant equal to that of the light reaction ($k_{obs} = k_I$), as strongly fluorescent PQ_A^- state is formed. (The $P^*Q_A^-$ excited state has higher probability to deactivate via fluorescence emission because the photochemistry is blocked). Reduced cytochrome c^{2+} in excess to the RC (>20 cyt c/RC) at low ionic strength (5 mM) assured orders of magnitudes faster re-reduction of P^+ ($k_D \approx 10^6$ s^{-1}) than by charge recombination ($k_{PA} \approx 12$ s^{-1}) under our conditions (data not shown).

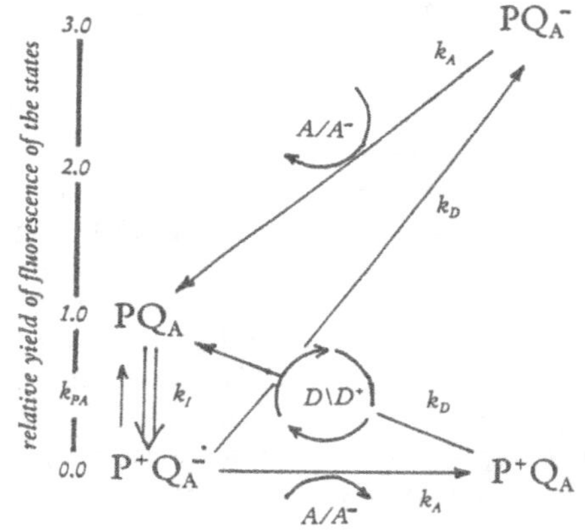

Figure 1 Reaction paths with rates and fluorescence levels of PQ_A redox states connected to donor (D/D^+) and acceptor (A/A^-) pools.

The rise and drop in fluorescence yield is limited either by the (acceptor and donor)pool sizes (Fig. 2) and/or by the rates of the kinetics (Fig. 3). The effects will be treated separately.

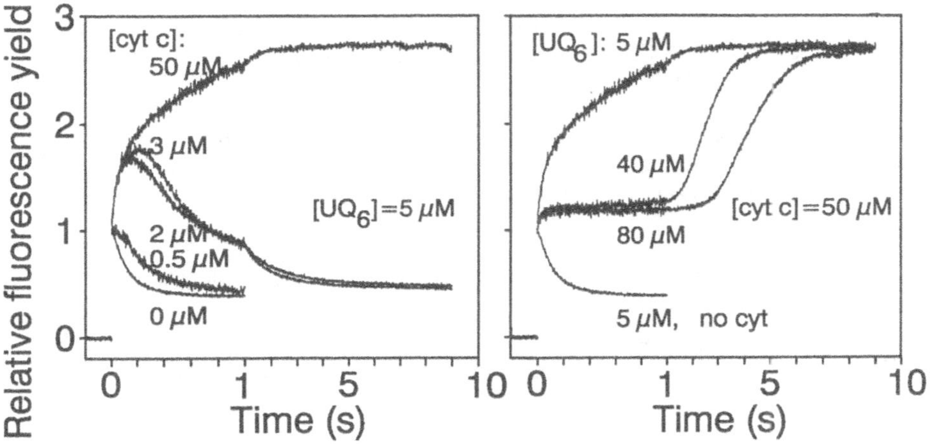

Figure 2 Traces of dimer fluorescence (910 nm) of 1 μM RC upon rectangular shape of illumination with laser diode (810 nm) attributed to donor (left) and acceptor (right) titration on dual time scale.

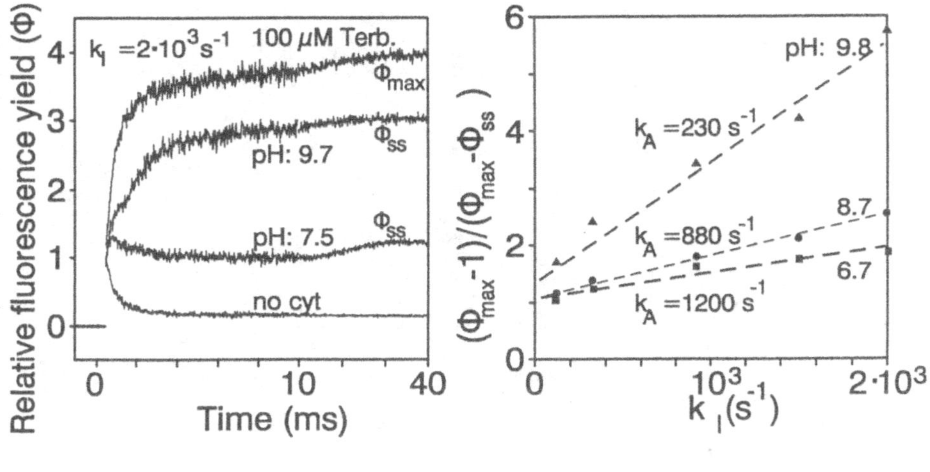

Figure 3 Induction of dimer fluorescence upon high illumination ($k_I = 2.10^3$ s^{-1}) (left). Steady-state level of fluorescence vs. attenuated k_I (right).

3.2 Competition of pool-sizes

Small acceptor pool-size was adjusted by addition of 5 μM UQ_6 to 1 μM RC at the beginning of donor (cyt c) titration (Fig. 2 left). Although the fluorescence started to increase (indicating that $k_D >> k_A$), the donor pool soon exhausted and the yield dropped. At higher donor concentration (50 μM), however, the donor pool became larger than the acceptor pool and the yield reached its maximum under steady state conditions.

The process was tried to be reversed by titrating now UQ_6 to the RC with 50 μM cyt c (Fig. 2 right) but as many as 80 μM UQ_6 could only compensate but not reverse the effect of the donor. The duration of the initial lag phase (where the yield is close to one) is the measure of the smaller pool-size and the steady-state level will show which pool is larger: the donor pool if the fluorescence is high and the acceptor pool if it is low.

3.3 Kinetic limitation

If the photochemical rate becomes not negligible to k_A (the smallest rate in our case), k_A will be the bottleneck in the induction of fluorescence. It is indicated by dramatic changes in fluorescence upon increase of pH (Fig. 3 left). The changes in k_A can be determined from the steady-state fluorescence values, ϕ_{ss} using the minimum model (Fig. 3 right).

4. Conclusions

The yield of dimer fluorescence from bacterial RCs is controlled by the redox states of P and Q_A. The induction of fluorescence depends on the pool-sizes and kinetic limitations both on the acceptor and the donor sides of the RC. Similarly to green plants, the Kautsky-effect of bacterial RC is a rapid and sensitive intrinsic probe of electron and proton transfer when combined with other parallel measurements.

Acknowledgements

This work was supported by grants from the Hungarian Science Foundation (OTKA 17362/95), Foundation for Hungarian Education and Research (FEFA III 1034), Balaton (NP-715/95-F) and International Human Frontier Science Program (RG-329/95 M).

References

1 Cramer, W.A. and Knaff, D.B. (1990) Energy Transduction in Biological Membranes, Springer-Verlag, New York
2 Shinkarev, V.P. and Wraight, C.A. (1993) in: The Photosynthetic Reaction Center, (Deisenhofer H. and Norris J., eds.) pp. 193-255, Academic Press
3 Wraight, C.A. and Clayton, R.K. (1974) Biochim. Biophys. Acta 333, 246-260
4 Okamura, M.Y. and Feher, G. (1992) Annu. Rev. Biochem. 61, 861-896
5 Clayton, R.K. (1972) Proc. Nat. Acad. Sci. USA 69, 44-49
6 Clayton, R.K. (1980) Photosynthesis: Physical Mechanisms and Chemical Patterns, Cambridge University Press
7 Schenck, C.C., Parson, W.W., Holten, D. and Windsor, M. (1981) Biochim. Biophys. Acta 635, 383-392
8 Godik, V.I. and Borisov, A.Yu. (1977) FEBS Lett. 82(2), 355-358
9 Amesz, J. and Vasmel, H. (1986) in: Light Emission by Plants and Bacteria, (Govindjee, Amesz J. and Fork D.Ch., eds.) pp 423-450, New York, Academic Press
10 Hoff, A.J. and Fischer, M.R. (1993) Molecular Physics 78(4), 799-819
11 Van Grondelle, R. Dekker, J.P., Gillbro, T. and Sundström, V. (1994) Biochim. Biophys. Acta 1187, 1-65
12 Zankel, K.L., Reed, D.W. and Clayton, R.C. (1968) Proc. Natl. Acad. Sci. USA 61, 1243-1249
13 Clayton, R.K. (1977) Photosynthetic Organelles, Special Issue of Plant @ Cell Physiol. 87-96
14 Stadnichuk, I.N. and Lukashev, E.P. (1982) Mol. Biol. (Russ.) 16(5), 991-996
15 Govindjee (1995) Australian Journal of Plant Physiology 22(2), 131-160
16 Maróti, P. and Wraight, C.A. (1988) Biochim. Biophys. Acta 934, 314-328
17 Osváth, Sz., Laczkó, G., Sebban, P. and Maróti, P. (submitted to Photosynth. Research)

ACID-BASE TITRATION OF ISOLATED REACTION CENTERS OF *RHODOBACTER SPHAEROIDES*

László Kálmán, Pierre Sebban* and Péter Maróti
JATE University, Institute of Biophysics, Egyetem utca 2, Szeged, Hungary H-6722
*Centre de Genetique Moleculaire, CNRS, Gif/Yvette, France

1. Introduction

The flash induced proton uptake by the isolated reaction center (RC) from *Rhodobacter sphaeroides* has been determined directly by a variety of methods such as spectrophotometry, conductimetry and pH electrode [1,2]. These studies offered similar results: the stoichiometry of light induced proton binding depended on the redox state of the RC and on the pH of the solution but was always substoichiometric i. e. less than one proton per RC. It was explained by pK shifts of the nearby protonatable amino acid residues exposed to the influence of the negative charge on the semiquinone. Due to the relatively short range of electrostatic interaction and screening effects of charged residues, the contribution of only a few protonatable residues in the process had to be assumed to explain the observed proton binding stoichiometry.

Another type of proton uptake, the protonation of the protein as a whole, upon lowering the pH in the dark, has not been the subject of much attention [3] although the acid-base titration of water soluble proteins has become a routine tool in the physico-chemistry of macromolecules [4]. The property of buffering capacity of RC is of great importance as the measured values can be compared to derived data from the structure (number and pKs of protonatable groups) and related to the availability of protonatable amino acids to the water phase. As not all of the protonatable residues of the protein are exposed to the aqueous phase, the relevance of exposed and buried groups can be studied with this method.

2. Materials and Methods

RCs from *Rhodobacter sphaeroides* were prepared and purified as described earlier [1]. The buffer (10 mM Tris) and (if required) the ionic detergent (LDAO) was removed and replaced by non-ionic detergents (Triton X-100) after long (48 hours) dialysis at 4°C. The buffering capacity of concentrated (10-20 μM) RC was measured by addition of known amounts of HCl with a syringe into a 1 cm rectangular air-tight cuvette equipped with glass/calomel pH electrode and stirring bar. Changes in pH were detected after complete mixing (about 1 s).

The buffering capacity is a differential ratio of

$$\beta = -\frac{\Delta [H^+]}{\Delta (pH)} \qquad (1)$$

P. Mathis (ed.), Photosynthesis: from Light to Biosphere, Vol. I, 799–802.

where $\Delta[H^+]$ is the increment of strong acid added to the solution and $\Delta(pH)$ the resultant increment in pH. Expressing $\Delta(pH)$ by the activities of the H^+-ions in the solution ($pH=-\log a_H$) and expanding the ratio by f_H, the activity coefficient of H^+ -ions, we obtain

$$\beta = \frac{2.3 a_H}{f_H} \cdot \frac{f_H . \Delta[H^+]}{\Delta a_H} \qquad (2)$$

Out of $\Delta[H^+]$ mole injected hydrogen ions, only a fraction of $f_H.\Delta[H^+]$ will be used to increase the H^+-ion activity of the aqueous solution (Δa_H) and to decrease the activities of OH^- (Δa_{OH}) and deprotonated buffer ions (Δa_B) i.e. partly to fill the proton sinks. The remaining part of injected hydrogen ions, $(1-f_H).\Delta[H^+]$, is deactivated and will contribute to the increase of buffering capacity. Replacing $f_H.\Delta[H^+] = \Delta a_H - \Delta a_{OH} - \Sigma\Delta a_B$ and introducing the equilibrium proton dissociation constants, K_i, we get

$$\beta = 2.3 \frac{10^{-pH}}{f_H} \left(1 + \frac{K_{water}}{a_H^2} + \Sigma_i [B]_i \frac{K_i}{(K_i + a_H)^2} \right) \qquad (3)$$

in units of mol H^+/pH. The first two terms in the bracket refer to the contribution of water ($K_{water} = 10^{-14}$ M^2 ionproduct of water and $[H^+] = 10^{-pH}$ concentration of hydrogen ions) and the summation should be extended to all protonatable groups (buffers, amino acids of the RC) of dissociation constants K_i in the solution. If the contribution of water is negligible, the buffering capacity is linearly proportional to the concentration of the buffering groups ($[B_i]$) and inversely proportional to the proton activity coefficient of the hydrogen ions (f_H).

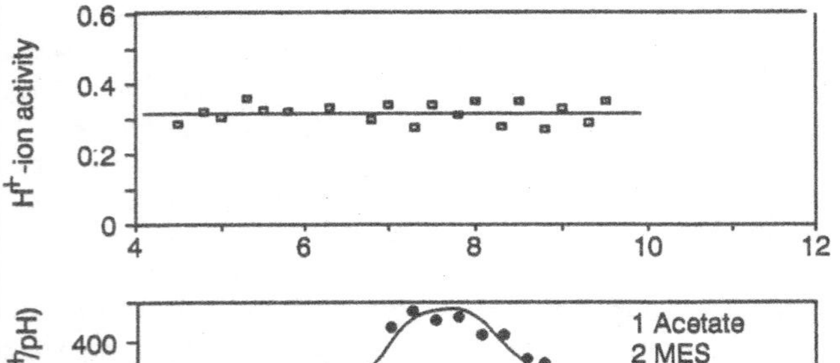

Figure 1. Determination of buffering capacity (β) and H^+-ion activity (f_H) in 0.03 % Triton X-100 detergent.

3. Results and Discussion

3.1. Determination of f_H. The proton activity coefficient in dilute aqueous solutions of amino acids (e.g. histidine) or water soluble protein of known structure (e.g. lysozyme) is unity, thus the acid-base titration experiments analyzed according to Eq. (3) directly offered the expected literature values (data not shown).

Micellular solutions, however, decreased the activity coefficients of H^+-ions significantly. The buffering capacity was monitored by a set of pH buffers as reporter molecules which covered wide pH range (Fig. 1.). $f_H=0.33$ was measured in the non-ionic detergent, Triton X-100

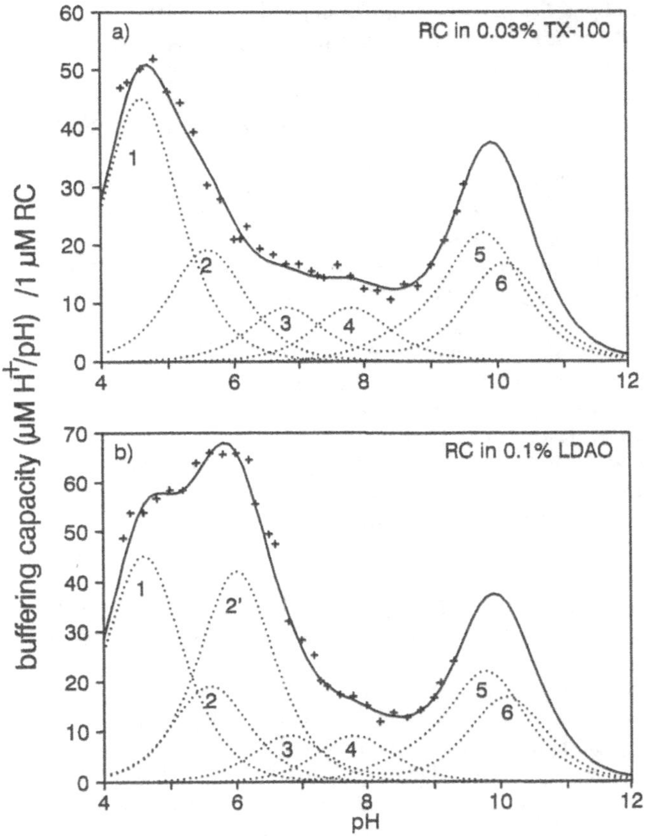

c)

#	pK	in LDAO	in TX-100	Group
1	4.5	59	59	Asp+Glu
2	5.6	25	25	LDAO
2'	6.0	55	-	LDAO
3	6.8	12	12	His
4	7.8	10	10	Tris
not shown	8.6	3	3	Cys
5	9.8	28	28	Tyr
6	10.1	22	22	Lys

Figure 2 PH-dependence of measured and calculated buffering capacities of RC in 0.03% Triton X-100 (a), 0.1% LDAO (b) and table of fitting parameters (c).

slightly above its critical micelle concentration, independently on pH. The decreased activity of H^+-ions can be rationalized by distorted hydrogen-bond structure of water: all water molecules continue to be hydrogen-bonded to their neighboring water molecules, but the intermolecular links can be bent and stretched to produce irregular and varied networks. The detergent molecules penetrate into and thus break the extended network of hydrogen bonds of water molecules.

3.2. Buffering capacity of RC. The buffering capacity of detergent solution was always subtracted from that of the RC solution of the same detergent concentration. The acid-base titration curve of RC was reversible up to pH 10.0, therefore it represented thermodynamic equilibrium. The pH-dependence of β of RC (referred to 1 μM concentration) in Triton (Fig. 2a) and LDAO (Fig. 2b) detergents were decomposed into protonatable groups (Fig. 2c) with assumptions of
- close to solutional pKs of the amino acids,
- $f_H=0.33$ (i.e. the RC in the micelle does not lower this value measured with detergents only),
- additional group (probably bound LDAO) contributes to the protonation around pH 5.6 and
- 10 μM Tris buffer remained (as trace) after dialysis.

Detergent (LDAO) molecules bound to the RC was observed in the RC crystals [5]: most of detergents covered the wide hydrophobic surface around the centre of the protein including the transmembrane helices of the subunits M, L and H and parts of the cofactors. They could be all replaced by Triton via dialysis in our experiments.

4. CONCLUSIONS

The acid-base titration of the RC protein offers new tool among routinely applied (spectrophotometric, conductimetric, electric, etc.) methods and opens new perspectives in study of protonation events.

The LDAO had two populations in RC solution - presumably bound and free with a small but significant pK shift between them (5.6 and 6.0, respectively).

All of the protonatable groups available in RC [1,2] were required to explain the relatively high buffering capacity observed in our acid-base titration studies. The pH-dependence of β could be well fitted by the sum of Henderson-Hasselbalch-type titration curves with solutional pKs of the titratable groups of the RC. It can be concluded, that even residues that from x-ray structures appear to be totally buried can exchange protons with the aqueous bulk phase on an extended time scale.

ACKNOWLEDGEMENTS: This work was supported by the Hungarian Science Foundation (OTKA 17362/95), Balaton project (NP-715/95-F), Foundation for Hungarian Education and Research (FEFA III 1034) and International Human Frontier Science Program (RG-329/95 M).

References

1 Maróti, P. and Wraight, C.A. (1988) Biochim. Biophys. Acta 934, 314-328 and 329-347
2 McPherson, P.H., Okamura, M.Y. and Feher G. (1988) Biochim. Biophys. Acta 934, 348-368
3 Fabian, M., Chamorovsky, S.K., Zakharova, N.I., Selivanov, V.A. and Kononenko, A.A. (1981) Mol. Biol. (USSR) 15 (2), 439-446
4 Tanford, Ch. (1966) Physical Chemistry of Macromolecules, John Wiley & Sons, Inc., New York, London, Sydney.
5 Roth, M., Lewit-Bentley, A., Michel, H., Deisenhofer, J., Huber, R. and Oesterhelt, D. (1989) Nature (London) 340, 659-662

REGULATION OF PHOTOSYSTEM II: EFFECTS OF pH AND LIGHT ON FLASH-INDUCED ABSORPTION CHANGES AT 820 nm

Anja Krieger

Section de Bioénergétique (CNRS URA 1290) DBCM, Bât. 532, C.E.A. Saclay, 91191 Gif-sur-Yvette, France
on leave from: Julius von Sachs Institut, Lehrstuhl für Botanik I, Mittl. Dallenbergweg 64, 97082 Würzburg, Germany

Introduction. When higher plants are exposed to excess light, processes are induced which lead to an increase in the dissipation of excitation energy by non-photochemical routes. These processes occur at high proton concentrations in the thylakoid lumen. It has been demonstrated that reversible inhibition of PS II can be induced by a pH-dependent release of a Ca^{2+} ion from the donor side of PS II. In Ca^{2+} depleted PS II the O_2 evolving complex is inhibited, such that the efficiency to donate electrons to P_{680}^+ is greatly reduced. In parallel the midpoint potential of Q_A is shifted from -80 mV to +65 mV [1].
In active centres, P_{680}^+ is reduced by the water-splitting complex via tyrosine Z with a polyphasic kinetic, the rate constant for the reduction depends on the S-state of the manganese cluster. In the states S_0 and S_1, P_{680}^+ is reduced to 85% with a $t_{1/2}$ = 20 ns, 10 % with 50 ns and a very small amount with a µs halftime. In S_2 and S_3, 40% of the total signal are reduced with a halftime of 50 ns, 40% with 260 ns and about 20% with a halftime of 30 µs [2]. Upon inactivation of the water-splitting complex by e.g. incubation of PS II at pH 4.0, the ns components of the P_{680}^+ reduction are slowed down, their amplitude is reduced drastically and the overall kinetic is dominated by an approximately 120-150 µs component which is thought to reflect a $Q_A^- P_{680}^+$ recombination reaction [3].

Measurements of fluorescence decay kinetics have shown that inactive PS II is characterized by a fluorescence behaviour similar to those of so-called 'open' centres in which Q_A is oxidised. Excitation energy is quenched by charge recombination in inactive reaction centres. Measurements of absorption changes at 820 nm, which reflect P_{680}^+ reduction kinetics, should answer the question whether a recombination reaction occurs in the reaction centre of PS II after inhibition of the water-splitting complex and whether such reactions allow further P_{680}^+ formation at high light intensities and thereby efficient energy dissipation.
We have measured P_{680}^+ reduction at 820 nm under repetetive excitation in thylakoid membranes in the presence of a proton gradient and in low pH-treated PS II enriched membrane fragments. The effect of additional background illumination during the measurement was studied in the presence and absence of electron acceptors.

Materials and Methods. PS II enriched membrane fragments [4] (PS II particles) and thylakoid membranes were prepared from spinach. PS II particles were treated at low pH as described in [1], in thylakoid membranes a proton gradient was established in the dark by hydrolysis of ATP. Measurements of flash-induced absorption changes at 820 nm were performed as described [5].

P. Mathis (ed.), Photosynthesis: from Light to Biosphere, Vol. I, 803–806.

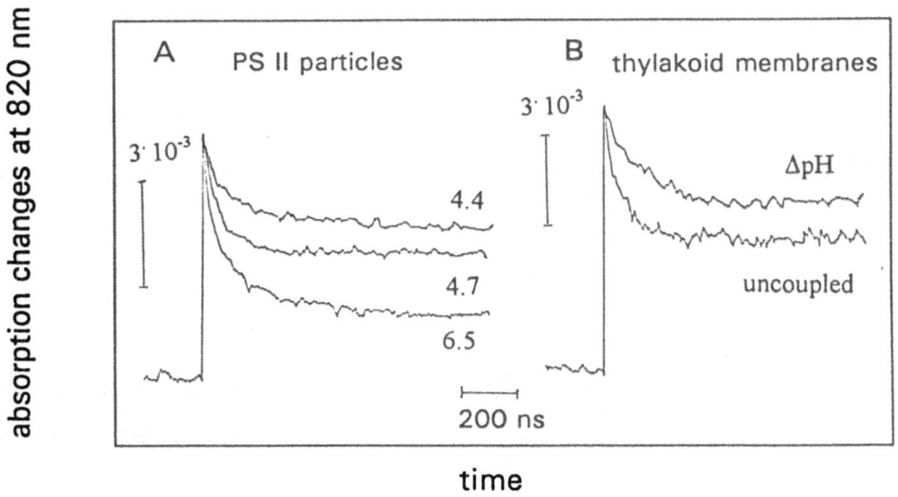

Fig. 1. Absorption changes at 820 nm induced by non-saturating laser flashes (300 ps puls width at half amplitude, excitation energy: 0.35 mJ/s, repetition rate 5Hz) as a function of pH in PS II particles and thylakoid membranes, bandwith of 20 MHz, average of 64 signals. **A:** PS II particles preincubated at pH 4.4, 4.7 and 6.5, respectively. All measurements were carried out at pH 6.5. The suspension contained 46 µg Chl/ml, 1 mM p-phenylbezoquinone (p-PBQ), 4 mM ferricyanide (Fecy) and 0.05 % dodecylmaltoside to reduce scattering. **B:** Thylakoid membranes in the presence of a ΔpH built by hydrolysis of ATP and uncoupled thylakoid membranes. The suspension contained 40 µg Chl/ml, 4 mM Fecy, 40 mM NH_4Cl, and 50 % glycerol to reduce scattering.

Results. Fig. 1 shows the time course of flash-induced absorption changes at 820 nm under repetitive illumination in PS II particles and thylakoid membranes measured in the presence of electron acceptors. In PS II particles incubated at pH 6.5 and uncoupled thylakoid membranes, a fast phase with $t_{1/2}$ in the ns-range contributes largely to the reduction kinetics. P_{700}, which contributes to absorption changes at 820 nm, was oxidised in thylakoid membranes by ferricyanide and so P700 does not contribute to the signal.
Low pH-treatment does not change the total amplitude of the signal but influences the reduction kinetics. The amplitude of the ns kinetics is reduced in PS II particles after incubation at pH 4.4 or 4.7 and in thylakoids in the presence of a ΔpH. The reduction kinetics are dominated by a slower component ($t_{1/2}$ = 120-150 µs, data not shown).
The effects of background illumination with continuous light on flash-induced absorption changes at 820 nm in PS II particles in the presence of Fecy and p-PBQ are shown in Fig. 2. For both, active and inactive samples, the ΔA820 decreased with increasing backgound light. In active samples (pH 6.5) the signal is suppressed at high intensity background light (I=4000 µmol quanta $m^{-2}s^{-1}$). In inactive centres, pretreated at pH 4.7 or 4.4, no such saturation was obtained. Even at high light intensities, the total amplitude of the signal only decreased to 55 % (pH 4.4) and 44 % (pH 4.7) of the value without background light. A similar behaviour was observed in thylakoid membranes in the presence of a proton gradient. Short term illumination with background light even at high intensities, as used for the measurements shown here, caused no irreversible changes in the ΔA820 kinetics neither in active nor in pH-treated samples.
Fig. 2B shows the light saturation behaviour of the amplitude of ΔA820 signals of low pH-treated PS II particles in the presence and absence of electron acceptors. In the absence of

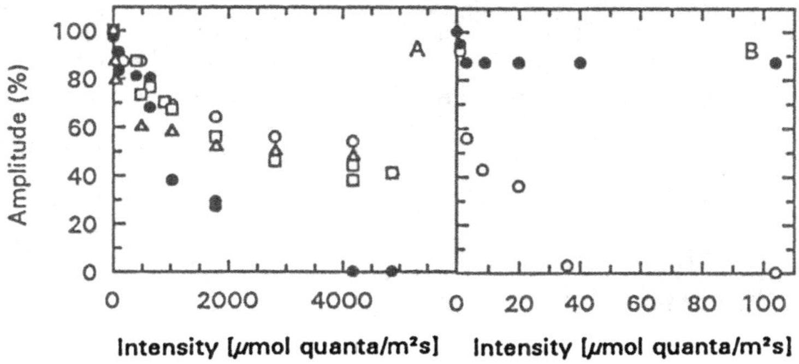

Fig. 2. Total initial signal amplitude of ΔA820 in PS II particles as a function of the intensity of background illumination. Samples were illuminated during the measurements of flash-induced absorption changes with continious light. **A**: PS II particles were pretreated at pH 4.4 (open circles) or 4.7 (squares) or directly incubated at pH 6.5 (closed circles). 4 mM ferricyanide and 1 mM p-PBQ were used as electron acceptors, Thylakoid membranes in the presence of a proton gradient without electron acceptor (triangle). **B**: PS II particles were pretreated at pH 4.4 and the measurements were performed in the absence (open circles) and in the presence (filled circles) of electron acceptors. All amplitudes are normalised to the control value (no background light).

electron acceptors, the flash-induced signal is suppressed. Absorption changes are no longer observed at a background light intensity of 40 µmol quanta $m^{-2}s^{-1}$. In the presence of electron acceptors, 15 % of the total amplitude disappeared. These 15 % reflect probably PS II reaction centres which have no contact to the artificial electron acceptors. In active samples in the absence of electron acceptors no signal could be seen implying that all centres are blocked in a 'closed' state within the first few flashes.

Discussion. Inactivation of the water-splitting complex by pH treatment of PS II particles or a proton gradient in thylakoid membranes is reflected in measurements of flash-induced absorption changes at 820 nm by a reduction and finally complete disappearence of the ns components of the P_{680}^{+} reduction (Fig. 1). The ns component is replaced by a ca. 120µs component which is thought to reflect the recombination of Q_A^{-} and P_{680}^{+} [3].
In active samples under background illumination, the ΔA820 signal disappears at high light intensities in the presence of electron acceptors (Fig. 2A). In these samples, the amplitude of the ns component is not affected over a broad range of intensities of the background light, indicating that the electron donation from the watersplitting complex functions and the electron transport is blocked due to acceptor side limitation. The yield of the primary charge separation in PS II is reduced if Q_A^{-} is present [5,6]. In inactivated PS II, flash-induced signals of ΔA820 is reduced by ca. 50 % and stay then constant during an increase of the light intensity. The reduction of the amplitude up to illumination of 700 µmol quanta $m^{-2}s^{-1}$ seems to reflect the same phenomenon in active and inactive PS II. It is known from previous measurements that in the presence of a proton gradient heterogenity of the reaction centres exists: a percentage of reaction centres gets inactivated while the other centres stay active indicating an all-or-none mechanism.
It can be concluded that in pH-inactivated PS II primary charge separation still occurs whereas this is inhibited in active PS II. The question arises which reactions could be responsible for the remaining 50 % of the signal found in pH treated samples. In the absence of electron acceptors (Fig. 2B), ΔA820 is already suppressed at low intensity background light.

This indicates that even without an additional electron acceptor an electron transport between acceptor- and donor side occurs which is not possible in active centres. First the electron donation from the water splitting complex and electron transfer from Tyr_Z to P_{680}^+ is inhibited and alternative reactions donating electrons to P_{680}^+ can occur. Second it had been demonstrated [1] that the midpoint potential of Q_A is more positive in inactive samples. In reaction centres with Q_A in such a "high potential form", a recombination between Q_A^- and P_{680}^+ could occur without creating an excited state of P_{680}.
In the presence of electron acceptors, recombination reactions in the PS II reaction centre are much more efficient (Fig. 2). Electrons are transported between acceptor and donor side probably mediated by p-PBQ. p-PBQ is reduced at the acceptor side and donates via an unidentified redox component electrons to P_{680}^+.

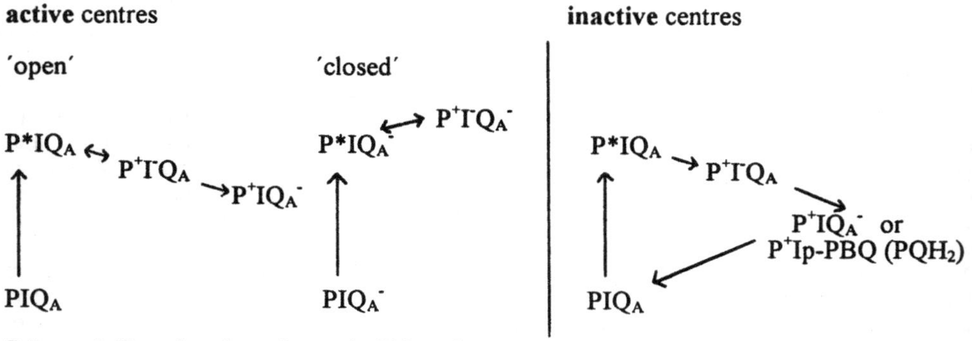

Scheme 1. Reactions in active and pH-inactivated photosystem II.
Absorption of a photon of light in active 'open' centres and inactive centres results in a high quantum yield of charge separation. This leads to charge stabilization by electron transfer from $Pheo^-$ to Q_A. In the case of inactive centres an efficient charge recombination reactions occurs between acceptor and donor side of PS II either directly ($P^+Q_A^-$ recombination) or via reduced plastoquinone or p-PBQ in the experimental system. In active centres in the presence of Q_A^- the yield of charge separation is suppressed during high intensity background illumination [5,6].

Conclusions. During high intensity illumination PS II inactivated at low pH is able to perform charge separation. In inhibited reaction centre the lifetime of P_{680}^+ is increased and an efficient recombination reaction of the charge pair $P_{680}^+Q_A^-$ occurs. Recombination via the state $P_{680}^+Pheo^-$ would be expected to give a high yield of triplet chlorophyll leading to harmful reactions involving singlet O_2. A shift in the midpoint potential of Q_A favours a direct recombination to the ground state of P_{680}, avoiding such reactions.

References.
[1] Krieger, A., Rutherford, A.W. and Johnson, G.N. (1995) Biochim. Biophys. Acta 1229, 193-201.
[2] Brettel, K., Schlodder, E. and Witt, H.T. (1984) Biochim.Biophys. Acta 766, 403-415.
[3] Haveman, J. and Mathis, P. (1976) Biochim. Biophys. Acta 440, 346-355.
[4] Berthold, D:A:, Babcock, G.T.and Yocum, C:F: (1981) FEBS Lett. 61, 231-234.
[5] v. Mieghem, F., Brettel, K., Hillmann, B., Kamlowski, A., Rutherford, A:W: and Schlodder, E. (1995) Biochem. 34, 4798-4813
[6] Schatz,G.H., Brock,H.,Holzwarth, A. (1987) Proc.Natl.Acad.Sci. USA 84, 8414-8418

PARALLEL LOW-TEMPERATURE FLUORESCENCE AND EPR MEASUREMENTS OF MN-DEPLETED PHOTOSYSTEM II.

Robert H. Schweitzer and Gary W. Brudvig,
Department of Chemistry, Yale University,
New Haven, CT 06511, USA.

1. Introduction

Fluorescence is widely used to probe photosynthetic systems ranging from purified pigment-protein complexes to intact plants.[1-3] Many factors such as the pH gradient across the thylakoid membrane, the temperature and the redox state of Q_A are known to affect fluorescence,[2] but the processes affecting fluorescence are still not well understood at the molecular level. In order to determine how the redox states of the donor-side species in photosystem II (PSII) affect its fluorescence, we have prepared PSII membranes in different stable charge-separated states at low temperatures. EPR spectroscopy has been used to ascertain the redox state of PSII and parallel fluorescence measurements have been made on the same samples. We find that Chl_Z^+, a chlorophyll species that mediates electron transfer from cytochrome b_{559} (cyt b_{559}) to $P680^+$,[4] is a potent quencher of fluorescence in PSII.

2. Procedures

BBY-type PSII membranes were prepared by a modified version[5] of the Berthold et al.[6] procedure. Mn depletion was done by treatment with 5 mM NH_2OH.[7] Cyt b_{559} and Q_A were oxidized by resuspending a pellet of PSII membranes into a buffer containing 1 mM potassium ferricyanide, and then the ferricyanide was removed by pelleting and resuspending the sample in buffer without ferricyanide. Prior to freezing the samples, a Tb^{3+}-chelate complex was added as an internal fluorescence standard (details on the synthesis and use of the Tb^{3+}-chelate complex will be described in a forthcoming paper) and 200 μM DCMU was added from a stock solution in ethanol. All manipulations of the PSII membranes were done under a green safe light. Fluorescence measurements were made with a Perkin Elmer 650-40 fluorescence spectrometer equipped with a home-built nitrogen-flow cryostat. EPR measurements were carried out on a home-built X-band spectrometer interfaced to a Macintosh IIx computer and equipped with an Oxford ESR-900 liquid helium cryostat.[8] Both fluorescence and EPR measurements were made on the same samples in EPR tubes. In order to allow fluorescence measurements to be made on optically dense samples, plugs were positioned in the EPR tubes which gave a short optical path at the top of the sample. The EPR signal intensities of different samples were

P. Mathis (ed.), Photosynthesis: from Light to Biosphere, Vol. I, 807–810.

scaled by the g=4.3 EPR signal from rhombic iron, which was used as an internal standard, and the fluorescence intensities from different samples were scaled by the emission at 547 nm from the added Tb^{3+}-chelate complex.

3. Results and Discussion

An initial indication that the redox state of the donor side of PSII affects its fluorescence is shown in *Figure 1*. When a sample of untreated, O_2-evolving PSII membranes was frozen in the dark and then illuminated at a temperature from 85 - 125 K, the fluorescence intensity initially increased and then slowly decreased. Further illumination by bright actinic light at the same temperature produced a further drop in the fluorescence intensity. The drop in intensity indicates the formation of a fluorescence quencher by the excitation light. Past work[9] has shown that, under these illumination conditions, a stable charge-separated state slowly forms over the time course of the illumination in which either cyt b_{559} or Chl_Z functions as the electron donor.

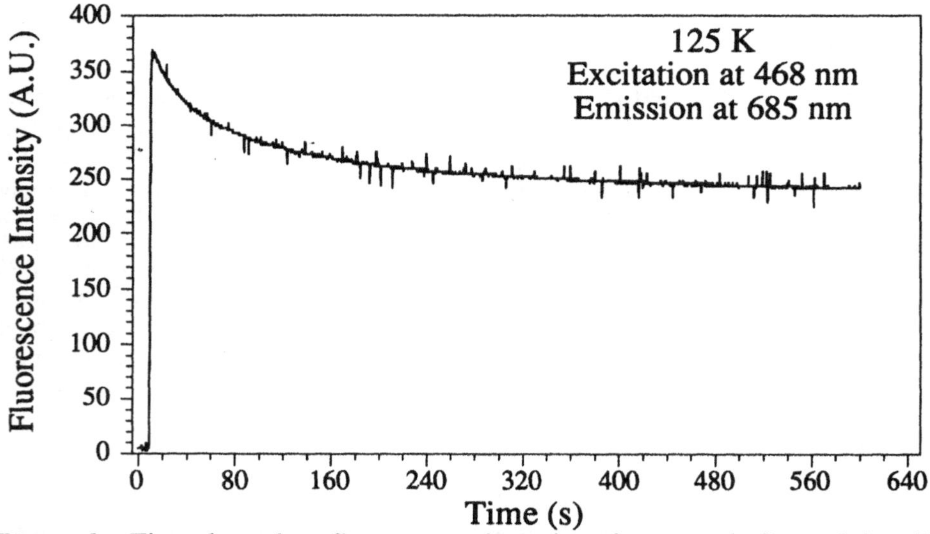

Figure 1. Time dependent fluorescence intensity of untreated, O_2-evolving PSII membranes during illumination at 125 K.

In order to determine the role of tyrosine D (Y_D), cyt b_{559} or Chl_Z in the observed fluorescence quenching, parallel fluorescence and EPR measurements were made on the same sample which was prepared in a defined charge-separated state. Mn-depleted PSII membranes were used to simplify the possible charge-separated states. On the basis of the work of Buser et al.,[7] conditions were chosen to maximize the formation of either the $Chl_Z^+ Q_A^-$ or the $Y_D^\bullet Q_A^-$ charge-separated state. First, the sample was treated with ferricyanide in order to oxidize cyt b_{559} and Q_A. Then, experiments were done where the amount of oxidized Chl_Z was varied by changing the illumination temperature. The sample that was preilluminated at 240 K was characterized by an EPR spectrum containing very little signal from Chl_Z^+ and a high fluorescence (*Figures 2A and 3A*).

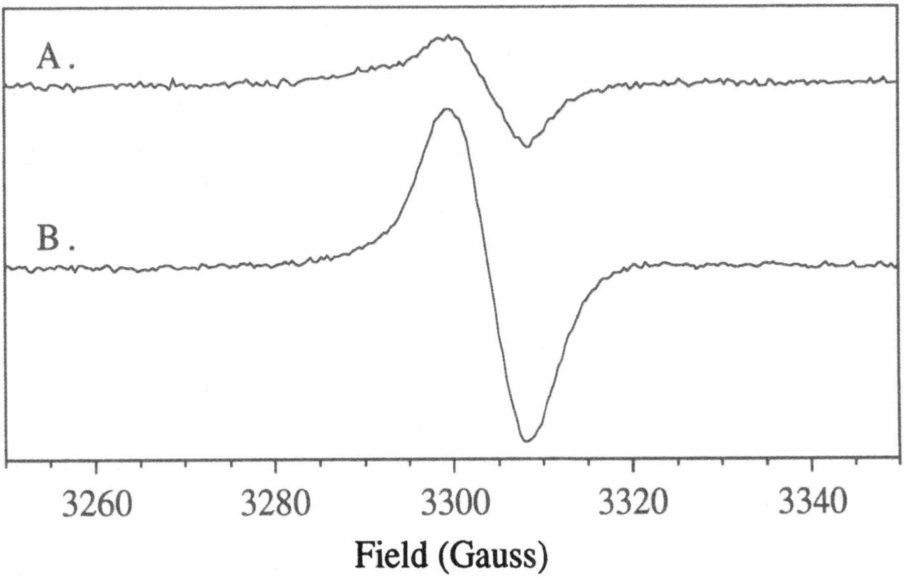

Figure 2. EPR spectra of ferricyanide-washed, DCMU-treated, Mn-depleted PSII membranes preilluminated at 240 K (A) or 176 K (B). After preillumination, the sample was illuminated for 15 min at 85 K to induce a stable charge-separated state in all centers.

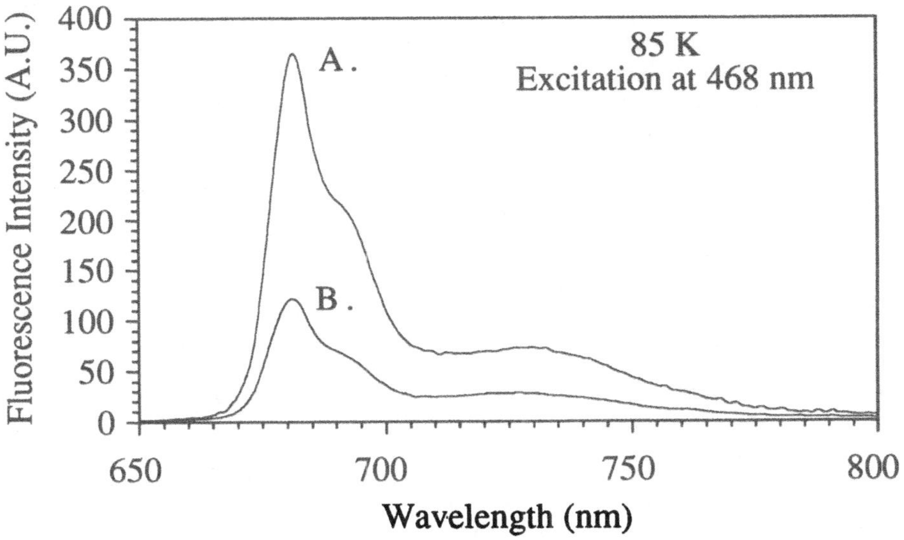

Figure 3 . Fluorescence spectra of the same samples used in *Figure 2*. The samples were preilluminated at 240 K (A) or 176 K (B) and subsequently illuminated at 85 K.

In contrast, the sample that was preilluminated at 176 K contained primarily the Chl_Z^+ Q_A^- charge-separated state and had a much lower fluorescence (*Figures 2B and 3B*).

Both samples A and B had the same state of the electron-acceptor side; Q_A was reduced and the Q_B site contained DCMU. In addition, cyt b_{559} was oxidized in both samples. Therefore, we conclude that the low fluorescence in sample B was due to the larger amount of Chl_Z^+ in this sample. These preliminary results indicate that Chl_Z^+ is a potent quencher of fluorescence in PSII. However, a number of other factors, such as the pH and buffer species, may also affect the low-temperature fluorescence yield of PSII. Studies are in progress in which the amount of Chl_Z^+ is changed systematically by varying the pretreatment and preillumination conditions in order to obtain a quantitative correlation between the amount of Chl_Z^+ and the extent of fluorescence quenching.

There are several important implications of these results. First, the fact that a species within the PSII reaction center produces significant fluorescence quenching shows that antenna chlorophylls are well coupled to Chl_Z in the PSII reaction center at low temperature. A quantitative analysis of the quenching may yield insight into energy transfer in PSII at low temperature. Second, fluorescence quenching by Chl_Z^+ needs to be considered in the interpretation of PSII fluorescence, especially in low-temperature studies when Chl_Z^+ can be stably formed.

Is quenching by Chl_Z^+ important under physiological conditions? Further studies are needed to answer this question. Based on past work on the electron-transfer reactions of PSII involving cyt b_{559} and Chl_Z,[7, 10] it is known that electron donation by cyt b_{559}/Chl_Z is very inefficient when the O_2-evolving center (OEC) is active. The steady-state amount of Chl_Z^+ may not be significant in active PSII. However, the yield of photooxidation of cyt b_{559}/Chl_Z increases markedly when the OEC cannot turn over, as is the case at low temperature or in Mn-depleted centers. Therefore, fluorescence quenching by Chl_Z^+ may be important under physiological conditions when the OEC is damaged or during the process of photoactivation when the Mn cluster is being assembled. In this way, cyt b_{559}/Chl_Z may function in the photoprotection of PSII.

References

1. Seely GR & Connolly JS (1986) In: Govindjee, Amesz J & Fork DC (ed) Light Emission by Plants and Bacteria, pp 99-133. Academic Press, New York.
2. Krause GH & Weis E (1991) Annu. Rev. Plant Physiol. 42: 313-349.
3. Evans EH & Brown RG (1994) J. Photochem. Photobiol. B: Biol. 22: 95-104.
4. Thompson LK & Brudvig GW (1988) Biochemistry 27: 6653-6658.
5. Beck WF, de Paula JC & Brudvig GW (1985) Biochemistry 24: 3035-3043.
6. Berthold DA, Babcock GT & Yocum CF (1981) FEBS Lett. 134: 231-234.
7. Buser CA, Thompson LK, Diner BA & Brudvig GW (1990) Biochemistry 29: 8977-8985.
8. Beck WF, Innes JB, Lynch JB & Brudvig GW (1991) J. Magn. Reson. 91: 12-29.
9. de Paula JC, Innes JB & Brudvig GW (1985) Biochemistry 24: 8114-8120.
10. Buser CA, Diner BA & Brudvig GW (1992) Biochemistry 31: 11449-11459.

EFFECT OF TRYPSIN ON BACTERIAL PHOTOSYNTHETIC REACTION CENTERS IN THE DARK AND UNDER ILLUMINATION: EVIDENCE OF LIGHT-INDUCED CONFORMATIONAL CHANGES

I.A.Smirnova[1], A.Blomberg[2], L.-E.Andréasson[1] & P.Brzezinski[1]
[1]Dept. of Biochemistry & Biophysics, and [2]Dept. of General & Marine Microbiology, Göteborg University and Chalmers University of Technology, Medicinaregatan 9C, S-41390 Göteborg, Sweden

1. Introduction

Bacterial photosynthetic reaction centers (RCs) from *Rhodobacter sphaeroides* consist of three protein subunits - H, M and L. The L and M subunits ligate all cofactors involved in photo-induced charge separation, including the primary electron donor, a bacteriochlorophyll dimer (P), and the primary and the secondary electron acceptors - bound ubiquinones Q_A and Q_B (for a review see [1]). Following light-induced excitation of P an electron is transferred sequentially from P to Q_A and Q_B with a quantum yield of about unity. One possible explanation for this high quantum yield is that structural changes follow charge separation, stabilizing the charge-separated state. Experimental results in support of this idea have been reported by Kleinfeld et al. [2], who compared rates of electron transfer in RCs frozen under illumination and in the dark. Light-induced structural changes upon forming PQ_A^- were resolved in time in measurements of voltage changes across planar RCs-lipid layers [3,4]. Recently, we also demonstrated differences in functionality of RCs treated with trypsin in the dark and under illumination [4]. To further investigate the light-induced structural changes, in this work we have compared degradation patterns of RCs after trypsinolysis using SDS-PAG electrophoresis and reverse-phase HPLC. Trypsination positions were identified by N-terminal sequence analysis. In particular, we focused on degradation fragments which were different in the samples treated in the dark and under illumination.

2. Procedure

RCs from *R. sphaeroides* R-26 were isolated in lauryl dimhetylamine N-oxide as described by Feher&Okamura [5].
Trypsin treatment was performed in a thermostated water bath at 25°C for 10 min with 5.5, 44, and 220 μg/ml trypsin (Sigma) under illumination or in the dark. The proteolysis was stopped by the addition of 4 to 8-times excess of soybean trypsin inhibitor (Merck), and the samples were then frozen in liquid nitrogen (see also [4].
The electrophoresis was performed on 4%/12% SDS polyacrylamide gels according to Laemmli [6]; Coomassie Brilliant Blue staining was used. The blotting was performed in Towbin buffer with PVDF-protein sequencing membrane (Bio-Rad).
For analysis of the short polypeptides (MW<10 kd) 0.1% trifluoroacetic acid (TFA) was added to the trypsinated samples and they were centrifuged. Supernatants were ultrafiltrated in a Microcon-10 device (Amicon). Filtrates were analyzed by reverse-

P. Mathis (ed.), Photosynthesis: from Light to Biosphere, Vol. I, 811–814.

phase HPLC on a μRPC C2/C18 SC (100×2.1 mm) column. The column was eluted with a linear gradient of acetonitrile, 0.1% TFA using a Smart[TM] system (Pharmacia). The N-terminal sequence analysis of blots and peaks was performed on a pulsed-liquid sequencer 473 A (Applied Biosystems, New York).
Flash-induced absorbance changes were measured as described previously [4]. The cuvette path length was 10 mm. Before measurements the samples were diluted 3-times and mixed with 5-7 molar excess of ubiquinone (Q_{10}, Sigma).

3. Results and Discussion

Degradation pattern analysis. Degradation patterns of RCs treated with trypsin in the dark and under illumination were analyzed. We varied trypsin concentration in a wide

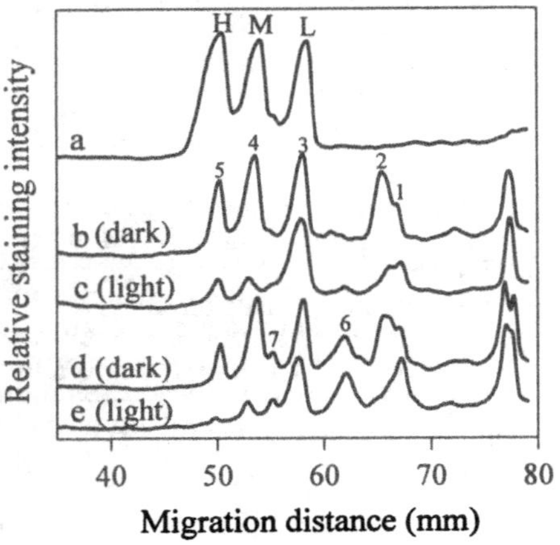

Figure 1. SDS-PAGE of RCs after trypsin treatment in the dark and under illumination. 15 μM RCs were treated with trypsin at 44 μg/ml (*b, c*) and 220 μg/ml (*d, e*) in the dark (*b, d*) and under illumination (*c, e*). Trace *a* - untreated RCs. 8 μg RC protein was loaded on each well. Bands *6* and *7* belong to trypsin inhibitor and trypsin, respectively.

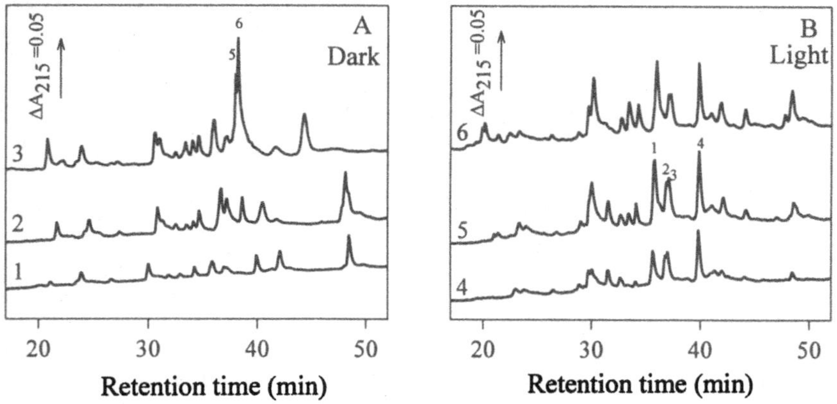

Figure 2. Reverse-phase HPLC separation of polypeptides with MW<10 kd. The samples after trypsin treatment in the dark (A) and under illumination (B) were prepared as described in Procedure section. Trypsin was used at concentrations: 5.5 μg/ml (curves 1, 4), 44 μg/ml (curves 2, 5) and 220 μg/ml (curves 3, 6). The linear gradient of acetonitril was from 7 to 28 %.

range (5.5 - 220 μg/ml) to be able to follow the degradation kinetics and to investigate differences at the initial degradation steps. The electrophoretical patterns are shown in Fig.1. Bands 3, 4 and 5 were attributed to the L, M and H subunits, respectively, owing to their N-terminal sequences. Trypsinolysis under illumination (Fig.1, *c* and *e*) gave rise to more degraded H and M subunits than in the dark (Fig.1, *b* and *d*). In the entire range of trypsin concentrations the bands positioned at #2 (MW~20 kd) were much more prominent in the samples treated in the dark. Sequence analysis of this band showed that it was a fragment of the H subunit starting from T90.
Small degradation products (MW<10 kd) were analyzed using reverse-phase HPLC. Fractions 1, 2, 3 and 4 (Fig.2, *5*) were characteristical to the samples treated under illumination, particularly on the first steps of degradation (compare samples obtained with 5.5 μg/ml, Fig.2, *4* and *1*). Fractions 5 and 6, characteristic to the samples treated in the dark, were observed at later degradation steps. All these fractions were sequenced, and results are summarized in a table:

fraction #	polypeptide, subunit	comment
1	the major: from G71, H	illumination, low trypsin concentration
	the minor: from L190, H	
2, 3	from S250, the N-terminal of H	
4	the N-terminal of M, 13 amino acids	
5	the N-terminal of M	dark, high trypsin concentration
6	the 1st: the N-terminal of M	
	the 2nd: from T90, H	

According to the three-dimensional RC structure [7] all identified cleavage points are positioned relatively close to the protein-globule surface. However, the degradation pathways in the dark and under illumination were different. Under illumination the bond at R89 (H subunit) is more protected from trypsinolysis, and the bonds at R70, R189, K249 (H subunit) and R13 (M subunit) are more accessible as compared to the treatment in the dark.
The effect of trypsin on the photochemical activity of the RCs. Absorbance changes associated with the recombination reactions $P^+(Q_AQ_B)^- \rightarrow PQ_AQ_B$ and the electron transfer from the primary to the secondary acceptor ($Q_A^-Q_B \rightarrow Q_AQ_B^-$) after a laser flash were recorded. As was found previously the charge-separation absorbance changes decreased after trypsin treatment [4]. The recombination reaction was biphasic as a sum of two reactions: $P^+Q_A^- \rightarrow PQ_A$ and $P^+Q_B^- \rightarrow PQ_B$ with an amplitude ratio ~1:1, indicating an ~50% occupation of the Q_B-binding sites. Trypsin treatment did not affect considerably this ratio. The electron transfer from Q_A^- to Q_B was monitored as an increase in the optical absorbance at 445 nm (Fig.3). Treatment in the dark did not influence dramatically the electron transfer from Q_A^- to Q_B (Fig3, traces 3, 5, 7). A 10-minute illumination by itself lead to a decrease of the Q_B-reduction amplitude (Fig.3, traces 1, 2) which may be caused by photo-damage of the Q_B binding sites. Treatment with increasing trypsin concentrations under illumination resulted in a decrease of the Q_B-reduction amplitude without any noticeable change in the electron-transfer rate (Fig.3, traces 4, 6, 8). At trypsin concentrations of 220 μg/ml the absorbance change was not resolved (trace 7). This effect may be due to an RC population with an about 100 times retarded $Q_A^-Q_B \rightarrow Q_AQ_B^-$ electron transfer. This population increases with increasing trypsin concentrations. Most likely this effect is caused by degradation of the H subunit which was shown to control electron transfer involving the Q_B [8].

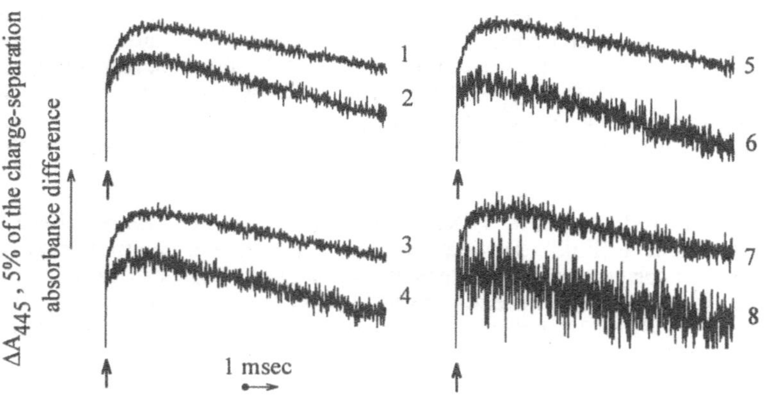

Figure 3. Kinetics of Q_B reduction after trypsin treatment of RCs. The absorbance changes were normalized to the PQ_A-$P^+Q_A^-$ absorbance differences. Short arrows indicate the moments of laser flashes. Traces 3, 5, 7 - after treatment in the dark; traces 4, 6, 8 - after treatment under illumination with trypsin concentrations 5.5 (3; 4), 44 (5; 6) and 220 (7; 8) µg/ml. Traces 1 and 2 - after 10 min incubation without trypsin in the dark and under illumination, respectively.

4. Concluding remarks

The analysis of degradation patterns of RCs treated with trypsin in the dark and under illumination shows marked differences in the degradation pathways indicating light-induced conformation changes in the RCs. Most of these differences were detected on the H subunit which may be related to its proposed function as a regulator of electron transfer in RC [8].

Acknowledgments
This study has been supported by grants from the Swedish Natural Science Research Council and fellowships (to I.S.) from Wenner-Grenska Samfundet and the Göteborg University.

References
1. Gunner, M. (1991) *Current Topics in Bioenergetics 16*, 319-367.
2. Kleinfeld, D., Okamura, M.Y., & Feher, G. (1984) *Biochemistry 23*, 5780-5786.
3. Brzezinski, P., Okamura, M.Y., & Feher, G. (1992) in *The Photosynthetic Bacterial Reaction Center II: Structure, Spectroscopy and Dynamics* (Breton J., & Verméglio, A., eds.) pp 321-330, Plenum Press, New York & London.
4. Brzezinski, P. & Andréasson, L.-E. (1995) *Biochemistry 34*, 7498-7506.
5. Feher,G. & Okamura, M.Y. (1978) in *The Photosynthetic bacteria* (Sistrom, W.R. & Clayton, R.K. eds.) pp.349-386, Plenum Press, New York.
6. Laemmli, U.K. (1970) *Nature 227*, 680-685.
7. Ermler U., Fritzsch G., Buchanan S.K. & Michel H. (1994) *Structure* 2, 925-936.
8. Debus, R.J., Feher, G. & Okamura, M.Y. (1985) *Biochemistry 24*, 2488-2500.

INVESTIGATIONS OF PHOTOINDUCED ELECTRON TRANSFER IN PYROPHEOPHYTIN-ANTHRAQUINONE MODEL SYSTEMS.

Andrei Y. Tauber[1], Juho Helaja[1], Nikolai V. Tkachenko[2], Helge Lemmetyinen[2], Ilkka Kilpeläinen[3] and Paavo H. Hynninen[1]
[1]Department of Chemistry, University of Helsinki, P.O. Box 55, FIN-00014 University of Helsinki, Finland; [2]Institute of Materials Chemistry, Tampere University of Technology, P.O. Box 589, FIN-33101 Tampere, Finland; [3]Institute of Biotechnology, University of Helsinki, P.O. Box 45, FIN-00014 University of Helsinki, Finland.

1. Introduction

The primary photosynthetic electron-transfer occurs in a series of donors and acceptors whose mutual distances and orientations are determined by the surrounding reaction-centre protein [1]. In this connection, model donor-acceptor (D-A) systems with well defined geometries are especially interesting for the elucidation of the mechanism of photoinduced electron transfer (PET) [2]. We have prepared a series of four pyropheophytin-anthraquinone linked molecules (*Fig. 1*) along with the reference compounds, pyropheophytin *a* (H_2P, pyro = 13^2-demethoxycarbonyl) and Zn(II)-pyropheophytin *a* (ZnP). Here we present some results from our conformational and photochemical studies on the model D-A compounds.

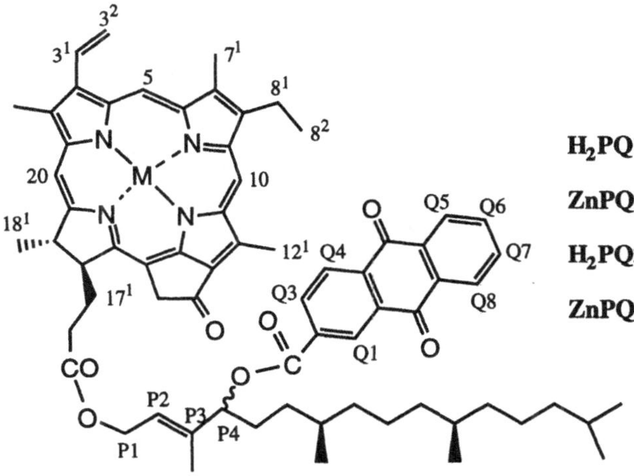

	M	P4 stereochemistry
H_2PQ1	2H	*S*
ZnPQ1	Zn	*S*
H_2PQ2	2H	*R*
ZnPQ2	Zn	*R*

Figure 1. Pyropheophytin-anthraquinone D-A model compounds.

P. Mathis (ed.), Photosynthesis: from Light to Biosphere, Vol. I, 815–818.

2. Experimental

The synthesis and complete chromatographic separation of the P4-diastereomers of P4-oxy-(9,10-anthraquinone-2-carbonyl)-pyropheophytin *a* (H_2PQ1 and H_2PQ2, *Fig. 1*) were accomplished as described previously [3]. The zinc(II)-complexes of H_2P (ZnP) and the pyropheophytin-anthraquinone derivatives (ZnPQ1 and ZnPQ2) were prepared according to the standard procedure [4] by refluxing the chloroform solution of the corresponding metal-free derivative with a five-fold excess of Zn(II)-acetate in methanol.

The ^1H and ^{13}C NMR spectra of the compounds were assigned using two-dimensional NMR techniques (HMQC, HMBC and ROESY). Measurements were done at 300 K (27 °C) in $CDCl_3$ on a Varian Unity 500 FT spectrometer. The experimental details have been described elsewhere [5,6]. The computer-assisted molecular modelling calculations were performed in the MM+ force-field with standard parameters under HyperChem program run on 486 PC.

The steady-state absorption spectra were recorded on a Unicam 8700 Series UV/Visible spectrometer (Unicam Ltd. UK) and the fluorescence spectra on a Shimadzu RF 5001 PC spectrofluorimeter (Japan). For the fluorescence measurements, samples were prepared in a 1 cm cuvette and showed an optical density of 0.1 at the Soret absorption maximum; the excitation wavelength was 410 nm for the H_2PQ compounds and 425 nm for their Zn-complexes.

The time-resolved fluorescence measurements were carried out on a time-correlated single photon counting instrument (Edinburgh Instrument Model 199) as described previously [7]. Pulses of about 15 ps (fwhm) at 590 nm were used for excitation; fluorescence decays were collected in the range from 620 to 780 nm with an instrumental response of about 120 ps (fwhm).

The transient absorption spectra were measured by means of a setup described elsewhere [7]. Coumarin 120 dye laser was used for excitation (wavelength 430 nm, pulse duration about 30 ns), and transient curves were collected in the range from 380 nm to 720 nm with the time resolution limited by the excitation pulse duration. Generally, 10 to 50 decays were averaged in order to improve the signal-to-noise ratio.

3. Result and Discussion

The results from the NMR, photochemical investigations and computer-assisted molecular modelling led us to the conclusion that all four D-A molecules studied occur in solution as two main types of conformers, the folded ones and the opened ones. In $CDCl_3$ or CH_3CN, the dominating conformer on the NMR time-scale is the folded one. The ^1H NMR spectra showed significant differences in the chemical shifts between diastereomeric pairs H_2PQ1 and H_2PQ2, ZnPQ1 and ZnPQ2. The chemical shift differences ($\Delta\delta$-values) were even larger between the Zn-complexes and the corresponding metal-free D-A compounds (ZnPQ1 and H_2PQ1; ZnPQ2 and H_2PQ2). The observed $\Delta\delta$-values were attributed to the alterations occurring in the ring-currents of the D and A π-systems as a result of intramolecular π-π interactions between the π-systems. A detailed report of the NMR results will be published in our forthcoming paper.

Here we report briefly the results obtained by ROESY and molecular modelling on the solution structures of the Zn-complexes (ZnPQ1 and ZnPQ2). The two-dimensional rotating-frame Overhouser enhancement spectroscopy (ROESY) is known to provide information of the distance between correlative protons. We have used ROESY to define the mutual orientation between the anthraquinone and pyropheophytin parts in our D-A compounds. The ZnPQ1 and ZnPQ2 showed several ROESY correlations between the anthraquinone and pyropheophytin protons. No similar ROESY correlations were observed in the case of the metal-free H_2PQ1 and H_2PQ2 compounds, which indicates that the mutual distance between the D and A planes exceeds 4-5 Å. Both Zn-complexes showed a clear ROESY correlation between the 18^1-CH_3 and Q1-**H** protons. This proves that these protons are located on the same side of the pyropheophytin macrocycle. The mutual orientation of the D and A parts could then be defined more specifically from the other ROESY correlations observed between the anthraquinone and pyropheophytin protons. This information was used to construct computer-assisted molecular models for ZnPQ1 and ZnPQ2.

The computer-assisted molecular modelling provided information of the energy differences between different conformers. The energy-minimized folded conformers of ZnPQ1(**I**) and ZnPQ2(**II**) are presented in *Fig. 2*. The distance between the π-system planes was found to be *ca.* 4 Å for both isomers. The ^{1}H-^{1}H distances in structures **I** and **II** provided by molecular modelling, were in good agreement with the ROESY results. The obtained minimized energies for the folded conformers were *ca.* 130 Kcal/mol, whereas for the open conformers the energies were 15-20 Kcal/mol higher. However, these energy differences and structures should not be accepted without certain reservations, since the MM+ force-field is not optimized to model π-π interactions between complicated aromatic systems.

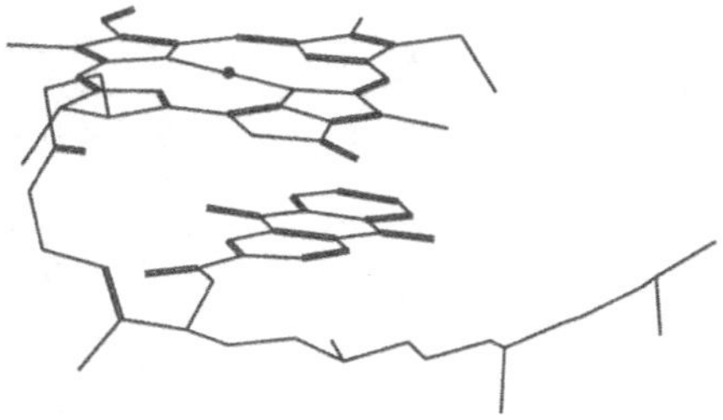

I

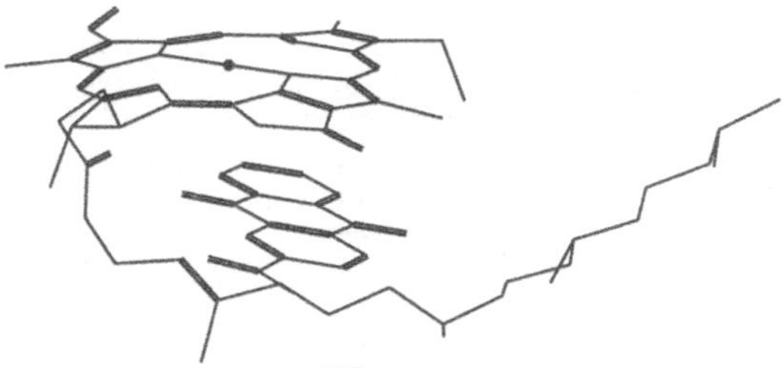

II

Figure 2. Molecular models of ZnPQ1 and ZnPQ2

The absorption spectra of H_2PQ1, H_2PQ2 and their Zn-complexes were found to be very similar to those of the reference compounds (H_2P and ZnP) in toluene and CH_3CN. A small red shift (< 2 nm) in the absorption maxima and broadening of the Q-band could be observed in the spectra of the D-A compounds.

Also the fluorescence spectra of the D-A compounds studied were almost identical with those of the reference compounds. Nevertheless, their fluorescence intensity was significantly quenched relative to the reference compounds. This supports the interpretation that two types of conformers were present. According to this interpretation, the folded conformers were predominating and defined the absorption spectra. However, they underwent fast intramolecular electron-transfer from the excited states of the pyropheophytin donor to the ground state of the anthraquinone acceptor and did not contribute to fluorescence intensity. Hence, the fluorescence intensity was determined by the open conformers, which possessed an A-D separation that was too long for efficient electron transfer.

Since the environment is known to affect substantially the rate of electron transfer, photochemical studies were performed in two solvents with very different polarities. Toluene (ε 2.38) was chosen as a representative of the most nonpolar solvents and acetonitrile (ε 37.5) as one of the most polar solvents. Fluorescence quenching was stronger in acetonitrile than in toluene for all D-A compounds studied. In both solvents, the fluorescence quenching was more efficient for H_2PQ1 and ZnPQ1 (diastereomers *S*) than for H_2PQ2 and ZnPQ2 (diastereomers *R*). The relative fluorescence intensities of the D-A compounds *vs.* the reference compounds (H_2P for the metal-free D-A compounds and ZnP for their Zn(II)-complexes) are summarized in the Table 1.

Table 1. Relative fluorescence intensities and triplet state yields.

Solvent	Compound	Relative fluorescence intensity	Triplet-state yield
Toluene	H_2PQ1	0.33	0.40
	H_2PQ2	0.40	0.50
	ZnPQ1		0.02
	ZnPQ2	0.21	0.12
Acetonitrile	H_2PQ1	0.11	0.10
	H_2PQ2	0.16	0.15
	ZnPQ1	0.05	0.02
	ZnPQ2	0.13	0.12

The transient absorption spectra were measured in the range from 380 to 720 nm with a time resolution of about 50 ns (excitation pulse duration). Prior to the measurements, samples were fluxed with N_2 to remove O_2. The transient absorption spectra obtained for the D-A compounds were similar to those of the reference compounds and also very close to the differential triplet-state spectra of chlorophyll *a* and pheophytin *a* [8]. The decays were almost mono-exponential with lifetimes of about 100 μs. Therefore, we conclude that the relaxation of the charge-separated state is faster than 50 ns for the compounds studied. The triplet-state yields of the D-A compounds *vs.* the reference molecules (Table 1) were calculated as the average reduction of the transient signal intensities throughout the spectrum. The observation that the singlet and the triplet states for the metal-free D-A compounds are quenched approximately equally, leads us to the conclusion that, for the metal-free D-A molecules, the PET occurs only from the singlet excited state of the donor. For the zinc D-A compounds, the quenching observed for the triplet state of the donor was more pronounced than that for its singlet excited state. Hence we conclude that, for the Zn-complexes, the intramolecular PET can also take place from the triplet state of the donor.

References

1 Deisenhofer, J.; Epp, O.; Miki, K.; Huber, R.; Michel, H. (1984) J. Molec. Biol. **180**, 385-398.
2 Wasielewski, M. R. (1992) Chem. Rev. **92**, 435-461.
3 Tauber, A. Y.; Kostiainen, R.; Hynninen, P. H. (1994) Tetrahedron **50**, 4723- 4732.
4 Boucher, L. J. and Katz, J. J. (1967) J. Am. Chem. Soc. **86**, 4703-4708.
5 Helaja, J.; Hyvärinen, K.; Heikkinen, S.; Kilpeläinen, I.; Hynninen, P. H. (1995) J. Mol. Struct. **354**, 71-79.
6 Hyvärinen, K; Helaja, J.; Kuronen, P.; Kilpeläinen, I.; Hynninen, P. H. (1995) Magn. Reson. Chem., **33**, 646-656.
7 Tkachenko, N. V.; Grandell, D.; Ikonen, M.; Jutila, A.; Moritz V.; Lemmetyinen H. (1993) Photochem. Photobiol. **58** (2), 284-289.
8 Chibisov A. K. (1969) Photochem. Photobiol. **10**, 331-347.

EARLY STEPS IN PHOTOSYNTHETIC REACTION CENTERS PROBED BY ACCUMULATED PHOTON ECHO EXPERIMENTS

P. Schellenberg, R.J.W. Louwe, S. Shochat,
P. Gast, A.J. Hoff, T.J. Aartsma
Department of Biophysics, University of Leiden
P.O. Box 9504, 2300 RA Leiden, The Netherlands

1. Introduction

During the past years, a large number of experiments has been performed to unravel the kinetics of energy and electron transfer in photosynthetic reaction centers (RCs). Most of these experiments are of the pump-probe type and provide the excited state lifetime T_1 of the special pair or the lifetime of later intermediates. At low temperature, complementary information can be obtained by measuring the total dephasing time T_2 of the excited state by coherent resonance techniques or by spectral hole-burning. Here we describe the results of accumulated photon echo measurements on RCs of *Rhodobacter (Rb.) sphaeroides* R26 and (M)Y210W at 1.5 K. The relationship between the homogeneous linewidth and the total dephasing time is given by

$$\Gamma_h = \frac{1}{\pi T_2} = \frac{1}{\pi}\left(\frac{1}{2T_1} + \frac{1}{T_2^*}\right)$$

with T_2^* being the pure dephasing time. Since the accumulated photon echo signal is just the Fourier transform of the homogeneous lineshape, hole burning and accumulated photon echo (APE) measurements give equivalent information.

The accumulated photon echo decay of RCs of *Rb. sphaeroides* R26 is characterized by multiple time constants. A very fast component is attributed to vibronic relaxation due to excitation of the phonon sideband, while the slower components reflect electron transfer. These results are discussed in relation to those of previous photon echo measurements [1] and of hole-burning experiments [2].

2. Experimental

The experimental set-up is the same as described before [3]. The laser pulse width was about 1.5 ps, as determined from the autocorrelation signal. The pulse shape was transform limited.

RCs from *Rb. sphaeroides* R26 and (M)Y210W were prepared as previously described [4]. To enhance the glass forming properties of the solution, two parts of glycerol per part buffer were added prior to cooling of the sample. The concentration was adjusted to an optical density of about 0.3 at the measuring wavelength. To enhance the efficiency of triplet state formation of the primary donor (^{3}P), the quinone (Q_A) was prereduced.

P. Mathis (ed.), Photosynthesis: from Light to Biosphere, Vol. I, 819–822.

3. Results and Discussion

The photon echo decay in RCs of *Rb. sphaeroides* (M)Y210W is shown in Fig. 1a for different excitation wavelengths. The dephasing kinetics is multiexponential with a fast component which decays in less than 1 ps, and two slower time constants of 20 and 120 ps, respectively. We attribute the fast component to rapid vibronic dephasing following excitation of the phonon sideband. The longer decay components are associated with electron transfer from the special pair to the primary acceptor, and are observed upon direct excitation of the 0-0 transition. Primary charge separation in the (M)Y210W mutant at 1.5 K is obviously much slower than that in native RCs (Fig. 1b). The time constants obtained from accumulated photon echo measurements are basically in agreement with those from incoherent pump-probe experiments [5].

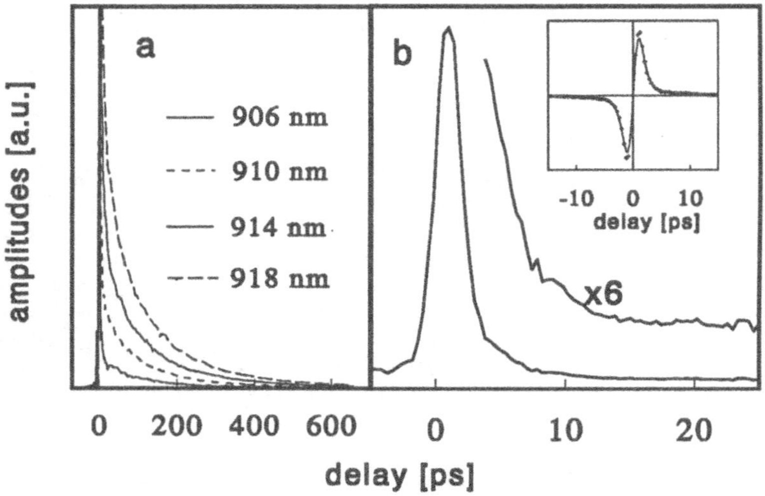

Figure 1 a: APE signal from Rb. sphaeroides M(Y)210W at different probing wavelengths. The peaks are normalized to equal intensity. The full height around t = 0 is not shown. The phonon contribution increases towards the blue wing of the absorption profile.
b: APE signal from Rb. sphaeroides R26. Insert: Antisymmetrized plot

The relative amplitude of the fast decay component increases as the excitation wavelength is shifted from the red to the blue side of the 890 nm band of the absorption spectrum, as expected from a phonon contribution. The relative amplitude of this component is determined by the strength of the electron-phonon coupling. The slower components will be observed with excitation in the long-wavelength edge of the absorption spectrum where the zero-phonon transitions are located. Indeed, the 20 and 120 ps components are only observed in the wavelength range of 910 – 920 nm (Fig. 2). Since the photon echo signal is just the Fourier transform of the homogeneous line shape spectrum, the Debye–Waller–factor can in principle be determined by calculating the ratio of the amplitude of the fast decaying component to the total amplitude [6]. In our case, the fast component cannot be resolved because of the limited time resolution of our experiment. Hence the amplitude of the fast component is subject to a large uncertainty and the Debye-Waller factor cannot be determined reliably from these measurements.

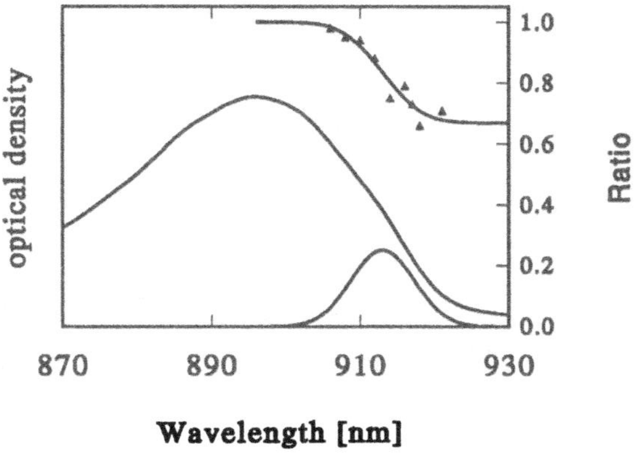

Figure 2: Amplitude ratio from APE-signals of short decay time to total decay time vs. absorption wavelength for M(Y)210W. The Gaussian distribution (with arbitrary height) for the zero–phonon lines is calculated from these datapoints, λ_{max} = 913 nm, σ = 6 nm. The absorption spectrum is also shown.

The electron transfer component is much faster in *Rb. sphaeroides* R26 compared to the mutant. It can be separated from that associated with dephasing of the phonon–coupled electronic states by an antisymmetrization procedure [7], as shown in Fig. 1b. The best fit of this antisymmetrized signal is obtained with a biexponential function, leading to decay times of 1.2 ps and 8 ps. The zero–phonon line width in hole burning experiments corresponds to the short lifetime component measured in APE.

The general pattern for the wavelength dependance resembles that in the mutant: The zero–phonon contribution vanishes as one goes to the blue wing of the inhomogeneous absorption profile. Apart from signal to noise consideration, this is probably the reason why the long decay component was not observed in previous APE–experiments [2]. In these measurements, only a very fast decay (< 150 fs) was observed, which we attribute to the contribution of the phonon side–band.

Another important observation is the striking dispersion of decay times in *Rb. sphaeroides* R26 and, on a longer timescale, in the mutant (M)Y210W. Such a behaviour was also previously reported for population lifetime measurements of the excited state [5], and several different models has been put forward to account for this multiexponential decay.

The explanation favoured by our work is the dispersion model [8], which assumes a broad distribution of the free energy difference between P^* and P^+I^-, I being the primary acceptor. This explanation is based on the assumption of an inhomogeneity of solvent cages due to the intrinsic disorder of the protein. In this case, the decay would be a stretched exponential as observed in kinetic measurements on disordered systems. Alternatively, a limited number of conformational substates of the protein giving rise to different free energy gaps can also be put forward to account for these data. Such a discrete substate model was first proposed to explain the reaction dynamics in myoglobin [9]. Different discrete mutual arrangements of the chromophores were also proposed for *Rps. viridis* [10]. Discrete substates would lead to a multiexponential behaviour with each exponent representing one substate. Both of these models are consistent with our data.

Another approach for explaining the kinetic pattern observed is the parking state model [11] where competing electron transfer to the primary acceptor in the inactive branch is included, and backtransfer to the special pair and electron transfer along the active branch takes place in a subsequent step. Such a process would lead to a loss of phase relation of the excited state following the first transfer step. Therefore the multiexponential kinetics observed in T_1 –experiments would not show up in APE–measurements. Therefore, this model can clearly be excluded on the basis of our measurements.

4. Conclusions

The APE–signal of the special pair (P890) in *Rb. sphaeroides* R26 and M(Y)210W consists of a very fast decay component due to excitation into the phonon side–band and a component, whose decay time is determined by the electron–tranfer time to the primary acceptor.
The special pair of bacterial RCs obviously exhibits a strong electron–phonon coupling, contrary to what is observed for monomeric chlorophylls, which are only weakly coupled to the matrix. The high Debye–Waller factor can be understood in the context of Stark– measurements [12] which show that the dimer state possesses a significant charge transfer character. The strong electron–phonon coupling is also confirmed by hole–burning experiments [2].
The short decay time of the signal in *Rb. sphaeroides* R26 is comparable to the linewidth observed in transient hole burning measurements. Such experiments for M(Y)210W have not been done yet. The decay patterns for the dephasing of the excited state are strongly dispersive, which is attributed to the amorphous character of the systems. The involvement of a parking state to explain the multiexponential decay can be excluded.

5. Acknowledgements

The investigation was supported by the Life Science Foundation (LSF) which is subsidized by the Netherlands Organization for Scientific Research (NWO). P.S. thanks the European Community for a grant from the Human Capital and Mobility Program (contract No. ERBCHBGCT930361)

References

1 Meech, S.R., Hoff, A.J. and Wiersma, D.A. (1986) Proc. Natl. Acad. Sci. USA 83, 9464-9468
2 Jankowiak, R., Tang, D., Small, G.J. and Seibert, M. (1989) J. Phys. Chem. 93, 1649-1654
3 Louwe, R.J.W. and Aartsma, T.J. (1994) J. Lumin. 58, 154-157
4 Shochat, S. et al (1994) Photosyn. Res. 40, 55-66
5 van Noort, P.I., (1994) Ph. Thesis, Leiden University
6 Saikan, S., (1990) Phys. Rev. B 41, 3185-3189
7 Engh, R.A., Petrich, J.W. and Fleming, G.R. (1985) J. Phys. Chem. 89, 618-621
8 Vos, M.H. et al (1991), Proc. Nat. Acad. Sci. USA 88, 8885-8889
9 Frauenfelder, H., Nienhaus, G.U., and Johnson, B.J. (1991) Ber. Bunsenges. Phys. Chem. 95, 272-278
10 Shkuropatov, A.Y., Ganago, A.O. and Shuvalov, V.A. (1991) FEBS Letters 287, 142-145
11 Hamm, P. et al (1993) Biochim. Biophys. Acta 1142, 99-105
12 Lösche, M., Feher, G., Okamura, M.Y. (1987) Proc. Nat. Acad. Sci. USA 84, 7537-7541

CYANIDE BINDING AT THE NON HEME IRON OF PSII AND EFFECTS ON CERTAIN EPR SIGNALS.

Y. Sanakis and V. Petrouleas
NCSR "Democritos", 15310 Aghia Paraskevi, Greece

1. Introduction

CN^- binds at or in the vicinity of the non-heme iron of PSII in a stepwise fashion [1,2]. At low concentrations, 10-20 μM cyanide competes with NO for binding to the iron [1]. At higher concentrations (K_d 0.1-0.2 mM) cyanide modifies the Q_A^- Fe^{2+} state inducing a pronounced EPR signal at g=1.98 [1]. The temperature dependence of this signal is consistent with an excited electronic state of the complex. At high concentrations (K_d ca 1.2 mM in PSII membrane fragments) cyanide converts the Fe^{2+} to a low spin state (S=0) [2]. We report here on a number of new EPR signals associated with the cyanide binding to the iron-quinone complex.

2. Materials and Methods

The cyanide treatment and the EPR experiments were as described in [1, 2]. The iron in the low spin state was oxidised with ferricyanide 10 mM. In order to minimise contributions from the oxidant in the EPR region of interest excess ferricyanide was washed away and the final concentration in the EPR sample was 0.5 mM.

3. Results and Discussion

3.1 *The $Q_A^-Fe^{2+}$ EPR signal at g=1.98 is accompanied by a ground state signal at g=2.0038*

Fig. 1 shows the g=2 region of the EPR spectrum at 4.2 K of a BBY sample treated with 340 mM NaCN at pH 6.5, and subsequently reduced with 40 mM dithionite. Background signals from a control sample (340 mM NaCl + dithionite) have been subtracted. The spectrum consists of an "absorption-shaped" resonance with a peak position at g=2.0038 and an apparent width of approx. 7 G. The g=1.98 is vanishingly small at this temperature and power. A detailed examination of the g=2.0038 signal (data to be presented elsewhere) shows that the signal has similar

P. Mathis (ed.), Photosynthesis: from Light to Biosphere, Vol. I, 823–826.

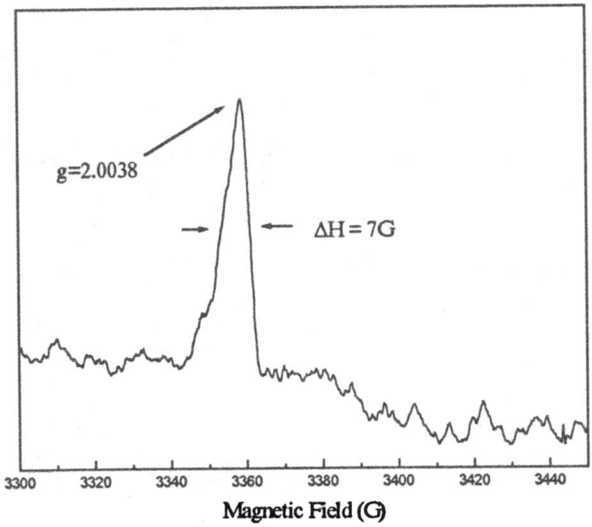

Fig. 1. The ground state EPR signal accompanying the g=1.98 of the state $Q_A^-Fe^{2+}$. A BBY sample was treated with 340mM NaCN at pH 6.5 and subsequently reduced with dithionite. The spectrum of a control sample (340mM NaCl + dithionite) has been subtracted.
EPR conditions: T=4.2K, m-w power 31mW, m-w freq., 9.42 GHz, mod app., 4Gpp.

microwave power dependence as well as similar variation with pH and cyanide concentration with the g=1.98 signal. What is distinctly different is the temperature dependence of the signal. The g=2.0038 signal decreases rapidly with increasing temperature above 4.2K, unlike the g=1.98 signal which shows a maximum at about 12 K [1]. These observations suggest that the g=2.0038 signal is associated with the same electronic configuration that gives rise to the g=1.98 signal but results from a different transition, probably, the ground doublet of the complex.

3.2 *Effects of cyanide on the split pheophytin signal*
A weak magnetic interaction between $Pheo^-$ and the $Q_A^-Fe^{2+}$ complex splits the $Pheo^-$ signal by about 50 Gauss [3]. Modifications of the $Q_A^-Fe^{2+}$ state such as the pH induced shift of the $Q_A^-Fe^{2+}$ signal from 1.82 to 1.9 induce small but distinct alterations to the split pheophytin signal [4]. We have examined the effects of the cyanide induced conversion of the $Q_A^-Fe^{2+}$ signal to g=1.98, 2.0038 on this signal. Fig. 2 compares the split pheophytin signal of an untreated sample characterised by the g=1.9 signal at pH 6.5 with a similar sample treated with 300 mM NaCN so as to produce a maximum g=1.98 signal in the state $Q_A^-Fe^{2+}$. Notably, the high field component broadens and shifts to the right while the low field component is smeared out. Qualitatively, these effects can be explained by a theoretical model [5] assuming that only the EPR detectable $Q_A^-Fe^{2+}$ resonance at g=1.8 - 2.0 contributes to the split spectrum and the interaction between the two radicals is of an exchange character. A more thorough analysis is currently in progress.

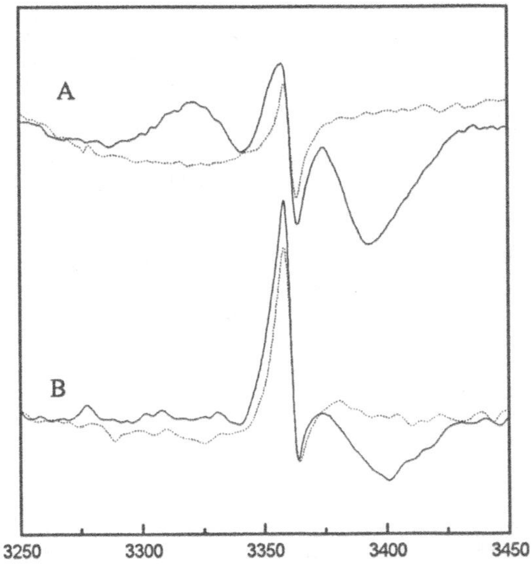

Fig. 2. Effect of the cyanide treatment at pH 6.5 on the split pheophytin signal. BBY membranes were treated with 300 mM NaCl (A) or 300 mM NaCN (B) at pH 6.5 and subsequently reduced with dithionite. Dotted lines:dark, solid lines : after illumination at 200K.
EPR conditions: T=4.2K, m-w power 31mW, m-w freq., 9.42 GHz, mod. App., 4 Gpp

3.3 *Oxidation of the low spin iron to Fe^{3+} (S=1/2)*
The untreated Fe^{2+}(S=2) in the oxidised state shows characteristic light sensitive high spin EPR signals in the g=5.5 to 8 region [6]. We have treated with ferricyanide the iron following conversion to low spin by the high pH cyanide treatment [2] and searched for low spin Fe^{3+} signals. Such signals were difficult to observe not only because of strong interference with the EPR signal of the oxidant but also because of unfavourable relaxation effects, which prohibited the detection at 4.2 K. Careful recording of dark - light spectra at about 17 K resulted in the spectrum of Fig. 3. The spectrum contains two weak (as expected due to the unfavourable recording conditions) but reproducible features at 2.60 and 2.33 attributable, respectively, to the g_x and g_y components of Fe^{3+}(S=1/2). These results offer additional support for the cyanide induced conversion of the iron to low-spin. Furthermore the light sensitivity of the signals provides alternative evidence to the observation that the iron remains in site following the cyanide treatment.

Fig.3. Low spin Fe^{3+} EPR signals following the high pH cyanide treatment. Difference (dark-illumination at 200K) spectrum of a BBY sample treated with 300 mM NaCN at pH=8.0 and oxidized with ferricyanide.
EPR conditions: T, 17K, m-w power, 31mW, m-w freq., 9.42 GHz, mod. app., 16Gpp.

References

1 Koulougliotis, D., Kostopoulos, Th., Petrouleas, V. and Diner, B.A. (1993) Biochim. Biophys. Acta 1141, 275-282.
2 Sanakis, Y., Petrouleas, V. and Diner, B. A. (1994) Biochemistry 33, 9922-9928.
3 Klimov, V.V., Dolan, Ed, Shaw, E.R. and Ke, B. (1980) Proc. Natl. Acad. Sci. USA 77, 7227-7231
4 Rutherford, A.W. and Zimmermann, J.L. (1984) Biochim. Biophys. Acta 767, 168-175.
5 Schepler, K.L., Dunham, W.R., Sands, R.H., Fee, J.A. and Abeles, R.H. (1975) Biochim. Biophys. Acta 397, 510-518.
6 Petrouleas, V. and Diner, B.A. (1986) Biochim. Biophys. Acta 849, 264-275.

ISOLATION AND PRELIMINARY CHARACTERIZATION OF A D1/D2/CYTOCHROME *b*-559 COMPLEX FROM THE CYANOBACTERIUM *PHORMIDIUM LAMINOSUM.*

M. Aured and R. Picorel
Department of Plant Nutrition, Estación Experimental Aula Dei (CSIC), Apdo. 202, 50080 Zaragoza, Spain

1. Introduction

The reaction center (RC) of phosystem II (PSII) is defined in terms of a protein unit capable of photochemical charge separation between the primary donor (P680) and the primary acceptor (Phe *a*). The isolation and characterization of D1/D2/cytochrome *b*-559 complex which constitutes the PSII RC is a useful step in the study of the details of molecular mechanism in PSII; also, comparisons of the PSII RC to the three-dimensional structure of the reaction centers of the purple non-sulfur photosynthetic bacteria has allowed to make predictions about the structure of the PSII RC. Along this line, we have isolated a D1/D2/cyt *b*-559 complex from a highly purified oxygen evolving core complex (OECC) preparation.

2. Procedure

Cells were grown at 45°C in 10 L bottle in the inorganic medium of Castenholz (1). The growth medium was bubbled with 5% CO_2 in air and continously illuminated from fluorescence tubes for 12 days. Cells were pelleted by centrifugation at 4°C for 10 min at 8,000xg. The pellet was washed once with buffer A [0.5 M sorbitol, 10 mM $MgCl_2$, 12.5 mM EDTA, 10 mM Hepes/NaOH, and 5 mM phosphate buffer (KH_2PO_4/K_2HPO_4), pH 7.5] plus 1 mM PMSF (2). The cells were sonicated for 20 min with 10 sec interval period at 4°C, and the homogenate was centrifuged at 200xg for 6 min. The supernatant was centrifuged again at 16,000xg for 150 min. The pellet was resuspended in buffer A and stored at -80°C.
The OECC preparation was isolated from thylakoid membranes. Membranes were washed twice with buffer C [10 mM $MgCl_2$, 10 mM Hepes/NaOH and 5 mM phosphate buffer (KH_2PO_4/K_2HPO_4), pH 7.5] to remove phycobiliproteins and then solubilized with 1% (w/v) LDAO (lauryl dimethylamine oxide) for 1 h, and the resulting mixture was centrifuged at 100,000xg for 1 h at 4°C. The solubilized material was loaded onto a 10-17.5% sucrose gradient containing 20% (v/v) glycerol, 0.05% (w/v) LDAO, 5 mM $MgCl_2$ and 20 mM Hepes/NaOH, pH 7.0 and centrifuged at 100,000xg for 24 h at 4°C (3). The PSII enriched fraction was carefully removed and loaded into the ion-exchange DEAE-Toyopearl column equilibrated with 20 mM Tris-HCl, pH 8.0 and 0.03% dodecyl β-maltoside (DM). The column was washed

P. Mathis (ed.), Photosynthesis: from Light to Biosphere, Vol. I, 827–830.

with the same buffer until the absorption at the red peak was aproximately 0.1-0.2, and then the OECC complex was eluted with 0.03% DM, 300 mM NaCl and 20 mM Tris-HCl, pH 8.0 (4). The OECC complex was incubated with 0.8 M Tris-HCl, pH 9.0, for 40 min, and centrifuged at 300,000xg for 3 h at 4°C to remove the 33 kDa protein, diluted in 200 mM sucrose, 0.03% DM, 1.5 % taurine and 20 mM Bis-Tris, pH 6.5. The core complex was solubilized with 4% Triton X-100 and the D1/D2/cyt *b*-559 complex was purified by Q-Sepharose Fast Flow chromatography. The complex was eluted with a 20-100 mM $MgSO_4$ gradient.

Secondary electron transport activity was undertaken at 4°C using 1 mM $MnCl_2$ and 200 μg/ml silicomolybdate (SiMo) as artificial electron donor and acceptor, respectively. The reduction of SiMo was monitored by the increase in absorption at 600 nm, shielding the phototube with a Scott BG18 cut-off filter. The PSII RC complex suspensions were illuminated from the top with a slide projector. The actinic light was filtered through a Scott RG665 and KG3 filters. The concentration of reduced SiMo over the period of illumination was calculated using the extinction coefficient of 4.8 mM^{-1} cm^{-1} (5,6). To determine the Chl/Phe ratio, PSII RC samples were extracted with 80% acetone, sonicated from 1-2 min to favor the release of pigments, and centrifugated for 5 min with a microfuge. Then, 1.5 mM HCl final concentration was added to the supernatant to induce total pheophytinization of the Chl. The concentration of Chl is calculated from the difference in the content of Phe between acetonic and pheophytinizated extracts (7). The amount of cyt *b*-559 was determined from the dithionite ***minus*** ferricyanide difference absorption spectra using the extinction coefficient, $\varepsilon^{1\,mM}_{559\,nm} = 21.0\ mM^{-1}\ cm^{-1}$ (7).

3. Results and Discussion

The absorption spectra of OECC and PSII RC are shown in Fig.1A and Fig.1B, respectively. Note the different 416/435 nm peak ratio between both preparations. The intensity of the 435 nm band mainly due to Chl is an indication of the purity of the PSII RC preparation.

Low temperature (77K) fluorescence spectra was measured in OECC and PSII RC samples upon excitation at 435 nm. The emission spectra of OECC and PSII RC complexes were characterized by an intense band at 683-684 and 682 nm, respectively (not shown).

Figure 2 shows a SDS-PAGE of the OECC samples isolated without (lane A) and with (lane B) PMSF. In lane A several bands appear at low molecular weight (20 kDa and lower), and the bands at 30 kDa and 34 kDa corresponding to D1 and D2, respectively were less marked. In contrast, the low molecular weigth bands were not found in lane B. Western blotting of both gels (data not shown) confirmed that one band at around 20 kDa was a D1 breakdown product. These data indicated that the presence of PMSF in the purification buffers, eliminate protein degradation. PSII RC obtained in the presence of PMSF is showed in the Fig.2C, the material contains several polypeptides at about 60 kDa (D1/D2 heterodimer), 34 kDa (D2), 30 kDa (D1), 20 kDa (probably D1 degradation product), and 9 kDa (α-subunit of cyt *b*-559). The photochemical activity of the PSII RC was about 9,500 μeq e^- μmol Chl^{-1} h^{-1}.

The Chl/Phe ratio was determined by spectroscopic methods. The total concentration of Chl plus Phe was calculated from the acetonic extract after its pheophytinization with 1.5 mM HCl, where Chl was completely converted to Phe, using the extinction coefficient of 49.3 $mM^{-1}cm^{-1}$ at 666 nm. The Chl concentration was calculated from the difference between this value and that calculated from the acetonic extract spectrum for Phe using an extinction coefficient of 6.504 $mM^{-1}cm^{-1}$ at 535 nm (7). From these data it was determined a pigment content of 5.5-6 Chl per 2 Phe. From the dithionite ***minus*** ferricyanide spectra it was calculated a ratio of 6 Chl /2 Phe /1 cyt ***b***-559. At this point we do not know the amount, if any, of CP47 present in this preparation.

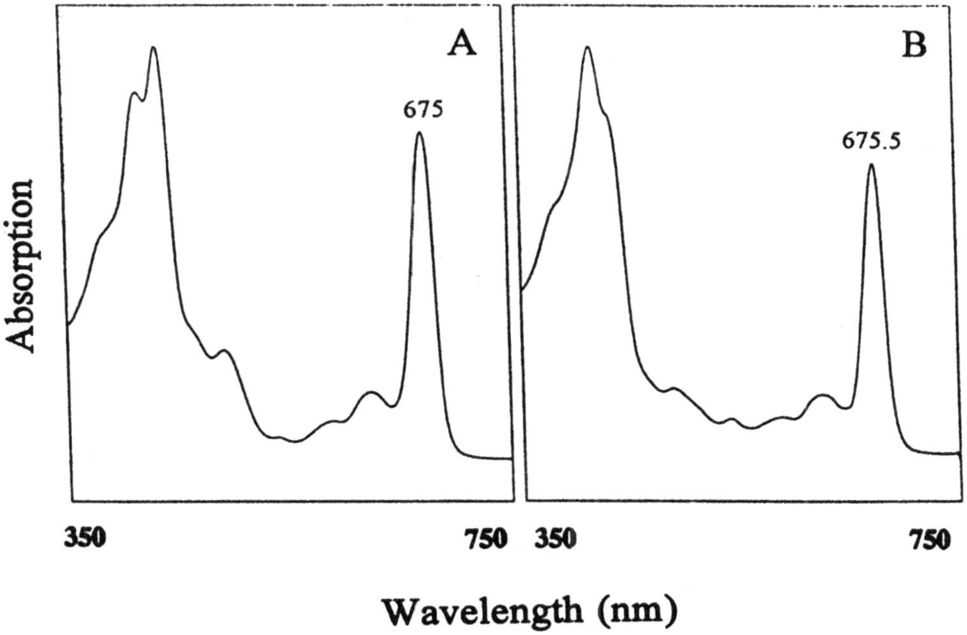

FIGURE 1: (A) Electronic absorption spectrum of the OECC complex preparation in 0.03% DM and 20 mM Tris-HCl, pH 8.0, (B) Electronic absorption spectrum of the PSII RC complex preparation in 200 mM sucrose, 0.03% DM, 1.5 % taurine, and 20 mM Bis-Tris, pH 6.5.

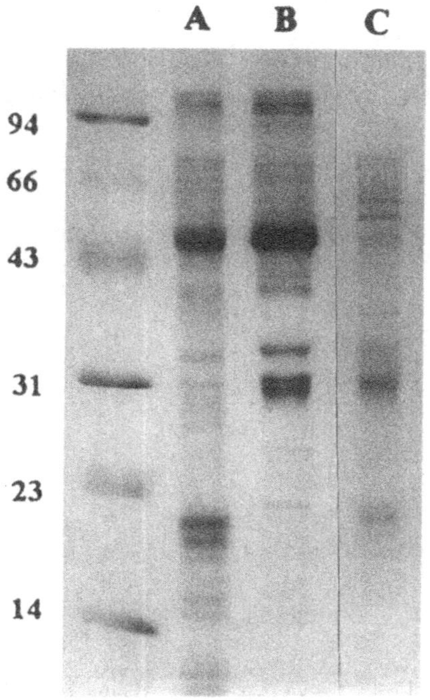

FIGURE 2: 10-17% gradient SDS-PAGE in 8 M urea of OECC isolated without (A), and with (B) 1mM PMSF, and PSII RC prepared using PMSF (C).

References

1 Castenholz, R.W. (1970) Schweiz. Z. Hydrol. 32, 538-551
2 Stewart, A.C. and Bendall, D.S. (1980) Biochem. J. 188, 351-361
3 Koike, H. and Inoue, Y. (1985) Biochim. Biophys. Acta 807, 64-73
4 de Vitry, C., Diner B.A. and Popot J.L. (1991) J. Biol. Chem. 266: 25, 16614-16621
5 McTavish, H., Picorel, R. and Seibert, M. (1988) Plant Physiol. 89, 452-456
6 Montoya, G., Cases, R., Rodríguez, R., Aured, M. and Picorel, R. (1994) Biochemistry 33, 11798-11804
7 Montoya, G., Yruela, I. and Picorel, R. (1991) FEBS Lett.283: 2, 255-258

DYNAMIC PHOTOACOUSTIC DIAGNOSTICS OF LASER INDUCED PROCESSES IN BACTERIAL REACTION CENTERS

Oleg V. Puchenkov, Zohar Kopf and Shmuel Malkin,
Biochemistry Department.The Weizmann Institute of Science,
Rehovot. 76100 Israel.

1. Introduction

Light induced pressure changes in photosynthetic materials result from heat emission, molecular volume changes and gas exchange [1]. When the light is pulsed or modulated transient pressure waves are formed and may be monitored by suitable detectors and recorded as photoacoustic (PA) signals. The relative weight of each contribution to the PA signal depends on the experimental arrangement. Thus, when a microphone is placed in the gas phase around an irradiated suspension, its signal monitors heat production from the sample. In the case where the sample is a leaf or a lichen and the modulation frequency is sufficiently low, the signal contains a strong contribution from oxygen evolution [1]. In this case the effect of molecular volume changes is negligible. On the other hand, in an experimental arrangement where a pressure sensor is coupled directly to the sample, both heat production and molecular volume changes will contribute. If the temperature is lowered to the maximum density point of an aqueous medium (3.98° C for pure water), heat production does not result in pressure changes and the resulting signal is entirely due to molecular volume changes.

Several studies used this second experimental arrangement type, employing short laser flashes for excitation [2-6]. Initially, the possibility of molecular volume changes was not regarded and the pressure changes were thought to represent only heat production. Calibration of the PA signal with a photochemically inert solution of identical light absorption and thermoelastic parameters, resulted in calorimetric information on the electron transfer process in the μs time range [2,3]. Recent studies [4-6], where experiments at the maximum density temperature were included, disclosed indeed notable molecular volume changes. Subtracting the signal due to molecular volume changes, in experiments at higher temperatures, enabled more exact calorimetric measurements.

The time dependence of the PA signal is however complex, consisting of several peaks and crests until the signal dies out. Part of this is due to the characteristics of the sensor, which tends to resonate. Part of this is due to the echoes produced by the sound pulse wave. The first peak and crest reflect the immediate pulse of pressure increase formed at the path of the laser beam, which propagates as a sound wave towards the sensor. Thus, a deconvolution of the signal to separate out the true time characteristics of the pressure changes is necessary. However, in most of the studies only the amplitude of the signal (the difference between the first maximum and minimum points) was used to represent the signal.

With this approach of using PA amplitudes only, a previous study [4] on reaction centers from *R. sphaeroides* R-26 gave the following results: A molecular volume change

P. Mathis (ed.), Photosynthesis: from Light to Biosphere, Vol. I, 831–834.

of about -12 Å^3 (a contraction) per reaction center was found to be induced by light absorption; the associated quantum yield of the photoreaction was very close to 1; the energy storage for the 590nm photons was 40%. Assuming that the photoreaction measured was the electron transfer between P_{870} and Q_A, which is faster than the time span in the PA observation (the electron transfer to Q_B is too slow for this type of PA detection), it follows that the enthalpy of the state $P_{870}^+ Q_A^-$, relative to the ground state, is about 0.83 eV.

While these numbers give an adequate idea on the magnitudes involved, they had to be reviewed by a more precise approach, for the following reasons: The molecular volume change per reaction center was obtained as the saturation value of the volume change at very high light pulse energies divided by the amount of reaction centers contained in the volume transversed by the laser beam. These values were uncertain: the saturation value could not actually be achieved and had to be found by extrapolation, and the volume that the laser beam transverses is not well defined, as the intensity profile continuously declines further from the beam center (having a Gaussian shape). Any deviation from these numbers can affect the quantum yield estimation (obtained by dividing the initial slope of the molecular volume change vs. the number of photons in a flash by the maximum volume change per reaction center) and hence also the estimation for the enthalpy of the charge separated state.

Therefore we undertook to re-evaluate these numbers by a refined approach, which will take into account all the above reservations. This report gives a concise account of the work. For a comprehensive account of experimental and theoretical details, see [7,8].

2. Experimental

The experimental system consisted of: (a) a pulsed, frequency doubled Nd:Yag laser (532nm, half width 35 ps), (b) an additional light source projector for delivering a continuous background light, (c) a spectrophotometric cuvette coupled to a pressure detector (ceramic PZT disc), which was enclosed in a stainless steel housing and pressed to the external wall together with vacuum grease for acoustic coupling. The laser beam passing through the cell was nearly parallel after optical conditioning by lenses and a pinhole. The cell was thermostated (± 0.2° C) using a temperature range 1-28° C. Data acquisition, on line monitoring and PA signal processing were done under the control of a virtual instrument written on Lab-View 3.0 software. The laser flash energy was adjusted by neutral density filters and measured by a pyroelectric radiometer. In each run of experiments requiring accumulation of PA waveforms at fixed laser energy we applied a program that rejected data from flashes that occasionally deviated from the set value of energy. For exact computations the intensity profile of the laser across the beam cross-section was measured precisely (conforming to a Gaussian shape).

Reaction centers of *R. Sphaeroides* (R-26 and wt) were prepared following the procedure in [9] and diluted in a TL buffer (10mM Tris, 1mM EDTA and 0.1% LDAO, pH 8.0). The maximum density point of the buffer was at about 3.5° C.

3. Data Analysis and Results

For the deconvolution of the PA signal we made a Fourier transformation to the frequency manifold, converting each time (t) dependence to a transformed equivalent frequency (ω) dependence. In the frequency domain the ultimate signal S(ω) is given as a product: $S(\omega) = \Gamma(\omega) \cdot K(\omega) \cdot [V_T(\omega) + V_{SR}(\omega)]$, where V_T and V_{SR} are the volume changes associated with the thermal and molecular structural rearrangement, respectively. K(ω) is the medium transfer function, describing the conversion of volume changes to pressure and its subsequent propagation, and Γ(ω) is the detector transfer function that takes care of the sensor and electronics response. The combined transfer function Γ(ω) ·

$K(\omega)$ can be calibrated by a reference solution in which the laser flash energy is totally and promptly converted to heat and where $V_{SR}=0$. For this, we used analytical expressions for $V_T(\omega)$ and $K(\omega)$ [7]. These functions depend on the thermoelastic properties of the medium. $V_T(\omega)$ is proportional to $(\beta\ E)/(\rho\ C_P)$, where β is the thermal expansion coefficient, C_P the heat capacity, ρ the density and E the absorbed laser flash energy, and depends also on the cross sectional distribution of the laser intensity and its time profile. Following this calibration we made experiments with reaction centers at the density inversion point, so that $V_T(\omega) = 0$. Here we considered a spatial distribution of volume changes dependent on the density of photon absorption, assuming Poisson distribution of photons to reaction centers, with adjustable parameters ϕ (the quantum yield of electron transport from P_{870} to Q_A) and v_o (the specific molecular volume change per reaction center for an electron transfer event). From this, one can calculate $V_{SR}(\omega)$ which depends on the values chosen for the free parameters, calculate $S(\omega)$, calculate its Fourier transform S(t) and compare with the experimental waveform. By adjusting the free parameters it was possible to calculate waveforms indistinguishable from the experimental ones, matching also the dependence of the PA amplitude on the total energy of the laser flash (Figs. 1, 2). This matching occurs for $\phi = 1$ and $v_o = -32$ Å^3 (± 3% standard deviation).

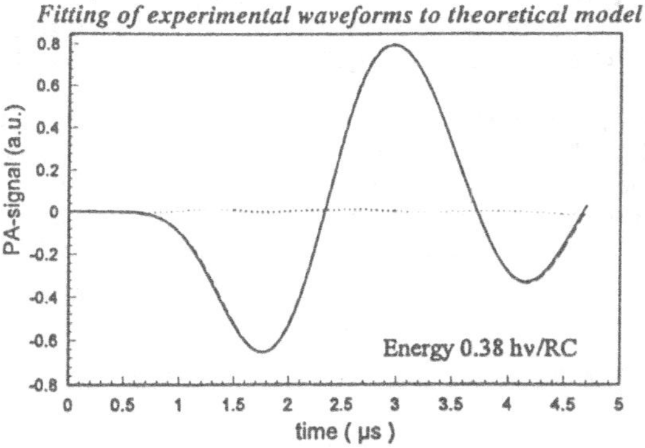

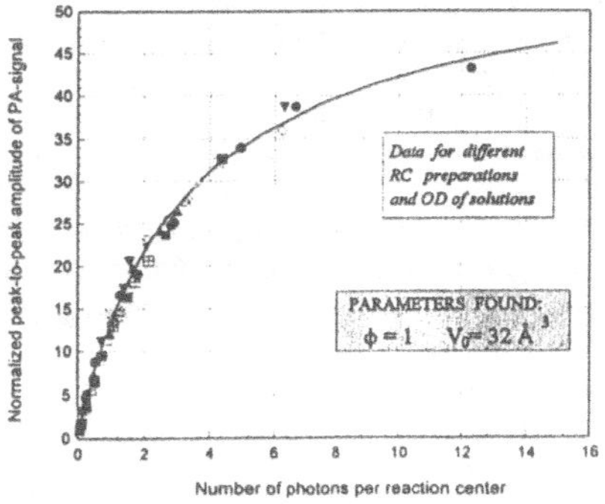

Fig. 1 (left). *Experimental (—) and theoretically fitted (----) PA signal waveform.*
Fig. 2 (right) *Experimental (points) and theoretically fitted curve of PA amplitude vs. no. of photons (in a single pulse) / Reaction-center*

It was noticed that the waveform of the total PA signal at temperatures higher than that of the maximum density point do not generally match the waveform of the molecular volume change. The contribution of the photothermal effect was therefore found by subtracting the waveform of the later from the total signal waveform. This photothermal signal was calibrated against a photochemically inert reference. With low laser energies (<< 1 hν / RC) there was an energy storage of 24% (± 2%), which for the wavelength used (532 nm) corresponds to an enthalpy gain of 0.56 (± 0.04) eV.

Analysing the PA photothermal signals at higher laser energies, there is a mixed response also from photons that hit closed reaction centers during the laser pulse time (≈ 35ps). Their contribution increases as the laser energy increases. In the range of low laser light energies the dynamics of the thermal and molecular volume changes was too fast to be discerned in the time window of the PA experiments. At high laser energies, however, the best analysis of the thermal contribution waveform from photons hitting closed reaction centers (i.e. in a state $P_{870}{}^+Phe_L{}^-$, hence these photons excite primarily Phe_M) indicates a mixture of a prompt part (≈85%), and a slow part (≈15%) with dynamics that corresponds to a time constant of about 600 (± 100ns). The nature of this long-lived state is yet unknown.

We also carried out experiments in the presence of continuos background light. We noticed the expected decrease of ϕ as the continuous light intensity increased, due to the closure of reaction centers (in a state $P_{870}{}^{+}Q_{B}{}^{-}$ - this state relaxes in about a second). Interestingly, a long exposure to a very strong background light (reducing ϕ to almost 0) induced some adaptive reversible changes in the reaction centers, which was noticed by a very slow relaxation of ϕ when the background light was switched off. There was an initial rapid relaxation to $\phi \approx 0.8$ within a minute (kinetically not resolved). But to reach the high dark relaxed value of 1 took about 15-20 minutes more.

4. Discussion

Our values for $v_o = -32$ Å^3 per reaction center is in good agreement with Arata and Parson [10], who used a hydrophone as a slow (ms) volume monitor. Our value of the enthalpy of the state $P_{870}{}^{+}Q_{A}{}^{-}$ (0.56 eV), while in good agreement with the delayed light measurements (0.63 eV) of Arata and Parson [11], does not agree at all with the hydrophone measurement ($\leq$ 0.13 eV). It is not clear to date what is the reason for this discrepancy. It remains to be seen whether in a time scale between 10μs and 1ms there is free energy transformation from enthalpic to entropic form (the standard free energy of $P_{870}{}^{+}$ and $Q_{A}{}^{-}$, from redox titrations, while each species is equilibrating with the external medium, is 0.6 eV [10]).

The molecular volume change does not amount to more than about 5% of the total volume of the reaction center protein. This shows the sensitivity of the method. It is not yet clear what is the origin for this change. It could be an electrorestriction effect, but it could also reflect small variation in the protein structure. In that respect the experiments of Mauzerall et al. [6] who manipulated the reaction center, artificially replacing the native quinone by a series of other ones, seem very attractive.

5. References

1. Malkin S & Canaani O. *Ann. Rev. Plant Physiol. Plant Mol. Biol.* 45: 493-526 (1994).
2. Nitsch C, Braslavsky SE & Schatz, GH. *Biochim. Biophys. Acta* 934: 201-212 (1988).
3. Nitsch C, Schatz GH & Braslavsky SE. *Biochim. Biophys. Acta* 975: 88-95 (1989).
4. Malkin S, Churio MS, Shochat S & Braslavsky SE. *J. Photochem.. Photobiol.* B 23: 79-86 (1994).
5. Delosme R, Béal D &Joliot P. *Biochim. Biophys. Acta* 1185: 56-64 (1994).
6. Mauzeral DC, Gunner MR & Zhang JW. *Biophys. J.* 68: 275-280 (1995).
7. Puchenkov OV. *Biophys. Chem.* in the press.
8. Puchenkov OV, Kopf Z & Malkin M. *Biochim. Biophys. Acta* in the press.
9. Fehr G & Okamura MY. In: "The photosynthetic bacteria" (Eds. Clayton RK & Sistrom WR), pp.349-386 Plenum. New York (1978)
10. Arata H & Parson WW. *Biochim. Biophys. Acta* 636: 70-81 (1981).
11. Arata H & Parson WW. *Biochim. Biophys. Acta* 638: 201-209 (1981).

Acknowledgement

This work was supported in part by the Israel Academy of Science and Humanities and by a postdoctoral fellowship to O.P. from the Feinberg Graduate School. Thanks are due to Dr. V. Brumfeld for his indispensable help and Dr. A. Scherz for his support. We acknowledge the gift of reaction centers from the Biophysics Dept, State University of Leiden, The Netherlands, by permission of Dr. A. Hoff and Dr. P.Gast.

NOVEL EFFECTS OF NITRIC OXIDE BINDING IN PHOTOSYSTEM II.

Charilaos Goussias, Yiannis Sanakis and Vasili Petrouleas.
Institute of Materials Science, N.C.S.R. "Demokritos", 15310 Aghia Paraskevi, Athens, Greece.

1. Introduction

Nitric oxide (NO) interacts with important PSII components. Notable is the competition of NO and CO_2/HCO_3^- for binding to the non-heme iron(II) of the acceptor side [1, 2], and the reversible interaction of NO with the tyrosyl radical Y_D^+ of the donor side [1]. We assign this latter effect to a nitrosation reaction. Furthermore, we report on two new EPR signals induced by NO treatment, coupled to the photosynthetic electron transfer.

2. Materials and methods.

BBY membranes at pH 6.5 were bubbled anaerobically with 3ml of a 2:3 gas mixture of $NO:N_2$ [1] and incubated in the dark at 10°C for 5 minutes (g=2.013 signal) or 4 hours (nitroxide radical signal). EPR spectra were recorded using a Bruker ER 200D- SRC spectrometer.

3. Results and discussion

3.1 *The reversible interaction of NO with Y_D^+ is attributed to a nitrosation reaction.*
It has been shown earlier that NO interacts reversibly with Y_D^+ to produce an EPR silent species [1]. The k_d for this effect is approximately 3μM. While NO interacts relatively quickly with Y_D^+ (within a few minutes at 4°C), it also dissociates easily by 2 minutes' evacuation at 0-4°C. Based on relatively recent work on the NO interaction with phenoxyl radicals [3] and with the tyrosyl radical of ribonucleotide reductase (M. Fontecave, personal communication), we assign the NO interaction with Y_D^+ to a nitrosation reaction.

3.2 *A photo-induced nitroxide radical signal*
In figure 1 we show the spectra of BBY membranes treated with 3ml of a 2/3 gas mixture NO/N_2 for 4h at 10 °C in the dark. As expected (see 3.1), the Signal II_{slow} is eliminated [1]. The broad signal at high fields represents free NO in solution. Subsequent illumination at 200K results in the formation of a new EPR signal. The

P. Mathis (ed.), Photosynthesis: from Light to Biosphere, Vol. I, 835–838.

features of this signal are typical of a nitroxide radical, as revealed after removal of free NO (fig. 2). From the spectrum of figure 2 the hyperfine tensor parameters are calculated to be $A_{\parallel}$=44 Gauss and $A_{\perp}$=22 Gauss leading to a value of 29.3 Gauss for the A_{iso} ($A_{iso}=(A_{\parallel}+2A_{\perp})/3$). The magnitude of the A_{iso} falls within the range usually observed for nitroxide radicals of the type >C=NO· (iminoxyl) [4].

Development of the maximum nitroxide radical signal at levels comparable to that of Signal II_{dark} requires long incubation times with NO (several hours) and appears to correlate with the release of Mn^{2+}. The nitroxide radical signal is vanishingly small in manganese-depleted preparations. If, however, the Tris-treatment is carried out after the NO-incubation a sizeable NR signal is produced by illumination.

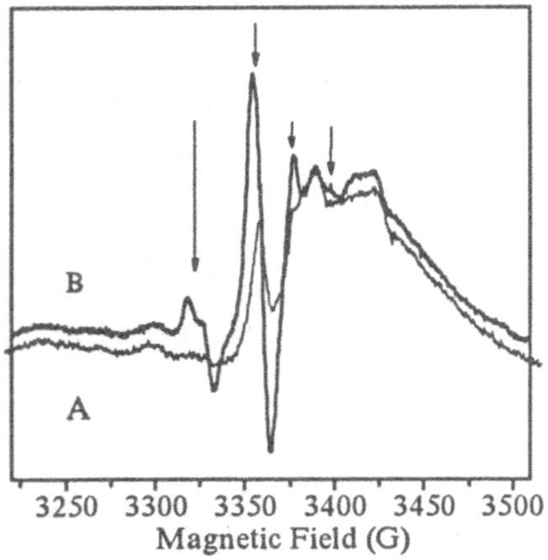

Figure 1. EPR spectra in the g=2.00 region of BBY membrane treated with NO in the dark. A, spectrum in the dark; B, after illumination at 200K. The arrows indicate the photoinduced signal. EPR conditions:T, 11K; m-w freq, 9.42 GHz; m-w power, 50μW; mod.amp.,4 Gpp

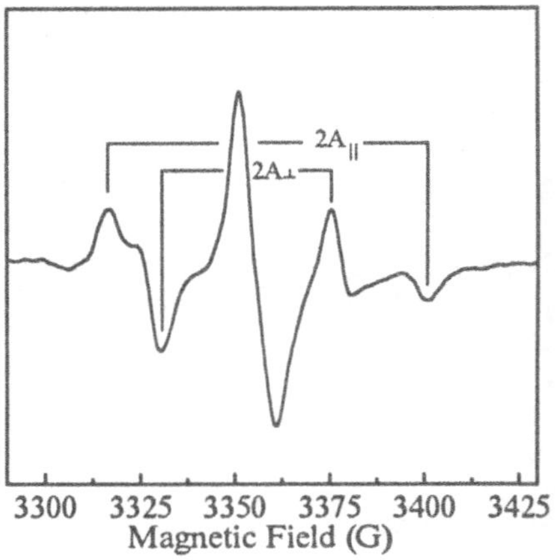

Figure 2. The nitroxide radical signal. EPR conditions as in fig. 1.

We assume that during the NO treatment a precursor species is formed slowly by interaction of NO with a PSII donor-side component. During light-induced charge separation this species either acts as an electron donor or interacts with an electron donor to form the nitroxide radical.

3.3 *A new EPR signal at g=2.013*

A new EPR signal (fig. 3) with an inflection point at g=2.013 is observed following NO treatment of BBY membranes, illumination at 200K and subsequent incubation in the dark at temperatures varying from -20°C to -50°C (half-time of signal evolution at -30°C is approximately 10 min). The signal is unstable at temperatures higher than -20°C.

Figure 3. The g=2.013 signal produced as described in the text. The feature at g=2.0016 is not related to the signal. Difference spectra (incubation at -30°C -prior illumination at 200K) recorded at A, 11K, 50μW and B, 200K, 200mW

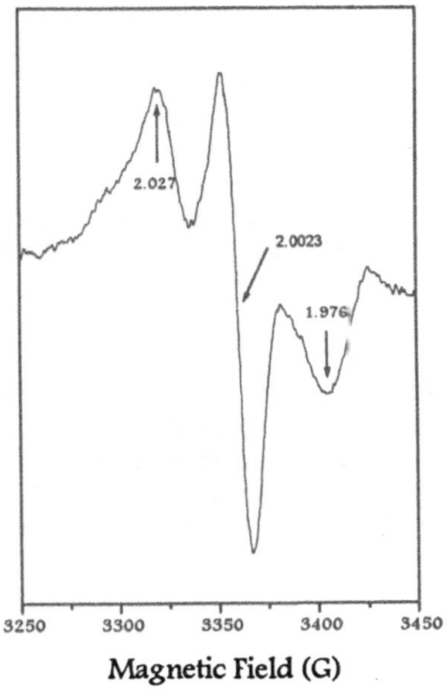

Figure 4. The NO-induced signal following illumination at 200K and incubation at -30°C in the presence of 80mM cyanide at pH 6.5. The derivative contribution at g=2.0023 is not part of the signal. EPR conditions: T, 11K; m-w power, 50μW; mod. amp., 4Gpp.

The signal does not correlate with the binding of NO to Tyr D^+ (see 3.1) or the formation of the nitroxide radical species (see 3.2). It is produced in both manganese-depleted and intact BBY membranes illuminated at 77K, 200K or 240K. The process of formation of the g=2.013 signal requires prior charge separation and formation of the $Q_A^-Fe^{2+}$-state before NO can react to form the EPR active species.

In cyanide-treated BBY membranes (under conditions that the $Q_A^-Fe^{2+}$-state exhibits the g=1.98 signal [5, 6]) a noticeable change on the shape of the g=2.013 signal takes place (Fig. 4). The microwave saturation properties at 11K also change from $P_{1/2}$=100μW, for the untreated sample, to $P_{1/2}$=10μW in the presence of cyanide.The identity of the signal is currently under investigation.

References

1 Petrouleas, V. And Diner, B.A. (1990) Biochim. Biophys. Acta, 1015, 131-140.
2 Diner, B.A. and Petrouleas, V. (1990) Biochim. Biophys. Acta, 1015, 141-149.
3 Janzen, E.G., Wilcox, A.L. and Manoharan, V (1993) J. Org. Chem 58, 3597-3599.
4 Berndt, A., & Neugebauer, F. A. (1987) Landolt-Börnstein Numerical Data and Functional Relationships in Science and Technology, Group II: Atomic and Molecular Physics (Fischer, H. Ed.), vol. 17c, Springer-Verlag.
5 Koulougliotis, D., Kostopoulos, T. Petrouleas, V. And Diner, B.A. (1993) Biochim. Biophys. Acta, 1141, 275-282.
6 Sanakis, Y., Petrouleas, V. And Diner, B.A., (1994) Biochenistry, 33, 9922-9928.

CONSTRUCTION AND INITIAL CHARACTERISATION OF A D2-LEU205TYR MUTANT OF *CHLAMYDOMONAS REINHARDTII*

C. Andronis[1], J.R. Durrant[2], B.A. Diner[3] S. Merry[2], D.R. Klug[2], and P.J. Nixon[1]
Photosynthesis[1] and Molecular Dynamics[2] Groups, Department of Biochemistry, Imperial College of Science, Technology and Medicine, London SW7 2AY, UK. [3]Central Research and Development Department, E.I. Du Pont de Nemours and Company, Wilmington, Delaware 19880-0173, USA.

1. Introduction

In the bacterial reaction centre (RC), absorption of light by the special pair P generates the excited state P*. Within approximately 3 ps an electron is transferred from P* to a bacteriopheophytin molecule on the active branch ($BPheo_A$) (1). Tyr210 of the M subunit (*Rhodobacter sphaeroides* numbering) has been suggested to facilitate electron transfer between the primary donor P and $BPheo_A$(2). Thus, site directed mutants of *Rb. sphaeroides* were constructed, where M-Tyr210 was replaced by phenylalanine (3, 4), isoleucine (4) or leucine (3). The replacement of M-Tyr210 with phenylalanine and leucine slowed down the formation of the radical pair $P^+BPheo_A^-$ from 3 ps to 16 ps and 22 ps, respectively (5), indicating the importance of this residue for primary electron transfer. The effect of the M-Tyr210Phe mutation was later found to originate from the molecular properties of the exchanged amino acids and not from secondary structural disorders (6).

Although the structure of the PS II reaction centre has not been determined, models have been consctructed (7) based on the sequence similarities between the L/M subunits of the bacterial RC and D1/D2, respectively. In the isolated PS II RC, the formation of the radical pair $P680^+$ $Pheo^-$ is thought to occur, at room temperature, with a lifetime of approximately 21 ps (8). This contrasts with the 3 ps lifetime that was reported for the bacterial RC. The amino acid residue in PSII thought to be analogous to M-Tyr210 is a leucine at position 205 of D2 (*Chlamydomonas reinhardtii* numbering, Leu206 in higher plants) (7). To test whether the difference detected between the PS II RC and the bacterial RC in terms of the rate of primary charge separation is because the analogous residue to M-Tyr210 in PS II is a leucine rather than a tyrosine, we have constructed the mutant D2-Leu205Tyr, in the unicellular green alga *Chlamydomonas reinhardtii.*

2. Materials and Methods

Chlamydomonas strain CC125 (wild type) was obtained from the Chlamydomonas Genetic Centre, Duke University, Durham, NC, USA. Cells were grown in Tris Acetate Phosphate (TAP) medium or High Salt minimal (HSM) medium and handled as described in (9). Plasmid pH3 was provided by J.-D. Rochaix. Site-directed mutagenesis was carried

P. Mathis (ed.), Photosynthesis: from Light to Biosphere, Vol. I, 839–842.

out according to (10) using synthetic oligonucleotides. Chloroplast transformation using a particle gun was performed essentially as described in (11).

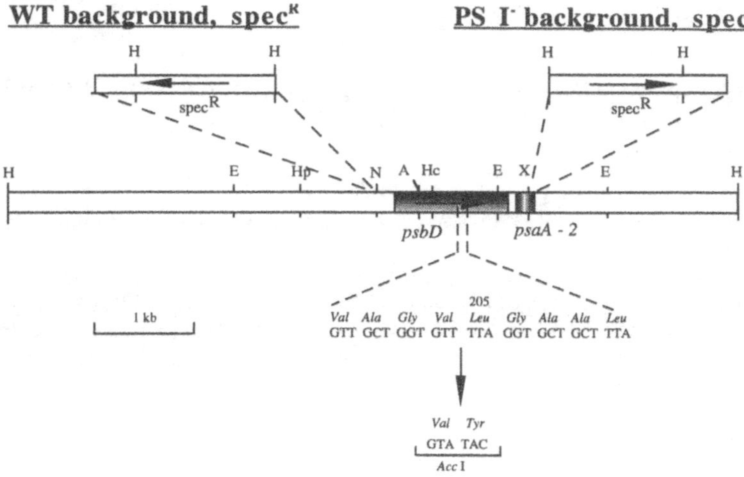

Figure 1: Composite map showing the mutations made to the *psbD* gene and flanking regions in the chloroplast genome of *C. reinhardtii*. The positions and orientations of *psbD, psaA-2* and the spec^R cassette in the various mutants are indicated. Restriction sites are: *Hind* III (H), *Eco* RI (E), *Nsi* I (N), *Acc* I (A), *Hinc* II (Hc) and *Xba* I (X).

Growth rates were determined in HSM medium at 25^{0}C and at an incident light intensity of 50-70 μE.m^{-2}.s^{-1}. The chlorophyll content of *Chamydomonas* cells and of various preparations was calculated using the equations of Arnon (12). Oxygen evolution of whole cells was measured in the presence of 1 mM potassium ferricyanide and 1 mM dichloro-*p*-benzoquinone (DCBQ) as electron acceptors. Measurements of the decay of variable fluorescence were carried out according to (13). PS II RC particles were isolated using a method described in (14). Femtosecond transient absorption measurements and data analyses were carried out according to (8, 15).

3. Results and discussion

A 4.7 kb *Hpa* I/*Hind*III fragment containing the *psbD* gene (which encodes D2) was subcloned from pH3 into the plasmid vector pTZ19U to generate plamid pCA1. A mutagenic oligonucleotide was designed to introduce the Leu205Tyr mutation together with a site for the restriction enzyme *Acc* I. In order to positively identify transformants, the *aadA* selectable marker, which confers resistance to spectinomycin (16), was placed either upstream or downstream of the *psbD* gene (Figure 1).This was done in order to generate two versions of the mutant in either a wild type or a PS I$^-$ background. As can be seen from Figure 1, immediately downstream of *psbD* lies *psa-A2*, the second exon of the *psaA* gene, which encodes one of the two main apoproteins of the Photosystem I RC. The selectable marker was introduced into an *Xba* I site inside the *psa-A2* exon in order to inactivate this gene. This would result in the generation of a mutant that would be PS I$^-$ in addition to carrying the D2-Leu205Tyr mutation. The creation of such a mutant would help us in the preparation of PS II RCs using the method of (14) where a mutant lacking the PS I and ATPase complexes was used for the isolation of PS II RCs.

Plasmids carrying the mutated *psbD* gene were introduced into the chloroplast genome of wild type (CC125) cells using a particle gun. Spectinomycin-resistant colonies were screened for the existence of the new *Acc* I restriction site by Southern blotting (Figure 2). Sequencing of DNA isolated from the transformants confirmed the presence of the Leu205Tyr mutation and revealed no differences in any other region of *psbD*.

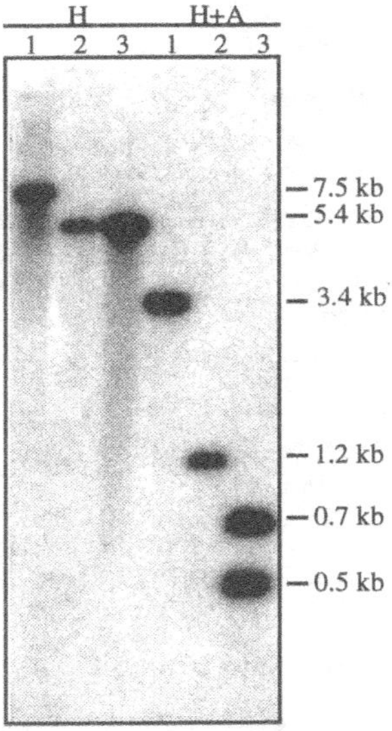

Figure 2: Southern blot hybridisation of DNA obtained from wild type (lanes 1); PS I$^-$ control (lanes 2) and D2-L205Y mutant in a PS I$^-$ background (lanes 3), following digestion with *Hind* III, *Hind* III/*Acc* I. The blot was probed with a 0.6 kb *psbD*-specific probe.

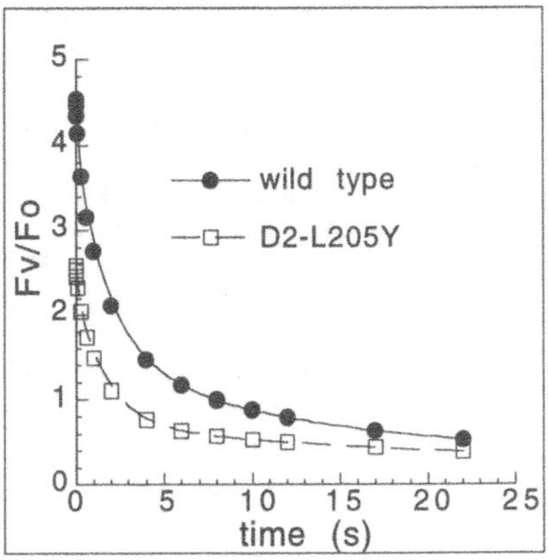

Figure 3: Relaxation of variable fluorescence (F_v) resulting from charge recombination between Q_A^- and the PS II donor side following a single saturating 2-μs flash excitation of whole cells of *C. reinhardtii*, in the presence of DCMU. The variable fluorescence F_v is normalised to F_0 [F_v/F_0]. The F_0 values at 50 μs (the earliest time point), are as follows: wild type, 7156; D2-L205Y, 11012.

The D2-Leu205Tyr mutant was able to grow photoautotrophically at similar growth rates to the wild type. Oxygen evolution measurements revealed that the mutant could evolve oxygen at approximately 60-70% of the wild type rates, although this could be attributed to a reduced number of functional PS II centres present in the mutant. Measurements using a control transformant that carried the *aadA* cassette upstream of the *psbD* gene but had an otherwise wild-type genotype, showed that the presence of the cassette alone did not affect the expression of *psbD* (data not shown).

Measurements of the relaxation of variable fluorescence in whole cells, did not show any serious disturbances in electron transfer between Q_A and Q_B in the mutant (not shown). Figure 3 shows that the mutation also has little effect on the donor side of PS II. However, a higher F_0 value was observed in the mutant. This result indicated a decrease in the photochemical quenching of fluorescence in the D2-Leu205Tyr mutant.

Femtosecond transient absorption measurements were performed on isolated PS II RCs isolated from the D2-Leu205Tyr and wild-type control strains. The rate of radical pair formation in the wild type occured with a lifetime of approximately 21 ps, the same as for pea PS II RCs (15). Preliminary measurements using PS II RC particles from the mutant, showed a considerable slowdown in the formation of the radical pair from 21 to approximately 38 ps. Moreover, a 3 ps component that has previously been associated with a small proportion of radical pair formation (17), does not appear in the mutant.

4. Conclusions

A genetic system has been developed for constructing site-directed mutants of the D2 protein in *Chlamydomonas reinhardtii* in either a wild-type or a PS I$^-$ background. The alteration of Leu205 to a Tyr resulted in reduced oxygen evolution rates as well as a decrease in the photochemical quenching of fluorescence in the mutant, although the latter was able to grow photoautotrophically at wild type rates. Moreover, femtosecond transient absorption measurements showed a slowdown in the primary electron transfer process. Therefore, it is evident that the difference observed between the bacterial reaction centre and the PS II RC in terms of the rate of primary charge separation, cannot be simply attributed to the difference of only this amino acid residue.

CA wishes to acknowledge the financial support of the State Scholarship Foundation of Greece (IKY). This work was funded by the Royal Society and BBSRC.

5. References

1. Kirmaier C. and Holten D. (1993) In: The Photosynthetic Reaction Centre, Deisenhofer J and Norris JR (eds) Vol. II, pp 49-70, Academic Press,
2. Allen J.P., Feher G., Yeates T.O., Komiya H. and Rees D.C. (1987). *Proc. Natl. Acad. Sci. USA* 84, 5730-5734
3. Gray K.A., Farchaus J.W., Wachtveitl J., Breton J. and Oesterhelt D. (1990). *EMBO J.* 9, 2061-2070
4. Nagarajan V., Parson W.W., Gaul D. and Schenk C. (1990). *Proc. Natl. Acad. Sci. USA* 87, 7888-7892
5. Finkele U., Lauterwasser C., Zinth W., Gray K.A. and Oesterhelt D. (1990). *Biochemistry* 29, 8517-8521
6. Chirino A.J., Lous E.J., Huber M., Allen J.P., Schenk C.C., Paddock M.L., Feher G. and Rees D.C. (1994). *Biochemistry* 33, 4584-4593
7. Ruffle S.V., Donnelly D., Blundell T.L. and Nugent J.H.A. (1992). *Photosynth. Res.* 34, 287-300
8. Hastings G., Durrant J.R., Hong Q., Barber J., Porter G. and Klug D.R. (1992). *Biochemistry* 31, 7638-7647
9. Harris H. (1989) The *Chlamydomonas* Sourcebook. Academic Press, New York
10. Kunkel T.A., Roberts J.D. and Zakour R.A. (1987). *Methods Enzymol.* 154, 367-382
11. Boynton J.E. and Gillham N.W. (1993). *Methods Enzymol.* 217, 510-536
12. Arnon D.I. (1949). *Plant Physiol.* 24, 1-15
13. Nixon P.J. and Diner B.A. (1992). *Biochemistry* 31, 942-948
14. Alizadeh S., Nixon P.J., Telfer A. and Barber J. (1995). *Photosynth. Res.* 43, 165-171
15. Giorgi L.B., Durrant J.R., Alizadeh S., Nixon P.J., Joseph D.M., Rech T., Barber J., Porter G. and Klug D.R. (1994). *Biochim. Biophys. Acta* 1186, 247-251
16. Goldschmidt-Clermont M. (1991). *Nucleic Acids Res.* 19, 4083-4089
17. Klug D.R., Rech T., Joseph D.M., Barber J., Durrant J.R. and Porter G. (1995). *Chem. Phys.* 194, 433-442

REACTION CENTER-PHOSPHOLIPID REVERSE MICELLES: KINETICS OF CHARGE RECOMBINATION

A. Mallardi[1], R. Angelico[1], M. Della Monica[2], M. Giustini[1], G. Palazzo[2] and G. Venturoli[3]
[1]CNR-CFILM, Via Orabona 4, 70126 Bari (Italy). [2]Dip. Chimica, Università di Bari (Italy). [3]Dip. Biologia, Università di Bologna (Italy).

1. Introduction

Phospholipids in apolar, organic solvents spontaneously form micelles. Such reverse (or inverted) micellar aggregates can be schematically described as consisting of water droplets (small water pools of about 20-200 Å diameter), surrounded by polar heads of the phospholipid molecules facing into the interior of the micelle while the hydrophobic chains are directed towards the bulk, continuous organic phase[1]. Although the existence and the possible role of such structures in native membranes remain a topic of research and discussion, reverse micelle (R.M.) solutions provide a powerful tool in investigating, *in vitro,* the behaviour of integral membrane proteins. The anisotropic and amphipatic nature of biological membrane is preserved in these organized microassemblies, as well as, the functionally relevant interface between phospholipid headgroups and water. More importantly, the presence of a continuous organic phase, in which hydrophobic compounds are soluble, allows a detailed study of the interaction of liposoluble molecules with membrane proteins, avoiding possible artefacts present in aqueous solutions (direct micelles of detergent and liposomes). It should be noted that the local concentration of liposoluble molecules in liposomes and in detergent dispersions is not easely controlled, being affected by the topological discontinuity of the hydrophobic domain. Here we report the preliminary results of a study on the effect of both the ubiquinone-10 (Q) concentration and the temperature on the rate of charge recombination in Reaction Centers (RC) from *Rhodobacter sphaeroides* R-26 solubilized in *n*-hexane R.M. Materials and methods are described elsewhere[2].

2. Results

The RCs were first incorporated into liposomes and the R.M. solution was formed by extracting into *n*-hexane the phospholipid-protein complex. In order to optimally preserve the RC

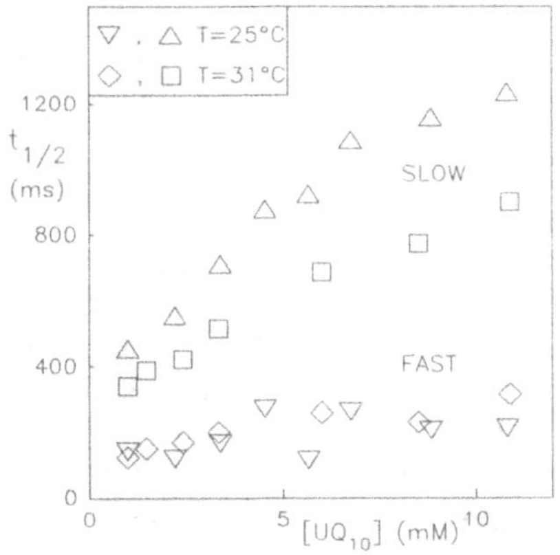

Figure 1 Q dependence of P^+ decay in R.M. at two temperatures.

P. Mathis (ed.), Photosynthesis: from Light to Biosphere, Vol. I, 843–846.

photoactivity it is necessary to use proteoliposomes made of a mixture of three phospholipids: phosphatidylcholine (PC), phosphatidylethanolamine (PE), and phosphatidylserine (PS) in the ratio PC/PE/PS = 1:1:2.

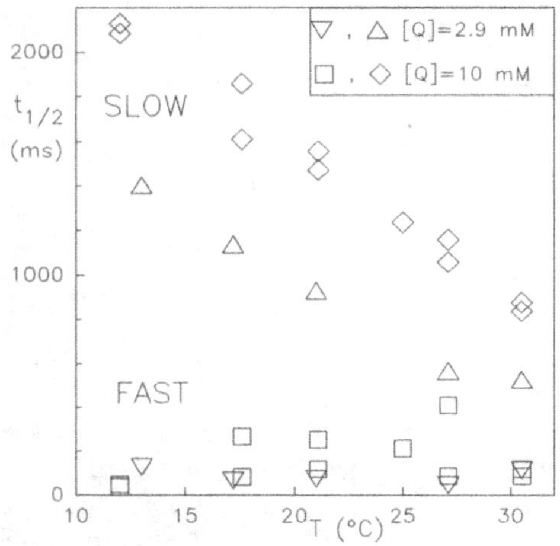

Figure 2 Temperature dependence of P^+ decay in R.M. at two different values of [Q].

The kinetic competence of RC's in performing normal photo-induced electron transfer was examined by flash spectrophotometry. In R.M. the RC lacks any exogenous electron donor to P^+ and the charge recombination of flash-generated states is observed. The recovery of P^+ was monitored at 605 nm at different temperatures and Q concentrations. Under all conditions tested, the decay can be accounted for by two exponential phases. Results of this deconvolution procedure are summarized in Figures 1-2. Fig. 1 shows the quinone dependence of the recombination kinetics, at two temperatures (25°C and 31°C). The half-time ($t_{1/2}$) of the fast phase is essentially unaffected by the concentration of Q and its amplitude remains constant accounting for approximately 10% of the total (not shown). On the other hand, the $t_{1/2}$ of the slow phase strongly depends on [Q], increasing at increasing quinone concentration. This effect is observed over the whole temperature range investigated (see Fig. 2).

In Fig. 2 the temperature dependence of $t_{1/2\ fast}$ and $t_{1/2\ slow}$ is presented at two different values of [Q]. The half-time and the amplitude of the fast phase are temperature independent. On the contrary, the $t_{1/2}$ of the slow phase shows a marked temperature dependence, being halved by a 20°C increase of the temperature.

3. Discussion

The decay of P^+ is accurately described by two exponential phases: a slow component, accounting for 90% of the recovery under all conditions and a fast minor phase, characterized by kinetic parameters which are unaffected by temperature and Q concentration. In wiew of this independence, we attribute the latter component to $P^+Q_A^-$ recombination, taking place in a fraction (approximately 10%) of RC's in which the Q_B site has been irreversibly damaged (most likely during the extraction in organic solvent). The half time of the slow phase, originating from RC's in which the photoinduced charge separation involves Q_B, increases at increasing Q concentrations.

This is the behaviour predicted when the binding equilibrium of quinone to the Q_B site is established rapidly as compared to the charge recombination process from $P^+Q_A^-$. Under this assumption, the overall recombination from $P^+Q_B^-$ is described by a single component, with an apparent rate constant $(k_p)^3$ given by eq. 1, where k_{AP} is the rate constant of $P^+Q_A^-$ recombination and K_{AB}^{app} is determined by the equation:

$$k_p = \frac{k_{AP}}{1 + K_{AB}^{app}} \quad (1)$$

$$K_{AB}^{app}=\frac{K_{AB}K_{bind}[Q]}{1+K_{bind}[Q]} \qquad (2)$$

in which K_{AB} is the equilibrium constant for electron transfer from Q_A^- to Q_B, and K_{bind} is the equilibrium constant for Q binding to the RC Q_B site. In terms of half-times eq.(1) can be written as:

$$\frac{t_{1/2\ P}}{t_{1/2\ AP}}-1=K_{AB}^{app} \qquad (3)$$

From the data of Fig. 1 K_{AB}^{app} can be calculated using eq.(3) and the value $t_{1/2\ AP}=80$ ms (measured in presence of terbutryne). The values of K_{AB}^{app} obtained as a function of Q concentration at 25° and 31° (Fig. 3) have been fitted to equation 2, yelding the K_{AB} and K_{bind} parameters listed in Tab.I.

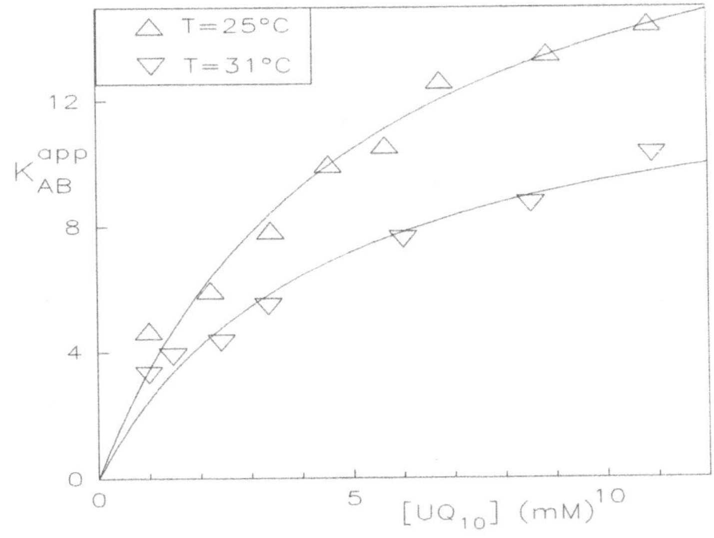

Figure 3 K_{AB}^{app} as a function of quinone concentration at 25°C and 31°C. The lines are fitting curves according to eq(2).

From the data of table I, ΔH° and ΔS° for the electron transfer and binding processes can be determined according to

$$-k_BT\ \ln K=\Delta G^\circ=\Delta H^\circ-T\Delta S^\circ \qquad (4)$$

Where k_B is the Boltzmann constant.

Table I

T (°C)	K_{AB}	K_{bind} (M^{-1})
25±0.2	21.4±2.0	191±38
31±0.2	13.8±1.4	219±50

The obtained values of ΔG° (at 298 K), ΔH° and ΔS°, are listed in Tab.II.

Table II Thermodynamic parameters for the electron transfer $Q_A^-Q_B \rightarrow Q_AQ_B^-$ and for the binding equilibria.

$\Delta G_{AB}°$ (meV)	-78.7	$\Delta G_{bind}°$ (meV)	-134.9
$\Delta H_{AB}°$ (meV)	-571	$\Delta H_{bind}°$ (meV)	+177
$\Delta S_{AB}°$ (meV K^{-1})	-1.65	$\Delta S_{bind}°$ (meV K^{-1})	+1.05

The data of Fig. 2 can now be used to test the consistency of the above reported procedure. Combining equations (2) and (4) we obtain:

$$K_{AB}^{app} = \frac{\exp\left[-\frac{\Delta H_{AB}°}{k_B T} + \frac{\Delta S_{AB}°}{k_B}\right] \exp\left[-\frac{\Delta H_{bind}°}{k_B T} + \frac{\Delta S_{bind}°}{k_B}\right] [Q]}{1 + \exp\left[-\frac{\Delta H_{bind}°}{k_B T} + \frac{\Delta S_{bind}°}{k_B}\right] [Q]} \tag{5}$$

Fig. 4 shows the experimental values of K_{AB}^{app} obtained using eq.(3) for two different values of [Q] and the correspondent temperature dependence of K_{AB}^{app} predicted by equation 5, when the thermodynamic parameters of Table II are assumed. The values of Tab.II reasonably account for the temperature dependence at different Q concentrations. The deviation from the calculated behaviour occurring at the higher Q concentrations below 20°C might be related to a decrease in the phospholipid mobility. An increase in the micellar wall rigidity could depress the binding equilibrium resulting in values of K_{AB}^{app} lower than predicted.

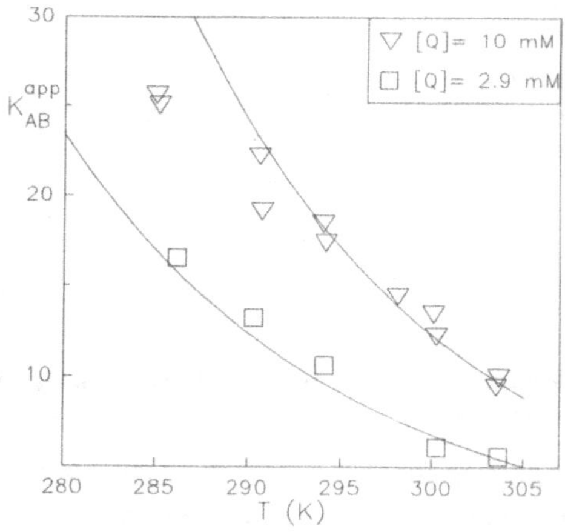

Figure 4 K_{AB}^{app} as a function of the temperature. Continuous lines are drawn according to eq.(5) (see text).

4.References

[1] Walde, P., Giuliani, A.M., Boicelli, C.A. and Luisi, P.L. (1990) Chem. Phys. Lipids 53, 265-288.

[2] Agostiano, A., Catucci, L., Della Monica, M., Mallardi, A., Palazzo, G. and Venturoli, G. (1995) Bioelectrochem. Bioenerg. in press.

[3] Shinkarev, V.P. and Wraight, C.A. (1994) in The Photosynthetic Reaction Center (Deisenhofer, J. and Norris, J.R., eds.) pp.193-255, Academic Press, New York, USA.

PHOTOACTIVITY OF REACTION CENTER IN DIFFERENT MEMBRANE MODELS

L. Catucci, A. Agostiano, G. Colafemmina and M. Della Monica
Dipartimento di Chimica, Università di Bari, Via Orabona 4, 70126 Bari, ITALY.

Introduction

The Reaction Center (RC) from *Rhodobacter sphaeroides* plays a prominent role in the field of bacterial photosynthesis. Its structure has been determined by high resolution X-ray cristallography and its photochemistry is probably better characterized than that of RCs from any other photosynthetic organism. Despite this widespread interest, little information concerning its photoactivity (in particular the charge recombination) in presence of phospholipids (e.g. in chromatophore or in membrane mimetic systems) has been reported.In this proceeding we report preliminary results concerning the influence of lipid environment on RC charge recombination kinetics. Preparation methods and investigative techniques have been described in a previous publication (1).

Results and Discussion

In Fig. 1 RC charge recombination kinetics (followed at 600 nm by flash-spectroscopy) in liposomes made of different phospholipids (PL) are reported. It is evident that in all the liposomes $P^+Q_B^-$ decay is strongly slowed down compared to that obtained in LDAO dispersion ($t_{1/2}$=800 ms), suggesting greater stabilization of the $P^+Q_B^-$ state in a lipidic environment. Furthermore, the chemical structure of the PL employed strongly affects the experimental half-time of P^+ decay, which increases from 1580 ms in liposomes made of phosphatidylcholine (PC), to 2560 ms in those of phosphatidylethanolamine (PE) and finally to 3325 ms in those made of phosphatidylserine (PS). This influence, already reported for other membrane proteins (2,3), is probably related to either specific structural features of the

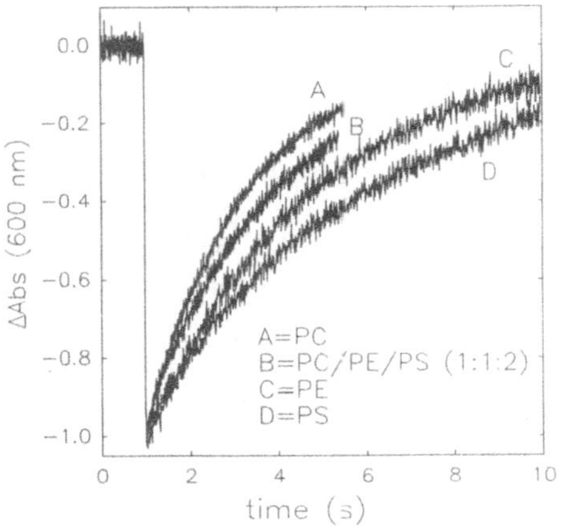

Figure 1 Charge recombination kinetics of the RC in different proteoliposomes. [RC]=5 μM, [Q]=250 μM; A, C, D: [PL]=10 mM; B: [PL]=25 mM.

P. Mathis (ed.), Photosynthesis: from Light to Biosphere, Vol. I, 847–850.

lipids or to their macroscopic organization.

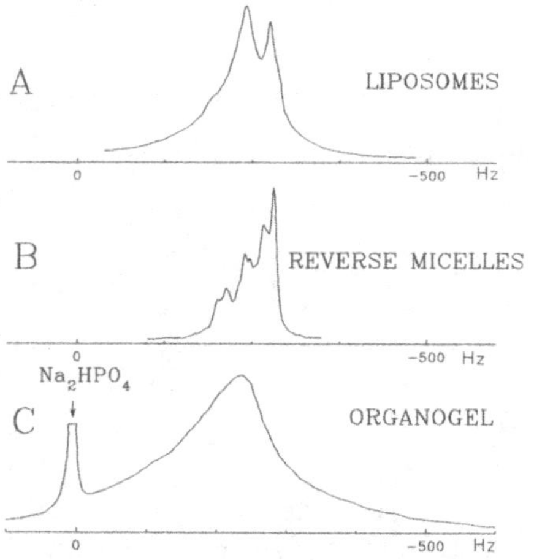

Figure 2 Phospholipid ^{31}P NMR spectra of different RC hosting systems.

Specific interactions between the RC and the polar head group of the lipid used can be involved. In fact, PS, characterized by a negatively charged polar head, can interact with the positively charged zones of the RC, located between the hydrophylic moieties of the protein, stabilizing the charge separate state more than the other two phospholipids. PC and PE, instead, are both zwitterionic molecules, but differ for their packing properties. In fact, PC prefers to organize in the bilayer phase, while PE prefers a reverse hexagonal (H_{II}) organization (4). These dissimilar molecular shapes could induce differences in the vescicle curvature and in the bilayer fluidity accounting for the different kinetic behaviour of the RC in PC and PE proteoliposome preparations.

In the case of liposomes made of a mixture of PL, whose main components PC, PE, PS are in a ratio of 1:1:2, the experimental half-time was about 2000 ms. In Fig. 2A, the ^{31}P-NMR spectrum relative to this latter liposome preparation is reported. The enlarged peaks of the two main components (PC and PS) of the mixture, covering that of PE, confirm reduced mobility of the lipid polar heads in the liposomal bilayer. The charge recombination rate decrease observed in proteoliposomes, could indicate that a decrease in the phospholipid mobility increases the stabilization of the charge separated state, in agreement with investigations carried out below and above the phospholipid phase transition temperature in RCs of *Rhodopseudomonas viridis* reconstituted in liposomes (5).

In order to evaluate the influence of lipid mobility on RC photoactivity,

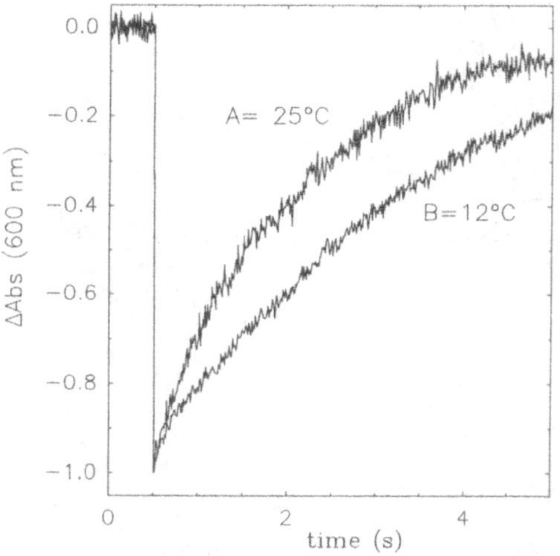

Figure 3 RC charge recombination kinetics in reverse micelles at two different temperatures. [RC]=3 μM, [Q]=10 mM.

different membrane mimetic systems have been used.
Proteoliposomes prepared using the mixture PC:PE:PS in the ratio 1:1:2, have been extracted with hexane. This composition allows the better yield of protein transfer into organic phase (6). The PL in the apolar solvent are organized as reverse micelles, characterized by a lipid mobility greater than that present in liposomes and acting as host system for the RC.
Recombination kinetic measured in this system (at 25 °C) results faster compared to that in proteoliposomes, with an observed half-time of about 1000 ms, as shown in Fig. 3 (trace A). This could probably be due to the increased lipid mobility, confirmed by the ^{31}P-NMR spectrum showed in Fig. 2B, where all resonance bands of the phospholipids are narrow and clearly resolved. The RC kinetic behaviour in this system is strongly influenced by the quinone concentration (as widely reported in the relative proceeding of Mallardi et al.(6)) but, keeping the quinone concentration constant, a dependence on temperature has been observed. The experimental half-time obtained at 12°C, at which a reduced lipid mobility is plausible, is slowed down to about 1900 ms, as evident in Fig. 3 (trace B).

Table I - Results of the deconvolution analysis of the kinetic traces -

Samples	Ampl. %	$t_{1/2}$ (ms)	Ampl. %	$t_{1/2}$ (ms)	Ampl. %	$t_{1/2}$ (ms)
liposomes PC	11	684	89	1810		
liposomes PE	2	350	98	2685		
liposomes PS	8	821	92	3910		
liposomes PC/PE/PS	31	1580	69	2460		
reverse micelles t=12°C	6	50	94	2085		
reverse micelles t=25°C	8	211	92	1236		
organogel [Q]=2 mM	42	114	58	1120		
organogel [Q]=6 mM	10	66	17	640	73	2160

In all systems examined the RC recombination kinetics never fit to a single exponential decay and any dependence on the measure wavelength was found in the range 540-610 nm. Two exponential phases are required to accurately fit the recovery of P^+. The results of the deconvolution analysis are summarized in Table I. In proteoliposomes the half-times relative to the two phases are always greater than the value reported in literature for $P^+Q_A^-$ recombination. Instead, in reverse micelles the half-time of the fast phase can be ascribed to a $P^+Q_A^-$ recovery.
In PL reverse micelle system viscosity is strongly affected by both the PL concentration and the ratio [water]/[lipid]. Increasing the lipid concentration the gelification of the

micellar system is obtained. A complex multiphasic kinetic is found for RCs in these organogels. In some cases three exponential phases are required to fit the P^+ decay. The fast phase, accounts for the $P^+Q_A^-$ recombination. In fact, after the addition of terbutryne (an inhibitor of electron transfer from Q_{A-} to Q_B) the whole kinetic becomes mono-exponential with the same half-time. In organogel the slow phase of the RC charge recombination kinetic (attributable to $P^+Q_B^-$) presents a prolonged half-time. This result is in agreement with the hypothesis that the charge separated state, $P^+Q_B^-$, is more stabilized in a system characterized by a reduced lipid mobility typical of the organogel and as shown in the Fig. 2C, where the ^{31}P-NMR is reported. The single broad NMR band suggests, in fact, a lesser conformation freedom of lipids in the organogel system.

The fast phase amplitude shows a dependence on the quinone and PL concentrations. In Fig. 4 the traces relative to the charge recombination of RC at two different quinone concentrations are reported. The amplitude of the fast phase in trace A ([Q]= 2 mM) is 42%, while in trace B ([Q]= 6 mM) is 10%. An increase of the fast phase amplitude with PL concentation rise has been also observed (data not shown).

The possibility to induce an increase in the extent of the fast phase in a micellar system, by means of dehydration, has been already reported by Kendall-Tobias et al. (7). Increasing the PL amount, keeping the water amount constant (as we have done), induces the same phenomenon obtained decreasing the amount of water at constant PL concentration, suggesting an influence of the ratio [water]/[PL] on the capability of the Q_B site to act as electron acceptor. However, further water addition does not decrease the fast phase amplitude. Currently the dependence of the extent of $P^+Q_A^-$ recombination on both the quinone concentration and the ratio [water]/[PL] is as yet unclear and further investigations are required.

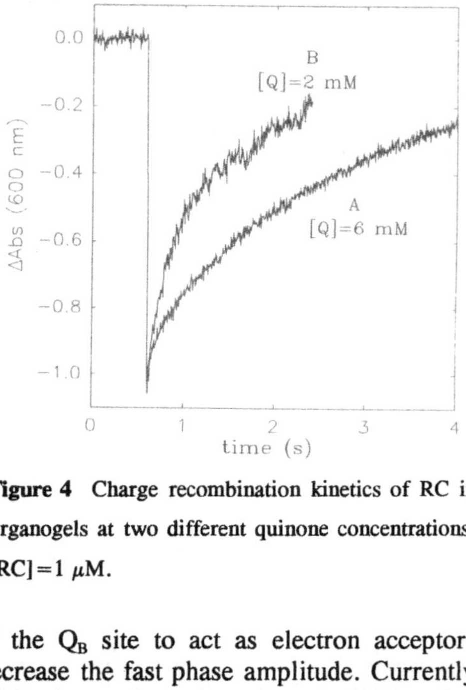

Figure 4 Charge recombination kinetics of RC in organogels at two different quinone concentrations. [RC]=1 μM.

References

1 Agostiano, A., Catucci, L., Della Monica, M., Mallardi, A., Palazzo, G. and Venturoli, G. Bioelectrochem. Bioenerg. 00(1995)BEB01789 T (In press).
2 Jensen, J.W. and Schutzbach, J.S. (1988) Biochemistry 27, 6315-6320.
3 Wiedmann, T.S., Pates, R.D., Beach, G.M., Salmon, A. and Brown, M.F. (1988) Biochemistry 27, 6469-6474.
4 De Kruijff, B., Kullis, P.R., Verkleij, A.J., Hope, M.G., Van Echteld, C.G.A. and Taraschi, T.F. (1985) in The Enzymes of Biological Membranes (Martonosi A.N., ed.) pp. 131-204, Plenum Press, Ney York.
5 Baciou, L., Gulik-Krzywicki, T. and Sebban, P. (1991) Biochemistry 30, 1298-1302.
6 Mallardi, A., Angelico, R., Della Monica, M., Giustini, M. and Palazzo, G. This volume.
7 Kendall-Tobias, M.W., Celis, H., Almanza De Celis, S. and Crofts, A.R. (1981) Biochim. Biophys. Acta 635, 585-601.

P* VIBRATIONAL COHERENCE IN MEMBRANE-BOUND REACTION CENTERS OF THE M210 TYR →LEU MUTANT OF R. SPHAEROIDES

Marten H. Vos[1], Michael R. Jones[2], C. Neil Hunter[2], Jacques Breton[3], Jean-Christophe Lambry[1] and Jean-Louis Martin[1]
[1] Laboratoire d'Optique Appliquée, INSERM U275, CNRS URA 1406, Ecole Polytechnique-ENSTA, 91120 Palaiseau, France; [2] Krebs Institute for Biomolecular Research and Robert Hill Institute for Photosynthesis, Department of Molecular Biology and Biotechnology, University of Sheffield, Western Bank, Sheffield, S10 2UH, United Kingdom;
[3] SBE/DBCM, CEN de Saclay, 91191 Gif-sur-Yvette Cedex, France

1. Introduction

For bacterial reaction centers (RC), the potential energy surface of the excited donor state P^* is strongly displaced with respect to the ground state for a set of low-frequency vibrational modes, the molecular characteristics of which are essentially unknown. We have previously studied these vibrational modes in the time domain using femtosecond spectroscopy to monitor oscillations in the stimulated emission band shape [1,2] and shown that phase correlations for those modes are maintained on the timescale of electron transfer (ET). A similar conclusion was recently obtained in a spontaneous emission study of bacterial reaction centers [3]. For wild type (WT) RCs at low temperature, the observed oscillatory features damp concomittantly with depletion of P^* by ET and set a *lower limit* of the timescale of the vibrational dephasing process by interactions with the bath. The relation between the damping and ET can be studied using RCs with modified ET properties. In this paper we present preliminary results on RCs of the M210 Tyr→Leu single-residu mutant of *R. sphaeroides*, for which it is known that ET slows down [4]. A comparison is made with WT RCs.

2. Materials and Methods

The construction of the mutant reaction centers where Tyr M210 is changed to Leu has been described previously [5]. The deletion and expression systems used to construct mutants with a reaction center-only phenotype are described in detail in Ref. 6. The growth of mutant strains of *R. sphaeroides* and the preparation of intracytoplasmic membranes for spectroscopy was as described in Refs. 2,5. The measurements were performed on RC-only membranes suspended in a glycerol/buffer glass at 10K in a convection cryostat. The arrangement of the femtosecond pump-probe spectrometer running at 30 Hz was essentially as described previously [2]. The ~30 fs pump pulses were centered at 880 nm. After dispersion in the monochromator, the test and reference probe beams were focussed

P. Mathis (ed.), Photosynthesis: from Light to Biosphere, Vol. I, 851–854.

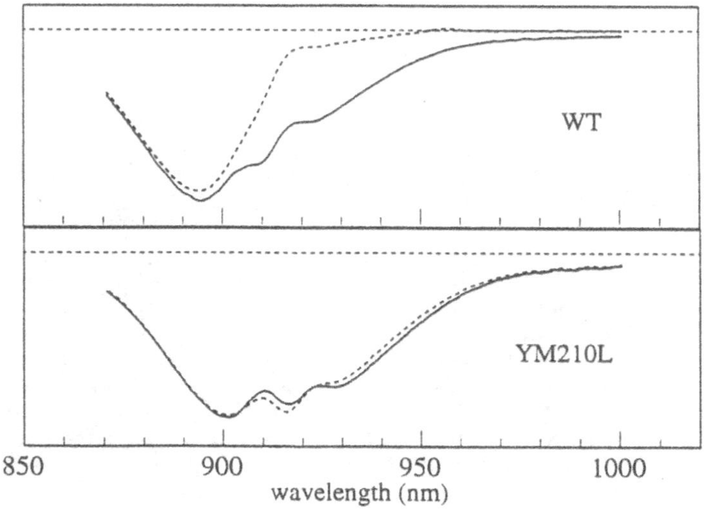

Figure 1 *Transient absorpton spectra at delay times of 500 fs (solid lines) and 3 ps (dashed lines).*

on a CCD camera, which was red out in a dual array configuration. The readout time of the camera was fast enough for shot-to-shot normalization at 30 Hz, of entire spectra with a resolution of ~1 nm.

3. Results and Discussion

3.1 Transient spectra

The shape of the 10K ground state electronic spectra of the P band are very similar for WT and Y(M210)→L (not shown), with the spectrum of the mutant slightly red-shifted (~3 nm). The similar shape indicates that the vibrational progression underlying the P band is similar. Transient absorption spectra in the P band absorption and P^* emission region are shown in Fig. 1 for the mutant and WT at delay times of 500 fs and 3 ps. The spectra are superpositions of the bleaching of the P band in the blue-most part of the spectra and stimulated emission, which is red-shifted with respect to the P band. The stimulated emission spectrum of the mutant is also qualitatively similar to that of WT. However, the overall spectrum does virtually not decay on the timescale of a few ps and in contrast to WT, the observed spectral evolution is only due to vibrational dynamics. In this mutant, at 10K, P^* decays on the timescale of hundreds of picoseconds.

The characteristic shape of the transient spectra with three maxima is the result of the superposition of the sharply rising edges of the absorption and emission spectra as is explained elsewhere [7]. The vibrational dynamics can best be studied in the kinetics at individual wavelengths at either edge of the emission band where the fundamental frequencies with which the spectrum oscillates are detected [1]. Below data are shown at the detection wavelength of 940 nm, at the red edge of the emission band both for WT and for the mutant.

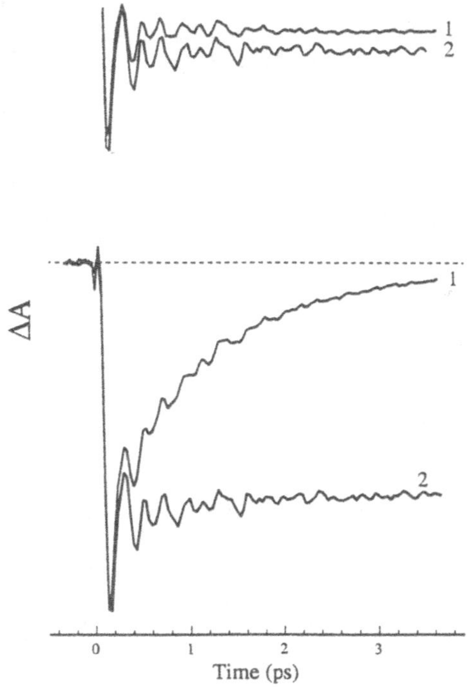

Figure 2:
Transient kinetics at 940 nm (lower) and their oscillory parts, obtained by subtracting an exponential fit from the kinetics (upper). 1: WT. 2: Y(M210)→L mutant.

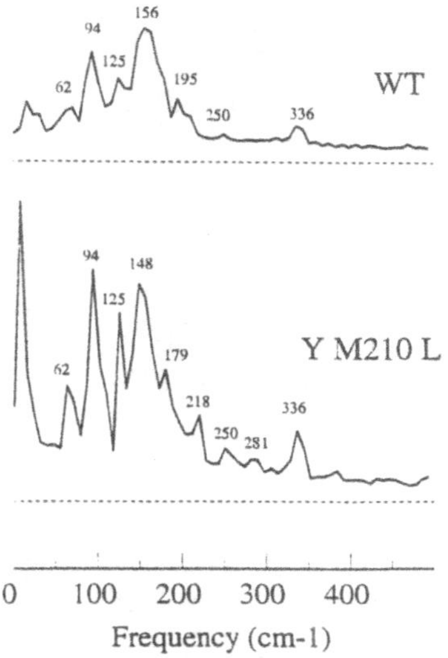

Figure 3:
Fourier transforms of the oscillatory parts of Fig. 2.

3.2 Transient kinetics

Fig. 2 shows the transient kinetics at 940 nm. Both kinetics are clearly modulated in a complicated manner. Comparison of the oscillatory parts shows that the modulations are very similar for the two types of RCs. However, in WT they damp out on the timescale of ~1 ps, whereas in the mutant oscillatory features are still observed after 3-4 ps. This strongly indicates that, for WT, damping of the oscillations is primarily due to depopulation of the P^* state, which also occurs in ~1 ps, and *not* to vibrational dephasing or relaxation by interactions with the bath.

Fig. 3 shows the Fourier transform (FT) spectra of the oscillatory parts. The very similar oscillatory patterns correspond to FT spectra with the main vibrational bands at the same positions. Minor differences are found in the very small bands in the relatively high frequency region (> 180 cm^{-1}). The bands of the mutant FT spectrum are substantially narrower, which corresponds to the observed longer damping times. A quantitative analysis shows that the damping times for most modes, assuming exponential damping, are ~2 ps or more. In addition to homogeneous broadening, the FT bands can be broadened inhomogeneously by frequency dispersion and by congestion of bands. Therefore this damping time should be regarded as a *lower limit* for the true dephasing time, and, with the usual assumption that dephasing precedes or is concomitant with energy exchange, the vibrational relaxation time.

In the Y(M210)→L mutant the ET properties are modified without strongly modifying the activated vibrational modes. We would like to stress that this does not necessarily imply that these modes are not involved in reaching the transition zone for ET. In the M210 mutants the energy levels of the involved electronic states can be substantially altered and also the electronic couplings may be modified. The probability of ET from the transition zone depends on those factors.

References

1 Vos, M.H., Rappaport, F., Lambry, J.-C., Breton, J. and Martin, J.-L. (1993) Nature 363, 320-325.
2 Vos, M.H., Jones, M.R., Hunter, C.N., Breton, J., Lambry, J.-C. and Martin, J.-L. (1994) Biochemistry 33, 6750-6757.
3 Stanley, R.J. and Boxer, S.G. (1995) J. Phys. Chem. 99, 859-863
4 Finkele, U., Lauterwasser, C., Zinth, W., Gray, K.A. and Oesterhelt, D. (1990) Biochemistry 29, 8517-8521.
5 Jones, M.R., Heer-Dawson, M., Mattioli, T.A., Hunter, C.N. and Robert, B. (1994) FEBS Lett. 339, 18-24.
6 Jones, M.R., Visschers, R.W., van Grondelle, R. and Hunter, C.N. (1992) Biochemistry 31, 4458-4465.
7 Vos, M.H., Jones, M.R., Hunter, C.N., Breton, J., Lambry, J.-C. and Martin, J.-L. (1995) in Reaction Centers of Photosynthetic Bacteria, Structure and Dynamics III, (Michel-Beyerle, M.E., ed.) Springer, Berlin, in press

EFFECTS OF CAVITIES IN THE BACTERIAL REACTION CENTER

M. Schiffer, Y.-L. Deng, A. Marrufo, & D.K. Hanson.
Center for Mechanistic Biology & Biotechnology,
Argonne National Laboratory, Argonne, IL 60439, USA

1. Introduction

A site-specific double mutant of *Rhodobacter capsulatus*, in which the large aromatic residues M208Tyr and L181Phe in the interior of the photosynthetic reaction center (RC) complex were replaced by smaller theonine residues, showed a dramatic reduction in the number of assembled complexes and was incapable of photosynthetic growth. The cavity created by the smaller side chains interferes mostly with the assembly of the complex. Phenotypic revertants were recovered in which a spontaneous second-site mutation restored photocompetence in the presence of the original site-specific mutations. In these strains, an Ala to Pro substitution in a neighboring transmembrane helix (at M271) resulted in an increased yield of RC complexes. To test the hypothesis that the original phenotype was due to a cavity, other mutants were constructed where L180Phe and M207Leu were replaced with alanines that created similar-sized voids at other positions in the membrane-spanning interior. The L180Ala-M207A mutant had the same phenotype. Coupling of the above proline substitution to these new cavity mutants also resulted in photocompetant strains that carry increased levels of RC complexes. Therefore, the proline substitution at M271 serves as a global suppressor of the phenotype caused by these internal cavities. The proline substitution slightly increases the thermal stability of the complex at higher temperatures, but both mutant and suppressor strains have about the same stability at the optimal culture temperature, where both are less stable than the wild-type strain. Therefore, the proline substitution may suppress the nonphotosynthetic phenotype of cavity mutants by facilitating folding of the nascent polypeptides as they assemble with cofactors to form the transmembranar RC complex. The proline replacement occurs at a pre-existing kink in a transmembrane helix where it can be accommodated without introducing a strain in the structure. The function of prolines in transmembrane helices might be to promote folding and/or assembly in general (1).

P. Mathis (ed.), Photosynthesis: from Light to Biosphere, Vol. I, 855–858.
Kluwer Academic Publishers.

2. Result and Discussion

To further our understanding of the effect of cavities, we studied phenotypic revertants of cavity generating mutants, and have created voids at additional sites in the RC with site-specific mutagenesis.

i. We have selected and genotypically characterized phenotypic revertants of the PS^- L180Ala - M207Ala double mutant. Surprisingly, none carried the M271Ala → Pro mutation, which we found previously to suppress its PS^- phenotype. Two independently isolated revertants carried the M207Ala → Glu mutation (L180Ala was still present), while two others carried a suppressor mutation, M211Ala → Val, and retained the L180Ala-M207Ala site-specific mutations. Both M207Glu and M211Val could be acting to suppress the PS^- phenotype by partially filling the cavity created by the original L180Ala-M207Ala mutations, suggesting that indeed the cavity was causing the problem with this mutant.

ii. The L187Phe → Ala and M214Leu → Ala mutations, when combined, created a cavity at the L-M subunit interface at the level of the bacteriopheophytins. This cavity does not significantly alter the function or assembly of the RC complex. Though this mutant is analogous to the previously described AA mutant (L180Phe → Ala + M207Leu → Ala), the cavity created (Figure 1) is not as deeply buried in the complex, and it is further away from the special pair.

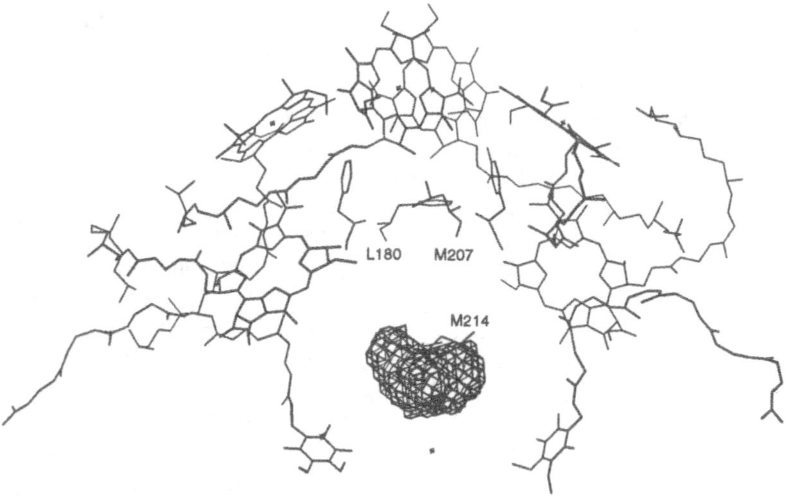

Figure 1. Cavity generated by the replacement of L187Phe and M214Leu by alanines, calculated and displayed by program VOIDOO (2) with the wild-type *Rhodobacter sphaeroides* reaction center structure (3). Side chains of wild-type residues are shown.

iii. We also replaced conserved tryptophan residues at the symmetry-related positions L156 and M183 (Figure 2), which are located near the special pair close to the periplasmic surface of the RC, with phenylalanines or alanines. The phenylalanine substitution slightly affected the growth phenotype, but more severe consequences resulted from substitution with the smaller alanines, especially the mutation at L156 on the active side.

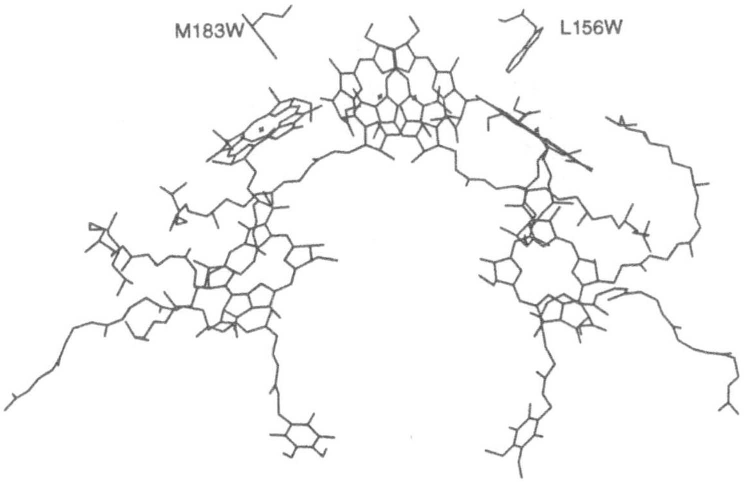

Figure 2. Location of the large M183 and L156 tryptophan residues near the periplasmic surface of the reaction center. Cavities which border the "special pair" bacteriochlorophyll dimer (primary electron donor) are generated by replacement of these residues with alanines.

The photocompetence of the family of alanine mutants is:
L156Ala-M183Ala < L156Ala < M183Ala < wild type. The L156Ala-M183Ala double mutant is incapable of photosynthetic growth and contains approximately 36% the amount of RC complexes as the wild type. We coupled the M271Ala → Pro global suppressor mutation with the L156Ala-M183Ala double mutant to see if it enhanced the yield of RCs in this strain, but no increase in the number of RC complexes was observed. This shows that the effect of the cavity created by the Trp to Ala substitution is different in these mutants than in the cavity generating mutants we described previously. Selection of phenotypic revertants is in progress.

3. Conclusions

Our results indicate that the effect of a cavity in the reaction center depends on its location. Cavities near the special pair interfere with photosynthetic growth, while similar cavities distant from the special pair do not. We have observed previously that lack of photosynthetic growth can depend on interference with RC function (4), or interference with assembly (1) which results in insufficient numbers of RCs in the photosynthetic membrane. The substitution of proline at M271, at a bend of the helix, appears to counteract a defect in folding and/or assembly of the complex (1). Since it does not restore photosynthetic growth lost by the Trp to Ala mutation at L156 and M183, therefore the mutations at M183 and especially at L156 are likely to influence the RC function. Spectroscopic studies of these mutants will be necessary to show the functional effects of these mutations. Since these mutations are located between the special pair of bacteriochlorophylls and the monomeric bacteriochlorophylls they might influence the electron transfer between them, and shed new light on the very fast 3ps electron transfer step.

4. Acknowledgement

We thank George Johnson for making the figures. This work was supported by the U.S. Department of Energy, Office of Health and Environmental Research, under Contract W-31-109-ENG-38 and by Public Health Service Grant GM36598.

5. References

1 Schiffer, M., Ainsworth, C.F., Deng, Y.-L., Johnson, G., Pascoe, F.H. and Hanson, D.K. (1995) J. Mol. Biol. in press

2 Kleywegt, G.J. and Jones, T.A. (1994) Acta Crystallogr. 50, 178-185

3 Chang, C.-H., El-Kabbani, O., Tiede, D.M., Norris, J. and Schiffer, M. (1991) Biochemistry 30, 5352-5360

4 Hanson, D.K., Tiede, D.M., Nance, S.L., Chang, C.-H. and Schiffer, M. (1993) Proc. Nat. Acad. Sci.USA 90, 8929-8933

COMPENSATION FOR L212GLU IN BACTERIAL REACTION CENTERS

D. K. Hanson[1], Y.-L. Deng[1], P. Sebban[2], and M. Schiffer[1]. [1]Center for Mechanistic Biology and Biotechnology, Argonne National Laboratory, Argonne, IL USA; [2]Centre de Génétique Moléculaire, CNRS, Gif/Yvette, France

1. Introduction

In wild-type bacterial reaction centers (RC), residue L212Glu, which is located about 5 Å away from Q_B, is involved in the delivery of the second proton to Q_B^{2-} [1-4]. We previously constructed the L212Glu-L213Asp → Ala-Ala double mutant of *Rhodobacter capsulatus*, and it is incapable of photosynthetic growth (PS^-) due to interruption of the proton transfer pathway to Q_B[3,4]. We have isolated several photocompetent (PS^+) phenotypic revertants of this L212-L213AA double mutant [3-7]. The compensatory mutations that restore function in these strains are diverse and show that neither L212Glu nor L213Asp is absolutely required for efficient light-induced electron or proton transfer. Genotypic revertant and second-site mutations, located within the Q_B binding pocket or at more distant sites, can compensate for mutations at L212 and L213 to restore photocompetence. One of the phenotypic revertants of the L212Ala-L213Ala double mutant carries a genotypic reversion of L213Ala to Asp; the Ala substitution at L212 remains. We were intrigued that this L212Glu → Ala mutant of *R. capsulatus* is photocompetent, while the L212Glu → Gln mutant of *R. sphaeroides* is not, particularly since the sequence identity in the Q_B site of these two strains is 90% [8]. To this end, we constructed the L212Glu → Gln mutant in *R. capsulatus*, and it is also PS^-. To determine the function that is lost in the L212Gln mutant but restored by Ala at that site, we selected four PS^+ revertants from the L212Gln strain.

2. Results and Discussion

None of the revertants carried the single base transversion that would restore the wild-type Glu codon; each carried a different second-site substitution in addition to the original L212Gln mutation. Complementation mapping [5] was used to determine that this single second-site mutation was responsible for restoration of the PS^+ phenotype. The compensatory mutations carried by the phenotypic revertants are: Rev2: L228Gly → Asp; Rev4: M231Arg → Cys; Rev5: L227Leu → Phe; Rev6: L231Arg → Cys. Among the suppressor mutations of the L212Ala-L213Ala double mutant were some mutations that are similar to Rev4, that have been shown to restore the PS^+ phenotype by increasing the negative environment surrounding Q_B, sometimes from a large distance [3,4,7]. Therefore, we have also constructed plasmids that coupled some of these other compensatory mutations with the L212Glu → Gln mutation: L212Gln + M5Asn → Asp; L212Gln + M43Asn → Asp; L212Gln + M231Arg → Leu. These substitutions also restored the PS^+ phenotype to varying extents. Their relative growth rates under photosynthetic conditions are summarized in Table 1.

P. Mathis (ed.), Photosynthesis: from Light to Biosphere, Vol. I, 859–862.
Kluwer Academic Publishers.

Table 1. Relative photosynthetic growth rates of strains carrying mutations which compensate for loss of L212Glu.

Strain	Growth
Wild-type	++++++
L212Glu → Gln[e]	–
Rev2 (L212Gln + L228Gly → Asp)[s]	++
Rev4 (L212Gln + M231Arg → Cys)[s]	++++
Rev5 (L212Gln + L227Leu → Phe)[s]	+++++
Rev6 (L212Gln + L231Arg → Cys)[s]	+++
L212Gln + M5Asn → Asp[c]	+++
L212Gln + M43Asn → Asp[c]	+
L212Gln + M231Arg → Leu[c]	++++

[e]site-specific mutation
[s]spontaneous phenotypic revertant
[c]constructed

It is likely that the photocompetence of the L212Glu → Ala strain, as opposed to the PS$^-$ phenotype of the L212Glu → Gln mutant, is due to the size difference between the side chains. The Ala substitution could accomodate a water molecule which could assume the role played by L212Glu in proton delivery in the wild type, similar to observations with the L223Ser → Ala (PS$^-$) and L223Ser → Gly (PS$^+$) mutants of *R. sphaeroides* [9,10].

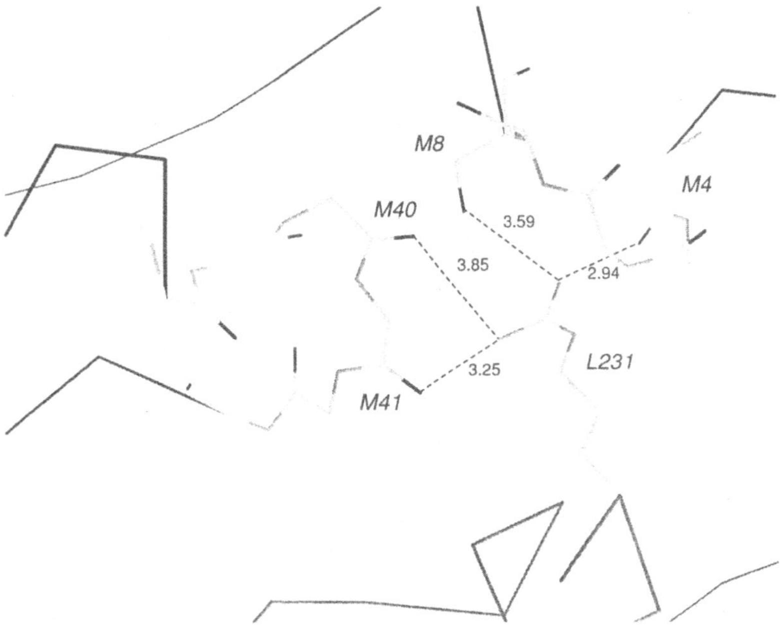

Figure 1. Molecular model, using the *R. sphaeroides* RC structure [11] of hydrogen bonds between L231Arg (→ Cys in Rev6) and residues in two loops of the M chain.

The mechanism(s) of suppression of these second-site mutations is not yet clear, but it may involve some displacement of the protein backbone to allow a water molecule to be accomodated near Q_B. Residues L227Leu and L228Gly are conserved in five of the six strains of photosynthetic bacteria whose RC genes sequences are known; L231Arg and M231Arg are strictly conserved. Modeling of these substitutions using the *R. sphaeroides* RC structure [11] shows that some of these residues lie at the interface of the L and M chains, thus the mutations might cause chain rearrangement as was observed for the L222Tyr → Phe mutant of *R. viridis* [12]. In particular, residue L231Arg [Fig. 1] forms hydrogen bonds to the backbone carbonyl oxygen atoms of residues M4, M8, M40, and M41, thus acting to tether this segment of the L chain near Q_B to two different loops of the M subunit that are part of the second sphere of residues that sequesters Q_B from the external aqueous environment. Loss of these hydrogen bonds could lead to a major rearrangement of this region of the RC, which could change the orientation of protonatable residues and the structure of internal water molecules near Q_B to generate a substitute proton delivery pathway.

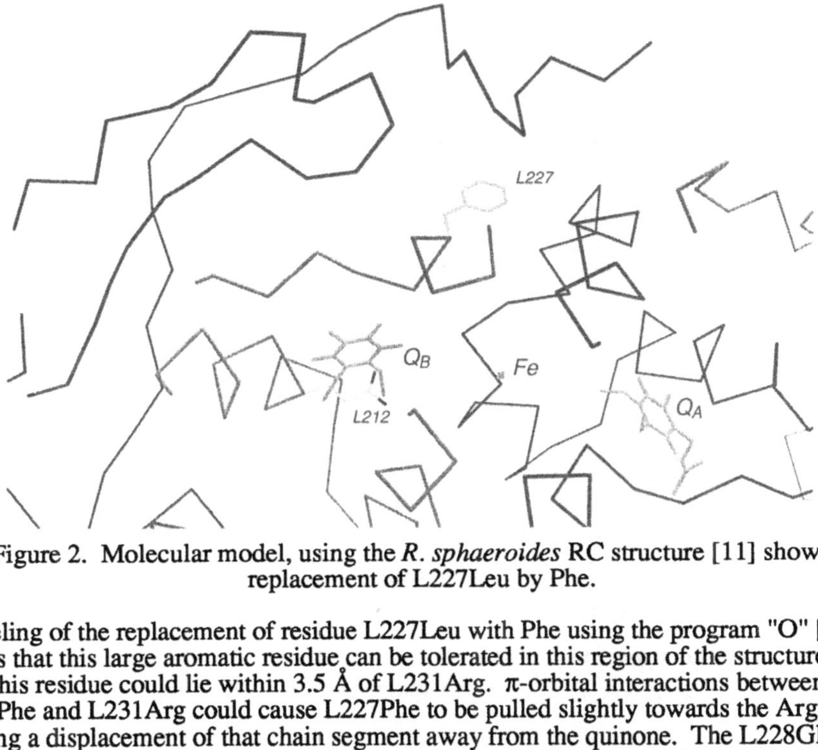

Figure 2. Molecular model, using the *R. sphaeroides* RC structure [11] showing replacement of L227Leu by Phe.

Modeling of the replacement of residue L227Leu with Phe using the program "O" [13] shows that this large aromatic residue can be tolerated in this region of the structure [Fig. 2]. This residue could lie within 3.5 Å of L231Arg. π-orbital interactions between L227Phe and L231Arg could cause L227Phe to be pulled slightly towards the Arg, thus causing a displacement of that chain segment away from the quinone. The L228Gly → Asp replacement is difficult to accomodate without rearrangement of the L chain backbone.

The replacement of M231Arg by a non-protonatable residue has been observed before as a mutation that compensates for the loss of L213Asp [3,6]. Loss of M231Arg breaks a salt bridge between that residue and two glutamic acid residues of the H chain. The pH dependence of K_2, the $Q_A^-Q_B \leftrightarrow Q_AQ_B^-$ equilibrium constant, is altered by this mutation, revealing differences in the pKs of titratable groups in this strain compared to the wild type

[3,7]. This residue is 15 Å away from Q_B, and we have proposed that the breakage of the salt bridge as a result of the mutation causes realignments of salt bridges within the network of charge-charge interactions in this region of the RC, thereby partially relocating the negative charge much closer to the quinone [7]. An analogous mechanism may be proposed to explain the electrostatic effects of the M5Asn → Asp (13 Å away from Q_B) and M43Asn → Asp (9 Å away from Q_B) substitutions in restoring proton transfer to Q_B in strains which contain Ala residues at L212 and L213 [7, and Hanson et al., unpublished]. When they are coupled with the L212Glu → Gln mutation, these substitutions must be acting in a comparable manner to restore proton transfer, thereby suppressing the PS^- phenotype. In the case of RCs carrying these electrostatic substitutions, the proton transfer pathways must be different than that of the wild type, or that of the L212Ala mutant where function of L212Glu might be restored by a water molecule.

We thank George Johnson for the figures. M. S., Y.-L. D, and D. K. H. were supported by U. S. Dept. of Energy, OHER Contract No. W-31-109-ENG-38 and by U. S. PHS Grant GM36598. P. S. was supported by NATO grant CRG920725 and NSF-CNRS grant CDP900350.

References

1 Paddock ML, Rongey SH, Feher G, Okamura MY (1989) *Proc. Natl. Acad. Sci. USA* 86, 6602-6606.
2 Takahashi E, Wraight CA (1992) *Biochemistry* 31, 3, 55-866.
3 Hanson DK, Baciou L, Tiede DM, Nance SL, Schiffer M, Sebban P (1992) *Biochim. Biophys. Acta* 1102, 260-265.
4 Maróti P, Hanson DK, Baciou L, Schiffer M, Sebban P (1994) *Proc. Natl. Acad. Sci. USA* 91, 5617-5621.
5 Hanson DK, Nance SL, Schiffer M (1992) *Photosyn. Res.* 32, 147-153.
6 Hanson DK, Tiede DM, Nance SL, Chang C-H, Schiffer M (1993) *Proc. Natl. Acad. Sci. USA* 90, 8929-8933.
7 Sebban P, Maróti P, Schiffer M, Hanson DK (1995) *Biochemistry* 34, 8390-8397.
8 Schiffer M, Chan C-K, Chang C-H, DiMagno TJ, Fleming GR, Nance S, Norris J, Snyder S, Thurnauer M, Tiede DM, Hanson DK. (1992) In: *The Photosynthetic Bacterial Reaction Center: Structure, Spectroscopy, and Dynamics* (J. Breton and A. Vermeglio, eds.), NATO ASI Series, Plenum Press, London, pp. 351-361.
9 Paddock ML, McPherson PH, Feher G, Okamura MY (1990) *Proc. Nat. Acad. Sci. USA* 87, 6803-6807.
10 Paddock ML, Feher G, Okamura MY (1995) *Biophys. J.* 68, A246.
11 Chang, C-H, El-Kabbani O, Tiede D, Norris J, Schiffer M (1991) *Biochemistry* 30, 5352-5360.
12 Sinning I (1992) *TIBS* 17, 150-154.
13 Kleywegt GJ, Jones TA (1994) *Acta Crystallogr.* 50, 178-185.

TRAPPING IN MUTANT PHOTOSYNTHETIC BACTERIA

A. Freiberg*, J. P. Allen, J. C. Williams, and N. W. Woodbury
Center for the Study of Early Events in Photosynthesis, Arizona State University, Tempe, USA and *Institute of Physics, Tartu, Estonia

1. Introduction

Despite its fundamental importance, the knowledge of how antennas are coupled (both structurally and functionally) to the reaction centre (RC) is still rather limited. Similarly, the origin of fast multiexponential kinetic components of antenna emission is poorly understood. This may be a reason why there is no general agreement between the kinetic data from different laboratories (1, 2). A modern and powerful approach to attack the problem is to design mutants with altered structure and energy or electron transfer parameters. In this study, chromatophores from mutant *Rhodobacter (Rb.) sphaeroides* were studied. The single, double and triple hydrogen bond RC mutants used have substantially altered midpoint potentials of the primary electron donor-the bacteriochlorophyll (Bchl) dimer, with no changes in the chemical identity of the dimer (3). The varied midpoint potential results, first of all, in a change in the free energy gap between the excited primary donor, P*, and the charge-separated state, P^+H^-. (H denotes the pheophytin primary acceptor of the RC.) It also results in altered charge separation and recombination rates in the RC. The primary charge separation time in the mutants used varied 17-fold (from 3.4 ps to 52 ps) compared to wild-type bacteria (3). All these changes should have implications in the kinetics of antenna emission. The mutants used are also simpler than the wild-type organisms, as they possess only a core antenna and lack the peripheral one. There are about 25 antenna Bchl molecules per RC in all mutant strains with their Q_y absorption band at 875 nm, like in wild type. For comparison, emission kinetics in the membranes of *Rb. sphaeroides* antenna-only mutant and of wild-type *Rhodospirillum (Rs.) rubrum* were measured. A time-correlated single photon counting system with a microchannel-plate photomultiplier with 50-70 ps instrument response function was used to monitor the decay kinetics at room temperature. Low excitation intensity (0.1 mW/cm^2 at 590 nm or 860 nm), the adding of 0.5 mM terbutryn, short-cutting of the electron transfer chain at the primary quinone acceptor site and preventing of the accumulation of long-lived intermediates in the RC, and intense stirring of the solvent combined with reduced pulse repetition rate (down to 380 kHz) were used to ensure that almost all of the RC were in the open, photoactive state during the measurements. The kinetics was followed over a 10 ns time interval.

P. Mathis (ed.), Photosynthesis: from Light to Biosphere, Vol. I, 863–866.

2.Results

The picosecond emission kinetics of all *Rb. sphaeroides* mutants can be approximated by a sum of three decaying exponents as shown in Table 1. The data in Table 1 are average numbers over about 10 separate measurements. The key of the labels reads, e. g. HF(L168): the histidine (H) residue has been replaced by the phenylalanine (F) residue at the L168 RC protein site. In mutants, the shortest decay time varies roughly linearly with the primary charge separation time, τ_{rc}, according to the empirical formula τ(in ps)$\approx 56+2\tau_{rc}$. (Fig.1).

Table 1. Picosecond emission decay times and relative amplitudes (both with ±15 % error) of mutant *Rb. sphaeroides* and-wild type *Rs. rubrum* chromatophores. The free energy gaps, ΔG, are in meV, the detrapping efficiency, $D=k_{-1}/(k_2+k_{-1})$, is given in % units.

Sample	τ_1/A_1	τ_2/A_2	τ_3/A_3	$\tau_{rc}(3)$	ΔG_{cal}	ΔG (3)	D
pseudo-WT	72/11.8	280/1	710/1	3.4	131	120	6
HF(L168)	61/27	230/2	650/1	3.6	133	140	6
LH(M160)	61/10.8	240/2	700/1	5.7	99	100	9
LH(L131)	82/3.8	250/2	640/1	12.2	54	65	18
LH(M160)+LH(M197)	99/2.4	250/2	690/1	24	25	25	30
LH(L131)+LH(M160) +FH(M197)	160/1	400/2	780/1	52	-16		49
antenna-only mutant		230/1	750/7				
Rs. rubrum	55/79	330/1		2.8	184		5

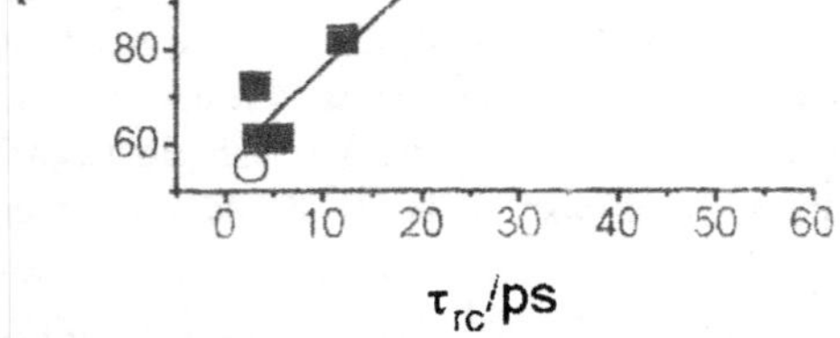

Figure 1. The shortest emission lifetime as a function of charge separation time in *Rb. sphaeroides* mutants. The hollow circle is for *Rs. rubrum.*

The two slower decay constants are about 250 ps and about 700 ps in all mutants (except the triple mutant where they are 400 ps and 780 ps, respectively). The ratio of the amplitudes of the shorter and the intermediate component decreases as τ_{rc} increases, but clearly in a non-linear fashion.

3. Discussion

A comparison of the 700 ps decay component with the data for antenna-only mutant (Table 1) and for the isolated core antenna complexes (4) reveals its possible origin as at least partly due to the emission of antenna complexes, that are not effectively coupled to the RC. Note that there is no such emission component in the reference *Rs. rubrum* membranes. A four-level kinetic scheme that takes into account the recombination of the RC P^+H^- state into the P* state and the subsequent detrapping of the energy from the RC was used to analyse the origin of the two other decay components:

$$A \underset{k_{-1}}{\overset{k_1}{\Leftrightarrow}} P^* \underset{k_{-2}}{\overset{k_2}{\Leftrightarrow}} P^+H^- \overset{k_3}{\Rightarrow} P^+Q_a^{-\cdot}$$

Here k_1 represents the rate of energy transfer from the antenna to the RC, k_{-1} is the total rate of energy detrapping, k_2 and k_{-2} are the forward and the backward electron-transfer rates in the RC, respectively, and k_3 is the secondary electron-transfer rate that stabilises the electron on the quinone acceptor. Under a very reasonable assumption that $k_2 >> k_{-2}$, k_3 the above kinetic system has three approximate solutions (5): $\lambda_1 = k_1k_2/(k_1+k_{-1}+k_2)$; $\lambda_2 = k_3+k_{-1}k_{-2}/(k_2+k_{-1})$; $\lambda_3 = k_1+k_{-1}+k_2 - k_1k_2/(k_1+k_{-1}+k_2)$. It is easy to see that λ_1^{-1} is linearly proportional to k_2^{-1}, or what is the same, to τ_{rc}: $\lambda_1^{-1} = k_1^{-1}+(1+k_{-1}/k_1)\tau_{rc}$ and may represent the experimental time constant τ_1. Then it immediately follows from Fig.1 that $k_{-1}/k_1 \approx 1$ and $k_1^{-1} \approx k_{-1}^{-1} = 56$ ps. λ_2 is limited by the lifetime of the charge separated state, P^+H^-. It may correspond to the experimental time constant τ_2. λ_3 represents a fast energy equilibration process which could hardly be resolved with our present time resolution. Not only the linear increase of the short lifetime on τ_{rc}, but also the non-linear decrease of the amplitude ratio, A_1/A_2, finds its explanation within this simple model. It can be shown that in a quasi-equilibrium condition $A_1/A_2 \approx k_1/k_{-2} = k_1\tau_{rc}\exp(\Delta G/kT)$. ΔG in this equation is the free energy gap between the RC states P* and $P^+ H^-$. The adequacy of the model was further assured by the realistic ΔG values calculated according to the above equation, ΔG_{cal}, and by comparing them to the existing estimates by other methods (Table 1).

However, the model also predicts a gradual shortening of τ_2 with τ_{rc} (about two times for the double mutant with respect to pseudo-WT), which was not clearly resolved in our data. This discrepancy can probably be explained with methodological reasons. In mutants with longer τ_{rc}, it is increasingly more difficult to separate the closing decay times on the background of the emission of uncoupled pigments. The ratio of the excitation energy of P* that transfers back to the antenna, $k_{-1}/(k_{-1}+k_2)$, was estimated to increase from .06 in the pseudo-wild-type *Rb. sphaeroides* membranes to .5 in the triple mutant (Table 1). In the case of *Rs. rubrum* D=5 %. This is a much smaller number than the 25±5 % reported earlier (6). One can notice also serious deviations when the ratio of the rates k_1 and k_{-1} ($k_1/k_{-1} \approx 1$ in this work, but ≈ 5.3 in (6)) as well as their absolute values are compared. These discrepancies need further analysis. Here we note only two essential assumptions used throughout this work the validity of which is not *a priori* evident: (i) Energy transfer rates between the antenna and the RC are invariant in different mutants; (ii) τ_{rc} in intact antenna-RC complexes is the same as in isolated RC. There are some other related problems. One is the absorption peak position of the RC in mutant membranes. Within error limits, after isolation and purification, the peaks are the same as in wild-type RC, except for HF(L168) that has a blue-shifted P absorption peak at 850 nm. We did not succeed yet an accurate measuring of the RC absorption spectra in chromatophores. Another problem we have ignored is the .5 quantum yield of charge separation in a triple mutant (3). This may result in the overestimation of the linear coefficient of τ_{rc} in the equation for λ_1^{-1}, and may mean that, in reality, $k_{-1}<k_1$.

There is only one recent paper on this subject to be compared with (7). Three *Rb. sphaeroides* mutants with up to six times altered τ_{rc} were studied by picosecond absorption technique. Although the observed trends have been qualitatively similar to ours, the measured lifetimes (including the wild-type species) have been much shorter, probably because of the rather intense excitation pulses used.

Acknowledgements

Our thanks are due to Dr. J. M. Peloquin and to Dr. K. Timpmann for many helpful discussions considering the experimental details and to Miss Xuan for technical assistance. This work was performed under the NAS/NRC COBASE grant.

References

1. Borisov, A. Yu., Freiberg, A. M., Godik, V. I., Rebane, K. K. and Timpmann, K. E. (1985) Biochim. Biophys. Acta 807, 221-229

2. Woodbury, N. and Bittersmann, E. (1990) in Current Research in Photosynthesis (Baltscheffsky, M., ed.) Vol. II, pp. 165-168, Kluwer Academic Publishers, Dordrecht, The Netherlands

3. Williams, J. C., Alden, R. G., Murchison, H. A., Peloquin, J. M., Woodbury, N. W. and Allen, J. P. (1992) Biochemistry 31, 11029-11037; Murchison, H. A., Alden, R. G., Allen, J. P., Peloquin, J. M., Taguchi, A. K. W., Woodbury, N. W. and Williams, J. C. (1993) ibid 32, 3498-3505; Peloquin, J. M., Williams, J. C., Lin, X., Alden, R. G., Taguchi, A. K. W., Allen, J. P. and Woodbury, N. W. (1994) ibid 33, 8089-8100

4. Freiberg, A. and Timpmann, K. (1992) J. Photochem. Photobiol. 15, 151-158

5. Xiao, W., Lin, S., Taguchi, K. W. and Woodbury, N. W. (1994) Biochemistry 33, 8313-8322

6. Timpmann, K., Zhang, Fu Geng, Freiberg, A. and Sundström, V. (1993) Biochim. Biophys. Acta 1183, 185-193

7. Beekman, L. M. P., van Mourik, F., Jones, M. R., Visser, H. M., Hunter, C. N. and van Grondelle, R. (1994) Biochemistry 33, 3143-3147

SITE-SPECIFIC MUTAGENESIS AROUND THE SPECIAL PAIR OF THE REACTION CENTER *RHODOPSEUDOMONAS VIRIDIS*

M. Bibikova[1], Th. Arlt[2], W. Zinth[2] and D. Oesterhelt[1],
[1]Max-Planck-Institut für Biochemie, D-82152 Martinsried,
[2]Institut für Medizinische Optik, Ludwig-Maximilians-Universität, D-80797 München

1.Introduction

The reaction center (RC) from the purple, non-sulfur photosynthetic bacterium *Rhodopseudomonas (Rps) viridis* has been crystallized and the X-ray structure solved to 2.3 A resolution (1, 2). A system for site-specific mutagenesis of the photosynthetic RC in *Rps. viridis* has been recently developed (3). By the means of in vitro mutagenesis it is possible to introduce specific alterations in the protein structure and to study the functional role of single amino acids in the electron transfer properties of the RC. Specific changes of amino acids in following positions were prepared (Tab.1):

<u>Positions M200 Histidine and L173 Histidine</u>

The side chains of histidines L173 and M200 are the ligands to the central Mg^{2+} ions of the special pair Bacteriochlorophylls (Bchls) in the RC. The mutagenic change of one of these amino acids results in *Rb. capsulatus* in the replacement of the corresponding Bchl with Bacteriophephytin (Bphe), providing a Bchl-Bphe heterodimer at the special pair (4).

<u>Positions M195 Tyrosine and L168 Histidine</u>

Tyrosine M195 and histidine L168 are localized near the primary electron donor of the RC from *Rps. viridis*. L168 His provides a hydrogen bond to one of the the Bchls of the special pair. Both residues play an important role in protein - cofactor interactions, establishing the redox-potentials needed for electron transfer in the RC (5).

<u>Positions M208 Tyrosine and L181 Phenylalanine</u>

Tyrosine M208 is situated in close proximity to the special pair, monomer $Bchl_A$ and $Bphe_A$. This residue lies directly in the path of electrons from the special pair along the active branch of the RC. Phenylalanine L181 lies at the corresponding position on the inactive branch. The interactions of Tyr M208 may favour electron transfer along the active branch (6).

2. Mutagenesis

All specific amino acid changes (Tab.1) were prepared by the Kunkel method of site-specific mutagenesis using uracil-containing DNA (7). Mutated fragments were cloned

P. Mathis (ed.), Photosynthesis: from Light to Biosphere, Vol. I, 867–870.

into the broad host range vector pRK404, which is stably maintained in Rps.viridis. The E.coli strain S17-1 was used to mobilize pRK404 derivatives into Rps.viridis using a conjugation procedure. The DNA of cells taken for RC isolation was resequenced to ascertain the engineered genetic background. The isolated RCs were used for biophysical characterization of mutants.

Position of the mutation	Name of the mutant	Change of the amino acids	Change of the nucleotides	Micro-aerobic growth	Photo-synthetic growth
L168 His	L168 HF	His -> Phe	CAC->TTC	+	+
L173 His	L173 HI	His -> Ile	CAC->ATC	+	-
	L173 HT	His -> Thr	CAC->ACC	+	-
	L173 HF	His -> Phe	CAC->TTC	+	-
L181 Phe	L181 FY	Phe -> Tyr	TTC->TAT	+	+
	L181 FL	Phe -> Leu	TTC->CTG	+	+
M195 Tyr	M195 YF	Tyr -> Phe	TAC->TTC	+	+
	M195 YF+ L168 HF			+	+
	M195 YH	Tyr -> His	TAC->CAC	+	+
M200 His	M200 HI	His -> Ile	CAC->ATC	+	-
	M200 HF	His -> Phe	CAC->TTC	+	-
	M200 HT	His -> Thr	CAC->ACC	+	-
M208 Tyr	M208 YF	Tyr -> Phe	TAT->TTC	+	+
	M208 YF+ L181 FY			+	+
	M208 YL	Tyr -> Leu	TAT->CTG	+	+

Table 1. Site-specific mutations in the RC of *Rps. viridis*.

Until now, large amounts of *Rps.viridis* cultures could be grown only under phototrophic conditions. The resulting selective pressure causes the formation of the revertants containing primary and/or secondary mutations. For this reason the possibility of alternative growth modes for the *Rps. viridis* site-specific mutants was studied. Microaerobic growth (0.1% O_2) of *Rps. viridis* in the dark in liquid cultures was performed successfully, but the amount of the RCs induced under this growing conditions is only 1-10% of the RCs, induced during phototrophic growth.

3.Results

In the primary processes of bacterial photosynthesis, light energy is converted into chemical energy by a charge separation in the RC. From the exited state P* , which exists for 3.5 ps in the wild type, an electron is transferred to the bacteriochlorophyll B_A. The corresponding radical-pair state $P^+B_A^-$ lives for just 0.65 ps, forming the radical pair $P^+H_A^-$ in the second electron transfer step (8).

The mutation L168 His to Phe causes the loss of a hydrogen bond to the acetyl group of ring I of the $Bchl_A$ of the special pair. The optical absorption spectrum of the L168 HF mutant reaction center shows a blue shift of 35 nm in the peak of the Bchl dimer (Tab.

2). The midpoint potential of the donor measured by redox titrations was found to decrease by 80 mV in the mutant compared to the wild type (WT)(Tab.2).

Sample	Absorption Maximum of Isolated RC (nm)	Em(P) , mV	Initial e-transfer, t_1 [ps]
Wild Type	970	520	3.5 + 0.5
L168HF	930	440	1.1 + 0.3
M195YF	947	475	3.1 + 0.4
M195YH	935	540	5.7 + 0.6
L168HF + M195YF	890(910)	430	0.8 + 0.2
L181FY	966	495	1.7 + 0.3
L181FL	972	520	n.d.
M208YF	966	535	20 + 4
M208YL	972	535	n.d.
L181FY + M208YF	966	495	8.2 + 1.0

Table 2. RCs of *Rps. viridis*: absorption maxima, redox potentials of the special pair and initial electron transfer constants (monoexponential fit of the measured P^*decay).

The decrease in the P/P^+ midpoint potential is related to a lowering of the energy levels of all P^+ containing intermediate states including the energy level of $P^+B_A^-$. If one assumes that the energy level of $P^+B_A^-$ in WT RC is close to that of P*, it can be concluded that in the mutant L168HF the energy level of $P^+B_A^-$ should be far below P*. According to the electron transfer theory of Marcus (9), this could lead to a significant increase in the rate of the initial electron transfer.

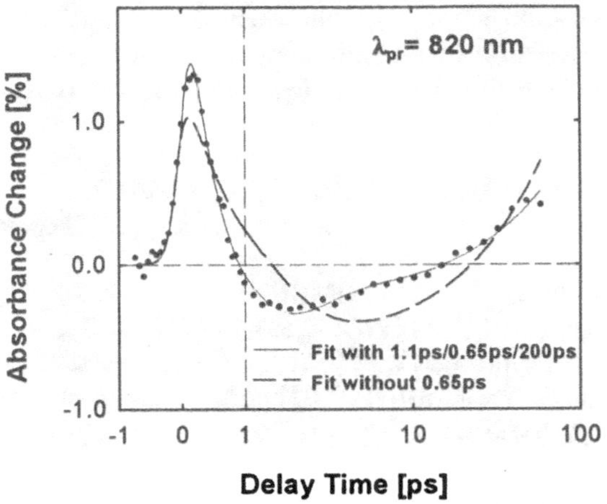

Fig. 1. Time-resolved absorption data for the mutant L168HF. The probing wavelength was taken in the Q_y absorption band of the monomeric bacteriochlorophyll. In the modified RC a strong bleaching of the $Q_y(B)$ band is observed at 820 nm. Solid and broken lines belong to model functions with and without the short 0.65 ps component, respectively.

Experimentally, the initial electron transfer time constant, measured in the L168HF mutant was determined to be 1.1 ps, compared to 3.5 ps measured for the wild type (Tab.2). As a result of the fast decay of P*, the state $P^+B_A^-$ is populated stronger than in WT RCs and reaches a population of about 30%. A strong bleaching of the Bchl monomer band was detected (Fig.1), which confirms the existence of $P^+B_A^-$ as a real intermediate state in bacterial photosynthesis

Tyrosine M210 plays an important role in the fast rate of the primary electron transfer in *Rb. sphaeroides* (10). The location of this tyrosine and its polarity lower the energy of the state $P^+B_A^-$. The exchange of tyrosine by the nonpolar Phe should slow down the rate of electron transfer. The corresponding mutation M208YF in the *Rps. viridis* RCs leads to a significant increase of the initial electron transfer time constant from 3.5 ps in the wild type to 20 ps in the mutated RCs (Tab.2).

4. Summary

Three groups of amino acids located close to the special pair of the reaction center have been mutagenized. The histidines M200 and L173 liganding the special pair from both sides have been exchanged for phenylalanine, isoleucine and threonine, but in no case photosynthetic growth or expression of the reaction centers could be observed. A second group comprises histidine L168 and tyrosine M195 which form hydrogen bridges to the special pair. Loss of these H-bonds causes blue shifts of the absorption maximum between 15 and 50 nm (Tab.2) and changes in the redox potential of the special pair which influence the initial electron transfer steps (Tab.2). Specifically the L168HF mutation shows acceleration of the initial electron transfer from 3.5 to 1.1 ps and allows the unambiguous demonstration of the accessory bacteriochlorophyll-b as the first intermediate. A third group of mutants concerns tyrosine M208 and the corresponding symmetric position L181 Phe which have a significant influence on the electron transfer kinetics. In *Rps. viridis* pronounced changes in kinetics were found and all effects observed could be discussed in terms of the Marcus electron transfer theory.

References

1. Deisenhofer, J., Michel, H. (1989) EMBO J. , **8,** 2149-2170.
2. Deisenhofer, J., Michel, H. (1987) Photobiochemistry and Photobiophysics, Suppl., 133-140.
3. Laußermair, E., Oesterhelt, D. (1992) EMBO J., **11**, 777-783.
4. Bylina, E.J., Youvan, D.C. (1988) PNAS USA, **85**, 7226-7230.
5. Woodbury, N.W., and Williams J.C. (1993), Biochemistry, **32**, 3498-3505.
6. Gray, K., Wachtveitl, J., Oesterhelt, D. (1992), Eur. J. Biochem., **207**, 723-731.
7. Kunkel, T.A. (1985), PNAS USA, **82**, 488-492.
8. Dressler K., Murchison, H.A., Alden, R.G., Allen, J.P., Peloquin, J.M., Taguchi, A.K.W., Umlauf E., Schmidt S., Hamm P., Zinth W., Buhanan S., and Michel H. (1991), Chem. Phys. Lett., **183**, 270-276.
9. Marcus R.A. (1988), Chem. Phys. Lett., **146**, 13-22.
10. Finkele U., Lauterwasser C., Zinth W., Gray K., Oesterhelt D. (1990), Biochemistry, **29**, 8517-8521.

CHARACTERISATION OF PSEUDOREVERTANTS FROM A DELETION-MUTANT IN THE LARGE HYDROPHILIC LOOP OF THE PHOTOSYSTEM II SUBUNIT CP 43 FROM *SYNECHOCYSTIS* sp. PCC 6803

Matthias G. Kuhn
Max-Volmer-Institut für Biophysikalische und Physikalische Chemie, Technische Universität Berlin, Straße des 17. Juni 135, 10623 Berlin, Germany.

1. Introduction

The Photosystem II (PSII) catalyses the light-induced oxidation of water to molecular oxygen. The smallest particle capable of water splitting isolated so far consisits of seven polypeptides. These are the CP43-, CP47- the D1- and the D2-proteins, the α- and β-subunits of cytochrome b559 and the PsbI protein (1). The structure of the D1/D2 reaction center is thought to be related to the bacterial reaction center (bRC) (2). In homology to the L and M subunit of the bRC D1 and D2 are predicted to harbour the cofactors (4 chlorophylls, 2 pheophytins, 2 quinones and 1 non-heme Fe). However, at the electron donor site a similarity between purple bacteria and PSII is missing. Therefore less information is available about the coordination of the Mn-cluster of the oxygen evolving complex in PSII

The hydropathy plots for CP47 and CP43 reveals interesting features. Both proteins are predicted to have a large extrinsic loop stretching out into the lumen (3). According to the most recent PSII model these loops are large enough to interact with the lumenal part of D1 and D2 where the oxygen evolving complex is located (4). The role of the large lumenal loop of CP 43 has been studied by site-directed mutagenesis. Different deletions have been introduced. All lead to an obligate photoheterotrophic phenotype indicating the functional importance of this unique CP43 domain (5). A similar situation is given for the CP47, but other than in CP43 some domains in the loop of CP47 allow deletions without affecting PSII function (6, 7).

In order to understand the functional and structural impact of the different deletions of CP43 we screened for spontaneous photoautotrophic (pseudo)revertants from the obligate photoheterotrophic deletion mutants. Any supression of the deletion may give an interesting insight into the function of this unique structural element. Here we report about the characterisation of two different intragenic supressions of a deletion in the large extrinsic loop of CP43.

2. Materials and Methods

Synechocystis sp. PCC 6803 (*Syn.* 6803) cells were grown in BG11 medium. Mutant cultures were maintained on plates in the presence of 25 μg/ml kanamycin. In order to guarentee PSII independent heterotrophic growth 10 mM glucose and 20 μM atrazine were added. Liquid cultures were grown in 500 ml flasks while bubbling with sterile air.
In order to increase the frequency of reversions we tried a rather 'naive' approach. Cells were spread over appr. 2 cm^2 as a thick layer on BG11 plates lacking glucose and PSII

P. Mathis (ed.), Photosynthesis: from Light to Biosphere, Vol. I, 871–874.

inhibitors. As *Syn*. 6803 is capable of taking up DNA from the environment this step may increase the occurence of revertants. Some cells may utilize DNA from dead and partially rotted cells. As a matter of fact, we observed about 5 times higher frequency of occurence of photoautotrophic cells.
Nucleic-acid manipulation were done as described earlier (5). The *psbC* fragment used to characterise the revertants was amplified by PCR using standard protocolls. Sequencing of *psbC* from revertants was performed by using the ΔTaq Cycle Sequencing Kit from USB.
PSII specific biophysical analysis often requires a PSII preparation with low contamination by PSI. Therefore, we developped a methode of a PSII preparation based on the procedure described by Tang and Diner (8). Characterisation of this PSII preparation will be published elsewhere.

3. Results and Discussion

Screening for revertants was as yet only succesfull for the deletion-mutant 8 (Δ8) (L364/370D). 10 revertant colonys (out of appr. 75) were picked (PR1 to PR10) and analysed.

Characterisation of pseudorevertants.

In order to analyse *psbC* directly a PCR product covering almost the second half of the *psbC* gene was amplified and characterised. All revertants still showed the unique *Eco*RV site which was originally generated by the deletion together with a silent mutation (see Fig. 1). The PCR fragment of the revertants showed appr. wildtype size indicating an insertion event in the second half of the *psbC* gene (the deletion of 21 nt is visible after separation of the PCR fragment on 2% agarose gels). In order to see whether this insertion event is responsible for the new photoautotrophic phenotype of the revertants the PCR fragments were used to transform the original Δ8 mutant. All PCR fragments from the different pseudorevertants could transform the Δ8 mutant into the phenotype of the revertant.
Pseudorevertants PR1 to PR10 were sequenced and two different were found. The sequence of PR1 and PR3 are shown in figure 1. The deduced amini-acid sequence is also shown in figure 1 and figure 3.
In both pseudorevertants the *psbC* insertion originated from an region upstream of the deletion. The sequence 5'-TGGACCCAACGGTTT-3' (nt 1107 to nt 1121, underlined in figure 2) was doubled in the revertant PR3 and trippled in the revertant PR1. These events resulted in the deduced amino acid sequence as shown in figure 2. Similar reversion events have been reported for the psbD2 gene but only for a single nt deletion (frameshift mutant) (10).
Preliminary characterisation of both pseudorevertants showed that the growthrate for R1 is quite similar to wildtype whereas R3 is somewhat reduced in the photoautotrophic capacity. Additionally, we have preliminary indications of Y_{Dox} beeing affected in R3. Further characterisations are currently being done.
In the amino-acid sequence of PR3 only aspartate at position 370 is conserved after the insertion event. The two positively charged residues lysine 366 and arginine 368 were replaced by proline and glycine, respectively. In order to test the effect of the additional proline 366 in PR3 on the location of two functionally interesting arginines at position 377 and 378 we modelled the structure of the loop region from residue 361 to 379 (not shown). The modelling showed that the insertion in PR3 has a significant impact on the position of these two arginines. A simple sketch of this result is shown in figure 2.

The data presented here undermines the functional importance of the large extrinsic lumenal loop of CP43. According to the latest PSII model CP43 is located far outside of the actual PSII reaction center (4). As the large extrinsic loop is long enough (130 aa-residues) to stretch out to functionally importent domains of the PSII donor site we still propose this loop as being involved in the coordination of an active oxygen evolving complex. Our current model sees the two arginines at position 377 and 378 (close to the position of the deletion in D8) as important for PSII funtion (see figure 3). In the CP47, analogue arginines (position 384 and 385) have been tested by site-directed mutagenesis. Conversion of both residues to glycine or aspartate yields about 50% loss of steady state oxygen evolution (9). A deletion covering two arginines at position 357 and 358 leads to an obligate photoheterotrophic phenotype (6).

Figure 1

WILDTYPE

CCT TGG TTG GAG CCC CTC CGT GGA CCC AAC GGT TTG GAT TTG GAC
Pro Trp Leu Glu Pro Leu Arg Gly Pro Asn Gly Leu Asp Leu Asp

AAA CTC CGC AAT GAC ATT CAG CCC TGG CAA GTA
Lys Leu Arg Asn Asp Ile Gln Pro Trp Gln Val

DELETIONMUTANT D8

CCT TGG TTG GAG CCC CTC CGT GGA CCC AAC GGT TTG GAT ATC CAG
Pro Trp Leu Glu Pro Leu Arg Gly Pro Asn Gly Leu Asp Ile Gln
↑ EcoRV

CCC TGG CAA GTA
Pro Trp Gln Val

PR1

CCT TGG TTG GAG CCC CTC CGT GGA CCC AAC GGT TTT GGA CCC AAC
Pro Trp Leu Glu Pro Leu Arg Gly Pro Asn Gly Phe Gly Pro Asn

GGT TTT GGA CCC AAC GGT TTG GAT ATC CAG CCC TGG CAA GTA
Gly Phe Gly Pro Asn Gly Leu Asp Ile Gln Pro Trp Gln Val

PR3

CCT TGG TTG GAG CCC CTC CGT GGA CCC AAC GGT TTT GGA CCC AAC
Pro Trp Leu Glu Pro Leu Arg Gly Pro Asn Gly Phe Gly Pro Asn

GGT TTG GAT ATC CAG CCC TGG CAA GTA
Gly Leu Asp Ile Gln Pro Trp Gln Val

Nucleotide and protein sequences of supressor mutants. Both pseudorevertants (PR1 and PR3) originated from deletion mutant D8. The deletion of the amino acid sequence LDKLRND (bold in wildtype) generated an *Eco*RV restriction site by an additional silent mutation (bold in the deletionmutant sequence). The insertion event causing the supression is shown for the pseudorevertants PR1 and PR3 (bold). The insertion was derived from a trippling, respectively doubling of an upstream region of the deletion (underlined).

Figure 2

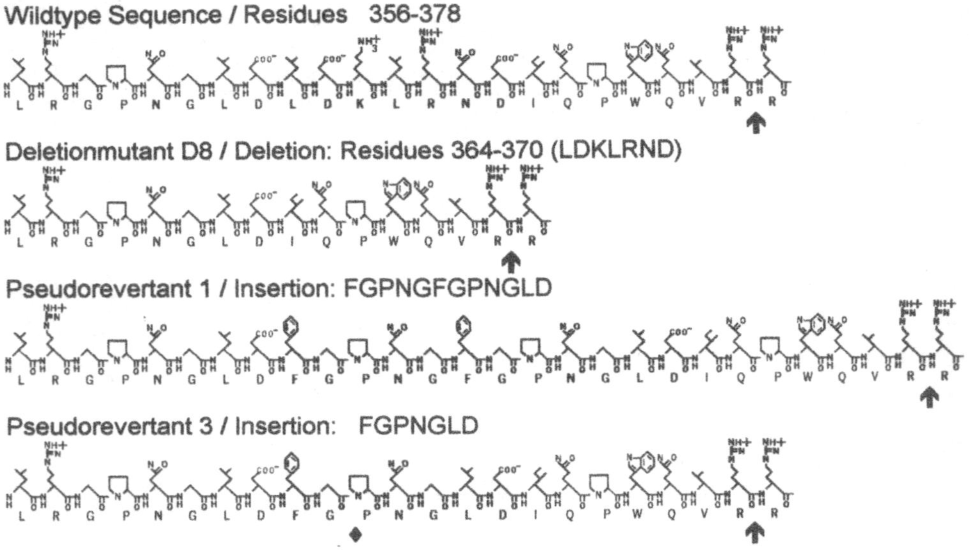

Comparison of the amino acid sequence of the wildtype, the deletion strain and the two different pseudorevertant strains. The deletion and the insertions are indicated in bold. The position of the double arginine motif is indicated in bold and by an arrow. The proline wich might influence the position of the 'double-arginine' in PR3 is indicated by an ♦.

The effect of the pseudoreversions on the function of PSII especially on the watersplitting mechanism will be studied more extensively in the future. Purified PSII membrane fragments (with active oxygen evolution) from wildtype and the pseudorevertants will be analysed regarding the S2 multilinesignal and the stabiliy (kinetics) of the Y_{Dox} radical using EPR and ENDOR techniques.

Acknowledgement: This work was supported by DFG (Sfb 312 TP A4)

References

1 Ghanotakis DF, Demetriou DM and Yocum CF (1987) Biochem Biophys Acta **891**, 15-21
2 Trebst A (1986) Z Naturforsch **41c**, 240-245
3 Bricker TM (1990) Photosynth Res **24**, 1-13
4 Boekema E, Hankamer B, Bald D, Kruip J, Nield J, Boonstra AF, Barber J and Rögner M (1995) Proc Natl Acad Sci (USA) **92**, 175-179
5 Kuhn M and Vermaas WF (1993) Plant Mol Biol **23**, 123-133
6 Eaton-Rye JJ and Vermaas WF (1991) Plant Mol Biol **17**, 1165-1177
7 Haag E, Eaton-Rye JJ, Renger G and Vermaas WF (1993) Biochem **32**, 4444-4454
8 Tang X-S And Diner B (1994) Biochem **33**, 4594-4603
9 Putnam-Evans C and Bricker TM (1992) Biochem **31**, 11482-11488
10 Klees H, Vermaas WF and Edelmann M (1992) Biochem **31**, 11065-11071

FTIR SPECTROSCOPY OF Q_B PHOTOREDUCTION IN *RB. SPHAEROIDES* REACTION CENTERS: EFFECTS OF SITE–DIRECTED REPLACEMENTS AT GLU L212, ASP L213, AND ASP L210 AND OF $^1H/^2H$ EXCHANGE.

E. Nabedryk[1], J. Breton[1], R. Hienerwadel[1], C. Fogel[2], W. Mäntele[2], M. L. Paddock[3], and M. Y. Okamura[3]; [1]SBE/DBCM, CEA Saclay, France; [2]Universität Erlangen, RFA; [3]U.C. San Diego, USA

1.Introduction

In order to investigate protonation or environmental changes of carboxylic acid residues occuring upon photoreduction of the secondary quinone acceptor Q_B in the reaction center (RC) of the photosynthetic bacteria *Rhodobacter sphaeroides* 2.4.1., we have performed light–induced FTIR spectroscopy on RCs from wild–type (Wt) and several site–directed mutants where carboxylic amino acid residues lying in the binding pocket of Q_B, i.e., Glu L212, Asp L213, and Asp L210 were diversely substituted (1,2). IR spectroscopy is well suited for investigating carboxylic acid sites since changes of protonation or of environment of Asp and Glu residues result in shifts and/or intensity changes of vibrational modes in the 1760–1700 cm^{-1} region where the C=O stretching mode from the COOH side chain group of protonated carboxylic acid residues is the main IR contributor. The single flash–induced FTIR Q_B^-/Q_B spectra have been obtained at pH 7 for native RCs and RC mutants containing either a single–site mutation, with Gln at L212 (EQ L212), Asn at L213 (DN L213) or Asn at L210 (DN L210), or a double–site mutation with both Gln at L212 and Asn at L213 (EQ L212 + DN L213). Additional residue replacements have been investigated at L 213, including Leu (DL L213), His (DH L213) and Ser (DS L213). The assignment of an IR band to the protonation/deprotonation of a particular carboxylic side chain was analyzed by combining the effects of site–directed mutagenesis and $^1H/^2H$ isotope exchange.

2.Methods

RC samples (≈1 mM, pH 7.0 –Tris–HCl, 50 mM–) contained a 10 fold excess of ubiquinone–10, 10 mM ascorbate and 20 mM diaminodurene (3). Light–induced FTIR difference spectra were acquired at 15°C with a Nicolet 60SX spectrometer upon single–turnover flash excitation by a Nd YAG laser (7 ns, 530 nm). $^1H/^2H$ exchange was performed by incubating RCs in ^2H_2O for 24h at 20°C.

3.Results and Discussion

Figure 1 depicts the light–induced FTIR difference spectra corresponding to the Q_B to Q_B^- transition in native 2.4.1 *Rb. sphaeroides* RCs, and in the mutants DN L210, DN L213, EQ L212, and EQ L212 + DN L213. In these Q_B^-/Q_B spectra, negative bands originate from the neutral Q_B state while positive bands arise from the Q_B^- state. The two main features common to the spectra of native and mutant RCs are the positive semiquinone anion band at 1479 cm^{-1} (this peak was normalized in all of the spectra) and the negative Q_B band at ≈1641 cm^{-1}. Figure 1 shows that the amplitude of the

P. Mathis (ed.), Photosynthesis: from Light to Biosphere, Vol. I, 875–878.

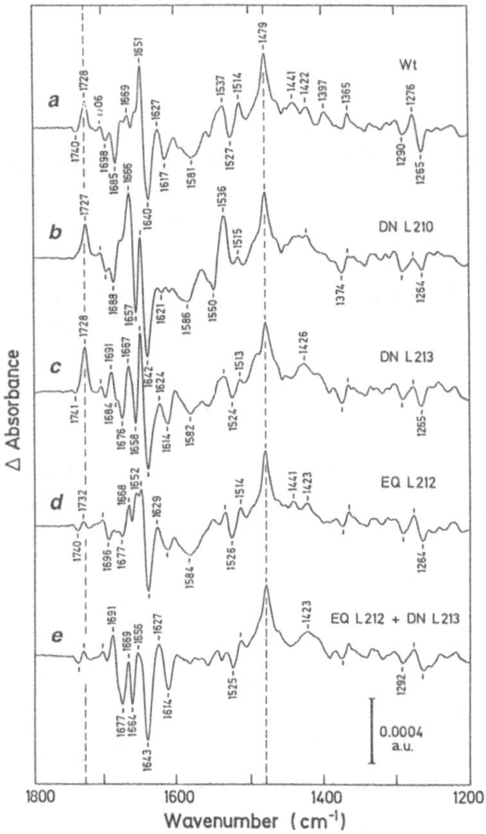

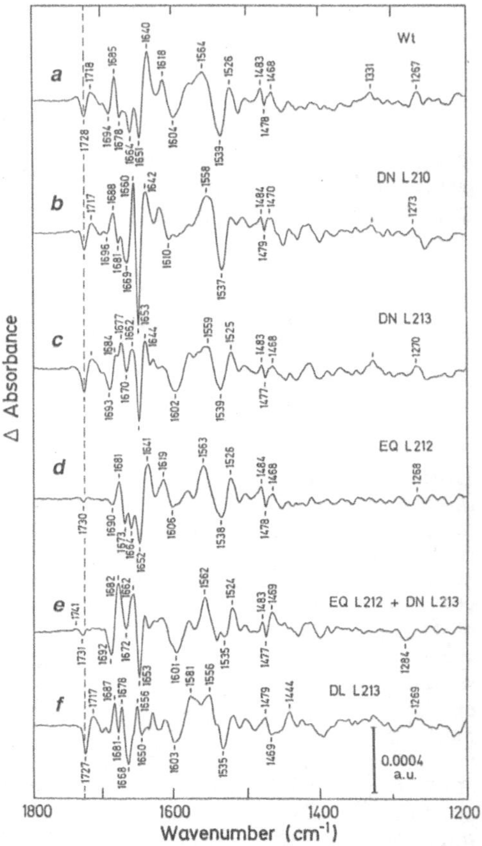

Figure 1 (left): Light–induced Q_B^-/Q_B FTIR difference spectra of *Rb. sphaeroides* RCs in ^1H_2O. a) wild type; b,c,d,e: mutant RCs. 4 cm^{-1} resolution. a. u., absorbance units.
Figure 2 (right): Calculated double–difference spectra between Q_B^-/Q_B spectra of *Rb. sphaeroides* RCs obtained in ^2H_2O and ^1H_2O: ^2H_2O–minus–^1H_2O.

absorbance changes in the amide I and amide II regions is not very different in Wt and in the mutants demonstrating the absence of large structural changes induced by the mutations in either the Q_B or Q_B^- states. These observations are taken to indicate that the differences found in the Q_B^-/Q_B spectra of Wt and of the mutants are localized to the mutation site.

Of particular interest is the positive 1728–cm^{-1} peak observed in Wt, DN L210 and DN L213 (Fig. 1a–c) and its absence in the mutants EQ L212 and EQ L212 + DN L213 (Fig. 1d,e). This signal lies in the typical absorption region of COOH groups from Asp and Glu residues. The increase of absorption at 1728 cm^{-1} is indicative of an increase in protonation of an Asp or Glu residue concomitant with the reduction of Q_B. The amplitude of the 1728–cm^{-1} band is similar in native and DN L210 RCs but is increased in DN L213 as well as in DL L213, DS L213 and DH L213 (see Fig. 3). The intensity of the 1728–cm^{-1} band is significantly reduced in ^2H_2O and a new feature appears at

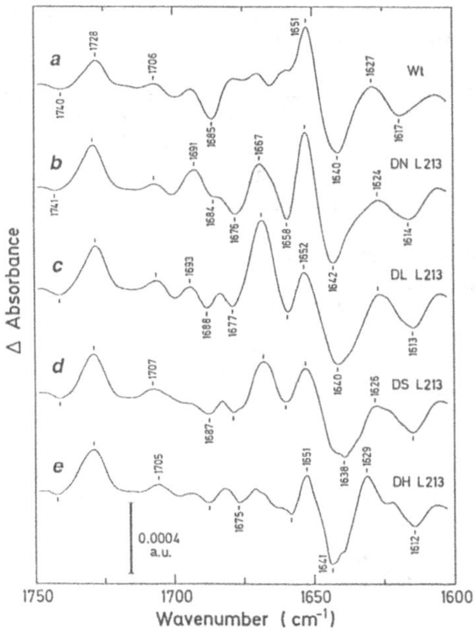

Figure 3: Comparison of Q_B^-/Q_B spectra in the 1800–1600–cm^{-1} region for the native RCs and the mutants DN L213, DL L213, DS L213 and DH L213.

1717 cm^{-1} (4). These changes in ^2H_2O are seen in the double–difference spectra (Fig. 2) calculated between Q_B^-/Q_B spectra obtained for a given RC in ^2H_2O and ^1H_2O (^2H_2O–minus–^1H_2O). In these double–difference spectra, only the residues affected by the photoreduction of Q_B **and** sensitive to $^1H/^2H$ exchange will give rise to differential and/or shifted bands. Note the negative peak at 1728 cm^{-1} and positive peak at ≈1717cm^{-1} common to the Wt, DN L210, DN L213 and DL L213 spectra (Fig. 2a–c,f). The frequency of the new feature observed at 1717 cm^{-1} is consistent with a $^1H/^2H$ isotope shift characteristic for a protonated carboxylic group accessible to solvent. The smaller intensity of the peak at 1717 cm^{-1} in ^2H_2O relative to that at 1728 cm^{-1} in ^1H_2O suggests that either COO^{2}H has a smaller extinction coefficient than COO^{1}H or the protonation of the carboxylic residue upon Q_B^- formation is smaller in ^2H_2O than in ^1H_2O. The comparison of the double–difference spectra obtained for Wt, DN L210, DN L213 and DL L213 (Fig. 2a–c,f) demonstrates that a common residue gives rise to the 1728–cm^{-1} signal and is similarly sensitive to $^1H/^2H$ exchange. On the other hand, the double–difference spectra ^2H_2O–minus–^1H_2O for EQ L212 and the double mutant (Fig. 2d, e) show only weak signals at ≈1731(–) and 1741(+) cm^{-1}. This is taken to indicate that Glu L212 is involved in the 1728–cm^{-1} peak seen in Wt, DN L210, and the various mutants at L213. Therefore, the 1728–cm^{-1} band is assigned to partial proton uptake by Glu L212. From kinetic IR data with μs time resolution, a transient signal at 1725 cm^{-1} present in Wt and absent in EQ L212 was proposed to arise from protonation of Glu L212 (5).

An estimation of the proton uptake by Glu L212 in Wt, DN L210 and DN L213, can be made based on molar extinction coefficients and half–widths for COOH (6) and semiquinone (7) bands. The IR signal at 1728 cm^{-1} in the Q_B^-/Q_B spectrum of Wt RCs has a half–width of ≈10 cm^{-1}, considerably less than for Glu in solution (56 cm^{-1}). The half–width of Q_B^- at 1479 cm^{-1} is ≈15 cm^{-1}, while the amplitude of the semiquinone peak is about twice that of the 1728–cm^{-1} band (Fig. 1a). The main sources of uncertainty are the values for the extinction coefficients and the determination of the baselines and half-widths of the IR bands, both of which may be in error by ≈25%. Nevertheless, based on these data, the integrated absorption at 1728 cm^{-1} corresponds to a proton uptake by Glu L212 upon Q_B^- formation of ≈0.4 H^+ in native RCs and DN L210 and ≈0.6 H^+ in DN L213. Thus, Glu L212 is partially ionized in the Q_B state becoming more protonated in the Q_B^- state. This directly confirms the earlier models in which a pK_a of > 10.5 for Glu L212 in the Q_B^- state was deduced (1,8). The proton on Glu L212 is thus available to rapidly protonate the fully reduced quinone following second electron transfer.

In contrast, the FTIR Q_B^-/Q_B spectra show no evidence for change of protonation or environment of Asp L213 upon Q_B^- formation. Apart from the increase of the amplitude of the 1728-cm^{-1} band, the other signals in the 1760–1700-cm^{-1} region are not significantly influenced by the various mutations at L213 (Fig. 3). This is consistent with the attributed pK_a< 5 for Asp L213 (8). The increased protonation of Glu L212 in DN L213 RCs is explained by a decreased Glu L212 pK_a value due to the loss of a negatively charged Asp L213. Thus we favor a model in which Asp L213 is essentially ionized both in the Q_B and Q_B^-states. The Asp L210 may contribute to the small differential signal at 1732(+)/1740(–) cm^{-1} (Fig. 1d,e) that is affected by $^1H/^2H$ exchange (Fig. 2d,e). Thus we suggest that Asp L210 is partially protonated at pH 7.

A negative signal at 1685 cm^{-1} in native RCs (Fig. 1a), shifted by a few wavenumbers in ^2H_2O, is proposed to arise from the absorption change of the amide I C=O mode of Glu L212. This band is absent in EQ L212 and EQ L212 + DN L213 spectra (Fig. 1d,e), whereas it is observed at 1688–1687 cm^{-1} in DN L210 (Fig. 1b,2b), DL L213 (Fig. 2f,3c), DS L213 (Fig. 3d), and DH L213 (Fig. 3e), and at 1684 cm^{-1} in DN L213 (Fig. 2c,3b). A more detailed discussion of the 1685-cm^{-1} band is given in (4). A signal at 1691(+)/1677(–) cm^{-1} observed in both DN L213 and the double-mutant (Fig. 1c,e) is tentatively assigned to an absorption change of the C=O side chain mode of Asn L213, since only weak signals in this spectral range are detected in the Q_B^-/Q_B spectra of Wt and mutant RCs lacking Asn L213 (Fig. 3).

In addition to the notion that mutations at the Q_B site induce only localized perturbations in the immediate vicinity of the mutated residue, the two most important conclusions from these FTIR studies are that after the first electron transfer step forming Q_B^- 1) Glu L212 increases its protonation state, consistent with its proposed function as a proton donor to the fully reduced quinone and 2) Asp L213 is proposed to be in the ionized state consistent with the increased proton uptake of Glu L212 in DN L213 RCs and with its function to facilitate proton transfer by creating a negative electrostatic potential near Q_B.

Acknowledgmemts: We thank G. Feher for his sustained interest in this FTIR work and E. Abresch for reaction center preparation. Part of this work was supported by an EEC grant (SCI* 0335-C) and a NIH grant (GM41637).

References

1–Okamura, M. Y., & Feher, G. (1992) *Ann. Rev. Biochem.* 61, 861–896.

2–Paddock, M. L., Rongey, S. H., McPherson, P. H., Juth, A., Feher, G., & Okamura, M. Y. (1994) *Biochemistry* 33, 734–745.

3–Breton, J., Berthomieu, C., Thibodeau, D. T., & Nabedryk, E. (1991) *FEBS Lett.* 288, 109–113.

4–Nabedryk, E., Breton, J., Hienerwadel, R., Fogel, C., Mäntele, W., Paddock, M. L., & Okamura, M. Y. *Biochemistry*, submittted.

5–Hienerwadel, R., Grzybek, S., Fogel, C., Kreutz, W., Okamura, M. Y., Paddock, M. L., Breton, J., Nabedryk, E., & Mäntele, W. (1995) *Biochemistry* 34, 2832–2843.

6–Venyaminov, S. Yu., & Kalnin, N. N. (1990) *Biopolymers* 30, 1243–1257.

7–Bauscher, M., Nabedryk, E., Bagley, K., Breton, J. & Mäntele, W. (1990) *FEBS Lett.* 261, 191–195.

8–Takahashi, E., & Wraight C. A. (1992) *Biochemistry* 31, 855–866.

DEGENERATE FOUR-WAVE MIXING ON PHOTOSYNTHETIC REACTION CENTERS

T. Arlt, R. Heinecke, H. Penzkofer and W. Zinth
Institut für Medizinische Optik, Universität München
Barbarastr. 16, D-80797 München

1. Introduction

Investigations of the ultrafast processes in photosynthetic reaction centers are one of the topics in femtosecond spectroscopy today. The photosynthetic reaction center (RC) is a membrane protein complex containing different chromophores. Optical excitation starts a stepwise electron transfer across this membrane protein via the electron carrying chromophores. In the first step the electron is transferred from a pair of excitonically coupled bacteriochlorophylls called special pair P to a single bacteriochlorophyll molecule B within 3 picoseconds (ps) [1]. Optical pump-probe spectroscopy of photosynthetic reaction centers has been used during the last decade to study the electron transfer reaction in detail (for an overview see [2]). However, pump-probe experiments are not able to deliver complete information. As a consequence additional techniques are required. Degenerate four-wave mixing is an interesting coherent technique to characterize the nonlinear response of a medium. In the present paper it is used to investigate Q_y transition moments (Q_y(P) at 865 nm and Q_y(B) at 802 nm) of chromophores in RCs of *Rhodobacter sphaeroides* (R26) in detail. The data are compared with results obtained from dye molecules.

2. Materials and Methods

The experimental setup is shown in Fig. 1. A lasersystem based on a CPM dye laser delivers femtosecond pulses in the range from 700 nm to 1000 nm (for details of the system see [1]). The pulses used for the experiments have a spectral width of about 20-25 nm and an energy of about 1 μJ. The instrumental response function of the four-wave mixing experiments shows a FWHM of about 90 fs. Two pulses with wavevectors $\underline{k}_1$ and $\underline{k}_2$ are used for the experiment. The first laser pulse $\underline{k}_1$ creates a nonlinear polarization in the sample which decays very rapidly in the case of an inhomogeneous broadened system treated here. The second pulse $\underline{k}_2$, which is delayed by the time t, rephases the differentconstituents of the inhomogeneously broadened line and the so-called photon echo is emitted in the direction $2\underline{k}_2 - \underline{k}_1$:

$$I_{2k_2-k_1}(t) = I(0)\cdot\exp(-t/\tau) \sim \exp(-t/(T_2/4)) \quad \text{Eq. (1)}$$

P. Mathis (ed.), Photosynthesis: from Light to Biosphere, Vol. I, 879–882.

τ is the measured decay time of the scattered signal and T_2 the optical dephasing time. In the case of a simple two-level system T_2 is given by:

$$1/T_2 = 1/T_{2PD} + 1/(2 \cdot T_1) \quad \text{Eq. (2)}$$

Here T_{2PD} represents the pure dephasing time and T_1 the energy- or population relaxation time.

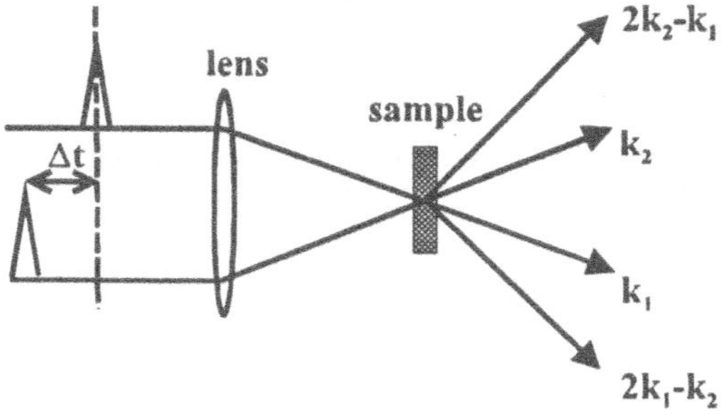

Fig. 1: Experimental setup used in the four-wave mixing experiments. The scattered signal (photon echo) in direction $2\underline{k}_2 - \underline{k}_1$ is observed as a function of the delay Δt.

3. Results and Discussion

Q_Y(B) band:

Fig. 2a shows the signal for reaction centers (R.26) scattered in direction $2\underline{k}_2$-$\underline{k}_1$ (λ=804 nm). The data are well described by a sech^2 instrumental response function and a monoexponential decay with a time constant of τ = 85 fs ±10 fs (solid line). In Fig. 2b the results on RC are plotted together with measurements on the dye molecules IR 140 (Lambda Physics) and Kodak 9860. For IR 140 a decay time of 75 fs ±10 fs could be detected, whereas for the Kodak dye the decay time is below the time resolution; i.e. τ < 40 fs. If we treat the RC as an inhomogeneous broadened system, the decay time τ = 85 fs for the RC corresponds to a dephasing time T_2 of about 340 fs (Eq. (1)).
From up-conversion experiments the time constant $\tau_{tra} \approx$ 150-200 fs for the energy transfer from Q_x(B) to Q_y(P) and/or Q_x(P) to Q_y(P) is known [3]. For an estimate we can consider a T_1-time constant in this range (τ = 150 fs) and neglect pure dephasing processes (see below): this leads to a dephasing time of T_2 = 300 fs (Eq. (2)). The decay of the coherence of the transition moment could be determined by such a T_1-process.

Q_Y(P) band:

In Fig. 3 the photon echo signal of reaction centers at 865 nm is plotted for 300 K and 17 K. The signals show a fast initial decay. At low temperatures a slow decaying background remains, which is probably due to an artefact of aggregating bacteriochlorophyll molecules. The observed time constant τ of about 60 fs at room temperature (Fig.3a) is weakly reduced to 40 fs at low temperatures (Fig.3b). At 100K and 150 K a similar time constant is observed (data not shown). In this context it must be noted, that time constants of 40 fs and faster are close to the resolution limit of the used experimental setup. The observation of a speed up of the decay at lower temperatures is surprising since a slowing down is expected for the temperature dependence of pure dephasing processes. As a consequence we conclude that the dephasing time T_2 should be determined by the energy relaxation time T_1; e.g. by some ultrafast initial reaction in P^*.

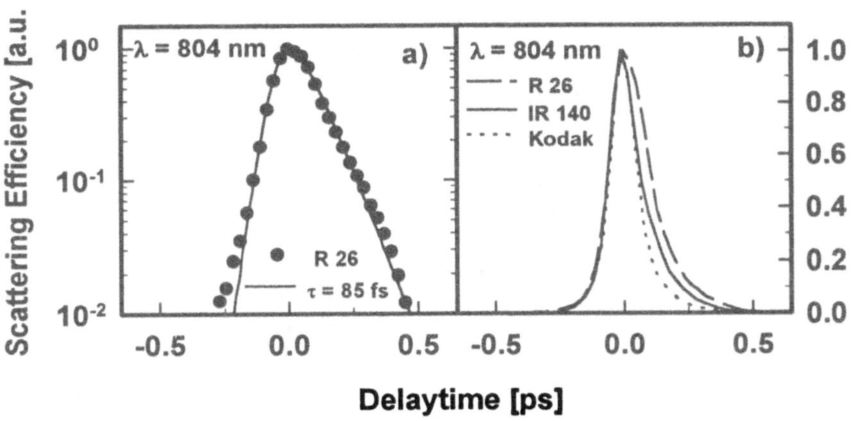

Fig.2: a) Scattered signal for reaction centers (R26) at 804 nm (points) and a model function using a $sech^2$-function and a monoexponetial decay with 85 fs (solid line). b) For comparison the signal of the dyes IR 140 and Kodak 9860 is shown. Note the different scales used in a) and b).

The observed coherent decay times compare well with results from other techniques.
In ref. [4] it was shown that the dephasing of BChl molecules is largely electronic in character at low temperatures. Femtosecond pump-probe experiments at different temperatures (10 K - 290 K) revealed oscillations within the decay of the excited electronic level P^* with a main damping time of 200 fs practically independent of temperature [5]. This also supports the finding of the present paper that pure dephasing does not contribute strongly to the decay of the coherence of excited level of $Q_y(P)$ in photosynthetic RCs. A remaining explanation for the observed dephasing time could refer to changes of the excited electronic state after excitation due to coupling of P^* to charge-transfer states within the special-pair [4], [6]. In this case the first electron transfer step probably starts from this modified state. The detected decay constant of 60 fs at room temperature corresponds to a time constant T_2 of 240 fs, if pure dephasing processes are neglected. Fast processes within the excited state P^* were found also very recently: time resolved infrared [6] and near-infrared [7] spectroscopy yielded transients with time constants of 200 fs and 250 fs, respectively.

4. Summary

We have presented photon-echo experiments on photosynthetic reaction centers resolving rapid processes on the 100 fs time scale. The loss of coherence is found to be completed on the 200 fs - 300 fs time scale. Experiments at low temperatures show that population-relaxation or reactive processes (T_1-processes) and not quasielastic (pure dephasing) processes dominate the measured dephasing times. The data in the Q_y (P) band support a fast dephasing due to coupling to charge-transfer states or other excited states within the special-pair.

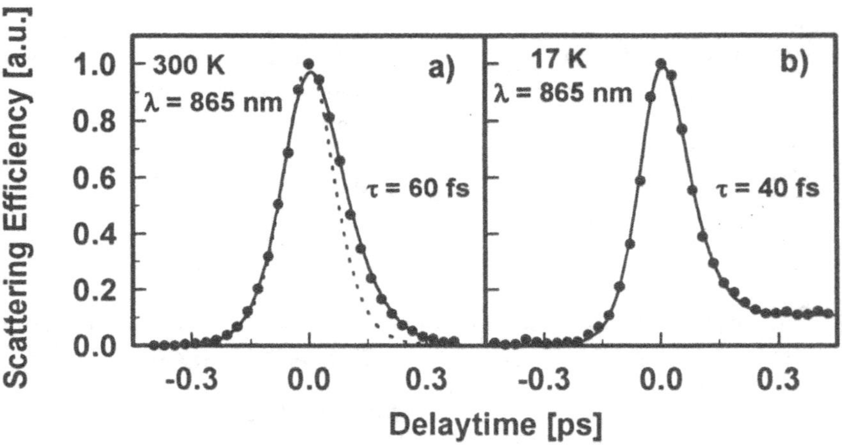

Fig.3: Scattered signal of reaction centers (R 26) in the Q_y (P) band at 865 nm for 300 K and 17 K. Points refer to the measured data, the dotted line represents the instrumental response function. The solid lines are model-functions using exponential decay processes.

We thank Prof. H. Scheer and M. Meyer (Botanisches Institut der Universität München) for the sample preparation.

References

[1] Arlt, T., Schmidt, S., Kaiser, W., Lauterwasser, C., Meyer, M., Scheer, H. and Zinth, W. (1993) Proc. Natl. Acad. Sci USA 90, 11757-11761

[2] The Photosynthetic Reaction Center (Deisenhofer, J. and Norris, J.R., eds.), Academic Press, Inc. (1993), San Diego, USA

[3] Du, M., Rosenthal, S.J., Xiaoliang, X., DiMagno, T.J., Schmidt, M., Hanson, D.K., Schiffer, M., Norris, J.R. and Fleming, G.R. (1992) Proc. Natl. Acad. Sci USA 89, 8517-8521

[4] Meech, S.R., Hoff, A.J. and Wiersma, A.J. (1985) Chem. Phys. Letters 121, 287-292 .

[5] Vos, M.H., Rappaport, F., Lambry J.-C., Breton, J. and Martin J.-L. (1993) Nature 363, 320-325

[6] Hamm, P., Zurek, M., Mäntele, W., Meyer, M., Scheer, H. and Zinth, W. (1995) Proc. Natl. Acad. Sci USA 99, 1826-1830

[7] Streltsov, A.M., Yakovlev, A.G., Shkuropatov, A.Y. and Shuvalov, V.A. (1995) FEBS Letters357, 239-241

Experimental observation of multiple trapping/charge separation steps in the isolated PS2 reaction centre.

Shigeichi Kumazaki[‡], D. Melissa Joseph, Ben Crystall, Yasuhiro Tachibana, James R. Durrant, James Barber, George Porter, Keitaro Yoshihara[‡] and David R. Klug[*].

Molecular Dynamics and Photosynthesis Research Groups, Centre for Photomolecular Sciences, Department of Biochemistry, Imperial College, London SW7 2AY, and [‡] Institute for Molecular Science, Myodaiji, Okazaki 444, Japan.

Introduction

There have been a number of studies which have addressed the kinetics of primary charge separation in the isolated PS2 reaction centre (reviewed by van Grondelle et al., 1994 and Babcock & Diner, 1995). The kinetics of the PSII RC have been the source of some controversy. In particular there is as yet no consensus regarding the dominant time-constant which is associated with primary charge separation. A number of researchers have found time constants of approximately 3 ps and 21 ps at room temperature (Hastings et al., 1992; Durrant et al., 1993; Wiederrecht et al., 1994;. Schelvis et al., 1994; McCauley et. al, 1992), but these studies do not agree on the extent to which these components are associated with trapping/charge separation.

Studies of spectrally resolved stimulated emission in PSII RC's have been restricted to the vibrational sideband of P680, as this is the only spectrally distinct section of the emission spectrum. These studies have concluded that when P680 is photoselectively excited, 70 ± 20 % of the loss of stimulated emission is associated with a 21ps time constant and the remaining 30 % associated with a 3ps component (Durrant et al., 1993; Klug et al., 1995). These transient absorption studies had a time resolution of 200fs in the vecinity of the emission sideband, and are broadly in agreement with a measurement of spontaneous emission loss in PSII RC's made using a streak camera with an instrument response function of 18ps (Freiberg et al., 1993). The streak camera measurements and the transient absorption studies are however not consistent with studies of spontaneous emission in PSII RC's made using time-correlated single photon counting in which the instrument response function was 30-50ps wide (Roelofs et al., 1991; Holzwarth et al. 1994) and in which the majority of spontaneous emission loss was associated with a 3ps component with the minority associated with an 18ps component.

P. Mathis (ed.), Photosynthesis: from Light to Biosphere, Vol. I, 883–886.

We attempt to resolve this discrepancy by making time resolved spontaneous emission measurements using fluorescence up-conversion with an instrument response of approximately 600fs and single photon counting measurements with an instrument response of 45ps. We show here that our up-conversion data are consistent with our time-correlated single photon counting data, and that in both studies the spontaneous emission kinetics are dominated by a time constant of approximately 20 ps. These spontaneous emission studies are *quantitatively* consistent with studies of the simulated emission loss in PSII RC's monitored by transient absorption spectroscopy. These new data also exhibit some interesting additional features which demonstrate a further distinction between electron and energy transfer kinetics in PSII RC's and those found in PB RC's.

Experimental results

The excitation conditions for all of the experiments were chosen to achieve photoselection of the C680 chlorin pool. Detection was at the peak of the emission sideband (approx. 730 nm) to reduce problems associated with scatter from the excitation pulse. The emission data, instrument response and fitted function are illustrated in Fig. 1a and 1b and the fitting parameters are presented in Table 1. Figure 2 shows transient absorption data at 545nm which also demonstrates the presence of a 100-200ps component.

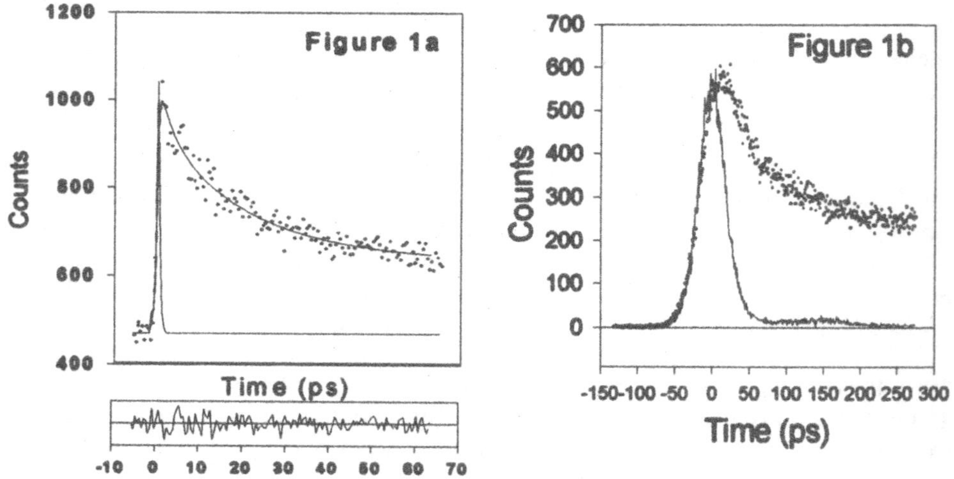

Figure 1. a) Shows fluorescence up-conversion data while b) shows and example of time-correlated single photon counting data. The results of the analysis of this data is given in Table 1.

Table 1: Lifetimes and Amplitudes of spontaneous Emission from PSII reaction centres measured by both fluorescence up-conversion and time-correlated single photon counting.

Up-Conversion Double exponential fit	1.6ps(0.26)	19.5ps(0.49)	∞(0.25)	
Up-Conversion Single exponential fit	15.7ps(0.67)		∞(0.33)	
Photon counting global analysis	15.6ps(0.74)		98ps(0.15)	>1ns(0.11)

Results are presented as lifetime(relative amplitude).

Discussion

Although the up-conversion data has considerably lower signal to noise ratio than the photon counting data, the former experiment successfully resloved the 21ps and 3ps components while the latter was unable to distinguish them, and demonstrated a complex χ^2 surface with poorly defined minima. This difference presumably reflects the practical difficulties associated with collecting and analysing multiexponential kinetics in which there are many components which are significantly shorter than the decay of the instrument response function.

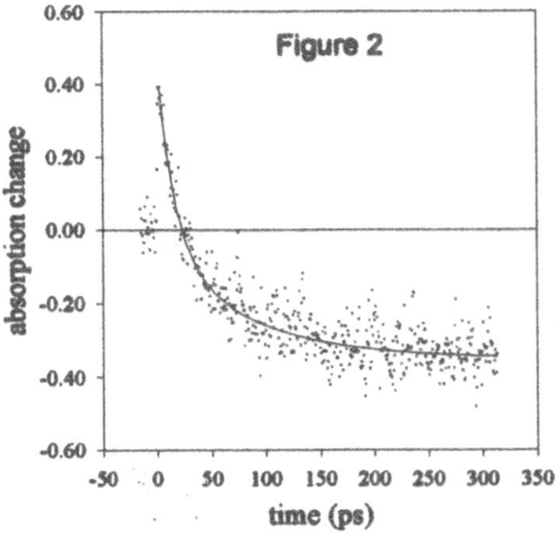

Overall we conclude as previously, that the majority of charge separation is associated with a lifetime of about 21 ps although there is also some loss of excited singlet states associated with a lifetime of about 3 ps.

One surprising aspect of both the up-conversion and the photon counting data, is the extremely high amplitude of the slower components. After 50 ps, the level of fluorescence is still approximately 30% of the initial amplitude. The ratio of the sum of the 3 ps and 21 ps components to the slower components is

3:1 in both the photon counting and the up-conversion experiments. This high background level is consistent with streak camera measurements which also found that components longer than approx. 20ps had a surprisingly large amplitude (Freiberg et al., 1993). The photon counting data shows that half of this background decays with a time constant of 100ps. This result is shown here to be consistent with transient absorption measurements, which also demonstrate the presence of a component of 100-200ps time constant and whose amplitude is consistent with a loss of 15% of the initial singlet state population (figure 2).

References

Diner, B.A. & Babcock, G.T., in : Oxygenic photosynthesis: the light reactions, eds. C.Yocum and D. Ort (Kluwer Dordrecht, 1995) In Press.

Durrant, J.R., Hastings, G., Joseph, D.M., Barber, J., Porter, G., & Klug, D.R. (1993) *Biochem. 32(32)*, 8259 - 8267.

Freiberg, A., Timpmann, K., Moskalenko, A.A., & Kuznetsova, N.Y. (1993) *Biochim. Biophys. Acta 1184(1)*, 45 - 53.

Hastings, G., Durrant, J.R., Barber, J., Porter, G., & Klug, D.R. (1992) *Biochem. 31*, 7638 - 7647.

Holzwarth, A.R., Müller, M.G., Gatzen, G., Hucke, M., & Griebenow, K. (1994) *J. Luminescence 60 & 61*, 497 - 502.

Klug, D.R., Rech, T., Joseph, D.M., Barber, J., Durrant, J.R., & Porter, G. (1995) *Chem. Phys.*, in press.

McCauley, S.W., Baronavski, A.P., Rice, J.K., Ghirardi, M.L., & Mattoo, A.K. (1992) *Chem. Phys. Lett.* 198(5), 437-442.

Roelofs, T.A., Gilbert, M., Shuvalov, V.A. & Holzwarth, A.R., (1991) *Biochim. Biophys. Acta* , 1060, 237-244.

Schelvis, J.P.M., van Noort., P.I., Aartsma, T.J. & van Gorkom, H.J., (1994) *Biochim. Biophys Acta,* 1184, 242-250.

van Grondelle, R., Dekker, J.P., Gillbro, T., & Sundström, V. (1994) *Biochim. Biophys. Acta* 1187, 1-65.

Wiederrecht, G.P., Seibert, M., Govindjee & Wasielewski, M.R., (1994) *Proc. Natl. Acad. Sci. USA,* 91, 8999-9003

Low-intensity fluorescence spectra from the photosystem II reaction centre measured using a picosecond synchroscan streak camera. Effects of sample degradation on the observed lifetimes.

I.H.M. van Stokkum, F. van Mourik, M.-L. Groot, and R. van Grondelle
Faculty of Physics and Astronomy, Vrije Universiteit, De Boelelaan 1081, 1081 HV Amsterdam, The Netherlands

1. Introduction

Optimal information is obtained when processes are monitored at many wavelengths simultaneously with the highest possible time resolution. The combination of a streak camera with a spectrograph enables a true spectrotemporal investigation of complex and vulnerable photoactive systems. For this pilot study we chose the PSII reaction center. The improved time resolution of the apparatus relative to SPT techniques (1) allows the unambigueus observation of picosecond processes. Due to the high sensitivity of the system short data accumulation times with low light intensity are possible which prevents systematic errors due to sample degradation.

2. Materials and Methods

The PSII reaction center sample was prepared as in (2,3), oxygen in the sample was removed using the glucose-oxidase/catalase as in (4) For the excitation of the sample the frequency doubled light from a Mira RegA (Coherent) oscillator/regenerative amplifier combination was used, giving pulses of about 200 fs at 400 nm, with a repetition rate of 200 kHz. The pulses were attenuated with neutral density filters to an energy of ~ 1 nJ per pulse, corresponding with an average energy of ~ 300 μW. The pulses were focused into the sample using an f=15 cm lens. The sample was contained in a rotating cell (plm. 300 rpm) to prevent excessive accumulation effects. Under the experimental conditions each spot is hit by < 10 pulses per transit through the focus. The light path of the cell was 2 mm, the (horizontal) focus of the laser was imaged onto the (vertical) slit of the spectrograph using f/8 optics.
The synchroscan streak camera (Hamamatsu C5680 and M5675) was sychronized to the Mira oscillator using a photodiode (Hamamatsu C1808–02). The full width at half maximum (FWHM) of the instrument response function (IRF) of the streak camera without spectrograph was typically 1.8 ps. The IRF of the total system including the Chromex IS250 spectrograph equipped with a 50 grooves/mm grating was 2.6–3 ps, this was checked using the white-light continuum pulses of the RegA to simulate a broad-band emission spectrum. Measurements were performed with the help of a data acquisition system including a C4880 Dual Scan Cooled CCD camera (Hamamatsu HPD TA). Accumulation of ~ 30 seconds per dataset (accumulation on the CCD chip) resulted in a 1024×1024 matrix of integrated fluorescence intensity. The full time and wavelength ranges were 200 ps and 300 nm, respectively. Experiments started after approximately 10 min adjustment of the set up. A total of 40 data sets was thus obtained over a period of

P. Mathis (ed.), Photosynthesis: from Light to Biosphere, Vol. I, 887–890.

40 minutes. Data were corrected for shading (6) and analysed off-line with the help of the Singular Value Decomposition and global analysis (7).

3. Results and Discussion

The time profile of the IRF of the total system as determined with pump light (centered at 400 nm) scattered from the sample is shown in Figure 1. The FWHM of the Gaussian

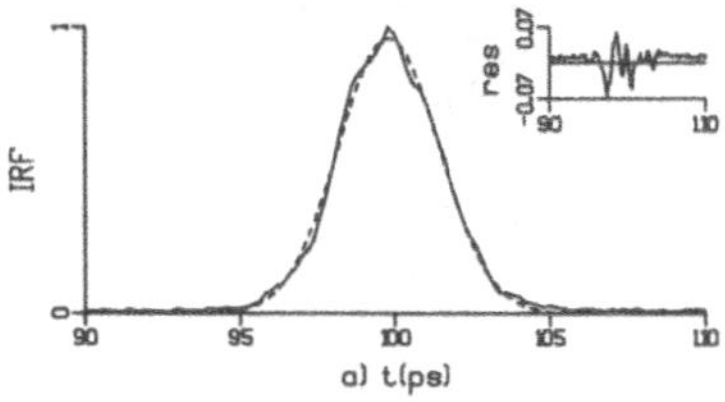

Figure 1. Time profile of the IRF of the total system fitted with a Gaussian (dotted). Inset shows residuals.

which fitted the IRF well was 3.5 ps.
The singular value decomposition of a matrix containing the fluorescence recorded from 625–725 nm and from 0–200 ps resulted in two singular values significantly different from the noise singular values (Figure 2c). Thus the rank of this matrix is two, implying

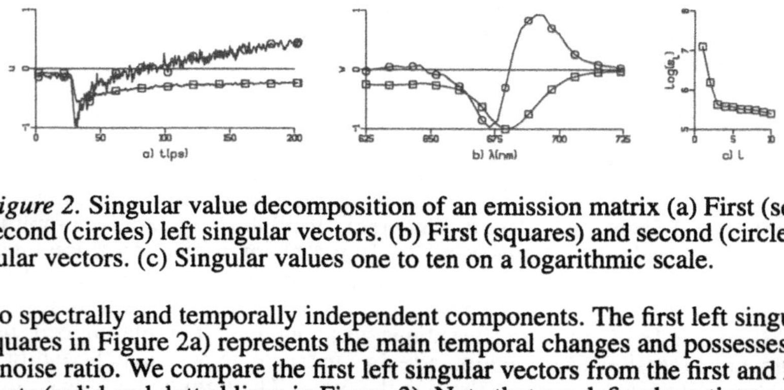

Figure 2. Singular value decomposition of an emission matrix (a) First (squares) and second (circles) left singular vectors. (b) First (squares) and second (circles) right singular vectors. (c) Singular values one to ten on a logarithmic scale.

two spectrally and temporally independent components. The first left singular vector (squares in Figure 2a) represents the main temporal changes and possesses a high signal to noise ratio. We compare the first left singular vectors from the first and last experiments (solid and dotted lines in Figure 3). Note that we define here time zero as the max-

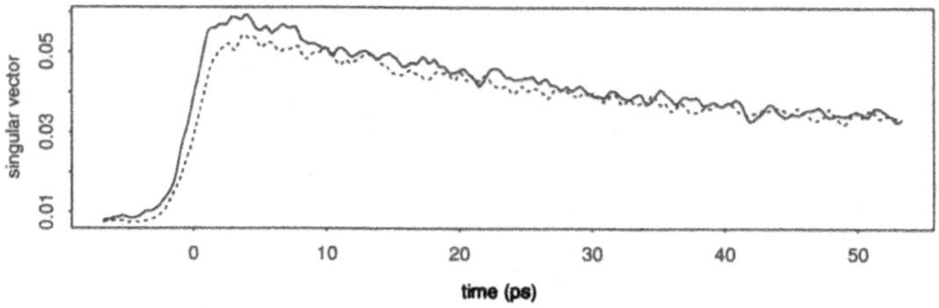

Figure 3. First left singular vectors from the first (solid) and last (dotted) experiment.

imum of the IRF, which is thus a result of the global analysis described below. The rise of the fluorescence emission is different. A fast phase is only present in the first experiment (solid line). This was confirmed by a global analysis of the data. We used a model with three independently decaying components. We analysed the 40 data sets measured consecutively in 20 adjacent pairs, with a delay parameter for each data set to allow for drift of the IRF. Decay Associated Spectra (DAS) of the first and last experiment are shown in Figure 4. Both the first and the second rate constant have become considerably slower in the last experiment, and the contribution of the third, slowest component (triangles) has increased. The accompanying DAS (triangles in Figure 4) appears to shift to

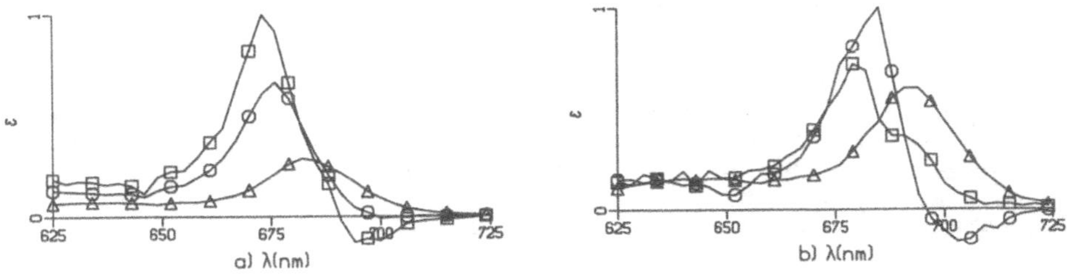

Figure 4. Decay Associated Spectra from first (a) and last (b) experiment. Life times: (a) 1.3 ps (squares), 43 ps (circles), 5.5 ns (triangles). (b) 17 ps, 87 ps, 5 ns.

the red.
The trends in the rate constants as a function of exposure time are depicted in Figure 5.

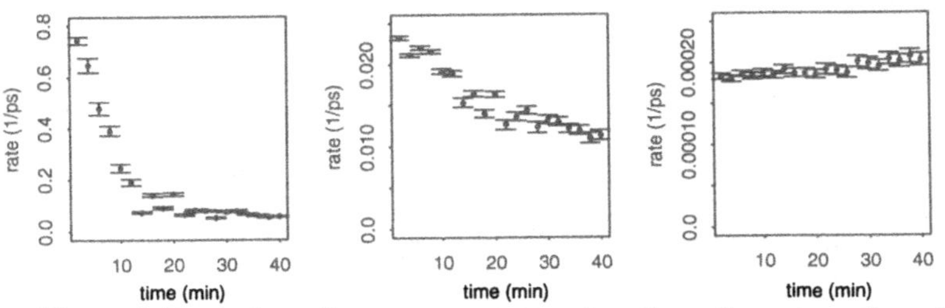

Figure 5. Evolution of rate constants as a function of exposure time. Horizontal bars indicate ±standard error.

One possible explanation for the trends observed in Figure 5 is that in the early experiments a fourth component is present, and that the very fast 1-2 ps component degrades gracefully. We checked this hypothesis by performing fits with more components. Five component fits were always unsatisfactory, whereas a four component fit was satisfactory with the first, but not with the last experiments. The DAS of the additional component of 25±3 ps is the smallest and most red (circles in Figure 6).
Roelofs et al. (1) found lifetimes of 2, 41 and 160 ps, and 1.09, 5.87 and 29.57 ns. The two shortest lifetimes are in accordance with the results from our first experiments. Regarding the range of longer lifetimes, a combined analysis of experiments on different time scales is necessary (1).

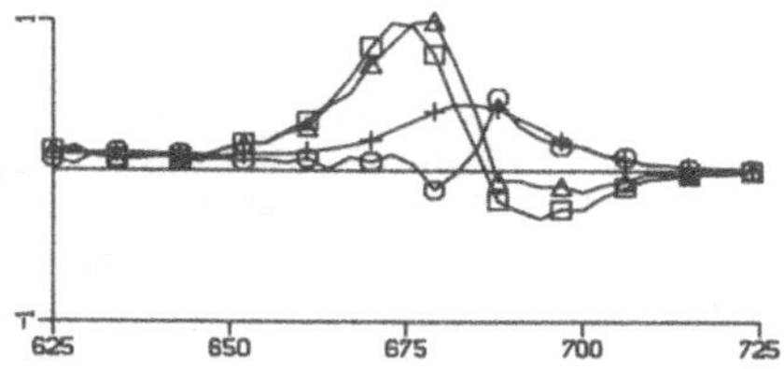

Figure 6. Decay Associated Spectra from first experiment. Life times: 2.0 ps (squares), 25 ps (circles), 47 ps (triangles), 5.3 ns (plus symbols).

Despite the preliminary nature of the dataset (no sets with less than 5 min. 'preillumination' were recorded), some interesting changes are occuring during the course of the experiment. The fast decay component decreases within ~10 min. which is quite surprising considering the low excitation density used here. The changes do not correspond with complete destruction of the sample since no long-lived 670 nm components are observed in the 'late' data sets. We note that the early and late rate constants correspond to those extracted from SPT (1) and pump-probe experiments (5) respectively.

Acknowledgements

We are grateful to Hamamatsu Photonics Deutschland GmbH for giving us the opportunity to use the spectrograph – streak camera system and to U. Denzér and T. Miyazaki for help with the set up of the instrument. We thank H. van Roon and F. Calkoen for the expert preparation of the PS II reaction centre particles. This research was supported by the "Nederlandse organisatie voor Wetenschappelijk Onderzoek (NWO)" via the "Stichting voor Fundamenteel Onderzoek der Materie (FOM)" and the "Foundation for Life Sciences (SLW)".

References

1 Roelofs, T.A., Gilbert, M., Shuvalov, V.A. and Holzwarth, A.R. (1991) *Biochim. Biophys. Acta* **1060**, 237-244
2 Kwa, S. L. S.; Newell, W. R.; van Grondelle, R.; and Dekker, J. P. (1992) *Biochim. Biophys. Acta* **1099**, 193-202
3 Eijckelhoff, C., and Dekker, J.P. (1995) *Biochim. Biophys. Acta*, **1231**, 21-28
4 Groot, M.-L., Peterman, E.J.G, van Kan, P.J.M.,van Stokkum, I.H.M., Dekker, J.P. and van Grondelle, R. (1994) Biophys. J. 67, 318-330
5 Durrant, J.R., Giorgi, L.B., Barber, J., Klug, D.R., Porter, G. (1990) *Biochim. Biophys. Acta* **1017**, 167
6 High Performance Digital Temporal Analyzer Instruction Manual (1995) Hamamatsu Photonics Deutschland GmbH
7 Van Stokkum, I.H.M., Brouwer, A.M., Van Ramesdonk, H.J. and Scherer, T. (1993) *Proc. Kon. Ned. Akad. v. Wetensch.* **96**, 43-68

MODIFICATION OF INACTIVE PSII CENTERS BY LIGHT.

Kevin Oxborough[1,2], Ladislav Nedbal[1,3] & John Whitmarsh[1]
[1]Department of Plant Biology, University of Illinois, [2]Photosynthesis Research Unit, ARS/USDA, 1201 W. Gregory Drive, Urbana, Illinois 61810. [3]Present address: Institute of Microbiology, Opatovicky mlyn, 37981 Trebon, Czech Republic. [4]Present address: Dept. Biology, University of Essex, Wivenhoe Park, Colchester, Essex CO4 3SQ

In dark-adapted leaves of 'healthy' plants, between one quarter and one third of photosystem II (PSII) reaction centers are impaired in their ability to transfer electrons from water to plastoquinone. Although these 'inactive' PSII centers are capable of reducing the primary quinone acceptor, Q_A, oxidation of Q_A^- occurs with a half-time of 1.5 - 2 s (Chylla and Whitmarsh 1989) compared to ~ 1 ms at active PSII centers.

Additionally, inactive PSII centers differ from active PSII centers in the following ways:

- While active PSII centers are located almost exclusively within the grana lamellae, inactive PSII centers are concentrated within the stromal regions of the thylakoid membrane (Henrysson and Sundby 1990, Chylla and Whitmarsh 1989).
- The absorption cross-section of inactive PSII centers is approximately half that of active PSII centers (Chylla and Whitmarsh 1990, Nedbal et al. 1991).

Current evidence suggests that little or no energy transfer occurs between the pigment beds associated with inactive and active centers. Consequently, the quantum yield of PSII photochemistry could be substantially reduced by the presence of inactive centers; by about 10% in spinach, for example (Nedbal et al. 1991).

It has been suggested that inactive PSII centers are modified by high light such that they become capable of electron transfer to the plastoquinone pool (Neale and Melis 1990); the proposal being that these centers act as a reserve pool which partially compensates for normally active PSII centers under photoinhibitory conditions.

We have examined the effect of high light treatment on PSII centers within spinach leaves through measurements of the electrochromic shift and chlorophyll fluorescence. While our data do indicate that inactive PSII centers are modified by light, this modification does not appear to involve conversion to an active form.

The concentration of inactive PSII centers within spinach leaves was assessed by comparing the amplitude of the absorbance change at 518 nm (ΔA_{518}) induced by two saturating, single-turnover flashes spaced 30 ms apart (Chylla and Whitmarsh 1990).

P. Mathis (ed.), Photosynthesis: from Light to Biosphere, Vol. I, 891–894.

The fast (< 1 μs) phase of ΔA_{518} induced by the first flash (ΔA_{FL1}) reflects charge separation at all centers and is directly proportional to the number of centers which undergo charge-separation (e.g. Chylla and Whitmarsh 1989). During the 30 ms interval before the second flash, the primary, stable electron-acceptors of active PSII centers and PSI centers are re-oxidized. Since inactive PSII centers are re-oxidized with a half-time of 1.5-2 s, most of these centers are "closed" at the time of the second flash. Consequently, the fast-phase of ΔA_{518} induced by the second flash (ΔA_{FL2}) only represents charge-separation at active PSII centers plus PSI centers. The difference between ΔA_{FL1} and ΔA_{FL2} can therefore be attributed to charge-separation at inactive PSII centers.

The data presented in Figure 1 show the change in the fraction of ΔA_{FL1} representing inactive PSII centers (ΔA_{FL1} - ΔA_{FL2}) which occurred during dark-adaptation of a spinach leaf after 5 min illumination with 900 μmol m^{-2} s^{-1} red light. What these data show is that, after light treatment, the contribution of inactive PSII centers to ΔA_{FL1} is less than half of that from a dark-adapted leaf. Virtually all of this light-induced decrease in ΔA_{FL1} - ΔA_{FL2} was due to a decrease in ΔA_{FL1}; ΔA_{FL2} remained within 2% of the dark-adapted level. Consequently, this lowering of ΔA_{FL1} - ΔA_{FL2} would appear to represent a large decrease in the number of inactive PSII centers which are capable of stable charge-separation and not the conversion of these centers to an active form, which would have led to a proportional increase in ΔA_{FL2} (of approximately 8%).

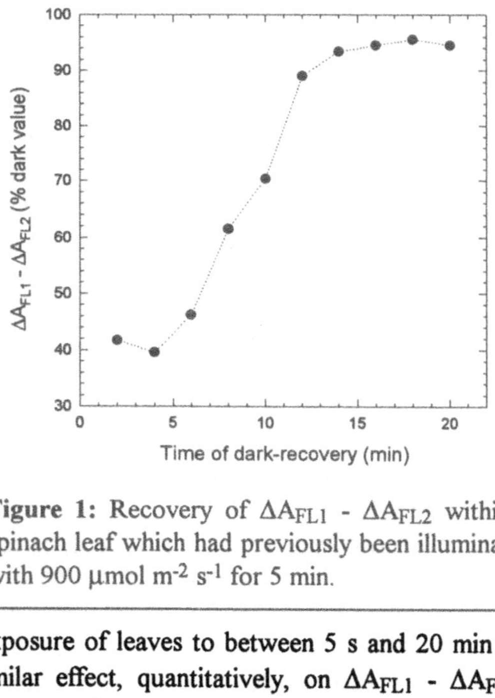

Figure 1: Recovery of ΔA_{FL1} - ΔA_{FL2} within a spinach leaf which had previously been illuminated with 900 μmol m^{-2} s^{-1} for 5 min.

Exposure of leaves to between 5 s and 20 min illumination at 900 μmol m^{-2} s^{-1} had a similar effect, quantitatively, on ΔA_{FL1} - ΔA_{FL2}, with minimal effect on ΔA_{FL2}; a decrease of up to 3% was observed after prolonged light treatment, which was reversed within 10 min.

The fluorescence yield from PSII centers is at a maximum when Q_A is reduced, for example, by a saturating light pulse. Following a saturating, single-turnover flash, the fluorescence yield from inactive centers remains very high for several hundred ms (since

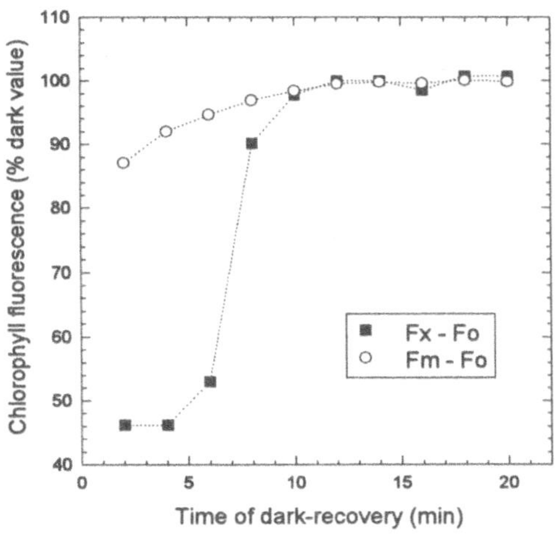

Figure 2: Recovery of variable fluorescence yield from inactive PSII centers (Fx - Fo) and all PSII centers (Fm - Fo) within a spinach leaf which had previously been illuminated with 900 μmol m^{-2} s^{-1} for 5 min.

Q_A^- is oxidized very slowly) while the fluorescence yield from active PSII centers decreases to a minimum value (Fo) within a few ms.

The data presented in Figure 2 show the difference between the maximum and minimum fluorescence yield from inactive PSII centers (Fx - Fo) and all PSII centers (Fm - Fo) within a spinach leaf following 5 min illumination with 900 μmol m^{-2} s^{-1} (i.e. the same treatment as was used to produce the ΔA_{518} data in Figure 1). Fo was determined at 0.6 μmol m^{-2} s^{-1}, Fx was determined after a saturating, single-turnover flash and Fm during a 250 ms pulse of white light (9000 μmol m^{-2} s^{-1}). The 5 min light treatment would be expected to decrease the fluorescence yield from PSII through the well documented 'non-photochemical' quenching associated with the formation of a ΔpH across the thylakoid membrane. This type of quenching is probably responsible for much of the decrease in Fm - Fo observed during the first few min after the light-treatment. In this instance, it would be quite reasonable to also attribute the decrease of Fx - Fo to ΔpH-dependent quenching. However, exposure to as little as 5 s illumination with 900 μmol m^{-2} s^{-1} also induced a decrease in Fx - Fo of approximately 50%, which was sustained for several minutes, whereas Fm - Fo recovered to 100% of the dark-adapted level within the first 2 min (data not shown). This was quite a surprising result since the greater than 50% decrease in Fx - Fo would be expected to decrease Fm - Fo by approximately 8%.

In summary, these data indicate that between 60% and 75% of inactive PSII centers within spinach leaves are modified by light such that a saturating, single-turnover flash does not induce stable charge separation. In addition, the variable chlorophyll fluorescence yield from inactive PSII centers is reduced by over 50% following light-treatment.

The light-dependent decrease in charge-separation within inactive PSII centers could be caused by stabilization of Q_A^-. However, this would be expected to decrease Fx - Fo through an increase the measured value of Fo. No such increase in Fo was observed after any of the light-treatments used.

The decrease in Fx - Fo, in the absence of any quenching of Fm - Fo would be expected if energy-transfer between inactive and active PSII centers were to occur in the light. Since inactive and active PSII centers are located within different regions of the membrane, this would obviously require the relocation of some centers.

The decreased capacity for charge-separation within inactive PSII centers could also be explained by the light-dependent formation of a dissipative electron cycle around these centers. Such a cycle would have to be fast enough to dissipate the charge within the response time of the system used for the measurements of ΔA_{518} (< 30 μs). A PSII electron cycle could also explain the quenching of Fx - Fo (measured after a saturating, single-turnover flash) in the absence of any quenching of Fm - Fo, if the cycled was saturated by the 250 ms pulse of 9000 μmol m^{-2} s^{-1} white light used to determine Fm.

Whatever the reason for the decreased contribution of inactive PSII centers to ΔA_{FL1} and the decrease in Fx - Fo, these data do not indicate that light-treated inactive PSII centers are capable of transferring electrons from water to plastoquinone and therefore represent strong evidence against light-induced activation of inactive PSII centers.

References:

Chylla, R.A. and Whitmarsh, J. (1989) Inactive Photosystem II complexes in leaves. Turnover rate and quantitation. Plant,Physiol. 90, 765-772

Chylla, R.A. and Whitmarsh, J. (1990) Light saturation response of inactive Photosystem II reaction centers in spinach. Photosynth. Res. 25, 39-48

Henrysson, T. and Sundby, C. (1990) Characterization of photosystem II in stroma thylakoid membranes. Photosynth. Res. 25, 107-117

Neale, P.J. and Melis, A. (1990) Activation of a reserve pool of photosystem II in *Chlamydomonas reinhardtii* counteracts photoinhibition. Plant Physiol. 92, 1196-1204.

Nedbal, L. Gibas, C. and Whitmarsh, J. (1991) Light saturation curves show competence of the water splitting complex in inactive photosystem II reaction centers. Photosynth. Res. 30, 85-94.

STOICHIOMETRY OF CYTOCHROME *b*-559 WITHIN THE ISOLATED PHOTOSYSTEM TWO REACTION CENTRE COMPLEX OF *CHLAMYDOMONAS REINHARDTII* AS DETERMINED BY ISOTOPIC LABELLING OF THE POLYPEPTIDE SUBUNITS

Sedigheh Alizadeh, James Barber and Peter Nixon
Photosynthesis Research Group
Department of Biochemistry
Imperial College of Science, Technology and Medicine,
London SW7 2AY, U.K.

1. Introduction

Cytochrome *b*-559 (Cyt *b*-559) is an integral component of all photosystem two (PSII) complexes and is composed of two subunits, α and β, which are encoded by the *psbE* and *psbF* genes, respectively. It is likely that the cytochrome is composed of an αβ heterodimer with a histidine residue within each subunit acting as a ligand to the haem. Its role in electron transport is still unclear although it is widely thought that it may function to protect PSII against photoinhibition [reviewed in 1].

There has been much controversy over the stoichiometry of this cytochrome within PSII with values of one and two being suggested [2]. For the PSII reaction centre complex (PSII RC) which is composed of D1, D2, PsbI as well as Cyt *b*-559, it is widely assumed, based mainly on the use of extinction coefficients at 559 nm for the reduced-oxidised cytochrome, that there is one Cyt *b*-559 per complex [3, 4]. However, Shuvalov and co-workers have argued that there are in fact 2 cytochromes per PSII RC [5].

Recently we have described the preparation of a PSII RC from the green alga *Chlamydomonas reinhardtii* [6]. We have shown that this preparation is very similar in terms of pigment content, cytochrome content [6] and photochemical activity [7] to the PSII RC isolated from pea. In order to explore further the stoichiometry of the polypeptide components within this complex, we report in this paper the isolation of a ^{14}C-labelled complex. Analysis of this preparation suggests that the polypeptide constituents are present in equimolar amounts and that there is, therefore, one Cyt *b*-559 per PSII RC.

2. Procedures

2.1 *Chlamydomonas* strain and growth conditions

Mutant F54-14 of *Chlamydomonas reinhardtii* which lacks the photosystem one and chloroplast ATP-ase complexes was used to isolate the PSII RC. In order to obtain radiolabelled material, the mutant was grown at 25°C in 2 l of Tris-acetate-phosphate medium containing [^{14}C]-sodium acetate (supplied at 59mCi.mmol^{-1}) at a final activity of 0.5 μCi.ml^{-1}. Cells were harvested after 4 days of growth when the $A_{750} \sim 0.8$.

2.2 Isolation of the ^{14}C-labelled PSII RC

^{14}C-labelled thylakoid membranes containing 5.8 mg of chlorophyll (Chl) were diluted two-fold with unlabelled thylakoids and the PSII RC isolated by anion-exchange

P. Mathis (ed.), Photosynthesis: from Light to Biosphere, Vol. I, 895–898.

chromatography of Triton X-100-solubilised thylakoid membranes as described in [6]. The pea PSII RC was prepared as described in [8].

2.3 SDS-PAGE and phosphorimage analysis
Polypeptides were resolved through 10-17% or 16-22% gradient polyacrylamide gels containing 6 M urea as described in [6]. After electrophoresis, protein was fixed in 40% methanol + 13.5% formaldehyde and stained with Coomassie blue R250. Gels were incubated in 5% glycerol and dried under vacuum before exposing to a phosphor screen. The phosphor screen was then scanned by a PhosphorImager and signals quantified using the ImageQuant software (Molecular Dynamics Ltd). In this method the integration volume of the signal is calculated from the sum of the pixel values within an object, such as a rectangle, containing the band, after subtraction of the background level. The number of carbon atoms in each polypeptide was estimated from the decoded DNA sequences of *psbA* [9], *psbD* [10], *psbE* and *psbF* [11] and *psbI* [12] assuming no N-terminal modifications.

3. Results and discussion

3.1 Isolation and characterisation of a ^{14}C-labelled PSII RC
PSII RCs radiolabelled with ^{14}C were isolated from the F54-14 mutant *of Chlamydomonas reinhardtii* which lacks the photosystem one complex and chloroplast ATP-ase. The absorption spectrum of the purified complex was indistinguishable to that published previously [6] confirming the authenticity of the preparation (data not shown). Dithionite-reduced minus ferricyanide-oxidised absorption spectra indicated that there was approximately 1.1 Cyt *b*-559 per 6 Chl *a* using an extinction coefficient of 23.4 $mM^{-1}cm^{-1}$ at 559 nm for Cyt *b*-559 [3] and 103 $mM^{-1}cm^{-1}$ at 676 nm for Chl *a* [13]. This stoichiometry is again similar to that published previously [6].

Figure 1 shows the Coomassie blue-stained gel (left-hand panel) and corresponding phosphorimage (right-hand panel) of the radiolabelled PSII RC following SDS-PAGE in a 10-17% polyacrylamide gel. Minor protein contamination can be observed in the *C. reinhardtii* PSII RC compared to the PSII RC from pea, probably because the small amount of radiolabelled thylakoid membranes used in the preparation prevented optimal washing of the anion-exchange column. These contaminating bands were excluded from subsequent analysis to determine the polypeptide stoichiometry. As reported previously [6] upon SDS-PAGE the PSII RC from *C. reinhardtii* gives rise to bands corresponding to a heterodimer of D1 and D2 (HD) at about 60 kDa, D1, D2, the α subunit of Cyt *b*-559 and two further low-molecular-weight subunits which are assumed to be PsbI and the β subunit. As illustrated in Figure 1, there is no evidence in the phosphorimage for the presence of an additional low-molecular-weight polypeptide component of the PSII RC that may have so far escaped detection because it stains poorly with Coomassie blue.

3.2 Stoichiometry of the protein subunits within the PSII RC
The radioactive signals were analysed using a Phosphorimager system. It was assumed that there were no variations in signal due to distortions in the thickness of the dried gel and that there was minimal aggregation of the low-molecular-weight subunits. For the purposes of quantification, the D1 and D2 bands migrating at about 30 kDa were grouped together as were the unresolved PsbI and β subunits. By using a 16-22% polyacrylamide gel the D1 and D2 proteins could be resolved (data not shown). Quantification of these two bands revealed a D1 : D2 stoichiometry of 1 : 0.99±0.04 (n=3); therefore the polypeptide stoichiometry was normalised to 2 copies of D1 and D2. Control experiments

were also performed to confirm that the signals were not saturating and that there were no D1 or D2 degradation products present in the PSII RC (data not shown).

Table I shows that when the contribution of the D1/D2 heterodimer was excluded, the subunits within the PSII RC were present in approximately equimolar amounts. If the heterodimer is considered to be part of the PSII RC then a ratio of approximately 0.82±0.06 copies of the α subunit per 2 copies of D1 and D2 per 1.28±0.20 copies of PsbI and the β subunit was obtained. However, we do not exclude the possibility that the D1/D2 heterodimer may in fact be a contaminant of the PSII RC and not a solubilisation artefact as often assumed [4]. The lower than expected value for PsbI and β could be the result of loss of protein during fixation of the gel. Other fixation conditions such as the use of 25% methanol and 10% acetic acid led to an underestimation of the stoichiometry of the low-molecular-weight subunits probably due to this reason. Alternatively, sub-stoichiometric amounts of PsbI may be present within the PSII RC because of extensive washing of the PSII RC on the column with 0.2% Triton X-100. Future work will aim to confirm these stoichiometries with other radiolabelled preparations.

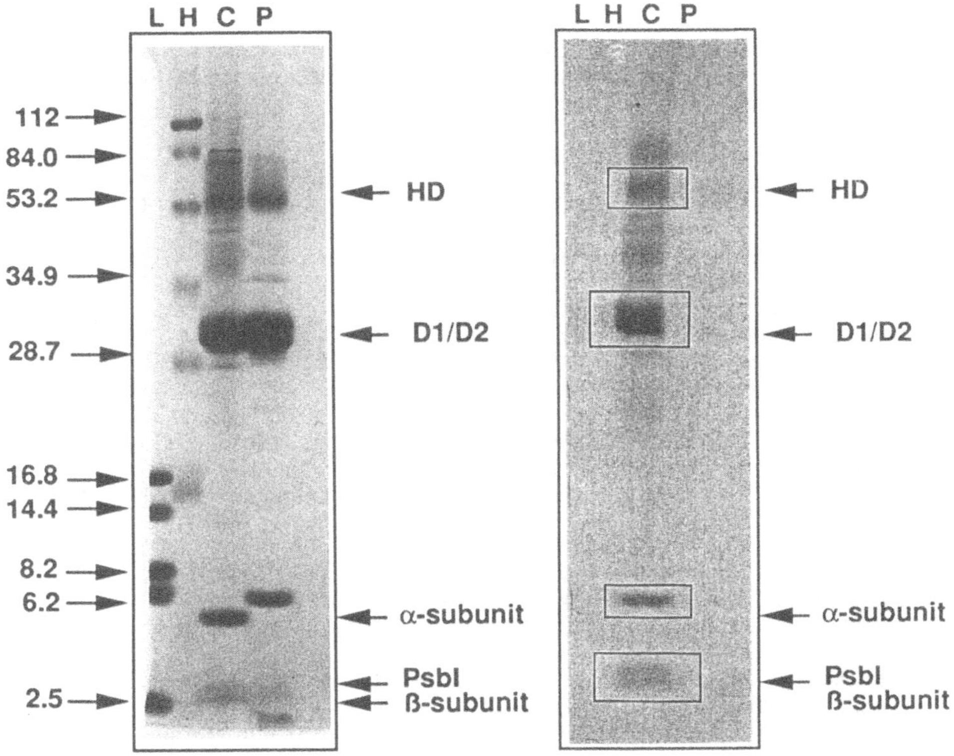

Figure 1: SDS-PAGE analysis of the pea PSII RC (P) and the ^{14}C-labelled PSII RC of *C. reinhardtii* (C). The gel was stained with Coomassie blue (left-hand panel) from which a phosphorimage was obtained (right-hand panel). Lanes L and H contain low and high molecular weight markers, respectively. HD indicates position of the D1/D2 heterodimer. Bands quantified by phosphorimaging are shown boxed in the right-hand panel.

Table I: Subunit stoichiometry within the *C. reinhardtii* PSII RC

Polypeptide band	Number of carbon atoms[a]	Ratio of subunits[b]
D1/D2 heterodimer (HD)	3599	0.26±0.04
D1/D2	3599	2
α subunit	430	1.03±0.06
PsbI + β subunit	444	1.61±0.24

[a]Predicted from decoded gene sequences
[b]Calculated by dividing the integration volume of the radioactive signal by the number of carbon atoms present within the polypeptide band, normalised to 1 copy each of D1 and D2. Average and standard error of 14 measurements, at three different exposures, of the gel shown in Figure 1.

4. Conclusion

Our data suggests that the five subunits within the PSII RC of *C. reinhardtii* (D1, D2, PsbI and the α and β subunits of Cyt *b*-559) are present in equimolar amounts. These results therefore support a stoichiometry of 1 Cyt *b*-559 per PSII RC assuming a homogeneous population of PSII RCs.

References

1 Cramer, W.A., Tae, G.S., Furbacher, P.N. and Böttger, M. (1993) Physiol. Plant. 88, 705-711
2 Buser, C.A., Diner, B.A. and Brudvig, G.W. (1992) Biochemistry 31, 11441-11448
3 Miyazaki, A., Shuna, T., Toyashima, Y., Gounaris, K. and Barber, J. (1989) Biochim. Biophys. Acta 975, 142-147
4 Nanba, O. and Satoh, K. (1987) Proc. Natl. Acad. USA 84, 109-112
5 Shuvalov, V.A., Schreiber, U. and Heber, U. (1994) FEBS Letts. 337, 226-230
6 Alizadeh, S., Nixon, P.J., Telfer, A. and Barber, J. (1995) Photosynth. Res. 43, 164-171
7 Giorgi, L.B., Durrant, J.R., Alizadeh, S., Nixon, P.J., Joseph, D.M., Rech, T., Barber, J., Porter, G. and Klug, D.R. (1994) Biochim. Biophys. Acta 1186, 247-251
8 Chapman, D.J., Gounaris, K. and Barber, J. (1988) Biochim. Biophys. Acta 933, 423-431
9 Erickson, J.M., Rahire, M. and Rochaix, J.-D. (1984) EMBO J. 3, 2753-2762
10 Erickson, J.M., Rahire, M., Malnoe, P., Girard-Bascou, J., Pierre, Y., Bennoun, P. and Rochaix, J.-D. (1986) EMBO J. 5, 1745-1754
11 Alizadeh, S., Nechushtai, R., Barber, J. and Nixon, P. (1994) Biochim. Biophys. Acta 1188, 439-442.
12 Kunstner, P., Guardiola, A., Takahashi, Y. and Rochaix, J.-D. (1995) J. Biol. Chem. 270, 9651-9654
13 De Las Rivas, J., Telfer, A. and Barber, J. (1993) Biochim. Biophys. Acta 1142, 155-164.

^{13}C MAS NMR RELAXATION STUDY OF THE Q_A BINDING IN *RHODOBACTER SPHAEROIDES*-R26 REACTION CENTERS.

B.-J. van Rossum[1], W.B.S. van Liemt[1], G.J. Boender[1], P.Gast[2], A.J. Hoff[2], J. Lugtenburg[1] and H.J.M. de Groot[1],
[1]Leiden Institute of Chemistry and [2]Department of Biophysics, University of Leiden, the Netherlands.

1. Abstract
The carbonyls of the primary quinone Q_A in the reaction centre (RC) of *Rhodobacter (Rb.) sphaeroides*-R26 reconstituted with 1 or 4-^{13}C ubiquinone-10 (UQ-10, Figure 1) have been characterized by a CP-MAS NMR relaxation study. The linewidth of the 1-^{13}C signal at room temperature (RT) is 150Hz and at T=230K it is 250Hz. The 4-^{13}C signal can only be observed below ~255K, at 190K the linewidth is ~220Hz. The observed linewidth strongly depends on the degree of hydration of the sample. The $T_1\rho$ of the 4-^{13}C signal (3.8±0.9 ms at 250K and 8±2 ms at 230K) seems to decrease as the temperature approaches 255K. This suggests dynamic Q_A-protein interactions which could explain the disappearance of the the 4-^{13}C signal for T>255K. The $T_1\rho$ of the 1-^{13}C signal is nearly temperature independent (8±1 ms at RT and 7.2±0.8 ms at 250K). In addition, T_1 and T_2 were checked, but they cannot account for the loss of the 4-^{13}C signal for T>255K. The ^{13}C T_1 is longer than 1 s for both carbonyls which provides evidence for a rigid protein environment. Both the 1-^{13}C T_2 (4.2±0.5 ms at RT and 4.3±0.2 ms at 230K) and 4-^{13}C T_2 (~7±3 ms at 190K) are long and the corresponding homogeneous linewidths are small (≤75Hz), showing that the signals are inhomogeneously broadened. For the 1-^{13}C signal additional heterogeneous broadening occurs at T=230K. This suggests structural heterogeneity of the quinone pocket. The RT T_2 of the protons in the neighbourhood of the 1-^{13}C label is very short (≤10μs), indicating a rigid environment of the label.

2. Introduction
The photosynthetic RC of *Rb. sphaeroides*-R26 contains two chemically identical quinones with a different function. Q_A is the primary electron acceptor and is asymmetrically bound to the protein at the two keto functionalities. Magic angle spinning (MAS) NMR studies using specifically labelled quinones revealed a remarkable difference in the NMR properties between the two carbonyl carbons, as the 4-^{13}C CP/MAS signal could only be detected at temperatures below T~255K. The purpose of the present investigation is to characterize the relaxation properties of the two carbonyl carbons using RCs reconstituted with [1-^{13}C] and [4-^{13}C]

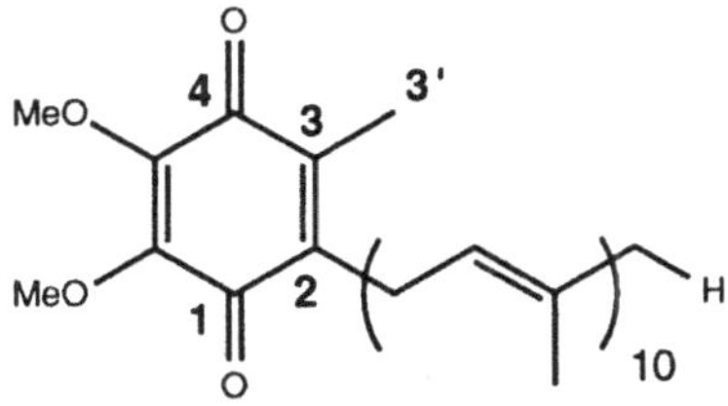

"Figure 1. Structure and partial IUPAC numbering of UQ-10"

P. Mathis (ed.), Photosynthesis: from Light to Biosphere, Vol. I, 899–902.

ubiquinon-10, in order to find an explanation for the peculiar temperature dependence of the CP/MAS signal intensity at the 4 C=O.

3. Experimental

Site-specific selective labeling with ^{13}C of UQ-10 is described elsewhere[1]. The Q_A reconstitution was performed in subdued light[2]. The (100 MHz ^{13}C CP/MAS) NMR spectra were obtained using a MSL-400 NMR spectrometer with a 4mm MAS probe (Bruker, Karlsruhe, Germany). In order to perform low-temperature measurements, the sample was frozen with liquid nitrogen cooled bearing gas[3]. A home-built spinning speed controller[4] was used to keep the spinning rate constant. Typical ^1H and ^{13}C 90° pulses were 4μs. The T_{CH} and T_1^ρ were obtained from CP-efficiency measurements, fitting the intensity $I(\tau)$ of the experimental data to the double exponential function

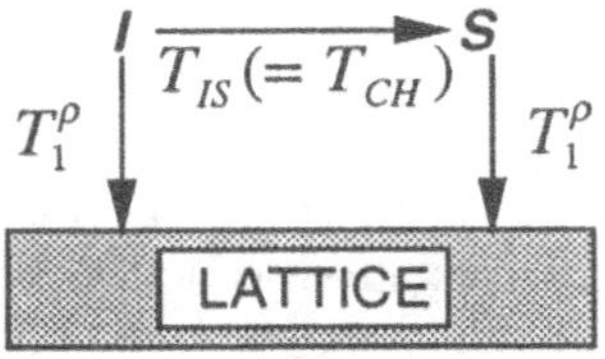

"Figure 2. Principles of cross-polarization"

$$I(\tau)=\frac{I_0}{T_{CH}}\frac{\left(\exp\frac{-\tau}{T_{1\rho}^H}\right)-\left(\exp\frac{-\tau}{T_{CH}}\right)}{\frac{1}{T_{CH}}-\frac{1}{T_{1\rho}^H}},$$

which takes into account the coherence transfer from the abundant (I) spins to the rare (S) spins, and the relaxation to the lattice under spin-lock conditions (Fig.2). The T_1 of the ^{13}C labels was probed with an inversion recovery pulse sequence at $\omega_r/2\pi$=7 kHz. The T_2 of the ^{13}C labels was measured with a rotor-synchronized Hahn-echo after CP with $\omega_r/2\pi$=6993 Hz yielding τ_r=143 μs. The T_2 of the protons in the vicinity of the 1-^{13}C label was roughly estimated with a Hahn-echo before CP (Figure 3).

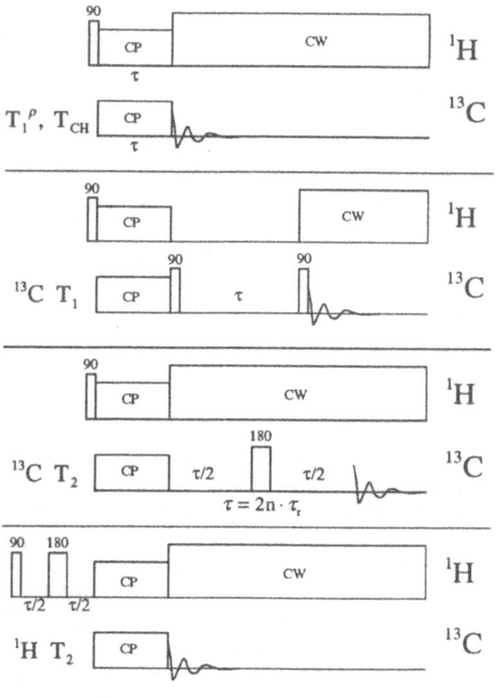

"Figure 3. Pulse sequences"

4. Results

The CP/MAS NMR response of the label in RCs reconstituted with [1-^{13}C] or [4-^{13}C] UQ-10 is found at 183.8 ppm or 183.1 ppm (190K) respectively, relative to TMS. The linewidth of the [1-^{13}C] UQ-10 signal is ~150Hz at RT and ~250Hz at T=230K. The linewidth of the [4-^{13}C] UQ-10 is ~220Hz at T=190K, and the line has only been observed below 255K (Fig.4). The linewidth of both labels was very sensitive to the water content of the sample.

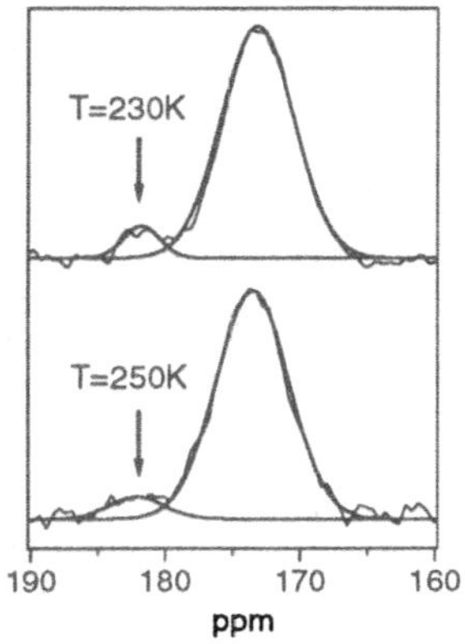

"Figure 4. [4-^{13}C] UQ-10 R26 signals (indicated by the arrows) at two temperatures close to 255K. As guide to the eye, the computer fits are added (solid lines)"

The different relaxation parameters are summarized in Table 1. T_{CH} and $T_{1\rho}$ of the [1-^{13}C] label (Fig.5) are essentially independent of the temperature. Also, the T_2 of the [1-^{13}C] label (Fig.6) is independent of temperature and the corresponding homogeneous linewidth is 75Hz. The T_{CH} of the [4-^{13}C] label at 230K is 1.5±0.5 ms (Fig.5) while the T_{CH} at 250K is difficult to fit, and was estimated and fixed at 1.5 ms. The $T_{1\rho}$ at 230K (8 ms) is comparable to the $T_{1\rho}$ of the [1-^{13}C] label. In contrast, the $T_{1\rho}$ at 250K is ~3.8 ms. The T_2 of the [4-^{13}C] label is ~7 ms at T=190K, which translates into a homogeneous linewidth of ~50 Hz.

5. Discussion and Conclusions

The relaxation phenomena can be characterized by different time constants. T_1 characterizes the depopulation of energy levels of the nuclear spin due to energy exchange with the lattice (spin-lattice relaxation). T_{CH} characterizes the rate of coherence-transfer during CP. The longitudinal proton relaxation in the rotating frame, $T_{1\rho}$, characterizes the depopulation of the energy levels created by the spin-lock field during CP. Finally, T_2 characterizes the decrease of the transverse coherence due to relaxation by fluctuations of the magnetic field at the nucleus (spin-spin relaxation). The $T_{1\rho}$ of protons close to the [1-^{13}C] UQ-10 label is long and almost temperature independent.

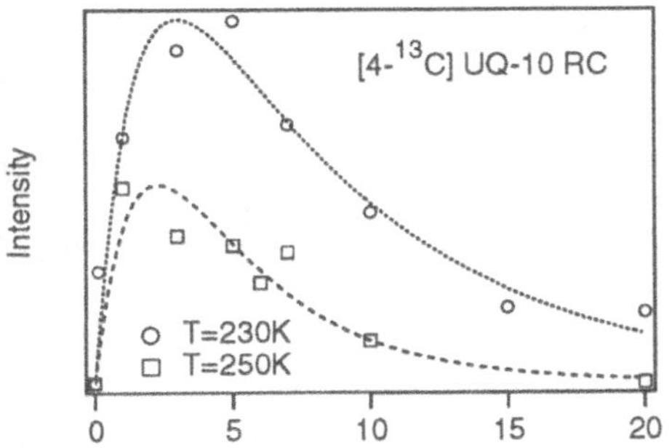

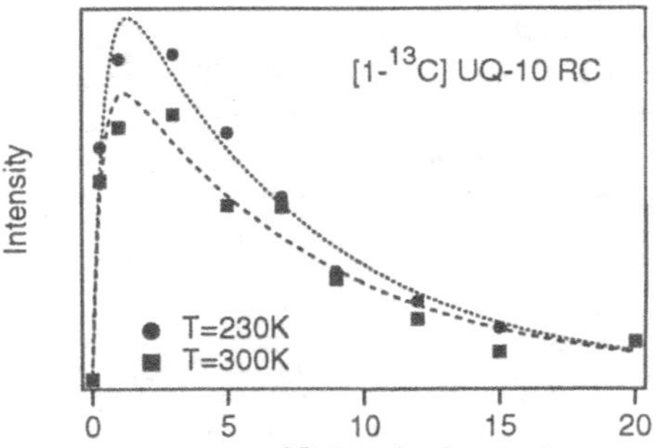

"Figure 5. Cross polarization efficiency measurements on the [1-^{13}C] and [4-^{13}C] UQ-10 label in RCs. Dashed lines are computer fits of the double exponential function $I(\tau)$ to the data. The theoretic point (0, 0) is added."

	$T_{1\rho}$	T_{CH}	$^{13}C\ T_1$	$^{13}C\ T_2$	$^1H\ T_2$
[1-^{13}C]UQ-10	8 ms (RT) 7.2 ms (250K)	0.4 ms (RT) 0.44 ms (250K)	> 1s (230K)	4.2 ms (RT) 4.3 ms (230K)	< 10μs (RT)
[4-^{13}C]UQ-10	3.8 ms (250K) 8 ms (230K)	1.5 ms (250K) 1.5 ms (230K)	> 1s (190K)	~7 ms (190K)	–

"Table 1. Relaxation parameters of the [1-^{13}C] and [4-^{13}C] UQ-10 labelled RCs."

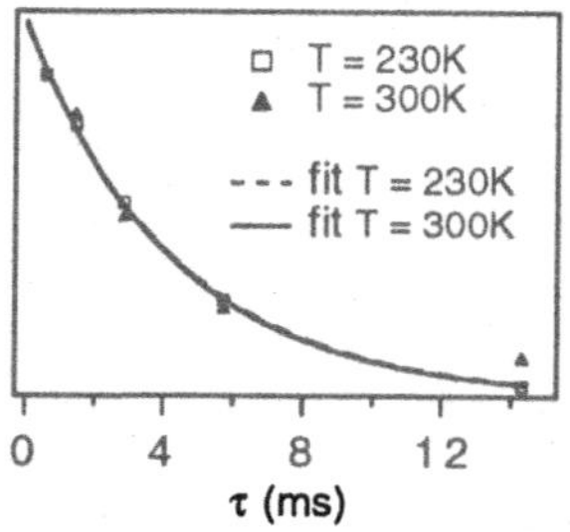

"Figure 6. ^{13}C T_2 relaxation curves of the [1-^{13}C] UQ-10 label in RCs at RT and at T=230K"

This corroborates the picture of a rigid protein environment. In contrast, the observed shortening of the $T_{1\rho}$ of protons close to the [4-^{13}C] UQ-10 label as the temperature approaches 255K suggests slow (~50 kHz) dynamic Q_A-protein interactions interfering destructively with the spin-locked protons; in this case, during CP magnetization is not transferred from the *I* spins to the *S* spins, but leaks away to the lattice (Fig.2). This process may also depend on the degree of hydration of the sample, and measurements on newly prepared [4-^{13}C] UQ-10 labelled RCs are in progress to confirm this interesting temperature dependence of $T_{1\rho}$.

The T_2 of the protons in the environment of the [1-^{13}C] UQ-10 label is short. Hence the homonuclear dipolar interaction is not (partially) averaged out by motion. This provides strong evidence that the protein surrounding of the label is rigid. The T_1 of both ^{13}C labels under strong heteronuclear decoupling is long (> 1s). Hence energy transfer to the lattice is difficult, which also corroborates the finding that the protein environment is rigid.

The linewidth $\Delta\upsilon$ is the sum of the homogeneous and the inhomogeneous linewidth. If no inhomogeneous linebroadening is present, the linewidth is dominated by ^{13}C T_2-processes ($\Delta\upsilon=1/\pi T_2$). The presence of additional inhomogeneous linebroadening provides evidence for structural heterogeneity of some kind.The T_2 of both ^{13}C labels is long and independent of temperature for the [1-^{13}C] UQ-10 label. The corresponding homogeneous linewidths are relatively small. Hence the variations in the magnetic field at the nucleus are small. This suggests that on the NMR time scale the electronic surrounding of the label is rather static. The substantial inhomogeneous linebroadening provides evidence for more than one electronically slightly different conformation of the quinone and/or its binding site. For the [1-^{13}C] labeled RC this disorder becomes more pronounced at low temperatures. This is in contrast with the extremely rigid, well defined nearly crystalline ordered structure probed with ^{13}C labels in the vicinity of the special pair[5].

1 Van Liemt, W.B.S., Steggerda, W.F., Esmeijer, R. & Lugtenburg, J. (1994) Rec. Trav. Chim. Pays-Bas 113, 153-161
2 Okamura, M.Y., Isaacson, R.A. & Feher, G. (1975) Proc. Nat. Acad. Sci. USA 72, 3491-3495.
3 Fischer, M.R., de Groot, H.J.M., Raap, J., Winkel, C., Hoff, A.J. & Lugtenburg, J. (1992) biochemistry 31, 11038-11049.
4 De Groot, H.J.M., Copié, V., Smith, S.O., Allen, P.J., Winkel, C., Lugtenburg, J., Herzfeld, J. & Griffin,R.G. (1988) J. Magn. Res. 77, 251-257.
5 Shochat, S., Gast, P., Hoff, A.J., Boender, G.J., van Leeuwen, S., van Liemt, W.B.S., Vijgenboom, E., Raap, J., Jugtenburg, J. & de Groot, H.J.M. (1995) Spectrochimica Acta 51A, 135-144.

THE COUPLING OF LIGHT-INDUCED ELECTRON TRANSFER AND PROTON UPTAKE: ELECTROSTATIC CALCULATIONS ON THE PHOTOSYNTHETIC REACTION CENTRE FROM *RHODOPSEUDOMONAS VIRIDIS*.

C. Roy D. Lancaster[‡], M. R. Gunner[§], and Hartmut Michel[‡], [‡]Max-Planck-Institut für Biophysik, Abteilung Molekulare Membranbiologie, Heinrich-Hoffmann-Str. 7, D-60528 Frankfurt am Main, Germany, and [§]City College of New York, Department of Physics, 138th Street and Convent Avenue, New York NY 10031, USA

1. Introduction

In photosynthetic reaction centres (RCs), light-induced electron transfer is coupled to proton uptake from the cytoplasm at the binding site of the secondary quinone, Q_B, where doubly reduced Q_B takes up two protons. The sequence of light-induced electron and proton transfer reactions that occur within the RC in the process of reducing Q_B to Q_BH2 involves the neutral redox state DQ_AQ_B, the radical states $DQ_A^- Q_B$ and $DQ_AQ_B^-$, the diradical state $DQ_A^- Q_B^-$, and, if transfer of the second electron to Q_B precedes transfer of the first proton, the doubly reduced state $DQ_AQ_B^{2-}$ [1]. The effect of changes in the quinone redox states on the titration of protonatable amino acid residues has been studied previously by electrostatic calculations on the RCs of *Rb. sphaeroides* [2,3,4] and *Rp. viridis* [5]. However, these calculations were based on structures with Q_B sites which were weakly defined [6,7]. A more reliable Q_B site structure has been obtained from X-ray analysis of crystals which have full occupancy of ubiquinone-2 in the Q_B site of the *Rp. viridis* RC (Lancaster and Michel, manuscript in preparation). Based on this new structure, we have used a continuum dielectric model and finite difference technique to identify clusters of electrostatically interacting ionisable residues and calculate the average protonation of titratable residues in the five redox states presented above.

2. Methods

The non-hydrogen atom co-ordinates of the *Rp. viridis* RC structure introduced above were used in the calculations. Protons were added with the programs HPP and PROTEUS (Joguine and Gunner, unpublished). Sulphate, detergent and water molecules were ignored after the placement of the protons. The program DelPhi was used to solve the Poisson equation to obtain the electrostatic potential in and around the RC essentially as described by Gunner and Honig [3], one difference being that the charge distribution of the cofactors was adapted from [8]. Arg, Asp, Glu, non-ligated His, Lys, Tyr, the heme propionates, the chain termini, and all cofactors but the carotenoid were treated as ionisable residues. All other residues were assumed to have no net charge. One residue of particular interest as a potential proton donor to Q_B is Ser L223, possibly *via* a transiently protonated state Ser L223-OH_2^+. Ser L223 was considered to be protonatable with a pK_{sol} of -1.7 [5] in a separate series of calculations with 293 ionisable sites considered. The average protonation of the titrating sites was determined by Monte Carlo sampling of the protonation states of the protein complex [3]. A detailed account of the computational methods and parameters will be given elsewhere.

P. Mathis (ed.), Photosynthesis: from Light to Biosphere, Vol. I, 903–906.

3. Results

Groups of residues can be identified where the ionisation state of these residues are interdependent because of the large electrostatic interaction energy between them. These will be referred to as clusters. Given a threshold criterion that charge-charge interactions change the pK of a residue by 2 pH units, a cluster of 47 titrating residues is identified which are directly or indirectly (*via* other titrating residues) coupled to Q_A and Q_B (cf. Figure 1). These can be subdivided into a group of 23 residues more strongly coupled to Q_B (the Q_B cluster) and another 24 which are more strongly coupled to Q_A (the Q_A cluster). The two clusters are linked *via* Glu M232, Asp H236, and Asp M238. The Q_B cluster extends for over 33 Å from His H72 to the N-terminus of the M subunit. Both residues are solvent-accessible, thus providing at least two possible entry points for cytoplasmic protons.

There are a number of significant differences between the Q_A and Q_B clusters. Within the Q_B cluster, there is a surplus of three acidic residues. This imbalance is maintained in the centre of the cluster where Glu L212, the only active residue to couple strongly to Q_B, interacts strongly with twice as many acidic than basic residues (Glu H177, Glu M234, Glu L210, Asp H174, Asp H125, and Asp M230 *versus* Lys H133, Arg H181, and Arg M231). The Q_A cluster differs from the Q_B cluster in four important characteristics. In addition to the relative lack of acidic residues, the Q_A cluster has fewer and smaller charge-charge interactions within the cluster, its residues are more solvated, and there is less positive electrostatic potential arising from the polypeptide backbone towards stabilising negative charges than in the Q_B cluster.

In the ground state (DQ_AQ_B) at pH 7, all residues in the Q_A cluster, despite the surplus of basic residues, are ionised, except for Glu L104 which, when neutral, can donate a hydrogen bond to the ring V carbonyl oxygen of ϕ_A. Within the Q_B cluster, residues are predominantly ionised despite the surplus of acidic residues and high desolvation penalties. The exceptions are the amino terminus of the M subunit, which is neutral, and the three residues Glu L212, Glu H177, and Glu M234. None of the three Glu residues exhibits a typical Henderson-Hasselbalch titration curve. However, the sum of the fractional protonation of these three residues (Glu cluster) is approximately described by such a curve, with a total protonation of 2 at low pH and of 1 at high pH. The pK_a'' value of the Glu cluster, corresponding to a summed fractional protonation of 1.5, is 6.4 in the neutral state DQ_AQ_B. These three Glu residues act as a tightly coupled cluster of interdependently titrating residues within the larger Q_B cluster. These will be seen to function as a buffer for the Q_B site.

A maximum of five residues within the Q_B cluster account for virtually all of the change in protein charge at neutral pH upon changing the quinone redox states. Considering the first four RC redox states (cf Figure 2), there is virtually no difference to the system, whether Ser L223 is considered protonatable or not. Almost all of the change in charge induced on the whole protein at neutral pH can be accounted for by the sum of the fractional protonation of the three acidic residues Glu L212, Glu H177, and Glu M234 (6.6 Å, 8.1 Å, and 13.8 Å away from Q_B, respectively). In the state $DQ_AQ_B^{2-}$, Asp M43 (8.3 Å away from Q_B) is calculated to be involved in proton uptake, when Ser L223 is considered non-ionisable. When considering protonation of Ser L223 in this redox state, Ser L223 is protonated at the expense of Asp M43 and, partially, Glu M234.

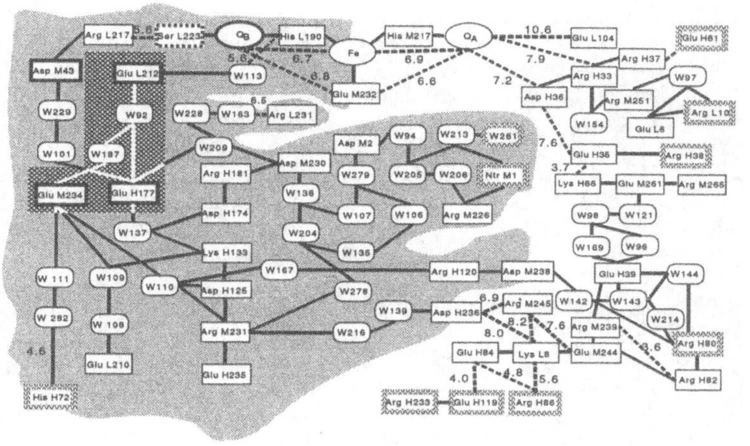

Figure 1. Schematic diagram of the Q_A cluster and the Q_B cluster. This diagram includes water molecules (W) identified in the X-ray structure of the ubiquinone-2-reconstituted RC (Lancaster and Michel, manuscript in preparation). Waters with numbers greater than 162 were not identified in the original 1PRC structure [9]. Solid lines (—) link residues and water molecules which are within hydrogen bonding distance (< 3.5 Å). Dashed lines (---) designate longer distances (explicitly given in Å). Bold frames highlight the five residues identified in Figure 2. Frames with wave patterns designate residues which are at least partially solvent exposed. The grey background defines the Q_B cluster, the white background the Q_A cluster, and the chequered background the Glu cluster.

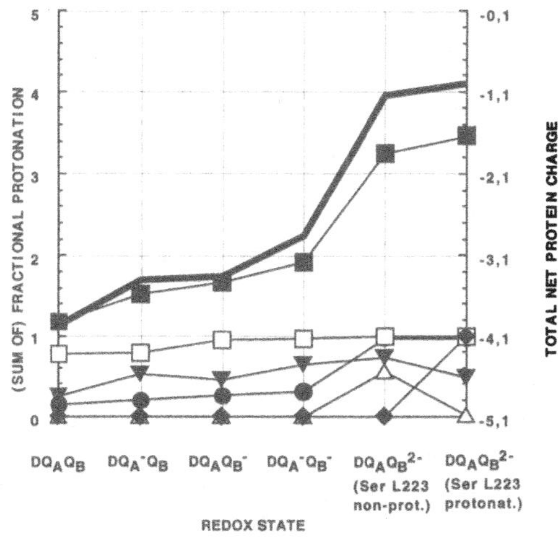

Figure 2. Fractional protonation at pH 7.0 as calculated from Monte Carlo sampling for the five residues Asp M43 (Δ), Glu H177 (●), Glu L212 (□), Glu M234 (▼), and Ser L223 (◆) in different quinone redox states. Also plotted is the sum of the protonation of all five residues (■), for the first four redox states this is equivalent to the protonation of the Glu cluster) and the total proton uptake calculated for the whole RC (—). The latter quantity is the only one plotted with respect to the right ordinate.

4. Discussion

At least some of the protonatable residues calculated to be important have been implicated to be functionally relevant in the proton transfer process. Based on the spectroscopic analysis of site-specific mutants of the reaction centre from *Rb. sphaeroides* (reviewed in [1]) it was suggested that the first proton to Q_B is transferred *via* Asp L213 and Ser L223, the second *via* Asp L213 and Glu L212. In *Rp. viridis*, residue L213 is an Asn, but its role can apparently be complemented [10] by Asp M43 (which is Asn M44 in *Rb. sphaeroides*). A recent mutation of Glu H173 to Gln [11] in *Rb. sphaeroides*, corresponding to Glu H177 in *Rp. viridis*, has been shown to give rise to a transfer rate for the second electron from Q_A to Q_B which is 100-fold reduced compared to the wild type. These experimental results corroborate our findings of important roles for Ser L223, Asp M43, Glu L212, and also for Glu H177. While the Glu cluster and possibly Asp M43 and Ser L223 account for the net change in charge, the other residues shown in Figure 1 (the Q_B cluster and the Q_A cluster) are important for establishing the electrostatic environment and thus setting the electrostatic potentials as is the arrangement of the polypeptide backbone.

The work presented here demonstrates how electrostatic calculations can help to identify amino acid residues possibly involved in proton uptake to the Q_B site. A strongly interacting cluster of the three Glu residues L212, H177, and M234 near Q_B accounts for the net effect of this charge compensation at neutral pH for all redox transitions up to and including the diradical state $DQ_A^-Q_B^-$. This Glu cluster is part of a larger interacting network of ionisable residues around Q_B (Q_B cluster) that extends to the solvent accessible surface. An analogous cluster of ionisable residues around Q_A (Q_A cluster) has different electrostatic properties and is apparently not involved in proton uptake at neutral pH. If the lifetime of Q_B^{2-} is long enough, then additionally Ser L223 and possibly Asp M43 may take up protonation. The calculated complex titration behaviour of the strongly interacting residues of the Q_B cluster may be a common feature in proton transferring membrane proteins.

5. Acknowledgements

We thank Anthony Nicholls and Barry Honig at Columbia University, New York, for timely and continuing modifications of DelPhi.

6. References

[1] Okamura, M.Y. and Feher, G. (1992) *Annu Rev Biochem 61,* 861-869.

[2] Beroza, P., Fredkin, D.R., Okamura, M.Y., and Feher, G. (1991) *Proc. Natl..Acad. Sci. USA 88,* 5804-5808.

[3] Gunner, M.R. and Honig, B. (1992) In: *The Photosynthetic Bacterial Reaction Center II* (Breton, J., Verméglio, A., eds) Plenum Press New York, 403-410.

[4] Beroza, P., Fredkin, D.R., Okamura, M.Y., and Feher, G. (1995) *Biophys. J. 68,* 2233-2250.

[5] Cometta-Morini, C., Scharnagl, C., and Fischer, S.F. (1993) *Int. J. Quant. Chem. Quant. Biol. Symp. 20,* 89-106.

[6] Lancaster, C.R.D., Ermler, U., and Michel, H. (1995) In: *Anoxygenic Photosynthetic Bacteria* (Blankenship, R.E., Madigan, M.T. and Bauer, C.E., eds.) Kluwer Academic Publishers, Dordrecht, The Netherlands, pp. 503-526.

[7] Lancaster, C.R.D., and Michel, H. (1995) In: *Reaction Centers of Photosynthetic Bacteria. Structure and Dynamics. Proceedings of the Feldafing-III-Workshop* . (Michel-Beyerle, M.E., ed.) Springer-Verlag, Berlin, in the press.

[8] Treutlein, H., Schulten, K., Brünger, A.T., Karplus, M., Deisenhofer, J., and Michel, H. (1992). *Proc.Natl Acad Sci USA 89,* 75-79.

[9] Deisenhofer, J., Epp, O., Sinning, I. and Michel, H. (1995) *J. Mol. Biol. 246,* 429-457.

[10] Rongey, S.H., Paddock, M.L., Feher, G., and Okamura, M.Y. (1993) *Proc. Natl. Acad. Sci. USA 90,* 1325-1329.

[11] Takahashi, E. and Wraight, C.A. (1995) *Biophys J 68,* A95.

WAVELENGTH AND TEMPERATURE DEPENDENCE OF THE $P^+Q_A^-$ RECOMBINATION REACTION IN THE M210W REACTION CENTRE ONLY MUTANT OF *RB. SPHEAROIDES.*

M.E. van Brederode[1], I.H.M van Stokkum[1], M.R. Jones[2] and R. van Grondelle[1].
[1]Faculty of Physics and Astronomy, Free University, Amsterdam
[2] Department of Molecular Biology and Biotechnology, University of Sheffield, United Kingdom.

Introduction.

In the photosynthetic bacterium *Rb. sphaeroides*, the reaction centre (RC) contains four molecules of bacteriochlorophyll, two molecules of bacteriopheophytin and two quinones and one non-heme iron. The pigments are aligned in two symmetric pigment chains. Photosynthesis in the RC starts with excitation of a bacteriochlorophyll dimer (P). From the excited dimer an electron is transferred via one of the bacteriopheophytin molecules (H_A) to one of the quinone molecules (Q_A). Blocking further electron transfer leads to recombination from Q_A^- to P^+ on a millisecond timescale. The rate of recombination from the $P^+Q_A^-$ state in the M210W mutant has been found to increase relative to the WT. As in the WT, the reaction rate increases with decreasing temperature in the M210W mutant [1]. In this work we present measurements on the wavelength and temperature dependence between 5 and 298 K of the multi-phasic $P^+Q_A^-$ recombination in the M20W RC only mutant of *Rb. sphaeroides*. At cryogenic temperatures we observe a strong wavelength dependence over the Q_y absorption band of P, with the 'red' reaction centres being faster than the 'blue' reaction centres. It is suggested that a variation of the electronic coupling between P and Q_A may explain this effect.

Materials and Methods.

Sample preparation and mutant construction.
Construction of the RC-only strain was carried out as described in [2]. The $P^+Q_A^-$ recombination measurements were performed in 20 mM 1,10 o-phenanthroline, 50 mM potasium phosphate buffer in 70% (v/v) glycerol.
$P^+Q_A^-$ recombination.
Transient absorbance measurements on a μs and ms timescale after a 8 ns 590 nm laserflash were performed on a set-up essentially as described in [3].

P. Mathis (ed.), Photosynthesis: from Light to Biosphere, Vol. I, 907–910.

Results.

Transient absorbance changes of the M210W mutant, after excitation with a 5 ns 590 nm light flash, were recorded at 59 wavelengths between 700 and 950 nm, on time scales from 0-500 μs, 0-20 ms and 0-200 ms. From global analysis [4] of all the kinetic traces onthe three time-scales, it followed that three independently decaying spectral components were needed to satisfactorily fit the data. The lifetimes of the spectral components were 100 μs, 4.7 ms and 13.9 ms, with relative contributions of 0.1, 0.3 and 0.6 respectively.The corresponding decay associated spectra (DAS) are shown in Figure 1.

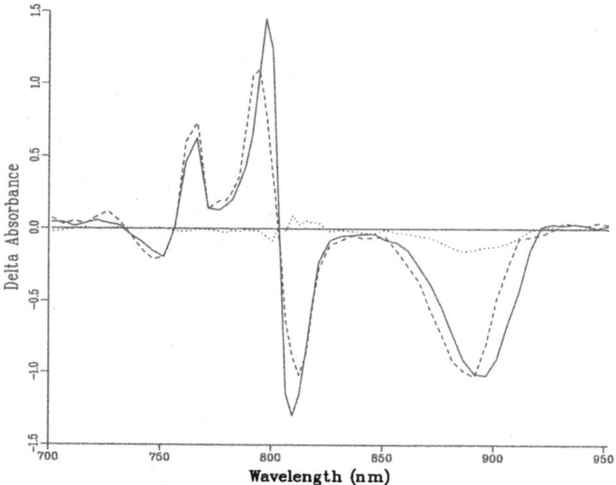

Figure 1. Decay associated spectra of the transient absorbance changes of the M210W mutant measured on a μs and ms timescale at 5 K. The accompaning lifetimes of the spectra are 100 μs, 4.7 ms and 13.9 ms for the dotted, solid and dashed spectrum, respectively. For comparison in the P band region the dashed spectrum (13.9 ms) was multiplied by 2.3.

The 100 μs DAS is mainly composed of a bleaching of P and shows only some minor features around 800 nm. In view of the characteristic lifetime and spectrum [5], this component is attributed to the transition $^3P \rightarrow P$. Most likely, these triplets are directly formed from P*, since the rate of primary charge separation in the M210W mutant at cryogenic temperatures is slowed down to several hundreds of picoseconds in open RC [1]. As a result, the formation of triplets by intersystem crossing directly from P* can efficiently compete with the charge separation process. The two spectra with the lifetimes of 4.7 ms and 13. 9 ms are associated to a multi-exponential decay of $P^+Q_A^-$. Two remarkable differences between these two decay associated spectra should be noted.

Firstly, the DAS show a large difference in the P band with the 4.7 ms DAS clearly red shifted relative to the 13.9 ms DAS. Secondly, the electrochromic bandshift around 800 nm has a significantly larger amplitude for the red 'fast' RCs than for the 'blue' slow RCs. At 77 K the $P^+Q_A^-$ recombination spectrum of the M210W mutant shows the same

characteristics in the P band region as at 5 K, with lifetimes of 6 and 20 ms (results not shown). Global analysis of the transient absorbance changes of the $P^+Q_A^-$ recombination in the M210W mutant at RT, using two spectral components yielded identical DAS in the P band, with lifetimes of 24 ms and 105 ms (results not shown). Also for $P^+Q_A^-$ recombination measured at 77 K in membrane bound WT and M210L RCs of *Rb. sphaeroides,* no spectral shift in the DAS in the P band was observed, although these reactions are clearly bi-exponential at that temperature (results not shown).
The temperature dependence of the $P^+Q_A^-$ recombination kinetics was probed at 885 nm and fitted to a bi-exponential decay (Figure 2). For comparison the temperature dependence of the $P^+Q_A^-$ recombination in the membrane bound WT RC of *Rb. sphaeroides* is also shown.

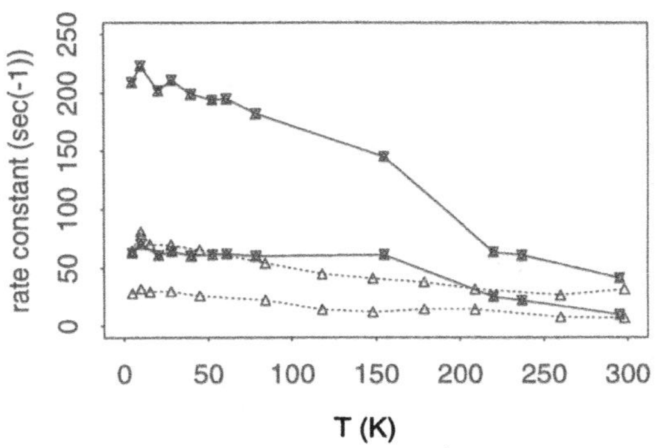

Figure 2. Temperature dependance of the $P^+Q_A^-$ recombination kinetics in the WT (dashed lines) and M210W mutant (solid lines).

The results agree well with the earlier measurements of Nagarajan et al. [10], who analysed the same process assuming a mono-exponential decay.
At all temperatures in the M210W RC the $P^+Q_A^-$ recombination rate has increased relative to the WT RC. Upon cooling from 293 K to temperatures lower than 40 K, the lifetime of the fast component in the $P^+Q_A^-$ recombination in the M210W mutant decreases from 24 ms to 4-5 ms, whereas the lifetime of the slow component decreases from 105 ms to 13-16 ms. The latter is about the same as the lifetime measured for the fast phase in $P^+Q_A^-$ recombination in WT RCs at low temperatures.

Discussion.

The rate of $P^+Q_A^-$ recombination in the M210W mutant at 5 K and 77 K shows a pronounced wavelength dependence, with the faster rates observed in the red wing of the P band. One possibility to explain these effects is that due to a structural variation a distribution of the electronic coupling exists, with the 'red' RCs having the stronger coupling. The difference in the amplitude of the electrochromic bandshift over the monomeric bacteriochlorophylls, with the fast 'red' RCs having the larger amplitude, supports the idea that a difference in electronic coupling underlies the difference in ET rate. Evidence for the presence of different structural $P^+Q_A^-$ states in WT RCs has already been obtained by [6], who have shown that back ground illumination during cooling reversibly slows down the $P^+Q_A^-$ reaction. At room temperature no wavelength dependence of the $P^+Q_A^-$ recombination rate was observed in the M210W mutant, which indicates that at room temperature the different conformations are in equilibrium. A relation between red shift and an increase in the $P^+Q_A^-$ recombination rate has also been observed by studying the effect of increasing pressure on the $P^+Q_A^-$ recombination rate [7].Furthermore, this correlation between red shift and charge recombination may also underlie the increase of the several electron transfer processes with decreasing temperature in WT RCs. However, for the WT and the M210L RCs the $P^+Q_A^-$ recombination at 77 K needed only a single spectral component in the P band region, although the reactions were multi-exponential. A wavelength dependence of the $P^+Q_A^-$ recombination in the WT RC of *Rb. sphaeroides* has been observed to be present in another spectral region, around the crossing points of the band shift over the H_A and B_A bands (801 nm and 757 nm). Possibly, the distribution of electronic couplings is not as manifest in the P -Band region of the WT RC and other RCs with mutations at the M210 positions as it is in the M210W mutant, for reasons which are not clear yet.

Acknowledgements.

This research was supported by the Dutch Foundation for Fundamental Research (NWO) through the Foundation for Life Sciences (SLW) and the EC contract CT930278. MRJ is a BBSRC Senior Research Fellow.

References

1.Nagarajan, V., Parson, W.W., Davis, D. and Schenck, C.C. (1993) Biochemistry 32, 12324-12336

2Jones, M.R., Heer-Dawson, M., Mattioli, T.A., Hunter, C. N. and Robert, B. (1994), FEBS Lett 339, 18-24.

3.Groot, M.-L., Peterman, E.J.G., Van Stokkum, I.H.M., Dekker, J.P. and Van Grondelle, R. (1995) Biophys. J. 68, 281-290.

4.Van Stokkum, I.H.M., Brouwer, A.M., Van Ramesdonk, H.J. and Scherer, T. (1993) Proc. Kon. Ned. Acad. v Wetensch. 96, 43-68.

5.Shochat, S., Arlt, T., Francke, C., Gast, P., Van Noort, P.I., Otte, S.C.M., Schelvis, H.P.M., Schmidt, S., Vijgeboom, E., Vrieze, J., Zinth, W. and Hoff, A.J. (1994) Photosynth. Res. 40, 55-66.

6.Kleinfeld, D., Okamura, M.Y. and Feher, G. (1984), Biochemistry 23, 5780-5786.

7.Hoganson, C.W., Windsor, M.W., Farkas, D.I. and Parson, W.W. (1987), Biochim. Biophys. Acta 892, 275-283.

ANGULAR AND TEMPERATURE DEPENDENCE OF LINEAR-DICHROIC MAGNETIC FIELD EFFECTS IN PHOTOSYNTHETIC REACTION CENTERS

B. van Dijk[1], J.K. Hicks[2], P.J. Hore[2] and A.J. Hoff[1]
(1) Department of Biophysics, Huygens laboratory, University of Leiden, P.O. Box 9504, 2300 RA Leiden, The Netherlands
(2) Physical Chemistry Laboratory, Oxford University, South Parks Road, Oxford OX1 3QZ, United Kingdom

1. Introduction

The steady state triplet concentration of a photosynthetic reaction center (RC) can be changed by applying an external magnetic field. This MFE (Magnetic field effect) arises due to both the RPM (Radical Pair Mechanism) and the MIMS (Magnetic field Induced Mixing of triplet Sublevels) effect [1, 2].

A modulation of the triplet concentration by an external magnetic field, $\vec{B}$, can be detected optically by monitoring the transmission of the sample at a wavelength where the extinction coefficients for triplet and singlet state differ. Probing the effect as a function of B at a fixed wavelength results in the so called MFE-curve; probing it as a function of wavelength results in the Triplet-minus-Singlet (T-S) difference spectrum. The latter technique was labelled MODS (Magneto-Optical Difference Spectroscopy) [3].

Due to dipolar and hyperfine interactions, the change in triplet concentration is also dependent on the orientation of the triplet with respect to $\vec{B}$. So, in general, the MODS signal will be different for light polarised parallel and perpendicular to the external field. Using polarised light we can record the linear-dichroic T-S spectrum or the LD-MFE curves [4]. The experimental set-up is described in [5]

Recently we introduced the Rota-MARY technique to measure LD-MODS spectra [6]. In this technique the magnitude of $\vec{B}$ is constant and $\vec{B}$ is rotating with respect to the polarisation direction of the excitation/probe light beam. The first results showed a small magnetic field dependency of the LD-MODS spectrum of RC's of the photosynthetic bacterium *Rhodobacter sphaeroides*. Because earlier measurement were performed using a rather large sinusoidal magnetic field modulation this effect was overlooked. In order to understand these effects the results should be compared to computer simulation since there is no direct analytical relation between the LD-MFE signal and the orientation of the optical transition moments.

2. The problems

After preliminary measurement we found that the exact shape of the LD-MFE curves is highly dependent on the modulation depth; therefore we improved the measurement by taking a much smaller and more homogeneous modulation field. It

P. Mathis (ed.), Photosynthesis: from Light to Biosphere, Vol. I, 911–914.

must be stressed however that even at this small modulation there is still distortion left. Going to even smaller modulation amplitudes would make the measurement too time-consuming and too sensitive to long-term drift of the detection circuits. The Rota-MARY technique would in principle remove the distortion of the measured MFE-curves since there is no field modulation but unfortunately our old set-up does not allow the high fields needed to investigate the interesting regions of the MFE-curves. A new set-up is in development and will be in operation soon.

Experiments at different modulation amplitudes (not shown) indicated that the zero-crossing around 70 Gauss and the low field features are severely disturbed with increasing modulation depth; the 190 Gauss zero-crossing is less sensitive to modulation.

3. Simulations

The simulations are performed by solving the stochastic Liouville equation of the Radical Pair including Zeeman, hyperfine, exchange and dipolar interactions, to model the RPM, and including the MIMS effect. We assume isotropic excitation in the simulated experiment, and average the result over all orientations of the RC. The parallel and perpendicular signals are calculated from the averaged triplet concentration for a given field and subtracted to obtain the LD MFE-curve. For the triplet decay rates k_x and k_y 8000 s^{-1} was used, 1200 s^{-1} for k_z, the zero-field splitting parameters $|D|$ and $|E|$ are resp. 200 and 35 G, and the exchange interaction J was assumed to be 7 G.

The dipolar coupling of the radical pair was set to zero because it makes little difference to the simulation compared to the more realistic figure of 4.8 G and it greatly decreases the time needed for calculations.

A weighting function is used to give the probability of different values of the hyperfine coupling:

$$f = \frac{1}{\sqrt{2\pi}}\frac{4}{A}\exp\frac{-8H^2}{A^2} \tag{1}$$

For each of the 3 spatial dimensions a 5-point Gaussian distribution was used.($H = \pm\frac{A}{4}$, $\pm\frac{A}{2}$, 0) For each orienation of the RC the average value over all ($5^3 = 125$) possible combinations of the hyperfine couplings is used. A is taken to be 16.3 G.

4. Results

In Fig. 1 the LD-MFE curves are shown for different detection wavelengths. The structure at B <10 Gauss is clearly visible only at the 821 nm band. Also one can see that the zero-crossing around 70 Gauss is noticeably lower when detected at 821 nm. The 900 and 809 nm LD-MFE curves are roughly the same. This is not very suprising since we now from LD-ADMR measurements [7] that the orientation of the 900 and 809 nm bands are more or less equal while the orientation of the 821 nm band is different.

Computer simulations where performed to check the LD-MFE dependence on the orientation of the transition moment used to record the curve. The angular dependence of the LD-MFE curves is quite strong and effects such as the shift of the zero-crossing around 70 Gauss can be reproduced. Unfortunately we are not able to reproduce the exact shape of the 821 nm LD-MFE curve without tilting the transition moment away from the triplet xy-plane. Since we know from LD-ADMR that

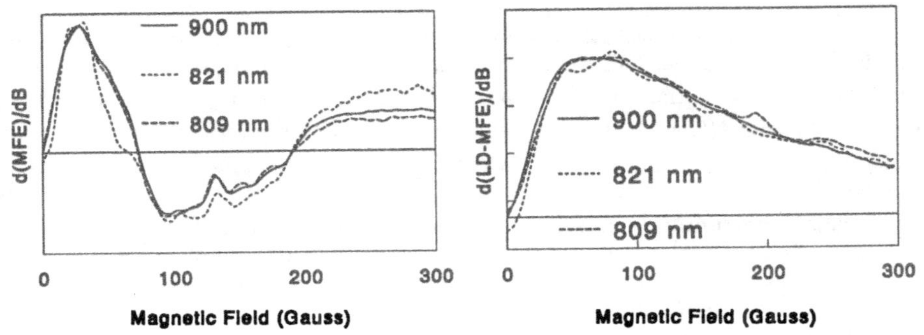

Figure 1: The LD-MFE curves of RC's of *Rhodobacter sphaeroides* R26 measured at 30 K (left) and 90 K (right) with a modulation amplitude of 5 Gauss

the angle is at most 10 degrees [7] this seems highly unlikely to be of any physical significance. More extensive simulations are in progress.

5. The LD-MFE curve

In Fig. 2 the MFE curves for light polarised parallel and perpendicular to $\vec{B}$ are shown at different temperatures. Below 50 K the curves originate mainly from the MIMS effect; above 50 K the RPM prevails.

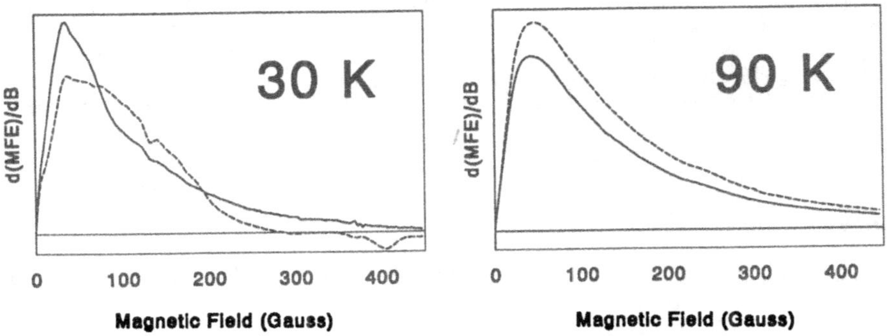

Figure 2: The LD-MFE curves measured at different temperatures (dotted line: perpendicular, solid line: parallel). All curves where recorded using a modulation amplitude of 6 Gauss, and a modulation frequency of 312 Hz.

In the low temperature curves it can be seen that there are sharp features at 135 and 400 Gauss. (Both these features disappear quickly when the modulation depth is increased.) The 135 Gauss dip arises presumably from an avoided level-crossing of the x and y triplet levels when $\vec{B}$ is almost parallel to the y-axis. In the simulation a very shallow, broad depression can be modelled which is not nearly as sharp as the experimentally observed dip [8].

At this moment we do not have a reasonable explanation for the dip at 400 Gauss, which is only present in the perpendicular signal. From this we can conclude that

RC's with there optical transition moment $\vec{\mu}$, close to or parallel to $\vec{B}$ do not show this structure. The optical transition moment for the Q_y band used to record the curves in Fig. 2 is polarised more or less along the triplet y-axis so for those RC's where $\vec{\mu} \parallel \vec{B}$, $\vec{B}$ is along the triplet y-axis. The structure around 400 Gauss must therefore arise from RC's with there $\vec{\mu}$ perpendicular to the field, thus having $\vec{B}$ in the triplet xz-plane. It is most likely that $\vec{B}$ along the x-axis does not give such structure since there are level-crossings for this orientation [8]. It can therefore be concluded that the structure around 400 Gauss is mainly caused by RC's with $\vec{B}$ along or close to the triplet z-axis, for those RC's an (avoided) level-crossing occurs at 200 Gauss, the zero-field splitting of the triplet. At this point it is unclear what the mechanism behind this feature can be.

On going to higher temperatures almost all structure disappears showing that both features can be attributed to ^3P since at high temperatures the MFE is mainly connected with $^3[P^+I^-]$ via the RPM.

The high temperature LD-MFE curves at different wavelength are shown in Fig. 1. As expected most of the structure is gone. Again the 821 nm band differs a little from the other wavelengths at very low fields. Computer simulations at higher temperature are in progress but not yet available

6. Conclusions

The wavelength and temperature dependence of the LD-MFE curves show more structure than previously recognised. We think it's worth the effort of trying to model these features since it can give information about the relative orientations of the optical transition moments composing the T-S spectrum which is still not completely understood. Since the differences are very small and highly sensitive to small offsets in the signal and the depth of the magnetic field modulation it is at this time very difficult to compare the experimental results with computer simulations in any meaningful way.

References

1 Schulten,K., Staerk,J., Weller,A., Werner,H.-J. and Nickel,B. (1976) Z. Phys. Chem. Neue Folge 101, 371

2 Michel-Beyerle,M.E., Haberkorn,R., Bube,W., Steffens,E., Schröder,H., Neuser,H.J., and Schlag,E.W. (1976) Chem. Phys. 17, 139

3 Hoff,A.J., Lous,E.J., Moehl,K.W. and Dijkman,J.A. (1985) Chem. Phys. Lett. 114, 39

4 Lous,E.J., and Hoff,A.J. in: *The Photosynthetic Bacterial Reaction Center* (J. Breton and A. Verméglio, eds.) p. 71/ Plenum, New York 1988

5 van der Vos,R., Franken,E.M., Sexton,S.J., Shochat,S., Gast,P., Hore,P.J., and Hoff,A.J., (1995) Biochim. Biophys. Acta 1230, 51

6 van Dijk,B., van der Vos,R. and Hoff,A.J. (1994) Chem. Phys. Lett. 226, 206

7 Vrieze,J., submitted to Biochim. Biophys. Acta

8 Hicks,J.K. and Hore,P.J., (1995) Chem. Phys. Lett. 237, 183

IRON DEFICIENCY AFFECTS THE DONOR SIDE OF PHOTOSYSTEM II

Namachevayam Nedunchezhian, Anunciación Abadía and Javier Abadía

Department of Plant Nutrition, Estación Experimental de Aula Dei,

CSIC, Apdo 202, 50080 Zaragoza, Spain.

1. Introduction

Based on the Fe content of the different complexes within the photosynthetic apparatus it has been proposed that Fe-deficiency should affect to different extents each complex, in the decreasing order PSI>Cyt_{b-f}>PSII>ATPase (1, 2). Measurements of the stoichiometry of different components of the photosynthetic apparatus of Fe-deficient plants generally support this order (3-6), although a specific effect of Fe-deficiency on Cyt_f has been reported recently in maize (7). However, Morales et al. (8) indicated that in moderately Fe-deficient leaves the quantum yield of O_2 evolution was more affected than photochemical efficiency estimated from Chl fluorescence parameters. This suggests that Fe-deficiency may cause a functional impairment in the O_2 evolving system in PSII.

2. Procedure

Sugar beet (Beta vulgaris L. cv. Monohil, Hilleshög, Landskröna, Sweden) was grown hydroponically in a growth chamber. Leaves used had Chl concentrations of 35-37 (control) and 6-8 $nmol{\cdot}cm^{-2}$ (Fe-deficient). Thylakoids and PS II membranes (PSIIm) were isolated from sugar beet leaves as described previously (9). Chlorophyll and protein were estimated according to Refs. (10) and (11), respectively. Light saturated rates of whole chain, PSI and PSII electron transport activity were estimated as O_2 consumption or evolution (12) at pH 6.8. PSII activity was also determined by following the reduction of DCPIP at 590 nm. MV, DCBQ and DPC were 1 mM and NH_2OH and $MnCl_2$ were 5 mM, respectively. 1-D LiDS-PAGE was done as in (13) using a gradient of 12-20% polyacrylamide. 2-D was made as in (14).

P. Mathis (ed.), Photosynthesis: from Light to Biosphere, Vol. I, 915–918.

3. Results and Discussion

3.1 Effect of iron deficiency on photochemical activities

Whole chain electron transport was decreased in the Fe-deficient plants (Table 1) by 65%. However, PSI electron transport activity was only inhibited by 25%. The PSII electron transport activity mediated by both DCBQ and DCPIP was inhibited by approximately 60% in the Fe-deficient thylakoids and PSIIm preparations (Tables 1 and 2).

Table 1. Whole chain, PSI and PSII electron transport activities in thylakoid preparations from leaves of control and Fe-deficient plants. Data are given in μmol O_2. (mg Chl)$^{-1}$. h^{-1} and are the mean ± SE of 3 replicates. Fe-deficient leaves had 6-8 nmol Chl·cm^{-2}.

Electron transport		Control	-Fe	% inhibition
Whole chain	H_2O-->MV	215±12	73±11	66
PSII	H_2O-->DCBQ	302±12	119±3	61
PSI	$DCPIPH_2$-->MV	461±16	328±20	25

Table 2. Effect of various exogenous electron donors on the PSII electron transport activity in thylakoid and PSII membrane preparations from leaves of control and Fe-deficient plants. DCPIP was used as electron acceptor. Data are given as μmol DCPIP reduced·(mg Chl)$^{-1}$·h^{-1} and are the mean ± SE of 3 replicate experiments. Fe-deficient leaves had approximately 6-8 nmol Chl·cm^{-2}.

Electron donor	Thylakoids			PSIIm		
	Control	-Fe	% inhibition	Control	-Fe	% inhibition
H_2O	245±10	101±7	59	348±14	122±11	65
$MnCl_2$	258±11	121±11	54	417±10	159±4	62
NH_2OH	271±7	194±10	29	497±13	338±9	33
DPC	284±6	215±8	25	532±11	379±9	30

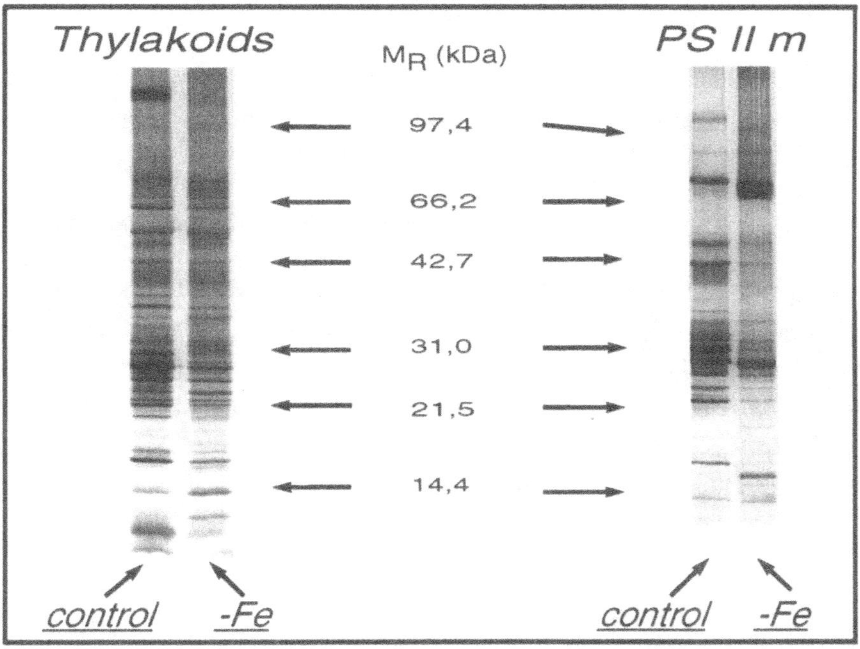

FIGURE 1.- 1-D polypeptide profile of thylakoids and PSIIm from control and -Fe plants.

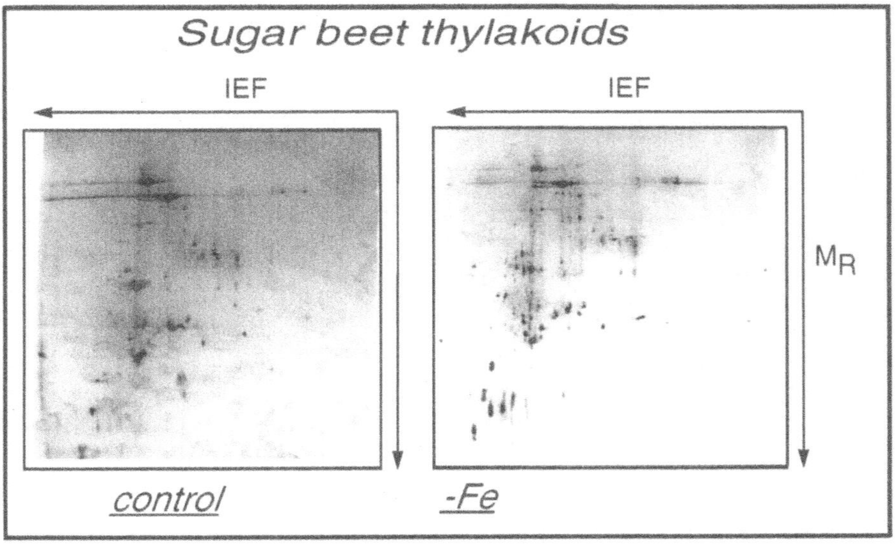

FIGURE 2.- 2-D polypeptide profile of thylakoids from control and -Fe plants.

3.2 Effects of Exogenous PSII electron donors

$MnCl_2$, which donates electrons close to water oxidation site, induced only a marginal restoration of the PSII activity in both thylakoid and PSII membranes. However, DPC and NH_2OH restored a significant part of the PSII electron transport (Table 2).

3.3 Iron deficiency induced changes in thylakoid and PSII membrane polypeptides

The 1-D LiDS-PAGE of thylakoid and PSII membrane polypeptide profiles of control and iron deficient plants is shown in Fig. 1. The relative amounts of some polypeptides were reduced in thylakoids and PSII membranes from Fe-deficient plants. Some of the polypeptides that are reduced, such as those at 49, 42, 33, 27-25, 24, and 17 kDa are part of the PSII complex. The 2-D IEF/SDS-PAGE of thylakoid profiles of control and Fe deficient plants (Fig. 2) also indicates differences in the polypeptide composition.

3.4 Possible causes for the decreases in PSII electron transport rates

The preliminary data presented here suggest that the donor side of PSII may be impaired in moderately Fe-deficient plants. The cause for this impairment may reside in the specific loss of any of the elements essential for the functioning of this complex. We are currently investigating whether the Fe-deficiency induced loss PSII electron transport rates could be correlated with the loss of PSII polypeptides and other co-factors.

Acknowledgements. The technical assistance of Aurora Poc is gratefully acknowledged. Supported by grants PB91-0057 and AIR3-CT94-1973 to J.A.

References

1 Terry N. and J. Abadía (1986) J. Plant Nutr. 9: 609-646.
2 Abadía J. (1992) J. Plant Nutr. 15: 1699-1713.
3 Spiller, S.C. and N. Terry. (1980). Plant Physiol. 65: 121-125.
4 Nishio, J.N. and N. Terry (1983) Plant Physiol. 71: 688-691.
5 Nishio, J.N. and N. Terry (1985) Plant Physiol. 78: 296-299.
6 Sandmann, G. (1985) Photosyn. Res. 6: 261-272.
7 Sharma S. and G.G. Sanwal (1992) J. Plant Physiol. 140: 527-530.
8 Morales, F., A. Abadia and J.Abadia. (1991). Plant Physiol. 97: 886-893.
9 Quílez, R., A. Abadia and J. Abadia. (1992). J. Plant Nutr. 15: 1808-1819.
10 Lichtenthaler, H.K. (1987). Meth. Enzymol. 148:350-382.
11 Markwell M., Haas S.M., N.E. Tolbert and L.L. Lieber (1981) Meth. Enzymol. 72:350-382.
12 Izawa,S. (1980). Meth. Enzymol. 69: 413-433.
13 Laemmli,U.K. (1970). Nature 227: 680-685.
14 Rémy R. and F. Ambard-Bretteville (1987) Meth Enzymol 148: 623-632.

REACTION CENTERS OF *RHODOBACTER SPHAEROIDES* R26 AFTER EXCHANGE WITH BACTERIAL- AND PLANT-TYPE PHEOPHYTINS AT $H_{A,B}$-SITES

M. Meyer[1], G. Fritzsch[2], U. Ermler[2] , C. Geskes[3], H. Michel[2], G. Heinze[3], H. Scheer[1]
[1] Botanisches Institut, Menzingerstr. 67, 80638 München, Germany
[2] MPI für Biophysik, Heinrich-Hoffmannstr. 7, 60528 Frankfurt/Main, Germany
[3] Institut für Physikalische Chemie, Albertstr. 21, 79104 Freiburg, Germany

1. Introduction

The native bacteriopheophytin a (BPhe a) at sites H_A and H_B in reaction centers (RC) of *Rhodobacter (Rb.) sphaeroides* R26 has been exchanged with several modified bacterial-type and plant-type pheophytins. They have modifications at C-3, C-7, C-13^2 and C-17^3. To investigate the three-dimensional structure of RC bevore and after exchange, native RC and RC in which BPhe a is exchanged with Phe a have been crystallized.

Abbreviations: $B_{A,B}$ = binding-sites for BChl's at the active and inactive branch, BPhe a = bacteriopheophytin, $H_{A,B}$ = binding-sites for BPhe's at the active and inactive branch, rsp., *Rb.* = *Rhodobacter*, RC = reaction center(s), P = primary donor, Phe a = pheophytin a, $Q_{A,B}$ = binding-sites for the quinones, THF = tetrahydrofurane

2. Procedure

RC and pigments were extracted by standard procedures (1,2). Modification of the pigments and exchanges were prepared as described previously (3). Characterization of modified RC was done according to (3).

3. Result and Discussion

3.1 *Pigment characterization*
Several bacterial-type and plant-type pheophytins were tested (Fig. 1).

P. Mathis (ed.), Photosynthesis: from Light to Biosphere, Vol. I, 919–922.

Figure 1.

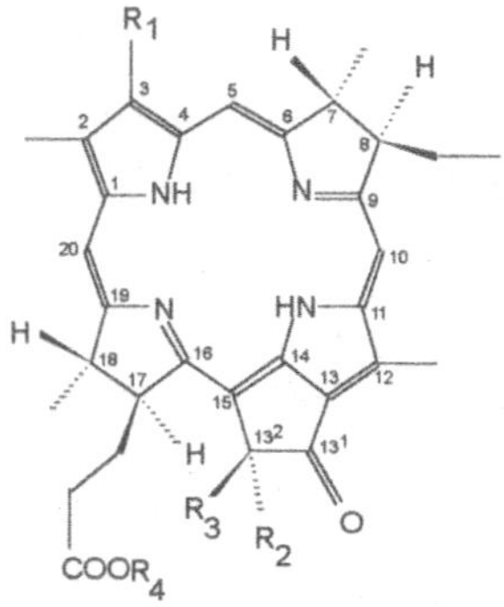

Pigment	R_1	R_2	R_3	R_4
BPhe a_p	$COCH_3$	$COOCH_3$	H	$C_{20}H_{39}$
BPhe a_p'	$COCH_3$	H	$COOCH_3$	$C_{20}H_{39}$
BPhe a_{gg}	$COCH_3$	$COOCH_3$	H	$C_{20}H_{33}$
[3^1-OH]-BPhe a	CH_2OH	$COOCH_3$	H	$C_{20}H_{39}$
[3-vinyl]-BPhe a	$CHCH_2$	$COOCH_3$	H	$C_{20}H_{39}$
13^2-OH-BPhe a	$COCH_3$	$COOCH_3$	OH	$C_{20}H_{39}$
Pyro-BPhe a	$COCH_3$	H	H	$C_{20}H_{39}$
[17^4-methyl]-BPhe a	$COCH_3$	$COOCH_3$	H	CH_3

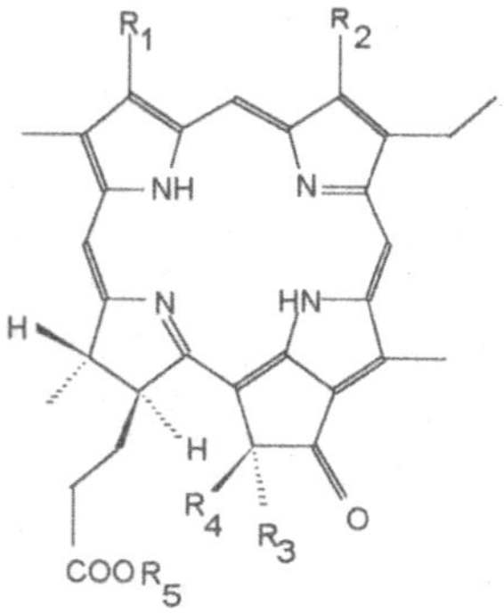

Pigment	R_1	R_2	R_3	R_4	R_5
Phe a	$CHCH_2$	CH_3	$COOCH_3$	H	$C_{20}H_{39}$
Phe a'	$CHCH_2$	CH_3	H	$COOCH_3$	$C_{20}H_{39}$
[3-acetyl]-Phe a	$COCH_3$	CH_3	$COOCH_3$	H	$C_{20}H_{39}$
13^2-OH-Phe a	$CHCH_2$	CH_3	$COOCH_3$	OH	$C_{20}H_{39}$
Pyro-Phe a	$CHCH_2$	CH_3	H	H	$C_{20}H_{39}$
[17^4-methyl]-Phe a	$CHCH_2$	CH_3	$COOCH_3$	H	CH_3
Phe b	$CHCH_2$	CHO	$COOCH_3$	H	$C_{20}H_{39}$
[7^1-OH]-Phe b	$CHCH_2$	CH_2OH	$COOCH_3$	H	$C_{20}H_{39}$

Structure of the bacterial- (top) and plant-type (bottom) pheophytins.

The redox potentials of all these pigments were measured in THF: differences in the hydrogenation state (chlorin *vs.* bacteriochlorin) changed the redox potential for oxidation, while the redox potentials for reduction were affected strongly by the β-pyrrolic substituents.
Strong changes in the absorption spectra of the modified pigments as compared to the native BPhe a are only seen after transformation of the bacteriochlorin to the chlorin and after substituation of the C-3 group by a vinyl- or a hydroxy-group.

3.2 *Characterization of the modified reaction centers*

In the case of [3-vinyl]-BPhe a the occupation of H_A and H_B can be followed spectroscopically (Fig. 2).

Figure 2.

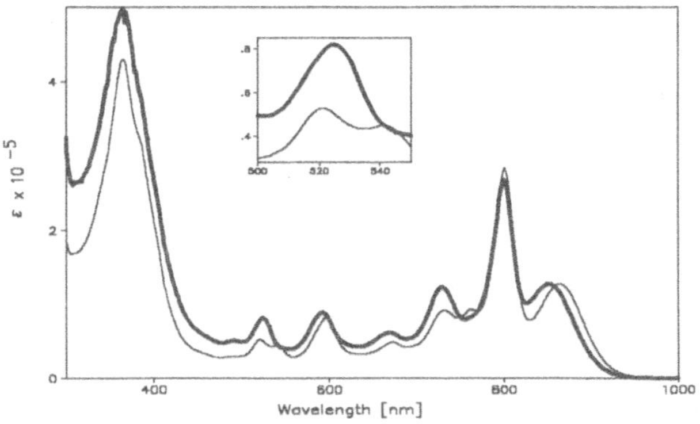

Absorption spectra of RC exchanged with [3-vinyl]-BPhe a after the first (thin line) and the second exchange (heavy line).

After the first exchange one can see that the band at 536 nm is split into one at 521 nm (= introduced [3-vinyl]-BPhe a at H_B) and one at 541 nm (= native BPhe a at H_A, 4), while after the second exchange it sharpens and shifts to 521 nm. This would suggest an easier exchange at the inactive site H_B.
The absorption spectra of RC exchanged with pigments which absorb at different wavelengths compared to the native BPhe a, show the characteristic features of the newly introduced pigments (Table 1).

Table 1.

pigments	Q_X ($H_{A,B}$)	Q_X ($B_{A,B}$; P)	Q_Y ($H_{A,B}$)	Q_Y ($B_{A,B}$)	Q_Y (P)
BPhe a_P	536 - 538	598	758	803	865
BPhe a_{gg}	539 - 540	591	758	802	865
[3^1-OH]-BPhe a	518 / 542	597	722	801	865
[3-vinyl]-BPhe a	525	592	729	800	851
13^2-OH-BPhe a	533	595	758	802	864
Pyro-BPhe a	538	592	757	803	862
[17^4-methyl]-BPhe a	541	594	760	803	859
Phe a	509 / 542	595	674	800	862
[3-acetyl]-Phe a	516 / 544	593	689	803	865
13^2-OH-Phe a	506 / 539	593	673	801	865
Pyro-Phe a	509 / 542	596	674	801	861
[17^4-methyl]-Phe a	539	594	672	802	862
Phe b	540	596	664	801	864
[7^1-OH]-Phe b	512 / 546	597	669	800	865

Absorption bands of the modified RC after the second exchange

It is noteworthy that all modified RC are capable of light induced electron transfer to $Q_B^{\bullet-}$; the different quantum yields and recombination times (τ_1, τ_2) are given in Table 2.

Table 2.

pigment at $H_{A,B}$	τ_1 [ms]	a_1	τ_2 [ms]	a_2	bleaching [%]
BPhe a	100 ± 2	0.30	730 ± 20	0.70	98
BPhe a_{gg}	90 ± 7	0.39	885 ± 85	0.61	86
[3^1-OH]-BPhe a	98 ± 7	0.47	709 ± 40	0.53	92
[3-vinyl]-BPhe a	110 ± 10	0.84	400 ± 50	0.16	60
Pyro-BPhe a	102 ± 5	0.83	500 ± 80	0.17	84
Phe a	120 ± 5	0.40	700 ± 35	0.60	77
[3-acetyl]-Phe a	100 ± 5	0.30	800 ± 34	0.70	64

Flash-induced electron transfer. Kinetics and relative yields of electron transfer to $Q_{A,B}$.

3.3 *Crystallography*

Comparing the crystal structure of RC of *Rb. sphaeroides* R26 and 2.4.1 the main difference between the two RC are observed at the carotenoid (spheroidene) binding-site. The space occupied in the wild-type RC by the carotenoid is replaced in the mutant by 1) a detergent molecule (containing 14 C-atoms), covering the same interaction sites as the carotenoid, and 2) the side-chain of the amino acid Phe M162, which changes its position to fill the space of the missing carotenoid tail. Comparing the other cofactors (BChl's, BPhe's, quinones) of both RC, their position and electron densities seem to be unchanged. This observation is in agreement with identical rate constants measured for the electron transfer in both RC (5).

References

1 Satoh, N. and Murata, N. (1978) Biochim. Biophys. Acta 501, 103-111
2 Feher, G. and Okamura, M.T. (1978) in: The Photosynthetic Bacteria (Clayton R.K. and Sistrom W.R., eds), pp. 349-386, Plenum Press, New York
3 Meyer, M. and Scheer, H. (1995) Photosyn. Res. 44, 55-65
4 Breton, J., Bylina, E.J. and Youvan, D.C. (1989) Biochemistry 28, 6423-6430
5 Arlt, T., Schmidt, S., Kaiser, W., Lauterwasser, C., Meyer, M., Scheer, H. and Zinth, W. (1993) Proc. Natl. Acad. Sci. 90, 11757-11761

MODELLING THE LOW TEMPERATURE FLUORESCENCE KINETICS OF THE PHOTOSYSTEM II REACTION CENTER: ENERGY TRANSFER AND RADICAL PAIR RELAXATION

Lars Konermann, Guido Gatzen and Alfred R. Holzwarth

Max-Planck-Institut für Strahlenchemie, Stiftstr. 34-36, D-45470 Mülheim a.d. Ruhr; Germany

Introduction

Since the reaction center (RC) of the photosystem II (PSII-RC, also called D_1-D_2-cyt-b_{559} complex) has been isolated for the first time [1] a variety of studies has been carried out to elucidate its structure and function. The interpretation of any spectroscopic data is strongly hampered by the severe congestion of the Q_y-bands of all chromophores contained in the PSII-RC. Thus the spectral positions of P_{680}, the primary electron donor, the pheophytins (Pheo) and the remaining chlorophylls (Chl) are still discussed controversially. In a recent study (Konermann & Holzwarth, *submitted to Biochemistry*) we presented a model to analyze the absorption spectrum of the PSII-RC and to describe its temperature dependence. This model is based on theoretical calculations of the temperature-dependent homogeneous lineshapes for the individual pigments [2,3] and it considers the inhomogeneous broadening of the spectra. As a result of that study a detailed decomposition of the absorption spectrum of the PSII-RC was obtained. The model assumes relatively weak coupling between the chromophores in the RC core such that with the exception of the putative P_{680} dimer a description in terms of contributions by individual chromophores is appropriate. It was concluded that typical samples consist of a mixture of RC particles which bind a varying number of Chl molecules, probably located outside the core RC. Thus the average Chl/2Pheo ratio varies between 6 and 7 in most preparations.
Now this spectral model is extended and is applied to the analysis of the low temperature (77 K) fluorescence kinetics of the PSII-RC. Important questions in this respect concern the influence of inhomogeneous broadening on the energy transfer kinetics and the mechanism of radical pair (RP) formation and relaxation.

P. Mathis (ed.), Photosynthesis: from Light to Biosphere, Vol. I, 923–926.

Materials and Methods

Samples: The PSII-RC samples for this study were prepared from spinach leaves according to the method of van Leeuven et al. [4]. These samples had a pigment ratio of 6.2 ± 0.2 Chl : 2 Pheo as determined by HPLC.
Time-resolved fluorescence measurements: Fluorescence decay curves were measured with single photon counting as described [5]. Excitation was at 650 nm with a repetition rate of 800 KHz. The full width at half maximum of the system response function was about 30 ps.
Kinetic Modelling: In order to describe the fluorescence kinetics of the PSII-RC, the appropriate master equation [6] has to be solved for many different spectral "arrangements". Each arrangement represents i) the spectral characteristics of one single PSII-RC out of the inhomogeneously broadened ensemble and ii) the pigment composition of this single RC (due to the heterogeneity in pigment composition). To obtain the fluorescence kinetics of the ensemble hundreds of such different arangements were averaged. The individual arrangements were chosen by a random process under consideration of the parameters obtained from the spectral model.
Energy transfer is described by using the Förster theory [7]. The energy transfer rates are proportional to the overlap integral in the Förster formula and to the term κ^2/R^6 (κ denoting the orientation factor and R the distance between the pigments). The overlap integral can be calculated from the homogeneous spectra of the pigments but the terms κ^2/R^6 had to be obtained by fitting to the kinetic data (for the pigments in the core of the PSII-RC these values were taken from the structure of the bacterial RC). Also the rate constants connected with RP formation had to be fitted. A genetic algorithm was used as fitting procedure. The data were analyzed globally for several detection wavelengths and for three time windows ($t < 150$ psec, $t < 3$ nsec, $t < 20$ nsec).

Results and Discussion

As experimental basis for this modelling study we used time resolved fluorescence data obtained from measurements of the PSII-RC at 77 K and an excitation wavelength of 650 nm. Fluorescence decays were recorded for several detection wavelengths between 675 nm and 690 nm (data not shown).
In a first step the kinetic model described above was fitted to these data, including a one-step mechanism for RP formation (see Fig.1a). With such a model no acceptable correspondence between the experimental data and the fitted decay curves could be obtained. In order to improve the quality of the fits, the model was modified in such a way that the free energy difference for charge separation ΔG_{CS} was assumed to be different for the individual PSII-RCs in the sample. Such a static distribution for ΔG_{CS} is illustrated in Fig.1b. We assumed a gaussian shaped

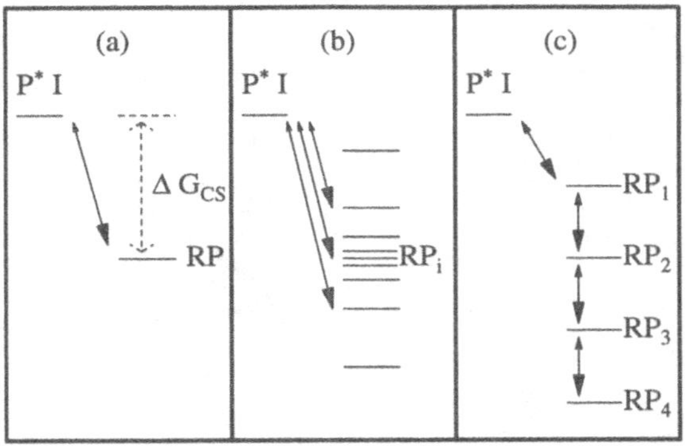

Figure 1. Models for RP formation in RCs. (a) one-step mechanism, (b) one-step mechanism with a distribution in free energy difference, (c) sequential RP relaxation.

distribution function whose width and maximum were fitting parameters. Also with such a mechanism of RP formation the low temperature fluorescence kinetics of the PSII-RC could not be described within the framework of our model. Furtheron a mechanism with several sequential RPs was tested. A primary RP is formed after charge separation which then relaxes in a series of successive steps, each one with a lower free energy than the previous one (Fig.1c). With five RP states this model finally resulted in an acceptable fit to the experimental data from the picosecond time range up to 20 nsec. This sequential RP relaxation presumably should not really be interpreted as a step by step reaction. It should rather be considered as a continuous relaxation process during which the protein surrounding and the pigments react to the electrostatic forces that are caused by the sudden creation of a pair of ions within the RC. This picture is in accordance with the "dynamic solvation model" that recently was proposed to explain the fluorescence kinetics of bacterial RCs [8].

Due to the inhomogeneous broadening of the pigment spectra the rate constants for energy transfer show a pronounced dispersion. Depending upon the average energy difference between two molecules participating in such a transfer reaction the distribution functions for these rates can be strongly asymmetric (not shown). Within the RC core (consisting of P_{680}, two Pheo and two accessory Chls) energy equilibration proceeds on the sub-picosecond timescale which is in line with the experimental findings of Durrant et al. [9]. Energy transfer from the Chls on the periphery of the complex is typically at least one order of magnitude slower than within the core. Interestingly none of these outer Chls transfers energy to P_{680} directly.

Our kinetic model predicts a characteristic temperature dependence of the stationary fluorescence. At low temperature part of the excitation energy gets trapped on pigments that absorb on the very red side of the spectrum so that they can no longer transfer their energy to P_{680}, which absorbs at slightly shorter wavelengths. At 10 K the above mentioned spectral model predicts an accessory Chl absorbing at around 682 nm and one of the outer Chls absorbing at around 683 nm which both can serve as such excitation traps. At higher temperatures the

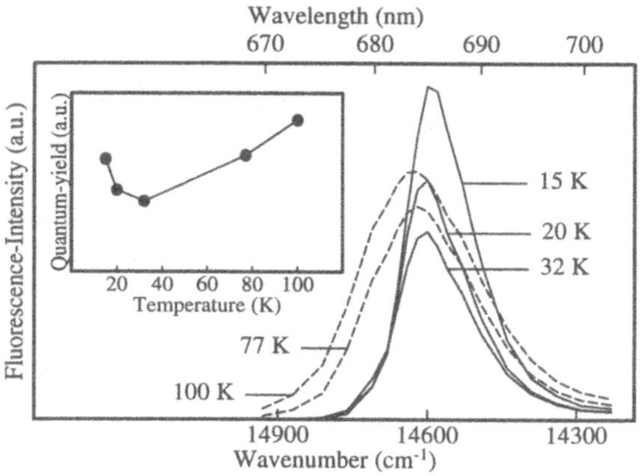

Figure 2. Temperature dependence of the stationary fluorescence spectrum as predicted by the kinetic model. The inset shows the fluorescence quantum yield as a function of temperature.

contribution of fluorescence from "blue" pigments in the complex increases. As a result of this the maximum of the stationary fluorescence shifts from 683 nm at 100 K to 685 nm at 15 K and the fluorescence quantum yield shows a minimum at about 35 K. These predictions from the model agree quite well with the experimental findings of Groot et al. [10]. The temperature dependence of the stationary fluorescence spectra together with the relative fluorescence quantum yield (calculated by integrating the spectra) are depicted in Fig.2.

References

1 Nanba, O. and Satoh, K. (1987) Proc. Natl. Acad. Sci. USA 84, 109-112.
2 Pullerits, T., van Mourik, F., Monshouwer, R., Visschers, R.W. and van Grondelle, R. (1994) J. Luminesc. 58, 168-171.
3 Rebane, K.K. (1970) Impurity Spectra of Solids. Elementary theory of vibrational structure, pp. 1-253, Plenum Press, New York.
4 van Leeuwen, P.J., Nieveen, M.C., van de Meent, E.J., Dekker, J.P. and van Gorkom, H.J. (1991) Photosynth. Res. 28, 149-153.
5 Holzwarth, A.R., Müller, M.G., Gatzen, G., Hucke, M. and Griebenow, K. (1994) J. Luminesc. 60&61, 497-502.
6 Beauregard, M., Martin, I. and Holzwarth, A.R. (1991) Biochim. Biophys. Acta 1060, 271-283.
7 Förster, T. (1965) in Modern Quantum Chemistry. Part III Action of Light and Organic Crystals (Sinanoglu, O., ed.), pp. 93-137, Academic Press, New York.
8 Peloquin, J.M., Williams, J.C., Lin, X., Alden, R.G., Taguchi, A.K.W., Allen, J.P. and Woodbury, N.W. (1994) Biochemistry 33, 8089-8100.
9 Durrant, J.R., Hastings, G., Joseph, D.M., Barber, J., Porter, G. and Klug, D.R. (1992) Proc. Natl. Acad. Sci. USA 89, 11632-11636.
10 Groot, M.L., Peterman, E.J., van Stokkum, I.H.M., Dekker, J.P. and van Grondelle, R. (1995) Biophys. J. 68, 281-290.

THE IMPACT OF THE F26S MUTATION IN THE β SUBUNIT OF CYTOCHROME b559 ON THE FUNCTION AND STABILITY OF PHOTOSYSTEM II IN TOBACCO

Tsafrir. S. Mor[1], Himadri B. Pakrasi[2] and Itzhak Ohad[1]
[1]Dept. of Biological Chemistry, The Hebrew University of Jerusalem, Jerusalem 91904, Israel; [2]Biology Dept., Box 1137, Washington University, St. Louis, MO 63130, USA

1. Introduction

Cytochrome b559 (cyt b559), an obligatory component of the photosystem II (PSII) core complex, is a heterodimeric protein with the heme moiety coordinated by histidine residues of the two subunits and can exist in two thermodynamic forms, one with a low midpoint potential (20-80 mv) the other with a high midpoint potential (330-400 mv) (reviewed in [1, 2]). The exact role of the cyt b559 is not yet known, but a widely accepted hypothesis suggests that it is involved, possibly by means of a putative molecular switch between the high and low redox potential forms, in protection of PSII from photodamage [3-5].

The 9 kDa α subunit and the 4 kDa β subunit of cyt b559 are the products of the *psbE* and *psbF genes* respectively [6]. Although the two polypeptides are well conserved in all species examined (see references cited in [7]), a certain degree of variance is observed at the gene structure level. In cyanobacteria and higher plants these genes are part of the *psbEFLJ* operon. However, in *Chlamydomonas reinhardtii* and related species these genes are transcribed separately from distant parts of the chloroplast genome [7, 8]. Furthermore, in certain higher plants such as spinach and pea, the transcript of the *psbF* gene is subjected to RNA editing resulting in conversion of a UCU triplet (serine) to a UUU triplet specifying a highly conserved phenylalanine residue (F26) which is encoded as such in other plant, algal and cyanobacterial species (e.g. tobacco and *C. reinhardtii*) [9]. This particular occurrence of post-transcriptional modification is most likely species-specific as was recently demonstrated [10] by transformation of tobacco with a mutated version of the gene that should direct the incorporation of the amino acid serine in the absence of RNA editing. The resultant mutant, S6, containes a mutated form of the β subunit with an F26S change [10]. The mutant is photosynthetic but its photoautotrophic growth was slow and its highly fluorescent leaves indicates functionally defective albeit operational PSII [10]. In this work we report further characterization of this mutant and describe progress in generation of cyt b559 site directed mutants in the green alga *C. reinhardtii*.

2. Procedure

The wild type (wt) and S6 mutant varieties of tobacco were grown aseptically in MS medium supplemented with 3% sucrose under low light (10/14 h light/dark regime) as described [9]. Whole plantlets (one month after sowing) were homogenized in the

P. Mathis (ed.), Photosynthesis: from Light to Biosphere, Vol. I, 927–930.

presence of buffer A (4 M sucrose, 50 mM MES pH 6.5, 5 mM $MgCl_2$ and 20 mM NaCl). Total shoot proteins were precipitated over-night in 90% acetone at -20°C, dried in vacuum and solubilized in electrophoresis sample buffer. SDS-PAGE, electrotransfer and immunodecoration were done by previously described methods [11]. Variable fluorescence was measured [11] either on whole leaves in buffer A or on total shoot homogenates. Thermoluminescence (TL) [11] was recorded after dark adaptation of the leaves for 5 min at 16-18°C, then cooling by a stream of liquid nitrogen and excitation by continuous white light during cooling from -2°C to -40°C [12].

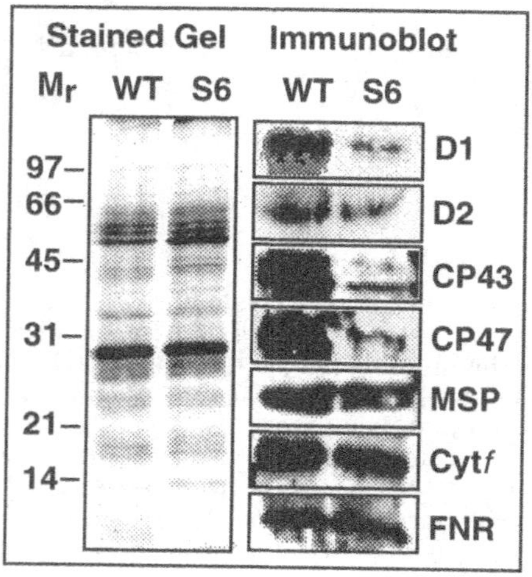

Figure 1. Comparison of thylakoid protein content in wt and S6 mutant. Proteins were separated by SDS-PAGE (left), electrotransfered to nitrocellulose filters and immunodecorated with specific antibodies (right).

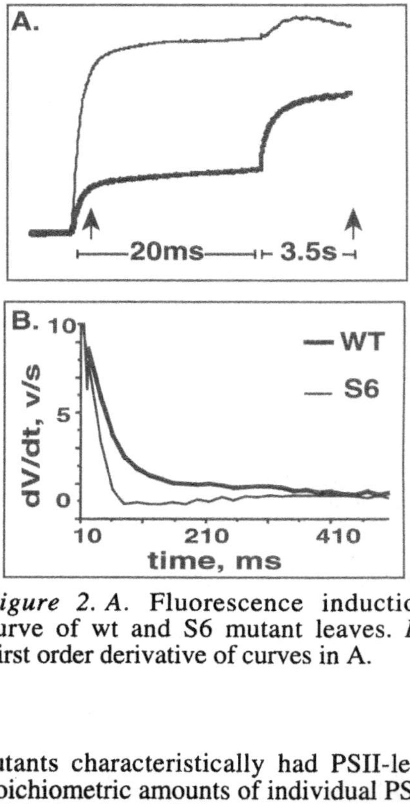

Figure 2. A. Fluorescence induction curve of wt and S6 mutant leaves. *B.* First order derivative of curves in A.

3. Results and Discussion

Many of the reviously reported cyt b559 mutants characteristically had PSII-less phenotypes and only residual and usually non-stoichiometric amounts of individual PSII subunits (e.g. [13-15]. Similarly, measurements of the levels of PSII core proteins present in homogenates of the S6 mutant and wt demonstrate a reduced content of the PSII D1, D2, CP43 and CP47 proteins in the S6 mutant relative to the wt strain as compared on an equal chlorophyll basis, (Fig. 1). However the major PSII antenna, the light harvesting complex II (LHCII) is present at slightly elevated amounts in the S6 mutant relative to the wt as judged by the Coomassie stained gel (Fig. 1). The increased levels of the chlorophyll *a/b* binding antenna proteins relative to the level of the PSII core proteins are also manifested by lower ratio of chlorophyll *a/b*. This ratio is 2.7 in the wt but only 1.9 in the mutant. While the relative content of the PSII intrinsic membrane proteins is drastically reduced in the mutant, the level of the peripheral membrane manganese stabilizing protein (MSP) of PSII is not significantly affected by the cyt b559 mutation (Fig. 1). Similar results have been were reported before for other PSII mutants [13, 16]

As a control the relative amounts of cytochrome *f* (cyt *f*) and ferredoxin-NADP reductase (FNR) were found to be nearly equal in the mutant and in the wt. Thus it seems that the F26S mutation results in a specific lowering of the PSII core intrinsic protein levels.

Because the S6 mutant is autotrophic [10] it was of interest to assess whether the residual amounts of the PSII subunits were assembled in a functional complex. Fluorescence induction measurements indicate that the S6 mutant has higher intrinsic fluorescence (F_o) and lower variable (F_v) fluorescence measured *in vivo* (Fig. 2A) as well as *in vitro* (data not shown). The F_v/F_o ratio of the mutant although low (10%) as compared with that of the wt indicate presence of functional PSII centers. This ratio may be low due to the high level of the initial fluorescence (F_o). Since an excess LHCII relative to the level of PSII subunits is evident in the mutant we interpret these results as an indication for disconnection of the LHCII from the PSII units. The high initial fluorescence F_o level can be explained by the chlorophyll fluorescence of the antennae. The slope of the fluorescence rise of the mutant during the initial 40 ms to 100 ms is significantly lower as compared with that of the wt leaves (Fig. 2B). This may be due to lower rate of plastoquinol reduction relative to its oxidation in the mutant as compared to the wt, in agreement with the S6 mutant having apparently fewer active PSII centers with the rest of the electron transport chain presumably intact (Fig 1). Alternatively these data may also indicate that the active PSII centers present in the mutant are functionally altered with lower rates of electron transport within the PSII reaction center.

Examination of the thermoluminescence glow curves (Fig. 3) reveals the presence of the TL B band ($Q_B^-/S_{2,3}$ charge recombination peak) in the S6 mutant as well as in the wt. However the B band of the mutant is down shifted by 7°C, from 26°C in the wt to 19°C. This substantiates the presence of functional but possibly photoinhibited PSII centers in the S6 mutant [12]. The shift in the TL B band, the high initial fluorescence and the possible disconnection of the LHCII from the PSII units in the mutant may indicate a chronic state of photoinhibiton. This may be in agreement with the low steady state levels of the PSII core polypeptides relative to the other thylakoid membrane proteins.

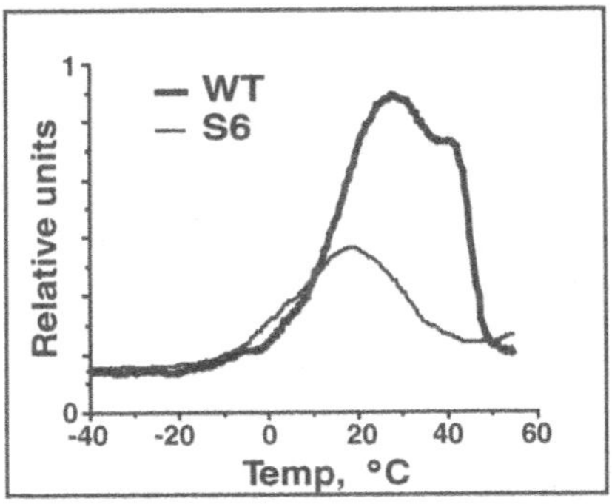

Figure 3. Thermoluminescence glow curves of wt and S6 mutant leaves.

The F26S mutation of the S6 mutant is the first described mutation in the hydrophobic domain of cyt b559 having an organized but functionally altered PSII. This study shows that the F26S mutation in the β subunit of cytochrome b559 of the S6 mutant has a substantial impact on the overall functionality of PSII and in particular may bring about its destabilization and hence the lower steady state level of the individual PSII subunits. However we cannot rule out the possibility that the cyt b559 defect has deleterious effects on the assembly process of PSII eventually inducing degradation of the misassembled PSII subunits resulting in fewer centers. We are currently investigating these hypotheses. The structural implications of the mutation are still obscure. The current model of cyt b559 [1, 2] suggests a polarity across the transmembrane α–helix of the β subunit with the surface of the helix opposite to the heme axial ligand histidine residue more hydrophilic than the other surface. The F26S mutation may disrupt the suggested directionality of the helix and possibly interfere with either one or both heme binding and interaction with adjacent PSII subunits. The aromatic side chain of F26 may also be involved as a conduit in the putative electron transfer between the cyt b559 heme and its redox partners, possibly the PSII primary electron donor, P680, or primary acceptor, Pheo [2]. The absence of this residue may reduce the rates of the electron transfer reactions making PSII more susceptible to photoinhibition.

In order to facilitate the understanding of the role that cyt b559 allegedly plays in the protection of PSII from photoinhibition, in parallel to the work on the tobacco S6 mutant we have created the same mutation in the *psbF* gene of the transformable eukariotic green alga *C. reinhardtii*. Several other site directed mutants of *C. reinhardtii*. are now in the process of investigation in our laboratory.

4. Acknowledgments

We are indebted to Dr. P. Maliga and Dr. R. Bock of Rutgers University for their kind gift of the S6 mutant and wt seeds. We thank Mr. Nir Keren for assisting in the TL measurements (supported by ISF grant A). This work was supported by a BARD grant awarded to H.B.P. and I.O.

5. References

1 Cramer, W.A., Tae, G.-S., Furbacher, P.N. and Böttger, M. (1993) Physiol. Plant. 88, 705-711
2 Whitmarsh, J. and Pakrasi, H.B. (1995) in Oxygenic Photosynthesis: The Light Reactions (Ort, D.R. and C.F. Yocum, ed), vol. 4, pp. In Press, Kluwer Academic, Dordrecht, The Netherlands
3 Babcock, G.T., Widger, W.R., Cramer, W.A., Oertling, W.A. and Metz, J.G. (1985) Biochemistry 24, 3638-3645
4 Nedbal, L., Samson, G. and Whitmarsh, J. (1992) Proc. Natl. Acad. Sci. USA 89, 7929-7933
5 Barber, J. and De Las Rivas, J. (1993) Proc. Natl. Acad. Sci. USA 90, 10942-10946
6 Herrmann, R.G., Alt, J., Schiller, Widger, W.R. and Cramer, W.A. (1984) FEBS Lett. 176, 239-244
7 Mor, T.S., Ohad, I., Hirschberg, J. and Pakrasi, H.B. (1995) Mol. Gen. Genet. 246, 600-604
8 Alizadeh, S., Nechushtai, R., Barber, J. and Nixon, P. (1994) Biochim. Biophys. Acta 1188, 439-442
9 Bock, R., Hagemann, R., Kössel, H. and Kundla, J. (1993) Mol. Gen. Genet. 240, 238-244
10 Bock, R., Kössel, H. and Maliga, P. (1994) EMBO J. 19, 4623-4628
11 Zer, H., Prasil, O. and Ohad, I. (1994) J. Biol. Chem. 269, 17670-17676
12 Ohad, I., Koike, H., Shochat, S. and Inoue, Y. (1988) Biochim Biophys Acta 933, 288-298
13 Pakrasi, H.B., De Ciechi, P. and Whitmarsh, J. (1991) EMBO J. 10, 1619-1627
14 Shukla, V.K., Stanbekova, G.E., Shestakov, S.V. and Pakrasi, H.B. (1992) MolMicrobiol 6, 947-956
15 Tae, G.S. and Cramer, W.A. (1992) Biochemistry 31, 4066-74
16 Nilson, F., Andersson, B. and Jansson, C. (1990) Plant Mol. Biol. 14, 1051-1054

3. Evolution of photosynthesis

ORIGIN AND EVOLUTION OF PHOSPHORYLATING AND NON-PHOSPHORYLATING GLYCERALDEHYDE-3-PHOSPHATE DEHYDROGENASES

Rüdiger Cerff
Institut für Genetik, Technische Universität Braunschweig,
Spielmannstr. 7, D-38106 Braunschweig, Germany

I. Plant glyceraldehyde-3-phosphate dehydrogenases (class I & III)

Chloroplast and cytosolic phosphorylating GAPDHs play key roles in the energy metabolism of plants. They are marker enzymes of two metabolic pathways, the photosynthetic Calvin cycle and glycolysis, sharing a number of reactions which are catalyzed by distinct (iso-) enzymes unique to each (Figure 1). In angiosperms Calvin cycle GAPDH (EC 1.2.1.13) is composed of two isoenzymes, an A_2B_2 heterotetramer and an A_4 homotetramer [1]. Subunit B (GapB) differs from subunit A (GapA) by about 20 % of its amino acid residues and has a slightly larger molecular weight owing to a flexible and highly charged C-terminal extension of about 30 additional amino acids conserved in all higher plants [2]. This C-terminal tail of unknown function is susceptible to proteolytic cleavage and is not essential for catalytic activity [3]. GapA and GapB are encoded in the nucleus and are subject to transcriptional activation by light [4, 5].

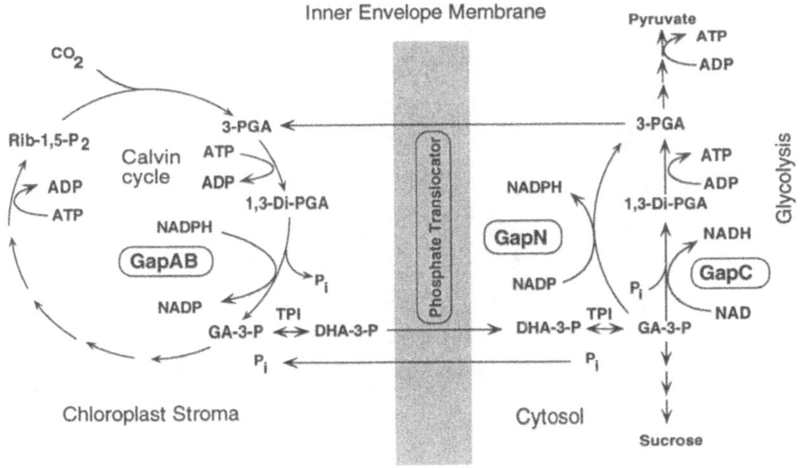

Figure 1. Compartmentalization and catalytic functions of plant GAPDHs. The reaction of non-phosphorylating GapN is irreversible. For phosphorylating GapAB and GapC, which catalyze the reversible oxidation of glyceraldehyde-3-phosphate (GA-3-P), only the directions dominating *in vivo* are indicated. Triose phosphates (GA-3-P and its isomer DHA-3-P generated by TPI) synthesized in the Calvin cycle or during starch degradation in plastids are exported to the cytosol in exchange for inorganic phosphate *via* the phosphate translocator [6]. Abbreviations used: DHA-3-P, dihydroxyacetone phosphate; 3-PGA, 3-phosphoglycerate; 1,3-Di-PGA, 1,3-bisphosphglycerate; Rib-1,5-P2, ribulose-1,5-bisphosphate; TPI, triosephosphate isomerase.

P. Mathis (ed.), Photosynthesis: from Light to Biosphere, Vol. I, 933–938.

Cytosolic GAPDH of glycolysis (GapC, EC 1.2.1.12) from angiosperms is a tetramer of identical or electrophoretically distinguishable subunits [1, 7] encoded by multiple genes which are expressed constitutively or stimulated by environmental stress factors such as anaerobiosis, heat shock and salinity [8-11].

All eukaryotic GapA and GapB proteins are chloroplast specific, while eukaryotic GapC enzymes usually function in the glycolytic pathway of the cytosol. Trypanosomes are exceptional in that they contain two largely different GapC enzymes, one located in the cytosol and the other in a microbody like organell called the glycosome harboring the major part of glycolysis from glucose to 3-phosphoglycerate in these organisms [12]. Recently, a new *GapC* gene was discovered in Scots pine (*Pinus sylvestris*) encoding a GapC protein with a transit peptide which is targeted to chloroplasts [13]. This nuclear *GapCp* gene is highly expressed and has a wide distribution in gymnosperms and ferns. It encodes a NAD-specific GAPDH of unknown function which, however, may play a potential role in carbohydrate catabolism of the chloroplast (chlororespiration, see [14]). Also entirely different functions may be envisaged, since GAPDH has been implicated in a number of other cellular activities unrelated to glycolysis and photosynthesis such as nuclear export of tRNAs [15] DNA repair [16] and protein phosphorylation [17]. *GapCp* is only marginally more closely related to *GapC* genes from green plants than to the *GapC* gene of red algae [18] suggesting that *GapCp* originated early in plant evolution, probably in green algae in striking parallel to the *GapA/GapB* gene duplication.

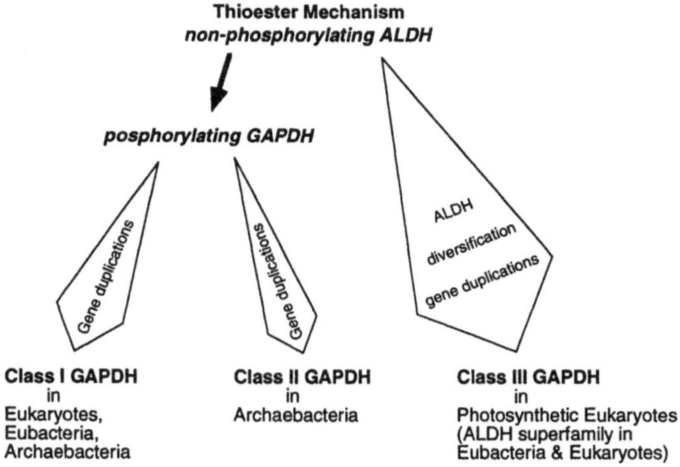

Figure 2. Schematic outline of enzyme relationships and potential evolutionary origins of glyceraldehyde-3-phosphate dehydrogenases. Three different types of GAPDH can be distinguished, one of which, class III, is a non-phosphorylating aldehyde dehydrogenase (ALDH) with no sequence similarity to its phosphorylating counterparts, class I and class II. Substrate specificity of class III GAPDH for glyceraldehyde-3-phosphate probably originated independently by convergent evolution, while its basic catalytic mechanism (hydride transfer and thioester formation) may possibly be homologous to that of phosphorylating GAPDH (see text).

Non-phosphorylating GAPDH (GapN) was discovered 40 years ago by Arnon et al. [19]. It catalyzes an irreversible reaction (see Figure 1) and seems to be restricted to the cytosol of photosynthetic eukaryotes including algae [20] where it generates NADPH for biosynthetic processes from triose phosphates exported from photosynthesizing chloroplasts [21]. The recent cloning and characterization of GapN cDNAs [22] led to the surprising discovery that non-phosphorylating GAPDH belongs to the aldehyde dehydrogenases (ALDH) which have no sequence homology to phosphorylating GAPDH. The ALDH superfamily comprises multiple homologous enzymes from eubacteria and eukaryotes which are specific or non-specific for particular aldehyde substrates. A phylogenetic analysis suggests that ALDH diversity reflects an early phase of intense metabolic radiation antedating the separation of pro- and eukaryotes [22].

These findings firmly establish that GAPDH originated at least twice independently in evolution leading to three separate classes of enzymes as depicted in Figure 2 (class I, class II and class III). Although unrelated at the sequence level phosphorylating and non-phosphorylating GAPDH resemble one another with respect to catalytic hydride transfer and covalent thioester formation. Hence, phosphorylating and non-phosphorylating thioester mechanisms may have a common evolutionary origin suggesting that GAPDH originated after ALDH function and prior to the separation into two distinct GAPDH classes I and II (Figure 2). The two classes share 10 to 15 % amino acid identity and were thought to be restricted to eubacteria/eukaryotes (class I) and archaebacteria (class II), respectively [23]. The recent discovery of a class I GAPDH in the archaebacterium *Haloarcula vallismortis* [24] and the characterization of the corresponding operon (unpublished) suggest that also in the case of class I GAPDH the diversification of enzymes may have occurred prior to the separation of primary life forms.

II. Chloroplast and cytosolic GAPDHs (class I) are of eubacterial origin.

Although plant GAPDH genes *GapA*, *GapB*, *GapC*, *GapCp* and *GapN* are all encoded in the nucleus, it was suggested already some time ago that genes *GapA/GapB* encoding the Calvin cycle enzyme may be derived from the endosymbiotic genome of the chloroplast ancestor [25-27]. The characterization of the first GAPDH genes from cyanobacteria confirmed this prediction and, in addition, provided unexpected evidence for an endosymbiotic (eubacterial) origin also of *GapC* [28].

On the GAPDH tree all eukaryotic *GapC* and *GapAB* lineages appear as terminal branches grafted on two major eubacterial branches corresponding to *gap1* (= *gapC*) and *gap2* (= *gapA*), respectively. The highly divergent genes *gap1* and *gap2* are present in all cyanobacteria so far investigated (*Anabaena*, *Synechocystis*, *Synechococcus*) and an apparent orthologous gene *gap1* is also present in *E. coli*. This clearly suggests that eukaryotic genes *GapC* and *GapAB* are of eubacterial origin and became fixed in nuclear genomes *via* endosymbiotic gene translocations. Since *gap1* of the purple bacterium *E. coli* is more closely related to *GapC* genes from higher eukaryotes than to cyanobacterial *gap1* and, furthermore, purple bacteria probably gave rise to mitochondria [29], we have previously suggested that *GapC* of plants, animals and fungi may be of mitochondrial origin [18, 28]. In light of the new GAPDH data from trypanosomes [12] and *Euglena* [30] this hypothesis may now be integrated into a more complex scenario suggesting a biphyletic origin of the modern eukaryotic cell containing plastids and/or mitochondria. This model supposes that *GapC* of kinetoplastids/*Euglena* and probably also *Euglena GapA* were acquired independently of *GapC** (+mitochondria) and *GapA** (+chloroplasts) of higher eukaryotes (see Figure 3 and review [31]).

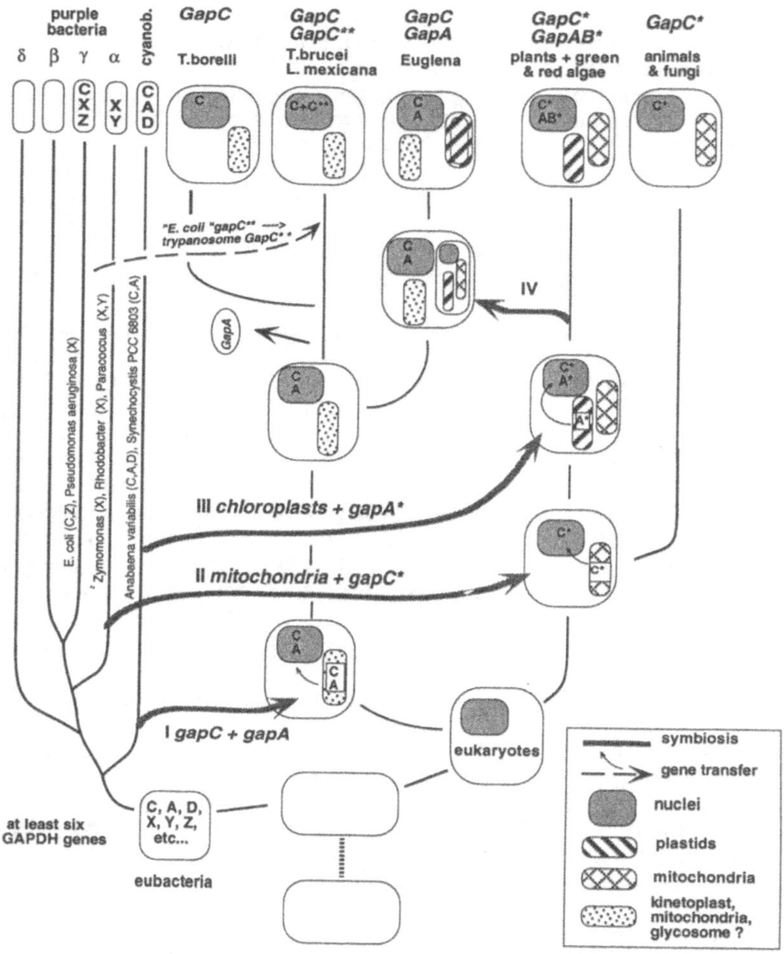

Figure 3. Evolutionary scenario explaining the distribution of GAPDH genes in present-day eubacteria and eukaryotes in terms of three independent primary endosymbioses giving rise to two separate lineages leading to trypanosomes/*Euglena* (genes *GapC* and *GapA*) and higher eukaryotes (genes *GapC** and *GapA**), respectively. Gene *GapC*** was acquired recently by trypanosomes from an *E. coli* like ancestor probably in an non-endosmybiotic context. Endosymbiosis IV: Origin of the *Euglena* chloroplast *via* incorporation of a eukaryotic endosymbiont possibly related to green algae [29, 32]. The schematic eubacterial tree on the left showing the branching order of purple bacteria (alpha, beta, gamma) and cyanobacteria is based on the 16S rRNA phylogeny of Woese [33]. There are six or more different paralogous GAPDH genes in present-day eubacteria showing between 50 and 60 % amino acid sequence divergence in pairwise comparisons. Cyanobacteria contain types C, A and D and share type C with *E. coli*. Types X and Y occur in alpha-purple bacteria and types C, X and Z in gamma-purple bacteria [18, 31].

III. References

1. Cerff R: Separation and purification of NAD- and NADP-linked glyceraldehyde-3-phosphate dehydrogenases from higher plants. In: Edelman M, Hallick RB, Chua NH (eds) Methods in Chloroplast Molecular Biology, 683-694. Elsevier Biomedical Press, Amsterdam (1982).
2. Brinkmann H, Cerff R, Salomon M, Soll J: Cloning and sequence analysis of cDNAs encoding the cytosolic precursors of subunits GapA and GapB of chloroplast glyceraldehyde-3-phosphate dehydrogenase from pea and spinach. Plant Mol Biol 13: 81-94 (1989).
3. Zapponi MC, Iadarola P, Stoppini M, Ferri G: Limited proteolysis of chloroplast glyceraldehyde-3-phosphate dehydrogenase (NADP) from *Spinacia oleracea.* Biol Chem Hoppe-Seyler 374: 395-402 (1993).
4. Dewdney J, Conley TR, Shih M-C, Goodman HM: Effects of blue and red light on expression of nuclear genes encoding chloroplast glyceraldehyde-3-phosphate dehydrogenase of *Arabidopsis thaliana.* Plant Physiol 103: 1115-1121 (1993).
5. Cerff R, Kloppstech K: Structural diversity and differential light control of mRNAs coding for angiosperm glyceraldehyde-3-phosphate dehydrogenases. Proc Natl Acad Sci USA 79: 7624-7628 (1982).
6. Stitt M: The flux of carbon between the chloroplast and the cytosol. In: Dennis DT, Turpin DH (eds) Plant Physiology, Biochemistry and Molecular Biology, 309-326. Longman, Singapore (1990).
7. Russell DA, Sachs MM: The maize cytosolic glyceraldehyde-3-phosphate dehydrogense gene family: organ-specific expression and genetic analysis. Mol Gen Genet 229: 219-228 (1991).
8. Ostrem JA, Vernon DM, Bohnert HJ: Increased expression of a gene coding for NAD:glyceraldehyde-phosphate dehydrogenaseduring the transition from C3 photosynthesis to crassulacean acid metabolism in *Mesembryanthemum crystallinum.* J Biol Chem 265: 3497-3502 (1990).
9. Yang Y, Kwon H-B, Peng H-P, Shih M-C: Stress responses and metabolic regulation of glyceraldehyde-3-phosphate dehydrogenase genes in *Arabidopsis*. Plant Physiol 101: 209-216 (1993).
10. Russell DA, Sachs MM: Differential expression and sequence analysis of the maize glyceraldehyde-3-phosphate dehydrogenase gene family. Plant Cell 1: 793-803 (1989).
11. Martinez P, Martin W, Cerff R: Structure, evolution and anaerobic regulation of a nuclear gene encoding cytosolic glyceraldehyde-3-phosphate dehydrogenase from maize. J Mol Biol 208: 551-565 (1989).
12. Michels PA, Marchand M, Kohl L, Allert S, Wierenga RK, Opperdoes FR: The cytosolic and glycosomal isoenzymes of glyceraldehyde-3-phosphate dehydrogenase in *Trypanosoma brucei* have a distant evolutionary relationship. Eur J Biochem 198: 421-428 (1991).
13. Meyer-Gauen G, Schnarrenberger C, Cerff R, Martin W: Molecular characterization of a novel, nuclear-encoded, NAD-dependent glyceraldehyde-3-phosphate dehydrogenase in plastids of the gymnosperm *Pinus sylvestris* L. Plant Mol Biol 26: 1155-1166 (1994).
14. Vermiglio A, Ravenel J, Peltier G: Chlororespiration: a respiratory activity in the thylakoid membrane of microalgae and higher plants. In: Wiessner W, Robinson DG, Starr RC (eds) Cell Walls and Surfaces, Reproduction, Photosynthesis, 188-205. Springer Verlag, Berlin (1990).
15. Singh R, Green MR: Sequence-specific binding of transfer RNA by glyceraldehyde-3-phosphate dehydrogenase. Science 259: 365-368 (1993).

16. Meyer-Siegler K, Mauro DJ, Seal G, Wurzer J, DeRiel JK, Sirover MA: A human nuclear uracil DNA glycosylase is the 37-kDa subunit of glyceraldehyde-3-phosphate dehydrogenase. Proceedings of the National Academy of Sciences USA 88: 8460-8464 (1991).
17. Kawamoto RM, Caswell AH: Autophosphorylation of glyceraldehyde-3-phosphate dehydrogenase and phosphorylation of protein from skeletal muscle microsomes. Biochemistry 25: 656-661 (1986).
18. Liaud M-F, Valentin C, Martin W, Bouget F-Y, Kloareg B, Cerff R: The evolutionary origin of red algae as deduced from the nuclear genes enconding cytosolic and chloroplast glyceraldehyde-3-phosphate dehydrogenases from *Chondrus crispus*. J Mol Evol 38: 319-327 (1994).
19. Arnon DI, Rosenberg LL, Whatley FR: A new glyceraldehyde phosphate dehydrogenase from photosynthetic tissues. Nature 173: 1132-1134 (1954).
20. Mateos MI, Serrano A: Occurence of phosphorylating and non-phosphorylating NADP-dependent glyceraldehyde-3-phosphate dehydrogenases in photosynthetic organisms. Plant Sci 163-170 (1992).
21. Kelly GJ, Gibbs M: Non-reversible D-glyceraldehyde 3-phosphate of plant tissue. Plant Physiol 52: 111-118 (1973).
22. Habenicht A, Hellman U, Cerff R: Non-phosphorylating GAPDH of higher plants is a member of the aldehyde dehydrogenase superfamily with no sequence homology to phosphorylating GAPDH. J Mol Biol 237: 165-171 (1994).
23. Hensel R, Zwickl P, Fabry S, Lang J, Palm P: Sequence comparison of glyceraldehyde-3-phosphate dehydrogenases from the three urkingdoms: evolutionary implication. Can J Microbiol 35: 81-85 (1989).
24. Prüß B, Meyer HE, Holldorf AW: Characterization of the glyceraldehyde 3-phosphate dehydrogenase from the extremely halophilic archaebacterium *Haloarcula vallismortis*. Arch Microbiol 160: 5-11 (1993).
25. Brinkmann H, Martinez P, Quigley F, Martin W, Cerff R: Endosymbiotic origin and codon bias of the nuclear gene for chloroplast Glyceraldehyde-3-phosphate dehydrogenase from maize. J Mol Evol 26: 320 - 328 (1987).
26. Shih M-C, Lazar G, Goodman HM: Evidence in favor of the symbiontic origin of chloroplasts: Primary structure and evolution of tobacco glyceraldehyde-3-phosphate dehydrogenase. Cell 47: 73-80 (1986).
27. Martin W, Cerff R: Prokaryotic features of a nucleus-encoded enzyme. cDNA sequences for chloroplast and cytosolic glyceraldehyde-3-phosphate dehydrogenases from mustard (*Sinapis alba*). Eur J Biochem 159: 323-331 (1986).
28. Martin W, Brinkmann H, Savona C, Cerff R: Evidence for a chimeric nature of nuclear genomes: Eubacterial origin of eukaryotic glyceraldehyde-3-phosphate dehydrogenase genes. Proc Natl Acad Sci USA 90: 8692-8696 (1993).
29. Gray MW: The endosymbiont hypothesis revisited. Int Rev Cytol 141: 233-357 (1992).
30. Henze K, Badr A, Wettern M, Cerff R, Martin W: A nuclear gene of eubacterial origin in *Euglena gracilis* reflects cryptic endosymbioses during protist evolution. Proc Natl Acad Sci USA in press (1995).
31. Cerff R: The chimaeric nature of nuclear genomes and the antiquity of introns as demonstrated by the GAPDH gene system. In: Go M (eds) *Tracing Biological Evolution in Protein and Gene Structures*. Proceedings of the 20th Taniguchi International Symposium (Nagoya, Japan), in press. Elsevier Science, Amsterdam (1995).
32. Martin W, Somerville CC, Loiseaux-DeGoër S: Molecular phylogenies of plastid origins and algal evolution. J Mol Evol 35: 385-404 (1992).
33. Woese CR: Bacterial evolution. Microbiol Rev 51: 221-271 (1987).

PLASTID ORIGINS

Christopher J. Howe[1], Adrian C. Barbrook[1] & Peter J. Lockhart[2]
[1] Department of Biochemistry, University of Cambridge, Tennis Court Road, Cambridge, CB2 1QW UK, [2]Molecular Genetics Unit, Massey University, Palmerston North, New Zealand

1. Introduction

There is now little disagreement that plastids arose from symbiosis between photosynthetic prokaryotes and non-photosynthetic hosts, although some present-day plastids may be derived subsequently by secondary endosymbiosis in which a non-photosynthetic host acquired a plastid from an existing photosynthetic eukaryote (1). However, there is still considerable disagreement over whether all plastids are derived from the same photosynthetic endosymbiotic ancestor (often referred to as a "monophyletic" origin) or from different ones (often referred to as a "polyphyletic" origin). There are three main groups of plastids whose origins must be considered. These are (i) the "chlorophyte" lineage, characterized by the possession of chlorophylls *a* and *b*, and including green algae and land plants; (ii) the "rhodophyte" lineage, characterized by the possession of chlorophyll *a* and phycobiliprotein, and including the red algae; and (iii) the "chromophyte" lineage, characterized by the possession of chlorophylls *a* and *c* (and in some cases phycobiliprotein), and including diverse groups of brown and other algae.

There is also disagreement over the evolutionary importance of the prochlorophytes, oxygenic photosynthetic bacteria containing chlorophylls *a* and *b* (and shown more recently to contain a chlorophyll *c*-like pigment [2]) and lacking phycobiliprotein. Some workers argue that the prochlorophytes do not share common ancestry with chlorophyte plastids to the exclusion of cyanobacteria (oxygenic photosynthetic prokaryotes containing chlorophyll *a* and phycobiliprotein) and are in fact polyphyletic (3,4); others dispute this (5).

Three main kinds of molecular data have been used in trying to resolve questions of plastid origins. These are (i) the presence or absence of particular proteins; (ii) gene organization; and (iii) DNA and protein sequences. However, there are difficulties in interpretation of the data that we believe have not been adequately addressed. We discuss these difficulties in this paper, and suggest possible solutions, where they exist. We concentrate on the first of the questions above, whether plastids are monophyletic or polyphyletic, although the same principles are relevant to the second.

P. Mathis (ed.), Photosynthesis: from Light to Biosphere, Vol. I, 939–944.

2. Molecular data

2.1 Presence or absence of proteins

The demonstration that the rhodophyte alga *Porphyridium cruentum* contained a light-harvesting chlorophyll-binding protein immunochemically related to the LHC proteins of green plants, and the failure to detect such a protein in cyanobacteria were taken by Wolfe *et al.* to indicate a monophyletic origin of rhodophyte and chlorophyte plastids (6). The occurrence of an immunochemically related protein had also been reported earlier from the chromophyte *Phaeodactylum* (7), and it had been proposed that homologues did exist in cyanobacteria (1). Dolganov *et al.* (1995) succeeded in demonstrating the existence of cyanobacterial homologues (8), and hence the presence of LHC homologues must therefore be regarded as equally consistent with a polyphyletic or a monophyletic origin of plastids.

2.2 Gene organization

The clustering of certain groups of genes in different plastids that are not clustered elsewhere does provide evidence in support of a monophyletic origin. Examples of this are the *rpo/rps2/atp* and *psbB/N/H* genes, which are clustered in rhodophyte and chlorophyte plastids, but not in cyanobacteria (9). However, in the absence of an explicit probabilistic model for plastid genome rearrangements, the significance of these data is difficult to quantify (10, 11), particularly since it is known that certain areas of chloroplast (and possibly other plastid) genomes act as "hotspots" of recombination and mutation (12, 13), which may increase the probability of similar groupings being generated independently. A further difficulty with analyses of these data is that for practical reasons they are generally based on only a few selected taxa.

Similarities of gene organization between plastids and cyanobacteria to the exclusion of non-oxygenic organisms, such as the presence of a group I intron in a tRNALeu(UAA) gene of plastids and cyanobacteria, have been taken by some authors to imply an origin of plastids within the cyanobacteria (e.g. 14). This represents an over-interpretation of the data, which would be equally consistent with plastids being a sister group to the cyanobacteria. The appropriate inference is that oxygenic photosynthesis evolved once and that oxygenic photosynthesisers constitute a monophyletic group.

2.3 Sequence data

Even so, some analyses of primary sequence data do not support even this suggestion. Inferences from analyses of *rbcLS* sequences have implied that the chromophyte and rhodophyte plastids originated in the beta-group of purple photosynthetic bacteria (e.g. 15) and that the prochlorophytes, higher plants and cyanelle have evolved from the gamma-group of purple photosynthetic bacteria (15, 16). This would imply that the ability to oxidize water and produce oxygen developed more than once in evolution. However, this analysis conflicts with the results from other sequences and also the demonstration that a polypeptide immunochemically related to the 33 kDa extrinsic "oxygen evolving" polypeptide of Photosystem II of higher plant plastids exists in rhodophytes, chromophytes, prochlorophytes and cyanobacteria (17). The anomalous results of the *rbcLS* analyses are probably caused by comparing genes that are in fact paralogous.

Most analyses of other genes have indicated an origin for plastids among oxygenic photosynthetic bacteria. However, while some imply a monophyletic origin for all plastids, others have implied polyphyly (1, 18, 19). To date, the largest collection of coding sequences analysed from photosynthetic taxa is of *tufA* sequences (14). When all the sequences were analysed using a maximum likelihood method, weak support was found for a monophyletic origin of plastids from within the cyanobacteria. By contrast, analyses using parsimony or distance matrix methods suggested a polyphyletic origin of plastids, with chloroplasts forming a sister group to the cyanobacteria (14).

Further analyses we have carried out of this and other datasets question the support they appear to provide for the monophyletic hypothesis. For example under Neighbor Joining (on observed differences at first and second codon positions) with the *tufA* dataset we find 87% bootstrap support for the existence of the edge separating the plastids *Cryptomonas, Cyclotella, Porphyra, Cyanophora, Euglena and Astasia* from the bacteria *Anacystis, Gloeobacter, Gloeothece, Phormidium, Plectonema, Prochlorothrix,* and *Spirulina*. However, when either the higher plant (nuclear encoded) *Arabidopsis* or *Glycine* sequence is included in the same analysis, support for the monophyletic topology drops dramatically (to 63% or 53% respectively). In contrast, support changes little if an extra (plastid encoded) sequence such as *Chlamydomonas* or *Codium* is included in the analysis.

The questions raised are why different analyses should support different hypotheses, and why we should find considerably less support for the monophyletic hypothesis, if it is correct, when one or more taxa are additionally sampled. This is contrary to expectations and indicates that the models of sequence evolution underlying the analysis may be inappropriate (20).

We believe that two important problems particularly interfere in sequence analyses used to infer plastid origins: non-random nucleotide composition ("substitutional bias") and variations in the patterns of the sites free to vary in eubacterial and plastid sequences. As we discuss, both need to be considered together.

3. Substitutional Bias

3.1 The problem

A feature of plastid genomes (of chlorophytes, rhodophytes and chromophytes) is that they are generally much more AT-rich than those of oxygenic prokaryotes. It can be inferred, therefore, that the process of nucleotide substitution over time is asymmetric, tending to substitute GC base pairs with AT ones. The models of sequence evolution that have been used until very recently (see below) do not take account of this, and sequences that have been subjected to biased substitution will seem artifactually similar and tend to be grouped together in a phylogenetic analysis. The problem this causes when analysing plastid origins was first recognized in analyses of the AT-rich cyanelle genome of *Cyanophora paradoxa* (18). It is important to note that the fraction of sites in a sequence that actually influence the tree topology may be small, so the problem may exist even if sequences have base compositions that are not observed to be greatly dissimilar when all positions are considered. When studying the effect at 1st+2nd codon positions we now typically examine the base composition at either the parsimony sites or variable sites in these positions (e.g. 21).

3.2 Dealing with substitutional bias

Few analyses of plastid origins have tried to take account of the problem of substitutional bias. One approach has been to use protein sequences rather than DNA sequences, on the grounds that they will not be affected by the bias (e.g. 14). However, it is clear that in some datasets, the first and second codon positions are affected and that bias can affect protein sequences (e.g.18, 22, 23). A second approach is to attempt to correct the branch lengths within trees for the effect of biases (24).

Another approach has been to compare the trees obtained with plastid encoded sequences with trees based on sequences which are plastid encoded in some lineages, but nuclear encoded in others (18). A change in the position of a taxon according to whether it is represented by a plastid or a nuclear sequence may indicate that it is being influenced by bias. An example of a sequence whose location varies is *secA*, which encodes a large protein involved in the translocation of polypeptides across the thylakoid membrane of photosynthetic bacteria and plastids. This gene is in the plastid of rhodophytes and chromophytes, but in the nucleus of higher plants. We have determined partial gene sequences for a number of oxygenic bacteria, and used these in conjunction with sequences published for plastid homologues to generate phylogenetic trees. Detailed analysis of the tree topologies generated using different methods is still in progress, but a representative tree is shown in Figure 1.

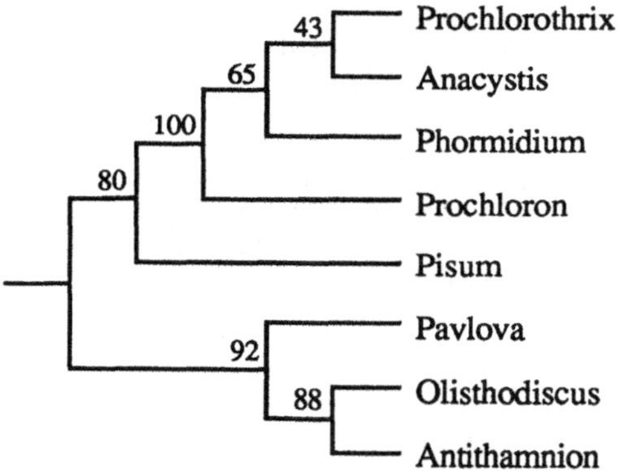

Figure 1. Majority rule consensus tree (Fitch Margoliash, 100 bootstraps) of *secA* sequences (227 amino acids) from prokaryotes, chlorophytes (*Pisum*), rhodophytes and chromophytes (*Pavlova, Olisthodiscus, Antithamnion*). ACB, unpublished.

This reconstruction places the chlorophyte (where the sequence is nuclear encoded) as a sister group to one containing the photosynthetic bacteria and rhodophyte/chromophyte plastids. A protein parsimony analysis gave a broadly similar result. This is comparable with some of the results from the *tufA* analysis, and adds support to the proposal that AT-bias is at least in part responsible for support given by some analyses to a monophyletic origin.

The fourth and most sophisticated approach is to transform the observed differences between all pairs of compared sequences by taking the logarithm of the determinant of the divergence matrix for each. This results in a set of path estimates which can, under different tree selection criteria, e.g. neighbor joining or split decomposition, identify the correct tree even when different lineages have been evolving under different substitutional biases (21, 25, 26). We have applied this approach to *atpB* and *tufA* datasets. Figure 2 shows our result with *atpB* sequences. It is similar in structure to that found with *tufA* sequences when higher plant sequences are omitted from the analysis and the data are untransformed. That is, there is still a small amount of apparent evidence for an edge which separates plastids and oxygenic prokaryotes. Determining whether this is actually evidence for monophyly or not requires an understanding of the pattern of sites free to vary in these sequences.

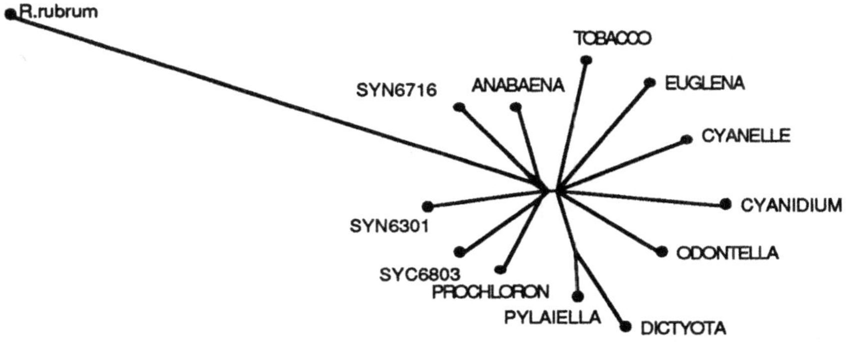

Figure 2 . SplitsTree graph obtained using LogDet on *atpB* sequences (1st & 2nd codon positions). SYN: *Synechococcus sp.*; SYC: *Synechocystis sp.* PCC numbers. Resolution between taxa with different LHCs was not improved by adding other available plastid sequences or by removing either taxa or constant sites in the analysis.

4. Sites free to vary

It is well recognised that the pattern of sites free to vary can differ in different taxonomic groups for the same coding sequences (27). Even without substitutional biases such patterns can mislead evolutionary tree reconstruction (Lockhart et al., submitted). The problem is illustrated below in a simple example with only two character states, G and C.

```
Outgroup      GGGGG
Cyano 1       CGGGG
Cyano 2       CGGGG
Cyano 3       CGGGG
Plastid 1     CGCGC
Plastid 2     CGCCG
```

Positions 3-5 are free to vary and site saturated (randomized) in plastids. They are not free to vary in cyanobacteria. By chance, patterns will arise (e.g. at position 3) that link the plastids to the exclusion of cyanobacteria. This is essentially an extreme example of the Felsenstein zone (10). The tendency for spurious patterns to arise linking the plastids will be greater if they also have a similar substitutional bias. With the *atpB* alignment we estimate 33% of codons are free to vary within the oxygenic prokaryotes, 43% within plastids and 48% within oxygenic prokaryotes and plastids. Similar patterns are found with *tufA* sequences. We are now seeking to determine whether such a difference is sufficent to account for the presence of the central edge between plastids and bacteria.

5. Conclusions

It is clear that phenomena such as substitutional bias and differences in the number and distribution of sites free to vary can in theory impose artificial patterns on phylogenetic trees based on sequence data. Support for the suggestion that these phenomena are responsible for the edges separating plastids from photosynthetic bacteria that are identified in some analyses comes from the application of the LogDet transformation, and the analysis of datasets such as *tufA* and *secA*, where the plastid sequences are not all plastid encoded. It remains to be seen whether there is in fact sufficient historical information left in the sequences to allow reliable trees to be constructed.

6. Acknowledgements

We thank A. Larkum, J. Packer and the FARSIDE systematics group for helpful discussions, C. Delwiche for providing his *tufA* alignment prior to publication, and D. Huson and R. Wetzel for providing SPLITSTREE V1.0.

References

1 Howe, C.J. *et al.* (1995) Trends Ecol. Evol. 7, 378-383
2 Larkum, A.W.D. (1994) Proc. Natl Acad. Sci. USA 91, 679-683
3 Palenik, B. and Haselkorn, R. (1992) Nature 355, 265-266
4 Urbach, E. *et al.* (1992) Nature 355, 267-270
5 Lockhart, P.J. *et al.* (1993) Photosynthesis Res. 37, 61-68
6 Wolfe, G.R. *et al.* (1994) Nature 367, 566-568
7 Grossman, A. *et al.* (1990) Mol. Gen. Genet. 224, 91-100
8 Dolganov *et al.* (1995) Proc. Natl Acad. Sci. USA 92, 636-640
9 Reith, M. and Munholland, J. (1993) Plant Cell 5, 465-475
10 Felsenstein J. (1988) Annu. Rev. Genet. 22, 521-565
11 Beanland, T.J. and Howe, C.J. (1992) Comp. Biochem. Physiol. 102, 643-659
12 Howe, C.J. *et al.* (1988) Curr. Genet. 13, 343-349
13 Clegg, M.T. *et al.* (1994) Proc. Natl Acad. Sci. USA 91, 6795-6801
14 Delwiche, C.F. *et al.* (1995) Mol. Phyl. Evol. 4, 110-128
15 Morden, C.W. and Golden, S.S. (1991) J. Mol. Evol. 32, 379-395
16 Shimada A. *et al.* (1995) J. Mol. Evol. 40, 671-677
17 Fairweather M.S. *et al.* (1994) Biochem. Biophys. Res. Commun. 205, 1497-1502
18 Lockhart, P.J. *et al.* (1992) J. Mol. Evol. 34, 153-162
19 Lockhart, P.J. *et al.* (1992) FEBS Letts 301, 127-131
20 Penny, D. *et al.* (1994) in "Models in Phylogeny" pp.211-230 OUP, Oxford, UK
21 Lockhart, P.J. *et al.* (1994) Mol. Biol. Evol. 11, 605-612
22 Olsen, G.J. *et al.* (1994) J. Bacteriol. 176, 1-6
23 Sueoka, N. (1961) Proc. Natl Acad. Sci. USA 47, 1141-1149
24 Steel, M.A. *et al.* (1993) Nature 364, 440-442
25 Lake, J.A. (1994) Proc. Natl Acad. Sci. USA 91, 1455-1459
26 Steel, M.A. (1994) Appl. Math. Lett. 7, 19-23
27 Miyamoto, M.M. and Fitch, W.M. (1995) Mol. Biol. Evol. 12, 503-513

FROM NAPHTHO- TO BENZOQUINONES - (R)EVOLUTIONARY REORGANISATIONS OF ELECTRON TRANSFER CHAINS

Nitschke, W.[1,5], Kramer, D.M.[1,2], Riedel, A.[3] & Liebl, U.[4]
[1] Biologie II, University of Freiburg, Schänzlestr. 1, 79104 Freiburg, FRG.
[2]Dept. Biophys., Univ. of Illinois/USA
[3]Inst. Biophysics and Phys. Biochem., 93040 Regensburg, FRG.
[4]LOA, Ecole Polytechnique-ENSTA, 91120 Palaiseau Cedex, France
[5]BIP-CNRS, 13402 Marseille Cedex 20, France

1. Introduction

During recent years, studies addressing evolution of bioenergetic electron transport chains mainly focussed on the evolutionary pathways of the enzymes involved, such as (a) ATPases (1), (b) photosynthetic RCs (2-5), (c) oxidases (6) or most recently cytochrome bc complexes (7-10).
However, in addition to the major events concerning molecular evolution of the enzymes mentioned above, a reorganisation of the thermodynamic properties of complete electron transport chains must have occured in response to switches in quinone content during evolution. During such changes, species apparently swapped between quinones that significantly differ from each other with respect to redox midpoint potential. The evolutionary aspects of the respective transition(s) will be discussed below.

2. Distribution of quinones within the kingdom of the Bacteria.

The phylogenetic relationships between species of the kingdom Bacteria as based on 16 S r-RNA comparisons (11) are shown in Fig.1a. The presence of a survey of quinone content in microorganisms (12) allows to label most branches of this phylogenetic tree by the quinone content of the species in the respective phylum (Fig.1b). The majority of branches consists of species containing exclusively or predominantly 2-methyl-1,4-naphthoquinone ("menaquinone"). Only proteobacteria (formerly referred to as purple bacteria and relatives) and cyanobacteria contain 2,3-dimethyl- or 2,3-dimethoxy-5-methyl- substituted 1,4-
benzoquinones, i.e. plastoquinone and ubiquinone. This pattern of distribution of quinones differs significantly from that obtained for other evolutionary markers, such as type of reaction centre (photosystem 1/RCI- or photosystem 2/RCII-like, ref.13, Fig. 1c) or connectivity of the cytochrome b subunit in cytochrome bc complexes (Fig.1d).
The evolutionary events affecting the mentioned three markers as depicted in Fig. 1b-d therefore appear to be independent.

P. Mathis (ed.), Photosynthesis: from Light to Biosphere, Vol. I, 945–950.

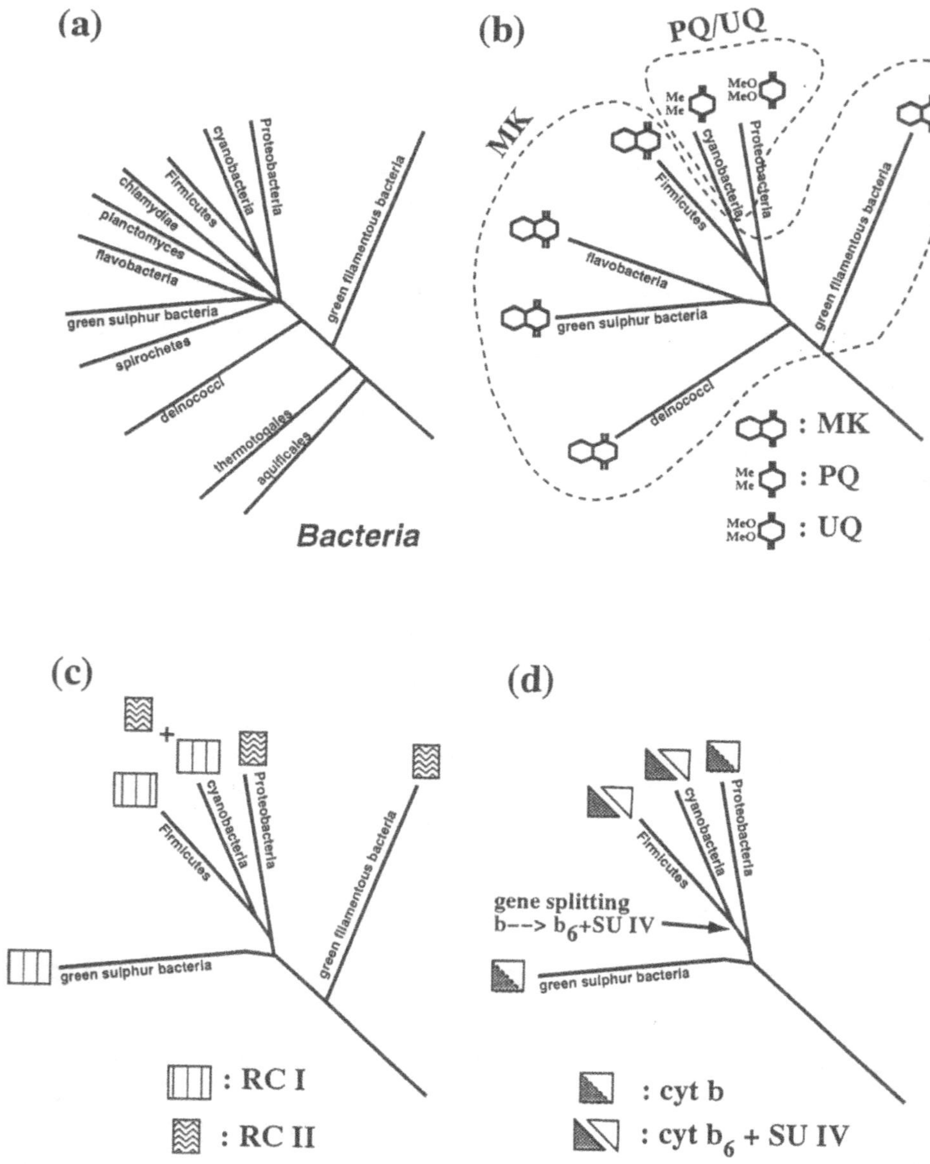

Figure 1. (a) Phylogenetic tree of the kingdom Bacteria based on 16 S r-RNA sequences (11). (b) Distribution of quinones over the phylogenetic tree according to the data reported in (12). (c) Distribution of RCI- and RCII-type photosystems (modified from (5)). (d) Distribution of "long" cytochrome b and split cytochrome b_6 + SU IV (according to (9,10)).

3. The transition naphtho- to benzoquinone occured twice and independently during evolution.

A closeup of the region of the phylogenetic tree under consideration (Fig.2) suggests that separate MK to PQ- and MK to UQ-transitions had occured. Within the phylum of the proteobacteria, the respective event took place only after the branching-off of the δ-subgroup. The use of differently substituted benzoquinones in cyano- and purple bacteria (and hence also in chloroplasts and mitochondria) therefore seems to be due to two independent "inventions" of higher potential quinones by two different bacterial phyla in order to substitute the "outfashioned" menaquinones, rather than to evolutionary diversification from a common benzoquinone-type ancestor.

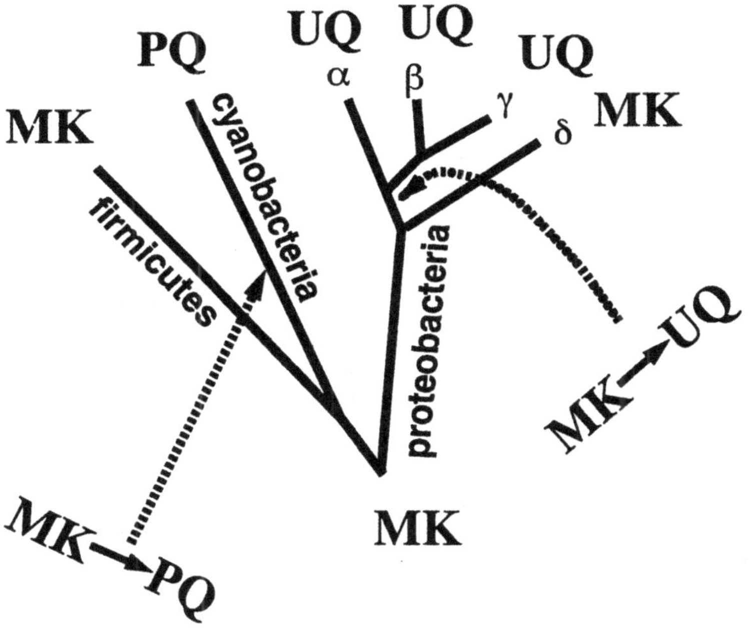

Figure 2. Region of the phylogenetic tree (11) relevant to the naphtho- to benzoquinone transition.

4. The transitions from naphtho- to benzoquinone conserved the functional properties of the respective electron tranport chains

Data on functional characteristics of electron transfer chains using MK are scarce. However, during recent years, a number of enzymes involved in such electron transfer chains have been purified and studied (notably from representatives of the firmicutes, i.e. the former Gram-positive bacteria, see refs 7 and 14). These enzymes appear to closely resemble their counterparts from electron transport chains based on UQ/PQ. An examination of the electrochemical parameters of redox components involved in MK-based electron transport suggested that all these compounds show E_m-values lower than those of their counterparts from UQ/PQ-based bioenergetic pathways thereby resulting in a similar redox equilibrium between electron transport components (15-17). We have recently studied photoinduced electron transport involving the RC and the cytochrome bc complex in whole cells of heliobacteria (considered to belong to the phylum of the firmicutes) as a model system for the functional parameters of MK-based bioenergetic electron transport (15; D.M. Kramer, B. Schoepp, U. Liebl and W. Nitschke, unpublished).

The results clearly show that no fundamental difference in the mode of functioning exists between the MK based heliobacterial light-induced electron transport and the formerly studied respective reactions in bioenergetic chains using UQ/PQ.

Fig. 3 shows a comparison of E_m-values and electron transport pathways (photosynthetic and respiratory) between UQ/PQ-based (left half) and MK-based (right half of Fig.3) systems. As can be seen from this figure, the “middle” region of redox centres (i.e. those at +/-300 mV from the potential of the respective quinone) is shifted compensating for the differences in E_m of the quinones. Otherwise, most electron transport pathways seem to be operative in both systems. A difference might be the role of the succinate/fumarate redox couple which is a good electron donor into benzoquinone-based chains. The respective electron donation into naphthoquinone-based chains is energetically uphill possibly indicating that the succinate-dehydrogenase related enzyme in firmicutes actually works as a fumarate-reductase (however, for an alternative interpretation see (14)).

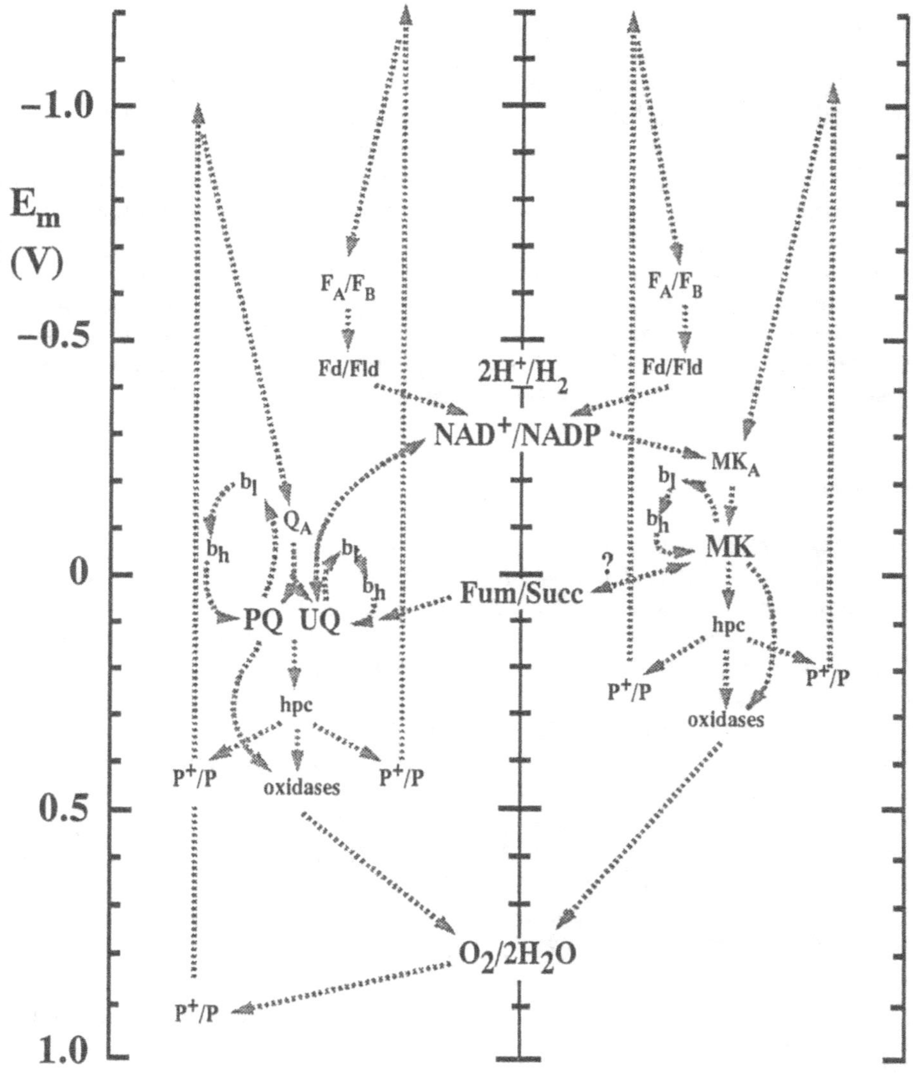

Figure 3. Redox components and their electron transfer reactions in UQ/PQ- (left half) and MK- (right half) based bioenergetic chains. No distinction between photosynthetic and respiratory pathways is made in this picture. hpc stands for "high potential chain". No references for the various redox reactions are cited since the mere list of references would exceed the total space of this contribution by far.

Why then?

If no change in function was brought about by the transition form MK to PQ/UQ, why then did this change occur? It is certainly intriguing that a high (yet not full, see below) correlation exists between O_2-tolerance of species and type of quinone. The high sensitivity of reduced MKH_2 to oxidation under aerobic conditions (as painfully known in every-day work on these systems) might have resulted in significant leakage of reducing equivalents out of the bioenergetically productive electron transfer chains once oxygen became abundant in earth's atmosphere. The respiring Gram-positive bacilli, however, stand out as quite healthy examples weakening this argument. The apparent organisation of their bioenergetic enzymes into supercomplexes as well as the relatively high amounts of these enzymes as compared to UQ-based respiration might be a consequence of efforts to counteract and compensate these kinds of losses.

References

1 Nelson, N. (1992) *Biochim.Biophys.Acta 1100*, 109-124
2 Blankenship, R.E. (1992) *Photosynth.Res. 33*, 91-111
3 Blankenship, R.E. (1994) *Antonie van Leeuwenhoek 65*, 311-329
4 Rutherford, A.W. and Nitschke, W. (1996) in *Origin & Evolution of Biological Energy Conservation* (Baltscheffsky, H., ed.), VCH, in press
5 Nitschke, W., Mattioli, T. and Rutherford, A.W. (1996) in *Origin & Evolution of Biological Energy Conservation* (Baltscheffsky, H., ed.), VCH, in press
6 Castresana, J.,Lübben M., Saraste, M. & Higgins, D.G. (1994) *EMBO J. 13*, 2516-2525
7 Kutoh, E. and Sone, N. (1988) *J.Biol.Chem. 263*, 9020-9026
8 Lübben, M., Kolmerer, B. and Saraste, M. (1992) *EMBO J. 11*, 805-812
9 Schütz, M., Zirngibl, S., leCoutre, J., Büttner, M., Xie, D.-L., Nelson, N., Deutzmann, R. and Hauska, G. (1994) *Photosynth.Res. 39*, 163-174
10 Sone, N., Sawa, G., Sone, T. and Noguchi, S. (1995) *J.Biol.Chem. 270*, 10612-10617
11 Woese, C.R. (1987) *Microbiol. Rev. 51*, 221-271
12 Collins, M.D. and Jones, D. (1981) *Microbiol. Rev. 45*, 316-354
13 Pierson, B.K. and Olson, J.M. (1989) in *Microbial Mats* (Cohen, Y. & Rosenberg, E., eds.) pp. 402-427
14 Hägerhäll, C., Aasa, R., von Wachenfeldt, C. and Hederstedt, L. (1992) *Biochemistry 31*, 7411-7421
15 Nitschke, W., Liebl, U., Matsuura, K. and Kramer, D.M. (1995) *Biochemistry*, in press
16 Liebl, U., Pezennec, S., Riedel, A., Kellner, E. and Nitschke, W. (1992) *J.Biol.Chem. 267*, 14068-14072
17 Riedel, A., Kellner, E., Grodzitzki, D., Liebl, U., Hauska, G., Müller, A., Rutherford, A.W. and Nitschke, W. (1993) *Biochim.Biophys.Acta 1183*, 263-268

SEQUENCE OF *atpI;atpH* GENES OF *Prochloron* sp. ATPASE GENE CLUSTER: IMPLICATIONS FOR CHLOROPLAST ORIGIN.

Scaramuzzi, Carol. D*+, Hiller, Roger. G+, Wrench, Pamela. M+, Lockhart, Peter. J # and Larkum, Anthony.W. D.*
*School of Biological Sciences, University of Sydney, NSW 2006.Aust.
+School of Biological Sciences, Macquarie University, NSW 2109. Aust.
Department of Microbiology and Molecular Genetics, Massey University, Palmerston North, New Zealand.

1. Introduction

A consensus of algal evolution views eubacterial ancestors with photosynthetic capacities proliferating to evolve the chloroplasts of present day algae via endosymbiotic events [1]. Debate persists however as to i) the phylogeny of the host organism's chloroplast and ii) whether the evolution of chloroplasts was mono or polyphyletic.

The photosynthetic prokaryote, *Prochloron* sp., is a phylogenetic enigma. This organism, a symbiont of several species of didemnid ascidians from tropical waters, contains a chlorophyll *c*-like pigment [2] in addition to chlorophyll *a* and *b*. Consequently, *Prochloron* has added stimulus to the debate on the origin of chloroplasts, as have two other prochlorophytes, the filamentous fresh water alga, *Prochlorothrix hollandica* [3] and the free-living unicellular alga, *Prochlorococcus marinus*. [4,5]. These prochlorophytes may represent an anciently diverged group widely separated from the cyanobacteria that are ancestral to the algal and higher plant chloroplast, or alternatively, they may represent an anciently diverged group, widely separated from the cyanobacteria, which have independently acquired chlorophylls *b* and *c* . Thus it is important to take into account the relationships of these three prochlorophytes to each other as well as their relationships to the cyanobacteria and green algae.

The ATPase operon is a conserved gene cluster that has been well characterised. We have used the polymerase chain reaction (PCR) to amplify a fragment of DNA from *Prochloron*, of approximately 1.1 kb which comprises much of the *atpI* and *atpH* genes and a non-coding intergenic region. These findings have been evaluated with regard to chloroplast evolution.

2. Materials and Methods.

2.1 Isolation of DNA.

Didemnid ascidians containing *Prochloron* were collected on One Tree Reef, Capricorn Section, Great Barrier Reef of Queensland, Australia. *Prochloron* cells were extracted by gently teasing the ascidians with a razor blade thereby allowing release of the *Prochloron* DNA was isolated and purified using a modified method of Rochaix (1982) [6].

P. Mathis (ed.), Photosynthesis: from Light to Biosphere, Vol. I, 951–954.

2.2 Polymerase Chain Reaction (PCR).

PCR experiments were conducted using oligo-primers (20 and 23 mer) designed from conserved regions of AtpI and AtpH (consensus sequence of higher plant data) incorporating inosine (i) where necessary. **Primer 1** (ggicaa/gcat/cttt/ctggca) was designed using the amino acid sequence, **GQHFYWQ** from a conserved region of AtpI near the amino terminus. **Primer 2** (cca/gtaiatigtiagigct/ctccat) was designed using the amino acid sequence, **MEALTIYG** from a conserved region of AtpH near the carboxyl terminus. Reactions contained 20 ng of target DNA, approximately 20 ng of oligo-primers, 2.5 mM $MgCl_2$ and 1.0 unit of thermostable polymerase (Advanced BioTechnology) in a total volume of 50 ul. The following cycle conditions were used:

Temperature (^oC)	Time (mins)	Cycle Number
94	5	1
94	1	2-36
45	1	2-36
72	2	2-36
72	10	37

PCR products were electrophoresed in an 0.7% agarose gel in Tris Acetate Buffer, excised from the gel, purified using Geneclean (BIO101) and ligated to the vector pGEM-T (Promega) in a 3:1 molar ratio (insert:vector). Ligations were performed at 4^oC for 16 hours and ligated DNA transformed into *Escherichia coli* strain JM101 and selected on LB plates containing 100 ug per ml ampicillin, 40 ug per ml 5-bromo-4 chloro-3-indolyl-B-D-galactosidase (X-gal) and 40 ug per ml isopropyl-B-D-thiogalactopyranoside (IPTG). Positive transformants were selected and the DNA was isolated and sequenced. Nucleotide sequences were obtained from database entries as listed below:

GROUP	ORGANISM	ACCESSION NO.
Eubacteria:	*Bacillus subtilus*	P37808
	Escherichia coli	ECUNC01
Anaerobic bacteria:	*Propionigenium modestum*	X53960
	Vibrio alginolyticus	X16050
Thermophilic bacteria:	PS3	X07084
Purple bacteria:	*Rhodospirilum rubrum*	M37308
Cyanobacteria:	*Anabaena* sp. PCC7120	M21659
	Synechococcus sp 6716	X70431
Prochlorophyta:	*Prochloron* sp	This work
Rhodophyta:	*Antithamnion* sp	X63382
	Galdieria sulphuraria	X67814
Chromophyta:	*Odentella sinensis*	X60752
	Pavlova lutherii	X64731
Euglenophyta:	*Euglena gracilis*	X70810
Higher plants:	*Marchantia polymorpha*	Y00686
	O. sativa	X15901
	Spinach	X03775

Deduced amino acid sequences were analysed using Staden [7] and GCG. Alignments and phylogenetic trees were generated with PILEUP (GCG) using the Dayhoff 250 PAM matrix , a gap weight of 3.0 and a gap length penalty of 0.2. Nucleotide sequences and phylogenetic trees were analysed using the program LogDet [8].

3. Results and Discussion

A 1.063 kb fragment of *Prochloron* DNA was amplified by PCR, purified, subcloned and sequenced. The DNA fragment is composed of 690 nucleotides that predict a protein sequence of 230 amino acids corresponding to part of the AtpI protein and 201 nucleotides that predict a protein sequence of 67 amino acids corresponding to part of the AtpH protein. The non-coding sequence between the 3' end of *atpI* and the 5' end of *atpH* is composed of 172 nucleotides. In higher plants the genes encoding the AtpI and AtpH proteins of the ATPase operon are separated by a non-coding region of 692 nucleotides (spinach) and 795 nucleotides (rice). This large intergenic region is not seen in the eubacteria, photosynthetic bacteria, Cyanobacteria, rhodophytes or the Chromophyta, eg. *Rhodospirillum rubrum*; 101 nucleotides, *Anabaena* sp PCC7120; 83 nucleotides, and *Odentella sinensis*; 70 nucleotides. The intergenic space in *Prochloron* for these two genes of 172 nucleotides more closely approximates the intergenic space of the prokaryotic Cyanobacteria and the eukaryotic algae. The predicted product of the *atpI* gene has a carboxyl terminus which resembles that of the Cyanobacteria. This extended carboxyl terminus is not seen in the higher plants, eubacteria or purple bacteria (Figure1) and is composed of a high percentage of charged amino acids (E, glutamic acid; D, aspartic acid; H, histidine) and the polar residue glycine.

B.subtilus	AFIFTMLTMV **Y**MSHKISHDH
E. coli	AFIFMVLTIV **Y**LSMASEEH
P. modestum	SFVFIMLTMV **Y**IQGSIGDAEYLED
V.alginolyticus	AFVFMMLTIV **Y**LSMAHEDPDH
PS3	AFIFTMLTMV **Y**MAHKVSHDH
R.rubrum	AYVFTILTCI **Y**LNDAINMH
Anabaena sp	ALIFATLAAA **Y**IGEAMEDH**HGEEHEEHH**
Synechococcus	ALVFATLAAT **Y**IGEAMEGH**GGEHEEAHS**
Prochloron sp	ALIFATLAAS **Y**IAASMEGH**GEEHD**
Antithamnion sp	ALISTLSAA- **Y**IGEALEGHGEE
G. sulphuraria	ALIFATLAAA **Y**IGEAIEEH
O. sinensis	ALIFSTLAAA **Y**IAEALE
P. lutherii	ALIFATLSAA **Y**IGEALE
Euglena	ALIFATLSGS **Y**IGEAMEGHH
M polymorpha	ALIFATLAAA **Y**IGESMEGHH
O. sativa	ALIFATLAAA **Y**IGESMEGHH
Spinach	ALIFATLAAA **Y**IGESLEGHH

Figure 1: Comparison of the Carboxyl termini of AtpI proteins. The carboxyl termini of the Cyanobacteria and *Prochloron* are indicated in bold type as is the invariant tyrosine residue

It is speculated that these differences are functionally significant in the Cyanobacteria and Prochlorophyta. Further, the predicted protein sequence immediately preceding the carboxyl terminus is remarkably conserved and contains an invariant tyrosine residue (Figure 1).

The translated sequences were aligned using PILEUP and phylogenetic trees were generated (data not shown). The trees for both AtpI and AtpH were similar and both indicated the most divergent form to be *Rhodospirilum rubrum*, along with the other bacteria (eg. *E.coli*, *B. subtilus*). *Prochloron* was most closely associated with the

Cyanobacteria. Chloroplasts from algae and higher plants also claded with this group but were more widely separated.Since it is known that analysis based on such limited information cannot yield robust trees, trees based on nucleotide sequences are preferable, providing that base compositional bias is taken into account when composing ancient lineages. Thus the LogDet method [8] has been used to determine phylogenetic trees for the nucleotide sequences of the *atpI*, *atpH* and *atpB* genes (Figure 2A) and the recently published data for the elongation factor genes, *EF-1* (Figure 2B). The results of LogDet transformation, together with Neighbor Joining [9] or Split composition [10] indicate insufficient resolution to have confidence in the root of the tree (Figure 2). Such starbursts occur even when outgroup sequences are not included in the analysis. We conclude that at present there is insufficient data to support any one hypothesis concerning the relationship of the cyanobacteria, the prochlorophytes and the photosynthetic bacteria

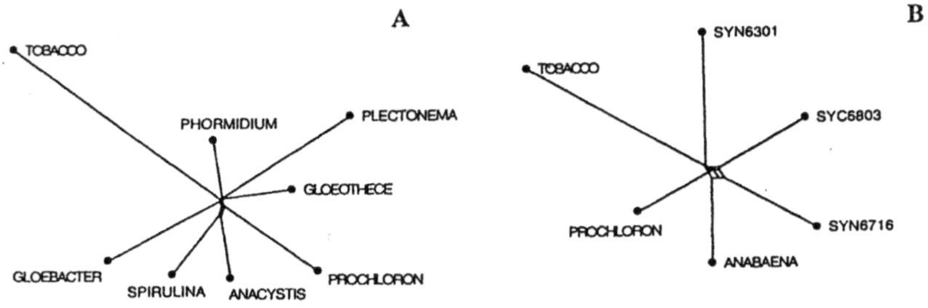

Figure 2. Majority consensus trees from 100 bootstrap runs /LogDet/invariant sites model/Neighbor Joining. Outgroups used for *atpI, atpH* and *atpB*, were *Bacillus subtilus, Propionigenium modestum*, and for *EF-1*, *Thermatoga maritima*, *Deinonema* sp.

References:

1 Cavalier-Smith, T (1987) Ann. N.Y. Acad. Sci 501,55-71
2 Larkum, A.W.D, Scaramuzzi, C.D,Cox, G.C, Hiller, R.G. and Turner, A.G (1994) Proc Natl Acad Sci (USA) 91,679-683
3 Burger-Wiersma, T, Veenhuis, M, Korthals, H.J, Van de Wiel, C.C.M.and Mur, R. (1986) Nature (London) 320,262-264
4 Chisolm, S.W., Olsen, R.J., Zettler, E.R.m Goericke,R., Waterbury, J.B and Welschmeyer, N.A. (1988) Nature (London) 334,340-343
5 Palmer, J. D. (1985) Ann. Rev. Genet. 19, 325-354
6 Scaramuzzi, C.D, Stokes, H.W. and Hiller, R.G. (1992) Plant Mol. Biol. 18,467-476
7 Gleeson, T.J and Staden, R. (1991) CABIOS. 7, 398
8 Lockhart, P.J, Steel, M.A,Hendy, M.D and Penny, D. (1994) Mol.Biol. Evol. 11,605-612
9 Saitou, N and Nei, M. (1987) Mol Biol Evol 4, 406-425
10 Bandelt, H-J and Dress, A.W.M. (1992) Adv Math 92, 47-105

PHYLOGENETIC ANALYSIS OF PHOTOSYSTEMS I AND II

Steven C Halls, Stephen R Downie, and Govindjee

University of Illinois, Urbana, IL 61801 U.S.A. E-mail: gov@uiuc.edu

Photosystem I (PS I) and Photosystem II (PS II) reaction centers include heterodimers of the gene products of *psa*A (PSI-A) and *psa*B (PSI-B) and of *psb*A (D1) and *psb*D (D2), respectively. Structural and sequence similarity exists among the heterodimer pair PSI-A and PSI-B, and D1 and D2 (Svensson et al., 1990; Golbeck and Bryant, 1991; Nitschke and Rutherford, 1991; Blankenship, 1992; Ruffle et al., 1992). Many sequences are now available for reaction center proteins of PS I and PS II. Golbeck and Bryant (1991) presented alignments of PSI-A and PSI-B for 7 taxa. Svensson et al. (1992) compiled 38 sequences of D1 (27 taxa) and 15 sequences of D2 (12 taxa) and showed that the sequences were strongly conserved. Ruffle et al. (1992) have shown relationships among the D1 and D2 proteins. In this paper, we present the first phylogenetic analysis of PSI-A and PSI-B, and extend the existing reaction center alignments of D1 and D2 to include several additional taxa.

Selected published protein sequences were retrieved from SWISS-PROT, PIR, and GenPept data bases using BLAST (Altschul et al., 1990). Additional sequences were obtained for *Porphyra purpurea* (M. Reith and J. Munholland, unpublished), *Cyanophora paradoxa* (V. L. Stirewelt and D. A. Bryant, unpublished), and *Zea mays* (Sopory et al., 1993, from Larrinua and McLaughlin, 1987). In total, 39 D1 and 11 D2 sequences of different genera, and 14 PSI-A and 14 PSI-B sequences, representing 13 genera each, were aligned. Alignments were done as described by Svensson et al. (1992), and Golbeck and Bryant (1991). The resulting data matrices were analyzed using PAUP (Phylogenetic Analysis Using Parsimony) version 3.1.1 (Swofford, 1993). All heuristics searches were replicated 100 times with random addition sequence and tree bisection-reconnection branch swapping. The options mulpars, steepest descent, collapse, and acctran optimization were selected. Gaps were treated as missing data. The amount of phylogenetic information in the parsimony analyses was estimated using the consistency, CI, (excluding uninformative characters) and retention, RI, indices. The trees were rooted by positioning the root along the branch connecting *Synechococcus* to the rest of the network.

Multiple alignments of D1, D2, PSI-A, and PSI-B sequences resulted in matrices of 360, 353, 804, and 743 positions, respectively; consensus sequences for each group of reaction center proteins are presented below (Complete data matrices are available upon request.).

D1 (Photosystem II)

```
PROKARYOTES   MTTTLERRES ASLWERFCSW ITSTDERLYV GWFGVLMIPT LLAATICFII AFIAAPPVDI DGIREPVAGS LLYGNNIISG AVVPSSNAIG LHFYPIWEAA
ALGAE           AI           A   E                          T  SV                      S           T    I T
MONOCOTS        AI       T   G   N                          T  SV                      S               II T A
DICOTS          AI       E   G   N                          T  SV                      S                I T A

PROKARYOTES   SLDEWLYNGG PYQLVVFHFL IGICCYMGRE WELSYRLGMR PWICVAYSAP VAAATAVFLI YPIGQGSFSD GMPLGISGTF NFMLVFQAEH NILMHPFHML
ALGAE                        I    F   V                      A  F        S    V                          I
MONOCOTS       V           E I      L VA                     A                                           I
DICOTS         V           E I      L VA                     A           F                               I

PROKARYOTES   GVAGVFGGSL FSAMHGSLVT SSLVRETTEN ESQNYGYKFG QEEETYNIVA AHGYFGRLIF QYASFNNSRS LHFFLAAWPV VGIWFTALGV STMAFNLNGF
ALGAE                                  I         A N
MONOCOTS                               I         A E  R                                                        I
DICOTS                                 I         A E  R                                                        I

PROKARYOTES   NFNQSVVDSQ GRVINTWADI INRANLGMEV MHERNAHNFP LDLASVESAP VALTAPSING
ALGAE                                                            --- ----
MONOCOTS                                                    A  --- ----
DICOTS                                                      A  --- ----
```


P. Mathis (ed.), Photosynthesis: from Light to Biosphere, Vol. I, 955–958.

D2 (Photosystem II)

```
PROKARYOTES    MTIAVGRAAA VERGWFDVLD DWLKRDRFVF IGWSGLLLFP CAFMALGGWL TGTTFVTSWY THGLASSYLE GANFLTVAVT SPADAFGHSL LFLWGPEAQN
ALGAE              I TTYK EK     DA                         SYF                       T      C    A  S T  NSMA      V
MONOCOTS           L  VPK E NDL  IM     R                    YF                              C    A  S T  NSLA     L
DICOTS             L  FTK D NDL  IM     R                    YF V                            C    A    T  NSLA     L

PROKARYOTES    NLTRWFQLGG LWFFVALHGA FGLIGFMLRQ FEIARLVGIR PYNAIAFSGP IAVFVSVFLM YPLGQSSWFF APSFGVAGIF RFILFLQGFH NWTLNPFHMM
ALGAE          DF   C I     A                          N           A           I                   A
MONOCOTS       DF   C       T         A           L    Q        S              I                   A
DICOTS         DF   C       T                     L    Q                       I                   A                     M

PROKARYOTES    GVAGILGGAL LCAIHGATVE NTLFEDGEDS NTFRAFEPTQ AEETYSMVTA NRFWSQIFGI AFSNKRWLHF FMLFVPVTGL WMSSVGIVGL ALNLRAYDFV
ALGAE                                       DGA       N    S                   V                          AI
MONOCOTS                                    DGA       N                        V                          AI V
DICOTS                                      DGA       N                        V                          AL V

PROKARYOTES    SQELRAAEDP EFETFYTKNI LLNEGIRAWM APQDQPHENF IFPEEVLPRG NAL
ALGAE             I                              A         V
MONOCOTS          I                              A       L
DICOTS            I                              A       L
```

PSI-A (Photosystem I)

```
PROKARYOTES    MTISPPEREA -KARVVVDND PVPTSFEKWA KPGHFDRTLA RGPKTTTWIW NLHALAHDFD THTSDLEDIS RKIFSAHFGH LAVVFIWLSG MYFHGAKFSN
ALGAE              T      K VKIA  K    E             S   S K  N           D      S            V      Q  GII
MONOCOTS        M RS  P-- -EVKI   R    K    E   R    S  L  K  D           D      S      E     V      Q  SII
DICOTS          I RS  P-- -EVKIL  R    K    E   R    S  I  K  D           D      S      E     V      Q  SII L

PROKARYOTES    YEAWLADPTG IKPSAQVVWP IVGQGILNGD VGGGFHGIQI TSGLFQLWRA SGITNSFQLY CTAIGGLVMA GLMLFAGWFH YHKRAPKLEW FQNVESMLNH
ALGAE               S                    E           Q        L           SEL    T          A  F          A
MONOCOTS            S      G             E           R     S  F  I        SEL         A IF  S             A    A  SHI
DICOTS              S      G             E           R        F  I        SEL         A  F  A             A    A    I

PROKARYOTES    HLAGLLGLGS LAWAGHQIHV SLPINKLLDA GVAAKDIPLP HEFILNPSLY PELDPGFASG LTPFFTLNWG AYSDFLTFKG GLNPVTGGLW LSDTAHHHLA
ALGAE            G                            S   DPQE           RD M AQ Y S SK   A       S E
MONOCOTS                   S               QF      DP E           RD L AQ Y S  ER A          K AE  S R      I         I
DICOTS                     S               QF N    DP-E       -   RD L AQ Y S  E  A          K A     R              T

PROKARYOTES    IAVLFIIAGH MYRTNWGIGH SLKEILEAHK GPFTGEGHKG LYEVLTTSWH AQLAINLAML GSLTIIVAQH MYAMPPYPYL ATDYPTQLSL FTHHMWIGGF
ALGAE            V  LV                M       R               I                F    S    H    S
MONOCOTS         I  L                G  D            QR        I          SL        T  V  H
DICOTS           I  L                G  D            Q         I           L           V  H

PROKARYOTES    LIVGGAAHGA IFMVRDYDPA MNQNNVLDRV LRHRDAIISH LNWVCIFLGF HSFGLYIHND TMRALGRPQD MFSDTAIQLQ PVFAQWVQNL HTLAPGGTAP
ALGAE          C   AG  A             N Y  L     I                                  S                         I  T  F    L
MONOCOTS           A   A           T TRY DL                                        S                               A   RL
DICOTS             A   A           T TRY DL                    A                   S                         I  T  A   SA

PROKARYOTES    NALA-TASYA FGGDVVAVGG KVAMMPIVLG TADFMVHHIH AFTIHVTVLI LLKGVLFARS SRLIPDKANL GFRFPCDGPG RGGTCQVSGW DHVFLGLFWM
ALGAE              -    W G- ELG H          S    S
MONOCOTS       G TTS SLTW G  EL         LL   P       L
DICOTS         G TAS SLTW G S L         LL   P       L

PROKARYOTES    YNSLSIVIFH FSWKMQSDVW GTVAPDGTVS HITGGNFAQS AITINGWLRD FLWAQASQVI NSYGSALSAY GLMFLGAHFV WAFSLMFLFS GRGYWQELIE
ALGAE               V                   TAS G                              S     Q            I
MONOCOTS          I V                  ISDQ V T            S                     Q    S       F
DICOTS            I V                 SISDQ V T            S                     Q    S       F

PROKARYOTES    SIVWAHNKLK VAPAIQPRAL SIIQGRAVGV AHYLLGGIAT TWAFFLARIL SVG
ALGAE                                  T                     S      I A
MONOCOTS                      T                 T                   I A
DICOTS                        T        V        T                   I A
```

PSI-B (Photosystem I)

```
PROKARYOTES    MATKFPKFSQ DLAQDPTTRR IWYGIATAHD FETHDGMTEE NLYQKIFASH FGHLAIIFLW TSGNLFHVAW QGNFEQWVKD PVNTRPIAHA IWDPHFGKAA
ALGAE                     G                       S                     Q                          IT   LHV               P
MONOCOTS        ELR  R    G            F          S  DI    R   N        Q                        S IQ                     P
DICOTS           LR  R    G            F          S  DI    R   N        Q                           Q                     P

PROKARYOTES    VEAFTQAGAS NPVNIAYSGV YHWWYTIGMR TNGELYQGAI FLLILASLAL FAGWLHLQPK FRPSLSWFKN AESRLNHHLA GLFGVSSLAW AGHLVHVAIP
ALGAE               RG    G           Q           QD T SV      VSA F              Q                 S             T
MONOCOTS            RG  A G           Q     L     ED T  L     F ST S  IG         WK                 S             T
DICOTS              RG  L             Q     L     ED T  L        AIS  I          WK  V              S             T

PROKARYOTES    ESRGQHVGWD NFLSTMPHPA GLAPFFTGNW GVYAQNPDTA SHVFGTSQGA GTAILTFLGG FHPQTESLWL TDIAHHHLAI AVLFIVAGHM YRTNFGIGHS
ALGAE              R          VL   Q       S    AA                E S  Q                      M                            R
MONOCOTS       A   EY R N    DVL   Q   G LL Q    NL      SS N L   T                                      FI LI
DICOTS         A   EY R N    DVL       G L  Q    NL      SS   L   A                                      FI L

PROKARYOTES    IKEILNAHTK VGGPFGAGHQ GLYDTYNNSL HFQLGLALAS LGVITSLVAQ HMYSLPPYAF IAQDHTTQAA LYTHHQYIAG FLMVGAFAHG AILLVRDYDP
ALGAE          L A  D  KP PS NL    K      I                                        I F                  I C           FI
MONOCOTS         DL E  TP P  RL R  K      I                                  S       F                  I T           FI
DICOTS         M DL D  IP P  RL R  K      I                                  A       F                                 I

PROKARYOTES    AQNKGNVLAR VLDHKEAIIS HLSWVSLFLG FHTLGLYVHN DVVVAFGTPE KQILIEPVFA QFIQAAHGKL LYGFDVLLSN PDSIASTAWP NYGNVWLPGW
ALGAE          E          I               T                  MQ                           A         S  STSV    GQ S---I
MONOCOTS       E  ED      M              A                   ML              I       S    T T    I  S  TSGP FN GR T---L
DICOTS         E  ED      M E            A                   ML              I       S    T S       S  TSGP FN GR S---I

PROKARYOTES    LDAINSGTNS LFLTIGPGDF LVHHAIALGL HTTTLILVKG ALDARGSKLM PDKKDFGYAF PCDGPGRGGT CDISAWDAFY LAMFWMLNTI GWVTFYWHWK
ALGAE               NNQ
MONOCOTS        N V ENS
DICOTS          N V ENS

PROKARYOTES    HLGVWEGNVA QFNESSTYLM GWFRDYLWLN SSQLINGYNP FGTNNLSVWA WMFLFGHLVW ATGFMFLISW RGYWQELIET LVWAHERTPL ANLVRWKDKP
ALGAE            TL                                          M S      F                                      K
MONOCOTS        ITL QG  S                                    M S                                        A            I
DICOTS           TL QG  S                                    M S                                        A            I

PROKARYOTES    VALSIVQARL VGLAHFSVGY ILTYAAFLIA STAGKFG
ALGAE                                 F           S
MONOCOTS                              F           S
DICOTS                                F           S
```

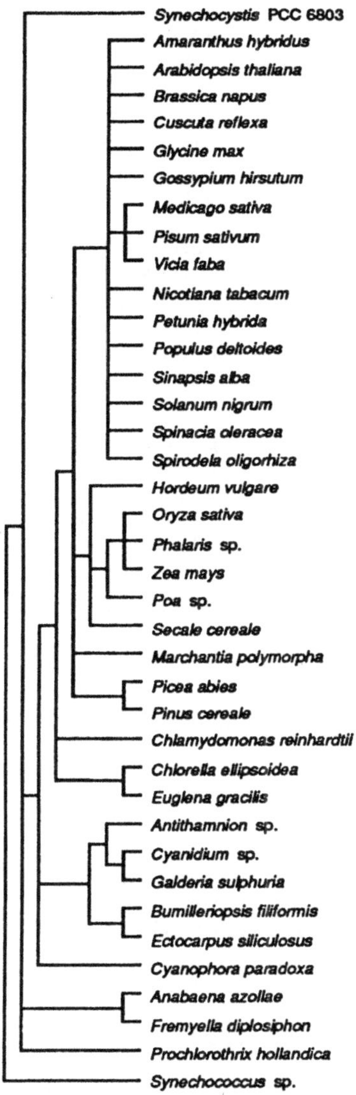

Fig. 1. Strict consensus of 1,660 minimal length 400-step trees derived from parsimony analysis of selected *psbA* (D1) sequences. CI=0.62, RI=0.69.

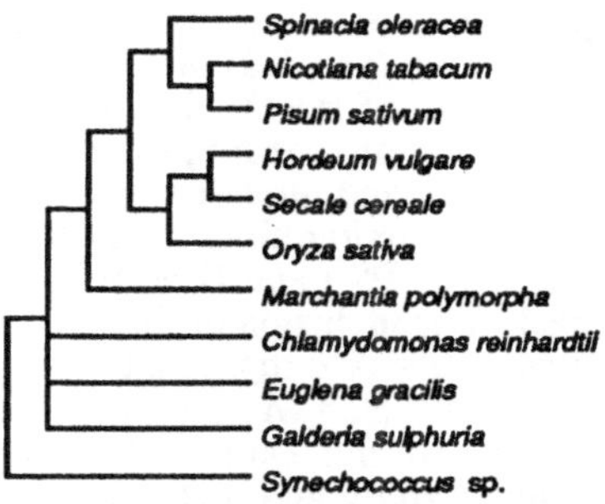

Fig. 2. Strict consensus of 4 minimal length 144-step trees derived from parsimony analysis of selected *psbD* (D2) sequences. CI=0.73, RI=0.62.

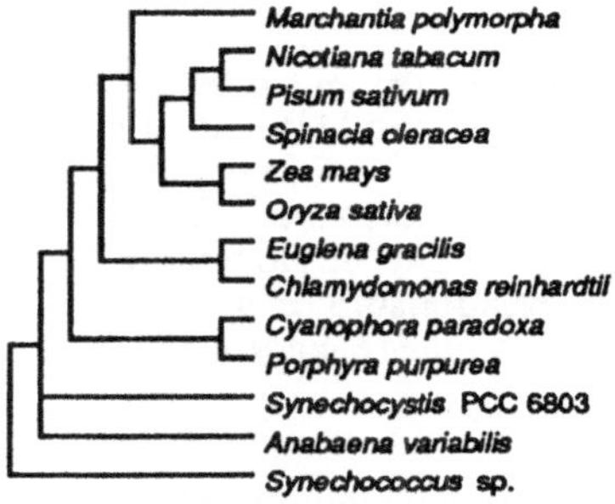

Fig. 3. Strict consensus of 2 minimal length 659-step trees derived from parsimony analysis of selected PSI-A sequences. CI=0.69, RI=0.69.

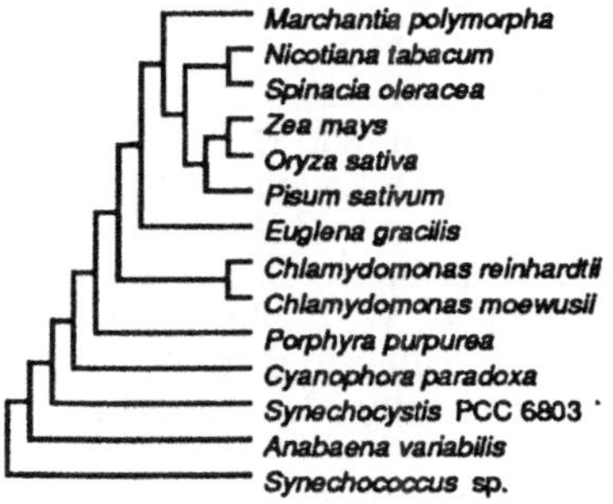

Fig. 4. Single maximally parsimonious 696-step tree derived from the analysis of selected PSI-B sequences. CI=0.72, RI=0.74.

The multiple alignment of each of the reaction center subunits was facilitated by the introduction of none (i.e., D2) or very few gaps (D1, one gap; PSI-B, three gaps; PSI-A, 16 gaps). Analyses of the D1, D2, and PSI-A data matrices resulted in 1,660, 4, and 2 maximally parsimonious topologies, whose strict consensus trees, with accompanying tree lengths, consistency and retention indices, are shown in Figs. 1, 2, and 3, respectively. A single maximally parsimonious tree obtained from the analysis of PSI-B sequences, is presented in Fig. 4. PS I and II reaction center proteins are highly conserved among the taxa examined. The inclusion of the *Pisum sativum* A2 PSI-A sequence (Lehmbeck et al., 1986) resulted in a single maximally-parsimonious tree of 1033 steps, 374 steps greater than the trees constructed without this sequence (CI excluding uninformative characters = 0.70; RI = 0.64). In this single PSI-A tree (not shown), this *Pisum* sequence arises as sister taxon to *Euglena* and away from the other *Pisum* and dicot sequences. In most cases, the evolutionary relationships inferred on the basis of PS I and II reaction center protein sequences and maximum parsimony are in agreement with traditional concepts of relationship. Generally the dicots and monocots each comprise a clade and are sistergroups.

Remarks: We have presented here the current consensus sequence of core PS II reaction center proteins (D1 and D2), and of core PS I reaction center proteins (PSI-A and PSI-B) based on published as well as unpublished sequences. Further, we have presented here the first phylogenetic analysis of PSI-A and PSI-B proteins and an extended analysis of D1 and D2 proteins. The phylogenetic trees, thus obtained, provide a hypothesis of the evolutionary relationship of the reaction centers of prokaryotes, cyanelle, rhodophytes, chlorophytes and other eukaryotes (gymnosperms, monocots and dicots).

Acknowledgments

Steven Halls acknowledges an IPR predoctoral training grant for financial support. We thank Drs. D. A. Bryant, V. L. Stirewelt, M. Reith and J. Munholland for providing us with unpublished sequences.

Altschul, S. F., Gish, W., Miller, W., Myers, E. W. and Lipman, D. J. (1990). *J. Mol. Biol.* **215**, 403-410.

Blankenship R. E. (1992). *Photosynth. Res.* **33**, 91-111.

Golbeck, J. H. and Bryant, D. A. (1991). *Current Topics in Bioenerg.* **16**, 83-176.

Larrinua, I. M. and McLaughlin, W. E. (1987). In: "*Progress in Photosynthesis Research*", edited by J Biggins, Vol 4, Nijhoff, The Hague, pp. 649-652.

Lehmbeck, J., Rasmusen, O. K., Bookjans, G. B., Jepsen, B. R., Stummann, B. M. and Henningsen, K. W. (1986). *Plant Mol. Biol.* **7**, 3-10.

Nitschke W. and Rutherford A. W. (1991). *TIBS* **17**, 241-245.

Ruffle, S. V., Donnelly, D., Blundell, T. and Nugent, J. H. A. (1992). *Photosynth. Res.* **34**, 287-300.

Sopory, S. K., Ghirardi, M. L., Greenberg, B. M., Elich, T., Edelman, M. and Mattoo, A. K. (1993). In: "*Photosynthesis: Photoreactions to Plant Productivity*", edited by Y. P. Abrol, P. Mohanty and Govindjee, Kluwer, Dordrecht, pp. 131-156.

Svensson, B., Vass, I. and Styring, S. (1990). *EMBO J.* **9**, 2051-2059.

Svensson, B., Vaas, I. and Styring, S. (1992). *Z. Naturforsch* **46c**, 765-776.

Swofford, D. L. (1993). PAUP: Phylogenetic analysis using parsimony, version 3.1.1 Computer program distributed by the Illinois Natural History Survey, Champaign, IL.

CLONING BY FUNCTIONAL COMPLEMENTATION IN *E. coli* OF THE *gap2* GENE OF *Synechocystis* sp. PCC 6803 SUPPORTS AN AMPHIBOLIC ROLE FOR CYANOBACTERIAL NAD(P)-DEPENDENT GLYCERALDEHYDE-3-PHOSPHATE DEHYDROGENASE

F. Valverde, M. Losada and A. Serrano
Instituto de Bioquímica Vegetal y Fotosíntesis, CSIC y Universidad de Sevilla, Apdo. 1113, E-41080 Sevilla, Spain

1. Introduction

Phosphorylating glyceraldehyde-3-phosphate dehydrogenase (GAPDH) is a well known glycolytic enzyme (EC 1.3.1.12) found in the cytosol of all organisms so far studied, while a second GAPDH (EC 1.3.1.13) is located in the chloroplast stroma being a key enzyme of the reductive pentose phosphate pathway [1,2]. In cyanobacteria only one enzyme with both NAD and NADP-linked activities has been isolated so far, a dual (amphibolic) role in both biosynthetic and catabolic pathways having been proposed from biochemical data [3]. However, the presence of three *gap* genes potencially encoding for GAPDHs with different pyridine nucleotide specifity has been recently reported in *Anabaena variabilis* [4] raising the question of their physiological roles. We have cloned the *gap2* gene from *Synechocystis* sp. PCC 6803, which encodes for NAD(P)-dependent GAPDH, by using as cloning strategy the complementation of an *E. coli gap*⁻ mutant with a genomic library. In this way, the glycolytic pathway was restored in the complementant *E. coli* clone, thus indicating that the enzyme is able to function with that capacity in *Synechocystis* as well. Since NAD(P)-dependent GAPDH is already known to be essential in the reductive pentose phosphate pathway, it seems that it displays both anabolic and catabolic roles in *Synechocystis*. Our present studies have moreover shown that although two different *gap* genes are present in the *Synechocystis* genome, only the NAD(P)-dependent enzyme is expressed under normal culture conditions in this cyanobacterium.

2. Procedure

2.1 *Bacterial strains and growth conditions*

Synechocystis sp. PCC 6803 and all other cyanobacteria were cultured in BG11 medium at 30°C. *Escherichia coli* K-12 and DH5α strains were cultured in Luria Broth medium supplemented with ampicilin (100 μg/ml) when necessary. *E. coli gap*⁻ mutans DF-221 (*E.coli* Genetic Stock Center, Yale University, USA) and W3CG [5] were cultured in M63 minimal medium supplemented with succinate/glycerol (20% w/v) or glucose (20% w/v) and when necessary with ampicilin and, only for W3CG, tetracyclin (15 μg/ml). All *E.coli* strains were cultured at 37°C with continuous agitation.

2.2 *Enzyme assays and protein techniques*

Enzyme assays were performed spectrophotometrically in the forward direction (oxidation of G3P) by monitoring the appearance of either NADH or NADPH in the presence of

P. Mathis (ed.), Photosynthesis: from Light to Biosphere, Vol. I, 959–962.

sodium arsenate [6]. One unit of enzyme is defined as the amount which catalyzes the formation of 1 μmol of NAD(P)H per min. Cell-free extracts were obtained in the cold by sonication (70W, 2 min) followed by centrifugation at 40,000 *g* for 25 min. The native and recombinant NAD(P)-dependent GAPDHs were purified to electrophoretic homogeneity by a method involving the following steps: $(NH_4)_2SO_4$ fractionation, and anionic exchange, hydrophobic and dye-affinity chromatographies. Monospecific antibodies were raised in rabbit against the purified *Synechocystis* NAD(P)-dependent GAPDH. The antiserum was used (1:100 dilution) for Western blot analysis after SDS-PAGE (12%) and transfer to nitrocellulose of both cell-free extracts and purified preparations from different cyanobacteria and *E. coli* strains. Protein was determined using the BioRad Protein Dye reagent.

2.3 *Cloning strategy and DNA manipulation*
Competent cells of the *E. coli gap*⁻ mutants DF-221 or W3CG were transformed with a genomic library from *Synechocystis* sp. PCC 6803 in BlueScript SK^+ plasmid. Recombinant plasmids were isolated by alkaline extraction from clones able to grow in selective media (M63 plus glucose and antibiotics). Cell-free extracts of these clones were checked for both NAD and NADP-linked GAPDH activities. A recombinant plasmid with a 10 kb insert harboring the putative *Synechocystis gap2* gene was selected and named pFV1. After subcloning, a plasmid with a 4 kb insert that still retained the capacity to complement the *E. coli gap*⁻ mutants was obtained, being then called pFV8. Southern blots were performed as described in [7]. *Synechocystis* genomic DNA was completely restricted with different enzymes, run on 0.7% agarose gels and blotted to nylon membranes. A 350 bp Bgl II-Ava I *gap2* fragment of pFV8 and a 650 bp Sau 3AI fragment from the *E.coli gapc* gene labelled with ^{32}P dCTP were used as probes in homologous (65°C) or heterologous (55°C) hybridization conditions respectively. For DNA sequencing exonuclease stepwise deletions (Erase-a-Base kit, Promega) of the pFV8 plasmid were sequenced by using a modified dideoxi chain termination method and the Taq polymerase of Promega (Taq Track kit). PCR amplifications were carried out with the Pharmacia Taq polymerase in a programmable Eppendorf thermocycler. Primers constructed from highly conserved amino acid sequences were: 5'GTC/TAAG/CCAG/CCAC/TGTT/CG/CA/TC/TCG3' and 5'ACCATG/ACTG/ATTG/ACTC/TACC/GCC3'. Annealing was performed at 45°C and 35 cycles followed. After electrophoresis on agarose gel the amplified 500 bp band was scised, cloned in pGEMt plasmid (Promega) and subsequently sequenced.

3. Results and Discussion

The functional complementation of *E. coli gap*⁻ mutants with a genomic library [8] has been used for the first time as a cloning strategy for a photosynthetic GAPDH. By using this method we isolated a gene of *Synechocystis* sp. PCC 6803 coding for a NAD(P)-dependent GAPDH presumably able to play both catabolic and anabolic roles. As is shown in Table 1, whereas only the NAD-linked activity is present in *E. coli*, both NAD- and NADP-linked activities were clearly overexpressed in *E. coli gap*⁻ mutants transformed with the recombinant plasmid pFV8 which harbored a 4 kb insert with only one complete ORF. Noteworthy, the ratio between both pyridine nucleotide-linked activities were about 1 in extracts of both *Synechocystis* and the transformed *E. coli* mutants that recovered the capacity to grow with glucose as carbon source. These results show that a gene encoding for a cyanobacterial NAD(P)-dependent GAPDH is able to functionally substitute the glycolytic enzyme and suggest that a single NAD(P)-dependent GAPDH enzyme should be present in *Synechocystis* cells. In agreement with this proposal we have found that during GAPDH purification from this cyanobacterium both pyridine nucleotide-linked activities overlapped in all chromatographic steps, only one peak being observed (not shown). On the other hand, GAPDH activity staining with either pyridine nucleotide gave a single identical band in native acrylamide gels of cell-free extracts from both the *E. coli* recombinant clone and *Synechocystis*. Furthermore, when the *Synechocystis* crude extract was incubated for 1 h with the anti-GAPDH antibody both NAD- and NADP-linked activities decreased in a parallel manner down to eventually reach undetectable levels. Similar results were obtained in extracts from *Anabaena* sp. PCC 7120, but the NAD-linked activity was not completely inhibited. Western blot analysis showed that this antibody recognizes the same 38 kDa protein in extracts of both transformed *E. coli* mutants and *Synechocystis* (Fig. 1), demonstrating that the GAPDH subunit is correctly expressed by the recombinant clone. Western analysis gave also a unique protein band of about 38 kDa in crude extracts from different cyanobacteria, but not with those from *E. coli* wild type and non-photosynthetic eukaryotes, indicating that the antibody does not cross-react with typical glycolytic GAPDHs (Fig. 1).

Table I. GAPDH activities in cell-free extracts of *Synechocystis* sp. PCC 6803 and the following *E. coli* strains: K-12 (wild type), the used *gap*$^-$ mutants and complemented mutans with recombinant plasmids harboring the *Synechocystis gap2* gene. The ability of the *E. coli* strains to grow with different carbon sources is also shown.

	GAPDH activity (U/mg protein)		growth[a)]	
	NAD-linked	NADP-linked	succ./glyc.	glucose
Synechocystis sp. PCC 6803	0.12	0.13		
E. coli K-12	0.86	<0.001[b)]	+++	++++
E. coli DF221	<0.001	<0.001	+	-
E. coli W3CG	<0.001	<0.001	++	-
E. coli DF221 pFV8	0.60	0.61	++	++
E. coli W3CG pFV8	6.6	6.5	+++	++++

a) Only for *E. coli* strains
b) Negligible values (below the detection limit)

The 1033 bp ORF which contained the insert of pFV8 was sequenced (accession number X83564, EMBL nucleotide sequence database), compared with sequence databases and found to have a high degree of identity to other *gapa,b* genes. The protein sequence deduced from the gene showed the higher degree of identity (about 80%) with *gapa* genes from higher plants and the *gap2* gene of *A. variabilis* [4], the only cyanobacterial *gapa*-like gene currently available in databases. It exhibits typical active-site sequences and a total of five Cys residues. Putative promotor and Shine-Dalgarno sequences were found in the upstream region of the coding sequence which exhibits a GTG as starting codon.

Using the same cloning strategy than with *Synechocystis* we have obtained several complementant clones with genomic libraries from *Anabaena* sp. PCC 7120 and *Calothrix* sp. PCC 7601. All of them presented NAD-linked activity and only a *Calothrix* clone also NADP-linked activity. The recombinant 38 kDa proteins are only slightly detected in Western blots by the anti-GAPDH (*gap2*) antibody and the corresponding genes only dimly recognized in DNA-DNA hybridization experiments with the *gap2* probe (Fig.2). The signal was much clear when a *gapc* probe was used (Fig. 3) thus suggesting they could be *gapc*-like clones which may express an active NAD-dependent GAPDH in these cyanobacteria.

On the other hand, isolation of DNA from strains representing the main cyanobacterial taxonomic groups has been done and PCR amplification experiments have been carried out using as primers conserved sequences present in all *gap* genes known (see Section 2.3.). *Synechococcus* sp. PCC 7942, *Gloeobacter* sp. PCC 7421; *Pseudanabaena* sp. PCC 6903; *Anabaena* sp. PCC 7120, *Calothrix* sp. PCC 7601 and *Fischerella* sp. PCC 7414 were tested, among others, for correct amplifications and in all cases the expected fragment of about 500 bp was obtained. Some of these fragments have been cloned in the pGEMt vector and are currently being sequenced, so far indicating that they are actually part of *gap* genes (data not shown). This would eventually lead us to better understanding of the cyanobacterial *gap* system as well as give clues for systematic and evolutionary questions.

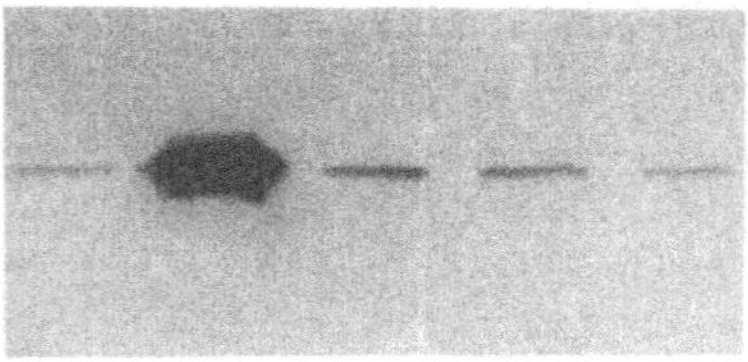

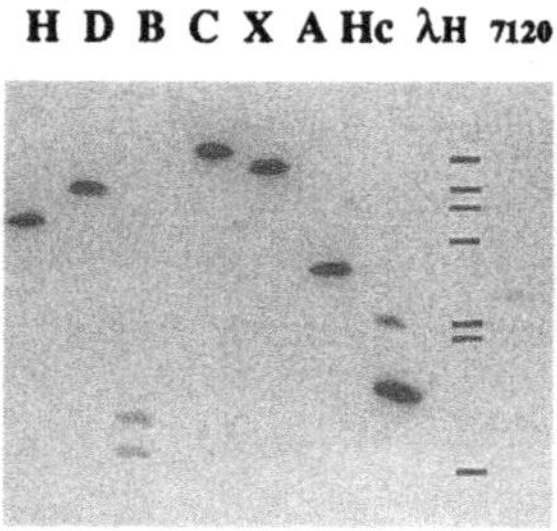

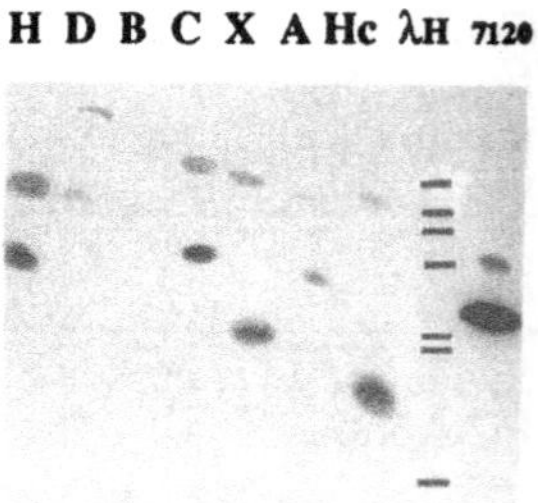

Figure 1 (left). Western blot analysis of crude extracts (70 μg of protein) of *Synechocystis* (lane a); *E. coli* W3CG transformed with pFV8 (lane b); strains PCC 7120, PCC 7119 and ATCC 29413 (lanes c-e, respectively). In all cases the anti-GAPDH antibody recognizes the same 38 kDa protein. ***Figure 2 (center).*** Homologous Southern blot of *Synechocystis* PCC 6803 genomic DNA restricted with different enzymes and the 350 bp BglII-AvaI probe from pFV8. H:HindII; D:DraI; B:BstI; C:ClaI; X:XbaI; A:AvaII; Hc:HincII; λH:marker (23.1, 9.4, 6.5, 4.3, 2.3, 2.0, 0.56 Kb); 7120:complementant clone from *Anabaena* PCC 7120 (see text). ***Figure 3 (right).*** Heterologous Southern blot of 650 bp Sau 3AI probe from *E. coli gapc* gene and the same filter used in Fig. 2.

Supported by grants PB91-0085 and PB94-0033 of DGICYT and Junta de Andalucía.

References

1. Fothergill-Gilmore, L.A. and Michels, P.A.M. (1993) Prog. Biophys. Molec. Biol. 59, 105-235
2. Cerff, R. (1982) *in* Methods in Chloroplast Molecular Biology (Edelman, M., Hallick, R.B. and Chua, N.H., eds.) pp. 683-694 , Elsevier Biomedical Press, Amsterdam
3. Hood, W and Carr, N.G. (1969) Planta 86, 250-258
4. Martin, W., Brinkmann, H., Savonna, C. and Cerff, R. (1993) Proc. Natl. Acad. Sci. USA 90, 8692-8696
5. Ganter, C. and Plückthun, A. (1990) Biochemistry 29, 9395-9402
6. Mateos, M.I. and Serrano, A. (1992) Plant Sci. 84, 163-170
7. Sambrook, J., Fritsch, E.F. and Maniatis, T. (1989) *in* Molecular Cloning, a Laboratory Manual. 2nd edition.(vol. I) Cold Spring Harbor Laboratory Press, New York
8. Conway, T; Sewell, G.W. and Ingram, L.O. (1987) J. Bacteriol. 169, 5653-5662.

CHARACTERIZATION OF A cDNA ENCODING A FUCOXANTHIN-CHLOROPHYLL PROTEIN FROM THE CHROMOPHYTE ALGA Heterosigma carterae

D.G. Durnford & B.R. Green
Dept. of Botany, University of British Columbia, Vancouver, B.C. Canada V6T 1Z4.

1. Introduction

Heterosigma carterae is a unicellular marine raphidophyte which, along with other chromophyte algae, is characterized by the presence of two additional membranes around the chloroplast and by the possession of both chlorophylls *a* and *c*. Its fucoxanthin-Chl *a/c* proteins (FCP) are immunologically related to those of diatoms and to the Chl *a/b* (CAB) polypeptides of green plants and chlorophyte algae (1). We have previously reported the isolation of a number of related antenna polypeptides of *Heterosigma* (1). We report here the cDNA sequence encoding the major FCP of *Heterosigma* and show that it is indeed a member of a complex multi-gene family.

2. Procedure

2.1 *Cloning, sequencing, blotting*

All methods were according to (2). A cDNA library was made from *Heterosigma* polyA+ RNA (isolated according to 3) using Lambda-Zap and Giga-Pak Gold kits (Stratagene) according to the manufacturer's recommendations. Several tryptic peptide sequences were obtained from the major FCP of *Heterosigma*, purified as previously described (1). They were used to design degenerate primers to PCR amplify a gene specific probe from genomic DNA, which was used to isolate a number of clones from the cDNA library. The largest clone obtained did not have the N-terminal end; this was obtained by the RACE technique (4). A ^{32}P labelled StyI/BssHII fragment from the *Fcp*1 cDNA was used to probe a blot of total RNA to determine transcript size. The same probe was used for a Southern blot of *Heterosigma* genomic DNA restricted with the enzymes SacI and DraI.

2.2 *Sequence Analysis*

Protein sequences were aligned using the program Macaw (Ver 2.03, (5)) with manual refinement. Pairwise distances between the sequences were calculated with the PROTDIST program (Dayhoff weighting option) and the tree constructed using the Neighbor Joining method (6) found in the PHYLIP package (7). Only the most reliably aligned parts of the sequences were included in the analysis.

3. Results

The nucleotide sequence of the complete *Fcp*1 cDNA is shown in Fig.1, with its deduced

P. Mathis (ed.), Photosynthesis: from Light to Biosphere, Vol. I, 963–966.

```
    •+1/1                                       •28/10
aaa atg tct ctc aag ctc gcc acc ctc gct gcc gcc ctc atg ggt gcc tcc gcc ttc gtg
    met ser leu lys leu ala thr leu ala ala ala leu met gly ala ser ala phe val

•58/20                                          •88/30
gcc ccc aac aag atg ggc gtg gcc aag tcc tcc tcc gcc ctg aag atg agc ttc gag aac
ala pro asn lys met gly val ala lys ser ser ser ala leu lys met ser phe glu asn
                                                                 Δ
•118/40                                         •148/50
gag atc ggc gtc cag gcc cct ctg ggc ttc tgg gac cct ctg ggc ctt ctg gat gag gcc
glu ile gly val gln ala pro leu gly phe trp asp pro leu gly leu leu asp glu ala

•178/60                                         •208/70
gac cag gag cgg ttc gac cgc ctc cgt acc gtc gag atc aag cac ggc cgc atc tcc atg
asp gln glu arg phe asp arg leu arg thr val glu ile lys his gly arg ile ser met
                                     T   V   E   I   K   X               S   M
•238/80                                         •268/90
ctt gcc atc ctt ggc cac ctg gtg acc acc gct ggt gtg cgt ctg cct ggc gct tac gac
leu ala ile leu gly his leu val thr thr ala gly val arg leu pro gly ala tyr asp
 L   A  [Y]  L   G   F?  L   V   T  [F]  A   G?          L   P   G   A   Y   D
•298/100                                        •328/110
ctg gct ggc gat cag ttc tcc agc ctg ccc acc ggc ctg aag gct ctg tct gct ctg cct
leu ala gly asp gln phe ser ser leu pro thr gly leu lys ala leu ser ala leu pro
 L   A   G   D   Q   F   S   S   L   P   T
•358/120                                        •388/130
gct gct ggt gtg gcc cag acc atc ggc ttc att ggt ctg att gag ctg ggt ttt gct cag
ala ala gly val ala gln thr ile gly phe ile gly leu ile glu leu gly phe ala gln

•418/140                                        •448/150
atc aag gag gag ctg gag gct gac tgc gag gcc cgc atg gac gct gcc ggg tgg gat gac
ile lys glu glu leu glu ala asp cys glu ala arg met asp ala ala gly trp asp asp
         E   E   L?  E?  G?  G?  X   E   A
•478/160                                        •508/170
gag aag aag gac tcc aag cgc gca att gag ctg aac aac ggc cgt gcc gcc cag atg ggt
glu lys lys asp ser lys arg ala ile glu leu asn asn gly arg ala ala gln met gly

•538/180                                        •568/190
atc ctt gcc ttg atg gtc cac gag cag ctg gac aac aac cct tac atc atc aac tct ctg
ile leu ala leu met val his glu gln leu asp asn asn pro tyr ile ile asn ser leu
                                                             I   I   N   S   L
•598/200                                        •628/210
ctg ggc tcc cct gtg gac ttc aac gct ggc ttc taa aca gtg ttt ttt ttt cct tct cca
leu gly ser pro val asp phe asn ala gly phe OCH
 L   G   S   P   V  E/D  F   N   A   G   F   H?

aaa att ttt gac ttt ctc ttg acc ttc tga ctt ttc tgg gcc cag gct ctt tct agt gtt
ggg ttg ctg tgc tgg gtt agt tcg cag ttt gta gcc tga tgc gtg ttt cat cat ata gac
tct ttt gtt ggt atc ttg act tga ctt ctt ttg ctt gac tgc tca caa tct ttc ata cac
ttt gaa cta tag tac cct tct ttc •858
```

Figure 1

Nucleotide sequence of the *Fcp*1 cDNA from *Heterosigma*. The nucleotide and amino acid positions are given above the first nucleotide of the codon (•). Tryptic fragment sequences are shown in bold below the appropriate amino acid. Ambiguous amino acids are indicated by a question mark (?). Amino acids not matching the derived gene sequence are indicated by square brackets []. A putative processing site is indicated by an open triangle (Δ).

amino acid sequence in 3-letter code, and the amino acid sequences of tryptic peptides in 1-letter code (bold). The close match between deduced sequence and tryptic peptides shows that this gene is encoding the major FCP of *Heterosigma*. This gene is clearly a member of a multigene family as shown by the Southern blot in Fig.2A. The blot of total genomic DNA was probed with a *Fcp*1 fragment at 65°C (1X SSC). It revealed the presence of multiple bands of 1.1 kb or higher. The intensity of hybridization varied among bands, suggesting that there are multiple gene copies on a fragment or that not all gene copies are equally related. This was confirmed by washing the blot at more stringent temperatures (data not shown). An RNA blot (Fig.2B) shows that the *Fcp*1 gene transcript is abundant and about 0.98 kb. The width of the band may be due to the presence of transcripts from of number of related but not identical genes.

Fig.3 shows an unrooted distance tree of sequences from the Chl *a/c* antenna proteins of several chromophyte algae: *Heterosigma* (raphidophyte), *Phaeodactylum* and *Odontella* (diatoms), *Macrocystis* (phaeophyte) and the dinoflagellate *Amphidinium*. These are compared with all members of the tomato CAB family (labelled according to gene type) and Lhca and Lhcb genes (no type known) from the chlorophyte algae *Dunaliella* and *Chlamydomonas*, and the euglenophyte *Euglena*.

This preliminary tree shows a clear separation of tomato LHCII (Lhcb1-3) and CP26 (Lhcb5) protein sequences from the LHCI (Lhca1-4), CP29 (Lhcb4) and CP24 (Lhcb6) sequences. The green algal and *Euglena* sequences group with the other CABs, but cannot be assigned more specifically to higher plant gene types. The FCP sequences from different chromophyte algae form a group apart from the intrinisic peridinin-Chl protein (*i*PCP) sequences of the dinoflagellates, but they are more related to them than to the CABs.

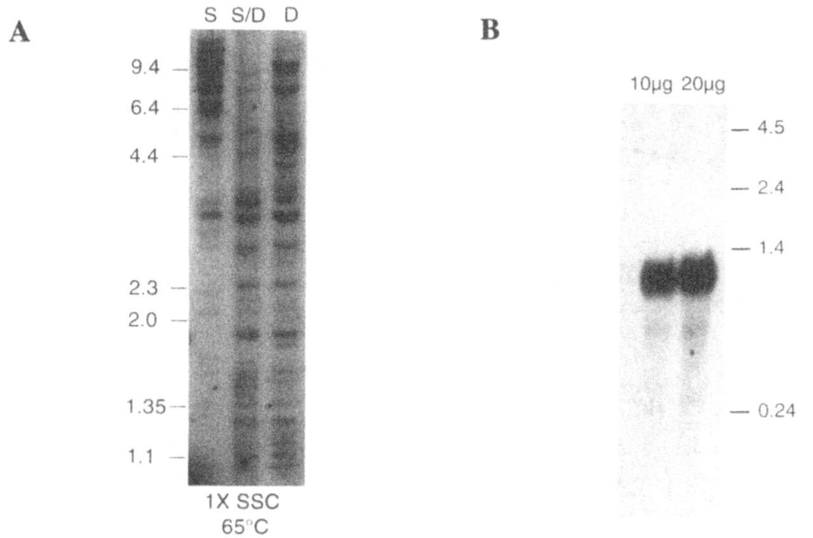

Figure 2

Southern blot of *Heterosigma* genomic DNA (A) and a Northern blot of total RNA (B) probed with a StyI/BssHII fragment from the *Fcp*1cDNA. Enzymes used in the genomic digests of the southern blot are: S, SacI and D, DraI.

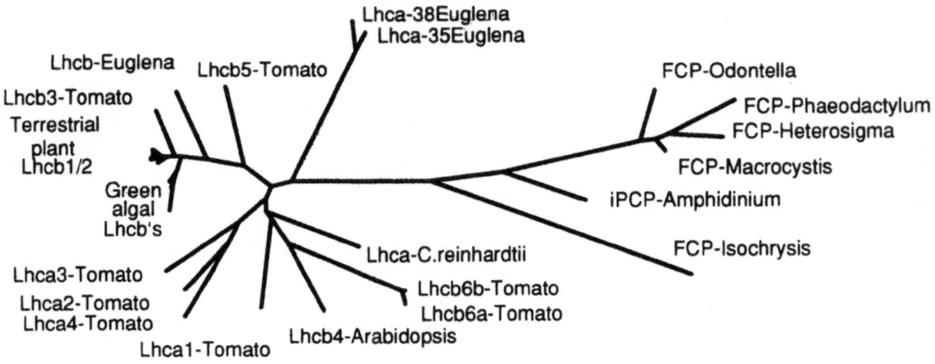

Figure 3

A Neighbor Joining tree showing the relationships between the CABS and the FCPs. Lengths of the branches are proportional to the distances between the sequences and as indicated by the scale.

4. Conclusions

The Southern blot analysis indicates that the *Fcp*1 gene is part of a large nuclear encoded family with over twenty members. This family would therefore appear to be just as complex as the Chl *a/b* family of higher plants (8). Both the Southern analysis and our previous immunological work (1) indicate that there are various degrees of relatedness within this family, as in the CAB family.

The FCPs and *i*PCPs are evolutionarily related to the CABs, although the relationship is somewhat distant. However, the central core residues that bind Chl *a* and those thought to participate in helix stabilizing ionic interactions (9). This means that the overall folding pattern of the FCPs should be similar to the CABs, and some of the same residues should be binding Chl *a* (9,10).

References

(1) Durnford, D.G. and Green, B.R. (1994) Biochim. Biophys. Acta 1184, 118-123
(2) Sambrook, J., Fritsch, E.F. and Maniatis, T. (1989) Molecular Cloning, a laboratory Manual. Cold Spring Harbour Press.
(3) Chromczynski, P. and Sacchi, N. (1987) Anal. Biochem. 162, 156-159
(4) Jain, R., Gomer, R.H., and Murtagh, J.J. (1992) Biotechniques 12(1), 58-59
(5) Schuler, D.M., Altschul, S.F. and Lipman, D.J. (1990) Proteins: structure, function and genetics 9, 180-190
(6) Saitou, N. and Nei, M. (1987) Mol. Biol. Evol. 4(4), 406-425
(7) Felsenstein, J. (1992) PHYLIP . Seattle, WA. University of Washington
(8) Green, B.R., Pichersky, E. and Kloppstech, K. (1991) Tr. Biochem. Sci. 16, 181-186
(9) Kühlbrandt, W., Wang, D.N. and Fujiyoshi, Y. (1994) Nature 367, 614-621
(10) Green, B.R. and Kühlbrandt, W. (1995) Photosyn. Res. In Press.

A STRUCTURAL COMPARISON OF THE PHOTOACTIVE YELLOW PROTEINS (PYP) FROM THREE HALOPHILIC PURPLE PHOTOTROPHIC BACTERIA

G. Van Driessche @, B. Samyn @, M. Koh @, G. Van den Broeck @, T.E. Meyer # and J. Van Beeumen @

@ Dept. of Biochemistry, Physiology and Microbiology, University of Gent, Ledeganckstraat 35, B-9000 Gent, Belgium
# Dept. of Biochemistry, University of Arizona, Tucson, AZ 85721

1. Introduction

Photoactive yellow protein (PYP) has been isolated from three halophilic purple bacteria: *Ectothiorhodospira halophila* [1], *Rhodospirillum salexigens* [2] and *Chromatium salexigens* (Meyer et al., unpublished). The photoactivity of PYP was determined to be remarkably similar to that of the bacterial rhodopsins [3, 4]. Bacteriorhodopsin is a membrane-bound protein of 33 kDa containing a Shiff based linked retinal chromophore [5], whereas PYP is small and completely water soluble. Amino acid sequence determination of *E. halophila* PYP showed that the polypeptide chain contains 125 amino acids [6]. The only cysteine residue, at position 69, covalently binds the chromophore [6]. It was recently proven by spectroscopic, mass spectrometric and X-ray crystallographic analyses that the chromophore is p-coumaric acid [7, 8]. The three-dimensional structure of *E. halophila* PYP, first determined by McRee et al. [9] and recently refined at 1.4 Å resolution [10], indicates that PYP contains a central beta sheet flanked on either side by short loops and helices while bacteriorhodopsins contain seven closely packed membrane-spanning α-helices [11]. Thus, it is clear that, with exception of the photocycle, bacteriorhodopsins and PYP's are structurally quite different. Little is known about the functional role of PYP. There is some evidence that it is involved as photoreceptor in negative bacterial phototaxis in the presence of blue light [12]. To gain some insight into the properties of PYP and to learn what residues are conserved and might therefore be structurally and functionally important, we have determined the sequences of *R. salexigens* and *C. salexigens* PYP.

2. Results and Discussion

The complete amino acid sequences of the photoactive yellow proteins from *Chromatium salexigens* and *Rhodospirillum salexigens* were determined by N-terminal Edman degradation of native and/or electroblotted proteins, and enzymatically generated peptides. The proposed sequences and the nature of the chromophore was further verified by electrospray ionisation mass spectrometry of the native and dithiothreitol treated native protein [13]. Both PYP's are 125 amino acids long and contain one cysteine at position 69.

P. Mathis (ed.), Photosynthesis: from Light to Biosphere, Vol. I, 967–970.

In Figure 1, we have aligned the three different PYP amino acid sequences. No insertions or deletions are required for optimal alignment. The PYP of *E. halophila* is 76.0% identical to PYP from *C. salexigens* and 71.2% to that from *R. salexigens*. There is 69.7% identity of *C. salexigens* PYP with *R. salexigenss* PYP. The high similarity is unexpected since these bacterial species, from strongly different phototrophic families, have very few electron transfer proteins in common [1, Meyer et al., unpublished results]. Moreover, the alignment reveals that cysteine 69, the site of covalent attachment with the chromophore p-coumaric acid as determined for the PYP from *E. halophila* [6, 7], is conserved.

					5					10					15					20					25
(1)	M	E	H	V	A	F	G	S	E	D	I	E	N	T	L	A	K	M	D	D	G	Q	L	D	G
(2)		D	I		H				D					S			N		S		Q	D			Q
(3)			M	I	K			Q	D					A	M		D		G		A		I		D

					30					35					40					45					50
(1)	L	A	F	G	A	I	Q	L	D	G	D	G	N	I	L	Q	Y	N	A	A	E	G	D	I	T
(2)										A	S		K	V									G		
(3)										E	T		T			A							E	L	

					55					60					65					70					75
(1)	G	R	D	P	K	E	V	I	G	K	N	F	F	K	D	V	A	P	C	T	D	S	P	E	F
(2)						S								E							K		Q		
(3)			S		Q	D										I						T	E		

					80					85					90					95					100
(1)	Y	G	K	F	K	E	G	V	A	S	G	N	L	N	T	M	F	E	Y	T	F	D	Y	Q	M
(2)	Q		R							N				A						V					
(3)	G		R		R					N		D			A					V					

					105					110					115					120					125
(1)	T	P	T	K	V	K	V	H	M	K	K	A	L	S	G	D	S	Y	W	V	F	V	K	R	V
(2)	K													V	D					I					L
(3)	Q										R		I	T						I					

Figure 1. Alignment of the photoactive yellow proteins isolated from *Ectothiorhodospira halophila* (1), *Chromatium salexigens* (2) and *Rhodospirillum salexigens* (3). Cysteine at position 69 is the site of the covalent attachment of the chromophore. For the latter two sequences, only residues that differ from the corresponding positions in (1) are given. The underlined residues consitute the six β-strands that constitute the central β-sheet [10].

A possible explanation for the high sequence similarity of PYP's from the three different purple phototrophic families is that these proteins exhibit an important and similar function which would be lost if there is greater sequence divergence. The N-terminal half of the three PYP sequences appears to be more variable than the second half. This may be related to the presence of the covalent binding site of the chromophore in the second half of the protein. The refined three-dimensional structure of *E. halophila* PYP [10] shows that the two halves of the protein are separated by a central beta sheet. The first twenty six amino acids are on the opposite side of the chromophore binding region and therefore contribute to a greater variability to the first half of the protein [10].

The three-dimensional structure of *E. halophila* PYP [10] indicates that the conserved amino acids Tyr98, Arg52, Thr50, Glu46 and Tyr42 are important for the protein-chromophore interaction in that the last two residues make a hydrogen bond with the phenolic oxygen of the chromophore during the dark state of the photocycle [10] (Figure 2). Further, recent spectroscopic experiments reveal that the primary step of the photocycle involves cis-trans isomerization of the acrylic double bond whereby the chromophore, Tyr98, Arg52 and likely also Val66, probably swing out so that the chromophore hydroxyl is subsequently exposed to solvent where it is protonated [10, 14] (Figure 2). This phenomenon would explain the overall observed conformational change of the protein during the light-induced chromophore bleaching [15,], the increased exposure of hydrophobic residues, and proton uptake [16, 17].

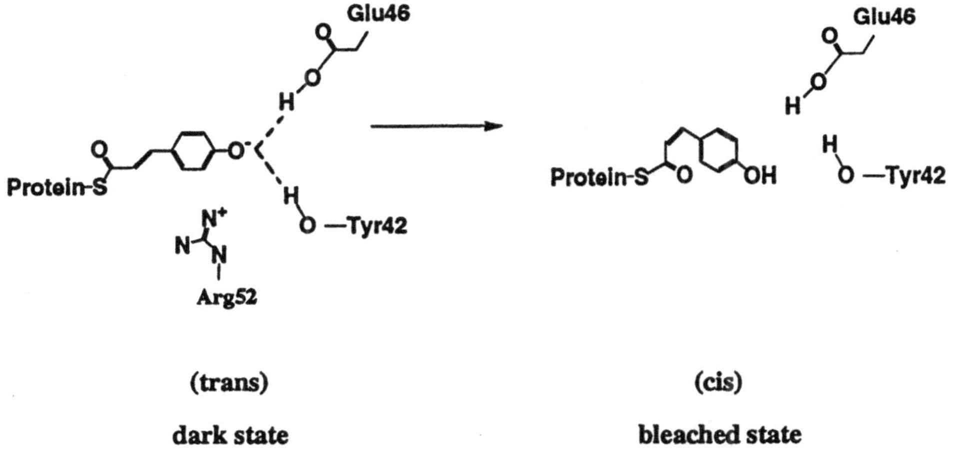

Figure 2. Model for interaction of the cysteine-chromophore with Tyr42 and Glu46 or Arg52 during cis-trans isomerisation of the acrylic double bond of the chromophore between the two different states of the photocycle

To locate a possible interaction site of PYP with another reaction partner, we have mapped the conservative residues of the three different PYP sequences onto the refined three-dimensional structure of *E. halophila* [10, 13]. We observe a conserved surface region in the direct spatial environment of cysteine 69 which binds the chromophore and is centered around tyrosine 98. Borgstahl et al. [10] proposed that arginine 52, which is located in a cavity of the protein surface just under the chromophore molecule and which shields the chromophore from solvents, functions as a gateway for transmitting the phototactic signal to the outside of the protein. We propose that the interaction site may be composed of either arginine 52 or tyrosine 98, in combination with the cysteine-chromophore complex.

Mapping of the conserved positively and negatively charged residues onto the *E. halophila* PYP three-dimensional model shows that the negatively charged patch, composed of the residues Glu2, Asp19, Asp20, Asp24, Asp48 and Asp53, described by Borgstahl et al. [10, 13] is not conserved. This negatively charged site is not close to Tyr98 and is even more highly charged by both positively and negatively charged residues in the new sequences: e.g. Lys5, Asp17 and Asp25 in *R. salexigens* PYP and Asp22 in *C. salexigens* PYP.

3. References

1 Meyer, T.E. (1985) Biochem. Biophys. Acta 806, 175-183
2 Meyer, T.E., Tollin, G., Causgrove, T.P., Cheng, P. and Blankenship, R.E. (1991) Biophys. J. 59, 988-991
3 Hoff, W.D., Kwa, S.L., Van Grondelle, R. and Hellingwerf, K.J. (1992) Photochem. Photobiol. 56, 529-539
4 Miller, A., Leigeber, H., Hoff, W.D. and Hellingwerf, K.J. (1993) Biochim. Biophys. Acta 1141, 190-196
5 Oesterhelt, D. and Stoeckenius, W. (1973) Proc. Natl. Acad. Sci. USA 70, 2853-2857
6 Van Beeumen, J.J., Devreese, B., Van Bun, S.M., Hoff, W.D., Hellingwerf, K.J., Meyer, T.E., McRee, D.E. and Cusanovich, M.A. (1993) Protein Sci. 2, 1114-1125
7 Hoff, W.D., Van Stokkum, I.H.M., Van Ramesdonk, H.J., Van Brederode, R., Brouwer, A.M., Fitch, J.C., Meyer, T.E., Van Grondelle, R. and Hellingwerf, K.J. (1994) Biophys. J. 67, 1691-1705
8 Baca, M., Borgstahl, G.E.O., Boissinot, M., Burke, P.M., Williams, D.R., Slater, K.A. and Getzoff, E.D. (1994) Biochem. 33, 14369-14377
9 McRee, D.E., Tainer, J.A., Meyer, T.E, Van Beeumen, J., Cusanovich, M.A. and Getzoff, E.D. (1989) Proc. Natl. Acad. Sci. USA 86, 6533-6537
10 Borgstahl, G.E.O., Williams, D.R. and Getzoff, E.D. (1995) Biochem. 34, 6278-6287
11 Henderson, R., Baldwin, J.M., Ceska, T.A., Zemlin, T.A., Beckman, E. and Downing, K.H. (1990) J. Mol. Biol. 213, 899-929
12 Miller, A., Leigeber, H., Hoff, W.D. and Hellingwerf, K.J. (1993) Biochim. Biophys. Acta 1141, 190-196
13 Koh, M., Van Driessche, G., Samyn, B., Van den Broeck, G., Meyer, T.E., Cusanovich, M.A. and Van Beeumen, J.J. (1995) Biochemistry, submitted
14 Van Brederode, M.E., Gensch, T., Hoff, W.D., Hellingwerf, K.J. and Braslavsky, S.E. (1995) Biophys. J. 68, 1101-1109
15 Meyer, T.E., Fitch, J.C., Bartsch, R.G., Tollin, G. and Cusanovich, M.A. (1990) Biochim. Biophys. Acta 1016, 364-370
16 Meyer, T.E., Tollin, G. Hazzard, J.H. and Cusanovich, M.A. (1989) Biophys. J. 56, 559-564
17 Meyer, T.E., Cusanovich, M.A. and Tollin, G. (1993) Arch. Biochem. Biophys. 306, 515-517

ORIGINS OF ORGANELLES IN PLANTS AND ALGAE AS INFERREDFROM COMPARISONS OF HIGHLY CONSERVED CHAPERONE PROTEINS.

Adrián K. Arakaki and Alejandro M. Viale.
Programa Multidisciplinario de Biología Experimental, Departamento de Microbiología, Facultad de Ciencias Bioquímicas y Farmacéuticas, Universidad Nacional de Rosario, Suipacha 531, 2000 Rosario, Argentina.

1. Introduction

Although a large amount of molecular evidence provides now an amply support to the endosymbiotic hypothesis, these very data also brought about controversial observations on the *specific* affiliation of the bacteria from which chloroplasts and mitochondria aroused (1-5,7). The arguments reside now over whether each of these organelles derive from a single event of endosymbiosis, or represent the aftermaths of several and independent events involving distinct prokaryotic lineages (1-5). In the case of chloroplasts, pronounced structural and biochemical similarities, patterns of gene organization and expression, sequence similarities, etc., advocate strongly for a cyanobacterial ancestry (1-5). Nevertheless, controversial phylogenies obtained from several molecules have raised doubts on whether non-green algae plastids (*i.e.*, rhodophytes, chromophytes, cryptophytes) derive from the same cyanobacterial lineage than green algae and land plants (1-5,7). Moreover, inferences based in ribulose bisphosphate carboxylase (Rubisco) genes (*rbc*) have been interpreted as indicating that plastid genomes in non-green algae derive (at least in part) from α- or β-proteobacterial ancestors (2,4).

The molecular chaperones represent distinct families of essential proteins, ubiquitously distributed among eubacteria, mitochondria and chloroplasts (6). Some of these highly conserved proteins, such as Hsp60 and Hsp70, have also proved to constitute valuable phylogenetic tools (7-9). We have drawn evolutionary trees based in these molecules, and these inferences support a common origin of all plastids from within the cyanobacterial lineage.

2. Procedure

Since comparisons of polypeptide sequences seem to provide some advantages over those of nucleotide sequences (10), we have used translated sequences in this work. Alignments were done as described (7,8), and refined by visual inspection. Evolutionary distance matrices were computed by using the amino acid conversion table compiled by Dayhoff *et al.* (7,8). Phylogenetic trees were drawn by the neighbor-joining (NJ)

P. Mathis (ed.), Photosynthesis: from Light to Biosphere, Vol. I, 971–974.

distance method, and confidence limits were placed by using the bootstrap procedure (7,8). Hsp sequences from plastids, mitochondria and bacteria used in this study were obtained from *Entrez* (Release 16.0) databases (National Center for Biotechnology Information). The sequences of *Chlamydomonas reinhardtii* and *Porphyra purpurea* Hsp60s were kindly provided to us by Drs. S. Gantt and M. Reith, respectively.

3. Results and Discussion

We have shown previously that evolutionary trees based in Hsp60 support an origin of plant and non-plant mitochondria among the α-proteobacterial *Ehrlichia/Rickettsia* group (7). In phylogenetic inferences based in the highly conserved chaperones Hsp60 (Fig. 1) or Hsp70 (9, A.M. Viale and A.K Arakaki, unpublished), which include both plastid and nuclear-encoded proteins (6-9), plant plastids as well as green and non-green algae plastids clustered with their homologs present in cyanobacterial species. These results, which are in agreement to rRNA-derived trees (1-5), supports the notion that all plastids originated among the cyanobacterial lineage. The presence of distinct Hsp60 in *C. reinhardtii* showing homology to the α- and β- subunits of plant chloroplast Hsp60s (11, Fig. 1) suggests that this gene duplication occurred in the lineage that led to land plants, as mentioned previously (7). It is worth mentioning the closer relationship between the operon-encoded Hsp60s from cyanobacteria and those of rhodophytes (Fig. 1), as noted previously (7). Interestingly, the second cyanobacterial Hsp60 (*Synechocystis*2) seems more related to its green algae/plant counterparts (Fig. 1).

In Rubisco-based phylogenetic trees, the non-green algae cluster with the α- and β-Proteobacteria, and separately from the cyanobacteria, green algae and plants which cluster with the γ-Proteobacteria (1-5, A.M. Viale and A.K. Arakaki, unpublished). To explain the differences between these inferences from those of other macromolecules (1-5,7-9, this work), lateral transfer of genes from a proteobacterial endosymbiont (including the protomitochondrion [4]) in the non-green algae lineage has been postulated (2-5). Nevertheless, no further evidence other than Rubisco comparisons support the latter hypotheses (1-5,7-9, this work) and, in fact, the closer affiliation in both rRNA and Hsp60 trees (8) of β- and γ-proteobacterial species (rather than α- and β-Proteobacteria) argues against it. Therefore, we favor the alternative explanation of a differential retention of duplicated Rubisco operons originally present in the common photosynthetic ancestor of cyanobacteria and Proteobacteria (2), which also agrees with our previous proposal of the presence of paralogous Rubisco operons in the common ancestor of the α-, β- and γ-proteobacterial lineages (12).

The common cyanobacterial ancestor to all plastids may thus have contained *multiple* gene duplications (including genes other than *rbc*, *e.g. hsp60*), which were differentially lost during the diversification that followed within the distinct photosynthetic lineages. This possible scenario is exemplified in Fig. 2, in which the latter step would represent the multiple copies of a given gene (in this case, *rbc*) in the common ancestor to cyanobacteria and Proteobacteria. Further data supporting this hypothesis include, among others, the observed clustering of Rubisco sequences

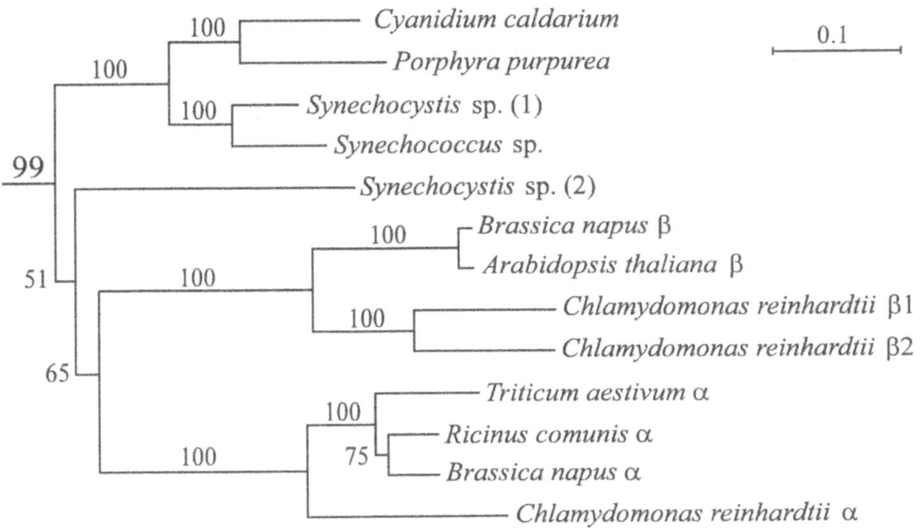

Figure 1. Evolutionary relationships between chloroplasts and (eu)Bacteria. Unrooted NJ phylogenetic trees were derived using Hsp60 protein sequences, and the cyanobacteria/plastids subtree is shown. The evolutionary distance scale (substitutions/site) and the bootstrap support values (percentage of trees) are indicated.

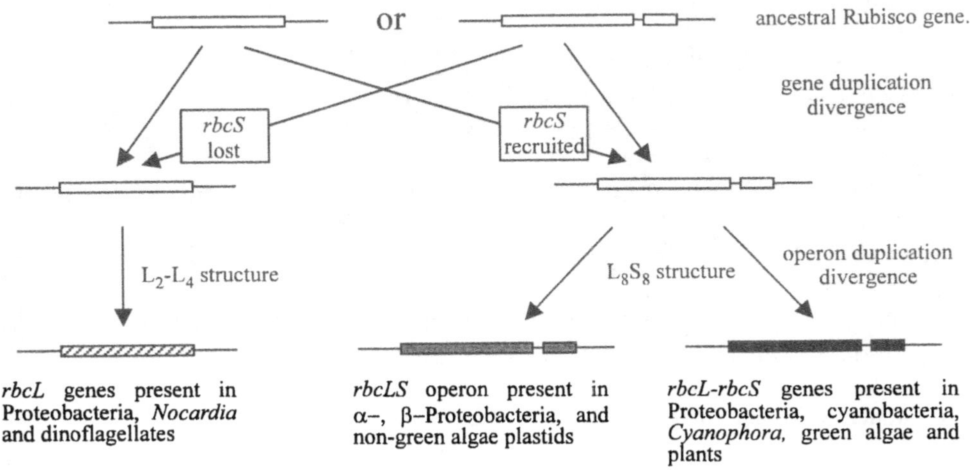

Figure 2. Proposed evolution of Rubisco genes in (eu)Bacteria and photosynthetic eukaryotes. The common ancestor to Proteobacteria and cyanobacteria possessed multiple Rubisco genes (lower part of the figure), that generated the distinct enzymes found in extant autotrophic organisms. Two possible scenarios are shown, in which the aboriginal Rubisco was either composed only by large subunits (upper left), or both large and small subunits (upper right).

belonging to α-Proteobacteria (*i.e.*, *Nitrobacter* sp.) with the cyanobacteria/ green algae/ plant cluster rather than to the non-green algae (A.M. Viale and A.K. Arakaki, unpublished), the presence of Rubiscos composed only by large-subunits in dinoflagellates (13) as well as in phototrophic and chemotrophic bacteria (14-17), and the immunological relatedness of antenna complex polypeptides from rhodophytes, chlorophytes and chromophytes (18), which supports the proposal that the ancestor to all plastids also contained different antenna systems (19).

Acknowledgments

Supported by grants from Fundación Antorchas (Argentina), the Third World Academy of Sciences (Italy) and the International Foundation for Science (Sweden).

References

1 Howe, C.J., Beanland, T.J., Larkum, A.W.D. and Lockhart, P.J. (1992) Trends Ecol. Evol. 7, 378-383.
2 Martin, W., Somerville, C.C. and Loiseaux de Goër, S. (1992) J. Mol. Evol. 35, 385-404.
3 Cavalier-Smith, T. (1992) BioSystems 28, 91-106
4 Palmer, J.D. (1993) Nature 364, 762-763.
5 Gray, M.W. (1993) Curr. Opin. Genet. Develop. 3, 884-890.
6 Ellis, R.J. (1994) Curr. Opin. Struct. Biol. 4, 117-122.
7 Viale, A.M. and Arakaki, A.K. (1994) FEBS Lett. 341, 146-151.
8 Viale, A.M., Arakaki, A.K., Soncini, F.C. and Ferreyra, R.G. (1994) Int. J. Syst. Bacteriol. 44, 527-533
9 Falah, M. and Gupta, R.S. (1994) J. Bacteriol. 176, 7748-7753.
10 Hasegawa, M. and Hashimoto, T. (1993) Nature 361, 23.
11 Thompson, M.D., Paavola, C.D., Lenvik, T.R. and Gantt, J.S. (1995) Plant Mol. Biol. 27, 1031-1036.
12 Kobayashi, H., Viale, A.M., Takabe, T., Akazawa, T., Wada, K., Shinozaki, K., Kobayashi, K. and Sugiura, M. (1991) Gene 97, 55-62.
13 Morse, D., Salois, P., Markovic, P. and Woodland Hastings, J. (1995) Science 268, 1622-1624.
14 Tabita, F.R., Gibson, J.L., Bowien, B., Dijkhuizen, L. and Meijer, W.G. (1992) FEMS Microb. Lett. 99, 107-110.
15 Kalkus, J., Reh, M. and Schlegel, H.G. (1990) J. Gen. Microbiol. 136, 1145-1151.
16 Stoner, M.T. and Shively, J.M. (1993) FEMS Microb. Lett. 107, 287-292.
17 Yaguchi, T., Chung, S.Y., Igarashi, Y. and Kodama, T. (1994) Biosci. Biotech. Biochem. 58, 1733-1737.
18 Wolfe, G.R., Cunningham, F.X., Durnford, D., Green, B.R. and Gantt, E. (1994) Nature 367, 566-568.
19 Bryant, D.A. (1992) Curr. Biol. 2, 240-242.

PHYLOGENETIC ANALYSIS OF PHOTOSYNTHETIC REACTION CENTERS OF PURPLE BACTERIA AND GREEN FILAMENTOUS BACTERIA

Kenji V.P. Nagashima[1], Satoshi Hanada[1], Akira Hiraishi[2], Keizo Shimada[1] and Katsumi Matsuura[1]
[1]Department of Biology, Tokyo Metropolitan University, Hachioji, Tokyo 192-03, Japan and [2]Laboratory of Environmental Biotechnology, Konishi Co. Ltd., Sumida, Tokyo 130, Japan

1. Introduction

Photosynthetic reaction center complexes of purple bacteria and green filamentous bacteria resemble each other and are the best characterized of all photosynthetic reaction centers [1,2]. The core part of the complex consists of two related integral membrane proteins, known as L and M subunits, with bacteriochlorophylls and other cofactors. The cytochrome subunit with four *c*-type hemes is attached to the LM core complex at the periplasmic side.

According to the phylogenetic analysis of bacteria based on 16S rRNA, the group of green filamentous bacteria forms a deep branch on the common line of eubacterial descent [3]. Four other groups of phototrophic bacteria, i.e., purple bacteria, green sulfur bacteria, heliobacteria and cyanobacteria, have radiated significantly later than the branching. The early branching of the green filamentous bacteria is supported by comparative sequence analysis of elongation factor Tu [4]. To explain the distant relationship in phylogeny and the similarity in the reaction center complex between the green filamentous bacteria and the purple bacteria, Blankenship [2] suggested a possibility of a horizontal transfer of photosynthetic genes between ancestor species of purple bacteria and *Chloroflexus*.

In purple bacteria, two variations of reaction center complexes are known in terms of the presence of the cytochrome subunit. Examining the phylogenetic distribution of the reaction centers without the cytochrome subunit, we suggested that the common ancestor of purple bacteria had the cytochrome subunit and the subunit was independently lost in at least four lines in each subgroup of the α subclass of purple bacteria [5,6]. On the other hand, we have also suggested a possibility of horizontal transfer of reaction center genes from an ancestral *Rhodospirillum* species to the ancestor of *Rubrivivax gelatinosus* [7,8].

To investigate the evolution of reaction centers in purple bacteria and green filamentous bacteria, especially in terms of the frequency of horizontal gene transfer, we determined nucleotide sequences of the reaction center genes in 10 species.

2. Materials and Methods

A part of *puf* operon which includes most of L and M genes was amplified by polymerase chain reaction (PCR) method using two primers as shown in Fig. 1. The primers were designed at 5' terminal region of *puf*L and 3' terminal region of *puf*M using

P. Mathis (ed.), Photosynthesis: from Light to Biosphere, Vol. I, 975–978.

consensus sequences among 6 purple bacterial species. From 60-70% of species examined, the expected fragments of 1.5 kbp were obtained. The fragments were cloned in plasmids pT7Blue(R) or pUC118, fractionated and sequenced using a Taq Dye Primer Cycle Sequencing Kit and a 373A DNA Sequencer (Applied BioSystems).

Alignment of the sequences and construction of phylogenetic trees were carried out using Clustal V and MEGA programs with the neighbor-joining method.

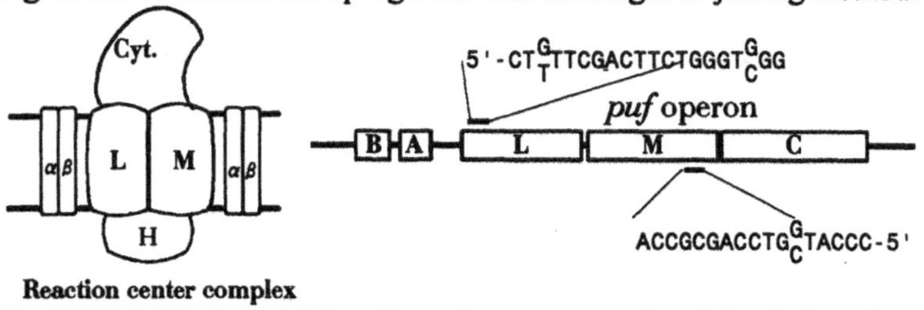

Fig. 1 Schematic representation of reaction center complex of purple bacteria and PCR primers for amplification of L and M subunit genes.

3. Results and Discussion

Figure 2 shows the phylogenetic tree based on the nucleotide sequence of 16S rRNA of purple bacteria and green filamentous bacteria. We determined the sequences in 9 species for the comparison with the reaction center genes. The phylum of purple bacteria is further divided into five subclasses. Three of these, α, β and γ subclasses, include photosynthetic bacteria and are shown in Fig 2. The phylogenetic tree based on the sequence of soluble cytochrome *c* shows high consistency with the tree based on 16S rRNA, and therefore the tree is generally accepted. A newly isolated species of green filamentous bacteria, *Chloroflexus aggregans*, is also included in Fig. 2 [9,10].

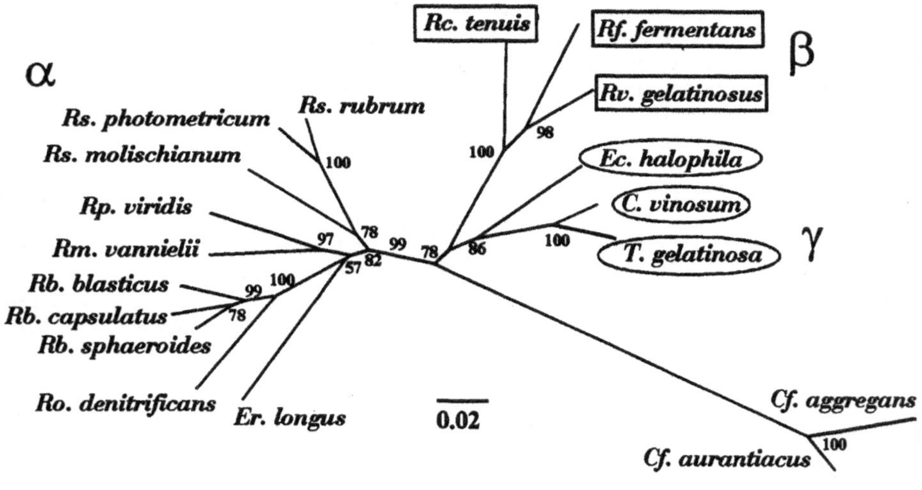

Fig. 2 Phylogenetic tree based on nucleotide sequence of 16S rRNA. Species belonging β and γ subclasses of purple bacteria are surrounded by rectangles and ovals, respectively.

The deduced amino acid sequences of L and M subunits were compared in 17 species. The motif of the primary structures and the important amino acids as the ligands for cofactors are well conserved. Phylogenetic trees based on the sequence alignments are drawn in Fig. 3 for M subunit and in Fig. 4 for L subunit. In both trees of the reaction center proteins, the species belonging to β and γ-subclasses of purple bacteria were all distributed within the cluster of α-subclass, especially near *Rhodospirillum* species. These positions in the phylogenetic trees are inconsistent with those in the trees based on rRNA (Fig. 2) and soluble cytochrome *c* (data not shown), in which each subclass forms a distinct cluster.

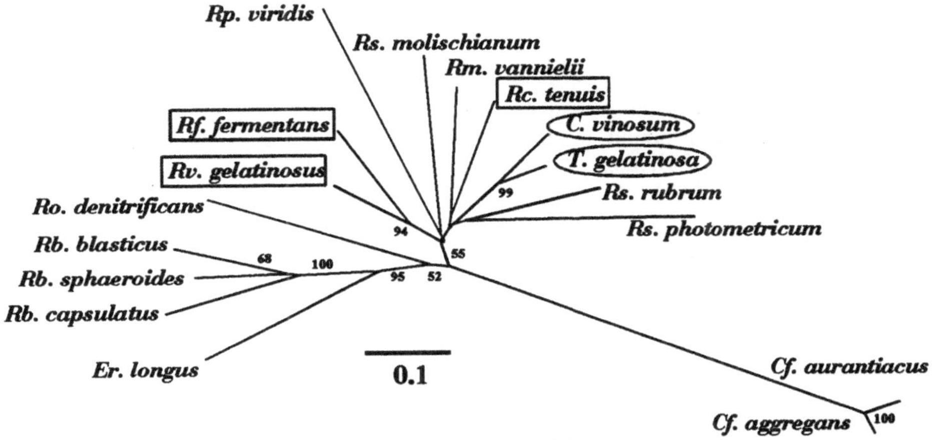

Fig. 3 Phylogenetic tree based on amino acid sequence of M subunit.

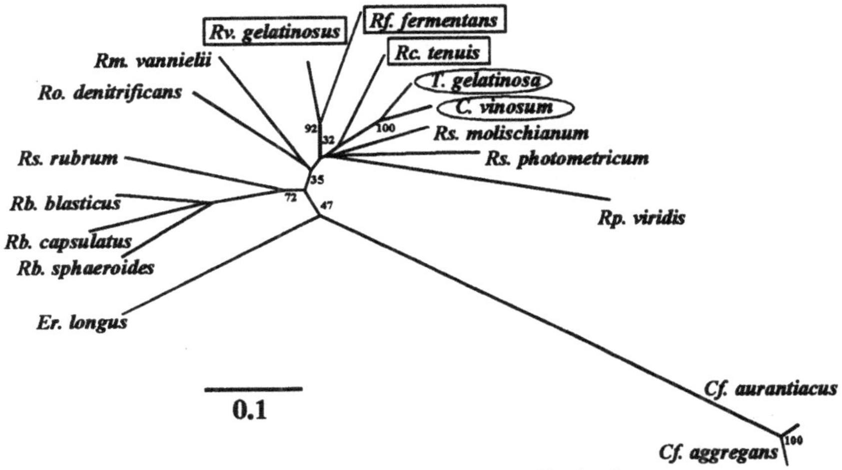

Fig. 4 Phylogenetic tree based on amino acid sequence of L subunit.

On the other hand, the tree topology of the reaction center proteins among β and γ subclasses is in principle similar to that based on rRNA and cytochrome *c*. The topology among species of α subclass after omitting species in β and γ subclasses also retains that based on rRNA with some exceptions which have different positions between L and M subunits. These results indicate that the genes for the reaction center primarily came from the line of descent in purple bacteria, and the major horizontal transfer of the photosynthetic genes took place once from the ancestral *Rhodospirillum* species to a common ancestor of species in β and γ subclasses.

The species which have the reaction center complex without the bound cytochrome subunit, i.e., *Rb. sphaeroides*, *Rb. capsulatus*, *Er. longus* and *Rs. rubrum*, do not form a single cluster in the trees shown in Figs. 3 and 4. This indicates that the reaction center without the cytochrome subunit has not a single origin (and has not spread by horizontal gene transfer) but appeared on different lines with independent mutations as has been suggested previously [5,6].

The *Chloroflexus* species forms a definite outgroup in the trees based on the reaction center genes (Figs. 3 and 4) as in that based on rRNA. Relative distances from the common ancestral point of purple bacteria to each species of purple bacteria and *Chloroflexus* are in principle retained both in the reaction center trees and in the ribosomal tree. This indicates that the suggested possibility of horizontal transfer of photosynthetic genes between *Chloroflexus* and purple bacteria [2] is not supported by the present analysis.

In summary, following conclusions were made. (1) The phylogenetic trees of the reaction center proteins in purple bacteria and green filamentous bacteria are primarily consistent with the tree of 16S rRNA, indicating that the reaction center genes generally came from the line of decent except a single major horizontal transfer of the genes within purple bacteria. (2) The reaction center genes of β and γ subclasses of purple bacteria have been originated from a horizontal gene transfer from an ancestral *Rhodospirillum* species in α subclass. (3) The reaction centers without the bound cytochrome subunit in purple bacteria have not a single origin, in terms of the deletion of the cytochrome subunit.

Acknowledgments: This work was supported in part by the grant-in-aid 05266215 (for Priority Areas), 05640734, and 07839014 from the Ministry of Education, Science, and Culture, Japan.

References

1 Deisenhofer, J., Epp, O., Miki, K., Huber, R. and Michel, H. (1985) Nature 318, 618-624
2 Blankenship, R.E. (1992) Photosynth. Res. 33, 91-111
3 Woese, C.R. (1987) Microbiol. Rev. 51, 221-271
4 Ludwig, W. et al. (1993) Antonie van Leeuwenhoek 64, 285-305
5 Matsuura, K. and Shimada, K. (1990) in Current Research in Photosynthesis (Baltscheffsky, K. Ed.) Vol. 1, pp 193-196, Klewer Academic Publishers, Dordrecht
6 Matsuura, K. (1994) J. Plant Res. 107, 191-200
7 Nagashima, K.V.P., Shimada, K. and Matsuura, K. (1993) Photosynth. Res. 36, 185-191
8 Nagashima, K.V.P., Matsuura, K., Ohyama, S. and Shimada, K. (1994) J. Biol. Chem. 269, 2477-2484
9 Hanada, S., Hiraishi, A., Shimada, K. and Matsuura, K. (1995) J. Gen. Appl. Microbiol. 41, 119-130
10 Hanada, S., Hiraishi, A., Shimada, K. and Matsuura, K. (1995) Int. J. System. Bacteriol. 45, in press

INDEX OF NAMES

GPSR Compliance

The European Union's (EU) General Product Safety Regulation (GPSR) is a set of rules that requires consumer products to be safe and our obligations to ensure this.

If you have any concerns about our products, you can contact us on ProductSafety@springernature.com

In case Publisher is established outside the EU, the EU authorized representative is:

Springer Nature Customer Service Center GmbH
Europaplatz 3
69115 Heidelberg, Germany

Batch number: 10761593

Printed by Printforce, the Netherlands

PHOTOSYNTHESIS: FROM LIGHT TO BIOSPHERE

Photosynthesis: from Light to Biosphere

Volume II

Proceedings of the Xth International Photosynthesis Congress, Montpellier, France, 20–25 August 1995

edited by

PAUL MATHIS

Section de Bioénergétique, CEA-Saclay, France

KLUWER ACADEMIC PUBLISHERS

DORDRECHT / BOSTON / LONDON

Library of Congress Cataloging-in-Publication Data

International Congress on Photosynthesis (10th : 1995 : Montpellier, France)
Photosynthesis : from light to biosphere : proceedings of the Xth International Photosynthesis Congress, Montpellier, France, 20-25 August 1995 / edited by Paul Mathis.
p. cm.
Includes indexes.
ISBN-13:978-94-010-6562-7
1. Photosynthesis--Congresses. I. Mathis, P. II. Title.
QK882.I56 1995
581.1'3342--dc20 95-47402

ISBN-13:978-94-010-6562-7 e-ISBN-13:978-94-009-0173-5
DOI: 10.1007/978-94-009-0173-5

Published by Kluwer Academic Publishers,
P.O. Box 17, 3300 AA Dordrecht, The Netherlands.

Kluwer Academic Publishers incorporates
the publishing programmes of
D. Reidel, Martinus Nijhoff, Dr W. Junk and MTP Press.

Sold and distributed in the U.S.A. and Canada
by Kluwer Academic Publishers,
101 Philip Drive, Norwell, MA 02061, U.S.A.

In all other countries, sold and distributed
by Kluwer Academic Publishers Group,
P.O. Box 322, 3300 AH Dordrecht, The Netherlands.

Printed on acid-free paper

GENERAL CONTENTS

CONTENTS TO VOLUME II

4. Reaction centers : green sulfur bacteria and PSI

PRESENT STATE OF THE CRYSTAL STRUCTURE ANALYSIS OF PHOTOSYSTEM I AT 4.5 Å RESOLUTION

W.D.Schubert*, O.Klukas*, N.Krauß*, W.Saenger*, P.Fromme, and H.T.Witt

* Institut f. Kristallographie, Freie Universität Berlin, Takustr.6, 14195 Berlin
Max-Volmer-Institut, Technische Universität Berlin, Straße des 17. Juni 135, 10623 Berlin, Germany

1. Introduction

In water-oxidizing photosynthesis of higher plants and cyanobacteria two photosystems, PS I and PS II, coupled in series, drive the overall process. The primary act starts with a light-induced charge separation, whereby both in PS I as well as in PS II an electron is channeled vectorially from excited chlorophylls near the lumenal side of the photosynthetic membrane via intermediate electron carriers to electron acceptors near the stromal side (1). Despite these two similar primary acts in both complexes, the function and structure of PS I and PS II are very different (2). In photosynthesis only the structure of the photosynthetic reaction center (RC) of purple bacteria is known at high resolution (3). Although this bacterial RC is not able to oxidize water, it displays important functional and structural similarities to those of PS II (4,5). However, no such indirect information is available for PS I, because it is evolutionarily related to the reaction center of green sulfur bacteria (6,7) of unknown structure. Recently, however, structural information on PS I was made available following the successful crystallization of PS I from the thermophilic cyanobacterium *Synechococcus elongatus*. The molecular mass of each of the 11 subunits constituting the complex in this bacterium was obtained from their genetically-deduced amino acid sequences (8). The two large membrane intrinsic subunits A and B (~ 83 kDa) bind the main parts of the electron transfer chain (9); the primary electron donor P700 (probably a Chl *a* dimer), A_0 (a Chl *a*), A_1 (vitamin K_1), and a 4Fe-4S cluster, F_X. The electron transfer chain continues to subunit C (8.8 kDa) located towards the stromal side, containing the terminal electron acceptors - two 4Fe-4S clusters, F_A and F_B. Subunit C is surrounded by subunit D (15.4 kDa) and E (8.4 kDa). Subunit F (15.1 kDa) is located at the lumenal side. A further five small membrane intrinsic subunits are present in PS I: I (4.6 kDa), J (4.5 kDa), K (8.5 kDa), L (15.5 kDa), and M (3.5 kDa). The total mass of one PS I monomer, including ~ 90 antenna chlorophylls, is ~ 340 kDa. PS I was isolated from *Synechococcus elongatus* in a trimeric form (10,11), a monomer of which represents the functional PS I unit (12). The trimer is assumed to be the active form *in vivo* (13,14). Crystals suitable for structure analysis were obtained by crystallization of the trimer, diffracting X-rays to a maximum resolution of 4 Å (15). The crystallization was achieved by decreasing the salt ($MgSO_4$) concentration of the protein solution to ~ 5 mM through dialysis (16). A first model of PS I was proposed on the basis of an electron density map calculated at 6 Å resolution

P. Mathis (ed.), Photosynthesis: from Light to Biosphere, Vol. II, 3–10.

(16,17). In the recent past progress was made regarding the reproducibility of protein isolation and crystallization permitting the collection of native and derivative data sets to higher resolution. Freshly grown crystals, together with high-intensity and highly-collimated synchrotron radiation, ensured successful data evaluation to 4.5 Å resolution despite a high mosaicity of 0.7°.

2. Overall structure

The space group of the crystals is $P6_3$. The crystallographic *a,b* plane is observed to be parallel to the membrane plane, while the crystallographic *c*-axis is both parallel to the membrane normal and to the crystallographic threefold (trimerization) axis respectively. Fig. 1 shows the arrangement of helices for one PS I monomer as deduced from the 4.5 Å map. The view is parallel to the membrane plane onto the plane defined by the threefold and the local, pseudo-twofold axis (see below). Essentially, the arrangement remains unchanged as compared to the 6 Å model. Yet, a number of differences are obvious. Thus, while most so-called transmembrane helices could not be observed to span the membrane in the 6 Å map, the improved resolution now reveals that all such helices span the membrane completely (see Fig. 1). Secondly, in the 6 Å map 21 transmembrane helices were observed; the corresponding number in the new map is 29. Furthermore, while two symmetry-related helices parallel to the membrane plane were identified in the 6 Å map, another pair near the stromal side of the core is visible in the new 4.5 Å map. The cubes indicate loci of high electron density which can unambiguously be identified as the positions of the three 4Fe-4S clusters, known to be located on the stromal side of PS I. The lone cluster still within the membrane plane may be clearly attributed to F_X.

3. Helices attributed to subunits A and B

Fig. 2 shows the top view from the stromal side onto the membrane plane. The shaded transmembrane helices in the indicated two regions A and B are related to each other by a local, pseudo-twofold axis. This axis, indicated by the black dot, is both parallel to the *c*-axis and the membrane normal, and runs through the center of the cluster F_X. Thus helix a corresponds to a', b to b', etc. On the basis of the known high homology of the amino acid sequence of subunits A and B (45%) the shaded helices are assigned to subunits A and B. The open helices, not symmetry-related, are attributed to the small membrane intrinsic subunits. Near the center of Fig. 2, five helices from each of the large subunits - e, f, g. h, i and e', f', g', h', i' - surrounding the twofold rotation axis are seen to form the core of the monomer. The peripheral helices a, b, c, d, and a', b', c', d' were already observed in the 6 Å map; whereas, the four helices u, t and u', t' were only identified in the 4.5 Å map. Each of the A and B subunit is therefore assigned a total of 11 transmembrane helices. This result is in satisfactory agreement with the amino acid sequence analysis (8) and its corresponding hydropathy plot (18) which predicts 11 transmembrane regions (18,19). There is one noticeably long helix indicated in the hydropathy plot of subunit A. Since helix g in the central region A matches this description, we assign this region to subunit A and, consequently region B to subunit B as already shown in the 6 Å map. Helices lying parallel to the membrane are indicated by dots: two near the lumenal side, n and n', detected in the 6 Å map, and two near the stromal side, s and s', identified in the 4.5 Å map. Both pairs, n, n'

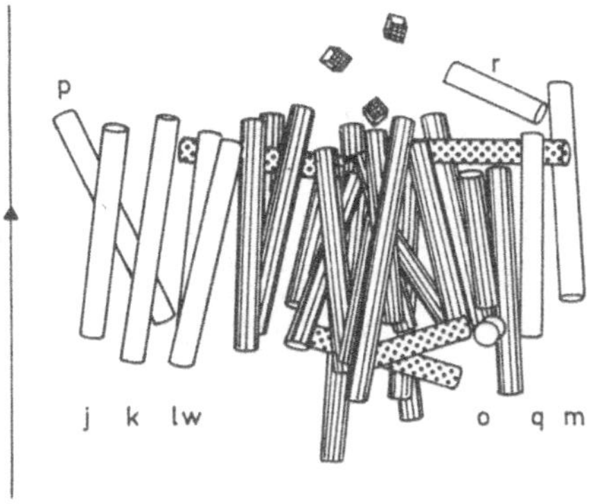

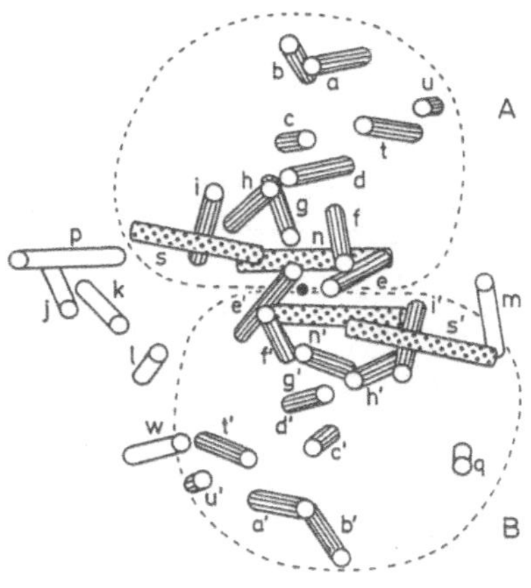

Fig. 1. (see left). Present model of the arrangement of the helices in the PS I monomer. The helices are depicted as columns. View direction is parallel to the membrane plane onto the plane defined by the crystallographic threefold and the pseudo-twofold axes (see Fig. 2). The threefold axis is indicated. Helices attributed to subunits A and B are shaded and dotted. The positions of the three 4Fe-4S clusters are depicted as cubes and indicate the stromal side.
Fig. 2. Helix arrangement with view direction from the stromal side onto the membrane plane. (Fig. 1 rotated by 90° about the horizontal.) The local twofold rotation axis is indicated by the black dot (details see text).

and s, s', occur as symmetry-related sets and are therefore also assigned to subunits A and B.

4. Helices belonging to the small membrane intrinsic subunits I, J, K, L and M

The 7 unshaded transmembrane helices in Figs. 1 and 2 lacking symmetry-related counterparts must, by exclusion, belong to these small subunits and to the membrane anchor of the lumenal subunit F. The helices j, k and a new helix p are located in close proximity to the threefold axis. Since PS I trimers do not exist in deletion mutants of subunit L (20), L must be assigned to the domain connecting the three monomers. L was predicted to contain 2 membrane-spanning helices (18); therefore, two of j, k or p must belong to L. Helices m and q are located on the opposite side of PS I. Since PS I is smaller in this region in mutants lacking subunits F and J (Boekema and Rögner, unpublished) and because both were predicted to contain one helix (18), m and q must each belong to either F or J. The remaining three helices l, w and one of j, k, p must belong to the subunits M, K or I. (I has as yet not been detected in the SDS gels of our PS I crystals.)

5. The electron transfer chain

P700: The overall position of the primary electron donor has remained unchanged. However, the broadly globular region of the density map assigned to the two chlorophylls of P700 now show clear features of large aromatic molecules. Fig. 3

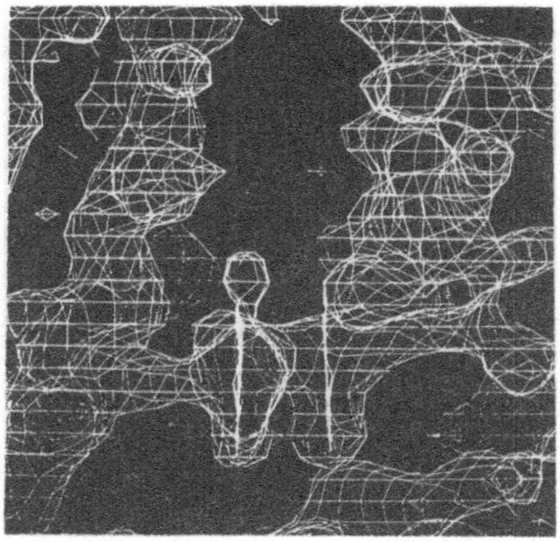

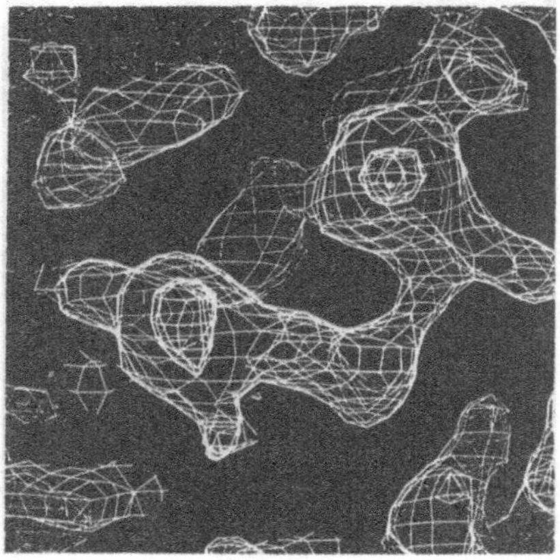

Fig. 3. (see left). Section of the electron-density map in the region surrounding P700. View direction is along the planes of the rings of the assigned two chlorophylls of P700 and along the plane of the membrane (see text).
Fig. 4. Section of the electron density map showing the region where the two 4Fe-4S clusters F_A and F_B are located. The heavier, spherical contours indicate high electron density and are assigned to the loci of F_A and F_B. The contours between F_A and F_B are attributed to two one-turn helices.

shows the contours of the two chlorophylls of P700. The electron density of the chlorophyll on the right is mainly behind the section chosen for this figure and is therefore only partially visible. Modeling the head groups of chlorophyll molecules into the pockets of electron density indicates that the chlorophyll rings are perpendicular to the membrane plane and parallel to each other with a perpendicular distance of at least 4.5 Å. The center-to-center distance is 9 Å. The left-hand chlorophyll is seen to be connected to helix e' by a broad segment of density. We interpret this to be a histidine side chain pointing towards the Mg atom of the chlorophyll. The same constellation with helix e is observed behind this section for the right-hand chlorophyll. His 685 and His 660 of subunits A and B, respectively, have been proposed to be involved in the coordination of P700 (19).
ENDOR experiments of the oxidized P_{700}^{+} relative to the orientation of the PS I crystal in a magnetic field confirm the orientation of P700. Furthermore, the circular orientation of one chlorophyll of P700 within its plane has been determined at atomic resolution by this technique (Käß et al., unpublished).
Fig. 4 shows a section of the electron density map in the region of the two terminal Fe-S clusters, F_A and F_B. In both the 6 Å map and in the 4.5 Å map the axis F_A-F_B is tilted by 54° with respect to the c axis. The centroid-to-centroid distance between F_A and F_B is 12 Å. Peaks of high electron density (heavier contours) indicate the space occupied by each cluster. Close to F_A and F_B two pockets of electron density resembling α-helices are observed in the 4.5 Å map (see Fig. 4). Their positions show a clear resemblance to two one-turn α-helices of the soluble ferredoxin from *Peptococccus aerogenes*, which also contains two 4Fe-4S clusters at 12 Å distance (21). ESR measurements showed that the relative orientation of the F_A and F_B clusters in PS I is the same as in the ferredoxin. This information was provided by measurements of the g-tensor axes of F_A^- and F_B^- in dependence on the

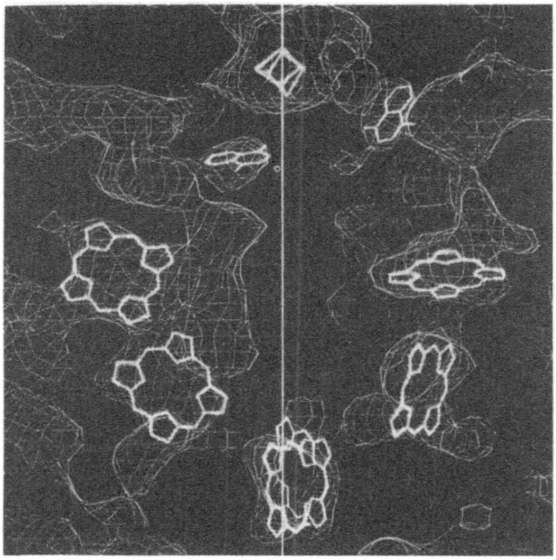

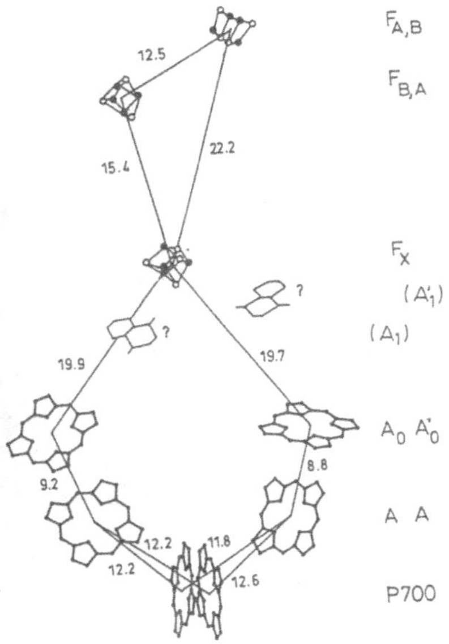

Fig. 5. (see left). Section of the electron density map in the region between P700 and F_X and present assignments to the electron carriers. The twofold rotation axis is indicated. The positions of the two vitamin K_1 molecules are tentative. View direction is not directly along the plane of the rings of the chlorophylls of P700 as in Fig. 3, but is rotated about the local twofold axis by 18° to make all electron carriers visible on one plane.

Fig. 6. Arrangement of the geometrical centers of the electron carriers (distances in Å). View direction is as in Fig. 5.

orientation of PS I crystals in a magnetic field (Kamlowski et al., in these proceedings). As the molecular structure of this ferredoxin is known (21), the structural similarity in combination with the crystallographic data allows to predict in subunit C the positions of the 16 atoms of the two 4Fe-4S clusters and those of the eight cysteins through which they are coordinated with the two helices. An assignement of F_A and F_B to the respective clusters is, however, not yet possible because of the pseudo-symmetry of both the ferredoxin and subunit C.

The section of the electron density map at 4.5 Å resolution between F_X and P700 is shown in Fig. 5. While the positions of the accessory chlorophylls, A and A', first observed in the 6 Å map remain unchanged in the 4.5 Å map, the inclinations of the planes have been modified. The most notable differences between the model derived from the maps at 6 Å and 4.5 Å resolution, respectively, entail the position of the electron carriers, A_0 (chlorophyll) and A_1 (vitamin K_1). The electron densities which were attributed to A_1 and A_0 at 6 Å now clearly constitute part of the extended helices, e and e', which appeared to end in the middle of the membrane in the 6 Å maü but are now completely membrane spanning. Instead, two new chlorophyll molecules could be identified at 4.5 Å resolution. Their positions are related to each other by the twofold symmetry within subunits A and B. They are

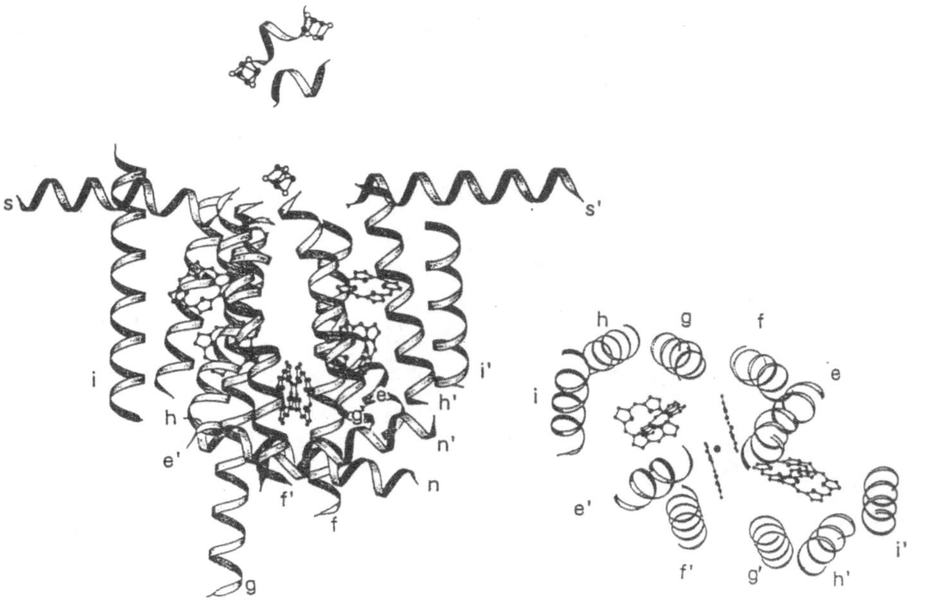

Fig. 7. Left: Present model of the architecture of the inner core of PS I. The view is as in Fig. 1. The helices are shown in form of spirals. Right: View direction as in Fig. 2 (Fig. 7, Left, rotated by 90° about the horizontal). For clarity, for this view, the helices have not been depicted in full length and horizontal helices have been omitted.

located at the same depth relative to the membrane surface as the A_o of the 6 Å resolution map but at a distance of 12 Å from the twofold symmetry axis. One of them should represent the redox active A_o. As it is not yet clear along which of the two branches the electron transfer proceeds, A_o cannot at present be identified. This arrangement of the electron transfer chain is now similar to that of the purple bacteria RC. Two contours of electron density in the new map could possibly accommodate the vitamin K_1 head groups. However, they are as yet indistinguishable from large protrusions from α-helices representing aromatic side chains.

Fig. 6 schematically depicts the distance between the centers of the electron carriers. Based on the new constellations of the chlorophylls between P700 and F_X, several new distances had to be evaluated.

6. The framework of the inner core of PS I

Fig. 7 shows the core of the 10 helices of subunits A and B surrounding and supporting the electron transfer chain as well as the two one-turn helices of subunit C. The arrangement of the helices e-i and e'-i' resembles that of the 10 helices of subunits L and M of the purple bacteria. The top view in Fig. 7, right, shows that the helices are all slightly tilted with respect to the membrane normal. Regarding the four horizontal helices s and s' near the stromal limit of subunits A and B, lie parallel to the membrane plane, while n and n' near the lumenal side are slightly inclined to this plane. Helix s extends nearly as far as the connecting domain (link for trimerization), while helix s' ends at the outer surface, where PS I is exposed to the lipid bilayer.

7. The antenna chlorophylls

A unique feature of PS I is that the two subunits A and B harbor both the electron transfer chain as well as the antenna chlorophylls. Recently, data sets have been collected to a resolution of 4.0 Å. Work on the best possible map is still in progress; we have though embarked on postulating a preliminary antenna model which presently includes 64 chlorophylls (see Fig. 8). Although the final overall concept of the antenna assembly cannot be described yet, some preliminary statements seem opportune. In the case of PS I there is no similarity with an arrangement of the chlorophylls as recently shown for the light-harvesting complex LH 2 of purple bacteria (22). There is also no evidence for a structural resemblance to the assembly of chlorophylls in the LHC II of higher plants (23). Clearly, an independent principle was developed in PS I. The distance of the chlorophyll centers is 8-15 Å, no preferable orientation being obvious. The chlorophylls are oriented close to the periphery of subunits A and B often with voids between the transmembrane helices. The overall arrangement of the chlorophylls is as if they are randomly positioned along the wall of an oval bowl with the bottom oriented towards the lumenal side and a large opening towards the stromal side. This "wall" may function as a storage device in which an excited state is rapidly delocalized, ready to be trapped by P700 located towards the bottom of the device; but this suggestion is

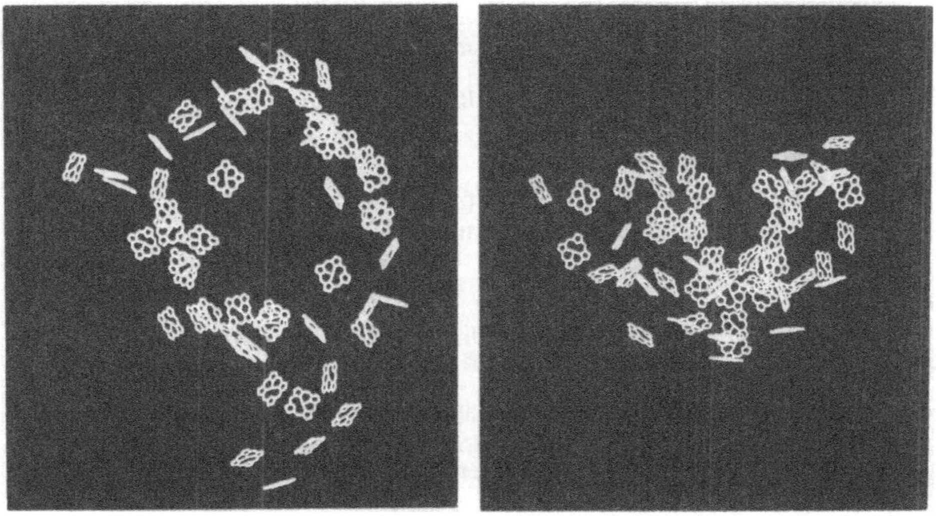

Fig. 8. Present model of the arrangement of 64 antenna chlorophylls of PS I. Left: top view as in Fig. 2. Right: side view as in Fig. 1.

open to debate. The shape of the antenna arrangement modeled primarily at 6 Å resolution (see top and side view in (16)) is similar to that in Fig. 8; however, the orientation of the chlorophylls was predominantly parallel to the membrane normal. This difference is obviously due to the lower resolution and the number of only 40 chlorophylls observed in the 6 Å map.

The studies were supported by the Deutsche Forschungsgemeinschaft, Sonderforschungsbereich 312 (A3, D1).

References

1 Witt, H.T. (1987) *New J. Chem.* 11, 91-101
2 The Photosystems (1992), Topics in Photosynthesis 11 (ed. Barber), Elsevier, Amsterdam
3 Deisenhofer, J., Epp, O, Miki, K., Huber, R. and Michel, H. (1985) Nature 318, 618-624
4 Trebst, A. (1986) *Z. Naturforsch.* 41c, 240-245
5 Deisenhofer, J. and Michel, H. (1985) *EMBO J.* 8, 2149-2170
6 Nitschke, W., Feiler, U., Lockau, W. and Hauska, G. (1987) *FEBS Lett.* 218, 283-286
7 Mathis, P. (1990) *Biochim. Biophys. Acta* 1018, 163-167
8 Mühlenhoff, U., Haehnel, W., Witt, H.T. and Herrmann, R.G. (1993) *Gene* 127(1), 71-78
9 Golbeck, J.H. (1992) *Ann. Rev. Mol. Biol.* 43, 293-324
10 Boekema, E.J., Dekker, J.P., v. Heel, M.G., Rögner, M., Saenger, W., Witt, I. and Witt, H.T. (1987) *FEBS Lett.* 217, 218-286
11 Ford, R.C. and Holzenburg, A. (1988) *EMBO J.* 7, 2287-2293
12 Rögner, M., Mühlenhoff, U., Boekema, E.J. and Witt, H.T. (1990) *Biochim. Biophys. Acta* 1015, 415-424
13 Hladik, J. and Sofrova, D. (1991) *Photosyn. Res.* 29, 171-175
14 Shubin, V.V., Tsuprun, V.L., Bezsmertnaya, I.N., Karapetyan, N.V. (1993) *FEBS Lett.* 334(1), 79-82
15 Witt, I., Witt, H.T., DiFiore, D., Rögner, M., Hinrichs, W., Saenger, W., Granzin, J., Betzel, C. and Dauter, Z. (1988) *Ber. Bunsenges. Phys. Chem.* 92, 1503-1506
16 Witt, H.T., Krauß, N., Hinrichs, W., Witt, I., Fromme, P., Pritzkow, W. and Saenger, W. (1992) Research in Photosynthesis I, Proc. IX Int. Congr. Photosynthesis, Nagoya, Japan (ed. Murata, N.), 521-528, Kluwer, Dordrecht
17 Krauß, N., Hinrichs, W., Witt, I., Fromme, P., Pritzkow, W., Dauter, Z. Betzel, C., Wilson, K.S., Witt, H.T. and Saenger, W. (1993) *Nature* 361, 326-331
18 Mühlenhoff, U., thesis, Technische Universität Berlin (1991)
19 Fromme, P., Schubert, W.-D. and Krauß, N. (1994) *Biochim. Biophys. Acta* 1187, 99-105
20 Chitnis, V.P., Xu, Q., Yu, L., Golbeck, J.H., Nakamoto, H., Xie, D.-L., Chitnis, P.R. (1993) *J. Biol. Chem.* 268, 11678-11684
21 Adman, E.T., Sieker, L.C. and Tensen, L.H. (1973) *J. Biol. Chem.* 248, 3987-3996
22 McDermott, G., Prince, S.M., Freer, A.A., Hawthornthwaite-Lawless, A.M., Papiz, M.Z., Cogdell, R.J. and Isaacs, N.W. (1995) *Nature* 374, 517-521
23 Kühlbrandt, W., Da Neng, W. and Fujiyoshi, Y. (1994) *Nature* 367, 614-621

BIOCHEMICAL ASPECTS OF THE REACTION CENTER IN GREEN SULFUR BACTERIA - COMPARISON WITH OTHER FeS-TYPES

G.Hauska, C.Hager-Braun, N. Schneebauer, M.Schütz, R.Zimmermann and N.Nelson*,
Lehrstuhl für Zellbiologie und Pflanzenphysiologie, Universität Regensburg, Germany and *Roche Institute of Molecular Biology, Nutley/N.J., USA

1. Introduction

Two prototypes of reaction centers (RC) in chlorophyll dependent photosynthesis are discerned according to the terminal electron acceptors. The Q-type reduces quinones and occurs in PSII of chloroplasts and cyanobacteria, in purple - and in green non-sulfur bacteria. The FeS-type reduces FeS-centers and occurs in PSI, in green sulfur bacteria and in heliobacteria (1). The structure of the Q-type RC has been solved to the atomic level for purple bacteria (2, 3). Structural resolution of the FeS-type in cyanobacterial PSI is under way (4, 5). With 11 subunits this is intrinsically rather complex. For this reason the simpler case of the FeS-type RC in green sulfur bacteria is investigated with rising intensity (6). Lately the RC of heliobacteria has been added, which in several aspects is even simpler (7). Rather intriguing is the conclusion that the RC in green sulfur - and heliobacteria have a homodimeric core structure, in contrast to the heterodimeric arrangement of all the other RC. For detailed references consult two recent reviews, one on green sulfur - (6), and one on heliobacteria (7). With respect to PS1 we refer to Golbeck and Bryant (8) and Chitnis et al. (9).

2. Common features and differences

FeS-type RC creates a reducing power strong enough for the direct reduction of $NAD(P)^+$. The electrons come from water via PSII in cyanobacteria and chloroplasts. In green sulfur bacteria they originate from reduced sulfur compounds like sulfide, as depicted in Fig.1. The electron transport from sulfide to the RC involves sulfide-quinone oxidoreductase (SQR), a putative flavoprotein, the menaquinone pool, and the cytochrome bc-complex, most likely via the Q-cycle (10). In heliobacteria electrons come from organic compounds. The reducing equivalents reach $NAD(P)^+$ via ferredoxin and ferredoxin-NADP oxidoreductase (FNR in Fig.1), a reaction which can be carried out with isolated FeS-type RC, as has been again demonstrated recently for *Chlorobium vibrioforme* in a heterologous system with spinach ferredoxin and FNR (11).

P. Mathis (ed.), Photosynthesis: from Light to Biosphere, Vol. II, 11–16.

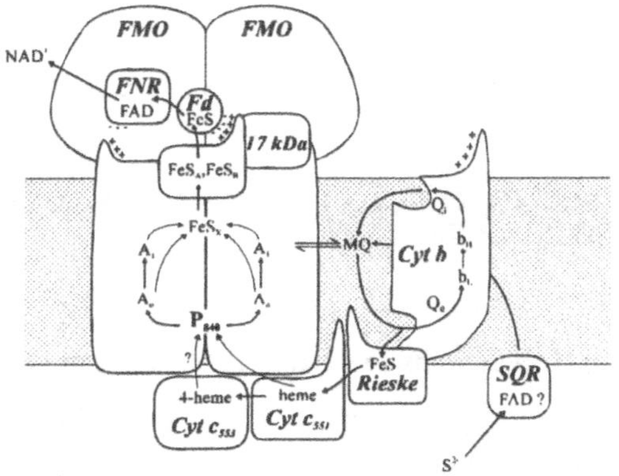

Fig. 1 - Organisation of Electron Transport in *Chlorobium*.
Q_O and Q_i stand for quinol-oxidizing - and quinol-reducing site, b_L and b_H for low - and high-potential heme-b in the bc-complex. Other terms are explained in the text.

The FeS-type RC consists of the primary electron donor P, the primary electron acceptor A_0, followed by A_1 and the three FeS-centers X, A and B. P is a special pair of chlorophylls, Chl *a* in P700 of PSI, BChl *a* in P840 of green sulfur bacteria, and BChl *g* or *g'* in P800 of heliobacteria. The midpoint potentials of P800 in heliobacteria and P840 in green sulfur bacteria with 240 mV is about 200 mV more negative than the one of P700. A_0 is a Chl *a*-monomer - interestingly not only in PSI, but also in green sulfur - and heliobacteria (6-8). More specifically, A_0 in *Chlorobia* is a Chl *a*-isomer with a phytol-side chain (12), and in heliobacteria it is 8'-OH-Chl *a* esterified with farnesol (13). The secondary electron acceptor A_1 in PSI is phylloquinone (Vit-K1), or OH-phyllochinone (8). For *Chlorobia* and heliobacteria Vit-K2 (menaquinone = MQ) is in discussion. However, doubts remain, because MQ is easily lost from the RC during isolation from *Chlorobium* (6), and total removal of MQ from membranes of *Heliobacterium chlorum* did not affect electron transfer in the RC (14). Thus A_1 may be MQ, but may not obligatorily function in electron transfer to FeS-center X. It may be responsible for cyclic electron flow (Fig. 1). Based on Mößbauer spectroscopy and EXAFS the three FeS-centers X, A and B in PSI have been identified as 4Fe-4S-clusters (8). Like P, center X is bound between the two central, large polypeptides, denoted PsaA and -B in PSI. Both contain an identical dodecapeptide with two cysteines. It was this X-binding peptide which identified the genes for the core subunits, PscA (s. ref. 15 for the nomenclature) in *Chlorobium* (16) and PshA in *Heliobacillus* (17). In both cases 9 of the 12 residues are conserved. Fig. 2 shows this in a folding model of the C-terminal part for 5 out of 11 putative transmembrane helices. It resembles the fold of the core proteins in the Q-type RC (2, 3). Further conserved residues shown in Fig. 2 are candidates to bind P, A_0 and A_1. The conserved histidine in helix VIII, originally envisaged to bind P in analogy to the Q-type RC (6), has been excluded by directed mutagenesis in *Chlamydomonas*, the histidine in helix X being a better candidate (18).

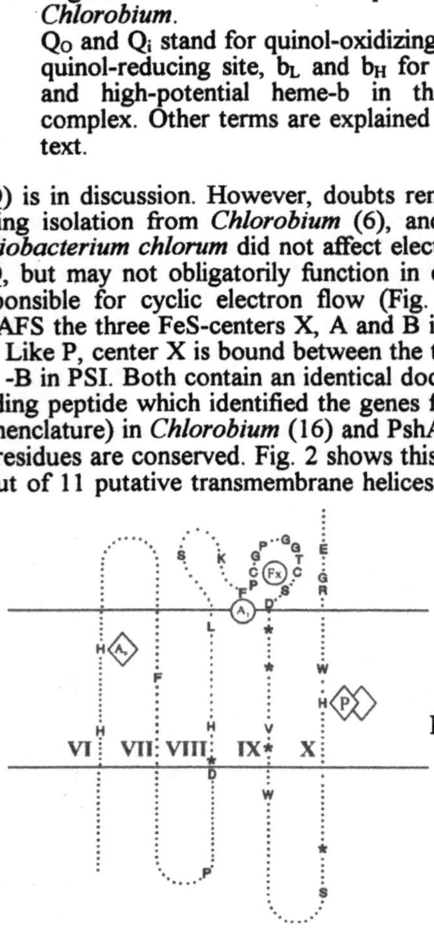

Fig.2 - Folding Model for the Redox Center Binding Part in the Core Subunit
Hydrophobic stretches are numbered in Roman. Conserved residues are given in single letter code, non-conserved ones are represented by dots. Aromats are indicated by asterisks.

FeS-centers A and B are located on the extra subunit PsaC in PSI, formerly called "bound ferredoxin" (8,9). The corresponding subunit PscB has been identified for *Chlorobium*, and contains a peculiar, positively charged N-terminal extension (19; Fig. 3). Such N-termini have also been found for PscA and cytochrome *b* (20; Fig. 1) and may dock negatively charged peripheral proteins in *Chlorobium*, like ferredoxin. In the C-terminal part PscB contains the significant 8 cysteines. Site directed mutations of PsaC in cyanobacteria indicate that the first 3 cysteines and the last bind center A, the others center B (21). The same has been suggested from sequence comparison of PscB and PsaC, to explain the turn around of the redox potentials of the *Chlorobium* FeS-centers A and B with respect to PSI (16). In *Chlorobium* both centers are significantly more negative than in PSI, where -530 and -580 mV for A and B are found, respectively (8), but the A-like center is the more negative (22). It is to say, however, that there are differences in the EPR spectra, especially in line widths, and that therefore the general assignment to centers X, - A and - B in all the FeS-type RC is not generally accepted (6, 7, 23; see contributions by Hager-Braun, Itoh, Scott to these proceedings). The FeS-centers A and B in an expression protein of *psa*C for PSI can be reconstituted, but show extremely broad EPR-lines, even when the protein is rebound to the core of the P700-RC, unless the subunit PsaD is also bound (24). The interaction with PsaD restores the natural shape of the EPR-spectra, probably by decreasing the flexibility in PsaC. A corresponding expression protein from *Chlorobium-psc*B was obtained and reconstituted, and again shows very broad lines (Golbeck and Nelson, these proceedings).

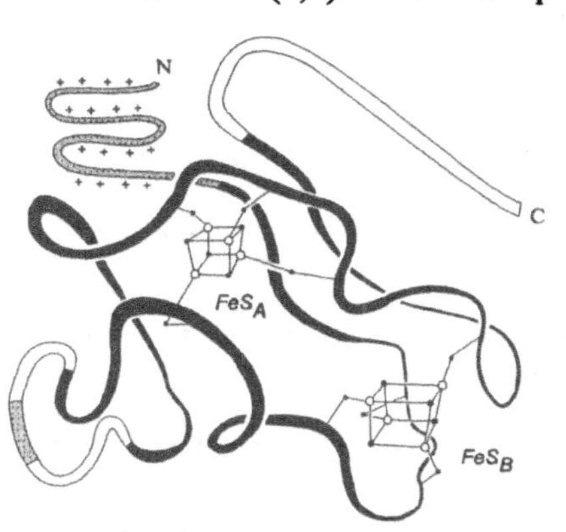

Fig.3 - Folding Model for PscB
The model is based on the structure of bacterial ferredoxin (35) and the model for PsaC (36).

The nature of the secondary electron donor to the P840-RC is controversial (6; question mark in Fig. 1). Spectroscopic *in vivo*-studies favor a cytochrome *c*-553, while in most of the reports on photoactive preparations a photooxidizable cytochrome *c*-551 is present (6, 25; s. contributions by Albouy, Hager-Braun, Itoh, Sakurai, Scheller to these proceedings). Only in the preparation by Feiler et al. (26) cyt *c*-553 is retained and cyt *c*-551 is lost, and recently a photoactive preparation without any cytochrome was achieved (25). Since cyt *c*-551 not only fractionates with part of the RC, but also copurifies with cytochrome *b*, it actually may represent the cytochrome *c*-component of the *bc*-complex (Fig. 1). An analog to cytochrome c_1 or *f* has not been found in *Chlorobium* (20). The sequence of the gene demonstrates that cyt *c*-551 represents a c-type of its own (27). Cyt *c*-553 is considered to be a tetraheme protein (26, Albouy et al. in these proceedings), in analogy to the secondary electron donor in certain Q-type RC (28). In this context the denotation PscC for cyt *c*-551 (15) seems premature - also plastocyanin is not regarded a subunit of the P700-RC!

For the subunits in the PSI-RC not less than 14 genes, *psa*A-N, have been identified, - 13, the first 12 + *psa*N, for plants and algae, and 11 (*psa*A-F and *psa*I-M) for cyanobacteria (9). The products of the later are detectable in crystals (29). Thus, next to the 2 subunits binding the redox centers, 9-11 additional subunits are present. The minimum polypeptide composition of photoactive P840-RC in *Chlorobium limicola* comprises only 4 subunits - PscA, PscB, the FMO-protein and a fourth subunit denoted PscD. The gene for PscD has been sequenced and represents a polar protein of 16.52 kDa ("17 kDa" in Fig. 1), with a positive net-charge (25). Chaotropic agents like urea, NaJ, or other halogenides/pseudohalogenides specifically remove this subunit and lead to inactivation of the RC (Hager-Braun, unpublished). This is very reminiscent of the removal of PsaD (9), and if it interacts with the FeS-protein PscB in analogy to PsaC-PsaD (Fig. 1) we expect that rebinding of both subunits to the core of the P840-RC should restore the original, narrower EPR-lines (s. above).

The N-terminal part of the PscA-protein represents a covalently linked antenna with about 20 BChl *a*-molecules per RC (6). Highly purified RC contain about 35 BChl *a*-molecules, suggesting the presence of 2 FMO-proteins (Fig. 1), each with 7 BChl *a*-molecules. It is uncertain, however, whether this remaining FMO is bound to the RC, and if so, whether it excerts a stabilizing effect in addition to PscD (25). About 2 carotenoid molecules are also retained in the isolated RC (6).

The P800-RC of heliobacteria may even be simpler, because photoactive preparations consist of the large core polypeptide PshA only (7). Neither the protein PshB with FeS-centers A and B, nor the electron donating cytochrome have been characterized beyond spectroscopic detection in membranes. They are not firmly bound. FeS-center X has not been detected yet by EPR, but its presence is suggested by the binding peptide in the PshA-sequence (17), and by optical spectroscopy (30).

In the case of the RC the transition to simplicity is not found from the eucaryotic organelle to the procaryote, but among procaryotes, from oxygenic photosynthesis with two RC, to photosynthesis with only one RC.

3. Homodimeric reaction centers

Why a dimer at all? A dimeric structure of the RC-core is called for by only half the binding requirements for P as well as for FeS_X in a single core protein (s. above). The homodimeric structure of the FeS-type RC in green sulfur - and heliobacteria is supported by genetic and biophysical evidence. Only one gene has been detected which represents all the previously determined peptide sequences in both cases. This raises the probability of only one core polypeptide practically to certainty (16, 17). Moreover, a comparatively more symmetric organisation of the RC pigments is supported by the zero-field-splitting of the P-triplet, by Raman spectroscopy of P and A_0, as well as by paramagnetic measurements of P^+ (6). These measurements demonstrate an even distribution of the unpaired electron over the two chlorophylls of P, and an equal probability of electron transfer in the two parallel branches (Fig. 1). The coincidence of homodimeric RC and anaerobic life suggest that atmospheric oxygen exerted the selection pressure towards gene duplication and heterodimeric RC. A possible reasoning is that the asymmetry with a firmly bound primary quinone in the preferred electron transport branch allowed for more efficient charge separation, lower steady state concentration of excited P, and thus a decreased damage by oxygen (6). At the same time efficient energy conservation was kept via the exchange of the secondary, losely bound quinone with the quinone pool.

However, two identical core proteins do not conclusively reflect a homodimeric structure of the P840-RC of *Chlorobia*. Asymmetry may be induced by asymmetric binding to another component, as has been demonstrated for the interaction of growth factors with their receptors (31), and could be the case for binding the subunits PscB and -D (Fig. 1). Indeed, PscB as well as PsaC are asymmetric (Fig. 3; s. 5), although they probably evolved by gene duplication. Is the P800-RC of heliobacteria, which does not require a bound FeS-A/B protein (PshB) for photoactivity, the actual case of a homodimeric RC?

4. A Note on Evolution

In view of the evidences that early conditions were anaerobic it is reasonable to assume that homodimeric RC preceded the heterodimers, and that the ancestor was of the FeS-type. The extents of conservation within the genes for the different core proteins suggest that gene duplication occured at least three times, twice in the Q-types and once in the FeS-types. This means that a homodimeric structure with the shorter core polypeptide of the Q-type RC existed. It can not be decided whether the ancestral homodimeric RC had short or long core polypeptides - in other words, whether they had the N-terminal antenna or not. Furthermore, it is plausible that the homodimeric RC evolved from a monomeric organisation (s. 1, 6, 28, 32 for more details).

Noteworthy chlorophyll-dependent photosynthesis is confined to the eubacterial descent. It does not occur in archaebacteria. On the other hand, key components of respiration like cytochrome oxidase, cytochrome *b*, the Rieske FeS-protein do occur in eubacteria and archaebacteria, and are related down to the level of functionally important amino acids (33, 34). The old textbook notion that photosynthesis preceded respiration is not plausible anymore.

References

1 Blankenship, R.E. (1992) Photosynth.Res. 33, 91-111
2 Deisenhofer, J., Epp, O., Miki, K., Huber R. and Michel, H. (1985) Nature 318, 618-624
3 Deisenhofer, J., Epp, O., Sinning, I. and Michel, H. (1995) J.Mol.Biol. 246, 429-457
4 Krauss, N., Hinrichs, W., Witt, I., Fromme, P., Pritzkow, W., Dauter, Z., Betzel, C., Wilson, K.S., Witt, H.T. and Saenger, W. (1993) Nature 361, 326-331
5 Fromme, P., Schubert, W.-D. and Krauß, N. (1994) Biochim.Biophys.Acta 1187, 99-105
6 Feiler, U. and Hauska, G. (1995) in Advances in Photosynthesis: Anoxygenic Photosynthetic Bacteria (Blankenship, R.E., Madigan, M.T. and Bauer, C.E., eds.) Kluwer Academic Publishers, Dordrecht, The Netherlands, in press
7 Amesz, I. (1995) ibid., in press
8 Golbeck, J.H. and Bryant, D.A. (1991) Curr.Top.Bioenerg. 16, 83-177
9 Chitnis, P.R., Xu, Q., Chitnis, V.P. and Nechushtai, R. (1995) Photosynth.Res. 44, 23-40
10 Klughammer, C., Hager, C., Padan, E., Schütz, M., Schreiber, U., Shahak, Y. and Hauska, G. (1995) Photosynth.Res. 43, 27-34
11 Kjaer, B. and Scheller, H.-V. (1995) Photosynth. Res. in press
12 van de Meent, E.J., Kobayashi, M., Erkelens, C., van Veelen, P.A., Otte, S.C.M., Inoue, K., Watanabe, T. and Amesz, I. (1992) Biochim.Biophys. Acta 1102, 371-378
13 van de Meent, E.J., Kobayashi, M., Erkelens, C., van Veelen, P.A. and Amesz, I. (1991) Biochim.Biophys.Acta 1058, 356-362
14 Kleinherenbrink, F.A.M., Ikegami, I., Hiraishi, A., Otte, S.C.M. and Amesz, I. (1993) Biochim.Biophys.Acta 1142, 69-73
15 Bryant, D.A. (1994) Photosynth.Res. 41, 27-28
16 Büttner, M., Xie, D.-L., Nelson, H., Pinther, W., Hauska, G. and Nelson, N. (1992) Proc.Natl.Acad.Sci.USA 89, 8135-8139
17 Liebl, U., Mockensturm-Wilson, M., Trost, J.T., Brune, D.C., Blankenship, R.E. and Vermaas, W. (1993) Proc.Natl.Acad.Sci.USA 90, 7124-7128

18 Cui, L., Bingham, S.E., Kuhn, M., Käß, H., Lubitz, W. and Webber A.N. (1995) Biochemistry 34, 1549-1558
19 Illinger, N., Xie, D.-L., Hauska, G. and Nelson, N. (1993) Photosynth.Res. 38, 111-114
20 Schütz, M., Zirngibl, S., le Coutre, J., Büttner, M., Xie, D.-L., Nelson, N., Deutzmann, R. and Hauska, G. (1994) Photosynth.Res. 39, 163-174
21 Zhao, J., Li, N., Warren, P.V., Golbeck, J.H. and Bryant, D.A. (1992) Biochemistry 31, 5093-5099
22 Nitschke, W., Feiler, U. and Rutherford, A.W. (1990) Biochemistry 29, 3834-3842
23 Kjaer, B., Jung, Y.-S., Golbeck, J.H. and Scheller, H.-V. (1994) Photosynth.Res. 41, 105-114
24 Li, N., Zhao, J., Warren, P.V., Warden, J.T., Bryant, D.A. and Golbeck, J.H. (1991) Biochemistry 30, 7863-7872
25 Hager-Braun, C., Xie, D.-L., Jarosch, U., Herold, E., Büttner, M., Zimmermann, R., Deutzmann, R., Hauska, G. and Nelson, N. (1995) Biochemistry 34, 9617-9624
26 Feiler, U., Nitschke, W. and Michel, H. (1992) Biochemistry 31, 2608-2614
27 Okkels, J.S., Kjaer, B., Hansson, Ö., Svendsen, I., Lindberg-Möller, B. and Scheller, H.-V. (1992) J.Biol.Chem. 267, 21139-21145
28 Nitschke, W. and Rutherford, A.W. (1991) Trends Biochem.Sci. 16, 241-245
29 Jekow, P., Fromme, P., Witt, H.T. and Saenger, W. (1995) Biochim.Biophys. Acta 1229, 115-120
30 Kleinheerenbrink, F.A.M., Chiou, H.-C., LoBrutto, R. and Blankenship, R.E. (1994) Photosynth.Res. 41, 115-123
31 De Vos, A.M., Ultsch, M. and Kossiakoff A.A. (1992) Science 255, 306-312
32 Vermaas, W.F.J. (1994) Photosynth.Res. 41, 285-294
33 Castresana, J., Lübben, M. and Saraste, M. (1995) J.Mol.Biol. 250, 202-210
34 Castresana, J. and Saraste, M. (1995) Trends Biochem.Sci., in press
35 Adman, E.T., Sicker, L.C. and Jensen, L.H. (1973) J.Biol.Chem. 248, 3987-3996
36 Oh-Oka, H., Takahashi, Y., Kariyama, K., Saeki, K. and Matsubara, H. (1988) J.Biochem.(Tokyo) 103, 962-968

MUTATIONAL ANALYSIS OF PHOTOSYSTEM I POLYPEPTIDES

P. R. Chitnis[1], V. P. Chitnis[1], Q. Xu[1], Y. S. Jung[2], L. Yu[2], J. H. Golbeck[2]
[1] Division of Biology, Kansas State University, Manhattan, Kansas, USA
[2] Department of Biochemistry, University of Nebraska, Lincoln, Nebraska, USA

1. Introduction

Photosystem I (PSI) in cyanobacteria and chloroplasts is a multisubunit membrane-protein complex that catalyzes electron transfer from reduced plastocyanin or cytochrome c_6 to oxidized ferredoxin or flavodoxin (see recent reviews 1, 2). The PsaA and PsaB subunits form a heterodimeric core that harbors ~100 antenna chlorophyll *a* molecules, the primary electron donor, P700, and a chain of electron acceptors A_0, A_1, and F_X. PsaC, PsaD, and PsaE are peripheral subunits that are located on the *n*-side (stromal in chloroplasts and cytoplasmic in cyanobacteria) of the photosynthetic membrane. PsaC binds the terminal electron acceptors, F_A and F_B, each a [4Fe-4S] cluster. PsaF, PsaL, PsaK, PsaI, PsaJ, and PsaM of cyanobacterial PSI are integral membrane proteins.

The cyanobacterium *Synechocystis* sp. PCC 6803 provides a suitable model system for studying the organization and function of PSI. Structure and function of its PSI complex is similar to that from higher plants, but its genome can be easily manipulated. *Synechocystis* sp. PCC 6803 can take up extraneous DNA and incorporate it into the genome by homologous recombination (3), thus enabling one to mutate specific genes. We have embarked on a program to identify functions of PSI subunits through targeted mutagenesis of the genes encoding these subunits in *Synechocystis* sp. PCC 6803. We cloned the genes that code for PsaD, PsaE, PsaF, PsaJ, PsaL, and PsaI from *Synechocystis* sp. PCC 6803 and subsequently generated mutants in which these genes have been interrupted or deleted (4-11). Study of these cyanobacterial mutants has helped us identify *in vivo* functions of PSI proteins that do not bind any redox centers. Among these accessory subunits, PsaE and PsaD are required for maximal rates of ferredoxin-mediated $NADP^+$ photoreduction (8). PsaL is required for the formation of PSI trimers (12, 7). PsaF can be chemically cross-linked *in vitro* with plastocyanin (13, 14). PsaJ and PsaI may be involved in proper assembly of PsaF and PsaL, respectively (9, 11). In this paper, we describe some examples to illustrate the versatile use of cyanobacterial mutant strains in physiological and structure-function studies.

P. Mathis (ed.), Photosynthesis: from Light to Biosphere, Vol. II, 17–22.

2. Results

2.1 *Functions of some accessory subunits are crucial under stress conditions.*

Growth of subunit-deficient mutant strains indicates the physiological role of the missing polypeptide. We have previously reported that many subunit-deficient mutant strains have wild-type growth rates under photoautotrophic conditions (2). Examples of such mutants include the strains lacking PsaL or PsaI (7, 11). When the PsaL-less (ALC7) or the PsaI-less (AIC9) strains are grown at 25°C, they are indistinguishable from the wild type in their photoautotrophic growth, but at high temperature (40°C), the mutants grow slower than the wild type (Table I), indicating that the defects caused by the absence of PsaL or PsaI become limiting for growth under stress conditions.

Table I. Photoautotrophic growth of some PSI mutants is slowed by high temperature stress

Mutant Strain	Growth Temperature	Growth rate - Doubling time (h)
WT	25°C	29
	40°C	26
ALC7	25°C	30
	40°C	36
AIC9	25°C	30
	40°C	40

2.2 *PsaE and the cyclic electron flow around PSI.*

In addition to its function in linear electron flow, PSI also participates in a cyclic process that transfers electrons through cytochrome b_6/f complex. Photoheterotrophic growth of a mutant strain in a mineral medium supplemented with an organic carbon source and a PSII inhibitor can indicate role of the missing PSI protein in cyclic electron transfer (15). When we examined growth of the PsaE-less mutant strain under different conditions, the strain was found to be impaired in photoheterotrophic growth (Fig. 1). Similar mutant strain of *Synechococcus* sp. PCC 7002 cannot grow under equivalent conditions (15). This prompted us to study cyclic electron flow in the PsaE-less strain.

Methods for direct examination of cyclic electron transfer are not available. Therefore the contribution from cyclic flow was indirectly assessed by monitoring $P700^+$ rereduction in the presence of different electron transfer inhibitors (Table II). Rates of $P700^+$ rereduction after a period of white light were measured using whole cells of *Synechocystis* sp. PCC 6803 as previously described (7, 15). The 1/e time for $P700^+$ rereduction was 23 and 25 ms in the wild-type and AEK2 cells, respectively (Table II). After addition of DCMU, electrons are supplied by the respiratory NAD(P)H dehydrogenase complex and by cyclic electron flow around PSI, and the 1/e time for $P700^+$ reduction increases to 160 ms in the wild-type strain, but 351 ms in the mutant

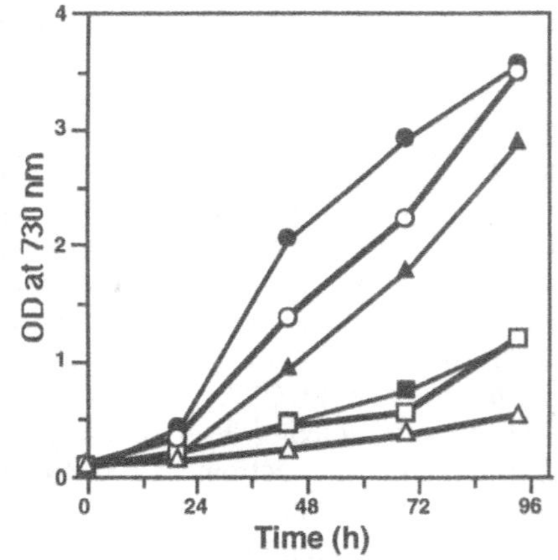

Figure 1. Growth of the wild-type (filled symbols, thin lines) **and PsaE-less** (open symbols, thick lines) **strains** under photoautotrophic (BG11; rectangles), mixotrophic (BG11+glucose; circles), or photoheterotrophic (BG11+glucose +DCMU; triangles) conditions.

Table II. 1/e time (ms) for P700$^+$ rereduction in the wild type (WT) and PsaE-less (AEK2) strains of *Synechocystis* sp. PCC 6803

Inhibitor	WT	AEK2	Comment
None	23	25	P700$^+$ rereduction after a block of light
DCMU	160	351	No electrons from PSII
DCMU, glucose	140	200	Additional electrons from glucose
DCMU, KCN	118	300	Oxidase blocked
DCMU, KCN, MV	300	300	MV precludes cyclic flow
DBMIB	660	653	No electrons enter P700$^+$ pathway
DBMIB, DCMU	655	665	No electrons from PSII
DBMIB, glucose	670	673	Additional electrons from glucose

Abbreviations: MV, methyl viologen; DBMIB, Dibromomethylisopropyl-benzoquinone

strain. Addition of KCN to the DCMU-poisoned sample accelerates the 1/e time for P700$^+$ reduction due to inhibition of cytochrome oxidase which competes with PSI for electrons. Further addition of methyl viologen removes electrons from the acceptor side of PSI, thereby inhibiting cyclic electron flow. When photosystem II and cyclic electron flow are inhibited, the 1/e time decreased to 300 ms in both wild-type and mutant strains. DBMIB inhibits electron donation to the cytochrome b_6/f complex from PSII, the respiratory NAD(P)H dehydrogenase complex and the cyclic pathway, and the rate of P700$^+$ reduction is slowed to a basal level. This rate probably represents a 'leak' due to the reduction of P700$^+$ from physiologically insignificant reductants. In contrast to the wild type, the PsaE-less cells showed 1/e time of 300 ms with or without cyclic electron flow, indicating an *in vivo* role of PsaE in cyclic electron flow around PSI.

2.3 *Deletions in the PsaD subunit*

As described in the previous section, subunit-deficient mutants can be used to identify functions of specific PSI proteins. The next logical step is to employ site-directed mutagenesis to understand structure-function relations. Here we describe site-directed deletion mutations in *psaD*. We used the ADC4 strain of *Synechocystis* sp. PCC 6803 as a recipient strain for *psaD* mutagenesis (8). This strain was generated by replacing *psaD* with a chloramphenicol-resistance cassette (Fig. 1). To introduce the wild-type and mutant *psaD* into the ADC4 strain, we generated the pK6 recombinant plasmid by cloning the 1.6 kb 5' flanking region (*Eco*RI-*Xba*I fragment), the wild-type *psaD* (*Xba*I-*Pst*I fragment), a kanamycin resistance gene cartridge (*Pst*I fragment), and 1040 bp 3' flanking region (*Pst*I-

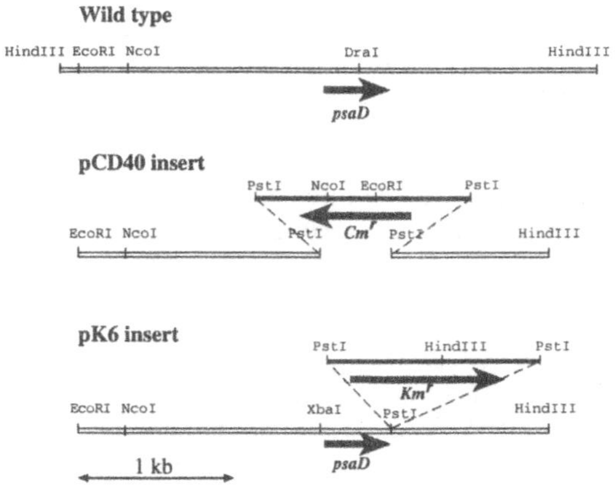

Figure 2. Scheme for introduction of mutated *psaD* into the PsaD-less strain. The restriction map of DNA region around *psaD* in the wild-type strain is shown on the top line. The restriction map of the insert in pCD40 that was used to generate the ADC4 strain (middle line) and of the insert in pK6 (bottom line) is also shown. Arrows show location, size and direction of open reading frames for *psaD*, *Km^r* (gene for kanamycin resistance), and *Cm^r* (gene for chloramphenicol resistance).

Table III. Deletions in the PsaD subunit of PSI

Strain	Deleted region	Size (# amino acids)	Molecular Mass kDa		Fd-mediated $NADP^+$ photoreduction rate µmoles/h/mg Chl
			Predicted	Apparent	
K6	None (Wild-type)	140	15.465	18.1	440
ADC4	no PsaD	-	-	-	0
K31	Ala34-Gly45	128	14.248	-	0
K33	Leu71-Phe75	135	14.820	16.8	0
K35	Asn108-Gly115	132	14.624	16.3	0
K36	Leu71-Val107	103	11.077	-	0
K74	Lys135-Val140	134	14.778	17.0	0
K75	Val128-Val140	127	14.046	16.3	0
K76	Lys117-Val140	116	12.770	14.4	0

*Hin*dIII fragment) into the polylinker of pGem3z (Fig. 2). Appropriate restriction sites were introduced by incorporating them in the oligonucleotide primers that were used for DNA amplification. ADC4 cells were transformed with pK6. Kanamycin-resistant transformants were selected and segregated for two generations. They were replica-plated to confirm the absence of chloramphenicol-resistance gene. The resulting K6 strain contains an active, wild-type *psaD* gene, as shown by PCR analysis of genomic DNA, Western blotting of thylakoid proteins, and from ESR analysis of the F_A/F_B centers (data not shown).

Deletion mutations in *psaD* were generated using PCR or appropriate restriction enzymes (16). The DNA fragments containing deletions were cloned into pK6 by replacing its *Xba*I-*Pst*I fragment. The amplified fragment was completely sequenced to confirm the presence of desired mutation and to ensure the fidelity of Taq polymerase. The recombinant DNAs containing mutations were used to transform the ADC4 strain and to obtain mutant strains according to the procedure described earlier for the K6 strain. We examined the accumulation of mutant PsaD proteins in the membranes using Western blotting (Fig. 3). PsaD could not be detected in the membranes from the K31 and K36 strains. *In vitro* reconstitutions have previously shown that these deletions disable the assembly of PsaD into thylakoids (16). K36 contains a relatively large deletion of 37 amino acids, while deletion in K31 removes amino acids between two putative α-helices in PsaD. Thylakoids from the other deletion strains contained PsaD mutant proteins. Apparent molecular weights of the mutant PsaD proteins were estimated from their migration during Tricine-urea-SDS-PAGE (Table III). As expected, the mutant proteins showed proportionate decrease in mass compared to the wild-type protein. On equal chlorophyll basis, many deletion strains showed less intense bands that were recognized by the anti-PsaD antibody (Fig. 3). This may have resulted from the loss of some epitopes in the deletion mutants or from

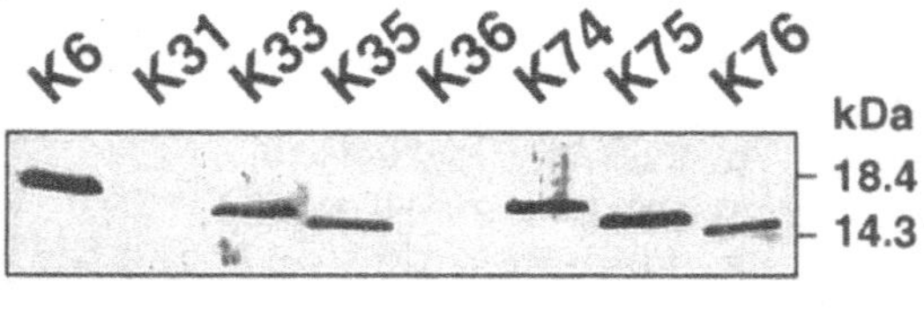

Figure 3. Western analysis of PsaD in the mutant membranes. Photosynthetic membranes were isolated from cyanobacterial strains, proteins were separated by Tricine-urea-SDS-PAGE, and transferred to an Immobilon membrane. Immunodetection was performed using an anti-PsaD primary antibody, horseradish peroxidase conjugated secondary antibody and enhanced chemiluminescence reagents.

the decreased accumulation of the mutant PsaD in the membranes. Among mutant PsaD proteins that accumulated in the membranes, were the proteins with C-terminal deletions of up to 24 residues. Therefore, the C-terminal domain of PsaD is not essential for its assembly into thylakoids. Interestingly, this region contains surface-exposed lysyl residues and protease cleavage sites (17).

Photosynthetic membranes from the PsaD mutant strains were also used to estimate the reductase activity of PSI in a coupled ferredoxin-mediated $NADP^+$ photoreduction assay (Table III). Rates of $NADP^+$ photoreduction were measured using ferredoxin: $NADP^+$ oxidoreductase, and by monitoring the change in the absorption of NADPH at 340 nm (8, 18). The rate of $NADP^+$ reduction was 440 μmoles per mg chlorophyll per h when K6 membranes were used. In contrast, membranes from the PsaD-less ADC4 strain and from the strains containing deletion mutations in PsaD were unable to reduce $NADP^+$. Therefore, the deletion of C-terminal amino acids in K74, K75, and K76 strains or the absence of residues near the central basic domain in PsaD in K33 may be detrimental to the function of PsaD in ferredoxin-mediated electron transfer. Alternatively, the deletions may affect overall conformation of PsaD, thereby causing the observed defect in ferredoxin-mediated $NADP^+$ photoreduction.

3. Discussion

Several researchers have effectively applied targeted mutagenesis to identify functions of photosynthetic proteins and then to correlate their structural features with functional characteristics (2). Subunit-deficient mutants of *Synechocystis* sp. PCC 6803 are available for all subunits, except PsaK and PsaM (2). Mutant strains for these proteins are available in another cyanobacterium, *Synechococcus* sp. PCC 7002 (DA Bryant, personal communication). These targeted mutants have been used to study how the lack of specific gene products affects *in vivo* function and organization of PSI.

Functions of three subunits of PSI are clearly known; PsaA, PsaB, and PsaC bind cofactors and redox centers of PSI. The remaining accessory subunits are conserved during evolution, as revealed from the remarkable sequence similarity among proteins from cyanobacteria and chloroplasts. Yet, many of these accessory subunits are dispensable in cyanobacteria. Mutant strains of *Synechocystis* sp. PCC 6803 that lack PsaE, PsaF, PsaL, PsaJ, and PsaI are similar to the wild type in their photoautotrophic growth. In some cases, metabolic flexibility in cyanobacteria can account for the photoautotrophic growth. For example, *Synechocystis* cells, that are grown in BG11, contain substantial pools of flavodoxin and ferredoxin (19). Although ferredoxin-mediated electron transfer is decreased in PsaE-less strain, flavodoxin reduction does not require PsaE, thus supporting normal photoautotrophic growth (8). Detailed analyses of PSI from many mutants have revealed only minor defects. How can one account for this disparity between evolutionary conservation of some PSI proteins and their dispensable function in cyanobacteria? We argue that functions of these subunits may be dispensable under laboratory conditions, but are important for survival in the real world. For example, the presence of PsaL or PsaI grants a significant growth benefit at high temperature (Table I). Apparently dispensable subunits may provide subtle improvements in PSI stability and function and offer significant evolutionary advantage. The proposed role of

PsaE in cyclic electron flow (Table II), which accounts for only a minor contribution to overall photosynthetic energy metabolism (15), can still be essential for survival during evolution.

Subunit-deficient cyanobacterial mutants can be used as recipient strains for introducing genes with site-directed mutations. Such structure-function studies have been performed for PsaB, PsaC, PsaD, and PsaE. Random mutagenesis of a gene can also be used to identify residues that are necessary for function of an essential subunit. Here we described generation of site-directed deletions in *psaD* and their use in dissecting functional domains in the protein (Table III). When coupled with biochemical and biophysical analyses, mutational analysis of cyanobacterial PSI proteins clearly provides a powerful tool in understanding intricacy and complexity of PSI.

5. Acknowledgments

This work was supported in part by grants from the National Science Foundation (MCB 9405325 to PRC and MCB 9202751 to JHG), and the United States Department of Agriculture-NRICGP (92-37306-7661 to PRC). This is contribution #95-37-A from the Kansas Agricultural Experiment Station.

6. References

1. Golbeck, J.H. (1994) in The Molecular Biology of Cyanobacteria (Bryant, D.A., ed), pp. 179-220, Kluwer Academic Publishers, Dordrecht, The Netherlands
2. Chitnis, P.R., Xu, Q., Chitnis, V.P., and Nechushtai, R. (1995) Photosynthesis Res. 44, 23-40
3. Williams, J.G.K. (1988) Meth. Enzymol. 167, 766-778
4. Chitnis, P.R., Reilly, P.A., Miedel, M.C., and Nelson, N. (1989) J. Biol. Chem. 264, 18374-18380
5. Chitnis, P.R., Reilly, P.A., and Nelson, N. (1989) J. Biol. Chem. 264, 18381-18385
6. Chitnis, P.R., Purvis, D., and Nelson, N. (1991) J. Biol. Chem. 266, 20146-20151
7. Chitnis, V.P., Xu, Q., Yu, L., Golbeck, J.H., Nakamoto, H., Xie, D.-L., and Chitnis, P.R. (1993) J. Biol. Chem. 268, 11678-11684
8. Xu, Q., Jung, Y.S., Chitnis, V.P., Guikema, J.A., Golbeck, J.H., and Chitnis, P.R. (1994) J. Biol. Chem. 269, 21512-21518
9. Xu, Q., Odom, W.R., Guikema, J.A., Chitnis, V.P., and Chitnis, P.R. (1994) Plant Mol Biol 26, 291-302
10. Xu, Q., Yu, L., Chitnis, V.P., and Chitnis, P.R. (1994) J. Biol. Chem. 269, 3205-3211
11. Xu, Q., Hoppe, D., Chitnis, V.P., Odom, W.R., Guikema, J.A., and Chitnis, P.R. (1995) J. Biol. Chem. 270, 16243-16250
12. Chitnis, V.P., and Chitnis, P.R. (1993) FEBS Lett. 336, 330-334
13. Wynn, R.M., and Malkin, R. (1988) Biochemistry 27, 5863-5869
14. Hippler, M., Ratajczak, R., and Haehnel, W. (1989) FEBS Lett. 250, 280-284
15. Yu, L., Zhao, J., Mühlenhoff, U., Bryant, D.A., and Golbeck, J.H. (1993) Plant Physiol. 103, 171-180
16. Chitnis, P.R., and Nelson, N. (1992) Plant Physiol. 99, 239-246
17. Xu, Q., Guikema, J.A., and Chitnis, P.R. (1994) Plant Physiol. 106, 617-624
18. Jung, Y.-S., Yu, L., and Golbeck, J.H. (1995) Photosynth. Res. In press
19. Bottin, H., and Lagoutte, B. (1992) Biochim. Biophys. Acta 1101, 48-56

FERREDOXIN REDUCTION BY PSI WITH WILD TYPE, DELETED AND SITE-DIRECTED MUTANTS FROM THE CYANOBACTERIUM *SYNECHOCYSTIS* SP. PCC 6803

P. Sétif, J. Hanley, P. Barth, H. Bottin and B. Lagoutte
CEA Saclay - DBCM/SBE and URA CNRS 1290, 91191 Gif sur Yvette, France

1. Introduction

Photosystem I (PSI) is a light-driven transmembrane oxidoreductase which catalyses the reduction of soluble ferredoxin (Fd) in oxygenic photosynthetic organisms. In cyanobacteria, it is composed of at least 11 polypeptides (1). Primary charge separation and the first steps of charge stabilization are occuring within the two large membrane-spanning subunits named PsaA and PsaB, which each provide two cysteine ligands to bind the [4Fe-4S] center F_X. The terminal acceptors of PSI are the [4Fe-4S] centers F_A and F_B which are both carried by subunit PsaC. The positions of the three iron-sulfur centers within the PSI complex were determined from the crystal structure of trimeric PSI from *Synechococcus* sp. (2). This study showed that, whereas F_X lies at the pseudo symmetry axis of the complex, symmetry is lost at the level of PsaC, as the F_A to F_B axis is tilted away from the membrane plane. Soluble Fd is a very acidic protein which interacts with PSI at its stromal side. Most of the stromal-protruding region of PSI is due to the hydrophilic subunits PsaC, PsaD and PsaE (3). The subunits PsaD and PsaE presumably do not carry any cofactor and were previously shown to be both important for the interaction between PSI and Fd (4-8). We will describe Fd reduction by PSI in the absence of either PsaD or PsaE and in site-directed mutants of these two subunits.

2. Materials and Methods

Flash-absorption spectroscopy measurements were performed as previously described (9,10). Genetic engineering of the PsaE deleted strain and of the site-directed mutants of PsaE were described in (8). Similar procedures were used for generating the PsaD mutants (Hanley et al., unpublished).

3. Results and Discussion

The kinetics of Fd reduction by wild type (WT) PSI, both prepared from *Synechocystis* 6803, have been recently studied by flash-absorption spectroscopy (9,10). At pH 8.0 in

P. Mathis (ed.), Photosynthesis: from Light to Biosphere, Vol. II, 23–28.

the presence of salts, the existence of three different first order phases was observed. These phases exhibit $t_{1/2}$ of about 500 ns, 15 μs and 110 μs respectively. They were ascribed to Fd reduction within PSI-Fd complexes that are preformed before flash excitation. The dissociation constants associated with each of these components are similar, with a global value ≈ 0.5 μM. Salts and pH effects pointed the importance of electrostatic interactions between PSI and Fd. It was particularly found that the affinity of Fd for PSI increases at lower pH values, with a dissociation constant as low as 0.05 μM at pH 5.8 in the absence of salts. Under these conditions, Fd reduction is faster than at pH 8.0 with only two first-order phases with $t_{1/2}$ of 500 ns and 9 μs (10).

Fig. 1 exhibits the kinetics of Fd reduction in WT PSI and in PSI complexes prepared from strains in which either the psaD gene or the psaE gene was deleted. The absence of the corresponding subunits in the PSI complex was checked by electrophoresis and immunoblotting.

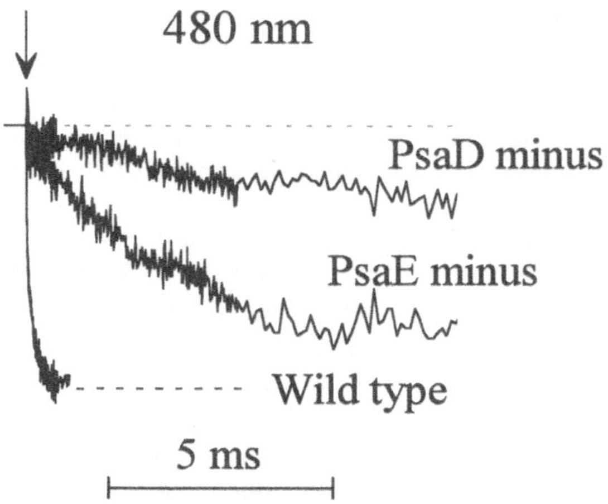

Figure 1: kinetics of Fd reduction in PSI complexes prepared from WT, psaD deleted and psaE deleted strains ([PSI] =0.2 μM; [Fd] = 2.0 μM).

With both deleted mutants, no fast first-order kinetics of Fd reduction could be observed, up to a concentration of ferrodoxin of 8 μM. This indicates that in both cases, Fd affinity for PSI is decreased at least 16-fold compared to WT PSI. However, a slower phase was observed in both cases, due to second order processes of Fd reduction. For the PsaE minus mutant, the observed rate increases linearly with the Fd concentration. The second-order rate constant that can be derived from this linear dependence is $k_2 = 1.5 \times 10^8\ M^{-1}\ s^{-1}$. This is only two to three times smaller than the corresponding rate observed with WT PSI ($k_2 = 2\text{-}5 \times 10^8\ M^{-1}\ s^{-1}$ (9)). For the PsaD minus mutant, Fd reduction is so slow that reoxidation of reduced Fd impedes the easy observation of Fd reduction by PSI under most conditions. Only an upper limit for the second order rate constant can be calculated: $k_2 < 0.5 \times 10^8\ M^{-1}\ s^{-1}$.

It has been previously reported that, after reconstitution of PSI isolated from the psaE deleted strain with isolated PsaE, fast kinetics of Fd can be recovered (8). The same kind of experiment was performed with PSI isolated from the psaD deleted strain and reconstituted with PsaD. First-order kinetics similar to those observed with WT PSI are obtained, with a maximum signal amplitude at saturating Fd concentrations corresponding to 80% of the signal observed with WT PSI. The presence of subunits PsaC and PsaE in most of the PSI complexes were observed by electrophoresis and immunoblotting. The presence of these two subunits in PSI of the PsaD-less mutant allowed to successfully reconstitute fast Fd reduction and is in accordance with a recent report (11).

After cross-linking PSI and Fd with EDC and NHS, it has been recently possible to identify the residues involved in the covalent bond between Fd and subunit PsaD of PSI (6) (only PSI polypeptide crosslinked to Fd, in accordance with previous studies: (4,5)): D93 in Fd and K106 in PsaD. Whereas D93 lies close to the flexible C-terminus of Fd, K106 is located in a region which was found to be partially solvent accessible (12). Moreover, several charged residues located close to K106 in the primary sequence of PsaD were recently found to be accessible to proteases (13). Different site-directed mutants were obtained by mutating (*) charged residues among all highly conserved residues (which are underlined) in the region of PsaD surrounding K106 (↓):

100 ↓ 110

YLHPADGVFPEKVNEGREAQGTK

* * * *

When changing one of the two conserved basic residues of this region for alanine (mutants K106A and R111A), Fd reduction by PSI is quite similar to the WT behavior (Table I). Only a 30% increase in the dissociation constant is found at pH 8.0 for both mutants and a pH dependence similar to that observed for the WT is obtained. In both cases, the affinity of Fd for PSI increases at lower pH ($K_d \approx 0.05\ \mu M$ for WT) and only the two fastest phases of Fd reduction are observed at pH 5.8, as in WT (10). These results indicate that K106 and R111 participate only weakly to the global energy of Fd binding on PSI. The observation that K106 is crosslinked to Fd could be due to the fact that this residue is the only lysine residue available for crosslinking with Fd. However it must be recalled that Fd can be photoreduced in the covalent complex. Moreover, fast kinetics of Fd reduction are observed which are very similar to those found in the non covalent complex (Lelong & Sétif, unpublished observations), thus showing that Fd is properly bound to PSI in the covalent complex.

More pronounced effects are observed when changing one of the conserved acidic residues of the above region for a neutral one (mutants D100S and E105Q). In both cases, the kinetics of Fd reduction become faster at pH 8.0, with only two phases ($t_{1/2} <$ 1 μs and ≈ 10 μs) and the affinity of Fd for PSI is increased about five times ($K_d \approx 0.1$ μM). These properties are similar to those observed with WT at pH 5.8 and the Fd reduction process becomes little pH dependent in any of these two mutants (Table I). In the most simple and direct interpretation of these results, we propose that, in WT PSI, either D100 or E105 becomes protonated between pH 5.8 and pH 8, and that this protonation results in a larger electrostatic interaction with negatively charged Fd. The close spatial proximity of residues D100 and E105 could drive the first protonation of one of these two groups to the neutral region. For both mutants in the pH range 5.8 to 8, the net charge of PsaD would thus be identical to that of WT at pH 5.8.

Table I. Fd reduction in site-directed mutants of PsaD.

	kinetic behavior at pH 8 (+ salts)		kinetic behavior at pH 5.8 (no salts)	
Syn. strain	number of 1st order phases ($t_{1/2}$)	dissociation constant K_d	number of 1st order phases ($t_{1/2}$)	dissociation constant K_d
wild type	3 (500 ns, 15 μs, 110 μs)	0.6 μM	2 (500 ns, 9 μs)	0.05 μM
K106A or R111A	idem WT	0.8-0.9 μM	idem WT	≤ 0.1 μM
D100S or E105Q	2 (< 1 μs, 10 μs)	0.1 μM	idem WT	≤ 0.1 μM

The primary sequence of the PsaE subunit in *Synechocystis* is shown on the next page with highly conserved residues being underlined. In a previous study (8), different site-directed mutants of the PsaE subunit allowed to derive two basic conclusions: the N and

C-termini are involved in PsaE integration in the PSI complex and all mutants in which the PsaE subunit was assembled in PSI exhibited kinetics of Fd reduction similar to that of WT. These mutants included W17A, Y16F/Y18F and K29A. This list has been recently extended with mutants R12A and R42Q behaving as the WT. However mutant **R39Q** is strongly affected in its interaction with Fd. In this mutant, first-order phases of Fd are completely absent, up to a Fd concentration of 10 μM, despite the fact that the PsaE subunit is present in the PSI complex, as probed by electrophoresis and immunoblotting. This indicates a more than 20-fold decrease in Fd affinity for PSI. A slower second order process of Fd reduction is observable with a rate constant similar to that of the PsaE deleted mutant (Table II).

```
1        10        20        30        40        50
ALNRGDKVRIKRTESYWYGDVGTVASVEKSGILYPVIVRFDRVNYNGFSG
           *    ***            *         *  *

         60        70
SASGVNTNNFAENELELVQAAAK
```

Table II. Fd reduction in R39Q of PsaE.

	kinetic behavior at pH 8 (+ salts)		
Syn. strain	first order phases ($t_{1/2}$)	dissociation constant K_d	second-order rate constants
wild type	3 (500 ns, 15 μs, 110 μs)	0.6 μM	$3\text{-}5 \times 10^8\ M^{-1}\ s^{-1}$
R39Q	no	>> 10 μM	$1.4 \times 10^8\ M^{-1}\ s^{-1}$

4. Conclusion

In the present study, much larger effects of site-directed mutagenesis on Fd reduction were found with mutant R39Q of PsaE than for the different mutants of PsaD studied yet. These observations suggest that, whereas both subunits are involved in Fd docking, the PsaE subunit is more essential for a tight binding of Fd, through one of its basic residues, R39. However, comparison of second-order rate constants show that Fd reduction is more altered in the absence of PsaD than in the absence of PsaE.

In addition to binding and fast reduction of Fd, other roles have been proposed for the stromal PSI subunits PsaD (stabilization and orientation of PsaC, shielding from oxygen) and PsaE (stabilization of PSI, cyclic electron transfer) (reviewed in (1)).

References

1. Golbeck, J.H. (1994) in The Molecular Biology of Cyanobacteria (Bryant, D.A., ed.) pp. 319-360, Kluwer Acad. Publ., Dordrecht
2. Krauss, N., Hinrichs, W., Witt, I., Fromme, P., Pritzkow, W., Dauter, Z., Betzel, C., Wilson, K.S., Witt, H.T. and Saenger, W. (1993) Nature 361, 326-330
3. Kruip, J., Boekema, E.J., Bald, D., Boonstra, A.F. and Rögner, M. (1993) J. Biol. Chem. 268, 23353-23360
4. Zanetti, G. and Merati, G. (1987) Eur. J. Biochem. 169, 143-146
5. Zilber, A.L. and Malkin, R. (1988) Plant Physiol. 88, 810-814
6. Lelong, C., Sétif, P., Lagoutte, B. and Bottin, H. (1994) J. Biol. Chem., 269, 10034-10039
7. Sonoike, K., Hatanaka, H. and Satoh, S. (1993) Biochim. Biophys. Acta 1141, 52-57
8. Rousseau,F., Sétif, P. and Lagoutte, B. (1993) EMBO J. 12, 1755-1765
9. Sétif, P. and Bottin, H. (1994) Biochemistry 33, 8495-8504
10. Sétif, P. and Bottin, H. (1995) Biochemistry 34, 9059-9070
11. Xu, Q., Armbrust, T.S., Guikema, J.A. & Chitnis, P.R. (1994) Plant Physiol. 106, 1057-1063
12. Lagoutte, B. and Vallon, O. (1992) Eur. J. Biochem. 205, 1175-1185
13. Xu, Q., Guikema, J.A. & Chitnis, P.R. (1994) Plant Physiol. 106, 617-624

STRUCTURAL ORGANIZATION OF THE ACCEPTORS A_1, F_X, F_A AND F_B IN PHOTOSYSTEM I FROM EPR IN SOLUTION AND SINGLE CRYSTALS

A. Kamlowski, A. van der Est, P. Fromme* and D. Stehlik
Phys. Dep., FU Berlin, Arnimallee 14, D-14195 Berlin and
*Max-Volmer Inst., TU Berlin, Germany

1. Introduction

Since the first X-ray structure of Photosystem I (PSI) at 6 Å resolution [1], steady progress has been made towards a more reliable arrangement of the cofactors within the membrane-bound proteins, as summarized earlier at this meeting [2]. All electron transfer cofactors bind to just three proteins (PsaA, B, C), most of them to the largest membrane spanning proteins, PsaA and PsaB; only the terminal iron-sulfur centers F_A and F_B are part of the PsaC protein located at the stroma side of the membrane. In the PSI reaction center six Chlorophyll-*a* (Chl) based entities (two of the primary donor dimer, P_{700}, two accessory Chls, A and A′, and the remaining two assigned to A_0 and A_0') are now positioned with confidence (see Fig. 6 of [2]). In addition, the average center position and space of highest electron density associated with each of the three (4Fe-4S) clusters F_X, F_A and F_B are well defined within a triangular arrangement. In contrast, the locations of the quinone acceptor A_1 and the inactive counterpart quinone A_1' remain obscure. Two possible positions in between A_0 (A_0') and F_X are tentatively offered in [2], quite different from those suggested before [1]. In addition to the uncertainty in the location of A_1 (A_1') the relative orientation of these quinone acceptors within the PS I reaction center is even less accessible with the available electron density maps. Precisely in that respect EPR-spectroscopy is able to provide its most reliable piece of structural information, namely the relative orientation within the reaction center for the charge carriers P_{700}^{+}, A_1^{-}, F_X^{-}, F_A^{-}, F_B^{-} in the respective functional charge-separated state. In this contribution we present the consequences for the orientation of these cofactors based on recent EPR-data.

2. Materials & Methods

PSI particles were prepared from the cyanobacteria *Synechococcus elongatus* as used for crystallization [1]. The charge-separated states $P_{700}^{+\cdot}A_1^{-\cdot}$ and $P_{700}^{+\cdot}F_X^{-}$ were characterized by transient EPR in solution as described in [3] and [4]. Extention to high-frequency EPR (95GHz) became possible by significant instrumental advancements by T. Prisner in the laboratory of K. Möbius (FU Berlin) and follows the procedures reported in the recent application to the reaction center (bRC) of the purple bacteria *Rb. sphaeroides* [5]. PSI single crystals were grown with water as a precipitation agent [1]. Rod-shaped crystals were used. The morphological axes are collinear with the crystal axes. The c-axis corresponds to the elongated rod-axis. Adjacent edges

P. Mathis (ed.), Photosynthesis: from Light to Biosphere, Vol. II, 29–34.

of the hexagonal cross-section determine the crystalline a- and b-axis. The main improvement with respect to our previous study on photoreduced F_A^- [6] is a significant increase in concentration of F_B^- (at least a factor of 3) through photoaccumulation at higher temperatures (up to 200K). Standard EPR-spectra were measured as a function of crystal orientation to the magnetic field at low temperature (18K).

3. Results

3.1 The quinone acceptor A_1 in the $P_{700}^{+\cdot}A_1^{-\cdot}$ state
Our previous results with time-resolved EPR at microwave frequencies of about 9 and 24 GHz [3] have established a significantly different orientation of the quinone acceptor A_1 (vitamin K_1, VK1) in PSI as compared to Q_A (ubiquinone, UQ_{10}) in bRC. This comparison could now be extended to 95 GHz (A. van der Est et al., to be published). The spin-polarized transient EPR-spectra of the $P^{+\cdot}Q^{-\cdot}$state of both reaction centers are compared in Fig. 1. The pattern of Zn-substituted bRC (broken line) has been discussed in detail [5] with the conclusion that the relative orientation of the quinone acceptor $Q^{-\cdot}$ in the charge-separated state $P_{865}^{+\cdot}Q_A^{-\cdot}$ does not change with respect to the ground state, at least to the accuracy of the X-ray data. Inspection of the $P_{700}^{+\cdot}A_1^{-\cdot}$ spectrum (solid line) in Fig. 1 yields the following comparisons: (i) considerable difference in the lower-field region (left) where only g-tensor components of the quinone contribute and (ii) highly similar polarisation pattern in the upper-field region (right), which is dominated by the primary donor, P.

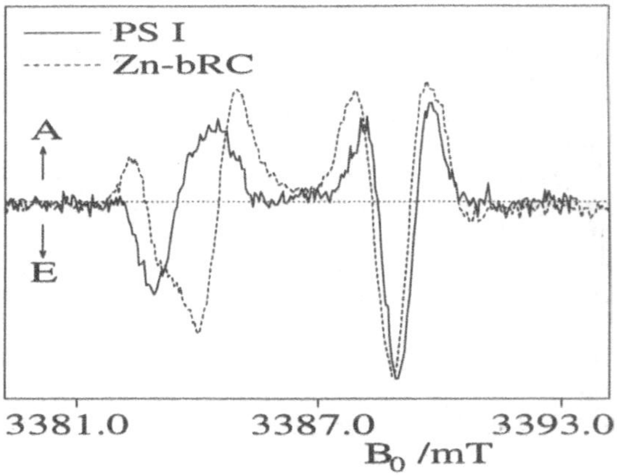

Fig. 1 Spin polarized time resolved EPR spectra of $P_{700}^{+\cdot}A_1^{-}$ in PSI (solid line) and $P_{865}^{+\cdot}Q_A^{-\cdot}$ in Zn-bRC of ***Rb. sphaeroides*** (dashed line) at a microwave frequency of 95 GHz. Experimental conditions as in Fig. 2 of ref. [5].

With some basic knowledge of the polarisation mechanism (see [3,5] and references given there) observation (ii) translates into similar orientations of the g-tensor axes and thus of the general molecular arrangement of the primary donors $P_{865}^{+\cdot}$ and $P_{700}^{+\cdot}$ with respect to the axis connecting P and Q in the respective reaction center. This is in line with results for the triplet state of $^3P_{700}$ [7] and ENDOR-studies on $P_{700}^{+\cdot}$ in PSI single crystals [8]. In contrast, observation (i) indicates that the relative orientations of the quinone acceptors $Q_A^{-\cdot}$ and $A_1^{-\cdot}$ are very different within the respective reaction center. For example, the $g_{xx}(Q)$ in-plane axis along the C=O double bonds is known to include an angle of about 62° with the PQ connecting axis in bRC [3,5] while these two axes are close to collinear in PSI. In any case, a rather precise orientation of the active A_1-quinone (VK_1) in PSI is predicted together with an approximate PQ-distance (25–30 Å) similar to that in bRC (29 Å). Note however, that the absolute position of A_1 cannot be obtained from EPR-data because only relative parameters enter the analysis. Another significant difference of the quinone sites in the two reaction centers should be mentioned. While the relative quinone orientation is hardly changed upon substitution of Q_A with quinone molecules in bRC, major orientation changes are observed for substitution of A_1 in PSI with the same quinones.

3.2 The first (4Fe-4S) iron sulfur center F_X

Transient EPR is able to measure the charge transfer step from A_1 to the first iron-sulfur center. Systematic removal of the iron-sulfur center F_A/ F_B and F_X establishes F_X as the first acceptor following A_1 [4]. Orientational information about F_X^- comes from EPR-studies on oriented PSI particles [11]. The minimal principal g-value of F_X^-, termed g_{xx} (or g_3), is aligned along the membrane normal. If g_{xx} is associated with the mixed-valence pair of the reduced 4Fe-4S cluster, the corresponding Fe-Fe diagonal of the distorted cube of F_X is aligned along the membrane normal.

3.3 Structural organization of the F_A/F_B PsaC-Protein

PSI single crystals proved to be well suited for low-temperature EPR-studies on the reduced terminal iron-sulfur centers F_A^- and F_B^- [6]. The full g-tensor orientation of both centers has to be measured in order to obtain structural information about the PsaC protein within the PSI unit. This in turn requires sufficient reduction of F_B which was achieved by high temperature (200K) photoaccumulation as mentioned in section 2. A full account of the orientation dependent EPR-studies and their evaluation will be published elsewhere. Here, we collect the essential results together with an outline of the strategy used to conclude the structural analogy of the PsaC core to that of bacterial ferredoxins of known structure. Hence, a prediction of the orientation of the PsaC protein with respect to the overall PSI unit is obtained.

3.3.1 Orientation of the **g** tensors of F_A^- and F_B^- with respect to the PSI crystal

The symmetry of the unit cell with six PSI units in a double trimer arrangement [1] with the c-axis as a six-fold screw axis helps to extract full tensor determination from a single rotation pattern in a plane which allows orientation of the **c**-axis and the **a**- (or **b**-) axis along the magnetic field B_0. The g-tensor parameters are collected in Tab. 1. Both the principal values and the angle between the principal axes and the crystalline **c**-axis (or membrane normal) agree well with previous studies on oriented PSI particles [11] except for the angle β_X. Its deviation from 90° could only be resolved in the single crystal study.

Table 1 The principal g-values and angles of principal axes of F_A^- and F_B^- with respect to the crystallographic **c**-axis or to membrane normal, $\vec{n}$ (ref. [11]).

	g-values of the FeS center			angles with c-axis, $\vec{n}$			ref.
	g_{xx}	g_{yy}	g_{zz}	β_X	β_Y	β_Z	
F_A^-	1.856	1.941	2.051	53°	48°	64°	this work
F_B^-	1.880	1.916	2.056	77°	60°	34°	±3°
F_A^-	1.855	1.942	2.050	52°	48°	65°	[11]
F_B^-	1.886	1.920	2.067	90°	(60°)	30°	[11]

3.3.2 Redox-State $(F_A^-) + (F_B^-)$ vs. $(F_A^- F_B^-)$

Because, the single crystals are not stable under strongly reducing conditions, the state $(F_A^-) + (F_B^-)$ in which individual PSI units exist with either F_A or F_B reduced is

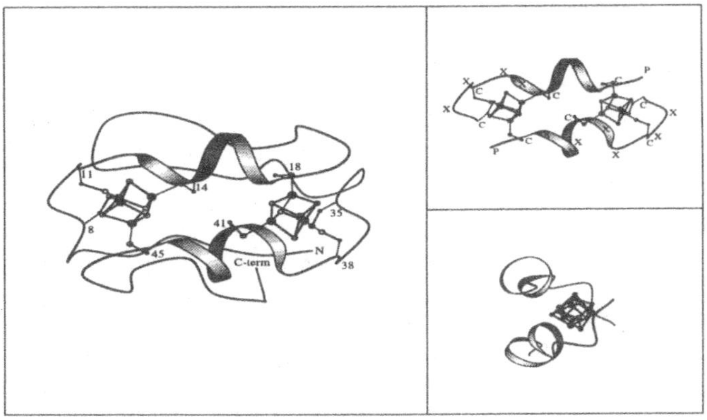

Fig. 2 PaFd structure [12] (see text). The numbering refers to the cystein residues ligating the Fe atoms. ***left***: view along the pseudo-C_2(Fd)-axis. ***upper right***: Only the cystein binding pattern is shown. ***lower right***: view along the center-to-center connecting line.

obtained in our experiments. None of the characteristic features of "interaction" EPR spectrum of the fully reduced state (F_A^- F_B^-) (see e.g. [11]) were observed. Therefore, pairs of EPR resonances, one from the F_A^- and one from the F_B^- manifold, have to be assigned to each of the six inequivalent PSI monomers per unit cell. This produces a set of transformation matrices relating the **g** tensors of both centers, F_A^- and F_B^-.

3.3.3 Transformation Matrix R(F_A^-+F_B^-)

The symmetry of the PSI unit cell and the EPR rotation patterns provide restrictions for the possible (in principle 36) transformation matrices relating the experimentally determined g-tensors of F_A^- and F_B^-. Although a unique choice cannot be reached based on experimental criteria alone, we have listed just one preferred transformation in Tab. 2 as characterized by the Euler angles relating the two g-tensor axis systems.

Table 2 Transformation matrices R(F_A^-+F_B^-) relating the axis systems of $\mathbf{g}(F_A^-)$ and $\mathbf{g}(F_B^-)$

	EULER angles		
R(F_A^-+F_B^-) obtained from	$\varphi/°$	$\theta/°$	$\psi/°$
(1) PSI single crystal data	48	94	57
(2) simulation of PSI single crystal data with PaFd structure	56	97.5	54

The choice is eventually done by comparison with transformation matrices calculated on the basis of the known structure of the two (4Fe-4S) cluster ferredoxin of *Peptococcus aerogenes* (PaFd) [12]. This structure is shown in Fig. 2 viewed along a local C_2-axis featured by this protein. The chosen representation emphasizes the distorted cubane-like (4Fe-4S) centers bound to the protein backbone by cysteine (cys) ligands to the Fe atoms. Also emphasized are two parallel one-turn helices which keep the two clusters at their characteristic center to center distance (~12Å). A truncated structure is also shown in Fig. 2 (right, with orthogonal projections), which is reduced to the two cys binding motifs (CXXCXXCXXXCP). These are also found in the sequence of PsaC in PSI [13]. Mutagenesis studies on PsaC [14] showed that

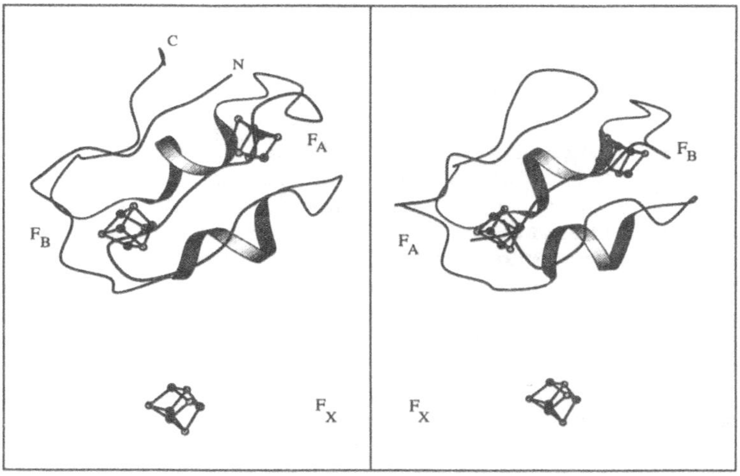

Fig. 3 Orientation of the PaFd structure within the F_X, F_A, F_B triangle. Note that the helices are behind the cluster-cluster connecting line, in contrast to Fig. 2. This representation corresponds to the upper part of the arrangement in Fig. 6 of [2].

Cluster I (cys 8, 11, 14, 45) of PaFd would correspond to F_B, Cluster II (cys 35, 38, 41, 18) to F_A. Based on this analogy transformation matrices can be calculated from the PaFd structure if the g-tensor axes assignment to the respective cubane structure is known. Reliable assignments have been obtained from EPR-studies on model compounds [15] which possibly differ between oxidized and reduced FeS centers. We have tested various assignment schemes and the best agreement with experiment-based transformation matrices was reached for the following assignment: the principal g-tensor axes x and y are collinear to the Fe-Fe distance vectors associated with the mixed-valence and equal-valence pair, respectively. As demonstrated in Tab. 2, a satisfactory correspondance is reached for the transformation matrices inferred from the experimentally determined g-tensors and from the PaFd structure with the above assignment. A different transformation matrix has been obtained from the simulation of the "interaction spectrum" of $(F_A^- F_B^-)$ in [11].

3.3.4 Orientation of the PsaC protein in PSI

The final interpretation step is the orientation of the PaFd model (Fig. 2) as a PsaC substitute with respect to the PSI unit. One way is to optimize the orientation for best agreement with the EPR-data. The other is to adjust it against the electron density map [1,2]. The optimal orientation is characterized by two parameters: (i) the angle between the **c**-axis and the distance vector $\overline{F_A F_B}$, which is accurately available from the X-ray data (see above), (ii) the angle by which the local C_2(Fd)-axis deviates from a reference plane made-up by the **c**-axis and the $\overline{F_A F_B}$ distance vector. While (i) is fixed with the X-ray value of 54°, (ii) is optimized in both ways to 115.....120° (65.....60°). This result leaves a two-fold ambiguity because by principle a 180° rotation around the C_2(Fd)-axis yields the same set of angles. Fig. 3 shows the PaFd structure oriented this way within the triangular arrangement F_X,F_A,F_B established for the PSI structure. For Fig. 3 (right) F_A is proximal to F_X vs. (left) with F_B closer to F_X (see also the different location of the C- and N-terminal of the PsaC model).

4. Discussion

The EPR-data obtained here for the redox-state $(F_A^-) + (F_B^-)$ in PSI single crystals give evidence for the structural analogy of the cores of PaFd and PsaC and allow the orientation of PsaC in PSI to be determined. The same result is obtained by inspection of the X-ray electron density map (see[2]). As outlined above, the results imply an assignment of g-tensor axes to the (4Fe-4S) cube structure and with it a prediction of the Fe-Fe pair associated with the mixed- or equal-valence pair. A completely independent assignment of the mixed-valence pair is reached from recent NMR-results on a PaFd analogous ferredoxin of *Clostridium acidi urici* (CauFd)[16].It is based on the Curie-type temperature dependence of the corresponding cysteine NMR-resonances. Interestingly, the same sequence specific mixed-valence pair assignment is concluded as implied in the EPR-analysis presented here.
The remaining two-fold ambiguity in the PsaC orientation (see Fig. 3) is expected to be removed soon by on-going studies pursuing biochemical, structural as well as spectroscopic approaches.

5. References

[1] Krauß, N., Hinrichs, W., Witt, I., Fromme, P., Pritzkow, W., Dauter, Z., Betzel, C., Wilson, K., Witt, H.T., and Saenger, W. (1993) Nature 362 , 326–331.
[2] H.T. Witt et al., (1995) Proc. Xth Int. Photosyn. Congr., Montpellier.
[3] Füchsle, G., Bittl, R., van der Est, A., Lubitz, W., Stehlik, D. (1993) Biochimica Biophysica Acta 1142, 23–35.
[4] van der Est, A., Bock, C., Golbeck, J., Brettel, K., Sétif, P., and Stehlik, D., (1994) Biochemistry 33, 11789–11797.
[5] Prisner, T.F., van der Est, A., Bittl, R., Lubitz, W., Stehlik, D., Möbius, K., (1995) Chem. Phys. 194, 349-360.
[6] Brettel, K., Sieckmann, I., Fromme, P., van der Est, A. and Stehlik, D., (1992) Biochimica Biophysica Acta 1098, 266–270.
[7] Sieckmann, I., Brettel, K., Bock, C., van der Est, A., Stehlik, D., (1993) Biochem. 32, 4842-4847.
[8] Käß, H., PhD thesis (1995), TU Berlin.
[9] van der Est, A., Sieckmann, I., Lubitz, W., Stehlik, D. (1995) 194, 349-360,
[10] van der Est, A., et al., (1995) Proc. Xth Int. Photosyn. Congr. Montpellier.
[11] Guigliarelli, B., Guillausier, J., More, C., Sétif, P., Bottin, H. and Bertrand, P. (1993) J. Biol. Chem. 268, 900-908.
[12] Adman, E.T., Sieker, L.C., and Jensen, L.H. (1976) J. Biol. Chem. 251, 3801.
[13] Muehlenhoff, U. et al., (1993) Gene 127, 71-78.
[14] Zhao, J., Li, N., Waren, P.V., Golbeck, J.H. and Bryant, D.A. (1992) Biochemistry 31, 5093-5099 and Yung et al., (1995) Proc. Xth Int. Photosyn. Congr.
[15] Ruis, G. and Lamotte, B. (1989) J. Am. Chem. Soc. 111, 2464-2469 and Gloux, J., Gloux, P., Lamotte, B. et al. (1994) J. Am. Chem. Soc. 116, 1953-1961.
[16] Bertini, I., Capozzi, F., Luchinat, C., Piccioli, M. and Vila, A.J. (1994) J. Am. Chem. Soc. 116, 651-660.

MULTIFREQUENCY ELECTRON SPIN ECHO STUDIES OF NITROGEN HYPERFINE COUPLING IN P700+: SPECTRAL SIMULATIONS AND THEIR IMPLICATIONS FOR ELECTRONIC STRUCTURE

Michelle Mac, John McCracken and Gerald T. Babcock, Michigan State University Department of Chemistry, E. Lansing , MI 48824 USA

1. Introduction

Oxygenic photosynthesis in higher plants and cyanobacteria requires the interplay of two pigment-protein reaction centers, Photosystem I and Photosystem II (PSI and PSII). In each of these reaction centers, the primary charge separation event proceeds via the light-induced generation of a chlorophyll a (chl a) cation radical (P700 in PSI and P680 in PSII). This photochemically active chlorophyll is distinguished from the light harvesting and energy transfer chlorophylls by its spectral and redox properties. After initial charge separation, subsequent electron transfer events occur on a timescale fast enough to prevent recombination. This highly effective, unidirectional electron transfer occurs with a quantum yield approaching unity. Because these events are controlled by the spatial and electronic arrangement of the co-factors involved, knowledge regarding the structure of these radicals is imperative to understand better the electron transfer process in photosynthetic systems. Monomeric[1] and dimeric[2] chl a structures have been suggested, for the primary donor in PSI, however, neither has been proven conclusively. The electronic structure of the radical remains elusive as well, however the presence of P700+ makes this system especially amenable to the application of electron magnetic resonance techniques. Previous 1H and 15N ENDOR[3,4] studies have indicated either a monomeric or highly asymmetric spin density distribution for P700+. Recent nitrogen ESEEM[5] and HYSCORE[4] (HYperfine Sublevel CORrelation) experiments have corroborated this hypothesis. Spectral simulations of these ESEEM data have not been successful in determining 14N hyperfine and quadrupole parameters conclusively. This precludes quantization of the number of 14N nuclei giving rise to a particular hyperfine interaction. Here we report results of multifrequency ESEEM experiments and corresponding spectral simulations of P700+ containing either natural abundance 14N or isotopically enriched with 15N.

P. Mathis (ed.), Photosynthesis: from Light to Biosphere, Vol. II, 35–38.

2. Materials and Methods

The PSI reaction centers were isolated from *Synechocystis 6803* by using the methods of Noren et al.[6]. Isotopic enrichment with 15N was accomplished by growing the cultures on a medium containing 15N (Cambridge Isotope Labs) potassium nitrate. Mass spectrometry was performed to ensure isotopic labeling (data not shown). P700+ was generated by illumination at 0° C for 30 seconds. The radical was trapped by freezing immediately to 77 K. The radical's presence was confirmed by continuous wave EPR (data not shown).

ESEEM data were collected on a home built spectrometer that has been described.[7] A three pulse ($\pi/2$-τ-$\pi/2$-T-$\pi/2$) pulse sequence was used for stimulated echo experiments. Dead time reconstruction was performed prior to Fourier transformation by the methods of Mims.[8] Computer simulation of the ESEEM data was performed on a Sun Sparcstation 2 computer by using Matlab (Mathworks, Nantick, MA) software. The experimental dead time was included in the simulations. An isotropic g- and an axial A-tensor were assumed.

3. Results and Discussion

Three pulse ESEEM spectra of P700+ containing natural abundance 14N and collected at multiple frequencies (A: 8.9 GHz; B: 9.7 GHz and C: 11.5 GHz) are shown in Figure 1. These spectra are dominated by the quadrupole coupling of the 14N nucleus (I=1). To determine the field position at which the hyperfine and nuclear Zeeman fields cancel (exact cancellation), it is essential that the experiment be performed at many microwave frequencies. If the exact cancellation condition is met, then the nuclear quadrupole resonances can be read directly from the ESEEM spectrum. The spectra are characterized by three sharp low frequency features (two of which add to

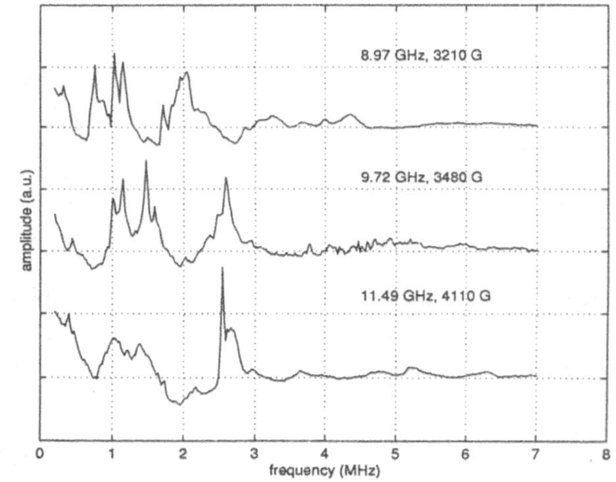

Figure 1: Multifrequency ESEEM spectra from P700+ containing natural abundance 14N. Spectrometer conditions: microwave frequency and magnetic field strength indicated on spectra; pulse power: 30 W (A), 80 W (B), 100 W (C); pulse width: 22 ns FWHM (all); pulse repetition rate: 30 Hz (A,B), 60 Hz (C); tau value: 240 ns (A), 250 ns (B), 300 ns (C); sample temperature: 4 K.

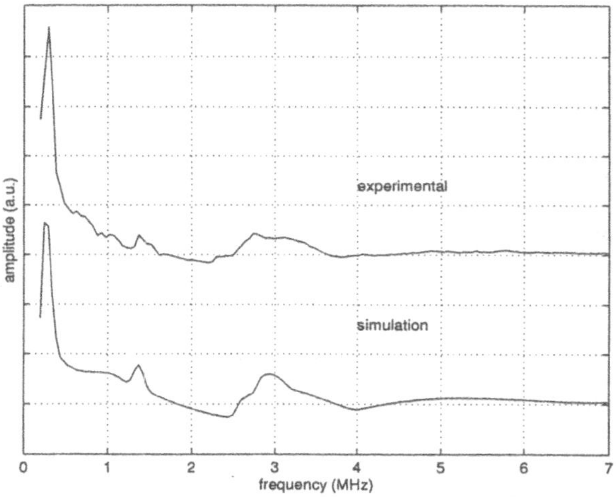

Figure 2: Experimental (top) and simulated (bottom) ESEEM spectra of 15N enriched P700+. Spectrometer conditions: microwave frequency: 8.8 GHz; magnetic field strength: 3167 G; pulse power: 30 W; pulse width: 22 ns FWHM; pulse repetition rate: 90 Hz; tau: 250 ns; sample temperature: 4 K. Simulation parameters: Set 1, single 15N nucleus: Aiso=3.60 MHz, r=2.15 Å; Set 2: two 15N nuclei : Aiso=2.75 MHz, r= 2.60 Å; all other parameters as in experimental.

the third) and a broad feature observed at a higher frequency (a $\Delta m_I=2$ or double quantum transition). The lineshape of this double quantum feature depends on the anisotropic portion of the hyperfine coupling. Figure 1B shows that the exact cancellation condition is met at H_o= 3480 G, 9.7 GHz. Two sets of quadrupole parameters were determined: 1) e^2qQ = 2.70 MHz, η= 0.85 and 2) e^2qQ = 2.81 MHz , η= 0.73, where e^2qQ is the quadrupole coupling constant and η the asymmetry parameter. To eliminate contributions from the quadrupole couplings and resolve the hyperfine couplings, the reaction centers were enriched with 15N (I=1/2).

The 15N data, Fig. 2 (top) show two sets of hyperfine parameters: ($A_\perp$,$A_\parallel$, Aiso) 1) 2.00 MHz, 3.71 MHz, 2.81 MHz (one nucleus contributing) and 2) 1.71 MHz, 2.14 MHz, 2.70 MHz (two nuclei contributing).. The experimental spectrum was simulated by using these parameters and also including a set of weakly coupled 15N nuclei. These hyperfine tensors (scaled to 14N) and the quadrupole parameters measured from Fig. 1 were then used to simulate the 14N ESEEM spectrum shown in Fig. 3 (the experimental spectrum (top) is shown for comparison). Only three nitrogens were included in the simulations. This is consistent with results from spin

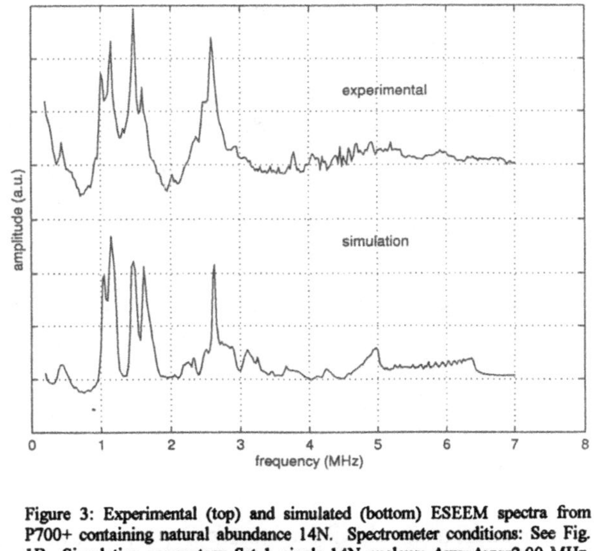

Figure 3: Experimental (top) and simulated (bottom) ESEEM spectra from P700+ containing natural abundance 14N. Spectrometer conditions: See Fig. 1B. Simulation parameters: Set 1, single 14N nucleus: Axx=Ayy=2.00 MHz, Azz=3.71 MHz, e2qQ=2.81 MHz, η=0.73, α=0°, β=90°, γ=0° ; Set 2, two 14N nuclei: Axx=Ayy=1,71 MHz, Azz=2.14 MHz, e2qQ=2.70 MHz, η=0.85, α=0°, β=90°, γ=0°; all other conditions as in experimental.

density calculations on chl a cation radical monomers with the unpaired electron residing in an a_{1u} type molecular orbital that show π spin density on only three nitrogens (one nitrogen with a large amount of unpaired spin density and two nitrogens with smaller, almost identical amounts of unpaired spin density), the remaining unpaired spin is delocalized asymmetrically over the carbon framework of the heterocycle.[9] The large dipolar contribution to the hyperfine coupling in set one is also consistent with a large π spin density. The two sets of 14N's show similar quadrupole parameters, these are expected as well. Numerous studies have shown little variation in the quadrupole parameters of a variety of chlorophyll and bacteriochlorophyll molecules.[5]

4. Conclusions

These data represent the first successful ESEEM simulations of nitrogen hyperfine and quadrupole coupling in P700+. The results of Resonance Raman and optical hole burning experiments have provided evidence that the ground state of P700 is that of a chl a dimer.[10] The magnitudes of the hyperfine coupling constants and the number of nuclei giving rise to these couplings indicate a highly asymmetric spin density distribution for the cation radical. Because there is a contribution from weakly coupled 15N nuclei included in the 15N simulation, a dimeric model for P700+ cannot be discarded at this time, although these weak couplings may arise from putative histidine ligands to the Mg of chl a.

5. References

1 O'Malley, P.J. and Babcock,G.T. (1984) Proc. Natl. Acad. Sci. USA 81, 1098
2 Norris, J.R., Uphaus, R.A., Crespi, H.L. and Katz, J.J. (1971) Proc. Natl. Acad. Sci. USA 68,625
3 Rigby, S.E.J.; Nugent, J.H.A.; and O'Malley, P.J. (1994) Biochemistry 33, 10043
4 Kab, H.; Bittersmann-Weidlich, E.; Andreasson, L.-E.; Bonigk, B.; Lubitz, W. (1995) Chem. Phys. 194, 419
5 Davis, I.H.; Heathcote, P.; MacLachlan, D.J.; Evans, M.C.W. (1993) Biochim. Biophys. Acta 1143, 183
6 Noren, G. H.; Boerner, R.J.; Barry, B.A. (1991) Biochemistry 30, 3943
7 McCracken, J.; Shin, D.-H.; Dye, J.L. (1992) Appl. Magn. Reson. 3, 30
8 Mims, W.B. (1984) J. Magn. Reson. 59, 291
9 Fajer, J.; Davis, M.S.; Brune, D.C.; Spaulding, L.D.; Borg, D.C.; Forman, A. (1977) Brookhaven Symp. Biol. 28, 74
10. Golbeck, J.H. (1992) Annu. Rev. Plant Physiol. Plant Mol. Biol. 43, 293

6. Acknowledgements

The authors would like to thank Dr. Neil Bowlby for assistance in sample preparation. Additionally, grant support from the GE Foundation (M.M), National Institutes of Health (J.M.) and United States Department of Agriculture (G.T.B.) is gratefully acknowledged.

TIME-RESOLVED FLUORESCENCE STUDIES OF PHOTOSYSTEM I ANTENNAE

N. Schneebauer and G. Hauska,
Universität Regensburg, Institut für Zellbiologie und Pflanzenphysiologie,
D-93053 Regensburg, Germany

1. Introduction

The antenna of photosystem (PS) I in higher plants and cyanobacteria consists of several chlorophyll (Chl) containing protein-complexes. We can distinguish between outer, light harvesting complexes (LHC) and the central antenna on the reaction-center (RC)-complex. Time resolved spectroscopy gives us a tool for investigating the energy transfer in the antenna system after a short excitation. Much work was done with emphasis on the kinetics of energy migration on a ps-time scale (for a review, see [1]), neglecting longer time domains, which also give valuable information. Therefore we present fluorescence characteristica of the central antenna of PS I reaction centers from *Anabaena variabilis* in the time region from sub-ns up to μs.

2. Materials and methods

PS I was isolated from the cyanobacterium *Anabaena variabilis* with Triton X-100. The P 700 reaction centers were derived from PS I by treatment with SDS followed by sucrose density gradient centrifugation [2]. The preparation contains the primary donor, P 700, the complete acceptor chain and 80 - 100 antenna Chl-a molecules. The samples were filled in thin capillary tubes with an inner diameter of 0.3 mm and put into a cryostate.
Fluorescence decay measurements were performed using time-correlated single photon counting (for a review see [3]). Excitation was provided by a mode-locked, cavity-dumped Ar-ion laser at 496.5 nm. The pulses had a duration with a FWHM of 200 ps. The excitation energy was $2 * 10^{-9}$ J ($5 * 10^{-9}$ photons/puls). The Poisson-equation was used to estimate the probability of the number of photon-hits at one reaction center. In combination with the Chl-a concentrations in our samples and a repetition rate of 4 kHz the time for a secondary hit at a reaction center lies in the range of seconds. The decay curves were the basis for the 3d-plots in figure 3 and 5. For a numerical analysis the decay curves were fitted with the convolution of the excitation pulse and a sum of exponential functions. Best fits were achieved with a sum of two exp-functions with decay-times τ_1 and τ_2.
For measurements in the μs-time-region the mode-locking-unit was switched off. This leeds to pulses with a variable FWHM from 15 ns to 50 ns. The fluorescence signal was collected in a multi-channel-analyzer.

P. Mathis (ed.), Photosynthesis: from Light to Biosphere, Vol. II, 39–42.

2. Results and discussion

Figure 1 shows in the lower trace the fluorescence spectrum of PS I reaction centers of *A. variabilis*, collected in the first 100 ns after a 15 ns-pulse.

The spectrum exhibits a broad maximum at 676 nm with a sharp rise at the short wavelength side. At the long wavelength flank From these findings it can be clearly deduced as a typical spectrum of chla-molecules in a protein matrix [4]. In agreement with our experiments in the ps-time-domain (see section below) we ascribe this maximum to prompt fluorescence from the antenna chla-molecules. Additional peaks, which would be a hint for long-living fluorescence components (or for delayed fluorescence), could not be detected. The prereduction of the probes with ascorbate or sodium dithionite did not alter these results (fig. 1, middle and upper trace). The fluorescence decay (e.g. in fig. 2 at 680 nm) in the observed wavelength region showed no contribution of long-living or delayed fluorescence. Within 100 ns after the pulse the fluorescence decreases to the dark rate. For testing the sensitivity of the detection unit 1.2 µs after the first pulse a weak secondary pulse was emitted by the laser (see fig. 2). The experiments were extended to a observation time scale of 1 ms (not shown here) without any change of the results: no longer living fluorescence component could be detected. Also decreasing the temperature down to 2 K (not shown here) did not alter this findings. To explain the absence of long-living fluorescence components we propose, in accordance to [5], a pathway for the rereduction of the primary donor, which involves the efficient formation of a ^{T}P 700. The spin-forbidden transition from a triplett state to the ground state has a much longer life-time and, in the case of chla-molecules, lies at wavelengths longer than 800 nm.

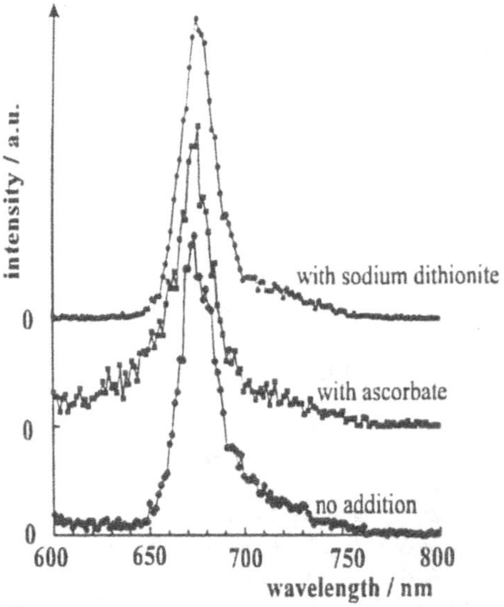

Fig. 1: Fluorescence spectra of PS I reaction centers (*A. variabilis*) without addition, with ascorbate and with sodium dithionite, resp., after a 15 ns pulse (Δt = 100 ns, T = 258 K)

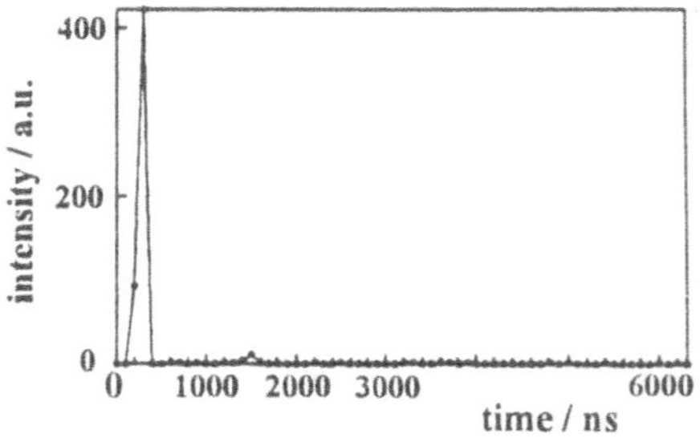

Fig. 2: Fluorescence decay collected at 680 nm from PS I RC (*A. variabilis*) after excitation with with a 15 ns pulse (channel width: 100 ns, T = 258 K)

The fluorescence decay of a PS I sample under highly reducing conditions (pH 10.5, addition of a sodium dithionite) at 258 K after a 200 ps pulse (fig. 3) shows a broad band at 680 nm and a plateau region until 740 nm. For PS I samples preoxidized with ferricyanide the plateau only reaches 710 nm (fig. 4).

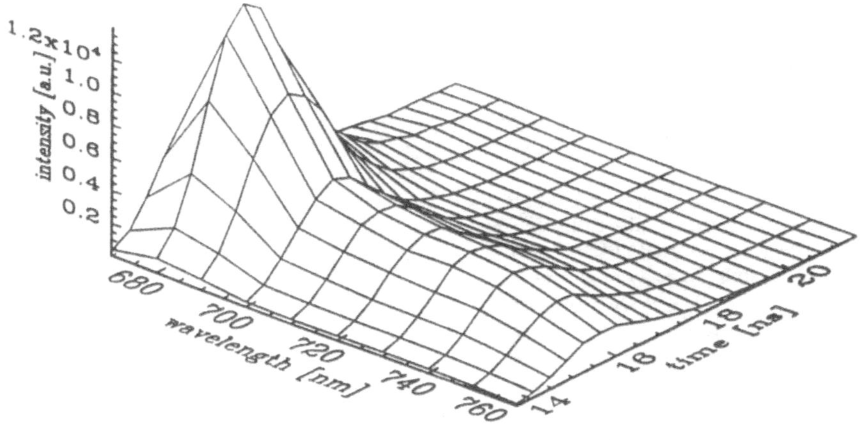

Fig: 3: Fluorescence decay of PS I RC (*A.variabilis*) under highly reducing conditions after a 200 ps excitation (T = 258 K)

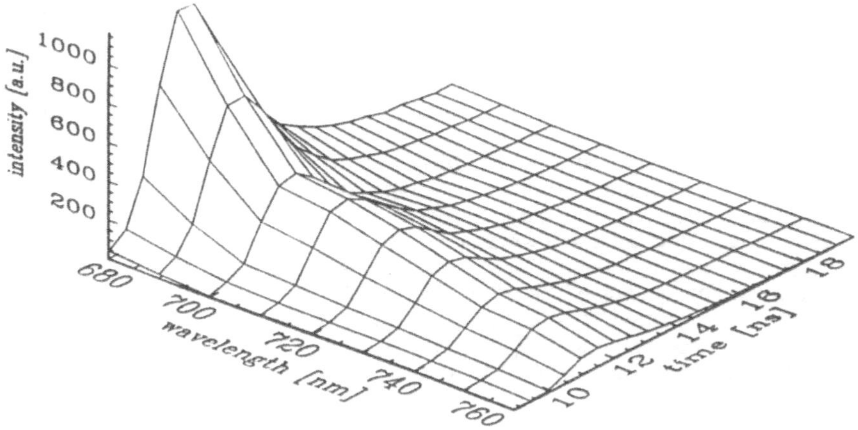

Fig: 4: Fluorescence decay of oxidized PS I RC (*A.variabilis*) after a 200 ps excitation (T = 242 K)

For all samples the longer decay-time, τ_2, is 3 - 4 ns with an uncertainty of $\pm$ 0.75 ns. J.P. Ide et al. [6] determined the lifetime of free Chl-a in detergent micelles with 5.7 ns. Therefore we interprete τ_2 in agreement with [7] as the fluorescence decay of functionally not connected antennae complexes.

The first decay-time τ_1 of the oxidized samples did not change with decreasing the temperature from 242 K to 113 K (not shown here). For both temperatures the values of τ_1 lie between 1 and 1.5 ns. The observed fluorescence signal cannot come from recombination of P 700^+ with reduced early electron acceptors, because in the preoxidized samples no charge separation can occur. It must be emitted from antenna complexes.

For the reduced samples τ_1 was nearly constant at 1 ns over the whole wavelength region (fig. 5, lower trace). A decrease of the temperature to 123 K leeds to a significant prolongation of τ_1 for wavelengths greater than 680 nm (fig. 5, upper trace). A maximum was reached at 710 nm with a τ_1 of 2 ns. In agreement with our results at longer time scales (see above) and the known recombination times (see ref. cited in [1]) for this samples contributions from recombination of P 700^+ with the early electron acceptors also must be excluded.

The value of τ_1 (1 - 2 ns) prohibits an interpretation as natural life-time of free chla-molecules (see discussion above). The rise of τ_1 when reducing the temperature can be explained in terms of the Förster-Dexter mechanism for energy transfer. At low temperatures the energy transfer within the antenna-complex is slowed down due to a much smaller spectral overlap between individual donor-acceptor molecules. Therefore τ_1 represents a fluorescence component arising from energy-transfer in an antenna-complex. Considering the sensitivity of the fluorescence to the redox-state of the primary donor P 700 (fig. 3 and 4) we can say that this decay-time of about 1 ns comes from intact antenna complexes connected to reaction centers. The distribution of the fluorescence intensities demonstrate that a majority of the Chl-a antenna molecules emit at 680 nm. Assuming a Stokes-shift of 5 - 7 nm, as in [8], the plateau up to 740 nm can be ascribed to fluorescence from antenna Chl a molecules with an absorbtion maximum higher than that of P 700.

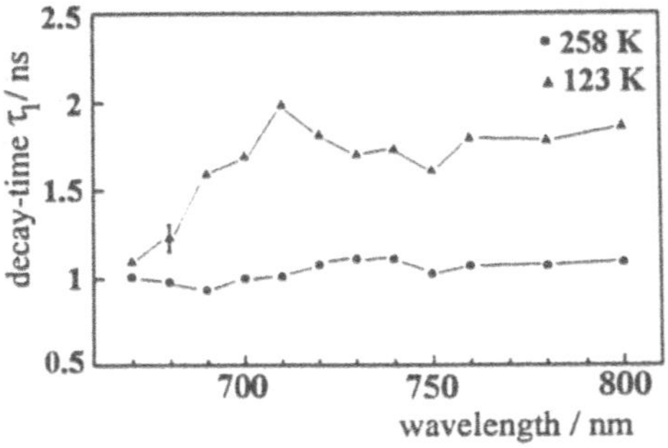

Fig. 5: Decay-time τ_1 of PS I RC (*A. variabilis*) at two different temperatures

The addition of an oxidant (ferricyanide) decreases the fluorescence intensity drastically. As can be seen in fig. 4 the relative fluorescence intensity of the oxidized sample is more than one magnitude smaller than in the case of the reduced one. This means that the oxidized primary donor quenches fluorescence much better than the reduced one. A similar result result was found by [9] for the primary donor P 798 in *Heliobacterium chlorum*. Furthermore this quenching effect of the oxidized donor leeds also to a decrease of the fluorescence between 700 nm and 740 nm.

References:

[1] Sétif, P. (1992), in: The Photosystems: Structure, Function and Molecular Biology, J. Barber (ed.), Elsevier Science Publishers, 471-499

[2] Schoeder, H.-U. and Lockau, W. (1986), FEBS Lett. 199, 23-27

[3] Holzwarth, A.R. (1995), Meth. Enzym., 246, 334-362

[4] Ide, J.P., Klug, D.R., Kuhlbrandt, W., Porter, G., and Barber, J. (1986), in: Ultrafast Phenomena V, Fleming, G.,R., and Siegman, A.E., (ed.), Springer Series in Chem. Phys.,46, 406-408

[5] Sétif, P. and Brettel, K. (1990), BBA,1020, 232 -238

[6] Ide, J.P., Klug, D.R., Kuhlbrandt, W., Giorgi, L.B., Porter, G., Gore, B., Doust, T., and Barber, J. (1986), Biochem. Soc. Trans., 14, 34

[7] Schatz, G.H. and Holzwarth, A.R. (1986), Photosynth. Res., 10, 309-318

[8] Zucchelli, G., Jennings, R.C., and Garlaschi, F.M. (1992), BBA, 1099, 163-169

[9] Deinum, G., Kramer, H., Aartsma, T.J., Kleinherenbrink, F.A.M., and Amesz, J. (1991), BBA, 1058, 339-344

SEARCH FOR THE PRESENCE AND ROLE OF A SECOND PHYLLOQUINONE IN THE CORE OF PHOTOSYSTEM I

Thomas Schwartz and Klaus Brettel
Section de Bioénergétique (CNRS-URA 1290), DBCM, CEA-Saclay,
91191 Gif-sur-Yvette Cedex, France

1. Introduction

Phylloquinone (PhQ), also called vitamin K_1, functions as secondary electron acceptor A_1 in photosystem I (PS I) (see Ref. 1 for a review on PS I). Two PhQ molecules are present in PS I. According to Refs 2 and 3, one of them is readily extracted with hexane without considerable loss of activity in electron transfer (ET) to extrinsic electron acceptors, indicating that only one PhQ molecule is involved in ET in PS I. This is in contrast to photosystem II and to purple bacteria where two quinone molecules function as sequential electron acceptors. Here, we investigate a possible participation of the second PhQ by studying directly ET at the level of the secondary acceptor A_1 prior to and after extraction with hexane. The PhQ content is quantified by HPLC.

2. Material and Methods

Biological Samples. PS I was isolated from the cyanobacterium *Synechocystis* sp. PCC 6803 and lyophilized according to Ref. 3. Lyophilized PS I was extracted 3 times with dry hexane or hexane containing 0.3% methanol. The 3 supernatants were collected and taken to dryness under vacuum. The pellet was dried under nitrogen gas flow and stored in evacuated tubes at -20 °C. Prior to spectroscopic experiments, dry samples were solubilized with 50 mM Tricine (pH 7.6), 10 mM potassium chloride and 1.2% octylglucoside - 0.6% sodium cholate, sonicated 3 times for 15 seconds and centrifuged to remove unsolubilized particles. The chlorophyll (Chl) content was determined according to Ref. 4.
Flash Absorption Spectroscopy. Flash induced absorbance changes at 820 nm were measured with a time resolution of approx. 5 ns using a set-up described in Ref. 5. For the UV and blue spectral region, we used a set-up similiar to the one described in Ref. 6. Its time resolution was approx. 10 ns.
Phylloquinone Analysis. Dry PS I preparations were extracted with chloroform - methanol (2:1). After mixing thoroughly, the extracts were recovered quantitatively and taken to dryness. These residues, as well as the residues of the hexane and hexane/methanol extracts (see above), were dissolved in a small volume of ethanol and analyzed by reversed-phase HPLC using ethanol containing 6% water as eluant (1ml/min). Two 30 cm analytical columns (µBondapak C18, Waters / Ultrasphere ODS, Beckman) were used in series in order to improve the separation. The elution profile was recorded between 220 and 790 nm using a photodiode array detector (Waters 996). A retention time of 19.5 min was observed for PhQ standards.The amount of PhQ present in biological samples was calculated from the area under the peak around 19.5 min in the

P. Mathis (ed.), Photosynthesis: from Light to Biosphere, Vol. II, 43–46.

elution profile at 270 nm. Analysis of the complete data sets revealed that PhQ coeluted with a species which had the same spectrum as Chl *a* (possibly Chl *a'*) and contributed to the absorption at 270 nm. This contribution was subtracted, using the elution profile at 431.5 nm (absorption maximum of Chl *a*) and the 270 to 431.5 nm absorbance ratio of Chl *a* (0.2935).

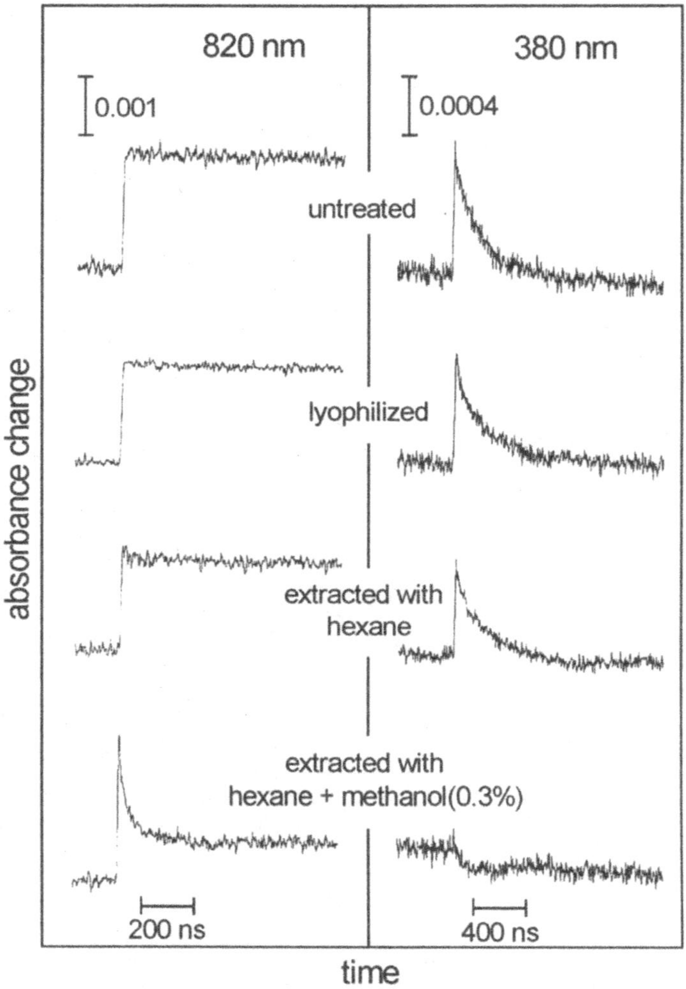

Fig. 1. Absorbance changes at 820 and 380 nm in various samples at non-saturating excitation. Left (820 nm): 50 μM Chl; 10 mm optical path; excitation by 300 ps laser pulses at 532 nm, approx. 0.5 mJ/cm^2. Right (380 nm): 100 μM Chl; 2 mm optical path; excitation by 7 ns laser pulses at 600 nm, approx. 0.3 mJ/cm^2.

P700 Analysis was performed by two independent methods : (a) Oxidized-minus-reduced difference spectra were measured by adapting the method described in Ref. 7 to a single beam UV-VIS spectrometer (Cary 5E). A first absorption spectrum was recorded 70 seconds after addition of ferricyanide (0.36 mM). Then dry sodium ascorbate (1.8 mM) was added and after 2 minutes a second absorption spectrum was recorded and subtracted from the first spectrum. The P700 content was calculated from the absorbance change at the maximum of bleaching (around 700 nm) using a differential extinction coefficient of 64000 $M^{-1}cm^{-1}$ for $P700^+$/P700. (b) Absorbance changes at 820 nm on a ms time scale were recorded under saturating excitation by two synchronuously fired Xe flashes (pulse duration, 3 μs) placed at opposite sides of the sample cuvette. The P700 content was calculated using a $P700^+$ extinction coefficient of 6500 M^{-1} cm^{-1} at 820 nm [8].

3. Results and Discussion

3.1 *Electron Transfer*

Fig. 1 shows flash induced absorbance changes at 820 nm (essentially due to $P700^+$) and at 380 nm (essentially due to A_1^-) after various treatments. The sample extracted with pure hexane behaved very similarly to the controls (untreated and lyophilized samples), showing long-lived $P700^+$ and transient formation of A_1^-, the latter being reoxidized with $t_{1/2} \approx 160$ ns, presumably due to ET to F_X [9]. Thus, extraction with pure hexane did not change the kinetics of ET at the level of A_1.

Extraction with hexane containing 0.3% methanol should remove both molecules of PhQ [3], and in line with this reference, we observed a short lived signal at 820 nm ($t_{1/2} \approx 30$ ns, reflecting charge recombination in the primary pair $P700^+A_0^-$) (Fig. 1, bottom, left). Formation of A_1^- as detected at 380 nm was largely suppressed (Fig. 1, bottom, right).

3.2 *Phylloquinone Content*

Assuming that pure hexane extracts one out of two PhQ molecules present in PS I [2, 3], we expected to recover the same amount of PhQ in the supernatants as in the final pellet after 3-fold extraction with hexane. However, we found approx. 5 times less PhQ in the

Table 1. Amount of phylloquinone recovered in the supernatants and in the pellet after 3-fold solvent extraction of 1 mg lyophilized PS I

	extraction with hexane	extraction with hexane + methanol(0.3%)
supernatants	0.3 nmol	2.0 nmol
pellet	1.6 nmol	0.1 nmol

Table 2. Quantification of phylloquinone and P700

preparation	PhQ/PS I (nmol/mg)	Chl *a*/PS I (nmol/mg)	Chl *a*/P700 (mol/mol)	PhQ/P700 (mol/mol)
lyophilized	2.4 ± 0.2	91 ± 4	109 ± 4[a] 102[b]	2.9[a] 2.7[b]
extracted with hexane	2.1 ± 0.2	93 ± 4	107 ± 1[a] 96[b]	2.4[a] 2.2[b]
extracted with hexane+methanol(0.3%)	0.2 ± 0.2	38 ± 3	76 ± 2[a]	0.4[a]

[a] P700 content determined from the oxidized-minus-reduced difference spectrum
[b] P700 content determined from flash induced absorbance changes at 820 nm

hexane supernatants than in the pellet (see Table 1). After extraction with hexane containing 0.3% methanol, we recovered the large majority of PhQ in the supernatant, as expected (see Table 1). The surprising result with pure hexane may indicate that the PhQ extractable with hexane in our PS I preparation is not an integral component of PS I, but rather associated with lipids attached to the PS I complex (cyanobacteria contain a considerable amount of PhQ in addition to that bound to PS I [10, 11]).
In order to check whether the PhQ remaining after extraction with pure hexane represents one or two molecules per PS I complex, we tried to relate the PhQ content to the P700 content in our samples. As detailed in Table 2, a PhQ to P700 ratio of approx. 2.3 was obtained for the sample extracted with pure hexane. We would like to mention some errors which may have entered into this ratio: scattering of data on PhQ, Chl *a* and P700 content between different experiments as indicated in Table 2; loss of PhQ during extraction and preparation of samples for HPLC (in the order of 10% according to controls using added PhQ as internal standard); errors in the extinction coefficients used; presence of inactive PS I complexes still containing PhQ (the data in Fig. 1 indicate a loss of approx. 20% in signal amplitudes due to the lyophilization). Taking into account these possible errors, it is still highly likely that the PhQ remaining after extraction with pure hexane represented more than one (most likely two) molecules per PS I complex. We have no consistent explanation for the deviation of our results from those obtained by Biggins and Mathis [3] who used virtually identical samples and extraction procedures. The deviation from the results obtained by Malkin [2] might be related to the use of different samples (PS I complexes from spinach containing 200 Chl/P700).

4. Conclusion

As, in our hands, samples extracted with pure hexane still contained approx. two PhQ molecules per PS I complex, our observation of unmodified kinetics at the level of the secondary electron aceptor A_1 does not exclude a participation of both PhQ molecules in electron transfer in PS I.

Acknowledgement

We would like to thank G. Berger for performing the HPLC, H. Bottin for discussion and initiation to the preparation methods and.B. Le Rest for the realisation of fast data transfer from the digitizing oscilloscope to a personal computer.

References

1 Golbeck, J.H. (1992) Annu. Rev. Plant Physiol. Plant Mol. Biol. 43, 293-324
2 Malkin, R. (1986) FEBS Lett. 208, 343-346
3 Biggins, J. and Mathis, P. (1988) Biochemistry 27, 1494-1500
4 Arnon, D.I., McSwain, B.D., Tsujimoto, H.Y. and Wada, K. (1974) Biochim. Biophys. Acta 357, 231-245
5 Brettel, K. and Golbeck, J.H. (1995) Photosynth. Res., in press
6 Gerken, S., Brettel, K., Schlodder, E. and Witt, H.T. (1987) FEBS Lett. 223, 376-380
7 Markwell, J.P., Thornber, J.P. and Skrdla, M.P. (1980) Biochim. Biophys. Acta 591, 391-399
8 Mathis, P. and Sétif, P. (1981) Isr. J. Chem 21, 316-320
9 Lüneberg, J., Fromme, P., Jekow, P. and Schlodder, E. (1994) FEBS Lett. 338, 197-202
10 Takahashi, Y., Hirota, K. and Katoh, S. (1985) Photosynth. Res. 6, 183-192
11 Schoeder, H.-U. and Lockau, W. (1986) FEBS Lett. 199, 23-27

THE ROLE OF LONG WAVELENGTH ABSORBING CHLOROPHYLLS IN TRIMERIC PS I ISOLATED FROM CYANOBACTERIUM *SYNECHOCOCCUS ELONGATUS*.

B. Sopko[1], D. Sofrová[1], J. Hladík[1], B.A. Gulyayev[2], V.L. Tetenkin[2]

[1]Dept. of Biochem., Faculty of Natural Sci., Charles Univ., Albertov 2030, 128 40 Praha 2, Czech Republic; [2]Fac.of Biology, Leninskie Gory, Moscow, 119899, GSP, Rossia

1.Introduction

Photosystem I (one of the two protein complexes responsible for light induced charge separation in cyanobacteria) transfers electrons from plastocyanin (or cytochrome C_6) via ferredoxin to $NADP^+$ reductase. However, despite its better stability during isolation, the PS I function and properties are not as well studied as the ones of PS II. Let us therefore summarise the known facts of PS I: at first there is no significant change in the fluorescence quantum yield when PS I becomes fully oxidised or reduced, at second charge separation does occur even when the acceptor A_1 is removed [1], at third the main emission fluorescence peak occurs at higher wavelengths than the reaction centre P_{700} absorbs, and at fourth PS I contains chlorophylls absorbing at 705 nm and higher [2]. The other studied phenomenon is the question of whether PS I occurs *in vivo* in the trimeric or monomeric form, however, there is more evidence for the trimeric one [3].

2.Materials and Methods

PS I in the trimeric form was prepared from thylakoid membranes using Triton X100 and further purified using mild SDS electrophoresis [4]. The temperature dependence of fluorescence spectra and quantum yields of fluorescence ere measured using an LS-5B spectrofluorometer (Perkin-Elmer, UK) with modified accessory for measurements at different temperatures from 80 K up to ambient temperature. Free chlorophyll, used as standard, was obtained by adding Triton X-100 to the sample used for the measurement to the concentration of 4%. Absorption spectra at 293 K and 77 K and LD spectra were obtained using an M40 spectrophotometer by Zeiss Jena (Jena, Germany) with home made accessory for measuring

P. Mathis (ed.), Photosynthesis: from Light to Biosphere, Vol. II, 47–50.

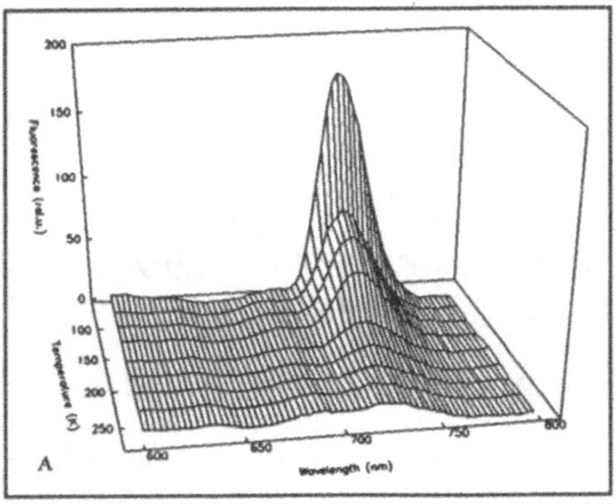

Figure 1a *Dependence of PS I fluorescence emission spectra on temperature*

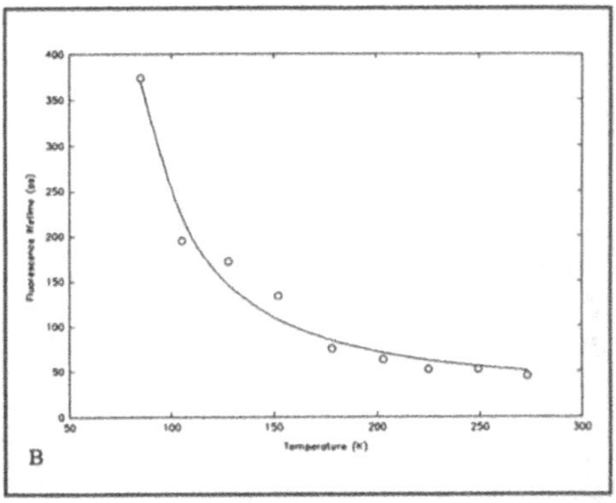

Figure 1b *The result of non-linear curve fit of Tusov's equation to fluoresccence lifetimes obtained from the emision spectra.*

at 77 K (in the case of LD spectra an LD dichroic filter was added, the sample was oriented by pressure in PAA gel). CD spectra at 77 K were measured using a Mark V dichrograph by Jobin-Yvon (Longjumeau, France).

3.Results and discussion

Figure 1a shows the temperature dependence of PS I fluorescence emission spectra. From these spectra, we determined the fluorescence quantum yields and subsequently calculated the corresponding theoretical lifetimes using a coefficient obtained from the quantum yield of fluorescence and fluorescence lifetime at room temperature. The validity of these calculations was confirmed by direct measurement of the fluorescence lifetime at 77 K; the results were within the limits of experimental error. We used the lifetimes thus calculated to determine the fluorescence activation energy according to Tusov's modification of Arhenius's equation (the result of nonlinear regression is shown in figure 1b) [5]. The result of this nonlinear regression was:

$$\tau\,(\text{ns}) = \frac{1}{\left(0.51 + 51.32.e^{-\frac{23.1\ meV}{k.t}}\right)}$$

Using this activation energy value (23.1 meV), we tried to estimate the energetical level of P_{700} charge separation. Taking into account the first excited state energy of the antenna (maximum of fluorescence emission) plus the above mentioned activation energy we concluded that the charge separation energy is 1742.7 meV, thus suggesting that the P_{700} recombination fluorescence should occur at 711.5 nm for room temperature (for 77K the energy is

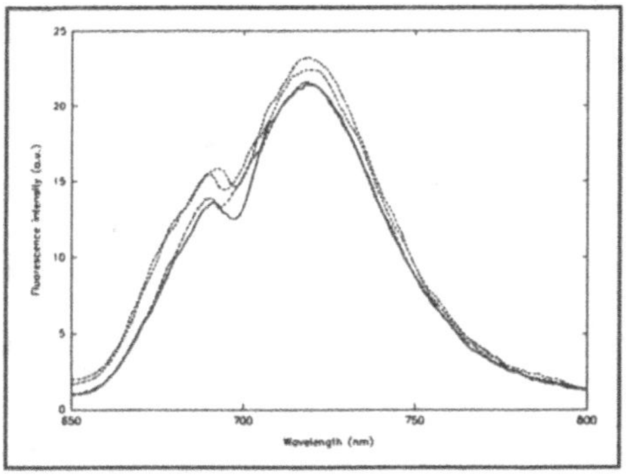

Figure 2a *Effect of different concentrations of $Na_2S_2O_4$ on fluorescence emission spectra of PS I. Full line - oxidised PS I, long dash - 0.33 mg/ml of sodium dithionite, medium dash - 0.66 mg/ml of $Na_2S_2O_4$, short dash - 1.00 mg/ml of $Na_2S_2O_4$.*

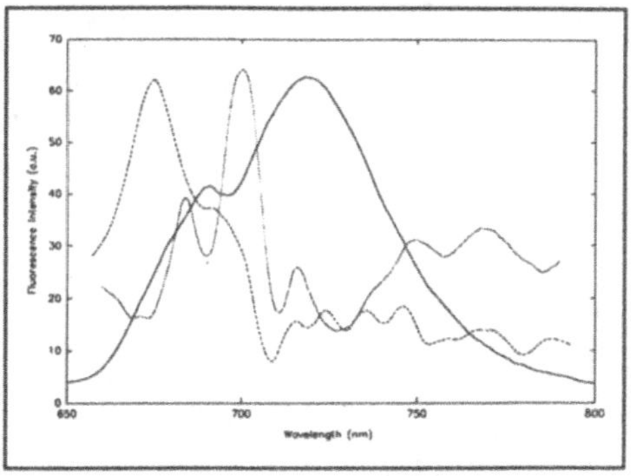

Figure 2b *Resulting spectra from factor analysis of fluorescence emission spectra with different concentration of $Na_2S_2O_4$ (see figure 4). Full line - spectrum associated with fluorescence emission of PS I antennal chlorophylls. Dashed line - spectrum associated with free chlorophyll fluorescence spectrum. Dotted line - spectrum associated with P_{700} emission spectrum.*

1706.8 meV and recombination fluorescence at ought to be at 725.4 nm).

To justify our results we have attempted to estimate the wavelength of recombination fluorescence of P_{700} using dithionite titration (figure 2a) and subsequent mathematical factor analysis of the spectra obtained (figure 2b). The first subspectrum can be described as the emission fluorescence of PS I antenna chlorophylls, the second one we attributed to the required P_{700} recombination fluorescence (reversible after subsequent oxidation) and we consider the third one to be a side effect of dithionite (setting free the outermost chlorophylls; this fluorescence is irreversible). However, the energy of the first excited state of P_{700} has the energy evidently corresponds to the emission wavelength of 703 nm and not at 711.5nm as calculated above. Considering that there are no differences in the fluorescence quantum yield of PS I in the fully oxidised or reduced state [1], we have decided that the energy transfer from antenna chlorophylls to P_{700} is carried out via special non-fluorescent chlorophylls absorbing at the suggested wavelengths. In figure 3 we can see the absorption spectrum

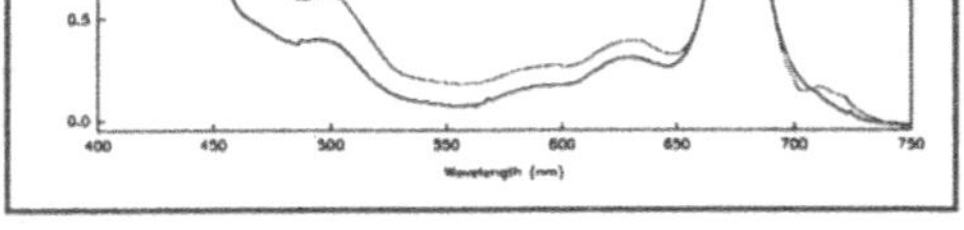

Figure 3 *The PS I absorption spectra at 290 K (full line) and 77 K (dashed line).*

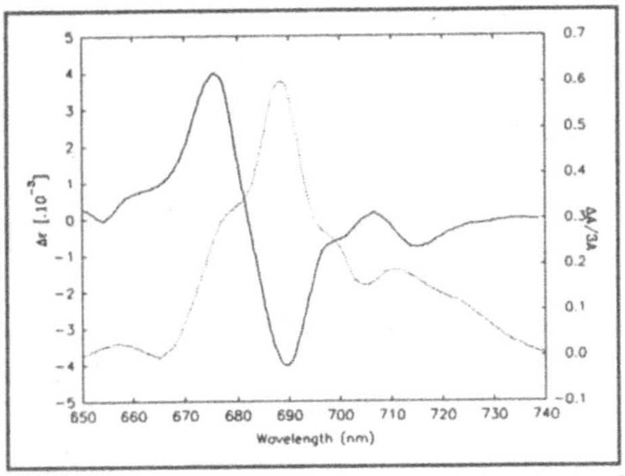

Figure 4 *CD (full line) and LD (dashed line) spectra of isolated PS I particles at 77 K.*

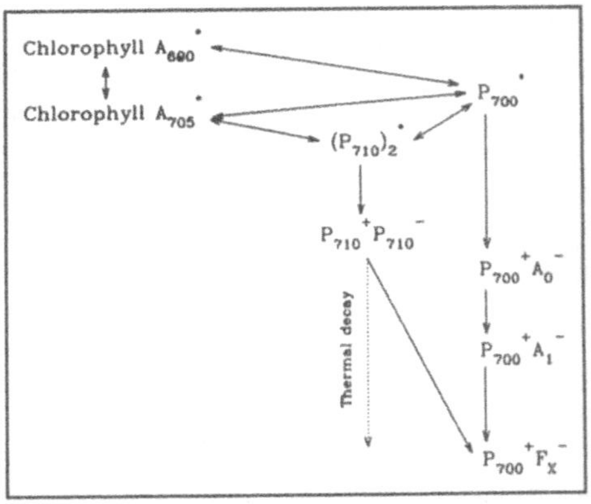

Figure 5 *Scheme of excitation energy transfers in PS I.*

of PS I at 77K and room temperature with the second derivative of this spectrum and indeed we can clearly spot an absorption peak at 710 nm; the same but for 77 K and does confirm our result for room temperature (however, the absorption maximum is at 724 nm).

Hence, we have found that there really are chlorophylls in PS I absorbing at the wavelengths, assessed by the measurement of fluorescence activation energy but without any knowledge of their conformations. Trying to resolve this problem we have carried out linear and circular dichroism measurements (figure 4) at 77 K, resulting in the conclusion that these chlorophylls are in dimeric form (see the CD spectrum at 724 nm) and oriented in parallel to P_{700} (see the LD spectrum). The LD result thus solves also the problem laid by Woolf et al. [6].

Figure 5 summarises all our results in one scheme of energy transfers in PS I, in which however, all energy levels are approximative only.

References

1 Golbeck, J.H. (1992) Annu. Rev. Plant Physiol. Plant Mol. Biol. 43, 293-324.

2 Shubin, V.V., Bezsmertnaya, I.N. and Karapetyan, N.V. (1992) FEBS Lett. 309, 340-342.

3 Hladík, J. and Sofrová, D. (1991) Photosynth. Res. 29, 171-175.

4 Komenda, J., Hladík, J. and Sofrová, D. (1989) J. Photochem. Photobiol. 3, 575-592.

5 Stahl, U., Tusov, V.B., Paschenko, V.Z. and Voigt, J. (1989) Biochim. Biophys. Acta 973, 198-204.

6 Woolf, V.M., Wittmershaus, B.P., Vermaas, W.F.J. and Tran, T.D. (1994) Photosynth. Res. 40, 21-34.

ON THE SPATIAL DISTRIBUTION OF CHLOROPHYLL SPECTRAL TYPES IN PHOTOSYSTEM I

G.Trinkunas[1] and A.R.Holzwarth[2]
[1]Institute of Physics, Vilnius, 2600, Lithuania
[2]Max-Planck-Institut für Strahlenchemie, D-45470, Mülheim (Ruhr), Germany

1. Introduction

It is now generally accepted that photosynthetic antenna are strongly energy disordered and heterogeneously broadened. This conclusion emerges from kinetic data for the energy transfer processes on time scales from femtoseconds to piscoseconds [1] and from hole burning experiments covering almost all photosynthetic systems. Due to the scarce knowledge about the antenna structure and the complexity of the processes involved the comparative analysis of experimental data and theoretical model predictions for the structure/function relationships is required in order to find the proper excitation energy transfer parameters. Even the antenna structural data at hand a major problems in determining the excitation energy delivery paths comes from the lack of the knowledge about the spatial arrangement of the pigment spectral types.

Two views in this respect are presently under discussion. The first one is based on the assumption of a complete random nature of the spatial-spectral correlations of the pigment properties. In the kinetic excitation dynamics modelling the spatial spectral type distributions on a given structure are generated at random and the kinetics is found as the result of an averaging procedure [2-6]. Such an approach practically excludes the possibility to access the structural antenna information from the kinetic data. The second one assumes that the specific spatial distribution of spectral types is responsible to a large part for the measured excitation decay kinetics in antenna [7]. This model can still involve a great deal of random spectral characteristics around one site. Certainly also the type of antenna will decide which model is the preferred one.

We have taken the latter model as the basis for simulating the excitation dynamics in cyanobacterial photosystem (PS) I core antenna/reaction center (RC) particles by the exact solution of the Pauli master equation for excitation motion on a 10x10 regular square lattice [8]. A striking feature of the system is the presence of pigments that absorb at longer wavelengths than the primary electron donor. Therefore, for describing the inhomogeneous spectral distribution by seven discrete spectral forms, we had to include two "red" spectral forms in order to properly describe the time-resolved fluorescence decays. The specific feature of that simulation consisted in the calculation of so-called decay-associated spectra (DAS) [7] and comparing them to the results of global analysis of experimental data [9]. The simulation of the two longest experimental lifetimes and their DAS as well as the steady state fluorescence spectrum shape resulted in a single arrangement of pigment spectral types with the "red" pigments clustered in the corner of the lattice with the P700 pigment at a distance

P. Mathis (ed.), Photosynthesis: from Light to Biosphere, Vol. II, 51–54.

of two lattice constants [8]. While providing new principal insight into the spatial arrangement of Chl spectral types in PS I this study suffered from several limitations. First, concerning the PS I structure, a two-dimensional model is hardly realistic given the present knowledge. A single regular lattice spacing could be far from reality also. One of the most stringent restrictions was however that we applied a manual search procedure to find the lattice arrangements. Given the enormous parameter space chances are high that good arrangements were simply missed.
We present here the results of the simulations of the excitation dynamics free from the above discussed limitations. While using the same criteria [8] to select the actual spectral type distributions on the model lattice we extend the number of used parameters (the lattice constant and the rate of charge separation from P700 pigment) to include two extra spatial scalings: the mean spacings between P700 and the "red" pigments and to their nearest neighbor pigments. Furthermore we use a three-dimensional lattice. For the automated search of the actual spectral lattice arrangements we employ genetic algorithms (GAs) [10].

2. Procedure

The arrangement of the pigment spectral types on the lattice can be defined as a string of spectral type names. Due to conservation of the spectral content of the system, the search space is formed by a multitude of string order permutations. The concatenated bit representations of four continuous parameters form a bit string which defines the multitude of possible bit combinations also. GAs are promising for this kind of random search problem since they can sample large spaces efficiently due to rapid convergence on solutions [10]. The idea behind the GAs is borrowed from the Darwinian evolution theory. The search starts with a random generation of a population of strings (solutions), the latter playing the role of chromosomes. When decoded, they represent the set of parameters of the target function under optimization. In this way every string is to be ranked according to its fitness value. The latter mimics the natural selection and defines whether the solution is included in the mating pool or "dies". After the mating pool is formed, pairs of parent strings are selected at random and crossed over, yielding a pair of new individuals. This particular operation combines the parent "genes" to produce the offspring "chromosome", thus forming an intelligent randomization strategy for the search of the "fittest" population. The offsprings, with a small predefined probability, are subjected to random mutation. In the optimization procedure this helps to avoid locking in a local minimum. With the evaluation of the fitness of new individuals the above genetic operations are repeated and the process proceeds till the desired solutions are found. The fitness of the order as well as the bit strings representing the particular solution of the problem is defined as the inverse power of the sum of dimensionless least squares deviation from experimental lifetimes as well as all the points in the DAS and steady state fluorescence spectrum. For more details we refer to [11] where the GAs has been applied for the excitation dynamics simulations for the first time.

3. Results and discussion

We performed searches on three-dimensional lattice models of PS I including nearest-neighbor and next to nearest neighbor excitation transfer approximations. In most of the search runs over 100 generations were needed for a 1000 string population to reach a good fit level. The procedure provided over 10% of the final generation population to be good solutions. About 10% of the latter differed in the "red" pigment locations. The common feature of these arrangements is in line with earlier findings for the two-dimensional PS I model [8]: red pigments are spread on the surface of the system without direct transfer contact of the "reddest" pigment to P700. However, the arrangements vary in the extent of the "red" pigment clustering. The detailed description of all the results will be published elsewhere. Here we concentrate on the discussion of the model with the P700 pigment site fixed in the center of a plane applying the next-to-nearest neighbor transfer approximation. Among the models studied this one is closest to the available PS I particle structure [12]. The continuous parameter distributions describing the good solutions in the final generation are presented in Figure 1.

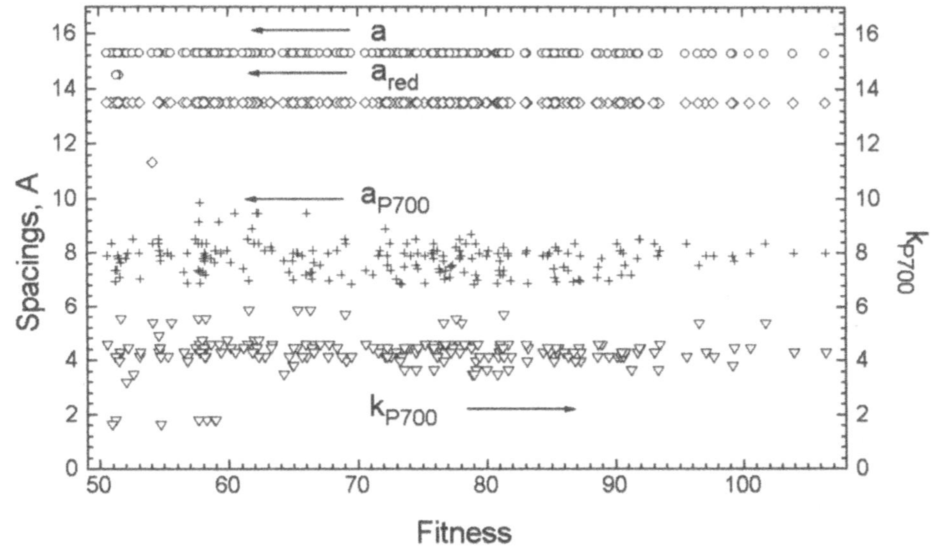

Figure 1

Open circles, diamonds and crosses stand for lattice constant (a), mean spacings from the "red" (a_{red}) or P700 pigment (a_{P700}) to the nearest neighbors, respectively. It is seen, that pigment density around the P700 pigment is higher than the average one in contrast to what is proposed in [1]. Due to the fact that the intrinsic charge separation rate k_{P700} scales the trapping-dependent part of the zero momentum of the excitation decay function, the distribution of its values (open triangles) provides a direct indication of the rate limiting tendency for the particular solution. k_{P700} covers values almost from the theoretical limit of the trapping-dependent excitation decay (k_{P700}=1.2ps^{-1}) up to k_{P700}=6ps^{-1} which is an indication for the close to migration-limited decay.

The corresponding spectral type distributions are presented in Figure 2 showing a view parallel to the membrane. Cubes represent the Chl molecules. Seven different spectral types [8] (5 pigments of spectral type 1 with absorption maximum at 654.1 nm, 23 - 2 - 667.3 nm, 38 - 3 - 678.3 nm, 22 - 4 - 686.7 nm, 7 (one of which is P700) - 5 - 700 nm, 4 - 6 - 712.2 nm, 1 - 7 - 724 nm) are distinguished by the grey levels indicated also by the spectral type numbers. The left lattice (a) stands for migration-limited model (k_{P700}=5ps^{-1}; a=13.5Å; a_{P700}=9.2Å; a_{red}=11.1Å) while the right one (b) stands for a trapping-limited model (k_{P700}=1.8ps^{-1}; a=14.7Å; a_{P700}=7.4Å; a_{red}=13.5Å). The (b) model as compared to (a) shows more compact "red" pigment distribution around the P700 pigment site while both models do not indicate the direct "reddest" pigment contact to the P700 pigment.

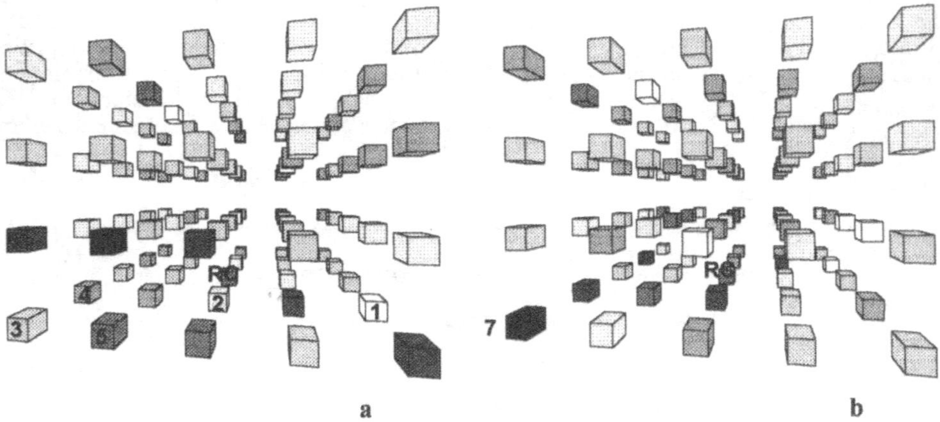

Figure 2

To conclude, we found that most of the spectral lattice arrangements correspond to the intermediate rate-limiting model (see k_{P700} value distribution in Figure 1) including several solutions for the two extremes also. More detailed high time resolution and temperature dependent absorption and/or fluorescence data are required in order to decide which of these situations is actually realized in the PS I core antenna.

Acknowledgments
Financial support by a travel grant to G.T. from the Sonderforschungsbereich 189, Heinrich-Heine-Universität Düsseldorf, is gratefully acknowledged.

References

1. Van Grondelle, R., Dekker, J.P., Gillbro, T. and Sundström, V. (1994) Biochim. Biophys. Acta 1187, 1-65
2. Jia, Y., Jean, J.M., Werst, M.M., Chan, C. and Fleming, G.R. (1992) Biophys.J. 63, 259-273
3. Werst, M., Jia, Y.W., Mets, L. and Fleming, G.R. (1992) Biophys.J. 61, 868-878
4. Pullerits, T., and Freiberg, A. (1992) Biophys. J. 63, 879-896
5. Laible, P.D., Zipfel, W. and Owens T.G. (1994). Biophys. J. 66, 844-860
6. Lin, S., Kleinherenbrink, F.A.M., Chiou, H.-C., and Blankenship, R.E. (1994) Biophys. J. 67, 2479-2489
7. Beauregard, M., Martin, I. and Holzwarth, A.R. (1991) Biochim.Biophys.Acta 1060, 271-283
8. Trinkunas, G. and Holzwarth, A.R. (1994) Biophys. J. 66, 415-429
9. Holzwarth, A.R., Schatz, G., Brock, H. and Bittersmann, E. (1992) Biophys.J. 64, 1813-1826
10. Goldberg, D.E. (1989) Genetic Algorithms in Search, Optimisation, and Machine Learning, pp. 1-412, Addison-Wesley Publishing Co
11. Trinkunas, G. and Holzwarth, A.R. (1994) Lith. J. Phys., 34, 287-292
12. Krauss, N., Hinrichs, W., Witt, I., Fromme, P., Pritzkow, W., Dauter, Z., Betzel, C., Wilson, K.S. and Saenger, W. (1993) Nature 361, 326-331

DYNAMICS OF PRIMARY PHOTOCHEMISTRY OF PSI IN MEMBRANES, MONOMERIC AND TRIMERIC COMPLEXES OF *SPIRULINA* AT 77K: CORRELATION OF P700 OXIDATION AND FLUORESCENCE QUENCHING AT 760 NM

N.V. Karapetyan[1], V.V. Shubin[1], I.N. Bezsmertnaya[1], Reto J. Strasser[2]
[1]Biochemistry Institute, RAS, Moscow 117071, RUSSIA; [2]Laboratory of Bioenergetics, University of Geneva, CH-1254,Jussy-Geneva, SWITZERLAND

1. Introduction

In contrast to the reaction center (RC) in bacteria or in photosystem II (PSII) of plants, the RC of PSI of *Spirulina* contains about 40 molecules of chlorophylls (Chl) which are tightly bound with P700 (1). Deletion of that Chls of PSI core antenna causes the inactivation of P700. At least 8-10 Chls per P700 are necessary to conserve the activity of the RC (1). The precise organization of Chls bound with P700, which are responsible for several fluorescence quench and dequench phenomena is not yet clear. Marked light-induced changes of the fluorescence intensity at 755-760 nm of PSI Chls accompanying the P700 oxidation were found for some cyanobacteria at 77K (2,3). As compared with PSI of higher plants and algae, the cyanobacterial PSI lacks a light-harvesting complex. About 90 Chls per P700 are located in a core antenna. It was shown that the fluorescence yield at 760 nm (F760), emanated by Chl absorbing at 735 nm, is at low temperature proportional to the amount of reduced P700. The 77K emission at 760 nm and its light-induced quenching were found only in membranes and PSI trimeric complexes of *Spirulina*, but not for monomeric complexes (4,5). Only F720, but no F760 was found in cyanobacteria *Synechocystis* or *Synechococcus* (6) which contain PSI trimeric complexes too. The trimeric complexes of PSI were not found in higher plants and algae which emit at 735 nm. The aim of this paper is to study the DYNAMICS and the ORGANIZATION of PSI units of *Spirulina*. The existence of energy migration between the antenna of several PSI RCs (grouping) is probed by the correlation of simultaneously measured light-induced changes of the intensities of fluorescence (F760) and reflectance (R820) at 77K in membranes, monomeric and trimeric complexes of *Spirulina*.

2.Procedure

The isolation of the membranes and the monomeric and trimeric complexes of PSI were done as described earlier (4,5). The samples (50 μM) were placed on a Millipore filter, and put in a metallic holder and immersed in liquid nitrogen. A multibranched fiberoptic system was used for the simultaneous optical measurements such as F690, F720, F760 and the reflectance at 820 nm. The electronic instruments were combination of different instruments of Hansatech Ltd. King's Lynn, Norfolk PE 32 1J1, GB. As excitation light a HeNe laser (633nm with a shutter opening of 2ms) or the head of the Hansatech Instrument PEA (time response better than 50 μs) were used. The measuring beam at 820nm was modulated with 5KHz.

P. Mathis (ed.), Photosynthesis: from Light to Biosphere, Vol. II, 55–58.

3.Results

In the following figures typical experimental traces are shown as examples. Detailed quantitative results will be published elsewhere.

Fig. 1 shows the time courses of light-induced changes of the fluorescence of PSII and PSI in *Spirulina*, measured simultaneously at 690 (F690), 720 (F720) and 760 nm (F760). The intensity of F690 (emitted by Chls of PSII core antenna) increased, F760 (Chls of PSI core antenna of trimeric complexes) decreased and the yield of F720 (emitted by some Chls of PSI antenna of trimeric complexes) remained constant as well as for membranes and PSI of monomeric complexes. No light induced changes of the PSI complexes (monomers or trimers) were found at room temperature, although the photo oxidation of P700 happens to be normal.

Fig. 2 shows the time course of the light-induced decrease at 760 nm (F760) of PSI trimers complexes as a function of excitation light intensities on a logarithmic time scale. The rate of the light induced decrease of F760 of PSI trimer complexes or membranes is proportional to the light intensity I (insert of Fig. 2 where 1/Half-decay-time versus I is shown).

Fig. 3 shows (from top to bottom) (A) the fluorescence kinetics for F760 and (B) the simultaneously measured photooxidation of P700 in PSI monomers complexes, membranes and PSI trimer complexes. In all cases the light induced decrease of F760 (in membranes and PSI trimer complexes) is faster than the photooxidation of P700.

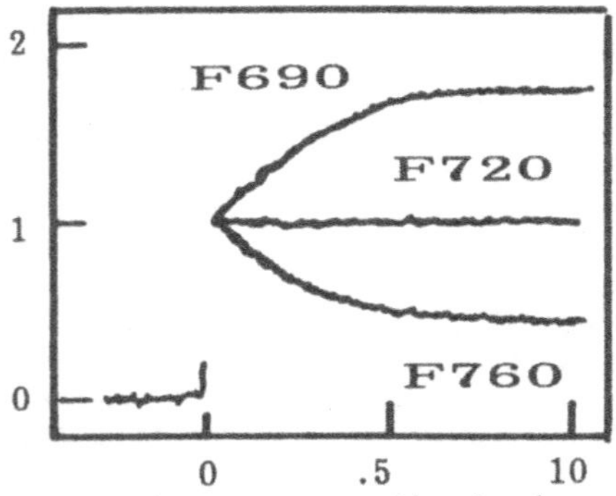

Fig.1

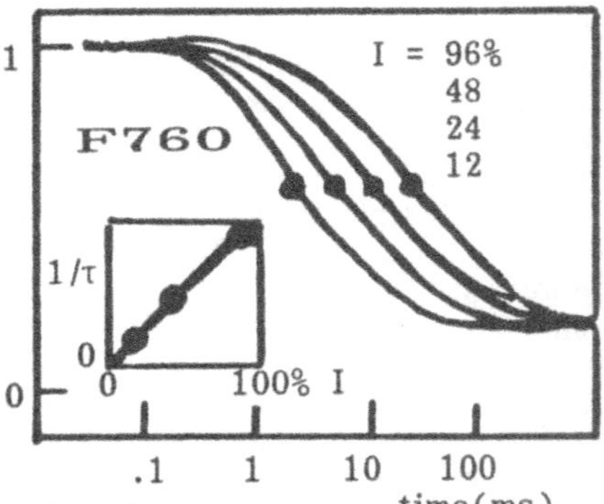

Fig.2

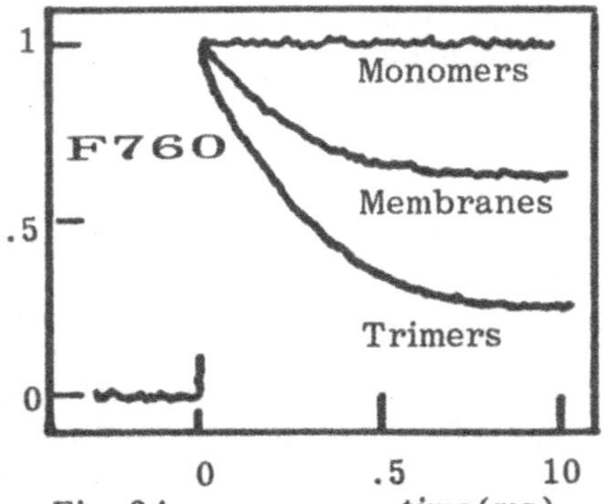

Fig.3A

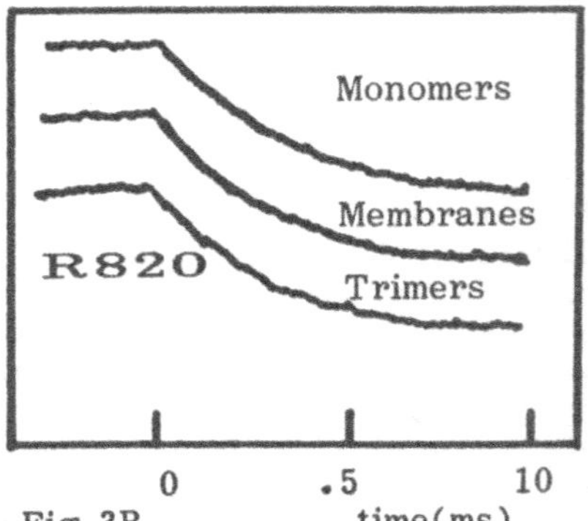

Fig.3B

4. Dynamics and Mechanistic of P700 oxidation and F760 quenching

It has been shown in leaves and chloroplasts of higher plants (7) that the kinetics of P700 photooxidation follows a second order type of kinetics. If this is the case for PSI in *Spirulina* too, then the reciprocal values of the fraction of reduced P700 (a=P700/P700(total)) and the fraction of the fluorescence decrease F760 (V=(Ft-Fs)/(Fo-Fs)) should plot as a straight line.

Fig. 4A shows the kinetics of the relative P700 photooxidation (a) and the relative fluorescence decrease at 760nm (V).

Fig. 4B shows the reciprocal values of a and V versus time. Both lines are straight, however with different slopes.

Fig. 4C shows the dynamics of the two signals V versus a which shows a highly non-linear behavior.

Fig. 4D shows the linear relationship of the plot 1/V versus 1/a. The slope is typically bigger than one. We assume that a PSI unit with an oxidized P700 can be called as closed and it exhibits a lower fluorescence yield (due to increased radiationless energy dissipation) than an open unit with a reduced RC. According to the grouping concept and by the use of the energy flux theory of Strasser (8,9) a general equation for the fluorescence behavior can be derived for PSI in membranes and trimeric complexes of *Spirulina*. This general equation is the same as it was derived by Strasser (8) for PSII in higher plants, and it has been reconfirmed recently twice(10,11).

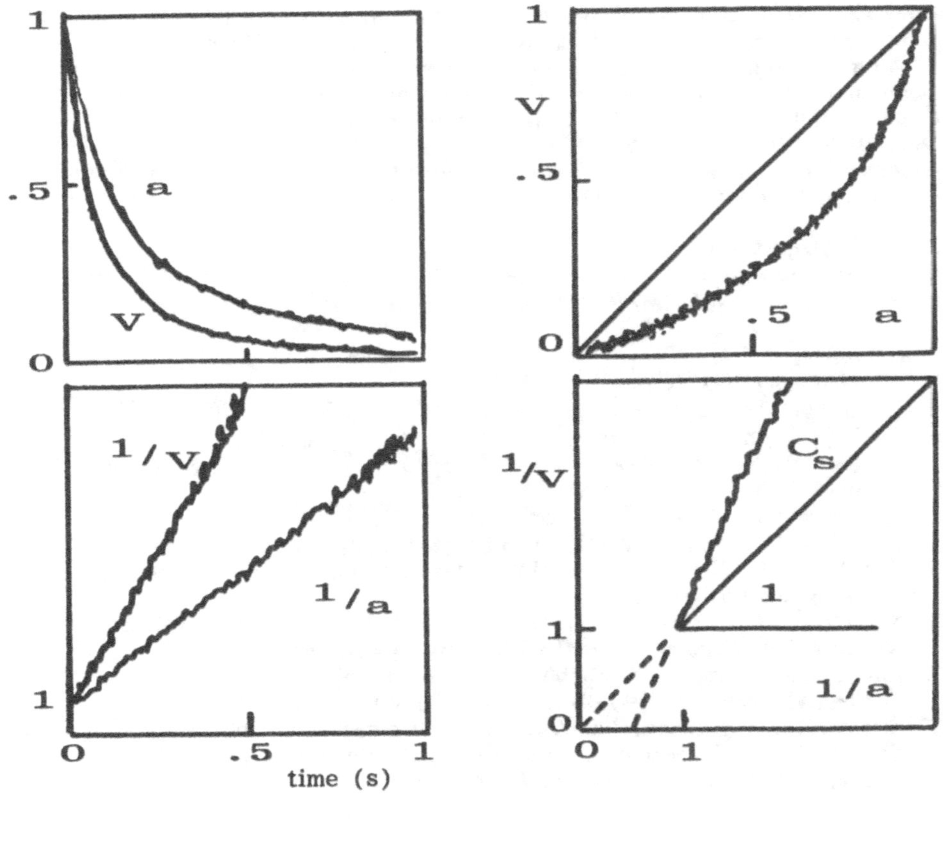

5. Discussion

Our aim is to show the qualitative behaviour only. Therefore we formulate the equations as a fraction of reduced P700 with the symbol *a*.

Eq. 1. $a = [1 + {}^2k.t]^{-1}$ Eq. 4. $1/a = 1 + {}^2k \, . \, t$

Eq. 2. $V = a / [1 + Cs(1 - a)]$ Eq. 5. $1/V = - Cs + (1 + Cs) \, . \, 1/a$

Eq. 3. $V = [1 + (1 + Cs) \, . \, {}^2k \, . \, t]^{-1}$ Eq. 6. $1/V = 1 + (1 + Cs) \, . \, {}^2k \, . \, t$

Equation 1 describes the kinetics of P700 photooxidation as a second order reaction. Equation 2 describes the general hyperbolic behavior of fluorescence versus the fraction of open (with P700 reduced) RCs. C is the only constant describing a hyperbola which goes through the two points (0/0) and (1/1). This constant C has different biophysical meanings according to different authors and they have been labeled as C(j), C(p) and C(s) (12) We use here the constant Cs derived by Strasser (8) and recently confirmed (10,11). This constant C(s) is proportional to the overall grouping probability (13), p(G), which indicates the probability that an exciton of one antenna excites the antenna Chl of a neighbor unit. Where for the trimers of PSI in *Spirulina* C(s) = p(G) * (Ft-Fs)/Fs where Fs is the stationary state in the light.
Equation 3 shows the relative fluorescence kinetics V derived from Eq. 1 and Eq. 2. Equation 4, 5 and 6 are the reciprocal presentation of the equations 1, 2 and 3.

With the findings reported here we conclude that the antenna of PSI units in trimeric complexes, isolated or in membranes are forming groups which are energetically connected. The comparison of the kinetics of trimeric PSI complexes with membranes at 77 K gives a probe to estimate the fraction of monomeric and trimeric PSI complexes in membranes. At room temperature the experimental conditions are as such that P700 oxidation follows a pseudo first order type which hides the information of energetic connectivity (grouping). *In PSII of plants connectivity provides a gain in photochemistry whereas the connectivity observed in PSI trimers of Spirulina provokes a gain in the radionless dissipation.*

6. Acknowledgments

This work was supported by Swiss Fund National grant #.31.33678.92 to RJS and by the Russian Foundation for Fundamental Research, grant 94-04-12252.

7. References

1. Krauss et al. (1993) Nature 361, 326-331
2. Duval,J.C.,Thomas, J.C.and Choquet,Y. (1986) Biochim. Biophys.Acta 848,352-358
3. Shubin V.V. et. al. (1991) Biochim. Biophys. Acta 1060, 28-36
4. Shubin,V.V.,Bezsmertnaya,I.N. and Karapetyan,N.V.(1992) FEBS Lett.309,340-342
5. Shubin V.V. et. al (1993) FEBS Lett. 334, 79-82
6. Holzwarth, A.R. et. al (1993) Biophys. J. 64,1813-1826
7. Strasser R.J. and Butler, W. (1976) Biochim. Biophys. Acta 449, 412-419.
8. Strasser R.J. (1981) In. V International Photosynthesis Congress Balaban (ed. Akoyunoglou) pp. 727-737.
9. Sironval C. R.J. Strasser and M. Browers (eds) (1984) In. Protochlorophyllide Reduction and Greening, 307-316, Martinus Nijhoff, The Hague.
10. Lavergne and Trissl (1995) Biophys. J. 68, 2474-2492.
11. Trissle and Lavergne (1994) Aust. J. Plant Physiol 22, 183-193.
12. Strasser, R.J., Eggenberg, Pfister K. and Govindjee (1992) Arch. Sci. 45, 207-224.
13. Strasser, R.J. and Greppin, H. (1981) In. III International Photosynthesis Congress, Balaban (ed. Akoyunoglou) pp. 717-726.

FERREDOXIN OXIDATION IN CYCLIC ELECTRON TRANSPORT IN THE CYANOBACTERIUM *Synechocystis* PCC 6803

Mercedes Roncel, José M. Ortega and Manuel Losada
Instituto de Bioquímica Vegetal y Fotosíntesis, Universidad de Sevilla y CSIC, Apdo 1113, E-41080 Sevilla, Spain

1. Introduction

It is generally accepted that cyclic electron transport involves components of the linear photosynthetic chain including plastoquinone (PQ), cytochrome *bf* complex, plastocyanin, photosystem I (PSI) and ferredoxin (Fd) (1). However, the question of how electrons pass from Fd to the PQ pool completing the cycle remains to be elucidated.

Different pathways for the oxidation of Fd in cyclic electron transport have been suggested. A direct pathway between Fd and cytochrome b_{563} of the cytochrome *bf* complex has been traditionally proposed, mainly based on the inhibition of cyclic phosphorylation by antimycin and on the observation of a PSI-dependent photoreduction of cytochrome b_{563} (2,3). However, experimental results obtained recently favours the view that electron from Fd reach the cytochrome *bf* complex via the PQ pool. Hosler and Yocum (4) haved suggested that a reduced form of ferredoxin-NADP$^+$ oxidoreductase (FNR) (probably the semiquinone form) could donate directly electron to PQ pool. Evidence in favour of an involvement of FNR is that cyclic electron flow is inhibited by disulfodisalicylidenepropane-1,2-diamine and maleimides which are considered to inhibit FNR specifically (5). On the other hand, in order to explain the specific inhibition of cyclic phosphorylation by antimycin, it has been postulated a new enzyme called ferredoxin-plastoquinone reductase (FQR) as the only mediator closing the cycle (6). The involvement of both FNR and the respiratory NAD(P)H dehydrogenase (NDH) has been also proposed (7-10), based on studies of inhibitors effect on the dark rereduction of P700$^+$ (7,8) and on the behaviour of mutant cells in which NDH genes had been inactivated (9). There is at present insufficient information to support or reject any of these suggestions conclusively (11).

In order to clarify the pathway(s) of Fd oxidation in cyclic electron transport and the possible involvement of FNR and NDH in such process, the effect of different inhibitors of these enzymes on the dark reoxidation of Fd after photoreduction by PSI has been investigated. An assay previously developed by Bendall and coworkers (12,13), which allows to measure both Fd photoreduction and dark reoxidation by PSI, was used. The results show that Fd oxidation by PSI is sensitive to FNR and NDH inhibitors, supporting the idea that both FNR and NDH might be implicated in cyclic electron flow around PSI.

P. Mathis (ed.), Photosynthesis: from Light to Biosphere, Vol. II, 59–62.

2. Procedure

Synechocystis PCC 6803 was grown photoautotrophycally at 30 °C in BG11 medium under aeration with 3% CO_2 in air. Continuous illumination (120 µE m^{-2} s^{-1}) was provided by fluorescent lamps. Thylakoid membranes were prepared and separated from cytoplasmic membranes by sucrose gradient centrifugation as decribed in (14).

Ferredoxin reduction and oxidation by PSI was assayed by using the spectrophotometric method described by Cleland and Bendall (12). NDH and FNR activities were assayed by monitoring NAD(P)H oxidation (15) or $NADP^+$ reduction (16) at 340 nm, respectively.

3. Results and Discussion

3.1 *Fd reduction and oxidation by PSI*

Under anaerobic conditions, when DCMU is present to block electron donation from PSII, cyclic electron transfer around PSI is the only possible form of photosynthetic electron flow. Illumination with selective PSI-light of *Synechocystis* thylakoids induces an absorbance decrease at 463 nm due to Fd reduction, followed by a steady-state situation where there was no net change in the redox state of Fd. After turning off the light, Fd reoxidation can be followed by an absorbance increase (Figure 1). This assay has been extensively used by Bendall and coworkers (11-13) as a measurement of cyclic electron flow around PSI. In a series of light-dark cycles no diminution of the absorbance changes amplitude was observed (data not shown), thus supporting the assumption that Fd reoxidation is driven by PSI and not by a final electron acceptor. The spectra of the absorbance changes above described were actually due to changes in the redox state of Fd (data not shown).

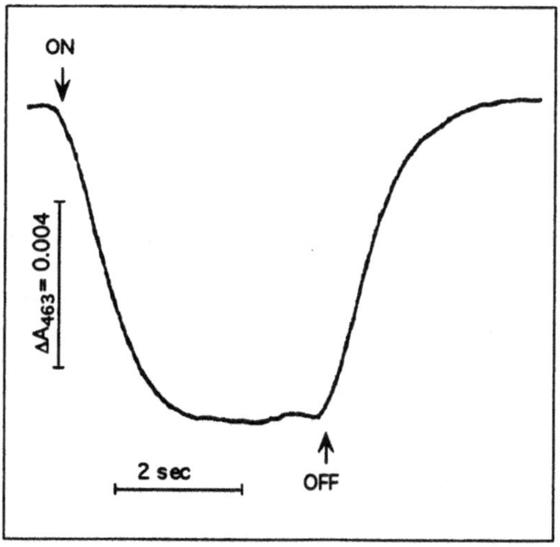

Figure 1. Typical kinetic trace of photoreduction and dark reoxidation of Fd induced by illumination with red light in *Synechocystis* thylakoid membranes.

3.2 *Effect of NADPH and different inhibitors on Fd oxidation by PSI*

In order to elucidate whether FNR and NDH participate in the electron pathway from Fd to PQ, the effect of different inhibitors of these enzymes on Fd oxidation by PSI described in Fig.1 was tested in *Synechocystis* thylakoids (Table I).

Table I
Effect of NADPH and inhibitors of FNR and NDH on Fd oxidation in *Synechocystis* thylakoid membranes

Addition	Fd oxidation activity	
	μmol Fd mg Chl^{-1} h^{-1}	%
No addition	118	100
Heparin	50	42
NADPH	62	53
Rotenone	54	46
Rotenone + NADPH	43	36

The inhibitors were used at the concentration where maximum percentage of inhibition was obtained, i.e., 25 μM heparin and 60 μM rotenone. 250 μM NADPH was used.

Heparin has been described as a non redox-active molecule which binds to FNR, thus inhibiting both Fd and $NADP^+$ binding to the enzyme (4). The addition of this inhibitor eficiently blocks FNR activity in *Synechocystis* thylakoid membranes (data not shown). Table I shows that Fd oxidation is partially sensitive to this inhibitor, thereby indicating a possible involvement of FNR in this reaction. Additional evidence for this interpretation came from inhibition obtained by *o*-phenylendimaleimide, an inhibitor which reacts with free sulphydryl groups on FNR, thus preventing also the binding of Fd (data not shown). The partial inhibitory effect induced by the addition of NADPH could be due to its continuous ocupation of the nucleotide binding site on FNR at the concentration used (12). The above results are in consequence consistent with a possible role of FNR on Fd oxidation by PSI.

NDH is also a candidate to be involved in Fd oxidation, rotenone having been described as a potent inhibitor of this enzyme in mammalian mitochondria (17) and *Anabaena variabilis* (18). We have confirmed that rotenone effectively inhibits NDH activity in *Synechocystis* thylakoid membranes (data not shown). As shown in Table I, rotenone inhibits also partially Fd oxidation by PSI. Notwithstanding, Manasse and Bendall (13) reached the opposite conclusion, but they used rotenone at a concentration of 5 μM, clearly below the effective concentration level employed in the above described papers (17, 18). Other NDH inhibitors, such as Hg^{2+} ions and 2-thenoyltrifuoroacetone, also affected Fd oxidation, but in a minor extension (data no shown). These results seems to indicate that NDH could be also implicated in Fd oxidation by PSI.

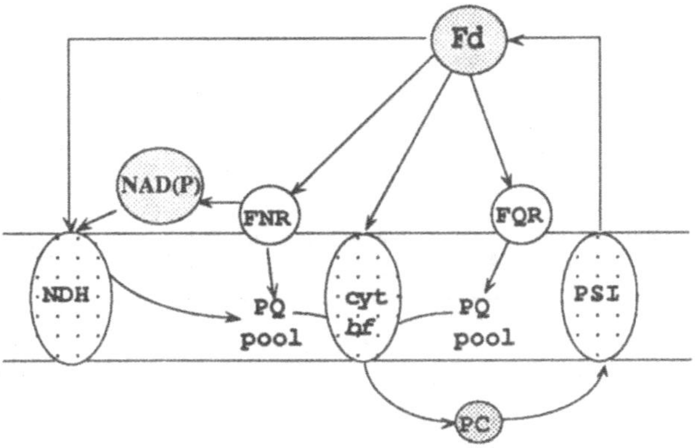

Figure 2. Scheme of possible pathways for ferredoxin oxidation by PSI.

The results presented here seem to indicate that both FNR and NDH might participate on Fd oxidation by PSI in the cyanobacterium *Synechocystis*, but as shown in the scheme of Figure 2, they do not exclude other possibilities.

This work was supported by DGICYT (Grants PB 91-0085, PB 94-0033).

References

1 Scherer, S. (1990) TIBS 15, 458-462
2 Arnon, D. I. (1992) in Trends in Photosynthesis Research (Barber, J., Guerrero, M.G. and Medrano, H. eds.) pp. 103-123, Intercept, Andover
3 Furbacher, P.N., Girvin, M.E. and Cramer, W.A. (1989) Biochemistry 23, 8990-8998
4 Hosler , J.P. and Yocum, C.F. (1985) Biochim. Biophys. Acta 808, 21-31
5 Shakak, Y., Crowther, D.K. and Hind, G. (1981) Biochim. Biophys. Acta 636, 234-243
6 Moss, D.A. and Bendall, D.S. (1984) Biochim. Biophys. Acta 767, 389-395
7 Mi, H., Endo, T., Schreiber, U. and Asada, K. (1992) Plant Cell Physiol. 33, 1099-1105
8 Mi, H., Endo, T., Schreiber, U., Ogawa, T. and Asada, K. (1992) Plant Cell Physiol. 33, 1233-1237
9 Mi, H., Endo, T., Schreiber, U., Ogawa, T. and Asada, K. (1994) Plant Cell Physiol. 35, 163-173
10 Friedrich, T., Steinmüller, K. and Weiss, H. (1995) FEBS Lett. 367, 107-111
11 D.S. Bendall and Manasse, R.S. (1995) Biochim. Biophys. Acta 1229, 23-38
12 Cleland, R.E. and Bendall, D.S. (1992) Photosynth. Res. 34, 409-418
13 Manasse, R.S. and Bendall, D.S. (1993) Biochim. Biophys. Acta 1183, 361-368
14 Murata, N. and Omata, T. (1988) in Methods in Enzymology (Packer, L. and Glazer, A.N. eds.) pp. 245-251, Academic Press Inc, San Diego
15 Berger, S., Ellersiek, U. and Steinmüller, K. (1991) FEBS Lett. 286, 129-132
16 Fillat, M. (1988) Ph. D. Thesis, University of Zaragoza
17 Slater, E.C. (1967) Methods in Enzymology (Estabrook, R.W. and Pullman, M.E., eds) pp. 48-57, Academic Press Inc, New York
18 Alpes, I., Scherer, S. and Böger, P. (1989) Biochim. Biophys. Acta 973, 41-46

MECHANISM OF ELECTRON TRANSFER FROM PLASTOCYANIN AND CYTOCHROME c_6 TO PHOTOSYSTEM I IN A NUMBER OF EVOLUTIONARILY DIFFERENTIATED ORGANISMS

M. Hervás[1], J.A. Navarro[1], A. Díaz[1], H. Bottin[2] and M.A. De la Rosa[1]
[1]Instituto de Bioquímica Vegetal y Fotosíntesis, Universidad de Sevilla y CSIC, Facultad de Biología, Sevilla, Spain; [2]Département de Biologie Cellulaire et Moléculaire, CEA Saclay, Gif-sur-Yvette, France

1. Introduction

In some cyanobacteria and green algae, plastocyanin (Pc) and cytochrome c_6 (Cyt) act as alternative electron donors to Photosystem I (PSI), depending on the relative availability of copper and iron in the culture medium. The two metalloproteins are acidic in green algae, as is Pc in higher plants, but can be either basic or acidic in cyanobacteria [1].
Although widely investigated, the mechanism of electron donation to PSI is still under debate. In this respect, the determination of the structures of Cyt and Pc from the same organism and subsequent comparison looking for reaction partner binding sites (e.g., hydrophobic or negative patches) would be of great interest for both functional and structural studies.
For the electron transfer from Pc to PSI, fast kinetic experiments have revealed the existence of two different kinetic phases in higher plants and some eukaryotic algae: a protein concentration independent so-called fast kinetic phase, and a second diffusion-controlled phase. Half times of 12 μs and 110 μs have respectively been determined for the two phases in spinach PSI [2].
This work is aimed at studying the factors which determine the appearance of the fast phase of electron transfer to oxidized P700 in the PSI complex. A comparative kinetic study by laser flash spectroscopy has been carried out using Pc, Cyt and PSI particles isolated from a number of evolutionarily differentiated organisms, i.e. higher plants, green algae and cyanobacteria. A more detailed analysis of the different kinetic models is presented in [3].

2. Procedure

2.1. *Obtention of metalloproteins and PSI particles*
Pc from spinach was isolated according to [4]. Pc and Cyt from the green alga *Monoraphidium braunii* were purified as previously described [5]. Pc and Cyt from the cyanobacterium *Anabaena* PCC 7119 were obtained according to [6]. Cyt and Pc from *Synechocystis* PCC 6803 were isolated following the method recently described [7]. Spinach and *Monoraphidium* PSI were obtained with digitonin following the method described in [8]. *Synechocystis* and *Anabaena* PSI particles were obtained by β-dodecyl maltoside solubilization as described in [9].

P. Mathis (ed.), Photosynthesis: from Light to Biosphere, Vol. II, 63–66.

2.2. *Laser flash absorption spectroscopy*
Kinetics of flash-induced absorption changes in PSI were followed at 820 nm after sample excitation by means of a ruby laser and using a silicon photodiode as measuring detector. Both the sample and photodiode were protected from actinic ligth by 820 nm narrow-band glass filters. The photodiode output was amplified and further recorded in a digital oscilloscope. For experiments with spinach PSI and *Synechocystis* 6803 donor proteins, the set-up used was that described in [9]. All experiments were carried out at room temperature (23°C) in a 1 mm pathlength cuvette. Exponential analyses were performed using the Marquardt method with the software devised by P. Sétif. Kinetic parameters were calculated according to the reaction mechanisms herein proposed (see scheme below).

3. Results and Discussion

The comparative kinetic analysis herein reported using a number of evolutionarily differentiated organisms has allowed us to infer interesting conclusions on the mechanism of electron transfer to PSI. From the experimental data obtained, three different kinetic models are proposed by assuming either an oriented collisional reaction mechanism (type I), a minimal two-step mechanism involving complex formation followed by intracomplex electron transfer (type II), or rearrangement of the reaction partners whitin the complex before electron transfer takes place (type III):

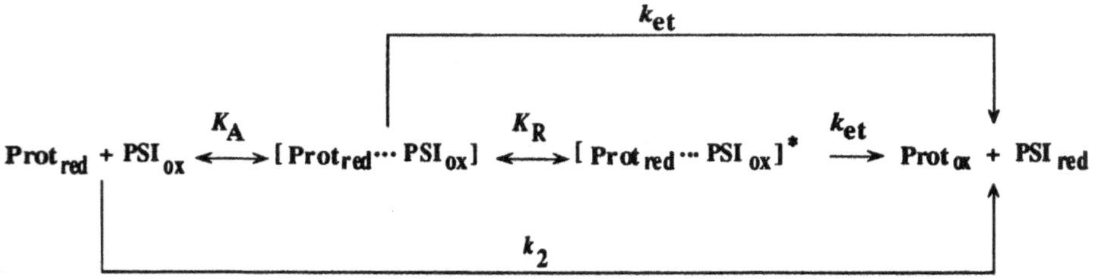

Table I shows the kinetic parameters obtained for the different protein/PSI systems by fitting the experimental data to the corresponding model. Proposal of a type III mechanism is only tentative for Pc and Cyt from *Monoraphidium*, which exhibit a fast kinetic phase but a linear donor protein concentration dependence of the observed rate constant for the slow phase.
It is interesting to note that the fast phase in *Anabaena* PSI ($t_{1/2}$=4 μs), which is observed only with Cyt but not with Pc, is significantly faster than in any other organism (Table I). Such findings indicate that PSI was able to first optimize its interaction with positively charged Cyt by means of long distance attractive electrostatic interactions between the positive electron donor protein and negative PSI particles. In addition, Cyt appears to be more efficient than Pc -mainly with regard to the fast kinetic phase- in photosynthetic organisms capable of using both Pc and Cyt as electron donors to PSI. The evolutionary pathway leading to the appearance of the fast phase would thus involve structural modifications in both PSI and donor proteins. An interesting question still remains to be answered on account of which Pc has been evolutionarily choosen over the ancient Cyt as the electron donor to PSI.

Table I. Kinetic parameters for PSI reduction by plastocyanin and cytochrome c_6 in evolutionarily differentiated organisms.

PSI	Donor protein	Type of mechanism[a)]	Fast phase $t_{1/2}$ (μs)	Fast phase Amplitude[b)] (%)
spinach	sp[c)] Pc	III	11.0	70
	Mb Pc	III	17.8	30
	Mb Cyt	III	21.0	55
	Ana Pc	II	--	--
	Ana Cyt	II	--	--
	Syn Pc	I	--	--
	Syn Cyt	I	--	--
Monoraphidium	sp Pc	III	15.1	<15
	Mb Pc	III(?)	nd[d)]	<10
	Mb Cyt	III(?)	8.0	50
Anabaena	sp Pc	I	--	--
	Ana Pc	I	--	--
	Ana Cyt	III	4.1	35
Synechocystis	Syn Pc	I	--	--
	Syn Cyt	I	--	--

[a)]see models in the scheme above; [b)]at 300 μM donor protein concentration; [c)]sp, spinach; Mb, *Monoraphidium braunii*; Ana, *Anabaena* 7119; Syn, *Synechocystis* 6803; [d)]nd, not detectable.

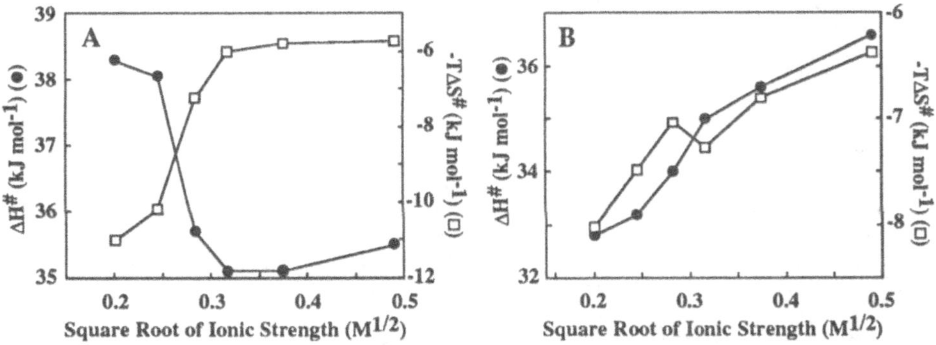

Figure 1. Effect of ionic strength on the enthalpic and entropic terms of the overall reaction of PSI reduction by Pc (A) or Cyt (B) in *Anabaena*. The thermodynamic parameters were calculated at 298 K. Samples were 30 μM in Pc or Cyt, and 2 μM in P700.

A comparative thermodynamic analysis of P700 reduction in organisms showing different reaction mechanisms, such as *Anabaena* and *Synechocystis*, has been carried out. The activation parameters for the reduction of P700 by Pc and Cyt were estimated from the Eyring plots of kinetic constants. When analysing the ionic strength dependence of the activation parameters, it was observed that the apparent activation free energy ($\Delta G^{\neq}$) decreases whit increasing ionic strength with *Synechocystis* Pc and Cyt but increases with ionic strength with the proteins of *Anabaena*. This is consistent with the previously proposed existence of attractive electrostatic forces in the interaction of PSI with the donor proteins in *Anabaena* and of repulsive interactions in *Synechocystis*. It is also most interesting to note that such a change in $\Delta G^{\neq}$ is the result of opposite changes in the enthalpic ($\Delta H^{\neq}$) and entropic ($-T\Delta S^{\neq}$) terms with Pc and Cyt from *Synechocystis* and Pc from *Anabaena*, but this compensation effect is not observed with *Anabaena* Cyt. This compensation effect can be attributed either to the removal of water molecules during partners association and/or to the burial of accesible hydrophobic protein groups upon complex formation. Moreover, the term $\Delta S^{\neq}$ is positive in the cyanobacterial systems, but negative in green algae. These findings could suggest differences in solvent reorganization during complex formation. Such an enthalpy/entropy compensation is more evident as the reaction mechanism becomes more complex, i.e. from the less evolved type I and II mechanisms (cyanobacteria) to the more evolved type III mechanism (spinach). This fact is indicative not only of the important role played by solvent reorganization (or other phenomena such as hydrogen bonding or hydrophobic interactions), but also of the decrease in the electrostatic interactions as the reaction mechanism is optimized. This is on line with the recently proposed role of water molecules in the control of electron transfer.

We have also analyzed the effect of temperature on the fast phase of the *Anabaena* Cyt/PSI system. Surprisingly, neither the kinetic rate constant nor the contribution of the fast phase to the total absorbance change depends on temperature. Moreover, increasing glycerol concentration drastically decreases the percentage of fast phase in this system, but does not alter the rate constant

References

1 Golbeck, J.H. (1992) Annu. Rev. Plant Physiol. Plant Mol. Biol. 43, 293-324

2 Bottin, H. and Mathis, P. (1985) Biochemistry 24, 6453-6460

3 Hervás, M., Navarro, J.A., Díaz., A., Bottin, H. and De la Rosa, M.A. (1995) Biochemistry (in press)

4 Yocum, C.F. (1982) in Methods in Chloroplast Molecular Biology (Edelman, M., Hallick, R.B. and Chua, N.H., eds.) pp. 973-981, Elsevier Biomedical Press, Amsterdam

5 Campos, A.P., Aguiar, A.P., Hervás, M., Regalla, M., Navarro, J.A., Ortega, J.M., Xavier, A.V., De la Rosa, M.A. and Teixeira, M. (1993) Eur. J. Biochem. 216, 329-341

6 Medina. M., Díaz, A., Hervás, M., Navarro, J.A., Gómez-Moreno, C., De la Rosa, M.A. and Tollin, G. (1993) Eur. J. Biochem. 213, 1133-1138

7 Hervás, M., Navarro, F., Navarro, J.A., Chávez, S., Díaz, A., Florencio, F.J. and De la Rosa, M.A. (1993) FEBS Lett. 319, 257-260

8 Boardman, N.K. (1971) Methods Enzymol. 23, 268-276

9 Hervás, M., Ortega, J.M., Navarro, J.A., De la Rosa, M.A. and Bottin, H. (1994) Biochim. Biophys. Acta 1184, 235-241

HIGH ENRICHMENT OF P700 BY ETHER-ACETALDEHYDE EXTRACTION OF ANTENNA PIGMENTS IN PS-I PARTICLES.

Isamu Ikegami[1], Shigeru Itoh[2] & Masayo Iwaki[2],
[1]Faculty of Pharmaceutical Sciences, Teikyo University, Sagamiko, Kanagawa 199-01, Japan; [2]National Institute for Basic Biology, Okazaki 444, Japan

1. Introduction

In the photosystem (PS) I reaction-center complexes of plants and cyanobacteria, more than 100 chlorophyll molecules, as well as the reaction-center chlorophyll-dimer (P700), are associated with the two PS I core polypeptides of 82 kDa each. The large number of antenna chlorophyll on this reaction center has, however, impeded precise optical measurements of the primary events. Ikegami and Katoh (1,2) showed that antenna pigments can be selectively extracted with diethylether, and that extraction becomes more efficient with increasing the amounts of water in the ether. A minimum molar ratio of 11 in spinach and 15 in cyanobacterial PS I particles has been achieved as the ratio of 'total' chlorophyll (Chl) *a* to P700 by treatment with wet ether, where 'total' refers to the total number of the chlorophyll *a* molecules that include the two constituent chlorophyll *a* molecules for P700. Most of the chlorophyll remaining in ether-treated particles still functions to transfer excitation energy to P700 and it seems to be located in the central part of the PS I reaction center (3). In this study, we attempted to prepare PS I particles with a smaller antenna size, and we succeeded in preparing PS I particles with a molar ratio of total Chl *a* to P700 below 9. The particles contain at least two fewer antenna chlorophylls than wet ether-extracted particles and, thus, they have the simplest chlorophyll composition among the PS I particles isolated to date.

2. Materials and Methods

The extraction of antenna pigments from spinach PS I particles was carried out essentially as described previously (1,2). Spectrophotometric assays of extracted particles were determined as described previously (3,4).

3. Results and Discussion

3.1 *Extraction of pigments with mixtures of ether and various organic solvents*
About 60% of the Chl *a* was removed without loss of P700 by the extraction of the PS I particles with dried ether (1,2). When we added various organic solvents at 8% (v/v) to the dried ether, more pigments were extracted. The amounts of extracted Chl and P700 increased with increasing the dielectric constant of the solvent added to the ether (Table 1). P700 was at least five times more resistant to the extraction than the antenna Chl, so that P700 was enriched by about five times relative to antenna pigments. However, it was

P. Mathis (ed.), Photosynthesis: from Light to Biosphere, Vol. II, 67–70.

Table 1. Chl *a* and P700 contents in spinach PS I particles extracted with dried diethylether containing 8% (v/v) each organic solvent. The dielectric constants of ether and each solvent added to ether are also shown. The Chl *a*/P700 ratio represents the molar ratio of total Chl *a* to P700.

Sample	Solvents added (8%, v/v)	Chl *a* (%)	P700 (%)	Chl *a*/P700 (mol/mol)	dielectric constants (ε)
PS-I complex		100	100	150	—
Extracted with dry ether	—	40.0	100	60	4.34
Extracted with dry ether and	chloroform	18.8	83	36	4.8
	isopropanol	12.5	82	25	18.3
	propionaldehyde	10.8	65	27	18.5
	n-propanol	7.5	54	23	20.1
	acetaldehyde	2.5	47	10	21.1
	ethanol	6.3	44	24	24.3
	methanol	1.8	9.3	31	32.6

Table 2. Chl *a* and P700 contents of spinach PS I particles extracted with wet diethylether that contained various amounts of acetaldehyde. Wet ether that contained water at 50% saturation was used. The ratio of Chl *a*/P700 represents the molar ratio of total Chl *a* to P700.

Sample	Acetaldehyde content (%, v/v)	Chl *a* (%)	P700 (%)	Chl *a*/P700 (mol/mol)
PS-I complex	-	100	100	150
Extracted with wet ether	0	12.0	100	20
	2	4.9	86	10.5
	3	4.3	80	10.1
	4	3.3	65	9.6
	5	2.5	54	8.9
	6	1.4	33	8.3
	8	1.1	25	8.5

difficult to obtain the ratio of Chl *a* to P700 below 20, suggesting that P700 is as sensitive as the remaining 20 antenna Chls to extraction with organic solvents. The ether-acetaldehyde mixture was the only solvent found to yield ratios of Chl *a* to P700 below 20. When we added various amounts of acetaldehyde to ether that contained water at 50% saturation (Table 2), the ratio of total Chl *a* to P700 decreased with increases in the acetaldehyde content of the ether, and finally reached 8-9 at the content of 5-6%. At the contents above 8%, P700 was denatured. The optimum content of acetaldehyde decreased when we increased the water content of the ether. These results indicate that the effect of acetaldehyde resembles that of water in the ether and that acetaldehyde is a little more effective in the enrichment for P700 than is water.

3.2 *Absorption spectra of the particles extracted with ether and acetaldehyde*

Figure 1A shows the absorption spectra of the ether plus acetaldehyde-treated particles with a ratio of total Chl *a* to P700 of 9. P700 was in an almost fully reduced state just after the preparation. In the presence of ferricyanide (dotted line), the spectrum in the red region had a peak at 676 nm with a distinct shoulder at 683 nm. The shoulder arose from the absorption of $P700^+$ (3). Upon addition of ascorbate (solid line), absorption increased at 675 nm and 683-710 nm, with a distinct shoulder at 695 nm. This is, therefore, the first reported preparation that shows a distinct peak due to P700 in the entire spectrum at room temperature. There were no absorption bands due to carotenoids or Chl *b* in the region around 450-500 nm (Fig. 1A inset). This contrasts with the wet ether-extraction, by which the residual Chl *b* always gave a shoulder at 480 nm (1,2). The difference between the reduced and the oxidized spectra (Fig. 1B) gave the difference spectrum of P700/$P700^+$. Curve analysis of the absorption spectra at 77K of the ether plus acetaldehyde-extracted particles showed that the ratio of areas of each Gaussian component was nearly 1:2:2:1:1:2 for forms of Chl with peak wavelengths at 661, 670, 675, 681, 686 (A_0) and 694 nm (P700) (Table 3). Thus, the particles had antennae with, at least, two fewer molecules of Chl than that prepared previously with ether that contained water at 80% saturation (3). The number of chlorophyll molecules responsible for the band at 670 nm decreased from three to two, and that for the band at 681 nm decreased from two to one.

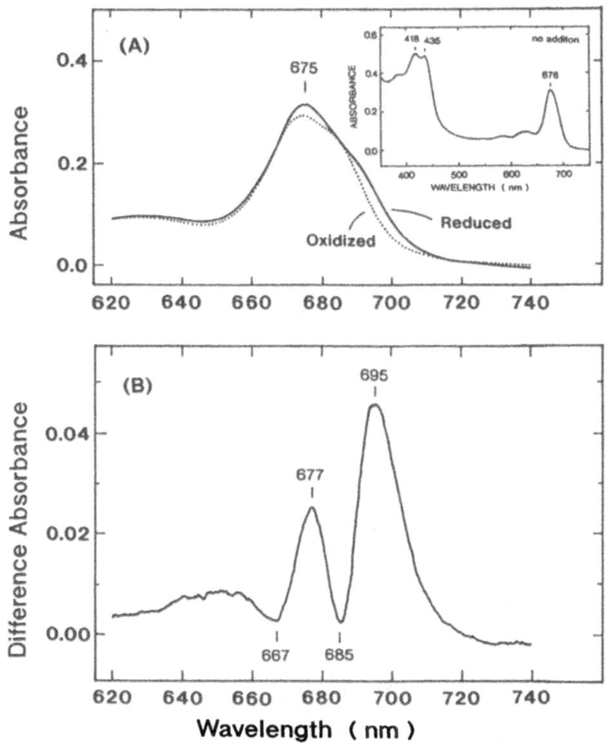

Fig. 1. Absorption (A) and reduced-minus-oxidized difference (B) spectra at 5°C of spinach PS I particles treated with an ether-acetaldehyde mixture, which has a molar ratio of total Chl *a* to P700 of 9. Potassium ferricyanide (0.3 mM) was added initially to generate the oxidized spectrum (broken line in A) and then ascorbate (5 mM) was added to the same sample to generate the reduced spectrum (solid line in A). The spectrum in the blue region, recorded without any redox reagents, is shown in the inset.

3.3 *Light-induced changes in absorption of P700*

The electron-transfer activity in the ether plus acetaldehyde-treated particles was examined by monitoring the flash-induced changes in absorption of P700 at 430 nm (Fig. 2). Without addition of quinone, only decay of the P700 triplet state was observed in the μs to ms time range upon flash excitation. In the absence of the acceptor phylloquinone (A_1), P700$^+$ seemed to be re-reduced in the ns time range by the back flow of electrons from A_0^-, as previously reported in ether-extracted particles (4), through which the triplet state of P700 is generated. After incubation with quinone, the initial amount of P700$^+$ increased, with a subsequent decay with an apparent half-time of 7 ms. This result indicates that the transfer of electrons from A_0^- to the added quinone occurred at a rate that was rapid enough to compete against the charge recombination. The reduced quinone then seemed to reduce the Fe-S centers that reduced P700$^+$ more slowly. The difference absorption spectrum generated 2 ms after the flash excitation was typical of the difference spectrum of P700/P700$^+$ (Fig. 2, inset). A change in absorption of P700 was also induced by continuous illumination of the ether plus acetaldehyde-treated particles. The rate of photooxidation of P700 was extremely low without addition of the quinone. The slow photooxidation must reflect a slow bypass flow of electrons from A_0 to Fe-S centers in the absence of the acceptor phylloquinone. The addition of quinone accelerated the photooxidation of P700. These features are essentially the same as those reported for wet ether-extracted particles that lacked phylloquinone (4).

Table 3. Parameters of the Gaussian components estimated by curve analysis of the absorption spectra at 77K of ether plus acetaldehyde-extracted particles. The data were taken from the reduced and oxidized (in parentheses) spectra, respectively.

Chl form	Peak position (nm)	Peak height (rel.unit)	Band width (1/e)(cm^{-1})	Area (rel.unit)
Chl-a-660	660.6	1.30	350	1.49
	(661.0)	(1.24)	(338)	(1.36)
Chl-a-670	669.3	2.35	288	2.42
	(669.8)	(2.32)	(287)	(2.31)
Chl-a-675	675.0	2.13	255	2.13
	(675.0)	(1.97)	(253)	(1.94)
Chl-a-681	680.7	1.39	232	1.25
	(680.7)	(1.33)	(225)	(1.16)
Chl-a-686	685.7	0.90	319	1.12
(A_0)	(686.0)	(0.85)	(350)	(1.16)
Chl-a-686	-	-	-	-
($P700^+$)	(686.1)	(0.82)	(210)	(0.67)
Chl-a-694	694.0	1.00	441	1.85
(P700)	(693.6)	(0.26)	(419)	(0.43)

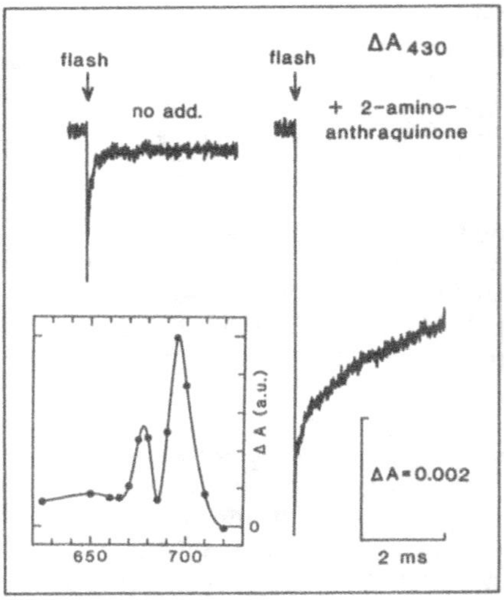

Fig. 2. Flash-induced absorption changes in the ether-acetaldehyde-extracted particles. Absorption changes were recorded at 430 nm in the absence (left) and in the presence (right) of 10 μM 2-amino-9,10-anthra-quinone, respectively, with the flash from the 532 nm Nd-YAG laser. The ratio of total Chl *a* to P700 of the particles was 9. Inset: the difference absorption spectrum determined 2 ms after flash excitation after reconstitution with 10 μM 2-amino-9,10-anthraquinone.

4. Conclusion

Particles that contained only 6 Chls in addition to P700 and the primary electron acceptor (A_0), in a single reaction-center unit, were prepared by extraction with a mixture of ether and acetaldehyde, which had the smallest antenna size of any reported particles. Chl *b*, which always retained in wet ether-extracted particles, was also absent from this preparation. The decrease in the antenna size did not affect the primary charge separation since the particles were still photoactive. In the recent X-ray crystallographic studies of PS I core particles (5), two monomeric chlorophyll *a* molecules might be located close to P700 and one of them might function as a mediator of primary electron transfer between P700 and A_0. If such is the case, these chlorophyll might be also as stable as P700 to the ether-acetaldehyde treatment. The chlorophyll forms with absorption at 670, 681 and/or 661 nm could be candidates for such chlorophyll. Thus, the five molecules of Chl *a*, namely, two in P700, one in A_0 and two in any of the forms absorbing at 661, 670, or 681 nm might be essential to the charge separation. The distinct bands of P700 and $P700^+$ in the present preparation should greatly facilitate optical measurements of the primary processes in PS I.

References

1 Ikegami, I. and Katoh, S. (1975) Biochim. Biophys. Acta 376, 588-592
2 Ikegami, I. and Katoh, S. (1989) Plant Cell Physiol. 30, 175-182
3 Ikegami, I. and Itoh, S. (1988) Biochim. Biophys. Acta 934, 39-46
4 Itoh, S., Iwaki, M. and Ikegami, I. (1987) Biochim. Biophys. Acta 893, 508-516
5 Krauss, N., Winfried, H., Witt, I., Fromme, P., Pritzkow, W., Dauter, Z., Betzel, C., Wilson, K.S., Witt, H.T. and Saenger, W. (1993) Nature 361, 326-331

PHOTOSYSTEM I / FERREDOXIN INTERACTION: SITE DIRECTED MUTAGENESIS OF PSI-E

Lagoutte, B., Barth, P. and Sétif, P., CNRS-URA 1290 /
CEA-Service de Bioénergétique, DBCM, CEN Saclay
191, Gif sur Yvette Cedex, France

1.Introduction

Photosystem I (PSI) is a multisubunit membrane complex catalyzing the reduction of ferredoxin (Fd) upon illumination. This reduction event is of fundamental importance for the full metabolism of the cell, as reduced Fd will then participate in the activation process of a large number of enzymes in most of the metabolic pathways. Concomitant with this reduction of Fd, the oxidation of plastocyanin takes place on the other side of the membrane, as a consequence of the reduction of the newly oxidized reaction center ($P700^+$).

The reduction of Fd is a fast , sub-microsecond process [1,2], which likely requires a very precise interaction with PSI. It is now established that two subunits of this photosystem, PSI-D and PSI-E, are central to this interaction. We more precisely define in this work the role of some PSI-E residues.

2.Procedure

2.1 *Materials*

PSI complex was isolated rom *Synechocystis* 6803 after breaking the cells in a French pressure cell at 10 000 psi. We then solubilized the membrane fragments in ß dodecyl maltoside and purified the monomeric form of PSI on sucrose gradients as described [3]. Ferredoxin was extracted from the same organism [4].

2.2 *Methods*

Protein composition of the different PSI particles were analyzed by classical SDS polyacrylamide gel electrophoresis [1]. The presence or absence of PSI-E subunit was tested on western blots using monospecific polyclonal antibodies. Flash-induced absorption changes were measured at 296K at 580 nm. Excitation of P700 at 694.3 nm was provided by a ruby laser; the flash duration was of 6 ns. The measuring light was obtained using a 10 nm bandwith filter and a 200 W tungsten lamp [2].

P. Mathis (ed.), Photosynthesis: from Light to Biosphere, Vol. II, 71–74.

2.3 *Genetic constructions*
Site directed mutants in *psa*E were essentially produced as already described [1]. Briefly, a 6803 strain deleted in *psa*E (FKE2) was first obtained by transformation of the wild type with a deleted construct. A second plasmid bearing the site mutated gene was then used for a second round of transformation of the deleted strain.

Results and Dicussion

When a preformed complex between PSI & Fd is briefly illuminated, a global decrease in the absorption of the different Fe-S centres is observable between 460 and 600 nm. This decrease partly reflects the reduction state of Fd, due to the fact that the reduction of oxidized ferredoxin gives a larger change in absorption than the concomitant but opposite change due to the transition from $(FeS_{A-B})^{red}$ to (FeS_{A-B}).
In these conditions, it is possible to follow the reduction kinetic of the PSI associated Fd, which exhibits 3 first order fast phases of $t_{1/2}$ 500 ns, 20 µs and 100 µs (Fig. 1A). The Kd of the reaction is 0.5 µM. The same characteristics are observed when using membranes as well as purified PSI. When the same reduction kinetic is followed on a E minus PSI, there is a complete disappearance of the first order fast phases (Fig. 1A). A second order reduction process still occurs (Fig. 1B), and a similar total amplitude is reached when using a higher amount of ferredoxin. A fine description of PSI-E mutants was thus possible, some being shortly described below (see Fig; 2). Of the numerous mutations we have designed, some have no effects on both PSI-E integration and ferredoxin reduction, others result in a complete failure of PSI-E integration and a third class leads to a mixture of wild type and E deleted PSI complex. We presently characterized just one mutant with a normal level of E subunit but with a E less phenotype, the result of a single arginine suppression (R39Q).

N-terminal sequence

A large number of modifications have been designed on this most charged sequence of the protein [1]. Multiple charge replacements often lead to the complete absence of the subunit. This is not the case for less drastic changes, like the R12A mutant, for which the observed modification is a decrease in the total amplitude of the first order kinetics to 30% of the wild type value (similar to W17A, see Fig. 1C). This likely reflects a structural problem as a lower amount of subunit seems to be stably associated to the PSI. We expect, also based on some labelling experiments [1], that this part of the sequence is directly required for the correct anchorage of PSI-E in the PSI complex.

C-terminal sequence

The deletion of the 8 terminal amino acids prevents integration of the subunit in PSI, whereas the simple deletion of the positive terminal lysine is of no effect. In agreement with this last point, it has been possible to label this residue using fluorescein isothiocyanate in mid basic conditions without disturbing the association of the subunit to PSI.

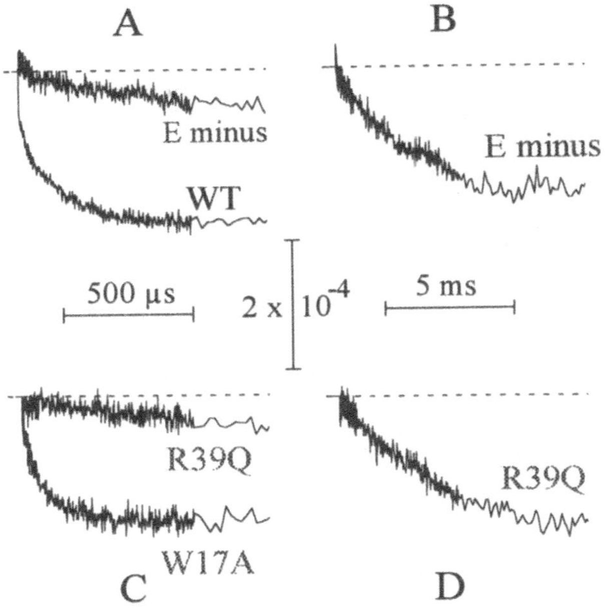

Figure 1: reduction kinetics of ferredoxin by PSI particles from wild type and some *psa*E mutants of *Synechocystis* 6803

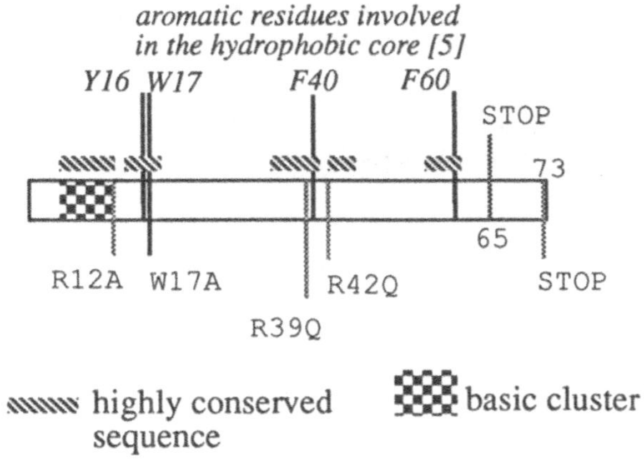

Figure 2: some basic features and mutations of PsaE from *Synechocystis* 6803

Aromatic core

The unique tryptophane of PSI-E is part of a hydrophobic core; this is obvious from the NMR structure [5]. The replacement of this residue could be predicted to affect the general structure of the polypeptide, and possibly its stable integration in PSI. Mutant W17A was designed to test this hypothesis. It has a similar behavior as mutant R12A, with a less dramatic effect as the remaining amplitude of the first order kinetics is then of 60% (Fig. 1C). This can be can be explained by the fact that the suppression of just one aromatic residue is not drastic enought to completly destroy the core of the protein, just leading to a less stable association.

Central arginines

Two arginines, R39 and R42 are present close to the middle part of the sequence, at the beginning of the mobile loop described in the NMR structure [5]. Both have been changed separately for glutamines. If the R42Q mutation is a silent one, R39Q has dramatic effects on the reduction of ferredoxin, with the suppression of all the fast first-order reactions, the second order rate constant being very similar to that observed for E minus (Fig. 1C-D).

4. Conclusion

From this brief discussion, it appears that PSI-E brings a very important positive charge at the rigth position in the PSI site for ferredoxin (R39). This likely occurs through a precise anchorage of this subunit, possibly involving charged sequences close to the N and C terminal ends of this protein.

References

1 Rousseau F., Sétif P., and Lagoutte B. (1993) EMBO J. 12, 1755-1765.
2 Sétif P.Q.Y. and Bottin H. (1994) Biochemistry, 33, 8495-8504.
3 Kruip J., Boekema E.J., Bald D., Boonstra A.F. and Rögner M. (1993) J. Biol. Chem., 268, 23353-23360.
4 Bottin H., and Lagoutte B. (1992) Biochim. Biophys. Acta, 1101, 48-56.
5 Falzone CJ., Kao Y-H., Zhao J., MacLaughlin KL., Bryant DA. and Lecomte JTJ. (1994) Biochemistry, 33, 6052-6062.

KINETICS OF PsaC REDUCTION IN PHOTOSYSTEM I

James E. Franke, Leszek Ciesla and Joseph T. Warden
Department of Chemistry, Rensselaer Polytechnic Institute,
Troy, NY 12180-3590

1. Introduction

A continuing focus of contemporary studies in photosystem I (PS I) is the elucidation of the electron transfer sequence, kinetics and molecular structures of the low-potential, iron-containing acceptors: F_X, F_A and F_B [1]. Recent advances in purification methodology for membrane proteins have enabled the isolation of highly purified PS I reaction centers either containing or deficient in the PsaC peptide, which binds the iron-sulfur clusters F_A and F_B [2]. The availability of such biochemically-defined preparations facilitates a determination of the electron transfer kinetics associated with the reduction and reoxidation of Center F_X, thus providing a temporal guide to the transit of the electron from P700 to its membrane-associated terminal acceptor.

We have proposed that in both *Synechococcus* sp. PCC 6301 and spinach PS I preparations the iron-sulfur cluster F_X is reduced within the 5 ns time constant of our flash spectrometer [3]. However, the kinetics of the reoxidation of this cluster, coincident with the reduction of the non-heme iron acceptor (either F_A or F_B) bound to the PsaC peptide, has not been determined. Experimental verification by transient absorption spectroscopy of the participation of F_X in forward electron transport is difficult due to the low, molar absorptivity of this component and also due to the similarity of the difference spectra of the PsaC electron acceptors (the iron-sulfur clusters, F_A and F_B, together designated, P430) to that for F_X. In this paper we present a comparison of the difference spectra for F_X and P430 and demonstrate that spectral differences between these species provide the capability for monitoring electron transfer from F_X to the PsaC-associated iron-sulfur clusters.

2. Materials and Methods

Photosystem I particles from *Synechococcus* sp. PCC 6301 containing (P700-F_A/F_B *complex*) or depleted of the F_A and F_B iron-sulfur clusters (P700-F_X *core protein*) were prepared according to modifications of a literature procedure [2,4]. For transient optical studies, the PS I particles were diluted to 10-25 μg/mL Chl in a 50 mM Tris buffer (pH 8.3) which contained 50 μM DCPIP and *ca.* 2 mg/ml dithiothreitol as reductant.

Transient absorbance changes in the sub-microsecond time-domain were measured with a laboratory-designed, flash-photolysis spectrometer. The spectrophotometer consisted of a 300 W. Eimac Xenon-arc monitoring source mated to a 200 mm monochromator (P.T.I.; bandwidth 1-4 nm). An EMI model 9816 photomultiplier coupled with a 55 MHz AC preamplifier (Analog Modules) was utilized to detect the optical transients. The minimum instrumental response time ($t_{1/2}$) was 60 ns. Actinic illumination (0.5 - 2 mJ, 601 nm, 0.5 Hz repetition rate) was provided by a TDL 51 dye laser/YG 581C Nd-YAG

P. Mathis (ed.), Photosynthesis: from Light to Biosphere, Vol. II, 75–78.

laser combination (Continuum). All measurements were performed at room temperature (19 °C). A LeCroy 9410 digital oscilloscope was used for data collection and signal averaging. Kinetic analysis of the experimental data was performed with the IGOR Pro graphing and analysis package (WaveMetrics).

3. Results

To establish the feasibility for monitoring the kinetics of electron transfer to the P430 cluster(s) detailed difference spectra for both F_X and P430 are required. Determination of the F_X difference spectrum followed the protocol of Parrett *et al.* [2], utilizing the *Synechococcus* P700-F_X core protein, which is deficient in the PsaC peptide. Similarly, the difference spectrum for P430 was obtained with the intact PS I P700-F_A/F_B complex from *Synechococcus* following the procedure of Ke [5].

The transient difference spectra for the F_X (open circles) and P430 (closed circles) centers are presented in Figure 1. These spectra are both characterized by minima in the 420-430 nm range and a bleaching extending from 390 nm to beyond 480 nm. Utilizing the differential molar absorptivity of $P700^+/P700$ at 430 nm as determined by Hiyama and Ke (45,000 M^{-1} cm^{-1} [6]) for calibration, the differential molar absorptivity for P430 can be estimated as 13,000±500 M^{-1} cm^{-1}, in accord with the reported value of 12,000 M^{-1} cm^{-1} [6]. Similarly, the photoreduced-minus-oxidized difference spectrum for the F_X cluster reveals a broad, absorption bleaching having a minimum at *ca.* 422 nm. This spectrum resembles the difference spectrum of the P430 cluster, but is distinguished from that of P430 by a discrete, narrow shoulder in the region of 410-418 nm as well as a diminished absorbance at wavelengths above 450 nm. The difference spectrum for the F_X cluster is consistent within experimental error at wavelengths greater than 400 nm with previous determinations [7,8]. The differential molar absorbance for the F_X cluster is calculated to be 15,000±500 M^{-1} cm^{-1} at 422 nm.

According to the difference spectra presented in Fig. 1, flash photolysis of the PS I complex should induce a transient positive absorbance in the wavelength region of 450 to 470 nm, diagnostic of the $P700^+F_X^-$ state, which relaxes to 10-25% of the initial amplitude concurrent with electron transfer from F_X to P430. Transient absorption kinetics at three wavelengths (430 nm, 416 nm and 453 nm) for the *Synechococcus* PS I complex are illustrated in Figure 2. The traces at 430 and 416 nm are dominated principally by the bleaching of P700 and the predicted absorbance increase signifying electron transfer from F_X to P430 is at the level of instrumental sensitivity for these two wavelengths. At 453 nm, however, the transient absorption increase is followed by a pronounced decay with a half-life of ca. 800 ns. The relative extent of this absorption change is consistent with the differences in molar absorptivity between F_X and P430 as detailed in Figure 1.

The attribution of the 800 ns kinetic component to the reduction of P430 by the reduced F_X cluster can be corroborated by comparison of the kinetics at 453 nm in the PS I complex with those obtained from the P700-F_X core protein. According to this kinetic assignment, removal of the terminal acceptor complex should result in the elimination of the 800 ns component as the electron is stabilized on the F_X cluster prior to recombination with $P700^+$. Figure 3 (upper trace) illustrates the kinetics observed at 453 nm in the P700-F_A/F_B complex. In contrast to these data the *Synechococcus* PS I core protein, which is depleted in PsaC, does not exhibit the 800 ns decay kinetics at 453 nm (Figure 3 lower trace). Instead, the charge separation in the $P700^+F_X^-$ couple relaxes in ca. 800 μs. Figure 3 demonstrates that observation of the 800 ns transient in photosystem I is contingent on the presence of a functional PsaC containing oxidized clusters. These kinetics are consistent with the recent suggestion by Sigfridsson *et al.* that the electrogenic phase associated

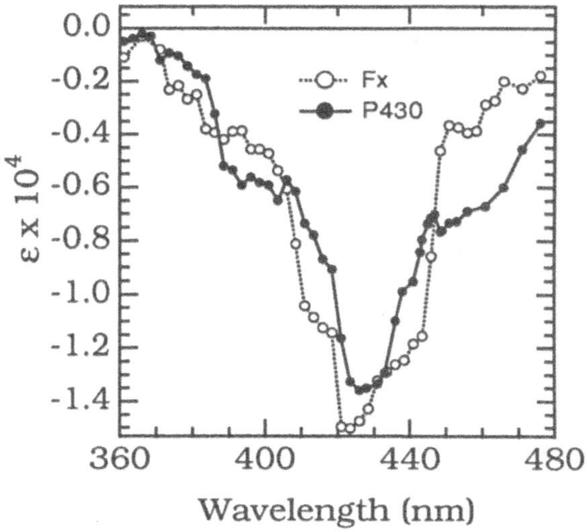

Figure 1. Transient absorption difference spectra for F_X and P430 in the photosystem I complex from *Synechococcus*. Each point is an average of 32 to 64 flashes. The scaling for the ordinate axis is expressed in differential molar absorptivity ($M^{-1}cm^{-1}$).

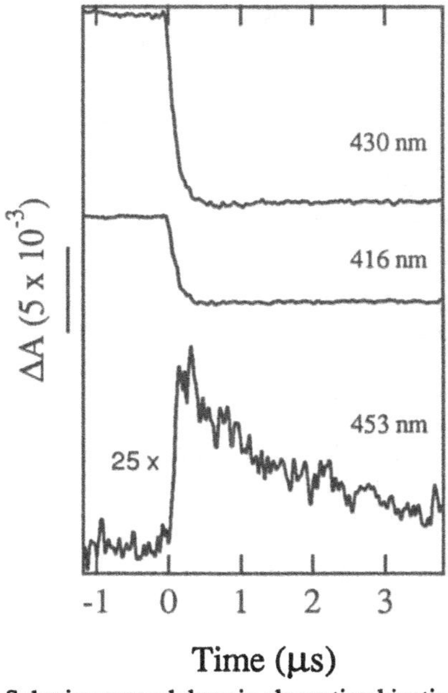

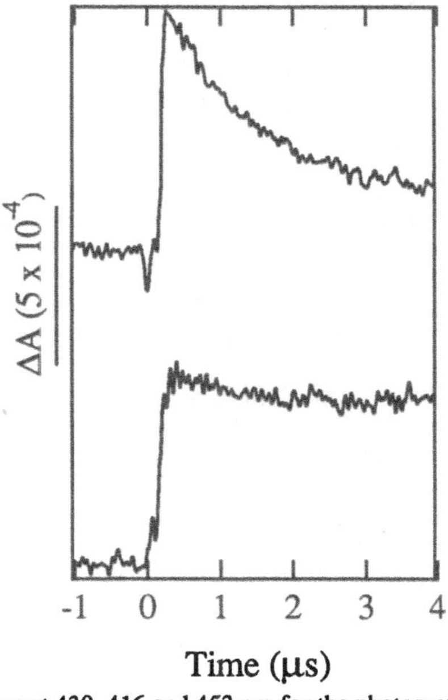

Figure 2. Submicrosecond domain absorption kinetic traces at 430, 416 and 453 nm for the photosystem I complex.
Figure 3. Submicrosecond domain absorption kinetics for the PS I complex (upper trace) and the P700-F_X core preparation (lower trace).

with reduction of PsaC occurs with a time constant of less than 1 μs [9].

The observation that the reduction of F_X in *Synechococcus* occurs in less than 5 ns [3], implies that the rate-controlling electron transfer step on the reducing side of PS I is the reduction of a cluster in the PsaC peptide. The preliminary structure of the PS I complex from *Synechocystis* [10] indicates that the two 4Fe-4S clusters of PsaC lie at a distance of 15 and 22 Å from the F_X cluster. Applying these distances to the relationship proposed by Dutton [11], which correlates the free-energy optimized rate for electron transfer with distance, predicts that electron transfer between the sites separated by 15 Å would exhibit a characteristic τ of *ca.* 900 ns. In contrast direct electron transfer between F_X and the cluster at 22 Å separation would require a τ of *ca.* 13 ms. The data presented in Figures 2 and 3 are consistent temporally with electron transport from F_X to the crystallographically-denoted closest cluster in PsaC, provisionally identified as F_A [12].

The assignment of the kinetics at 453 nm to electron transfer solely between F_X and the F_A cluster of the PsaC protein is admittedly circumstantial and cannot be demonstrated conclusively with the current data. However, PS I particles depleted of F_B by incubation with $HgCl_2$ [13] retain the ca. 800 ns kinetics at 453 nm, consistent with our assignment of F_A as the entry point for electrons into the PsaC protein.

Acknowledgments

This work was supported in part by the National Institutes of Health (GM26133) and by a Biomedical Research Support Grant.

References

1. Golbeck, J.H. and Bryant, D.A. (1991) Curr. Topics Bioenerg. 16, 83-177.
2. Parrett, K.G., Mehari, T., Warren, P. and Golbeck, J.H. (1989) Biochim. Biophys. Acta 973, 324-332.
3. J.T. Warden (1990) Current Research in Photosynthesis vol. II, 635-638.
4. Golbeck, J.H., Parrett, K.G., Mehari, T., Jones, K.L. and Brand, J.J. (1988) FEBS Lett. 228, 268-272.
5. Ke, B. (1972) Arch. Biochem. Biophys. 152, 70-77.
6. Hiyama, T. and Ke, B. (1972) Biochim. Biophys. Acta 267, 160-171.
7. Sauer, K., Mathis, P., Acker, S. and Van Best, J.A. (1978) Biochim. Biophys. Acta 503, 120-134.
8. Koike, H. & Katoh, S. (1982) Photochem. Photobiol. **35**, 527-531.
9. Sigfridson, K. Hansson, O. and Brzezinski, P. (1995) Proc. Natl. Acad. Sci., USA 92, 3458-3462.
10. Krauss, N., Hinrichs, W., Witt, I., Fromme, P., Pritzkow, W., Dauter, Z., Betzel, C., Wilson, K.S., Witt, H.T. and Saenger, W. (1993) Nature 361, 326-331.
11. Moser, C.C., Keske, J.M., Warncke, K., Farid, R.S. and Dutton, P.L. (1992) Nature 355, 796-802.
12. Zhao, J.D., Li, N., Warren, P.V., Golbeck, J.H. and Bryant, D.A. (1992) Biochemistry 31, 5093-5099.
13. Kojima, Y., Niinmi, Y., Tsuboi, S., Hiyama, T. and Sakurai, H. (1987) Bot. Mag. Tokyo 100, 243-253.

ELECTROGENIC LIGHT REACTIONS IN PHOTOSYSTEM I: EFFECTS OF PLASTOCYANIN, METHYL VIOLOGEN AND DITHIONITE ON A 100 NANOSECOND TO 100 MILLISECOND TIME SCALE.

Kalle Sigfridsson, Örjan Hansson and Peter Brzezinski
Department of Biochemistry and Biophysics, Göteborg University and Chalmers University of Technology, Lundberg Laboratory, Medicinaregatan 9C, S-413 90 GÖTEBORG, Sweden

Introduction

The electron-transfer (ET) components of the Photosystem I (PSI) reaction center include a chlorophyll primary electron donor, P700, and a number of electron acceptors: a chlorophyll (A_0), a quinone (A_1) and three 4Fe-4S clusters (F_X, F_A and F_B) (see refs.1,2). The sequence and kinetics of the light-induced ET reactions through PSI have been partly investigated using a number of different techniques, including flash-induced optical and EPR spectroscopy. However, these measurements have provided little information about the ET reactions between the Fe-S centers at room temperature, because their optical spectra are nearly identical.

The three-dimensional structure of a cyanobacterial PSI complex has been determined at 6 Å resolution (3). The structure suggests that all ET steps, from P700 to the F_X, F_A and F_B centers should be electrogenic in nature, i.e. they should be accompanied by an electric-field component parallel to the membrane normal. We have previously studied light-induced voltage changes in PSI complexes oriented in phospholipid layers, adsorbed on a Teflon film (4). Three phases (< 1 μs, 30 μs and 200 μs) were discerned in the charge separation and attributed to forward ET involving the iron-sulfur clusters in PSI. The fastest component, with a relative amplitude of 50%, was not possible to resolve in time. We have now improved the time resolution of the experimental setup, and determined the lifetime of a part of this component to 250/500 ns (with/without dithionite). In addition, we have investigated light-induced electrogenic events after addition of plastocyanin (Pc) as an electron donor and methyl viologen (MV) as an electron acceptor.

Material and Methods

Digitonin-solubilized PSI particles (D-144) from spinach with about 240 chlorophyll per P700 were used. The preparation of lipid vesicles from soybean lecithin, incorporation of PSI and formation of a lipid layer was as before (4). The experimental set-up is similar to the one described previously (4, and references therein).

Results and Discussion

Pulsed illumination of the PSI-containing lipid layer leads to a rapid voltage increase, followed by a 50 ms decay (Fig. 1). From the similarity with optical measurements on PSI particles in solution (4), we ascribe this signal to a charge separation from P700 to centers F_A and/or F_B followed by a charge recombination. A small fraction of the voltage (≈ 10%) decreases with a time constant of ≥ 1 s, limited by the setup time constant.

P. Mathis (ed.), Photosynthesis: from Light to Biosphere, Vol. II, 79–82.

Addition of ferricyanide (to oxidize P700) abolishes all signals, as expected from signals originating from PSI, and showing that there is no contribution from PSII in this preparation.

The charge-separation process was studied at a higher time-resolution (~ 50 ns). We found three components in the voltage increase with time-constants of about 500 ns, 30 μs and 200 μs (Fig. 1). Faster components were obscured by an artifact in the reverse direction. The 30 and 200 μs phases are not present when dithionite is added at pH 10. However, the major part of the ns-phase is still preserved, but with a lifetime of 250 ns (Fig. 1). Addition of dithionite at pH 10 is known to reduce F_A and F_B and to result in a more rapid charge recombination from center F_X. This is also observed for the electrical signal (Fig. 1) and the time constant obtained, 330 μs, is similar to what we obtained optically with PSI particles in solution (4) and to literature values during chemically and photoreduced conditions at room temperature (see ref. 2). There is also a contribution (≈ 30%) of a 2 ms kinetic phase in the electrical signal.

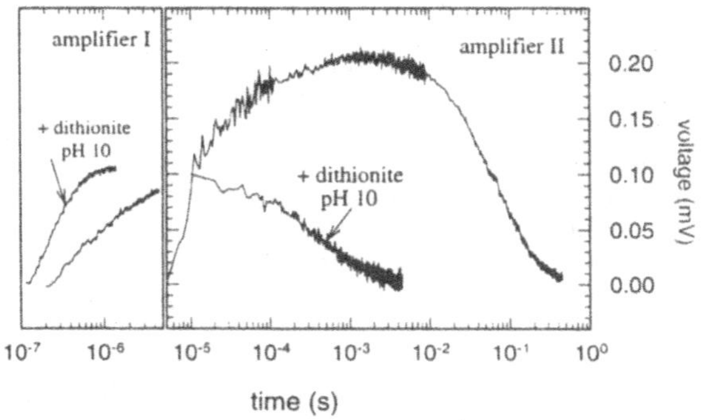

Fig. 1. A logarithmical illustration of the light-induced voltage changes in PSI complexes oriented in a lipid layer. The left part of the figure shows the nanosecond phases and the right part illustrates the microsecond and millisecond components. The buffer solution consisted of 10 mM Tris-HCl (pH 8.0), 10 mM KCl, 10 mM $CaCl_2$ and 2 mM ascorbate or 200 mM glycine (pH 10.0), 10 mM KCl, 10 mM $CaCl_2$ and 40 mM sodium dithionite. Amplifier risetime, ~ 50 nanoseconds (I) and ~ 1 microsecond (II).

In a further attempt to verify the origin of the above signals, reduced Pc or oxidized MV were added to the PSI-containing cell compartment. Neither Pc nor MV are expected to penetrate the lipid layer. Following pulsed illumination, at the first flash, an increase in voltage was followed by a decrease with a time constant of 50 ms, below the baseline. At the ensuing flashes an increased charge-separation amplitude was observed. The explanation is as follows. Two different populations of PSI are in principal possible in the lipid layer (fig. 2), one with P700 oriented towards the Teflon film (T-population) and one with P700 exposed to solution (S-population). Charge displacements in the two populations will result in voltage changes of opposite signs and a net signal can be obtained if $N_T \neq N_S$, where N_T and N_S are the number of PSI complexes in each population. The sign of the charge-separation voltage change shows that $N_T > N_S$. In the presence of Pc, after the first flash, charge recombinaiton is prevented in the S-population by reduction of $P700^+$ by Pc and occurs only in the T-population. On the following flashes charge separation and recombination is only observed in the T-population and any reactions in the S-population are expected to occur on time-scales shorter than the resolution of the setup. The back-reaction to $P700^+$ is much faster than ET from Pc to P700. Consequently, only the T-population gives rise to detectable charge separation and recombination signals. That no ET from Pc to P700 can be discerned in single flash experiments is explained by the fact that this reaction takes place in the S-population with time constants similar to the 30 and 200 μs ET reactions on the acceptor side.

When MV (E_m = -440 mV) is added (20-30 μM), the sub-ms part of the voltage increase is preserved, without any significant changes in time constants or amplitude ratios. But now the charge separation is followed by a further 50 ms rising component (Fig. 3, trace A), instead of a decay. What we observe here, is mainly the 50 ms recombination from the S-population. Only the T-population can donate electrons to the added acceptor and no electrons can recombine to $P700^+$ in this population. However, the recombination is unaffected in the S-population and follows immediately after the net charge separation originating from the T-population.

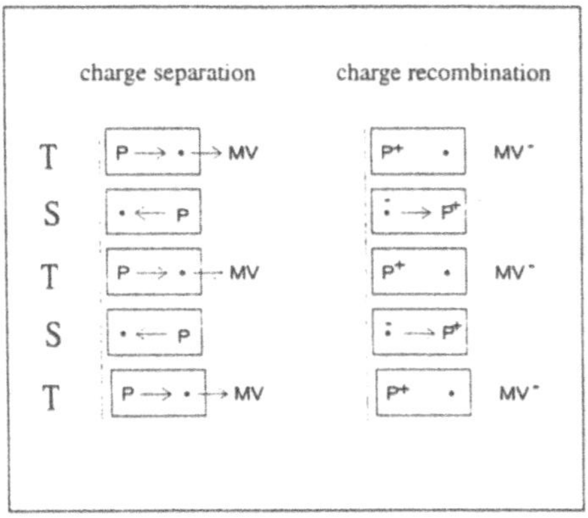

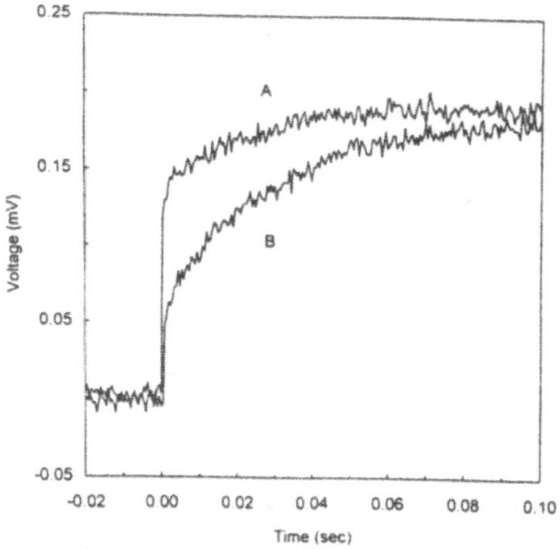

Fig 2. Schematic representation of the PSI-lipid layer. Two orientations with P700 (P) towards the Teflon film (T) and P towards the solution (S) are indicated. The dotted pattern symbolizes the Teflon film. The arrows indicate the direction of ET.
Fig 3. Light-induced voltage changes after methyl viologen additions. The MV additions are 25 μM (A) and 50 μM (B). The lower amplitude of the charge separation preceding the 7 millisecond component in trace B is artificial and due to dilution.

In an attempt to separate a component originating from MV reduction we increased the concentration of the acceptor. At higher concentrations (>30 μM), the initial charge separation is followed by a ms phase that precedes the 50 ms component (Fig. 3, trace B). This phase is attributed to the reduction of MV by the T-population. The time constant decreases with concentration. It is 7 ms at 50 μM MV. A similar time constant for MV reduction has earlier been obtained by (5), where a ~ 2 ms component was found at this concentration.

From the experiments with Pc and MV it is possible to quantitate the degree of orientation, defined as $(N_T-N_S)/(N_T+N_S)$. We found that ~ 70 % of PSI are oriented with the donor side towards the Teflon film.

The results reported here and in (4) indicate an involvement of F_A and F_B in the 30 μs and 200 μs phases of the electrical signal. The reduction potentials of F_A and F_B are -520 and -580 mV, respectively, as determined by low-temperature EPR spectroscopy under equilibrium conditions (see ref.1). Assuming that the potentials are not too different during the charge-separation process, an ET between F_A and F_B, away from P700, can be exergonic only if F_B is closest to P700. This leaves us with two possible models, assuming a serial ET through the PSI reaction center (4).

> In model I, the ns-component is assigned to the reduction of F_B, followed by an ET to F_A with a time constant of 30 μs. There are then three possibilities for the 200 μs phase: an ET to a soluble electron acceptor, a proton transfer in the opposite direction or a conformation change.
>
> In model II, all three components are assigned to ET reactions within PSI. The sub-microsecond voltage increase is attributed to reduction of F_X and the following phases to ET from F_X to F_B (30 μs) and from F_B to F_A (200 μs).

A choice between the two models could, in principle, be obtained from a comparison of the amplitudes of the electrical signals with those expected from the three-dimensional structure (3). Unfortunately, the uncertainties in the experimental amplitudes are, at present, too large to unambiguously rule out one of the two models.

We have previously estimated the time constant of the ET beween F_B and F_A to be of the order of 30 μs (4), which would argue in favor of model I. This model is also compatible with the suggestion that F_B is reduced within a time shorter than 500 ns (5).

On the other hand, we observe a 250 ns component of significant amplitude under conditions when F_B/F_A and F_X are mainly reduced and oxidized, respectively. This lifetime, as well as the 330 μs lifetime of the subsequent charge recombination, is similar to those obtained by others (6) and has been assigned to the reduction of F_X. This would leave us with model II.

Acknowledgements
This work was supported by the Swedish Natural Science Research Council.

References
1. Golbeck, J.H. (1993) Curr. Opin. Struct. Biol. 3, 508-514.
2. Golbeck, J.H. (1987) Biochim. Biophys. Acta 895, 167-204.
3. Krauss, N., Hinrichs, W., Witt, I., Fromme, P., Pritzkow, W., Dauter, Z., Betzel, C., Wilson, K.S., Witt, H.T. and Saenger, W. (1993) Nature 361, 326-331.
4. Sigfridsson, K., Hansson, Ö. and Brzezinski, P. (1995) Proc. Natl. Acad. Sci. USA 92, 3458-3462.
5. Setif, P. and Bottin, H (1994) Biochemistry 33, 8495-8504.
6. Luneberg, J., Fromme, P., Jekow, P and Schlodder, E. (1994) FEBS Lett. 338, 197-202.

Photoactive Reaction Center (PscA/Cytochrome c_{551})$_2$ Complex of *Chlorobium limicola*

Hirozo Oh-oka[1], Saki Kakutani[1], Shoichiro Kamei[1], Hiroshi Matsubara[1], Masayo Iwaki[2], & Shigeru Itoh[2], Dept. of Biol., Fac. of Sci., Osaka Univ.[1], Osaka 560, & National Inst. Basic Biology[2], Okazaki 444, Japan

1. Introduction

Photosynthetic reaction center (RC) of green sulfur bacteria is structurally and functionally related to the photosystem I RC of plants and cyanobacteria, especially in that they both have Fe-S centers on the acceptor side (1,2,3). We previously reported that the partially purified RC complex of *Chlorobium limicola* showed light-induced oxidation of cytochrome c_{551} as well as EPR signals of three distinct Fe-S centers (4). We have further purified the crude RC complex in order to get the detailed information concerning the minimal requirement for the photoactive P840. The finally obtained complex (CRC-2) is made up of only two types of subunits, the core and cytochrome c_{551}, respectively, and exhibits the light-induced primary charge separation state at 77 K. This complex will be express as (PscA/cytochrome c_{551})$_2$, as described below.

2. Procedure

Isolation of Reaction Center Complex: The crude photosynthetic RC complex prepared as reported previously (4) was further purified with a series of column chromatography according to the scheme in Fig. 1. The detailed procedure will be described in elsewhere (a manuscript has been submitted).

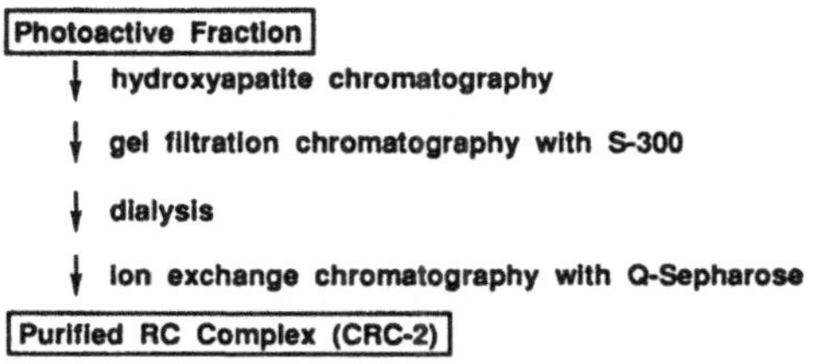

Figure 1. Procedure to prepare highly purified RC complex (CRC-2).

Cloning and Sequencing of the cytochrome c_{551} gene: The N-terminal sequence of cytochrome c_{551} was determined with a gas-phase sequencer (Applied Biosystems, model 473A) after transferring on PVDF membrane from SDS-PAGE. The C-terminal peptide fragment recovered after BrCN-cleavage was purified with a Waters HPLC system (600E) and sequenced. Two oligonucleotides were synthesized according to these sequences to amplify *C. limicola* chromosomal DNA by PCR. A single PCR product was obtained and used to identify the cytochrome c_{551} gene by Southern blotting. The appropriate size DNA fragments were extracted from agarose gel, ligated into the plasmid pUC18 and transformed into E. coli HB101 cells. Positive clones were obtained by colony

P. Mathis (ed.), *Photosynthesis: from Light to Biosphere, Vol. II*, 83–86.

hybridization with the PCR product as a probe. The plasmid DNA of these clones was mapped and subcloned into pUC118 for sequencing by the Sanger dideoxy chain-termination method using a DNA sequencer (Applied Biosystems, model 373A).

Measurement of Transient Difference Spectra: Time-resolved transient absorption spectra of CRC-2 were obtained at 77 K using a combination of a multichannel photodiode array spectrophotometer and a 532-nm, 10-ns hfwm excitation laser flash which are assembled in the NIBB laboratory. The reaction mixture contained 60% (V/V) glycerol in 50 mM Tris-HCl (pH8.0), 2 mM sucrose-monolurate, 10 mM cysteine and 1 mM EDTA with a glucose/catalase system.

3. Result and Discussion

The SDS-PAGE analysis of the purified CRC-2 complex indicates only two bands at 68 and 21 kDa, respectively (Fig. 2). The 68-kDa subunit is a core protein. The 21-kDa subunit is cytochrome c_{551} that is stainable with TMBZ (4).

The absorption spectrum of the CRC-2 complex shows a Qy-absorption band of BChl *a* at 817 nm with a prominent shoulder at 830 nm (Fig. 3a). The 671-nm peak can be assigned to the Qy-band of BChl-671 (formerly BChl 663). One molecule of BChl-671 seems to function as a primary acceptor. From the chemically oxidized - minus - reduced difference spectrum (Fig. 3 b), the ratio of heme *c* to P840 is calculated to be 1.6 - 1.7 ($\Delta\varepsilon_{830}$ = 100 mM^{-1}cm^{-1} for P840 and $\Delta\varepsilon_{551-540}$ = 20 mM^{-1}cm^{-1} for heme *c*, respectively).

The nucleotide and deduced amino acid sequences of the

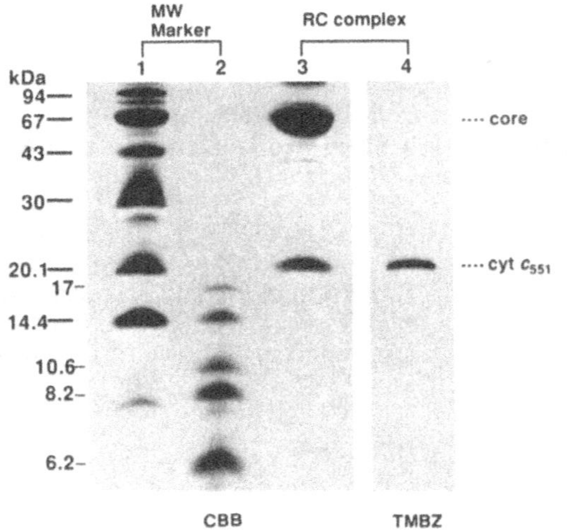

Figure 2. SDS-PAGE analysis of the CRC-2 complex. Lanes 1 - 3 were stained with CBB, and lane 4 by TMBZ. Lanes 1 and 2, molecular-weight markers; lanes 3 and 4, the CRC-2 complex.

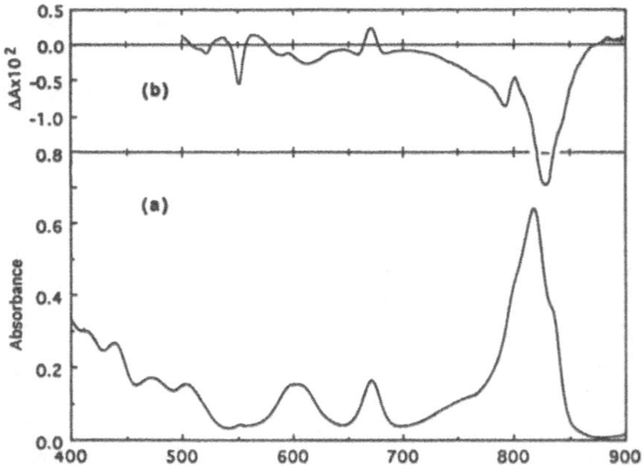

Figure 3. Absorption spectrum (a) and chemically oxidized-minus-reduced difference spectrum (b) of the CRC-2 complex at 293 K. The desalted sample (A_{810}=0.56) was oxidized by addition of a small amount of ferricyanide and then reduced by an excess amount of ascorbate.

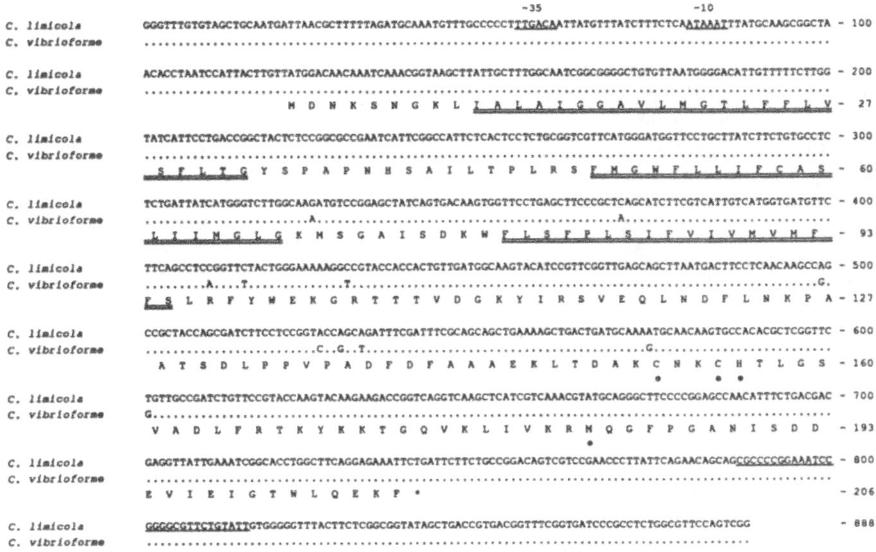

Figure 4. Alignment of nucleotide and amino acid sequences of cytochrome c_{551} in the reaction center complex of *C. limicola* and *C. vibrioforme*. The predicted membrane-spanning regions (double underline) and the heme binding site (•) are indicated.

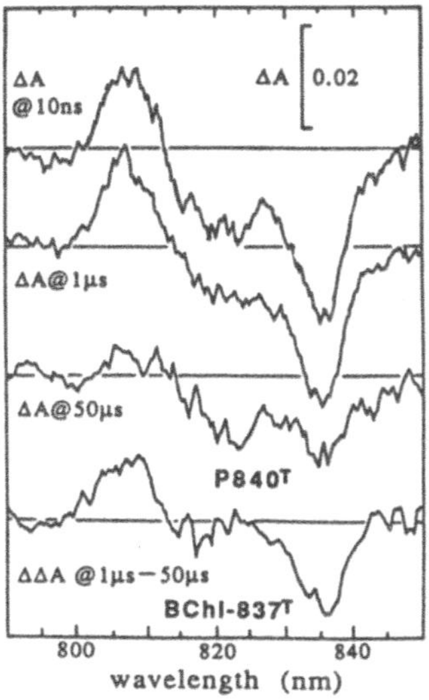

Figure 5. Time resolved difference spectrum of the CRC-2 complex at 10 ns, 1 μs and 50 μs after the laser flash at 77 K and the difference between 1- and 50- μs spectra.

cytochrome c_{551} gene are compared with those from *C. vibrioforme* (5) in Fig.4. Their amino acid sequences are completely identical and show typical alignment of Cys-X-X-Cys-His to bind a single heme *c*. On the nucleotide level, they show only 11 replacements, exhibiting 98% homology to each other. This strongly suggest the existence of a pair of cytochrome c_{551} in our *C. limicola* RC complex, although Okkels et al. (5) reported that *C. vibrioforme* RC complex contains 0.9 - 1.2 hemes/P840. In fact, we have recently demonstrated that two hemes *c* function as the electron donos to P840 in the *C. tepidum* RC (6).

Time-resolved difference spectra at 77 K are shown in Fig. 5, although no absorption change of cytochrome c_{551} is observed either at 77 K or at room temperature. At 10 ns , a narrow bleach at 836 nm with a shoulder around 820 - 830 nm and a positive change at 808 nm are detected. The extent and spectral shape do not significantly change until 1 µs. The spectrum afterward decays in two phases with $t_{1/2}$ of 20 and 100 µs, respectivley (data not shown). These components seem to correspond to BChl-837^T (7) and P840^T, respectivley, whose typical spectra are shown in Fig. 5.
In conclusion, the present RC complex contains two sets of the core and cytochrome c_{551} subunits and can be expressed as (PscA/cytochrome c_{551})$_2$ complex. Since this complex undergoes the primary charge separatrion followed by triplet formation, the minimal requirement for the photoactivity of the so-called "homodimeric" *Chlorobium* RC complex is the two identical core subunits that bear P840 and other pigments.

References

1. Blankenship, R.E. (1992) Photosynth. Res. 33, 91-111.
2. Feiler, U., Nitschke, W. and Michel, H. (1992) Biochemistry 31, 2608-2614.
3. Büttner, M., Xie, D.L., Nelson, H., Pinther, W., Hauska, G. and Nelson, N. (1992) Proc. Natl. Acad. Sci. U.S.A. 89, 8135-8139.
4. Oh-oka, H., Kakutani, S., Matsubara, H., Malkin, R. and Itoh, S. (1993) Plant Cell Physiol. 34, 93-101.
5. Okkels, J.S., Kjær, B., Hansson, Ö., Svendsen, I., Møller, B.L. and Scheller, H.V. (1992) J. Biol. Chem. 267, 21139-21145.
6. Oh-oka, H., Kamei, S., Matsubara, H., Iwaki, M. and Itoh, S. (1995) FEBS Lett. 365, 30-34.
7. Hoff, A.J., Vasmel, H., Lous, E.J. and Amesz, J. (1988) in Green Photosynthetic Bacteria (Olson, J.M. et al., Eds.) pp 119-125, Plenum Press, New York.

STRUCTURAL ORGANIZATION OF PHOTOSYSTEM I POLYPEPTIDES

Q. Xu, V. P. Chitnis, K. An, P. R. Chitnis
Division of Biology, Kansas State University, Manhattan, KS 66506, USA

1. Introduction

Photosystem I (PSI) in cyanobacteria and chloroplasts is a multisubunit membrane-protein complex that catalyzes electron transfer from reduced plastocyanin (or cytochrome c_6) to oxidized ferredoxin (or flavodoxin) (1, 2). PsaA and PsaB form a heterodimeric core that harbors ~100 chlorophyll *a* molecules, the primary electron donor, P700, and the electron acceptors A_0, A_1, and F_X. PsaC, PsaD, and PsaE are peripheral subunits that are located on the reducing side of PSI. PsaC binds the terminal electron acceptors, F_A and F_B. The remaining proteins of cyanobacterial PSI are integral membrane-proteins. Among them, PsaL is required for the formation of PSI trimers (3, 4). PsaF can be chemically cross-linked *in vitro* with plastocyanin (5, 6). PsaJ and PsaI are involved in proper assembly of PsaF and PsaL, respectively (7, 8).

Despite the recent X-ray crystallographic analysis of the PSI complex at 6-Å resolution, detailed organization of PSI polypeptides remain to be explored (9). We are studying architecture of PSI using biochemical approaches and mutational analyses of the PSI polypeptides (2). These studies have revealed several details about the topography and subunit organization of the PSI complex.

2. Results

2.1 Accumulation of the PSI proteins in subunit-deficient cyanobacterial mutants

The absence or decreased accumulation of specific polypeptides in the subunit-deficient mutants may indicate assembly requirements and subunit interactions (2). Table I summarizes accumulation of PSI subunits in the photosynthetic membranes of different mutants. These data suggested structural interactions among PsaE, PsaF, and/or PsaJ, between PsaD and PsaL, or between PsaI and PsaL in the organization of PSI.

2.2 Chemical cross-linking

Subunit interactions in PSI may be identified using chemical cross-linking, followed by immunodetection (Fig. 1A). Analyses of cross-linked products in subunit-deficient PSI complexes help in identification of comigrating species. When the wild-type and PsaE-less PSI were treated with glutaraldehyde and the proteins were used for

P. Mathis (ed.), Photosynthesis: from Light to Biosphere, Vol. II, 87–90.

Table I: Accumulation of various PSI subunits in cyanobacterial mutants

Strain [a]	Missing or inactive genes	Subunit in the photosynthetic membranes[b]							
		PsaA-B	PsaC	PsaD	PsaE	PsaF	PsaL	PsaI	PsaK
WT	none	+	+	+	+	+	+	+	+
PSI-less	*psaA-B*	-	-	-	-	-	-	-	-
PsaC-less[c]	*psaC2*	+	-	-	-	+	+	+	+
ADC4	*psaD*	+	+	-	+	+	+[e]	+	+
AEK2	*psaE*	+	+	+	-	+	+	+	+
AFK6	*psaF-psaJ*	+	+	+	+[d]	-	+	+	+
AJC8	*psaJ*	+	+	+	+	+[e]	+	+	+
ALC7-3	*psaL*	+	+	+	+	+	-	+	+
AIC9	*psaI*	+	+	+	+	+	+[e]	-	+

[a] See ref. 2 for original references to the mutant strains

[b] The cyanobacterial PSI contains at least eleven subunits (2). However, only the subunits for which antibodies are available, are presented. + denotes the presence whereas - denotes the absence of a certain subunit.

[c] L. McIntosh, personal communication.

[d] Easily extracted by chaotropes.

[e] Absence or depletion in the Triton X-100 solubilized PSI complex.

Western blot analysis, three major cross-linked products with apparent molecular masses of 21, 25, and 29 kDa were cross-reacted with antibodies against PsaE, PsaC, and PsaD (Fig. 1A). The 29-kDa species results from cross-linking between PsaD and PsaL (10). The 25-kDa species was absent in the PsaE-less PSI, but was recognized by antibodies against PsaE, PsaC, and PsaD in wild type, indicating formation of cross-linking between PsaD and PsaE or between PsaD and PsaC. Because of the electrophoretic comigration of the PsaD-PsaC and PsaD-PsaE species, it is advantageous to isolate the PsaD-PsaC species using the cross-linked PsaE-less PSI. We used this strategy to purify the PsaD-PsaC species (Fig. 1B). Such PsaD-PsaC preparation can be used to identify the cross-linked domains.

2.3 Treatment with proteases and NHS-biotin

Treatment of assembled and functional PSI complexes with surface-labeling probes and proteases, followed by immunodetection and N-terminal amino acid sequencing, provide insights regarding the topography of PSI subunits. We have used thermolysin, Glu-C protease, and *N*-hydrosuccinimidobiotin (NHS-biotin) in topographical investigations. Thermolysin cleaves peptide bonds on the amino side of hydrophobic residues. Under appropriate reaction conditions, Glu-C protease specifically hydrolyzes the peptide bond at the carboxyl side of glutamyl residues. NHS-biotin reacts with unblocked N-termini and ε-amino groups of lysyl residues. Most PSI subunits have numerous sites available for reaction with these probes, but only the surface-exposed sites can be accessed by them (Table II). We have mapped the regions in PsaE and PsaD of *Synechocystis* sp. PCC 6803 that are accessible to *in vitro* proteolysis and NHS-biotin (11). Lys^{74} at the C-terminus of PsaE was accessible for biotinylation. Lys^{107}, or Lys^{118}, or both in PsaD could be modified by NHS-biotin. The accessible protease cleavage sites in PsaE and PsaD also map in the C-terminal domains of these proteins. Subunit-deficient PSI complexes can be used to probe for the new protease

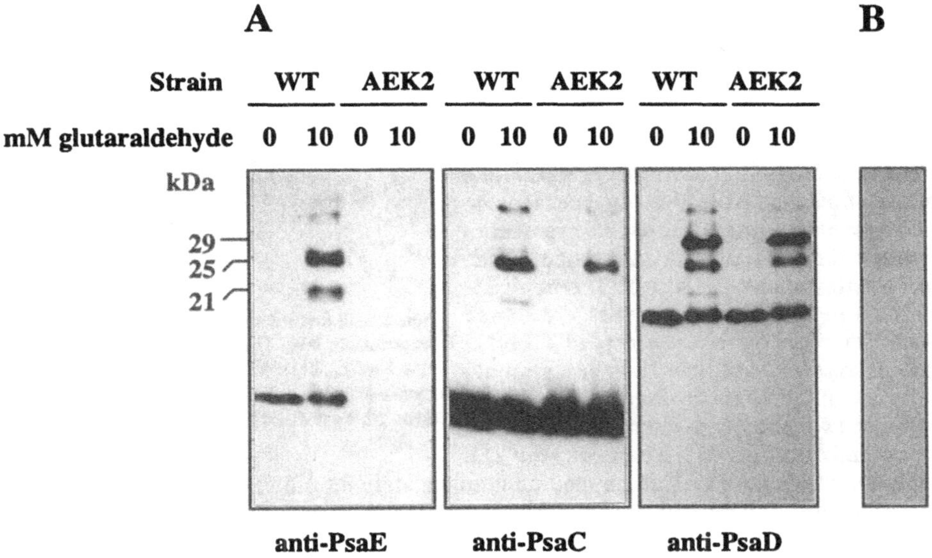

Figure 1. Cross-linking of PSI subunits and purification of the cross-linked species. Panel A, The Triton X-100 solubilized wild-type (WT) and PsaE-less (AEK2) PSI complexes were treated with 10 mM glutaraldehyde (10). The proteins were separated by Tricine-urea-SDS-PAGE, and transferred to Immobilon-P membranes. Immunodetection was performed using horseradish peroxidase conjugated secondary antibody and enhanced chemiluminescence reagents. Panel B, The PsaD-PsaC species in the cross-linked PsaE-less PSI complex was separated by electrophoresis, isolated, and electroeluted. The purify of such species was examined using Tricine-urea-SDS-PAGE, followed by staining with Coomassie blue.

sites in the PSI subunits that may be exposed by the absence of a specific subunit. We used this approach to identify the PsaB domains that may interact with PsaD (12). Incubation of the wild-type PSI with thermolysin yields 22-kDa C-terminal fragments of PsaB. The absence of PsaE, PsaF, PsaJ, or PsaL does not alter their resistance to further proteolysis. When the PsaD-less PSI is treated with thermolysin, the 22-kDa C-terminal fragments of PsaB are rapidly cleaved, with concomitant accumulation of a 16-kDa fragment and then a 3.4-kDa fragment. We have proposed that these cleavage sites are in the extramembrane loops on the cytoplasmic side of the thylakoids and are shielded by PsaD (12).

Table II. Use of surface probes to study topography of PSI. Sites in PSI proteins from *Synechocystis* sp. PCC 6803 that are available (Avl.) and accessible (Acc.) for reaction with surface probes

Subunit	NHS-biotin		Thermo-lysin		Glu-C protease[a]	
	Avl.	Acc.	Avl.	Acc.	Avl.	Acc.
PsaA	24	Yes	173	Yes	17	Yes
PsaB	21	Yes	164	Yes	20	Yes
PsaD	13	Yes	23	Yes	14	Yes
PsaF	9	Yes	32	Yes	9	Yes
PsaL	2	Yes	38	Yes[b]	5	No
PsaE	5	Yes	16	Yes	5	Yes
PsaC	4	Yes	15	No	4	Yes
PsaK[c]	≥4	Yes	≥10	Yes	≥0	No
PsaI	1	No	11	No	3	No
PsaJ	2	No	11	No	1	No
PsaM	2	No	11	No	1	No

[a] Under the Glu-hydrolysis condition.
[b] In PSI monomers, but not in trimers.
[c] Based on the partial amino acid sequence.

3. Discussion

Topographical investigations of PSI by protein chemistry techniques complement the structural analysis of PSI by crystallography. Surface probes, domain-specific antibodies, chemical cross-linking, protease accessibility, chaotropic extraction, and *in vitro* reconstitution can be used to study protein interactions and surface-exposed domains. We have extended these analyses to subunit-deficient PSI complexes from the mutant strains of the cyanobacterium *Synechocystis* sp. PCC 6803 and discovered additional organizational features of PSI complexes (2). These findings have allowed us to propose a model for organization of PSI subunits (Fig. 2). Such model would be refined by the mutational analyses of the PSI subunits and X-ray crystallographic analyses of PSI crystals. Use of cyanobacterial mutants in topographical studies is indeed a powerful tool for understanding structural organization of PSI.

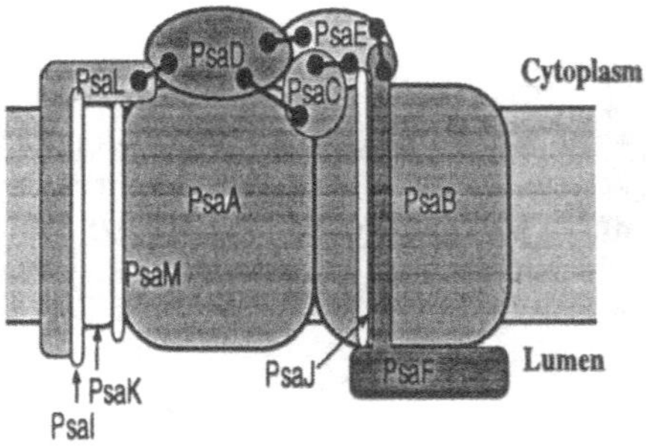

Figure 2. Subunit interactions and topography in cyanobacterial PSI. The subunits that can be cross-linked with glutaraldehyde are connected with a dumbbell line. The subunits with surface-exposed ε-amino groups are shaded.

4. Acknowledgments

This work was supported in part by grants from the US National Science Foundation (MCB 9405325), the US Department of Agriculture-NRICGP (92-37306-7661), and the US National Institute of Health (R01GM53104). This is contribution #95-38-A from the Kansas Agricultural Experiment Station.

5. References

1. Golbeck, J.H. (1994) in The Molecular Biology of Cyanobacteria (Bryant, D.A., ed), pp. 179-220, Kluwer Academic Publishers, Dordrecht, The Netherlands
2. Chitnis, P.R., Xu, Q., Chitnis, V.P. and Nechushtai, R. (1995) Photosynth. Res. 44, 23-40.
3. Chitnis, V.P. and Chitnis, P.R. (1993) FEBS Lett. 336, 330-334
4. Chitnis, V.P., Xu, Q., Yu, L., Golbeck, J.H., Nakamoto, H., Xie, D.-L. and Chitnis, P.R. (1993) J. Biol. Chem. 268, 11678-11684
5. Wynn, R.M. and Malkin, R. (1988) Biochemistry 27, 5863-5869
6. Hippler, M., Ratajczak, R. and Haehnel, W. (1989) FEBS Lett. 250, 280-284
7. Xu, Q., Odom, W.R., Guikema, J.A., Chitnis, V.P. and Chitnis, P.R. (1994) Plant Mol. Biol. 26, 291-302
8. Xu, Q., Hoppe, D., Chitnis, V.P., Odom, W.R., Guikema, J.A. and Chitnis, P.R. (1995) J. Biol. Chem. 270, 16243-16250
9. Krauss, N., Hinrichs, W., Witt, I., Fromme, P., Pritzkow, W., Dauter, Z., Betzel, C., Wilson, K.S., Witt, H.T. and Saenger, W. (1993) Nature 361, 326-331
10. Xu, Q., Armbrust, T.S., Guikema, J.A. and Chitnis, P.R. (1994) Plant Physiol. 106, 1057-1063
11. Xu, Q., Guikema, J.A. and Chitnis, P.R. (1994) Plant Physiol. 106, 617-624
12. Xu, Q. and Chitnis, P.R. (1995) Plant Physiol. 108, 1067-1075

EPR Studies Of Site-Directed Mutants Of The PsaC Protein Of Photosystem I In *Anabaena variabilis* ATCC 29413

W. -Z. He[1], R. M. Mannan[2], S. Metzger[3], J. Whitmarsh[3], H. B. Pakrasi[2] and R. Malkin[1]

[1]Department of Plant Biology, University of California, Berkeley, CA 94720-3102. [2]Department of Biology, Washington University, St. Louis, MO 63130. [3]Center of Biophysics and Computational Biology, University of Illinois, Urbana, IL 61801

Introduction

The electron transfer carriers of photosystem I (PS I) include the primary donor P700, A_0, A_1, F_X, F_A and F_B. The latter three are known [4Fe-4S] clusters [1]. One of the unresolved questions in PS I function is the passage of electrons through the terminal acceptors, F_A and F_B. Based on their redox potentials and photoreduction behaviors at low temperature, it was proposed that F_B preceded F_A in the electron transport route [2]. However, when F_B was selectively denatured [3, 4], the room- and low-temperature photoreduction kinetics of F_A were hardly affected, suggesting a parallel electron transfer pathway through F_A and F_B.

F_A and F_B are bound to a 9-kD PsaC protein. Amino acid sequences of PsaC, available from a variety of cyanobacteria, algae and higher plants, are highly conserved, with two CxxCxxCxxxCP motifs providing ligands to two [4Fe-4S] clusters. More recently, site-directed mutagenesis of the PsaC protein of a cyanobacterial *Synechococcus* sp. was carried out and the resulting gene was overexpressed in *E. coli* [5]. *In vitro* reconstitution with mutant proteins identified the cysteine residues liganding F_A and F_B. Earlier reports suggested a [3Fe-4S] center was assembled at the modified site after changing one of the cysteine ligands to an aspartate, leading to the conclusion that photoreduction of F_A is independent of the presence of F_B. However, in a recent report, the same group has concluded that a mixed ligand [4Fe-4S] cluster, rather than a [3Fe-4S] was present at the modified site [6]. Therefore, the question of whether F_A or F_B is required for PS I electron transfer reactions remains open.

We have used an *in vivo* approach to study the function of PsaC. This involved the incorporation of altered PsaC into PS I. A number of mutants have been constructed using the filamentous cyanobacterium *Anabaena variabilis* ATCC 29413 [7], and in this paper, we present the EPR characterization of two mutants, C13D and C50D, where either the C-13 or the C-50 residue of PsaC was changed to an aspartate residue.

P. Mathis (ed.), Photosynthesis: from Light to Biosphere, Vol. II, 91–94.

Results

The C13D and C50D mutants can grow photoautotrophically or heterotrophically in the dark when supplemented with fructose. The C13D mutant showed a slower rate of autotrophic growth. The doubling times are given in Table I, together with the rate of $NADP^+$ photoreduction. When Fd was not saturating, the C13D mutant showed a slower rate in this *in vitro* assay.

Table I. Growth and electron transfer rates of wildtype and mutant strains.

	Wildtype	Deletion	C13D	C50D
Doubling Time (h)				
Dark-grown	38 ± 3	36 ± 3	37 ± 4	35 ± 5
Light-grown	36 ± 3	-	62 ± 5	40 ± 5
DCIP/Asc ⟶ $NADP^+$ (μM [mg chl]$^{-1}$ hr^{-1})				
1.6 μM Fd	54 ± 8	-8	33 ± 6	43 ± 4
3.2 μM Fd	74 ± 8	-	72 ± 5	72 ± 5

When thylakoid membranes from the wildtype strain were frozen with dithionite under illumination, both F_A and F_B were reduced (Fig. 1). The wildtype showed the typical interacting spectrum for F_A and F_B (g=2.05, 1.94, 1.92, 1.89). The deletion strain shows an axial signal of a [4Fe-4S] center (g=2.05, 1.92). This species, attributed to a reduced NADH/NADPH dehydrogenase (NDH), has been found in a variety of cyanobacteria [8]. The C13D mutant shows the rhombic F_A signal (g=2.05, 1.94, 1.86) in addition to NDH. In contrast, the C50D mutant shows the F_B signal (g=2.07, 1.93, 1.88). The EPR characteristics of F_A in C13D and F_B in C50D are retained. This confirms the assignment of C-13 and C-50 as ligands to the cubane iron of F_B and F_A, respectively [5].

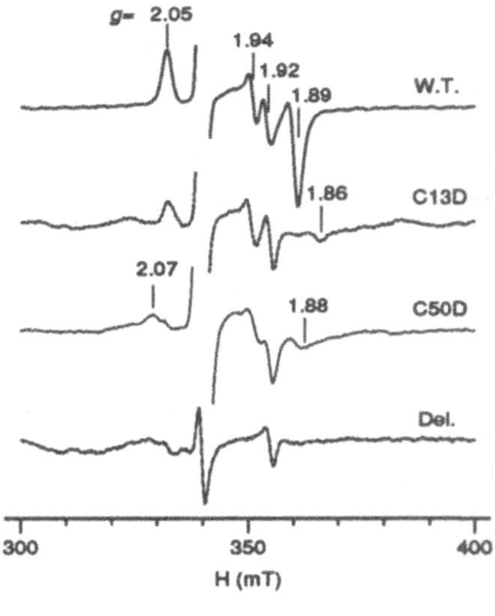

Fig 1. EPR spectra of fully reduced thylakoid membranes from wildtype (1.4 mg chl ml^{-1}), deletion strain (1.3 mg chl ml^{-1}) and C13D (1.0 mg chl ml^{-1}) at 18 K and C50D (1.5 mg chl ml^{-1}) at 25 K. Samples were frozen with dithionite under illumination. Spectrometer settings: gain, 2.0×10^5; modulation 1.25 mT at 100 kHz; microwave 9.440 GHz at 20 mW.

The question remains as to the nature of the modified cluster. Initially it appeared that the clusters at the affected sites were lost. There is no [3Fe-4S] cluster found in either C13D or C50D mutants. To examine the presence of F_X, the EPR spectra of samples frozen under

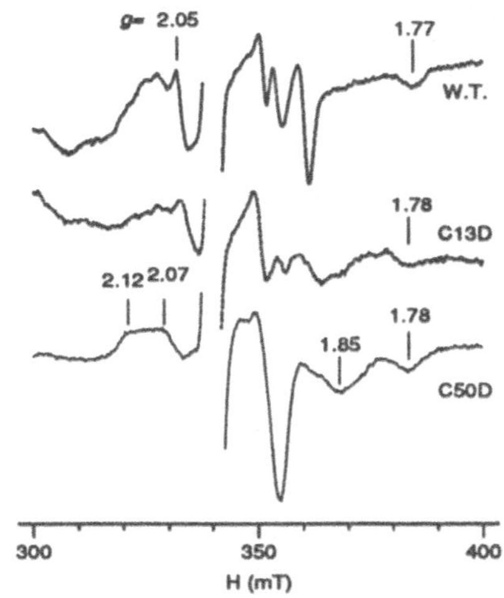

Fig. 2 EPR spectra of fully reduced thylakoids from WT, C13D and C50D at 12K. Samples were frozen in the light with dithionite and PMS (0.1 mM). Instrument settings are as in Fig 1.

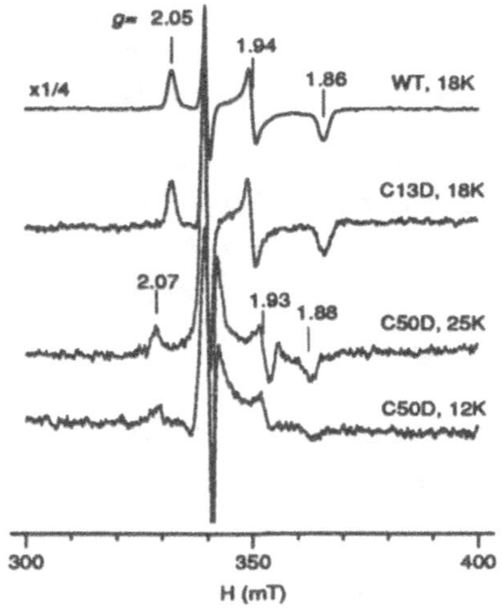

Fig. 3 Light-minus-dark EPR spectra. Samples were frozen after incubation in dark with 5mM ascorbate. Light spectra were recorded while illuminated. Instrument settings are as in Fig. 1.

illumination with dithionite and PMS added were recorded at 12 K (Fig. 2). In addition to the F_X signal (g=1.78), we also found a set of new resonances in C50D mutant (g=2.12, 1.93, 1.85). We believe that these arise from modified F_A, similar to the one recently described by Yu et al. [6] and denoted as F_A'. However, in contrast to their *in vitro* reconstituted system, we find no evidence for a spin-spin interaction between this F_A' and F_B.

There is no comparable F_B' EPR signal found in the C13D mutant. Although it is possible that the mixed ligand [4Fe-4S] may adopt an S=3/2 state leading to an extensively broadened EPR spectrum which is difficult to observe, triple flash experiments showed only one terminal electron acceptor in C13D (identified as F_A based on its EPR properties), contrasting with two acceptors in C50D, F_A' and F_B (data not shown).

Either F_A or F_B can be irreversibly photoreduced when illuminating the sample at low temperature. As the light-minus-dark (L-D) spectra in Fig 3 show, the characteristic F_A^- signal (g=2.05, 1.94, 1.86) appears along with the $P700^+$ radical for both the wildtype and the C13D mutant. This suggests that the point-mutation does not affect the functional assembly of F_A. The result also confirms the earlier finding that F_A can be photoreduced in the absence of F_B [3,4,9]. The g=1.92 signal of NDH is sometimes seen in the dark spectrum, but not in the L-D difference spectrum. The L-D spectrum of C50D shows only F_B (g=2.07, 1.93, 1.88) if the sample was frozen with ascorbate in the dark, even at 12K. However, if C50D samples are frozen with dithionite, an axial signal (g=2.05, 1.92) appears under illumination at 12K and is reversible at low temperature. The nature and origin of this signal is under investigation.

Redox titrations of F_A and F_B were carried out for the wildtype and the C13D mutant strain. Mid-potentials of -505 and -540 mV were obtained for F_A and F_B respectively in wildtype. These values are comparable to those reported for higher plant F_A and F_B centers [2]. F_A in C13D has a mid-potential of -510 mV, close to that in the wildtype.

Conclusions

A targeted change of a cysteine ligand to [4Fe-4S] in C13D of PsaC renders F_B undetectable by EPR, but gives a mixed ligand cluster in C50D. In both cases, the untargeted [4Fe-4S] cluster retained its EPR characteristics and electron transfer functionality. Surprisingly, the C13D mutant, which seems to be lacking F_B, can still grow photoautotrophically, and *in vitro* can photoreduce spinach Fd and $NADP^+$. Therefore, F_B appears not to be absolutely required for forward electron transfer from F_X to Fd. This finding is in contrast to our earlier *in vitro* study with mercurial treated thylakoids from spinach [9]. Furthermore, the fact that PsaC protein can functionally assemble without F_B in C13D may imply that it can adopt a conformation similar to a monocluster ferredoxin, such as that from *Bacillus thermoproteolyticus*, where a rigid α-helix forms in place of the second cluster [10]. The mixed ligand F_A' in C50D is not irreversibly photoreduced under normal conditions. Moreover, when F_B and F_A' are both reduced, they do not show line-shifting due to spin-spin interaction as seen in wildtype.

Acknowledgments

This work is supported in part by grants from NSF (to R.M.), NIH (to H.B.P.) and USDA-ARS to (J.W.). We thank Mr. Richard Chain for his gifts of spinach Fd and FNR used in this study.

References

1. Golbeck, J. H. & Bryant, D. A. (1991) In: Current Topics In Bioenergetics. (C. P. Lee ed.) Vol. 16, pp83-177. Academic Press, New York
2. Heathcote, P., Williams-Smith, D. L., Sihra, C. K. & Evans, M. C. W. (1978) Biochim. Biophys. Acta. 503, 333-342
3. Golbeck, J. H. & Warden, J. T. (1982) Biochim. Biophys. Acta. 681, 77-84
4. Fujii, T., Yokoyama, E. -I., Inoue, K. & Sakurai, H. (1990) Biochim. Biophys. Acta. 1015, 41-48
5. Zhao, J., Li, N., Warren, P. V., Golbeck, J. H. & Bryant, D. A. (1992) Biochemistry. 31, 5093-5099
6. Yu, L., Bryant, D. & Golbeck, J. H. (1995) Biochemistry. 34, 7861-7868
7. Sandmann, G. & Malkin, R. (1983) Arch. Microbiol. 136, 49-53
8. Mannan, R. M., Whitmarsh, J., Nyman, P. and Pakrasi, H. B. (1991) Proc. Natl. Acad. Sci. USA. 88, 10168-10172
9. He, W. -Z. & Malkin, R. (1994) Photosynthesis Research. 41, 381-388
10. Fukuyama, K., Nagahara, Y., Tsukihara, T., Katsube, Y., Hase, T. & Matsubara, H. (1988) J. Mol. Biol. 199, 183-193

CHARACTERIZATION OF THE PRIMARY ELECTRON DONOR OF PHOTOSYSTEM I, P700, BY ELECTROCHEMISTRY AND FOURIER TRANSFORM INFRARED (FTIR) DIFFERENCE SPECTROSCOPY

E. Hamacher*, J. Kruip#, M. Rögner# and W. Mäntele*
*Inst. f. Phys. u. Theoret. Chemie, Egerlandstr. 3, 91058 Erlangen, Germany
#Institut für Botanik, Schloßgarten 3, 48149 Münster, Germany

1. Introduction

Information about dynamic behaviour of a protein "at work" can principally be obtained by vibrational spectroscopy [1]. If a protein reaction can be triggered specifically and spectra of two functionally relevant states can be taken at sufficient precision, the vibrational difference spectra shows characteristic changes at the level of individual bonds which are connected to this specific protein reaction [2,3]. Photosynthetic proteins like photosystem I perform a light dependent reaction which can be interpreted as a chain of redox reactions: absorption of light energy brings about photochemical charge separation between the primary donor P700 and the terminal acceptors F_AF_B along a chain of redox cofactors. This electron transport leads to characteristic changes in the vibrational spectra of P700 and its surrounding protein [4,5]. Redox triggering by the use of electrochemistry allows to adjust defined redox states of PS I, and, in combination with vis/IR spectroscopy, information about the electronic and molecular changes of **one** cofactor and its protein environment during the photosynthetic activity of PS I is available.
Here we present an approach for the study of P700 from PS I of the cyanobacterium *Synechocystis* PCC 6803. Two different aggregation states of PS I -monomeric and trimeric complexes- were investigated. Reversible electrochemical oxidation and rereduction of P700 is used for the spectroscopic investigation (vis, IR) of the selective P700 cation formation. Midpoint potentials are determined by titration of the 700 nm band as well as by titration of the flash induced absorbance changes of P700. In addition, FTIR difference spectra with redox dependent absorbance changes and midpoint potentials according to those in the visible region were obtained.

2. Results

Spectroelectrochemistry. For the monomeric and the trimeric complex of PS I, application of an oxidizing potential (+460 mV) leads to the disappearence of the 700 nm absorption of neutral P700 (overlapped by ca. 100 Chl antenna Chls absorbing at 680 nm; see Fig. 1, *upper inset*) and to the rise of the 820 nm cation band of $P700^+$. Equilibration of the sample at intermediate potentials leads to titration of the 700 nm band (see Fig. 1). The dependence of the amplitude on the applied potential can be approximated with a Nernst curve with n = 1 transferred electrons and a midpoint potential of E_m = +260 mV ± 10 mV (see Fig. 1, *lower inset*).

P. Mathis (ed.), Photosynthesis: from Light to Biosphere, Vol. II, 95–98.

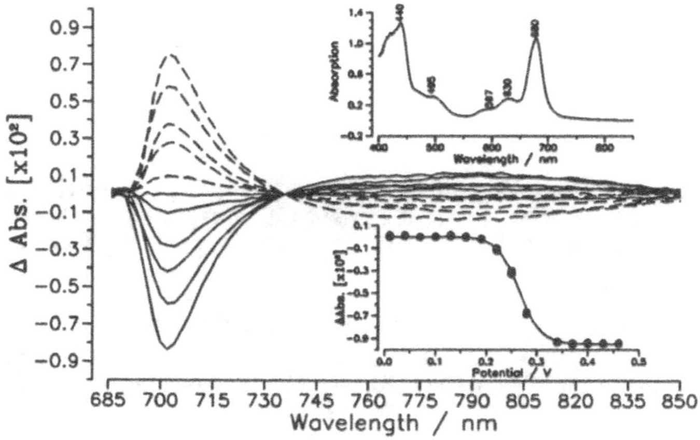

Fig. 1. Electrochemically-induced difference spectra obtained from a PS I monomeric sample shown in the *upper inset* (spectroelectrochemical cell, 10 μm pathlength, ~ 1.6 mM P700 in MES-buffer, pH 6.6 at T= 5 °C); differences calculated with respect to the reduced state at 0 mV (oxidation) and 460 mV (reduction). *Lower inset*: titration curve for the static difference spectra; *solid line*: Nernst curve with E_m= 260 mV and n=1 transferred electrons; *full circles*: oxidative titration; *open circles*: reductive titration.

Flash induced charge separation. The kinetic signal of flash induced charge separation and recombination (P700 ⇒ $P700^+$ ⇒ P700) for the monomeric complex is shown in Fig. 2 (*inset*). Application of increasingly oxidizing potentials leads to electrochemical oxidation of P700 and thus to a decrease of the photochemically induced signal. A plot of the amplitude of the flash induced potential at 707 nm *vs.* the applied potential (Fig. 2) was approximated with a Nernst curve with E_m= 260 mV ± 10 mV. Within the limits of error, identical titration curves were obtained for the monomeric and the trimeric complex (data not shown).

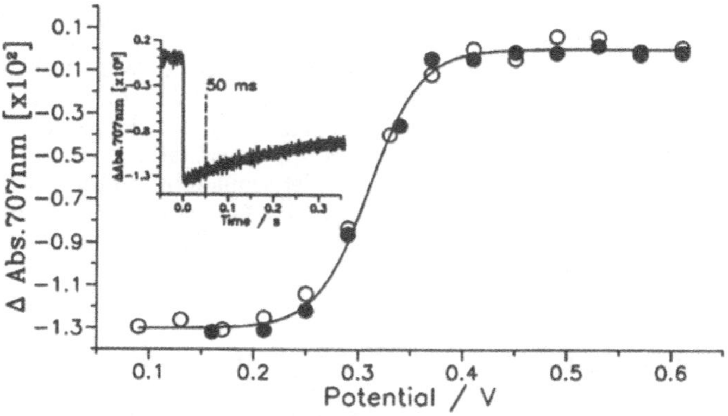

Fig. 2. Titration curve obtained for the flash induced kinetic traces from 0 to 520 mV; amplitudes were taken after 50 ms. *Full circles*: oxidative titration, *closed circles*: reductive titration; *solid line*: Nernst curve with E_m= 260 mV ± 10 mV, n= 1 transferred electrons. Conditions ca. 1.6 mM P700, MES buffer pH 6.6, T= 15 °C. *Inset*: Flash induced kinetic trace at 707 nm for fully reduced P700 at 0 mV.

<u>FTIR difference spectra.</u> The FTIR difference spectrum for the oxidation provides the most pertinent informationon changes upon redox transition in the 1750 to 1550 cm^{-1} region (see Fig. 3).

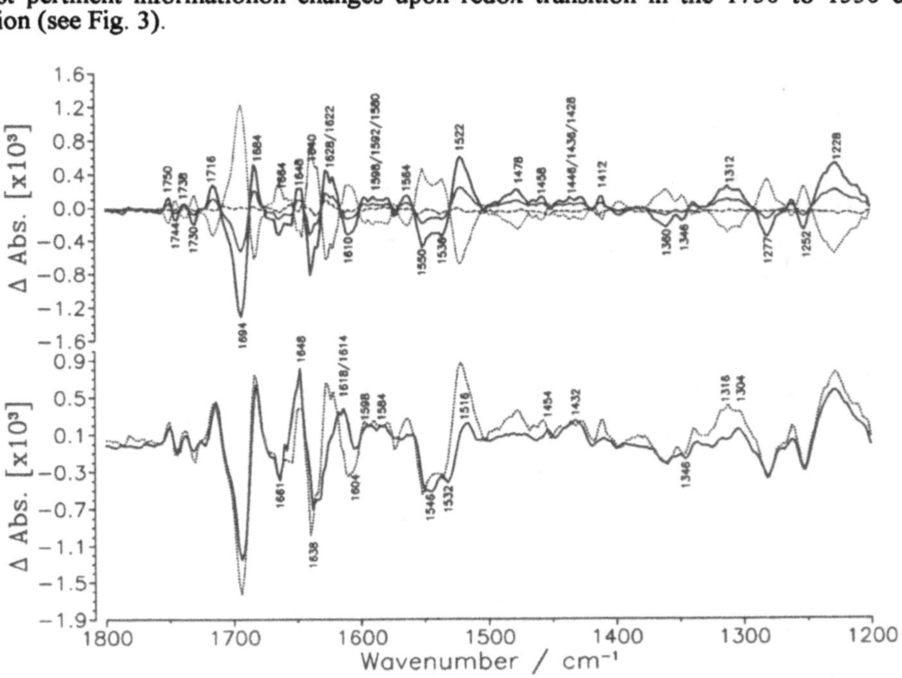

Fig. 3. FTIR difference spectra for the oxidation (*solid lines*) and the rereduction (*dotted lines*) of P700 from PS I monomeric complex (*upper Fig.)*; conditions as in Fig. 1, and the PS I trimeric complex (*lower Fig.)* in ^2H_2O-buffer (*solid line*) and ^1H_2O-buffer (*dashed line*); numbers refer to the band positions of the *solid line* spectra.

In the carbonyl stretching region, three sharp difference bands were obtained; in line with [5] the difference bands at 1750(+)/1744(-) cm^{-1} and 1738(+)/1730(-)cm^{-1} can be assigned to the two 10a ester C=O groups of the two Chl molecules comprising P700. Different spectral positions are indicating different protein environment: the 10 a ester C=O group at 1738(+)/1730(-)cm^{-1} has a weak interaction with its environment in the neutral and in the oxidized state, whereas the 10 a ester group at 1750(+)/1744(-) cm^{-1} seems to be unbonded in both cases. The difference band at 1716(+)/1694(-) cm^{-1} can be assigned to the 9 keto C=O group(s) of both Chl a molecules [6,7]. An alternative interpretation would be that both 9 keto C=O groups of P700 absorb at 1694 cm^{-1}, and that upon cation formation this band is split into a higher and a lower wavenumber component at 1716 and 1684 cm^{-1} [7], suggesting that the two Chl a molecules of P700 are energetically inequivalent but interacting in the ground state, while in $P700^{\bullet+}$ the electronical state is localized on either one of the two Chl molecules.

The region between ~1670 and 1635 cm^{-1} difference signals can be attributed to the amide I modes of the protein backbone; as expected for amide I modes $^1H_2O \rightarrow {}^2H_2O$ exchange leads to small shifts between 0 - 10 cm^{-1} [8] in this region (see Fig. 3, *lower part*).

In the region of the amide II modes at approx. 1550 cm^{-1} [8] is difficult to interpret because, in addition, absorption changes of the macrocycle (C=C/C-C modes) are expected [9]. Comparing the difference spectra of primary donor oxidation of PS I and bacterial reaction centers with chlorophyll model spectra in the 1600 to 1500 cm^{-1} region of macrocycle C=C/C-C vibrations leads to significant insight concerning the localization of the oxidized state. The difference spectrum of cation formation for monomeric Chl a in solution exhibits three bands at 1597(-), 1570(+) and 1549(-) cm^{-1} which show up at nearly identical positions in the P700/P700$^{\bullet+}$ spectra of PS I monomeric complex (see Fig. 3, *upper part*), suggesting the Chl monomer to be a useful model for P700$^{\bullet+}$ oxidation. However, in the case of P/P$^{\bullet+}$ spectra of the purple bacteria *Rps. viridis* and *Rb. sphaeroides* [3], the absorption of the C=C/C-C vibrations appears to be upshifted by approximately 30 cm^{-1} relative to the model spectra of monomeric BChl oxidation (BChl a$^{\bullet+}$/BChl a and BChl b$^{\bullet+}$/BChl b) in solution [10]. This was ascribed to an oxidized state delocalized on both macrocycles which form the primary donor [11], and suggests that monomeric Chl may be a poor model for description of the primary donor oxidation in this spectral region. Altogether, the positions of C=C/C-C bands which are sensitive to the distribution of the positive charge in the dimer suggest that the positive charge of the oxidized P700 is localized predominantly on only one of the two Chl molecules composing the primary donor of PS I.
The predomonant localization of the positive charge of P700$^{\bullet+}$ is also supported by FTIR difference spectra in the 3000 to 1800 cm^{-1} spectral range. In this region, a broad and strong low-energy electronic transition was detected for bacterial RC [11]. This broad absorbance feature was found to be characteristic for the native primary electron donor, and absent in the heterodimer mutant of RC. More general, the absence of this broad absorbance feature in the P700$^{\bullet+}$/P700 difference spectra can be taken as an additional proof for an asymmetric localization of the hole on P700.

3.References

[1] W. Mäntele, *Trends Biochem. Sci* **18**, 197 (1993).
[2] W. Mäntele, M. Leonhard, M. Bauscher, E. Nabedryk., J. Breton. and D. A. Moss, in *Reaction Centers of Photosynthetic Bacteria* (Edited by M.E. Michel-Beyerle), Vol. 6, p. 31, Springer-Verlag, Berlin (1990).
[3] M. Leonhard and W. Mäntele, *Biochemistry* **32**, 4532 (1993).
[4] B. A. Tavitian, E. Nabedryk, W. Mäntele and J. Breton, *FEBS Lett.* **201**, 151 (1986).
[5] E. Nabedryk, M. Leonhard, W. Mäntele and J. Breton, *Biochemistry* **29**, 3242 (1990).
[6] K. Okada, K. Uehara and Y. Ozaki, *Photochem. Photobiol.* **57,** 958 (1993).
[7] P. Moënne-Loccoz, B. Robert and M. Lutz, in *Techniques and New Developments in Photosynthetic Research* (Edited by J. Barber) NATO ASI Series, Plenum, New York (1989).
[8] J. L. R. Arrondo, A. Muga, J. Castresana and F. M. Goni, *Prog. Biophys. molec. Biol.* **59**, 23 (1993).
[9] M. Leonhard, PhD thesis, University of Freiburg, Germany (1993).
[10] W. Mäntele, in *The Photosynthetic Reaction Center* (Edited by J. Deisenhofer and J. R. Norris), Chap. 10, p. 239, Vol. II. Academic Press, New York (1993).
[11] J. Breton, E. Nabedryk and W. W. Parson, *Biochemistry* **31**, 7503 (1992).

THE OXIDIZING SITE OF PHOTOSYSTEM I MODULATES THE ELECTRON TRANSFER FROM PLASTOCYANIN TO P700$^+$

Michael Hippler[1], Friedel Drepper[2] and Wolfgang Haehnel.

Lehrstuhl für Biochemie der Pflanzen, Institut für Biologie II, Albert-Ludwigs-Universität, Schänzlestr. 1, D-79104, Freiburg, Germany; Present adress: [1]Département de Biologie Moléculaire, Université Genève, 30 Quai Ernest Ansemet; [2]Département de Biologie Cellulaire et Moléculaire, CEA-Saclay, 91191 Gif sur Yvette Cedex, France;

1. Introduction

The reaction between soluble plastocyanin (Pc) and the photosystem I (PSI) complex from spinach appears to be an attractive system for studying the mechanismen of interprotein electron transfer within complexes. Cross-linking of Pc to the PsaF subunit of PSI (1) in a functional conformation enabled a fast 13 - 15 µs-reduction of P700$^+$ (2). This halftime is comparable to that found in intact thylakoids (3). Recently Drepper et al. (4) analyzed the reaction between Pc and PSI in detail. They showed that the increase in redox potential by 50 mV of bound as compared to soluble Pc is consistent with the preferential binding of reduced as compared to oxidized Pc to PSI (4). Here we focuse on the effect of pH and temperature on the redox properties of soluble and bound Pc, repectively.

2. Material and Methods

Preparation of the materials. Solubilized PSI-particles (PSI-200) were isolated from spinach leaves according to Wynn and Malkin (1). Chlorophyll (Chl) concentrations were determined according to (5). Pc was isolated from spinach leaves as described by Ratajczak et al. (6). The concentration of the Pc solutions were determined spectroscopially using an extinction coefficient of 4.9 $mM^{-1}cm^{-1}$ at 597 nm for the oxidized form (7) and had an absorbance ratio A_{278}/A_{597} of 1.2-1.4. Cross-linked complexes between PSI-particles and Pc were obtained as described (8). *Kinetic measurements.* The absorbance change of P700 were measured on a single beam spectrophotometer essentially as described by Haehnel et al. (9). The buffers used were: Mops, pH 7-6.5; Mes, pH 6-5.5; Acetat, pH 5.2-4.5. The pH value of the samples with acetat buffers was finally adjusted with 1 N HCL. Flash-redoxtitration with Pc Cross-linked to PSI were done with following reaction mixture: PSI particles 95 µg Chl/ml, 100 mM buffer, 5 mM $MgCl_2$, 1 mM sodium ascorbate, 0.2 mM methylviologen, 0.1 mM diaminodurene (DAD), 0.05 % (w/v) Triton-X-100. *Redox potentiometry.* Anaeroic redox titration were carried out under argon gas in a cuvette containig 4 ml of sample with an optical pathlength of 10 mm. The following redox mediators were used: N,N,N',N'-tetramethy-p-phenylendiamine TMPD and parabenzoquinone PBQ at 5 µM, Ferri- and Ferrocyanid at 10 µM. For measurements of the redox potential a platinum electrode and a Ag/AgCl reference electrode (saturated KCL) were used. The electrode was calibrated using solutions of quinhydrone at different pH at 25°C.

3. Results and Discussion

pH-Dependence of the midpoint potential of Pc cross-linked to PSI

The determination of the redox midpoint potential of Pc bound to PSI was performed on PSI particles cross-linked to PSI. The amplitude of the fast reduction kinetic with a half time of 15 µs was measured at different redox potentials. The amplitude as a function of the redox potential was fitted to the Nernst's equation for the transfer of one electron. Fig. 1 shows the midpoint potential of cross-linked Pc and P700 as a function of pH. The differences between the Em values of PSI and Cross-linked Pc, remain constant until pH 4.75, which corresponds to a similar driving force for electron transfer, expressed by ΔG = -55 mV. The Em value of cross-linked Pc seems to increases sligthly from 421 mV to 438 mV at pH 4.5,

P. Mathis (ed.), Photosynthesis: from Light to Biosphere, Vol. II, 99–102.

while the redox potential of PSI is unaffected by the pH. For comparison we have also determined the redox potenial of soluble Pc as a function of pH. Consistent with previous data (7) the midpoint potential of Pc increases from 360 mV at pH 7 to 430 mV at pH 4.5. The intercept of the constant portion and the rising midpoint potential with the slope of about 58 mV/pH indicates a pK value of soluble Pc of 5.5.

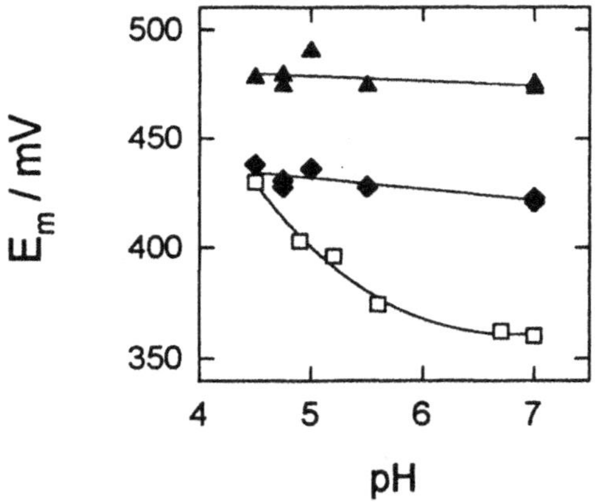

Figure 1: The dependence of the redox midpoint potentials of P700 (▲), Pc Cross-linked to PSI (♦) and Pc (□) on the pH. Pc was titrated according to its absorbance change at 597 nm.

The protonation of Pc has been shown to change the distorted tertahedral structure of the Cu(I) center in reduced Pc to a trigonal structure with a pK value of 5.6 or 4.9 as detected by NMR, because of protonation of the liganding N of His87. Such a orientation of the ligands is extremely unfavourable for Cu(II) and should lead to a redox-inactive Pc (10). The fact that the midpoint redox potential of Pc cross-linked to PSI and P700 remains almost constant until pH 4.5 indicates a shift of the pK value of Pc from pH 5.5 in soluble Pc down to 4.5 or even lower, when bound to PSI. Thus a binding of Pc has to result in a deprotonation of protonated soluble Pc at low pH values. We suggest that the underlying mechanismen could be described as a pK shift of His87 of Pc, preventing the formation of redox inactive Pc at these pH values. We further conclude that the interaction of the hydrophobic northern face involving His87 of Pc with a hydrophobic patch of PSI shifts its pK to a lower value (9).

Temperature dependence of the the reaction rate of Pc cross-linked to PSI

Fig. 2 shows the reaction rate of the electron transfer within the complex between Pc and PSI as a function of the temperature. Pc Cross-linked to PSI donates electrons to PSI with a halftime of about 14 µs at 304 K and 92 µs at 202 K. Apparently the reaction rate is not strongly dependend on temperature. The slope of the Arrhenius plot gives an activation energy of $E_A = 9$ kJ/mol. The standard change of activation enthalpy, obtained from Eyring analysis, gives a value of about 69 meV.

Furthermore the temperature dependences of the midpoint potential of soluble and bound Pc have been studied as shown in Fig. 3. The data are corrected for the temperature dependence of the reference electrode. While the slope of bound Pc is similar to that of P700 the slope with soluble Pc is considerably higher. The Em of P700 increased with decreasing temperature, resulting a standard entropy of reduction ($\Delta S°$) of -0.75 meV per degree,

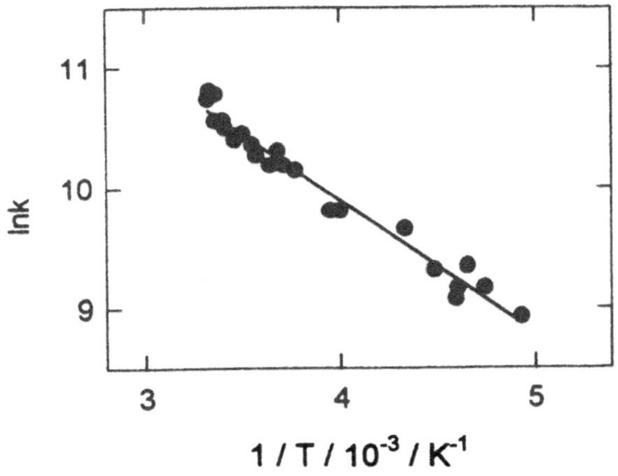

Figure 2: Temperature dependence of the fast reduction phase of Pc Cross-linked to PSI. The flash induced absorbance change was was detected at 703 nm. Every point is the result of one flash. The cross-linked PSI was suspended in measuring buffer and additionally 60 % (v/v) glycerol.

determined from the slope (Calculation of these values is straightforward from the relation $\Delta G° = \Delta H° - T\Delta S°$), and a standard enthalpy ($\Delta H°$) of -698 meV, given by the intercept (Em as T - 0 K). The Em of Pc cross-linked to PSI increased lineraly with decreasing temperature, yielding $\Delta S°$ of 0.64 meV per degree and $\Delta H°$ =-612 meV.

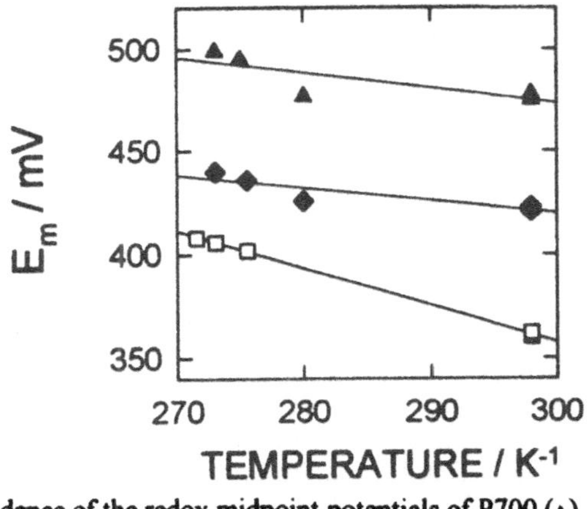

Figure 3: The dependence of the redox midpoint potentials of P700 (▲), Pc cross-linked to PSI (◆) and Pc (□) of temperature.

In contrast the Em of soluble Pc exhibited a more pronounced temperature dependence, resulting in thermodynamic parameters of $\Delta S°$ = 1.81 meV/K and $\Delta H°$ = -900 meV (Fig. 3). It is remarkable that the standard entropy of soluble Pc is more than 2 times higher than that of bound Pc.

The standard enthalpy change $\Delta H°$ of the electron transfer reaction of cross-linked Pc to PSI yields a value of about $\Delta H° = -86$ meV, if we imply that ΔH and ΔS are not temperature dependent. The reorganisation energy λ can be calculated by following equation (11)

$$E_A \approx \Delta H^{*\prime} \approx \frac{\lambda}{4} + \frac{\Delta H^{\circ\prime}}{2}\left[1 + \frac{\Delta G^{\circ\prime}}{\lambda}\right] - \frac{(\Delta G^{\circ\prime})^2}{4\lambda} \qquad 1$$

taking the values of ΔG=-55meV; $\Delta H°$=-86 meV and a change in activation enthalpy of about 69 meV a reorganisation energy of about λ = 418 meV can be estimated. This is a rather small value indicating that the complex provides conditions where a change in the oxidation state is coupled to a small reorganization of solvent molecules. This is consistent with a hydrophobic contact site in the complexes (9, 13) and supports the fast inter-protein electron transfer in spite of the relatively low driving force.

Moser and Dutton (12) developed an empirical approximation for intraprotein electron transfer that include a log relationship between the optimal rate log k_{et} and the edge-to-edge distance R with log k_{et}= 15-0.6R-3.1$(\Delta G-\lambda)^2/\lambda$; where ΔG is the free energy change and λ the reorganisation energy. Taking the reorganisation energy of 418 meV and a rate of log k_{et} = 4.66 an edge-to-edge distance of R = 14.5 Å can be calculated.

The nearest atom of the π-electron system of Pc to P700 is expected to be the Cu liganding $N\delta_1$ of His87, which is involved in the electron transfer route from the copper to P700 (9). Based on the structure of cyanobacterial PSI (13) and a structure of an algal Pc, Fromme et al. (14) proposed a docking site for Pc with a distance from the copper center of Pc to the center of the chlorophylls of P700 of about 20 Å (14). Since the distance of $N\delta_1$ of His87 to the copper is about 4.4 Å, and the distance of the last atom of the π-electron system of the chlorophyl dimer of P700 (measured from P of Rsp. *viridis*) to its center is about 4.4 Å our estimated edge-to-edge distance of 14.5 Å increases to about 23 Å, which is slightley greater than the distance obtained by Fromme et al. (14).

References

1 Wynn, R.M. and Malkin, R. (1988) Biochemistry 27, 5863-5869
2 Hippler, M., Ratajczak, R. and Haehnel W. (1989) FEBS Lett. 250, 280-284
3 Haehnel, W., Ratajczak, R. and Robenek, H. (1989) J. Cell Biol., 108, 1397-1405
4 Drepper, F., Hippler, M., Nitschke, W. and Haehnel, W. (1995) submitted.
5 Porra, R.J., Thompson, W.A. and Kriedemann, P.E. (1989) Biochim. Biophys. Acta, 975, 384-394
6 Ratajczak, R., Mitchell, R. and Haehnel, W. (1988) Biochim. Biophys. Acta 933, 306-318
7 Katoh, S., Shiratori I. and Takamija A. (1962) J. Biochem. 51, 32 - 40
8 Hippler (1994) Dissertation, Universität Freiburg, Germany
9 Haehnel, W., Jansen, T., Gause, K., Klösgen, R.B., Stahl, B., Michl, D., Huvermann, B., Karas, M. and Herrmann, R.G. (1994) EMBO J. 13, 1028 -1038
10 Guss, J.M., Harrowell, P.R., Norris, V.A. and Freemann H.C. (1986) J. Mol. Biol., 192, 361-387
11 Marcus and Suttin (1985) Biochim. Biophys. Acta, 811, 265-322
12 Moser and Dutton (1992) Biochim. Biophys. Acta, 1101, 171-176
13 Krauss N., Hinrichs W., Witt I., Fromme P., Pritzkow W., Dauter Z., Betzel C., Wilson K S., Witt H.T. and Saenger W. (1993) Nature 361, 326 - 331
14 Fromme, P.,Schubert, W.D and Krauss, N. (1994) Biochim. Biophys. Acta, 1187, 99-105

PERFUSION CHROMATOGRAPHY AS A MEANS FOR VERY QUICK ISOLATION OF THE PHOTOSYSTEM I CORE COMPLEX AND LIGHT-HARVESTING I SUBCOMPLEXES

Staffan E. Tjus[1], Margrit Roobol-Boza[1], Lars-Olof Pålsson[2] and Bertil Andersson[1]
[1]Dept. of Biochemistry, Stockholm University, S-10691 Sweden.
[2]Dept. of Physical Chemistry, University of Umeå, S-90187 Sweden.

1. Introduction.

Photosystem I is a supra-molecular enzyme complex mediating electron transport from Plastocyanin to NADP. It has in higher plants and cyanobacteria a core complex constituted by at least 14 polypeptides ranging between 84 and 3 kDa (PsaA-N) while higher plant photosystem I also contains a peripheral light-harvesting complex (LHC I) (1-2). The core complex carries all the cofactors needed for electron transport and the structure of a cyanobacterial photosystem I core complex was recently determined at 6 Å resolution (3). The core complex also harbours an inner chlorophyll *a* antenna that in higher plants together with the LHC I *a/b* antenna very rapidly and efficiently captures solar energy (4). LHC I is constituted by four polypeptides, Lhca1-4 with molecular weights around 20-24 kDa (5).

Native photosystem I of higher plants has been fractionated into the core complex and LHC I (6). Further subfractionation of LHC I has also been performed yielding LHC I-680 (Lhca2+3) and LHC I-730 (Lhca1+4) (7-8) and LHC I-680A (Lhca3) and LHC I-680B (Lhca2) (9). However, all these traditional methods employ either very time-consuming sucrose gradient centrifugation or the use of highly denaturing SDS during "green gel electrophoresis".

To overcome these earlier obstacles we have designed a very rapid method to isolate the photosystem I core complex and various LHC I subcomplexes (10) by the use of anion-exchange perfusion chromatography (11-15). This technique takes advantage of a new type of chromatographic bead with pores inside (POROS) that at once both enlarges the reactive surface area and permits a highly alleviated elution flow.

2. Materials and Methods

2.1 *Isolation procedures*

Spinach was used to isolate thylakoid membranes as in (16) and from these, native photosystem I used for further fractionation was isolated by solubilisation with Triton X-100 and sucrose density gradient centrifugation as in (17).

Native photosystem I in a buffer containing 50 mM sorbitol and 5 mM EDTA (pH 7.5) at 0.5 mg chlorophyll/ml was so solubilised with ß-D-dodecyl maltoside (2.5 mg/ml) and Zwittergent-16 (2.8 mg/ml) for 80 min at 10°C and slow stirring. Solubilised material corresponding to 1 mg chlorophyll was then loaded to a 1 ml anion-exchange 20 POROS Q column (PerSeptive Biosystems Inc., USA) combined with a FPLC system (Pharmacia, Sweden). The column was equilibrated and washed with 50 mM Tricine (pH

P. Mathis (ed.), Photosynthesis: from Light to Biosphere, Vol. II, 103–106.

8.1), 0.03% dodecyl maltoside and 0.02% Zwittergent and eluted with a 0-400 mM NaCl gradient.

2.2 *Analytical methods*
proteins were identified by SDS-PAGE combined with western blotting. Assay of 77 K fluorescence emission and determination of fluorescence yield was performed as in (4). Photosystem I electron transport from ascorbate/DCIP to methylviologen was measured as oxygen consumption with an oxygen electrode. Chlorophyll was determined as in (18)

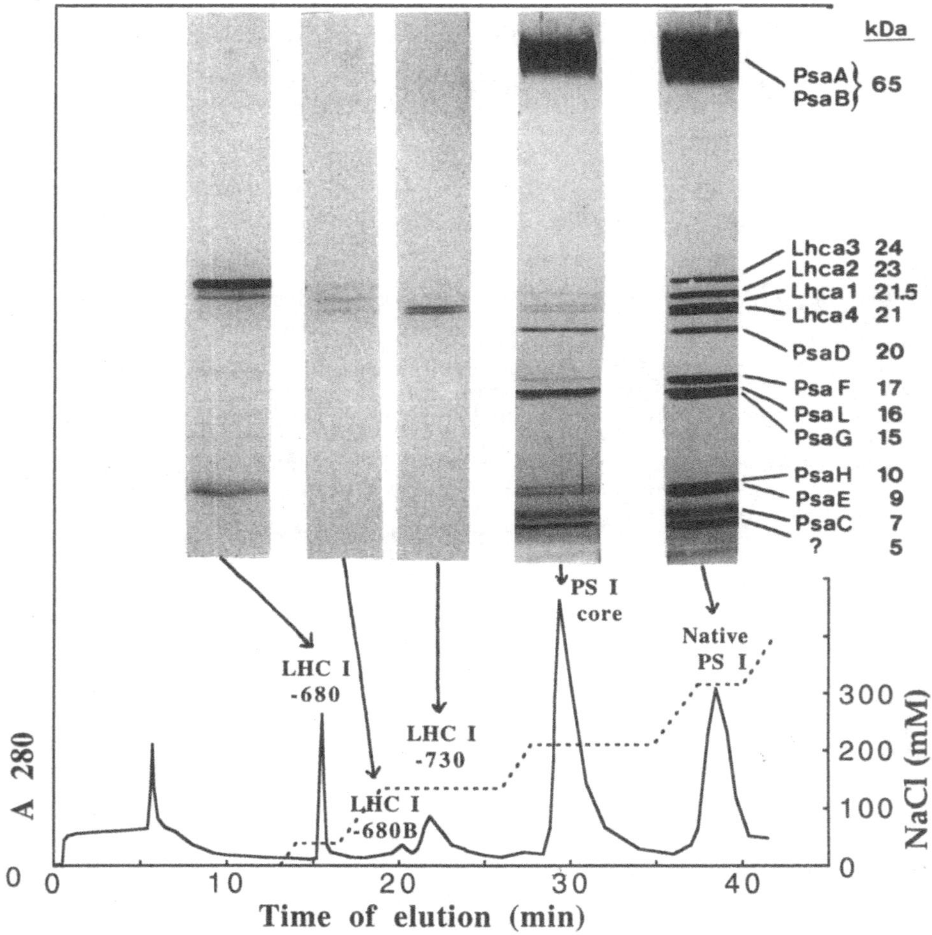

Figure 1. Elution profile of native photosystem I solubilised witt dodecyl maltoside and Zwittergent and subjected to anion-exchange perfusion chromatography. Inset is the SDS-PAGE profile of the five chlorophyll-protein fractions.

3. Results

NaCl-gradient elution of the POROS Q anion-exchange column, loaded with native photosystem I solubilised with dodecyl maltoside and Zwittergent, rendered five chlorophyll-protein peaks within 40 min (Fig.1).

SDS-PAGE and 77 K fluorescence emission spectra (Fig.1) identified these fractions as;

(1) **LHC I-680** isolated after only 15 min at 40 mM NaCl.
(2) **LHC I-730** eluted after 22 min at 135 mM NaCl.
(3) In between, also at 135 mM salt, a small peak enriched in **LHC I-680B**.
(4) The **photosystem I core complex** after 30 min at 210 mM NaCl.
(5) and **unsolubilised native photosystem I** after 40 min at 315 mM NaCl.

The LHC I subfractions all showed a chlorophyll *a/b* ratio around 2-3 while the core fraction was nearly devoid of chlorophyll *b* and the native photosystem I displayed a chlorophyll *a/b* ratio of 7.4.

The isolated native photosystem I consumed oxygen at 915 µmoles O_2 mg (Chl h)$^{-1}$ and the core complex at 1535 µmoles O_2 mg (Chl h)$^{-1}$ while their fluorescence quantum yields were 0.6% and 0.4% respectively.

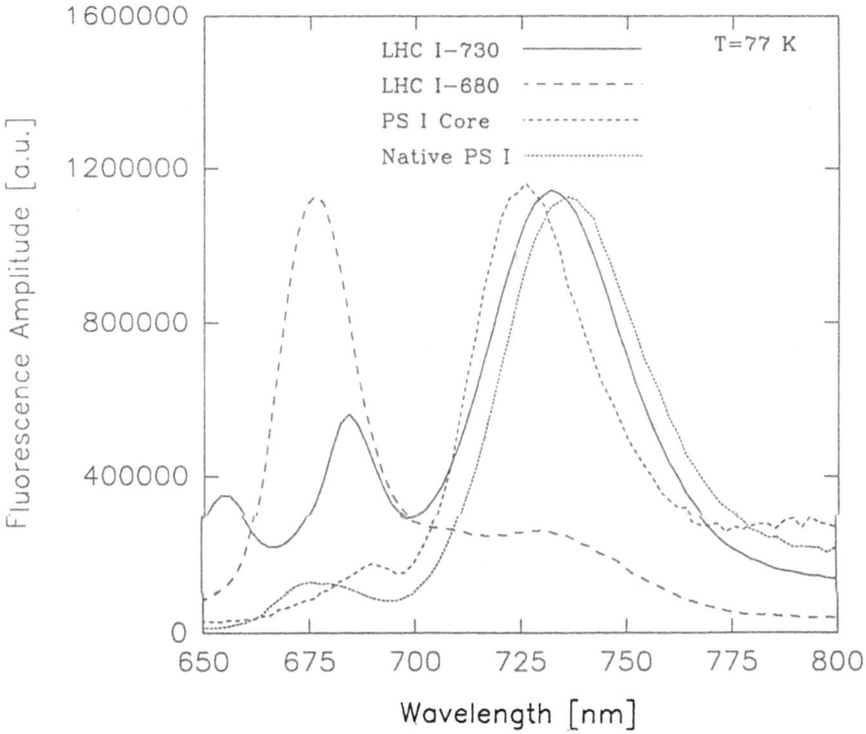

Figure 2. Fluorescence emission spectra at 77 K of eluted chlorophyll-protein fractions shown in figure 1. Excitation was at 435 nm and samples were kept in 60% glycerol

4. Discussion

The above subfractionation pattern show of a very rapid isolation; The photosystem I core complex was obtained in 30 minutes compared to ≈20 h using density gradient centrifugation and even more significant the LHC I subfractions were obtained in as short as ≈15-20 min instead of traditional ≈50 h ultracentrifugation.

All fractions showed typical low temperature fluorescence maxima and the reaction centre containing fractions high electron transport and low fluorescence yields. This altogether demonstrates a very swift isolation procedure yielding chlorophyll-binding photosystem I fractions in highly intact states

This novel method should be generally applicable for fast separation of photosynthetic membrane proteins and specifically useful for further studies of photosystem I. For example our groups have recently identified ultrafast energy hopping of ≈200 femtoseconds between chlorophylls in LHC I (18) which findings are due to be further consolidated with these intact prepararations. Furthermore, experiments are in advanced progress to rapidly isolate also our present starting material, native photosystem I, directly from solubilised thylakoids.

References

1 Golbeck, J.H. (1992) Annu. Rev. Plant Physiol. Plant Mol. Biol. 43, 293-324
2 Andersen, B. and Scheller, H.V. (1993). in Pigment-Protein Complexes in Plastids: Synthesis and Assembly (Sundqvist, C. and Ryberg, M., eds.), pp. 383-418
3 Krauss, N., Hinrichs, W., Witt, I., Fromme, P., Pritzkow, W., Dauter, Z., Betzel, C., Wilson,K.S., Witt, H.T. and Saenger, W. (1993) Nature 361, 326-331
4 Pålsson, L.O., Tjus, S.E., Andersson, B. and Gillbro, T. (1995) Chem. Phys.194, 291-302
5 Jansson, S. (1994) Biochim. Biophys. Acta 1184, 1-19
6 Haworth, P., Watson, J.L. and Arntzen, C.J. (1983) Biochim. Biophys. Acta 724, 151-158
7 Lam, E., Ortiz, W. and Malkin, R. (1984) FEBS Lett. 168, 10-14
8 Bassi, R. and Simpson, D. (1987) Eur. J. Biochem. 163, 221-230
9 Knoetzel, J., Svendsen, I. and Simpson, D.J. (1992) Eur. J. Biochem. 206, 209-215
10 Tjus, S.E., Roobol-Boza, M., Pålsson, L.O. and Andersson, B. (1995) Photosynth. Res., in press
11 Afeyan, N.B., Fulton, S.P., Gordon, N.F., Mazsaroff, I., Varady, L. and Regnier, F.E. (1990) BIO/TECHNOLOGY 8, 203-206
12 Afeyan, N.B. and Fulton, S.P. (1991) J. Chromatogr. 544, 267-279
13 Fulton, S.P., Shahidi, A.J., Gordon, N.F. and Afeyan, N.B. (1992) BIO/TECH.NOLOGY 10, 635-639
14 Lehman, E.D., Joyce, J.G., Freymeyer, D.K., Bailey, J., Herber, W.K. and Miller, W.J. (1993) BIO/TECHNOLOGY 11, 207-212
15 Regnier, FE. (1991) Nature 350, 634-635
16 Andersson, B., Åkerlund, H.-E. and Albertsson, P.Å. (1976) Biochim. Biophys. Acta 423, 122-132
17 Mullet, J.-E., Burke, J.J. and Arntzen, C.J. (1980) Plant Physiol. 65, 814 -822
18 Arnon, D.I. (1949) Plant Physiol. 24, 1-14
19 Pålsson, L.O., Tjus, S.E., Andersson, B. and Gillbro, T. (1995) Biochim. Biophys. Acta 1230, 1-9

Temperature Dependence of the Reoxidation Kinetics of A_1^- in PS I

E. Schlodder[1], K. Brettel[2], K. Falkenberg[1] and M. Gergeleit[1]
[1]Max-Volmer-Institut für Biophys. Chemie, TU-Berlin, Str. des 17. Juni 135, 10623 Berlin, Germany, [2]Section de Bioénergétique (CNRS, URA 1290), DBCM, CEA-Saclay, 91191 Gif-sur-Yvette Cedex,

Introduction

Photosystem I (PS I) is a membrane protein complex, consisting of at least 11 subunits. PSI mediates the light-driven electron transfer from reduced plastocyanin or cytochrome c_6 to ferredoxin. The PS I complex contains two large subunits (A, B of about 83 kDa), which bind about 100 antenna chlorophyll *a* molecules and the following redox centers: P700 (presumably a chlorophyll a dimer), A_o (a monomeric chlorophyll a), A_1 (a phylloquinone) and F_X (a [4Fe-4S] iron-sulfur-cluster). The terminal electron acceptors F_A and F_B (two further [4Fe-4S] iron-sulfur-clusters) are both coordinated by subunit C. The structure of PS I has been determined at 4.5 Å resolution in a recent X-ray crystallographic study using crystals of trimeric PS I of Synechococcus elongatus [1].
After excitation of PS I by light, the primary radical pair $P700^+A_o^-$ is formed. Subsequent charge stabilization is achieved by electron transfer from A_o^- to the secondary acceptor A_1. At room temperature, the reoxidation of A_1^- occurs with a half-life of 200 ns in PS I complexes from Synechococcus [2], whereas in PS I from spinach A_1^- decays biphasically with $t_{1/2} \approx 25$ ns and 150 ns and relative amplitudes depending on the preparation [3]. Recent studies provided direct evidence that the electron from A_1^- is transferred to F_X [4-6]. At 10 K, electron transfer to the iron-sulfur-clusters has been reported to be blocked and replaced by $P700^+A_1^-$ charge recombination in a fraction of PS I [7]. This indicates a dramatic change in the electron transfer kinetics and the appearance of heterogeneity with decreasing temperature..
In this work, we studied the temperature dependence of the decay kinetics of A_1^- and $P700^+$ throughout the full range from 300 to 77 K.

Materials and Methods

Trimeric PS I complexes with about 100 Chl/P700 were isolated from thermophilic cyanobacteria Synechococcus elongatus according to Ref. [8]. The PS I complexes were purified by sucrose density gradient centrifugation and by anion exchange chromatography. For low temperature measurements, the samples were diluted with glycerol to a final glycerol concentration of 65% (v/v) unless otherwise stated. Before freezing in a variable temperature cryostat, 5 mM sodium ascorbate and 500 µM DPIP were added. Glycerol was omitted at room temperature and 100 µM benzyl viologen was additionally added.
Absorbance changes were measured as described previously [2,4]. The samples were excited by laser flashes of 3 ns pulse duration at 532 nm or by Xe-flashes (15 µs duration). The time course of the flash-induced absorbance changes was fitted to a (multi)exponential decay using an algorithm that minimizes the sum of the unweighted least squares.

Results and Discussion

Flash-induced absorbance changes at 385 nm have been measured at different temperatures between 300 K and 77 K in order to analyze the temperature dependence of the

P. Mathis (ed.), Photosynthesis: from Light to Biosphere, Vol. II, 107–110.

reoxidation kinetics of A_1^-. At this wavelength, the absorbance changes are predominantly due to the formation and the decay of A_1^- [2]. Fig. 1 shows the time course of the absorbance changes at selected temperatures. At room temperature we observe a mono-exponential decay with a half-life of 180 ns (upper trace in Fig. 1), in good agreement with our previous measurements [2,4]. The rate of A_1^- reoxidation is markedly slowed down with decreasing temperature. The lower the temperature, the longer the time scale that has to be used in order to follow the decay of the flash-induced absorbance increase. At 220 K, the single exponential fit (solid line) yields a half-life of 4.5 µs. Below 200 K, two exponentials are required for a satisfactory fit. This is evident from the decay at 180 K. The following half-lives and relative amplitudes were obtained by the fit: 27 µs (46%) and 211 µs (54%). If the temperature is lowered further from 180 K to 77 K, the relative amplitude of the fast phase decreases to zero, whereas that of the slow phase increases. The half-life of the slow phase changes only slightly. At 77 K, the signal decays with a half-life of about 170 µs. Based on the difference spectrum of the slower phase in the range between 300 and 500 nm (not shown) and measurements at 826 nm monitoring the decay of $P700^+$ (see Fig. 3), we attribute the slow phase to the charge recombination of $P700^+A_1^-$ in PS I complexes, in which the electron transfer to F_X is blocked. The fast phase is due to PS I centers, in which the forward electron transfer (rate k_f) competes with the charge recombination (rate k_r). In Fig. 2, the rate of the fast phase of A_1^- reoxidation is plotted vs. the reciprocal of temperature. In the range between 310 and 190 K, the observed rate ($=k_f + k_r$) is considerably larger than k_r ($\Rightarrow k_f >> k_r$) so that the data represent essentially forward electron transfer.

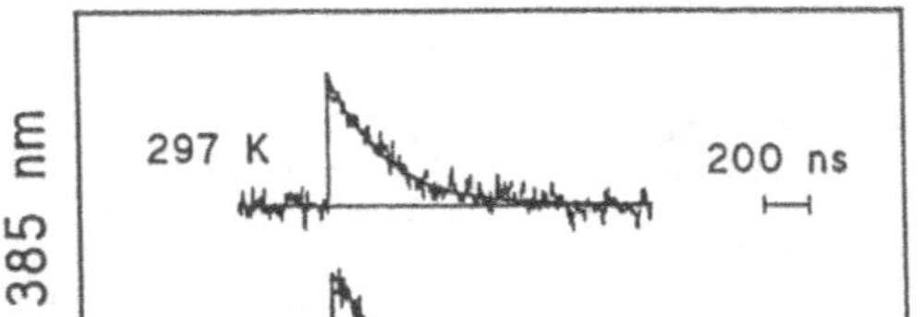

Fig.1: Flash-induced absorbance changes at 385 nm attributed to the formation and decay of A_1^- in PS I complexes. The signals (between 64 and 512 averages) are normalized to their initial amplitudes. At 297 K, the initial amplitude corresponds to a differential molar extinction coefficient of ≈8000 $M^{-1}cm^{-1}$. At lower temperatures, the amplitude depends on the excitation rate, since the state $P700^+F_{A/B}^-$ accumulates to some extent due to its long life-time.

By using the Arrhenius equation, an activation energy of 220 meV ± 20 meV has been determined is this range for the electron transfer from A_1^- to F_X. In terms of nonadiabatic electron transfer theory, the relatively large activation energy can be explained when the standard Gibbs free energy of reaction ΔG^0 is considerably smaller than the reorganization energy λ. As a further consequence of the large activation energy (i.e. small Franck-Condon factor), the electronic coupling has to be large in order to explain the high rates for the electron transfer from A_1^- to F_X, thus the distance between A_1 and F_X has to be small. Using an expression for the rate with a single vibration coupled to the electron transfer (eqn. 85 in Ref. [9]), an activated temperature dependence is expected in a range, where 2kT is larger or comparable to the energy $\hbar\omega$ of the characteristic frequency. Therefore, $\hbar\omega$ should be <50 meV. The experimental data can be described e.g. with the following parameters: ΔG^o = -54 meV; $\hbar\omega$ = 38 meV; λ = 1.2 eV and $2\pi/\hbar \cdot V_R^2 = 6 \cdot 10^{10}$ eVs^{-1} (V_R= electronic coupling). Although the data do not allow an unequivocal determination of the parameters, possible values of λ were in all cases > 1 eV.

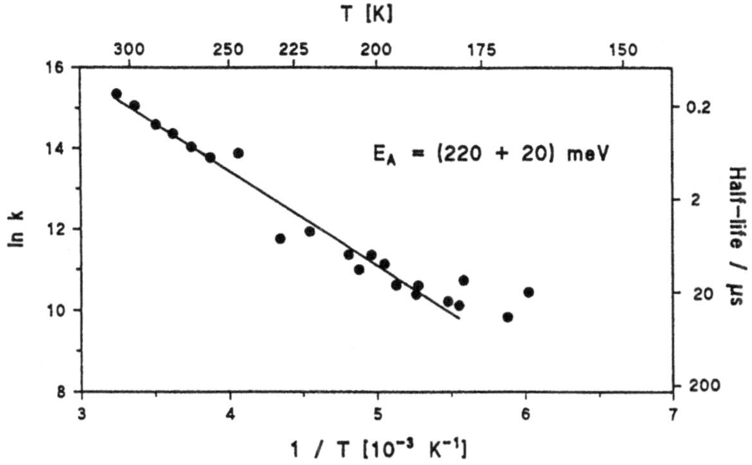

Fig. 2: Arrhenius plot of the decay constant of flash-induced absorbance changes at 385 nm attributed to the forward electron transfer from A_1^- to F_X.

In order to estimate the distance between A_1 and F_X, an empirical relation has been used that gives the free-energy optimized rate k_{op} as a function of the edge-to-edge distance R: $k_{op} = 10^{13} s^{-1} \cdot \exp(-\beta(R-R_o)$ with β=1.4 Å^{-1} and R_o = 3.6 Å representing van der Waals contact [10]. With $k_{op} = 9 \cdot 10^{10}$ s^{-1} (calculated for $-\Delta G^o = \lambda$ = 1.2 eV, T = 295 K and the other parameters as above) we obtain an edge-to-edge distance of R ≈ 7 Å. Unfortunately, the position of A_1 can not be identified unambigiously at the present state of the structure analysis of PS I [1].

Fig.3a shows absorbance changes at 826 nm induced by the first flash given to PS I complexes frozen in the dark. In accordance with the 385 nm measurements, a decay component with $t_{1/2}$ = (190 ± 30) µs appeares below 200 K, due to the charge recombination of $P700^+A_1^-$. In paralell, the amplitude of the slow component, indicating the yield of $P700^+FeS^-$ - formation, decreases. This can not be explained by the competition between forward electron transfer and charge recombination (see above), but is obviously caused by an inhibition of forward electron transfer in a fraction of the PS I complexes (≈ 50%). This inhibition occurs in a narrow temperature interval centered around a transition temperature $T_{50\%}$. For 65% glycerol $T_{50\%}$ = 168 K was found (see Fig.3b). Interestingly enough, the use of different glass-forming solvents or different concentrations of the cryoprotectants results in significant changes of the transition temperature (e.g. 153 K for50% and 200 K for 85% glycerol).

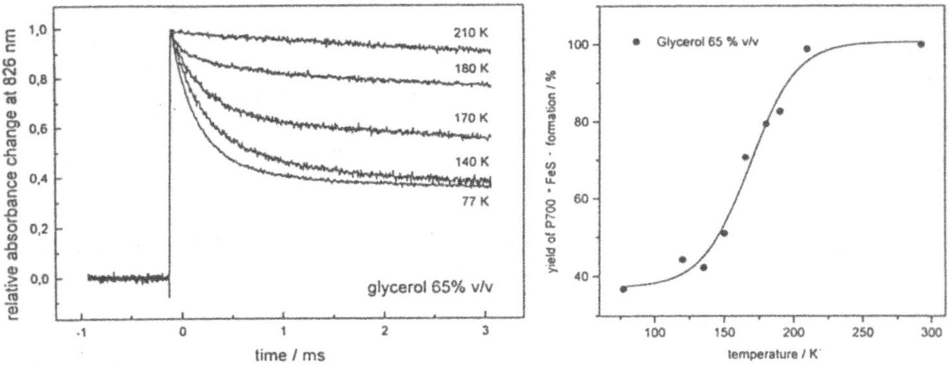

Fig.3: left) Flash-induced absorbance changes at 826 nm measured with PS I complexes in the presence of 65% v/v glycerol and 5 mM ascorbate. The initial amplitude corresponds to $\Delta A = 1.4 \cdot 10^{-3}$. **right)** Efficiency of $P700^{+}FeS^{-}$-formation as function of temperature. The solid line represents the fit using a logistic sigmoid.

In all cases, three different fractions of PS I complexes can be distinguished at low temperature (T < 100 K) according to their capability to perform charge separation up to $F_{A/B}$, F_X or only up to A_1 as described above. The state $P700^{+}F_{A/B}^{-}$, wich is stable at low temperatures, can be induced in 35% ± 10% of the PS I centers by illumination with several flashes.

The heterogeneity of the low temperature photochemistry may reflect PS I complexes frozen in different conformational substates. The solvent effects suggest that the inhibition of the forward electron transfer to F_X is caused by the extreme increase of the viscosity up to > 10^{13} Poise due to the glass transition. This can be explained by the assumption that the dynamics of the protein is strongly influenced by the solvent.

Acknowledgement

This work was supported by grants from the DFG (Sfb 312, A5) and from the DAAD (Procope).

References

1 Schubert, W.D., Klukas, O.,Krauß, N., Saenger, W., Fromme, P. and Witt, H.T. (1995) these proceedings

2 Brettel, K. (1988) FEBS Lett. 239, 93-98

3 Sétif, P. and Brettel, K. (1993) Biochemistry 32, 7846-7854

4 Lüneberg, J., Fromme, P. ,Jekow, P. and Schlodder, E. (1994) FEBS Lett. 338, 197- 202

5 van der Est, A., Bock, C., Golbeck, J., Brettel, K., Sétif, P. and Stehlik, D. (1994) Biochemistry 33, 11789--11797

6 Moënne-Loccoz, P., Heathcote, P., Maclachlan, D.J., Berry, M.C., Davis, I.H. and Evans, M.C.W. (1994) Biochemistry 33, 10037-10042

7 Sétif, P., Mathis, P. and Vänngård,T. (1984) Biochim. Biophys. Acta 767, 404-414

8 Witt, I., Witt, H.T., Gerken, S., Saenger, W., Dekker, J.P. and Rögner, M., (1987) FEBS Lett. 221, 260-264

9 Devault, D. (1980) Q. Rev. Biophys. 13, 387-978

10 Moser, C.C., Keske, J.M., Warncke, K., Farid, R.S. and Dutton, P.L. (1992) Nature 355, 796-802

INTERACTION OF THE SUBUNIT PsaC WITH ITS BINDING SITE ON THE PS I CORE HETERODIMER

J. Biggins, S. M. Rodday and L. Do.
Department of Molecular Biology, Cell Biology and Biochemistry
Brown University, Providence, RI 02912, USA.

1. Introduction

The terminal steps of electron transfer in PS I are mediated by the three 4Fe4S clusters F_X, F_A and F_B (1). We have proposed that a surface-exposed domain in the immediate vicinity of the F_X cluster on the PS I core heterodimer is involved in the binding of the PsaC subunit bearing the clusters F_A and F_B (2). The putative binding site comprises the two interhelical loops -PCDGPGRGGTC- contributed by PsaA and PsaB and contains the four cysteine ligands of the F_X cluster. The model has been tested experimentally by chemical modification and proteolysis of the F_X-core (2), and site directed mutagenesis of residues in the interhelical loop of PsaB in both *Synechocystis* (3) and *Chlamydomonas* (4). The results were consistent with the proposed model and indicated the importance of electrostatic interactions between one or both arginine residues in the putative core binding site and negatively charged domains on the PsaC subunit. We now extend our experiments to identify the specific carboxylate-containing residues on the surface of PsaC that are implicated in the subunit interaction with the core heterodimer .

2. Model for the PsaC-core interaction

Computer-generated models of the binding of PsaC on the PS I core were constructed using the 1.8 Å crystal structure of *Peptococcus aerogenes* ferredoxin (5) as a simulator for PsaC. Although the *P. aerogenes* Fd has only modest amino acid residue homology with PsaC it does possess two 4Fe4S clusters separated by 12 Å with ligand coordination identical to the F_A F_B clusters on the PS I subunit (1, 6).

9 --**F_B**-- 24 27 32
PsaC MSHSVKIY**D**TCIGCTQCVRACPL**D**VL**E**MVPW**D**GCKAG
P.aer AYVIN**D**SCIACGACLP**E**CPVNIQQGSIYAI**D**A**D**S
6 17 31 33

4647 61 72
QIAASPRT**ED**CVGCKRCETACPT**D**FLSIRVYLGA**E**TTRSMGLAY
CI**D**CGSCASVCPVGAPNP**ED**
37 5354
--**F_A**--

P. Mathis (ed.), Photosynthesis: from Light to Biosphere, Vol. II, 111–114.

The crystal structure of the simulator Fd (5) was aligned into the model of the putative core binding site (2) using the coordinates for the three FeS centers from the 6 Å PS I crystal structure (6) in two orientations corresponding to either F_A or F_B proximal to F_X. Rotations of the Fd were then performed within the binding site around the 22 Å and 12 Å axes and an exhaustive search was conducted for alignment between the two interhelical loop arginine residues on the binding site and negatively charged domains on the simulator subunit. *Figure 1* shows an example of one favorable interaction in the case of the cluster F_B proximal to F_X. *Table 1* below lists all subunit domains predicted to interact favorably for both F_A or F_B proximal to F_X.

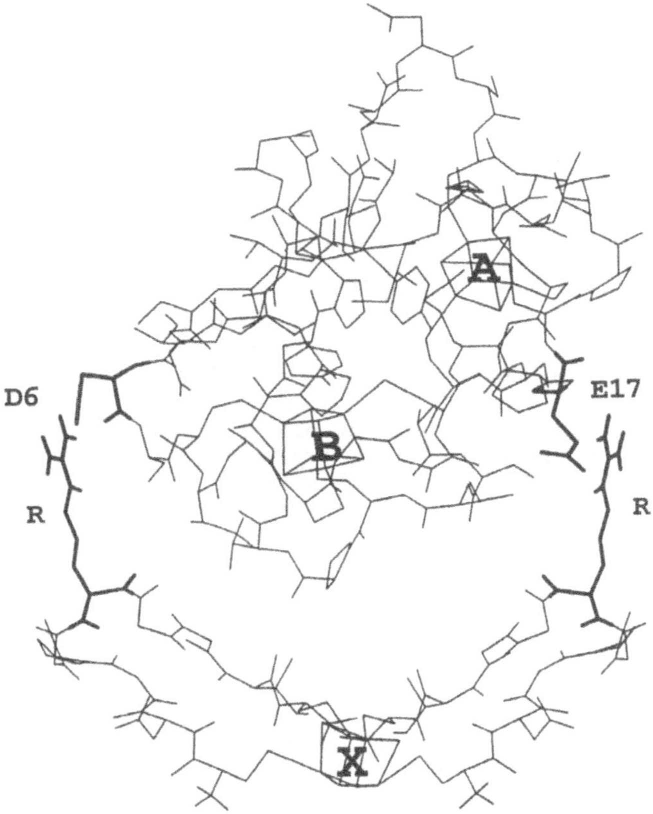

Figure 1: Molecular model showing the *P. aerogenes* Fd (5) as a simulator for PsaC in the putative core binding site in the F_X domain (2). The coordinates for the three FeS clusters were obtained from the 6 Å PS I crystal structure (6). The example shows F_B proximal to F_X and the Fd residues D6 and E17 aligning within 4 Å with the two arginine residues on the putative core binding site. All combinations are presented in *Table 1*.

Table 1: Results of molecular modeling *P. aerogenes* Fd (5) in the putative PS 1 core binding site (2) as shown in *Figure 1*. The degree of interaction for all combinations of the five negatively charged domains on the subunit aligning with the two positively charged arginines in the binding site are presented for both F_A and F_B proximal to F_X. Refer to the sequence alignment above for equivalent domains in PsaC.

Subunit domains			Interaction with loop arginines: Proximity to F_X — F_A	F_B
D6	-	E17	--	++
D6	-	D31D33	--	++
D6	-	D37	-	+
D6	-	E53D54	--	+
E17	-	D31D33	++	+
E17	-	D37	++	++
E17	-	E53D54	+	++
D31D33	-	E53D54	-	+
D31D33	-	D37	--	--
D37	-	E53D54	--	--

[++, very favorable; +, favorable; -, unlikely; --, very unlikely]

3. Experimental

Seven structural changes were introduced into PsaC to test the functional interaction of the subunit with the PS 1 core by in vitro cluster insertion of F_A and F_B, and reconstitution of electron transfer. The residues D9, D24, E27, D32, E46/E47, D61 and E72 were all changed to arginines by oligonucleotide-directed mutagenesis and the seven mutants were cloned into the *Afl* 111-*Bam H* 1 site in pEt-3d in the *E. coli* BL21DE3 plysS expression host for overproduction of the mutant apoproteins. Wild type PsaC and mutant apoproteins were purified from inclusion bodies (7) solubilized in 7 M urea, 50 mM tris-Cl pH 8.3 and 10 mM dithiothreitol using a Sephadex G-75 sizing column to remove large molecular mass contaminants and target protein aggregates from PsaC monomers.

4. Results

Preliminary results are shown in *Figure 2* below where the D9R and E72R mutant polypeptides are compared with wild type PsaC in their ability to reconstitute PS 1-F_X cores in vitro. The data presented show the % reconstitution of electron transfer to the FeS centers F_A and F_B, measured by flash transient spectrophotometry (2), as a function of protein concentration (3, 4). It can be seen that the wild type PsaC and E72R mutant are virtually identical and 100 % reconstitution of the PS 1-FX cores can be achieved at a ratio of ca 50 mole protein/mole P700. In contrast the D9R mutant protein shows a much lower level of reconstitution and no greater than 70 % could be achieved even at very high protein concentrations. This indicates that the D9R mutant PsaC is either incapable of insertion of functional F_A and F_B redox centers or, more likely, has an impaired interaction with its binding site on the core heterodimer. The former possibility is presently under investigation.

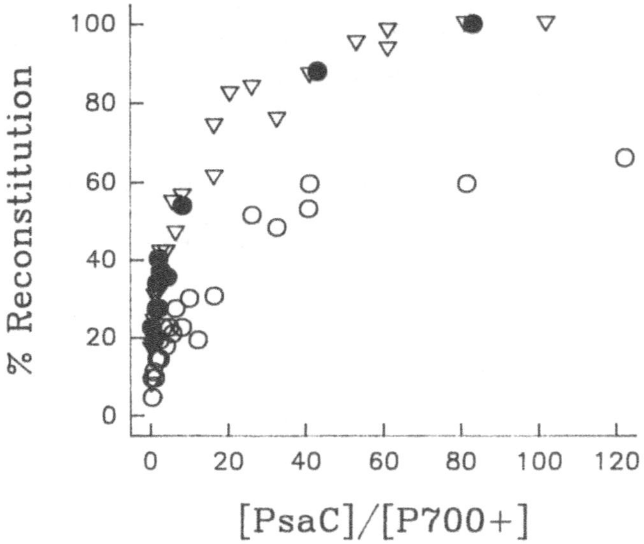

Figure 2: In vitro reconstitution of PS I-F_X cores by wild type PsaC (∇), mutant D9R (O) and mutant E72R (●). Restoration of full electron transfer in the reaction center was measured by flash spectrophotometry (2).

5. Conclusion

Mutant E72R is functionally identical to wild type PsaC whereas mutant D9R fails to fully reconstitute the F_X core. Therefore, we provisionally suggest that the PsaC domain containing the residue 9R is unable to interact favorably with its contact zone on the binding site on the PS I core heterodimer. We conclude that the D9 domain of PsaC interacts with a binding site loop arginine of PsaA or PsaB in support of the model proposed.

6. References

1 Golbeck, J. H. (1992) Annu. Rev. Plant Physiol. Plant Mol. Biol. 43, 293-324
2 Rodday, S. M., S-S Jun and Biggins, J. (1993) Photosynth. Res. 36, 1-9
3 Rodday, S. M., Schulz, R., McIntosh, L and Biggins, J. (1994) Photosynth. Res. 42, 185-190
4 Rodday, S. M., Webber, A., N., Bingham, S., E. and Biggins, J. (1995) Biochemistry 34, 6328-6334
5 Adman, E., T., Sieker, L., C. and Jensen, L., H. (1976) J. Biol. Chem. 251, 3801-3805
6 Krauss, N., Hinrichs, W., Witt, I., Fromme, P., Pritzkow, W., Dauter, Z., Betzel, C., Wilson, K., S., Witt, H., T. and Saenger, W. (1993) Nature 361, 326-331
7 Zhao, J., Warren, P., V., Li, N., Bryant, D., A. and Golbeck, J., H. FEBS Lett. 276, 175-180

QUINONE EXCHANGE AT THE A_1 SITE IN PHOTOSYSTEM I [PSI]

A. Barkoff[1], N. Brunkan[1], S. W. Snyder[1], A. Ostafin[1], M. Werst[1], J. Biggins[2] and M. C. Thurnauer[1], [1]Argonne National Laboratory, Argonne, IL, 60439, USA; [2]Brown University, Providence, RI, 02912, USA

1. Introduction

Quinones play an essential role in light-induced electron transport in photosynthetic reaction centers (RC). Study of quinone binding within the protein matrix of the RC is a focal point of understanding the biological optimization of photosynthesis. In plant and cyanobacterial PSI, phylloquinone (K_1) is believed to be the secondary electron acceptor, A_1, similar to Q_a in the purple bacterial RC (1). Photoinduced electron transfer is initiated by reduction of the electron acceptor (A_0), a chlorophyll species, by the photoexcited primary donor $^*P_{700}$. A_1 acts as a transient redox intermediate between A_0 and the iron-sulfur centers (FeS).

We have examined the characteristic PSI electron spin polarized (ESP) electron paramagnetic resonance (EPR) signal as a marker of the interacting radical pairs developed during electron transfer (2-5). The ESP EPR signal was assigned to the interacting radical pair, $P_{700}^{+}A_1^{-}$. The ESP EPR signal from PSI isolated from spinach was destroyed by K_1 extraction and restored by chemical reconstitution of K_1, either protonated, $^HK_1$ or perdeuterated, $^DK_1$ (2). When quinone-depleted PSI was reconstituted with $^DK_1$, the ESP EPR signal exhibited linewidth narrowing and significant intensity changes indicating that K_1 was directly involved in the transient redox process (2). Further, the spectral shifts observed after reconstitution with $^DK_1$ could be duplicated by incubation of PSI particles with $^DK_1$ without prior extraction of the native K_1 (4). Attempts to exchange quinones lacking a hydrophobic phytyl group such as 1,10-naphthoquinone (NQ) or 1,4-Cl_2NQ were unsuccessful (4).

Here we examine the conditions required for K_1 exchange in PSI isolated from the thermophilic cyanobacterium, *Synechococcus lividus*, and compare them with those required for PSI isolated from spinach. The advantage of the former system is that it is possible to study the exchange of $^HK_1$ into deuterated PSI particles in addition to the exchange of $^DK_1$ into protonated PSI particles, providing a complete data set that can be used for modeling. We have used the correlated radical pair polarization (CRPP) ESP model (6) to reproduce the observed lineshapes. Simulations allow us to determine whether the 'exchanged' quinone occupies the A_1 binding site in its native orientation, a perturbed orientation, or an alternate site generated as a consequence of the isolation procedure.

P. Mathis (ed.), Photosynthesis: from Light to Biosphere, Vol. II, 115–118.
Kluwer Academic Publishers.

2. Materials and Methods

PSI particles, (^{H}PSI, ^{D}PSI), were isolated from *S. lividus* (7). $^{H}K_1$ was purchased from Aldrich and used without further purification. $^{D}K_1$ was extracted from *S. lividus* (99.6 % deuterated). PSI preparations were incubated with K_1 dissolved in ethanol (100 µM, 2% ethanol by volume). Control samples have PSI, 2% ethanol and no added quinones. ESP EPR signals were monitored by a light modulation, phase sensitive detection method as previously described (4). Light modulated signals are collected at 500 Hz, 0.01 mW microwave power and 1.25 G modulation amplitude.

3. **Results and Discussion**

The shape and intensity of the ESP EPR signal exhibited by a radical pair is dependent upon several factors, including: the time-dependence of electron transfer, spatial orientation, and isotopic composition of the involved species. Disruption of K_1 binding or other quinones in the A_1 site will perturb the ESP EPR signal (4,6,8).

The rates of quinone exchange depend upon incubation time and temperature. Figure 1 shows the time dependence of the exchange of $^{H}K_1$ into ^{D}PSI at 50°C. The control sample exhibits an ESP EPR signal characteristic for the perdeuterated PSI (^{D}PSI). Quinone exchange in these samples produces an ESP EPR signal arising from $^{D}P^{+}_{700}\ ^{H}K^{-}_{1}$. The protonated component of this isotopically mixed case gives rise to a severely attenuated ESP EPR signal relative to the deuterated component. Thus, the exchange process is manifest by a diminution of the overall signal, particularly the low field part arising predominantly from quinone resonances (5,8,9). By this measure, 0.5 hr. incubation time at 50°C is sufficient for complete quinone exchange. At 30°C quinone exchange is partial even after 16 hr. incubation (data not shown).

The temperature dependence (0-50°C) for $^{H}K_1$ incubated with ^{D}PSI is presented in Figure 2a. For each temperature an incubation period of 0.5 hr. is used. A transition from slow to fast exchange at a temperature between 40°C and 45°C is observed. Figure 2b shows exchange of $^{D}K_1$ into ^{H}PSI. Incubation at 45°C for 0.5 hr. is required to produce the intense narrow signal characteristic of $^{D}K_1$ in the site. We observed that the transition temperature for PSI particles derived from spinach is in the range 30-40°C. As a thermophilic bacterium, *S. lividus* would be expected to have a membrane composition optimized for survival at temperatures higher than for plant PSI. The primary adaptive mechanism of membranes to temperature is to shift the lipid composition to poise the gel-liquid crystal melting to slightly above ambient temperatures. We suggest that the difference in temperature-dependent transition observed for cyanobacterial and plant PSI is controlled by differences in the gel-liquid crystal melting temperature for the two cases. Further examination of this issue, as well as an investigation of the effect of detergent and solvent (10) on the exchange, are in progress.

The correlated radical pair polarization model (CRPP) for ESP can be used to approximate the observed lineshapes (6,11). Figure 3 shows the simulated spectra obtained for all isotopic variants. The electronic coupling parameters and the geometry for $P^{+}_{700}A^{-}_{1}$ used for the simulations have been previously described (6). The upper and lower inset show the contribution from 'P^{+}_{700}' and 'K^{-}_{1}', respectively. The changes observed in the experimental EPR spectrum of ^{D}PSI following incubation with $^{H}K_1$, Figure 1, and ^{H}PSI incubated with $^{D}K_1$, Figure 2b, are consistent with the differences

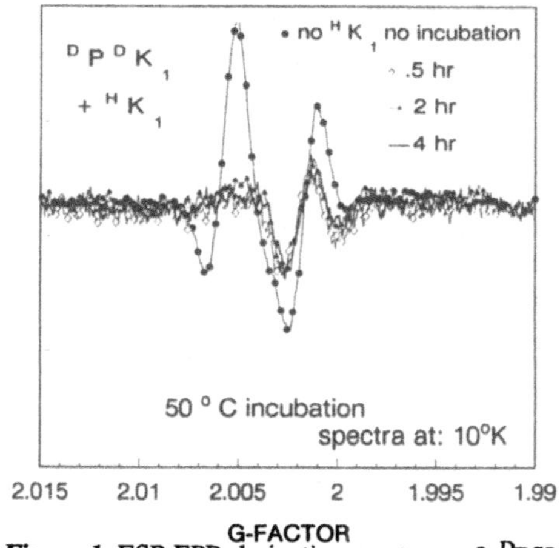

Figure 1. ESP EPR derivative spectrum of ^{D}PSI from *S. lividus* incubated with $^{H}K_1$ at 50°C: 0.5 hr (◇), 2 hrs (▲), 4 hrs (—). Control (●).

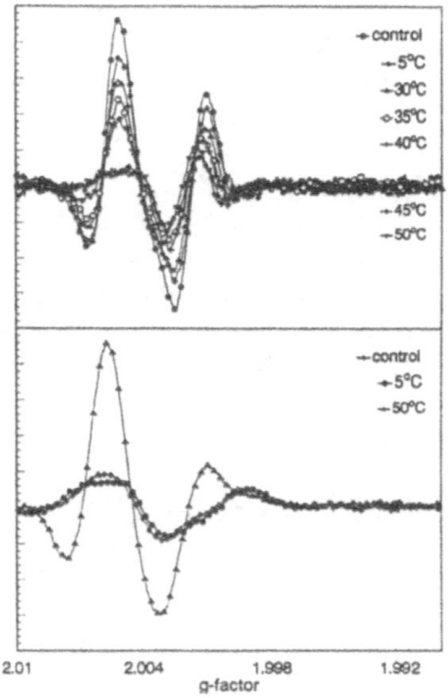

Figure 2. ESP EPR derivative spectrum of: (a) ^{D}PSI from *S. lividus* incubated with $^{H}K_1$ for 0.5 hr: 5°C (◇), 30°C (▲), 35°C (O), 40°C (+), 45°C (♦), 50°C (▽). Control (●). (b) ^{H}PSI incubated with $^{D}K_1$ for 0.5 hr: 5°C (●) and 50°C (▲). Control (◇).

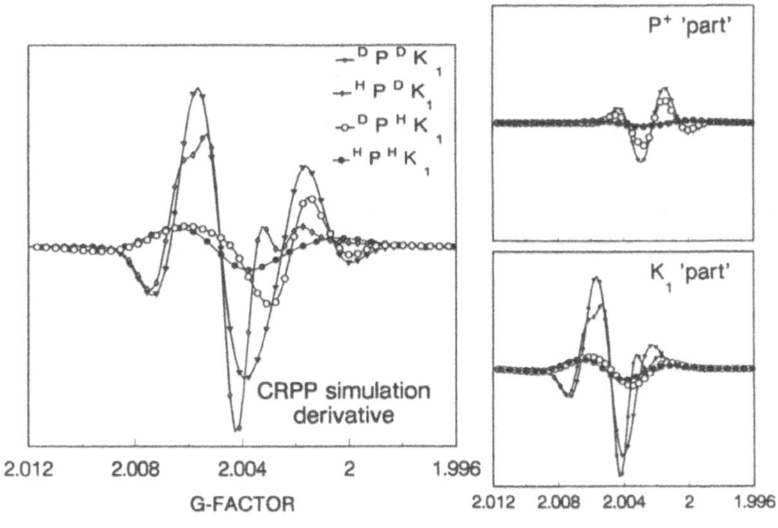

Figure 3. Simulated derivative ESP EPR spectra for all isotopic variants ($^{D}P^{D}K_1$, ▼/ $^{H}P^{D}K_1$,◇/ $^{D}P^{H}K_1$, O/ $^{H}P^{H}K_1$, ●). Insets: contribution from 'P^{+}_{700}' and 'K_1'.

observed in the simulated spectra with K_1 in the site. Variations in the dipolar angles, (not shown) indicate that the EPR spectrum is very sensitive to the orientation of the radicals in the site and suggest that the exchanged quinone is indeed in the site occupied by A_1.

4. Acknowledgments

The authors acknowledge J. R. Norris, G. Kothe, S. Weber, and E. Ohmes for useful discussions and for the modeling program. A. Barkoff and N. Brunkan, Undergraduate Research Participants, were supported by ANL's Division of Educational Programs. This work was supported by the U.S. DOE, Office of Basic Energy Sciences, Division of Chemical Sciences, under contract W-31-109-Eng-38.

References

1. Golbeck, J.H. and Bryant, D.A. (1991) in Current Topics in Bioenergetics (Lee, C.P., ed.), Vol. 16, pp. 83-177, Academic Press, New York.
2. Rustandi, R.R., Snyder, S.W., Feezel, L.L., Michalski, T.J., Norris, J.R. and Thurnauer, M.C. (1990) Biochemistry 29, 8030-8032.
3. Snyder, S.W., Rustandi, R.R., Biggins, J., Norris, J.R. and Thurnauer, M.C. (1991) Proc. Natl. Acad. Sci., USA 88, 9895-9896.
4. Rustandi, R.R., Snyder, S.W., Biggins, J., Norris, J.R. and Thurnauer, M.C. (1992) Biochim. Biophys. Acta 1101, 311-320.
5. Snyder, S. and Thurnauer, M.C. (1993) in The Photosynthetic Reaction Center (Deisenhofer, J. and Norris, J.R., eds.), Vol. II, Chapter 11, pp. 285-330, Academic Press, San Diego.
6. Kothe, G., Weber, S., Ohmes, E., Thurnauer, M.C. and Norris, J.R. (1994) J. Phys. Chem. 98, 2706-2712.
7. Biggins, J. and Mathis, P. (1988) Biochemistry 27, 1494-1500.
8. Van der Est, A., Sieckmann, I., Lubitz, W. And Stehlik, D. (1995) J. Chem. Phys. 194, 349-359.
9. Feezel, L.L., Gast, P., Smith, U. and Thurnauer, M.C. (1989) Biochim. Biophys. Acta 974, 149-155.
10. Warren, P.A., Golbeck, J.H. and Warden, J.T. (1993) Biochemistry 32, 849-857.
11. Morris, A.L., Snyder, S.W., Zhang, Z., Dutton, P.L., Robertson, D.E., Gunner, M.R. and Thurnauer, M.C. (1995) J. Phys. Chem. 99, 3854-3866.

A NEW SET OF SITE-DIRECTED MUTATIONS IN PHOTOSYSTEM I CORE REACTION CENTER FROM *SYNECHOCYSTIS* SP. PCC 6803

R. Schulz[1], L.B. Smart[1], J. Yu[1] and L. McIntosh[1,2]
[1]DOE Plant Research Laboratory and [2]Department of Biochemistry, Michigan State University, East Lansing, MI 48824-1312, USA

1. Introduction

The photosystem I (PS I) core reaction center is comprised of a heterodimer of homologous 82-83- kDa polypeptides, PsaA and PsaB, which bind four of the six electron transfer components of PS I (P700, A_0, A_1, F_X). The remaining two components (F_A and F_B) are carried by PsaC, a 8.9 kDa subunit of PS I [1, 2]. The corresponding genes of these three PS I subunits were inactivated by interposon mutagenesis and selection under light-activated heterotrophic growth (LAHG) conditions [3, 4, 5, 6, 7 and J. Yu and L. McIntosh, unpublished], indicating that the PS I reaction center is dispensable for heterotrophic growth of *Synechocystis* 6803. We are employing site-directed amino acid substitutions and deletions in *Synechocystis* 6803 to investigate the ligands of F_X and amino acid residues which may interact with A_0 and A_1. This technique is also being used to study the assembly of the PS I reaction center; specifically, the proposed leucine "zipper" between PsaA and PsaB and residues which may be required for interaction between the PsaA/PsaB core and PsaC (Fig. 1, Tab. 1).

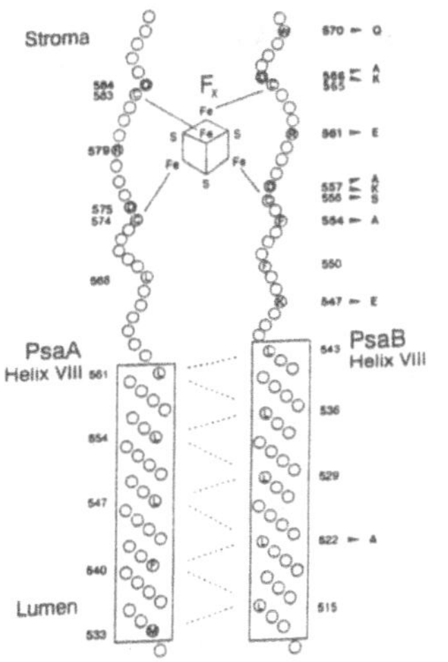

Tab. 1. *psa*B mutants in *Synechocystis* sp. PCC 6803 and presumed function of mutated amino acids in PS I.

Mutant Strain	Presumed function of amino acids in PS I
W570Q	Interaction with A_1
D566A	Redox enviroment for F_X
D566K	Redox enviroment for F_X
R561E	Interaction with PsaC
D557A	Redox enviroment for F_X and interaction with PsaC
D557K	Redox enviroment for F_X and interaction with PsaC
C556S	One ligand of F_X
F554A	Interaction with A_1
K547E	Interaction with PsaC
ΔL522	Part of leucine zipper

Fig. 1. Schematic diagram of portions of the PsaA and PsaB proteins from *Synechocystis* sp. PCC 6803 and of the [4Fe-4S] center F_X. Site-directed amino acid changes are indicated. Regions predicted to form helices are bounded by rectangles. Dashed lines represent the possible leucine zipper interaction. Figure adapted from [8].

P. Mathis (ed.), Photosynthesis: from Light to Biosphere, Vol. II, 119–122.

2. Procedure and Results

Oligonucleotides for site-directed mutagenesis were designed to effect the desired change in the coding sequence of *psa*B while also destroying or adding a restriction site. Site-directed mutagenesis was performed with an oligonucleotide-directed *in vitro* mutagenesis kit as directed by the manufacturer (Amersham). DNA sequencing was performed as described in [9]. Construction of mutated recombinant plasmids for transformation of *Synechocystis* 6803 and segregation of homozygous mutants under LAHG-conditions were performed essentially as described [8, 10]. Incorporation of mutations in the recombinant plasmids were verified by restriction mapping and by DNA sequencing. These constructs were used to transform the strain ΔB-RCPT, which allowed a rapid segregation of the site-directed mutations free of contamination from wild-type *psa*B DNA. Spectinomycin resistant colonies (Sp^R) were selected under LAHG conditions. Selection procedure was controlled by Southern blotting (Fig. 2).

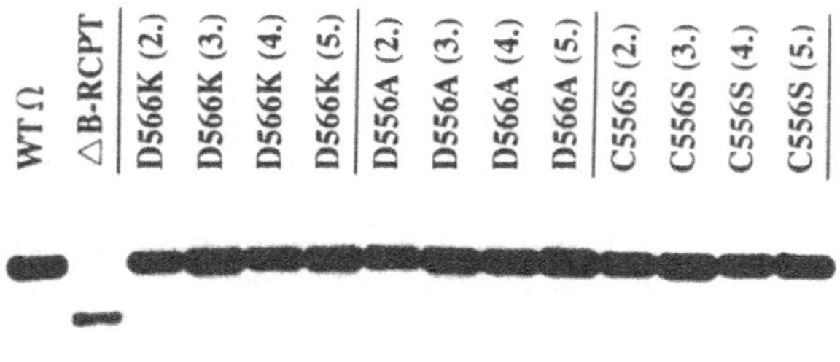

Fig. 2. Southern analysis of genomic DNA of mutant strains D566K, D566A and C556S, isolated from cells after the second (2.), third (3.), forth (4.) and fifth (5.) selection step. DNA was digested with HindIII and hybridized with a digoxigenin labelled HindIII fragment of *psa*B as a probe.

Comparision with DNA of WT Ω and the strain ΔB-RCPT confirmed complete segregation even after the second selection step. The presence of the desired mutations in Sp^R colonies was verified by two different methods: Firstly, genomic DNA of all mutants after at least five rounds of selection was digested with restriction enzymes in the region of recombination and was subjected to Southern analysis. The hybridization pattern confirmed correct recombination and complete segregation of all mutant strains. Secondly small-scale preparations of genomic DNA from all mutant strains were used as templates for PCR amplification of the mutated region of the *psa*B gene. The resulting PCR products were digested with the informative restriction enzymes. Agarose gel electrophoresis showed correct incorporation of all site-directed mutations (Fig. 3).

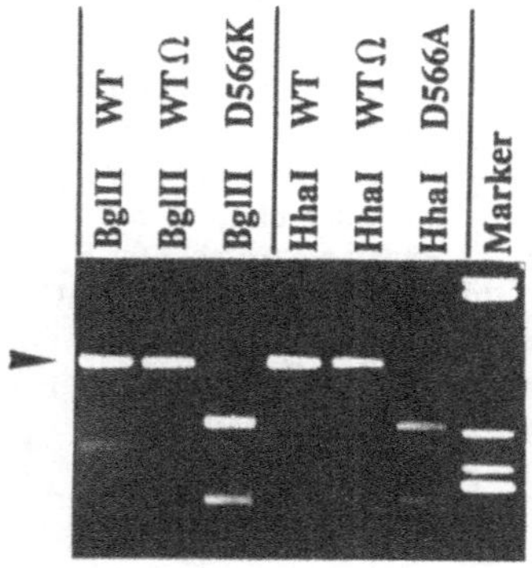

Fig. 3. Two examples of PCR analysis of *Synechocystis* transformants are shown. Genomic DNA, isolated from WT 6803, WT Ω and mutant strains D566K and D566A was amplified by PCR using primers complementary to *psa*B. The PCR products (766 bp in length, marked by an arrowhead) were digested with the informative restriction enzyme (BglII, HhaI) and separated by agarose gel electrophoresis. As expected, only the PCR fragments amplified from mutant DNA show the restriction site which was added by site-directed mutagenesis. Marker: pBR322 TaqI-fragments.

Characterization of mutants was performed as shown and described in Fig. 4.

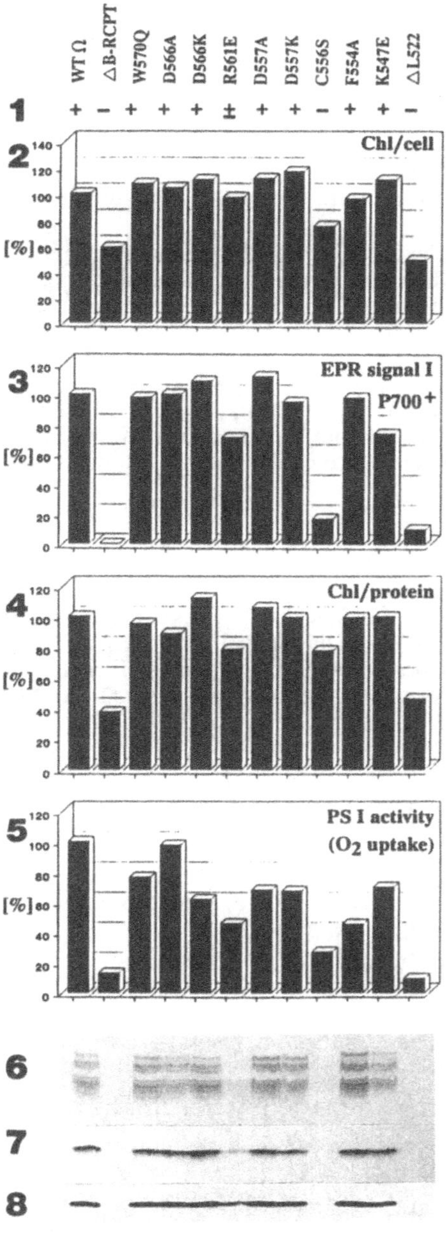

Fig. 4. Characterization of mutants.
(1) Autotrophic growth was tested on BG-11 plates without supplemental glucose in the light [8].
(2) Chlorophyll (Chl) was extracted from whole cells with methanol and quantified by using published extinction coefficients [11]. Data were normalized to equal optical density of the cultures at 730 nm. An average of at least seven chlorophyll assays from at least three cultures was used.
(3) $P700^+$-signal was obtained by room temperature EPR (Varian E-4 EPR Spectrometer, Palo Alto, CA, USA) of whole cells, essentially as described [4]. $P700^+$-signal intensity was calculated as the maximum signal in the light minus dark signal at the same H-value. Data were normalized to equal chlorophyll concentration of the EPR samples.
(4-8) Characterization of thylakoid membrane preparations. Cells were broken with a bead beater (Biospec Products, Bartelville, OK, USA) and isolation of thylakoid membranes was performed as described [4].
(4) Chlorophyll (Chl) measurements were normalized to equal protein content of the preparations. An average of three or six chlorophyll assays from one or two membrane preparations was used.
(5) For each mutant three PS I activity measurements were performed as previously described [8].
(6-8) Immunoblots of thylakoid membrane proteins, probed with antibodies recognizing (6) PsaA/B (66 kDa), (7) PsaC (9 kDa) or (8) PsaD (15 kDa), respectively. SDS-PAGE and immunoblotting were performed as described [4]. Equal amounts of protein (150 µg) were loaded in each lane. Molecular masses were estimated by comparison to migration of prestained molecular weight standards.

3. Discussion

(1) All mutant strains under investigation are genetically verified.
(2) The poor ability of mutant strain R561E to grow autotrophically, reduced PS I

activity, EPR signal for $P700^+$ and protein contents of PsaA/B, PsaC and PsaD seems to be caused by an altered interaction between PsaA/PsaB core and PsaC [12].
(3) Data shown for mutant C556S, in parallel to mutant C565S [8, 13], together with EPR data and absorbance change measurements (results not shown) indicate that this cysteine most likely play a crucial role in coordinating the [4Fe-4S] center F_X and in attaining high efficiency of electron transfer from $P700^+$ to F_A and F_B.
(4) Mutants W570Q and F554A exhibit in general wild type characteristics. Thus the presumed function of these amino acids to interact with A_1 could not be proved by the approach of this work.
(5) Very weak signals in all characterization experiments of the mutant ΔL522 make the hypothesis most likely that this mutation disrupted the proposed leucine zipper.
(6) Reduced $P700^+$ signal and PS I activity in mutant K547E and a little decrease in all protein contents investigated by Western blotting may indicate a role of the changed amino acid in interaction between PsaB and PsaC.
(7) While being located in PsaB on comparable positions, the mutants D566A/D566K and D557A/D557K, with presumed function of the aspartates to play a role in the redox enviroment for F_X, react in a different manner. All four mutant stains show in general wild type characteristics, but the EPR signals for F_X are changed for D566 mutants only (data not shown).

4. Acknowledgements
We are grateful to Dr. N.R. Bowlby (DOE-PRL, MSU) for performing room temperature EPR spectroscopy and for helpful discussions and Dr. G.T. Babcock (Dept. of Chemistry, MSU) for use of his spectrometer. We thank Dr. J. Guikema (Kansas State University) and Dr. D.A. Bryant (Pennsylvania State University) for providing PsaA/B, PsaC and PsaD antibodies. This work was supported by the Department of Energy (DE-FG02-90-ER 20021 to L.M.) and by Michigan State University Research Excellence Funds (to L.M.).

5. References
[1] Golbeck, J.H. (1992) Annu. Rev. Plant Physiol. Plant Mol.Biol. 43, 293-324
[2] Chitnis, P.R., Xu, Q., Chitnis, V.P. and Nechushtai, R. (1995) Photosynth. Res. 44, 23-40
[3] Anderson, S.L. and McIntosh, L. (1991) J. Bacteriol. 173, 2761-2767
[4] Smart, L.B., Anderson, S.L. and McIntosh, L. (1991) EMBO J. 10, 3289-3296
[5] Smart, L.B. and McIntosh, L. (1991) Plant Mol. Biol. 17, 959-971
[6] Smart, L.B. and McIntosh, L. (1993) Plant Mol. Biol. 21, 177-180
[7] Smart, L.B., Bowlby, N.R., Anderson, S.L., Sithole, I. and McIntosh, L. (1994) Plant Physiol. 104, 349-354
[8] Smart, L.B., Warren, P.V., Golbeck, J.H. and McIntosh, L. (1993) Proc. Natl. Acad. Sci. USA 90, 1132-1136
[9] Newman, T., de Bruijn, F.J., Green, P., Keegstra, K., Kende, H., McIntosh, L., Ohlrogge, J., Raikhel, N., Somerville, S., Thomashow, M., Retzel, E. and Somerville, C. (1994) Plant Physiol. 106, 1241-1255
[10] Williams, J.G.K. (1988) Methods Enzymol. 167, 766-778
[11] Lichtenthaler, H.K. (1987) Methods Enzymol. 148, 350-382
[12] Rodday, S.M., Schulz, R., McIntosh, L. and Biggins, J. (1994) Photosynth. Res. 42, 185-190
[13] Warren, P.V., Smart, L.B., McIntosh, L. and Golbeck, J.H. (1993) Biochemistry 32, 4411-4419

SITE-DIRECTED MUTATIONS OF *PsaB* FOR THE STUDY OF COFACTOR-PROTEIN AND PROTEIN-PROTEIN INTERACTIONS OF PHOTOSYSTEM I

L. Krabben, H. Käß, E. Schlodder, M. Kuhn & W. Lubitz.
H. Xu[#], S. Bingham[#] & A. Webber[#]
Max-Volmer-Institut für Biophysikalische und Physikalische Chemie, Technische Universität Berlin, Straße des 17. Juni 135, D-10623 Berlin, Germany.
[#]*Department of Botany, Arizona State University, Tempe, AZ 85787-1601, U.S.A.*

1. Introduction

The multiprotein complex Photosystem I (PSI) of higher plants, algae and cyanobacteria functions as a light driven plastocyanin/ferredoxin oxidoreductase. The terminal electron acceptor is $NAD(P)^+$ which is reduced by ferredoxin. The PSI core is a heterodimer of the PsaA and PsaB subunits which harbour the redox cofactors. These cofactors are the primary donor $\mathbf{P_{700}}$ (chlorophyll *a* dimer) and the electron acceptors $\mathbf{A_0}$ (chlorophyll *a* monomer), $\mathbf{A_1}$ (phylloquinone) and $\mathbf{F_X}$ (4Fe-4S cluster) [1]. The two terminal iron-sulfur clusters $\mathbf{F_A}$ and $\mathbf{F_B}$ are bound to the stromal PsaC subunit. The X-ray crystallographic structure of PSI is presently resolved at a resolution of 4.5 Å (see Schubert et al., in these proceedings and [2]). In this structure the two core subunits of PSI, PsaA and PsaB, are represented by eleven α-helices each which follow a local twofold symmetry. At the current resolution it is not possible to determinate amino acid residues involved in the coordination of cofactors. For the assignment of functional substructures and for the understanding of the function of specific amino acid residues site directed mutagenesis combined with comparative spectroscopical and biochemical analyses of the wild type and mutants is necessary. For a better understanding of light driven electron transfer processes our interest is focused especially on the cofactor-protein interactions and the (redox)-protein binding domains of PsaA and PsaB. *Chlamydomonas* (*C.*) *reinhardtii* is the most suitable species allowing genetic manipulation of chloroplast DNA and purification of the PSI for different spectroscopic and biochemical studies [3].

2. Materials and Methods

C. reinhardtii wild type and mutants were grown in HSA medium [4, 5]. For preparation of PSI cells were broken by passage through a French pressure cell. Thylakoids were washed twice and solubilized with 0.83% N-dodecyl-β-D maltoside. The thylakoid extract was separated on a sucrose gradient. The lower fraction of this gradient was loaded on a DEAE (TSK 650S) column. PSI enriched membrane fragments were eluted from the column by a Mg_2SO_4 gradient pooled and concentrated.
The PSI preparations were analysed by SDS-PAGE and by different spectroscopic techniques. The chlorophyll (Chl)/P_{700} ratio was calculated from the flash induced absorbance difference at 703 nm. The redoxpotential of P_{700} was determined by measuring the magnitude of flash-induced absorbance changes as a function of the redox potential poised in the reaction medium. The spectra of absorbance changes between 550 and 750 nm were recorded as described earlier [6]. Electron Nuclear Double Resonance (ENDOR) spectroscopy was performed as in [8].

P. Mathis (ed.), Photosynthesis: from Light to Biosphere, Vol. II, 123–126.

3. Results and Discussion

3.1 PSI Preparation

The preparation described here yields high quality PSI complexes which are suitable for all biochemical and spectroscopic studies. The Chl/P_{700} ratio is about 100:1 and the electron transport chain is active up to F_A/F_B. The protein pattern of a typical PSI preparation is shown in figure 1A. Almost no PSII proteins are detectable. Spectroscopic analyses of PSI isolated from *C. reinhardtii* shows similar properties as reported for PSI isolated from other species. The $P_{700}^{+\bullet}$ ENDOR spectrum of wildtyp (WT) is shown in figure 2. Assignment of line pairs could be done in analogy to *Synechococcus elongatus* [8] (data not shown). No significant structural differences of $P_{700}^{+\bullet}$ exist between *C. reinhardtii* and *Synechococcus.* Optical difference spectra of P_{700} show the typical pattern with the prominent band at 698 nm attributed to oxidation of P_{700} (see figure 3).

3.2 Protein-Protein Interactions

For the protein-protein interactions we are currently testing different domains in the predicted lumenal and stromal loops of PsaB for existence of structural elements involved in subunit binding. One possible domain for the docking of cytochrome c_6/plastocyanin is a region in PsaA and PsaB which shows a homology to an α-helix of the bacterial reaction center. In the reaction centre of *Rb. sphaeroides* this helix is oriented parallel to the membrane on the periplasmic site and is possibly involved in binding the electron-carrier cytochrome c_2 [7]. We raised an antiserum against an oligopeptide resembling this epitope in PsaB. As shown in figure 1B the antiserum crossreacts specifically with the PsaA/B subunits. This antiserum will be very useful for the analysis of directed mutants in this region of PsaB. We already introduced different mutations in this region which are currently under investigation. These studies will also help to understand electron transfer from plastocyanin to $P_{700}^{+\bullet}$.

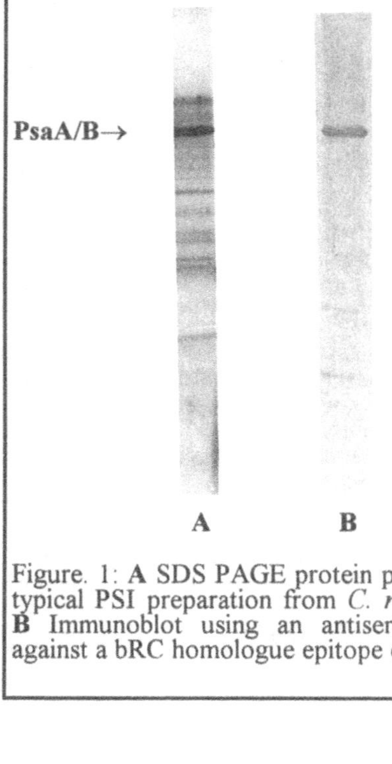

Figure. 1: **A** SDS PAGE protein pattern of a typical PSI preparation from *C. reinhardtii.* **B** Immunoblot using an antiserum raised against a bRC homologue epitope of PsaB.

3.3 Cofactor-Protein Interactions

According to the current model of the PSI reaction center one of the P_{700} (dimer)-chlorophylls is predicted to be liganded by a conserved histidine of the transmembrane helix VIII or helix X of the PsaB subunit. Site-conversion of PsaB His 523 (helix VIII) to Leu results in an impaired assembly of PSI but does not alter ENDOR and ESEEM characteristics of $P_{700}^{+\bullet}$ [8]. PsaB His 656 (helix X) has been altered to Leu and Asn. H656L does not stably accumulate PSI whereas mutant H656N yields considerable levels of PSI.

Figure 2 shows the ^{1}H-ENDOR spectra of $P_{700}^{+\bullet}$ from the WT and the mutant H656N. Assignment of hyperfine couplings (hfc's) derived from the spectrum to molecular positions in the spin carrying half of $P_{700}^{+\bullet}$ was done as previously described [8]. The ENDOR spectrum of the mutant is distinctly different from that of wild type, in particular the hfc of line pair (3, 3') is increased by 20 % in this mutant, whereas some of the smaller hfc's are reduced. This clearly indicates an altered interaction of this chlorophyll with its local protein surrounding indicating that the $P_{700}^{+\bullet}$ environment has been modified by this mutation.

Further characterisation of this mutant by redox-titration and flash induced optical absorption difference spectroscopy showed additional evidence for an altered chlorophyll-protein interaction. The difference spectra are dominated by the bleaching at 696 nm attributed to the photooxidation of P_{700}. This band shows a minor blueshift of about 2 nm in the mutant. More interestingly, the mutant spectrum exhibits a new light induced band appearing at 668 nm (Figure 3) which is at present not understood. Furthermore the redox potential of P_{700} shifted from approx. 440 mV to 490 mV (data not shown).

These results provide first experimental evidence that histidine at position 656 in PsaB most probably directly interacts with P_{700} and changes its electronic properties.

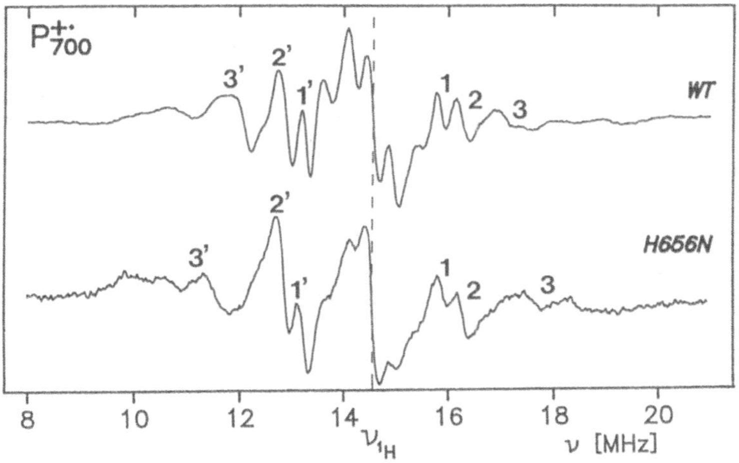

Figure 2: *C. reinhardtii* wild type and mutant H656N ^{1}H-ENDOR spectra of $P_{700}^{+\bullet}$ in PSI particles in frozen solution at 130 K. The ENDOR measurements were performed on a Bruker ESP 300E X-band EPR spectrometer with home built ENDOR accessories. Experimental conditions: microwave power 5.1 mW, RF power 200W, time constant 640 ms, total averaging time 60 min, frequency modulation amplitude (FM) 100 kHz.

Additional mutants carrying different substitutions at residues of interest are already under investigation in our laboratories. We are currently also working on an improved structural model of PSI in order to better understand the effects of these mutations on the spectroscopic and functional properties of $P_{700}^{+\bullet}$.

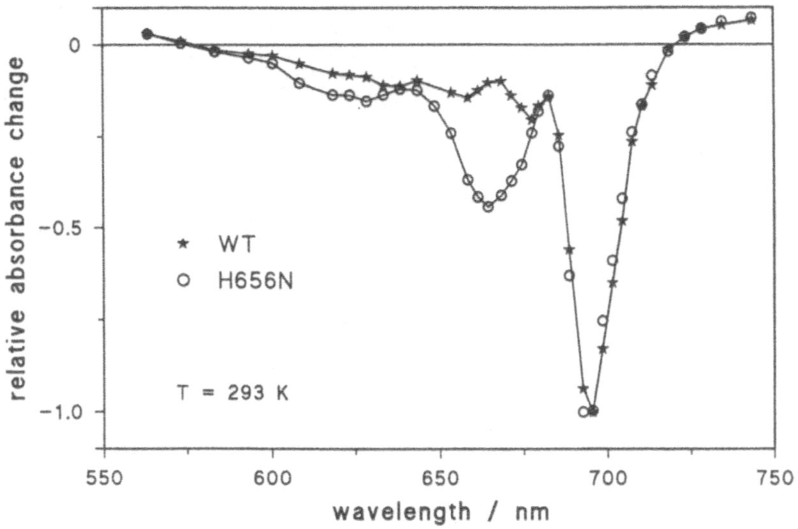

Figure 3: Flash-induced absorbance difference spectrum of $P_{700}^{+\bullet}F_{A/B}^{-}$ vs $P_{700}\ F_{A/B}$ of PS I complexes from *C. reinhardtii* measured at room temperature. The sample was excited by saturating flashes of about 15 µs duration. In the mutant H656N a new band appears at about 667 nm. The main bleaching band attributed to the oxidation of P_{700} is slightly blueshifted (2 nm) in the mutant H656N.

We thank Irene Geisenheimer (TU Berlin) for technical assistance. L. Krabben is grateful for a grant from DAAD. This work was supported by the National Research Initiative Competitive Grants Program of the USDA (award no. 93-37306-6473), DFG (Sfb 312 TPA4) and NATO (CRG 950759).

References

1. Golbeck, J. H. (1992) Ann. Rev. Plant Mol. Biol. 43, 293-324
2. Krauss, N., Hinrichs, W., Witt, I., Fromme, P., Pritzkow, W., Dauter, Z., Betzel, C., Wilson, K., Witt, H. & Saenger, W. (1993) Nature **361**, 326-331
3. Webber, A. N., Bingham, S. E., Gibbs, P. B., Misra, L. M. & Warden, B. (1992) Research in Photosynthesis **1**, 561-564
4. Sueoka, N. (1960) Proc. Natl. Acad. Sci. U.S.A. **46**, 83-91
5. Harris, E. H. (1989) The Chlamydomonas Sourcebook, Academic Press, San Diego and New York
6. Lüneberg, J., Fromme, P., Jekow, P. & Schlodder, E. (1994) FEBS Lett. **338**, 197-102
7. Kuhn, M., Fromme, P. & Krabben, L. (1994) Trends in Biochem. Sci. **19**, 401-402
8. Cui, L., Bingham, S. E., Kuhn, M., Käß, H., Lubitz, W. and Webber, A. N. (1995) Biochem. **34**, 1549-1558

IN VIVO SITE-DIRECTED MUTATIONS OF THE CYSTEINE LIGANDS TO F_A AND F_B IN *SYNECHOCYSTIS* sp. PCC 6803: A COMPARISON WITH *IN VITRO* RECONSTITUTED PHOTOSYSTEM I COMPLEXES

Yean-Sung Jung[1], Jianping Yu[2], Lian Yu[1], Jindong Zhao[3], Donald A. Bryant[3], Lee McIntosh[2], and John H. Golbeck[1]
[1]Department of Biochemistry, Center for Biological Chemistry, University of Nebraska, Lincoln, NE 68588-0664 USA
[2]Department of Energy Plant Research Laboratory, Michigan State University, East Lansing, MI 48824-1312 USA
[3]Department of Biochemistry and Molecular Biology, Pennsylvania State University, University Park, PA 16802 USA

1. Introduction

The F_A and F_B iron-sulfur clusters are located on a tightly bound 8.9-kDa polypeptide in Photosystem I (PS I). This 81 amino acid protein, known as PsaC, has sequence homology with 2[4Fe-4S] ferredoxins, especially around the two CxxCxxCxxxCP iron-sulfur binding motifs which bind the F_A and F_B clusters. Previous work with *Escherichia coli*-expressed PsaC mutants and *in vitro* reconstituted Photosystem I complexes implied that cysteines 11, 14, 17 and 58 ligate F_B, while cysteines 48, 51, 54 and 21 ligate F_A [1]. An unexpected finding was that mixed-ligand [4Fe-4S] iron-sulfur clusters were located in the C14D-PS I complex [2] and in the C51D-PS I complex [3]. In both instances, electrons could be transferred to the mixed-ligand iron-sulfur cluster at 15 K and at room temperature, thus supporting $NADP^+$ photoreduction rates similar to the wild-type. In this work, *in vivo* mutations in C14D and C51D were incorporated into PsaC in *Synechocystis* sp. PCC 6803 [4]. Also, cysteine to serine mutations (C14S and C51S) and cysteine to alanine mutations (C14A and C51A) were engineered in the same sites. The Photosystem I complex was isolated and analyzed by EPR spectroscopy. This article summarizes the structural and functional properties of *in vivo* aspartate, serine and alanine mutants in the C14 and C51 positions of PsaC in *Synechocystis* sp. PCC 6803, and offers a comparison with *in vitro* reconstituted reaction centers generated by rebinding *E. coli*-expressed PsaC proteins to P700-F_X cores.

2. Materials and Methods

In vivo site-directed mutants C14D, C14S, C14A, C51D, C51S and C51A were generated in *Synechocystis* sp. PCC 6803 and the organisms grown in the presence of glucose under light-activated heterotrophic growth conditions in the dark. Membranes were isolated by breaking the cells with a Bead-Beater and centrifugation; PS I complexes were isolated from the membranes with ß-D-dodecyl maltoside and sucrose density ultracentrifugation. *In vitro* site-directed mutants were assembled from *Escherichia-coli* expressed proteins and purified P700-F_X cores as described in refs. [1, 2 and 3]. EPR studies were performed with a Bruker ECS-106 X-band spectrometer equipped with a liquid helium cryostat and a xenon arc lamp. Odd-integer spin systems ($S \geq 1/2$) from reduced [4Fe-4S] clusters and integer spin systems ($S = 2$) from reduced [3Fe-4S] clusters were distinguished using a dual-mode resonator (Bruker DM-4116).

P. Mathis (ed.), Photosynthesis: from Light to Biosphere, Vol. II, 127–130.

3. Results

3.1. Cysteine->Aspartic Acid

3.1.1. Cysteine->Aspartic Acid (C14D)

When frozen in darkness and illuminated at 15 K, the EPR spectrum of the *in vivo* C14D-PS I complex depicts resonances characteristic of F_A, with *g* values of 2.046 , 1.943 and 1.850. The maximum signal intensity occurs at 15 K, a value similar to that of F_A in the wild-type. When the C14D-PS I complex is frozen during illumination, only resonances characteristic of F_A are visible; however, the maximum signal intensity occurs at ≤4.2 K, indicating relief of power saturation by a nearby spin system. While this could be either F_B or F_X, we rule out the latter because the maximum signal intensity of F_A occurs at 15 K in Hg-treated PS I complex where the F_B cluster has been destroyed [5]. Hence, the magnetic interaction is attributed to F_B, which we surmise is present as a high-spin (S ≥ 3/2) [4Fe-4S] cluster. We are seeking evidence for its existence in the region around *g* = 5.5. The *in vitro* C14D-PS I complex behaved differently, where a S = 1/2 ground state F_B cluster was detected in the mutated site [2]. The spectrum of the F_B cluster in the *in vitro* C14D-PS I complex has *g*-values of 2.118, 1.911, 1.883 and linewidths of 75, 60 and 85 MHz.

3.1.2. Cysteine->Aspartic Acid (C51D)

When frozen in darkness and illuminated at 15 K, the EPR spectrum of the *in vivo* C51D-PS I complex depicts a weak set of resonances characteristic of F_B, with *g* values of 2.066, 1.930 and 1.880. However, a significant amount of $P700^+$ is formed which is irreversible at 18 K, implying charges separation with F_B or F_A (but not F_X). We surmise that the F_A cluster is reduced, but that it is in a high-spin (S ≥ 3/2) state when F_B is oxidized in the same reaction center. The maximum signal intensity occurs at 18 K, a value similar to that of F_B in the wild-type complex. When the C51D-PS I complex is frozen during illumination, a strong set of resonances characteristic F_B are found, but also a new set of resonances are seen with *g*-values of 2.084, 1.922, and 1.846, which are attributed to F_A in the modified site. The maximum signal intensity of F_B occurs at 18 K and the modified F_A at ≤ 4.2 K, a temperature lower than that of F_A in the wild-type complex. The lack of a significant interaction spectrum indicates that there is weak magnetic coupling between F_B and F_A in the mutant C51D-PS I complex. The *in vitro* C51D-PS I complex has a S = 1/2 cluster in the F_A site, with *g*-values of 2.044, 1.943 and 1.853 and a temperature optimum of ≤ 4.2 K when the sample is frozen in darkness and illuminated at 15 K. The spectral *g*-values and lineshapes, the temperature dependencies, and the responses to light at 298 K are identical to an *in vitro* C51D-PS I complex [3].

3.2. Cysteine->Serine

3.2.1. Cysteine->Serine (C14S)

When frozen in darkness and illuminated at 15 K, the EPR spectrum of the *in vivo* C14S-PS I complex depicts resonances characteristic of F_A, with *g* values of 2.048 , 1.943 and 1.855. The maximum signal intensity occurs at 9 K, a value lower than that of F_A in the wild-type, indicating enhanced spin relaxation properties at the non-mutated site. When the C14S-PS I complex is frozen during illumination, resonances characteristic of F_A appear along with a set of resonances from F_B. The maximum signal intensity of F_A and F_B occurs at ≤4.2 K, and the complex spectrum indicates significant magnetic interaction between the two clusters. The *in vitro* C14S-PS I complex behaved differently, where a S = 1/2 ground state F_B cluster was detected in the mutated site. The extracted spectrum of the F_B cluster in the *in vitro* C14S-PS I complex has *g*-values of 2.118, 1.911, 1.883 and linewidths of 75, 60 and 85 MHz.

3.2.1. Cysteine->Serine (C51S)

When frozen in darkness and illuminated at 15 K, the EPR spectrum of the C51S mutant depicts a weak set of resonances characteristic of F_B, with g values of 2.066, 1.930 and 1.877. Similar to C51D, there is no indication of reduced F_A, and we again surmise that the F_A cluster is reduced, but that it is in a high-spin ($S \geq 3/2$) state when F_B is oxidized in the same reaction center. The maximum signal intensity occurs at 15 K, a value similar to that of F_B in the wild-type complex. When the C51S-PS I complex was frozen during illumination, a strong set of resonances characteristic F_B are observed in addition to a new set of resonances with g-values of 2.113, 1.916, and 1.837, which are attributed to to F_A in the modified site. The maximum signal intensity of the modified F_A cluster occurs at $\leq$ 4.2 K, a temperature lower than that of F_A in the wild-type PS I complex. There is a noticeable lack of an interaction spectrum, indicating that magnetic coupling between F_A and F_B is weak. The *in vitro* C51D-PS I complex has a $S = 1/2$ cluster in the F_A site, with g-values of 2.044, 1.943 and 1.853 and a temperature optimum of $\leq$ 4.2 K when the sample is frozen in darkness and illuminated at 15 K. The spectral g-values and lineshapes, the temperature dependencies, and the responses to light at 298 K are identical to an *in vitro* C51D-PS I complex.

3.3. Cysteine -> Alanine

3.3.1. Cysteine -> Alanine (C14A)

When frozen in darkness and illuminated at 9 K, the EPR spectra of the C14A-PS I complex shows a set of broad resonances characteristic of F_X, with g-values of 2.04, 1.87 and 1.78.When the C14A-PS I complex is frozen under illumination, the same F_X resonances are present at a higher spin concentration (ca. 2 to 2.5 times). This data agrees with Western blots which show the lack of PsaC (as well as PsaD and PsaE) protein(s) on the C14A-PS I membranes (see Yu, this volume). In contrast, the *in vitro* C14A protein was able to rebind to the P700-F_X core. When frozen in darkness and illuminated at 15 K, the EPR spectrum of the *in vitro* C14A-PS I complex showed a set of resonances characteristic of F_A at g = 2.046, 1.943 and 1.853, and an equally strong set of resonances with g-values of 2.118, 1.911, 1.883 and linewidths of 75, 60 and 85 MHz in the mutated F_B site. When frozen during illumination, the C14A-PS I complex shows an interaction spectrum remarkably similar to that found in the wild-type PS I complex. The *in vitro* C14A-PS I complex supported rates of $NADP^+$ reduction 41% that of a PS I complex reconstituted with wild-type PsaC.

3.3.2. Cysteine -> Alanine (C51A)

When frozen in darkness and illuminated at 9 K, the EPR spectra of the C51A-PS I complex also shows a set of broad resonances characteristic of F_X, with g-values of 2.068, 1.865 and 1.780.When the C14A-PS I complex is frozen under illumination, the same F_X resonances are present at a higher spin concentration (ca. 2 to 2.5 times). This data also agrees with Western blots which show the lack the PsaC (as well as PsaD and PsaE) protein(s) on the C51A-PS I membranes (see Yu, this volume). Similar to C14A, the C51A protein was able to rebind to the P700-F_X core. When frozen in darkness and illuminated at 15 K, the EPR spectrum of the *in vitro* C51A-PS I complex showed a set of resonances characteristic of F_B at g = 2.070, 1.930 and 1.880 at a temperature optimum of 15 K, and an equally strong set of resonances characteristic of F_A at g = 2.045, 1.942, 1.854 at a temperature optimum $\leq$4.1 K. When frozen during illumination, the C51A-PS I complex shows an interaction spectrum similar to that found in the wild-type PS I complex, but with a temperature optimum $\leq$ 6 K. The *in vitro* C51A-PS I complex supported rates of $NADP^+$ reduction 59% of that in a PS I complex reconstituted with wild-type PsaC.

4. Discussion

The original intent in this work was to engineer [3Fe-4S] clusters into the F_A and F_B sites of PsaC. The rationale was to place a high-potential redox trap into the F_A and F_B sites to determine whether electron transfer can occur without a properly functioning (i.e. low potential) F_A or F_B cluster. The choices of aspartic acid, serine and alanine were made with because aspartate supports [3Fe-4S] clusters in naturally-occurring ferredoxins, that serine can support a [3Fe-4S] cluster in the F_X binding site and that alanine, by the absence of a suitable ligand on its side group, should support only [3Fe-4S] clusters. To summarize:

1) Only [4Fe-4S] clusters are found in the mutant PS I complexes. The premise is that the P700-F_X core is unable to bind a PsaC protein containg a [3Fe-4S] cluster; instead, it binds PsaC containing a normal [4Fe-4S] and a mixed-ligand [4Fe-4S] cluster.
2) All mutant PS I complexes containing a mixed-ligand [4Fe-4S] cluster are capable of electron transfer to F_A and F_B at 15 K and room temperature, and all support high rates of electron throughput to $NADP^+$.
3) The similarities in the *in vitro* reassembled and *in vivo* engineered C51D-PS I and C51S-PS I complexes is likely because oxygen from aspartate and serine (alternatively, water or OH^-) provide the ligands to F_A in the modified site.
4) The differences in the *in vitro* reassembled and *in vivo* engineered C14D-PS I and C14S-PS I complexes is a spin state difference between the F_B cluster in the two sets of mutants. Perhaps different ligands are functioning *in vivo* and *in vitro*.
5) The differences in the *in vitro* and *in vivo* C14A-PS I and C51A-PS I complexes are profound; only the *in vitro* mutant supports mixed-ligand [4Fe-4S] clusters.

The similar EPR characteristics of the F_B cluster in the mixed-ligand sites of *in vitro* C14D-PS I, C14S-PS I and C14A-PS I complexes is particularly striking. Alanine lacks a suitable ligand, which may be provided by the carboxylate oxygen in aspartic acid and hydroxy/alkoxide in serine (or water, or OH^-), and this is undoubtedly the reason behind the inability of the C14A-PS I complex to support an [4Fe-4S] cluster *in vivo*. We propose that ß-mercaptoethanol, present in the reaction mixture, has 'rescued' a [4Fe-4S] cluster in the *in vitro* C14A-PsaC complex, thereby allowing a protein containing a mixed-ligand F_B cluster to rebind to the P700-F_X core. The similarity to the F_B spectrum in the *in vitro* C14D-PsaC and C14S-PsaC complexes implies that ß-mercaptoethanol provides the ligand in those mutated sites; the implication is that the sulfur thiolate as good a ligand as the oxygen assumed in aspartate or serine mutants. We also propose that ß-mercaptoethanol has 'rescued' a [4Fe-4S] cluster in the *in vitro* C51A-PS I complex. Indeed, the similarity to the F_A cluster in the mixed-ligand sites of *in vitro* C51D-PS I, C51S-PS I complexes suggests that they are also derived from a thiolate ligand provided by ß-mercaptoethanol. The PsaC protein does not assemble in the *in vivo* mutants of C14A-PS I and C51A-PS I, most likely because in the absence of a suitable ligand, a [4Fe-4S] cluster does not form.

5. References

1. Zhao, J.D., Li, N., Warren, P.V., Golbeck, J.H. and Bryant, D.A. (1992) Biochemistry 31, 5093-5099.
2. Yu, L., Bryant, D. A, and Golbeck, J.H.(1995) Biochemistry 34, 7861-7868.
3. Yu, L., Vassiliev, I. R., Jung, Y-S., Bryant, D. A. and Golbeck, J.H. (1995) J. Biol. Chem. (in press).
4. Yu, J., Smart, L. B., Jung, Y-S., Golbeck, J. H. and McIntosh, L. (1995) Plant Mol. Biol. (in press).
5. Jung, Y-S., Yu, L. and Golbeck, J.H. (1995) Photosynth. Res. (in press).

ELECTRON TRANSFER VIA THE F_X CLUSTER OF PS I IN PsaB MUTANTS OF *SYNECHOCYSTIS* SP. PCC 6803

Ilya R. Vassiliev[1], Yean-Sung Jung[1], Rüdiger Schulz[2], Lawrence B. Smart[2], Lee McIntosh[2] and John H. Golbeck[1]
[1]Department of Biochemistry and Center for Biological Chemistry, University of Nebraska, Lincoln, NE 68588-0664, USA
[2]Department of Energy Plant Research Laboratory, Michigan State University, East Lansing, MI 48824-1312, USA

F_X is a unique interpolypeptide [4Fe-4S] cluster in PS I, in which two of four cysteine ligands are provided by PsaA, and the other two by PsaB polypeptide of the reaction center heterodimer. The role of F_X for forward electron transfer at room temperature is not fully understood, although optical studies on the kinetics of forward electron transfer [1] and studies of electrogenic reactions [2] suggest that F_X is an obligatory component of PS I. While assessing the PS I function by EPR and kinetic optical spectroscopy, we introduced a genetic component in addressing the issue of F_X role. The cysteine ligands to F_X in PsaB (C556 and C565) were changed to serines, with the expectation of altering electron transfer through the cluster. Also, we probed whether the substitution of adjacent aspartates (D566 and D557) to alanine or lysine exert any influence on F_X properties.

Materials and Methods

Mutants of *Synechocystis* sp. PCC 6803 with substitutions in the F_X biding niche were generated by protocol described previously [3] for one of the mutants (C565) used in this study. The organism grown in the presence of glucose under light-activated heterotrophic conditions in the dark. PS I complexes were isolated from thylakoid membranes using β-dodecylmaltoside. Room temperature absorbance difference changes at 820 nm (ΔA_{820}) were measured with a laboratory-built double-beam spectrometer using frequency-doubled Nd:YAG laser flash (532 nm, 10 ns) excitation [4]. Kinetic traces were acquired on a 10 μs to 5 s time scale and decomposed to a sum of exponentials and a baseline. EPR studies were performed with a Bruker ECS-106 X-band spectrometer equipped with a liquid helium cryostat and a xenon arc light source. For single flash excitation the Nd:YAG laser beam was diverted directly into the EPR cavity.

Results and Discussion

The ΔA_{820} kinetics (Fig. 1) were used to identify the relative contributions of various electron acceptors to P_{700}^+ decay in the wild type (WT) and mutant PS I complexes. Figure 2 shows the results of kinetics decomposition as the amplitudes of individual exponentials plotted against their life times. We carried out a verification protocol on WT PS I complex [4] to determine whether the computer-fitted exponentials correlate with known back reactions of P_{700}^+ (for review, see [5]). The major contribution to ΔA_{820} kinetics in the WT PS I complex is brought about by a ca. 80-100 ms component. This

P. Mathis (ed.), Photosynthesis: from Light to Biosphere, Vol. II, 131–134.

component, along with a 10-20 ms component should be attributed to the F_A^-/F_B^- back reaction which was resolved earlier as a single component with a 30 ms half-time [5]. These two components were eliminated upon chemical reduction of F_A/F_B with dithionite or upon removal of the PsaC subunit with urea. The slower components of ca. 300 ms and 3-4 s are probably a result of P_{700}^+ reduction from reduced DCPIP and oxygen. The 1-ms component ascribed to the F_X^- back reaction, has a minor contribution to P_{700}^+ reduction in the WT PS I complex. Two faster components with life times of ca. 10 and 100 μs, corresponding to that reported for A_1^- back reaction [6] can be also resolved in the WT complex.

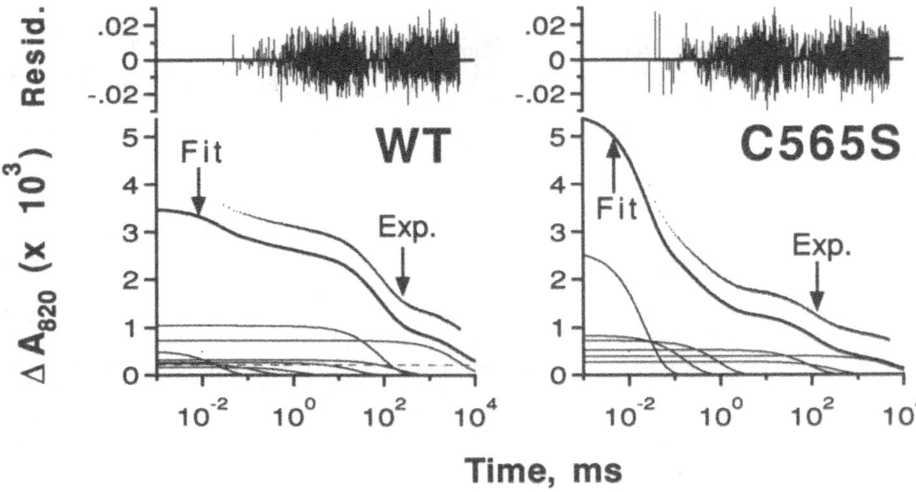

Figure 1. Kinetics of absorbance changes at 820 nm measured in PS I complexes from the wild type (WT) and C565S mutant of *Synechocystis* sp. PCC 6803. The reaction media is 25 mM Tris buffer, pH 8.3 with 0.03% ß-dodecylmaltoside, 10 mM sodium ascorbate and 4 μM DCPIP; the chlorophyll concentration is 50 μg ml^{-1}. Excitation flash intensity, 58 mJ. Each experimental trace (Exp.) is an average of 20 measurements taken at 50 s intervals and is shown with a vertical offset relative to the multiexponential fit (Fit).

Figure 1 (right) shows the ΔA_{820} decay kinetics of the C565S mutant complex, and the multiexponential fit of the data. In the wild type, around 40% of P_{700}^+ is reduced via a back reaction from F_A^- and/or F_B^- whereas in the C565S this back reaction contributes only around 10% to P_{700}^+. The major contribution to ΔA_{820} in C565S comes from 24 μs and 176 μs components (48 and 15%), which should correspond to A_1^- (and probably chlorophyll triplet; see below), and a 1-ms component (13%) arising from F_X^-.

To check for possible contribution of the chlorophyll triplet to ΔA_{820} in the tens of microseconds domain, we analyzed the flash intensity dependency of WT and C565S PS I complexes kinetics. As seen from the decomposition results (Fig. 2), in both the WT and C565S, the F_A/F_B (ca. 10 and 80 ms) and F_X (ca. 1 ms) components tend to be saturated upon increasing the flash intensity, whereas the 10-μs (and to a lesser extent, the ca. 100-μs component) is not saturated at high flash intensities. This implies a contribution of the chlorophyll triplet at high flash intensities and might explain a higher calculated initial amplitude of ΔA_{820} in the C565S mutant than in WT at equal chlorophyll concentration (Fig. 1). However, the lower contribution of F_A/F_B components in the C565S mutant relative to WT is obvious throughout the whole range of flash intensities tested.

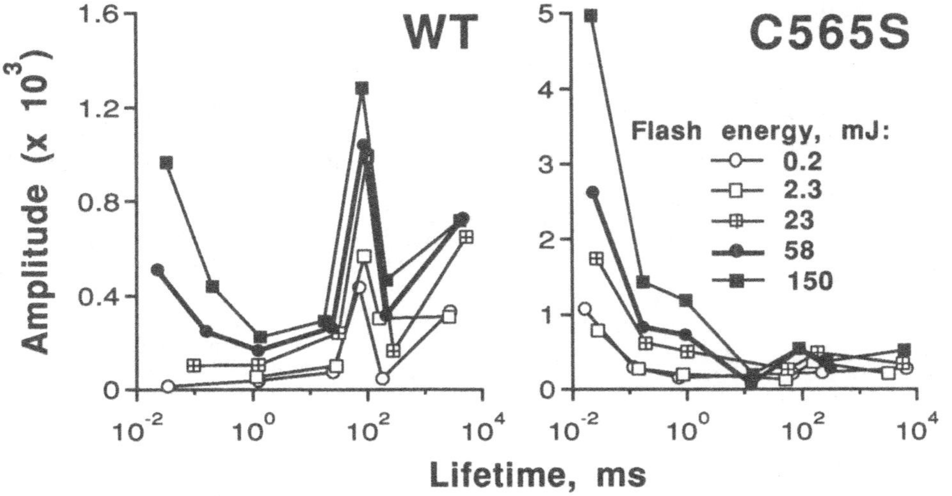

Figure 2. The results of multiexponential decompositions ΔA_{820} kinetics of PS I complexes from WT and C565S mutant at different excitation flash intensities plotted as the amplitudes of individual exponentials versus their life times

The other mutant with a modified ligand to F_X, C556S, has a ΔA_{820} kinetics remarkably similar to that of C565S, with a relatively low contribution of the F_A^-/F_B^- back reaction, which is resolved as a single 53.4-ms component. The aspartate mutants (D566A, D557A, D566K and D557K) have ΔA_{820} kinetics essentially similar to that of WT, with a major contribution to ΔA_{820} from a 80-100-ms component. The increase in the contribution of F_X^- and A_1^- to the overall back reaction with P_{700}^+ at the expense of the back reaction from F_A^- and F_B^- in C556S and C565S suggests a decrease in quantum efficiency of F_A/F_B photoreduction due to substitution of the F_X ligand. Substitutions of the aspartates adjacent to F_X-binding cysteines induce minor or almost no changes in the P_{700}^+ decay kinetics.

The properties of the iron-sulfur clusters in mutant and WT PS I complexes were studied by low-temperature EPR spectroscopy. The *g*-value and linewidths of F_X resonances differed slightly from the wild type in the aspartate mutants, but the optimum temperature and microwave power saturation of F_X were nearly identical to the WT, indicating similar relaxation properties. The EPR spectra of the modified F_X' cluster of the cysteine mutants showed *g*-values of 2.04, 1.94 and 1.81 and linewidths relatively narrower than in WT complex. The $g = 1.81$ resonance also had a narrower temperature optimum and different power saturation characteristics relative to the WT. However, F_A and F_B could be quantitatively reduced in C565S and C556S and in all four aspartate mutants upon freezing during continuous illumination.

To investigate the relative quantum efficiency of electron transfer at cryogenic temperatures, we compared the electron transport efficiencies in cysteine mutants with the wild type by measuring the $g = 1.944$ midfield resonance of F_A^- under illumination of dark-frozen samples at 15 K with a train of single-turnover saturating laser flashes. The higher number of quanta required for saturation of F_A^- (Fig. 3) strongly suggests that in these mutants, the photoreduction of F_A occurs at all relevant temperatures with a decreased quantum efficiency.

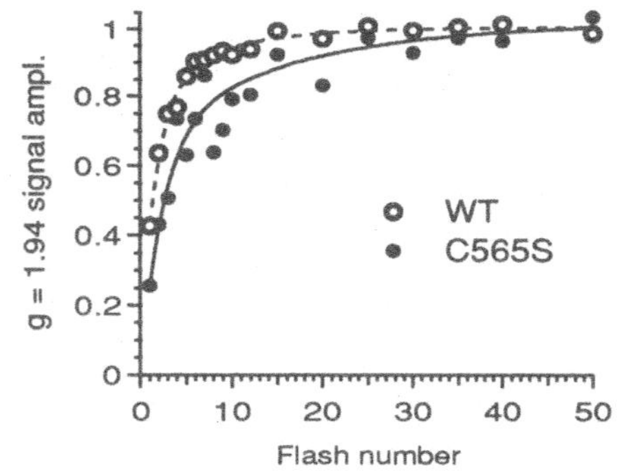

Figure 3. Flash saturation dependency of amplitude of $g = 1.944$ resonance of WT and C565S PS I complex. The samples were suspended at a chlorophyll concentration of 200 μg ml^{-1} in 25 mM Tris buffer, pH 8.3 with 1 mM sodium ascorbate, 30 μM DCPIP and 60% (v/v) glycerol, frozen to 15 K in darkness, and illuminated in the EPR cavity with train of 10-ns flashes from a Nd:YAG laser. The data are normalized to the average value of the last two flashes points.

Conclusion

Cysteines in positions 556 and 565 in PsaB play a crucial role in attaining highly efficient electron transfer within PS I complex. Although in these mutants majority of electrons do not pass beyond F_X and back react to P_{700}^+ from F_X^- and A_1^-, a low efficiency leak to F_A/F_B still exists. On the other hand, substitutions of adjacent aspartates in F_X binding niche, which yields mutants capable of photoautotrophic growth, exerts much less or almost no effect on electron transfer in PS I. Based on this investigation, we conclude that a functional F_X is required for photoreduction of the F_A and F_B clusters at both room and cryogenic temperatures.

Acknowledgments. This work was supported by grants from the National Science Foundation (MCB-9205756) to J.H.G. and the Department of Energy (DE-FG02-90-ER20021) and M.S.U.-R.E.F. to L.M.

References

1. Lüneberg, J., Fromme, P. Jekow, P.and Schlodder, E. (1994) FEBS Lett. 338, 197-202
2. Sigfridsson, K., Hansson, Ö. and Brzezinski, P. (1995) Proc. Natl. Acad. Sci. USA. 92, 3458-3462
3. Smart, L.B., Warren, P.V., Golbeck, J.H. and McIntosh, L. (1993) Proc Natl Acad Sci USA. 90, 1132-1136
4. Vassiliev, I.R., Jung, Y.-S., Smart, L.B., Schulz, R, McIntosh, L. and Golbeck, J.H. (1995) Biophys. J. 69 (in press)
5. Golbeck, J. H. (1987) Biochim. Biophys. Acta. 895, 167-204
6. Brettel, K., and Golbeck, J.H. (1995) Photosynth. Res. (in press)

IDENTIFICATION OF THREE PHOTOACTIVE [4FE-4S] CLUSTERS IN PHOTOSYNTHETIC REACTON CENTER PREPARATIONS OF THE GREEN SULFUR BACTERIA *CHLOROBIUM VIBRIOFORME*

M. Paul Scott[1], Bodil Kjœr[2], Henrik Vibe Scheller[2] and John H. Golbeck[1]
1 University of Nebraska, Lincoln, Department of Biochemisty and 2 Royal Veterinary and Agricultural University, Department of Plant Biology, Copenhagen, Denmark

1. Introduction

The photosynthetic reaction center (RC) of green sulfur bacteria has been identified as an FeS type RC (1,2), related to photosystem I (PS I) of cyanobacteria and higher plants. PS I is characterized by the presence of three [4Fe-4S] clusters, designated F_A, F_B and F_X, which serve as terminal electron acceptors and play a key role in stabilizing the charge separated state. In the green sulfur bacterium *Chlorobium*, proteins predicted to contain contain iron sulfur clusters have been identified. Thus PscB contains the two clusters analagous to F_A and F_B in PS I (3), while a homodimer of PscA is considered to contain the cluster analogous to F_X (4). Many groups (5-11) have reported EPR spectra of *Chlorobium* RC or membrane preparations and have identified signals representing one or more of these iron sulfur clusters. A consensus has developed in which one EPR spectrum with a high-field depression near g=1.85 is seen on illumination of RCs at cryogenic temperatures (5,6,8-11), and another spectrum is seen in photoaccumulated or chemically reduced conditions with an inflection at about g=1.94 and a trough at about g=1.88 (5-9,11). Some groups have identified a third very small resonnance with a trough at about g=1.78 (8,9). We present evidence that the EPR signal with an inflection at about g=1.94 and a trough at about g=1.88 is composed of two signals overlaid and therefore contains a heretofore unidentified resonance in the*Chlorobium* RC. The two clusters yielding the composite spectrum plus the one giving rise to the g=1.85 signal are the F_A, F_B and F_X analogs of *Chlorobium*.

2. Results

2.1 *Preparation of the C. vibrioforme RC.*
By maintaining strictly anaerobic conditions throughout the preparation of the RC, we have prepared a high quality RC capable of photoreducing $NADP^+$ at a rate of 333 μmols NADPH mg bacteriochlorophyll $a^{-1}hr^{-1}$ (Kjær *et al.*, manuscript in review). SDS-PAGE

P. Mathis (ed.), Photosynthesis: from Light to Biosphere, Vol. II, 135–138.

(11) shows the RC contains PscA, PscB and PscC as well as polypeptides of 9 and 15 kDa and the Fenna-Matthews-Olson protein.

2.2 *Center I is reduced by light at 10K*
Illumination of the sample at 10K results in transfer of a single electron from P840 because in the frozen sample, $P840^+$ cannot be re-reduced by an exogenous electron. In this condition, the *Chlorobium* RC gives a rhombic spectrum with *g* values of 2.08, 1.90 and 1.85 (figure 1a). The cluster which this signal is derived from has been designated Center I (11) and similar spectra have been reported by other groups (5,6,8-11). The resonance did not dissappear when the light was turned off as shown in figure 1b, indicating that the charge separation between $P840^+$ and Center I is stable at 10 K.

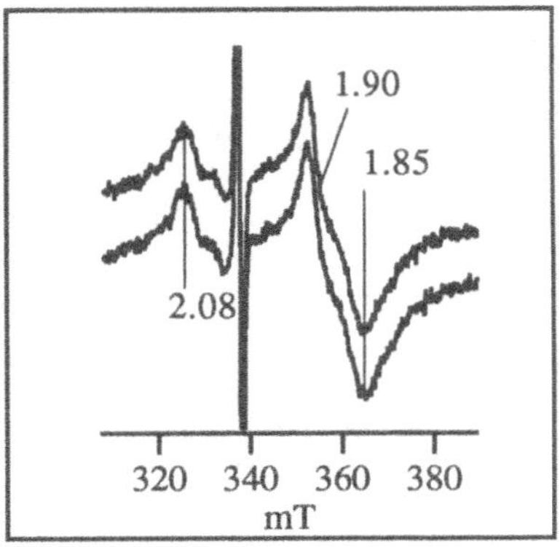

Figure 1. EPR spectra of the RC illuminated at 15K. A. Light minus dark. B. Light off minus dark.

2.3 *Mild chemical reduction reduces Center II*
In an effort to identify further electron carriers, we chemically reduced the RC by adding dithionite to 1.7 mM at pH 10.0. In a sample which was frozen and measured in darkness, we observed a rhombic spectrum with *g* values of 2.08, 1.92, and 1.88 (figure 2) and a temperature optimum of 12 K. This spectrum can be simulated using line widths of 70, 100, and 100 Mhz and *g*-values of 2.083, 1.915, and 1.877, consistant with the hypothesis that this signal is derived from a single [4Fe-4S] cluster. In keeping with the nomenclature of reference 11, we designate this cluster Center II. Because this is the only cluster chemically reduced in these conditions, its midpoint potential must be the highest of the clusters in theRC complex. At or below 6K an additional small trough at about g=1.94 could be seen in the chemically reduced sample, suggesting the presence of another cluster with FAster relaxation properties than Center II. The small size of this trough probably indicates that this center is partially reduced under these conditions and therefore has a lower mid-point potential than Center II.

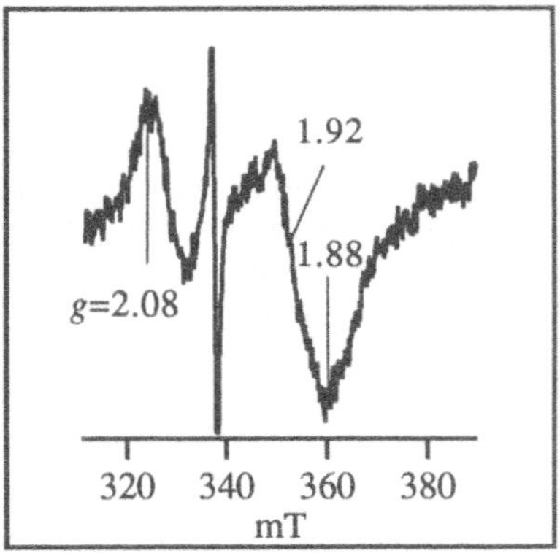

Figure 2. EPR spectra of the chemically reduced RC.

2.4 *Photoaccumulation of the reduced sample results in reduction of two [4Fe-4S] clusters*
We next froze the reduced sample under illumination in order to accumulate electrons in the acceptor system. The resulting spectrum (figure 3) is similar to previously reported ones (5-9,11). The spectrum has an inflection point at g=1.94 and a trough at g=1.88. Since this sample contains a chemically reduced center which gives a trough at g=1.88 (Center II), it is likely that this signal represents two spectra overlaid. Also, the trough of the inflection (g=1.91) corresponds to the position of the Fast relaxing cluster observed in the chemically reduced sample. We therefore examined the relaxation properties of the photoaccumulated spectrum in an attempt to de-convolute the two signals contained within it. Above 18K, the$g$=1.88 trough was the prominant feature, while below 6K it was not detectable, and the g=1.91 trough was prominant. Since it is not possible for the high-field trough to vary in intensity independantly of the mid-field inflection in a single EPR signal, this signal must be derived from two signals overlaid. The clusters differ in their spin relaxation properties and hence exhibit different temperature behavior. Below 10 K, the trough at g=1.91 is more prominant with a maximum signal intensity of 8 K. The g=1.88 trough is prominant above 10 K with a maximum signal intensity of 12K.

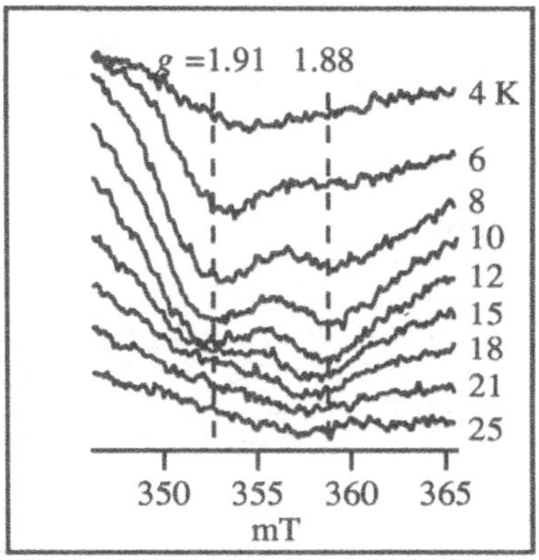

Figure 3. EPR spectra of the RC following chemical reduction and freezing during illumination.

Because the spectrum of Center II contains a trough at g= 1.88 and has a temperature optimum of 12K like the g=1.88 trough in figure 3, we conclude that Center II contributes the g=1.88 trough to the spectrum shown in figure 3. The signal intensity of Center II is higher in figure 3 than in the chemically reduced sample. This indicates that this center is photoactive and that the chemical reduction of both clusters was incomplete. We conclude that the g=1.91 trough in the photoaccumulated spectrum is actually a feature of the spectrum of a heretofore unidentified electron acceptor which we designate Center III. While we have never observed a spectrum of this component individually, from the composite spectrum (figure 3) we know that it should undergo rapid spin relaxation, having a temperature optimum of about 8 K. Further, becuase the Center III signal is smaller than the Center II signal when reduced chemically, it has a mid-piont potential lower than Center II.

3. Discussion

3.1 *Identification of three [4Fe-4S] clusters*
While the primary structures of the subunits PscA and PscB indicate the presence of three [4Fe-4S] clusters in the *Chlorobium* RC, the EPR spectra of these clusters have been a subject of controversy. Our RC preparations give EPR signals similar to others in the

literature (5-11). We demonstrate that two EPR spectra (figures 1 and 3) contain contributions from three [4Fe-4S] clusters of the *C. limicola* RC. Two observations support our conclusion that the spectrum in figure 3 is derived from two clusters. First, the temperature dependence of the signal shows that the two troughs belong to different resonances. Second, we have obtained a spectrum from the same sample (figure 2) which represents a single component (Center II) of the complex spectrum shown in figure 3.

3.2 *Analogy with PS I*

It is important to assign each *Chlorobium* cluster to its counterpart in PS I. F_A, F_B and FX were originally identified by the *g*-values of their EPR signals (12). None of the [4Fe-4S] clusters of *C. vibrioforme* resemble any of the PS I clusters on this basis. This is not surprising given the large structural differences between the *Chlorobium* RC and PS I. The assignment can better be based on the mid-point potentials of the clusters. F_A and F_B are reported to have mid-point potentials of -530 and -580 mV, respectively (13), while the F_X potential is -705 mV (14). We have determined that Center II has a higher mid-point potential than Center III, establishing that it is not the F_X analog. Also, Center III is chemically reduced when Center I is not, indicating that the mid-point potential of Center III is between those of Centers II and I. This makes Center III an unlikely F_X analog as well. Our data therefore indicates that Center I is the most likely F_X analog. This makes the*Chlorobium* RC is different from PS I in that electrons accumulate at F_X under illumination at 10 K. More data are needed to verify the assignment of the Chlorobium EPR signals to their corresponding iron-sulfur clusters. Correlation of these clusters with the presence of the PscB polypeptide or determination of their redox potentials may solve this problem.

References

1. Lokau,W., and Nitschke, W., (1993) Physiol. Plantarum, 88:371-381.
2. Blankenship, R. (1994) Anton Leeuwenhoek Int J Gen M, 65: 311-329.
3. Illinger, N., Xie, D-L., Hauska, G. and Nelson, N. (1993) Photosynthesis Res. 38:111-114.
4. Büttner, M., Xie, D-L., Nelson, H., Pinther, W., Hauska, G., and Nelson, N. (1992) Proc. Natl. Acad. Sci. USA 89:8135-8139.
5. Kusumoto, N., Inoue, K., Hidekazu, N., and Sakurai, H. (1994) Plant Cell Physiol. 35:17-25.
6. Feiler, U., Nitschke, W., and Michel, H. (1992) Biochemistry 31:2608-2614.
7. Miller, M. Liu, X., Snyder, S., Thurnauer, M. and Biggins, J. (1992) Biochemistry 31:4354-4363.
8. Oh-oka, H., Kakutani, S., Matsubara, H., Malkin, R., and Shigeru, I. (1993) Plant Cell Physiol. 34:93-101.
9. Nitschke, W., Feiler, U., and Rutherford, A. W. (1990) Biochemistry 29:3834-3842.
10.Jennings, J., and Evans, M. (1977) FEBS Lett. 75:33-36.
11. Kjœr, B, Jung, Y-S, Yu L, Golbeck JH, Scheller HV (1994) Photosynth. Res. 41: 105-114.
12. Evans, M., Reeves, S. and Cammack, R. (1974) FEBS Lett. 49:111-114.
13. Ke, B., Hansen, R., and Beinert, H., (1973) Proc. Natl. Acad. Sci. USA, 70:2941-2945.
14. Chamorovsky, S., and Cammack, R. (1982) Photobiochem. Photobiophys. 4, 195-200.

HEAT-INDUCED PIGMENT ALTERATION IN THE PHOTOSYSTEM I AND II REACTION CENTERS

Masami Kobayashi[1], Masayo Iwaki[2], Tatsuya Tomo[2,3], Hideo Kise[1], Kimiyuki Satoh[2,3], Shigeru Itoh[2]

[1] Inst. Materials. Sci., Univ. Tsukuba, Tsukuba, Ibaraki 305, [2] Natl. Inst. Basic Biol., Okazaki, Aichi 444, [3] Fac. Sci., Okayama Univ., Okayama, Okayama 700, JAPAN

1. INTRODUCTION

Primary donor of photosystem (PS)1, P700, can be concentrated by Ikegami's ether extraction method [1,2]. In this sample, ca. 10 molecules of chlorophyll (Chl) *a*, one molecule of Chl *a'* and one molecule of Chl *b* per P700 are present. Recently we found a heat-induced drop in P700 photoactivity in ether-treated PS1 particles in a short incubation even at 30 ℃, although ether-untreated PS1 particles are very tolerant to high temperature. Heat-induced inactivation has been also observed in D1/D2/cyt b559 RC2 complex without ether-treatment. However, little is known about the reason. Here we report on the heat-induced pigment alteration in PS1 particles and in PS2 RC.

2. MATERIALS AND METHODS

The ether extraction of PS1 particles from spinach was carried out as reported before [1]. D1/D2/cyt b559 RC complex was prepared essentially as described before [3,4]. The samples dissolved in a buffer were incubated for 10 min at various temperature under air. Pigments were then extracted by acetone and analyzed by normal-phase HPLC as performed before [4,5]. P700 activities were measured by chemical

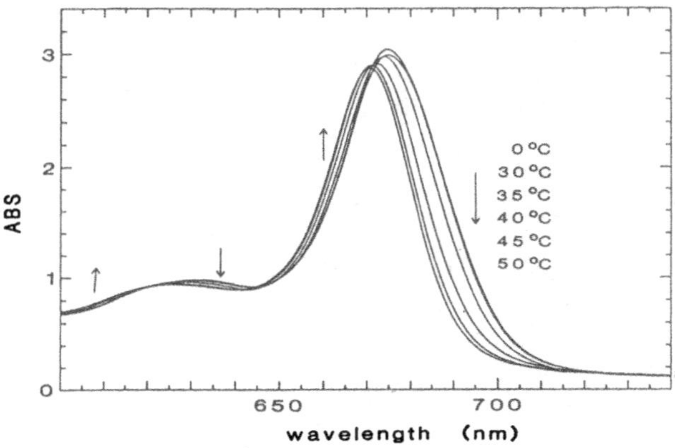

Fig. 1 Absorption spectral change of ether-treated PS1 particles after incubation at several temperature for 10 min.

P. Mathis (ed.), Photosynthesis: from Light to Biosphere, Vol. II, 139–142.

oxidation methods.

3.1 Heat-induced absorption spectral change of PS1

Fig. 1 shows the absorption spectra in the red band region of ether-treated PS1 particles after heat-incubation. Absorption in the longer wavelength was clearly decreased and shorter one was slightly increased with increasing the temperature. Approximately 20 % reduction of absorption area in the region of 640nm~ 730nm was observed at 50℃, indicating some pigments might get out of the original position.

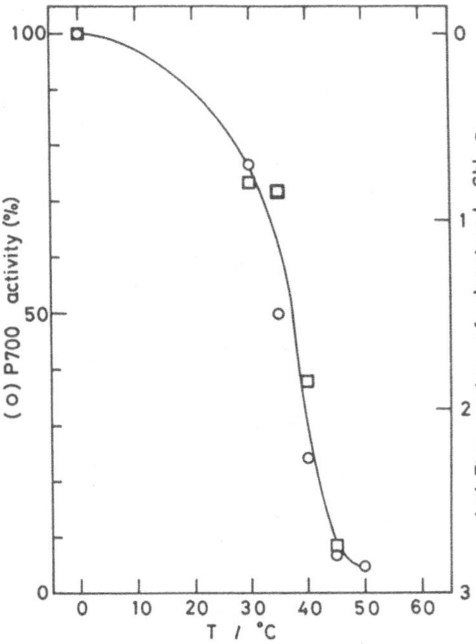

Fig. 2 Heat-induced reduction of P700 activity of ether-extracted PS1 particles and the number of denatured Chl *a*.

3.2 Heat-induced reduction of P700 activity

P700 activities were decreased with increasing incubation temperature. Relative activity is 100 at 0℃, 75 at 30℃, 50 at 35℃, 25 at 40 ℃, 7 at 45℃ and 5 at 50℃ (Fig. 2).

3.3 HPLC analysis of heat-treated PS1 particles

Fig. 3 shows typical HPLC traces for acetone extract of PSI particles incubated at 0℃ and 45℃ for 10 min. No change has been observed before and after incubation at 0℃ : pigment composition is Chl *a* : Chl *a'* : Chl *b* = 10 : 1 : 1. However, PSI particles incubated at 45℃ have shown the production of significant amount of artifacts (X, Y and Z). Main artifact, Z, has been identified as an oxidized Chl *a*, C10-OH-Chl *a* (Fig. 3), and other artifacts, Y and Z, also seem to be allomers on the basis of their retention times.

Results of HPLC analyses are summarized in Fig. 4. Some Chls *a* were oxidized to yield C10-OH-Chl *a* during heating: one molecule of Chl *a* at 35℃, two molecules at 40℃ and three molecules at 45℃ were oxidized. No alteration of Chl *a'* and Chl *b*, however, has not been observed. Under unaerobic condition, C10-OH-Chl *a* was not produced even at high temperature (data not shown), suggesting some molecules of Chl *a* in P700-enriched PSI particles are very sensitive to oxidation under air even at room temperature.

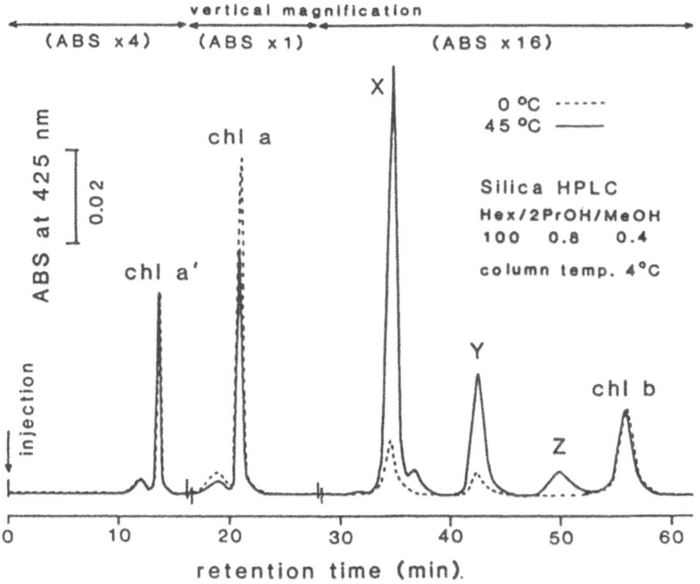

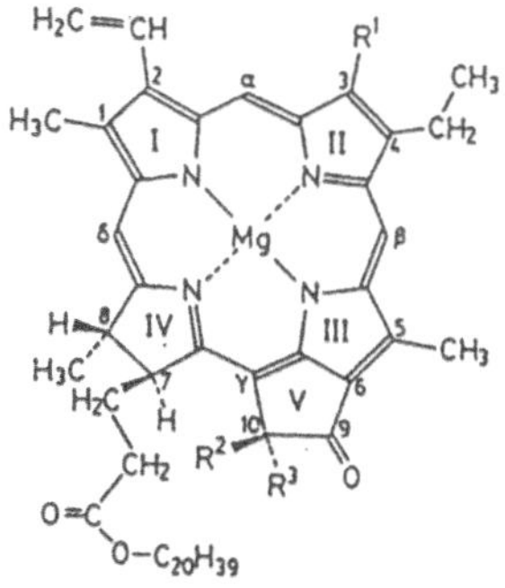

	R^1	R^2	R^3
Chlorophyll a	CH_3	H	$COOCH_3$
Chlorophyll a'	CH_3	$COOCH_3$	H
Chlorophyll b	CHO	H	$COOCH_3$
X	CH_3	OH	$COOCH_3$

Fig. 3 HPLC traces of acetone extract of heat-treated PS1 particles (Top) and molecular structure of chlorophylls (Bottom).

3.4 HPLC analysis of heat-induced D1/D2/cyt b559 complex

Fig. 5 exhibits the pigment composition of D1/D2/cyt b559 RC2 complex incubated at several temperatures. No alteration of Chl *a*, Pheo *a* and β -Car (6:2:2) in D1/D2/cyt b559 reaction center complex has been observed even at 70℃, indicating the high tolerance of pigments in the RC2 complex to oxygen.

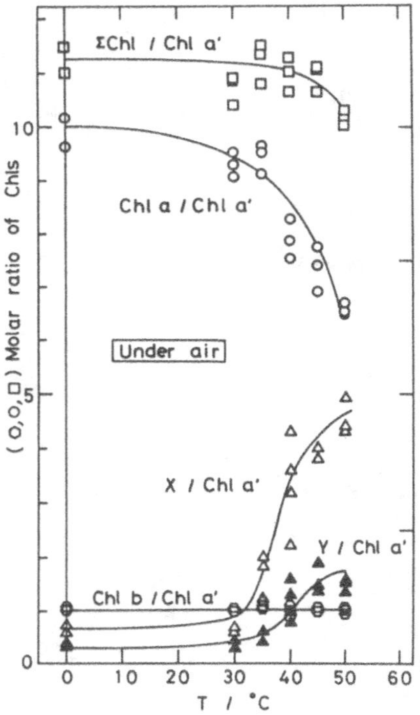

Fig. 4 Pigment composition of ether-treated PS1 particles after incubation at several temperatures for ten min determined by HPLC analyses.

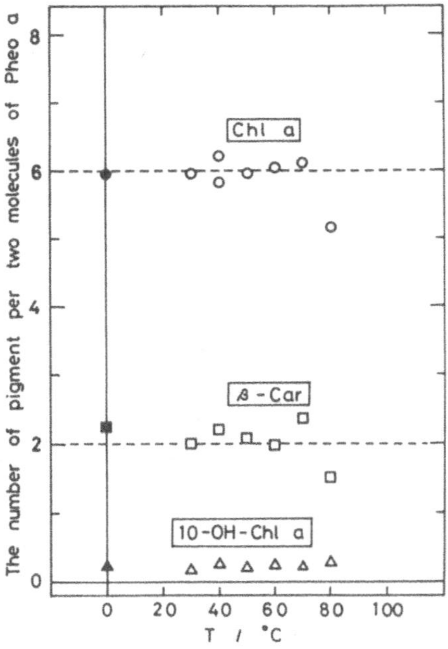

Fig. 5 Pigment composition of D1/D2/cyt b559 complex incubated at several temperatures for ten min determined by HPLC analyses.

Summary

We thus demonstrated that P700 activity of ether-treated PS1 particles are easily affected by heat and some Chls *a* in this particles are quite sensitive to oxygen even at room temperature, although pigments in D1/D2/cyt b559 RC2 complex are very tolerant to oxygen even at high temperature.

REFERECES

[1]I. Ikegami and S. Katoh (1975) *Biochim. Biophis. Acta*, **376**, 585-592.
[2]I. Ikegami et al., (1995) *Plant Cell Physiol.*, **36**, 857-864.
[3]O. Nanba and K. Satoh (1987) *Natl. Acad. Sci. USA*, **84**, 109-112.
[4]M. Kobayashi et al. (1990) *FEBS Lett.*, **260**, 138-140.
[5]M. Kobayashi et al. (1988) *Biochim. Biophys. Acta*, **936**, 81-89.

QUINONE BINDING IN *Rhodobacter Sphaeroides* AND PHOTOSYSTEM I STUDIED BY TRANSIENT EPR.

A. van der Est[1], I. Sieckmann[1], W. Lubitz[2] and D. Stehlik[1]
[1]Fachbereich Physik, Freie Universität Berlin,
Arnimallee 14, 14195 Berlin (Germany).
[2]Max Volmer Institut, Technische Universität Berlin,
Straße des 17. Juni 135, 10623 Berlin (Germany)

1. Introduction

In reaction centres of purple bacteria (bRC's) and Photosystem I (PS I), light excitation leads to a transmembrane electron transfer from the excited singlet state of the primary donor, P, to a quinone acceptor, Q, on a picosecond timescale. In bRC's of *Rhodobacter sphaeroides,* P is a bacteriochlorophyll-a dimer, P_{865}, and the primary quinone acceptor, Q_A, is ubiquinone-10 (UQ_{10}). In PS I the donor, P_{700}, is probably a chlorophyll-a dimer and the acceptor A_1 is phylloquinone also known as vitamin K_1 (VK1). In contrast to the charge separation from P to Q, which is similar in the two systems, the subsequent steps are different. In bRC's, electron transfer from $Q_A^{-\cdot}$ to the secondary quinone acceptor Q_B occurs on a timescale of 0.1 ms [1]. In PS I the iron-sulphur centre F_X accepts an electron from $A_1^{-\cdot}$ with a rate constant k^{-1} = 280 ns [2, 3]. One possibility for investigating the relationship between the function and nature of the quinones is to systematically vary their properties by exchanging the native species with artificial ones having the desired properties. Several such studies have been carried out on bRC's and PS I (see [4] and references therein). The lifetimes of the states $P_{700}^{+\cdot}A_1^{-\cdot}$ and $P_{865}^{+\cdot}Q_A^{-\cdot}$ allow them to be studied conveniently using transient EPR. Their EPR spectra, which are described by the correlated, coupled radical pair (CCRP) mechanism (see [5, 6] for reviews and [7] with references therein), show characteristic spin polarization patterns as a result of the sudden light induced charge separation from a singlet precursor. Calculations of such spectra at various microwave frequencies [7–9] show that they are particularly sensitive to the orientation of the quinone relative to the dipolar coupling vector, z_d, which joins $P^{+\cdot}$ and $Q^{-\cdot}$. A comparison of the results for PS I of cyanobacteria and Zn-substituted RC's of *Rb. sphaeroides* (Zn-bRC's) [4, 7] indicates that the quinone orientation is

P. Mathis (ed.), Photosynthesis: from Light to Biosphere, Vol. II, 143–146.

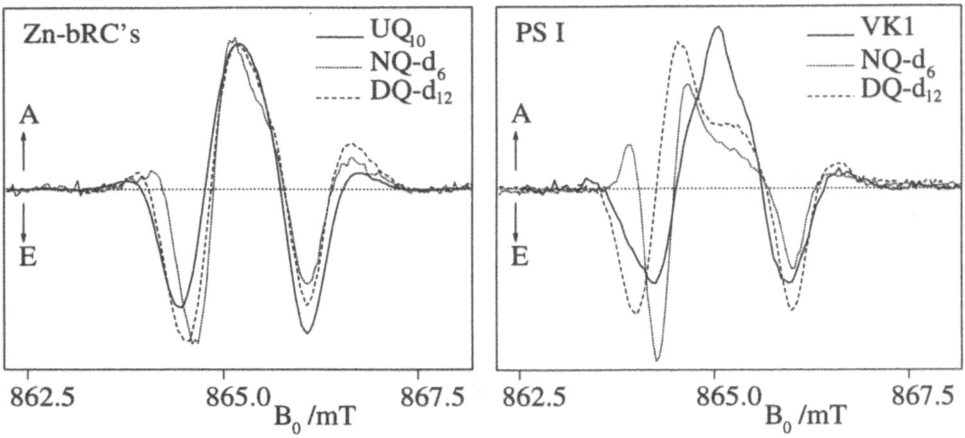

Figure 1 Comparison of K-band transient EPR spectra at 50 K of Zn-bRC's and PS I containing various quinones. Zn-bRC spectra(left): time window 50-300 ns, microwave power/frequency: 0.32 mW, 24.2695 GHz. PS I spectra (right): time window 0.2-1.2 μs, microwave power/frequency: 6 mW, 24.227 GHz. The magnetic field values for the PS I spectra have been adjusted to correspond to a microwave frequency of 24.2695 GHz. A linear baseline correction has been performed to remove weak signals from $^3P_{865}$ and $^3P_{700}$.

different in the two systems. Here, we will report transient EPR results on PS I and Zn-bRC's in which the native quinones have been replaced with fully deuterated naphthoquinone (NQ-d_6) and duroquinone (DQ-d_{12}).

2. Materials and Methods

The quinone exchange procedures and methods of sample preparation are described in [4]. The transient EPR spectrometer and the methods used to extract the spectra from the data sets are given in [10].

3. Results and Discussion

Fig. 1 shows the K-band transient EPR spectra of Zn-bRC's and PS I containing the respective native quinones, UQ_{10} and VK1, as well as fully deuterated naphthoquinone (NQ-d_6) and duroquinone (DQ-d_{12}). It is immediately apparent that the spectra of $P^{+\cdot}Q^{-\cdot}$ in PS I differ from one another whereas the Zn-bRC spectra are very similar. Simulations of the spectra [4] show that NQ-d_6, DQ-d_{12} and UQ_{10} all have the same orientation in Zn-bRC's. This is consistent with independent experimental information that $Q^{-\cdot}{}_A$ is hydrogen bonded to the protein. Our results indicate that this bonding is similar for the three quinones and that it determines their orientation. In PS I, on the other hand, NQ-d_6, DQ-d_{12} and VK1 are all oriented differently and none of the quinones has the orientation found in Zn-bRC's. The polarization pattern on the low field side of the spectra can be used to estimate the relative orientation of z_d and the quinone x and y-axes which lie in the molecular plane with x parallel to

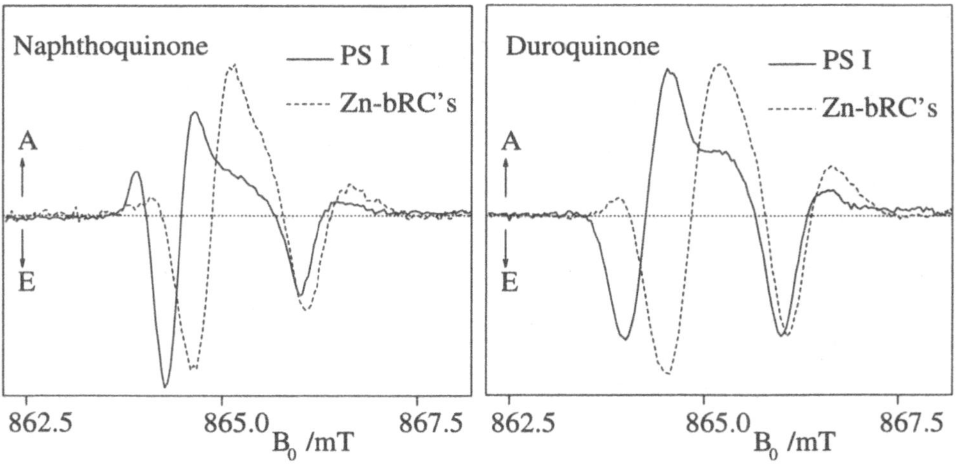

Figure 2 Comparison of K-band transient EPR spectra at 50 K of Zn-bRC's and PS I containing NQ-d_6 (left) and DQ-d_{12} (right). The experimental conditions are given in fig. 1.

the carbonyl bonds. An analysis of the pattern observed with fully deuterated PS I samples [7, 8] (not shown) indicates that VK1 is oriented with its x-axis parallel to the dipolar vector, z_d. For NQ-d_6 (Fig. 1 right, dotted curve), the low field A/E/A polarization pattern shows that in this case the quinone y-axis is parallel to z_d. For DQ-d_{12}, z_d lies in the xy-plane roughly half way between x and y. These differences clearly point to a different type of bonding in PS I. Additional evidence for this comes from the width of the spectra in the two RC's. As can be seen in Fig. 2 the width of the spectra in the two species is different although they contain the same quinone. From these spectra it is obvious that the anisotropy in the quinone g-tensor is larger in PS I. In semiquinone radicals, the g-anisotropy is due to deviations of the x and y components from the free electron value. These deviations are given [11–13] by

$$\Delta g_{xx} = \frac{2\xi\rho_o^\pi C_y^2}{\Delta E_{n\pi^*}} \qquad \Delta g_{yy} = \frac{2\xi\rho_o^\pi C_x^2}{\Delta E_{n\pi^*}}. \tag{1}$$

The values of Δg_{xx} and Δg_{yy} are determined largely by the energy difference, $\Delta E_{n\pi^*}$, between the lone pair n-orbitals on the oxygen atoms and the half filled π^* orbital. In the case of H-bonding, the stabilization of the n-orbitals leads to an increase in $\Delta E_{n\pi^*}$ and thus to reduction in the anisotropy. In contrast, $\pi - \pi^*$ interactions, for example with a neighbouring amino acid residue, stabilize the π^* orbitals which leads to a reduction of the energy gap and thus to an increase in the g-anisotropy. The larger g-anisotropy in PS I therefore suggests that A_1 may be bound by $\pi - \pi^*$ interactions and not hydrogen bonds. This is also consistent with the fact that the orientation of DQ-d_{12}, NQ-d_6 and VK1 in PS I differ by a rotation about the normal to the quinone plane [4].

4. Conclusions

The differences between the transient EPR spectra of $P^{+\cdot}Q^{-\cdot}$ in quinone exchanged bRC's and PS I indicate that: (i) All quinones studied have the same orientation in bRC's whereas in PS I they all have different orientations. (ii) The anisotropy of the quinone g-tensors is significantly larger in PS I than in bRC's. The nature of the differences suggest that in bRC's the quinones are hydrogen bonded to the protein whereas in PS I $\pi - \pi^*$ interactions appear to hold the quinone in place.

5. Acknowledgements

This work was supported by grants from the Deutsche Forschungsgemeinschaft (Sfb 312 and Sfb 337).

References

1 Kleinfeld, D., Okamura, M. Y., and Feher, G. (1984) Biochemistry 23, 5780–5786 .
2 Brettel, K. (1989) FEBS Lett. 239, 93–98.
3 Bock, C. H., van der Est, A. J., Brettel, K., and Stehlik, D. (1989) FEBS Lett. 247, 91–96.
4 van der Est, A., Sieckmann, I., Lubitz, W., and Stehlik, D. (1995) Chem. Phys. 194, 349–359.
5 Hore, P. J. (1989) in Advanced EPR, Applications in Biology and Biochemistry (Hoff, A. ed.), pp. 405–440, Elsevier, Amsterdam.
6 Snyder, S. W. and Thurnauer, M. C. (1993) in The Photosynthetic Reaction Center (Norris, J. R. and Deisenhofer, J.eds.), volume II, Academic Press, New York.
7 Füchsle, G., Bittl, R., van der Est, A., Lubitz, W., and Stehlik, D. (1993) Biochim. Biophys. Acta 1142, 23–35.
8 Stehlik, D., Bock, C. H., and Petersen, J. (1989) J. Phys. Chem. 93, 1612–1619.
9 van der Est, A. and Stehlik, D. (1995) in Reaction Centers of Photosynthetic Bacteria (Michel-Beyerle, M. ed.), Springer, Berlin, in press.
10 Stehlik, D., Bock, C. H., and Thurnauer, M. C. (1989) in Advanced EPR Applications in Biology and Biochemistry (Hoff, A. ed.), pp. 370–403, Elsevier, Amsterdam.
11 Burghaus, O., Lubitz, W., Plato, M., MacMillan, F., Rohrer, M., and Möbius, K. (1993) J. Phys. Chem. 97, 7639–7647.
12 Stone, A. J. (1963) Proc. Roy. Soc. A271, 424.
13 Stone, A. J. (1963) Mol. Phys. 6, 509.

ΔG DEPENDENCE OF THE RATE CONSTANT OF $P700^+A_0^- \rightarrow P700^+Q^-$ REACTION IN THE QUINONE-RECONSTITUTED PSI REACTION CENTER.

Masayo Iwaki[1], Shigeichi Kumazaki[2], Keitaro Yoshihara[2], Tatsuo Erabi[3] and Shigeru Itoh[1], [1]Natl. Inst. Basic Biol., [2]Inst. Mol. Sci., Okazaki 444 and [3]Tottori Univ., Tottori 680 (Japan)

1. Introduction

The time constants (1/k) of the charge separation between the donor chlorophyll (P700) and the acceptor chlorophyll (A_0), and the electron transfer from A_0^- to the phylloquinone (PhyQ) in photosystem (PS) I reaction center (RC) has been shown to be 6-28 ps and 21-200 ps, respectively [1-4]. We measured the 1/k values of $A_0^-Q \rightarrow A_0Q^-$ to be 23, 23 and 34 ps with PhyQ, menaquinone and methylnapthoquinone reconstituted as Q [4]. The values are shorter than the time constant (200 ps) of the reaction from the primary acceptor bacteriopheophytin (H) to ubiquinone (Q_A) in the RC of purple bacterium *Rb. sphaeroides* [5]. In this study, the rates (k values) of $A_0^-Q \rightarrow A_0Q^-$ reaction were measured by ps- and ns-laser spectroscopy at 280 K. The energy gap (ΔG) was varied between -1.1 and +0.2 eV by reconstitution of 13 different quinone/quinonoid molecules (a-m in Fig. 1) to PS I PhyQ site.

2. Materials and Methods

Lyophylized spinach PS I particles were extracted with diethyl ether to obtain 30 Chl/P700 particles that lack PhyQ and carotenoids but retains A_0, F_X and F_A/F_B. The reaction mixture contained 50 mM Tris-Cl (pH 7.5), 0.05% Sucrose Monolaurate and 0.5 μM P700. Quinones dissolved in dimethyl sulfoxide were added and incubated overnight at 0 ^oC [6].

The ps pump-probe spectra were measured using a dye laser (1 ps, 605 nm, 10 Hz) with 2-ps and 1.5-nm resolution, at 4 ^oC. The ns spectra were measured with a Nd-YAG laser (15 ns, 532 nm) with 3.5-ns time resolution [7].

In each measurement, transient spectra were measured in the presence of 0.2 mM ferricyanide (ΔA under P700-preoxidized condition) firstly, and then with 10 mM ascorbate and 2 μM phenazine methosulfate (ΔA under P700-prereduced condition). The difference between ΔA (P700-prereduced) minus ΔA(P700 preoxidized) at the same delay time was calculated to extract the electron transfer process by eliminating the absorption changes of excited antenna chlorophylls [4].

P. Mathis (ed.), Photosynthesis: from Light to Biosphere, Vol. II, 147–150.

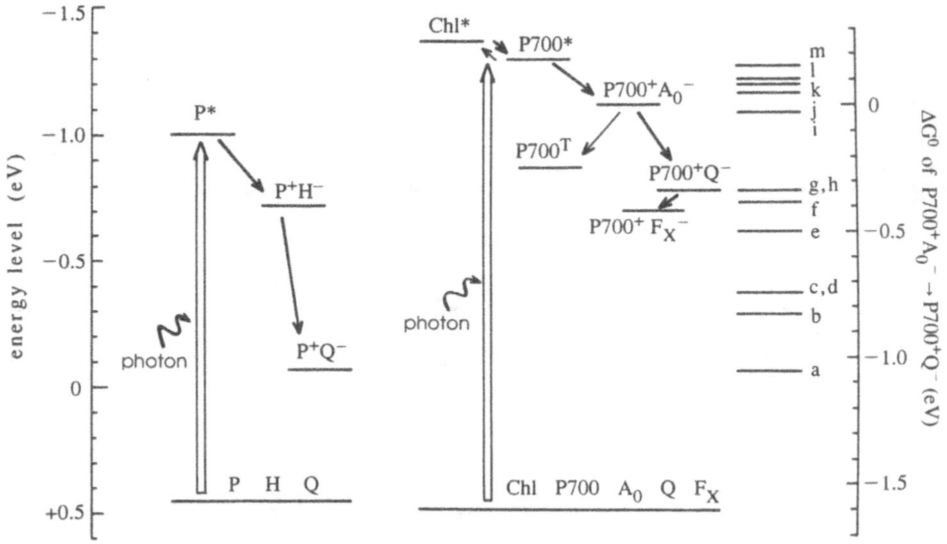

Fig. 1. Energy diagram and electron transfer path in RCs of PS I (right) and purple bacteria (*Rb. sphaeroides*) (left). Energy levels of P700$^+$Q$^-$ in the Q-reconstituted PS I are also shown. a)2,4,5,7-tetranitro-FL, b)2,4,7trinitro-FL, c)2,3Br$_2$-NQ, d)2,3Cl$_2$-NQ, e)3anilino-2chloro-NQ, f)2methyl-NQ, g)menaquinone, h)phylloquinone, i)2amino-AQ, j)1,2-diamino-AQ, k)anthraldehyde, l)FL, m)2,6diamino-AQ. NQ=1,4-naphthoquinone, AQ=9,10-anthraquinone, Fl=9-fluorenone

3. Estimation of ΔG values of the P700$^+$A$_0^-$Q→P700$^+$A$_0$Q$^-$ reaction.

ΔG between A$_0^-$Q and A$_0$Q$^-$ states was calculated from the $E_m(Q)$ and $E_m(A_0)$. We first estimated the E_m in situ at the PhyQ site of Q_{e-i}. Redox state of P700, Q and FeS centers, which were monitored after the flash light, were simulated by assuming the dynamic equilibration of electrons among the components. Then, ΔE_m between $E_m(F_X)$ and each $E_m(Q)$ was estimated. With $E_m(Fx)$=-700 mV as a standard, we calculated $E_m(Q)$ values. $E_m(Q_{e-i})$ values thus estimated were found to be shifted by -310 mV from their $E_{1/2}$ values determined polarographically in dimethylformamide. The E_m of other molecules that are more positive than $E_m(Q_e)$ or more negative than $E_m(Q_i)$, respectively, were also estimated to have the same -310 mV shift to their $E_{1/2}$s in DMF. The method seems fairly reliable since the equilibrium between A$_0$ and Q_{j-m} after the flash was also reasonably interpreted by these E_m values [see ref. 6 for details]. The $E_m(A_0)$ was estimated to be -1120 mV (180 mV lower than the P700* level with $E_m(P700)$= +480 mV). Horizontal error bars for ΔG in Fig. 2 represent the range of E_m(Q) that reflect different $E_m(Fx)$ (-650 mV to -730 mV) and $E_m(A_0)$ values in literature [6,7].

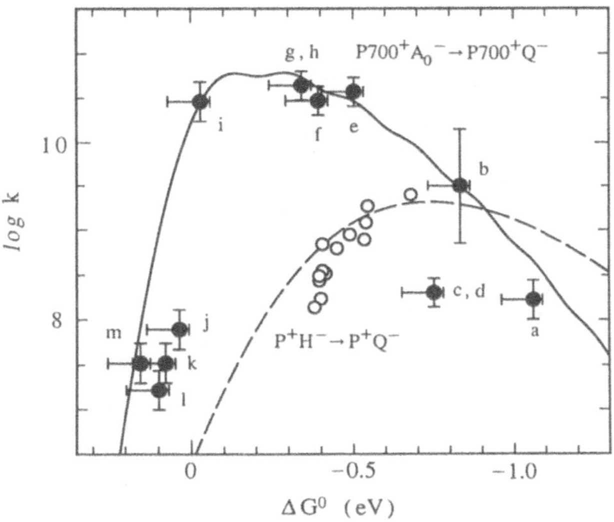

Fig.2. Plot of log k versus ΔG of the $P700^+A_0^-Q \rightarrow P700^+A_0Q^-$ reaction in the PS I RC (●). Similar plot (○) of the $H^-Q_A \rightarrow HQ_A^-$ reaction in the Rb. sphaeroides *RC [after ref.5]. The curves were drawn by Eqn 1.*

$$k_{ET} = \frac{2\pi}{\hbar} H_{rp}^2 (4\pi\lambda_{out} k_B T)^{-1/2} \times \sum_{m=0}^{\infty} \left[\frac{e^{-S} S^m}{m!}\right] \exp\left[-\frac{(\lambda_{out} + \Delta G^0 + mh\nu)^2}{4\lambda_{out} k_B T}\right] \quad (1)$$

$$S = \frac{\lambda_{in}}{h\nu}$$

$$H_{rp} = H_{rp}^0 \exp\left(\frac{-\beta(r - r_0)}{2}\right) \quad (2)$$

$$\lambda_{out} = \frac{(\Delta e)^2}{4\pi\epsilon_0}\left(\frac{1}{2a_D} + \frac{1}{2a_A} - \frac{1}{r}\right)\left(\frac{1}{\epsilon_{op}} - \frac{1}{\epsilon_s}\right) \quad (3)$$

4. Results and Discussion

4-1. Rate of the $P700^+A_0^-Q \rightarrow P700^+A_0Q^-$ reaction in Q-reconstituted RC.

In the PS I RC depleted of PhyQ, we observed a rise of bleach of P700 (formation of P700*) within 1-ps, and a 8±2-ps rise time of A_0^- signal as the bleach at 685-nm peak [4] followed by a very slow decay of >10 ns. After reconstitution of Q_{e-i} whose E_m's are not very different from that of PhyQ=Q_h, A_0^- decayed with 22-30 ps. Quinones with more negative (j-m) or more positive (a-d) estimated E_m values gave k values significantly smaller [7]. The rate constant seemed to be mainly determined by $E_m(Q)$ rather than the molecular structure.

4-2. Energy gap dependence of the rate of $A_0^-Q \rightarrow A_0Q^-$ reaction.

Fig. 2 shows a plot of log k vs ΔG of $A_0^-Q \rightarrow A_0Q^-$ reaction. Each point corresponds to a-m quinone/quinonoid in Fig.1. The data were fitted with a curve calculated by the quantum mechanical expression (Eqn. 1) of the Marcus theory [8]. Similar plot in the $H^-Q_A \rightarrow HQ_A^-$ reaction [5] was also shown. The different parameters indicate the wide capacity of protein matrix in controlling the quinone reactions [7].

4-3. Reorganization energy, λ, of $A_0^-Q \rightarrow A_0Q^-$ reaction.
1) λ_{total} of 0.3 eV is about a half that in $H^-Q_A \rightarrow HQ_A^-$ reaction.
2) The low λ_{total} gives a low medium reorganization energy λ_{out} if λ_{in}s of Q and A_0 are similar to that of H and Q_A [4].
3) The low λ_{out} cannot be explained by the difference in distance and suggests the low dielectric constant (ε_s) of protein (Eqn. 3).
4) The low ε_s also explain the extremely low $E_m(PhyQ)$ of -0.8 V.

4-4. Optimization of the reaction rate with PhyQ.
5) Intrinsic PhyQ (ΔG=-0.34eV) gives the highest rate indicating the optimization of the reaction, i.e. at the $\lambda = -\Delta G$ condition.
6) The optimization with the small ΔG is accomplished by the low λ_{out}.
7) Electronic coupling (H_{rp}) is one order higher than that in $H \rightarrow Q_A$ reaction. If H_{rp}^{o}s of both reactions are the same, an empirical relation by Moser et al.[9] gives a 0.78 nm edge-to-edge distance between A_0 and PhyQ, shorter than that (1.0 nm) between H and Q_A.
8) The k decreases at low ΔG (normal region) and less sharply at high ΔG (inverted region). This suggests the coupling of reaction with the vibrational modes.

5. Conclusion

Because of the small λ_{out}, the $A_0 \rightarrow$PhyQ reaction in PS I is set to be very fast with a low energy loss of 0.3 eV that is about a half that in the $H \rightarrow Q_A$ reaction in *Rb. sphaeroides*. The small λ_{out} seems to be accomplished by the low ε of the proteinaceous environments [7]. The PS I RC thus is highly optimized to produce the strong reducing power necessary for the reduction of ferredoxin in fixing CO_2.

References

1. Wasielewski,M.R., Fenton,J.M., Govindjee (1987) *Photosynth. Res.* 12, 181-190.
2. Shuvalov,V.A., Nuijis,A.M., van Gorkom,H.J., Smit,H.W.J., Duysens,L.N.M. (1986) *Biochim. Biophys. Acta* 850, 319-323.
3. Hastings,G., Kleinherenbrink,F.A.M., Lin,S., McHugh,T.J., Blankenship,R.E. (1994) *Biochemistry* 33, 3193-3200.
4. Kumazaki,S., Iwaki,M., Ikegami,I., Kandori,H., Yoshihara,K., Itoh,S. (1994) *J. Phys. Chem.* 98, 11220-11225.
5. Gunner,M.R., Dutton,P.L.(1989) *J. Am. Chem. Soc.* 111, 3400-3412.
6. Iwaki,M., Itoh,S. (1994) *Plant Cell. Physiol.* 35, 983-993.
7. Iwaki,M., Itoh,S. (1995) submitted.
8. Marcus,R.A., Sutin,N. (1985) *Biochim. Biophys. Acta* 811, 265-322.
9. Moser,C.C., Dutton,P.L. (1992) *Biochim. Biophys. Acta* 1101, 171-176.

ANTISENSE REPRESSION OF *PsaE* mRNA IN TRANSGENIC BARLEY (*Hordeum vulgare* L.)

Søren Kjærulff, Birger Lindberg Møller and Henrik Vibe Scheller, Plant Biochemistry Laboratory, Department of Plant Biology, Royal Veterinary and Agricultural University, 40 Thorvaldsensvej, DK-1871 Frederiksberg C, Copenhagen, Denmark.

1. Introduction

The photosystem I (PSI) complex of barley (*Hordeum vulgare* L.)is composed of at least 13 different polypeptides (1) and catalyses the light driven electron transport from plastocyanin to ferredoxin through the five electron acceptors A_0, A_1, F_X, F_A, F_B (2). The PSI-E subunit is an extrinsic polypeptide located on stromal surface of PSI partly covering PSI-C (3). A cDNA encoding PSI-E was first isolated and characterized from barley (4). Subsequently *PsaE* cDNAs and genes have been characterized from several species of plants and cyanobacteria (2). PSI-E is not essential for PSI function but is facilitating ferredoxin reduction (5). Studies of a *Synechococcus* sp. PCC 7002 *PsaE* inactivation mutant suggested that PSI-E is involved in cyclic electron transport around PSI (6). Targeted gene transfer by homologous recombination has not been achieved in the nuclear genome of plants. Inactivation of genes encoding components of the photosynthetic apparatus has thus been carried out in prokaryotes or using chloroplast genes. Consequently most *in vivo* functional studies of PSI have been carried out in cyanobacteria whereas functional studies of plant PSI have mostly been carried out *in vitro*. In order to investigate the function of PSI-E *in vivo* in higher plants we have established a system for transforming barley using the antisense approach.

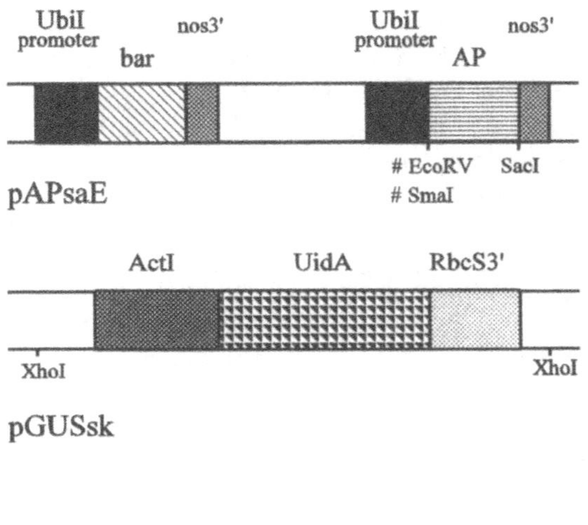

Figure 1. Constructs for transformation of immature barley embryos. Panel A: The full length *PsaE* cDNA was driven in reverse orientation by the *Ubi 1* maize promoter and terminated by the nos polyadenylation signal. Panel B: The *UidA* reporter gene was driven by the actin 5′ and terminated by RbcS 3′.

P. Mathis (ed.), Photosynthesis: from Light to Biosphere, Vol. II, 151–154.

The method is based on introduction of the DNA by particle bombardment of dissected immature embryos (7). We report here the generation of numerous independent lines of transgenic plants with functional expression of the *bar* and *UidA* genes. PCR analysis show integration of the *bar* gene into the genome of barley. Transgenic plants with reduced level of *PsaE* mRNA were identified by Northern blotting.

2. Procedure

2.1 *Plasmid*
The first plasmid pAPsaE was constructed from plasmid pAHC25 (8) kindly provided by Dr. P. Quail (Plant Gene Expression Center, Albany, CA). The plasmid contains the *UidA* reporter gene encoding GUS and the *bar* gene as a selective marker. Each are driven by the maize *Ubi1* promoter and first intron and terminated by the polyadenylation signal of *nos* 3′ gene from *Agrobacterium tumefaciens.* The *UidA* gene was digested from the pAHC25 plasmid with *Sma*I and *Sac*I and the vector fragment was ligated with a *Sac*I-*EcoR*V fragment of 0.5 kb containing a full length *PsaE* cDNA (3) so that the *Ubi1* promoter drives the *PsaE* cDNA in reverse orientation (Fig. 1). The *UidA* gene fused to the strong promoter of *Act1* and first intron, followed by the polyadenylation signal from the *RbcS* was digested from the plasmid pDM803 (kindly provided by Dr. D. McElroy, Plant Gene Expression Center, Albany, CA) with *Xho*I and ligated with pBluescript SK+ digested with *Xho*I. The obtained plasmid designated pGUSsk (Fig. 1) was used for transformation of immature barley embryos. Standard molecular biological methods were preformed as described (9).

2.2 *Plant materials*
Barley (*Hordeum vulgare L.*) cultivar Golden Promise was used throughout. The plants were grown in growth chambers using a 16 h light/8 dark period at 12°-16° C.

2.3 *Transformation and regeneration*
The two plasmids constructed were introduced separately or combined into immature barley embryos dissected from young caryopses using the particle bombardment method (7). Bialaphos resistant callus lines were regenerated to plants as described (7).

2.4 *GUS activity*
GUS expression in callus tissue was detected histochemically by staining with 5-bromo-4-chloro-3 indolyl-β-glucuronide (X-gluc) (10).

2.5 *PCR analysis*
The first oligomer was designed to correspond to the promoter of the maize ubiquitin gene and the second oligomer was designed to correspond to the opposite strand of the *bar* gene located after the ubiquitin promoter. Standard PCR conditions were used (9). The thermocycling parameters were 1 min at 94°C, 2 min at 63°C and 2 min at 72°C for 40 cycles.

2.8 *Northern blotting analysis*
Total RNA was isolated (11) from young top leaves of transgenic plants and from a wild type plant grown under identical conditions. Blotting, prehybridization, hybridization were carried out as described previously (1). A 300 bp *Sma*I of the *PsaE* cDNA was labelled with [α-^{32}P]dCTP and used as a probe. The filter was reprobed with a [α-^{32}P]dCTP labelled 28S gene from wheat. Bands on the autoradiograms were quantified using a gel scanner (Pharmacia, Denmark).

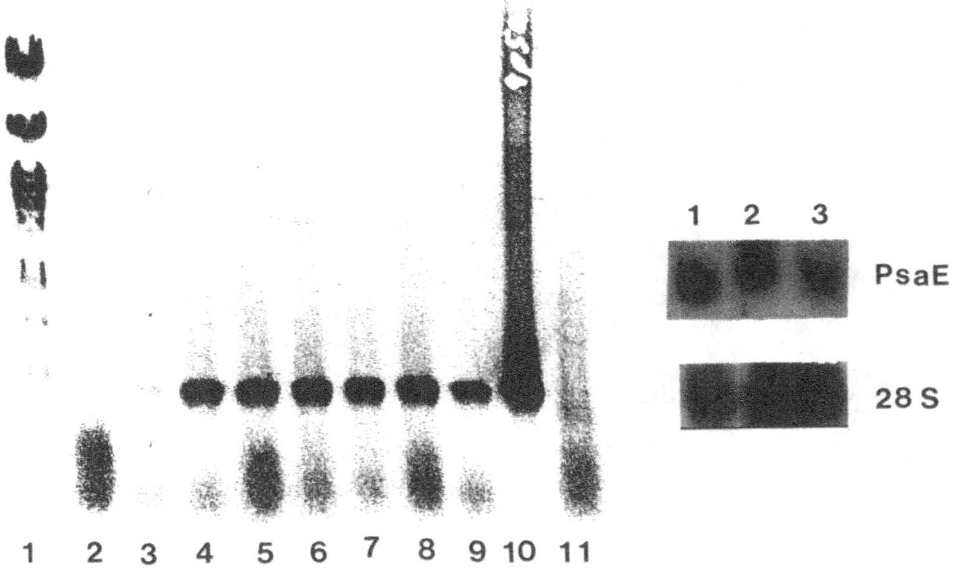

Figure 2: PCR and Northern blotting analysis of independent transgenic plants. Panel A: PCR analysis of seven independent lines (lane 3-9) and a positive pAPsaE plasmid control (lane 10) and wild type (lane 2 and 11) with corresponding ubiquitin and *bar* primers. Panel B: Lane 1 contained total RNA from wild type and lanes 2 and 3 contained total RNA from two independent transgenic plants.

3. Results and discussion

The plasmid pAPsaE was introduced either alone or together with the pGUSsk plasmid into immature barley embryos by particle bombardment. Several bialaphos resistent callus lines were obtained. After 8 weeks, resistant cells were transferred to shoot induction medium. Green plants were regenerated from several different callus lines (Fig 3).The presence of the pAPsaE plasmid was monitored by PCR analysis (Fig.2). In all transgenic lines and in the positive control plasmid, the expected 350 bp band was detected (Fig. 2). In the wild type plant cells, the band could not be amplified. Southern blotting revealed that the antisense gene was integrated into the genome of barley (data not shown). Transgenic callus lines were stained GUS positive demonstrating cotransformation of the pGUSsk and pAPsaE plasmids (Fig.3). Northern blotting of two independent transgenic lines and one wild type plant showed marked differences in the steady state level of *PsaE* mRNA (Fig.2). The reduction was calculated from the ratio of *PsaE* mRNA to 28S rRNA. One transgenic line showed a reduction by 62% compared to the wild type whereas another line showed no reduction. The line with a reduced transcript level showed normal growth in a greenhouse at 12°C. The promoter used to drive the antisense gene is the maize ubiquitin promoter. Histochemical studies of expressed GUS activity reveal that the promoter is most active in rapidly dividing cells and almost silent in mature leaves (12). Similarly, GUS was not expressed in all tissues of transgenic rice (12). A constitutive promoter like the 5′region of the rice actin gene (13), may give a stronger antisense expression in all cells and at every developing state of the plant.

Figure 3. Histochemical analysis of independent resistent callus lines six weeks after particle bombardment.

The data show that transgenic barley plants can be obtained by particle bombardment of immature embryos and that antisense repression of *PsaE* mRNA to a significantly reduced level is possible in barley.

4. References

1. Kjærulff, S., Andersen. B., Nielsen, V.S., Møller B.L. and Okkels. J.S. (1993) J. Biol. Chem. 268, 18912-18916
2. Bryant,D.A (1992) in The Photosystems: Sturcture, Function and Molecular Biology (Barber J., ed) pp.501-549, Elsevier, Amsterdam, The Netherlands
3. Li, N., Zhao, J., Warren, P.V.,Warden, J.T., Bryant, D.A. and Golbeck, J.H. (1991) Biochemistry 30, 7863-7872
4. Okkels., J.S., Jepsen, L.B., Hønberg, L.S., Lehmbeck, J., Scheller, H.V., Brandt, P., Høyer-Hansen, G., Stummann, B., Henningsen, K.W., Wettstein, D. and Møller, B.L. (1988) FEBS Lett. V 237, 1,2, 108-112
5. Rousseau, F., Setif, P. and Lagoutte, B. (1993) EMBO J. 12,1755-1765
6. Yu, L., Zhao, J., Muhlenhoff, U., Bryant, D.A. and Golbeck, J.H. (1993) Plant. Physiol. 103, 171-180
7. Wan, Y. and Lemaux, P.G. (1994) Plant Physiol. 104, 37-48
8. Christensen. A.H., Sharrock, R.A. and Quail, P.H. (1992) Plant Mol. Bio.l 18, 675-689
9. Sambrook, J., Fritsch E.F. and Maniatis T. (1992) Cold Spring Harbor Laboratry Press
10. Jefferson, R.A., Kavanagh, T.A., Bevan, M.W. (1987) EMBO J 6, 3901-3907
11. Okkels, J,S., Scheller, H.V. and Møller, BL (1992) N. Murata (ed), Research in Photosynthesis, Vol. I, 473-476. Kluwer Academic Publishers
12. Cornejo, M.-J. Luth, D., Blankenship, K.M., Anderson, O.D. and Blechl, AE. (1993) Plant Mol. Biol. 23, 567-581
13. Zhang, W., McElroy, D. and Wu, R. (1991) The Plant Cell 3, 1155-1165

CHARACTERIZATION OF *E. coli* EXPRESSED PSI-C MUTANTS

Helle Naver[1], M. Paul Scott[2], John H. Golbeck[2], Birger Lindberg Møller[1] and Henrik Vibe Scheller[1]
[1]Department of Plant Biology, Plant Biochemistry Laboratory, Royal Veterinary and Agricultural University, DK 1871 Frederiksberg C, Copenhagen, Denmark.
[2]Department of Biochemistry, University of Nebraska, Lincoln, NE 68588-0664.

1.Introduction

The PSI-C subunit of PSI is an iron-sulfur protein which contains the terminal acceptors centers A/B. The amino acid sequence of the PSI-C subunit is highly conserved among species and contains a repeated CXXCXXCXXXCP motif that is characteristic for 2[4Fe-4S] proteins. Alignment analysis clearly shows that PSI-C is similar to the soluble bacterial 2[4Fe-4S] ferredoxins apart from an 8 residue internal loop and a 15 residue C-terminus not present in the ferredoxins (1,2). Since the ferredoxins are soluble and the PSI-C polypeptide is membrane bound, the additional amino acids in the loop and in the C-terminus may be responsible for the binding of PSI-C to the PSI core (2). We decided to make two *Eschericia coli* expressed PSI-C deletion mutants, one lacking the 8 residue loop (PSI-$C_{Hor}\Delta L$) and one lacking the 10 C-terminal residues (PSI-$C_{Syn}\Delta C$), and investigate their ability to fold and form [4Fe-4S] clusters. On the basis of the structure of *Peptococcus aerogenes* ferredoxin (3) it has been predicted that neither the internal loop nor the C-terminal extension interact with the iron-sulfur clusters (4,5,6). However deletion of either the loop or the C-terminus may result in proteins unable to refold and form [4Fe-4S] clusters. If the folding properties of the PSI-C deletion mutants are intact then the interaction between PSI and the PSI-C deletion mutants can be investigated. A PSI core preparation specifically devoid of PSI-C, -D and -E can be used for reconstitution by addition of *E. coli* expressed PSI-C and -D in the presence of the reagents Na_2S, $FeCl_3$ and 2-mercaptoethanol to ensure rebuilding of the FeS clusters of PSI-C (7,8,9,10). Using this reconstitution system we show that the mutant forms of PSI-C both share properties known to be characteristic for iron-sulfur proteins and both have retained their function as electron acceptors in PSI.

2. Procedure

P. Mathis (ed.), Photosynthesis: from Light to Biosphere, Vol. II, 155–158.

The PSI-$C_{Hor}\Delta L$ and PSI-$C_{Syn}\Delta C$ were designed as shown in Fig. 1 and made as described (7,11). The PSI-$C_{Hor}\Delta L$ was made using the barley PSI-C as template (Okkels unpublished) and the PSI-$C_{Syn}\Delta C$ was made using the *Synechococcus* sp. 6301 PSI-C (PSI-C_{Syn}) as template (13). All the expressed proteins were recovered in inclusion bodies and were isolated as described (11). The iron-sulfur clusters of the *E.coli* expressed proteins were rebuilt as described (8,10). A barley PSI core specifically devoid of PSI-C,-D and -E was prepared by urea treatment of barley PSI isolated with the use of the detergent β-decylmaltoside (7, 12). Reconstituting the photosystem I core lacking PSI-C, -D and -E with either of the PSI-C deletion mutants was performed as described (7,8). Flash induced absorbance changes were measured as described (7).

```
                      1       10        20        30       40
Barley PSI-CBor       SHSVKIYDTCIGCTQCVRACPTDVLEMIPWDGCKAKQIAS
Synecho. PSI-CSyn     SHSVKIYDTCIGCTQCVRACPLDVLEMVPWDGCKAGQIAA
Synecho.PSI-CSynΔC    SHSVKIYDTCIGCTQCVRACPLDVLEMVPWDGCKAGQIAA
Barley PSI-CBorΔL     SHSVKIYDTCIGCTQCVRACPTDVL--------KAKQIAS
P. aerogenes fd         AYVINDSCIACGACKPECPVNI---------QQG-SIY

                              50        60        70        80
Barley PSI-CBor       APRTEDCVGCKRCESACPTDFLSVRVYLGPETTRSMALSY
Synecho. PSI-CSyn     SPRTEDCVGCKRCETACPTDFLSIRVYLGAETTRSMGLAY
Synecho.PSI-CSynΔC    SPRTEDCVGCKRCETACPTDFLSIRVYLGA
Barley PSI-CBorΔL     APRTEDCVGCKRCESACPTDFLSVRVYLGPETTRSMALSY
P.aerogenes fd        AIDADSCIDCGSCASVCPVGANPED
```

Figure 1. Amino acid sequence of mutant and wild type PSI-C proteins. The underlined sequences in the wild type PSI-C proteins were deleted resulting in two mutant proteins PSI-$C_{Hor}\Delta L$ and PSI-$C_{Syn}\Delta C$. The residues differing between the barley PSI-C_{Hor} and the *Synechoccus* sp. 6301 PSI-C_{Syn} are shown in bold. The sequence of *Peptococcus aerogenes* is shown for comparison.

3.Results and Discussion

The *E.coli* ex-pressed and pu-rified PSI-C_{Hor}, PSI-$C_{Hor}\Delta L$, PSI-C_{Syn} and PSI-$C_{Syn}\Delta C$ were analyzed by SDS PAGE (Fig. 2). The PSI-$C_{Syn}\Delta C$ migrates as a smaller protein than PSI-C_{Syn} as expected. PSI-$C_{Hor}\Delta L$ migrates like the full-length PSI-C_{Hor}. However sequencing both the plasmid construct by the dideoxy reaction (14) and the expressed protein by Edman degradation confirmed the expected sequence of PSI-$C_{Hor}\Delta L$. We rebuilt the [4Fe-4S] clusters of the mutant and wild type proteins and determined their absorption spectra under anaerobic conditions. Subsequently the samples were oxidized by exposing them to air whereby the spectra changed. The spectrum of PSI-

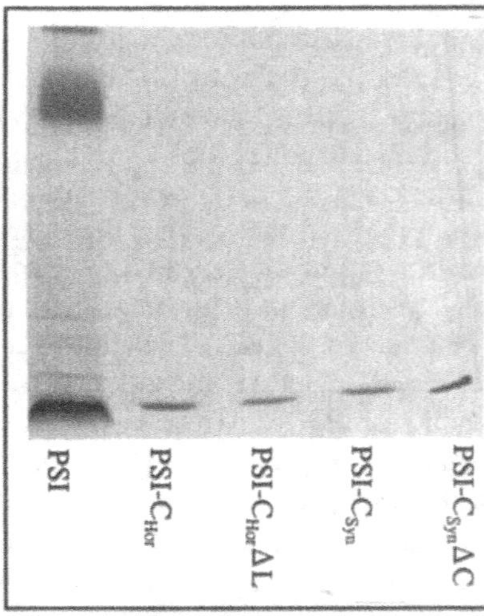

Figure 2. SDS PAGE of isolated wild type and mutant PSI-C.

$C_{Syn}\Delta C$ is shown in Fig. 3. Similar spectras were obtained for the PSI-C_{Hor} and PSI-$C_{Hor}\Delta L$ (data not shown). Thus both PSI-$C_{Hor}\Delta L$ and PSI-$C_{Syn}\Delta C$ can be folded into iron-sulfur proteins. As is the case for native PSI-C they become denature upon exposure to air (15). To examine the ability of the iron-sulfur proteins PSI-$C_{Hor}\Delta L$ and PSI-$C_{Syn}\Delta C$ to accept electrons from center X, they were used to reconstitute a barley PSI core. The reconstituted complexes were characterized by flash induced absorption change at 834 nm in the ms time range. The extent of the reconstitution was monitored by observing the change from the 1 ms backreaction from center X to P700$^+$ to the slower backreaction from center A/B (Fig. 4). Both the PSI-C_{Hor} and the PSI-$C_{Syn}\Delta C$ were able to rebuild the core in the presence of PSI-D. The PSI-$C_{Hor}\Delta L$ reconstituted only 55% of the core. The equivalent experiment was repeated using a *Synechococcus* sp. 6301 PSI core and similar results obtained (data not shown).

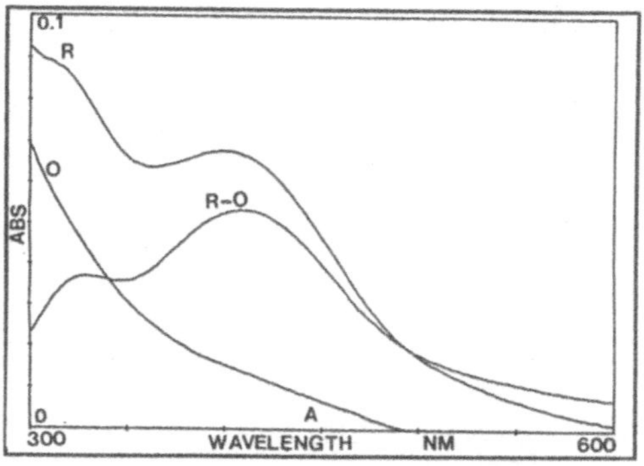

Figure 3. Absorption spectra of the PSI-$C_{Syn}\Delta C$. The PSI-$C_{Syn}\Delta C$ was reconstituted according to Parrett et al. (10) and desalted. The spectrum of the reconstituted protein was recorded under anaerobic conditions (R). The sample was subsequently exposed to air and the spectrum recorded (O). The difference spectrum was calculated (R-O).

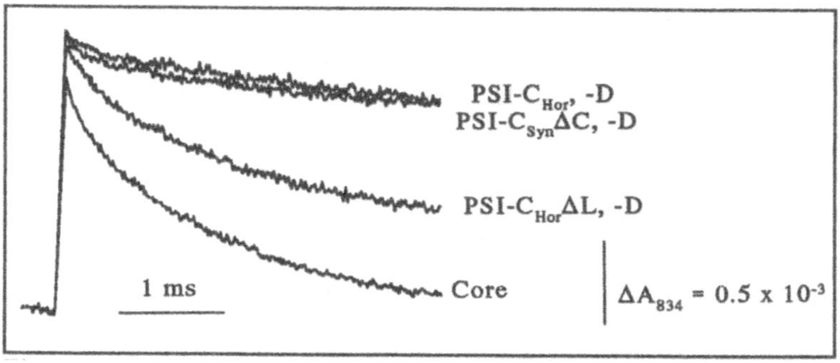

Figure 4. Flash induced absorption changes of P700 in a barley core reconstituted with wild type and mutant PSI-C . A saturating amount of barley PSI-D was included in all reconstitution experiments.

In conclusion, both PSI-$C_{Hor}\Delta L$ and PSI-$C_{Syn}\Delta C$ could be purified form *E.coli* inclusion bodies in high amounts and could be folded into iron-sulfur proteins in vitro. Further both were able to interact with the PSI core as confirmed by their ability to accept electrons from center X. The importance of the C-terminal extension as well as of the loop region of PSI-C with respect to the ability of PSI-C to bind to the PSI core can thus be studied using the modified PSI-C proteins.

1. Høj, P., Svendsen, I., Scheller, H.V. and Møller, B.L. (1987) J. Biol. Chem. 262, 12676-12684.
2. Scheller, H.V., Svendsen, I. and Møller, B.L. (1989) Carlsberg Res. Commun. 54, 11-15.
3. Adman, E.T., Sieker, L.C. and Jensen, J.H. (1973) J. Biol. Chem. 248, 3987-3996.
4. Dunn, P.P.J. and Gray, J.C. (1988) Plant Mol. Biol. 11, 311-319.
5. Oh-oka, H., Takahashi, Y., Kuriyama, K., Saeki. K. and Matsubara, H. (1988) J. Biochem 103, 962-968.
6. Andersen, B., and Scheller, H.V. (1993) in: Pigment Protein complexes in Plastids. Synthesis and Assembly (Sundqvist, C and Ryberg, M., eds) pp. 383-418, Acedemic Press, Orlando, FL
7. Naver, H., Scott, M.P., Andersen, B., Møller, B.L. and Scheller, H.V. (1995) Physiol. Plant. In press.
8. Zhao, J. Warren, P.V., Li, N., Bryant, D.A. and Golbeck, J.D. (1990) FEBS Lett., 276, 175-180.
9. Li, N, Zhao, J., Warren, P., Warden, J.T., Bryant, D.A. and Golbeck, J.D. (1991) Biochemistry 30, 7863-7872.
10. Scott, P. Kjærulff, S., Scheller, H.V. and Okkels, J.S. (1992) in Research in Photosynthesis (N. Murata, ed) I, pp 593-596.
11. Parrett, K.G., Mehari, T. and Golbeck, J.H. (1990) Biochem. Biophys. Acta. 1015, 341-352.
12. Andersen, B., Scheller, H.V. and Møller , B.L. (1992) FEBS Lett. 311, 169-173.
13. Herman, P., Adiwilaga, K., Golbeck, J.H. and Weeks, D.P. (1994) Plant Physiol. 104, 1459-1461.
14. Sanger, F., Nicklen, S. and Coulsen, A.R. (1977) DNA sequencing with chain-terminating inhibitors. PNAS US. 74:5463-5467.
15. Oh-oka, H., Itoh, S., Saeki, K., Takahashi, Y. and Matsubara, H. (1991) Plant Cell Physiol. 32, 11-17.

ELECTRON-NUCLEAR-NUCLEAR SPECIAL TRIPLE RESONANCE STUDIES OF A QUINONE ELECTRON ACCEPTOR IN THE PHOTOSYNTHETIC REACTION CENTRES OF GREEN SULPHUR PHOTOSYNTHETIC BACTERIA.

I.P. Muhiuddin[1], S.E.J. Rigby[2], M.C.W. Evans[2] and P. Heathcote[1].
[1]School of Biological Sciences, Queen Mary and Westfield College, London E1 4NS and [2]Department of Biology, University College London Gower Street, London WC1E 6BT.

1. INTRODUCTION

The reaction centres of green sulphur photosynthetic bacteria and *Heliobacteria* are thought to resemble Photosystem I (see [1] for a review).These reaction centres are attracting interest because of the recent evidence suggesting they are homodimeric, because only a single gene resembling *psaA* or psaB can be found [2,3].ENDOR studies of $P840^{+\bullet}$ have shown a symmetrical distribution of the unpaired electron between the two bacteriochlorophyll *a* molecules of this electron donor in green sulphur bacteria, suggesting a functional symmetry [4].

Although these bacterial reaction centres are thought to resemble Photosystem I (PSI), there are contradictory reports as to whether they contain a functional secondary electron acceptor equivalent to the phylloquinone (A_1) in PSI [1]. Nitschke *et al* [5] had previously reported photoaccumulation of an electron paramagnetic resonance (EPR) signal in green sulphur bacteria that resembled the EPR signal of $A_1^{\bullet-}$ photoaccumulated in PSI. We have repeated these experiments with membranes isolated from *Chlorobium limicola* f sp. *thiosulphatophilum* grown in the presence of acetylene to inhibit bacteriochlorophyll *c* synthesis [6]. We have photoaccumulated an EPR signal that resembles that arising from $A_1^{\bullet-}$ in spinach PSI [7]. We have shown that this signal arises from a napthoquinone by using Special Triple (ST) spectroscopy, which also gives information on the electronic structure of the quinone electron acceptor *in vivo* in the reaction centre of green sulphur bacteria.

2. MATERIALS AND METHODS.

Chlorobium limicola f. sp. *thiosulphatophilum* (strain Tassajara) was grown as in [8] in the presence of acetylene. Typically ca. 20ml of air space per liter was left over a freshly innoculated culture and gassed briefy with acetylene (personal communication, J.G.Ormerod). After two days growth, cells were harvested by centrifugation and stored in liquid nitrogen until use.

Preparation of a membrane fraction was performed as far as possible under argon. This involved breaking thawed cells in a French pressure cell. Following this, the cells were

P. Mathis (ed.), Photosynthesis: from Light to Biosphere, Vol. II, 159–162.

incubated at 30°C (45 minutes) with lysozyme and EDTA at respective concentrations of 50ug/ml and 1mM. The resulting cell lystate was centrifuged at 12 000g (12 min, 4^0C), and the membranes pelleted from the supernantant by centrifugation at 140 000g (2 hr). The pellet obtained was diluted with sparged buffer (10mM Tris-Cl,pH 8.0) and centrifuged again at 140 000g (2hr). A minimal volume of buffer was used to resuspend the washed membranes which were then aliquoted (250µl) into 3 mm internal diameter EPR tubes for immediate preparation of samples for spectroscopy.
EPR samples were prepared, and EPR signals photoaccumulated by 2 minutes illumination at 205K, as detailed in [7]. During sample preparation, anaerobic conditions were maintained with a stream of oxygen free argon. EPR and ST resonance spectra were obtained using a Bruker ESP300 EPR spectrometer as described previously[9].

3. RESULTS AND DISCUSSION

Illumination of *C. thiosulphatophilum* membranes at 205K with dithionite produces an EPR spectrum (Figure 1a) with g=2.0048 and ΔHptp=9.5G. This spectrum is very similar to the spectrum of A1●— photoaccumulated in PSI under the same conditions (Figure 1b)[7].

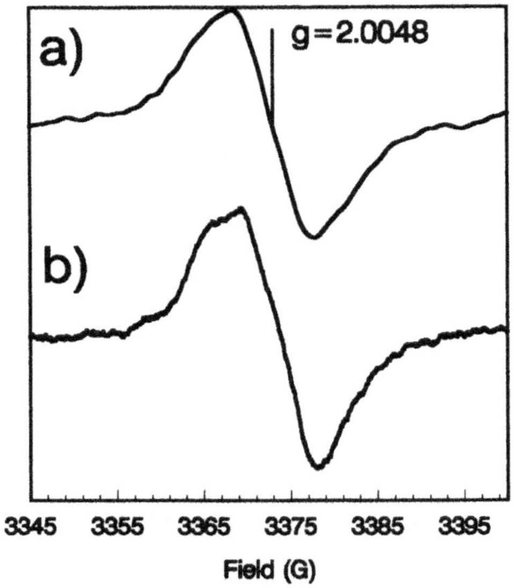

Figure 1. EPR spectra of (a) *C. thiosulphatophilum* membranes and (b) spinach PSI digitonin particles, each illuminated for 3 minutes at 205K.

The *C. thiosulphatophilum* membrane preparation allows for only relatively low (≤10µM) concentrations of photosynthetic reaction centres to be attained, leading to low spectroscopic signal to noise ratios. We have offset this problem to some extent through the use of Electron-Nuclear-Nuclear Special TRIPLE (ST) resonance spectroscopy which offers greater signal to noise than the standard ENDOR technique. The most intense features of such frozen solution spectra are expected to arise from the $A_{\perp}$

components of axial hyperfine coupling constants (hfcs) to hydrogen bonded protons [10] and protons from methyl and methylene protons β to the aromatic π system bearing the unpaired electron [11].

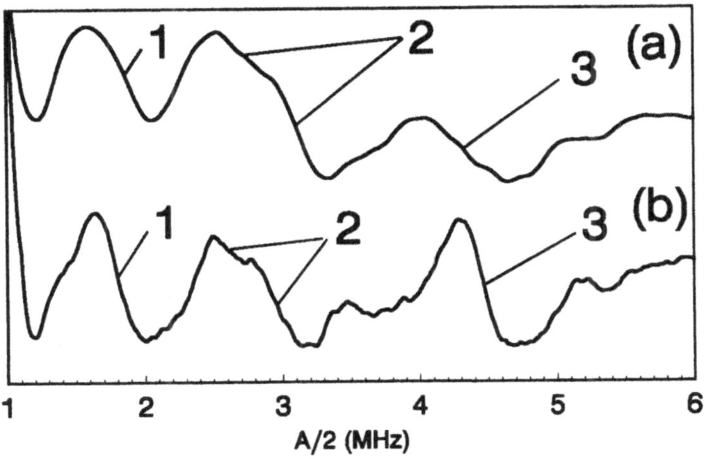

Figure 2. Special TRIPLE spectra of (a) "$A_1^{\bullet-}$" in *C. thiosulphatophilum* membranes and (b) $A_1^{\bullet-}$ in spinach digitonin PSI. Conditions: (a) m.w. power 10mW; r.f. power 200W; r.f. mod. depth 230KHz; temp. 120K: (b) as (a) but r.f. mod. depth 158KHz; temp. 60K.

Comparison of the ST spectrum of the *C. thiosulphatophilum* radical, Figure 2a, with that of $A_1^{\bullet-}$ photoaccumulated in spinach PS1 digitonin particles, Figure 2b, reveals features 1, 2 and 3 as common to both spectra. These features have been assigned in spinach $A_1^{\bullet-}$ using biosynthetic labelling and comparison with model compounds [12]. Feature 1 is assigned as the $A_\perp$ component of the hfc to a β-methylene proton at position 3, while feature 2 arises from the overlapped $A_\perp$ components of hfcs to two protons hydrogen bonded to the quinone oxygens. Feature 3 in PS1 $A_1^{\bullet-}$ is assigned as the $A_\perp$ component of the hfc to the 2-methyl group of phylloquinone. Due to the signal to noise constraints mentioned earlier, $A_\parallel$ components for the *C. thiosulphatophilum* hfcs have not been detected. Hfcs and assignments for both the *C. thiosulphatophilum* $A_1^{\bullet-}$-like species and for spinach $A_1^{\bullet-}$ are collected in Table 1.

The precise chemical nature of the quinone in *C. thiosulphatophilum* photosynthetic reaction centres is unclear. Menaquinone-7, 1'hydroxymenaquinone and 1'-oxo menaquinone ("Chlorobium quinone") have been isolated from membrane preparations of this organism but so far no report has appeared on the nature of the quinone in the reaction centre. All the aforementioned quinones are 2-methyl naphthoquinones. Menaquinone and 1'-hydroxymenaquinone will therefore give ST spectra very similar to phylloquinone. The carbonyl group of 1'-oxomenaquinone makes such similarity less likely for this quinone.

The data collected in Table 1 suggest that the $A_1^{\bullet-}$-like *C. thiosulphatophilum* radical and the $A_1^{\bullet-}$ phylloquinone anion radical of PSI have very similar electronic structures. The small differences in the hfcs of the two radicals is probably as much a function of the large r.f. modulation depths used in the acquisition of the *C. thiosulphato-*

Table 1.
Hyperfine coupling constants and resonance assignments for spinach $A_1^{\bullet-}$ and the $A_1^{\bullet-}$-like radical of *C. thiosulphatophilum*.

Feature	Hyperfine coupling constant (MHz)		Assignment
	C. thiosulphatophilum	spinach[a]	
1	3.8	3.6	3-β-CH_2 $A_\perp$
2	5.4	5.2	H-bond $A_\perp$
	6.2	6.0	H-bond $A_\perp$
3	8.8	9.0	2-Methyl $A_\perp$

Notes: a, data from [12]. Accuracy of hfc measurement estimated at ±0.1MHz for spinach $A_1^{\bullet-}$ and ±0.2MHz for *C. thiosulphatophilum*.

philum data as of any real differences in electronic structure. We have argued previously [12] that the characteristic hfcs measured for $A_1^{\bullet-}$ reflect a protein induced perturbation of the electronic structure of the bound phylloquinone molecule which enables it to function in PS1 at the required redox potential (circa -800mv). This ST study suggests that in *C. thiosulphatophilum* a similar quinone molecule functions in a similar way to the phylloquinone of PSI.

4. CONCLUSIONS

1) A semiquinone anion radical can be photoaccumulated in the *C. thiosulphatophilum* reaction centre at 205K.
2) Special TRIPLE spectroscopy suggests that the radical formed is a 2-methyl naphthoquinone with a distorted electronic structure.
3) The affect of the *C. thiosulphatophilum* reaction centre on the protein bound quinone is very similar to that exerted by Photosystem I on the bound phyllosemiquinone $A_1^{\bullet-}$.

We thank the BBSRC (UK) for financial support.

References

[1]Lockau, W. and Nitschke, W. (1993) Physiol. Plant. 88, 372-381.
[2]Büttner, M., Xie, D-L., Nelson, H., Pinther, W., Hauska, G. and Nelson, N. (1992) Proc. Natl. Acad. Sci USA 89, 8135-8139.
[3]Liebl, U., Mockensturm-Wilson, M., Trost, J.T., Brune, D.C., Blankenship, R.M. and Vermaas, W. (1993) Proc. Natl. Acad Sci USA 90, 7172-7128.
[4] Rigby S.E.J., Evans, M.C.W. and Heathcote, P. (1994) FEBS Lett. 350, 24-28.
[5]Nitschke, W., Feiler, U., Lockau, W. and Hauska, G. (1987) FEBS Lett. 218, 283-286.
[6]Ormerod, J.G., Nesbakken, T. and Beal, S.I. (1990) J. Bacteriol. 172, 1352-1360.
[7]Heathcote, P., Hanley, J.A. and Evans, M.C.W. (1993) Biochim. Biophys. Acta 1144, 54-61.
[8]Jennings, J.V. and Evans, M.C.W. (1977) FEBS Lett. 75, 33-36.
[9]Rigby, S.E.J., Nugent, J.H.A. and O'Malley, P.J. (1994) Biochemistry 33, 10043-10050.
[10]O'Malley, P.J. and Babcock, G.T. (1986) J. Am. Chem. Soc. 108, 3995-4001.
[11] O'Malley, P.J. and Babcock, G.T. (1984) J. Chem. Phys. 80, 3912-3913.
[12]Rigby, S.E.J., Evans, M.C.W. and Heathcote, P. (1995) Biochemistry, submitted.

ELECTRON NUCLEAR DOUBLE RESONANCE (ENDOR) STUDIES OF THE PHYLLOQUINONE ELECTRON ACCEPTOR (A_1) IN THE PHOTOSYSTEM 1 REACTION CENTRE.

P. Heathcote[1], S.E.J. Rigby[2] and M.C.W. Evans[2].
[1]School of Biological Sciences, Queen Mary and Westfield College, Mile End Road, London E1 4NS and [2] Department of Biology, University College London, Gower Street, London WC1E 6BT.

1. INTRODUCTION

The Photosystem 1 reaction centre contains a primary electron donor, a dimer of chlorophyll *a* molecules (P700), and five electron acceptors. The primary electron acceptor A_0 is thought to be a chlorophyll *a* monomer, and to donate electrons to a phylloquinone (vitamin K_1) secondary electron acceptor A_1. Subsequently electrons are transferred through three [4Fe-4S] centres called Fe-S_X, Fe-S_A and Fe-S_B. In order to transfer electrons from A_0 to Fe-S_X, the redox couple $A_1/A_1^{\bullet-}$ must have an E_m Of *circa* -800mV, whereas phylloquinone *in vitro* has a much more positive E_m (*circa* -100mV). This suggests that the protein binding pocket of the phylloquinone in Photosystem 1 is effecting a large change in the electronic structure of the bound quinone.

In order to examine the electronic structure of the bound phylloquinone, we have studied the semiquinone anion radical in Photosystem 1 in frozen solution using electron nuclear double resonance (ENDOR) and electron nuclear nuclear Special TRIPLE (ST) resonance spectroscopies in conjunction with biosynthetic deuteration *in vivo*. The semiquinone anion of phylloquinone in both protonated and deuterated alkaline ethanol was also studied. The results of these studies reveal protein-quinone hydrogen bonding in Photosystem 1, and a perturbation of the electronic structure of the bound phylloquinone.

2. MATERIALS AND METHODS

Photosystem 1 particles were isolated from spinach using digitonin. The methionine auxotroph of *Anabaena variabilis* strain met 27 [1] was grown on media supplemented with protonated or methyl-d_3 methionine ($CD_3S(CH_2)_2CHNH_2CO_2H$), and Photosystem 1 particles isolated using lauryl maltoside [2]. The growth on methyl-d_3 methionine results in deuteration of the phylloquinone (mass increases from 451 to 467 indicating an average of 16 protons out of a total of 42 are labelled). This treatment results in a narrowing of the electron paramagnetic resonance (EPR) spectrum from *A. variabilis* Photosystem 1 demonstrating that this EPR signal arises from the phylloquinone.

The EPR signal of $A_1^{\bullet-}$ was generated by photoaccumulation at 205K in the presence of sodium dithionite at pH8.0 [3]. The phylloquinone semiquinone anion was

P. Mathis (ed.), Photosynthesis: from Light to Biosphere, Vol. II, 163–166.

produced by stoichiometric sodium borohydride reduction of alkaline ethanolic solutions of the quinone under a stream of argon [4].

ENDOR and Special TRIPLE resonance spectra were obtained using a Bruker ESP300 EPR spectrometer as described previously [5].

3. RESULTS AND DISCUSSION

Hyperfine anisotropy limits frozen solution ENDOR spectroscopy of semiquinone anion radicals to the observation of hyperfine couplings (hfcs) to methyl (and occasionally methylene) protons β to the aromatic system and hyperfine couplings to solvent protons hydrogen bonded to the carbonyl oxygens. Hfcs to both types of proton have axial ENDOR lineshapes, with the anisotropy ($A_\parallel$-$A_\perp$) being 3-3.5MHz for methyl protons and the hydrogen bond hfcs being essentially traceless ($A_\parallel$-2$A_\perp$ ≈0) [6,7]. The two types of proton may be distinguished either by using deuterated solvent, in which case the hydrogen bond hfcs disappear, or conversely by biosynthetic deuteration of the quinone. Here the former method was used to assign hfcs in spectra of phyllosemiquinone *in vitro*, while the latter method is used in the study of $A_1^{\bullet-}$.

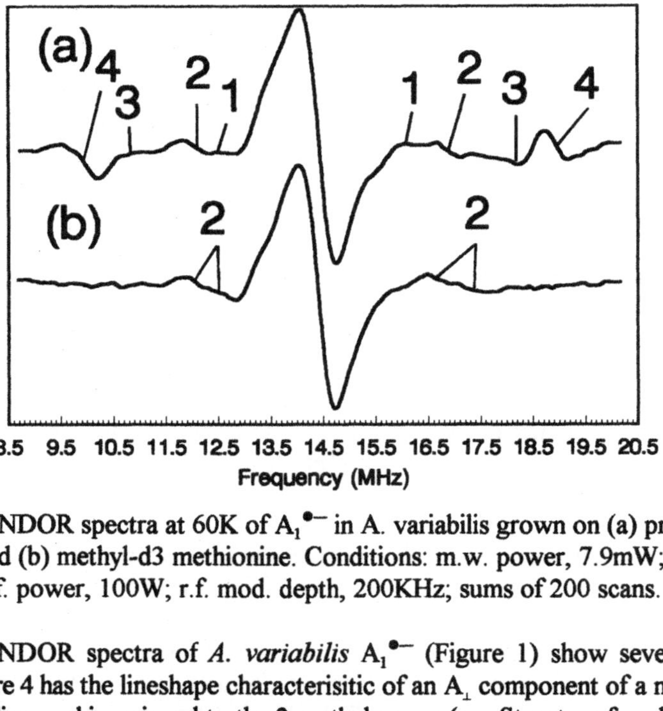

Figure 1. ENDOR spectra at 60K of $A_1^{\bullet-}$ in A. variabilis grown on (a) protonated and (b) methyl-d3 methionine. Conditions: m.w. power, 7.9mW; r.f. power, 100W; r.f. mod. depth, 200KHz; sums of 200 scans.

The ENDOR spectra of *A. variabilis* $A_1^{\bullet-}$ (Figure 1) show several resolved features. Feature 4 has the lineshape characterisitic of an $A_\perp$ component of a methyl group hyperfine coupling, and is assigned to the 2-methyl group (see Structure for phylloquinone carbon atom numbering scheme). The broader features 1 and 3 may arise from the $A_\perp$ and $A_\parallel$ components respectively of an hyperfine coupling to a 3-methylene proton. This leaves feature 2 (which appears to be split into two in the spectrum of deuterated material, Figure

1b) which, since it is insensitive to deuteration, is assigned to hyperfine couplings to two protein protons H-bonded to the quinone oxygens. Features 1 and 3 are better resolved in the Special TRIPLE (ST) spectra in Figure 2, and their ST lineshapes confirm the above assignments. Furthermore a new feature, 5, is observed in the ST spectra. The lineshape and coupling of feature 5 assign it as the overlapped $A_{\parallel}$ features of the 2-methyl and H-bond hyperfine couplings.

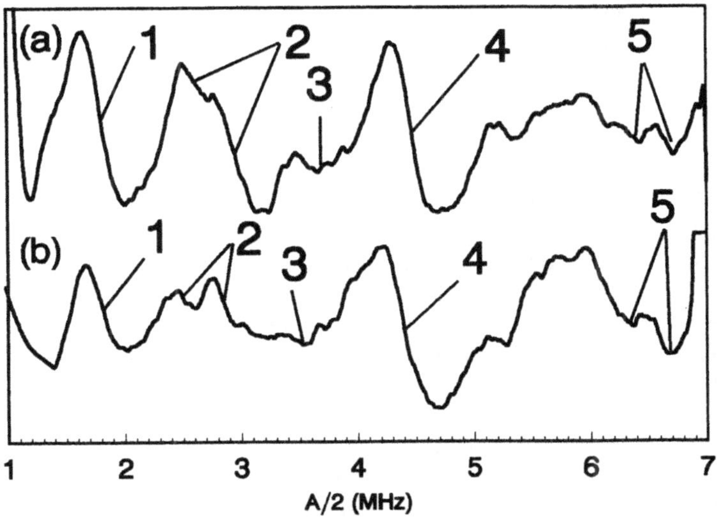

Figure 2. Electron-nuclear-nuclear Special TRIPLE spectra at 60K of $A_1^{\bullet-}$ in (a) spinach and (b) A. variabilis Photosystem 1. Conditions: m.w. power, 10mW; r.f. power, 200W; r.f. mod. depth 158KHz; sums of 300 scans.

Table 1.

Hyperfine coupling constants and resonance assignments for $A_1^{\bullet-}$ and for the phylloquinone anion radical ($PhQ^{\bullet-}$) in alkaline ethanol.

Feature	Hyperfine coupling constant (MHz)			Assignment
	$A_1^{\bullet-}$		$PhQ^{\bullet-}$	
	A. variabilis	Spinach		
1	3.6	3.6	3.0	?3-β-CH_2 $A_{\perp}$
2	-5.0	-5.2	-2.2	H-bond $A_{\perp}$
	-5.8	-6.0		H-bond $A_{\perp}$
3	7.6	7.7	-----	?3-β-CH_2 $A_{\parallel}$
4	9.0	9.0	6.8	2-Methyl $A_{\perp}$
5	12.8	12.6	10.0	2-Methyl $A_{\parallel}$
	13.4	13.4	5.2	H-bond $A_{\parallel}$

Comparison of the $A_1^{\bullet-}$ and phylloquinone anion radical ($PhQ^{\bullet-}$) hfcs (see Table 1) shows an increase in both 2-methyl, hydrogen bond and possibly 3-methylene proton hfcs occuring on binding to the protein. This suggests a major change in the spin density distribution within the quinone, with $A_1^{\bullet-}$ having more unpaired spin at C(2), oxygen and possibly C(3) than $PhQ^{\bullet-}$. Shortening of the O—H distance in $A_1^{\bullet-}$ could also contribute to the increased hydrogen bond hfcs. The ST spectra of Figure 2 and hfcs of Table 1 show that the spin density distribution in *A. variabilis* and spinach $A_1^{\bullet-}$ is very similar, indicative of similar A_1 binding sites in both species.

4. CONCLUSIONS

1) $A_1^{\bullet-}$ is hydrogen bonded at both carbonyl oxygens. This suggests that A_1 does not reside in a hydrophobic region of Photosystem 1 that mimics an hydrocarbon solvent which would be a straightforward way to lower the E_m of the $A_1/A_1^{\bullet-}$ redox couple.

2) Hydrogen bond hyperfine couplings are increased in $A_1^{\bullet-}$ relative to the phyllosemiquinone radical *in vitro* ($PhQ^{\bullet-}$). This reflects increased spin density at the carbonyl oxygens and/or shortening of the O—H distance in $A_1^{\bullet-}$

3) The 2-methyl group hyperfine coupling is increased by 30% in $A_1^{\bullet-}$ relative to $PhQ^{\bullet-}$. This represents a large, protein induced distortion in the electronic structure of the quinone which may be responsible for the unusual redox potential of A_1.

We thank the BBSRC (UK) for financial support.

References.

[1] Currier, T.C., Haury, J.F., and Wolk, C.P. (1977) J. Bacteriol. 129, 1556-1562.
[2] Barry, B.A., Bender, C.J., McIntosh, L., Ferguson-Miller, S. and Babcock, G.T. (1988) Isr. J. Chem. 28, 129-132.
[3]Heathcote, P., Hanley, J.A., and Evans, M.C.W. (1993) Biochim. Biophys. Acta 1144, 54-61.
[4] Hales, B.J., and Case, E.E. (1981) Biochim. Biophys. Acta 637, 291-302.
[5] Rigby, S.E.J., Nugent, J.H.A., and O'Malley, P.J. (1994) Biochemistry 33, 10043-10050.
[6] O'Malley, P.J., and Babcock, G.T. (1984) J. Phys. Chem. 80, 3912-3913.
[7] O'Malley, P.J., and Babcock, G.T. (1986) J. Am. Chem. Soc. 108, 3995-4001.

TEMPERATURE-DEPENDENT STUDIES OF CHARGE RECOMBINATION IN *HELIOBACILLUS MOBILIS*

Hung-Cheng Chiou and Robert E. Blankenship
Department of Chemistry and Biochemistry, Center for the Study of Early Events in Photosynthesis, Arizona State University, Tempe, Arizona, 85287-1604, U.S.A.

1. Introduction

Heliobacteria are newly discovered non-oxygen evolving photosynthetic bacteria (1,2). These organisms contain a very simple apparatus of a single pigment-protein complex with about 35-40 bacteriochlorophyll *g* per reaction center (RC) (3,4,5). Studies have shown that the RC of heliobacteria is similar in many ways to that of the RC of green sulfur bacteria and Photosystem I (PS I) of higher plants (6,7). Therefore, heliobacteria may be the bacterial models for PS I in the same way that purple bacteria are models for Photosystem II (8).

Electron transfer processes in the RC of PS I have been extensively studied and well established (9) while the charge separated states of heliobacteria are not yet well resolved, in particular whether or not a quinone is involved in the electron transfer processes as one of the secondary electron acceptors. The kinetics of oxidized P798 from the purified RC of *Heliobacillus mobilis* (*Hc. mobilis*) without F_A/F_B showed a recombination time constant of 14 ms between $P798^+$ and F_X^- at room temperature (4). Another recombination process that has been reported is between $P798^+$ and A_0^-, which has a time constant of 17 ns under reducing condition at room temperature from membranes of *Heliobacterium chlorum* (*H. chlorum*) (10). At 5 K, a mono-exponential decay with time constant of 2.3 ms was observed in membranes of *H. chlorum* (11). The identity of this 2.3 ms lifetime state is not yet clear. The two best candidates would be $P798^+F_X^-$ and $P798^+A_1^-$ since $P798^+A_0^-$ recombined with a time constant of 55 ns at 6 K (10).

In order to investigate the charge-separated states in heliobacteria, we have performed temperature-dependent studies of charge recombination using millisecond transient absorption difference spectroscopy for both membranes and purified RC samples. A temperature range of 4-29°C was used to study the activation energy. A broad range of 60-298 K was used to explore the possible identity of the 2.3 ms charge recombination state.

2. Materials and Methods

Membranes and RCs of *Hc. mobilis* were prepared as described previously (4). Both samples were stored in 50 mM Tris, 10 mM ascorbate, pH8, 15% glycerol under nitrogen at -20°C.

Transient absorption difference measurements on the millisecond timescale were performed using a single spectrophotometer of local design as described by Kleinherenbrink et al.(12). For the temperature range from 4 to 29°C, a thermostatted circulator was used to regulate the temperature of the cuvette holder. An Iron-Constantan thermocouple was connected to the junction of the cuvette holder and cuvette for the actual temperature reading. A typical optical density at 788 nm for both samples was 0.5 in a 5 mm pathlength cuvette. For the temperature range of 60-298 K, 67% glycerol was added to both samples to form a clear glass at low temperature. Samples were contained in a low temperature cell with an optical pathlength of 1.2 mm. The typical optical density at 788 nm

P. Mathis (ed.), Photosynthesis: from Light to Biosphere, Vol. II, 167–170.

of the sample in the measuring cell was 0.3-0.5 at room temperature. The cell was attached to the cooled finger of an Air-Products closed circulated helium refrigerator to cool the sample to the desired temperature.

3. Results and Discussion

The RC samples used here only contain the $P798F_X$ core complex without $F_{A/B}$ acceptors and cyt *c* which was confirmed by EPR measurements and heme staining of SDS-PAGE samples (4). Kinetics of flash-induced absorbance changes at 798 nm at various temperatures for both membrane and RC samples are shown in Fig 1.

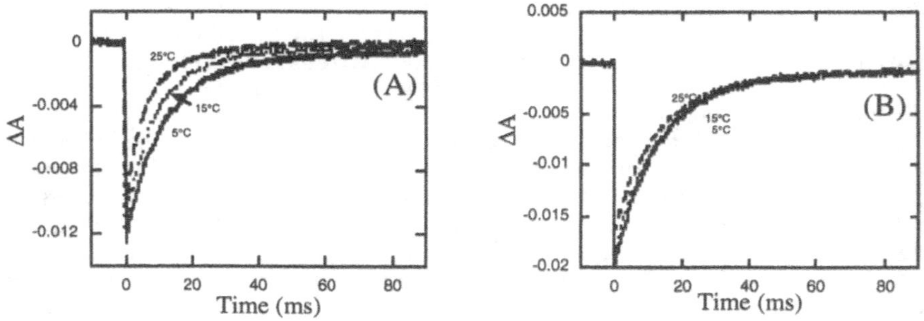

Fig. 1. Absorbance difference kinetics at 798 nm at 25°C, 15°C and 5°C for (A) membrane and (B) RC samples.

The decay of charge recombination becomes slower in membrane samples as the temperature decreases, while the decay kinetics for RC remains the same in the temperature range of 4-29°C. The amplitude of initial photobleaching increases a little upon cooling for both samples which is due to the slightly narrowing of the Q_y absorption band.

Arrhenius plots of activation energy for both samples are shown in Fig 2.

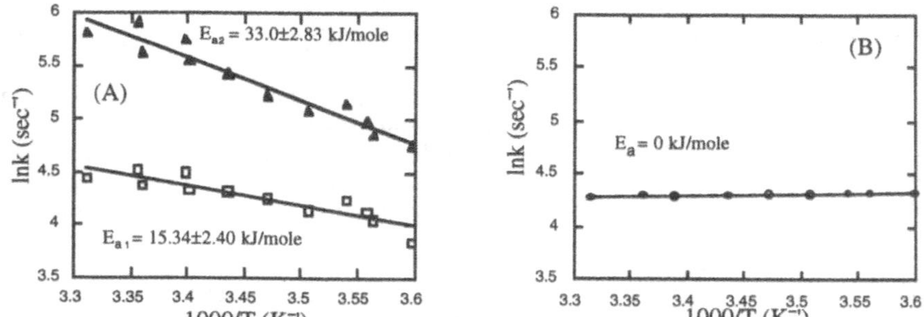

Fig. 2. Arrhenius plots of activation energies for (A) membrane and (B) RC samples. The temperature range is 4-29°C. Solid triangles: 3-8.6 ms component; open squares: 12-22 ms component and open circles: 14 ms component.

Two components were obtained from the fitting of kinetic traces in the membrane samples in the temperature range of 4-29°C. A 3-8.6 ms component representing the cyt *c* oxidation has an activation energy of 33.0±2.8 kJ/mole (344.4±29.0 meV/mole). A 12-22 ms component represents the recombination of the P798+F_X^- state or both P798+F_X^- and P798+$F_{A/B}^-$ states. The Arrhenius plot shows an activation energy of 15.34±2.40 kJ/mole (159.1±24.9 meV/mole) in membranes. Only one 14 ms component representing P798+F_X^- recombination was obtained for RC samples. The Arrhenius plot shows that the recombination rate of this state is temperature-independent (Fig. 2B) in this temperature range.

Kinetic traces of RC samples at 290 K, 190 K (both at 798 nm) and 90 K (at 794 nm) are shown in Fig. 3.

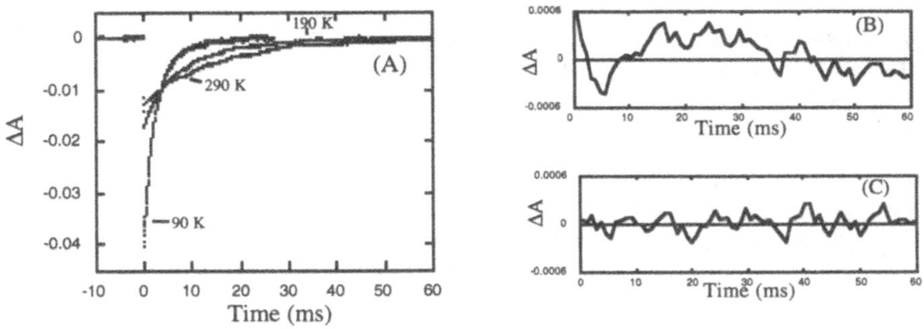

Fig. 3. (A) Absorbance difference kinetics of RC at 290 K, 190 K and 90 K. The residuals of the kinetic trace at 190 K for (B) one component plus a constant and (C) two components plus a constant.

One kinetic component is sufficient to fit the traces at 290 K and 90 K with the recombination time of 14 ms and 2.7 ms, respectively. The kinetic trace at 190 K is better fitted with two kinetic components as indicated by residuals (Fig. 3B and 3C) consisting of 59% of a 12.3 ms and 41% of a 3.8 ms component.

The decay components and the relative percentage of amplitude change for each component at various temperatures for RC samples are shown in Fig. 4.

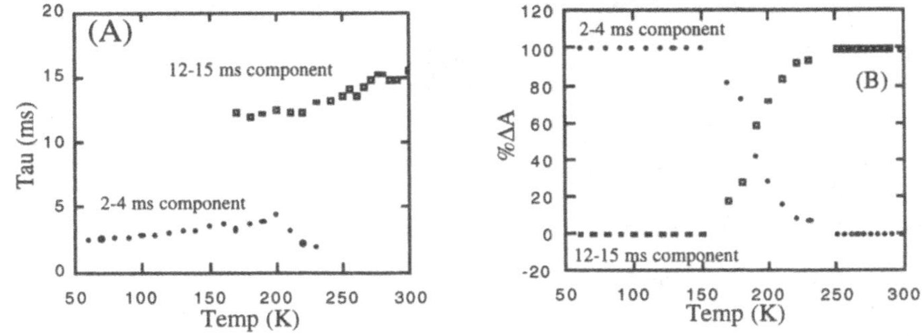

Fig. 4. (A) The charge recombination rates and (B) the relative percentage of amplitude distributions for both 12-15 ms component and 2-4 ms component over a broad temperature range for RC samples.

When the temperature is greater than 240 K, the 12-15 ms component is dominant representing the charge recombination state of the P798+F_X^- state. Two components (12-15 ms and 2-4 ms) were found between 230 K and 170 K with various amplitude distributions. When the temperature is lower than 160 K, the 2-4 ms component becomes the dominant component. The membrane samples also showed the same kind of decay behavior upon cooling (data not shown). The switch over gradually from 12-15 ms to 2-4 ms component upon cooling may indicate the shift of the $P798^+F_X^-$ recombination state to a state that is prior to $P798^+F_X^-$. This 2-4 ms decay component may represent an intermediate state between the $P798^+F_X^-$ and the $P798^+A_0^-$ states since the $P798^+A_0^-$ state decays with time constant of 55 ns at low temperature. It is also possible that the 2-4 ms decay component still represents recombination of the $P798^+F_X^-$ state due to some kind of structural change in RC upon cooling. Whether or not this 2-4 ms component is analogous to the $P700^+A_1^-$ state of PS I remains to be tested. However, it is worth noting that this 2.3 ms recombination state is about 10X slower than the $P700^+A_1^-$ state of PS I with a time constant of 200 µs at low temperature. The 12-15 ms recombination representing the $P798^+F_X^-$ in heliobacteria is also about 10X slower than the $P700^+F_X^-$ state of PS I with a lifetime of 1.2 ms at room temperature.

4. Acknowledgments

We thank Dr. Su Lin for helpful discussions. This work was supported by Grant MCB 9418415 from the National Science Foundation. This is publication No. 250 from Arizona State University Center for the Study of Early Events in Photosynthesis.

5. References

1. Gest, H. and Favinger, J.L. (1983) *Arch. Microbiol.* **136,** 11-16.
2. Madigan, M.T. and Ormerod, J.G. (1995) in *Anoxygenic Photosynthetic Bacteria* (Blankenship, R.E., Madigan, M.T. and Bauer, C.E., eds.) pp. 17-30, Kluwer Academic Publishers. Printed in The Netherlands.
3. Michalski, T.J., Hunt, J.E., Bowman, M.K., Smith, U., Bardeen, K., Gest, H., Norris, J.R. and Kate, J.J. (1987) *Proc. Natl. Acad. Sci. USA* **84**, 2570-2574.
4. Trost, J.T., and Blankenship, R.E. (1989) *Biochemistry* **28**, 9898-9904.
5. Van de Meent, E.J., Kobayashi, M., Erkelens, C., van Veelen, P.A., Amesz, J. and Watanabe, T. (1991) *Biochim. Biophys. Acta* **1058**, 356-362.
6. Büttner, M., Xie, D.L., Nelson, H., Pinther, W., Hauska, G. and Nelson, N. (1992) *Proc. Natl. Acad. Sci. USA* **89**, 8135-8139.
7. Liebl, U., Mockensturm-Wilson, M., Trost, J.T., Brune, D.C., Blankenship, R.E., and Vermaas, W. (1993) *Proc. Natl. Acad. Sci. USA* **90**, 7124-7128.
8. Blankenship, R.E. (1994) *Antonie van Leeuwenhoek* **65**, 311-329.
9. Golbeck, J.H., and Bryant, D.A. (1991) *Current Topics in Bioenergetics* **16**, 83-177.
10. Kleinherenbrink, F.A.M., Aartsma, T.J. and Amesz, J. (1991) *Biochim. Biophys. Acta* **1057**, 346-352.
11. Smit, H.W.T., Van Dorssen, R.J., and Amesz, J. (1989) *Biochim. Biophys. Acta* **973**, 212-219.
12. Kleinherenbrink, F.A.M., Chiou, H-C., LoBrutto, R. and Blankenship, R.E. (1994) *Photosynth. Res.* **41,** 115-123.

CHARACTERIZATION OF THE PHOTOSYNTHETIC REACTION CENTRE FROM *CHLOROBIA*

Christine Hager-Braun[1], Astrid Riedel[2], Dian-Lin Xie[3], Rainer Zimmermann[1], Nathan Nelson[3] and Günter Hauska[1]

[1]Department of Cell Biology and Plant Physiology, University of Regensburg, Germany; [2]Institute of Biophysics, University of Regensburg, Germany; [3]Roche Institute of Molecular Biology, Nutley, New Jersey, USA

1. Introduction

Bacterial photosynthetic reaction centres (RC) are of great interest as they are less complex than PSI and PSII and therefore can serve as models. The structure of the Q-type-RC from purple bacteria has been elucidated by X-ray crystallography, but characterization of the FeS-type-RC from green sulfur bacteria and Heliobacteria is lagging behind.
We have purified a RC-complex from *Chlorobium tepidum* and *C. limicola* (1) and characterized it by SDS-PAGE, absorption spectroscopy and EPR-spectroscopy (2). Here we give a brief summary of the results.

2. Materials and Methods

Chlorosome-depleted membranes of *Chlorobium tepidum, C. limicola* and *C. vibrioforme* were prepared according to (3). RC of *C. tepidum* and *C. limicola* was isolated by solubilization of chlorosome-containing membranes with Triton X-100, anion exchange chromatography and a sucrose density gradient (SDG) as described in (1). Proteins were separated by SDS-PAGE and analyzed with a Hirschmann Elscript 400 AT densitometer. Absorption spectra were recorded with a Kontron Uvikon 860 spectrophotometer, photobleaching in continuous light was measured with an Aminco dualwavelength DW2 spectrophotometer. The EPR-spectra were recorded on a Bruker 200 X band spectrometer with a helium cryostat and Oxford Instruments temperature control system.

3. Results and Discussion

3.1. Characterization of the RC by absorption spectroscopy, SDS-PAGE and photobleaching

The final sucrose density gradient separates three coloured zones as shown schematically in Figure 1A. In the top there is a brownish to green zone where bacteriachlorophyll *c* and bacteriapheophytin are accumulated in varying contents. The middle band is a brownish-green zone and the lower band is bluish-green; both fractions contain RC, but with different compositions of subunits (for details see (1)).
The absorption spectrum and the polypeptide composition of the bluish-green zone for *C. tepidum* are shown in Figure 1(B) and (C), respectively.

P. Mathis (ed.), Photosynthesis: from Light to Biosphere, Vol. II, 171–174.

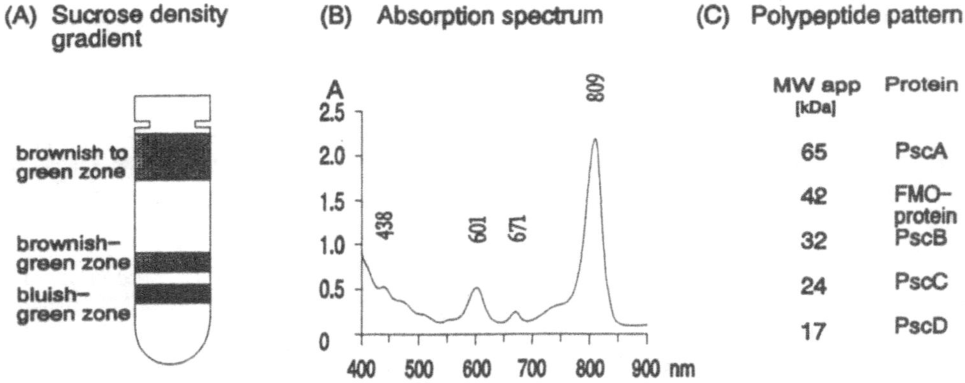

Figure 1: A sucrose density gradient separates a brownish-green zone and a bluish-green zone (A). The absorption spectrum of the bluish-green fraction is depicted in (B), whereas the polypeptide composition of the RC-complex is shown in (C) with the apparent molecular weights of the subunits in SDS-PAGE (nomenclature according to (4)).

In this fraction of *C. tepidum* pigments absorb at 400-500 nm, 551 nm, 601 nm, 671 nm and 809-810 nm.
Five proteins are found, namely the RC-core-protein PscA, the BChl *a*-binding protein (=FMO-protein), PscB (carrying the FeS-centres F_A and F_B), cytochrome c_{551} (=PscC) and PscD. PscD is a positively charged polypeptide (1) with so far unclear function. Incubation of photoactive RC with chaotropic agents results in a decrease of photoactivity combined with a separation of PscD. This 17 kDa-subunit probably stabilizes the RC-complex. In *C. limicola* the bluish-green fraction consists of PscA, FMO-protein, PscB and PscD, but lacks cytochrome *c*.
Photobleaching of P_{840} in continuous light is confined to the bluish-green zone while the brownish-green is inactive. This photoactivity is very stable. For *C. tepidum* in samples kept anaerobically at 4°C, or frozen at -20°C, after one week nearly no decrease was observed and even after 7 weeks about 70% of photobleaching still remained, although the samples have been tested at room temperature weekly.

3.2. Characterization of the RC by EPR-spectroscopy
The purified RC-complexes (bluish-green zone) from *C. tepidum* and *C. limicola* were characterized by EPR-spectroscopy and compared with the data obtained on chlorosome-depleted membranes (2).
In the membranes of *C. tepidum* and *C. limicola* the FeS-clusters F_B, F_A and F_X were photoreduced. The signals show narrow linewidths in contrast to the data for *C. vibrioforme* by Kjaer et al. (5). Therefore we tested membranes from *C. vibrioforme* (cells were kindly donated by H.-V. Scheller) and found also narrow linewidths, but they are still broader than in PSI. Table 1 summarizes the linewidths of the g_y signal of F_B and F_A of membranes and isolated RC-complexes.

	Membranes			isolated RC			PS I
g_y	*C. tepidum*	*C. limicola*	*C. vibrio-forme*	*C. tepidum*	*C. limicola*	*C. vibrio-forme*	
1.91	4	5/3[6]	3	4.5	5/3[7]	9[5]	1.6[8]
1.94	7.5	5-6/5.3[6]	-	6	8-10/5.5[7]	7.5[5]	2.2[8]

Table 1: Linewidth ΔH_{pp} in mT of the g_y-signal of F_A and F_B derived from difference spectra. Without reference numbers: our measurements

In purified RC all three FeS-centres were also detected as depicted in Figure 2 for *C. tepidum*. The dark spectrum shows no contamination with any paramagnetic substance. F_B was photoreduced at 4K, F_A was observed upon illumination at 200 K. Further illumination at 200 K reduced F_X, seen as a broad trough at $g_x = 1.79$.

Figure 2

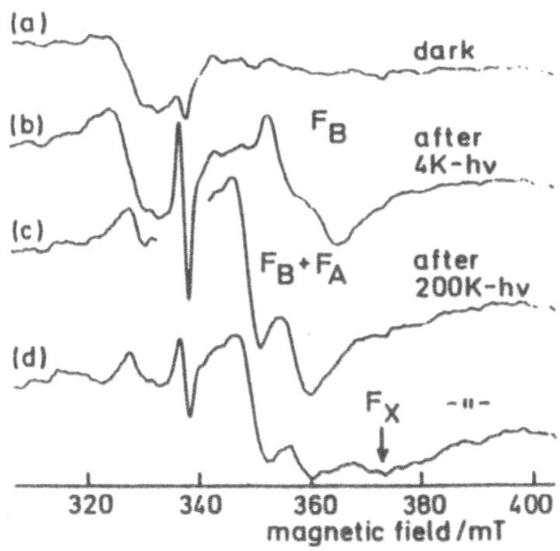

Figure 3

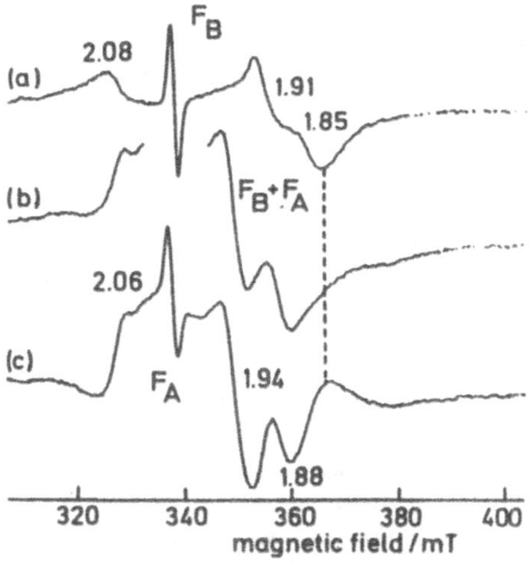

Figure 2: EPR-spectra of the FeS-centres in the bluish-green RC-fraction of *C. tepidum* at pH 8 in the presence of 50 mM dithionite. EPR-conditions: temperature 12 K; microwave frequency 9.46 GHz; microwave power 12.6 mW; modulation amplitude 1.6 mT; F_X was detected at 8 K, 32 mW, 2.0 mT.

Figure 3: Difference spectra showing the FeS-centres F_A and F_B: (a): Fig.2b minus a; (b): Fig.2c minus a; (c): Fig.2c minus b.

The following g-values can be derived from the difference spectra in Figure 3:
$g_z = 2.08$, $g_y = 1.91$ and $g_x = 1.85$ for F_A and $g_z = 2.06$, $g_y = 1.94$ and $g_x = 1.88$ for F_B.
The magnetic interaction of reduced F_B and F_A results in a shift of the g_x-signal of F_B to 1.88, which causes a peak in the difference spectum of F_A, as indicated by a dashed line. Chemical reduction of F_A and F_B in the presence of 50 mM dithionite at pH 8 could not be achieved, at pH 10.5 it could only be obtained partly; about 50 % of FeS-centres were reduced chemically compared to photoreduction and the yield varied amongst different preparations.

Low temperature EPR-spectroscopy offers the possibility to estimate the extent of damage of the RC-complex due to the preparation procedure. The spin-polarized triplet signal of P_{840} indicates how many RC-complexes are blocked in electron transfer beyond the primary electron acceptor A_0. Comparing the amplitude of the spontaneously formed triplet signal during illumination to the 100%-value (resulting by photoaccumulation at pH 10.5 after addition of 50 μM methylviologen) documents that only about 10-15% of RC can not transfer electrons beyond A_0 (Figure 4a and b, respectively). A g=2 signal arising from P_{840}^+, recorded under continuous illumination, represents the total amount of intact P_{840} involved in electron transfer to FeS-centres. The signal remaining after switching off the light shows that about 50% of RC-complexes (in some preparations up to 70%) were capable of stable charge separation (Figure 5).

Figure 4

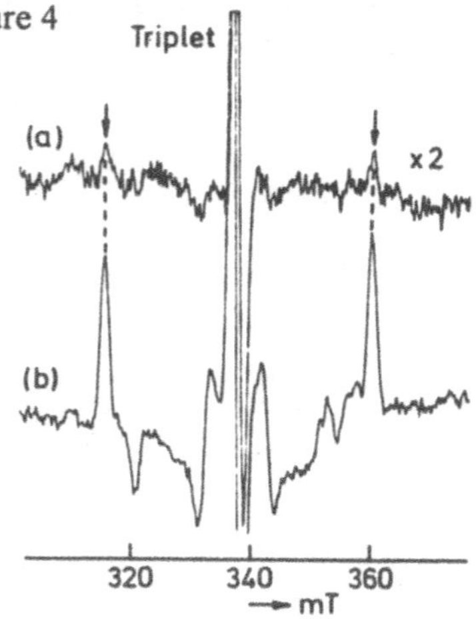

Figure 5

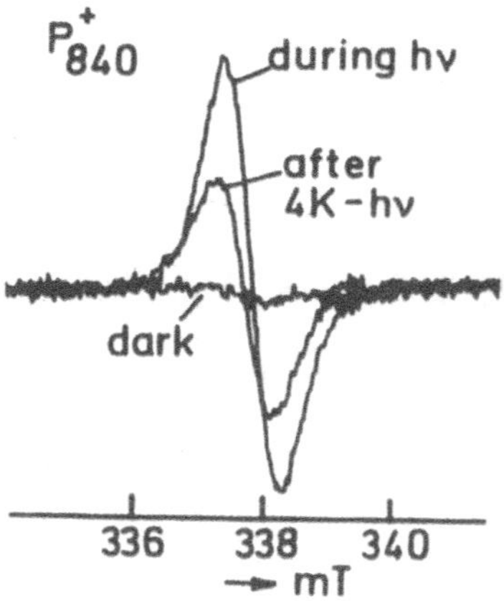

Figure 4: Spin-polarized triplet signal of P_{840}. The photoactive RC-complex was reduced with 50 mM dithionite at pH 8 (spontaneously formed triplet, (a)). Photoaccumulation occured at pH 10.5 after addition of 50 μM methylviologen (100% triplet, (b)).
Figure 5: EPR-spectrum of the P_{840}^+ radical of purified RC from *C. tepidum* at pH 8 with 50 mM dithionite. The spectra were taken in the dark, after 15 min illumination at 4K in the cavity and during illumination.

In future we want to find the minimal composition that is required for an active RC-complex using EPR-spectroscopy for characterization .

References:

(1) Hager-Braun et al. (1995) Biochemistry 34: 9617-9624
(2) Riedel et al., in preparation
(3) Klughammer et al. (1995) Photosynth. Res. 43: 27-34
(4) Bryant (1994) Photosynth. Res. 41:27-28
(5) Kjaer et al. (1994) Photosynth. Res. 41: 105-114
(6) Nitschke et al. (1990) Biochemistry 29: 3834-3842
(7) Feiler et al. (1992) Biochemistry 31: 2608-2614
(8) Nugent et al. (1981) Biochim. Biophys. Acta 634: 249-255

Evidence for the Existence of trimeric Photosystem I in the Cyanobacterial Thylakoid Membranes by recombinant Antibodies Fragments

Georgios Tsiotis[1], Winfried Haase[2], Shirley Müller[1] and Andreas Engel[1]
[1]) M. Müller Institute for Microscopical Structural Biology, Biozentrum, University of Basel, CH-4056 Basel, Switzerland
[2]) Max-Planck-Instute für Biophysik, D-60528 Frankfurt am Main, Germany

1. Introduction

Photosystem I (PSI) is a pigment protein complex of the thylakoid membrane which mediates light-driven electron transfer from plastocyanin (or cytochrome c_{553}) to ferredoxin. The PSI complex in cyanobacteria is comprised of at least twelve polypeptides, which are products of the *psa*A to *psa*L-genes (see [1] for review). The major component of the photoreactioncenter is a heterodimer composed of the *psa*A/*psa*B gene products each of which has a molecular mass of 80 kDa. This heterodimer carries the antenna chlorophylls and the prosthetic groups that are involved in the electron transport. Remaining protein subunits have masses between 4 kDa and 20 kDa and can be divided into three topologically diverent groups. The *Psa*C-*psa*E gene products are located on the stromal side, the *psa*F subunit is located on the lumenal side of the thylakoid membrane, and the subunits derived from the *psa*J-*psa*M genes are hydrophobic and are predicted to possess *trans*-membrane *a*-helices. Although monomeric and trimeric PSI complexes have been purified from the mesophilic cyanobacteria *Synechocystis* PCC 6803 and *Synechococcus* PCC 7002 [2-4], little information is available about the *in vivo* form of the PSI [5-7].

Here we describe monoclonal antibodies (mAb) that were raised against the native whole trimeric PSI complex purified from *Synechococcus* PCC 7002. The recombinant antibody fragments were used for the *in situ* labeling of PSI in thylakoid membranes.

2. Procedures

The trimeric PSI complex was solubilized and purified as described previously [3]. Generation of monoclonal antibodies was achieved following published procedures. The molecular interactions between the mAbs and the PSI were measured using the BIAcore system (Pharmacia). Amplification, cloning, expression and purification of the V_H and V_L gene domains were performed according to [8]. Mass measurements were made using a scanning transmission electron microscope (STEM) according to [9]. Experimental details are presented in the figures legends.

3. Results and Discussion

ELISA documented the binding of the mAbs to the native trimeric PSI. In addition, the kinetic constants $K_{1E7}=8.84*10^8\ M^{-1}$ and $K_{5B11}=1.7*10^6\ M^{-1}$ for the interaction between PSI and mAbs were assessed by surface plasmon resonance (BIAcore).

P. Mathis (ed.), Photosynthesis: from Light to Biosphere, Vol. II, 175–178.

Reactivity of the mAb with the denatured PSI complex was determinated by immunoblot analysis of the purified PSI complex of which was first separated into subunits by SDS-PAGE. While the mAb 5B11 reacted with the *psa*D protein, the mAb 1E7 showed no reaction with any part of the denatured PSI complex (fig. 1). Thus, the mAb 1E7 recognizes a conformational epitope of the trimeric PSI. The following observations corroborate this hypothesis: (i) The reactivity of the mAb 1E7 decreased upon urea induced conformational change of the PSI complex. (ii) The mAb only recognized the trimeric PSI from *Synechococcus* PCC 7002 while the mAb 5B11 also crossreacted with the *psa*D subunit of other strains.

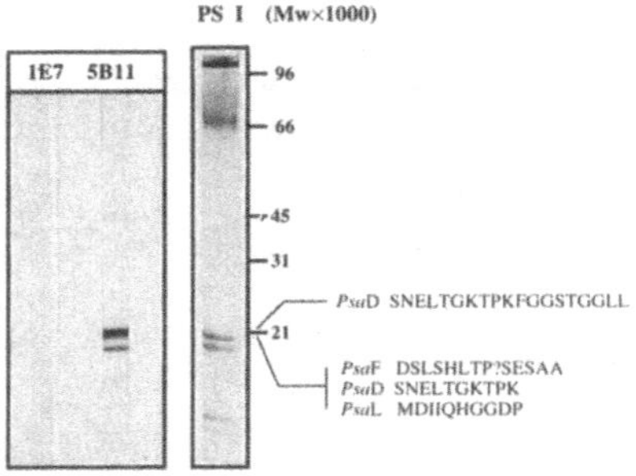

Figure 1. Immunoblot showing reactivity of mAbs with the PSI subunits. Characterization of the immunoreactive PSI subunit by N-terminal sequencing (PSI lane).

The antigen-binding affinity of the generated mAb was investigated by immunoelectron microscopy. As shown in fig. 2A,B the mAbs only labeled one side of the thylakoid membrane. This side was identified by performing labeling experiments with a mixture of both mAbs (fig. 2C). This mixture of mAbs also only reacted with one membrane side which means that the mAb 1E7 labeled the same surface as the mAb 5B1, i.e., the stromal side of the thylakoid membrane.

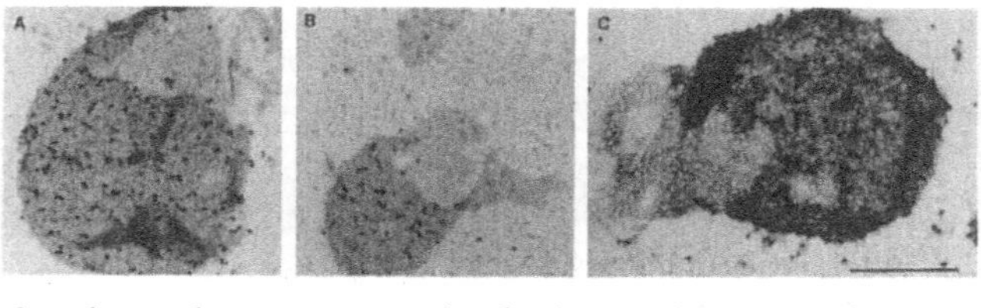

Figure 2. Isolated membranes were adsorbed to carbon coated grids and incubated with monoclonal antibodies (60 min, RT). Binding sites of the primary antibodies were detected with secondary gold coupled antibodies (goat anti-mouse IgG+IgM). Afterwards samples were negatively stained with uranyl acetate and viewed in the electron microscope. 1E7 (A), 5B11 (B) and with a mixture of mAbs (1E7, 5B11) directed against PSI (C). Bar represents 250 nm.

The Fv fragments expressed in *E.coli* were purified by streptavidin-affinity chromatography using the binding affinity of the strep tag which is attached to the V_H chain. The antigen-binding affinity of the purified Fv fragments was detected by a direct antigen Fv fragment binding ELISA, and by immunoelectron microscopy. Results obtained by ELISA revealed binding ability of the purified Fv fragments with the trimeric

PSI complex. Immunogold labeling and electron microscopy was used to visualize the reaction of the purified Fv fragments with the PSI complex on the thylakoid membranes. They were found to label specifically only one side of the thylakoid membrane (fig. 3). These results are in agreement with the immmunoelectron microscopic results of the whole mAbs.

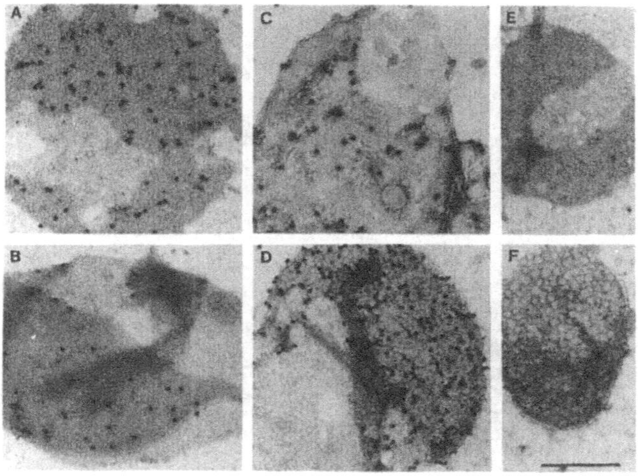

Figure 3. On-grid immunogold labeling of isolated thylakoid membranes from *Synechococcus* PCC 7002 with fragments from mAbs fused to Myc or Strep tag peptides. Detection of binding sites was either by the anti-Myc antibody 9E10 and a gold-coupled goat anti-mouse antibody (A and C) or by strep-Av-gold (B and D). Fv-1E7 (A and B), Fv-5B11 (C and D). In controls (E and F) the fragments were omitted using either only the ant-Myc tag peptide antibody 9E10 and the gold-coupled goat anti-mouse antibody (E) or the gold-coupled streptavidimarker (F). Bar represents 250 nm.

Incubation of recombinant antibody fragments with thylakoid membrane extracts and subsequent use of a streptavidin affinity column allowed the isolation of a Chla-containing protein complex. SDS gels showed that the isolated complex contained the same proteins as the PSI complex. Mass measurement by STEM yielded a mass peak at 985±50 kDa (fig. 4). This value is similar to the molecular mass of the trimeric PSI complexes estimated by gelfiltration and sedimentation centrifugation [3, 10], suggesting that the eluted complex is a trimeric PSI. In addition, the information contained in the trimeric organization of the complex contributes to the epitope since the trimer was selectively removed from the thylakoid extracts by streptavidin-affinity chromatography. The binding of the Fv fragment to thylakoid membranes in *situ* and the recognition of only the trimeric PSI form from the thylakoid extracts provides evidence that PSI exists as a trimer in the thylakoid membranes.

In summary, we have generated mAbs against the native PSI complex from the cyanobacterium *Synechococcus* PCC 7002. The epitope of one mAb is located on the the *psa*D protein. The second recognizes a conformational epitope. Both mAbs react on the same thylakoid membrane side with the PSI. The variable regions of the mAbs were cloned, sequenced and expressed in *E.coli* with different expression levels. The expressed

Fv fragments were purified by streptavidin affinity chromatography and were functionally intact. The 1E7 recombinant antibody recognizes specifically the trimeric form of the PSI complex in thylakoid extract and the immunogold labeling experiments show that this form exists*in vivo*.

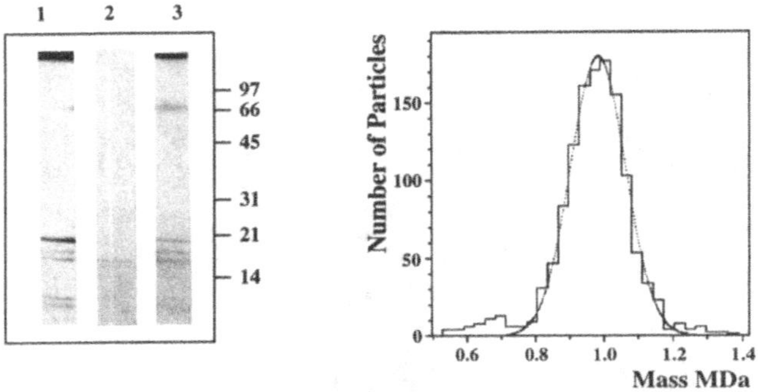

Figure 4. (Left) SDS-gel electrophoresis of PSI control (lane 1), of purified Fv fragment (lane 2) and of the Chl*a*-binding complex stained with Coomassie (lane 3). (Right) Mass analysis of the Chl*a*-binding complex which was specifically eluted from a streptavidin-Sepharose column by addition of diaminobiotin after immobilization of the thylakoid extracts with Fv fragments on the column.

Further, the bifunctionality of engineered Fv fragments to bind the antigen and steptavidin permit us to use this as a tool for one step purification of trimeric PSI complexes from thylakoid extracts by affinity chromatography [11]. Finally, the binding of the Fv fragment to the PSI complex increases the hydrophilic moiety in PSI complex and so may facilitate crystallization of membrane proteins [12].

References

1.Goldbeck,J. and Bryant,D. (1991) In Current Topics in Bioenergetics, Vol. 16 (Lee C.P. ed) pp.83-177, Academic Press, New York
2.Rögner, M., P.J. Nixon, and B. Diner (1990) J. Biol. Chem. 265, 6189-6196
3.Tsiotis, G., W. Nitschke, W. Haase, and H. Michel (1993) Photosynth Res 35, 285-297
4.Kruip, J., E. Boekema, D. Bald, A. Boonstra, and M. Rögner (1993) J. Biol. Chem. 268, 23353-23360
5.Hefti, A., C.R. Ford, M. Miller, R.P. Cox, and A. Engel (1992) FEBS Lett 296, 29-32
6.Hladik, R. and D. Sovrova (1991) Photosynth. Res. 29, 171-175
7.Kruip, J., D. Bald, E. Boekema, and M. Rögner (1994) Photosynth Res 40, 279-286
8.Kleymann, G., C. Ostermeier, A. Skerra, B. Ludwig, and H. Michel (1995b) Bio/Technology 13, 155-160
9.Müller, S.A., K.N. Goldie, R. Buerki, R. Häring, and A. Engel (1992) Ultramicrosscopy, 46, 317-334
10.Tziatzos, C., P. Schuck, D. Schubert, and G. Tsiotis (1994) Z. Naturforsch. 49c, 220-222
11.Tsiotis, G., W. Haase, A. Engel, and H. Michel (1995) Eur. J. Biochem. In press
12.Ostermeir,C., S. Iwata, B. Ludwig, and Michel,H. Submitted

DEGRADATION OF PHOTOSYSTEM I REACTION CENTER PROTEINS DURING PHOTOINHIBITION IN VITRO

Kyoko Baba[1], Shigeru Itoh[2] and Satoshi Hoshina[1]: [1]Department of Biology, Faculty of Science, Kanazawa University, Kakuma, Kanazawa 920-11 and [2]National Institute for Basic Biology, Myodaiji, Okazaki 444, Japan

1. Introduction

Strong light irradiation inhibits photosynthetic activities. This process is called photoinhibition. The photoinhibition of photosystem II (PSII) electron transport has been well studied (1-3). On the other hand, there are a few reports about the inactivation of photosystem I (PSI) electron transport by the light treatment in vitro and in vivo. Illumination of isolated spinach chloroplasts under aerobic conditions caused the destruction of three iron-sulfur centers of photosystem I (PSI), while the electron transfer between A_0 and F_X was inhibited by the illumination under extremely reducing conditions (4, 5). Light treatment of PSI complex caused simultaneous P700 degradation and chlorophyll bleaching (6). Sonoike and Terashima (7) and Havaux and Davaud (8) demonstrated the PSI photoinhibition in vivo. The component on the acceptor side of the PSI, A_1 or F_X was the first site of inactivation and that inactivation of the electron transfer led to the degradation of the PSI reaction center subunit (s) (7). However, the mechanism of PSI photoinhibition has not been clarified yet. In this study, we focused on the effect of histidine, a singlet oxygen scavenger, on the photoinhibition of spinach PSI core complex. Histidine provided a significant protection of the PSI reaction center proteins against degradation.

2. Materials and Methods

PSI particles (PSI-180) were prepared by solubilizing spinach thylakoid membranes with 0.8% (w/v) Triton X-100 as previously described (9). PSI core complex with Chl/P700 ratio of 100 (PSI-100) was prepared from PSI-180 by the treatment with Triton X-100, followed by sucrose density gradient ultracentrifugation. After sucrose density gradient ultracentrifugation, the PSI-100 band was diluted with distilled water and stirred with Bio-Beads SM-2 for 3 h to remove Triton X-100. The mixture was filtered through a nylon mesh (82 μ) before centrifugation. PSI-100 was resuspended at 100 μg Chl/ml, in 50 mM Tricine-NaOH (pH 7.8) containing 0.1M sorbitol, 10 mM NaCl, and 0.05% (w/v) Triton X-100 and illuminated with white light (PFD=1.1×10^4 μE m^{-2} s^{-1}) at 25°C for 2h with stirring. PSI-mediated electron transfer from reduced dichlorophenol indophenol (DCIP) to methyl viologen (MV) was measured by

P. Mathis (ed.), Photosynthesis: from Light to Biosphere, Vol. II, 179–182.

O_2 consumption using a Clark type oxygen electrode (Rank Brothers) at 25°C as described in (10). The P700 content was determined chemically by measuring the 50 μM ferricyanide-oxidized minus 1 mM ascorbate-reduced difference spectrum as described in (10). The amount of amino group was calculated from the absorbance at 345 nm after PSI-100 was allowed to react with trinitrobenzenesulfonate (TNBS). Polypeptide analysis was carried out by Tricine-SDS-polyacrylamide gel electrophoresis (PAGE) in 12.6% T and 3% C gel (11).

3. Results and Discussion

Fig. 1 shows the absorption spectra of PSI-100 treated after in the light with or without histidine. Chlorophyll content in PSI-100 without histidine decreased about 40% after photoinhibitory treatment for 2h. On the other hand, chemically determined P700 content stayed constant even when the electron transfer activity decreased by 45% after photoinhibitory treatment (Table 1).

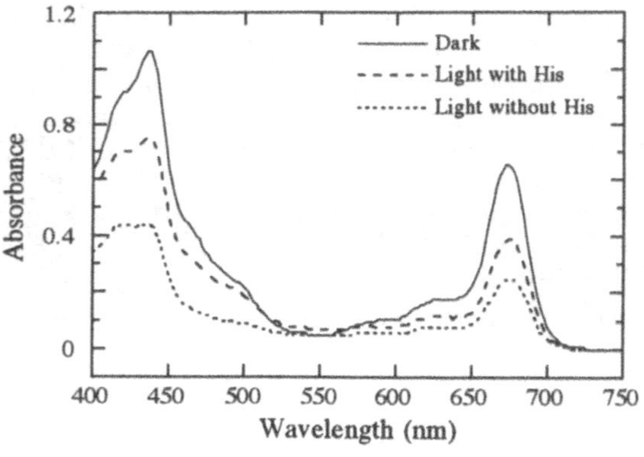

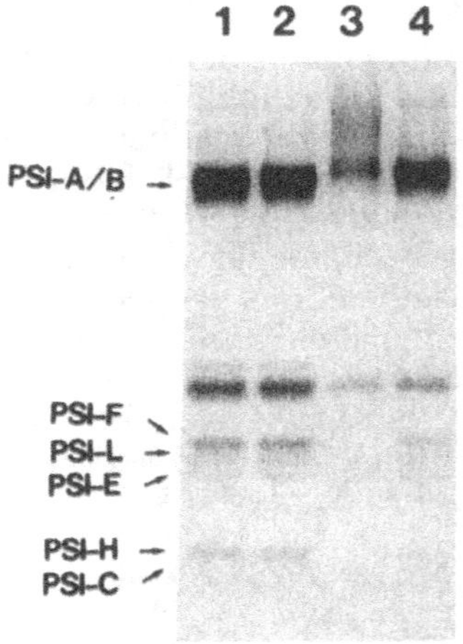

Fig. 1. Room temperature absorption spectra of PSI-100 after photoinhibitory treatment with or without histidine. PSI-100 was incubated for 2 h in the dark (solid line), in the light without histidine (dotted line) and in the light with 25 mM histidine (broken line).

Fig. 2. Polypeptide compositions of PSI-100 treated with the light in the presence or absence of histidine. Dark without histidine (lane 1), dark with 25 mM histidine (lane 2), light without histidine (lane 3), light with 25 mM histidine (lane 4).

Irradiation of spinach PSI-180 with the strong white light also caused a bleaching of Chl and a decrease in the MV photoreduction activity with almost no loss of P700. Measurements of the flash-induced absorbance change indicated that A1 and/or Fx was destroyed by the photoinhibitory treatment. Neither inactivation nor Chl bleaching in PSI-180 was induced by the light treatment in the absence of oxygen (K. Baba, S. Itoh, G. Hastings and S. Hoshina, unpublished). As shown in Fig. 1 and Table 1, Chl bleaching and the inhibition of the electron transport activity were partially overcome by the addition of histidine that is known to be a singlet oxygen scavenger (12).

PSI-100 was treated with the light in the presence or absence of histidine and its polypeptide composition was analyzed by SDS-PAGE (Fig. 2). The contents of the reaction center proteins, PSI-A and PSI-B decreased dramatically by the photoinhibitory treatment (Fig. 2, lane 3). The fragment of the reaction center proteins was not detected in this gel system. For high resolution of low molecular weight of proteins, the PSI-100 treated with the light was analyzed by SDS-PAGE in 16.5% T and 3% C gel. However, no fragment of the reaction center proteins was observed. We could not also detect any fragment of the reaction center proteins using Western blot analysis of the light-treated PSI-100 probed with antibody against CPIe (PSI-A and PSI-B) from *Synechococcus elongatus* (13). PSI-F, L, H and C were also destroyed by the photoinhibitory treatment. When the PSI-100 was treated with the light, histidine provided a significant protection of the reaction center proteins, and slight proteiction of PSI-F, L, H and C against the degradation (Fig. 2, lane 4).

The amount of α- and ε-amino groups in PSI-100 approximately doubled after the PSI complex was treated with the light in the absence of histidine (data not shown), suggesting that the proteins are hydrolyzed into small fragments and the amount of TNBS-reactive N-terminus increases.

Table 1. Effect of histidine during photoinhibitory treatment on the MV photoreduction activity and P700 content in spinach PSI-100.

Incubation	MV photoreduction	P700 content[a]
	% of control	
light without histidine	45	100
dark with histidine	100	96
light with histidine	59	nd[b]

a) chemically determinedP700 content.
b) not determined.

Sonoike and Terashima (7) reported that the photoinhibitory treatment of cucumber leaves led to the slight degradation of the PSI reaction center proteins and a new band appeared just below the proteins. Strong irradiation of the isolated PSI-100 seems to result in the drastic degradation of PSI proteins.

It is noteworthy that no loss of P700 was detected with the degradation of the reaction center proteins. This means that the structure around the site of P700 in the light-treated PSI-100 may be actively maintained before the dissociation of the proteins for SDS-PAGE (dissociation medium; 2 M urea, 2% (w/v) SDS, 2% (v/v) 2-mercaptoethanol, 35 mM sucrose, and 60 mM Tris-HCl, pH 6..8).

These results suggest that toxic oxygen molecule generated in the PSI reaction center complex during photoinhibitory treatment is responsible for the in vitro degradation of the proteins and the inactivation of PSI.

4. Acknowledgements

We would like to thank Professor K. Wada for his support and helpful comments and Dr. I. Enami of Science University of Tokyo for the gift of the antibody against CPIe. This study was supported in part by a Grant-in-Aid (05640728) from the Ministry of Education, Science and Culture, Japan to SH.

References

1 Powles, S.B. (1984) Ann. Rev. Plant Physiol. 35, 15-44.
2 Prasil, O., Adir, N. and Ohad, I. (1992) in Topics in Photosynthesis. Vol. 11 (Barber J., ed.), pp. 295-348. Elsevier, Amsterdam.
3 Aro, E.M., Virgin, I. and Andersson, B. (1993) Biochim. Biophys. Acta 1143, 113-134.
4 Inoue, K., Sakurai. H. and Hiyama, T. (1986) Plant Cell Physiol. 27, 961-968.
5 Inoue, K., Fujii, T., Yokoyama, E., Matsuura, K., Hiyama, T. and Sakurai, H. (1989) Plant Cell Physiol. 30, 65-71.
6 Purcell, M. and Carpentier, R. (1994) Photochem. Photobiol. 59, 215-218.
7 Sonoike, K. and Terashima, I. (1994) Planta 194, 287-293.
8 Havaux, M. and Davaud, A. (1994) Photosynth. Res. 40, 75-92.
9 Hoshina, S. and Itoh, S. (1987) Plant Cell Physiol. 28, 599-609.
10 Hoshina, S., Sakurai, R., Kunishima, N., Wada, K. and Itoh, S. (1990) Biochim. Biophys. Acta 1015, 61-68.
11 Schagger, H. and von Jagow, G. (1987) Anal. Biochem. 166, 368-379.
12 Mishra, N.P. and Ghanotakis, D.F. (1994) Biochim. Biophya. Acta 1187, 296-300.
13 Kashino, Y., Enami, I., Satoh, Ka. and Katoh, S. (1990) Plant Cell Physiol. 31, 479-488.

TWO DIMENSIONAL FOURIER TRANSFORM ELECTRON SPIN ECHO ENVELOPE MODULATION SPECTROSCOPY OF THE ^{14}N PRIMARY DONORS, $P700^{+\cdot}$ AND $P870^{+\cdot}$

M.C.W. Evans, P.J. Bratt, P. Heathcote and P. Möenne Loccoz

University College London, Department of Biology,
Gower Street, London,WC1E 6BT, UK

Introduction

The initial electron transfer event in photosynthesis is the photooxidation of the reaction center chlorophyll electron donor and the generation of the cation radical form of this special bacteriochlorophyll or chlorophyll reaction center species and the consequent reduction of an electron acceptor. In purple bacteria, *Rhodobacter sphaeroides* the reaction center bacteriochlorophyll is termed P870, whilst in green plant photosystem 1 the reaction center chlorophyll is denoted as P700. The photochemically active primary donor is only distinguishable from the bulk chlorophylls in the reaction center by its redox and spectroscopic properties.

A recent paper[1] by Käß et al using 2D ESEEM and spectral simulation on ^{15}N labelled *Rhodobacter sphaeroides* has suggested that although the primary donor is dimer of bacteriochlorophyll a molecules with an electron distribution of 2:1 at room temperature, at 4K there is a conformational change resulting in a 5:1 electron distribution. As this is a somewhat surprising result, a study was initiated using 1D and 2D ESEEM techniques on ^{14}N *Rhodobacter sphaeroides* at 4K to investigate the proposed conformational change.

Of greater interest is the primary donor in green plant photosystem 1, $P700^{+\cdot}$ which is thought to be a dimer of chlorophyll a molecules with a very asymetric electron distribution ratio[2,3] of about 3:1, although the possibility of P700 being a perturbed monomer of chlorophyll a has not been excluded. This work presents convincing spectroscopic evidence which indicates the presence of the minor half of the chlorophyll a dimer as well as the major half. All of the isotropic hyperfine coupling constants, (a_{iso}) which are directly indicative of the electron density are given for the high electron density half of the dimer.

Experimental Section

Photosystem 1 was prepared from spinach chloroplasts by two procedures. Digitonin photosystem 1 particles were prepared by the procedure of Boardman[4]. The final photosystem 1 pellet was suspended in 50 mM Tris-HCl buffer pH 8.0, containing

P. Mathis (ed.), Photosynthesis: from Light to Biosphere, Vol. II, 183–186.

5 mM $MgCl_2$ and 2mM EDTA. The preparation normally had a P700/Chlorophyll ratio of 1/150-200. Triton photosystem 1 particles were prepared by solubilisation with Triton-X100 followed by hydroxyl-apatite chromatography. After concentration by ultrafiltration the particles were stored in a buffer containing 20 mM Tris-HCl pH 8.0, 100 mM NaCl, 0.1% Triton-X100. These preparations have a P700/Chlorophyll ratio in the range of 1/40-60.

Reaction centers were prepared from wild type *Rb. sphaeroides* strain WS8 by standard procedures using lauryldiethylamine N-oxide, (LDAO) as a detergent). Iron free reaction centers were prepared by the method of Feher and Okamura[5]. The final preparations were suspended in a buffer containing 50 mM Tris, PH 8.0, 0.05% LDAO.

Samples in which the oxidised reaction center chlorophyll radicals were obtained by illumination were prepared in EPR tubes containing 0.3ml of a solution of 0.5 - 4.0 mg/Chlorophyll/ml in the case of Photosystem 1, and 50μM reaction centers for the bacterial samples. The tubes were illuminated for 1 minute at room temperature and frozen in liquid nitrogen under illumination to oxidise the reaction center chlorophylls. No cryoprotectants were used with the Photosystem 1 preparations as we have previously shown that the properties of the electron acceptor complex are altered by glycerol[6]. Samples of bacterial reaction centers were prepared with and without 60% glycerol, no effect of the use of this cryoprotectant was observed on the spectra reported. Rautter et al[7] reported variations in the electron distribution in bacterial reaction centers depending on the detergent used for reaction center isolation and also following damage caused by freezing and thawing the preparations. We have so far as possible avoided thawing and refreezing samples and have consistently observed constant values in our preparations, which are essentially the same as those reported by Rautter et al[7] in chromatophores and LDAO preparations.

ESEEM spectra were recorded using a Bruker ESP 380E pulsed EPR spectrometer. Where shown, all 1D ESEEM spectra were generated using the three pulse stimulated ECHO sequence, ($\pi/2$-τ-$\pi/2$-T-$\pi/2$-Acq). Two Dimensional ESEEM spectroscopy was performed using the four pulse HYSCORE pulse sequence. Both of these pulse sequences suffer from a "blind spot" depending upon the value of τ, therefore many spectra were acquired using differing values of τ, the spectra shown in this abstract are representative of a series of experiments. All data was analysed using WINEPR software on a 33MHz IBM compatible computer.

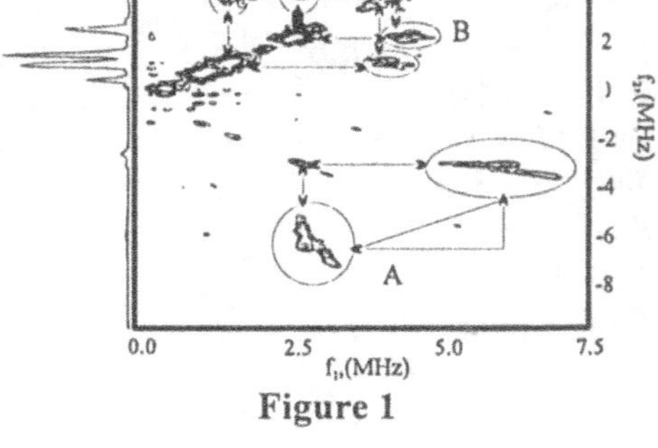

Figure 1

Results

The HYSCORE spectrum of $P700^{+\cdot}$ Is shown in Figure 1. The HYSCORE experiment consists of two sectors, along the diagonal of each sector is the 1 D ESEEM spectrum and peaks which are hyperfine coupled will be labelled by off diagonal peaks. Therefore the HYSCORE spectrum can be used to locate the double quantum transitions which are crucial in calculating a_{iso} . In addition to this the size of the hyperfine interaction detrmines which sector the off diagonal coupling appears in. In figure 1 double quantum transitions may be seen in the lower sector,(A) and connect the region of the 1D ESEEM spectrum at approximately 2.5 MHz with the region between 5-6 MHz. These double quantum transitions arise from the four nitrogens in the high electron density half of the proposed P700 dimer. Furthermore, in the top sector of the HYSCORE spectrum, smaller couplings, (B,C) may be seen labelling the region of the spectrum at approximately 3.6-4.2 MHz. It is suggested that these couplings are due to the double quantum transitions of the four nitrogens in the low electron density half of the P700 dimer. In order to accurately locate the double quantum transitions it is necessary to examine the 1D ESEEM spectrum, as it has a far higher digital resolution. In figure 2 is shown the stimulated echo spectrum of $P700^{+\cdot}$ (A) acquired with an interpulse delay time, (τ) of 112 ns. As indicated by the HYSCORE spectrum a cluster of peaks in the 5.0 - 6.0 MHz region is quite clearly seen. An expansion of this region may be seen in the inset, (B). This cluster may be resolved into four peaks which are due to the four nitrogens in the high electron density half of the P700 dimer. From inspection two sets of quadrupolar parameters may be extracted, (e^2Qq= 2.62 ,2.83 MHz and η= 0.75,0.7). Using the protocol of Britt et al[8] good estimates of a_{iso} may be calculated. These values are shown in Table 1. It can be seen that the a_{iso}'s of the high electron density half of the dimer are reduced from the monomeric values. Using these values it is possible to calculate the a_{iso}'s of the low electron density half of the dimer and calculate the frequencies in the ESEEM spectrum at which these nitrogens should appear, which is in the 3.6-4.2 MHz region of the spectrum. As can be seen in figure 2 a relatively large peak appears in this region and is also labelled by off diagonal peaks in the HYSCORE spectrum,(B and C).

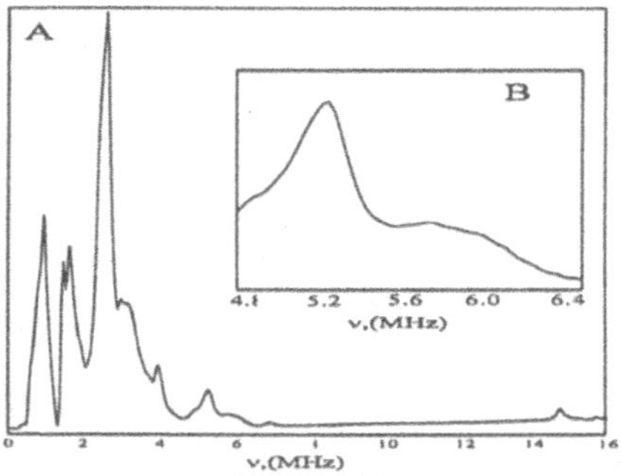

Figure 2

	Chl $a^{+\cdot}$[9]	$P700^{+\cdot}$(L)
N1	3.10	2.89
N2	3.10	2.41
N3	2.52	2.27
N4	2.23	1.97

Table 1: A comparison between the values of ^{14}N a_{iso}'s obtained in this study of $P700^{-\cdot}$ with Chl $a^{+\cdot}$ In MHz

In table 1 it is possible to compare the results obtained in this study with those of Chlorophyll a. It is clear that the results obtained in this study are reduced in comparison with those of Chl $a^{+\cdot}$ indicating a 4:1 electron distribution ratio.

The bacterial reaction center chlorophyll of *Rhodobacter sphaeroides*, $P870^{+\cdot}$ was studied by 1D and 2D ESEEM methods at 4K. The reaction centers were studied with and without cryoprotectant and no differences were observed. The methodology was the same as that used for $P700^{+\cdot}$. 2D ESEEM utilising the HYSCORE pulse sequence was used to locate the double quantum transitions and 1D ESEEM using the stimulated echo sequence was used to accurately determine the double quantum frequencies which are crucial for the estimation of a_{iso}. Figure 3 shows the 1D ESEEM spectrum of $P870^{+\cdot}$ with expansions of the double quantum region for various τ values and the HYSCORE projection. It can be clearly seen by comparison with Figure 2 that the double quantum transitions are at a lower frequency than $P700^{+\cdot}$ implying a less asymmetric electronic distribution. The a_{iso}'s of the high electron density half of the dimer may be calculated and are 2.52, 2.06, 1.90, and 1.64 MHz. These values are comparable to those obtained by Lendzian et al at room temperature using ENDOR spectroscopy. We have so far as is possible avoided thawing and re-freezing the samples and have consistently observed constant values in our experiments. These results indicate a 2:1 electronic distribution at 4K.

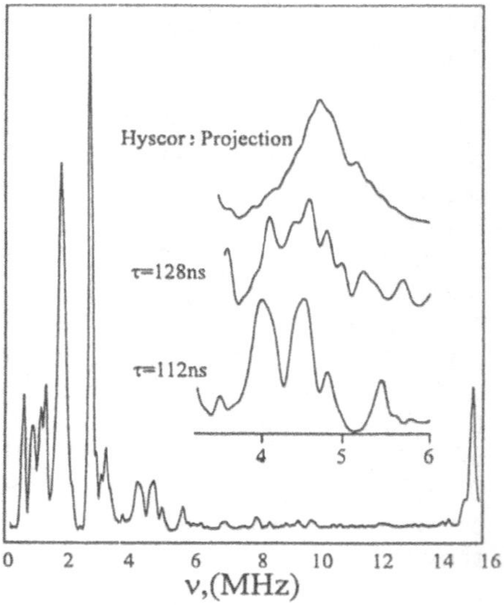

Figure 3

References

1)Käß, H., Rautter, J. Bönigk, B., Höfer, P. And Lubitz, W. (1995), *J. Phys. Chem. 99*, 436-448.

2)Davis, I. H.; Heathcote, P.; Maclachlan, D. J.; and Evans, M. C. W., *.Biochim Biophys. Acta.*, (1993), *1143*, 183-189.

3)Rigby, S.E.R.; Nugent, J. H. A.; and O'Malley, P.J.(1994), *Biochemistry,*, *33*, 10043-10050.

4)Boardman, N. H., *Meth. In Enzymol.,* (1971), *23*, 268-276.

5)Feher, G.; and Okamura, M. Y. (1978) in The Photosynthetic Bacteria, Clayton, R. K. and Sistrom W. R., eds., Plenum Press, New York, pp 349-396.

6 Evans, M.C.W. and Heathcote, P, (1980), *Biochim. Biophys. Acta,*590, 89-96

7)Rautter, J., Lendzian, F., Lubitz, W., Wang, S., and Allen (1994) J. P., *Biochemistry,*, *33,* 12077-12084.

8)Britt, R.D.; Zimmerman, J.L.; Sauer, K.; and Klein, M. P.(1989), *J. Am. Chem. Soc.,*

9)Astashkin, A.V., Dikanov, S.A.,Tsvetkov, Yu. D.,and Goldfeld, M.G. (1987),

A MOLECULAR ANALYSIS OF PHOTOSYSTEM 1 MUTANTS OF *CHLAMYDOMONAS REINHARDTII.*

V. Patel, D. W. Wright, B. J. Hallahan, S. Purton and M.C.W. Evans.
Biology, Darwin Building, University College London, Gower St., London WC1E 6BT, UK.

1. Introduction

Photosystem 1 (PS1) can be considered to be a light driven plastocyanin:ferredoxin oxidoreductase . The core complex of PS1 contains at least twelve polypeptides, designated PsaA - PsaL. Of these, the homologous, chloroplast-encoded subunits PsaA and PsaB are the largest (approximately 83kDa each).and coordinate the redox components P_{700}, A_O (The primary electron acceptor), A_1 (the phylloquinone acceptor), and a number of chlorophyll and β-carotene molecules. One of the three [4Fe-4S] centres of PS1, Fe-S_X is also believed to be coordinated by these two proteins.
[4Fe-4S] iron sulphur centres require four cysteine residues for coordination. Both PsaA and PsaB contain a twelve residue region of highly conserved amino acid residues. Within this region, there are two invariant cysteine residues, and it is generally believed that these provide the ligands for Fe-S_X; the centre being coordinated between the two polypeptides. This arrangement is very unusual and may impose some constraints on the interactions of the other subunits of Photosystem 1.We have created several mutants of *C. reinhardtii* with amino acid substitution mutants in the conserved region of PsaA to examine the role that these amino acids may have in the coordination of Fe-S_X and their possible effect on the assembly and function of PS1. In addition, we have developed a strategy to isolate mutants carrying functional changes within this region. This approach relies on the selection for PSI function (and hence phototrophic growth) in mutants carrying random changes to chosen residues.

2. Procedure

2.1 Mutants of *C. reinhardtii.*

In the chloroplast genome of *C. reinhardtii*, PsaA is encoded on three widely separated exons. Exon 3 (*psaA*-3) codes for the region containing the conserved cysteine residues and is located on *Bam*HI fragment 3. This fragment was isolated and cloned into a Bluescript vector and the resulting plasmid used to create several mutants. A disruption mutant of PsaA was created by inserting a selectable marker for chloroplast transformation (the '*aadA* cassette') within the coding sequence of *psaA*-3 and introducing the construct into the chloroplast genome by homologous replacement of the endogenous psaA-3 DNA using the Biolistic transformation procedure (1). Site-directed changes made within *psaA*-3 were linked to the *aadA* cassette by cloning it immediately downstream of the gene and screening transformants for the introduction of the site-directed change. The changes made are to the

P. Mathis (ed.), Photosynthesis: from Light to Biosphere, Vol. II, 187–190.

proximal cysteine (C575) and to the adjacent aspartate (D576) as follows: C575D, C575H, C575S and D576L.(figure 1).Additionally, we have created a control transformant carrying the *aadA* cassette linked to the wild-type *psaA*-3 to test whether the presence of the cassette affects PS1 function. These mutants have been characterised using various biochemical and biophysical techniques as described below.

2.2 A search for functional PsaA mutants

Two approaches have been adopted to identify permissible changes to the conserved region of PsaA. Firstly, we have isolated spontaneous suppressors of the non-functional D576L mutant capable of phototrophic growth. We are currently sequencing the psaA-3 DNA in four such isolates to identify the molecular basis of the suppression phenotype. Secondly, we are using two pools of degenerate oligonucleotides in combination with a three step PCR strategy to create a library encoding all possible amino acid combinations for the three residues (RFP) immediately upstream of C575 (figure 1). This library will be introduced into the C575D mutant and functional transformants selected by their ability to grow phototrophically on minimal medium.

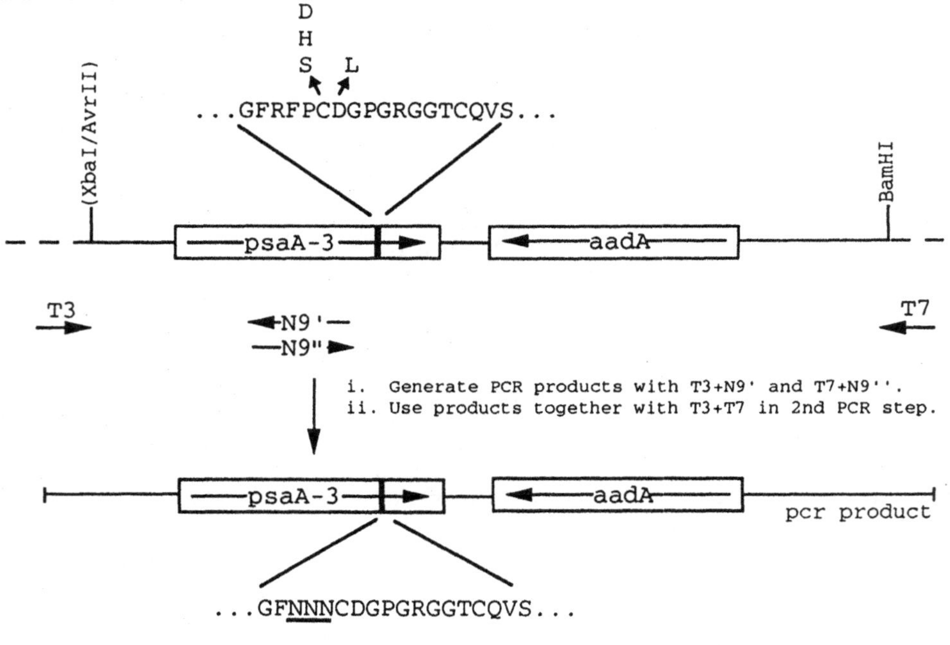

3. Results and Discussion

We have created four mutants: C575D, C575H, C575S and D576L and confirmed that these strains are homoplasmic for the introduced mutation. For each, at least two separate transformant lines were obtained and the site directed changes confirmed by DNA sequencing. We also created a transformant strain carrying a wild-type *psaA*-3 but with the *aad*A cassette downstream of the gene. This transformant, called 2137::*aad*A, serves as a control to demonstrate that the insertion of the *aad*A cassette downstream of *psaA*-3 has no effect on the expression of the gene or the functioning of PS1.

Analysis of the growth characteristics of the mutants showed that while the control strain grew normally, the three C575 mutants and the D576 mutants were unable to grow

Table 1

	NADP	Methyl Viologen
	µmole /mg Chlorophyll/hr	
Wild type	2.37	217.4
C575H	0.99	1.7
C575D	0.90	n/a
D576L	0.90	128.2
Boiled	0.57	47.6

NADP reduction measured as light induced change in OD340. Methyl viologen reduction as O_2 uptake.

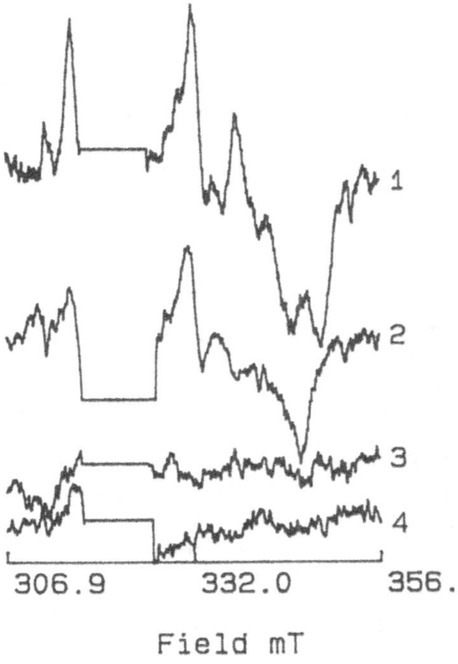

Figure 2 EPR spectra in the g=2.00 region of *C. reinhardtii* unfractionated thylakoid membrane fragment reduced with ascorbate. The spectra shown are light minus dark difference spectra. 1) Wild type. 2) D576L. 3) C575H, 4) C575D. Chlorophyll concentration 6 mg/ml. EPR conditions Temp. 15K, microwave power 10mW, Modulation width 1mT.

autotrophically and showed decreased growth rates with acetate. Non-denaturing "green" gel analysis showed that the control mutant has normal amounts of the PS1 reaction centre (CP1), C575H and D576L contained reduced amounts of CP1 but the C575D and C575S mutants showed no CP1 band.

The ability of *C. reinhardtii* membrane fragments to catalyse electron transfer through PS1 was determined by measuring NADP photoreduction and oxygen uptake coupled to Methyl Viologen reduction (Table 1). Methyl viologen reduction rates by wild type membranes are similar to those seen with spinach chloroplasts, but rates of NADP reduction are very low, perhaps indicating a need for the use of specific ferredoxin or flavodoxin.

None of the mutants catalyse significant rates of NADP reduction. The C575H mutant was also lacking in the ability to reduce methyl viologen. However the D576L mutant showed electron transfer to methyl viologen at about half the rate of the wild type.

Spectroscopic analysis allows a more detailed investigation of the components of the PS1 reaction centre. EPR analysis at room temperature and at cryogenic temperatures can be used to identify the components of the reaction centre; light induced spectral changes indicating the presence of photochemically active reaction centres. Analysis of the mutants shows that a light induced signal with the properties of the reaction centre chlorophyll, P700, can be detected in wild type and the control mutant, and also in C575H, ~30% of wild type, and in D576L, ~50% of wild type. No light inducible signals which can be assigned to PS1 are seen in C575D. The C575S mutant has not been analysed In wild-type PS1, illumination at low temperature results in electron transfer through a chain of redox components to the iron-sulphur centres Fe-$S_{A/B}$. These are detected by characteristic EPR spectra at 15K. No photoreducible iron-sulphur spectra can be detected in C576D, as would be expected from the complete absence of the CP1 complex. We have also been unable to detect photoreduction of PS1 iron-sulphur centres in C575H

preparations, even though these contain CP1 and apparently catalyse the low temperature photooxidation of P700. The presence of large amounts of other iron-sulphur centres makes it difficult to detect the PS1 centres by chemical reduction.
The photoreduction of the PS1 iron-sulphur centres is seen in chloroplast membranes from wild type, the control mutant and D576L (Fig 2). The spectrum of the Fe-$S_{A/B}$ centres seen in *C. reinhardtii* preparations varies somewhat from that classically seen in spinach preparations in that the spectrum suggests that in some centres the electron goes to Centre B rather than Centre A. However in the D576L mutant membranes, the spectrum is further modified. The spectrum suggests that in all centres the electron goes to the same centre, but the spectrum of that centre is modified compared to wild type. g_y is similar to that of Centre A, but g_Z is shifted to about g=1.87, from g=1.86 in wild type, and g_x cannot be detected as it is concealed by large none photosystem 1 signals around g=2.02, suggesting an overall narrowing of the spectrum.
Changes in C575 and the adjacent D576 would be expected to affect the Fe-S_X centre. It is difficult to obtain a complete spectrum of Fe-S_X as it is largely concealed below the spectrum of the other centres so that only the g_Z region is readily observed. Initial measurements of the g_Z region of the Fe-S_X spectrum show that the spectrum in D576L is unaltered compared to that of the wild type. No spectrum has been detected in the C575H or C575D mutants. Formation of the $P700^+/A_1^-$ radical pair can be detected by fast time resolved EPR measurements. Applying this technique to the mutants we again find that D576L and wild type are essentially the same while no signals can be detected in C575H or C575D mutants.

Conclusion.
The inactivation of PS1 by mutation of C575 in PsaA indicates the importance of this residue to the formation of an active reaction centre. However, conclusive evidence that the conserved cysteine region is involved in formation of Fe-S_X will depend on the preparation of mutants, in which a modified centre or activity can be detected, as in the PsaB cysteine to serine mutant prepared by Smart et al.(2)
The result from the modification of D576 is surprising in that the effect observed is not on Fe-S_X but on the Fe-$S_{A/B}$ complex. Rodday et al(3) have suggested that this aspartate residue is involved in forming the binding site for PsaC. Our results support that proposal: modification of this residue altering the binding of the PsaC protein, changing the EPR spectrum and inhibiting the ability of the modified PS1 to reduce NADP.
In order to advance our understanding of the mechanism of electron transfer in PS1, our aim is not to completely disrupt PS1 function, but to create subtle changes that can provide valuable information about the function of the conserved region of amino acids that are believed to coordinate Fe-S_X. The mutagenesis strategy based on the isolation of functional suppressors or the selection of functional site-directed changes should provide a portfolio of such mutants.

1) Goldschmidt-Clermont, M. (1991) Nucl. Acid. Res. 19, 4083-4089
2)Smart, L.B., Warren, P.V., Golbeck, J.H. and McIntosh, L. (1993) Proc. Natl. Acad. Sci. USA 90: 1132-1136
3)Rodday ,S.M., Webber, A.N., Bingham, S.E. and Biggins, J. (1995). Biochemistry 34, 6328-6334

PRIMARY ELECTRON TRANSFER REACTIONS IN HELIOBACTERIA AND PHOTOSYSTEM I.

Winfried Leibl, Bruno Toupance, and Jacques Breton

CEA Saclay, Section de Bioénergétique, DBCM, 91191 Gif-sur-Yvette, France

1. Introduction

During the last years, evidence has accumulated that the photosynthetic reaction centers (RCs) of different organisms can be clustered into two main classes according to the nature of their secondary electron acceptors (1–3). Purple photosynthetic bacteria and photosystem II (PS II) from oxygenic organisms belong to the family of quinone-type RCs with a quinone–iron acceptor complex working as a two-electron gate. The RCs of green sulfur bacteria, Heliobacteria, and of photosystem I (PS I) of higher plants and cyanobacteria belong to the group of Fe–S type RCs with low-potential iron-sulfur clusters as electron acceptors. The availability of high-resolution X-ray structures for the RCs of the purple bacteria *Rp. viridis* and *Rb. sphaeroides* led to a considerable understanding of this type of RCs and therefore these bacterial systems serve as a good working model for the PS II reaction center. In contrast, for the Fe–S type RCs only low-resolution structural information is available on a PS I RC from *Synechococcus* (4). As the RC of PS I is intensively investigated and rather well understood, there is a tendency in the literature to take PS I as a model for other bacterial Fe–S type RCs.

The reaction center of PS I contains 3 iron-sulfur centers (F_X, F_A, and F_B) as terminal electron acceptors. Two further acceptors mediate electron transfer from the primary donor, P700, to the Fe–S centers; a chlorophyll *a* functions as the primary acceptor (A_O) and a quinone (phylloquinone) as secondary acceptor (A_1). EPR and flash absorption spectroscopy have provided ample evidence for an analogy of the acceptor side in Heliobacteria, green sulfur bacteria and PS I, at least with respect to the Fe–S centers. Whether the analogy also holds for the earlier acceptors, especially the quinone A_1, is still controversial (see 5 and references therein).

Upon light-induced charge separation, an electron is transferred across the membrane, building up a membrane potential. Each electron transfer step, which has a component perpendicular to the membrane is called electrogenic and contributes to this potential. Photoelectric techniques based on macroscopic electrodes allow to measure not only the kinetics of electron transfer but also the dielectrically weighted transmembrane distances between the cofactors. We applied a photoelectric technique to study the different steps of electron transfer in Heliobacteria and in PS I on a picosecond and nanosecond time scale. The comparison of different systems aims at the elucidation of the structure–function relationship of photosynthetic energy conversion in this family of RCs.

P. Mathis (ed.), Photosynthesis: from Light to Biosphere, Vol. II, 191–194.

2. Procedure

Cells of *Synechocystis* PCC 6803 were grown in liquid media of BG 11 and PS I particles were prepared as described (6). Extraction of F_A and F_B was performed by urea treatment (7) and extraction was verified by absorbance change measurements at 820 nm. Cultures of *Heliobacterium chlorum* and *Heliobacillus mobilis* were grown in an anaerobic hood in Medium 112 of the American Type Culture Collection containing 2 mM ascorbate. Membrane fragments were prepared by a french press treatment or a short sonication and were purified by centrifugation steps. For electrical measurements, membranes were washed in 2 mM MES pH 6 and reversibly or irreversibly oriented in the measuring cell (see Fig. 1). Care was taken to minimize the exposure to oxygen of the Heliobacteria samples. Photoelectric measurements were carried out on oriented membrane fragments in a microcoaxial measuring cell (Fig. 1). Details can be found elsewhere (8).

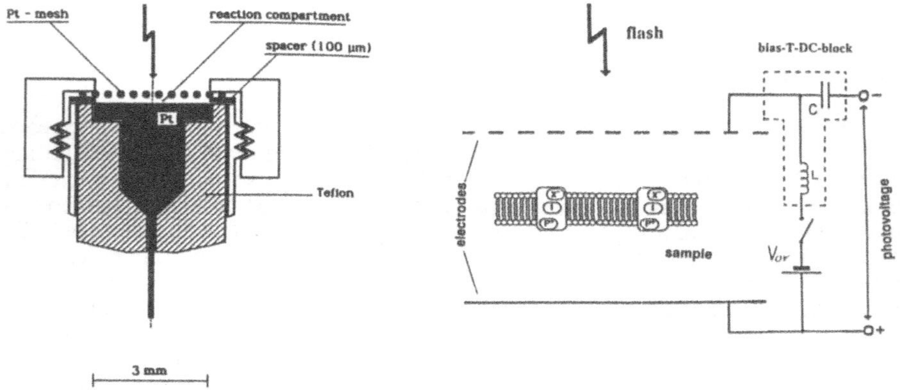

Fig. 1: (a) Drawing of the microcoaxial cell for photovoltage measurements. The sample is the dielectric of the capacitor formed by the electrodes. (b) Orientation of membrane fragments. The membranes are reversibly aligned by the electric field of a short DC–pulse (300 V/cm, 20 ms) preceding the excitation flash. Irreversible orientation can be achieved by electrophoretic sedimentation of the membranes onto the lower electrode.

3. Results and Discussion

In Heliobacteria as in PS I, excitation with a picosecond flash leads to an initial fast rise of the photovoltage signal with a time constant of 10–30 ps. The kinetics of this phase depend on the energy of the excitation flash. It is therefore attributed to trapping of the excitation energy by the very first photochemical step of charge separation between the primary electron donor (P) and the primary acceptor (A_O). In the following, the amplitudes of further electrogenic phases will be given relative to this phase.

A recent study of the picosecond photovoltage in PS I membranes from *Synechocystis* revealed a second step of charge separation with a kinetics of 50±15 ps (9). It was attributed to subsequent electron transfer to A_1 and its relative amplitude (0.25±0.13) is a measure for the dielectrically weighted transmembrane distance A_O–A_1 relative to P700–A_O.

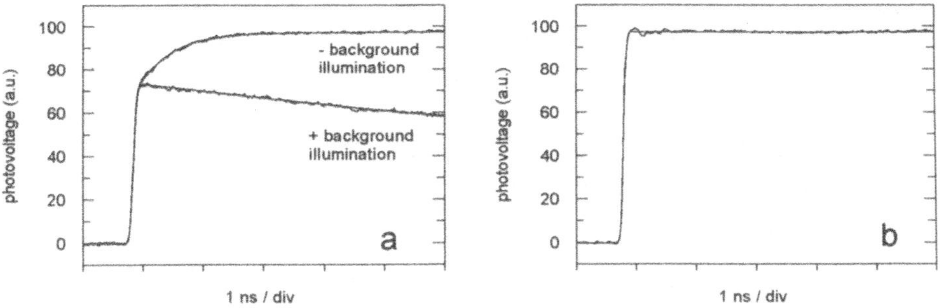

Fig. 2: Photovoltage kinetics from (a) Heliobacteria and (b) PS I. A 700–ps phase present in Heliobacteria but not in PS I disappears upon background illumination under reducing conditions and is replaced by a negative phase with a time constant of about 20 ns. The traces in (a) are normalized to equal amplitude of the fast rising phase.

Within the first nanosecond after the flash, picosecond photovoltage signals are clearly different in Heliobacteria and PS I (Fig. 2). In Heliobacteria a second electrogenic phase is characterized by a time constant of 600–700 ps and a much larger relative amplitude of about 0.6. Together with kinetic data from absorption spectroscopy (10,11) this phase can securely be attributed to the reoxidation of the primary acceptor (A_o) by electron transfer to a further acceptor, X. This interpretation is supported by the observation that the second phase disappears when the reduced state of the secondary acceptors is photoaccumulated. Under these conditions a recombination of the primary radical pair is observed with a time constant of 20 ns (Fig. 2a). A similar back reaction can be induced in PS I (not shown). However, in PS I much more reducing conditions are necessary and the process is only slowly reversible, a behaviour which is characteristic for the double reduction of the quinone acceptor A_1. The fact that photoaccumulation of the acceptor X in the reduced state in Heliobacteria is easily achieved and rapidly reversible suggests that no double reduction of a quinone acceptor is involved.

Extending the electric measurements to the nanosecond time scale, a third electrogenic phase was detected in PS I. It showed a kinetics of 220 ns and a relative amplitude of 0.76. Gradual removal of the extrinsic protein carrying F_A and F_B by urea treatment shows that the amplitude of the 220–ns phase is linearly correlated with the fraction of $F_{A,B}$ present, but that the kinetics of this phase is not significantly changed. Extrapolation gives a relative amplitude of 0.32 in the completely $F_{A,B}$–extracted system. We attribute the remaining phase to electron transfer from A_1 to F_X. A more detailed report on these experiments will be presented elsewhere (12).

There is no evidence for a corresponding third electrogenic phase in Heliobacteria. Double–flash measurements indicate that the acceptor which is reduced in 700 ps is reoxidized only after about 500 μs. This is outside the present time window of the electrical measurements.

The results of photoelectric experiments on both systems are summarized in Fig. 3. It shows the evolution of electrogenicity (transmembrane electron transfer) on a logarithmic time scale. The dielectrically weighted transmembrane location of PS I cofactors as deduced from the photoelectric measurements is indicated on the right. Comparison of the different electrogenic phases in Heliobacteria and PS I shows that the amplitude of the 700–ps phase in Heliobacteria is nearly identical to the dielectric distance A_O–F_X in PS I. A possible interpretation of this results is that a quinone acceptor corresponding to A_1 is absent in Heliobacteria. Another, but unlikely possibility is that A_1 is reduced more slowly (700 ps) but reoxidized much more rapidly (<<700 ps) than its equivalent in PS I (220 ns). The fact, that the reoxidation kinetics of A_o in Heliobacteria is slower by a factor of about 20 compared to PS I is in line with a larger distance between A_o and the next acceptor in this system. Further investigation of the energetics of the different electron transfer reactions and of the further electron transfer to F_A and F_B in Heliobacteria will hopefully allow to understand the significant differences in the structure/function of the cofactors in the Fe–S type reaction centers.

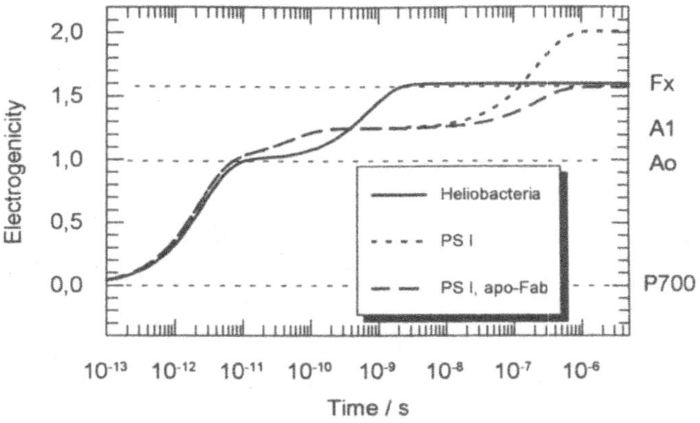

Fig. 3: Electron transfer across the membrane as a function of time in Heliobacteria, PS I, and a PS I after extraction of $F_{A,B}$. The curves are calculated from the exponential time constants and relative amplitudes obtained from best fits of kinetic photovoltage traces obtained on different time scales. For clarity, as the kinetics of the first phase is energy-dependent in the intact system, the estimated intrinsic rate of charge separation (as it would be measured in an isolated RC, about 2 ps) is used.

References

1 Mathis, P. (1990) Biochim. Biophys. Acta 1018, 163–167
2 Nitschke, W. & Rutherford, A.W. (1991) Trends Biochem. Sci. 16, 241–245
3 Blankenship, R.E. (1992) Photosynth. Res. 33, 91–111.
4 Krauss, N., Hinrichs, W., Witt, I., Fromme, P., Pritzkow, W., Dauter, Z., Betzel, C., Wilson, K.S., Witt, H.T., & Saenger, W. (1993) Nature 361, 326–331..
5 Amesz, J. (1993) Romanian J. Biiophys. 3, 195–204
6 Bottin, H. & Sétif, P. (1991) Biochim. Biophys. Acta 1057, 331–336
7 Golbeck, J.H., Parrett, K.G., Mehari, T., Jones, K.L., & Brand, J.J. (1988) FEBS Lett. 228, 268–272
8 Trissl, H.-W. & Wulf, K. (1995) Biospectroscopy 1, 70
9 Hecks, B., Wulff, K., Breton, J., Leibl, W., & Trissl, H.-W. (1994) Biochemistry 33, 8619–8625
10 Nuijs, A. M., Van Dorssen, R.J., Duysens, L.N.M., & Amesz, J. (1985) Proc. Natl. Acad. Sci. USA 82, 6865–6868
11 Van Kan, P.J.M., Aartsma, T.J., & Amesz, J. (1989) Photosynth. Res. 22, 61–68
12 Leibl, W., Toupance, B., & Breton, J. (1995) Biochemistry in press.

PROPERTIES OF THE LONG-WAVELENGTH EMITTING CHLOROPHYLLS IN ISOLATED PHOTOSYSTEM I PARTICLES OF *SYNECHOCOCCUS* sp.

Lars-Olof Pålsson[1], Eberhard Schlodder[2], Rienk van Grondelle[1] and Jan P. Dekker[1] 1). Department of Physics and Astronomy, Vrije Universteit, De Boelelaan 1081, 1081 HV Amsterdam, The Netherlands. 2.) Max-Volmer Institut für Biophysikalische und Physikalische Chemie, Technische Universität Berlin, Strasse des 17. Juni 135, 10623 Berlin, Germany.

1. Introduction.

Most photosynthetic systems contain one or more chromophores with a lower energy level than the level of the primary donor [1]. These chromophores could have a focusing function to concentrate excitations in the vicinity of the reaction centre but it is also possible that they serve to increase the absorption cross section and thereby enhance the efficiency of the photosynthetic unit [2]. The low-energy traps of photosystem I (PS I) are among the deepest considering the various photosynthetic systems. The spectral bandwidth of these chromophores in PS I of *Synechocystis* was found to be large, which is probably due to specific pigment-protein interactions [3]. This results in a large spectral overlap between P700 and the long-wavelength chromophores, which explains the efficient uphill energy transfer at higher (physiological) temperatures.

In a previous study the properties of the long-wavelength emission of PS I core preparations of *Synechocystis* were investigated [3]. This study showed that there is basically only one long-wavelength pigment within this organism which was designated C-708 and had the properties of a chlorophyll *a* dimer. In the present study we continue to investigate the long-wavelength emission in PS I and discuss the properties of the long-wavelength emission of PS I core complexes of *Synechococcus* sp. by using site-selective fluorescence and absorption spectroscopy at cryogenic temperatures. We conclude that the PS I complexes of *Synechococcus* and *Synechocystis* show considerable differences in the amounts and energies of the long-wavelength chlorophylls.

2. Materials and methods.

Trimeric complexes (~100 Chl/P700) were isolated from the thermophilic cyanobacterium *Synechococcus elongatus* as described in [4], except that PS I was extracted with 0.6 % (w/w) n-dodecyl-β,D-maltoside and purified by the use of a Mono Q column (Pharmacia). For elution a $MgSO_4$ gradient (25 - 200 mM) in 20 mM Mes pH 6.4 and 0.02 % n-dodecyl-β,D-maltoside was applied. The absorption spectrum was recorded on a Cary 219 spectrophotometer with an Oxford flow cryostat. The site-selective fluorescence experiments were performed using a CW Ti:Saph. laser (Coherent 890) pumped by an CW Ar^+-laser (Coherent Innova 300). The detection system consisted of an imaging CCD camera (Chromex). The anisotropy was measured using a Rhochon polarizer which separated the vertical and horizontal polarizations in space. The images where then focused to different parts of the CCD camera. The anisotropy

P. Mathis (ed.), Photosynthesis: from Light to Biosphere, Vol. II, 195–198.

was calculated according to $r = (I_{\|}-GI_{\perp})/(I_{\|}+2GI_{\perp})$. G is the correction factor ($G = 1.05$). In the fluorescence experiments an Utreks liquid Helium cryostat was used. The sample material was diluted in 20 mM Mes pH 6.5, 10 mM $CaCl_2$, 10 mM $MgCl_2$, 0.05 % n-dodecyl-β,D-maltoside and 70 % w/v glycerol and confined in an acrylic kuvette.

3. Results.

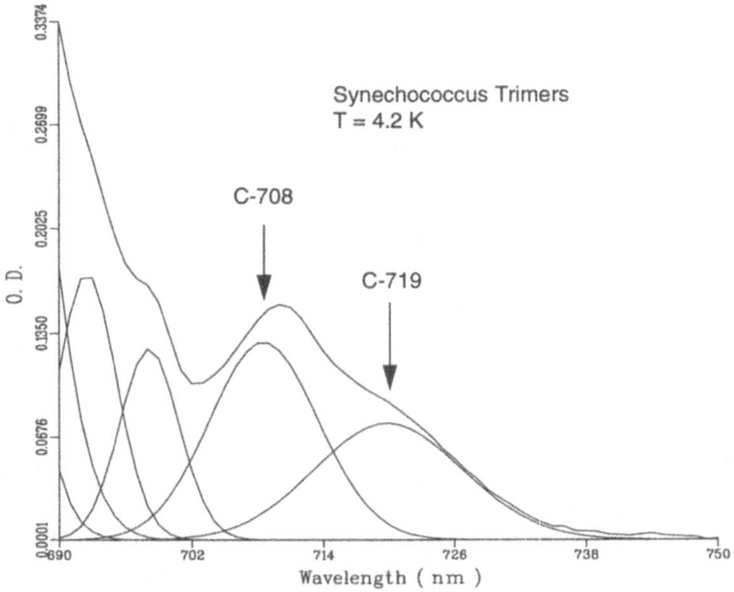

Figure 1. *The absorption spectrum of Synechococcus trimers at 4.2 K in the 690-750 nm intervall. The results of the fit with Gaussian components .*

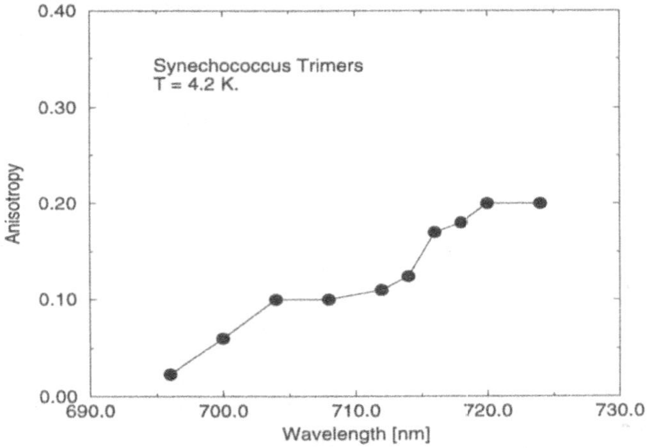

Figure 2. *The fluorescence anisotropy in the emission maximum of Synechococcus trimers as a function of excitation wavelength at 4.2 K.*

Figure 1 displays the low temperature absorption spectrum. Below 150 K a distinct peak is observable at 714 nm which disappears into the strong 680 nm absorption profile at T > 150 K (data not shown). Deconvoluting the 4 K absorption spectrum with Gaussian spectral components resulted in six different spectral forms (Fig. 2 and Tab. 1). The results obtained here are markedly different from those obtained in *Synechocystis* [3]. The redmost spectral form in *Synechocystis*, denoted C-708 [3], corresponds very well to the 707 nm component obtained in this study, both with respect to position and bandwidth. But in *Synechococcus* there is another spectral component located at 719 nm and has a bandwidth of almost 20 nm. From here on we denote this component C-719.

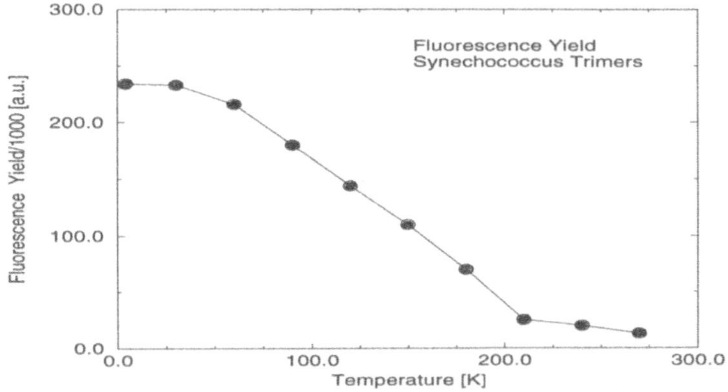

***Figure 3**. The temperature dependence of the fluorescence yield of Synechococcus trimers Excitation at 610 nm.*

Band nr	Position [nm]	Bandwidth [nm]	Amp. [a.u.]
1	680.0	8.96	0.81
2	684.9	10.00	0.11
3	689.6	7.70	0.23
4	697.6	7.70	0.13
5	707.1	12.30	0.11
6	719.1	18.70	0.08

***Table 1**. Results of the fitting of the absorption spectrum with Gaussian band.*

The fluorescence maximum of *Synechococcus* trimers at 4 K and at excitation wavelength 690 nm is located at 731 nm (data not shown). The anisotropy was calculated at the wavelength maximum. In Figure 2 the anisotropy as function of excitation wavelength is shown. At the blue-most excitation there is very little anisotropy, while in the 704-712 nm region there is an anisotropy of 0.10. The maximum anisotropy is 0.20 and this level is reached at excitation at 720 nm and longer excitation wavelengths. It should be noted that the anisotropy as a function of excitation wavelength for *Synechococcus* exhibits a markedly different behavior from that of the *Synechocystis* preparations and spinach PS I-200 particles [3]. Furthermore, the maximum anisotropy only reaches 0.20 in *Synechococcus* as compared to 0.30-0.40 for the *Synechoccystis* preparations and the spinach PS I-200 particles [3].

4. Discussion.

The maximal anisotropy that could be obtained for *Synechococcus* was 0.20. This is clearly less then what was observed for the *Synechocystis* preparations and spinach PS I-200 [3]. However, a very high anisotropy is not to be expected in this case since more than one redshifted pigment per monomer has been detected. A preliminay estimate of the oscillator strengths reveals that there are ~4 C-708 and ~4 C-719 pigments per P700 dimer. These numbers are reasonable inreagreement with conclusions at 300 K of [5]. Another interesting feature is the rise of the anisotropy as a function of excitation wavelength. There is no continous rise of the anisotropy like it was observed in the *Synechocystis* particles and in PS I-200 of spinach [3]. This could, however, be explained by the results of the fit of the absorption spectrum with Gaussian components. There are, in the wavelength region 696-720 nm, overlapping spectral components. It can not be expected that these are alligned parallel. Furthermore, downhill energy transfer to C-719 will also result in a depolarization process and a low steady-state anisotropy in this wavelength region. Selective excitation of the redmost spectral form can accordingly only be achieved at wavelengths longer than 720 nm. The residual anisotropy of $r = 0.20$, for excitation at 720 nm, is therefore the anisotropy of the long-wavelength emitting chromophores of *Synechococcus*. This anisotropy is also a reflection of the alignment of all (connected) long-wavelength chromophores of the *Synechococcus* trimers. The possibility of energy transfer between the long-wavelength pigments, donor-donor type from monomer to monomer, can not be totally excluded. However, this will be an energy transfer process over relatively long distances and it will therefore be inefficient. The uphill energy transfer process at 4 K must also be considered to be inefficient. The energy difference between C-708 and C-719 is $\Delta E = 240\ cm^{-1}$. Accordingly, at 4 K $\Delta E \gg k_BT$. Even at 100 K $\Delta E > k_BT$, but as can be seen in figure 3, the quantum yield is clearly less than at 4 K. This indicates that the photochemical energy conversion in the reaction centre is more efficient and there must therefore be some uphill energy transfer to the reaction centre from the low-energy pigments. Eventhough $\Delta E > k_BT$ this energy transfer process is possible if one considers the width of the inhomogenous distribution. At higher temperatures the overlap between the red part of the 707 nm spectral component and the blue part C-719 will increase and this will promote the uphill energy transfer.

5. Acknowledgements.

This project is supported by the Netherlands Foundations for Chemical Research (SON) and Life Sciences (SLW). LOP is supported by a fellowship within the HCM program project ERBCHBGCT 930361 supported by the EC.

6. References.

1. van Grondelle, R., Dekker, J.P., Gillbro, T. and Sundström, V. (1994) Biochim. Biophys. Acta **1187**, 1-65.
2. Trissl, H.-W. and Wilhelm, C. (1993) Trends Biochem. Sci. **18**, 415-419.
3. Gobets, B., van Amerongen, H., Monshouwer, R., Kruip, J., Rögner, M., van Grondelle, R. and Dekker, J.P. (1994) Biochim. Biophys. Acta **1184**, 75-85.
4. Witt, I., Witt, H.T., Gerken, S., Saenger, W., Dekker, J.P. and Rögner, M. (1987) FEBS Lett. **221**, 260-264.
5. Holzwarth, A.R., Schatz, G., Brock, H. and Bittersmann, E. (1993) Biophys J. **64**,1813-1826

PREPARATION AND CHARACTERIZATION OF A PHOTOACTIVE REACTION CENTER COMPLEX FROM THE HELIOBACTERIUM *HELIOBACILLUS MOBILIS*

Kazuhito Inoue[1], Mitsuhiro Akaike[1], Akihiko Hattori[1], Noriaki Kusumoto[2] and Hidehiro Sakurai[2]
[1]Dept. Biol., Fac. Sci., Kanagawa Univ., Hiratsuka, Kanagawa 259-12, Japan, and [2]Dept. Biol., Sch. Educ., Waseda Univ., Tokyo 169-50, Japan

1. Introduction

Several reports indicate that the reaction center (RC) of heliobacteria is similar to that of green sulfur bacteria and PSI of higher plants and cyanobacteria (1,2). They belong to the Fe-S type RC which contains very low-potential Fe-S centers as the secondary electron acceptors. Recently, Kusumoto et al. (3,4) successfully purified an RC complex from the green sulfur bacterium *Chlorobium tepidum* using Triton X-100, which contains photoreducible Fe-S centers and photooxidizable cytochrome *c*. This RC preparation consisted of 5 subunits, namely, 65- (RC core), 41- (FMO), 32- (Fe-S), 23- (Heme), and 18-kDa (unkown) polypeptide. As for heliobacteria, there are few reports on purification of RCs retaining photoactive Fe-S centers. RC core protein is the only peptide of the RC complex in heliobacteria whose DNA sequence has been sequenced (2). Thus it seems to be important to prepare a photoactive RC complex for elucidating the structure and function of RCs of heliobacteria. We solubilized RC particles from *Heliobacillus mobilis* membranes with Triton X-100, and purified them by sucrose-density gradient centrifugation. The RC-particles contained photoreducible Fe-S center(s) as determined by EPR. The RC-preparation was further purified by DEAE Bio-Gel A chromatography with respect to peptides.

2. Materials and Methods

Cells of *H. mobilis* (ATCC43427) were grown anaerobically (5) in a 1.5-l glass bottles for 2 days at 37°C, harvested by centrifugation, and stored at -80°C untill use. The thawed cells were disrupted with a BioNebulizer (Glas-Col, Indiana, USA) three times at 1.0 MPa of N_2 gas. After centrifugation at 36,000xg for 15 min, the supernatant was further centrifuged at 160,000xg for 60 min to precipitate membranes, from which RC particles were extracted with 3.0 % Triton X-100 in a buffer containing 50 mM Tris-HCl (pH7.8), 5 mM DTT, 100 mM NaCl, 0.1 mM EDTA, 10 mM sodium ascorbate and 0.5 mM sodium dithionite with stirring at 0°C for 60 min. After centrifugation at 250,000xg for 1h, the supernatant obtained was layered on a step-gradient of sucrose (30, 32.5, 35, 37.5, 40% (w/v)) in the above buffer containing 0.1% Triton X-100, and centrifuged at 250,000xg for 24h. The RC complex was recovered in a dark green band just above 35% sucrose layer. The RC-rich fraction was subjected to DEAE Bio-Gel A column (1.8x10cm) equilibrated with 20 mM Tris-HCl (pH7.8), 2.5 mM $MgCl_2$, 0.1 mM EDTA, 0.1% Triton X-100, 2 mM DTT, and 0.5 mM sodium dithionite. The RC complex was barely retained on the column and eluted from the column with the same buffer. Above

P. Mathis (ed.), Photosynthesis: from Light to Biosphere, Vol. II, 199–202.

procedures were carried out under N_2 gas, and all the media used had been purged with pure Ar gas in order to remove O_2.

EPR spectra were obtained with an X-band EPR spectrometer (model JES-RE2X, JEOL), equipped with a liquid-helium cryostat (model JES-LTR5X, JEOL). Room temperature absorbance spectra were recorded with a Hitachi U-3210 spectrophotometer.

3. Results and Discussion

We studied on two types of RC preparation; (I) RC-particle obtained by sucrose-gradient centrifugation, and (II) a preparation obtained by further purifying preparation I through DEAE Bio-Gel A chromatography.

Fig. 1 shows the absorption spectra at room temperature of RC-preparation I and II, respectively. Both of the spectra were very similar in shape, and had peaks at 786, 718, 672, 577, 493 and 411nm.

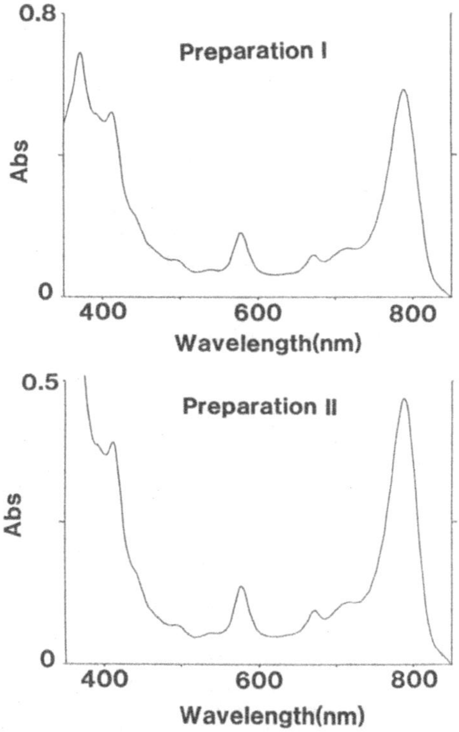

Fig. 1. Absorption spectra of RC preparation I and II.

When the preparation I was frozen in the dark at pH 8, no EPR signals attributable to iron-sulfur centers were observed. When the frozen sample was illuminated for 2 min at 298K, light-induced Fe-S center signal (g=1.89, 2.05) was observed. The profile of the spectrum resembled to those of Fe-S center(s) in membranes of *H. chlorum* (1,6,7,8) and

H. mobilis (9). Trost and Blankenship (9) purified an RC of *H. mobilis* solubilized with the detergent Deriphat 160c followed by sucrose-density gradient centrifugation. With respect to peptide, the preparation had only two peptides, a major component at 47 kD and a minor one at 40 kD, but, it contained no Fe-S center with a midpoint potential of -590 mV which was present in membranes nor light-indued Fe-S center signals. The magnitude of the light-induced Fe-S center spectum at g=1.89 of our preparation I (Fig. 2) was about 20% of that obtained by chemical reduction with dithionite at pH 10. When the sample was illuminated at 4K, a slight increase in the magnitude of signals of Fe-S center was observed. The fact that RC preparation I showed no Fe-S center signal when it was fozen in the dark at pH 8.0 in a medium containing 0.5 mM dithionite and 2 mM DTT, indicates that the preparation I was free from a Rieske-type Fe-S center present in heliobacteria (10). When preparation was frozen in the dark at pH 8, large signals were detected at g=2.03, 1.94, 1.87, and EPR signals were intensified by dithionite (data not shown). The identity of these signals are not certain at present, but these may arise from 3Fe-4S type Fe-S centers.

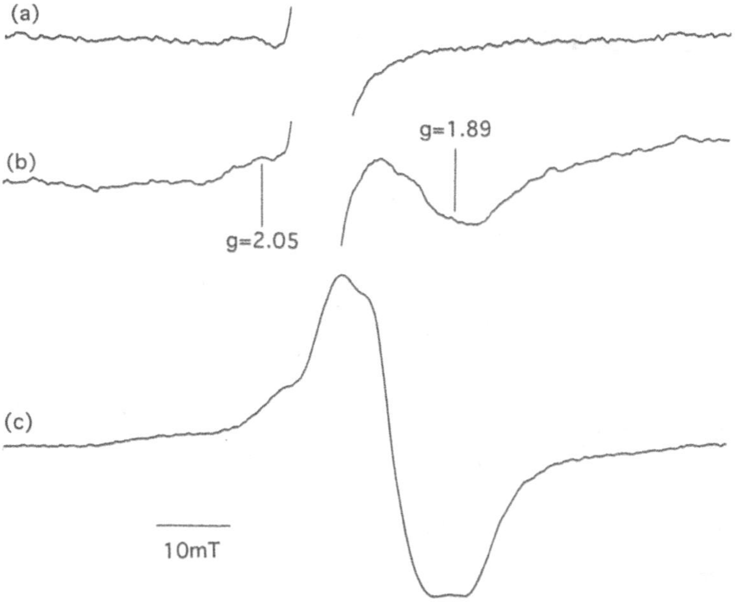

Fig. 2. Light-induced EPR spectra of RC preparation I.
(a) The sample was incubated in a medium containing 20 mM Tris-HCl (pH 8.0), 5 mM $MgCl_2$, 0.1 mM EDTA, 0.1% Triton X-100, 2 mM DTT, and 0.5 mM sodium dithionite and frozen in the dark. (b) After recording a dark spectrum (a), the sample was thawed, illuminated for 2 min at 298K, then frozen in liquid nitrogen with illumination during freezing, and EPR spectra measured. (c) The sample was incubated in a medium containing 20 mM glycine, 20 mM 2-amino-2-methyl-1,3-propanediol-NaOH (pH 10), 5 mM $MgCl_2$, 0.1 mM EDTA, 0.1% Triton X-100, 2 mM DTT, and 20 mM sodium dithionite and frozen in the dark. EPR conditions:temperature, 7K; microwave power, 20 mW; frequency, 9.07 GHz; modulation amplitude, 1 mT. Accumulation, three times.

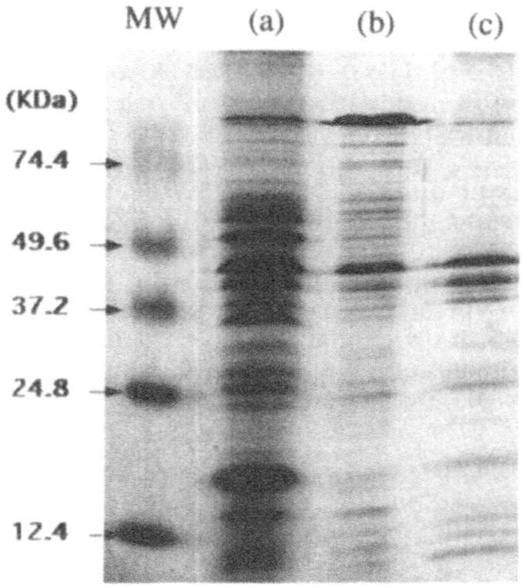

Fig. 3. SDS-PAGE (15-25% linear gradient) of *H. mobilis* membrane (a), RC preparation I (b), and RC preparation II (c). The gel was stained with CBB.

Fig. 3 shows polypeptide analysis by SDS-PAGE of RC preparation I and II. The preparation I contained about 20 peptide bands. The preparation II contained 8 (46.5, 42.3, 38.8, 26.3, 18, 13.6, 12.6, 11.5 kD) polypeptides as major bands (b). These results indicate that DEAE BioGel A was very effective for purifying RC of heliobacteria with respect to the peptide composition as found in *Chlorobium tepidum* (3), and suggest that some peptide in the preparation II (c) carries the putative F_A/F_B centers.

4. Acknowledgements

This work supported in part by Grant-in-Aid for Scientific Research (06640852) from the Ministry of Education, Science and Culture, Japan to H.S.

REFFERENCES

1 Prince, R.C. Gest, H. and Blankenship, R.E.(1985) Biochim. Biophys. Acta 810, 377-384
2 Liebl, U., Mockensturm-Wilson, M., Trost, J. T., Brune, D. C., Blankenship, R. E. and Vermaas, W.(1993) Proc. Natl. Acad. Sci. USA 90, 7124-7128
3 Kusumoto, N., Inoue, K., Nasu, H. and Sakurai, H. (1994) Plant Cell Physiol. 35, 17-25
4 Kusumoto, N., Inoue, K. and Sakurai, H. (1995) Photosyn. Res. 43, 107-112
5 Beer-Romero, P. and Gest, H. (1987) FEMS Microbiol. Lett. 41, 109-114
6 Brok, M., Vasmel, H., Horikx, J. T. G. and Hoff, A. J. (1986) FEBS Lett.194, 322-326
7 Nitschke, W., Setif, P., Liebl, U., Feiler, U. and Rutherford, A. W. (1990) Biochemistry 29, 11079-11088
8 Fischer, M. R. (1990) Biochim. Biophys. Acta 1015, 471-481
9 Trost, J. T. and Blankenship, R. E. (1989) Biochemistry 28, 9898-9904
10 Liebl, U., Rutherford, A. W. and Nitschke, W. (1990) FEBS Lett. 261, 427-430

BIOCHEMICAL AND OPTICAL STUDIES OF PHOTOACTIVE REACTION CENTER COMPLEX FROM THE GREEN SULFUR BACTERIUM *Chlorobium tepidum*

N. Kusumoto, *K. Inoue, H. Nasu, A. Tomioka
and H. Sakurai
Dept. of Biol., Sch. of Educ., Waseda Univ., Nishiwaseda, Tokyo 169-50, Japan; *Dept. of Biol., Fac. of Sci., Kanagawa Univ.,Hiratsuka, Kanagawa 259-12, Japan.

1. Introduction

By taking great care to maintain anaerobic conditions throughout the course of preparation, we succeeded in preparing an reaction center (RC) complex from the green sulfur bacterium *Chlorobium tepidum* which could photooxidize ~60% of the bound cytocrome (Cyt) c_{551}. The RC complex contained about 50-70 BChl molecules and 1.2-1.7 photooxidizable Cyt per P840, three bound Fe-S centers (CF_A, CF_B, CF_X; C for *Chlorobium*), and RC was composed of 5-6 peptides (1). We also measured reduction kinetics of photooxidized Cyt c_{551} and oxidation kinetics of a photoreduced Fe-S center, and obtained their optical oxidation-reduction difference spectra (2). We report here preparation and the properties of RC complex in more detail.

2. Materials and methods

The RC complex was purified as described (1): solubilization with Triton X-100, sucrose density-gradient centrifugation, DEAE chromatography, hydroxyapatite (HA) chromatography. Xenon flash-induced absorbance changes in a purified RC complex were measured by a single beam spectrophotometer as previously described (3). For amino acid sequencing, each peptide band on the gel after SDS-PAGE was transfered to PVDF membrane, and directly sequenced with a gas phase protein sequencer (Model 477A, Applied Biosystems).

3. Results and discussion

3.1 *Biochemical studies of RC complex*

Fig. 1A shows the absorption spectra of the preparations during purification. The membranes precipitated at 160,000 x g centrifugation (a) contained many chlorosomes absorbing at 746 nm, which were largely removed from an RC-rich supernatant fraction (b) obtained by solubilization of the membranes with Triton followed by differential centrifugation at 250,000 x g. After sucrose-density gradient centrifugation, the absorption spectrum of the RC-rich fraction (c) indicated that the peak at about 810 nm was greatly enriched in comparison with the absorption peaks at about 670 nm and 430 nm. The peak fractions were further purified by DEAE Bio-Gel

P. Mathis (ed.), Photosynthesis: from Light to Biosphere, Vol. II, 203–206.

A column chromatography, and by HA column chromatography, yielding a pale green colored fraction containing RC complex (e). Compared with (c), the shoulders at ~800 and ~830 nm were more prominent in (e). Fig. 1B shows SDS-PAGE analysis of the peptide composition during purification and Fig. 1C shows analysis of heme proteins. The peak fraction from sucrose-gradient centrifugation (c) contained more than 20 polypeptide bands, of which 4 were positive in heme staining. DEAE chromatography was very effective in purifying RC complex in terms of polypeptide and heme protein composition (d). HA chromatography removed some portion of the 41-kDa protein (BChl *a*-binding protein or FMO protein) yielding the final preparation (e) which contained 5 principal bands with apparent molecular masses of 65, 41, 32, 23, and 18 kDa. The 65-kDa band was assigned to the RC core peptide. The 41-kDa band was

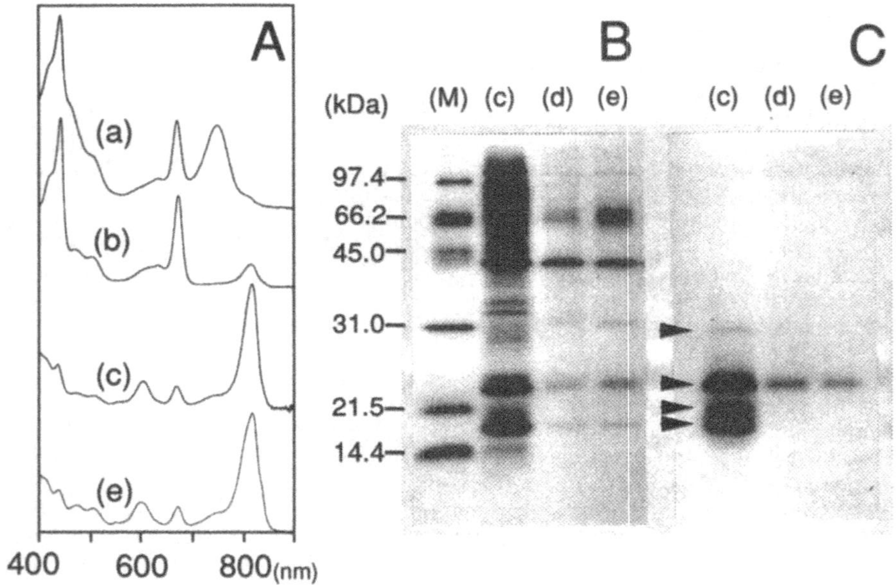

Fig.1. Absorption spectra (A), CBB staining for peptides (B) and TMBZ staining for heme (C) of preparations during the course of purification. (a), membrane fragments; (b), supernatant of Triton extract; (c) after sucrose-density gradient centrifugation; (d), after DEAE chromatography; (e), after HA chromatography; (M), moleculer weight markers.

assigned to be FMO protein from N-terminal amino acid sequencing: ALFGSNXVTT, which agreed with the sequence reported by Dracheva et al. (4). The 32-kDa band was rich in cysteine, and assigned to Fe-S protein. The 23-kDa band was assigned to Cyt both from heme-staining and from N-terminal amino acid sequencing:

MXKNSNGKLIALAVGGAVKMGALFFXVXFLXXY. The partial sequence completely agreed with that of *C. vibrioforme* reported by Okkels et al. (5). More recently, Oh-oka et al. (6) reported similar conclusion on the amino acid sequence of the peptides of *C. tepidum* RC complex. The identity of 18-kDa peptide is under investigation.

3.2 *Optical studies*

In a previous study, we reported flash-induced optical kinetics at room temperature of Cyt c_{551} and an Fe-S center (CF_A/CF_B) bound to a purified RC complex (2). At 551 nm in the absence of mPMS, the light-induced absorbance change decayed with kinetics approximating a single exponentially decaying component with a $t_{1/2}$ of ~200 ms. The $t_{1/2}$ usually ranged from ~150 to 400 ms, and might be ascribed to reductants (ascorbate, dithiothreitol etc.) present in the reaction mixture. The decay was accelerated by mPMS, and, at 0.5 μM mPMS, the flash-induced absorbance change returned to the original level in about 25 ms. At 430 nm, a large part of the absorbance change decayed with kinetics which depended on mPMS concentration in the same way as at 551 nm. However, the decay of a small part of the absorbance change was not significantly accelerated by low concentrations (<100 nM) of mPMS and the change decayed exponentially with a $t_{1/2}$ of ~300 ms at 0.5 μM mPMS. This spectral component was asigned to reduced Fe-S centers (2).

The flash-induced absorbance change at 605 nm, which was attributed to P840 oxidation, returned to almost the original level in less than 2 ms (2). Recently, we reexamined the absorbance change more carefully, and found that ~90% of the flash-induced absorbance change returned to the original level in less than 2 ms, and that ~10 % decayed slowly with kinetics similar to those of A551 absorbance change (Fig 2). These results can be explained as follows: ~90% of $P840^+$ was quickly reduced by Cyt *c* in less than 2 ms, and 10% was reduced with the same kinetics to those of Cyt c^+, indicating that oxidation-reduction of P840 was in equilibrium with that of Cyt *c* as reported by Oh-oka et al.(6).

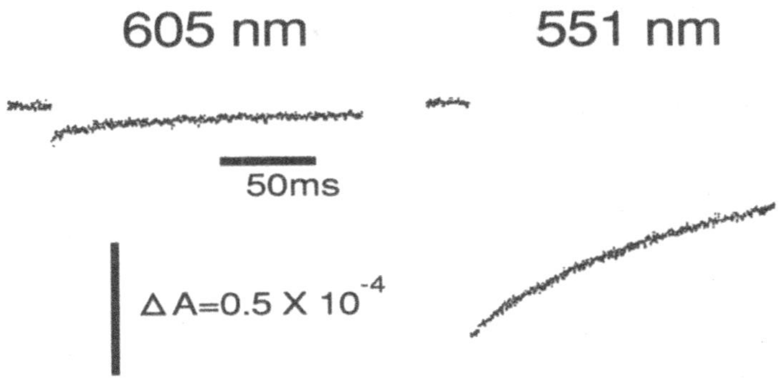

Fig.2 Flash-induced absorbance change kinetics in RC complex. Each trace is the mean of 64 measurements. Other conditions were as in (2).

Our tentative scheme representing the electron transfer kinetics in a purified RC of *C. tepidum* at room temperature is shown in Fig. 3.

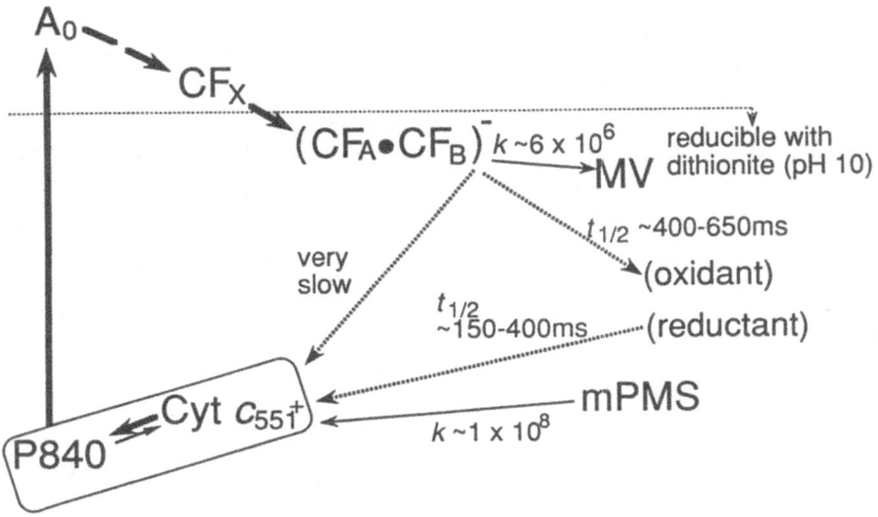

Fig. 3 Schematic representation of the electron transfer in a purified RC of *C. tepidum* at room temperature. The value of k is expressed in units of $M^{-1}s^{-1}$.

ACKNOWLEDGMENTS

Chlorobium tepidum was kindly donated to us by Professor M. T. Madigan. Triquat and mPMS were generous gifts from Dr. T. Erabi of Tottori University and from Prof. T. Yagi of Shizuoka University, respectively.

This work was supported in part by a Grant-in-aid for Scientific Research from the Ministry of Education, Science and Culture, Japan (06640852) to H.S.

REFERENCES

1. Kusumoto N., Inoue K., Nasu H. & Sakurai H. (1994) Plant Cell Physiol. 35, 17-25
2. Kusumoto N., Inoue K. & Sakurai H. (1995) Photosyn. Res. 43, 107-112
3. Inoue K., Sakurai H.& Hiyama T. (1986) Plant Cell Physiol. 27, 961-968
4. Dracheva. S., Williams J.C. & Blankenship R.E (1992) in Research in Photosynthesis (Murata, N) Vol I, pp. 465-468, Kluwer Academic Publishers, Dordrecht, The netherlands
5. Okkels J.S., Kjær B., Hansson Ö., Svendson I., Møller B.L. & Scheller H.V. (1992) J. Biol. Chem. 267, 21139-21145
6. Oh-oka H., Kamei S., Matsubara H., Iwaki M. & Itoh S. (1995) FEBS Lett. 365, 30-34

5. Oxygen evolution

THE OXYGEN-EVOLVING COMPLEX IN PHOTOSYSTEM II AS A METALLO-RADICAL ENZYME.

Gerald T. Babcock, Department of Chemistry, Michigan State University, East Lansing, Michigan 48824

1. The Appearance of O_2 in the Atmosphere

Substantial amounts of oxygen (≥15% of the atmosphere) appeared approximately two billion years ago as the result of the evolutionary development of oxygenic photosynthesis. The H_2O/O_2 couple is a uniquely well-suited biological redox couple and enzymatic mechanisms by which to exploit its unique physico-chemical properties developed. Because of its high reduction potential, O_2 is an ideal oxidant - its use in substrate oxidation maximizes the free energy made available in bioenergetic and respiratory processes. Similarly, water is thermodynamically stable and ubiquitous. Its use as a substrate allowed photosynthesis to proliferate. In addition to its thermodynamic properties, O_2 is a ground state electronic triplet, which introduces spin restrictions on its reactivity and provides the opportunity for biological control in enzyme active sites. Transition metals are well-suited for this purpose and, in general, O_2/H_2O metabolism occurs in metalloproteins.

At present, molecular understanding of the means by which the O_2/H_2O couple is manipulated enzymatically is emerging. In this article, the current status of our insight into the mechanism of water oxidation in Photosystem II is reviewed, recent results are introduced and discussed, lessons that may be applicable to PSII from O_2 - metabolizing enzymes are considered, and a new model for the water splitting process is introduced.

2. Water Oxidation: Conventional Views and Counterindications

Figure 1 (top) summarizes the prevailing view of the Photosystem II and its oxidizing side. Light absorption generates $P680^+$, which is rereduced by the redox active tyrosine Y_Z (D1 - Y160) in times fast compared to the deleterious charge recombination reaction between $P680^+$ and Q_A^-. The $(Mn)_4$ complex rereduces $Y_Z^\bullet$ with S-state dependent kinetics and stores oxidizing equivalents that are used ultimately for the concerted oxidation of water on the $S_3 \rightarrow S_4 \rightarrow S_0$ transition. Ca^{2+} and Cl^- are essential cofactors in the process.

Key to this view is that Y_Z serves a simple electron transfer function and is remote from the $(Mn)_4/Ca^{2+}/Cl^-$ oxygen-evolving complex.

P. Mathis (ed.), Photosynthesis: from Light to Biosphere, Vol. II, 209–215.

As summarized in the lower portion of Figure 1, there are a number of observations now reported that are inconsistent with the conventional view of water-splitting (1-8 and references therein). Thus, Y_Z is rotationally mobile and involved in a disordered hydrogen bond; its immediate environment appears to be hydrophilic. These properties are counter to what one would expect for a pure election transfer cofactor, as each operates to increase the reorganization energy, which will slow the election transfer process. Given the low free energy difference between $P680^+$ and H_2O, such a situation (high reorganization energy, low driving force) is disadvantageous. Gilchrist *et al* (6) have shown recently that Y_Z is within 5 Å of the $(Mn)_4$ ensemble, further contradicting the conventional view. Finally, kinetic data, particularly the facts that the $Y_Z^{\bullet}$ reduction rate limits water oxidation on the $S_3 \rightarrow S_4 \rightarrow S_0$ transition and that proton release in PSII generally follows the oxidation of Y_Z, not its reduction, are not easily accommodated within the conventional view.

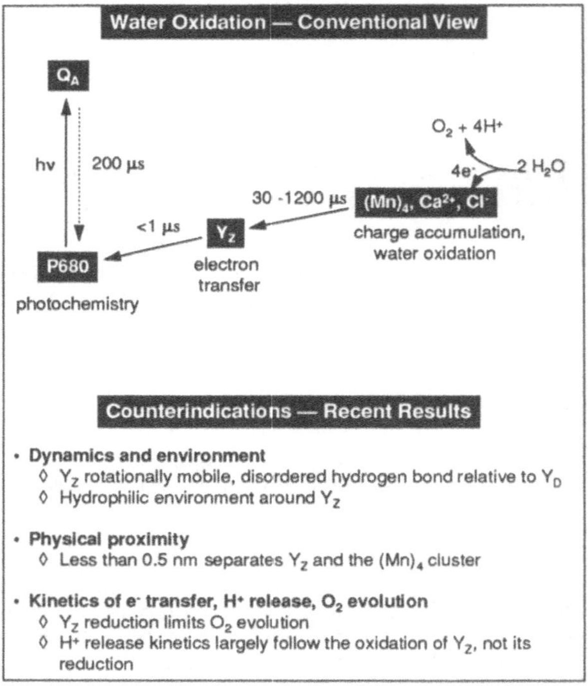

Figure 1

3. *Metallo-Radical Enzymes - Relevance to the PSII/OEC*

Confronted with these counterindications to the conventional view of water oxidation, we can turn to other amino acid radical-containing metalloenzymes for potential analogies and insights into water oxidation. Figure 2 summarizes the current status of work on representatives in this class of proteins (see 9 for a review). In broad outline, there appear to be three distinct families, as shown in the Figure. The underlying operational and structural principles for all three families are similar: (a) the metal center catalyzes radical formation and the radical subsequently initiates catalysis by abstracting a hydrogen atom from substrate; (b) the metal center, the radical, and the substrate binding site are usually in close physical proximity. The O_2-dependent family is especially illustrative, as shown by the example in the Figure. Dioxygen activation at the metal generates sufficient oxidizing power to produce the radical, which, in turn, is sufficiently oxidizing to abstract a proton and electron from substrate.

Viewed in this light, the PSII/OEC can be considered to have properties analogous to those of the O_2-dependent radical enzymes. All that is necessary is to postulate that, in PSII, the thermodynamic sink is $P680^+$ and that the water/oxygen couple serves as substrate, i.e., a simple reversal of the direction of the H atom current in the O_2-

dependent radical enzymes. This formulation is summarized in Figure 3. In this model, $(Mn)_4$ acts to bind substrate and to delocalize the four oxidizing equivalents required for water oxidation. $Y_Z^{\bullet}$ acts as the H^+/e^- abstractor on each S-state transition and its rereduction and protonation on $S_3 \rightarrow S_4 \rightarrow S_0$ rate limits O = O bond formation. Overall, the mechanism predicts that water oxidation is a charge neutral process: oxidation of the $(Mn)_4$ complex on each S-state transition is accompanied by the removal of a proton from the bound substrate.

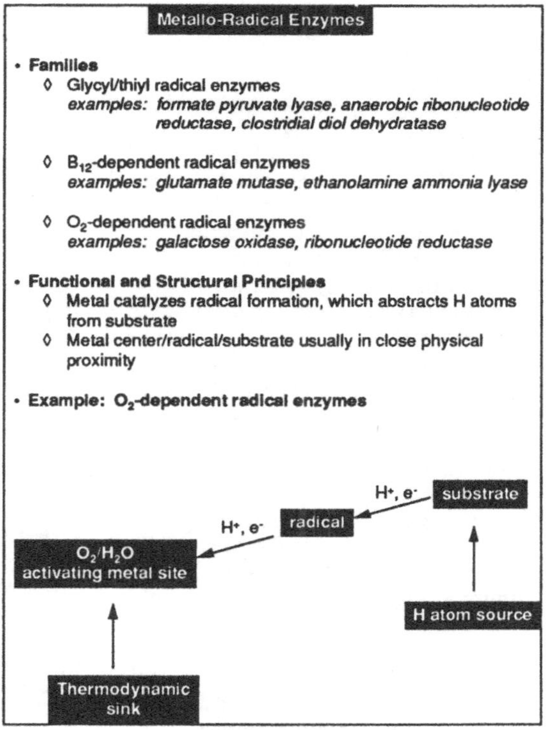

Figure 2

4. *Model Construction*

Figure 4 summarizes this view of water oxidation and folds Y_Z into the OEC. Considerable data are available now that provide information that can be used to develop the structural and functional details of the model. These results provide insight into the structure of the $(Mn)_4$/Ca components of the OEC and the protein-bound ligands to the metals, the likely binding sites for substrate water, the location of Y_Z relative to the metal centers, the details of the proton release pathways, and the roles of Ca^{2+} and Cl^- in the water oxidation process. Many of the groups that have contributed data are indicated in the Figure; more detail and full references can be found in References 3, 4, 10, and 11. Summarizing the results from this work, the following points emerge. (A) $(Mn)_4$, Ca^{2+}, and Cl^- are all integral components of the active site. The $(Mn)_4$ center forms an open, C-shaped structure with carboxylate and oxo bridges on the closed end and a 5.5 Å distance separating the two Mn on the open end. Within this C-shaped structure, we postulate that substrate water binds terminally to the two Mn on the open end. (B) Ca^{2+} appears to be associated with the $(Mn)_4$ cluster and linked to one of the Mn on the open end of the C-shaped structures through a carboxylate bridge, analogous to the Mn/Ca structure in concanavalin A. (C) Cl^- ligates one of the Mn on the open end of the structure in the lower S-states. In agreement with an original proposal by Rutherford (12), we suggest that Cl^- functions to prevent premature oxidation of water by acting as a shuttling ligand between Ca^{2+} and Mn binding sites, as summarized below (see also 13). (D) The $(Mn)_4$ cluster in the S_0 state has valences of +4, +3, +3 and +3 for the four metals; along with Ca^{2+}, the overall positive charge on the five metals is +15 (14). In this model, we postulate that this positive charge is neutralized by five oxo bridges (-10 total), Cl^-, and four carboxylate ligands, some of which are bridging. Recent directed mutagenesis implicates D1 - D170, D1 - E333, D1 - D342, and the carboxy terminus of D1 - A344 as the residues donating carboxy ligands (11 and refs. therein). Two histidines (D1 - H332 and D1 - H337) are also likely cluster ligands. (E) Y_Z is in close physical proximity to the $(Mn)_4/Ca$, analogous to the placement of Y_{122} in ribonucleotide reductase relative to its binuclear Fe center (15). We suggest further that D1 - H190 is hydrogen-bonded to the Y_Z phenol proton in the Y_Z reduced state and, in conjunction with D1 - E189, forms a proton conduction pathway to remove H^+ from the oxygen-evolving complex. The removal of the phenol proton through the H190/E189 pathway upon oxidation of Y_Z is key to the subsequent reprotonation of the phenol by substrate water protons upon $Y_Z^{\bullet}$ reduction.

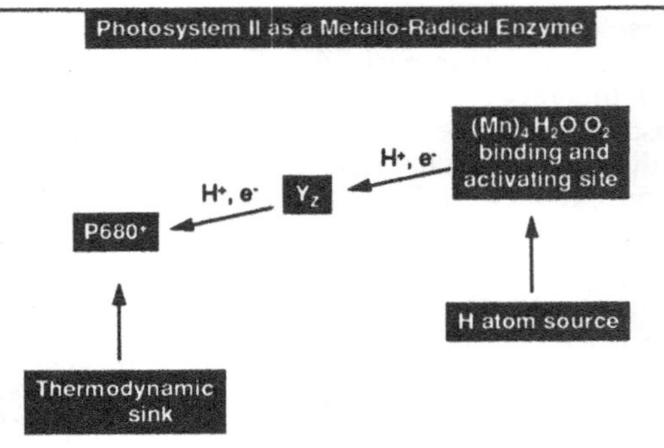

- **Simple reversal of H atom current from that in O_2-dependent radical enzymes**
- **$(Mn)_4$ acts as substrate binding site and sink for oxidizing equivalents**
- **H^+, e^- abstraction from $(Mn)_4$-bound H_2O or OH^- on each S-state transition; rate-limits O=O bond formation on $S_3/S_4/S_0$**
- **Overall water-splitting process at $Y_Z/(Mn)_4$ active site is charge neutral; analogous to O_2 reduction in cytochrome oxidase**
- **Rotational mobility in Y_Z phenol head group necessary for directional H^+ uptake and release**

Figure 3

Taken together, these results suggest the proposed structure for the $Y_Z/(Mn)_4/Ca^{2+}/Cl^-$ OEC shown in Figure 5. Y_Z is oriented toward the open end of the C-shaped $(Mn)_4$ cluster. Relative to the $Y_Z/(Mn)_4$ moieties, the COO^--bridged Ca^{2+} and D1 - H190 are oriented toward the inner aqueous phase and act to facilitate water and H^+ currents for the $(Mn)_4$ cluster and Y_Z, respectively. $P680^+$ is not shown in the Figure, but is expected to be located to the right of Y_Z as the scheme is drawn.

5. *Proposed Reaction Cycle for H_2O Oxidation*

Within the context of the structure proposed in Figure 5, a water oxidation scheme that involves the repetitive abstraction of H^+ and e^- from bound H_2O and OH^- substrates on each S-state transition can be easily constructed (see also Ref. 6). The postulated reaction cycle is shown in Figure 6. The S_0 state initiates the cycle and has H_2O and Cl^- ligated to the two substrate binding Mn on the open end of the C-shaped structure. The valence on the Mn ions of the cluster is +3, +3, +3, and +4 in S_0 (14); we suggest that the upper, dioxo-bridged $(Mn)_4$ dimer assumes the +3, +3 state, whereas the lower dimer, with Cl^- bound to the substrate binding Mn, assumes the +3, +4 state. Photon absorption generates $Y_Z^{\bullet}$, which abstracts H^+ and e^- as shown to produce S_1. The bound OH^- in this state is formally hydroxide and the upper dimer stores the added oxidizing equivalent in its +3, +4 state. This S-state advance, as is each of the following, is charge neutral. The $S_1 \rightarrow S_2$ transition occurs in the upper dimer and produces a bound oxo and a +4, +4, valence on the Mn dimer. Generation of the oxo triggers a ligand shuttling process so that the lower substrate binding site and Ca^{2+} exchange H_2O and Cl^-. The halide had been bound to the lower dimer in the lower S states to prevent formation of two Mn-bound OH^- species in S_2, which could collapse to form H_2O_2 and prematurely discharge the two oxidizing equivalents stored in the $(Mn)_4$ cluster (see also 12). We note that this H_2O/Cl^- ligand exchange process could generate a substantial dipole, which may be responsible for both the Bohr effects noted for proton release and the electrochromic shift that occur on the $S_1 \rightarrow S_2$ transition (8). Upon the $S_2 \rightarrow S_3$ and $S_3 \rightarrow S_4$ transitions, the $Y_Z^{\bullet}$-driven, H^+/e^- abstraction process repeats itself for the substrate molecule in the lower dimer pair. At the conclusion of the process, the S_4 state is produced, which has two Mn bound oxos, held in close physical proximity - the distance separating the two substrate-binding, terminal Mn ions is 5.5 Å in S_1 (10), and we expect that this distance does not vary greatly as the S-state advance process proceeds. These two oxos collapse to form O_2 in a process that is kinetically limited only by their rate of formation by $Y_Z^{\bullet}$. Upon formation of the O = O bond and expulsion of the O_2 product, the $(Mn)_4/Ca^{2+}$ cluster reloads with H_2O and Cl^- to reproduce the S_0 state. We propose that the reloading occurs on the same time scale as the discharge of O_2 and, depending on the details of the H_2O and OH^- binding, is likely to account for the slow pH-dependent proton release that occurs on the $S_3 \rightarrow S_4 \rightarrow S_0$ process (we note, in agreement with (8), that proton release is equivalent to OH^- uptake experimentally). In this mechanism, all other proton release processes (with the possible exception of a residual substrate loading process that may occur on $S_0 \rightarrow S_1$) are coupled to the oxidation of Y_Z and account for the fact that most of the proton releases that accompany water oxidation are fast relative to the kinetics of S-state advance (8).

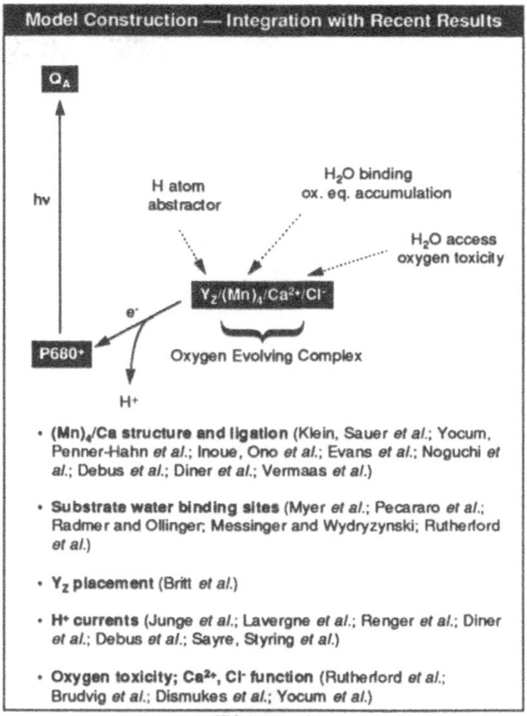

Figure 4

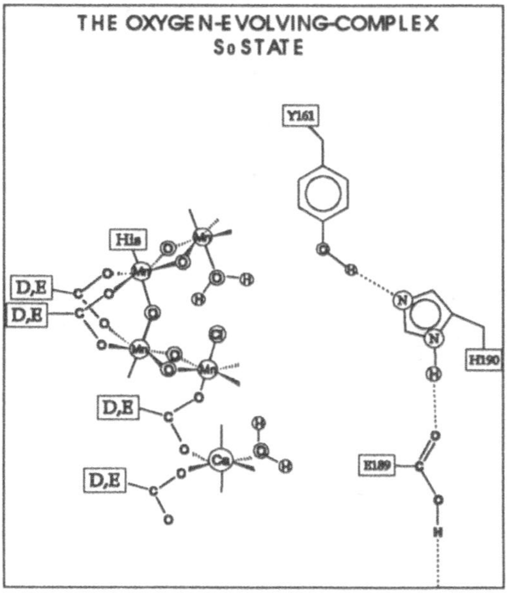

Figure 5

An attractive feature of the scheme in Figure 6 is that the $(Mn)_4/Ca^{2+}$ cluster remains relatively rigid during the

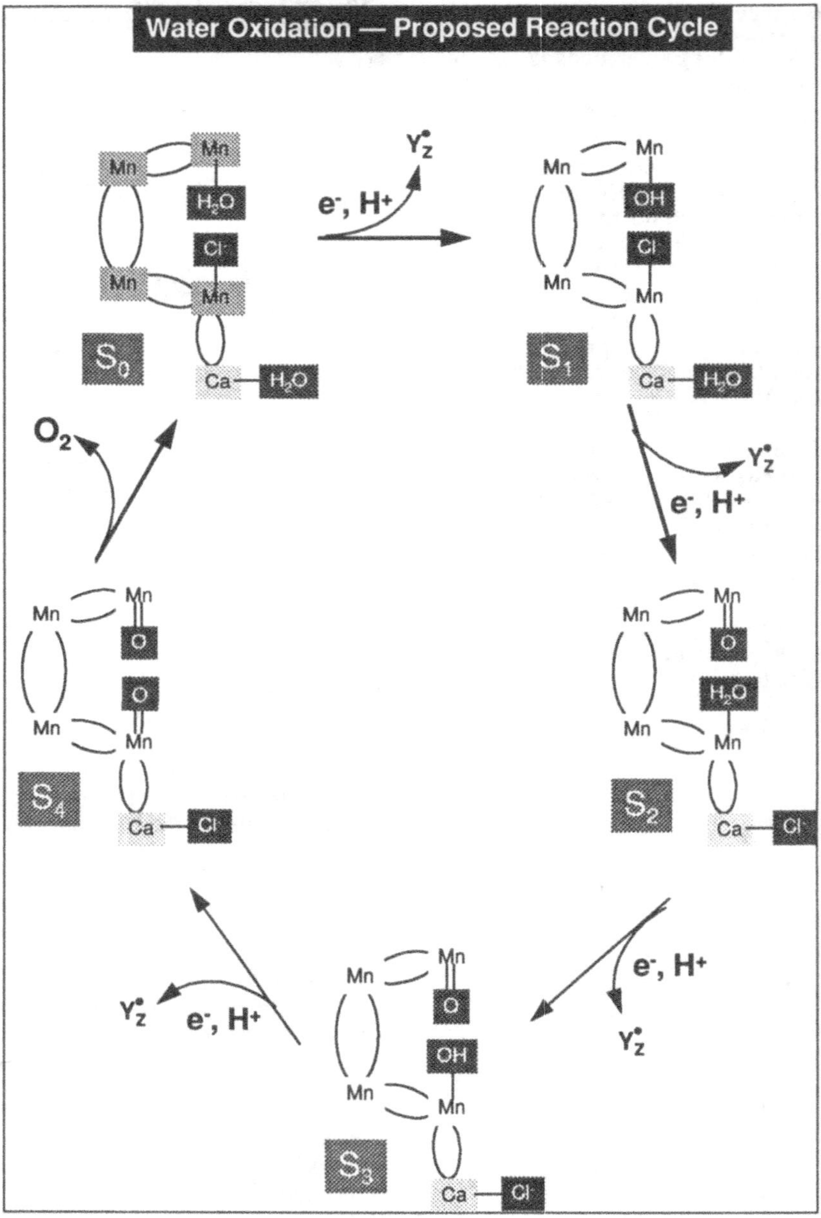
Water Oxidation — Proposed Reaction Cycle
S0
S1
S2
S3
S4
Mn
H2O
Cl-
Ca
OH
O
e-, H+
Yz•
O2

Figure 6

S-state cycle, which is consistent with the main body of XAFS results that have accumulated (10, 14, and refs. therein). The mechanism is conservative in terms of the nuclear rearrangments, and only the Mn-bound substrate water molecules undergo significant nuclear motion.

6. Conclusions

The model developed here postulates that the OEC in PSII is a unit that consists of Y_Z, $(Mn)_4$, Ca^{2+} and Cl^- and that each has specific functions during catalysis. $Y_Z^{\cdot}$ acts to abstract H^+ and e^- from $(Mn)_4$-bound substrate water; $(Mn)_4$ provides water binding sites and delocalizes oxidizing equivalents; and Ca^{2+} and Cl^- operate in a ligand exchange mode with $(Mn)_4$ and H_2O to pacify potentially toxic and/or reactive intermediates. Electroneutrality is maintained on each S-state transition, and water oxidation occurs as a concerted process as the $S_3 \rightarrow S_4 \rightarrow S_0$ transition takes place.

Acknowledgments

The ideas summarized here have grown out of extensive discussions with the following coworkers and collaborators: Matthew Gardner, Nykos Lydakis-Simantiris, Cecilia Tommos, Curt Hoganson, Xiao-Song Tang, Kurt Warncke, Bruce Diner, John McCracken, and Stenbjörn Styring. Very helpful conversations also occurred with Richard Debus, David Britt, Jerome Lavergne and Charles Yocum. This work is supported by the USDA/CRGO Photosynthesis Program and by NIH.

References

(1) Mino, H.; Kawamori, A. (1994) Biochem. Biophys. Acta 1185, 213-220
(2) Tang, X.-S.; Zheng, M.; Chisholm, D. A.; Dismukes, G. C.; Diner, B. A. (1995) Biochem., submitted
(3) Tommos, C.; Tang, X.-S.; Warncke, K.; Hoganson, C.W.; Styring, S.; McCracken, J.; Diner, B.A.; Babcock, G.T. (1995) J. Am. Chem. Soc., submitted
(4) Hoganson, C. W.; Lydakis-Simantiris, N.; Tang, X.-S.; Tommos, C.; Warncke, K.; Babcock, G. T.; Diner, B. A.; McCracken, J.; Styring, S. (1995) Photosynth. Res., in press
(5) Svensson, B.; Vass, I.; Cedergren, E.; Styring, S. (1990) EMBO J. 9, 2051-2059
(6) Gilchrist, Jr. M. L.; Ball, J. A.; Randall, D. W.; Britt, R. D. (1995) Proc. Natl. Acad. Sci. U.S.A., in press
(7) Babcock, G. T.; Blankenship, R. E.; Sauer, K. (1976) FEBS Letters 61, 286-289
(8) Lavergne, J.; Junge, W. (1993) Photosyn. Res. 38, 279-296
(9) Metal Ions in Biological Systems; Sigel, H.; Sigel, A., Eds.; Marcel Dekker: New York, 1994; Vol. 30
(10) Dau, H.; Andrews, J.C.; Roelofs, T.A.; Latimer, M.J.; Liang, W.; Yachandra, V.K.; Sauer, K.; Klein, M.P. (1995) Biochem. 34, 5274-5287
(11) Chu, H.-A.; Nguyen, A.P.; Debus, R.J.; (1995) Biochem. 34, 5859-5882
(12) Rutherford, A.W. (1989) Trends in Biochem. Sci. 14, 227-232
(13) Noguchi, T.; Ono, T.-a.; Inoue, Y. (1995) Biochim. Biophys. Acta 1228, 189-200
(14) Riggs, P.J.; Mei, R.; Yocum, C.F.; Penner-Hahn, J.E. (1992) J. Am. Chem. Soc. 114, 10630-10631
(15) Feig, J.; Lippard, S. (1994) Chem. Rev. 94, 759-810

MANGANESE MODELS FOR PHOTOSYNTHESIS

J.-J. Girerd
Laboratoire de Chimie Inorganique, URA CNRS 420, Université Paris-Sud, 91405 Orsay, France

A cluster of 4 Mn located in the reaction centre of PS-II probably acts both as the active site and as a charge accumulating device of the water-splitting enzyme in photosystem II.[1] The observation of the participation of Mn in such an important biological process, stimulated resarch on coordination chemistry of this element, specially in its +III and +IV oxidation states.[2] We studied different aspects of this chemistry and we will focus here on red-ox properties, magnetic properties and structural properties. We will try at the end to apply to the natural system what we learnt from the study of model complexes.

Red-ox properties. We prepared the complexes $[Mn_2^{III,IV}O_2(bisimMe_2en)_2]^{3+}$ [3] and $[Mn_2^{III,IV}O_2(bispicMe_2en)_2]^{3+}$.[4] They contain the famous dimanganese di-μ-oxo unit with is certainly a component of the OEC. The structure of $[Mn_2^{III,IV}O_2(bisimMe_2en)_2]^{3+}$ is represented in Figure 1. The ligand bispicMe$_2$en is deduced from bisimMe$_2$en by replacement of each of the two imidazole groups by a pyridine group.

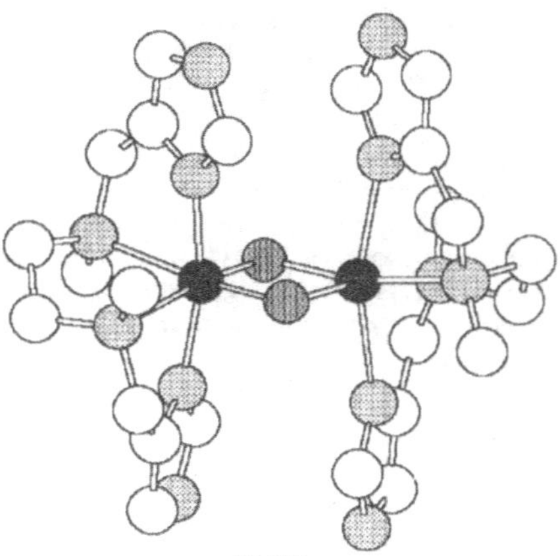

Figure 1. Structure of $[Mn_2^{III,IV}O_2(bisimMe_2en)_2]^{3+}$ cation.

Although the imidazole ring is linked to Mn by the N atom which is not in general involved in metal coordination by histidine, this complex can be useful as a model for the OEC. As usual in these complexes, the valences are trapped with a Mn-imidazole distance of 2.208 Å for Mn^{III} and 2.004 Å for Mn^{IV}. The complexes

P. Mathis (ed.), Photosynthesis: from Light to Biosphere, Vol. II, 217–222.

$[Mn_2^{III,IV}O_2(bispicMe_2en)_2]^{3+}$ and $Mn_2^{III,IV}O_2(bisimMe_2en)_2]^{3+}$ have very similar UV-vis and cw-EPR spectra. On the opposite, the results in cyclic voltammetry in AcN, are quite different. Voltammogram of $[Mn_2^{III,IV}O_2(bispicMe_2en)_2]^{3+}$ consists of two reversible red-ox couples: a [III,IV]/[III,III] with $E_{1/2}$ = 0.5 V/NHE and a [IV,IV]/[III,IV] with $E_{1/2}$ = 1.4 V/NHE. The voltammogram of $[Mn_2^{III,IV}O_2(bisimMe_2en)_2]^{3+}$ in the high potential region exhibits two couples with $E_{1/2}$ = 1.4 V/NHE and $E_{1/2}$ = 1.04 V/NHE. Detailed analysis of these couples has shown that they are due to an equilibrium (with a kinetics slow versus the characteristic time of CV) between two different structural isomers of $[Mn_2^{III,IV}O_2(bisimMe_2en)_2]^{3+}$. These isomers are likely related to the possibility for a ligand such as those used here to lead to two different geometries around each Mn ion: the cis α and the cis β.

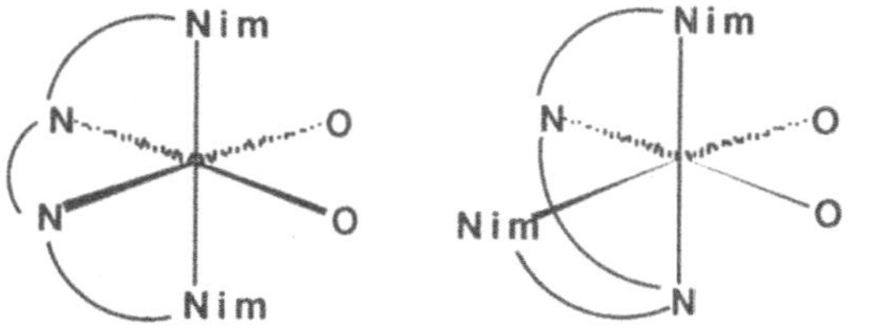

cisα　　　　cisβ

The species characterized by X-ray diffraction (figure 1) is the αα one. The other one in solution could be the ββ or the αβ. For $[Mn_2^{III,IV}O_2(bispicMe_2en)_2]^{3+}$, only one isomer would exist in solution (very likely the αα which is observed by X-ray diffraction[4]). Preliminary analysis shows that the potentials to be compared are 1.4 V for $[Mn_2^{III,IV}O_2(bispicMe_2en)_2]^{3+}$ and 1.04 V for $[Mn_2^{III,IV}O_2(bisimMe_2en)_2]^{3+}$: relatively to py, the imidazole makes the Mn^{IV} more accessible. This is in agreement with the stronger basicity of imidazole. This study shows also that a change in the configuration of the terminal ligands can modify the red-ox properties. As it is proposed that the majority of the ligands in the OEC are of the carboxylate type, the Mn^{IV} must be still easier to achieve.

Nuclearity of the cluster can also strongly affect the red-ox potential. We have a nice example with the bpy Mn-oxo clusters. The $[Mn_2^{III,IV}O_2(bpy)_4]^{3+}$ complex[5] has been recognized since a long time as a model of the OEC.[6] Its CV presents two red-ox couples: $[Mn_2^{III,IV}O_2(bpy)_4]^{3+}$ is reduced to the [III,III] form at $E_{1/2}$ = 0.53 V/NHE (irreversible) and the $[Mn_2^{IV}O_2(bpy)_4]^{4+}$ complex to the [III,IV] form at $E_{1/2}$ = 1.49 V/NHE.[6] A trinuclear cluster $[Mn_3^{IV}O_4(bpy)_4(H_2O)_2]^{4+}$ has been prepared;[7] it can be reduced only at 0.5 V/NHE and in an irreversible manner. A tetranuclear cluster $[Mn_4^{IV}O_6(bpy)_6]^{4+}$ has been prepared.[8] Its structure (Figure 2) is reminiscent of the model proposed by Klein et al. for the OEC.[9]

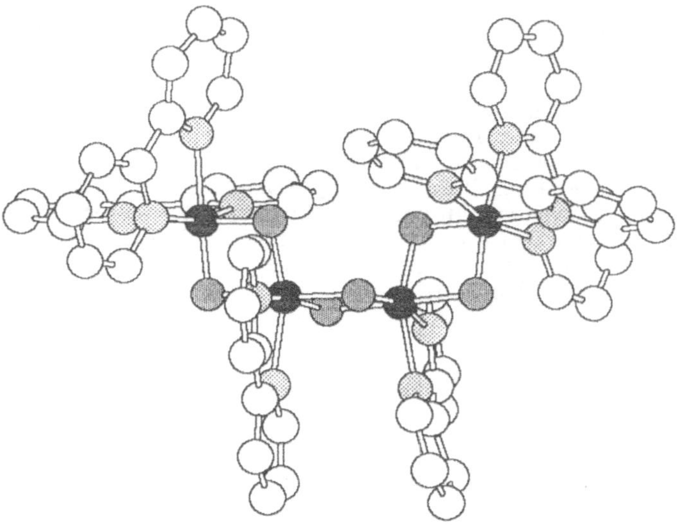

Figure 2. Structure of the $[Mn_4^{IV}O_6(bpy)_6]^{4+}$ cation.

Again, this cation can be reduced only around 0.5 V/NHE and in an irreversible manner. The decrease in oxidizing power from $[Mn_2^{IV}O_2(bpy)_4]^{4+}$ to $[Mn_3^{IV}O_4(bpy)_4(H_2O)_2]^{4+}$ and $[Mn_4^{IV}O_6(bpy)_6]^{4+}$ is spectacular: the global charge is the same and each cation presents the $Mn^{IV}O_2(bpy)_2$ motif. In fact this decrease in potential is due to the spread of the charge on the cluster. Another way to say the same thing, is to state that introduction of oxo groups decrease the oxidizing power of Mn^{IV}.

The magnetic properties. We will focus on the magnetic properties of the tetranuclear $[Mn_4^{IV}O_6(bpy)_6]^{4+}$ cation. Each Mn^{IV} ion (d^3) has a spin S = 3/2. Strong antiferromagnetic interactions were detected between nearest neighbours: $J_{ab} = J_{cd} = -176$ cm^{-1} and $J_{bc} = -268$ cm^{-1}. This leads to a S = 0 ground state by cancellation of the spin as illustrated in Figure 3a.

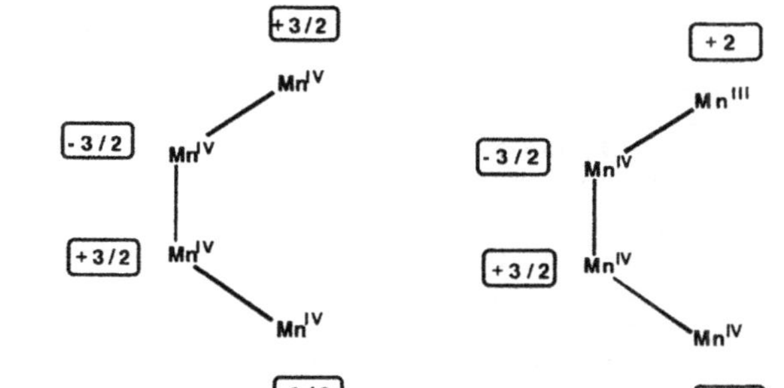

Figure 3. Magnetic properties of Mn tetramers with a linear topology. (a) The [IV,IV,IV,IV] case (S = 0). (b) The [III,IV,IV,IV] case (S = 1/2).

In the S_2 state the OEC, it has been proposed that the distribution of oxidation states is IV_3III.[9] Keeping antiferromagnetic couplings between neighbours, it is easy to see that this coupling scheme can explain qualitatively that the natural cluster has a ground state with S = 1/2.[10] In fact, it is important to realize that with such a linear topology, the two terminal spins can be relatively uncoupled as soon as the AF J_{bc} coupling is strong enough (the bc pair becomes a "diamagnetic " link). This suggests a possible origin for the g = 4.1 ground state of the cluster in the S_2 step: a ferromagnetic J_{ad} coupling can tilt the terminal spins to achieve a ground state for the cluster with S > 1/2 (Figure 4 a). Calculation shows that in fact a moderately strong ferro interaction is needed. This could imply the existence in the g = 4.1 state of the cluster, of an efficient monoatomic coupler between the terminal Mn (Figure 4b).

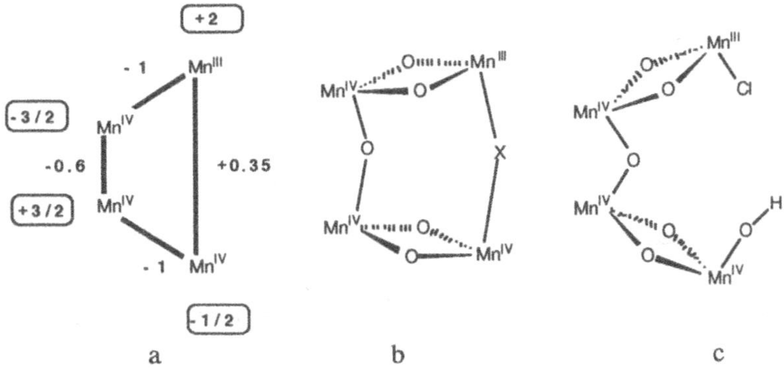

Figure 4. (a) A ferromagnetic coupling between Mn terminal atoms can lead to the g = 4.1 ground state of the cluster in the S_2 step (coupling parameters indicated in arbitrary unit). (b) A monoatomic bridge could achieve a F coupling strong enough. (c) Relaxation to an "open" conformation will give a S = 1/2 ground state.

Relaxation (Figure 4c) to a more "open" conformation will give a S = 1/2 ground state.

Chemical and Structural informations on the Mn-chloride interaction. Several examples of the $[Mn_2^{III}O(RCO_2)_2]^{2+}$ core are known.[11] In particular we studied the hydrolysis in aqueous solution of $Mn_2^{III}O(AcO)_2(bpy)_2Cl_2$. The substitution of chloride by water on Mn^{III} is possible but slow. We crystallized the different steps of the hydrolysis: $[Mn_2^{III}O(AcO)_2(bpy)_2Cl(H_2O)]^+$ and $[Mn_2^{III}O(AcO)_2(bpy)_2(H_2O)_2]^{2+}$.[12] The Mn^{III}-OH_2 distance is on the average 2.2-2.3 Å and 2.5-2.6 Å for Mn^{III}-Cl.

In a previous study[13], we had prepared in organic medium $[Mn_3^{IV}O_4(bpy)_4Cl_2]^{2+}$. In acidic aqueous solutions, the only complex obtained in presence of chloride, is the aquo complex, $[Mn_3^{IV}O_4(bpy)_4(H_2O)_2]^{4+}$. The hydrolysis of the Mn^{IV}-Cl bond seems much faster than in the Mn^{III} case. The Mn^{IV}-OH_2 distance is on the average 2.0 Å and 2.3 Å for Mn^{IV}-Cl. These complexes may allow to access quantitatively to the corresponding thermodynamic and kinetic data.

A model. At this point it is possible to elaborate on the biochemical and chemical data to propose a very simple model for oxygen evolution, based on two remarks: i) oxidation takes place preferentially on a Mn atom surrounded by oxo groups; ii) the access of water to the oxidation site must be controlled such that chloride can compete with water on Mn^{III}; on Mn^{IV} the substitution of water by chloride is fast.

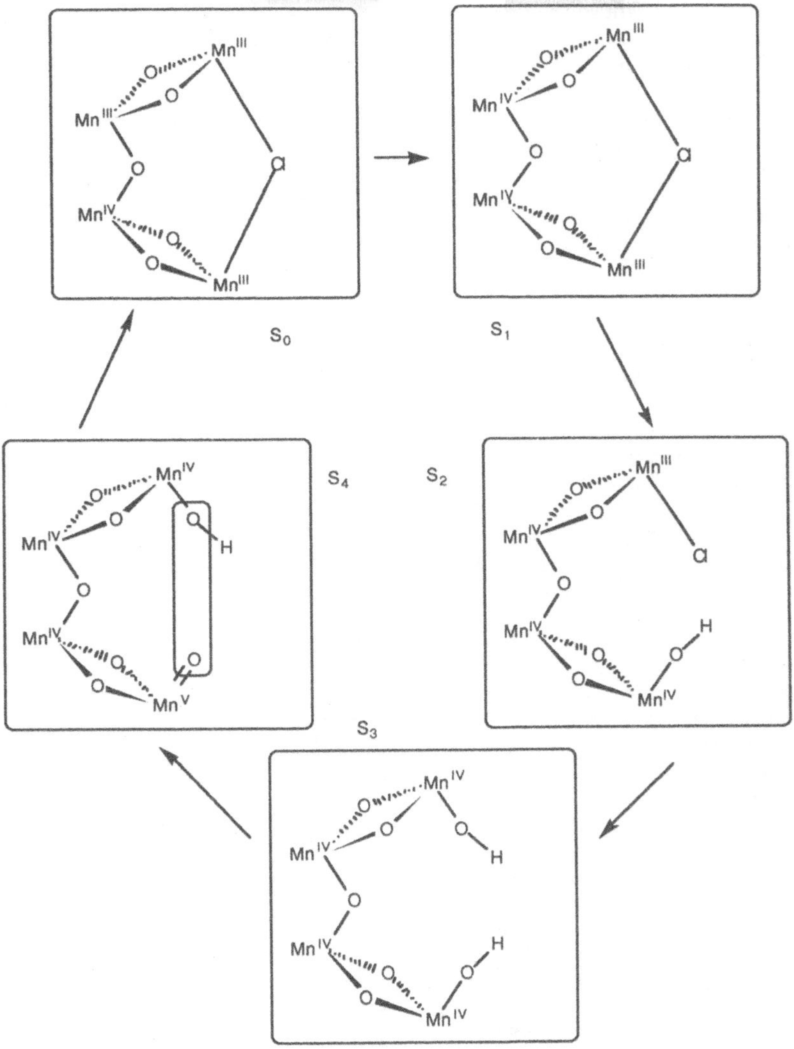

Let us comment step by step this scheme. The S_0 ---> S_1 transition: in S_0, the Mn^{IV} is in a oxo rich position, let us say c; the new Mn^{IV} is created in a position equally rich in oxo groups, i. e. position b. Using the long Mn^{III}-Cl distance found in models, the presence of a bridging chloride is found possible.[14] The S_1 ---> S_2 transition: the Mn^{IV} is now created in a terminal position, d. This induces a change in coordination since the Mn^{IV}-Cl distance is too short and water has strong affinity with Mn^{IV}. This rationalizes the observation of two configurations for the cluster: a low temperature one with a ferromagnetic interaction between the terminal Mn atoms leading to a g = 4.1 ground state and a relaxed one giving the multiline. The S_2 ---> S_3 transition: Mn^{IV} is created on position a and chloride is substituted by water. The S_3 ---> S_4 transition: Mn^{V} may now appear in a or d since these positions can be stabilized by deprotonation of the hydroxyl groups. The Mn^{V}=O group can create the OO bond with the hydroxo group linked to position a. Surely this is a very simple model and even if it contains some piece of truth, several features remain to be explained as deprotonation steps and role of calcium ion.

Acknowledgements. I deeply thank all the colleagues with whom I had the pleasure to work and whose names appear in our joint publications: R. Albach, E. Anxolabéhère-Mallart, N. Auger, G. Blondin, A. Boussac, M. Cesario, M. Corbella, M. Delroisse, Y. Frapart, A. Gleizes, J. Guilhem, D. Lexa, G. Matouzenko, S. Ménage, C. Pascard, C. Philouze, L. Tchertanova, J.-B. Verlhac, J.-L. Zimmermann. My special thanks go to J.-L. Seris.

References

1 For reviews, see Debus, R. J. (1992) Biochem. Biophys. Acta 1102, 269-352 and Rutherford, A. W. (1989)Trends Biochem. Sci. 14, 227-232.

2 Wieghardt, K. (1989) Angew. Chem. Int. Ed. Engl. 28, 1153-1172. Vincent, J. B. and Christou, G. (1989) Adv. Inorg. Chem. 33, 197-257. See also Manganese Red-Ox Enzymes (1992) (Pecoraro, V. L., ed.), VCH, New-York.

3 Boussac, A.; Albach, R.; Anxolabéhère-Mallart, E.; Delroisse, M.; Verlhac, J.-B.; Blondin, G.; Girerd, J.-J.; Guilhem, J.; Cesario, M.; Rutherford, A. W. and Lexa, D. submitted for publication.

4 Glerup, J.; Goodson, P. A.; Hazell, A.; Hazell, R.; Hodgson, D. J.; McKenzie, C.J.; Michelsen, K; Rychlewska, U.; Toftlund, H. (1994) Inorg. Chem. 33, 4105-4111.

5 Plaskin, P. M.; Stoufer, R.C.; Mathew, M.; Palenik, G. J. (1972) J. Am. Chem. Soc. 94, 2121-2122.

6 Cooper, S. R. and Calvin, M. (1977)J. Am. Chem. Soc. 99, 6623-6630.

7 Sarneski, J. E.; Thorp, H. H.; Brudvig, G. W.; Crabtree, R. H. and Schulte, G. K. (1990) J. Am. Chem. Soc. 112, 7255-7260.

8 Philouze, C.; Blondin, G.; Girerd, J.-J.; Guilhem, J.; Pascard, C.; Lexa, D. (1994) J. Am. Chem. Soc. 116, 8557-8565.

9 Yachandra, V. K.; DeRose, V. J.; Latimer, M. J.; Mukerji, I.; Sauer, K.; Klein, M. P. (1993) Science 260, 675-679.

10 Dismukes, G. C. and Siderer, Y. (1981) Proc. Natl. Acad. Sci. USA 78, 274-278. Britt, R. D.; Lorigan, G. A.; Sauer, K.; Klein, M. P.; Zimmermann, J.-L. (1992) Biochim. Biophys. Acta 1040, 95-101. Lorigan, G. A.; Britt, R. D. (1994) Biochemistry 33, 12072-12076.

11 Wieghardt, K.; Bossek, U.; Ventur, D., Weiss, J. (1985) J. Chem. Soc. Chem. Commun. 347. Sheats, J. E.; Czernuszewics, R. S.; Dismukes, G. C.; Rheingold, A. L.; Petrouleas, V.; Stubbe, J.; Armstrong, W. H.; Beer, R. H.; Lippard, S. J. (1987) J. Am. Chem. Soc. 109, 1435. Ménage, S.; Girerd, J.-J.; Geizes, A. (1988) J. Chem. Soc. Chem. Commun. 431-432. Vincent, J. B.; Tsai, H.-L.; Blackman, A. G.; Wang, S.; Boyd, P. D. W.; Folting, K.; Huffman, J. C.; Lobkowsky, E. B.; Hendrickson, D. N.; Christou, G. (1993) J. Am. Chem. Soc. 115, 12353- 12361.

12 Albach, R.; Matouzenko, G.; Tchertanova, L.; Girerd, J.-J. in preparation.

13 Auger, N.; Girerd, J.-J.; Corbella, M.; Gleizes, A.; Zimmermann, J.-L. (1990) J. Am. Chem. Soc. 112, 448-450.

14 Critchley, C. (1985) Biochem. Biophys. Acta 811, 33-46.

ELECTRON SPIN ECHO - ENDOR STUDIES OF THE TYROSINE RADICALS AND THE MANGANESE CLUSTER OF PHOTOSYSTEM II

R. David Britt, David W. Randall, James A. Ball, M. Lane Gilchrist, Jr., Dee Ann Force, Bradley E. Sturgeon, Gary A. Lorigan,
Department of Chemistry University of California, Davis, CA
Xiao-Song Tang, Bruce A. Diner,
Du Pont de Nemours and Company, Wilmington, DE
Melvin P. Klein,
Structural Biology Division, Lawrence Berkeley Laboratory, University of California, Berkeley, CA
Michael K. Chan,
Department of Chemistry and Biochemistry, Ohio State University, Columbus, OH
and William H. Armstrong
Department of Chemistry, Boston College, Chestnut Hill, MA

In this paper we describe recent electron spin echo - electron nuclear double resonance (ESE-ENDOR) experiments performed to characterize PSII radicals and the manganese cluster of the PSII oxygen-evolving complex (OEC). ENDOR spectroscopy provides sensitive detection of nuclear spin transitions in radicals and paramagnetic transition metal complexes. ENDOR spectroscopy is particularly useful in the study of large complex biological systems such as the PSII reaction center because the data obtained are unambiguously specific to the paramagnetic species irradiated in the electron spin component of the double resonance experiment. The ENDOR spectra displayed in this paper were obtained with the Davies ESE-ENDOR sequence[1-3] implemented on our laboratory-built pulsed EPR spectrometer.[4-5]

We have recently utilized ESE-ENDOR to study the broad radical EPR signal[6] photogenerated in Ca^{2+}-depleted PSII preparations. Several suggestions have been made as to the molecular origin of this radical signal, including an oxidized histidine,[6] the redox-active tyrosine Y_Z•,[7] and a water-derived radical.[8] Our ESE-ENDOR study provides strong evidence that the broad radical arises from a fraction of Y_Z• trapped by illumination and quick freezing in these non-oxygen-evolving samples.[9] Figure 1 displays new ESE-ENDOR data supporting this assignment. Trace (A) is the ESE-ENDOR spectrum for Y_Z• cryotrapped under illumination in a Mn-depleted PSII preparation from a *Synechocystis* 6803 mutant where the dark-stable Y_D• radical is removed by the conversion of the D2 tyrosine-160 to a phenylalanine, allowing the Y_Z• EPR signal to be observed without interference.[10-12] Trace (B) is the ESE-ENDOR spectrum of the Ca^{2+}-depletion radical cryotrapped in similar fashion.[9] The ENDOR spectra are quite similar, and the transitions can be assigned to those expected for a tyrosine radical.[9,11,13] Focusing on frequencies above the 15 MHz ^1H Larmor frequency, we assign overlapping transitions in the 17-20 MHz frequency range to the 2 and 6 ring protons and

P. Mathis (ed.), Photosynthesis: from Light to Biosphere, Vol. II, 223–228.

Figure 1:

(A) The ESE-ENDOR spectrum of $Y_Z\bullet$ in Mn-depleted PSII particles from the D2-Tyr160Phe mutant of *Synechocystis* 6803 lacking the $Y_D\bullet$ radical signal. (The PSII membranes were exchanged in ^2H_2O to examine hydrogen bonding to this tyrosine,[10–12] but ^1H ENDOR transitions from the non-exchangeable ring and β-methylene protons remain unaffected.)

(B) The ESE-ENDOR spectrum of the broad radical photogenerated in Ca^{2+}-depleted spinach PSII membranes (from ref 9).

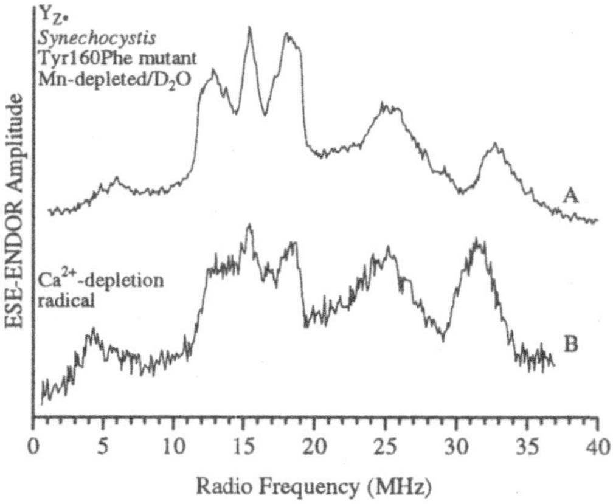

a weakly-coupled β-methylene proton. The broad peaks centered at 25 MHz are due to the 3 and 5 ring protons. The highest frequency peaks, in the 30-35 MHz range, arise from the strongly coupled β-methylene proton of each radical. The frequencies are slightly different for the two radicals, but the isotropic hyperfine coupling for the β-methylene proton is very sensitive to the dihedral angle θ between the tyrosine ring normal and the methylene C-H bond direction, varying as $\cos^2\theta$. Only a 2° decrease in θ or a 0.04 increase in the C_1 spin density for the *Synechocystis* $Y_Z\bullet$ relative to that of $Y_Z\bullet$ trapped in the Ca^{2+}-depleted sample would account for this frequency shift.

Thus in both Mn-depleted and Ca^{2+}-depleted PS II preparations some fraction of $Y_Z\bullet$ appears to be trapped by rapid freezing following illumination. However, the resultant radical EPR signal is not broadened in the Mn-depleted sample, indicating that the broadening is indeed due to a magnetic interaction between $Y_Z\bullet$ and the Mn cluster in the Ca^{2+}-depleted preparation. Using a dipolar coupling approximation, we have simulated the linewidth of the broadened $Y_Z\bullet$ signal in the Ca^{2+}-depleted preparation with a Mn-$Y_Z\bullet$ distance of only 4.5 Å.[9] It is difficult to imagine that the Y_Z tyrosine would be so close to the Mn cluster if its function were solely that of electron transfer. We therefore consider the possibility that Y_Z is directly involved in water oxidation. In this view, this redox-active tyrosine is as much a part of the oxygen-evolving complex as the Mn cluster itself. Figure 2 displays our model of water oxidation starting with the formation of P_{680}^+ (a). Electron transfer to re-reduce P_{680}^+ leaves Y_Z in its neutral radical form $Y_Z\bullet$ as the phenolic proton is transferred to a proximal base B (b). This proton can be quickly shuttled into the lumenal phase. $Y_Z\bullet$ is then re-reduced as an electron is transferred from the Mn cluster, advancing the S-state clock from S_n to S_{n+1} (c). The resulting deprotonated phenoxy group of Y_Z is a strong base which can abstract a proton from a water molecule acidified by its ligation to a high valence Mn ion of the Mn cluster (d). One could also consider a variant of this model which replaces the sequential electron and proton transfers in steps (b-d) with a single H atom transfer.

The number of coupled Mn ions in the Mn cluster of the OEC continues to be debated. Based on details of spin state interconversions between g=2 and g=4.1 Mn signal forms[14,15] and EPR simulations of the g=2 "multiline" EPR signal,[16–18] the current

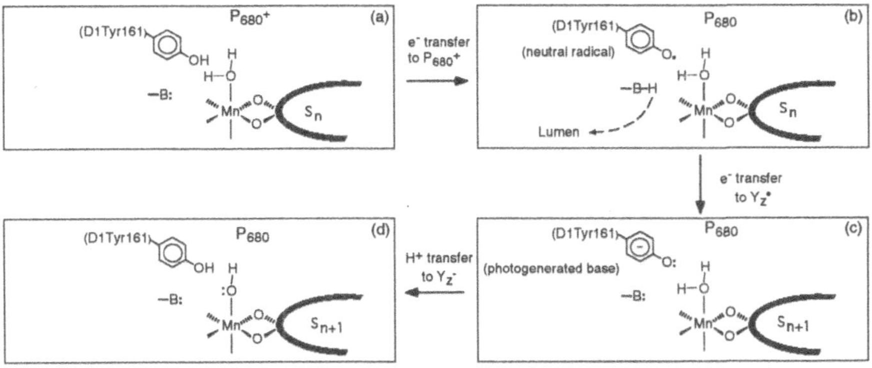

Figure 2: The proposed role of $Y_Z\bullet$ in water oxidation (adapted from ref. 9).

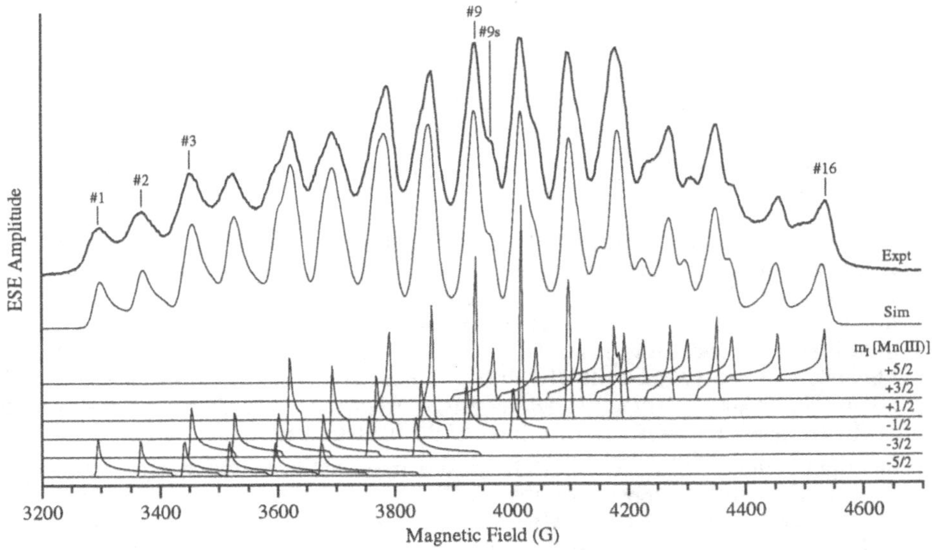

Figure 3: The di-μ-oxo-tetrakis(2,2′-bipyridine)dimanganese(III,IV) model compound field-swept ESE spectrum, simulation, and decomposition of simulation into its constituent 36 powder patterns. For each m_I[Mn(III)] set the m_I[Mn(IV)] values decrease from +5/2 to -5/2 from left to right.

predominant model places four Mn ions in a single magnetically-coupled tetranuclear cluster. However, based on multifrequency EPR simulations of the g=2 multiline signal using a dinuclear Mn cluster model with large ^{55}Mn nuclear quadrupole interactions and large anisotropic hyperfine couplings, Åhrling and Pace proposed that such a dinuclear Mn cluster exists magnetically isolated within the OEC.[19]

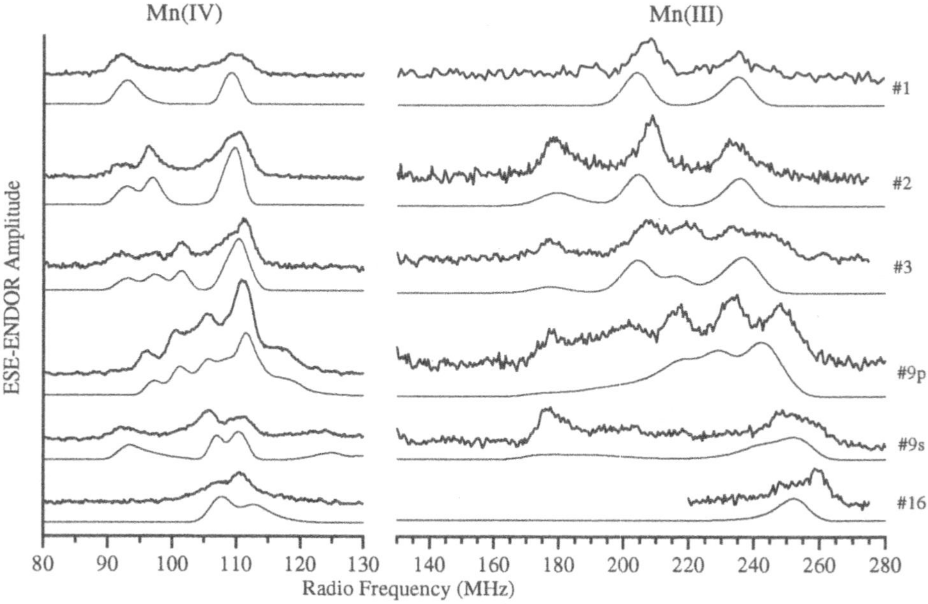

Figure 4: Mn(III)Mn(IV) model compound ESE-ENDOR spectra and simulations.

The high spectral density of EPR transitions in multinuclear Mn clusters combined with the effects of inhomogeneous broadening lead to EPR spectra with limited resolution. For example, an $S = 1/2$ state of a tetranuclear Mn complex would present 1296 transitions in the EPR spectrum. However only about 19 lines are resolved in the g=2 multiline EPR spectrum of the OEC. Figure 3 shows that even the EPR spectrum of a dinuclear mixed valence Mn(III)Mn(IV) model compound [di-μ-oxo-tetrakis(2,2$'$-bipyridine)dimanganese(III,IV)] is deleteriously affected by significant spectral overlap, with only 16 peaks resolved out of a total of 36 EPR transitions.

We have performed ^{55}Mn ESE-ENDOR in order to overcome the spectral congestion problem of CW-EPR in these Mn systems.[20] Figure 4 displays representative ESE-ENDOR spectra of the dinuclear Mn(III)Mn(IV) model compound. This is an interesting system for analysis because the EPR lineshape is largely determined by ^{55}Mn hyperfine interactions which also dominate the ENDOR spectrum. The ENDOR transitions for ^{55}Mn nuclei of the Mn(III) and Mn(IV) ions are clearly resolved. The ENDOR spectra obtained at the outermost pairs of lines in the EPR spectrum are the simplest to interpret because the ENDOR transitions originate from unique ^{55}Mn nuclear spin states. In contrast, ENDOR spectra obtained in the central region of the EPR spectrum contain multiple ^{55}Mn nuclear spin transitions due to the degeneracy of states giving rise to EPR transitions in this region. The superposition of multiple ENDOR transitions results in appreciable structure in the ENDOR features obtained in this region. Because of hyperfine anisotropy for both ions, particularly for Mn(III), there are significant "angle selection" effects which must be included in ENDOR simulations as well as the selective

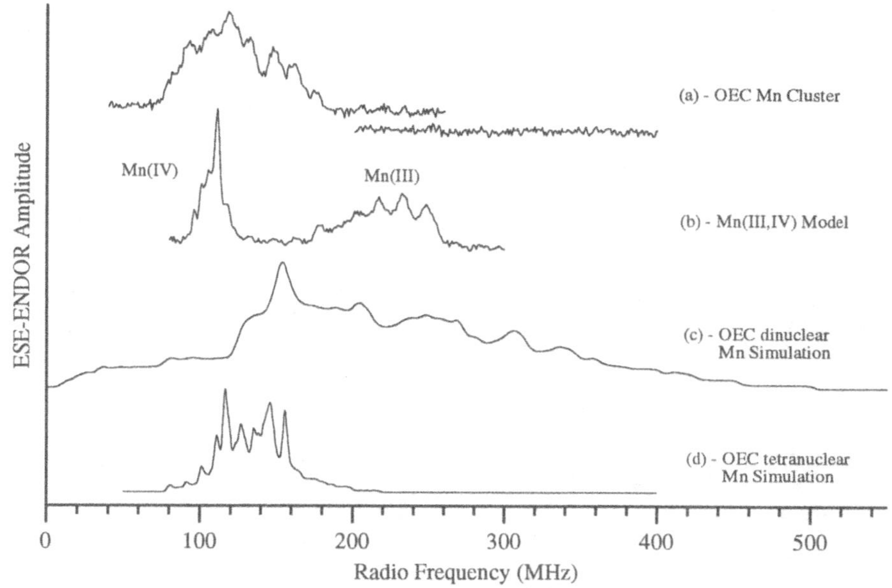

Figure 5: (a) ESE-ENDOR of the OEC Mn cluster. (b) ESE-ENDOR of the Mn(III)Mn(IV) model. (c) ESE-ENDOR simulation using the dinuclear Mn simulation parameters utilized by Åhrling and Pace[19] (d) ESE-ENDOR simulation using tetranuclear Mn simulation parameters adapted from Zheng and Dismukes[17] (Hyperfine values of $A_{\perp}$= +211 MHz, $A_{\parallel}$= +288 MHz instead of $A_{\perp}$= +226 MHz, $A_{\parallel}$= +280 MHz are used for one Mn ion). Quadrupole parameters are $P_{\parallel}$ = +7.0 MHz and η = 0.1 for each Mn.

inclusion of initial nuclear spin states. Powder pattern simulations performed across the EPR spectrum give good fits to the experimental data with the following parameters: for Mn(IV), $A_{\perp}$ = +212 MHz, $A_{\parallel}$ = +231 MHz $P_{\parallel} = 3e^2Qq/40$= +2.0 MHz, η = 0.3; and for Mn(III), $A_{\perp}$ = -480 MHz, $A_{\parallel}$ = -360 MHz , $P_{\parallel}$ = -4.5 MHz, η = 0.1.

^{55}Mn ENDOR data were also recorded at several field positions near the center of the g=2 multiline EPR signal arising from Mn in the S_2 state of the PS II oxygen-evolving complex. Figure 5a shows a representative spectrum taken over two frequency bands. The peaks are observed to be more closely spaced in frequency than the well-separated Mn(IV) and Mn(III) peaks of the model compound (Figure 5b). A dinuclear Mn cluster ENDOR simulation using the spin hamiltonian parameters of Åhrling and Pace[19] dramatically fails to simulate the ^{55}Mn ENDOR spectrum of the multiline EPR signal (Figure 5c). Dinuclear Mn EPR spectral simulations based on parameters that give a good simulation of the ENDOR are of insufficient spectral width to match the multiline signal (not shown). Therefore we consider that the constraints of simulating both the EPR and ENDOR spectra rule out a dinuclear Mn origin for the multiline signal. In contrast, tetranuclear model spin hamiltonian parameters provide reasonable simulations of both the EPR spectrum[17] and the ^{55}Mn ENDOR spectrum (Figure 5d). The observed ENDOR patterns indicate the hyperfine couplings to each of the four Mn nuclei are of comparable magnitudes.

References

1 Davies, E. R. (1974) Physics Lett. **47A**, 1-2

2 Hoffman, B. M., DeRose, V. J., Doan, P. E., Gurbiel, R. J., Houseman, A. L. P. and Telser, J. (1993) in Biological Magnetic Resonance, V 13 (Berliner L. J. and Reuben J., eds.) pp. 151-218, Plenum, New York

3 Britt, R. D. (1993) Curr. Opin. Struct. Bio. **3**, 774-779

4 Sturgeon, B. E. and Britt, R. D. (1992) Rev. Sci. Instrum. **63**, 2187-2192

5 Sturgeon, B. E., Ball, J. A., Randall, D. W. and Britt, R. D. (1994) J. Phys. Chem. **98**, 12871-12883

6 Boussac, A., Zimmermann, J.-L., Rutherford, A. W. and Lavergne, J. (1990) Nature **347**, 303-306

7 Hallahan, B. J., Nugent, J. H. A., Warden, J. T. and Evans, M. C. W. (1992) Biochemistry **31**, 4562-4573

8 Kusunoki, M. (1995) Chem. Phys. Lett. **239**, 148-157

9 Gilchrist, M. L., Ball, J. A., Randall, D. W. and Britt, R. D. (1995) Proc. Natl. Acad. Sci. USA *in press*

10 Tang, X.-S., Zheng, M., Chisholm, D. A., Dismukes, G. C. and Diner, B. A. (1995) *manuscript submitted to Biochemistry*

11 Tommos, C., Tang, X.-S., Warncke, K., Hoganson, C. W., Styring, S., McCracken, J., Diner, B. A. and Babcock, G. T. (1995) J. Am. Chem. Soc., *in press*

12 Diner, B. A, Tang, X.-S., Zheng, M., Dismukes, G. C., Force, D. A., Randall, D. W., Britt, R. D. (1995) Proceedings of the Xth International Photosynthesis Congress. (Mathis P., ed.) *this volume* Kluwer, Dordrecht

13 Rigby, S. E. J., Nugent, J. H. A. and O'Malley, P. J. (1994) Biochemistry, **33**, 1734-1742

14 Kim, D. H., Britt, R. D., Klein, M. P. and Sauer, K. (1990) J. Am. Chem. Soc. **112**, 9389-9391

15 Brudvig, G. W. (1989) in Advanced EPR: Applications in Biology and Biochemistry, (Hoff A. J., ed.) pp. 839-863, Elsevier, Amsterdam

16 Kusunoki, M. (1992) Chem. Phys. Lett. **197**, 108-116

17 Zheng, M. and Dismukes, G. C. (1992) in Research in Photosynthesis, (Murata N., ed.) pp. 305-308, Kluwer Academic Publishers, Dordrecht, The Netherlands

18 Bonvoisin, J., Blondin, G., Girerd, J.-J. and Zimmermann, J.-L. (1992) Biophys. J. **61**, 1076-1086

19 Åhrling, K. A. and Pace, R. J. (1995) Biophys. J. **68**, 2081-2090

20 Randall, D. W., Sturgeon, B. E., Ball, J. A., Lorigan, G. A., Chan, M. K., Klein, M. P., Armstrong, W. H. and Britt, R. D. (1995) *manuscript submitted to J. Am. Chem. Soc.*

ENVIRONMENT AND FUNCTION OF THE REDOX ACTIVE TYROSINES OF PHOTOSYSTEM II

Bruce A. Diner, Xiao-Song Tang - Central Research and Development Dept., E. I. Du Pont de Nemours & Co., Wilmington, DE, USA
Ming Zheng, G. Charles Dismukes - Department of Chemistry, Princeton University, Princeton, NJ, USA
Dee Ann Force, David W. Randall, R. David Britt - Department of Chemistry, University of California, Davis, CA, USA

Introduction

Recent experimental evidence is increasingly supportive of a role for the redox active tyrosine, Y_Z, of Photosystem II (PSII) to be a direct participant in the redox chemistry associated with water oxidation. The "S3-split EPR signal" observed upon illumination of Ca^{2+}-extracted PSII centers with intact Mn clusters has been interpreted as indicating the presence of Y_Z• as close as 4.5Å to the Mn cluster [1]. Kinetic evidence [2] has shown a structural change (probably a deprotonation) to precede the reduction of Y_Z• in the reaction $S3Y_Z$• -> [$S4Y_Z$] -> $S0Y_Z + O_2$, pointing to a more complex role for Y_Z• than that of an electron acceptor alone. Finally, a comparison of the redox chemistry of tyrosines in other enzyme systems has led to the proposal that Y_Z• is involved in hydrogen atom abstraction in advancing the S-states of the oxygen-evolving complex [3].

In an effort to better understand the function of the redox active tyrosines in Photosystem II, we have developed a system for studying in detail the spectroscopic properties of Y_Z• and Y_D•. We [4] and Mino et al. [5] have recently learned, using somehat different techniques, how to trap Y_Z• at low temperature in PSII centers depleted of manganese. In trapping Y_Z• in mutant PSII core complexes that lack Y_D (D2-Tyr160Phe) we [4] have been able to study Y_Z• free of contamination by Y_D• and chlorophyll cation radicals.

Experimental Results and Discussion

Except where otherwise indicated (Fig. 4), all of the experiments presented in this paper were carried out using Mn-depleted PSII core complexes isolated from the cyanobacterium, *Synechocystis* 6803, according to Tang et al. [4]. Fig. 1 shows an EPR spectrum preceding illumination of PSII core complexes from the mutant D2-Tyr160Phe. Upon freezing under illumination, the Y_Z• radical is trapped, disappearing reversibly on warming the sample above 250K. That this signal indeed comes from tyrosine is shown in Fig. 2, where specifically deuterated tyrosine, incorporated into the reaction centers, results in the expected partial loss of hyperfine structure of the radical, consistent with odd-alternant spin density localized on the ring carbons C1, C3 and C5 [6,7].

We have examined the dipolar coupling between the non-heme iron of the reaction center and Y_D• and Y_Z• [9]. A similar temperature dependence and rate for the accelerated spin-lattice relaxation of each of the two tyrosines and the non-heme iron indicates that the distance between each of the tyrosyl radicals and the non-heme iron is equivalent (within ± 1Å), supporting the location of the tyrosines equidistant from the C2 symmetry axis on which the non-heme iron is assumed to sit, as in the reaction centers of

P. Mathis (ed.), Photosynthesis: from Light to Biosphere, Vol. II, 229–234.

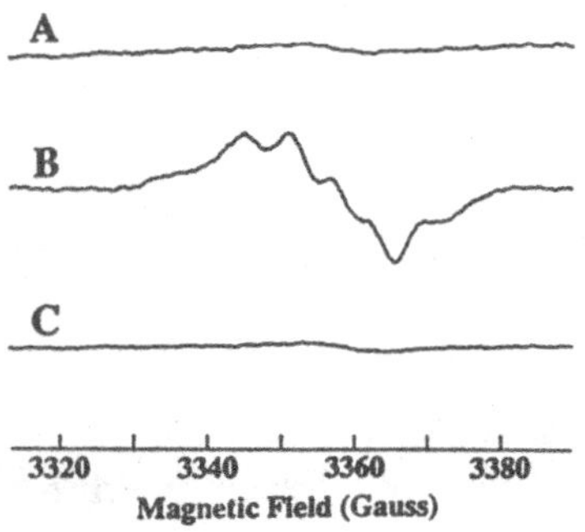

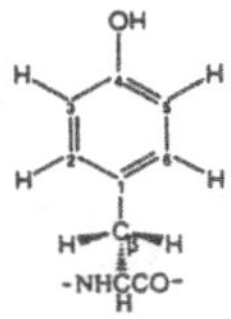

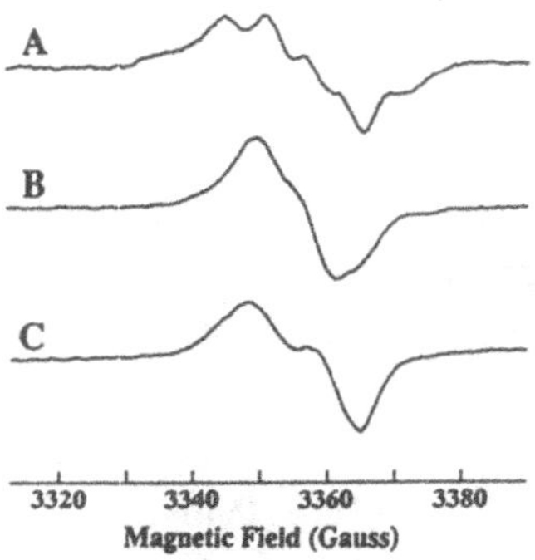

Figure 1. (left). EPR spectra of a PSII core complex isolated from the mutant D2-Tyr160Phe lacking Y_D. A) dark-adapted, B) frozen under illumination, C) thawed and refrozen in the dark.
Figure 2. (right) EPR spectra of Y_Z•, trapped as in Fig. 1 in PSII core complexes isolated from cells labeled as described by Barry and Babcock [8]. A) grown on normal BG-11 growth medium, B) growth medium containing β-methylene -3',3'-deuterated tyrosine, C) growth medium containing ring-3,5-deuterated tyrosine.

the purple non-sulfur photosynthetic bacteria [10]. Using the dipolar coupling of the homologous non-heme iron and the oxidized primary donor $P870^+$ as a molecular ruler, the distance between the non-heme iron of PSII and each of the tyrosines comes to 37±5Å, consistent with the identification of Y_Z• and Y_D• with D1-Tyr161 and D2-Tyr160, respectively, the PSII homologues of *Rb. sphaeroides* L-Arg135 and M-Arg165 (M-His162 in *Rs. viridis*). Y_Z is oxidized by $P680^+$, 100 to 10^5 times faster than Y_D, depending on whether the manganese cluster is absent or present, respectively (see [11] for review). This much faster oxidation rate for Y_Z most likely reflects a closer proximity of Y_Z to $P680^+$ than for Y_D• [12, 13] and a location for $P680^+$ on the Y_Z side of the C2 symmetry axis.

The reduction potentials of the redox couples, Y_Z•/Y_Z and Y_D•/Y_D, differ by ~250mV [14,15]. Possible explanations are differences in hydrogen bonding strength in the reduced and/or oxidized forms, differences in the pKa of the base that accepts the proton released upon tyrosine oxidation, and/or stabilization of the plus charge produced upon oxidation. We examine here two of these properties with regard to Y_Z• and Y_D•.

First we show that both are hydrogen-bonded. A hydrogen bond to Y_D• has been detected by both ESEEM and cw-ENDOR using H_2O/D_2O exchange [16,17,18]. The H_2O-D_2O difference of Fig. 3c, shows this exchangeable proton with 3.1MHz coupling and an axial line shape (Fig. 3c). Using a point dipole approximation this coupling would place a proton at a distance of 1.87Å from the tyrosyl radical oxygen (spin density = 0.25, [19]). This distance is really only consistent with a hydrogen bonded proton. We have done a similar H_2O/D_2O exchange experiment for Y_Z• in the Y_D-less mutant and find by TRIPLE a much broader exchangeable feature at ~2.7MHz (Fig. 3f).

Deuterium pulsed ENDOR on the same sample (Fig. 4) shows deuterium (^2H) coupling that can only come from deuterium hydrogen-bonding to the phenolic oxygen of the tyrosyl radical. The symmetric splitting of the perpendicular turning points of the ENDOR spectrum of Fig. 4b, around the frequency of the scaled proton peak of Fig. 4a, demonstrates that this splitting arises from the electric quadrupolar interaction for the I=1 deuterium nucleus. The Y_Z• and Y_D• spectra (Fig. 4c,d) have a spectral envelope of

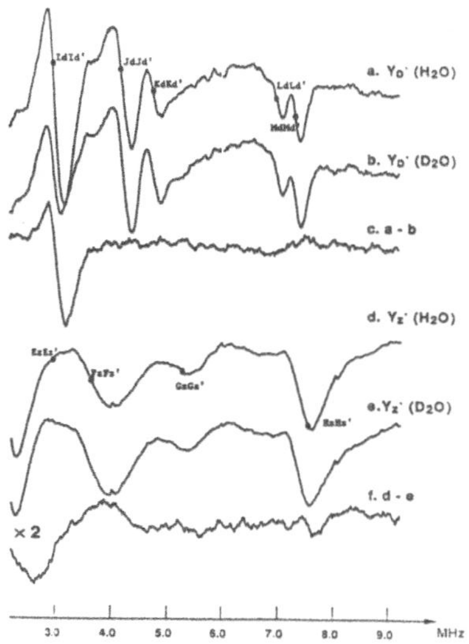

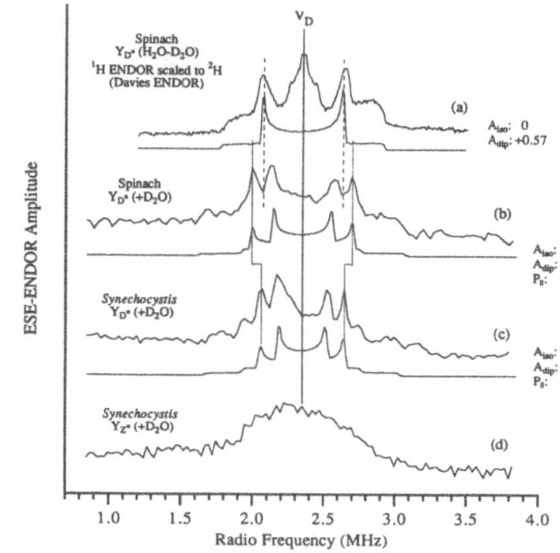

Figure 3. (left) ^{1}H Special TRIPLE spectrum of Y_Z• and Y_D• radicals from Mn-depleted wild type and D2-Tyr160Phe mutant PSII core complexes incubated for 14-16h in H_2O buffer or D_2O buffer (pH = pH = 7.5). a) Y_D• with H_2O buffer; b) Y_D• with D_2O buffer; c) a-b; d) Y_Z• with H_2O buffer; e) Y_Z• with D_2O buffer; f) d-e. Temperature 135K. The frequency axis has been converted to hyperfine frequency.

Figure 4. (right) Y_Z• and Y_D• ESE-ENDOR spectra and simulations. a) scaled ^1H_2O-^2H_2O difference Davies ^{1}H ESE-ENDOR spectrum of Y_D• in Mn-depleted spinach BBY membranes. The spectrum is scaled in frequency by the ratio $\gamma(^2H)/\gamma(^1H)$ to allow comparison to the subsequent ^{2}H ENDOR traces. Simulation parameters are in MHz units; b) ^{2}H Mims ENDOR spectrum of Y_D• in Mn-depleted ^2H_2O-exchanged spinach BBY membranes; c) ^{2}H Mims ENDOR spectrum of Y_D• in Mn-depleted ^2H_2O-exchanged *Synechocystis* PSII core complexes prepared as in Fig. 3. The dipolar coupling is slightly reduced compared to the spinach BBY membranes. d) ^{2}H Mims ENDOR spectrum of Y_Z• cryotrapped after illumination of Mn-depleted ^2H_2O-exchanged *Synechocystis* PSII core complexes from mutant D2-Tyr160Phe prepared as in Fig. 3.

approximately the same width. However, while the ^{2}H peaks are sharp and clearly defined for Y_D• (Fig. 4b,c), they are smeared out for Y_Z• (Fig. 4d). The complete lack of resolved hyperfine and quadrupolar structure indicates that there is considerable disorder in the ^{2}H couplings of Y_Z• relative to Y_D• and indicating a dispersion of hydrogen bond strengths in the case of Y_Z•. Measurement of the g_x-component of the anisotropic g-tensor for Y_Z• and Y_D•, sensitive to environmental perturbations to the non-bonded pairs of electrons on the tyrosyl radical phenolic oxygen, has lead to a similar conclusion [20], with the strength of the hydrogen bonds ranging from being both stronger and weaker than in the case of Y_D•. Thus Y_Z• and Y_D• are clearly both hydrogen bonded and we conclude that at least hydrogen bonding in the oxidized state is probably not responsible for the differences in the redox properties of the tyrosyl radicals.

Another explanation for the difference in reduction potential could be the way in which the proton released upon tyrosine oxidation is dealt with by the environments of Y_Z and Y_D. If, for example, the proton released by Y_D upon oxidation were to be

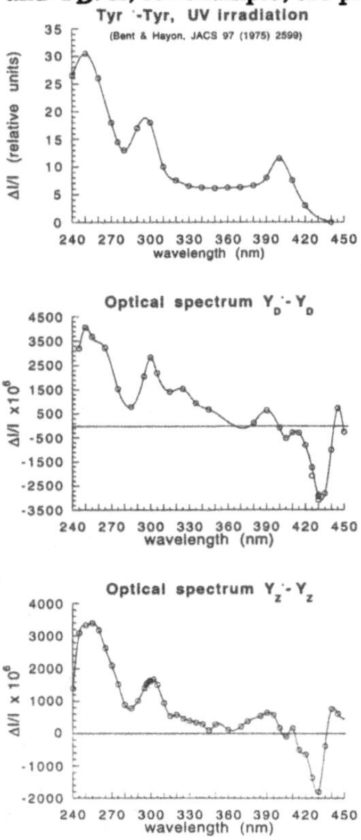

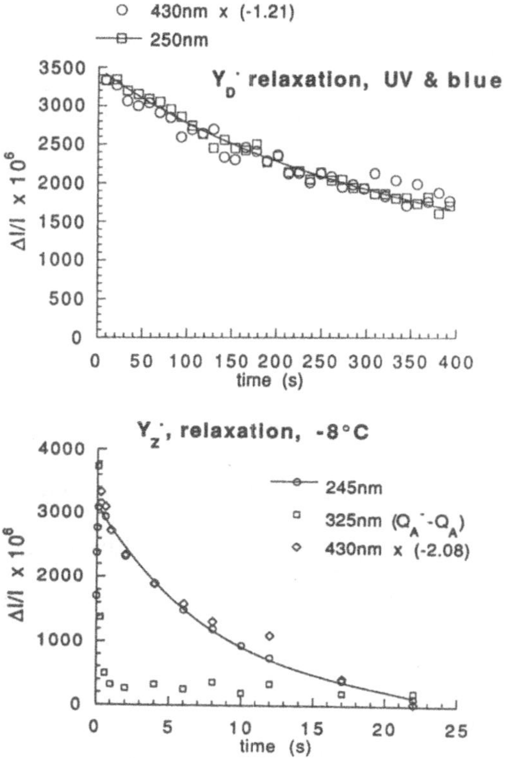

Figure 5. (left) - Comparison of optical difference spectra of model Tyr•-Tyr in aqueous solution following UV irradiation [21], with Y_D•-Y_D and Y_Z•-Y_Z in *Synechocystis* Mn-depleted PSII core complexes. These spectra were measured at pH 6.0 at 600ms after a single saturating flash and between 2 and 18s after the last of a train of 5 saturating flashes (0.33Hz), respectively for Y_Z• and Y_D•. Y_Z•-Y_Z and Y_D•-Y_D were measured at -8°C and 25°C, respectively, under conditions where Q_A^- had been completely reoxidized by ferricyanide. Y_Z•-Y_Z was measured in Mn-depleted D2-Tyr160Phe mutant PSII core complexes the presence of 600µM $K_3Fe(CN)_6$ and Y_D•-Y_D in wild type Mn-depleted PSII core complexes in the presence of 100uM and 100µM $K_4Fe(CN)_6$. In the case of both Y_Z•-Y_Z and Y_D•-Y_D the difference spectra are the equivalent of all PSII centers at a reaction center concentration of 20-25nM giving rise to the absorbance changes observed. Corrections have been made for absorbance changes arising from ferri- and ferrocyanide.
Figure 6. (right top) - Relaxation of Y_D•-Y_D measured at 250 and 430nm after a train of 5 saturating actinic flashes (0.33Hz) under the same conditions as in Fig. 5. The 430nm measurement has been multiplied by -1.21 to allow plotting on the same scale.
Figure 6. (right bottom) Relaxation of Y_Z•-Y_Z measured at 250, 325 and 430nm after a single saturating flash under the same conditions as in Fig. 5. The 430nm measurement has been multiplied by -2.08 to allow plotting on the same scale. The 325nm relaxation is entirely from Q_A^--Q_A, except for the residual contribution from Y_Z•-Y_Z at long times.

transported quickly to the membrane surface from the hydrophobic interior, then Y_D• would be more stabilized were the same not true of Y_Z•. Fig. 5 shows the optical difference spectra for Tyr•-Tyr in water [21], Y_D•-Y_D. and Y_Z•-Y_Z. The spectra are obviously all quite similar in the UV, where the absorption spectra reflect the radical alone. Both Y_Z• and Y_D• differ in the blue from the model tyrosine showing a derivative shaped spectrum with a minimum at ~429 and 432nm and maxima at 440 and 445nm for Y_Z• and Y_D•, respectively. This band shift most likely arises from a localized electrochromic effect on nearby chlorophylls produced by the plus charge associated with tyrosine oxidation. We have found (in collaboration with Jérôme Lavergne) that the mutation D1-His198Gln, which causes a blue shift in the optical difference spectrum of $P680^+$-P680, also produces a comparable shift in the optical difference spectrum of Y_Z•-Y_Z, indicating that the detecting chromophore is P680 itself. The band shifts observed in the blue for Y_Z•-Y_Z and Y_D•-Y_D are fairly similar in both cases, though the intensity with respect to the UV portion of the spectrum and the width appear to be very slightly larger in the case of Y_D•. The kinetics of relaxation were compared in the UV (245 and 250nm) and in the blue (430nm) (Fig. 6). Relaxation occurs with the same kinetics indicating that the positive charge is retained during the lifetime of the tyrosyl radical. The charge is then lost as the tyrosine is reprotonated upon reduction.

The spectrum measured here for Y_Z•-Y_Z shows the same ratio of signal amplitude in the UV and blue as observed 0.5ms after an actinic flash at room temperature[22], indicating that any adjustments to the band shift have already occurred at this time. Such adjustments could arise either from a fraction of centers that undergoes an internal proton loss to the solvent phase or from some or all centers where there is just an internal displacement of charge. Proton release has been observed on millisecond and submillisecond time scales associated with oxidation of Y_Z [23,24], though such protons are most likely Bohr protons released at the surface, with little influence on the electrochromic effect. It is likely that the proton released by Y_Z and Y_D upon oxidation is fully retained at least in Mn-depleted PSII centers.

The spectra of Fig. 5 for Y_Z•-Y_Z and Y_D•-Y_D are consistent with conditions under which their reduction potentials have been estimated - i.e. following equilibration of the donor and acceptor sides of the reaction center. Assuming that the chlorophylls of the reaction center are distributed with C2 symmetry and that the dielectric constants of the two sides of the reaction center are not drastically different, it appears that the fraction of centers retaining plus charge in the immediate environments of Y_Z and Y_D is probably not very different. Thus retention or not of the proton released upon tyrosine oxidation is also not the explanation for the difference in reduction potential of the two tyrosines.

As the proton released upon tyrosine oxidation appears to remain close by during the lifetime of the tyrosyl radical, we are left with the pKa of the proton acceptor (higher for Y_D) or the solvation of the plus charge by permanent or induced dipoles (greater for Y_D) as the remaining reasons for the difference in reduction potential of the Y_Z•/Y_Z and Y_D•/Y_D redox couples.

We have recently found that in Mn-depleted PSII core complexes the β-methylene protons of Y_Z• show a dispersion of angles with respect to the plane of the aromatic ring [25]. Whether this dispersion is a reflection of a heterogeneity in the structure of the pocket in which Y_Z is located or a variable twisting of the ring arising from of a dispersion in the angles of the hydrogen bonds to the non-bonded pairs of electrons of the phenolic oxygen of the radical is still an open question. It should be kept in mind that the system we are studying has lost the manganese cluster. However, recent pulsed ENDOR measurements of Gilchrist et al. [1] on the "S3 split signal", observed in Ca^{2+}-depleted but manganese containing PSII core complexes and attributed to Y_Z• [1], shows a very similar broadened signal for the strongly coupled β-methylene proton . This observation implies that the dispersion in the structure of Y_Z• exists even where the manganese cluster is still intact. It is thus possible that the heterogeneity in the structure of Y_Z, not observed for Y_D, reflects a mobility associated with its function in water oxidation. Y_Z is oxidized and reduced by different species, $P680^+$ and the manganese cluster, respectively. In the case of Y_D, however, the oxidant and the reductant are the same ($P680^+/P680$). As in the oxidation of the tyrosines, electron and proton transfer are coupled (possibly only concerted for the oxidation), the mobility of Y_Z may also allow the proton released upon Y_Z oxidation to be delivered to one site, and the proton acquired upon $Y_Z\bullet$ reduction provided by a different site. Y_Z would then flip back and forth between two conformations like a "toggle switch", with the preferred conformation possibly influenced by the presence of a nearby plus charge.

Acknowledgements:
Support is gratefully acknowedged from the NRI/CRGP of the USDA to B.A.D. and R.D.B.and from NIH to G.C.D. and R.D.B.

References
1 Gilchrist, M.L., Ball, J.A., Randall, D.W. and Britt, R.D. (1995) Proc. Natl. Acad. Sci. USA, in press
2 Rappaport F., Blanchard-Desce, M. and Lavergne, J. (1994) Biochim. Biophys. Acta 1184, 178-192
3 Hoganson C.W., Lydakis-Simantiris N., Tang, X.-S., Tommos C., Warncke, K., Babcock G.T., Diner, B.A., McCracken, J. and Styring, S. (1995) Photosyn. Res., in press
4 Tang, X.-S., Zheng, M., Chisholm, D.A., Dismukes, G.C. and Diner, B.A. (1995) Biochemistry, submitted
5 Mino, H. and Kawamori A. (1994) Biochim. Biophys. Acta 1185, 213-220
6 Barry, B.A., El-Deeb, M.K., Sandusky, P.O. and Babcock, G.T. (1990) J. Biol. Chem. 265, 20139-20143
7 Boerner, R.J. and Barry, B.A. (1993) J. Biol. Chem. 268, 17151-17154
8 Barry, B.A. and Babcock, G.T. (1987) Proc. Natl. Acad. Sci. USA 84, 7099-7103
9 Koulougliotis, D., Tang, X.-S., Diner, B.A. and Brudvig, G.W. (1995) Biochem. 34, 2850-2856
10 Ermler, U. Fritzsch G, Buchanan S.K. and Michel, H. (1994) Structure 2, 925-936
11 Diner, B.A. and Babcock, G.T. (1996) In Oxygenic Photosynthesis: The Light Reactions (Ort, D. and Yocum, C. eds.) Kluwer, Dordrecht, The Netherlands, in press
12 Hoganson , C.W. and Babcock, G.T. (1988) Biochem. 27, 5848-5855
13 Bock, C.H., Gerken, S., Stehlik, D. and Witt, H.T. (1988) FEBS Lett. 227, 141-146
14 Boussac, A. and Etienne, A.-L. Biochim. Biophys. Acta 766, 576-581
15 Vass, I. and Styring, S. (1991) FEBS Lett. 226, 347-351
16 Rodriguez, I.D., Chandrashekar, T.K., and Babcock, G.T. (1987) In PRogress in Photosynthesis, (Biggins, J. ed.) Martinus Nijhoff Publishers, Dordrecht, Vol. I pp 119-133
17 Evelo, R.G., Dikanov, S.A., Hoff, A.J. Chem. Phys Lett. 157, 25-30
18 Tang, X.-S., Chisholm, D.A., Dismukes, G.C., Brudvig, G.W. and Diner, B.A. (1993) 32, 13742-13748
19 Warncke, K., Babcock, G.T., McCracken J. (1994) J. Am. Chem. Soc. 116, 7332-7340
20 Un, S., Tang, X.-S. and Diner, B.A. (1995) Submitted
21 Bent, D.V. and Hayon E. (1975) J. Am. Chem. Soc. 97, 2599-2606
22 Metz, J.G., Nixon, P.J., Rögner, M, Brudvig, G.W. and Diner, B.A. (1989) Biochem. 28, 6960-6969
23 Renger, G and Voelker, M. (1982) FEBS Lett. 149, 203-207
24 Bögershausen, O. and Junge W. (1995) Biochim. Biophys. Acta 1230, 177-185
25 Tommos, C, Tang, X.-S. Warncke K., Hoganson, C.W., Styring, S., McCracken, J., Diner, B.A. and Babcock, G.T. (1995) J. Am. Chem. Soc., in press

FTIR STUDIES ON THE STRUCTURE AND REACTIONS OF THE OXYGEN-EVOLVING CENTER IN PHOTOSYSTEM II

Takumi Noguchi*, Taka-aki Ono and Yorinao Inoue
Photosynthesis Research Laboratory, Solar Energy Research Group,
The Institute of Physical and Chemical Research (RIKEN),
Wako, Saitama 351-01, Japan

1. Introduction

Molecular oxygen is evolved in plants as a result of oxidation of water, which is a terminal electron donor in photosystem II (PS II). It has been known that the catalyst of this reaction is the oxygen-evolving center (OEC) that consists of four Mn ions bound to the PS II proteins. In spite of extensive spectroscopic investigations, the structure of OEC and the reaction mechanism are still unresolved [1].

We have been applying Fourier transform infrared (FTIR) spectroscopy to studies on the structure of OEC [2-5]. In large PS II proteins, the FTIR spectrum specific to OEC has been obtained as a light-induced difference spectrum due to structural changes of OEC upon S_1-to-S_2 transition. The detected FTIR signals include information on the structures and molecular interactions of ligands, proteins, and substrate water. In this study, the FTIR spectrum of OEC was analyzed by using PS II membranes perturbed in various ways such as Ca^{2+} depletion, Cl^- depletion, H-D exchange, and ^{15}N-labeling. These analyses provided a new insight into the structure and reaction of OEC.

2. Materials and Methods

BBY-type PS II membranes were prepared from spinach according to [6]. ^{15}N-labeled spinach was cultured with a medium containing $K^{15}NO_3$ and $Ca(^{15}NO_3)_2$ as nitrogen sources. Ca^{2+} depletion was done by low-pH treatment [7]. Cl^- depletion was performed according to [8]. FTIR spectra were measured on a JEOL JIR-6500 spectrophotometer equipped with an MCT detector (EG&G JUDSON IR-DET101). The sample temperature was adjusted to 250 K in a cryostat (Oxford DN1704). Light-induced difference spectra were obtained by subtraction between the two spectra taken before and after single-flash illumination supplied by a Nd-YAG laser (532 nm, 7 ns).

3. Results and Discussion

3.1. *Detection of an FTIR difference spectrum of OEC between the S_1 and S_2 states*

Figure 1A (solid line) shows an FTIR difference spectrum of untreated PS II membranes upon single-flash illumination. The sample includes a mixture of 2 mM ferri- and 18 mM ferrocyanide in Mes buffer at pH 5.5. Under this condition of redox potential and pH, the non-heme iron, which could work as an endogenous electron acceptor if it has been preoxidized, is kept reduced and an electron is abstracted from the protein by an exogenous electron acceptor, ferricyanide [9]. On the donor side, upon

P. Mathis (ed.), Photosynthesis: from Light to Biosphere, Vol. II, 235–240.

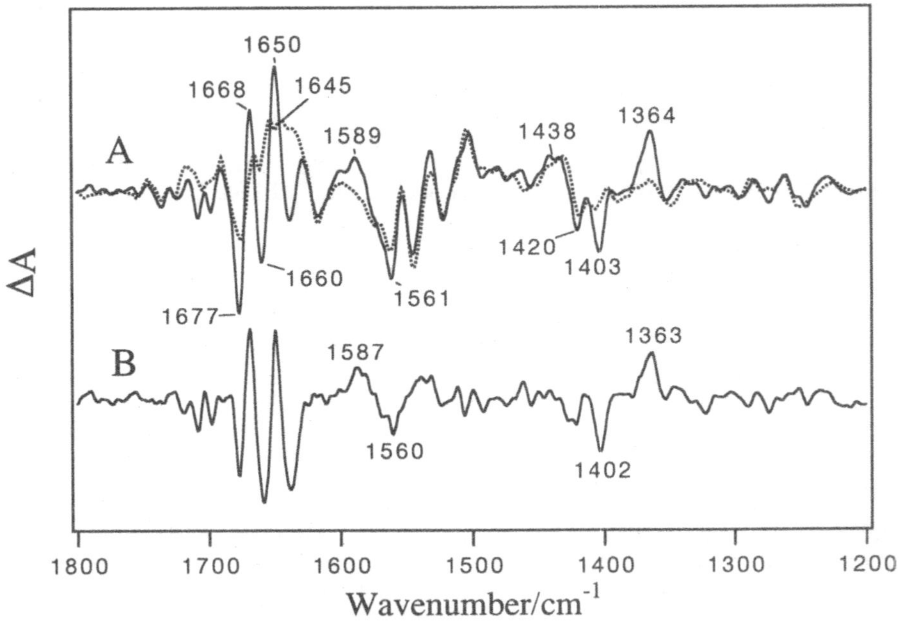

Fig. 1. (A) S_2/S_1 FTIR difference spectra of untreated (solid line) and Ca^{2+}-depleted (dotted line) PS II membranes. (B) Double difference spectrum between Ca^{2+}-depleted and untreated S_2/S_1 spectra

Table 1. Ca^{2+}-sensitive COO^- bands and the coordination structures

		$\nu_a(cm^{-1})$	$\nu_s(cm^{-1})$	$\Delta\nu(cm^{-1})$	Coordination
Untreated PSII	S_1	1560	1403	157	bridging
	S_2	1587	1364	223	unidentate
Ca^{2+}-depleted PSII		bands are absent			

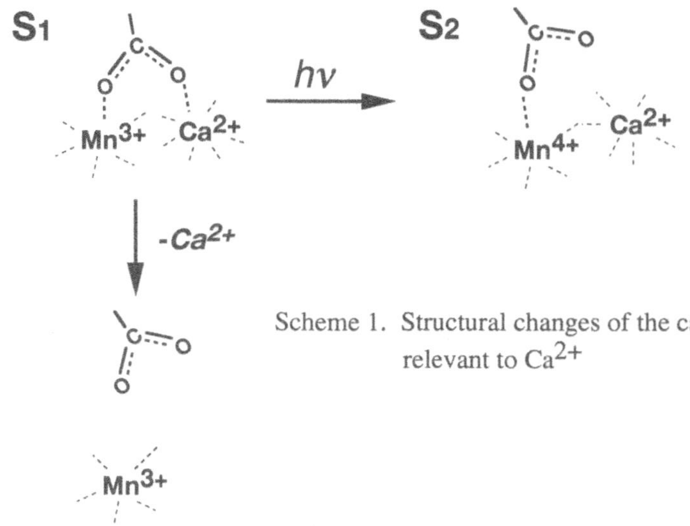

Scheme 1. Structural changes of the carboxylate ligand relevant to Ca^{2+}

single-flash illumination, the Mn-cluster is oxidized to be converted from S_1 to S_2 state. Since ferri- and ferrocyanide do not show any bands in 1800-1000 cm^{-1}, the spectrum in Fig. 1A (solid line) solely represents the difference between the S_1 and S_2 states of OEC. This S_2/S_1 difference spectrum was verified by good agreement with the calculated S_2/S_1 spectrum by subtraction of the Q_A^-/Q_A spectrum from the $S_2Q_A^-/S_1Q_A$ spectrum [4], and also by the fact that the temperature dependence of the typical S_2/S_1 band at 1404 cm^{-1} followed that of the S_2 formation [3].

3.2. *A carboxylate ligand connecting Mn and Ca^{2+} [4]*

Ca^{2+} is a cofactor indispensable for oxygen evolution. To investigate the structural relevance of Ca^{2+} to the Mn-cluster, the S_2/S_1 difference spectrum of Ca^{2+}-depleted PS II membranes (Fig. 1A dotted line) was compared with that of untreated membranes (solid line). The double difference spectrum (Fig. 1B) clearly showed that the negative bands at 1560 and 1403 cm^{-1} (S_1) and the positive bands at 1587 and 1364 cm^{-1} (S_2) were lost upon Ca^{2+} depletion. These bands were basically unaffected by ^{15}N substitution, indicating that they were due to nitrogen-free vibrations. Taking into consideration the frequencies of vibrational modes of amino acids and proteins as well as their infrared intensities [10], the above bands were assigned to the asymmetric (1560 and 1587 cm^{-1} bands) and symmetric (1403 and 1364 cm^{-1} bands) COO^- stretching modes of a carboxylate group of Asp, Glu or the C-termini. It has been known that there is a relationship between the coordination structure of a COO^- group and the frequency difference between its symmetric and asymmetric modes [11]. According to this criterion, the frequency differences of the above COO^- bands, i.e., 157 cm^{-1} for S_1 and 223 cm^{-1} for S_2, indicate that this COO^- group possesses a bridging structure in the S_1 state and an unidentate structure in the S_2 state (Table 1). Together with the fact that the appearance of the above COO^- bands depends on the presence of Ca^{2+}, the following conclusions on the structure and reaction of this COO- group were obtained (Scheme 1): 1) In the S_1 state, the COO^- group forms a carboxylate bridge between redox-active Mn and Ca^{2+}. 2) In the S_2 state, the coordination bond to Ca^{2+} breaks and the COO^- takes an unidentate structure. 3) Upon Ca^{2+} depletion, this COO^- ligand is liberated from the Mn ion and the bands become undetectable in the S_2/S_1 spectrum.

In Fig. 2, the S_2/S_1 difference spectrum of Cl^--depleted PS II membranes (solid line) was compared with that of untreated ones (dotted line). This S_2/S_1 difference spectrum was obtained as a double difference spectrum of Q_A^-/Q_A from $S_2Q_A^-/S_1Q_A$ measured with DCMU, to avoid the presence of anions other than OH^- in the PS II sample. The stretching bands of the above bridging carboxylate ligand were still observed at 1595/1564 and 1363/1402 cm^{-1}, indicating that the absence of Cl^- little affect the structure of the carboxylate bridge between Mn and Ca^{2+}.

3.3. *A carboxylate ligand interacting with substrate water [5]*

To investigate the molecular interactions of substrate water in OEC, the effect of H-D exchange on the S_2/S_1 spectrum was studied (Fig. 3A). Upon deuteration, the band at 1561 cm^{-1} upshifted by 18 cm^{-1} to 1579 cm^{-1}. This upshifting band was unaffected by ^{15}N substitution and hence assigned to the asymmetric COO^- stretching mode. FTIR measurements of model carboxylate compounds also showed upshifts of the asymmetric stretching bands upon deuteration (Table 2). When carboxylate ligands form hydrogen bonds with water molecules coordinated to the metal ion, the upshift was as much as 20 cm^{-1}, whereas carboxylate groups interacting with bulk water showed much smaller upshifts of 7 cm^{-1}. Thus, the large upshift of the COO^- band upon deuteration in the S_2/S_1 spectrum can be interpreted that there is a COO^- ligand of the

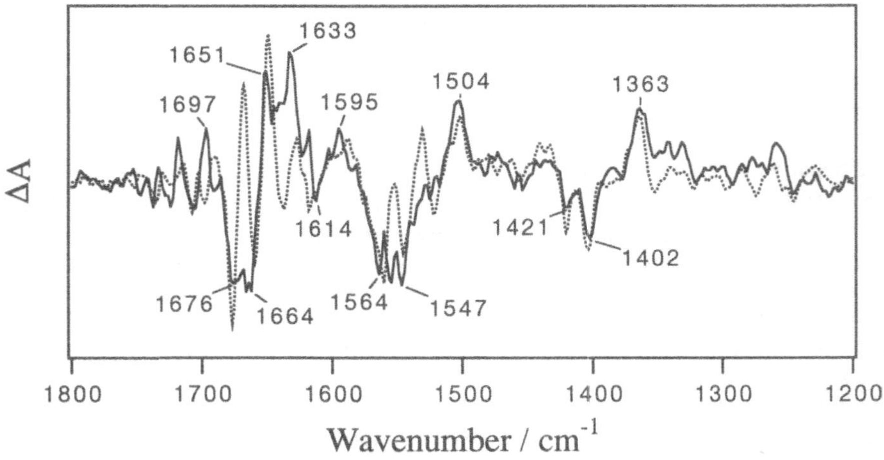

Fig.2. S_2/S_1 FTIR difference spectrum of Cl^--depleted PS II membranes (solid line) compared with that of untreated PS II membranes (dotted line)

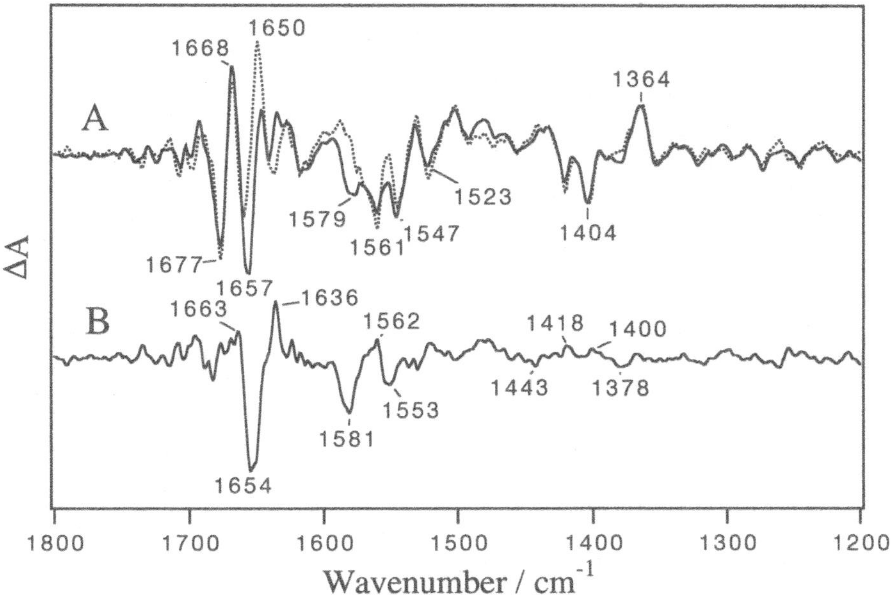

Fig. 3. (A) S_2/S_1 FTIR difference spectrum of PS II membranes in D_2O buffer (solid line) compared with that in H_2O buffer (dotted line). (B) Double difference spectrum between the spectra in H_2O and in D_2O

Table 2. Effects of H-D exchange on the frequencies of asymmetric COO^- stretching modes of model carboxylate compounds and the O_2-evolving center of PS II

	ν_{as} (cm^{-1})		$\Delta\nu$ (cm^{-1})	Structure
	H_2O	D_2O		
CH_3COO^-	1557	1564	+7	HOH-OCO-HOH
Mn-EDTA	1585	1592	+7	M-OCO-HOH
$(CH_3COO)_2Ni \cdot 4H(D)_2O$	1534	1554	+20	M-OCO-HO(H)-M
OEC (S_1 state)	1561	1579	+18	M-OCO-HO(H)-M

Mn ion strongly hydrogen-bonded with a water molecule that is also bound to Mn. Note that this COO^- ligand is a different one from the COO^- bridge connecting Mn and Ca^{2+}, because 1) the three peaks at 1581/1562/1553 cm^{-1} in the double difference spectrum (Fig. 3B) is considered to be a result of an upshift of a differential signal at ~1550(+)/~1560(-) cm^{-1}, whereas the bridging COO^- shows a differential signal at 1587(+)/1560(-), and 2) the symmetric bands of the bridging COO^- group at 1364/1404 cm^{-1} was not affected by deuteration at all. We speculate that the COO^- ligand hydrogen-bonded to substrate water could work as a primary proton acceptor in the water-cleavage reaction.

3.4. *Conformational changes of the proteins in OEC*

In the amide I region of 1700-1600 cm^{-1}, several, relatively strong bands were observed in the S_2/S_1 spectrum (Fig. 1A solid line). This implies that some conformational changes occur around the Mn cluster upon the S_2 formation. When Ca^{2+} is removed, most of these bands were lost (Fig. 1A). Upon Cl^- depletion, similarly, the band structures were mostly reduced (Fig. 2). These results indicate that both Ca^{2+} and Cl^- are necessary for conformational changes of OEC upon S_1-to-S_2 transition. The spectral changes of the amide I bands upon H-D exchange (Fig. 3) were interpreted as the large downshifts of the bands around 1659 cm^{-1} to those around 1645 cm^{-1}. This kind of large downshift upon deuteration is typical of a random coil conformation [12], and hence the result indicates that hydrophilic regions with random coil conformations are present around the Mn-cluster. Since only the part of protein in OEC can be deuterated by incubation in D_2O buffer, it is suggested that accessibility of substrate water is controlled by the arrangements of the hydrophilic and hydrophobic domains in OEC.

Acknowledgments

This work was supported by a grant for Photosynthetic Sciences and a Special Grant for Promotion of Research (to T.N.) at the Institute of Physical and Chemical Research (RIKEN) given by the Science and Technology Agency (STA) of Japan, and by a Grant-in-Aid for Cooperative Research (No.05304006) from the Ministry of Education, Science and Culture of Japan.

References

1. Debus, R. J. (1992) Biochim. Biophys. Acta 1102, 269-352
2. Noguchi, T., Ono, T. and Inoue, Y. (1992) Biochemistry 31, 5953-5956
3. Noguchi, T., Ono, T. and Inoue, Y. (1993) Biochim. Biophys. Acta 1143, 333-336
4. Noguchi, T., Ono, T. and Inoue, Y. (1995) Biochim. Biophys. Acta 1228, 189-200
5. Noguchi, T., Ono, T. and Inoue, Y. Biochim. Biophys. Acta, in press
6. Ono, T. and Inoue, Y. (1986) Biochim. Biophys. Acta 850, 380-389
7. Ono, T., Izawa, S. and Inoue, Y. (1992) Biochemistry 31, 7648-7655
8. Ono, T., Noguchi, T., Inoue, Y. , Kusunoki, M., Yamaguchi, H. and Oyanagi, H. (1995) J. Am. Chem. Soc. 117, 6386-6387
9. Noguchi, T. and Inoue, Y. (1995) J. Biochem. 118, 9-12
10. Venyaminov, S. Yu. and Kalnin, N. N. (1990) Biopolymers 30, 1243-1257.
11. Nakamoto, K. (1986) in Infrared and Raman Spectra of Inorganic and Coordination Compounds, pp. 231-233, Wiley, New York
12. Surewicz, W. K. and Mantsch, H. H. (1988) Biochim. Biophys. Acta 952, 115-130

STRUCTURAL CHARACTERIZATION OF THE MANGANESE CLUSTER IN THE PHOTOSYNTHETIC OXYGEN EVOLVING COMPLEX USING X-RAY ABSORPTION SPECTROSCOPY

J.E. Penner-Hahn[1], P.J. Riggs-Gelasco[1], E. Yu[3], P. DeMarois[3], and C.F. Yocum[2], Departments of [1]Chemistry and [2]Biology, and [3]The Biophysics Research Division, The University of Michigan, Ann Arbor, Michigan, USA 48109-1055

1. Introduction

The oxygen evolving complex (OEC) in Photosystem II requires Mn, Ca^{2+}, and Cl^- for activity. The role of Ca^{2+} in the OEC is not clearly understood. In the absence of the 23 and 17 kDa extrinsic polypeptides, non-physiological concentrations of both Ca^{2+} and Cl^- are required for maximal activity.[1] Most studies suggest that there are two distinct binding sites present in higher plants, a low affinity site with $K_m \approx$ 1-2 mM and a more tightly bound site with $K_m \approx$ 50-100 μM.[2,3] Initial studies of Ca^{2+} focused on its replacement by other metals. Of the divalent cations, only Sr^{2+} is able to substitute for Ca^{2+} to give an active complex. Mn^{2+} itself can compete for the calcium binding site(s) in photoactivation and cation reconstitution studies, giving inactive complex.[4-6] The lanthanides (Ln^{3+}) are very effective competitive inhibitors of the Ca^{2+} site, despite the increased charge on Ln^{3+} (the K_i for La^{3+} was estimated to be 0.05 mM compared to the K_m for Ca^{2+} of 0.5]). Monovalent cations (Na^+, K^+ and Cs^+) weakly inhibit calcium activation of oxygen evolution ($K_i \approx$ 8 mM[7])

The affinity of Ca^{2+} for its site(s) depends both on the redox state of the Mn cluster and on the presence of extrinsic polypeptides. In the absence of the 23 kDa polypeptide, the Ca^{2+} responsible for OEC activation has a low affinity for its binding site and can be readily replaced by competing cations.[8-10] The binding affinity of calcium is highest in S_1 and lowest in S_3, [11] a finding consistent with the observation that illumination facilitates removal of Ca^{2+} in the presence of chelators.[9,12]

Replacement of Ca^{2+} and/or depletion of Ca^{2+} from the Ca site causes changes in the EPR signal attributed to the Mn cluster.[13,14] This, together with the effect of Ca^{2+} on activity and the dependence of Ca^{2+} affinity on S state, suggests that the Ca^{2+} site that is responsible for activation of H_2O oxidation is near the Mn cluster. If so, it may be detectable by x-ray absorption spectroscopy (XAS).

In general, XAS has proven to be a useful technique for probing the structure of the Mn site in the OEC. There is now agreement that EXAFS spectra of the OEC show Mn•••metal scattering peaks at approximately 2.7 Å and 3.3-3.6 Å.[15-18] The 2.7 Å interaction is attributed to Mn•••Mn scattering from $Mn(\mu\text{-}O)_2Mn$ units. However, interpretation of the more distant interaction has proven more controversial, with

P. Mathis (ed.), Photosynthesis: from Light to Biosphere, Vol. II, 241–246.

possible contributions coming from Ca and/or Mn scattering, in addition to likely contributions from C next nearest neighbors to the Mn [16,17,19,20]

In order to determine whether Ca^{2+} contributes to the 3.3 Å feature, we have measured EXAFS data for Sr^{2+}-, Dy^{3+}-, and La^{3+}-substituted PSII reaction center complexes. If the Ca^{2+} is bound in the vicinity (≤ 4 Å) of the Mn cluster, then changes in the EXAFS may be detected when the Ca^{2+} is removed or when different metals replace the Ca^{2+}.

2. Experimental Procedures

Highly active (1000-1400μg O_2/ mg chl/ hr) reaction center complexes were isolated from spinach as previously described.[21,22] and frozen at -80°C prior to treatment. Ca^{2+}-substituted were prepared by first concentrating the complex by centrifugation and then resuspending the pellet in 50 mM MES (pH 6) containing the appropriate metal salt ($SrCl_2$ or $LnCl_3$). The complex was then reconcentrated and the pellet was packed into a lucite EXAFS, with a portion of the pellet saved for EPR. A calcium depleted sample was prepared by subjecting the pellet to 2 cycles of washing in a Ca^{2+}-free buffer (50 mM MES). Activity assays were performed in 50 mM MES (pH 6) containing either 10 mM $CaCl_2$ or 10 mM NaCl using 0.31 mM 2,6-dichlorobenzoquinone as an acceptor.

Mn K edge x-ray absorption spectra were measured at Stanford Synchrotron Radiation Laboratory on beamline 7-3 using Si[220] crystals. All measurements were made at 10 K and spectra were measured using a Canberra 13-element solid state Ge detector to monitor Mn Kα fluorescence. Spectra were calibrated by simultaneously measuring the absorption spectrum of $KMnO_4$; the distinctive $KMnO_4$ preedge feature was assigned an energy of 6543.3 eV. EXAFS data reduction and data analysis followed standard procedures.[18]

3. Results

Sr^{2+} substitution results in decreased activity (≈ 40% of control activity can be reconstituted by Sr^{2+} incubation) that is reversed by incubation in $CaCl_2$.[4,11,23] Lanthanide substitution results in irreversible inactivation.[24] Additional evidence that Ca^{2+} was successfully replaced is provided by characteristic changes in the EPR spectra for the S_2 state following Sr^{2+} substitution. There is an increase in the g ≈ 4.1 EPR intensity relative to g ≈ 2 multiline signal and a decrease in multiline hyperfine splitting from ≈ 89 G (control) to 77 G (Sr^{2+} substituted), with altered intensity patterns. The lanthanide substituted samples have no activity and do not give an S_2 EPR signal. Lanthanide binding is additionally indicated by the broadening of the hyperfine features of the EPR signal of the stable tyrosine radical, Y_D.[25] when Dy^{3+} is substituted. Finally, there was no detectable Ca x-ray fluorescence in the substituted samples, indicating that substitution was > 90% complete.

Replacement of Ca^{2+} by another cation causes no significant change in the x-ray absorption near edge structure (XANES) region, as illustrated in Fig. 1. In contrast, calcium depletion results in a significant shift in the edge to lower energy. Edge fits[26] to this spectrum are consistent with the presence of ca. 15±10% Mn(II). Previous reports of shifts in the edge energy following Ca^{2+} depletion have attributed this to small conformational changes. For the present samples, however, the shift in edge energy appears to result from the production of Mn(II). As shown in Fig. 2, this Mn, which does not contribute a detectable 6-line EPR signal, can be removed by treatment with chelex. This suggests that, at least for the present samples, Ca^{2+} depletion destabilizes the Mn cluster, leading eventually to decomposition, with loss of Mn(II) to solution. The Mn(II) that is released can bind at a variety of site (including, perhaps, the Ca^{2+} site). The low symmetry of these binding sites renders the Mn(II) EPR non-detectable.[27] Based on the XANES data (Figs. 1 and 2) there is no indication that replacement of Ca^{2+} with another cation causes any significant change in the Mn cluster structure.

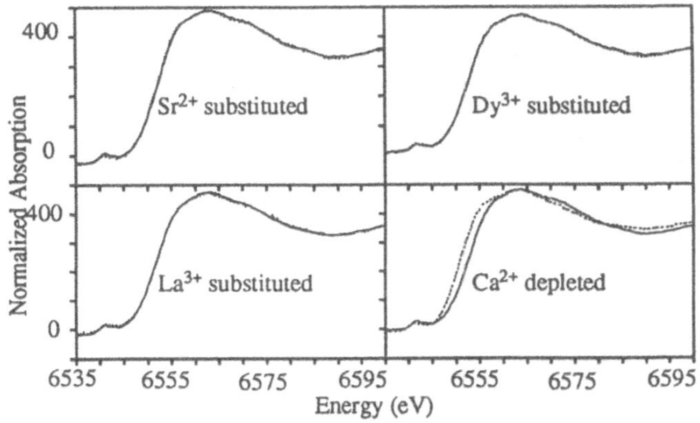

Figure 1. XANES spectra for Ca^{2+} modified samples. Dashed line = modified sample, solid = control. Only Ca^{2+} *depletion* gives a significant change in the edge energy.

A more sensitive probe of cluster structure is provided by the EXAFS spectra (Fourier transforms are shown in Fig. 3). Once again, there is no significant change in structure following Ca^{2+} substitution. In particular, we find no significant change in the ≈ 3.3 Å peak in any of the substituted samples.

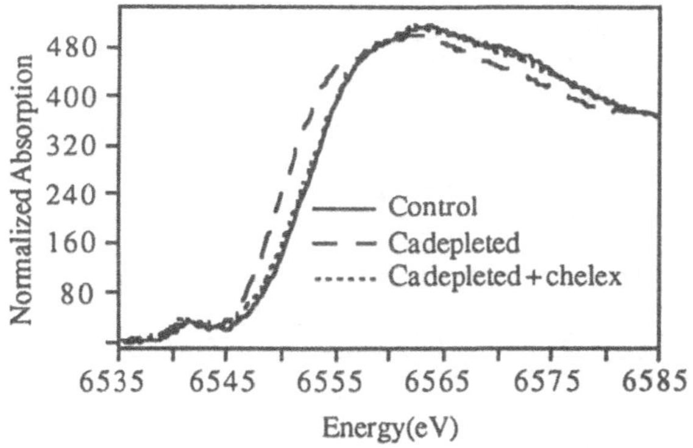

Figure 2. XANES spectra for Ca^{2+} depleted samples, with and without the addition of chelex to remove labile Mn(II).

It is important to note, however, that the absence of a signal in EXAFS does not indicate the absence of a scatterer. There are numerous examples of atoms within the coordination sphere of an absorber that are not detectable by EXAFS due to disorder. In particular, Fe–Fe interaction in dinuclear Fe proteins is often not detectable, particularly when only a single atom bridges between the absorber and the scatterer. Thus, the absence of a significant change in the Fourier transform does *not* mean that there the Ca^{2+} is > 3.5 Å from the Mn but does mean that the Mn–Ca interaction, if present, must be weak.

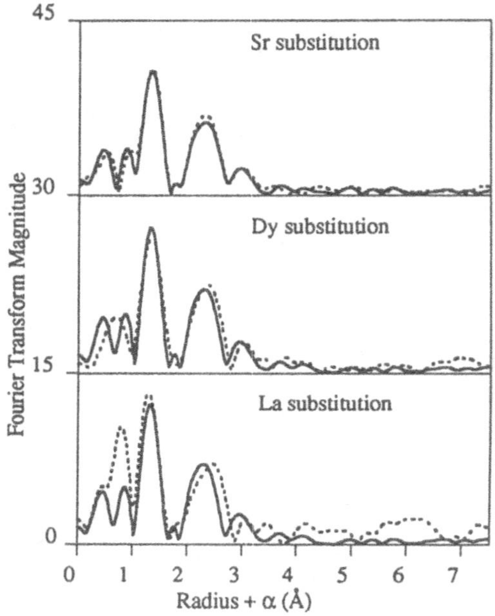

Figure 3. FT of EXAFS spectra for Ca^{2+} substituted samples. For each set, dashed line is modified sample, solid line is control.

An important question is the extent to which Mn•••M (M=Ca, Dy, Sr) *should* be distinguishable. Although Ca, Dy, and Sr have very different EXAFS scattering envelopes, the limited *k* range over which protein EXAFS can be measured (k_{max}≈11.8 $Å^{-1}$) limits ones ability to distinguish between the amplitude envelopes for these scatterers. Fortunately, the phase shift (i.e., the apparent bond length) is quite sensitive to the nature of the scatterer. Thus, if a scattering peak that is due to Sr is fit with Ca (or Mn) the apparent distance will be in error by ca. 0.1 Å. A comparable error (with the opposite sign) occurs if a Mn•••Dy peak is fit with Mn-Ca (or Mn-Mn). We observe no change in any of the structural parameters for the OEC when Ca^{2+} is replaced, and in particular, there is no change in the apparent Mn•••Mn distance at 3.3 Å, as summarized in Table 1. This strongly suggests that there is no significant Ca/Sr/Dy contribution to the OEC EXAFS in these samples.

Table 1. Best fit results for a three shell model of the EXAFS for control and Ca^{2+} substituted OEC.

	Mn-O/N			Mn•••Mn			Mn•••M		
Sample	R(Å)	CN	σ^2	R(Å)	CN	σ^2	R(Å)	CN	σ^2
Control	1.855 (1)	3	7 (1)	2.726 (2)	1	1 (1)	3.25 (1)	1	6 (4)
Sr	1.852 (8)	3	7 (1)	2.712 (1)	1	2 (1)	3.26 (3)	1	9 (0)
Dy	1.864 (1)	3	5 (1)	2.734 (1)	1	2 (1)	3.26 (2)	0.75	9 (5)

Although there is no change in the ≈3.3 Å Mn–Mn distance, there *is* a change in the ≈2.7 Å Mn-Mn distance. The variation in the ≈2.7 Å Mn-Mn distance between native (Ca^{2+}) and Sr^{2+}/Dy^{3+} substituted samples are given in Table 2 for replicate samples. Based on comparison of the data for independent samples measured at different times, the precision of the Mn-Mn distance is ca. 0.004

Table 2. Change in ≈ 2.7 Å Mn-Mn distance following Ca^{2+} substitution.

Sample	$\Delta R_{Mn•••Mn}$
Sr #1	-0.015
Sr #2	-0.012
Dy #1	+0.011
Dy #2	+0.013

Å. Thus, although the accuracy with which the precise Mn-Mn distance can be determined is on ca. 0.02 Å, the ability to detect changes in Mn-Mn distance is much better. Although the changes in Table 2 are small, they are substantially larger than the precision of the measurements and are thus likely to represent real structural changes.

4. Discussion

There is substantial variability in published protocols for calcium depletion. Most of the reported protocols utilize some combination of chelator, pH treatment, light, and high salt concentrations. In this study, we sought to isolate the structural perturbations due to cation substitution from those which might arise from chelator binding to the Mn cluster,[28] and, therefore, employed a simple wash/incubation protocol for metal substitution. EPR, activity and elemental analysis data all point to efficient Ca^{2+} replacement in these samples using this procedure.

On the basis of analogous Sr^{2+} substitution experiments, Yachandra et al. have assigned the 3.3 Å interaction to Mn•••Ca scattering.[29] In this work, replacement of Ca^{2+} by Sr^{2+}-substitution gave a significant increase in the amplitude of the third peak in the Fourier transform (i.e. the 3.3 Å interaction); a mixture of Sr and Mn produced the best fit for this interaction. Similarly, Nugent and co-workers have reported that outer shell interaction (at 3.6 Å in their data) decreases in amplitude when Ca^{2+} is replaced by Na^{+}. These results suggest that at least some of the outer shell scattering is due to a Mn•••Ca interaction.

The absence of significant changes in either the EXAFS or the XANES in our samples when Ca^{2+} is replaced demonstrates that any Mn•••Ca interaction in these samples is too weak to be detected by EXAFS. One possible explanation for the differences between our results and those found by other workers is that the 17 and 23 kDa extrinsic polypeptides, which are absent in our samples, provide bridging ligands which make a Mn•••Ca interaction sufficiently rigid to be EXAFS detectable. However, this possibility is difficult to reconcile with the similarity between our EPR spectra and those reported for other Ca-substituted samples. Our results thus suggest that the heavy scatterer(s) at ≈3.3 Å are most likely Mn atoms. This is consistent with the disappearance of the 3.3 Å feature following reduction of Mn by hydroquinone in Ca^{2+}-saturated samples.[18]

The observation of a small but reproducible change in the 2.7 Å interaction following Ca^{2+} substitution demonstrates that Ca^{2+} replacement does cause a minor structural perturbation of the Mn cluster. The sign of the change is not correlated with the size of the ion in the Ca site. However, there is an inverse correlation between the Lewis acidity of the metals used for Ca^{2+}-replacement and the average Mn•••Mn distance observed by EXAFS. One mechanism that could account for these changes is hydrogen bonding between a water bound to the Ca^{2+} and a $Mn_2(\mu\text{-}O)_2$ unit. When Ca^{2+} is replaced with Sr^{2+}, the water will become less acidic and thus will form a weaker hydrogen bond to the oxo bridge. Conversely, when Ca^{2+} is replaced with Dy^{3+}, the water will be more acidic and thus will form a stronger hydrogen bond. This model, while speculative, is consistent with the known effects of Ca replacement. Experiments are in progress to test the validity of this structural model.

5. Acknowledgments

Supported in part by grant USDA-NRICGP-92-37306-7662 to CFY and grant GM-45205 from the US NIH to JPH. SSRL and NSLS are supported by the US Department of Energy with additional support from the NIH Research Resource program.

6. References

1. Ghanotakis, D. F.; Babcock, G. T.; Yocum, C. F. *FEBS Lett.* **1985**, *192*, 1-3.
2. Cammarata, K. V.; Cheniae, G. M. *Plant Physiology* **1987**, *84*, 587-589.
3. Kalosaka, K.; Beck, W. F.; Brudvig, G.; Cheniae, G. *Current Research in Photosynthesis* **1990**, *1*, 721-724.
4. Ghanotakis, D. F.; Babcock, G. T.; Yocum, C. F. *FEBS Lett.* **1984**, *167*, 127-130.
5. Ono, T.-A.; Inoue, Y. *Biochim. Biophys. Acta* **1983**, *723*, 191-201.
6. Tamura, N.; Cheniae, G. M. *FEBS Lett.* **1986**, *200*, 231-236.
7. Waggoner, C. M. Ph.D. Thesis, Univsersity of Michigan, 1989.
8. Ghanotakis, D. F.; Topper, J. N.; Babcock, G. T.; Yocum, C. F. *FEBS Lett.* **1984**, *170*, 169.
9. Boussac, A.; Maison-Peteri, B.; Etienne, A.-L.; Vernotte, C. *Biochim. Biophys. Acta* **1985**, *808*, 231-234.
10. Miyao, M.; Murata, N. *FEBS Lett.* **1984**, *168*, 118-120.
11. Boussac, A.; Rutherford, A. W. *FEBS Lett.* **1988**, *236*, 432-436.
12. Miyao, M.; Murata, N. *Photosynthesis Research* **1986**, *10*, 489-496.
13. Boussac, A.; Zimmerman, J.-L.; Rutherford, A. W. *Biochemistry* **1989**, *28*, 8984-8989.
14. Ono, T.-A.; Inoue, Y. *Biochim. Biophys. Acta* **1990**, *1015*, 373-377.
15. George, G. N.; Prince, R. C.; Cramer, S. P. *Science* **1989**, *243*, 789-791.
16. MacLachlan, D. J.; Hallahan, B. J.; Ruffle, S. V.; Nugent, J. H. A.; Evans, M. C. W.; Strange, R. W.; Hasnain, S. S. *Biochem. J.* **1992**, *285*, 569-576.
17. DeRose, V. J.; Mukerji, I.; Latimer, M. J.; Yachandra, V. K.; Sauer, K.; Klein, M. P. *J. Am. Chem. Soc.* **1994**, *116*, 5239-5249.
18. Riggs-Gelasco, P.; Mei, R.; Yocum, C. F.; Penner-Hahn, J. E. *J. Am. Chem. Soc.* **1995**, submitted.
19. Penner-Hahn, J. E.; Fronko, R. H.; Pecoraro, V. L.; Yocum, C. F.; Betts, S. D.; Bowlby, N. R. *J. Am. Chem. Soc.* **1990**, ***112***, 2549-2557.
20. MacLachlan, D. J.; Nugent, J. H. A.; Bratt, P. J.; Evans, M. C. W. *Biochim. Biophys. Acta* **1994**, *1186*, 186-200.
21. Ghanotakis, D. F.; Demetriou, D. M.; Yocum, C. F. *Biochim. Biophys. Acta* **1987**, *891*, 15-21.
22. Berthold, D. A.; Babcock, G. T.; Yocum, C. F. *FEBS Lett.* **1981**, *134*, 231-234.
23. Sivaraja, M.; Tso, J.; Dismukes, G. C. *Biochemistry* **1989**, *28*, 9459-9464.
24. Bakou, A., Ghanotakis, Demetrios F. *Biochim. Biophys. Acta* **1993**, *1141*, 303-308.
25. Bakou, A.; Buser, C.; Dandulakis, G.; Brudvig, G.; Ghanotakis, D. *Biochim. Biophys. Acta* **1992**, *1099*, 131-136.
26. Riggs, P. J.; Mei, R.; Yocum, C. F.; Penner-Hahn, J. E. *J. Am. Chem. Soc.* **1992**, *114*, 10650-51.
27. Yocum, C. F.; Yerkes, C. T.; Blankenship, R. E.; Sharp, R. R.; Babcock, G. T. *Proc. Natl. Acad. Sci. U. S. A* **1981**, *78*, 7507-11.
28. Boussac, A.; Zimmerman, J.-L.; Rutherford, A. W. *FEBS Lett.* **1990**, *277*, 69-74.
29. Yachandra, V. K.; DeRose, V. J.; Latimer, M. J.; Mukerji, I.; Sauer, K.; Klein, M. P. *Science* **1993**, *260*, 675-679.

RECONSTITUTION OF THE WATER-OXIDIZING COMPLEX IN Mn-DEPLETED PHOTOSYSTEM II COMPLEXES BY USING SYNTHETIC BINUCLEAR Mn-COMPLEXES

S.I.Allakhverdiev[1], M.S.Karacan[2], G.Somer[2], N.Karacan[2], E.M.Khan[3], S.Y.Rane[3], S.Padhye[3], V.V.Klimov[1], G.Renger[4]
[1]Inst.Soil Sci.& Photosynth.RAS,Pushchino, Russia; [2]Depart.Chem.,Univ.Gazi,Ankara,Turkey; [3]Depart.Chem.,Univ. Poona,India. [4]Inst.Biophys. & Phys.Chem.,TU,Berlin,Germany

1. Introduction

Although significal progress has been achieved in understanding the photosynthetic water oxidation, especially with respect to the kinetic pattern /1,2/, key mechanistic questions still remain to be answered. Likewise, the structure of the water oxidase and the nature of its protein matrix are not yet resolved /1/. With respect to the latter problem, it now seems clear that a functionally competent water-oxidizing complex (WOC) contain 4 Mn /2-5/. In thylakoids and photosystem II (PS II) membrane fragments deprived of their Mn, the oxygen evolution capacity can be restored by a process called photoactivation /1,2/. Recently, various Mn complexes have been synthesized and analyzed as possible models for the Mn cluster in the WOC /6-11/ and reconstitution of oxygen evolution by using some of them has been demonstrated /9-11/. Many of these complexes contain binuclear or tetranuclear Mn centers. It therefore appeared interesting to use synthetic binuclear Mn complexes for reconstitution experiments on the oxygen evolution capacity of PS II preparations lacking a functionally competent WOC. The present study describes the reconstitution of the WOC with the binuclear Mn complexes $Mn_2(LwOx)_4(OAc)_2$ and $Mn_2O(LwOx)_2(OAc)_2(H_2O)_2$, symbolized by M-2 and M-3, respectively, where LwOx is the ligand 2-hydroxy-1,4-naphtoquinone monoxime and Ac is acetate. It is shown that these compounds are more efficient than $MnCl_2$ in reconstituting O_2 evolution.

2. Materials and Methods

Subchloroplast PS II membrane fragments (DT-20) isolated as described earlier /3-5/. These preparations, refferred to as DT-20, exhibited O_2 evolution rates of 250 µmol/ (mg of Chl)h under saturating light and in the presence of 0.3 mM $K_3[Fe(CN)_6]$ as electron acceptor. Removal of both Mn and the three extrinsic regulatory proteins with apparent m.m. of 33,24 and 18 kDa, was carried out by N,N,N',N'-tetramethyl-

P. Mathis (ed.), Photosynthesis: from Light to Biosphere, Vol. II, 247–250.

ethylendiamine (TEMED) and 0.5M $MgCl_2$ /12/. The rate of oxygen evolution was measured with a Clark type electrode. The assay mixture contained DT-20 preparations suspended in a buffer solution of 25 mM MES-NaOH(pH 6.5)/10 mM NaCl/5 mM $CaCl_2$/300 mM sucrose. The binuclear complexes used in this study was prepared according to the method /13/.

3. Results and Discussion

In TEMED/$MgCl_2$-treated DT-20 preparations the oxygen evolution capacity is virtually completely eliminated as shown in the bottom trace of Fig.1. A more pronounced effect was obtained in the presence of the binuclear Mn complexes M-2 and M-3. A comparison of the traces in Fig.1 reveals that the initial rate of the oxygen evolution in the presence of 2 molecules of M-3 per PS II, i.e., 4 Mn/PS II, is 5-6 times higher than that of samples reconstituted with 10 $MnCl_2$/PS II. The rate can be further enhanced if in addition the 33 kDa protein is used for the reconstitution of the O_2 evolution activity. The effect of compound M-3 and the 33 kDa protein on the reconstitution is presented in Fig.2. The date reveal two interesting phenomena: (i) the maximum effect (in terms of the Mn stoichiometry) is already achieved at a ratio of 4 Mn/PS II, i.e. a further increase of the M-3 concentration does not lead to higher O_2 evolution rates in the reconstituted samples, and (ii) addition of the 33 kDa protein leads to a marked stimulatory effect at all concentrations of M-3. Under optimum conditions more than 80% of the control activity can be restored in the presence of M-3 and the 33 kDa protein. Another cofactor of a functionally competent WOC is Ca^{2+} /1,2,9/. Ca^{2+} is not required for the photoligation of the dimeric Mn complex M-3. Normally, the assay medium for the measuring the oxygen evolution contained Ca^{2+} at concentration of 5 mM. This might be sufficient to satisfy of a Ca^{2+} demand in O_2 evolution under these conditions. In order to check for this possibility, the same experiment was performed with the M-3-reconstituted sample in the absence of Ca^{2+}. No significant differences were observed, as shown by the dotted curve in Fig.1. Accordingly, a comparatively high Ca^{2+} concentration is required for O_2 evolution in samples photoactivated with M-3. However, it has to be empahasized that a special treatment for Ca^{2+} depletion was not applied. Therefore, a small amount of Ca^{2+} bound to the PS II in the preparation could satisfy a low Ca^{2+} demand.

The results of the present study reveal that the binuclear Mn complexes are powerful substances for establishing inside the protein matrix a Mn cluster that is fully competent for photosynthetic water oxidation. The protein matrix also seems to be surprisingly flexible in accepting comparatively bulky complexes like M-2 and M-3 as Mn source for photoligation and establishing a functional competent water oxidase. These compounds are markedly more efficient than Mn^{2+} in photoactivation of the oxygen evolution capacity (Fig.1). Therefore, it is inferred that the complex M-3 (also M-2) is taken up by PS II as a preformed binuclear cluster. The aromatic ligand probably favors the transport into the neighborhood of the redox-active tyrosine Y_Z and the binding domain of

FIGURE 1. Poloarographic traces of oxygen evolution in DT-20 preparations deprived of their Mn and extrinsic polypeptides and reactivated under different conditions. Trace 1, Mn-depleted DT-20 sample; 2, Mn-depleted DT-20 sample reconstituted with 2 μM $MnCl_2$ (corresponding to 10 Mn/PS II); 3, Mn-depleted DT-20 sample reconstituted with 0.4 μM complex M-3 (corresponding to 4 Mn/PS II); 3', the same as trace 3 but without $CaCl_2$ in the assay medium; 4, Mn-depleted DT-20 sample reconstituted with 0.4 μM M-3 and in the presence of the 33 kDa protein; 5, untreated control DT-20 sample. Chl, 20 μg/ml.

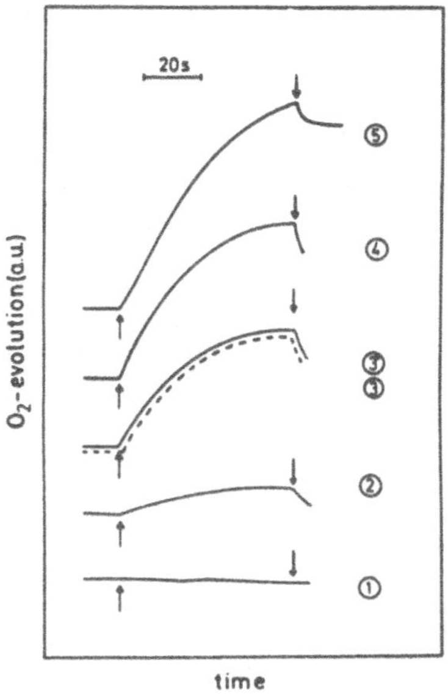

the WOC. This idea is supported by previous findings /14/. It is inferred that the ligand LwOx facilitates the interaction with the PS II donor side. The aromatic ligands (LwOx) are probably stripped off only inside the PS II comlex and substituted by amino acid residues of the matrix. It seems likely that the binuclear core is not disrupted and becomes incorporated as an entity. Furthemore, the Mn(III) valence state of the Mn in complex M-3 in probably also in support of a suitable ligation that eventually leads to a functionally competent WOC. In this respect it is interesting to note that in the intact WOC of the dark-relaxed redox state S_1 one of the two binuclear groups is assumed to attain the oxidation state Mn(III)Mn(III) /15/. The titration curve presented in Fig.2 indicates that a maximum level of restoration is achieved at 4 Mn per PS II. This finding provides an independent line of evidence for a tetranuclear Mn cluster forming the core of an intact WOC. In addition, the data of Fig.2 also indicate that the extrinsic 33 kDa protein is essential in achieving maximum activity of O_2 evolution.

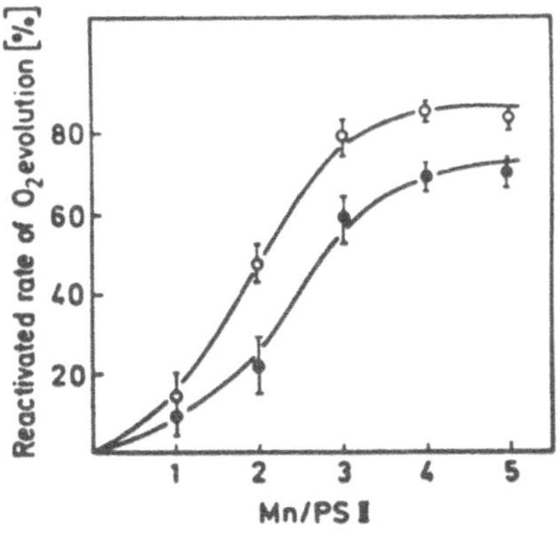

FIGURE 2. Oxygen evolution rate of Mn-depleted DT-20 samples as a function of the concentration of complex M-3 in the reconstitution medium in the absence (•) and presence (o) of the 33 kDa protein.

References

1. Debus,R.J. (1992) Biochim.Biophys.Acta 1112, 269-353
2. Renger, G. (1993) Photosynth.Res. 38, 229-247
3. Klimov,V.V., Allakhverdiev,S.I., Shuvalov,V.A. and Krasnovsky,A. (1982) FEBS Lett. 148, 307-312
4. Allakhverdiev,S.I., Shafiev,M.A. and Klimov,V.V. (1986) Photobiochem. Photobiophys. 12, 61-65
5. Allakhverdiev,S.I. and Klimov,V.V. (1992) Z.Naturforsch. 47C, 57-62
6. Vincent,J.B. and Christou,G. (1989) Adv.Inorg.Chem. 33, 197-257
7. Pecoraro, V.L. (1992) in Manganese Redox Enzymes (Pecoraro,V.L., ed.) pp. 197-231, VCH Publishers, New York
8. Armstrong,W.H. (1992) in Manganese Redox Enzymes (Pecoraro, V.L.,ed.) pp.261-286, VCN Publishers, New York
9. Klimov,V.V., Ananyev,G.M., Allakhverdiev,S.I. Zharmukhamedov,S.K., Mulay,M., Hedge,U. and Padhye,S. (1990) in Current Research Photosynthesis (Baltscheffsky, M., ed.) v.1, pp.247-254, Kluwer, Dordrecht
10. Allakhverdiev,S.I., Karacan,M.S., Somer,G., Karacan,N., Khan,E.M., Rane,S.Y., Padhye,S., Klimov,V.V. and Renger,G. (1994) Biochemistry 33, 12210-12214
11. Allakhverdiev,S.I., Karacan,M.S., Somer,G., Karacan,N., Khan,E.M., Rane,S.Y., Padhye,S., Klimov,V.V. and Renger,G. (1994) Z.Naturforsch. 49c, 587-592
12. Ananyev,G.M., Wydrzynski,T., Renger,G. and Klimov,V.V. (1992) Biochim. Biophys.Acta 1100, 303-311
13. Rane,S.Y., Dhavale,D.D., Mulay,M.P. and Khan,E.M. (1990) Spectrochim. Acta A46, 113
14. Erixon,K. and Renger,G. (1973) Biochim.Biophys.Acta 333, 95-106
15. Yachandra,V.K., de Rose,V.J., Latimer,M.J., Mukerji,I., Sauer,K. and Klein,M.P. (1993) Science 260, 675-679

ADVANCED EXAFS STUDIES OF THE S_1-STATE MANGANESE CLUSTER IN PLANT PHOTOSYSTEM II

M. Kusunoki[1], T. Takano[1], T. Ono[2], T. Noguchi[2], Y. Yamaguchi[3], H. Oyanagi[3] & Y. Inoue[2]: [1]*Dep. Phys., Meiji Univ., Kawasaki 214, Japan;* [2]*Photosynthesis Science , RIKEN, Wako 351-01, Japan;* [3]*Electrotechnical Lab., Tsukuba 305, Japan.*

1. Introduction

Plant photosystem II (PSII) is a light-driven enzyme to oxidize water to produce molecular oxygen's and physiological protons inside the thylakoide membranes. In a cycle of this enzyme reaction, four electron transfers through a reaction center P680 take place one by one to result in five redox states of the oxygen-evolving complex (OEC), called "Kok's S_i-states (i=0-4)", in which S_1 is the only dark stable state and S_4 a transient state before emitting O_2 during a spontaneous $S_3 \rightarrow S_0$ turnover [1]. The catalytic center of the OEC is believed to be a tetranuclear Mn cluster incorporating cofactors of 1 Ca^{2+} and some CL^- ions [2]. Hitherto, two Mn-selective spectroscopies of low-temperature EPR and X-ray absorption fine structure (XAFS) have been utilized to study the molecular and electronic structures of Mn [3]. A celebrating 'multiline' EPR signal due to S_2 [4] was clearly shown to arise from the superexchange-coupled Mn tetramers specified by four non-zero effective hyperfine coupling constants [5,6]. More direct structural information on the Mn complex has been provided by fluorescence EXAFS (extended x.a.f.s.) studies on any treated sample of PSII membranes, prepared in S_1 [7-10], S_2[7,8], Ca^{2+}/Sr^{2+}-substituted S_1 [11], Ca^{2+}-depleted S_1 [12] states and so on.

In this report, we present a high-resolution EXAFS spectrum for the extensively studied S_1 state of PSII, which was abstracted from our own raw XAFS data using an improved method of EXAFS data analysis [13]. This method can generate a non-wavy smooth background spectrum and a properly normalized EXAFS spectrum via stepwise least-square procedures. Furthermore, in order to determine many highly correlated EXAFS parameters pertinent to a heterogeneous Mn-tetramer and estimate upper and lower standard deviations for them, we will introduce the statistical χ^2 function [13] and abstract well-characterized structural information on Mn in PSII.

2. Materials and Methods

BBY type PSII membranes capable of O_2 evolution (600-800 mmoles O_2/mg Chl/ml) were prepared from spinach, and relaxed in darkness for 6 hr at 0 °C. The dark adapted membranes were washed once with 400 mM sucrose, 20 mM NaCl, 0.5 mM EDTA.2Na and 40 mM Mes-NaOH (pH 6.5), and resuspended in the same buffer medium lacking EDTA. After centrifugation at 35,000 × g for 35 min, the resulting pellet was applied onto

P. Mathis (ed.), Photosynthesis: from Light to Biosphere, Vol. II, 251–254.

membrane filter in a Cu holder of 0.5 mm in thickness. Manganese K-edge X-ray absorption spectra have been measured for condensed spinach PSII membranes in the S_1 state at 30 K at the Tsukuba Photon Factory BL-4C station with use of a single-element Si(Li) solid state (SSD) detector placed 1.2 cm apart from the sample.[14]

Recently, we have shown that, in the case of fluorescence EXAFS, the absorbance μ defined by I_D/I_F with I_D and I_F being an SSD count and a front-chamber count, respectively can be related with a theoretical EXAFS spectrum $\chi(k)$ by $\mu=\mu_{BG}(k) + \Delta\mu(k)\{1+\chi(k)\}$, in which $\Delta\mu(k)$, an absorbance contribution from central atoms when which are isolated from the surroundings, depends on the wave number k of a photoelectron as $\Delta\mu(k) \approx \Delta\mu_0 \exp(2\sigma_0^2 k^2)$, but this dependence is negligible since σ_0^2 is typically of order $1\text{x}10^{-4}$ Å^2, and $\mu_{BG}(k)$ is a k-dependent background spectrum [13]. This finding lead to a sufficiently accurate definition of an experimental EXAFS spectrum, $y(k)\equiv\{\mu-\mu_{BG}(k) - \Delta\mu_0\}/\Delta\mu_0$, to be compared with $\chi(k)$. We have used a new powerful method based on stepwise least-square procedures [13] to determine not only $\mu_{BG}(k)$ as a linear combination of three kinds of "non-integer power functions" but also $\Delta\mu_0$. Significantly, this method is free from uncertainties accompanied by the conventional methods using cubic splines, a high-order polynomial and a moving-smooth [15]. Furthermore, for curve-fitting and error estimation of EXAFS parameters, we have used the statistical χ^2 function defined by $\chi^2=\sum_{i=1}^{N}\{y_i-\chi(k_i)\}^2/\sigma_i^2$ with dispersion $\sigma_i^2=\overline{\delta y_i^2}$ at the i-th data point (i=1-N)[16], rather than a widely-used fit-index using a standard deviation σ_i incorrectly assumed to be proportional to $1/k_i^n$ (n=3 for k^3 weighted data). Here, $\chi(k_i)$ is the theoretical EXAFS spectrum [17] involving N_{shell} sets of EXAFS parameters, i.e. radius R_j, coordination number N_j, Debye-Waller factor σ_j^2, a shell-dependent part of the energy threshold ΔE_{0j}; $j=1-N_{shell}$. To get a good initial guess for the non-linear curve-fitting, we calculate a Fourier transform of $k^n y(k)$, designated $\Phi_n(R)$, because whose absolute value $|\Phi_n(R)|$ represents a 'radial structure function'. In general, there are a number of local minima on the χ^2 hypersurface in the p-dimensional parameter space. At the true minimal point $\mathbf{a}\equiv(a_1, a_2,\cdots, a_p)$, the minimal χ^2 must satisfy an equation, $\chi^2(\mathbf{a})/(N-p) \approx 1$[16]. This statistical principle can be generalized to define the standard deviations for these parameters. When there is a strong correlation between two parameters, say, a_1 and a_2, a contour equation, $\chi^2(a_1+\Delta a_1, a_2+\Delta a_2) - \chi^2(a_1, a_2) \approx 2$, can be used to define the upper and lower standard deviations of a_1 and a_2, by $\Delta_+=\max(\Delta a_i)$ and $\Delta_-=-\min(\Delta a_i)$, respectively.

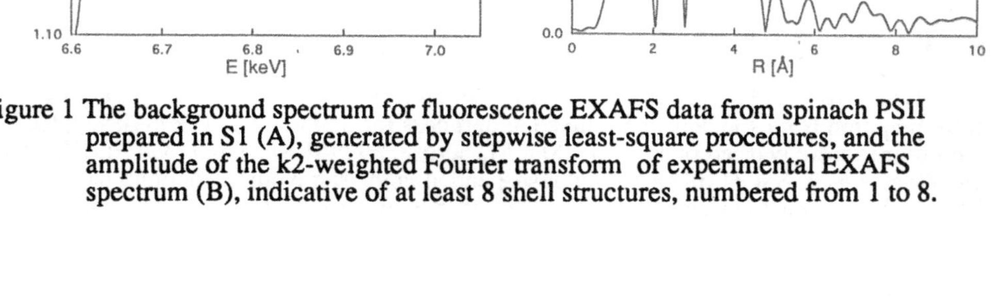

Figure 1 The background spectrum for fluorescence EXAFS data from spinach PSII prepared in S1 (A), generated by stepwise least-square procedures, and the amplitude of the k2-weighted Fourier transform of experimental EXAFS spectrum (B), indicative of at least 8 shell structures, numbered from 1 to 8.

Table 1 The best curve-fitting results for EXAFS from PSII membranes in S_1, at the statistical minimum of $\chi^2/(N-p) \approx 1.08910$. The upper and lower standard deviations (Δ_+ and Δ_-, respectively) of 32 EXAFS parameters were calculated as described in text in the pair-correlation approximation in each shell.

Shell I(atom)	$N_i-\Delta_-/N_i/N_i+\Delta_+$	$R_i-\Delta_-/R_i/R_i+\Delta_+$ [Å]	$\sigma_i^2-\Delta_-/\sigma_i^2/\sigma_i^2+\Delta_+\times10^3$ [Å^2]	$\Delta E_{0i}-\Delta_-/\Delta E_{0i}/\Delta E_{0i}+\Delta_+$[eV]
1(O)	0.59[a] / 0.63 / 0.71	1.798[b] / 1.805 / 1.815	1.00 / 1.00/ 2.55	-20.0 / -19.8 / -19.3
2(O)	6.92 / 7.23 / 8.71	2.253[b] / 2.259 / 2.266	25.7 / 26.5 / 31.8	-20.0 / -19.8 / -19.2
3(N)	0.23[a]/ 0.34 / 1.20[c]	2.320[b]/ 2.381 / 2.511	1.00 / 3.95 / 35.0	-20.0 / -15.9 / -1.19
4(Mn)	1.07 / 1.19 / 1.48	2.741 / 2.750 / 2.761	4.23 / 5.54 / 8.64	-2.97 / -2.31 / -1.50
5(Mn)	0.57[a] / 0.68 / 0.95	3.417 / 3.424 / 3.440	1.00 / 1.12 / 5.74	-2.77 / -2.31 / -1.73
6(Ca)	0.66[a] / 0.78 / 0.99	3.693 / 3.715 / 3.753	1.00 / 1.00 / 4.52	-7.40 / -5.98 / -3.46
7(Mn)	0.56[a] / 0.78 / .1.07	4.395 / 4.415 / 4.436	1.00 / 1.00 / 4.96	-2.72 / -2.31 / -1.80
8(Mn)	0.48[a] / 0.73 / 1.08	4.656 / 4.680 / 4.706	1.00 / 1.00 / 6.10	-2.70 / -2.31 / -1.81

([a]The restricted value at σ_i^2=0.001[Å^2], [b]The restricted value at ΔE_{0i}= -20[ev], [c]The restricted value at σ_i^2=0.035 [Å^2])

3. Results, Discussions and Conclusions

In Fig.1A, we present a resultant EXAFS background spectrum, $\mu_{BG}(k)/\Delta\mu_0+1$, for the Mn XAFS from spinach PSII prepared in S_1. We would emphasize that this background was routinely determined without allowance. A difference between the raw XAFS (a wavy line and EXAFS background (a smooth line) spectra yields the experimental EXAFS spectrum $\{y_i$ or $y(k_i)$; $i=1$–$N\}$. The k^2-weighted radial structure function $|\Phi_2(R')|$, plotted in Fig.1B, is clearly seen to consist of at least 6 peaks above a noise level. The 1st, 2nd and 3rd peaks at R' ~ 1.4 Å, 2.1 Å and 3.0 Å, respectively, look to coincide with past observations[7-10]. But, the present $|\Phi_2(R')|$ significantly reveals some new features: 1) the 1st broad peak around R' ~ 1.4 Å looks to contain 3 shells, numbered as 1,2,3 from left to right, in which the 3rd one is new; 2) the 3rd peak at R' ~ 3.0 Å is more intense than the past observations[7-10]; 3) the 4th, 5th and 6th peaks at R' ~ 3.5 Å, 4.0 Å and 4.5 Å, respectively, have never been observed above noise level. In particular, the 3rd feature is extremely important because it must arise from heavy atom scatterers, such as Mn and, less probably, Ca. Based on these features, we adopted a reasonable 8-shell model (see Table 1 to simulate the experimental EXAFS spectrum y(k). After many trial-and-error's for curve fitting, we have found the best local minimum with $\chi^2/(N-p) \approx 1.08910$, which is significantly close to 1!! (N=166, and p=28 because the ΔE_0 values for the same atoms were assumed to be equal). In Fig.2, we compare the simulated EXAFS spectrum (a dashed line) with the experimental one (a solid line). And, in Table 1, we have summarized the resultant EXAFS parameters. It should be noted that, since any Fourier-filtering technique was not used here, our method is free from a severe restriction on p, i.e. $p < N_{idp}$,

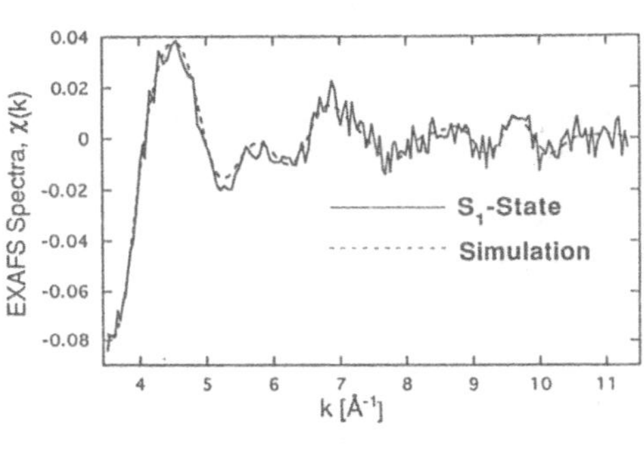

Figure 2. An EXAFS spectrum of a spinach PSII preparation in S_1-state (a solid line) and a simulated EXAFS spectrum (a dashed line).

where N_{idp} is the number of independent data points given by $2(R_{max}-R_{min})(k_{max}-k_{min})/\pi$ [15] ($\approx$ 21 for full filtering). Furthermore, we have estimated the upper and lower standard deviations for these EXAFS parameters, by taking into account the strong correlations between N_i and σ_i^2 and between R_i and ΔE_{0i} for all the shells (i=1–8), which are summarized in Table 1 (Note that we imposed two restrictions, $\sigma_i^2 > 0.001$ Å^2 and $\Delta E_{0i} >$ -20 eV). On the other hand, a 7-shell model, in which the 1st peak in $|\Phi_2(R')|$ is assumed to contain 2 shells of (O,N)-ligands, lead to a similar minimum point with $\chi^2/(N-p') \approx$ 1.18955, indicating that the 8-shell model is much more probable than the 7-shell's. Thus, we find: (1) a new shell of scatterers at R≈2.38 Å must exist, representing a few (N,O)-ligands (His, Tyr and H_2O are probable candidates); (2) a dominant 2nd shell of scatterers at R≈2.26 Å are quite diffusive, suggesting a heterogeneous Mn tetramer, and much more distant than [7-8]; (3) a minor 1st shell of scatterers are located at the same distance as [7-8], representative of di-μ-oxo bridges in Mn model compounds; (4) a prominent 4th shell of Mn scatterers at R≈2.75 Å have the coordination number 1.07-1.48, in agreement with [7-10]; (5) five new shells of heavy-atom scatterers at more distant R≈3.4 Å, 3.7 Å, 4.4 Å and 4.7 Å were observed with an average bond number of 1-2 per a Mn tetramer, suggesting a distorted Mn tetramer possibly incorporating Ca^{2+} ion.

Acknowledgment Thanks are due to the staff of the Photon Factory for their kind help during our experiment.

References

1 Joliot, P. and Kok, B. (1975) in Bioenergetics of Photosynthesis: Oxygen Evolution in Photosynthesis (Govindjee, ed.) pp. 387-412, Academic Press, New York
2 Debus, R.J. (1992) Biochim. Biophys. Acta, 1102, 269-352
3 Sauer, K., Yachandra, V.K., Britt, R.D. and Klein, M.P. (1992) in Manganese Redox Enzymes (Pecoraro, V.L., ed.) pp. 141-175, VCH, New York
4 Dismukes, G.C. and Sidler, Y. (1981) Proc. Natl. Acad. Sci. USA, 78, 274-278
5 Kusunoki, M. (1992) Chem. Phys. Lett. 197, 108-116
6 Bonvoisin, J., Blondin, G., Girerd, J.-J. and Zimmermann, J.-L. (1992) Biophys. J., 61, 1076-1086
7 DeRose, V.J., Mukerji, I., Latimer, M.J., Yachandra, V.K., Sauer, K. and Klein, M.P. (1994) J. Am. Chem. Soc. 116, 5239-5249
8 MacLachlan, D.J., Hallahan, B.J., Ruffle, S.V., Nugent, J.H.A., Evans, M.C.W., Strange, R.W. and Hasnain, S.S. (1992) Biochem. J., 285, 569-576
9 George, G., Prince, R.C. and Cramer, S.P. (1989) Science, 243, 789-791
10 Penner-Hahn, J., Fronko, R.M., Pecoraro, V.L., Yocum, C.F., Betts, S.D. and Bowlby, N.R. (1990) J. Am. Chem. Soc., 112, 2549-2557
11 Yachandra, V.K., DeRose, V.J., Latimer, M.J., Mukerji, I., Sauer, K. and Klein M.P. (1993) Science, 260, 675-679
12 MacLachlan, D.J., Nugent, J.H.A., Bratt, P.J. and Evans, M.C.W. (1994) Biochim. Biophys. Acta, 1186, 186-200
13 Kusunoki, M. and Takano, T. (1995) SR Sci. & Techn. Information, Vol.5, No.9, 1-6
14 Kusunoki, M., Ono, T., Noguchi, T., Inoue, Y. and Oyanagi, H. (1993) Photosynth. Res., 38, 331-339
15 Lee, P.A., Citrin, P.H., Eisenberger, P. and Kincaid, B.M. (1981) Rev. Mod. Phys., 53, 769-806
16 Bevington, P.R. (1969) Data Reduction and Error Analysis for the Physical Sciences, pp. 204-246, McGraw-Hill, New York.
17 Rehr, J.J., Mustre de Leon, J., Zabinsky, S.I. and Albers, R.C. (1991) J. Am. Chem. Soc., 113, 5135-*5140

SELECTIVELY ^{13}C-LABELLED TYROSINE RADICALS STUDIED BY ELECTRON PARAMAGNETIC RESONANCE SPECTROSCOPY.

R.J. Hulsebosch, J.S. van den Brink, P. Gast, A.J. Hoff, Department of Biophysics, Huygens Laboratory, Leiden University, P.O. Box 9504, 2300 RA Leiden, The Netherlands
S.A.M. Nieuwenhuis, J. Raap, J. Lugtenburg, Leiden Institute of Chemistry, Gorlaeus Laboratories, Leiden University, P.O. Box 9502, 2300 RA Leiden.

1. Introduction

Research during the past decade has shown that tyrosine radicals fulfil an important role in several biological processes. To obtain more information about the structure and spin density distribution of these *in vivo* radicals, comparison with the unperturbed model system, the neutral π-radical without protein environment, is essential. In this presentation we shall give a complete characterization of the tyrosine model radical using selective ^{13}C-labelling and EPR spectroscopy at two frequencies: 9.2 GHz and 34.8 GHz. A comprehensive study of selectively ^{13}C-enriched organic radicals in frozen solution offers new insights in isotropic and anisotropic ^{13}C-hyperfine interactions (hfi). Since ^{13}C-spectroscopy is very sensitive to structural changes, it provides, with respect to proton hyperfine spectroscopy, more detailed information about the spin density distribution and geometrical structure of the radical. Using ^{13}C-labelling, it is also possible to probe the spin density at so-called "blind" positions: molecular carbon positions that are not coupled to a proton. This property has enabled us to determine accurately the spin density on the phenoxyl oxygen and its neighbouring carbon.

2. Materials and Methodes

^{13}C-labelled L-tyrosines were synthesized as described [1], to obtain selectively enriched singly-labelled molecules with labels at position 1', 2', 3', 4' and the doubly-labelled 3', 5' molecule. An additional proton/deuteron exchange was also performed to decrease the amount of proton hfi and bring out more clearly small ^{13}C-couplings. For all samples, NMR results show that more than 95% of the protons were exchanged and 95% of the carbons were enriched. Solutions (5 mM) of a specific tyrosine were prepared in alkaline (pH 11) NaOH/H_2O, loaded into EPR quartz sample tubes and deoxygenated by bubbling with argon. The tubes were closed under argon pressure with a rubber stopper and were slowly frozen in liquid nitrogen. A stable neutral tyrosine radical was created by irradiation of the sample for two minutes with the full spectrum (275-335 nm) of a 100 Watt Oriel Mercury lamp, at liquid nitrogen temperature.
X-band (9.2 GHz) EPR spectra were obtained using a Varian E-9 spectrometer with 100 kHz field modulation. The spectrometer was equiped with a multipurpose TE_{102} cavity and a liquid nitrogen cooled gas flow cryostat. The X-band measurements were performed at about 90 K. Q-band (34.8 GHz) EPR spectra were obtained using a home-built spectrometer described in [2]. All measurements were performed in a home-built

P. Mathis (ed.), Photosynthesis: from Light to Biosphere, Vol. II, 255–258.

bath cryostat filled with cold N_2 gas (T~100 K). Spectral simulations were done with a Fortran program based on the "POWFUN" EPR simulation program provided by C. Hoganson.

3. Results and Discussion

3.1 The g-tensor

An important indicator of the nature of a neutral tyrosine π-radical is its *g*-tensor. The orientation of the *g*-axes in the molecular system is defined as follows: g_{xx} is parallel to the C1-C4-O axis, g_{zz} is perpendicular to the aromatic ring and the g_{yy}-axis is perpendicular to the other two axes. The axes are graphically represented in Figure 1. Using the Euler angle ϕ, one can transform the hyperfine components, defined in the molecular system, to the *g*-axes system. Simulation of the spectrum recorded at Q-band, which is more sensitive to *g*-anisotropy than the X-band spectrum, of a perdeuterated tyrosine species, where the hfi are negligibly small, enabled us to determine the principal *g*-tensor components: g_{xx}=2.0068±0.0002, g_{yy}=2.0043±0.0002 and g_{zz}=2.0023±0.0002. These values are in good agreement with those reported earlier for single crystal measurements [3]. Only the neutral tyrosine radical was created, and, as distinct from room temperature experiments [4], no other radicals, such as dopa-derived radicals (g_{iso}=2.0021) or a CO_2^- radical (g_{iso}=2.0001), were observed.

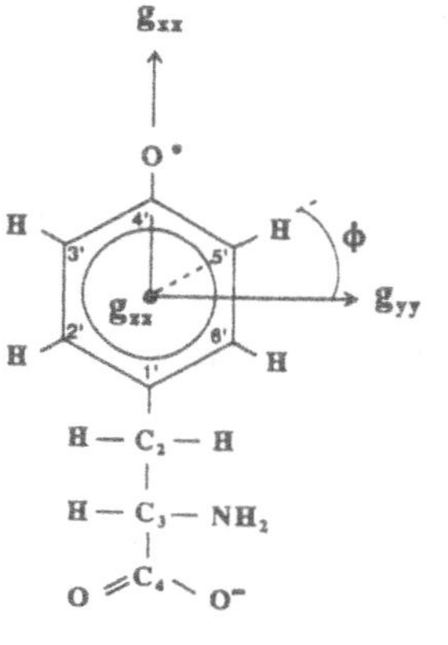

Figure 1: The tyrosine radical.

3.2 ¹H-hfi

Several X-band EPR studies have been performed on the tyrosine radical model system [4,5]. All major proton hfi were determined and with the McConnell relations [6], the corresponding spin densities were determined. The spin density is distributed over the phenoxyl ring and follows an odd alternant pattern: high spin density is found on the phenoxyl oxygen and on the carbons C1, C3 and C5, low spin density on the other ring carbons. The spin density on the protons and other molecular positions can be neglected. Simulating our additional Q-band spectra of selectively deuterated tyrosines helped us to determine the proton hfi more accurately. All proton hfi are tabulated in Table I. Knowing the two methylene β-proton hfi and assuming that the angle between the β-protons is 120°, an ideal sp^3-hybridization, the dihedral angles, θ_1 and θ_2, and spin density on C1 can be determined. Solving the resulting equations for θ results in *two* possible solutions: θ_1=75° or θ_1=111° and thus also in *two* possible spin densities on C1, namely 0.40 or 0.20. The spin densities following from proton hfi are summarized in Table II. Note that the spin density on C4 and the phenoxyl oxygen cannot be determined from proton hfi.

3.3 ¹³C-hfi

In comparison with proton hfi, the theory of ¹³C-hfi is more complicated, although the basic principles are the same. Magnitude and sign of ¹³C-hfi are sensitive functions of the spin distribution. In particular, they strongly depend on the spin densities of neighbouring atoms and on structural changes of the molecular framework. Dipolar ¹³C-hyperfine couplings are determined by the spin density on the labelled carbon and by the spin density on the neigbouring atoms [7]. The role of the neighbouring atoms becomes

Table I : Tyrosine radical ^1H- and ^{13}C-hyperfine interactions and Euler angles

hfi	A_{xx} (mT)	A_{yy} (mT)	A_{zz} (mT)	A_{iso} (mT)	$A_{K\text{-}F}$ (mT)	ϕ	θ	ψ
C_2-H_2	0.17	0.27	0.04	0.16	-	23	0	0
C_3-H_3	-0.99	-0.28	-0.70	-0.66	-	-23	0	0
C_5-H_5	-0.99	-0.28	-0.70	-0.66	-	23	0	0
C_6-H_6	0.17	0.27	0.04	0.16	-	-23	0	0
C_1-$H_{\beta 1}$	0.15	0.15	0.15	0.15	-	0	0	0
C_1-$H_{\beta 2}$	1.25	1.13	1.13	1.17	-	0	0	0
1'-^{13}C	**-0.20**	**-0.20**	**3.40**	**1.00**	**1.18**	**0**	**0**	**0**
2'-^{13}C	**-0.65**	**-0.85**	**-1.05**	**-0.85**	**-0.93**	**23**	**0**	**0**
3'-^{13}C	**-0.50**	**-0.50**	**1.90**	**0.30**	**0.86**	**-23**	**0**	**0**
4'-^{13}C	**-1.00**	**-0.98**	**-0.68**	**-0.89**	**-0.94**	**0**	**0**	**0**
5'-^{13}C	**-0.50**	**-0.50**	**1.90**	**0.30**	**0.86**	**23**	**0**	**0**
g_{xx} = 2.0068		g_{yy} = 2.0043		g_{zz} = 2.0023				

more important if the spin density at these atoms is high, or when it is low at the labelled carbon. Figure 2 shows by way of example the set of experimental and simulated EPR spectra of a [1'-^{13}C]-tyrosine radical. Spectral simulation yields $A^C_{xx}=A^C_{yy}=-0.20$ mT and $A^C_{zz}=3.40$ mT for the principal values of the ^{13}C-hyperfine tensor. A spin density of 0.37 is deduced from the observed dipolar, anisotropic splittings. This result immediately excludes the second orientation of the methylene β-protons, $\theta_1=111°$, $\theta_2=9°$ with spin density 0.20, derived from the proton hfi. In all cases, the complete data set consisting of EPR spectra recorded at 9.2 GHz and 34.8 GHz, of protonated, deuterated and ^{13}C-labelled tyrosine radicals, could be simulated very well with one unique set of parameters.

The observed isotropic and anisotropic ^{13}C-hfi for all labelled positions are given in Table I. Comparison with isotropic ^{13}C-hfi determined with the Karplus-Fraenkel theory [8], which are also tabulated in Table I, shows a large discrepancy for the positions C3 and C5. A possible explanation for this difference can be a change in hybridization of the atomic orbitals as suggested by the observed non-collinearity of the proton hyperfine tensor with the C-H bond which is possibly due to a slight distortion of the aromatic ring or to a change in bond angle of the C-H proton.

For the [4'-^{13}C]-labelled tyrosine radical, the observed ^{13}C-hfi are: $A^C_{xx}=-1.00$ mT, $A^C_{yy}=-0.98$ mT and $A^C_{zz}=-0.68$ mT. The sign is chosen negative on basis of the

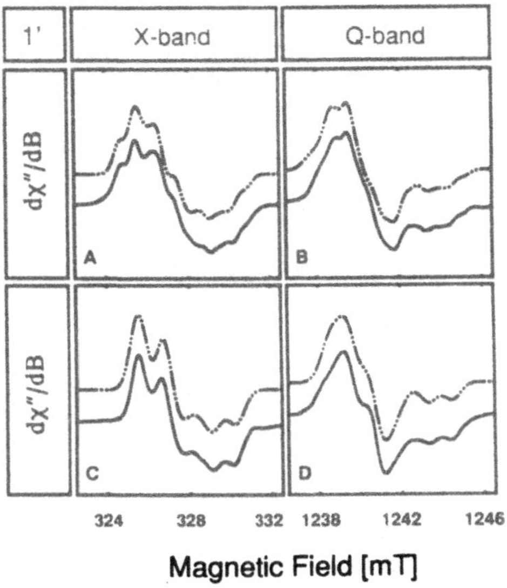

Figure 2: Examples of experimental (solid line) and simulated (dashed line) EPR spectra of selectively ^{13}C-labelled tyrosine radicals, in this case: [1'-^{13}C$_1$]-tyrosine (A, B) and [1'-^{13}C$_1$; 3',5'-^{2}H$_2$]-tyrosine (C, D).

Karplus-Fraenkel theory. From the observed anisotropy we were able to determine the spin density on C4 and on the phenoxyl oxygen. All spin densities are tabulated in Table II. There is good agreement with theoretically determined spin densities and those obtained from proton hfi

Table II : Experimental and theoretical derived spin densities for the tyrosine radical.

Atom	Experimental		Calculated		
	(^{1}H)	(^{13}C)	SVWN[9]	Hückel[3]	SCF[3]
C_1	0.20/0.40	0.37	0.29	0.24	0.31
C_2	±0.06	−0.02	−0.06	0.01	−0.07
C_3	0.26	0.25	0.21	0.20	0.26
C_4	?	0.04	−0.01	0.07	−0.03
C_5	0.26	0.25	0.24	0.20	0.26
C_6	±0.06	−0.02	−0.06	0.01	−0.02
O_4	?	0.13	0.36	0.26	0.28

4 Conclusions

The total spin density distribution in the UV-generated neutral tyrosine radical in frozen solution has been determined from isotropic proton and anisotropic ^{13}C-hyperfine couplings obtained from EPR measurements at two frequencies. ^{13}C-labels form an ideal probe to reveal the spin density located at, for proton hfi spectroscopy, "blind" positions such as the carbon C4. The relatively low spin density on the phenoxyl oxygen indicates that there is a large amount of delocalisation of the unpaired electron, which can be explained by a short C-O bond length implicating some double-bond character. A ^{13}C-label at C1 enabled us to determine the spin density more direct, resolving the uncertainty of the orientation of the methylene β-protons. The observed isotropic ^{13}C-hfi at C3 and C5 indicate that the aromatic ring might be somewhat distorted at these positions.
The present characterization of the *in vitro* tyrosine radical and the interpretation of ^{13}C-hfi forms a solid base for comparison with ^{13}C-labelled tyrosine radicals incorporated in *in vivo* systems. Our ability to probe the spin density with atomic selectivity at all positions, even at the phenoxyl oxygen, makes future *in vivo* experiments very promising. Incorporation of ^{13}C-labelled tyrosines in Photosystem II of cyanobacteria is in progress.

Acknowledgements

This research was supported by the Netherlands Foundation for Life Sciences (SLW) and financed by the Netherlands Organisation for Scientific Research (NWO).

References

[1] Winkel, C. (1989) Thesis, University of Leiden.
[2] Van den Brink, J.S. (1995) Thesis, University of Leiden. Brok, M. (1986) *ibid.*
[3] Fasanella, E.L. & Gordy, W. (1969) Proc. Natl. Acad. Sci. U.S.A. 62, 299.
[4] Tomkiewicz, M., McAlpine, R.D. & Cocivera, M. (1972) Can. J. Chem. 50, 3848.
[5] Hoganson, C.W. & Babcock, G.T. (1992) Biochemistry 31, 11874.
[6] McConnell, H.M. (1956) J. Chem. Phys. 24, 764.
[7] Atherton, N.M. (1993) Principles of Electron Spin Resonance, Ellis Horwood N.Y.
[8] Karplus, M. & Fraenkel, G.K. (1961) J. Chem. Phys. 35, 1312.
[9] Qin, Y. & Wheeler, R.A. (1995) J. Am. Chem. Soc. 117, 6083.

THE EXTRINSIC 33 KDA PROTEIN OF PHOTOSYSTEM II: IMPROVED EXPRESSION PLASMIDS AND PROGRESS IN THE MUTATIONAL ANALYSIS

A. Seidler[1,2], H. Michel[2] and A.W. Rutherford[1]
[1]Section de Bioénergétique, CNRS URA 1290, DBCM, CEA Saclay, 91191 Gif-sur-Yvette, France
[2]Max-Planck-Institut für Biophysik, Heinrich-Hoffmann-Str. 7, 60528 Frankfurt, Germany

1. Introduction

Three extrinsic proteins with an apparent molecular weight of 33, 23 and 16 kDa are located on the lumenal side of Photosystem II. Their function in oxygen-evolution has been established by dissociation and reconstitution experiments (1). The closest relationship to oxygen-evolution has the so-called 33 kDa protein. After removal from Photosystem II (PSII) the activity decreased to 20-45% (1). Deletion of the *psbO* gene encoding the extrinsic 33 kDa protein in cyanobacteria led to a decrease in oxygen-evolving activity to 30-40% (2, 3, 4, 5) wheras a mutant of *Chlamydomonas reinhardtii* lacking the synthesis of this protein showed no oxygen-evolving activity at all (6). In PSII-enriched membranes from higher plants or thylakoid membranes of the cyanobacteria lacking the 33 kDa protein the manganese cluster, at which oxygen evolution takes place, is only stable in the presence of high concentrations of Cl^- (1, 10). Site-directed mutagenesis on the 33 kDa protein was carried out by us and Burnap and co-workers (7, 8, 9, 10) in order to understand the function of the 33 kDa protein and to identify residues involved in binding of the protein to PSII. Some mutations led to a decrease in oxygen-evolution. We have extended our studies and present data on the protein variants in this communication.

2. Procedures

Over-expression in *E.coli* and purification of the 33 kDa protein, isolation of PSII and reconstitution experiments were carried out as described in (11) with the modifications described in (8).

3. Results and Discussion

3.1 Construction of improved expression plasmids

3.1.1 Construction of the expression plasmid pAS3

In order to facilitate site-directed mutagenesis we have generated the new expression plasmid pAS3 (fig. 1). This plasmid is based on the general expression plasmid pASK40 (12). The expression is controlled by the regulatory elements of the *E.coli lac*-operon followed by the ribosome binding site of the expression plasmid pDS12. As in the expression plasmid pDS1233 we used the thylakoid transfer domain of the transfer peptide of the 33 kDa protein as signal peptide in order to direct the protein into the periplasm of *E.coli*. In addition, the plasmid carries the ampicillin resistance gene, the

P. Mathis (ed.), Photosynthesis: from Light to Biosphere, Vol. II, 259–262.

gene encoding for the Lac repressor, an origin of replication and the intergenic region of the f1 phage. The latter allowed preparation of single stranded DNA and therefore site-directed mutagenesis directly in the expression plasmid. The expression level in strain JM83 was 20% of the total periplasmic protein (not shown) which is comparable with the result obtained with the expression plasmid pDS1233 (11). A slightly higher level (25%) can be obtained with the strain TG1 carrying the expression plasmid pDS1233 (fig. 2).

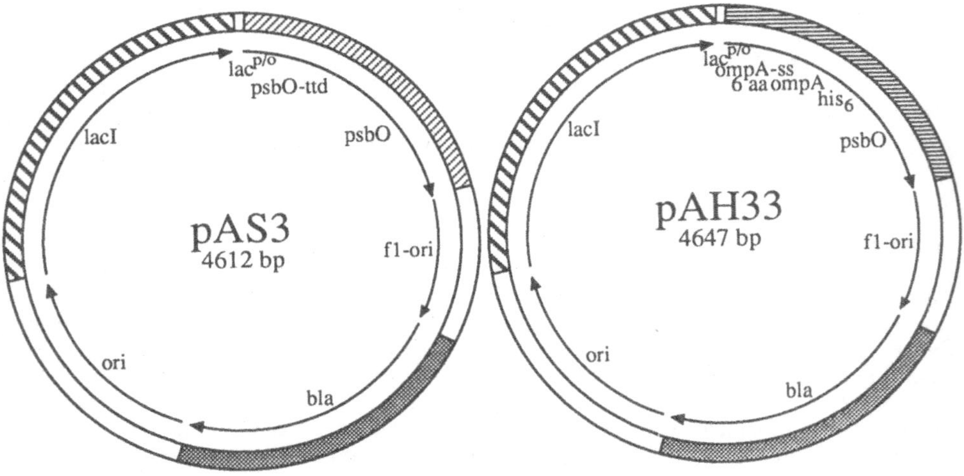

Fig. 1: Expression plasmids pAS3 and pAH33
$lac^{p/o}$ = lac promoter/operator, psbO-ttd = thylakoid transfer domain of the transit peptide of the 33 kDa protein, ompA-ss = signal sequence of the outer membrane protein A, f1-ori = intergenic region of the f1-phage, bla = β-lactamase gene, ori = origin of replication, lacI = Lac repressor gene.

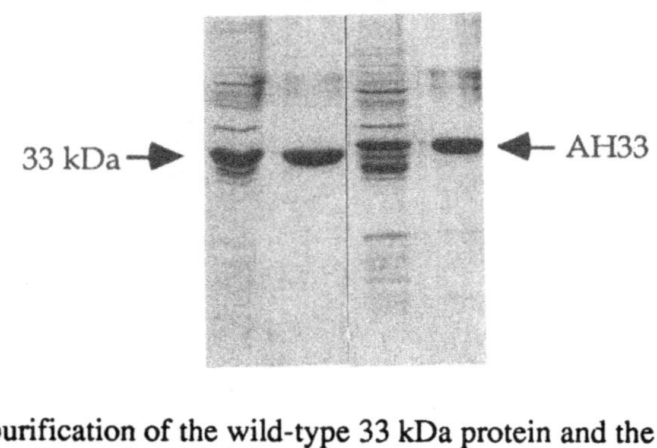

Expression and purification of the wild-type 33 kDa protein and the 33 kDa protein with either C-terminal or N-terminal histidine tail.
Lane 1 and 3 : periplasmic protein fractions; lane 2 and 4: purified proteins; lane 1 and 2: wild-type 33 kDa protein; lane 3 and 4: 33 kDa protein with N-terminal histidine tail. (The faint bands around 70 kDa are due to contaminations in the buffer.)

3.1.2 Construction of the expression plasmid pAH33

The expression plasmid pAH33 was constructed in order to facilitate the purification of poorly expressed variants of the 33 kDa protein (for example D212N and D212K)(8) and of crosslinked products (A.S., submitted). The 33 kDa protein is expressed from this plasmid with an N-terminal histidine tail which allows the purification via immobilised metal affinity chromatography. The histidine tail was introduced at the N-terminus since its introduction at the C-terminus led to a lowered affinity of the protein to PSII (13). The plasmid pAH33 is based on the expression plasmid pASK60 (14). The expression is under control of the regulatory elements of the *lac* operon. Efficient translation initiation is assured by a tandem ribosome binding site. The chimeric protein expressed from this expression plasmid is accumulating to 16% of the periplasmic proteins and could be purified by IMAC (fig. 2).

3.2 Site-directed mutagenesis on the 33 kDa protein

All conserved acidic amino acid residues of the 33 kDa protein were subjected to site-directed mutagenesis. Aspartic acid residues were exchanged with asparagin and lysin and glutamic acid residues were exchanged with glutamin and lysin. In (8) we have reported the activity of PSII reconstituted with a 10fold excess of 33 kDa protein variant over PSII. However, with the wild-type protein a 3 fold excess is sufficient for maximal restoration of oxygen evolution (13). We have used this excess in order to look for changes in the water splitting activity and not for minor changes in the affinity of the 33 kDa protein. Mutations of residues E104, E109 and D157 led to a significant decrease in activity (table 1)[1] . We have recorded light saturation curves with PSII reconstituted with one of the 33 kDa protein variants E104K, D109K and D157K led to no change in the "K_M" value, indicating that the decrease in the activity was due to a incomplete reconstitution due to lower affinity of the variants of the 33 kDa protein to PSII rather than to a slow down of the S-state turnover.

protein variant	activity
wild-type	100%
E104Q	103%
E104K	74%
D109N	98%
D109K	80%
D157N	95%
D157K	85%

Table 1: Oxygen-evolving activity of $CaCl_2$-washed PSII reconstituted with a 10fold excess of a protein of the 33 kDa protein.

We have also expressed the variant D9K in *E.coli* strain SF110 (15). This variant did not accumulate in the periplasm in strains like DH5α, TG1 or JM83, probably due to rapid

[1] In (8) we reported that a 10fold excess of the variants D97N and E104Q can not restore the activity as the wild-type protein. These results we are not able to reproduce. The likely reason for our initial observation is that the two variants were the first expressed and purified. At this time we used PMSF instead of DIFP as protease inhibitor in order to inhibit an *E.coli* serine protease which copurifies with the 33 kDa protein. Although PMSF was added to all buffers during purification we observed sometimes degradation of the purified protein. The obseved reduction in activity was probably due to proteolytic damage of the acceptor side of PSII since SDS-PAGE revealed binding of the variant to the same extent as the wild-type protein.

degradation (7). Reconstitution of various amounts of this protein variant to PSII revealed that its binding constant was identical to that of the wild-type protein which is constistent with ref. (10). This excludes the possibility that aspartic acid residue 9 is involved in binding to PSII. This result is in agreement with our recent findings that this residue might be involved in an intramolecular salt bridge (unpublished).
In our previous studies we were unable to purify the variants D212N and D212K because they accumulated poorly in *E.coli* in a soluble form and were eluted from a DEAE column over a wide range of the salt gradient indicating that the variants have several conformations. Expression in the protease deficient *E.coli* strain SF110 did not led to an increased accumulation. Therefore, we expressed the variant D212N with a histidine tail and purified the protein via immobilised metal ion affinity chromatography. The purified protein could be bound to $CaCl_2$-washed PSII, although with a somewhat lower affinity. The mutation had no influence on the oxygen-evolving activity.

4. References

1. Murata, N. and Miyao, M. (1985) Trends Biol. Sci. 10, 122-124
2. Burnap, R.L. and Sherman, L.A. (1991) Biochemistry 30, 440-446
3. Philbrick, J.B., Diner, B. and Zilinskas, B.A. (1991) J. Biol. Chem. 266, 13370-13376
4. Mayes, S.R., Cook, K.M., Self, S.J., Zhang, Z. and Barber, J. (1991) Biochim. Biophys. Acta 1060, 1-12
5. Bockholt, R., Masepohl, B. and Pistorius, E.K. (1991) FEBS-Lett. 294, 59-63
6. Mayfield, S.P., Bennoun, P. and Rochaix, J.-D. (1987) EMBO J. 6, 313-318
7. Seidler, A. and Michel, H. (1991) in: Trends in Photosynthesis Research (Barber, J., Medrano, M. and Guerrero, M.G., eds.) pp. 149-156, Intercept, Andover
8. Seidler, A., Röll, K. and Michel, H. (1992) in Current Research in Photosynthesis (Murata, N., ed.), Vol. 2, pp. 409-412, Kluwer, Dordrecht
9. Burnap, R.L., Shen, J.-R., Jursinic, P.A., Qian, M., Inoue, Y. and Sherman, L.A. (1992) in Research in Photosynthesis (Murata, N., ed.) Vol. 2, pp. 179-182 Kluwer, Dordrecht
10. Burnap, R.L., Quian, M., Shen, J.-R., Inoue, Y. and Sherman, L.A. (1994) Biochemistry 33, 13712-13718
11. Seidler, A. and Michel, H. (1990) EMBO-J. 9, 1743-1748
12. Skerra, A., Pfitzinger, I. and Plückthun, A. (1991) Bio/Technology 9, 273-278
13. Seidler, A. (1994) Protein Engineering 7, 1277-1280
14. Schmidt, T.G.M. and Skerra, A. (1993) Protein Engineering 6, 109-122
15. Baneyx, F. and Georgiou, G. (1991) J. Bacteriol. 173, 2696-1703

Protolytic Reactions at the donor and the acceptor side of Photosystem II core particles

Oliver Bögershausen and Wolfgang Junge

FB Biologie/ Chemie, Abt. Biophysik, Babarastr. 11 , Universität Osnabrück, D-49069 Osnabrück, Germany

1. Introduction

One molecule of dioxygen and four protons are the products of water cleavage in thylakoids. Photosystem II uses four quanta of light to drive its catalytic manganese cluster through a series of increasingly oxidizing states $S_0 \Rightarrow S_1 ... S_4$. The highest state S_4 spontaneously decays to S_0 [1,2]. Oxygen liberation only occurs in transition $S_3 \Rightarrow S_{4,0}$, whereas protons are released in all four transition with different extents depending on preparation and pH [3]. With core particles this pattern of proton release is featureless, nearly one proton pro electron is released in every transition. We used core particles [4] to investigate proton release at high time resolution. The rate of proton release was affected by three factors: 1) the detergent concentration, 2) added electron acceptor, and 3) the pH. The first effect reflects the transition from aggregation to monodispersity, and the second one the superimposition of protolytic reactions at both sides of PS II. As the rates of proton transfer over the four redox transitions was constant, under certain conditions (low pH) even faster than the one of the reduction of Tyr_Z^+ the measurable proton release was entirely attributable to the electrostatic response of several amino acids to charge displacements at the donor side.

2. Material and methods

Photosystem II core particles were prepared from 12 days old pea seedlings according to Van Leeuwen et al. [4] with modifications described in [5]. According to literature [4] and own data (not documented), the second quinone acceptor, Q_B, is absent from this preparation. We checked the activity of core particles by measuring oxygen evolution under continuous light on a Clark-type electrode. Flash spectroscopy was performed with a setup as described in [6]. The sample was excited by saturating flashes of a Xenon lamp (10 µs FWHM, Schott RG 610). Buffer was removed by gel filtration (PD-10, Pharmacia). The measuring medium contained, if not otherwise indicated, 2 µM Chl., 5 mM $CaCl_2$, 5 mM $MgCl_2$, and 0.01 % (w/v) *n*-dodecyl ß-D-maltoside (ß-DM). 2,5-dichloro-*p*-benzoquinone (DCBQ) and hexacyanoferrate (III) (FeCy) served as artificial electron acceptors. Proton release was measured by absorption changes of added dyes (all at 30 µM) namely Methyl red (MR), Bromocresol purple (BCR), Phenol red (PR), and Cresol red (CR) depending on pH. The optical pathway was 2 cm. Absorption

P. Mathis (ed.), Photosynthesis: from Light to Biosphere, Vol. II, 263–266.

transients in the UV were recorded with 8 μM Chl. and 1 cm optical pathway. The set of three Kok parameters (population of S_1, misses, and double hits) was determined at 295 nm as described in [5].

3. Results

The oxygen-evolving activity of core particles varied between 1200 and 1800 μmol O_2 /(mg Chl. · h). We determined the half-rise times of electron transfer out of the oxygen evolving complex from absorption transients at 360 nm they were 75 μs, 270 μs, and 4.2 ms for $S_1 \Rightarrow S_2 \Rightarrow S_3 \Rightarrow S_0$ respectively (see also [7]). Upon the 4th flash it was 86 μs, however, here the attribution to $S_0 \Rightarrow S_1$ was less reliable due to the high proportion of misses (20%).

Aiming at a kinetic correlation between the electron transfer from the oxygen evolving complex to Tyr_Z with proton release, we had to overcome kinetic limitations due to the aggregation of core particles. Increasing the detergent concentration from 0 to 0.05 % (w/v) accelerated the pH-transients reaching a plateau at 0.01% (w/v) (for a residual slow component, see further down). The half-rise time of the rapid component was then strictly dependent on the pH but independent of the chemical nature or the concentration of the indicator dye. We determined half-rise times of 250 μs at pH 7.5, 70 μs at pH 6.0, and 12 μs at pH 5.2 [5]. They were the same in every oxidation step and depending on pH faster or slower than the respective electron transfer to Tyr_Z. The oscillatory pattern of water oxidation and the rate constants of Tyr_Z^+- reduction were not likewise affected by pH

The residual slow component of the acidification could be attributed to superimposition of protolytic events at the acceptor side of Photosystem II. This was evident from the dependence on DCBQ and FeCy as electron acceptors. At pH 5.5 about 50% of the acidification occurred slowly. When using only DCBQ the slow component of the net acidification turned into an alkalinisation. The rate of the slow component of the net acidification varied depending on the DCBQ concentration (see Fig.1). When describing the acidification phenomenologically with the sum of two exponentials there was a linear dependence between the rate of the slow phase and the acceptor concentration ($k = m \cdot [DCBQ]$ with $m \approx 1$, if k was in ms^{-1}, and [DCBQ] in μM)

Because of these results we assigned the kinetic components to the following reactions: a) The fast component represents proton release at the donor side. It is an electrostatic answer of the protein boundary to the new positive charge on Tyr_Z (for a detailed discussion see [3,5]). This fast proton release is partly counteracted (20% at pH 6.5 and 50 % at pH 5.5) by a fast proton uptake into the protein moiety upon the reduction of Q_A. b) The slow component reflects the release of protons which have been rapidly bound at the acceptor side. This release kinetically follows the reoxidation of Q_A^- by added DCBQ. c) The very slow alkalinisation is caused by proton uptake which follows the dismutation reaction of the semiquinone form of DCBQ [8] as described by reaction (3) and (4). These reactions are prevented when FeCy is added, which oxidises both Q_A^- and $DCBQ^-$. We describe the events at the acceptor side as follows:

$$(1) \quad Q_A^- \cdot B^- + H^+ \underset{k_{-1}}{\overset{k_1}{\rightleftharpoons}} Q_A^- \cdot BH \qquad (2) \quad Q_A^- \cdot B^- + DCBQ \xrightarrow{k_2} Q_A \cdot B^- + DCBQ^-$$

$$(3) \quad DCBQ^- + DCBQ^- \xrightarrow{k_3} DCBQ^{2-} + DCBQ \qquad (4) \quad DCBQ^{2-} + 2H^+ \xrightarrow{k_4} DCBQH_2$$

therein B^- denotes an intrinsic buffering amino acid residue in the vicinity of Q_A.

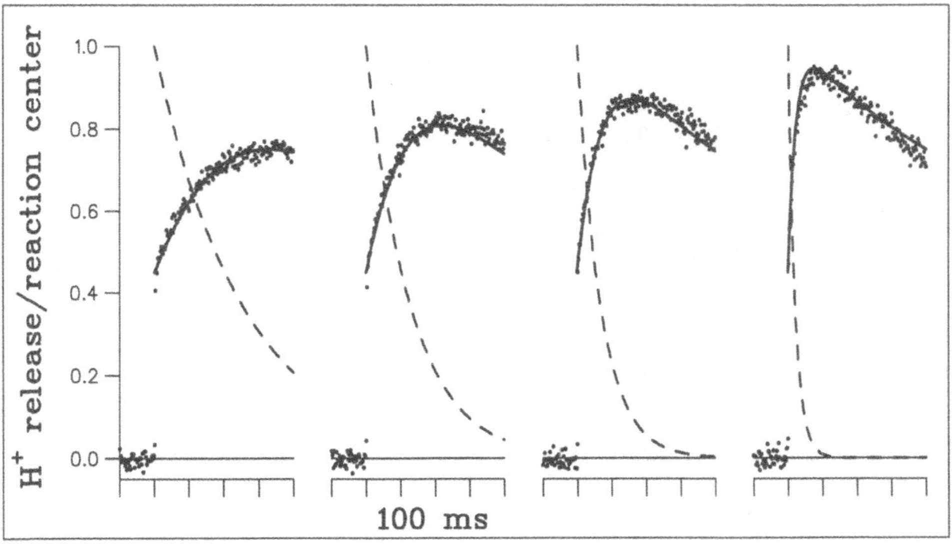

Fig.1: Acidification following flash excitation of core particles at different concentrations of DCBQ. The acidification of the medium was monitored by absorption changes at 560 nm of the dye MR (30 μM). The DCBQ concentrations were 25 μM, 50 μM, 100 μM, and 300 μM (left → right). The original traces (dots) were normalised in terms of released H^+ per reaction center. The solid line represents the fit (equ. (9)) using the kinetic parameters given in the text. The dashed line shows the reoxidation of Q_A^- by DCBQ (sum of equations (5) and (6)). pH 5.5, 50 μs per address (only every 10th data point is displayed), 15 transients averaged.

We fitted the original data with following system of differential equations (B^- was omitted for clarity):

$$\frac{d[Q_A^-]}{dt} = -(k_2 \cdot [DCBQ] \cdot [Q_A^-]) - (k_1 \cdot [H^+] \cdot [Q_A^-]) + (k_{-1} \cdot [Q_A^- H^+]) \quad (5)$$

$$\frac{d[Q_A^- H^+]}{dt} = (k_1 \cdot [H^+] \cdot [Q_A^-]) - (k_{-1} \cdot [Q_A^- H^+]) \quad (6)$$

$$\frac{d[DCBQ^-]}{dt} = (k_2 \cdot [DCBQ] \cdot [Q_A^-]) - (k_3 \cdot [DCBQ^-]^2) \quad (7)$$

$$\frac{d[DCBQ^{2-}]}{dt} = k_3 \cdot [DCBQ^-]^2 \cdot 0.5 \quad (8)$$

$$\frac{d[H^+]}{dt} = -(k_1 \cdot [H^+] \cdot [Q_A^-]) + (k_{-1} \cdot [Q_A^- H^+]) - (k_3 \cdot [DCBQ^-]^2) \quad (9)$$

In this system the fast release of 1 H^+ at the donor side and the fast uptake of n (0<n<1) protons at the acceptor side were treated as an unresolved initial jump. The fit starts at 1-n H^+. The solid lines in Fig. 1 represents the fit with the following results:

$k_1 = 10^{10}\ M^{-1}\ s^{-1}$ (fixed, assuming the diffusion controlled protonation of a base in contact with bulk water [9]),

$k_{-1}= 2.6 \cdot 10^4\ M^{-1}\ s^{-1}$, $k_2= 1.7 \cdot 10^6\ M^{-1}\ s^{-1}$, and $k_3= 3.2 \cdot 10^8\ M^{-1}\ s^{-1}$
with $k_{-1}/ (k_1 \cdot [H^+])= [Q_A^-] / [Q_A^- H^+] = 0.45 / 0.55 \rightarrow n = 0.55$

Further we assumed again a diffusion controlled protonation of $DCBQ^{2-}$ ($k_4= 10^{10}\ M^{-1}\ s^{-1}$), which meant that the rate of alkalinisation depended only on k_3.

Summary

We determined a detergent concentration which prevented the kinetical damping of pH-transients as caused by the aggregation of the hydrophobic core particles in aqueous solution (see also [10]). The apparent half-rise time of the net acidification was shortened upon lowering the pH. At acid pH it was faster than electron transfer from the oxygen evolving complex into Tyr_Z^+. This kinetic behaviour did not depend on the nature and the concentration of the added pH-indicator. We conclude that it reflects the protolysis of several electrostatically interacting acids at the surface of the protein in response to a new positive charge on Tyr_Z. The extent of deprotonation persisted upon electron transfer from the manganese cluster to Tyr_Z^+. This electrostatically triggered proton release totally covered the chemical produced protons from water cleavage in core particles (for thylakoids see [11]).
The superimposition of acid-base reactions at both sides of PSII was unravelled by characterising the protonation / deprotonation of the intrinsic (Q_A) and extrinsic electron acceptors. Our experiments revealed a rate constant of the dismutation of $DCBQ^-$, almost diffusion controlled, which was of the same order of as given in literatur ($6.3 \cdot 10^8\ M^{-1}\ s^{-1}$ [8], $2.6 \cdot 10^8\ M^{-1}\ s^{-1}$ [12]). The acceleration of the slow component of acidification when increasing the acceptor concentration can be described by a bimolecular reaction between Q_A^- and the artifical electron acceptor DCBQ with $k = 1.7 \cdot 10^6\ M^{-1}s^{-1}$. The deprotonation of the protein moiety of Q_A follows the reoxidation without delay.

Acknowledgements

Stimulating discussions with our other collegues in the EU-HCM Network (EU-DGXII-HCM-CHRX-CT94-0524) are gratefully acknowledged.

References

[1] Renger, G. (1993) Photosynth. Res. 38, 229-247.
[2] Debus, R.J. (1992) Biochim. Biophys. Acta 1102, 269-352.
[3] Lavergne, J. and Junge, W. (1993) Photosynth. Res. 38, 279-296.
[4] Van Leeuwen, P.J., Nieveen, M.C., Van de Meent, E.J., Dekker, J.P., and Van Gorkom, H.J. (1991) Photosynth. Res. 28, 149-153.
[5] Bögershausen, O. and Junge, W. (1995) Biochim. Biophys. Acta 1230, 177-185
[6] Junge, W. (1976) in Chemistry and Biochemistry of Plant Pigments (Goodwin, T. W., ed), pp. 233-333, Academic Press, New York
[7] Haumann, M., Bögershausen, O., and Junge, W. (1994) FEBS Lett. 355, 101-105
[8] Rich, P.R. and Bendall, D.S. (1980) Biochim. Biophys. Acta 592, 506-518.
[9] Eigen, M. (1963) Angew. Chem. 12s, 489-588.
[10] Junge, W. and Polle, A. (1986) Biochim. Biophys. Acta 848, 265-273
[11] Haumann, M. and Junge, W. (1994) Biochemistry 33, 864-872
[12] Yamazaki, I. and Ohnishi, T. (1966) Biochim. Biophys. Acta 112, 469-481

ADRY AGENT-INDUCED CYCLIC AND NON-CYCLIC ELECTRON TRANSFER AROUND PHOTOSYSTEM II

Vitaly D. Samuilov, Eugene L. Barsky and Andrei V. Kitashov, Department of Cell Physiology and Immunology, School of Biology, Moscow State University, Moscow 119899, Russian Federation

1. Introduction

Deactivation of the photosynthetic water-oxidizing complex in the S_2 and S_3 states is accelerated by compounds known as ADRY (acceleration of the deactivation reactions of the water-splitting system Y) reagents. These reagents are oxidized by Mn cluster and re-reduced by an endogenous donor [1, 2], apparently by reduced membrane pool plastoquinone Q_P [3]: artificial cyclic electron flow involving ADRY reagents proceeds around photosystem II (PSII). Anilinothiophene and phenylhydrazone derivatives (ANT2p, 2-(3-chloro-4-trifluoromethyl)anilino-3,5-dinitrothiophene; CCCP, carbonyl cyanide *m*-chlorophenylhydrazone) known as uncouplers of the photosynthetic and oxidative phosphorylation are powerful ADRY reagents [4].

Here, we provide data that ADRY effect exhibit also other uncouplers (TTFB, 4,5,6,7-tetrachloro-2-trifluoromethylbenzimidazole, and PCP, 2,3,4,5,6-pentachlorophenol). The capacity fer H^+ transfer (not merely for electron transfer) is apparently a common property of tested ADRY reagents. At low concentrations, ADRY reagents are catalysts of the cyclic electron transfer around PSII. With an increase in concentrations, they serve as electron donors for non-cyclic electron transfer in PSII.

2. Materials and Methods

Pea chloroplasts were isolated as described earlier [3]. O_2 evolution by chloroplasts was recorded using a Clark-type closed Pt electrode at saturating intensity of the white light of ~0.1 $W \cdot cm^{-2}$. Photoreduction of ferricyanide (FeCy) at saturating red light (λ>600 nm) was measured spectrophotometrically at 420 nm. Photoreduction of silico-molybdate (SiMo) was measured as ferrocyanide formation in the reaction: $SiMo_{red}+FeCy_{ox}\rightarrow SiMo_{ox}+FeCy_{red}$ [5]. The incubation medium contained 0.4 M sucrose, 3.5 mM NaCl, 50 mM Tricine-NaOH (pH 7.8), 4 μM gramicidin D and chloroplasts at a chlorophyll concentration of 5-6 μg/ml in the spectrophotometric measurements and 12-15 μg/ml in the oxymetric measurements. FeCy and SiMo concentrations were 2 mM and 50 μM in the oxymetric measurements, 0.5 mM and 50 μM in the spectrophotometric measurements, respectively. The incubation medium was supplemented with 1 μM 3-(3,4-diclorophenyl)-1,1-dimethylurea (DCMU) in the experiments with SiMo.

P. Mathis (ed.), Photosynthesis: from Light to Biosphere, Vol. II, 267–270.

3. Results and Discussion

Fig. 1 shows that TTFB inhibits the electron transfer from H_2O to FeCy with I_{50} ~0.15 μM. O_2 evolution insensitive to DCMU in the presence of SiMo+FeCy was inhibited by TTFB at higher concentrations. At the same time the SiMo photoreduction rate measured by ferrocyanide formation was slightly hindered by TTFB (Fig. 1).

Similar data were obtained with other tested reagents. Table 1 shows that I_{50} values for TTFB, ANT2p, CCCP and PCP in the $H_2O \rightarrow$SiMo electron transfer were approx. 7-20 times more than in the $H_2O \rightarrow$FeCy electron transfer.

SiMo is apparently reduced by PSII acceptor-side non-heme iron [6]. 2,5-Dibromo-3-methyl-6-isopropyl-*p*-benzoquinone (2 μM) inhibiting plastoquinol Q_ZH_2 oxidation in the cytochrome *b/f* complex [7] decreased the SiMo photoreduction rate by no more than 5% (data not shown).

Sayre and Homann [8] suggested that the interaction of ADRY reagents with the oxidizing side of PSII can lead to H_2O_2 formation. Wydrzynsky et al. [9] reported on H_2O_2 production by intact PSII preparations without the addition of ADRY reagents. This H_2O_2 production was visualized by an exogenous catalase-induced enhancement of O_2 evolution. Different lines of evidence have been presented for H_2O_2 formation at the PSII donor side [10, 11]. Light-induced superoxide formation located at the PSII acceptor side also leads to H_2O_2 production [12]. H_2O_2 produced by PSII was responsible for photoinhibition of PSII reaction centres [13]. The addition of catalase to thylakoids decreased the rate of photoinhibition [13].

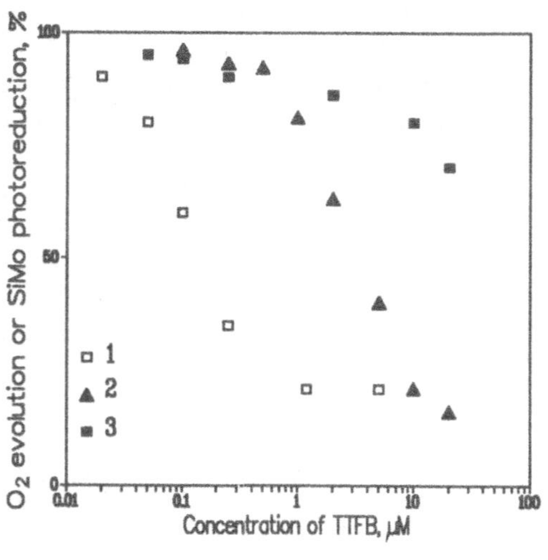

Figure 1. Effect of TTFB on light-induced O_2 evolution with FeCy (1) or SiMo (2) as Hill reagents and SiMo reduction (3) by pea chloroplasts incubated with gramicidin D. The 100% rates of O_2 evolution and FeCy reduction through the $H_2O \rightarrow$SiMo+FeCy electron transfer were ~105 and 400 μmol/mg Chl per h, respectively.

The experiment carried out with CCCP indicated that the inhibition of O_2 evolution in the $H_2O \rightarrow$SiMo electron transfer reaction was unaltered upon the addition of superoxide dismutase, catalase or their combination (Table 2). Therefore, the suggestion of H_2O oxidation to H_2O_2 or of superoxide production seems an unconvincing explanation of the inhibition of O_2 evolution by ADRY reagents. The above results imply that these reagents serve as electron donors for SiMo photoreduction occurring in the absence of O_2 evolution.

The effect of the tested reagents was compared with the effect of NH_2OH, a known inhibitor of

Table 1. I_{50} values for TTFB, ANT2p, CCCP and PCP in the $H_2O \rightarrow FeCy$ and $H_2O \rightarrow SiMo$ electron transfer reactions in pea chloroplasts

Addition	I_{50} (µM) for	
	FeCy reduction ($H_2O \rightarrow FeCy$)	O_2 evolution ($H_2O \rightarrow SiMo+FeCy$)
TTFB	0.15	3
ANT2p	1.0	17
CCCP	2.0	20
PCP	6.2	45

Experimental conditions were as in Fig. 1.

photosynthetic water oxidation and an efficient electron donor to PSII. Table 3 shows that TTFB, ANT2p, CCCP and PCP produced results similar to NH_2OH: they inhibited light-induced O_2 evolution and displayed a high electron-donor activity (assuming the electron-donor activity of H_2O to be 100%) in the reaction of SiMo pho-toreduction. As shown by cyclic voltammetry, redox potentials for TTFB, CCCP and PCP are equal to +1.18, +1.17 and +1.09 V, respectively [5].

Thus, the ADRY reagents can maintain both cyclic and non-cyclic electron flows around PSII. The ADRY reagent-dependent Hill reaction with SiMo is a manifestation of the non-cyclic electron flow, the ADRY reagent-dependent inhibition of the Hill reaction with FeCy is a manifestation of the cyclic electron flow. It was shown [14] that FeCy is reduced in PSII by plastoquinone Q_P at neutral pH values. The tested ADRY reagents oxidized on the PSII donor side are also reduced by this component [14]. The ADRY reagent-induced cyclic and non-cyclic electron flow are illustrated by a scheme:

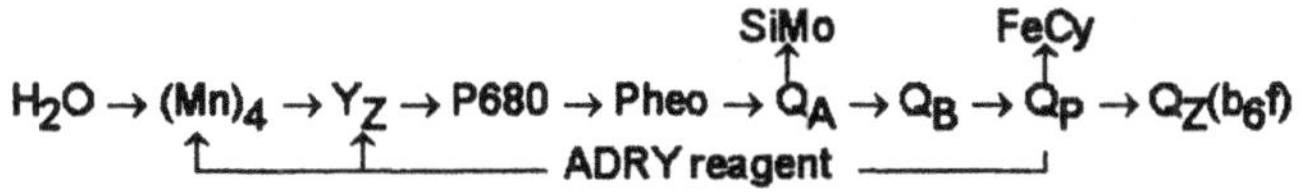

Inhibition of the FeCy Hill reaction by TTFB and CCCP was maximum at the pH values

Table 2. Effect of CCCP, catalase of bovine liver and CuZn-superoxide dismutase (SOD) of human erythrocytes on O_2 evolution in the $H_2O \rightarrow SiMo+FeCy$ electron transfer reaction in pea chloroplasts

Additions	Rate of O_2 evolution (%)
—	100
CCCP	21
CCCP+SOD	19
CCCP+SOD+catalase	18
CCCP+catalase	19

Experimental conditions were as in Fig. 1. The CCCP, SOD and catalase concentrations were 50 µM, 200 and 100 µg/ml, respectively. The 100% rate of O_2 evolution in the absence of additions was 95 µmol/mg Chl per h.

Table 3. Rates of O_2 evolution and SiMo photoreduction in the $H_2O \rightarrow$SiMo+FeCy electron transfer reaction in pea chloroplasts incubated with different reagents [5]

Reagent	Concentration (mM)	Rate of electron transport (%)	
		O_2 evolution	SiMo reduction
NH_2OH	1.0	5	75
TTFB	0.02	10	70
ANT2p	0.1	10	80
CCCP	0.1	9	86
PCP	0.1	12	78

Experimental conditions were as in Fig. 1. The 100% rates of O_2 evolution and SiMo reduction in the absence of added reagents were 100-110 and 380-400 μmol/mg Chl per h, respectively.

corresponding to the pK of these reagents [14]. The rested uncouplers serve apparently as H^+ donors, and not merely as electron donors on the oxidizing side of PSII [14].

Acknowledgements. This work was supported by research grant No ML7300 from the International Science Foundation and Russian Government and by grant No 94-04-12227a from the Russian Foundation for Fundamental Research.

References

1. Renger, G., Bouges-Bocquet, B. and Delosme, R. (1973) Biochim. Biophys. Acta 292, 796-807.
2. Hanssum, B., Dohnt, G.and Renger, G. (1985) Biochim. Biophys. Acta 806, 210-220.
3. Samuilov, V.D. and Barsky, E.L. (1993) FEBS Lett. 320, 118-120.
4. Renger, G. (1972) Biochim. Biophys. Acta 256, 428-439.
5. Samuilov, V.D., Barsky, E.L., Gubanova, O.N., Klimov, V.V. and Kozlov, Yu.N. (1995) FEBS Lett. 357, 55-57.
6. Schansker, G. and Van Rensen, J.J.S. (1993) Photosynth. Res. 37, 165-175.
7. Jones, R.W. and Whitmarsh, J. (1988) Biochim. Biophys. Acta 933, 258-268.
8. Sayre, R.T. and Homann, P.H. (1979) Arch. Biochem. Biophys. 196, 525-533.
9. Wydzynski, T. Ångström, J. and Vänngård,T. (1989) Biochim. Biophys. Acta 973, 23-28.
10. Ananyev, G.M., Wydrzynski, T. Renger, G. and Klimov, V.V. (1992) Biochim. Biophys. Acta 1100, 303-311.
11. Klimov, V.V., Ananyev, G.M. Zastryzhnaya, O.M., Wydrzynski, T. and Renger, G. (1993) Photosynth. Res. 38, 409-416.
12. Ananyev, G., Renger, G, Wacker, U. and Klimov, V. (1994) Photosynth. Res. 41, 327-338.
13. Bradley, R.L., Long, K.M. and Frash, W.D. (1991) FEBS Lett. 286, 209-213.
14. Samuilov, V.D., Renger, G., Paschenko, V.Z., Oleskin, A.V., Gusev, M.V., Gubanova, O.N., Vasil'ev, S.S. and Barsky, E.L. (1995) Photosynth. Res. (in press).

X-RAY ABSORPTION LINEAR DICHROISM SPECTROSCOPY (XALDS) ON THE PS II MANGANESE COMPLEX - METHODICAL ASPECTS

H. Dau[1], J.C. Andrews[2], T.A. Roelofs[2], M.J. Latimer[2], W. Liang[2], V.K. Yachandra[2], K. Sauer[2] and M.P. Klein[2]
[1]FB Biologie/Botanik, Univ. Marburg, Lahnberge, D-35032 Marburg, Germany; [2]Structural Biology Division, Lawrence Berkeley Laboratory, UC Berkeley, 1 Cyclotron Rd., CA 94720, U.S.A.

1. Introduction

Extended X-ray Absorption Fine Structure (EXAFS) spectroscopy has been employed to investigate the structure of the PS II manganese complex. It allows the high-accuracy determination of the distances between the absorbing manganese atom and backscattering atoms of the first, second and third coordination shell. But usually binding angles can not be derived from EXAFS data. Angular information becomes accessible by X-ray absorption spectroscopy on (partially) oriented preparations [1-3], a technique which in the following is denoted as X-ray Absorption Linear Dichroism Spectroscopy (XALDS).

2. Methods, Results and Discussion

Orientation procedure. Orientation of PS II membrane particles was obtained as shown in Fig. 1. To increase the amount of sample, the paint-and-dry procedure was done several times (2 to 8 times).

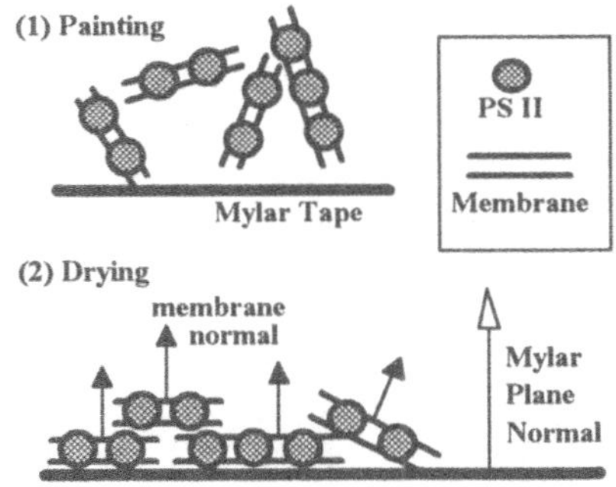

Figure 1. Mylar tape was glued onto a supporting frame. The PS II-membrane suspension was painted with a fine brush onto the the Mylar tape and dried for 20 min under a gentle stream of nitrogen gas (in the dark at 4°C).

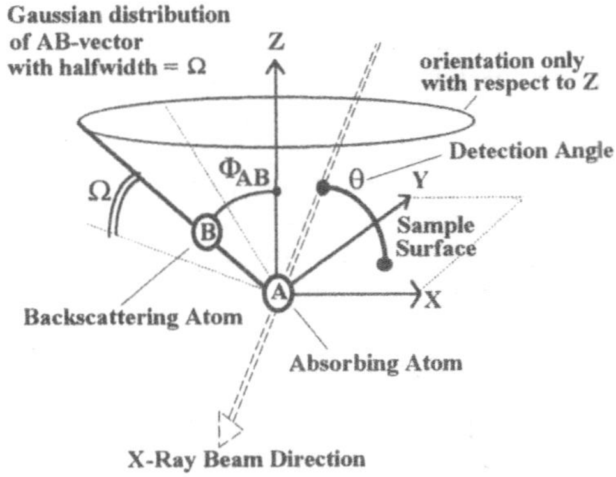

Figure 2. The probability that the AB-vector has the orientation ϕ is assumed to be $P_{AB}(\phi) = \sin\phi\ \exp[-\ln 2\ (\phi - \phi_{AB})/\Omega^2]$.

P. Mathis (ed.), Photosynthesis: from Light to Biosphere, Vol. II, 271–274.

Extent of orientation. By the paint-and-dry procedure samples are obtained with a preferential orientation of the PS II-membrane normal perpendicular to the surface of the substrate (an imperfect one-dimensional orientation, not comparable to two-dimensional crystals). Presumably, the orientation process is driven by gravity force and determined by the sheet-like geometry of the membrane particles. It is unknown whether specific surface interactions contribute. Using the oriented membrane particles, dichroism of spectroscopic signals is observable (Fig. 3-5). Because of its relative simplicity (lack of hyperfine structure) the angle dependence of the cyt b_{559} EPR signals can be easily simulated under the assumption of 'Gaussian disorder' [1]. Comparison with experimental results reveals (see Fig. 4) that, for 3 repetitions of the paint-and-dry procedure, the 'mosaic spread angle' (which corresponds to Ω in Fig. 2) is about 20°.

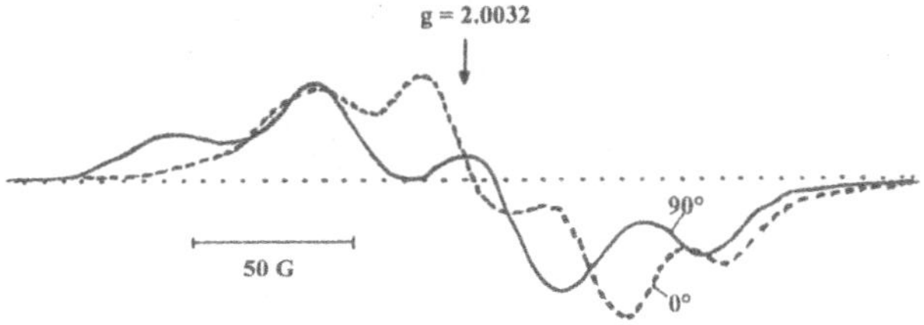

Figure 3. Tyr-D EPR signal of oriented PS II membrane particles. The angle between the sample and the magnetic field was either 0° or 90° (T = 20 K, P = 50 μW, ΔG = 2.5 G).

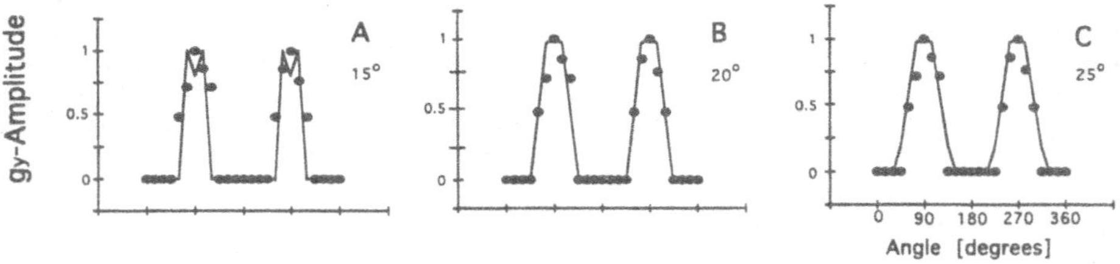

Figure 4. Orientation dependence of the g_y-EPR signal of oxidized cyt b_{559}. The normalized amplitude of the first derivative of the g_y signal is shown versus the detection angle. The filled circles give the experimentally determined g_y amplitude, and the solid line represents the amplitudes simulated for different disorder angles (Ω = 15°, 20°, 25°). The paint-and-dry procedure had been repeated three times.

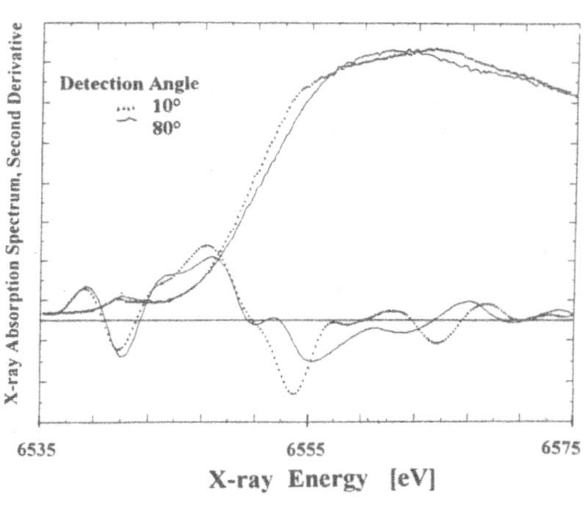

Figure 5. Dichroism of the X-ray absorption edge of samples in the ammonia-modified S_2-state. For experimental details see [1].

Angle dependence of EXAFS. We assume that curved-wave terms in the EXAFS amplitude and phase functions are of minor importance. (It should be noted that for short interatomic distances and for small values of the photoelectron wavenumber, k, this assumption may not hold.) Then, for a distinct atom pair (A, absorbing atom; B, backscattering atom) the EXAFS x-ray absorption cross-section, μ_{AB}, is proportional to $\cos^2 \gamma$ where γ is the the angle between the electric field vector (of the x-rays) and the individual AB-vector. Assuming that the imperfect one-dimensional orientation is reasonable well described by $P_{AB}(\phi)$ as defined in Fig. 2, we obtain

$$\mu_{AB}(\theta, k) = \mu_{AB}{}^{iso}(k)\ F_{AB}(\theta) \quad (1)$$

$$\text{with } F_{AB}(\theta) = \frac{\{\int^{\pi} (3/2 \sin^2\theta \sin^2\phi + 3 \cos^2\theta \cos^2\phi)\quad P_{AB}(\phi)\, d\phi\}}{\{\int P(\phi)\, d\phi\}} \quad (2)$$

$$= 1 + C_{AB}\ \{\ 1 + 3 \cos(2\,\theta)\} \quad (3)$$

$$\text{with} \quad C_{AB} = 1/8 + 3/8\ I_{AB} \quad (4)$$

$$\text{and } I_{AB} = \int (P_{AB}(\phi)\ \cos(2\,\phi))\, d\phi \ / \int (P_{AB}(\phi))\, d\phi\ . \quad (5)$$

In Eq. 1, $\mu_{AB}{}^{iso}(k)$ represents the EXAFS spectrum of a non-oriented (isotropic) sample, which usually is assumed to be of known functional form [1- 4].

Under favorable conditions (see below), curve-fitting of XALDS-data by Eq. 1-5 allows the determination of the angle ϕ_{AB} (provided $\mu_{AB}{}^{iso}$ and the mosaic spread angle (Ω) are known).

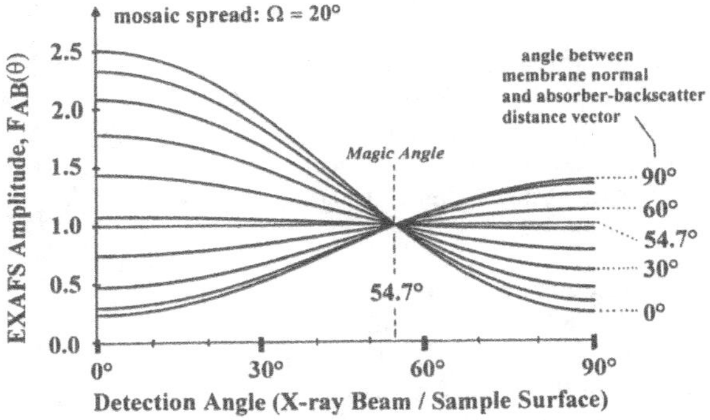

Figure 6. Calculated angle dependence of the amplitude of EXAFS oscillations (for a single backscattering atom per absorbing atom). It was assumed that the disorder of the partially oriented membrane particles is characterized by a mosaic spread angle of 20°.

Orientation of the PS II manganese cluster. For the PS II-Mn complex in its S_1- or S_2-state, two distinct Mn pairs with almost identical AB-distances of 2.7 Å (distance degeneracy) seem to determine the EXAFS oscillations which give rise to a peak of the μ(k)-Fourier-transform at about 2.2 Å (second Fourier peak). Due to the distance degeneracy the $\mu_{AB}(k)$ of both Mn pairs are indistinguishable, and the individual $\mu_{AB}(\theta, k)$ are not resolvable by XALDS. In this case, application of Eq. 1-5 results in an average angle [2,3]. However, by treatment of the PS II with ammonia (ammonia-modified S_2-state) a more favorable situation can be created. One Mn-Mn distance is lengthened from 2.72 to 2.87 Å. Consequently, the distance degeneracy is removed and the apparent coordination numbers for the individual angles can be determined [1]. A fit of the angle dependence of the individual apparent coordination numbers by Eq. 3-5 is shown in Fig. 7. Based on the structural model of Yachandra et al. (1993) and on the angles determined by XALDS on ammonia-treated PS II, a model for the orientation of the PS II manganese complex is proposed (Fig. 8).

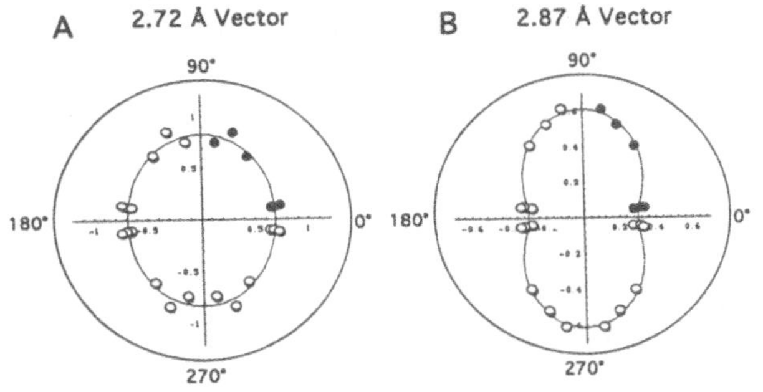

Figure 7. Angle dependence of the apparent coordination number. The apparent coordination numbers, N(θ), are plotted versus the detection angle in the form of a polar diagram. The following angles result from the fit of the calculated $F_{AB}(\theta)$ to the data points:
(A) 55° ± 4°, (B) 67° ± 3°

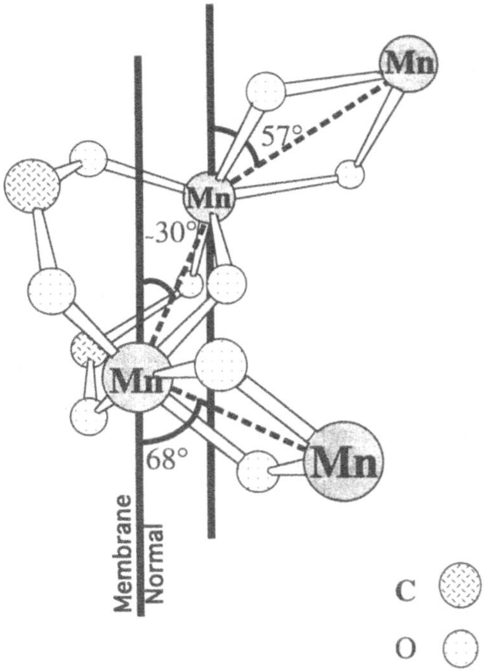

Figure 8. Proposed structure and orientation of the PS II manganese complex in its S_2-state. All manganese atoms are octaedrically coordinated; only the bridging ligands are shown.

Acknowledgement. Support by the NATO (Collaborative Research Grant to H.D. and V.K.Y.) and by the BMBF (Erforschung kondensierter Materie, 48.1 Dau) is gratefully acknowledged. For further acknowledgements see [1].

References.
[1] Dau, H., Andrews, J.C., Roelofs, T.A., Latimer, M.J., Liang, W., Yachandra, V.K., Sauer, K. and Klein, M.P. (1995) Biochemistry 34, 5274-5287.
[2] George, G.N., Prince, R.C. and Cramer, S.P. (1989) Science 243, 789-791.
[3] Mukerji, I., Andrews, J.C., DeRose, V.J., Latimer, M., Yachandra, V.K., Sauer, K. and Klein, M.P. (1994) Biochemistry 33, 9712-9721.
[4] Yachandra, V.K., DeRose, V.J., Latimer, M.J., Mukerji, I., Sauer, K. and Klein, M.P. (1993) Science 260, 675-679.

A TOPOLOGICAL ANALYSIS OF THE PSB B AND PSB O GENE PRODUCTS OF PHOTOSYSTEM II USING EPITOPE-DIRECTED ANTIBODIES

Klaus-D. Irrgang
Max-Volmer-Institute for Biophys. & Phys. Chem.,
TU Berlin, Strasse des 17. Juni 135, 10623 Berlin, FRG

INTRODUCTION

Photosynthetic water oxidation and plastoquinone reduction is catalysed by a manganese-containing thylakoid membrane protein complex referred to as water plastoquinone oxidoreductase. This enzyme was localised within Photosystem II (PS II) on the lumenal side of the thylakoids in cyanobacteria, lower and higher plants and found to consist of 15-20 different subunits. It is still a matter of debate which of these polypeptides are essential for its catalytic activity. Among them the D1-, D2-polypeptides, the $cytb_{559}$, CP47 and CP43 were shown to be functionally important by analysing various deletion and single site mutants whereas the roles of the extrinsic 33 kDa protein (the psb O gene product) and several low molecular mass polypeptides remain to be clarified (for reviews see ref. 1,2). Besides these functional aspects a detailed structural analysis of the water plastoquinone oxidoreductase is still lacking until today. Using a combination of chemical cross-linking and immunological screening with site-specific antibodies the topology and nearest neighbourhood of some of its subunits has been elucidated (1). Electron microscopical studies are available for PS II membrane fragments, PS II core (PS II CC) and core subcomplexes and the supramolecular structure was deduced using high resolution EM and image analysis (3-5). In this study the structure and function of CP47 and the psb O gene product were investigated using epitope-directed antibodies. These IgGs recognize specific sequence domains either nearby the disulfide bridge of the 33 kDa protein or a peptide region within the lumenal loop E of CP47.

MATERIALS AND METHODS

PS II membrane fragments have been purified from spinach leaves according to Berthold et al. (6) with modifications described as in (7). Tris-washed BBY were obtained after extracting extrinsic polypeptides with 0.8 M Tris-HCl at pH 8.35. Oxygen-evolving PS II CC were isolated as described by Haag et al. (3). $CaCl_2$-washed PS II was prepared essentially as described in (4). Light-induced oxygen yield per flash has been measured using a Clark-type electrode as previously reported by Renger (8). The Chl concentration was 50 μM and $K_3[Fe(CN)_6]$ and/or PPBQ (final concentrations 1 mM and 0.2 mM) were used as electron acceptors. On the average PS II membrane fragments contained 220 Chl (a+b) molecules/ centre. PS II polypeptides were electrophoretically analysed as

P. Mathis (ed.), Photosynthesis: from Light to Biosphere, Vol. II, 275–278.

previously described (3,4). Electroblotting was carried out according to Towbin (9) using PVDF membranes. Immunodecorations were screened with alkaline phosphatase conjugated goat anti rabbit IgG in combination with Fast Red /Naphtol AS MX phosphate. Peptide-directed antibodies were elicited against an internal domain of the psb O gene product *(L17-T30)* and a peptide region of the lumenal loop E of CP47 *(V417-E428)*. The peptides were synthesized by a Fmoc-technique using an automated solid phase peptide synthesizer (Rainin, model PS3). The sequences of the peptides were analysed by liquid pulse sequencing (Applied Biosystems sequencer, model 477a) coupled to a PTH-amino acid analyser (model 120) and the masses by ion spray mass spectrometry. The theoretically and practically determined masses were identical (M_r = 1417 Da). For immunization the peptides have been coupled to keyhole limpet hemocyanin by EDC. IgGs and peptide-directed IgGs were purified from the antisera via protein A Sepharose and peptide affinity chromatography using standard procedures. Immunotitrations and topological investigations on various types of PS II preparations were carried out by ELISA in combination with goat anti rabbit IgG alkaline phosphatase and p-nitro-phenylphosphate as a substrate.

RESULTS AND DISCUSSION

Two peptide-directed antibodies have been elicited against two PS II core complex polypeptides designated as psb B - (CP47) and psb O gene products (extrinsic 33 kDa polypeptide) to investigate the structure and function of these subunits within the water oxidizing enzyme. The anti psb O gene product was raised against an internal peptide region of the extrinsic subunit around the first cysteine residue (*C 28*) supposed to form a disulfide bridge (*C28-C51*) in vitro (10,11). Using purified peptide-directed IgGs western blot experiments showed that the antigenic determinant could only be recognized in O_2-evolving PS II core complexes in vitro under reductive conditions with DTE in the presence or absence of urea (see Fig. 1). These results clearly show that 1. the epitope is buried within the interior of the polypeptide in situ and in vitro; 2. the antibodies are both specifically reacting with the sequence domain and also recognizing a definite unfolded protein conformation.

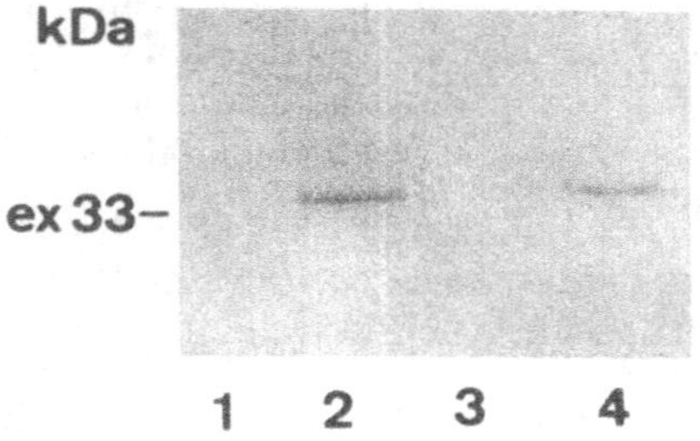

Fig.1 Western blotting of PS II (CC) polypeptides challenged with peptide-directed anti psbO protein lane 1: PS II CC proteins: + 5 M urea, - DTE; lane 2: PS II CC : - urea, + 20 mM DTE; lane 3: PS II CC proteins: - urea, - DTE; lane 4: PS II CC proteins: + 5 M urea, + 20 mM DTE; 5 µg Chl were loaded / lane; IgG dilution 1/500; SDS/PAGE was carried out without urea and DTE

The epitope was predicted to reveal a rather low antigenicity based on hydrophilicity plots using the approach of Hopp and Woods (12). It is worthwile to notify that the immunodecorations on western blots and ELISA plates could be completely blocked in the presence of surplus oligopeptide (*L17-T30*) (not depicted). These results directly support those previously obtained by Tanaka et al. (10,11) on the isolated extrinsic subunit. The cystine was shown to be essential for stabilising the secondary structure of the protein which could not rebind to PS II membrane fragments in its reduced state (10). In addition to that a closed disulfide bridge of the psb O protein is also functionally important for the water oxidizing activity. It could be previously shown that DTE inactivates the water oxidase. The opening of cystines was demonstrated by site-specific modifications of the cysteines with NEM or p-hydroxymercuribenzoate which prevented a reactivation of the enzyme by reoxidation (13). The peptide-directed antibodies against CP 47 were elicited against a domain of loop E *(V417-E428)* previously suggested to be a part of the binding domain (E364-D440) of the psb O protein in spinach PS II (14). Additionally it has been shown by Haag et al. (15) that deletions in the psb B gene in this region (so-called Δ10 (D416-F420)-, Δ11(R422-E428)-mutants) from Synechocystis sp. PCC 6803 slightly affect photoautotrophic capacity and did not lead to an increased requirement for Ca^{2+}. Theoretically the epitope *(V417-E428)* was predicted to have an intermediate antigenicity (12). The specificity of the anti CP47 IgGs has been again investigated by western blotting. The results obtained are demonstrated in Fig.2:

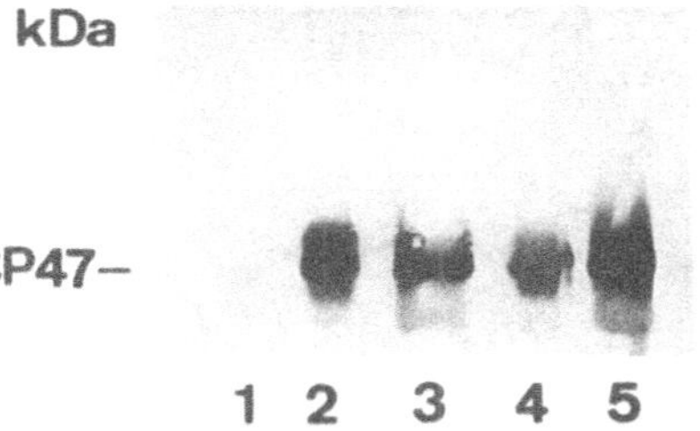

Fig.2: Western blotting of polypeptides of various PS II preparations challenged with anti CP47
lane 1: marker proteins; lane 2: PS II CC (-33 kDa protein), 5 μg Chl; lane 3: Tris-washed PS II membrane fragments 20 μg Chl; PS II CC, 5 μg Chl; PS II membrane fragments, 20 μg Chl;

Fig. 2 shows that in all PS II preparations analysed so far the epitope is accessible to the peptide-directed anti CP47 under denaturing conditions. Under more native conditions using an ELISA the following results were obtained (see Tab. 1): the strongest immunoresponse was observed in Tris-washed PS II membrane fragments, followed by BBY-type PS II membrane fragments whereas PS II CC showed only a very weak immunoreaction and PS II reaction centre complexes did not react. These results can be evaluated as follows: in Tris-washed PS II the extrinsic subunits including the psb O gene product have been removed. The selected peptide-region within the loop E of CP47 is supposed to be located at the border of the binding domain for the extrinsic 33 kDa protein which became accessible after its removal. This supports and extends the work reported earlier (14). The immunoresponse in PS II membrane fragments can be explained by the fact that some of the oxygen evolving complexes lost the 33 kDa protein and are therefore labelled with the IgGs, but it is also not excluded that this region of loop E is not buried by the extrinsic subunit in situ. The rather weak reactivity of the IgGs with their antigenic determinant in PS II CC can be explained by the fact that in the detergent solubilised complexes the binding region is modified in an unknown manner or occupied by detergent molecules. PS II reaction centre complexes were used as a negative control, because they do not contain any CP47. Since anti CP47 IgGs were bound to PS II membrane fragments it was interesting to investigate whether they would be able to affect

the enzymatic activity of the water oxidase. Therefore the oxygen yield per flash has been measured after incubating PS II membrane fragments at a constant Chl concentration with various amounts of IgG. Remarkably the number of Chl molecules per oxygen evolving centre decreased by about 20-25 % (170 Chl/oxygen evolving centre; n=3). These observations are interpreted as an increase of the number of active enzyme centres being able to oxidize water to O_2 and H^+ (not shown). Obviously the binding of the IgGs to the peptide-region induced a conformation favouring a more appropriate conformation of the catalytic site. This implies that either CP47 is directly involved in providing ligands for the manganese cluster or is indirectly altering the structure of a neighbouring subunit (e.g. the D1 protein) that itself is a catalytic subunit of the water oxidase. Summarising the results it can be established that the described epitope-directed antibodies are useful tools to analyse structure-function relationships within the water oxidase. Work is in progress to investigate the activation in more detail and to visualize the binding regions of the IgGs in PS II samples by immune electron microscopy.

Tab. I Epitope-analysis of the region *V417-E428* of loop E of CP47 in various PS II preparations by ELISA using an peptide-directed IgG population

sample	relative immunoresponse
PS II	++
Tris-washed PS II	++++
PS II core complexes	+/-
PS II reaction centre complexes	-

The immunoresponse was photometrically measured at 405 nm. Values of 6 different measurements have been averaged; background absorption $A_{405} = 0.026$

Acknowledgements: The author thanks Prof. G. Renger for his continous support and interest in this work, Dr. C. Weise for sequence analysis of the peptides and Dr. P. Franke for mass spectroscopical analysis, Prof. H. Diringer for providing facilities for the immunisation experiments.

REFERENCES

1. Debus, R. J. (1992) Biochim. Biophys. Acta 1102, 269-352
2. Vermaas, W. F. J., Styring, S., Schröder, W. P. & Andersson, B. (1993) Photosynth. Res. 38, 249-263
3. Haag, E., Irrgang, K.-D., Boekema, E. J. & Renger, G. (1990) Eur. J. Biochem. 189, 47-53
4. Irrgang, K-D., Boekema, E. J., Vater, J. & Renger, G. (1988) Eur. J. Biochem. 178, 209-217
5. Boekema, E. J., Boonstra, A. F., Dekker, J. P. & Rögner, M. (1994) J. Bioenerg. Biomembr. 26, 17-29
6. Berthold, A. D., Babcock, G. T. & Yocum, C. F. (1981) FEBS Lett. 134, 231-234
7. Völker, M., Ono, T., Inoue, Y. & Renger, G. (1985) Biochim Biophys. Acta 806, 25-34
8. Renger, G. (1972) Biochim. Biophys. Acta 256, 428-439
9. Towbin, H., Staehelin, T. & Gordon, J. (1979) Proc. Natl. Acad. Sci. U.S.A. 76, 4350-4354
10. Tanaka, S. & Wada, K. (1988) Photosynth. Res. 17, 255-266
11. Tanaka, S., Kawata, Y., Wada, K. & Hamaguchi, K. (1989) Biochemistry 28, 7188-7193
12. Hopp T. P. & Woods, K. R. (1981) Proc. Natl. Acad. U. S. A. 78, 3824-3828
13. Irrgang, K.-D. Geiken, B., Lange, B. & Renger, G. (1992) Res. in Photosynth. (Murata, N. ed.) Vol. II, 417-420, Kluwer Acad. Publ., Dordrecht, The Netherlands
14. Odom, W. R. & Bricker, T. M. (1992) Biochemistry 31, 5616-5620
15. Haag, E., Eaton-Rye, J.J., Renger, G. & Vermaas, W. F. J. (1993) Biochemistry 4444-4454

DEUTERIUM ISOTOPE EFFECTS ON THE KINETICS OF $Y_Z^{\bullet}$ REDUCTION IN OXYGEN EVOLVING PHOTOSYSTEM II MEMBRANES

N. Lydakis-Simantiris[1,2], C. W. Hoganson[2], D. F. Ghanotakis[1] and G. T. Babcock[2].
Departments of Chemistry, [1]University of Crete, Iraklion 71409, Crete, Greece, and [2]Michigan State University, East Lansing, MI 48824, USA.

INTRODUCTION

In Photosystem II, (PSII), two redox active tyrosines, Y_Z and Y_D, have been discovered [1]. The role of each of these radicals has been the subject of intense research in recent years. Of these two, only Y_Z, identified as Tyr-161 of the D1 protein, participates in the electron transfer chain from the manganese cluster to the reaction center chlorophyll complex, P680 [2,3]. However, the exact mechanism by which this electron transfer occurs is not clear yet. Based on recent spectroscopic data, a model has been proposed that predicts that Y_Z acts as a hydrogen atom extraction agent in a $(Mn_4)/Y_Z^{\bullet}$ active site, rather than as a simple electron transfer mediator [4]. This model is tested in the work reported here by time-resolved electron paramagnetic resonance spectroscopy on PSII samples suspended in either ^1H_2O or ^2H_2O buffers. Experimental data showing kinetic isotope effects on the reduction kinetics of $Y_Z^{\bullet}$ are presented and discussed.

MATERIALS AND METHODS

Oxygen-evolving BBY membranes from spinach were prepared according to [5]. These membranes were washed twice with 25mM MES (pH 6.0), 15mM NaCl, either in ^1H_2O (H-MN buffer) or in ^2H_2O (D-MN buffer). Room temperature time-resolved electron paramagnetic resonance experiments were carried out on a BRUKER ER-200 X-band instrument by using a home-built continuous flow system. Samples were relaxed to the S_1 state in the dark before undergoing flash excitation. Experimental conditions are described in the figure captions. Saturating light flashes (16-20μs full width at half maximum) from a home-built xenon flash lamp were used to trigger the S-states transitions. Analysis of the data was carried out with Origin program (Microcal. v.35). Oxygen evolution activity of the samples was measured by a Clark-type oxygen electrode. BBY membranes in H-MN buffer revealed activity of ~800μM O_2 x mg Chl^{-1} h^{-1}, while those in D-MN buffer had 40-50% less activity (400-500μM O_2 x mg Chl^{-1} h^{-1}). However, BBY membranes in D-MN gained all the activity back when they were assayed with H-MN buffer. Since it has been well established that the rate determining step for steady-state oxygen evolution activity is among the reactions in the acceptor side of PSII, we assume that the loss of ~50% of activity is due to a reversible slow down of one or more of these reactions. All samples were tested by EPR for PSI contamination, and an insignificant amount of PSI was found (data not shown).

RESULTS AND DISCUSSION

Proton release during the S-state transitions has been shown to be dependent on the sub-chloroplast system in which it has been examined. More specifically, a pH dependent 1:0:1:2 pattern has been determined for BBY membranes, while a pH

P. Mathis (ed.), Photosynthesis: from Light to Biosphere, Vol. II, 279–282.

independent 1:1:1:1 is followed in reaction center complexes (RCCs), (for a recent review see [6]). A number of redox active amino acids that are in the vicinity of the oxygen evolving complex, which might buffer the proton release process in the BBY membranes, are apparently absent in LHC and extrinsic polypeptide-depleted RCCs. These data suggest that a proton transfer coupled electron transfer process may be operational in the water-splitting reaction in Photosysten II. Although a considerable amount of work has been done on proton release during the S-state transitions, deuterium isotope effects on the proton-electron transfer have not been examined thoroughly. Renger and coworkers have reported a small isotope effect (k_H/k_D 1.39) on the $S_3 \rightarrow [S_4] \rightarrow S_0$ transition [7], while Sinclair and coworkers have reported that there is no isotope effect [8]. Here, we present time-resolved EPR data showing S-state transition dependent kinetic isotope effects.

Figure 1. EPR kinetic traces and their best fits for the YZ reduction. BBY membranes, (2.2mg Chl/mL), were suspended either in H-MN (a) or in D-MN (b) buffer. 0.75mM DCBQ was used as electron acceptor. 6000 traces were averaged. Experimental conditions: Time constant 35µs, modulation amplitude 4 Gpp, microwave power 7dB, frequency 9.4027 GHz, magnetic field locked at low field peak (3340 G), temperature 25oC, flash frequency 0.66 Hz.

Figure 1 shows kinetic traces and their best fits for the reduction of $Y_Z^•$ by the $(Mn)_4$ cluster in BBY membranes in ^1H_2O (trace a) and in ^2H_2O (trace b) buffers under steady-state conditions. Kinetic analysis of trace a (see below), resulted in lifetimes and signal amplitudes that are in a close agreement with those reported in [9]. Moreover, our experimental conditions allowed us to consider at least three S-state transtitions taking place during the steady state experiment. Our data show a biphasic decay of $Y_Z^•$ in both steady-state traces. The $S_1Y_Z^• \rightarrow S_2Y_Z$ and $S_2Y_Z^• \rightarrow S_3Y_Z$ transitions participate in the fast phase and the $S_3Y_Z^• \rightarrow [S_4] \rightarrow S_0Y_Z$ in the slower process. The half-life times of the fast decay are 140µs and 330µs for traces a and b, respectively. These half-life times correspond to 60% and 56% of the total amplitudes of the traces. The slow parts of traces a and b showed half-life times of 1.4ms and 2.3ms, with amplitudes of 40% and 44% of the total amplitude, respectively. From the above data we obtain a k_H/k_D ratio of 2.36 for the convoluted $S_1Y_Z^• \rightarrow S_2Y_Z$ and $S_2Y_Z^• \rightarrow S_3Y_Z$ transitions and an isotope value of 1.64 for the $S_3Y_Z^• \rightarrow [S_4] \rightarrow S_0Y_Z$.

The deconvoluted kinetic traces and their best fits for the $S_1Y_Z^• \rightarrow S_2Y_Z$ and $S_2Y_Z^• \rightarrow S_3Y_Z$ transitions are shown in figures 2 and 3. We used a continous flow techique to accumulate data from 6000 pairs of flashes while the sample was given enough time (more than 15min) to relax to the S_1 state in the dark before the next set of flashes. Analysis of the traces in figures 2 and 3 gave the following results: The half-life times of the $S_1Y_Z^• \rightarrow S_2Y_Z$ transition were 80-90µs and 230-250µs (figure 2), while the half-life times of the $S_2Y_Z^• \rightarrow S_3Y_Z$ were 350-370µs and 450-470µs (figure 3) for the non-deuterated and deuterated samples, respectively. These data exhibit is presented on table 1.

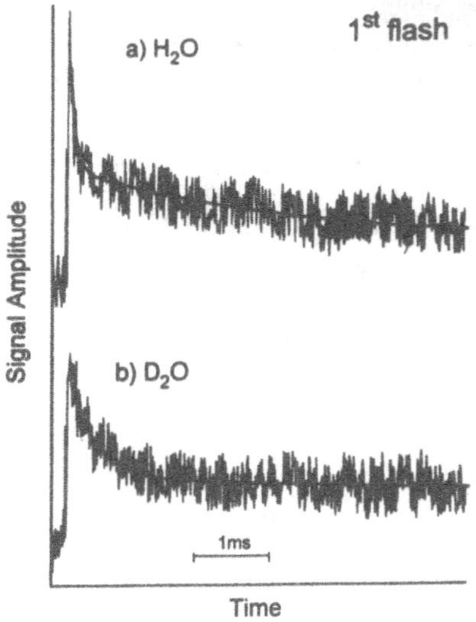

Figure 2. EPR kinetic traces and their best fits for the $S_1Y_Z^\bullet \rightarrow S_2Y_Z$ transition triggered by the first flash in a two flashes experiment. 6000 sets of flashes were averaged. Dark adapted BBY membranes, (2.0mg Chl/mL), suspended either in H-MN or in D-MN were circulated in a continuous flow system. The two flashes were 10ms apart and the sample was replaced before the next set of flashes. Other experimental conditions as in figure 1.

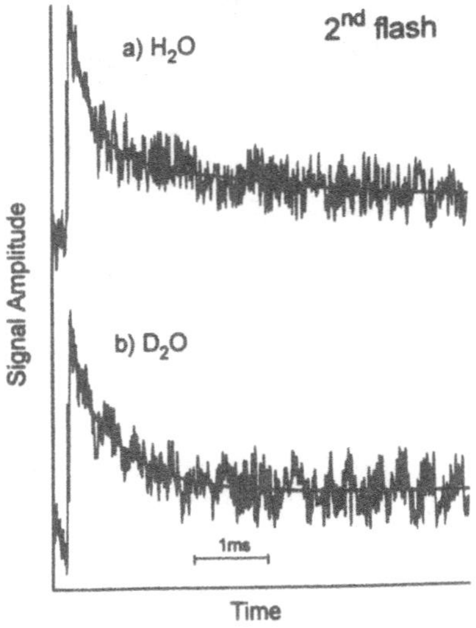

Figure 3. $S_2Y_Z^\bullet \rightarrow S_3Y_Z$ transition kinetic traces derived from the second flash of the experiment descrived in figure 2. All the condition were the same as in figure 2.

	steady state expt.				
	$t_{1/2}$ fast (µs)	$t_{1/2}$ slow (ms)	1st flash $t_{1/2}$ (µs)	2nd flash $t_{1/2}$ (µs)	O_2-evolution µM O_2 mg Chl^{-1} h^{-1}
BBYs in H-MN	140 (60%)	1.4 (40%)	80-90	350-370	800
BBYs in D-MN	330 (56%)	2.3 (44%)	230-250	450-470	400-500
k_H/k_D	2.36	1.64	2.9	1.3	

Table 1. Decay times and, in parentheses, the corresponding amplitudes derived from the time-resolved EPR experiments described in the text. The oxygen evolution activity of the samples is shown as well.

From the experiments described above, some interesting conclusions can be made. First, an isotope effect can be observed in all three of the three transitions we have studied Second, the isotope effect is dependent on the specific S-state transition. These observations can be explained if it is assumed that these S-state dependent isotope effects reflect different electron-proton transfer reactions on the oxidizing side of PSII, and/or differences in the structure/distance of the donor-acceptor system.

A model has been proposed recently [4] in which Y_Z funtions as a hydrogen-atom abstracting agent, extracting both an electron and a proton in a concerted or a sequential mechanism from the $(Mn_4)/(H_2O)$ ensemble on each S-state transition. A schematic view of that model is shown in figure 4.

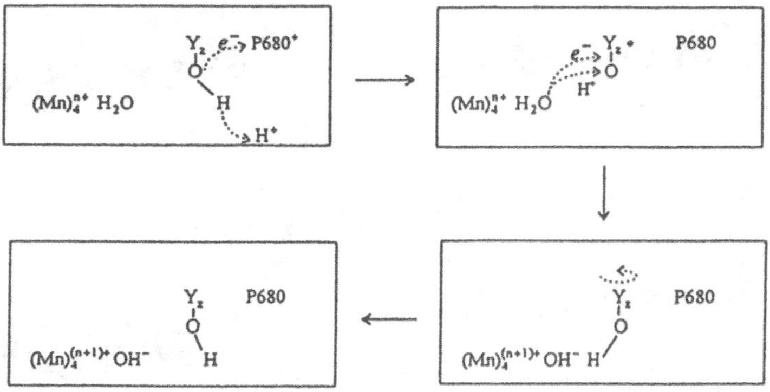

Figure 4. Schematic representation of a proposed model for the function of Y_Z. On each S-state transition hydrogen atom abstraction from the $(Mn)_4/H_2O$ cluster occurs.

According to this model, one would expect different isotope effects for each transition since the O-H bond dissociation energies are expected to differ according to the initial S-state. Moreover, this model and our data so far are in good agreement with the proton release measurements on RCCs, which have shown that one proton is released when Y_Z is oxidized [6].

We are working on refining the conditions in four flashes experiments that will determine precisely the S-state dependency of the deuterium isotope effects and the implications that these effects have on the electron transfer reactions that occur on the oxidizing side of PSII.

Acknowledgments

This work was supported by USDA competitive research grands office, Photosynthesis program.

REFERENCES

1. Barry, B. A., Babcock, G. T., (1987) Proc. Natl. Acad. Sci. USA, 84,7099-7103.
2. Babcock, G. T., Barry, B, A., Debus, R. J., Hoganson, C. W., Atamian, M., McIntosh, L., Sithole, I., Yocum, C. F., (1989) Biochemistry 28, 9557-9565.
3. Barry, B., (1993) Photochem. Photobiol. 57: 179-188.
4. a) Hoganson, C. W., Lydakis-Simantiris, N., Tang, X-S., Tommos, C., Warncke, K., Babcock, G. T., Diner, B. A., McCracken, J., Styring, S., (1995) Photos. Res. submitted, b) Tommos, C., Tang, X-S., Warncke, K., Hoganson, C. W., Styring, S., McCracken, J., Diner, B. A., and Babcock, G. T., (1995) J. Am. Chem. Soc. submitted
5. Ghanotakis,D.F., Topper, J, Babcock, G.T., Yocum, C.F. (1984) Biochim. Biophys. Acta 767, 524-531.
6. Lavergne, J., Junge, W., (1993) Photosynth. Res. 38, 279-296.
7. Renger, G., Bittner, T., Messinger, J., (1994) Biochem. Soc. Trans. 22, 318-322.
8. Sinclair, J., Sarai, T., Arnason, T., Garland, S., (1978) in Photosynthetic Oxygen Evolution, H. Metzner, ed., Academic, London, pp.295-320.
9. Hoganson, C. W., Babcock, G. T., (1988) Biochemistry 27, 5848-5855.

HETEROGENEITY IN SUBSTRATE WATER BINDING TO PHOTOSYSTEM II

J. Messinger, W. Hillier, M. Badger and T. Wydrzynski
Research School of Biological Sciences, The Australian National University, Canberra, ACT 0200, AUSTRALIA

1. Introduction

In higher plants water oxidation is performed by the membrane bound chlorophyll-protein complex photosystem II (PS II) which is mainly located in the grana stacks of the thylakoid membrane. With regard to water oxidation each PS II operates independently and cycles through five different redox states termed S_0, S_1 ,..., S_4 [1]. The higher S-states, S_2 and S_3, are reduced in the dark to S_1 through reduction by electrons from the acceptor side or by the reduced form of the tyrosine residue 161, Y_D , of the reaction center polypeptide D2, while S_0 is slowly oxidized to the dark stable S_1 state by Y_D^{ox}. The S_4 state represents the redox state where four electrons have been abstracted from the catalytic site of water oxidation. This state converts within a few milliseconds upon oxygen release into the lowest, natural S-state, S_0. The catalytic site of water oxidation is formed by four electronically coupled manganese ions, which are probably organized as two μ-oxo bridged dimers, with perhaps different functions. The univalent oxidation steps, $S_i \rightarrow S_{i+1}$ (i = 0, 1, 2, 3), are each driven by the energy of an absorbed photon which leads to a charge separation in the reaction center of PS II that is formed by a specially bound chlorophyll a dimer (or monomer), termed P680, and a pheophytin molecule, Pheo. $Pheo^-$ transfers its electron to the acceptor side quinones Q_A and Q_B , while $P680^+$ is reduced by the tyrosine residue 161, Y_Z , of the reaction center protein D1. In turn, Y_Z abstracts the electrons from the manganese cluster with rate constants that depend on the S-state.
Although a considerable amount of information is available about the organization of the four manganese ions and their redox states [2], only little is known about the actual chemistry during water oxidation, i.e.: a) In which S-state(s) do the two water molecules bind to the catalytic site and what are their binding affinities? b) Is the access of water to the catalytic site regulated? c) Is the substrate water partially oxidised, e.g. to a peroxidic intermediate, before the S_4 state? d) Do both substrate water molecules bind to the manganese ion(s) prior the O-O bond formation ? e) What is the protonation state of the (bound) substrate water molecule(s) in the different S-states? f) What triggers the O-O bound formation? The lack of information concerning these questions is due to the fact that in this reaction water is not only the substrate for PS II but also the solvent. To circumvent this problem attempts have been made to use various water analogs, i.e. H_2S, HOR (e.g. R = CH_3, C_2H_5), NH_3 (and organic derivates), hydrogen peroxide and its analogs N_2H_4 and NH_2OH, and water isotopes (D_2O, $H_2{}^{17}O$ and $H_2{}^{18}O$) in various activity measurements like oxygen yield measurements, EPR and mass spectrometry. With the latter technique it is, in principle, possible to measure the kinetics of the substrate water

P. Mathis (ed.), Photosynthesis: from Light to Biosphere, Vol. II, 283–286.

exchange reactions after rapid injection of $H_2{}^{18}O$ into a preflashed thylakoid suspension. The rate and activation energy of the exchange provide information about the mode of binding and possible diffusion limitations in a particular S-state. Through a significant improvement of time resolution we were recently able to measure for the first time the exchange kinetics of the substrate water molecules in the S_3 state of PS II. A surprising hetereogeneity was found with one slow (t_s = 500 ms) and one fast (t_f < 40 ms) exchanging molecule [3]. Here we report on some further improvement of the technique and on our first results concerning the S_2 state.

2. Procedure

Thylakoid samples were prepared from spinach according to the following procedure: two batches (200 g) of leaves were ground each in 500 ml 'SHMN-buffer' (400 mM sucrose, 50 mM Hepes/NaOH, pH 7.0/4°C, 5 mM $MgCl_2$ and 15 mM NaCl) and filtered through two layers of mira-cloth separated by one layer of cotton wool and immediately centrifuged for 5 min at 5000 rpm at 4°C (4 tubes, Beckman, GSA rotor). The pellets were carefully resuspended in 400 ml SHMN-buffer and centrifuged again for 10 min at 7500 rpm (2 tubes, Beckman, GSA rotor). The pellets were resuspended in 160 ml SHMN-buffer, thoroughly homogenised and centrifuged for 10 min at 7500 rpm (4 tubes, Beckman, SS34 rotor). After a further wash in 80 ml SHNM-buffer, the thylakoids were resuspended to about 3 mg Chl/ ml in SHNM-buffer, frozen as small aliquots in liquid nitrogen and then stored at - 80°C until used. Before the measurements, the samples were thawed on ice. SHMN-buffer adjusted to pH 6.8 at 10°C or 20°C was also used to dilute the thylakoids to the measuring concentration of 500 µM chlorophyll.
The exchange of $H_2{}^{16}O$ against $H_2{}^{18}O$ in the S_2 state was measured at 10°C using the following flash/$H_2{}^{18}O$-injection protocol (Scheme 1):

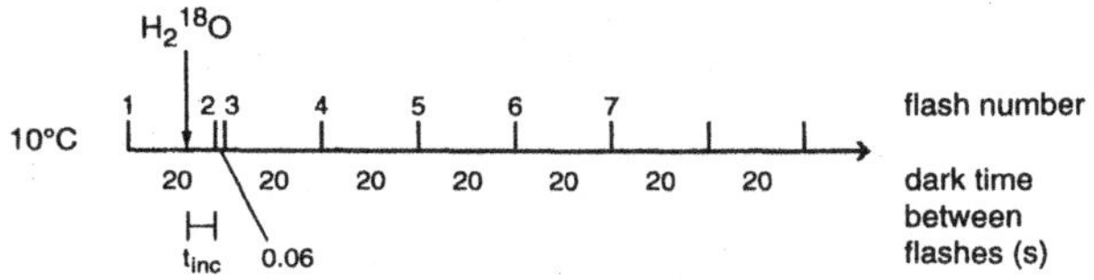

Scheme 1

At 20°C, a dark time of 1 s was used between the firstflash and the double flash.The samples were illuminated by saturating, single turnover flashes produced by a Xe-flash lamp (EG&G, 4 µFd, 1kV capacitor). The flash lamp and the syringe were triggered from a personal computer. One preflash and a 30 min darktime in the cuvette were used to preset the sample to the state $S_1Y_D{}^{ox}$. As the CR 700-200 syringe has a dead volume of 8 µl, the syringe was washed after each measurement first in distilled water and than in 15 µl $H_2{}^{18}O$, before the final volume of $H_2{}^{18}O$ was taken up under further mixing. This procedure led to a significant increase in the final $H_2{}^{18}O$ concentration in the sample chamber. The flash induced oxygen signals were measured with a mass spectrometer as described earlier [3]. At times up to about 10 s between injection and the 3rd flash it is impossible with our current setup to resolve the signals due to the background oxygen content of the added $H_2{}^{18}O$ from the flash-induced oxygen evolution by PS II. Therefore, at m/e = 34 and 36 the background signals were determined by several injections of $H_2{}^{18}O$ and the mean of the corresponding peak heights was subtracted from the observed oxygen signal of the double flash. The corrected signal amplitudes were then normalized to the sum of the respective yields of the flashes 4 to 7 in the same train of flashes and within one time course to the (normalized) steady state peak height of the double flash that is reached at long incubation times with $H_2{}^{18}O$ in the S_2 state.

3. Result and Discussion

For the S_2 state measurements a similar set up and protocol was used as described in [3]. However, a new sample chamber was built with a volume of 165 µl compared to previously 300 µl. With the smaller sample chamber, twice the chlorophyll concentration (500 µM) and less $H_2{}^{18}O$ (30 µl at m/e = 34 and 60 µl at m/e = 36) could be used and a better signal to noise ratio was achieved. The amplitude of the injection artifact was reduced to about 75 % (m/e = 34) and 45 % (m/e = 36) of the equilibrium amplitude of the oxygen yield in the detecting double flash, so that the addition of glucose, glucose oxidase and catalase to the $H_2{}^{18}O$ could be avoided (previously it was necessary for measurements at m/e = 34). The injection/mixing times were measured with a PAM 101 Chlorophyll Fluorometer (Walz, Germany) by injecting thylakoids into the buffer-filled sample chamber as described in [3]. Complete mixing was achieved after 30 ms and 40 - 50 ms for 30 µl and 60 µl injection volume, respectively (data not shown). The dead time before the injection was reduced to 10 ms by increasing the air pressure against the solenoid switch at the injection button (see Fig. 1 in [3]). Since our thylakoids have a carbonic anhydrase activity (data not shown) the final $H_2{}^{18}O$ enrichment could be calculated from the CO_2 mass peaks (m/e = 44, 46, 48) to be 17% and 33% at m/e = 34 and 36, respectively.

3.1 *Substrate water exchange in the S_2 state*

The substrate water exchange in the S_2 state was measured using a rapid double flash technique (see Scheme I). Figure 1 displays the resulting oxygen yields of the double flash (❑) as a function of the time between $H_2{}^{18}O$ injection and the double flash excitation as measured at 10°C (A,B) and 20°C (C,D). The time between the first flash and the double flash was kept constant to minimize effects of S_2 and S_3 decay on the reults. A comparison

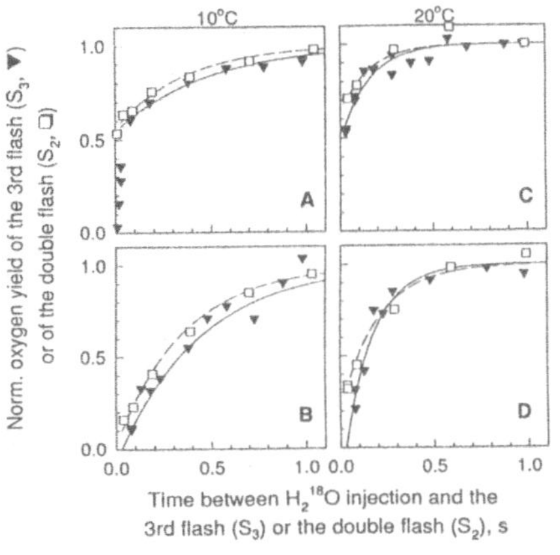

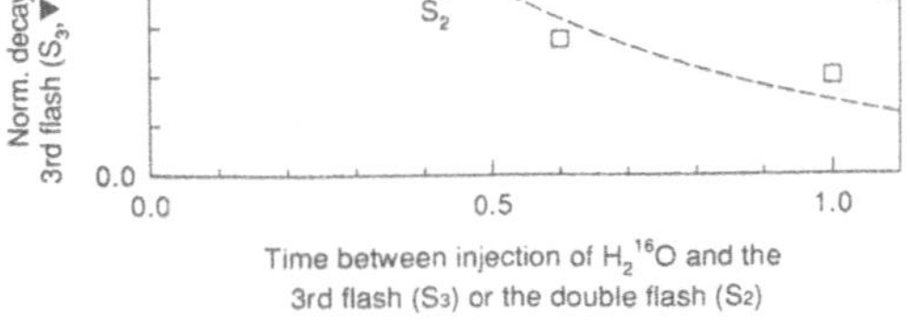

left Figure 1: Substrate water exchange in the S_2 (❑) and S_3 (▼) state of PS II at 10°C/pH 6.8 and 20°C/pH 6.8 after injection of $H_2{}^{18}O$ into the spinach thylakoid suspension. The oxygen yields were measured at m/e = 34 (A, C) and m/e = 36 (B, D).
right Figure 2: Substrate water exchange in the S_2 (❑) and S_3 (▼) state of PS II at 10°C/pH 6.8 after injection of $H_2{}^{16}O$ into a 63% $H_2{}^{18}O$-enriched spinach thylakoid suspension. The oxygen yields were measured at m/e = 36.

with the data obtained previously [3] for the S_3 state (▼) reveals only minor differences in the exchange properties of these to S-states. As a complete equilibration of the fast exchanging substrate water molecule occurs during the 60 ms darktime of the double flash, no conclusions can be drawn at this stage about a fast exchange in the S_2 state.The data for the slow exchange in S_2 were simulated by the following equations:

Eqs. 1a,b (m/e 34): 10°C $y = 0.59 + 0.41 * (1 - e^{-kt})$, 20°C $y = 0.7 + 0.3 * (1 - e^{-kt})$

Eqs. 2a,b (m/e 36): 10°C $y = 0.10 + 0.90 * (1 - e^{-kt})$, 20°C $y = 0.3 + 0.7 * (1 - e^{-kt})$

The offsets in Eqs. 1a,b account mainly for the fast exchange in S_3 (and S_2?), while those in Eqs. 2a,b take the small amount of slow exchange during the double flash into account. As for the S_3 data, an offset of 30 ms was used to compensate for the injection. The following rate constants have been used for: $k_s^{34} = 2.8\ s^{-1}$, $k^{36} = 2.8\ s^{-1}$ (10°C) and $k_s^{34} = 6.5\ s^{-1}$, $k^{36} = 5.0\ s^{-1}$ (20°C). Due to the large discrepancy in the rates at 20°C no reliable estimation of the activation energy for the S_2 state is possible at present. However, it seems to be in the same range as for S_3, possibly somewhat smaller. Overall, the exchange rates are very similar in S_2 and S_3.Therefore, a direct binding of the slowly exchanging substrate water molecule to a Mn ion that changes its formal redox state from Mn^{3+} to Mn^{4+} during the $S_2 \rightarrow S_3$ transition seems to be unlikely.

3.2 *Exchange of bound $H_2{}^{18}O$ by injected $H_2{}^{16}O$*

To obtain further evidence for this surprising result we measured also the exchange of bound substrate $H_2{}^{18}O$ against $H_2{}^{16}O$ by injecting 50 µl SHMN-buffer into an about 63% $H_2{}^{18}O$ enriched thylakoid suspension at 10°C. This led to an rapid decrease of the $H_2{}^{18}O$ concentration to 48%. In this case, the decay of the oxygen yield at m/e = 36 reflects the exchange reaction. This approach has the advantage of a much better signal to noise ratio in the flash pattern at m/e = 36; however, the amplitude of change during the time course is much smaller. The results are shown in Figure 2. The decay of the 3rd flash (S_3) or the double flash (S_2) is shown after subtraction of the corresponding equilibrium signal amplitudes obtained at long times after injection. For comparison, the data are also normalized to values obtained by extrapolated to $t_{inc} = 0$ (these are 'theoretical' values, as up to about 40 ms a fast decay occurs due to the dilution of the sample caused by the injection). Qualitatively, a similar result is obtained as in Fig. 1. Again a slightly faster substrate exchange is found for S_2 than for S_3. However, a simulation of the data (see lines in Fig. 2) using formulas for simple pseudo first order decays reveals rate constants of $k^{36}_{18\rightarrow16}(S_3, 10°C) = 0.95\ s^{-1}$ and $k^{36}_{18\rightarrow16}(S_2, 10°C) = 1.9\ s^{-1}$, which are significantly smaller than those obtained for the exchange of bound '$H_2{}^{16}O$' against $H_2{}^{18}O$. The magnitude of this difference seems rather large to be accounted solely to differences in zero point energies. However, together with the high activation energies this result indicates a very strong binding of the slowly exchanging substrate water molecule in S_3 and S_2. As proposed earlier [3] this could be realized by a terminal manganese oxo ligand. The slightly higher exchange rate in S_2 and the smaller relative difference between the $k_{16\rightarrow18}$ and $k_{18\rightarrow16}$ rate constants for S_2 than S_3 indicate a somewhat weaker binding of the slowly exchanging substrate molecule in S_2. A more detailed discussion will be presented after further refinement of the data in a future publication.

[1] Kok, B., Forbush, B. & McGloin, M. (1970) Photochem. Photobiol. 11, 457 - 475
[2] Debus, R. (1992) Biochim. Biophys. Acta 1102, 269 - 352
[3] Messinger, J., Badger, M. & Wydrzynski, T. (1995) Proc. Natl. Acad. Sci. USA 92, 3209 - 3213

UNCOUPLING OF DONOR SIDE ACTIVITY - EVIDENCE FOR CYCLIC ELECTRON FLOW AROUND PHOTOSYSTEM II *IN VIVO* ?

[1,2]O.Prášil, [1]Z.Kolber, [3]J.A,Berry, [1]P.G.Falkowski
[1]Oceanographic and Atmospheric Sciences Division, Department of Applied Science, Brookhaven National Laboratory, Upton, NY, 11973-5000; [2]Institute of Microbiology, Academy of Sciences of Czech Republic, Třeboň, 379 81, Czech Republic; [3]Carnegie Institution of Washington, Department of Plant Biology, Stanford, CA, 94305

1. Introduction

Over a wide range of light intensities the relationship between the changes in variable fluorescence yield ($\Phi=F_v/F_m$) and oxygen flash yields (Y_{O2}) is linear, implying that the reduction of Q_A is closely coupled to the oxidation of water and that all electrons delivered to Q_A originated from water [1]. At high irradiances, however, Φ often appears to be greater than Y_{O2}, suggesting an uncoupling between the oxidation of water and the reduction of PSII acceptors. One interpretation consistent with the apparent uncoupling is cyclic electron flow around PSII; i.e. Q_A can be photochemically reduced without the electron originating from water [1].

We have studied the conditions under which such uncoupling might occur *in vivo*. Simultaneous measurements of the changes in Φ and Y_{O2} were obtained in intact *Chlorella* cells on a bare platinum oxygen rate electrode. Our results show that uncoupling between yields of variable fluorescence and oxygen evolution also occurs transiently during light/dark transitions when the acceptor side of PSII is over-reduced. The decline in the number of higher S states (and the loss of Y_{O2}) in some of the reaction centers of PSII is not caused by recombination between $S_{2/3}$ states and the acceptor side [2,3], but rather by competitive donation of electrons by an alternative donor.

2.Procedure

Chlorella pyrenoidosa was grown as previously described [1]. The cells were harvested by centrifugation, resuspended in fresh medium and placed on the platinum button of a rate-measuring oxygen electrode [1]. The repetitive (0.4 Hz) saturating and single-turnover flash excitation for Φ and Y_{O2} determination was provided by the xenon flash lamp of a fast-repetition-rate (FRR) fluorometer [4]. The blue (Corning 4-96) continuous background light was provided by 500W Oriel high-pressure xenon arc lamp and its intensity was adjusted using neutral density filters. Thermoluminescence was measured with a custom-built instrument. Cells on the sample stage were pre-illuminated by blue LEDs of desired intensity and then excited by a xenon flash lamp at 2°C. Thermally stimulated emission was subsequently recorded during heating to +70°C.

P. Mathis (ed.), Photosynthesis: from Light to Biosphere, Vol. II, 287–290.

3. Results and Discussion

The flash-induced quantum yield of fluorescence (Φ) and the yield of oxygen (Y_{O2}) are coupled in dark adapted samples (*Figure 1.*, Dark period 1) and during non-saturating continuous background light (Actinic on). A significant hysteresis occurs after the actinic light is switched off (Dark period 2). Under these conditions, Φ recovers to its previous dark level within 5-10s, while the oxygen flash yield does not attain its pre-illumination steady state value until about 60s later.

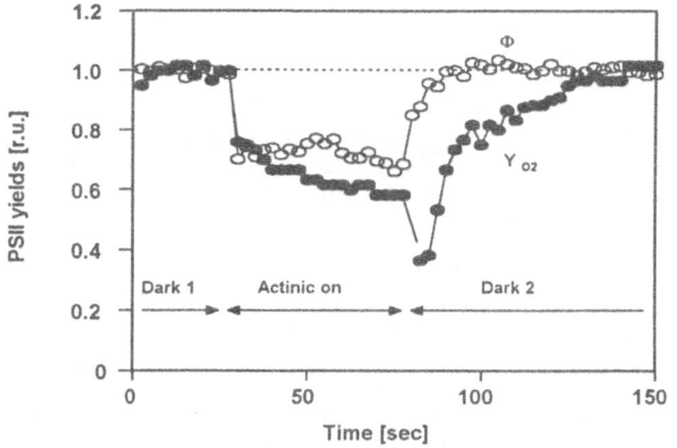

Figure 1. Changes in flash-induced yields Φ (O) and Y_{O2} (●) prior to, during, and following a 50s exposure to continuous blue-green actinic light at 60 μmol quanta$\cdot m^{-2} \cdot s^{-1}$.

The rise of Φ reflects relaxation of photochemical quenching mechanisms. We have observed a changes in functional cross-section of PSII occurring in parallel to changes in Φ (not shown). Contrary to our expectations, Φ (calculated as $(F_m' - F_i)/F_m'$) can attain its maximal value *before* fluorescence yield F_i (i.e. the fluorescence yield before the actinic flash) has fully relaxed towards F_o. At the same time, the maximal flash-induced fluorescence yield F_m' becomes *higher* than F_m measured in dark adapted state. If we assume that Φ obtained with single-turnover saturating flashes is a measure of the photochemical yield of PSII, and is proportional to the number of open PSII reaction centers, the changes in Φ and F_i suggest that i) PSII reaction centers are reoxidized within 5-10s after cessation of actinic light; and ii) under these conditions another non-photochemical quencher influences the fluorescence yields. Though the identity of this quencher and the mechanism of its function is not fully understood, this quenching of variable fluorescence is strikingly similar to "thermal" quenching or "R" quenching that are related to the redox state of the plastoquinone pool [5,6,7].

The recovery of Y_{O2} has several phases. The amplitude and kinetics of these phases depend on the intensity and duration of actinic light and treatment after transition from light (summarized in Table 1). There is a strong correlation between the kinetics of Y_{O2} and the relaxation of F_i. In general, the recovery of Y_{O2} is significantly faster when the acceptor side of PSII is oxidized, e.g. if weak far red light is present during dark recovery. The recovery of Y_{O2} can be slowed down by anaerobiosis. The effect of these treatments on Φ is much less pronounced.

Table 1 Recovery kinetics (shown as halftimes τ for recovery from light adapted state to the steady state in dark) of flash induced fluorescence yield Φ, fluorescence yield in the dark F_i and oxygen flash yield Y_{O2} under different conditions. In all cases, the cells were first exposed to 2min of continuous blue actinic light of 100 μmol quanta·$m^{-2}\cdot s^{-1}$

Transition:	Halftimes in [s] for recovery of:		
	Φ (F_v'/F_m')	F_i	Y_{O2}
actinic → dark	5	16	20
actinic → weak far red (720nm)	5	7	10
actinic → weak blue (450nm)	5	12	15
actinic → dark + DCMU	2	2	-
actinic → dark + anaerobiosis	25	40	40

These results show that there is a transient uncoupling between the yields of fluorescence and of oxygen release. Though, to our knowledge such direct measurements of the hysteresis between Y_{O2} and Φ have not been reported, several previous reports have suggested fast deactivation of S_3 states after light/dark transition due to direct recombination between higher S states and charges stored at the acceptor side (Q_A^-, Q_B^-) [3,8]. The acceleration of S states deactivation was hypothesized to be due to: i) over-reduction of the PSII acceptor side, and/or ii) shortening of the lifetime of recombination. An alternative hypothesis to this direct recombination mechanism postulates the existence of a cyclic pathway around PSII that would shuttle electrons from quinone acceptors to the donor side of PSII, temporally disconnecting the water-splitting complex. The signal that would open such escape pathway might be over-reduction of the plastoquinone pool or an extensive ΔpH that cannot be dissipated via ATP synthesis.

In order to check the validity of the recombination hypothesis, we have measured thermoluminescence (TL) signatures at various time intervals following continuous actinic illumination by blue light. At these intervals, cells were rapidly cooled to low temperature and excited by 0,1,2 or 3 consecutive flashes. In this way, assuming that number of misses is low, we were able to estimate the total number of charge pairs active in linear transport from water to plastoquinone (*Figure 2.*) and the lifetimes of the higher S states (*Figure 3.*). We have found that: i) after continuous illumination the total number of reaction centers active in linear electron transport decreases by about 40%, and ii) the halftime of higher S states decreases slightly (from 15-20s in dark adapted state to approx. 10s after continuous illumination). Since the interval between flashes was 2.5s, such an increase in the rate of $S_3Q_B^-$ recombination would not be sufficient to account for sharp decline in concentration of S_3 states as observed in *Figure 1.* Furthermore, the absence of Q ($S_2Q_A^-$) and A ($S_3Q_A^-$) bands from thermoluminescence glow curves proves that the fast decline of higher S states is not due to recombination with Q_A^- that might accumulate due to over-reduction of the plastoquinone pool. Thermoluminescence experiments therefore do not support the hypothesis that the recombination pathway is a major reason for the observed transient decline in Y_{O2} after light/dark transition.

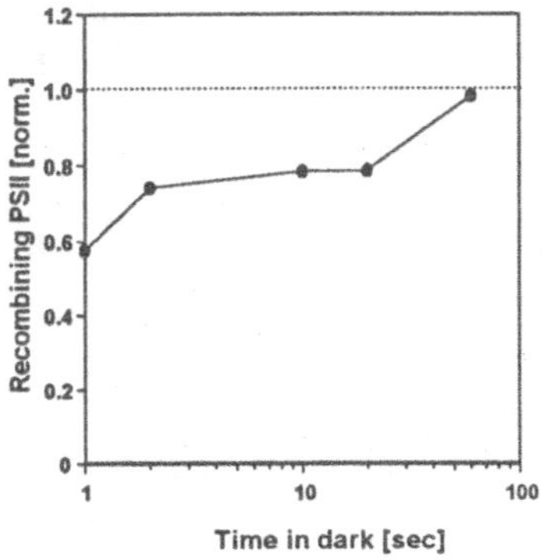

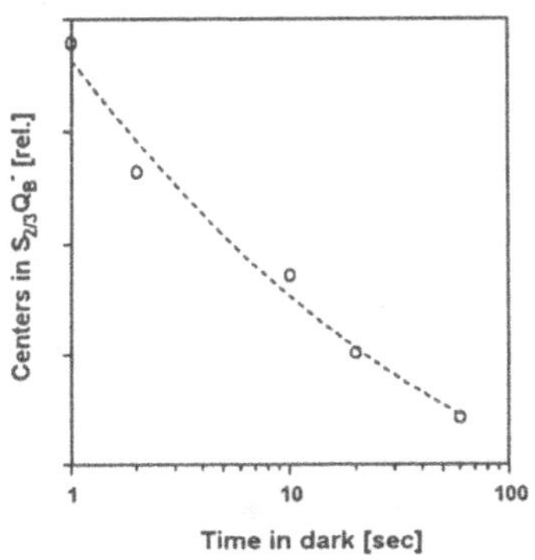

Figure 2. (left) Fraction of PSII reaction centers active in linear electron flow from H_2O to plastoquinone after transition from blue (450 nm, 100μmol quanta·m^{-2}·s^{-1}) continuous light to dark. Number of active reaction centers was estimated from thermoluminescence measurements as the sum of areas under TL glow curves after 0,1,2 and 3 flashes. The dotted line gives number of active reaction centers in dark adapted samples.

Figure 3. (right) Concentration of $S_2Q_B^-$ and $S_3Q_B^-$ charge pairs in dark, after transition from blue continuous light to dark. The number of centers in S_2 and S_3 states was estimated from areas under TL curves measured at given time intervals after transition, with no further excitation.

Though we are still lacking direct proof for existence of a cyclic electron pathway around PSII, our data strongly suggest that his pathway exists. As has been suggested several times, it might involve transfer of electrons from Q_A or Q_B to the donor side through cyt b_{559} [9]. Though such a pathway would probably not have high rates of electron transport, it might still be useful under fluctuating light conditions, when temporal over-reduction of the plastoquinone pool together with low lumenal pH makes the adjustment of linear operation of Photosystem II difficult.

References

1 Falkowski, P.G., Wyman, K., Ley, A.C. and Mauzerall, D.C. (1986) Biochim. Biophys. Acta 849, 183-192
2 Lemasson, C. and Barbieri, G. (1971) Biochim. Biophys. Acta 245, 386-397
3 Franck, F. and Schmid, G.H. (1993) Z. Naturforsch. c 48, 603-608
4 Kolber, Z., Barber, R. T. et al. (1994) Nature 371, 145-149
5 Delosme, R. (1971) C. R. Acad. Sc. Paris 272, 2828-2831
6 Vernotte, C., Etienne, A.L. and Briantais, J.-M. (1979) Biochim. Biophys. Acta 545, 519-527
7 Neubauer, C. and Schreiber, U. (1987) Z. Naturforsch. c 42, 1246-1254
8 Sundblad, L.G., Schröder, W.P. and Åkerlund, H.E. (1988) Biochim. Biophys. Acta 973, 47-52
9 Nedbal, L., Samson, G. and Whitmarsh, J. (1992) Proc. Natl. Acad. Sci. USA 89: 7929-7933

BALANCE BETWEEN PHOTOACTIVATION AND PHOTOINHIBITION OF PHOTOSYSTEM II IN *CHLAMYDOMONAS REINHARDTII* LACKING THE 23 kDa EXTRINSIC SUBUNIT

M. Rova[1,2], B. Mc Ewen[2], P.-O. Fredriksson[2] and S. Styring[1]
[1]Dept. of Biochemistry, Arrheniuslaboratories, Stockholm Univ., S-106 91 Stockholm, Sweden.
[2]Dept. of Chemistry, University of Karlstad, 651 88 Karlstad, Sweden.

1. Introduction

Photoactivation (1) is the process in which the Mn cluster of photosystem II (PS II) is assembled and O_2-evolution is acquired. The oxygen-evolving complex consists of four Mn ions, requires Cl^- and Ca^{2+} as cofactors and is stabilized by three extrinsic subunits of 16, 23 and 33 kDa. Photoactivation proceeds via two successive photoreactions, each proposed to involve photooxidation of one Mn^{2+} to Mn^{3+} (2). The product of the first photoreaction has to be rearranged in darkness before the second photoact and the intermediates decay if photoactivation does not proceed. Cl^- is essential for photoactivation and has been proposed to stabilize the intermediate state, thereby increasing the yield of photoactivation (3). Also Ca^{2+} and the 33 kDa subunit of PS II increase the efficiency of photoactivation (4, 5).

PS II is susceptible to donor-side photoinhibition if the water oxidizing capacity is impaired (6). This is due to oxidizing events in PS II connected with the appearance of unusually long-lived radicals of the strongly oxidizing primary donor $P680^+$ and/or $TyrZ^+$ and leads to degradation of the D1 reaction centre protein. Donor-side photoinhibition occurs at low light intensities in for example chloride-depleted thylakoids (7) or in a mutant lacking the 23 kDa extrinsic subunit of PS II (8).

Chlamydomonas reinhardtii is able to grow heterotrophically and PS II is completely assembled and fully functional in the dark except for the Mn cluster that has to be photoligated to the reaction centre before O_2-evolution can occur. We have studied a mutant, FUD 39, of *C. reinhardtii* lacking the 23 kDa extrinsic subunit. It contains wild type level of PS II but shows a decreased O_2-evolving activity that is strongly chloride-dependent (8). The fraction of centres active in water splitting in thylakoid membranes was 1.0 for the wild type but only 0.6 for the FUD 39 mutant at saturating Cl^- concentration (8). This leads to the proposal that the photoactivation process is less efficient in the mutant due to a lower affinity for Cl^- and that photoinhibition might compete with photoactivation. Here we have tested this hypothesis by photoactivation of dark grown cells initially lacking oxygen evolution.

2. Materials and Methods

C. reinhardtii, wild type and the mutant FUD 39, was grown on Tris-acetate-phosphate

P. Mathis (ed.), Photosynthesis: from Light to Biosphere, Vol. II, 291–294.

medium, pH 7.0, in complete darkness. In the early exponential phase, the cells were exposed to light of different intensities for various periods of time to accomplish photoactivation/photoinhibition. Thylakoid membranes were prepared in the dark by sonication (8) immediately after the light treatment. Electron transfer from water or 2,2′-diphenylcarbonic dihydrazide (DPC) was measured as 2,6-dichlorophenolindophenol (DCIP) reduction at 590 nm as described previously (8) at 5-10 µg Chl ml^{-1}, in a buffer containing 25 mM HEPES-NaOH (pH 7.0), 50 mM NaCl, 400 mM sucrose, 6 µg ml^{-1} gramicidin and 35 µM DCIP. When used, DPC was added to a final concentration of 1 mM.

3. Results and Discussion

3.1 *In vivo photoactivation is more efficient in wild type than in FUD 39*
Relative photoactivation is measured as the ratio between electron transfer from H_2O or DPC to DCIP in thylakoid membranes from cells treated as described above. Electron transfer from DPC occurs in all centres while water oxidation is possible only in photoactivated centres. Fig. 1 shows that wild type is photoactivated efficiently in about 10 minutes. Furthermore, there is little difference between the two light intensities used. Photoactivation of the mutant however is much less efficient. It takes longer times and the level of photoactivation differs between the two light intensities, indicating stronger light-dependence in the mutant. To test our mutant strand we have run SDS-PAGE that shows that FUD 39 lacks the 23 kDa subunit but has wild type level of the 33 kDa subunit (results not shown). This shows that the 23 kDa protein is not absolutely essential for photoactivation but its presence enhances the yield and efficiency.

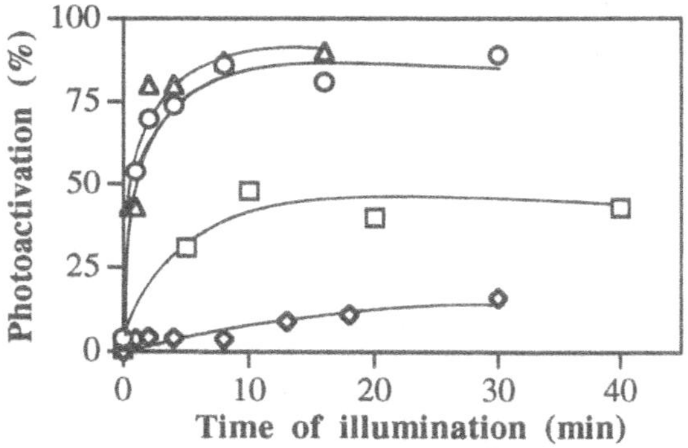

Figure 1. *Photoactivation of wild type (triangles and circles) and FUD 39 (squares and diamonds) cells at low (1.5 µE m^{-2} s^{-1}) (triangles and squares) and high (17 µE m^{-2} s^{-1}) (circles and diamonds) light intensity.*

3.2 *Photoactivation yield is very sensitive to light intensity in FUD 39*
Fig. 2A shows the time course for photoactivation of FUD 39 at various light intensities. Below 0.2 µE m^{-2} s^{-1} no photoactivation occurs. At higher intensities the efficiency increases to around 1 µE m^{-2} s^{-1}, where 70-75% of the maximal oxygen evolving activity is reached if all centres were functional. The optimal light intensity for photoactivation occurs over a narrow interval (Fig 2B). At higher light intensities the efficiency decreases again. At high light intensities (> 5 µE m^{-2} s^{-1}) as little as 40% of the centres are photoactivated while even fewer (15-20%) show oxygen evolution when photoactivation is performed at 60 µE m^{-2} s^{-1}. It should be noted that the total DPC mediated electron transfer (100%) was constant during the whole period of illumination irrespective of illumination regime (0-60 µE m^{-2} s^{-1}). Together with the results in Fig. 1 this indicates that the yield of photoactivation in FUD 39 is much more sensitive to the light intensity regime than the yield in the wild type.

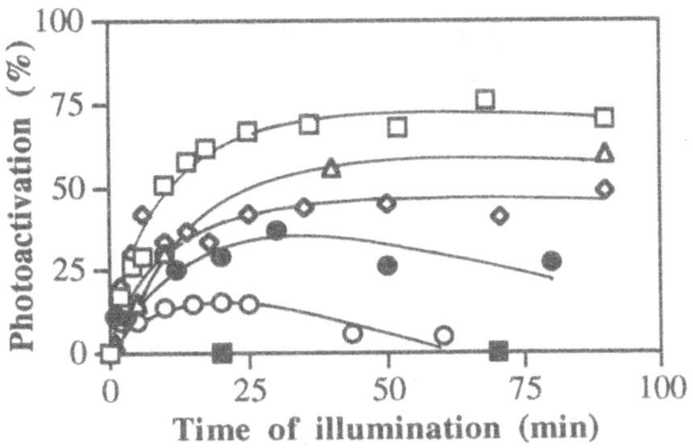

Figure 2A. The time course of photoactivation of FUD 39 mutant cells at different light intensities. Filled squares 0.2 μE m-2 s-1, triangles 0.8 μE m-2 s-1, open squares 1 μE m-2 s-1, diamonds 2 μE m-2 s-1, filled circles 17 μE m-2 s-1 and open circles 60 μE m-2 s-1.

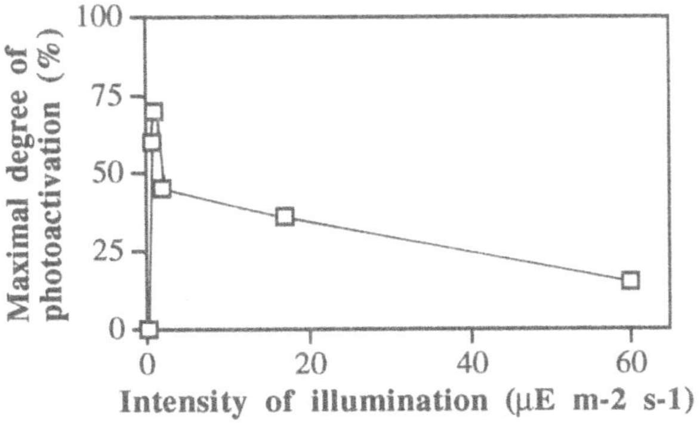

Figure 2B. Maximal water splitting activity in FUD 39 cells after photoactivation at different light intensities.

3.3 *The photoactivating capacity is irreversibly photoinhibitedat higher light intensities*
Fig. 2 shows that there are two effects of light intensity on photoactivation. First, the maximum yield decreases, at constant illumination time, with light intensities > 2 μE m-2 s-1 although DPC mediated electron transfer do not decrease. Second, the yield decreases with illumination time for the highest light intensity. We hypothesise that this reflects our earlier results, when we found that FUD 39 is susceptible to donor-side photoinhibition (8) and that this first involves the oxygen evolving complex. We therefore characterized the capacity for photoactivation in cells that had suffered various degree of photoinhibition (Fig. 3).

After various times of illumination, at a light intensity high enough to induce a decrease in photoactivation after about 30 minutes, the cells were transferred to the optimal light intensity for photoactivation and further activated for 40 minutes to reach the maximal level of activation in this particular sample. After short illumination at the high light intensity the second, optimal illumination resulted in complete photoactivation. Thus, no photoinhibition had occured. However, after longer exposure to the high illumination it became impossible to photoactivate the cells to the full extent. This reflects irreversible photoinhibition of the ability to perform photoactivation. The fraction of photoinhibited centres increases with time and eventually dominates the pool of cells.

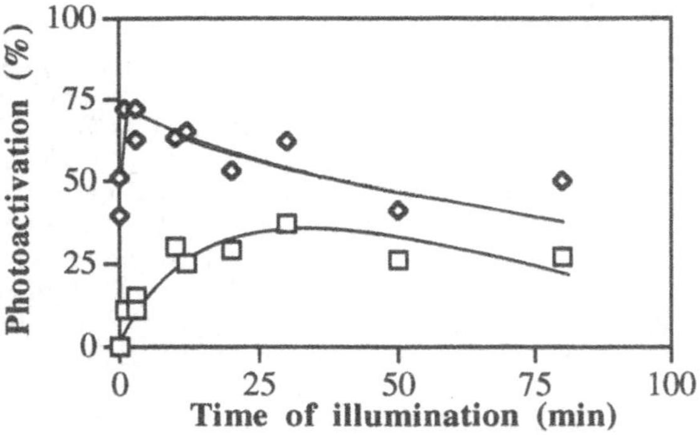

Figure 3. FUD 39 cells photoactivated at 17 μE m-2 s-1 (squares) for different times. For each time there are two samples. One is transferred directly to activity measurements. The other sample is transferred, after the illumination at 17 μE m-2 s-1 to 0.8 μE m-2 s-1 (diamonds) for 40 minutes to achieve optimal photoactivation before it is taken to the assay.

4. Conclusions

We have earlier shown that a fraction of PS II centres in FUD 39 seems to have no water splitting activity. Here we have compared the efficiency of the photoactivating process *in vivo*, in wild type and FUD 39 and we found that the mutant is less efficiently photoactivated even at an optimal light intensity. This could be explained, if we assume that the 23 kDa subunit has a Cl^- concentrating role, by the results of others (3) which shows that Cl^- has a positive effect on the yield of photoactivation. Presumably in the cells of FUD 39 the Cl^- concentration is not sufficient to accomplish full photoactivation.

We also found that the photoactivation yield in FUD 39 decreases with high light intensities. However, cells in the earlier part of the time course of photoactivation with strong light (Fig. 3) still have capacity to photoactivate if they are transferred to a lower light intensity. We suggest that this reflects the dark period necessary for photoactivation and that our results indicate that FUD 39 requires a longer dark period than wild type. However, Fig. 3 also shows that the cells will loose their photoactivating capacity irreversibly at prolonged illumination with high light. This reflects a counteracting effect of irreversible photoinhibition as FUD 39 is very susceptible to donor side photoinhibition. Thus, the final level of water splitting seems to be set by a balance between photoactivation and photoinhibition, which becomes easily discernible in the mutant.

Acknowledgements
This work was supported by the Swedish Natural Science Research Council.

References

1 Radmer, R. & Cheniae, G. (1977) in Primary Processes of Photosynthesis (Barber, J., ed) Vol. 2, pp 303-348, Elsevier, Amsterdam
2 Miller, A-F. & Brudvig, G.W. (1990) Biochemistry 29, 1385-1392
3 Miyao, M. & Inoue, Y. (1991) Biochemistry 30, 5379-5387
4 Tamura, N. & Cheniae, G. (1987) Biochim. Biophys. Acta 890, 179-194
5 Tamura, N., Inoue, Y. & Cheniae, G.M. (1989) Biochim. Biophys. Acta 976, 173-181
6 Styring, S. & Jegerschöld, C. (1994) in Photoinhibition of Photosynthesis from molecular mechanisms to the field (Baker, N.R. & Bowyer, J. R., eds) pp.51-73, BIOS Scientific Publishers, Oxford, UK
7 Jegerschöld, C., Virgin, I. & Styring, S. (1990) Biochemistry 29, 6179-6186
8 Rova,M.,Franzén,L.-G.,Fredriksson,P.-O.&Styring,S (1994) Photosynth.Res.39, 75-83

SECONDARY STRUCTURE ANALYSIS OF THE 33 kDA PROTEIN FROM PHOTOSYSTEM II AND ITS DEPENDENCE ON PH

T.V. Shutova[1], K.-D. Irrgang[2], V.K. Opanasenko[1], V.V. Klimov[1] and G. Renger[2]
[1]Institute of Soil Science and Photosynthesis, RAS, Pushchino, Moscow Region, 142292 Russia; [2]Max-Volmer-Institute for Biophysical und Physical Chemistry, Technical University of Berlin, Straße des 17. Juni 135, D-10623 Berlin, Germany

1. Introduction

Photosynthetic water oxidation into O_2 and 4 H^+ comprises a sequence of four univalent electron abstraction steps from a manganese-containing functional unit, referred to as water-oxidizing complex (WOC), that is part of Photosystem II (PS II) (for a review see [1]). With respect to the functional and/or structural role of particular polypeptides in the WOC, a lysine-rich extrinsic protein of 33 kDa has been thoroughly analyzed (for a review see [2]). Although this protein does not contain direct ligands to the manganese cluster, it is an important constituent for the functional and structural integrity of the WOC. In addition to stabilizing the manganese cluster and binding of the cofactor Ca^{2+} [3] it probably modifies the reaction pathway of water oxidation [4] and might also be involved in pH-induced changes of the functional properties of the WOC ([5] and references therein). Only limited information is available on the structure of the extrinsic 33 kDa protein: (i) the N-terminus is important for binding to the membrane [6]; (ii) a conserved S-S bridge is essential for the overall structure [7], and (iii) a significant fraction of the protein attains a β-sheet structure as deduced from different approaches [8,9]. Recently, a method applied earlier for determination of the buffer capacity of thylakoids and PS II preparations [10-13] was used to analyze pH-induced structural changes of the isolated 33 kDa protein in buffer solution. A marked hysteresis was observed in acid-base versus base-acid titrations [10]. More detailed analyses (Shutova, Irrgang, Klimov and Renger, in preparation) revealed that this feature reflects a structural rearrangement of the protein due to deprotonation/protonation of dicarbonic amino acid residues (Asp, Glu). The present report is an attempt to analyze the secondary structure of the isolated 33 kDa protein in buffer solutions by using a theoretical approach based on the method developed by Ptitzyn and Finkelstein [14].

2. Methods

The numerical method for secondary structure predictions is based on the resolved amino acid sequences of 33 kDa proteins [15] and on a molecular theory which takes into account local interactions inside each chain region and long-range interactions between different regions. Both modes of interaction are incorporated into a single Ising-type model. The local interactions are evaluated on the basis of the stereochemical theory, describing the relative stability of α- and β-structures for different residues in synthetic polypeptides, while the long-range effects are approximated by the interactions of each chain region with a suitable template [14]. The computational algorithm emerging from

P. Mathis (ed.), Photosynthesis: from Light to Biosphere, Vol. II, 295–299.

this theoretical approach referred to as ABL program provides a satisfactory description of the content of secondary structural elements. In addition, this procedure also permits to determine the probability of location and stability of these elements as a function of environmental parameters like pH, temperature, dielectric constant, and ionic strength.

3. Results and Discussion

The fractions of secondary structure elements calculated in this study for the extrinsic 33 kDa protein at 300K, ionic strength of 0.1 M, and dielectric constant of 78.5, are compiled in Table 1 and are compared with data reported by other groups [8,9,16]. Values of 13%, 28% and 17% were obtained for the fractions of α-helices, β-sheets and turns, respectively. In general, the calculations show a qualitative fit with the features reported in the literature. Interestingly, the content of α-helices is predicted to be comparatively small in correspondence with a recent analysis of CD data [8], but in contrast to other theoretical calculations [16] and FTIR measurements [9]. The origin of this marked difference still needs to be clarified.

Table 1: Secondary structure analysis of the extrinsic 33 kDa protein from spinach

Secondary structure element	Chou-Fasman [16]	Xu et al. [8]	Ahmed et al. [9]	this study
α-helix	28	9	27	13
β-sheet	25	38	36	28
turn	12	17	24	17

A more detailed picture of the secondary structure predictions is presented in Fig. 1A, where the structural elements are depicted along the peptide chain. A total of four α-helices and eleven β-strands is expected to exist within the extrinsic 33 kDa protein. A striking feature of this structure is a kind of clustering of adjacent β-sheets in the middle of the sequence, while the α-helices are predicted to be located near the more hydrophilic termini of the protein. A closer inspection of the β-sheet cluster reveals that there exist adjacent anti-parallel strands that probably stabilize each other and form an extended domain. It is also possible to include in this domain the adjacent β-sheets (β_2 and β_3). This assignment would comprise the intramolecular S-S-bridge. It is conceivable that this large β-sheet domain represents a buried region not readily accessible from the outer phase. Interestingly, the β_6-sheet contains dicarbonic amino acid residues which could be protonated in the acidic range, thus giving rise to marked structural changes.

Fig. 1B shows the secondary structure predicted for the 33 kDa protein at pH 2.8. A qualitative inspection of the data readily shows that the α-helix content markedly decreases while the content of β-sheets remains mainly unaffected. This acid-induced conformational change is ascribed to protonation of dicarbonic amino acid residues (Asp, Glu). It remains to be shown how these structural changes affect the observed hysteresis effect in the titration curve [see [10] and Shutova, Irrgang, Klimov and Renger, in preparation).

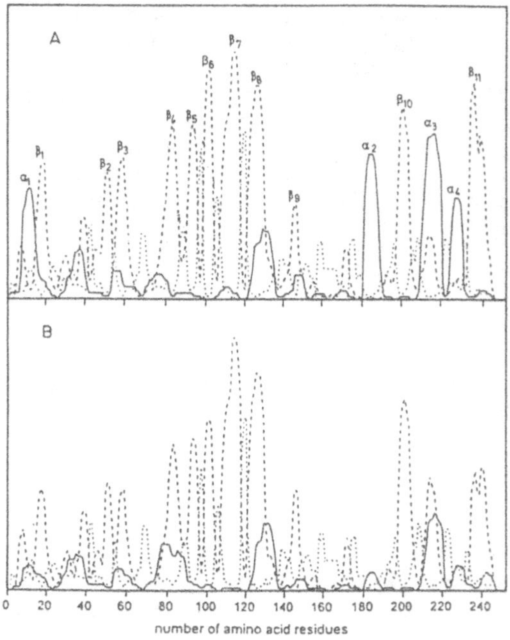

Fig. 1

Predicted secondary structural elements along the polypeptide chain of the extrinsic 33 kDa protein at pH 6.8 (A) and 2.8 (B); α-helices are symbolized by full line, β-sheets by dashed line, and turns by dotted line. The α-helices and β-sheets are labeled by numbers (α_1-α_4 and β_1-β_{11}, respectively), turns are not explicitly marked.

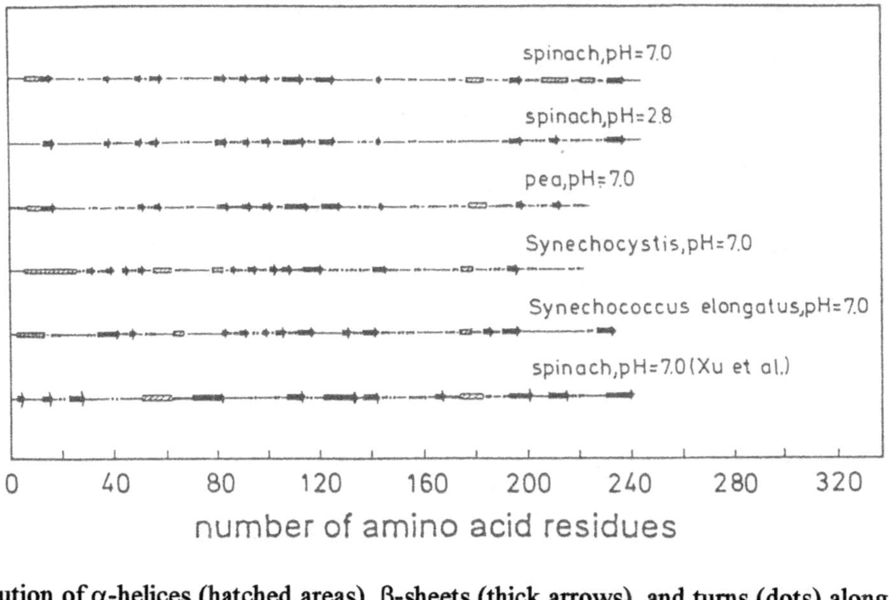

Fig. 2
Distribution of α-helices (hatched areas), β-sheets (thick arrows), and turns (dots) along the polypeptide chain of the extrinsic 33 kDa proteins from different species.

Fig. 2 compiles the general features of the secondary structure elements predicted by the ABL program on the basis of sequence data on 33 kDa proteins from different species [15,18-20]. A qualitative inspection of the data reveals the existence of a large β-sheet domain in the middle of the protein. The bottom part of Fig. 2 shows that a similar feature was obtained by Xu et al. [8] by the method described in [17]. However, differences exist for the N- and C-terminal regions. Based on the data presented here, the extended β-sheet region in the central part is inferred to be a common feature of all 33 kDa proteins. The possible relevance of the buried β-sheet domain for the structural and functional role has to be investigated in future studies.

Acknowledgement

Financial support by Russian Foundation of Basic Research and Deutsche Forschungsgemeinschaft is gratefully acknowledged.

References

1 Renger, G. (1993) Photosynth. Res. 38, 229-247
2 Vermaas, W.F.J., Styring, S., Schröder, W.P., and Andersson, B. (1993) Photosynth. Res. 38, 249-263
3 Ädelroth, P., Lindberg, K., and Andreasson, L.-E. (1995) Biochemistry 34, 9021-9027
4 Fiege, R., Zweygart, W., Irrgang, K.-D., Adir, N., Geiken, B., Renger, G., and Lubitz, W. (1995) these proceedings
5 Kebekus, U., Messinger, J., and Renger, G. (1995) Biochemistry 34, 6175-6182
6 Eaton-Rye, J.J. and Murata, N. (1989) Biochim. Biophys. Acta 977, 19-26
7 Tanaka, S. and Wada, K. (1988) Photosynth. Res. 17, 255-266
8 Xu, Q., Nelson, J., and Bricker, T.M. (1994) Biochim. Biophys. Acta 1188, 427-431
9 Ahmed, A., Tajmir-Riahi, H.A., and Carpentier, R. (1995) FEBS Lett. 363, 65-68
10 Opanasenko, V., Shutova, T., and Klimov, V. (1992) Biol. Mem. (Moscow) 5, 881-890
11 Shutova, T., Opanasenko, V., Ananyev, G., and Klimov, V. (1992) Biol. Mem. (Moscow) 5, 891-898
12 Shutova, T., Khristin, M., Opanasenko, V., Ananyev, G., and Klimov, V. (1993) Biol. Mem. (Moscow) 6, 1115-1128
13 Shutova, T., Opanasenko, V., and Klimov, V. (1994) Biochemistry (Moscow) 59, 733-737
14 Ptitzyn, O.B. and Finkelstein, A.V. (1983) Biopolymers 2, 15-22
15 Oh-Oka, H., Tanaka, S., Wada, K., Kuwabara, T. and Murata, N. (1986) FEBS Lett. 197, 63-66
16 Chou, P.Y. and Fasman, G.D. (1978) J. Adv. Enzym. 47, 95-148
17 Teeter, M.M. and Whitlow (1988) Proteins 4, 262-273
18 Watanabe, A., Minami, E.I., Murase, M., Shinohara, K., Kuwabara, T. and Murata, N. (1987) Progr. Photosynth. Res. (Biggins, J. ed.) Vol. IV, 629-636, Martinus Nijhoff, Dordrecht

19 Philbrick, J.B. and Zilinskas, B.A. (1988) Mol. Gen. Genet. 212, 418-425
20 Motoki, A., Miura, K., Shimazu, T., Kanai, S., Hirano, M. and Katho, S. (1992) Res. Photosynth. (Murata, N. ed.) Vol II, 413-416, Kluwer, Dordrecht

X-RAY ABSORPTION SPECTROSCOPY OF THE WATER OXIDATION COMPLEX OF PS II

Sandra Turconi[1], Colin P. Horwitz[2], S.T. Weintraub[2], Joseph T. Warden[3], Jonathan H.A. Nugent[1] and Michael C.W. Evans[1]
1Biology Dept, Univ. College London, Gower st, London WC1E 6BT, UK; [2]Chemistry Dept, Carnegie Mellon University, 4400 Fifth Ave, Pittsburgh, PA 15213-2683, USA; [3]Chemistry Dept, Rensselaer Polytechnic Institute, Troy, NY 12180-3590, USA

1. Introduction

Water oxidation in higher plants, algae and cyanobacteria is performed at a manganese containing active site within Photosystem II (PS II). The driving force for this reaction comes from $P680^+$ created by the charge separation in the reaction centre of PS II. $P680^+$ abstracts the electrons from water via an intermediate tyrosine, Tyr Z. Four electrons are released per oxidation of a single water molecule, requiring the manganese complex to cycle between five oxidation states (termed S_{0-i}). These states are thought to be different oxidation states of the manganese atoms. It is known that inorganic cofactors, like Cl^- and Ca^{2+} are present and are necessary for maximal oxygen evolving activity. The above described complex, consisting of four manganese atoms, ligated amino-acids and cofactors is termed Water Oxidation Complex, WOC.
The structure of the WOC is not determined to date and little is understood about the precise mechanism of water oxidation and the exact role of the cofactors.
Currently, structural information on the manganese site can be obtained only by X-ray absorption spectroscopy. Two methods are available: XANES (X-ray Absorption Near Edge Structure) which yields information on the oxidation state of the absorbing metal, and EXAFS (Extended X-ray Absorption Fine Structure) which gives information on the radial distribution of atoms in the coordination sphere of the central atom.
In this paper, we present results of X-ray absorption spectroscopy study of PS II and a series of inorganic manganese compounds.

2. Experimental

Experiments were carried out on PS II enriched thylakoid membrane particles known as BBYs. X-ray spectroscopy was performed at the Synchrotron Radiation Source at Daresbury, UK. Pulsed EPR measurements were carried out using a Bruker ESP380E X-band pulsed spectrometer based at UCL. Further experimental details are described in [1].

P. Mathis (ed.), Photosynthesis: from Light to Biosphere, Vol. II, 301–304.

3. Results and Discussion

3.1 *XANES of Photosystem II*

Positions of the rising edge of manganese absorption were determined in:

- untreated BBY particles, dark adapted at 273 K for different periods [2]: 10 minutes ("active") and 4 hours ("resting"). These S_1 samples were illuminated at 200 K to obtain respective S_2 states.
- Ca^{2+} - depleted particles, using high NaCl concentrations, as described in [1], with the extrinsic polypeptides reconstituted.
- PS II treated with sodium acetate, as described in [3] with 3-(3,4-Dichlorophenyl) 1,1-dimethylurea (DCMU) added in the final resuspending buffer.
- BBY in which (dimethylureidophenyl)phenylnitroxide (PNPDU) was added after the oxidation of the non-heme iron with $K_3Fe(CN)_6$ to limit the WOC to two electron turnovers [4]. To achieve the S_2 state, samples were illuminated at 200K. A subsequent illumination at 250 K leads to the S_3 state [4].

The results are presented in Table 1.:

sample \ state	edge position (eV) S_1	S_2	S_3
active	6551.8	6553.0	-
resting	6551.8	6553.6	-
Ca^{2+} - depleted	6550.0	6550.3	6550.7
acetate -treated + DCMU	6550.2	6550.9	-
$K_3Fe(CN)_6$ + PNPDU	-	6551.5	6552.0

Note: edge positions are accurate to $\leq \pm 0.2$ eV

These values are a probe of the oxidation state of the manganese. Therefore, following observations can be made:

- The oxidation state of Mn does not change during the 4h dark adaptation period. However, in the "active" BBYs the oxidation of the Mn seems to be less complete in the S_2 state.
- In all treated PS II samples there are significant, though small shifts in edge energies between different states. This indicates that changes in the oxidation state of manganese may occur upon transition between different S-states, though probably only in a fraction of centres.
- The 0.7 eV upward shift in edge energy in the acetate treated BBY indicates that the g=4.1 EPR signal (the only EPR signal present) is associated with the S_2 state.

3.2 *Model Compounds*

Coordination complexes of manganese are a very useful additional source of information. The structure of the complexes are known and the study of their properties in correlation to their known structural features could lead, through comparison, to better understanding of the both structure and function of the manganese site in PS II. Using EXAFS and XANES spectroscopies, we have studied the structure of a group of dinuclear manganese coordination compounds bridged by two oxygen atoms (bis-μ-oxo). An important characteristic of this compounds is that they contain two crown-ether moieties per Mn-dimer which bind one alkaline or alkaline-earth cation each and brings them in close neighbourhood of the manganese atoms. This is of special interest in relation to the problem of the location and role of a Ca^{2+} ion known to be a cofactor of OEC necessary for oxygen evolution.

The experiments were carried out on powdered compounds at room temperature, in transmission mode. The Mn bis-μ-oxo compounds were synthesised as described in [5].
The bis-μ-oxo compound [(3,3'-17-crown-6-SALPN)Mn(IV)(μ-o)]$_2$ and its derivatives containing Na^+, K^+, Ca^{2+} and Ba^{2+} within the crown-ether moiety were studied.

Fourier Transformed EXAFS spectra of the bis-μ-oxo compounds are shown in Fig. 1.

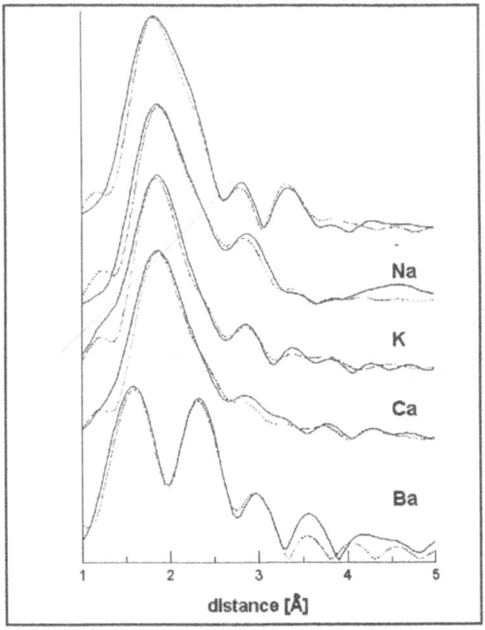

Fig.1 (right) The upper most trace is the compound without any cation in the crown ether moiety, other traces correspond to compounds containing cations as labelled. The continuous line represents experimental data and the broken line the theoretical fit.

The main features of the structural data obtained can be summarised as follows:
- The first shell of atoms surrounding Mn contains two sub-shells, the first of which consists of 4 oxygens at a distance of 1.84-1.9 Å, the second of 2 nitrogens at 2.25-2.36 Å from the manganese ($\pm$ 2% is the estimated accuracy for all distances).
- The Mn - Mn distances increase slightly upon the introduction of cations, from 2.66 Å for bis-μ-oxo to 2.75-2.79 Å for the compounds containing metal cations. There is no apparent dependence of these distances on the charge of the introduced cation.
- The cations in the crown-ether moiety are at a distance of 3.56-3.70 Å from the Mn.

- The only exception to the above observations is the Ba-containing bis-μ-oxo compound, where the Mn-Ba distance is found to be only 2.66 Å. Moreover, an additional shell of light atoms is present at the distance of 2.25 Å, making the combined O-N shell more compact and closer to Mn (1.69 Å and 1.90 Å, respectively).
- The values of the inflection points of the rising absorption edges of the studied model compounds are within 0.5 eV (6553.2 - 6553.7 eV).

4. Conclusions

A striking feature of the model compounds EXAFS is the effect of the introduction of Ba^{2+} in the crown-ether moiety. This raises concern over the substitution of Ca^{2+} by larger cations in the WOC.
The Mn-cation distances are about 3.6 Å, which corresponds to the distances reported for a metal-containing shell in the WOC [5,6,7]. In PS II data it is not possible to distinguish between Mn and Ca for this shell since the difference in their atomic number is judged to be too small (given the low the signal/noise level). However, for bis-μ-oxo Ca, a fit of the 3.63 Å shell with Mn instead of Ca was significantly worse, suggesting that Ca is clearly distinguishable from Mn.
The presence or absence of the cation alone does not alter the position of the absorption edge of manganese for more than ± 0.3 eV. This is true even for the Ba-containing compound with its drastically changed Mn-environment, suggesting that the position of the edge, though reflecting to some extent the covalency of Mn, is dominated by the oxidation state of the metal.
The values presented here for the edge positions of some of the treated PS II samples are somewhat different from those reported previously [3].The new values, which we consider more accurate, show still significant differences between various S-states.

Aknowledgements
We are thankfull to Prof. Gary W. Brudvig for supplying PNPDU. Financial support from the Biotechnology and United Kingdom BBSRC is gratefully aknowledged.

References
1 MacLachlan, D.J., Nugent, J.H.A., Bratt, P.J., Evans, M.C.W. (1994) Biochim. Biophys Acta 1186, 186-200
2 Koulougliotis, D., Hirsh, D.J., Brudvig, G.W. (1992) J. Am. Chem. Soc. 114, 8322-8323
3 MacLachlan, D.J., Nugent, J.H.A., Evans, M.C.W. (1994) Biochim. Biophys. Acta 1185, 103-111
4 Bocarsly, J.R. and Brudvig, G.W. (1992) J. Am. Chem. Soc. 114, 9672-9767
5 Horwitz, C.P., Warden, J.T., Weintraub, S.T. (1995?) Inorg. Chim. Acta, submitted
6 Penner-Hahn, J.E., Fronko, R.M., Pecoraro, V.L., Yocum, C.F., Bretts, S.D., Bowlby, N.R.(1990) J.Am.Chem.Soc. 112, 2549-2557
7 Yachandra, V.K., DeRose, V.J, Latimer, M.J., Mukerji, I., Sauer, K. and Klein, M.P. (1993) Science 260, 675-679

Effects of mild trypsinization on the properties of a (17, 23 and 33 kDa)-retaining PSII core complex

E. Kouimtzoglou[1], R. K. Mishra[1], M. Haumann[2], W. Junge[2] and D. F. Ghanotakis[1]
[1]Department of Chemistry, University of Crete, Iraklion, Crete, GREECE, [2]Biophysics, University of Osnabrueck, 49069 Osnabrueck, GERMANY

Abstract

Solubilization of Photosystem II (PS II) membranes with the non-ionic detergent 6-O-(N-heptylcarbamoyl)-methyl-α-D-glucopyranoside (HECAMEG) allowed the isolation of an oxygen evolving PS II core complex which retained the 23 and 17 kDa extrinsic polypeptides. The pattern of proton release as a function of flash number in this PSII complex, which retained all three extrinsic polypeptides, was found to be close to 1:1:1:1. Mild trypsinization of the PSII complex resulted in proteolysis of the D1 protein, while all three extrinsic proteins (17, 23 and 33 kDa) and the manganese complex remained intact; under these conditions the proton release pattern remained featureless, close to 1:1:1:1.

Introduction

Photosystem II complexes, depleted of the light harvesting complex provide an excellent system for the study of the role of the various polypeptides in the structural and catalytic properties of Photosystem II. In the present communication we report a simple procedure for the isolation of PSII core complex, which retains all three extrinsic polypeptides. The PSII core complex was characterized by various biophysical tecniques, and a series of limited proteolysis experiments.

Materials and Methods

Photosystem II enriched thylakoid membranes were prepared from spinach as described earlier (1). In order to maintain the same concentration of residual Triton X-100 during solubilization of the PS II membranes with HECAMEG (HECAMEG, 6-O-(N-heptylcarbamoyl)-methyl-α-D-glucopyranoside, was purchased from VEGATEC, France), the membranes, obtained after treatment of thylakoids with Triton X-100, were washed only once with a specific volume of SMN (the chlorophyll concentration during the wash was 2 mg Chl/ml). Isolation of an oxygen evolving PS II core complex (HECAMEG-PS II core complex) retaining the 23 and 17 kDa extrinsic polypeptides was achieved by solubilization of the PS II membranes with HECAMEG. For the isolation of this complex, V ml of PS II membranes, in SMN (0.4 M sucrose, 50 mM Mes, pH 6.0 and 10 mM NaCl) at 0.75

P. Mathis (ed.), Photosynthesis: from Light to Biosphere, Vol. II, 305–308.

mg Chl ml^{-1}, were mixed with V ml of 60 mM HECAMEG in SMN. The mixture was incubated for 10 min on ice in the dark and was subsequently centrifuged at 40,000 x g for 30 min. The green supernatant was diluted with V ml of SMN and 10 mM $MgCl_2$ was added to the mixture to facilitate the aggregation of the light harvesting complex, which was recovered by a second centrifugation at 40,000 x g for 20 min. The green supernatant, which contained the PS II complex, was mixed with 1.5xV ml of a solution containing 40 % (w/v) polyethylene glycol (M_r 6000), 50 mM Mes, pH 6.0 and 10 mM NaCl, and was centrifuged at 40,000 x g for 30 min. The pellet was washed once with SMN and was resuspended in the same medium. Approximately 20 % of the total chlorphyll in the PS II membranes was recovered in the final pellet.

The antibody "anti-$D1_N$" was raised against a synthetic peptide homologous to the following N-terminus of the D1 protein: acetyl-TAILERRESESLWGRFC-amide, and was a kind gift from Dr. Autar K. Mattoo. "anti-D1" was a polyclonal antibody raised against D1.

Results and Discussion

The polypeptide composition of the HECAMEG-PSII core was examined by SDS page and a series of antibodies. The major polypeptides of the complex are the 47 and 43 kDa PS II Chl binding proteins, the D1 and D2 reaction center polypeptides, Cyt b_{559}, the 17, 23 and 33 kDa extrinsic polypeptides, and the CP26, CP29 and CP24 Chl binding proteins.

The PSII core complex which retains all three extrinsic proteins was used in a series of trypsinization experiments in order to get information about the structural and functional topography of Photosystem II. Treatment of the PSII core complex, at 4^oC, with a low concentration of trypsin (trypsin/Chl = 0.1) resulted in the trypsinization of the D1 (Fig. 1) and D2 (data not shown) polypeptides; under these condition though the extrinsic 17, 23 and 33 kDa proteins remained almost intact. Since a polyclonal antiserum raised against a synthetic peptide specific to the N-terminus (anti-$D1_N$) of the D1 protein failed to detect any D1 fragments, while another antibody (anti-D1) reacted with three species (Fig. 2), we suggest that, under our conditions, trypsin cleaves the D1 polypeptide at two sites one close to the N terminus and the other between helices II and III.

The results of the characterization of control and trypsinized PSII complexes are summarized in Table I. Treatment with a low concentration of trypsin did not result in any major modification of the reducing side, at least as far as DCMU sensitivity is concerned. Moreover, an EPR study of the oxidizing side of PSII revealed that electron transport from the Mn complex to $TyrZ^+$ was not blocked by trypsin treatment (data not shown). The fact that no stimulation of oxygen evolution activity was observed upon addition of $CaCl_2$ in the trypsinized complex is consistent with the observation that, under our conditions, the 17 and 23 kDa polypeptides are not affected by trypsin (see Fig. 1A).

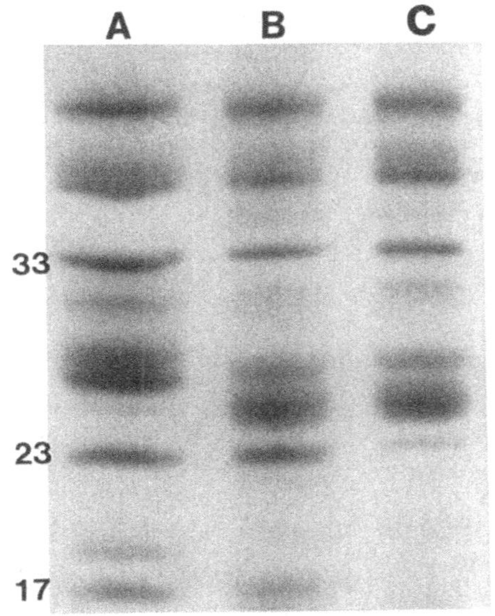

Figure 1: Polypeptide pattern of A) HECAMEG-PSII core complex, B) trypsinized PSII core complex and C) 2M NaCl-treated trypsinized PSII core complex.

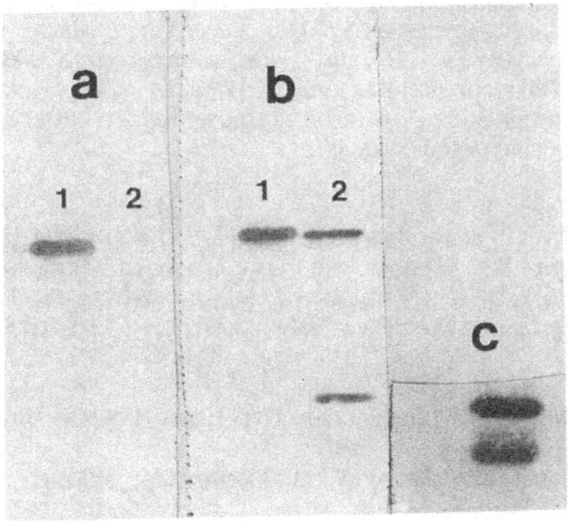

Figure 2: Western blot analyses of the D1 protein. The antibodies (a) "anti-$D1_N$", and (b) "anti-D1" (see text for details) were used in (1) control and (2) trypsinized HECAMEG-PSII core complexes. (c) shows the two D1 fragments on an overloaded gel.

Table I: Characterization of control and trypsinized PSII core complexes

	PREPARATION	
	Control PSII core	Trypsinized PSII core
i) Mn-content (determined by EPR)	4 Mn / $TyrD^+$	3.9 Mn / $TyrD^+$
ii) VO_2 (μmoles O_2 per mg Chl per hr)		
Additions		
DCBQ	1070	660
DCBQ + $CaCl_2$	1080	680
DCBQ + $CaCl_2$ + DCMU	250	180
$Fe(CN)_6^{3-}$	380	220
$Fe(CN)_6^{3-}$ + $CaCl_2$	580 *	490*
$Fe(CN)_6^{3-}$ + $CaCl_2$ + DCMU	130	180
iii) Proton release (see ref. 2)	0.96:0.96:0.96:1.12	1.0:0.92:1.08:1.0

*The same effect was observed in the presence of $MgCl_2$.

Since previous measurements on proton release were carried out with a PSII core complex depleted of the 17 and 23 kDa polypeptides (2), we studied the proton release pattern in this extrinsic protein retaining PSII complex. As shown in Table I, the pattern of proton release in this PSII core complex, which retains all three extrinsic proteins was close to 1:1:1:1, and it was not altered by mild trypsinization. Since the trypsinized PSII core retains only 62% of the control oxygen evolution activity, further experiments are now in progress in order to better characterize the oxidizing and reducing side of the complex.

Acknowledgements:

The kind gifts of antibodies from Prof. Roberto Bassi, Prof. Bertil Andersson, Dr. Autar K. Mattoo and Dr. Georgios Tsiotis are gratefully acknowledged. This work has been supported by a grant from the European Union (DG-XII / Human Capital and Mobility - Network CHRX-CT94-0524).

References

1. Ghanotakis DF, Babcock GT and Yocum CF (1984) Biochim Biophys Acta 765: 388-398.
2. Lubbers K, Haumann M. and Junge W (1993) Biochim. Biophys. Acta 1183: 210-214.

THEORETICAL AND EXPERIMENTAL STUDIES OF THE ORIENTATION DEPENDENCE OF THE S2 STATE MULTILINE SIGNAL OF PHOTOSYSTEM II

Karin Åhrling* & R.J. Pace, Department of Chemistry, Australian National University; *CRC for Plant Science, Canberra, Australia

1.INTRODUCTION

Simulation of the S_2 oxidation state multiline EPR signal of photosystem II (PSII) at three frequencies have shown (1) that its features can be explained in terms of a magnetically isolated MnIII-MnVI dimer. The multiline, although qualitatively similar to EPR spectra of synthetic di-μ–oxo bridged MnIII-IV dimers with antiferromagnetic coupling (2), has additional features. It has more peaks, in particular peaks of variable low intensity at the edges, which together with greater spacings between peaks gives a broader spectrum. In addition, the multiline spectrum shows superhyperfine structure. These additional features cannot be attributed to ligand hyperfine interactions nor to interactions between 3 or more Mn. Work in this group (Smith and Pace, this publication) shows that the Mn are organised as 2 non-interacting pairs, one of which gives rise to the multiline. The interactions giving rise to the multiline can be explained by assuming an unusual ligand geometry at both Mn, which would give rise to normally forbidden transitions from quadrupole interactions. For a quadrupole interaction of the magnitude indicated here, the Mn ions must be 5 coordinate or have a 6th, weakly bound ligand.The parameter set resulting from the simulations at three frequencies have allowed us to propose a geometry for the Mn-dimer (Fig 1). The simulations assume that with alcohol present in the buffer (4% EtOH) the net S=1/2 state is a well isolated ground state ($|J|>10\ cm^{-1}$) (3), see also Pace and Smith (this publication).

We report here preliminary studies of the multiline of one dimensionally ordered PSII particles together with simulations of their orientation dependence. Successful simulation of these spectra will allow us to predict an orientation of the Mn-dimer in the membrane plane. The parameter set used in the simulations are the same as used in the unoriented powder pattern multiline simulations (1) and shown in *Table 1.*

Previous orientation dependent studies on S_2 state signals in this group (4) have resulted in spectra of the non-alcohol form of the multiline, which has a nearby first excited S=3/2 state, the g=4.1 signal (3). This is because the ethanol (added to the PSII sample) has a higher vapour pressure than water and evaporates when drying the samples onto mylar strips. The vapour pressure of propanol on the other hand is lower. The addition of 4% propanol to the sample resulted in the alcohol form of the multiline as evidenced by the

P. Mathis (ed.), Photosynthesis: from Light to Biosphere, Vol. II, 309–312.

Param	x	y	z
a1	-9.8	-11.6	-3.1
q1	-0.76	-1.00	+1.76
a2	-9.4	-13.1	-5.0
q2	+1.83	-0.96	-0.87
g	1.972	1.974	2.009
lw	1.6	1.6	1.6

Table 1. True single ion parameters for multiline simulation a1/q1(MnIII), a2/q2(MnIV). Units: mT

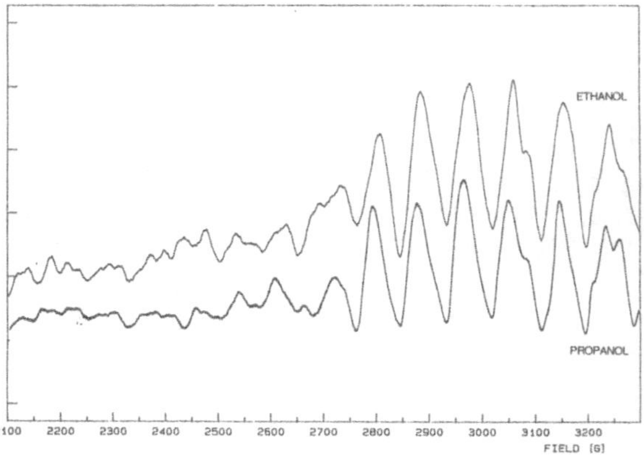

Figure 1. Powder pattern ethanol (top) and propanol (bottom) spectra. Downfield region only.

lack of 4.1 signal. There is no significant difference between the multiline signal of conventional PSII preparations with 4% ethanol or 4% propanol added. The signal with propanol added may however be somewhat better resolved. See *Figure 1.*

2. PROCEDURE

2.1*Experimental*:
PSII particles were prepared from freshly picked spinach according to the method of Pace et al (3). The particles showed an activity of 250-300 μmol O_2 per mg chl h^{-1} and were used immediately for oriented sample preparation.

Preparation of Oriented Samples. The method is derived from that of Rutherford (5). All work was carried out at 4° C, under nitrogen atmosphere and dim green light. To the freshly prepared particles in a buffer containing 20 mM MES, 10 mM $MgCl_2$, 15 mM NaCl, 400 mM Sucrose were added 4% Propanol, 50 μM DCMU. The concentration of the sample was about 12 mg Chl mL^{-1}. Three coats of the material were painted onto mylar strips, with 30 min between coats. The strips were dried overnight under N_2 atmosphere at 90% RH with propanol present. The strips (~12) were packed in a N_2-flushed quartz capillary tube. The tubes were flushed with N_2, sealed and allowed to dark-adapt at room temperature for 10 minutes. The samples were illuminated for 3 min at 195K using yellow light, intensity ~600 W m^{-2}. Samples were cooled to 77K prior to loading them into the EPR cavity.

X-band (9.425 GHz) EPR spectroscopy was performed using a Bruker ESP300E. Three orientations of the mylar strips to the applied field (reported as the angle between the normal to the membrane plane and the applied magnetic field) were taken.

2.2 *Theory:*
The model Hamiltonian assumes that the multiline arises from a well isolated, antiferromagnetically coupled MnIII-MnIV dimer in a S=1/2 ground state and therefore takes the form:

$$H=\beta \mathbf{H_o}\cdot \mathbf{g}\cdot \mathbf{S} - g_n\beta_n \mathbf{H_o}\cdot \mathbf{I_1} - g_n\beta_n \mathbf{H_o}\cdot \mathbf{I_2} + \mathbf{I_1}\cdot \mathbf{A_1}\cdot \mathbf{S_T} + \mathbf{I_1}\cdot \mathbf{Q_1}\cdot \mathbf{I_1} + \mathbf{I_2}\cdot \mathbf{A_2}\cdot \mathbf{S_T} + \mathbf{I_2}\cdot \mathbf{Q_2}\cdot \mathbf{I_2}$$

Here $\mathbf{H_o}$ is the applied magnetic field vector, **g** the g-tensor, β the Bohr magneton, g_n the nuclear g-value (assumed scalar) and β_n the nuclear Bohr magneton. The subscripts 1 and 2 refer to the MnIII and MnIV, respectively.

$\mathbf{A_1}$ and $\mathbf{A_2}$ are the effective nuclear hyperfine interaction tensors including the spin projection factors for each individual ion spin. In the strong coupling limit, this gives A_1effective (MnIII) = $2a_1$ and A_2 effective (MnIV) = $-a_2$, where a_1 and a_2 are the true ion parameters.

$\mathbf{Q_1}$ and $\mathbf{Q_2}$ are the quadrupole interaction tensors and $\mathbf{I_1}$, $\mathbf{I_2}$ are the nuclear spin vectors (I=5/2).

The applied magnetic field is taken to be along the laboratory z-axis.

For a detailed development of the Hamiltonian see Åhrling and Pace (1).

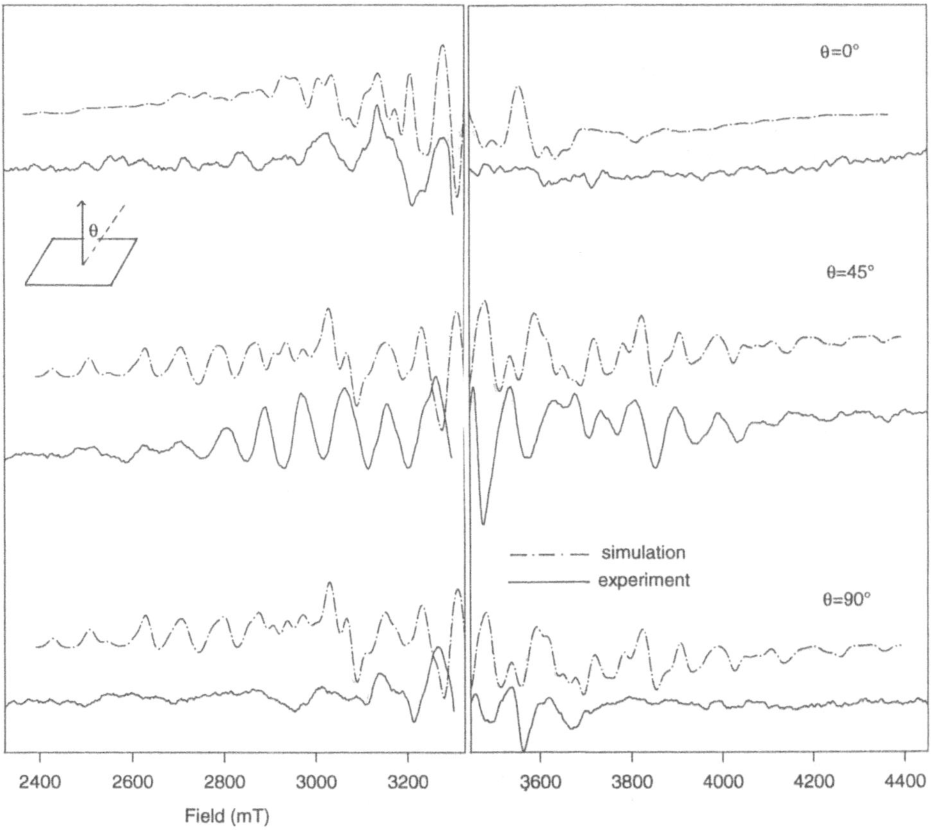

Figure 2. Experimental and simulated spectra of oriented PSII particles. Spectrometer conditions: Fr 9.425, Modulation Amplitude 20 G, Power 6.35 mW, Temperature 8K.

To simulate the spectra of one dimensionally ordered particles two rotations are performed: One rotates the principal axes (assumed coincident) of the molecule into the membrane plane and the other rotates the membrane plane to the applied field. The statistical variation of the alignment of the particles at each orientation has been calculated using a gaussian distribution with a standard deviation of 15°. The simulations of the 'oriented' multiline spectra are preliminary and have not been subjected to a minimisation. The only parameters fitted here were the two rotation angles of the molecular (principal axes) to the laboratory axis and the cone angle. All other parameters were those for the powder pattern spectra (*Table 1*).

3. RESULTS

Figure 2 shows the experimental oriented spectra together with the simulations. That the spectra were well oriented could be seen by the presence of g_X component of Cyt b-559 in the parallel orientation (g~2.22) and g_Z in the perpendicular (g~2.97). A cubic base line has been subtracted to remove such contributions.

The spectra show a high degree of orientation dependence not previously reported. The spectra show most of the intensity in the 45° orientation, with the other orientations almost devoid of intensity except in the central region. The most plausible reason for this is that the x,y & z principal axes of the molecule must all be oriented approximately within the 45° cone. Indeed the simulation results indicate such an orientation with α=45°and β=45°, see *Figure 3*.

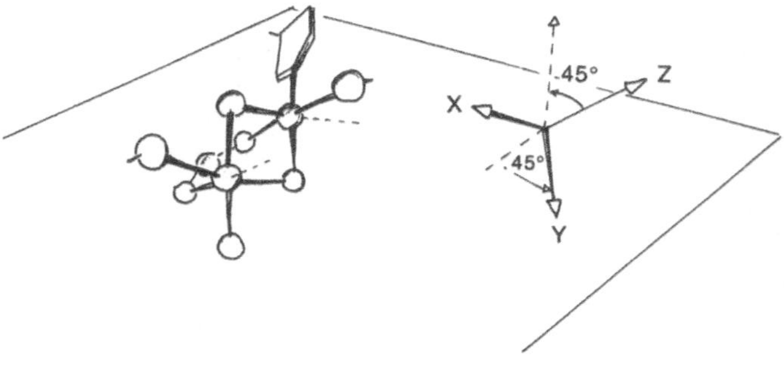

Figure 3. Result of the preliminary simulations indicate an orientation of the principal axes 45° off the membrane plane.

4. REFERENCES

1 Åhrling, K. A. and Pace, R. J. (1995) Biophys. J. 68, 2081-2090
2 Dismukes, G. C. and Siderer, Y. (1981) Proc. Natl. Acad. Sci. USA 78, 274-278
3 Pace, R. J., Smith, P., Bramley, R. and Stehlik, D. (1991) Biochim. Biophys. Acta 1058, 161-170
4 Smith, P. J., Åhrling, K. A. and Pace, R. J. (1993) J. Chem. Soc. Faraday Trans. 89, 2863-2868
5 Rutherford, A. W. (1985) Biochim. Biophys. Acta 807, 189-201

SYNTHESIS OF STEREO-SPECIFICALLY β-DEUTERATED L-TYROSINES

S.A.M. Nieuwenhuis, C. Mul, N.J.van Belle, J. Lugtenburg and J. Raap
Leiden Institute of Chemistry, Gorlaeus Laboratories, Leiden University,
P.O. Box 9502, 2300 RA Leiden.

1. Introduction

Tyrosine radicals are known to play an important role in many biological systems, for instance in ribonucleoside diphosphate reductase(1), prostaglandin H synthase(2) and photosystem II (PS II)(3). The two β-protons of L-tyrosine are magnetically inequivalent and can be denoted as (pro-)H_R and (pro-)H_S. The ESR lineshape of the *in vitro* tyrosyl radical is determined mainly by hyperfine interactions (hfi) to one of these two β-protons (the proton hfi of these two protons being very different: 11-13 G and <2.2 G(4)) and to the two protons at the 3' and 5' ring positions(4). In order to assign which β-proton, H_R or H_S, has the largest interaction, we have synthesized two specifically β-methylene deuterated L-(2S)-tyrosines (**1a** and **1b,** fig. 1). Since the hyperfine interaction of deuterium with a radical is one-sixth of the proton hfi, these two tyrosines will enable us to determine exactly the proton hyperfine interaction of H_R and H_S. The 3',5'-protons of the aromatic ring have been deuterated to decrease the proton hfi of these positions, facilitating the analysis of the effect of the specific deuteration at C-3.

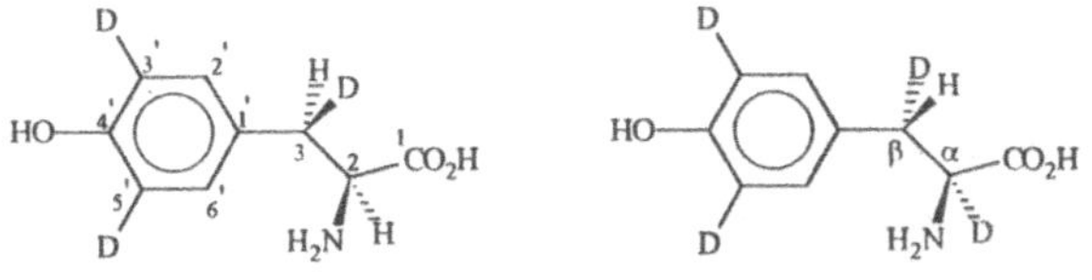

1a **1b**

Figure 1: The two synthesized specifically labeled L-(2S)-tyrosines (D=²H).

2. Synthesis

α-Amino acids can be prepared by catalytically reducing α-acylaminoacrylic acids with H_2. With a chiral catalyst, optically active amino acids can be obtained in high yields(5). Most of the catalysts that have been developped, however, produce D-(2R)-amino acids(6-8). D-tyrosine, for instance, has been prepared in 92% enantiomeric excess from α-benzoylamino-4-hydroxy-cinnamic acid with the chiral catalyst Rhodium(norbornadiene)((S,S)-bis-(diphenylphosphino)butane)(6).

P. Mathis (ed.), Photosynthesis: from Light to Biosphere, Vol. II, 313–316.

For incorporation in biological systems, L-amino acids are required. We therefore prepared the enantiomer of the aforementioned catalyst, expecting that the use of Rhodium(norbornadiene)((R,R)-bis(diphenylphosphino)butane) (abbr. Rh(NBD)((R,R)-chiraphos)) would lead to the required isomer.
The synthesis of (2S,3S)[2,3,3',5'-$^2H_4$]-tyrosine (**1b**) is depicted in figure 2. p-Methoxybenzaldehyde **2b** was condensed with N-benzoylglycine, yielding Z-oxazolinone **3b** in 80% yield by heating for 15 min at 80°C with 1 eq. sodium acetate in acetic anhydride(9). Oxazolinone **3b** was subsequently hydrolysed to Z-α-benzoylamino-4-methoxycinnamic acid **4b** by refluxing with 1.2 eq potassium hydroxide in water; acidification with HCl and double recrystallization (75% ethanol) yielded 85% pure **4b**. Then, compound **4b** was reduced with deuterium gas with the catalyst Rh(NBD)((R,R,)-chiraphos), indeed yielding stereoselectively the protected tyrosine **5b** (83%). The methoxy- and benzoyl protecting groups were removed by refluxing for 1 hour with 48% HBr(10); affording (2S,3S)[2,3-$^2H_2$]-tyrosine. Subsequent deuteration of the 3',5'-positions of the aromatic ring by refluxing overnight in 10% ^{2}HCl in $^2H_2$0 yielded the specifically labeled tyrosine **1b** in 30% overall yield; **1b** was recrystallized by dissolving in 2N NaOH and subsequent neutralization by HCl.
(2S,3R)[3,3',5'-$^2H_3$]-tyrosine (**1a**) was prepared analogously to tyrosine **1b** (see fig. 2); the chiral reduction was now performed with hydrogen gas. The required deuterated p-methoxybenzaldehyde **2a** was synthesized from methyl *p*-methoxybenzoate by reduction with LiAl$^2H_4$ and subsequent oxidition with pyridinium chlorochromate to [1-$^2H_1$]-p-methoxybenzaldehyde **2a** (72% overall yield).

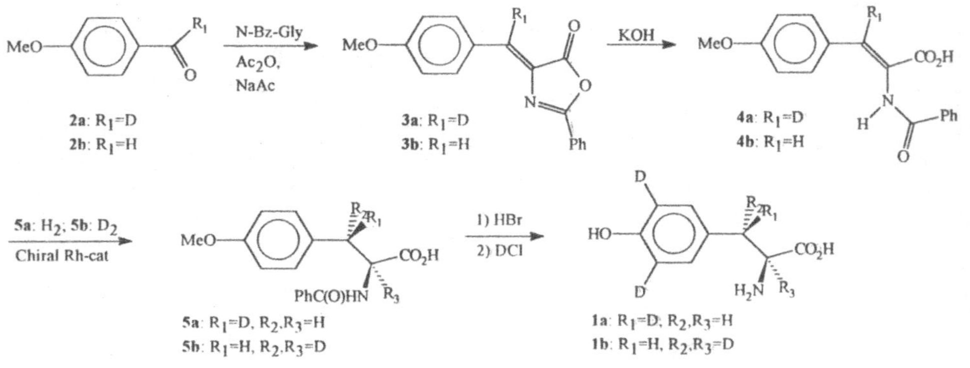

*Figure 2: Synthesis of deuterated L-(2S)-tyrosines **1a** and **1b** (D=^{2}H).*

3. Results and discussion

Enantiomeric purity at C-2 - identification of **1a** and **1b** by HPLC

In order to determine the enantiomeric purity of the synthesized tyrosines at carbon atom 2, we have measured HPLC of the synthesized compounds. Since L-(2S)-tyrosine and D-(2R)-tyrosine have the same chemical properties, a chiral derivatization is necessary; the formed diastereomers are separable by HPLC. Figure 3 shows the spectra of **1a** (fig 3c) and **1b** (fig 3d); integration gives an enantiomeric purity at C-2 of >98%. Reference samples of L-(2S)-tyrosine (fig 3a) and a mixture of D/L-tyrosine (fig 3b) were used to identify the peaks; furthermore, a mixture of (reference) L-tyrosine and either of tyrosines **1a** and **1b** shows only one peak (not shown).

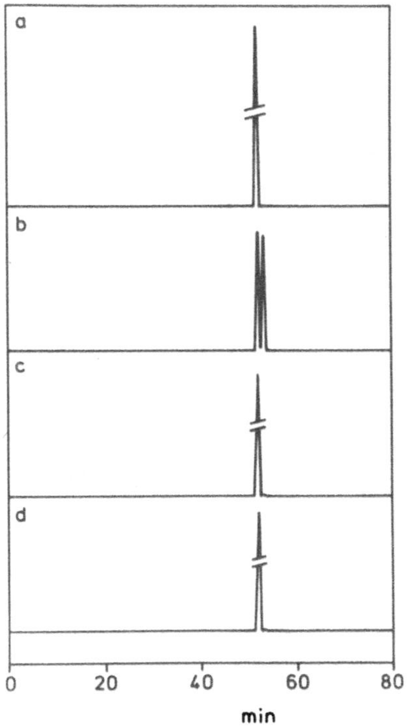

*Figure 3: HPLC diagram of L-(2S)-tyrosine(a), a mixture of D- and L-tyrosine(b), synthesized tyrosine **1a** (c) and **1b** (d). Chiral derivatization was performed with N-acetyl-L-cysteine and 1,2-phthalicdicarboxaldehyde in basic solution. The samples were eluted with 5% acetonitril in 30 mM NaOAc (pH=4), which was substituted by 30 mM NaOAc (pH=7.6) in acetonitril (25:75) in 2 hours.*

Enantiomeric purity at C-3 - identification of **1a** and **1b** by NMR

^{1}H NMR spectra of compounds **1a** and **1b** have been measured to determine the degree of deuterium incorporation and the stereochemistry at C-3. Both compounds show nearly complete incorporation of ^{2}H at C-3 (see fig. 4) and, within experimental error, complete stereospecificity at C-3. The chemical shifts of 3R-H (δ 3.19 ppm) and 3S-H (δ 3.30 ppm) have previously been assigned[11,12]. Normal (2S)-tyrosine shows an ABX spectrum for the methylene protons, as shown in figure 4c. In (2S,3R)-[3,3',5'-$^2H_3$]-tyrosine (**1a**), the 3S-H remains, and we detect an AB spectrum ($^3J_{H\text{-}H}$=5.13 Hz). Furthermore, only the 2',6'-protons of the aromatic ring can be seen (δ 7.22 ppm, s; not shown). The ^{1}H NMR spectrum of (2S,3S)[2,3,3',5'-$^2H_4$]-tyrosine (**1b**) shows the 3R-H (fig. 4b) and the 2',6'-protons (not shown) as singlets.

^{13}C NMR spectra of compounds **1a** and **1b** confirm the results obtained by ^{1}H NMR. Both spectra are identical to the ^{13}C NMR spectrum of natural L-(2s)-tyrosine; the only differences are the signals of the deuterated carbons, which appear as triplets ($^1J_{C\text{-}^2H}$ 19-25 Hz).

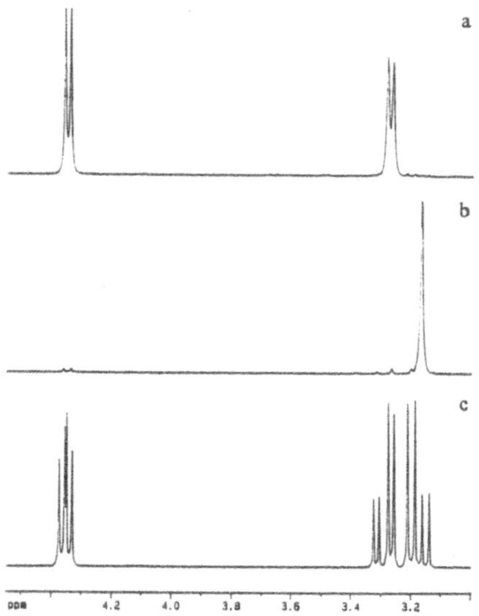

*Figure 4: Aliphatic region of the ^1H NMR spectra of (2S,3R)[3,3',5'-$^2H_3$]-tyrosine **1a** (a), (2S,3S)[2,3,3',5'-$^2H_4$]-tyrosine **1b** and L-(2S)-tyrosine (c). δ 3.17: 3R-H; δ 3.27: 3S-H; δ 4.34: 2-H.*

4. Conclusion

(2S,3R)[3,3',5'-$^2H_3$]-tyrosine (**1a**) and (2S,3S)[2,3,3',5'-$^2H_4$]-tyrosine (**1b**) have been synthesized in very high (>98%) enantiomeric purity and with high deuterium incorporation, as has been determined by HPLC and NMR analysis. These tyrosines are of considerable interest for the study of the role of tyrosine residues in biological systems, for instance in the study of the exact conformation at the β-methylene position of Tyr_D in photosystem II.

References

1) Sjöberg, B.M. and Gräslund, A. (1983) Adv. Inorg. Biochem. 5, 87-110.
2) Smith, W.L., Eling, T.E., Kulmacz, R.J., Marnett, L.J. and Tsai, A.L. (1992) Biochem. 31, 3-7.
3) Andersson, B. and Styring, S. (1991) Current Topics in Bioenergetics 16, 1.
4) Hoganson, C.W. and Babcock, G.T. (1992) Biochem. 31,11874-11880.
5) Knowles, W.S., Sabacky, M.J. and Vineyard, B.D. (1972) J. Chem. Soc. Chem. Comm. 10-11.
6) Fryzuk, M.D. and Bosnich, B. (1977) J. Am. Chem. Soc. 99 (19), 6262-6267.
7) Ojima, I., Kogure, T. and Yoda, N. (1980) 45, 4728-4739.
8) Achiwa, K. (1976) J. Am. Chem. Soc. 98 (25), 82265-8266.
9) Baltazzi, E. and Phil, D. Quart. Rev (1955) 9, 150.
10) Kirby, G.W. and Michael, J. (1973) J. Chem. Soc. P.T. I, 115-120.
11) Winkel, C. (1989) Ph.D. Thesis, University of Leiden, the Netherlands.
12) Kainosho, M. and Ajisaka, K. (1975) J. Am. Chem. Soc. 97 (19), 5630-5631.

A TWO DIMER MODEL OF Mn ORGANISATION IN PS-II - OBSERVATION OF TWO FORMS OF THE G=4.1 SIGNAL.

P. J. Smith* and R. J. Pace, Dept of Chemistry, Australian National University, Canberra 0200 Australia. * CRC for Plant Science, Canberra ACT 0200 .

Introduction

Oxidation of water to molecular oxygen is catalysed by the manganese containing Oxygen Evolving Complx (OEC) of photosystem (PS) II in higher plants and algae. A minimum of 4 Mn ions per reaction centre is required for functional water oxidation. The location of the Mn in the protein complex, as well as the mechanism operating at the catalytic site are as yet unresolved (review 1).

Continuous low temperature illumination advances the OEC from the diamagnetic, dark stable S_1 state, to the paramagnetic, S_2 state. Two characteristic EPR signals arising from the Mn cluster are generated under various conditions. First there is the multiline signal (2), characterised by 18-22 hyperfine features covering a width of ~190 mT and centred at g=2. In addition there is a signal centred near g=4.1, which is a broad, relatively featureless resonance under normal conditions (3). These signals may be generated individually or co-generated, depending upon the buffer, cryoprotectant and illumination conditions used with the PS-II sample. When alcohols (methanol, ethanol ≤ 3%) are present in the buffer medium, only the multiline is observed after 200K sample illumination, and its temperature dependence is that of an isolated, spin 1/2 ground state (4). In the absence of alcohols and with sucrose cryoprotectant, 200K illumination co-generates the multiline and 4.1 signals. These signals then behave quantitatively (4) like the spin 1/2 ground (multiline) and spin 3/2 , first excited state, of a weakly exchange coupled Mn heterodimer (III-II or III-IV). However, when 30% ethylene glycol or glycerol are used as cryoprotectant, the 4.1 signal alone is generated at ~ 130K illumination, and its temperature behaviour is consistent with that of a ground state form (5). Here we report that two distinct types of '4.1' signals may occur in the formal S_2 state of the OEC. One is clearly the first excited state of an antiferromagnetically coupled Mn dimer, and we observe the next excited state (spin 5/2) at g~6 with the expected temperature dependence. The other form of 4.1 signal we confirm to be a ground state species. While both types of 4.1 signals are very similar at X-band, they are readily distinquishable at Q-band.

Procedure

PS-II material was prepared from spinach and stored as previously described (4). Either 0.4 M sucrose or 30% v:v ethylene glycol was used as cryoprotectant. Samples for ESR

P. Mathis (ed.), Photosynthesis: from Light to Biosphere, Vol. II, 317–320.

(~ 10g/l chlorophyll) were thawed on ice in darkness for 1 hour before use. EDTA (2mM final) was added, but no electron acceptors, DCMU or alcohol (to allow development of the 4.1 signal). Samples were illuminated (3min) with strong yellow light (~600 Wm^{-2}) at 130 K or green light (~300 Wm^{-2}) at 200K in a N_2 flow cryostat. EPR measurements were peformed on Varian V-4502 (Q-band) and Bruker ESP300E (X-band) spectrometers. Signal intensities were estimated from extrapolations to zero microwave power as previously described (4) and all spectra are illuminated minus dark (S_1 state) background.

Results and Discussion

Fig 1a) shows the temperature developtment of signals in the g3-g10 region produced by 200K illumination of sucrose cryoprotected PS-II. At low temperature only the 4.1 signal is seen. This spectrum shows strong axiality, with a peak to trough width of 32 mT and g_{av} of 4.10. No splitting of $g_\perp$ components is resolved. As the temperature is increased, intensity builds in the g~6 region. This appears not to be a line broadening effect, certainly on the downfield edge of the 4.1 signal whose position does not shift significantly with temperature. The shape of the g~6 signal is estimated by substracting appropriately scaled amounts of the 5K spectrum from the spectra at higher temperature (fig 1b)). Fig 1c) shows the relative temperature dependences of the multiline, g=4.1 and g=6 signals. The curves are fits to a Boltzmann model (4) of the intensities, assuming the multiline, g~4.1 and g~6 signals arise from the spin1/2, 3/2 and 5/2 respectively states of an axially symmetric exchange coupled MnIII-IV dimer, with J/k=-3.5K. This value of the exchange coupling constant, J, is very similar to that inferred earlier (4) from the multiline and 4.1 signals only (J/k=-3.0K) The data indicate that the multiline signal arises from a MnIII-IV dimer, with approximate axial symmetry, supported by recent multi-frequency simulation studies from this laboratory on the multiline signal (6).

In contrast with the above, fig 2a) shows the temperature development of the g=4.1 signal formed in ethylene glycol cryoprotected PS-II particles, following illulmination at 130K. Little if any multiline is apparent nor developed. The signal shape is constant up to ~ 20K and then broadens at higher temperature. No obvious additional intensity develops in the g~6 region, nor any other region of the spectrum (g~2.6 to g~10).

Fig 2b) shows the temperature dependence of the doubly integrated signal amplitude. The signal is strongly sugggested to be a ground state (or very near one), with a possible upward deviation from Curie behaviour suggesting a next excited state within ~ 20 cm^{-1}. No obvious candidate for such a state is evident however.

That these two '4.1' signals arise from different species is clear when the spectra are compared at Q-band. Fig 3 shows the Q-band signals for the excited state (sucrose, 200K illumination) and ground state (ethylene glycol, 130K illumination) froms. Both now exhibit a partially resolved splitting of g=4 components. For the excited state species, the $g_\perp$ components are estimated to be at $g_{\perp1}$ =4.35 and $g_{\perp2}$ = 4.14 (±0.02) with, g_{av} ~4.25, ie at least as high as the X-band estimate. The signal appears highly axial suggesting $E/D \leq 0.02$ and D> 1cm^{-1} for the fine structure terms of the 3/2 state (7). The ground state signal shows a slight drop in g_{av} from the X-band value to g_{av}~3.98 (±0.05). The wings of this spectrum are much broader at Q-band than those of the excited state form. From the X- and Q-band data, this signal could also arise from a near axial

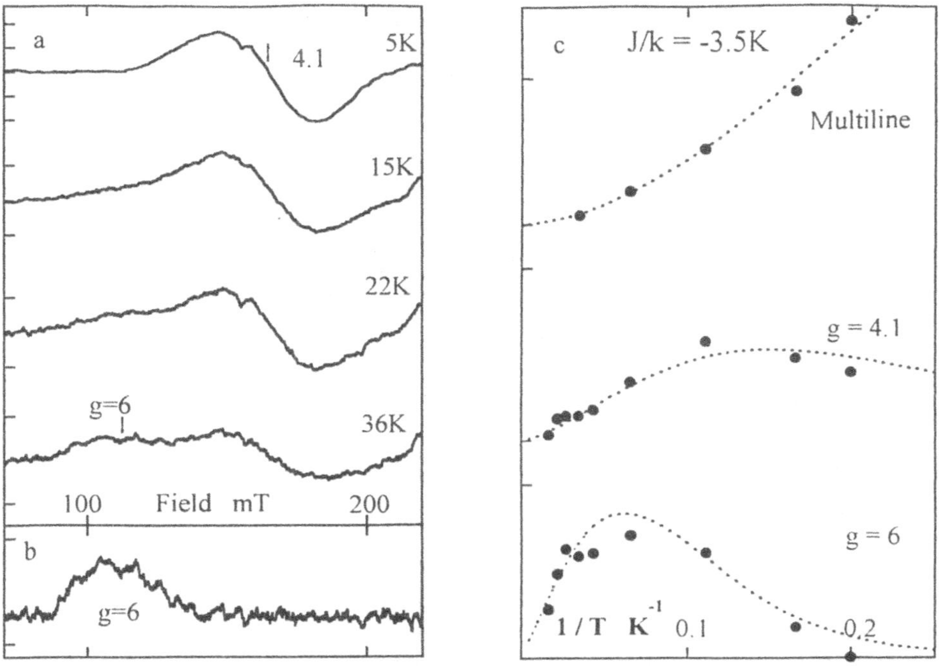

Fig. 1. a) Temperature development of low field signals produced by 200K illumination of 0.4M sucrose medium PS-II. b) shape of g~6 signal developed at temperatures above 20K, estimated as described in text. Spectrometer conditions: microwave frequency 9.423GHz, power 6.3mW, modulation frequency 100KHz, modulation amplitude 1.5mT average of 6 scans. c) Relative temperature dependences of multiline, g=4.1 and g~6 signal intensities of material in a). Intensity scale is arbitrary due to different methods of signal quantitation (4). Curves are fit to model (see text).

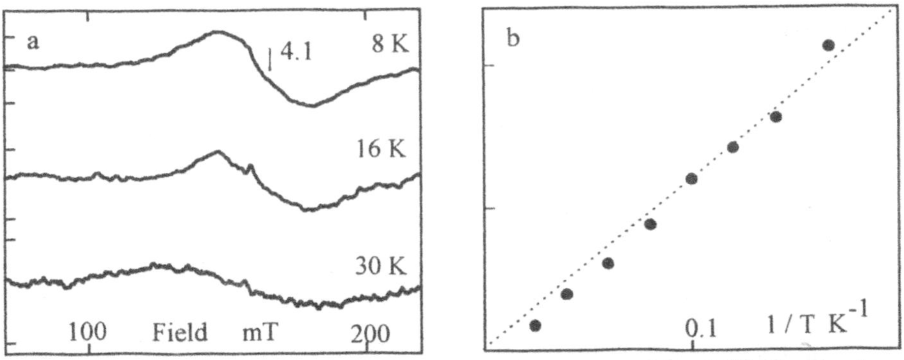

Fig. 2. a) Temperature development of low field signal formed by 130K illumination of 30% ethylene glycol medium PS-II. Spectrometer conditions as in fig 1. b) Temperature dependence of g~4 signal intensity (from double integrations) of material in a).

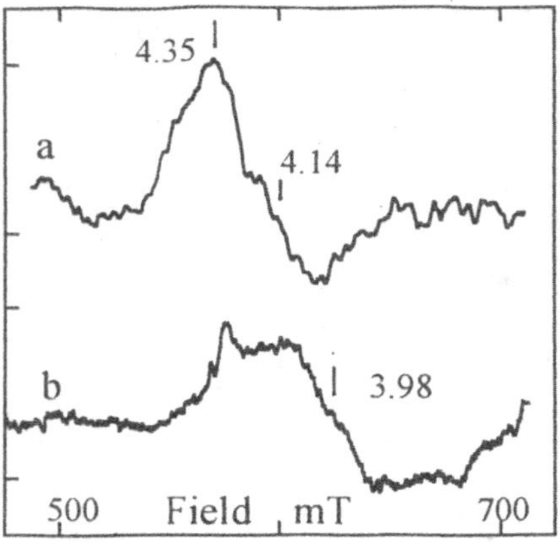

Fig. 3. Q band spectra of the excited state (0.4M sucrose, 200K illumination), (a), and the ground state (30% ethylene glycol, 130K illumination), (b), forms of the 4.1 signal. Partially resolved components are indicated as shown. Spectrometer conditions: temperature 9K, microwave frequency 34.75GHz, Power 1.5mW, modulation frequency 100KHz, amplitude 1mT.

spin 3/2 state, but with a D value less than that inferred for the excited state form (ie < 1 cm^{-1}).

In summary the evidence is strong that the multiline, excited state g~4.1 and g~6 signals arise from an isolated antiferromagnetically coupled MnIII-IV dimer (or a spin system formally equivalent to this). The coupling is variable, being weak (J~2 cm^{-1}) in the absence of alcohol and stronger (>10 cm^{-1},(4)) in the presence of alcohol in the buffer medium. This might reflect μ-oxo bridge protonation effects (8). The origin of the ground state 4.1 species is not so clear. The signal behaves as if it arose from an electron transfer intermediate between the multiline centre (concluded here to be an isolated dimer) and the z-tyrosine(9). However a Mn pair (ie the other two Mn) in plausible oxidation states would not give a spin 3/2 ground state. At least 3 spin centres would be required for a 3/2 ground state. One possibility is a Mn pair interacting with an oxidised protein side chain. possibly histidine. The presence of the latter has been suggested from studies on Ca depleted systems (10), although the precise manner of its possible involvement with the Mn is unclear.

References

1 Debus, R.J. (1992) Biochim. Biophys. Acta 1102, 269-352.
2 Dismukes, G.C. and Siderer, Y. (1981) Proc. Natl. Acad. Sci. USA 78, 274-278.
3 Zimmermann, J.-L. and Rutherford, A.W. (1986) Biochem. 25, 4609-4615.
4 Pace, R.J., Smith, P., Bramley, R. and Stehlik, D. (1991) Biochim. Biophys. Acta 1058, 161-170.
5 de Paula, J.C., Beck, W.F. and Brudvig, G.W. (1986) J. Am. Chem. Soc. 108, 4002-4009.
6 Åhrling, K.A. and Pace, R.J. (1995) Biophys. J. 68, 2081-2090.
7 Smith, P.J., Åhrling, K.A. and Pace, R.J. (1993) J. Chem. Soc. Faraday Trans. 89, 2863-2868.
8 Larson, E.J., Riggs, P.J., Penner-Hahn, J.E. and Pecoraro, V.L. (1992) J. Chem. Soc. Chem. Commun. 39, 102-103.
9 Casey, J.L and Sauer, K. (1984) Biochim. Biophys. Acta 767, 21-28.
10 Boussac, A., Zimmermann, J.-L. and Rutherford, A.W. (1990) Science 347, 303-306.

ELECTRON DONATION FROM EXOGENOUS DONORS TO PHOTOSYSTEM II STUDIED IN *CHLAMYDOMONAS REINHARDTII* MUTANTS.

C. Kullander[1,5], P-O. Fredriksson[1], R.T. Sayre[2], J. Minagava[3,4] A.R. Crofts[4] and S.Styring[5]
[1]Dept. of Natural Science, Univ. of Karlstad, Sweden. [2]Dept. of Plant Biology and Biochemistry, Ohio State University, Columbus, Ohio 43210 USA [3] Photosynthesis Research Lab. RIKEN, 2-1 Hirosawa, Wako Saitama 351-01 Japan [4]Dept. of Physiology and Biophysics, Univ.of Illinois, Urbana, IL 61801 USA. [5]Dept. of Biochemistry, Arrhenius Lab. Stockholm Univ. Sweden

1. Introduction

Photosystem II (PSII) extracts electrons from water and transports them to plastoquinone via a series of redox components on the D1 and D2 proteins (reviewed in 1). The water splitting complex consists of a cluster of four manganese ions on the lumenal side of PSII. Assembly of the manganese cluster is a light dependent process called photoactivation. During photoactivation Mn^{2+} is hypothesized to bind to its site in the water splitting complex (2). With the arrival of a photon, Mn^{2+} donates an electron to Tyr_Z^{ox} on D1 forming an intermediate with bound Mn^{3+}, this intermediate is stabilised by Ca^{2+} and Cl^-. Photooxidation of a second Mn^{2+} stabilises the first Mn^{3+}. This is followed by the ligation of two other Mn^{2+} ions, forming the functional $(Mn)_4$-cluster. Tyr_Z is oxidised by the primary electron donor P_{680}^+ and is normally reduced by the Mn-cluster. Another redox active tyrosine, Tyr_D on D2, may compete with Tyr_Z in reducing P_{680}^{ox}. However, Tyr_D is a much poorer electron donor.
There exist several potential electron donors to PSII centers that lack the Mn-complex. Mn^{2+} and DPC (diphenyl carbazide) are efficient exogenous electron donors. Both Mn^{2+} and DPC donate electrons to hydroxylamine extracted PSII (lacks the Mn-cluster) in a bifunctional manner showing one high and one low affinity site. The high affinity site most likely reflects electron donation to Tyr_Z while the origin for the low affinity site is less clear although donation to Tyr_D has been proposed (3).
We have studied Mn^{2+} and DPC electron donation in dark grown *Chlamydomonas reinhardtii* wild type and a series of site directed point mutants in the D1 protein. Dark grown *C. reinhardtii* shows normal PSII levels but does not assemble the Mn-cluster which facilitates studies of electron donation from added donors without any perturbing washing procedures to first remove the Mn-cluster.
The mutants studied are changed in the conserved amino acids Pro162, His190 and Phe186 on the D1 protein. Structural modelling (4) locates these residues near Tyr_Z where they might be involved in electron transfer to and from Tyr_Z. Tyr_Z itself has also been exchanged (D1-Tyr 161) and electron donation has been studied in this mutant.

2. Material and methods

C. reinhardtii wild type and mutants were grown heterotrophically on Tris-acetate-phosphate medium in complete darkness for 5-20 days. Some mutants are very slow growing. Thylakoid membranes were prepared as described (5). The preparation was

P. Mathis (ed.), Photosynthesis: from Light to Biosphere, Vol. II, 321–324.

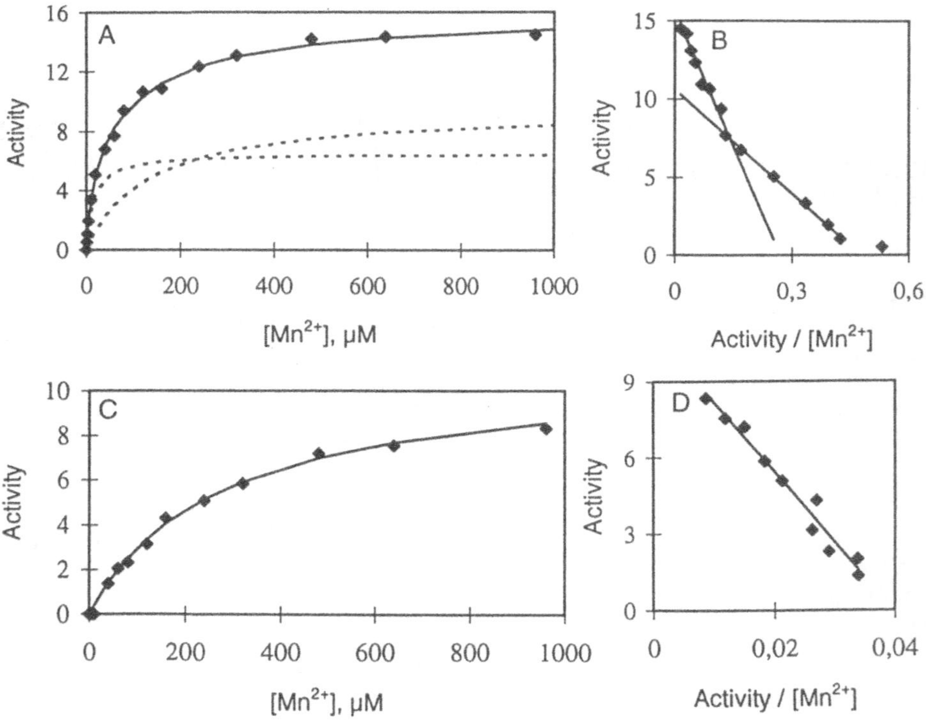

Figure 1. **A.** Mn^{2+} electron donation to wild type *C. reinhardtii*. The dotted lines represent binding curves to two sites with K_m and V_{max} that were calculated by successive approximation (6). **B.** An Eadie-Hofstee plot of the data in A indicating two sites. **C.** Mn^{2+} electron donation to Phe186Leu. **D.** An Eadie-Hofstee plot of the data in C indicating one site. Activity is given as µmol DCIP red/(mgChl*h).

modified with an EDTA-wash (1 mM) to remove divalent metal ions and two additional washes to remove the EDTA. Electron transfer in the thylakoids was measured spectrophotometrically as DCIP (2,6-dichlorophenol-indophenol, 35 µM) reduction. Assay conditions; 25 mM HEPES-NaOH, pH = 7.0, 15 mM NaCl, 400 mM sucrose, 6 µg/ml gramicidine (buffer 1) and 10 µg Chl/ml. Ca^{2+} was omitted to prevent photoactivation during the measurement. Electron donation to PS I was inhibited by DBMIB (2,5-dibromo-3-methyl-6-isopropyl-p-benzoquinone, 6 µg/ml). The background activity was recorded 1/2-1 min prior to illumination, the initial activity after illumination is reported. Mn^{2+} (as $MnCl_2$) and DPC were added from 20 and 50 times stock solutions respectively. DPC was dissolved in methanol. To measure O_2-evolution, dark grown wild type and mutants were illuminated for 20 min in low light, 10-15 µE/(m^2*s), prior to preparation of thylakoids to allow photoactivation (some mutants are very sensitive to light). The O_2-evolution was measured as described (5) in buffer 1 with PpBQ as electron acceptor.

	Mn^{2+} donation				DPC donation				O_2- evolution
	Site 1		Site 2		Site 1		Site 2		
	K_m	V_{max}	K_m	V_{max}	K_m	V_{max}	K_m	V_{max}	
Wildtype	15	6.5	129	9.5	28	6	470	37	63
Tyr161Phe	820	10	-	-	630	44	-	-	0
Pro162Leu	73	21	-	-	197	79	-	-	53
Pro186Leu	274	11	-	-	500	68	-	-	11
His190Phe	12	2	335	12	7	7	368	46	5
His190Tyr	294	15	-	-	11	6	890	80	0

Table 1. Mn^{2+} and DPC electron donation and O_2-evolution in *C. reinhardtii*. Activities are given as µmol DCIP red/(mg Chl*h) and concentrations in µM. The parameters are simulated as described by Spears et al. (6). (-) indicates that no second site was observed. Oxygen evolution is given as µmol O_2/(mg Chl*h). All activities are related to chlorophyll concentration, the PSII/ PSI ratio is not yet measured and might vary between the different mutants.

3. Results and Discussion

In the wild type Mn^{2+} and DPC electron donation are bifunctional, implying two sites (Fig. 1.A, 1B and Table 1). These results are in agreement with the results in NH_2OH washed PSII membranes of wheat (3). The high affinity site for Mn^{2+} and DPC represent about 40% and 15% of the total V_{max} respectively. The total V_{max} of DPC electron donation is 2,5-3 times that of Mn^{2+}. We note that to be detectable a site must show a V_{max} of 2-3 µmol DCIP red/(mg Chl*h), sites with lower V_{max} might not be observable. This also holds for the O_2-evolution.

Table 1 shows that Mn^{2+} and DPC electron donation are drastically altered in the mutants. Tyr161Phe, Pro162Leu and Phe186Leu (Fig 1 C och D) show only one site for Mn^{2+} and DPC electron donation. In Tyr161Phe an important structural change is introduced with the removal of Tyr_Z and its exchange into Phe. Interestingly, this severe alteration still allows electron transfer from both Mn^{2+} and DPC. Tyr161Phe has a very low affinity for Mn^{2+}, the K_m is more than six times higher for Tyr161Phe than for the low affinity site of the wild type. K_m for DPC donation is modified and about two times higher than in the wild type. Surprisingly high activity can be reached. V_{max} for Mn^{2+} donation is about 75% of the total V_{max} for the wild type while V_{max} for DPC donation is approximately the same as for the wild type. Tyr161Phe does not evolve any oxygen and is very sensitive to light.

Pro162Leu, a mutation in the residue next to Tyr_Z, shows a single affinity for Mn^{2+}, K_m = 73 µM, is an intermediate between the high and the low affinity sites in the wild type. The activities for Mn^{2+} and DPC electron donation are higher than in the wild type, which simply could reflect a higher PSII ratio. Pro162Leu evolves oxygen at a diminished rate compared to the wild type.
Phe186Leu is quite similar to Tyr161Phe. It has a single site both for Mn^{2+} and DPC donation. The K_m is higher for both Mn^{2+} and DPC as compared to the low affinity site of the wild type. V_{max} of the different sites is 70% and 160% of the total V_{max} for wild type in

Mn^{2+} and DPC electron donation, respectively. Oxygen evolution in this mutant can be photoactivated to some extent.

We have also studied two mutants in His190 which closely interacts with Tyr_Z (4, 7, 8). His190Phe has two electron donation sites for both Mn^{2+} and DPC. V_{max} of the low affinity site has increased at the expense of the high affinity site for both Mn^{2+} and DPC. Interestingly, we managed to observe a small O_2-evolution (7% of wild type) in His190Phe. This is different to earlier results in a similar mutant in *Synechosystis* sp. PCC 6803 (9). Presumably our results reflect our mild photoactivation procedures. We hypothesize that the low oxygen evolution is related to the existence of a small fraction of a high affinity site for Mn^{2+} and DPC in this mutant. His190Tyr, shows two donation sites for DPC but only one donation site for Mn^{2+}. V_{max} of the Mn^{2+} donation site is about the same as for wild type while the affinity is very low. The low affinity site of the DPC site is severely altered, both K_m and V_{max} has doubled. His190Tyr does not evolve any O_2. In His 190 Tyr_Z is possble but difficult to oxidize (7). It is likely that this makes Mn^{2+} photooxidation and photoactivation inefficient leading to absence of oxygen evolution.

4. Conclusions

The data presented here support the existence of two electron donation sites for both Mn^{2+} and DPC, one with a high and one with a low affinity. Site directed point mutations on the D1 protein alters this electron donation pattern. Changes made in Tyr161, Pro162 and Phe186 alters the electron donation for both Mn^{2+} and DPC to single low affinity sites. In Tyr161Phe, Phe186Leu and His190Tyr the affinity for Mn^{2+} is significantly lowered. This seems to be related to the absence of, or low rates of O_2-evolution in the mutants.

It is interesting that in two of this mutants, Tyr161Phe and His190Tyr, Tyr_Z is either absent or very difficult to induce (7). In Phe186Leu Tyr_Z has not been studied. For Pro162Leu, the affinity for Mn^{2+} is less altered and a quite high O_2-evolution is maintained suggesting that the normal Mn^{2+} site is little affected.

Acknowledgements -This work was supported by the Swedish National Science Research Council.

References

1 Andersson, B. & Franzén, L-G. (1992). L. Ernster (Ed.) *Molecular Mechanisms in Bioenergetics*. Elsevier Science Publisher B.V. Vol. 23 121-143
2 Tamura, N. & Cheniae, G. (1987) Biochim. Biophys. Acta 890, 179-194
3 Blubaugh, D. &Cheniae, G. (1990) Biochemistry 29, 5109-5118
4 Svensson, B., Vass, I., Cedergren, E. & Styring, S (1990) EMBO J. 9, 2051-2059
5 Rova,M., Franzén,L-G., Fredriksson,PO., & Styring,S.(1994)Photosynth.Res.39,75-83
6 Spears, G., Sneyd, J.G.T. & Loton, E.G. (1971) Biochem. J. 125, 1149-1151.
7 Roffey,R., van Wijk,K., Sayre,R, & Styring,S. (1994) J.Biol. Chem. 269 5115-5121
8 Diner, B. R., Nixson, P.J & Farchaus, J.W (1991) Curr. Opin. Struct.. Biol. 1, 546-554
9 Debus, R. (1992) Biochim. Biophys. Acta 1102, 269-352

REBINDING OF THE 33 kDa EXTRINSIC PROTEIN TO PHOTOSYSTEM II MEMBRANES WHICH CONTAIN FOUR, TWO OR ZERO MANGANESE PER PS II REACTION CENTER

Carola Leuschner and Terry M. Bricker
Department of Microbiology
Louisiana State University
Baton Rouge, LA 70803

1. Introduction

The light-driven oxidation of water to molecular oxygen and the reduction of plastoquinone to plastoquinol is catalysed by Photosystem II (PS II). In higher plants, three extrinsic proteins (33, 24 and 17 kDa) are associated with this oxygen-evolving complex of PS II. The 33 kDa extrinsic protein is present in all oxygenic organisms and is required for binding the 24 and 17 kDa extrinsic proteins (1). PS II membranes depleted of the 33 kDa extrinsic protein lose two manganese and the ability to evolve oxygen at low chloride concentrations (<100 mM) (2). Based on these results the 33 kDa extrinsic protein is considered to be a manganese-stabilizing protein. While the 33, 24 and 17 kDa extrinsic proteins appear to be present in equimolar amounts (3), their stoichiometry with respect to PS II remains controversial. Previous studies indicated 1-3 molecules of each extrinsic protein per PS II reaction center (3-5). Recently, results from our laboratory (6) indicated the presence of two copies of the 33 kDa extrinsic protein per PS II reaction center in both PS II membranes and oxygen-evolving PS II core complexes. Here we present data examining the rebinding of the 33 kDa extrinsic protein to PS II membrane preparations containing four, two or zero manganese per PS II reaction center.

2. Procedure

Oxygen-evolving PS II membranes were isolated as described in (7) and (8). PS II membranes depleted of the 33 kDa extrinsic protein and containing four manganese per PS II reaction center were prepared by treatment with NaCl-urea (9). These membranes were resuspended in a low chloride buffer to prepare membranes containing two manganese per PS II reaction center. PS II membranes containing zero manganese per PS II reaction center were prepared by alkaline Tris treatment according to (3). Isolation of the 33 kDa extrinsic protein was performed as described in (6). For

P. Mathis (ed.), Photosynthesis: from Light to Biosphere, Vol. II, 325–328.

the reconstitution experiments, aliquots of the three different types of PS II membranes were incubated for 1 hour in the dark with purified 33 kDa extrinsic protein (0 and 1400 nM) in 300 mM sucrose, 100 mM NaCl, 10 mM $MgCl_2$, and 50 mM Mes-NaOH, pH 6.0 (incubation buffer). The amount of rebound 33 kDa extrinsic protein in PS II membranes was determined immunologically essentially as described in (6).

3. Results and Discussion

Fig. 1A. shows a typical "Western" blot which was used in this study. These blots contained both standard concentrations of the 33 kDa extrinsic protein (within a background of PS II membranes (6)) and membrane fractions to which the 33 kDa extrinsic protein had been rebound. Quantification was performed from these "Western" blots and analyzed using the Quantiscan 2 program (Biosoft Inc.). A plot of

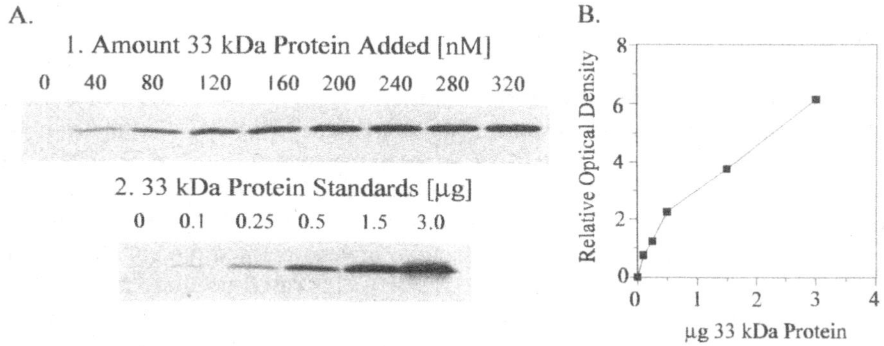

Figure 1: Analysis of the Binding of the 33 kDa Extrinsic Protein to PS II Membranes. A1 shows a "Western" blot of samples from PS II membranes containing four manganese per PS II reaction center which were incubated at given concentrations of 33 kDa extrinsic protein (0 - 320 nM); A2 shows purified 33 kDa protein (with 33 kDa-depleted PS II membranes) applied at the indicated concentrations as standards. The amounts of bound 33 kDa extrinsic protein in each PS II membrane preparation was calculated from the calibration curve (B), which had been analysed via the Quantiscan 2 program.

the standard curve generated from analysis of the "Western" blot shown in Fig. 1A is shown in Fig. 1B. Figure 2 shows a direct plot of the amount of the 33 kDa extrinsic protein bound to four, two and zero manganese-containing PS II membranes with increasing concentrations of added 33 kDa extrinsic protein. For four manganese-containing PS II membranes a saturation level was reached at a concentration of 360 nM of added 33 kDa extrinsic protein, while the two and zero manganese-containing PS II membranes became saturated at a concentration of 500 nM.

Figure 2: **Rebinding of the 33 kDa Extrinsic Protein to PS II Membranes Containing Four, Two and Zero Manganese per PS II Reaction Center**

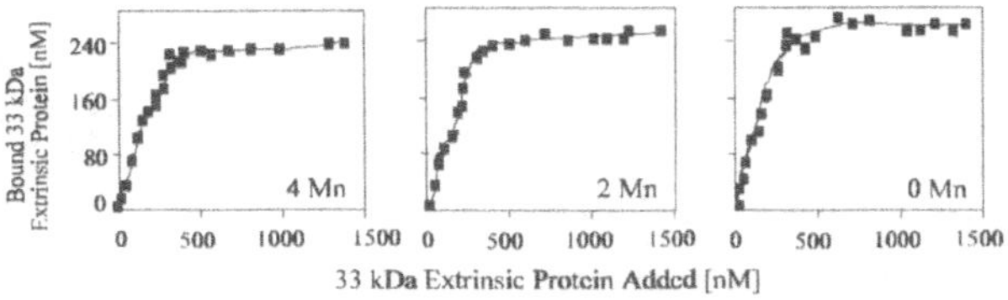

The stoichiometry was calculated to be 2.0-2.1 moles of the 33 kDa extrinsic protein per mole of PS II centers for all three types of PS II membranes. These data are in agreement with results from Xu and Bricker (6). No release of bound 33 kDa extrinsic was detected during the washing of reconstituted PS II membranes indicating that the two molecules are tightly bound to the PS II membrane preparations. Our results differ significantly from those obtained by Miyao and Murata (3), Enami et al. (10) and Andersson and Styring (11), who proposed the existence of one molecule of 33 kDa protein per PS II reaction center. This discrepancy might be due to different techniques in analyzing the amounts of bound 33 kDa protein. Underestimation of bound 33 kDa protein can also be caused by applying an incorrect extinction coefficient for 33 kDa protein (12). Binding constants for the two copies of the 33 kDa protein were determined as described by Klotz (13). A graphical determination of the first stoichiometric binding constant was performed on the basis of the following equation: $\lim S_b/E_t \times S_f = K_{a1}$ $[M^{-1}]$ in which S_b is the concentration of bound substrate, E_t the enzyme or receptor concentration, S_f the amount of free substrate and K_{a1} the first stoichiometric binding constant. These binding constants (Table 1) indicate that there is a four-fold decrease in K_{a1} upon release of the chloride-sensitive manganese. No

Table 1

Type of PS II Membranes	Hill Plot	Dissociation constant K_s [nM]	Binding constant K_a [nM^{-1}]	Stoichiometry [mol 33 kDa protein/ mol PS II]
4 Mn	h = 1.74	K_{s1} 3.6 ± 0.1 K_{s2} 1.4 ± 0.2	K_{a1} 0.28 ± 0.03 K_{a2} 0.72 ± 0.12	2.01 ± 0.08
2 Mn	h = 1.64	K_{s1} 15.4 ± 0.2 K_{s2} 5.0 ± 1.6	K_{a2} 0.07 ± 0.00 K_{a2} 0.23 ± 0.08	2.03 ± 0.06
0 Mn	h = 2.16	K_{s1} 15.2 ± 0.1 K_{s2} 4.4 ± 1.4	K_{a1} 0.07 ± 0.00 K_{a2} 0.25 ± 0.05	2.14 ± 0.13

further decrease in K_{a1} was observed upon removal of the chloride-insensitive manganese. The second stoichiometric binding constant (K_{a2}) was then calculated algebraically (13). The values observed for K_{a2} are higher than those observed for K_{a1} for all PS II membrane types. Again, release of the chloride-sensitive manganese led to about a three-fold decrease in the strength of binding of the 33 kDa extrinsic protein and no further decrease in the K_{a2} was observed upon release of the chloride-insensitive manganese. The analysis of our data using a Hill plot gave strong evidence that the two copies of the 33 kDa extrinsic proteins bind with positive cooperativity, with Hill coefficients ranging from 1.7 to 2.2 (Figure 3). This indicates that these molecules

Figure 3: Hill Plot Analysis for PS II Membranes Containing Four, Two and Zero Manganese per PS II Reaction Center. The data for binding of the 33 kDa protein were applied to a Hill plot. B is the amount of bound 33 kDa protein, Bmax is the amount of maximal binding. The slope h of each linear graph represents the Hill coefficient.

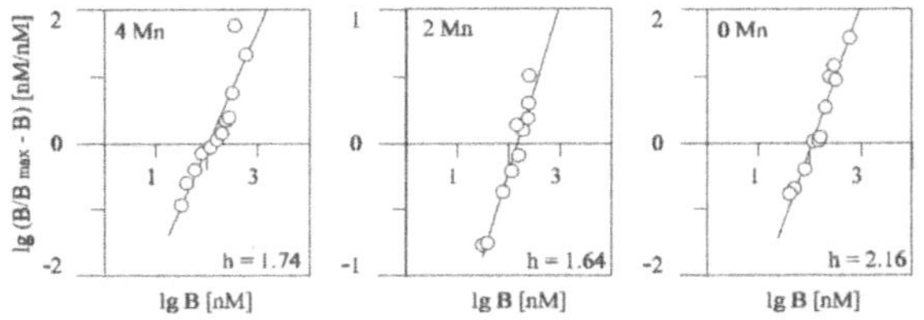

must interact either directly or indirectly during their binding to PS II. Our results clearly indicate that the binding of the 33 kDa extrinsic protein to PS II is affected by the tetranuclear manganese cluster. In particular, the presence of the chloride-sensitive manganese modulates the strength of binding of both copies of the 33 kDa extrinsic protein.

4. Acknowledgments

This work was supported by NSF Grant 93-04955. Special thanks to Ms. Laurie K. Frankel for her help in preparing this manuscript.

5. References

1. Kavelaki K. & Ghanotakis D.F (1991) *Photosynth. Res.*, 29, 149 - 155
2. Mavankal G., McGain D., Bricker T.M. (1986) *FEBS Lett.* 202, 235 - 239
3. Miyao M. & Murata N. (1989) *Biochim. Biophys Acta* 977, 315 - 321
4. Millner P.A., Gogal G., Barber J. (1987) *Photosynth Res.* 13, 185 - 198
5. Enami I., Ohta S., Mitsuhashi S., Takahashi S., Ikeuchi M., Katoh S. (1992) *Plant Cell Physiol.* 33, 291 - 297
6. Xu Q. & Bricker T.M. (1992) *J. Biol. Chem.* 267, 25816 - 25821
7. Berthold D.A., Babcock G.I., Yocum C.F. (1981) *FEBS Lett.* 153, 231 - 234
8. Ghanotakis D.F. & Babcock G.T. (1983) *FEBS Lett.* 153, 231 - 234
9. Miyao M. & Murata N. (1984) *FEBS Lett.* 170, 350 - 354
10. Enami I., Kaneko M., Kitamura N., Koike H., Somoike K., Inoue Y., Katoh S. (1991) *Biochim. Biophys Acta* 1060, 224 - 232
11. Andersson B. & Styring S. (1991) *Curr. Top. Bionerg.* 16, 1 - 81
12. Kuwabara T. & Murata N. (1984) in *Advances in Photosynthesis Research* (Sybesma, C. ed.) Vol .1. pp 371 - 374, Martinus Nijhoff Publishers/The Hague, The Netherlands
13. Klotz I.M. (1990) in: *Protein function - a practical approach* (Rickwood B., Hames B.D. Creighton T.E., eds)pp: 24 - 54 Irl Press Oxford University Press

HIGH AFFINITY BINDING OF THE Ca^{2+} ESSENTIAL FOR O_2 EVOLUTION IS DEPENDENT ON THE EXISTENCE OF THE Mn-CLUSTER BUT IS INDEPENDENT OF THE EXTRINSIC PROTEINS.

Changguo Chen and G. M. Cheniae
U. Kentucky, Lexington, KY, USA

1. INTRODUCTION

Treatment of PSII membranes with high concentrations of NaCl plus EGTA or low pH results in inhibition of O_2 evolution blocking the $S_3 \rightarrow S_0$ transition which can be overcome by addition of Ca^{2+} (1-3). Efforts to quantify the effects of such treatments on PSII Ca^{2+} abundance have not yielded consensus conclusions. More recently, evidence has appeared suggesting that 1 of the 2 Ca^{2+}/PSII is associated with LHCII while the other is linked to OEC (4,5). Measurements of $^{45}Ca^{2+}$ binding to Ca^{2+}-depleted PSII membranes (6) and Ca^{2+} dissociation from PSII membranes by acid pH (7) have lent support to this idea. Here, the above premise is analyzed using Ca^{2+}-depleted PSII membranes, PSII core complex particles, and photoactivated PSII membranes. Measurements also were made of the effects of pH and extrinsic polypeptide composition on the K_m value(s) of Ca^{2+} for the reconstitution of O_2 evolution by Ca^{2+} in the absence of Ca^{2+} antagonistic cations.

2. PROCEDURES

BBY-type PSII membranes were prepared from 7-10 day old wheat seedlings. Oxygen evolving PSII core complex particles (70 Chl/P680) were prepared as in Ref. 8. NaCl/NH_2OH-PSII was prepared and photoactivated as described in Ref. 9. Ca^{2+}-depleted PSII membranes were prepared by extraction with 1.5 M NaCl/1 mM EGTA/20 μM A23187/40 mM Mes-NaOH, pH 5.0 at 0.05 $\mu E \cdot m^{-2} \cdot s^{-1}$. Treatment of these membranes with 2.6 M urea/500 mM $(CH_3)_4NCl$/25 mM Mes-Tris/0.2 mM EGTA removed the 33 kDa protein. pH 3.0 extraction of PSII membranes was done essentially as described in Ref. 10. Reconstitutions of Ca^{2+}-depleted membranes were made in the absence of Ca^{2+} antagonistic ions with preincubations in 0.8 M sucrose/50 mM Mes-Tris/30 mM $(CH_3)_4NCl$/0.5 mM PBQ at the pH values and Ca^{2+} concentrations indicated before measurements of rates of O_2 evolution.

P. Mathis (ed.), Photosynthesis: from Light to Biosphere, Vol. II, 329–332.

3. RESULTS AND DISCUSSION

Figure 1 A shows the effects of NaCl-pH 5 treatment of PSII membranes in near darkness on Ca^{2+} abundance and V_{O2} in the absence (Δ) and presence (○) of 15 mM Ca^{2+}. The data reveal a) the initial rate of decrease of V_{O2} in the absence of Ca^{2+} is faster than the decrease of Ca^{2+} abundance, and b) with time, the minus Ca^{2+} V_{O2} rates approached zero and the Ca^{2+} abundance diminished from ~2 to 0.5 Ca^{2+}/PSII. The faster decline of V_{O2} than decline of Ca^{2+} abundance could reflect dissociation of Ca^{2+} from its functional binding site but not from the membrane (4), or some perturbation of OEC in addition to dissociation and loss of Ca^{2+}. In either scenario, Ca^{2+} addition to the extracted membranes almost completely reversed the inhibition of V_{O2} (○). The ratio of the rate of V_{O2} measured in the presence and absence of Ca^{2+}, is an indicator of the capability of the membranes to respond to added Ca^{2+}, irrespective of the above two scenarios. Figure 1 B shows a plot of this ratio vs membrane Ca^{2+} abundances.

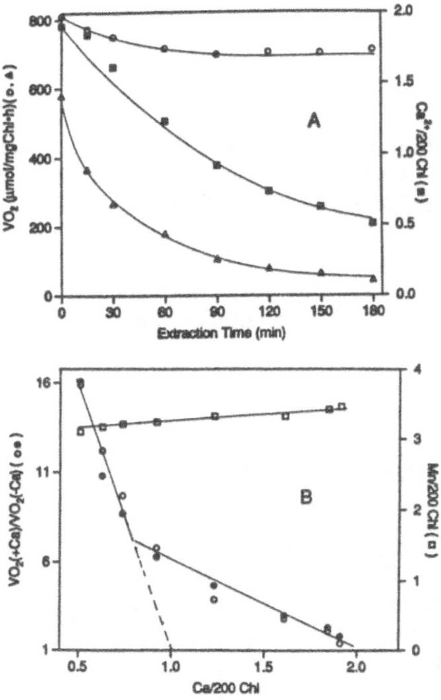

FIGURE 1

Inspection of Figure 1 B reveals a) the Mn abundance was unaffected by the extraction, b) the response of the membranes to added Ca^{2+} increased linearly between the limits of ~ 2 to 1 Ca^{2+}/PSII, and c) at <~1 Ca^{2+}/PSII, the slope of the response to Ca^{2+} increased sharply, a result of very small decreases of V_{O2} measured in the absence of Ca^{2+}. Thus, the capability of the PSII membranes to respond to added Ca^{2+} is determined primarily by removal of 1 Ca^{2+}/PSII. Extraction of control PSII membranes containing ~ 4 Mn and 2 Ca^{2+}/PSII with NH_2OH or Tris abolished V_{O2} and the Mn-cluster and dissociated 1 Ca^{2+}/PSII. NH_2OH or Tris treatment of PSII Core Complex particles lacking LHCII, containing CP 28-30, ~ 4 Mn, and ~ 0.44 Ca^{2+}/PSII, and showing a strong response of V_{O2} to added Ca^{2+}, also abolished V_{O2} and dissociated all of PSII Mn and Ca^{2+}. Such data support the idea that the 2 Ca^{2+}/PSII of PSII membranes are distributed as 1 Ca^{2+}/OEC and 1 Ca^{2+}/LHCII, and they show that dissociation of the Mn-cluster results in dissociation of the 1 Ca^{2+}/OEC.

Figure 2 shows that illumination of NaCl/NH_2OH-PSII membranes containing 1 Ca^{2+}/LHCII in a Mn^{2+}/Ca^{2+}/DCIP mixture leads to parallel photoassembly of the Mn-cluster and the 1 Ca^{2+}/OEC. As shown (Figure 2 A), V_{O2}, Mn, and Ca^{2+} abundance all increase in parallel with time, a result supporting an idea that the photoassembly of the Mn-cluster leads to creation of the binding site for the 1 Ca^{2+}/OEC (11; see however 12). In the experiment of Figure 2 B, membranes were photoactivated (20 min) at the pH values shown then V_{O_2} and Mn/Ca^{2+} abundances were determined. Over a wide pH range, V_{O_2} and Mn/Ca^{2+} abundances were affected similarly. A replot (Figure 2 C) of the data of Figs. 2 A and B, in which the ratio of the abundance of photoassembled Mn to Ca^{2+} is plotted vs. the various pH values or time of photoactivation, shows that good fit to the data is obtained with the dashed line showing 4 Mn and 1 Ca^{2+} per OEC.

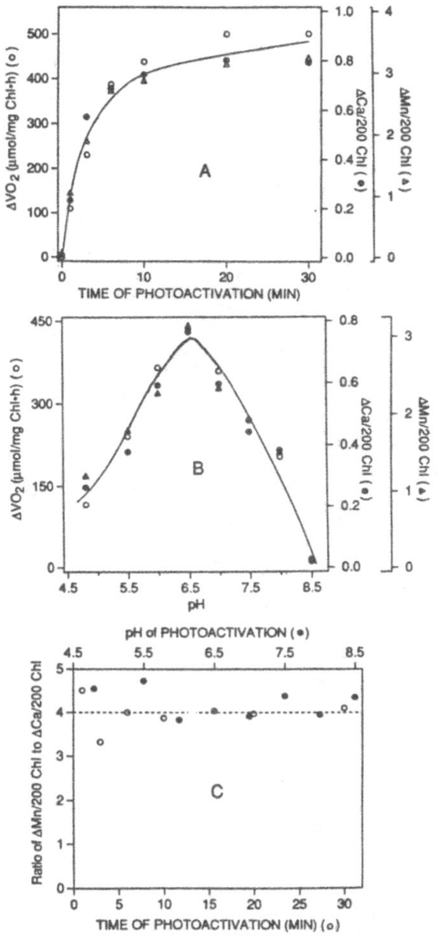

FIGURE 2

The dissociation constant(s) of the Ca^{2+} required to reactivate V_{O_2} by Ca^{2+}-depleted PSII membranes is heterogeneous at pH ~ 6.2 with the majority of the reports showing two constants (~ 50 μM and 1 mM) (1-3). Such values are inflated, since the measurements were made in the presence of Na^+ which competes for the Ca^{2+} site (3). Moreover, the two binding sites are interconvertible with the high affinity site dominating at higher pH values (11, 13), a result suggesting that the two constants indicate different conformations of the same binding site. We preincubated and assayed Ca^{2+}-depleted PSII membranes at various Ca^{2+} concentrations in the absence of competing Na^+/K^+ ions then K_m and V_{max} values were determined from Eadie-Scatchard plots for the site(s). These measurements were made at different pH values with Ca^{2+}-depleted PSII membranes containing different populations of the extrinsic polypeptides. Inspection of Table I reveals: 1) at pH values < pH 7.2, heterogeneity of Ca^{2+}-binding is observed irrespective of the membrane/extrinsic polypeptide composition; 2) with increasing pH, the K_m of the high-affinity site decreases and at pH 7.2 reaches a value of only 2-3 μM, irrespective of the membrane/extrinsic polypeptide composition; and 3) at pH 7.2, only one binding site is observed. This 2-3 μM K_m value is comparable to value(s) reported for some EF-hand

Ca^{2+}-binding proteins (3). We estimate from data obtained with membranes containing the 33 kDa protein an apparent pK-value of ~ 6.2 for the conversion of the K_m of the high-affinity site to its minimum value. This contrasts to the apparent pK-value of ~ 4.7 for Ca^{2+}-release and inhibition of O_2 evolution (7).

Table I Effects of pH and Extrinsic Polypeptide Composition on Ca^{2+} Km and V_{max} Values Obtained From Reconstitution of Oxygen Evolution by PSII Membranes in the Absence of Ca^{2+} Antagonistic Cations

Extrinsic Polypeptide Composition	pH	Km_1	Km_2	V_{max1}	V_{max2}
		(μM)		(μMols O_2/mg Chl•h)	
17 kDa/23 kDa/33 kDa	6.2	11.3	109.7	274.7	136.2
	6.7	8.8	967.1	368.3	105.5
	7.2	3.1	-	369.0	-
33 kDa	5.2	23.7	428.9	55.3	186.0
	6.2	10.0	118.2	106.8	341.5
	6.7	3.0	22.0	115.9	346.6
	7.2	2.1	-	341.5	-
None	6.2	4.9	160.7	69.7	222.6
	7.2	3.2	-	308.6	-

REFERENCES

1 Debus, R.J. (1992) Biochim. Biophys. Acta 1002, 269-352
2 Rutherford, A.W., Zimmerman, J.-L. and Boussac, A. (1992) in The Photosystems: Structure, Function, and Molecular Biology (Barber, J., ed.) pp 179-229 Elsevier Science Publishers, Amsterdam
3 Yocum, C.F. (1991) Biochim. Biophys. Acta 1059, 1-15
4 Han, K.-C. and Katoh, S. (1992) in Research in Photosynthesis (Murata, N., ed.) Vol II, pp 365-368. Kluwer Academic Publishers, Dordrecht, The Netherlands
5 Han, K.-C. and Katoh, S. (1993) Plant Cell Physiol. 34, 585-593
6 Andreasson, L.-E., Ädelroth, P. and Lindberg, K. (1994) Biochem. Soc. Trans. 22, 347-352
7 Krieger, A. and Weis, E. (1993) Photosynth. Res. 37, 117-130
8 Ghanotakis, D.F., Demetriou, D.M. and Yocum, C.F. (1987) Biochim. Biophys. Acta 891, 15-21
9 Miyao, M. and Inoue, Y. (1991) Biochemistry 30, 5379-5387
10 Ono, T.-A. and Inoue, Y. (1988) FEBS Lett. 227, 147-152
11 Tamura, N. and Cheniae, G.M. (1988) in Light Energy Transduction in Photosynthesis: Higher Plant and Bacterial Models (Stevens, S.E., Jr. and Bryant, D.A., eds) The American Society of Plant Physiologists
12 Shinohara, K., Ono, T-A. and Inoue, Y. (1992) Plant Cell Physiol. 33, 281-289
13 Homann, P.H. (1988) Biochim. Biophys. Acta 934, 1-13

ELECTROGENICITY OF ELECTRON AND PROTON TRANSFER IN PHOTOSYSTEM II OF GREEN PLANTS

Michael Haumann, Monika Hundelt, Wolfgang Drevenstedt and Wolfgang Junge

FB Biologie/ Chemie, Abt. Biophysik, Barbarastr. 11, Universität Osnabrück, D-49069 Osnabrück, Germany

1. Introduction

Photosystem II (PSII) produces dioxygen from water in a reaction cycle using four quanta of light. The primary product is the charge pair P+ and Q_A^-. P+ is reduced via Tyr-161 on D1 (Y_Z) by the manganese (and histidine?) containing oxygen evolving complex (OEC) [1], and Q_A^- reduces either the non-heme iron (Fe(III)) or Q_B, depending on the ambient redox potential [2]. Of the five increasingly oxidized states of the OEC, S_1 is stable in the dark and O_2 appears on the third flash of light causing $S_3 \Rightarrow S_4 \rightarrow S_0$ [3].

It is generally agreed on that the 1st flash oxidizes one of the Mn atoms (hence Mn(III)→Mn(IV)) [1]. Studying samples depleted of the cofactor Cl^-, we recently found that His is oxidized instead, whereas in control samples this amino acid is oxidized only on the second flash (causing $S_2 \Rightarrow S_3$) [4]. Using these properties we scrutinized the electrogenicity of the redox reactions involving Mn / Y_Z^+, His / Y_Z^+, and Q_A^- / Fe(III). Unlike other authors who studied the electrogenicity of Y_Z / P+ [5] and Q_A^- / Q_B [6] by photovoltaic techniques we used the electrochromic response of carotenoids [7].

Compared to the electrogenicity of the reaction producing Y_Z^+ and Q_A^- (100% = 28.5Å [5]) we found the following relative electrogenicities: [MnH_2O] to Y_Z^+ during $S_4 \rightarrow S_0$, 7%; His to Y_Z^+ (during $S_2 \Rightarrow S_3$ in controls or $S_1^* \rightarrow S_2^*$ in Cl^--depleted material), 12%, attributable to proton release from His+; Q_A^- to Fe(III), 30%. In the latter case 10% were attributable to electron transfer and 20% to proton uptake by Fe(II).

2. Materials and Methods

Unstacked thylakoids were prepared from pea seedlings as in [4] and stored at -80°C. Cl^--depletion was performed as in [4]. After thawing, thylakoids were suspended at 30µM chlorophyll in 20mM NaCl (10mM Na_2SO_4, Cl^--depleted), 2.6g/l BSA, 10µM DNP-INT, 3mM hexacyanoferrate(III) and 100µM DCBQ. Samples were dark-adapted for 30 min. Flash-spectrophotometry was performed as in [3] with saturating flashes from a Q-switched Ruby laser (1st flash) (694nm, 50ns FWHM) and a Xenon-lamp (flashes 2-9) (>610nm, 10µs FWHM). The optical pathlength was 1cm. Up to 500 transients were averaged (Nicolet Pro30) for signal-to-noise improvement. The transmembrane voltage was recorded by electrochromic transients at 522nm [7] subtracting background transients at this wavelenght as recorded in the presence of gramicidin (0.5µM) [9] which grounds the transmembrane potential in less than 10µs. By recording absorption transients at 360nm [8] we checked that gramicidin did not interfere with water oxidation. With DNP-INT and hexacyanoferrate(III) added, photosystem I was silent (checked at 700nm).

3. Results and Discussion

a) Fig. 1A shows transients at 522nm obtained with dark-adapted thylakoids at pH 7.2 on flashes 3, 4, 5 and 7 out of a series of 9 flashes. All transients were normalized on the initial extent of the 3rd flash for comparison. The transient of flash no. 3 (mainly inducing transition $S_3 \Rightarrow S_0$) was also plotted for comparison (upper transients). All transients

P. Mathis (ed.), Photosynthesis: from Light to Biosphere, Vol. II, 333–336.

revealed an unresolved jump due to Y_Z^+/Q_A^- formation in $\leq$ 1μs. It was followed by a decay due to the discharge of the membrane capacitor by ionic currents with about 25ms half-decay time. The transient on flash no. 3 decayed more slowly than the ones on flashes 4 and 5. The decay on flash no. 7 (congruent to the 3rd flash of the 2nd cycle) was nearly the same as the one on flash no 3. When this experiment was repeated at pH 6 the difference between flashes 3-5 was still the same. We subtracted the transient on flash no. 5 from the one on flash no. 3. The resulting transient (Fig. 1B) revealed a half-rise time of 1.2ms typical for transition $S_3 \Rightarrow S_0$ [8] (smooth line) and an extent of 5% of the initial extent of flash no. 3. With: misses (α) = 5%, double hits (β) = 0 and initial populations of S_0 = 20% and S_1 = 80% [9] about 70% and 0% of centers were in S_3 prior to the 3rd and 5th flash. Thus, the true relative extent of the electrogenic reaction on transition $S_4 \rightarrow S_0$ was 7% of the one of Y_Z^+-Q_A^-. Taking their distance as 28.5Å [5], the normal component of electron transfer [MnH_2O] in S_4 to Y_Z^+ covered a distance of at least 2Å.

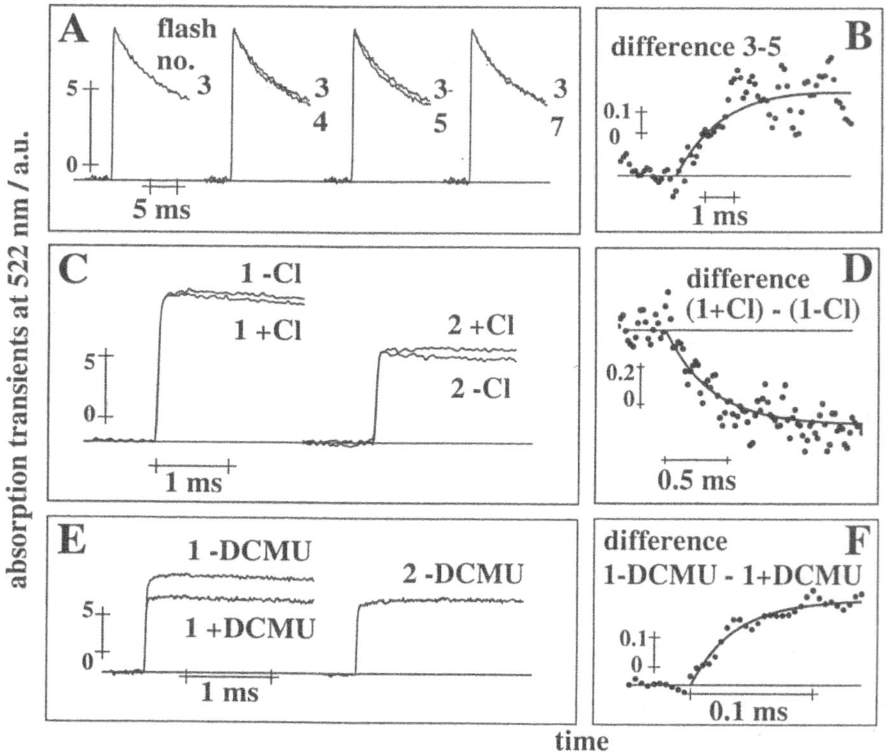

FIGURE 1: Electrochromic absorption transients at 522 nm in dark-adapted thylakoids (+/-Gramicidin). A), B) Time resolution 100 μs per address, flash spacing 0.5 s. C), D) 20 μs per address, flash spacing 100 ms. E), F) 5 μs per address, flash spacing 1 s. Note different time scales. For details see text.

Was this electrogenicity due to electron transfer or to proton release? i) The extent of the component was pH independent, which indicated that it was not coupled to the strongly pH dependent proton release on transition $S_3 \Rightarrow S_0$ [9]. ii. Without normalization the extent of the transient on flash 3 at about 5ms equaled the one on flash 5. This behaviour probably indicated a similar electrogenic reaction on both transitions, $S_1 \Rightarrow S_2$ and $S_4 \Rightarrow S_0$, with no indication for an *additional* (e.g. protonic) component on the latter. We conclude: The electrogenicity is probably due to electron transfer from [MnH_2O] to Y_Z^+.

b) Fig. 1C shows transients induced by the first 2 flashes in dark-adapted control (+Cl = Cl^--repleted) and Cl^--depleted (-Cl) thylakoids. On the 1st flash the jump was followed by a decay in the presence of Cl^- (lower trace) but by a small rise to an extent exceeding the former one in the absence of Cl^- (upper trace). On the 2nd flash the transient with Cl^- showed a small rising component (upper trace) and was larger than the one without Cl^- (lower trace). In the absence of Cl^- from the 3rd flash on (not shown) transmembrane charge recombination between Q_A^- and $P680^+$ occured in about 60% of centers [10]. We subtracted the curves in the absence from the ones in the presence of Cl^-. The result for flash no. 1 (Fig. 1D) revealed a half-rise time of 250µs (smooth line) and an extent of 8% of the initial jump on flash no. 2. An oppositely directed transient with similar time course was obtained on flash no. 2 (not shown). This transient was slightly larger (10%) which was probably due to a contribution (≤3%) of charge recombination in centers without Cl^-.

We conclude: An *additional* rising component with the typical half-rise time of transition $S_2 \Rightarrow S_3$ of 250 µs [8] appears on the 2nd flash in controls and on the 1st flash in Cl^--depleted thylakoids. With regard to the amount of still active centers in Cl^--depleted material (40%) the electrogenicity of this component was 12% (3.5Å) of Y_Z^+-Q_A^-. The additional electrogenic reaction was coupled to the oxidation of the putative His [4]. We have previously shown, that one proton was released on the latter reaction [4]. Therefore, we attribute the electrogenicity to proton release from His^+ into the lumen.

c) In Fig. 1E we compare transients induced by the 1st flash without and with DCMU (30µM) to block $Q_A^- \rightarrow Q_B$. Even after prolonged dark-incubation with hexacyanoferrate(III) the extent of the transient without DCMU remained about 30% larger than the one on the 2nd flash. Photosystem I did not contribute to the 1st flash by more than 5% because of its oxidation (≥95%) in the dark (evident from transients at 700nm, not shown). On the other hand, the oxidation of the non-heme iron by hexacyanoferrate(III) in the dark was evident from rapid proton uptake upon its reduction on the 1st flash (not shown, see [11]). In the presence of DCMU the transient on the 1st flash (Fig. 1E, 1 +DCMU) was about as large as the one on the 2nd flash without DCMU. Then, proton uptake was absent on the first flash [11] which probably indicated that the iron was not oxidized [2]. The formation of the difference 1st flash (-DCMU) - 1st flash (+DCMU) yielded the transient in Fig. 1F. Its half-rise time was 25µs (smooth line) typical for electron transfer from Q_A^- to Fe(III) [2]. We conclude: The reduction of Fe(III) by Q_A^- is coupled to an electrogenic reaction of 30% (8.5Å) of Y_Z^+-Q_A^-.

We have recently proposed, that Q_A may be closer to P by about 3Å in PSII than in the BRC [13]. If the non-heme iron remains in the same position as in the BRC this implies an electrogenicity of $Q_A^- \rightarrow$Fe(III) of 10% of Y_Z^+-Q_A^-. Electron transfer from Q_A^- to Q_B contributes only 5% [6]. We have previously shown, that proton uptake followed the reduction of Fe(III) [12]. Based on these results we attribute 1/3 of the electrogenicity of Fe(III) reduction to electron transfer and 2/3 to proton uptake.

4. Conclusions

In Scheme 1 we summarized relative electrogenicities as obtained in this work and combine them with distances from the literature. We have separated $Y_{Z/D}$ by 40Å like their bacterial counterparts [13]. If we adopt the distance of 30Å from [14] than these cofactors would be located as indicated by the dashed square for Y_D. The other cofactors would be rearranged accordingly. The relative electrogenicities (Scheme 1; c, d, f, g, h, i) directly translate into relative depth along the membrane normal, as plotted in Scheme 1, only if one assumes a homogeneous dielectric. However, the local dielectric strength likely increases towards both boundaries [15]. The respective distances in the Y_Z / Mn-His / Fe regions are then expected to expand accordingly, say, threefold. We conclude: The Mn-cluster is located between 2 and 6Å below Y_Z^+ closer to the protein-water boundary. This position differs strongly from the one proposed in [16] and implies that the cluster is far out of the plane formed by Y_Z, P_{680} and Q_A [13].

SCHEME 1: Tentative arrangement of redox cofactors in PSII. Contributions of electron and proton transfers from this work are indicated by arrows. The center-to-center distances (in Å) were taken from: a, b [16]; c, e [5] and references therein; d [6]. The distances obtained in this work, f, g, h, i, are lower limits, expected for a homogeneous dielectric.

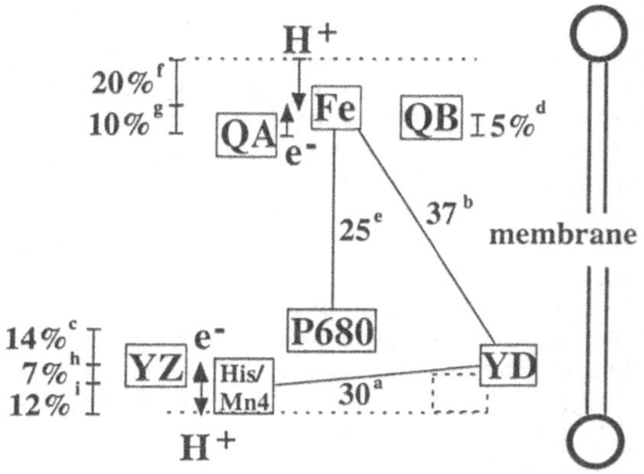

Acknowledgements

Stimulating discussions with our other collegues in the EU-HCM Network (EU-DGXII-HCM-CHRX-CT94-0524) are greatefully acknowledged.

References

1 Debus, R.J. (1992) Biochim. Biophys. Acta 1102, 269-352.
2 Diner, B.A., Petrouleas, V. and Wendoloski, J.J. (1991) Physiol. Plant. 81, 423-436.
3 Forbush, B., Kok, B. and McGloin, M.P. (1971) Photochem. Photobiol. 14, 307-321.
4 Haumann, M., Drevenstedt, W., Hundelt, M. and Junge, W. (1995) Biochim. Biophys. Acta, submitted.
5 Pokorny, A., Wulf, K. and Trissl, H.W. (1994) Biochim. Biophys. Acta 1184, 65-70.
6 Hook, F. and Brzezinski, P. (1994) Biophys. J. (1994) 66, 2066-72.
7 Emrich, H.M., Junge, W. and Witt, H.T. (1969) Z. Naturforsch. B24, 1144-1146.
8 Renger, G. and Hanssum, B. (1992) FEBS Lett. 299, 28-32.
9 Haumann, M. and Junge, W. (1994) Biochemistry 33, 864-872.
10 Lübbers, K., Drevenstedt, W. and Junge, W. (1993) FEBS Lett. 336, 304-308.
11 Haumann, M. and Junge, W. (1994) FEBS Lett. 347(1), 45-50.
12 Bögershausen, O. and Junge, W. (1995) Biochim. Biophys. Acta 1230, 177-185.
13 Mulkidjanian, A., Cherepanov, D., Haumann, M. and Junge, W. (1995) Biochemistry, submitted.
14 Astashkin, A.V., Kodera, Y. and Kawamori, A. (1994) Biochim. Biophys. Acta 1187, 89-93.
15 Svensson, B., Vass, I. and Styring, S. (1991) Z. Naturforsch. C. 46, 765-776.
16 Un, S., Brill, T.M. and Zimmermann, J.L. (1994) Proc. Natl. Acad. Sci. USA 91, 5262-5266.

EVIDENCE ON THE SPECIFIC INTERACTION BETWEEN MANGANESE AND CYT *b-559* ON THE PS II OXIDIZING SIDE.

Noriaki Tamura[1], Ikuko Iwasaki[2], Sayo Shibano[1], Izumi Oka[1] and Shigeki Okayama[2].

Dept. Environ. Sci., Fukuoka Women's Univ., Fukuoka 813[1] and Dept. Biol., Fac. Sci., Kyushu Univ., Fukuoka 810[2], Japan.

1. Introduction

Cytochrome *b-559*. a protoheme protein, is an essential component for structural integrity of PS II in oxygenic photosynthesis, and exhibits mainly three types of redox potential forms such as high potential- (HP-), intermediate potential- (IP-) and low potential- (LP-) forms. Although it is unclear how a variety of redox potential forms of cytochrome *b-559* function in PS II, these positively functional roles in photosynthetic electron transfer are suggested as follows: 1) the participation in cyclic electron transfer around PS II (1, 2), 2) the involvement in the light-induced assembly (photoactivation) of the apo-water-oxidizing complex (2, 3), and 3) a protection of photoinhibitory damage of PS II (4 - 7).

Recently, an interesting model was proposed on the protective function of cytochrome *b-559* against photoinhibition on both sides of PS II (7). In this model, if the acceptor side is over-reduced by strong light illumination, the LP form, which would exist as an oxidized form before the illumination, should accept electron(s) from the over-reduced quinones. On the other hand, if the donor side lacking the water-oxidizing complex is over-oxidized by exposure of strong or weak light, the HP form should donate electron(s) to the over-oxidized tyrosines of Z and D. In these situations, the IP form is expected to function as a reservoir of the HP- or LP- form. Accordingly, cytochrome *b-559* may function in attenuation of photoinhibitory damage of both sides of PS II by sensing the redox potential environment around PS II and tuning its redox potential forms.

We have reported that the HP-form is converted to the LP form via two intermediate potential forms, the IP form and a novel intermediate potential form with E_m of +350 mV, during photoinhibition on the donor side of PS II (8). This transition of the redox potential forms may be irreversible. In this report, we have studied effects of Mn-binding or Mn-liganding on the transition of redox potential forms of cytochrome *b-559*. The partially reversible transition between them is suggested to be dependent on the Mn-binding or Mn-liganding on the PS II oxidizing side.

2. Materials and Methods

The O_2-evolving PS II membranes were prepared from spinach leaves by the procedure of Ref. 9. NH_2OH-treated PS II membranes were obtained by incubation of the O_2-evolving PS II membranes in buffer A (50 mM Mes-NaOH, pH 6.5, 20 mM NaCl, 0.4 M sucrose) with 3 mM NH_2OH for 30 min at 4 °C in darkness. These preparations

P. Mathis (ed.), Photosynthesis: from Light to Biosphere, Vol. II, 337–340.

were used after storage at -80 °C. To delete the bound Mn from the membranes, PSII preparations were incubated with 2 mM EDTA in buffer A twice, and then washed with buffer A twice. Chemical modification of PS II membranes with 1-ethyl-3-(3-dimethyl-aminopropyl)carbodiimide (EDC) in the presence of 1 M glycine was carried out at 20 °C as previously reported (10). Photoactivation of the apo-water-oxidizing complex was done in buffer A containing 2mM $MnCl_2$, 50 mM $CaCl_2$, 100 μM 2, 6-dichloro-phenolindophenol and PS II extrinsic proteins as previously reported (13).

The O_2 evolution was assayed polarographically (11). Redox titrations of cytochrome *b-559* in PS II membranes were anaerobically carried out at 10-15 °C as described in Refs. 8 and 12. The absorption spectra were recorded with Shimadzu UV-300 spectrophotometer. From analysis of reduced-minus-oxidized difference spectra in the range of E_h examined, the obtained spectra had little contribution from cytochromes *b-563* and *f*. The values and relative amounts of each redox potential form of cytochrome *b-559* were estimated from the best fit of curves by analyzing the titration data with a nonlinear curve-fitting Igor program (Wavemetrics, U.S.A.), as reported earlier (8). The titration curves depicted in Figures in Results and Discussion were fitted to a linear combination of multiple n=1 Nernst equation curves.

Results and Discussion

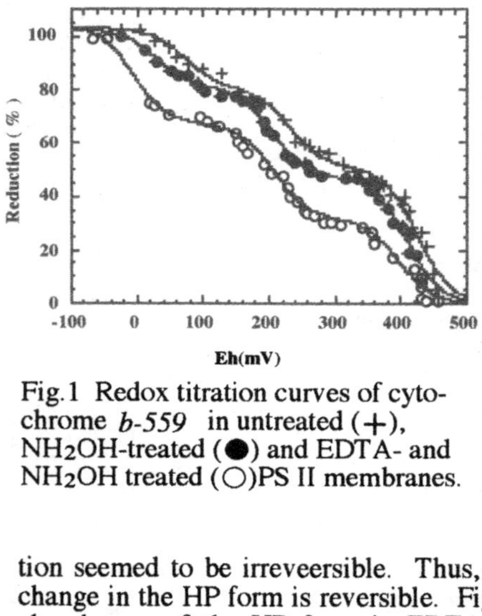

Fig.1 Redox titration curves of cytochrome *b-559* in untreated (+), NH2OH-treated (●) and EDTA- and NH2OH treated (○)PS II membranes.

Fig. 1 shows redox titration curves cytochrome *b-559* in three types of PS II membranes. O_2-evolving PS II membranes exhibit three redox components of cytochrome *b-559* with E_m values and relative amounts of +425, +233 and +55 mV, and 51, 30 and 19 %, respectively. NH_2OH-treatment, which induced a drastic decrease in O_2 evolution activity and almost depletion of functional Mn (13), caused little change in these redox parameters of cytochrome *b-559*. Further treatment of NH_2OH-treated PS II membranes with 2 mM EDTA resulted in decrease in the HP form by 16 % of the total cytochrome *b-559*, but increase in the LP form by 15 %. However, we did not obtain a complete loss of the HP form by EDTA treatment.

Recently , we observed the same type of phenomena, in which the HP form was converted to redox forms with lower E_m's by photoinhibitory treatment (8). The conversion by photoinhibition seemed to be irreveersible. Thus, it has been examined whether EDTA-sensitive change in the HP form is reversible. Fig. 2 shows the effects of exogenous Mn^{2+} on the abundances of the HP form in EDTA- and NH_2OH-treated PSII membranes. After 10min-dark-incubation of EDTA- and NH_2OH-treated PS II membranes with 1mM Mn^{2+}, the membranes were twice washed with buffer A and then subjected to redox titrations. Addition of Mn^{2+} to the membranes increase the HP form by 8 % of the total cytochrome *b-559*, equivalent to about a half of the EDTA-sensitive HP form. In contrast, addition of other divalent cations such as Ca^{2+}, Mg^{2+} and Zn^{2+} induced little increase in abundances of the HP form (data not shown). Fig. 3 shows the effects of EDC on the Mn^{2+}-induced increase in the HP form. EDC, a chemical modifier of carboxyl groups, suppresses the binding of Mn^{2+} to the high affinity Mn-binding sites, resulting in loss of capability of photoactivation of the apo-water-oxidizing complex (10, 14). NH_2OH-treated PS II

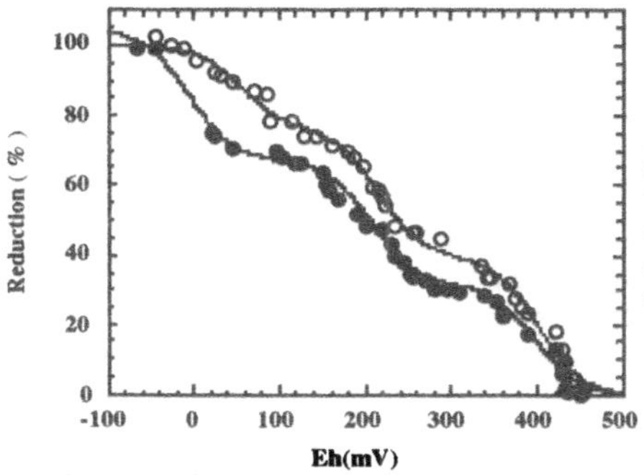

Fig.2 Redox titration curves of cytochrome *b-559* in EDTA- and NH_2OH-treated PS II membranes incubated with (○) or without (●) Mn^{2+}.

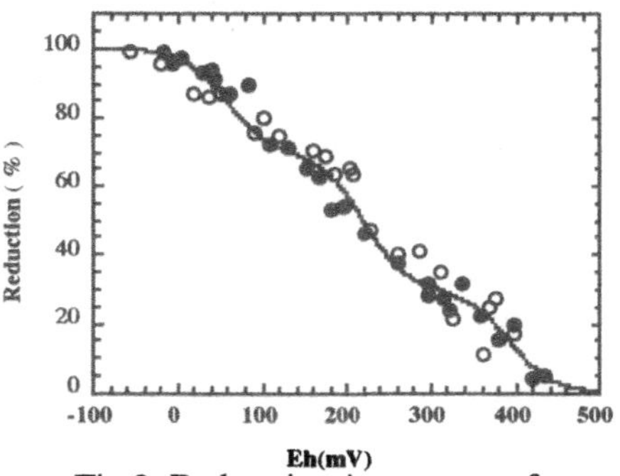

Fig.3 Redox titration curves of cytochrome *b-559* in EDC-modified PS II membranes incubated with (○) or without (●) of Mn^{2+}

membranes that were subjected to EDC-modification and EDTA-treatment were employed. The E_m values and relative amplitudes of cytochrome *b-559* in chemically-modified membranes were almost the same as EDTA- and NH_2OH-treated ones (Fig.1, open circles). Addition of Mn^{2+} to the membranes induced little change in the redox parameters of cytochrome *b-559*, compared with those in the absence of Mn^{2+}. Results in Figs. 2 and 3 suggest that the binding of Mn^{2+} to the high-affinity Mn-binding sites on the PS II oxidizing side is probably contributed to the stability of the HP form.

We have studied whether reassembly of the Mn cluster affects the redox parameters of cytochrome *b-559* (Fig.4). The relative amounts of HP-, IP- and LP-forms were 31, 45 and 24 % in non-photoactivated preparations. Photoligation of Mn^{2+} resulted in increase of the HP form and decrease in the IP form by 14 and 13 % of the total cytochrome *b-559*, respectively, but did not affect the LP form drastically. The similar result have already been reported with Tris-treated chloroplasts (3). These results indicate that the transition of the IP form (possibly the LP form) to the HP form occurs, accompanied with the light-induced formation of Mn cluster. In our experimental conditions, the O_2 evolution activity was restored by ≈ 35 % of the original activity. The fact that a number of reassembled Mn cluster is not proportional to that of the newly converted HP form by photoactivation is probably due to photoinhibition concomitantly occurring during photoactivation.

Unique properties of redox potentials of cytochrome *b-559* may be derived from the hydrophobicity of the environment around the heme pocket (15, 16). The existence of redox potentials with discrete and restricted E_m values probably results from a discrete structural change in cytochrome *b-559*, which includes tilting of the α-helices of the α and β subunits of the cytochrome, the non-coplanar orientation of imidazole ligands of two subunits and dipole interactions between α-helices of the cytochrome and those of proteins in the neighbor of the cytochrome (15). Our results suggest that the binding of Mn^{2+} to the high-affinity Mn-binding sites, probably on the D1 and D2 proteins, is one of determinants of such change. Furthermore, photoligation of Mn^{2+} also increased abundances of the HP form. Interestingly, Mizusawa and Yamashita (3) reported that the conversion of redox potential forms with lower E_m's to the HP form is much faster than the assembly of the Mn cluster. Taken together, the HP form, which is regenerated by specific binding of Mn^{2+} or formation of a Mn dimer of the basal component of the Mn cluster (11), may physiologically function in a protection against photoinhibition on

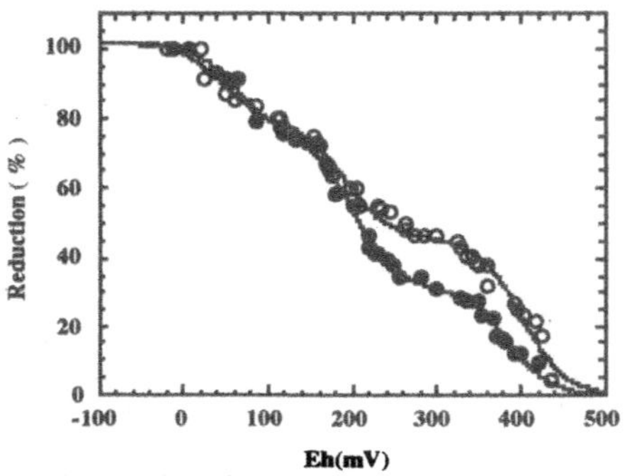

Fig. 4 Redox titration curves of cytochrome *b-559* photo-activated (○) and non-photoactivated(●) PS II membranes. NH_2OH-treated PS II membranes were photoactivated and subsequently incubated with 2 mM EDTA twice. Non-photoactivated ones were prepared by dark-incubating NH_2OH-treated PS II membranes with a photoactivation medium followed by EDTA treatment.

the donor side during the reassembly of PS II reaction center complexes.

Acknowledgments

This study is supported in part by Grants-in-Aid for Scientific Research from the Ministry of Education, Science and Culture of Japan. We thank Ms.'s M. Kuwahara and Y. Sasaki for their helpful technical supports.

References

1 Heber, U., Kirk, M.R. and Boardman, N.K. (1979) Biochim. Biophys. Acta 546, 292-306
2 Cramer, W.A., Theg, S.M. and Widger, W.R. (1986) Photosynth. Res. 10, 393-403
3 Mizusawa, N. and Yamashita, T. (1992) in Research in Photosynthesis (Murata, N. ed.) Vol II. pp.425-428, Kluwer Academic Publishers, Dordrecht, The Netherlands
4 Thompson, L.K. and Brudvig, G.W. (1988) Biochemistry 27, 6653-6658
5 Buser, C.A., Diner, B.A. and Brudvig, G.W. (1992) Biochemistry 31, 11449-11459
6 McNamara. V. and Gounaris, K. (1992) in Research in Photosynthesis (Murata, N. ed.) Vol II. pp.89-92, Kluwer Academic Publishers, Dordrecht, The Netherlands
7 Barber, J. and Rivas, J. De L. (1993) Proc. Natl. Acad. Sci. USA 90, 10942-10946
8 Iwasaki, I., Tamura, N. and Okayama, S. (1995) Plant Cell Physiol. 36, 583-589
9 Radmer, R., Cammarata, K., Tamura, N., Ollinger, O. and Cheniae, G. (1986) Biochim. Biophys. Acta 850, 21-32
10 Tamura, N., Tanaka,T., Wakamatsu, K., Inoue, H. and Wada, K. (1992) in Research in Photosynthesis (Murata, N. ed.) Vol II. pp.405-408, Kluwer Academic Publishers, Dordrecht, The Netherlands
11 Tamura, N. and Cheniae, G. (1987) Biochim. Biophys. Acta 890, 179-194
12 Takamiya, K. and Hanada, H. (1980) Plant Cell Physiol. 21, 979-988
13 Tamura, N. and Cheniae, G. (1985) 809, 245-259
14 Preston, C. and Seibert, M. (1991) Biochemistry 30, 9615-9624
15 Klishtalik, L.I., Tae, G.-S., Cherepanov, D.A. and Cramer, W.A. (1993) Biophys. J. 65, 184-195
16 Tae, G.-S. and Cramer, W.A. (1994) Biochemistry 33, 10060-10068

STRUCTURAL STUDIES OF THE S_2-STATE MANGANESE CLUSTER IN PLANT PHOTOSYSTEM II. THEORETICAL ANALYSIS OF "MULTILINE" EPR SIGNALS.

Koji Hasegawa and Masami Kusunoki,
Department of Physics, School of Science & Technology, Meiji University, 1-1-1 Higashimita, Tama-ku, Kawasaki 214, Japan

1. Introduction

The water-splitting and oxygen-evolving reaction in plant photosystem II (PSII) occurs through five intermediates, designated S_0 to S_4, in a cycle of four electron abstractions by successively generated oxidants P680+ at the reaction center [1]. The first direct evidence for involving Mn as positive charge accumulation place came from a discovery of the 18-20 multiline EPR signal, centered at g=2, in the S_2 state at low temperatures [2]. A second broad EPR signal was soon found around g=4.1-4.2, in the same S_2 state[3,4]. And a third one, centered at g=4.8, was discovered in the dark stable S_1 state in parallel polarization EPR experiments [5]. Among these EPR signals, the multiline signal is most informative on the nuclearity of superexchange-coupled Mn spins and hence has been most extensively studied in several laboratories [6]. A remarkable sensitivity of this signal on sample treatments was reasonably interpreted as arising from the sensititivity of projection factors $\{P_i, i=1\text{-}4\}$ in the effective nuclear hyperfine coupling constants (hcc)$\{\mathbf{A_i}\}$ rather than the monomeric hcc $\{\mathbf{a_i}\}$[7]. An isotropic model for g-tensor and $\mathbf{a_i}$-tesors was found [7] to be able to fairly well simulate the multiline EPR spectra at both X-band [8,9] and S-band [10] frequencies, with use of

$$\mathbf{A_i}=P_i\mathbf{a_i};\ (|A_1|, |A_2|, |A_3|, |A_4|) \Leftrightarrow (37.6\text{ G}, 87.0\text{ G}, 93.7\text{ G}, 122.5\text{ G}),\quad (1)$$

where the symbol $\Leftrightarrow$ implies that the vector elements on the l.h.s are equal to those on the r.h.s in any order of rearrangement. This clearly shows that four Mn spins in the S_2 state must be superexchange-coupled to be summed up to a total spin $\mathbf{S=S_1+S_2+S_3+S_4}$, involving no EPR-silent pair. Moreover, it was shown [7] that, if the Mn tetramer is in the Mn(3,4,3,3) or Mn(3,4,4,4) valence state and contains only 1 or 2 strong antiferromagnetic coupling(s) and the other weak ones, eqs 1 would be difficult to be satisfied by any model except {a dimeric Mn(3,4) plus two monomeric Mn(4) spins}, although the preference of the higher valence state is partly due to the assumed ranges of |a(3)|=95-80 G and |a(4)|=85-70 G for Mn(3) and Mn(4) ions, respectively. In this paper, eqs 1 and the other facts that the multiline signal arises from the S=1/2 ground state [6] and the first excited state lies ca. 30 cm^{-1} above it [11] will be used to determine what values of six super exchange interactions in $H_{exch}=\Sigma_{i>j} J_{ij}\mathbf{S_i}\cdot\mathbf{S_j}$ can reproduce the multiline signal with the monomeric hcc being allowed to take any value in the reported ranges of |a(3)|=90-70 G and |a(4)|=90-70 G [12]. Furthermore, we analyze what values of the monomeric hcc must be assumed to reproduce the multiline signal.

P. Mathis (ed.), Photosynthesis: from Light to Biosphere, Vol. II, 341–344.

2. Guide Lines for Theoretical Analysis

The Schrödinger's equation to determine spin state vectors for the Mn tetramer, $|\Phi_n(SM)>$, are expanded in terms of the composite-spin state vectors, $|S_{12}S_{34}SM>$, as

$$\{H_{exch}-E_n(S)\}|\Phi_n(SM)> = \sum_{S_{12},S_{34}} C_n(S_{12},S_{34},S)\{H_{exch}-E_n(S)\}|S_{12}S_{34}SM>=0 , \quad (2)$$

with

$$|S_{12}S_{34}SM> = \sum_{M_{ij},M_i} C^{SM}_{S_{12}M_{12}S_{34}M_{34}} C^{S_{12}M_{12}}_{S_1M_1S_2M_2} C^{S_{34}M_{34}}_{S_3M_3S_4M_4} \prod_{i=1}^{4} |S_iM_i> , \quad (3)$$

where $C^{c\gamma}_{a\alpha b\beta}$ is the Clebsh-Gordan coefficient, $|S_iM_i>$ is the single-spin state vector and the sums are taken according to the vector coupling rule [13]. In the isolated ground state $\Phi_0(SM)$, the projection factors $\{P_i\}$ are calculated according to the formulae:

$$P_i=<\Phi_0(SM)|S\bullet S_i|\Phi_0(SM)>/[S(S+1)]; \; i=1\text{-}4, \quad (4)$$

In both cases of Mn(3,4,3,3) and Mn(3,4,4,4), we assume that the first two spins, $S_1=2$ and $S_2=3/2$, are strong-antiferromagnetically bridged by a fixed $J_{12}\approx250$ cm^{-1}, corresponding to a di-μ-oxo bridged Mn(3)-Mn(4) pair, because the other choices of Mn(3)-Mn(3) and Mn(4)-Mn(4) pairs will lead to a failure of interpretation. Then, using Jacobi-Rutishauser's method, we can numerically solve eq 2 for given sets of five J_{ij}-values, which are meshed as equally spaced by 3 cm^{-1} or 10 cm^{-1} depending on the situation. The resultant spin states are checked with respect to three points, i.e. (a) whether the ground state is $S=1/2$ or not, (b) whether the first excited state lies 20-50 cm^{-1} above the $S=1/2$ ground state or not, and (c) whether eqs 1 are satisfied or not when the conditions for $|a_i|$ (i=1-4) mentioned above are employed. When these conditions are all satisfied, we can say such a Mn tetramer can reproduce the multiline signal, and then we add 1 to a "reproduction count (R)". In this way, we can draw a contour map of the total R in a two dimensional space of any pair of J_{ij}'s, where the remaining three J_{ij}'s are assumed to be either weak antiferro- or weak ferro-magnetic in the reported range of -20 $cm^{-1} < J_{ij} < 40$ cm^{-1}[14,15]. This assumption is equivalent to saying that the Mn cluster may contain 1 or 2, strong ($J > 150$ cm^{-1}) or intermediate (40 $cm^{-1} < J < 150$ cm^{-1}), antiferromagnetic interactions in addition to $J_{12}=250$ cm^{-1}. Among 10 possible pairs, only 6 pairs becomes independent by the symmetry.

3. Results, Discussions and Conclusions

In Fig.1a, we plot the contour map of the R-distribution in the J_{13}-J_{34} (J_{14}-J_{34}) plane for Mn(3,4,4,4), in which numbers, 1,5,10, max=28, imply that only the contours at R=1, 5, 10 and the maximum point at R=28 are explicitly drawn. The counted Mn tetramers are seen to belong to the 1dimer+2monomers ('DM^2') model, because only J_{12} is strong-antiferromagnetic. It should be noted that the R-distribution extends up to $J_{34}\approx130$ cm^{-1} along a line $J_{34}\approx2J_{13}$, but the total count is not high (If the condition of the first excited state is removed, the R-distribution will be doubly elongated along the line $J_{34}\approx2J_{13}$ and much more intensified to make the 2dimers('D^2')-model possible, e.g., an exceptional bond of $J_{13}\approx80$ cm^{-1} for a 3.27 Å Mn-Mn interaction [14] and J_{34}= 174-288 cm^{-1} [14,15]). As regard the intermediate and strong antiferromagnetic couplings,

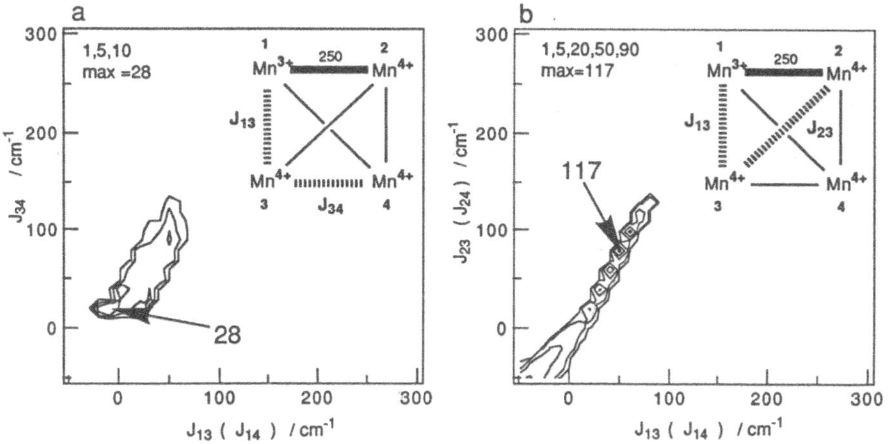

Figure 1 The contour maps of the reproduction counts R for the multiline signal as functions of J_{13} and J_{34} (J_{14} and J_{34}) in panel a, and J_{13} and J_{23} (J_{14} and J_{24}) in panel b, both for Mn(3,4,4,4). For a detail, see text.

the DM^2 tetramers can be classified into 'I'-type (J_{12}), 'L'-type (J_{12}/J_{13} et c.), 'T'-type ($J_{12}/J_{13}/J_{14}$ et c.), 'U'-type, 'Z'-type ($J_{12}/J_{23}/J_{34}$) and 'Δ'-type. The Mn tetramers counted in Fig.1a are I-type near the origin, L-type or U-type outside the I-type region. Next, the most interesting case, where J_{12}, J_{13} and J_{23} make a triangle, is plotted in Fig.2b. The R-distribution consists of a series of quite sharp peaks running along a line $J_{23} \approx 1.4 J_{13}$ and a low plateau around the origin. The corresponding Mn(3,4,4,4) tetramers belonging to Δ-type(majority) and I-type(minority) DM^2 models, respectively, are found to have the first excited state with spin S=3/2 and S=1/2, respectively.

To compare between the two valence-states and see the mountain shape more clearly for the case of Fig.1a, we have plotted, in Fig.2, the integrated R along the J_{13}-axis for the Mn(3,4,3,3) and Mn(3,4,4,4) clusters. Notably, the R-distributions are quite similar between two valence-states, and clearly shows that the J_{34} must be weak- or intermediate-antiferromagnetic, and can not be strong like di-μ-oxo bridged Mn(3)-Mn(3) and Mn(4)-Mn(4) interactions with $J > 170$ cm^{-1}[14,15]. Inspecting the calculated eigen states, we found that the R-distributions corresponding to two peaks around $J_{34} \approx 25$ cm^{-1} and ≈ 50 cm^{-1} mainly arise from Mn-tetramers having the first excited state with S=3/2 and S=1/2, respectively. Although the R-distributions are in general almost indistinguishable between Mn(3,4,3,3) and Mn(3,4,4,4), the distributions of monomeric hcc $\{a_i, i=1\text{-}4\}$ are seen, in Fig.3, to be quite different between them. A clear-cut difference appears in that both the $|a_3|$ and $|a_4|$ values must be smaller than 80 G for Mn(3,4,3,3), while these inequalities are opposite for Mn(3,4,4,4). This result seems to be satisfied by the hcc for Mn(3) and Mn(4) ions in manganese catalase(3,4) but not satisfied by those in di-μ-oxo bridged Mn(3,3) and Mn(4,4) dimers [12]. Moreover, in the case of Mn(3,4,3,3), $|a_1|$ and $|a_2|$ almost satisfy an inequality $|a_1| < |a_2|$ and are widely distributed, but, in the case of Mn(3,4,4,4), $|a_1|$ (> 80 G) must be larger than $|a_2|$ (< 85 G). Significantly, the reported hcc values for di-μ-oxo bridged Mn(3,4) dimers [12] are entirely consistent with the latter. [Another set of the effective hcc, (20.4 G, 86.3 G, 93.3 G, 122.5 G), were obtained by a computer simulation of a multiline signal by Bonvoisin et al. [16]. We carried out the same theoretical analysis using these values in eqs 1, revealing that the R's are vanishingly small in all the six cases, mostly due to a too small value of 20.4 G.]

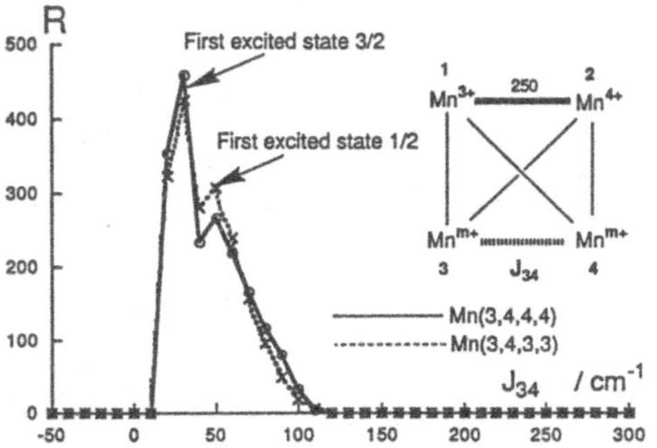

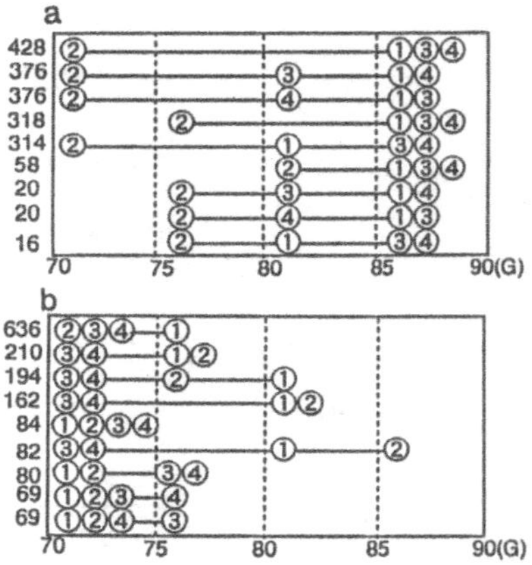

Figure 2 (left) The reproduction counts R for the mutiline signal as functions of J_{34} to compare between Mn(3,4,3,3) and Mn(3,4,4,4) (see text). These consist of two components with the S=3/2 and S=1/2 first excited states.

Figure 3 (right) The distributions of the monomeric hyperfine coupling constants (hcc), $|a_1|$ through $|a_4|$, in the cases of Mn(3,4,3,3) (panel b) and Mn(3,4,4,4) (panel a) as shown in Fig.2. See text for discussions.

In conclusions, (1) the number of strong antiferromagnetic couplings ($J > 150$ cm^{-1}) in the S_2-state Mn tetramer can be neither 2 or 3, but probably 1, supporting the DM^2 model; (2) among the 6 types of DM^2 models, Z-type and T-type can be excluded from candidates; (3) the Mn(3,4,4,4) valence state is more probable than Mn(3,4,3,3); (4) the total spin S in the first excited state is mostly 1/2 in the U-type DM^2 models, mostly 3/2 in Δ-type and a mixture between them in I-type.

References

1 Joliot, P. and Kok, B. (1975) in Bioenergetics of Photosynthesis: Oxygen Evolution in Photosynthesis (Govindjee, ed.) pp. 387-412, Academic Press, New York
2 Dismukes, G.C. and Sidler, K. (1981) Proc. Natl. Acad. Sci. USA, 78, 274-278
3 Casey, J.L. and Sauer, K. (1984) Biochim. Biophys. Acta, 767, 21-28
4 Zimmermann, J.-L. and Ratherford, A.W. (1984) Biochim. Biophys. Acta, 767, 160-167
5 Dexheimer, S.L. and Klein, M.P. (1992) J. Am. Chem. Soc., 114, 2821-2826
6 Debus, R.J. (1992) Biochim Biophys. Acta, 1102, 269-352
7 Kusunoki, M. (1992) Chem. Phys. Lett., 197, 108-116
8 Pace, R.J., Smith, P., Bramley, R. and Stehlik, D. (1991) Biochim. Biophys. Acta, 1058, 161-170
9 Kim, D.H., Britt, R.D., Klein, M.P. and Sauer, K. (1992) Biochemistry, 31, 541-547
10 Haddy, A., Aasa, R. and Andréasson, L.-E. (1989) Biochemistry, 28, 6954-6959
11 Hannson, Ö., Aasa, R. and Vänngård, T. (1987) Biophys. J., 51, 825-832
12 Zheng, M., Khanglov, S.V., Dismukes, G.C. and Barynin, V.V. (1994) J. Am. Chem. Soc., 33, 382-387
13 Varshalovich, D.A., Moskalev, A.N. and Khersonskii, V.K. (1988) in Quantum Theory of Angular Momentum, pp.235-361, World Scientific, Singapore
14 Wieghardt, K. (1989) Angew. Chem. Int. Ed. Engl., 28, 1153-1172
15 Thorp, H.H. and Brudvig, G.W. (1991) New. J. Chem., 15, 479-490
16 Bonvoisin, J., Blondin, G., Girerd, J.-J. and Zimmeremann, J.L. (1992) Biophys. J., 61, 1076-1086

REINVESTIGATION OF THE PHASES OF REDUCTION OF P_{680}^+ IN THE MICROSECOND TIME DOMAIN

F. Rappaport[a,b], G. Porter[b], J. Barber [c], D. Klug [b], J. Lavergne [a].
a: Institut de Biologie Physico-Chimique, 13 rue Pierre et Marie Curie, 75005 PARIS (France)
b: Centre for Photomolecular Sciences, and c: Photosynthesis Research group, Department of Biochemistry, Imperial College of Sciences Technology and Medicine, London SW7 2AZ (UK)

1.Introduction

The comparison of the kinetics of electron transfer monitored from the absorption changes at 295 nm and of the electrochromic changes in the Soret band of chlorophyll has suggested the occurence of an electrostatic relaxation phase with a half time of 35 µs, faster than the reduction kinetics of Y_Z^+ [1]. This electrostatic event, that was proposed to reflect a proton release, was observed on state $Y_Z^+S_3$, and also on state $Y_Z^+S_0$ at pH>7 It was also reported that the absorption changes in the blue region immediately after the formation of Y_Z^+ displayed an S-dependent amplitude, confirming earlier results [2]. This feature, analysed under more details in the present paper, was proposed to reflect the presence of a fraction of oxidized P_{680} in equilibrium with Y_Z^+ in the $Y_Z^+S_2$ and $Y_Z^+S_3$ sates, due to the higher net charge of the catalytic centre in states S_2 and S_3. The electrostatic field present in these states is expected to decrease the $[P_{680}Y_Z^+]/[P_{680}^+Y_Z]$ equilibrium constant, consistent with the slower rate of the reaction $P_{680}^+Y_Z \rightarrow P_{680}Y_Z^+$, reported in [3]. In the present paper two questions are examined: (i) what are the initial equilibrium constants between P_{680} and Y_Z immediately after the formation of Y_Z^+ in the presence of the various S-states, and (ii), how does this initial equilibrium evolve, depending on electrostatic relaxation phases presumably linked with Y_Z^+-triggered proton release.

2. Materials and methods

2.1 *Measurements in the Soret band region.*
Whole cells of the strain S-56 of the alga *Chlorella sorokiniana* were used after a treatment with p-benzoquinone (10^{-4} M), as described in[2]. This strain, isolated by P. Bennoun, is a double mutant lacking the PSI reaction centre and part of the LHCII complement. The algae were resuspended in a medium containing 50 mM KCL and 50 mM Mes pH6.5 at a chlorophyll concentration of about 10 µM. Absorption changes were measured with a Joliot-type spectrophotometer, described in [1,2].
2.2 *Measurement at 820 nm.*
Measurements were performed using a laser diode (30 mW, 820 nm, 10 ns-10µs range) or a continuous beam from a tungsten lamp transmitted through a monochromator (1-200 µs range). The fluorescence from the sample was cut off using a second monochromator or spatial and glass filters. The excitation flashes were provided by a frequency doubled-YAG laser with the pulse energy decreased to 0.5 mJ/cm^2. Granal membranes (BBY) prepared

P. Mathis (ed.), Photosynthesis: from Light to Biosphere, Vol. II, 345–348.

according to [4] were used at a chlorophyll concentration of 40 µM in a buffer containing 0.3 M Sucrose, 0.003% Triton X100, NaCl 10 mM, 10mM Mes pH 6.5 and 2,6 DCBQ 25 µM as an electron acceptor. 50 µM potassium octocyanotungstate (WCN, a kind gift of Dr. A. Stemler, midpoint potential of +570 mV) was added to the medium to preoxidize entirely the contamining P_{700}.

3. Results

3.1 *Dependence of the amount of P_{680}^+ present in the µs-domain on the S-states.*

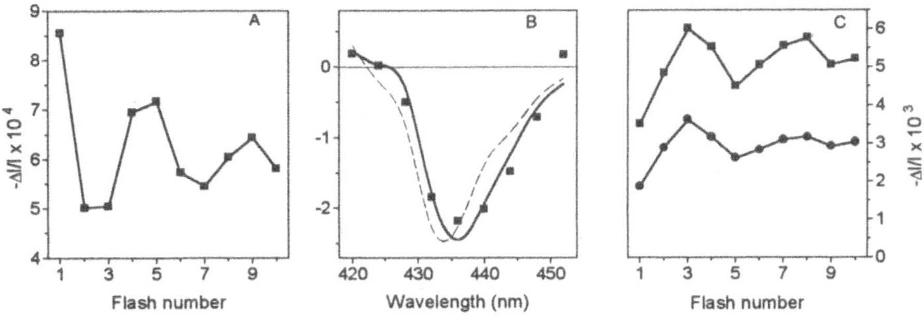

Figure 1: A, Difference of absorption changes (440 - 424 nm) induced at 15 µs by each flash of a series. B, Spectrum of the oscillating component defined as $(F_2+F_3-F_2-F_5)$ where F_i stands for the absorption changes after the ith flash. The curves are the best fits using the spectrum of $(P_{680}^+ - P_{680})$ (dashed line, 0.09 reaction centre units) and the spectrum of $(P_{680}^+ - P_{680}) - (Y_Z^+ - Y_Z)$ (solid line, 0.08 reaction centre units). C: absorption changes at 820 nm induced by each flash of a series at 1.5 µs (square) and 10 µs (circle).

As shown in Fig.1A the amplitudes of the absorption changes recorded at 15 µs after each flash of a series for the difference (440 - 424 nm) are markedly dependent on the S-states. The oscillating pattern suggests that the signal has a roughly similar amplitude in the presence of S_0 or S_1 on the one hand and of S_2 or S_3 on the other hand. The same feature is observed at all wavelengths in the blue region. The spectrum of the oscillating component ressembles that of $(P_{680}^+ - P_{680})$ as shown by the dashed curve in Fig 1B (redrawn from [5]). The occurence of a variable amount of P_{680}^+ in a time domain much longer than the P_{680}^+ reduction half time may reflect a modulation by the S-states of the equilibrium constant $[Y_Z^+P_{680}]/[Y_ZP_{680}^+]$. In this case, a decreased amount of Y_Z^+ should accompany the increase in the amount of P_{680}^+. Indeed the quality of the fit is significantly improved when using the spectrum of $(P_{680}^+ - P_{680}) - (Y_Z^+ - Y_Z)$ (see the solid line in Fig.1B). Using a calibration based on a comparison of the $(Q_A^- - Q_A)$ spectrum with litterature spectra where the absolute values of the extinction coefficient were indicated, we reckon the additional amount of P_{680}^+ present a few µs after a flash given in states S_2 or S_3 with respect to S_0 or S_1 to be about 0.08 reaction centre units.
The information obtained in the Soret band (Fig.1B) is appropriate to estimate the extent of the amount of P_{680}^+ present in excess in states $Y_Z^+S_2$ and $Y_Z^+S_3$ with respect to states $Y_Z^+S_0$ and $Y_Z^+S_1$. It gives no clue, however, concerning the absolute amount of P_{680}^+ in each S-states because too many spectral contributions interfere in this region (e.g. Y_Z^+, P_{680}^+, Q_A^-). More discriminating informations can be obtained around 820 nm where the cationic form of chlorophyll has a specific absorption band. As shown in Fig.1C, the 820 nm absorption changes present after a few µs depends on the S-states and the oscillating

pattern is similar to the one observed in the Soret band region. A contamination of P_{700}^+ was eliminated by using the strong oxydant WCN. On the other hand a P_{680}^+ contribution from a small fraction of centres inactive in oxygen evolution had to be substracted, based on the following rationale. In the presence of DCMU (20 μM), and varying the frequency of the actinic excitation from 0.05 Hz to 5 Hz, we could discrimate two components: about 90 % of the centres have a reactivation half time of 2.9 s, while the remaining 10% have a reactivation half time shorter than 200 ms. The half time of 2.9 s is consistent with the recombination reaction $S_2Q_A^- \rightarrow S_1Q_A$ at pH 6.5 [6] and can be ascribed to active oxygen evolving centres. The other component arises from fast recombining centres inactive in oxygen evolution: it is suppressed by the addition of benzidine (50 μM), which is an efficient electron donor to Y_Z in Mn-extracted centres, but which does not react with intact PSII. This contribution measured in the presence of DCMU under a flashing frequency of 5 Hz was substracted from the decays observed after each flash of a series in the absence of inhibitor. We thus obtained for the amount of P_{680}^+ present at a few μs after a flash figures of: 0.12, 0.10, 0.16, 0.19 (in reaction centres units) in the presence of S_0, S_1, S_2, S_3 respectively. The increase in states S_2 or S_3 with respect to S_0 or S_1 is in good agreement with the data obtained in the electrochromic region.

3.2 *Kinetics of reduction of P_{680}^+ in the μs time domain.*

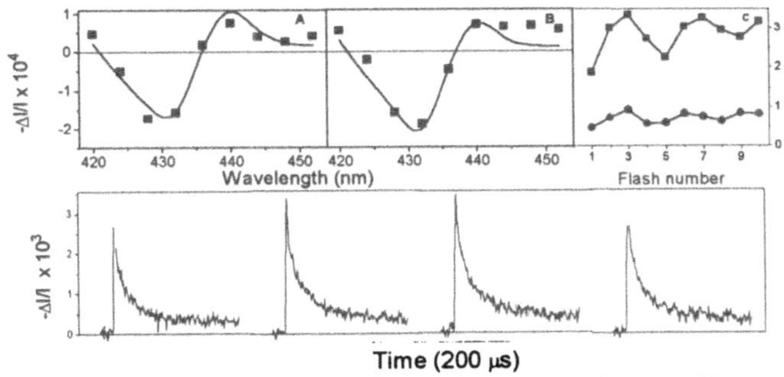

Figure 2: Top: Panels A, B: spectrum of the absorption changes between 15 μs and 100 μs after the second and third flash respectively and fits by a combination of the spectra of (P_{680}^+ - P_{680}) and (Y_Z^+ - Y_Z) (for the 2nd and 3rd flash respectively: 0.05 and 0.07 for the former spectrum and 0.2 for the latter in both cases). Panel C: absorption changes at 15 μs (■) and 100 μs (●) at 820 nm after each flash of a series. Bottom: Absorption changes at 820 nm after the first, second, third and fourth flash (from left to right).

We measured the absorption changes occuring in the 100 μs time range after each flash of a series both in the Soret region (Fig. 2A-2B) and at 820 nm (Fig. 2C and bottom). As previously observed by Schlodder et al. [7] the decays observed at 820 nm may be fitted with two components with half times of 5μs and 30-40 μs irrespective of the S-states; only the amplitude of the 35 μs transient appears to be dependent on the redox state of the OEC. Although their interpretation is more complex due to the overlap of several spectral contributions, the results obtained in the Soret region are in qualitative agreement with the 820 nm information. As shown in Fig. 2A-2B, the absorption changes observed after the second and third flashes between 15 and 100 μs reflect the disappearance of a similar amount of P_{680}^+.

4. Discussion

The present results show that a significant fraction of P_{680}^+ is present at a few µs in the presence of all the S-states, confirming earlier work [7,8]. In these works [7,8], the modulation by the S-states was interpreted as reflecting equilibria preexisting *before* the flash in the chain of secondary donors. Thus, in a fraction of centres, P_{680}^+ could not be reduced with a rapid sub-µs kinetics, because a secondary donor is already in its oxidized state. This class of model has become unlikely, however, since it was demonstrated that Y_Z is the direct donor to P_{680} [9] and most probably the only intermediate between P_{680} and the OEC. If a significant fraction of Y_Z^+ were present in equilibrium with the stabilized S_2 or S_3 , this should appear as an oscillation of the EPR Signal II, which could have hardly escaped detection. Furthermore, such models cannot account for the fact that a significant fraction of P_{680}^+ is also found in the µs-range after a flash given in the S_0 or S_1 states. Clearly, the equilibrium constant between Y_Z and the Mn-cluster is expected to be very large in these states. Thus, we favour the alernative interpretation that the presence of P_{680}^+ in the few µs-time-range reflects a low equilibrium constant between P_{680} and Y_Z rather than between Y_Z and the Mn-cluster. The modulation by the S-states is then explained by the electrostatic influence of the OEC which bears a positive net charge in states S_2 and S_3. If this charge is located closer to Y_Z than to P_{680}, it causes an increase of the midpoint potential of Y_Z with respect to P_{680}. In this framework, the figures that we obtained for the amount of P_{680}^+ imply that the $[Y_Z^+P_{680}]/[Y_ZP_{680}^+]$ equilibrium constant is about 7-9 in states S_0 and S_1 and 5-4 in states S_2 or S_3. This shift of the equilibrium constant requires only a small gradient of the electrostatic potential (≈ 20 mV) between P_{680} and Y_Z in the presence of the charged states S_2 or S_3. If the above interpretation is correct the P_{680}^+ present at a few µs is expected to decay concomitantly with the reduction of its equilibrium partner Y_Z^+, or possibly faster if a stabilization phase occurs that increases the $[Y_Z^+P_{680}]/[Y_ZP_{680}^+]$ equilibrium constant. From previous result [1], we expected both processes to take place, depending on the S-states. For instance, in the presence of S_2 , the decay of P_{680}^+ would accompany that of Y_Z^+ (half-time ≈200 µs), whereas in the presence of S_3 , it would be controlled by a 35 µs relaxation phase, much faster than the reduction of Y_Z^+ (1ms). At variance with these predictions but in agreement with the previous work of Schlodder et al. [7], we observed similar kinetics on all S-states, dominated by a 30-40 µs phase. It thus seems that a stabilization phase occurs in the 30-40,µs range on the Y_Z^+ state, irrespective of the S-states. This conclusion is somewhat puzzling, since the other kinetics events occuring in this region are all markedly influenced by the S-states.

References

1 Rappaport, F., Blanchard-Desce, M. and Lavergne, J. (1994) Biochim. Biophys. Acta 1184, 178-182.
2 Lavergne, J. (1984) FEBS Lett. 173, 9-14
3 Brettel, K., Schlodder, E. and Witt, HT (1984) Biochim. Biophys. Acta 766, 403-415.
4 Berthold, D.A., Babcock, G.T. and Yocum, C.F. (1981) FEBS Lett. 134, 231-234.
5 Gerken, S., Dekker, J.P., Schlodder, E. and Witt, H.T. (1989) Biochim. Biophys. Acta 977, 52-61.
6 Buser, C.A, Diner, B.A. and Brudvig, G.W. (1992) Biochemistry 31, 11449-11459.
7 Schlodder, E., Brettel, K. and Witt, H.T. (1985) Biochim. Biophys. Acta 808, 123-131.
8 Bowes, J.M., Crofts, A.R. and Itoh, S. (1979) Biochim. Biophys. Acta 547, 336-346.
9. Gerken, S., Brettel, K., Schlodder, E. and Witt H.T. (1988) FEBS Lett. 237, 69-75.

THE CHARACTERIZATION OF CHLORIDE-DEPLETED PHOTOSYSTEM II BY UV-VIS SPECTROSCOPY.

Hanna Wincencjusz, Hans J. van Gorkom
Department of Biophysics, Huygens Laboratory of the State University, P.O.Box 9504, 2300 RA Leiden

1. Introduction

Chloride is an obligatory cofactor for water oxidation by Photosystem II in plants. There is some controversy whether inhibition of the Oxygen Evolving Complex in the absence of chloride occurs after one or two electron transfer events. The character of the oxidized species is also uncertain: is it manganese or an amino acid residue?
Absorption difference spectroscopy in the UV-VIS region is one of the methods to distinguish between species being oxidized upon successive flashes in the PSII membrane fragments. Our results prompt us to suggest manganese oxidation upon the first and Tyr_Z oxidation upon the second and subsequent flashes.

2. Materials and Methods

Sample preparation. Photosystem II membrane fragments from spinach were prepared according to Berthold et al.[1] with minor modifications. NaCl was avoided in washing steps after the solubilization with Triton X-100. The particles were stored at pH 6.0 (5mM MES, 0.3M sucrose) and diluted for measurements with buffer containing:50 mM Hepes pH 7.5, 0.3M sucrose, 50mM Na_2SO_4. To complete the chloride depletion each sample was incubated at pH 7.5 for 15 min (just before measurement). The oxygen evolving activity was restored by addition of 50mM NaCl. The rates of oxygen evolution were nihil in the absence of chloride and 350-450 µmol $O_2.h^{-1}.mgChl^{-1}$ after addition of 50mM NaCl.
UV-VIS spectra. PSII membrane fragments were diluted for measurements to 0.2mg Chl/ml. Artificial electron acceptors: 50µM DCBQ and 100µM ferricyanide were added to prevent limitations at the PSII acceptor side. Absorption difference measurements were carried out at room temperature in a single beam apparatus described previously [2] using a tungsten halogen lamp for wavelengths above 290nm and a deuterium lamp below 290nm. Saturating laser flashes were spaced at 1s. The depicted spectra correspond to the initial absorption changes (time constant 0.2ms).

3. Results

At 295nm the absorption changes originating from the donor side can be conveniently measured (isosbestic point for the acceptor side contribution). Fig.1 shows typical traces obtained at 295nm for Cl^--depleted (A) and Cl^--reconstituted (B) PSII preparations. The

P. Mathis (ed.), Photosynthesis: from Light to Biosphere, Vol. II, 349–352.

absence of chloride did not seem to affect the first electron transfer event in Photosystem II. In contrast significant differences between the Cl^--depleted and Cl^--reconstituted particles were noted upon the second and following flashes.

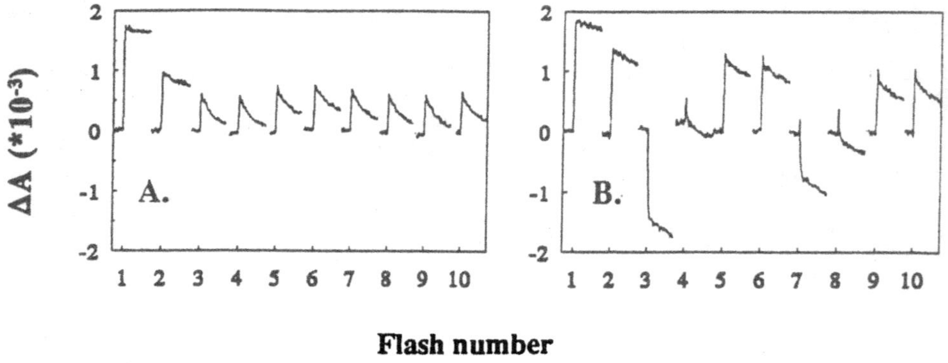

Fig.1 The donor side absorption changes recorded at 295nm for Cl^--depleted (A) and Cl^--reconstituted (B) PSII. The flashes were spaced at 1s, for each flash a trace of 300ms has been recorded.

To learn more about the identity of the involved components UV-VIS spectra have been measured. The absorption difference spectra induced by the first saturating flash are essentially the same for Cl^--depleted and Cl^--reconstituted samples. This similarity suggests oxidation of the same component regardless of the presence or absence of chloride. Since it is widely agreed that manganese is oxidized by the first flash in native PSII, we conclude that manganese is oxidized in Cl^--free PSII as well and that the presence of chloride is not required for the first electron transfer step in the Oxygen Evolving Complex. However something appears to happen during the first few miliseconds after the flash. In Cl^--depleted PSII the absorbance increase induced by the first flash was followed by a small decrease in a few miliseconds (Fig.3). This effect was observed in the region 300nm-350nm (Fig.2). It was insensitive to the

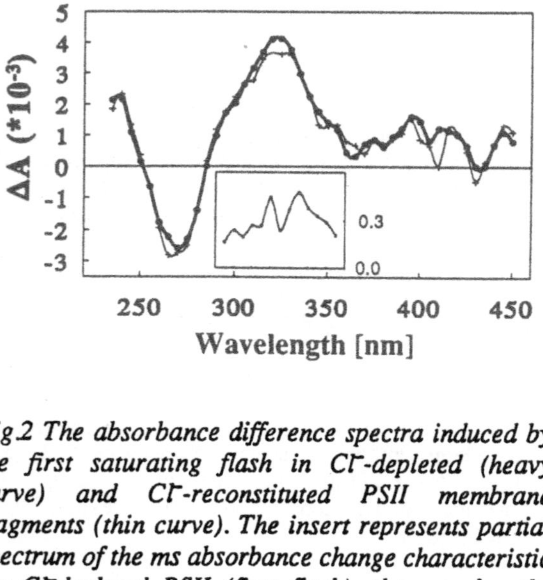

Fig.2 The absorbance difference spectra induced by the first saturating flash in Cl^--depleted (heavy curve) and Cl^--reconstituted PSII membrane fragments (thin curve). The insert represents partial spectrum of the ms absorbance change characteristic for Cl^-depleted PSII (first flash), the wavelengths correspond to values given under the Figure, amplitudes of the signals are magnified 5:1.

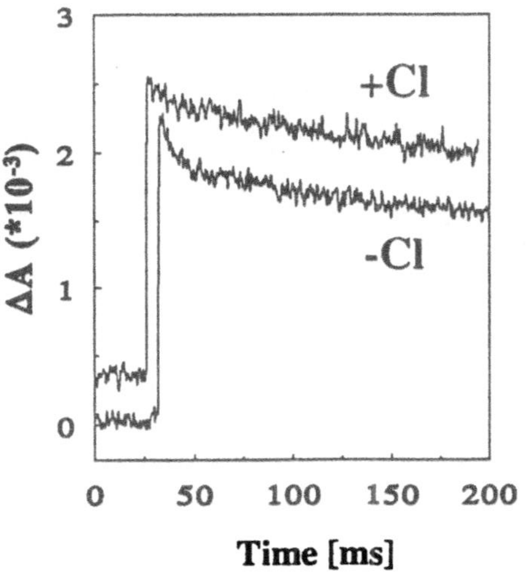

Fig.3 The kinetic traces recorded upon the first flash at 340nm - an example showing the small ms decay phase in Cl⁻-depleted PSII.

addition of DCMU, artificial donors and acceptors and we do not know if this spectral change takes place at the donor or acceptor side of the reaction center, nor if it reflects an electron transfer reaction. Electron transfer from a His residue to manganese is not excluded.

Fig.4 shows difference spectra obtained upon the second saturating flash. Here substantially smaller signals have been measured in the absence of chloride. To resolve the nature of the donor side component oxidized upon the second flash the contribution of the acceptor side has been substracted ($Q_A \rightarrow Q_A^-$spectrum from [2]). As expected the spectrum obtained for Cl^--reconstituted PSII is the normal $S_2 \rightarrow S_3$ spectrum (not shown). The resulting spectrum for Cl^--depleted material is shown on Fig.4 (heavy line). The two positive bands peaking at around 260nm and 300nm have been earlier reported [2] as characteristic for the $Tyr_Z \rightarrow Tyr_Z^+$ spectrum. Similarly the band shift at 435nm has been attributed to the $Tyr_Z \rightarrow Tyr_Z^+$ difference spectrum. From the 3rd flash on the absorbance changes had the same amplitude and decay kinetics (see Fig.1). The spectra depicted for the 3rd and following flashes indicate Tyr_Z oxidation also in this case (see Fig.4,thin line).

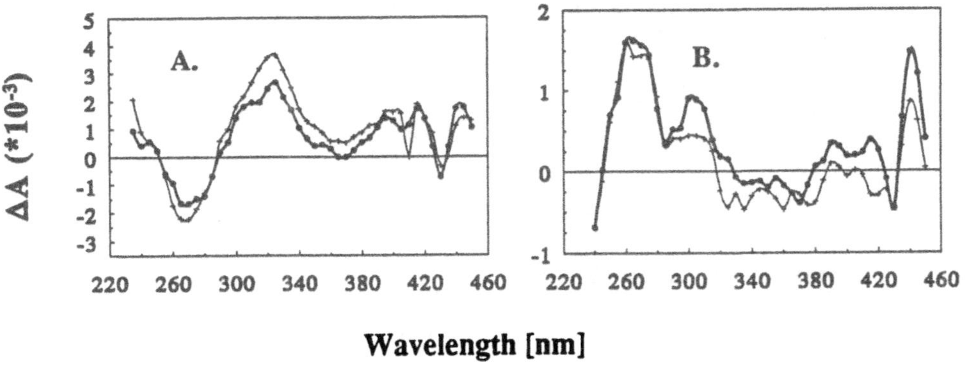

Fig.4 A. The absorption difference spectra induced by the second flash in Cl⁻-depleted (heavy curve) and Cl⁻-reconstituted (thin line) PSII membrane fragments. B. The donor side contribution in the spectra induced by the second (heavy curve) and subsequent flashes (thin curve) in Cl⁻-depleted PSII (these spectra were obtained after substracting the $Q_A \rightarrow Q_A^-$ spectrum).

4. Conclusions

The results presented in this study demonstrate that only one oxidizing equivalent can be accumulated by Oxygen Evolving Complex in Cl^--depleted Photosystem II. On the second and subsequent flashes the absorbance changes appeared to be due to Q_A^- Tyr_Z^+ formation and recombination. The detailed similarity between the UV-VIS spectra recorded for Cl^--depleted and Cl^--reconstituted samples upon the first flash indicated a regular $S_1 \rightarrow S_2$ transition, which means manganese oxidation in chloride free Photosystem II. If the small absorbance decrease in the region 300-350nm can be attributed to the donor side, our measurements are consistent with the EPR data indicating appearance of a modified S_2 state upon the first flash and suggest that the modification takes place in a few miliseconds after the flash. Chloride may be required for the stabilization of the S_2 state in its native form, not for its initial formation.

In contrast to our measurements several authors reported S-state advancement up to the S_3 state in chloride depleted particles [3,4,5]. Based on the EPR data Boussac et al. attributed the first electron transfer event to the oxidation of a His residue and the second to the oxidation of manganese [6]. According to the absorption spectra published in [7] it should be rather easy to distinguish between the oxidation of manganese and His residue by measuring absorption changes around 350nm, where the signal arising from His (or His OH-adduct) should be about half the signal arising from manganese oxidation. We did not find such a difference in our measurements.

Our results do agree with the finding that P_{680}^+ reduction, measured by the µs rise of chlorophyll fluorescence, remains fast only one flash beyond the S_2 state, which was interpreted to indicate the formation of a long lived Tyr_Z^+ on that flash [8]. Also a recent study on redox interactions of Tyr_D with the S-states of the OEC [9] indicated the formation of a substantial amount of long-lived Z^+ from the second flash on.

References

1 Berthold, D.A., Babcock, G.T. and Yocum, C.F. (1981) FEBS Lett. 134, 231-234

2 Dekker, J.P. and van Gorkom, H.J., Brok,M., Ouwehand, L. (1984) Biochim. Biophys. Acta 764, 301-309

3 Boussac, A. and Rutherford, A.W. (1992) Biochemistry 31, 7441-7445

4 Baumgarten, M., Philo, J.S. and Dismukes, G.C. (1990) Biochemistry 29, 10814-10822

5 Homann, P.H., Gleiter, H., Ono, T. and Inoue, Y. (1986) Biochim. Biophys. Acta 850, 10-20

6 Boussac, A. and Rutherford, A.W. (1994) J. Biol. Chem. 269, 12462-12467

7 Boussac, A., Zimmermann, J., Rutherford, A.W. and Lavergne, J. (1990) Nature 347, 303-306

8 Itoh, S., Yerkes, C.T., Koike, H., Robinson, H.H. and Crofts, A.R. (1984) Biochim. Biophys. Acta 766, 612-622

9 Deak, Z., Vass, I. and Styring, S. (1994) Biochim. Biophys. Acta 1185, 65-74

AMINO ACID RESIDUES OF THE EXTRINSIC 33 kDa PROTEIN INVOLVED IN ELECTROSTATIC INTERACTION WITH INTRINSIC PROTEINS OF PS II

Taro Miura[1,3], Seitaro Takahashi[1], Masaharu Kamo[2], Eriko Nakamura[1], Hisataka Ohta[1], Yasunori Inoue[3] and Isao Enami[1]
[1]Department of Biology, Faculty of Science, Science University of Tokyo, Kagurazaka, Shinjuku-ku, Tokyo 162, Japan; [2]Research Institute for Biosciences, Science University of Tokyo, Yamazaki, Noda, Chiba 278, Japan; [3]Department of Applied Biological Science, Faculty of Science and Technology, Science University of Tokyo, Yamazaki, Noda, Chiba 278, Japan

1. Introduction

The extrinsic 33 kDa protein of oxygen-evolving PS II complex is associated with lumenal surface of the thylakoid membrane and plays an important role in maintaining functional binding of the Mn cluster. Crosslinking of the protein with PS II intrinsic components by a water-soluble carbodiimide which directly links amino and carboxyl groups in contact, prevents release of the protein upon $CaCl_2$- or urea/NaCl-wash or heat-treatments, thereby stabilizing binding of the Mn-cluster and oxygen evolution against these treatments [1-3]. Based on these studies, the association of the 33 kDa protein with PS II has been proposed to involve electrostatic interactions between different charges on the extrinsic protein and some PS II intrinsic subunits as well as hydrogen bonding. The 33 kDa protein has been also demonstrated to be associated with, or is in close proximity to, virtually all of the intrinsic PS II components [4-9]. There is, however, only very limited information concerning the residues responsible for binding of the protein to PS II. Chemical modification of charged amino acid residues can easily alter surface charges on proteins and therefore is a useful technique to identify residues involved in the electrostatic interaction between the 33 kDa protein and intrinsic PS II proteins. Here, either positive or negative charges on the extrinsic 33 kDa protein were eliminated using chemical modification techniques, and the abilities of the resultant modified protein to rebind to PS II and to reactivate oxygen evolution were examined. Data are present showing that positive charges on the 33 kDa protein are essential for its electrostatic interaction with intrinsic PS II proteins. Identification of positively charged residues on the 33 kDa protein involved in the interaction was also attempted, and Lys4, Lys101, Lys105, Lys233 and Lys236 of the protein were found to be responsible for its binding with intrinsic PS II proteins.

2. Materials and Methods

Oxygen-evolving PS II membranes were prepared from spinach chloroplasts with Triton X-100. The extrinsic 33 kDa protein was extracted from the PS II membranes by 1 M $CaCl_2$ treatment and purified by DEAE-Sepharose CL-6B column chromatography. Chemical modification of the 33 kDa protein was carried out as described in Results and Discussion. Isoelectric focusing was performed in a 5.5% polyacrylamide gel covering a

P. Mathis (ed.), Photosynthesis: from Light to Biosphere, Vol. II, 353–356.

pH range of 3.5-10 or 4.0-6.0. For reconstitution, the PS II membranes were washed with 2.6 M urea/0.2 M NaCl to remove the extrinsic proteins. The urea/NaCl-washed PS II membranes were incubated with the modified proteins in medium A that contained 40 mM Mes-NaOH (pH 6.5), 0.4 M sucrose, 5 mM $MgCl_2$ and 10 mM NaCl at 0°C in the dark at a Chl concentration of 0.5 mg/ml. The modified proteins were supplemented at a protein to Chl ratio of 0.6 (w/w). The reconstituted PS II membranes were collected by centrifugation, then washed once with and resuspended in medium A. The amounts of the 33 kDa protein rebound by the reconstitution were determined by SDS-polyacrylamide gel electrophoresis and immunoblotting analysis with an antibody raised against the 33 kDa protein from spinach. Oxygen evolution was measured at 25°C with a Clark-type oxygen electrode in medium A, to which 5 mM $CaCl_2$ and 0.4 mM phenyl-*p*-benzoquinone were supplemented.

3. Results and Discussion

Chemical modification of negatively charged carboxyl groups on the 33 kDa protein was carried out in distilled water containing 37.75 μM of the protein, 100 mM glycine methyl ester (GME) and 2 mM 1-ethyl-3-(3-dimethylaminopropyl)-carbodiimide (EDC) at 25°C for 12 hr in the dark. In this reaction (see eq 1), a negatively charged carboxyl group is replaced by an uncharged methyl group. As a result, the isoelectric pH (pI) of GME-modified protein shifted from 5.2 of unmodified protein to a value above 8.5 (Table I). The GME-modified protein was able to rebind to urea/NaCl-washed PS II membranes (data not shown) and to reactivate the oxygen evolution to an extent comparable to that achieved by rebinding of the unmodified proteins (Table I). This indicates that surface negative charges on the 33 kDa protein do not participate in its functional binding to intrinsic PS II proteins.

Chemical modification of positively charged amino groups on the 33 kDa protein was performed in 20 mM phosphate buffer (pH 6.5) containing 48 μM of the protein and 8 mM of N-succinimidyl propionate (NSP) at 25°C for 90 min in the dark. In this reaction (see eq 2), positively charged amino groups are replaced by uncharged methyl groups. Upon this modification, the pI of the protein shifted from 5.2 to 4.4-4.7 (Table I). The NSP-modified protein with pI of 4.7 cannot rebind to PS II at all (data not shown) which resulted in a complete loss of the reactivating capability (Table I). A pI shift from 5.2 to 4.7 was calculated to correspond to elimination of about 7 positive charges on the 33 kDa protein. These results indicate that upon elimination of 7 positive charges, the protein completely lost its abilities to rebind to PS II and to reactivate oxygen evolution. Since NSP specifically

$$\text{Protein-COO}^- + NH_2CH_2COOCH_3 \xrightarrow{\text{EDC}} \text{Protein-CONHCH}_2\text{COOCH}_3 \quad (1)$$

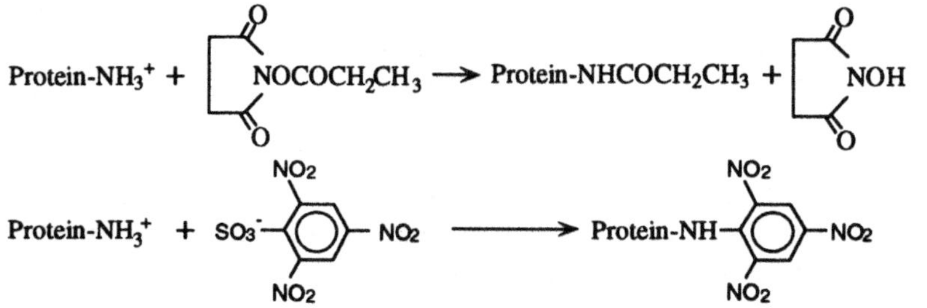

(2)

(3)

Table I. Restoration of oxygen evolution after reconstitution of the 33 kDa protein modified with GME, NSP or TNBS

		Oxygen evolution (μmol O_2/mg Chl/h)	Reactivation (%)
Urea/NaCl-washed PS II membrane		71.4	
+ unmodified 33 kDa protein	(pI=5.2)	371.3	100
+ GME-modified 33 kDa protein	(pI>8.5)	373.3	101
+ NSP-modified 33 kDa protein	(pI=4.4-4.7)	72.4	0.5
+TNBS-modified 33 kDa protein I	(pI=4.5-4.8)	71.0	0
+TNBS-modified 33 kDa protein II	(pI=5.1)	375.2	101

modifies the amino group of Lys residues, we concluded that positively charged Lys residues of the 33 kDa protein participate in its binding with PS II intrinsic proteins.

In order to identify Lys residues of the 33 kDa protein involved in the electrostatic interaction with intrinsic PS II proteins, we used trinitrobenzene-sulfonic acid (TNBS) to modify Lys residues, since TNBS-modified amino acid shows an absorbance at 345 nm which can be easily detected. The isolated 33 kDa protein (4.4 μM) and NaCl-washed PS II membranes (1 mg Chl/ml) which retained the 33 kDa protein were treated with 3 mM and 4 mM TNBS, respectively, in medium A at 25°C for 15 hr. After modification, NaCl-washed PS II membranes were precipitated and once washed with medium A and then the 33 kDa protein was extracted from the PS II membranes by 1 M $CaCl_2$ treatment. The 33 kDa proteins modified with TNBS in the isolated and the bound forms were referred here to as TNBS-modified 33 kDa protein I and II, respectively. TNBS specifically reacts with amino groups and eliminates positive charges on proteins (see eq 3). The TNBS-modified 33 kDa protein I (pI=4.5-4.8) completely lost its abilities to rebind to PS II and to reactivate oxygen evolution, while the TNBS-modified 33 kDa protein II (pI = 5.1) retained its rebinding and reactivating capabilities (Table I). For identification of Lys residues modified with TNBS, these two TNBS-modified 33 kDa proteins (2.5 μM) were digested with V-8 protease at 2.5 μg/ml in 0.1 M Mes-NaOH (pH 6.5), 0.1 mM $CaCl_2$ and 2 M urea after denaturing in 8 M urea and 25 mM Mes-NaOH (pH 6.5) for 16 hr at 37°C. After proteolysis at 37°C for 2 days, the samples were subjected to reversed-phase HPLC using a Waters RP-C4 column. The column was developed by using a 0-75% (v/v) gradient of acetonitrile in 0.1% (v/v) trifluoroacetic acid. When peptide fragments were monitored at 345 nm, the TNBS-midified 33 kDa protein I showed twenty peaks in the chromatogram of reversed-phase HPLC, while only one peak was detected in the TNBS-modified 33 kDa protein II (data not shown). This indicates that most of amino groups reactive with TNBS on the 33 kDa protein exist in the region interacting with PS II intrinsic proteins. Mass-spectroscopic analysis and N-terminal amino acid sequencing of the TNBS-modified peptide fragments revealed that the modified amino acid residues in the TNBS-modified 33 kDa protein I are N-terminus, Lys 4, Lys 101, Lys 105, Lys 233, Lys 236 and some undetected Lys residues. On the other hand, the modified amino group in the TNBS-modified 33 kDa protein II was only N-

```
          10         20         30         40         50
(NH2)-EGGKRLTYDE IQSKTYLEVK GTGTANQCPT VEGGVDSFAF KPGKYTAKKF

          60         70         80         90        100
CLEPTKFAVK AEGISKNSGP DFQNTKLMTR LTYTLDEIEG PFEVSSDGTV

         110        120        130        140        150
KFEEKDGIDY AAVTVQLPGG ERVPFLFTIK QLVASGKPES FSGDFLVPSY

         160        170        180        190        200
RGSSFLDPKG RGGSTGYDNA VALPAGGRGD EEELQKENNK NVASSKGTIT

         210        220        230        240
LSVTSSKPET GEVIGVFQSL QPSDTDLGAK VPKDVKIEGV WYAQLEQ
```

Fig. 1 Assignment of TNBS-modified amino acid residues to the primary structure of the 33 kDa protein. The circle represents N-terminal amino group modified with TNBS both in the isolated and the bound forms of the 33 kDa protein to PSII. The boxed amino acids show the Lys modified with TNBS in the isolated form but not in the 33kDa protein bound form.

terminus. These results were assigned in the primary structure of the 33 kDa protein (Fig. 1). Thus, we concluded that N-terminus of the 33 kDa protein exposes to lumenal surface but a number of Lys residues, at least Lys 4, Lys 101, Lys 105, Lys 233 and Lys 236, are present in the region interacting with PS II intrinsic proteins and these Lys residues participate in electrostatic interaction with PS II intrinsic proteins.

This work was supported in part by a Grant-in-Aid for Co-operative Research (05304006) to I.E. from the Ministry for Education, Science and Culture of Japan.

References

1. Enami, I., Kaneko, M., Kitamura, N., Koike, H., Sonoike, K., Inoue, Y. and Katoh, S.(1991) Biochim. Biophys. Acta 1060, 224-232.
2. Enami, I., Tomo, T., Kitamura, M. and Katoh, S.(1994) Biochim. Biophys. Acta 1185, 75-80.
3. Enami, I., Kitamura, M., Tomo, T., Isokawa, Y., Ohta, H. and Katoh, S.(1994) Biochim. Biophys. Acta 1186, 52-58.
4. Isogai, Y., Yamamoto, Y. and Nishimura, M.(1985) FEBS Lett. 187, 240-244.
5. Enami, I., Satoh, K. and Katoh, S.(1987) FEBS Lett. 226, 161-165.
6. Enami, I., Miyaoka, T., Mochizuki, Y., Shen, J.-R., Satoh, K. and Katoh, S.(1989) Biochim. Biophys. Acta 973, 35-40.
7. Enami, I., Mochizuki, Y., Takahashi, S., Kakuno, T., Horio, T., Satoh, K. and Katoh, S.(1990) Plant Cell Physiol. 31, 725-729.
8. Mei, R., Green, J.P., Sayre, R.T. and Frasch, W.D.(1989) Biochemistry 28, 5560-5567.
9. Enami, I., Ohta, S., Mitsuhashi, S., Takahashi, S., Ikeuchi, M. and Katoh, S.(1992) Plant Cell Physiol. 33, 291-297.

ISOLATION AND CHARACTERIZATION OF A HIGHLY PURIFIED OXYGEN-EVOLVING PHOTOSYSTEM II CORE COMPLEX FROM AN ACIDOPHILIC AND THERMOPHILIC RED ALGA, *Cyanidium caldarium*

Isao Enami[1], Hiroshi Murayama[1], Masahasru Kamo[2], Hisataka Ohta[1] and Katsuyoshi Nakazato[3,4]
[1]Department of Biology, Faculty of Science, Science University of Tokyo, Kagurazaka, Shinjuku-ku, Tokyo 162, Japan; [2]Research Institute for Biosciences, Science University of Tokyo, Yamazaki, Noda, Chiba 278, Japan; [3]PRESTO, JRDC; [4]Solar Energy Research Group, The Institute of Physical and Chemical Research (RIKEN), Wako, Saitama 351-01, Japan

1. Introduction

The essential components of oxygen-evolving PSII complex are identical between cynanobacteria and higher plants; they include the D1, D2 reaction center proteins, the intrinsic Chl-binding proteins CP47 and CP43, the α and β subunits of cyt *b*-559, the *psb*I gene product, the extrinsic 33 kDa protein and perhaps some other low molecular weight polypeptides. In contrast to these similarities, however, there are two extrinsic proteins that are significantly different between PSII complexes from higher plants and cyanobacteria. In higher plant PSII, the extrinsic 23 and 17 kDa proteins serve in maintaining maximum oxygen-evolving activity under physiological ionic condition. In cyanobactrial PSII, on the other hand, these two extrinsic proteins are absent but a low potential cyt *c*-550 and a 12 kDa protein have been found as the alternative extrinsic components in maintaining oxygen evolution and stability of PSII in vivo [1-3]. This suggests a functional replacement of cyt *c*-550 and the 12 kDa protein by the 23 and 17 kDa proteins during evolution from cyanobacteria to higher plants. In order to determine whether red algal PS II involves cyt *c*-550 and the 12 kDa protein or the 23 and 17 kDa proteins and to compare the PSII complex of red algae with that of cyanobacteria and higher plants, we purified an intact PSII complex highly competent in oxygen evolution from an acidophilic and thermophilic red alga, *Cyanidium caldarium*. Our results revealed that the PSII from the red alga contained cyt c-550 and the 12 kDa protein but not the 23 and 17 kDa proteins as extrinsic components. This indicates that the replacement of PSII extrinsic cyt *c*-550 and the 12 kDa protein by the extrinsic 23 and 17 kDa proteins occurred during evolution from red algae to green algae and higher plants.

P. Mathis (ed.), Photosynthesis: from Light to Biosphere, Vol. II, 357–360.

2. Materials and Methods

Cyanidium caldarium was photoautotrophically cultured at 40°C and pH 2-3 as described in [4]. The algal cells grown for one weak were collected by centrifugation, washed three times with distilled water and then suspended in medium A that contained 25% glycerol, 10 mM $CaCl_2$ and 50 mM Mes (pH 6). After the cells were disrupted by agitation with glass beads on ice in the dark, the thylakoid membranes were collected by centrifugation at 100,000 x g for 20 min and suspended in medium A. The thylakoid membranes were solubilized with 2% n-heptyl-β-D-thioglucoside (HTG) and then loaded onto a layer of 1 M sucrose in medium A followed by centrifugation at 410,000 x g for 1 hr. This resulted in a green band at the surface of the 1 M sucrose layer and a dark blue precipitate. The precipitate showed a high oxygen-evolving activity and was referred to as crude PSII particles, whereas the green band was rich in PSI and had only a very low oxygen-evolving activity. The crude PSII particles still retained a large amount of phycobiliproteins and some PSI components. To further purify the PSII complex, the crude PSII particles were solubilized with 2% n-dodecyl β-D-maltoside (DM) and then loaded onto a DEAE Sepharose CL-6B column. After washing the column with 0.05% DM, 25% glycerol, 0.2 M NaCl and 50 mM Mes (pH 6) by which phycobiliproteins and PSI components were eluted, the purified PSII complexes were eluted with the same medium containing 0.33 M NaCl.

3. Results and Discussion

Polypeptide compositions of the PSII complex purified from the red alga *C. caldarium* were shown in Fig. 1 (lane 1). The PSII complex was essentially devoid of phycobiliproteins and other contaminating components. Protein components of the PSII complex were

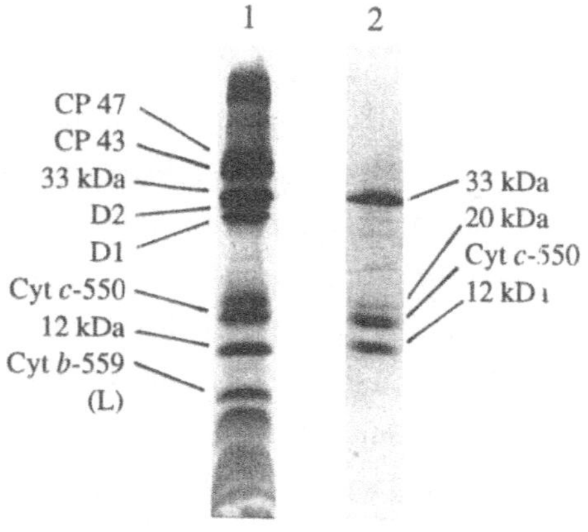

Fig. 1 SDS-polyacrylamide gel electrophoresis of the purified PSII complex of *Cyanidium caldarium* (lane 1) and of the extrinsic proteins extracted by Tris-wash from the PSII complex (lane 2). SDS gel electrophoresis was carried out with a gradient gel of 16-22% containing 7.5 M urea in the buffer system of Laemmli [5].

identified as shown in Fig. 1 by immunoblotting analysis with antibodies raised against major PSII polypeptides of spinach as well as against two extrinsic proteins, cyt *c*-550 and the 12 kDa protein, from a thermophilic cyanobacterium, *Synechococcus vulcanus*. No polypeptides in the red algal PSII complex were found to crossreact with antibodies against the extrinsic 23 and 17 kDa proteins, suggesting that these two extrinsic components are not present in the complex. On the other hand, the antibody against *S. vulcanus* cyt *c*-550 crossreacted with a polypeptide with an apparent molecular weight of 17 kDa, indicating that this band is cyt *c*-550. This was also confirmed by heme-staining of the gel and N-terminal amino acid sequencing of the 17 kDa protein [6]. The dense band with an apparent molecular weight of 12 kDa did not crossreact with antibody against the 12 kDa protein of *S. vulcanus*. The N-terminal sequence of this polypeptide, however, showed a moderate homology with the 12 kDa protein of PSII from *S. vulcanus* and *Phormidium laminosum* [6].

Cyt *c*-550 and the 12 kDa protein in cyanobacterial PSII can be released upon treatment with alkaline Tris or high concentrations of divalent cations [1]. In order to check whether a similar situation occurs with cyt *c*-550 and the 12 kDa protein present in the red algal PSII, we treated the red algal PSII with 1 M Tris (pH 8.5). As shown in lane 2 of Fig. 1, red algal cyt *c*-550 and the 12 kDa protein were released, together with the extrinsic 33 kDa protein, by 1 M Tris. This indicates that both cyt *c*-550 and the 12 kDa protein are real extrinsic components of the red algal PSII. On the CBB-stained gel of the extract of Tris-wash (lane 2 in Fig. 1), we also noticed the presence of a band at 20 kDa which appeared slightly above the band of cyt *c*-550, although this band was more weakly stained by CBB as compared with other extrinsic proteins. Its N-terminal amino acid sequence indicated that this 20 kDa polypeptide has no significant homology with any known PSII polypeptides [6], suggesting that it may be a new extrinsic component of PSII from the red alga.

Table I shows the oxygen-evolving activity of the purified red algal PSII complex. The PSII showed a high activity of 2375 μ moles O_2/mg Chl/hr when phenyl-*p*-benzoquinone (PBQ) was used as an electron acceptor. The activity was almost completely inhibited by 3-(3,4-dichlorophenyl)-1,1-dimethylurea (DCMU), and also decreased significantly when potassium ferricyanide (Fecy) was used as the acceptor, indicative of the retention of an intact Q_B-site in the purified PSII complex. These activities were not enhanced by the addition of exogenous $CaCl_2$. Presence of 2 mM EDTA in the assay medium, however, decreased the activity to 63% of the control. The suppressed activity was nearly completely restored not only by Ca^{2+} but also by Mg^{2+}. A similar enhancing effect of Mg^{2+} has been

Table I Effects of electron acceptors and divalent cations on oxygen evolution of the PS II complexes purified from *Cyanidium caldarium.*

Acceptors	Oxygen evolution* (μmol O_2/mg Chl/hr)
0.6 mM PBQ	2375
2.0 mM Fecy	742
0.6 mM PBQ+2 mM Fecy	2523
0.6 mM PBQ+10 μM DCMU	74

Divalent cations	Oxygen evolution* (μmol O_2/mg Chl/hr)
Control (PBQ)	2375
+10 mM $CaCl_2$	2424
+2 mM EDTA	1485
+10 mM $CaCl_2$	2573
+10 mM $MgCl_2$	2226

*Oxygen evolution was measured at 25°C in 50 mM Mes (pH 6.5), 25% glycerol and 20 mM NaCl.

observed in cyanobacterial PSII but not in higher plant PSII.

In the present work, we described a simple procedure for purification of oxygen-evolving PS II complex from a thermophilic and acidophilic red alga, *Cyanidium caldarium.* The purified PS II complex contained cyt *c*-550 and the 12 kDa protein but not the 23 kDa and 17 kDa proteins as extrinsic components. Based on the polypeptide compositions and characteristics of oxygen-evolving activity of the PS II complex, we concluded that the red algal PSII is evolutionarily more closely related with cyanobacterial PSII rather than with higher plant PSII.

We thank Dr. Jian-Ren Shen of the Institute of Physical and Chemical Research for his critical reading of the manuscript. We also thank Drs. T. Horio, T. Kakuno, M. Ikeuchi and J.-R. Shen for providing the antibodies against spinach PS II polypeptides and against cyt *c*-550 and the 12 kDa protein from *S. vulcanus.* This work was supported in part by a Grant-in-Aid for Co-operative Research (No. 05304006) from the Ministry for Education, Science and Culture of Japan.

References

1. Shen, J.-R., Ikeuchi, M. and Inoue, Y.(1992) FEBS Lett.301, 145-149.
2. Shen, J.-R. and Inoue, Y.(1993) Biochemistry 32, 1825-1832.
3. Shen, J.-R., Vermaas, W. and Inoue, Y.(1995) J. Biol. Chem. 270, 6901-6907.
4. Enami, I. and Fukuda, I.(1975) Plant Cell Physiol.16, 211-220.
5. Ikeuchi, M. and Inoue, Y.(1988) Plant Cell Physiol.29, 1233-1239.
6. Enami, I., Murayama, H., Ohta, H., Kamo, M., Nakazato, K. and Shen, J.-R. (1995) Biochim. Biophys. Acta in press.

EVIDENCE FOR ELECTROSTATIC INTERACTION OF THE LOOP A ON CP 47 WITH THE EXTRINSIC 33 kDa PROTEIN

Hisataka Ohta[1], Naoshi Yoshida[1], Masanori Sano[1], Masahiko Hirano[2], Katsuyoshi Nakazato[3,4] and Isao Enami[1]
[1]Department of Biology, Faculty of Science, Science University of Tokyo, Kagurazaka, Shinjuku-ku, Tokyo 162, Japan;[2]Biological Sciences Department, Toray Research Center, Inc., Kamakura 248, Japan;[3]PRESTO, JRDC; [4]Solar Energy Research Group, The Institute of Physical and Chemical Research (RIKEN), Wako, Saitama 351-01, Japan

1. Introduction

The extrinsic 33 kDa protein associated with lumenal surface of thylakoid membranes plays an important role in stabilizing binding and maintaining functional conformation of the Mn cluster present at the catalytic center of the H_2O-splitting enzyme. A chlorophyll a-binding protein CP 47 which is one of the core antenna functioning in transfer of excitation energy to the reaction center has been recently suggested to structurally interact with the oxygen-evolving site of PS II [1-4]. Based on hydropathy analysis of the derived amino acid sequence of CP 47, Bricker proposed that the protein possesses 6 transmembrane α helixes with three extrinsic loops, loops A, C and E, exposed to lumenal surface and two extrinsic loops, loops B and D, exposed to stromal surface, of the thylakoid membrane (see Fig. 2) [5]. We have reported that all the 33 kDa protein can be covalently and stoichiometrically linked to CP 47, without any significant effects on the oxygen-evolving activity, by treatment of oxygen-evolving PS II membranes with 1-ethyl-3-(3-dimethylaminopropyl)-carbodiimide (EDC) [6]. The result suggests that the extrinsic 33 kDa protein attaches directly to the intrinsic CP 47 protein through an electrostatic interaction. In the present paper, we attempted to identify domains on the 33 kDa protein and CP 47 that are crosslinked by EDC, by means of protease digestion of the EDC-crosslinked product followed by N-terminal sequence analysis. The results obtained showed that, in addition to crosslinking of the 33 kDa protein with the extrinsic loop E of CP 47 [7], a domain of Lys 60-Lys 137 on the 33 kDa protein is crosslinked to the extrinsic loop A of CP 47, most probably Asp 87 and/or Glu 94 in loop A. In addition, an intramolecular crosslinkage was also found in the 33 kDa protein between carboxyl groups of Glu 246 and/or C-terminus and ε-amino group(s) of Lys(s) present in the domain from Lys 186 to Lys 233.

2. Materials and Methods

Preparation of oxygen-evolving PS II membranes from spinach and crosslinking of PS II membranes with EDC were carried out as described previously [6]. Preparative quantities of CP 47, the 33 kDa protein and the crosslinked product of these two proteins were isolated by SDS polyacryamide gel electrophoresis followed by electroelution. The

P. Mathis (ed.), Photosynthesis: from Light to Biosphere, Vol. II, 361–364.

bands of CP 47 and the 33 kDa protein resolved from control PS II membranes and the band of the crosslinked product resolved from 0.5% EDC-treated PS II membranes were excised from gels, incubated for 30 min in 10% SDS, 5% 2-mercaptoethanol and 8 M urea, and then electroeluted in 25 mM Tris-glycine buffer (pH 8.8) and 0.1% SDS. The electroeluted samples were dialyzed against 0.1% SDS to remove urea and glycine and then precipitated by adding 80% acetone to remove SDS. The pellets were dissolved in distilled water and then treated with lysyl endopeptidase (LEP) at 0.05 mg/ml in 100 mM Tris-HCl buffer (pH9.0) at 37°C for 24 hr. The protease-digested samples were subjected to reversed-phase HPLC with a Waters RP-C4 column. Peptides were eluted with a 0-75% gradient of acetonitrile in 0.1% trifluoroacetic acid and monitored at 210 nm. Peptides containing putative crosslinking sites were selected and sequenced with a protein sequencer (Applied Biosystems, model 477A).

3. Results and Discussion

Fig. 1 shows peptide map of the crosslinked product which was resolved by the reversed-phase HPLC after digestion with LEP at 37°C for 24 hr. The peptide pattern did not change upon further prolonged treatment up to 48 hr with the protease. When the peptide pattern of the crosslinked product was compared with that of either CP 47 or the 33 kDa protein, most of the peaks in the chromatogram were correlated with peaks in the chromatogram of each protein (data not shown), but four new peptide peaks that were not found in the chromatogram of each protein appeared in the crosslinked product (peaks A-D in Fig. 1). These peaks are expected to be crosslinked fragments containing both CP 47 and the 33 kDa protein. N-terminal amino acid sequences of these four peptide peaks were determined.

N-terminal sequencing of peak A yielded two sequences corresponding to Arg 5-Tyr 16 of the 33 kDa protein and Glu 353-Phe 362 of CP 47, respectively (Table I). Since the sequence from Glu 353 to Phe 362 is located in loop E of CP 47, the result indicates that the large extrinsic loop E of CP 47 was crosslinked with N-terminal region of the 33 kDa protein (see Fig. 2), which is consistent with the results reported by Odom and Bricker [7].

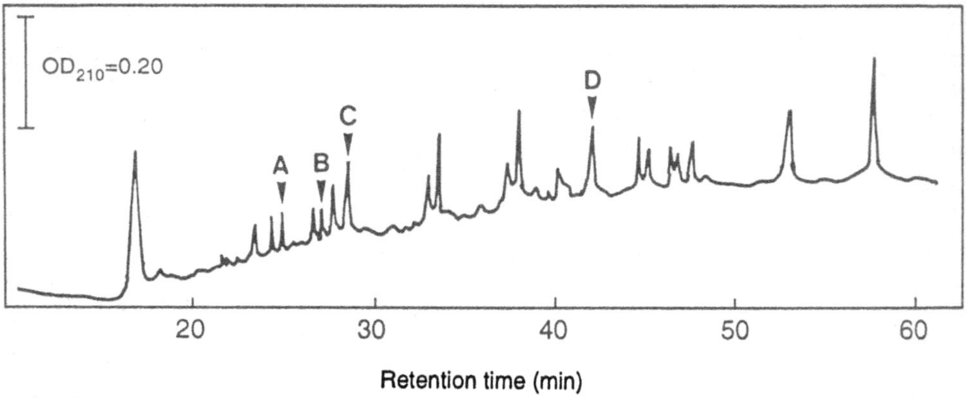

Fig. 1 HPLC peptide mapping of LEP digest of the crosslinked product (CP 47-33 kDa protein adduct).

Table I N-Terminal sequence analysis of peaks A, B and C obtained by LEP-digestion of crosslinked product between the 33 kDa protein and CP 47

Peak A	Arg	Leu	Thr	Tyr	Asp	Glu	Ile	Gln	Ser	Lys	Thr	Tyr	(33kDa protein R5 - Y16)
	Glu	Leu	Phe	Val	nd	nd	Met	Pro	nd	Phe	(CP47 E353 - F362)		
Peak B	Phe	nd	Leu	Glu	Pro	Thr	Lys	Phe	Ala	Val	(33kDa protein F50 - V59)		
	Pro	Val	Leu	Asp	Pro	Met	nd	nd	Gln	Gly	(CP47 P50 - G59)		
Peak C	Gly	Arg	Gly	Gly	Ser	Thr	Gly	Tyr	Asp	Asn	Ala	(33kDa protein G160 - A170)	
	Ile	Glu	Gly	Val	Trp	Tyr	Ala	Gln	Leu	(33 kDa protein I237 - L245)			

N-terminal sequencing of peptide peak B yielded two sequences corresponding to Phe 50-Val 59 of the 33 kDa protein and Pro 50-Gly 59 of CP 47 (Table I). Since the sequence from Pro 50 to Gly 59 is located in loop A of CP 47 and one fragment of the peak C starts from Gly 160 of the 33 kDa protein (see below), the result indicates that the extrinsic loop A of CP 47 exposed to lumenal surface of the thylakoid membrane was crosslinked to a residue(s) within the domain of Phe 50-Lys 159 of the 33 kDa protein. This is the first report showing a direct interaction of the 33 kDa protein with loop A of CP 47. Since amino acid residues crosslinked by EDC can not be released by the Edman degradation, Asp 53 of CP 47 and Glu 53 and Lys 56 of the 33 kDa protein which were detected by the sequencing protocol (Table I) cannot be the residues crosslinked by EDC. Thus, amino acids crosslinked by EDC should be present in the domain from Met 60 to Val 96 on loop A of CP 47 and in the domain from Lys 60 to Lys 159 of the 33 kDa protein. In this domain of CP 47, only two amino acids, Asp 87 and Glu 94, are present as charged residues (see Fig. 2). On the other hand, the domain from Lys 60 to Lys 159 of the 33 kDa protein contains 7 Lys residues. Among these Lys residues, Lys 60, 66, 76 and 130 are conserved in the sequences from cyanobacteria to higher plants. It is, therefore, most likely that the ε-amino group(s) of one or several of these conserved Lys residues on the 33 kDa protein was crosslinked with carboxyl group(s) of Asp 87 and /or Glu 94 on loop A of CP 47 (see Fig. 2).

N-terminal sequencing of peak C resulted in two sequences corresponding to Gly 160-Ala 170 and Ile 237-Leu 245 of the 33 kDa protein (Table I). This indicated that this peptide fragment is an intramolecular crosslinking product of the 33 kDa protein. A comparison of these sequences with primary sequence of the protein indicated that the crosslinking occurred between carboxyl group(s) of Glu 246 and/or C-terminus and ε-amino group(s) of Lys among six Lys residues present in the sequence from Lys 186 to Lys 233. Among these six Lys residues, only Lys 190 is conserved in the sequences from cyanobacteria to higher plants, whereas the other five Lys residues are not. It is, therefore, most likely that the ε-amino group of Lys 190 is the crosslinking partner of the carboxyl group of Glu 246 or C-terminus. Peptide of peak D showed a single sequence corresponding to Try 59-Pro 68 of LHC II apoprotein, suggesting that this peptide is contamination from LHC II apoprotein.

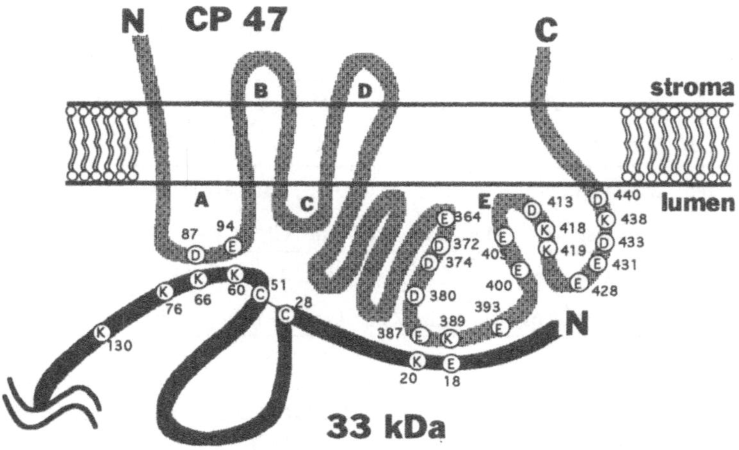

Fig. 2 Schematic diagram of electrostatic interaction sites between the extrinsic 33 kDa protein and the intrinsic CP 47 protein.

Based on these results obtained here, we proposed a model for electrostatic interaction sites between the extrinsic 33 kDa protein and the intrinsic CP 47 protein in Fig. 2. The 33 kDa protein interacts with two domains of loops A and E exposed to lumenal surface of CP 47. The negatively charged carboxyl group(s) of Asp 87 and/or Glu 94 on loop A of CP 47 electrostatically interacts with the positively charged amino group(s) of Lys 60 and/or Lys 66 (or Lys 76, Lys 130) in the 33 kDa protein. The negatively charged carboxyl group of Glu 387 and/or the positively charged amino group of Lys 389 (or other charged amino acids) on loop E of CP 47 interact with the positively charged amino group of Lys 20 and/or the negatively charged carboxyl group of Glu 18 in the 33 kDa protein, respectively.

This work was supported in part by a Grant-in-Aid for Co-operative Research (05304006) to I.E. from the Ministry for Education, Science and Culture, Japan.

References

1. Enami, I., Satoh, K. and Katoh, S.(1987) FEBS Lett. 226, 161-165.
2. Enami, I., Miyaoka, T., Mochizuki, Y., Shen, J.-R., Satoh, K. and Katoh, S.(1989) Biochim. Biophys. Acta 973, 35-40.
3. Enami, I., Tomo, T., Kitamura, M. and Katoh, S.(1994) Biochim. Biophys. Acta 1185, 75-80.
4. Gleiter, H.M., Haag, E., Shen, J.-R., Eaton-Rye, J.J., Seeliger, A.G., Inoue, Y., Vermaas, W.F.J. and Renger, G.(1995) Biochemistry 34, 6847-6856.
5. Bricker, T.M.(1990) Photosynth. Res. 24, 1-13.
6. Enami, I., Kaneko, M., Kitamura, N., Koike, H., Sonoike, K., Inoue, Y. and Katoh, S.(1991) Biochim. Biophys. Acta 1060, 224-232.
7. Odom, W.R. and Bricker, T.M.(1992) Biochemistry 31, 5616-5620.

PROPERTIES OF THE Mn-CLUSTER OF PHOTOSYNTHETIC OXYGEN EVOLVING CENTER IN Ca^{2+} AND/OR Cl^--DEPLETED PHOTOSYSTEM II

Taka-aki Ono[1], Takumi Noguchi[1], Masami Kusunoki[2], Yorinao Inoue[1]
Hirotaka Yamaguchi[3] and Hiroyuki Oyanagi[3]

[1]Solar Energy Research Group, Photosynthesis Laboratory, The Institute of Physical and Chemical Research (RIKEN), Wako, Saitama 351-01, Japan
[2]School of Science and Technology, Meiji University, Kawasaki, Kanagawa 214, Japan
[3]Electromechanical Laboratory, Tsukuba, Ibaraki 305, Japan

1. Introduction

Photosynthetic water oxidation is catalyzed by the tetranuclear Mn-cluster located in the photosystem (PS) II and proceeds through four stable intermediate reaction states, denoted as Si (i = 0 to 3), by use of the energy of four photons [1,2]. In this system, both Ca^{2+} and Cl^- are stoichiometrically present and function as indispensable inorganic cofactors for the normal cycling of the S-state mechanism and the structural stability of the Mn-cluster. Capability of oxygen evolution is lost when either Ca^{2+} or Cl^- is depleted from PS II complex [3]. In the absence of either of these ions, only a limited number of oxidizing equivalents for water oxidation could be accumulated on the donor side of PS II, but the chemical properties responsible for accumulating charge are still a matter of debate [4,5]. In order to understand the mode of action of these two ions on the oxygen evolution, it is necessary to learn the characteristic changes in the structure and function of the Mn-cluster in the absence of these ions. Here, we present results of thermoluminescence (TL) and Mn K-edge X-ray absorption near edge structure (XANES) studies on the oxygen evolving center (OEC) depleted of Ca^{2+} and/or Cl^-.

2. Materials and Methods

PS II membranes with 600-800 μmoles O_2/mg Chl/h were prepared from spinach and dark adapted for overnight at 0 °C. All the following procedures were carried out under complete darkness or dim green safe light unless otherwise noted. For Ca^{2+} depletion, the membranes were treated with a citrate buffer at pH 3.0 for 5 min as in [6], and resuspended in 0.4 M sucrose/20 mM Mes-NaOH (pH 6.5) (Buffer A) supplemented with 20 mM Cl^- and 0.5 mM EDTA•Na. For Cl^- depletion, the membranes were washed with 2M NaCl buffer (pH 6.5) [7], and resuspended in Buffer A supplemented with 20 mM Ca^{2+} after three wash with the same buffer. For Ca^{2+} and Cl^- depletion, the membranes were treated at pH 3.0 for the first time for depletion of Ca^{2+} and then followed by 2 M NaCl wash for Cl^- depletion, and resuspended in Buffer A after three wash with the same buffer. Ca^{2+} was added as $Ca(OH)_2$/Mes (pH 6.5) and Cl^- was

P. Mathis (ed.), Photosynthesis: from Light to Biosphere, Vol. II, 365–368.

added as NaCl. It is noted that these methods enable us to prepare Ca^{2+} and/or Cl^- depleted OEC retained the dark adapted S_1-state.

For TL measurements, samples were excited with a Xe flash at 5 °C and cooled quickly in liquid N_2, and then light emission during warming was recorded against the sample temperature as described [7]. For XANES measurements, samples mounted on a membrane filter were illuminated with a saturating Nd-YAG laser flash at 5~10 °C, and then immediately cooled to 77K in liquid N_2 as described [8,9]. XANES spectra were measured at the Photon Factory of the National Laboratory for High Energy Physics beam line 4B using Si (111) monochrometor. Fluorescence X-ray was detected with a modified silicon (lithium) solid-state detector placed 10 mm from the sample surface. Sample was kept 30 K during the measurement of spectrum, which was an average of seven to ten scans. Energy was calibrated by the sharp pre-edge peak of the $KMnO_4$ spectrum at 6543.3 eV.

3. Results and Discussion

Figure 1 shows the effects of Ca^{2+} and/or Cl^- depletion on TL glow curves measured in the presence of DCMU. Upon single flash illumination, the Ca^{2+}/Cl^--depleted (low pH/NaCl/Cl^--free washed) membranes showed TL-band peaking at 38 °C. The TL peak temperature further upshifted to 41 °C by the addition of Ca^{2+}. These upshifted peak temperatures are quite similar to that reported in PS II of which Cl^- is replaced by SO_4^{2-}, suggesting that Cl^- depletion by this protocol leads a similar type of modification in the Mn-cluster to that induced by SO_4^{2-} substitution. The elevated peak temperature in Ca^{2+}/Cl^--depleted OEC was downshifted to 10 °C by the addition of Cl^- only, but did not change much any more by further addition of Ca^{2+}. The resulted peak temperature after addition of both Ca^{2+} and Cl^- is almost the same as found in the TL-band arising from $S_2Q_A^-$ charge recombination in untreated control OEC. The result indicates that Cl^- depletion causes the changes in the redox properties of the Mn-cluster irrespective of the presence of Ca^{2+}. This may imply that Ca^{2+} and Cl^- function rather independently each other. In the Cl^--depleted (NaCl/Cl^--free washed) membranes, a

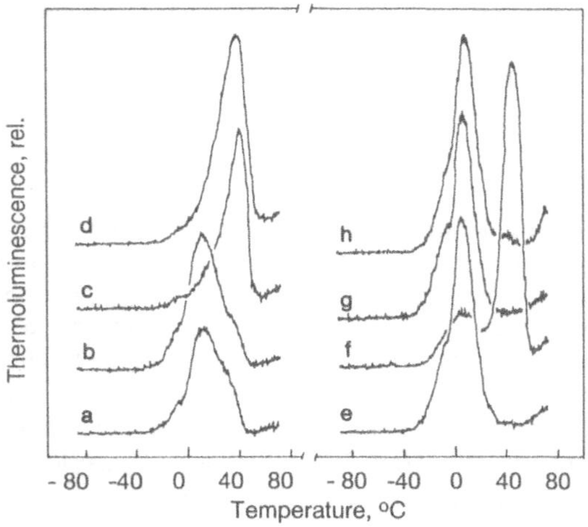

Figure 1. Effects of Ca^{2+} and/or Cl^- depletion on TL glow curves. Sample membranes (0.35 mg Chl/ml) were illuminated with a flash at 5 °C in the presence of 10 µM DCMU. Low pH / NaCl /Cl^-- free washed membranes with no addition (d), 20 mM Ca^{2+} (c), 20 mM Cl^- (b) and 20 mM Ca^{2+} plus 20 mM Cl^- (a). Untreated control membranes (e), low pH treated membranes (f), 2M NaCl washed (g) and low pH/NaCl washed membranes (h) with 20 mM Cl^-.

flash excitation induced a TL-band peaking at 42 °C (data not shown) in consistent with the result in the low pH/NaCl/Cl^--free washed membranes. In contrast, the Ca^{2+}-depleted (low pH treated) membranes showed a upshifted TL-band at 43 °C presumably due to some modification of the redox properties of the Mn-cluster by the binding of the 24 kDa extrinsic protein [7].

Effects of Ca^{2+} or Cl^- depletion on the S-state transition and the structure and/or electronic states of the Mn-cluster have been studied by means of XANES measurements [10,11]. Figure 2 summarizes the flash dependent changes of the Mn K-edge energy in the untreated control, Ca^{2+}-depleted (low pH treated) and Cl^--depleted (NaCl/Cl^--free washed) membranes. The flash number dependent period-four-change in the Mn K-edge found in the control OEC was not observed any more in the Ca^{2+} or Cl^--depleted OEC. It is of note that the manifestation of the depletion is considerably different between Ca^{2+} and Cl^-. The half-height energy of the Ca^{2+} -depleted S_1-state was lower by 1.5-1.7 eV than that of the normal S_1-state, and upshifted by about 1eV after the first flash followed by further upshift after subsequent flashes, and then reached the energy that was higher by 0.9 eV than that after the first flash. The marked downshift of the K-edge energy can be interpreted to mean that the carboxylate ligand is liberated from the Mn-cluster by the depletion of Ca^{2+} [12]. In the Cl^--depleted membranes, the K-edge energy in the S_1-state was almost the same as that of the normal S_1-state and upshifted after the first

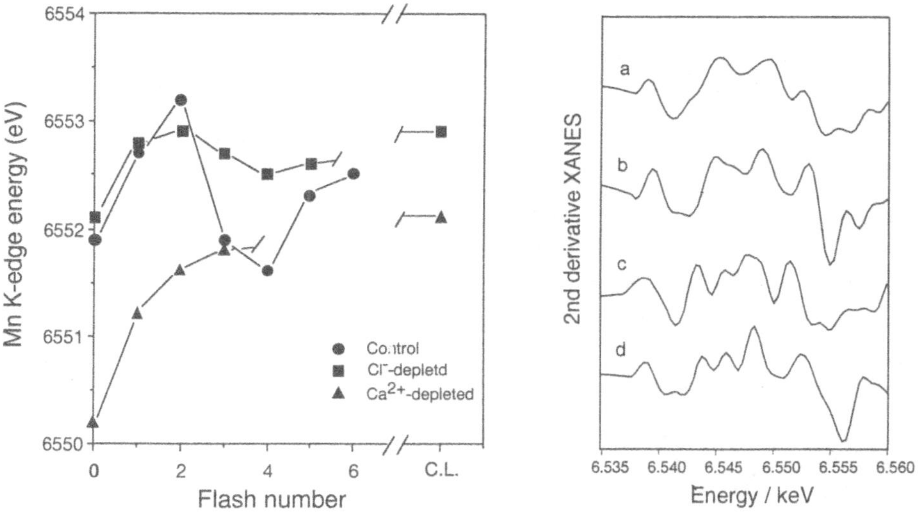

Figure 2. Changes of half-height energy of Mn K-edge after a series of flashes and continuous illumination (C.L.) in untreated control (●), Ca^{2+}-depleted (▲) and Cl^--depleted (■) PS II membranes.

Figure 3. Second derivatives of Mn K-edge spectra. Untreated PS II membranes after 0 (a) and 1 (b) flashes, and Cl^--depleted PS II membranes after 0 (c) and 1 (d) flashes.

flash, but did not undergo significant change after subsequent flashes. The result may suggest that much smaller structural modification is induced by Cl^- depletion than by Ca^{2+} depletion.

The possible structural changes of the Mn-cluster induced by Cl^--depletion could be emphasized by the second derivatives of the K-edge spectra as shown in figure 3. The overall features of the second derivative of the spectrum of the S_1- and S_2-states in the control OEC was in good agreement with that reported [13,14] even though sample concentration was very low in this study. The characteristic profiles found in the control S_1-state spectra were also preserved in the Cl^--depleted S_1 spectra, but some pronounced changes were induced by Cl^- depletion. We may, therefore, conclude that the structure of the Mn-cluster is subtly modified by Cl^- depletion. A flash illumination induced similar changes in the spectra both in the control and the Cl^--depleted OEC. It has been indicated that the oxidation of the Mn atom from Mn (III) to Mn (IV) upon S_1-to-S_2 transition is responsible for the change in the profiles of the derivative spectra. This implies that the Mn atom is oxidized from Mn (III) to Mn (IV) after one flash in Cl^--depleted OEC in consistent with the upshift of the K-edge energy after one flash.

Acknowledgements: This research was supported by STA (Photosynthesis Science and MR science) and MESC (Nos., 07839021, 07308050).

References

1 Joliot, P., Barbieri, G. and Chaud, R. (1969) Photochem. Photobiol. 10, 309-329
2 Kok, B., Forbush, B., McGloin, M. (1970) Photochem. Photobiol. 11, 457-475
3 Debus, R. J. (1992) Biochim. Biophys. Acta 1102, 269-352
4 Boussac, A. and Rutherford, A. W. (1994) J. Biol. Chem. 269, 12462-12467
5 Vliet, van P., Boussac, A. and Rutherford, A. W. (1994) Biochemistry 33, 12998-13004
6 Ono, T. and Inoue, Y. (1988) FEBS Lett. 227, 147-152
7 Ono, T. and Inoue, Y. (1992) Biochemistry 31, 5953-5956
8 Ono, T., Noguchi, T., Inoue, Y., Kusunoki, M., Matsushita, T. and Oyanagi, H. (1992) Science 258, 1335-1337
9 Ono, T., Noguchi, T., Inoue, Y., Kusunoki, M., Yamaguchi, H. and Oyanagi, H. (1994) Biochemical Society Transaction 22, 331-335
10 Ono, T., Noguchi, T., Inoue, Y., Kusunoki, M., Yamaguchi, H. and Oyanagi, H. (1993) FEBS Lett. 330, 28-30
11 Ono, T., Noguchi, T., Inoue, Y., Kusunoki, M., Yamaguchi, H. and Oyanagi, H. (1995) Journal of American Chemical Society 117, 239-248
12 Noguchi, T., Ono, T. and Inoue, Y. (1995) Biochim. Biophys. Acta 1228, 189-200
13 Yachandra, V. K., DeRose, V. J., Latimer, M. J., Mukerji, I., Sauer, K. and Klein, M. P. (1993) Science 260, 675-679
14 DeRose, V. J., Latimer, M. J., Zimmermann, J.-L, Mukerji, I., Yachandra, V. K., Sauer, K. and Klein, M. P. (1995) Chemical Physics 194, 443-459

EPR/ENDOR STUDIES OF THE WATER OXIDIZING COMPLEX IN PHOTOSYSTEM II

R. Fiege[1], W. Zweygart[1], K.-D. Irrgang[1], N. Adir[2], B. Geiken[1], G. Renger[1] and W. Lubitz[1]
[1]Max-Volmer-Institut für Biophysikalische und Physikalische Chemie, Technische Universität Berlin, Str. d. 17. Juni 135, D-10623 Berlin, Germany, [2] Department of Physics, 0319, University of California, San Diego, La Jolla, CA. 92093-0319, USA

1. Introduction

Photosynthetic water oxidation to dioxygen takes place via a sequence of four univalent redox steps at a manganese containing complex (see [1] and references therein). Among the different intermediary redox states S_i of the system only S_2 is fairly well characterized by EPR spectroscopy. Several changes of the structural and/or functional integrity induced by specific treatments (Ca^{2+} depletion, amine addition) give rise to characteristic changes of the S_2 signals [2]. In the present study EPR spectra have been measured to analyze possible effects of the extrinsic regulatory proteins on the water oxidizing complex (WOC) in redox state S_2. Another problem of mechanistic relevance is the binding of substrate water. In principle ^1H ENDOR spectroscopy should provide information on the interaction of H_2O with the EPR detectable functional manganese center. So far only few studies are reported in the literature [3-5]. As an extension of our earlier ENDOR work on PS II membrane fragments [3] results are presented here for PS II core complexes.

2. Material and Methods

PS II membrane fragments were isolated according to the procedure of [6] with modifications as described in [7]. The 33 kDa protein was stepwise extracted by salt washing as in [8] and concentrated via Centripep centrifugation (Amicon). SDS/urea/PAGE analysis showed that the purity of the protein was > 95% [8]. Light saturated rates of oxygen evolution and the oxygen yields per flash have been measured with a Clark-electrode in suspensions containing the sample , buffer A (50 mM MES-NaOH, pH 6.5, 15 mM NaCl, 5 mM $MgCl_2$, 400 mM sucrose) and either $K_3[Fe(CN)_6]$/PPBQ (1 mM/0.2 mM final conc.) or 2,6 DCBQ (final conc. 0.5 mM) as electron acceptors [9]. PS II membrane fragments revealed oxygen evolution rates of 400±50 μmol O_2 /mg Chl h. Peroxide production has been detected photometrically as reported by Geiken [10]. The manganese content was determined by atomic absorption spectroscopy at $\lambda = 279.5$ nm using a graphite furnace technique with $MnCl_2$ as a standard. PS II core complexes have been purified as described in [11]. The EPR spectra were monitored on a Bruker ESP 300E and on a Varian E-9 spectrometer using a standard Bruker TE_{102} cavity with an Oxford ESR 9 cryostat. For the ENDOR measurements a high sensitivity TM_{110} cavity was developed [12] attached to an Oxford ESR 910 cryostat.

P. Mathis (ed.), Photosynthesis: from Light to Biosphere, Vol. II, 369–372.

3. EPR Results and Discussion

The EPR spectra of PS II membrane fragments exhibit both the $g = 4.1$ and the multiline signal (MLS). Continuous illumination in the frozen state (Fig. 1 top) leads to the same spectrum as a single light flash at 0°C followed by rapid freezing (Fig. 1 center). Both methods limit PS II to a single turnover.

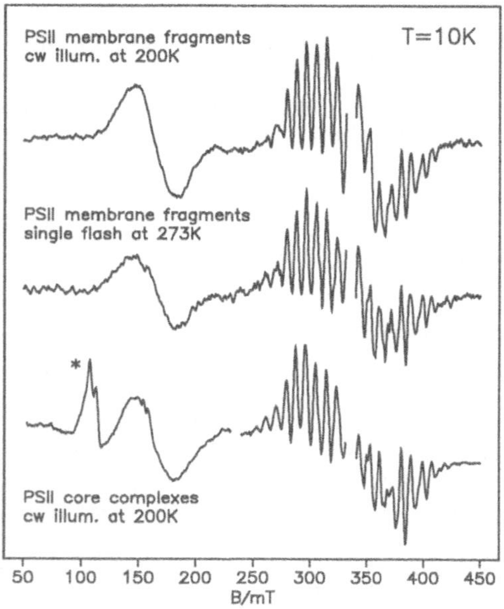

Figure 1: Light minus dark difference spectra measured in suspensions of either PS II membrane fragments or PS II core complexes in buffer A containing 400 mM mannitol instead of sucrose (buffer B)
exp. cond. : mw frequency = 9.44 GHz, mw power = 10.0 mW, field mod.: = 2.0 mT_{pp}

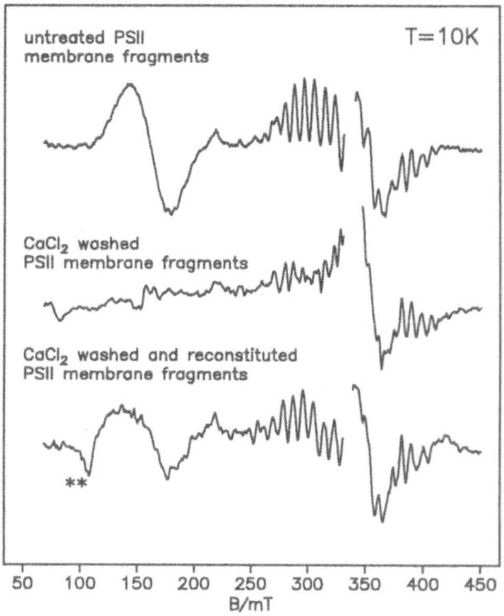

Figure 2: Light minus dark difference spectra of untreated, $CaCl_2$ washed and reconstituted PS II membrane fragments suspended in buffer B
exp. cond. as in Fig. 1

* cyt b-559 high spin signal
** artifact of subtraction

In PS II core complexes lacking the 17 and 24 kDa proteins virtually the same signals are observed as in PS II membrane fragments. Therefore, the 17 and 24 kDa proteins are inferred neither to provide ligands to the Mn cluster nor to affect its structure.

Markedly different features are observed after removal of all extrinsic proteins from PS II membrane fragments. In these samples the oxygen evolution activity is drastically decreased (about 10% compared with untreated control) accompanied by a significant increase of peroxide production from 5 (control) to 40 μmol H_2O_2 / mg Chl h. Likewise, the EPR spectra exhibit different patterns. The dark spectrum (not shown) reveals that part of the manganese becomes paramagnetically decoupled and $Mn^{2+}(H_2O)_6$ is formed. This decoupled manganese appears to remain rather weakly bound to the protein because only two manganese are detected after a further washing step that is required for atomic absorption measurements. Upon illumination of salt washed samples that retain the weakly bound manganese at 200K a multiline like signal is observed (Fig. 2 center), while the $g = 4.1$ signal is not detected in this preparation (it remains to be analyzed if the latter effect is a general feature of this sample type). In comparison with the untreated control (Fig. 2 top) the total amplitude of the multiline signal is diminished to about 50% and the line spacing is reduced from 8 - 9 mT to 6 - 7 mT. Reconstitution with the purified 33 kDa protein in the presence of Ca^{2+} led to partial recovery of the oxygen evolution rate (up to 50 % of the control) and concomitant decrease of H_2O_2 formation (<20 μmol H_2O_2 / mg Chl h). In these reconstituted samples an S_2 state MLS is detected (Fig. 2 bottom) which closely resembles that of the control. It has to be emphasized that this recovery can be observed only in the presence of both, the 33 kDa protein and Ca^{2+}. The data support the idea that the S_2 multiline signal originates from a binuclear cluster. This is in line with a recent proposal based multifrequency EPR simulations of the MLS spectrum [13].

4. ENDOR Results and Discussion

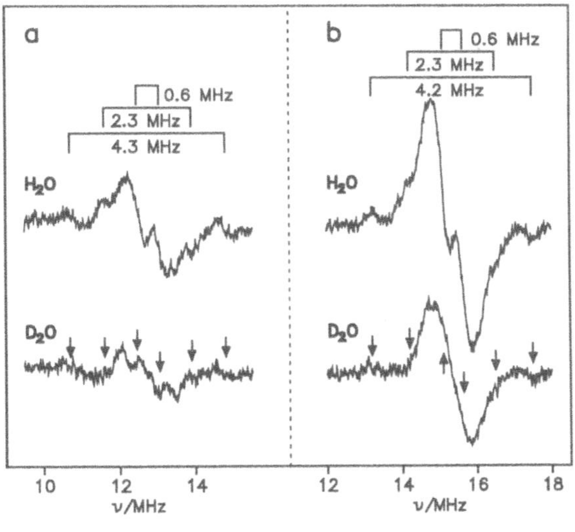

Figure 3: Spectra of S_2 - S_1 ENDOR at different field positions of PS II core complexes suspended either in H_2O or D_2O. The amplitude of the signals is not scaled, thus the signal amplitudes can directly be compared. The arrows (bottom) indicate the positions of lines changed by the D_2O treatment.
experimental conditions: mw frequency 9.5 GHz, mw power 20 mW, rf power 150 W, frequency mod. 12.5 kHz with ± 100 kHz dev, T=2 K;
a magn. field 298 mT at $g = 2.28$,
b magn. field 359 mT at $g = 1.89$

^{1}H-ENDOR studies have been performed with PS II core complexes at two different field positions. To check for non-specific signals, ENDOR spectra were measured in samples where the WOC is kept in the EPR silent S_1 state. The spectra (not shown) exhibit 3 to 4 hyperfine couplings (Hfc's), the largest one at 3.1 - 3.3 MHz. These signals that occur in the same spectral region as the S_2-MLS are not affected by D_2O exchange and are very likely due to EPR signals not associated with the redox states S_i of the WOC. Upon generation of the S_2 state by cw illumination at 200K new Hfc's appear. The S_2 *minus* S_1 difference spectra exhibit Hfc's of 0.6 MHz, 2.3 MHz and 4.2 - 4.3 MHz. These ENDOR signals have been assigned to the S_2 state. These three hyperfine components were also found in *Synechococcus elongatus* [4]. This finding suggests that the proton environment is very similar in the S_2 state of both organisms. Our results also agree with the ^1H hyperfine data reported in [3] and [5] for PS II membrane fragments (spinach). Therefore it seems reasonable to conclude that the proton environment is not essentially altered in different samples with a fully intact WOC. Larger Hfc's could not be detected probably due to the low signal-to-noise ratio and to the smaller intensity of the large couplings. Within the framework of the dipole-dipole model the Hfc's can be used to estimate the distance between the electron spin of the Mn cluster and surrounding protons. With the data so far available this model leads to the conclusion that protons from substrate water or exchangeable protons from the protein matrix are at a distance larger than 2.7 Å apart from the electron spin of the coupled manganese.

Acknowledgement

Financial support by Deutsche Forschungsgemeinschaft: "Schwerpunktprogramm Bioanorganische Chemie" is gratefully acknowledged, work at UCSD was supported by a grant from USDA (93-37306-9579).

References

1 Renger, G. (1993) Photosynth. Research 38, 229–247
2 Debus, R.J. (1992) Biophys. Biochim. Acta 1102, 269–352
3 Zweygart, W., Weyhermüller, T., Renger, G., Wieghardt, K., and Lubitz, W. (1992) in Research in Photosynthesis (Murata, N., ed.) pp. 289–292, Kluwer Academic Publishers, Dordrecht, The Netherlands
4 Tang, X.-S., Sivaraja, M., and Dismukes, G.C. (1993) J. Am. Chem. Soc. 115, 2382–2389
5 Kawamori, A., Inui, T., Ono, T., and Inoue, Y. (1989) FEBS Lett. 254, 219–224
6 Berthold, D.A., Babcock, G.T., and Yocum, C.F. (1981) FEBS Lett. 134, 231–234
7 Völker, M., Ono, T., Inoue, Y., and Renger, G. (1985) Biochim. Biophys. Acta 806, 25–34
8 Irrgang, K.-D., Boekema, E. J., Vater, J., and Renger, G. (1988) Eur. J. Biochem. 178, 178–209
9 Haag, E., Irrgang, K.-D., Boekema, E. J., and Renger, G. (1990) Eur. J. Biochem. 189, 47–53
10 Geiken, B., PhD thesis, Technische Universität, Berlin, 1993
11 Adir, N., Okamura, M.Y., and Feher, G. (1992) in Research in Photosynthesis (Murata, N., ed.) pp. 195–198, Kluwer Academic Publishers, Dordrecht, The Netherlands
12 Zweygart, W., Thanner, R., and Lubitz, W. (1994) J. Magn. Res., Series A 109, 172–176
13 Åhrling, K.A. and Pace, R.J. (1995) Biophys. J. 68, 2081–2090

ILLUMINATION RESPONSE OF THE S_{-1} STATE INVESTIGATED BY X-RAY ABSORPTION SPECTROSCOPY

P.DeMarois[1], P.Riggs-Gelasco[2], J. Penner-Hahn[2], C.Yocum[2,3]
[1]Biophysics Research Division, [2]Department of Chemistry, [3]Department of Biology, University of Michigan, Ann Arbor, MI 48109, USA

1. Introduction

X-ray absorption spectroscopy (XAS) monitors oxidation state changes in Mn in the O_2 evolving complex (OEC) of PSII; oxidation has been demonstrated by shifts to increased K-edge energies for $S_1 \rightarrow S_2 \rightarrow S_3$ and a shift to lower edge energy, indicating Mn reduction, for $S_3 \rightarrow [S_4] \rightarrow S_0$ (1). Dark-adapted samples treated with low concentrations of NH_2OH show a two flash delay in O_2 evolution, which corresponds to the S_{-1} state and to a 2-electron reduction of the Mn complex. However, it has been reported that illumination is required for S_{-1} formation (3). We have found that, using Reaction Center Core complexes, NH_2OH treatment in the dark gives K-edge energy lower than that of S_1 consistant with formation of S_{-1}(2); the EXAFS spectrum is unaffected and little O_2 evolution activity is lost. The corresponding XANES spectra for PSII samples reduced by NH_2OH, NH_2NH_2 and hydroquinone (H_2Q) also show reduction in the dark. H_2Q treated samples, have previously been shown by XANES and EPR to contain 2 Mn (II)/reaction center (2,4). The present data addresses the effect of low temperature illumination on these reduced samples.

2. Procedure

2.1 *Sample Preparation*

The PSII reaction center complex preparation lacking 17 and 23 kDa extrinsic polypeptides and LHC was used in all experiments. Reductant incubations were carried out in darkness; all samples contained 10 mM $CaCl_2$. The reductants and times of incubation were NH_2OH (100 µM, 3min.), NH_2NH_2 (200 µM, 35 min) and H_2Q (200 µM, 30 min.). Reduction was terminated by dilution/washing or with $Fe(CN)_6^{3-}$ (for H_2Q). EDTA sensitivity was confirmed for the NH_2OH treated samples. Centrifuged samples were immediately packed into lucite cells and stored in the dark at 77K. Illuminations (20 min.) at 200K in ethanol/dry ice bath in an optical dewar utilized a 300 watt light source.

P. Mathis (ed.), Photosynthesis: from Light to Biosphere, Vol. II, 373–376.

2.2 *Measurements*

X-ray absorption data were collected at 10K at the Stanford Synchrotron Radiation Laboratory (SSRL) beamline 7-3 using a Si(220) double crystal monochromator. Incident intensity was detuned by 50% and Mn Kα fluorescence was monitored by a Canberra 13-element Ge solid state flourescence detector. The data were normalized as previously reported and energy values determined as the first inflection point of K-edge.

3. Results and Discussion

Previous reports have discussed the difficulty in assigning oxidation states to Mn (III) and Mn(IV) complexes from XANES spectra (5). Edge structure is dependent on coordination number, bond length and Mn oxidation state. The energy shifts between the edge spectra shown in Fig. 1 were estimated using the first inflection point of the Mn K-edge. The values for the NH_2OH-induced S_{-1} and S_0 states are 6550.9 eV and 6551.6 eV respectively. The untreated S_1 and S_2 states reported here have K-edge values of 6552.1 and 6553.0 eV respectively, giving shifts of 0.7 eV ($S_{-1} \rightarrow S_0$), 0.5 eV ($S_0 \rightarrow S_1$) and 0.9 eV ($S_1 \rightarrow S_2$).

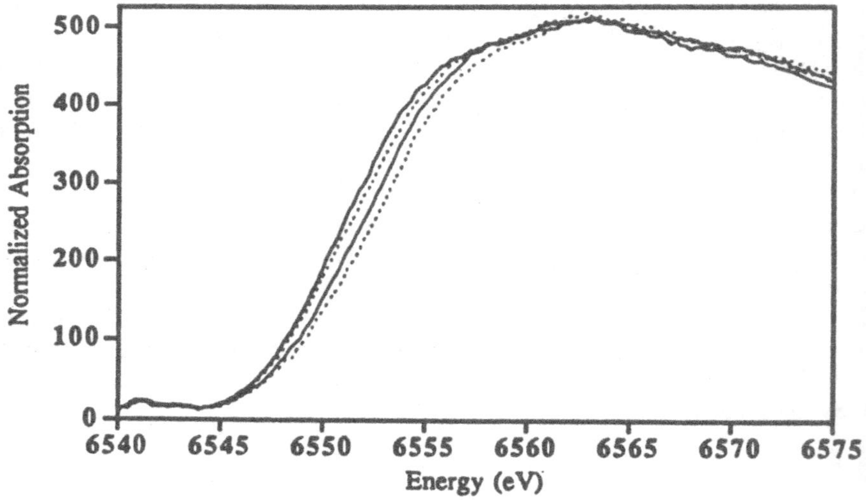

Figure 1. XANES spectra of the NH_2OH treated OEC. From lowest to highest edge energy: a) NH_2OH, dark, b) NH_2OH, 20 minute illumination at 200K. c) Untreated control, dark. d) Control, 20 minute illumination at 200K. Spectra are normalized to a constant edge jump.

This is the first evidence that the NH_2OH-reduced complex undergoes Mn oxidation upon low temperature illumination, and contrasts with the report that the combination of NH_2OH and low-temperature illumination reduces Mn (3). The NH_2OH treated samples retain 75% of the control activity which, in conjunction with the complete photoreversibility of the XANES changes, provide evidence that reduced Mn centers are retained at native binding sites. Photo oxidized NH_2OH treated samples show no detectable S_2 EPR multiline signal, thus demonstrating that the shift in Fig 1 cannot arise from unreduced centers advancing to the S_2 state

The corresponding spectra for OEC derivatives reduced by H_2Q or NH_2NH_2 are shown in Fig. 2. Hydrazine treatment results in an edge shift similar to that observed with NH_2OH, and similarly shows evidence for oxidation after 200K illumination. The K-edge energy values for the NH_2NH_2 treated samples are 6549.6 eV and 6550.6 eV for the dark and illuminated samples respectively, giving a 1.0 eV shift upon illumination As reported previously, H_2Q-treated samples show dramatically different edges, consistent with formation of ca. 50% Mn (II)(2). Hydroquinone reduction is reversed by continuous illumination at 4°C followed by dark adaptation. However, in contrast to the other reduced samples, the H_2Q reduced species does not show any oxidation under continuous illumination at 200K, as shown in Fig. 2.

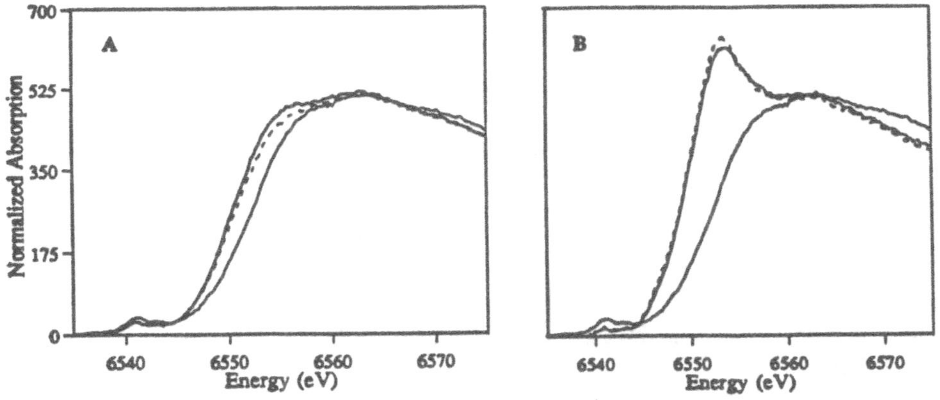

Figure 2: XANES spectra of a) NH_2NH_2 treated samples. Spectra are reduced, reduced and illuminated 20 minutes at 200K, and control S_1. b) H_2Q treated samples, reduced, illuminated, and control.

The edge advances detected after low temperature illumination of NH_2OH- and NH_2NH_2-reduced samples suggest that the reduced Mn in these samples occupies native binding sites in a ligand environment that permits oxidation by 200 K illumination. In contrast, the reduced Mn in H_2Q treated samples cannot be oxidized by illumination at 200 K. This is consistant with previous suggestions (2,4-6) that the reduced Mn in H_2Q treated samples is present in a non-native configuration. Previous studies showed that H_2Q and NH_2OH exhibit a synergestic inhibition of O_2 evolution (4), and we have suggested that this is due to the existence of 2 separate reduction sites. The present data provide further support for the presence of different reduced species following NH_2OH or NH_2NH_2 and H_2Q treatment.

The results presented here provide new insights into the nature of Mn reduction. First, our data support the conclusion that no bound reductant is associated with S_{-1} (7), since edge advances are clearly detected for samples reduced with either NH_2OH or NH_2NH_2. Second, the likely first target for NH_2OH and NH_2NH_2 reduction are the Mn(IV) atoms present in S_1. There is general agreement that the Mn oxidation states for S_1 are IV, IV, III, III; we assign oxidation states of III, III, III, III to S_{-1}, so the Mn(III) in this state must arise from reduction of Mn (IV). H_2Q treatment produces an S_{-1} state (IV, IV, II, II) as well, but failure to observe an increase in edge energy after 200 K illumination shows that the state of Mn ligation is substantially different from that observed in the other reduced derivatives of the OEC. Our results support the following model for reductant-induced conversions among the lower S-states in PSII:

$$\underset{(S_1)}{\text{Mn [IV, IV, III, III]}} + 2\ NH_2OH \text{ or } NH_2NH_2 \rightarrow \underset{(S_{-1})}{\text{Mn (III, III, III, III)}} + N_2$$

$$\underset{(S_1)}{\text{Mn[IV, IV, III, III]}} + H_2Q \rightarrow \underset{(S_{-1})}{\text{Mn[IV, IV, II, II]}} + Q$$

4. Addendum

This research was supported by grants from NCGRIP (USDA) to CFY and NIH (JEPH).

References

1 Ono, T., Noguchi, T., Inoue, Y., Kusunoki, M., Matsushita, T., and Oyanagi, H. (1992) Science, 258, 1335-1337
2 Riggs, P., Mei, R., Penner-Hahn, J., and Yocum, C. (1992) J. Am. Chem. Soc. 114, 10650-10651
3 Guiles, R., Yachandra, V., McDermott, A., Cole, J., Dexheimer, S., Britt, R., Sauer, K. and Klein, M. (1990) Biochemistry 29, 486-496
4 Mei, R. and Yocum, C. (1992) Biochemistry 31, 8449-8454
5 Riggs-Gelasco, P., Mei, R., Penner-Hahn, J. in "Mechanistic Bioinorganic Chemistry," (in press)
6 Riggs-Gelasco, P. (1995) Ph.D. Thesis, University of Michigan, Ann Arbor, MI
7 Kretschmann, H., Pauly, S. and Witt, H.T. (1991) Biochim. Biophys. Acta 1059, 208-214

A MODELLING APPROACH FOR UNDERSTANDING WATER OXIDATION IN PHOTOSYNTHESIS

Vincent L. Pecoraro*, Michael J. Baldwin, M. Tyler Caudle, Neil A. Law, Jeff W. Kampf, and Paul G. Rasmussen, Chemistry Department, The University of Michigan, Ann Arbor, MI, USA 48109-1055

The reducing equivalents required in photosynthesis are supplied by the oxygen evolving complex (OEC)[1] that catalyzes the four-electron oxidation of two water molecules to dioxygen and four protons, as shown in equation 1.

$$2\,H_2O \longrightarrow O_2 + 4\,H^+ + 4\,e^- \qquad (1)$$

The OEC contains a tetranuclear manganese cluster that is responsible for water oxidation that occurs as soon as the enzyme has accumulated four oxidizing equivalents. While the precise manganese ligation has not been deduced, there is evidence for μ_2-O^{2-} bridges and imidazole moieties as Mn ligands.[2]

The kinetic scheme for this reaction has been described by Kok's S-state cycle in which the states S_0 to S_4 represent progressively more oxidized forms of the OEC. There is some controversy regarding the nature of the site of oxidation in the highest S-state transitions, due in part to the magnitude of the shift in Mn X-ray absorption edge energy upon the S_2 to S_3 transition,[3] the observation of non-manganese associated epr signatures in S_3*[4] and the very high oxidation state required for S_4. We are currently using modelling chemistry to address three different possibilities for the final distribution of oxidation equivalents in S_4. These possiblities include (1) a Mn(IV)$_3$Mn(V) cluster, (2) a Mn(IV)$_4$ cluster with oxidation of a nearby protein residue (including possibly Y_Z) or (3) a Mn(IV)$_4$ cluster with oxidation of a substrate molecule (i.e., water). Imidazole has been suggested[4] to be oxidized in S_3* and has been implicated as a Mn ligand from pulsed epr experiments.[2] Furthermore, the possibility that histidine acts as a bridge between metals has been considered. In the first half of this contribution we will consider mechanisms for triggering redox activity from Mn clusters using point 2 above. The second portion of this work will examine the properties of models that contain coordinated imidazole.

As the EXAFS data on the OEC indicate the presence of bis(μ-oxo) Mn dimers in the OEC,[5] our model studies have concentrated primarily on the redox reactions of dimeric Mn units. Recently we have evaluated the role of protons as mediators of the physical and chemical properties of bis(μ-oxo) Mn dimers using the complexes [Mn(X-Salpn)(μ_2-O)]$_2$ where X-Salpn= 1,3-bis(X-salicyliminato)propane with X= H, Cl, OMe

P. Mathis (ed.), Photosynthesis: from Light to Biosphere, Vol. II, 377–380.

or NO_2.[6] From these studies we concluded that protonation of oxo bridges in manganese cluster can profoundly perturb structure and, more importantly, the reactivity. In particular, we have noted that the $Mn(IV)_2$ to Mn(III)Mn(IV) reduction potential could be shifted more positive by 500 mV per protonation of the dimer. As it is well known that proteins can masterfully create environments that can fine tune the magnitude of hydrogen bonds and orchestrate complex proton relay systems over tens of Ångstroms, we suggested[6] that the trigger for water oxidation could be proton transfer as illustrated in figure 1. In this model Y_z or another residue could donate a proton to the Mn cluster causing a shift of the center to more oxidizing potentials and subsequently induce or "trigger" the production of dioxygen.

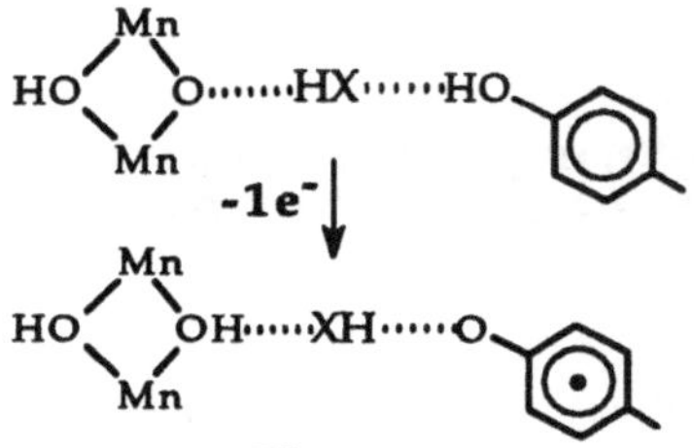

Figure 1

We have observed that the addition of two protons to any of the $[Mn(IV)(X\text{-}Salpn)(\mu_2\text{-}O)]_2$ complexes in dichloromethane, acetonitrile or trimethylacetonitrile leads nearly quantitatively to the generation of $Mn(III)(X\text{-}Salpn)(H_2O)_2^+$. This is an exciting reaction as it illustrates how protonation of oxo bridges in a high valent manganese dimer leads to oxidative reactivity. Furthermore, the presence of small amounts of water is essential for the complex reduction. Mechanisitically we know that the first steps of the reaction proceed through protonation of the dimer forming the purple $[Mn(IV)(X\text{-}Salpn)]_2(\mu_2\text{-}O)(\mu_2\text{-}OH)^+$ followed by the green $[Mn(IV)(X\text{-}Salpn)(\mu_2\text{-}OH)]_2^{2+}$. We also know that the product of the reaction is $Mn(III)(X\text{-}Salpn)(H_2O)_2^+$. However, the identity of the reductant remains mysterious. The fact that $[Mn(IV)(Salpn)(\mu_2\text{-}OH)]_2^{2+}$ reduces to $Mn(III)(Salpn)(H_2O)_2^+$ in the solid state and that identical chemistry is seen in the oxidatively resistant solvent trimethylacetonitrile supports the contention that solvent is not the source of reducing equivalents. We are presently evaluating whether a small amount of the ligand is being oxidized or, as originally suggested by Boucher and Coe,[8] that H_2O_2 is the product. If peroxide is being generated, then each of the steps in figure 2 is precedented based on model chemistry from our laboratory.

In an effort to understand this reactivity better and also to define the properties of $[Mn(IV)(X\text{-}Salpn)]_2(\mu_2\text{-}O)(\mu_2\text{-}OH)^+$ more fully, we evaluated whether the oxo bridges of these compounds are sufficiently nucleophilic to be alkylated. The reaction of methyl triflate with $[Mn(IV)(Salpn)(\mu_2\text{-}O)]_2$ succesfully leads to the isolation of $[Mn(IV)(Salpn)]_2(\mu_2\text{-}O)(\mu_2\text{-}OCH_3)^+$ demonstrating that oxo bridges in manganese clusters can react as nucleophiles. Mass spectral analysis of $[Mn(IV)(Salpn)]_2(\mu_2\text{-}O)(\mu_2\text{-}OCH_3)^+$ confirms that the starting dimer was alkylated only once and the similarity in UV-vis spectra, magnetic coupling, and Mn-Mn distance (from EXAFS) with that of $[Mn(IV)(Salpn)]_2(\mu_2\text{-}O)(\mu_2\text{-}OH)^+$ assure us that the methyl transfer occurred at the oxo bridge. The enhanced solubility of methylated derivative has allowed us to probe the redox behavior of this molecule and confirms the previous observation that substantial shifts towards oxidizing potentials result when positive charge is placed on the bridging oxo ligand. The addition of one equivalent of proton to $[Mn(IV)(Salpn)]_2(\mu_2\text{-}O)(\mu_2\text{-}OCH_3)^+$ first yields the green $[Mn(IV)(Salpn)]_2(\mu_2\text{-}OH)(\mu_2\text{-}OCH_3)^{2+}$ which subsequently decomposes to $Mn(III)(Salpn)(H_2O)_2^+$. Thus, while both oxo bridges must carry positive charge for the reaction to occur, at most only one dissociable proton site is required. Stopped flow kinetic studies of this proton triggered redox reaction demonstrate that there is a transient intermediate that forms prior to the product Mn(III) species. We believe that we have isolated this material since the decomposition of this intermediate is slow using the alkylated derivative. The identity of this compound may provide important information that will allow us to identify the species being oxidized in this system.

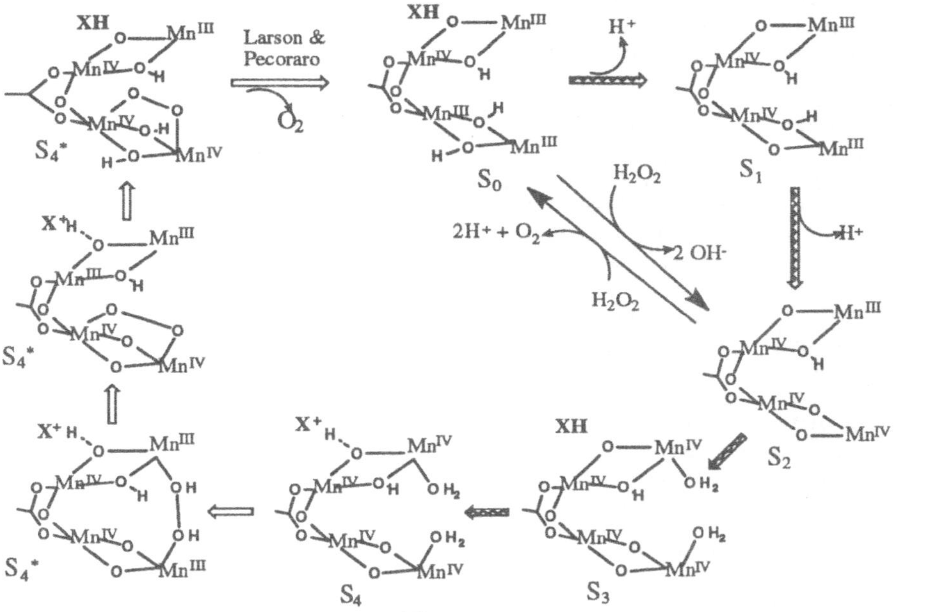

Figure 2

Recent pulsed ENDOR and ESEEM studies[2] point to the liklihood of imidazole as a ligand to the manganese cluster in the OEC and suggest the possibility that imidazole might act as a bridge between manganese ions through both nitrogens in the manner that histidine behaves in the CuZn superoxide dismutase.[9] The number of Mn(III) and Mn(IV) complexes that have coordinated imidazole is small and there are none that employ an imidazole group as a bridge between metals. We felt that the preparation of higher-valent manganese complexes with N-bound and N,N'-bridging imidazole would be a prudent strategy to evaluate this structural proposition. However, the strong oxophilicity of manganese, especially in the +3 to +5 oxidation levels, makes this a difficult objective. To achieve this goal we have utilized the bidentate and bis-bidentate chelating agents, H_2DCBI and H_2(N-Me)DCBI, illustrated in figure 3. The N-methylated ligand 4,5-dicarboxy-N-methylimidazole (N-MeDCBIH_2) can serve as a bidentate chelate using the carboxylate moiety and the imidazole nitrogen. Reaction of this ligand with Mn(III)(Salpn)(acetonylacetonate) affords the Mn(III)(Salpn)(H(N-Me)DCBI) complex that has the structure illustrated in figure 4. There is an

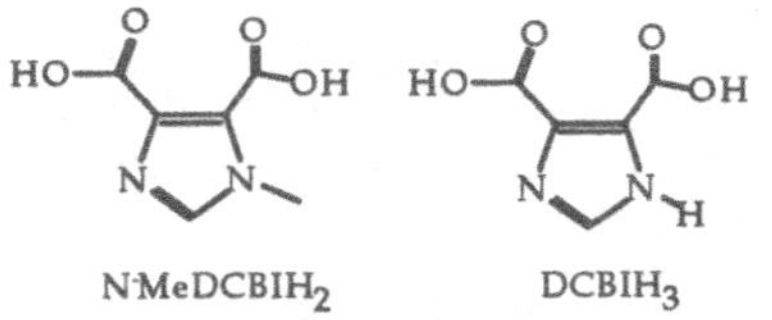

Figure 3

exceptionally long axis along the imine nitrogen-Mn(III)-imidazole nitrogen axis, although the Mn-O(carboxylate) is also considerably longer than a typical Mn-carboxylate bond so that it is unclear whether there is a definite Jahn-Teller axis.

Cyclic voltammetry indicates that the complex can be reversibly oxidized at +890 mV vs. SCE in CH_3CN as compared to Mn(III)(Salpn)(acetonylacetonate) that shows an oxidation at +520 mV. This 370 mV shift to more positive potential for the oxidation of the Mn(III)(Salpn)(H(N-Me)DCBI) species illustrates the destabilization of high valent manganese by nitrogen donors. Nevertheless, the addition of thianthryl perchlorate to Mn(III)(Salpn)(N-MeDCBI) yields the one electron oxidized species $Mn(IV)(Salpn)(H(N-Me)DCBI)^+$. The EPR spectrum exhibits the low field signal split by manganese hyperfine that is the hallmark of the Mn(IV) (S=3/2) systems.

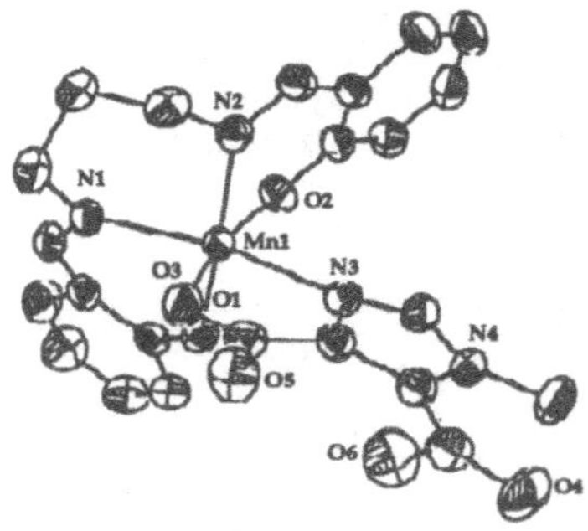

Figure 4

The addition of the unalkylated H_3DCBI to Mn(III)(Salpn) (acetonylacetonate) allows us to easily prepare $Mn(III)(Salpn)(H_2DCBI)$. However, this material is extremely insoluble in most solvents and is difficult to characterize in solution. However, it is soluble in pyridine and addition of a second equivalent of Mn(III)(Salpn)(acetonylacetonate) gives the imidazolate bridged dimer $(Salpn)Mn(III)(DCBI)Mn(III)(Salpn)^-$, as shown by mass spectral data. The most likely structure for the imidizole-bridged dimer is illustrated in figure 5.

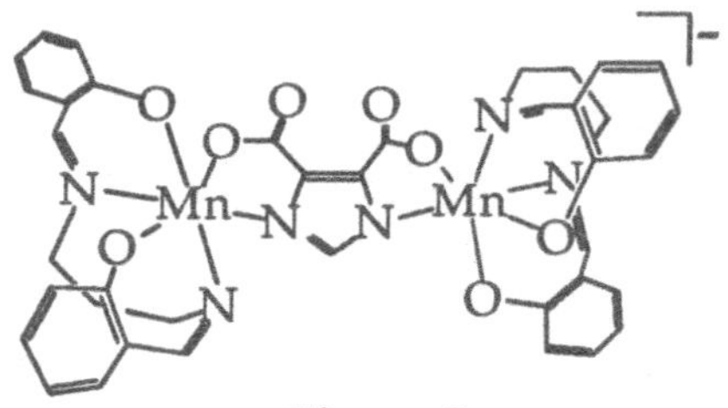

Figure 5

Acknowledgments

MTC (GM17348) and MJB (GM15102) wish to thank the NIH for postdoctoral fellowships. VLP wishes to thank the NIH (GM39406) for financial support of this work. The authors appreciate the helpful conversations over the years from Professor Jim Penner-Hahn and his students, and Professors Charlie Yocum and Jerry Babcock.

References

1) Debus, R. J. *Biochim. Biophys. Acta* **1992**, *1102*, 269-352.
2) Tang, X.-S.; Diner, B. A.; Larsen, B. S.; Gilchrist, M. L.; Lorigan, G. A.; Britt, R. D. *Proc. Nat. Acad. Sci., USA* **1994**, *91*, 704-708.
3) Ono, T.; Noguchi, T.; Inoue, Y.; Kusunoki, M.; Matsushita, T.; Oyanagi, H. *Science* **1992**, *258*, 1335-1337.
4) Boussac, A.; Zimmerman, J.-L.; Rutherford, A. W.; Lavergne, J. *Nature* **1990**, *347*, 303-306.
5) Klein, M. P.; Sauer, K.; Yachandra, V. K. *Photosyn. Res.* **1993**, *38*, 265-277.
6) Baldwin, M. J.; Stemmler, T. L.; Riggs-Gelasco, P. J.; Kirk, M. L.; Penner-Hahn, J. E.; Pecoraro, V. L. *J. Am. Chem. Soc.* **1994**, *116*, 11349-11356.
7) Baldwin, M. J.; Gelasco, A.; Pecoraro, V. L. Photosynth. Res. **1993**, 38, 303-308.
8) Boucher, L. J.; Coe, C. G. *Inorg. Chem.* **1975**, *14*, 1289-1294.
9) Tainer, J. A.; Getzoff, E. D.; Beem, K. M.; Richardson, J. S.; Richardson, D. C. *J. Mol. Biol.* **1982**, *160*, 181.

CHARACTERIZATION OF A COLD-SENSITIVE ASSEMBLY MUTANT OF MANGANESE-STABILIZING PROTEIN

Scott D. Betts, Jeannine R. Ross,
Eran Pichersky, and Charles F. Yocum,
Departments of Biology and Chemistry,
The University of Michigan,
Ann Arbor, MI 48109, USA

1. Introduction

Two of four manganese atoms are released from the O_2-evolving complex (OEC) following extraction of the extrinsic 33-kDa Manganese-Stabilizing Protein (MSP). Non-physiological concentrations of Cl^- (200 mM) stabilize the Mn cluster in the absence of MSP [1]; however, O_2 evolution activity drops below 50% and decays more rapidly under continuous long-term illumination. These defects are reversed upon reassembly of MSP into the OEC. We have targeted conserved residues in MSP by site-directed mutagenesis in order to identify amino acids that are essential in its folding, assembly, and function. Here we report the analysis of one such mutant. Substitution of alanine for the conserved valine at residue 235 destabilizes MSP and causes a cold-sensitive assembly defect. Once assembled into the OEC, however, the mutant protein binds tightly and functions normally. Based on our results from functional analyses of mutant and wild-type MSP, we propose a model in which the two copies of MSP in the OEC serve separate functions.

2. Procedures

2.1. *Assay systems*

The three extrinsic OEC proteins were extracted sequentially from PSII membranes [2]. The 17- and 23-kDa extrinsic subunits were released from PSII by exposure to 2 M NaCl ("salt-washed PSII"). Native MSP was extracted from salt-washed PSII by incubation in 2.6 M urea, 200 mM NaCl ("urea-washed PSII").

P. Mathis (ed.), Photosynthesis: from Light to Biosphere, Vol. II, 381–384.

2.2 *Purification of recombinant MSP*
Description of production and analysis of mutant V235A-MSP will be described elsewhere. MSP was purified from the supernatant fraction of urea-solubilized inclusion bodies and folded as described in the accompanying paper in this volume [Betts, S.D., Ross, J.R., Pichersky, E., and Yocum, C.F.]. The concentration of MSP was determined by the Lowry assay [3] using BSA as a standard.

2.3 *Reconstitution of the OEC*
Manganese-stabilizing protein was assembled into the OEC by mixing urea-washed PSII with MSP [2]. The concentration of PSII was determined from chlorophyll concentrations assuming a stoichiometry of 250 chlorophyll per PSII reaction center [4]. O_2 evolution was assayed in the presence of 20 mM Ca^{2+} and 100 mM Cl^-. For low-salt treatment, reconstitution mixtures were diluted to 10 mM Cl^-, incubated for 1 h at 22°C and assayed. Bound MSP was measured using a laser scanner with an internal digital integrator.

3. Results and Discussion

3.1 *Alanine at position 235 destabilizes unassembled MSP*
Val235 is conserved in 10 of 12 MSP sequences; the exceptions are *Euglena* and *Chlamydomonas*, where it is replaced by Ile. Substitution of Val235 by Ala inhibits assembly of MSP into PSII at 4°C (Figs.1&2). Hydrophobic interactions are weakened at low temperatures; thus, the loss of two methyl groups from the side chain at residue 235 (Val to Ala) likely weakens hydrophobic packing either within MSP or between MSP and PSII. The mutant protein migrates more slowly through partially denaturing polyacrylamide gels than does wild-type protein, indicating that it is more unfolded than wild-type MSP. However, bound V235A-MSP is stable: it is not released during prolonged incubation at low temperature. These results demonstrate that Val235 in wild-type MSP stabilizes the assembly-competent monomer, but is less critical for stability of the

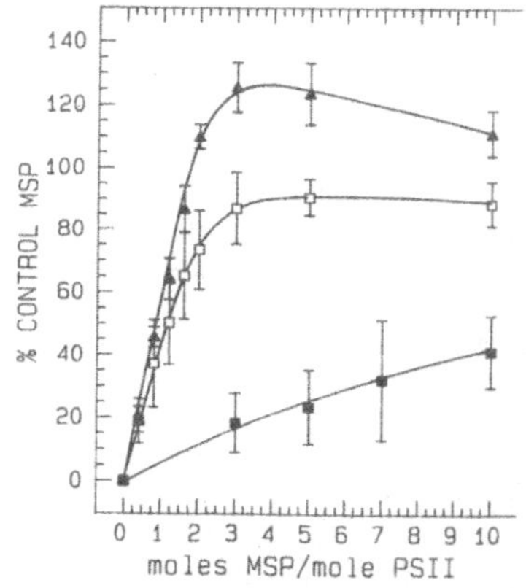

Figure 1. Densitometric analysis of MSP assembled into PSII. Unassembled MSP was removed from reconstitution mixtures by centrifugation, and washed PSII pellets were analyzed by SDS-PAGE. Triangles, wild-type MSP assembled at 4°C and 22°C; open squares, V235A-MSP assembled at 22°C; solid squares, V235A-MSP assembled on ice.

assembled protein. We conclude that Val235 functions primarily to lower the activation energy required for assembly of MSP into PSII.

3.2 One mole MSP per mole PSII is required for O_2 evolution, but two moles are required to stabilize this activity

Rates of O_2 evolution reach maximum values at lower ratios of added MSP to PSII than are required both for maximum stability and for saturation of MSP binding sites. The rate of O_2 evolution reaches a maximum value between 1.2 and 1.6 moles wild-type MSP per mole PSII (Fig.2A), while maximum stability of O_2 evolution requires addition of 2 moles wild-type MSP per mole PSII (Fig.2B). This difference between the level of added MSP required for maximum rates and stability of O_2 evolution increase proportionately for mutant MSP. The rate of O_2 evolution is nearly maximal when 1.6 moles V235A-MSP is added per mole PSII (Fig.2A), while 3 moles of the mutant subunit must be added to fully stabilize O_2 evolution (Fig.2B). The stability curves (Fig.2B) closely track the binding curves (Fig.1), which show that the level of bound MSP reaches a maximum value between additions of 2 and 3 moles of either wild-type or V235A-MSP per mole PSII.

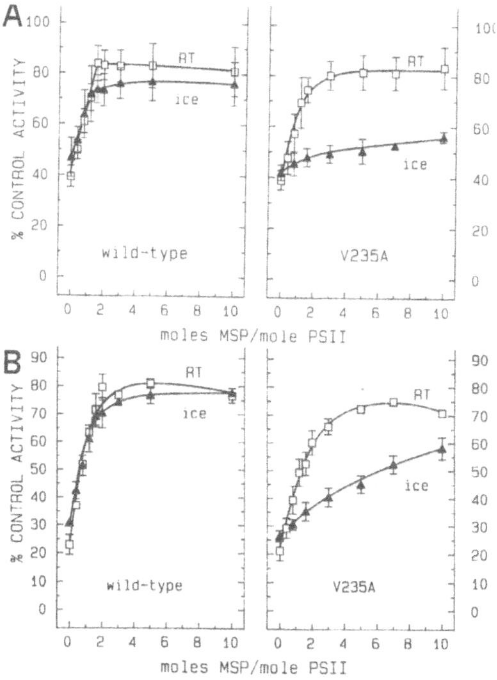

Figure 2. O_2 evolution activity of reconstitution mixtures. **A.** Steady-state rates of O_2 evolution. **B.** Stability of O_2 evolution activity (total O_2 produced during 4 min assay).

Following assembly of MSP into PSII in high salt, the ability of assembled protein to stabilize the Mn cluster can be tested directly by incubating reconstitution mixtures in low salt concentrations (10 mM Cl^-). Following room temperature incubation at low salt, rates of O_2 evolution reach maximum values between 1.2 and 1.6 moles MSP per mole PSII for both wild-type and mutant subunits (Fig.3). The corresponding half-maximal values estimated from these curves are 0.65 and 0.75 moles MSP per mole PSII,

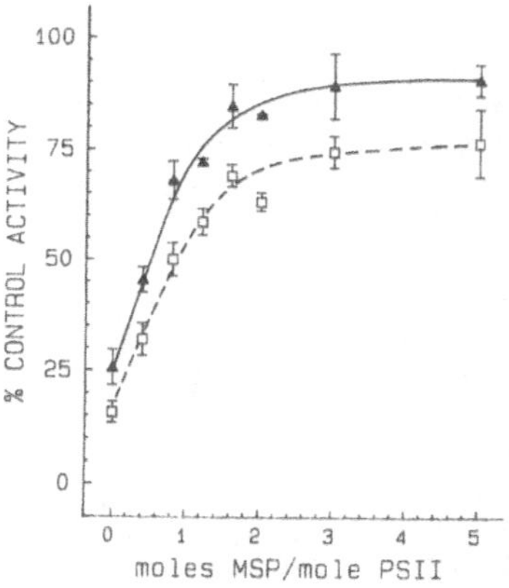

Figure 3. Rates of O_2 evolution following incubation in low salt at 22°C. Triangles, wild-type; squares, V235A.

respectively. The curve obtained with V235A-MSP is shifted almost uniformly downward 10-15% relative to the wild-type curve, indicating that the shift is most likely a result of differences between urea-washed PSII preparations. Most importantly, Fig. 3 shows that in low salt at room temperature V235A-MSP will assemble into PSII with the same efficiency as the wild-type protein, and that less than two moles of added MSP are required to stabilize the Mn cluster and restore maximum activity.

3.3 *Model*

Our results (Figs.2&3) indicate that only one copy of MSP is required to *activate* O_2 evolution, but that a second copy is required to fully *stabilize* O_2 evolution. We propose a sequential-binding model for assembly of MSP into the OEC. The first molecule of MSP in the OEC stabilizes the Mn cluster and activates O_2 evolution (Figs.2A&3). Binding of the second copy of MSP is dependent on occupation of the first binding site; MSP at the second site provides structural support that may help to reduce the decay of O_2 evolution due to photoinhibition (Fig.2B).

4. Addendum

This work was supported by grants from NSF (C.F.Y.) and USDA (E.P.). S.D.B. was supported in part by the Cellular Biotechnology Training Program (CST32GM08353).

References

1 Miyao, M., and Murata, N. (1984) FEBS Lett. 170, 350-354
2 Betts SD, Hachigian TM, Pichersky E, Yocum CF (1994) Plant Mol. Biol. 26, 117-130
3 Lowry, O.H., Rosebough, N.J., Farr, A., and Randall R.L. (1951) J. Biol. Chem. 193, 265-275
4 Berthold, D.A., Babcock, G.T., & Yocum, C.F. (1981) FEBS Lett. 134, 231-234

FUNCTIONAL IDENTITY BETWEEN SPINACH AND *ARABIDOPSIS* MANGANESE-STABILIZING PROTEIN IN SPINACH PHOTOSYSTEM II

Scott D. Betts, Jeannine R. Ross,
Eran Pichersky, and Charles F. Yocum,
Departments of Biology and Chemistry,
The University of Michigan,
Ann Arbor, MI 48109, USA

1. Introduction

The Mn cluster of PSII is stable in the absence of the 33-kDa subunit (also called the Manganese-Stabilizing Protein [MSP]) only when chloride is present at concentrations greater than 100 mM [1]. In the absence of MSP, Cl^--stabilized PSII evolves O_2 at low rates, and this activity is unstable under continuous saturating illumination. MSP can rebind to PSII *in vitro,* restore high rates of O_2 evolution, and decrease the susceptibility of PSII to photoinhibition. We have devised an approach to characterizing site-directed mutants of MSP purified and renatured from bacterial inclusion bodies [2], and have also compared the ability of recombinant *Arabidopsis* MSP and native spinach MSP to assemble into the spinach OEC *in vitro* and reconstitute O_2 evolution activity. Initially, *Arabidopsis* MSP restored activity to nearly the same level as the native protein, but it assembled into the spinach OEC with about half the native affinity. We have now optimized the folding protocol for recovering soluble MSP from inclusion bodies. Both recombinant *Arabidopsis* and spinach MSP assemble into the spinach OEC with native affinity and function. Our reconstitution results also indicate a stoichiometry of two molecules of MSP per PSII reaction center, in agreement with results reported by Xu and Bricker [3].

2. Procedures

2.1 *Assay systems*

The three extrinsic OEC proteins were extracted from PSII membranes as described previously [2]. The 17- and 23-kDa extrinsic subunits were released from PSII by exposure to 2 M NaCl ("salt-washed PSII"). Native MSP was extracted from

P. Mathis (ed.), Photosynthesis: from Light to Biosphere, Vol. II, 385–388.

salt-washed PSII by incubation in 2.6 M urea/200 mM NaCl ("urea-washed PSII"). The concentration of PSII reaction centers was determined from chlorophyll concentrations assuming a stoichiometry of 250 chlorophyll per PSII [4].

2.2 *Isolation of recombinant spinach MSP*

The *psbO* gene encoding preMSP was amplified by PCR from a spinach cDNA library (a gift from J. Norita and W. Gruissem), and cloned into the expression vector pET8c as described for *Arabidopsis psbO* [2]; the 3′ cloning site was *Eco*R1 instead of *Bgl*II. Expression of *psbO* was induced as described previously [2]; the resulting inclusion bodies were incubated in urea at room temperature; MSP was purified from the supernatant fraction by anion-exchange chromatography under partially denaturing conditions. Urea was removed from chromatographic fractions by dialysis, and MSP was purified from the dialysis retentate by anion-exchange chromatography.

2.3 *Reconstitution of the OEC and other methods*

Recombinant MSP was assembled into the OEC by incubation on ice or at room temperature [2]. Oxygen evolution was measured using a Clark-type O_2 electrode at 25°C. The concentration of MSP was determined by the Lowry assay [5] using BSA as a standard. Densitometric analysis was performed using a laser scanner with an internal digital integrator.

3. Results and Discussion

3.1 *Summary of folding protocol*

MSP was purified for *in vitro* folding by anion-exchange chromatography under partially denaturing conditions. This step yields relatively pure and concentrated MSP in urea, which was removed by dialysis. Two cysteine residues are conserved in all published MSP sequences; a disulfide bridge between these two residues in spinach MSP is required for binding to PSII [6]. Sulfhydryl assays indicate that the intramolecular disulfide is reduced in up to 60% of MSP in inclusion bodies (data not shown). Air oxidation of MSP sulfhydryls has been shown to go to completion in just a few hours at pH 8 [6]. We observe similar kinetics for disulfide formation. Finally, MSP was prepared for functional analysis by anion-exchange chromatography under non-denaturing conditions. This second chromatographic step separates monomeric, oxidized MSP from traces of degradation products and cross-linked species.

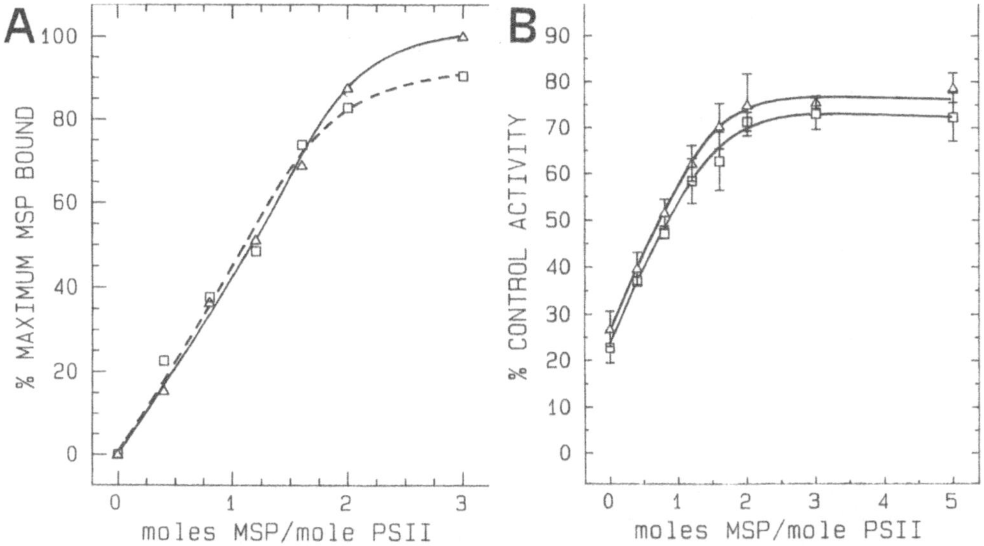

Figure 1. Reconstitution of the spinach OEC with recombinant spinach (triangles) and *Arabidopsis* MSP (squares). A. Densitometric analysis of MSP bound to PSII after washing to remove unassembled MSP. B. O_2 evolution activity.

3.2 *Functional identity of spinach and Arabidopsis MSP*

The concentration-dependence of recombinant spinach MSP in restoring O_2 evolution activity to spinach PSII was indistinguishable from the native protein (Fig.1B & Ref.2). An identical result was obtained with *Arabidopsis* MSP (Fig.1). The curves in Fig. 1B are shifted apart about 5% on the y-axis; this is most likely due to differences between urea-washed PSII preparations in maximum recoverable activity. Fig. 2 presents a sequence comparison. Spinach and *Arabidopsis* MSP differ at 41 out of 247 amino acid residues, including 25 non-conservative substitutions. Our data demonstrate that these substitutions do not affect assembly of MSP or its ability to function normally in the spinach OEC.

3.3 *Two moles MSP per mole PSII*

The stoichiometry of MSP in PSII has been reported as 1, 2, or 3 moles per mole reaction center [see Ref.3]. Under our reconstitution conditions, assembly proceeded linearly and reached a maximum between two and three moles MSP per mole PSII (Fig.1A). The half maximal values for activity and binding for the curves in Fig. 1 occurred at 1-1.2 moles MSP per mole PSII. Analysis of supernatant fractions of reconstitution mixtures indicated that assembly is highly

```
So   EGG.KRLTYDEIQSKTYLEVKGTGTANQCPTVEGGVDSFAFKPGKYTAKKFCLEP
At   --AP-------------M-------------ID--SET-S-------A-----F--

So   TKFAVKAEGISKNSGPDFQNTKLMTRLTYTLDEIEGPFEVSSDGTVKFEEKDGID
At   -S-T¬--DSV---AP-E----------------------A---S-N-K-E----

So   YAAVTVQLPGGERVPFLFTIKQLVASGKPESFSGDFLVPSYRGSSFLDPKGRGGS
At   -------------------V---D-----D--T-K-------------------S

So   TGYDNAVALPAGGRGDEEELQKENVKNVASSKGTITLSVTSSKPETGEVIGVFQS
At   ---------------------V------T-A-V-E---K--K------------E-

So   LQPSDTDLGAKVPKDVKIEGVWYAQLEQ
At   ------------------Q----G---
```

Figure 2. MSP sequences from spinach (So) and *Arabidopsis* (At). Dash indicates identical residue.

efficient. When the molar ratio of MSP to PSII in reconstitution mixtures was increased from two to three, the ratio of assembled to unassembled MSP decreased dramatically from about 10:1 to 2:1 (data not shown). Our results are consistent with a stoichiometry of two molecules of MSP per PSII reaction center and provide independent evidence supporting the result reported by Xu and Bricker [3].

4. Addendum

This work was supported by grants from NSF (C.F.Y.) and USDA (E.P.). S.D.B. was supported in part by the Cellular Biotechnology Training Program (CST32GM08353).

References

1 Miyao, M., and Murata, N. (1984) FEBS Lett. 170, 350-354
2 Betts, S.D., Hachigian, T.M., Pichersky, E., and Yocum, C.F. (1994) Plant Mol. Biol. 26, 117-130
3 Xu, Q., and Bricker, T.M. (1992) J. Biol. Chem. 267, 25816-25821
4 Berthold, D.A., Babcock, G.T., and Yocum, C.F. (1981) FEBS Lett. 134, 231-234
5 Lowry, O.H., Rosebough, N.J., Farr, A., and Randall R.L. (1951) J. Biol. Chem. 193, 265-275
6 Tanaka, S., and Wada, K. (1988) Photosyn. Res. 17, 255-266

REGULATION OF THE O_2-EVOLVING MECHANISM DURING N_2-FIXATION IN THE DIAZOTROPHIC CYANOBACTERIUM *CYANOTHECE* SP. ATCC 51142.

Pascal C. Meunier, James W. Watters, Milagros S. Colon-Lopez and L. A. Sherman, Department of Biological Sciences, Purdue University, West Lafayette, IN 47907, USA

1. Introduction

The cyanobacterium *Cyanothece* sp. ATCC 51142 fixes N_2 in short, intense bursts of activity which are spaced 24 hours apart (1). This behavior, which occurs under light/dark or continuous light conditions, suggests the control of N_2 fixation by a circadian rhythm, which could also control other cell functions. The capacity and the properties of O_2 production by the S-state mechanism in cultures subjected to cycles of 12 hours light and 12 hours of darkness were also determined, and the peak of O_2 evolution is 12 h out of phase with N_2 fixation. Our results suggested that the stability of Mn centers in the dark, their sensitivity to the redox state of quinones, the capacity for O_2 production, the photoreactivation capacity, and the presence of super-reduced S-states (S_{-1} and S_{-2}) were all modulated in *Cyanothece*.

2. Materials and methods

2.1 Experimental

Cyanothece sp ATCC 51142 was cultivated in 25L carboys of ASP_2^- medium (N deficient) bubbled with air. N_2 fixation was determined by the acetylene method (1). Oxygen production was measured by a bare Pt electrode having an amplifier with a high-pass filter time constant of 1.2 s. The amplitudes of O_2 production were evaluated using a modified form of the correction for respiratory transients (2). The ratio used to divide the respiratory transient (the constant "x") was uniformly fixed at 3.0 instead of being determined from the first flash for each experiment (this ratio is the average in *Synechocystis* sp. PCC 6803). This prevented problems with the yields after photoactivation, since the S_3 state did not have sufficient time to deactivate and PSII produced O_2 under the first flash. In all cases, the resulting corrections were smaller than that which would have been applied on *Cyanothece* yields by the original method. This resulted in substantial O_2 yields under the first flash after 15 min. of dark adaptation in *Cyanothece* sp. ATCC 51142. Nevertheless, we found that the small correction resulted in lower residual standard deviations than no correction at all or a full one, and significantly improved the consistency of the results.

2.2 Analysis of results

The amplitudes were analyzed by a generic equation of the form (here given for a 5-step analysis):

P. Mathis (ed.), Photosynthesis: from Light to Biosphere, Vol. II, 389–392.

$$Y_{n+5} + a_1Y_{n+4} + a_2Y_{n+3} + a_3Y_{n+2} + a_4Y_{n+1} + a_5Y_n = 0 \quad (1)$$

where the a_i are the traces (a.k.a. "sigma coefficients") of the 5-step S-state transition matrix (e.g., the determinant is the 5th trace). Note that the eigenvalues, but not the traces, retain the same meaning. The only assumption is that the properties of O_2 production are constant during the measurement. The traces were fitted by Singular Value Decomposition (3), without which the tests described below are considerably more difficult to implement. Singular Value Decomposition (SVD) is a form of multivariable linear regression that allows to discriminate against poorly correlated or poorly defined variables. The number of variables discriminated against is called the "nullity" of the fitting. Thus, a fitting with a nullity greater than zero basically indicated that the data should be fitted by a simpler model. The threshold for this discrimination is somewhat arbitrary, but is linked to the accuracy of the measurements. We used a fixed threshold of 0.005, by opposition to the relative threshold suggested in (3). Our fixed threshold translates to a relative threshold of approximately 0.001. However, the use of the fixed threshold improved the reliability of the discrimination.

In order to determine the number of S-states necessary to describe the data, we performed three sets of fittings: a) with one more flash than the minimum necessary; b) with two extra flashes; and, c) using flashes 1-15, where the minimum necessary number of flashes is 2N (N=4, 5, 6 or 7). Within each set, the best fit was determined by using the residual standard deviation as the selection criterion. Fits that had nullities greater than zero were discarded. When two standard deviations were within 50% of the smallest one, the simplest analysis (i.e., the smallest N) was selected. Thus, a significant improvement in the fittings was necessary to consider 5-, 6- or 7-step fittings over a 4-step one. Then, the best fits of each set were compared; if the same result was observed in at least two sets, then this result was kept. When all three sets differed, the simplest analysis (i.e., the smallest N) was selected.

3.Results and Discussion

The capacity for whole chain electron transport measured with a Clark electrode is shown in Fig. 1. This capacity varies by more than a factor of two and cycles with a 24 hour period, and is maximal at about mid-day. Note that the rates start declining before the light/dark transition. They also increase before the dark/light transition, which implies a clocked regulation by the cell (in opposition to a simple, direct response to environmental conditions). N_2 fixation also occurs at narrowly defined intervals, at 4 h into the dark (Fig.1). The addition of 400 µM phenol-*p*-benzoquinone (PpBQ) resulted in an inhibition of O_2 production rates even in the presence of 2 mM ferricyanide (Fig.1). However, no inhibition was seen in the presence of 2 mM PpBQ (not shown), which indicates that PpBQ is not inhibiting by blocking an electron transport site. The concentration dependence (not shown) seems to indicate that PpBQ had difficulty penetrating the cell membrane and thus could not oxidize the PSII electron acceptors at a sufficient rate. We observed a similar effect in the cyanobacterium *Synechocystis* sp. PCC 6803 (not shown). The inhibition was least in the dark phase; however, there was certainly no lack of available reducing power in the dark since nitrogen was reduced at that time. We believe that this phenomenon may indicate a variable sensitivity to reduced quinones; the lower sensitivity would be useful during N_2 fixation.

After dark-adapting samples on a bare platinum electrode, we found that the O_2 yields (averaged over the first 10 flashes) were variable and sometimes negligible (Fig.2).

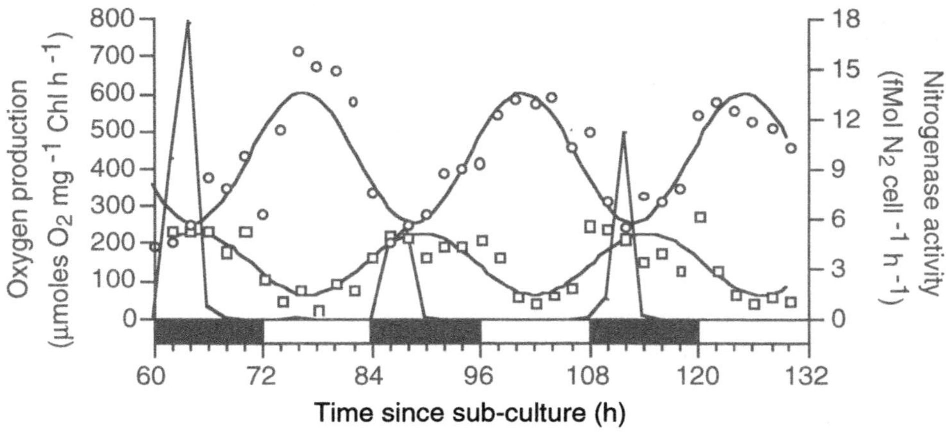

Fig. 1. Rates of O_2 production in the presence of 10 mM $NaHCO_3$ (O) and in the presence of 400 μM PpBQ (□), and rates of N_2 fixation (no symbol) in light/dark cultures. The bottom bar represents when the cultures were in the dark.

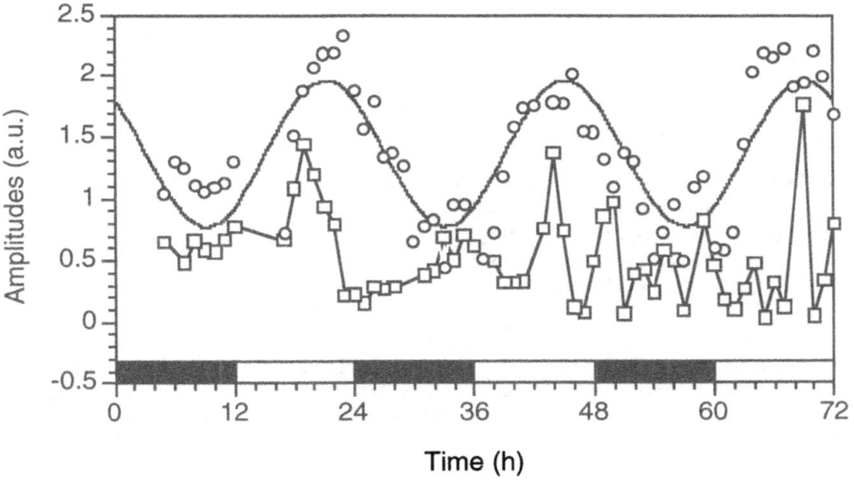

Fig.2. Amplitudes of O_2 production in light/dark cultures. The bar represents when the cultures were in the dark. The initial amplitudes after dark adaptation are shown by squares (□) and the amplitudes after photoactivation are shown by circles (O).

This did not happen when using *Synechocystis* sp. PCC 6803. Thus, the number of centers that stayed active after dark adaptation was indicative of Mn stability. The stability of Mn usually reached a peak earlier than the PSII capacity for O_2 production. More than 10-fold increases in the amplitudes of the O_2 yields were obtained by pretreating the samples with 10 Hz flashes (through a red filter similar to RG-610) for 3 min. This PSII capacity for O_2 production clearly cycled with a period of 24 hours. The decline in PSII Mn stability (Fig.2) may be the consequence of the decline in the capacity for whole chain electron transport (Fig.1) resulting in the reduction of the PQ pool (not shown). Thus, PSII may be downregulated as a consequence of the downregulation of other components.

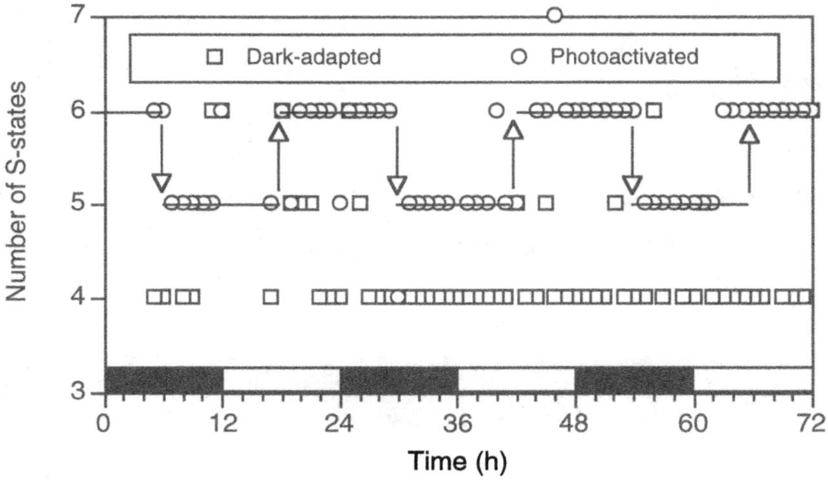

Fig. 3. Number of S-states required to describe O_2 production. The arrows represent the switch between 5 and 6 step mechanisms, and do not represent experimental events. Gaps were left when experiments could not be analyzed.

The oxygen yields were analyzed with 4, 5, 6 and 7 step models in order to determine the involvement of super-reduced S-states (Fig.3). After the dark-adaptation period, a 4-step model was found to be satisfactory most of the time. This indicates that the centers stable in the dark did not deactivate below S_0 or S_1. However, after the photoactivation treatment, the presence of significant quantities of S_{-1} and S_{-2} states could be detected. The number of S-states was mostly 5 or 6 in alternating 12-hour periods. The transitions between 5 and 6-steps mechanisms were 6 hours out of phase with the light/dark transitions. We interpret this as indicating different properties of the centers stable in the dark compared to those that have been freshly activated. Freshly activated centers were unstable and quickly populated the S_{-1} and S_{-2} states. The S_{-2} state was populated most quickly when PSII capacity was in excess and the plastoquinone (PQ) pool was reduced (not shown). After the end of N_2 fixation (at about midnight) the PQ pool became oxidized again and thus the S_{-2} state was slower to form. Thus, a component of the observed PSII behavior is a response to the redox state of the PQ pool; however, we hypothesize that the extent of this response is regulated by the circadian rhythm. Evidence for this is the greater resistance of PSII Mn to reduced quinones during the dark phase (Fig.1), and the increase in PSII activity that anticipates the dark/light transition (Figs. 1 and 2).

References

1 Schneegurt, M.A., Sherman, D.M., Nayar, S. and Sherman, L.A. (1994) J.Bacteriol.176, 1586-1597
2 Meunier, P.C., Burnap, R.L. and Sherman, L.A. (1995) Photosynth. Res. (in press)
3 Press, W.H., Flannery B.P., Teukolsky, S.A. and Vetterling, W.T. (1989) in Numerical Recipes in Pascal, pp.61-74, Cambridge University Press, New York

PARTIAL CHARACTERIZATION OF TWO DIFFERENT MANGANESE BINDING SITES IN PSII: BINDING AFFINITIES AND pH DEPENDENCIES.

A. Magnuson and L.-E. Andréasson
Dept. of Biochemistry and Biophysics,
Göteborg University and Chalmers University of Technology
Medicinaregatan 9C, S-413 90 Göteborg, Sweden

1. Introduction

The nature of the amino acids involved in the water splitting mechanism on the donor side of PSII is still a matter of speculation. Histidine and carboxylic acid residues have been suggested as ligands to manganese in the water oxidizing complex, on the basis of spectroscopic data and site-directed mutagenesis (see ref. 1 for a review).

Extraneous Mn^{2+} acts as an electron donor to PSII, depleted of extrinsic regulatory proteins and native manganese, in a process which may include ligation of Mn^{2+} to native sites in the protein matrix. In earlier studies manganese affinities have been estimated as ranging from < 1 μM to about 200 μM, depending on the method used (a compilation of results can be found in ref. 2).

In this study the binding of manganese to PSII has been investigated by
a) light-induced, steady state electron transfer from Mn^{2+} to DCIP [3], on the minutes time scale, and
b) time resolved μs kinetics of the re-reduction of $P680^+$ by tyrosine Y_Z . The latter has been shown to be affected by the photoligation of Mn^{2+}, possibly due to electrostatic interactions [4]. The effect on both of these types of assays, after chemical modification of histidine and carboxylates, has been analyzed.

2. Experimental procedures

2.1 *Photosynthetic material*

PSII-membranes were prepared by partial solubilization of thylakoid membranes with Triton X-100. To dislodge the three extrinsic proteins with masses of 16, 24 and 33 kDa, the particles were washed in 1M $CaCl_2$. Manganese was extracted either by treatment with hydroquinone (1,4-benzoquinol) and EDTA, or by Tris washing.

2.2 *Chemical modification*

Amino acid residues were chemically modified at pH 6.5 prior to or after manganese extraction, respectively. In the former case, manganese was extracted after the modification was completed. Histidyl residues were modified with diethylpyrocarbonate (DEPC) [5], and carboxylic acid residues with 1-ethyl-3-[(3-dimethyl-amino)propyl] - carbodiimide (EDC) [6].

P. Mathis (ed.), Photosynthesis: from Light to Biosphere, Vol. II, 393–396.

2.3 *Steady state measurements*
The affinity for Mn^{2+} was assayed after removal of bound manganese in chemically modified and un-modified PSII membranes. The rate of photo-reduction of 2,6-dichlorophenolindophenol (DCIP) was measured at different concentrations of added $MnCl_2$, in 20 mM Mes or Hepes buffer, at pH values from 5.6 to 8.0. Apparent K_M values for Mn^{2+} were calculated by least-square fitting of the kinetic data, plotted as $[Mn^{2+}]/v$ *vs.* $[Mn^{2+}]$ (i.e. Hanes plot).

2.4 *Time resolved measurements*
In PSII membranes modified with DEPC, manganese binding was monitored at pH 7.4, by observing the effect of added Mn^{2+}, on the reduction kinetics of $P680^+$ by tyrosine Y_Z [4]. 30 mM Hepes buffer, PSII membranes, 2,6-dichloro-p-benzoquinone (DCBQ) and $MnCl_2$ was added to a measuring cuvette. The concentration of free Mn^{2+} was buffered with citrate. Flash-induced absorption transients from $P680^+$, were measured at 830 nm after an excitation pulse from a Nd/YAG laser (532 nm).

3. Results and discussion

3.1. *Steady-state measurements*
The steady state measurements of DCIP reduction as a result of electron donation by Mn^{2+} revealed that the apparent K_M value for Mn^{2+} is strongly dependent of pH. The affinities for Mn^{2+} binding were shown to be decreasing with decreasing pH, leveling out to about 10 mM below pH 6.2 in unmodified PSII membranes. This behavior indicates the existance of Mn^{2+} binding amino acids that become protonated in this pH range (Fig:s 1 and 2).

3.1.1. *Histidine modification*
In samples modified with DEPC after removal of native manganese, the K_M for Mn^{2+} was roughly equal to that of the unmodified material, at pH values between 6.2 and 7.2 (Fig.1). Below pH 6.2 and above pH 7.2, the affinity was lower than in the unmodified material. Apparently the DEPC modification affects a histidine residue which binds manganese tightly at high pH, but also interferes with manganese binding at low pH. In the modified samples, we could not observe any group titrating at low pH.
However, the K_M values in material modified in the presence of native manganese, was significantly lower than in both the unmodified material and in that modified in the absence of bound manganese. One may suggest a preliminary model, where chemical modification of histidine preserves or induces a conformation with high affinity for Mn^{2+} when it is done prior to removal of native manganese, a conformation that would otherwise be lost upon manganese removal.

3.1.2. *Carboxylate modification*
The EDC-modified PSII samples exhibited generally lower Mn^{2+} affinities than unmodified material. Furthermore, the manganese affinities after overall modification in the absence of native manganese, was clearly lower than in the samples where modification was conducted in presence of Mn (Fig.2).
Our conclusion here is that at least one carboxylic acid residue is shielded from EDC modification by originally bound manganese, possibly functioning as a native ligand. The affinity for Mn^{2+} below pH 6.2, is lower in PSII modified in the presence of manganese, than in unmodifed PSII. This indicates the modification of a Mn^{2+} binding group with a pK value in this range, that is not protected by native manganese, and that

carboxylate residues other than at the native binding sites, contribute to the binding of Mn^{2+} in connection with its function as an artificial electron donor to PSII.

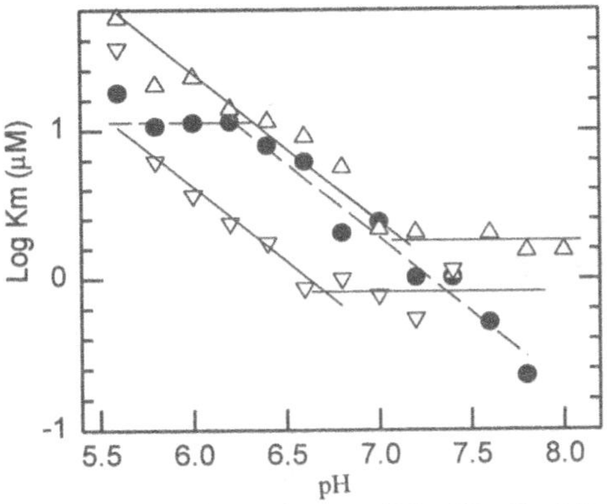

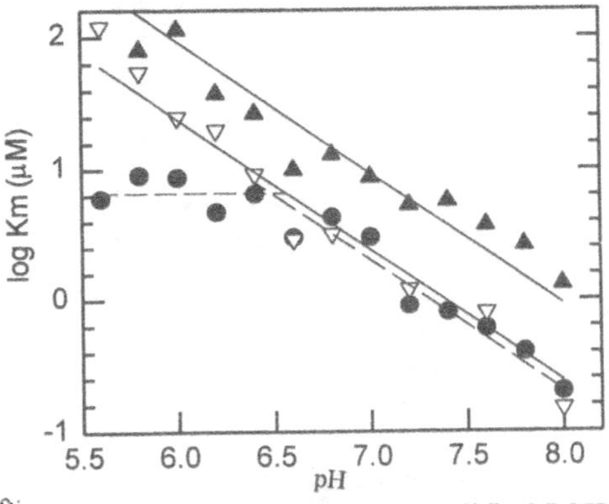

Figure 1. pH dependence of the K_M for Mn^{2+}, in unmodified and DEPC-modified PSII membranes. (●), unmodified PSII membranes; (∇), PSII membranes modified in the presence of manganese; (Δ), PSII modified after manganese removal.

Figure 2. pH dependence of the K_M for Mn^{2+}, in unmodified and EDC-modified PSII membranes. (●), unmodified PSII membranes; (∇), PSII membranes modified in the presence of manganese; (Δ), PSII modified after manganese removal.

3.2. *Time-resolved kinetics of P680$^+$ re-reduction*

In both types of DEPC-modified samples the amplitude of the 10 μs decay phase of P680$^+$, associated with the oxidation of Y_Z, decreased with added Mn^{2+} as observed earlier [4] in unmodified material (Fig.3). Two smaller, slower decay transients, with time constants of 40 μs and 0.2 - 1 ms respectively, were rather unaffected by changes in the Mn^{2+} concentration. The 10 μs re-reduction phase was less sensitive to added Mn^{2+} in the overall modified material, than in the samples modified in the presence of native manganese, as the apparent dissociation constant at pH 7.4 increased from 90 μM to 500 μM, approximately (Fig.4). This clearly suggests that the histidine involved in these measurements is shielded by the native manganese during modification.

The results from the steady state measurements are in accordance with other evidence of the involvment of histidine and carboxylic acids as manganese ligands. Our investigations show that there are two binding sites for added Mn^{2+} involving histidine, indicated by the difference between the values for dissociation constants for the sites modified by DEPC. The dissociation constant at pH 7.5 is in the order of 1 μM in the steady state measurements, which is to be compared to 100 μM in the time-resolved experiments. These results confirm earlier conjectures that there are two sites for manganese re-binding, and that the low-affinity site influences the photooxidation of tyrosine Y_Z [4,7]. Although the modification of both sites by DEPC is affected by natively bound manganese, it remains to be investigated whether or not these sites are in fact identical to binding sites for native manganese.

Acknowledgement

The authors wish to thank Dr Örjan Hansson for valuable assistance with the flash photolysis experiments.

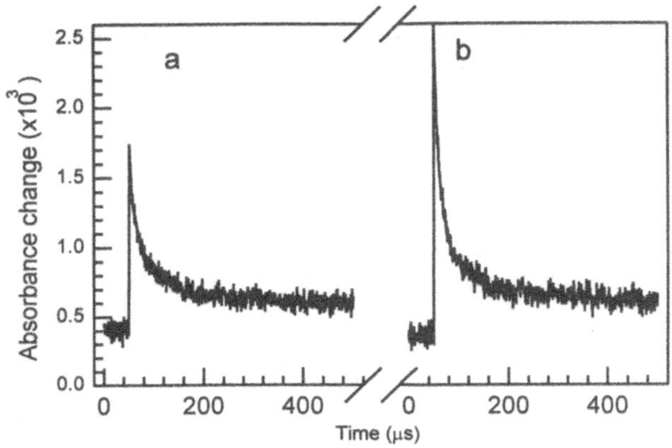

Figure 3. Kinetic traces of absorption trancients at a free [Mn^{2+}] of 0.3 mM. (a), DEPC-modified PSII membranes in presence of manganese; (b), PSII membranes modified after after manganese removal.

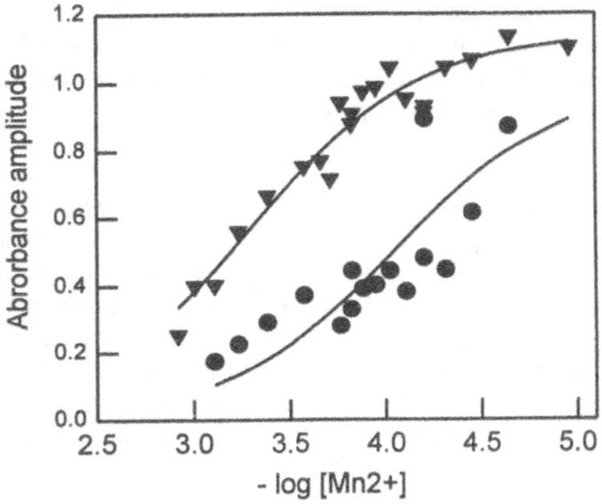

Figure 4. Normalized absorbance amplitudes of the 10 μs phase in DEPC-modified PSII membranes. (●), PSII membranes modified in the presence of manganese; (∇) PSII modified after manganese removal.

References

1 Debus, R.J. (1992) Biochim. Biophys. Acta 1102, 269-352
2 Ghirardi, M.L. and Seibert M. (1992) in Research in Photosynthesis (Murata, N. ed.), Vol II, pp 357-360, Kluwer Academic Publishers, Dortrecht, The Netherlands
3 Blubaugh, D.J. and Cheniae, G.M. (1990) Biochemistry 29, 5109-5118
4 Hoganson, C.W., Casey, P.A., and Hansson, Ö. (1990) Biochim. Biophys. Acta 1057, 399-406
5 Tamura, N., Ikeuchi, M. and Inoue, Y. (1989) Biochim. Biophys. Acta 973, 281-289
6 Preston, C. and Seibert, M. (1991) Biochemistry 30, 1915-9624
7 Miller, A.F. and Brudvig, G.W. (1990) Biochemistry 29, 1385-1392

ISOLATION AND CHARACTERIZATION OF PHOTOSYSTEM II PARTICLES WHICH HAVE INTACT Q_B SITES BUT HAVE LITTLE Q_B-PLASTOQUINONE

Kazuhiko Satoh, Hirokazu Egashira, Yasuhiro Kashino, Hiroyuki Koike and [1]Kouichi Yoneyama
Department of Life Science, Faculty of Science, Himeji Institute of Technology, Harima Science Garden City, Hyogo 678-12, and [1]Weed Center, Utunomiya University, Utunomiya, Tochigi 321, Japan.

1. Introduction

The most difficult thing in studying the functions of Q_B in photosystem II (PS II) is that there have been no report on the PS II preparation which has no Q_B-plastoquinone (PQ) but still has the intact Q_B site. Purified PS II core complexes so far reported have lost the Q_B-PQ, but the binding affinities of the Q_B site have already shown to be modified [1]. Therefore, we have to use crude PS II complexes which still have Q_B-PQ and have to introduce several indirect methods in order to measure binding affinities of various benzoquinone derivatives (BQs) to the Q_B site [2-4].

From the experiments measuring the binding affinities of the Q_B site for BQs and from the studies how various BQs accept electrons at the donor side of PS II, we found that many BQs accept electrons at the two site; the Q_B site and PQ site [4]. BQs bind to the Q_B site and accept an electron directly from reduced Q_A (the Q_B site) or BQs are reduced via PQ which functions as Q_B (the PQ site). From these results, we proposed that the PQ molecule at the Q_B site is not replaced by another PQ molecule, but transfers electrons and protons to a pool PQ molecule by moving its head quinone ring in and out the Q_B pocket while anchoring to the PS II particles by its isoprene side chain [5]. Of course, at a slow rate, the PQ molecule at the Q_B site might be replaced by another PQ molecule. This hypothesis also explain why crude PS II preparations which were not further treated with excessive detergents still retain Q_B although they had no free PQ molecules [6].

Here, we will report that addition of 3-(3,4-dichorophenyl)-1,1-dimethylurea (DCMU) during the isolation of PS II particles by a detergent treatment reduced the amount of Q_B-PQ but retaining the Q_B site intact. Using these PS II particles, we found that the isoprene side chain longer than 3 units might be enough to anchor the PQ molecule to function as Q_B as mentioned above.

2. Materials and Methods

P. Mathis (ed.), Photosynthesis: from Light to Biosphere, Vol. II, 397–400.

PS II particles were prepared from the thermophilic cyanobacterium, *Synechococcus vulcanus* as reported by Egashira et al. [7] except that DCMU was added when the thylakoid membranes were treated with n-heptyl- β -D-thiogulcoside. The PS II particles were washed twice by centrifugation to remove the DCMU, and were stored at -80°C before use.

Rates of oxygen evolution were measured using a Clark-type oxygen electrode at 30°C in a reaction medium containing 1.0 M sucrose, 50 mM Mes-NaOH (pH 6.7), 10 mM NaCl and 5.0 mM $MgCl_2$. Absorbance changes at 290 nm due to photoreduction of tetramethyl-*p*-benzoquinone (duroquinone, DQ) were measured at 25°C using a Hitachi double beam spectrophotometer type 356 [8]. The reaction mixture was the same as mentioned above except 0.2 mM DQ and/or indicated concentrations of PQ with various lengths of the isoprene side chain (PQs) were added. The concentration of PS II particles corresponded to 9.8 μ g Chl/ml.

PQ_1, PQ_2, PQ_3 and PQ_4 (for structure, see right, n=1~4) were synthesized using 2,3-dimethyl-*p*-benzoquinone and alcohols which have corresponding isoprene chains, and were purified by high performance liquid chromatography.

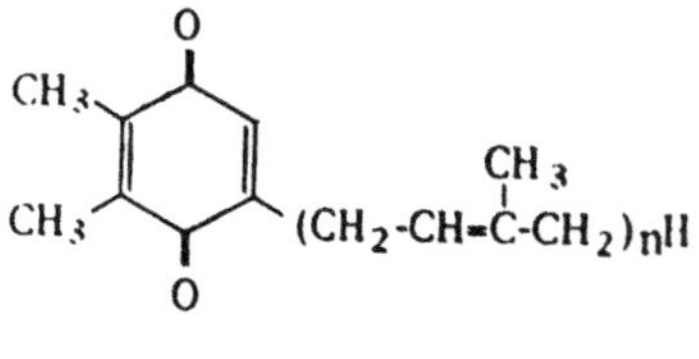

3. Results and Discussion

3.1 *Characteristics of the PS II particles isolated in the presence of DCMU*

Because DQ has almost no affinity to the Q_B site [2,4] and photoreduction of DQ was not observed in PS II core complexes which have no Q_B [1], rates of photoreduction of DQ must be a good measure for the amount of remaining Q_B-PQ. Table 1 shows rates of oxygen evolution in the presence of 2,5-dichloro-*p*-benzoquinone (2,5-DCBQ) ,DQ and ferricyanid. The rate of oxygen evolution in the presence of 2,5-DCBQ decreased to about

Table 1. . The rate of oxygen evolution in the presence of 2,5-DCBQ decreased to about
Oxygen evolution in PS II particles prepared with and without DCMU

Electron acceptor	Control PS II	PS II particles prepared with DCMU	
		Preparation 1	Preparation 2
2,5-DCBQ	2870	1240	440
DQ	1730	357	0
Ferricyanide	—	473	178
Ferricyanide +DCMU	—	0	0

(Rates; μ moles O_2/mg Chl/h)

a half level in the PS II particles prepared in the presence of DCMU (the central column). This might be because 2,5-DCBQ was reduced at both the Q_B and PQ sites [4] and the reduction of 2,5-DCBQ at the PQ site becomes low due to the decrease of electron flow through the Q_B site as a result of reduced amounts of Q_B-PQ. Oxygen evolution in the presence of DQ,
however, decreased much more greatly; down to about one-fifth of the control PS II particles (preparation 1 represents a typical case). In one example, we could not detect oxygen evolution with DQ used as an electron acceptor although a relatively good electron flow was observed with 2,5-DCBQ (Table 1, preparation 2).

Oxygen evolution in the presence of ferricyanide was not so large and was completely inhibited by DCMU, implying the intactness of the Q_B site again (see Table 1). In the case with DQ and 2,5-DCBQ, DCMU inhibited completely or about 90%, respectively, (data not shown).

The results mentioned above suggest that the addition of DCMU during the isolation of PS II particles largely decrease the amount of PQ bound to the Q_B site.

3.2 *Effects of PQ with various lengths of isoprene chain on the rate of DQ reduction*

As can be imagined from the data shown in Table 1, photoreduction of DQ measured by light-induced increases of the absorbance at 290 nm was slow but it continued for several minutes. When 10 μM PQ_3 was used as an electron acceptor, the quinone was reduced rapidly and the reduction was completed within 10 to 20 sec after switching on the actinic light (data not shown). When DQ photoreduction was measured in the presence of both 0.2 mM DQ and 10 μM PQ_3, a rapid absorbance increase due to the reduction of PQ_3 occurred for 10 to 20 sec, and then continuous increases were observed attributable to DQ reduction which were faster than those with DQ alone. Table 2 shows the relative rates of duroquinone reduction in the presence of 10 μM PQ_1, PQ_2, PQ_3 or PQ_4.

Table 2. Relative rates of DQ reduction in the presence of various PQs.

	No PQ	PQ_1	PQ_2	PQ_3	PQ_4
Rate of DQ reduction	13	16	28	33	28

(Rates; relative units. For detail, see the Text and Materials and Methods).

PQ_1 had little effects on the rate of DQ reduction, but PQ_2 and PQ_3 greatly increased the rate, implying that the two PQs are mediating electron flow from PS II to DQ. The rapid reduction of PQs also supports the idea that the Q_B site still remains to be intact without the intrinsic Q_B-PQ. Fig. 1 shows the effects of the incubation time on the rate of DQ photoreduction in the presence of 10 μM PQ_3. The rate gradually increased and reached a maximum after 5-min incubation of the PS II particles with PQ_3. This result suggests that binding and rearrangements of PQ_3 to and in the PS II particles are necessary for the acceleration of the DQ reduction. The concentration dependency of acceleration of the DQ reduction showed that about 10 μM of PQs were enough to induce the maximum effect (data not shown).

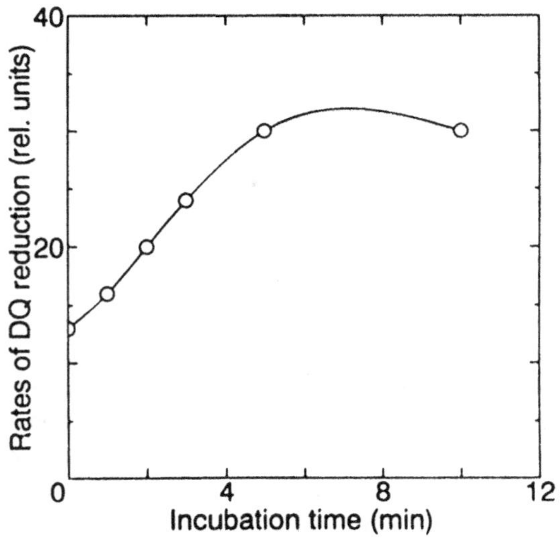

Fig. 1. Effects of incubation time on the rate of DQ reduction.

All the results show that the presence of DCMU during the isolation of PS II particles actually decreased the amount of PQ bound to the Q_B site, and the PS II particles thus obtained can be used to study the functions of Q_B. From the data presented here, we propose that the isoprene side chain longer than 3 units is needed to keep the PQ molecule at the Q_B site and function as Q_B moving its head quinone ring to transfer electrons and protons to another free plastoquinone molecule anchoring the isoprene chain to the PS II complexes.

References

1 Ikeuchi, M. and Inoue, Y.(1986) Arch. Biochem. Biophys.247, 97-107
2 Satoh, K., Koike,H., Ichimura, T. and Katoh,S. (1992) Biochim. Biophys. Acta 1102, 45-52
3 Satoh, K., Katoh, S., Donner, A. and Oettmeier, W .(1994) Plant Cell Physiol.35, 461-468
4 Satoh, K., Oh-hashi, M., Kashino,Y. and Koike, H. (1995) Plant Cell Physiol. 36, 597-605
5 Satoh, K., Kashino, Y. and Koike, H.(1993) Z. Naturforsch. 48C, 168-173
6 Ohno,T., Satoh, K. and Katoh, S.(1986) Biochim. Biophys. Acta 852, 1-8
7 Egashira,K.,Kashino,Y.,Koike.H. and Satoh.K.(1995) The accompanying paper in this book
8 Yamagishi, A. and Katoh, S.(1985) Biochim. Biophys. Acta 807, 74-80

ON THE BINDING AND FUNCTION OF CHLORIDE IN PHOTOSYSTEM II

K. Lindberg & L.-E. Andréasson
Dept. of Biochemistry and Biophysics,
Göteborg University and Chalmers University of Technology,
Medicinaregatan 9C, S-413 90 Göteborg, Sweden

1. Introduction

We have in our previous work shown that there exists a site in PS II where bound Cl^- exchanges slowly with Cl^- in the surrounding medium [1]. The exchange is characterized by a $t_{½}$ of about 1 h and a K_d of 20 μM [2]. Complete removal of all Cl^-, including the slowly exchanging Cl^-, by dialysis against Cl^--free medium (<1μM Cl^-) for 24 h decreases the oxygen evolving activity to about 25-40 % of the control, suggesting that Cl^- is not an absolute requirement for oxygen evolution but might be necessary for full activity. Our further studies have shown that O_2-evolution activity, lost by incubation in a Cl^--free medium, could rapidly (in 15 s) be recovered by fairly high concentrations of Cl^- (K_d =0.5 mM). Complete removal of Cl^- led to a gradual increase of the g=4 EPR signal at the expense of the multiline signal. The original relation between the signal amplitudes was immediately restored upon the addition of Cl^-. This indicates the presence of a Cl^--binding site with properties (affinity, kinetics) different from those of the high-affinity, slowly exchanging site. The significance of these observations in relation to possible models for binding and function of Cl^- is discussed.

2. Materials and Methods

Cl^--free PS II membranes were prepared by diluting the membranes 70-fold in Cl^--free buffer containing 20 mM Mes-NaOH (pH 6.3) and 400 mM sucrose in glass-distilled, deionized water (<1μM Cl^-) and centrifuging at 12 100 x g for 2.5 min. This washing procedure is repeated three times. This removes all unbound Cl^- (=washed membranes). These membranes were further dialyzed against Cl^--free medium for 24 h in order to remove the slowly exchanging Cl^- (=dialyzed membranes).

$^{36}Cl^-$-labeled PS II membranes were prepared by suspending washed membranes in buffer with 200 μM NaCl (42 % $Na^{36}Cl$) included. After incubation at 0°C for 10-20 h in the dark the PS II membranes were washed free from unbound $^{36}Cl^-$ with Cl^--free buffer as above.

Oxygen-evolution activity was measured with a Clark-type electrode (Hansatec) with 1 mM PPBQ as an electron acceptor and other conditions as indicated in the figures.

X-band EPR signals from the S_2 state, induced by illumination of PS II samples at 200 K, were measured at a microwave power of 20 mW and a temperature of 11 K.

P. Mathis (ed.), Photosynthesis: from Light to Biosphere, Vol. II, 401–404.

3. Results and Discussion

To study the effects of removal of Cl^-, PS II membranes were first labeled with $^{36}Cl^-$ for 20 h as described in materials and methods. We have shown before that this procedure selectively labels a slowly exchanging binding site for Cl^- characterized by a dissociation constant of 20 μM, and that the oxygen-evolving activity of the labeled material was practically the same in a Cl^--free medium as in 25 mM Cl (Fig. 1 at time zero). In addition, the distribution of the EPR signals of the S_2 state in the labeled and washed PS II membranes was essentially the same as in the presence of 5 mM Cl^- (Fig. 2a, b). This shows that the single Cl^-, introduced during the labeling procedure and located in the slowly exchanging site, can maintain full activity and normal distribution of EPR signals.

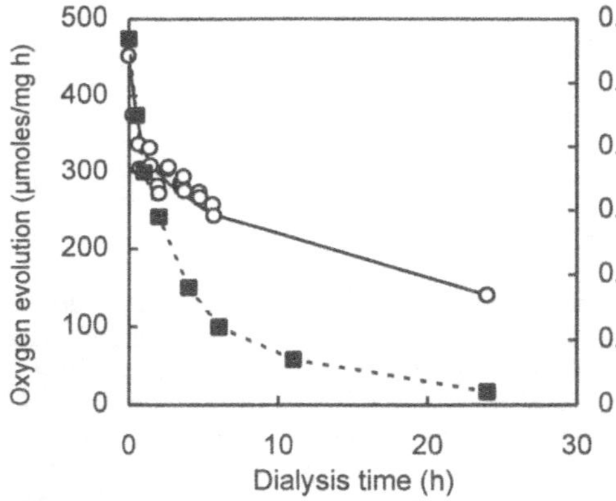

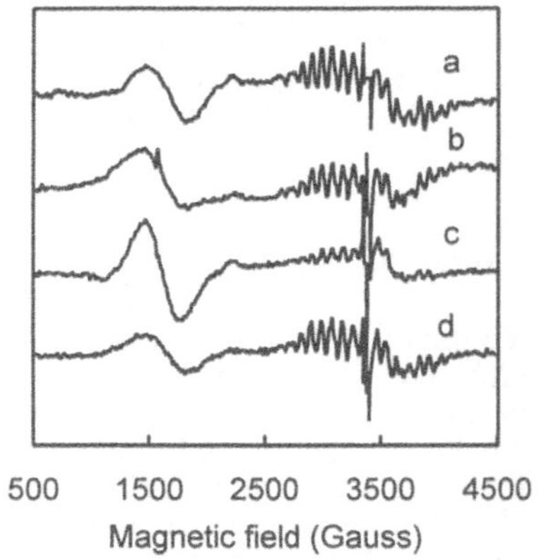

Figure 1. O_2-evolution rate and slowly exchanging Cl^- in a Cl^--release experiment. $^{36}Cl^-$-labeled PS II membranes were labeled as in materials and methods and dialyzed against Cl^--free buffer. Samples were taken out for measurement of bound ^{36}Cl (■) and the O_2-evolution rate (○) in Cl^--free buffer.

Figure 2. EPR difference spectra, (illuminated minus dark-adapted) of (a) untreated PS II membranes in 5 mM Cl^-, (b) washed membranes, (c) dialyzed membranes and (d) 15 s after addition of 5 mM Cl^- to dialyzed membranes.

Fig. 1 shows the dissociation of Cl^- from PS II membranes, measured by following the loss of $^{36}Cl^-$ from labeled membranes. The loss of bound Cl^-, which had a half-time of about 1 h, was accompanied by a simultaneous decrease in oxygen-evolution activity. After 24 h, when all bound radioactivity had been lost, approximately 30% of the original activity remained.

The intensities of the EPR signals of the S_2 state are different from those with saturating concentrations of Cl^- present, with a shift in intensity towards the g=4 signal at the expense of the multiline signal (Fig. 2d).

The binding of Cl^- to PS II was studied by measuring the reactivation of oxygen evolution 15 s after the suspension of dialyzed membranes at different concentrations of Cl^- in the measuring cell. From these measurements a K_d for Cl^- of 0.5 mM was determined (data not shown), in agreement with several earlier reports. This value is, however, much higher than that of the slowly exchanging Cl^- (20 μM) which was shown above to be able to sustain high oxygen-evolution activity in PS II. It appears, therefore, that Cl,$^-$ bound in either of two different environments with high and low affinity and different rates of exchange, respectively can maintain high activity.

Not only was oxygen evolution rapidly restored by addition of Cl^- to the depleted PS II membranes but also the original distribution of the S_2 state EPR signals (Fig. 2c). Thus, it appears that the rapidly exchanging, low-affinity Cl^- site also in this respect behaves as slowly exchanging, high-affinity site. The binding of Cl^- to washed and dialyzed membranes, respectively, was also measured using radioactive labeling with $^{36}Cl^-$. In both cases this procedure labeled the slowly exchanging binding site for Cl^- in PS II with a dissociation constant for Cl^- of 20 μM. In the former case, the labeling can be characterized as an exchange reaction between Cl^- present in the site and added $^{36}Cl^-$ since quick washing results in the removal of weakly associated Cl^- but will leave the slowly exchanging, strongly bound Cl^- intact. In the latter case, on the other hand, the binding takes place at an empty site. In the washed membranes the half-time of labeling in 100 μM Cl^- was about 1 h whereas in the dialyzed membranes the halftime was only about 10 min. (Fig. 3).The rate of labeling of the washed membranes is determined by the low rate of dissociation of the bound Cl^-. This is supported by the identical halftimes of dissociation and binding of Cl^- in Fig. 3. In the dialyzed Cl^--free membranes, on the other hand, the rate of binding is not limited by the dissociation reaction but by other but obviously faster processes.

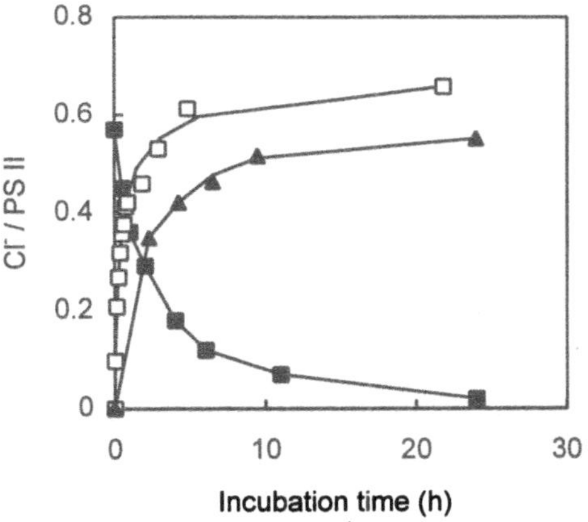

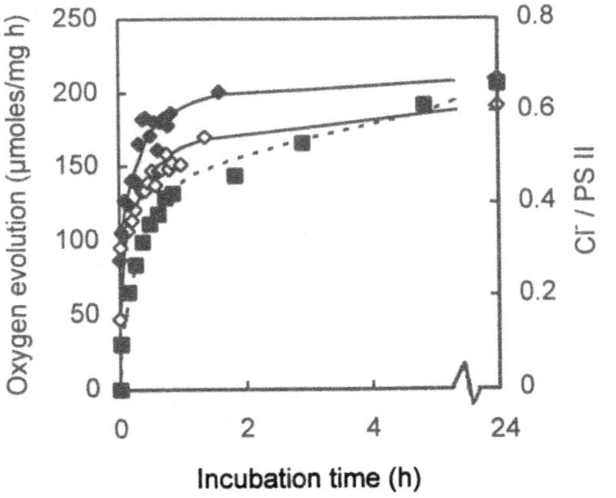

Figure 3. Binding and release of Cl^- in PS II. Dialyzed (□) and washed (•) PS II membranes were incubated at 100 μM Cl^- (42 % $^{36}Cl^-$). Samples were washed with Cl^--free buffer to remove unbound Cl^- and the amount of bound Cl^- was determined. (■) Dissociation of Cl^- from $^{36}Cl^-$-labeled PS II membranes dialyzed against Cl^--free buffer as in Fig.1.

Figure 4. Effects of Cl^--binding on O_2-evolution activity. Dialyzed membranes were incubated in 100 μM Cl^- (42 % $^{36}Cl^-$). At the times indicated samples were withdrawn, treated as in Fig. 3 and the amount of bound $^{36}Cl^-$ determined (■). Parallel to this O_2-evolution activity in 100 μM Cl^- (◆) and in Cl^--free buffer (◊) was determined.

Our results indicate that reactivation of PS II by Cl^- may occur either at two sites with different binding properties where, depending on the conditions, either site is responsible for the activity or, alternatively, at one site with changing properties. To investigate these possibilities further we studied the time dependence of the reactivation process in dialyzed membranes incubated in the presence of 100 μM Cl^-. To measure the total activation involving Cl^- bound to both types of sites an assay medium with 100 μM Cl^- was used. To measure the activation due to binding to the high-affinity, slowly exchanging site, the PS II membranes were diluted in an Cl^--free assay medium (the final concentration of Cl^-, 1 μM, derives from dilution of the PS II membranes). The dilution should lead to dissociation of the low-affinity, rapidly exchanging Cl^- (K_d=0.5 mM) and leave only the slowly exchanging site occupied.

In both assay media a clear time-dependence of the activation process was observed (Fig. 4) but the activity in 100 μM Cl^- was significantly higher than in the Cl^--free medium. Our interpretation is that the transfer of the PS II membranes from 100 μM Cl^- to the Cl^--free environment caused the dissociation of activating Cl^- from the low-affinity site. The residual activation is caused by Cl^- in the slowly exchanging site. However, as the time of incubation in 100 μM Cl^- increased, an increasing fraction of the restored activity was retained when the membranes were diluted in the Cl^--free medium (Table 1). The increase followed the occupation of the slowly exchanging Cl^- site (Fig. 4). This indicates that the activating role which is initially held by Cl^- bound to the low-affinity, rapidly exchanging site is gradually taken over by Cl^- in the high-affinity, slowly exchanging site. One may also view this as a time-dependent, structural alteration in PS II which gradually changes the low-affinity, rapidly exchanging site into a high-affinity, slowly exchanging site.

Table 1. Activation of oxygen evolution in Cl^--depleted PS II by Cl^-

Incubation in 100 μM Cl^- (h)	Activation measured in 0 μM Cl^- / Activation measured in 100 μM Cl^-
0.5	0.63
1	0.78
2	0.80
24	0.85

In summary, Cl^- binding, responsible for high oxygen-evolution activity in PS II and normal EPR properties of the S_2 state, may occur either as high-affinity (K_d= 20 μM) and slowly exchanging (several minutes to hours) or as low affinity (K_d= 0.5 mM) and rapidly exchanging (seconds). We suggest that the high-affinity state is the normal state of binding. On dissociation of Cl^- from PS II, a conversion to the low-affinity state occurs but, following ligation of Cl^- to this state, PS II slowly returns to the high-affinity state. The conversion between these states may be associated with structural changes. We also propose that methods for rapid depletion of Cl^-, e.g. high pH in the presence of sulfate, may lead to an irreversible conversion to the low-affinity state. The dissociation constants after such treatments are generally found to be around 1 mM. These methods usually lead to the dissociation of the 16 and 23 kDa extrinsic subunits. We have earlier found that after removal of these subunits the capability of high-affinity, slow-exchange binding of Cl^- is lost.

References

1 Lindberg, K., Wydrzynski, T., Vänngård, T. and Andréasson, L.-E. (1990) FEBS Lett. 264: 153-155

2 Lindberg, K, Vänngård, T. and Andréasson, L.-E. (1993) Photosynth. Res. 38: 401-408

Low-Frequency Modes in the Light-Induced FTIR Difference Spectrum of Oxygen-Evolving Photosystem II Particles

Matthew T. Gardner and G.T. Babcock
Department of Chemistry, Michigan State University, East Lansing, MI 48824, U.S.A.

ABSTRACT:

The low frequency region of the infrared spectrum of oxygen evolving Photosystem II membranes was investigated using light induced FTIR difference spectroscopy and isotopic labeling of substrate. Preliminary results are reported here. What is being assigned tentatively as the $S_2Q_A^-/S_1Q_A$ difference spectrum exhibits several low frequency infrared active vibrations. Two peaks, a positive going feature appearing around 800cm^{-1}, and one difference peak appearing around 669/546 cm^{-1} are seen. When ^{18}O labeled H_2O is used, the 800cm^{-1} peaks shows no shift, while the 669cm^{-1} difference peak may show isotopic dependence. Control experiments with manganese-depleted samples do not exhibit these peaks. These peaks may represent signals arising from the manganese cluster of Photosystem II. The frequencies at which these peaks appear are consistent with them possibly arising from bridging di-μ-oxo structures, bridging μ-oxo, di-μ-carboxylato, or possibly terminal oxo groups. This is one of the first observations of light-induced difference peaks in this region of the infrared spectrum of Photosystem II.

EXPERIMENTAL:

BBY-type particles capable of oxygen evolution were prepared as described in (1). Samples for FTIR and EPR were resuspended in 1 mM MES (pH=6.0), 5 mM CaCl2, 5 mM NaCl. The particles were assayed in a Clark-type electrode and found to evolve oxygen at a rate of 650 to 700 μmoles O_2/mg chl/ hour (400 μmol DCBQ used as the acceptor). Tris-treatment was carried out as described in (2) For EPR assay, the samples, 300μg chlorophyll, were partially dehydrated onto thin strips of mylar. For FTIR measurements, the sample was dried onto a AgCl window, sealed with another AgCl window, and placed in a homebuilt N_2 gas flow cryostat. A germanium filter blocked the He-Ne reference laser from entering the sample compartment. FTIR spectra were recorded at -80 C, 4 cm^{-1} resolution, on a Nicolet 520P FTIR spectrometer equipped with a modified Nicolet MCT-B detector. 500 scans per sample were recorded, and 3 samples were averaged for each spectrum. EPR controls were carried out on a Varian E-4 spectrometer equipped with a TE cavity. Illumination of the sample in the microwave cavity and the FTIR spectrometer was carried out with a fiber-optic illuminator. The actinic light was passed through a long-pass filter(λ>650 nm) and a 1 cm cuvette filled

P. Mathis (ed.), Photosynthesis: from Light to Biosphere, Vol. II, 405–408.

with 5% copper sulfate solution (to filter IR). Incident power on the sample in both cases was 100 mW/cm^2.

RESULTS:

Figure #1 demonstrates the ability of the current instrumental set up and detector to detect signals that are calculated to have similar intensities as those expected from the OEC of Photosystem II. The top trace is the average of 3000 scans taken with 6 different samples of a di-μ-oxo bridged manganese dimer. The calculated sample in each sample was 6.25 x 10^{-11} moles of sample. Due to large baseline fluctuations relative to the size of the sample absorbances, a square function was subtracted from the baseline of the average spectrum in order to make comparison to the reference spectrum easier. It can be clearly seen, that the features present in the top spectrum are in fact attributable to actual sample absorbances. The features of particular interest are the features from 600-700 cm^{-1}. This is the region of the IR spectrum of PSII in which metal-oxo bonds will absorb.

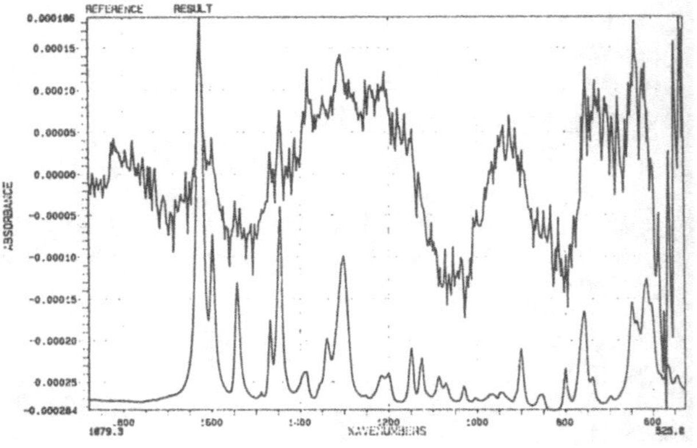

Figure #1 top: FTIR spectrum of $6x10^{-11}$ moles Mn_2O_2 complex. 3000 scans, 2 cm^{-1} res. bottom: reference of same compound. 128 scans, 2 cm^{-1}

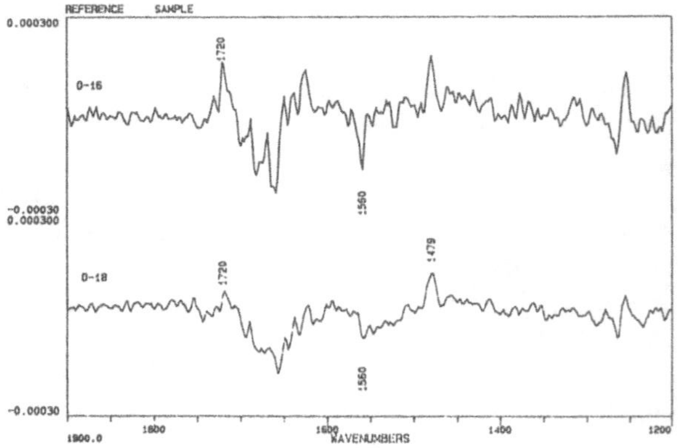

Figure#2 top: Light induced difference spectrum of O_2 evolving BBY particles. bottom: Spectrum of same, with ^{18}O buffer See experimental section for spectral acquisition details.

FTIR SPECTROSCOPY:

Figure #2, top, shows the light-induced difference spectrum of BBY particles. The spectrum is very similar to that reported for $S_2Q_A^-/S_1Q_A$. (3). It displays the positive going feature at 1479cm^{-1}. attributed by some to be characteristic of the Q_A^- species (4). The origin of this peak is currently a hotly disputed issue. A potentially important observation in this spectrum is the lack of the 1403/1364 cm^{-1} difference peak assigned to a calcium ligating carboxylate residue (3c). The fact that these samples are fully competent in oxygen evolution (650-700 μmoles O_2/hour/mg chl) and that the remaining peaks generally duplicate those reported for the $S_2Q_A^-/S_1Q_A$. difference spectrum leads to assigning this spectrum as $S_2Q_A^-/S_1Q_A$. Figure #3, bottom, is a spectrum showing samples treated identically, but resuspended in 66-70% ^{18}O-labeled

H_2O containing buffer. A puzzling aspect of this spectrum is the reduced intensity of the signals. The samples were exposed to identical conditions regarding light and drying conditions. Upon close examination, the spectra are basically identical in terms of peak positions and relative intensities of peaks. At this modest level of signal-to-noise, it appears as though isotopic labeling of substrate has no effect on the infrared spectrum between 1200 and 1900 cm^{-1}

In the top spectrum of figure #3, one notices several areas of interest. There are several positive and negative going features in the area around 1050cm^{-1}. This region may be slightly higher than one would expect metal-oxo species to absorb and is free from contributions from the lower frequency protein amide backbone vibrations (5). Upon isotopic substitution, these peaks showed no apparent shifts in frequency. The positive going feature at 800 cm^{-1} in Figure #3, top,is intriguing. Of particular interest is the fact that this feature shows no associated negative going partner. Usually in lightinduced FTIR of proteins the difference peaks appear as a result of small perturbations on existing molecules. However when a feature which is entirely positive or negative with no opposite going partner, is observed it can be indicative of an entirely new chemical species being formed or eliminated. In the case of PSII, the water oxidationpathway may include formation of terminal oxo groups (6). These terminal oxo groups depending upon the oxidation state of the manganese to which it is bound, would absorb in the 700-950 cm^{-1} region of the spectrum. If ^{18}O was substituted into a terminal oxo group, a diatomic harmonic oscillator would predict a shift of approximately 30 cm^{-1} (7). Figure #3, bottom, shows that upon substitution with labeled water, the 800 cm^{-1} peak showed no such shift. While this may seem to eliminate it as a possible teminal-oxo species, more extensive isotopic labeling studies are planned to

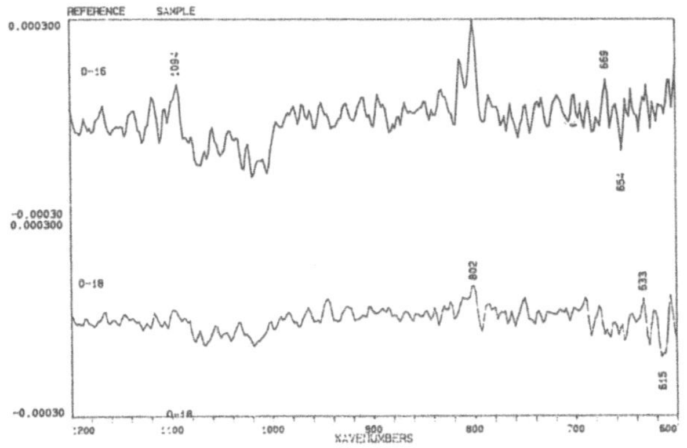

Figure #3:Low frequency region of same samples in figure #2. top: ^{16}O, bottom ^{18}O.

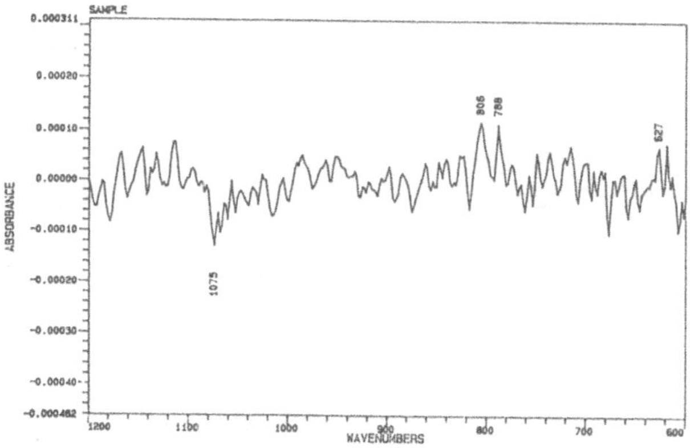

Figure #4:Light induced difference spectrum of tris-treated BBY's. Low frequency region. See experimental section for spectral acquisition details.

completely rule it out. Another interesting result is the appearance of a difference peak at 669/654 cm^{-1}. This peak, while of modest amplitude and signal-to-noise, proved reproducible, and when ^{18}O was present, it showed a shift to 633/615 cm^{-1}. This is the region of the spectrum where di-μ-oxo linked, or mono-μ-oxo,di-μ-carboxylato linked dimers of manganese absorb (8). Additionally, the shift observed is appropriate,and consistent with that previously observed for structures of this type. The light-induced 15 cm^{-1} shift (669 to 654 cm^{-1}) could be due to a one electron oxidation of one of the two proposed di-μ-oxo linked manganese dimers present in the OEC. No detailed study of the effect of one electron oxidations of these types of dimers on their characteristic infrared bands has been published to date.

Tris washed BBY's, when treated in the same way, showed a complete loss of all of the low-frequency signals observed in the Mn-containing samples (Figure #4). This last result is a necessary but not sufficient result on it's own to show that the low-frequency signals originate from the Mn-ensemble. There is a residual peak present at 806 cm^{-1}, the origin of which is unknown. The high frequency region of the spectrum showed substantial differences from that of oxygen-evolving particles. Upon Tris treatment the 33 kDa protein is removed (2), possibly resulting in relatively large local rearrangements around the Mn cluster.The fact that the 669/654 cm^{-1} pair is not present in the Tris treated spectrum, and shows an apparent isotope dependence is good evidence for it originating from the oxygen-evolving-cluster, and it's frequency and isotopic shift supports it's assignment as possibly vibration of a di-μ-oxo linked dimer or mono-μ-oxo,di-μ-carboxylato bridged manganese atoms.

REFERENCES:

(1) Berthold, D.A., Babcock, G.T., Yocum, C.F. (1981) FEBS Lett. 134, 231-234
(2) Tamura, N., Cheniae, G.M. (1987) Biochim. Biophys. Acta. 890, 179-194
(3) a) Noguchi, T., Ono, T., Inoue, Y. (1992) Biochemistry 31, 5953-5956 b) Noguchi, T., Ono, T., Inoue, Y. (1992) in Research in Photosynthesis (Murata, N., ed.) Vol. II, 309-312, Kluwer, Dordrecht c) Noguchi, T., Ono, T., Inoue, Y. (1995) Biochim. Biophys. Acta 1228, 189-200
(4) a) Noguchi, T., Ono, T., Inoue, Y. (1993) Biochim. Biophys. Acta 1143, 333-336 b) Berthomieu, C., Nabedryk, E., Mantele, W., Breton, J. (1990) FEBS Lett. 269, 363-367
(5) Arrondo, J.L.R., Muga, A., Castresana, J., Goni, F.M. (1993) Prog. Biophys. Mol. Biol. 59, 23-56
(6) a) Messinger, J., Badger, M., Wydrzinski, T. (1995) Proc. Natl. Acad. Sci. 92, 3209-3213 b) Hoganson, C.W., Lydakis-Simantiris, N., Tang, X.-S., Tommos, C., Warncke, K., Babcock, G.T., Diner, B.A., McCracken, J., Styring, S. Photosynthesis Research, in press.
(7) a) Workman,J.M., Powell, R.D., Procyk, A.D., Collins, T.J., Bocian, D.F. (1992) Inorg. Chem. 31, 1550-1551 b) Collins, T.J., Gordon-Wylie, S.W., (1989) J. Am. Chem. Soc. 111, 4511-4513 c) Czernuszewicz, R.S., Su, Y.O., Stern., M.K., Macor, K.A., Kim, D., Groves, J.T., Spiro, T.G. (1988) J. Am. Chem. Soc. 110, 4158-4165
(8) a) Cooper, S.R., Calvin, M. (1977) J. Am. Chem. Soc. 99, 6623-6630 b) Sheats, J.E., Czernuszewicz, R.S., Dismukes, G.C., Rheingold, A.L., Petrouleas, V., Stubbe, J., Armstrong, W.H., Beer, R.H., Lippard, S.J. (1987) J. Am. Chem. Soc. 109, 1435-1440

SELECTIVE EFFECT OF CARBOXYL AND HISTIDYL MODIFIERS ON BINDING AND SINGLE TURNOVER PHOTOOXIDATION OF Mn^{2+} BY PHOTOSYSTEM II

Maria L. Ghirardi, Thomas W. Lutton and Michael Seibert
National Renewable Energy Laboratory, Golden CO 80401, U.S.A.

1. Introduction

Photosynthetic water-splitting is catalyzed by a tetrameric manganese cluster bound to amino acid residues on the oxidizing side of the photosystem II (PSII) reaction center proteins. The cluster can be removed by Tris or NH_2OH treatments that eliminate O_2 evolution but not PSII electron transport function. His337 on the D1 protein was the first residue experimentally identified as a PSII Mn ligand (1,2). A number of ligands to Mn were also identified by molecular biology techniques, such as Asp170, Asp342, and His332 on D1 (3,4), Glu69 on D2 (5), and, more recently, His337 and Glu333 on D1 (6).

Our approach to identify ligands to Mn involves assessing the effects of bound Mn on DPC photooxidation by PSII (i.e., the DPC-inhibition assay) following specific chemical modifications. For this purpose, we use Tris-treated PSII membranes that are devoid of the Mn cluster and measure the inhibition of PSII-mediated steady-state electron transfer from DPC to DCIP by added $MnCl_2$ (2,7-9). High affinity Mn ligands thus identified may be involved in photoactivation of O_2 evolution or reflect final ligands to the Mn cluster. We have previously shown that high affinity Mn inhibits up to 50% of the initial rate of DCIP photoreduction in the presence of 200 μM DPC (2,7,8) by non-competitively (2,9) blocking steady-state electron transfer from DPC to Y_Z^+ (10,11) without affecting DPC photooxidation by Y_D^+ (12). The latter accounts for the remaining 50% of the rate of DCIP reduction (9,11). The inhibition of DPC photooxidation by high affinity Mn was also deconvoluted into 4 independent components (2,8), two associated with carboxyl and two with histidyl residues. Each component corresponds to one Mn ion bound to a specific amino acid residue.

Recently, we found that one of the 4 Mn-binding components binds a photooxidizable Mn (13) as suggested earlier (10). In this report we characterize the residue that binds the photooxidizable Mn using flash-induced fluorescence decay measurements.

2. Procedure

PSII membrane fragments were isolated from market spinach and Tris-treated (8).

P. Mathis (ed.), Photosynthesis: from Light to Biosphere, Vol. II, 409–412.

EDC and Pt terpyridinium modifications were done as described (8,14). Initial rates of DCIP photoreduction were determined spectrophotometrically (8). Flash-induced fluorescence decay was measured with a home-built instrument (13). Actinic flashes were provided by a Xenon flash lamp (3 μs HWHM) and filtered through a Corion LS-650 bandpass filter. Monitoring flashes were supplied by an array of 621-nm emitting LEDs at a flash rate of 100 Hz. Fluorescence emission, detected at 90° from actinic and probe flashes by a silicon photodiode protected by a Corion RG-715 cut-on filter, was amplified by a home-built circuit. The detection system was gated for 100 μs following the actinic flash. Data were collected with a DT2819 A/D board using Global Lab data acquisition software. Dark-adapted PSII samples were diluted to 20 μg Chl/ml, treated with 50 μM DCMU, and placed in 4 mm x 10 mm cuvettes.

3. Results and Discussion

Manganese-binding components detected by the DPC-inhibition assay can be specifically inactivated by either the carboxyl modifier EDC (8) or by the histidine-modifier Pt terpyridinium (14; Preston & Seibert, in preparation). Table I shows that either EDC or Pt separately inactivate about half of the Mn-binding components detected by the DPC-inhibition assay. In order to determine which treatment removes the ligand to the photooxidizable Mn, we applied the technique of flash-induced fluorescence decay (4) to assess the effects of each treatment on the accumulation of Q_A^- (after a single flash) upon addition of $MnCl_2$ to Tris-treated membranes. Under these conditions, an actinic flash will generate the species $Y_Z^+ Q_A^-$ (a high fluorescent state) in about 20-40 μs (4). In the absence of an electron donor to Y_Z^+, the $Y_Z^+ Q_A^-$ state recombines in 15-100 ms giving rise to a low fluorescent state (F_{final}). Addition of increasing concentrations of PSII donors results in a raise in the F_{final} fluorescence level. It is possible to estimate the binding affinity of the donor to PSII by titrating the amount of accumulated Q_A^- as a function of the concentration of added donor

Membrane treatment	Rate of DCIP reduction with 200 μM DPC (+Mn/-Mn)	Number of components modified
none	0.62	0
10 mM EDC	0.83	2.2
50 mM EDC	0.87	2.6
50 μM Pt	0.84	2.3

Table I. Effect of EDC and Pt-terpyridinium modification on the number of high affinity Mn-binding components out of a possible 4 that are not detectable by the DPC-inhibition assay (steady-state light).

(4). We used F_{final}/F_{max} as a measure of accumulated Q_A^-. In order to eliminate the non-linearity between fluorescence yield and Q_A^- concentration, PSII membranes used for these measurements were resuspended in the absence of divalent cations. The data were analyzed as Scatchard plots, where F_{final}/F_{max} was plotted as a function of (F_{final}/F_{max})/[donor]. The slope of the line in the Scatchard plot is an estimate of the binding affinity of the donor (K_d); the Y-intercept determines the relative amount of donor-binding PSII centers (DBC).

Table II shows that either EDC or Pt treatment eliminates about 30% of the DBC for DPC; however, neither treatment affects the binding affinity (K_d) of the remaining sites for DPC. On the other hand, the effect of the two chemical modifiers on Mn binding to PSII membranes is very different. The histidyl modifier hardly affects the amount of DBC and causes only a small increase in the K_d for Mn as compared to untreated membranes. Treatment with the carboxyl modifier, though, results in elimination of about 80% of the high-affinity Mn DBC and uncovers DBC with low affinity for Mn. If the concentration of EDC during the treatment is increased from 10 mM to 50 mM (which does not result in inactivation of significantly more Mn binding components as detected by the DPC-inhibition assay, see Table I), the high affinity Mn DBC are completely eliminated, but the uncovered low affinity DBC stays fairly constant. The selective effect of EDC on high affinity Mn clearly indicates that the ligand to the high affinity photooxidizable Mn in Tris-treated PSII membranes is a carboxyl residue. Similar results were presented by Nixon & Diner (4) regarding the effects of site-directed mutations of Asp170 D1 in *Synechocystis* on binding of photooxidizable Mn. The lack of effect of Pt on high affinity Mn binding suggests that histidyl residues do not bind a photooxidizable Mn.

The nature of the ligand to the low affinity Mn uncovered by the EDC treatment is not known. Nixon & Diner (4), who observed the same phenomenon in *Synechocystis* attributed the low affinity component to Mn binding with second-order to the membranes. The low affinity component may be a histidyl residue that binds a Mn that is not photooxidized by a single flash if Mn is also bound to the carboxyl residue. However, if the carboxyl is eliminated by chemical modification with EDC, then the Mn bound to the histidyl may become photooxidizable, albeit with a lower K_d. This interpretation agrees with the observation by Blubaugh & Cheniae (15) that the K_m for $MnCl_2$ photooxidation by Tris-treated membranes is substantially higher in EDC-treated samples due to binding of Mn^{2+} to a histidine residue. We will next attempt a double modification of Tris-treated membranes to investigate whether Pt treatment further removes the low affinity, Mn-binding component from EDC-treated samples. Our results confirm previous reports on the involvement of a carboxyl residue in the binding of photooxidizable Mn (4) and correlate that function with a ligand detected by the DPC-inhibition assay. This validates the assay as a technique to study ligands that may bind Mn of physiological importance.

Membrane treatment	DPC titration		$MnCl_2$ titration			
			High affinity		Low affinity	
	DBC	K_d, μM	DBC	K_d, μM	DBC	K_d, μM
none	0.90	49-58	0.93	0.94	0	-
10 mM EDC	0.67	64	0.22	0.75	0.39	58.5
50 mM EDC	-	-	0	-	0.37	25
50 μM Pt	0.70	54	0.90	1.47	0	-

Table II. Effect of EDC and of Pt terpyridinium modification on the relative number of donor binding centers (DBC) and binding affinity (K_d) for DPC and Mn in Tris-treated PSII membranes (flash-induced fluorescence decay).

4. References

1 Preston, C. and Seibert, M. (1989) *Current Research in Photosynthesis* (Baltscheffsky, M., ed.) Vol I, pp. 925-929, Kluwer Academic Publishers, The Netherlands.
2 Preston, C. and Seibert, M. (1991) Biochemistry 30, 9625-9633.
3 Boerner, R.J., Nguyen, A.P., Barry, B.A. and Debus, R.J. (1992) Biochemistry 31, 6660-6672.
4 Nixon, P.J., Trost, J.T. and Diner, B.A. (1992) Biochemistry 31, 10859-10871..
5 Vermaas, W.F.J., Charite, J. and Shen, G. (1990) Biochemistry 29, 5325-5332.
6 Chu, H-A., Nguyen, A.P. and Debus, R.J. (1995) Biochemistry 34, 5859-5882.7
7 Seibert, M., Tamura, N. and Inoue, Y. (1989) Biochim. Biophys. Acta 974, 185-191.
8 Preston, C. and Seibert, M. (1991) Biochemistry 30, 9611-9624.
9 Ghirardi, M.L. and Seibert, M. (1992) *Research in Photosynthesis* (Murata, N., ed) Vol.2, pp. 357-360, Kluwer Academic Publishers, The Netherlands.
10 Hoganson, C.W., Ghanotakis, D.F., Babcock, G.T. and Yocum, C.F. (1989) Photosynth. Res. 22, 285-293.
11 Blubaugh, D.J. and Cheniae, G.M. (1990) Biochemistry 29, 5109-5118.
12 Blubaugh, D.J., Atamian, M., Babcock, G.T., Golbeck, J.H. and Cheniae, G.M. (1989) Biochemistry 30, 7856-7597.
13 Ghirardi, M.L., Lutton, T.W. and Seibert, M., submitted.
14 Ratilla, E.M.A., Brothers, H.M., II and Kostic, N.M. (1987) J. Am. Chem. Soc. 109, 4592-4599.
15 Blubaugh, D.J. and Cheniae, G.M. (1992) *Research in Photosynthesis* (Murata, N., ed) Vol 2, pp.361-364, Kluwer Academic Publishers, The Netherlands.

S_3 STATE OF THE O_2-EVOLVING COMPLEX: STRUCTURE OF THE MN COMPLEX OF PHOTOSYSTEM II DETERMINED BY X-RAY ABSORPTION SPECTROSCOPY

Wenchuan Liang, Theo A. Roelofs, Gary T. Olsen, Matthew J. Latimer, Roehl M. Cinco, Annette Rompel, Kenneth Sauer, Vittal K. Yachandra, and Melvin P. Klein. Structural Biology Division, Lawrence Berkeley National Laboratory, and Department of Chemistry, University of California, Berkeley, CA 94720, USA.

1. Introduction

The oxygen-evolving complex (OEC) of photosystem II (PS II) in plants and cyanobacteria catalyzes the oxidation of two water molecules to one molecule of dioxygen [1]. A Mn_4 complex in the OEC is believed to cycle through five intermediate states (S_0-S_4) to couple the four-electron oxidation of water with the one-electron photochemistry occurring at the PS II reaction center. After long-term dark adaptation (tens of minutes), the S_1 state predominates; the S_4 state is a transient intermediate that spontaneously returns to the S_0 state upon release of oxygen [2].

Investigation of the structures of the Mn complex at each S-state is essential to a better understanding of the catalytic mechanism of photosynthetic oxygen evolution. We have used X-ray absorption spectroscopy (XAS) to study the local structure of the Mn complex, and have proposed a model for it [3] based on studies of the Mn K-edges and the extended X-ray absorption fine structure (EXAFS) of the S_1 and S_2 states. EXAFS experiments on the S_3 and S_0 states have been difficult because of the dilute nature of samples that can be saturated with single flashes. To explore the structures beyond the S_2 state, we have conducted EXAFS studies on chemically treated S_3 and S_0 states accumulated by continuous illumination at cryogenic temperatures [1]. Recently, however, with a much improved detection system, we have been able to collect a series of Mn K-edge and EXAFS spectra of flash-induced S-state samples prepared under physiological conditions. The K-edge results from these "native" S-states confirm our previous result from the study of the chemically treated S_3 state that the oxidative equivalent accumulated during the S_2 to S_3 transition is not stored on the Mn atoms [T. A. Roelofs et al., submitted].

In this report we present the first EXAFS results for S_3 state samples created under physiological conditions with saturating flashes. Pure S_3 EXAFS spectra were generated from linear combinations of the flash-induced EXAFS spectra, with each coefficient obtained from the best fit of the flash-induced multiline EPR signal (MLS) oscillation pattern to Kok's model [2]. The Fourier transforms and some preliminary fitting of the EXAFS spectra show a shift to longer distance of the second Fourier peak relative to the S_1 and S_2 states. This represents a significant structural change in the complex. In contrast, we find that calcium-depleted preparations advanced to a modified S_3 state do not exhibit such a shift in the second peak.

P. Mathis (ed.), Photosynthesis: from Light to Biosphere, Vol. II, 413–416.

2. Materials and Methods

See accompanying paper by T. A. Roelofs et al. for a description of the PS II-enriched membrane preparation, of the protocol for flash illumination, of the EPR and XAS collection methods, and of the procedure for extracting pure S-state spectra from the spectra of samples illuminated by multiple flashes. EXAFS fitting procedures are described in Ref. 4. Calcium-depleted sample preparation used a low pH/citrate treatment [5].

3. Results

3.1 Fourier Transforms of S_3 Spectra

In Figure 1 we show Fourier transforms (FT's) of EXAFS data for pure S_2 and S_3 states. The FT's for the four samples from which we have extracted the pure S_3 spectrum all show peaks I and II that are significantly smaller than the corresponding peaks of the S_2 spectrum. For the S_3 state, the amplitude of peak II is consistently less than that of peak I, although to varying degrees. As shown by the vertical line in Figure 1, peak II of the S_3 spectra is shifted to a longer distance compared to that observed in S_2 data. Fits to the EXAFS equation (discussed below) confirm this as a lengthening of Mn—Mn distances.

Peak III appears substantially above the noise level in one of the four pure S_3 FT's, and is readily distinguishable from the noise in two others. Due to its small size, the amplitude of peak III is largely affected by differences in noise components. Whether this peak in the S_3 spectra is larger or smaller than that in the S_2 data is therefore uncertain. Furthermore, although all discernible third peaks seem shifted to a higher "apparent distance" than that observed in the S_2 spectrum, to report a definite shift is premature.

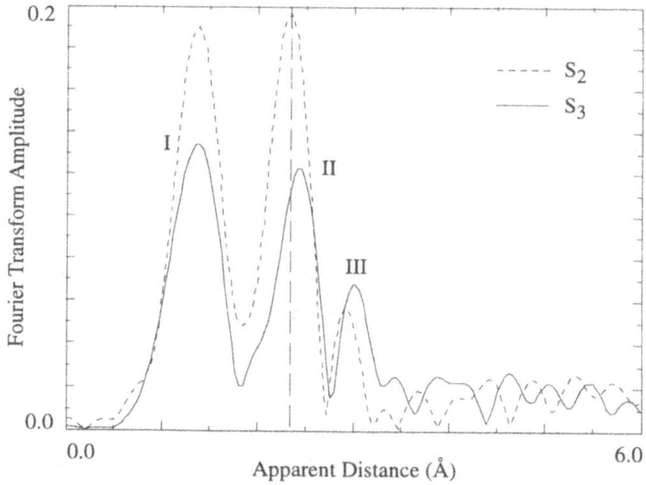

Figure 1. Fourier transforms of the k^3-weighted EXAFS data (3.5-11.5Å^{-1}) of PS II-enriched membranes in pure S_2 [6] and S_3 states. The S_3 spectrum is derived from the averaged k-space data for two samples prepared from the same stock. The vertical line is aligned with the maximum of peak II of the S_2 transform, and it is clear that peak II for the S_3 state is shifted to a longer distance. (Preliminary fits to peak II confirm a lengthening of Mn—Mn interactions.)

In Figure 2 we show the FT's of EXAFS data from a calcium-depleted sample in a modified S_3 state overlaid on a native S_2 state spectrum. Although there are similar reductions in amplitude for peaks II and III as compared to the case for a native S_3 state sample, there is no shift in distance for peak II. Peak III is dramatically reduced in amplitude.

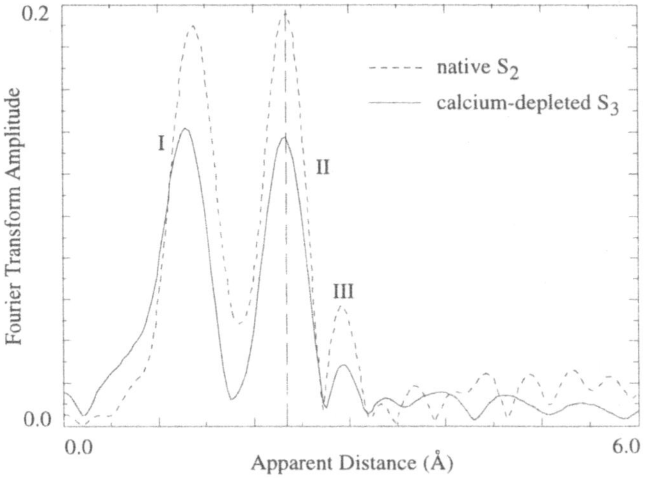

Figure 2. Fourier transforms of the k^3-weighted EXAFS data (3.5-11.5Å^{-1}) of PS II-enriched membranes in native S_2 [6] and calcium-depleted S_3 states. Fits to peak II show no change in distance compared to native S_2 state samples, but do show an increased disorder which accounts for the lower amplitude. (The vertical line is aligned with the maximum of peak II of the S_2 transform.)

3.2 Preliminary Fits of the S_3 Spectra to the EXAFS Equation

S_3 data analysis which is still in progress has led to some preliminary fits for peak II. For the four pure S_3 spectra we have considered, a single shell fit to Mn gives a distance of 2.82 ± 0.04Å with a coordination number of 0.9 ± 0.1 (implying 2 interactions) and a Debye-Waller factor of 0.006 ± 0.001Å^2. (Fit parameters are reported for a 90% confidence interval.) If we instead fit to two Mn—Mn distances, we find separations of 2.82 ± 0.02 and 2.97 ± 0.03Å and coordination numbers of 0.8 ± 0.1 and 0.5 ± 0.1, respectively, with Debye-Waller factors fixed at 0.002Å^2.

Fitting to peak III is more difficult due to its large noise components, but we have averaged the *k*-space data for two samples taken from the same preparation, therefore reducing the noise level and allowing for some tentative conclusions. Peak III can be fit to one Mn—Mn distance, two Mn—Mn distances, or to one Mn—Mn and one Mn—Ca distance; all are around 0.1 to 0.3Å longer than those found in corresponding fits for the S_1 and S_2 states. The two shell fit incorporating Ca has slightly less error than the fit with only Mn. The tentative nature of these fits, however, does not allow for any definite conclusions regarding peak III. The lower amplitude of Fourier peak III in the calcium-depleted S_3 state spectrum is close to the noise level of the data, making fitting yet more difficult, but this behavior agrees well with our model where Ca contributes to peak III [3,4].

4. Discussion

When the Mn complex undergoes transition from the S_1 state to the S_2 state, there is little change observed in the fit parameters for peak II: the average Mn—Mn distance for two interactions changes from 2.74Å to 2.73Å [6]. On transition to S_3, however, we find an increase in single shell fit distance to ~2.8Å. A two shell fit gives one interaction at ~2.8Å and one at ~3.0Å. This pattern of little or no structural change (S_1 to S_2) followed by a structural perturbation (S_2 to S_3) is consistent with evidence from the Mn K-edge spectra of pure S-states [T. A. Roelofs et al., submitted]. Edges from calcium-depleted samples show a similar behavior in that little or no shift is observed in the S_2 to S_3 transition [4], but here the EXAFS spectra do not exhibit a lengthening of the ~2.7Å interaction as seen in the native S_3 state samples. This difference in structure between the native and calcium-depleted S_3 state samples indicates that these two states are quite dissimilar in the core di-μ-oxo bridged structure of the complex.

It is interesting to note that the structural perturbation involved in the S_3 state is also different from the perturbation observed in the S_2 states for ammonia-treated samples [7], fluoride-treated samples [8], and samples illuminated at 130K ($g = 4.1$ EPR signal) [9]. Each of these altered S_2 states gives a two shell Mn—Mn fit of around 2.71 and 2.85Å that is consistent with the alteration of one of the two di-μ-oxo-bridged Mn binuclear units proposed in our model [3]. Our preliminary fits for the S_3 data, however, show an elongation of both bridges. Although each of the three altered S_2 states may have very similar structures, the S_3 state's structure is distinctly different.

5. Conclusion

EXAFS results show that there is a significant change in the Mn—Mn distances on advancing from the S_2 to the S_3 state. This indicates a change in the di-μ-oxo bridged structure proposed for the S_1 and S_2 states. We speculate that this can be due to protonation of the oxo bridges, loss of one or more of the oxo bridges, or the formation of a peroxo intermediate.

This research was supported by the Director, Office of Basic Energy Sciences, Division of Energy Biosciences of the U.S. Department of Energy, under Contract No. DE-AC03-76SF00098, and by a grant from the National Science Foundation (DMB91-04104). We acknowledge Dr. Holger Dau for his insightful discussion of early results and his help with preliminary data collection. We thank Joy C. Andrews for her help at the X-ray beamlines.

References

1 Sauer, K., Yachandra, V. K., Britt, R. D., and Klein, M. P. (1992) in Manganese Redox Enzymes (Pecoraro, V. L., ed.) pp. 141-175, VCH Publishers, New York
2 Kok, B., Forbush, B., and McGloin, M. (1970) Photochem. Photobiol. 11, 457-475
3 Yachandra, V. K., DeRose, V. J., Latimer, M. J., Mukerji, I., Sauer, K., and Klein, M. P. (1993) Science 260, 675-679
4 Latimer, M. J. (1995) Ph.D. Thesis, University of California, Berkeley, CA
5 Ono, T. and Inoue, Y. (1988) FEBS Lett. 227, 147-152
6 DeRose, V. J., Mukerji, I., Latimer, M. J., Yachandra, V. K., Sauer, K., and Klein, M. P. (1994) J. Am. Chem. Soc. 116, 5239-5249
7 Dau, H., Andrews, J. C., Roelofs, T. A., Latimer, M. J., Liang, W., Yachandra, V. K., Sauer, K., and Klein, M. P. (1995) Biochemistry 34, 5274-5287
8 DeRose, V. J., Latimer, M. J., Zimmermann, J.-L., Mukerji, I., Yachandra, V. K., Sauer, K., and Klein, M. P. (1995) Chem. Phys. 194, 443-459
9 Liang, W., Latimer, M. J., Dau, H., Roelofs, T. A., Yachandra, V. K., Sauer, K., and Klein, M. P. (1995) Biochemistry 34, 5274-5287

RECENT ADVANCES TOWARD A STRUCTURAL MODEL FOR THE PHOTOSYNTHETIC OXYGEN-EVOLVING MANGANESE CLUSTER

Matthew J. Latimer, Holger Dau‡, Wenchuan Liang, Joy C. Andrews, Theo A. Roelofs, Roehl M. Cinco, Annette Rompel, Kenneth Sauer, Vittal K. Yachandra and Melvin P. Klein. Structural Biology Division, Lawrence Berkeley National Laboratory and Department of Chemistry, University of California, Berkeley, CA 94720, USA, and ‡FB Biologie/Botanik, Phillips-Universität Marburg, D-35032, Germany.

1. Introduction

Photosynthetic water oxidation occurs in the oxygen evolving complex (OEC) of photosystem II (PS II). One-electron photo-oxidations in the reaction center of PS II are coupled to the four-electron oxidation of water in the OEC. As the PS II reaction center sequentially extracts electrons, the OEC cycles through five intermediate oxidation states (S_0-S_4) where S_0 is the least oxidized and S_4 is a transient state which decays to S_0 with the release of a dioxygen molecule. A complex of four manganese atoms has been shown to function in charge accumulation and is thought to form the catalytic site for the water oxidation reaction. The structure of this manganese complex and the mechanism of water oxidation have been the subject of vigorous inquiry by a great many research groups [1].

We have proposed a model [2] for the Mn complex based largely upon extended X-ray absorption fine structure (EXAFS) spectroscopy of Mn in PS II preparations. The core feature of this model is two di-μ-oxo bridged Mn pairs yielding two Mn-Mn interactions of ~2.7Å. The Mn pairs are connected by a single bridging oxygen which results in another Mn-Mn distance of ~3.3Å. These distances and structural interpretations arose from a comparison of the Mn-EXAFS fitting results of PS II compared to known structural motifs from a large number of inorganic Mn complexes [3]. Experiments on partially oriented preparations indicated that the scattering vectors giving rise to the ~2.7Å interaction are oriented close to the plane of the membrane (averaging 60°±7° from the membrane normal), while the vectors contributing to the ~3.3Å interaction are aligned closer to the membrane normal (43°±10°) [4]. This angular information was included in the model to indicate the position of the complex within the membrane. A closely bound Ca was added to the model, based on EXAFS from Sr-substituted preparations, and a directly-ligated chloride was also included, based on a large body of indirect evidence for a chloride binding site in PS II. Finally, a histidine ligand to Mn was included based on electron spin-echo envelope modulation experiments.

In this paper we describe a revised structural model for the manganese cluster in PS II based on the results of recent studies on PS II preparations subjected to various treatments (Sr-substitution, ammonia or fluoride inhibition, or low temperature illumination to produce the S_2-state exhibiting the g=4.1 EPR signal). The information gained from these studies has helped confirm and extend our structural model.

P. Mathis (ed.), Photosynthesis: from Light to Biosphere, Vol. II, 417–420.

2. Results and Discussion

2.1 The Mn-Mn Interaction at ~2.7Å

The second Fourier peak in the EXAFS spectrum of PS II is generally agreed to arise from 2-to-3 di-μ-oxo bridged Mn pairs. Although there is certainly more than one Mn-Mn interaction, spectra from unmodified PS II preparations in the S_1-state and in the S_2-state where the multiline EPR signal is exhibited show only a single resolved distance [3]. In recent work we have shown that this distance degeneracy is lifted in the S_2-state where the g=4.1 EPR signal is exhibited [5] and in preparations which have been inhibited with ammonia [6] or with fluoride [7]. The Fourier transforms of the EXAFS from these preparations are shown in Figure 1. In each of these preparations, the second Fourier peak (peak II) is observed to have a smaller amplitude than in control samples. Also, in each case, curve-fitting results have indicated increased disorder and that the data can be fit with two Mn-Mn interactions, one at approximately 2.69-2.73Å and a second interaction at ~2.78-2.87Å.

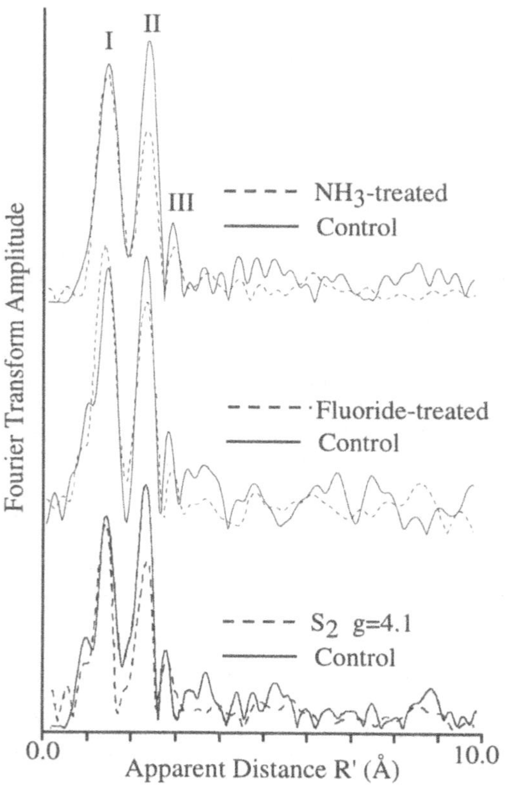

Figure 1. Fourier transforms of the EXAFS from ammonia or fluoride treated PS II samples in the S_2-state, and from an untreated sample illuminated at 140K to generate the S_2-state exhibiting the g=4.1 EPR signal. For comparison each trace is plotted with a control sample in the S_2-state exhibiting the multiline EPR signal.

A variety of models can be employed to explain the increased distance heterogeneity including direct binding of ammonia to the Mn complex, ligand exchange reactions which are affected by temperature, protonation of oxo-bridges, or changes in oxidation state of the individual dimers dependent upon ligand environment. Fundamentally, however, the resolution of two different ~2.7Å distances (one relatively unchanged and the other lengthened from that found in untreated/unilluminated preparations) implies two relatively independent Mn-Mn dimeric units.

EXAFS spectra from partially oriented ammonia-treated membranes has yielded further information about the relative angles of the individual manganese vectors [6]. EXAFS curve fitting revealed Mn-Mn interactions at 2.72 and 2.87Å, and changes with angle of the apparent coordination number of each vector allowed determination of angles of 55° (2.72Å interaction) and 67° (2.87Å interaction) relative to the membrane normal. Angles determined for the ammonia-treated complex are likely to be similar to the angles found in the untreated complex.

2.2 Ca/Sr experiments

Ca is an essential co-factor for oxygen evolution and a great deal of effort in recent years has centered on elucidating its role in PS II. We have shown, through EXAFS experiments on PS II preparations which have had Sr substituted for Ca that the amplitude of the third Fourier peak is enhanced relative to Ca-containing preparations (Fig. 2) [2,8]. Detailed fitting of the EXAFS data from these preparations [8] has indicated that this change in amplitude can be explained by an exchange of strontium for calcium in a binding site ~3.45Å from a manganese atom. By comparison with compounds of known structure which contain calcium-metal distances, we proposed that calcium and manganese may be bridged by one or more single oxygen atoms derived from carboxylate or phenolate protein residues, protein backbone carbonyls, or possibly water or hydroxide. Although the data can be explained with just calcium contributing to peak III, a careful consideration of chemically reasonable structural models indicates the most likely explanation is that both Mn-Mn and Mn-Ca interactions contribute to peak III.

Spectra from Ca-depleted preparations appear to be further modified in the amplitude of all three fourier peaks. Peak III is definitely decreased in these preparations and there appears to be increased heterogeneity in peak II as well. The relevance of the structure of the Ca-depleted preparations to that in untreated preparations is unclear.

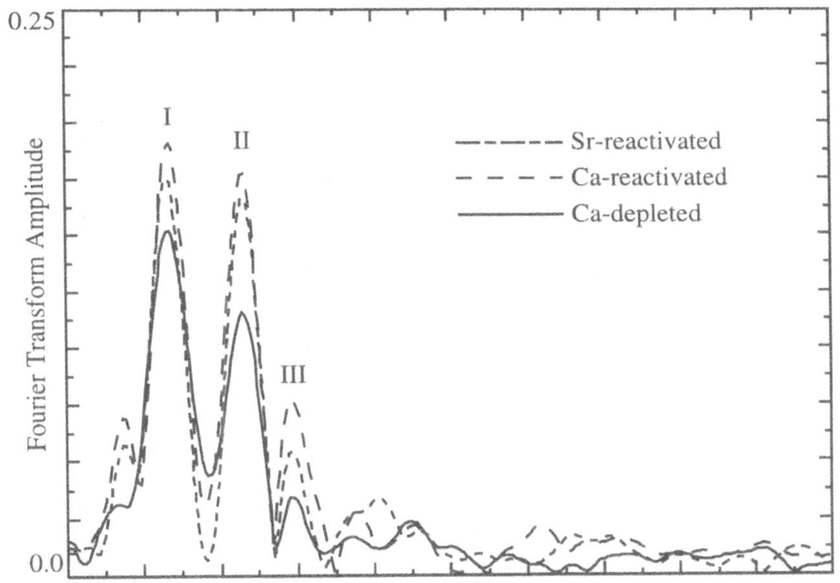

Figure 2. Fourier transforms of EXAFS spectra of Ca-depleted preparations in the S_1-state which have been reactivated with Sr, Ca, or with no reactivation. The major difference between Ca- and Sr-reactivated preparations is seen in the amplitude of peak III. In the Ca-depleted spectrum, changes are observed in amplitude of all three peaks relative to the samples containg Ca and Sr. Calcium depletion of PS II preparations was achieved by use of a low pH/citrate method [9]. Calcium and strontium reactivation procedures and sample characterization are described in references 8 and 10.

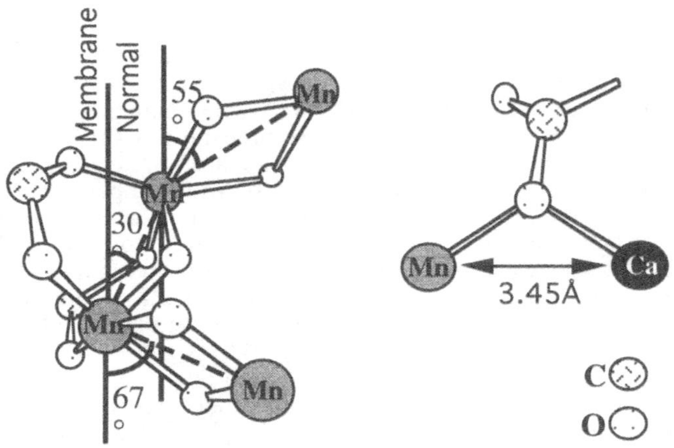

Figure 3. Revised structural model for the Mn cluster of PS II. One possible positioning of the cluster within the membrane is shown (left). Also shown is a proposal for a calcium binding site (right), although which Mn atom of the complex is involved in this site is undetermined at this time.

The revised model.
Our revised model for the Mn cluster in PS II is presented in Figure 3 [6]. The histidine and chloride ligands are not shown, although at least one histidine ligand has been proven and a chloride ligand is very likely. The model we propose here for the manganese cluster in PS II is a combination of structural elements and the exact positions of these elements relative to each other is unknown. In addition, we note that there appear to be significant structural changes occurring in the S_2-to-S_3 transition (see Liang et al., this volume), thus the structural model presented here is at present applicable to the S_1 and S_2 states.

This research was supported by the Director, Office of Basic Energy Sciences, Division of Energy Biosciences of the U.S. Department of Energy, under Contract No. DE-AC03-76SF00098, and by a grant from the National Science Foundation (DMB91-04104).

References

1 Sauer, K., Yachandra, V. K., Britt, R. D., and Klein, M. P. (1992) in Manganese Redox Enzymes (Pecoraro, V. L., ed.) pp. 141-175, VCH Publishers, New York
2 Yachandra, V. K., DeRose, V. J., Latimer, M. J., Mukerji, I., Sauer, K., and Klein, M. P. (1993) Science 260, 675-679
3 DeRose, V. J., Mukerji, I., Latimer, M. J., Yachandra, V. K., Sauer, K., and Klein, M. P. (1994) J. Am. Chem. Soc. 116, 5239-5249
4 Mukerji, I., Andrews, J. C., DeRose, V. J., Latimer, M. J., Yachandra, V. K., Sauer, K., and Klein, M. P. (1994) Biochemistry 33, 9712-9721.
5 Liang, W., Latimer, M. J., Dau, H., Roelofs, T. A., Yachandra, V. K., Sauer, K., and Klein, M. P. (1994) Biochemistry 34, 5274-5287
6 Dau, H., Andrews, J. C., Roelofs, T. A., Latimer, M. J., Liang, W., Yachandra, V. K., Sauer, K., and Klein, M. P. (1995) Biochemistry 34, 5274-5287
7 DeRose, V. J., Latimer, M. J., Zimmermann, J.-L., Mukerji, I., Yachandra, V. K., Sauer, K., and Klein, M. P. (1995) Chem. Phys. 194, 443-459
8 Latimer, M. J., DeRose, V. J., Mukerji, I., Yachandra, V. K., Sauer, K., and Klein, M. P. (1995) Biochemistry in press.
9 Ono, T. and Inoue, Y. (1988) FEBS Lett. 227, 147-152
10 Latimer, M. J. (1995) Ph.D. Thesis, University of California, Berkeley, CA

INORGANIC COFACTOR RELATIONS AND PROPERTIES OF PHOTOSYSTEM II IN THYLAKOIDS OF *PROCHLOROTHRIX HOLLANDICA*.

Jun Wang and Peter H. Homann, Dept. of Biological Science,
Florida State Univ., Tallahassee, Florida 32306-3015, USA.

1. Introduction

In general, the principal proteinaceaous and inorganic constituents of the photosynthetic water oxidizing complex (WOC) of Photosystem II (PSII) are highly conserved among organisms capable of oxygenic photosynthesis. However, differences appear to exist between the eukaryotic and prokaryotic WOCs regarding aspects of their requirements for the inorganic cofactors Cl^- and Ca^{2+} (1). Important determinants for the binding of the two inorganic cofactors in the WOC of green plants are the extrinsic polypeptides of approximately 33, 23 and 17 kD molecular mass (the *psbO*, *psbP* and *psbQ* gene products, resp.). The 23 and 17 kD polypeptides form a shield that sequesters the inorganic ions at their binding sites and, in the case of Cl^-, greatly increases its apparent binding constant. Since the prokaryotic WOC lacks these two polypeptides, or analogs thereof, its cofactor relations must necessarily be different from that of the eukaryotic complexes (1).

With the above considerations in mind, we have investigated aspects of function and cofactor requirement of the WOC in thylakoid preparations from the prokaryotic alga *Prochlorothrix hollandica*. Among the cyanobacteria, *Prochlorothrix* is attractive for comparative studies of a WOC that constitutively lacks the 23 and 17 kD extrinsic polypeptides because this organism uses a Chl *a/b* light-harvesting protein and in this respect resembles green algae and plants (2). It was found that functional Cl^- is tightly bound in the PSII of *Prochlorothrix* in spite of the absence of the 17 and 23 kD polypeptides. This must be due to compensatory structural properties. On the other hand, the cofactor depleted WOC of *Prochlorothrix* was much more unstable than that of higher plants even when its 17 and 23 kD polypeptides had been removed.

2. Materials and Methods

Highly active thylakoid preparations from *Prochlorothrix* (kindly provided by Dr. G.S. Bullerjahn) were prepared by modifying published procedures (2-4): cells were broken by a combination of sonication and *Yeda* Press treatment while suspended 0.4 M sucrose, 0.5 M betaine, 40 mM Na-Tricine pH 7.5 and 10 mM $MgCl_2$. The thylakoid fraction used in this study was collected by centrifugation for 20 min at 100,000 g after a precentrifugation at 3,500 g and was suspended in the isolation medium. Its photosystem

P. Mathis (ed.), Photosynthesis: from Light to Biosphere, Vol. II, 421–424.

II activity varied (250 - >500 μmol O_2 (mg Chl ·h)$^{-1}$) and remained almost unchanged during storage at -80 C. All assays were performed as described elsewhere (5).

3. Results

The requirements for optimal water oxidizing activities of *Prochlorothrix* thylakoids resembled those reported for preparations from typical cyanobacteria (3,4,7) in that relatively high concentrations of some protein-structure stabilizing solute as well as Mg^{2+} or Ca^{2+} had to be present in the assay medium. A unique property of PSII in the *Prochlorothrix* thylakoids was detected in thermoluminescence measurements: After a 90 min incubation in the dark at room temperature, the emission was not maximal after an illumination with 1 or 2 light flashes as is normally the case, but after three flashes instead (Fig. 1A). Apparently, one of the usually stable oxidants in the WOC, i.e. Tyr_D^+ or S_1, had become reduced in the dark.

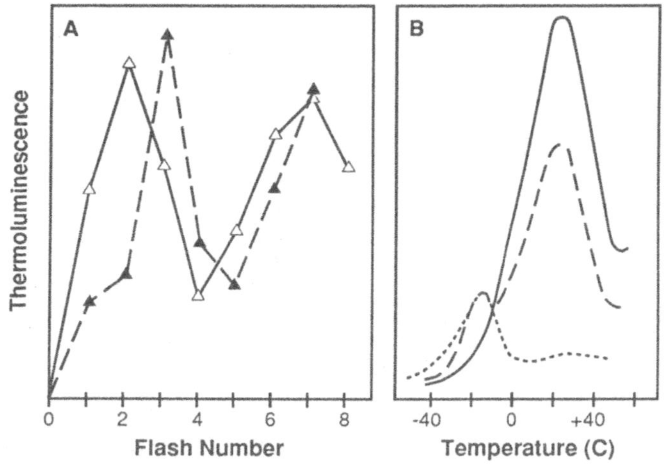

Fig.1. Thermoluminesence of *Prochlorothrix* thylakoids. A. Peak intensities after illumination with single flashes at pH 7.5. Δ: after 10 min dark adaptation at 25C; ▲: after 90 min in the dark at 25C. B. Emissions after illuminations at -30C in the presence of DCMU: —— : control; ---- : Cl^--depleted after readdition of 20 mM NaCl; ····· : NH_2OH treated control to remove functional Mn for demonstration of A_T emission band.

When *Prochlorothrix* thylakoids were suspended in Cl^--free media, they did not lose activity as fast is known for PSII preparations from higher plants after removal of their Cl^- sequestering 17 and 23 kD polypeptides. This could be attributed to rather low apparent dissociation constants for activating Cl^- of about 100 μM at pH 6.0 and around 1 mM at pH 7.5 which resembled those determined for thylakoids and PSII membranes from green plants carrying their 17 and 23 kD polypeptides (Table 1). As in other systems, reactivation of the Cl^- depleted PSII was not specific for Cl^- but could be achieved with other anions with an effectiveness $Cl^- > Br^- > I^- > NO_3^-$. The WOC in Cl^- depleted *Prochlorothrix* thylakoids was, however, unusually unstable, causing the

Table 1. Comparison of the apparent dissociation constants of activating Cl^- for PSII of various types of Cl^- depleted photosynthetic membranes. Values given in mM. -extr. pp and +extr. pp refer to the extrinsic 17 and 23 kD polypeptides.

pH	*Spinacia oleracea* thylakoids[1)]	*Phytolacca americana* PSII membranes[2)] - extr. pp	+ extr. pp	*Prochlorothrix hollandica* thylakoids
6.0	-	1.1	0.12	0.1
7.2	0.9	8	1.5	1[3)]

[1)] from Kelley and Izawa; Biochim. Biophys. Acta 502 (1978), p. 205
[2)] from Homann, Biochim. Biophys. Acta 934 (1988), p. 6
[3)] value determined for pH 7.5

activity after optimal reactivation to always remain significantly below that of Cl^--sufficient controls (Table 2), especially when incubated at 25C (not shown). An appearance of the A_T-thermoluminescence band at -18 C in the emission of the Cl^--reactivated thylakoids (Fig. 1B) suggested loss of functional Mn (8) , but attempts to restore high activities through published photoreactivation procedures were not successful.

To obtain a significant loss of activity in the absence of Ca^{2+}, stabilizer deficient media had to be used that lacked Mg^{2+} as well (Table 2). However, any treatments designed to deplete Ca^{2+} were even more detrimental to the WOC than Cl^- depletion. None of the lost activity could be recovered by added Ca^{2+}, or by a combination of Ca^{2+} and Mn^{2+} which had been found to be effective with preparations from typical cyanobacteria under similar conditions (4,9).

Table 2. Photosystem II activities of *Prochlorothrix* thylakoids after treatments allowing Cl^- or Ca^{2+} dissociations. Unless indicated, assays and incubations (but not the $-Cl^-$ wash) were at pH 7.5 in the absence of stabilizing 0.5 M betaine + 0.4 M sucrose. Salts provided 20 mM Cl^-. "Controls" are "washed $+Cl^-$" or "before incubation at 25C" resp.; "Reactivated" refers to assay + 20 mM NaCl or 10 mM $CaCl_2$, resp.

Assay condition	Wash with Cl^- free medium pH 6.0	pH 7.5	20 min incubation with salts NaCl	$MgCl_2$	$CaCl_2$
	$\mu mol\ O_2\ (mgChl \cdot h)^{-1}$				
Control (+bet./sucr.)	-	-	255	320	300
Control	204	240	140	175	210
After treatment	40	40	65	115	180
"Reactivated"	140	144	65	-	-

4. Discussion

The goal of the present investigation was to better understand the cofactor relations of a PSII that constitutively lacks the 17 and 23 kD extrinsic polypeptides which appear to be critical for the PSII of green plants to be able to protect its functional Cl^- and Ca^{2+}. It was shown that the design of the PSII of *Prochlorothrix* allowed functional Cl^- to be bound as effectively as it is in the green plant PSII when its 17 and 23 kD polypeptides are in place. Perhaps the more hydrophobic *psbO* protein (10) of *Prochlorothrix* is responsible for this compensatory structural property. The observed stabilizing effect of Mg^{2+} in general and with respect to functional Ca^{2+} is a known but apparently yet unexplored feature of the cyanobacterial WOC (c.f. refs. 3,7). Unexpected, however, was the considerable degree of irreversibility of the activity loss of the *Prochlorothrix* WOC after removal of Cl^- which presumably was a consequence of a labilization of the functional Mn_4-cluster at the catalytic site of the WOC. The problem of irreversible inactivation was even more pronounced after treatments designed to remove functional Ca^{2+}. In fact, all attempts to recover any activity failed. Obviously, the integrity of the microenvironment of the catalytic site in the *Prochlorothrix* WOC depends much more strongly on the presence of its inorganic cofactors than it does in green plants. Another indication for a structural difference at the catalytic centre was a relatively rapid reduction of one of the usually dark-stable oxidants for which apparently there is precedence, at least qualitatively and *in vivo*, in other cyanobacterial PSIIs (11). Yet, all these features do not reflect fundamentally different properties of the WOC of *Prochlorothrix* and that of green plants: a reduction in the dark of a stable oxidant in the WOC has been reported for PSII membranes from higher plants at a higher pH than used in this study (12), and with the same system it has been found that Cl^- deficiency facilitated Mn-dissociation when the temperature was raised above 35 C (13).

References

1 Debus, T.J. (1992) Biochim. Biophys. Acta 1102, 269
2 Miller,K.R., Jacob, J.S. Burger-Wiersma,T. and Matthijs, H.C.P (1988) J. Cell Sci. 91, 577
3 Stewart, A.C and Bendall, D.S. (1981) Biochem. J. 194, 877
4 Satoh,K and Katoh,S. (1985) Biochim. Biophys. Acta 806, 221
5 Rashid, A. and Homann, P.H. (1992) Biochim. Biophys. Acta 1101, 303
6 Mamedov, M., Hayashi, H. and Murata, N. (1993) Biochim. Biophys. Acta 1142, 1
7 Stewart, A.C., Ljungberg, U., Akerlund, H.E. and Andersson, B. (1985) Biochim. Biophys. Acta 808, 353
8 Vass, I. and Inoue, Y. (1992) in The Photosystems: Structure, Function and Molecular Biology (Barber, A., ed.) pp.259-294, Elsevier, Amsterdam
9 Pistorius, E.K and Schmid, G.H. (1984) FEBS Lett. 171, 173
10 Mor, T.S., Post, A.F. and Ohad, I. (1993) Biochim. Biophys. Acta 1141, 206
11 Engels, D.H., Lott, A., Schmid, G.H. and Pistorius, E.K. (1994) Photosynth. Res. 42, 237
12 Plijter, J.J., De Groot, A., Van Dijk, M.A. and Van Gorkom, H.J. (1986) FEBS. Lett. 195, 313
13 Nash, D., Miyao, M. and Murata, N. (1985) Biochim. Biophys. Acta 807, 127

TERBIUM SUBSTITUTION OF THE CALCIUM-BINDING SITES IN THE OXYGEN-EVOLVING COMPLEX OF PHOTOSYSTEM II.

C. Hatch§, M. Grush†, R. Bradley§, R. LoBrutto§, and S. Cramer†, and W. Frasch
§Center for the Study of Early Events in Photosynthesis, Department of Botany, Arizona State University, Tempe, AZ 85281-1601, and †The Department of Applied Science, University of California-Davis, Davis, CA

Introduction

The oxygen-evolving complex (OEC) advances sequentially through a series of oxidation states (S_n-states) to supply single electrons to the photosystem II reaction center. In most of the S-state transitions, the electron donor is a cluster of four manganese, which gives rise to a multiline EPR signal at g=2 or an EPR signal at g=4 in the S_2-state (1). Analysis of the manganese cluster by EXAFS spectroscopy has provided evidence that the cluster is composed of two μ-oxo-bridged Mn dimers with a Mn-Mn distance of 2.7 Å that are separated by 3.3 Å via a carboxylato bridge (2).

The OEC requires Ca^{2+} as an essential cofactor (1). A citrate (pH 4) wash depletes the OEC of Ca^{2+} and inhibits the S_3-S_0 transition (3) which can be partially reversed by the addition of Ca^{2+} or to a lesser extent by Sr^{2+}. When depleted of Ca^{2+} in this manner, the OEC does not give rise to the multiline signal in the S_2-state. However, a split EPR signal centered at g=2 with a line width of about 160 G appears in the S_3-state (4). From Mn-EXAFS studies, a scatterer 3.3-3.4 Å away from the Mn cluster has been attributed to Mn, Ca, or a combination of both (2, 5-7).

Lanthanides serve as probes of Ca^{2+}-binding sites due to the close similarity in their ionic radii, and should clearly be distinguishable from Mn by EXAFS spectroscopy. Addition of lanthanides to Ca^{2+}-depleted PSII preparations prevented advancement from the S_1-state, and was postulated to be close to tyrosine Y_D (8).

Results presented here demonstrate that Tb^{3+} can displace Ca^{2+} from its binding site in the OEC without prior depletion of the cation. The Mn-cluster of the Tb^{3+}-substituted preparations remains structurally intact and can advance to the S_2-state. Although the Tb^{3+} must be very close to the manganese cluster, Tb^{3+} is not needed to fit the 3.4 Å distance in the Mn-EXAFS spectrum of these preparations.

Procedure

To prepare Tb^{3+}-substituted PSII membranes, the PSII preparations (3) were washed in 10 mM NaCl and 25 mM Mes, pH 6.5 and resuspended to 0.2 mg chlorophyll/ml in the same buffer also containing 30 μM $TbCl_3$. The sample was washed repeatedly with 1 mM EDTA to obtain 8 Tb/4Mn after a final resuspension in SMN. Some samples were washed with 70 μM NTA, pH 6.5 until approximately 2 Tb/4 Mn was obtained after a final resuspension in SMN. Continuous wave EPR spectroscopy was performed using an X-band Bruker ESP 300E EPR spectrometer with an Oxford Instruments liquid He cryostat.

P. Mathis (ed.), Photosynthesis: from Light to Biosphere, Vol. II, 425–429.

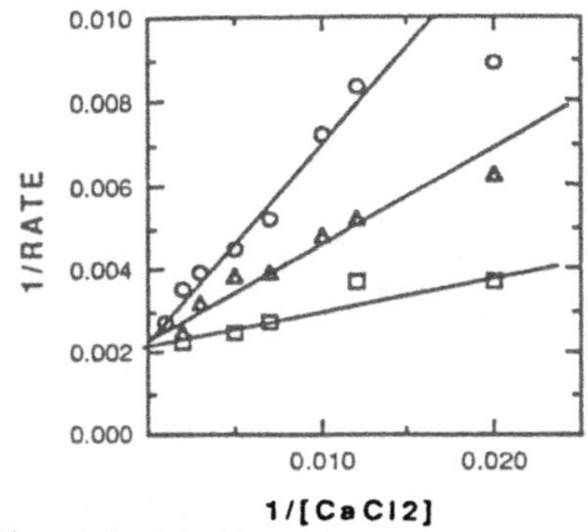

Figure 1. Double reciprocal plot of the rate of water oxidation catalyzed by PSII preparations at 30 (p), 60 (Δ), and 120 (m) μM $TbCl_3$.

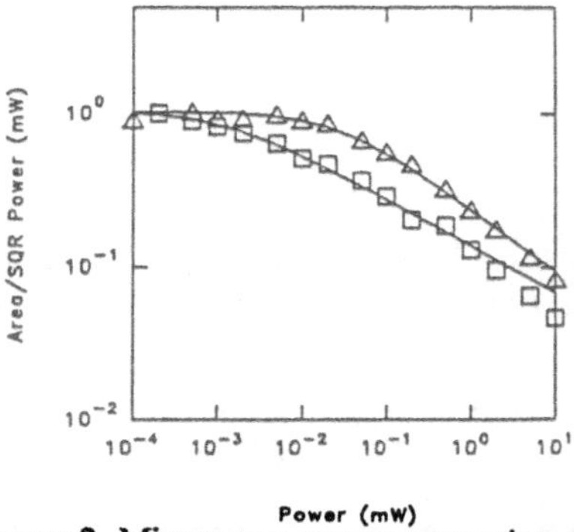

Figure 2. Microwave power saturation of Y_D°. Tb^{3+}-substituted PSII preparations containing 1.7 Tb/4 Mn were dark adapted 10 m at 25 C (p) then illuminated at 200 K for 5 m (Δ).

Mn K-edge EXAFS data were taken on beamlines X19A and X10C at the National Synchrotron Light Source using Si(220) and Si(111) monochromator crystals, respectively, and a vertical entrance slit of 2 mm. Samples were kept at 12-18K with an Oxford line and the data sets were averaged together. Data were analyzed using the EXAFSPAK software as obtained from the Biotechnology Group at the Stanford Synchrotron Radiation Laboratory in addition to our own data reduction software. A single scattering approximation to the EXAFS equation was used with tabulated phase and amplitude functions to fit the data over a *k*-range of 3.5-11.5 $Å^{-1}$. The Fourier-filtered data were back-transformed over several different ranges and analyzed before fitting the raw data. Coordination numbers are reported on a per Mn basis.

Results and Discussion

Figure 1 shows a double-reciprocal plot of the rate of water oxidation catalyzed by PSII preparations as a function of the Ca^{2+} concentration at various fixed concentrations of $TbCl_3$. These results indicate that Tb^{3+} is a linear competitive inhibitor of the binding of Ca^{2+} to the OEC. By replotting the slopes of Fig. 1 as a function of the Tb^{3+} concentration, Tb^{3+} was found to have a K_D = 5.05 μM for the Ca^{2+}-binding sites. Although the Ca^{2+} binding site(s) have been measured to have a K_D = 1 mM under conditions similar to those of Fig. 1, Ca^{2+} binds in a rapid equilibrium ordered manner such that multiple catalytic turnovers of the enzyme typically occur before Ca^{2+} can dissociate from these sites (9). Thus, addition of $CaCl_2$ to the assay medium is normally not needed under these conditions to obtain maximal rates of water oxidation. The fact that it is not necessary to deplete PSII preparations of Ca^{2+} prior to Tb^{3+} substitution probably results from the 200 fold increase in site affinity for Tb^{3+} as compared to Ca^{2+}.

The addition of lanthanides to Ca^{2+}-depleted PSII preparations was previously found to inhibit the S_1-S_2 state transition (7). A sensitive measure of this transition is the increase in the microwave power needed to saturate the Y_D° signal in the S_2-state from that in S_1 due to a change in the relaxation properties of the fast-relaxing Mn-cluster (10). Figure 2 shows power saturation curves of Y_D° in Tb^{3+}-substituted PSII preparations before and after illumination at 200 K to induce formation of the S_2-state. The relaxation enhancement of Y_D° observed upon illumination shows that after Tb^{3+} substitution into the Ca^{2+}-sites by the methods reported here, the OEC can still advance to the S_2-state.

Figure 3 shows the light-minus-dark difference EPR spectra of PSII preparations

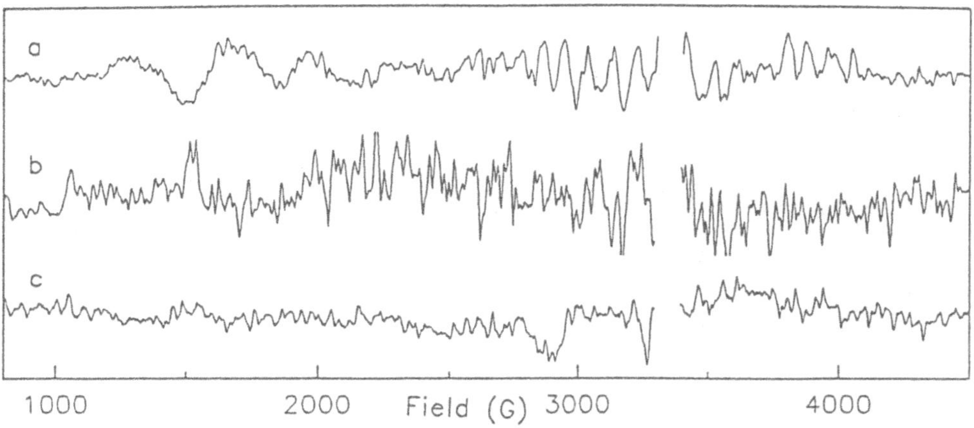

Figure 3. The effect of Tb^{3+} substitution on the S_2-state EPR signals from the Mn-cluster in the OEC. Light minus dark spectra of control PSII preparations (a), and PSII preparations that contain 1.7 Tb/4 Mn (b), and 7.6 Tb/4 Mn (c).

before and after substitution of Ca^{2+} with Tb^{3+}. After addition of Tb^{3+} sufficient to inhibit 90 % of water oxidation activity, washing the PSII preparations with EDTA results in a preparation that contains about 8 Tb/4 Mn as determined by atomic spectroscopy. As shown in spectrum c, neither the g=4 nor the multiline EPR signals were visible in these Tb^{3+}-substituted samples. Washing the Tb^{3+}-substituted PSII preparations with NTA, resulted in samples that contain about 2 Tb/4 Mn. As shown in spectrum b, these remaining two Tb^{3+} are sufficient to prevent the observation of the EPR signals from the manganese in the OEC. Since the data of Fig. 2 established that illumination at 200 K induced the S_2-state, the absence of these EPR signals must result from manganese cluster relaxation caused by the faster relaxing Tb^{3+}. The complete loss of these EPR signals suggests that the sites at which Tb^{3+} is bound are close to the manganese cluster.

The split-EPR signal formed in the S_3-state was also not observed in these Tb^{3+}-substituted samples, which suggests that the species that gives rise to this radical is proximal to the bound Tb^{3+} as well. Although it was still possible to observe the EPR signal from $Y_D°$ in the presence of the bound Tb^{3+} (Fig. 3), it is noteworthy that the relaxation enhancement of this signal resulting from the S_1-S_2 transition is significantly larger than that for Ca^{2+}-containing samples (9). Thus, while the faster relaxing Tb^{3+} contributes significantly to the relaxation enhancement of $Y_D°$, it is too far away from the radical to eliminate this signal.

The Mn K-edge EXAFS of PSII samples which have had Tb^{3+} substituted for Ca^{2+} are presented in Figure 4 along with the Fourier transform of the data. Also shown are the best fit to the data and the residual resulting when the fit is subtracted from the raw data. Three main peaks are observable in the Fourier transform. The first peak can be fit with a shell of O at 1.87 Å and a shell of N at 2.23 Å. Adding a third component results in a shell of O at 1.88 Å and shells of N at 2.21 Å and 2.44 Å. This reduces the fit error by a factor of 4 with very little change in the components of the second and third peak. We attempted to fit the second and third peaks with a variety of backscatterers: C, N, O, S, Cl, Ca, Mn, and Tb. The second peak was nicely simulated by 0.97 Mn at a distance of 2.75 Å. The third peak was best fit using 0.47 Mn at a distance of 3.36 Å. These

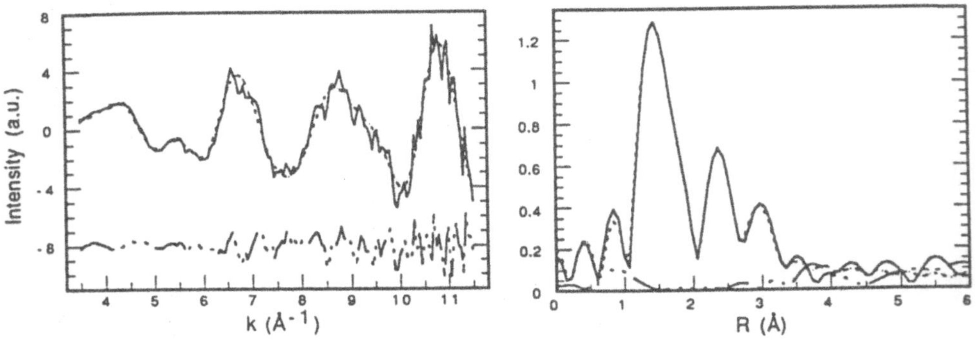

Figure 4. Left: Mn EXAFS on PSII preparations containing about 8 Tb/4 Mn. Right: Fourier Transforms of the Mn EXAFS, data has not been phase corrected so true distances will be ~0.3 Å longer. Raw data (solid line), best fit (dotted line), and residual (dotted and dashed line).

coordination numbers and distances are consistent with those that have been reported for control S_1 preparations (5, 6). Neither peak requires another component for the fit, as can be seen upon examination of Figure 4, especially noting the residual. This suggests that the Ca^{2+} binding site is most likely located greater than 3.6 Å away from the Mn cluster.

A search profile for the most likely Mn-Tb distance yielded a shell of 0.6 Tb at 4.05Å. However, it must be cautioned that upon addition of this component our fit becomes over determined, the quality of the fit is only increased by 7%, and the corresponding peak in the fourier transform is barely above the noise level of the data. Fitting of our data with Tb or Ca instead of Mn resulted in a fit error which was worse by more than a factor of five. We also note that attempts to include a shell of Tb at either 3.41 Å or 4.36 Å changed the quality of the fit by less than 2%.

From the data presented here we conclude that the Ca^{2+} binding site is most likely located at least 4 Å away from the Mn cluster.

References

1 Debus, R. (1992) *Biochimica et Biophysica Acta* 1102, 269-352.
2 Yachandra, V., DeRose, V., Latimer, M., Mukerji, I., Sauer, K., & Klein, M. (1993) *Science* 260, 675-678.
3 Boussac, A. & Rutherford, A. W. (1988) *Biochemistry* 27, 3476-3483.
4 Boussac, A., Zimmermann, J., & Rutherford, A. (1989) Biochemistry 28, 8984-8989.
5 DeRose, V., Mukerji, I., Latimer, M., Yachandra, V., Sauer, K., & Klein, M. (1994) *J. Am. Chem. Soc.* 116, 5239-5249.
6 Penner-Hahn, J., Fronko, R., Pecoraro, V., Yocum, C., Betts, S., & Bowlby, N. (1990) *J. Am. Chem. Soc.* 112, 2549-2557.
7 Bakou, A., Buser, C., Dandulakis, G., Brudvig, G., & Ghanotakis, D. (1992) *Biochimica et Biophysica Acta* 1099, 131-136.
8 Nugent, J. H. A., MacLachlan, D. J., Rigby, S. E. J. and Evans, C. W. (1993) *Photosynthesis Research* 38, 341-346.
9 Frasch, W., and Mei, R. (1987) *Biochemistry* 26, 7325-7332.
10 Boussac, A., & Rutherford, A. W. (1994) *Biochemistry* 269, 12462-12467.

Acknowledgments
Thanks to T. Mattioli, A. Boussac, and J. Christiansen for helpful scientific discussions and J. Christiansen for assistance with data acquisition. Center for the Study of Early Events in Photosynthesis publication #246.

PHOTOASSEMBLY OF THE TETRA-Mn SITE OF PHOTOSYNTHETIC WATER OXIDATION: EPR EVIDENCE FOR DARK LIGATION OF A BINUCLEAR Mn INTERMEDIATE.

Gennady M. Ananyev & G. Charles Dismukes

Princeton University, Departament of Chemistry,
H. Hoyt Laboratory, Princeton, New Jersey, 08544

1. Introduction

Unlike several stable inorganic clusters observed in biology which undergo spontaneous self-assembly like iron-sulfur proteins and dimetal-oxo-carboxylates, the water oxidizing complex (WOC) possesses an intrinsically unstable core. The tetra-Mn-Ca active center of the WOC of PSII can be reversibly reassembled in a process called photoactivation [1]. We have determined the experimental conditions needed to optimize both the yield and the kinetics of photoactivation using two improvements over previous methods: (a) An ultra-sensitive Clark-type O_2 microcell which exhibits a rapid response time as short as 100 ms, utilizes a tiny volume 5 µl, and is equipped with bright LEDs (λ_{max}=660 nm) that are capable of controlling both the light and dark intervals, while minimizing photoinhibition that occurs under continuous white illumination; (b) A new highly efficient chemical chelator, an aromatic diamino tetracarboxylate, for complete extraction of Mn and Ca ions under mild conditions.

We use these new tools to kinetically resolve the first light-induced intermediate in the photoactivation process and provide initial characterization, using EPR spectroscopy, suggesting a possible binuclear Mn_2(II,II) precursor forms initially in the dark.

2. Procedure

BBY membranes were prepared from spinach and stored in 10% glycerol at 77 K. O_2 evolving activity was 400-500 µmoles O_2/mg Chl · h. The complete removal of Mn was done at 0.25-1.0 mg Chl/ml in basic medium (mM): 300 sucrose/35 NaCl/25 MES-NaOH buffer pH 6.0 with 1 mM ascorbate and 25-35 mM tetrapropionato-1,3-di(aminomethyl)benzene (TPDBA). After 90 s incubation at 20°C solution was cooled immediately to 2°C and washed 3 times with basic medium. The apo-PSII in the absence of $MnCl_2$ showed no O_2 evolving activity. The time-course of O_2 recovery was measured directly in microcell (V=5 µl) equipped with a silicone membrane (1-5 µm thickness) and coupling to DC/AC-amplifiers with adjustable low- and high-frequency roll-offs 0.3 and 3.0 Hz, respectively. Illumination from ultra-bright LEDs (HLMP-8102, Hewlett-Packard) permitting variable duration of light (0.1-60 ms) and dark (0.1-10 s) pulses, and intensities 8-80 mW/cm^2. EPR spectra were obtained at 9.3 GHz on a Bruker ESP 300 spectrometer operating with TE_{102} cavity Q=8000; 100 kHz modulation frequency; 5 G modulation amplitude and Oxford Instruments ESR 900 liquid helium cryostat.

P. Mathis (ed.), Photosynthesis: from Light to Biosphere, Vol. II, 431–434.

3. Results and Discussion

The PSII protein complex, depleted of inorganic cofactors, can be photoactivated by addition of Mn^{2+}, Ca^{2+}, Cl^-, K_3FeCN_6 and pulse-light illumination (Fig.1,A). Fig.2 shows a semilogarithmic plot of the population of the inactive apo-PSII centers (which are photoactivatable) vs. total time of pulse-light illumination. The break in the curve shows that recovery of O_2 evolution is a biphasic process with either two kinetic phases or two populations of PSII centers having different kinetics.

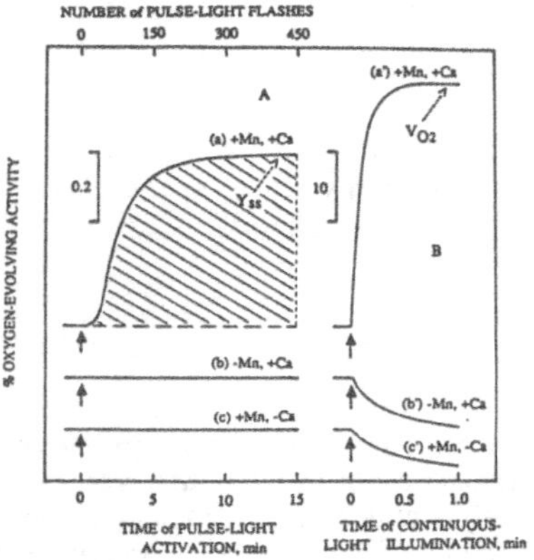

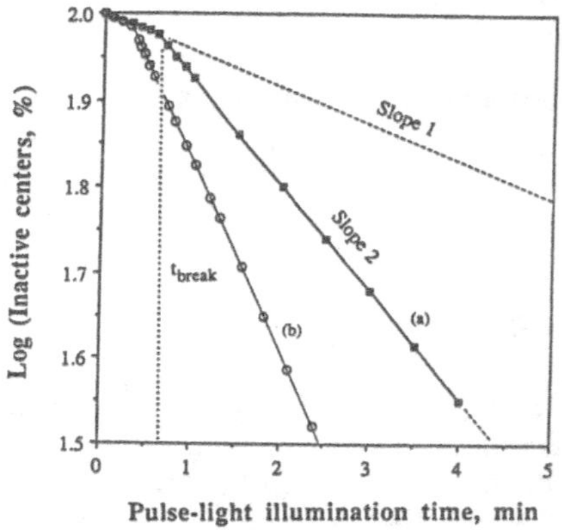

Fig.1. Kinetics of recovery of O_2 evolution by apo-PSII using pulse-light illumination at t_{light}=20 ms, t_{dark}=2 s, I=80 mW/cm^2 (A); and following pulse-light photoactivation O_2 activity was monitored using saturated continuous light at 8 mW/cm^2 (B). 8 atoms Mn^{2+}/RC, 8 mM $CaCl_2$ and 0.8 mM K_3FeCN_6 are present, as noted.
Fig.2. Log of % inactive centers vs. integrated pulse-light illumination time: (a) at 8 atoms Mn^{2+}/RC; (b) at 40 atoms Mn^{2+}/RC.

To achieve maximal recovery of the O_2 evolution rate (Y_{ss}) by photoactivation both the duration of the light pulses (30 ms<t_{light}<60 ms at t_{dark}=2 s) and the dark interval between light pulses (2 s<t_{dark}<3 s at t_{light}=40 ms) must be optimized (Fig.3A,B).

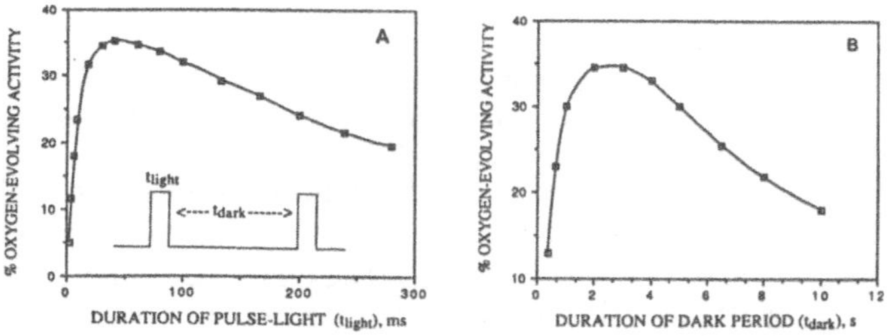

Fig.3. Dependence of the maximal yield O_2 evolution (as shown in Fig.1,B) on light and dark intervals: (A) variable t_{light} at t_{dark}=2 s; (B) variable t_{dark} at t_{light}=40 ms.

The half-time of photoactivation accelerates by four-fold from $t_{1/2}$=250 s to 60 s and the yield increases by three-fold when the duration is increased from t_{light}=10 ms to 180 ms (data not shown). To achive photoactivation a dark process must occur after the light step[2]; maximum Y_{ss} is observed at $t_{dark}//t_{light}$=50-70(Fig.3).

Calcium is absolutely required for photoactivation [2-4]. However, the $t_{1/2}$ increases with the concentration of accessible $CaCl_2$, $MgCl_2$, NaCl and H^+. The slopes of these plots show the relative ability of each cation to slow the photoactivation process, as follows: 13 s/mM $CaCl_2$, 4.7 s/mM $MgCl_2$, 0.14 s/mM NaCl and 130 s/unit pH.

As shown in Fig.4, the presence of 2.0 Mn/RC leads to resolved regular structure with Δ=44-45 G, a coupling which is found in binuclear $^{55}Mn_2$(II,II) complexes and proteins [5,6]. This EPR signal decreases after illumination at 160 K, suggesting proximity to tyrosine Z^+. In the presence of Ca^{2+} or Mg^{2+} the splitting increases to Δ=60-65 G, suggesting disruption of the binuclear Mn site by divalent cations.

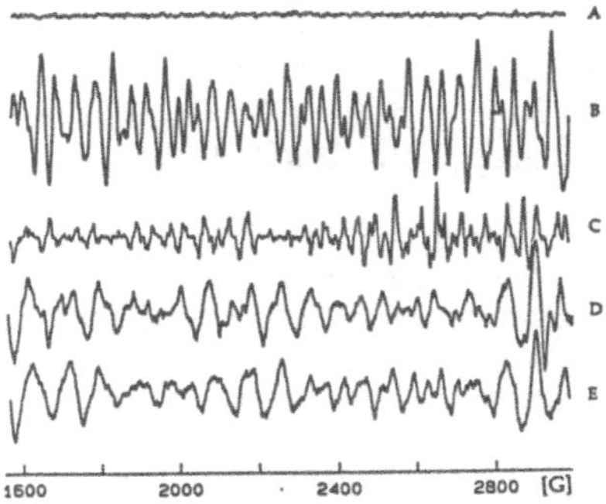

Fig.4. EPR spectra of apo-PSII after Fourier-filtering to remove low frequency components: (A) apo-PSII; (B) apo-PSII + 2 Mn/RC, dark incubation; (C) difference spectrum dark-minus-light /illumination at 160 K, 10 min, λ_{max}=660 nm/; (D) same as A+10 mM $CaCl_2$; (E) same as A+10 mM $MgCl_2$. Measurements were done at 25 K.

A maximum yield of photoactivation is observed at E_m=380 mV by controlling the redox-potential of the medium using ferri/ferrocyanide. A sharp maximum occurs in the half-time for photoactivation at 330 s when E_m=340 mV. These effects appear to be due to a true redox-potential dependence, since ferricyanide exhibits little non-specific influence on photoactivation yield above 100 μM concentration . Below this concentration a rapid decrease in photoactivation yield occurs simply because an electron acceptor is required.

TPDBA-extraction of PSII membranes provides effective chelating of the Mn^{n+} and Ca^{2+} ions at low concentration of chelator, while not damaging the centers as seen by the high photoactivation yield (Fig.1). Using TPDBA to deplete Mn and Ca from PSII membranes we find that better than 95% of the maximum O_2 evolution rate could be recovered with 4.0 Mn/P680 (Fig.5A). From the monophasic rising portion of the titration

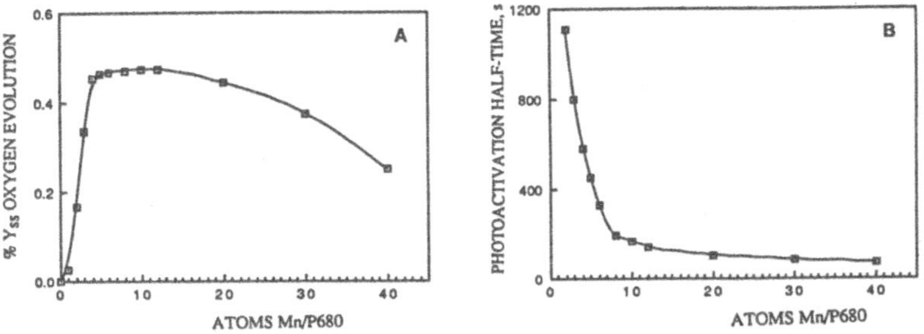

Fig.5. Dependence of the yield Y_{ss} (A) and half-time $t_{1/2}$ (B) on concentration of added $MnCl_2$.

curve a single Michaelis constant for photoactivation is evident with K_m=2.6 μM Mn^{2+}. Increasing the amount of Mn^{2+} from 2 to 40 Mn/P680 accelerates the rate of photoactivation (Fig.5,B) with clear evidence for at least two populations of sites, a high activity subset (< 5 Mn/P680) having the strongest differential influence on acceleration of photoactivation, and a low activity subset having at least a 100-fold weaker influence on the rate of photoactivation. The initial slope of $t_{1/2}$ on Mn concentration reveals a power law in $[Mn]^n$ with n=1.1-1.2. These data indicate that the rate-limiting step in photoactivation (slope 2,Fig.2) requires uptake of 1Mn atom/RC and that a total of 4 atoms of Mn are required for full recovery of O_2 activity. Additional Mn^{2+} ions accelerate the rate of O_2 evolution without increasing the yield. At low $[Mn^{2+}]$ an earler step in photoactivation (slope 1, Fig.2) occurs before $t_{1/2}$ is reached, and this step also requires 1 Mn.

Model for the photoactivation process.

After complete removal of Mn^{n+}, Ca^{2+} and extrinsic proteins, the apo-PSII state is generated. Addition of Mn^{2+} yields dark intermediate IM_1 and eventually IM_2 with the latter state detected by EPR as a spin-coupled binuclear Mn_2(II,II) precursor. The data reveal two sequential light dependent steps, each requiring 1 atom of Mn(II) and a dark step in-between. We propose these represent the first two Mn atoms to photoligate via the sequence as shown in the scheme. Divalent cations, including Ca^{2+}, Na^+ and H^+ each inhibit one or more of these steps. The slope 1 process in Fig.2 is assigned to the first light step I, while the slope 2 process in Fig.2 is assigned to the second light step 2. Two additional Mn(II) ions will be required in subsequent steps. The step(s) where the functional Ca^{2+} and Cl^- ions interact are not yet known, but are required for photoactivation. Competition between Ca^{2+} binding at a Mn binding site is in agreement with earler studies [2-4].

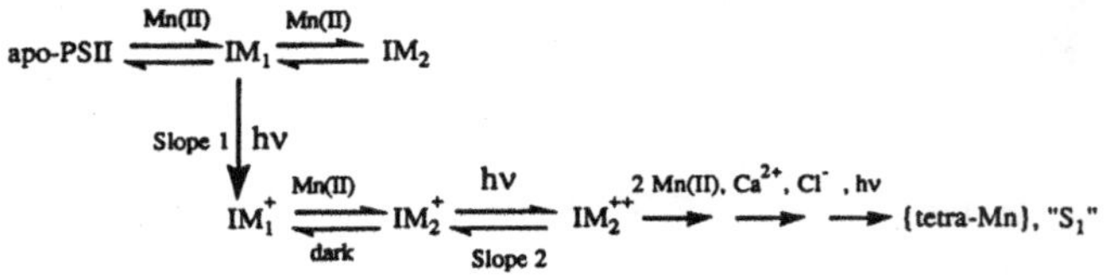

4. Conclusions

(1) A new pulse-light method allows greatly expanded kinetic studies of photoactivation in the *ms-min* time range. Photoactivation is shown to be a biphasic kinetic process.
(2) Divalent cations, Na^+ and H^+ slow the rate of photoactivation of O_2 evolution.
(3) Titration results show that the WOC active center is saturated by 4 Mn/P680.
(4) EPR spectra reveal that apo-PSII forms a Mn_2II dimer in the dark.
(5) The steady-state rate-limiting step of photoactivation in excess Mn depends on $[Mn]^n$, n=1.1-1.2.

Acknowledgments. We thank Dr. Sergei V. Khangulov for discussion, Peter J. Pessiki for synthesis Mn-Ca-chelator TPDBA and the NIH for financial support, GM39932.

References

1. Radmer, R. & Cheniae, G.M.(1977) in Primary Process of Photosynthesis (Barber, J., ed.) Vol. 2, pp 303-348, Elsevier, Amsterdam
2. Tamura, N. & Cheniae, G.M. (1987) Biochim. Biophys. Acta 890, 179-194
3. Ono, T. and Inoue, Y. (1983) Biochim. Biophys. Acta 723, 191-201
4. Miller, A.-F. and Brudvig, G. (1989) Biochemistry 28, 8181-8190.
5. Khangulov, S.V., Pessiki, P.J., Barynin, V.V., Ash, D.E. and Dismukes, G.C. (1995) Biochemistry 34, 2015-2025
6. Reczkowsky, R.S. and Ash,D.E. (1992) J. Am. Chem. Soc. 114, 10992-10994

SPECTROSCOPICAL CHARACTERISATION AND OXYGEN MEASUREMENTS OF WHOLE CELLS FROM WILDTYPE AND MUTANTS OF SYNECHOCYSTIS sp.PCC 6803

H.Kretschmann

Max-Volmer-Institut fuer Biophysikalische und Physikalische Chemie,Technische Universitaet Berlin,Strasse des 17.Juni 135, 10623 Berlin,Germany.Present adress: Bogazici University,Institut of Environmental Sciences,80815 Bebek,Istanbul,Turkey

1.Introduction

In Cyanobacteria , some algae and higher plants the light induced water-oxidation to molecular oxygen is catalysed by a Mangan containing enzym-complex (OEC) of Photosystem II (PS II). In the water-splitting reaction the oxidation states S0 to S4 are running succesively with a oxygen-evolution in the S3 to S0 transition. The location of the Mn-cluster in PS II was assumed between the lumenal site of the D1,D2 polypeptides and the PsbI protein [1]. The structural involvement of the chrophyll containing polypeptide CP43 and CP47 which both have a large extrinsic loop in the lumen is a matter of discussion [2-6].
We applied stable absorption-changes- , flourescence- and oxygen-evolution-measurements on whole cells of Synechocystis sp.PCC 6803 wild type and revertants of a CP43 (loop E) deletion mutant (see:M.Kuhn,this Proceedings),to characterize the possible role of CP43 on the water-splitting.

2.Materials and Methods

Whole cells from Synechocystis sp. PCC 6803 were cultivate in BG11. For the measurements we used cells of wild type and two different revertants R1 and R3. The difference for the revertants result from different length of the inserted sequence in the CP43 loop:

WT Deletion (AA 364-370): seq. LDKLRND
R1 Insertion: seq. FGPNGFGPNGLD
R3 Insertion: seq. FGPNGLD

(see M.Kuhn,this proceeding).
Absorption-change in the UV and red were measured as described in [7]. The flourescence-oszillation pattern were measured with a equipment as described in [8]. Oxygen-evolution were measured with a Zircon-dioxyde electrode [9].The puffer solution contain 0.3 M Mannitol,20 mM MES,20 mM CaCl and 10 mM MgCl at pH 6.5.For all measurements 600 µm DCBQ as acceptor were used.

P. Mathis (ed.), Photosynthesis: from Light to Biosphere, Vol. II, 435–438.

3. Result and Discussion

Oxygen-evolution pattern from whole cells of wild type (WT),R1 and R3 are shown in Fig.1. The WT and R1 shows a maximal oxygen evolution after the fourth flash,which indicates a dark stable S0-state.Fitting of the pattern with the Kok-Model give 100% S0,0% S1,0% misses and 2% double-hits for WT and 100% S0,0%S1,0% misses and 2% double-hits for R1. Therefore the change of the sequence in the CP43 loop of R1 shows no effect on the oxygen-evolving process.For WT and R1 we get a "ideal" behaviour of the oxygen-evolution pattern with practical no misses and double-hits.In contrast , R3 shows a maximal oxygen-evolution after the third flash which indicates a dark stable S1-state. Fitting led to 11% S0,89% S1,4% misses and 7% d.-hits. The pattern looks like the known pattern from higher plants or preparation from cyanobacteria [10,11]. The upward shiftig of the pattern from R3 can be also observed using stable absorption-changes at 367nm (Mn-oxidation [12]),Electrochromic-absorption-changes at 694 (charge of the OEC [13] and flourescence oscillation (charge of OEC [14]) as shown in Fig.2. The cause of the change of the oxidation state of one Mn result from the difference in the sequence of the CP43 loop E from R3. Despite the known different aminoacid-sequenz, over the details only speculations are possible.Therefore the oxidation state or the junction to the Mn-cluster of the stable Tyr D on the D2 polypeptid may be change through the structural change of the CP43 loop.The alterate state of Tyr D led then to a oxidation of one Mn.Besides from such assumption the main result is the structural involvment of the hydrophilic loop from CP43 in the environment of the water-splitting-complex. A further interesting result is the opposite sign of the electrochromic signals for R3 without change the approx. 0:0:1:1 charge pattern.The opposite pattern can also result from structural change in the loop E of the CP43. The reverse sign of the signal at 694nm indicate in R3 a red shift of the antenna Chl-band in contrast to the known blue

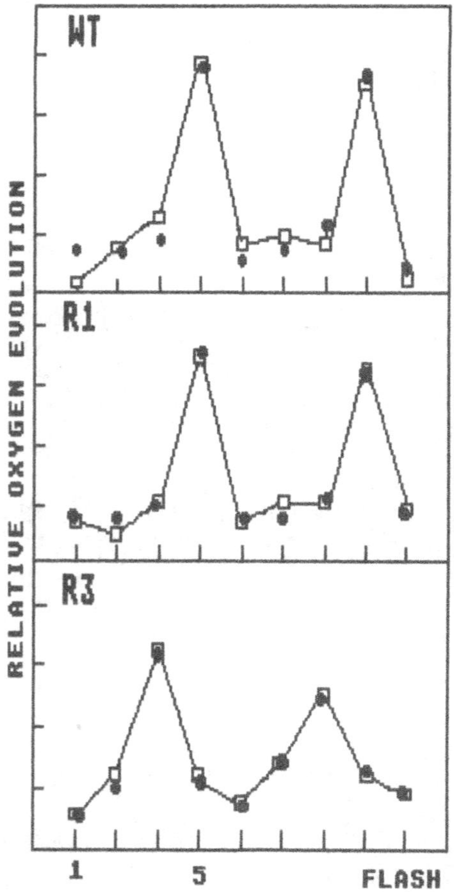

Fig.1: Flash induced oxygen evolution of wild type and revertants R1,R3;□:Exp. Data, ●:Fitting-data,see Text.

shift [13].The flourescence-oscillation pattern which result from charges on the OEC shows also a approx. 0:0:1:1 pattern as the electrochromic signals. Besides the shifting the Signals are equal for R1 and R3 (Fig. 2B).After 24 h dark adaptation of R1 and R3 a further difference in the oxygen-evolution pattern can be observed (Fig.3).R3 shows a normal oscillation pattern with a slight maximum in the third flash and a strong

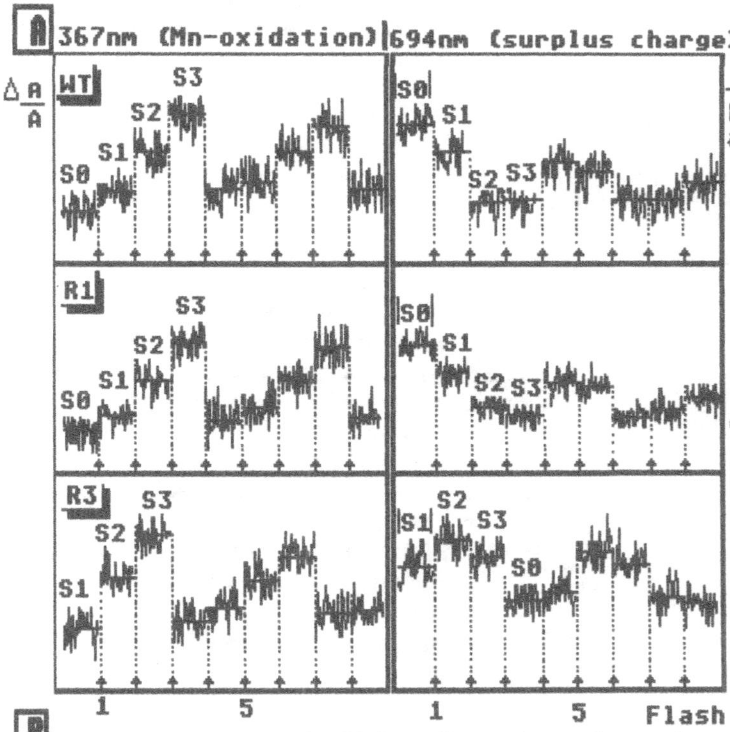

Fig.2A: Stable absorption changes at 367 nm (left) and 694 nm (right) for whole cells of wild type and revertants R1 and R3, At 694nm the first signal is superposed by an artefact which is indicated by brackets

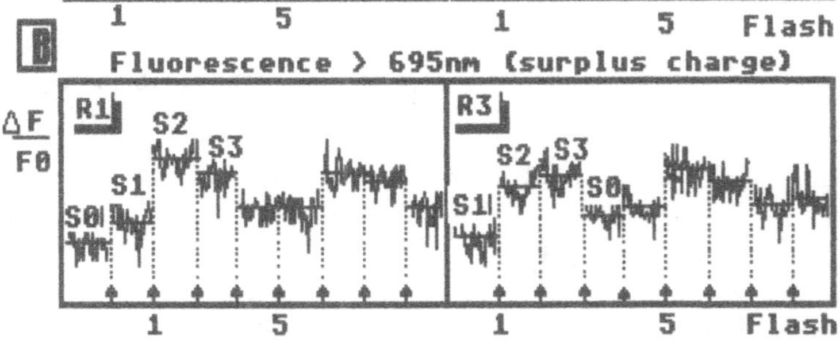

Fig.2B: Flourescence change 200ms after a flash for the revertants R1 and R3.

damping. In contrast,R1 shows in each flash a signal,but no oscillation like pattern as with R3 (fig.3B).Only after 200 preflashes a slight oscillation pattern on the top of a high offset in each flash can be observed (Fig.3C).Whereas the R1 with a dark stable S0-state shows a photoactivation behaviour [15,16],the R3 need no preflashes to show normal oxygen evolution.R3 shows a behaviour as higher plants (dark stable S1,no photoactivation) whereas R1 look like cyanobacteria with a lower dark stable S0-state and necessity for photoactivation.On the basis of the here present results it seem's that structural change

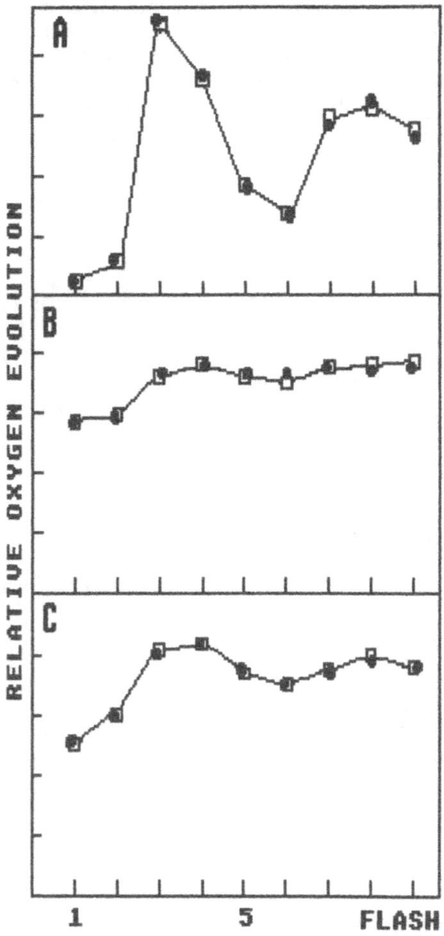

in the extrinsic loop of CP43 have an impact. In a more unspecific way,a shifting of the dark stable S0-state of whole cells from cyanobacteria to a stable S1-state occur after preparation of PS II complexes from them [12]. This may be a result from a treatment with detergents and gradient methods.Therefore caution are necessary in interpretation of measuring data from such artificial created systems.

Fig.3: Flash induced oxygen evolution of revertants R1 and R3 after 24 h dark adaption;A:R3,B:R1,C:R1 with 200 preflashes.
□ : Exp.data
• : Fitting data:
[A] R3: 29% S0,71% S1,17% misses 4% d.-hits
[B] R1: 100% S0,26% misses,20% d. d.hits
[C] R1: 100% S0,18% misses,21% d.hits

Acknowledgement:This work was supported by DFG Sfb 312 TP A4

References

1 Debus,R.J.;Biochim.Biophys.Acta 1102,p.269-352,(1992)
2 Bricker,T.M;Photosynth. Res. 24,p.1-13,(1990)
3 Boekema,E. et al.;Proc.Natl.Acad.Sci.(USA) 92,p.175-179,(1995)
4 Kuhn,M. and Vermaas,W.F.;Plant. Mol. Biol. 23,123-133,(1993)
5 Eaton-Rye,J.J. et al.;Plant. Mol. Biol. 17,p.1165-1177,(1991)
6 Haag,E. et al.;Biochem. 32,p.4444-4454,(1993)
7 Kretschmann,H. et al.;Biochim.Biophys.Acta 1059,p.208-214,(1991)
8 Aksoy,N. and Saygin,O.;Photochem.Photobiol. submitted,(1995)
9 Meyer,B.;Thesis,TU-Berlin,(1987)
10 Joliot,B. et al.;Photochem.Photobiol. 10,p.309-329,(1969)
11 Messinger,J. et al.;Z.Naturforsch. 460,p.1033-1037,(1991)
12 Saygin,O. and Witt,H.T.;Biochim.Biophys.Acta 893,452-469,(1987)
13 Saygin,O. and Witt,H.T.;FEBS 187,p.224-226,(1985)

AN IMPROVED HOST STRAIN OF *SYNECHOCYSTIS* SP. PCC 6803 FOR INTRODUCING SITE-DIRECTED MUTATIONS INTO THE D1 PROTEIN OF PHOTOSYSTEM II.

Hsiu-An Chu, Anh P. Nguyen, and Richard J. Debus*
Department of Biochemistry, University of California at Riverside
Riverside, CA 92521-0129 U.S.A.

1. Introduction

To identify amino acid residues that ligate the manganese and calcium ions of Photosystem II, or that are otherwise crucial for water oxidation, we have employed site-directed mutagenesis to individually target every conserved lumenal carboxylate, histidine, and tyrosine residue in the D1 polypeptide of the cyanobacterium *Synechocystis* sp. PCC 6803 [1-3]. All mutations were constructed in this organism's *psb*A-2 gene and were introduced into a host stain that lacks all three *psb*A genes [4]. A number of procedures for isolating oxygen-evolving PSII core complexes from *Synechocystis* 6803 are available [5-8]. However, because PSII in *Synechocystis* tends to elute near PSI in ion-exchange chromatography [5,6,8] and because *Synechocystis* 6803 typically contains 5-9-fold more PSI that PSII (*e.g.*, [8-10]), the isolation of mutant PSII particles would be greatly facilitated if the mutations were introduced into a *Synechocystis* host strain that lacks PSI.

Photosystem I was first deleted from *Synechocystis* 6803 by McIntosh and coworkers [11,12]. These authors segregated mutants with inactivated *psa*A [11] or *psa*B [12] genes under light-activated heterotrophic conditions (complete darkness except for 5 min of light every 24 hrs). Unfortunately, the resulting strains were very sensitive to light. Vermaas and coworkers later showed that similar strains could be segregated more conveniently in continuous, dim light (5 $\mu E\ m^{-2}\ s^{-1}$) and that the light-sensitivity of such strains could be alleviated by additionally inactivating the *apc*E gene that encodes the polypeptide that anchors the phycobilisome to the thylakoid membrane and PSII [13]. These authors have used the resulting PSI^-/*apc*E$^-$ strains to examine aspects of PSII function [14,15] and respiratory electron transport pathways [16,17].

In the present communication we describe the construction of an improved host strain of *Synechocystis* sp. PCC 6803 that lacks PSI and *apc*E function in addition to lacking all three *psb*A genes. This strain is transformable and will facilitate the isolation and characterization of mutant PSII particles, particularly those from mutants with labile water oxidizing complexes and with low PSII contents.

2. Materials and Methods

Cell Growth and DNA manipulations. Cells were propagated as described

P. Mathis (ed.), Photosynthesis: from Light to Biosphere, Vol. II, 439–442.

previously [1]. Where indicated, media were supplemented with 2.5 μg chloramphenicol (Cm) per mL, 0.1 μg erythromycin (Em) per mL, 5 μg kanamycin (Km) per mL, 20 μg spectinomycin (Sm) per mL, 20 μg gentamycin (Gm) per mL, or 15 μg tetracycline (Tet) per mL. Isolations of genomic DNA, PCR amplifications, and Southern analyses were performed as described previously [1,4]. The construction of the original Sm^{r}/Em^{r}/Cm^{r} host strain lacking all three *psb*A genes was described previously [4].

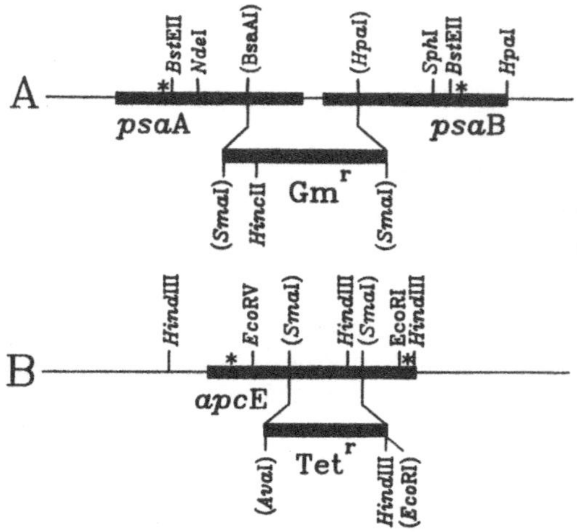

FIGURE 1. The *psa*A, *psa*B, and *apc*E genes of *Synechocystis* 6803. (A) Strategy for replacing the 3' 28% of *psa*A and the 5' 15% of *psa*B with DNA encoding resistance to Gm. Sites for key restriction endonucleases are shown. Parentheses indicate sites lost during construction. Asterisks show the position of the oligonucleotides employed for PCR amplification during cloning. (B) Strategy for replacing part of *apc*E with DNA encoding resistance to Tet.

Preparation of host strain lacking all three psbA genes and PSI. A 3.4 kb fragment of *Synechocystis* 6803 genomic DNA bearing the 3' 70% of *psa*A and the 5' 70% of *psa*B was amplified with oligonucleotides designed on the basis of the sequences of the *Synechocystis psa*A and *psa*B genes [18]. The oligonucleotides employed were 5'-TTTGTTGGGCTTAGGC-TCC-3' and 5'-AAGAACAGAGA-GTTGGTGCCG-3'. The 2.8 kb *Nde*I/*Sph*I fragment of this PCR product (see Figure 1A) was substituted for the 258 bp *Nde*I/*Sph*I fragment of pUC19 [19], creating the plasmid pANPSI. To delete the 3' 28% of *psa*A and the 5' 15% of *psa*B from the *Synechocystis* chromosome, the 1.2 kb *Hpa*I/*Bsa*AI fragment of pANPSI (see Figure 1A) was replaced with the 1.9 kb Gm^{r} fragment of pRZ1107 [20]. The resulting plasmid (pANPSIGm^{r}) was transformed into the original Sm^{r}/Em^{r}/Cm^{r} host strain and single colony transformants were selected for ability to grow on Gm at 5-6 μE m^{-2} s^{-1}. (To facilitate the use of the Gm^{r} cassette, the 1.9 kb *Bam*HI/*Bam*HI Gm^{r} fragment of pRZ1107 was substituted for the Ω fragment of pHP45Ω [21] to creating the plasmid pHP45Gm^{r}.)

Preparation of host strain lacking all three psbA genes, PSI, and apcE. A 2.2 kb fragment of *Synechocystis* 6803 genomic DNA bearing most of the *apc*E gene [13] was amplified with degenerate oligonucleotides designed on the basis of conserved regions in the amino acid sequences of the *apc*E gene-products of four cyanobacteria [22]. The oligonucleotides employed were 5'-GCYAACCGKATYTTCGTNGGTGG-ATCCCC-3' and 5'-GTAMCGTTCGTAGGGTACCGTRTCTTCACC-3'. The 1.6 kb *Eco*RV/*Eco*RI fragment of this PCR product (see Figure 1B) was ligated into the *Hin*cII/*Eco*RI sites of pUC119 [23] to create the plasmid pANapcE. The 940 bp *Sma*I/*Sma*I fragment of this plasmid (see Figure 1B) was replaced with the 1.4 kb Tet^{r} *Ava*I/*Eco*RI fragment of pBR322 to create the plasmid pANapcETet^{r}. This plasmid was transformed into the fully segregated Sm^{r}/Em^{r}/Cm^{r}/Gm^{r} strain described above and single

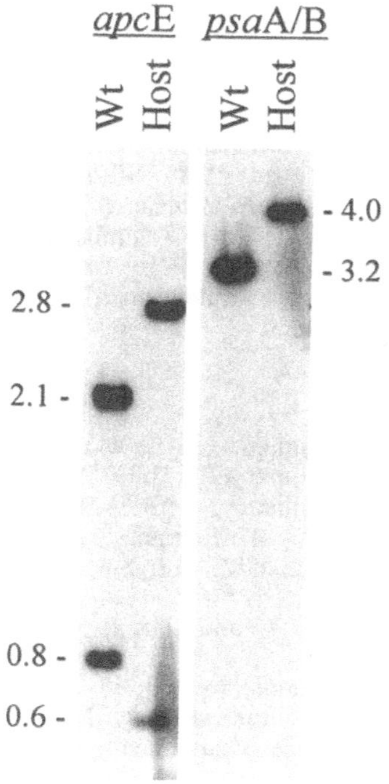

FIGURE 2. Southern-blot analyses of total chromosomal DNA from wild-type *Synechocystis* 6803 "Wt" and from a $Sm^r/Em^r/Cm^r/Gm^r/Tet^r$ transformant (the $PSII^-/PSI^-/apcE^-$ "Host"). Fragment sizes expressed in kb are indicated. *Right lanes*: DNA was digested with *Bst*EII and hybridized with [^{32}P]-labeled plasmids carrying *psa*A/*psa*B and the Gm^r gene. In the Host strain, the 3.2 kb fragment is larger, indicating substitution of part of *psa*A/*psa*B by the larger Gm^r gene (see Figure 1A). *Left lanes*: DNA was digested with *Hin*dIII and hybridized with [^{32}P]-labeled plasmids carrying *apc*E and Tet^r. In the Host strain, the 2.1 kb fragment is larger and the 0.8 kb fragment is smaller, indicating substitution of part of *apc*E by the larger Tet^r gene (see Figure 1B).

colony transformants were selected for ability to grow on Tet at 50-60 μE m^{-2} s^{-1}. Because Tet is light-sensitive, filters were transferred to fresh solid media every 10 days. (The original $Sm^r/Em^r/Cm^r$ host was also transformed with pANapcETet^r but, in this case, the filters were transferred to fresh media every 3-4 days.)

3. Results and Discussion

The $Sm^r/Em^r/Cm^r/Gm^r$ transformants were light-sensitive and turquoise-blue in color, as expected for *Synechocystis* cells lacking PSI [11-13]. Southern-blot analysis of total chromosomal DNA extracted from a $Sm^r/Em^r/Cm^r/Gm^r$ transformant (Figure 2) shows that the wild-type 3.2 kb *Bst*EII/*Bst*EII fragment bearing psaA and psaB (Figure 1A) is larger in the transformant (4.0 kb), reflecting the substitution of part of these genes by the larger fragment encoding resistance to Gm. The absence of the wild-type 3.2 kb fragment in the transformant's DNA confirms that the transformant has fully segregated. Segregation was also confirmed by PCR amplifications conducted with a different set of oligonucleotides; a 1.4 kb fragment amplified in wild-type cells increased to 2.1 kb in the Gm^r transformants (not shown).

The $Sm^r/Em^r/Cm^r/Gm^r/Tet^r$ transformants were not sensitive to light. Their color was light olive-green on solid media and yellow with a tinge of green in liquid media. This coloration is consistent with the $PSI^-/apcE^-$ mutant described by Vermaas and coworkers [13]. (In contrast, $Sm^r/Em^r/Cm^r/Tet^r$ transformants were dark olive-green, as expected for a strain that contains PSI but lacks phycobilisomes [9].) Southern-blot analysis of total chromosomal DNA extracted from a $Sm^r/Em^r/Cm^r/Gm^r/Tet^r$ transformant (Figure 2) shows that the wild-type 2.1 kb and 0.8 kb *Hin*dIII/*Hin*dIII fragments that together bear *apc*E (Figure 1B) have increased to 2.8 kb and decreased to 0.6 kb, respectively, in the transformant, reflecting the substitution of part of *apc*E by the larger fragment encoding resistance to Tet. The absence of the wild-type fragments in the transformant's DNA confirms that the transformant has fully segregated, forming the desired $PSII^-/PSI^-/apcE^-$ host strain. Interestingly, this strain appears to be much more sensitive to Tet than the $PSII^-/apcE^-$ strain that still contains PSI. Despite repeated transformation attempts, no Sm^r/Em^r/Cm^r/Gm^r/Tet^r transformants were obtained until the filters were transferred to fresh Tet-containing media every 10 days instead of every 3-4 days.

The new PSII$^-$/PSI$^-$/*apc*E$^-$ host strain is transformable with our mutation-bearing *psb*A-2 plasmids [4], yielding single-colony Km^r transformants. The resulting Sm^r/Cm^r/Gm^r/Tet^r/Km^r cells exhibit the same coloration as the PSII$^-$/PSI$^-$/*apc*E$^-$ host strain. However, their light-saturated rates of oxygen-evolution are at least 6-fold higher than when examined previously in the presence of PSI and *apc*E [1-3]. A similar increase in oxygen evolving activity in the original PSI$^-$/*apc*E$^-$ strain was reported by Vermaas and coworkers [13]. We are in the process of introducing our large collection of mutation-bearing *psb*A-2 plasmids [1-3] into the new PSII$^-$/PSI$^-$/*apc*E$^-$ host.

4. References

1 Chu, H.-A., Nguyen, A.P. and Debus, R.J. (1994) Biochemistry 33, 6137-6149.
2 Chu, H.-A., Nguyen, A.P. and Debus, R.J. (1995) Biochemistry 34, 5839-5858.
3 Chu, H.-A., Nguyen, A.P. and Debus, R.J. (1995) Biochemistry 34, 5859-5882.
4 Debus, R.J., Nguyen, A.P. and Conway, A.B. (1990) in Current Research in Photosynthesis, Vol. I (Baltscheffsky, M. ed.), pp. 829-832, Kluwer Academic Publishers, Dordrecht.
5 Burnap, R.L., Koike, H., Sotiropoulou, G., Sherman, L.A. and Inoue, Y. (1989) Photosynth. Res. 22, 123-130.
6 Noren, G.H., Boerner, R.J. and Barry, B.A. (1991) Biochemistry 30, 3943-3950.
7 Kirilovsky, D.L., Boussac, A.G.P., van Mieghem, F.J.E., Ducruet, J.-M.R.C., Sétif, P.R., Yu, J., Vermaas, W.F.J. and Rutherford, A.W. (1992) Biochemistry 31, 2099-2107.
8 Tang, X.-S. and Diner, B.A. (1994) Biochemistry 33, 4594-4603.
9 Rögner, M., Nixon, P.J. and Diner, B.A. (1990) J. Biol. Chem. 265, 6189-6196.
10 Aizawa, K., Shimizu, T., Hiyama, T., Satoh, K., Nakamura, Y. and Fujita, Y. (1992) Photosynth. Res. 32, 131-138.
11 Smart, L.B., Anderson, S.L. and McIntosh, L. (1991) EMBO J. 10, 3289-3296.
12 Smart, L.B. and McIntosh, L. (1993) Plant Mol. Biol. 21, 177-180.
13 Shen, G., Boussiba, S. and Vermaas, W.F.J. (1993) The Plant Cell 5, 1853-1863.
14 Shen, G. and Vermaas, W.F.J. (1994) J. Biol. Chem. 269, 13904-13910.
15 Shen, G. and Vermaas, W.F.J. (1994) Biochemistry 33, 7379-7388.
16 Vermaas, W.F.J., Shen, G. and Styring, S. (1994) FEBS Lett. 337, 103-108.
17 Vermaas, W.F.J. (1994) Biochim. Biophys. Acta 1187, 181-186.
18 Smart, L.B. and McIntosh, L. (1991) Plant Mol. Biol. 17, 959-971.
19 Yanisch-Perron, C., Vieira, J. and Messing, J. (1985) Gene 33, 103-119.
20 Yin, J.C.P., Krebs, M.P. and Reznikoff, W.S. (1988) J. Mol. Biol. 199, 33-45.
21 Prentki, P. and Krisch, H.M. (1984) Gene 29, 303-313.
22 Capuano, V., Braux, A.-S., Tandeau de Marsac, N. and Houmard, J. (1991) J. Biol. Chem. 266, 7239-7247.
23 Vieira, J. and Messing, J. (1987) Methods Enzymol. 153, 3-11.

Photoactivation and S-state cycling kinetics in photosystem II mutants in *Synechocystis* sp. PCC 6803

Robert L. Burnap, Ming Qian, Sufian Al-Khladi, and Coy Pierce
Department of Microbiology and Molecular Genetics
Oklahoma State University, Stillwater, Oklahoma 74078 USA

1. Introduction The core of the oxygen-evolving photosystem II complex is minimally composed of several highly conserved intrinsic proteins (D1, D2, cyt. b_{559}, CP47, CP43, & *psbI* product) and one moderately conserved extrinsic protein (*psbO* gene product). Eukaryotes possess the *psbP* and *psbQ* gene products, which modulate Ca^{+2}, Cl^-, requirements, respectively, of the H_2O-oxidation reaction. Cyanobacteria do not have *psbP* and *psbQ* , but have different polypeptides including a cytochrome (cyt. c_{550}), that together with the ubiquitous manganese-stabilizing protein (MSP, *psbO* gene product), form the extrinsic portion of the H_2O-oxidation complex. Recent analyses of various H_2O-oxidation mutations affecting the binding and function of extrinsic polypeptides of the cyanobacteria H_2O-oxidation complex are presented. The focus is on alterations in the kinetics of photoactivation (light-dependent assembly of the Mn cluster) and alterations in the kinetics of S-state cycling.

2. Materials and Methods *Synechocystis* sp. PCC6803 wild type and mutant cells were maintained in BG-11 medium supplemented with 5 mM glucose. Site-directed mutations were introduced into MSP as described in [1]. Mutations in the D1 protein were constructed according to Chu et al [2]. Hydroxylamine (HA) extraction of PSII Mn and photoactivation (flash and continuous illumination) of HA-extracted cells were performed according to [3]. Flash O_2-yield measurements were performed using a bare platinum electrode that permits the centrifugal deposition of samples upon the electrode surface. Estimates of the PSII turn-over time performed according to the double flash methods described in [4].

3.0 Results and Discussion
3.1 *Photoactivation* of ΔpsbO cells (which lack MSP) under continuous illumination occurs at an intrinsically faster rate (about 4-fold) than the wild-type **(Fig. 1)**. Virtually all of the increase in the rate of photoactivation can be accounted for by a 4-fold increase relative quantum yield of photoactivation as indicated by the yield of photoactivation as a function of flash number **(Fig. 2)**. Experiments measuring the flash frequency dependence of photoactivation indicate a multiquantum process in the mutant resembling the wild-type, but with significant increases in yields at all flash frequencies examined **(Fig. 3)**. The characteristics of these curves [5] suggest that the higher quantum yield of photoactivation observed for ΔpsbO cells occurs in the absence of large changes in the rate-limiting dark rearrangement, and is probably due to increased accessibility or affinity

P. Mathis (ed.), Photosynthesis: from Light to Biosphere, Vol. II, 443–446.

(and photooxidation) of Mn^{2+} at one or both of the two binding sites involved in the initial stages of the process.

Figure 1

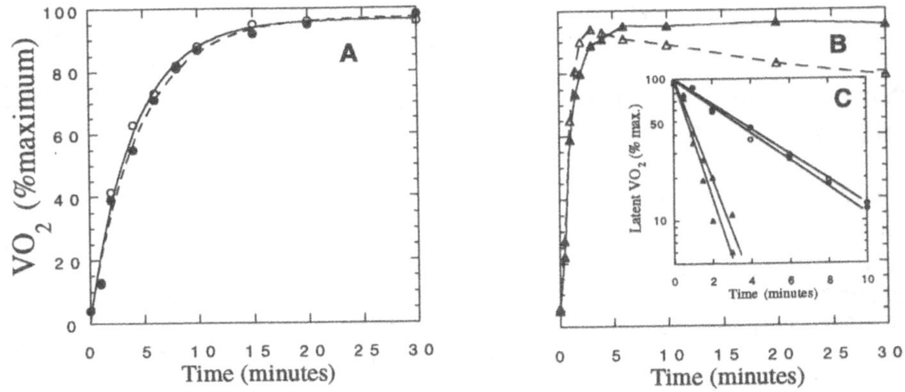

Figure 1. Photoactivation of hydroxylamine (HA)-extracted wild-type (circles, Panel A, open and closed circles) and ΔpsbO cells (Panel B, open and closed triangles) exposed to continuous illumination at two different light intensities: 15 (open symbols) and 65 (solid symbols) μmoles photons m^{-2} sec^{-1}. Panel C plots the decline in the percent of un-photoactivated PSII centers during the initial course of the photoactivation treatment, with symbols according to Panels A and B.

Figure 2.

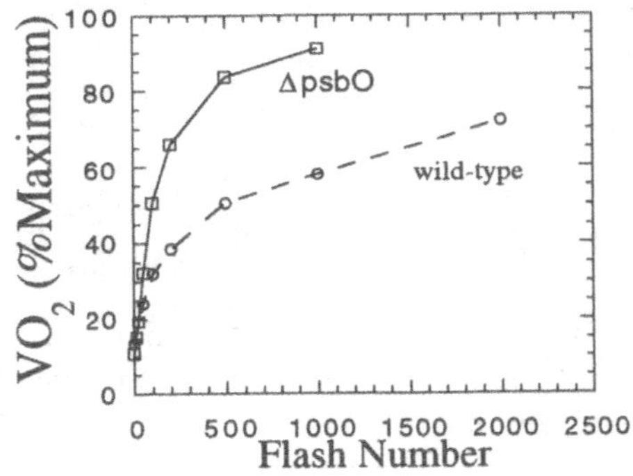

Figure 3.

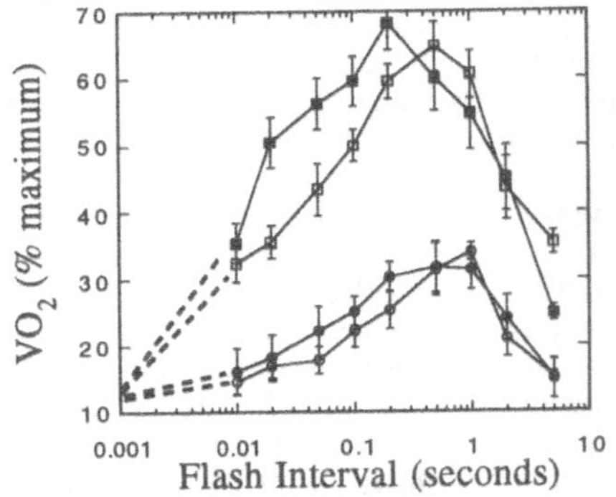

Figure 2. Photoactivation of HA-extracted wild-type and ΔpsbO cells as a function of flash number.

Cells were subjected various numbers of flashes given at 2 Hz.

Figure 3. Photoactivation of HA-extracted wild-type and mutant cells under flash illumination given at different intervals. HA-extracted in Hepes-NaCl buffer with 5 mM glucose (closed symbols) and without glucose (open symbols) were subjected to 100 flashes and the rate of light-saturated DCBQ-supported O_2 evolution determined. The dashed lines

3.2. *PSII catalytic activity* The H_2O-splitting enzyme acts as a charge accumulator and passes through a series of progressively more oxidized states (S-states): $S_0 \rightarrow S_1 \rightarrow S \rightarrow S_3 \rightarrow S_4$, with S_4 an unstable intermediate coinciding with the release of dioxygen and return to S_0. Bare platinum electrode experiments indicate that the S_3-->S_0 transition is markedly slowed in whole cells of the ΔpsbO strain [6], consistent with experiments with spinach PSII particles biochemically depleted of MSP [7]. However, the ΔpsbO strain [6] conclusions were based on the kinetic rise time of O_2 appearance, which could only be resolved for the mutant and could also be explained by differences in diffusion rates rather than the postulated alteration in enzyme kinetics. To obtain a comparison of the mutant and the wild type, double flash experiments [4] were performed to estimate PSII turnover times **(Fig. 4)**. The results are in very good agreement with previous measurements and put the half-time of the O_2-yielding $S_3 \rightarrow S_0$ transition at approximately 1.2 ms for the wild-type as opposed to a remarkably slow 8-12 ms for the ΔpsbO strain (variation in ΔpsbO strain appears to be due to physiological differences depending on culture state of cells). Preliminary fluorescence decay measurements indicate Q_A^- reoxidation rates dominated by decays in the sub-millisecond range for both cell types, indicating that the limitation is localized on the donor side. Also shown are two D1 mutants R64V and R64Q (see Table I also).

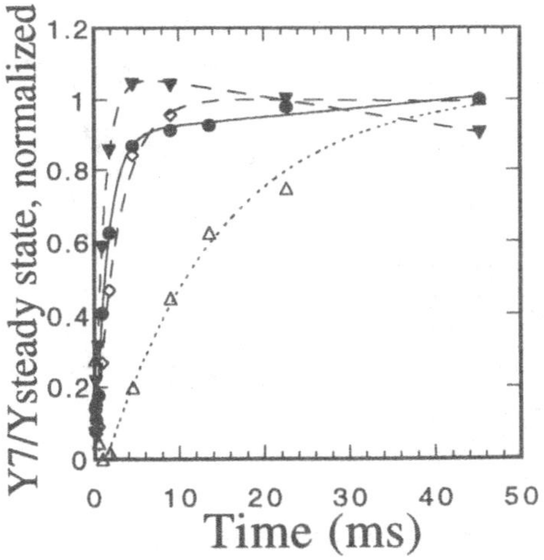

Figure 4. PSII turn-over times of cells undergoing primarily the $S_3 \rightarrow S_0$ transition were estimated by monitoring the increase in the O_2 yield on the 7th flash as a function of increasing the time interval between the 3rd and 4th flash. Wild-type, —●—; ΔpsbO,----Δ----; D1-R64V, —▼— ;R64Q, —◊—.

3.3 *Site-Directed mutagenesis of PSII* At this point we have constructed approximately 35 different amino acid substitution mutations at 19 distinct and highly conserved amino acid positions along the MSP amino acid sequence. Some of this work is published [1]. Mutations were also produced at two highly conserved arginine residues on the putative lumenal domain of the D1 protein. These were targeted since MSP probably interacts with basic residues on the intrinsic portion of the reaction center [8]. Table I summarizes some results of obtained upon analysis of several mutants.

MSP has been shown to modulate the Ca^{2+} requirement of PSII in vivo [9]. The majority of MSP mutants are able to grow autotrophically at low concentrations of Ca^{2+} (1/100 normal), yet many of these appear to only bind MSP weakly suggesting that even weakly bound MSP may be sufficient to stabilize the Ca^{2+} site. An exception is mutant, N138K, which is unable to grow 1/100 Ca^{2+} concentrations, despite the accumulation of normal levels of MSP. All the other mutants which do not grow at reduced Ca^{2+} concentrations (e.g. Δ*psbO* , C20S) and share the common property of being unable to accumulate MSP [1]. N138K exhibits substantial rates of O_2 evolution when (and only when) the cells are grown with sufficient Ca^{2+} , and the fact that O_2 evolution activity is not easily lost during membrane isolation (Qian & Burnap, unpublished) suggests MSP probably has a moderate binding affinity. Perhaps one of the most interesting observations is that this mutant undergoes very rapid (Δ*psbO* -like) photoactivation following hydroxylamine extraction. No other MSP accumulating mutant has yet been identified sharing this property. It will be important to investigate this result in view of findings that Ca^{2+} controls access to the active site. N138 is located in a stretch of 8 amino acids that is highly conserved in cyanobacteria, but has been deleted during the evolution of higher plant MSP. Since higher plants have additional extrinsic polypeptides thought to be involved in the retention of Ca^{2+} near the active site, we hypothesized that this region may be the primordial version of this functionality that was later replaced by the extrinsic 23 kDa protein and the N138K mutation was designed to introduce positive charge density into this region.

Table I
MSP & D1 mutants

Strain	Low Ca^{2+} growth	V_{max} O_2 evol., % wt	S_2-lifetime seconds	S_3-->S_0 ms	rapid photo-activation
wild-type	yes	100	23	1.2	no
ΔpsbO	no	44	110	8.9	yes
MSP-N138K	no	49	29	1.3	yes
MSP-D159K	yes	40	80	3.8	no
MSP-D159N	yes	64	44	2.2	no
D1-R64V	nd	70	15	0.8	yes
D1- R64Q	nd	55	35	1.8	nd
D1-R64E	nd	15	nd	nd	nd
D1-R334V	nd	51	95	8.5	nd

We thank Dr. George Cheniae and Dr. Rick Debus for valuable discussions and clones (R.D.). We gratefully acknowledge support from the U.S. Department of Agriculture.

4. References

[1] Burnap, R.L., et al . (1994) Biochemistry 33, 13712-13718.
[2] Chu, H.-A., Nguyen, A.P. and Debus, R.A. (1994) Biochemistry 33, 6150-6157.
[3] Cheniae, G.M. and Martin, I.F. (1972) Plant Physiology 47, 568-575.
[4] Bouges-Bocquet, B. (1973) Biochimica et Biophysica Acta 292, 772-785.
[5] Cheniae, G.M. and Martin, I.F. (1972) Plant Physiology 50, 87-94.
[6] Vass, I., Ono, T.A. and Inoue, Y. (1987) Biochim Biophys Acta 892, 224-235.
[7] Miyao, et al (1987) Biochimica et Biophysica Acta 890, 151-159.
[8] Shen, J.-R. and Inoue, Y. (1991) Plant and Cell Physiology 32, 453-457.
[9] Philbrick, J.B., Diner, B.A. and Zilinskas, B.A. (1991) JBC 266, 13370-13376.

AN FTIR STUDY OF CALCIUM INTERACTIONS IN PHOTOSYSTEM II

H. M. Zhang, G. Fischer* and T. Wydrzynski
Research School of Biological Sciences, *Department of Chemistry,
The Australian National University, Canberra, ACT 0200, AUSTRALIA

1. Introduction

The specific role which calcium plays in the O_2-evolving reactions of Photosystem II (PS II) is unknown even though calcium is generally considered to be an essential cofactor [1]. At least two tightly bound Ca^{2+} ions are associated with each PS II unit, one of which is located in a light harvesting component and the other in the core complex itself [2,3]. There are many treatments, usually involving 1-2M NaCl or 1M $CaCl_2$ salt washes to variously remove the 16, 23 and 33kDa extrinsic proteins, which induce a strong calcium dependence in O_2 evolution. However, there are controversies as to the functional role of calcium under these conditions [4,5]. In this communication, we report on the O_2-evolving activity, Ca and Mn contents and FTIR-detectable protein conformations of salt washed spinach PS II samples. We find that even though a large calcium demand in O_2 evolution can be created under these conditions, the two tightly bound Ca^{2+} ions in PS II remain intact. Rather, from the FTIR analysis, it appears that the removal of 16, 23kDa extrinsic proteins by 2M NaCl treatment induces a conformational change in PS II which is associated with the loss in activity. This conformational change is then restored by the superficial binding of calcium to PS II, most likely to the 33kDa extrinsic protein. In contrast, of the treatments which do release one of the tightly bound Ca^{2+} ions from PS II, we find no further increase in the fraction of Ca^{2+}-restorable O_2 activity. We conclude that many of the Ca-induced effects on O_2 evolution are mediated by conformational changes produced by the experimental conditions and are not related to the tightly bound calcium.

2.Methods

PS II samples were prepared from hydroponically grown spinach with Triton X-100 treatment according to the method of Berthold et al [6]. The excess detergent was removed by washing four times with the standard buffer medium containing 0.4M sucrose/15mM NaCl/5mM $MgCl_2$/40mM Mes-NaOH, pH 6.5. The 33kDa extrinsic protein was isolated from freshly prepared PS II membranes by salt washes as described earlier [7]. The isolated protein was concentrated by Amicon ultrafiltration to give a final protein concentration of 20-50mg/ml. Infrared spectra were recorded at 2 cm^{-1} resolution using a Bruker IFS 66 spectrometer with a MCT detector and OPUS data station. 20μl of PS II or isolated 33kDa protein sample was loaded between two CaF_2 windows using a 6-μm aluminium spacer. The sample temperature was kept at 4°C and the sample compartment was continuously purged with nitrogen to eliminate absorption by water vapour in the spectral region of interest. The quantitative analysis of protein secondary structure was calculated by the method of factor analysis [8]. Rates of O_2 evolution were

P. Mathis (ed.), Photosynthesis: from Light to Biosphere, Vol. II, 447–450.

measured at 25°C with a Clark-type oxygen electrode. The Ca and Mn contents were determined by atomic absorption spectroscopy after removal of contaminating Ca^{2+} with Chelex 100 [4].

3. Results and Discussion

The removal of the 16 and 23kDa extrinsic proteins from PS II samples by treatment with 2M NaCl inactivates O_2 evolution. This is shown in Table 1 where the activity of this sample drops to 27% of the untreated control. The expected Mn content does not change after the NaCl treatment nor does the Ca content, as reported earlier [4]. However, the O_2 evolving activity can be reactivated to 68% of the control by the specific addition of excess $CaCl_2$ (20mM) during the O_2 assay measurement. It was initially suggested that the loss of activity by NaCl treatment is due to the release of one of the two tightly bound Ca^{2+} ions. The discrepancy in results, however, was clarified by showing that it is the presence of small amounts of detergent during the NaCl treatment which induces the loss of the tightly bound Ca^{2+} ion [3]. This effect is demonstrated in Table 1 for our samples which were treated with NaCl plus 0.05% Triton X-100. In this case, the activity drops to 2% of the control. A more gentle treatment to release one of the tightly bound Ca^{2+} ions employs EGTA and the ionophore A23187 during the NaCl wash. This is also shown in Table 1. Here the activity drops to 29% of the control. When a tightly bound Ca^{2+} ion is removed, the addition of 20mM $CaCl_2$ also increases the O_2 evolving activity, but to no more than what is observed for simple NaCl-washed samples in which none of the tightly bound Ca^{2+} ions is removed.

Table 1. O_2 activity and metal contents of NaCl-washed PS II samples

Sample	O_2 rate[a] (μmol/mg chl/h)				Mn/200chl	Ca/200chl
	-Ca[b]		+Ca[c]			
1. Control PS II	620	(100)	632	(102)	3.9±0.2	2.0±0.1
2. 16 and 23kDa-depleted PS II[d]						
NaCl treated	167	(27)	421	(68)	4.0±0.3	1.9±0.4
NaCl/Triton treated	12	(2)	161	(26)	4.1±0.2	1.1±0.1
NaCl/EGTA/A23187 treated	180	(29)	372	(60)	4.2±0.2	0.94±0.2
3. Ca^{2+}-reconstituted PS II[e]						
NaCl treated	165	(27)	421	(68)	—	2.0±0.2
NaCl/Triton treated	13	(2)	167	(27)	—	1.3±0.2
NaCl/EGTA/A23187 treated	182	(29)	322	(52)	—	2.1±0.3

[a] The values in parentheses are percentages of the control. [b] Assay medium contained 1mM EGTA. [c] Assay medium contained 20mM $CaCl_2$. [d] 16 and 23kDa depletion was carried out by incubation of PS II samples in 2M NaCl for 30min in the dark, 2M NaCl plus 0.05% Triton X-100 under room light for 30min, or 2M NaCl /1mM EGTA/40mM Mes-NaOH, pH 6.5 in the presence of 20μM A23187 ionophore in the dark for 1hr, respectively. [e] Ca^{2+}-reconstitution was performed by incubation of 16 and 23kDa-depleted samples in 20mM $CaCl_2$ in the dark for 1 hr followed by washing with the standard buffers without $MgCl_2$

In order to test whether the tightly bound Ca^{2+} ion that can be removed may be related to O_2 evolution, the Ca^{2+}-depleted samples were incubated with 20mM $CaCl_2$, thoroughly washed with Chelex 100 treated buffers and reassayed for the Ca content and O_2 activity. For NaCl/EGTA/A23187 treated sample the lost tightly bound Ca^{2+} ion could be reconstituted but not for the NaCl/Triton X-100 treated sample. Either the different treatments remove Ca^{2+} ions from different environments or the detergent causes irreversible damage. Regardless, the levels of O_2 activity in all cases in the absence and presence of 20mM $CaCl_2$ are nearly the same as before reconstitution. Thus, the Ca^{2+}-restorable O_2 activity in NaCl-treated (16 and 23kDa protein-depleted) samples is independent of the two tightly bound Ca^{2+} ions.

The extrinsic 33kDa protein is known to be essential for stabilizing and optimizing the O_2-evolving activity. It can be removed by further treatment of NaCl-treated samples with 1M $CaCl_2$ [7]. As shown in Table 2, the Mn and Ca contents of our PS II samples also remain unaffected by this salt treatment although the activity drops further to 15% of the control. The restoration of O_2 evolution by the readdition of 20mM $CaCl_2$ is much less than that in the NaCl-treated sample alone. When the isolated 33kDa protein is reconstituted along with 20mM $CaCl_2$, the O_2 evolution activity approaches the level to what is observed for the NaCl-treated sample. Interestingly, since $CaCl_2$ is used to prepare the 33kDa protein, in its isolated form considerable calcium binds to it, some of which is removed by EGTA wash. These results are also shown in Table 2. It is clear, however, that the 33kDa protein in the intact PS II samples does not bind Ca. Thus, it seems that the binding of Ca^{2+} to the 33kDa protein is related to the Ca^{2+}-restorable O_2 activity created by the loss of the 16 and 23kDa proteins after NaCl treatment.

Table 2. O_2 activity and metal contents of 33kDa protein depleted and reconstituted PS II samples.

Sample	Mn/200chl	Ca/200chl	O_2 rate (μmol/mg chl/h)			
			no addition		20mM $CaCl_2$	
1. Control PS II	3.9±0.2	2.0±0.1	620	(100)	632	(102)
2. $CaCl_2$ treated	4.1±0.2	2.6±0.3	93	(15)	149	(24)
3. Reconstituted PS II[a]			279	(45)	347	(56)

Ca content in isolated 33kDa proteins

Sample	Ca/33kDa protein[b]
untreated	40
EGTA treated	12

[a] Reconstitution was carried out by incubation of the $CaCl_2$-treated PS II with the isolated 33kDa protein (protein/chl=1.5, w/w) in the standard buffer medium for 20min in darkness at 0°C. The reconstituted PS II sample was collected by centrifugation at 35,000×g and the unbound 33kDa was removed by resuspension in the buffer. [b] The concentration of 33kDa protein was determined according to [7]

Fig. 1 shows the FTIR spectra for the 33kDa protein before and after EGTA and heat denaturation treatment. Spectral variations in the range of the amide I band (1600-1700cm^{-1}) reflect protein conformational differences under these conditions. Using a multivariate analysis based on a set of 17 protein standards of known conformation, the α helix, β sheet, turn and random coil contents were quantified. The results for the 33kDa protein are shown in Table 3 and compared with a similar analysis for NaCl-treated PS II samples. Clearly the addition of $CaCl_2$ to the EGTA-treated 33kDa protein increases the α helix structure at the expense of the turns. For NaCl-treated PS II samples, the addition of $CaCl_2$ also increases in the α helix structure.

We conclude that the Ca^{2+}-restorable O_2 activity of NaCl-treated PS II is related to the renaturing of the 33kDa protein upon Ca^{2+} binding after the loss of the 16 and 23kDa

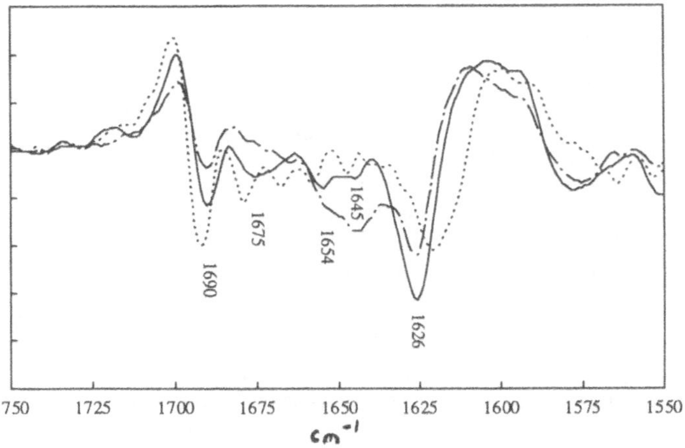

Fig.1 Second derivative spectra of the isolated 33kDa protein; (—) EGTA treated (peak positions shown), (—·—) in 20mM $CaCl_2$, (· · ·) denatured.

Table 3. Secondary structure of the isolated 33kDa protein and NaCl-washed PS II samples

Type of secondary structure	PS II sample			isolated 33kDa protein		
	control	NaCl washed				
		-Ca	+Ca[b]	-Ca[a]	+Ca[b]	denat[c]
α-helix%	54.3	49.0	51.3	20.2	29.3	12.0
β-sheet%	14.3	16.0	15.5	42.5	41.4	40.0
turn%	17.9	19.2	18.5	24.9	18.5	31.7
other%	13.5	15.8	14.9	12.4	10.8	16.3

[a] The isolated proteins were treated with 2mM EGTA. [b] In 20mM $CaCl_2$. [c] Denatured by heating at 95°C for 5min

proteins and that this effect is not related to the two tightly bound Ca^{2+} ions inherent to the PS II complex. The tightly bound Ca^{2+} ion that can not be removed by the treatments reported here may still be functionally related to the O_2-evolving reactions.

References

1. Debus, R. (1992) Biochim.Biophys.Acta 1102, 269-352
2 Cammarata, K.V. and Cheniae, G. (1987) Plant Physiol. 84, 587-595
3. Han, K.C. and Katoh, S. (1992) in Research in Photosynthesis (Murata, N., ed.) Vol. II, pp.6.365-368
4. Shen, J.R. and Katoh, S. Plant Cell Physiol. (1991) 32, 439-446
5. Pauly, S., Schlodder, E. and Witt, H.T. (1993) Biochim. Biophys. Acta 1099, 203-210
6. Berthold, D.A., Babcock, G.T. and Yocum, C.F. (1981) FEBS Lett. 134, 231-234
7. Xu, Q. and Bricker, T.M. (1992) J. Biol, Chem. 267, 25816-25821
8. Lee, D.C., Haris, P.I., Chapman, D. and Mitchell, R.C.(1990) Biochemistry 29, 9185-9193

EFFECTS OF FORMATE ON THE OXYGEN-EVOLVING MECHANISM

Alan Stemler [1,2] Jérôme Lavergne[1]
[1]Institut de Biologie Physico-Chimique, 13 rue Pierre et Marie Curie, 75005, Paris, France. [2]Permanent address: Section of Plant Biology, University of California, Davis, CA 95616 USA

1. Introduction

Formate ions have multiple effects on photosystem II, both on the acceptor side and donor side of the reaction center. We used flash-induced absorbance changes at 295nm to study the effects of formate on the oxygen-evolving mechanism. We find that treating BBY membranes and spinach thylakoids with formate and reduced 2,6 dichlorobenzoquinone (DCBQ) destabilizes the S_1 state in the dark, and can result in a population of S_0 near 100%.

2. Procedure

BBY membranes were prepared as in [1], omitting the second triton incubation. Spinach thylakoids were prepared as in [2]. Absorption changes were measured with a Joliot-Béal spectrophotometer [3]. The first saturating actinic flash of a series was a 50 ns pulse from a ruby laser. Subsequent flashes were from a xenon lamp (2μs at half-peak). Flash-induced O_2 yields were measured on a bare-platinum electrode with xenon lamp excitation (also 2μs at half-peak).

3. Results and Discussion

1. Absorption changes: BBY membranes incubated in control buffered solution that included DCBQ, showed typical period-four oscillations in UV absorption changes when given saturating flashes (Fig.1,A) [4,5]. Minima (which correspond to oxygen evolution maxima) occur on the 3rd,7th, and 11th flash. In the presence of formate, the minima are shifted one flash, occurring on the 4th, 8th, and 12th flash. Similar results were seen with spinach thylakoids (Fig. 1,B). The plots of Figs. 1 and 3 show the stable changes induced by each flash in the 100ms range. Alternatively, one can plot the extent of the kinetic phase with $t_{½}$ ~1ms [5,6]. This phase is specific to the $S_3Y_Z^+ \rightarrow S_0Y_Z + O_2$ reaction, so that this type of plot is roughly equivalent to an O_2 yield sequence. As shown in Fig. 4, the formate-induced shift is also evident in this representation. We interpret this shift as due to a destabilization of the S_1 state by formate. From a curve fitting procedure [5],based on the extinction coefficients of the S state transitions at

P. Mathis (ed.), Photosynthesis: from Light to Biosphere, Vol. II, 451–454.

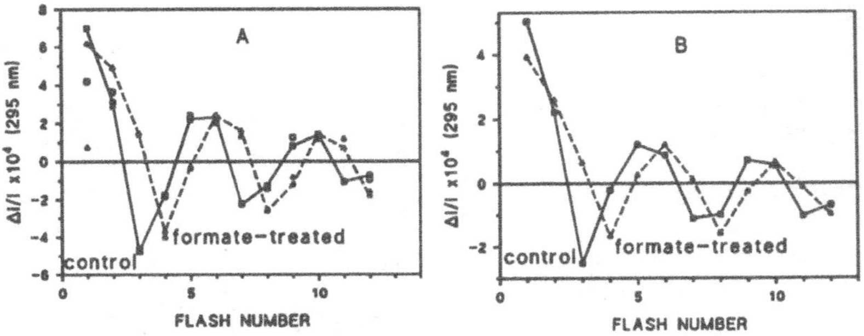

Fig. 1 The pattern of UV absorption changes following flash excitation of BBY membranes (A) and thylakoids (B) treated with formate and DCBQ. Controls were in 0.05M Na phosphate, pH 6.4, and 0.02M NaCl. Formate-pretreated samples were in 0.05M Na phosphate, pH 6.4, 0.1M NaCl, and 0.05M Na formate. All samples were given 25μM DCBQ and 5μM Na ascorbate, and were incubated in the dark for 10 min at 34°C. The open symbols are from a fitting procedure based on extinction coefficients determined by Lavergne [5]. The first two flashes were excluded from the fitting procedure.

295nm, the calculated S_1 population was 75% in the control and 12% in the formate-treated samples. The shifted pattern could be observed again after a 5-10min dark period provided both formate and DCBQ were both present. To obtain the maximum shift $S_1 \rightarrow S_0$, some *reduced* DCBQ was required, most likely to act as an electron donor to S_1. It was also helpful to warm the sample mildly (33-35°C) to maximize the S_0 population. Under optimum conditions, the level of S_0 approached 100%. This particular formate effect on PS II was neither reversed nor prevented by bicarbonate added either before or after the formate-DCBQ treatment. On the other hand, bicarbonate added alone did not itself cause a shift in the pattern as did formate (data omitted).

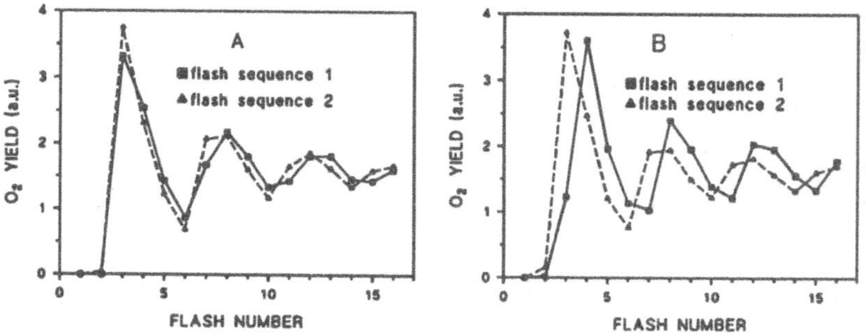

Fig 2 Flash yields of oxygen in thylakoids pretreated with DCBQ and either without (A), or with (B) formate. The thylakoids were pretreated as described in Fig. 1, then washed twice to remove DCBQ. Bicarbonate was added to relieve acceptor-side limitation due to formate. The thylakoids, still kept dark, were placed on the surface of

a platinum electrode. A sequence of saturating flashes from a xenon lamp was given then, after a 10 min dark period, a second sequence was given.

2. Oxygen flash yields: We confirmed that formate and DCBQ caused an $S_1 \rightarrow S_0$ shift by measuring oxygen evolution on a bare-platinum electrode (Fig.2, A and B). In this case, thylakoids were pretreated in the presence of DCBQ and either with or without formate. The samples were then washed in their respective buffered solutions in order to remove DCBQ, which can interfere with the measurement. The controls, (Fig. 2A), showed a typical maximum on the 3rd flash and period-four oscillations thereafter. This was true both for the first flash sequence given and on a second sequence given 10min later. Formate-treated samples, in contrast, showed maxima on the 4th, 8th,12th,and 16th flash for the first flash sequence, and a normal pattern on the second flash sequence. The disappearance of the $S_1 \rightarrow S_0$ shift in the second sequence is specific to this procedure where DCBQ was washed away after the formate treatment.

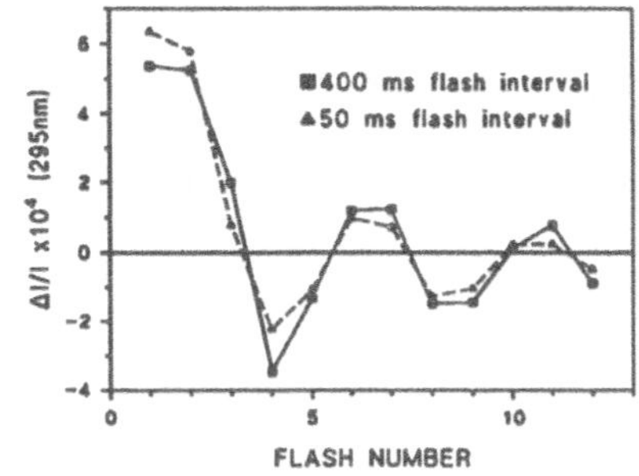

Fig. 3 The pattern of UV absorption changes in formate and DCBQ-treated BBY membranes given flashes at either 400 or 50 ms intervals.

3. Effect of flash rate: The pattern of absorption changes induced by formate-DCBQ is almost the same whether flashes are given at 400 or 50ms intervals (Fig. 3), except for a larger damping in the latter case due to acceptor-side limitation. It is therefore unlikely that the shifted pattern observed after formate treatment could be due to reduction of S_2 and S_3 states by reduced tyrosine Y_D, occuring during the flash sequence, since this reaction has a $t_{1/2}$ of about 1s. Moreover, Vermaas et al [7] report that formate blocked electron transfer from Y_D to the higher S-states and that the transfer was restored by bicarbonate.

From the results presented here, we draw the following conclusions: 1. The presence of formate and reduced DCBQ destabilizes the S_1 state to form S_0. Tentatively, we propose that formate binding causes a structural change in the Mn cluster that allows its redox equilibration with DCBQ. It should be noted, however, that the absorption change observed at 295nm on the first flash in formate-treated material is larger than expected for the $S_0 \rightarrow S_1$ transition[6,8]. It appears that the modified S_0 state induced by formate treatment has specific spectral properties. 2. At the present time we cannot completely rule out an alternative hypothesis. Formate treatment may not really affect the S_0,S_1 distribution, but may act on Y_D, (i) causing its full reduction and (ii) accelerating dramatically the $S_2Y_D \rightarrow S_1Y_D^+$ reaction. As the phase shifted sequence can still be observed with a 50ms flash interval, the reaction with Y_D would have to be completed in less than 50ms whereas, under normal conditions, it occurs in the 1s range. As previously mentioned, this hypothesis is also inconsistent with the conclusions of Vermaas et al [7].

References

1 Berthold, D.A., Babcock, G.T., and Yocum, C.F. (1981) FEBS Lett. 134,231-234
2 Joliot, P., Lavergne, J., and Béal, D. (1992) Biochim. Biophys. Acta 1101,1-12
3 Joliot, P., Béal, D., and Frilley, B. (1980) J. Chim. Phys. 77,210-218
4 Dekker, J.B., van Gorkom, H.J., Wensink, J., and Ouwehand, L. (1984) Biochim. Biophys. Acta 767,1-9
5 Lavergne, J. (1991) Biochim. Biophys. Acta 1060,175-188
6 Velthuys, B.R. (1981) in Photosynthesis (Akoyunoglou, G., ed.) Vol II pp 75-85 Balaban International Science Services, Philadelphia
7 Vermaas, W.F.J., Renger, G., and Dohnt, G. (1984) Biochim. Biophys. Acta 76,194-202
8 van Leeuwen, P.J., Heimann, C., and van Gorkum, H.J. (1993) Photosynthesis Res.38, 323-330

NEW PROBES OF NATIVE PSII: TRAPPING OF S STATE INTERMEDIATES

Jonathan H.A. Nugent and Sandra Turconi
Department of Biology, University College London,
Gower Street, London WC1E 6BT, U.K.

1. **Introduction**

In PSII, electron donation from water is probably located on the heterodimer of D1/D2 polypeptides. Tyrosine Y_Z (located on D1) acts as an intermediate electron carrier between $P680^+$, the oxidised primary chlorophyll donor (bound to D1/D2), and the water oxidising complex (WOC) which includes a manganese cluster. The WOC cycles through states termed S states due to the removal of electrons via Y_Z. The S_2 state is the best understood, mostly from studies involving two EPR signals near g=2 and g=4, termed the multiline and g=4.1 signal [1]. The S_3 and S_4 states are difficult to obtain in high yield, especially at the high concentrations required for EPR.
In inhibited samples, following ammonia treatment or where the cofactors Ca^{++} or Cl^- are removed or replaced, the cycle appears to be inhibited at S_3. This allows the formation of a split g=2 signal (Figure 2) termed the "S3" signal and this is attributed to the interaction of the S_2 state with a radical from histidine, tyrosine or an intermediate of water oxidation, $S_2 X^+$ [2-9]. The native S_3 state produced during flash turnover of PSII does not give this signal. A recent paper [10] has indicated an oxidation of manganese on the native S_2 to S_3 step, so that the relevance of data from inhibited systems is not clear. It is therefore uncertain whether Mn, a Mn ligand or an amino acid is oxidised when S_3 is formed *in vivo*.
We have performed a series of experiments to investigate this question. We have found that "S3" type g=2 split signals can be found in untreated preparations and that a new EPR signal near g=5 provides a further probe of PSII electron transfer.

2. Materials and Methods

Chloroplast PSII membranes (Chl a / Chl b ratio 2:1) were prepared using Triton X-

P. Mathis (ed.), Photosynthesis: from Light to Biosphere, Vol. II, 455–458.

100 by the method based on that of [11]. Chloroplast PSII membranes were washed in a buffer containing EDTA to remove adventitiously bound Mn^{2+}, followed by washing under room light in a buffer containing $CaCl_2$ and NaCl to prevent cofactor depletion. Both pea and spinach PSII with high rates of oxygen evolution {400-1000 ug O_2/mg Chl/ml} were used.
For EPR, samples were either untreated or treated with electron acceptors such as ferricyanide or quinones such as phenyl benzoquinone (PPBQ) and 2,6 dichlorobenzoquinone (DCBQ). Samples were illuminated at a variety of temperatures from 4-295K and examined by EPR at cryogenic temperatures using a Jeol RE1X spectrometer fitted with an Oxford Instruments cryostat. EPR conditions: Temperature 9K and microwave power 10mW.

3. Results and Discussion

New g=2 split signals were found in PSII membranes (Figures 1 and 2). They have an asymmetric lineshape and are attributed to an interaction between a S=1/2 g=2 radical (oxidised amino acid or intermediate of water oxidation) and the Mn complex. The signal is not formed in all PSII centers due to the inefficiency of the procedures for trapping these states.
The characteristics of the signals show them to be different from those found in cofactor-depleted or ammonia-treated reaction centers. No "S3" g=2 split signal is detected in untreated PSII membranes using the illumination conditions required for production of "S3" signals in inhibited samples [2-9] or in the native S_3 state.
A new 20-25mT broad g=5 EPR signal was also found (Figure 1). Although it is possible this originates from the non-haem iron of PSII, this signal, which shows Curie behaviour down to 4.2 K and a rapid spin relaxation rate, probably arises from the Mn complex. Characteristics indicate a spin state of S=3/2 (or a middle Kramer's doublet of a S=5/2 state). Some of the g=5 signal is lost when illumination forms the g=2 split signal (Figure 1). This indicates an electron donor/electron acceptor relationship with the g=2 split signal or that the g=5 signal arises from a precursor state of that giving the g=2 split signal. Similar yields of the new signals were obtained in several PSII preparations from both spinach and pea.

The generation of a g=2 split signal indicates that Y_Z can operate at cryogenic temperatures and that interactions of a radical and the manganese cluster can be observed in untreated PSII samples. Recent experiments have shown that the "S3" signal in calcium-depleted samples arises from an interaction between oxidised Y_Z [12] and the Mn cluster, confirming the our earlier ideas [6-7]. Therefore in the inhibited preparations [2-9] the formation of a modified S_3 by manganese oxidation via Y_Z^+ may occur but at a rate that allows a more significant population of S_2 X^+, possibly S_2 Y_Z^+ to be trapped than in untreated samples.

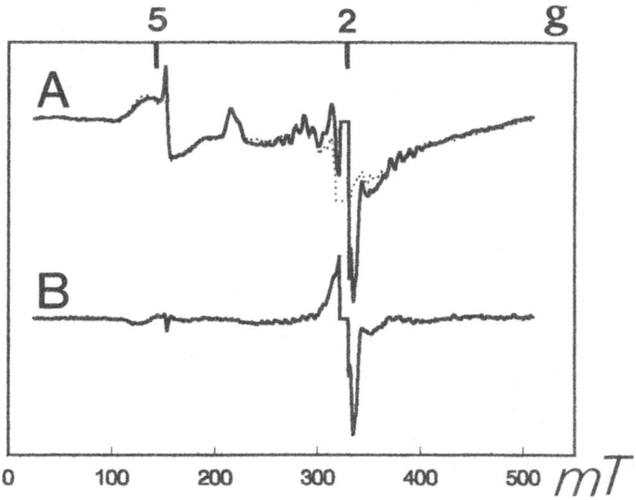

Figure 1. New EPR signals in untreated PSII membranes.
A. Spectrum of PSII sample illuminated at cryogenic temperatures superimposed on that of the dark adapted g=5 containing spectrum.
B. Light-dark difference spectrum showing partial loss of the g=5 signal and appearance of a split signal near g=2 following illumination.

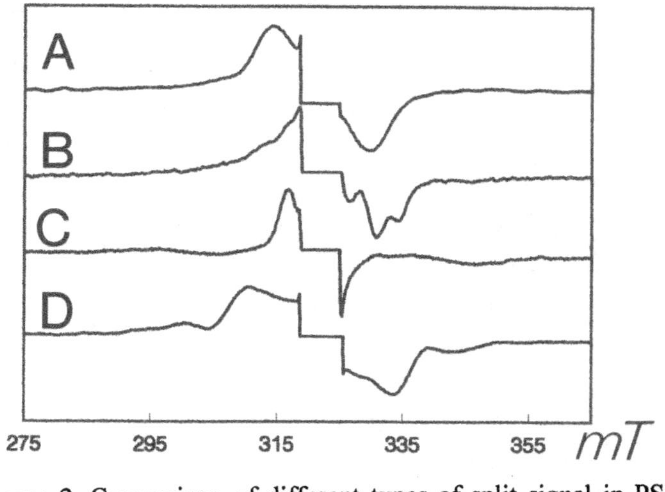

Figure 2. Comparison of different types of split signal in PSII.
A. "S3" type signal in calcium-depleted PSII.(g=2 radical removed)
B,C. Two new forms found in native PSII membranes in this study.
D. "S3" type signal in acetate-treated PSII.

The g=5 signal may be related to g=4.1 type S_2 signals. Analysis of the conditions required to obtain g=4.1 signals suggest that changes in the ligands to Mn, perhaps at the water substrate binding site, control the appearance of this form of S_2. It is tempting therefore to suggest that the g=5 signal represents the Mn cluster with an intermediate of water oxidation trapped as a ligand or that the cluster is trapped in a higher S state conformation. The native g=2 split signal would then be a trapped intermediate state in S state turnover.

4.Conclusions

We have found three new EPR signals in untreated PSII membranes. Our results indicate that Y_Z operates close to and interacts with the Mn cluster in native PSII. These new signals will be useful probes to further investigate PSII electron transfer.

Acknowledgements

We acknowledge support from the U.K. Biotechnology and Biological Science Research Council.

References

1. Debus, R.J.(1992) Biochim. Biophys. Acta 1102, 269-352.
2. Boussac, A., Zimmermann, J.L. and Rutherford, A.W. (1989) Biochemistry 28, 8984-8989.
3. Boussac, A., Zimmermann, J.L., Rutherford, A.W. and Lavergne, J. (1990) Nature 347, 303-306.
4. Baumgarten, M., Philo, J.S. and Dismukes, G.C. (1990) Biochemistry 29, 10814-10822.
5. Ono, T., Kusunoki, M., Matsushita, T., Oyanagi, H. and Inoue, Y. (1991) Biochemistry 30, 6836-6841.
6. Hallahan, B.J., Nugent, J.H.A., Warden, J.T. and Evans, M.C.W. (1992) Biochemistry 31, 4562-4573.
7. Maclachlan, D.J. and Nugent, J.H.A. (1993) Biochemistry 32, 9772-9780.
8. van Vliet, P., Boussac, A. and Rutherford, A.W. (1994) Biochemistry 33, 12998-13004.
9. Kusunoki, M. Chem. Phys. Lett. (1995) 239, 148-157.
10. Ono, T., Noguchi, T., Inoue, Y., Kusunoki, M., Matsushita, T. and Oyanagi, H. (1992) Science 258, 1335-1337.
11. Ford, R.C. and Evans, M.C.W. (1983) FEBS Lett. 160, 159-163.
12. Gilchrist, M.L., Ball, J.A., Randall, D.W. and Britt, R.D. (1995) Proc. Natl. Acad. Sci. USA in press

MANGANESE OXIDATION STATES OF THE FLASH-INDUCED S-STATES OF PHOTOSYSTEM II

T.A. Roelofs[†‡§], W. Liang[†‡¶], M.J. Latimer[†‡], R. Cinco[†‡], A. Rompel[†‡], J.C. Andrews[†‡], V.K. Yachandra[†], K. Sauer[†‡] and M.P. Klein[†].
[†]Structural Biology Division, Lawrence Berkeley Laboratory,
[‡]Department of Chemistry, University of California at Berkeley, CA 94720.
[§]Present address: Dept. of Physics, Free University Berlin, Germany.
[¶]Present address: Dept. of Biochemistry, Stanford University, California.

Introduction

The oxygen-evolving complex (OEC) of Photosystem II (PS II) contains a tetranuclear manganese cluster that has been implicated as the catalytic site for photosynthetic water oxidation (1). The OEC cycles through five intermediate states, denoted S-states S_i, where the index i indicates the number of positive charges accumulated at the OEC and runs from 0 to 4. The mode of charge accumulation during the S-state cycling can be studied by X-ray Absorption Near-Edge Structure (XANES-) spectroscopy and, with regard to the $S_2 \rightarrow S_3$ transition, has been controversial. Results on an S_3-like state, populated by low temperature continuous illumination of chemically treated PS II, suggest the absence of direct Mn-oxidation during the $S_2 \rightarrow S_3$ transition (2). On the other hand, XANES experiments on flash-induced S-states (3,4) revealed similar shifts in the edge position upon the first and second flash and led the authors to conclude that direct Mn oxidation takes place during the $S_2 \rightarrow S_3$ transition. However, in the latter studies no independent information was available on the actual S-state distributions in the flash-treated samples.

Because of the apparent discrepancy between the results of Ono et al. (3) and our previous studies (2), we deceided to combine XANES-experiments on flash-induced S-states with a detailed EPR characterization of the flash-induced S-state advance in the same samples. We monitored the S_2-population by measuring its EPR multiline signal (MLS) amplitude and have fit the damped oscillation pattern with the Kok model, using misses and double hits as the only variables (5). This model description allowed for the calculation of the flash-induced S-state distributions, which are crucial for the accurate extraction of the edge spectra of the *pure* S-states from the experimentally acquired flash-induced spectra.

Materials and Methods

PS II-enriched membranes were isolated from spinach according to a modified BBY-procedure (6), and had a oxygen evolution rate of 400-550 μmol O_2/ (h • mg Chl). Samples (4-5 mg Chl / ml, with 500 μM. phenyl-*p*-benzoquinone) were dark-adapted for 1 h at 4 °C. A double-preflash protocol synchronized almost all PS II centers into the S_1-Y_D^+-state (7). Each sample was then given 0, 1, 2, 3, 4, or 5 flashes, with 1.5 s intervals (Xe flash lamp system giving white light flashes of 14 μs full width at half maximum), before being immersed in liquid nitrogen within 1-2 s.

The EPR spectra, both the Y_D^+ signal and the MLS, were recorded as (6). Mn K-edge spectra were acquired at SSRL and NSLS, as described elsewhere (8). Typically 16-20

P. Mathis (ed.), Photosynthesis: from Light to Biosphere, Vol. II, 459–462.

scans were averaged for each sample. A linear background, fit to the signal before the pre-edge, was subtracted, and the spectra were normalized at the energy of maximal absorption. The position of the main edge was defined by its first inflection-point energy (IPE), determined from the zero-crossing of the second derivative of the spectrum. Analytical differentiation of a third-order polynomial fit to the data over an interval of 3.0 eV on each side of a data point produced the derivative spectra. The estimated error in the IPE is ± 0.1 eV for individual samples. The second derivative signal was also used to characterize the shape of the edge spectrum (9).

Results and Discussion

Flash-induced changes

The flash-induced MLS-oscillation pattern is shown in Fig. 1. Samples dark-adapted after the preflash treatment did not exhibit any detectable amount of MLS. The first flash induced the maximal amount of MLS. The second flash led to a reduction in MLS to ~25% of the maximal value. The MLS nearly disappeared (~5%) after the third flash. The fourth flash led to a slight recovery in MLS-amplitude (~15%), whereas upon the fifth flash the MLS recovered to ~50% of the maximal value.

The application of 0 - 5 flashes after the preflash protocol did not lead to an increase of the Y_D^+-signal, indicating that at that point the recombination between reduced Y_D and the higher S-states did not take place to any significant extent. Flash-induced S-state advance can thus be described in terms of only two parameters (misses and double hits), neglecting recombination processes involving reduced Y_D. Assuming that after the preflash protocol the samples have a 100% S_1-population, the Kok model describes the damped period-four MLS oscillation well. The best fit (dashed line) was found with 12% misses and 5% double hits. Using these fit values, the S-state distributions in samples given 0, 1, 2, 3, 4, or 5 flashes were calculated.

In Fig. 2 we show the inflection-point energy (IPE) of the Mn K-edge spectra of the PS II membranes, as a function of the number of flashes applied. Samples not given any flash (after the preflash treatment) have their Mn K-edge around 6551.8 eV. The first

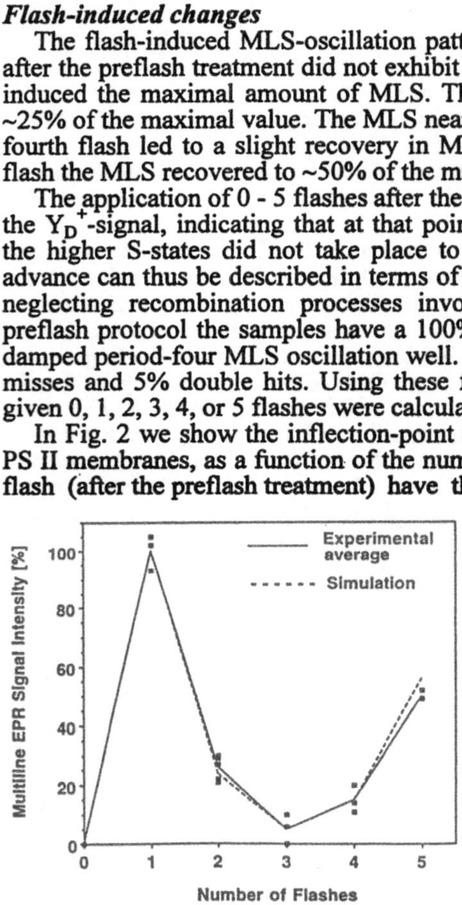

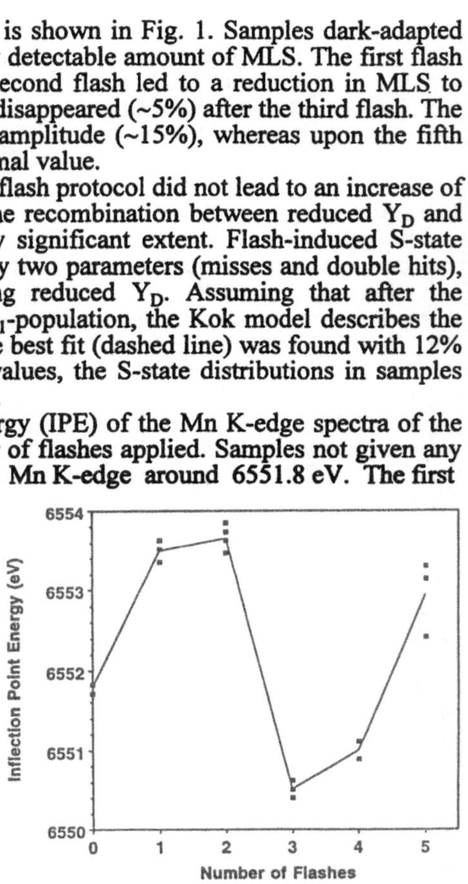

Fig. 1 (left): Flash-induced EPR MLS-oscillation pattern of PS II membranes. The MLS-amplitudes were normalized to the amplitude after the first flash. The solid line shows the average of 3 - 4 independent and identical series; the dashed line is the best fit to the MLS-oscillation pattern, using Kok's model with 12% misses and 5% double-hits.

Fig. 2 (right): Inflection-point energy (IPE, in eV) of the Mn K-edge of PS II membranes as a function of the number of applied flashes.

flash induced a shift of 1.5 - 2.0 eV to higher energy, with the edge at approx. 6553.5 eV. Upon the second flash, only a small shift of 0.3 eV to higher energy was observed. Application of the third flash led to a large shift of the edge to lower energy, resulting in an edge position around 6550.5 eV, well below that of a 0-flash sample. The fourth and fifth flashes returned the edge to higher energies.

S-state edge spectra

The edge spectrum of a sample given any number of flashes is a superposition of the edge spectra of the four S-states, each of these S-state spectra weighted by its population in that particular sample. Combining the flash-induced S-state distributions, calculated from the MLS-oscillation pattern, with the spectra of samples given 0, 1, 2 or 3 flashes, the edge spectra of the pure states S_0 through S_3 were extracted. The resulting S-state edge spectra are shown in Fig. 3.

The IPE found for the pure S-state spectra are 6550.1, 6551.7, 6553.5 and 6553.8 eV for S_0, S_1, S_2, and S_3, respectively. Each of the $S_0 \rightarrow S_1$ and the $S_1 \rightarrow S_2$ transitions is accompanied by a shift of the edge to higher energies by approx. 1.6 - 1.8 eV, indicating that during those transitions the Mn-cluster undergoes oxidation. The $S_2 \rightarrow S_3$ transition, however, leads to a small edge shift of only 0.3 eV, suggestive of the absence of a direct Mn-oxidation step.

We previously showed (9) that the shape of the Mn-edge in the spectral region around the inflection point, between 6545 and 6555 eV, reflects predominantly oxidation state changes, whereas in the region between 6555 and 6575 eV changes in the coordination environment of the Mn become more pronounced. In Fig. 4 we show the second derivatives of the normalized S-state edge spectra. Several features change reproducibly with S-state. During the $S_1 \rightarrow S_2$ transition a pronounced positive feature appears at 6552

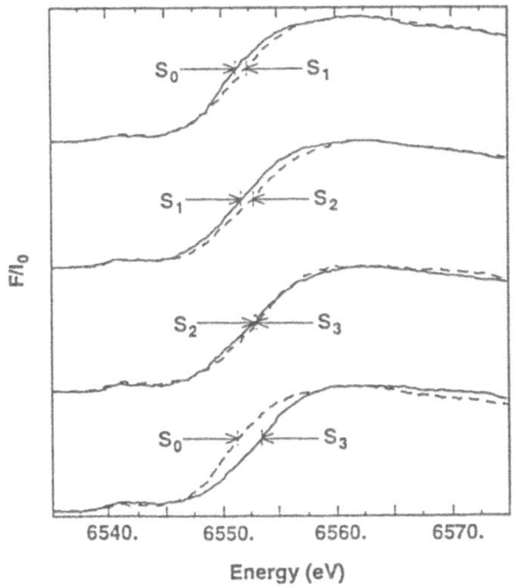

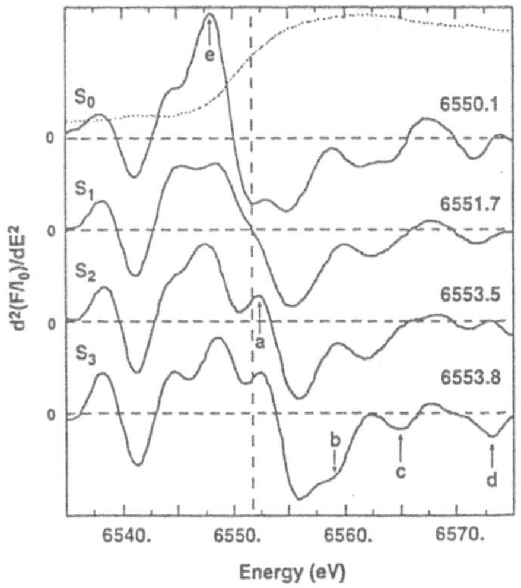

Fig. 3 (left): Normalized Mn K-edge spectra for the pure S-states of the Mn cluster of PS II, as calculated from the flash-induced edge spectra of samples given 0, 1, 2 or 3 flashes. **Fig. 4** (right): Second derivatives of the normalized pure S-state edge spectra of the Mn-cluster of PS II. For clarity a vertical dashed line has been drawn at the inflection-point energy (IPE) of the S_1-state. The IPE (in eV) for each S-state is given at the right. For reference the spectrum of the S_0-state is plotted as well (dashed line).

eV, marked with **a**, whereas above 6555 eV the edge remains virtually unchanged. The shape of the edge in the spectral range between 6545 and 6555 eV remains mostly unchanged at the $S_2 \rightarrow S_3$ transition. However, reproducible changes occur between 6555 and 6575 eV (indicated with **b**, **c** and **d**) These results suggest that changes occur in the coordination environment of the Mn, rather than in the oxidation state of the Mn. The spectrum of the S_0-state lacks feature **a** around 6552 eV. A positive feature around 6548 eV (marked **e**) is present, having a much larger amplitude and a significantly narrower width than in the S_1-spectrum. This intense and narrow feature possibly suggests the presence of Mn(II) or a greater amount of Mn(III) in the complex (9).

Oxidation state of the Mn-cluster

Based on our analysis of the Mn edge spectra of the flash-induced S-state, we propose the following scheme for the oxidation states of the Mn-cluster during the S-state cycle: S_0: (II, III, IV, IV) or (III, III, III, IV), S_1: (III, III, IV, IV), S_2: (III, IV, IV, IV) and S_3: (III, IV, IV, IV, *). In a previous paper (9) we showed that the S_1-state is more likely to consist of Mn_4 (III, III, IV, IV) than $Mn(III)_4$. Upon the formation of S_2 one of the Mn(III) is oxidized to Mn(IV), shifting the edge position to higher energy. During the $S_2 \rightarrow S_3$ transition no direct Mn oxidation occurs, because both the position of the edge and its shape in the 6545 - 6555 eV region change very little. The oxidation equivalent resulting from the reaction center turnover, indicated in the scheme as *, is likely stored in the proximity of the Mn-cluster, because the MLS disappears. Upon the $S_3 \rightarrow S_0$ transition the cluster is reset to the most reduced state of the Kok-cycle. The S_0-state possibly contains Mn(II) or a greater amount of Mn(III), as suggested by feature **e** in the second derivative of the edge. The transition to the S_1-state involves Mn oxidation, the disappearance of the feature **e** in the second derivative of the edge, and a shift in the edge position to higher energy.

In summary, we have shown that during the $S_2 \rightarrow S_3$ transition, the edge position shifts very little (0.3 eV) and the shape of the edge between 6545 and 6555 eV shows only subtle changes. Therefore, we conclude that no direct Mn oxidation is involved in this transition. However, at present we cannot rule out the possibility that the expected edge shift of a putative Mn oxidation step is partially compensated by structural rearrangements in the coordination environment of the Mn-cluster.

References

1 Debus, R.J. (1992) Biochim. Biophys. Acta 1102, 269-351
2 Guiles, R.D., Zimmermann, J.-L., McDermott, A.E., Yachandra, V.K., Cole, J.E., Dexheimer, S.L., Britt, R.D., Wieghardt, K., Bossek, U., Sauer, K. and Klein, M.P. (1990) Biochemistry 29, 471-485
3 Ono, T., Noguchi, T., Inoue, Y., Kusunoki, M., Matsushita, T. and Oyanagi, H. (1992) Science 258, 1335-1337
4 Kusunoki, M., Ono, T., Noguchi, T., Inoue, Y. & Oyanagi, H. (1993) Photosynth. Res. 38, 331-339
5 Kok, B. and Joliot, P. (1975) in Bioenergetics of Photosynthesis (Govindjee, ed.) pp. 387-412, Academic Press, New York
6 Liang, W. (1994) Ph.D.-Thesis, University of California at Berkeley (LBL Report 36632, Lawrence Berkeley Laboratory)
7 Evelo, R.G., Styring, S., Rutherford, A.W. and Hoff, A.J. (1989) Biochim. Biophys. Acta 973, 428-442
8 Liang, W., Latimer, M.J., Dau, H., Roelofs, T.A., Yachandra, V.K., Sauer, K. and Klein, M.P. (1994) Biochemistry 33, 4923-4932
9 Yachandra, V.K., DeRose, V.J., Latimer, M.J., Mukerji, I., Sauer, K. and Klein, M.P. (1993) Science 260, 675-679

COMPARISON OF STRUCTURE AND FUNCTION OF THE PHOTOSYNTHETIC OXYGEN-EVOLVING COMPLEX OF SPINACH AND THE GREEN ALGA *SCENEDESMUS OBLIQUUS*

H. Schiller, S. Klingelhöfer, W. Dörner, H. Senger and H. Dau
FB Biologie/Botanik, Philipps-Universität, 35032 Marburg, Germany

1. Introduction

This investigation is part of a project aimed at the comparison of the structure and function of the oxygen evolving complex of evolutionary diverse organisms.

2. Materials and Methods

PS II membrane preparations of spinach were obtained using a Triton X-100 purification method described elsewhere (1). *Scenedesmus* preparations were carried out by a modified procedure. Oxygen evolution measurements were measured polarographically using 5 mM $K_3[Fe(CN)_6]$ and 0.25 mM 2,6-dichlorobenzoquinone (DCBQ) as electron acceptors. EPR spectra were collected with a X-band EPR spectrometer (ESP 300E, Bruker, Oxford, UK). Measuring conditions were (if not otherwise stated): temperature, 8 K; microwave power, 30 mW; modulation amplitude, 20 G. Samples were illuminated for five minutes at 200 K. Ammonia treated samples were obtained by suspending in 100 mM NH_4Cl buffer, illuminating for five minutes at 200 K and subsequent annealing for one minute at room temperature in the dark.

3. Results and Discussion

3.1 Influence of NH_4Cl and $(NH_4)_2SO_4$ on PS II activity. The concentrations of NH_4Cl and $(NH_4)_2SO_4$ were determined which caused a inhibition of oxygen evolution by 40 % at pH 7.5 (30 mM for NH_4Cl and 10 mM for $(NH_4)_2SO_4$). Adding these concentrations of ammonia salts, the enzyme activity was measured over a pH range from 3 to 9. The actual NH_3 concentration at each pH value was calculated using the Henderson-Hasselbach equation. In *Fig. 1* the PS II activity is plotted as a function of

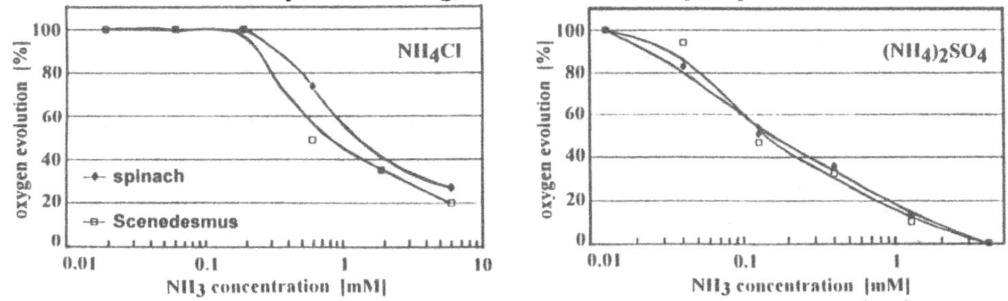

Fig. 1: Inhibition of PS II activity by NH_4Cl or $(NH_4)_2SO_4$. Activity of samples is plotted against the calculated NH_3 concentration.

P. Mathis (ed.), Photosynthesis: from Light to Biosphere, Vol. II, 463–466.

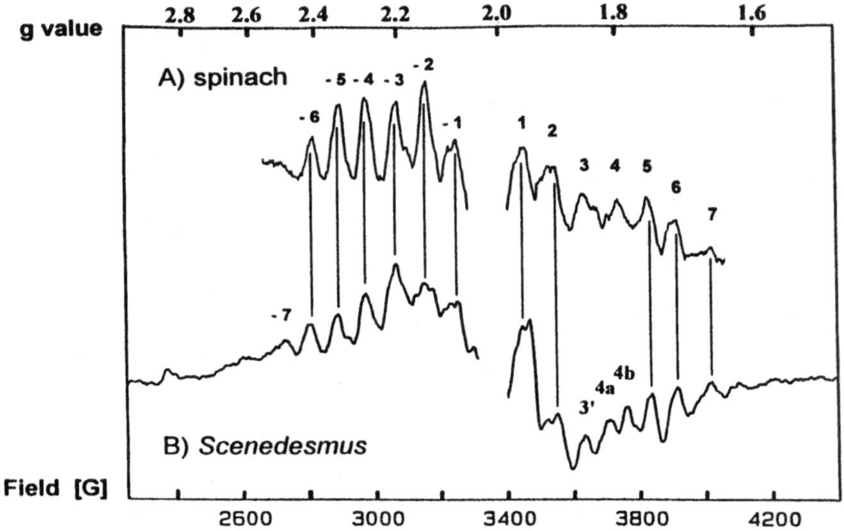

Fig. 2: Comparison of the g = 2 multiline region in EPR spectra of **A)** spinach and **B)** *Scenedesmus*. Measuring conditions as in Materials and Methods.

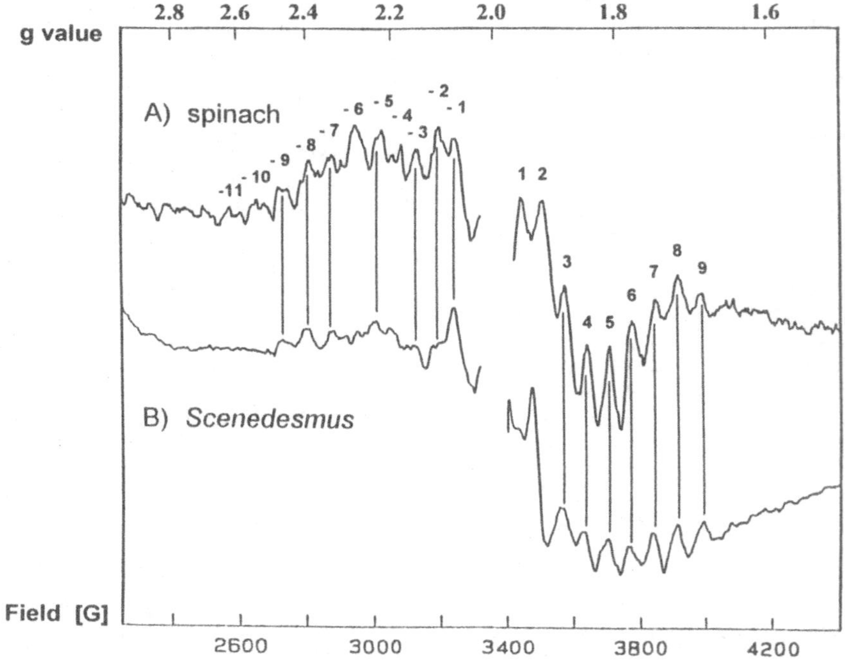

Fig. 3: Comparison of the EPR spectra of NH_3-modified and annealed **A)** spinach and **B)** Scenedesmus preparations. Measuring conditions as in Materials and Methods.

the actual NH_3 concentration. For each data point the enzyme activity without NH_3 at the respective pH value was referred to as 100 %.
Both ammonia salts inhibit oxygen evolution by the action of NH_3, which competes with Cl^- and H_2O for binding sites within the PS II donor side (2). For NH_4Cl the inhibition begins at 1.2 mM NH_3, whereas for $(NH_4)_2SO_4$ it starts immediately. Therefore the Cl^- binding site may be the first target for ammonia and Cl^- seems to prevent its action by competitive inhibition. The graphs of both organisms are very similar, indicating that there is no difference of spinach and *Scenedesmus* with respect to the inhibition with ammonia.

3.2 EPR spectroscopy. In *Fig. 2*, g = 2 multiline spectra of spinach and *Scenedesmus* PS II membrane preparations are shown. The signals of both organisms have a similar line spacing. Some peaks of the *Scenedesmus* spectrum, however, show reproducibly a different finestructure, especially in the right part of the spectrum (peaks 3 and 4) and at peak -2.

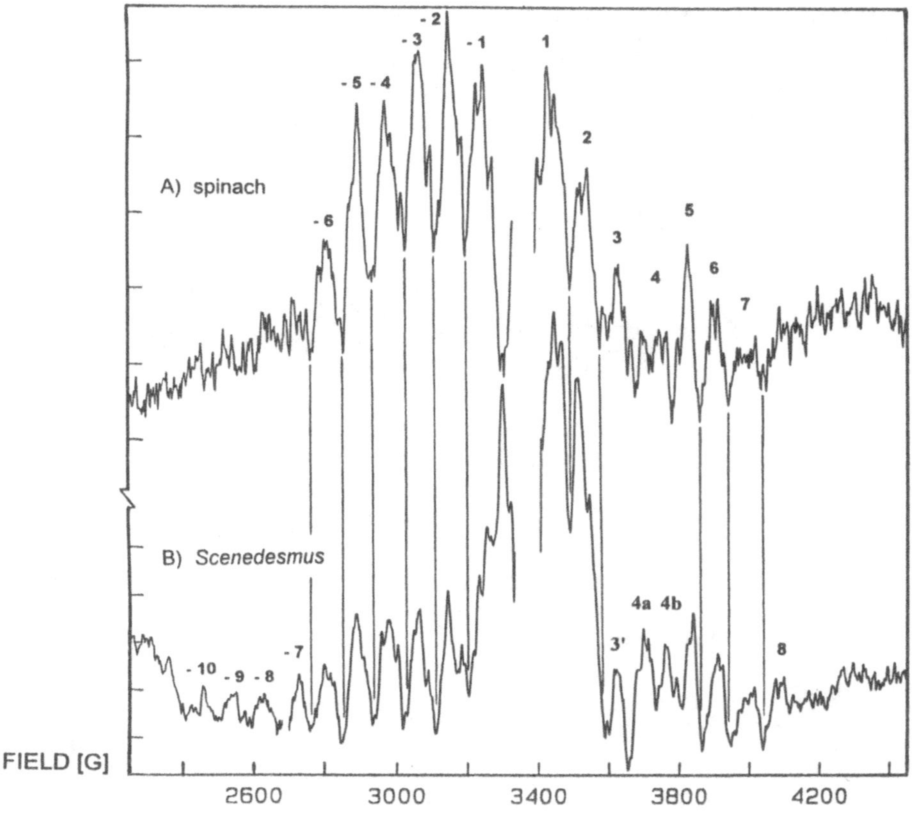

Fig. 4: EPR spectra of A) spinach and B) *Scenedesmus* PS II membrane particle preparations measured with a modulation amplitude of 5 G.

Upon ammonia treatment the shape of the multiline signal changes because of binding of NH_3 to the manganese cluster (3,1). In both, spinach and *Scenedesmus*, the spectra changed upon NH_3 binding (*Fig. 3*). The NH_3 spectra look similar, but the ratio of the peak heights in the right part of the spectra differs slightly. In *Scenedesmus* peaks 3-9 are of similiar magnitude, whereas the height of the spinach peaks decreases strongly from peak 5 to 9.
In *Fig. 4* high-resolution spectra of untreated preparations are compared. The subdivision of the single peaks is comparable in the left part, but differs as in *Fig. 2* in the right part of the spectra, especially in the region of the peaks 3 and 4. Like in Fig. 3, the difference between the peak intensities is smaller in *Scenedesmus* than in spinach. In the alga spectrum, the difference in the heigth of peaks -6 and -7 is much smaller than in the corresponding spinach region.
Conclusion. We interpret the differences in the EPR spectra as due to small differences in the protein environment of the manganese cluster. Therefore not only higher plants and cyanobacteria have not fully identical PS II donor sides (4), but also more closely related groups like green alga and higher plants.

3.3 X-ray absorption spectroscopy. XAS measurements (K-edge and EXAFS spectra) were carried out with preparations of spinach and *Scenedesmus* in S_1 and S_2 states at the synchrotron radiation source of the Daresbury Laboratories (UK). Preliminary results show that the edge inflection point energy and therefore the oxidation states of the oxygen evolving complex manganese atoms are comparable in spinach and the green alga *Scenedesmus obliquus*.

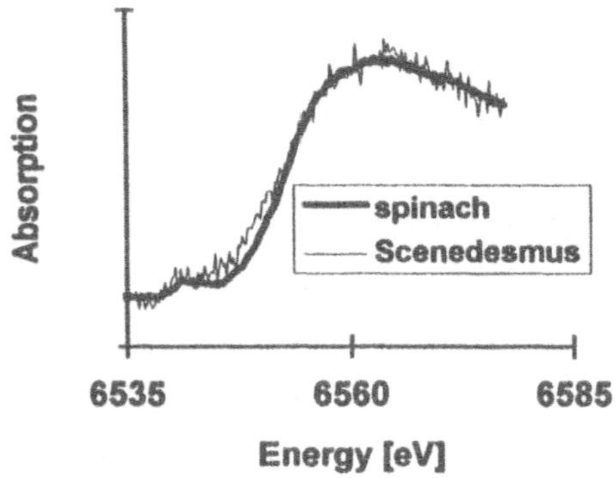

Fig. 5: X-ray absorption edge spectra of spinach and *Scenedesmus* preparations in the S2-state. The edge shapes and the pre-edge heights are similar, but possibly not fully identical.

5. Acknowledgement

Support by the BMBF (Erforschung kondensierter Materie, 48.1 Dau) and a Darsbury grant for european users is gratefully acknowledged. HS was supported by a scholarship of the Deutsche Studienstiftung.

6. References

1 Dau, H., Andrews, J. C., Roelofs, T. A., Latimer, M. J., Liang, W., Yachandra, V. K., Sauer, K. and Klein, M. P. (1995) Biochemistry 34, 5274-5287
2 Sandusky, P. O. and Yocum, C. F. (1984) Biochim. Biophys. Acta 766, 603-611
3 Britt, R. D., Zimmermann, J., Sauer, K. and Klein, M. P. (1989) J. Am. Chem. Soc. 111, 3522-3532
4 Barry, B. A., Boerner, R. J. and de Paula, J. C. (1994) in The Molecular Biology of Cyanobacteria (Bryant D. A., ed.) pp. 217-257, Kluwer Academic Publishers, Dordrecht, The Netherlands

IS PHOTOSYSTEM II KINETICALLY AND ENERGETICALLY HETEROGENEOUS IN *SYNECHOCHYSTIS* 6803?

Zsuzsanna Deák, László Sass and Imre Vass
Institute of Plant Biology, Biological Research Center, P.O. Box 521, Temesvári krt. 62, H-6701 Szeged, Hungary

1. Introduction

A useful method to study the function of photosystem II (PSII) in the naturally transformable cyanobacterium, *Synechocystis* 6803 - a wide-spread plant material for the molecular analysis of electron transport in the PSII reaction center - is thermoluminescence [for recent review see Ref. 1]. Thermoluminescence (TL) is light emission at characteristic temperatures arising from preilluminated samples upon warming up in the dark. In the case of photosynthetic samples this light emission arises from the recombination of positive and negative charges stabilized on the redox components of PSII. The TL intensity is proportional to the amount of recombining charges, whereas the peak temperature is characteristic for the free energy loss during charge stabilization: the higher the peak position the deeper the stabilization of the charge pair [2]. With dark adapted thylakoids or with PSII enriched membranes from higher plants, illumination with one flash results in a single TL band at around 30-40 °C, called the B-band, which arises from the $S_2Q_B^-$ recombination [3, 4]. If the electron transfer between Q_A and Q_B is blocked, e.g., by DCMU, the B-band is replaced by the so called Q-band arising from the $S_2Q_A^-$ recombination at around 0-10 °C [3, 5].

In intact *Synechocystis* 6803 cells the situation is more complicated and illumination with one flash leads to the appearance of three TL bands. In the present study we characterized these bands and the results indicate that, contrary to spinach thylakoids, PSII is kinetically and energetically heterogeneous in *Synechocystis* 6803.

Abbreviations: Chl, chlorophyll; DCMU, 3-(3,4-dichlorophenyl)-1,1'-dimethylurea; PSII, photosystem II; Q_A and Q_B, primary and secondary plastoquinone electron acceptors of PSII; S_0-S_4, charge storage states of the water-oxidizing complex; TL, thermoluminescence.

2. Materials and Methods

2.1 The strain of *Synechocystis* sp. PCC 6803, used in our experiments is the glucose-tolerant strain originally isolated by J.G.K. Williams [6]. It was routinely grown in BG-11 medium supplemented with 5 mM glucose [6]. The cells were resuspended in 25 mM Hepes, pH 7.0 for the measurements.
2.2 *Thylakoid membranes* were isolated from spinach [7] and suspended in 50 mM Hepes, pH 7.5, 0.4 M sucrose, 15 mM NaCl and 5 mM $MgCl_2$.

P. Mathis (ed.), Photosynthesis: from Light to Biosphere, Vol. II, 467–470.

2.3 Thermoluminescence was measured with a setup similar to that described previously [2], but equipped with computer data acquisition. Measuring conditions were the same as earlier presented [8]. Deconvolution of the TL curves to individual components was performed as previously described [2].

3. Results and Discussion

Single-flash excitation of dark adapted *Synechocystis* 6803 cells resulted in three TL components with peak temperatures at 20-25 °C, 35-40 °C and 55-60 °C (Fig. 1A, upper curve). This is in contrast with PSII membrane preparations from higher plants, where only a single TL component is observed after one flash from the $S_2Q_B^-$ recombination at 30-40 °C (Fig. 1B, upper curve). In the presence of DCMU, the B band is abolished with the concomitant appearance of the Q band (Fig. 1B, lower curve). Similarly, in *Synechocystis* 6803 cells the 20-25 °C component completely disappeared, and the 35-40 °C component markedly decreased (Fig. 1A, lower curve). Because the addition of DCMU blocks electron transport between Q_A and Q_B, the conclusion can be drawn that Q_B^- was involved in the charge recombination that created both the 20-25 °C and the 35-40°C components in *Synechocystis* 6803.

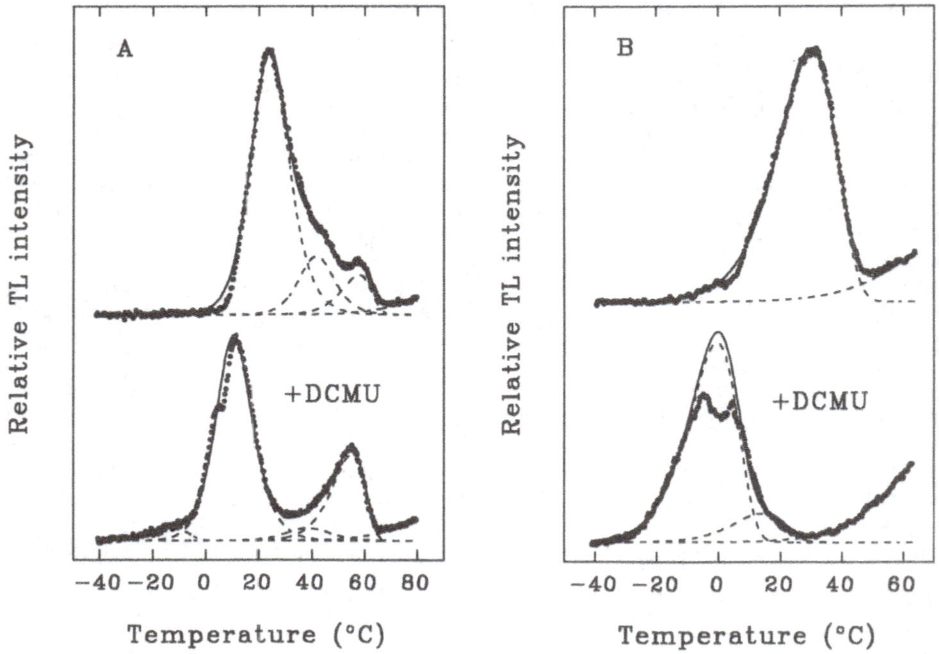

Figure 1. Single-flash induced thermoluminescence of *Synechocystis* 6803 (A) and thylakoid membranes (B). Thermoluminescence was measured after excitation with one flash at 5 °C in the absence (upper curve) or at -10 °C in the presence of 10 μM DCMU (lower curve). Besides the measured thermoluminescence signal (dots) the deconvoluted components (dashed lines) and the summary of these (solid line) are also shown. The artifactual shoulder at 0 °C, which is caused by the solid-liquid phase transition of the samples, was omitted from the curve fitting procedure.

For the origin of the two components in the main TL band an obvious assumption could be the participation of different S states in the charge recombination. In order to obtain more information about the positive charges involved in the recombination, the flash-induced oscillation of the main TL band was also measured. A characteristic period-four oscillation in the intensity was observed with maxima after two and six flashes in the case of both the 20-25 °C and the 35-40 °C components (Fig. 2). These oscillatory patterns indicate that both of the 20-25 °C and the 35-40 °C components arise from the $S_2Q_B^-$ recombination after the first flash, and from the $S_{2(3)}Q_B^-$ after two or more flashes in *Synechocystis* 6803, as in the case for the B band of thylakoids from higher plants [3, 4].

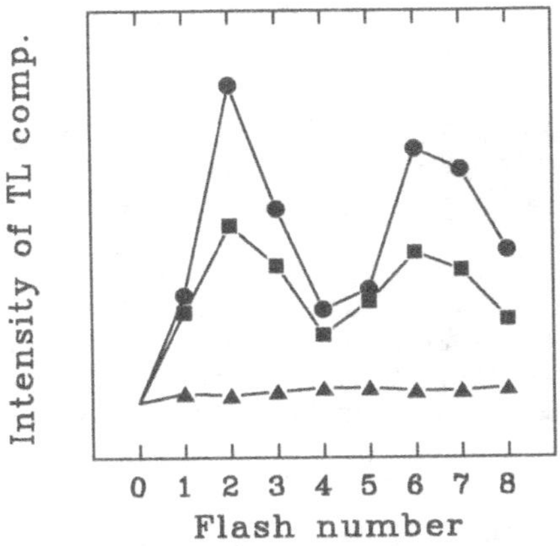

Figure 2. The flash-induced oscillation of the deconvoluted components of the main thermoluminescence band in *Synechocystis* 6803. Thermoluminescence was excited with one to eight flashes, 1 Hz frequency, at 5 °C. The measured curves were deconvoluted and the intensity of the components are plotted as a function of flash number (20-25 °C component: circle, 35-40 °C component: square, 55-60 °C component: triangle).

In order to characterize the two components kinetically, TL was measured in samples exited by one flash at 5 °C then dark incubated for various time intervals at the same temperature before TL measurement. Supporting the heterogeneous origin of the B band, there are differences in the decay kinetics of the two components deconvoluted from the TL curves obtained after various dark adaptation times (Fig. 3). The half time for the 20-25 °C component is about 2 min at 5 °C, while the lifetime of the 35-40 °C component is two times longer, it is about 4 min.

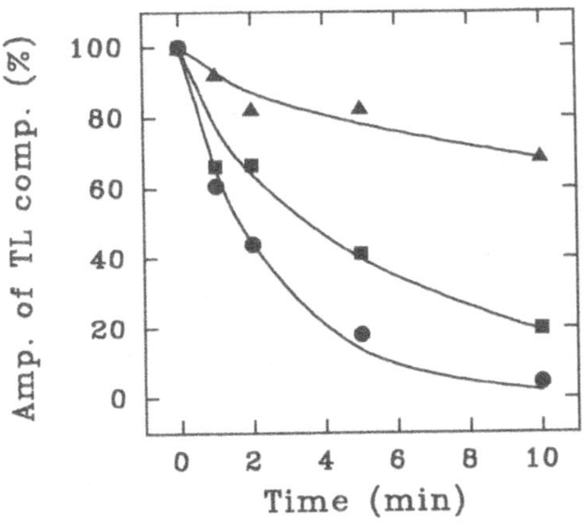

Figure 3. The decay of the three thermoluminescence components of *Synechocystis* 6803. The sample was excited with one flash at 5 °C followed by dark adaptation for various periods at the same temperature, then cooled down to -40 °C. The data are shown as percentage of the values at 0 min dark adaptation (20-25 °C component: circle, 35-40 °C component: square, 55-60 °C component: triangle).

The properties of the 55-60 °C component are very similar to those of the C band in higher plants, which is assigned to the Tyr-D$^+$$Q_A^-$ recombination [9]. The fact that in *Synechocystis* 6803 cells the 55-60 °C component exists even in the absence of DCMU (Fig. 1A, upper curve) suggests that in these cells the Q_A to Q_B electron transport is blocked in part of the PSII centers, which also lack functional S_2 and S_3 states and therefore capable of Tyr-D$^+$$Q_A^-$ charge stabilization. Supporting this idea, the 55-60 °C band is enhanced by DCMU (Fig. 1A, lower curve) as it is in the case of higher plants [10, 11]. There was no significant change in the intensity of the 55-60 °C component upon using various numbers of excitation flashes (Fig. 2). The halftime of this component is about 20 min at 5 °C (Fig. 3) which means that Tyr-D$^+$$Q_A^-$ charge pair is really very stable.

4. Conclusions

Single flash excitation of *Synechocystis* 6803 cells results in three TL components with peak temperatures at 20-25 °C, 35-40 °C and 55-60 °C. Our results indicate that the 20-25 °C and 35-40 °C components arise from the $S_{2(3)}Q_B^-$, while the 55-60 °C component from the Tyr-D$^+$$Q_A^-$ recombination.

We conclude that PSII kinetically and energetically heterogeneous in cyanobacterium *Synechocystis* 6803. First, in part of the PSII centers the electron transport is blocked between Q_A and Q_B. In addition, these centers lack available S_2 and S_3 states and therefore Q_A^- recombines with Tyr-D$^+$. Second, the functional centers appear to be heterogeneous in respect to the energetic stability of the $S_{2(3)}Q_B^-$ charge stabilization state.

Acknowledgment

This work was supported by OTKA T/017049 and OTKA F/017454, grants from the Hungarian Scientific Research Fund.

References

1 Vass, I. and Inoue, Y. (1992) in Topics in Photosynthesis, Vol. 11: The Photosystems: Structure, Function and Molecular Biology (Barber, J., ed.) pp. 259-294, Elsevier, Amsterdam, The Netherlands
2 Vass, I., Horváth, G., Herczeg, T. and Demeter, S. (1981) Biochim. Biophys. Acta 634, 140-152
3 Rutherford, A.W., Crofts, A.R. and Inoue, Y. (1982) Biochim. Biophys. Acta 682, 457-465
4 Demeter, S. and Vass, I. (1984) Biochim. Biophys. Acta 764, 24-32
5 Demeter, S., Droppa, M., Vass, I. and Horváth, G. (1982) Photobiochem. Photobiophys. 4, 163-168
6 Williams, J.G.K. (1988) Methods Enzymol. 167, 766-778.
7 Takahashi, M. and Asada, K. (1982) Plant Cell Physiol. 23, 1457-1461
8 Vass, I., Cook, K.M., Deák, Zs., Mayes, S.R. and Barber, J. (1992) Biochim. Biophys. Acta 1102, 195-201
9 Sane, P.V. and Rutherford, A.W. (1986) in Light Emission by Plants and Bacteria (Govindjee, J. Amesz, and D.C. Fork, eds.) pp. 329-361, Academic Press, New York
10 Desai, T.S., Sane, P.V. and Tatake, V.G. (1975) Photochem. Photobiol. 21, 345-350
11 Ichikawa, T., Inoue, Y. and Shibata, K. (1975) Biochim. Biophys. Acta 408, 228-239

THE REACT MECHANISM OF CHLORADION IN PHOTOSYN-THETIC O_2 EVOLUTION

Yu, Hong qi (Dept. of Biology, Inner Mongolia Education College, Huhhot, Inner Mongolia 010010 P. R. China)

It's known that[3] the water—splitting reaction or photosynthetic O_2 evolution is carried out on the internal surface of the thylakoid membrane on the oxidizing side of PS Ⅱ. Cl^- is employed at the oxidizing side of PS Ⅱ, and the symptom of Chlorisis[11]is :

a). leaves wither;

b). roots become into wooden club shape.

In this paper, the mechanism of the role of Cl^- in O_2 evolution and the two light enhancement effect in one photosystem have been presented.

1. **Material and Method**

Fresh spinach obtain at local market

Isolation medium [24]: containing 0. 4M sucrose, 0. 03M KCL and 0. 02M Tricine buffer (PH7. 5)

$K_3Fe(CN)_6$ 2mM in Tricine buffer

DBMIB 10_{-5}M in react solution

Incandescent lamp 100 watt

Monocolour glass of 650nm & 690nm

Oxygen Electrode (made in Xin Hua instrument factory of China)

LKB, BROMMA 2210 Recorder 2—channel 10mv

1). Add DBMIB the inhibitor of PSI to 10^{-5}M and the $K_3Fe(CN)_6$ to the isolated chloraplast solution sign as No1;

2). Add alcohol 95% volume equal to DBMIB of No1's and the same $K_3Fe(CN)_6$ of No1's sign as No2……

3)Light separately on 650nm, 690nm and white light

4) measure the change of O_2 evolution in react solution, Take down by LKB recorder as follow:

2. **Results**

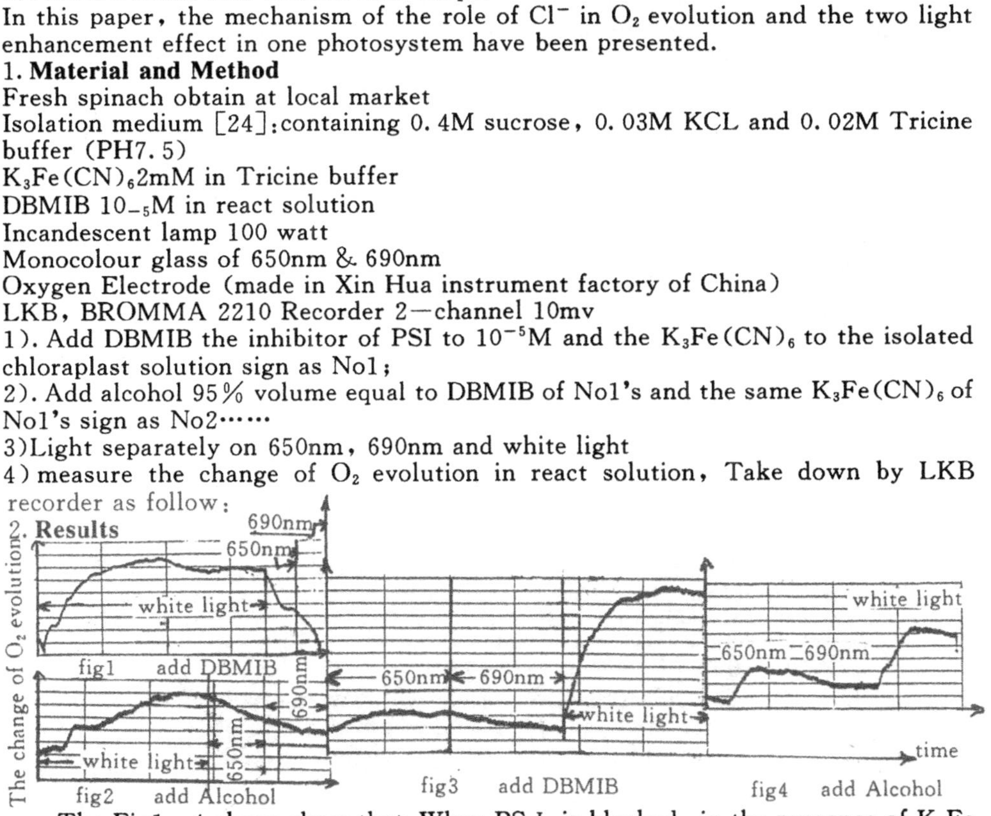

The Fig1—4 above show that: When PS Ⅰ is blocked, in the presence of $K_3Fe(CN)_6$, the O_2 evolution of PS Ⅱ is equal to or larger than the O_2 evolution of PS Ⅰ

P. Mathis (ed.), Photosynthesis: from Light to Biosphere, Vol. II, 471–474.

working together with PS Ⅱ. The two light enhancement effect happened in one photosystem. The role of PS I is just replenishing the energy to e^- formed by PS Ⅱ to raise potential to finish the process of electric energy to ATP & $NADPH_2$ or make photosynthetic electron transfer chain run in normal cycle. And Cl^- is the primary electron donor of $P680^+$ or the key spot where light energy is absorbed by PS Ⅱ jions water—splitting reaction.

3. **Discussion**

By comparing thylakoid with dye laser[2], I found that the thylakoid can be a natural, high effect, tunable, controllable by P680 & P700, pumped by visible sunlight and perfect dye laser.

Both PS I and PS Ⅱ consist of more than two kinds of pigments, one is acting pigment, eg: chlolorophll$_a$; the other is accessory pigment, eg: chlorophll$_b$, carotenoid etc. All of the pigments have the structure of four energy levels. Among them, the first singlet is converting energy level. The triplet or metastable state is belong to all of pigments, but it's not used directly in energy converting. It's known that [5] there is few chlorophll being in this energy state in normal photosynthesis.

In dye laser, [2][23]the condition of generating laser is single pass gain larger than single pass lose. The farther light is from central frequence of laser, the more difficult it generates laser. 680nm is the central wavelenth of PS Ⅱ. Light longer than 680nm is far from central frequence of PS Ⅱ, the light can't pump the PS Ⅱ to the state of population inversion and can't emit laser 680nm. So red drop occured. When the light longer than 680nm is supplemented by a light of 650nm, the two light simultaneously excite the acting and accessory pigments of every photosystem. The two kinds of pigments promote with each other in their near energy level.

The excited accessory pigment induce the chlorophll$_a$ into the state of population inversion so that the two lights can excite the PS Ⅱ into the state of population inversion more easily than single light such as the mechanism of He—Ne laser[14]. Therefore: two light enhancement effect happened in one photosystem. We shoud explain "TWO LIGHT ENHANCEMENT EFFECT" as "two pigment enhancement effect" not as "two photosystem enhancement effect". Let's look back at the work done by Duysens and Amesz in 1962.

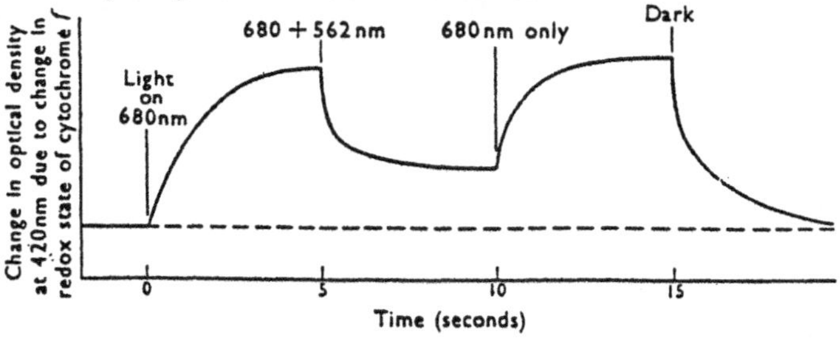

Fig. 5–3 Oxidation and reduction of cytochrome *f* in the red alga *Porphyridium*. Increase in OD_{420} is due to oxidation and decrease in OD_{420} is due to reduction of cytochrome *f*. Light of 680 nm is absorbed by Photosystem I (chlorophyll *a*) and at 562 nm is absorbed by Photosystem II (phycoerythrin). (After Duysens and Amesz, 1962, *Biochim. Biophys Acta*, 64, 243.)

This fig show: Plant photosynthesis involves two light reaction; one sencitized by chlorophll$_a$(680nm light), and one sencitized by chlorophll$_b$ or other accessory pigments(562nm light). Both of the two kinds of pigments exist in one photosystem.

And the work done by Myers and French in 1960[3] show that PS I and PS II need't be applied simultaneously to get optimum photosynthesis but that they may be given alternately with a short dark period of a few seconds between them. This indicated that the two—light reaction could store their photochemical products for a short time before reacting with the electron transfer chain.
In an word, PS I and PS II were put in the order of Z—scheme. The water—splitting reaction or photosynthetic O_2 evolution depend upon PS I and PS II. Both of them can't carry out Hill reaction without the other though their effects are different from each other. PS I can turn light energy into electric energy just as PS II and replenish the energy to e^- formed by PS II to raise potential to finish the process of electric energy to ATP& $NADPH_2$, so that the photosynthetic transfer chain run in normal cycle; just as water is pumped step by step from the bottom to the top of a mountain.
Then the photosynthetic O_2 evolution process can be express as :
Chlorophll$_a$ and accessory pigment in thylakoid membrane pumped by sunlight and emit laser 680nm

$P680 \xrightarrow{hv} P680^+$

The Cl^- lose e^- in two possible way

$E_{0Cl_2/Cl^-} = 1.36V \quad E_{0MnO_4^-/Mn^{2+}} = 1.49V$

1). $10Cl^- + 2MnO_4^- + 16H^+ = 5CL_2 + 2Mn^{2+} + 8H_2O$

if $E_{0\,P680^+/P680} > 1.49$, $P680^+$ can take an e^- from Cl^-

as: $P680^+ + Cl^- = P680 + Cl$

2). Laser 680nm excite Cl^- (which is near the PS II) to Rydberg atom[2]
And the Cl^- (which excited into Rydberg atom) emits an electron in the electric field of $P680^+$.

$Cl^- \xrightarrow{hv} Cl^{-*}$

$Cl^{-*} + P680^+ = Cl + P680$

$2Cl = Cl_2$

$Cl_2 + H_2O = HClO + HCL$

$HClO \xrightarrow{hv} HCl + O$ (the O first conbine with Carotenoid or Mn—Protein)

In the same way:

$4P680 \xrightarrow{4hv} 4P680^+$

$4Cl^- - - - 4Cl^{-*}$

$4Cl^{-*} + 4P680^+ = 4Cl + 4P680$

$4Cl = 2Cl_2$

$2Cl_2 + 2H_2O \xrightarrow{4hv} 4Cl^- + 4e^- + O_2 + 4H^+$

Based on the results above: When Cl^- is lack in plant, there is no Cl^- giving an e^- to the $P680^+$ to reduce it to P680. The photosynthesis cease here. H_2O can't be photolysised anymore, but the root pressure still exists and the plant don't stop absorbing H_2O while H_2O can't be consumed physically, except a little evaparation. In this condition, the PS II is blocked and the photosynthesis stops; the plant can only consume itself till it withers and dies; meanwhile the roots just expands its space to accommodate the "surplus" H_2O in order to moderate the crisis of surplus H_2O for a short time.
For instance:
The spitting water of plant: when there is no light shining at night, the photosynthesis is so weak that the plant must regulate H_2O circulation to keep its physical balance.

The wound bleeding of plant is also a way by which plant regulates H_2O circulation to keep physical balance by itself when plant is wounded by accident.

4. **Summary**

The water—splitting reaction starts at the lase680nm pumped by light. Photoexcited P680 produce a strong Oxidant $P680^+$ Which promotes the Oxidation of the primary electron donor Cl^- to Cl.

Usually laser 680nm or thylakoid is in the state of population inversion. When it absorbs a photon, the photon is amplified by stimulated emission of readiation and result in laser 680nm emission.

Excited P680 emits an e^- then turn into $P680^+$; excited Cl^- free in thylakoid give an e^- to $p680^+$ to reduce it to P680. One Cl_2 produce one Oxygen atom. In the same way, photosystem releasing one O_2 need $4Cl^- + 2H_2O + 4hv + 4P680$. Photoexcitation of P700 generates a strong oxidant $p700^+$ which replenishes the potential energy to e^- which come from PS Ⅱ to finish the accumulation of assimilatary power—ATP & $NADPH_2$. When the primary electron donor is H_2S and so on, the mechanism is as same as H_2O's eg; $H_2S + Cl_2 = 2HCl + S$

$4Cl_2 + 2Na_2SO_3 + 5H_2O = 2H_2SO_4 + 2NaCl + 6HCL$

So the equation of light primary reaction is

$Cl_2 + H_2O = 2HCl + A$

5. **Literature cited**

1. W. M. Yen and M. D. Levenson Springer Series in Optical Science'Vol. 54
2. Guo, Guang can; High Physics Selected Series — section of optics nnd atom phvsics. High education pub. co
3. D. O. Hall; K. K. Rao. The institute of Biology Studies BIology, NO. 37 Photosynthesis Thrid edition Edward Arnold
4. The proceedings of Chinese the mechanism and physioligy of photosyntheisis congress Oct. 1992 Jiu Jiang
5. G. Richter; Plant Metablism — Physiology and Biochemistry of Primary metabolism, Geory Thieme Publishers Stuttgart, 1978
6. Chen, rongsan; Inorganic & Chemistry, people pub. co
7. China, Photosynthesis study's development Acta Vol. Ⅰ. Ⅱ, Ⅲ.
8. China, Plant Physiology Communication Acta Vol. (1—6) 1984—1994.
9. China, Plant Physiolgy and Biochemistry Development Acta Ⅰ, Ⅱ, Ⅲ。
10. Xing, qiyi; Basic Organic Chemistry, High education pub, co
11. Salisbury and C Ross; Plant Physiolgy
12. W. Sweater, Cell Biology, One evolving path P251—268, Sciencepub, co 1983
13. M. Sargent Ⅲ M. o. SCully W. E. Lamb; LASER PHYSICS, Addison— wesley 1974
14. Chu, shenglin; Atom physics, peope education c0.
15. Cao, zongxing; Plant phvsiolgy, P75—123,1984, High education pub. co
16. Jolit, P. Chim Phys. Physicochim, Biol,58 584 1961
17. Joliot, P Biochim Bophys Acta 102:116—134 1965
18. Joliot. P Barbieri. G:& Chabaud. R Photochem, PhotoBiol 10:309+329 1969
19. KOK,B Forbush. B& Mcgloin, M. Photochem, Photobiol, 11:457—475,1970
20. Fowler. CF & KOK. B;Biochim, Biophys Acta 357:299—307 1974
21. Slater, E. C, in Mechanism of Energy Conversion, FEBS 8th conf, proc Vol. 28 Berghetallleg P133
22. USA,Science Acta 10:9—15,1987(in Chinese)
23. Xu, rongpu; Laser instrument and technicality Summary, Beijing Industry COllege CO.
24. Thomas. C. Moore; Research Experimence in Plant Phvsiologgy, second edition.

Y_Z as a Hydrogen-Atom Abstractor in the Water-Oxidizing Chemistry of Photosystem II

Cecilia Tommos and Stenbjörn Styring
Department of Biochemistry,
Arrhenius Laboratories for Natural Sciences,
Stockholm University, S-106 91 Stockholm, Sweden

The Photosystem II/Oxygen-Evolving Complex (PSII/OEC) catalyzes the light-driven oxidation of water to dioxygen in oxygenic organisms. Scheme 1 conveys the conventional view of the redox-components and chemistry involved (1). The water-splitting reactions are thermodynamically driven by the ~ 1.3 V generated by the light-induced oxidation of the PSII reaction-center chlorophyll complex P_{680}. The oxidizing equivalents created at P_{680}^+ are transferred, one by one, to a tetra-nuclear Mn-cluster, identified as the catalytic core of the OEC, via the tyrosyl residue Y_Z (tyrosine 161 on the D1 protein). Y_Z re-reduces P_{680}^+ in the sub-µs time range, which prevents wasteful charge-recombination reactions and provides a quantum yield close to one. The $(Mn)_4$-cluster binds substrate water and, by cycling through five redox states, denoted S_0-S_4, accumulates the four oxidizing equivalents required for the water splitting reaction. Oxidation of water occurs upon formation of the S_4 state, dioxygen is released, and the system relaxes back to the S_0 state. The two inorganic cofactors, chloride and calcium, are also essential for the process (2). PSII contains a second redox-active tyrosine, Y_D, positioned at 161 on the D2 protein. The Y_D tyrosine functions as an auxiliary electron donor not directly involved in the water-splitting chemistry of PSII (1).

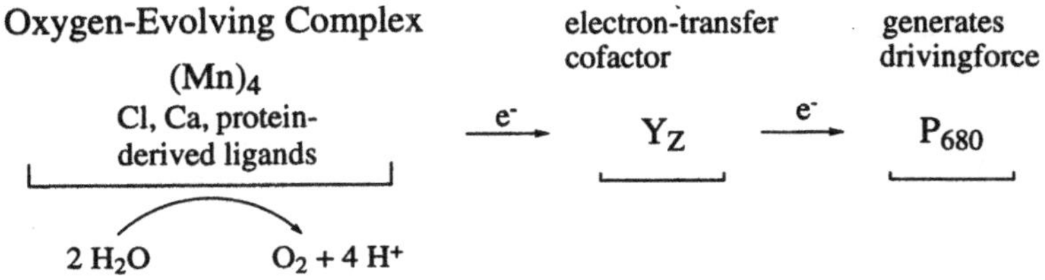

Scheme 1

Properties of the two redox-active tyrosines

On the basis of $^1H_2O\backslash^2H_2O$-exchange data from ENDOR and ESEEM studies, $Y_D^{\bullet}$ was proposed to be a neutral, deprotonated radical with its phenol oxygen involved in a hydrogen-bond (1). The exchange rate was extremely slow, indicating a shielded site for the Y_D radical. From modeling studies, it was suggested that Y_Z and Y_D form hydrogen-bonds to nearby D1-His190 and D2-His190 (D2-His189 in *Synechocystis*), respectively (3). The prediction made for the D2 protein was later experimentally supported by site-directed mutagenesis studies. Upon replacement of D2-His189, both the normal EPR spectrum of

P. Mathis (ed.), Photosynthesis: from Light to Biosphere, Vol. II, 475–478.

$Y_D^{\bullet}$ (4, 5), as well as the ENDOR detected hydrogen-bond (5), were removed. These results support the proposed 'proton-rocking' model for the $Y_D\backslash Y_D^{\bullet}$ redox chemistry where, upon oxidation, the proton migrates to the base (D2-His189) and, upon re-reduction, returns to the tyrosine (1).

In contrast, no pronounced $^1H_2O\backslash ^2H_2O$-exchangeable features could be observed in the $Y_Z^{\bullet}$ ENDOR spectrum (6, 7). These results indicate that $Y_Z^{\bullet}$ is not involved in a hydrogen-bonding interaction equivalent to the well-ordered hydrogen-bond detected for the $Y_D^{\bullet}$ radical. This rationalizes the observation that point-mutations at D1-His190 do not significantly alter the $Y_Z^{\bullet}$ EPR spectrum (8, 9). However, the D1-190 histidine is essential for the oxygen-evolving reactions of PSII (5, 9, 10). Point-mutations at this position impede the electron transfer between Y_Z and P_{680}^{+} dramatically. Moreover, Debus *et al.* report oxygen-evolution activety of the D1-H190 mutants only upon changing the histidine to another relatively large, basic residue (10). These observations have been rationalized by postulating the following events (6, 11). The Y_Z tyrosine is hydrogen-bonded to D1-His190 in its reduced form. Upon oxidation, the phenol O-H bond breaks, the proton becomes fully associated with the histidine, and the neutral, deprotonated $Y_Z^{\bullet}$ radical is formed. The histidine deprotonates rapidly, and the proton is transported out of the radical site towards the lumenal bulk medium. The suggestion that the proton resides only transiently on the histidine is in agreement with FTIR data (8). In addition, proton ejection studies on PSII core preparations show a stoichoimetric release pattern upon each S-state transition (12), consistent with the proposed mechanism (6, 11). Finally, a proton-transfer mechanism, initiated by oxidation of Y_Z and mediated by D1-His190, may provide an explanation for the positive trap involved in the A_T thermoluminescence band, but, without invoking a *redox-active* histidine. Upon charge-recombination, the equilibrium shifts, and the ejected proton returns to the tyrosine via the D1-190 histidine residue (for a discussion of the A_T band, see 13 and refs. therein).

In addition to the observed difference in hydrogen-bonding status of the two tyrosyl radicals, the motional dynamics of the redox centers differ. A detailed ESEEM study rendered evidence of low rotational mobility (< 4°) about the C_{para}-C_β bond for $Y_D^{\bullet}$ in *Synechocystis* (14). This is in contrast to the Y_Z radical in the same organism, where both ENDOR and ESEEM line shapes show evidence of motional broadening (6). Quantitative information could be derived from the ESEEM data, which showed a ~ 14° variation in the dihedral angle about the C_{para}-C_β bond of $Y_Z^{\bullet}$ (6). These results are in agreement with observations made by Kawamori and co-workers, who proposed higher mobility at the Y_Z site, as compared to the Y_D site, from temperature-dependent ENDOR studies in spinach (15). Another distinction between Y_Z and Y_D is suggested from modeling studies that predict that the Y_Z site is more hydrophilic than is the protein environment around Y_D (16).

To summarize, for the Y_D tyrosine the phenol oxygen is proposed to be hydrogen-bonded to nearby D2-His189, the site is hydrophobic and shielded from the bulk medium. In addition, there is evidence of low motional disorder in the radical site. In contrast, Y_Z is not involved in a well-ordered hydrogen-bond in its oxidized form and is proposed to slough its phenol proton upon oxidation. Moreover, the protein environment of the Y_Z radical is suggested to be more hydrophilic with a higher degree of mobility in comparison to the Y_D site. These observations appear to be inconsistent with Y_Z functioning as a pure electron-transfer cofactor; in order to lower reorganization during redox change and maximize the rate, one expects that the electron-transfer process will occur in a hydrophobic environment and that changes in structure upon oxidation and reduction will be minimized. The $Y_Z\backslash Y_Z^{\bullet}$ couple shows several properties that clearly deviate from these principles that govern simple and rapid electron transfer. The site is hydrophilic, and the de- and reprotonation of Y_Z upon redox change are expected to involve considerably more nuclear motion as compared to the proposed 'proton-rocking' motion of the $Y_D\backslash Y_D^{\bullet}$ couple.

A mechanistic model for the function of Y_Z in PSII

Recently, a hydrogen-atom abstraction function for Y_Z in the water-splitting chemistry of PSII has been proposed (6, 11). The overall reaction proposed in this model is shown in Scheme 2. In this mechanism, the $(Mn)_4\backslash Y_Z$ unit forms the catalytic center of the Oxygen-Evolving Complex; the $(Mn)_4$-cluster provides binding sites for substrate and delocalizes oxidizing equivalents, and $Y_Z^{\bullet}$ acts by abstracting hydrogen-atoms from substrate aquo- or hydroxy-species, ligated to the Mn-ions in the cluster, on each S-state transition. The proposed hydrogen-abstraction function accounts for the differences observed between Y_Z and Y_D, as described above, and the paradoxical behavior of Y_Z relative to that expected from a pure electron-transfer cofactor. Moreover, studies of other radical-containing metalloenzymes show a growing number of systems where the radical performs a hydrogen-atom abstraction function (see 17 and refs. therein). In general terms, a metal-containing cofactor generates a nearby radical species, by reacting with oxygen or by facilitating the homolytic cleavage of a bond. The radical, in turn, initiates the catalytic cycle by abstracting a hydrogen-atom from substrate. Examples include the prostaglandin H synthase enzyme in which the heme reacts with peroxide to generate a $Fe^{IV}/Y^{\bullet}$ species; the tyrosyl radical initiates the catalytic reaction by abstracting a hydrogen-atom from bound substrate. A second example is provided by ribonucleotide reductase. Upon activation of oxygen at its di-iron active site, a tyrosyl radical is generated, which, directly or indirectly, participates in hydrogen-atom abstraction at the ribonucleotide substrate.

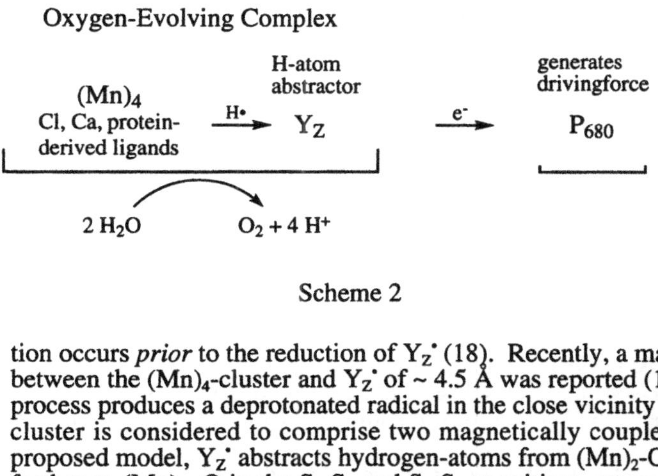

Scheme 2

The complete S-state cycle of the proposed mechanistic model occurs as postulated in the following steps. Upon each turnover, Y_Z is oxidized by P_{680}^{+} and its phenol proton is sloughed from the radical site towards the lumen. This is consistent with the observation that a large part of the proton release that accompanies water oxidation occurs *prior* to the reduction of $Y_Z^{\bullet}$ (18). Recently, a magnetic dipole-dipole distance between the $(Mn)_4$-cluster and $Y_Z^{\bullet}$ of ~ 4.5 Å was reported (19). Thus, the proton ejection process produces a deprotonated radical in the close vicinity of the metal center. The Mn-cluster is considered to comprise two magnetically coupled $(Mn)_2$ dimers (20). In the proposed model, $Y_Z^{\bullet}$ abstracts hydrogen-atoms from $(Mn)_2$-OH_2 to produce $(Mn)_2$-OH and further to $(Mn)_2$=O in the S_0-S_1 and S_1-S_2 transitions, respectively. The same procedure occurs in the other dimer, whose substrate binding site was protected by Cl until after S_2 formation (21), in the higher S-state transitions to produce two $(Mn)_2$=O species upon S_4 formation. The two oxo-ligands (each of which carries a formal charge of -2) become oxidized in a concerted fashion, the oxygen-oxygen bond is formed, and dioxygen is released as the system resets to the S_0 state. This model accounts for the report by Messinger *et al.* which shows one fast and one slowly exchanging substrate water molecule in the S_3 state, and for the observation that Y_Z reduction in the S_3-S_4-S_0 transition rate-limits oxygen evolution (23). Moreover, the model in Scheme 2 recasts the distinction between chemical and electrostatic protons (18), and shows how the release of chemically produced protons can occur upon oxidation of Y_Z, rather than upon its reduction.

Acknowledgements
The work and ideas presented in this manuscript is the result of a collaborative effort and we gratefully acknowledge G. T. Babcock, J. McCracken, K. Warncke, C. W. Hoganson and N. Lydakis-Simantiris at Michigan State University, and B. A. Diner and X.-S. Tang at Du Pont de Nemours and Co. C.T. acknowledge support by the Foundation BLANCEFLOR Boncompagni-Luovisi, née Bildt during a research visit 1994 at Michigan State Univesity. S.S. acknowledge support by the Swedish Natural Science Research Council.

References

1 Diner, B. A. and Babcock, G. T. (1995) in Oxygenic Photosynthesis: The Light Reactions (Ort, D., Yocum, C., eds.) Kluwer Academic Publishers, Dordrecht, in press
2 Debus, R. J. (1992) Biochim. Biophys. Acta 1102, 269-352
3 Svensson, B., Vass, I., Cedergren, E. and Styring, S. (1990) EMBO J. 9, 2051-2059
4 Tommos, C., Davidsson, L., Svensson, B., Madsen, C., Vermaas, W. and Styring, S. (1993) Biochemistry 32, 5436-5441
5 Tang, X.-S., Chisholm, D. A., Dismukes, G. C., Brudvig, G. W. and Diner, B. A. (1993) Biochemistry 32, 13742-13748
6 Tommos, C., Tang, X.-S., Warncke, K., Hoganson, C. W., Styring, S., McCracken, J., Diner, B. A. and Babcock, G. T. (submitted) J. Am. Chem. Soc.
7 Tang, X.-S., Zheng, M., Chisholm, D. A., Dismukes, G. C., Diner, B. A. (submitted) Biochemistry
8 Bernard, M. T., MacDonald, G. M., Nguyen, A. P., Debus, R. J. and Barry, B. A. (1995) J. Biol. Chem. 270, 1589-1594
9 Roffey, R. A., van Wijk, K. J., Sayer, R. T. and Styring, S. (1994) J. Biol. Chem. 269, 5115-5121
10 Chu, H.-A., Nguyen, A. P., Debus, R. J. (1995) Biochemistry 34, 5839-5858
11 Hoganson, C. W., Lydakis-Simantiris, N., Tang, X.-S., Tommos, C., Warncke, K., Babcock, G. T., Diner, B. A., McCracken, J. and Styring, S. (1995) Photosyn. Res., in press
12 Lübbers, K., Haumann, M. and Junge, W. (1993) Biochim. Biophys. Acta 1183, 210-214
13 Kramer, D. M., Roffey, R. A., Govindjee and Sayre, R. T (1994) Biochim. Biophys. Acta 1185, 228-237
14 Warncke, K., Babcock, G. T. and McCracken, J. (1994) J. Am. Chem. Soc. 116, 7332-7340
15 Mino, H. and Kawamori, A. (1994) Biochim. Biophys. Acta 1185, 213-220
16 Svensson, B., Vass, I. and Styring, S. (1991) Z. Naturforsch 46, 765-776
17 Sigel, H. and Sigel, A. eds. (1994) Metal Ions in Biological Systems Vol. 30, Marcel Dekker, New York, USA
18 Lavergne, J. and Junge, W. (1993) Photosyn. Res. 38, 279-296
19 Gilchrist, Jr., M. L., Ball, J. A., Randall, D. W. and Britt, R. D. (1995) Proc. Natl. Acad. Sci. USA, in press
20 Yachandra, V. K., DeRose, V. J., Latimer, M. J., Mukerji, I., Sauer, K. and Klein, M. P. (1993) Science 260, 675-679
21 Rutherford, A. W., Zimmermann, J.-L. and Boussac, A. (1992) in The Photosystems: Structure, Function and Molecular Biology (Barber, J., ed.) pp. 179-229, Elsevier, Amsterdam, The Netherlands
22 Messinger, J., Badger, M. and Wydrzynski, T. (1995) Proc. Natl. Acad. Soc. USA 92, 3209-3213
23 Babcock, G. T., Blankenship, R. E. and Sauer, K. (1976), FEBS Lett. 61, 286-289

MODIFICATIONS OF PS II-PROTEIN DOMAINS BY TRYPSIN INDUCE A HIGH VARIABILITY OF H^+-RELEASE STOICHIOMETRY WITHOUT AFFECTING THE ENZYMATIC ACTIVITY OF THE WATER OXIDASE

Wacker, U., Irrgang, K.-D. and Renger, G., Max-Volmer Institut für Biophysikalische und Physikalische Chemie, TU-Berlin, Straße des 17. Juni 135, 10623 Berlin, Germany

Introduction

Photosynthetic water oxidation involves a sequence of univalent redox steps which split the substrate water into molecular oxygen and four protons. At a first glance, the amount of protons released into the lumen during each redox step might provide information on the mechanism of water oxidation. However, it has to be considered that the redox active Mn-complex is embedded into a protein matrix (Renger, 1978). Therefore, the detected H^+-release does not necessarily reflect the oxidation of the substrate water. Instead additional interactions have to be considered (Renger, 1987): a) pK-shifts of membrane bound buffer groups due to formation of charged species (Y_Z^+, $Pheo^-$, Q_A^-, etc.), b) kinetic limitations of H^+-exchange on either side of the membrane owing to bound buffer groups, c) conformational changes which could alter: 1) the number of interacting buffer groups, 2) the accessibility of a H^+-exchanging redox center and 3) the electrostatic interaction between redox centers and buffer groups.
The present study analyses effects on H^+-release induced by tryptic modifications of protein domains, which are not of immediate necessity for water oxidation. PS II-proteins which are most likely to contain interacting buffer groups or H^+-shielding domains were selected for this investigation.

Materials and Methods

PS II membrane fragments (PS II m. f.) were prepared according to Berthold et al. (1981) with minor modifications described in Wacker (1993). Bovine pancreas trypsin from Sigma (<0.1 BTEE Chymotrypsin, 12500 BAEE/mg activity) was dissolved in demineralized water to prepare stock solutions of 10 mM, 1 mM, 100 µM and 10 µM trypsin, which were stored at -20°C until use. Suspensions of PS II m. f. (c_{Chl} = 6.5 mM) were mixed with trypsin stock solution on a Vortex for 5 s and incubated in the dark for two hours at 6°C. The incubation was stopped by addition of 20 mM benzamidine and freezing in liquid nitrogen. All samples were stored at -78°C until use.
Digestion patterns were electrophoretically analysed by SDS/Urea-PAGE as described by Irrgang et al. (1986) using the buffer system of Laemmli (1970). Samples of 20 µg Chl were loaded into each well. The gels were run over night at 10-20 mA/gel and 5°C. Western blotting was performed according to Towbin et al. (1979) using PVDF-membranes. For better illustration the staining pattern was redrawn on a transparency.

P. Mathis (ed.), Photosynthesis: from Light to Biosphere, Vol. II, 479–482.

Since the samples used for manganese determination originated from a second tryptic proteolysis, a Western blot with α-33 was repeated to ensure comparability. The staining pattern was in line with the results depicted in fig. 1 (data not shown).
Absorption changes of bromocresol purple (20 μM) at 580 nm were recorded at pH=6.8, in a suspension containing dark adapted PS II m. f. (10 μM Chl, 10mM NaCl, 100 μM $K_3[Fe(CN)_6]$. A home built flash spectrometer (Wacker, 1993) was used for detecting and averaging of 25 traces comprising eight signals induced by saturating, red (Schott, RG630/2) Xenon-flashes (≈10 μs FWHM).
Oxygen evolution was measured with a Clark-type electrode as described in Renger (1972) in a suspension containing PS II m. f. (50 μM Chl, 10 mM MES/NaOH pH=6.80, 10 mM NaCl, 1 mM $K_3[Fe(CN)_6]$). After correcting for kinetical limitations a value of 210 Chl per oxygen evolving PS II-reaction center (RC II) was obtained. A light saturated rate of 540 μmol O_2 mg Chl^{-1} h^{-1} was found with 500 μM phenyl-p-benzoquinone as an additional exogenous acceptor.
Manganese was determined at 279.5nm in a Perkin Elmer 301 atom absorption spectrometer. 10 μl of sample were atomized at 1200°C in a graphit tube.

Results

Fig. 2 shows the H^+-release patterns of trypsin modified, dark adapted PS II m. f.. The untreated control exhibits a damped oscillation of the extent of H^+-release as well as a striking H^+-uptake after the first flash and minor uptake within the 50ms time domain after the 4th, 5th and 8th flash. The comparatively small oscillation is mainly due to an extra double hit in the first flash caused by the rapid electron transfer from Q_A^- to the non-haem iron center that has been preoxidized by $K_3[Fe(CN)_6]$ (for discussion, see Renger et al., 1987). Likewise, the pronounced H^+-uptake after the first flash is to a large extent related to a reduction of the preoxidized non-haem iron center (Renger et al., 1987). On the other hand, the oscillating uptake in the subsequent flashes has to be ascribed to processes that could depend on the redox state of the water oxidase.
The further evaluation and comparison of the data reveals:
a.) trypsin treatments with 0.01 trypsin/RC II (T/RC II) and 0.1 T/RC II induce a decrease of signal amplitude and slower H^+-release kinetics. This result can be explained by an increase of the membrane buffer capacity and a lengthening of the diffusion pathway, respectively, due to the exposure of formerly buried buffer groups after the cleavage of several peptide bonds. A correlation with the beginning of the degradation of various proteins can be found (fig. 1). The most susceptible proteins are the 22 kDa-protein and the D1. The extent of the H^+-uptake after the first flash is markedly diminished at 0.1 T/RC II while H^+-uptake after the 4th, 5th and 8th flash completely disappears. This phenomenon indicates that the H^+-uptake after the first flash originates from at least two effects.
b.) At 1 T/RC II the period four oscillation of H^+-release starts to fade and at 10T/RC II it almost completely disappears. A considerable decrease of buffer capacity is observed. Under these conditions an advanced

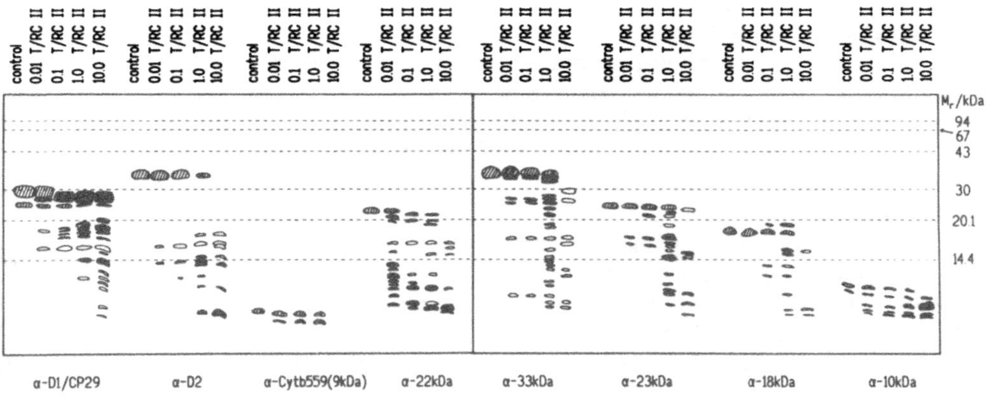

Fig. 1: Diagram of immunodecorated Western blots of PS II-proteins and peptide fragments after limited tryptic proteolysis. Hatched spot: intact protein; full spot: major fragment; open spot: minor fragment or unspecifity.

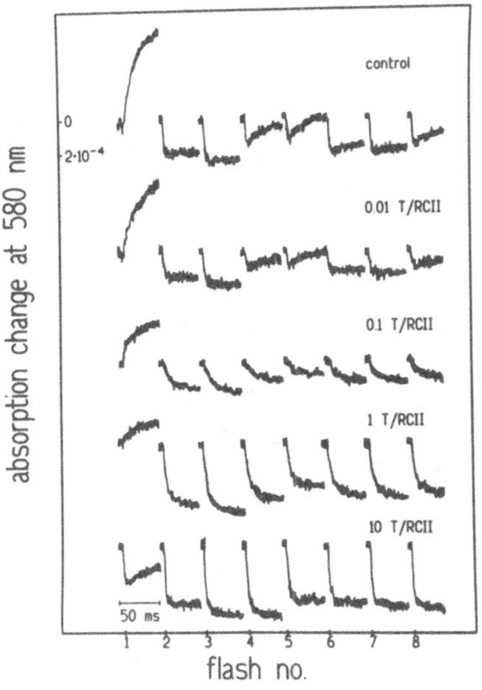

Fig. 2: H^+-release patterns of trypsin modified, dark adapted PS II m. f.

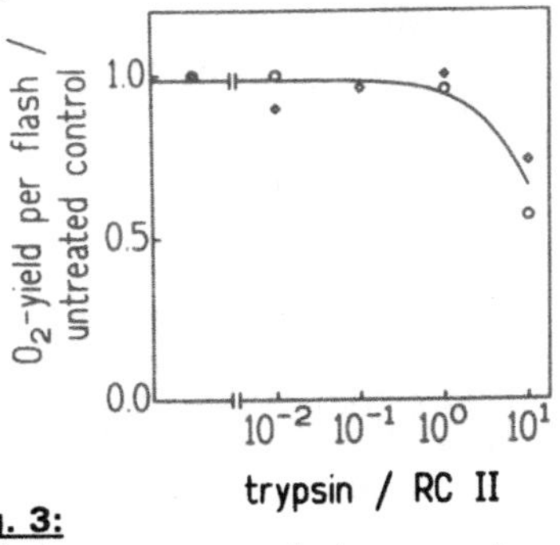

Fig. 3:
Relative oxygen evolution capacity of trypsin modified PS II m. f.

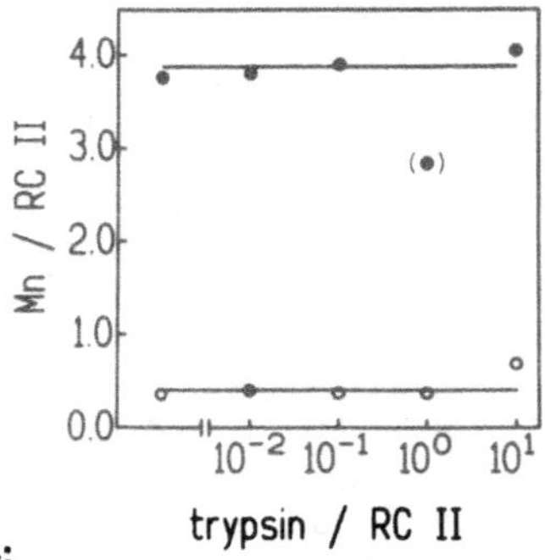

Fig. 4:
Manganese content of trypsin modified PS II m. f. Open symbols: supernatant. Closed symbols: pellet.

degradation of protein domains shielding the catalytic center of the water oxidase is obvious from fig. 1.
c.) The oxygen evolving capacity remains invariant up to 1T/RC II and declines by 30-40% at 10T/RC II (fig. 3).
d.) The manganese content (4 Mn/RC II) remains virtually unaffected even at the highest trypsin concentration used (fig. 4).

Discussion

The results of this study reinforce the evidence of our preceding investigation (Wacker et al., 1990). They clearly show that the H^+-release pattern is altered by selective proteolytic cleavage of PS II-protein domains that are not of immediate necessity for water oxidation. Therefore the resulting structural modifications have to be considered as minor in terms of their influence on the enzymatic activity of the water oxidase. This finding provides further proof for the previous conclusion that the experimentally detected H^+-release pattern does not necessarily reflect the protolytic events at the catalytic site of water oxidation (Renger 1978, 1987). Furthermore, structural changes were found to lead to significant modulations of the buffer capacity. As an implication of this result, it has to be considered that transient conformational changes of the protein matrix might affect the extent as well as the kinetics of experimentally determined H^+-release.

Acknowledgements

The authors would like to thank B. Andersson, G.-S. Tae and W. Cramer for antibodies, S. Hohm-Veit for oxygen measurements and K. Scharf for Mn-determinations by atomic absorption spectroscopy. The financial support by Deutsche Forschungsgemeinschaft and Fonds der Chemischen Industrie is gratefully acknowledged.

References

Berthold, D. A., Babcock, G. T. and Yocum, C. F. (1981) FEBS Lett. 134, 231-234
Irrgang, K.-D., Renger, G. and Vater, J. (1986), FEBS Lett. 204(1), 67-75
Laemmli, U. K. (1970), Nature (London) 227, 680-683
Renger G. (1972), Biochim. Biophys. Acta 256, 428-439
Renger, G. (1978) in: Photosynthetic Oxygen Evolution (Metzner, H. (ed.)), 229-248, Academic Press, New York
Renger, G. (1987) Photosynthetica 21, 203-224
Renger, G., Wacker, U. and Völker, M. (1987), Photosynthesis Res. 13, 167-189
Towbin, H., Staehelin, T. and Gordon, J. (1979), Proc. Natl. Acad. Sci. U.S.A. 76, 4350-4354
Wacker, U., Haag, E. and Renger, G. (1990), in: Current Research in Photosynthesis, Vol. I, 869-872, Baltscheffsky, M. (ed.), Kluwer Academic Publishers, Dordrecht, The Netherlands
Wacker, U. (1993), Dissertation TU-Berlin, D83

PROTON RELEASE ASSOCIATED WITH PHOTOSYSTEM II IN A MUTANT OF *ARABIDOPSIS THALIANA* LACKING *TRANS*-HEXADECENOIC ACID

Marie José Delrieu and Antoine Trémolières
Institut de recherches sur les plantes, URA 1128, Bat. 630, Université Paris XI, 91405 Orsay, France.

1. Introduction

The light-harvesting chlorophyll *a/b* complex of photosystem II (LHC II), the most abundant chlorophyll-complex in higher plants, is organized in trimers. Synthetic LHC II monomers trimerize *in vitro* only in the presence of a lipid fraction isolated from pea thylakoids [1]. Lipids, particularly phosphatidyl glycerol (PG), appear to be involved in the stabilization of trimers [2] by interacting, directly or indirectly, with the N-terminal domain of LHC II [3]. PG also appears to be an integral part of the monomeric complex [3]. A specific role for the unusual fatty acyl group, Δ3, trans-hexadecenoic acid (*trans*-$C_{16:1}$) in photosynthesis, has frequently been proposed because *trans*-$C_{16:1}$-PG is present only in chloroplast membranes [4]. This fatty acid is atypical because of the *trans* configuration, and because of the double bond near the carboxyl rather than the methyl end of the fatty acid. However, in a preliminary analysis of LHC II structure and function in a mutant of *Arabidopsis thaliana* devoid of *trans*-$C_{16:1}$ in PG (but having a normal content of PG), McCourt et al [5] were unable to observe any effect of the mutation on the efficiency of energy transfer between LHC II and PS II and between the photosystems. Oligomeric LHC II did not appear to be impaired in this mutant as one might expect if LHC II structure was substantially altered.The present work shows that *trans*-$C_{16:1}$-PG is indispensable in LHC II for proton conduction from the water-oxidizing complex to the lumen.

2. Material and Methods

The mutant was isolated in the lab. of J. Browse. The mutant carries a defective allele of a locus, designated *fad4*, which is required for the desaturation of palmitic acid at position two of PG. Isolated thylakoids were prepared as described previously [6] from *Arabidopsis thaliana*, and suspended in 300 mM sorbitol, 10 mM NaCl, 5 mM $MgCl_2$, 40 mM Mes-NaCl (pH 6.5). The oscillation patterns of O_2 evolution and fluorescence yield induced by a series of flashes in dark-adapted thylakoids were measured as described in [7]. pH-transients in the suspending medium were determined by absorption changes of the dye bromocresol purple (BCP) at 580 nm [8]. Thylakoids equivalent to 10 µM Chlorophyll were suspended in a medium containing 10 mM NaCl, 0.5 mM hexacyanoferrate III (Fecy), 4 µM nigericin, 7 µM bromocresol purple. From the traces obtained in the presence of the indicator, curves recorded in the absence of the indicator were subtracted.

P. Mathis (ed.), Photosynthesis: from Light to Biosphere, Vol. II, 483–486.

3. Results and Discussion

Figure 1 shows in the wild-type and the *fad4* mutant thylakoids, the same flash-induced fluorescence yield measured 80 ms after each flash of a series, wich oscillates with a periodicity of four. The oscillatory pattern of fluorecence yield has been assumed to reflect in a population of PS II centers, the existence of a local electric field on the donor side, induced by an oxidized manganese when oxidation of S_1 to S_2 occurs [7]. As shown in Figure 2, the oscillation pattern of the O_2 flash yield was also almost identical. This indicates that there is an apparent absence of an effect of the *fad4* mutation on PS II activity and O_2 evolving properties.The fluorescence and O_2 yield oscillations were obtained from mutant and wild-type leaves gathered at two different stages of plant development, at an early stage (a), and at the first appearance of flowering (b). In the latter case, the period 4-fluorescence oscillations were found inexistent, even though the O_2 evolving capacity was intact. This shows that the O_2 yield and fluorescence oscillations are intrinsically not the same, and cannot be described with the same properties for the S_i states [7,9].

In thylakoids, the protonation equilibrium is under the influence of the oxidation of water on the PS II donor side, the reduction of the acceptor quinones Q_A/Q_A^-, Q_B/Q_B^-, and the reoxidation of plastoquinol by cyt b_6/f, dependent on PS I activity. The release of protons from water oxidation is a composite of the chemical production (1 H^+/e^-) and a more indirect electrostatic response, dependent on the properties of the protein environment of PS II [10,8]. The uptake of protons at the acceptor quinone, about 1 H^+/e^- [11] cancels the chemical proton release originating from water oxidation. Figure 3 shows the transient absorption changes of bromocresol purple at 580 nm after the first height flashes of a series given to thoroughly dark adapted samples (WT: wild type, M: mutant) from leaves

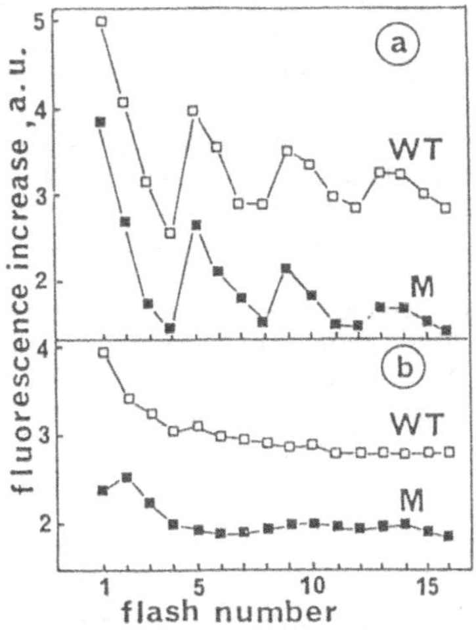

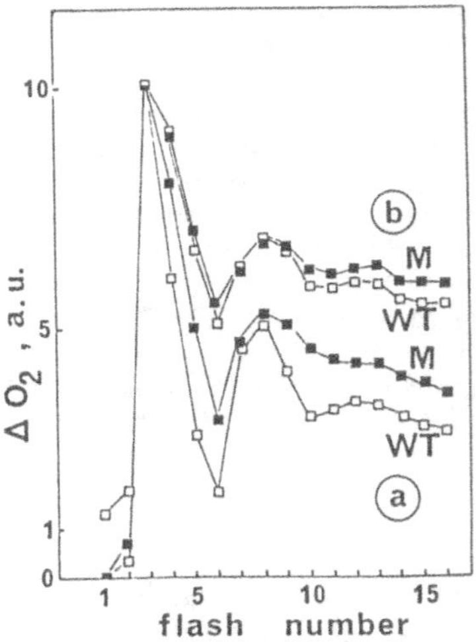

Figure 1. Flash number-dependent fluorescence yield measured 80 ms after each flash in wild-type (WT) and *fad4* mutant (M) *Arabidopsis thaliana* thylakoids at pH 6.5, in the presence of 0.5 mM FeCy, from plants at an early stage of development (a),at the first appearance of flowering (b). The sample was dark-adapted for 6 min. The flash interval was 730 ms. The fluorescence yield after the first flash for M was the same as for WT; the curves have been offset for presentation.

Figure 2. Flash number-dependent O_2 yield in wild-type (WT) and *fad4* mutant (M) thylakoids, from plants at an early stage of development (a), at the first appearance of flowering (b). Same conditions as in Figure 1.

gathered at an early stage of development (a). The downward-directed transient reflects acidification. pH transients in M- thylakoids and in control are compared at pH 6.1 and pH 6.9. A pH-dependent oscillation of proton release was only found in WT-thylakoids. M-thylakoids did not reveal any oscillations in the pH-range from 5.8 to 7.5. The lower trace is the difference between the WT and M-thylakoid pH transients at pH 6.9 (WT-M). This represents the extents of electrostatically induced proton release and uptake in WT thylakoids, since no oscillations were detected in M-thylakoids. For some other measurements, 5 μM DBMIB was added, in order to inhibit plastoquinol reoxidation by the cyt b_6/f complex and to abolish proton release from this process. In this case, the oscillations of the electrostatically proton release/uptake are exclusively observed. Assuming that the PS II/ PS I ratio is about 1:1 in the WT sample [12], the proton release extents over the first four flashes was then 1.5:0:-1.5:0 at pH 6.1, and -1.5:0:1.5:0 at pH 6.9. The average extent over all pH was 0:0:0:0, found in the M-thylakoids. This shows that the electrostatic effects induced the release of 1.5 protons on $S_1 \rightarrow S_2$ and the recapture of 1.5 protons on $S_3 \rightarrow S_0$ at acidic pH, and that the reverse event (or the same event with OH^-) occurred at basic pH in the WT sample. The crossover point where no proton was released was at pH 6.65 after a thorough dark adaptation (as already found in Ref.10), but this point shifted to more acidic pH values after several illuminations, as shown in Figure 4. The pH dependent curves may be explained if the catalytic site of water oxidation in the dark is polarized with an equal amount of positive and negative charges that attract H_2O molecules, H^+, OH^-: more H^+ are attracted at acidic pH, and more OH^- at basic pH [8]. According to this model, the steep transition with increasing pH, indicates a change of the extra H^+ release into extra OH^- release [8]. This

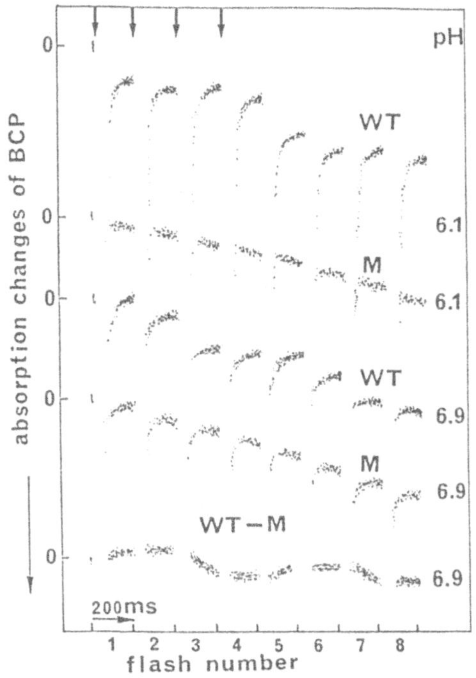

Figure 3. pH transients in the external medium induced by a series of height flashes in dark-adapted wild-type (WT) and *fad4* mutant (M) thylakoids (a) at pH 6.1 and pH 6.9.The time per sweep was 200 ms. The flashes were spaced 1.3 s apart.

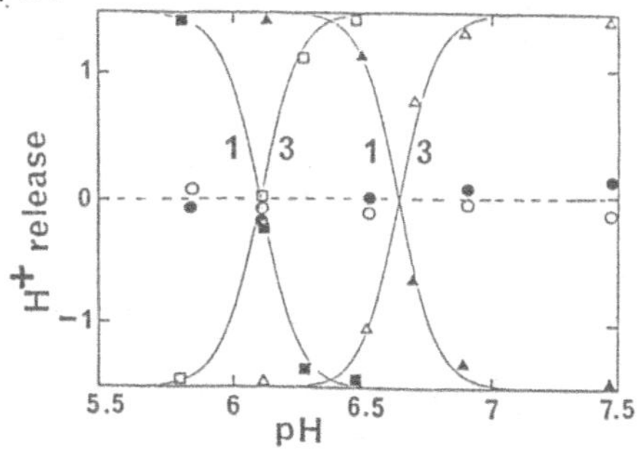

Figure 4. Extent of pH changes induced by the first (1) and the third (3) flash in a series of flashes, as a function of pH in the presence of 5 μM DBMIB, in wild-type thylakoids after a long dark adaptation (trianges), after 5 series of 16 flashes separated by 3 min dark adaptation (squares); in mutant thylakoids (circles).

mechanism could be related to regulation of violaxanthin deepoxidation in LHC II [13]. The pH dependence of violaxanthin deepoxidation exhibits the same strong cooperativity (with an inflexion point at pH 6) [13], as observed in the pH dependence curves of proton release associated with water oxidation. This could indicate that the same polymeric complexes in LHC II are involved in both associated events.

The absence of oscillations of proton release in M-thylakoids resembles the featureless and pH independent proton release stoichiometry in thylakoids from plants grown under intermittent light [14] and PS II core particles [15]. In the latter cases, the non-oscillating behaviour was caused by the absence of LHC II proteins. In the *fad4* mutant thylakoids (a), important oscillatory variations in fluorescence yield were observed, which indicates the presence of the electric events generally responsible for electrostatic proton relaxation at the oxidizing side of PS II. Our results suggest that in mutant thylakoids (as in PS II core complexes), the electrostatically induced protons are not translocated from PS II to the lumen.The coupling of electric field and proton movements is corroborated by the fact that in the absence of period 4 fluorescence oscillations in the control (b) in Figure 1, no significant oscillating pattern of proton release was observed (not shown). LHC II has been proposed to shield the site of proton production and to channel protons from the water oxidation site to the lumen [16]. Our results suggest a possible role for *trans*-$C_{16:1}$-PG, in the conductive mechanism of protons through LHC II. This could indicate that if the proton-translocating amino-acid side chains are to do their job, they must be free to move within the membrane. The molecules of the phospholipid bilayer, while necessarily aligned, must not be locked into a rigid crystalline lattice, as they are when all the fatty acid chains are saturated.

References

1 Hobe, S., Prytulla, S., Kühlbrandt, W. and Paulsen H. (1994) The EMBO J. 13, 3423-3429

2 Rémy, R., Trémolières, A. and Ambart-Bretteville, F. (1984) Photobiochem. Photobiophys. 7, 267-276

3 Nussberger, S.,Dörr, K., Wang, D. and Kühlbrandt, W. (1993) J. Mol. Biol. 234, 347-356

4 Dubacq, J.P. and Trémolières, A. (1983) Physiol. Veg. 2, 293-312

5 MC Court, P., Browse, J., Watson, J., Arntzen, C. and Somerville, C.R. (1985) Plant Physiol. 78, 853-858

6 Delrieu, M.J. (1981) Photobiochem. Photobiophys. 3, 137-144

7 Delrieu, M.J. and Rosengard, F. (1993) Photosynt. Res. 37, 205-215

8 Delrieu, M.J. (1995) Biochim. Biophys. Acta (in press)

9 Delrieu, M.J. and Rosengard, F. (1991) Biochim. Biophys. Acta 1057, 78-88

10 Haumann, M. and Junge, W. (1994) Biochemistry 33, 864-872

11 Haumann, M. and Junge, W. (1994) FEBS Lett. 347, 45-50

12 Jahns, P. and Junge, W. (1992) Biochemistry 31, 7390-7397

13 Pfündel, E. and Dilley, R.A. (1993) Plant Physiol. 101, 65-71

14 Jahns, P. and Junge, W. (1992) Biochemistry 31, 7398-7403

15 Jahns, P., Haumann, M., Bögershausen, O. and Junge, W. (1992) in Research in Photosynthesis (Murata N., ed.) pp.333-336, Kluwer Academic Publishers,Dordrecht.

16 Jahns, P. and Junge, W. (1990) Eur. J. Biochem. 193, 731-736

LIMITED PROTEOLYSIS OF RECOMBINANT MANGANESE STABILIZING PROTEIN OF *SYNECHOCOCCUS ELONGATUS*

Akihiro Motoki[1], Tsuneo Shimazu[1], Masahiko Hirano[1], Sakae Katoh[2]
[1]Toray Research Center, Inc., Kamakura 248 and [2]Department of Biology, Toho University, Funabashi 274, Japan

1. Introduction

The extrinsic 33 kDa protein or Mn-stabilizing protein (MSP) that is associated with the lumenal surface of photosystem II (PSII) complexes plays an important role in photosynthetic oxygen evolution in plants, algae and cyanobacteria (1). Oxygen evolution is strongly suppressed by release of MSP and restored by rebinding of the protein (2). Removal of the protein also induces gradual loss of two Mn atoms (2,3). Mutants of cyanobacteria which lack MSP evolve oxygen in reduced rates and are highly susceptible to photoinhibition (4-6). Thus, the binding of MSP to PSII complexes is essential for the maximum oxygen evolution and stabilization of the Mn cluster. Recently, we have cloned and sequenced the *psbO* gene that encodes MSP from the thermophilic cyanobacterium *Synechococcus elongatus* and discussed the thermostability of the protein in comparison with the corresponding proteins from other organisms (7,8). In this communication, we report the expression of the *S. elongatus psbO* gene and limited proteolysis of the recombinant protein by chymotrypsin. The results suggest that a loop structure of the MSP polypeptide involving Phe^{156} has an important role in the binding of the protein to PSII complexes.

2. Materials and Methods

The *psbO* gene of *S. elongatus* encoding the mature MSP was amplified by the polymerase chain reaction, inserted into *NcoI/Bam*HI site of the pET 8c expression vector and expressed in *E.coli* BL21 cells. A site directed mutagenesis of the *psbO* gene was carried out according to the procedures described in (9-11). Oxygen-evolving PSII complexes were isolated from the cyanobacterial thylakoids by the method of Ichimura et al. (12). The reconstitution experiments were done according to the procedure described in (13). Oxygen-evolving activities were measured with a Clark-type oxygen electrode (12, 14). Limited proteolysis of MSP by chymotrypsin (1: 100 w/w) was performed in 50 mM $K\text{-}PO_4$ (pH9.0) and 150 mM NaCl and stopped by addition of 2.5 mM PMSF.

3. Result and Discussion

The recombinant MSP (rMSP) was abundantly expressed in the *E.coli* cells. Western blotting and CBB stained gel showed that more than 20 % of the total soluble cell proteins was rMSP. Taking advantage of the thermostability of the *S. elongatus* MSP (7,8), a major part of *E. coli* proteins was removed by treating cell extracts at 70 ℃ for 20 min, then rMSP was highly purified by gel filtration and ion exchange column chromatography. The yield of the purified protein was about 20 mg per liter of the culture medium. The N-

P. Mathis (ed.), Photosynthesis: from Light to Biosphere, Vol. II, 487–490.

terminal amino acid sequence, amino acid composition and molecular mass (35 kDa) of rMSP were identical to those of the native mature protein with no transit peptide. The recombinant protein restored the oxygen-evolving activity of $CaCl_2$-washed *Synechococcus* PSII complexes as effectively as the native protein (data not shown). These results show that rMSP is structurally and functionally identical to the native protein.

Fig. 1 shows the polypeptide patterns of rMSP that was treated with chymotrypsin for varied periods of time. Proteolysis of the protein for 1 h yielded several bands including two major fragments of about 26 kDa and 22 kDa and two minor ones of about 15 kDa and 11 kDa. The protein totally lost its ability to restore oxygen evolution in $CaCl_2$-washed complexes after 2 h of chymotrypsin-treatment. The N-terminal amino acid sequences and molecular masses of the fragments suggested two cleavage sites between Phe^{156} and Leu^{157} and between Phe^{190} and Ser^{191} (Fig. 2). Thus, the 26 kDa, 22 kDa, 15 kDa and 11 kDa fragments are Ala^{1} - Phe^{190}, Ala^{1} - Phe^{156}, Leu^{157} - Ala^{246}, and Ser^{191} - Ala^{246}. The Phe^{156} and Leu^{157} pair is conserved among 10 different MSPs from higher plants and cyanobacteria and present in a region (Val^{148} - Gly^{163}) of the MSP sequence which is predicted to form a loop structure exposed to the outer space (Fig. 3). Thus, a possibility arises that the Phe^{156} and Leu^{157} pair is involved in the binding of MSP to PSII complexes. To examine this possibility, a site directed mutagenesis was carried out to construct a rMSP molecule with a nick between Phe^{156} and Leu^{157} and a Met residue inserted before Leu^{157}. Fig. 4 shows that the protein migrated as a single band in none-denaturing PAGE but split into two fragments of about 22 and 15 kDa in SDS-PAGE. The modified MSP (156/157) failed to bind to MSP-free PSII complexes. The loss of the binding capacity is, therefore, ascribed to either introduction of a nick or insertion of a Met residue in the loop region of the polypeptide. The results strongly suggest that the loop region play an important role in binding of MSP to PSII complexes.

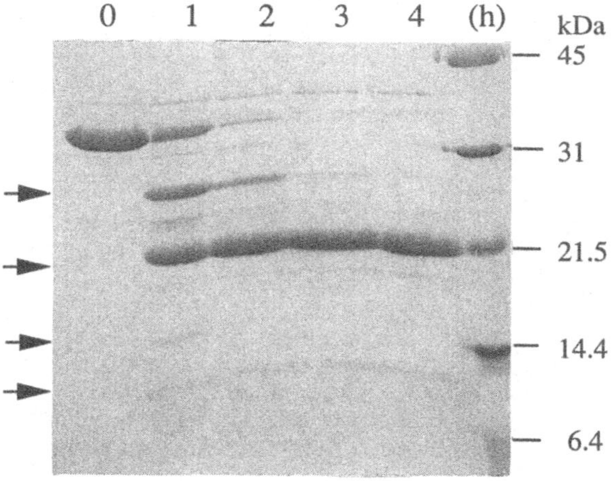

Fig. 1. Limited proteolysis of rMSP by chymotrypsin. The protein was incubated with chymotrypsin for indicated periods of time and then analyzed by SDS-PAGE. Arrows indicate the four fragments mentioned in text.

	150 160 170 180 190 197
Synechococcus elongatus	NVPSYRTANFLDPKGRGLASGYDSAIALPQ...AKEEELARANVKRFSLTKGQI
Anacystis nidulans R2	R------S------A---TT--ES-V-I-SA..GDA-D-TKE-----VTGQ-E-
Synechocystis sp. PCC 6803	------G-G-----A---YT---N-V---SA..-DK..F.-T-K-ETP-G--TL
Anabaena sp. PCC 7120	K-----GSA--------VV----N-V---AQ..-DD-D-T-T----AEILN-K-
Chlamydomonas reinhardtii	L-----GSS--------GST---N-V---A..R-DA---LKE---ITKAL--SA
Spinach (Spinacia oleracea)	L-----GSS--------GST---N-V---AGGRGD----QKE-N-NVASS--T-
Garden pea (Pisum sativum)	L-----GSS--------AST---N-V---AGGRGD----GKE-N-SAASS--K-
Cress (Arabidopsis thaliana)	L-----GSS--------GST---N-V---AGGRGD----VKE---NTAASV-E-
Wheat (Triticum aestivum)	L-----GSS--------GST---N-G---RGGRGD-----KE---NA-SST-N-
Potato (Solanum tuberosum)	L-----GSS--------GST---N-V---AGGRGD----QKE---NTASLT-K-

Fig. 2. Comparison of the amino acid sequences of the MSPs and the deduced sites of cleavage by chymotrypsin. Dash indicates an amino acid that is the same as that of the *S. elongatus* protein; Dot, an amino acid deleted; ↓, cleavage sites.

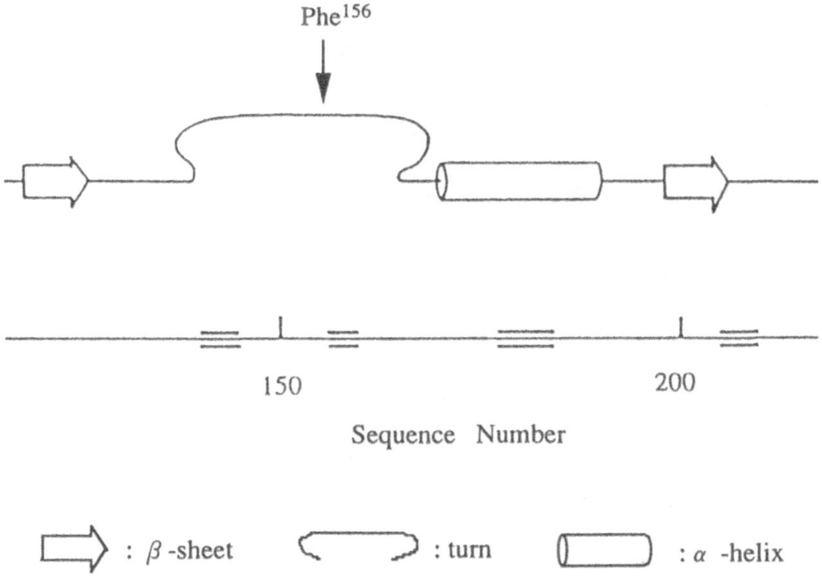

Fig. 3. Secondary structure of the MSP predicted by the method of Gascuel and Golmard (15) .

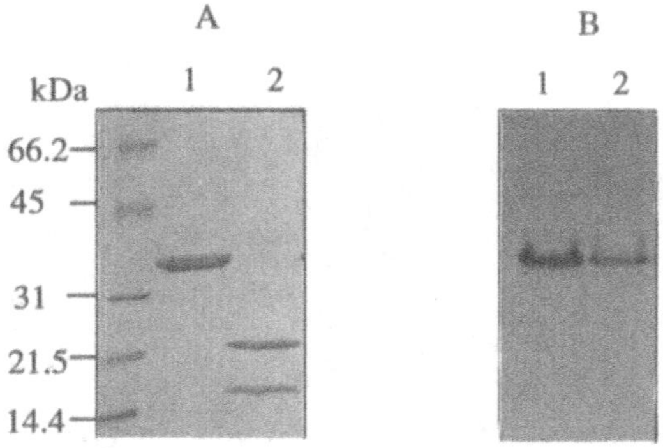

Fig. 4. The SDS-PAGE (A) and non-denaturing PAGE (B) of MSP (156/157).
Lane1, native MSP; lane2, MSP (156/157)

Acknowledgements

This work was performed under the management of Research Association for Biotechnology as a part of the R & D Project of Basic Technology for Future Industries supported by NEDO (New Energy & Industrial Technology Development Organization).

References

1 Vermaas, W. F. J. and Ikeuchi, M. (1991) in the Photosynthetic Apparatus: Molecular Biology and Operation (Bogorad, L. and Vasil, I.K., eds.) vol. 7B, pp.25-111, Academic Press, San Diego
2 Kuwabara, T., Miyao, M., Murata, T. and Murata, N. (1985) Biochim. Biophys. Acta 806, 283-289
3 Miyao, M. and Murata, N. (1984) FEBS Lett. 170, 350-354
4 Burnap, R. L. and Sherman, L. A. (1991) Biochemistry 30, 440-446
5 Mayes, S. R., Cook, K. M., Shelf, S. J., Zhang, Z. and Barber, J. (1991) Biochim. Biophys. Acta 1060, 1-12
6 Philbrick, J. B., Diner, B. A. and Zilinskas, B. A. (1991) J. Biol. Chem. 266, 13370-13376
7 Motoki, A., Miura, K., Shimazu, T., Kanai, S., Hirano, M. and Katoh, S. (1992) in Research in Photosynthesis (Murata, N. ed.) vol. II, pp. 413-416, Kluwer Academic Publishers, Dordrecht
8 Miura, K., Shimazu, T., Motoki, A., Kanai, S., Hirano, M. and Katoh, S. (1993) Biochim. Biophys. Acta 1172, 357-360
9 Taylar, J. W., Ott, J. and Eckstein, F. (1985) Nucleic Acids Res. 13, 8764-8785
10 Nakayama, K. and Eckstein, F. (1986) Nucleic Acids Res. 14, 9679-9698
11 Sayer, J. R., Krekel, C. and Eckstein, F. (1992) Biotechniques 13, 592-596
12 Ichimura, T., Miyairi, S., Satoh, K. and Katoh, S. (1992) Plant Cell Physiol. 33, 299-305
13 Koike, H. and Inoue, Y. (1985) Biochim. Biophys. Acta 807, 64-73
14 Katoh, S., Satoh, K., Yamagishi, A. and Yamaoka, T. (1975) Plant Cell Physiol. 16, 1093-1099
15 Gascuel O. and Golmard J. L. (1988) CABIOS. 4, 357-365

6. Cytochrome complexes

STRUCTURE AND FUNCTION IN THE BC$_1$-COMPLEX OF *RHODOBACTER SPHAEROIDES*.

A.R. Crofts[#*], Blanca Barquera[%], Georg Bechmann[#], Mariana Guergova[#], R. Salcedo-Hernandez[%], Beth Hacker[%], Sangjin Hong[#] and R.B. Gennis[#%]
Program in Biophysics and Computational Biology[#], Dept. of Microbiology[*], and Dept. of Biochemistry[%], University of Illinois, Urbana IL 61801

1. Introduction.

The photosynthetic chain of *Rb. sphaeroides* contains two major enzymes, the reaction center and the bc$_1$-complex, which transfer electrons between catalytic sites through redox reactions of ubiquinone-10 (Q) in the lipid phase and cytochrome (cyt) c$_2$ in the periplasmic phase. The turn-over of the chain is well described by a modified Q-cycle (1-3). With the exception of the bc$_1$-complex, structures of the proteins involved in light-harvesting and photosynthetic turn-over are available from X-ray crystallography (4-6), or from the solved structures of closely related proteins (7, 8). A secondary structure for cyt *b* has been proposed based on profile analysis of aligned sequences (8-11); hydropathy plots suggest that cyt b is organized in eight transmembrane spans (A-H). Spans in the N-terminal stretch (a helix) and between helices A and B (ab helix), C and D (cd helix) and possibly between E and F (ef helix) have been tentatively identified as amphipathic surface helices. We have suggested a tertiary folding pattern based on the location of inhibitor resistance lesions, site-directed mutagenesis to identify heme ligands, molecular engineering of conserved residues and of residues predicted to impinge on the quinone binding sites (12). A crystallographic structure for cyt *f* of the related b$_6$f-complex from chloroplasts is available (13), but the very weak sequence homology makes this a dubious template for prediction of a structure for cyt c_1. A number of laboratories have reported crystals of the mitochondrial complex or its subunit, with diffraction to atomic resolution, and it seems likely that a structure will be available in the near future. The three main subunits are highly conserved, so the mitochondrial complex will likely "contain" the bacterial structure. Our mutant strains are summarized in Table I.

2. Mechanism of ubiquinone reduction at the Q$_i$-site.

Redox titration of the bc$_1$-complex from *Rhodobacter sphaeroides* shows three cyt *b* components, titrating with $E_{m,7}$ values of -90 (b_L), 50 (b_H) and 150 (b_{150}) mV (2, 14). The isolated complex contains two *b*-type hemes, associated with the cyt *b* subunit (15), and the sequence and molecular engineering studies suggests only two conserved heme binding sites (9-11, 16). When titrations are performed in the presence of antimycin A (AA), the cyt b_{150} component is lost, and when AA is added to preparations poised so that cyt b_{150} is reduced but cyt b_H is oxidized (at $E_{h,7}$ ~ 100 mV), AA induces an oxidation of the cyt b_{150} form (17-19). The higher potential component (cyt b_{150}) was seen in the isolated complex only if the preparation was supplemented with quinone (15, 19), but

P. Mathis (ed.), Photosynthesis: from Light to Biosphere, Vol. II, 493–500.

Table I

Mutation	ΔE_m	Pho. Growth	Kinetic effects
Q_o-site (Wild type has V_{max} ~500 mol e^-/mol cyt b_H/s ($t_{1/2}$ ~1 ms), K_m ~1.66 mM)			
F144S	None	++++	Slowed QH_2 oxidation, myxothiazol resistant.
T160A,V	None	++++	Conserved polar residue to a neutral, little effect on function.
N279Y	None	++++	Mimics myx^r strain in yeast; V_{max} 174, Km 3.2 mM
I292F	None	-	V. slow QH_2 oxidation; V_{max} 4, K_m 2.9 mM
P294S	None	++++	QH_2 oxidation slowed, otherwise normal; K_m 3.45 mM
E295D	Minor	++++	QH_2 oxidation slowed 2-fold; K_m 2.63 mM
E295G	Minor	+++	QH_2 oxidation slowed 9-fold; K_m 4 mM
E295Q	b_L -30	++	QH_2 oxidation slowed 50-fold; no reliable K_m value
W296F	b_L -20	+++	QH_2 oxidation weakly inhibited; K_m 3.3 mM
W296L	b_L -20	+++	QH_2 oxidation slowed 2-fold; K_m 4 mM
Y297F	None	++++	QH_2 oxidation weakly inhibited; K_m 1.05 mM
Y297S	Minor	++	QH_2 oxidation slowed 25-fold; K_m 3.4 mM
L305D	ND	++	Slow QH_2 oxidation (V_{max} 8.4); K_m 1.55 mM
K329G	None	+++	Slow QH_2 oxidation ($t_{1/2}$ ~10 ms); K_m 1.54 mM
G332S	None	+++	Slow QH_2 oxidation ($t_{1/2}$ ~5 ms); K_m 2.7 mM
G332I	None	-	Q_o-site blocked, but complex assembled normally
G332D	None	+++	Slow QH_2 oxidation (V_{max} 15.4); K_m 3.2 mM
V333M	None	++++	Stigmatellin resistant, otherwise normal kinetics; K_m 2.5 mM
Q_i-site (Wild-type has E_m values: b_H 43 ±5, b_L -90 mV. Cyt b_{150} is 20-30% total cyt b_H)			
G48V	None	-	Q_i-site blocked, but AA-induce oxid. still present
A52V	None	++++	Normal electron transfer, weak AA resistance
R114A	Large	-	Spectra odd, complex assmbl. but dead, no AA-induced oxid.
R114K	b_H 12	++++	Kinetics and AA titre normal, AA-induced oxid. low (0.1)*
W128F	None	+++	bc_1-complex low but assembled, AA-induced oxid. 0.75*
W128A	b_H 16	+++	Kinetics fairly normal, AA-induced oxid. low (0.33)*
V209A	None	++++	Kinetics normal, funiculosin sensitive at Q_o-site
V209T	b_H 10	++++	Kinetics normal, no funiculosin sensitivity, b_{150} low
I213V	None	++++	E-transfer from cyt b_H slightly slowed when pool oxidized
V209A/I213V	None	++++	Kinetics, redox centers normal, funiculosin sens. at Q_o-site
H217A	Minor	-	The bc-complex assembled, but blocked in cyt b oxidation.
S218A	None	++++	Normal kinetics, AA-induced oxid. 0.85
T219A	b_L -31	+++	Slowed b_H-oxidation, AA-induced oxid. much greater (1.9)*
G220V	None	++++	Near normal kinetics of Q_i-site, AA-induced oxid. low (0.2)*
N221I	b_H 20	+++	Slowed b_H-oxidation, AA-induced oxid. much greater (2.3)*
N222A	b_H 20	++++	Near normal kinetics of Q_i-site, AA-induced oxid. low (0.4)*
N223Q	b_H 22	++++	Normal kinetics, AA-titre, AA-induced oxid.
N223V	b_H 10	+++	Kinetics slightly slowed. AA-induced oxid. low (<0.1)*
K251M	None	++++	Ant^r, e-transfer from cyt b_H slowed, esp. with oxidized pool
K251I	None	++++	Ant^r, e-transfer from cyt b_H slowed, esp. with oxidized pool
D252A	None	-	Normal complexed assembled, but oxidation of b_H blocked
D252N	None	+++	Inhbited Q_i-site, but weak turn-over detectable.
K251D/D252K (see below)			Ant^r, e-transfer from cyt b_H slowed

*AA-inucded oxid. values are normalized with respect to wild-type.

(P294L, L305A, K329E,A, G48D, R114Q, H217Q, F248L No bc_1-complex assembled. G48A, N221S, N223A, N223D, P224A Normal kinetics, AA-titre, AA-induced oxidtn.)

isolation the bacterial complex using a $(His)_6$ tagged cyt *b* anu a Ni-affinity column gives a complex showing the cyt b_{150} effects (20). The two hemes of cyt *b* are assumed to represent cyt b_L and cyt b_H; as discussed below, cyt b_{150} is thought to be a form of cyt b_H.

The cyt b_{150} phenomena reflect the mechanism of reduction of Q at the Q_i-site of the complex. In the modified Q-cycle, this is thought to involve two sequential one-electron transfers from cyt b_H to Q bound at the site, with formation of an intermediate, relatively stable, semiquinone (1-3). Recent attempts to account for cyt b_{150} have invoked changes in the mid-point potential of cyt b_H such that the cyt has different E_m values when Q, Q^-, or QH_2 are bound at the site (18, 19). We suggest here an alternative model based on schemes previously proposed to account for AA-induced oxidation of cyt *b* (17), and the reduction of cyt b_H observed following excitation of chromatophores in the presence of myxothiazol (21), in which the E_m values for all components have single values, and the reduction of cyt b_H to give cyt b_{150} occurs by the reactions at the Q_i-site which have already been demonstrated. No *ad hoc* assumptions are needed, and the equilibrium constants are provided by direct measurement from existing data. We propose that formation of cyt b_{150} occurs through the reverse reaction at the Q_i-site which is demonstrated by cyt b_H reduction following flash illumination in the presence of myxothiazol (21):

$$QH_2(\text{pool}) + \text{cyt } b_H . Q \Leftrightarrow Q(\text{pool}) + \text{cyt } b_H^- . Q^-(2H^+) \qquad (1)$$

We have previously proposed a mechanism for AA induced oxidation of cyt b_{150} (or the elimination of cyt b_{150} from the redox titrations in the presence of AA) based on this reaction (17), which is similar to the well established mechanism by which reduction of Q_A by Q_B^- is induced on addition of DCMU to photosystem II following formation of the semiquinone at the Q_B-site (cf. 22). In the reaction below, the semiquinone at the Q_i-site <u>accepts</u> an electron from ferrocyt b_H^-, and leaves as the quinol, so that ferricyt b_H is stabilized by AA binding.

$$\text{cyt } b_H^- . Q^- (2H^+) \Leftrightarrow \text{cyt } b_H . QH_2 \Leftrightarrow \overset{\nearrow QH_2}{\text{cyt } b_H.\text{vacant}} \Leftrightarrow \overset{AA \searrow}{\text{cyt } b_H.AA} \qquad (2)$$

Partial reactions giving rise to the overall process of eq (1) are:

$$QH_2(\text{pool}) + \text{cyt } b_H . Q \Leftrightarrow Q(\text{pool}) + \text{cyt } b_H . QH_2 \qquad (1a)$$

$$\text{cyt } b_H . QH_2 \Leftrightarrow \text{cyt } b_H^- . Q^-(2H^+) \qquad (1b)$$

At any defined pH, the equilibrium constant for formation of $[\text{cyt } b_H^- . Q^-(2H^+)]$ (the cyt b_{150} form) according to equation (2) is given by:

$$K_{eq} = K_{(1a)} . K_{(1b)}$$
$$= K_{a(QH2)}/K_{a(Q)} . \exp\{(E_{m(\text{cyt } bH)} - E_{m(QH^{\cdot}/QH2)}).F/RT)\} \qquad (3)$$

where K_a values are association (binding) constants for QH_2 and Q, and $E_{m(\text{cyt } bH)}$ and $E_{m(QH^{\cdot}/QH2)}$ refer to the mid-point potentials of cyt b_H, and of the semiquinone/quinol couple bound at the site, respectively.

Examination of equation (3) shows that an increased value of K_{eq} could be expected if $E_{m(cytbH)}$ is increased, if $E_{m(QH^{\cdot}/QH2)}$ is decreased, if the binding of quinol is increased, or if the binding of quinone is decreased. However, the implied importance of the latter pair of terms is misleading. When expressed with respect to the potential of the quinone pool, the ratio of binding constants for Q and QH_2 contributes to the value for $E_{m(QH^{\cdot}/QH2)}$ in such a way as to cancel the pre-exponential term. This gives:

$$K_{eq} = \exp\{(E_{(\text{cyt } bH)} + E_{m(Q/QH^{\cdot})} - 2\,E_{m(Qpool)}).F/RT)\} \qquad (4)$$

(where $E_{m(Q/QH^{\cdot})}$ refers to the bound couple), which can also be derived directly from eq. 1. The relation between these two expressions for K_{eq} reflects the dependence of the value of $E_{m(Q/QH^{\cdot})(bound)}$ on the stability constant for the bound reactants in the disproportionation reaction, and on the value for $E_{m(Q/QH2)(bound)}$, which itself depends on the relative affinities of Q and QH_2. The important factors in defining a value for K_{eq} are the E_m value

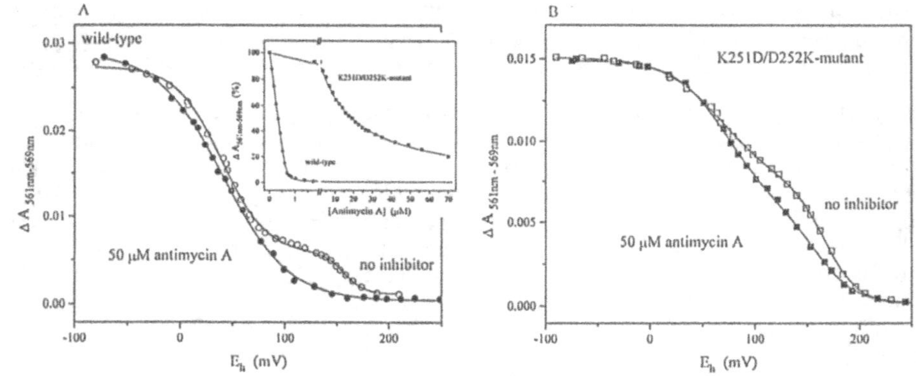

Figure 1. Redox titrations of cyt b_H in wild-type and K251D/D252K strains. Insert: Titrations with the antimycin showing inhibition of cyt b_H oxidation in the two strains.

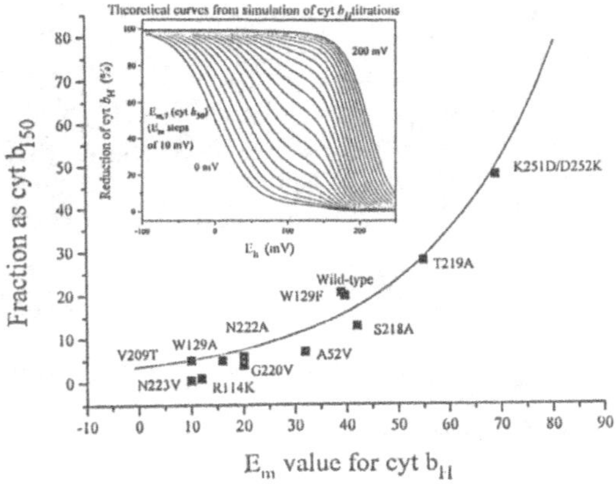

Figure 2. Dependence on the E_m value for cytb b_H of fraction in the cyt b_{150} form in mutant strains. The line draw through the points is from values calculated from a computer model based on the hypothesis. Insert: Set of theoretical titration curves.

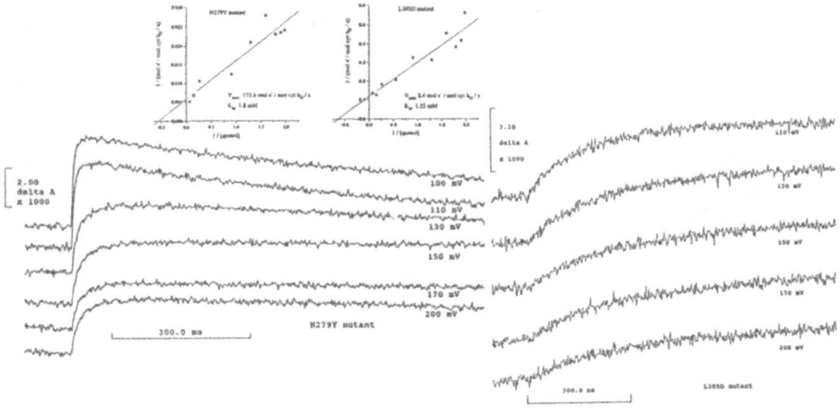

Figure 3. Traces showing the reduction of cyt b_H in N279Y and L305D strains. Inserts: Plots (1 / rate v. 1 / [quinol]) from which K_m and V_{max} were calculated.

for cyt b_H, and the stability of the semiquinone, while the relative affinities of Q and QH_2 determine the apparent mid-point of the cyt b_{150} form.

Support for the above mechanisms comes from analysis of strains with mutations in the Q_i-site (23, 24, Fig. 1 and Table I). In these mutants we measured the fraction of cyt b_H which titrates in the high potential cyt b_{150} form, the extent of AA induced oxidation of cyt b_{150} and the correlation between these factors and the E_m value for cyt b_H. Several mutants have a lower fraction of cyt in the b_{150} form, and a reduced extent of cyt b_{150} oxidation on addition of AA; most dramatic is the low level of the cyt b_{150} component in strain N223V, but mutants V209T, G220V, N222A, R114K and W129A all follow this trend. In all of these strains, the $E_{m,7}$ for cyt b_H was lower than in wild-type. Conversely, strain T219A, and the double mutant K251D/D252K, both of which have a higher $E_{m,7}$ for cyt b_H, both showed a higher amplitude of AA induced oxidation, and a higher ratio of cyt b_{150}. Note that in mutants in which the E_m of cyt b_H was changed the E_m values for cyt b_{150} were relatively little changed, in line with the above mechanism, suggesting that these mutations did not much change the relative affinities of Q and QH_2.

Computer simulations show that the redox titration curves for cyt b_{150} and for AA induced oxidation (showing [cyt b_H^- . $Q^-(2H^+)$]), can be well modeled in wild-type strains using the assumption that reaction (1) is the mechanism for formation of cyt b_{150}, by using the following values: $E_{m(cyt\ bH)} = 45 \pm 5$ mV; $E_{m(QH^{\cdot}/QH2)} = 205 \pm 5$ mV; $K_{a(QH2)}/K_{a(Q)} = 100$; $E_{m(Qpool)} = 90$ mV; pK_{bH} 7.8; pK_{QH}. 12.5. With these parameters, $E_{m(cyt\ b150)}$ ~146 mV, and the component accounts for ~ 20-30% of the total cyt b_H at pH 7.0. We simulated the titration curves to be expected when E_m values for cyt b_H were changed. Substitution of the measured E_m values for cyt b_H from the mutant strains, keeping other values constant, then provided a good fit for the amount of cyt b_{150} found in these strains (Fig. 2). In addition, the stable level of reduction seen in kinetic experiments in the absence of inhibitors, at E_h values where the pool was oxidized before flash, could also be accounted for by the modeling parameters in terms of the K_{eq} for reduction of Q_i by cyt b_H.

Several mutants with a change in E_m for cyt b_H do not follow the pattern expected from the hypothesis. We would suggest that in these strains, the other terms in eqs. 3 or 4 which govern the stability of the $Q^{\cdot-}$ or binding of Q or QH_2 will be of importance

3. Role of Lysine 251 and aspartate 252 in the function of the Q_I-site of the bc_1-complex of *Rb. sphaeroides*.

Two highly conserved charged residues, lysine 251 and aspartate 252 of subunit *b* in the bc_1 complex of *Rhodobacter sphaeroides*, have previously been implicated in the Q_i-site on the basis of AA resistant mutations. In order to test the role of the charge contribution from this pair, the sequence has been inverted by site-directed mutagenesis to give strain K251D/D252K. The inversion of these charged residues has the following effects on the bc_1-complex:

1) AA has approximately 2000 times lower affinity than in the wild type (Fig. 1).
2) The E_m of cyt b_H is increased by 30-40 mV, the highest increase so far introduced by mutagenesis (Fig. 1).
3) The mutant grows photosynthetically at rates close to those of wild-type. The steady-state DBH:cyt c oxidoreductase activity of the bc_1 complex is about 3 times slower than in the wild type. The electrogenic processes of the complex, measured from the slow phase of the carotenoid change in the absence of inhibitor, are also slowed about three-fold. The maximal rate of reduction of cyt b_H following flash excitation in the presence

of AA at 140 μM is somewhat slowed, but direct comparison is difficult because the change in E_m precludes measurement at the same relative concentrations.
4) The reduction of cyt b_H by quinol in the presence of myxothiazol at pH 9.0 by a reversal of the quinone reducing site, is 6 times slower, but has a higher amplitude.
5) The formation of cyt b_{150} at 2 x the level in wild-type, the stable reduction of cyt b_H following a flash with the pool oxidized, and the increased cyt b_H reduction through reversal of the Q_i-site are accounted for by the increased E_m of cyt b_H (hypothesis above).

These results suggest that the slowed kinetics reflect changes in rate constant(s) at the Q_i-site. Although there is some decrease in steady-state activity in this mutant, K251 and D252 do not seem to be specifically required for catalysis because the turnover is not dramatically altered. Since the increased concentration of cyt b_{150} is well explained by the change in E_m for cyt b_H, it does not seem likely that these residues are involved in the stabilization of the semiquinone at the Q_i-site. In addition, the change in E_m value for cyt b_{150} is as predicted, suggesting that the affinities of Q and QH_2 are unaltered. Loss of charge at D252 has a strong inhibitory effect, but the charge on K251 does not seem to be important (Table I). This pattern of effects is similar to those seen when residues involved in the protonation pathway of the Q_B-site are changed ( ), suggesting a similar role. Nevertheless, K251 is obviously important for AA binding and may interact directly with the inhibitor. The 30-40 mV increase of the midpoint potential of the cyt b_H may reflect a differential coulombic effect due to a longer distance from the heme charge of cyt b_H to the positive charge of K251 than to the negative charge of D252 in the wild-type, and a reversal of these distances in the mutant.

4. The contribution of the ef-loop to the function and structure of the Q_O-site of the bc_1-complex of *Rb. sphaeroides*.

We have used site directed mutagenesis and biophysical assay to probe the structure-function interface at the Q_o-site. A number of inhibitor resistance mutations have mapped to the span between helices E and F, and in addition, the span contains an almost totally conserved - PEWY- region, P294, E295, W296, Y297. Mutational replacement in this span shows that none of the residues is essential to function, but replacement of any lowers activity. Since resistance mutations occur on either side of the -PEWY- span, it seems likely that this region is near to the Q_o-site (25). Many residue changes in this region show marked effects on the site, as reflected in modified kinetics of quinol oxidation, and binding of inhibitors (see Table I). Several of these have not previously been identified as contributing to the site. In general, although the maximal rates vary >200-fold, the mutants with modified kinetics all show the increase in rate of QH_2 oxidation as the pool becomes reduced over the range $E_h < 180$ mV, with an approach to saturation at potentials where the pool is half-reduced, which in wild-type has been interpreted as showing a site at which QH_2 binds in preference to Q by a factor 10-100 (3, and Fig. 3). This behavior in the mutant strains shows that the relative affinities of Q and QH_2, and K_m for QH_2, are not greatly affected. The main effects of mutation appear to be on k_{cat}.

Mechanism of quinol oxidation.

Although there is a consensus that QH_2 oxidation follows the pattern expected from the modified Q-cycle, details of the mechanism at the structural level, and interpretation of mutational effects, are more controversial. In addition to ligands for quinone, the Q_o-site has contributions from the Fe.S subunit, and interactions with cyt c_1 (26, and Daldal, F., personal communication), which might effect function if perturbed. Following earlier work showing an effect of the redox state of the quinone pool on the EPR spectrum of the 2Fe.2S center of the complex (27), Ding et al. (28), using wild-type and mutants strains from *Rb. capsulatus*, have investigated the effects of extraction of ubiquinone, and concluded that different spectral changes can be detected at different local concentrations. They suggested that these reflect two different populations of bound quinone at the Q_o-site, called $Q_{s,o}$ and $Q_{w,o}$ for strongly and weakly binding species. They have extended these observations to mutant strains, and kinetic studies of turn-over and inhibitor binding, concluding that the mechanism might reflect a double occupancy of the Q_o-site, with both sites showing equal binding for Q and QH_2. More recently, they have discussed possible mechanisms and atomic details of ligation (29, 30). In general, the mechanisms recognize the need for a rapidly exchanging site required by the kinetic data (1-3); it would seem likely that $Q_{w,o}$ provides this function, and that $Q_{s,o}$ acts as a prosthetic group regulating the delivery of electrons to the metallic acceptors (2Fe.2S and cyt b_L) of the complex. In collaboration with Dutton's lab, we have looked at similar effects in some of the strains from *Rb. sphaeroides* summarized in the Table above.

The double-occupancy mechanism requires that the catalytic site must contain two distinct sets of ligands to provide for the differential binding of the two species. If the protein provided these, it would be expected that ligands specific for one or other species would be detected by mutation, but no such differential effects have been seen among the many mutants generated by our two groups (25). Furthermore, there is no clear correlation between effects of mutation on the spectral changes attributed to $Q_{w,o}$, and interpreted as showing substantial changes in affinity, and the small changes in K_m for QH_2 oxidation. Myxothiazol, stigmatellin and UHDBT eliminate the spectral changes associated with <u>both</u> strong and weak binding, with the same titration ($\leq$ 1mol/mol cyt c_1 for stigmatellin) as that giving rise to inhibition of electron flow (28). These inhibitors compete for a common site (30, 31); however, stigmatellin and UHDBT, but not myxothiazol, directly effect the redox potential and spectra of the 2Fe.2S center, indicating differences in the liganding, and these differences <u>are</u> reflected in a differential resistance to the inhibitors in different mutant strains (25, 30). These effects are difficult to explain in the double-occupancy model. The failure to detect differential effects of mutation on Q-binding could mean that the ligands for $Q_{w,o}$ are provided by $Q_{s,o}$; in this context, the titration effects could be explained by suggesting that the inhibitors displace $Q_{s,o}$ (29), and the relative lack of effect of mutation on $K_m(QH_2)$ by the fact that binding does not involve protein ligands. Still unexplained is the discrepancy between the tighter binding of quinone indicated by kinetic experiments, and the equal binding of Q and QH_2 from the EPR data. If the structure shows that the Q_o-site has only a single quinone, alternative explanations of the binding effects on the EPR spectrum of 2Fe.2S will have to be found.

Acknowledgments. We thank Les Dutton and Fevzi Daldal for extensive helpful discussion and access to unpublished material, and NIH for support though a grant RO1 GM35438.

References

1. Crofts, A. R., Meinhardt, S. W., Jones, K. R. and Snozzi, M. (1983). *Biochim. Biophys. Acta*, 723, 202-218.
2. Crofts, A. R. (1985). In *The Enzymes of Biological Membranes*, (Martonosi, A.N., ed.), Vol. 4, pp. 347-382, Plenum Publ. Corp., New York.
3. Crofts, A.R. and Wang, Z. (1989) *Photosynth. Res.* 22, 69-87.
4. Komiya, H., Yeates, T.O., Rees, D.C., Allen, J.P. and Feher, G. (1988) *Proc. Natl.*

Acad. Sci. USA, **85**, 9012-9016
5. Chang, C.-H., El-Kabbani, O., Tiede, D., Norris, J. and Schiffer, M. (1991) *Biochemistry* 30, 5352-5364
6. Ermler, U., Fritzsch, G., Buchanan, S.K. and Michel, H. (1995) *Structure* **2**, 925-936
7. Benning, M.M., Wesenberg, G., Caffrey, M.S., Bartsch, R.G., Meyer, T.E., Cusanovich, M.A., Rayment, I. and Holden, H.M. (1991) *J. Mol. Biol.* 220, 673
8. McDermott, G., Prince, S.M., Freer, A.A., Hawthornethwaite-Lawless, A.M., Papiz, M.Z., Gogdell, R.J. and Isaacs, N.W. (1995) *Nature*, 374, 517-521.
9. Widger, W.R., Cramer, W.A., Herrmann, R.G. and Trebst, A. (1984) *Proc. Natl. Acad. Sci. 81,* 674-678.
10. Saraste, M. (1984) *FEBS Lett.* 166, 367-372.
11. Crofts, A.R., Robinson, H.H., Andrews, K., Van Doren, S. and Berry, E. (1987) In *Cytochrome Systems: Molecular Biology and Bioenergetics* (Papa, S., Chance, B. and Ernster, L., eds.) pp. 617-624, Plenum Publ., New York.
12. Crofts, A.R., Hacker, B., Barquera, B., Yun, C.-H. and Gennis, R.B. (1992) In *Research in Photosynthesis*, (Murata, N., ed.), Vol. II, pp. 463-470. Kluwer Academic Publishers, Dordrecht.
13. Martinez, S.E., Smith, J.L., Huang, D., Szczepaniak, A. and Cramer, W.A. (1993) In *Research in Photosynthesis* (Murata, N., ed.) Vol. II, pp. 495-498. Kluwer Academic Publ., Dordrecht
14. Dutton, P.L. (1978) *Methods. Enzymol.* 54, 411-435
15. Andrews, K.M., Crofts, A.R. and Gennis, R.B. (1990) *Biochemistry*, 29, 2645-2651.
16. Yun, C.-H., Crofts, A.R. and Gennis, R.B. (1991) *Biochemistry*, 30, 6747-6754
17. Meinhardt, S. W. and Crofts, A. R. (1984). In *Advances in Photosynthesis Research* (Sybesma, C., ed.), Vol. 1, pp. 649-652, Martinus Nijhoff/Dr. W. Junk Publishers, The Hague.
18. Rich, P.R., Jeal, A.E., Madgwick, S.A. and Moody, A.J. (1990) *Biochim. Biophys. Acta* 1018, 29-40
19. Salerno, J.C., Xu, Y., Osgood, P. Kim. C.H. and King, T.E. (1989) *J. Biol. Chem.* 264, 15398-15403
20. Bechmann, G., Salcedo-Hernandez, R., Guergova, M., Crofts, A.R. and R.B. Gennis (1995) These proceedings.
21. Glaser, E.G., Meinhardt, S.W. and Crofts, A.R. (1984) *FEBS Lett.* 178, 336-342.
22. Crofts, A. R. and Wraight, C. A. (1983) *Biochim. Biophys. Acta* 726, 149-186.
23. Hacker, B., Barquera, B., Crofts, A.R. and Gennis, R.B. (1993) *Biochemistry* 32, 4403-4410
24. Hacker, B., Barquera, B. Gennis, R.B. and Crofts A.R. (1994) *Biochemistry*, 33, 13022-13031
25. Gennis, R.B., Barquera, B., Hacker, B., Van Doren, S.R., Arnaud, S., Crofts, A.R., Davidson, E., Gray, K.A. and Daldal, F. (1993) *J. Bioenerg. Biomemb.* 25, 195-209
26. Van Doren, S.R., Gennis, R.B., Barquera, B. and Crofts, A.R. (1993) *Biochemistry*, 32, 8083-8091.
27. Robertson, D.E., Prince, R.C., Bowyer, J.R., Matsuura, K., Dutton, P.L. and Ohnishi, T. (1984) *J. Biol. Chem.* **259**, 1758-1763
28. Ding, H.G., Robertson, D.E., Daldal, F., & Dutton, P.L. (1992) *Biochemistry* **31**, 3144-3158
29. Ding, H.G., Robertson, D.E., Daldal, F., & Dutton, P.L. (1995) In press
30. Ding, Moser,C., Robertson, D.E. Tokito, M.K., Daldal, F. and Dutton, P.L. (1995) In press
31. Meinhardt, S.W. and Crofts, A.R. (1982). *FEBS Lett.*, **149**, 217-222.

Q_o SITE MUTANTS OF THE CYTOCHROME *bc*$_1$ COMPLEX OF *Rhodobacter capsulatus*

A. Sami Sarıbaş and Fevzi Daldal, Department of Biology, Plant Science Institute, University of Pennsylvania, Philadelphia, PA 19104 U.S.A.

1.Introduction

Ubiquinone cytochrome *c* oxidoreductase (*bc*$_1$ complex and its plant counterpart *b*$_6$*f* complex) is a central component in energy transduction systems. It catalyzes electron transfer from ubihydroquinone to cyt *c* and mediates vectorial proton translocation across the membrane, thereby generating a proton motive force used to produce ATP (for recent reviews see 1, 2).

The *bc*$_1$ complex from non-sulfur purple bacterium *Rhodobacter capsulatus* encoded by the *fbcFBC* (*petABC*) operon has three subunits; cyt *b*, cyt *c*$_1$ and Rieske FeS protein. Cyt *b* is the most conserved of the three subunits (3), and it coordinates two heme groups, named b_H and b_L after their high and low redox midpoint potentials (E_{m7}), respectively. The Rieske FeS protein has a [2Fe2S] cluster which can be detected by EPR spectroscopy. There are two distinct quinone binding sites in the complex, Q_o where the oxidation of ubihydroquinone takes place, and Q_i where ubiquinone is reduced. The Q_o site is formed of the cyt *b* and the Rieske FeS subunits at the positive site of the membrane, while the Q_i site is confined only to the cyt *b* subunit at the negative side of the membrane. Although the *bc*$_1$ complex has been crystallized (4,5), its high resolution crystal structure is not yet available. However, as cyt *b* subunit is an integral membrane protein, modelling studies were used to depict it as being made of eight transmembrane (A to H), and at least one transversal, helices (*cd*) (Figure 1). Studies with spontaneous inhibitor resistance (Inh^R) mutants have revealed two portions of cyt *b* (Q_oI and Q_oII) involved in binding the Q_o site inhibitors (6). The Q_oI subdomain extending from the residues M140 to T163 has been the focus of our work. In this region, first the role of the Inh^R residues M140, F144, G152, G158 and T163 has been examined by site-directed saturation mutagenesis (7,8) and biochemical/biophysical studies (9,10) (Figure 2). Results indicated that these residues are important for Q_o site catalysis by influencing the binding of Q/QH_2 at the Q_o site. Next, considering that mutations at functionally essential residues will yield inactive *bc*$_1$ complexes and therefore be missed as Inh^R mutants, we have chosen several highly conserved residues, including G146 and Y147 (Figure 1 and 2), as targets for further studies to elucidate their role in the Q_o site structure and function. G146 is one of the four invariant glycine residues of cyt *b* which have been implicated in the heme groups (11). In addition, it has been shown that a spontaneous mutation at the equivalent position, G131S in yeast mitochondrial cyt *b*, yields an unassembled *bc*$_1$ complex (12). Y147 is also highly conserved among the *bc*$_1$ and *b*$_6$*f*

P. Mathis (ed.), Photosynthesis: from Light to Biosphere, Vol. II, 501–506.

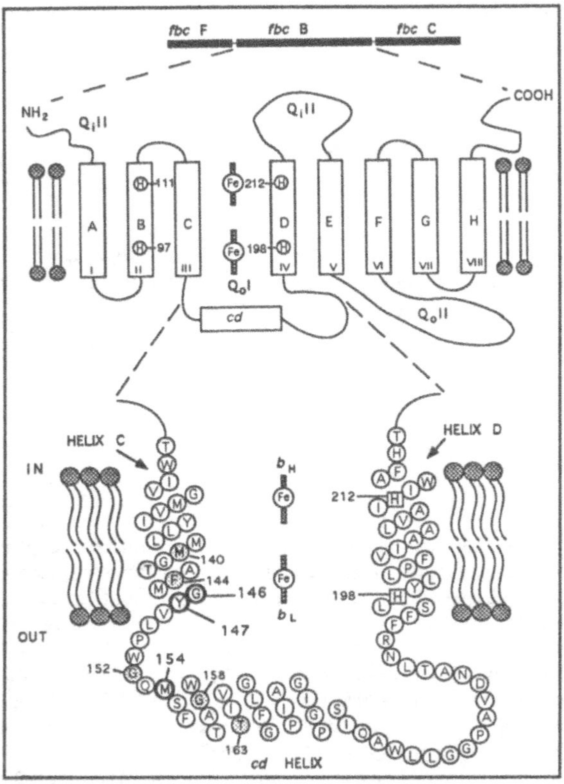

Figure 1. The fbc FBC operon, a model for membrane spanning helices of cyt b subunit and its QoI region. Residues studied in this work are shown in thick circles.

complexes (3), is located on the same face of helix C as the residues M140 and F144, which interfere with the Q_o site functions (9,10) (Figure 1). In addition, a mutation at the equivalent position, Y132C in *C. reinhartii* cyt *b*, was found to confer resistance to myxothiazol (13). In this work the properties of various mutations obtained at these two sites are described. It appears that introduction of residues even slightly larger than glycine at position 146 disturbs the conformation of cyt b_L heme, leads to an empty Q_o site and consequently, hinders electron transfer and disables Ps^+ growth. In contrast, Y147A and S define a new class of Q_o site mutations which lower dramatically the single turnover Q_o site kinetics and bc_1 complex activity without affecting the binding of Q/QH_2. These findings indicate that they must affect a step subsequent to the binding of the substrate to Q_o site. Interestingly, the effect of these mutations can be restored by a second mutation (M->V or I) at position 154 of cyt *b*, only six amino acids away. Considering

	140	146 147		154	163
Bacteria					
R. capsulatus	M G T A F M	**G Y**	V L P W G Q	**M**	S F W G A T V I T
R. rubrum	M G T A F M	**G Y**	V L P W G Q	**M**	S F W G A T V I T
Mitochondria					
S. cerevisiae	I A T A F L	**G Y**	C C V Y G Q	**M**	S H W G A T V I T
H. sapiens	M A T A F W	**G Y**	V L P W G Q	**M**	S F W G A T V I T
M. musculus	M A T A F W	**G Y**	V L P W G Q	**M**	S F W G A T V I T
L. smithii	M A T A F V	**G T**	M L P - - -	-	- - - - - - - - -
P. aurelle	Q V V A F F	**G L**	V L C C T H	**L**	E S I T L T I A A
Chloroplast (b_6)					
S. oleracea	A S F G V T	**G Y**	S L P W D Q	**I**	G Y W A V K I V T
C. protothecoides	V S F G V T	**G Y**	S L P W D Q	**I**	G Y W A V K I V T
Nostoc	V S F G V T	**G Y**	S L P W D Q	**V**	G Y W A V K I V S

Figure 2. Cyt b sequences from selected species between the positions 140 and 163 of Q_oI region.

that a single substitution at position 154 from M to V has no significant effect on the bc_1 complex activity, the role of a large residue at position 147 may be to control the hydrophilicity of the Q_o site to allow maximum rates of electron transfer during ubihydroquinone oxidation.

2.Procedures

2.1 *Genetic Methods*

Site-directed mutants were constructed as described in (7). After mutagenesis, a 500 bp fragment was excised from the RF form of the appropriate phage M13, and subcloned into an expression vector (pMTS1) carrying the entire *fbcFBC* operon. Then plasmids with the appropriate mutations were conjugated into a bc_1-deficient *R. capsulatus* host strain (MT-RBC1) and mutations were reconfirmed by sequencing these pMTS1/MT-RBC1 derivatives. Ps^+ spontaneous revertants from the Ps^- mutants Y147A and S were isolated from chemoheterotropically grown cultures, and mutations were defined by DNA sequencing after transformation into appropriate strains (14).

2.2 *Biochemical and Biophysical Methods*

Chromatophore membranes were prepared in 50 mM MOPS buffer pH 7.0 containing 100 mM KCl using a French pressure cell as described earlier (7), except that they were washed three times. Protein measurements, SDS-PAGE and reduced *minus* oxidized optical difference spectra were performed as described in (7). Single-turnover kinetics of cyt b_H reduction and cyt *c* oxidation and determination of reaction center concentration were performed using a dual-beam flash-photolysis spectrophotometer (Biomedical Instrumentation Group, University of Pennsylvania) as described in (15). EPR spectroscopy was carried out using a Bruker model ESP-300E equipped with a helium cryostat as described earlier (16).

3.Results and Discussion

The G146A and V substitutions yielded Ps^- mutants of which only the latter reverted to Ps^+ at a low frequency, however no supressor mutations were encountered among them. Initially G146V produced a mixture of small (5 mm) and large (12 mm) colonies under respiratory growth conditions, whereas G146A always yielded large colonies. Upon restreaking the small colonies of G146V again yielded a mixture of small and large colonies. Furthermore, isolation and recloning of the *fbcFBC* operon from these small and large colonies indicated that the mutation affecting the colony size was confined to the *fbc* operon. However, G146A and V were virtually identical in respect to their spectral and kinetic properties, thus these mutations which probably have no major effect on the bc_1 complex, were not further pursued. In the case of Y147 substitutions, while Y147A and S were Ps^-, Y147F and V were Ps^+. The latter mutants grew slightly slower than a wild type strain, while the former reverted to Ps^+ growth with an approximate frequency of 10^{-7} to 10^{-8}. DNA sequence analysis revealed that these revertants contained an additional substitution changing M154 to V or I. Note that, these latter substitutions are naturally present in the b_6f complexes, while Y147 is very highly conserved (Figure 2). Finally, based on the data obtained using SDS-PAGE, western blot analyses and optical difference spectra all of the G146 and Y147 substitutions yielded properly assembled bc_1 complexes at a wild type level (not shown). Steady-state DBH: cyt *c* reductase activities of the G146A and V, and Y147A and S mutants were low (approximately 10 % of that of a wild type strain). All of the others, including the Ps^+

mutants Y147F and V, had approximately 20 to 40 % of the wild type activity (Table 1). Potentiometric titrations of the chromatophore membranes isolated from these mutants indicated that the redox mid-point potentials ($E_{m,7}$) of the cyt b_H and b_L hemes remained essentially unaffected (Table 1). Further, the EPR spectrum of the [2Fe2S] cluster of the Ps^- mutants Y147A and S, as well as the Ps^+ mutants Y147F and Y147V were identical to that of a wild type strain, indicating that these mutations did not affect significantly the binding of QH_2/Q to the Q_o site (Figure 3).

Table 1. Phenotypic Analysis of the G146, Y147 and M154 mutants.

Strain	Phenotype	Doubling Time (min^{-1})	% DBH Activity	b_L (mV)	b_H (mV)	Inhibitor resistance
pMTS1	Ps^+	150	100 [a]	-126	71	Myx^S
MT-RBC1	Ps^-	na	<1	na	na	na
pBG146A	Ps^-	na	10	-126	56	na
pBG146V	Ps^-	na	10	-120	73	na
pBY147F	Ps^+	180	43	-100	66	Myx^S
pBY147V	Ps^+	200	23	-108	71	Myx^R [b]
pBY147S	Ps^-	na	14	-110	68	na
pBY147A	Ps^-	na	13	-132	63	na
pBY147S+ M154I	Ps^+	nd	nd	nd	nd	Myx^R
pBY147A+ M154V	Ps^+	nd	nd	nd	nd	Myx^R
pBM154V	Ps^+	nd	nd	nd	nd	Myx^R [b]

[a] Based on the value of 4222 nmol cyt *c* reduced per mg membrane protein per min.
[b] Resistance is not strong. na= not applicable; nd=not determined.

On the other hand, flash-induced kinetic studies indicated that in these mutants single turnover electron transfer rates were dramatically lowered (20-30 fold lower than that of a wild type strain) (Figure 4 and Table 2).

Table 2. Single turnover kinetics of the G146, Y147 and M154 mutants.

Strain	Q_o->b_H (s^{-1})	E_h (mV)	cyt *c* red. (s^{-1})	E_h (mV)
pMTS1	553	*110*	338	*106*
pBG146A	15	*100*	21	*100*
pBG146V	16	*100*	16	*100*
pBY147F	233	*100*	108	*100*
pBY147V	200	*100*	87	*100*
pBY147S	25	*102*	14	*102*
pBY147A	31	*103*	20	*103*
pBY147S+ M154I	240	*100*	117	*103*
pBY147A+ M154V	210	*106*	98	*106*
pBM154V	475	*106*	181	*108*

red.=reduction

Further analyses with the bc_1 complex inhibitors antimycin, myxothiazol and stigmatellin indicated that electron transfer still takes place from the [2Fe2S] cluster to cyt c_1, whereas

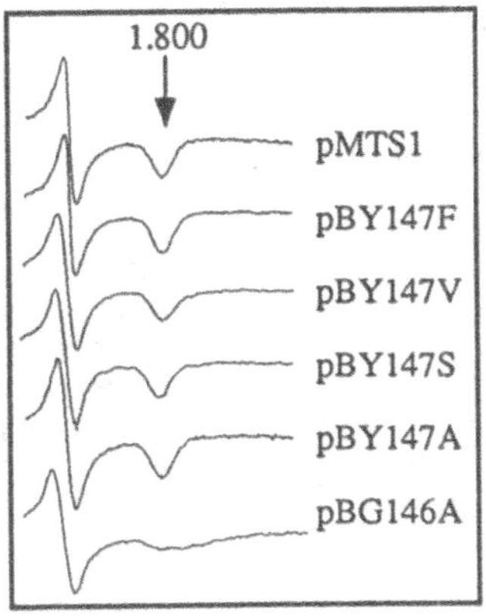

Figure 3. EPR spectra of the [2Fe2S] cluster of Y147 and G146 mutants (Eh=200 mV)

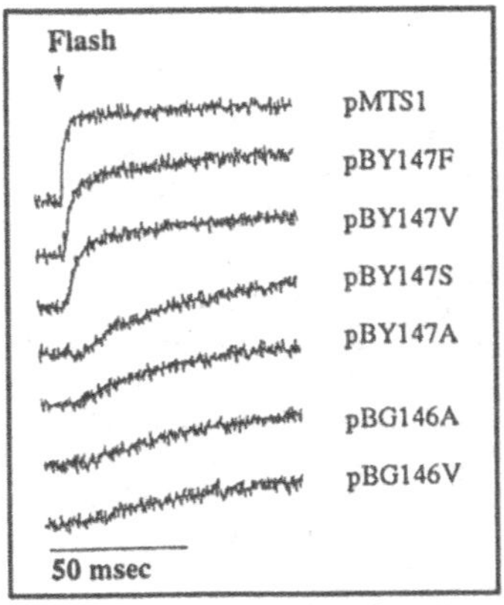

Figure 4. Qo site kinetics of Y147 and G146 mutant in the presence of antimycin (Eh=100 mV)

those from Q_o to the [2Fe2S] cluster and Q_o to cyt b_L were impaired (Figure 5). Thus, small amino acid side chains at position 147 affect Q_o site catalysis without perturbing the binding of QH_2/Q to the Q_o site. Interestingly, these effects were compensated by substituting the position M154 with V or I, only six residues away. These additional substitutions restored electron transfer rates to approximately 40-50% of that of a wild type strain (Table 2). Unlike the Y147A and S mutants the G146A substitution has an EPR spectrum of the [2Fe2S] cluster with a broad g_x signal, indicating a decreased Q_o site occupancy by QH_2/Q (Figure 3). Flash-induced kinetics obtained with this latter mutant also indicated that electron transfer was dramatically lower than that of a wild type strain (Table 2).

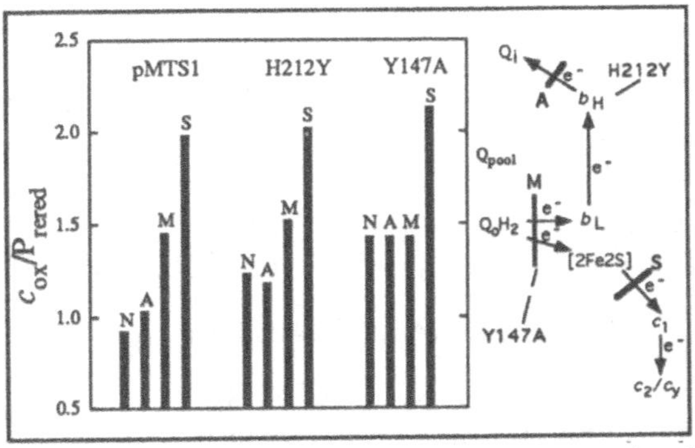

Figure 5. Ratio of Cyt c oxidized / reaction center rereduced in the presence of selected inhibitors. N, no inhibitor; A, antimycin; M, myxothiazol; S, stigmatellin; pMTS1, wt; H212Y, cyt bH deficient strain (Daldal and Gray, pers. comm.) and Y147A (Saribas et al., 1995).

Finally, EPR spectroscopy of the cyt *b* hemes revealed that only the G146A and V mutations affected them (Figure 6). In particular, the sharp signal with a g value of 3.791 seen in the wild type strain was modified in both G146A and V mutants and shifted to approximately g=3.713 and 3.722, respectively. Disturbances in the cyt b_H signal were rather minor, indicating that these mutations affect predominantly the conformation and packing of the cyt b_L heme.

In summary, the two different classes of mutants described in this study revealed invaluable information about the role of the amino acid side chains contributing to the Q_o site functions. Clearly, G146 is located near the cyt b_L heme binding pocket, and a glycine at this position provides optimal spatial comfort to the porphyrin ring of this heme group. Introduction of even an additional methyl group to this position affects the orientation of the heme group as evidenced by EPR spectra. Once the proper three dimensional conformation of this redox center is disturbed, and the Q/QH_2 binding to the Q_o site is affected, then electron transfer

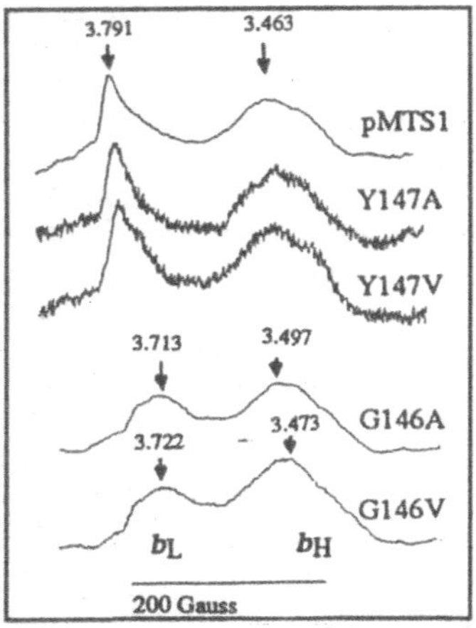

Figure 6. EPR spectra of the cyt bH and bL hemes of the Y147 and G146 substitutions.

no longer takes place efficiently, leading to a defective bc_1 complex. On the other hand, Y147 affects some step(s) of electron transfer from Q_o to the [2Fe2S] cluster and to cyt b_L without perturbing the binding of Q/QH_2 to the Q_o site. This position requires a large amino acid side chain to possibly form a bridge on the electron transfer path from the Q_o site to the nearest redox center. When this side chain is substituted with a smaller one then this bridge is perturbed causing a dramatic decrease in the rate of electron transfer, consequently leading to a defective bc_1 complex. Future analysis of additional residues with similar properties should further contribute to our understanding of the functional mechanism of the Q_o site.

Acknowledgments

This work was supported by NIH grant GM 38237.

References

1 Cramer, W.A., Martinez, S.E., Huang, D., Tae, G.-S., Everly, R.M., Heyman, J.B., Cheng, R.H., Baker, T.S. and Smith, J.L. (1994) J. Bioenerg. Biomembr. 26, 31-47
2 Gray, K.A. and Daldal, F. (1995) in Anoxygenic Photosynthetic Bacteria (Blankenship, R.E., Madigan, M.T. and Bauer, C., Eds.) pp.747-774, Kluwer Academic Publishers, Dordrecht, The Netherlands
3 Degli-Esposti, M.D., De Vries, S., Crimi, M., Gelli, A., Patarnello, T. and Meyer, A. (1993) Biochim. Biophys. Acta 1143, 243-271
4 Berry, E.A., Huang, L.-S., Earnest, T.N. and Jap, B.K. (1992) J. Mol. Biol. 224, 1161-1166
5 Yu, C.A. and Yu, L. (1993) J. Bioenerg. Biomembr. 23, 259-273
6 Daldal, F., Tokito, M.K., Davidson, E. and Faham, M. (1989) EMBO J. 8, 3951-3961
7 Atta-Asafo-Adjei, E. and Daldal, F. (1991) Proc. Nat. Acad. Sci. U.S.A. 88, 492-496
8 Tokito, M.K., and Daldal, F. (1993) Mol. Microbiol. 9, 965-978
9 Robertson, D.E., Daldal, F. and Dutton, P.L. (1990) Biochemistry 29,, 11249-11260
10 Ding H., Daldal, F. and Dutton, P.L. (1995) submitted.
11 Tron, T., Infossi, P., Coppee, J-Y. and Colson, A.M. (1991) FEBS Letters, 278, 26-30
12 Lemesle-Meunier, D., Brivet-Chevillotte, P., Di Rago, J.P., Slonimski, P.P., Bruel, C., Tron, T. and Forget, N. (1993) J. Biol. Chem. 268, 15626-15632
13 Bennoun, P., Delosme, M., Godehart, I. and Kück, U. (1992) Mol. Gen. Genet. 234, 147-154
14 Saribas, A.S., Ding, H., Dutton, P.L. and Daldal, F. (1995) submitted
15 Gray, K.A., Dutton, P.L. and Daldal, F. (1994) Biochemistry 33, 723-733
16 Robertson, D.E., Ding, H., Chelminski, P.R., Slaughter, C., Hsu, J., Moonraw, C., Tokito, M., Daldal, F. and Dutton, P.L. (1993) Biochemistry 32, 1310-1317

PURIFICATION AND COMPOSITION OF THE CYTOCHROME b_6f COMPLEX FROM *CHLAMYDOMONAS REINHARDTII*

Jean-Luc Popot[1], Yves Pierre[1], Cécile Breyton[1], Yves Lemoine[2], Yuichiro Takahashi[3] and Jean-David Rochaix[4]

[1]Institut de Biologie Physico-Chimique and Collège de France, C.N.R.S. URA 1187, 11 rue Pierre et Marie Curie, F-75005 Paris, France. [2]Université de Lille & Ecole Normale Supérieure, 46 rue d'Ulm, F-75230 Paris Cedex 05, France. [3]Department of Biology, Okayama University, Okayama, Japan. [4]Département de Biologie Moléculaire & de Biologie des Plantes, Université de Genève, CH-12111 Genève, Suisse.

1. Introduction

Most structural studies of the b_6f complex to date have been carried out either on higher plants or on cyanobacteria (for reviews, see refs. 1-4). Unicellular algae, such as *Chlorella sorokiniana* and *Chlamydomonas reinhardtii*, have long been recognized as convenient organisms for combined spectroscopic, biochemical and genetic studies of the b_6f complex structure and function (5). In the present article, we describe the purification and composition of the b_6f complex from *C. reinhardtii*.

2. Results and Discussion

2.1 Purification protocol

Treatment of thylakoid membrane preparations from *C. reinhardtii* with the neutral detergent Hecameg (HG) preferentially solubilizes the CF_0CF_1 ATP synthase and the cytochrome b_6f complex, resulting in a *ca.* 3-fold increase of the b_6f/protein ratio (Fig. 1, lane 2). Assessment of the distribution of the b_6f complex in this and subsequent fractionation experiments was based on that of two heme-stainable proteins recognized previously as cytochromes f and b_6 (6, 7). Fractionation of the supernatant on a 10-30% sucrose gradient in the presence of HG and lipids

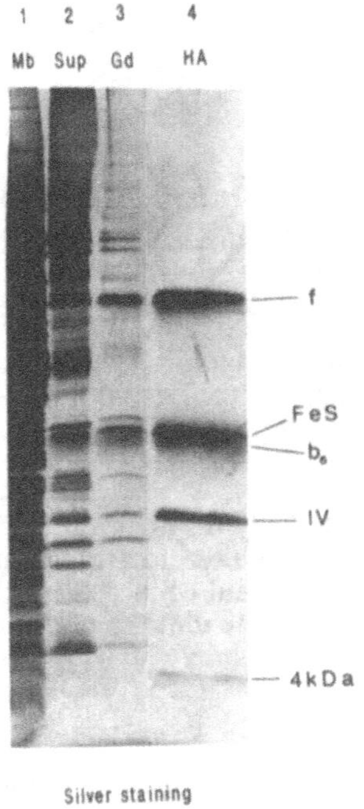

FIG. 1. **Urea/SDS-PAGE and immunoblot analysis of *Chlamydomonas reinhardtii* b_6f complex preparations.** Polypeptide composition of **(1)** thylakoid membranes prepared according to refs. 22-23, **(2)** supernatant following solubilization of thylakoid membranes (1.5 g/l chlorophyll) with 25 mM HG, **(3)** b_6f-enriched fraction from sucrose gradient, **(4)** purified b_6f complex eluted from a hydroxylapatite column. From ref. 8.

P. Mathis (ed.), Photosynthesis: from Light to Biosphere, Vol. II, 507–512.

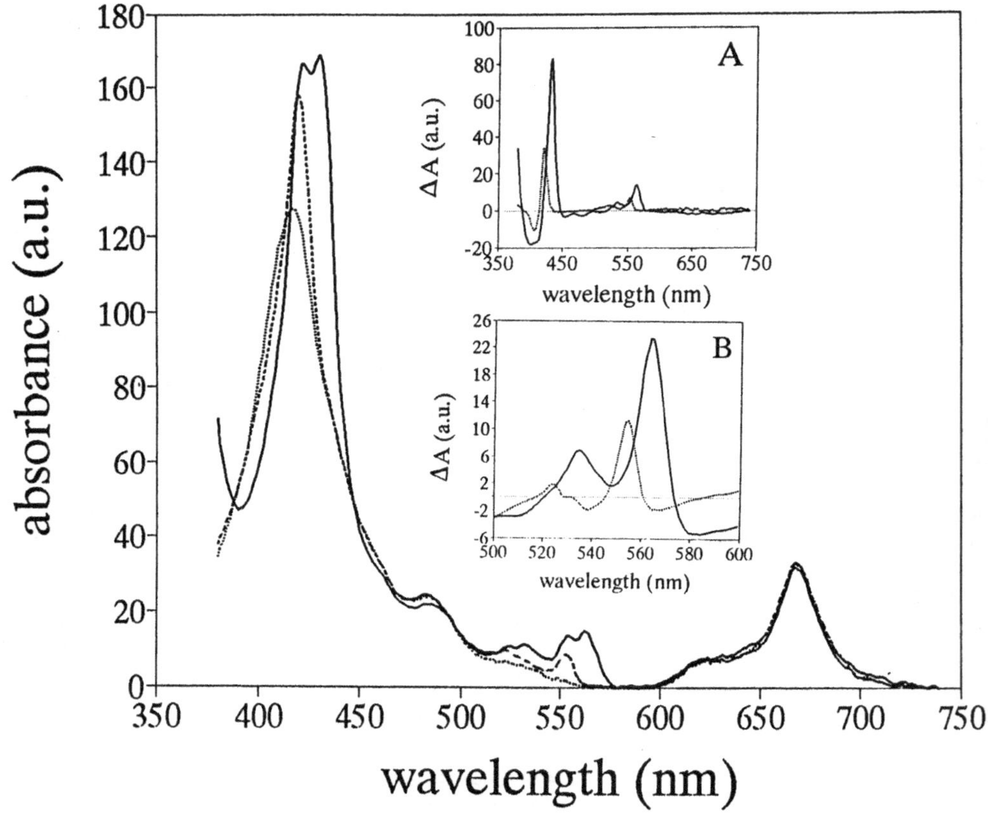

FIG. 2. **Absorbance spectra of purified b_6f preparations.** Oxidized (....), ascorbate-reduced (----) and dithionite-reduced (——) near UV/visible spectra of a purified preparation. The spectrum of oxidized b_6f was calculated by subtracting from the spectrum of air-equilibrated b_6f the contribution of reduced cytochrome *f*. *Inset A* : near UV/visible difference spectra : ascorbate-reduced minus air-oxidized (....) and dithionite-reduced minus ascorbate-reduced (——). *Inset B* : visible difference spectra (from a different b_6f preparation) : ascorbate-reduced minus ferricyanide-oxidized (----) and dithionite-reduced minus ascorbate-reduced (——). From ref. 8.

revealed that 18.5, 12.5 and ~4 kDa proteins comigrate with the two cytochromes (Fig. 1, lane 3). Further purification on a hydroxylapatite column resulted in a preparation devoid of any other polypeptides (Fig. 1, lane 4). The overall yield of the purification was close to 50%, a 20-l growth of wild type *C. reinhardtii* cells yielding 5-7 mg of pure complex (8).

2.2 Spectroscopic properties

The near UV/visible spectrum of a purified b_6f preparation is shown in Fig. 2. Absorbance peaks in the Soret, α and β regions of the spectrum are characteristic of the presence of cytochrome *f*. Additional bands at ~460 and 483 nm indicate the presence of carotenoids and a band at 667-668 nm that of chlorophyll *a*. Reduction with dithionite or borohydride reveals the presence of cytochrome b_6. Difference spectra (Fig. 2, insets) show absorption maxima at 523 and 554 nm for cytochrome *f* and 534 and 564 nm for cytochrome b_6, and isosbestic points in the Soret region at 414 and 420 nm, respectively. The b_6 to *f* ratio, determined from redox titration curves, is ~1.9:1, indicating that none of the two *b*-type hemes has been lost during the preparation (8).

2.3 Electron transfer activity

The purified preparation actively catalyzed electron transfer from decylplastoquinol to spinach plastocyanin (5 μM). Under our experimental conditions, one molecule of cytochrome *f* transfered 270 ± 60 electrons per second (8). This rate is >3-fold higher than the highest rates hitherto reported for purified higher plant b_6f preparation under comparable conditions (see refs. 9, 10, and references therein). It is similar to that estimated *in vivo* (see refs. 2, 11). Electron transfer was totally inhibited by the addition of 1 μM C_{13}-stigmatellin (8).

2.4 Polypeptide composition

The subunit composition of the purified complex was studied by SDS-PAGE using either gradient gels in the presence of urea (12) or a 3-step gel system devoid of urea (13). Staining with silver revealed five bands in the first system, with apparent M_r 39.5, 18.5, 18, 12.5 and *ca.* 4 kDa (Fig. 1, lane 4). The 4-kDa band often appeared heterogeneous. In the discontinuous, 3-step gel system, the 4-kDa band was resolved into 2, sometimes 3 or 4 bands of variable intensity (14). The pattern observed depended on currently

TABLE

Polypeptide composition of purified *C. reinhardtii* b_6f complex

Apparent M_r (kDa) (SDS-PAGE)[a] Urea	3-step	Protein identity	Actual M_r (kDa)[b]	Criteria for identification: TMBZ staining	N-terminal sequencing	Immunolabeling[c] N-term.	C-term.	other
39.5	44	cyt. *f*	31.8	+	+[d]	+[d]	+[d]	
18.5	22	Rieske	18.6	-	+[d]	+[e]	*n.d.*	
18	26	cyt. b_6	25.4	+	blocked	+[d]	+[d]	
12.5	17	sub. IV	17.4	-	blocked	+[d]	+[d]	
~4	4.5-5.5	PetX	4.0	-	+[f]	+[f]	+[f]	
"	"	PetG	4.0	-	blocked[f,g]	*n.d.*	+[f]	
"	"	PetL	3.4	-	blocked[h]	*n.d.*	*n.d.*	+[h]

[a] Apparent M_r on 12-18% polyacrylamide/urea gels (12) or on 3-step polyacrylamide gels (13); the pattern of migration of the small subunits in the latter system was variable, and no definitive assignments are proposed; [b] from nucleic acid sequence data, taking into account known posttranslational modifications (processing and cofactors); [c] immunolabeling of SDS-PAGE replicas with antipeptide antisera, unless otherwise indicated; peptide sequences are given in ref. 8; [d] data from ref. 8; [e] data from ref. 15; [f] data from ref. 14; [g] found unblocked in ref. 17; [h] antiserum to recombinant PetL; data from ref. 16; *n.d.* : not determined.

incompletely identified factors that affect the migration and staining of the subunits.

Identification of the bands was based on their apparent M_r, heme content as revealed by TMBZ staining, N-terminal sequencing and cross-reaction with antisera. It is summarized in the TABLE (data from refs. 8, 14-16). The high-MW bands correspond to cytochrome *f*, the Rieske iron-sulfur protein, cytochrome b_6 and subunit IV, in this order, as established previously (6, 7). The 4-kDa band comprises three small polypeptides, PetG (14, 17), PetX (14, 17, 18) and PetL (16). Immunoblotting of SDS-PAGE gels shows that all of these seven proteins are absent in thylakoid membranes prepared from mutant FuD4, that lacks the b_6f complex (8, 14-16).

2.5 Subunit stoichiometry

In order to examine subunit stoichiometry, b_6f complex was purified from *C. reinhardtii* cells grown on [$^{14}C_1$]acetate. The ratio of ^{14}C activity in the bands containing cytochrome *f*, Rieske protein + cytochrome b_6, subunit IV and the three ~4-kDa subunits, normalized to the number of carbon atoms per protein(s), was 1.02 ± 0.08, 1.03 ± 0.04, 0.89 ± 0.05 and 0.99 ± 0.08, respectively. These data are compatible with all seven subunits being present in equimolecular amount. The protein content of the purified preparations (112 ± 9 mg per mole of cytochrome *f*) is very close to that expected assuming this stoichiometry (103 mg per mole). All seven subunits (15, 16, 18) but the Rieske protein (15) are integral subunits. Our current view of the b_6f complex polypeptide composition and transmembrane topology is summarized in Fig. 3.

As described elsewhere, the purified b_6f complex from *C. reinhardtii* is a dimer whose experimentally determined molecular weight (300 ± 25 kDa including bound detergent) is close to that expected from its composition (344 ± 10 kDa) (8, 19). Upon delipidation, the dimer monomerizes and loses the Rieske protein (8, 19).

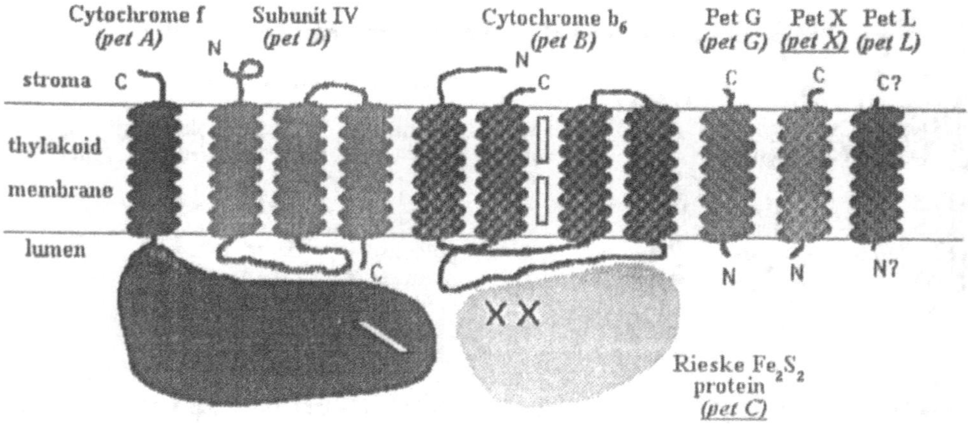

FIG. 3. **Subunit composition and topology of *C. reinhardtii* cytochrome b_6f.** Subunit transmembrane topologies are based on earlier results obtained on b_6f complexes from other sources (cytochromes *f* and b_6 and subunit IV; *cf.* ref. 4), on our analysis of the mode of membrane association of the Rieske protein (15), and on the sensitivity to proteolysis and mode of membrane association of PetG (15, 24), PetL (16) and PetX (18).

2.6 Origin of the chlorophyll a *molecule*
Purified preparations of b_6f complex from *C. reinhardtii* contain about one molecule of chlorophyll *a* per cytochrome *f*, as previously observed in *Synechocystis* PCC6803 (20) and spinach (10) b_6f preparations. There are ~900-1000 chlorophyll molecules per cytochrome b_6f complex in thylakoid membranes from *C. reinhardtii* (8), 99.9 % of which are removed during the purification. The single chlorophyll found associated with the purified complex therefore may well be a contaminant. A number of indirect arguments however run against this view, among which : *i)* the chlorophyll *a*/cytochrome *f* ratio, although somewhat variable from one preparation to another, is generally very close to 1 (0.90 ± 0.15 over 10 preparations), suggesting the presence of an unique high-affinity site per complex; *ii)* very little free chlorophyll is found in the supernatant following solubilization of the membranes with HG (<0.2 molecule per cytochrome *f*); *iii)* b_6f complex purified from strongly etiolated *C. reinhardtii* cells also contains chlorophyll (YP and JLP, unpublished observations).

In order to directly examine whether the b_6f-bound chlorophyll originates from free chlorophyll released upon membrane solubilization, *C. reinhardtii* thylakoid membranes were solubilized using HG micelles containing [^{3}H]chlorophyll *a*. The supernatant was fractionated and the b_6f complex purified. The specific activity of b_6f-bound chlorophyll was found to be 10-40-fold lower than that of free, micelle-bound chlorophyll. While preliminary, these observations further support the view that chlorophyll *a* is associated with the b_6f complex prior to solubilization and that little association of or exchange with the radioactive chlorophyll occurs in the course of the purification. They strongly suggest that a molecule of chlorophyll *a* is an authentic component of the b_6f complex.

It is interesting to note that, *in vivo*, the turnover of the b_6f complex in *Chlorella sorokiniana* and in *C. reinhardtii* cells generates a spectral shift typical of the electrostatic perturbation of a chlorophyll (21). This observation has been interpreted as reflecting the presence of a chlorophyll molecule in the immediate vicinity of the site of proton release at the Q_o site of cytochrome b_6f (21).

3. Conclusion

The purification protocol developed for the present work provides preparations of the b_6f complex from *C. reinhardtii* that are both highly pure and highly active in electron transfer. Biochemical and spectroscopic characterization of the purified complex asserts its great similarity to higher plant b_6f, and, therefore, its usefulness as an experimental model. Some of the conclusions from the present work presumably can be extended to higher plant b_6f. This most likely applies to the presence of three ~4 kDa integral subunits, PetG, PetL and PetX, and of a molecule of chlorophyll *a*. The latter seems to preexist to solubilization and therefore is likely an authentic component of the b_6f complex. The purity and yield of the preparations have made it possible to undertake 2D and 3D crystallogenesis experiments.

4. Acknowledgments

This research has been supported by the Centre National de la Recherche Scientifique, the Collège de France, the Commissariat à l'Energie Atomique, and by grants from the EEC (BIO2-CT93-0076), from the Ministère de la Recherche et de la Technologie (87.C.0385) and from IMABIO to JLP.

5. References

1. Malkin, R. (1992) *Photosynth. Res.* **33**, 121-136.
2. Hope, A.B. (1993) *Biochim. Biophys. Acta* **1143**, 1-22.
3. Trumpower, B.L., & Gennis, R.B. (1994) *Annu. Rev. Biochem.* **63**, 675-716.
4. Cramer, W.A., Martinez, S.E., Furbacher, P.N., Huang, D., & Smith, J.L. (1994) *Curr. Opin. Struct. Biol.* **4**, 536-544.
5. Harris, E.H. (1989) *The Chlamydomonas sourcebook. A comprehensive guide to biology and laboratory use.* (First ed.). Academic Press, San Diego.
6. Lemaire, C., Girard-Bascou, J., Wollman, F.-A., & Bennoun, P. (1986) *Biochim. Biophys. Acta* **851**, 229-238.
7. Wynn, R.M., Bertsch, J., Bruce, B.D., & Malkin, R. (1988) *Biochim. Biophys. Acta* **935**, 115-122.
8. Pierre, Y., Breyton, C., Tribet, C., Kramer, D., Olive, J., & Popot, J.-L. (1995) *Submitted for publication.*
9. Black, N.T., Widger, W.R., & Cramer, W.A. (1987) *Arch. Biochem. Biophys.* **252**, 655-661.
10. Huang, D., Everly, R.M., Cheng, R.H., Hetmann, J.B., Schägger, H., Sled, V., Ohnishi, T., Baker, T.S., & Cramer, W.A. (1994) *Biochemistry* **33**, 4401-4409.
11. Cramer, W.A., Black, M.T., Widger, W.R., & Girvin, M.E. (1987) In *The Light Reactions* (Barber, J., ed.), pp.447-493, Elsevier, Amsterdam.
12. Piccioni, R.G., Bennoun, P., & Chua, N.-H. (1981) *Eur. J. Biochem.* **117**, 93-102.
13. Schägger, H., & von Jagow, G. (1987) *Anal. Biochem.* **166**, 368-379.
14. Pierre, Y., & Popot, J.-L. (1993) *C. R. Acad. Sci. (Paris), Ser. III* **316**, 1404-1409.
15. Breyton, C., de Vitry, C., & Popot, J.-L. (1994) *J. Biol. Chem.* **269**, 7597-7602.
16. Takahashi, Y., Rahire, M., Breyton, C., Popot, J.-L., Joliot, P., & Rochaix, J.-D. (1995) *Submitted for publication.*
17. Schmidt, C.L., & Malkin, R. (1993) *Photosynth. Res.* **38**, 73-81.
18. de Vitry, C., Breyton, C., Pierre, Y., & Popot, J.-L. (1995) *These proceedings.*
19. Breyton, C., Tribet, C., Olive, J., Recouvreur, M., & Popot, J.-L. (1995) *These proceedings.*
20. Bald, D., Kruip, J., Boekema, E.J., & Rögner, M. (1992) In *Research in Photosynthesis. Proceedings of the IXth International Congress on Photosynthesis* (Murata, N., ed.), Kluwer Academic Publishers, Nagoya, pp. 629-632.
21. Joliot, A., & Joliot, P. (1995) *These proceedings.*
22. Chua, N.H., & Bennoun, P. (1975) *Proc. Natl. Acad. Sci. USA* **72**, 2175-2179.
23. Atteia, A., de Vitry, C., Pierre, Y., & Popot, J.-L. (1992) *J. Biol. Chem.* **267**, 226-234.
24. Haley, J., & Bogorad, L. (1989) *Proc. Natl. Acad. Sci. USA* **86**, 1534-1538.

ELECTROGENIC REACTIONS IN THE HIGH-POTENTIAL SEGMENT OF THE CYTOCHROME bc_1 - COMPLEX

A.Yu. Semenov, D.A. Bloch, O.A. Gopta and N.B. Ugulava
A.N. Belozersky Institute of Physico-Chemical Biology, Moscow State University, 119899 Moscow, Russia

1. Introduction

The electron transfer function of the cytochrome bc_1-complex is accompanied by generation of the transmembrane electric potential difference ($\Delta\psi$) as part of the proton pumping activity. Our knowledge of electrogenic events within the bc_1-complex of purple bacteria is derived essentially from measurements of the carotenoid electrochromic absorption changes **(1,2)**. By means of this method it has been shown that electrogenic reactions in the bc_1- complex are partly suppressed by antimycin A, an inhibitor of the quinone- reducing site (Q_c or Q_i), and completely by myxothiazol, an inhibitor of the quinol-oxidizing site (Q_z or Q_o) **(2,3)**. In the presence of myxothiazol under alkaline conditions a negative electrogenic reaction is observed **(3-5)**.

The results obtained by carotenoid bandshift assay have led to the proposal of a general model of electrogenic reactions in the bc_1-complex **(3)**. According to the model proposed by Robertson and Dutton, there are two main electrogenic reactions, namely: (i) electron transfer between the low (b_l) and the high-potential (b_h) cytochrome b hemes, and (ii) a subsequent electron transfer between reduced b_h and the ubiquinone molecule at the quinone- reducing Q_i-site. In accordance with this scheme both the Q_i- and Q_o-sites should be located at or close to the borders of the membrane dielectric. Consequently, both protonation of the doubly reduced Q^{2-} at the Q_i-site, and deprotonation of QH_2 at the Q_o- site should be non- electrogenic reactions.

The alternative model **(6)**, which is mainly based on the analysis of accessibility of b hemes to membrane-impermeable redox agents **(7)**, predicts a different arrangement of redox-centres in the bc_1-complex. According to this model, the total electrogenesis within the bc_1-complex is due to (i) electron transfer between b_l and b_h and (ii) antimycin- sensitive deprotonation of QH_2 at the Q_o-site. Correspondingly, Q_i and b_h should be located at the outer surface of the chromatophore membrane, whereas Q_o and b_l should reside approximately in the middle of the membrane dielectric.

Electrogenic events during electron transfer via the cytochrome bc_1-complex of *Rhodobacter sphaeroides* chromatophores have been studied electrometrically and the data obtained suggest that the total electrogenesis within the bc_1-complex includes: (i) electron transfer between the low- and high- potential cytochrome b hemes; (ii) proton

P. Mathis (ed.), Photosynthesis: from Light to Biosphere, Vol. II, 513–518.

binding on reduction of quinone at the Q_i- site and (iii) proton release on oxidation of QH_2 at the Q_o-site **(8-10)**.

The electrogenic nature of proton release on oxidation of QH_2 predicts that Q_o-site is somewhat buried in the protein dielectric. Hence, the electron transfer via Rieske FeS centre and cyt c_1 to cyt c_2 should compensate the electric potential increase due to the release of one of the two protons.

In the present work, the possible electrogenic reactions in the high-potential branch of the bc_1-complex have been investigated by the direct electrometry in *Rb. sphaeroides* chromatophores and in proteoliposomes containing purified mitochondrial bc_1-complexes. In the latter case, the flash-induced chemical photoreduction (FIRE) technique has been applied.

2. Procedure

Chromatophores were isolated by French press disruption from the cells of *Rb. sphaeroides* grown anaerobically according to **(8)**. The incubation mixture contained 20 mM Hepes, pH 7.5, 50 mM KCl, 2 mM potassium ferrocyanide, 2 mM potassium cyanide, 2 μM methylene blue, 2 μM antimycin A, 1 μM myxothiazol, 50 μM terbutrine. The ambient redox potential varied in the range +180 - +320 mV.

Cytochrome bc_1-complex was isolated from bovine heart mitochondria essentially according to **(11)** and stored in liquid nitrogen until used. The bc_1-complex containing proteoliposomes were prepared by the cholate dialysis method. The incubation mixture contained 25 mM HEPES and 2 mM potassium EDTA (pH 7.0). Catalytic amounts of cyt c (0.1 μM) and mitochondrial cytochrome c oxidase (0.1 μM) were present to maintain the cyt c_1 in the oxidised state in the dark before the flash.

Measurements of membrane potential generation by either chromatophores or cytochrome bc_1-complex containing proteoliposomes adsorbed onto the surface of asolectin-impregnated collodion film were done and the kinetic data were processed as in **(8,12)**.

3. Results and Discussion

3.1. Cytochrome c_1 photoreduction in the mitochondrial bc_1-complex

The flash-induced chemical photoreduction (FIRE) technique was developed by Tollin and co-workers **(13)** to investigate the cyt c photoreduction by the flash-induced flavin semiquinone generation. Another FIRE system based on flash photolysis of phenazine methosulphate (PMS) was developed later **(14,15)**. This FIRE system has been used to study the mitochondrial cytochrome bc_1-complex, paticularly in relation to the electron transfer mechanism in the FeS Rieske - cyt c_1 part of the complex. The photochemistry of PMS is complicated but well understood **(16,17)**. On the basis of these data and the well-defined absorption spectra of PMS and its photoproducts **(18)** the reductant of cyt c has been identified as the neutral semiquinoid form of PMS (PMS-SQ) **(14)**.

In our recent work a similar approach has been used to investigate the flash-induced reduction of the initially oxidised cyt c_1 in the mitochondrial cytochrome bc_1- complex **(A.Yu. Semenov et al., unpublished)**. It has been shown that the laser flash induces the equimolar one-electron reduction of cyt c_1 by PMS-SQ. Similar results were obtained with the cytochrome bc_1-complex containing proteoliposomes.

In order to check the possible electrogenicity accompanying the reduction of cyt c_1 in the collodion film - bc_1-complex containing proteoliposomes system the direct electrometrical method has been used **(8,9)**. 100 µM of PMS was added into both compartments of the experimental cell, since this concentration was defined as optimal for our purposes. The photoelectric signal was smaller at a lower concentration of the redox dye, since this concentration was unsaturated. At higher PMS concentration the photoelectric response has been also diminished because of absorbance of the triggering light flash by the redox dye.

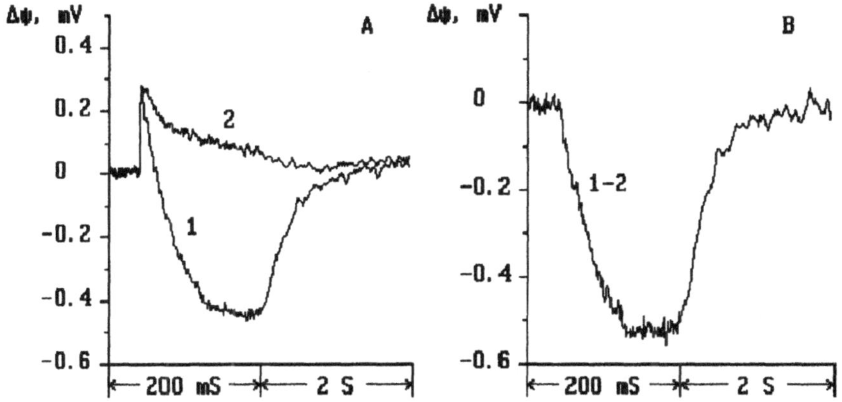

Fig. 1. Kinetics of the membrane potential generation by the FIRE of cyt bc_1-complex containing proteoliposomes. Incubation mixture:25 mM Hepes, 2 mM potassium EDTA, pH 7.0, 0.1 µM cyt c, 0.1 µM cytochrome-c oxidase, and 100 µM PMS. 1 - without ferrocyanide, 2 - after addition of 2 mM ferrocyanide.

Fig. 1 illustrates the membrane potential generation under the conditions described. The laser flash induced fast $\Delta\psi$ generation, positive inside proteoliposomes, followed by a slower phase of the opposite polarity (minus inside vesicles) (Fig.1A, trace 1). The addition of ferrocyanide reducing cyt c_1 resulted in disappearance of the slower negative phase without alteration of the fast positive one (Fig. 1A, trace 2). Fig. 1B presents the difference between the traces 1 and 2 from Fig. 1A (± ferrocyanide). The kinetics of $\Delta\psi$ generation presented in Fig. 1B is similar to the kinetics of the flash-induced cyt c_1 reduction measured by flash-spectrometry (not shown). The data obtained show that the fast positive phase of the $\Delta\psi$ generation which is still observable when cyt c_1 is reduced before the flash (Fig. 1A, trace 2) is due to some flash-induced PMS artifact, while the difference kinetic curve (Fig. 1B) reflects the electrogenesis due to the reduction of cyt c_1. Unfortunately, we were not able to calculate the amplitude of this phase in relation to the other phases.

3.2. FeS Rieske centre photooxidation in Rhodobacter sphaeroides chromatophores.

It was previously shown by the flash-induced carotenoid bandshift that there is no electrogenic reaction between the FeS Rieske centre, cyt c_1 and cyt c_2 **(3)**. This conclusion was made on the basis of the failure to observe any $\Delta\psi$ generation by the cytochrome bc_1 -complex in the presence of antimycin and myxothiazol.

It is noteworthy that the expected electrogenic stage in the presence of both inhibitors should be characterized by the opposite polarity to the other stages, since the only reaction that might proceed under these conditions is the electron transfer between FeS centre, cyt c_1 and cyt c_2 . Interestingly, the small negative phase was observed in the kinetics of the flash-induced carotenoid bandshift in the presence of antimycin and myxothiazol (see Fig. 2, Ref. **(3)**).

In order to study the possible electrogenicity of the electron transfer through the high-potential segment of the bc_1-complex, we applied the direct electrometrical assay. In the presence of antimycin and myxothyazol the charge transfer reactions in the bc_1 -complex are inhibited. The reaction between FeS centre, cyt c_1 and cyt c_2 is possible, if these redox centres are in the reduced state before the laser flash (i.e. at E_h lower, than ~+240 mV). In this case the electron transfer is initiated by the oxidation of a special pair bacteriochlorophyll P870.

As it is shown previously by both carotenoid bandshift **(1)** and direct electrometry (for review see **(19)**) the rereduction of $P870^+$ by cyt c_2 is electrogenic and compiles about 20% of the phase due to the charge separation between $P870^+$ and Q_A^-. On the other hand, the electrogenic reaction accompanying the release of two protons on oxidation of QH_2 at centre "o" compiles approx. 10 % of the primary charge separation **(10)**. Since the release of protons is accompanied by the electron transfer in the same direction, the latter should neutralize the electrogenesis provided by one of the two protons released. Consequently, the negative electrogenic step through the high-potential segment of the bc_1 -complex should also compile 10% of the primary charge separation. Thus, taking into account all these considerations and similar kinetics of electron transfer between FeS centre and cyt c_1 ($t_{1/2}$ = 200 μs), cyt c_1 and cyt c_2 ($t_{1/2}$ = 150 μs]), cyt c_2 and P870 ($t_{1/2}$ = 3 - 300 μs), depending on certain conditions (for review see **20**), this negative phase should be disguised by the positive one ascribed to the reduction of $P870^+$ by cyt c_2.

The E_m values of all the three redox centres are close. Therefore, it is difficult to switch off the centres by the successive redox titrations. The only possible way to influence the redox state of the FeS Rieske centre without any alteration of that of cyt c_1 and cyt c_2 is the action of chromone inhibitor stigmatellin, which is known to significantly increase the FeS Rieske midpoint potential **(21)**. It was also shown recently that stigmatellin addition results in complete reduction of the Rieske centre and disturbance of rapid equilibrium between the FeS and cyt c_1 **(A.Yu. Semenov et al., unpublished)**. Thus under certain conditions stigmatellin which prevents the FeS-----> cyt c_1 electron transfer seems to be a useful tool for clarification of the possible electrogenicity of this reaction.

The experiments have shown that at E_h = +200 mV in the presence of antimycin and myxothiazol, i.e. when QH_2 oxidation by the bc_1-complex was completely inhibited, stigmatellin addition did not influence the $\Delta\psi$ generation measured electrometrically. Thus, in accordance with the previous observation **(3)** the electron transfer from the FeS Rieske centre to cyt c_1 seems to be electrically silent.

3.3 Arrangement of redox centres.

The question arises: how to explain this result in the framework of the model **(6-10)** which places the Q_o - site in the hydrophobic region of the protein? It is worth while to consider in this connection the partial reactions of the electron transfer in the high-potential branch on oxidation of QH_2 at centre "o". The overall process includes the following reactions:

(1) oxidation of QH_2 by FeS;
(2) oxidation of FeS by cyt c_1;
(3) oxidation of cyt c_1 by cyt c_2.

The result obtained only excludes reaction (2) from consideration. The electrogenic nature of reaction (3) follows from the experiment presented in Fig. 1. As to reaction (1), it seems impossible to reveal this partial reaction from the others. In conclusion: according to the data obtained there are two alternative arrangements of the redox centres in the high-potential branch of the cytochrome bc_1 - complex which are presented in Fig. 2.

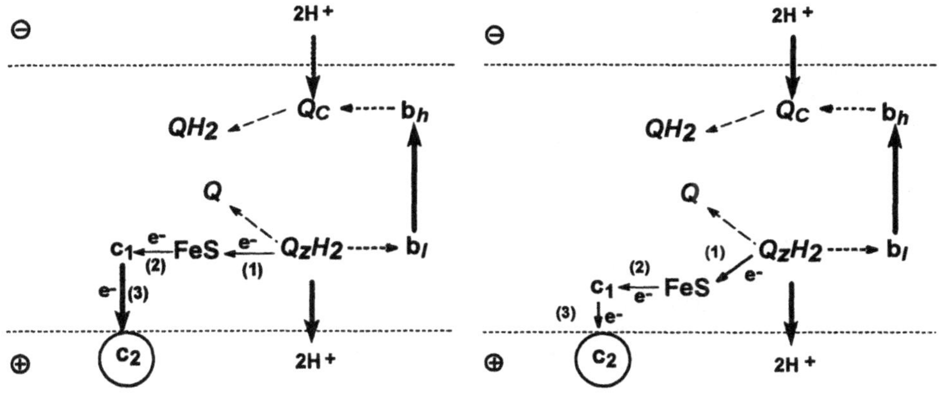

Fig. 2. A general scheme of charge transfer reactions and possible arrangement of redox centres in cyt bc_1-complex. The bold arrows indicate main electrogenic reactions. Figures in brackets refer to the numberred reactions (see the text). The dotted lines indicate the conditional borders of the protein/membrane dielectric.

The first model (Fig. 2, left panel) assumes that the only electrogenic reaction is reaction (3). In accordance with this model, the Q_o site, FeS Rieske centre and cyt c_1 are buried into the protein dielectric and the electron transfer between these centres is parallel to the membrane plane. While the electron transfer between cyt c_1 and cyt c_2 is electrogenic and is characterized by the opposite polarity compared to the overall electrogenic process.

The second model (Fig. 2, right panel) predicts that both reactions (1) and (3) are electrogenic, while reaction (2) is electrically silent. In this case, both cyt c_1 and FeS Rieske centre are equally immersed into the protein dielectric, while Q_o - site is buried somewhat deeper.

Further experiments are needed to distinguish between these two alternative models.

Acknowledgements: This work is supported by the Wellcome Trust Interlaboratory Collaboration Grant, by the International Science Foundation (grants No MLM300 and M5T300) and by the Russian Foundation for the Fundamental Research (grant No 94-04-13094). We wish to thank Drs. M.D. Mamedov, A.J. Moody, P.R. Rich and W. Nitschke for stimulating and useful discussions. The mitochondrial cytochrome c oxidase was a generous gift of Dr. A.J. Moody.

REFERENCES

1 Jackson, J.B. and Dutton, P.L. (1973) Biochim. Biophys. Acta 325, 102-113.
2 Glaser, E.C. and Crofts, A.R. (1984) Biochim. Biophys. Acta 766, 322-333.
3 Robertson, D.E. and Dutton, P.L. (1988) Biochim. Biophys. Acta 935, 273- 291.
4 Glaser, E.C. and Crofts, A.R. (1986) Proceedings of the Fourth European Bioenergetics Conference, Prague, p. 122.
5 Glaser, E.C. and Crofts, A.R. (1987) in: Cytochrome Systems: Molecular Biology and Bioenergetics (Papa, S., Chance, B., Ernster, L. eds.) pp. 625- 631, Plenum, New York.
6 Konstantinov, A.A. (1990) Biochim. Biophys. Acta 1018, 138-141.
7 Popova, E.Yu. and Konstantinov, A.A. (1987) in Cytochrome Systems: Molecular Biology and Bioenergetics (Papa, S., Chance, B. and Ernster, L., eds.), pp. 751-765, Plenum, New York.
8 Drachev, L.A., Kaurov, B.S., Mamedov, M.D., Mulkidjanian, A.Ya., Semenov, A.Yu., Shinkarev, V.P., Skulachev, V.P. and Verkhovsky, M.I. (1989) Biochim. Biophys. Acta 973, 189-197.
9 Mulkidjanian, A.Ya., Mamedov, M.D., Semenov, A.Yu., Shinkarev, V.P., Verkhovsky, M.I. and Drachev, L.A. (1990) FEBS Lett. 277, 127- 130.
10 Semenov, A.Yu., Bloch, D.A., Crofts, A.R., Drachev. L.A., Gennis, R.B., Mulkidjanian, A.Ya. and Yun, C.-H. (1992) Biochim. Biophys. Acta 1101, 166- 167.
11 Rieske, J.S. (1967) Methods Enzymol. 10, 239-245.
12 Drachev, L.A., Mamedov, M.D. and Semenov, A.Yu. (1987) FEBS Lett. 213, 128-132.
13 Ahmad, I., Cusanovich, M.A. and Tollin, G. (1981) Proc. Natl. Acad. Sci. USA 78, 6724-6728.
14 Moody, A.J. and Rich, P.R. (1988) EBEC Rep. 5, 69.
15 Moody, A.J., Brandt, U. and Rich, P.R. (1991) FEBS Lett. 293, 101- 105.
16 Chew, V.S.F. and Bolton, J.R. (1980) J. Phys. Chem. 84, 1903-1908.
17 Chew, V.S.F., Bolton, J.R., Brown, R.G. and Porter, G. (1980) J. Phys. Chem. 84, 1909-1916.
18 Zaugg, W.S. (1964) J. Biol. Chem. 239, 3964-3969.
19 Semenov, A.Yu. (1991) Sov. Sci. Rev., Section D. Physicochem. Biol. 10, 45-75.
20 Crofts, A.R. and Wang, Z. (1989) Photosynth. Res. 22, 69-87.
21 von Jagow, G. and Ohnishi, T. (1985) FEBS Lett. 185, 311-315.

THE CYTOCHROME B6F COMPLEX AND STATE TRANSITIONS: A BIOCHEMICAL EVALUATION IN THE CHLOROPHYTE ALGA *CHLAMYDOMONAS REINHARDTII*

Claire S. Ting, Francis-André Wollman, and Jean-Luc Popot, Institut de Biologie Physico-Chimique, Service de Photosynthèse, 75005 Paris, France

1. Introduction

The cytochrome b_6f complex has a critical role in oxygenic photosynthetic reactions and is involved in both linear electron transport between Photosystem II (PSII) and Photosystem I (PSI) and cyclic electron transport reactions with PSI. In wild-type *C. reinhardtii*, lateral migration of cytochrome b_6f complexes from the grana to the stromal lamellae have been demonstrated to accompany transitions from State I to State II which are induced by cellular ATP depletion (1-5). This may facilitate cyclic photophosphorylation reactions between cytochrome b_6f complexes and PSI, and lead to increased ATP production in State II (1-4). An important step towards understanding the function underlying state transition-associated lateral migration of cytochrome complexes involves determining whether changes in the cytochrome b_6f complex and/or its association with other membrane proteins actually accompany these migrations. We addressed this in our study and also developed a protocol which minimized the possibility that cells would undergo unspecified state transitions during the isolation of thylakoid membranes.

2. Procedure

C. reinhardtii cells (wild-type strain, 137c) grown in TAP medium at 25°C and 1000 lux (continuous) were placed in State I by incubation with 10 μM DCMU under high light conditions (4000 lux, 20 min) or in State II by incubation with 20 μM Antimycin A and 4 mM salicylhydroxamic acid in darkness (20 min). Cells in State I or II were then frozen in liquid nitrogen. Prior to cell disruption by sonication, each frozen block of cells was manually broken into smaller fragments and suspended in HEPES buffer (25 mM HEPES-KOH (pH 7.5), 5 mM $MgCl_2$, 0.3 M sucrose, final concentrations). The frozen slurry of cells in buffer was then sonicated for 2 x 80 sec; the majority of cells were broken as they came out of the ice phase. This procedure minimized the chance cells would undergo unspecified state transitions during the initial membrane purification steps involving cell centrifugation and breakage. Although cells can be fixed in State I or II using compounds such as *p*-benzoquinone (6), we avoided this method because of the possible multiple effects of this compound. Thylakoid-

P. Mathis (ed.), Photosynthesis: from Light to Biosphere, Vol. II, 519–522.

enriched membrane fractions were then isolated according to established protocols (7).

Dimeric cytochrome b_6f complexes were purified from thylakoid membranes using the detergent Hecameg and a protocol recently developed in our laboratory (8). In experiments where the stability of these complexes was examined, Hecameg was exchanged for dodecylmaltoside. Purified complexes were loaded into Spectra/Por dialysis tubing and dialyzed overnight at 4°C against 0.2 mM dodecylmaltoside and 20 mM Tricine. Proteins were characterized using analytical SDS-PAGE performed with gradient gels (20 cm x 30 cm, width = 0.1 cm) containing 12%-18% acrylamide and 8 M urea. Gels were stained with tetramethylbenzidine (TMBZ) in order to visualize heme-containing polypeptides and silver nitrate.

3. Results and Discussion

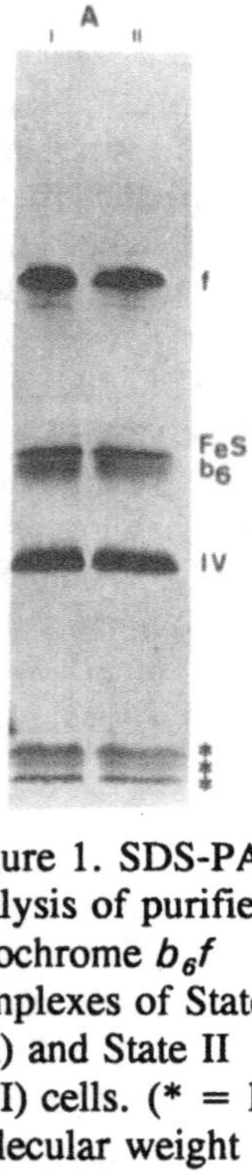

Figure 1. SDS-PAGE analysis of purified cytochrome b_6f complexes of State I (AI) and State II (AII) cells. (* = low molecular weight subunits ≤ 4 kDa)

Cytochrome b_6f complexes purified from cells in State I and State II did not exhibit major biochemical differences and were identical in terms of cytochrome content and polypeptide composition. Figure 1 indicates that both cytochromes *f* and b_6 were associated with the purified complexes and additional heme-containing polypeptides were not isolated with the complex in either state (Fig.1, lanes AI, AII). In addition, polypeptides corresponding to the Rieske FeS protein, subunit IV, and low molecular weight subunits (≤ 4 kDa) were also isolated with the complex in both States I and II (Fig.1, lanes AI, AII). The absence of major differences between cytochrome complexes isolated from State I and State II cells was further supported by the similarity in their absorption spectra and their ascorbate-reduced minus ferricyanide oxidized difference spectra (data not shown).

In the absence of added lipids, dimeric cytochrome b_6f complexes contained in detergent micelles are destabilized; this destabilization is reflected in the loss of the Rieske FeS protein and in the conversion of some dimeric cytochrome complexes into monomers. We observed that cytochrome complexes purified from State I and State II cells exhibited similar dissociation characteristics and the same propensity to monomerize. Following centrifugation on a sucrose gradient, the Rieske FeS protein dissociated first from both State I and State II complexes (Fig.2, lanes 1 to 6, shown only for State II). Towards the middle-bottom of the gradient (lanes 7 to 10), cytochrome complexes were present which were associated with cytochrome *f*, cytochrome b_6, subunit IV, low molecular weight (≤ 4 kDa) subunits, and the Rieske FeS protein. However, the majority of the Rieske protein had dissociated from these complexes and migrated at the gradient top. Only one or two fractions collected near the gradient bottom (lanes 11 and 12) stained strongly for the Rieske protein and may have contained complexes associated with bound Rieske. Based on past measurements of cytochrome complex sedimentation coefficients under the same experimental conditions as in our study (C. Breyton, personal communication), it was concluded that the majority of complexes from which the Rieske FeS protein had dissociated were monomeric. This would tend to suggest that in our study, dimeric cytochrome complexes of both State I and State II cells dissociated primarily into monomers following this destabilization treatment.

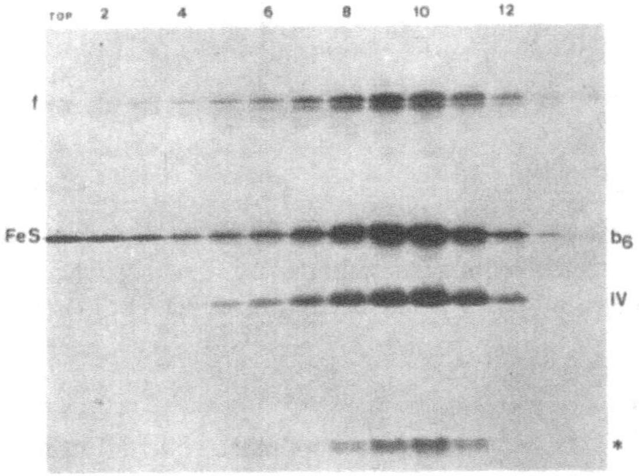

Figure 2. SDS-PAGE analysis of gradient fractions collected from a 5% to 20% sucrose density gradient loaded with cytochrome b_6f complexes purified from State II cells. Gradients were centrifuged at 55,000 rpm for 3 hrs (4°C). Silver nitrate staining was used in visualizing proteins. (* = low molecular weight subunits $\leq$ 4 kDa)

We next examined whether differences exist in protein interactions between cytochrome complexes and PSII and/or PSI polypeptides in these states. A prerequisite for this study involved developing an efficient detergent solubilization system which did not result in the selective extraction of particular proteins. We observed that the extraction of reaction center and cytochrome complex proteins was not uniform with Mega 9, octyl glucoside, dodecylmaltoside, and Triton X-100/Digitonin. In each case, cytochrome complex proteins were selectively extracted at low detergent concentrations. Zwittergent 3-12 was finally chosen for use in our studies because the extent of solubilization of these polypeptides was comparable at a given detergent concentration and protein extraction was efficient. Following Zwittergent solubilization of State I and State II thylakoid membranes, proteins were separated on a sucrose density gradient; gradient fractions were then examined using analytical SDS-PAGE. Our preliminary results suggested that in State II, there was an increase in the proportion of cytochrome b_6f complexes comigrating in PSI (66 kDa, 60-64 kDa, 20 kDa, 17-18 kDa)-containing fractions; in addition, there was an increased comigration of proteins associated with the PSII light harvesting complex (mobile fraction, 25 kDa, 26.5 kDa, 30.5 kDa) and with the PSII reaction center (47 kDa core polypeptide) in these PSI-containing fractions (Figs. 3A and 3B). This may reflect genuine changes in the interactions among some or possibly all of these proteins during state transitions.

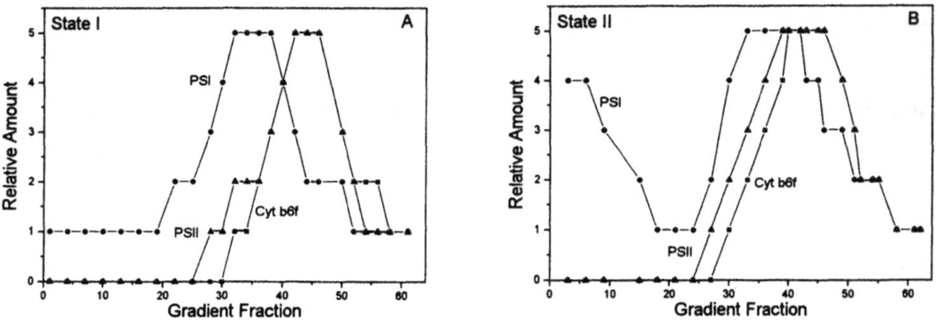

Figure 3. Relative proportions of cytochrome b_6f, PSI, and PSII polypeptides in gradient fractions collected from a sucrose density gradient loaded with Zwittergent-solubilized thylakoid membranes isolated from State I (A) State II (B) cells.

4. Acknowledgements

This material is based on work supported by the North Atlantic Treaty Organization under a Grant awarded in 1994 to C.S. Ting (NSF/NATO Postdoctoral Research Fellow). Any opinions, findings, and conclusions or recommendations expressed in this publication are those of the author and do not necessarily reflect the views of NATO. C.S. Ting would like to thank Y. Pierre, C. Breyton, C. de Vitry, and D. Picot for their kind assistance and helpful discussions.

5. References

1. Horton, P. (1985) in Photosynthetic Mechanisms and the Environment (Barber J. and Baker N.R., eds) pp. 135-187, Elsevier Press, Amsterdam, The Netherlands
2. Horton, P., Lee, P., Fernyhough, P. (1990) Biochim. Biophys. Acta 1017:160-166
3. Bulté, L., Gans, P., Rebéille, F., Wollman, F-A (1990) Biochim. Biophys. Acta 1020:72-80
4. Wollman, F-A, Bulté, L. (1990) in Photoconversion Processes for Energy and Chemicals (Hall D.O. and Grassi, G., eds) pp.198-207, Elsevier Press, Amsterdam, The Netherlands
5. Vallon, O., Bulté, L., Dainese, P., Olive, J., Bassi, R., Wollman, F-A (1991) Proc. Natl. Acad. Sci. USA 88:8262-8266
6. Bulté, L., Wollman, F-A (1990) Biochim. Biophys. Acta 1016:253-258
7. Chua, N.H., Bennoun, P. (1975) Proc. Natl. Acad. Sci. USA 72:2175-2179
8. Pierre, Y., Breyton, C., Tribet, C., Kramer, D., Olive, J., Popot, J.L. (1995), submitted for publication

A MEMBRANE BOUND CYTOCHROME C 552 COMPLEX IN THE ANAEROBIC PHOTOSYNTHETIC BACTERIUM *HELIOBACTERIUM GESTII*

Ingrid Albert and Hartmut Michel, Max-Planck-Institut für Biophysik, Heinrich Hoffmann-Straße 7, D-60528 Frankfurt/Main

1.Introduction

The family of heliobacteria is a group of strictly anaerobic, photosynthetic bacteria with the unique bacteriochlorophyll g as their main photosynthetic pigment. Being classified as Gram-positive according to the analysis of their 16S rRNA [1] the heliobacteriaceae are the only photosynthetic representative of this group. Unlike most photosynthetic bacteria the heliobacteria do not posess highly developed intracytoplasmic membranes or chlorosomes [2]. The whole photosynthetic aparatus as well as membrane associated metabolic and catabolic processes have to be harboured in the cytoplasmic membrane. Judging from spectroscopic experiments and from molecular biology studies the reaction centre of the heliobacteria seems to be the bacterial counterpart to the photosystem I of the oxygenic organisms [3,4,5].
Since very little is known about secondary events in the energy converting system of the heliobacteria we focussed our interest on the cytochromes c of *Heliobacterium gestii.*

2.Results and Discussion

Membrane preparations of *Heliobacterium gestii* (*H. gestii*) show three major heme staining bands at 63, 30 and 19 kDa when analysed by SDS-PAGE. As in the other known heliobacteria no soluble cytochromes could be detected. The 19 kd heme staining band corresponds to a c-type cytochrome with absorption maxima at 421, 522 and 552 nm in the reduced state.
Purification attempts of the 19 kd cytochrome c lead to the iolation of this cytochrome together with two other polypeptides with apparent molecular weights of 32 and 42 kd, respectively, forming a stable complex even in the presence of chaotropic reagents.
When subjected to molecular sieve chromatography on an HPLC system the solubilized cytochrome c 552 complex eluted in a homogenous fraction with a molecular weight of approximately 200 kd. As indicated in figure 2 the three polypeptides are present in a constant stoichiometry when rechromatographed. The association of these three proteins therefore is unlikely to be due to accidental copurification but rather a functional association. Considering that about one third of the molecular weight of detergent

P. Mathis (ed.), Photosynthesis: from Light to Biosphere, Vol. II, 523–526.

solubilized proteins account for bound detergent molecules we suggest that the three polypeptides are complexed with a stoichiometry of 1:1:1.
Comparison of N-terminal sequences from both the 31 and 42 kd peptide with known protein sequences did not reveal any significant homologies

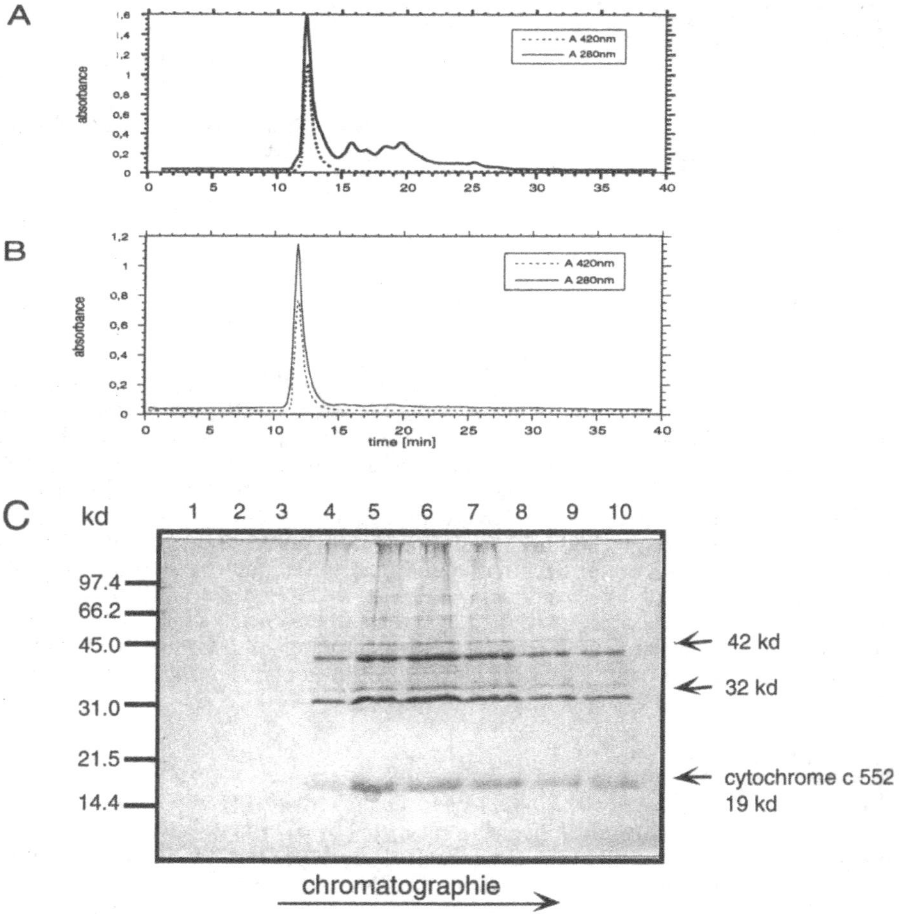

Figure 1: Characterization of the cytochrome c 552 complex by HPLC on a TSKgel G3000SW column (50mM potassium-phosphate pH 6.9, 0.3% Mega 9, flow rate: 1ml/min, fraction size: 0.5ml).

A: Chromatogram of an enriched cytochrome c 552 sample obtained after hydroxalapatite chromatography.

B: A sample of the fraction containing the cytochrome c 552 complex was rechromatographed under the same conditions as above.

C: Aliquots from each fraction of chromatography B were subjected to SDS -PAGE, the gel was stained with Coomassie Brilliant Blue.

Figure 2 shows redox titrations carried out on the enriched 19 kd cytochrome. The data can be fitted with a single Nernst equation for a one electron transition with E_m(pH7.5)=+213±10mV. This titration behavior may be taken as an indication for the 19 kd cytochrome beeing a monohaem cytochrome. But since we cannot exclude that this cytochrome contains several haems with identical redoxpotential, this question can only be answered definitively by analyzing the primary structure.
In membranes from *Heliobacterium chlorum* a reaction centre associated cytochrome c 552 with a midpoint potential of E_m(pH7.09=+170 mV) was observed to donate electrones to the photooxidized prímary donor P798+ with E_m=+225 mV [6]. Since the E_m determined for the19 kd cytochrome in *H. gestii* is only a little more negative than that of P798+ the 19 kd cytochrome may not function as a donor to P798+. Nevertheless the physiological function of this cytochrome remains to be investigated.

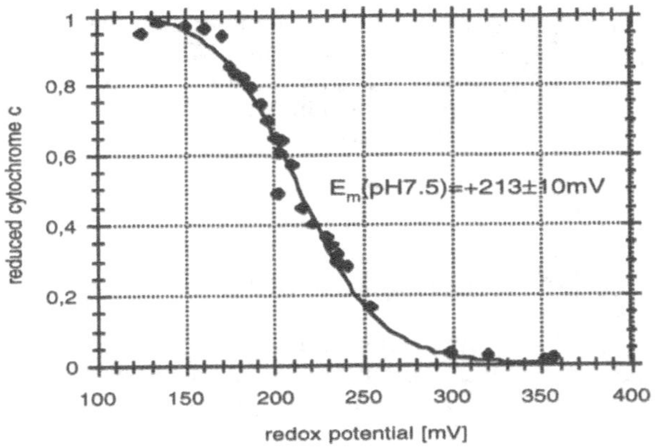

Figure 2: Redox titration on the enriched cytochrome c 552. The following mediators were present at 10µM concentration each: 1,4-Benzoquinone, 5-OH-1,4-Naphtoquinone, Duroquinone, Variamine Blue, Benzylviologen, Methylviologen, Neutral Red. Titrations were carried out at 20°C, in 50mM MOPS pH 7.5, 0.3% Mega 9, with ferricyanid for the oxidizing direction and with dithionite for the reducing direction. The amount of reduced cytochrome *c*-552 was measured by monitoring the absorption difference A_{536}-A_{552}.
The data obtained are fitted with a Nernst curve for a one electron transition with E_m=+213±10 mV.

Using a protocol usually employed to remove peripheral proteins from membranes [7] the 30 kd cytochrome c was released from the cytoplasmic membranes as the pH was increased from 10 to 12. The 19 kd cytochrome apparenty was not affected by the alkaline treatment while the 30kd cytochrome was extracted to about 70% at pH 12.

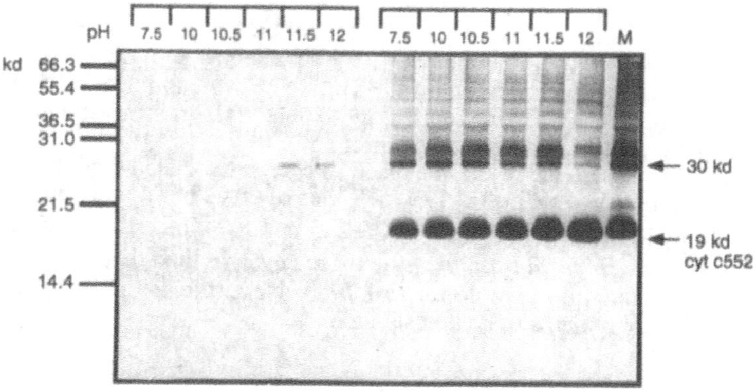

Figure 3: SDS-PAGE on the polypeptides extracted from cytoplasmatic membranes at alkaline pH (10-12). The gel was stained for hemes with TMBZ.
Membranes were incubated at pH values 10, 10.5, 11, 11.5, 12 for 30 min on ice. After centrifugation the supernatant and the sediments were analyzed for the presence of cytochrome c.

3.References

[1] Woese, C.R., Debrunner-Vossbrinck, B.A. and Oyaizu, H. (1985), *Science* **229**, 762-765.

[2] Beer-Romero, P. (1986) Comparative studies on *Heliospirillum gestii* and *Heliobacillus mobilis*. M.A. thesis, Indiana University, Bloomington.

3] Nitschke,W., Setif, P.,Liebl, U.and Rutherford, A.W. (1990) *Biochemistry* **29**, 11079-11088.

[4] Nitschke, W., Feiler, U. and Rutherford, A.W. (1990) *Biochemistry* ***29***, 3834-3842.

[5] Liebl, U., Mockensturm-Wilson, M., Trost, J., Brune, J., Blankenship, R.E. and Vermaas, W. (1993) *Proc. Nat. Sciences (USA)* **90**, 7124-7128.

[6] Prince, R. C., Gest H. and Blankenship R. E., Biochim. Biophys. Acta **810** (1985) 377-384.

[7] Singer, S. J. (1971) The molecular organization of biological membranes. In Structure and Function of Biological Membranes (L. I. Rothfield, ed.) pp145-222, Academic Press, New York.

STABILIZATION OF THE HIGH POTENTIAL FORM OF CYTOCHROME b-559 OF PHOTOSYSTEM II BY CALCIUM OR LACK OF OXYGEN

V. P.. McNamara[1], K. Gounaris[2] and J. Whitmarsh[1,3]
[1]Dept. of Plant Biology & [3]Photosynthesis Research Unit, USDA/ARS, University of Illinois at Urbana-Champaign, Urbana, IL 61801, U.S.A.;
[2]Dept. of Biochemistry, Imperial College, London SW7 2AY, U.K.

1. Introduction

Cytochrome b-559 is an essential component of the reaction center of Photosystem II (PS II) although its role remains unresolved. In addition to facilitating the assembly of PS II complexes [1], recent suggestions for its function have included participation in an electron cycle [2,3] or in a side path for electron transport through PS II [4,5] which provide protection against photoinhibition, photoactivation of oxygen evolution [6,7] and scavenging of photogenerated superoxide radicals [8].

In the dark, this cytochrome is found to occur in two interconvertible redox forms, a predominant high redox potential form (cyt b-559HP) with an E_m of 380mV and a low potential one (cyt b-559LP), with an E_m of around 60mV [1]. The interconvertibility of cyt b-559 may be significant in controlling its function since recent evidence has suggested that the minor LP form can accept electrons from reduced pheophytin following accumulation of the potentially damaging state P680Pheo$^-$Q$_A$$^{-}$·[5]. The high mid-point potential results from an extremely hydrophobic microdomain which excludes bulk solvent molecules from interacting with the heme group, in conjunction with the positive dipole potential arising from the parallel α-helices of the two subunit polypeptides which combine to ligate the heme [9,10]. Disruption to a layer of intermediate dielectric constant at the interface between the non-polar interior and polar exterior of the thylakoid membrane increases heme exposure to the bulk aqueous phase, thus converting the cytochrome to lower potential forms. Release of extrinsic polypeptides from the lumenal surface of the complex and exposure to detergents are among treatments which have been suggested to induce such a conversion [10].

2. Results and Discussion

2.1 Calcium is required to stabilize Cyt b-559HP

Photosystem II enriched membrane fragments were isolated according to the protocol of Berthold et al [11] with the added precaution of maintaining mildly reducing conditions thoughout the procedure by including 5mM ascorbate in each solution. Cyt b-559 was assayed by redox potentiometry, as shown in Figure 1. As we have shown

P. Mathis (ed.), Photosynthesis: from Light to Biosphere, Vol. II, 527–530.

previously [12], providing that the heme remains in its reduced state at all times, the membrane fragment preparations contain almost exclusively cyt b-559HP. Oxidation of the heme, as occurs prior to a reductive titration, results in the conversion of 50% of the cytochrome to lower potential forms. Destabilization of the oxidized cytochrome has been noted before [12,13], and probably arises from electrostatic effects resulting from the placement of additional positive charge in an extremely hydrophobic domain.

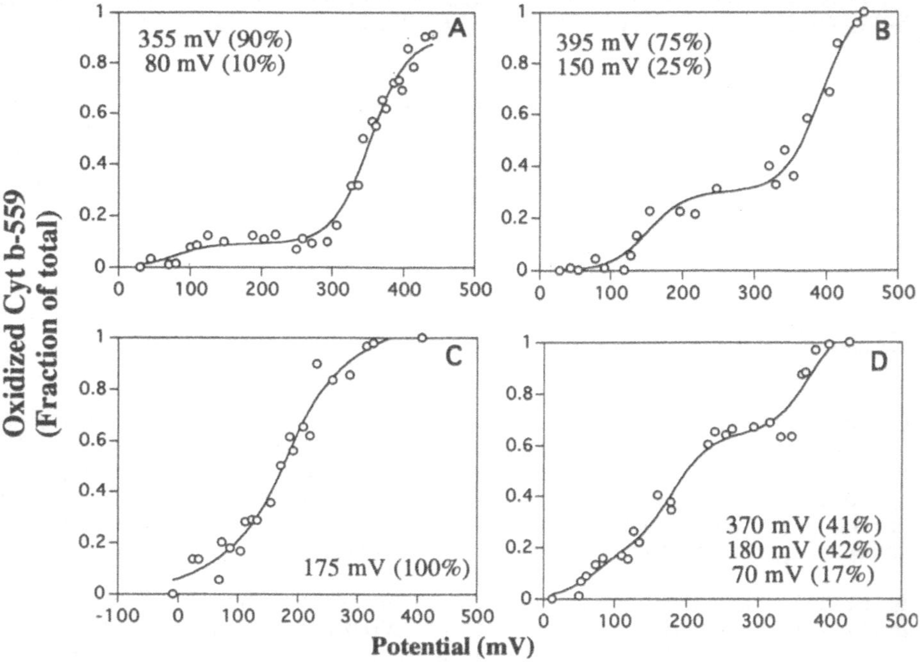

Figure 1. Oxidative redox titrations of cyt b-559 in PS II membrane fragments. Titrations were performed as described in [12]. A) Untreated PS II; B) PS II washed with 1M NaCl, 10 mM $CaCl_2$ in the dark; C) as in B) except 50μM EGTA was present and washing performed under low light; D) PS II washed with 10mM citrate (pH3.0)

To investigate the effect of components located on the lumenal surface of PS II on cyt b-559 stability, membranes fragments were washed with 1M NaCl, which is known to cause dissociation of the 17 and 23 kDa extrinsic polypeptides that stabilize water oxidation. Membranes that were washed in the dark, in the presence of millimolar concentrations of calcium, retained mostly the HP form of cyt b-559 (Fig.1B). This result is in contrast to the reported requirement for binding of the 23 kDa protein to stabilize cyt b-559HP [10]. These treated membranes were subsequently washed with 50μM EGTA under low levels of light (approx. 50μ moles photons $\cdot m^{-2} \cdot s^{-1}$), a treatment which causes the further release of calcium from the PSII reaction center [14]. As a result of the loss of calcium, all cyt b-559 was converted to an intermediate form with an E_m of 175 mV (Fig. 1C). An alternative method to deplete calcium from PS II involves washing the membranes in a solution buffered at pH3 with citrate. Both calcium and the extrinsic polypeptides are released from the complex at this pH. Upon return to physiological pHs, the extrinsic proteins rapidly re-associate with the membrane, leaving the complexes devoid of calcium [15]. Citrate-washed PS II membranes were titrated and found to have a complex mixture of at least three different cyt b-559 redox components. However, despite the retention of the 23 kDa polypeptide, the majority of cyt b-559HP had been converted to lower potential forms.

We conclude that the binding of calcium to the PS II reaction center, but not of the 23 kDa extrinsic polypeptide, is required to stabilize cyt b-559HP. It should be noted that each treatment was undertaken with ascorbate present to maintain reducing conditions and that only oxidative titrations are shown. We expect more significant shifts towards lower potential forms to be apparent if the heme is allowed to become oxidized.

2.2 Oxygen Destabilizes Cyt b-559HP

Exposure of the photosynthetic apparatus to strong light leads to the photoinactivation of PS II, resulting in, amongst other events, irreversible loss of cyt b-559HP [16]. However, recently it was shown that thylakoid membranes which were photoinhibited under strictly anaerobic conditions did not suffer from a loss of this HP component [5]. We therefore tested the possibility that oxygen was involved in the conversion to cyt b-559LP. PS II membrane fragments were suspended in a medium which was made anaerobic by the addition of a glucose oxidase/catalase trap and subjected to ten minutes of illumination with red light of 5000μ moles photons $\cdot m^{-2} \cdot s^{-1}$. As shown previously [5], this treatment resulted in the accumulation of some photoreduced cyt b-559LP. The proportion of cyt b-559HP was then tested by three-point titrations using ferricyanide, hydroquinone and dithionite, both for membranes maintained under anaerobic conditions (Fig 2A) and after re-oxygenation by bubbling with air for several minutes (Fig 2B). The proportion of cyt b-559HP was unchanged after exposure to the light under anaerobic conditions. After introducing oxygen into the sample, however, some 20% of the total cytochrome was converted to a non-HQ-reducible form.

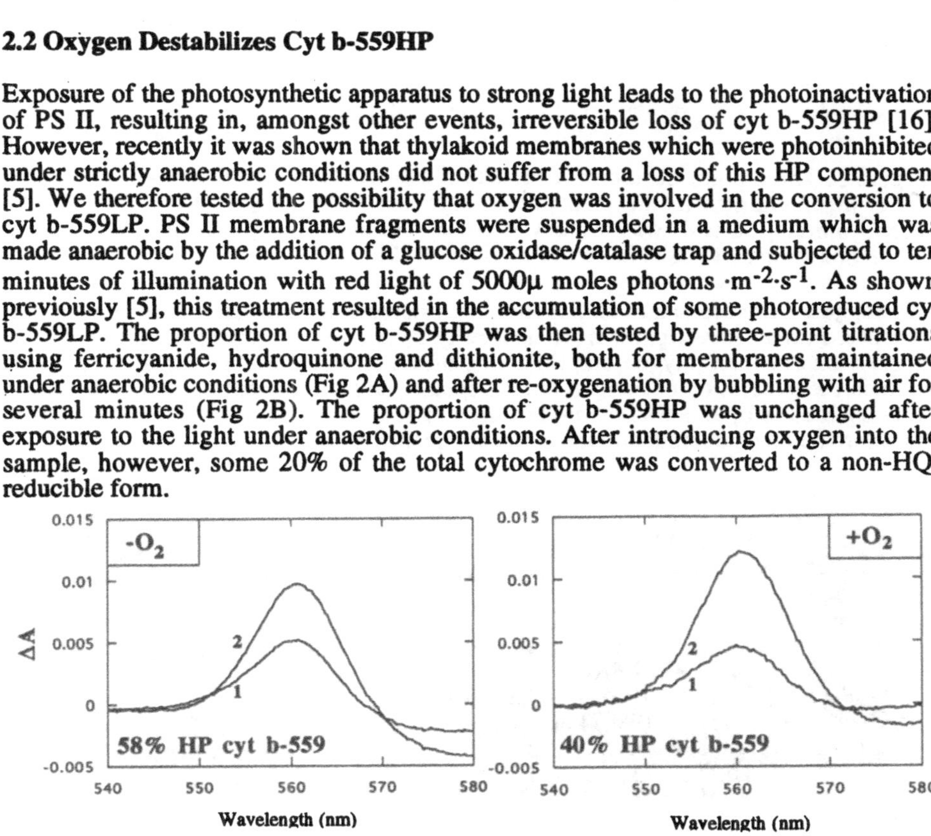

Figure 2. Three-point titrations of cyt b-559 in photoinhibited PS II membranes. The membranes were illuminated with 5,000 μ moles photons $\cdot m^{-2} \cdot s^{-1}$ of red light for ten minutes in the presence or absence of a glucose oxidase/catalase oxygen scavenging system. (1) HQ- FeCy spectrum ; (2) Dithionite-FeCy spectrum.

This oxygen dependent irreversible loss of cyt b-559HP has also been observed using other treatments to destabilize the cytochrome. In Fig 3, thylakoid membranes were incubated with ferricyanide for 7 minutes at 40°C in order to denature oxidized cyt b-559HP. Following the treatment, the proportion of cyt b-559HP fell from 68% to less than 10% (Fig 3A). However, the thylakoid membranes which were heat-treated under anaerobic conditions, did not suffer a loss of this form of the cytochrome. They were found to retain 63% of cyt b-559 in its HP form (Fig 3B).

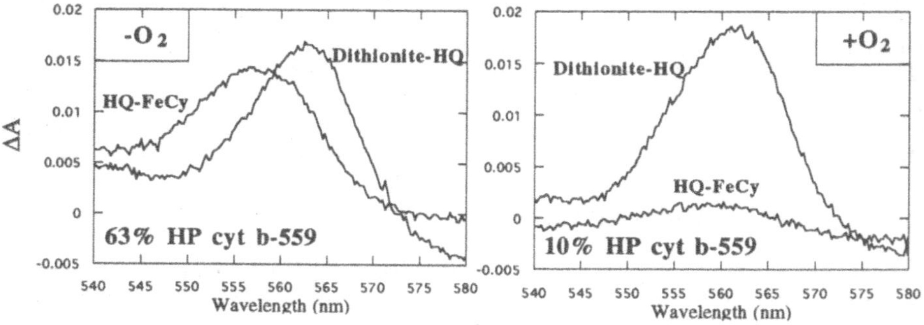

Figure 3. Three-point titrations of thylakoid membranes subjected to a heat shock of 40°C for 7 minutes in the presence or absence of an oxygen scavenging system. Ferricyanide (200 μM) was present in order to oxidize the heme. The ratio of cyt b-559 redox forms was calculated as described in [5].

We suggest therefore that cyt b-559HP is less stable when oxygen is present than in its absence. Further investigations are underway to examine the mechanistic basis of the interaction between oxygen and the heme of cyt b-559. Possible explanations could include the generation of oxygen radical species and subsequent oxidative damage to polypeptides or lipid peroxidation, leading to a reduction in the integrity of the barrier between the heme domain and the external aqueous phase.

References

1 Whitmarsh, J. and Pakrasi, H.B. (1995) in Advances in Photosynthesis Vol.4, Oxygenic Photosynthesis: The Light Reactions (Ort, D.R. and Yocum, C.F., eds.) in press
2 Thompson, L.K. and Brudvig, G.W. (1988) Biochemistry 27, 6653-6658
3 Barber, J. and DeLas Rivas, J. (1993) Proc. Natl. Acad. Sci. USA 90, 10942-10946
4 Nedbal, L. *et al.*, (1992) Proc. Natl. Acad. Sci. USA 89, 7929-7933
5 Poulson, M.E. *et al.*, (1995) Biochemistry, in press
6 Cramer, W.A. *et al.*, (1986) Photosynth. Res. 10, 393-403
7 Mizusawa, N. and Yamashita, T. (1990) in Research in Photosynthesis - Vol II (Murata, N., ed.) pp 425-428. Kluwer Academic Publishers, Dordrecht
8 Ananyev, G. *et al.*, (1994) Photosynth. Res. 41, 327-338
9 Krishtalik, L.I. *et al.*, (1993) Biophys. J. 65, 1-12
10 Thompson, L.K. *et al.*, (1989) Biochemistry 28, 8048-4056
11 Berthold, D.A. *et al.*, (1981) FEBS Lett. 143, 231-234
12 McNamara, V.P. and Gounaris, K., (1995) Biochim. Biophys. Acta, in press
13 Ortega, J.M. *et al.*, (1989) Plant Science 68, 71-75
14 Boussac, A. and Rutherford, A.W. (1988) FEBS Lett. 236, 432-436
15 Ono, T.-A. and Inoue, Y. (1988) FEBS Lett. 227, 147-152
16 Styring *et al.*, (1990) Biochim. Biophys. Acta 1015, 269-278

THE ROLE OF CYTOCHROME *b*559 IN PHOTOREACTIVATION OF OXYGEN-EVOLVING CENTER OF PHOTOSYSTEM II MEMBRANES

Naoki Mizusawa, Mitsue Miyao[1] & Takashi Yamashita,
Inst. Biol. Sci., Univ. of Tsukuba, Ibaraki 305, Japan,
[1]Natl. Inst. Agrobiol. Resources, Ibaraki 305, Japan

1. INTRODUCTION

Cytochrome *b*559 (Cyt *b*559) is one of the integral components of the photosystem II (PSII) complex. In untreated chloroplasts, 0.5 mol to 0.7 mol of Cyt *b*559 occurs in a high potential (HP) form (E_m7= +370 mV) per 400 Chl molecules, which is reducible with hydroquinone (HQ). However, the HP form is unstable that the inactivation treatments of O_2-evolving center, such as heat or 0.8 M Tris-treatment, convert it to a low potential (LP) form (E_m7= +60 mV) which is reduced by ascorbate but not by HQ. We recently reported [1] that the LP form of Cyt *b*559 was brought back to the HP form by the photoreactivation of O_2-evolving center in the Tris-treated and Mn-depleted thylakoids and that the recovery of the HP form was a more rapid reaction than the recovery of the O_2-evolving activity. So, the recovery of the HP form of Cyt *b*559 might be one of the primary reaction in photoreactivation. The reassembly of Mn to the O_2-evolving center [2] and the rebinding of 23- and 18-kDa proteins to the center [3] were raised as the candidates to trigger the recovery of the HP form in the primary process of reactivation, that the effects of Mn^{2+} and 23- and 18-kDa proteins on Cyt *b*559 should be also studied with our Tris-inactivated thylakoids.

In this study, we employed the photoreactivation system of the Tris-inactivated PSII membranes depleted of Mn and extrinsic proteins completely which were more suitable than the Tris-inactivated thylakoids for the study of Mn^{2+}, Ca^{2+} and 23- and 18-kDa proteins effects on the recovery of the HP form in details. The results show that the restoration of the HP form rapidly occurs by photoactivation in the absence of the extrinsic proteins and the recovered HP form functions as an electron acceptor, an essential factor for the photooxidation of Mn^{2+} to reconstitute Mn-cluster.

2. MATERIALS AND METHODS

PSII membranes were prepared from spinach thylakoids basically according to the method of Kuwabara and Hashimoto [4]. The PSII mem-

P. Mathis (ed.), Photosynthesis: from Light to Biosphere, Vol. II, 531–534.

branes were incubated with 0.8 M Tris–HCl (pH9.1)/5 mM EDTA/2 mM ascorbate for 2 hr at 4°C. The inactivated PSII membranes were collected by centrifugation and washed twice with SMN (0.4 M sucrose/ 50 mM Mes–Na(pH6.5)/20 mM NaCl).

The Tris–inactivated PSII membranes were first incubated with SM–1 M NaCl (0.4 M sucrose/50 mM Mes–Na (pH6.5)/1 M NaCl) for 5 min in the dark at 25°C. Next, 0.1 mM $Mn(CH_3COO)_2$ and 50 mM $CaCl_2$ were added to the suspension and it was further incubated for 5 min. Then, the membrane suspension was photoreactivated with continuous light (33 $\mu E \cdot m^{-2} \cdot s^{-1}$) or with 10 μs xenon flashes at 25 °C. A portion of photoactivated membranes was used for the determination of Cyt *b*559 redox properties. Another fraction of the membranes was diluted 20–fold with SMN containing 33–, 23– and 18–kDa proteins and the O_2–evolving activity of the suspension was measured by adding 0.5 mM phenyl–p–benzoquinone with a Clark–type O_2–electrode at 20°C. The content of HP form of Cyt *b*559 was determined by measuring the (2 mM HQ) minus (0.5 mM FeCN) difference spectra of the PSII membranes suspended with 2 mM HQ in the sample cell and with 0.5 mM ferricyanide (FeCN) in the reference cell. A spectrophotometer UV–2200 (Shimadzu, Japan) was used for the determination. The total content of Cyt *b*559 was determined from (dithionite) minus (FeCN) spectra. As the extinction coefficient for Cyt *b*559, $\varepsilon^{\Delta}(559\text{–}570)=15 mM^{-1} \cdot cm^{-1}$ was used [5].

3. RESULTS AND DISCUSSION

In our PSII membranes with the O_2–evolving activity of 350–500 μmol mg $Chl^{-1} \cdot hr^{-1}$, the total content of Cyt *b*559 was about 2 mol per PSII center, assuming a molar ratio of 220 Chl per reaction center. In untreated PSII membranes, 1.0–1.2 mol/220 Chl of Cyt *b*559 was present in the HP form, and 1.0–0.8 mol/220 Chl in the (LP + very low potential) forms. In contrast to the Tris–inactivated thylakoids,

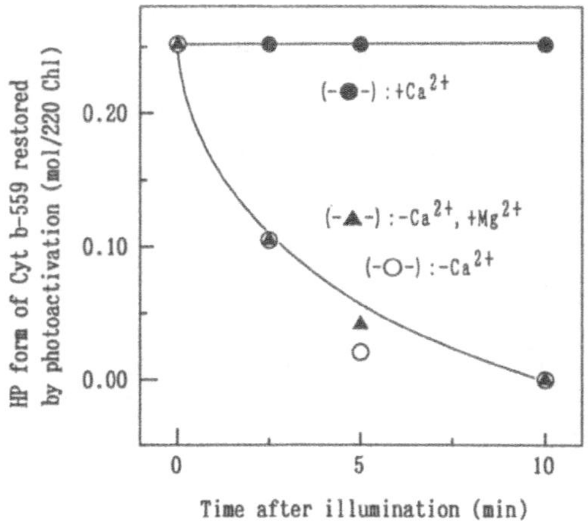

Fig. 1. Effect of Ca^{2+} on the dark–decay of the HP form of Cyt *b*559 which was restored by photoreactivation. Tris–inactivated PSII membranes were photoreactivated in the presence of 0.1 mM Mn^{2+} and 1 M NaCl under weak light for 10 min. After reactivation, 50 mM Ca^{2+} (–●–), 50mM Mg^{2+} (–▲–) or none (–O–) was added to the suspension of photoreactivated membranes and the suspension was incubated in darkness. At a designated time, the content of the HP form of Cyt *b*559 was determined.

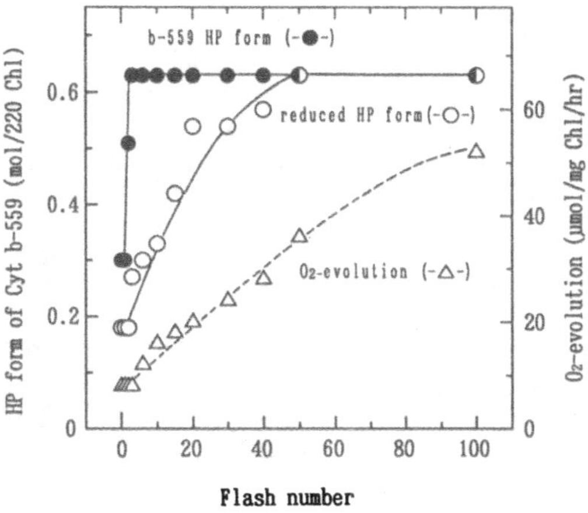

Fig. 2. Effects of repetitive flashes on the increases of the HP form of Cyt *b*559, photoreduced HP form and the O_2-evolving activity. Tris-inactivated PSII membranes were photoreactivated in the presence of 0.1 mM Mn^{2+}, 50 mM Ca^{2+} and 1 M NaCl with repetitive flashes at 1 s interval. (- ● -); the content of the HP form of Cyt *b*559, (-○-); the photoreduced HP form of Cyt *b*559, (-▲-); O_2-evolving activity.

where all Cyt *b*559 was converted to the LP form, Tris-inactivated PSII membranes still kept the HP form by 0.6–0.7 mol Cyt *b*559 per 220 Chl unit. Two third of the HP form was reduced state in darkness. Addition of 1 M Cl^- in the photoreactivation of the (Mn- and 33-kDa protein)-depleted PSII membranes enhances the recovery level of the O_2-evolving activity [6]. Thus, we usually carried out the photoreactivation in the presence of 1 M NaCl. Incubation of the inactivated membranes with 1 M NaCl caused the oxidation of the HP form in the dark and concomitantly decreased the content of the HP form from 0.6–0.7 mol/220 Chl to 0.3–0.6. After the photoreactivation of the Tris-inactivated PSII membranes in the presence of 0.1 mM Mn^{2+}, 50 mM Ca^{2+} and 1 M NaCl with continuous light for 10 min, the accumulation of the photoreduced HP form and the increase of the HP form (ΔHP form of Cyt *b*559, 0.2–0.25 mol/220 Chl) were observed. Even in the photoreactivation in the absence of artificial electron acceptor (DCPIP or PBQ), 90–120 μmol mg $Chl^{-1}\cdot hr^{-1}$ of O_2-evolving activity was restored as previously reported [7]. In the absence of Ca^{2+} from the reactivation mixture, the recovery of the HP form was 0.13–0.25 mol/220 Chl. Although Ca^{2+} was not essential for the recovery of the HP form of Cyt *b*559, Ca^{2+} stabilized the reduced HP form of Cyt *b*559 (Fig. 1). In the absence of Ca^{2+}, the photoreduced HP form was unstable and readily reoxidized by the following dark-incubation exponentially, but it was stabilized by the addition of Ca^{2+} at least for 10 min. The Ca^{2+} effect could not be substituted for Mg^{2+}. In the primary process of the reconstitution of Mn-cluster by photoreactivation, it is important to photooxidize Mn^{2+} by PSII and to stabilize Mn^{3+} or to keep Mn^{3+} away from the activated electron in PSII acceptor side or in quinones [8]. Our results suggested the possibility that the HP form of Cyt *b*559 was brought to the oxidized form in the presence of 1 M NaCl and then accepted an electron to oxidize Mn^{2+} to Mn^{3+} to reconstitute Mn-cluster. Ca^{2+} appears to stabilize Mn^{3+} formed by the photochemical reaction of PSII by suppressing the electron flow from the reduced HP form of Cyt *b*559 to Mn^{3+}.

To analyze the order of reaction process between the change in the HP form of Cyt *b*559 and the recovery of O_2-evolving activity, the photoreactivation was carried out with repetitive flashes at 1 s interval and the number of required flashes were analyzed (Fig. 2). The recovery of the HP form was a rapid reaction and occurred by dosing 2-flashes (60 % of the maximum) and saturated by 3-flashes. Subsequently, the HP form was photoreduced and the reduced HP form was accumulated. In contrast, the recovery of the O_2-evolving activity was a slow reaction and the recovery level was not saturated even by 100 flashes. On the photoreduction of HP form and the recovery of O_2-evolving activity, the flash interval dependence was examined at 20 flashes. On the both reactions, bell shaped curves were observed and the maximum values were obtained at the intervals between 0.2 and 2s. The results suggested that the recovery of the HP form was the primary process in the reconstitution of Mn-cluster and the oxidized HP form might be utilized as the electron acceptor for the photooxidation of Mn^{2+} to reconstitute Mn-cluster.

Next, we examined the effects of 10 μM DCMU on the recovery of the HP form of Cyt *b*559 by the photoreactivation with repetitive flashes. Interestingly, the recovery of the HP form was little affected by DCMU within 10 flashes. This suggested that the partial electron transport from Mn^{2+} to Q_A^- or nonheme iron was sufficient enough for the recovery of the HP form of Cyt *b*559. However, the photoreduction of the HP form and the recovery of the O_2-evolving activity were inhibited by DCMU. Addition of an artificial electron acceptor, 2 μM DCPIP, in the reactivation mixture also suppressed the photoreduction of the HP form while the recovery of the O_2-evolving activity was the same as in the absence of DCPIP. These results suggested that the HP form of Cyt *b*559 accepted an electron via Q_B or plastoquinone pool to reconstitute Mn-cluster.

REFERENCES

1 Mizusawa, N., Ebina, M. and Yamashita, T. (1995) Photosynth. Res. (in press)
2 Cramer, W.A., Theg, S.M. and Widger, W.R. (1986) Photosynth. Res. 10, 393–493.
3 Briantais, J.-M., Vernotte, C., Miyao, M., Murata, N. and Picaud, M. 808, 348–351.
4 Kuwabara, T. and Hashimoto, Y. (1990) Plant Cell Physiol. 31, 581–589
5 Cramer, W.A. and Whitmarsh, J. (1977) Ann. Rev. Plant Physiol. 28, 133–172.
6 Miyao, M. and Inoue, Y. (1991) Biochemistry 30, 5379–5387.
7 Tamura, N. and Cheniae, G.M. (1987) Biochim. Biophys. Acta 890, 179–194
8 Miyao-Tokutomi, M. and Inoue, Y. (1992) Biochemistry 31, 526–532.

BIOCHEMICAL AND GENETIC ANALYSIS OF THE CYTOCHROME C_Y OF *RHODOBACTER CAPSULATUS*

Hannu Myllykallio, Francis E. Jenney, Jr. and Fevzi Daldal
University of Pennsylvania, Department of Biology, Plant Science Institute, Philadelphia, PA 19104-6018, USA

1. Introduction

In *Rhodobacter capsulatus* photosynthetic growth depends on the cyclic electron transport between the photosynthetic reaction center (RC) and the cytochrome (cyt) bc_1 complex generating a membrane potential across the bacterial plasma membrane (1). Earlier models indicated that ubiquinone and soluble cyt c_2 were the sole electron carriers between these complexes. However, when the structural gene *cycA* encoding cyt c_2 was inactivated (2), the *R. capsulatus* strain thus obtained was still able to grow photosynthetically (Ps^+). This finding, combined with spectroscopic analyses of strains lacking cyt c_2 (3), demonstrated that in *R. capsulatus* cyclic electron transport continues in the absence of the cyt c_2 via a membrane bound component. Using interspecies complementation a gene responsible for this phenomenon was isolated and characterized (4). This gene - called *cycY* - encodes a c-type cytochrome named cyt c_y, which based on genetic and spectroscopic criteria, can donate electrons to the RC of *R. capsulatus in vivo* (4,5). On the other hand, a *R. capsulatus* double mutant, cyt c_2^- c_y^- (FJ2), is unable to grow photosynthetically, demonstrating that these two cytochromes are the sole electron carriers capable of connecting the RC and the cyt bc_1 complex. Sequence analysis of chromosomal DNA surrounding *cycY* indicated that this gene appears to be monocistronic (Figure 1, S. Lang, personal communication). Recently, a homologue of *cycY* has also been found in *R. sphaeroides* (J. Zeilstra and S. Kaplan, personal communication), but why this cytochrome is not active in photosynthetic electron transport *in vivo* is currently unknown. Using flash induced absorbance spectroscopy and chromatophores from various deletion strains lacking the cyts bc_1, c_2 and c_y (or a combination of them) Jenney *et al.* (5) have demonstrated that the membrane-bound cyt c_y is capable of reducing the RC more rapidly than the soluble cyt c_2. Furthermore, flash induced oxidation kinetics of cyt c_y is dependent on the presence or absence of the cyt bc_1 complex while that of cyt c_2 is independent of it. Altogether, these findings indicate that photosynthetic electron transport properties of the cyts c_y and c_2 are markedly different. One of the goals of this work is to define the molecular basis of these observed functional differences. As a first step toward this goal we here describe the purification and characterization of cyt c_y.

P. Mathis (ed.), Photosynthesis: from Light to Biosphere, Vol. II, 535–538.

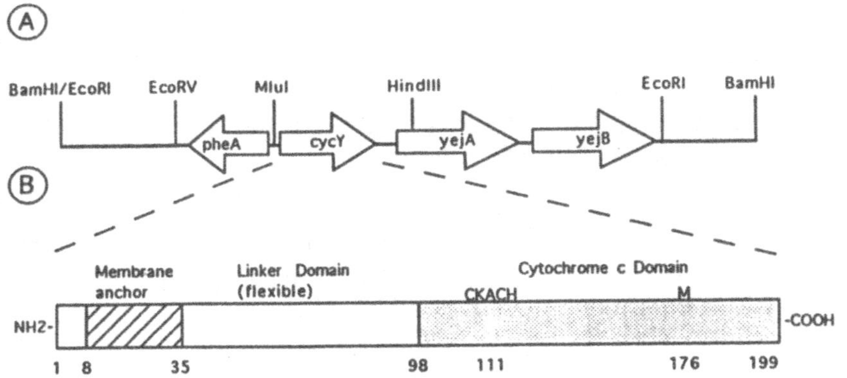

Figure 1. A) DNA sequence around cycY suggests that this gene is monocistronic, since ORFs surrounding it are not related to known functions of cyt c_y (yejA/B are a part of putative transport operon in *E. coli* (unpublished sequence, Accession No: U00008), and pheA is involved in biosynthesis of aromatic amino acids (not shown)). B) A domain structure of cyt c_y. For details, see the text.

2. Procedures

Molecular biological techniques are described elsewhere (3,4). Growth conditions (pHMFLAG/FJ2), solubilization of membranes and anion exchange chromatography were as described earlier (6), except that solubilization and binding of a sample to DEAE-Sepharose (Pharmacia, Sweden) were done using buffers containing 30 mM NaCl at pH 9.0. Also, during solubilization a protease inhibitor cocktail containing Pefabloc SC, leupeptin and pepstatin A (Boehringer Mannheim) was used at the recommended concentrations. After extensive washes with the buffer containing 30 mM NaCl in the original buffer, pH of the washing buffer was changed to pH 8.0, and cyt c_y eluted in this buffer using 100 mM NaCl. Eluted proteins were dialyzed against 50 mM Tris-Cl, pH 7.4, 150 mM NaCl, 2 mM EDTA (TBS). Affinity chromatography with anti-FLAG M2 agarose (IBI, Kodak) was done according to the manufacturer's instructions including 0.01 % dodecylmaltoside in all buffers. Western blots were done according to the same instructions except that 3 μg/ml of anti-FLAG M2 antibody was used. Spectroscopic analyses were performed according to (6).

3. Results and Discussion

3.1 Identification and purification of cyt c_y

To identify the gene product of *cycY* (cyt c_y) and to aid its purification the highly antigenic FLAG-epitope (N-AspTyrLysAspAspAspAspLys-C) was added to the carboxy-terminus of cyt c_y using site-directed mutagenesis. The FLAG'ed cyt c_y thus obtained was able to complement a *R. capsulatus* strain lacking both the cyts c_2 and c_y

(FJ2) and a *R. sphaeroides* strain lacking cyt c_2 (Gadc$_2$) to Ps^+. These findings indicated that the addition of this epitope to cyt c_y does not render it non-functional *in vivo*. Therefore, the biophysical and biochemical properties of FLAG'ed cyt c_y must be close to those of the native protein. Using monoclonal antibodies against the FLAG-epitope, cyt c_y was recognized on a Western blot as a polypeptide of ≈ 30 kDa. TMBZ staining of SDS-polyacrylamide gels also indicated that addition of the FLAG epitope modified the migration pattern of only one c-type cytochrome, which was also absent in FJ2. Furthermore, a Western blot analysis of a single colony of pHMFLAG/FJ2 (FJ2 containing FLAG'ed cyt c_y *in trans*) taken from a plate and suspended directly in the SDS loading buffer yielded only one band corresponding in size with that observed in chromatophore membranes. These findings indicate that this band must correspond to the mature and functional form of cyt c_y.

To study how cyt c_y is associated with cytoplasmic membranes, membrane sheets of *R. capsulatus* strain MT1131 (wild-type in terms of its cytochrome content) was isolated by osmotic lysis, and treated using either alkaline conditions or chaotropic agents. Treatments with 0.1 M CAPS-NaOH buffer, pH 10.5 or 1 M NaBr did not remove cyt c_y from these membranes indicating that this cytochrome is indeed an integral membrane protein (7), with its cyt c domain located at the periplasmic side.

Using a combination of DEAE-ion exchange and anti-FLAG affinity columns FLAG'ed cyt c_y was purified to homogeneity starting with membranes of the strain pHMFLAG/FJ2 as described in Procedures. The densitometric analysis of Coomassie blue stained SDS-polyacrylamide gels indicated that purified cyt c_y was over 95 % pure. The protein band present on these gels can be recognized either by the anti-FLAG antibodies, or stained using TMBZ, indicating that it is indeed a c-type cytochrome (not shown). Determination of the N-terminal amino acid sequence of the purified protein indicated that its putative signal sequence is not cleaved (C. Moomaw and C. Slaughter, personal communication), suggesting that cyt c_y is anchored to the membrane via its uncleaved signal sequence. Internal protein sequences from a tryptic digest of purified protein confirmed that deduced from the DNA sequence of *cycY*. The translational start site determined by the N-terminal amino acid sequence now obtained confirmed the initiating Met residue suggested by previous site-directed mutagenesis studies.

3.2 Spectral characteristics of cyt c_y

The ascorbate reduced spectrum of purified FLAG'ed cyt c_y shows maxima at 417, 521 and 551 nm, while the ferricyanide oxidized spectrum has a maximum at 409 nm. Potentiometric equilibrium redox titration of the purified protein is shown in Figure 2, and indicates a redox midpoint potential ($E_{m,7}$) for cyt c_y of 365 mV at 50 mM MOPS, pH 7.0 - 100 mM KCl. This value is in agreement with that obtained using flash induced absorbance spectroscopy (5). It appears that the addition of a FLAG epitope has no major effect on spectroscopic properties of cyt c_y.

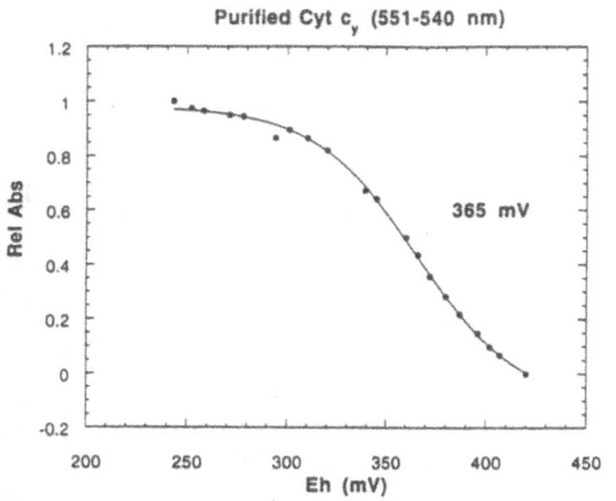

Figure 2. A potentiometric equilibrium redox titration of the cyt c_y. For details, see text.

3.3 Conclusions

The cyt c_y is now the best characterized member of a new subgroup of c-type cytochromes in which a cytochrome c-cassette appears to be fused to a ≈100 amino acids long N-terminal extension. The latter domain consists of a putative membrane anchor (a signal sequence like hydrophobic stretch long enough to span the membrane), and a linker region connecting the membrane anchor and the cyt c domain. The linker region connecting the two domains is fairly long (≈60 amino acids), and predicted to be relatively flexible (Figure 1), perhaps to facilitate cyt c_y's interactions with various components of the energy transducing machinery of *R. capsulatus*.

Electron transport reactions between the cyt bc_1 complex and the RC being a prototype of biological electron transfer reaction, it is worthwhile to study further whether the observed differences in the oxidation/reduction kinetics of the cyts c_y and c_2 are due solely to electrostatic interactions in the periplasm, or they are caused at least partially by interactions within the lipid bilayer (and/or the cytoplasm). To study the precise role of the N-terminal extension of cyt c_y in determining its characteristics, the construction of chimera between the cyts c_y and c_2 are in progress.

4. References

1. Daldal, F. (1990) in The Molecular biology of Membrane-bound Complexes in Phototrophic Bacteria (Drews, G and Dawes, E.A., eds.) pp. 85-93, Plenum Press
2. Daldal, F., Cheng, S., Applebaum, J., Davidson, E. and Prince, R.C. (1986) Proc. Natl. Sci. 83: 2012-2016
3. Prince, R.C., Davidson, E., Haith, C.E. and Daldal, F. (1986) Biochem. 25: 5208-5214
4. Jenney, F.E. and Daldal, F. (1993) EMBO J. 12: 1283-1292
5. Jenney, F.E., Prince, R.C. and Daldal, F. (1994) Biochem. 33: 2496-2502
6. Gray, K.A., Grooms, M., Myllykallio, H., Moomaw, C., Slaughter, C. and Daldal, F. (1994) Biochem. 33, 3120-3127
7. Jennings, M. (1989) Annu. Rev. Biochem. 58:999-1027

CRYSTAL STRUCTURE OF CYTOCHROME c_6 FROM THE GREEN ALGA *Monoraphidium braunii*. COMPARISON WITH PLASTOCYANIN

M.A. De la Rosa[1], C. Frazão[2], C. M. Soares[2], M.A. Carrondo[2,3], E. Pohl[4], Z. Dauter[5], K.S. Wilson[5], M. Hervás[1], J.A. Navarro[1] and G. Sheldrick[4]
[1]Instituto de Bioquímica Vegetal y Fotosíntesis, Sevilla, Spain
[2]Instituto de Tecnologia Química e Biológica, Oeiras, Portugal
[3]Instituto Superior Técnico, Oeiras, Portugal
[4]Institut für Anorganische Chemie, Universität Göttingen, Germany
[5]European Molecular Biology Laboratory, Hamburg, Germany

1. Introduction

Cytochrome (Cyt) c_6 is a soluble electron transport protein which is synthesized by some species of green algae and cyanobacteria to replace plastocyanin (Pc) when grown in the absence of copper and/or at a high iron concentration (*cf.* Moore and Pettigrew [1] and Sandmann [2]). These two proteins play the same physiological role in photosynthesis, i.e. the transfer of electrons from Cyt f to photosystem I (PSI). PSI reduction by Pc or Cyt c_6 exhibits an almost identical reaction mechanism when the two proteins are isolated from the same organism [3], a fact that has to be compulsorily explained by assuming a high degree of structural homology between the two metalloproteins. A great deal of structural information exists for Pc from higher plants and green algae [4], but up to now little was known about the heme-protein structure. Recently, Campos *et al.* [5] reported that Cyt c_6 from the green alga *Monoraphidium braunii* exhibits unusual EPR and Mössbauer spectra, which led to the proposal of two interchangeable acid-base forms with a pK_a of 5.0. Even more recently, Frazão *et al.* [6] reported the crystallization and preliminary X-ray diffraction studies of this same Cyt c_6, which have just culminated in the determination of its crystal structure at almost atomic resolution [7]. This work provides the first high resolution structure of Cyt c_6 from any organism, as well as a comparative structural analysis of Cyt c_6 and Pc from eukaryotic algae.

2. Procedure

The crystal structure of Cyt c_6 from the green alga *Monoraphidium braunii* has been solved by *ab initio* methods from single wavelength native data alone. No precise structural information was available for any Cyt c_6 prior to this study. The key to success was the collection of synchrotron data with an effective resolution of 1.2 Å and the presence of one iron and three sulfur atoms in the Cyt molecule. Extensive description of synchrotron diffraction data collection and processing, *ab initio* methods used (similar to those used for small molecules) and final structure refinement is presented in Frazão *et al.* [7].

P. Mathis (ed.), Photosynthesis: from Light to Biosphere, Vol. II, 539–542.

3. Results and Discussion

As shown in Fig. 1, the secondary structure of *Monoraphidium* Cyt c_6 follows the typical topology of class I *c*-type Cyts. Four α-helices and three loops, which contain six reverse turns, envelope the heme group in a hydrophobic pocket but leave exposed the edge of pyrrole ring C and D propionate group. The heme group is covalently attached to the polypeptide chain *via* two thioether bonds, one with Cys15 and the other with Cys18. The octahedral coordination of the iron atom is completed by the axial ligands His19 and Met61. The only other histidine residue, His30, is at the molecular surface, close to the exposed heme propionate group. The low isoelectric point of this Cyt (p*I* = 3.6) is a direct consequence of the high ratio of acidic to basic amino acids in the protein surface. Actually, an acidic negative region formed by residues Glu1, Asp69, Glu70, Asp71 and Glu72 becomes evident. Further details of the Cyt c_6 structure are given in [7].

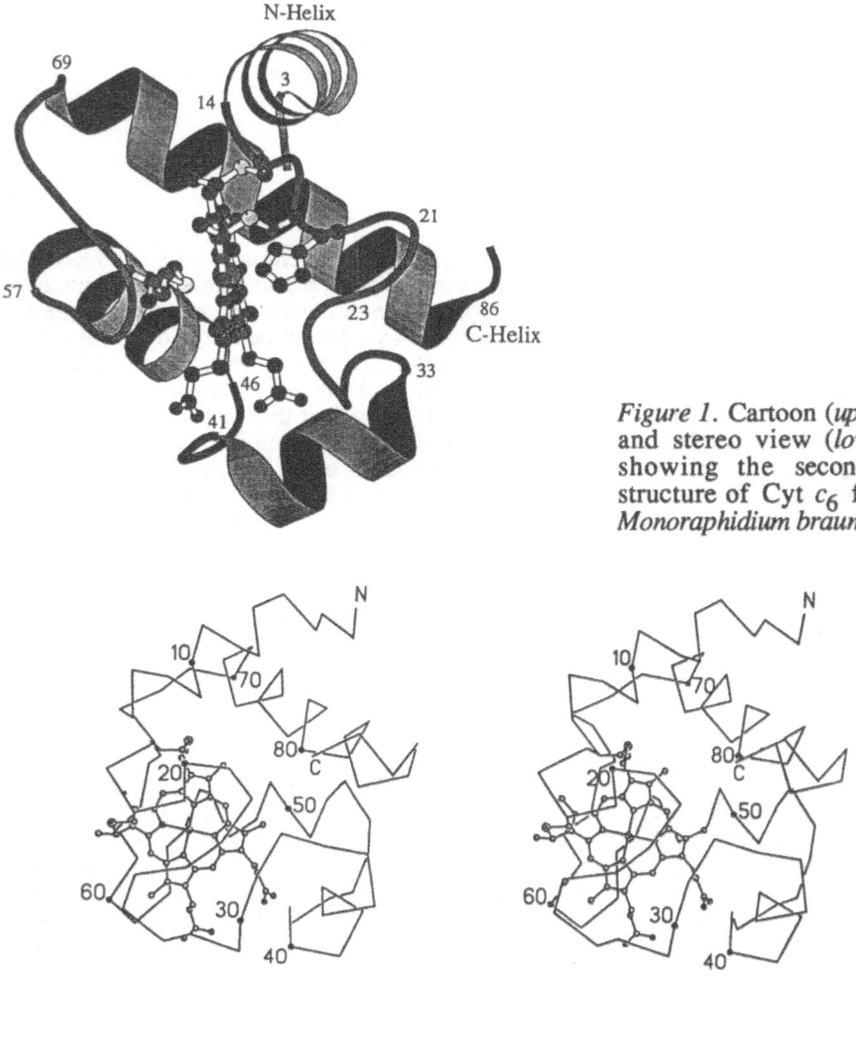

Figure 1. Cartoon (*upper*) and stereo view (*lower*) showing the secondary structure of Cyt c_6 from *Monoraphidium braunii.*

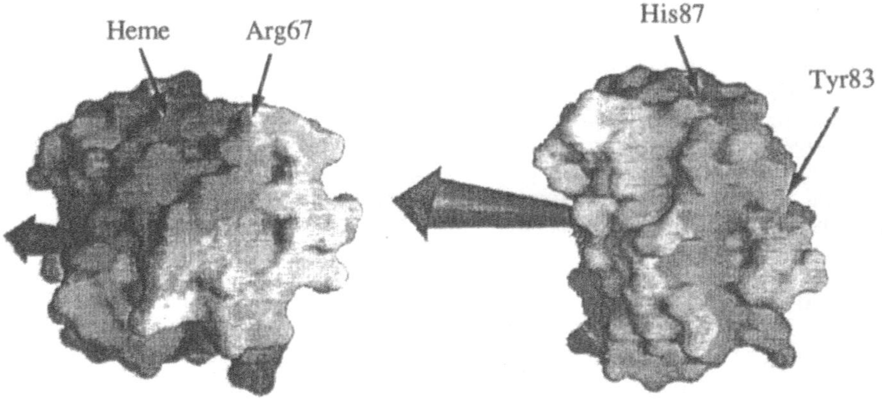

Figure 2. Comparison of the structures of *Monoraphidium* Cyt c_6 (*left*) and *Chlamydomonas* Pc (*right*) with molecular dipole moments aligned. The pictures show the electron transfer decay coupling mapping at the molecular surface. Darker zones correspond to high coupling values.

The structure of *Monoraphidium* Cyt c_6 herein described has been compared with that of Pc from the green alga *Chlamydomonas reinhardtii*, which has been recently solved at 1.5 Å resolution by Redinbo *et al.* [8]. Even though the heme- and copper-proteins have different topologies (the first is a hemic, α-helix based structure; the second, a β-sheet with β-strands), their main electrostatic characteristics are comparable. In fact, the two proteins exhibit a similar asymmetry, which is created by negatively charged areas and results in a net dipole moment in each molecule (Fig. 2). This allowed us to identify similar regions in the heme- and copper-proteins. The Cyt c_6 molecule contains not only an acidic region equivalent to the well-known "east negative face" of the Pc molecule, but also an electron transfer surface around the heme group which could be closely related to the "hydrophobic north pole" surrounding His87 of the copper-protein [4]. His87 has been proposed to be involved in the electron transfer pathway from the copper atom in Pc to the photooxidized chlorophyll molecule P700 in PSI. On the basis of these relative orientations, the heme group (or vicinal amino acid residues) might play the same role in Cyt c_6 as His87 in Pc regarding the PSI-protein redox interactions.

The equivalence between Cyt c_6 and Pc is, however, scarcely found with regard to their putative electron transfer sites for Cyt *f*. Actually, a critical difference becomes evident when one fails to find in the Cyt molecule an aromatic residue in its acidic patch equivalent to Tyr83 in the copper-protein. In principle, such a failure would be not so critical as Tyr83 in Pc does not appear to be involved in electron transfer to P700 [9]. Notwithstanding, Tyr83 (or alternatively another aromatic residue) is required for efficient electron transfer from Cyt *f* to Pc, thereby indicating that electron transfer from Cyt *f* and to P700 could follow different routes in the Pc molecule. All these findings can be explained by assuming that the Cyt c_6 molecule contains just one active site for electron transfer, which is used for interaction with both Cyt *f* and PSI, whereas the Pc molecule reacts with its redox partners in two different areas. A detailed analysis of this hypothesis would include site-directed mutagenesis and rapid kinetic studies.

We have already cloned and properly expressed in *E. coli* the genes coding for Cyt c_6 (*petJ*) and Pc (*petE*) from the cyanobacterium *Synechocystis* sp. PCC 6803 [10,11]. This is an adequate tool to address in a parallel way structural and functional studies with the two metalloproteins isolated from the same organism. The interaction of wild and mutant proteins, both with PSI and Cyt *f*, will thus shed new light on the role played by a number of specific amino acids in two different processes: the binding between the donor and acceptor partners, i.e. the placement of the two molecules into physical contact to prepare for the second phase, the actual process of electron migration from the donor to the acceptor centre.

Acknowledgements

This work has been supported by the JNICT, Portugal (Grant PBIC/C/BIO/1277/92), the DGICYT, Spain (Grants PB90-0099 and PB93-0922), the European Union (Networks CHRX-CT92-0072, CHRX-CT93-0143 and CHRX-CT94-0540, and project BIO2-CT94-2052), and the Fonds der Chemischen Industrie.

References

1 Moore, G.R. and Pettigrew, G.W. (1990) Cytochromes c. Evolutionary, Structural and Physicochemical Aspects, Springer, Berlin, Heidelberg
2 Sandmann, G. (1986) Arch. Microbiol. 145, 76-79
3 Hervás, M., Navarro, J.A., Díaz, A., Bottin, H. and De la Rosa, M.A. (1995) Biochemistry, in press
4 Gross, E.L. (1993) Photosynth. Res. 37, 103-116
5 Campos, A.P., Aguiar, A.P., Hervás, M., Regalla, M., Navarro, J.A., Ortega, J.M., Xavier, A.V., De la Rosa, M.A. and Teixeira, M. (1993) Eur. J. Biochem. 216, 329-341
6 Frazão, C., Dias, J.M., Matias, P.M., Romao, M.J., Carrondo, M.A., Hervás, M., Navarro, J.A., De la Rosa, M.A. and Sheldrick, G.M. (1995) Acta Crystallogr. D51, 232-234
7 Frazão, C., Soares, C.M., Carrondo, M.A., Pohl, E., Dauter, Z., Wilson, K.S., Hervás, M., Navarro, J.A., De la Rosa, M.A. and Sheldrick, G.M. (1995) Structure, submitted
8 Redinbo, M.R., Cascio, D., Choukair, M.K., Rice, D., Merchant, S. and Yeates, T.O. (1993) Biochemistry 32, 10560-10567
9 Haehnel, W., Jansen, T., Gause, K., Klösgen, R.B., Stahl, B., Michl, D., Huvermann, B., Karas, M. and Herrmann, R.G. (1994) EMBO J. 13, 1028-1038
10 Hervás, M., Navarro, F., Navarro, J.A., Chávez, S., Díaz, A., Florencio, F.J. and De la Rosa, M.A. (1993) FEBS Lett. 319, 257-260
11 Díaz, A., Navarro, F., Hervás, M., Navarro, J.A., Chávez, S., Florencio, F.J. and De la Rosa, M.A. (1994) FEBS Lett. 347-173-177

Electron transfer carriers in obligately aerobic photosynthetic bacteria from genera *Roseococcus and Erythromicrobium.*

Vladimir Yurkov, Barbara Schoepp and André Verméglio
CEA-DEVM/LBC- C.E. Cadarache, 13108 St. Paul lez Durance, France

1. Introduction

Obligately aerobic bacteriochlorophyll (Bchl) a-containing bacteria have been found at several places in different ecological niches. They are presented by marine species of genera *Erythrobacter* and *Roseobacter* (1,2), and freshwater species of genera *Erythromicrobium, Roseococcus* (3,4) and *Porphyrobacter* (5). The main features of the group are: unability to grow photosynthetically under anaerobic conditions despite of Bchl a production; low level of photosynthetic units in cells; strong inhibition by light of Bchl synthesis (6,7).

Several photoheterotrophic bacteria have the capability to transform energy for life processes by both photosynthesis and respiration. These electron (e^-) transfer systems share several different carriers like soluble cytochromes (cyt), quinone molecules and cyt *bc1* complex. In addition, the photosynthetic apparatus of all purple bacteria contains the photochemical reaction center (RC) and the antenna light-harvesting (LH) complexes (8).

Among obligately aerobic phototrophes cyt and RCs have been isolated from two species only - *Erythrobacter longus* and *Roseobacter denitrificans* (9,10). Investigation of the cyt patterns and their functions in newly described photosynthetic representatives will extend our knowleges about diversity of energetic pathways and may be used together with genetic data to specify a classification of freshwater aerobic bacteria containing Bchl a.

2. Materials and Methods

The following species were investigated: *Roseococcus thiosulfatophilus,* RB3; from genus *Erythromicrobium*: *E. ramosum,* E5; *E. sibiricum,* RB16-17; *E. ursincola,* KR99; *E. ezovicum,* E1; *E. hydrolyticum,* E4(1). All strains were cultivated in an airlift with a strong bubbling of sterile air in the dark at 30° C and pH 7.6 to 7.8 in medium described in (4).

Fractionation of cells and isolation of LHI-RC enriched fractions has been described previuosly (11).

The cyt in soluble, membrane and LHI-RC enriched fractions were assayed by determining reduced minus oxidized spectra at room temperature.

The presence of *c*-type cyt and their molecular weights were analysed by SDS-PAGE on an 11.5-16.5% linear gradient of acrylamide and were stained using 3,3',5,5'-tetramethylbenzidine (TMBZ) (12). Spectrophotometric measurements were performed with an apparatus similar to that one described in (13).

P. Mathis (ed.), Photosynthesis: from Light to Biosphere, Vol. II, 543–546.

3. Results and Discussion

We have analyzed the light-induced absorbance changes in whole cells of six aerobic photosynthetic bacteria in order to characterize the e^- donation to the RC and component involved in the cyclic e^- transfer. The experiments have been performed under both anaerobic and aerobic conditions. Under anaerobic conditions no light-induced absorbance changes can be observed for all species studied (data not shown). Such situation has already been reported for *R. denitrificans* (10,14) and explained by the rather high value of the mid-point potential of the primary e^- acceptor Q_A.

The analysis of the light-induced difference spectra in the presence of oxygen lead us to distinguish two different behaviours depending on the considered species. Very fast cyt photooxidation observed after flash exitation in the cells of *E. ursincola, E. sibiricum* and *R. thiosulfatophilus* indicates that immediate e^- donor to the RC in these species is a bound cyt. The light-induced difference spectra detected at 50 µs and 5 ms after flash excitation in the case of *E. ursincola* are depicted in Fig.1A. The difference spectrum measured at 50 µs is centered at 556 nm in agreement with the wavelength position of high potential hemes of RC bound cyt described for other purple bacteria. At 5 ms, the wavelength of maximum bleaching is centered at 551 nm and certainly corresponds to the photooxidation of a soluble cyt *c* as already observed for whole cells of *R. viridis* and *R. denitrificans* (10,15). The rate of e^- transfer between the high potential heme and the soluble fraction can be followed at 547 nm (Fig.1B). The half-time of this reaction is found to be equal to 250 µs. The soluble cyt *c* is then rereduced with an half-time of 10 ms. A very similar bahaviour is observed for *R. thiosulfatophilus* and *E. sibiricum* (not shown).

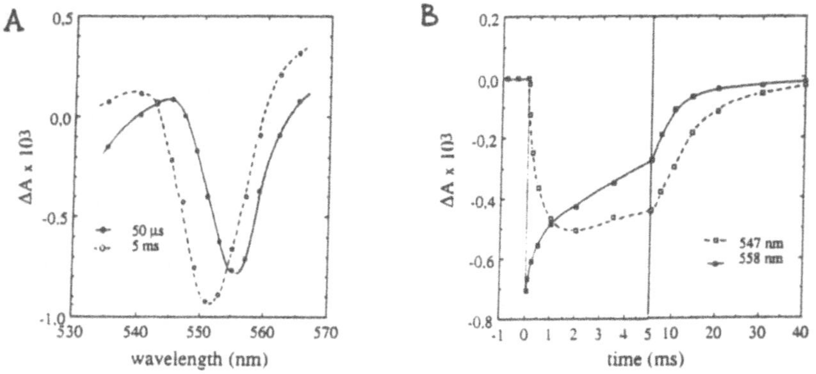

Figure1.Difference absorption spectra determined at 50µs and 5ms after a saturating flash (A) and kinetics of oxidation and re-reduction of cyts measured at 547 and 558 nm for *E. ursincola* whole cells (B).

The species *E. ramosum, E. hydrolyticum* and *E. ezovicum* do not possess a RC bound cyt. The light-induced difference spectrum detected at 50 µs corresponds to the c_2 type cyt photooxidation with an absorption band centered at 550 nm (Fig.2A;data for *E. hydrolyticum)*. There is no detectable spectral shift between 50 µs and 5 ms. Like it was observed in the cells of *R. sphaeroides*, the cyt photooxidation detected at 551-542 nm presents two phases: a fast phase (less than 50 µs) not resolved in our experiments and a slower phase with an half-time of 250 µs. This slow phase can also be found in the kinetics of P^+ rereduction followed at 542 nm or 602.5 nm (Fig.2B). The half-time of the cyt rereduction is about 40 ms.

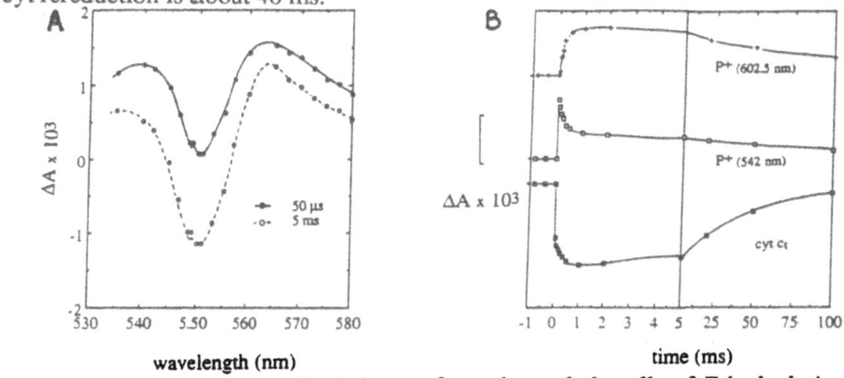

Figure 2. The same as Fig.1, but performed on whole cells of *E.hydrolyticum*.

The experiments described for whole cells of six species suggest that only *R. thiosulfatophilus, E. sibiricum* and *E. ursincola* contain a RC bound tetrahemic cyt *c*. To further support of that hypothesis and characterize the cyt content in studied species we have performed redox titrations and gel electrophoresis of their soluble and membrane fractions and LHI-RC particules. The results are summarized in Table 1.

Species	Soluble cyt				Membrane bound cyt			
	λ (nm)	Em (mV)	M.w. (kDa)	%	λ (nm)	Em (mV)	M.w. (kDa)	%
R.thiosulfatophilus	549	210	6,5	55	557	270		
	552	30	4,0	45	555	240		
					553	40	44,0	60
					553	-60		
					?	160	26,0	10
					?	100	23,0	15
					?	0	21,5	15
E.ursincola	549	312	9,0	45	557	330		
	550	292	6,5	48	556	305		
	c'	60	14,3	?	551	40	40,0	50
					551	-40		
					?	280	24,0	28
					?	180	21,5	10
					?	0	14,3	8
E.sibiricum	550	296	14,0	100	553	330		
					553	300		
					551	30	44,0	70
					551	-40		
					?	190	30,0	30
E.ramosum	549	330	14,3	90	551	325	30,0	30
	549	282	8,0	10	552	300	26,0	30
					550	215	8,0	30
					560	30	?	?
					563	-110	?	?
E.hydrolyticum	550	295	14,3	100	550	352	21,0	50
					552	260	30,0	
					?	80	?	50
					563	-150	?	
E.ezovicum	549	340	14,3	90	552	350	34,0	50
	549	315	8,0	10	550	290	30,0	
					560	45	?	50
					?	-110	?	

Table 1. Some cyt properties, isolated from dark-aerobic grown cells. (Em values are in a range of ±10 mV; ? means not estimated).

PAGE of LHI-RC particules of *R. thiosulfatophilus, E. sibiricum* and *E. ursincola* confirmed the presence of RC bound tetrahemic cyt *c*. The apparent molecular weights of

RC bound cyt (44.0 kDa in *R. thiosulfatophilus* and *E. sibiricum* and 40.0 kDa in *E. ursincola*) are quite similar to those described for another photosynthetic bacteria.

In addition to the RC bound tetrahemic cyt *c*, several membrane bound cyt are revealed by redox titration. *bc1* complex can be evidenced without ambiguities from the titrations and PAGE analysis in *E. hydrolyticum, E. ramosum* and *E. ezovicum*. The Em values for the high and low potential cyt *b* range from 70 to 30 mV and from 20 to -150 mV, respectively. In the other species, due to the presence of RC bound tetrahemic cyt, determination of the *bc1* hemes redox properties is render difficult. A membrane bound cyt *c* with Em around 350 mV is observed in *E. hydrolyticum, E. ramosum* and *E. ezovicum*. We have observed no cyt photooxidation for isolated membranes of *E. hydrolyticum, E. ramosum* and *E. ezovicum* (data not shown), which implied that, at least in these species, reduction of the photooxidized primary donor involves a soluble cyt *c* and not this membrane bound e^- carrier as postulated in *Protaminobacter ruber* (16).

One soluble cyt *c* with a mid-point potential equal to 295 mV is present in the soluble fractions of *E. sibiricum* and *E. hydrolyticum,* which is certainly involved in both the respiratory and photosynthetic activities. Only one soluble cyt *c* possess a redox potential high enough (Em 210 mV) to sustain respiration and photosynthesis in *R. thiosulfatophilus*. Several soluble *c*-type cyt are found in *E. ursincola, E. ramosum* and *E. ezovicum*. Purification and reconstitution experiments with isolated membranes will be necessary to assign a precise role of these cyt in electron transfer chains.

In conclusion, we have demonstrated that an efficient photoinduced e^- transfer is operative under aerobic conditions in intact cells of freshwater aerobic bacteria - *R. thiosulfatophilus, E. ursincola, E. sibiricum, E. hydrolyticum, E. ramosum* and *E. ezovicum.*

References

1. Shiba, T., and Simidu, U. (1982) Int. J. Syst. Bacteriol. 32, 211-217.
2. Shiba, T. (1991) System. Appl. Microbiol. 14, 140-145.
3. Yurkov, V., Lysenko, A. M., and Gorlenko, V. M. (1991) Microbiology 60, 362-366.
4. Yurkov, V., Stackebrandt, E., Holmes, A., Fuerst, J.A., Hugenholtz, P., Golecki, J., Gad'on, N., Gorlenko, V. M., Kompantseva, E. I., and Drews, G. (1994) Int. J. Syst. Bacteriol. 44, 427-434.
5. Fuerst, J. A., Hawkins, J. A., Holmes, A., Sly, L. I., Moore, C. I., and Stackebrandt, E. (1993) Int. J. Syst. Bacteriol. 43, 125-134.
6. Harashima, K., Kawazoe, K., Yoshida, I., and Kamata, H. (1987) Plant Cell Physiol. 28, 365-374.
7. Yurkov, V., and van Gemerden, H. (1993) Arch. Microbiol. 159, 84-89.
8. Drews, G. (1989) in Biology of autotrophic bacteria (Schlegel H. G. and Bowien B. ed.) pp 461-480, Science Tech. Publ. Madison.
9. Takamiya, K. (1989) in Aerobic photosynthetic bacteria (Harashima K., Shiba T., and Murata N. eds) pp 73-95, Springer- Verlag, Berlin.
10. Garcia, D., Richaud, P., Breton, J., and Vermeglio, A. (1994) Biochimie. 76, 666-673.
11. Yurkov, V., Gad'on, N., Angerhofer, A., and Drews, G. (1994) Z. Naturforsch. (in Engl.) 49(c), 579-586.
12. Thomas, P. E., Ryan, D., and Lewin, W. (1976). Analyt. Biochem. 75, 168-176.
13. Joliot, P., Beal, D., and Frilley, B. (1980) J. Chim. Phys. 77, 209-216.
14. Okamura, K., Mitsumori, F., Ito, O., Takamiya, K., Nishimura, M. (1986) J. Bacteriol. 168, 1142-1146.
15. Garcia,D., Richaud,P., Vermeglio,A. (1993). Biochim. Biophys. Acta. 1144, 295- 301.
16. Takamiya, K. (1995) Arch. Microbiol. 143, 15-19.

ELECTROGENIC PROTON DISPLACEMENTS IN THE CYTOCHROME bc_1-COMPLEX OF *Rhodobacter capsulatus*.

A. Mulkidjanian and W. Junge
Dep. of Biophysics, University of Osnabrück, D-49069 Osnabrück, Germany

1. Introduction

The cytochrome bc_1-complex, an electrogenic ubiquinol:cytochrome *c*-oxidoreductase, is formed from cytochrome *b* with low- and high-potential protohemes (b_l and b_h, respectively), cytochrome c_1 and FeS protein [1]. Robertson and Dutton [2] have attributed the electrogenicity exclusively to the transmembrane electron transfer: first from quinone-oxidizing center *P* and heme b_l on the periplasmic *p*-side to heme b_h in the middle of the membrane and secondly from heme b_h to the quinone reducing center *N* on the cytoplasmic *n*-side. Other workers, from the other hand, advocate the electrogenicity of proton displacements in the bc_1-complex [3-6].

In this work we monitored the kinetics of electron transfer, $\Delta\psi$ formation (measured by the electrochromic bandshift of carotenoids) and proton release into the lumen (measured with neutral red as pH-indicator) in *Rb. capsulatus* chromatophores under various conditions. Our data support the electrogenicity of proton displacements in the bc_1-complex and we discuss a model which allosterically couples the quinone-binding sites.

2. Materials and Methods

The preparation of chromatophores and the spectrophotometric measurements were conducted as described in [5,6]. The incubation medium contained 2 mM potassium ferrocyanide, 5 µM N,N,N',N'-tetramethyl-*p*-phenylendiamine (TMPD), 1 mM KCN; E_h=250-260 mV. Bovine serum albumin (BSA, 2 mg/ml) was used as a pH-buffer.

3. Results and Discussion

In the absence of inhibitors of the bc_1-complex heme b_h was reduced with $t_{1/2}$ of about 2.5 ms (Fig. 1a), whereas the onset of (b) the voltage transient, (c) the lumenal acidification (the time-unresolved fast component is unrelevant to the bc_1-complex [6]) and (d) cytochrome *c* reduction (cytochromes c_1 and c_2 were not distinguished) was slower ($t_{1/2}$=8-10 ms). In the presence of antimycin A (inhibitor of center *N*) the onsets of the voltage transient (f), of the lumenal acidification (g) and of the cytochrome *c* reduction (h) were again similar but faster than in the absence of antimycin A ($t_{1/2}$ = 4-5 ms). The heme b_h was reduced (e) somewhat slower ($t_{1/2}$ = 6-7 ms). In the presence of myxothiazol (inhibitor of center *P*) the reduction of heme b_h (i) was coupled with a reverse electrogenic phase (*j*) and with an alkalinization of the lumen (k).

P. Mathis (ed.), Photosynthesis: from Light to Biosphere, Vol. II, 547–550.

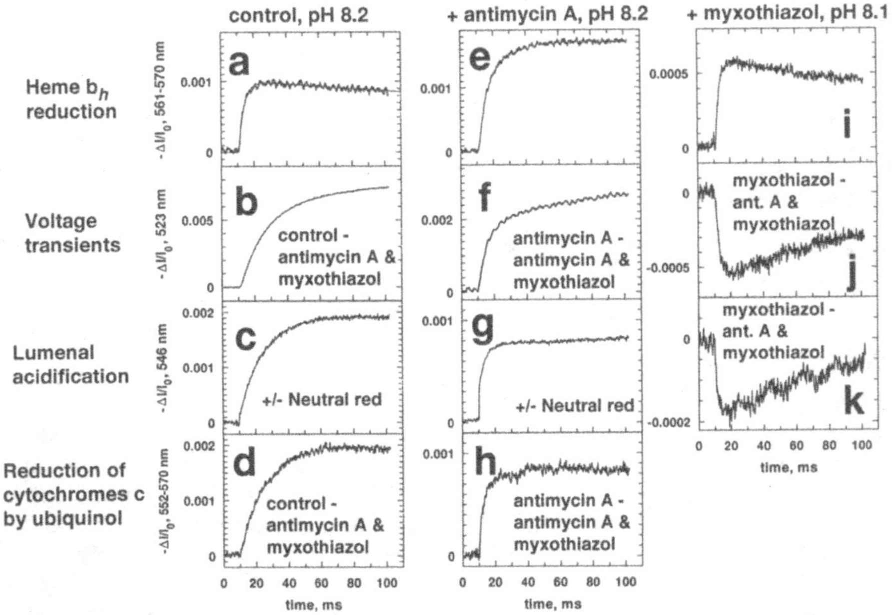

Fig. 1. Flash-induced absorbance transients in chromatophores of *Rb. capsulatus*. Incubation medium as in Materials and Methods plus 100 nM venturicidin. Additions where indicated: 5 μM antimycin A, 3 μM myxothiazol, 13 μM (*c*) or 39 μM (*g*, *k*) neutral red. (See text for details).

To check whether the rate of heme b_h reduction (Fig. 1a) was *apparently* fast due to a kinetic truncation by its oxidation via center *N* we used the sensitivity of the latter reaction to $\Delta\mu H^+$. Varying $\Delta\mu H^+$ by ionophores or inhibitors of ATP-ase it was possible to increase the observable fraction of reduced hemes b_h at pH 8.2 from one third (with valinomycin, not shown) to more than half of the total amount (with oligomycin, Fig. 2, upper trace). The rate of heme b_h reduction had not changed (the τ value varied around 3 ms). We conclude that the oxidation of heme b_h is limited thermodynamically but not kinetically and that the observed rate of heme b_h reduction reflected the *genuine* rate of electron delivery from ubiquinol in center *P*.

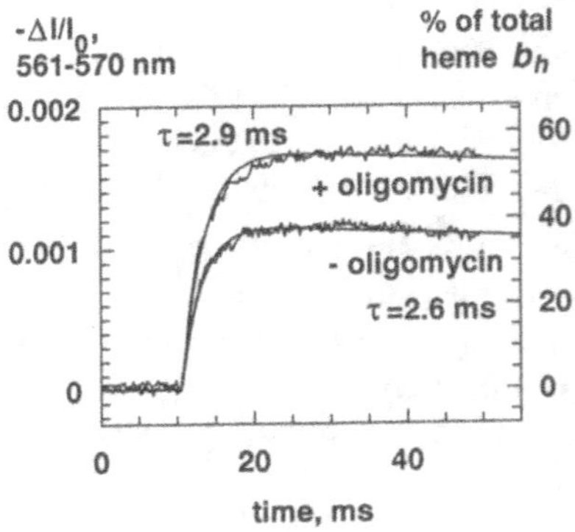

Fig. 2. Flash-induced reduction of heme b_h in the absence (lowere trace) and in the presence of oligomycin (1 μM). Incubation medium as in Materials and Methods.

The lower rate of the electrogenic reaction compared to the heme b_h reduction (Fig. 1a and 1b) could be caused by the contribution of two or more turnovers of the bc_1-complex. To check this we applied a flash that excited only 12% of RCs and thus provided ubiquinol for only one turnover of some bc_1-complexes. Fig. 3a shows that in this case (dashed trace) the electrogenic reaction was not faster than after a saturating flash (solid trace) but rather slower. The $\Delta\psi$ transients measured in the presence of antimycin A and reflecting the rate of ubiquinol oxidation were independent on the intensity of the flash (Fig 3b) thus ruling out the slowing of the latter reaction after weak flash.

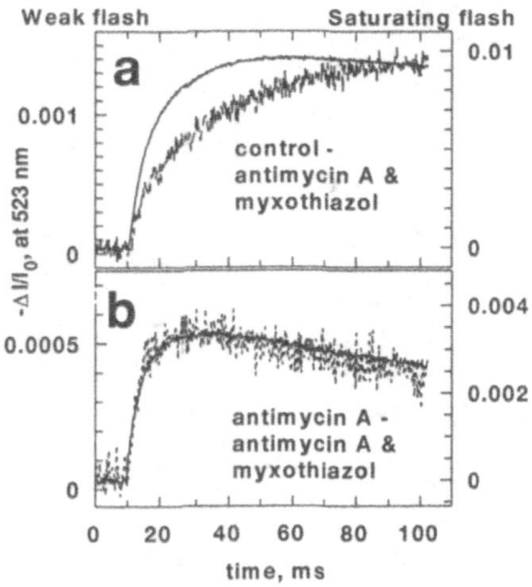

Fig. 3. Dependence of the electrogenicity in the bc_1-complex on the intensity of the exciting flash (note different scales). Conditions as in Fig. 1, pH 6.9.

We explain the data by a Q-cycle-type scheme [7,8] which is shown below. We assume that: 1) FeS and both hemes *b* are in protonic equilibrium with periplasmic phase *p*; 2) the quinone-binding centers *P* and *N* oscillate between *open* conformational substates which permit the exchange with the membrane quinone pool and the *closed* ones containing a bound and stabilised semiquinone. 3) We hypothesise, that the quinone-binding centers are conformationally coupled: *closing* of center *N* prevents *closing* of center *P* (shown by a "*spring*").

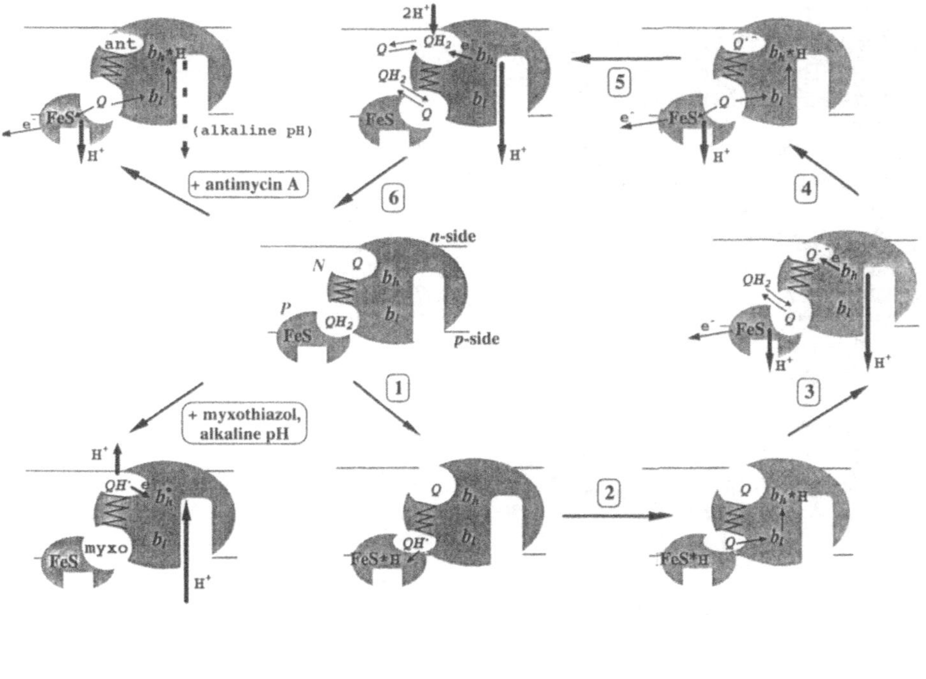

In the absence of inhibitors the first electron (e_1^-) did not appear on cytochrome *c* faster than e_2^- on heme b_h (compare Fig. 1a and 1d). We hypothesise that after the formation of semiquinone and *closing* of center *P* the E_m value of FeS is positevely shifted and e_1^- is trapped. The two protons that are released stay bound with FeS (H_1^+) and heme b_h (H_2^+) to compensate electrostatically the delivered electrons. Therefore the reduction of heme b_h is electrically silent (compare Fig. 1a and 1b). The stabilization of semiquinone in center *N* and its *closing* is suggested to promote the *opening* of center *P*. This causes decrease in the E_m value of FeS and enables the delivery of e_1^- to cytochromes *c* (Fig. 1d). The electrogenic release of H_1^+ follows the oxidation of FeS whereas the electrogenic release of H_2^+ follows the disapperance of the negative charge from the cytochrome *b* moiety. Therefore the initial rates of electrogenesis, lumenal acidification and re-reduction of cytochromes *c* in Fig. 1b-1d are similar.
In the presence of antimycin A e_1^- is not trapped by FeS and appears on cytochromes *c* faster (Fig. 1h). We suggest that antimycin A tightly bound in center *N* prevents *closing* of center *P* and the shift in the midpoint potential of FeS. From data in Fig. 1e-h we assume that the electrogenic reaction in the presence of antimycin A is coupled with the release of H_1^+ into the lumen. The attribution of this reaction to heme b_h reduction [9] is unlikely as the voltage transient (Fig. 1f) develops faster than heme b_h reduction (Fig. 1e) and follows the onsets of proton release (Fig. 1g) and cytochrome *c* reduction (Fig. 1h). As the oxidation of heme b_h is blocked, H_2^+ stays bound with reduced heme b_h at pH < pK of heme b_h (estimated as 7.5 in [10]) but is slowly electrogenically released at pH values higher than this pK (see [6,10] and the slower component in Fig. 1f).

In the presence of myxothiazol and under alkaline conditions the reduction of heme b_h by ubiquinol via center *N* takes place (Fig. 1i) and is accompanied with a reverse electrogenic phase (Fig. 1j). We assume that this electrogenesis is not limited to the electron transfer from center *N* to heme b_h as suggested in [2] but is additionaly contributed by proton release from center *N* [3,11] and by an electrostatic attraction of protons from phase *p* (Fig 1k).

Acknowlegements. A.M. is grateful to the Alexander von Humboldt Stiftung for a Research Fellowship. This research was supported by a grant to W.J. from the Deutsche Forschungsgemeinschaft (SFB 171--A2).

References

1 Trumpower, B.L. & Gennis, R.B. (1994) *Annu. Rev. Biochem. 63*, 675-716.
2 Robertson, D.E. & Dutton, P.L. (1988) *Biochim. Biophys. Acta 935*, 273-291.
3 Drachev, L.A., Kaurov, B.S., Mamedov, M.D., Mulkidjanian, A.Y., Semenov, A.Y., Shinkarev, V.P., Skulachev, V.P. & Verkhovsky, M.I. (1989) *Biochim. Biophys. Acta 973*, 189-197.
4. Konstantinov, A.A. (1990) *Biochim. Biophys. Acta 1018*, 138-141.
5 Mulkidjanian, A.Y., Mamedov, M.D., & Drachev, L.A. (1991) *FEBS Lett. 284*, 227-231.
6 Mulkidjanian, A.Y. & Junge, W. (1994) *FEBS Lett. 353*, 189-193.
7 Mitchell, P. (1976) *J. Theor. Biol. 62*, 327-367.
8 Crofts, A.R., Meinhardt, S.W., Jones, K.R., & Snozzi, M. (1983) *Biochim. Biophys. Acta 723*, 202-218.
9 Glaser, E. & Crofts, A.R. (1984) *Biochim. Biophys. Acta 766*, 322-333.
10 Petty, K.M. & Dutton, P.L. (1976) *Arch. Biochem. Biophys. 172*, 346-353.
11 Mulkidjanian, A.Y., Mamedov, M.D., Semenov, A.Y., Shinkarev, V.P., Verkhovsky, M.I., & Drachev, L.A. (1990) *FEBS Lett. 277*, 127-130.

A MÖSSBAUER STUDY OF THE CYTOCHROME BF COMPLEX FROM SPINACH

Jürgen Illerhaus[a], Y. Sanakis[b], Vasili Petrouleas[b], Wolfgang Haehnel*[a]
[a]Institut für Biologie II, Biochemie der Pflanzen, Universität Freiburg, Schänzlestr. 1, 79104 Freiburg, Germany
[b]Institute of Materials Science, NCSR "Democritos" 15310 Aghia Paraskevi, Athens, Greece

1. Introduction

The function of the cytochrome (cyt) bf complex depends on three redox-center carrying subunits of which each contains at least one iron. These are cyt f with one hem, cyt b with two hemes and the Rieske-2Fe2S protein. The ligands and the midpoint potential of these redox-centers are quite different from each other. The Mössbauer spectroscopy is exclusively sensitive to the iron centers. Its specificity makes it useful in predicting the structural and electronic features of iron centers in proteins. To our knowledge the only related Mössbauer study is that of a Rieske protein isolated from *Thermus thermophilus*. (1)

2. Materials and Methods

Hydrocultures of spinach were grown in a growth-chamber with 20mM 95% ^{57}Fe in the nutriant. The cyt bf complex was isolated using octylglucoside / cholate as detergents and characterized by SDS-Page and spectrophotometrically as described in (2). Oxidoreductase activity was measured with plastoquinone as reductant and horse heart cytochrome c as oxidant. Mössbauer spectra were obtained using a constant acceleration spectrometer and a ^{57}Co(Rh) source (3). The sample was 200 μM in cyt f and the pH was 6.5.

3. Results

Our isolated cyt bf - complex is composed from the four majour subunits, cyt f, cyt b, the Rieske-FeS protein, and subunit IV and one or more of the small subunits in the

* Support by the EC contract BC02CT - 930077 is gratefully acknowledged.

P. Mathis (ed.), Photosynthesis: from Light to Biosphere, Vol. II, 551–554.

range of 4 kDa. The cyt b to cyt f ratio was 2. We detected also a chlorophyll molecule of unknown function at a molar ratio of cyt f / chlorophyll of 0.9. The oxidoreductase activity was 17 s^{-1} .

In the untreated state of the complex all these components are oxidized to a great extent as detected by optical measurements (data not shown). While the b cytochromes were fully oxidized about 20% of the cyt f was found in the reduced state. The EPR spectrum (data not shown) shows a weak contribution from reduced Rieske, which represents not more than 5% of the total. These relative amounts of the reduced components are compatible with the respective midpoint potentials of the Rieske-FeS protein and cyt f of 280 - 320 and 340 - 360 mV respectively (4).

Representative Mössbauer spectra at 220K and 4.2K are shown in Fig. 1A and B. The spectrum of the untreated sample (upper trace) shows a well defined outer doublet with peak positions at -0.8 and +1.0 mm/s. These represent the oxidized cyt b and cyt f. An asymmetric inner doublet with unresolved structure represents the oxidized Rieske and the 20% of reduced cyt f. The fully reduced sample (Fig. 1A, bottom) shows a prominent doublet of the reduced cyt b and cyt f. The small peaks at -0.8, 0.4 and 2.0 mm/s in addition to a contribution overlapping with the left cyt peak correspond to the reduced Rieske.

The ascorbate reduced sample (middle trace) shows an intermediate spectrum. Cyt f and the Rieske are reduced, cyt b is oxidized. The peak at -0.3 mm/s corresponds to the reduced cyt f. The other half of this doublet at 1.0 mm/s interferes with one peak of the oxidized cyt b. The other half of the doublet of ox. cyt b interferes with the reduced Rieske at -0.8 mm/s. At lower temperatures the oxidized cytochromes (S=1/2) undergoes progressive relaxation broadening and at 4.2K (Fig. 1B) the spectrum is fully split into a magnetic spectrum with weakly defined features at about -1.8 mm/s and 3.0 mm/s. The Rieske doublet is pronounced and has peaks at -0.25 and 0.6 mm/s. In the reduced state (data not shown) the Rieske shows relaxation broadening in

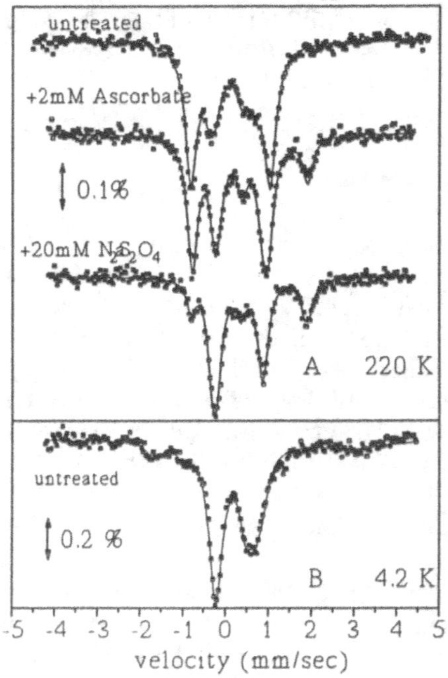

Figure 1: Representative Mössbauer spectra at 220K at various stages of reduction (A). Spectrum of the untreated sample at 4.2K (B).

contrast to the cytochromes.
The Mössbauer parameters of the isomeric shift (IS) the quadrupole splitting (QS) as well as the linewidth (Γ/2) are summarized in Table 1.
Preliminary computer fits of the spectra yield the data summarized in Table I.
An accurate determination of the abundance of the various components is currently complicated by the presence of an unidentified component representing 11% of the total iron. Following dithionite reduction this component appears as high spin Fe^{2+}. An approximate quantitation of the main contributions to the spectra show that 56% of the total iron is of cytochrome origin (cyt b and f) and 33% Rieske iron. Repeated freezing and thawing of the concentrated complex results in a partial denaturation of the Rieske and possibly of cyt b.

	T(K)	QS(mm/sec)	IS(Fe) (mm/sec)	Γ/2 (mm/sec)
Rieske oxid. [a]	4.2	0.66 (0.52)	0.24 (0.24)	0.19 (0.13)
	"	0.90 (0.91)	0.33 (0.32)	0.20 (0.13)
Rieske red. [b]	220	0.65 (0.61)	0.14 (0.22)	0.17
	"	2.76 (2.81)	0.68 (0.55)	0.13
cyt b + f oxid. [a]	220	1.85	0.22	0.15
cyt b + f red [c]	"	1.06	0.45	0.15

Table I: Mössbauer parameters of the isolated cyt bf - complex. QS: quadrupole splitting, IS: isomeric shift, Γ: linewidth. Values in parenthesis: Published data for Rieske isolated from *T. thermophilus*. The data were derived from measurements with [a] untreated sample, [b] ascorbatereduced, and [c] dithionite reduced sample.

4. Conclusions

The present Mössbauer data show that the components of the cyt bf complex can be distinguished either based on their different redox properties or their Mössbauer parameters. The Rieske parameters show similarities to those of the isolated protein from *T. thermopilus* (1) but small differences, not evaluated yet, appear to exist. The Rieske center and the cytochromes in the intact complex show faster relaxation properties than the isolated Rieske from *T. thermophilus* and models of cyt b (6). It is also interesting to note the relative small QS values of the cyt, which for the cyt b would suggest a nearly perpendicular orientation of the imidiazole planes (6). The distinct difference of the 5th. and 6th. ligands of cyt b and cyt f suggest that Mössbauer spectroscopy may reveal details of the iron centers.
Further analysis of the hyperfine parameters and binding studies with quinones or inhibitors may allow to discriminate the relative orientation of the 2Fe2S-cluster of the Rieske and the electronic structure of the cyt b hemes.

5. References

1 Fee, J. A., Findling, K. L, Yoshida, T., Hille, R., Tarr, G. E., Haershen, D. O., Dunham, W.R., Day, E.P., Kent, T. A., and E. Münck (1984) J. Biol. Chem. 259, 124-133

2 Hauska, E. (1986) Methods Enzym. 126, 271 - 285

3 Sanakis, Y., Petrouleas, V., and Diner, B.A. (1994) Biochemistry 33, 9922-9928

4 Hurt, E., and Hauska, G. (1982) J. Bioenerg. Biomemb. 14, 405-424

5 Martinez, S. E., Huang, D., Szczepaniak, A., Cramer, W. A., and Smith, J. L. (1994) Structure 2, 95-105

6 Medhi, O. K., and Silver, J. (1990) Chem. Soc. Dalton Trans. 263-270

THE MECHANISM OF CYT b6 PHOTOOXIDATION IN *Chlamydomonas reinhardtii* - THE POSSIBILITY OF A SEMIQUINONE CYCLE.[1]

Javier G. Fernández-Velasco, Deborah A. Berthold and Richard Malkin
Dept. Plant Biology. University of California. Berkeley, CA 94720-3102, USA

Introduction

The operation of (plasto)quinol:(plastocyanin)cyt c oxidoreductases (b6f complex in oxygenic systems) is generally explained by Mitchell's "Q-cycle" (1). In this model an oxidized quinone is released from the Q oxidizing site (Qo) after the concerted reduction of the high potential (Rieske-cyt f/c) and low potential (cyt bl, bh hemes) branches.The cyt bh is reoxidized at the Q-reducing site (Qr) by a quinone, either preexisting in the pool or the one released by the Qo site. The semiquinone formed would oxidize cyt bh in a second turnover, closing the catalytic cycle.The alternative "semiquinone-cycle" model proposed by Wikström and Krab (2) is similar in its first turnover, although without the reoxidation of cyt bh and formation of a semiquinone. In the completing second turnover, a semiquinone generated at Qo would move to Qr and oxidize cyt bh.

A mechanism called "reductant-induced-oxidation", has been previously put forth by Chain (3) to explain the photooxidation reaction of prereduced cyt b in spinach chloroplasts. This involves the generation by photoreduced ferredoxin of a semiquinone that interacts at Qr. Recently, Joliot and Joliot (4) have proposed that the light induced oxidant of prereduced cyt b in *Chlorella* is also a semiquinone, but in this case generated at the Qo site by the oxidized Rieske. After displacement, the semiquinone reacts at the Qr site oxidizing cyt bh.The main evidence for this is the development of a slow electrochromic signal (the putative semiquinone moving from Qo to Qr) that is faster than the oxidation of cyt b that follows after a 1-2 ms lag. This was taken as evidence that a semiquinone-type of reaction is, in principle, possible at low redox potentials.

Here we test both photooxidation models in *Chlamydomonas*. The former mechanism requires the reducing side of PS1 and must be insensitive to inhibitors of the Qo site. The latter requires the oxidizing side of PS1 and must be sensitive to inhibitors of Qo. Our results with a plastocyanin deficient (Pc^-) mutant of *Chlamydomonas* and inhibitors of the Qo and Qr sites, are consistent with the "semiquinone-cycle" mechanism at low Eh (2,4). Also we show the resolution of HQNO insensitive and sensitive phases of the slow electrochromic signal in agreement with the "semiquinone cycle".

Materials and Methods

Cultures of both wild type (CC-125) and Pc^- mutant (CC-1729) were grown autoheterotrophically (30 $\mu E.m^{-2}.s^{-1}$) at either 20 or 31°C. The mutant expresses the cyt c-552 carrier, even in the presence of copper, if grown at low but not at high temperatures (e.g.20 vs. 31°C) (5).Cells were permeabilized by nebulization and a fraction having a high rate of P700 rereduction was purified on a Percoll gradient. Flash induced redox reactions were studied at pH 7.0 in the presence of redox mediators and 10 µM of each valinomycin, nigericin and gramicidin at 25°C. They were monitored at 702-730 nm for P700, 554-547 nm for cyt f and 563-574 nm for cyt b. The carotenoid band shift was measured at 515 nm.

(1) Supported by a grant from the Nat. Institute of Health

P. Mathis (ed.), Photosynthesis: from Light to Biosphere, Vol. II, 555–558.

Results and Discussion

Fig. 1 shows that the wt strain, grown at either 20 or 31°C, contains very similar amounts of plastocyanin but no cyt c-552. Instead, the Pc$^-$ mutant can express cyt c-552 only at 20°C; no traces of plastocyanin are observed at high or low growth temperatures. Cyt f and cyt b (protein and spectrally resolved hemes) are present at comparable levels in both strains under all conditions. The wt preparation conserves plastocyanin that interacts rapidly with PS1-RC with sub-milisecond kinetics and with cyt f (t1/2 200 µs) (Fig.2). The stoichiometry of photoreactive P700:Pc:b6f is ca. 1:1:1. The preparations of wt cells grown at either 20 or 31°C show the same activity of the b6f complex, as concluded from the redox reactions of cyt f and b and the slow phase of electrochromic signal both at Eh 0 or -220 mV, the sensitivity to inhibitors and the rates of whole electron transfer chain (e. g. P700 rereduction after a train of flashes, sensitive to DCMU). It is concluded that growing cells at 31°C does not affect *per se* any of the functions studied here.

At Eh < -200 mV a single flash photooxidizes cyt b(s) after a 2-3 ms lag (t1/2 10-30 ms) (Fig.2). This reaction is sensitive to stigmatellin (2 µM) or DBMIB (10 µM). HQNO (10-30 µM) inhibits the oxidation as well (Fig. 2), indicating that the reaction

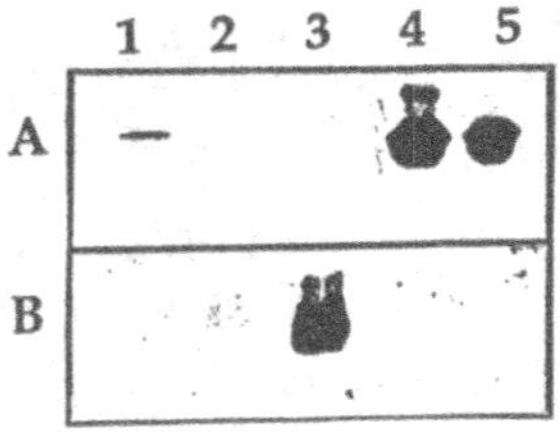

Fig.1 Western blot against plastocyanin (A) and cyt c-552 (B) of wt (1,4,5) and Pc$^-$ mutant (2,3) preparations. Cells grown at 20°C (3,5) and 31°C (2,4). Anti-plastocyanin and anti-cyt c-552 were purified from *C. reinhardtii* (S. Merchant); 10 µg Chl per lane. Lane 1 is a 1:50 dil of the load in lane 5.

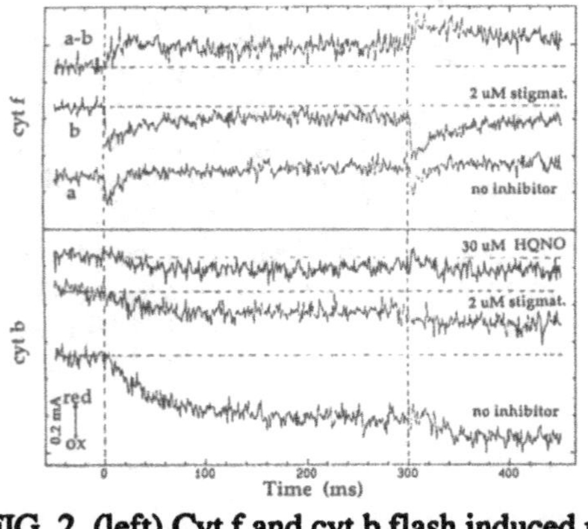

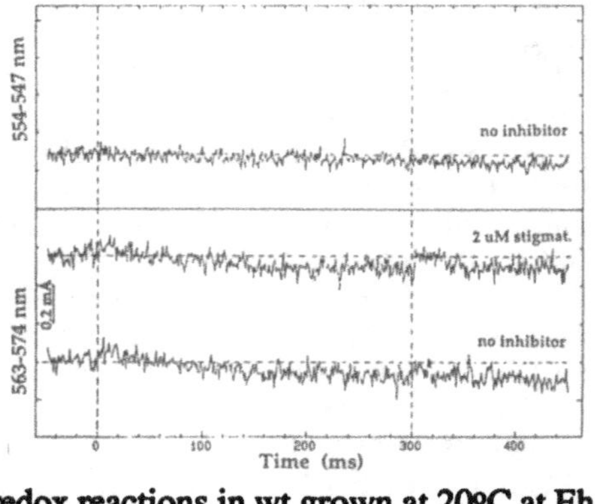

FIG. 2 (left) Cyt f and cyt b flash induced redox reactions in wt grown at 20°C at Eh -220 mV.Train of 2 flashes 300 ms apart. The stigmatellin sensitive cyt f rereduction (t 1/2 ca. 3 ms) is shown as trace a minus trace b (a-b). After a lag, see Fig. 6, one cyt b is photooxidized with the first flash (t 1/2 20-30 ms) and transiently reduced with the second flash. Preparations of wt cells grown at 31°C show an identical behavior. RC 500 µs.

FIG. 3 (right) Light induced signal changes for the Pc$^-$ mutant grown at 31°C at Eh -220 mV. Train of 2 flashes 300 ms apart. Only a very small signal, eventually identifiable as cyt b oxidation, can be recognized in a time resolved spectrum.

probably occurs at the Qc site, as considered by both photooxidation models. In the Pc⁻ mutant, specifically when grown at 20 °C, and due to the presence of the alternative electron carrier cyt c-552 expressed at this temperature, the b6f complex is also catalytically competent, as shown by the flash induced redox reactions of cyt f and cyt b at Eh 0 or -220 mV, as well by the sensitivity to inhibitors. Time resolved spectra after a single flash (2-300 ms) at Eh -220 mV confirms the signal specificity for both wt (20 and 31°C) and Pc⁻ (20 °C) strains (cyt b peak at 564 nm), including the lag in the cyt b photooxidation and the residual reaction in the presence of stigmatellin. This last is interpreted as a subpopulation of cells not being inhibited. In contrast, when grown at 31°C, conditions where cyt c-552 is not expressed and, therefore, in the absence of any carrier between PSI and cyt f, the b_6f complex is impaired for all redox reactions (Fig.3). Nevertheless photoreduction of ferredoxin is normal in this preparation (Fig.4).

As the wt grown at 31°C shows a normal system, we conclude that in the mutant grown at 31°C the lack of function of the b6f complex, including the cyt b photooxidation, is only due to the lack of plastocyanin or cyt c-552. This demonstrates that the photooxidation of cyt b is triggered by the appearance of an oxidizing equivalent at the high potential chain (probably at Qo), a fact not required in (3), and it is probably independent of the reducing side of PSI that operates normally in the mutant (i.e. photoreduction of ferredoxin), even at the growth temperature of 31°C. The inhibition of the cyt b photooxidation by stigmatellin, and DBMIB independently leads to the same conclusion, although involving the risk that the inhibitors could affect allosterically the Qr site (where the photooxidation of cyt b would occur).These results disprove for *Chlamydomonas* the mechanism of "reductant-induced-oxidation" (3). Conversely, they are consistent with the possibility that a semiquinone generated after a flash at the Qo site is the oxidant, at the Qr site, of the prereduced low potential chain, thus favoring a "semiquinone-cycle" (4).

Fig. 5 shows the redox transitions of cyt b(s) upon a train of flashes at Eh -220 mV. After the first flash, two phases of oxidation are seen. The fastest would represent the first turnover. Due to the difference in Em's, cyt bl is expected to be the heme finally oxidized. The slow phase is probably due to residual second turnovers elicited by either actinic effect or partial excess of PS1. The second flash induces a transient reduction (expected to be cyt bl) followed by a further oxidation, interpreted as the lost of the second electron equivalent present before the first flash, see below. After the 3rd flash a pseudo-steady state is stablished, similar to the typical at Eh 0 mV (both b hemes oxidized).

At E_h <-200 mV a slow phase of the electrochromic signal develops much faster (t1/2 ca. 3 ms) than the photooxidation of cyt b (t 1/2 10-30 ms), as already shown for *Chlorella* (4), (Fig. 6 and 2). The signal, that is inhibited as well by stigmatellin or DBMIB, is biphasic and can be deconvoluted into an HQNO-insensitive phase (without a significant lag) and a sensitive one (after a lag of ca.1-2 ms) (Fig.6). The preparation of wt grown at 31°C has the same behavior as the one at 20°C. On the contrary, the Pc⁻ strain grown at 31°C shows only a fast electrochromic signal (μs rise kinetics) with a decay that is insensitive to inhibitors, confirming the b6f origin of all the slow phases described. As noted before (4), a membrane potential originating negative charge crosses the membrane span faster than any electron movement at the level of the cyt b hemes chain. Probably, the HQNO insensitive relates with the movement of a semiquinone from the Qo, where it would be generated, to the Qr site, as proposed in (4). The kinetics of this phase (t1/2 3ms) are similar to the one of cyt f rereduction that is sensitive to Qo inhibitors (Fig. 2), as expected for a reaction at the Qo site. The HQNO-sensitive phase could be assigned to the movement of an electron from cyt bl to the Qr site leaving the chain. Although also with a lag, the expected kinetics parallelling cyt b oxidation are not seen. The apparent t1/2 is faster and may arise from the known leakeage of the HQNO inhibition at the Qr site. An

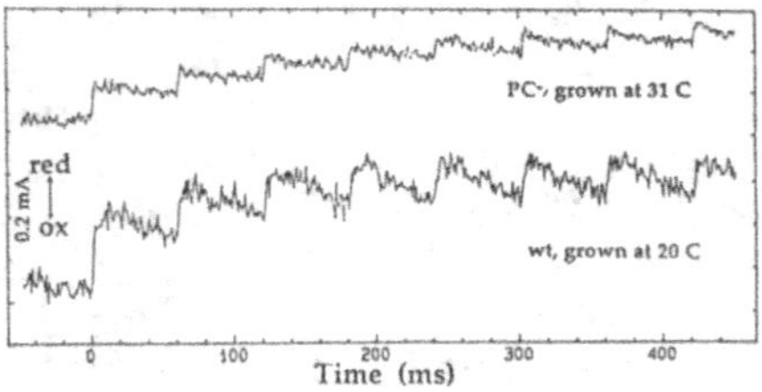

FIG.4 Optically deconvoluted signal, as in (6) for ferredoxin in the wt grown at 20°C and Pc- mutant grown at 31°C. Eh -220 mV.

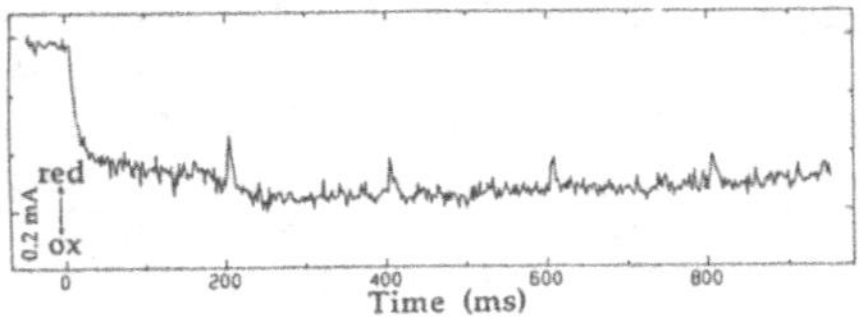

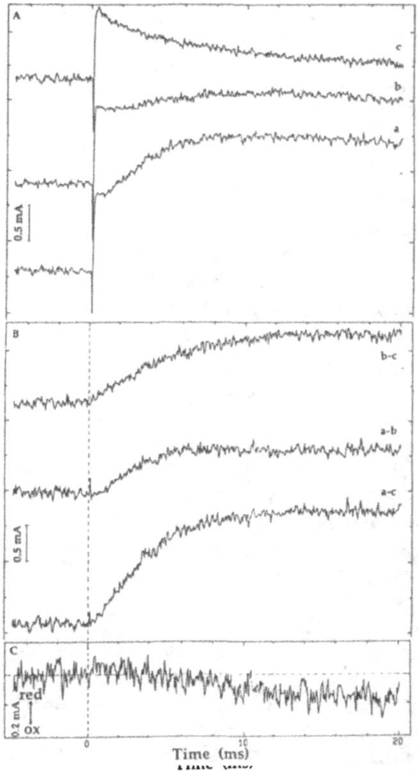

FIG.5 (above) Cyt b photooxidation and subsequent photoreduction events in a train of flashes in wt grown at 20°C.If considered only the fast phase after the first flash, the total cyt b complement is oxidized in almost two equal consecutive steps; i.e. bl and bh. Eh -220mV.

FIG.6 (right) The electrochromic signal phases (515 nm) (A,B) and the onset of cyt b photooxidation (C) at Eh -220 mV for wt grown at 20°C No ionophores. A) The total signal (fast and slow phases) no inhibitors (a), with 30 μM HQNO (b) and with 2 μM stigmatellin. B) The slow phases sensitive to stigmatellin (a-c), sensitive to HQNO (a-b) and HQNO-insensitive but stigmatellin sensitive (b-c). RC 20 μs.

identical analysis of the signal at 0 mV shows no ms lag in the HQNO sensitive phase.

After one turnover of this photooxidative reaction, the b6f complex would be left with its cyt bl oxidized but cyt bh still reduced. A second turnover of the Qo site will be able reduce cyt bl generating a plastoquinone, that in turn will oxidize both b hemes through the Qc site. Thereafter, typical turnovers of a "Q-cycle" would be allowed. Fig. 5 shows a train of flashes were these various events would take place. Physiologically, the photooxidation at low potential, probably involving a "semiquinone cycle",would occur as an unfrequent reaction, and for only one turnover, in any physiological condition were the plastoquinone pool and both hemes are reduced at the time that the Rieske center is photooxidized. The reaction would act as a "primer" to allow Q-cycle turnovers that otherwise would be blocked at low redox potentials.

References

1)Mitchell, P (1975) FEBS Lett.57,135. 2)Wikström, M and Krab, K (1986) J. Bioenerg. Biomembr. 18, 181. 3) Chain, R. (1982) FEBS Lett. 143, 273. 4) Joliot, P and Joliot, A (1994) PNAS 91, 1034. 5) Merchant, S. and Bogorad, L. (1987) Embo J. 6, 2531. 6) De Wolf, F, Krab, K, Visschers, R, De Waard, J and Kraayenhof, R (1988) BBA 936, 487.

MUTATIONAL ANALYSIS OF THE ROLE OF EXTRINSIC CYTOCHROME *C*-550 IN CYANOBACTERIAL PHOTOSYSTEM II

Jian-Ren Shen[1], Robert L. Burnap[2] and Yorinao Inoue[1]
[1]Solar Energy Research Group, The Institute of Physical and Chemical Research (RIKEN), Wako, Saitama 351-01, Japan; [2]Department of Microbiology and Molecular Genetics, Oklahoma State University, Stillwater, Oklahoma 74078.

1. Introduction

Cytochrome (cyt) *c*-550 is one of the major *c*-type cyts present in cyanobacterial cells [1]. Its cellular location and function, however, have not been well understood until recently. We have purified an intact PSII complex from a thermophilic cyanobacterium, *Synechococcus vulcanus*, and found that the PSII contained cyt *c*-550 as one of the extrinsic components in an amount stoichiometrically comparable with other intrinsic membrane proteins or with the extrinsic 33 kDa proteins [2]. In line with this, we have shown that most of the cyt is present in PSII, with no significant amount of the cyt detectable in any other soluble or membrane fractions upon systematic fractionation of the thermophilic cyanobacterial cells [1]. Release-reconstitution studies revealed that cyt *c*-550 is required not only for optimum oxygen-evolving activity but also for, in combination with the 33 kDa extrinsic protein, the effective binding to PSII of another extrinsic component, the 12 kDa protein [3]. These results suggested that the three cyanobacterial PSII extrinsic components are closely related with each other. This is confirmed by chemical crosslinking studies of the purified PSII complex, which showed that the three cyanobacterial extrinsic components are crosslinked together by 1-ethyl-3-(3-dimethylaminopropyl)carbodimide [4]. These results suggested that like the 33 and 12 kDa proteins, cyt *c*-550 is located at the lumenal side of thylakoids. This is supported by the fact that the gene encoding cyt *c*-550 in *Synechocystis* sp. PCC 6803 has a leader sequence corresponding to a transit peptide of 25 residues which has typical features of prokaryotic type transit peptide [5]. Moreover, this leader sequence has a significantly high homology with that of the 33 kDa extrinsic protein of the same cyanobacterium, strongly suggesting that cyt *c*-550 is transported into the lumen after its synthesis. Based on these results, we named the gene for cyt *c*-550 as *psb*V.

Here we report the effects of deleting either the *psb*V gene alone or both *psb*V and *psb*O genes on growth and photosynthetic electron transfer activities in *Synechocystis* sp. PCC 6803. The results obtained showed that deletion of cyt *c*-550 affected not only the PSII oxygen evolution but also the PSII stability in vivo; furthermore, simultaneous deletion of the *psb*V and *psb*O genes resulted in a mutant that cannot grow photoautotrophically due to loss of PSII activity and stability. Based on these and our previous results, a possible evolutionary relationship among extrinsic components of the donor side of PSII-type reaction centers is proposed.

2. Deletion mutagenesis of cyt *c*-550

Mutants lacking cyt *c*-550 was constructed either by disrupting the *psb*V gene with a kanamycin-resistant cassette or deleting the gene by replacing it with an erythromycin-resistant cassette in *Synechocystis* sp. PCC 6803 [5]. The resulted mutants were

P. Mathis (ed.), Photosynthesis: from Light to Biosphere, Vol. II, 559–562.

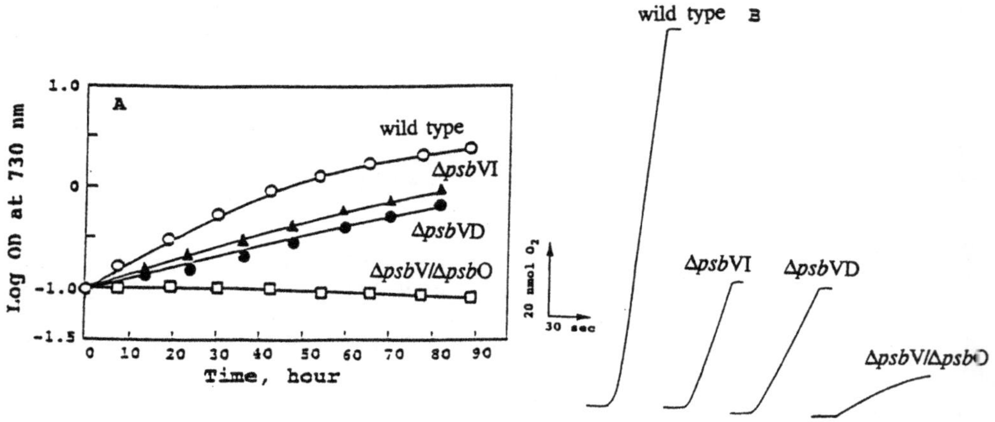

Fig. 1. A. Growth curves of *Synechocystis* sp. PCC 6803 wild type, Δ*psb*VI, Δ*psb*VD and Δ*psb*V/Δ*psb*O mutant strains in BG11 medium at 30°C. B. Traces of oxygen evolution measured under saturating light with 0.6 mM 2,6-dichloro-*p*-benzoquinone and 1 mM potassium ferricyanide as acceptors at 30° C in BG11.

Table 1 Growth rate, PSII content and photosynthetic electron transfer activities of *Synechocystis* sp. PCC 6803 wild type, Δ*psb*VI, Δ*psb*VD and Δ*psb*V/*psb*O mutant strains[a].

Strains	Wild type	Δ*psb*VI	Δ*psb*VD	Δ*psb*V/Δ*psb*O
Doubling time in BG11 (h)[b]	13	28	31	NG
Chl/PSII ratio[c]	580	860	1100	3300
	(100)	(67)	(53)	(18)
Electron transfer rates[d]				
H_2O to CO_2 (whole chain)	120	80	72	13
	(100)	(67)	(60)	(11)
H_2O to DCBQ/FeCN (PSII)	500	197	217	40
	(100)	(39)	(43)	(8)

[a]Numbers in parentheses are percentages relative to the wild type stain. Abbreviations used in the Table are: NG, no growth; DCBQ, 2, 6-dichloro-*p*-benzoquinone; FeCN, potassium ferricyanide.
[b]Doubling time was determined by measuring light scattering at 730 nm.
[c]Determined with the herbicide-binding method using ^{14}C-labeled atrazine.
[d]Expressed as μmol O_2 per mg Chl per hour.

designated Δ*psb*VI and Δ*psb*VD, respectively, for the insertional inactivation and deletion mutant.

Fig. 1 shows the photoautotrophic growth rates and traces of PSII oxygen evolution measured with 2,-6-dichloro-*p*-benzoquinone and potassium ferricyanide as acceptors, of the wild type and mutant strains. In the presence of 5 mM glucose in the growth medium, the two mutant strains grew at a same rate as that of the wild type strain [5], indicating that lack of cyt *c*-550 did not affect the photomixotrophic and photoheterotrophic growth of the cyanobacterium. In the absence of glucose, however, both Δ*psb*VI and Δ*psb*VD mutant strains grew at a rate less than half of the wild type (Table 1). This is caused by a 60% decrease in the PSII oxygen-evolving activity (Fig. 1, Table 1), since no effect on the PSI activity was observed in the mutant strains [5]. Determination of PSII abundance by means of herbicide-binding and Western-blot analysis revealed that deletion of cyt *c*-550 decreased the PSII content to 53-67% of that of the wild type (Table 1), implying a role of cyt *c*-550 in stabilization of the PSII complex in vivo. Since the decrease in PSII oxygen evolution was larger than the decrease in the PSII abundance, we conclude that cyt *c*-550 is required for both optimal PSII activity and stability in vivo. These effects were caused directly by inactivation of the *psb*V gene because insertion of a kanamycin resistance cassette just beyond the coding region did not show any effects on the growth and photosynthetic activities [5], which excluded the possibility that insertion of an antibiotic-resistant cassette may affect genes in the same operon or downstream of the *psb*V gene.

3. Double deletion of the *psb*V and *psb*O genes inhibits photoautotrophic growth of *Synechocystis* sp. PCC 6803

Since cyt *c*-550 is able to bind to PSII essentially independent of the 33 kDa protein, it is important to determine whether it can function without the 33 kDa protein in vivo. To address this question, we constructed a double deletion mutant lacking both cyt *c*-550 and the 33 kDa protein, by inactivating the *psb*V and *psb*O genes simultaneously in *Synechocystis* sp. PCC 6803 [6,7]. The resultant double deletion mutant, Δ*psb*V/*psb*O, evolved oxygen at a rate only 8% of the wild type strain, and this activity lost very rapidly under illumination during measurement (Fig. 1, Table 1). This is consistent with a role of cyt *c*-550 in oxygen evolution and also suggests a markedly increased sensitivity toward photoinhibition upon simultaneous loss of the 33 kDa protein and cyt *c*-550, a phenomenon often observed in systems with damaged PSII donor side. Moreover, the double deletion mutant contained only about 18% PSII as compared with that of the wild type strain (Table 1). As a result, the mutant was unable to grow photoautotrophically (Fig. 1). Since mutants lacking either the 33 kDa protein or cyt *c*-550 alone can still grow photoautotrophically [5,6], the present results clearly indicate that cyt *c*-550 can function independent of the 33 kDa protein in vivo, and that both cyt *c*-550 and the 33 kDa protein are required for maximum oxygen evolution and PSII stability in vivo. In other words, the oxygen evolution and photoautotrophic growth in the Δ*psb*O mutant is supported by cyt *c*-550 and vice versa.

4. Possible evolutionary relationship of PSII donor side components

The reaction center and acceptor side of PSII are homologous to those of photosynthetic purple bacteria. No apparent similarities, however, have been found at the donor side between purple bacterial reaction center and cyanobacterial and higher plant PSII. With the finding of cyt *c*-550 being present and functioning at the donor side of cyanobacterial PSII, we may propose a possible evolutionary linkage at the donor side from bacteria to higher plants, as shown in Fig. 2. At the donor side of purple bacterial reaction center, cyt *c*-2 functions as the electron donor, whereas at the donor side of cyanobacterial PSII, cyt *c*-550 functions in maintaining PSII oxygen evolution and stability. Although cyt *c*-2 and cyt *c*-550 have no significant sequence homology, both cyts are soluble, *c*-type, monoheme cyts and are located at the donor side of the reaction center, suggesting that cyt *c*-550 may be a "molecular appendix" remained during the evolution of PSII from purple bacterial reaction center. Furthermore, cyt *c*-550 is disappeared in higher plants and its function is possibly replaced to some extent by the 23 kDa extrinsic protein. This change from heme-protein to a non-heme protein may be a result of plants to adopt to their environment with less availability of soluble iron as compared with the aquatic environment where cyanobacteria live. A similar loose but apparent evolutionary linkage could also be seen at the donor side of PSI. The reaction center and acceptor side of PSI of both cyanobacteria and higher plants are homologous to the reaction center of green sulfur bacteria. In green sulfur bacteria, a membrane-bound *c*-type cyt, cyt *c*-551, functions as the electron donor to the reaction center, whereas in cyanobacteria either cyt *c*-550 or plastocyanin is the functional donor to the PSI reaction center. Cyt *c*-551 in green bacteria and cyt *c*-553 in cyanobacteria are not homologous in their sequence but have similar function. Interestingly, cyt *c*-553 also disappeared in higher plants and its function is replaced by plastocyanin, a copper-containing protein.

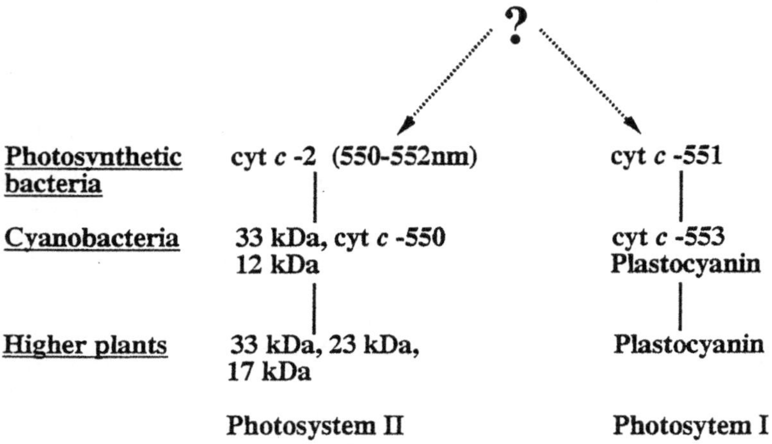

Fig. 2. A possible evolutionary linkage among the donor side components of PSII-type reaction centers from photosynthetic bacteria to higher plants. For comparison, the PSI donor side components from bacteria to higher plants were also presented.

References

1 Shen, J.-R. and Inoue, Y. (1993) J. Biol. Chem. 268, 20408-20413
2 Shen, J.-R., Ikeuchi, M. and Inoue, Y. (1992) FEBS Lett. 301, 145-149
3 Shen, J.-R. and Inoue, Y. (1993) Biochemistry 32, 1825-1832
4 Han, K.-C., Shen, J.-R., Ikeuchi, M. and Inoue, Y. (1994) FEBS Lett. 355, 121-124
5 Shen, J.-R., Vermaas, W. and Inoue, Y. (1995) J. Biol. Chem. 270, 6901-6907
6 Burnap, R.L. and Sherman, L. (1991) Biochemistry 30, 440-446
7 Shen, J.-R., Burnap, R.L. and Inoue, Y. (1995) Biochemistry in press

CYTOCHROME C_6 FROM THE THERMOPHILIC *SYNECHOCOCCUS ELONGATUS* *

M. Sutter[1], H. Sticht[2], R. Schmid[3], P. Hörth[1], P. Rösch[2], and W. Haehnel[1]
[1]Biochemie der Pflanzen, Schänzlestr. 1, 79104 Freiburg, Germany;
[2]Lehrstuhl Biopolymere, Universitätsstr. 30, 95440 Bayreuth, Germany;
[3]Fachbereich Biologie/Chemie, Postfach 4469, 49076 Osnabrück, Germany

1. Introduction

The electron transfer between the cytochrome (cyt) b_6f complex and Photosystem I (PSI) in oxygenic photosynthetic organisms is mediated by small soluble redox proteins. In higher plants it is exclusively plastocyanin (Pc), in green algae and cyanobacteria Pc can be replaced by cyt c_6.The alternate expression is regulated in some species by the availability of copper and iron [1]. Very recently the X-ray crystallographic structure of cyt c_6 from the green alga *Chlamydomonas reinhardtii* has been solved with a resolution of 1.9 Å [2]. The structure of PSI from the thermophilic cyanobacterium *Synechococcus elongatus* (*S. elong.*) has been solved at a resolution of 6 Å by Krauss et al. [3]. A Pc-docking site in this PSI structure has been discussed by Fromme et al. [4]. An improvement of the structure of PSI may reveal details of the binding site of cyt c_6. Therefore, we have investigated the structure of cyt c_6 from this organism and its reaction with PSI.

2. Procedure

Cells of *S. elong.* were grown at 57°C in the inorganic medium D of Castenholz according to [5]. For ^{15}N-,^{13}C-NMR the cells were grown with $Na^{15}NO_3$ and $NaH^{13}CO_3$. The isolation of cyt c_6 was carried out essentially as described in [6]. Cyt c_6 was further purified by preparative isoelectric focussing using immobilines to adjust a narrow gradient between pH 4.5 and 5.5. Cyt c_6 focused in the reduced state at pH 4.8 and in the oxidized state at pH 5.2. The amino acid sequence was determined by sequencing the N-terminus of the intact protein and a set of BrCN-fragments, which had been separated by HPLC and identified with electrospray mass spectrometry and UV/VIS-spectroscopy. For NMR measurements we used wild type and ^{15}N-,^{13}C-labelled cyt c_6. The results of ^1H-NMR measurements were not sufficient for a complete determination of the three-dimensional structure, mainly due to the degeneracy of the

*Support by Deutsche Forschungsgemeinschaft SFB 388/A1 is gratefully acknowledged.

P. Mathis (ed.), Photosynthesis: from Light to Biosphere, Vol. II, 563–566.

chemical shifts. Crystals of PSI trimer were a kind gift from Petra Fromme (Berlin). The kinetics of the laser flash induced absorption changes of PSI were followed at 820 nm as described previously [7]. The reaction mixture contained PSI trimer crystals dissolved to a final chlorophyll concentration of 0.1 mg/ml, 20 mM buffer given in the text, 1 mM sodium ascorbate, 0.1 mM diaminodurene, 0.1 mM methylviologen, 0.03% β-dodecyl maltoside and 10 μM cyt c_6.

3.Result and Discussion

In *S. elong.* we found no copper- or iron-mediated regulation of cyt c_6 expression. In cells grown in the presence of 0-20 μM $CuSO_4$ at a reduction of the iron concentration to 20% we could not detect any effect on the amount of cyt c_6 and did not find any evidence for the expression of Pc. In other organisms at 5 μM Cu^{2+} Pc would completely replace cyt c_6 [1]. We conclude that cyt c_6 is the only electron carrier in *S. elong.*

The cyt c_6 shows the typical absorption spectra of a c-type cytochrome with an α-band at 552.3 nm in the reduced state. In concentrated solutions a band at 693 nm in the oxidized state indicates that Met is the sixth ligand of the iron. The redox midpoint potential is 340 mV at pH 7.0/25°C (data not shown).

The complete amino acid sequence of cyt c_6 has been determined. It is nearly identical with that of cyt c_6 from the thermophilic cyanobacterium *Synechococcus lividus* (*S. liv.*) and has high homology with other c_6 cytochromes :

```
                    10         20         30         40         50         60         70         80
S. elong.:  ADLANGAKVF SGNCAACHMG GGNVVMANKT LKKEALEQFG MYSEDAIIYQ VQHGKNAMPA FAGRLTDEQI QDVAAYVLDQ AAKGWAG
S. liv.:    ADLANGAKVF SGNCAACHMG GGNVVMANKT LKKEALEQFG MNSEDAIIYQ VQHGKNAMPA FAGRLTDEQI QDVAAYVLDQ AAKGWAG
A. var.:    ADSVNGAKIF SANCASCHAG GKNLVQAQKT LKKEDLEKFG MYSAEAIIAQ VTNGKNAMPA FKGRLKPDQI EDVAAYVLGQ ADKSW
S. 79042:   ADLAHGGQVF SANCAACHLG GRNVVNPAKT LQKADLDQYG MASIEAITTQ VTNGKGAMPA FGSKLSADDI ADVASYVLDQ SEKGWQG
S. 6803:    ADLAHGKAIF AGNCAACHNG GLNAINPSKT LKMADLEANG KNSVAAIVAQ ITNGNGAMPG FKGRISDSDM EDVAAYVLDQ AEKGW
A. nid.:    ADLAHGGQVF SANCDSCHLG GRNVVNPAKT LEKADLDEYG MASIEAITTQ VTNGKGAMPA FGAKLSADDI EGVASYALDQ SGKEW
S. 6312:    ADIADGAKVF SANCAACHMG GGNVVMANKT LKKEALEQFG MNSADAIMYQ VQNGKNAMPA FGGRLSEAQI ENVAAYVLDQ SSNKWAG
C. reinh.:  ADLALGAQVF NGNCAACHMG GRNSVMPEKT LDKAALEQYLDGGFKVEIIYQ VENGKGAMPA WADRLSEEEI QAVAEYVFKQATDAAWKY
```

(*A. var.*: *Anabaena variabilis* , *S.* 79042: *Synechococcus* PCC79042, *S.* 6803: *Synechocystis* PCC6803, *A. nid.*: *Anacystis nidulans*, *S.* 6312: *Synechococcus* PCC6312, *C. reinh.*: *Chlamydomonas reinhardtii*, sequences from [8].) The heme binding site is clearly identified by the typical sequence motif -CXXCH-. His18 is the fifth heme-iron ligand. ^{1}H-NMR data indicate that the sixth ligand is Met58 (not shown). Cys14 and Cys17, which cannot be detected by N-terminal sequencing, could be verified by the results of mass spectrometry (*Fig. 1*). The native cyt c_6 has a mass of 9795 $\pm$ 1.1 Da including the covalently linked heme group and the iron. This is consistent with our results from the amino acid sequencing. The mass of 10252.6 of the labelled cyt c_6 corresponds to a ^{15}N-,^{13}C-content of 85% in this protein. A complete NMR data set has been collected of this labelled protein. One data set is shown in *Fig. 2*. The HNCO-experiment shows the correlation between the amide proton and the carbonyl carbon of

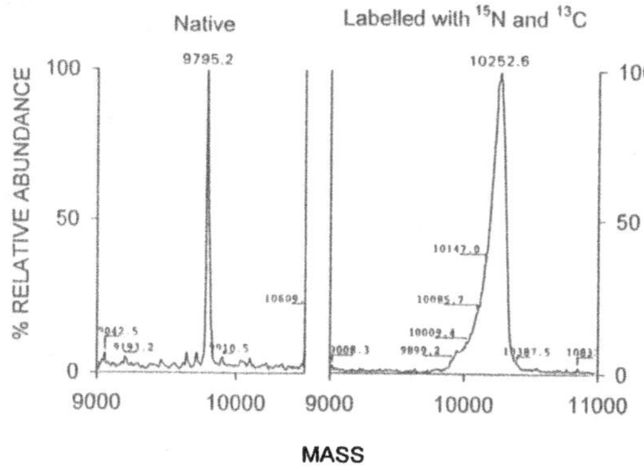

Figure 1. Electrospray mass spectrometry of cyt c_6 recorded on a TSQ700 (Finnigan). About 1 nMol protein dissolved in 1 μl 50% (v/v) methanol/1% (v/v) acetic acid was injected.

the peptide bonds. The x- and y-axes show the resonance frequencies of the amide protons (ω_2) and the correlated ^{13}C-carbonyl groups (ω_1), respectively. Therefore, each peptide bond that contains an amide proton should give one signal. Most of the expected 85 resonances can be seen in this spectrum and are well separated. A further enhancement of the resolution is obtained due to the additional spread of the peaks by the ^{15}N chemical shift of the amide nitrogens. (This third dimension cannot be seen in *Fig. 2*). The data analysis will be reported elsewhere.

Kinetics: The effect of $MgSO_4$ on the second-order rate constant of the reaction between cyt c_6 and dissolved PSI trimer crystals at different pH values is shown in *Fig. 3*. At the isoelectric point of reduced cyt c_6 (pH 4.8) and at a pH below this value $MgSO_4$ does not affect the electron transfer between P700$^+$ and cyt c_6, whereas the reaction is accelerated at pH values higher than the isoelectric point. The decreasing rate constants at increasing pH values suggest an increasing repulsion by negative charges of the reacting proteins. First attempts of cocrystallization of PSI and cyt c_6 at pH 6.5 were not successful, although cyt c_6, in the reduced state, had an effect on the growth of the PSI crystals (P. Fromme, personal communication).

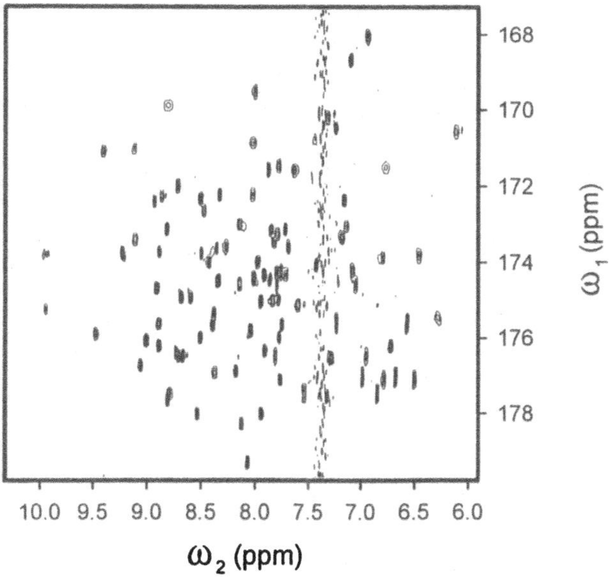

Figure 2. Two-dimensional projection of a heteronuclear HNCO NMR-experiment. The spectrum was obtained from a uniformly ^{15}N-,^{13}C-labelled sample of cyt c_6 from *S. elong*. at 298K, pH 6.3, recorded on a Bruker AMX 600 spectrometer.

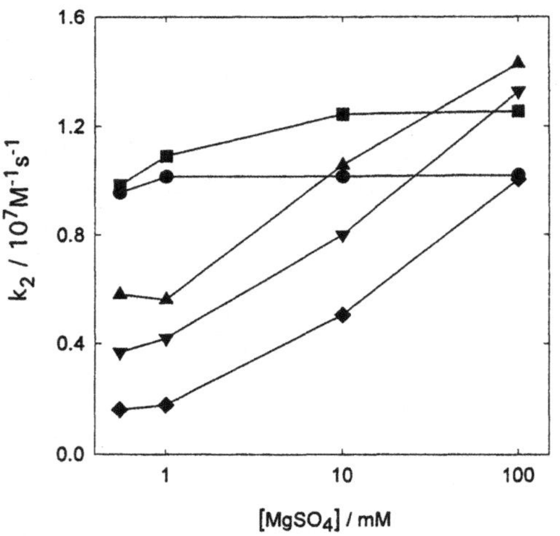

Figure 3. Effect of $MgSO_4$ on the second order rate constants between cyt c_6 and dissolved PSI crystals at different pH (● 20 mM sodium acetate pH 4.1, ■ 20 mM sodium acetate pH 4.8, ▲ 20 mM MES pH 5.7, ▼ 20 mM MES pH 6.5, ◆ 20 mM HEPES pH 7.5)

For crystallization the $MgSO_4$ concentration has to be lowered [5]. However, the decreasing second-order rate constants suggest an increase of the repulsion of cyt c_6 by PSI. During the electron transfer from reduced cyt c_6 to $P700^+$ the two proteins do not form a stable complex as found with Pc and PSI from spinach [7]. This is similar in other cyanobacteria [9]. In cyanobacteria PSI is not the only acceptor of the electrons from Pc/cyt c_6. The cytochrome oxidase can act as an alternative acceptor [1]. A stable complex between PSI and the soluble electron transfer protein could be of disadvantage for other electron pathways and the flexibility of the electron metabolism in the thylakoid of cyanobacteria. In chloroplasts of plants PSI is the only acceptor of the electrons from Pc/cyt c_6 and a stable complex with reduced Pc may accelerate the electron transfer [7].

References:

1 Sandmann, G., Reck, H., Kessler, E. & Böger, P. (1983) Arch. Microbiol. 134, 23-27
2 Kerfeld, C.A., Anwart, H.P., Merchant, S. & Yeates, T.O. (1995) J. Mol. Biol. 250, 627-647
3 Krauss, N., Hinrichs, W., Witt, I., Fromme, P., Pritzkow, W., Dauter, Z., Betzel, C., Wilson, K.S., Witt, H.T. & Saenger, W. (1993) Nature 361, 326-331
4 Fromme, P., Schubert, W.D., Krauß, N. (1994) Biochim. Biophys. Acta 1187, 99-105
5 Schatz, G.H. & Witt, H.T. (1984) Photobiochem. Photobiophys. 7, 1-14
6 Shin, M., Sakihama, N., Koike, H. & Inoue, Y. (1984) Plant Cell Physiol. 25, 1575-1578
7 Haehnel, W., Jansen, T., Gause, K., Klösgen, R.B., Stahl, B., Michl, D., Huvermann, B., Karas, M. & Herrmann, R.G. (1994) EMBO J. 13, 1028-1038
8 Moore, G.R., Pettigrew, G.W. eds. (1990) Cytochromes c, Springer Verlag, pp. 143-145
9 Medina, M., Diaz, A., Hervás, M., Navarro, J.A., Gómez-Moreno, C., De La Rosa, M. & Tollin, G. (1993) Eur. J. Biochem. 213, 1133-1138

REVERSIBLE ACTIVATION OF LHCII KINASE BY TRANSIENT LOW pH TREATMENT OF THYLAKOIDS IN THE DARK.

A. V. Vener*, P. J. M. van Kan*, A. Gal#, I. Ohad*#, and B. Andersson*
*Department of Biochemistry, Stockholm University, Stockholm, Sweden, and #Department of Biological Chemistry, The Hebrew University of Jerusalem, Jerusalem, Israel

1. Introduction

A unique light mediated redox-controlled phosphorylation of a number of proteins associated with the thylakoid membrane has evolved in chloroplasts [1]. Phosphorylation of the major light-harvesting chlorophyll *a/b* protein complex (LHCII) regulates the balance of excitation energy between the two photosystems [2,3]. Activation of LHCII phosphorylation is dependent on the redox state of the plastoquinone pool [2,3] and on cytochrome *bf* complex [4]. However, the specific mechanism involved in the redox mediated activation of the LHCII kinase is not yet understood. Elucidation of this regulatory mechanism has been hampered by the fact that the kinase activation process was so far achieved only under conditions in which all thylakoid electron carriers were reduced following either illumination or excess addition of reducing agents in the dark. Here we introduce a new experimental system in which LHCII kinase is activated in the dark by a short preincubation of thylakoid membranes at low pH in the absence of added reducing agents. Our studies indicate that activation and deactivation of LHCII phosphorylation is specifically related to a plastoquinol bound to the cytochrome *bf* complex in vicinity of the reduced Rieske Fe-S centre.

2. Experimental Procedures

Spinach (*Spinacia oleracea L*) thylakoids were isolated according to Andersson et al. [5]. Thylakoids from *Lemna perpusilla* strain 6746 (wild type) and mutant strain 1073 were prepared according to Gal et al. [4]

To phosphorylate LHCII the dark adapted thylakoids were diluted with 50 mM NaH_2PO_4, pH 4.3, 100 mM sorbitol and 5 mM $MgCl_2$ to 0.1 mg Chl./ml. The preincubation at low pH was performed in the dark at room temperature for 2-20 min and terminated by centrifugation in a microcentrifuge for 30 s. The supernatant was removed and the thylakoid membranes were resuspended (0.4 mg Chl./ml) in the phosphorylation medium containing 50 mM sodium phosphate, pH 7.4 or 8.0, 100 mM sorbitol, 5 mM $MgCl_2$, 10 mM NaF and 0.1 mM [γ-^{32}P]ATP (1 μCi per 30 μl). Phosphorylation was performed for the indicated times and was terminated by addition of electrophoresis sample buffer. The samples were subjected to electrophoresis using 15% acrylamide slab gels. To activate the kinase by light, thylakoids were preincubated in 50 mM sodium phosphate, pH 7.4, 100 mM sorbitol and 5 mM $MgCl_2$. Illumination was provided by a

P. Mathis (ed.), Photosynthesis: from Light to Biosphere, Vol. II, 567–570.

light projector (white light, 100 µmol photons m^{-2} s^{-1}).

For determination of Plastoquinol/Plastoquinone ratio lipid extraction of spinach thylakoids was performed according to Åberg et al. [6]. All procedures were carried out in the dark. Reverse phase HPLC was performed on Hewlett-Packard Hypersil ODS 3-mm column with a linear gradient of methanol : H_2O (9:1 v/v) to methanol : 2-propanol (2:1 v/v). The absorbance of the eluate was monitored at 210 nm. The elution times of plastoquinone and plastoquinol were determined using reference compounds.

For low temperature EPR spectroscopy samples of dark-adapted spinach thylakoids containing 3.8 mg Chl./ml were frozen in liquid nitrogen at stages corresponding to various states of activation or deactivation of the LHCII kinase. Directly afterwards, the X-band EPR spectra were measured at 15 K on a Bruker ESP 300 spectrometer equipped with a helium flow cryostat and temperature controller. EPR conditions were: microwave power 6.3 mW, modulation 0.12 mT.

3. Results and Discussion

Preincubation of spinach thylakoid membranes at pH 4.3 and subsequent transfer to the phosphorylation buffer, pH 7.4 or 8.0, results in a significant activation of the LHCII kinase in the dark (Fig. 1A). Notably, the level of the phosphorylation induced in the dark by preincubation at acidic pH is hundred fold higher than that of the basal phosphorylation at pH 4.3, or that at pH 7.4 without prior preincubation at low pH as judged by densitometry of autoradiograms. Considering the molar ratio of the two LHCII subunits (1:4), the relative phosphorylation of the mobile component, the 25 kDa protein is significantly higher than that of the stationary 27 kDa protein subunit [7]. Nigericin, NH_4Cl or FCCP have no effect on the extent or rate of the low pH induced phosphorylation. Thus, the activation of the kinase by the above procedure is not due to the formation or dissipation of a transmembrane pH gradient.

The activation of LHCII phosphorylation by low pH preincubation is related to a change in the membrane redox potential. Ferricyanide inhibits this phosphorylation completely (Fig. 1B). To test directly whether the reduction of the plastoquinone pool is

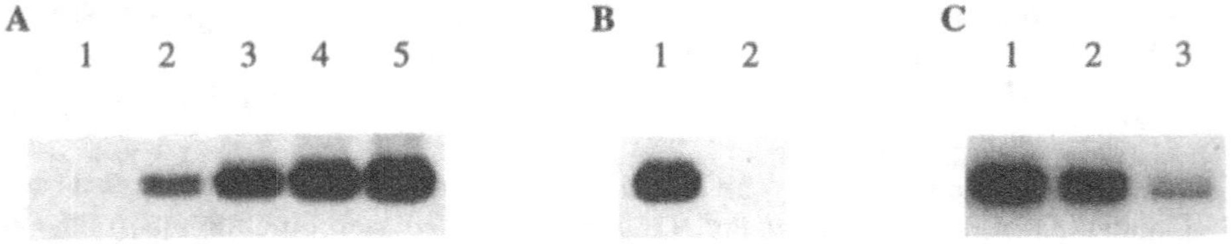

Fig. 1. Activation of the LHCII kinase in the dark by transient low pH preincubation of spinach thylakoids. Autoradiograms of the gels. **A**: 1 - phosphorylation of LHCII at pH 7.4 without preincubation at pH 4.3 (control); 2,3,4 - phosphorylation of LHCII at pH 7.4 in 1, 3 and 5 min (respectively) after transfer of the membranes from pH 4.3; 5 - phosphorylation of LHCII in the light (5 min). **B**: Phosphorylation of LHCII in the dark after low pH preincubation in the absence (1) and presence (2) of 10 mM ferricyanide. **C**: Phosphorylation of LHCII in the dark after low pH preincubation in the absence (1) and presence of 1 µM and 5 µM DBMIB (2,3).

involved in the low pH activation process, we have determined the relative levels of plastoquinol and plastoquinone in thylakoids by total quinone extraction and HPLC quantification. The ratio PQH_2/PQ in thylakoids incubated in the dark at pH 7.4 was in the range of 0.04-0.07. After 2 min of incubation at pH 4.3 this ratio increased to 0.20-0.27. However, transfer of thylakoids back to pH 7.4 in phosphorylation medium resulted in reoxidation of plastoquinol in only 1 min, in contrast to the slow LHCII kinase deactivation. The deactivation of the kinase was assayed by addition of [γ-^{32}P]ATP in different times after the transfer of thylakoids from pH 4.3 to pH 7.4 in the dark. The rate of deactivation after pH-shift was slow in comparison with the deactivation of LHCII kinase in the dark after pre-illumination of thylakoids ($t_{1/2} = 4$ and 1 min, respectively). The kinase activity was maximal in 1 min after the transfer of thylakoids from pH 4.3 to pH 7.4 when the plastoquinone pool was already reoxidised to a level incompatible with the activation process. These results therefore strongly suggest that the active state of the kinase once achieved is not related to the redox state of the "free pool" of plastoquinone and thus an additional intermediate should participate in the redox control of the kinase.

A possible redox sensor could reside within the cytochrome *bf* complex [3,4]. To explore this possibility we have assayed the effect of DBMIB, an inhibitor of cytochrome *bf* reduction [8], on the low pH activation of LHCII phosphorylation. As shown in Figure 1C, DBMIB inhibits the low pH induced phosphorylation of LHCII to about 50% at concentration of 1 μM and almost complete inhibition is obtained at 5 μM. These inhibition results provide evidence supporting the involvement of reduced cytochrome *bf* complex in activation of the LHCII kinase by transient low pH treatment. This conclusion was further supported by analysis performed on the thylakoids of the cytochrome *bf* deficient mutant 1073 of *Lemna* [4]. In the wild type *Lemna*, low pH preincubation in the dark induce the phosphorylation of LHCII polypeptides. However, the phosphorylation of LHCII could not be induced in the dark by the transient low pH treatment of the cytochrome *bf*-deficient thylakoids.

Taking advantage of slow LHCII kinase deactivation after low pH treatment, EPR spectroscopy was used to analyse the changes in the redox state of the cytochrome *bf* complex during the activation and deactivation of LHCII phosphorylation. In the dark-adapted thylakoids, initially incubated at pH 7.4, cytochrome *f* as well as plastocyanin and Rieske Fe-S centre are oxidised. Incubation of the thylakoids for 2 min in the dark at pH 4.3 reduces cytochrome *f* and plastocyanin. Transfer of the thylakoids from pH 4.3 back to pH 7.4 results in appearance of the EPR signal at g=1.90 (Fig. 2A), characteristic for a reduced Rieske Fe-S centre [9]. During the time of the LHCII deactivation (10 min) in darkness no signal of cytochrome *f* oxidation appear at g=3.5-3.7 (not shown). The Rieske Fe-S centre and plastocyanin remain reduced as well. However, the spectra show a decrease of the EPR signal around g=2.03 occurring as a function of incubation time (Fig. 2B), which might reflect a decrease of the g_z signal of the reduced Rieske Fe-S centre [8,9]. The g_z signal is known to be affected by displacement of bound plastoquinol from the reduced Rieske Fe-S centre when DBMIB and other quinone analogs compete for this binding site [8]. We have shown above (Fig. 1C) that DBMIB inhibits the activation of the kinase induced by low pH treatment. To further characterise the effect of DBMIB we performed EPR measurements in the similar experimental conditions and observed that this quinone analog indeed displaces plastoquinol and shifts the g_y and g_z signals of the Rieske Fe-S centre (Fig. 2A). Thus, the correlation of the changes in the EPR g_z signal of the reduced Rieske Fe-S centre with the kinase deactivation in the dark and inhibition by DBMIB indicate a connection between the kinase active state and occupancy of the quinol binding site in the vicinity of the reduced Rieske Fe-S centre.

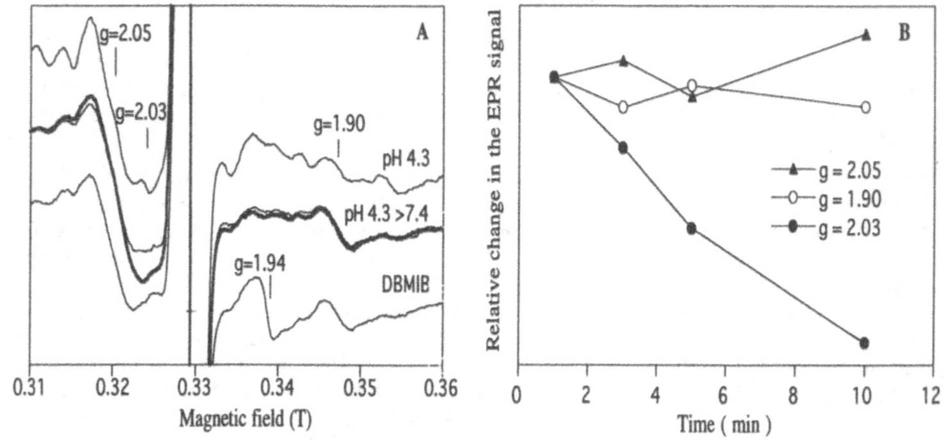

Fig. 2. Changes in the low temperature EPR spectra of Rieske Fe-S centre during activation and deactivation of the LHCII kinase. A: Spectra of dark-adapted thylakoids after preincubation at pH 4.3 for 2 min (upper trace); after the transfer of the membranes from pH 4.3 to pH 7.4 (middle traces: thin line - 1 min, thick line - 10 min after the transfer); after the transfer of the membranes from pH 4.3 to pH 7.4 in the presence of 5 μM DBMIB (lower trace). **B**: Changes in the signal of plastocyanin (g = 2.05) and g_y and g_z signals of Rieske Fe-S centre (g = 1.90 and 2.03, respectively) during the time of LHCII kinase deactivation.

We conclude that the activation of LHCII kinase depends on the plastoquinol bound at the quinol oxidising site of cytochrome *bf* complex. The deactivation of the kinase is consequently due to the release or oxidation of this bound plastoquinol.

Acknowledgements: This project was supported by the Swedish Natural Science Research Council and the Swedish Research Council for Agriculture and Forestry. A.V. Vener was supported by the Royal Swedish Academy of Sciences.

References

1 Bennett, J. (1977) Nature 269, 344-346
2 Allen, J. F., Bennett, J., Steinback, K. E., and Arntzen, C. J. (1981) Nature 291, 25-29
3 Allen, J. F. (1992) Biochim. Biophys. Acta 1098, 275-335
4 Gal, A., Shahak, Y., Schuster, G., and Ohad, I. (1987) FEBS Lett. 221, 205-210
5 Andersson, B., Åkerlund, H.-E., and Albertsson, P.-Å. (1976) Biochim. Biophys. Acta 423, 122-132
6 Åberg, F., Appelkvist, E.-L., Dallner, G., and Ernster, L. (1992) Arch. Biochem. Biophys. 295, 230-234
7 Larsson, U. K., Sundby, C., and Andersson, B. (1987) Biochim. Biophys. Acta 894, 59-68
8 Malkin, R. (1982) Biochemistry 21, 2945-2950
9 Huang, D., Everly, R. M., Cheng, R. H., Heymann, J. B., Schägger, H., Sled, V., Ohnishi, T., Baker, T. S., and Cramer, W. A. (1994) Biochemistry 33, 4401-4409

THE DELETION OF *petG* IN *Chlamydomonas reinhardtii* DISRUPTS THE CYTOCHROME *bf* COMPLEX

Deborah A. Berthold, Christian L. Schmidt, and Richard Malkin,
Dept. of Plant Biology, University of California, Berkeley, CA

1. Introduction

The cytochrome *bf* complex catalyzes the transfer of electrons from plastoquinol to plastocyanin with concomitant proton translocation across the membrane. It is comprised of four large subunits, cytochrome b_6, subunit IV, the Rieske iron-sulfur protein, and cytochrome *f* (1). In addition, several small proteins co-purify with the complex. The existence of a 4 kDa fifth subunit of the cytochrome *bf* complex (the *petG* protein) was reported based on its co-purification with the complex in a spinach preparation (2). In *Chlamydomonas reinhardtii*, the *petG* protein has also been found to co-purify with the cytochrome *bf* complex (3,4). We have chosen to investigate the role of *petG* in the complex by deleting *petG* from the *C. reinhardtii* chloroplast genome. As a control, we have also disrupted the reading frame that is found only 25 bp downstream of *petG* (ORF56). In this work we report that the *petG* protein has a role in either the assembly or the stability of the cytochrome *bf* complex.

2. Materials and Methods

2.1 *Chlamydomonas* strains
C. reinhardtii wildtype strains CC-124, CC-125, and CC-1928 were obtained from the *Chlamydomonas* Culture Collection (Duke University, North Carolina), and wildtype strain 7d+ was obtained from M. Goldschmidt-Clermont (Geneva).

2.2 Construction of the *petG*- and ORF56-deletions
A 1.5 kb spectinomycin resistance cassette containing the promoter region of *C. reinhardtii* chloroplast *atpA* and the bacterial *aadA* gene was cut out of pUC-atpX-aadA (5) for insertion into pG35, the 3.6 kb *Pst*I-*Hin*dIII fragment containing *petG* (Fig. 1). *petG* was excised at its flanking *Ssp*I sites, and replaced with this cassette. The *Taq*I-*Afl*III fragment at the 3' end of ORF56 was likewise deleted, and replaced with the spectinomycin-resistance cassette. In both cases, the coding direction of *aadA* is the same as the original gene.

2.3 Transformation of *Chlamydomonas*
C. reinhardtii cells were grown and plated as described in Boynton et al. (6). Plasmid DNA was precipitated onto tungsten or gold particles by standard methods, and transformation utilized a PDS-1000/He biolistic device (Dupont NEN). Cells were replated onto TAP plates containing 150 µg/ml spectinomycin and 50 µg/ml ampicillin.

P. Mathis (ed.), Photosynthesis: from Light to Biosphere, Vol. II, 571–574.

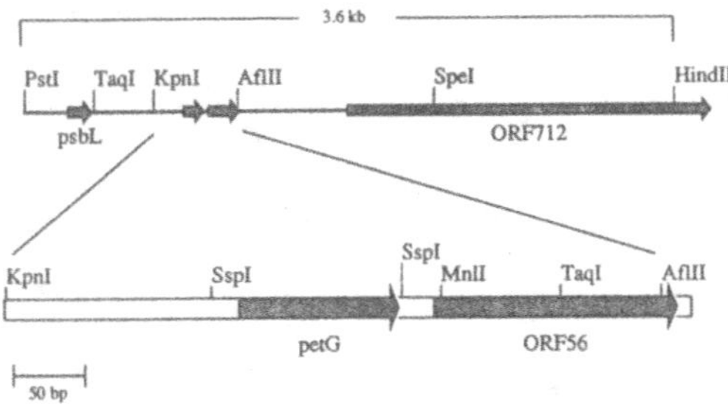

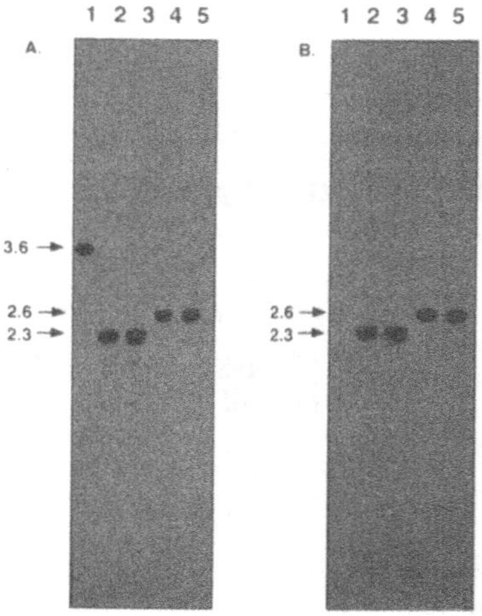

Figure 1. (above) Map of the *C. reinhardtii* chloroplast DNA fragment containing *petG* cloned into pBluescript as plasmid pG35. The restriction sites depicted are not necessarily unique.

Figure 2. (right) Southern analysis of *petG*- and ORF56-deletion strains. Total DNA from each strain was digested with *Pst*I and *Hin*dIII, run on an agarose gel, transferred, and hybridized with A. a *psbL* probe or B. an *aadA* probe. Lane 1, wildtype CC-1928; lane 2, *petG* deletion strain GF-3; lane 3, *petG* deletion strain GF-6; lane 4, ORF56 deletion strain 56-2A; lane 5, ORF56 deletion strain 56-6A.

2.4 Immunoblots and nucleic acid hybridization

Protein samples were run on denaturing 15% polyacrylamide gels, and transferred to nitrocellulose following electrophoresis. Binding of antibodies was detected by enhanced chemiluminescence. For Southern or northern hybridizations, DNA fragments for probes were labelled with ^{32}P using the random primer method. DNA or RNA was isolated by a minor modification of conventional methods.

3. Results

C. reinhardtii petG was cloned from chloroplast DNA of strain CC-125 using an oligonucleotide probe designed from the N-terminal sequence of the *petG* protein (3). Sequencing of *petG* revealed a transposition of guanosine and thymidine in codon 30 relative to a subsequently published sequence (7), resulting in an arginine rather than a leucine at this position. Sequencing of *petG* in three other *C. reinhardtii* strains (CC-124, CC-1928, 7d+) using PCR confirmed the transposition. The arginine is conserved in *Chlamydomonas eugametos* (8), *Cyanophora paradoxa* (9), and all organisms for which the *petG* sequence is known, with the exception of *Euglena gracilis* (10).

In order to determine whether the *petG* protein is essential for the functioning of the cytochrome *bf* complex, *petG* was deleted and replaced with a spectinomycin resistance cassette (5). As a control, a portion of ORF56 near the 3' end was also deleted and replaced with the spectinomycin resistance cassette. These constructs were transformed into *C. reinhardtii*, and the transformant colonies were recloned until homoplasmic as determined by Southern hybridization (Fig. 2). Two independent transformants of each were selected for characterization; the *petG* deletion strains GF-3 and GF-6, and the ORF56-deletion strains, 56-2A and 56-6A. The 3.6 kb band present in the wildtype lane of Fig. 2A is absent in the deletion strains, indicating no wildtype chloroplast genome remains. The spectinomycin cassette contains a *Pst*I restriction site, yielding a 2.3 kb band for the *petG*-deletion strains, and a 2.6 kb band for the ORF56 deletion strains (Fig. 2A), which are also visible when probed with the *aadA* cassette (Fig. 2B).

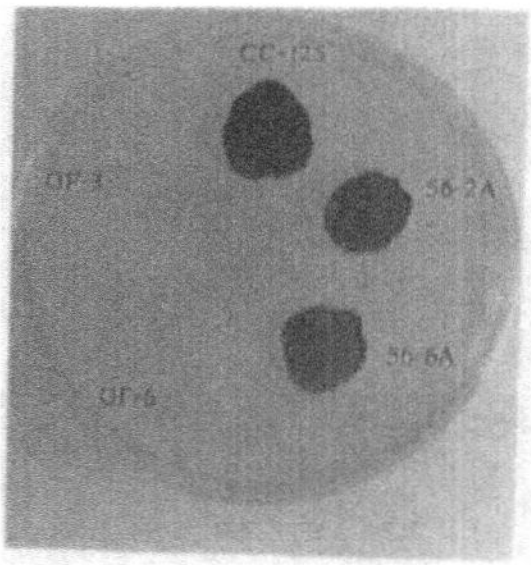

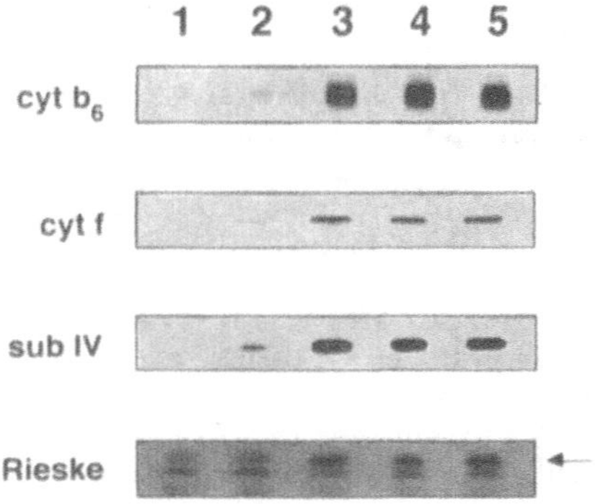

Figure 3. (above) Growth of the *petG*- and ORF56- deletion strains on HS media. Suspended *C. reinhardtii* cells (20 µl) of the indicated strains were spotted on the plate and incubated in bright light for 10 days.

Figure 4. (right) Immunoblots of the *petG*- and ORF56-deletion strains. Samples were loaded on a basis of constant chlorophyll. Rabbit sera generated against spinach cytochrome *f*, spinach subunit IV, maize Rieske iron-sulfur protein (from A. Barkan, University of Oregon) or a synthetic peptide conjugate (corresponding to *C. reinhardtii* cytochrome b_6 sequence) were used at a 1:10,000 dilution. The arrow indicates the band corresponding to the Rieske iron-sulfur protein; the lower band in this blot is a nonspecific cross-reaction of the anti-maize antibodies with an unidentified *Chlamydomonas* protein. Lane 1, GF-3; lane 2, GF-6; lane 3, 56-2A; lane 4, 56-6A; lane 5, wildtype CC-1928.

The ability of the deletion strains to grow photosynthetically was determined by growing the strains on HS minimal media in bright light. The *petG* deletion strains were unable to grow under these conditions (Fig. 3), but grew as normal on acetate-containing media in low light. The ORF56 deletion strains grew as well as the wildtype under both conditions.

Immunoblot analysis of the *petG*-deletion strains showed substantially reduced levels of cytochrome b_6, subunit IV, cytochrome *f*, and the Rieske iron-sulfur protein relative to wildtype levels (Fig. 4). In contrast, the ORF56-deletion strains showed normal levels of these proteins.

The reduced levels of the other subunits of the cytochrome *bf* complex in the *petG* deletion strains could be due to problems with assembly or stability of the complex in the absence of *petG*, or may be due to a problem in transcription. In order to rule out the latter possibility, total RNA was isolated from each of the strains and analyzed by northern hybridization. The level of RNA transcripts for *petB*, *PetC*, and *petD* were found to be normal in the *petG* deletion strains, and the *petA* transcript was found at a diminished level (Fig. 5). Among the photosynthetic ORF56 deletion strains, 56-2A showed decreased levels of every transcript, whereas 56-6A was similar to wildtype.

In addition to these genes, we analyzed transcripts from open reading frames in the vicinity of *petG*. The *psbL* gene (upstream from *petG*; see Fig. 1) showed variable levels of RNA in the *petG*- and ORF56-deletion strains (Fig. 4). Probing wildtype strain CC-1928 for ORF56 and ORF712 did not reveal a single defined transcript for either, but instead, very low levels of several large, possibly unprocessed transcripts (not shown). In the *petG*- and ORF56-deletion strains, the size of some of these large transcripts increased, suggesting that there was initiation at the *atpA* promoter region of the inserted *aadA* cassette. The *aadA* probe hybridized to a number of large transcripts in these strains (not shown).

4. Discussion

Deletion of *petG* from the chloroplast genome of *C. reinhardtii* results in a loss of photosynthetic function, which can be attributed to highly diminished levels of the cytochrome *bf* subunits. As a control we also disrupted ORF56, which is found 25 bases downstream of *petG*. This had no discernible effect on photosynthesis, or on the amount of the cytochrome *bf* complex as determined by immunoblots. We find normal levels of RNA transcripts for *petB*, *PetC*, and *petD*, and a reduced level of *petA* in the *petG* deletion strains. This lower level of *petA* transcript is unlikely to be responsible for the nonphotosynthetic phenotype of the *petG* deletion strains, however, because in the ORF56 deletion strain 56-2A, an even lower level of *petA* transcript has no effect on the ability of the strain to grow photosynthetically.

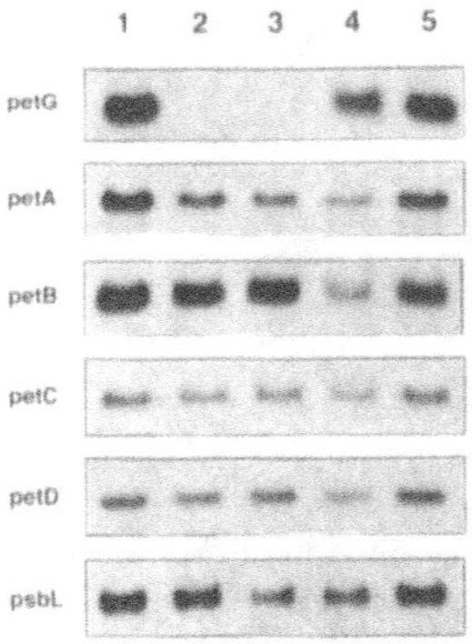

Figure 5. Northern analysis of the *petG*- and ORF56- deletion strains. Total RNA was electrophoresed, transferred to Hybond-N membrane, and probed with the labeled DNA as indicated. Each probe yielded a single band on the blot, and the following sizes were interpolated from RNA standards: *petG*, 0.7 kb; *petA*, 1.1 kb; *PetC*, 1.0 kb; *psbL*, 1.1 kb. Lane 1, wildtype CC-1928; lane 2, GF-3; lane 3, GF-6; lane 4, 56-2A; lane 5, 56-6A.

In the cytochrome *bf* complex, it has been observed that a single deletion of either *petA*, *petB*, or *petD* causes a large decrease in the level of all the other subunits (11). In these three deletion strains, the level of the *petG* protein is undetectable on immunoblots (11). Therefore, it is not surprising to find that in the *petG* deletion strains, the remaining subunits of the cytochrome *bf* complex are found at a markedly decreased level.

We have shown that the presence of the *petG* polypeptide is essential to the *C. reinhardtii* cytochrome *bf* complex for either assembly or stability of the complex. Work is underway to determine whether, in addition, it may have a function in catalysis.

References

1 Malkin, R. (1992) Photosyn. Res. 33, 121-136.
2 Haley, J. and Bogorad, L. (1989) Proc. Natl. Acad. Sci. USA 86, 1534-1538.
3 Schmidt, C.L. and Malkin, R. (1993) Photosynth. Res. 38, 73-81.
4 Pierre, Y., and Popot, J.-L. (1993) C.R. Acad. Sci. (Paris) Ser. 3, 316, 1404-1409.
5 Goldschmidt-Clermont, M. (1991) Nucl. Acids Res. 19, 4083-4089.
6 Boynton, J.E., Gillham, N., Harris, E., Hosler, J., Johnson, A., Jones, A., Randolph, B., Robertson, D., Klein, T., Shark, K., and Sanford, J. (1988) Science 240, 1534-1538.
7 Fong, S.E. and Surzycki, S.J. (1992) Curr. Gen. 21, 527-530.
8 Turmel, M. and Otis, C. (1994) Curr. Genet. 27, 54-61.
9 Stirevalt, V.L. and Bryant, D.A. (1989) Nuc. Acids Res. 17, 10095.
10 Hallick, R.B., Hong, L., Drager, R.G., Favreau, M.R., Monfort, A., Orsat, B., Spielmann, A., and Stutz, E. (1993) Nuc. Acid Res. 21, 3537-3544.
11 Kuras, R. and Wollman, F.-A. (1994) EMBO J. 13, 1019-1027.

A CHAIN REACTION MECHANISM FOR OXIDATION OF THE FULLY REDUCED CYTOCHROME b_6f COMPLEX.

David M. Kramer and Antony R. Crofts. Dept. of Plant Biology, Mirobiology and Program in Biophysics and Computational Biology. 265 Morrill Hall 505 S. Goodwin Ave. Urbana, IL 61801 U.S.A.

The Q-cycle (1) has provided a powerful framework for understanding the function of the cytochrome bc and b_6f complexes, accounting for the vast majority of available data. It is widely believed that the Q-cycle mechanism does not allow for turnover of the bc or b_6f complexes from conditions where both *b* cytochromes are pre-reduced. In the normal Q-cycle, quinol is oxidized at the quinol oxidase (Qo) site by oxidized Rieske iron-sulfur center (FeS), and the product semiquinone (SQ) reduces cyt b_L (the low potential cyt *b*). When both *b* cytochromes are pre-reduced, there is no apparent acceptor for the Qo site SQ. However, several laboratories have shown that turnover at low E_H does occur, although at diminished rates. After exciting photosystem I (PSI) with a single-turnover flash, cyt *b* is generally observed to go oxidized after a lag of 2-3 ms with a half time of about 20-30 ms. This is compared to a $t_{1/2}$ of 2-3 ms at higher potentials. These data have lead some to propose that the b_6f complex operates with a SQ-cycle (2,3), while others have suggested that alternative mechanisms, including a SQ-cycle may operate at low E_H (e.g. 4,5). In the version of the SQ-cycle favored by Joliot and Joliot (2,3), the SQ produced at the Qo site escapes and migrates as a SQ anion to the quinone reductase (Qi) site where it is reduced by cyt b_H (cyt b_L, then reduces cyt b_H). On the other hand, we have favored the operation of a Q-cycle under all but reducing conditions (6), partly because we failed to observe slippage reactions that are expected from SQ disproportionation. In this work, we further explored the turnover of the complex under conditions where cyt *b* is completely reduced prior to actinic illumination.

Materials and Methods.

Spinach thylakoids were prepared as in (6,7) and diluted in suspension medium at pH 7.5 to a concentration of 25 µg/ml in the presence of 1 µM gramicidin and approx. 5 mM Na dithionite. Spectroscopic measurements were made as described in (6). When very small signals were measured, up to 24 averages were collected. At least 20 minutes was allowed between measurements, and the samples were renewed after 80 minutes. No difference was noted between the first and final traces of a series. Actinic illumination was provided by a xenon flashlamp, blocked with an RG695 filter (Schott). The range of actinic intensities was obtained by inserting metal screens and/or neutral density filters into the actinic light path.

Results and Discussion.

Figure 1a shows the kinetics cyt *b* redox changes after an approximately 33% saturating flash. The re-reduction kinetics are distinctly biphasic. It has been previously

P. Mathis (ed.), Photosynthesis: from Light to Biosphere, Vol. II, 575–578.

shown that such biphasic re-reduction kinetics reflect the re-reduction of the cyt b_H and cyt b_L, respectively (7,8). This is confirmed in Fig. 1b, where the spectral changes of the fast and slow phases were found to have α-bands at about 563 565 nm, respectively, consistent with cyt b_H and b_L (7). The difference in the α-bands peaks is too small to reliably distinguish between the two forms, but we were able to measure the yields of cyt b_H and b_L oxidation by the extents of their respective re-reduction phases. Figure 2 shows that approximately 55-65% of the re-reduction kinetics were slow, even when the actinic flash intensity was lowered to the point where only 20% of one cyt b was oxidized per PSI center. This indicates that a significant fraction of cyt b_H was oxidized, even after flashes of very low intensities. We note that the absolute re-reduction rates varied by as much as 50% with conditions (probably reflecting the amount of added dithionite and its oxidation state), but the ratio of the fast and slow phases was always between 10 and 20, consistent with the relative E_m values of the two cyt b hemes (7).

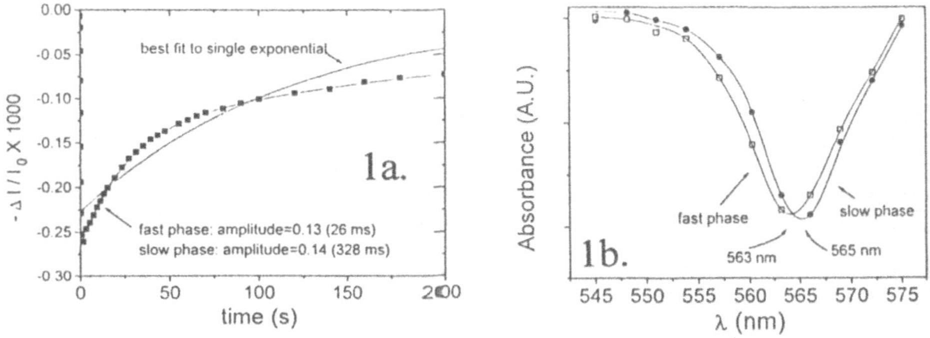

Starting from a fully reduced cyt b chain, the SQ cycle predicts that, if only one turnover of the Qo site occurs, only one cyt b heme will be oxidized per complex (i.e. one SQ produced at Qo, one cyt b oxidized at Qi). Given the relative E_m values of the b hemes in thylakoids (-50 and -150, see 8), we expect nearly all of the oxidized b heme to be the low potential form. Therefore, the data in Figs. 1 and 2, which suggest that nearly half of the oxidized cyt b is of the high potential form, provides strong evidence against the operation of a SQ-cycle at low E_H. Instead, a mechanism whereby both cyt b_H and b_L are oxidized in nearly equal proportions should be considered.

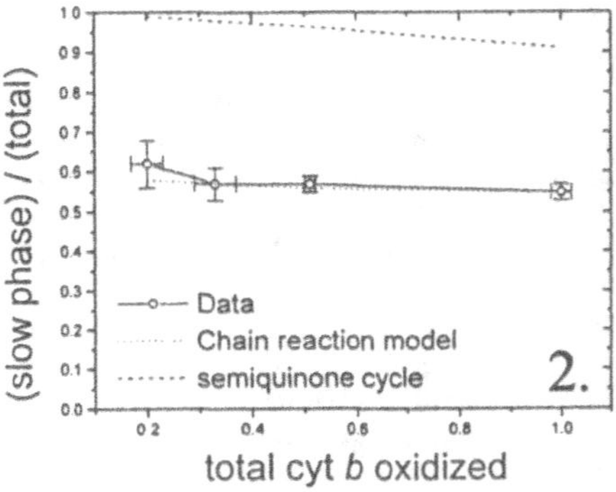

We propose that a "chain reaction" type model can account for this behavior within the framework of the Q-cycle model we developed for the b_6f complex (9). The scheme relies on two processes: a slow reaction, that produces small numbers of "open" b_6f complexes (with oxidized cyt b) and the normal Q-cycle turnover, which proceeds much more rapidly, producing plastoquinone which, in turn, opens more centers. Thus, the slow reaction "nucleates" a chain reaction that exponentially opens complexes. The Q-cycle produces PQ at the Qo site, and we expect that both cyt bH and bL would be oxidized upon reduction of PQ at the Qi site. This is illustrated in Fig. 3 a-c.

One possible candidate for the slow process is illustrated in Fig. 4 a-e. SQ is produced at the Qi site through oxidation by the Rieske iron-sulfur center (FeS). Cyt b_L then reduces the SQ producing PQH_2 at Qo. After another oxidizing equivalent migrates to FeS, via rapid redistribution by plastocyanin (PC), allowing PQH_2 at Qo to be fully oxidized by FeS and cyt b_L. PQ is then released into the membrane where it can oxidize both cyt b_L and b_H of a nearby b_6f complex. When opened, the complex can turn over at its normal, rapid rate; with each Qo site turnover it will produce 1 PQ, while consuming only 0.5 PQ at the Qi site. The excess PQ will be available for opening more complexes, which, in turn, will open still more (see above).

We performed numerical simulations using such a model. Starting with our previous kinetic model (9), we added a finite rate for oxidation of cyt b_L by SQ at the Qo site. We found that the extents of cyt b_H and b_L oxidation could be well modeled with a value for this reaction of 10s-1 or less (dotted line in Fig. 2). The simulated kinetics after a saturating flash were also reasonably similar to those found in our experiments, with a lag of approximately 3 ms, followed by a cyt *b* oxidation phase with $t_{1/2}$ of about 30 ms (not shown). Rate constants were similar to those in ref. 9, except for the oxidation of cyt bL by SQ at Qo. We also simulated the cyt b_H and b_L oxidation extents if a SQ-cycle was essentially the only pathway for cyt *b* oxidation. This yielded the dashed line in Fig. 2, where nearly 100% of oxidized cyt *b* was of the low potential form, except at high intensity flashes, where multiple-turnovers resulted in some cyt b_H oxidation.

The numerical simulations show that the observed behavior can be simulated, simply by adding a slow nucleating step, without changing the basic Q-cycle model. Indeed, the existence of a pathway for oxidation of cyt b by SQ at Qo can be surmised by the predicted complex geometry. Although reasonable fits could be obtained with a range of parameters, two additional constraints were important for obtaining a good fit with the chain reaction model. It was necessary to make the interaction of PC with cyt *f* relatively rapid, because the redistribution of oxidizing equivalents is required for opened complexes to turn over multiple times. In separate experiments, we obtained half times for cyt *f*

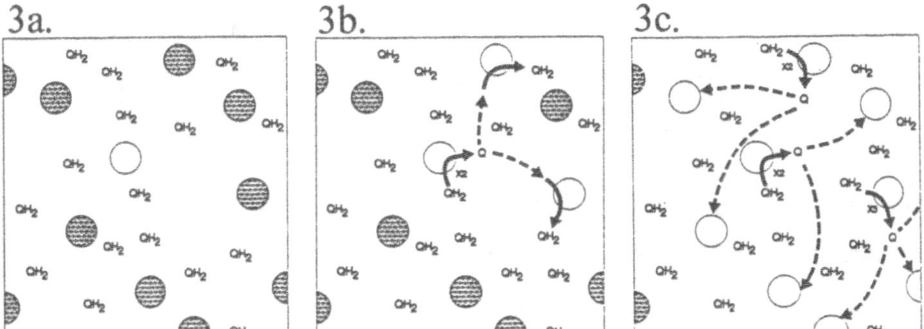

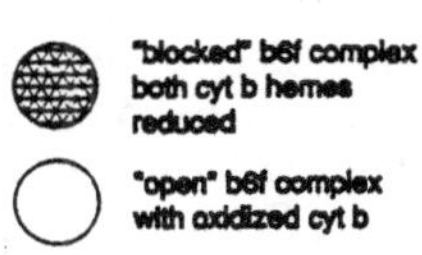

Figure 3. a) Complexes start with both cyt *b* hemes reduced. A slow nucleating reaction (about 10 s^{-1}) oxidizes both cyt *b* hemes in a small fraction of centers, "opening" them for normal turnover. b) "Open" complexes can turn over normally, producing PQ, which migrates to other b_6f complexes, opening them as well. c) Each of these opened complexes also produces PQ, opening still more centers. This proceeds until all oxidant is consumed.

oxidation after flashes of between 135 and 150 μs, regardless of the amplitude of the flash excitation or the degree of cyt f oxidation (data not shown). This is very near the expected half time for unbinding of oxidized PC from PSI (F. Drepper, personal communication), and we conclude that oxidation of cyt *f* by PC is limited by the release of oxidized PC from PSI. Therefore, we surmise that the migration of PC between cyt *f* docking sites, and its interaction with the complex are far more rapid than 135 μs, making it likely that PC is kinetically competent to catalyze the chain reaction. It was also necessary to make PQ unbinding from Qo and its interaction with the Qi site relatively rapid (less than about 0.5 ms). This is not unreasonable considering that PQH_2 dissociates from the Q_B site of PSII and migrates to the Qo site--a much larger distance than that between the Qo and Qi sites--in about 2 ms. The sensitivity of the model to these constraints suggests experiments to test it. For example, the model predicts that is the rate of PC migration were artificially slowed or constrained, the chain reaction would be severely hindered.

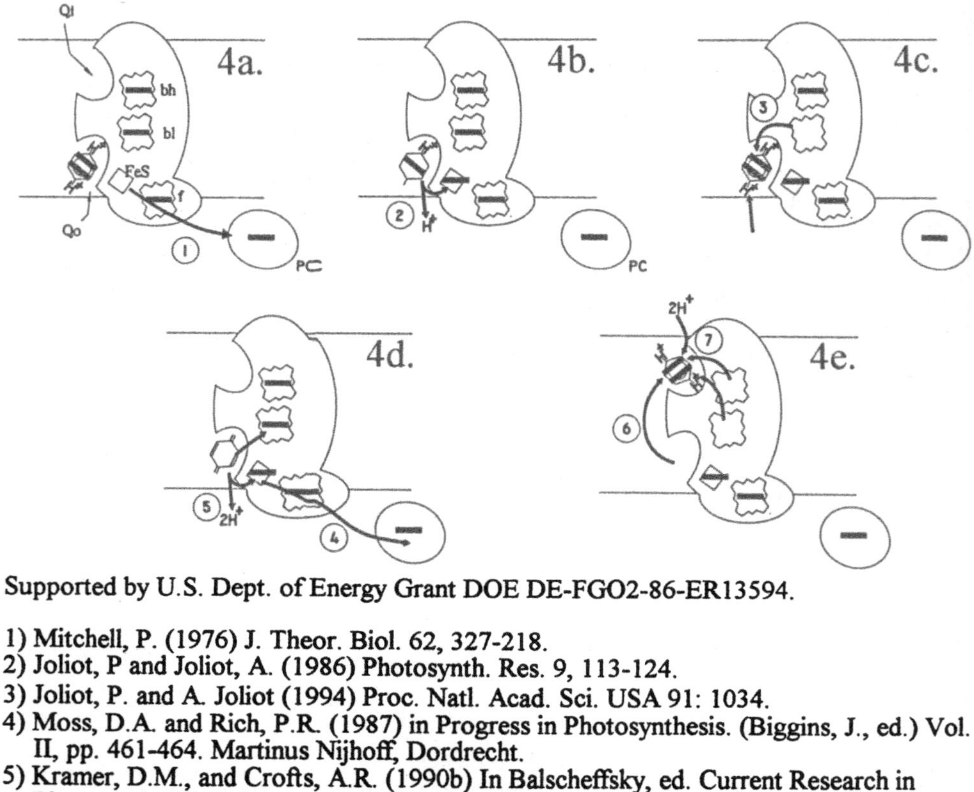

Supported by U.S. Dept. of Energy Grant DOE DE-FGO2-86-ER13594.

1) Mitchell, P. (1976) J. Theor. Biol. 62, 327-218.
2) Joliot, P and Joliot, A. (1986) Photosynth. Res. 9, 113-124.
3) Joliot, P. and A. Joliot (1994) Proc. Natl. Acad. Sci. USA 91: 1034.
4) Moss, D.A. and Rich, P.R. (1987) in Progress in Photosynthesis. (Biggins, J., ed.) Vol. II, pp. 461-464. Martinus Nijhoff, Dordrecht.
5) Kramer, D.M., and Crofts, A.R. (1990b) In Balscheffsky, ed. Current Research in Photosynthesis, Vol. 3, pp 381-384 Kluwer Academic, Dordrecht.
6) Kramer, D.M. and Crofts, A.R. (1993) Biochim. Biophys. Acta 1183, 72-84.
7) Kramer, D.M. and Crofts, A.R. (1994) Biochim. Biophys. Acta 1184, 193-201.
8) Nitschke, W., Berthold, I. and Hauska, G. (1987) in Progress in Photosynthesis. (Biggins, J., ed.) Vol. II, pp. 477-480. Martinus Nijhoff, Dordrecht.
9) Kramer, D.M. and Crofts, A.R. (1992) In Research in Photosynthesis (Murata, N., ed.) Vol. II, pp. 491-494, Kluwer, Dordrecht.

CYTOCHROME b559 WITHIN PHOTOSYSTEM II: STUDIES ON ITS PHOTOREDUCTION AND PHOTOOXIDATION IN PEA AND TARGETED MUTAGENESIS OF ITS GENES IN *CHLAMYDOMONAS*.

CLUES FOR A FUNCTIONAL MODEL OF THIS HAEM-PROTEIN RELATED TO PHOTOINHIBITION

Javier De Las Rivas [1], Francisco Morais [2], Peter Nixon [2] & James Barber [2]

[1] Department of Biochemistry & Molecular Biology. Faculty of Science. University of the Basque Country. P. O. Box 644 - Bilbao E48080 - SPAIN.

[2] Photosynthesis Research Group. Wolfson Laboratories. Department of Biochemistry. Imperial College of Science, Technology & Medicine - London SW7 2AY - UK.

1. Introduction

Cytochrome b559 within PSII. Cytochrome (cyt) b559, an integral membrane protein, is an intimate component of the photosystem II (PSII) multienzymatic complex in the thylakoid membranes of oxygenic photosynthetic organisms. Cyt b559 has two subunits, α and β, each with one predicted transmembrane α-helix domain. The haem cofactor is believed to be coordinated between two histidine residues one from each transmembrane α-helix. The role of this cyt in PSII function has been an enigma for many years (1), but recently the idea that it helps to protect PSII against photoinduced damage has emerged (2,3). PSII suffers more readily than other photosystems from irreversible damage due to excess illumination. This susceptibility of PSII to photo-induced damage underlies the physiological phenomenon of photoinhibition in plants (4). We have conducted a series of experiments on the photoreduction and photooxidation of cyt b559 within isolated PSII reaction centres (RC) (3,5) and cores from pea (6). Combined with these functional experiments, we have undertaken the construction of mutants in the *psbE* and *psbF* genes in the unicellular green algae *Chlamydomonas reinhardtii*.

Chlamydomonas reinhardtii, an ideal system to study cyt b559 by targeted mutagenesis. The unicellular green algae *Chlamydomonas reinhardtii* is a widely used organism to study photosynthetic processes, particularly structure/function relationships. Not only does it have a photosynthetic apparatus which is very similar to the one in higher plants, but its chloroplast genome can be transformed using biolistic technology (7) and it can grow heterotrophically in the dark allowing the isolation of non-photosynthetic mutants. The construction and study of the photosynthetic activity of cyt b559 mutants may be a useful approach to clarify the role of this important haem-protein in photosynthesis. The *psbE* gene encodes the α-subunit of the cyt b559 and the *psbF* gene encodes the β-subunit. The *psbE* and *psbF* genes have recently been sequenced from *C. reinhardtii* (8) and are very similar to their higher plant homologues. However, in contrast to all the plant and cynobacterial species so far examined the *psbE* and *psbF* genes are not co-transcribed in *C. reinhardtii*. This makes *C. reinhardtii* a more attractive system than the other model system, the cyanobacterium *Synechocystis sp* PCC 6803, for the construction of mutants.

2. Procedures

Biochemical studies on cyt b559 from pea were carried out as described in (3,5,6). DNA manipulation of the *psbE* and the *psbF* genes in *Chlamydomonas reinhardtii* were carried out as described in (9). Plasmid p78 (10) was used for the generation of plasmid constructs and was kindly provided by Prof. E. H. Harris.

P. Mathis (ed.), Photosynthesis: from Light to Biosphere, Vol. II, 579–582.

3. Results and Discussion

A proposed model for alternative electron transport in PSII linked to photoinhibition. We have reported a series of studies on the photo-redox activity of cyt b559 within isolated PSII complexes (3,5,6). In the first of these papers (3) we postulated a functional model for the role of this cyt. Our more recent results with PSII RCs (5) and core complexes (6) indicate that both photoreduction and photooxidation of cyt b559 can occur in PSII preparations under certain conditions and that these reactions are related to the amount of HP and LP redox forms of this cyt. We also provided some evidence that the sizes of the pools of HP and LP are related to pH (5).

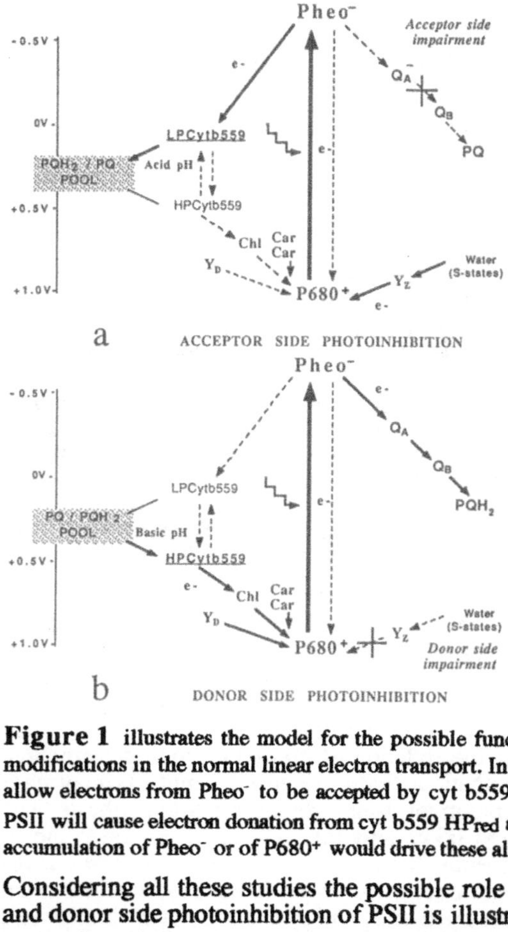

These data are relevant to our first model that proposed that cyt b559 acts to protect against donor and acceptor side photoinhibition (3). Under normal conditions linear electron transport within PSII from the water splitting system to the plastoquinone pool will occur, and neither the HP nor the LP forms of cyt b559 will take part in it. During photoinhibition, however, parts of the normal electron transport pathway may be disturbed. In fact, two mechanisms seem to be responsible for the photoinduced inactivation of PSII (4). One originated in an acceptor side impairment of electron transfer due to the over reduction of the secondary plastoquinone acceptors, mainly Q_A. And the other originated in a donor side impairment of the electron transfer from the water splitting system of PSII leading to an increased lifetime of $P680^+$ and other highly oxidising species.

Figure 1 illustrates the model for the possible function of cyt b559 within PSII. Each part shows modifications in the normal linear electron transport. In ***a*** the impairment of the acceptor side of PSII will allow electrons from $Pheo^-$ to be accepted by cyt b559 LP_{ox}. In ***b*** the impairment of the donor side of PSII will cause electron donation from cyt b559 HP_{red} and other auxiliary donors to $P680^+$. The eventual accumulation of $Pheo^-$ or of $P680^+$ would drive these alternative routes for electron transfer.

Considering all these studies the possible role of cyt b559 in protecting against acceptor and donor side photoinhibition of PSII is illustrated in **Figure 1 a & b**. Primary charge separation within PSII gives rise to $P680^+$ which is very oxidising (E_m=+1170 mV) and $Pheo^-$ which is very reducing (E_m=–600 mV). These radicals will be damaging when their lifetimes are not properly controlled. The '*specific mechanism*' of protection by cyt b559 might involve the photooxidation of cyt b559 HP_{red} by $P680^+$ to prevent secondary oxidations, and the direct reduction of cyt b559 LP_{ox} by $Pheo^-$ to compete with the further reduction of Q_A^-. In this way, cyt b559 could reduce the probability of damage due to the cation and anion radicals generated by primary charge separation.

Both alternative reactions, illustrated in **Figure 1**, should have a very low quantum yield compared to normal electron transport to ensure that they only occur when there is an impairment in PSII and, therefore, avoid any competition with linear electron flow. In this model, relaxation of the redox states of the two forms of cyt b559 in the dark involves electron exchange with the surroundings which has an intermediate redox potential (around +100 mV). This ambient redox potential lies conveniently between the potential of cyt b559 HP and LP and it is likely to change as a consequence of the reduction in the level of the plastoquinone pool. The reversibility of photooxidation in the dark upon addition of hydroquinone in RCs (5) supports this suggestion, as this quinone could provide a replacement for the missing PQH_2 pool in this isolated system. In contrast, the dark oxidation of photoreduced cyt b559 LP by addition of an oxidised quinone (equivalent to PQ pool) has also been demonstrated in isolated RCs (3).

The model correlates the light-induced redox changes of cyt b559 with its diverse redox potential forms, and we can postulate a pH dependent mechanism which could implicate a protonation / deprotonation step of a key amino acid to switch between the HP and LP forms. This mechanism could be in some way related to the function of this cyt proposed in (11) and it assumes that there is one cyt b559 per PSII. The key amino acid might be an acidic amino acid embedded in the proteinic matrix close to the haem, for example E7 or D12 from PsbE, or E130 from PsbA. The mechanism might also involve some of the several arginines present in the stromal side of cyt b559 (i. e. R8 and R18 in PsbE, and R5 and R13 in PsbF). The unusually positive redox potential value of cyt b559 HP (E_m = +0.4 V) requires a specially protected hydrophobic niche for the haem (1) . In this niche, near the N-termini of the two parallel α-helices, the haem is ligated to two histidines and the charged groups around them (like the ones in Glu, Asp or Arg) will drastically influence the positive dipole potential of the haem (1).

Strategies and clues for targeted mutagenesis within cyt b559. Our molecular biology work on the cyt b559 genes is aimed in the first step to see how essential the cyt is for the function of PSII. For this purpose we have undertaken the inactivation of the genes by insertion of a spectinomycin-resistance cassette (made available by Dr M. Goldschmidt-Clermont). The results corresponding to the construction of a *psbE*-null mutant of *C. reinhardtii* by insertional inactivation are presented in the paper by Morais *et al.* in these proceedings (9). For the production of a similar *psbF* mutant we have constructed a plasmid by insertion of the spectinomycin resistance cassette into an unique *Sca I* site of the *psbF* gene in the pF1 plasmid (9) (see **Figure 2**). This construct (called pF1aadF) has been used to transform *C. reinhardtii* wild type cells (CC-125) with the particle gun transformation method (7). We obtained several spectinomycin resistant colonies. Southern analyses of these colonies did not show any positive mutant with *psbF* correctly interrupted. This could be presumably due to a low frequency recombination because of the small size of the 3' end flanking region of the gene in the construct (12). In current experiments we are repeating this transformation. In any case, looking at the results obtained with *psbE* (9) we can infer that cyt b559 is an essential component of PSII needed for its structural integrity. In fact, the absence of the α-subunit of cyt b559 provokes the incapacity for the algae to grow photoautotrophically and the lack of functional oxygen evolving complex in the thylakoid membranes (9). This result contrast with a mutant strain of *C. reinhardtii* lacking the chloroplast *psbI* gene which is able to grow photoautotrophically under dim light (13). The PsbI protein is also intimately associated to PSII RCs but this *psbI*-deficient mutant did not show a complete inactivation of PSII (13). This could indicate that the cyt b559 is essential even at low light conditions or that there is a more drastic regulation of the expression of the *psbE* and *psbF* genes together to the *psbA* and *psbD* which encode the central proteins of PSII, D1 and D2. In fact, deletion of the individual genes, *psbE* or *psbF*, of cyt b559 in *Synechocystes 6803* leads to complete destabilization of the PSII RC (14). The importance of *psbF* in the photosynthetic growth of plants is also shown in the behavior of a *psbF* -mutant recently obtained in tobacco that only can grow in low light (15).

We have constructed another plasmid (called pKS5.7, see **Figure 2**) using pBluescript II KS as a vector. We cloned in it a 4.2 kb *KpnI/PstI* restriction fragment (from the *C. reinhardtii* ctDNA in p78) together to a 1.5 kb *PstI/BamHI* fragment (from *C. reinhardtii* chloroplast genome). This second fragment was obtained by PCR using purified chloroplast genome as a template. The next step is to use this plasmid for insertional inactivation and site-directed mutagenesis on specific sites of *psb F*. In this respect it is necessary to identify some key amino acids which may be involved in the function of the cyt b559. Based on the model presented in **Fig.1** charged negative residues around the haem would be important targets. From our pH experiments (5) we can postulate the involvement of a acidic group with a pKa between 6 and 8. It is clear that the pK's of amino acid residues in proteins depend on the environment around each group. This environment is modulated by the protein structure through the interaction with other residues and with water. The site-directed mutagenesis of D12 from PsbE and D4 from PsbF could be a good starting point to see the influence of these aspartic groups on the redox function of the haem.

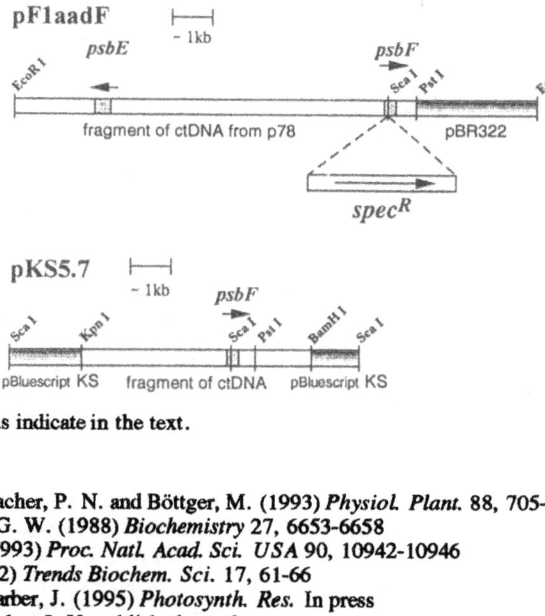

Figure 2 . Plasmids constructed as indicate in the text.

References

1. Cramer, W. A., Tae, G.-S., Furbacher, P. N. and Böttger, M. (1993) *Physiol. Plant.* 88, 705-711
2. Thompsom, L. K., and Brudvig, G. W. (1988) *Biochemistry* 27, 6653-6658
3. Barber, J. and De Las Rivas, J. (1993) *Proc. Natl. Acad. Sci. USA* 90, 10942-10946
4. Barber, J. and Andersson, B. (1992) *Trends Biochem. Sci.* 17, 61-66
5. De Las Rivas, J., Klein, J. and Barber, J. (1995) *Photosynth. Res.* In press
6. Klein, J., De Las Rivas, J.and Barber, J. Unpublished resutlts.
7. Boynton, J. E. and Gillham, N. W. (1993) *Methods in Enzymology* 217, 367-382
8. Alizadeh, S., Nechustai, R., Barber, J. & Nixon, P. (1994) *Biochim. Biophys. Acta* 1188, 439-442
9. Morais, F., Barber, J. and Nixon, P. (1995) in this Proceedings
10. Fong, S. E. and Surzycki, S. J. (1992) Curr. Genet. 21, 485-497.
11. Ortega, J.M., Hervas, M., and Losada, M. (1988) *Eur. J. Biochem.* 171, 449-455
12. Takahashi, Y. *et al.* (1991) *EMBO J.* 10, 2033-2040.
13. Künstner, P. Guardiola, A., Takahashi, Y. and Rochaix, J.-D. (1995) *J. Biol. Chem.* 270, 9651-9654
14. Pakrasi, H. B., Nyhus, K. J. and Granok, H. (1990) *Z. Naturforsch* 45c, 67-73.
15. Bock., R., Kossel, H. and Maliga, P. (1994) *EMBO J.* 13, 4623-4628.

CLONING AND SEQUENCING OF *PETB* AND *PETD* GENES; OVEREXPRESSION AND CHARACTERIZATION OF SUBUNIT IV FROM A CYANOBACTERIAL CYTOCHROME B_6F COMPLEX.

U. Boronowsky, J. Kruip and M. Rögner
Institute of Botany, University of Münster, Germany

1. Introduction

In the thylakoid membranes of cyanobacteria, the cytochrome b_6f complex is of central importance both for photosynthetic and respiratory electron transport. However, the yield of cyanobacterial preparations that could be obtained up to now has been too low for a detailed structural analysis. Overexpression and purification of the isolated subunits of this complex could help considerably for the understanding of structure/function relationships; it may also allow the reconstitution of a whole, functional complex.

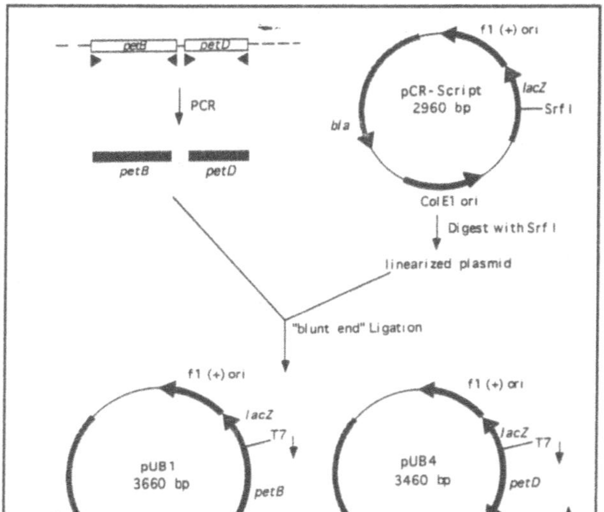

Fig. 1: Cloning of *petB* and *petD* PCR fragments into sequencing vectors

Site-directed mutations within these subunits may help in studying the function of highly conserved residues either *in vivo* or in isolated complexes. Similar approaches have already proven useful for the Rieske FeS protein [1] and the soluble domain of the cytochrome f protein [2].

In this work, sequences of the *petB* and *petD* genes from *Synechocystis* PCC 6803, encoding cyt b_6 apoprotein and subunit IV, respectively, could be determined by non-radioactive methods. The genes could be amplified by PCR and cloned into expression vectors under control of a T7 promoter (Fig.2). Overexpression of subunit IV in *E. coli* yielded up to 5% of the total cell protein as characterized by immunoblotting.

P. Mathis (ed.), Photosynthesis: from Light to Biosphere, Vol. II, 583–586.

2. Procedure

Total DNA was isolated from *Synechocystis* PCC 6803 wild type cells grown in BG11 media. Amplification of the *petB* and *petD* genes was achieved by PCR. All molecular biology techniques were performed as described in [5]. The PCR fragments were cloned blunt-end into sequen-cing vectors (Fig.1). Biotiny-lated sequencing primers allowed the detection of bands with alkaline-phosphatase-coupled streptavidin molecules (ECL). Subcloning into pQE-vectors (Qiagen) yielded positive clones, but over-production of *petB/petD*-His fusion genes was not successful. This was probably due to background expression of protein having a toxic effect on the cell. For this reason, both genes were sub-cloned into pT7-7 vectors permit-ting gene expression under strict control of a T7 RNA-polymerase promoter (Fig. 2). Restriction analysis and non-radioactive sequencing were used to verify correct insertion of the PCR fragments. Expression vector pT7-*petD* was transformed into *E. coli* BL21(DE3) cells carrying the T7 RNA-polymerase gene under control of a lac promoter. Cells were grown to mid-log phase and induced by addition of IPTG. Cell lysates were analyzed by SDS-PAGE according to [6]. Protein bands were identified by immuno-blotting using polyclonal antibodies raised against subunit IV from spinach (kind gift of Dr. W. Cramer)

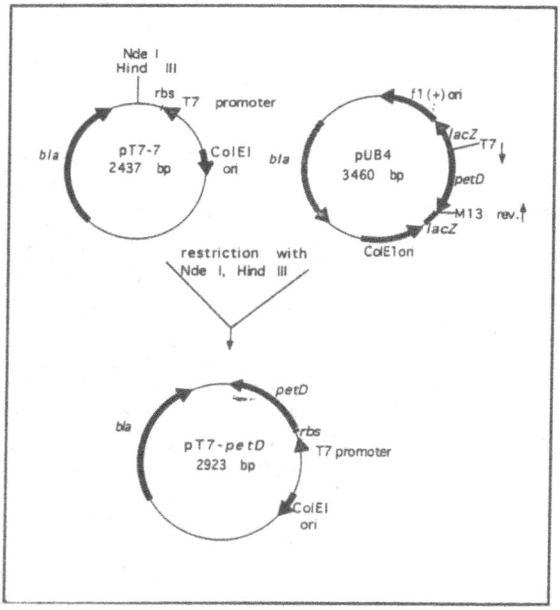

Fig. 2: Subcloning of *petD* into pT7 vector after digestion with NdeI and HindIII

3. Results and discussion

3.1. Non-radioactive sequencing of *petB* and *petD*

The genes *petB* and *petD* were amplified by PCR using total *Synechocystis* DNA preparations as template. Complete sequences of both genes could be determined with non-radioactive techniques and were found to be identical to the corresponding sequences obtained by partly radioactive techniques that have already been published [7,8]. Deduced amino acid sequences show a high degree of homology (75-85 %) between different species of cyanobacteria and chloroplasts of algae and higher plants. In contrast, homology between cytochrome b_6/subunit IV and cytochrome b from mitochondrial bc_1-

complexes on one hand and those from photosynthetic bacteria on the other does not exceed 25-40 % [9]. A seven amino acid-extension at the N-terminus of PetB can be found in sequences of *Synechocystis* and *Synechococcus* as well as in *Prochlorothrix*, but not in those of other cyanobacteria such as *Nostoc* and *Anabaena*. This may suggest that the relationship between unicellular cyanobacteria and *Prochlorothrix* is closer than estimated from the pigment composition of both species [7]. Hydropathy analyses predict four transmembrane helices and one amphiphilic helix at the lumenal (n-) side of the membrane for cyt b_6 and three transmembrane helices for SU IV. According to the "positive-inside-rule" [10] N- and C-termini of PetB and the N-terminus of PetD are located at the p-side of the membrane.

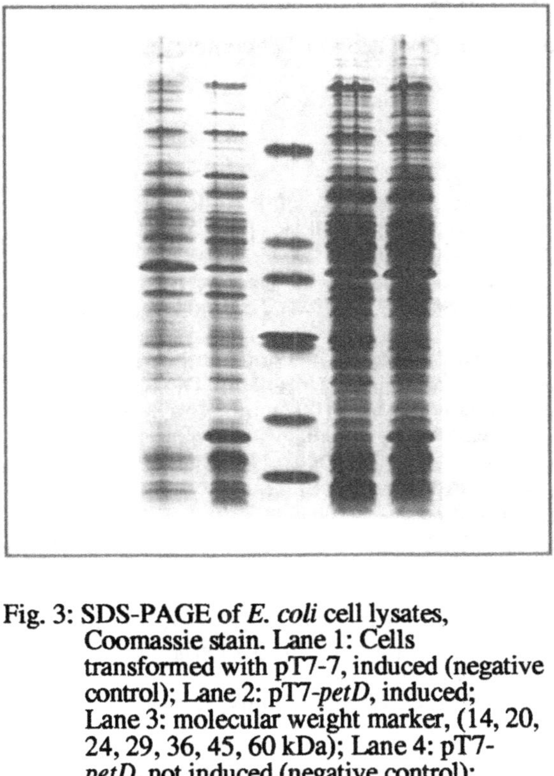

Fig. 3: SDS-PAGE of *E. coli* cell lysates, Coomassie stain. Lane 1: Cells transformed with pT7-7, induced (negative control); Lane 2: pT7-*petD*, induced; Lane 3: molecular weight marker, (14, 20, 24, 29, 36, 45, 60 kDa); Lane 4: pT7-*petD*, not induced (negative control); Lane 5: pT7-*petD*, induced.

3.2. Overexpression

Both genes could be subcloned into pT7-7 expression vectors; the new plasmids were named pT7-*petB* and pT7-*petD*, respectively (Fig.2). After induction of BL21(DE3) transformed with pT7-*petD*, an additional band could be detected by SDS-PAGE (Fig.3). Its apparent molecular weight of 17,8 kDa is in accordance with published data for SU IV (17,5 kDa for *Nostoc*, [9]). The protein could be further characterized by immunoblotting and specific recognition by antibodies raised against SU IV from spinach (data not shown). The high expression level (about 5 % of total cell protein) suggests that the protein is mainly localized in inclusion bodies. If this is the case, purification methods for subunit IV would have to include refolding steps in order to study the native protein. Concerning the cytochrome b6 apoprotein, a suitable expression system is yet to be found; maybe a fusion protein system, which would also permit purification by affinity chromatography, could be the right choice. Overproduction of both subunit IV and cytochrome b_6 apoprotein would offer the possibility for genetical modification of the overexpressed subunit in order to investigate the function of strictly conserved residues. Since both *Synechocystis* PCC 6803 and *Synechococcus* PCC 6301 are naturally transformable, more complicated procedures like conjugation and electroporation (which are necessary for the genetic manipulation of other cyanobacteria such as *Nostoc* [1] and *Anabaena* [3]) could be avoided. Recombinant, whole complexes could be isolated from these organisms for further investigations according to preparation procedures already established in our group [4]. For instance, the role of the four conserved histidines suggested to serve as ligands for the two b-haemes which are thought to be oriented parallel to the surface at each side of the membrane, similarly to the b-haemes of mitochondrial and bacterial bc_1-complexes [9,11], could be probed. To verify this hypothesis, the effect of site-directed deletions or mutations of these histidines could be investigated by spectroscopical analysis. Similar approaches could serve to localize possible binding sites for oxidized and reduced plastoquinone, thus supplying more data about the route of the cyclic electron flow involving the cytochrome b_6f complex. Due to the fission of mitochondrial cyt b into cyt b_6 and SU VI in chloroplasts and cyanobacteria, the postulated quinone binding site would have to be formed by the C- and N-termini of cyt b_6 and SU IV, respectively. Studies on a synthetic, overexpressed fusion protein could provide information about the implications of this fission on quinone binding. This could also help to elucidate structural and functional differences between b_6f complexes and bacterial bc_1-complexes.

4. References

1 Holton, B., Middleton, K., Zarka, D., Alsaadi, R. and Kallas, T. (1992) in: "Research in Photosynthesis", (Murata, ed.), Vol.II, pp. 567-570, Kluwer Academic Publishers, Dordrecht, The Netherlands

2 Cramer, W.A., Martinez, S.E., Huang, D., Tae, G.-S., Everly, R.M., Heymann, J.B., Cheng, R.H., Baker, T.S. and Smith, J.L. (1994) J. Bioenerg. Biomembr. 26, 31-47

3 Pakrasi, H.B. and Nyhus, K.J.(1992) in: "Research in Photosynthesis", Vol.I, pp. 569-572, (Murata, ed.)Kluwer Academic Publishers, Dordrecht, The Netherlands

4 Bald, D., Boekema, E.J., Kruip, J., Rögner, M. (1992) in "Research in Photosynthesis" (Murata, N., ed.) Kluwer Academic Publishers Vol. I , pp. 629-632

5 Sambrook, J., Fritsch, E.F. and Maniatis, T. (1989) "Molecular cloning - a laboratory manual" 2nd edition, Cold Spring Harbour Laboratory Press

6 Laemmli, U.K. (1970) Nature 227, 680-685

7 Kruip, J., Nixon, P., Osiewacz, H. D., Rögner, M. (1994) Biochim. Biophys. Acta, 1188, 443-446

8 Osiewacz, H.D. (1992) Arch. Microbiol. 157, 336-342

9 Kallas, T. (1994) in: The Molecular Biology of Cyanobacteria, (Bryant D.A., ed.), pp.259-317, Kluwer Academic Publishers, Dordrecht, The Netherlands

10 von Heijne, G. (1992) Nature 341, 456-458

11 Crofts, A., Hacker, B., Barquera, B., Yun, C.-H. and Gennis, R. (1992) Biochim. Biophys. Acta, 1101, 162-165

TOPOGRAPHICAL STUDIES ON THE CYTOCHROME B_6F COMPLEX FROM SPINACH BY CROSSLINKING WITH EDC AND OPA

A. Blokesch[1], K.-D. Irrgang[2], J. Salnikow[1], and J. Vater[1]
Technische Universität Berlin: [1]Institut für Biochemie und Molekulare Biologie, Franklinstraße.29, 10587 Berlin, and [2]Max-Volmer-Institut für Physikalische und Biophysikalische Chemie; Straße des 17. Juni 135, 10623 Berlin, Germany

1. Introduction

The cytochrome b_6f complex in the thylakoid membrane of cyanobacteria and chloroplasts plays a key role in the electron transfer between the two photosystems. This quinol-plastocyanin oxidoreductase consists of at least six protein subunits. For five of these the primary structures have been resolved for a variety of organisms widely distributed on the evolutionary scale (for reviews, see (1,2)). So far known only three subunits carry prosthetic groups: the cytochromes b_6 (two hemes) and f (one heme), and the Rieske protein (an iron-sulfur cluster). Further components are subunit IV (probably derived from an evolutionary split gene which originally linked this polypeptide to the C-terminus of cyt b_6), and subunit V which is encoded by petE, occasionally also referred to as petG (3,4). Subunit VI, along with the Rieske protein the only nuclear encoded protein, has been identified by immunodetection using peptide-directed antibodies against the petX-gene product (4), and partially sequenced at the N-terminus (5).
Recently Martinez et al. reported the X-ray-crystallography of isolated cytochrome f revealing the hydrophobic segments to consist predominantly of β-sheets and the heme iron to be ligated by the N-terminal amino group (6). Since attempts have failed so far to crystallize the entire complex, topological studies have been initiated by a variety of methods and achieved considerable progress. Membrane spanning segments have been predicted by hydropathy plots and their position and orientation confirmed by limited proteolysis (for reviews, see (1,2)). Crosslinking with bifunctional reagents is the method of choice to elucidate the adjacency of proteins in the native complex (7). In addition to previous studies with the short range (0.3-0.5 nm) crosslinker *ortho*-phthaldialdehyde (OPA) (8), we here report modifications using the zero-length crosslinker 1-ethyl-3-[3-(dimethylamino)propyl]carbodiimide (EDC) and discuss structure-function relationships.

2. Materials and Methods

Cyt b_6f complex from spinach was purified according to Hurt and Hauska (9) as modified by Shallan et al. (7). For immunostaining experiments the following protease inhibitors were added to all solutions throughout the preparation: 2 mM benzamidine, 5 mM ε-amino-n-caproic acid, and 50 μM 4-(2-aminoethyl)-phenylsulfonylfluoride. They were removed by dialysis prior to crosslinking and completely omitted in preparations assigned to the study of electron transport activity. OPA-crosslinking, electron transport measurements, cytochrome titration, protein

P. Mathis (ed.), Photosynthesis: from Light to Biosphere, Vol. II, 587–590.

determination, SDS-PAGE, heme staining and Western blot experiments were performed as described by Bhagwat et al. ((8) and references therein). In addition to the Laemmli system, gels with a smaller pore size were used for SDS-PAGE in order to obtain a higher resolution in the molecular mass range of 5-25 kDa. Resolving gels contained 5 M urea, 17.5 (w/v) % acrylamide, and 0.47 (w/v) % bisacrylamide. For immunostaining, the ECL/peroxidase system (Amersham) was used alternatively to the Fast Red/naphthol AS MX phosphate/alkaline phosphatase system (Sigma). Polyclonal antibodies against both cytochromes, subunit IV, the Rieske protein (10) and CP 14 (vide infra) (11) were raised in rabbits by injecting electrophoretically or chromatographically (10) purified proteins. Anti-subunit V was raised against a tridecapeptide corresponding to residues 25 to 37 of the mature petE gene product (3,4) coupled to *keyhole limpet* hemocyanin. For modifications using EDC, the purified cyt b_6f complex was dialyzed against 50 mM Mes-NaOH buffer (pH 6.5) containing 15 mM NaCl, 5 mM $MgCl_2$, 10 mM nonaoyl-N-methylglucamide, and 0.5 (w/v) % sodium cholate at 4°C for 2 hours. EDC was dissolved in water and immediately added to the samples. The optimum EDC-protein ratio (w/w) for crosslinking was found to be in the range of 1 to 10. Samples were incubated at 30°C for 30 minutes. Reactions were quenched with 250 mM sodium acetate. Electron transport was measured prior to quenching.

3. Results and Discussion

It was previously reported that treatment of the cyt b_6f complex with OPA mainly yielded conjugates between subunit IV with each of the two cytochromes, respectively (8). Below the subunit IV-cyt b_6 conjugate band (39 kDa), a 35 kDa band was only immunodecorated with anti-subunit IV (*Fig.1*). Immunostaining and apparent molecular mass led us to suggest that a dimer of subunit IV (2 x 17 kDa) was formed. This result is consistent with the current view that the active form of the cyt b_6f complex is a dimer ((1,2) and references cited there).

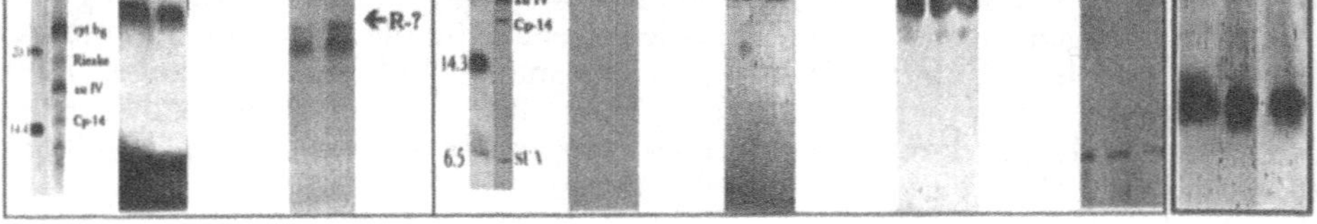

Fig. 1. **OPA-crosslinking:** Blots of purified (if not otherwise indicated) cyt b_6f complex were incubated without (-) and with (+, ++) OPA. The heme stained SDS-gel and immunodecorated blots with the respective antibodies are indicated. Abbreviations for conjugates (←) are *f* and b_6 for the respective cytochromes, *R* (the Rieske protein), *IV* (subunit IV) and *?* for yet unknown components. The titles "crude" and "purified" refer to preparations of the cytochrome b_6f complex.

An additional band of about 22 kDa molecular mass appeared involving subunit IV. The Rieske protein (20 kDa) was identified in a conjugate of 24 kDa apparent molecular mass.

The small differences in molecular mass between these conjugates and one of the components taking part in their formation led us to assume one of the small subunits V and/or VI as possible partners in crosslinking. Blot membranes hence were screened with anti-subunit V (anti-subunit VI being not yet available). The presence of its original 4.4 kDa band (3-5) was confirmed for our preparations, diminishing with higher concentrations of OPA. Occasionally, crossreaction with cytochrome f occured, as has been noted by Pierre et al. (5). None of the mentioned conjugates, however, showed a positive immunodecoration with anti-subunit V.

Purified preparations of the cyt b_6f complex frequently exhibited bands below subunit IV, in the range of 10 to 15 kDa. Therefore, cyt b_6f complex was assayed with antibodies against CP 14, which presumably is an antenna component of the adjacent photosystem II with yet undetermined primary structure (11). CP 14 was identified in both crude and purified preparations of the cyt b_6f complex with the same intensity, disappearing at higher OPA-concentrations without formation of detectable conjugates. Similar to photosystem II preparations, partial oligomerisation to aggregates of 66-70 kDa was observed (11). The copurification of CP 14 with the cyt b_6f complex is a hint for the location of this protein at the margin of PS II, adjoining to the neighbouring quinol-plastocyanin oxidoreductase. The lack of crosslinking of the pigment protein complex with any of the cyt b_6f subunits suggests that it is either located at a larger distance or amino acids to be conjugated are not surface-exposed,

Further studies were performed with the *zero*-length crosslinker EDC which had to be applied in concentrations about a 100fold higher as compared to OPA. OPA used in the concentrations appropriate for crosslinking attenuated 40 to 56 % of the initial electron transport activity (8); EDC-treatment even inhibited 70 to 90 % (*Fig. 2*).

Heme staining after EDC-treatment revealed the formation of conjugates similar to those obtained after OPA-treatment: conjugate bands exhibited apparent molecular

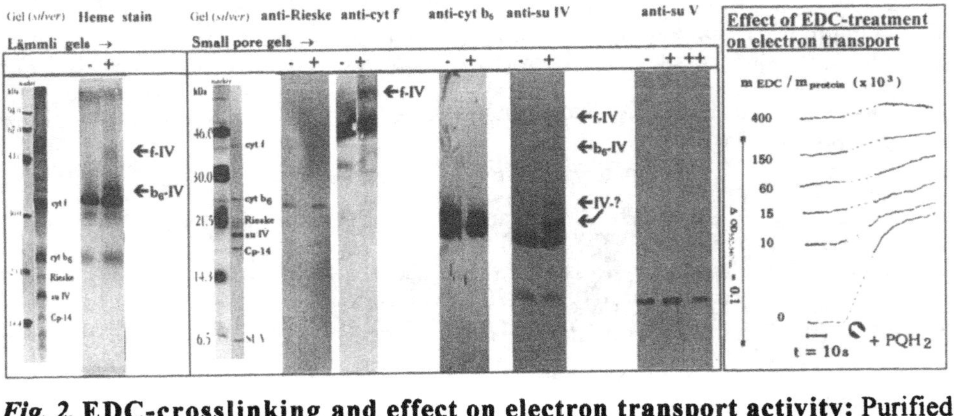

***Fig. 2.* EDC-crosslinking and effect on electron transport activity:** Purified cyt b_6f complex was incubated without (-) and with (+, ++) EDC. The heme stained SDS-gel and blots immunodecorated with the respective antibodies are indicated. Abbreviations for conjugates (←) are *f* and b_6 for the respective cytochromes, *IV* (subunit IV) and *?* for yet unknown components. *PQH_2*: plastoquinol.

masses of 39 and 51 kDa (*Fig. 2*). Immunoscreening unequivocally confirmed cytochrome f as a component of the 51 kDa conjugate. After incubation with anti-sub--unit IV, two weakly stained bands appeared corresponding to the conjugates detected by heme staining. Immunodecoration with anti-cytochrome b_6 showed no conjugate band in addition to the original protein band. The experimental data, however, are sufficient to infer that EDC like OPA predominantly links subunit IV to each of the two cytochromes, respectively. The weakness or absence of immunostained bands for two of the proteins taking part in conjugate formation may possibly be explained by a shielding of epitopes for antibody recognition after EDC-modification.
Two strong bands of 22 and 25 kDa proved the presence of subunit IV in conjugates with yet undetected smaller partners (*Fig. 2, lanes anti-su* IV). As in the case of the OPA-conjugates involving smaller proteins we were not able to confirm the petE gene product as the second component of the conjugates.
As a summary of our studies we note that both OPA and EDC linked subunit IV to each of the two cytochromes, respectively. It can be deduced that subunit IV is located in close distance to these two heme carrying proteins which directly participate in electron transport. Only OPA, the crosslinker with a slightly longer spacer length, was able to induce the formation of subunit IV-dimers.
In addition, conjugates involving smaller proteins and the Rieske protein (after OPA-treatment) or subunit IV (after OPA- and EDC-treatment) were observed. Subunit V could neither be proved to participate in their formation nor could it be excluded, since epitopes for antibody recognition might have become unaccessible after crosslinking. Work is in progress to raise antibodies against an undecapeptide taken from the petX gene product sequence and to examine its possible role in conjugate formation.

REFERENCES

1 Hauska, G., Nitschke, W., and Herrmann, R.G. (1988) J. Bioenerg. Biomembr. 20, 211-227
2 Cramer, W.A., Martinez, S.E., Huang, D., Tae, G..-S., Everly, R.M., Heymann, J.B., Cheng, R.H., Baker, T.S., and Smith, J.L. (1994) J. Bioenerg. Biomembr. 26, 31-47
3 Haley, J., and Bogorad, L. (1989) Proc. Natl. Acad. Sci. USA 86, 1534-1538
4 Schmidt, C.L., and Malkin, R. (1993) Photosynth. Res. 38, 73-81
5 Pierre, Y., and Popot, J.-L. (1993) C. R. Acad. Sci. Paris, Sciences de la vie 316, 1404-1409
6 Martinez, S.E., Huan, D., Szczepaniak, A., Cramer, W.A., and Smith, J.L. (1994) Structure 2, 95-101.
7 Shallan, M.A.-A.M., Radau, B., Salnikow, J., and Vater, J. (1991) Biochim. Biophys. Acta 1057, 64-68
8 Bhagwat, A.S., Blokesch, A., Irrgang, K.-D., Salnikow, J., and Vater, J. (1993) Arch. Biochem. Biophys. 304, 38-44
9 Hurt, E., and Hauska, G. (1981) Eur. J. Biochem. 117, 591-599
10 Hurt, E., Hauska, G., and Malkin, R. (1981) FEBS Lett. 134, 1-5
11 Irrgang, K.-D., Kablitz, B., Vater, J., and Renger, G. (1993) Biochim. Biophys. Acta 1143, 173-182

ACKNOWLEDGEMENT

This work is supported by the Deutsche Forschungsgemeinschaft (Sa 267/7-4). We are indebted to Dr. R. Malkin (Berkeley, CA) for providing us antibodies against the four main components of the cytochrome b_6f complex.

MOLECULAR WEIGHT AND STATE OF ASSOCIATION OF THE PURIFIED b_6f COMPLEX FROM *CHLAMYDOMONAS REINHARDTII.*

Cécile Breyton[1], Christophe Tribet[2], Jacqueline Olive[3], Michel Recouvreur[3], and Jean-Luc Popot[1]

[1]Institut de Biologie Physico-Chimique, C.N.R.S. URA 1187, 13, rue Pierre et Marie Curie, F-75005 Paris, France.[2]Ecole Supérieure de Physique et de Chimie Industrielles de la Ville de Paris, C.N.R.S. URA 278, 10 rue Vauquelin, F-75231 Paris cedex 05, France.[3]Institut Jacques Monod, C.N.R.S. UMR 9922, Université Denis Diderot, Tour 43, 2 place Jussieu, 75251 Paris cedex 05, France.

1. Introduction

A protocol is available to purify the b_6f complex from *C. reinhardtii* (1). It comprises three steps : specific solubilisation of thylakoid membranes with the neutral detergent Hecameg (HG), fractionation of the supernatant on a sucrose gradient, and a hydroxylapatite chromatography. The last two steps are performed in the presence of lipids and near the cmc of the detergent. The purified complex is enzymaticaly active, and comprises 7 subunits (1-2), only one of which, the Rieske protein, is not transmembrane (3-6).

2. Results and discussion

2.1.Molecular weight determination : the purified b_6 f *complex is a dimer.*
The molecular weight of the purified b_6f complex was determined using an estimate of its specific volume ($\bar{v}$, calculated from its composition (1)), its diffusion coefficient ($D_{20,w}$, determined by molecular sieving), and its sedimentation coefficient ($s_{20,w}$, determined by rate-zonal centrifugation on sucrose gradients) according to the Svedberg equation : $M = s_{20,w}RT/D_{20,w}(1-\bar{v}\rho)$, were ρ is the density of water at 20°C (Table 1). In order to make the measurement of bound detergent possible, the molecular weight determination was performed after transfering the purified b_6f complex into 0.2 mM laurylmaltoside (LM), either by molecular sieving or by dialysis. Detergent binding was estimated by rate-zonal centrifugation in the presence of [^{14}C]LM (fig. 1).

	Monomer	Dimer
$s_{20,w}$ (S)	6.3±0.2	10.4±0.4
$D_{20,w}$ (x10^{11})($m^2.s^{-1}$)	5.56±0.20	3.88±0.20
Bound LM (mol./part)	[130±10][a]	260±20
Isopycnic density (g/cm^3)[b]	*n.d.*	1.197
M (calculated) (kDa)[c]	[153±5][a]	344±10
M (measured) (kDa)[d]	128±10	300±25

TABLE 1. [a]assumes detergent binding to the monomer to be half that to the dimer (likely underestimated);[b]includes bound water;[c,d]including bound detergent, excluding bound water; *n.d.* : not determined.

P. Mathis (ed.), Photosynthesis: from Light to Biosphere, Vol. II, 591–594.

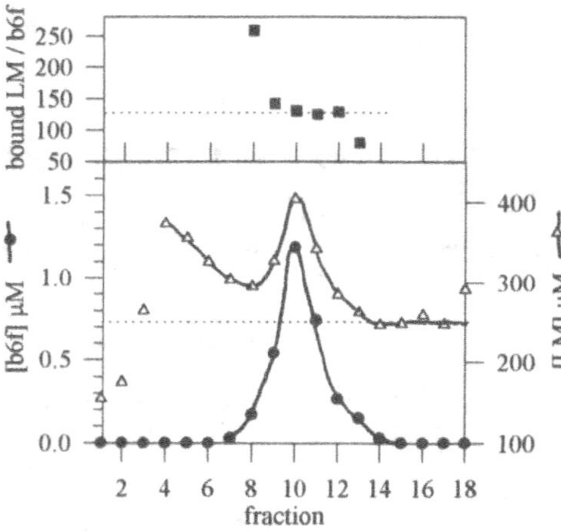

FIG. 1. **Determination of bound LM per b_6f particle.** Bottom : concentration pattern of detergent and cytochrome *f* along a 5-20% sucrose gradient prepared with 0.25 mM LM and 20 mM Tricine-NaOH, pH 8.0. b_6f/LM particles were prepared from b_6f purified in HG, dialysed against 0.25 mM LM, 20 mM Tricine-NaOH, pH 8.0, centrifuged on a 5-20% sucrose gradient prepared with the same buffer containing a mixture of ^{14}C-labelled/non labelled LM (specific activity : 3.1 x 10^{12} cpm/mole), and dialysed against the same radioactive buffer. ● : cytochrome *f* concentration determined spectroscopically, △ : total LM concentration calculated from the ^{14}C activity. Top : ■ : ratio of bound LM to cytochrome *f* (bound LM = total LM - 0.25 mM).

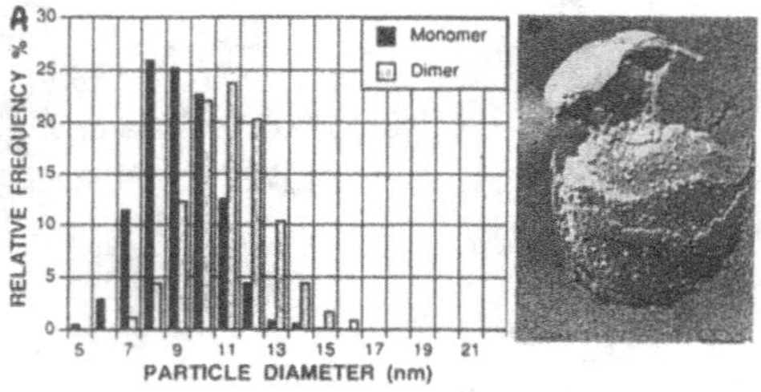

FIG. 2. **Electron microscopy analysis of reconstituted b_6f complex.** A. Distribution of the diameter of particles measured on freeze-fractured vesicles reconstituted from dimers (D) or monomers (M). B. Egg PC vesicule containing purified b_6f dimer observed after freeze-fracturing and Pt/C shadowing. bar : 0.1 μm.

The molecular weight determined experimentally (300±25 kDa) is in good agreement with a dimeric form of the complex (calculated MW = 344±10 kDa). When reconstituted into lipid vesicules and analysed by electron microscopy following freeze-fracturing, the complex appears as particles with a diameter of 10-11 nm (fig. 2A,B).

2.2.An excess of detergent leads to a light form corresponding to the monomer.

When the purified b_6f complex is transfered into 3 mM LM and layered onto a sucrose gradient, the dimer dissociates into a lighter form which has lost the Rieske protein (fig. 3c,d). The measured MW (128±10 kDa, Table 1) of this lighter form is in good agreement with a monomeric form of the complex (calculated MW = 153±5 kDa). Except for the Rieske protein, the monomer contains all the subunits found in the dimer (fig. 4). Having lost the Rieske protein, the monomer is not active, but there is no change in the cytochrome spectra. When reconstituted into lipid vesicules and analysed by freeze-fracturing, the particules show a diameter of 8 nm (fig. 2A; this result, obtained only once, needs to be confirmed), suggesting that the reconstituted particles are monomers.

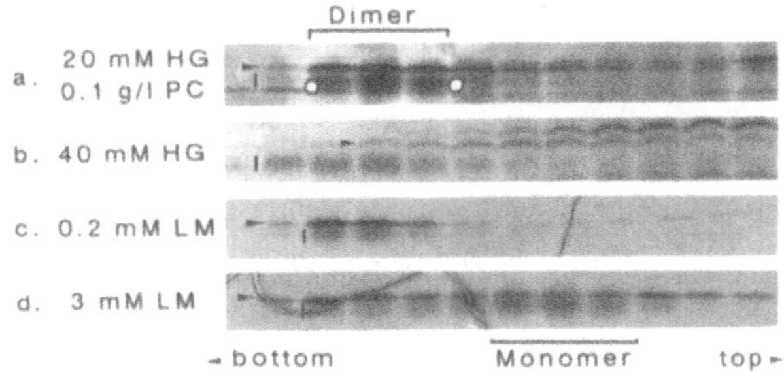

FIG. 3. **Sedimentation velocity analysis of the association state and composition of the b_6f complex under delipidating conditions.** Samples of the supernatant from a Hecameg solubilisation of thylakoid membranes (*a,b*) or of purified b_6f complex in Tricine-NaOH buffer, pH 8.0, containing either 0.2 mM LM (*c*) or 3 mM LM (*d*) were centrifuged on non-delipidating (*a,c*) or delipidating (*b,d*) sucrose gradients. Fractions were collected and analysed by SDS-PAGE. The areas shown correspond to the Rieske protein (▸) and to cytochrome b_6 (|). Gradient buffers : *a :* 20 mM Tricine-NaOH buffer, pH 8.0, containing 3 mM KCl, 3 mM $MgCl_2$, 20 mM HG and 0.1 g/l egg PC; *b, c, d :* 20 mM Tricine-NaOH buffer, pH 8.0, containing either 40 mM HG (*b*), 0.2 mM LM (*c*), or 3 mM LM (*d*). Centrifugation at 4°C. From Ref. (1).

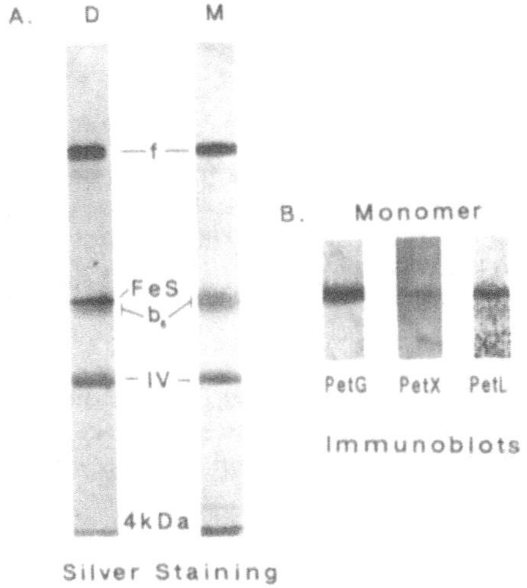

FIG. 4. **Subunit composition of the dimeric and monomeric forms of the b_6f complex.** *A.* SDS-PAGE analysis of purified dimer (D) and monomer (M). The dimer was obtained after centrifugation of the purified b_6f complex on a 0.2 mM LM gradient in 20 mM Tricine-NaOH, pH 8.0. The monomer was obtained after a first centrifugation of the purified b_6f complex on a 3 mM LM gradient in the same buffer, the upper band being submitted to a second centrifugation on a 0.2 mM LM gradient in the same buffer. *B.* Immunoblot analysis of the monomer with antisera against PetG, PetX and PetL. From Ref. (1).

2.3.Dimer to monomer transition and loss of the Rieske protein : a consequence of delipidation.

When the second step of the purification is performed with an excess of HG and without lipids (*i.e.* 25-40 mM HG instead of 20 mM HG and 0.1 g/l lipids), the Rieske protein dissociates from the complex (fig. 3a,b). In the same way, if the purified complex is layered onto a gradient prepared with 3-5 mM LM, the dimer normally present in 0.2

mM LM dissociates into monomers, losing the Rieske protein (fig. 3c,d). Note that although the concentration of detergent is quite different, the concentration of micelles is approximately the same in 25 mM HG and in 5 mM LM (cmc_{HG} = 19.5 mM, cmc_{LM} = 0.17 mM). At these high micellar concentrations, loss of the Rieske protein and monomerisation can be prevented by the addition of lipids in the gradients (data not shown). Those observations lead us to conclude that lipids are associated with the b_6f complex, stabilising the dimer. An excess of micelles (hydrophobic phase) tends to dissociate and dilute the lipids, disturbing the transmembrane region and thereby the interaction between the Rieske protein and the complex. The lipids may directly stabilise the complex, or may protect the transmembrane region from denaturation by detergent molecules.

Apparently, the loss of the Rieske protein and monomerisation, although induced by the same cause, are independent. Indeed, when the purification is performed in the presence of 25 mM HG and without lipids, the b_6f complex remains a dimer even though it has lost the Rieske protein. On the other hand, we observed a monomeric species retaining the Rieske protein while purifying the b_6f complex of a mutant lacking the PetL protein (6).

Monomerisation of the b_6f complex seems irreversible. Indeed, when the monomer is transfered into a medium containing lipids, it does not dimerise : a monomer layered onto a gradient prepared with 20 mM HG, 0.1 g/l lipid, remains a monomer; and preliminary results show that particules reconstituted into lipid vesicules from monomers remain so (fig. 2). Either the lipids associated with the dimer cannot be substituted with egg PC, or the change of conformation induced by delipidation is irreversible. The nature of lipids associated with the purified b_6f is under investigation.

3. Conclusion

These data show that the active purified b_6f complex from *C. reinhardtii* is a dimer. Delipidation brings about monomerisation and loss of the Rieske protein. While both events occur concomitantly, they can be observed independently one from another. Stability of the complex is ensured by working close to the cmc of the detergent and in the presence of lipids.

4. Acknowledgments

This research has been supported by the C.N.R.S., the Collège de France, the C.E.A. and by grants from the EEC (BIO-CT93-0076), from the Ministère de la Recherche et de la Technologie (87.C.0385) and from IMABIO to JLP. Particular thanks are due to Y. Takahashi and J.-D. Rochaix for a gift of PetL antiserum.

5. References

1. Pierre, Y., Breyton, C., Tribet, C., Kramer, D., Olive, J., & Popot, J.-L. (1995) *submitted for publication.*

2. Popot, J.-L., Pierre, Y., Breyton, C., Lemoine, Y., Takahashi, Y., & Rochaix, J.-D. (1995) *These proceedings.*

3. Breyton, C., de Vitry, C., & Popot, J.-L. (1994) *J. Biol. Chem.* **269**, 7597-7602.

4. de Vitry, C., Breyton, C., Pierre, Y., & Popot, J.-L. (1995) *in preparation.*

5. de Vitry, C., Breyton, C., Pierre, Y., & Popot, J.-L. (1995) *These proceedings.*

6. Takahashi, Y., Rahire, M., Breyton, C., Popot, J.-L., Joliot, P., & Rochaix, J.-D. (1995) *in preparation.*

THE 4-kDa CHLOROPLAST POLYPEPTIDE OF CYTOCHROME b_6f COMPLEX ENCODED BY THE NUCLEAR *petX* GENE: NUCLEIC AND PROTEIN SEQUENCES, TARGETING SIGNALS, AND MEMBRANE TOPOLOGY.

Catherine de Vitry, Cécile Breyton, Yves Pierre and Jean-Luc Popot

Laboratoire de Photosynthèse, C.N.R.S. URA 1187, Institut de Biologie Physico-Chimique, 13 rue Pierre et Marie Curie, 75005 Paris, FRANCE.

1. Introduction

We have previously identified by protein sequencing and immunodetection a nuclear-encoded 4-kDa polypeptide that is present in purified b_6f complexes and absent in b_6f-deficient mutants from *Chlamydomonas reinhardtii* and that we therefore considered as a novel b_6f subunit (1,2). This subunit is also present in higher plants (3). We have now sequenced the cDNA of the *petX* gene encoding this 4-kDa subunit and the mature protein by Edman degradation in *C. reinhardtii* and studied its membrane topology.

2. Results and discussion

2.1 Sequence and messenger RNA

Two oligonucleotides (Fig. 1) were used to screen by plaque hybridization a *C. reinhardtii* cDNA library cloned into λ-GT10 (4) and replicated in *Escherichia coli* C600. The EcoRI insert was isolated from one of the phage isolates, subcloned into Bluescript KS^-, and sequenced. The size of the transcript for *petX* gene is of about 0.6 kb (Fig. 2), which corresponds to the size of the insert.

3'-CCGAAGCAC/GGACGACGCGCA-5'(GFVLLRV)
3'-GACCAC/GCTCCTCCCGTTC-5' (LVEEGK)

FIG. 1. **Oligonucleotides and corresponding amino acid sequences.**

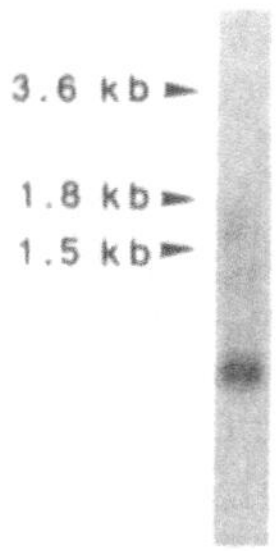

FIG. 2. **Northern blot of total RNA from *C. reinhardtii* probed with PetX cDNA.** Ribosomal RNA are taken as molecular weight markers.

2.2 The transit peptide : a long peptide rich in basic residues

The site of processing between the transit peptide and the mature protein was determined by comparing the N-terminal sequence of the protein and that deduced from the cDNA sequence. The transit peptide is 60 residue long and has basic residues typical for the transfer through the chloroplast envelope, but no hydrophobic segment at the C-terminal end as is found in most proteins transfered through the thylakoid membrane (Figs. 3,4).

P. Mathis (ed.), Photosynthesis: from Light to Biosphere, Vol. II, 595–598.

MAMSIAARSACCGVAAPRSSTVRVAAARPAVRPSLRTAGQKAAPGRGVATKA
VNELAMIA ▾ GEAEFIAGTALTMVGMTLVGLAIGFVLLRVESLVEEGKI

FIG. 3. **Amino acid sequence of PetX from *C. reinhardtii*.** ▾, processing site.

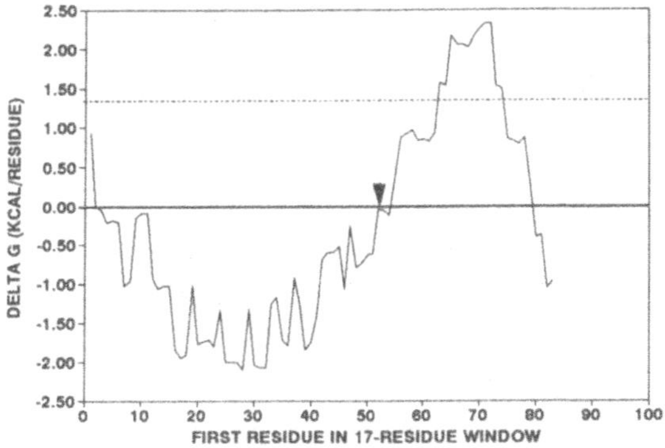

FIG. 4. **Hydrophobicity profile of PetX precursor.** The free energy of transfer from lipids to water calculated using the GES scale (5) has been averaged over a 17-residue window. ▾, processing site. Above the dotted line, sequence segments correspond generally to transmembrane helices in integral membrane proteins (6).

2.3 The mature protein : a 4-kDa hydrophobic protein

The mature protein is 39 residue long with a 4-kDa molecular weight. The sequence of the protein is identical to that deduced from the cDNA sequence but lacks the last residue (Fig. 3), probably due to the low yield for the last amino acid in protein sequencing. PetX is conserved in higher plants (Fig. 5).

C. reinhardtii	GEAEFIAGTALTMVGMTLVGLAIGFVLLRVESLVEEGKI
Spinach (3)	NAAAEIFXIAAVMNGLTLVG...

FIG. 5. ***C. reinhardtii* PetX mature protein aligned with spinach PetX N-terminus.**

The mature protein has a 24 residue long hydrophobic segment, which could form a transmembrane α-helix (Fig. 4). The average hydrophobicity of the most hydrophobic 17 amino acid residue stretch of PetX using the GES scale is 2.33 which is higher than the transmembrane α-helices of the other intrinsic $b_6 f$ subunits (*f* 2.28 (1 α-hel), b_6 1.93 (average 4 α-hel), IV 2.22 (average 3 α-hel), PetG 2.15 (1 α-hel), PetL 2.19 (1 α-hel)). The lateral amphipathy of this α-helix is low.

2.4 Membrane topology : a single α-helix with short extramembrane segments

Upon dissociating treatments, PetX was not extracted (Fig. 6). Thylakoid membranes were incubated for 10 min in the dissociating solution, subjected to 2 freeze/thaw cycles, to centrifugation; pellet and supernatant were analysed by urea-SDS/PAGE; proteins were identified by immunoblot. PetX was detected by a specific polyclonal antibody raised against a peptide corresponding to the N-terminus of PetX (Fig. 7A). These data further suggest that PetX forms a single transmembrane α-helix with short extramembrane segments.

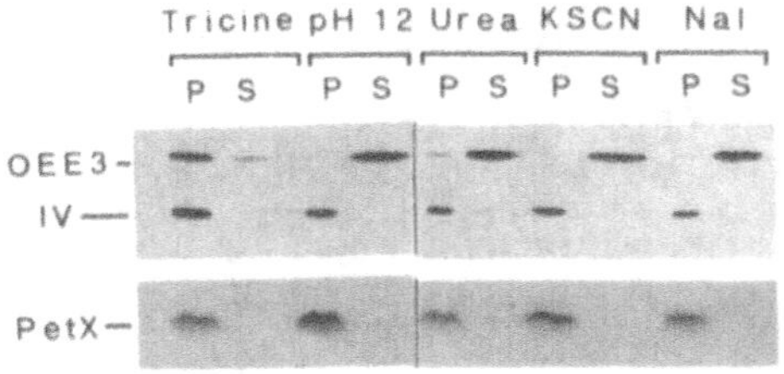

FIG. 6. **Extraction of thylakoid proteins by dissociating treatments.** 20mM tricine pH8.0; 6.8M urea, 2M KSCN, 1.5M NaI in tricine; CAPS pH12.0. S supernatant, P pellet. Immunoblots labeled with antisera-extrinsic 0EE3,-intrinsic subunit IV and -PetX.

2.5 C-terminus : indications for a localization on the stromal side of the membrane

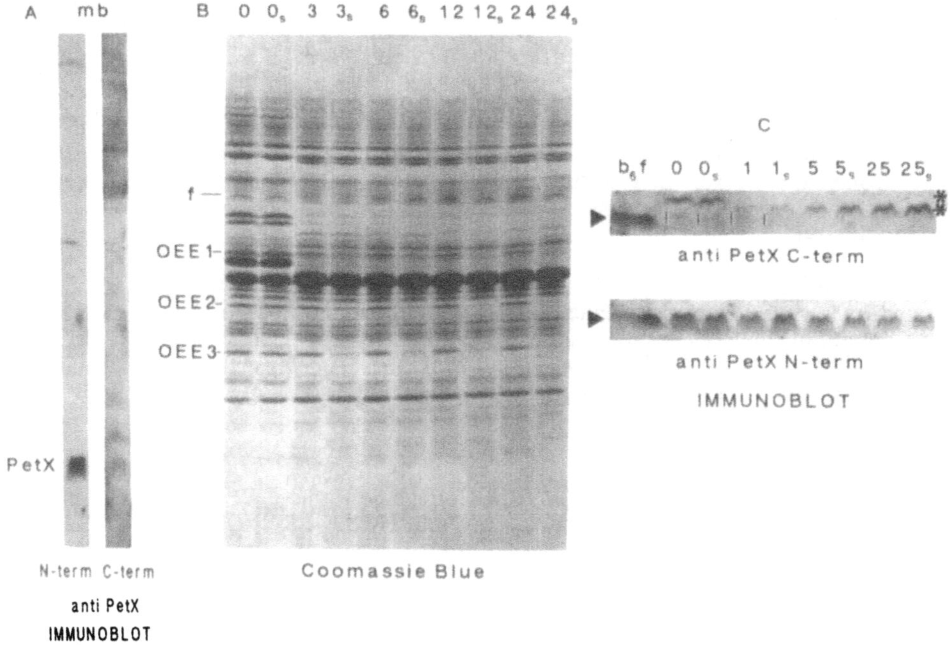

FIG. 7. **Proteolysis of thylakoid membrane proteins by trypsin.** Analysis by urea/SDS-PAGE. A, PetX immunodetection with antisera raised against the N-terminus (N-term) and the C-terminus (C-term), mb membranes. B, polypeptide pattern after incubation with increasing trypsin concentrations (μg/ml), when sonicated (s), Coomassie brillant blue staining. C, PetX immunodectection after incubation with trypsin (μg/ml). b_6f, isolated cytochrome b_6f; ▸ PetX; * crossreactions in membranes.

Proteolysis experiments suggest that PetX C-terminus would be localized in the stroma. Thylakoid membranes in presence of trypsine, were subjected or not to sonication and incubated for 1 hour at room temperature; pigments and lipids were extracted by methanol/ether prior urea-SDS/PAGE and immunoblot. Trypsine has two potential cleavage sites in PetX C-terminus (R_{29} and K_{38}). Antibodies were raised against peptides, GEAEFIAGTALTMY and YLLRVESLVEEGK, corresponding to PetX N-terminus and C-terminus (Fig. 7A). PetX was similarly cleaved by trypsin in intact membrane and sonicated vesicles (Fig. 7C) in contrast to lumenal proteins which are more cleaved when vesicles are open by sonication (OEE1,OEE2,OEE3 in Fig. 7B).

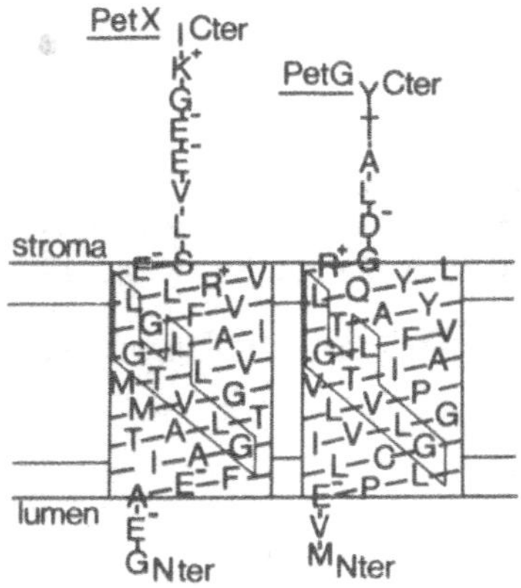

PetG is another $b_6 f$ subunit formed by a single α-helix with short extramembrane segments, whose transmembrane orientation is known (7). Residue identities between PetX and PetG are limited (29% in the putative transmembrane region). However, when PetX and PetG are folded into α-helices, a stripe of identical residues becomes apparent, suggesting either a common ancestry or convergent evolution. Proteolysis data, charge distribution and homology with PetG are consistent with a stromal localization of the C-terminus of PetX (Fig. 8).

FIG. 8. **Putative transmembrane arrangement of PetX and PetG of *C. reinhardtii*.** PetG sequence from ref. 8. Residue identities between PetX and PetG are boxed.

3. Conclusion

We characterized a new $b_6 f$ subunit, PetX, which belongs to the class of proteins formed by a single transmembrane α-helix with short extramembrane segments; three $b_6 f$ subunits of this class have now been identified, one is nuclear-encoded, PetX, and two are chloroplast-encoded, PetG (7) and PetL (9). The present results establish the primary sequence, targeting signals and transmembrane topology of PetX.

4. Acknowledgments

This research was supported by the CNRS, Collège de France, CEA and by grants from the EEC (BIO2CT930076), MRT (87C0385) and HC&MEEC(ERBCHRXCT920045).

5. References

1. Pierre, Y. & Popot, J.-L. (1993) *C. R. Acad. Sci. Paris* **316**, 1404-1409.
2. Popot, J.-L., Pierre, Y., Breyton, C., Lemoine, Y., Takahashi, Y. & Rochaix, J.-D. (1995) *These proceedings*
3. Schmidt, C.L. & Malkin, R. (1993) *Photosynth. Res.* **38**, 73-81.
4. Goldschmidt-Clermont, M. & Rahire, M. (1986) *J. Mol. Biol.* **191**, 421-432.
5. Engelman, D. M., Steitz, T.A., and Goldman, A. (1986) *Annu. Rev. Biophys. Biophys. Chem.* **15**, 321-353.
6. Popot, J.-L. & de Vitry, C. (1990) *Ann. Rev. Biophys. Biophys. Chem.* **19**, 369-403.
7. Haley, J. & Bogorad, L. (1989) *Proc. Natl. Acad. Sci. USA* **86**, 1534-1538.
8. Fong, E.S. & Surzycki, S.J. (1992) *Curr. Gen.* **21**, 527-530.
9. Takahashi, Y., Rahire, M., Breyton, C., Popot, J.-L., Joliot, P. & Rochaix, J.-D. (1995) *Submitted for publication.*

EFFECTS OF INACTIVATION OF GENES CODING FOR THE REACTION CENTER-BOUND CYTOCHROME SUBUNIT ON GROWTH AND ELECTRON TRANSFER IN PURPLE PHOTOSYNTHETIC BACTERIUM, *RUBRIVIVAX GELATINOSUS*

Kenji V.P. Nagashima, Keizo Shimada and Katsumi Matsuura
Dept. Biol., Tokyo Metropolitan Univ., Minamiohsawa 1-1, Hachioji, Tokyo 192-03, Japan

1. Introduction

Photosynthetic reaction center complexes of purple bacteria are composed of four subunits called L, M, H and cytochrome. The L and M subunits form a heterodimer in the membrane, in which a special pair of bacteriochlorophylls, monomeric bacteriochlorophylls, bacteriopheophytins and quinones are contained [1]. Electrons are sequentially transferred from the photo-oxidized special pair at the periplasmic side to the quinones at the cytoplasmic side through other chromophores. The H subunit is located on the cytoplasmic side of these two subunits. The cytochrome subunit serves as the immediate electron donor to the photo-oxidized special pair and re-reduced by a soluble cytochrome *c*. This subunit protrudes to the periplasmic side and has four *c*-type hemes. Some species of purple bacteria do not have the cytochrome subunit. In the species lacking the cytochrome subunit, the photo-oxidized special pair is directly reduced by the soluble cytochrome *c*.

Previously, we have shown that the purified reaction center complex of the purple bacterium, *Rubrivivax gelatinosus*, in which the cytochrome subunit has been removed by detergent treatment is directly re-reduced by the soluble cytochrome *c* following flash-induced photo-oxidation. The rate of oxidation of the soluble cytochrome *c* was comparable with that in the native form of the reaction center with the tetra-heme cytochrome [2, 3]. In this work, the gene encoding the cytochrome subunit of *R. gelatinosus* was inactivated mutationally to clarify the physiological role of the cytochrome subunit. Growth of the mutant was analyzed under various conditions.

2. Procedure

The genes encoding the reaction center proteins containing the cytochrome subunit of *R. gelatinosus* were cloned and sequenced previously [4, 5]. The plasmid, pGI7, derived from pUC plasmid (Amp^r) and containing the genes for the M and cytochrome subunits was digested at an unique *Bgl*II site within the cytochrome gene and kanamycin resistant cartridge was inserted (Fig. 1). The constructed plasmid was introduced to *R. gelatinosus* cells by electropolation. A kanamycin resistant and ampicilin sensitive clone was picked up and named as strain C244.

P. Mathis (ed.), Photosynthesis: from Light to Biosphere, Vol. II, 599–602.

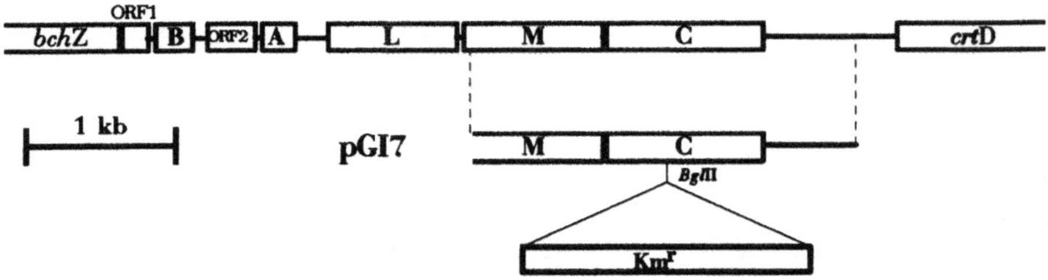

Fig. 1. Restriction map of photosynthetic genes (puf operon) of *R. gelatinosus* and construction of a plasmid for inactivation of cytochrome subunit gene. A kanamycin resistant cartridge was generous offer from Dr. Saeki (Osaka Univ.).

3. Results and Discussion

The strain, C244, in which the cytochrome subunit gene was genetically inactivated by the insertion of the kanamycin resistant cartridge was capable of growing under both aerobic respiratory and anaerobic photosynthetic conditions. This shows that the cytochrome subunit is not indispensable in *R. gelatinosus*. The finding is consistent with our previous observation that the rate of the electron transfer from the soluble cytochrome *c* to the reaction center core complex, in which cytochrome subunit had been removed by detergent treatment, was comparable to that to the native reaction center complex *in vitro* [2, 3]. The photosynthetic growth of the mutant suggests that the short-cut electron transfer also takes place in the growing cells.

Most purple bacteria have the cytochrome subunit. However, several species located at different lineage in the phylogenetic tree of purple bacteria do not have this subunit. From the phylogenetic distribution we have previously suggested that the cytochrome subunit had become less necessary in some species and had been lost during the course of evolution at least in four independent lines [6]. The present results in the mutant strain, C244, support this idea. The effect of the deletion of the cytochrome subunit in C244 is shown in Figure 2 as redox difference spectra of cytochromes in the membranes isolated from the wild type and the mutant strain grown photosynthetically. In C244, the difference α-absorption band of *c*-type hemes peaking around 552 nm was about 10% of that in the wild type, showing that the C244 does not synthesize the cytochrome subunit since the major *c*-type cytochrome in the membrane is the reaction center-bound cytochrome subunit. The insertion of the kanamycin resistant cartridge was further confirmed by Southern hybridization using kanamycin resistant gene and pUC gene as probes. On the other hand, no significant differences were observed between absorption spectra of C244 and wild type with regard to bacteriochlorophylls and carotenoids (data not shown).

Figure 3 shows the kinetics of the changes of the membrane potential in the membranes of wild type and C244 measured with carotenoid band shift. The extent of the membrane potential in the membrane of C244 induced by continuous light was nearly equal to that of the wild type, showing that the content and the function of the photosynthetic electron transfer system were not changed by the mutation except for the cytochrome subunit. These results indicate that deletion of the cytochrome subunit does not significantly affect constructions of the other photosynthetic proteins.

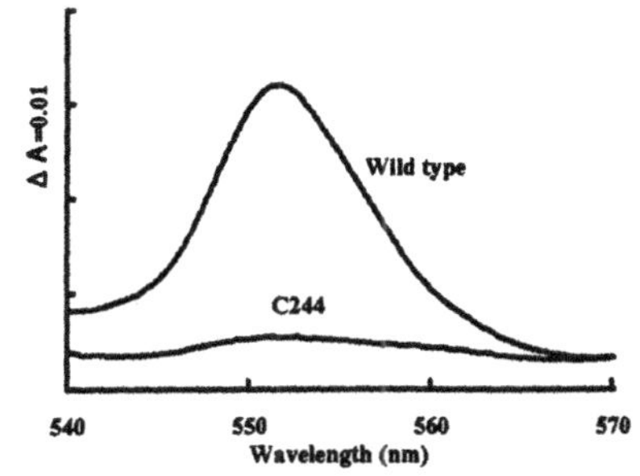

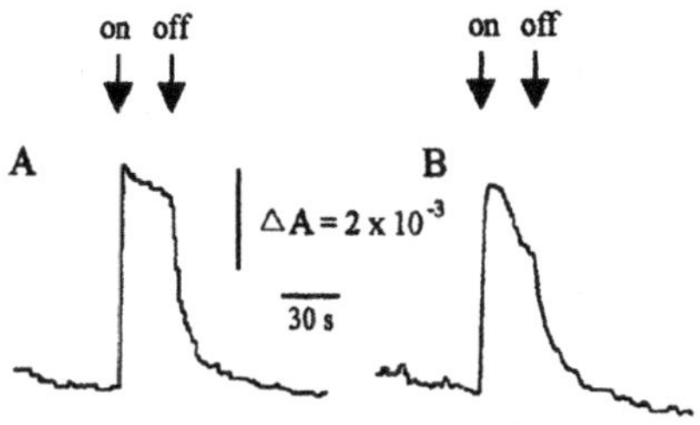

Fig. 2. Reduced-minus-oxidized difference spectra of membranes of wild type and C244 of *R. gelatinosus*. Cells were broken by French pressure cell. Membranes were prepared by differential ultracentrifugation and suspended in buffer containing 50 mM KCl and 50 mM K-phosphate, pH 7.6. Reduced minus oxidized difference spectra were obtained by subtracting spectra of the samples added 3.3 mM potassium ferricianide from those added 1 mM sodium dithionite.

Fig. 3. Kinetics of the changes of membrane potentials in the membranes of the wild type (A) and C244 (B) of *R. gelatinosus* induced by continuos light. The membrane potential was measured using the carotenoid band shift. The measurements were performed by recording the time course of the difference absorbance change at 495 minus 511 nm.

The growth rate in C244 under the anaerobic photosynthetic conditions was slower than that in the wild type as shown in Fig. 4. In the wild type, the doubling time under the photosynthetic conditions was about 140 minutes. The doubling time of C244 grown photosynthetically was nearly twice as long as that of the wild type, 270 minutes, showing that the photosynthetic growth rate of the cytochrome-subunit-less mutant of *R. gelatinosus* is significantly reduced. On the other hand, the aerobic respiratory growth rates showed no differences between C244 and the wild type (the doubling time was 130 min). The reduced photosynthetic growth rate in C244 indicates that the absence of the cytochrome subunit is somewhat disadvantageous for the photosynthetic growth under the conditions used. This may be caused by an increase in the rate of the back electron transfer to the photo-oxidized special pair due to the absence of the fast re-reduction by the hemes in the bound cytochrome subunit. Another possibility is that the cytochrome subunit is an essential factor in a pathway of electrons from a component other than soluble cytochrome *c*.

In the growth experiments shown in Fig. 4, aerobically grown cells with little pigmentation were transferred to the fresh medium in screw-capped test tubes. After the initial aerobic growth using residual oxygen, lag time which is presumably due to the synthesis of photosynthetic apparatus was observed. It was about 4 hours in the wild type and about 20 hours in C244. It is difficult to explain why the growth curve of C244 showed such long lag time before photosynthetic growth. The cytochrome subunit may be essential for the quick adaptation of the photosynthetic apparatus to the change of the ambient redox condition or may play important roles in higher order assembly of photosynthetic apparatus to form the electron transfer pathway.

All species without the cytochrome subunit, e. g., *Rhodobacter sphaeroides* and *Rhodospirillum rubrum*, are capable of significant aerobic growth which is

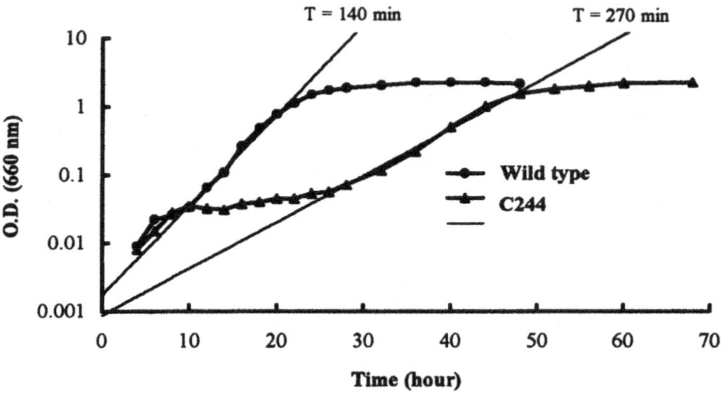

Fig. 4. Growth curves of wild type and C244 of *R. gelatinosus* under photosynthetic conditions. Growth medium contained 0.5% polypepton, 0.1% yeast extract, 0.5% sodium succinate and minerals. The measurements were started by inoculation of 1/1000 volume of late log phase cells grown aerobically in the same medium. Culture was grown at 30 °C in screw-capped tubes filled with the medium in the light supplemented with 60 W tungsten lamp placed 20 cm apart from the cultures. The absorbance at 660 nm was measured in the tube, the diameter of which was 18 mm. The average of three out of five independent measurements omitting the minimum and the maximum was plotted against time.

scarcely seen among the purple bacterial species having the cytochrome subunit, e. g., *Rhodopseudomonas viridis* and *Rhodospirillum molischianum*. In this study, it was shown that the cytochrome subunit is not essential for the photosynthetic growth in *R. gelatinosus*. *R. gelatinosus* is able to grow aerobically and might be in an evolutional process to delete the cytochrome subunit. It may be interesting to examine deletion mutants of the cytochrome subunit in other purple bacteria which grow only photosynthetically. More detailed physiological and photo-kinetic experiments using C244 and other site-directed mutagenesis will lead to clarification of the roles and evolution of the cytochrome subunit.

References

1 Deisenhofer, J., Epp, O., Miki, K., Huber, R. and Michel, H. (1985) Nature 318, 618-624

2 Fukushima, A., Matsuura, K., Shimada, K. and Satoh, T. (1988) Biochim. Biophys. Acta 933, 399-495

3 Matsuura, K., Fukushima, A., Shimada, K. and Satoh, T. (1988) FEBS lett. 237, 21-25

4 Nagashima, K. V. P., Shimada, K. and Matsuura, K. (1993) Photosynth. Res. 36, 185-191

5 Nagashima, K. V. P., Matsuura, K., Ohyama, S. and Shimada, K. (1994) J. Biol. Chem. 269, 2477-2484

6 Matsuura, K. and Shimada, K. (1990) in Current Research in Photosynthesis (Baltscheffsky, K. Ed.) Vol. 1, pp193-196, Klewer Academic Publishers, Dordrecht, The Netherlands

INHIBITOR BINDING TO ISOLATED CYTOCHROME *bf* COMPLEX.

A. B. Hope and P. Valente, School of Biological Sciences, Flinders University, G.P.O. Box 2100, Adelaide, S.A. 5001, Australia

1. Introduction

In current concepts the chloroplast cytochrome *bf* complex has sites specialised for the oxidation of plastoquinol (*p*-sites, Q_o-sites) and for the reduction of plastoquinone (*n*-sites, Q_r-sites). The inhibitors of quinol oxidation dibromothymoquinone (DBMIB) [1], stigmatellin [2] and tridecyl-stigmatellin (TDS; P. R. Rich, personal communication) are inferred to act by binding at or near *p*-sites. Use of a system including isolated *bf* complex [3-5] gives the opportunity to study the binding of these inhibitors in the presence or absence of quinol or quinone.

Binding of stigmatellin makes the reduction potential of all haem groups of the *bf* complex more negative [6], despite the *b* cytochromes being well removed in space from the putative binding site near the Rieske centre, in turn on a separate subunit from cyt *f*. The Rieske (FeS) centre in chloroplasts is made more redox positive by stigmatellin [7]. DBMIB binding apparently causes cyt *f* to be more redox negative in chloroplasts [8], but interpretation of the data requires knowledge of the relative potentials of the couples FeS/FeS^+, cyt f^{II}/cyt f^{III} and PC^I/PC^{II}. PC = plastocyanin. It is of importance to confirm that these changes are reflected kinetically during turnovers in the *bf* complex.

Quinols and quinones compete with DBMIB at *p*-sites. Either quinol tends to displace DBMIB or reduces it [9, 10]; reduced DBMIB is a much less potent inhibitor. Stigmatellin and TDS do not suffer from this ambiguity.

The disagreement, see [11], between reports that DBMIB causes inhibition of electron throughput at a ratio of one DBMIB per two cyt *f* [12], or at a ratio of 1:1 [10] has not been resolved. The use of the isolated system enables this to be further probed. When possible errors are accounted for, we conclude that DBMIB inhibits totally at a ratio of one to each monomer of *bf* complex.

2. Procedure

Isolated cyt *bf* complexes, photosystem (PS) 1 particles and PC were prepared from pea leaves [4], and suspended in a buffer containing (mM) sorbitol 150, $MgCl_2$ 10, KCl 5, ascorbate 1, HEPES/KOH 10, anthraquinone sulphonate 0.01, Mega-9 15, pH 7.5. Cuvette concentrations were typically: PS 1, 0.25-0.3 μM, PC 2-3 μM and cyt *bf* 0.4-0.8 μM in cyt *f*. Flash-induced absorbancy changes due to cyt b_{563} at 564 nm, cyt *f* at 554 nm, PC at 575 nm and P700 at 543 nm were deconvoluted from responses obtained at

P. Mathis (ed.), Photosynthesis: from Light to Biosphere, Vol. II, 603–606.

these four wavelengths [3, 4]. The signals obtained were the average of 8-16 flashes at each wavelength, at usually 0.1 Hz. In DBMIB experiments, preflashes were given to oxidise any bound $DBMIBH_2$ before recording the average of the next 8-16 responses.

Sources of error carefully considered included those in estimations of oxidised cyt *f* and PC as functions of time, in the volumes added to a cuvette to give the desired concentrations of inhibitor and of cyt *bf* complex, in the estimation of stock cyt *bf* concentration using an appropriate $\Delta\varepsilon$ for reduced minus oxidised cyt *f*, and in the proportion of total cyt *bf* complex that has the normal complement of all subunits; if FeS centres were missing, binding sites could be absent or modified. All syringes were calibrated for volume gravimetrically using a microbalance or spectrophotometrically. Solutions of inhibitors were based on weighing and checked using accepted molar extinction coefficients. Cyt *bf* concentration was based on an estimation of cyt f^{II} content using A_{554} - $(A_{540}+A_{560})/2$ and $\Delta\varepsilon = 20.0\ mM^{-1}\ cm^{-1}$ [13]. This agrees with that estimated from apparently accurate data for cyt *c* [14] after an appropriate shift of the wavelength scale by 3.5 nm: this haem and that in cyt *f* should have comparable absorbancy, and ε 's for cyt f^{II} and cyt f^{III} at 553.75 nm [13, 15] agree with those of cyt *c* at 550.25 nm [14]. With regard to the Rieske content of the cyt *bf* complex, our preparation [16] is dimeric [17, 4, 18] and active enzymically whereas its monomeric derivative is inactive and lacks the Rieske subunit [18]; there is no reason to suppose that conversion of dimer to monomer occurs in the course of these experiments.

3. Results and Discussion

All three inhibitors increased flash-induced [cyt f^{III}] and decreased [PC^{II}]; [$P700^+$] was relatively unchanged. In all experiments, cyt *bf* complex was present, but there was no added quinol. From the data two kinetic coefficients were obtained from fitting rate equations for the second-order reactions between PC and cyt *f* [4]: k_{FP} for the reduction of PC^{II} by cyt f^{II}, and k_{PF} for the reverse reaction; as well, from the quasi-constant values of [PC^{II}] and [cyt f^{III}] at 60-90 ms after the flash, the value of K_{eq} (= k_{FP}/k_{PF}) could be calculated. These independent data gave similar values for K_{eq}, which suggests the deconvolution and curve-fitting procedures were valid. k_{FP} was increased by inhibitors while k_{PF} was almost unchanged.

Fig. 1 shows the relation between the above parameters and stigmatellin concentration (relative to cyt *bf* concentration). It is possible to show that the observed K_{eq} (and k_{FP} in an analogous way) is given by:

$$K_{eq} = \frac{K_{eq}(\max)}{K_{eq}(0)}\left\{(1+[\mathrm{I}]/[\mathrm{E}]+K_B/[\mathrm{E}]) - \left[(1+[\mathrm{I}]/[\mathrm{E}]+K_B/[\mathrm{E}])^2 - 4[\mathrm{I}]/[\mathrm{E}]\right]^{1/2}\right\}/2$$

where [I] and [E] are the total concentrations of inhibitor and *bf* complex monomer, $K_{eq}(0)$ refers to cyt *bf* without I, and $K_{eq}(\max)$ to that when I is bound. A similar treatment was given in ref. 10. The curve through the data points is the result of fitting this equation. The K_B of stigmatellin was about $10^7\ M^{-1}$. The corresponding curve for the reputedly more strongly-binding TDS is shown in Fig. 2. The points lie close to a binding curve for very large K_B, suggesting that the errors referred to earlier were sufficiently small, or tended to cancell. Figure 3 illustrates the effect of DBMIB on the equilibrium coefficient. The maximum effect, an approximate doubling of K_{eq} from 1 to 2, was similar to that produced by stigmatellin and TDS. The K_B values for DBMIB and TDS could not be accurately determined because they were too great (~ 10^8).

Under control conditions, only a small amount of electron donation to cyt *f* by the

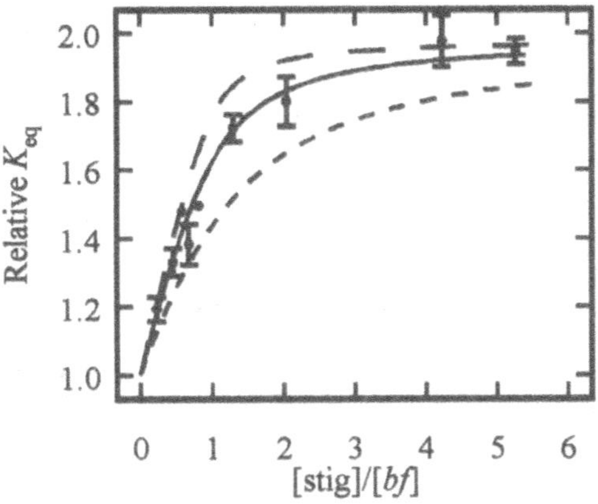

Figure 1. Stigmatellin binding. Mean and SEM of 3-7 measurements. The full line indicates a K_B of 10^7 M^{-1}, the long-dashed, $3x10^7$, the short-dashed, $3x10^6$

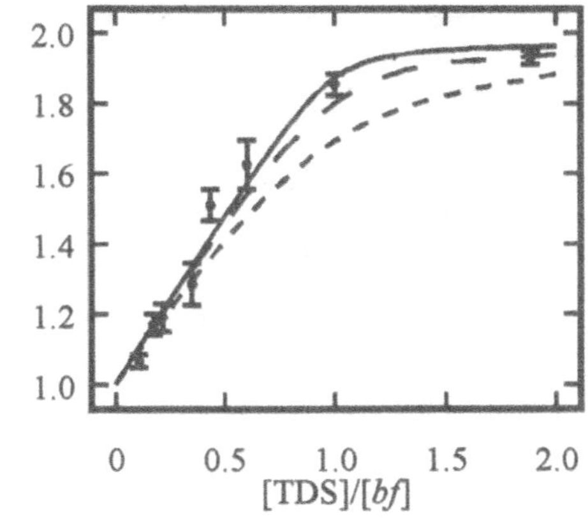

Figure 2. Tridecyl stigmatellin. Mean and SEM of 3-7 measurements. The full line indicates a K_B of 10^8 M^{-1}, the long-dashed, $3x10^7$, the short-dashed, $1x10^7$

Relative K_{eq} is plotted throughout to compare different experiments. $K_{eq}(0)$ varied somewhat but was around 1.0.

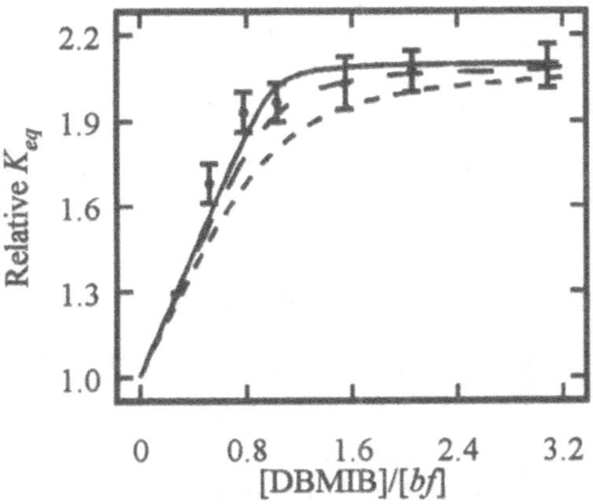

Figure 3. DBMIB binding. Mean and SEM of 2-6 measurements. The full line indicates a K_B of 10^8 M^{-1}, the long-dashed, $3x10^7$, the short-dashed, $1x10^7$.

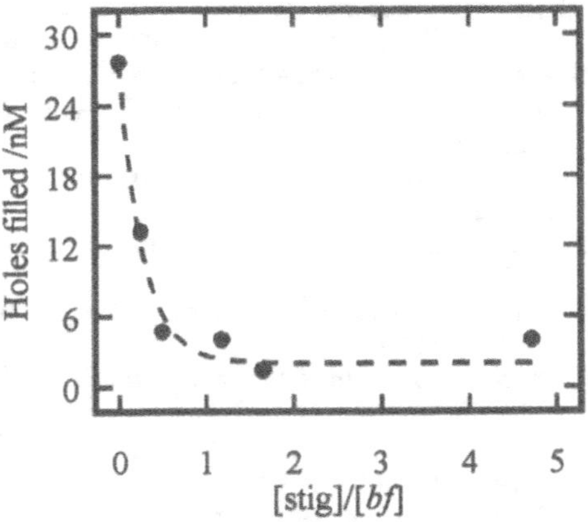

Figure 4. Stigmatellin-sensitive reduction of cyt f^{III}, PC^{II} and $P700^+$ In the absence of quinol, electrons are inferred to come from the Reiske centre.

Rieske centre in occurred the absence of quinol [4]. Stigmatellin inhibited this electron transfer (Figure 4), consistent with the reported change in the reduction potential of the FeS centre to a much more positive value [7].

The maximum effect of the inhibitors studied on the redox equilibrium between cyt *f* and PC corresponded to a change in reduction potential of 18 mV (a factor of two change in K_{eq} or k_{FP}), which agrees reasonably well with the reported change in cyt *f* due to stigmatellin, when measured by redox titration, of 25 mV [6]. While in some experiments occasional points fell above the binding curve for infinitely high K_B, the average data are consistent with DBMIB fully affecting K_{eq} with one bound per cyt *bf* monomer. Data from [2] on electron throughput (ditertiary butyl hydroquinone to PC) and unpublished data of ours are also easily fitted by the 1:1, "monomer model" with a K_B of 3×10^7 M^{-1}. The effect of all three inhibitors is argued elsewhere to be on the "on coefficient" of the encounter complex between cyt *f* and PC, and not on the inherent electron transfer rate (see [5] for a discussion of pH effects on this reaction).

The project was supported by the Australian Research Council and Flinders University. Thanks go to Prof. G. Höfle and Dr P. R. Rich for generously supplying stigmatellin and tridecyl-stigmatellin respectively, and the latter also for helpful discussions, to West Hiscock for essential help with experimental apparatus, and Western Mining Corporation for the use of a YAG laser.

References

1 Trebst, A. (1980) in Methods in Enzymology (San Pietro A. ed.), Vol. 69, pp. 675-715, Academic Press, New York
2 Oettmeier, W., Godde, D., Kunze, B. and Höfle, G. (1985) Biochim. Biophys. Acta 807, 216-219
3 Rich, P.R., Heathcote, P. and Moss, D.A. (1987) Biochim. Biophys. Acta 892, 138-51
4 Hope, A.B., Matthews, D.B. and Valente, P. (1994) Photosynth. Res. 40, 199-206
5 Hope, A.B., Matthews, D.B. and Valente, P. (1994) Photosynth. Res. 42, 111-120
6 Hauska, G., Herold, E., Huber, C., Nitschke, W. and Sofrova, D. (1989) Z. Naturforsch. 44c, 462-467
7 Nitschke, W., Hauska, G. and Rutherford, A. W. (1989) Biochim. Biophys. Acta 974, 223-226
8 Whitmarsh, J., Bowyer, J.R. and Crofts, A.R. (1982) Biochim. Biophys. Acta 682, 404-412
9 Kramer, D.A., Joliot, A. Joliot, P. and Crofts, A.R. (1994) Biochim. Biophys. Acta 1184, 251-262
10 Rich, P.R., Madgwick, S.A. and Moss, D.A. (1991) Biochim. Biophys. Acta 1058, 312-328
11 Hope, A.B. (1993) Biochim. Biophys. Acta 1143, 1-22
12 Graan, T. and Ort, D.R. (1986) Arch. Biochem. Biophys. 248, 445-451
13 Rigby, S.E.J., Moore, G.R., Gray, J.C., Gadsby, P.M.A., George, S.J. and Thomson, A.J. (1988) Biochem. J. 256, 571-577
14 Margoliash, E. and Frohwirt, N. (1959) Biochem. J. 71, 570-572
15 He, S., Modi, S., Bendall, D.S. and Gray, J.C. (1991) EMBO J. 10, 4011-4016
16 Hauska, G. (1986) Methods Enzymol. 126, 271-279
17 Chain, R.K. and Malkin, R. (1991) Photosynth. Res. 28, 59-68
18 Huang, D., Everly, R.M., Cheng, R.H., Heymann, J.B., Schägger, H., Sled, V., Ohnishi, T., Baker, T.S. and Cramer, W.A. (1994) Biochemistry 33, 4401-4409

THE CYTOCHROME bf COMPLEXES FROM GRANA AND STROMA LAMELLAE OF SPINACH CHLOROPLASTS ARE DIFFERENT

Elzbieta Romanowska[1] and Per-Åke Albertsson[2]
1 Dept. of Plant Physiology II, University of Warsaw, Poland
2 Dept. of Biochemistry, University of Lund, 221 00 Lund, Sweden

1. Introduction

The cytochrome bf (cyt. bf) complex operates in photosynthetic electron transport either in linear electron flow from PS II to PS I, as in cyclic flow around PS I.
The cytochrome bf complexes isolated from grana (B3) and stroma (T3) membranes are in dimeric form and contain six polypeptides with molecular masses of 34, 23, 20, 17, 4 kDa as well plastocyanin (PC) (1). As we show earlier (1) there is significantly more PC and 4 kDa protein in cyt. bf complex from stroma than from grana lamellae. The both complexes contain 64 kDa protein claimed to be kinase. We found (2), that purified cyt. bf complex from whole thylakoids contains two types of Rieske Fe-S protein having the isoelectric points of 5.1 and 5.4. This paper describes further differences in the isolated cyt. bf complexes from grana and stroma membranes.

2. Procedure

2.1 Thylakoids were prepared as described in (3). Grana derived vesicles (B3) and stroma membrane derived vesicles (T3) were obtained after sonication of stacked thylakoids present in an aqueous two-phase system, followed by three batch partitioning steps (3).
2.2 Isolation of cyt.bf complex from B3 and T3 vesicles was carried out by the method described in (1) and cytochrome f was determined by a spectroscopic method (4).
2.3 SDS-PAGE was used as reported in (5).
2.4 Two dimentional electrophoresis of purified cyt.bf complexes was used as reported in (6).
2.5 Native isoelectric focusing technique (IEF) was carried out by the significant modification of the method of Radola (7)

3. Result and Discussion

As shown earlier (1, 8) cyt.bf complex purified from grana (B3) membrane is different from the complex isolated from stroma (T3) lamellae. The content of the cyt. f and b is almost the same for both complexes. Both complexes contain copper, somewhat more in T3 than in B3, in agreement with the higher plastocyanin (PC) content in the T3 fraction. The polypeptide composition of the purified cyt. bf complex from grana and stroma membranes (Fig. 1a) shows the four major polypeptides: 34, 23, 20 and 17 kDa and

P. Mathis (ed.), Photosynthesis: from Light to Biosphere, Vol. II, 607–610.

additionaly PC band (subunit V, 4 kDa protein, is not visible on this gel). In addition, there is also a band of about 15 kDa which is more abundant in the B3 fraction.
The autors in Ref. 9 found this protein in cyt. bf preparation from Chlamydomonas and they suggest that it played role in the assembly of the cyt. bf complex. Similar polypeptide was observed in the PS II preparation from spinach, but chlorophyll was detected bound in this protein; it is proposed that it is an additional antena complex (CP14-15) within PS II (10). As seen in Fig. 1b the 15 kDa bands from both cyt. bf complexes clearly react with antibodies raised against a 14-15 kDa protein from PS II. The presence of this protein in isolated PS II (10) and also in our complexes (mainly from B3), may suggest that this protein can anchors the grana cyt.bf complex to the grana partitions, for example by binding to PS II α.

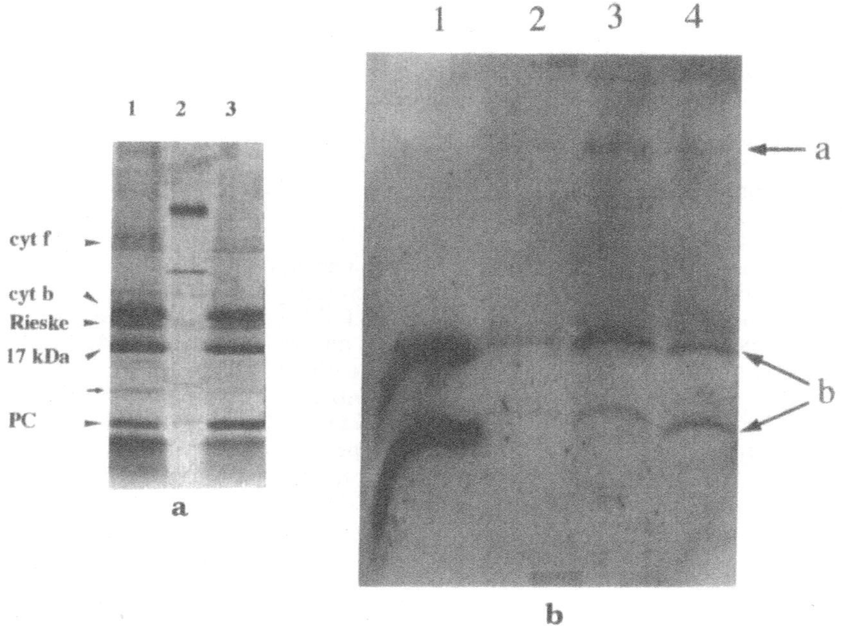

Figure 1. a. SDS-PAGE of the purified cyt.bf complex from grana (B3) and stroma (T3) membranes. Linear gradient gel from 12-22% polyacrylamide. Lane 1, cyt.bf complex from B3; lane 2, protein standards; lane 3, cyt. bf complex from T3. Silver staining.
b. The chemiluminescent detection of 14-15 kDa protein in cyt.bf complex from B3 and T3 membranes. Lane 1, PS II sample; lane 2-4, purified cyt.bf complex from T3, B3 and whole thylakoids, respectively. Lane 1-4, were probed with antibody against the 14-15 kDa protein. a. oligomeric form; b. 14-15 kDa double band.

Using purified cyt. bf complexes from grana and stroma membranes we obtained by two dimensional gel electrophoresis analyses that Rieske Fe-S protein was separated into two polypeptide spots having isoelectric points of 5.1 and 5.4, respectively (Fig. 2 ab). Earlier (6), we identified this two spots in isolated cyt. bf complex from thylakoids using antibody raised against isolated Rieske Fe-S protein. Physiological significance of the

presence of this two isoform of Rieske protein is difficult to explain. It is presumably necessary for dimeric function of the complex.
We applied the native IEF method for chacking the differences between monomeric and dimeric form of the cyt. bf complexes from grana and stroma membranes (Fig. 3ab). Using octylglucoside (OG) as detergent in both, the solubilization (2 and 4%) and separation (1%) media the polypeptide composition of the cyt.bf complex obtained by IEF is the same as previously shown by SDS-PAGE, monomeric and dimeric form of the cyt.bf complex are also visible (Fig. 3). Using O.3% Triton X-100 we did not observed mono and dimeric form of this complex. The results as for cytochrome bf complex from whole thylakoids we also obtained for cyt. bf complex from grana and stroma membranes (not shown). It seems that differences in the composition of the cyt.bf complex from grana and stroma membranes are important in the regulation of cyclic and non cyclic electron flow.

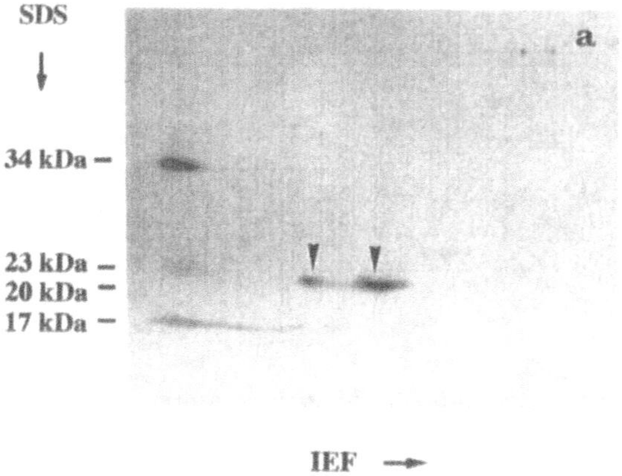

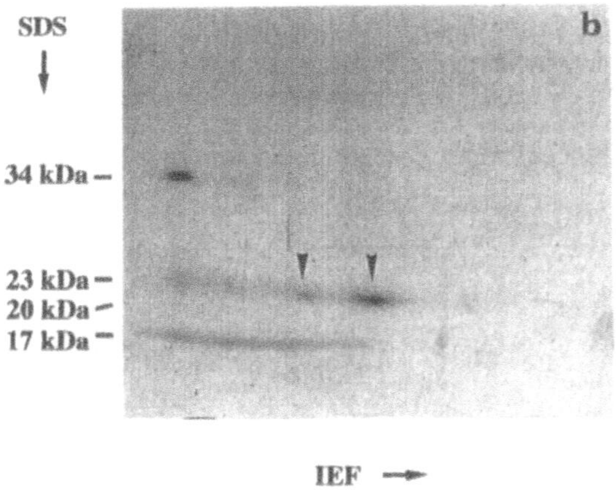

Figure 2. Two dimensional electrophoresis pattern of the purified cyt. bf complex from (a). B3 and (b). T3 membranes. The arrows indicate the Rieske Fe-S protein.

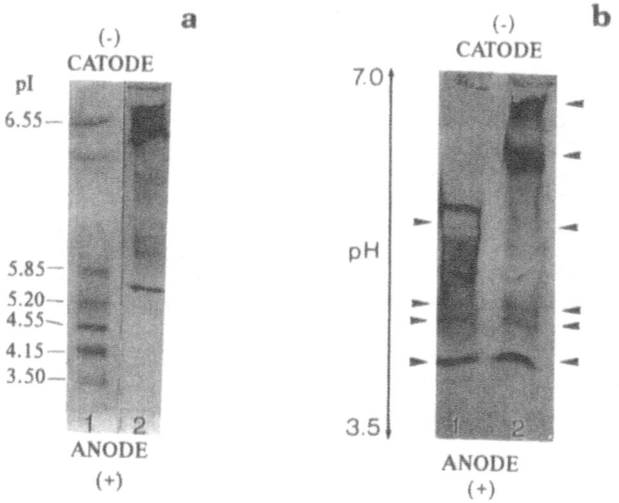

Figure 3. Native IEF in polyacrylamide Mini-Slab Gel (2% ampholine pH 3.5-7.0, 1% OG, 5% polyacrylamid) of cyt. bf complex from whole thylakoids. (a) Line 1, pI standards; lane 2, cyt. bf complex in 4% OG. (b) Lane 1, cyt. bf complex in 0.3% Triton X-100; lane 2, cyt.bf complex in 2% OG.

Acknowledgment

The authors thank Dr K-D. Irragang for determination 14-15 kDa protein in cytochrome bf complexes by chemiluminescent method.

References

1 Romanowska, E. and Albertsson, P-A. (1994) Plant Cell Physiol. 35, 557-568
2 Yu, S-G., Romanowska, E., Xue Z. and Albertsson, P-A. (1994) BBA 1185, 239-242
3 Andreasson, E., Svensson, P., Weibul, C. and Alberdsson, P-A. (1988) BBA 936, 339-350
4 Hauska, G. (1986) in Meth. in Enzymology (Fleischer, S., Fleischer, B. ed.), Vol.126, pp. 271-285, Academic Press, INC
5 Laemmli, U.K. (1970) Nature (London), 227, 680-685
6 Yu, S-G, Stefansson, H., Romanowska, E. and Albertsson, P-A. (1994) Photosynthesis Res. 41, 475-486
7 Radola, J. (1969) BBA 194, 335-338
8 Romanowska, E. and Albertsson P-A. (1992) in Res. in Photosynthesis (Murata N., ed)Vol. II, pp. 507-510, Kluwer Academic Publishers, Dordrecht, The Netherlands
9 Lamaire, C., Girard-Bascou, J., Wollman, F.A. and Bennoun, P. (1986) BBA, 851, 229-238
10 Irragang, K-D., Kablitz, B., Vater, J. and Renger, G. (1993) BBA 1143, 173-182

The cytochrome bc_1 -complex of the purple bacterium *Rhodobacter capsulatus*: spectroscopical and electrochemical investigations

F. Baymann, Institut für Biophysik, Albertstr. 23, 79104 Freiburg, Germany
D.E. Robertson, Recombinant Biocatalysis Inc., 512 Elmswood Av., Sharon Hill, PA 19070, USA
W. Mäntele, Institut für Physikalische und Theoretische Chemie, Egerlandstr. 3, 91058 Erlangen, Germany

Cytochrome *bc*-complexes act as proton pumps in respiration of mitochondria and bacteria and in photosynthesis of chloroplasts, cyanobacteria, heliobacteria and purple bacteria. *Rhodobacter capsulatus* has a bc_1-complex composed of three protein subunits: one of them contains a covalently bound *c*-heme (cytochrome c_1), the second an iron-sulfur cluster of the Rieske-type, and the third two coordinatively bound *b*-hemes (cytochrome b_L and cytochrome b_H).

The complex has two quinone binding sites: a quinol oxidizing site wich can bind up to two quinols [1], and a quinone reducing site, where a seminquinone anion can be stabilized. For the co-transfer of protons and electrons the Q-cycle model has been proposed [2].

In the present work a combination of protein electrochemistry and spectroscopy has been used to determine the midpoint potentials of the cofactors and the redox-induced difference spectra of each compound of the complex [3,4]. Difference spectra in the mid-infrared spectral region indicate changes in the bond geometry of the protein and its prostetic groups caused by the redox transition of the cofactors. This allows the conformational changes of the polypeptide backbone, the amino acid side chains, the cofactors as well as changes in the protonation state of these groups to be detected. The infrared difference signals in this spectra can lead to information about the molecular mechanism of the complex.

The midpoint potentials of the cofactors of the complex were determined electrochemically at different pH-values (Table 1), and the midpoint potential of the Rieske-protein for the first time could be measured at physiological temperature. The same value (+294 mV at pH 8.7) was found as previously determined by EPR spectroscopy [1].

Extinction coefficients of the α-bands of the cytochromes were calculated (Table 2) and IR-difference spectra were obtained, which characterize the redox reaction of each of the cofactors (Figure 1a - e). In addition a method has been developed to calculate the amount of protons picked up or released from the complex during the redox transition of the cofactors, evaluating the buffer signals in the mid-infrared region (Table 3).

It has been shown that the variability of the midpoint potentials of the three hemes and a small splitting of the α-band of cytochrome b_H are intrinsic properties of the complex. The

P. Mathis (ed.), Photosynthesis: from Light to Biosphere, Vol. II, 611–614.

extinction coefficient of cytochrome c_1 was calculated to 11.9 $mM^{-1}cm^{-1}$, i.e. much smaller as supposed so far [5,6]. At the conditions used its midpoint potential is 50 mV more positive than the values reported [7]. Unlike the quinone bound to Q_i-binding site, the quinones bound to Q_o-binding site were not able to react under the present conditions.

	pH 6.5	pH 8.7
cytochrome c_1	+354 mV ± 21 mV	+335 mV ± 25 mV
Rieske-protein	+337 mV	+294 mV
cytochrome b_H	+72 mV ± 21 mV	-11 mV ± 36 mV
cytochrome b_L	-83 mV ± 28 mV	ox: -145 ± 52 mV
cytochrome b_{150}	+161 mV ± 23 mV	+91 mV ± 32 mV
quinone Q_i	+170mV	+44 mV

Table 1: Midpoint potentials of the cofactors of cytochrome bc_1-complex; The midpoint potentials of the cytochromes were calculated from 16 redox titrations at pH 8.7 and 19 titrations at pH 6.5, monitored by difference spectra in the visible range; average values and standard deviations are shown. The midpoint potentials of the quinone bound to Q_i-binding site (quinone Q_i) and of the Rieske-protein were calculated from a single potential titration, monitored in the infrared region

cofactor	cytochrome c_1	cytochrome b_H	cytochrome b_L
extinction coefficient	11.9 $mM^{-1}cm^{-1}$ 553 nm -575 nm	21.9 $mM^{-1}cm^{-1}$ 561 nm - 575 nm	6.7 $mM^{-1}cm^{-1}$ 566 nm - 575 nm

Table 2: Differential extinction coefficients of the α-band of the cytochromes. The values were calculated assuming a value of 28.6 $mM^{-1}cm^{-1}$ [8] for the α-band of the cytochrome bc_1-complex (all cofactors oxidized minus all cofactors reduced)

The IR-difference spectra of all compounds of the complex exhibit distinct signals in the amide I and amide II region (1690-1610 cm^{-1} and 1575-1480 cm^{-1}, resp.), indicating redox-induced conformational changes of the protein backbone. They have the same size as the bands observed in the spectra obtained from soluble *c*-type cytochromes [3,9], indicating that the conformational changes accompanying the redox transitions are of comparable extent in both proteins. In contrast to this, the bands which have been assigned to vibrational changes of the porphyrin are more pronounced in the spectra of the complex than in spectra obtained from soluble cytochromes. This may be interpreted in terms of a more rigid protein structure of the cytochrome bc_1-complex, which force a strictly defined, different structure of the heme in its two redox states to ensure an efficient and fast electron transfer.

In addition to this, IR-difference spectra show some characteristic band features which can preliminarily be assigned to amino acid side chains involved in the conformational transitions of the complex: in particular we presume that an apartic or glutamic acid is involved in the redox reaction of cytochrome c_1, a proline or tryptophane in that of the Rieske-protein, a lysine with cytochrome b_L and two arginines, each with one of the *b*-hemes.

The Rieske-protein denatures if an external potential of +500 mV is applied to the protein solution at basic pH-values. The IR-difference spectra characterizing this reaction show a signal which is tentativly assigned to GLU 99. The structural importance of this amino acid [10] could thus be confirmed.

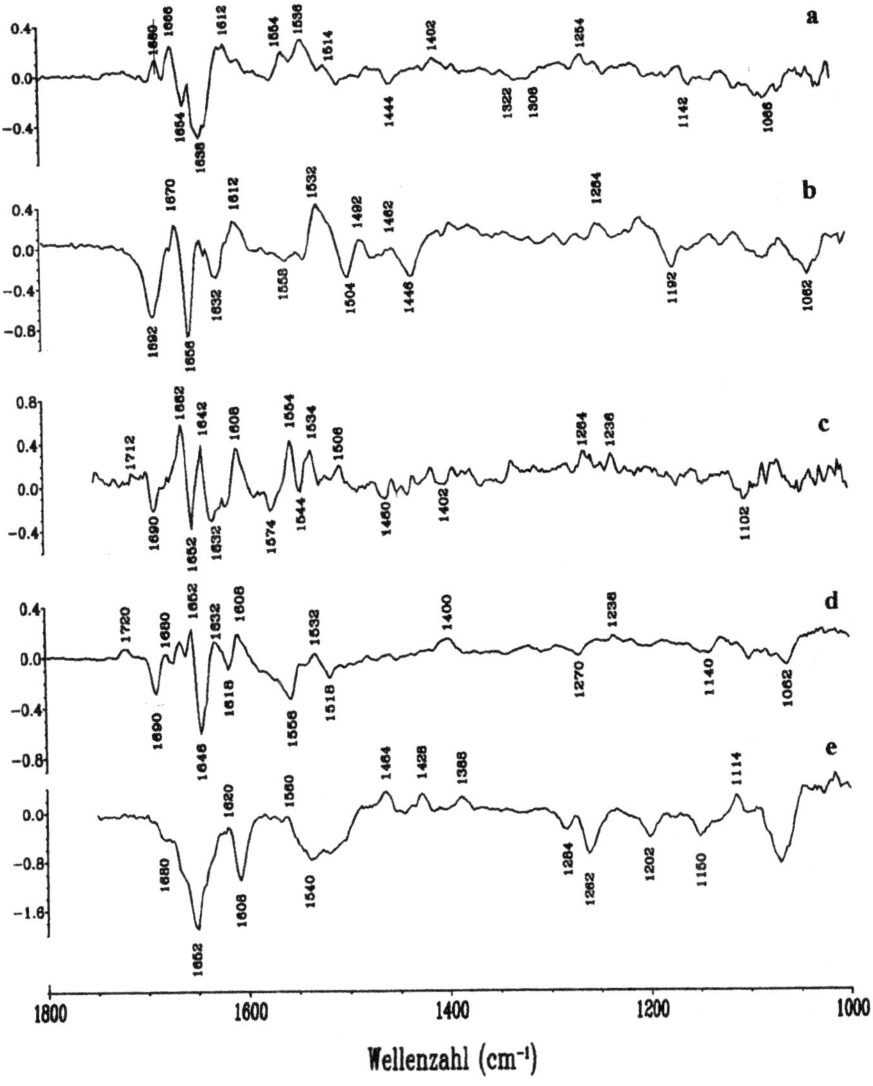

Figure 1: IR-difference spectra (reduced minus oxidized) of the cytochrome bc_1-complex. a) cytochrome c_1; b) Rieske-protein; c) cytochrome b_H; d) cytochrome b_L; e) quinone bound to Q_i-binding site. All measurements were done at pH 8.7. Signals have been normalised to a protein concentration of 1 mM and a pathlength of 1 cm.

compound	proton uptake			
	pH 6.5		pH 8.7	
cytochrome c_1		0		0.2
Rieske-protein	0.6 ± 0.4	0.6	0.5 ± 0.7	0.5
cytochrome b_H		0.6		0
quinone Q_i	2.0 ± 1.2	1.2	2.0 ± 0.4	1.9
cytochrome b_L	1.5 ± 1.7		0.6 ± 0.8	
cytochrome bc_1-complex	5.0 ± 1.1		4.1 ± 1.3	

Table 3. Amount of protons involved in the redox reactions of cytochrome bc_1-complex. Column 2 and 4: average values and standard deviations calculated from 57 experiments (40 in phosphate buffer, pH 6.5 and 17 in TRIS buffer, pH 8.7. Column 3 and 5: values calculated from the separated IR-difference spectra of the cofactors.

The reactions of the quinones bound to the complex contribute to the protonation reactions. In addition, redox induced protonation reactions of protein groups seem to balance the changes in the charge composition. This leads to a pH-dependence of the midpoint potentials of the cofactors.

It has been shown that the pH-dependence of the heme midpoint potentials is not caused by protonation reactions of the heme propionic acids. For the two *b*-hemes, aspartic or glutamic acids could tentatively be identified, probably contributing to this feature. It could be excluded that these two amino acids are involved in protonation reactions of cytochrome c_1 and the Rieske protein. Protonation reactions of additional aspartic or glutamic acids, causing differential features in the spectra of the cytochromes, are not reflected by the buffer signals. This may indicate internal proton exchange in the protein.

References

1: Ding, H., Robertson, D., Daldal, F., Dutton, P. (1992) Biochem. 31, 3144-3158
2: Mitchell, P. (1976) Journal of theoretical Biology 62, 327-267
3: Moss, D., Nabedryk, E., Breton, J., Mäntele, W. (1990),Eur. J. Biochem. 187, 565-572
4: Baymann,F., Moss, D.A., Mäntele,W. (1991) Anal. Biochem. 199, 269-274
5: West, J.C., Mitchell, P., Rich, P.R. (1988), Biochim. Biophys. Acta 933, 35-41
6: Garcia A.F., Venturoli, G., Gad'on, N., Fernandez-Velasco, J.G., Melandri, S.A., Drews, G. (1987) Biochim. Biophys. Acta 89,: 335-345
7: Robertson, D.E., Ding, H., Chelminski, P.R., Slaughter, C., Hsu,J., Moomaco, C., Tokito, M., Daldal, F., Dutton, P.L. (1993), Biochem. 32, 1310-1317
8: Yu, C.A., Yu, L., King,T.E. (1973) J. Biol. Chem. 249, 4905-4910
9: Baymann, F., Moss, D.A., Mäntele, W. (1991) in: Spectroscopy of Biological Molecules; eds: Hester.R.E., Girling,R.B., S. 79-80, The Royal Society, Cambridge
10: Graham L.A., Brandt, U., Sargent, J.S., Trumpower,B.L. (1993), J. of Bioenerg. and Biomembr. 25, 245-257

A SHIFT IN A CHLOROPHYLL SPECTRUM ASSOCIATED WITH ELECTRON TRANSFER WITHIN THE CYTOCHROME b/f COMPLEX.

Anne Joliot and Pierre Joliot
Institut de Biologie Physico-Chimique
13, rue Pierre & Marie Curie, 75005 Paris, France

The cytochrome b/f is a transmembrane complex which catalyzes the oxidation of plastoquinol and the reduction of plastocyanin within the thylacoid membrane of the chloroplast. In this work, we have investigated spectral changes in the red region of the spectrum associated with the electron or proton transfer reactions within the cyt b/f complex. The experiments were performed in anaerobic conditions with *Chlorella sorokiniana*, in the mutant strain S52, which lacks PSII and a large fraction of the light-harvesting complex (1). The illumination of mutant S52 in anaerobic conditions leads to an efficient cyclic process, involving PSI and the cyt b/f complex.

Figure 1 shows the flash-induced spectral changes measured in the presence of 0.4 mM dicyclohexyl-18-*crown*-6 (DCHC), which collapses the light-induced membrane potential (2), with a $t_{1/2} \approx 10$ ms. Spectrum 1, measured 1 ms after a saturating flash, displays several electrochromic shifts. The addition of DCHC - or of carbonylcyanide 4-trifluoromethoxyphenylhydrazone (FCCP) - accelerates the decay of these spectral changes, as it acts on the 515 nm absorption change. One can thus ascribe Spectrum 1 to electrochromic effects induced by the delocalized membrane potential. Spectrum 2 is measured after the complete dissipation of the membrane potential; a new spectral change is then observed, with a minimum at 652 nm and a maximum at 680 nm. This so-called 'red signal' (RS) corresponds to a red shift of a chlorophyll peaking around 669 nm. The same spectrum is observed in *Chlamydomonas reinhardtii*, BF_4 strain, which also lacks most of the chlorophyll antenna (3). The red signal is not observed in the double mutant strain BF_4 ac206 (4), also devoid of cyt b/f complex. Spectrum 3 is measured in the presence of 30 μM dinitrophenyl ether of iodonitrothymol (DI), a specific inhibitor of the quinol-oxidizing site (Q_o site) of the cyt b/f complex (5). Spectrum 3 displays an (ox-red) spectrum of plastocyanin with no contribution of the red signal. One can thus conclude that the red signal is associated with the electron transfer process within the cyt b/f complex.

In Figure 2 are shown the time-courses of the red signal and of the cyt f redox changes, following each flash of a series of oversaturating xenon flashes given 17 s apart. The amplitude of the red signal is measured as the difference $\Delta I/I$ (680 nm - 658 nm). These wavelenghts were chosen to eliminate the contribution of the electrochromic effect due to the delocalized membrane potential. The red signal has not been corrected for the contribution of plastocyanin spectral changes, responsive for the small negative signal observed at time zero. The half-time rise ($\approx$ 9.9 ms) of the red signal is close to that of cyt f reduction ($t_{1/2} \approx 8.1$ ms), which reflects the process of plastoquinol oxidation. One can exclude that the red signal is due to a local electrochromic effect

P. Mathis (ed.), Photosynthesis: from Light to Biosphere, Vol. II, 615–618.

dependent upon the redox state of cyt f, since no red signal change arises during cyt f oxidation and that no change in the cyt f redox state is observed during the decay of the red signal. The half-time of the red signal decay is 2.3 s. This value can vary from 2 to 5 s, depending upon the cultures and the physiological conditions. The addition of DI induces a parallel slowdown of cyt f reduction and of the red signal rise, which clearly demonstrates the correlation of these two processes.

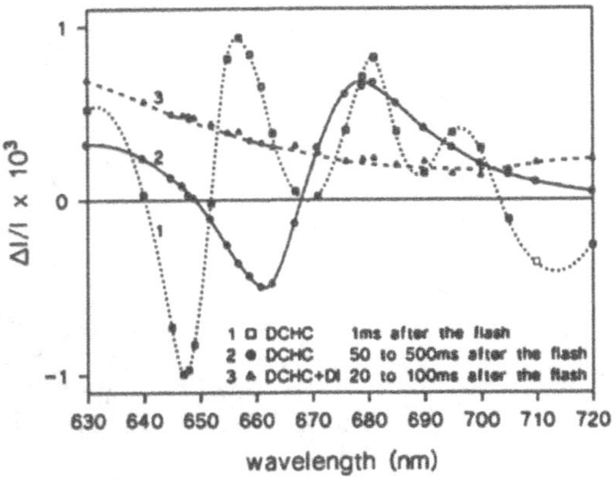

Figure 1.

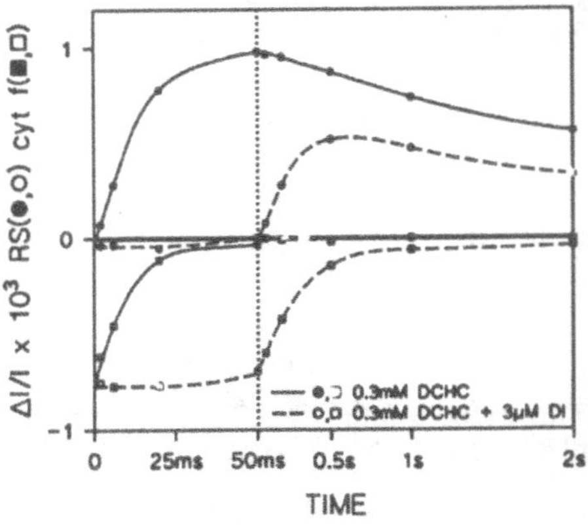

Figure 2.

Figure 1. Spectral changes induced by saturating flashes 10 s apart. 0.4 mM DCHC. Spectrum 3. 30 μM DI.

Figure 2. Effect of DI on the time-course of the red signal (RS) and of cyt f redox changes, following saturating xenon flashes given 17 s apart.

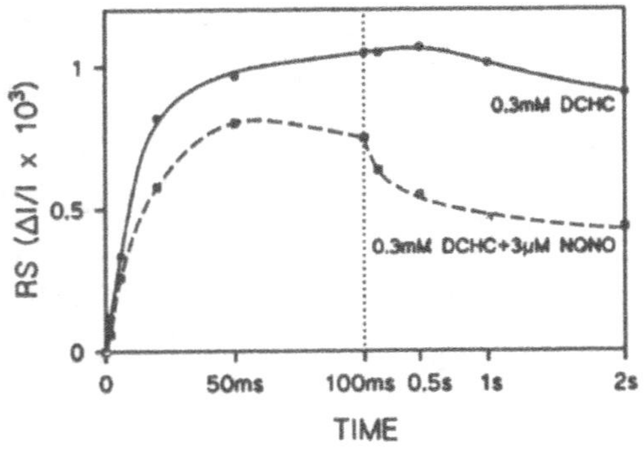

Figure 3.

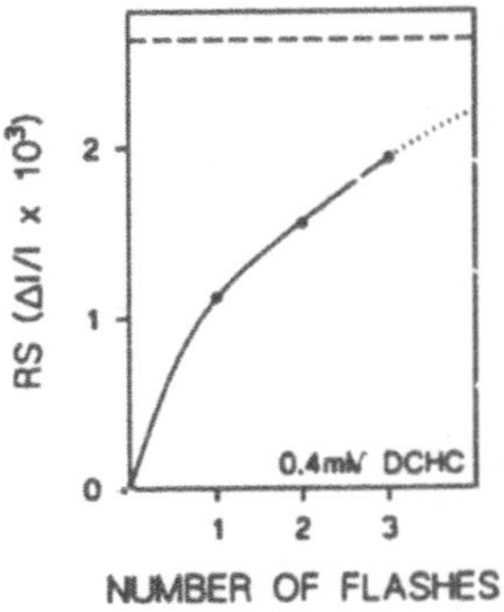

Figure 4.

Figure 3. Effect of NQNO on the time-course of the red signal (RS), following saturating flashes given 20 s apart (DCHC) or 15 s apart (DCHC + NQNO).

Figure 4. Amplitude of the red signal (RS) as a function of the number of saturating flashes (see text). Dashed line, maximum amplitude of the red signal after a strong continuous illumination.

In Figure 3 is shown the effect on the time-course of the red signal of 2-*n*-nonyl-4-hydroxyquinoline *N*-oxide (NQNO), a specific inhibitor of the quinone-reducing site (Q_i site) of the cyt b/f complex (6). The addition of NQNO does not inhibit the red signal but induces the acceleration of its decay, which displays a two-phase kinetics. From experiments Figs. 2 and 3, one can conclude that the red signal is associated with the electron transfer reactions or proton release occurring on the lumenal side of the membrane.

Figure 4 shows the amplitude of the red signal measured 50 ms after the last flash of a series of 1 to 3 saturating flashes given 30 ms apart. The maximum amplitude of the red signal was measured 50 ms after a 40 ms high-intensity continuous illumination (dashed line). Taking into account that there is about 0.6 cyt b/f complex per PSI center and that a single oversaturating flash induces about 1.3 charge separation, one can conclude that a large number of turnovers of the cyt b/f complex is required to reach a maximum amplitude of the red signal.

The amplitude of the red signal increases with DCHC concentration and is maximum for 0.3 mM. In the absence of DCHC, but in the presence of FCCP which collapses the electrochemical proton gradient (not shown), a red signal with a spectrum similar to that shown in spectrum 2, Figure 1 but of smaller amplitude, is observed. Its maximum amplitude, measured after a 20 ms strong continuous illumination, is about 0.25 of that measured in the presence of DCHC, while its decay ($t_{1/2} \approx 150$ ms) is about 20 times faster. Under the same conditions of illumination, but when both uncouplers are present (0.4 mM DCHC and 20 μM FCCP), the amplitude of the red signal is close to that observed in the presence of DCHC only, but with a faster multiphasic decay ($t_{1/2} \approx 150$ ms).

Conclusions. 1. The red shift, with an isobestic point around 669 nm, is related to the process of oxidation of plastoquinol and reduction of cyt f, occurring at site Q_o of the cyt b/f complex.

2. It has been reported that purified cyt b/f includes a single chlorophyll molecule (7-9). Popot et al. (10) have shown that this chlorophyll, which peaks at 667 nm - 668 nm, is an authentic component of the cyt b/f complex. The derivative of its spectrum is very close to the red shift we observed in vivo. We therefore conclude that the red shift can very likely be ascribed to a local electrochromic shift of the chlorophyll molecule bound to the cyt b/f complex.

3. As the time-course of the red signal does not correlate with redox changes of the carriers within the cyt b/f complex, we propose that this signal is an electrochromic shift, induced by the binding to the cyt b/f complex of one or several protons released during the course of plastoquinol oxidation. Proton binding to aminoacids belonging to cyt b/f polypeptides could perturb the electrostatic environment of the chlorophyll probe. In this hypothesis, FCCP, which is a proton carrier, would facilitate the equilibration between the proton-binding site and the lumenal space, thus accelerating the red signal decay.

4. The presence of DCHC, which is a specific K^+ carrier, is required to observe a red signal of maximum amplitude. One can assume that a charged DCHC binds in the vicinity of the chlorophyll, resulting in change in the local electrostatic field. Since the amplitude of an electrochromic shift is proportionnal to the square of the electrostatic field, the polarization of the chlorophyll by DCHC would rend this probe more sensitive to a perturbation of its electrostatic environment by proton binding. In favor of this hypothesis, we observed that the addition of DCHC, in the presence of FCCP which already collapsed the permanent delocalized membrane potential, induces a 2- to 3-times slowdown of the rate of cyt f reduction, following a saturating flash excitation. A similar effect of another K^+ carrier (valinomycin) has been reported in isolated chloroplasts

(11). Valinomycin induces a slowdown of the electron transfer reactions within the cyt b/f complex and a red shift of a b-cytochrome. The authors interpreted these effects, assuming that valinomycin/K^+ binds to the cyt b/f complex in the hydrophobic region of the membrane and perturbs the local electrostatic field within the complex.

5. The hypothesis we proposed implies that the binding of protons released from site Q_o is a fast process (< 2 ms) while the equilibration between the binding sites and the lumen is slow (several s.). This paradoxical conclusion could be interpreted if one assumes that protons are released at the Q_o site, in a pocket or a channel including a few water molecules. The transient proton concentration associated with plastoquinol oxidation would be extremely high ($\approx$ 10 M) i.e. $\approx 10^8$ times larger than the proton concentration at pH 7. This large transient proton concentration could induce a fast proton binding, even if the equilibration rate between the binding site and the bulk water is a slow process.

6. Figure 4 shows that the amplitude of the red signal does not linearly increase with the number of flashes, but tends to the maximum, following an exponential function. This suggests that there is a competition between the trapping of the protons at the protein-binding site and their release in the lumenal space. The rate constant for the trapping of the protons will be proportionnal to the fraction of open binding sites, which explains the quasi-exponential behaviour shown in Figure 4. Due to this competition, the number of binding sites involved in this process cannot be evaluated since the higher the number of binding sites, the higher the probability of trapping by these sites, compared to the direct proton release in the lumen. Whatever the number of binding sites, our results suggest that a significant number of protons formed during the process of plastoquinol oxidation can be transiently stored in the protein matrix before their release in the lumenal space. In the case of bacteriorhodopsin, a delayed proton transfer from the surface of the protein to the bulk water has been reported (12).

References

1. Lavergne, J., Delosme, R., Larsen, U. & Bennoun, P. (1984) Photobiochem. Photobiophys. 8, 207-219

2. Ovchinikov, Y.A, Ivanov, V.T. & Skrob, A.M. (1974) Membrane Actives Complexones, Elsevier, Amsterdam

3. Olive, J., Wollman, F-A., Bennoun, P. & Recouvreur, M. (1981) Arch. Biochem. Biophys. 208, 2, 456-467

4. Gorman, D.S. & Levine, R.P. (1965) Proc. Natl. Acad. Sci. USA 54, 1665-1669

5. Trebst, A. (1980) Methods Enzymol. 69, 675-715

6. Jones, R.W. & Whitmarsh, J. (1985) Photobiochem. Photobiophys. 9, 119-127

7. Bald, D., Kruip, J., Boekema, E.J. & Rögner, M. (1992) in Research in Photosynthesis. Proc. IXth Int. Cong. on Photosynthesis (Murata, N. ed.) Kluwer Academic Publishers, Nagoya, pp. 629-632

8. Huang, D., Everly, R.M., Cheng, R.H., Hetmann, J.B., Schägger, H., Sled, V., Ohnishi, T., Baker, T.S. & Cramer, W.A. (1994) Biochemistry 33, 4401-4409

9. Pierre, Y., Breyton, C., Tribet, C., Kramer, D., Olive, J. & Popot, J-L (1995) submitted

10. Popot, J-L., Pierre, Y., Breyton, C., Lemoine, Y., Takahashi, Y. & Rochaix, J-D. (1995) Proc. Xth Int. Cong. on Photosynthesis (Mathis, P. ed.) Kluwer Academic Publishers

11. Klughammer, C. & Schreiber, U. (1993) FEBS Lett. 336, 491-495

12. Heberle, J., Riesle, J., Thiedemann, G., Oesterhelt, D. & Dencher, N.A. (1994) Nature 370, 379-382

Acknowledgments. This research has been supported by the Centre National de la Recherche Scientifique (URA1187) and by EEC (Contract BIO2-CT93-0076).

THE PufX PROTEIN OF *RHODOBACTER SPHAEROIDES* IS REQUIRED FOR EFFICIENT UBIQUINONE / UBIQUINOL EXCHANGE BETWEEN THE REACTION CENTER AND THE CYTOCHROME BC$_1$ COMPLEX

Wolfgang P. Barz [1,4], Giovanni Venturoli [2], Francesco Francia [2], B. Andrea Melandri [2], André Verméglio [3] and Dieter Oesterhelt [1]

1 Max-Planck-Institute for Biochemistry, 82152 Martinsried, Germany
2 University of Bologna, Department of Biology, 40126 Bologna, Italy
3 CEA Centre d'Etudes de Cadarache, 13108 Saint Paul lez Durance, France
4 Present address: Department of Biochemistry and Biophysics, University of California, San Francisco, CA 94143-0448, USA

1. Abstract

The *pufX* gene is essential for photoheterotrophic growth of the purple bacterium *Rhodobacter sphaeroides*. It codes for a 9 kDa integral membrane protein which is associated with the RC–LH1 core complex. We constructed a chromosomal *pufX* deletion strain and studied the role of PufX in facilitating the efficient interaction between the membrane complexes required for light-driven cyclic electron transport. We demonstrate that the Q/QH$_2$ exchange rate of the RC is reduced approximately 500-fold by the deletion of *pufX* when the Q pool is largely reduced demonstrating an essential role of PufX for the access of ubiquinone to the Q_B site. The fast Q/QH$_2$ exchange is partially restored by suppressor mutations altering the macromolecular antenna structure. These results suggest an indirect role of PufX in structurally organizing a functional photosynthetic apparatus.

2. Introduction

Rhodobacter (R.) sphaeroides is a purple non-sulfur bacterium that is able to grow photosynthetically by using a light-driven, cyclic electron transport pathway. When grown either anaerobically in the light (photosynthetically) or aerobically in the dark under low oxygen partial pressure (semi-aerobically), *R. sphaeroides* produces an extensive system of intracytoplasmic membranes (ICM) which is functionally differentiated to capture light and catalyze photosynthetic energy conversion [1]. Two light-harvesting complexes, LH1 and LH2, absorb the light energy and transfer it to the reaction center (RC), where the primary photochemistry occurs. The RC molecules are surrounded by a ring of LH1 complexes [2]. The levels of LH2 antenna, which are arranged peripherally around the RC-LH1 complexes, vary with irradiance and oxygen tension during growth [3].

Two membrane-bound complexes form a light-driven, cyclic electron transfer pathway in the ICM: the RC and the cytochrome (cyt) bc$_1$ complex (RC). Photooxidation of the primary electron donor, a bacteriochlorophyll dimer, results in charge separation across the membrane [4]. Two subsequent turnovers of the RC lead to the reduction of Q_B, a ubiquinone (Q) molecule bound to the RC. Following binding of two protons, an ubiquinol (QH$_2$) molecule leaves the RC and enters a pool of Q/QH$_2$ molecules in the hydrophobic part of the membrane. Cyclic electron flow is completed by the cyt bc$_1$ membrane

P. Mathis (ed.), Photosynthesis: from Light to Biosphere, Vol. II, 619–624.

complex utilizing QH_2 as an electron donor. The electrons are transferred to cyt c_2, which is located in the periplasm and rereduces the photooxidized bacteriochlorophyll dimer of the RC. The redox reactions of the electron cycle are coupled with vectorial translocation of protons across the membrane, forming an electrochemical gradient which is used for ATP synthesis.

The genes for the antenna, the RC, and the cyt bc_1 complex have been cloned and sequenced [1]. The genes coding for LH1-a and -b (*pufB* and *pufA*) and two of the RC subunits (*pufM* and *pufL*) are located adjacently on the chromosome. These genes form the so-called *puf* operon that is transcribed as a polycistronic mRNA transcript [5]. An additional open reading frame downstream of *pufM*, termed *pufX*, was originally identified in *R. capsulatus* [6]. In *R. sphaeroides*, a similar open reading frame was found to be transcribed as part of the polycistronic *puf* mRNA [7]. Recent studies in *R. sphaeroides* [8, 9, 10] and *R. capsulatus* [11] have shown that the PufX protein is required for anaerobic, photosynthetic growth although PufX is not required for *puf* operon expression or primary charge separation in the RC. The precise role of PufX is unknown, but it has been observed that cyclic electron flow (induced by a train of flashes) is impaired in the absence of PufX [8, 11].

In this article, we summarize our recent studies [8, 12, 13, 14] on the molecular role of *pufX* in photosynthetic energy transduction. Using a novel, chromosomal *pufX* deletion mutant, a control strain containing *pufX* and two photocompetent suppressor mutants (carrying point mutations in *pufB* or *pufA*), we studied which step of energy transduction is impaired in the absence of PufX and can be restored by suitable point mutations in the LH1 genes. We provide *in vivo* evidence that PufX is essential for photophosphorylation under anaerobic conditions because it plays a key role in multiple turnover cyclic electron flow, although it has no apparent role following single flash excitation [13]. Detailed spectroscopic studies *in vivo* revealed that PufX is involved in the efficient interaction between the RC Q_B site and the Q pool [14].

3. Results and Discussion

3.1 Phenotypic characterization of pufX

Complementation of a *pufLMX* deletion strain with plasmid-borne *puf* operon constructs (lacking or containing *pufX*) demonstrated that *pufX* is strictly required for photoheterotrophic growth of *R. sphaeroides* [8, 15]. While during chemoheterotrophic growth (at high or low oxygen tensions) identical growth rates were observed in the presence or absence of *pufX*, this gene was strictly required for photosynthetic growth [8]. The impairment of photosynthetic growth was observed in a succinate or malate minimal medium as well as in rich media.

To determine whether the lack of *pufX* was lethal for illuminated *R. sphaeroides* cells or whether it only impaired the ability of photosynthetic growth (i.e. cell division), a *pufX*$^-$ culture was illuminated for 50 hours under anaerobic conditions, while another *pufX*$^-$ culture was incubated anaerobically in the dark. Aliquots were taken from both cultures, diluted, spread on agar plates and incubated aerobically. Similar numbers of colonies were obtained from the differently incubated cultures giving evidence that the photosynthetic incompetence of *pufX*$^-$ cells was not caused by light-induced death of the *pufX*$^-$ cells. Instead, photosynthetic growth (cell division) was impaired in the absence of *pufX*.

A possible reason for the impairment of photosynthetic growth in the absence of PufX could be the light-induced degradation of the RC. To study a potential role of PufX in protecting anaerobic cells against photodamage, the stability of the RC was determined during prolonged illumination under anaerobic conditions. For an illumination period of about 10 h, the concentration of photobleachable RC was similar in cells containing or lacking *pufX* [13]. Since the levels of RC/cell did not decrease in the absence of PufX, light-induced degradation of the RC could be excluded.

3.2 Expression of PufX and its localization
Northern blot analysis and mapping experiments have shown that the *pufX* open reading frame is transcribed as part of the 2.7 kb *puf* mRNA [11]. To test whether *pufX* is translated, polyclonal antibodies were raised against a synthetic peptide corresponding to a portion of the PufX amino acid sequence [8]. Western blot analysis showed that, under photosynthetic or semi-aerobic conditions, *pufX* is expressed as an integral membrane protein (termed PufX) of approximately 8–10 kDa, a value close to the predicted mass of 9.0 kDa. Cell fractionation experiments revealed the localization of the PufX protein in the ICM and its association with the RC–LH1 core complex [8]. Interestingly, even in the absence of *pufX* an intact RC-LH1 complex could be solubilized from the membrane (using octyl glycoside) demonstrating that PufX is not required for the binding of LH1 complexes to the RC.

3.3 Genetic characterization of second-site suppressor mutants
The requirement of *pufX* for photosynthetic growth of *R. sphaeroides* can be suppressed by spontaneous second-site mutations. We isolated and characterized 65 of these suppressor mutants to mechanistically understand the intergenic suppression of the non-photosynthetic phenotype [12]. In these strains (containing both chromosomal and plasmid-borne copies of *pufB* and *pufA*), we found single point mutations in the plasmid-borne *pufB* or *pufA* genes altering highly conserved residues of the LH1 complex. In 89% of the independently isolated suppressor mutants, specific base substitutions were observed in one of three tryptophan codons which are highly conserved among all purple bacterial LH1 polypeptides [12]. Spectral characterization and Western blot analysis were used to examine the expression of mutant LH1 polypeptides and LH1 assembly. Based on these studies, different groups of suppressor mutants could be distinguished [12]. Interestingly, all suppressor mutants had drastically reduced amounts of antenna complexes (20 - 35%), indicating severe changes in the antenna structure of these strains [12, 13]. These results suggest that the differently altered LH1 polypeptides do not substitute for PufX directly but lead to structural rearrangements in the macromolecular membrane organization, thus allowing photosynthetic growth in the absence of PufX.

3.4 Construction of a chromosomal pufX deletion strain
Until recently, all work on the PufX protein was performed using partial *puf* deletion strains which were complemented with plasmids carrying *puf* operon constructs [8, 9, 15]. As tools for the quantitative determination of the PufX function, these plasmid-complemented strains were unfavorable because of copy number effects [15], because of the presence of chromosomal copies of *pufB* and *pufA* and because of cis-acting elements regulating the expression of the *puf* operon [13, 16]. For these reasons, we constructed a novel *pufX* deletion mutant carrying one copy of a truncated *puf* operon in the correct chromosomal position [13]. The resulting mutant (termed PUFΔX) was compared with a control strain containing *pufX* (termed PUFC). In addition, green carotenoid mutants were isolated and termed PUFΔX/g and PUFC/g, respectively. Phenotypic characterization of these strains confirmed the requirement of *pufX* for photosynthetic growth [13].

3.5 Photophosphorylation in vivo
The requirement of *pufX* for photosynthetic growth suggests that the PufX protein is involved in the conversion of light into chemical energy. To study the importance of PufX

for photosynthetic energy transduction, light-induced ATP synthesis (photophosphorylation) was investigated in intact cells of PUFC and PUFΔX [13]. In addition to these strains, the photocompetent suppressor mutants Sup101 and Sup102 [12] were studied. In vivo measurements of the light-induced ATP synthesis revealed that PufX is indeed required for photosynthetic energy transduction [13]. While no photophosphorylation was observed in PUFΔX cells, light-driven ATP synthesis was restored in the suppressor mutants showing that specific point mutations in the LH1 genes can suppress the impairment of photophosphorylation caused by the deletion of *pufX*.

3.6 Flash-induced electrochromic carotenoid bandshift

The flash-induced generation of a membrane potential was analyzed spectroscopically to examine the role of PufX for cyclic electron transfer. When chromatophores isolated from semi-aerobically grown cells of PUFΔX/g were subjected to single-turnover flashes, no carotenoid bandshift could be detected [13]. Since the primary RC photochemistry was not impaired in these strains (see below), the lack of the carotenoid bandshift suggests that the PUFΔX/g chromatophores were unsealed vesicles. This interpretation was directly supported by the observation that the addition of exogenous cyt c to these chromatophores increased the rate and extent of cyt c photooxidation. In addition, potassium diffusion potentials did not induce any carotenoid in valinomycin-treated PUFΔX/g chromatophores, in contrast to PUFC/g chromatophores (data not shown). We conclude that the deletion of *pufX* changes the membrane structure of *R. sphaeroides* in a way that isolated chromatophores are no longer sealed vesicles.

For this reason, flash-induced electron flow was studied *in vivo* [13]. Following single flash excitation, the rise kinetics of the carotenoid bandshift were indistinguishable in anaerobically incubated cells of PUFC/g and PUFΔX/g, demonstrating that PufX is not required for single turnover of the RC and the cyt bc_1 complex.

3.7 Cyclic electron transfer induced by continuous illumination

Since single turnover experiments did not reveal any role of PufX in cyclic electron transfer, the light-induced generation of a membrane potential was measured *in vivo* using continuous illumination [13]. In anaerobically incubated cells, PufX was found to be required for multiple turnover electron flow: While continuous illumination induced an intense carotenoid bandshift in PUFC/g cells, only a very small signal was observed in PUFΔX/g cells. Interestingly, this small signal closely resembled the trace observed during very weak illumination (0.5 W/m^2) of PUFC/g. This low irradiance does not allow photosynthetic growth of WT demonstrating that the light-induced membrane potential observed during strong illumination of PUFΔX/g is not sufficient to sustain photosynthetic growth. These results suggest an important role of PufX in facilitating multiple turnover cyclic electron transport under anaerobic conditions [13].

Under aerobic conditions, however, the light-driven generation of the membrane potential was very similar in continuously illuminated cells of PUFC/g and PUFΔX/g [14]. In contrast to anaerobic conditions (where the membrane Q pool is completely reduced [17]), the absence of PufX has apparently no consequences on multiple turnover electron flow under aerobic conditions (where the Q pool is partially oxidized [17]). Similar results were obtained when the Q pool was oxidized *in vivo* using electron acceptors like dimethylsulfoxide (DMSO) or trimethylamine N-oxide (TMAO) [14]. These results suggest that (at least partial) oxidation of the Q pool effectively restores cyclic electron flow in the absence of PufX. Indeed, the addition of DMSO or TMAO restored photosynthetic growth in the absence of PufX [14].

3.8 The Q/QH_2 exchange rate at the reaction center
Multiple flash experiments *in vivo* (using flashes spaced 20 ms apart) revealed that the RC photochemistry does not depend on PufX for the first two flash excitations [14]. Following the third and subsequent flashes, however, efficient charge separation required PufX under anaerobic conditions (indicative of a reduced Q pool), suggesting an important role of PufX in efficiently replacing Q_BH_2 (formed by the first two flashes) by Q. Under more oxidizing conditions (cells treated with oxygen or TMAO), however, multiple turnover of the RC did not require PufX [14]. Therefore, in the absence of PufX, primary RC photochemistry following more than two flashes is only impaired when the Q pool is largely reduced. These results suggest that, under reducing conditions, PufX might be involved in the efficient access of Q to the Q_B site of the RC.
By varying the frequency of flash excitations, the importance of PufX for fast QH_2/Q exchange could be measured quantitatively [14]. When the Q pool was completely reduced (but not under more oxidizing conditions), the Q/QH_2 exchange rate was drastically reduced (about 500-fold) by the absence of PufX demonstrating an essential role of PufX for the access of ubiquinone to the Q_B site [14]. Interestingly, the fast Q/QH_2 exchange was partially restored in the suppressor mutants Sup101 and Sup102 demonstrating that the requirement of PufX for fast RC turnover (under reducing conditions) can be suppressed by suitable point mutations in the LH1 genes.

3.9 Kinetics of cyt b_{561} reduction induced by a single turnover flash
While multiple turnover of the RC was found to required PufX only under reducing conditions, the cyt bc_1 complex showed an opposite redox dependence: When the kinetics of flash-induced reduction of cyt b_{561} were monitored in chromatophores inhibited with antimycin A, PufX was found to be required for turnover of the cyt bc_1 complex only at high redox potential ($E_h > 140$ mV) [14]. At such redox poise, all the QH_2 molecules reacting at the cyt bc_1 complex are those released by the RC [18]. Thus, the availability of QH_2 molecules coming from the RC seems to be drastically reduced in the absence of PufX, suggesting a role of PufX in the lateral diffusion of QH_2 (formed at the RC) to the cyt bc_1 complex.

4. Conclusions

The different redox dependence observed for RC and cyt bc_1 activity might seem contradictory at first sight, but in fact demonstrate that PufX is important for both the access of Q to the RC (observed under reducing conditions when the concentration of Q is limiting) and for the transfer of QH_2 (revealed under oxidizing conditions) [14]. These results strongly suggest an indirect role of PufX in lateral, intramembrane transfer of Q and QH_2 (e.g. by structurally organizing the photosynthetic membrane). The results of the suppressor analysis [12] also point to a central role of PufX in organizing the macromolecular antenna structure around the RC. We suggest that PufX structurally arranges the RC surrounding thus enabling efficient access of Q to the Q_B site under reducing conditions where Q is limiting. The drastically reduced LH1 levels in the suppressor mutants suggest a role of PufX in preventing LH1 molecules from physically blocking the Q_B site. This interpretation is clearly supported by a recent study showing that PufX has no significant effect on the photosynthetic growth rates of *R. sphaeroides* strains lacking both types of antenna complex [10]. The fact that photosynthetic growth of such strains has no absolute requirement of PufX strongly supports our conclusion that PufX facilitates Q/QH_2 exchange between the RC and the cyt bc_1 complex under conditions where this Q/QH_2 exchange is limited by a wildtype photosynthetic unit structure.

References

1 Kiley, P.J. and Kaplan, S. (1988) Microbiol. Rev. 52, 50-69
2 Monger, T.G. and Parson, W.W. (1977) Biochim. Biophys. Acta 460, 393-407
3 Aagaard, J. and Sistrom, W.R. (1972) Photochem. Photobiol. 15, 209-225
4 Parson, W.W. (1987) in Photosynthesis (Amesz, J., ed.) pp.43-61, Elsevier Science Publisher B.V., Amsterdam
5 Zhu, Y.S. and Kaplan, S. (1985) J. Bacteriol. 162, 925-932
6 Youvan, D.C., Bylina, E.J., Alberti, M., Begusch, H. and Hearst, J.E. (1984) Cell 37, 609-619
7 Lee, J.K., Dehoff, B.S., Donohue, T.J., Gumport, R.I. and Kaplan, S. (1989) J. Biol. Chem 264, 19354-19365
8 Farchaus, J.W., Barz, W.P., Grünberg, H. and Oesterhelt, D. (1992) EMBO J. 11, 2779-2788
9 Westerhuis, W.H.J., Farchaus, J.W. and Niederman, R.A. (1993) Photochem. Photobiol. 58, 460-463
10 McGlynn, P., Hunter, C.N. and Jones, M.R. (1994) FEBS Lett. 349, 349-353
11 Lilburn, T.G., Haith, C.E., Prince, R.C. and Beatty, J.T. (1992) Biochim. Biophys. Acta 1100, 160-170
12 Barz, W.P. and Oesterhelt, D. (1994) Biochemistry 33, 9741-9752
13 Barz, W.P., Francia, F., Venturoli, G., Melandri, B.A., Verméglio, A. and Oesterhelt, D. (1995) Biochemistry 34, in press
14 Barz, W.P., Verméglio, A., Francia, F., Venturoli, G., Melandri, B.A. and Oesterhelt, D. (1995) Biochemistry 34, in press
15 Farchaus, J.W., Grünberg, H. & Oesterhelt, D. (1990) J. Bacteriol. 172, 977-985
16 Farchaus, J.W. and Oesterhelt, D. (1989) EMBO J. 8, 47-54
17 Jones, M.R., Richardson, D.J., McEvan, A.G., Ferguson, S.J. and Jackson, J.B. (1990) Biochim. Biophys. Acta 1017, 209-216
18 Crofts, A.R., Meinhardt, S.W., Jones K.R. and Snozzi, M. (1983) Biochim. Biophys. Acta 723, 202-218

7. Electron transfer proteins

MOLECULAR RECOGNITION IN ELECTRON TRANSFER PROTEINS

Carlos Gómez-Moreno, Marta Martínez-Júlvez, María F. Fillat
Departamento de Bioquímica y Biología Molecular y Celular
Universidad de Zaragoza, Spain

John Hurley and Gordon Tollin
Department of Biochemistry
University of Arizona, AZ, USA

1. Introduction

During the last few years the understanding of the mechanisms involved in photosynthesis has experienced a great advance. This is due, mainly, to the fact that a large number of structures of the different components of photosynthetic organisms have been known allowing the proposal of mechanisms at the atomic level.
This is the case for the mechanism of electron transport that is critical for the process of energy transduction in photosynthesis. Proteins participating in electron transfer processes are usually embedded into membranous structures so that the exchange of electrons with partner proteins has to be carried out by means of the smaller size electron transport proteins, which are more mobile and communicate the components of electron transport chains.
Electron transfer reactions between proteins require the formation of a transient complex of the two proteins involved. This complex allows the redox centres to lye at the adequate position for the reaction to take place. There is a great interest in deciphering the parameters that control electron transfer reactions since oxidation-reduction reactions are present in all types of organisms and participate in all kinds of processes. The importance of designing artificial systems, which could be able to support redox reactions in productive processes, adds interest to this area of research.
Among the different electron transport system involved in photosynthesis we have worked with that including ferredoxin and ferredoxin-NADP$^+$ reductase (FNR) from the cyanobacterium Anabaena PCC 7119, which participates in the reduction of NADP$^+$ by equivalents produced by photosystem I. Ferredoxin from Anabaena has a 11 kDa polypeptide chain that belongs to the plant-type [2Fe-2S] ferredoxins and has a very low midpoint redox potential of -430 mV [1]. The structure of Anabaena Fd has been recently solved at high resolution and exhibits a large amount of acidic amino acid residues in the surface of the protein [2].
FNR is a 36 kDa protein having a non-covalently bound FAD group with a midpoint redox potential of -344 mV [3]. The three-dimensional structure of the Anabaena protein has been determined at 1.8 Å resolution and shows two separated domains

P. Mathis (ed.), Photosynthesis: from Light to Biosphere, Vol. II, 627–632.

where the cofactor FAD and the substrate $NADP^+$ binds [4]. Different indirect experiments indicate that the second substrate, the electron transfer protein ferredoxin, which has a much larger size, binds to a cavity which occupies both domains in a position that would allow the close interaction between the iron sulfur centre of Fd and the FAD and $NADP^+$ in the enzyme [5,6].
In this work we describe kinetic experiments carried out with mutants of FNR where specific amino acid residues located in the surface of the enzyme and carrying positive charges have been replaced by negatively charged residues. The results indicate the specificity of the interaction between the two proteins for electron exchanging reactions.

Material and Methods

Oligonucleotide-directed mutagenesis of Anabaena PCC 7119 FNR. The FNR mutants were prepared using as template a construct from the complete *pet*H gene [7] made as follows. The AGT codon coding for Ser 84 was mutated to AGA using the polymerase chain reaction (PCR) in order to create a *Xba* I restriction site. A *Hind* III site was also introduced by PCR at the 3´ side after the stop codon of the *petH* gene and the *Xba* I/*Hind* III digested amplification product ligated into the expression vector *pTrc* 99a. This construct was used to transform the *E. coli* PC 0225 strain and the expression of the protein induced with IPTG as described in [8]. The K75E, K138E, K290E and K294E amino acid substitutions were carried out using the Transformer™ Site-directed mutagenesis kit from CLONTECH in combination with the following oligonucleotides: 5´-GAGTCTCAATTCTTCCGGCTTGCC-3´ for K75E, 5´-CAGCATTTCTTCACC CACAGGCCC-3´ for K138E, 5´-CTTGAGGTCTTCTTGGTAATCAC-3´ for K290E, 5´-GCGACCAGCTTCCTTGAGGTC-3´ for K294E and the trans oligo *Nde* I/*Sac* II 5´-AGTGCACCATCCGCGGTGTGA-3´ as selection primer.
Purification of the FNR mutants. Cells from 3.5 l of IPTG induced cultures were harvested, resuspended in 50 mM Tris/HCl, 0.1 mM EDTA, 0.1 mM mercaptoethanol, pH 8 and 1 mM PMSF (buffer A) and broken by ultrasonic treatment. The crude extract was separated from the cell membranes by centrifugation at 25.000 x g for 30 min at 4°C. The supernatant was poured onto a DE-52 column. After washing the column, FNR was eluted using buffer A with a NaCl gradient of 0 to 0.2 M. Analysis of the eluted fractions showed the presence of two peaks containing FNR with different N-terminal sequences. The tubes from the peak eluting at higher ionic strength (N-terminal MTQAKAKH) were pooled, diluted 1:1 in buffer A and poured onto a Cibacron Blue-Sepharose column. Proteins fixed to the column were eluted with a gradient of 0.1 to 0.5 M NaCl in buffer A and the fractions containing homogeneous FNR pooled and dialysed versus buffer A. UV-visible spectra and SDS-PAGE electrophoresis were used as purity criteria. Typical yield was 2 mg FNR/litre culture.
Steady-state kinetics. The diaphorase activity was assayed using DCPIP as the artificial electron acceptor and the ferredoxin-dependent NADPH-cytochrome **c** reductase activities were measured as described previously [9]. T.N. for the diaphorase activity were determined using a final concentration of 40 μM NADPH. T.N. for the ferredoxin-dependent NADPH-cytochrome **c** reductase activity were determined using 5-fold concentrations of Fd K_m values.
Laser flash photolysis experiments were performed at the University of Arizona with a nitrogen laser-pumped dye laser (BBQ 2A368 dye, 396-nm wavelength maximum,

0.1-mJ energy, pumped by a PRA Model LN100 nitrogen laser, 300-ps pulse duration as described in [10].

Results and Discussion

Our previous results on the interaction between Anabaena FNR and ferredoxin indicated the absolute requirement of an acidic, as well as an aromatic, amino acid residues for electron transfer between the two proteins when forming a complex in solution. Kinetic experiments using the laser flash photolysis technique had shown that E94 in ferredoxin is essential for electron transfer to FNR since its replacement by another amino acid not having a negative charge (Lys or Gln) decreased the rate constant by 50,000-fold, while replacement by another amino acid having negative charge (Asp) maintained a similar rate [10]. This technique has shown to be very appropriate for the determination of the first order rate constants of the electron transfer reactions which means the direct measurement of interprotein electron transfer rates in the absence of other catalytic steps. Similar changes were observed when F65 in Anabaena ferredoxin was replaced by a non-aromatic amino acid residue, while the replacement by another aromatic amino acid (Tyr or Trp) did not show any change in the rate [11,12].

These results indicated the high degree of specificity in the interaction between the reductase and ferredoxin for the electron transfer reaction. Moreover, the role of these important amino acid residues was suggested to be somewhat different to just maintaining intact the complex, since the measurement of the binding constant of the complex indicates differences of 2 to 3-fold in the mutant proteins with respect to the wild type ferredoxin, a change several orders of magnitude smaller than the change in rate. The midpoint redox potentials of the ferredoxin mutants were also similar to those of the wild type protein indicating that the difference in rate was not due to a major structural change in the electron transfer protein.

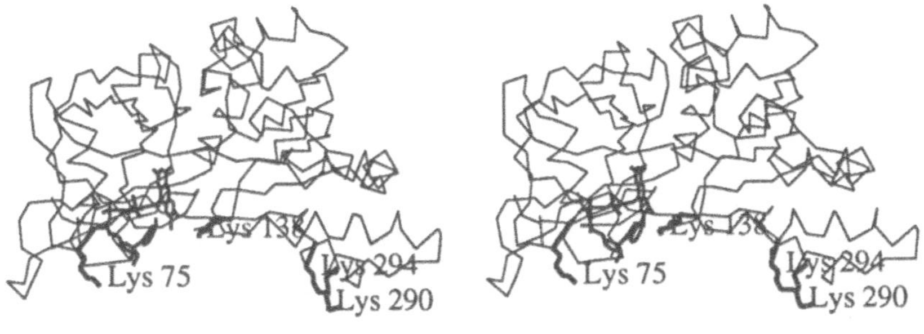

Figure 1: Stereo-view of the FNR molecule showing the positions of the amino acid residues which have been mutated.

Kinetic experiments with FNR mutants

The above mentioned results prompted us to design mutants of the reductase which could have hampered their ability to interact with ferredoxin. Previous chemical

modification experiments with the enzyme from Anabaena had shown the involvement of certain basic amino acid residues in FNR which could be important for the interaction with the electron transfer protein [13,14]. A close examination of the three-dimensional structure of the
enzyme points directly to a concave cavity formed by the two domains of the reductase which suggests that this is the area where ferredoxin binds to the enzyme.Many basic amino acid residues are located in the surface of that cavity pointing to the outside and these could also be considered as candidates to support the interaction with ferredoxin.
Having developed the methodology for the expression of Anabaena FNR in E. coli at high yields a systematic mutagenesis program was started in order to evaluate the effect of the breaking electrostatic interactions on the rate of electron transfer.
Four FNR mutants have been prepared which correspond to lysines in positions 75, 138, 290 and 294 in FNR. In Figure 1 the relative positions of the mutated residues in the FNR structure are shown. The purification of the mutant proteins from E. coli cell extract is quite simple as indicated in the Material and Methods section and corresponds essentially to the procedure described for the wild type protein [15]. It consists in the sequential use of an ionic exchange and affinity chromatographic steps. The fact that all the purified proteins show similar absorption spectra to the wild type protein also supports the proposal that no major structural changes have taken place by the mutation.

Table I: Steady-state kinetic constants for the reactions catalyzed by wild type and FNR mutants.

	Diaphorase[1] (NADPH-FNR-DCPIP) **T.N.**(min^{-1})	**Fd-NADP$^+$ reductase**[2] (NADPH-FNR-cytc) **T.N.**(min^{-1})	**Km$_{Fd}$** (μM)
wild-type FNR	3400	8700	17.8
mutant K75E	1900	<200	---
mutant K138E	3200	9700	346
mutant K290E	3300	10800	218
mutant K294E	3500	15100	207

Steady state kinetics.
The enzymatic activities of the FNR mutants were measured using two different enzymatic assays: the diaphorase activity, where the transfer of electrons between NADPH and an artificial dye is measured and the NADPH-cytochrome **c** reductase assay in which the transfer of electrons between FNR and the cytochrome **c** requires the presence of ferredoxin interacting with the enzyme.
Table 1 shows the turnover numbers obtained for both enzymatic activities. It can be observed that, while the diaphorase activities of the mutants in positions 138, 290 and 294 show similar rates to the wild type protein, the mutant in position 75 shows a decreased enzymatic activity which is still higher than 50 % of the native protein.
The ferredoxin-dependent cytochrome **c** reductase activity shows (Table 1) that all the mutants prepared, except K75E, exhibit even higher rates than the wild type protein. In agreement with the very high K_m values for ferredoxin obtained with these mutants

the concentration of ferredoxin required to reach saturation with the mutant proteins is, nevertheless, much higher than with the wild type. This is an indication of the partial impairment in the interaction of the enzyme with ferredoxin which has occurred upon the replacement of the charges in the surface of FNR. The mutant in position 75, which exhibited lower diaphorase activity does not show any appreciable cytochrome c reductase activity, indicating the lack of interaction between FNR and ferredoxin when the positive charge in position 75 is changed to a negative one.

Rapid kinetics.
The ability to exchange electrons with ferredoxin by the FNR mutants was also checked using the laser flash photolysis technique. Fig 2 shows the effect of the ionic strength in the assay on the rate constants in comparison to those with the wild type protein.

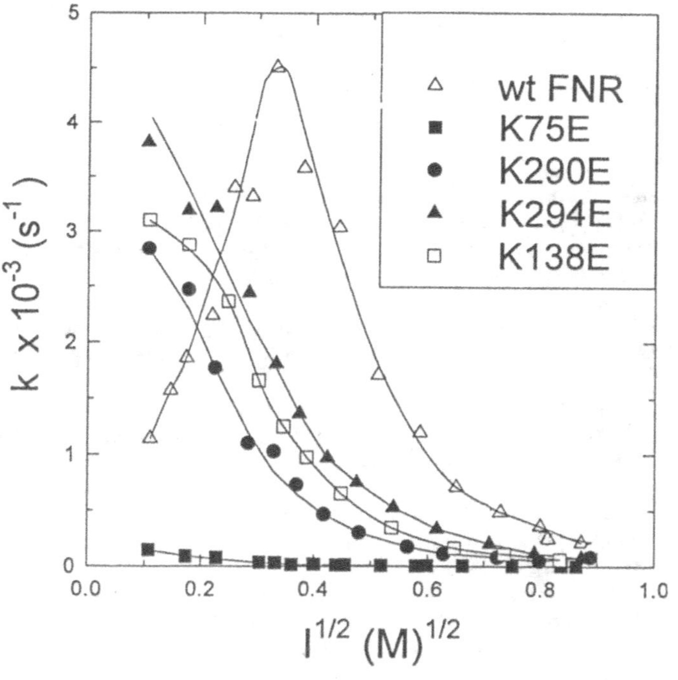

Figure 2: Ionic strength dependence for the reaction of mutant and wild type FNRs with wild type ferredoxin.

It is obvious the different behaviour of the wild type and of the mutant proteins with respect to the presence of salt in the assay. The wild type protein has a low rate constant at low salt concentrations that increases with the ionic strength. This is indicative of a mechanism where the protein-protein complex is stabilised by electrostatic interactions but, once it is formed, it has to rearrange and this occurs better at lower ionic strength values, that is, when the complex is more flexible. Mutants K138E, K290E and K294E show highest rate constants at the lowest ionic strength. This is consistent with the model proposed where the complex with the mutant proteins is more loose and becomes more unstable as the salt concentration increases. It is not fully understood, nevertheless, why the rates observed at low salt concentration in the mutant proteins have similar values to that of the wild type at optimum salt concentration.

The most striking result is that corresponding to the replacement of Lys in position 75 by a Glu. This mutant, in which a reversal of the charge has been made, has lost completely the ability to exchange electrons with ferredoxin and this occurs at all ionic strengths assayed. A more thorough characterization of this mutant is currently being made in order to reveal its function in ferredoxin-dependent electron transfer reactions supported by FNR.

Acknowledgements

This work was supported in part by BIO94-0621-C02-01 to C.G.-M. and DK15057 to G.T. M.M.-J. is the recipient of a fellowship from the Diputación General de Aragón. We acknowledge Juan Fernández-Recio for his help on the computer-graphic model presentation.

Bibliography

1. Salamon, Z. and Tollin, G. (1992) Bioelectrochem. Bioenerg. 27, 381-391
2. Rypniesky, W.R., Breite, D.R., Benning, M.M., Wesemberg, G., Oh, B.H., Markley, J.L. Rayment, I. and Holden, H.M. (1991) Biochemistry 30, 4126-4131
3. Pueyo, J.J., Gómez-Moreno, C. and Mayhew, S. G. (1991) Eur. J. Biochem. 202, 1065-1071
4. Serre, L., Vellieux, F., Fontecilla-Camps, J., Frey, M., Medina, M. and Gómez-Moreno, C., (1994) in Flavins and Flavoproteins (K. Yagi, ed.) 431-434. Walter de Gruyter, Berlin
5. Karplus, P.A. and Bruns, C.M. (1994) J. Bioenerg. Biomem. 26, 89-99
6. Apley, E.C. and Wagner, R. (1988) Biochim. Biophys. Acta 936, 269-275
7. Fillat, M.F., Flores, E. and Gómez-Moreno, C. (1993) Plant. Mol. Biol. 22, 725-729
8. Fillat, M.F., Borrias, W.E. , Weisbeek, P.J. (1991) Biochem. J. 280, 187-191
9. Sancho, J., Peleato, M.L., Gómez-Moreno, C. and Edmondson, D.E. (1988) Arch.Biochem.Biophys. 260, 200-207.
10. Hurley, J.K., Salamon, Z., Meyer, T.E., Fitch, J.C., Cusanovich, M.A., Markley, J.L., Cheng, H., Xia, B., Chae, Y.K., Medina, M., Gómez-Moreno, C. and Tollin, G. (1993) Biochemistry 32, 9346-9354
11. Hurley, J.K, Chen, H., Xia, B., Markley, J.L. Medina M., Gomez-Moreno C, and Tollin G. (1993) J.Am.Chem.Soc. 115, 11698-11701
12. Hurley, J.K., Medina, M., Gómez-Moreno, C. and Tollin, G. (1994) Arch. Biochim. Biophys. 312, 480-486
13. Medina, M., Mendez. E. and Gómez-Moreno, C. (1992) FEBS Letters 298, 25-28
14. Medina, M., Mendez. E. and Gómez-Moreno, C. (1992) Arch. Biochem. Biophys. 299, 281-286
15. Fillat, M.F., Pacheco, M.C, Peleato, M.L. and Gómez-Moreno, C in Flavins and Flavoprtoeins (k.Yagi, ed.) (1994) 447-450, Walter de Gruyter, Berlin

MUTAGENESIS STUDIES OF *ANABAENA 7120* [2Fe-2S] FERREDOXINS: PHYSICAL AND CHEMICAL CHARACTERISTICS OF HYBRID VEGETATIVE/HETEROCYST SEQUENCES

Young Kee Chae and John L. Markley, Department of Biochemistry, University of Wisconsin-Madison, 420 Henry Mall, Madison WI 53706, USA

1. Introduction

The photosynthetic cyanobacterium *Anabaena 7120* produces two [2Fe-2S] ferredoxins whose structures have been investigated by x-ray crystallography[1-3] and NMR spectroscopy[4-10]. One is the vegetative ferredoxin (VFd), which is involved in photosynthesis, and the other is the heterocyst ferredoxin (HFd), which is involved in nitrogen fixation. Although the two ferredoxins share 51% sequence identity[11] and are presumed to have evolved from a common ancestral gene, they have distinctly different physical and chemical properties, such as the isoelectric points, EPR spectra, UV-visible spectra, and immunochemical properties,[11] and NMR spectral characteristics.[8-9] VFd, which is expressed constitutively, plays a role in photosynthesis by transferring an electron from photosystem II to ferredoxin-$NADP^+$-reductase (FNR).[12] HFd, which is expressed only under conditions of nitrogen fixation when the nitrogen source in the liquid medium is depleted, acts as the immediate electron donor to component II of nitrogenase (the Fe protein).[12,13] Functional studies have shown that HFd can transfer an electron both to FNR and to the Fe protein; however, VFd can donate an electron only to FNR.[13] The three-dimensional structures of these two ferredoxins[1-3] are very similar to one another, with overall RMSDs less than 1 Å; this suggests that subtle differences in the region close to the cluster may be responsible the different functional properties of the two ferredoxins. It was proposed that VFd reacts poorly with the Fe protein because it does not bind optimally for electron transfer.[13] Such an idea is supported by evidence that mutations of surface residues can influence the electron transfer rate.[14] NMR studies, on the other hand,[8] revealed that a major difference between the two proteins is in their dynamics, with VFd more rigid than HFd as revealed by hydrogen exchange rates and helix stabilities. The objective of this study was to determine whether one can interchange the spectral properties characteristic for the vegetative and heterocyst ferredoxin by swapping a minimal number of conserved residues from the sequence of one ferredoxin to the other. Four complementary pairs of such "hybrid mutants" of *Anabaena 7120* vegetative and heterocyst ferredoxins were prepared by site-directed mutagenesis, and their UV-visible and ^{15}N NMR spectral properties were compared. All four of the mutants of vegetative ferredoxin containing a few residues from the heterocyst ferredoxin sequence, termed "VFd → HFd hybrids", showed UV-visible $A_{422\ nm}/A_{276\ nm}$ ratios much closer to that of HFd than to that of VFd. However, none of the hybrid heterocyst mutants, termed "HFd → VFd hybrids", showed the expected inverse; instead, their UV-visible spectra were all similar to that HFd. Similar results were obtained from ^{15}N NMR spectra of this series of proteins. Whereas the ^{15}N NMR patterns exhibited by the VFd → HFd hybrids were similar to that of HFd, those of the HFd → VFd hybrids, although changed, did not

P. Mathis (ed.), Photosynthesis: from Light to Biosphere, Vol. II, 633–638.

resemble that of VFd. We conclude that, whereas there appear to be several ways of changing the sequence of VFd so as to achieve the spectral "phenotype" HFd, the reverse is not so easily possible. It will be interesting to see whether studies of the physiological activities of these mutants will show trends similar to their spectral properties. If so, this may shed some light on the origin of specificity and evolution of their action in photosynthesis and nitrogen fixation.

2. Procedures used

Comparison of residues close to the [2Fe-2S] cluster of VFd and HFd (Figure 1) reveal that contact residues in the first half of the primary sequence of VFd are more hydrophobic than those of HFd, whereas those in the second half of the sequence of VFd are more hydrophhilic than those of HFd. Hybrid mutants (Table 1) were designed with residue swaps between several residues (42, 43, 45, 48, 50) in the first half, one residue (42) in the first half, one residue (78) in the second, and residues from both halves.

	1	11	21	31			
VFd	ATFKV	TLINE	AEGTK	HEIEV	PDDEY	ILDAA	EEQGY
HFd	ASYQV	RLINK	KQDID	TTIEI	DEETT	ILDGA	EENGI
	36	46		56	66		
VFd	DLP**FS**	**CRAGA**	**CSTCA**	GKLVS	GTVDQ	SDQSF	LDDDQ
HFd	ELP**FS**	**CHSGS**	**CSSCV**	GKVVE	GEVKQ	SDQIF	LDDEQ
	71	81		91			
VFd	IEAGY	V***LTCV***	AYPTS	DVVIQ	THKEE	DLY	
HFd	MGKGF	A***LLCV***	TYPRS	NCTIK	THQEP	YLA	

Figure 1. Sequence comparisons of *Anabaena 7120* vegetative (VFd)[15] and heterocyst (HFd)[16] ferredoxins. Residues that contact the iron-sulfur cluster[1-3] are highlighted: those in the first half of the sequence in bold and those in the second half in bold italic.

Table 1. Nomenclature of reciprocal hybrid mutants.

VFd → HFd hybrids		HFd → VFd hybrids	
name of mutant	substitutions to VFd	name of mutant	substitutions to HFd
HV	R42H, A43S, A45S T48S, A50V	VH	H42R, S43A, S45A, S48T, V50A
T78L	T78L	L78T	L78T
CHV	R42H, A43S, A45S T48S, A50V, T78L	CVH	H42R, S43A, S45A, S48T, V50A, L78T
R42H	R42H	H42R	H42R

Site-directed mutagenesis was performed by using the single-site elimination method.[17] The structural genes *petF* (vegetative ferredoxin) and *fdxH* (heterocyst ferredoxin) were excised from their overexpression vectors, pET9a/F and pKID, respectively, by two restriction enzymes, XbaI and EcoRV. These two DNA fragments were inserted into pBluescript KS II (+) (Stratagene, La Jolla, CA), which had been cut by the same enzymes, and the resulting plasmids were named pART and pIRIS2 which had *petF* and *fdxH*, respectively. The plasmids were heat denatured, and then annealed with proper mutagenic primers and selection primer. After these steps, the second strand was synthesized, and the resulting DNA was transformed into BMH 71-16 *mutS*, which cannot repair mismatch mutations due to the insertion of Tn10 into the chromosome. The plasmids were purified, cut by the selection enzyme, and then transformed into DH5a. The correct mutants, as determined by DNA sequencing,[18] were excised from pART and pIRIS2 and inserted into the pET9a(-) vector (a slightly modified form of pET9a; Novagen, Madison, WI). Proteins were overexpressed in BL21(DE3) cells with the pLysS system.[19] Previously described methods were used to reconstitute[2] and purify[8] the proteins. UV-visible spectra were taken with a Hewlett-Packard 8452A diode array spectrophotometer at room temperature. The scanning range was from 200 nm to 700 nm. The sample concentrations were adjusted so that the maximum absorbances were less than 1.2 in order to avoid nonlinear effects. For comparison purposes, the intensities of the spectra were normalized by the absorption maximum at 276 nm. All NMR spectra were taken on Bruker DMX 600 spectrometer. Concentrations varied with each sample, but were higher than 2 mM. Temperature was set to 298 K. The spectral range was 20,000 Hz (300 ppm spectral width). The carrier frequency was set to 120 ppm. To suppress the diamagnetic resonances, the overall recycling time (relaxation delay + acquisition time) was adjusted to about 300 ms. The total number of scans per sample was 40,000. The raw spectra were processed on Silicon Graphics workstation by using FELIX program (Biosym, CA). The raw data acquired on the UXNMR system were left-shifted by 69 data points before processing. Linear prediction of the first point was followed by the application of a line broadening factor of 30 Hz. The resulting spectra were Fourier transformed, and phase corrected. The FACELIFT software package[20] was used to correct the baseline. Spectra were referenced by assuming a the frequency of 120 ppm for the ^{15}N resonance of [^{15}N]-acetyl glycine.

3. Results

Hybrid mutants HV, VH, T78L, L78T, CHV, and CVH (Table 1) resembled HFd in terms of their behavior on the ion-exchange and gel filtration columns. They eluted from the Q-Sepharose Fast Flow (Pharmacia, Piscataway, NJ) column (3 cm × 10 cm) when the salt gradient reached 0.3 – 0.4 M NaCl, whereas VFd eluted from the same column at a higher salt concentration (about 0.4 – 0.5 M). These hybrid mutants eluted the Sephacryl S-100 High Resolution (Pharmacia, Piscataway, NJ) column (3 cm × 100 cm) in about the same time as HFd and all earlier than VFd.

The $A_{422\ nm}/A_{276\ nm}$ ratio is 0.56 for VFd and 0.76 for HFd.[12] $A_{422\ nm}/A_{276\ nm}$ ratios determined for the VFd → HFd hybrid mutants with the exception of R42H (Table 2) were similar to that of HFd (all higher than 0.74); the single-site mutation T78L was sufficient to effect the change. By contrast, $A_{422\ nm}/A_{276\ nm}$ ratios for the HFd → VFd hybrid mutants did not show the reciprocal trend; their absorbance ratios were all higher than 0.69, which is unchanged from that of HFd within the experimental error associated with protein purity.

Table 2. Absorbance ratios ($A_{422\ nm}/A_{276\ nm}$) of hybrid mutants.

mutants	ratio[a]	change (%)[b]	mutants	ratio[a]	change (%)[b]
VFd	0.56[c]	-	HFd	0.76[c]	-
CHV	0.745	+ 33.0	CVH	0.694	- 8.7
HV	0.743	+ 32.7	VH	0.721	- 5.1
T78L	0.744	+ 32.9	L78T	0.706	- 7.1
R42H	0.592	+ 5.7	H42R	0.720	-5.3

[a] Highest value of all preparations.
[b] Calculated by using values of 0.56 and 0.76 for VFd and HFd, respectively.
[c] Reported by Böhme & Schrautemeier.[11]

The pattern of ^{15}N signals from backbone amide nitrogens that show interactions with unpaired electron density in the reduced ferredoxins was compared in VFd, HFd, and the hybrid mutants (Figure 2). The patterns of the VFd → HFd hybrid mutants (with the exception of R42H) are clearly different from that of VFd and resemble, but are not identical to that of HFd. Again, the single substitution T78L produced nearly as large a

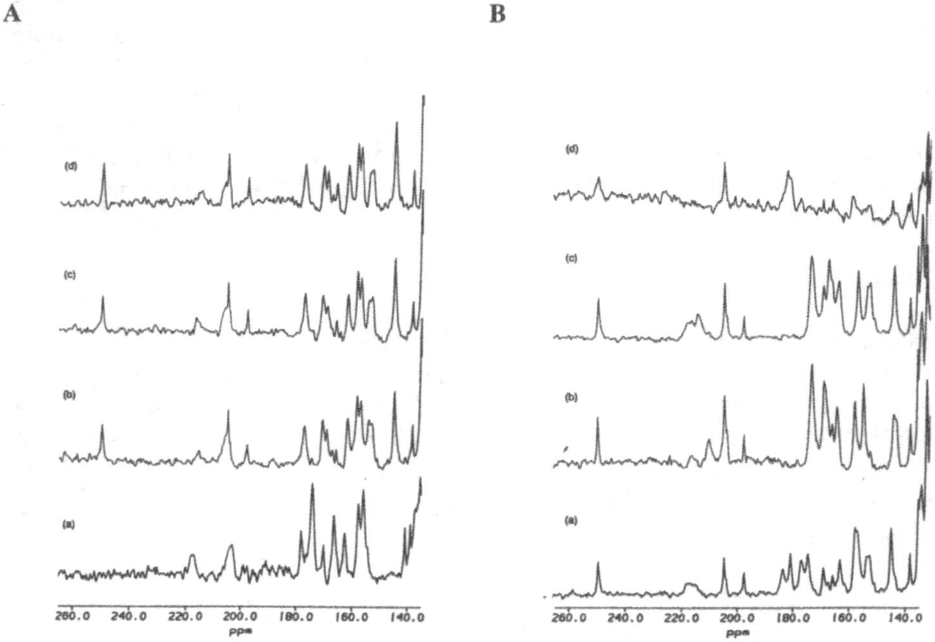

Figure 2. ^{15}N spectra of wild-type and hybrid *Anabaena 7120* ferredoxins. (A) From bottom, wild-type vegetative ferredoxin, CHV, HV, and T78L, (B) from bottom, wild-type heterocyst ferredoxin, CVH, VH, and L78T. See Figure 1 for sequences and Table 1 for definitions of the mutants.

change as the multiple-site mutations. Although the reciprocal HFd → VFd hybrid mutants showed hyperfine ^{15}N spectra that were somewhat different from the parent protein HFd, the changes did not make the spectra similar to that of VFd. The single-mutant, L78T, proved to be unstable and denatured during the NMR experiment leading to line broadening and degradation of signal intense.

4. Discussion

Because the x-ray structures of the two ferredoxins (VFd[1,3] and HFd[2]) are very similar to one other, their comparison has not led to definitive insights about the origins of their different physical properties and physiological roles. Spectroscopic comparisons, on the other hand, have revealed differences between the two ferredoxins that point to altered electronic properties of the region close to the cluster. The eight hybrid mutants studied here were designed to investigate the structural basis for these properties. The initial expectation was that the reciprocal mutations would lead to reciprocal changes in the spectral properties. The intriguing result, however, is that the changes appear to go only one way. The change from the VFd phenotype to the HFd phenotype can be effected by swapping a single amino acid residue *or* by swapping five other residues in a very different part of the sequence. By contrast, the reciprocal mutations in the HFd background, or even both sets together, were not sufficient to convert the spectral properties to those of VFd. The results appear to indicate that the structure of VFd, which has been shown by NMR studies to be more rigid than that of HFd, has very particular sequence requirements. Presumably other residue differences between the two proteins (Figure 1) are critical to establishing its structure. Perhaps the most interesting mutant is T78L, the single-site mutation that changes the spectral properties from VFd-like to HFd-like. Hydrophobicity analysis[21] of the residues close to the cluster showed differences between HFd (partial hydrophobicity surrounding the cluster) and VFd (more uniform hydrophobicity surrounding the cluster). This coupled with the striking spectral changes caused by T78L, a mutation that alters the hydrophobicity near the cluster, suggests that hydrophobicity around the cluster is closely related to electronic properties such as charge transfer from sulfur to iron. The change in hydrophobicity manifests itself in changes in the UV-visible spectrum, which can be thought of as a delocalized property, and also in changes in ^{15}N signals which are much more localized. The results suggest that VFd and HFd may use different electron-guiding paths for physiological electron transport. The antiferromagnetic coupling between the two iron atoms can be enhanced or suppressed by the characteristics of the surrounding residues. Whereas the T78L mutant serves to increase hydrophobicity in the region of one iron atom of the cluster (Fe_2), the HV mutations decrease the hydrophobicity around the other iron atom of the cluster (Fe_1).
The L78T mutant was found to be much less stable than other hybrid mutants. The stability of CVH, however, appeared to be close to that of wild type; the destabilization by L78T thus appears to be counteracted by the other mutations. The only symmetric changes in properties exhibited by this set of hybrid mutants is for R42H and H42R, which previously have been shown to lead reciprocal changes in stability: R42H showed lower stability than wild-type vegetative ferredoxin, and H42R showed higher stability than wild-

type heterocyst ferredoxin (Hurley et al., 1993). Furthermore, the amount of stabilization/destabilization is almost the same, 2 kcal/mole. The optical spectra of these two mutants resemble those of the respective wild-type ferredoxins (Table 2).

Acknowledgment. We thank Dr. Hong Cheng for kindly providing data on the R42H mutant. This research was supported by the Photosynthesis/Respiration Division, Competitive Grants Program, United States Department of Agriculture

References

1 Rypniewski, W.R., Breiter, D.R., Benning, M.M., Wesenberg, G., Oh, B.-H., Markley, J.L., Rayment, I. and Holden, H.M. (1991) *Biochemistry 30*, 4126-4131
2 Jacobson, B.L., Chae, Y.K., Markley, J.L., Rayment, I. and Holden, H.M. (1993) *Biochemistry 32*, 6788-6793
3 Holden, H.M., Jacobson, B.L., Hurley, J.K., Tollin, G., Oh, B.-H., Skjeldal, L., Chae, Y.K., Cheng, H., Xia, B. and Markley, J.L. (1994) *J. Bioenerg. Biomemb. 26(1)*, 67-88
4 Oh, B.-H. and Markley, J.L. (1990) *Biochemistry 29*, 3993-4004
5 Oh, B.-H., Mooberry, E.S. and Markley, J.L. (1990) *Biochemistry 29*, 4004-4011
6 Oh, B.-H., Westler, W.M., Darba, P. and Markley, J.L. (1988) *Science 240*, 908-911
7 Oh, B.-H., Westler, W.M. and Markley, J.L. (1989) *J. Am. Chem. Soc. 111*, 3083-3085
8 Chae, Y.K., Abildgaard, F., Mooberry, E.S. and Markley, J.L. (1994) *Biochemistry 33*, 3287-3295
9 Chae, Y.K. and Markley, J.L. (1995) *Biochemistry 34*, 188-193
10 Chae, Y.K., Xia, B., Cheng, H., Oh, B.-H., Skjeldal, L., Westler, W.M. and Markley, J.L. (1995) in Nuclear Magnetic Resonance of Paramagnetic Macromolecules (G.N. La Mar, ed.) pp. 297-317, Kluwer Academic Publishers, Dordrecht, The Netherlands
11 Böhme, H. and Schrautemeier, B. (1987) *Biochim. Biophys. Acta 891*, 1-7
12. Böhme, H. and Schrautemeier, B. (1987) *Biochim. Biophys. Acta 891*, 115-120
13 Schrautemeier, B. and Böhme, H. (1985) *FEBS Lett. 184(2)*, 304-308
14 Hurley, J.K., Salamon, Z., Meyer, T.E., Fitch, J.C., Cusanovich, M.A., Markley, J.L., Cheng, H., Xia, B., Chae, Y.K., Medina, M., Gomez-Moreno, C. and Tollin, G. (1993) *Biochemistry 32*, 9346-9354
15 Alam, J., Whitaker, R.A., Krogmann, D.W. and Curtis, S.E. (1986) *J. Bacteriol. 168*, 1265-1271
16 Böhme, H. and Hazelkorn, R. (1988) *Mol. Gen. Genet. 214*, 278-285
17 Deng, W.P. and Nickoloff, J.A. (1992) *Anal. Biochem. 200*, 81-88
18 Sanger, F., Niklen, S. and Coulson, A.R. (1977) *Proc. Natl. Acad. Sci. USA 74*, 5463-5467
19 Studier, F.M., Rosenberg, A.H., Dunn, J.J. and Dubendorff, J.W. (1990) *Methods. Enzymol. 185*, 60-89.
20 Chylla, R. A. and Markley, J.L. (1994) *J. Magn. Reson., Ser. B, 102,* 148-154
21 Kyte, J. and Doolittle, R.F. (1982) *J. Mol. Biol. 157*, 105-132

STRUCTURAL AND FUNCTIONAL ANALYSES OF SPINACH FERREDOXIN:THIOREDOXIN REDUCTASE

P.Schürmann[1], L.Gardet-Salvi[1], A.-L.Stritt-Etter[1], J.Li[1], L.-P.Chow[2], H.Iwadate[2], K.Yano[2], M.Kamo[2], A.Tsugita[2], M.Hirasawa[3], D.B.Knaff[3], Z.Salamon[4], G.Tollin[4], C.R.Staples[5], E.Ameyibor[5], W.Fu[5], M.K.Johnson[5]
[1]Laboratoire de Biochimie végétale, Université de Neuchâtel, CH-2007 Neuchâtel, Switzerland, [2]Research Institute for Biosciences, Science University of Tokyo, Japan, [3]Department of Chemistry and Biochemistry, Texas Tech University, Lubbock, TX 79409-1061, [4]Department of Biochemistry, University of Arizona, Tucson, AZ 85721 and [5]Department of Chemistry and Center for Metalloenzyme Studies, University of Georgia, Athens Georgia 30602, USA

1. Introduction

Ferredoxin:thioredoxin reductase (FTR) is one of the ferredoxin-dependent enzymes present in the chloroplast stroma (1) where it plays an important role in the ferredoxin/thioredoxin system of enzyme regulation. This light-dependent system regulates the activities of certain key chloroplast enzymes. Photoreduced ferredoxin reduces the FTR, which in turn reduces thioredoxins, theirby transforming an electron signal into a thiol signal. The reduced thioredoxins further interact with certain target enzymes which, when reduced, become activated or deactivated. Enzymes involved in carbon assimilation are activated by reduction whereas an enzyme of carbohydrate breakdown in the stroma becomes deactivated by reduction (2).

The FTR is a nucleus encoded chloroplast protein with an apparent molecular mass of approximately 30 kDa. It is composed of two dissimilar subunits one of which contains the active site, a redox-active disulfide bridge. The native FTR has a yellowish-brown color due to the presence of a [4Fe-4S] cluster (3) whose function is not yet established. FTR has been isolated and characterized from different organisms like spinach, corn, *Nostoc* and *Chlamydomonas* (3,4). The comparison of the molecular properties of FTR isolated from different organisms suggests, that the subunit containing the active site disulfide (subunit B, catalytic subunit) is rather well conserved, whereas the other subunit with no known catalytic function is variable in size (subunit A, variable subunit). In spinach the variable subunit appears larger (16 - 17 kDa) than the catalytic subunit (13 - 14 kDa) (3,5) as estimated by SDS-PAGE. A similar situation is found in corn, whereas in *Nostoc* and *Chlamydomonas* the variable subunit appears smaller than the catalytic subunit. The catalytic subunits from different organisms must have similar antigenic sites since they are recognized by heterologous antibodies in contrast to the variable subunits which are not.

In a series of elegant experiments it has been shown that the light signal is transmitted as a thiol-signal from FTR through thioredoxins to the target enzymes (6,7). These experiments have revealed that the redox active disulfide/dithiol is located in subunit B of FTR, from

P. Mathis (ed.), Photosynthesis: from Light to Biosphere, Vol. II, 639–644.

thereon called catalytic subunit. Quantitation of the non-heme iron and inorganic sulfide content as well as the visible absorption spectrum have suggested that FTR contains a [4Fe-4S] cluster. However, there is as yet no clear evidence for an involvement of this cluster in catalysis. Previous studies have indicated that the cluster is EPR silent in the protein as isolated or after reduction. When oxidized with ferricyanide the protein exhibited EPR signals indicative of an oxidized HIPIP-type [4Fe-4S] cluster. Based on oxidation-reduction titrations a midpoint potential of +410 mV was reported for the cluster, a value much too positive to make this redox process physiologically relevant for the transfer of electrons from ferredoxin to the active site disulfide (8).
FTR interacts very specifically with its natural electron donor, ferredoxin, with which it forms a tight 1:1 complex at low ionic strength (9,10). Negatively charged residues involved in this interaction have been mapped on the ferredoxin surface (10). However, the corresponding positively charged conterparts of the FTR are not yet known.
Recently the amino acid sequences of both subunits of the spinach FTR have been determined by conventional protein sequencing methods (11,12). In addition cDNA clones coding for the spinach FTR (13,14) and for the catalytic subunit of the corn FTR (14) as well as for the variable subunit of the *Anacystis nidulans* (15) have been isolated and characterized. In the present paper we present the primary structures of the spinach FTR and the functional roles of the different cysteine residues. In addition we report on some results indicating an involvement of the Fe-S cluster in electron transfer from ferredoxin to the active site disulfide.

2. Material and Methods

FTR has been isolated as described elsewhere (16). The detailed procedures for the amino acid sequence analyses of subunit A (11) and subunit B (12) as well as for the chemical modifications have been published. Details on the redox potential determinations have been reported (17) and the biophysical methods used for the functional analysis of the Fe-S cluster will be described elsewhere.

3. Results and Discussion

The two subunits of FTR have been separated by SDS-PAGE and the individual peptides sequenced by conventional protein sequencing techniques. In Figure 1 we show the amino acid sequences of the two subunits of the spinach enzyme. It can be seen that both subunits contain practically the same number of residues, 112 for subunit A and 113 for subunit B. Also their caluculated molecular masses, 12669 Da for subunit A and 12959 for subunit B are virtually identical. Whereas the calculated Mr for subunit B corresponds very well with estimates obtained by SDS-PAGE, the calculated Mr for subunit A is quite lower than expected from its electrophoretic migration which predicts a peptide of 16 kDa. We speculate that this difference is due to the rather particular N-terminal part of the subunit which is redundant and has clustered Ser, Pro and Glu residues as we have shown earlier (5). Based on the amino acid sequence information we have constructed degenerate oligonucleotides that were used to retrieve cDNA clones coding for the two subunits from spinach and for the catalytic subunit from corn (14,18). The protein sequences obtained by translation of the isolated cDNA clones correspond exactly to the primary structures determined for the isolated protein. In addition they show for each subunit a typical N-terminal transit peptide necessary for the translocation of this nucleus encoded protein into the chloroplast stroma.

Subunit A 1 EVALKSDSSTGFDSSSSSPPEEDEELKKNLEKVG**C**KVKVKSPLKVYHVPK 50
51 LPEVELTPDMVGVIKQYVGFWKGKYISPNYPFKVEYRIDVPDRGSVKLVV 100
101 HLKEEEFEIIAE 112

Subunit B 1 ADPSDKSMEVMRKFSEQF**C**RKSDTYF**C**VDKSVTAVVIKGLADHRDTLGAP 50
51 L**C**P**C**RHYDDKEAEAKQGFWN**C**P**C**VPMRERKE**C**H**C**MLFLTPDNDFAGKEQT 100
101 TITLDEIREVTSNM 114

Figure 1. Amino acid sequences of both subunits of spinach FTR.

Thiol reagents are efficient inhibitors of reduced FTR, whereas they are ineffective on oxidized FTR. Covalent modification of active site cysteines with N-ethylmaleimide (NEM) not only totally blocks the activity of FTR it also changes the absorption spectrum (19). The visible absorption of the protein increases and a new peak at 330 nm appears (Fig. 2). These observations suggest that the active site disulfide of the FTR is spatially in close contact with the Fe-S cluster. The nature and the coordination of this cluster have been probed by Resonance Raman spectroscopy. These studies provide unambiguous evidence for a $[4Fe-4S]^{2+}$ cluster that is coordinated exclusively by cysteinyl residues. It is therefore of special interest to identify the cysteine residues constituting the active site disulfide/dithiol and those liganding the [4Fe-4S] cluster.

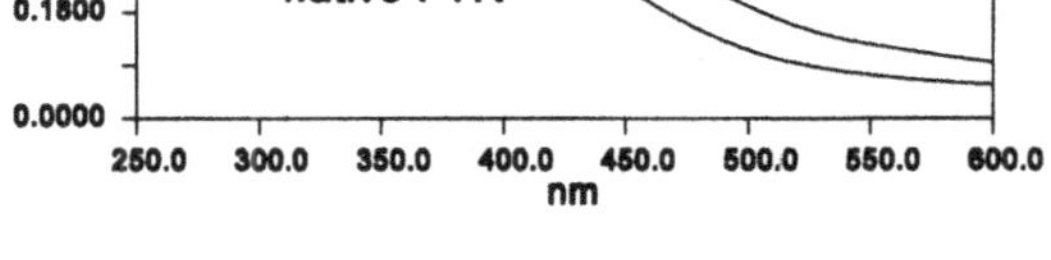

Figure 2. Absorption spectra of native and NEM-modified spinach FTR (19)

FTR from spinach contains nine cysteine residues, one in subunit A, the variable subunit, and eight in subunit B, the catalytic subunit. In order to find out which function each individual cysteine has we have elaborated conditions that allowed preferential modification of the different cysteines with radioactively labelled NEM or iodoacetate. The results of these experiments are summarized in Table 1.

Table 1. Functional roles of the individual cysteine residues

	Variable subunit
Cys35	accessible free thiol, no catalytic function
	Catalytic subunit
Cys19	Inaccessible, free thiol, no catalytic function
Cys27	accessible free thiol, no catalytic function
Cys52	liganding [4Fe-4S] cluster
Cys54	accessible partner of the active site disulfide/dithiol
Cys71	liganding [4Fe-4S] cluster
Cys73	liganding [4Fe-4S] cluster
Cys82	liganding [4Fe-4S] cluster
Cys84	inaccessible partner of the active site disulfide/dithiol

When native, oxidized FTR is treated with [^{14}C]NEM, the label is evenly distributed between the two subunits, indicating the presence of one accessible thiol per subunit. Since the variable subunit contains but one cysteine (Fig. 1) this must be the modified residue. It also allows the conclusion that the variable subunit is not involved in catalysis nor in Fe-S cluster ligation. The accessible thiol on subunit B is Cys27, a residue found also at the corresponding position in the corn enzyme. Under denaturing conditions a third thiol, Cys19, is detected. This residue seems to be inaccessible and is not conserved in the corn enzyme (14).

Modification of the reduced active site disulfide/dithiol seems to be somewhat variable. Whereas the inhibition is always complete, the number of modified cysteines varies between 1.2 and 1.8 thiols. The localisation of the modified residues has now shown, that only one of the two active site cysteines, Cys54, is accessible under non-denaturing conditions. The second cysteine, Cys84, labelled after reduction and under denaturing conditions, seems to be protected from the solvent and inaccessible to the reagent under native conditions. The labelling results suggest that in FTR we obtain only monoalkylation of the active site, a situation reported for most of the disulfide reductases. The variable numbers of modified residues we observed may be due to incomplete saturation of free thiols and/or incorporation of radioactivity into thiols made accessible due to the modification conditions.

The four remaining cysteines, Cys52, Cys71, Cys73 and Cys82 are liganding the [4Fe-4S] cluster. These results confirm the close proximity of active site disulfide and [4Fe-4S] cluster previously suspected based on the spectral changes upon NEM modification (19). The observed ligation and the proximity of the disulfide bridge to the cluster may confer particular functional properties to the Fe-S cluster.

Using cyclic voltammetry, a technique which has been demonstrated to be effective for the oxidation-reduction measurements of thiol groups (20), two midpoint potential values have been obtained for FTR, a negative value centered at -230 mV and a positive value at +340 mV. The measured negative potential at -230 mV can be attributed to the disulfide/dithiol redox couple of the active site, since its covalent modification with NEM abolishes the low potential signal. However, the high-potential signal remains present after alkylation of the active site. This observation and the similarity of the Em value for the electropositive component detected in spinach FTR to that reported previously for the [4Fe-4S] cluster in Nostoc FTR (8), make it seem highly likely that the Em = +340 mV component is the [4Fe-4S] cluster of FTR. The fact that this Em value is so much more positive than those of the dithiol/disulfide couples of FTR and thioredoxin makes it seem highly unlikely that the $[4Fe\text{-}4S]^{2+,3+}$ redox transition is involved in electron transfer from ferredoxin to the active site disulfide of FTR (17). The midpoint potential of the active site disulfide/dithiol of FTR, which is substantially more positive than the midpoint potential of ferredoxin (-420 mV) and slightly more negative than those of the two chloroplast thioredoxins (-210 mV) indicate that a transfer of electrons from ferredoxin to the thioredoxins is thermodynamically favorable.

In order to get further insights into a possible involvement of the [4Fe-4S] cluster of FTR in the transfer of electrons from ferredoxin to the active site disulfide, we tested artificial electron donors for their capacity to reduce FTR. Na-dithionite by itself is incapable of reducing the active site disulfide. However, if catalytic amounts of ferredoxin are present, FTR is efficiently reduced. Ferredoxin can be efficiently replaced by viologens, either methyl- or benzyl-viologen, both of which have midpoint potentials below -300 mV. Mediators with more positive redox potentials, like anthraquinone 2-sulphonate with a redox potential in the order of that of the active site disulfide/dithiol, are ineffective even at 500 fold excess over FTR (21). Although these results provide no direct experimental evidence it seems unlikely that the viologens reduce the disulfide bridge directly. They rather suggest that a group with a more negative potential than the disulfide bridge is the primary electron acceptor in the FTR. This group could be the Fe-S cluster.

The properties of the Fe-S cluster were further investigated using the combination of resonance Raman, variable temperature magnetic circular dichroism (VTMCD) and EPR spectroscopies. The cluster of the native protein can not be reduced by dithionite or by deazaflavin-mediated photochemical reduction, but it can be oxidized to the $[4Fe\text{-}4S]^{3+}$ oxidation state by ferricyanide. After monoalkylation of one of the active site cysteines with NEM the cluster exhibits different properties. It exhibits a slow relaxing EPR signal, g = 2.11, 1.99, 1.98, accounting for 1 spin/molecule, that is visible without significant broadening at temperatures up to 150 K and exhibits very weak temperature-dependent VTMCD bands in the UV-visible region. Resonance Raman studies indicate that NEM modification has a direct affect on the Fe-S vibrational modes of the [4Fe-4S] cluster. This species is tentatively attributed to a cysteinyl disulfide radical species that is formed by oxidative attachement of one of the active site cysteines to one of the bridging sulfurs of the [4Fe-4S] cluster. The present results can be summarized in a mechanistic scheme in which the [4Fe-4S] cluster replaces the role of the flavin present in other disulfide reductases and participates directly in the cleavage of the active site disulfide.

References

1 Knaff, D.B. (1995) in Oxygenic Photosynthesis: The Light Reactions (Yocum C.F. & Ort D.R. eds.), Kluwer Academic Publishers, Dordrecht, The Netherlands
2 Buchanan, B.B., Schürmann, P., Decottignies, P. and Lozano, R.M. (1994) Arch. Biochem. Biophys. 314, 257-260
3 Droux, M., Jacquot, J.-P., Miginiac-Maslow, M., Gadal, P., Huet, J.C., Crawford, N.A., Yee, B.C. and Buchanan, B.B. (1987) Arch. Biochem. Biophys. 252, 426-439
4 Huppe, H.C., Lamotte-Guéry, F.d., Jacquot, J.-P. and Buchanan, B.B. (1990) Planta 180, 341-351
5 Tsugita, A., Yano, K., Gardet-Salvi, L. and Schürmann, P. (1991) Protein Seq. Data Anal. 4, 9-13
6 Droux, M., Miginiac-Maslow, M., Jacquot, J.-P., Gadal, P., Crawford, N.A., Kosower, N.S. and Buchanan, B.B. (1987) Arch. Biochem. Biophys. 256, 372-380
7 Crawford, N.A., Droux, M., Kosower, N.S. and Buchanan, B.B. (1989) Arch. Biochem. Biophys. 271, 223-239
8 De La Torre, A., Lara, C., Yee, B.C., Malkin, R. and Buchanan, B.B. (1982) Arch. Biochem. Biophys. 213, 545-550
9 Hirasawa, M., Droux, M., Gray, K.A., Boyer, J.M., Davis, D.J., Buchanan, B.B. and Knaff, D.B. (1988) Biochim. Biophys. Acta 935, 1-8
10 De Pascalis, A.R., Schürmann, P. and Bosshard, H.R. (1994) FEBS Lett. 337, 217-220
11 Iwadate, H., Yano, K., Kamo, M., Gardet-Salvi, L., Schürmann, P. and Tsugita, A. (1994) Eur. J. Biochem. 223, 465-471
12 Chow, L.-P., Iwadate, H., Yano, K., Kamo, M., Tsugita, A., Gardet-Salvi, L., Stritt-Etter, A.-L. and Schürmann, P. (1995) Eur. J. Biochem. 231, 149-156
13 Falkenstein, E., von Schaewen, A. and Scheibe, R. (1994) Biochim. Biophys. Acta Bio-Energetics 1185, 252-254
14 Marc-Martin, S., Spielmann, A., Stutz, E. and Schürmann, P. (1993) Biochim. Biophys. Acta Bio-Energetics 1183, 207-209
15 Szekeres, M., Droux, M. and Buchanan, B.B. (1991) J. Bacteriol. 173, 1821-1823
16 Schürmann, P. (1995) in Methods in Enzymology, Biothiols (Packer L. ed.), pp.in press Academic Press, Orlando, Florida 32887
17 Salamon, Z., Tollin, G., Hirasawa, M., Knaff, D.B. and Schürmann, P. (1995) Biochim. Biophys. Acta Bio-Energetics 1230, 114-118
18 Gaymard, E. and Schürmann, P. (1995) These Proceedings
19 Schürmann, P. and Gardet-Salvi, L. (1993) Chimia 47, 245-246
20 Salamon, Z., Gleason, F.K. and Tollin, G. (1992) Arch. Biochem. Biophys. 299, 193-198
21 Schürmann, P., Stritt-Etter, A.-L. and Li, J. (1995) Photosynth. Res. (In Press)

DESIGN OF SYNTHETIC IRON-SULFUR PROTEINS

Brian R. Gibney, Stephen E. Mulholland, Francesc Rabanal
and P. Leslie Dutton*, Johnson Research Foundation,
Department of Biochemistry and Biophysics,
University of Pennsylvania, Philadelphia, PA 19104 U.S.A.

1. Introduction

Iron-Sulfur clusters have long been recognized to be components of the electron transfer systems of photosynthetic and respiratory energy conversion (1). More recent investigations find them increasingly in enzymes with functions as diverse as intermediary metabolism, nitrogen fixation and gene regulation. The proteins that accommodate the iron-sulfur clusters are complex and the basis for their structural properties as well as for their assembly and control of their catalytic properties are at present uncertain (2). This uncertainty is augmented by the inevitability that any protein structure must accommodate a multiplicity of biological requirements other than that of catalysis. Our approach to the study of complex natural redox enzymes is to design and synthesize minimal structures that test the assembly of the component peptides with the incorporation of a particular chosen cofactor. A further goal is to generate in protein structures simplified working versions of complex native enzymes, molecular maquettes with the view to provide systems which can be exhaustively studied without interference from protein domains which may be present for unrelated reasons (3).

The architectural basis for maquette construction is a well defined scaffold, here the four alpha helix bundle. The *de novo* design concepts for construction of stable four helix bundles are becoming well understood (4-8). The helices of our previously reported prototypes for a heme protein maquette were proven to be stable with respect to denaturants and incorporation of hemes increases their stability (2). In addition, the four helix bundle has been demonstrated to induce the assembly of a model for the (bacterio)chlorophyll dimer special pair of photosynthetic reaction centers (9,10). The designed alpha helix structure also provides for incorporation of loop regions between helices which offer a site for incorporation of cofactors not generally located in helical proteins, iron-sulfur clusters are one example.

This report focuses on the feasibility of synthesizing iron-sulfur cluster proteins. Apart from the iron-sulfur clusters in the respiratory complexes, complex I (several) and complex III (Rieske 2Fe2S cluster), the PS I reaction center(11,12) has three iron-sulfur clusters, designated F_X, F_A and F_B, which provide part of an efficient electron transfer pathway between P700* and ferredoxin. The complexity of PS I typifies the difficulty in

P. Mathis (ed.), Photosynthesis: from Light to Biosphere, Vol. II, 645–648.

approaching the physical and chemical fundamentals that govern the properties of the individual redox cofactors.

2. Procedures

2.1 *Peptide Design*

Our desire to incorporate an iron-sulfur cluster of distorted cubane geometry (13,14) using cysteine thiolate ligation into our prototype four helix bundle scaffold required several alterations in the basic design. First, the use of disulfides to covalently link helices had to be avoided due to the need to utilize cysteine thiolate ligation to the iron-sulfur site. Second, the lack of observed iron-sulfur clusters within helical domains suggested that the helix scaffold might not provide the optimal location for this multinuclear cluster. Given the above engineering constraints, a 67 amino acid helix-loop-helix peptide was designed (Figure 1) which retains the helical regions of the prototype but replaces the N-terminal disulfide linkage region with a loop region extracted from the consensus sequence of natural ferredoxins. In particular, we chose the binding domain from *P. aerogenes* which offered a crystallographically characterized loop which simplified modeling during the design stage (15). Three of the four cysteine thiolate ligands are derived from the -Cys-$(X)_2$-Cys-$(X)_2$-Cys- motif from *P. aerogenes*, with the fourth remote cysteine placed at the top of one of the helical regions. We retained the histidines at analogous positions (8, 22, 46 and 60) which were hoped to ligate hemes.

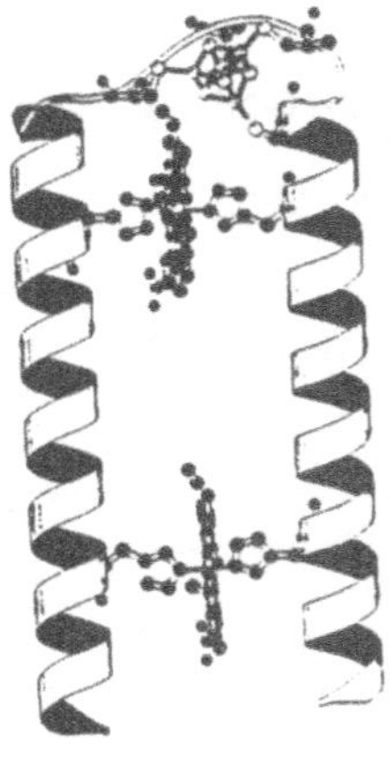

Figure 1: Model of Iron-Sulfur/ Heme Peptide

2.2 *Synthetic Methods*

The peptide was synthesized on a Milligen 9050 synthesizer using standard Fmoc/^{t}Bu solid phase synthetic methodologies and purified to homogeneity via reversed phase HPLC. The identity of the resultant peptide was confirmed by laser desorption mass spectrometry. The helical content (›60%) of the peptide was determined using CD spectrometry on an Aviv 62DS spectrometer. Peptide concentrations were determined spectrophotometrically using the Trp absorbance at 280nm (ε=5700 $M^{-1}cm^{-1}$). Iron-sulfur cluster assembly into purified apo-peptide were attempted using modified biochemical protocols under strictly anaerobic conditions at ambient temperatures (2). Hemes were incorporated into peptide both in the presence and absences of bound iron-sulfur clusters using standard literature methods under anaerobic conditions (3).

3. Results and Discussion

3.1 *Heme Cofactor Incorporation*

In the presence of bound iron-sulfur clusters, a heme titration typical of bis-His ligation to the hemes was observed as identified by the Soret absorbance at 413 nm, as is shown for a typical sample in Figure 2. After addition of two equivalents of hemin, dithionite reduction resulted in alpha and beta bands indicative of bis-His ligated Fe(II)protoporphyrin IX, λ_{max} at 529 and 560nm.

In the absence of iron-sulfur clusters, addition of less than one equivalent of hemin resulted in an absorption spectrum inconsistent with bis-His ligation to the heme.

Interestingly, the initial spectra (λ_{max} at 358, 418, 539 and 565 nm), representative spectra shown in Figure 2, were reminiscent of the UV-vis spectrum of ferric cytochrome P-450 (λ_{max} at 356, 417, 536 and 569 nm) which possess axial thiolate ligation to the iron (16). Additions of hemin, beyond one equivalent, result in UV-vis spectra typical of bis-His ligated heme. These data confirm the incorporation of the iron-sulfur cluster. Additionally, they suggest that heme-thiolate ligation may be possible within the alpha helical framework.

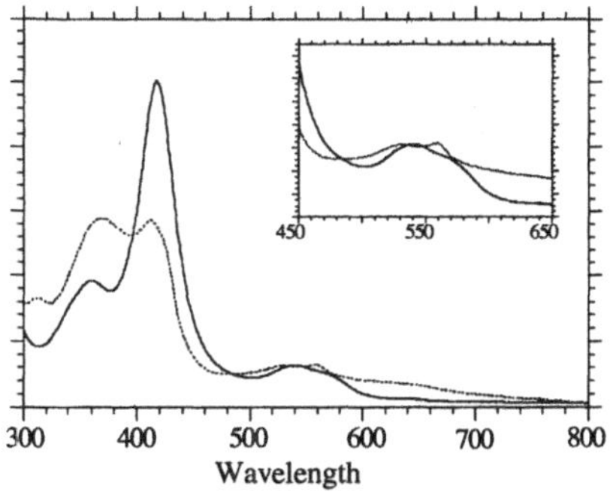

Figure 2:UV-vis of Bound Heme in [4Fe4S] containing peptide (dotted) and apopeptide (solid).

3.2 *EPR Spectroscopy*

Direct evidence for the successful assembly of an iron-sulfur cluster was obtained using EPR spectroscopy of the peptides. The initial iron-sulfur incorporation yielded an EPR silent species in the oxidized form (top spectrum, Figure 3). Reduction of this sample with dithionite results in observation of a rhombic spectrum typical of a paramagnetic (S=1/2) $[4Fe4S]^{1+}$ center and characterized by resonances at g= 2.05, 1.93 and 1.89 (middle two spectra, Figure 3). This EPR signature is lost upon warming of the sample to 40K (bottom spectrum, Figure 3), consistent with an exchange coupled multinuclear site. Variable temperature studies indicate that the signal has maximal intensity at 12K. The spectral features compare favorably with those of F_A isolated from the other components of PsaC (g-values of 2.04, 1.94 and 1.85 with maximal signal at 15K) (2,17,18).

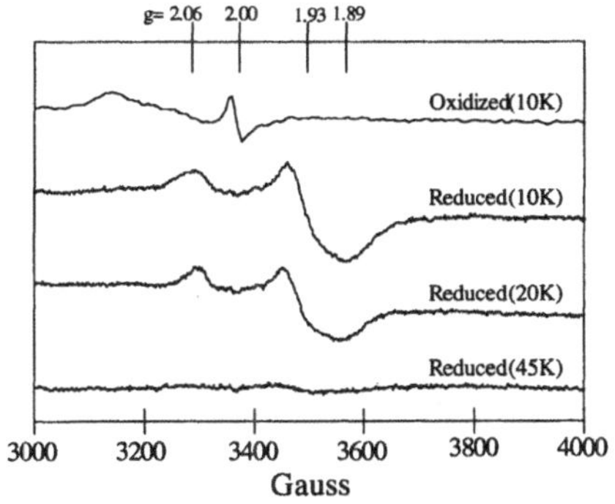

Figure 3:EPR spectra of [4Fe4S] Peptide.

Some biochemical reconstitution attempts resulted in data consistent with incorporation of a $[3Fe4S]^{1+}$ site into the peptide. In the oxidized state, this sample yielded an EPR spectrum with distinctly different characteristics than previously observed. The less resolved rhombic spectrum had rather high g-values (g= 2.15, 2.07 and 2.01), a maximal signal intensity at 20K. Reduction of this sample with dithionite yielded an apparently EPR silent species, consistent with the assigment to a $[3Fe4S]^0$ cluster.

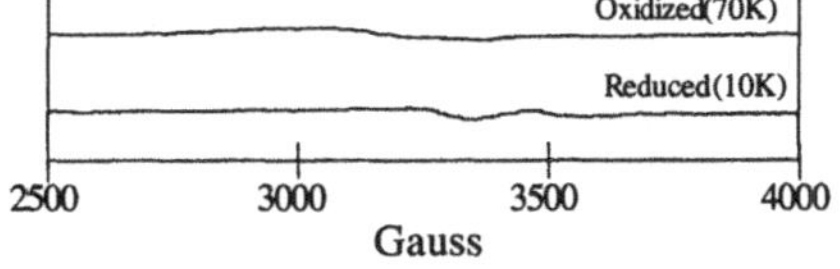

Figure 4:EPR spectra of [3Fe4S] Peptide.

4. Conclusions

Extracting minimalistic motifs from natural proteins, we have designed and successfully synthesized a mulitcentered redox protein containing an iron-sulfur cluster and two hemes. The peptide provides the iron-sulfur binding site which can accommodate a [4Fe4S] and, under ill-defined conditions, perhaps a [3Fe4S] center using biochemical reconstitution methods. We have not been able to ascertain the conditions which resulted in the alteration of the preferred iron-sulfur cluster nuclearity but factors implicated in natural systems include oxidative damage and low sulfur concentrations during reconstitution (19). With the successful introduction of hemes, photoactive porphyrin dimers and iron-sulfur clusters into synthetic four helix bundles, the production of functional molecular maquettes of oxidoreductase and/or photosynthetic proteins lies in the realm of possibility.

Acknowledgments
This work was supported by NIH grants GM 41048 and GM 48130. B.R.G. and F.R. wish to acknowledge reciept of postdoctoral fellowships from the National Institutes of Health and the European Molecular Biology Organization (EMBO), respectively. We thank Kim Sharp for aid in preparing Figure 1.

References

1 Cammack, R. (1992) in Advances in Inorganic Chemistry (Sykes, G., ed.), pp. 281-322, Academic Press, San Diego, CA.
2 Yu, L.; Zhao, J.; Lu, W.; Bryant, D.A.; Golbeck, J.H. (1993) Biochemistry 32, 8251-8258.
3 Robertson, D. E.; Farid, R. S.; Moser, C. C.; Urbauer, J. L.; Mulholland, S. E.; Pidikiti, R.; Lear, J. D.; Wand, A. J.; DeGrado, W. F.; Dutton, P. L. (1994) Nature 368, 425-32.
4 DeGrado, W.F.; Wasserman, Z.R.; Lear, J.D. (1989) Science 243, 622-628.
5 Regan, L.; DeGrado, W.F. (1988) Science 241, 976-978.
6 Ho, S.W.; DeGrado, W.F. (1987) J. Am. Chem. Soc. 109, 6751-6758.
7 Raliegh, D.P.; DeGrado, W.F. (1992) J. Am. Chem. Soc. 114, 10079-10081
8 Choma, C.T.; Lear, J.D.; Nelson, M.J.; Dutton, P. L.; Robertson, D. E.; DeGrado, W.F. (1994) J. Am. Chem. Soc. 116, 856-865.
9 Rabanal, F.; DeGrado, W. F.; Dutton. P. L., (1995) J. Am. Chem. Soc., submitted.
10 Rabanal, F.; Gibney, B. R.; DeGrado, W. F.; Moser, C.C.; Dutton, P.L. (1995) Inorg. Chim. Acta., submitted.
11 Fromme, P.; Schubert, W.D.; Krauss, N. (1994) Biochim. Biophys. Acta 894, 99-105.
12 Krauss, N; Hinrichs, W.; Witt, I.; Fromme, P.; Pritzkow, W.; Dauter, Z.; Betzel, C.; Wilson, K.S.; Witt, H.T.; Saenger, W. (1993) Nature 361, 326-331.
13 Holm, R.H.; Ciurli, S.; Weigel, J.A. (1990) Prog. Inorg. Chem. 38, 1.
14 Carnet, M.J.; Papaefthymiou, G.C.; Spartalian, K.; Frankel, R.B.; Holm, R.H. (1988) J. Am. Chem. Soc. 110, 6084.
15 (a) Adman, E.T.; Sieker, L.C.; Jensen, L.H.(1973) J. Biol. Chem. 248, 3987. (b) Adman, E.T.; Sieker, L.C.; Jensen, L.H.(1976) J. Biol. Chem. 251, 3806.
16 Dawson, J.H.; Sono, M. (1987) Chem. Rev. 87, 1255-1276.
17 Zhao, J.; Li, N.; Warren, P.V.; Golbeck, J.H.; Bryant, D.A. (1992) Biochemistry 31, 5093-5099.
18 Yu, L.; Bryant, D.A.; Golbeck, J.H. (1995) Biochemistry 34, 7861-7868.
19 Johnson, M.K.; Spiro, T. G.; Mortenson, L.E. (1982) J. Biol. Chem. 257, 2447-2452.

PRIMARY STRUCTURE AND POST-TRANSLATIONAL MODIFICATION OF FERREDOXIN-NADP REDUCTASE FROM *Chlamydomonas reinhardtii.*

Paulette Decottignies*, Pierre Le Maréchal*, Jean-Marie Schmitter§, Valérie Flesch* and Jean-Pierre Jacquot*.
*Institut de Biotechnologie des Plantes, Université Paris-Sud, 91405 Orsay Cedex, France. §Laboratoire de Biochimie, Ecole Polytechnique, 91128 Palaiseau Cedex, France.

1. Introduction

Ferredoxin-NADP reductase (FNR, E.C. 1.18.1.2) is a chloroplastic flavoprotein which transfers electrons from ferredoxin, an iron-sulfur protein, to NADP during photosynthesis. FNR was isolated first from spinach leaves (1). It is a monomeric enzyme of 35 kDa, which contains non covalently bound FAD and is able to form a 1:1 complex with ferredoxin (2). Its kinetic properties (2) as well as its aminoacid (3) have been determined, and its three-dimensional structure is now available (4). Spinach FNR has two stuctural domains, the stretch from residues 19 to 161 contains a β barrel structure which binds the flavin and the second domain, residues 162 to 314, binds the NADPH. Because of the key position of FNR in the photosynthetic electron flow, it was of interest to know its primary structure. However, the sequences of only a few FNR have been elucidated so far, especially by direct amino acid sequencing. We have purified, characterized and sequenced FNR from *Chlamydomonas reinhardtii*, a unicellular photosynthetic alga.

2. Procedure

Chlamydomonas reinhardtii cells (strain cw 15) were grown photoheterotrophically on acetate. Purification was achieved by chromatographic methods, as previously described (5). Cytochrome c reductase activity was measured as described (2).

Sequence determination
FNR was cleaved with trypsin or SV8 protease, and the generated peptides separated by RP-HPLC as previously described ((5) then submitted to sequence analysis. Automated Edman degradation was performed using an Applied Biosystems Model 476 A sequencer equipped with an on-line analyzer of phenylthiohydantoin amino acids. Mass spectrometry was performed as previously described (5).

PCR cloning
4 oligonucleotides were derived from the protein sequence, taking into account the coding bias of *C. reinhardtii* and corresponding : (F187-WRRC-F192)) (K66-IPFW-E71) (D256-QIYM-K251) (Y320-VEVHW-Q314). PCR cloning of the cDNA fragments was realized as described in (6) using a cDNA library in pUC as template.The fragments with the proper size were gel purified and cloned blunt-end in a Bluescript SK+ vector at the EcoRV site.

P. Mathis (ed.), Photosynthesis: from Light to Biosphere, Vol. II, 649–652.

3. Results and Discussion

3.1 *Purification and properties*

C. reinhardtii FNR was purified with a good yield in three chromatographic steps, including anion-exchange, gel filtration and affinity on a ferredoxine-Sepharose column. The homogeneity of the preparation was checked by SDS-PAGE and N-terminal sequencing (see below). Table 1 summarizes the physical and kinetic properties of *C. reinhardtii* FNR in comparison with its spinach counterpart. The algal FNR is a monomeric protein of 36 kDa, containing one FAD and exhibiting diaphorase activity and ferredoxin-dependent cytochrome c reduction activity. Laser desorption mass spectrometry allowed us to determine the molecular masses of both the apoenzyme (after chromatography on a reverse phase column in acidic conditions) and the holoenzyme (Table 1). The difference between the two values is 775 ± 10, which corresponds to the mass value of FAD (785.5 Da).

Table 1 : Structural and kinetic properties of FNR from *Chlamydomonas reinhardtii* and Spinach leaves. nd. = not determined

		Ferredoxin-NADP Reductase (FNR)	
		Chlamydomonas reinhardtii	Spinach
Subunit MM	SDS PAGE :	36 kDa	35 kDa
	Mass Spectrometry :	35685 Da (apoenzyme) 36460 Da (holoenzyme)	nd.
Native enzyme		Monomer	Monomer
Coenzyme		FAD	FAD
Cyt C reduction	kcat (s^{-1}) :	12	13
	K_m Fd (μM) :	0.8	0.8

3.2 *Protein sequence*

The following strategy was used to obtain the complete primary structure of *C. reinhardtii* FNR (Fig.1). In a first series of experiments, the protein was reduced with dithiothreitol and carboxymethylated with ^{14}C iodoacetate under denaturating conditions, then an aliquot was submitted to N-terminal sequencing, allowing identification of the first 32 residues. The remaining part of the derivatized protein was cleaved with SV8 protease, which generated sixteen significant peptides which were purified and sequenced. The two longest peptides were further cleaved with chymotrypsin and/or trypsin to identify the remaining residues. In a second series of experiments, native FNR (i.e. not reduced nor carboxymethylated) was digested with trypsin. In those conditions, only a few potential sites were actually cleaved, which facilitated the alignment of the peptides. Incubation of FNR with trypsin mostly generated three long peptides, and two short peptides corresponding to the amino and carboxy termini of the protein. The three long peptides were partially sequenced and an aliquot of each of them was submitted to a second tryptic cleavage which provided the missing stretches. Moreover, several peptides, especially those containing cysteinyl residues, were submitted to mass spectrometry analysis. All the masses measured were in total agreement with the ones calculated from the corresponding sequence (data not shown).

Comparison of the primary structure of *C. reinhardtii* FNR with the known sequences showed 41-46% identity.

```
                  10                  20                  30                  40
                  |                   |                   |                   |
A T K A S T A V T T D M S K R T V P T K L E E G E M P L N T Y S N K A P F K A K
                  50                  60                  70                  80
                  |                   |                   |                   |
V R S V E K I T G P K A T G E T C H I I I E T E G K I P F W E G Q S Y G V I P P
                  90                  100                 110                 120
                  |                   |                   |                   |
G T K^T I N S K G K^T E V P H G T R L Y S I A S S R Y G D D F D G Q T A S L C V R R
                  130                 140                 150                 160
                  |                   |                   |                   |
A V Y V D P E T G K E D P A K^D K G L C S N F L C D A T P G T E I S M T G P T G K
                  170                 180                 190                 200
                  |                   |                   |                   |
V L L L P A D A N A P L I C V A T G T G I A P F R S F W R R C F I E N V P S Y K
                  210                 220                 230                 240
                  |                   |                   |                   |
F T G L F W L F M G V A N S D A K L Y D E E L Q A I A K A Y P G Q F R L D Y A L
                  250                 260                 270                 280
                  |                   |                   |                   |
S R E Q N N R K G G K M Y I Q D K V E E Y A D E I F D L L D N G A H M Y F C G L
                  290                 300                 310                 320
                  |                   |                   |                   |
K G M M P G I Q D M L E R V A K E K G L N Y E E W V E G L K H K N Q W H V E V Y
```

Fig. 1 Sequence of *Chlamydomonas reinhardtii* FNR
The protein was cleaved and sequenced as described in the text. K^T and K^D respectively refer to trimethyllysine and dimethyllysine.

3.3 *Identification of three post-translationally modified residues*

During the course of sequence analysis, we observed an unusual aminoacid located in positions 83 and 89 of the primary structure. The corresponding peak of the phenylthiohydantoin (PTH) derivative eluted differently from any standard PTH aminoacid both in shape and in retention time and could correspond to a derivative (acetyl- or methyl-) of lysine or arginine. Mass spectrometry analysis of peptides containing this unusual amino acid indicated that its mass was 170. Thus, only three modified residues fulfilled this additional condition : NG-monomethylarginine, Nε-acetyllysine or Nε-trimethyl-lysine. When the three candidates were directly tested on our sequencer, in the same conditions as the unknown PTH, only PTH-Nε-trimethyllysine fitted the unknown PTH, both for its retention time and for the unusual broad shape of the peak. A third amino acid located at position 135 seemed to be also modified. The same approach was used to demonstrate that this residue was a dimethyllysine. The occurrence of modified lysines was definitely confirmed by cDNA sequencing (see below). Moreover, the mass precedently obtained by laser desorption mass spectrometry (35,680 ± 14) was in total agreement with the value calculated from the sequence (35,685)
Therefore, we can conclude that in *C. reinhardtii* FNR three specific residues of lysine have undergone a post-translational modification, the ε amino group being substituted with three methyl groups for K83 and K89 and with two methyl groups for K135.
Trimethyllysine, or other methylated residues, have never been observed in any FNR. Tri- and di-methyllysine have already been found in different proteins such as histones, myosin, cytochrome c, citrate synthase and heat shock proteins. In the plant kingdom, the occurrence of N-methylated residues is poorly documented, excepted for cytochrome c

from several higher plants, which was reported to contain residue(s) of Nε-trimethyllysine. The presence of a lysine methyltransferase which specifically methylates the ε-amino group of protein-bound lysines, using S-adenosyl L-methionine as the methyl donor, has been described in animal and bacterial organisms (7), but the role of this post-translational modification remains a matter of speculation.

3.4 *cDNA sequence and overexpression of the recombinant protein in* E. coli
Based on the amino acid sequence and taking into account the coding bias of *C. reinhardtii*, oligonucleotides were constructed in order to obtain cDNA fragments coding for FNR by realizing PCR reactions on a cDNA library as described in (6). Basically two overlapping fragments coding for the regions K66-D256 and F187-Y 320 were obtained, the whole deduced sequence being completely in agreement with the amino acid sequence. Most important, the cDNA sequence featured the presence of codons for Lys in the positions K83, K89 and K135, confirming that the post-translational modification involved lysine residues.
In order to obtain an expression system for FNR, it was necessary to obtain a cDNA coding for the whole polypeptide. This could not be achieved by PCR or RACE experiments and we decided thus to reconstruct a "synthetic gene" covering the sequence M1 to E71 by the use of four 70 nt long oligonucleotides (2 sens and 2 antisens) overlapping each over on 18 nt long sequence. A combination of PCR reactions yielded a 200 bp long fragment which could be fused to the earlier fragments covering the sequence K66-Y320 by PCR. The resulting 980 bp fragment which contained the NcoI and Bam HI sites present in the oligonucleotides was introduced into plasmid pET-3d. After induction with IPTG, recombinant FNR was purified with a good yield (2 mg/l). The N-terminus of the protein was sequenced up to 10th residue and was shown to be identical to the one of the algal protein. Recombinant FNR was active in the cytochrome c reduction test but its kinetic parameters have not been studied yet. Its methylation status is also under current investigation.

4. Perspectives

The recombinant protein, which is expected not to contain methylated lysines, could be used as a tool to further study the role of the methyllysines in FNR. First, it will compared with the native protein from *Chlamydomonas* : structural and enzymatic properties, ability to bind ferredoxin and thylakoid membranes. Then the recombinant protein will be used as a substrate to detect and purify the methyltransferase involved in the methylation process.

BIBLIOGRAPHY

1 Shin, M., Tagawa, K. and Arnon, D.I. (1963) Biochem. Z. **338**, 84-96.
2 Shin, M. (1971) Methods in Enzymol. **23**, 440-447.
3 Karplus, P.A., Walsh, K.A. and Herriot, J. R. (1984) Biochemistry, **23**, 6576-6583.
4 Karplus, P.A., Daniels, M.J. and Herriot, J. R. (1991) Science **251**, 60-66.
5 Decottignies, P., Le Maréchal, P., Jacquot, J-P., Schmitter, J-M and Gadal, P. (1995) Arch. Biochem. Biophys. **316**, 249-259.
6 Jacquot, J.P., Stein, M., Hodges, M. and Miginiac-Maslow. (1992) Nucleic Acid Res. **20**, 617.
7 Lee, H. W., Paik, W. K. and Borun, T.W. (1973) J. Biol. Chem. **248**, 4194-4199.

SPINACH FERREDOXIN-NADP$^+$ REDUCTASE: STRUCTURE-FUNCTION RELATIONSHIP AS STUDIED BY SITE-DIRECTED MUTAGENESIS

A. Aliverti[1], V.E. Pandini[1], F.A. Sternieri[1], M.E. Corrado[1], P.A. Karplus[2], and G. Zanetti[1]. [1]Dipartimento di Fisiologia e Biochimica Generali, Università di Milano, via Celoria 26, I-20133 Milano, Italy; [2]Division of Biological Sciences, Cornell University, Ithaca, NY 14853, USA

1. Introduction

In plants and cyanobacteria, ferredoxin-NADP$^+$ reductase (FNR) is the flavoenzyme that catalyses the last step in the photosynthetic electron transport chain, transferring reduction equivalents from ferredoxin (Fd) to NADP$^+$. The amino acid sequence of FNR is known from nine sources and the three-dimensional structure of the spinach enzyme has been refined to 1.7 Å (1). A role in the catalytic mechanism for a few residues has been suggested in the past by chemical modification studies (2). More recently, the powerful tool of site-directed mutagenesis in conjunction with X-ray crystallography has allowed us to probe more directly the bearing of protein structure on enzymatic catalysis. Four residues located in the isoalloxazine microenvironment of the FAD prosthetic group were individually replaced by site-directed mutagenesis: Tyr95, Ser96, Cys272 and Glu312. Tyr95 interacts with the *si* face of the flavin and makes a H-bond to the ribityl 4' OH group. Ser96 makes contacts through its amide nitrogen and its side chain with the N_5 position of the isoalloxazine, while Glu312 side chain makes a H-bond with the Ser96 hydroxyl group. Cys272 is one of the best conserved residue in the FNR family and, in a NADP-binding model recently proposed (1), it contacts by its side-chain the nicotinamide moiety of the substrate.

2. Procedure

Site-directed mutagenesis was carried out using the phosphorothioate-based method (3). FNR forms were overproduced in *E. coli*, purified and characterised essentially as described (4).

3. Result and Discussion

Five mutant FNR forms, *i.e.* Y95F, S96G, S96V, C272S and E312L, were produced and purified to homogeneity. All mutants incorporated FAD and displayed protein and FAD fluorescence properties identical to those of the wild-type FNR. The mutants Y95F, C272S and E312L showed essentially the same absorbance spectrum as the wild-type enzyme, whereas mutations at position 96 resulted in a substantial red-shift of the absorption bands, indicating a significant change in the microenvironment of the flavin (Fig. 1). As shown in Table I, FNR-Y95F maintained a substantial amount of the diaphorase activity of the wild-type enzyme, whereas it was more impaired as a Fd-

P. Mathis (ed.), Photosynthesis: from Light to Biosphere, Vol. II, 653–656.

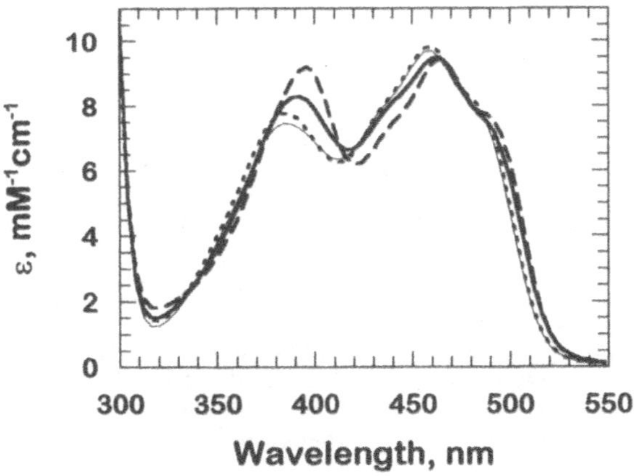

Figure 1. Absorption spectra in the visible region of wild-type (thin line), S96G (solid line), S96V (long dashes) and E312L (short dashes) FNR forms. The spectra of the Y95F and C272S FNR mutants are not shown for clarity.

dependent cytochrome *c* reductase. Amino acid replacements at positions 96, 272 and 312 caused more dramatic effects, FNR-S96V being almost inactive. Kinetic parameters for the diaphorase reaction using ferricyanide as acceptor are shown in Table II. FNR-Y95F showed a lower k_{cat} but essentially the same k_{cat}/K_m^{NADPH} ratio as the wild-type FNR, indicating that the reductive half-reaction (*i.e.* reduction of FNR by NADPH) was not affected by this mutation. This possibly holds also for the reactions with other dye acceptors, although it was not checked directly. Thus, the lower activity of FNR-Y95F could be due to decreased reoxidation rates of reduced flavin by the various electron acceptors.

The major effect of mutations at positions 96, 272 and 312 was on k_{cat} (Table II). In these mutant forms, it was the reductive half-reaction to be impaired as shown by the large decreases in k_{cat}/K_m^{NADPH}. This was confirmed for the Cys272 and Ser96 mutants by studying the reduction of bound FAD by NADPH through stopped-flow spectrophotometry. Rate constants of 105, 7.2, 2.1 and 0.04 s^{-1} were measured for wild-

Table I. Specific Activities with Various Electron Acceptors of the FNR Forms

enzyme form	electron acceptor INT[a] (U/FAD)	$K_3Fe(CN)_6$ (U/FAD)	cyt *c*[b] (U/FAD)
wild type	2800	20000	3900
Y95F	1100	13000	650
S96G	174	280	234
S96V	4.5	6.2	5.9
C272S	350	1000	550
E312L	110	174	48

[a]2-(*p*-iodophenyl)-3-(*p*-nitrophenyl)-5-phenyltetrazolium chloride
[b]Fd-dependent cytochrome *c* reductase activity.

Table II. Kinetic Parameters of the FNR Forms for the Diaphorase Activity Assayed with $K_3Fe(CN)_6$ as Electron Acceptor

enzyme form	k_{cat} (s^{-1})	K_m^{NADPH} (μM)	k_{cat}/K_m^{NADPH} ($s^{-1}M^{-1}$)
wild type	500	35	1.43×10^7
Y95F	270	14	1.93×10^7
S96G	13.7	144	9.51×10^4
S96V	0.26	106	2.47×10^3
C272S	31	77	4.03×10^5

type FNR, FNR-C272S, FNR-S96G and FNR-S96V, respectively. Details about rapid kinetic studies on these FNR forms have been published elsewhere (4,5).

Since the productive interaction between NADPH and FNR seemed to be impaired by replacing specific amino acids at the active site of the enzyme, it was of interest to study the interaction of these mutant enzymes with $NADP^+$. Upon binding $NADP^+$, the visible spectrum of wild-type FNR undergoes a perturbation, yielding the difference absorption spectrum shown in Fig. 2. The difference spectrum of the $NADP^+$/FNR-Y95F complex was very similar to that of the wild-type enzyme (not shown). The difference spectrum of the $NADP^+$/FNR-C272S complex, although similar in shape to that involving wild-type FNR, was much less intense. More interestingly, the difference spectra observed upon $NADP^+$ binding to S96G, S96V or E312L mutant forms displayed a completely different shape (for clarity, only the spectrum of the FNR-S96G mutant is shown in Fig. 2). These spectra strongly resembled those induced by binding of 2'-P-ADP-ribose, a substrate analogue lacking the nicotinamide ring, to wild type as well as mutant FNR forms (not shown). The K_d values for $NADP^+$ were found unchanged in the FNR forms carrying mutations at positions 96, 272, or 312. Thus, although all these mutants were fully competent in $NADP^+$ binding, the interaction between nicotinamide and isoalloxazine rings was partially impaired in the case of the C272S mutant and possibly, completely

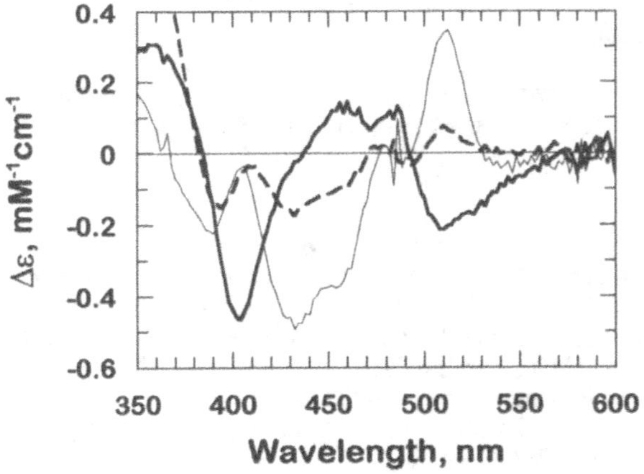

Figure 2. Difference absorption spectra of the complexes between $NADP^+$ and the FNR forms. Symbols are: thin line, wild-type FNR; solid line, FNR-S96G; dashed line, FNR-C272S.

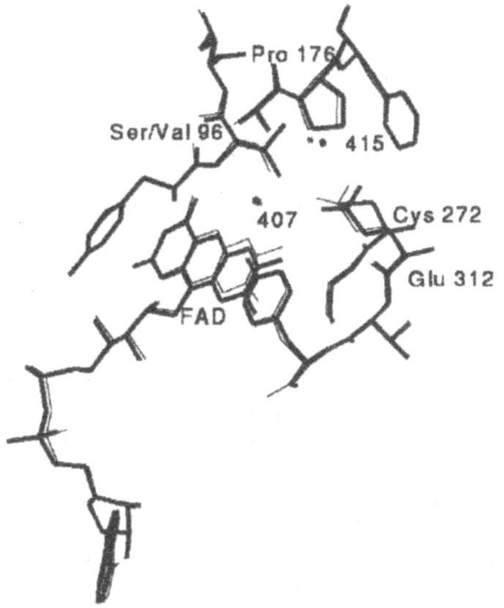

Figure 3. Structure of FNR-S96V (thick bonds) compared to that of the wild-type FNR (thin bonds). Two active-site water molecules (407 and 415) are also shown.

abolished when residues Ser96 or Glu312 were mutated. The three-dimensional structure of the most impaired mutant form, *i.e.* FNR-S96V, as determined by X-ray crystallography (Fig. 3), allowed us to rule out gross structural alterations and to assess that conformational modifications are confined to the surrounding of the mutated residue. In conclusion, a scheme of FNR catalysis is emerging, in which Ser96, Cys272 and Glu312 have a definite role in the positioning of the nicotinamide ring of NADP(H) for hydride transfer to/from the flavin to occur at high rate.

4. Acknowledgements

This work was supported from grants from Ministero della Ricerca Scientifica e Tecnologica (60%) and CNR target project "Biotechnologies and Bioinstrumentations".

References

1. Bruns, C.M. and Karplus, P.A. (1995) J. Mol. Biol. 247, 125-145
2. Zanetti, G. and Aliverti, A. (1991) in Chemistry and Biochemistry of Flavoenzymes (Müller F., ed.) Vol. 2, pp. 305-315, CRC Press, Boca Raton, FL
3. Sayers, J. R., Schmidt, W. and Eckstein, F. (1988) Nucl. Acids Res. 16, 791-802
4. Aliverti, A., Bruns, C.M., Pandini, V.E., Karplus, P.A., Vanoni, M.A., Curti, B. and Zanetti, G. (1995) Biochemistry 34, 8371-8379
5. Aliverti, A., Piubelli, L., Zanetti, G., Lübberstedt, T., Herrmann, R.G. and Curti, B. (1993) Biochemistry 32, 6374-6380

CHARACTERIZATION OF RECOMBINANT WILD-TYPE AND MUTANT FORMS OF FERREDOXIN I

L. Piubelli, F. Bellintani, L. Pirola, A. Aliverti and G. Zanetti. Dipartimento di Fisiologia e Biochimica Generali, Università degli Studi di Milano, Via Celoria 26, I-20133 Milano, Italy

1. Introduction

In oxygenic photosynthesis, ferredoxin I (Fd I) acts as a soluble, stromal shuttle carrying single electrons, by means of its [2Fe-2S] cluster, from photoreduced photosystem I (PSI) to the several enzymes that are ferredoxin-dependent. Although the molecular details of the interactions between Fd I and its electron donor and acceptors are still to be clarified, a critical role has been ascribed to carboxylate groups at the Fd I molecular surface. Electrostatic interactions between oppositely charged side-chains have been proposed to be involved in complex formation with ferredoxin-$NADP^+$ reductase (FNR) (1), with thioredoxin reductase (2) and with PSI (3,4). Moreover, a specific glutamate residue of ferredoxin has been recently shown to be essential in the electron-transfer to FNR, by site-directed mutagenesis of the *Anabaena* ferredoxin (5). Here, we describe the set up of a system for the expression of spinach leaves Fd I in *Escherichia coli* as well as of a procedure for its purification. This system has been used for the production of four Fd I mutant forms, in which two residues possibly involved in the interaction with FNR, *i.e.* Glu92 and 93, were replaced with different amino acids through site-directed mutagenesis.

2. Procedure

Plasmid p6SocFd/1, which carries the spinach cDNA encoding the precursor form of Fd I, was a gift from Prof. R.G. Herrmann. Site-directed mutagenesis was carried out using the phosphorothioate-based method. DNA sequencing was performed according to chain termination method. *E. coli* strain BL21 (DE3) was used as a host for expression of Fd I forms. The concentration of Fd I forms was determined spectrophotometrically using an ε_{420} of 9680 $mM^{-1}cm^{-1}$. NADPH-cytochrome *c* reductase activity was assayed essentially as published (6) using 6 nM recombinant FNR as a catalyst. Dissociation constants of the complexes between Fd I forms and FNR were determined by titrating the reductase protein fluorescence with Fd I forms. Cytochrome *c* (cyt *c*) and $NADP^+$ photoreductions were performed with hypotonically washed thylakoids reconstituted with the recombinant Fd I forms (3).

3. Results and Discussion

A gene construct encoding the mature form of spinach Fd I was obtained by recloning part of the cDNA insert excised from p6SocFd/1 in pET-11d with the aid of a synthetic

P. Mathis (ed.), Photosynthesis: from Light to Biosphere, Vol. II, 657–660.

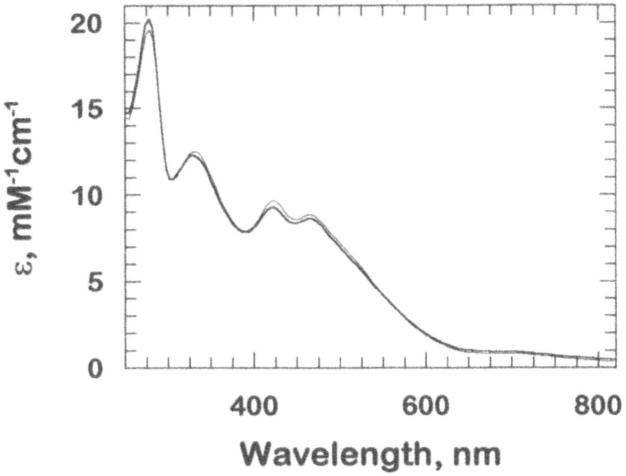

Figure 1. Absorption spectra of wild type Fd I (thin line) and Fd I-E92K (bold line).

DNA adaptor. Details of the cloning procedure, yielding plasmid pETFdI, have been reported elsewhere (6). The insert of pETFdI was subjected to site-directed mutagenesis in order to obtain mutant Fd I genes, where the codons for Glu92 and Glu93 were individually replaced with triplets coding for either Ala, Gln, or Lys, at position 92, and for Ala at position 93. Plasmids carrying wild-type and mutant Fd I genes were transferred to *E. coli* for overexpression. In crude extracts of induced cells, Fd I forms accounted for about 2.5% of the soluble proteins. The wild-type and the four mutant Fd I forms were isolated to homogeneity by a procedure comprising salt-promoted adsorption on Sepharose 4B, ion-exchange on Q-Sepharose in FPLC, and hydrophobic interaction chromatography on Phenyl-Sepharose in FPLC. The N-terminal residue of the recombinant protein was shown to be alanine by partial sequencing, confirming that methionine was removed *in vivo* from the N-terminus of the neosynthesised protein. Moreover, the mass spectrum of wild-type recombinant Fd I, obtained by matrix-assisted laser desorption/ionisation mass spectrometry, allowed the determination of a molecular mass of 10,478 for the apoprotein, which is in close agreement with the value of 10,483 calculated for the 97 residue mature polypeptide.
All mutant Fd I forms displayed fluorescence and absorption spectra very similar to those of the wild-type protein. The absorption spectrum of Fd I-E92K, in comparison to that of the wild-type protein, is shown in Fig. 1.
The effect of mutations at positions 92 and 93 of Fd I on its electron-transfer properties was initially studied using wild-type and mutant proteins as substrates in the NADPH-cyt *c* reductase reaction catalysed by soluble FNR. In this assay, Fd I-E93A showed the same activity as the wild-type protein. Mutants at position 92 instead supported a substantially higher electron-transfer rate. Cyt *c* reductase activity measured with E92Q, E92A and E92K forms were 190, 220 and 240% of that measured with wild-type Fd I, respectively.
Since the acidic cluster comprising residues 92-94 of Fd I was proposed to be involved in the interaction with FNR (1), the affinity of the Fd I mutants for the reductase was studied. The K_d values of the complexes between soluble FNR and Fd I mutants at

Table I. Properties of the mutant Fd I forms as compared with those of the wild-type protein

Fd I form	thylakoid-catalyzed activities[a]			E'_m[b]
	photoreduction		NADPH-cyt *c*	
	NADP$^+$	cyt *c*		
	μeq (mg Chl)$^{-1}$h^{-1}			mV
wild type	570	110	278	-401
E92Q	310	100	531	-328
E92A	310	90	651	-323
E92K	170	110	716	-308

[a] Fd I forms were 8 μM in NADP$^+$ photoreduction and NADPH-cyt *c* reductase assays, and 2 μM in cyt *c* photoreduction.
[b] Redox potential values are taken from (7).

position 92 were slightly higher (up to three-fold) as compared with the wild-type protein, indicating that Glu92 side-chain has a limited role in the interaction between the two proteins.
The effect of mutations at position 92 on the interaction of Fd I with both PSI and membrane-bound FNR was studied by performing various assays on thylakoid suspensions hypotonically washed to remove endogenous ferredoxin. Thylakoids were reconstituted with recombinant Fd I forms and NADPH-cyt *c* reductase activity as well as photoreduction of NADP$^+$ and cyt *c* were measured. As shown in Table I, Glu92 mutants showed a reduced ability to transfer electrons from PSI to FNR with respect to wild-type Fd I, whereas their functionality in cyt *c* photoreduction was unchanged, and their activity in NADPH-cyt *c* reduction even increased.

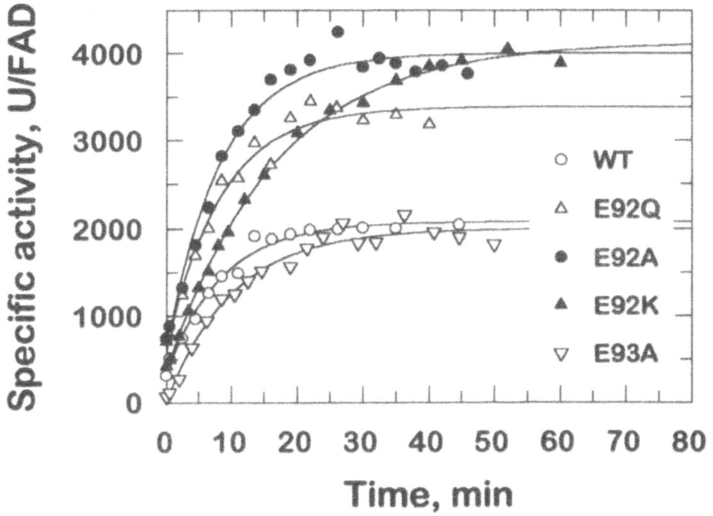

Figure 2. Time course of cross-linking of Fd I forms to FNR as monitored by cyt *c* reductase activity measured in the absence of added ferredoxin to the assay. Mixtures of 8 μM FNR and 40 μM Fd I forms were incubated with 5 mM EDC in the presence of 2 mM NADP$^+$. 1 μl-aliquots were assayed at each time.

The behaviour of the Fd I forms where Glu92 was replaced with uncharged or positively charged residues can be rationalised in terms of the effect of the side-chain replacement on the electrochemical properties of the [2Fe-2S]. Indeed, a correlation exists between shift in redox potential (Table I) and either increase in NADPH-cyt *c* reductase activity or decrease in $NADP^+$ photoreduction rate. Replacement of Glu92 in Fd I resulted in a positive shift ranging from 73 to 93 mV in the redox potential of the [2Fe-2S] cluster, which places it in between the redox potentials of the reductase and $NADP^+$. This results in the impairment of the electron transfer from PSI to FNR but in the enhancement of the electron transfer from NADPH to cyt *c*, which proceeds in the direction opposite to the physiological one.

Since the 92-94 glutamate cluster of Fd I has been proposed as a major site for cross-linking to lysine side-chains of FNR by carbodiimide treatment (1), it was of interest to perform cross-linking studies with Fd I mutants lacking possible modifier targets. Cross-linking between Fd I and FNR can be conveniently monitored by assaying modification mixtures for cyt *c* reductase activity. Indeed, as Fd becomes covalently bound to FNR, the latter acquires the ability to catalyse NADPH-cyt *c* reduction in the absence of free Fd added to the assay. As shown in Fig. 2, and by SDS-PAGE, elimination of a carboxylate either at position 92 or 93 of Fd I did not prevent cross-linking of Fd I to FNR. The effect of the most drastic mutation, *i.e* E92K, was just to slow down the cross-linking rate. It should be noted that even when cross-linked to FNR, Fd I mutants at position 92 supported enhanced electron transfer from FNR to cyt *c* (Fig. 2).

In conclusions, replacement of Glu93 side-chain of Fd I seems to have no effect on the interaction and electron transfer to FNR. Glu92 instead seems to have a minor role in complex formation with FNR. Replacement of Glu92 moderately affected the electron-transfer properties of Fd I probably as a consequence of the modification of the redox potential of the [2Fe-2S]. The modulation of the properties of the iron-sulfur cluster by Glu92 could be mediated through hydrogen bonding of its side-chain to Ser45, which is part of the [2Fe-2S] binding loop, as shown for the *Anabaena* ferredoxin (8).

4. Acknowledgements

This work was supported by grants from C.N.R. and MURST (40%) of Italy.

5. References

1. Zanetti, G., Morelli, D., Ronchi, S., Negri, A., Aliverti, A. and Curti, B. (1988) Biochemistry 27, 3753-3759
2. De Pascalis, A.R., Schürmann, P. and Bosshard, H.R. (1994) FEBS letters 337, 217-220
3. Zanetti, G. and Merati, G. (1987) Eur. J. Biochem. 169, 143-146
4. Lelong, C., Sètif, P., Lagoutte, B. and Bottin, H. (1994) J. Biol. Chem. 269, 10034-10039
5. Hurley, K.J., Medina, M., Gomez-Moreno, C. and Tollin, G. (1994) Arch. Biochem. Biophys. 312, 480-486
6. Piubelli, L., Aliverti, A., Bellintani, F. and Zanetti, G. (1995) Protein Express. Purif. 6, 298-304
7. Aliverti, A., Hagen ,W.R. and Zanetti, G. (1995) FEBS Lett., in press
8. Holden, H.M., Jacobson, B.L., Hurley, J.K., Tollin, G., Oh, B.-H., Skjeldal, L., Chae, Y.K., Cheng, H., Xia, B. and Markley, J.L. (1994) J. Bioenerg. Biomembr. 26, 67-87

INVOLVEMENT OF THE FAD DOMAIN RESIDUE TYROSINE 308 ON STABILITY AND FUNCTION OF FERREDOXIN-NADP$^+$ REDUCTASE.

Nora B. Calcaterra, Guillermo A. Picó, Néstor Carrillo, Jorgelina Ottado, Elena G. Orellano and Eduardo A. Ceccarelli. PROMUBIE, CONICET, and Facultad de Cs. Bioquímicas, Universidad Nacional de Rosario, Suipacha 531. (2000) Rosario, Argentina.

1. Introduction

Ferredoxin-NADP$^+$ reductases (FNR, EC 1.18.1.2) represent an important class of flavoenzymes that utilize a noncovalently bound FAD cofactor as the only redox center. In chloroplasts and cyanobacteria, FNR catalyzes the reversible electron transfer between NADP(H) and electron carrier proteins like ferredoxin or flavodoxin (1, 2). In addition to this physiological reaction, the enzyme is able to mediate *in vitro* the oxidation of NADPH by artificial electron acceptors (diaphorase activity) or the ferredoxin-cytochrome *c* couple (cytochrome *c* reductase). Plant FNRs are hydrophilic proteins of ~35 kDa, containing two structural domains (2, 3). The three dimensional structure determined at 0.17 nm (3) indicates that the prosthetic group is bound external to an antiparallel β-barrel that makes up the core of the FAD binding domain at the NH_2-terminal region of the protein. The C-terminal domain provides for NADP$^+$ binding, with the last 19 amino acids forming a distinct α-helix/β-strand region that folds back into the FAD site (2, 3). The terminal, conserved tyrosine (Tyr 308 in pea FNR) is stacked coplanar to the *re*-face of the flavin ring and makes extensive π-orbital contact with it, partially shielding the FAD group from solvent (Fig. 1). Results obtained after site-directed mutagenesis of Tyr 308 indicated that *i*) aromaticity was the most important factor to the function of this residue in FNR activity, and *ii*) all mutations introduced barriers to catalysis, without affecting FNR interaction with substrates (4). This effect might be related to a direct involvement of the tyrosine in catalysis, or to a structural role in FNR folding, FAD binding or both. In this study we use FAD fluorescence measurements and unfolding experiments to demonstrate that Tyr 308 plays a key role in FNR holoprotein stability.

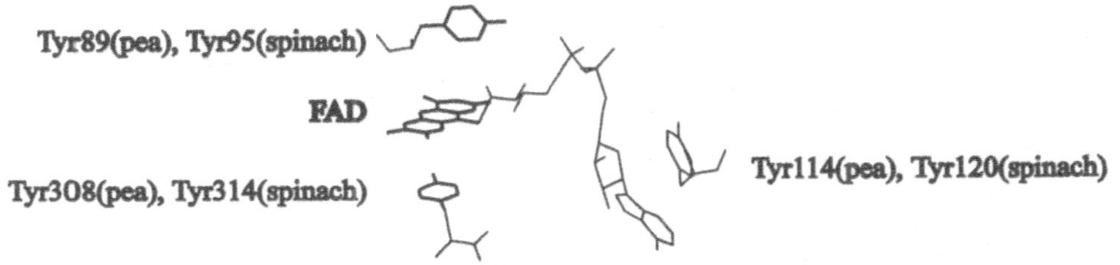

Fig. 1. Computer graphics model of the FAD region in pea FNR showing relative positions of Tyr 308, Tyr 89 and Tyr 114. Taken from x-ray diffraction data of the spinach enzyme (2).

P. Mathis (ed.), Photosynthesis: from Light to Biosphere, Vol. II, 661–664.

2. Experimental Procedures

FNR cDNA clones were kindly provided by Dr. J.C. Gray, University of Cambridge, UK. Sequences encoding the mature region of pea FNR (5) were ligated in-frame to the 3' end of the glutathione *S*-transferase gene in pGEX-3X (6). The procedures employed for cloning, mutagenesis, expression in *Escherichia coli* and purification of the FNR mutants Y308W, Y308F, Y308S and Y308G have been reported (4, 6). Concentrations of the different FNR variants were estimated by measuring the flavin released in 5 % (w/v) trichloroacetic acid, using HPLC and fluorescence detection. FAD emission spectra and relative quantum yields were determined by setting the excitation wavelength at 456 nm. Energy transfer was measured at 526 nm after excitation of the FNR samples at 295 nm. Thermal denaturations were determined by measuring intrinsic tryptophan fluorescence upon increasing the temperature of the sample at a rate of 0.7°C/min. Transitions were analyzed on the basis of the two-state model (7). Urea gradient gels were prepared and stained according to Creighton (8).

3. Results and Discussion

The fluorescence of bound FAD is known to be largely quenched in the native oxidized form of wild-type FNR (1). Accordingly, we observed that the emission of the pea enzyme at 526 nm was only 0.75 % of that of free FAD (Fig. 2A). Mutants containing an aromatic residue at position 308 (Y308W and Y308F) displayed small increases of the relative quantum yields, whereas removal of the aromatic side chain in the Y308G and Y308S variants caused a large fluorescent rise (Fig. 2A). These results indicate that the interaction between Tyr 308 and the isoalloxazine ring in wild-type FNR accounts for a significant fraction of the total FAD fluorescence quenching. It is likely that the phenol behaves as a static quencher of the flavin by enhancing nonradiative deactivation of the first excited singlet. The behavior of the tryptophan and phenylalanine substitutions suggests that aromaticity is the key factor involved in quenching, and that the aromatic residues introduced in the mutants adopt a conformation similar to that of Tyr 308 in wild-type FNR. Indeed, the Y308W mutant displays significant energy transfer between the indole and isoalloxazine moieties (Fig. 2B), suggesting that the two fluorophores are still close to each other in this FNR variant.

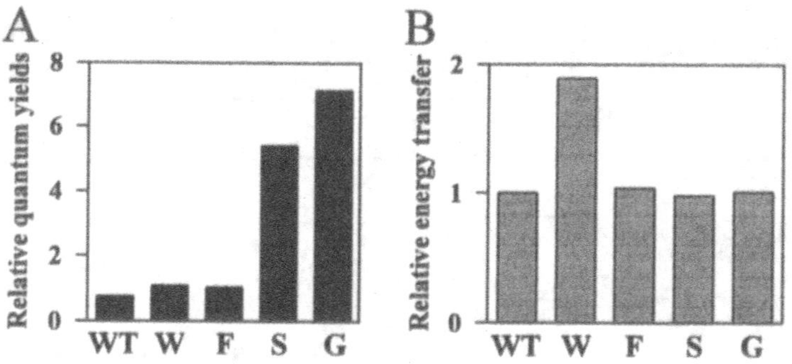

Fig. 2. FAD fluorescent properties of FNR variants. Emissions were monitored at 526 nm. (A) Quantum yields are given as percent of emission relative to free FAD, following excitation at 456 nm. (B) Energy transfers are expressed relative to the emission of wild-type FNR, after excitation at 295 nm. Residues introduced in place of Tyr 308 are indicated in a one-letter code.

Thermal unfoldings of wild-type and mutant FNR proteins followed a sigmoidal behavior and were irreversible, presumably due to FAD release above the transition midpoint temperature T_m (not shown). Denaturation curves were analyzed on the basis of the two-state model (7), and the corresponding $\Delta\Delta G_u$ values (changes in thermodynamic stability relative to the wild-type) are given in Fig. 3 for the four FNR mutants. All replacements led to destabilization, with the nonaromatic mutants Y308G and Y308S showing the lowest stability (Fig. 3).

The effect of Tyr 308 substitutions on FNR stability was also analyzed using transverse urea gradient gel electrophoresis. In this method, unfolding is detected by the reduced mobility of the denatured protein (8). A typical result obtained with the Y308F mutant is shown in Fig. 4. The urea concentrations at the transition midpoints (C_m) were used to calculate the changes in stability $\Delta\Delta G_u$ introduced by the mutations (Fig. 3). In agreement with the thermal denaturation profiles, substitution of Tyr 308 resulted in lower C_m values, with nonaromatic replacements leading to more extensively destabilized forms (Fig. 3). Therefore, the structural consequences of single substitutions at the C-terminal position of FNR, measured by both thermal- and urea-denaturation, followed a similar trend as that observed for enzyme activity (4, see also Fig. 3).

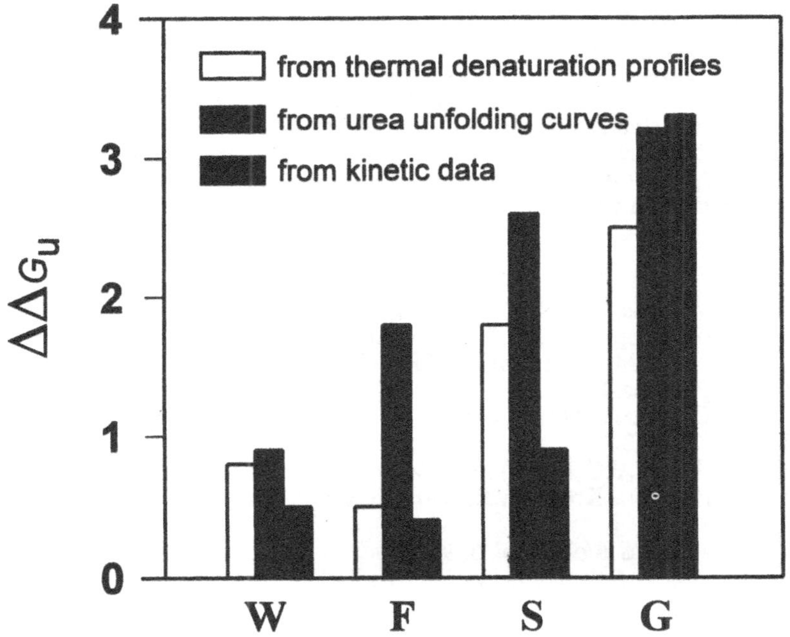

Fig. 3. Thermodynamic parameters characterizing the changes in stability and in catalytic efficiency of pea FNR, introduced by site-directed mutagenesis of Tyr 308. The $\Delta\Delta G_u$ values were calculated using established procedures (4, 7, 8).

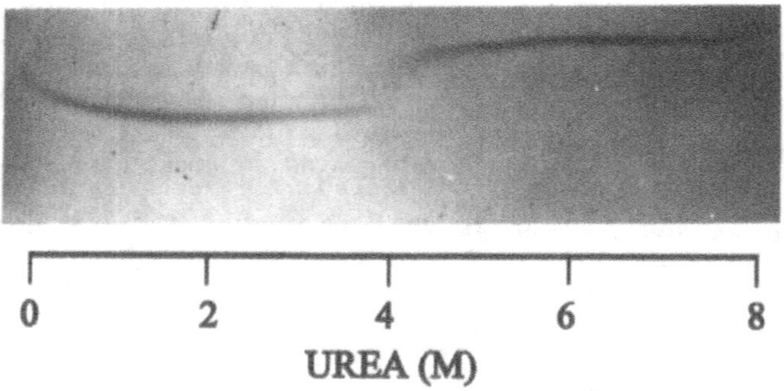

Fig. 4. Urea gradient gel electrophoresis pattern of the Y308F mutant version of FNR. The native enzyme was electrophoresed for 35 min at 250 V essentially as described by Creighton (8), and the gel was stained with Coommassie Brilliant Blue.

In toto, our results indicate that FNR holoenzyme stability is sensitive to the presence of an aromatic side chain (preferably tyrosine) at the C-terminal position. The increases in the free energy of activation introduced by mutagenesis of Tyr 308 are largely accounted for by reductions in the thermodynamic stability of the corresponding holoproteins. Although the tryptophan and phenylalanine mutants were designed to be conservative substitutions for Tyr 308, the two aromatic amino acids were not able to fully replace the functions of the wild-type residue, leading to FNR variants with decreased activity and stability (Fig. 3). This might explain why aromatic substitutions of the C-terminal tyrosine are not observed in the dozen of FNR sequences already available. It is likely that other FNR mutants with replacements at the FAD binding site will help to understand how these proteins modulate the properties of the versatile flavin cofactor.

References

1 Carrillo, N. and Vallejos, R. (1987) in *The Light Reactions. Topics in Photosynthesis.* (Barber, J., ed.) Vol. 8, pp. 527-560. Elsevier, Amsterdam.
2 Karplus, P.A., Daniels, M.J. and Herriott, J.R. (1991) *Science* 251, 60-66.
3 Bruns, C.M. and Karplus, P.A. (1995) *J. Mol. Biol.* 247, 125-145.
4 Orellano, E.G., Calcaterra, N.B., Carrillo, N. and Ceccarelli, E.A. (1993) *J. Biol. Chem.* 268, 19267-19273.
5 Newman, B.J. and Gray, J.C. (1988) *Plant Mol. Biol.* 10, 511-520.
6 Serra, E.C., Carrillo, N., Krapp, A.R. and Ceccarelli, E.A. (1993) *Protein Expres. Purif.* 4, 539-546.
7 Pace, C.N. (1990) *Trends Biotechnol.* 8, 93-98.
8 Creighton, T.E. (1986) *Methods Enzymol.* 131, 156-172.

PURIFICATION AND PROPERTIES OF PEA THIOREDOXIN m EXPRESSED IN *E. coli*.

J.López Jaramillo, A. Chueca and J. López Gorgé
Plant Biochemistry Department, Estación Experimental del Zaidín (CSIC), Granada. Spain.

1. INTRODUCTION

Thioredoxins (Td) are ubiquitous small M_r proteins with a redox fuction in different processes; this occurs through a reversible reduction of an S-S bridge formed between two cysteines existing in a conserved active site: Cys-Gly-Pro-Cys (1). A single Td has been found both in bacteria and mammals, being the 3-dimensional structures of E. coli and human Tds very well solved now. Plants represent an interesting challenge in that they contain a variety of Tds at least of three types (m, *f* and *h*), acording to their subcellular localization, biochemical reactivity and amino acid sequence. In chloroplasts two types of Tds are present, which display a high interaction with their target enzyme: Td *f* modulates fructose-1,6-bisphosphatase (FBPase)(2), while Td *m* preferentially activates NADP-malate dehydrogenase (NADP-MDH)(3).

The 3-dimensional structure of plant Tds has not been solved so far, being of high interest to obtain structural models which can help to understand the peculiarities of these proteins in plants. We have isolated a cDNA clone coding for pea Td *m*, whose nucleotide sequence and deduced primary structure has been determined (4) and subcloned into a pET expression vector. In order to get a more general view of the properties and specificity of Td *m*, this clone has been overexpressed in *E. coli* cells. The expressed pea Td m was purified and characterized, and its binding features with pea FBPase studied.

2. MATERIALS AND METHODS

Construction of recombinant pET-Td m plasmid. A cDNA coding for pea Td *m* (4) was amplified by PCR, using as primers two specific oligonucleotides complementary to the 5'and 3' ends of the coding region; these hold, respectively, Nde I and Bam HI restriction sites. Ala-65 was chosen as the first amino acid for the recombinant protein, and an initiation methionine added immediately upstream of the chosen alanine via the ATG codon present in the NdeI restriction site. Nde I and Bam HI sites were used to clone the fragment into the expression vector pET 12a.The resulting construction was used to transform *E. coli* strain BL 21(DE3).

P. Mathis (ed.), Photosynthesis: from Light to Biosphere, Vol. II, 665–668.

Expression and purification of Td m . Transformed cells were grown at 37° C in LB medium containing ampicillin. The expression was induced by 100 µM IPTG, and the protein production was allowed to occur for 4h at 37° C. The expressed protein appears basically in the soluble fraction of the *E. coli* cells, which were broken by three passages in a French pressure cell at 60 MPa. After centrifugation, the supernatant was heated for 10 min. at 80° C, the denatured proteins removed and the supernatant fractionated with ammonium sulfate (35-80% sat.). The pelleted proteins were dissolved in 30 mM Tris-HCl pH 7.9, 300 mM NaCl, 1,4 mM 2-mercaptoethanol and desalted by Sephadex G-50 filtration with the same buffer; the active fractions were pooled, concentrated and stored at -70° C.

Biochemical and inmunological methods. SDS-PAGE was perfomed as described in (5), and Western blotting according to (6). FBPase activity was determined in the presence of 10mM Mg^{2+} (7). In addition, FBPase and NADP-MDH activities were also measured after reduction of the enzyme with the DTT-Td *m* (8, 9). Sequence analysis of the N-ends was carried out by Edman degradation in an Applied Biosystems 470 A sequencer.

Interaction between the FBPase and Td m. Pea FBPase was preincubated, with or without expressed Td *m* in 100 mM Tris-HCl pH 7.9, 2.5 mM DTT, 0.4 mM fructose-1.6-bisphosphate, and 0.05 mM $CaCl_2$. The mixture was filtered through a Superose 12 column, eluted with the above buffer, and the fractions assayed by FBPase activity in the presence of excess Mg^{2+}, or after reduction with DTT-Td without addition of exogen Td *m*.

3. RESULTS AND DISCUSSION

From a pea cDNA library in λgt11 we have isolated several different positive clones to antibodies raised against pea Td *m*. One of them was subcloned in the EcoRI site of pTZ18R (4), and the isolated recombinant sequenced by the deoxynucleotide chain termination method. Our clone codes for a full 60 amino acids transit peptide, and for a Td m with 112 amino acids and 12,503 kDa molecular mass. As occurs with the transit peptide of pea Td *f*, that of Td *m* from pea starts with a MA motif in the N-terminus. A controversial point was the initial amino acid of the Td *m* amino-terminal end, since the deduced amino acid sequence shows downstream of the 60 residues of the procesing peptide the four amino acid sequence His-Ala-Arg-Glu. Native Td *m* appears by SDS-PAGE as a single polypeptide of about 16 kDa molecular mass, which shows by Edman degradation the following sequence in the N-terminus: AVNEVQVVND...Because of its total identity with the deduced sequence from our clone downstream of amino acid 65, we can state the existence of a post-translational modification during or after the transit of the precursor form into the chloroplast, with one step or sequential loss of residues His-Ala-Arg-Glu from the amino-terminal end. Td *m* from pea appears then to have 108 amino acids, with a 12 kDa molecular mass.

To study the properties and specificity of Td *m*, the gene was expressed in *E.coli*, for the production of the mature form of the protein. Because of the above mentioned identity with the N-terminus of native pea Td *m*, Ala-65 was chosen as the first amino acid for the recombinant pea Td *m*. The NdeI and Bam HI sites induced in the PCR product were used to clone the fragment into the expression vector

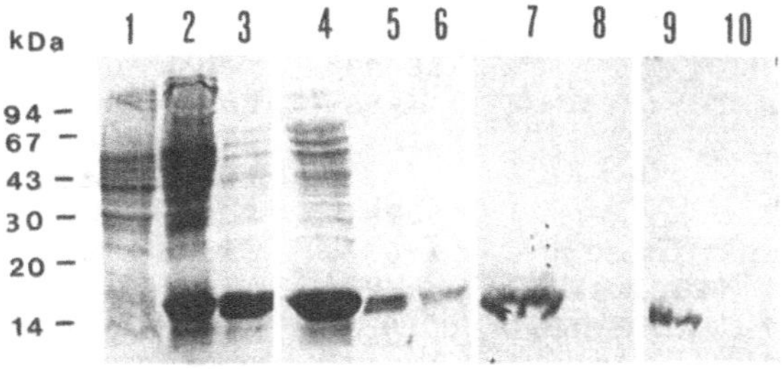

Fig 1. SDS-PAGE of the purification steps of recombinant pea Td *m*. **1.**pET 12a transformed *E. coli* lysate. **2.**pET-Td transformed *E. coli* lysate. **3.**supernatant after heat treatment. **4.**35-80% $(NH_4)_2SO_4$ sat. **5.**pooled active fractions after Sephadex G-50. **6.**pea Td *m*. Lanes **7,10** Western blot of Td *m* and Td *f* (lanes **8,9**). Lanes 7,8 were developed with antiserum against pea Td m. lanes 9,10 against anti Td *f*.

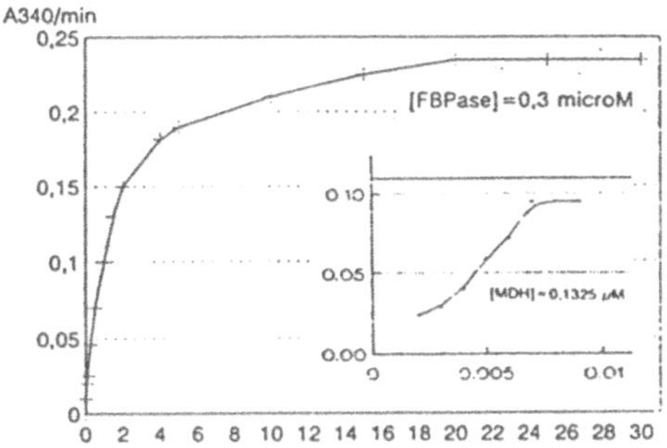

Fig 2. Saturation curves of pea FBPase and NADP-MDH by pea recombinant Td *m*

coli strain BL 21 (DE3). Small scale production was carried out to see if the recombinant protein was located in the soluble or insoluble fraction of the bacterial cells. After centrifugation of cell lysates, Western blots and activity assays showed the presence of high amounts of Td m in the soluble fraction. The recombinant pea Td *m* was purified from *E. coli* easily with a high yield of pure protein (20 mg/litre), as confirmed by SDS-PAGE (Fig.1). Polyclonal antibodies raised against pea Td *m* strongly reacted with both the native Td *m* and the recombinant protein indicating that the expressed protein had the correct primary structure. However, both proteins did not display positive reaction with anti pea Td *f* (Fig. 1). No reaction was found with lysates from control *E. coli* cells. The ability of the recombinant pea Td m to activate NADP-MDH and chloroplastic FBPase, in the presence of DTT or by Ca++, FBP and DTT, respectively, was investigated as a function of Td concentration (Fig.2).Table I shows the kinetic, immunological and physico-chemical characteristics of recombinant pea Td *m*. The saturation curves of

TABLE I. Kinetic, inmunological and physico-chemical characteristics of recombinant pea Td *m*.

Molecular mass	12 kDa
Length	108 aa
Isoelectric point	4,96
Charge	-2
anti Td *f*	--
anti Td *m*	100%
FBPase:Td m stoichiometry	1:60
Vmax ($A_{340}\cdot min^{-1}$) (FBPase)	0,235
Vmax ($A_{340}\cdot min^{-1}$) (NADPH-MDH)	0,095
$S_{0,5}$ (μM) (FBPase)	1
$S_{0,5}$ (μM) (NADP-MDH)	0,004
Hill coefficient (FBPase)	0,94
Hill coefficient (NADPH-MDH)	2,86

pea FBPase and NADP-MDH whith pea Td *m* show different kinetics (Fig.2); on the contrary that occurs with Td *f*, the saturation curve of FBPase-Td m shows hiperbolic kinetic. This behaviour can be explained on the assumption that chloroplast FBPase can bind the reduced and oxidized forms of Td with similar effectiveness (10); both are in an equilibium which depends on the redox state of system. The reduced form acts as activator, and the oxidazed one as a non competitive inhibitor of the calytic activity of FBPase (11). Assuming M_r of 160 kDa for pea chloroplast FBPase (12), and 12 kDa for pea Td *m* (4), a 60:1 stoichiomry could be determined from the saturation of pea FBPase with pea Td *m*. By contrast, the saturation for pea FBPase with pea Td *f* shows a 4:1 stoichiometry (13), a value lower than that reported for Td *m*.

When Td *m* is incubated with FBPase at pH 7.9 and 2.5 mM DTT, the bulk of FBPase elutes bound to Td *m*. Both can be detected by enzyme activity without addition of exogeneus FBPase and extra Td *m*, at an elution volume lower than that observed when FBPase was filtered alone. Interpolation in the plots obtained with M_r standars provides for the FBPase-Td complex a molecular mass about 250 kDa.

4. REFERENCES

1 Eklund H, Gleason F, Holmgren A. (1991) PROTEINS 11:13-28
2 Schürman P, Maeda K, Tsugita A. (1981) Eur. J. Biochem. 116:37-45
3 Jacquot J, Vidal J, Gadal P, Schürman P.(1978) FEBS Lett.96:243-246
4 López Jaramillo J, Chueca A, Sahrawy M, Hermoso R, Lázaro JJ, Prado FE, López Gorgé J. (1994) Plant Physiol. 105:1021-1022
5 Laemmli UK. (1970) Nature 227:680-685
6 Towbin H, Staehelin T, Gordon J. (1979) Proc Natl. Acad. Sci. USA 76:4350-4354
7 Lázaro JJ, Chueca A, López Gorgé J, Mayor F. (1974) Phytochemistry 13:2455-2461
8 Herting C, Wolosiuk RA (1980) Biochem. Biophys.Res.Commun.97:325
9 Johnson TC, Quiang R, Kung J, Buchanan B. (1987) Planta 171:321-331
10 Plá A, López Gorgé J. (1981) Biochem. Biophys. Acta 636:113-118
11 Soulié JM, Riviére M, Buc J, Goutero B, Richard J. (1987) Eur. J. Biochem. 162:271-274
12 Carrasco JL, Chueca A, Prado F, Hermoso R, Lázaro JJ, Ramos JL, Sahrawy M, López Gorgé J. (1994) Planta 193:494-501
13 Prado FE, Lázaro JJ, Hermoso R, Chueca A, López Gorgé J. (1992) Planta 188:345-353

SPECTROSCOPIC AND KINETIC CHARACTERIZATION OF SITE-SPECIFIC MUTANTS OF PLASTOCYANIN

S. Young, M. Ejdebäck, K. Sigfridsson, A. Samuelsson and Ö. Hansson,
Department of Biochemistry and Biophysics, Lundberg Laboratory,
Göteborg University and Chalmers University of Technology,
Medicinaregatan 9C, S-413 90 Göteborg, Sweden

1. Introduction

A series of mutants of spinach plastocyanin have been constructed by site-directed mutagenesis and expressed in *Escherichia coli* to elucidate the interaction between plastocyanin and photosystem 1. The proteins have been characterized by optical absorption and EPR spectroscopy, isoelectric focusing and redox titrations and the kinetics in their reaction with PS1 has been determined.

2. Results and Discussion

2.1 Hydrophobic patch; L12A, L12E, L12K, L12N, F35Y, Q88E

Table 1 summarizes pI, E^0 and spectroscopic parameters for the mutants. There are small changes in the EPR spectrum of the Leu-12 and Y83L mutants and the absorption maximum of the former is red-shifted while that of Y83L is blue-shifted (see also [1]). The changes are attributed to slight perturbations of the Cu site geometry but major structural changes can be excluded. The lack of a shift in the pI for L12E is consistent with the high pK_a of 6.6 for the Glu-12 residue (see below). The Leu-12 mutations induce shifts in E^0. The increase for L12K and the decrease for L12E can be explained by electrostatic interactions [2]. The decrease in E^0 for L12A and L12N is possibly caused by the Cu site perturbation.

The Leu-12 mutants display a reduced reactivity towards PS1 which is too drastic to be related to changes in E^0 or spectral properties (Fig. 1, [3]). A curve-fitting analysis shows that these mutants lack the fast, 70 000 s^{-1}, component present in the $P700^{ox}$ decay kinetics of WT Pc. On the other hand, a fast phase is present in the kinetics of F35Y (Table 2) and Q88E (not shown). The reactivity of L12E increases as the pH is lowered (Fig. 1) with an apparent pK_a of 6.6. WT, L12A or L12N do not exhibit such a pH dependence and we attribute it accordingly to the protonation of the Glu-12 residue [2]. There is a rate saturation in the kinetics at high Pc concentration (Fig. 2, [2]) which can not be explained by a simple binding model, since there is no fast phase. Instead we favor a model in which a rate-limiting conformational change occurs in the Pc-PS1 complex before the electron transfer (P = P700) [4,5]:

$$Pc^{red} + P^{ox} \underset{k_{off}}{\overset{k_{on}[Pc]}{\rightleftharpoons}} \underset{\text{distal}}{Pc^{red}...P^{ox}} \underset{k_c}{\overset{k_d}{\rightleftharpoons}} \underset{\text{close}}{Pc^{red}.P^{ox}} \xrightarrow{k_{et}} Pc^{ox}.P^{red}$$

P. Mathis (ed.), Photosynthesis: from Light to Biosphere, Vol. II, 669–672.

The lack of a fast phase for the Leu-12 mutants is attributed to a shift in the equilibrium towards the non-competent "distal" conformation ($K_{dc} = k_d/k_c << 1$). Thus, it can be concluded that Leu-12 is part of a critical region around the Cu ligand His–87, whose conserved structure is very important for the binding to PS1. The region also includes residues 10 and 90 [1], but not 35 and 88 (this work).

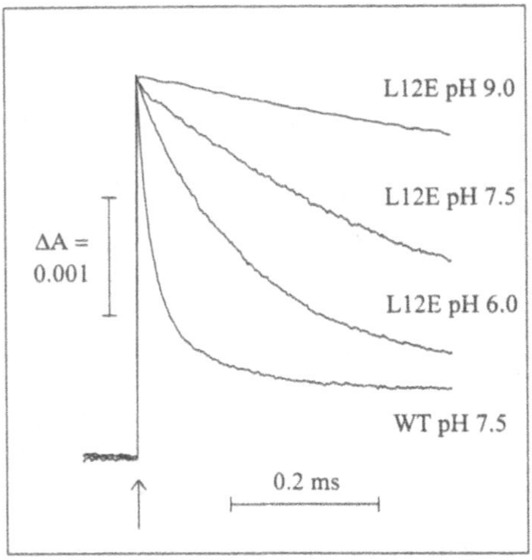

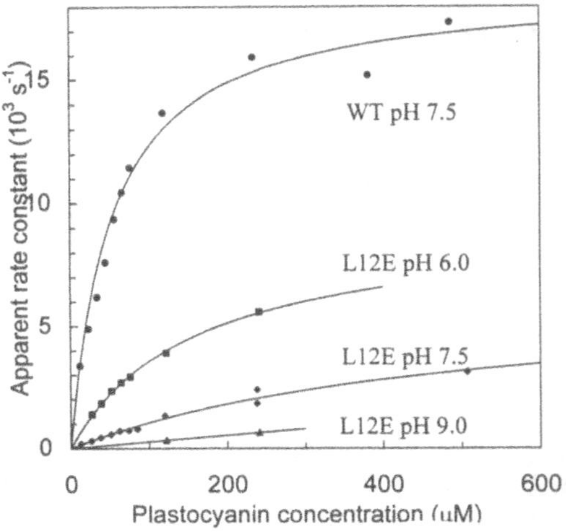

Figure 1. Flash-induced absorption transients at 830 nm of PS1 particles with WT Pc and L12E mutant at three different pH. Samples contained spinach D-144 PS1 particles (0.9 mg chl/ml), 2 mM Na-ascorbate, 0.1 mM methyl viologen, 0.24 mM Pc, 20 mM Tris (pH 7.5 or 9.0) or Mes (pH 6.0) and $MgCl_2$ (final ionic strength, 38 mM). Average of 16 flashes spaced 20 seconds apart.

Figure 2. Effect of Pc concentration on the apparent rate constant (k_2) of the intermediary phase in the decay kinetics of oxidized P700 for WT Pc compared to that of the fastest phase for the L12E mutant at three different pH.

Table 1. Isoelectric point (pI), reduction potential (E^0) and spectroscopic parameters for plastocyanin mutants. The latter properties were determined at pH 7.5.

Pc	pI	λ_{max} (nm)	$g_{\|\|}$	$A_{\|\|}$ (mT)	E^0 (mV)
WT	3.82	597	2.24	6.4	384
L12A	3.82	599	2.25	5.5	356
L12E	3.81	597	2.24	5.9	354
L12K	3.90	600	2.24	5.9	411
L12N	3.83	596	2.24	6.2	362
F35Y	3.80	597	2.24	6.3	385
D42N	3.87	597	2.24	6.4	387
Y83F	3.84	597	2.24	6.4	379
Y83H	3.88	597	2.24	6.4	419
Y83L	3.84	595	2.23	6.3	383
H87G	3.80	597(weak)	2.22/2.21	6.2/17.8	nd
H87Q	3.76	non-blue	2.22/2.24	17.5/17.5	nd
H87N	3.77	non-blue	2.22	18.2	nd
H87D	3.63	non-blue	2.22	17.8	nd

2.2 Acidic patch; D42N, Y83F, Y83H, Y83L

The spectral properties of these mutants are similar to WT (except Y83L, see above). For some of the mutants the pI shifts (Table 1), in accordance with the changes in the number of titratable groups. There is an increase in E^0 for Y83H at pH 7.5 (Table 1, [5,6]), but at higher pH it is close to WT [5], suggesting an electrostatic influence from His-83.

The decay kinetics of $P700^{ox}$ are biphasic for these mutants and for WT, with a concentration independent fast phase (k_1) and a slower phase with a rate constant (k_2) that saturates at high [Pc]. The relative amplitude of the fast phase saturates at a value (R_{max}) well below 1 (Table 2). (At 830 nm there is an additional (20%) slow phase due to Pc^{ox}.) This behavior can be accounted for by the above reaction scheme but not by a simpler binding model [2,5]. Table 2 summarizes the values obtained for k_c, k_d, k_{et} and K_{dc} (= k_d/k_c). K_{dc} is approx. 4 for WT but 6 for Y83L. Evidently, the smaller Leu residue favours a "close" complex with PS1, leading to the increased reactivity previously observed [1]. In contrast, K_{dc} is significantly lower for Y83H and D42N. Raising the pH results in an increase in K_{dc} for Y83H, suggesting that the protonation state of His-83 influences the equilibrium [5]. The effect of the D42N mutation can be interpreted similarily: An increase in charge of the acidic patch results in less of the reactive "close" complex. The conclusion is that both the acidic and hydrophobic patches make contact with PS1.

Table 2. Parameters obtained from analysis of flash-induced absorbance changes of P700, describing the electron transfer to PS1 at pH 7.5. Rate constants are in 10^3 s^{-1}

Pc	k_1	k_{2max}	R_{max}	k_{et}	k_d	k_c	K_{dc}
WT	74	19	0.74	69	20	5.2	3.9
F35Y	82	13	0.64	76	14	6.2	2.3
D42N	66	6.9	0.56	61	7.4	4.9	1.5
Y83H	41	7.0	0.37	31	9.2	9.7	0.9
Y83F	70	12	0.67	65	13	5.0	2.6
Y83L	75	15	0.82	72	16	2.7	5.7

2.3 The copper ligand His-87; H87G, H87A, H87N, H87E, H87Q, H87R

His-87 is a Cu ligand and it is likely that the ET to P700 occurs through this residue. In order to elucidate the role of this residue in the coordination of the copper and in the ET, His-87 has been mutated to a Gly, Asp, Asn, Glu, Gln and an Arg. The mutant proteins were difficult to express and none of the mutants H87R and H87E could be obtained. The absorbance at 597 nm in the UV-Vis spectrum of the WT protein disappeared in the mutants H87Q, H87D and H87N, while the type 1 EPR signal, characteristic of WT, changed into a type 2 (Table 1, Fig. 3). The ET to PS1 is impaired for these mutants due to the conformational change.

The mutant H87G as prepared has properties similar to the mutants above. If ferricyanide is added to fully oxidize the sample, a weak absorption appears at 597 nm in the UV-Vis spectra and a type 1 EPR signal can be seen in addition to the type 2 signal (Fig. 3). Our current hypothesis is that one population is locked in a type 1 conformation, possibly by buffer molecules, the other is of type 2. In summary, substitution of the His–87 results in a conformational change of the copper site, unabling the protein to transfer electrons to PS1. There are two different populations of H87G with different spectroscopic properties.

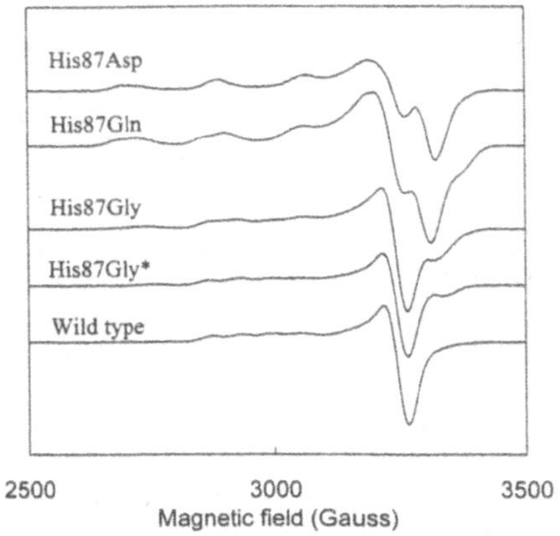

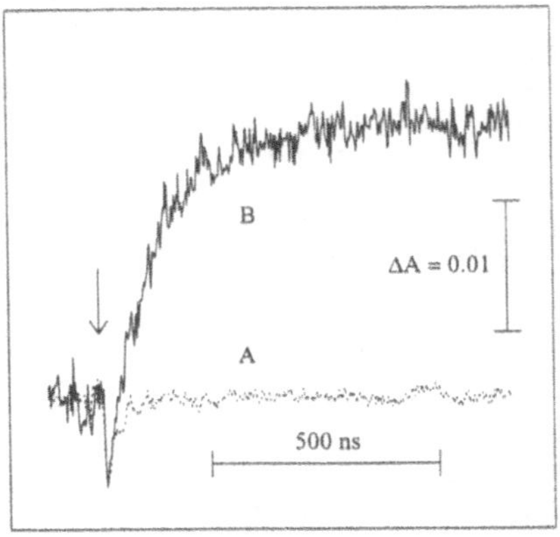

Figure 3. X-band EPR-spectra at 77 K of some His-87 mutants. *A spectrum obtained from a sample without ferricyanide has been subtracted.
Figure 4. Flash-induced absorption transient of the Tyr83His-Ru complex at 600 nm, before (A) and after (B) the addition of the external quencher $Ru(NH_3)_6Cl_2$.

2.4 Ruthenium-modification of Y83H
Intramolecular ET in Y83H has been investigated by attaching a photoactive ruthenium (Ru) complex to the His-residue [7]. The residue is surface exposed and located 10–12 Å from the copper ion at the entrance of a proposed natural ET pathway from cythocrome *f*.

$Ru(bpy)_2(CO_3)$ (bpy = 2,2`-bipyridine) has been attached to the residue to yield, after addition of excess imidazole (im), $Ru(bpy)_2(im)(His)^{2+}$. The protein-bound Ru complex is luminescent, with a 20 ns lifetime for the excited (Ru*) state. An ET quencher ($Ru(NH_3)_6^{3+}$) was added, to oxidize a small fraction of the bound Ru* complex, yielding Ru^{3+}. With copper reduced, before excitation, the copper oxidation was monitored by transient absorption spectroscopy at 600 nm (Fig. 4). The obtained k_{ET} value for the reaction was $9.5 \cdot 10^6$ s^{-1}. From an estimated driving force of 0.56 eV and an assumed reorganization energy of 0.87 eV, a strong electronic coupling (H_{DA} = 0.32 cm^{-1}) is suggested along the Ru-His83-Cys84-Cu pathway [7], stronger than predicted by theoretical calculations (H_{DA} = 0.06 cm^{-1}) [8].

References
1. Haehnel, W., Jansen, T., Gause, K., Klösgen, R. B., Stahl, B., Michl, D., Huvermann, B., Karas, M. and Herrmann, R. G. (1994) EMBO J. 13, 1028-1038
2. Sigfridsson, K., Young, S. and Hansson, Ö. (1995) Submitted.
3. Nordling, M., Sigfridsson, K., Young, S., Lundberg, L. G. and Hansson, Ö (1991) FEBS Lett. 291, 327-330
4. Bottin, H. and Mathis, P. (1985) Biochemistry 24, 6453-6460
5. Sigfridsson, K., Hansson, Ö., Karlsson, B. G., Baltzer, L., Nordling, M. and Lundberg, L. G. (1995) Biochim. Biophys. Acta 1228, 28-36
6. Kyritsis, P., Dennison, C., McFarlane, W., Nordling, M., Vänngård, T., Young, S. and Sykes, A. G. (1993) J. Chem. Soc. Dalton Trans. 2289-2296
7. Sigfridsson, K., Sundahl, M., Bjerrum, M. J. and Hansson, Ö. (1995) Submitted.
8. Christensen, H. E. M., Conrad, L. S., Mikkelsen, K. V., Nielsen, M. K. and Ulstrup, J. (1990) Inorg. Chem. 29, 2808–2816

PURIFICATION AND CHARACTERIZATION OF THE SULFIDE-QUINONE REDUCTASE (SQR) OF *RHODOBACTER CAPSULATUS* DSM 155

M. Schütz[§], Y. Shahak[&], E. Padan[$] and G. Hauska[§]

[§]Lehrstuhl für Zellbiologie und Pflanzenphysiologie, Universität Regensburg, Germany; [&]Institute of Horticulture, the Volcani Center, Agricultural Research Organization, Israel; [$]Division of Microbial and Molecular Ecology, Institute of Life Science, the Hebrew University, Israel

1. Introduction

Sulfide photooxidation by anoxygenic photosynthetic bacteria is well known and has been extensivly reviewed by Brune (1). The sulfide oxidizing entity in these organisms has been considered to be a flavocytochrome *c* (2). However, not all sulfide oxidizing species contain this component and alternatives have been discussed (1). Oxidation of sulfide by quinone has first been described for *Rhodobacter sulfidophilus* (3), later for the green sulfur bacteria *Chlorobium limicola f. thiosulfatophilum* (4) and *Chlorobium tepidum* (5) and for the non sulfur purple bacterium *Rhodobacter capsulatus* (6). For the last three organisms it was demonstrated that sulfide oxidation is connected to the reduction of the cytochrome-bc complex via the quinone pool. Sulfide dependent photosynthetic electron transport also has been observed in the oxygenic photosynthetic cyanobacterium *Oscillatoria limnetica* (7) that shifts into anoxygenic, bacteria like photosynthesis with sulfide as electron donor to photosystem-I. The inducible factor was characterized as sulfide-quinone reductase (SQR). This enzyme was isolated as a monomeric membrane bound 57kDa flavoprotein (8). Here we describe the purification and characterization of a similar enzyme from the non sulfur purple bacterium *Rhodobacter capsulatus* DSM 155 and give the first sequence information of the *sqr*-gene.

2. Materials and Methods

Purification of the SQR: Screwed batch cultures of *Rhodobacter capsulatus* were grown photoheterotrophically in RCV medium (9) supplemented with 150µM Na_2S. Chromatophores were prepared and their bacteriochlorophyll-a content was measured as described (10). Frozen chromatophores were resuspended in a buffer containing 50mM glycylglycin pH7.0 and 20mM EDTA at 0.8mg bacteriochloropyll-a per ml and homogenized. Thesit (Boehringer Mannheim) was added up to 10mM. After stirring on ice for 30 min the mixture was centrifuged for 2h at 200.000xg. The supernatant was applied on a DEAE-Cellulose column equilibrated with a buffer containing 10mM BisTris HCl pH6.5, 2mM EDTA and 0.4mM Thesit (BITE). Elution was performed with a linear gradient from 0 to 0.4M NaCl in BITE. The most active fractions were pooled and concentrated with Centricon-100 (Amicon). The sample was loaded on a density gradient from 10-40% sucrose in BITE and centrifuged for 14h at 120.000xg. The most active fractions were again combined, concentrated and applied on a Mono-Q column (Pharmacia, HR5/5). A linear gradient from 0-0.4M NaCl in BITE was performed. After SDS-PAGE the pure fractions were combined. *Spectroscopic Assays:* SQR activity was determined as sulfide-dependent decyl-UQ (Sigma) reduction in a spectrophotometer

P. Mathis (ed.), Photosynthesis: from Light to Biosphere, Vol. II, 673–676.

(Kontron Uvikon 860) recorded at 275nm under N_2 atmosphere using a milli-molar extinction coefficient of 15. Reaction mixture contained 10mM BisTris HCl pH6.5, 20μM UQ. The reaction was started by addition of 40μM Na_2S. Fluorescence spectra were measured in a Spex Fluoromax spectrofluorimeter at room temperature under air. *Other Assays:* Sulfide concentration was measured by the methylene-blue method (11). Genetical engineering procedures were done according to standard procedures.

3. Results and Discussion

Purification of SQR: Various methods were tested to detach the SQR from washed chromatophores. The most effective condition was the solubilization with Thesit. Using this detergent nearly 50% of the total activity of the chromatophores could be obtained (Table 1). The first step of the purification procedure was a chromatography with DEAE-cellulose. The fractions with the highest specific activity were eluted at low salt concentration. Further purification attempts with cation exchange chromatography, phenylsepharose column and ammonium sulfate precipitation resulted in a dramatic loss of activity. The concentrated fractions from the DEAE-cellulose column were loaded on a density gradient of 10-40% sucrose. The most active samples were recovered from 25 to 32% sucrose. These were applied on a Mono-Q column. By elution with a gradient from 0 to 0.4M NaCl in BITE pure SQR was obtained at low salt concetration. The resulting total purification factor of 27 (Table I) indicates a partial loss of activity during the purification procedure. A comparison of the intensities of the silver stained 55kDa bands in the SDS-PAGE at different purification steps (Figure 1) to the activity of the loaded samples (every lane represents the same activity) point out a great loss of activity during the first purification steps. This might be due to a loss of the prosthetic group from the SQR .

Table 1. Purification steps of the SQR

Purific. Fraction	Protein	Activity	Specific Activity	Purification Factor
	mg	nmol UQ/min	nmol UQ /mgProt min	
Chromatoph.	825	105600	128	1
Thesit extract	145	48000	331	2.6
DEAE-Column	6	7900	1320	10
Filtration	4	7200	1800	14
Suc. Gradient	1.2	3800	3120	25
MQ-Column	0.5	1800	3500	27

Figure 1. 12% SDS-PAGE at different purification steps. Every lane represents the same activity. (silver-stained)

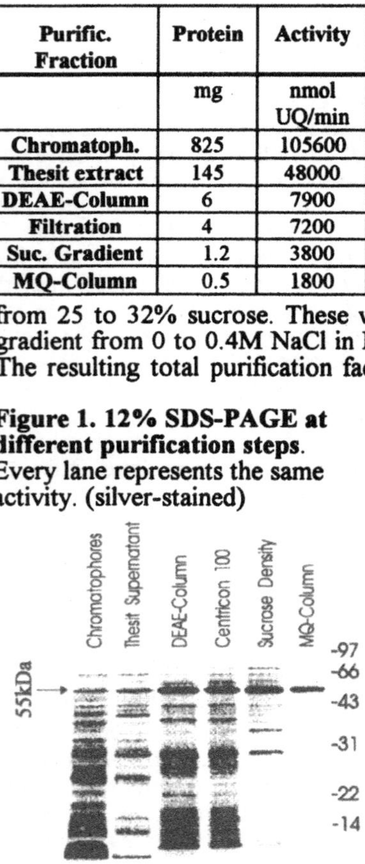

Properties of the purified SQR: The pH-optimum of the purified SQR is around pH6.3. The dependence on the concentration of its substrates was determined (Figure 2). The apparent K_M for sulfide is about 2μM and about 2μM for decyl-UQ. This K_M for sulfide agrees with the one determined for reduction of added UQ and of cytochromes (6) in chromatophores of *Rb.capsulatus*. It also correlates with the K_S of 2.3μM for sulfide dependent growth of *Rb.capsulatus*. (12) So the affinity for sulfide is four times higher and for the quinone substrate about 15 times higher than that of *O.limnetica* SQR. (8).

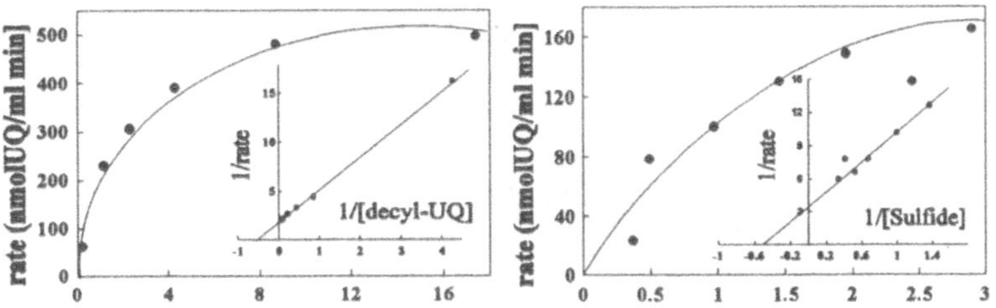

Figure 2. Dependence on the concentration of the substrates. A, Decyl-UQ conc. dependence. B, Sulfide conc. dependence. Rates were corrected for the noncatalytic reaction (about 1%)

The prosthetic group: Figure 3 shows the excitation and emission spectra of the purified SQR. The excitation maxima around 375nm and 458nm with a predominant peak at 280nm are typical for a flavoprotein. The maximum of the fluorescence is around 518nm. Adding sulfide to a final concentration of 160μM leads to an almost total decrease of the excitation maxima caused by the reduction of the FAD. Comparing the NH_2-terminal sequence (Figure 4) with the FAD-fingerprint of the FAD-binding domain of many FAD containing oxidoreductases (13) all the conserved amino-acids within the first thirty residues were found. So like the *O.limnetica* SQR the SQR of *Rb.capsulatus* is a flavoprotein.

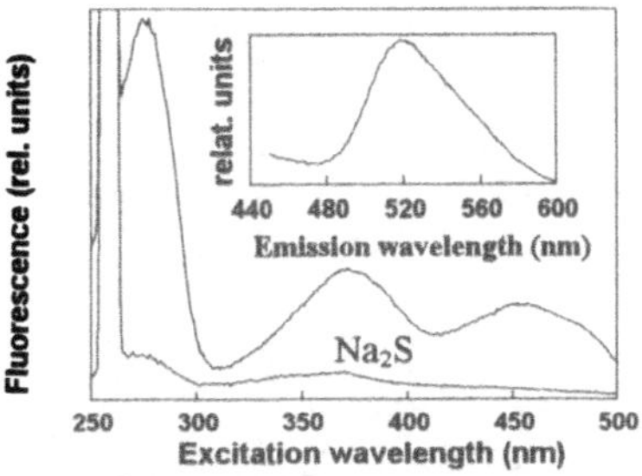

Figure 3. Fluorescence spectra of the SQR. Excitation spectrum (Emission at 518nm) *Inset*, Emission spectrum (Excitation at 375nm)

First sequence information: After tryptic digestion four peptides were sequenced. PCR with oligonucleotides deduced from these sequences and the NH_2-terminal sequence produced two predominant DNA-fragments of about 800bp and 1100bp (Figure 5). Three peptides and the NH_2-terminal sequence were found in the amino-acid sequence deduced from the PCR-fragments. Thus, this fragments represents about two thirds of the *sqr*-gene of *Rb. capsulatus* DSM 155. Comparison to the amino-acid sequences of other FAD containing oxido-reductases the highest identity was about 20% and the similarity about 40%. The hydrophobicity profile did not show any segment that satisfies the criteria for prediction of a membrane-spanning region. So either the SQR is anchored in the membrane with a still not seqenced region in the C-terminal part of the protein, or it is not an intrinsic membrane protein.

```
     1        10        20        30
R.c  AHIVVLGAGLGGAIMAYELREQVRKEDKVTVITKD
     ||:.|:||||:|   |||||. :.:::.|
O.l  AHVAVIGAGLAGLPTAYELRHILPRQQHV
      ** * * *  *   *  *        * *   *
FAD   KA A G G  G   A  A        A A   D
      RI I          I  I        I I   E
      HL L          L  L        L L
      SV V          V  V        V V
      TM M          M  M        M M
      QC C          C  C        C C
```

Figure 4. NH_2-terminus sequences of SQR of *Rb.capsulatus* and *O.limnetica*. Identical (|) or similar (:) amino-acids are indicated. The fingerprint of the FAD-binding domain of FAD enzymes taken from (13) is marked by asterisk (*). *O.limnetica* sequence published in (8)

Figure 5. Sequence of the PCR-Fragment and deduced amino-acid sequence

```
                                                 .........................
                                                 A  H  I  V  V  G  A  G  L
   1   GGGGGGGCGATCATGGCGTATGAGTTGCGCGAACAGGTGCGCAAGGAAGACAAGGTCACC
       G  G  A  I  M  A  Y  E  L  R  E  Q  V  R  K  E  D  K  V  T
  61   GTCATCACCAAGGACCCGATGTATCACTTCGTGCCCTCGAACCCCTGGGTGGCGGTGGGC
       V  I  T  K  D  P  M  Y  H  F  V  P  S  N  P  W  V  A  V  G
 121   TGGCGCGACCGCAAGGAGATCACCGTCGATCTGGCACCGACGATGGCCAGGAAGAACATC
       W  R  D  R  K  E  I  T  V  D  L  A  P  T  M  A  R  K  N  I
 181   GACTTCATCCCGGTGGCGCGAAACGCCTGCATCCGGCCGAAAACCGGGTCGAGCTGGAAA
       D  F  I  P  V  A  R  N  A  C  I  R  P  K  T  G  S  S  W  K
 241   ACGGCCAGTCGGTCAGTTGTGACCAGATCGTCATCGCCACGGCCCGGAACTGGCCTTCGA
       T  A  S  R  S  V  V  T  R  S  S  S  P  R  P  G  T  G  L  R
 301   ACGAGAATCGAGGGCTTCGGCCCGAGGGGCACACCCAGTCGAATCTGCCATATCGACCAT
       T  R  I  E  G  F  G  P  R  G  T  P  S  R  I  C  H  I  D  H
 361   GCCGAGGCGGCAGGAGCGGCCTTTGACCGCTTCTGCGAAAACCCCGGCCCGATCATCATC
       A  E  A  A  G  A  A  F  D  R  F  C  E  N  P  G  P  I  I  I
 421   GGCGCGCAAGGCGCCAGCTGCTTCGGCCCGGCTTACGAATTCACCTTCATCCTCGACACC
       G  A  Q  G  A  S  C  F  G  P  A  Y  E  F  T  F  I  L  D  T
 481   GCGCTGCGCAAGCGCAAGATCCGCGACAAGGTGCCGATGACCTTCGTCACCTCGGAACCC
       A  L  R  K  R  K  I  R  D  K  V  P  M  T  F  V  T  S  E  P
 541   TATGTCGGGCATCTGGGGCTGGACGGGGTGGGCGACACCAAGGGCCTGCTGGAAGGNAAC
       Y  V  G  H  L  G  L  D  G  V  G  D  T  K  G  L  L  E  G  N
 601   CTGCGCGACAAGCACATCAAATGGATGACCTCGTACCCGGATCAAGCGCGTCGAGCCCGG
       L  R  D  K  H  I  K  W  M  T  S  Y  P  D  Q  A  R  R  A  R
 661   CAAGATGGTGGTCGAAGAGGAGACCGAAGACGGCACGGTGAAGCCCGAGAAAGAGCTGCC
       Q  D  G  G  R  R  G  D  R  R  R  H  G  E  A  R  E  R  A  A
 721   TTCGGCTATGCGATGATGCTGCCCCTTTCGCGGATCAAGGCGCTGATGGGGATCGAGGG
       F  G  Y  A  M  M  L  P  L  S  R  D  Q  G  A  D  G  D  R  G
 781   GCTGGTCAATCGCGCGGCTTCGTCATCGTCGATCAGCACCAGCAGAACCCCACCTTCAAG
       A  G  Q  S  R  G  F  V  I  V  D  Q  H  Q  Q  N  P  T  F  K
 841   AATGTCTTCGCGGTGGGGGTCTGCGTGGCGATCCCGCCGGTCGGGCCCGGGCCGGTGCCC
       N  V  F  A  V  G  V  C  V  A  I  P  P  V  G  P  G  P  V  P
 901   TGCGGTGTGCCGAAGACCGGCTTCATGATCGAAAGCATGGTCACTGCCACCGCCCACAAC
       C  G  V  P  K  T  G  F  M  I  E  S  M  V  T  A  T  A  H  N
 961   ATTGGCCGGATCGTGCGCGGCCTTGAGCCCGACGAGGTCGGCAGCTGGAACGCCGTCTGT
       I  G  R  I  V  R  G  L  E  P  D  E  V  G  S  W  N  A  V  C
1021   CTGGCGGATTTCGGCGACAAGGGCATCGCCTTCGTGGCACAGCCGCAGATCCCGCCGCGC
       L  A  D  F  G  D  K  G  I  A  F  V  A  Q  P  Q  I  P  P  R
1081   AACGTGAACTGGTCCA.........
       N  V  N  W  S  S  Q  G  K
```

References

1 Brune, D.C. (1998) Biochim Biophys Acta 975, 189-221
2 Kusai, K. and Yamanaka, T. (1973) Biochim Biophys Acta 325, 304-314
3 Brune, D.C. and Trüper, H.G. (1986) Arch Microbiol 145, 295-301
4 Shahak, Y. et al. (1992) FEBS Lett 299, 127-130
5 Klughammer, C. et al. (1995) Photosynth Res 43, 27-34
6 Shahak, Y. et al. (1994) Photosynth Res 39, 175-181
7 Padan, E. (1979) Annu Rev Plant Physiol 30, 27-40
8 Arieli, B. et al. (1994) JBiol Chem 269, 5705-5711
9 Weaver, P.F. et al. (1975) Arch Microbiol 105, 207-216
10 Baccarini-Melandri, A. and Melandri, B.A. (1972) Methods Enzymol 23, 556-561
11 Cline, J.D. (1969) Limnol Oceanogr 14, 454-458
12 Van Gemerden, H. (1984) Arch Microbiol 139, 289-294
13 Wierenga, R.K. et al. (1986) J Mol Biol 187, 101-107

LOCALIZATION OF FERREDOXIN-NADP REDUCTASE WITH AND WITHOUT N-TERMINAL BLOCK IN SPINACH THYLAKOID MEMBRANES

Masateru Shin[1], Kozo Sakai[2], Shigehiro Obata[3], Izumi Nishimura[4], and Naoko Sakihama[5], [1]Kobe Yamate College, Chuo-ku, Kobe, 650; [2]Sekisui Plastics Co., Tenri-shi, Nara 632; [3]Kishida Chemicals, Co., Sanda-shi, Hyogo 669-13; [4]Faculty of Science, Kobe University, Nada-ku, Kobe, 657; and [5]Osaka Institute of Technology, Asahi-ku, Osaka 535, Japan

1. Introduction

Resent studies on binding of ferredoxin-NADP reductase (FNR) to spinach thylakoids provide evidence for existence of two FNRs that differ in binding affinity towards thylakoid membranes (1, 2). One is the loosely bound FNR (lFNR), located on the stromal surface of thylakoids in the form of a protein complex being associated with a connective protein named "connectein" to participate in the non-cyclic photosynthetic electron flow. The other is the tightly bound FNR (tFNR), localized in the more hydrophobic region of thylakoids. Its physiological function is not clear but it may be involved in the cyclic part of photosynthetic electron flow.

In this work, our interest was focused on whether the two FNRs differently localized have different primary structures. The primary structure of FNR has been determined with the mature protein (3) and that of the precursor peptide has been deduced from cDNA sequence (4). These two structures differ in amino acids at the following two positions. 1) N-terminus, γ-pyroglutamic acid (<Glu) for the mature protein and Gln for the precursor; and 2) the 269th residue, Val for the former and Phe for the latter. This may suggest that there is more than one FNR variant with difference at these two positions in spinach.

2. Procedure

2.1 *Extraction of FNRs from spinach chloroplasts*

Whole spinach chloroplasts were isolated by the method of Cockburn et al. (5) and suspended in 20 mM Tris-HCl buffer, pH 7.8, containing 5 mM $MgCl_2$.

Total FNR was extracted from the chloroplast suspension with 35% acetone, precipitated with 75% acetone, and air-dried.

lFNR was extracted from the chloroplasts by incubating in 20 mM Tris-HCl buffer, pH 7.8, containing 5 mM $MgCl_2$ and 40% saturated ammonium sulfate at 0° C for 15 min. The extracted lFNR was precipitated with ammonium sulfate at 75% saturation. The precipitate was dissolved with 20 mM Tris-HCl buffer, pH 9.3, and passed through a Sephadex G-25 column equilibrated with the same buffer to remove ammonium sulfate. The eluate was fractionated with acetone between 35 and 75%, and air-dried.

tFNR was extracted with 35% acetone from lFNR-depleted chloroplasts, precipitated with 75% acetone, and air-dried. The lFNR-depleted chloroplasts were prepared by removing lFNR from the whole chloroplasts with 20 mM Tris-HCl buffer, pH 7.8, containing 5

P. Mathis (ed.), Photosynthesis: from Light to Biosphere, Vol. II, 677–680.

mM $MgCl_2$ and 40% saturated ammonium sulfate at 0° C for 2 h. The lFNR-depleted chloroplasts were washed once to remove ammonium sulfate with an isotonic medium (10 mM Na-pyrophosphate buffer, pH 6.5, containing 5 mM $MgCl_2$ and 0.33 M sorbitol).

2.2 *N-terminal analysis*
Acetone precipitate of each FNR fraction was dissolved in an SDS-PAGE sample buffer (0.25 M Tris-HCl buffer, pH 6.8, containing 8% SDS and 8% β-mercaptoethanol) and heated at 60° C for 1 h. Proteins were separated on SDS-gel (acrylamide concentrations: 10% for resolving gel and 5% for stacking gel), transferred onto PVDF membrane (Transblot, Bio-Rad Laboratories, Hercules, CA, pore size 0.22 μm), and stained with Coomassie Brilliant Blue (CBB) R-250. The FNR peptide band, which was immunologically identified using a rabbit anti-spinach FNR antiserum, was cut out and directly analyzed for N-terminal sequence up to the 5th residue with a Tosoh amino acid sequence analyzer (Tosoh Corp., Tokyo). The N-terminus blocked with <Glu was deblocked by the γ-pyroglutamyl peptidase method (6).

2.3 *Identification of the 269th amino acid residue*
The FNR samples (lFNR and tFNR) were further purified by affinity chromatography on a Matrex Red A and a ferredoxin-Sepharose columns. Then they were carboxymethylated, and digested with 1/50 amount (in moles) of arginyl endopeptidase (Takara Shuzo, Kyoto) in 50 mM Tris-HCl buffer, pH 8.0, containing 2 M urea at 37° C for 16 h, with subsequent addition of the same amount of enzyme and further incubation for 8 h (7). The peptide fragments were separated by Tricine-SDS-PAGE (8). A 4% T, 3% C gel was used as a stacking gel and a 16.2% T, 6% C gel was a resolving gel. No spacer gel was used. Separated peptides were then electroblotted onto PVDF membrane, and stained with CBB. The peptide band (R-9), which included the 269th amino acid residue, was cut off from the PVDF membrane for sequence analysis.

2.4 *N-terminal analysis of PSI-bound FNR*
Spinach chloroplasts were isolated by using an isolation medium (50 mM Tris-HCl buffer, pH 8.0, containing 0.35 M NaCl, 5 mM $MgCl_2$ and 1 mM potassium ferricyanide). The chloroplasts were broken in a hypotonic medium (20 mM Tris-HCl buffer, pH 8.0, containing 5 mM $MgCl_2$ and 1 mM potassium ferricyanide). The thylakoids were mixed with 5% Triton X-100 in 50 mM Tris-HCl buffer, pH 8.3, containing 5 mM $MgCl_2$ and 1 mM potassium ferricyanide to make a detergent/chlorophyll ratio (w/w) of 25, and the mixture was incubated for 15 min with gentle stirring at 4° C in the dark. PSI particles in the mixture were adsorbed on a DEAE-Toyopearl 650M column equilibrated with 20 mM Tris-HCl buffer, pH 8.3, containing 5 mM $MgCl_2$. The column was washed with 20 mM Tris-HCl buffer, pH 8.3, containing 5 mM $MgCl_2$, 50 mM NaCl and 0.07% Triton X-100. The PSI particles adsorbed were then eluted by increasing NaCl concentration to 200 mM. Further purification was done by chromatography on a Sephadex G-150 column equilibrated with 20 mM Tris-HCl buffer, pH 8.3, containing 5 mM $MgCl_2$, 200 mM NaCl and 0.07% Triton X-100.
FNR bound on the PSI particles was extracted with 35% acetone using an ultrasonic disruptor for 10 min, precipitated with 75% acetone, and air-dried. The FNR fraction was separated by SDS-PAGE/electroblotting. The portion of PVDF membrane adsorbing FNR was immunologically identified and cut out for N-terminal sequence analysis.

3. Results and Discussion

3.1 *N-terminal structure of FNR*

Table 1. N-terminal analysis of spinach FNR

Preparation	Deblocking	N-terminal structure	Content %	Ratio Q:<Q
Total FNR	-	QIASD--	100	
	+	QIASD--	69	69:31
		IASDV--	31	
tFNR	-	QIASD--	100	
	+	QIASD--	100	100:0
lFNR	-	----------	0	
	+	----------	0	0:?
PSI-FNR	-	QIASD--	84	
		IASDV--	16	

The N-terminal amino acid of mature spinach FNR is reported to be <Glu (3), while Gln is deduced at this position from nucleotide sequence of its precursor cDNA (4). This suggests that the FNR precursor synthesized in cytosol undergoes processing in chloroplasts to generate an FNR peptide with Gln at the N-terminus (Q-FNR), and then Q-FNR might be converted to <Q-FNR, an FNR peptide with <Glu at the N-terminus, by cyclization of N-terminal Gln. However, we have recently found that Q-FNR is actually present in chloroplasts (9, 10). Since then, we are interested in physiological significance of the N-terminal block of FNR related to function and localization on thylakoids. We intended to clarify distribution of blocked and unblocked FNRs in different regions of thylakoids. However, the N-terminal moiety of FNR is extremely susceptible to degradation by endogeneous proteases during isolation (11). So we took special care to prevent the proteolytic attack: rapid extraction of FNR with acetone, separation by SDS-PAGE/electroblotting, and direct sequencing of immobilized FNR peptide on PVDF membrane.
First, N-terminal structure of total FNR was analyzed to estimate the content of blocked and unblocked FNRs in chloroplasts. Table 1 shows that total FNR released Gln at the first cycle before removal of N-terminal <Glu with γ-pyroglutamyl peptidase, while it released Gln and Ile after the removal. From the ratio of liberated Gln to Ile, Q-FNR and <Q-FNR were supposed to exist in chloroplasts in a ratio of 69:31. Next, lFNR and tFNR were separately extracted and analyzed. As for tFNR, Gln was detected as the sole N-terminal amino acid regardless of deblocking. On the other hand, lFNR released no amino acid without deblocking, indicating that its N-terminus was blocked. Unexpectedly, however, no Ile was liberated from lFNR even after deblocking. The reason why γ-pyroglutamyl peptidase effective in deblocking <Q-FNR in the total FNR preparation was inactive with the isolated lFNR preparation is under investigation.

3.2 *The 269th amino acid residue in lFNR and tFNR*
Digestion of spinach FNR with arginyl endopeptidase generates ten peptide fragments, R-1 to R-10, and the 269th amino acid is located on the peptide R-9 with a molecular mass of 6 kDa including the 251st to 301st amino acids (3). Therefore, a peptide band with a molecular mass closest to this value was analyzed. Both lFNR and tFNR had Val-269, and their partial sequence from the residues 251 to 270 matched perfectly with that reported by Karplus (3).

3.3 *N-terminal structure of PSI-bound FNR*
Triton X-100 used for isolation of the PSI particles liberated the bulk of membrane-bound

FNR during disruption of thylakoids. The liberated FNR could be separated from PSI fraction by column chromatography. Nevertheless, the isolated PSI particles still retained 1.5 mol FNR per mol P700. Such firm binding of FNR to PSI, surviving the detergent treatment, suggests that the PSI-bound FNR belongs not to the lFNR, but to the tFNR category. As we expected, the PSI-bound FNR released Gln as shown in Table 1. However, Ile was also detected without deblocking to give a ratio of 84:16. The minor FNR peptide might be generated from Q-FNR by proteolytic attack during isolation.

4. Conclusion

The two FNRs, tFNR and lFNR, with different binding affinity to thylakoids had different N-terminal amino acids. tFNR had Gln, an unblocked N-terminus. lFNR had a blocked N-terminus, and its N-terminal amino acid might most probably be <Glu. Since no other difference in primary structure was found so far between the two FNRs, their structural difference must be generated post-translationally. PSI-bound FNR had Gln at the N-terminus, suggesting that it belongs to the tFNR category.

References

1 Matthijs, H.C.P., Coughlan, S.J. and Hind, G. (1986) J. Biol. Chem. 261, 12154-12158

2 Shin, M., Nishikawa, T., Sekido, T. and Sakihama, N. (1994) in Flavins and Flavoproteins 1993 (Yagi, K., ed.) pp.413-421, Walter de Gruyter, Berlin, Germany

3 Karplus, P.A., Walsh, K.A. and Herriott, J.R. (1984) Biochemistry 23, 6576-6583

4 Jansen, T., Reiländer, H., Steppuhn, J. and Herrmann, R.G. (1988) Curr. Genet. 13, 517-522

5 Cockburn, W., Walker, D.A. and Baldry, C.W. (1968) Plant Physiol. 43, 1415-1418

6 Hirano, H., Komatsu, S., Nakamura, A., Kikuchi, F., Kajiwara, H.,Tsunasawa, S. and Sakiyama, F. (1991) Theor. Appl. Genet. 83, 153-158

7 Matsushita, H., Kato, I., Aoyama, H., Tsunasawa, S. and Sakiyama, F. (1989) Tanpakushitu Kakusan Koso 34, 374-379 (in Japanese)

8 Schägger, H. and von Jagow, G. (1987) Anal. Biochem. 166, 368-379

9 Nishikawa, T., Sakai, K., Sakihama, N. and Shin, M. (1992) in Research in Photosynthesis (Murata, N., ed.) pp.543-546, Kluwer Academic Publishers, Dordrecht, The Netherlands

10 Sakihama, N., Shin, M. and Obata, S. (1994) in Flavins and Flavoproteins 1993 (Yagi, K., ed.) pp.439-442, Walter de Gruyter, Berlin, Germany

11 Shin, M., Tsujita, M., Tomizawa, H., Sakihama, N., Kamei, K. and Oshino, R. (1990) Arch. Biochem. Biophys. 279, 97-103

NUCLEOTIDE BINDING TO THE PROTON-TRANSLOCATING NICOTINAMIDE NUCLEOTIDE TRANSHYDROGENASE FROM *Rhodospirillum rubrum.*

Tania Bizouarn[1], Christine Diggle[1], Rachel L.Grimley[1], Nick P.J.Cotton[1], Philip G.Quirk[1], Christopher M.Thomas[2] and J.Baz Jackson[1]

Schools of Biochemistry[1] and Biological Sciences[2], University of Birmingham, Edgbaston, Birmingham, B15 2TT, UK.

Abstract

Th_s is the soluble, NAD(H)-binding subunit of the proton-translocating transhydrogenase from *R.rubrum*. Recombinant Th_s expressed in *E.coli* reconstitutes transhydrogenation activity to chromatophore membranes depleted of the native protein. Using NMR spectroscopy the binding of nucleotide analogues to Th_s is shown to affect the conformational dynamics of a mobile loop that protrudes from the surface of the protein. Mutation of the invariant Tyr residue in the mobile loop leads to aberrant closure of the nucleotide loop during nucleotide binding and to loss of catalytic activity in the reconstituted enzyme.

1. Introduction

Transhydrogenase couples the transfer of reducing equivalents (hydride ion equivalents) between NAD(H) and NADP(H) to the translocation of protons across a membrane.

$$NADH + NADP^{+} + H^{+}_{out} \Leftrightarrow NAD^{+} + NADPH + H^{+}_{in}$$

The enzyme is located in the cytoplasmic membrane of bacteria, where it is responsible for NADPH production during cellular biosynthesis, and in the inner mitochondrial membrane of animal cells, where it may serve in the regulation of the tricarboxylic acid cycle by the protonmotive force [1]. In photosynthetic bacteria the enzyme resides in the intracytoplasmic membrane. The generation of a protonmotive force by the action of the photosynthetic electron transport chain drives the reaction from left to right. Transhydrogenase has separate catalytic binding sites for NAD(H) and for NADP(H) [2]. These are thought to reside in relatively hydrophilic domains of the protein that protrude into the bacterial cytoplasm from the membrane. The NAD(H) site resides in domain I and the NADP(H) site in domain III [3]. The strongly hydrophobic domain II spans the membrane. Uniquely, in *Rhodospirillum rubrum* transhydrogenase the NAD(H)-binding

P. Mathis (ed.), Photosynthesis: from Light to Biosphere, Vol. II, 681–684.

domain I exists as a separate polypeptide (called Th_s) [4-6]. This polypeptide can be easily dissociated from the domain II/III components of the enzyme. Separated Th_s and the domain II/III proteins are devoid of transhydrogenation activity but reconstitution leads to full recovery.

Recombinant Th_s from *R.rubrum* was expressed at high levels in *Escherichia coli* [7]. The purified recombinant protein completely restored transhydrogenation activity to membranes of *R.rubrum* that had been washed to remove their native domain I polypeptide. The thermodynamic and kinetic characteristics of NADH binding to Th_s were investigated by monitoring the fluorescence quenching of a sole Trp residue at position 72 in the recombinant protein. The K_D for NADH binding was approx. 20μM, the k_{on} approx. $10^7M^{-1}s^{-1}$, and the k_{off} approx. $150s^{-1}$.

On the basis of an analysis of the one-dimensional ^{1}H-NMR spectrum of Th_s it was proposed that a region from approximately V221 to A250, which straddles sites that are highly sensitive to cleavage by proteases, has a significantly greater segmental flexibility than is available to the remainder of the molecule [8]. This segment is evidently a mobile loop which emanates from the surface of the protein. Resonances in the NMR spectrum were provisionally assigned to specific amino acid residues in the segment V221 to A250. The narrow resonances that were attributed to amino acid residues within the mobile loop were considerably broadened and some were slightly shifted when NADH was added to recombinant Th_s [8]. This indicates a loss of segmental mobility of the loop and the induction of chemical exchange processes as the protons sampled a range of environments on the NMR time scale as the NADH became bound. In titrations with NADH, the resonances that broadened appeared to do so in parallel. However, titration with NAD^+ revealed a two-step binding reaction. At low concentrations of nucleotide only one resonance, the upfield Met at 1.97ppm, was specifically broadened. At high concentrations the NMR spectrum came to resemble that observed in the presence of NADH.

In this report we describe (a) the effects of binding of nucleotide analogues on the conformational dynamics of the mobile loop, and (b) the effect of mutation of an invariant Tyr residue in the mobile loop on the loop dynamics and on the catalytic properties of the reconstituted protein.

2. Procedures

Biochemical procedures, the expression and purification of Th_s, and the NMR and fluorescence spectroscopy were carried out as described [7,8]. Mutagenesis was carried out by the gapped duplex method [9].

3. Results

3.1 The effect of nucleotide analogues on the conformational dynamics of the mobile loop in Th_s

Low concentrations (<1mM) of NADP(H) had very little effect on the NMR spectrum of Th_s. Higher concentrations led to some broadening of the resonances deriving from the amino acid residues in the mobile loop. This is consistent with the view that NADP(H) binds to domain III of transhydrogenase. At high concentrations there appears to be some non-specific binding into domain I.

Acetylpyridine adenine dinucleotide ($AcPdAD^+$) is often used as an analogue for NAD^+. It had a very similar effect to that of NAD^+ on the conformational dynamics of the mobile loop in Th_s as revealed by NMR [8]. Thus, low concentrations led to specific broadening of the upfield Met resonance at 1.97ppm, whereas higher concentrations caused broadening of resonances attributable to Thr, Ala, a Met at 2.04ppm, and Tyr. There was also a slight change in the chemical shift of some of these resonances. Likely candidates for the residues within the mobile loop were discussed in [8].

Addition of 5^{l}-AMP had similar effects on the NMR spectrum to those of NAD^+ and $AcPdAD^+$ except that the broadening of the Thr, Ala, upfield Met and the Tyr were only evident at higher nucleotide concentrations.

3.2 Changing the invariant Tyr residue in the mobile loop did not affect NADH binding but it did alter the conformational dynamics of loop closure and result in considerable loss of transhydrogenation activity

Replacement of the invariant Tyr residue in the mobile loop with Asn gave rise to a mutant Th_s (Y235N) with the following properties:

1. Upon reconstitution with depleted chromatophore membranes, the k_{cat} for reduction of $AcPdAD^+$ by NADPH was decreased by 80% and the Km for $AcPdAD^+$ was increased 10 fold relative to the wild-type protein.
2. In contrast, the k_{cat} for light-driven "forward" transhydrogenation was decreased by only 50% in the reconstituted mutant protein.
3. The mutant docked with the depleted chromatophores with undiminished affinity.
4. The fluorescence properties of Trp72 were similar to those in wild-type Th_s. The K_D for NADH binding was similar to that in wild-type.
5. The NMR spectrum of the mutant was similar to that of the wild-type, except that (a) the Tyr resonances at 6.82 and 7.12ppm were lost (this unequivocally confirms the assignment of these resonances), and (b) the resonances attributed to Met residues were perturbed; most noticably the resonance at 1.97ppm was shifted downfield by approximately 0.05ppm. Upon titration with NADH, this resonance was specifically shifted and broadened at lower concentrations of nucleotide than those required to broaden other resonances in the spectrum. Thus in the mutant, a *two-step* binding process became evident which resembled that observed in wild-type protein during titrations only with the *oxidised* nucleotide.

Note that the equivalent of Tyr235 has been mutated also in *E.coli* transhydrogenase. [10]. It was shown that activity in the bacterial membrane fraction was down by 67-50% in these mutants.

4. Discussion

(A) The model that emerges from these results is that the adenosine moiety of the substrate nucleotide is recognised by, and binds into, the putative βαβ Rossman fold about 60 residues upstream from the mobile loop. This leads to a conformation of Th_s

that is characterised by the broadening and shifting of the upfield Met resonance. If the incoming nucleotide also has a nicotinamide moiety, and particularly if the nicotinamide is reduced, then the V221 - A250 loop immobilises on the surface of the protein.
(B) Substitution of Tyr235 with Asn does not change the fold of the protein, its ability to dock with domains II/III of transhydrogenase, or its affinity for NADH. However, the mutation does lead to an alteration in the dynamics of folding of the mobile loop upon upon NADH binding and this has a profound inhibitory effect on the transhydrogenation reactions.

5. Acknowledgments
We are grateful to Dr.B.A.Levine for discussions about the NMR experiments and to Drs.Ken Jakeman and John Hobman for advice on the mutagenesis.

References

1. Sazanov, L. A., and Jackson, J. B. (1994) *FEBS Lett.* **344,** 109-116.
2. Jackson, J. B. (1991) *J.Bioenergetics and Biomembranes* **23,** 715-741.
3. Yamaguchi, M., and Hatefi, Y. (1993) *J.Biol.Chem.* **268,** 17871-17877.
4. Cunningham, I. J., Williams, R., Palmer, T., Thomas, C. M., and Jackson, J. B. (1992) *Biochim.Biophys.Acta* **1100,** 332-338.
5. Williams, R., Cotton, N. P. J., Thomas, C. M., and Jackson, J. B. (1994) *Microbiology* **140,** 1595-1604.
6. Yamaguchi, M., and Hatefi, Y. (1994) *J.Bioenerg.Biomembr.* **26,** 435-445.
7. Diggle, C., Hutton, M., Jones, G. R., Thomas, C. M., and Jackson, J. B. (1995) *Eur.J.Biochem.* **228,** 719-726.
8. Diggle, C., Cotton, N. P. J., Grimley, R. L., Quirk, P. G., Thomas, C. M., and Jackson, J. B. (1995) *Eur.J.Biochem.* , in press.
9. Kramer, W., Drutsa, V., Jansen, H.-W., Kramer, B., Plugfelder, M., and Fritz, H.-J. (1984) *Nucl.Acids Res.* **12,** 9441-9456.
10. Olausson, T., Hultman, T., Holmberg, E., Rydstrom, J., Ahmad, S., Glavas, N. A., and Bragg, P. D. (1993) *Biochemistry* **32,** 13237-13244.

A NEW METHOD TO IDENTIFY FERREDOXIN-INTERACTING PROTEINS.

M. Isabel García-Sánchez, A. Javier Vigara, Cecilia Gotor, José M. Vega
Instituto de Bioquímica Vegetal y Fotosíntesis, CSIC y Universidad de Sevilla, Facultad de Química, Apartado 553, 41080 Sevilla, Spain

1. Introduction

Ferredoxin (Fd) is a central protein for transferring electrons from the photosynthetic chain to Fd-dependent enzymes in the chloroplasts. This protein is involved in NADP photoreduction, nitrite, ammonium and sulfite reduction, lipid biosynthesis and light regulation of chloroplastic enzymes. Fd is a monomeric protein of around 12 kD and contains a [2Fe-2S] cluster as prosthetic group. It is a highly acidic protein, and these acidic residues are involved in the interaction of Ferredoxin with Fd-dependent enzymes (1). Thus, Ferredoxin forms electrostatically stabilized complexes with these enzymes, being able to form covalent complexes by cross-linking with 1-ethyl-3-(3-dimethylaminopropyl) carbodiimide (EDC) (2-4). Recently, we have described the formation of a functional covalent complex between Fd and Fd-dependent glutamate synthase (GOGAT) from green alga (5). Furthermore, we have shown that different Fd-dependent enzymes, such as GOGAT and nitrite reductase, have a similar binding active site for Fd (6). In this report, we describe a detection system of Fd-interacting proteins in crude extracts of the alga *Chlamydomonas reinhardtii*, using GOGAT as a positive control of Fd-interacting protein. This system is based in the use of biotin-labeled Fd as probe and the posterior detection of biotin with avidin-alkaline phosphatase conjugate.

2. Procedure

2.1 Crude extract preparation

Chlamydomonas reinhardtii wild-type strain 21 gr was grown at 25°C in Sueoka liquid medium containing 5 mM NH_4Cl. Cultures were flushed with air supplemented with 5% CO_2 and continuously illuminated. Cells were broken by freezing in liquid nitrogen and thawing in 10 mM phosphate buffer, pH 7.5 (standard buffer). The homogenate was centrifuged and the resulting supernantant used as crude extract.

2.2 Biotinylation of C. reinhardtii *Fd*

Fd from *C. reinhardtii* was purified to homogeneity as previously described (7). A 25-fold molar excess of Sulfo-NHS-Biotin (Pierce) was mixed with Fd (2 mg/ml) in

P. Mathis (ed.), Photosynthesis: from Light to Biosphere, Vol. II, 685–688.

standard buffer overnight at 4°C. To remove unreacted biotin, the mixture was then dialyzed for at least 24 h at 4°C against standard buffer. Biotinylated Fd (^{Bt}Fd) was stored at 4°C until use.

2.3 Complex formation between Fd and GOGAT

Both proteins were covalently cross-linked by incubation in standard buffer for 3 h at 4°C in the presence of 5 mM EDC.

2.4 SDS-PAGE and Western blotting

SDS-PAGE was performed using 10% and 4% acrylamide separating and stacking gels, respectively (8). Following SDS-PAGE, the gel proteins were electrotransferred to polyvinylidene difluoride (PVDF) membranes (Millipore) using the Nova-Blot 2117 Multiphor semidry system (Pharmacia). Transfer was performed at room temperature for 45 min at a constant current of 0.8 mA·cm^{-2} filter surface, with 25 mM Tris-HCl (pH 10.4) as anode buffer and 40 mM 6-amino-n-hexanoic acid (pH 7.6) as cathode buffer. The membranes were rinsed in 0.1% Tween 20 in TBS and blocked in 5% nonfat dry milk, 0.1% Tween 20 in TBS for 45 min at room temperature. The membranes were then probed with rabbit polyclonal anti-*M. braunii* GOGAT (9) at a 1/5000 dilution in blocking buffer for 90 min at 4°C. The membranes were further washed and blocked for 20 min as above, and reprobed with goat anti-rabbit IgG alkaline phosphatase conjugate (Sigma) at a 1/1000 dilution in blocking buffer for 45 min at 4°C. After several washes, alkaline phosphatase activity was detected with a solution of nitrobluetetrazolium and 5-bromo-4-chloro-3-indolyl-phosphate at room temperature for 2-5 min.

2.5 Detection with ^{Bt}Fd

After transferring of gel proteins to PVDF membranes as described above, the membranes were rinsed in 0.05% Tween 20, 0.3% Triton X-100 in PBS and blocked in 5% BSA, 0.25% gelatin in PBS overnight at 4°C. The membranes were then incubated with ^{Bt}Fd (0.33 mg/ml) and 10 mM EDC in 5% BSA, 0.25% gelatin, 0.05% Tween 20, 0.3% Triton X-100 in PBS (incubation buffer) for 6 h at room temperature and gently shaking. After several washes, the membranes were blocked again overnight at 4°C. Avidin-alkaline phosphatase conjugate (Bio-Rad) at a 1/1000 dilution in incubation buffer was added and kept for 45 min at 4°C and gently shaking. After new washes, alkaline phosphatase activity was detected as described above.

3. Result and Discussion

3.1 Biotinylated Fd binds to GOGAT

The first step in the optimization of the new detection system was the suitable biotinylation of Fd and the cheking of non-interference of biotin in the Fd-binding to interacting proteins. For biotinylation, we selected the derivative Sulfo-N-hydroxysulfosuccinimide(NHS) ester because it was water-soluble and therefore useable in physiological buffers. Another important feature was that this derivative reacted with primary amines of the protein, with no alteration of the carboxyl functional groups for Fd-binding (5). In the standard biotinylation procedure we used a molar ratio of Sulfo-

NHS-biotin to Fd of 25:1, and extended dialysis to remove excess of reagent was necessary to get the best results (data not shown).

To check that [Bt]Fd was functional for protein-interaction, we allowed the formation of covalent complexes between [Bt]Fd and Fd-dependent enzymes in *C. reinhardtii* crude extracts. Previously, we have shown that rabbit polyclonal anti-*M. braunii* GOGAT recognized *C. reinhardtii* GOGAT and its covalent complex with Fd (5,9). Thus, these antibodies might be used to localize in Western blots protein bands corresponding to this complex (Fig. 1A, lane 1). When we used [Bt]Fd, the anti-GOGAT antibodies were able to detect the specific complex protein band, demonstrating the non-interference of biotin labeling in the Fd-interaction (Fig. 1A, lane 2). Besides the antibody detection, we used avidin-alkaline phosphatase conjugate to detect the [Bt]Fd-complex. We observed a specific band with the same mobility to that recognized by the antibody (Fig. 1A, lane 3). An extra protein band was also detected, that could correspond to other Fd-interacting protein or any biotin-dependent enzyme present in the crude extract as it is shown next. At this point, we can not rule out any of both posibilities. Because the conditions of the electrotransfer were optimal for transferring high molecular wight proteins, we did not detect protein bands corresponding to Fd-dependent enzymes of lower molecular weights.

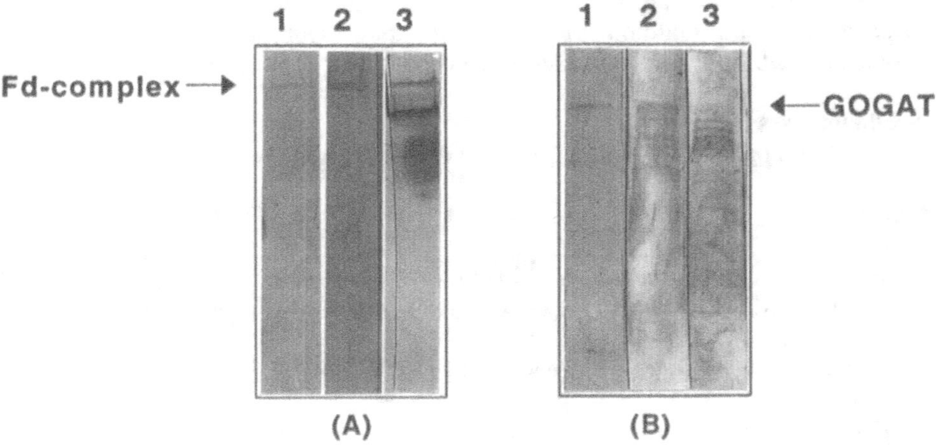

Figure 1. (A) Detection of Fd-complex. 50 μg of *C. reinhardtii* crude extract were incubated with Fd (lane 1) or [Bt]Fd (lanes 2 and 3) in the presence of EDC to allow the formation of Fd-complexes. The mixtures were subjected to SDS-PAGE followed by electrotransfer to membranes. Antibody detection was carried out on lines 1 and 2. Detection with avidin-alkaline phosphatase conjugate was used on lane 3.
(B) Detection of GOGAT. 50 μg of *C. reinhardtii* crude extract were subejected to SDS-PAGE followed by electrotransfer to membranes. Antibody detection was carried out on lane 1. Incubation with [Bt]Fd following by detection with avidin-alkaline phosphatase conjugate was performed on lane 2. Detection with avidin conjugate with no previous [Bt]Fd incubation was carried out on lane 3.

3.2 Optimization of the detection system using ^{Bt}Fd as probe

To get the highest sensitivity and best ratio signal/background noise, different variables were checked. When blocking conditions were evaluated, we observed the lowest nonspecific binding with 5% BSA and 0.25% gelatin instead of the traditional 5% nonfat dry milk. Longer blocking periods were also necessary, and the best results were obtained with overnight blocking at 4°C (data not shown). ^{Bt}Fd concentration and time and temperature of the incubation step were examined next. The observed optimal conditions were between 0.2 and 0.3 mg/ml for ^{Bt}Fd concentration and an incubation for at least 6 h at room temperature (data not shown). The results obtanied from a representative experiment of GOGAT detection in *C. reinhardtii* crude extract using ^{Bt}Fd as probe is shown in Fig. 1B. The detection conditions are described in the Procedure section. We detected the protein band corresponding to GOGAT with a sensitivity similar to the one obtained by antibody detection (Fig. 1B, compare lanes 1 and 2). However, we also observed extra bands that appeared when no previous ^{Bt}Fd incubation was performed (Fig 1B, lane 3). These bands could correspond to biotin-containing enzymes present in the crude extract as several carboxylases found in the plant kingdom (10).
In summary, we have design a detection system of Fd-interacting proteins in crude extracts that is simple, sensitive and reproducible. We think that this procedure can be applyed to the screening of expression libraries, allowing the cloning of cDNAs encoding Fd-dependent enzymes. Until now, no much information about these genes is available, and this method will be useful to isolate genes coding for novel proteins.

Acknowledgements
This work was supported by funding from DGICYT (Spain), Grant PB93-0735

References

1 Knaff, D.B. and Hirasawa, M. (1991) Biochim. Biophys. Acta 1056, 93-125
2 Privalle, L.S., Privalle, C.T., Leonardy, N.J., Kamin, H. (1985) J. Biol. Chem. 260, 14344-14350
3 Hirasawa, M., Chang, K-T., Knaff, D.B. (1991) Arch. Biochem. Biophys. 286, 171-177
4 Pueyo, J.J., Revilla, C., Mayhew, S.G., Gómez-Moreno, C. (1992) Arch. Biochem. Biophys. 294, 367-372
5 Vigara, A.J., García-Sánchez, M.I., Gómez-Moreno, C., Vega, J.M. (1994) Biol. Plant. 36(suppl), S170
6 Romero, L.C., Gotor, C., Márquez, A.J., Forde, B.G., Vega, J.M. (1988) Biochim. Biophys. Acta 957, 152-157
7 Galván, F., Márquez, A.J., Fernández, E. (1984) Z. Naturforsch. 40, 373-378
8 Laemmli, U.K. (1970) Nature 227, 680-685
9 Kozielski, F., Vigara, A.J., Vega, J.M. (1995) in Avances en el Metabolismo del Nitrógeno (Guerrero M.G.,ed) pp.89-94, Servicio de Publicaciones Universidad de Sevilla, Sevilla, Spain.
10 Wurtele, E.S., Nikolau, B.J. (1990) Arch. Biochem. Biophys. 278, 179-186

Site-Directed Mutagenesis of Acidic Patches of Plastocyanin

Byung Hyun Lee*, Takashi Hibino**, Tetsuko Takabe* & Teruhiro Takabe**, ***
*BioScience Center, School of Agricultural Science, Nagoya Univ., Chikusa-ku, Nagoya 464-01, Japan, **Department of Chemistry, Faculty of Science & Technology, Meijo University, Tenpaku-ku, Nagoya 468, Japan, and ***Research Institute of Meijo University, Tenpaku-ku, Nagoya 468, Japan

1. INTRODUCTION

Plastocyanin (PC) is a copper protein (10.5 kDa) that functions as a mobile electron carrier in the thylakoid lumen. It is reduced by cytochrome *f* (Cyt. *f*) in the Cyt. *b6/f* complex and oxidized by the reaction center chlorophyll ($P700^+$) in the photosystem I (PSI) complex within the electron transport chain of oxygen-evolving photosynthetic organisms. Whereas higher plants utilize only PC as an electron carrier between the two complexes, the reaction between PSI and Cyt. *b6/f* complexes in algae and cyanobacteria is mediated by either PC or Cyt. *c6*, water-soluble *c*-type Cyt. In some algae and cyanobacteria, the function of PC can be replaced by Cyt. *c6* in cells grown under conditions of copper deprivation. When PC and Cyt. *c6* are present in the same organisms, the isoelectric points of these proteins are always similar. This linked variation suggests the importance of total net charge on PC and Cyt. *c6* for the interactions with Cyt. *b6/f* or PSI. Although higher plant PC is an acidic protein, the isoelectric points of other photosynthetic organisms vary from 3.8 to 9.5. Plant PC has two highly conserved negative patches, and two possible reaction sites have been suggested on this protein. One (Site 1) is the "proximal" site close to the solvent-accessible imidazole ring of His87 which lies in the center of the northern hydrophobic patch. The other (Site 2) is the "remote" site of Tyr83 located at the center of two negatively charged patches of acidic residues (#42-45 and #59-61). Most previous studies have suggested the involvement of negative patches on PC in its reaction with Cyt. *f*, although the participation of Site 1 as the binding site has also been reported. For the interaction of PC with PSI, participation of a hydrophobic surface and/or negative patches on PC has been suggested. The role of negative patches on PC in photosynthetic electron transfer chains is still largely unclear. Site-directed mutagenesis can be an important tool to address the above questions. Since, we have recently developed an expression vector to produce mature plant PC from silene encoding the intermediate precursor PC (1,2), we decided to introduce systematically positive charges into the negatively charged patches by site-directed mutagenesis. It will be shown that modifications of residues (#42-45) caused a dramatic decrease in the electron transfer rate with Cyt. *f* and PSI, whereas those of residues (#59-62) had no effect. Electron transfer mechanisms of PC are discussed in relation to the role of the negative patches

2. PROCEDURE

P. Mathis (ed.), Photosynthesis: from Light to Biosphere, Vol. II, 689–692.

Site-Directed Mutagenesis of PC---The intermediate *silene* PC gene was subcloned into the EcoRI and Hind III digested pBluescript KS- plasmid (pBKSPC). The NcoI and BamHI fragment of pBKSPC was inserted into the same site of the pET3d expression vector to yield pETiPC. The pETiPC plasmid was introduced into *E. coli* BL21(DE3). The *in vitro* mutagenesis was performed on pBKSPC using a Mutan-K kit (Takara, Japan). The mutagenic oligonucleotides are designed systematically to incorporate systematically positive charges into the negative patches. Oligonucleotides were synthesized with a DNA synthesizer (Applied Biosystems, model 381A), phosphorylated with T4 poly-nucleotide kinase, annealed to the uridine-containing, single-strand DNA template, extended with T4 DNA polymerase, and ligated with T4 DNA ligase. They were transformed into *E. coli* MV1184, and colonies containing the mutated cDNA were identified by DNA sequencing using the dye-primer chain-termination method with a DNA sequencer (Applied Biosystems, model 373A). Each mutated cDNA was inserted into the NcoI/BamHI site of the pET3d expression vector and transformed into *E. coli* BL21(DE3).

Expression, Purification and Characterization of Mature PC---Expression and purifi-ation of mature PC, SDS-PAGE, immuno-blotting, and N-terminal amino acid sequencing were conducted as previously described (1,2). Acrylamide concentrations of native gel electrophoresis was 10%. The contents of polypeptide and copper were determined as previously described (1,2).

Kinetics---Cyt. *f* and PSI complex (PS I-200) were purified from *Brassica komatsuna* and cucumber, respectively, as previously described (1,2). Kinetic experiments on the electron transfer of PC with Cyt. *f* and PSI were performed using a Union Giken single-beam spectrometer equipped with a Union RA-401 stopped flow instrument and a flash lamp for excitation. The reaction of Cyt. *f* was monitored by measuring the absorbance at 422 nm. The flash-induced absorbance change of P700 was measured at 697 nm. The redox potentials of PC were obtained by a combined spectrophotometric/ platinum-calomel electrode method. All the experiments were done at 25°C.

3. RESULTS and DISCUSSION

Preparation of Negative Patch Mutant PC---The mutants were designed systematically to incorporate positive charges into the negative patches. The calculated net charges on the negative patch (#42-45) of wild-type, M1(Asp42Asn), M2(Glu43Lys), M3(Asp42Asn + Glu43Lys), and M4(Glu43Lys + Asp44Lys) mutant PCs were estimated as -4, -3, -2, -1, and 0, repectively, and those on the negative patch (#59-61) of wild-type, M5(Glu59Lys), and M6(Glu59Lys +Glu60Lys) mutant PCs were estimated as -3, -1, and +1, respectively (Table I). Compared to the net charge on a basic PC from *A. variabilis*, the mutants M3, M4, and M6 have the same, more positive (+1), and slightly negative (-0.5) net charges on the respective negative patches. Upon the expression of intermediate precursor PC from silene, the mature size PC accumulated in the periplasmic space of *E. coli.* The correct processing to the mature protein was confirmed by N-terminal amino acid sequencing of expressed PC (data not shown). Modifications of net charge on the protein molecules of mutants were detected by native PAGE, which separates the PC molecules on the basis of their total net charges.

Spectroscopic Properties of Mutant PC---All mutants (M1-M6) could be purified as holo-proteins, which is in sharp contrast to the case of Met92 mutant of silene PC that accumulated as apo-protein (2). Their absorption spectra are similar to that of the wild-type PC: absorption maximum, 597 nm; absorption coefficient, 4500 $cm^{-1} \times M^{-1}$; absorbance ratio A_{278}/A_{597}, 1.2. The EPR spectra of the six mutants were indistinguishable from that of wild PC (data not shown). The midpoint potentials of the mutants (M1-M6) were similar to that of the wild type within the range of 9 mV. These results indicate that the structural change of the copper binding site upon the modification of negative patches is minor and can be regarded as negligible.

Electron Transfer from Cyt. f to Negative Patch Mutants---The oxidation kinetics of Cyt. *f* by the six mutant PCs was examined using stopped-flow spectrophotometry at constant ionic strength (I = 0.1 M) and pH 8.0. The oxidation of Cyt. *f* by the Asp42Asn mutant (M1) decreased to 69 % of that by the wild-type PC, although in the case of spinach PC, no effect of this mutation has been reported (3). The rate constants decreased further in the following order; M1 > M2 > M3 > M4. Interestingly, we also found that the modification of residues #59-61 (M5 and M6) did not change the reaction rate. These results indicate that the negative patch (#42-45) on PC is important for the oxidation of Cyt. *f*, whereas the other one (#59-61) is not involved in the interaction with Cyt. *f*.

It has been shown that the effects of ionic strength on interacting macromolecules can be described by the following equations.

$$\ln k(I) = \ln k_\infty - V_{ii}\,(1 + KR)^{-1} \exp(-KR) \qquad (1)$$

$$V_{ii} = AD^{-1} Z_{cyt}\, Z_{pc}\, r_{12}\, R^{-2} \qquad (2)$$

where k(I) and k_∞ are the rate constant at a given (I) and at infinite ionic strength, respectively. R and K are the radius of the interacting domains (taken as 7.25 Å) and 0.33 $I^{1/2}$, respectively. The term V_{ii} is the electrostatic energy, defined by: A, a constant equal to 128.5 at 25°C; D, the effective dielectric constant (taken as 40); r_{12}, the distance between the interacting domains in the transition state complex (taken as 3.5 Å); Z_{cyt} and Z_{pc}, the charges on the interacting domains of Cyt. *f* and PC, respectively. The above equations suggest that at constant ionic strength, a linear relationship between ln k and Z_{pc} can be anticipated. The plots, ln k versus. Z_{pc} indicated that the rate constant (ln k) decreases linearly, to a first approximation, when the calculated net charges on the mutants (M1-M4) are plotted, whereas a constant value is obtained when the calculated net charges on the mutants (M5 and M6) are plotted. The rate constants for the wild-type and the six mutant PCs are plotted as a function of ionic strength, with the solid lines showing the fits of the experimental data to Equations 1 and 2. The rate constants of all mutants decreased with increasing ionic strength in a manner similar to that of the wild-type PC. The calculated parameters k_∞ and V_{ii} are shown in Table II. It can be seen that the values of k_∞ did not change significantly among the wild-type and the six mutant PCs, whereas the electrostatic attraction energies decreased upon the incorporation of positive charges into the negative patch (#42-45) on PC, but not into the other negative patch (#59-61).

Electron Transfer from Negative Patch Mutant PC to Photooxidized P700---The reduction of photooxidized P700 by PC was examined using flash photolysis. Compared to the wild-type, the reduction of photooxidized P700 by the mutant PCs occurred in the following order; wild-type > M1 > M2 > M3 > M4. The mutants M5 and M6 had only minor effect on the reduction of photooxidized P700. As was the case for the reduction of PC by Cyt. *f*, no first-order reactions were observed for the oxidation of PC by P700 under our experimental conditions. Plots of second-order rate constants (ln k) versus the calculated net charge on PC showed that the rate constant (ln k) decreased approximately linearly with increasing calculated net charge on the mutants (M1-M4). In this case, the mutants M5 and M6 showed slight decreases but their decreases were much smaller than those of the mutants M1-M4.

Effects of $MgCl_2$ on the oxidation of PC by $P700^+$ were also examined. Essentially similar results were obtained when KCl was used in place of $MgCl_2$. At concentrations of

$MgCl_2$ higher than 3 mM, both the wild-type and the mutants (M1-M4) exhibited decreased rate constants with increasing ionic strength, in a manner similar to that in the reduction of PC by Cyt. *f*. The experimental data at concentrations higher than 3 mM $MgCl_2$ were fitted to Equations 1 and 2, and the parameters k_∞ and V_{ii} were calculated. The values of k_∞ changed within three-fold upon the modification of acidic residues (#42-45). The observation that the parameter V_{ii} for the mutants M1-M4 increases with decreasing net charge on PC indicates that the acid residues (#42-45) are located in the interaction domains of PC for PSI.

The increase of rate constant with increasing ionic strength at low concentrations of $MgCl_2$ was previously interpreted as due to electrostatically stabilized complex formation between PC and PSI, although very large association constants 50-170 mM have been reported. The present results indicate that the increase of the rate constant produced by addition of low concentrations of $MgCl_2$ becomes smaller in the following order; wild > M1 > M2 > M3 >M4, which suggests that the substitution of negatively charged amino acids (#42-45) by neutral or positively charged ones decreased the long-lived complex formation between PC and PSI complex.

4. REFERENCES

1. Hibino,T., Lee,B.H., and Takabe,T. (1994) *J. Biochem.* **116**, 826-832
2. Hibino,T., Lee,B.H., Takabe,T., and Takabe,T. (1995) *J. Biochem.* **117**, 101-106
3. Modi,S., Nordling,M., Lundberg,L.G., Hansson,O., and Bendall, D.S. (1991) *Biochim. Biophys. Acta* **1102**, 85-90

TABLE I. Amino acid sequences of acidic patch region of mutant PC.

species	40	41	42	43	44	45	46	47	net charge of (#42-45)
Spinach	Val	Phe	Asp	Glu	Asp	Glu	Ile	Pro	-4
Silene wild-type	Leu	Phe	Asp	Glu	Asp	Glu	Val	Pro	-4
M1	Leu	Phe	<u>Asn</u>	Glu	Asp	Glu	Val	Pro	-3
M2	Leu	Phe	Asp	<u>Lys</u>	Asp	Glu	Val	Pro	-2
M3	Leu	Phe	<u>Asn</u>	<u>Lys</u>	Asp	Glu	Val	Pro	-1
M4	Leu	Phe	Asp	<u>Lys</u>	<u>Lys</u>	Glu	Val	Pro	0
Anabaena variabilis	Val	Phe	Asp	Ala	Ala	Leu	Asn	Pro	-1

species	57	58	59	60	61	62	63	net charge of (#59-61)
Spinach	Met	Ser	Glu	Glu	Asp	Leu	Leu	-3
Silene wild-type	Met	Pro	Glu	Glu	Asp	Leu	Leu	-3
M5	Met	Pro	<u>Lys</u>	Glu	Asp	Leu	Leu	-1
M6	Met	Pro	<u>Lys</u>	<u>Lys</u>	Asp	Leu	Leu	+1
Anabaena variabilis	Leu	Ser	His	Lys	Gln	Leu	Leu	+1.5

TABLE II. Electrostatic parameters for the electron transfer from Cyt. *f* to PC.

species	$k_\infty (M^{-1} s^{-1})$	V_{ii} (kcal / mol)	Z_{cyt}	Z_{pc}
wild-type	2.7×10^6	-12.5	7.6	-7.6
M1 Asp42Asn	2.2	-9.3	7.6	-5.7
M2 Glu43Lys	2.4	-5.7	7.6	-3.5
M3 Asp42Asn +Glu43Lys	2.1	-5.4	7.6	-3.3
M4 Glu43Lys +Asp44Lys	1.5	-4.5	7.6	-2.8
M5 Glu59Lys	2.9	-11.6	7.6	-7.1
M6 Glu59Lys +Glu60Lys	2.7	-11.7	7.6	-7.2

INTERACTION OF SPINACH PLASTOCYANIN WITH PHOTOSYSTEM I [#]

J.Reichert, L.Altschmied[*], R.B.Klösgen[*], R.G.Herrmann[*] & W.Haehnel.
Biochemie der Pflanzen, Albert-Ludwigs-Universität, Freiburg, Germany.
[*]Botanisches Institut, Ludwig-Maximilians-Universität, München, Germany.

1. Introduction

Plastocyanin (Pc) functions as the electron carrier between the cytochrome *bf* complex and photosystem I (PSI) in plant photosynthesis. Remarkable features of this ß-barrel copper protein are a flat hydrophobic north pole around the copper ligand H87 and two acidic patches on one side of the molecule. These two conserved patches are located south (42-45) and north (59-61) of Y83, a residue just next to the copper ligand C84. While Y83 has been proposed to play a role in the interaction with cytochrome f (1), the hydrophobic northern surface around H87 has been shown to be the docking site for PSI (2). Chemical modifications and cross-linking studies suggested an important function of the acidic patches in the interaction with PSI (for review see 3). In a very recent paper, neutralizing the negative charges by introducing lysine residues has suggested a dominant role of the southern acidic patch (4). A potential involvement of an internal hydrophobic channel in electron transfer was discussed when the crystal structure of poplar Pc was solved (5).

We studied the participation of these possible intra-molecular electron transfer routes, using site-directed mutated Pc expressed in *E. coli*. With the A90L mutant, which causes a deformation of the initially flat hydrophobic surface (2), a stable complex between A90L and PSI is still obtained. But the kinetics of the intra-complex electron transfer was altered. This revealed further evidence for the electron transfer to PSI *via* H87. To analyze which of the negative patches is involved in complex formation with PSI we substituted the acidic residues with the corresponding amides. Mutations in the aromatic core of Pc were used to test experimentally the hydrophobic channel hypothesis.

2. Methods

The generation of new spinach Pc mutants (DEDE42-45NQNQ, E59Q, EED59-61QQN. In the following we will use the number of the first altered residue followed by the new residues in the sequence for abbreviation of the acidic patch mutant plastocyanins. The hydrophobic core mutants are abbreviated as usual: Y80F, F82L), expres-

---

[#] Support by Deutsche Forschungsgemeinschaft SFB 388/A1 is gratefully acknowledged.

P. Mathis (ed.), Photosynthesis: from Light to Biosphere, Vol. II, 693–696.

sion, purification, isoelectric focusing (IEF), and laser flash kinetic measurements were performed as described (2). Electrospray mass spectrometry (MS) was carried out with a Finnigan TSQ700. 60 pmol Pc were injected in 3 μl 50% methanol, 1% acetic acid. 1-Ethyl-3-(3-Dimethylaminopropyl)-carbodiimid (EDC) cross-linking between Pc and PSI has been described previously (6).

3. Results

The Pc mutants 42NQNQ, 59Q, 59QQN, Y80F and F82L were expressed as truncated precursor and processed in *E. coli*. The identity of the isolated proteins was verified by MS as shown in Table I.

Since the amide differs from the carboxyl group by only 1 Da, the acidic patch mutants are difficult to verify by MS, but can easily be distinguished by IEF (Fig. 1). The replacement of Y80 could be verified by the UV spectrum. An alteration of the bands in the visible spectrum of oxidized Pc was not detectable (data not shown). The measured redox potentials are given in Table I. We conclude that our mutations did not affect the overall structure of Pc.

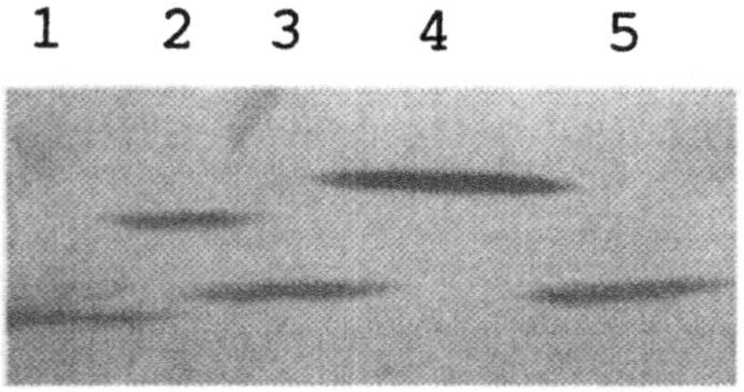

Fig. 1: IEF with the acidic patch mutants. Lane 1, recombinant spinach Pc (WT); lane 2, 59QQN; lane 3, 59Q; lane 4, 42NQNQ; lane 5, spinach Pc.

Table I: Properties of the new mutant Pc molecules. The measured molecular weight and the difference to the calculated values are shown. The midpoint potentials where obtained by cyclic titration with fericyanide and ascorbate at 25 °C. The buffer was 50 mM MOPS pH 7, 5 mM $MgCl_2$, 50 μM DAD.

Pc	MW Δ	E_m
42NQNQ	10404.5±.9-4.1	375 mV
59Q	10409.8±1.11.8	381 mV
59QQN	10407.3±1.32.3	396 mV
F82L	10377.2±.5-1.4	366 mV
Y80F	1397.4±1.9 0.8	371 mV

The A90L mutation introduces a hump at the flat north pole but still formes a complex with PSI. Up to now, only an A90L depending second-order flash-induced P700 kinetics was obtained with this mutant (2). But at high concentrations of A90L, the first order kinetics increased to a clearly resolved amplitude (Fig. 2). The electron transfer in the A90L-PSI complex has a decay with a halftime of 70 μs. Because WT Pc could be cross-linked to PSI200 particles, and this complex resembles the native dynamic complex in its electron transfer properties, we searched for the first-order halftime of $P700^+$ reduction with A90L cross-linked to PSI. As shown in Fig. 3, the A90L mutant could be cross-linked to PSI with EDC. This cross-linked complex is active in $P700^+$ reduction, and the cross-linked product shows indeed the 70 μs component.

The acidic patches have very different functions in complex formation with PSI. The

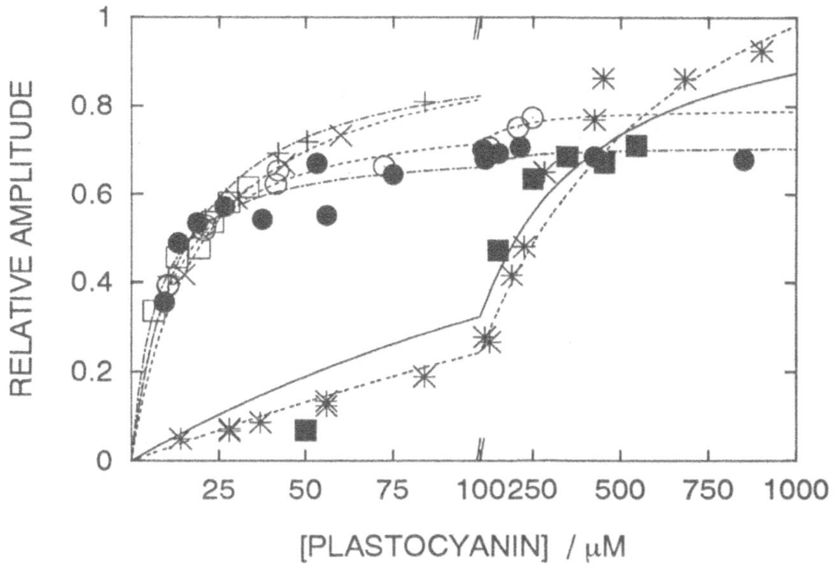

Fig. 2: Plots of the relative amplitude of the intra-complex kinetics varies with the Pc concentration. Data for WT (□), 42NQNQ (■), 59Q(○), 59QQN(●), Y80F (x), F82L (+) and A90L (*) are shown.

binding affinity decreased drastically with the 42NQNQ mutation (Fig. 2). The constant halftime was estimated as 160 μs. The minimal halftime of the second-order kinetics at infinite 42NQNQ concentration was extrapolated to be 10 times the value obtained with WT Pc (data not shown). However, attempts to cross-link 42NQNQ to PSI failed, expected for a negligible amount of a product. In contrast, with the northern acidic patch mutations, we find a fast kinetic component with a halftime of 11 μs, as in wild-type (WT) spinach Pc (Fig. 2). But the second-order halftime extrapolated to infinite 59Q concentration was 30% higher compared to that of the WT. The complete loss of the negative charges at the northern patch in the 59QQN mutant lead to a minimum second-order halftime, which was 2.5 times that measured with the WT. The value of the dissociation constant K^I_D for the complex between reduced Pc and PSI, derived from the first order amplitude, dropped to 70% and 30% for 59Q and 59QQN, respectively.

The mutations at the hydrophobic core, Y80F and F82L, had no effect on the kinetic properties of spinach Pc (Fig. 2).

4. Discussion

The assumed model for the interaction between Pc and PSI is a snugly fit between the flat northern surface of Pc and a complementary surface of PSI. Molecular modelling of the Pc surface with A90L revealed that the hump of L90 could increase the distance between the edge of H87 and the PSI surface by 1.3 Å. This increase is likely the reason for the decrease in electron transfer rate in the complex, as detected by the halftime of the fast P700$^+$ reduction kinetic component. The increase from 11 μs to 70 μs resulting from an elongated through space distance of 1.3 Å is consistent with the slope if the logarithm of the rate constant is plotted as a function of edge to edge distance as published by Moser et al. (7).

From our data, we can identify the southern acidic patch (42-45) as the most essential patch for complex formation of Pc with PSI. Such a drastic effect cannot be seen with the northern acidic patch. However, 59QQN does show a lower value of the dissociation constant K^I_D, i.e. a higher binding affinity in the reduced state as compared to the WT. The halftime of the second-order reaction extrapolated to infinite Pc concentrations, i.e. the minimal halftime, increased relative to the WT. The northern acidic patch (59-61) seems to tune the PSI binding for a rapid release of oxidized Pc after electron transfer. The mutations of the aromatic residues extending from the copper ligand M92 to the south pole of Pc had no effect on the electron transfer. Y80 is a conserved residue in all known Pc molecules, which makes a hydrogen-bond to the surface exposed residue 76 at the south pole. However, its mutation had no effect on the P700$^+$ reduction kinetics.

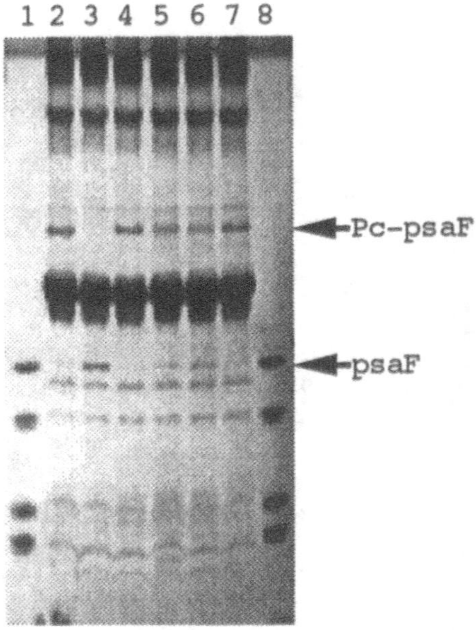

Fig. 3: EDC cross-linking products between PSI200 particles and Pc mutants. Lane 3: without Pc and with WT (4), G10L (5), A90L (6), Y83L (7); Schägger SDS-PAGE, stained with coomassie.

1. Modi, S., Nordling, M., Lundberg, L.G., Hansson, Ö. & Bendall, D.S. (1992) Biochim. Biophys. Acta **1102**, 85-90.
2. Haehnel, W., Jansen, T., Gause, K., Klösgen, R.B., Stahl, B., Michl, D., Huvermann, B., Karas, M. & Herrmann (1994) EMBO J. **13**, 1028-1038.
3. Redinbo, M.R., Yeates, T.O. & Merchant, S. J. (1994) Bioen. Biomem. **26**, 49-66.
4. Lee, B.H., Hibino, T., Takabe, T., Weisbeek, P.J. & Takabe, T. (1995) J. Biochem. **117**, 1209-1217.
5. Colman, P.M., Freeman, H.C., Guss, J.M., Murata, M., Norris, V.A., Ramshow, J.A.M. & Venkatappa, M.P. (1978) Nature **272**, 319-324.
6. Hippler, M., Ratajczak, R. & Haehnel, W. (1989) FEBS Lett. **250**, 280-284.
7. Moser, C.C., Keske, J.M., Warncke, K., Farid, R.S. & Dutton, P.L. (1992) Nature **355**, 796-802.

DYNAMIC INTERACTION BETWEEN PLASTOCYANIN AND PHOTOSYSTEM I

Friedel Drepper[1], Michael Hippler[2], and Wolfgang Haehnel
Biochemie der Pflanzen, Albert-Ludwigs-Universität, Freiburg, Germany

1. Introduction

Inter-protein electron transfer involves the diffusion of mobile proteins and a binding process during which a properly oriented pair of redox partners has to be formed to optimize the electron transfer between the active sites. A detailed knowledge of the binding mechanism and of the redox midpoint potential of the reaction partners in the transient complex is prerequisite for an understanding of the electron transfer reactions. We have studied this process between the soluble protein plastocyanin (Pc) and the photosystem I (PSI) complex from spinach. Two diagnostic parameters have been used to analyze several aspects of the biphasic reaction kinetics of $P700^+$ with Pc after a laser flash, i) the amplitude of a fast component with a constant halftime of 11 μs which originates from a first-order electron transfer within a complex between the two electron carriers formed in the dark before the flash (1) and ii) the halftime of the slower component which is largely due to the second-order binding of reduced Pc to PSI with oxidized P700. We have investigated the dynamics and binding of reduced and oxidized Pc, respectively, with respect to an efficient turn-over in the electron transport chain at PSI.

2. Materials

PSI particles (PSI-200) were isolated from spinach (2). Isolated plastocyanin was cross-linked to PSI particles with ethylenediamine-carbodiimide (3) at pH 6.5. Kinetic measurements of P700 at 703 nm induced by a frequency-doubled Nd-YAG laser have been described (4). The time course was analyzed by a fit to a sum of 2 or 3 exponentials using Marquardt's least squares algorithm.

Present address: [1]Département de Biologie, C.E.N. de Saclay, Service BPH, F-91190 Gif-Sur-Yvette (France); [2]Département de Biologie Moleculaire, Université de Geneve, 30 Quai Ernest Ansermet, CH-1211 Geneve, Switzerland

P. Mathis (ed.), Photosynthesis: from Light to Biosphere, Vol. II, 697–700.

3. Results and Discussion

1. *Competitive binding of oxidized and reduced plastocyanin to PSI*

Fig. 1, upper trace, shows kinetics of P700 in the presence of 22 μM reduced Pc (Pc^I) (superscript I and II indicate reduced and oxidized Pc, respectively, and + oxidized P700). The fast component with a halftime of 11 μs shows a relative amplitude A(1) of 0.65 of the total Pc dependent components. At increasing Pc^I concentrations this amplitude extrapolates to a maximal value of $f = 0.92$ (not shown). This limit of the complex formation allows to estimate the dissociation constant K_D^I of Pc^I and P700 from Eq. (1) as 9 μM.

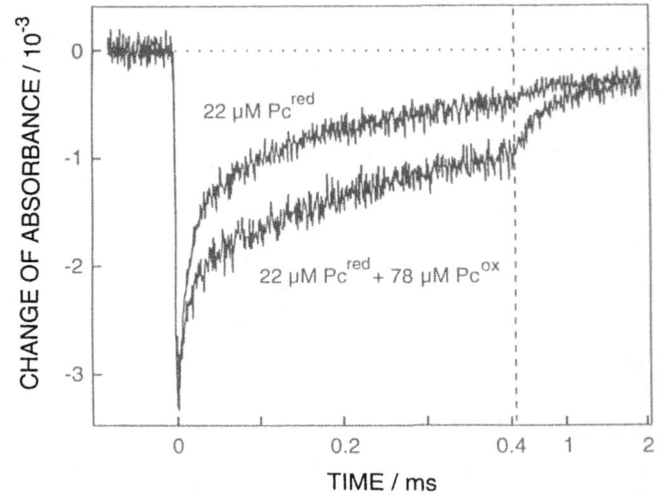

Figure 1. P700⁺ reduction kinetics of PSI particles induced by a laser flash in the presence of Pc. (Details see text)

$$K_D^I=(\frac{f}{A(1)}-1)[Pc^I] \qquad (1)$$

For the measurement of the lower trace in Fig. 1 78 μM oxidized Pc was added in addition to 22 μM Pc^I. The fast component decreased to a relative amplitude A(1) of 0.44 indicating some competitive binding of Pc^{II}. The dissociation constant K_D^{II} of oxidized Pc at reduced P700 can be derived, Eq. (2). With the values of f and K_D^I we estimate for K_D^{II} a value of 48 μM. This has been verified by measurements at other concentrations. Thus, we find that the binding of Pc^I to PSI with reduced P700 is about 5-7 times stronger than that of Pc^{II}.

$$K_D^{II}=[Pc^{II}]/\{\frac{[Pc^I]}{K_D^I}(\frac{f}{A(1)}-1)-1\} \qquad (2)$$

The slow component of $P700^+$ reduction is more difficult to analyse. In Fig. 1 it shows a halftime of 134 and about 280 μs in the absence and presence of Pc^{II}, respectively. It is largely due to the fast electron transfer with the rate constant k_{et} preceded by the second-order binding of the free Pc^I with a rate constant $k_{on}^{I,+}$. At a high excess of Pc^{II} over Pc^I we can estimate the dissociation constant $K_D^{II,+}$ of the complex between Pc^{II} and $P700^+$. If the concentration of Pc^I is the same in the two experiments (as in Fig. 1) we derive Eq. (3), where $t_{1/2}$ and $t_{1/2}^{II}$ are the respective halftimes in the absence and presence of the concentration $[Pc^{II}]$. Thus, we estimate a value of about 72 μM for $K_D^{II,+}$. If compared with K_D^{II} this indicates

$$K_D^{II,+}=[Pc^{II}]\frac{t_{1/2}}{t_{1/2}^{II}-t_{1/2}} \qquad (3)$$

that oxidation of P700 facilitates the release of Pc^{II} from PSI. These data provide experimental evidence for the effect of the charges in the redox centers on the binding between the proteins. In addition to electrostatic repulsion and attraction also the solvation energy may be a major contribution. Reduced Pc is preferentially bound by PSI as compared to oxidized Pc which is the opposite for the release in particular when P700 becomes oxidized. This shows an excellent adaptation to a rapid turnover of the electron transport through PSI.

3.2. Redox potential of bound plastocyanin

The different dissociation constants of reduced and oxidized Pc indicate a difference between the midpoint redox potential of Pc in the free and the bound state, E_m^{fr} and E_m^{bd}, respectively, as given by Eq. (4). The ratio of the dissociation constants from Fig. 1 corresponds to an increase of the midpoint potential of soluble Pc by about 40 mV when it is bound to PSI. Eq. (4) applies also to the effect of bound oxidized Pc on the redox potential of P700. Using the values of $K_D^{II,+}$ and K_D^{II} we estimate an increase of the redox potential of $P700^+$ by about 25 mV. The shift is likely to originate from the electrostatic repulsion between the positive charge at the Cu^{2+} center of Pc^{II} and $P700^+$. Except for the latter the other values have been directly determined by redox titration and are summarized in the table. E_m^{bd} has been determined with Pc cross-linked to PSI. The increased redox potential of Pc cross-linked to PSI in the active state (3) is in excellent agreement with the independent determination of the dissociation constants of Pc. The small difference in the redox potential between bound Pc and P700 of about 65 mV should limit the fast turn-over in the complex and is consistent with the value of f. The decrease in the driving force of the electron transfer from Pc^I to $P700^+$ is at the expense of an increased dissociation constant of oxidized Pc.

$$E_m^{bd} - E_m^{fr} = 59mV \cdot \log\left(\frac{K_D^{II}}{K_D^{I}}\right) \quad (4)$$

$Pc(Cu^{II}/Cu^{I})$		$P700^+/P700$	
E_m^{fr}/mV	E_m^{bd}/mV	E_m/mV	$E_m(Pc^{II})$/mV
360±5	410±10	475±10	ca. 500

3.3. How fast is oxidized Pc released from PSI after the electron transfer?

After the electron transfer oxidized Pc has to leave PSI. We have probed this release and the successive binding of another reduced Pc to PSI by a flash given at time intervals in the μs-range after a preceding laser flash. The results are shown in Fig. 2, left. After the first flash (trace a) $P700^+$ reduction was similar to that in Fig. 1, upper trace, with a halftime of 10 μs and A(1) of 0.76. A flash after 60 μs, (trace b1) induced no fast component. A delay in the single exponential decay with a halftime of 72 μs is seen from the deviation in the residuals (bottom). A flash after 400 μs (trace b2) shows that the fast component of P700 reduction was recovered. The amplitude of the fast component as a function of the time between the two flashes is shown in Fig. 2, right. This time course shows a rise with a halftime of 180 μs and a delay of 50 μs. In a first approximation we have analyzed this time course by three consecutive first-order steps, the electron transfer

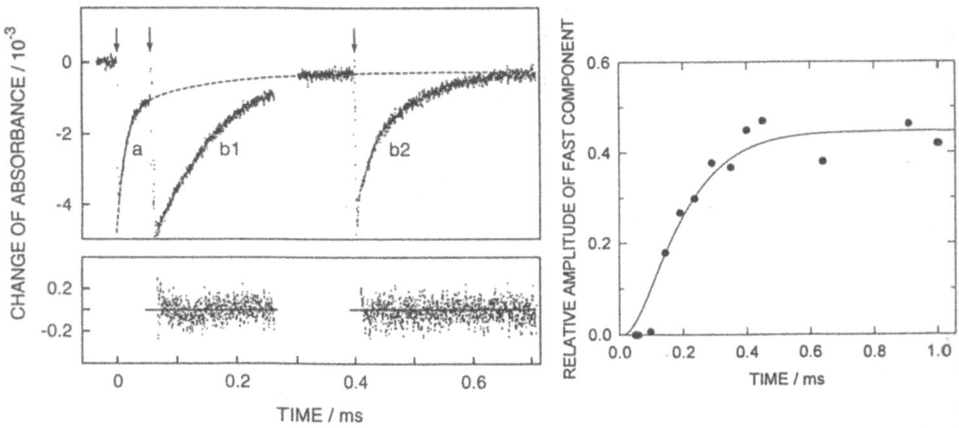

Figure 2. P700$^+$ reduction kinetics of PSI particles in the presence of 250 μM Pc^I induced by two flashes spaced at time intervals of 60 μs (trace **b1**) or 400 μs (trace **b2**). The time course after the first flash (trace **a**) without a second flash is indicated by the dashed line.
Right: Amplitude of the fast component as a function of the dark time between the two flashes. (The PSI particles with a chl a/b ratio of 4.3 were different from those used for Fig.1 with chl a/b=10.)

in the complex, the release of Pc^{II} from the complex, and the pseudo first-order binding of a new Pc^I to PSI ($[Pc^I] >> [P700]_{tot}$)

$$Pc^I + Pc^I{\cdot}P700^+ \rightarrow Pc^I + Pc^{II}{\cdot}P700 \rightarrow Pc^I + Pc^{II} + P700 \rightarrow Pc^{II} + Pc^I{\cdot}P700.$$

The rate constants of these steps are estimated as $k_{et} = 6\,3{\cdot}10^3\ s^{-1}$, $k_{off}{}^{II} = 9{\cdot}2\ 10^3\ s^{-1}$, and $k_{on}{}^{I} = 5{\cdot}10^7\ M^{-1}s^{-1}$, respectively. It should be pointed out, that our data (not included) show that the release of Pc^{II} is the rate limiting step of the electron transfer cycle between the cyt *bf* complex and PSI *in situ*. We do not find any evidence for a second binding site of Pc involved in electron transport (5). A more detailed analysis will be done by a complete set of differential equations.

Acknowledgement
Support by Deutsche Forschungsgemeinschaft Ha1084/5-3 and SFB 388 is gratefully acknowledged.

References

1 Haehnel, W., Pröpper, A. and Krause, H. (1980) Biochim. Biophys. Acta 593, 384-399
2 Wynn, R.M. and Malkin, R. (1988) Biochemistry 27, 5863-5869
3 Hippler, M., Ratajczak, R., and Haehnel, W. (1989) FEBS Lett. 250, 280-284
4 Haehnel, W., Jansen, T., Gause, K., Klösgen, R.B., Stahl, B., Michl, D., Huvermann, B., Karas, M., and Herrmann, R.G. (1994) EMBO J. 13, 1028-1038
5 Bottin, H. and Mathis, P. (1985) Biochemistry 24, 6453-6460

CHARACTERIZATION OF THE NAD(P)H-PLASTOQUINONE-OXIDOREDUCTASE FROM MAIZE THYLAKOID MEMBRANES

Edgar Funk and Klaus Steinmüller
Institut für Entwicklungs- und Molekularbiologie der Pflanzen,
Heinrich-Heine-Universität, Universitätsstr. 1, 40225 Düsseldorf, FRG

1. Introduction

The first complete sequences of plant plastid genomes revealed 11 open reading frames homologous to genes for mitochondrial complex I (NADH-ubiquinone-oxidoreductase). These open reading frames were named *ndh*-genes (*ndhA-K*) and it was proposed that they encode subunits of a plastidial enzyme complex with a putative NAD(P)H-plastoquinone-oxidoreductase activity [1].
Complex I is the first enzyme of the respiratory chain and occurs in mitochondria as well as in eubacteria. The enzyme of *E. coli* is made up of 14 different subunits and can be divided into three subfragments (Table 1). The NADH-dehydrogenase fragment consists of three subunits bearing the NADH-oxidation site. The membrane fragment is composed of seven proteins and contains the quinone-binding site. Both fragments are connected by a third subcomplex of four proteins [2]. This subcomplex arrangement is conserved in the mitochondrial complex I enzymes, that are equipped with 20 to 30 additional subunits [3]. In plastids and also in cyanobacteria the 11 genes *ndhA-K* encode subunits of the membrane and the connecting fragment (Table 1)[4]. These enzymes seem to lack a subcomplex homologous to the NADH-dehydrogenase fragment since no genes coding for the corresponding subunits have been found. It is possible that the enzymes of plastids and cyanobacteria are equipped with a different subcomplex for the input of electrons.

Table 1. Assignment of *E. coli* complex I subunits to homologous proteins of plastids and cyanobacteria

Subcomplex	*E. coli* subunits	Homologs in plastids and cyanobacteria
NADH-DH fragment	NUO-E,-F,-G	none known
connecting fragment	NUO-B,-C,-D,-I	NDH-K,-J,-H,-I
membrane fragment	NUO-A,-H,-J,-K,-L,-M,-N	NDH-C,-A,-G,-E,-F,-D,-B

P. Mathis (ed.), Photosynthesis: from Light to Biosphere, Vol. II, 701–704.

2. Materials and Methods

2.1. Production of antibodies against NDH-proteins
Fragments of different *ndh*-genes from tobacco and rice plastid DNA were cloned into pGEMEX 1 and pGEMEX 2 expression vectors (Promega) and used for the production of recombinant proteins. The constructs were expressed in *E.coli*. The recombinant fusion proteins were purified by preparative SDS-PAGE. We constructed fusions with fragments of NDH-A,-H,-J and -K. Antisera against the recombinant proteins were produced by Eurogentec Corp., Belgium.

2.2. Solubilization of the enzyme complex
Thylakoid membranes were isolated from fresh maize leaves according to Ghanotakis et al. [5]. All steps were carried out at 0-4 °C. The membranes were suspended in 50 mM MES/NaOH (pH 6.0), 15 mM NaCl, 5 mM $MgCl_2$, 1 mM Na-ascorbate, (buffer I). The thylakoid membranes were incubated with different concentrations of Triton X-100 and the technical alkylglucoside APG 225 (Henkel) at a chlorophyll concentration of 2 mg/ml for 1 h under agitation. Centrifugation at 100000 x g for 1 h separated the solubilized material from the remaining membranes.

2.3. Immunoaffinity purification of the enzyme complex
Antibodies against NDH-A or NDH-H were covalently coupled to protein-A sepharose (Pharmacia) according to standard coupling procedures [6]. The matrix-coupled antibodies were incubated with a 0.5% Triton extract from maize thylakoid membranes at 4 °C for 16 h. The matrix material was filled into chromatography columns and washed with buffer I. The bound proteins were then eluted with 0.1 M triethylamin pH 11.5.

2.4. Miscellaneous procedures
SDS-PAGE and western-blot analysis were carried out as described in Berger et al. [4]. The antigen-antibody complexes were detected with the ECL western-blot analysis system (Amersham Buchler).

3. Results and Discussion

3.1. Specificity of the antibodies
The antibodies against the recombinant NDH-H,-J and -K proteins were able to detect the corresponding NDH-proteins in the chloroplasts of several higher plant species in western-blot experiments. The apparent molecular masses of the proteins were 46 kDa (NDH-H), 18 kDa (NDH-J) and 28 kDa (NDH-K). These values are consistent with the molecular masses calculated from the NDH-H, NDH-J and NDH-K amino acid sequences. The antibody raised against the recombinant NDH-A protein failed to detect the corresponding chloroplast protein in the same experiments (data not shown).

3.2. Solubilization of the enzyme complex
Initially different concentrations of Triton X-100 and APG 225 were tested for solubilization of the thylakoid membranes. The best results were obtained with 0.5% Triton X-100, which quantitatively solubilized NDH-H (Fig. 1). Other enzymes brought into solution were the cytochrome b_6/f complex and the chloroplast ATPase while photosystem I and photosystem II were essentially not solubilized (data not shown).

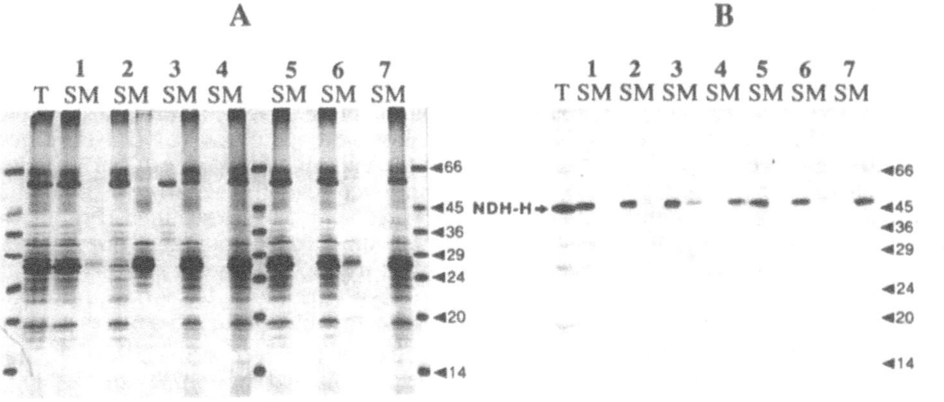

Figure 1. Solubilization of maize thylakoid membranes with Triton X-100 and APG 225. (A) SDS-PAGE analysis. (B) Western-blot analysis with the antibody against NDH-H. Thylakoid membranes were solubilized under the following conditions: 5% Triton (1), 2% Triton (2), 0.5% Triton (3), 0.1% Triton (4), 5% APG (5), 2% APG (6), 0.5% APG (7). T: total thylakoid proteins, S: solubilized proteins, M: residual membrane bound proteins.

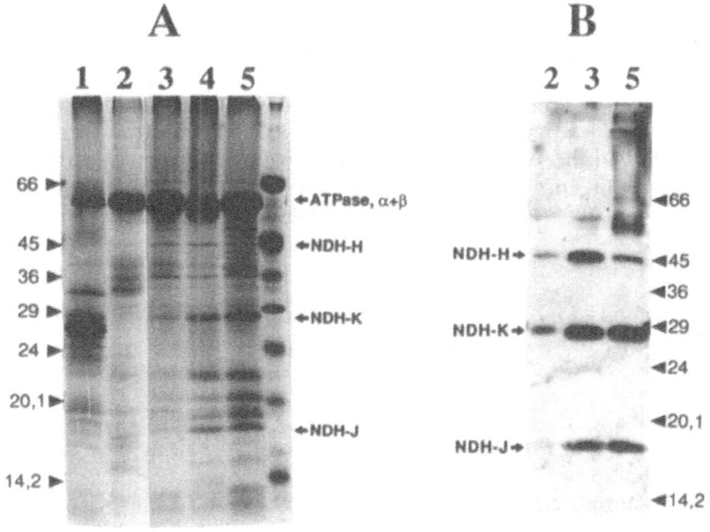

Figure 2. Immunoaffinity purification of the enzyme complex. (A) SDS-PAGE analysis, 60 μg protein was loaded per lane. (B) Western-blot analysis with the antibodies against NDH-H,-K,-J; 25 μg protein in lane 2, 5 μg protein in lanes 3 and 5. 1: total thylakoid proteins, 2: 0.5% Triton extract, 3: purified fraction of anti-NDH-A column, 4: purified fraction of the anti-NDH-H column after washing with buffer I containing 200 mM NaCl, 5: purified fraction of the anti-NDH-H column.

3.3. Immunoaffinity purification of the enzyme complex
The enzyme was further purified by chromatography on immunoaffinity columns carrying antibodies against NDH-H and NDH-A. The eluate of the anti-NDH-H column contained two prominent proteins at about 60 kDa and 12 other major proteins (Fig. 2A, lane 5). The two prominent proteins are the α and β subunits of the chloroplast ATPase visible as one band in Fig. 2A. Three of the 12 major proteins were identified as NDH-H,-K and -J by western-blot analysis (Fig. 2B). Washing of the column with high salt buffer reduced the contamination by the ATPase subunits and removed several other proteins. Four major proteins at 36, 22, 20.5 and 19 kDa and NDH-H,-K and -J were retained (Fig.2A, lane 4). Wether these four proteins are subunits of the complex I-homologous enzyme has to be determined by N-terminal sequencing. The proteins removed by the high salt treatment are either contaminations or also subunits of the complex. Immunoaffinity chromatography with the anti-NDH-A column gave a similar protein pattern as compared to the eluate from the anti-NDH-H column (Fig.2A, lane 3). The NDH-H,-K and -J subunits are also present in the preparation (Fig. 2B). This result demonstrates that the NDH-A protein must be present in the solubilized enzyme complex even though the NDH-A antibody failed to detect the protein in western-blot analysis.

4. Conclusions

Our results show that NDH-A,-H,-K and -J form a complex which can be solubilized from thylakoid membranes with Triton X-100. This is the first biochemical evidence that higher plant thylakoid membranes contain an enzyme complex homologous to mitochondrial complex I. A similar fragment of this enzyme, that contains NDH-H,-I,-J and -K has been isolated from the cyanobacterium *Synechocystis* sp. PCC6803 [4]. These four proteins constitute the connecting fragment of the *E. coli* complex I (Table 1). However no indication for the presence of a subunit of the membrane fragment was found in this preparation. In contrast, our preparation of the enzyme from maize also seems to contain the membrane fragment of the enzyme. This can be inferred from the presence of the NDH-A subunit in the solubilized complex. However, like with the cyanobacterial enzyme there is no indication for the presence of a NAD(P)H-dehydrogenase fragment in our preparation. This supports the view that the plastid enzyme lacks this subfragment.

Acknowledgements

This work was supported by a grant to K. Steinmüller from the DFG (SFB 189).

References

1 Ohyama, K., Kohchi, T., Sano, S. and Yamada, Y. (1988) TIBS 13, 19-22
2 Leif, H., Sled, V.D., Ohnishi, T., Weiss, H. and Friedrich, T. (1995) Eur. J. Biochem. 230, 538-548
3 Friedrich, T., Steinmüller, K. and Weiss, H. (1995) FEBS Lett. 367, 107-111
4 Berger, S., Ellersiek, U., Kienzelt, D. and Steinmüller, K. (1993) FEBS Lett. 326, 246-250
5 Ghanotakis, D.F., Babcock, G.T. and Yocum, C.F. (1984) Biochim. Biophys. Acta 765, 388-398
6 Harlow, E. and Lane, D. (1988) Antibodies: A Laboratory Manual pp. 522-523, Cold Spring Harbor Press, Cold Spring Harbor, USA

Presence of a large protein complex containing the *ndhK* gene product and possessing NADH-specific dehydrogenase activity in thylakoid membranes of higher plant chloroplasts

L. A. Sazanov, P. Burrows and P. J. Nixon
Photosynthesis Research Group
Department of Biochemistry
Imperial College of Science, Technology and Medicine,
London SW7 2AY, U.K.

1. Introduction

Plastid DNA of certain higher plants contains eleven open reading frames (*ndhA-K* genes) that are homologous to genes encoding subunits of the mitochondrial Complex I [1]. By analogy it was proposed that chloroplasts may contain an NAD(P)H-plastoquinone oxidoreductase [2,3] which may function in cyclic electron transport around the photosystem one complex or in chlororespiration [1-4]. Expression of the *ndh* genes may also be important in other types of plastid or at certain stages of chloroplast development, such as senescence [4,5].

The complex itself has not yet been purified from plants. Thus, the function of the *ndh* genes in chloroplast metabolism or indeed if they encode a NADH (or NADPH) dehydrogenase (DH) is not known. Major problems in isolating such a complex are potential mitochondrial contamination of the chloroplast preparation, as well as the presence of other possible NADH DH activities such as Ferredoxin-$NADP^+$ reductase (FNR), which has both NADPH and NADH DH activities [6].

The presence of a large complex with NADH DH activity in barley chloroplasts was suggested recently by native PAGE [5]. Our approach has been to fractionate the thylakoid membrane and to assay for NADH and NADPH DH activities and compare their distribution to that of the NdhK protein for which antibodies are available [3]. We show here that an NADH DH complex is present in both pea and spinach chloroplasts and is located mainly in the stromal lamellae membranes. In addition this activity co-purifies with the NdhK protein using both anion-exchange chromatography and sucrose density-gradient centrifugation.

2. Procedures

2.1 *Preparation of Plant Material*

Chloroplasts, thylakoid membranes including granal and stromal membrane fractions were obtained from greenhouse-grown peas or from spinach as described in [3]. Young and mature pea leaves were obtained 10 days and 16 days, respectively, after germination. Senescent leaves were obtained by incubation of detached spinach leaves in a dark wet box for 6 days at 20°C, as suggested in [7]. Intact chloroplasts were purified by centrifugation through percoll [8].

2.2 *Antibodies*

Antibodies specific for the NdhK protein [3] were affinity-purified using purified NdhK protein that was overexpressed in *E. coli.*. Plant samples containing 5 μg chlorophyll were immunoblotted as described [3]. Concentration of chlorophyll was assayed as described in [3] and protein by the Bradford method.

P. Mathis (ed.), Photosynthesis: from Light to Biosphere, Vol. II, 705–708.

2.3 *Assay of NADH and NADPH dehydrogenase activity*
NADH or NADPH oxidation was monitored at 340 nm in 50 mM Tris (pH 7.5), 0.5 mM EDTA buffer, containing 0.05% dodecyl-ß-D-maltoside (DM), 100µM of either NADH or NADPH and 250µM potassium ferricyanide. Succinate dehydrogenase activity was assayed as in [9]. Purified mitochondria from *Arum Maculatum* were generously provided by Prof. J. Palmer (Department of Biology, Imperial College, London).
2.4 *Anion-exchange chromatography*
Stromal lamellae membranes were solubilised by incubation with 2% DM (detergent/protein ratio = 5) for 30 min on ice. Solubilised material was applied to 1x5 cm DEAE-Fractogel column equilibrated with TB buffer (10mM Tris/pH 8.0, 0.5mM EDTA, 0.03%DM) and eluted with 0-0.5M NaCl gradient.

3. Results and Discussion

3.1 *Pure intact chloroplasts possess membrane-bound NAD(P)H dehydrogenase activity*
Pure intact spinach chloroplasts, prepared by centrifugation through a Percoll gradient, showed significant membrane-bound NAD(P)H DH activity (0.5-2 $\mu mol \cdot min^{-1} \cdot mg\ prot^{-1}$). No succinate DH activity, which is a marker enzyme for mitochondrial membranes, was detected. By comparison with the succinate DH activity shown by mitochondria isolated from *Arum maculatum*, we estimate the level of mitochondrial contamination in the chloroplast preparations to be less than 0.005% on a protein basis. We therefore exclude mitochondrial contamination as the source of the NAD(P)H DH activity. Chloroplasts prepared by a conventional centrifugation method [3] had very similar specific dehydrogenase activities and NdhK content and thus were used in subsequent work.

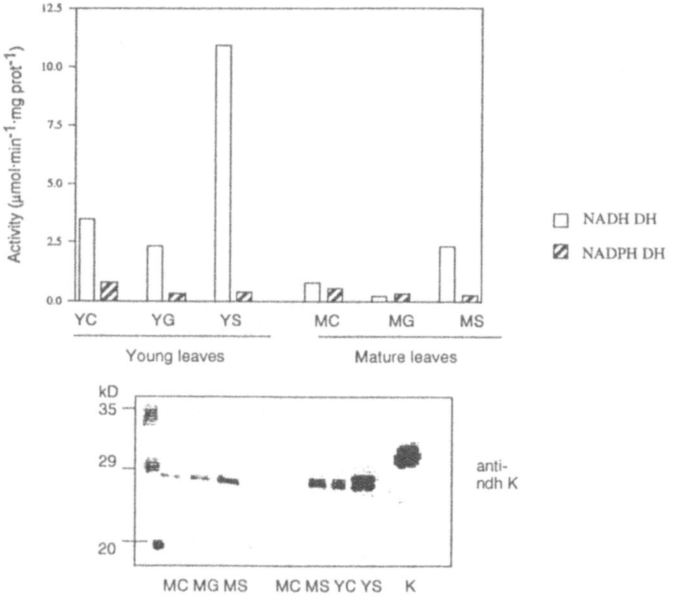

Fig.1. Distribution of NAD(P)H DH activity and NdhK subunit between chloroplasts (C), granal (G) and stromal lamellae membranes (S) prepared from young (Y) and mature (M) pea leaves. Preparations were assayed for NAD(P)H DH activity and probed with antibody specific for NdhK. 20 ng of purified NdhK (K) was immunoblotted for comparison.

3.2 *Distribution of NAD(P)H DH activity between granal and stromal lamellae fractions*
Thylakoid membranes prepared from pea chloroplasts by washing in hypotonic medium retain as membrane-bound about 50% of the NADH-specific activity and about 45% of the NADPH-specific activity, the rest of activity being soluble. Preparations of granal (PSII-enriched) and stromal lamellae (PSI-enriched) membrane fractions demonstrate significant enrichment of NdhK in the stromal lamellae fraction, as shown previously for this subunit [3] and NdhH [10]. Our results show that NADH- (but not NADPH-) specific activity is enriched to a similar extent (Fig.1).

NADH DH activity was not inhibited by rotenone, which is common when using ferricyanide as an electron acceptor. Activities with duroquinone and plastoquinone as substrates were too low to be measured reliably in the presence of highly absorbing chlorophyll and thus may require a more pure complex.

3.3 *NADH dehydrogenase activity and NdhK content are higher in young leaves*
Our results (Fig.1) show that both NADH-specific activity and NdhK content are significantly higher in chloroplasts and in stromal lamellae membranes of young pea leaves as compared to mature leaves, when compared on an equal chlorophyll and protein basis. This may indicate that the content of the putative chloroplast Complex I is higher at earlier stages of development, providing energy for plastids when the photosynthetic apparatus is not fully developed. Contrary to the suggestions in [5], we have not found an increase in NAD(P)H DH activity or NdhK content in senescent spinach chloroplasts (not shown). It should be noted, that the total content of the NdhK subunit, as determined immunochemically, is very low (<0.1% of total protein) even in chloroplasts from young leaves (Fig.1), indicating that the putative Complex I is probably a minor component of the thylakoid membranes. This result is consistent with the estimation from the rate of reduction of the PQ pool by NADH that less than 1 NADH dehydrogenase is present per 100 photosynthetic electron transfer chains in pea thylakoids [11].

3.4 *Co-elution of NADH dehydrogenase activity and NdhK subunit by anion-exchange chromatography*

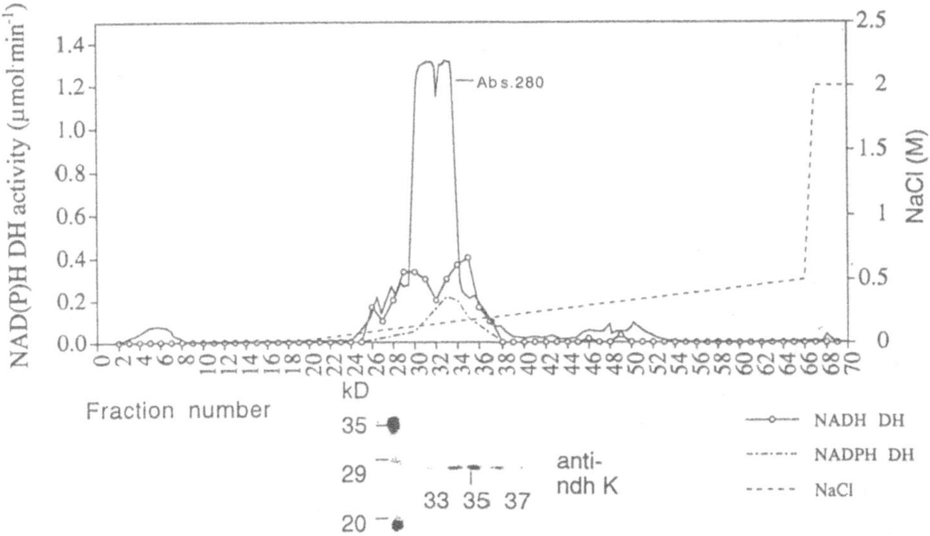

Fig.2. Anion-exchange chromatography of stromal lamellae membrane proteins. 2-ml fractions were assayed for NAD(P)H DH and immunoblotted with anti-NdhK antibody.

Stromal lamellae membranes were solubilised with dodecyl-ß-D-maltoside and applied to an anion-exchange column. Two NADH- (peaks in fractions 30 and 35) and one NADPH- (peak in fraction 33) specific activities were separated. Only the second of the NADH DH activities (peaking in fraction 35) coincided with elution of NdhK (Fig.2). This NADH DH activity also co-sedimented with NdhK in sucrose gradients (data not shown).

3.5 *NADH-specific activity and NdhK subunit can be attributed to a large protein complex*
Active fractions (34-36) from the anion-exchange column were filtered through an Amicon 100 (100 kD cut-off) concentrator. All the NADH DH activity as well as all the *ndhK* gene product were retained, while about 80% of the NADPH DH activity came through the filter (not shown). This indicates that NdhK (27 kD) and possibly other Ndh subunits are part of a larger complex (>100 kDa), while the NADPH-specific enzyme is much smaller, probably FNR (30-40 kD [6]).
In accordance with this prediction, the size of the NADH DH complex as determined by size-exclusion HPLC is approx. 400-800 kD (not shown).

4. Conclusions

1. Pure intact chloroplasts contain both membrane-bound NADH as well as NADPH dehydrogenase activities when assayed in the presence of ferricyanide.
2. The NADH (but not NADPH) dehydrogenase activity co-purifies with the *ndhK* gene product (Fig.2) and is enriched in stromal lamellae membranes (Fig.1). The NADH dehydrogenase activity is attributed to a large (400-800 kD) complex by HPLC size-exclusion analysis (data not shown).
3. Chloroplasts from young leaves appear to have higher NADH dehydrogenase activity and NdhK subunit content than chloroplasts from mature and senescent leaves (Fig.1). This may indicate the importance of this protein complex at earlier stages of plastid development.
4. This evidence supports the hypothesis that the *ndh* genes encode subunits of a plastidial NADH dehydrogenase.

References

1. Shinozaki, K. et al. (1986) EMBO J., 5, 2043-2049.
2. Ohyama, K., Kohchi, T., Sano, T. and Yamada, Y. (1988) Trends Biochem. Sci. 13, 19-22.
3. Nixon, P.J., Gounaris, K., Coomber, S.A., Hunter, C.N., Dyer, T.A. and Barber, J. (1989) J. Biol. Chem. 257, 10766-10769.
4. Berger, S., Ellersiek, U., Kinzelt, D. and Steinmuller, K. (1993) FEBS Lett. 326, 246-250.
5. Cuello, J., Quiles, M.J., Albacete, M.E. and Sabater, B. (1994) Plant Cell Physiol. 36, 265-271.
6. Suss, K.-H. (1982) in Methods in chloroplast molecular biology (Edelman, M., Hallick, R.B. and Chua, N.-H., Eds.), Elsevier Biomedical, p.957-952.
7. Kawasaki, S. and Takeuchi, J. (1989) Plant Physiol. 90, 338-344.
8. Douce, R. and Joyard, J. (1982) in Methods in chloroplast molecular biology (Edelman, M., Hallick, R.B. and Chua, N.-H., Eds.), Elsevier Biomedical, p.244-245.
9. Baginsky, M.L. and Hatefi, Y. (1970) J. Biol. Chem. 244, p.5313-5319.
10. Berger, S., Ellersiek, U., Westhoff, P. and Steinmuller, K. (1993) Planta 190, 25-31.
11. Rich, P.R. (1991) in 37th Harden Conference, Wye College, Kent, U.K.; Biochem.Soc.Publ., U.K., p.37.

NON-PHOTOSYNTHETIC TYPE FERREDOXIN FROM MAIZE RETARDS PHOTOAUTOTROPHIC GROWTH OF THE CYANOBACTERIUM *PLECTONEMA BORYANUM*

T. Matsumura, Y. Fujita and T. Hase
Institute for Protein Research, Osaka University, 3-2 Yamadaoka, Suita, Osaka 565, Japan

1. Introduction

Ferredoxins (Fds) are a group of iron-sulfur proteins which act as electron carriers in a number of biochemical processes. In higher plants, Fd is localized in plastids such as chloroplasts, leucoplasts and others. A photosynthetic-type Fd localized in chloroplasts is mainly reduced by photosystem I, while a nonphotosynthetic-type Fd localized in non-photosynthetic plastids is reduced by NADPH via Fd-$NADP^+$ reductase (FNR). Then reduced Fds donate electrons to a variety of Fd-dependent enzymes. It has been pointed out that the two types of Fd may play respective physiological roles in each plastid.

Maize (*Zea mays* L.) has at least two types of Fd, FdI and FdIII. These Fds differ in the manner of distribution and expression; FdI is a leaf-specific and light-inducible, whereas FdIII is mainly distributed in roots and its expression is constitutive (1). They also differ in the redox potential and electron transport abilities to FNR.

In this report, we have constructed a heterologous Fd-expression system in a cyanobacterium *Plectonema boryanum* to investigate the functional differentiations of the two Fds *in vivo*, and some physiological differences in electron partitioning of these Fds are discussed.

2. Procedure

2.1 *Cyanobacterial strains and culture conditions*

A cyanobacterium *Plectonema boryanum* IAM-M101 strain *dg5*, an dark-adapted strain, was used as the wild type. The cyanobacterial strains carrying a Fd-expression vector were grown in BG-11 medium containing 20mM Hepes-KOH pH8.0, 30mM glucose and 25μg/ml chloramphenicol in the light at 30 °C. The medium without glucose was used for the photoautotrophic growth of strains.

2.2 *The plasmids for expression of maize Fd cDNAs*

To express maize Fd cDNAs (FdI and FdIII) in the cyanobacterium, a series of expression plasmid was constructed from a shuttle vector pPBH201 between *P. boryanum* and *E. coli* (2) and an *E. coli* expression vector pTrc99A (Pharmacia). The mature parts of the Fd cDNAs were inserted under the *trc* promoter to express in *E. coli*. A segment carrying the promoter and the Fd coding region of the *E. coli*-expression vector was introduced

P. Mathis (ed.), Photosynthesis: from Light to Biosphere, Vol. II, 709–712.

into a shuttle vector pSVM30 which had been constructed by insertion of a *lacI*^{q} segment into pPBH201. Thus constructed plasmids consist of a Fd-expression part derived from pTrc99A, the *lacI*^{q} gene to control the *trc* promoter, a cyanobacterial cryptic plasmid pGL3 carrying a replication origin in the cyanobacterium, and an *E. coli* replication origin and *cat* (chloramphenicol acetyltransferase) gene for a selective marker. In addition to FdI and FdIII, FdIII* was also introduced into the shuttle vector. FdIII* encodes a polypeptide identical to that of FdIII but has some altered codons in order to raise the expression level of FdIII in *E. coli.*

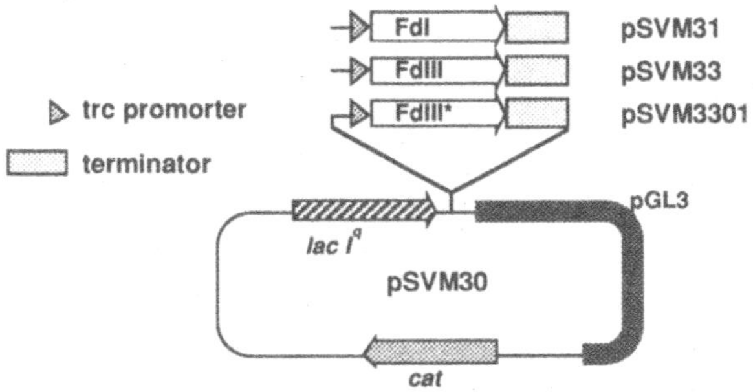

Fig. 1 Structure of the plasmids for expression of maize Fd cDNAs

2.3 *Electrotransformation of* P. boryanum

The plasmid DNAs (pSVM30, etc.) were introduced into the cells of *P. boryanum* strain *dg5* by electroporation (3). Cells applied with a single electric pulse (14.0 kV cm^{-1}, 8 ms) were plated onto a selective BG-11 agar plate containing 25 µg/ml chloramphenicol, and strains carrying the plasmid appeared in one to two weeks.

2.4 *Electron transfer activities of FdI and FdIII*

The electron transfer activities of FdI and FdIII were measured in the assay of photochemical reduction of $NADP^+$ by broken spinach chloroplasts and cytochrome *c* reduction by maize Fd-$NADP^+$ reductase (FNR).

3. Result and discussion

3.1 In vitro *electron transfer abilities of FdI and FdIII*

To examine whether the two Fds, FdI and FdIII, have different characteristics of electron transfer reaction, photoreduction of $NADP^+$ by photosystem I and cytochrome *c* reduction by NADPH in the presence of FNR were measured (Fig. 2). FdIII mediated the electron transfer from photosystem I to $NADP^+$ at a half rate of FdI. In contrast, FdIII did from NADPH to cytochrome *c* at a rate about two times higher than FdI. These results suggested that these Fds may differentiate in functioning as an electron carrier in the photosynthetic and non-photosynthetic plastids.

3.2 *Heterologous expression of maize Fds in the cyanobacterium*

A series of Fd-expression vector was introduced into cyanobacterial cells and the expression of Fd was examined by nondenaturing PAGE and Western-blot analysis using an antiserum against maize Fds. As shown in Fig. 3, both FdI and FdIII were expressed in the functional form with the iron sulfur cluster in the cyanobacterium. IPTG addition raised the expression levels of all Fds while all of them were expressed at lower levels in the absence of IPTG. The expression level of FdIII in the transformant with pSVM3301 carrying FdIII* gene, a FdIII over expression plasmid in *E. coli*, was significantly higher than that in the transformant with pSVM33. A combination of the two transformants and addition of IPTG allowed us to analyze the cyanobacterial growth according to various levels of FdIII.

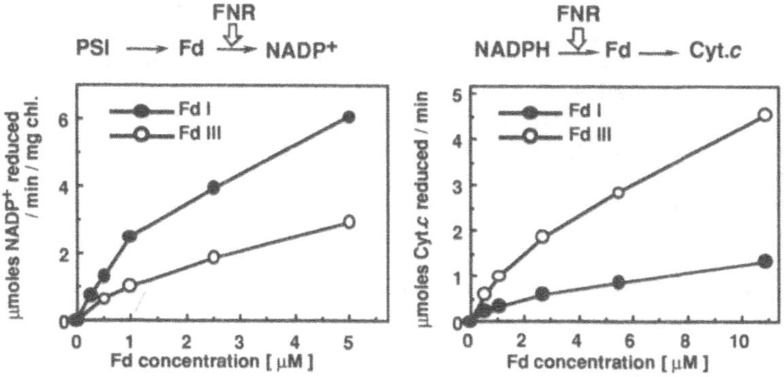

Fig. 2 Electron transfer activities of maize Fds
The assay systems are photoreduction of $NADP^+$ (left) and NADPH-cytochrome *c* reduction (right).

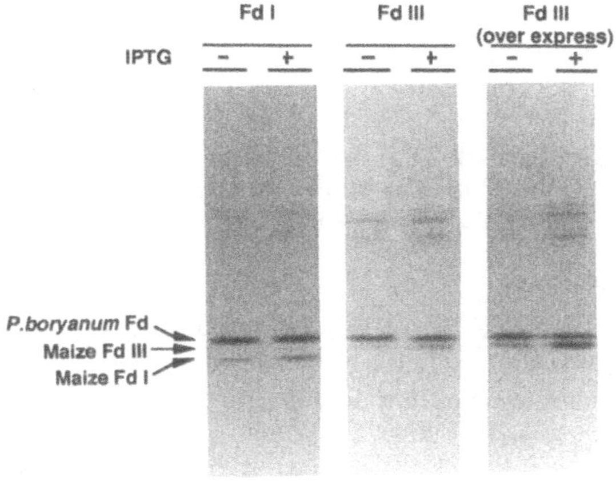

Fig. 3 Expression of maize Fds in *P. boryanum*
Crude extracts from transformants grown under photomixotrophic conditions with or without IPTG (1mM) were subjected to nondenaturing PAGE and analyzed by Western blotting using anti-maize Fds antiserum. Each lane contained 20µg of total soluble protein.

3.3 In vivo *effects of non-photosynthetic Fd*
To examine whether the expression of maize Fds causes any effects on growth of the cyanobacterium, the transformants were cultivated in two different conditions. As shown

in Fig. 4, the transformants expressing FdIII grew significantly slower than the control strains under photoautotrophic conditions, whereas the transformants expressing FdI showed no remarkable difference. The transformants carrying FdIII* gene (FdIII-overexpressing strains) grew much slower than the control strains even in the absence of IPTG. Furthermore, in the presence of IPTG, the FdIII-overexpressing strains no longer grow. This retardation of growth seems to be correlated with the expression level of FdIII. When the transformants expressing FdIII were photomixotrophically cultivated in the medium supplemented with glucose in the light, such growth retardation did not occurred. These effects of FdIII on the cyanobacterial growth could be interpreted as follows: the maize FdIII might function as an competitor for the NADPH photoreduction mediated by the cyanobacterial own Fd, causing the growth retardation in the photoautotrophic conditions, and FdI did not give such an effect in the cyanobacterial cells. These *in vivo* observations suggest that two types of Fd, photosynthetic and non-photosynthetic, functionally differentiate and are thus strictly controlled in their distribution and expression level according to the physiological conditions of each plastid in maize.

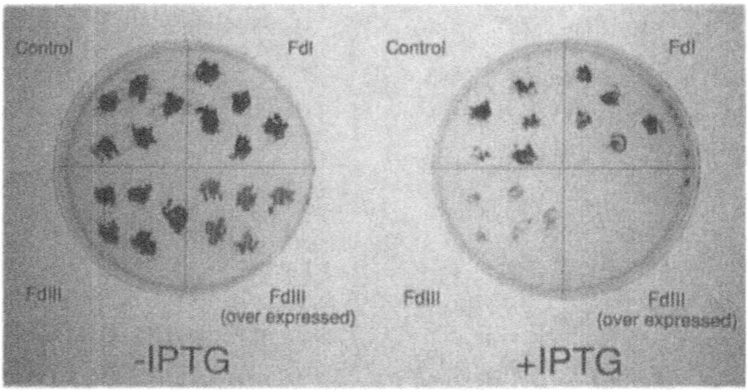

Fig. 4 Effect of expression of maize Fds on the photoautotrophic growth of *P. boryanum* Five transformants were independently isolated and they were cultivated in the photoautotrophic conditions with or without IPTG (1mM).

4. Acknowledgment

We are grateful to Dr. D. K. Walton (College of St. Scholastica) for the gift of the shuttle vector pPBH201.
T.M. is grateful for a fellowship from the JSPS for Young Scientists.

References

1. Hase, T. et al (1991) Plant Physiol. 97, 77-83
2. Walton, D.K. et al (1993) Nucleic Acids Res. 21, 746
3. Fujita, Y. et al (1992) Plant Cell Physiol. 33, 81-92

cDNA CLONING AND FUNCTIONAL EXPRESSION OF FERREDOXIN-DEPENDENT SULFITE REDUCTASE FROM MAIZE IN *E.coli* CELLS

Takashi Ideguchi, Tetsuyuki Akashi, Yayoi Onda and Toshiharu Hase
Institute for Protein Research, Osaka University, Suita, Osaka 565, Japan

1. Introduction

Sulfite reductase (SiR) catalyses reduction of SO_3^{2-} to S^{2-} through 6 electron transfer. The enzyme has [4Fe-4S] cluster and siroheme as prosthetic groups. In plants, SiR is localized in chloroplast stroma using ferredoxin (Fd) as a physiological electron donor (1). Maize SiR reported here is a monomeric protein (65kDa). *E.coli* has SiR with a heterooligomeric structure of $\alpha_8\beta_4$ (α:66kDa, β:64kDa) which utilizes NADPH as an electron donor. Four FADs and four FMNs are contained in the α_8 complex, and [4Fe-4S] cluster and siroheme are contained in the β subunit, which are bridged by S-atom of a cysteine residue in the presented model (2, 3, 4). Overproduction of the *E.coli* enzyme required co-expression with siroheme synthase (5).

Cloning and expression in *E.coli* of maize SiR was attempted so as to allow us to compare both types of the enzymes on primary structure and to obtain enough material for biochemical analyses taking advantages of its simple structure.

2. Materials and Methods

SiR extracted from maize leaves was purified by ammonium sulfate fractionation and following column chromatographies of anion exchange, gel filtration, hydrophobic, and affinity to Blue-Sepharose and Fd. The enzyme was further purified by preparative SDS-PAGE and the electroeluted polypeptide was employed as an antigen for immunization of a rabbit.

SiR activity was assayed by measuring the cysteine formed from S^{2-} and added O-acetyl-L-serine by the coupling reaction with cysteine synthase. The reaction was started by addition of sodium dithionite to the assay mixture containing methylviologen (MV) or Fd as an electron donor.

Maize cDNA library was constructed and screened essentially as described previously (6). The library constructed in an expression vector pUEX1 (Amersham) was primarily screened using the antiserum. The obtained cDNA was employed as a probe for secondary screening.

The nucleotide sequence was determined by the chain termination method (7) using equipments (DNA Sequencer ABI Model 373A) of Research Center for Protein Engineering, Institute for Protein Research, Osaka University.

For *in vitro* transcription, pGEM4Z (Promega) vector and SP6 RNA polymerase were used. The transcripts were translated in wheat germ extract in the presence of

P. Mathis (ed.), Photosynthesis: from Light to Biosphere, Vol. II, 713–716.

^{3}H-leucine. The labeled polypeptides were used for import analysis in chloroplasts.

An expression vector pTrc99A (Pharmacia) was employed to produce recombinant proteins in *E.coli*. Co-expression of SiR and siroheme synthase was achieved by cloning both genes into the identical vector with each own promoter. Recombinant SiR produced in *E.coli* was extracted by sonication and purified by the methods essentially same as described above.

Mutation by base substitution was introduced using synthetic oligonucleotides as PCR primers.

3. Results and Discussion

The 2.4kb DNA fragment was cloned from maize cDNA library and sequenced as described above. The complete nucleotide sequence will appear in the GSDB, DDBJ, EMBL and NCBI DNA databases with the following accession number D50679. It contained an open reading frame encoding a 70kDa polypeptide composed of 635 amino acid (aa) residues. The deduced amino acid sequence exhibited similarity to the β subunit of *E.coli* NADPH-SiR (32% aa identical), and an N-terminal extension absent in the β subunit indicated that the protein coded by this cDNA is to be a precursor of maize SiR. Four cysteine residues regarded as ligands for the prosthetic groups were conserved among other siroheme-containing enzymes. Remarkable similarity was exhibited in the flanking region (49% aa identical to *E.coli*, Fig.1).

```
                        *     *                                     *   *
Maize   (492-556): MACPALPLCPLAQTEAERGILPILKRIRAVFNKVGIKDSESVVVRITGCPNGCARPYMAELGFVG
E.coli  (421-484): MACVSFPTCPLAMAEAERFLPSFIDNIDNLMAKHGVSD-EHIVMRVTGCPNGCGRAMLAEVGLVG
```

Fig. 1. Comparison of amino acid sequences of maize SiR and the β subunit of *E.coli* SiR (3). Highly conserved regions including 4 ligand cysteines (*) are represented.

In fact, the precursor synthesized *in vitro* was translocated into chloroplasts to be processed to the mature form with a molecular mass of 65kDa. Although the processing site was not yet determined, a putative mature SiR was attempted to be expressed in *E.coli*. An ATG codon was introduced to a corresponding position close to the N-terminus of *E.coli* SiR. The transformed cells produced a 65kDa polypeptide as expected, which was cross-reacted with the antiserum against the authentic SiR, and SiR activity was found in a crude extract. It was reported that the level of enzymatically active SiR was limited by availability of siroheme in *E.coli* cells when SiR was overproduced (5). Co-expression of siroheme synthase also resulted in a remarkable increase of the activity in this case (Fig.2, Table 1). Insoluble products were significantly decreased when cells were grown at lower temperature.

Table 1. MV-SiR activity in crude extract. The activity was normalized for total soluble protein.

Co-expression of siroheme synthase	-	-	+	+
Growth temperature	37°C	27°C	37°C	27°C
Relative activity (%)	9.7	19	26	100

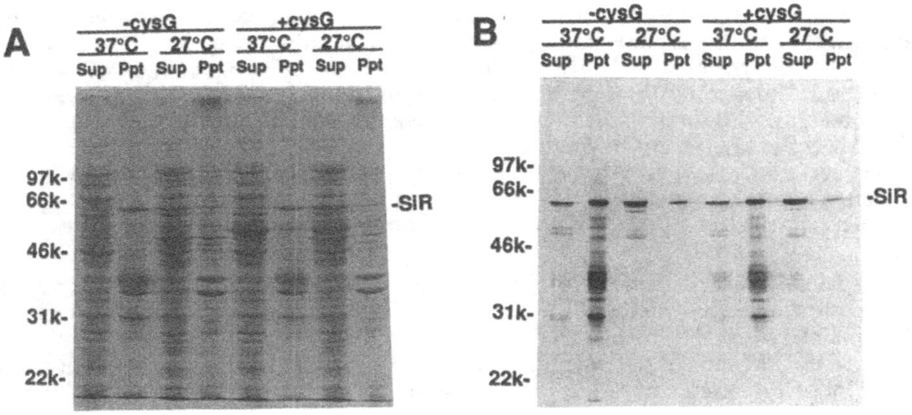

Fig. 2. Expression of maize SiR in *E.coli* cells. Sonicated cells were centrifuged to obtain soluble (Sup) and insoluble (Ppt) fractions. Siroheme synthase was co-expressed as described in the text (+cysG). Soluble (40 µg of total protein) and corresponding insoluble fractions were applied on SDS-PAGE and stained by CBB (A). One fourth samples (10 µg protein for Sup) were simultaneously detected by Western analysis (B).

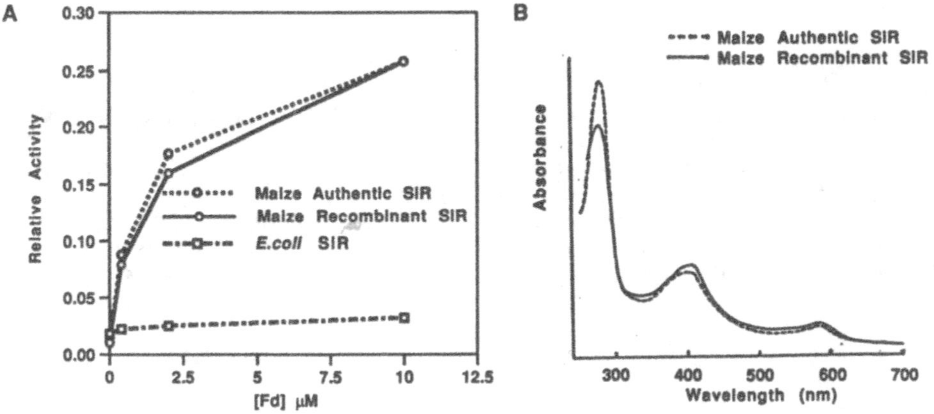

Fig. 3. Comparison of authentic and recombinant maize SiR. A). Fd-SiR activity. The enzymes of equivalent units on MV-SiR activity were used. Authentic SiR was purified as described above, while recombinant one is crude extract. B). UV-vis absorption spectra of the purified authentic SiR and recombinant SiR.

Fd-SiR activity of the recombinant SiR was in good agreement with the authentic one purified from maize leaves on various Fd concentration (Fig. 3A). The equivalent units of both enzymes on MV-SiR activity, which contained comparable amounts of the polypeptide detected by Western analysis, were subjected to Fd-SiR activity assay. Thus the recombinant SiR were indistinguishable from the authentic one in regards of Km for Fd and specific activity for MV and Fd.

UV-vis absorption spectrum of the purified recombinant SiR was almost same as that of the authentic one (Fig. 3B). They were characteristic of siroheme containing enzymes. The difference at 280nm was probably attributed to a slight difference in purity.

Consequently, no significant difference was observed between the recombinant SiR and the authentic one, although their exact N-termini were not yet determined. Several recombinants with different N-termini were constructed. They were also expressed in *E.coli* with SiR activity, suggesting that the N-terminal structure seems not to be crucial for enzymatic activity. Further investigation on the purified proteins is now in progress.

The expression system was applied to examine roles of ligand cysteine residues (C494, C500, C540, C544) in the enzymatic activity. By altering one of those cysteine residues to serine, four mutant SiRs were constructed. All of the mutants were produced in *E.coli* as well as the recombinant one, but no activity was detected. Each of the cysteine residues was confirmed to be critical for the enzymatic activity. The active recombinant SiR and the authentic one were eluted from Mono Q column by the same salt concentration with absorption in the wavelength characteristic for the prosthetic group. This spectroscopic property was not found in the cases of the cysteine mutants, suggesting that substitution of the cysteines caused a considerable deficiency in assembly to the prosthetic group.

References

1 Aketagawa, J. and Tamura, G. (1980) Agric. Biol. Chem. 44, 2371-2378
2 Siegel, L.M. and Davis, P.S. (1974) J. Biol. Chem. 249, 1587-1598
3 Ostrowski, J., Wu, J.Y., Rueger, D.C., Miller, B.E., Siegel, L.M. and Kredich, N.M. (1989) J. Biol. Chem. 264, 15726-15737
4 Ostrowski, J., Barber, M.J., Rueger, D.C., Miller, B.E., Siegel, L.M. and Kredich, N.M. (1989) J. Biol. Chem. 264, 15796-15808
5 Wu, J.Y., Siegel, L.M. and Kredich, N.M. (1991) J. Bacteriol. 173, 325-333
6 Sakakibara, H., Watanabe M., et al. (1991) J. Biol. Chem. 266, 2028-2035
7 Sanger, F., Nicklen S., et al. (1977). Proc. Natl. Acad. Sci. U.S.A. 74, 5463-5467

MOLECULAR PROPERTIES AND LOCALIZATION OF FERREDOXIN-NADP+-OXIDOREDUCTASE IN THE THYLAKOID MEMBRANE

Shi-Gui Yu , Hreinn Stefánsson and Per-Åke Albertsson
Department of Biochemistry, Chemical center,
University of Lund, P.O.Box 124, S-221 00 Lund Sweden

Introduction

Ferredoxin-$NADP^+$-oxidoreductase (FNR) plays a key role in the linear electron transport in photosynthesis. This enzyme catalyzes oxidation of ferredoxin and reduction of $NADP^+$, between the one-electron-carrier ferredoxin and the two-electron-acceptor $NADP^+$[1]. It is also involved in cyclic electron transport around PSI. It has been reported that there are two different pools of FNR, the loosely membrane-bound and the tightly membrane-bound. The loosely-bound FNR seems to be located at the surface of the thylakoid membranes and can easily be removed by extensive wash while the tightly-binding type is located in the inner part of the membranes [2]. N-terminal sequences of amino acids of these two types of FNR display some difference in amino acid residue [3] . The ratio between bound FNR and P700 has been estimated to be 0.4, using a non-ionic detergent n-decyl-β-D-maltopyranoside for the isolation procedure [4].
In this communication we characterize further these two FNR-fractions with respect to their molecular properties and the localization in the thylakoid membrane.

Materials and Methods

The thylakoid membranes were prepared as previously described [5].
The grana- and the stroma lamellae-derived thylakoid membrane fractions were isolated by sonication and an aqueous two-phase partitioning procedure [5-6].
The washing procedure for the removing of loosely-bound FNR was as described by Matthijs et al [2].
2-D electrophoresis, immunoblotting and microsequencing analyses were performed as earlier reported [5].
Protein phosphorylation was carried out principally as described in [5]. After light-induced protein phosphorylation, the membrane suspension was spun down and the pellets and supernatant were collected for 2-D analysis.

Results and Discussion

Two groups of polypeptides of FNR identified by antibodies raised against purified FNR and microsequence analysis are present in the 2-D protein pattern of the thylakoid membranes. One which has an apparent molecular mass of 35 kDa is localized at pH range 6.70-7.15 and another with a slightly lower molecular weight is localized at pH range

P. Mathis (ed.), Photosynthesis: from Light to Biosphere, Vol. II, 717–720.

6.15-6.70 (Figure 1).

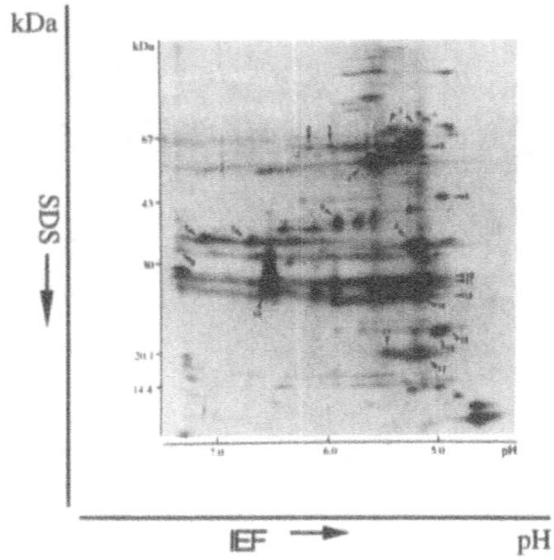

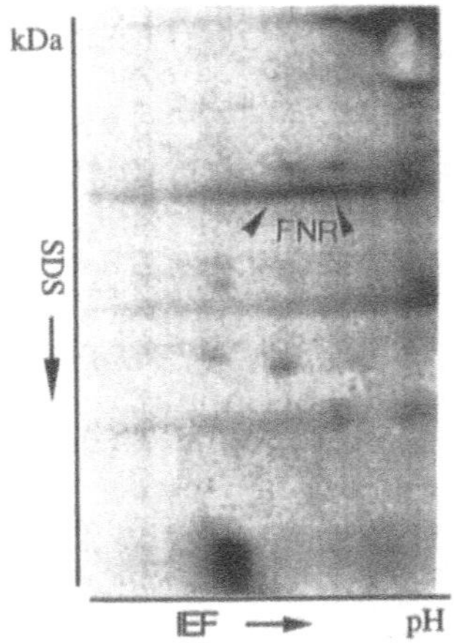

Fig. 1 2-D protein pattern of thylakoid membranes.
Total 100 µg of the polypeptides solubilized in lysis buffer was loaded into the IEF gel. The second dimensional gel was Silver-stained. Numerals represent identified polypeptides: [1] PSI core; [2] 64 kDa protein; [3] α subunits of CF1 ATPase; [4] β subunits of CF1 ATPase; [5] 45 kDa protein; [6] Aldolase; [7]; FNR; [8] 33 kDa polypeptide of the oxygen evolving complex; [9] cytochrome f; [10] CP29; [11] CP27; [12] CP25; [13] 23 kDa polypeptide of oxygen evolving complex; [14] cytochrome b; [15] CP24; [16] δ subunit of CF1 ATPase; [17] Rieske iron- sulfur protein .

Fig. 2 Phosphorimager of 2-D protein pattern of the dark-adapted thylakoids.
The thylakoid membranes were adapted in the dark for 60 min then subjected to protein phosphorylation that was induced by adding [γ-^{32}P] ATP. The membrane proteins were precipitated using precooled acetone to final concentration of 80%. The protein pellets were then treated applying phenol-extraction as described in ref. [5].

The phosphorylation of these two groups of FNR polypeptides was studied . The dark adapted thylakoid membranes have an intensive band in the phosphorimager of 2-D protein pattern under the conditions that protein phosphorylation is initiated by adding [γ-^{32}P] ATP without light illumination (Figure 2). However, after light- induced protein phosphorylation, the labelled FNR leaves the membrane and becomes soluble. This has been proved by the results from 2-D electrophoresis of the membrane pellets and the supernatant after light-induced protein phosphorylation (Figure 3-4). As can be seen, FNR shows up in the supernatant (Figure 4a-4b), rather than in the pellets (Figure 3).
In order to estimate the distribution of the different types of FNR in the thylakoid membrane, the freshly prepared thylakoid membranes of spinach chloroplasts were fractionated by the combination of sonication and aqueous two phase partitioning. It was observed that FNR is distributed all over the thylakoid membrane, the amount per chlorophyll being more in the stroma lamellae than in the grana stacks of the thylakoid membrane (Figure 5a-6a), the concentration of FNR per P700, however, is higher in the grana than that in the stroma

lamellae. If it is assumed that the ratio between FNR and P700 in the grana stacks is 1, then, this ratio would be about 0.6 in the stroma lamellae-originating membrane fraction.

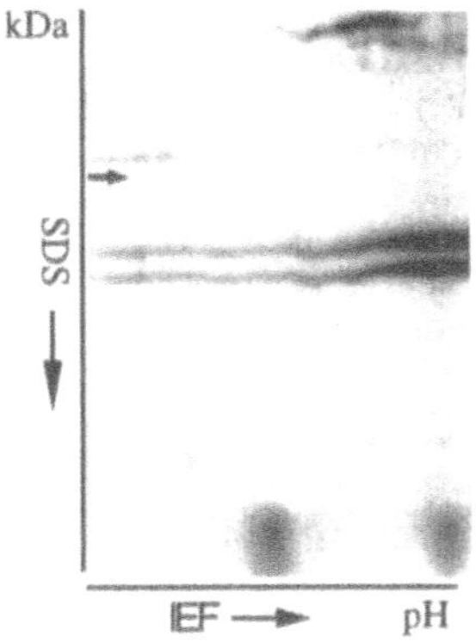

Fig. 3 Phosphorimager of 2-D protein pattern of the membrane pellets obtained after light-induced protein phosphorylation of thylakoid membranes. the FNR polypeptides have disappeared from their positions.

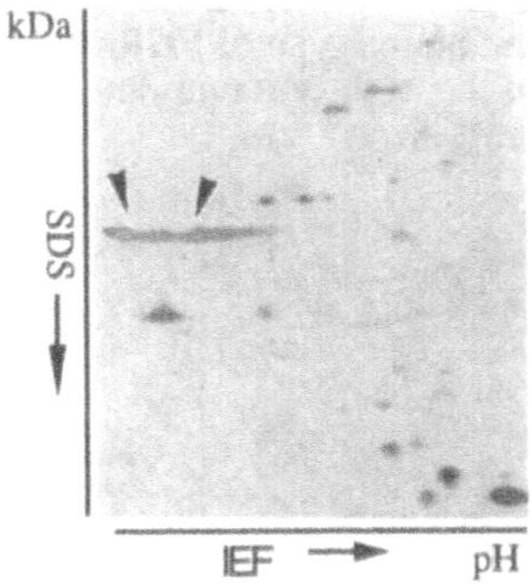

Fig. 4a 2-D protein pattern of the supernatant obtained after light-induced protein phosphorylation of thylakoid membranes. The arrowhead indicates the FNR polypeptides.

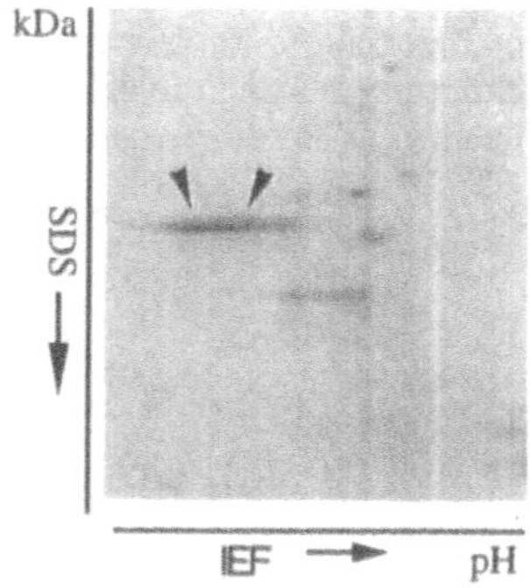

Fig. 4b Phosphorimager of 2-D protein pattern of the supernatant obtained after light-induced protein phosphorylation of thylakoid membranes. The arrowhead indicates the FNR polypeptides.

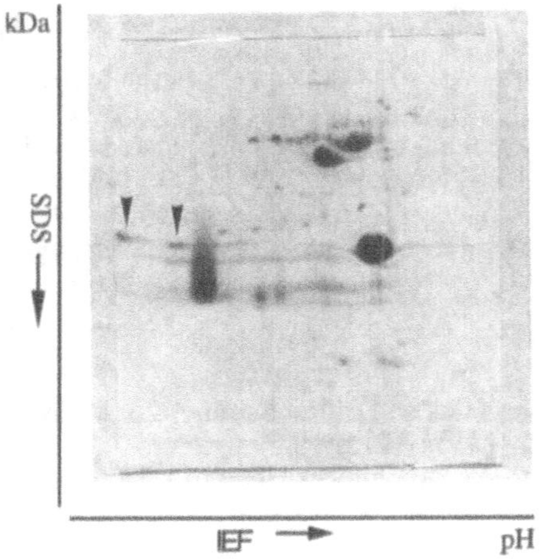

Fig. 5a 2-D protein pattern of the grana-derived membrane fraction isolated from the control thylakoids. The arrowheads indicate the localization of FNR polypeptides.

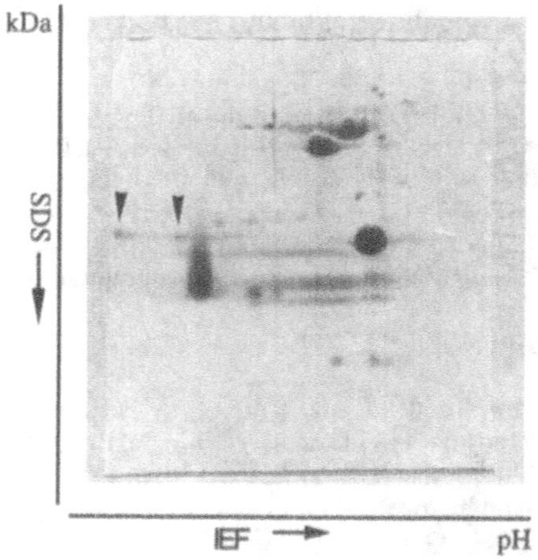

Fig. 5b 2-D protein pattern of the grana-derived membrane fraction isolated from the washed thylakoids.

For the purpose of removing the loosely-bound part of FNR the thylakoids were washed 5 times in 5 mM NaCl /10mM Tricine-NaOH pH 7.8, then fractionated in the same way as mentioned above. It was found that the amount of FNR removed from the grana-derived membrane fraction was much more significant than that removed from the stroma lamellae. About a half was removed from the grana but only a one fifth from the stroma lamellae, indicating that the loosely-membrane bound type of FNR is relatively more abundant in the grana region of the thylakoids while the tightly-membrane bound type of FNR is enriched in the stroma lamellae region (Figure 5-6).

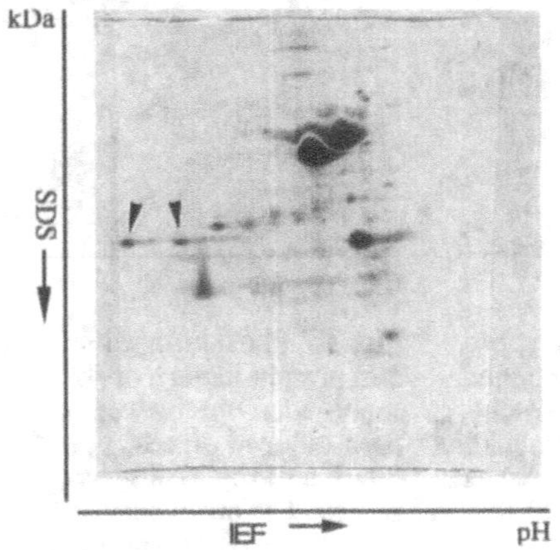

Fig. 6a 2-D protein pattern of the stroma lamellae-derived membrane fraction, isolated from the control thylakoids. The arrowheads indicate the localization of FNR polypeptides.

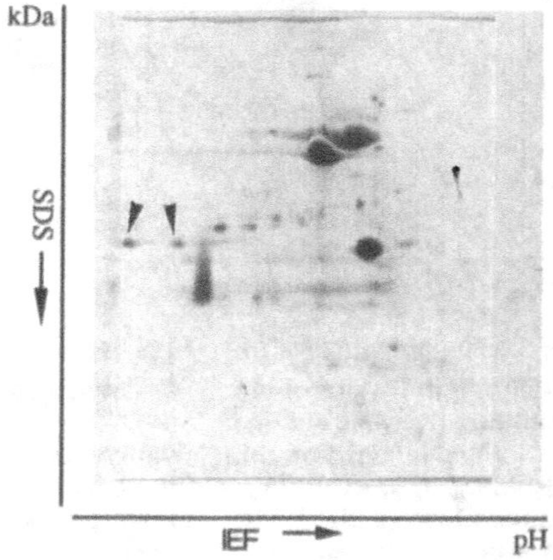

Fig. 6b 2-D protein pattern of the stroma lamellae-derived membrane fraction isolated from the washed thylakoids.

These data would be in support of a proposition that the tightly binding type of the FNR, which is mostly located in the stroma lamellae region, is involved in the cyclic electron transport around PSIβ while the loosely binding type, which is preferentially located in the grana region of the thylakoids, mediates the linear electron transport from H_2O to $NADP^+$ in this region [7-9].

We thank Mrs Ingun Sundén for skillful technical assistance. This work was supported by the Swedish Natural Science Research Council.

REFERENCES

[1] Karplus, P. A. and Bruns C. M. (1994) J. Bioenergetics and Biomembranes 26: 89-99.
[2] Matthijs, H.C.P. et al. (1986) J. Biol. Chem. 261, 12154-12159.
[3] Sakihama, N. et al. (1994) Plant Cell Physiol.vol. 35, supplement, 366 (1a H01).
[4] Andersen, B. et al. (1992) FEBS 311, 169-173.
[5] Yu, S. G. et al. (1994) Photosyn. Res.41:475-486.
[6] Albertsson, P-Å. et al. (1994) Methods in Enzymology (edited by Walter, H. and Johansson, G.), 228, pp 469-482.
[7] Albertsson, P-Å. et al. (1992) in Trends in Photosyn Res (edited by J. Barber, M.G.Guerrero and H. Medrano) pp 46-57.
[8] Albertsson, P-Å. et al. (1990) FEBS lett 273, 36-40.
[9] Albertsson, P-Å. (1995) Photosyn. Res. in press.

THE FERREDOXIN-BINDING SITE OF SPINACH NITRITE REDUCTASE.

Michelle M. Dose [1], Masakazu Hirasawa [1], Susan Kleis-SanFrancisco [2], Elaine L. Lew [1] and David B. Knaff [1,2]
[1]Department of Chemistry & Biochemistry, Texas Tech University
[2]Institute for Biotechnology, Texas Tech University
Lubbock, Texas 79409-1061, U.S.A.

1. Introduction

Ferredoxin:nitrite oxidoreductases (EC 1.7.7.1, hereafter referred to as nitrite reductase) are soluble enzymes, located in the chloroplast stroma in higher plants and algae, and also found in cyanobacteria. These enzymes catalyze the 6-electron reduction of nitrite to ammonia using reduced ferredoxin as the electron donor (1). Nitrite reductases are monomeric proteins, with molecular masses in the 60-65 kDa range, that contain a single siroheme and a single [4Fe-4S] cluster as prosthetic groups (1). Amino acid sequences, deduced from cDNA sequences, are known (1) for six nitrite reductases (including spinach - Reference 2). E_m values of -290 mV and -365 mV have been measured for the siroheme, the site of nitrite binding and reduction (1), and the [4Fe-4S] cluster, respectively, of spinach nitrite reductase (3). Kinetic measurements have established that electron flow occurs between the [4Fe-4S] cluster and the siroheme during turnover (3). Ferredoxin forms a high-affinity, electrostatically-stabilized 1:1 complex with nitrite reductase, with ferredoxin supplying the negative groups involved in complex formation and nitrite reductase the positive groups (1,4). It appears likely that a tryptophan is present at the ferredoxin-binding site (5) and that both lysine and arginine residues on nitrite reductase are directly involved in binding ferredoxin (4), at a site that is antigenically similar to the ferredoxin-binding sites on other ferredoxin-dependent enzymes (1).

2. Procedure

Spinach leaf nitrite reductase and ferredoxin were isolated and purified and nitrite reductase activity was assayed as described previously (4). Modification of nitrite reductase with [^{14}C] phenylglyoxal was carried out under conditions (4) where treatment affects only ferredoxin binding by the enzyme. Treatment of the enzyme with pyridoxal-5'- phosphate, followed by $NaBH_4$, was carried out by a modification of the procedure developed by Medina et al. (6). Amino acid sequencing was accomplished using a Porton Model PI 202E automatic sequenator.

3. Results and Discussion

We had previously demonstrated that treatment of nitrite reductase with the arginine-

P. Mathis (ed.), Photosynthesis: from Light to Biosphere, Vol. II, 721–724.

modifying reagent phenylglyoxal inhibits ferredoxin binding by nitrite reductase and enzymatic activity when reduced ferredoxin is used as the electron donor (4). No detectable changes in other properties of nitrite reductase were observed and phenylglyoxal treatment had no inhibitory effect on activity when the non-physiological reductant, reduced methyl viologen, was used as the electron donor (4). Furthermore, formation of the nitrite reductase/ferredoxin complex prior to treating the enzyme with phenylglyoxal completely protected nitrite reductase against inhibition (4). These results were interpreted as indicating the presence of at least one arginine at the ferredoxin-binding site of spinach nitrite reductase (4).

Treatment of nitrite reductase with [^{14}C] phenylglyoxal under conditions where ferredoxin-binding is specifically affected results in the incorporation of ca. 4 mol of modifying reagent per mol of enzyme, corresponding to 2 mol of arginine modified per mol of enzyme. Figure 1 shows the results of a reverse-phase HPLC separation of peptides produced when [^{14}C] phenylglyoxal-modified nitrite reductase was treated with *Staphylococcus aureus* V8 protease. Five radioactive peptides were located (indicated by asterisks), but three were equally radioactive regardless of whether nitrite reductase alone or the ferredoxin/enzyme complex was treated with phenylglyoxal. In contrast, two of these peptides (indicated by the arrows) were protected against labeling as a result of complex formation with ferredoxin.

As complex formation with ferredoxin also protects nitrite reductase against the inhibition in ferredoxin binding that results from treatment of the enzyme with phenylglyoxal (see above), it seems reasonable to conclude that the two protected peptides each contain at least one arginine residue that contributes to ferredoxin binding. The two protected peptides were sequenced and, for peptides derived from nitrite

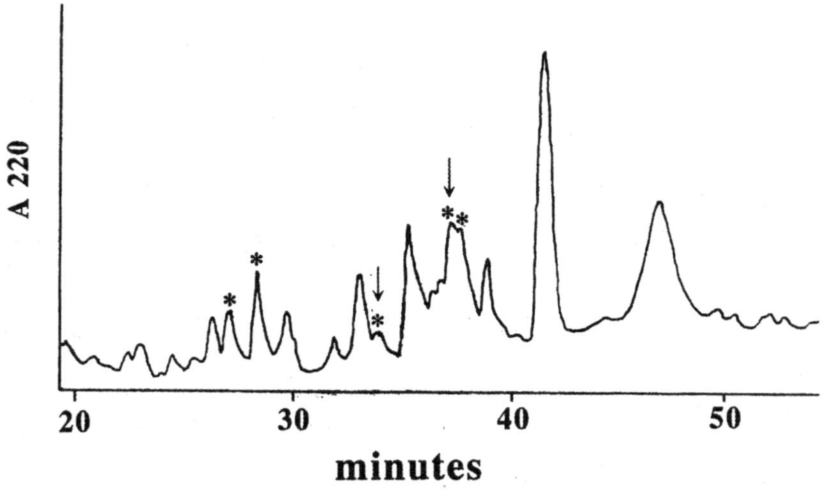

Figure 1. HPLC separation of nitrite reductase peptides after digestion with *S. aureus* V8 protease. Peptides were eluted from a Hamilton PRP-3 column (7 x 305mm) using a 60 minute linear gradient from 0 to 70% acetonitrile. Peptides were monitored at 220 nm and 0.1 ml aliquots were collected and analyzed for radioactivity.

reductase treated with phenylglyoxal in the absence of ferredoxin, gave a sequence that corresponded exactly to known portions of the sequence of the enzyme, except at the position of the single arginine (indicated by the asterisks in the sequences shown below) in each peptide, where a blank cycle was observed because the phenyl- glyoxal derivative of arginine does not exhibit the same retention time as that characteristic of any unmodified amino acids. The sequences of the two peptides are:

*
YLGVHPQKQQGLSFVGLHIPVGRLQADE

and

*
ILINQFGAVPRE

The first peptide, which begins at Y353 and ends at E380 (the numbers refer to the sequence of the mature spinach enzyme), contains a single arginine at position 375. The second peptide, which extends from I547 to E557, contains a single arginine at position 556. No obvious sequence homology could be found between either of these two nitrite reductase peptides and the ferredoxin-binding domain that has been proposed for $NADP^+$ reductase (7). These data have provided the first information about specific amino acids likely to be present at the ferredoxin-binding domain of a ferredoxin-dependent nitrite reductase.

We had previously demonstrated that, like the case described above for phenylglyoxal, treatment of spinach nitrite reductase with the lysine-modifying reagent *N*-acetylsuccinimide produces an enzyme that is indistinguishable from the native enzyme with respect to many properties, including its ability to reduce nitrite to ammonia when supplied with reduced methyl viologen as an electron donor, but is greatly inhibited in its ability to bind ferredoxin or use reduced ferredoxin as an electron donor (4). Unfortunately, *N*-acetylsuccinimide is not available in radioactively-labeled form, so another lysine-modifying reagent had to be substituted in experiments designed to identify specific lysines involved in ferredoxin binding. Figure 2 shows that treatment of nitrite reductase with the lysine-modifying reagent pyridoxal-5'-phosphate results in significant inhibition of the ferredoxin-linked activity of the enzyme, while actually stimulating the methyl viologen-linked activity. As is the case for *N*-acetylsuccinimide (4), complex formation with ferredoxin completely protects nitrite reductase against the inhibition of ferredoxin-linked activity caused by pyridoxal-5'-phosphate (Figure 2). Results similar to those of Fig. 2 were obtained if, subsequent to treatment with pyridoxal-5'-phosphate, $NaBH_4$ were added to reduce the Schiff-base formed by the reaction of pyridoxal-5-phosphate with the side-chain amino group of lysine.

The fact that two lysine-modifying reagents both specifically inhibit ferredoxin binding by nitrite reductase strongly suggests the presence of at least one lysine at the ferredoxin-binding site of the enzyme. In order to quantify this observation, and ultimately identify the lysine(s) involved, spinach nitrite reductase was treated first with pyridoxal-5'-phosphate and then with $[^3H]NaBH_4$. Approximately 4.5 mol of lysines were labeled per mol of enzyme, when nitrite reductase alone was modified. In experiments in which the ferredoxin/nitrite reductase complex was treated with pyridoxal-5'-phosphate/$NaBH_4$, ca. 2.5 mol of lysine per mol of enzyme were labeled. These results are consistent with the presence of 2 lysine residues at the ferredoxin-binding site of nitrite reductase.

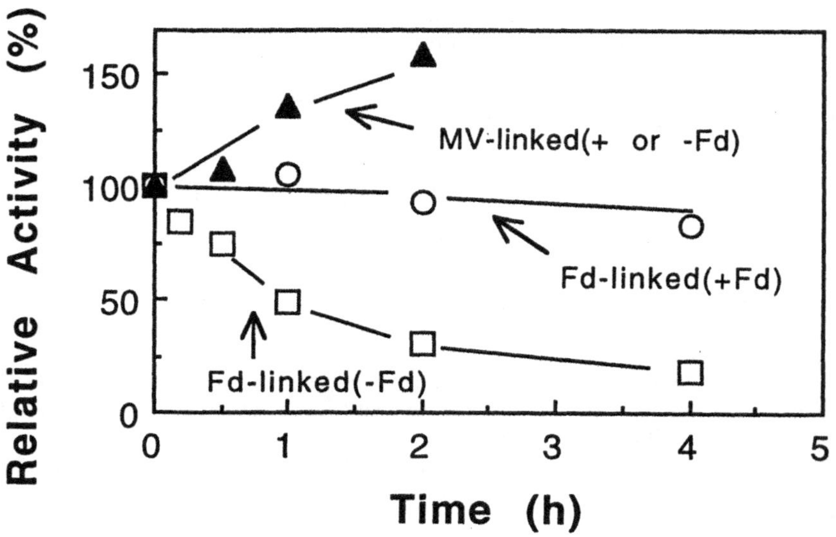

Figure 2. Effect of pyridoxal-5'-phosphate treatment of spinach nitrite reductase. Nitrite reductase (0.005mM) was incubated with a 8-fold excess of pyridoxal-5'-phosphate in 10 mM postassium phosphate buffer (pH 7.7). At the indicated times, aliquots were diluted 1,000-fold and assayed for activity.

Acknowledgements
This research was supported by a grant (DE-FG03-92ER20125 to D. B. K.) from the U.S. Department of Energy.

References

1. Knaff, D. B. and Hirasawa, M. (1991) Biochim. Biophys. Acta 1056, 93-125
2. Back, E., Burkhart, W., Moyer, M., Privalle, L. and Rothstein, S. (1988) Mol. Gen. Genet. 212, 20-26.
3. Hirasawa, M., Tollin, G., Salamon, Z. and Knaff, D. B. (1994) Biochim. Biophys. Acta 1185, 336-345.
4. Hirasawa, M., De Best, J. H. and Knaff, D. B. (1993) Biochim. Biophys. Acta 1140, 304-312.
5. Hirasawa, M., Proske, P. A. and Knaff, D. B. (1994) Biochim. Biophys. Acta 1187, 80-88.
6. Medina, M., Mendez, E. and Gómez-Moreno, C. (1992) FEBS Lett. 298, 25-28.
7. Karplus, P. A. and Bruns, C. M. (1994) J. Bioenerg. Biomembr. 26, 89-99.

Structure of the Thioredoxin 2 from *Anabaena* sp. pcc7120

Markku Saarinen[1], Florence K. Gleason[2] and Hans Eklund[1]

[1]Department of Molecular Biology
Swedish University of Agricultural Sciences
Box 590, BMC, S-751 24 Uppsala, Sweden
[2]Department of Plant Biology, University of Minnesota
St. Paul, Minnesota, 55108, USA

INTRODUCTION

Thioredoxins are small stable redox proteins with the active site sequence, Trp-Cys-Gly-Pro-Cys. For review see [1 and 8]. The first thioredoxin was detected in *Escherichia coli* as a reducing agent for ribonucleotide reductase, an enzyme which is essential for DNA synthesis [2]. Thioredoxin serves as a general protein reductant and can regulate the activity of other enzymes by reducing disulfides [3].

In contrast to *E. coli* and higher animals, multiple thioredoxins have been found in plants. The f- and m-type thioredoxins (Trx-f and Trx-m) are located in chloroplasts. Trx-f activates enzymes in the Calvin cycle, specifically fructose-1,6-bisphosphatase (FBPase) and phosphoribulokinase [4,5]. Trx-m is a general protein reductase with some preference for activation of NADP-malate dehydrogenase (MDH) [6].

The cyanobacterium, *Anabaena* sp. PCC7120, contains two different thioredoxins. One of these (thioredoxin-1) is very similar to spinach chloroplast Trx-m and *E. coli* thioredoxin in primary structure and activity. It serves as an efficient reducing agent for *Anabaena* ribonucleotide reductase. Low levels of an unusual thioredoxin, which is distantly related to other thioredoxins, was also found. In contrast to other thioredoxins, this protein (Trx-2) can be reduced by glutathione and has some functional similarity to spinach chloroplast thioredoxin f (Trx-f) [7].

Although Trx-2 has reasonable sequence homology with other thioredoxins, a closer look at the highly conserved residues in these proteins shows that Trx-2 is one of the most different. Twelve, out of a total of 38 highly conserved residues, are altered in Trx-2.

The three-dimensional structure of oxidized *E. coli* thioredoxin has been determined by X-ray diffraction [9,10]. The molecule has a central core of five strands of β-sheet enclosed by four α-helices (Fig. 1). The active site disulfide is located at the amino end of α-helix 2 in a short segment which is separated from the rest of the helix by a kink due to the conserved Pro40. The 14-membered disulfide ring is located on the surface of the protein, but the disulfide is located on the inside of the protrusion in a shallow pocket. Cys32 is exposed to solvent while Cys35 is recessed. Adjacent to the disulfide there is a flat hydrophobic surface, which presumably facilitates interaction with other proteins [11].

Here we describe the three-dimensional structure of *Anabaena* Trx-2, and discuss possible interactions of thioredoxins with their chloroplastic target proteins. Details of the structure determination are reported elsewhere [16].

P. Mathis (ed.), Photosynthesis: from Light to Biosphere, Vol. II, 725–728.

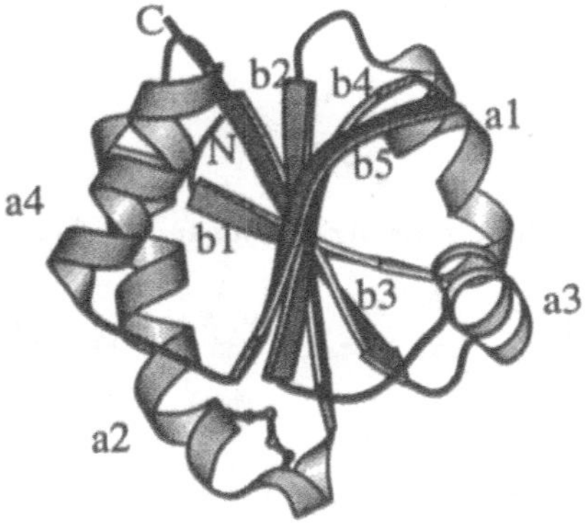

Figure 1. Ribbon diagram of Thioredoxin-2, drawn with program MOLSCRIPT [15]

RESULTS AND DISCUSSION

The two crystal forms of *Anabaena* Trx-2

Anabaena Trx-2 was crystallized in two related orthorhombic space groups, both $P2_12_12_1$, and the structure has been refined using data to high resolution. Extensive contacts are found in the direction of same crystallographic axis in both crystal forms. The contacts are here made by N-terminal residues 2 to 6, to an area close to the redox active disulfide of a symmetry related molecule. The carbonyl oxygen of Val4 and the amide nitrogen of Thr6 form hydrogen bonds to the amide nitrogen and carbonyl oxygen, respectively, of the symmetry related Val75. This valine has a cis-peptide bond to the conserved Pro76. The Cb-Cb-distance between Ile5 and the symmetry related Cys32, the more exposed and reactive of the cysteines, is 5 Å. Side chains of Lys2 and Gly3 are in van der Waals contact with Trp31 in a symmetry related molecule.

Structural comparison with *E. coli* thioredoxin

The overall structure of *Anabaena* Trx-2 is very similar to the previously determined x-ray structure of *E. coli* thioredoxin. The largest differences occur in the N-terminus, at a loop around residues 19 to 22, at a turn between β-strands 4 and 5, and at residues 62 to 70. The hydrogen bond system in the pleated sheet is exactly the same in the *Anabaena,* human and *E. coli* thioredoxins. In the helical regions there are however, slight differences. The largest differences are found around α-helix 3. In *Anabaena* Trx-2, there is a proper α-helix between residues 63-72. In *E. coli,* there is a 3_{10}-helix between residues 66-71.

In thioredoxins the active site helix α–2, residues 32-50, is bent due to the presence of an invariant Pro at position 40. In *E. coli,* the first part of this helix has regular α-helix hydrogen bonding for residues 32-39, while in *Anabaena* thioredoxin this part starts with three 3_{10}-helix bonds. One reason for the difference may be that the unusual Tyr at position 26 perturbs this part and forms a hydrogen bond to the carbonyl of Cys35. Additional differences are found in β turns 49-52 and 82-85.

The active site

Most conserved residues in 39 known thioredoxin sequences are located around the active site. Tyr26 and Gln36 are two major changes in Trx-2. The highly conserved Asp26 is almost buried in *E. coli* thioredoxin, and at position 36 there is usually a positively charged residue. The larger tyrosine side chain of Trx-2 occupies more space at the backside of the disulfide, and causes shifts in the peptide back bone of residues 27 to 41 and surrounding side chains.

The side chain of Gln36 does not interact with the main chain behind the active site as NZ of Lys36 in *E. coli* thioredoxin does. One water molecule has occupied this position. There are total of four water molecules found here in the Trx-2, instead of two in *E.coli* thioredoxin

The 14 membered disulfide ring of *Anabaena* and *E. coli* thioredoxins has the same conformation and the disulfide geometry is very similar in spite of the differences in main chain hydrogen bonding within this region. The residues involved in the disulfide bridge have moved further away from the side chain of residue 26. Due to this, Trp31 at the tip of the active site protrusion has shifted position by 1.5 Å. These differences may affect Trx-2-protein interactions.

Protein interaction site

Important features of the interaction area with other proteins are in addition to the active site, residues 31-36, two extended chains which contain mainly hydrophobic side chains, residues 74-76, and 92-94 are important in thioredoxin-protein interactions [11]. One invariant feature of all thioredoxins is a proline at position 76 with the *cis* configuration, which is located in a loop preceding the β4 strand. This residue is positioned in van der Waals contact with the active site cysteines and is important for both stability and function.

Around the equivalent *cis* peptide residues extensive protein-protein interactions are observed in the *Anabaena* Trx-2 crystals. These crystal contacts involve a major part of the proposed protein interaction area, and include two parallel hydrogen bonds to the main chain of residue 75 before the conserved *cis*-proline 76. Many of the proteins which interact with thioredoxin may also bind in a parallel manner to residues 74 - 76 and place the reactive cysteines in close proximity for reduction of the substrate. In the complex between human thioredoxin and its target peptide from the transcription factor NFκB, the peptide binds parallel to residues 71-75 with main chain - main chain hydrogen bonds to N74 and O74 equivalent to those observed by us [12].

Thioredoxins specificities

Interactions between thioredoxins and their target peptides can be visualized, if the regulatory peptides from FBPase, and MDH [13, 14] are aligned such, that the cysteine to be reduced has same position as Ile5 of the symmetry related N-terminal peptide in the crystal structure. See figure 2.

The modeled target peptides can all form two hydrogen bonds to the main chain at thioredoxin residue 75. In each case, the target cysteine to be reduced is positioned suitably for reduction, and residues on the N-terminal side of it can favor the correct interaction or discriminate against the incorrect one. With the observed crystal contacts as a starting point, the binding which gives the most reasonable pattern of interactions is when the target polypeptide is modeled in parallel to Trx at residues 75. Residue 1 of the peptide provides specificity for Trx-f towards FBPase. Similarly Lys68 in Trx-f and Trx-2 repel Lys3 in both N and C terminal peptides of MDH (in 'contact' with residue 3 in both, and with residue 4 only in C terminal peptide). Furthermore, Glu73 of Trx-2 can interact with Arg in position 4 of FBPase, but repel Asp in the corresponding position in MDH N terminal peptide.

Conclusions

Anabaena Trx-2 structure is very similar to *E. coli* thioredoxin, most of the major differences are found on the surface of the protein. Many of the in Trx-2 altered residues don't have any major effect on the structure. Two such residues, Tyr26 and Gln36, however, affect the active site structure, and can thus influence the redox properties and the protein interactions of *Anabaena* Trx-2. Thioredoxins can bind to their target proteins wih perallel hydrogen bonds at the *cis*-proline 76, this in contrast with the well established anti-parallel binding to glutathione by glutaredoxins.Based on the crystal contacts and alignement of the thioredoxins' chloroplastic target peptides, some determinants of thioredoxin specificities against MDH and towards FBPase can be identified.

a)

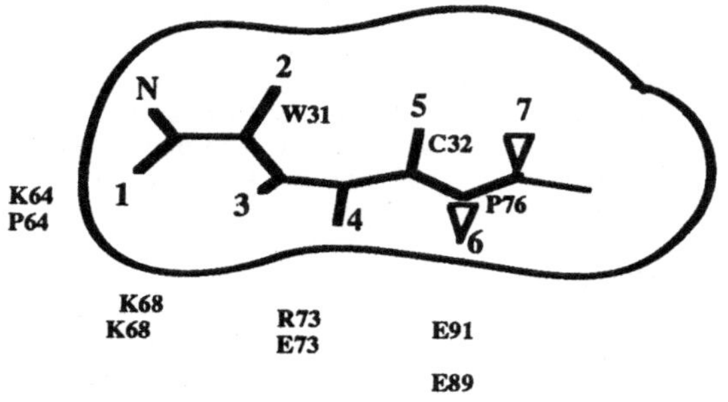

b)

	1				5					10				14
FBPase	E	E	Q	R	C	V	V	N	Y	C	G	P	G	D
MDH N-term.	T	R	K	D	C	F	Q	V	F	C	T	T	Y	D
C-1	A	E	K	K	C	V	A ...							
C-2							...	A	Y	C	D	V	P	E

Figure 2. a) Schematic view of the thioredoxin- protein interaction area.
b) Target peptides of chloroplastic thioredoxins.

References.

1. Holmgren, A. (1985). Ann. Rev. Biochem. 54, 237-271.
2. Laurent, T.C., Moore, E.C., Reichard, P. (1964) J. Biol. Chem. 239, 3436-3444.
3. Buchanan, B.B. (1980) Ann. Rev. Plant. Physiol. 31, 341-374.
4. Wolosiuk R. A.and Buchanan B. B. (1977) Nature 266, 565-567.
5. Wolosiuk R. A.and Buchanan B. B. (1978) Arch. Biochem. Biophys. 189, 97-101.
6. Jacquot, J.-P., Vidal, J., Gadal, P. & Schürmann, P. (1978) FEBS Lett. 96, 243-246.
7. Gleason, F.K. (1992a) J. Bacteriol. 174, 2592-2598.
8. Eklund H., Gleason, F. & Holmgren, A. (1991) Proteins 11, 13-28.
9. Holmgren, A., Söderberg, B.-O., Eklund, H. & Brändén, C.-I. (1975) Proc. Natl. Acad. Sci. USA 72, 2305-2309.
10. Katti, S.K., LeMaster, D.M. & Eklund, H.(1990) J. Mol. Biol. 212, 167-184.
11. Eklund, H., Cambillau, C., Sjöberg, B.-M., Holmgren, A., Jörnvall, H., Höög, J.-O., & Brändén, C.-I. (1984) EMBO J. 3, 1443-1449.
12. Qin, J., Clore,G.M., Kennedy, W.M.P, Huth, J.R. & Gronenborn, A.M (1995) Structure 3, 289-297.
13. Marcus, F.,Moberley, L., & P., L. S. (1988). Proc. Natl. Acad. Sci. USA, 85, 5379-5383.
14. Issakidis, E., Saarinen, M.,Decottignies, P.,Jacquot, J. P.,Cretin, C.,Gadal, P., & Miginiac- Maslow, M. (1994). J Biol Chem, 269, 3511-3517.
15. Kraulis, P. (1991) J. Appl. Crystallogr. 24, 946-950.
16. Saarinen, M., Gleason F.K. and Eklund, H (1995) Structure, in press.

THE DOCKING OF CYTOCHROME *f* WITH PLASTOCYANIN: THREE POSSIBLE COMPLEXES

Douglas C. Pearson Jr. and Elizabeth L. Gross
Department of Biochemistry and Biophysics Program, The Ohio State University
484 West 12th Avenue, Columbus, Ohio 43202 USA

1. Introduction

Plastocyanin (PC) and cytochrome *f* (cyt *f*) are reaction partners in photosynthetic electron transport. PC shuttles electrons from cyt *f* to P700 in the photosystem I complex (1). The specific mechanism for the electron transfer reaction between PC and cyt *f* has not yet been established; however, with the recent turnip cyt *f* structure determination by Martinez et. al. (2, 3), it is now possible to present a first hypothesis of this mechanism. Evidence shows that: (a) the initial attraction between the two molecules is electrostatic in nature (1, 4); (b) PC has two possible reaction sites, about H87 and Y83 (1); (c) cyt *f* has two possible reaction sites; Y1 and a positive patch of residues at the junction of cyt *f*'s two domains. These considerations were used to create the manual docks as presented here.

2. Procedure

2.1 *Atomic coordinates*
The atomic coordinates of oxidized poplar PC (5) were obtained from the Brookhaven Protein Data Bank (6). The coordinates of reduced cyt *f* were obtained from Martinez et. al. (2, 3).

2.2 *Docking configurations*
Simple molecular manipulations, including those necessary to create the manual docks, were performed using the HyperChem molecular modeling package, distributed by AutoDesk for use on Silicon Graphics IRIS (SGI) workstations running the IRIX operating system v4.x. The establishing criteria for a "dock" were: key residues of interest (i.e., electron transfer sites, observed covalent cross-links, etc.) were within 5Å, and severe steric overlap (more than 5 residues of within 4Å of each other) was eliminated. Two residues on different molecules were defined as interacting when at least one atom on each residue was within 6Å of the other; two residues were defined as overlapping when this distance was less than 4Å. Note that docking was performed with static molecular structures; no dynamic analysis has been performed at this time.

2.3 *Electrostatic analysis*
When useful, electrostatic analysis was performed on individual molecules and dock configurations using the package GRASP (7), distributed by Anthony Nicholls of Columbia University for use on SGI workstations running IRIX v4.0 or above. GRASP utilizes a double-grid algorithm to solve the finite-difference Poisson-Boltzmann for electrostatic using a continuum philosophy (8). (GRASP was also used to prepare the figures for presentation.)

P. Mathis (ed.), Photosynthesis: from Light to Biosphere, Vol. II, 729–732.

3. Results and Discussion

3.1 *Dock 1*

Dock 1 between poplar PC and turnip cyt *f* (Fig. 1) is the most electrostatically favorable dock between PC and cyt *f* of the three; the negatively charged residues on PC (D44, E59, E60, others) face a positive patch of residues on cyt *f* (K58, K65, K66, K187, others). When the two proteins are docked, the negative patches on PC align well with the positive patch on cyt *f* with no steric hinderance - in particular, D44 is situated next to K187, E59 and E60 with K65, and E60 alone with K58. This reflects the results of Morand et. al. (9) where covalent cross-links were observed between D44 (PC) and K187 (cyt *f*), and between E59 or E60 (PC) and an unidentified residue on cyt *f*. These results suggest K65 (or possibly K58) as that unknown residue on cyt *f*.

Dock 1 may constitute either a true electron-transfer-capable dock or a transient predocking complex. If it is electron-transfer capable, then an electron donation site on cyt *f* must be adjacent to an electron acceptor site on PC. However, the heme on cyt *f* is distant from PC in this dock (the Cu-Fe distance is 31.4Å). The most possible electron transfer pathway is the chain of water molecules (6) extending from the heme to the vicinity of K66; however, this is not likely. With respect to PC, however, Y83 is in a favorable position for electron transfer, and there is evidence for Y83's involvement in electron transfer from site-directed mutagenesis studies of Y83 (10). Interestingly, D42 is *not* involved in this dock, which would agree with observations of Modi et. al. (10) that mutation of D42 does not affect the rate of electron transfer.

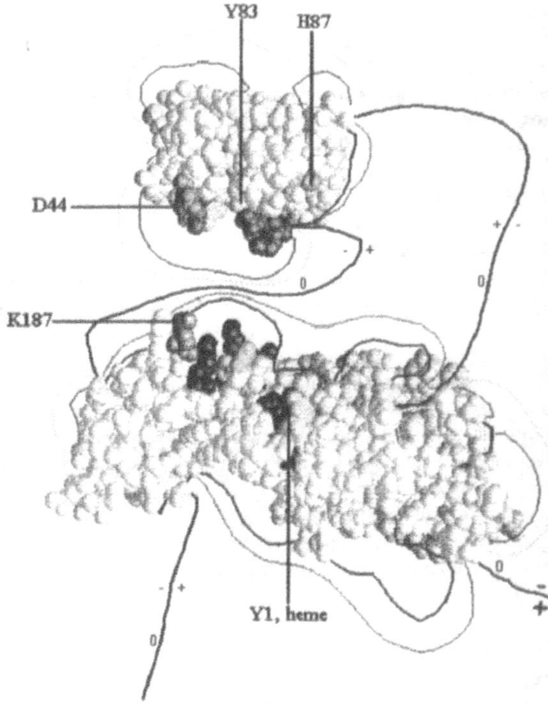

Figure 1. Dock 1, representing electrostatic attraction between the positive patch on cyt *f* and negative patches on PC. Plastocyanin was backed off a distance of 15Å to show the electrostatic interactions at 100 mM ionic strength, displayed here as two-dimensional interpolation plane contours at values of -1.0, -0.5, 0, 0.5, and 1.0 kT/e. The 0 contour and the positive and negative sides of that countour are marked, as are key residues in the interaction. The electrostatics calculation was performed by GRASP as described in the Procedure.

3.2 *Dock 2*

Fig. 2 presents a designed dock (Dock 2) with H87 (PC) next to Y1 (cyt *f*), thus providing a direct path of electron transfer from a ligand of the heme of cyt *f* to a ligand of the copper in PC. Note three important features: the H87/Y1 interaction as mentioned; the D44/K187 link from Dock 1 remains; and hydrophobic residues in the region of H87 and Y1 which would provide some attraction between the two sites (among them G9, L12, Q88, G89, and A90 on PC; Y1, F4, A62, Y160, and P161 on cyt *f*). Hydrophobic stabilization would be important since electrostatic repulsion exists (Fig. 2) due to the closeness of the copper atom and histidines on PC to the iron atom and associated positive charges on cyt *f*. Note also that L12 is involved in this dock. Modi et al. (10) observed that a L12N mutation increased the interaction of PC with cyt *f*. Tyrosines in close proximity could be hydrogen-bonding partners for a mutated asparagine 12.

Dock 2 is a conceivable electron-transfer capable dock due to the low Cu-Fe distance (15.3 Å) and the fact that H87 and Y1 are both ligands to the respective metal atoms. The co-operation of the D44/K187 link and the hydrophobic regions about H87 and Y1 may be essential. Once the electrostatic complex is formed, PC could rotate about the D44-K187 link into a position to form Dock 2, aided by the removal of waters from hydrophobic residues close to H87 and Y1. The question (1) as to whether the cross-linked complex of Morand et al. (8) is electron-transfer active could be answered by understanding the flexibility of the covalent cross-linkage between the two molecules (i.e, if there is enough flexibility for H87 to reach Y1).

There is some steric hindrance in Dock 2 particularly caused by G89 and A90 on PC. However, G89 and A90 are part of a loop which has two conserved glycines (G89, G91) and is likely quite flexible. Molecular dynamics analysis should give a better view of this loop's flexibility.

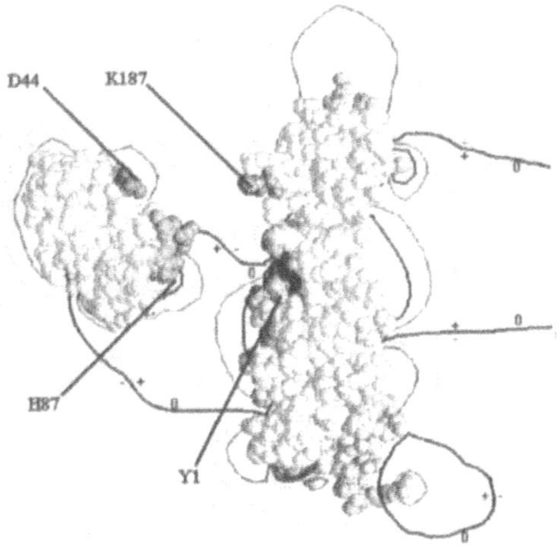

Figure 2. Dock 2, showing matched residues H87 (PC) with Y1 (cyt *f*) and D44 (PC) with K187 (cyt *f*). All conditions are as shown in Fig. 1.

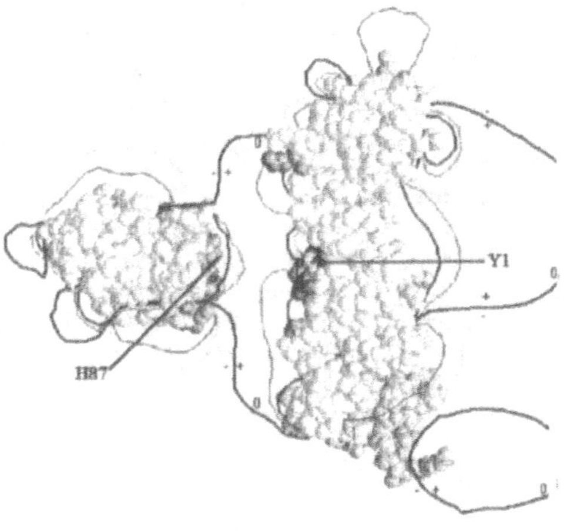

Figure 3. Dock 3, showing the H87/Y1 dock with optimized hydrophobic interactions. All conditions are as shown in Fig. 1.

3.3 *Dock 3*

A second H87/Y1 dock (Dock 3) is shown in Fig. 3. The cross-link between D44 on PC and K187 on cyt *f* present in docks 1 and 2 is broken here to allow a greater degree of hydrophobic interaction - particularly between F35, P36, L62, and P86 on PC, and I3, F4, Q7, Q103, and P117 on cyt *f*. Hydrophobic attractions may draw PC from a Dock 2 orientation to a Dock 3 orientation prior to or during electron transfer. The ultimate issue is optimizing the position of H87 on PC with respect to Y1 on cyt *f* for electron transfer (assuming Y1 to H87 is the electron transfer pathway). This may be achieved in Dock 2, or in Dock 3, or in a hybrid of the two docks. There are several hydrophobic residues in the region between H87 and Y1 which could serve in several different ways to give a dock between PC and cyt *f*; i.e., there may be several variations on Docks 2 and 3 which may be electron-transfer competent.

There is a reasonable degree of electrostatic repulsion in Dock 3; this may prove essential since, unlike other protein-protein interaction systems such as antigen-antibody, the molecules must separate once the reaction is completed so PC can leave and deliver its electron to P700.

4. Conclusion

We propose that PC and cyt f initially form an electrostatic pre-docking complex (Dock 1) after which it migrates to form an electron-transfer complex (Dock 2, Dock 3, or hybrid) which is primarily held together by hydrophobic interactions. We also postulate that the rate-limiting step in the electron-transfer process is the formation of the electrostatic pre-docking complex.

References

1 Gross, E. L. (1993) Photosynth. Res. 37, 103-116; Gross, E. L. (1996) in Adv. in Photosynthesis. D. Ort and C. Yokum, editors. In press.

2 Martinez, S. E., Huang, D., Szczepaniak, A., Cramer, W. A., and Smith, J. L. (1995) Structure. 2, 95-105.

3 Martinez, S. E., Cramer, W. A., and Smith, J. L. (1995) Biophys. J. 68, A246.

4 Durell, S. R., Labanowski, J. K., and Gross, E. L. (1990) Arch. Biochem. Biophys. 277, 241-254.

5 Guss, J. M., Bartunik, H. D., and Freeman, H. C. (1992) Acta Crystallogr., Sect B. 48, 790-811.

6 Bernstein, F. C., Koetzle, T. F., Williams, G. J. B., Meyer, E. F., Brice, M. D., Rogers, J. R., Kennard, O., Shimanouchi, T., and Tatsumi, M. (1977) J. Mol. Biol. 112, 535-542.

7 Nicholls, A., Sharp, K. A., and Honig, B. A. (1991) Proteins 11, 281-296.

8 Sharp, K. A., and Honig, B. A. (1990) Ann. Rev. Biophys. Biophys. Chem. 19, 301-332.

9 Morand, L. Z., Frame, M. K., Colvert, K. K., Johnson, D. A., Krogmann, D. W., and Davis, D. J. (1989) Biochemistry 28, 8039-8047.

10 Modi, S., Nordling, M., Lundberg, L.G., Hannson, O., and Bendell, D. S. (1992) Biochim. Biophys. Acta. 1102, 85-90.

THE HIPIP FROM *RHODOFERAX FERMENTANS* RAPIDLY REDUCES THE CYTOCHROME C-556 ASSOCIATED WITH THE PHOTOCHEMICAL REACTION CENTER

A. HOCHKOEPPLER[a], S. CIURLI[b], G. VENTUROLI[a], D. ZANNONI[a]
[a]Dept. of Biology, Bologna University, Irnerio 42, 40126 Bologna (I)
[b]Inst. of Agric. Chem., Bologna University, Berti-Pichat 10, 40127 Bologna (I)

1. INTRODUCTION

In purple photosynthetic bacteria a number of different soluble cytochromes and iron-sulfur proteins have been isolated and purified. Among the iron-sulfur proteins the ferredoxins are characterized by reduction potentials lying in the range from -250 to -650 mV and involve an $[Fe4S4]^{2+/1+}$ redox couple, while the high-potential iron-sulfur proteins (HiPIP) have a much higher reduction potential (from +90 to +450 mV) and utilize the $[Fe4S4]^{3+/2+}$ couple. Although many physico-chemical properties of HiPIPs are well known, their functional role is still a matter for discussion. In *Chromatium vinosum*, HiPIP was shown to deliver electrons to the membrane-bound cytochrome that re-reduces the photo-oxidized primary donor (P^+) of the reaction center (1). On the other hand, it was also shown that *Chromatium* HiPIP is not re-reduced by the homologous bc_1 complex (2). Recently, the kinetics of electron transfer bewteen HiPIPs from three different sources and the reaction center (RC) from *Rhodopseudomonas viridis* was studied (3). The involvement of HiPIP in the respiratory chain of *Rhodotermus marinus* was also reported (4).

Rhodoferax fermentans is a purple nonsulfur photoheterotrophic bacterium, exhibiting some interesting features, such as fructose fermentation, quinone composition, and the presence of a RC associated tetraheme cytochrome. The tetraheme from *Rhodoferax* has distinguishable α bands: three hems have been resolved at 551, 556, and 560 nm, with mid-point potentilas of 79, 294, and 354 mV, respectively (5,6).

We have previously reported that the HiPIP from *Rhodoferax* is able, upon RC photo-excitation, to re-reduce, on the millisecond time scale, cytochrome c-556 (cyt c-556), the direct electron donor to P^+ (7).

Here the identification of oxidized HiPIP as the product of RC photo-excitation is confirmed by direct spectroscopic detection of flash-induced HiPIP oxidation. The results of a more detailed kinetic analysis of the electron transfer between HiPIP and cyt c-556 are also discussed.

P. Mathis (ed.), Photosynthesis: from Light to Biosphere, Vol. II, 733–735.

2. MATERIALS AND METHODS

2.1 Cell growth and membrane isolation

Rhodoferax fermentans was grown photoheterotrophically in completelly-filled screw capped bottles as previously described (5). Cells were harvested at late log-phase, and membrane fractions isolated as reported (5).

2.2 HiPIP purification

After cell lysis, cell debris and membrane fragments were separated by differential centrifugation at 20,000 and 200,000g, respectively. The supernatant was concentrated, dyalized, and loaded onto a DEAE-cellulose column: the unretained HiPIP was washed out of the column, and then concentrated and loaded onto a CM-cellulose column developed with buffer only (5 mM phosphate, pH 6), and then with a step gradient (30, 60, 80 mM NaCl). The HiPIP containing fractions (eluted at 60 mM NaCl) were collected, concentrated, and loaded onto a Superdex 75 gel filtration column. Finally, an FPLC cation exchange chromatographic step (Mono S column) was performed. The homogeneity of the protein after this last chromatographic step was checked by SDS PAGE with 20% acrylamide gels.

2.3 Redox potentiometry

The mid-point potential of HiPIP was determined as previosly described (5) except that the absorbance changes related to HiPIP oxidation or reduction were recorded over the 450-600 nm wavelength interval. The redox mediators used (10 μM each) were: p-benzoquinone, 1,2-naphtoquinone, 1,4-naphtoquinone, and diaminodurene.

2.4 Kinetic spectrophotometry

The absorbance changes induced in *Rhodoferax* membranes upon RC photo-excitation by a single μs flash of light were measured as previosly described (6). All the measurements were performed in the absence of any reductant and redox mediator. The concentration of cytochromes was determined assuming an extinction coefficient of 19.0 mM^{-1} cm^{-1} (8).

3. RESULTS AND DISCUSSION

The electronic absorption spectrum of the HiPIP purified from *Rhodoferax* is very similar to that of other HiPIPs (9). In particular, the reduced protein has an absorption maximum at 388 nm, and the 280/388 nm absorbance ratio is 2.44. Upon oxidation (by potassium ferricyanide) of the protein, the absorption maximum undergoes quite a large red shift (69 nm) without an appreciable change in the extinction coefficient. Isosbestic points are detected at 258, 319, 388, and 416 nm. At wavelengths above 416 nm (and up to 700 nm) an increase in the extinction coefficient is observed upon HiPIP oxidation. Redox titrations of the purified HiPIP yield a mid-point reduction potential equal to 351 ± 16 mV. This value, lying in the range observed for other HiPIPs (10), coincides with the mid-point potential of the RC associated cyt c-556 (354 mV).

The role of HiPIP in the photosynthetic electron transfer chain of *Rhodoferax* has been investigated by directly monitoring the flash-induced oxidation of the HiPIP, and by

comparing its kinetics with that of the cyt c-556 redox changes observed under the same conditions. To do this, we have studied the absorbance changes detected after RC photo-excitation at three different wavelengths: 540, 556, and 586 nm, indicative of P, cyt c-556, and HiPIP photo-oxidation, respectively.
In the absence of HiPIP, when a suspension of *Rhodoferax* membranes is excited by a single flash of light, the photo-oxidation of the primary donor P of the RC can be observed. Upon addition of purified HiPIP in a large stoichiometric excess over the RC, a fast absorbance change is detected at 556 nm, corresponding to the sub-ms oxidation of one cyt c-556 per RC. The flash-induced oxidation of cyt c-556 is followed, under these conditions, by its dark re-reduction, which fits a single exponential yielding a pseudo first-order rate constant of k_{obs} = 130 s^{-1}. When the monitoring wavelength is 586 nm no appreciable light-induced absorbance changes are detected in the absence of HiPIP, showing that 586 nm is an isosbestic point for RC and cyt c-556 spectral contributions. When the same measurements are performed in the presence of HiPIP, a light-induced absorbance increase is detected at 586 nm, indicating oxidized HiPIP as the product of RC photo-excitation, in agreement with the difference spectrum of the purified protein. Thus, the flash-induced oxidation of the HiPIP and the re-reduction of cyt c-556 appear to be concomitant. The kinetics of the two reactions are, in fact, reasonably described by a single exponential, with comparable pseudo first-order rate constants. Moreover, by assuming $\Delta\varepsilon_{586}$ = 4.03 mM^{-1} cm^{-1} (an exctinction coefficient determined from the difference spectrum of the purified protein) a stoichiometry of approximately one oxidized HiPIP per reduced cyt c-556 is obtained.
Accordingly, the rapid re-reduction of the photo-oxidized cyt c-556 seems to be the role for HiPIP in the photosynthetic electron transfer chain of *Rhodoferax*.

REFERENCES

1 Kennel, S.J., Bartsch, R.G. and Kamen, M.D. (1972) Biophys. J. 12, 882-896.
2 Tan, J., Corson, G.E., Lane Chen, Y., Garcia, M.C., Guner, S. and Knaff, D.B. (1993)Biochim. Biophys. Acta 1144, 69-76.
3 Meyer, T.E., Bartsch, R.G., Cusanovich, M.A. and Tollin, G. (1993) Biochemistry 32, 4719-4726.
4 Pereira, M.M., Antunes, A.M., Nunes, O.C., da Costa, M.S. and Teixeira, M. (1994), FEBS Letters 352, 327-330.
5 Hochkoeppler, A., Moschettini, G, and Zannoni, D. (1995) Biochim. Biophys. Acta 1229, 73-80.
6 Hochoeppler, A., Zannoni, D. and Venturoli, G. (1995) Biochim. Biophys. Acta 1229, 81-88.
7 Hochkoeppler, A. Ciurli, S., Venturoli, G. and Zannoni, G. (1995) FEBS Letters 357, 70-74.
8 Dutton, P.L., Petty, K.M., Bonner, H.S. and Morse, S.D. (1975) Biochim. Biophys. Acta 387, 536-556.
9 Dus, K. de Klerk, H., Sletten, K. and Bartsch, R.G. (1967) Biochim. Biophys. Acta 140, 291.
10 Luchinat, C., Capozzi, F., Borsari, M. Battistuzzi, G. and Sola, M. (1994) Biochem: Biophys. Res. Commun. 203, 436-442.

CHARACTERISATION OF THE ELECTRON TRANSFER COMPLEX BETWEEN HiPIP AND THE PHOTOSYNTHETIC REACTION CENTRE FROM *RUBRIVIVAX GELATINOSUS*

Schoepp, B., Nitschke, W., Biologisches Institut II, University of Freiburg, Schänzlestr. 1, 79104 Freiburg, FRG.
Riedel, A., Institut für Biophysik und Physikalische Biochemie, University of Regensburg, 93040 Regensburg, FRG.

1. Introduction

Electron transport between *bc* complex and Reaction Centre (RC) is usually considered to be mediated by soluble cytochrome c_2 (for review, see 1). In many species of purple bacteria, a further soluble electron carrier was found the function of which was enigmatic: the High Potential Iron-sulfur Protein (HiPIP) (2). It was proposed that this HiPIP can substitute cytochrome c_2 in photosynthetic electron transfer (3). Recently, this role was clearly demonstrated in whole cells of *Rubrivivax gelatinosus* (4). It was shown that for electron transfer, the HiPIP forms a stable complex with the RC. In this contribution we further investigate the caracteristics of this complex. The spatial organisation of the complex between HiPIP and RC's tetraheme subunit was studied by EPR measurements on dried membranes. A comparison of the redox potentials of HiPIP in solution and linked to the RC, allows conclusions with respect to preferential binding of one of the redox states.

2. Material and Methodes

Membrane fragments and isolated HiPIP were prepared as published (4). LH_{870}-RC complex was purified according to Agalidis et al. (5). The orientation of membrane samples, adjustment of their redox states and EPR measuring conditions are described in (6). Cross-linking between purified HiPIP and isolated LH_{870}-RC was performed as in (7). The redox-poising and titration were performed essentially as described by Dutton (8) using ferricyanide and 1,4 - benzoquinone as redox mediators.

3. Results and Discussion

As shown in Figure 1 the typical signal of HiPIP (with $g \sim 2.11$ and $g \sim 2.03$) could be seen in dried, chemically oxidised membranes (solid line) and disappeared when the membranes were reduced by ascorbate in the dark (open triangles). When examined in more detail, the EPR spectra present three different g values : 2.11, 2.07 and 2.028. This spectrum differs from the one mesured by Beinert and Thomson (9) in that these authors describe only two components at $g = 2.11$ (g_z) and $g = 2.029$ (g_y and g_x) for the soluble HiPIP. The third component at $g = 2.07$ in our spectrum most probably represents a

P. Mathis (ed.), Photosynthesis: from Light to Biosphere, Vol. II, 737–740.

heterogeneous population of HiPIP molecules, i.e containing a dominant species with *g*-values at ~2.11 and ~2.03 (axial species), and a minor rhombic species with *g*-values at ~2.11, 2.07 and ~2.03. Such phenomena have been described for *Chromatium vinosum* (10) and *Rhodothermus marinus* (11) HiPIP's. The origin of this heterogeneity may be due to an intrinsic and unexplained protein heterogeneity and/or to protein dimerization (as proposed by Dunham et al. (10)) or to chemical equilibrium between two distinct species, differing in the electron distribution within the cluster (as proposed by Banci et al. (12)). Further experiments are needed to clarify this question.

The fact that the spectrum of the oxidised membranes showed a high amount of HiPIP bound to the membrane indicates a strong affinity of this soluble protein for one or several binding sites on the surface of the membrane. When reduced membranes were photooxidised (by 10 min illumination at room temperature and during freezing), the HiPIP signal reappeared. The ratio of photooxidisable to chemically oxidisable HiPIP was found to vary with preparations but was at least 25 %. A fraction of the membrane-associated HiPIP is therefore competent in electron transfer to the photooxidised RC probably via the tetraheme subunit (4). Since diffusion of electron carriers is prevented by the dried state of the sample, this fraction must have been bound to the RC in a conformation favorable to electron donation already in the ascorbate reduced dark state. We therefore investigated the spatial organisation of this complex.

Illumination of an oriented membrane sample that was seen to be in a partially oxidised, partially reduced state with respect to the HiPIP, resulted in a sizeable increase of the oxidised HiPIP. Both the chemically oxidised signal observed prior to illumination and the photooxidised part were seen to be dependent on orientation (Fig. 2). Therefore, in this sample a direct comparison of the orientations of photo-vs chemically oxidised HiPIP was possible.

Since a large part of membrane-associated HiPIP was poorly oriented, we eliminated the isotropic contributions by substracting a reference spectrum at a given orientation (130° in that case). The resulting spectra (Fig. 3) represent the well-oriented part of HiPIP

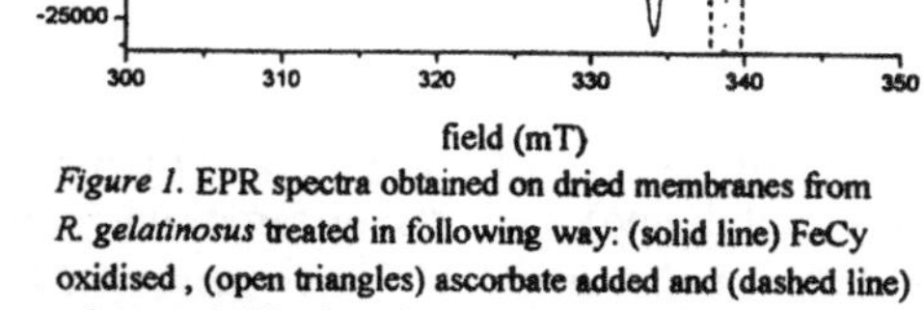

Figure 1. EPR spectra obtained on dried membranes from *R. gelatinosus* treated in following way: (solid line) FeCy oxidised, (open triangles) ascorbate added and (dashed line) subsequently illuminated.

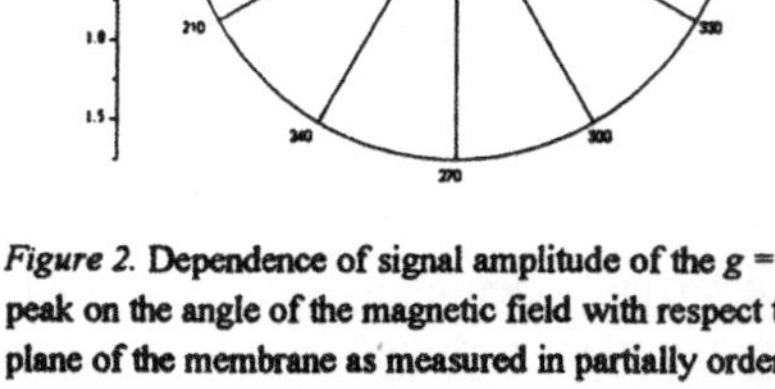

Figure 2. Dependence of signal amplitude of the $g = 2.02$ peak on the angle of the magnetic field with respect to the plane of the membrane as measured in partially ordered membrane multilayers. Closed symbols represent the signals in the partially oxidised, dark adapted sample and open symbols represent the signals induced by the illumination.

molecules. In these spectra, the high-field trough was found to shift upon rotation of the sample indicating that (a) the g_y and g_x values usually considered to coincide (9), are in fact slightly different and (b) that g_y and g_x are differently oriented. From the turning-point of this shift, the values of g = 2.024 and g = 2.012 were obtained for g_y and g_x respectively. The g_z direction also showed orientational anisotropy (Fig. 3). Furthermore, the position of the g_z peak also depends on orientation as seen in Figure 3. The orientation dependencies of the signals at g_z (at 2.103 and at 2.096) , g_y and g_x in photo- and chemically oxidised membranes are shown in Figure 4 A, B and C, respectively (signals in the polar plots were normalised to similar amplitudes). The following conclusions can be drawn : (a) The orientation dependencies of the photo-induced signals differ significantly from those of the chemically oxidised HiPIP indicating that the chemically oxidised fraction contains contributions from HiPIP that are associated to the membrane in a well-ordered way albeit incompetent in electron donation to P+. This fraction may represent HiPIP molecules bound to the RC in a conformation unfavorable for electron transfer or alternatively those that are bound to other membrane-integral enzymes, e.g the cytochromes *bc* complex or the terminal oxidase. In view of functional data (4), we favour the latter possibility. Furthermore, the relatively high background of isotropic signal suggest the presence of a fraction that is non-specifically attached to the membrane. (b) Two slightly different conformations of the complex between HiPIP and RC are able to perform light induced electron transfer as evidenced from Fig. 3 and 4. Although the noise is high, the different orientations confirm that the two measured g_z values (at g = 2.103 and g = 2.096) correspond to two structurally different HiPIP. This double-g_z phenomenon has been seen for *Chromatium vinosum* and *Rhodothermus marinus* (9, 10). This is probably in relation to the existance of the g = 2.07 species discussed before.

The fact that the HiPIP can be cross-linked with the RC (Fig. 5) confirms that these two proteins form a complex. Since the above described results showed that HiPIP could bind to several membrane integral enzymes, isolated LH-RC complex and HiPIP were used for cross-linking. First results from redox titrations of the HiPIP suggest a significant shift of the midpoint potential from 315 mV (determined by optical titration) for the soluble carrier to 365 mV (determined by EPR titration) for the RC-linked form (Fig. 6). This shift indicates a seven times higher affinity of the reduced form of the HiPIP for the RC, as compared to the oxidised form.

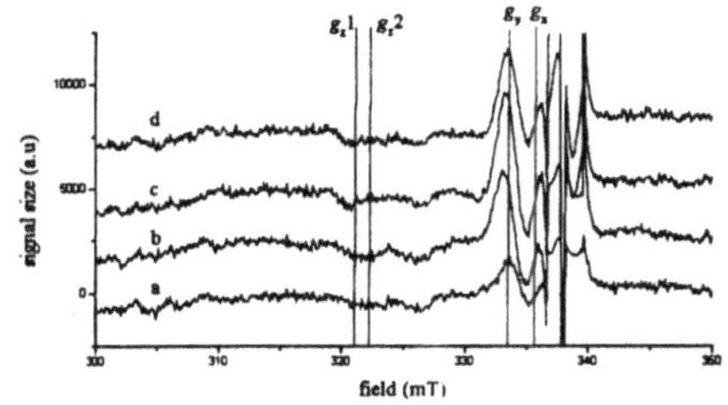

Figure 3. Dependence of the EPR spectra from photooxidised HiPIP bound to the membrane, on the angle of the magnetic field with respect to the plane of the membrane. The spectra at 100° (a), 70° (b), 40° (c) and 10° (d) were obtained by substracting the spectra recorded at 130° from each spectrum.

In conclusion, the results presented above demonstrate that HiPIP forms a well-ordered complex with the RC (and probably with other membrane integrated enzymes). Further studies on the cross-linked sample are in progress. With calculated g_z, g_y and g_x orientations, we will be able to propose an orientation for the 4Fe-4S cluster with respect to the membrane.

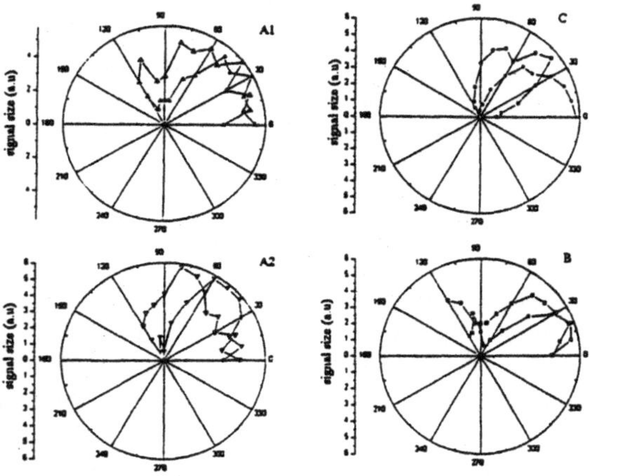

Figure 4. Dependence of signal amplitude of the g_z =2.11, g_z = 2.1, g_y and g_x values (respectively in panels A1. A2, B and C), on the angle of the magnetic field with respect to the plan of the membrane as in Figure 2. "Dark adapted" (closed symbols) and "added by illumination" (open symbols) signals are defined as in Fig. 2.

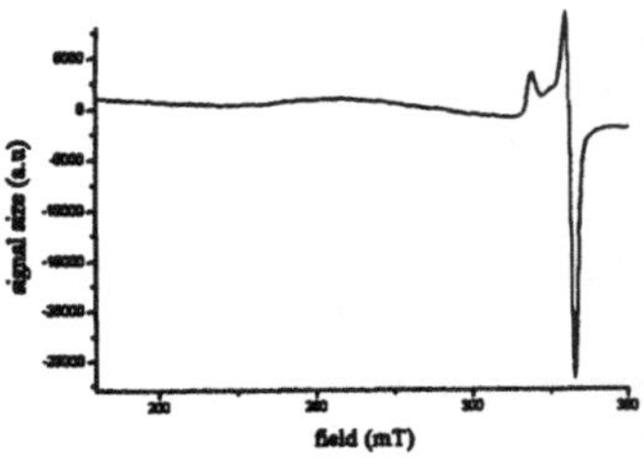

Figure 5. EPR spectrum from oxidized HiPIP cross-linked to the LH_{870}-RC complex.

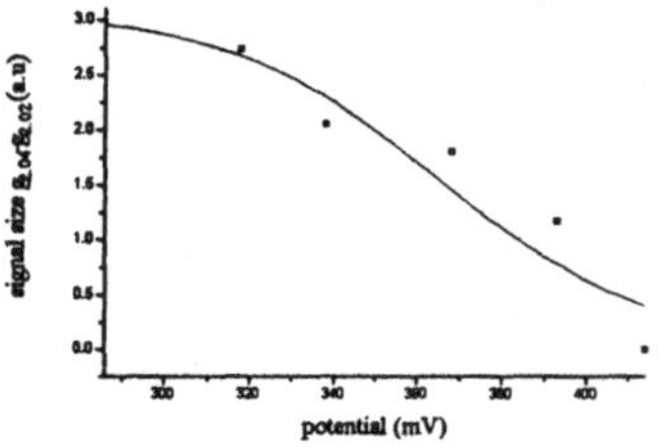

Figure 6. Midpoint redox potential of cross-linked HiPIP determinated by EPR mesurement as the g = 2.04- g = 2.02 amplitude. Curve is least-square fited of the experimental data to a Nernst equation with E_0 = + 365 mV.

References

1 Meyer, T.E. and Donohue T.J. (1995) in Anoxygenic Photosynthetic Bacteria (Blankenship, R.E., Madigan, M.T. and Bauer, C.E., Eds.), pp. 725-745, Kluwer Acad. Publ., Dordrecht, in press

2 Bartsch R.G. (1991) Biochim. Biophys. Acta 1058, 28-30

3 Kennel, S.J., Bartsch R.G. and Kamen M.D. (1972) Biophys. J. 12, 882-896

4 Schoepp, B., Parot, P., Menin, L., Gaillard, J., Richaud, P. and Verméglio, A. (1995) Biochemistry, in press

5 Agalidis, I. and Reiss-Husson, F. (1992) Biochim. Biophys. Acta 1098, 201-208

6 Nitschke, W., Jubault-Bregler, M. and Rutherford, A.W. (1993) Biochemistry 32, 8871-8879

7 Mühlenhoff, U., Kruip, J., Rögner, M., Bryant, D.A., Sétif, P. and Boekema, E. (1995) EMBO J., in press

8 Dutton, P.L. (1971) Biochim. Biophys. Acta 226, 63-80

9 Beinert, H. and Thomson, A.J. (1983) Arch. Biochem. Biophys. 222, 333-361

10 Dunham, W.R., Hagen, W.R., Fee, J.A., Sands, R.H., Dunbar, J.B. and Humblet, C. (1991) Biochim. Biophys. Acta 1079, 253-262

11 Pereira, M.M., Antunes, A.M., Nunes, O.C., da Costa, M.S. and Teixeira, M. (1994) FEBS Lett. 352, 327-330

12 Banci, L., Bertini, I., Ciurli, S., Ferreti, S., Luchinat, L. and Piccioli, H. (1993) Biochemistry 32, 9387-9397

SPECTROSCOPIC EVIDENCE FOR THE FLEXIBILITY OF PSAD IN SOLUTION

Z. Xia[a], R.W. Broadhurst[a], E.D. Laue[a], D.A. Bryant[b], J.H. Golbeck[c] and D.S. Bendall[a]
[a]Department of Biochemistry, University of Cambridge, Tennis Court Road, Cambridge CB2 1QW, U.K.; [b]Department of Molecular and Cell Biology, Pennsylvania State University, University Park, PA 16802; [c]Department of Biochemistry, University of Nebraska, Lincoln, NE 68583, USA.

1. Introduction

The photosystem I (PS I) reaction centre is a membrane-bound multiprotein complex which acts as a light driven oxidoreductase in photosynthesis. It utilises the energy of a single red photon to drive the energetically unfavourable reduction of oxidized ferredoxin by reduced plastocyanin or cytochrome c_6, with a remarkable quantum yield for charge separation across the thylakoid membrane (~100%) and an efficient conversion of the photon energy to chemical free energy (~40%)[1]. In cyanobacteria, it contains two large intrinsic subunits (PsaA, PsaB, M_r *ca* 83 kDa), five small intrinsic subunits (PsaI, PsaJ, PsaK, PsaL, PsaM of 3 - 15 kDa) and four small extrinsic subunits, three on the stromal side (PsaC, PsaD and PsaE of 8 - 16 kDa) and one on the lumenal side (PsaF, 15 kDa)[2]. While the biological functions of these proteins are gradually being uncovered, their structures remain largely unknown, despite the recent report of the X-ray structure at a resolution of 6 Å[3,4].

PsaD may play an important role in the transfer of electrons from PsaC, the terminal electron acceptor of the reaction centre, to ferredoxin[5-7]. It is also found to be involved in stabilizing the peripheral proteins at the stromal side[8,9]. Overexpression in *E. coli*[8] has made it possible to study the solution structure by using modern spectroscopic techniques, including UV-CD spectroscopy and heteronuclear NMR.

2. Materials and Methods

2.1 *Sample preparation*
The *psaD* gene from *Nostoc* sp. PCC 8009 was expressed in *E. coli* and purified as described previously[8]. ^{15}N-labelled PsaD was expressed by growing cells on a minimal medium containing 1g/l $(^{15}NH_4)_2SO_4$ supplemented with trace metal ions. Prior to CD and NMR experiments the sample was exhaustively dialysed in a Centricon against 50 mM phosphate, pH 5.5, and concentrated to about 1 mM. APMSF (1 mM) was added to stabilize the protein and 5% D_2O to lock the magnetic field.

P. Mathis (ed.), Photosynthesis: from Light to Biosphere, Vol. II, 741–744.

2.2 *NMR spectroscopy*
2D [^1H, ^{15}N]-HSQC-NOESY and HSQC-TOCSY spectra were recorded at 14.09T with a Bruker AMX-600 at 308K. Such spectra have the advantage that the dispersion of the ^{15}N resonances results in the separation of overlapping proton-proton J- and NOE-connectivities. The excellent water suppression achieved by the field gradient makes it possible to place the proton NH region in one of the indirectly detected dimensions, thus making best use of the limited resolution available. Data were acquired into 2048 data points in t2 (1024 real) and into 128 points (all real) in t1. Quadrature detection was achieved using the States procedures[10]. The spectral width was 8064.51 Hz in t2 and 1177.8 Hz in the indirectly detected dimension, with 32 transients recorded for each of the 128 t1 experiments.

All spectral processing was done on a Silicon Graphics Iris Indy workstation, using the Azara suite of programs provided by Wayne Boucher and the Department of Biochemistry, University of Cambridge. The code may be obtained via anonymous ftp to www.bio.cam.ac.uk in the directory ~ftp/pub/azara. The residual water peak was removed by convolution of the signal, then the spectra were zero-filled once in the proton dimension, linear predicted to 256 points in the nitrogen dimension and zero-filled to 512 points. The spectra were then apodized using a sine-bell squared window shifted by $\pi/2$, Fourier transformed and phase and baseline corrected. Chemical shifts were measured relative to internal TSP in the proton dimension and relative to carrier frequency (116 ppm) in the nitrogen dimension.

2.3 *CD spectroscopy*
CD spectra were recorded with a Jobin Yvon CD6 spectrometer at room temperature between 200 nm and 240 nm. A two second time constant was used during all the measurements. The spectra were an average of 4 repetitive scans, baseline corrected by subtraction of the buffer signal obtained under the same conditions. The protein concentration was estimated using ε_{280}=10.4 cm^{-1} mM^{-1}, calculated from total amino acid analysis.

3. Results and Discussion

Recombinant *Nostoc* PsaD was found to be unusually prone to proteolytic degradation. The protein was reasonably stable for 24 h at room temperature in the presence of the protease inhibitor APMSF, but more prolonged NMR experiments were not feasible. Attempts to improve the stability with other protease inhibitors, or by more extensive purification by a variety of procedures, failed. These observation suggested that parts of the protein are highly exposed to solvent, and probably lack regular secondary structure.

Direct evidence of secondary structure was sought from measurement of the CD spectrum in the far UV, which showed a low molecular ellipticity compared with some other proteins (see Fig. 1). A broad minimum was observed at around 212 nm which clearly indicates that PsaD possesses only a little secondary structure. The lack of a double minimum (208-210 and 222 nm) suggested a low helical content. An attempt to quantitate the secondary structure using the program k2d[11] failed, possibly in part because the ellipticity values were low, and partly because the protein is too flexible to have a single conformation (see below). Maximum values for the content of α-helix and β-strand were calculated from the four idealized spectra quoted by Yang *et al.*[12], which were based on the spectra of five proteins of known structure. At 206 nm, the ellipticity values of β-strand and β-turns are negligible compared to those of α-helix and random coil. On this basis the maximum helix content of PsaD was calculated to be 8%. Similarly, we obtained a maximum content of β-strand of 20% from the molecular ellipticity at 233 nm., where the contributions from β-turns and random coil are negligible. The real values may well be below these estimated maxima.

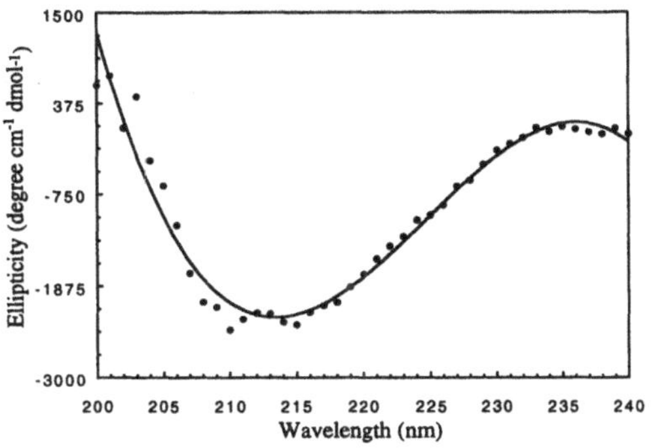

Fig. 1. CD spectrum of PsaD.

Part of a [^{1}H, ^{15}N]-HSQC-NOESY spectrum is shown in Fig.2. This experiment establishes the NOE connectivities between the Cα and sidechain protons and the amide protons labelled with ^{15}N. Only a few crosspeaks were observed in this spectrum. This strongly suggests that a large portion of the protein is in such an extended form that both the medium-range and the long range NOEs are unobservable. Lack of sequential NOE connectivities suggests that in a major part of the protein fast backbone internal motion quenches the NOE. Extremely long ^{15}N T_2 values (about 400 ms) have been found for some resonances (D.S. Bendall *et al.*, unpublished). Many inter-residue connectivities in the CαH region appear either downfield shifted (4.9 - 5.6 ppm) or upfield shifted (3.4-3.9 ppm) compared with random coil. These regions are normally associated with β-sheet and α-helix, respectively[13]. However, the proportion of these resonances is considerably less than would be expected for most proteins.

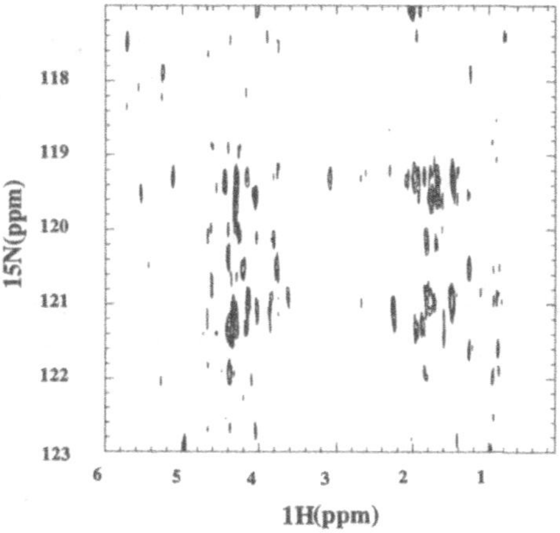

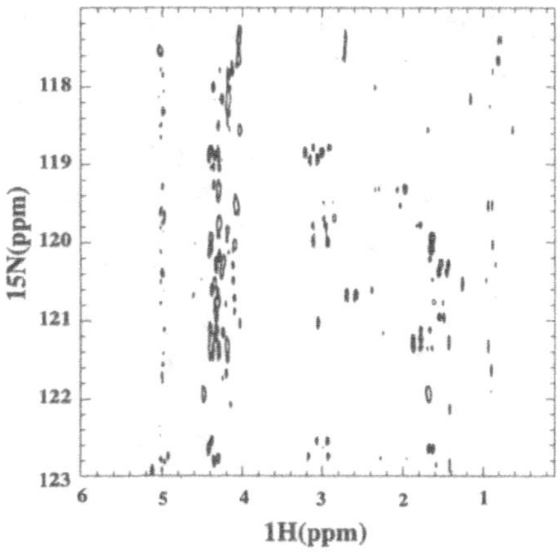

Fig. 2. [^{1}H, ^{15}N]-HSQC-NOESY spectrum. Fig. 3. [^{1}H, ^{15}N]-HSQC-TOCSY spectrum.

In contrast to the poor NOESY spectrum, a good TOCSY spectrum was obtained with efficient magnetization transfer and very sharp resonances for many residues (Fig. 3). This provides an additional proof of flexibility since longer relaxation times usually result in more efficient magnetization transfer because of favourable coupling and reduced dissipation during transfer.

PsaD is unlikely to be as flexible and unstructured *in vivo* as it is in solution. This suggests that a region of the protein associated with components of the PSI complex or with the thylakoid membrane, looses its regular structure in solution. Glucagon[14] and calcitonin[15], both of which associate with membranes, are unstructured in aqueous solution but can fold readily into predominantly helical amphiphilic strucutures at a membrane surface or in solvents of low polarity. In a similar manner, some organic solvents can increase the helical content of PsaD (Z. Xia *et al.*, unpublished).

4. Conclusion

CD and NMR spectroscopy suggest that a cyanobacterial PsaD is flexible, with little regular secondary structure, in aqueous solution.

Acknowledgement

This work was support the Engineering and Physical Sciences Research Council.

References

1 Golbeck, J.H. and Bryant, D.A. (1991) Current Topics in Bioenergetics 16, 83-177
2 Golbeck, J.H. (1993) Curr. Opin. Struct. Biol. 3, 508-514
3 Krauss, N., Hinrichs, W., Witt, I., Fromme, P., Pritzkow, W., Dauter, Z., Betzel, C., Wilson, K.S., Witt, H.T. and Saenger, W. (1993) Nature 361, 326-331
4 Fromme, P., Schubert, W.-D. and Krauss, N. (1994) Biochim. Biophys. Acta 1187, 99-105
5 Zilber, A.L. and Malkin, R. (1988) Plant Physiol. 88, 810-814
6 Zanetti, G. and Merati, G. (1987) Eur. J. Biochem. 169, 143-146
7 Lelong, C., Sétif, P., Lagoutte, B. and Bottin, H.(1994) J. Biol. Chem. 269, 10034-10039
8 Li, N., Zhao, J., Warren, P.V., Warden, J.T., Bryant, D.A. and Golbeck, J.H. (1991) Biochemistry 30, 7863-7872
9 Xu, Q., Armbrust, T.S., Guikema, J.A. and Chitnis, P.R. (1994) Plant Physiol. 106, 1057-1063
10 Novotny, J. and Sharp, K. (1992) Prog. Biophys. Mol. Biol. 58, 203-224
11 Andrade, M.A., Chacon, P., Meret, J.J. and Moran, F.(1993) Prot. Engng. 6, 383-390
12 Yang, J.T., Wu, C-S.C and Martinez, H.M. (1986) Meth. Enzymol. 130, 208-269
13 Wishart, D.S., Sykes, B.D. and Richards, F.M. (1991) J. Mol. Biol. 222, 311-333
14 Braun, W., Piotto, M.E., Pak, Y.M., Weiner, H. and Wüthrich, K. (1983) J. Mol. Biol. 169, 921
15 Motta, A., Pastore, A., Goud, N.A., Castiglione, M.A. and Morelli, G. (1991) Biochemistry 30, 10444

SOME PROPERTIES OF CYTOCHROME F IN THE CYANOBACTERIUM PHORMIDIUM LAMINOSUM

M.J. Wagner, J.C.L. Packer, C.J. Howe and D.S. Bendall
Department of Biochemistry, University of Cambridge,
Tennis Court Road, Cambridge CB2 1QW, U.K.

1. Introduction

Cytochrome *f* and plastocyanin are part of the electron transport chain between Photosystems II and I. The former is a component of the cytochrome *bf* complex, which mediates electron transfer between plastoquinol and plastocyanin. The crystal structure of the lumenal portion of turnip cytochrome *f* shows that cytochrome *f* is a long, thin, flat molecule. It is composed of two domains, the larger being attached to a membrane spanning polypeptide in the holoprotein, containing the haem group[1].
Plastocyanin, a blue copper protein, is located in the thylakoid lumen. The tertiary structure of a number of higher plant and algal proteins is well conserved throughout the proteins investigated[2]. Many of the higher-plant proteins feature two acidic patches at the eastern face of the molecule, and electron transfer involves the side chain of Tyr-83, which is located between the acidic patches.
Knowledge of the tertiary structure of cytochrome *f* has made it possible to develop a molecular model for the interaction. The electrostatic features of the interaction are probably very similar throughout the higher plants because all higher-plant plastocyanins are strongly acidic (isoelectric points cluster around pH 4). This is not true in the cyanobacteria. Cyanobacterial plastocyanins have isoelectric points varying from pH 5.5 in *Synechocystis* [3] to pH 8 in *Anacystis nidulans*.[4] We therefore believe that a study of the cyanobacterial proteins helps to determine residues involved in the interaction of the proteins.

2. Materials and Methods

2.1 *Protein purification*
Phormidium laminosum (strain OH-1-p) was grown under standard conditions[5]. For cytochrome *f* preparation, thylakoid fragments were isolated from 5-7 day old cultures[6]. Cytochrome *f* was isolated from the thylakoids by ethyl acetate extraction[7]. This yielded 52 nmoles of oligomeric cytochrome *f* per mg chlorophyll *a*. Spinach plastocyanin was prepared as described elsewhere[8]. *Phormidium* cytochrome *bf* complex was prepared with slight modifications of the method of Hurt and Hauska[9].

2.2 *Redox titration*
Rexox titration of cytochrome *f* was carried out in 10 mM MES pH 6.2 at 293 K and in the presence of ferri-/ferrocyanide as redox buffer ($E_0' = 410$ mV)[11]. The sample was

P. Mathis (ed.), Photosynthesis: from Light to Biosphere, Vol. II, 745–748.

made up to a final concentration of 0.5 mM ferricyanide and titrated over a range of 270 to 400 mV by addition of ferrocyanide. The extent of reduction of the cytochrome was monitored at 556 nm using a Cary 219 spectrophotometer, with an ascorbate reduced sample reference. Horse heart cytochrome *c* (Sigma C-2506) was used as a control for the titrations.

2.3 *Assay of plastoquinol:plastocyanin oxidoreductase activity*

The initial rate of reduction of spinach plastocyanin and horse heart cytochrome *c* were assayed in a Cary 219 spectrophotometer at 598 nm and 550 nm respectively. The assay buffer contained 0.1% (w/v) digitonin, 20 mM MES pH 6.2, 4 nM cytochrome *bf* complex, 0-800 mM NaCl and 8 μM plastocyanin or cytochrome *c* . The assay was initiated by the addition of 200 μM plastoquinol-1 for cytochrome *c* and 40 μM plastoquinol-1 for plastocyanin and the initial rate of reduction determined. Plastoquinol-1 was synthesised folowing the procedure given by Wood and Bendall[12]. Molar extinction coefficients of 27700 $M^{-1}cm^{-1}$ at 550 nm for cytochrome *c* and 4700 for plastoyanin at 598 nm were used. The rates were corrected for the uncatalysed rates of reduction of plastocyanin and cytochrome *c*.

3. Results and discussion

3.1 *Ionic strength dependence*

At low ionic strength horse heart cytochrome *c* was found to be a good electron acceptor for the *Phormidium* cytochrome *bf* complex, unlike higher plant plastocyanin. As the ionic strength was increased the rate of reduction with cytochrome *c* decreased while the opposite occured with spinach plastocyanin (Figure 1). This can only be explained if *Phormidium* cytochrome *f* is assigned a negative charge at the interaction site for the two electron

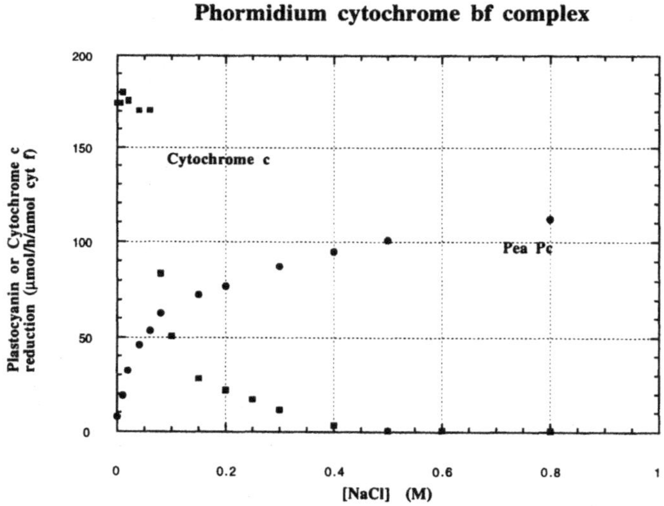

Figure1: Ionic strength dependence of cytochrome *c*/plastocyanin reduction by *Phormidium* cytochrome *bf* complex at pH 6.2

acceptors, bearing in mind that in horse heart cytochrome *c* positively charged residues cluster around the exposed haem edge and that Tyr-83 in most higer plants plastocyanin is located between 2 negatively charged clusters. This is supported by the fact that a number of *Phormidium* cytochrome *f* residues are neutralized or show charge reversal if compared with the turnip protein (Table 1). Similarly the*Phormidium* plastocyanin sequence shows a number of significant changes in the protein sequence when compared to the higher plants.[13] The 'small acidic patch' (residues 59-61) has disappeared. The 'large acidic patch' (residues 42-45) retains acidic residues at positions 42 and 43 but Glu-45 has been neutralized by becoming Gln and Asp-44 has undergone charge inversion by conversion to Lys.

Table 1: Non-conserved Basic Residues of Turnip Cytochrome *f*

Residue	Turnip	*Phormidium*
58	Lys	Gln
65	Lys	Ser
96	Lys	Val
106	Arg	Ala
110	Lys	Gln
122	Lys	Glu
185	Lys	Ala
187	Lys	Asp
201	Arg	Thr
209	Arg	Ala
222	Lys	Ala

3.2 *Midpoint Potential*

The midpoint potential of cytochrome *f* isolated from *Phormidium* thylakoids was determined to be approximately 276 mV using the ferri-/ferrocyanide couple (Figure 3). The potential of horse heart cytochrome *c* was determined as a control and found to be 250 mV. The midpoint potential is considerably lower for *Phormidium* cytochrome *f* than the values previously reported for other species, the midpoint potential of *Spirulina maxima* is reported to be 334 mV[14]. The potential usually observed for higher plants is 365 mV. This indicates that this part of the electron transport chain around the *Phormidium* cytochrome *bf* complex has a markedly lower mid-point potential than in higher plants and some cyanobacteria.

4. Conclusion

The results presented show that the charge properties and mid-point potential of *Phormidium* cytochrome *f* differ significantly from those of it's higher plant counter parts.

Acknowledgements

M.W. wishes to acknowledge the BBSRC for a studentship and thank M. Taylor for the preparation of *Phormidium* cytochrome *bf* complex and plastoquinol-1.

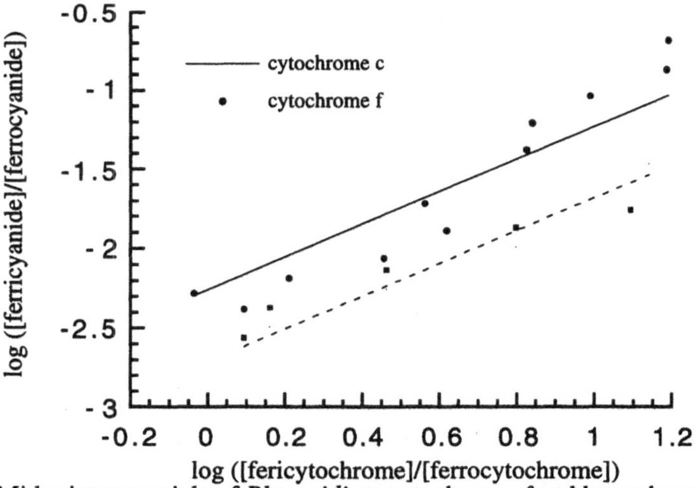

Figure 3: Midpoint potentials of *Phormidium* cytochrome *f* and horse heart cytochromec

5. References

1 Martinez, S.E., Huang, D., Szepaniak, A., Cramer, W.A. and Smith, J.L. (1994) Structure 2, 95-105
2 Redinbo, M.R., Yeates, T.O., and Merchant, S. (1994) J. Bioenerg. Biomembr. 26, 49-66
3 Hervas, M., Navarro, F., Navarro, J.A., Chavez, S., Diaz, A., Florencio, F.J. and De la Rosa, M.(1993) FEBS Lett. 319, 257-260
4 Widger, W.R., (1991) Photosynthesis Research 30, 71-84
5 Castenholz, R.W. (1970) J. Schweiz. Z. Hydrol. 32 538-551
6 Bendall, D.S., Bowes, J.M., Stewart, A.C. and Taylor, M.E. (1988) Meth. Enzymol. 167, 272-280
7 Bendall, D.S., Davenport, H.E., Hill, R. (1971) Meth. Enzymol. 23, 327-344
8 Plesceniar, M. and Bendall, D.S. (1970) Biochim. Biophys. Acta 216, 192-199
9 Hurt, E. and Hauska, G. (1981) Eur. J. Biochem. 117, 591-599
10 Hill, R. and Scarsbrick, R. (1951) New Phytol. 50, 98-111
11 O'Reilly, J.E. (1973) Biochim. Biophys. Acta 282, 509-515
12 Wood, P. and Bendall, D.S. (1976) Eur. J. Biochem. 61, 337-344
13 Varley, J.P.A., Moehrle, J.J., Manasse, R.S., Bendall, D.S. and Howe, C.J. (1995) Plant Mol. Biol. 27, 179-190
14 Ho, K. and Krogman, D.W. (1980) J. Biol. Chem. 255, 3855-3861

CONSERVED NON LIGANDING RESIDUES OF *RHODOBACTER (R.) CAPSULATUS* 'RIESKE' PROTEIN ARE ESSENTIAL FOR [2Fe-2S] CLUSTER PROPERTIES AND COMMUNICATION WITH THE QUINONE POOL

Ursula Liebl[1§], Vladimir Sled[2], Tomoko Ohnishi[2], and Fevzi Daldal[1]
[1]Dept. of Biology, and [2]Dept. of Biochemistry and Biophysics, University of Pennsylvania, Philadelphia, PA 19104, USA
[§]present address: Lab. Optique Appliquée, Ecole Polytechnique-ENSTA, 91120 Palaiseau Cedex, France

1. Introduction

The 'Rieske' iron-sulfur protein is one of three redox-active components in ubiquinol cytochrome (cyt) *c* oxidoreductases (cyt bc_1-complexes) (reviewed in 1,2). Its function is to oxidize ubiquinol (UQH_2) at the Q_o site, and subsequently reduce cyt c_1. The reduced form of the 'Rieske' [2Fe-2S] cluster has a distinct electron paramagnetic resonance (EPR) spectrum with a characteristic g_{av}-value of 1.91, and a relatively high redox midpoint potential (E_{m7}) [+310mV in *R. capsulatus*] (3). Spectroscopic studies indicate that the [2Fe-2S] cluster is bound to the protein by two nitrogen and two sulfur ligands (4,5), and results of site-directed mutagenesis asign C133 and H135 as well as C153 and H156 as cluster ligands in *R. capsulatus* (6).

In order to test the hypothesis that a network of charges as well as hydrogen bonds in the cluster environment are the major forces defining the redox properties of the [2Fe-2S] cluster, and to contribute to a structural model of the 'Rieske' protein, site-directed mutagenesis and biochemical and biophysical methods were used to study the effect of amino acid substitutions directly adjacent to the designated cluster ligands on properties of the [2Fe-2S] cluster. For that purpose the polar amino acid T134 and the hydrophobic residue L136 were replaced with the charged residues R and D, and with H as well as G. These substitutions were selected in order to probe the effect of positively and negatively charged amino acids on the [2Fe-2S] cluster, as well as to assess the role of size and hydrophobic nature of a residue at the respective position. These replacements created two distinctly different types of mutant strains. None of these substitutions affect the assembly of the cyt bc_1 complex in *R. capsulatus* although they perturb, in varying degrees, its function.

2. Procedures

R. capsulatus was grown as described in (7). Site-directed mutagenesis, DNA sequencing, preparation of chromatophores and DBH : cyt *c* reductase assays were performed as in (8). EPR spectra were recorded using a Bruker ESP-300E spectrometer equipped with a helium cryostat, and redox titrations were performed as described in (9). Redox titrations were carried out in the presence of 50 μM indigotetrasulfonate, duroquinone, phenazine methosulphate, 1,2-naphthoquinone-4-sulfonate, N,N,N',N'-tetramethyl-p-phenylene-diamine-dihydrochloride, tetrachloro hydroquinone, 40μM 1,2-naphthoquinone, and 10μM pyocyanine.

P. Mathis (ed.), Photosynthesis: from Light to Biosphere, Vol. II, 749–752.

3. Results and Discussion

Phenotypical characteristics of the mutant strains at position T134 and L136 of the 'Rieske' iron sulfur protein in *R. capsulatus* are presented in Table 1.

Strain	Phenotype	DBH Activity [%]	Doubling Time [min]	Reversion Frequency
pMTSI (WT)	Ps^+ $Stig^S$; Myx^S	100	170	
pFT134R	Ps^+ $Stig^S$; Myx^S	24	170	
pFT134G	Ps^+ $Stig^S$; Myx^S	10	320	
pFT134H	Ps^+ $Stig^S$; Myx^S	10	240	
pFT134D	Ps^-	2		$1x10^{-6}$
pFL136R	Ps^-	5		$1x10^{-5}$
pFL136G	Ps^-	4		$1x10^{-7}$
pFL136H	Ps^-	4		$7x10^{-8}$
pFL136D	Ps^-	2		$6x10^{-8}$

Table 1. Characteristics of T134 and L136 mutant strains.
Ps^+ and Ps^- indicate photosynthetic competence or incompetence; $Stig^S$ and Myx^S refer to stigmatellin and myxothiazol sensitivity; DBH:cyt c reductase activity is measured as nmol of horse heart cyt c reduced min^{-1} mg^{-1} chromatophores and is expressed as % of the activity of the wild type (WT) overproducing strain pMTSI; Approximate reversion frequencies of the mutant strains were measured by the ratio of Ps^+ to Ps^- colonies.

All T134 and L136 mutant strains overexpress cyt *b* and c_1 to WT levels as shown by optical absorption difference spectroscopy. Intracytoplasmic membrane preparations separated on SDS-PAGE contain all three protein subunits of the cyt bc_1 complex. Immunoblotting and detection with polyclonal antibodies raised against the 'Rieske' protein from purified *R. capsulatus* cyt bc_1 complex show the presence of 'Rieske' apoprotein in all mutant strains at levels similar to the WT overproducing strain pMTSI (data not shown).

T134 substitutions

EPR characteristics and occupancy of the Q_o site are not affected in the Ps^+ mutants T134R, H, G, although the detectable [2Fe-2S] cluster content is reduced compared to WT. All E_{m7} values are lowered compared to pMTS1: T134G<T134H<T134R : +210<+230<+260 mV (Figure 1). These data show that the substitution of T134, which has a small aliphatic side chain with a polar hydroxyl group, by charged amino acid residues or even by the apolar, much smaller G, does not affect its ability to bind UQ and to respond to the redox state of the UQ_{pool} and indicates a significant plasticity at this position of the protein. The moderate changes in the redox midpoint potential of the [2Fe-2S] clusters (at best 100 mV lower) suggest that T134 contributes to fine tuning of the [2Fe-2S] cluster E_{m7} but does not play a crucial role in its redox control. DBH:cyt *c* reductase activities measured under steady-state conditions in T134R, H, and G chromatophores range between 10% to 24% of the WT activity when cultures are grown anaerobically under photosynthetic conditions and chromatophores are prepared in the presence of 10-20 mM sodium ascorbate. Virtually no cluster signal or cyt *c* reductase activity can be detected when cells are grown (semi-)aerobically or preparation of chromatophores is done in the absence of ascorbate. These data correlate the loss of cluster stability, and of measurable DBH activity, with an exposure to more oxidising conditions. Possibly the substituted residues at position T134 cause structural changes in the immediate protein environment of the [2Fe-2S] cluster that enhance accessibility of oxygen and thus result in cluster destabilization.

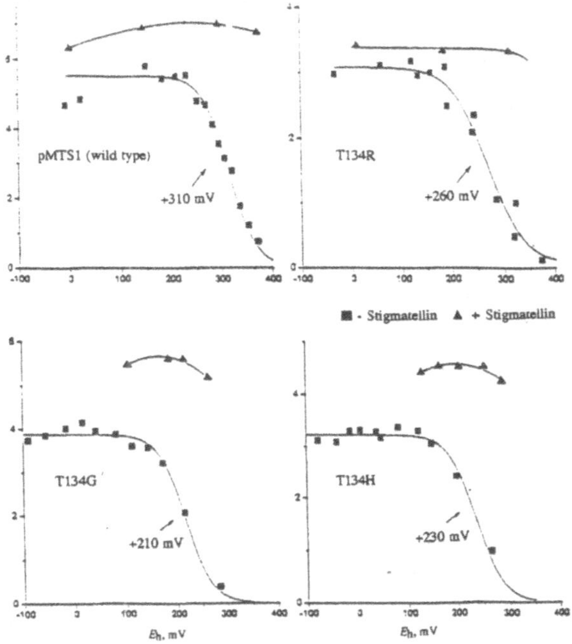

Figure 1. EPR redox titrations of pMTS1 (WT) and T134R, H and G, without inhibitor and in the presence of stigmatellin. E_{m7}s are indicated. Experimental conditions are described in the text.

T134D is Ps⁻ but can revert to Ps+ with a frequency of approximately 10^{-6}. A frequently observed T134D to N reversion indicates that the photosynthetic incompetence in the T134D mutant strain is due to a negative charge, rather than to a size effect. The fact that no second site T134D suppressor mutant was readily found suggests that a negative charge at this position is too destructive to be functionally compensated by a secondary amino acid substitution.

L136 mutant strains

All L136 substitutions result in Ps⁻ strains (Table 1). Their EPR spectra are altered (Figure 2) and no response to the redox state of the UQ_{pool} is observed, indicating perturbation of the cluster geometry and impaired UQ binding. The near WT redox midpoint potentials of L136R and L136H (+280 mV) indicate no significant charge effect on the cluster. Thus, for appropriate communication between the [2Fe-2S] cluster and the Q_{pool} hydrophobicity appears to be a primary requirement for an amino acid residue at position 136.

Intragenic suppressor mutation of L136H

An intragenic suppressor of L136H was identified as V44L which is a substitution naturally present in the 'Rieske' protein of spinach *b6f* complex. L136H+V44L is the only known intragenic suppressor in *R. capsulatus* 'Rieske' protein, and suggests a spatial interaction between these two residues. L136H+V44L is resistant to stigmatellin concentrations of $5x10^{-6}$M *in vivo*. Steady state DBH:cyt *c* reductase activity is approximately 15% of the WT rate indicating only partial suppression. While the presence of another suppressor mutation cannot yet be completely excluded in this mutant, the fact that the single mutant strain V44L is Ps+ argues that solely a size increase (V44L) is sufficient for a partial compensation.

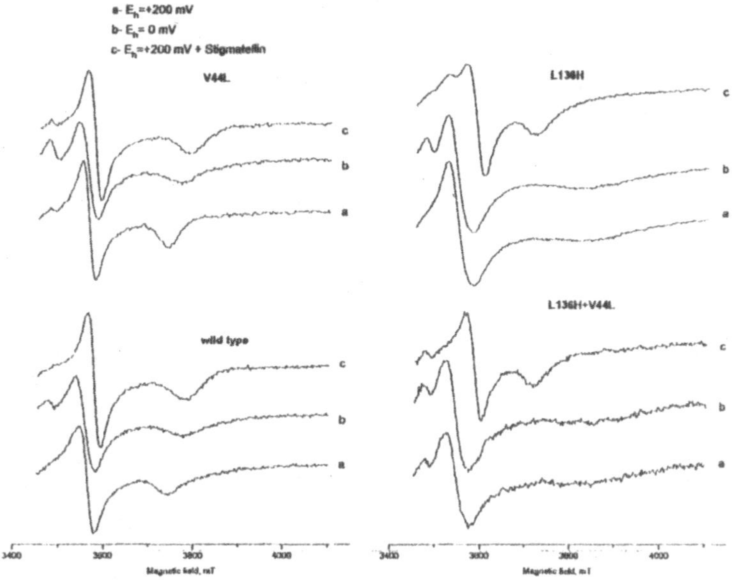

Figure 2. EPR spectra of pMTS1 (WT) and the mutant strains L136H, L136H+V44L, and V44L. E_h and absence or presence of inhibitor, respectively, are indicated.

Acknowledgement

This work was supported by a HFSPO Long Term Fellowship to UL, by NIH grant GM 38237 to FD, and by NSF grant MCB-94-18694 to TO. UL would like to thank Mimi Benach for technical help.

References

1. Knaff, D.B. (1993) Photosynth. Res. 35, 117-133
2. Gennis, R.B., Barquera, B., Hacker, B., van Doren, S.R., Arnaud, S., Crofts, A.R., Davidson, E., Gray, K.A. and Daldal, F. (1993) J. Bioenerg. Biomembr. 25, 195-209
3. Ding, H., Robertson, D.E., Daldal, F. and Dutton, P.L. (1992) Biochemistry 31, 3144-3158
4. Britt, R.D., Sauer, K., Klein, M.P., Knaff, D.B., Kraiuciunas, A., Yu, C.-A., Yu, L. and Malkin, R. (1991) Biochemistry 30, 1892-1901
5. Gurbiel, R.J., Ohnishi, T., Robertson, D.E., Daldal, F. and Hoffman, B.M. (1991) Biochemistry 30, 11579-11584
6. Davidson, E., Ohnishi, T., Atta-Asafo-Adjei, E., & Daldal, F. (1992) Biochemistry 31, 3342-3351
7. Gray, K.A., Dutton, P.L. and Daldal, F. (1994) Biochemistry 33, 723-733
8. Atta-Asafo-Adjei, E. and Daldal, F. (1991) Proc.Natl.Acad.Sci.U.S.A. 88, 492-496
9. Dutton, P.L. (1978) Methods Enzymol. 5, 411-435

MODIFIED LIGANDS TO F_A AND F_B IN PHOTOSYSTEM I:
Structural Constraints for the formation of Iron-Sulfur Clusters in the Free PsaC Protein.

Tetemke Mehari[1], Fengyu Qiao[2], M.Paul Scott[2], Donald A. Bryant[3] and John H. Golbeck[2]

[1]Department of Chemistry, Addis Ababa University, Addis Ababa, Ethiopia,
[2]Department of Biochemistry, and Center for Biological Chemistry, University of Nebraska, Lincoln, Nebraska 68588-0664
[3]Department of Biochemistry and Molecular Biology, and Center for Biomolecular Structure and Function, Penn State University, University Park, PA 16802

1. INTRODUCTION

PsaC in Photosystem I (PS I) shares similarities with a number of soluble 2[4Fe-4S] ferredoxins in terms of molecular mass, amino acid sequence, and the presence of two CxxCxxCxxxCP iron-sulfur cluster binding motifs. There is enough sequence homology with the 54-amino acid 2[4Fe-4S] ferredoxin from *Peptococcus aerogenes* (1) and the first 58 residues of the 106-amino acid [3Fe-4S]•[4Fe-4S] ferredoxin from *Azotobacter vinelandi* (2) to make predictions of tertiary structure based on the known 3-dimensional structure of these proteins. The sequences of naturally occurring ferredoxins suggest modifications that may be tolerated in PsaC. For example, several naturally-occurring ferredoxins contain aspartic acid at the position analogous to cysteines 14 in PsaC. These proteins are capable of supporting a mixed-spin ($S = 3/2$ and $1/2$) [4Fe-4S•3Cys1Asp] cluster in the modified site (3,4). Using site-directed mutagenesis, mutant PsaC proteins have been constructed with similar mutations (PsaC-C14D and PsaC-C51D), and the free proteins contain [3Fe-4S] clusters at the modified sites and the [4Fe-4S] clusters at the unmodified sites (5,6).

For further studies mutant PsaC proteins were constructed in which one of the cysteine ligands to an iron-sulfur cluster was changed to either the polar amino acid serine: PsaC-C14S and PsaC-C51S; or the neutral amino acid alanine: PsaC-C14A and PsaC-C51A.

2. MATERIALS AND METHODS

Mutant PsaC proteins were engineered in *E. coli*, synthesized as inclusion bodies and purified according to the protocol in Yu, L., et al. (6 and references therein). Purified PsaC was refolded with iron-sulfur clusters as described (7), and purified

P. Mathis (ed.), Photosynthesis: from Light to Biosphere, Vol. II, 753–756.

by ultrafiltration in an amicon cell over a YM-5 membrane (Amicon, Beverly, MA). For electrochemistry, the PsaC holoproteins were desalted and the ß-mercaptoethanol was removed by gel filtration chromatography over a Sephadex G-25 (pharmacia Biotech, Piscataway, NJ) column under anaerobic conditions in a Coy controlled environment chamber (Grass Lake, MI).

X-band EPR spectroscopy was performed using a Bruker ECS 106 spectrometer operating with a normal-mode (ST 8615) or a dual-mode (DM 4116) resonator. Temperatures were controlled using an Oxford Intelligent temperature controller with liquid nitrogen as the reference. Samples were approximately 2.5 mg mL^{-1}. Chemical reduction was carried out by adjusting the pH to 10.5 with 1 M glycine-NaOH and approximately 1 mg sodium dithionite per 300 µl sample. All manipulations of the PsaC holoprotein were carried out under anaerobic conditions. A matrix of temperature and microwave power levels were determined for each mutant protein to optimize the signal intensity (and S/N) and to determine the half-saturation parameter, $P_{1/2}$ (at a given temperature). Spin Quantitations were determined at power levels an order of magnitude lower than the measured $P_{1/2}$. The midpoint potentials of the mutant PsaC proteins were determined using electrochemical poising followed by EPR detection.

3. RESULTS

The EPR spectra of free PsaC-C14S, PsaC-C51S, PsaC-C14A and PsaC-C51A proteins were found to be similar to those observed previously for the PsaC-C14D and PsaC-C51D mutant proteins (6). With no additions, the EPR spectra show a peak at g = 2.020 and trough at g = 1.998 characteristic of oxidized [3Fe-4S] clusters (Fig. 1 doted line). The resonances disappear after mild chemical reduction with sodium dithionite at pH 8.0 (E_m = -480 mV), but not with ß-mercaptoethanol. When the samples are more strongly reduced by addition of sodium dithionite at pH 10.5 (E_m = -630 mV), the EPR spectra show a set of resonances with broad line shapes and a g_{av} of 1.94 characteristic of reduced [4Fe-4S] clusters (Fig 1 solid line). The line shapes of the merged midfield and highfield resonances of PsaC-C14S and PsaC-C14A are slightly broader than the corresponding features in PsaC-C51S and PsaC-C51A, a pattern observed earlier for PsaC-C14D and PsaC-C51D (6). These features are consistent with the presence of a S = 1/2 ground state $[4Fe\text{-}4S]^{2+/1+}$ cluster, identified as F_A in PS I-rebound PsaC-C14D, and by the presence of a S = 1/2 ground state $[4Fe\text{-}4S]^{2+/1+}$ cluster, identified as F_B in PS I-rebound PsaC-C51D.

There are a new set of broad resonances around g = 5.5 which are difficult to microwave-saturate. This rapidly relaxing spin system is attributed to the lowfield peak(s) of one or two of the Kramers doublets derived from a S = 3/2 [4Fe-4S] cluster. It is proposed that the resonance(s) around g = 5.5 arise from a high spin, mixed-ligand [4Fe-4S] cluster resident in the modified sites of mutant PsaC-C14S (Fig. 2 solid line) and PsaC-C51S . One likely origin of the high-spin [4Fe-4S] cluster is the conversion of a reduced [3Fe-4S] cluster (Fig. 2 dotted line) as in *Peptococcus furiosus* (3) and *Desulfovibrio africanus* (4) ferredoxins which contain similar CxxDxxCxxxCP motifs.

Charged and polar side groups represented by PsaC-C51D (8), PsaC-C14S and PsaC-C51S therefore appear capable of supporting a high-spin [4Fe-4S] cluster.

However, this signal may also be present, albeit at a reduced intensity, when the replacement amino acid contains a hydrophobic side group such as alanine in PsaC-C14A and PsaC-C51A.

To further characterize the properties of the mutant proteins a redox titration was performed on the $[4Fe\text{-}4S]^{2+/1+}$ clusters on free WT PsaC and the $[3Fe\text{-}4S]^{1+/0}$ and $[4Fe\text{-}4S]^{2+/1+}$ clusters on free PsaC-C14S, PsaC-C14A, PsaC-C51S and PsaC-C51A (Table I). Because of the low S/N, we didn't titrate the high-spin clusters in the modified site.

Table I

Midpoint Potentials of [3Fe-4S] and [4Fe-4S] Clusters in Mutant PsaC Proteins

Mutation	$[3Fe\text{-}4S]^{1+/0}$	$[4Fe\text{-}4S]^{2+/1+}$
C14D[1]	-98 mV	-520 mV
C14A	-171 mV	-551 mV
C14S	-132 mV	-520 mV
C51D[1]	-98 mV	-580 mV
C51A	-167 mV	-575 mV
C51S	-145 mV	-495 mV

[1]From Ref.6

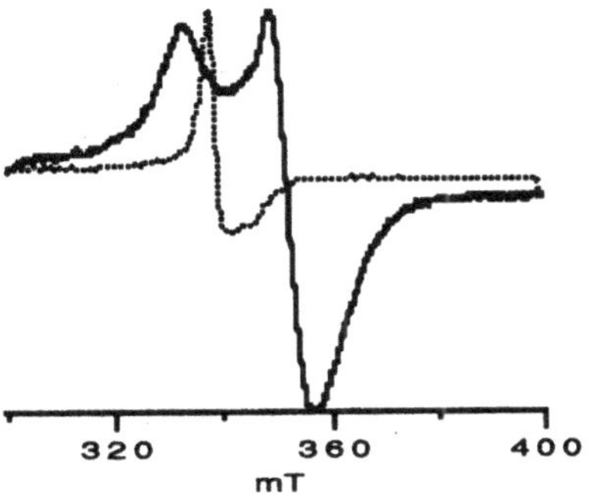

Fig. 1. EPR spectra of the oxidized [3Fe-4S] (dotted line) & reduced [4Fe-4S] (solid line) clusters in free PsaC-C14S

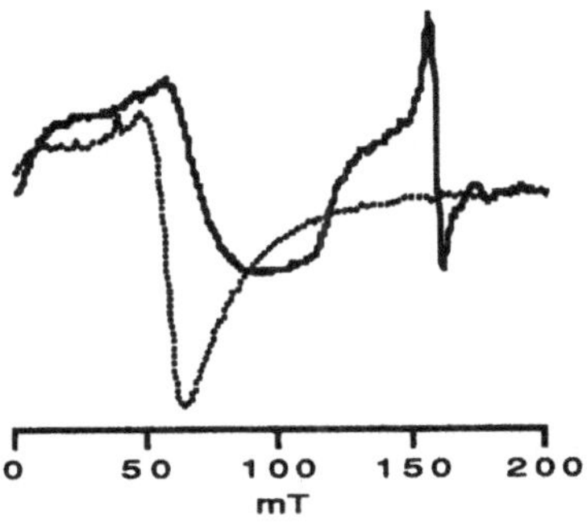

Fig. 2. Perpendicular (solid line) & parallel-mode (dotted line) EPR studies of reduced [3Fe-4S] in equilibrium with [4Fe-4S] clusters in free PsaC-C14S

4. DISCUSSION

The EPR spectroscopic properties of the iron-sulfur clusters in free PsaC proteins were determined after the introduction of charged (aspartate), polar (serine) and hydrophobic (alanine) amino acids in the second position. The following generalizations hold when the second cysteine of the motif is changed. Firstly, a

[3Fe-4S] cluster can be found in the oxidized protein and is assumed resident in the modified site. The midpoint potentials of the [3Fe-4S] clusters are ~400 mV more oxidizing than the wild-type [4Fe-4S] clusters. Secondly, a low-spin [4Fe-4S] cluster is found in the reduced protein and is assumed resident in the unmodified site. The replacement amino acids, which include those with charged, polar and a hydrophobic side groups, show no consistent pattern in affecting the reduction potential of the cluster in the unmodified site (Table I). Thirdly, high-spin [4Fe-4S] clusters are tentatively identified in the PsaC-C14S, PsaC-C51S, PsaC-C14A and PsaC-C51A reduced proteins and are assumed resident in the modified site as observed in PsaC-C14D (8) and PsaC-C51D (9). While the presence of a high-spin [4Fe-4S] cluster was not surprising in the aspartate (or serine) mutants, it was not expected in the alanine mutants.The issue is whether the aspartate and/or serine oxygen provide ligands to the cubane iron as does aspartate in *P. furiosus* (10,11), or whether water, hydroxide, or the thiolate from ß-mercaptoethanol, present in the reconstitution mixture, has contributed the fourth ligand. In this respect, it is interesting that only those mutant proteins which show evidence for a high-spin [4Fe-4S] cluster in the mutant site are capable of reconstituting onto P700-F_X cores.

In summary, these studies show that PsaC refolds only in the presence of two iron-sulfur clusters when cysteine is replaced in positions 14 and 51 by aspartate, serine and alanine. Since there was no occurrence of a mutant PsaC which contained only one [3Fe-4S] cluster or only one [4Fe-4S] cluster, a corollary to the above premise is that a stable 3-dimensional structure may require two conditions be met: 1) the presence of two iron-sulfur clusters, and 2) one of the two clusters is a cubane.

ACKNOWLEDGEMENTS

This research was supported by a Biotechnology Career Fellowship from the Rockefeller Foundation (to T.M.) and by National Science Foundation grants MCB-9205756 (to J.H.G.) and MCB-9206851 (to D.A.B.).

REFERENCES

1. Adman, E. T., Sieker, L. C. and Jensen, L. H. (1973) *J. Biol. Chem.* **248**, 3987-3996
2. Stout, C. D. (1989) *J. Mol. Biol.* **205**, 545-555
3. Conover, P. R., Kowal, A. T., Fu, W., Park, J. B., Aono, S., Adams, M. W. W., and Johnson, M. K. (1990) *J. Biol. Chem.* **265**, 8533-8541
4. Thomson, A., Breton, J., Butt, J., Hatchikian, E., and Armstrong, F. (1992) J. *Inorganic Biochem.* **47**, 197-207
5. Zhao, J., Li, N., Warren, P., Golbeck, J., and Bryant, D. (1992) *Biochemistry* **31**, 5093-5099
6. Yu, L., Zhao, J. D., Lu, W. P., Bryant, D. A., and Golbeck, J. H. (1993) *Biochemistry* **32**, 8351-8258
7. Li, N., Zhao, J., Warren, P., Warden, J., Bryant, D., and Golbeck, J. (1991) *Biochemistry* **30**, 7863-7872
8. Yu, L., Bryant, D., and Golbeck, J.H. (1995) *Biochemistry* **34**, 7861-7868
9. Yu, L., Vassilev, I., Jung, Y-S., Bryant, D., and Golbeck, J.H. (1995) *J. Biol. Chem.* (in press)
10. Telser, J., Smith, E. T., Adams, M. W. W., Conover, R. C., Johnson, M. K., and Hoffman, B. M. (1995) *J. Am. Chem. Soc.* **117**, 5133-5140
11. Park, J., Fan, C., Hoffman, B., and Adams, M. (1991) *J. Biol. Chem.* **266**, 19351-19356

BIOCHEMICAL CHARACTERIZATION AND THREE-DIMENSIONAL STRUCTURE OF DIMERIC HIGH POTENTIAL IRON SULFUR PROTEIN FROM *CHROMATIUM PURPURATUM*

Cheryl A. Kerfeld, Cheryl Chan and Todd O. Yeates
Molecular Biology Institute, University of California, Los Angeles

1. Introduction

Chromatium purpuratum is an anaerobic, marine species of purple-sulfur photosynthetic bacteria. Members of the genus *Chromatium* use reduced sulfur compounds as an electron source in photosynthesis and are thought to be more ancient than their non-sulfur counterparts, the members of the Rhodospirillaceae. In the Rhodospirillaceae, cyclic electron transfer between the membrane-bound reaction center (RC) and cytochrome bc1 complex in photosynthesis is mediated by a soluble, periplasmic, high potential c2-type cytochrome. In contrast, in the Chromatiaceae, there is no evidence of a c2-type cytochrome and the identity of the soluble mediator of electron transfer is uncertain. Very recently two groups have acquired data suggesting that High-Potential Iron-Sulfur proteins (HiPIPs) may perform this function (1; Andre Vermeglio, personal communication). In our investigation of the soluble fraction of *C. purpuratum* we have found no evidence of a c2-type cytochrome, but we have isolated large quantities of HiPIP.

HiPIPs are small, soluble proteins (62-85 amino acids) in which a single 4Fe-4S cluster is covalently linked to the polypeptide chain through Fe-Cys bonds. They are found in many species of phototrophic bacteria (all of these, with one exception, contain a tetraheme cytochrome subunit in the photochemical RC). The RC of *C. purpuratum* has been isolated and characterized with its associated antenna; it contains a tetraheme cytochrome and is thought to occur as an oligomer (2).

Several lines of evidence suggest that HiPIP forms dimers that are functionally relevant. Although HiPIPs from most bacterial species behave as monomers in gel filtration columns, one noted exception is HiPIP from *Rhodospirillum salinarum* which appears to be a tetramer or hexamer (3). In studies of self-exchange electron transfer kinetics, HiPIPs show very fast rates, indicative of a close association of HiPIP molecules in solution (4). Molecules in the *Ectothiorhodospira vacuolata* HiPIP crystals pack as dimers (5), as do those in the *Rhodocyclus tenuis* HiPIP crystals (6). Crystals of HiPIP from *Chromatium vinosum* did not exhibit dimerization within the lattice (7), however EPR studies of *C. vinosum* HiPIP indicate it dimerizes after freezing in 0.1-0.2 M NaCl (8). The putative dimerization interface in *C. vinosum* has been modeled (9) and is similar to that observed in the *R. tenuis* and the *E. vacuolata* crystals. Here we describe the isolation, characterization and X-ray structure determination of *C. purpuratum* HiPIP, the first HiPIP isolated as a dimer in solution.

P. Mathis (ed.), Photosynthesis: from Light to Biosphere, Vol. II, 757–760.

2. Results and Discussion

2.1) *Isolation and Characterization*

HiPIP was isolated from the soluble fraction of *C. purpuratum* cells by a combination of anion exchange and gel filtration chromatography. It is the most abundant protein present in this fraction. The isoelectric point was determined to be 4.0 (oxidized) and 4.3 (reduced) and the midpoint potential is +390 mV (pH 7.0), one of the highest in this class of proteins. The purity index (A_{283}:A_{388}) for the reduced form is 2.71. The purified protein migrates as a dimer in size exclusion chromatography on either HPLC (Figure 1) or FPLC gel filtration columns by comparison to elution of gel filtration standards and other small proteins such as cytochrome c551 or horse heart cytochrome c. Surprisingly the dimer is stable during SDS gel electrophoresis, even after treatment with beta-mercaptoethanol; only boiling *C. purpuratum* HiPIP in SDS sample buffer prior to electrophoresis abolishes the dimerization (data not shown). The molecular weight of *C. purpuratum* HiPIP, as determined by ion spray mass spectroscopy, is 9372 Da; there is no evidence of HiPIP isoforms as have been observed in other species of photosynthetic bacteria. Dissolved crystals were used for N-terminal amino acid sequencing; the 30 residues obtained show a high degree of homology to the primary structures of HiPIPs from *Chromatium gracile* and *Chromatium tepidum* and a slightly more distant relationship to *C. vinosum* (data not shown).

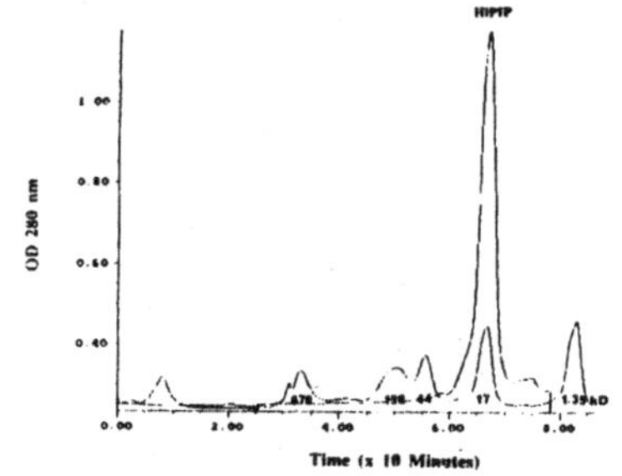

Figure 1. HPLC size-exclusion chromatography of HiPIP with BioRad gel filtration standards superimposed and labeled. Column: Toso Haas TSK3000. Mobile Phase: 25 mM $NaPO_4$, pH 7.0/100mM $NaSO_4$.

2.2) *Structure Determination and Refinement*

Crystals of *C. purpuratum* HiPIP were readily obtained in ammonium sulfate in the pH range of 4.5 and 8.0. The bulkiest crystals, obtained at pH 5.0, were subjected to X-ray analysis. The space group was determined to be primitive orthorhombic with cell dimension of 55.5, 108.8 and 37.2 Å. An examination of the axial reflections suggested the space group $P2_12_12$. Data from two crystals were processed and merged to obtain a 98% complete data set to 2.7 Å resolution. Molecular replacement using the program AMORE was successful with the HiPIP from *C. vinosum* as a search model. At the time there was no amino acid sequence data available for the *C. purpuratum* HiPIP so the search model was truncated to polyalanine with the exception of the four cysteines that ligate the 4Fe-4S cluster and invariant Tyr 19. Determination of the amino acid sequence and manual model building is underway using 3-fold averaged electron density maps generated by

RAVE. The model is being refined with the XPLOR positional refinement protocol, with imposed strict non-crystallographic symmetry constraints.
A phased anomalous difference fourier was calculated from phases derived from the present model after removal of the 4Fe-4S clusters and all atoms within 3.0 Å and following a cycle of the simulated annealing in XPLOR (Figure 2). Electron density for the 4Fe-4S clusters was observable to the 5.0 σ contour level, confirming the positioning of the *C. vinosum* search model.

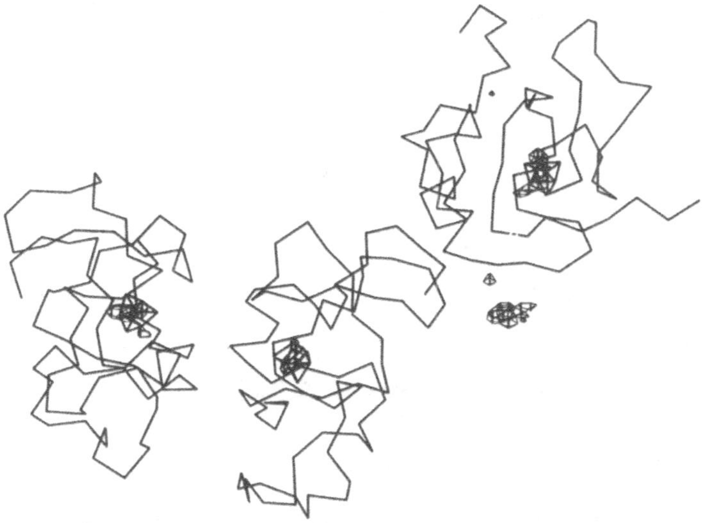

Figure 2. The contents of the asymmetric unit with a phased anomalous difference fourier map showing the electron density for the 4Fe-4S clusters. The electron density outside of the three molecules corresponds to the 4Fe-4S cluster of a molecule that forms a "dimer" with the third molecule in the asymmetric unit across the pure crystallographic two-fold symmetry axis.

2.3) *Preliminary Structure Analysis*
The current model of *C. purpuratum* HiPIP (R-factor of 26.3%) is shown in Figure 3. Approximately 70 % of the protein is now modeled. Like other HiPIPs, it contains little secondary structure and is a series of turns that envelop the 4Fe-4S cluster. Within the asymmetric unit (Figure 2) two molecules form a dimer (rotation between these molecules is 175° with negligible translation along the dyad axis). The distance between the two 4Fe-4S clusters is approximately 13.8 Å, similar to the observed distance between these atoms in other HiPIP dimers and the deduced distance from the EPR data. The third molecule in the asymmetric unit contacts a second HiPIP molecule across the crystallographic 2-fold symmetry axis; the 4Fe-4S clusters in this pair of molecules are likewise separated by ~13.0 Å. Thus the three molecules in asymmetric unit comprise one and one-half dimers. The two types of dimers are not completely superimposable, yet within them the distance between the 4Fe-4S clusters is maintained. EPR data on the *C. vinosum* dimeric form of the protein suggests that HiPIP forms two different types of dimers, with inter cluster distances of 13 and 16 Å.

Figure 3. Overall fold of *C. purpuratum* HiPIP

3. References

1. Hochkoeppler, A., Ciurli, S., Venturoli, G. & Zannoni, D. (1995). FEBS Lett. 357, 70-74.

2. Kerfeld, C.A., Yeates, T.O. & Thornber, J.P. (1993) BBA, 1185, 193-202.

3. Meyer, T.E., Fitch, J., Bartsch, R.G., Tollin, D. & Cusanovich, M.A. (1990). Biochem Biophys. Acta 1017, 118-124.

4. Bertini, I., Gaudemer, A., Luchinat, C. & Piccioli, M. (1993). Biochemistry 32, 12887-12893.

5. Benning, M.W., Meyer, T.E., Rayment, I. & Holden, H.M. (1994). Biochemistry 33, 2476-2483.

6. Rayment, I., Wesenberg, G., Meyer, T.E., Cusanovich, M.A. & Holden, H.M. (1992). J. Mol. Biol. 228, 672-686.

7. Carter, C.W., Kraut, J., Freer, S.T., Alden, R.A., Sieker, L.C., Adman, E. & Jensen, L.H. (1974). Proc. Natl. Acad. Sci. USA 69, 3526-3529.

8. Dunham, W.R., Hagen, W.R., Fee, J.A., Sands, R.H., Dunbar, J.B. & Humblet, C. (1991). BBA 1079, 253-262.

9. Adman, E.T., Mather, M.W. & Fee, J.A. (1993) BBA 1142, 93-98.

ACKNOWLEDGEMENTS

We would like to thank Dr. David Knaff of Texas Tech University for determining the midpoint potential of *C. purpuratum* HiPIP. This work was supported by the National Science Foundation's Postdoctoral Fellowship program in Plant Molecular Biology (C.A.K.); NSF-PYI and USPHS (T.O.Y.)

CLONING AND EXPRESSION OF cDNAs CODING FOR THE SPINACH FERREDOXIN:THIOREDOXIN REDUCTASE

E. Gaymard and P. Schürmann, Laboratoire de Biochimie végétale, Université de Neuchâtel, Rue Emile Argand 11, CH-2007 Neuchâtel, Switzerland

1. Introduction

Ferredoxin:thioredoxin reductase (FTR) is the key enzyme of the ferredoxin/thioredoxin system, the light-dependent enzyme regulatory system in oxygenic photosynthesis (1). This nucleus encoded chloroplast enzyme reduces thioredoxins with electrons supplied by illuminated thylakoids. It is an iron-sulfur protein, composed of two dissimilar subunits of about equal size, one of which contains a [4Fe-4S] cluster and a redox-active disulfide bridge functional in the reduction of thioredoxins. This subunit is therefore called the catalytic subunit. It has been proposed to be rather well conserved since it is readily detected by heterologous antibodies (2,3). The other subunit is of variable size when comparing the FTR from different plant species and has no known catalytic function. We have sequenced both subunits of the spinach FTR on the level of the protein (4,5) and also isolated and characterized a cDNA coding for the catalytic subunit of the enzyme from corn (6). Here we report the isolation and characterization of full-length cDNAs encoding both subunits of FTR from spinach and the expression of a functional recombinant FTR in *E.coli*. The isolation cDNAs coding for spinach FTR, however different from ours, has been reported (7).

2. Materials and Methods

A 159 bp fragment isolated by PCR from a spinach cDNA library (6) and coding for part of the catalytic subunit of FTR was used as a probe to screen the spinach cDNA library (Gift of Prof.W.Gruissem, Berkeley) using standard procedures. Positive plaques were purified and one cDNA clone of 738 bp was isolated and cloned in pBluescript SK+ (Stratagene) and then characterized by sequencing using the chain termination technique with Sequenase (U.S. Biochemical Corp.). In a similar manner, we isolated a 286bp fragment coding for part of the variable subunit of FTR. The cDNA library was screened and a clone of 804bp was isolated and characterized. Using PCR we have made a di-cistronic construction carrying, in phase and on the same DNA strand, the coding parts of the genes for both subunits. The construction (pET(FTR)) cloned in PET-3a-d expression vector (8) was used to transform *E.coli* BL21(DE3). Transformed cells were grown in LB medium to a OD_{600nm} of 0.8 and expression induced with 2mM isopropyl thio-β-D-galactoside (IPTG) for 2.5 hours.
To test FTR expression bacterial extracts were prepared for SDS-PAGE (9): cultured cells were resuspended in water (0.1 $OD_{600nm}/4\mu l$) and lysed by freeze-thawing (-70°C, 15 min.). One volume of sample buffer was added and the lysate vigourosly vortexed for 5 min. Proteins were separated electrophoretically on a 15% gel and probed after membrane transfer with antibodies raised against total FTR and against each subunit. Secondary antibodies

P. Mathis (ed.), Photosynthesis: from Light to Biosphere, Vol. II, 761–764.

conjugated to horse-radish peroxydase were used together with H_2O_2 and 4-chloro-1-naphtol as chromogenic substrate to visualize the FTR .
Protein samples for testing FTR activity were prepared as follows: 200 ml of transformed BL21(DE3) cells grown as described above were harvested by centrifugation. After resuspension in 3 ml of 100 mM NaKPi pH 7.0 containing 0.1 mM phenylmethylsulfonyl fluoride, 14 mM β-mercaptoethanol they were lysed using lysozyme or sonication. Lysates were centrifuged and 10 µl supernatant tested for FTR activity as described elsewhere (10).

3. Results and Discussion

Figure 1 shows the nucleotide sequence and the translation of the entire cDNA coding for a protein of 174 residues. A comparison with the known protein sequence (4) indicates that residues 63 to 174 represent the mature variable subunit, whereas the first 62 residues apparently constitute the transit peptide which displays some typical features of chloroplast transit peptides (11). It starts with Met-Thr-Thr-, a sequence reported in ferredoxin:NADP reductase from spinach (12). The N-terminal part contains no Pro nor Gly, the central region is rich in Ser (30%) with a total Ser content of 20%. The C-terminal 10 amino acids exhibit a high probability for β-sheet formation. The cleavage site sequence, Ile-Cys-Cys↓Glu, follows the proposed consensus.

```
             .         .         .         .         .         .
  1 ggaattcggcacgaggaacaaagaaagtaaaattatctgagacacagaaatccagtacct  60
 61 cttcacttcttctattccggataggaaaatcctcatctccatctcattattctcatctaa 120
121 aacaaatctaatctaaactcctccgatgacaacaggtgtggcagtaatgtcatcagcaac 180
                              M  T  T  G  V  A  V  M  S  S  A  T
181 agcagcatcaaccgcaaccgcaacggcggcggcgacggcgagaataccactcttcctaag 240
     A  A  S  T  A  T  A  T  A  A  A  T  A  R  I  P  L  F  L  S
241 ccgcaacaactcatcggccaccgtctgcagcaccctgaggtgcagaacaataacacgaac 300
     R  N  N  S  S  A  T  V  C  S  T  L  R  C  R  T  I  T  R  T
301 aagaaccagagccagactagcaatatgctgtgaagtagctttgaaatccgattcttcaac 360
     R  T  R  A  R  L  A  I  C  C  E  V  A  L  K  S  D  S  S  T
361 cgggttcgattcatcatcatcatcacctccagaagaagacgaggaattgaagaagaatct 420
     G  F  D  S  S  S  S  S  P  P  E  E  D  E  E  L  K  K  N  L
421 ggagaaagttggatgcaaggttaaggtgaaatcccctcttaaagtctaccatgttcctaa 480
     E  K  V  G  C  K  V  K  V  K  S  P  L  K  V  Y  H  V  P  K
481 attacctgaagttgaattaaccccagatatggttggggttattaagcagtatgttggatt 540
     L  P  E  V  E  L  T  P  D  M  V  G  V  I  K  Q  Y  V  G  F
541 ctggaaagggaaatacatttcccctaattatcctttcaaagttgagtaccgcatcgatgt 600
     W  K  G  K  Y  I  S  P  N  Y  P  F  K  V  E  Y  R  I  D  V
601 ccctgaccgtggttctgtcaaattagttgttcatcttaaagaagaagaattcgaaatcat 660
     P  D  R  G  S  V  K  L  V  V  H  L  K  E  E  E  F  E  I  I
661 agcagagtagaagtctaattatttctattttattggtattgatttatgtatattgtct 720
     A  E  *
721 ttcttcttccttgttgtctatacatctgattactatggcagattggcaggatgtatcaaa 780
781 aaaaaaaaaaaaaaaaaactcgag 804
```

Figure 1. Nucleotide sequence and deduced protein sequence of the cDNA clone coding for the precursor peptide of spinach FTR subunit A. The putative transit peptide is given in italics.

Figure 2 shows the nucleotide sequence and the translation of the entire cDNA which codes for a protein of 144 residues. Residues 31 to 144 correspond to the mature catalytic subunit (5) and the first 31 residues apparently constitute the transit peptide, which again displays some typical features of a chloroplast transit peptides. It starts with Met-Arg-Ala, a sequence sofar never found in spinach transit peptides, but reported for *Chlamydomonas reinhardii* pre-apoplastocyanin (13). The 10 N-terminal residues contain one charge, but no Pro nor Gly. The central part is rich in Ser (40%), the total Ser content being 20%. The C-terminal 10 amino acids exhibit a high probability for beta sheet formation, but contrary to the consensus there are 4 charged amino acids (Arg) present. The cleavage site Ile-Arg-Ala-Gln↓Ala again follows closely the proposed consensus (Val/Ile)-X-(Ala/Cys)↓Ala (14). The translated sequences of both subunits match exactly the protein sequences.

```
              .         .         .         .         .         .
  1 gaattcggcacgagatcaatttacaacaaaattaatataccatgaaagctcttcaagc  60
                                               M  K  A  L  Q  A
 61 ttcaatagcttacagcttccccatttcttctcctgcagcttctcccagacgtttctcccg 120
     S  I  A  Y  S  F  P  I  S  S  P  A  A  S  P  R  R  F  S  R
121 tgtaattcgtgctcaagcggatccttctgacaaatctatggaagtaatgaggaaatttc 180
     V  I  R  A  Q  A  D  P  S  D  K  S  M  E  V  M  R  K  F  S
181 tgagcagttctgtcgtaagtcagatacatactttgtgttgataaaagtgtcactgccgt 240
     E  Q  F  C  R  K  S  D  T  Y  F  C  V  D  K  S  V  T  A  V
241 tgttatcaagggattagcagatcacagggacacattaggtgcacccctttgcccatgtcg 300
     V  I  K  G  L  A  D  H  R  D  T  L  G  A  P  L  C  P  C  R
301 gcattatgatgacaaagaagctgaagcaaaacagggttttggaattgcccatgtgtgcc 360
     H  Y  D  D  K  E  A  E  A  K  Q  G  F  W  N  C  P  C  V  P
361 aatgagggagaggaaggagtgccactgtatgctgtttctgacccctgacaatgatttgc 420
     M  R  E  R  K  E  C  H  C  M  L. F  L  T  P  D  N  D  F  A
421 tggcaaggagcagactatcacattggatgaaattcgagaagttacatcaaatatgtaaaa 480
     G  K  E  Q  T  I  T  L  D  E  I  R  E  V  T  S  N  M  *
481 ttgtgatctgaagcctcttatgcgtagtaacagtaggcctgtacattacgtactatcacg 540
541 aggatgccaataactgttgtactgttatatgttgtatatactcttattctggagagcttt 600
601 gtcatctctaactcggctttatcaaggtatatagccatactcgtatctgagtggtgtatt 660
661 tggttttatgatactccgtatcatataatactgagtgttaaagaaagaaaaggtcaagaa 720
721 tgaagtactacaaaaaaa 738
```

Figure 2. Nucleotide sequence and deduced protein sequence of the cDNA clone coding for the precursor peptide of spinach FTR catalytic subunit. The putative transit peptide is given in italics.

Figure 3 shows by Western blot analysis of bacterial extracts the expression of recombinant FTR. These results clearly demonstrate that both subunits are expressed and recognized by the specific antibodies. In order to verify that functional recombinant FTR has been produced, bacterial extracts were tested for their capacity to reductively activate FBPase in the ferredoxin/thioredoxin system. Whereas in a control extract, from bacteria transformed with the expression vector carrying no FTR construct, there was background activity (4 nmol substrate hydrolized/min), the extracts from bacteria transformed with the FTR construct displayed very good activity (414 nmol substrate hydrolized/min). After partial purification the recombinant protein reveals the typical spectral properties known from the native FTR.

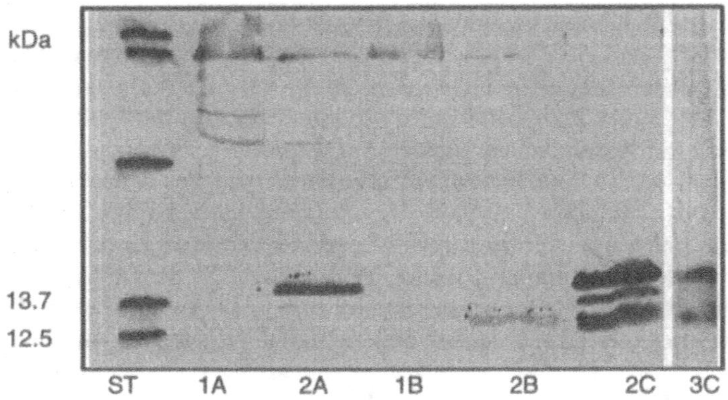

Figure 3. Western blot analysis of baterial extracts after expression of recombinant FTR. ST standard proteins; 1: extract from bacteria transformed with pET; 2: extract from bacteria transformed with pET(FTR); 3: native spinach FTR. The following antibodies were used: A - anti variable subunit, B - anti catalytic subunit, C - anti holoprotein.

References

1 Buchanan, B.B. (1992) Photosynth. Res. 33, 147-162
2 Droux, M., Jacquot, J.-P., Miginiac-Maslow, M., Gadal, P., Huet, J.C., Crawford, N.A., Yee, B.C. and Buchanan, B.B. (1987) Arch. Biochem. Biophys. 252, 426-439
3 Huppe, H.C., Lamotte-Guéry, F.d., Jacquot, J.-P. and Buchanan, B.B. (1990) Planta 180, 341-351
4 Iwadate, H., Yano, K., Kamo, M., Gardet-Salvi, L., Schürmann, P. and Tsugita, A. (1994) Eur. J. Biochem. 223, 465-471
5 Chow, L.-P., Iwadate, H., Yano, K., Kamo, M., Tsugita, A., Gardet-Salvi, L., Stritt-Etter, A.-L. and Schürmann, P. (1995) Eur. J. Biochem. 231, 149-156
6 Marc-Martin, S., Spielmann, A., Stutz, E. and Schürmann, P. (1993) Biochim. Biophys. Acta Bio-Energetics 1183, 207-209
7 Falkenstein, E., von Schaewen, A. and Scheibe, R. (1994) Biochim. Biophys. Acta Bio-Energetics 1185, 252-254
8 Studier, F.W., Rosenberg, A.H., Dunn, J.J. and Dubendorff, J.W. (1990) in Methods in Enzymology, Vol.185 pp.60-89, Academic Press, London
9 Laemmli, U.K. (1970) Nature 227, 680-685
10 Schürmann, P., Stritt-Etter, A.-L. and Li, J. (1995) Photosynth. Res. (In Press)
11 Von Heijne, G., Steppuhn, J. and Herrmann, R.G. (1989) Eur. J. Biochem. 180, 535-545
12 Jansen, T., Reiländer, H., Steppuhn, J. and Herrmann, R.G. (1988) Curr. Genet. 13, 517-522
13 Merchant, S., Hill, K., Kim, J.H., Thompson, J., Zaitlin, D. and Bogorad, L. (1990) J. Biol. Chem. 265, 12372-12379
14 Gavel, Y. and Von Heijne, G. (1990) FEBS Lett. 261, 455-458

The Role of PSI-D in Fast Phase Reduction of Soluble Ferredoxin

Jonathan Hanley, Pierre Sétif, Hervé Bottin and Bernard Lagoutte

Introduction

PSI is a multi-subunit pigment-protein complex of approximately 300kDa which functions as a light driven plastocyanin-ferredoxin oxido-reductase. The three dimensional X-ray crystal structure of PSI indicates that it is mostly embedded in the photosynthetic membrane with several peripheral subunits protruding into the aqueous environment (Krauss *et al.* 1993). Two of these subunits, PSI-D and PSI-E, have been proposed previously as candidates for the binding and orientation of soluble ferredoxin. In the present work we have used mutagenesis to delete PSI-D and have characterised the resultant PSI-D minus mutant using flash spectroscopy and electron paramagnetic resonance spectroscopy.

Experimental Procedures

Plasmid Construction.

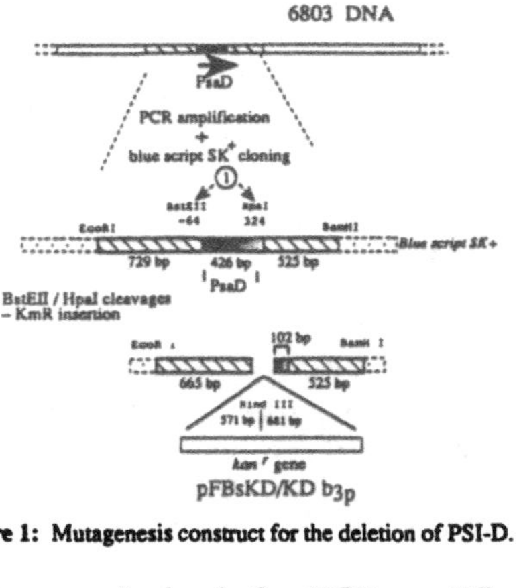

Figure 1: Mutagenesis construct for the deletion of PSI-D.

A DNA fragment of 1680bp was obtained after PCR amplification based on the previously published flanking sequences of the *psa*D gene in *Synechocystis* 6803 (). Unique restriction sites for EcoRI and BamHI, respectively, were introduced at either end of the DNA fragment using the PCR primers. This allowed easy subcloning of the

P. Mathis (ed.), Photosynthesis: from Light to Biosphere, Vol. II, 765–768.

fragment into a Blue script SK$^+$ plasmid (Stratagene) (fig. 1). The cloned *psa*D gene was completely sequenced including approximately 200bp of each flanking sequence. The Bluescript construct, labelled BSpsaD, was then further engineered in order to gain a plasmid aimed at suppressing the *psa*D gene in the Cyanobacterium. Firstly it was restricted at the unique Bste II and Hpa I sites then the the Hinc II Km^r cassette from transposon Tn 903 (Pharmacia) was then blunt end ligated into it. The resulting plasmid, labelled pFBsKD, was used to delete *psa*D from the *Synechocystis* genome (fig. 1). Transformant selection started by establishing an antibiotic gradient through the agar. Transformed cells were progressively selected as single colonies on BG11 agar plates containing 5 mM glucose and kanamycin to a final concentration of 50µg/ml.The selection procedure was carried out under very low light conditions.
Biological Samples. Ferredoxin was isolated from *Synechocystis* sp. PCC 6803 and purified according to Bottin and Lagoutte (1992). Thylakoid membranes (1mg Chl/ml) were obtained by French pressing followed by solubilisation with 1% (w/v) ß-dodecyl maltoside (Rögner et al. ,1990). Flash absorption spectroscopy was carried out according to Sétif and Bottin (1994)

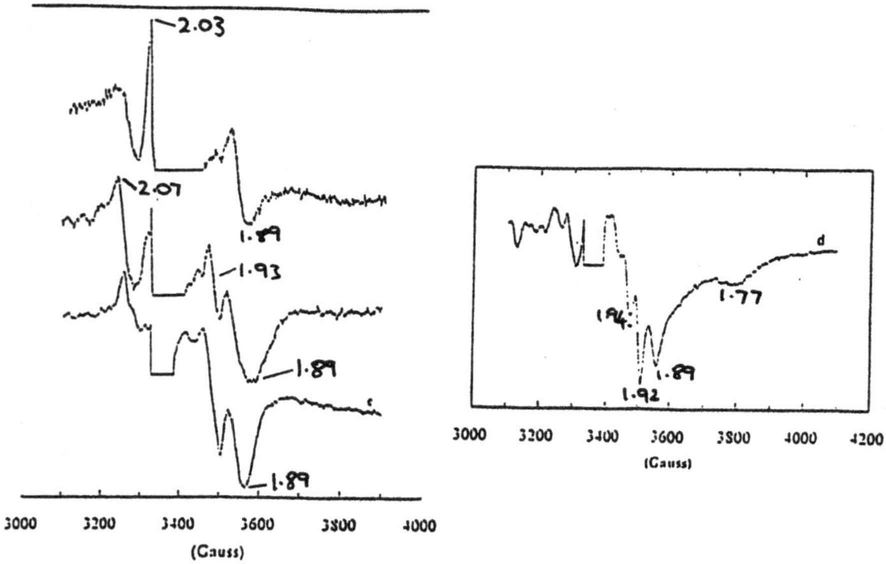

Figure 2: EPR spectra of the PSI-D deleted mutant. A) ascorbate/DCPIP, pH 8.0, 20 min dark adaptation frozen in darkness. B) As above, illuminated at 15K. C) ascorbate/DCPIP, pH 8.0, illuminated at room temperature and frozen under illumination. D) Dithionite pH 10.0, illuminated at room temperature and frozen under illumination. All samples were at 1 mg/ml chlorophyll.

EPR. Low temperature EPR measurements were performed 15 K with a Bruker ESR 200 X-band spectrometer. Samples were prepared at pH 8 (50 mM Tricine/NaOH) in the presence of 5 mM sodium ascorbate and 30 mM DPIP. Samples were then either illuminated at room temperature for 1 minute before freezing under illumination or dark adapted for 20 minutes and frozen in the dark. Dark adapted samples were illuminated at 15K. Samples were also prepared at pH 10.0 in the presence of dithionite then illuminated at room temperature before freezing under illumination .

Results

Molecular biology. The deletion of the *psa*D was monitored by Southern blot analysis and in addition, western blot analysis confirmed the absence of any immunologically related PSI-D polypeptide. Western probes using antibodies raised against PSI-C and PSI-E demonstrated that these two polypeptides are incorporated into the membrane system in the absence of PSI-D..

Electron Paramagnetic Resonance Spectroscopy (EPR) Figure 2 clearly shows that in the PSI-D deleted mutant iron sulphur centres are still present, however, they are not in a native conformation . In the dark adapted sample there are peaks at g=2.03 and 1.89 (fig. 2a). Under these conditions one would normally expect the 4Fe-4S centres to be oxidised and thus EPR silent. The peak at g=2.03 is lost on illumination of the sample(fig. 2b). Illumination at 15K or room temperature results in very similar spectra with peaks at g=2.07, 1.93, and 1.89 (fig 2b and 2c), this is consistent with the reduction of Fe-S_B alone. At pH 10.0 in the presence of dithionite with illumination at room temperature and freezing under illumination results in a spectra with peaks at g=1.94, 1.92, 1.89 and 1.77 consistent with the reduction of Fe-S_X, Fe-S_A and Fe-S_B, although the relative proportions of the peaks are altered as compared to the wild type (fig. 2d).

Flash-absorption spectroscopy In isolated PS-I, light excitation at room temperature leads to charge separation between $P700^+$ and the terminal acceptor either Fe-S_A or Fe-S_B. This is followed by a charge recombination to $P700^+$ from one of these two centres which can be observed by monitoring the decay of $P700^+$ at 820 nm. The addition of soluble ferredoxin leads to a much slower decay of $P700^+$ as electron transfer proceeds from $(F_A, F_B)^-$ to ferredoxin and P700+ is re-reduced much more slowly by DCPIP(fig. 3a). In the PSI-D deleted mutant, in the absence of soluble ferredoxin, there are two decay components of $P700^+$ with a $t_{1/2}$ of 30 ms and 1ms, respectively (fig. 3b). The 1ms decay can most probably be ascribed to recombination between Fe-SX^- and $P700^+$. In the presence of ferredoxin the decay of $P700^+$ becomes much slower indicating, in the absence of PSI-D, PS-I still retains some ability to reduce soluble ferredoxin. Sétif and Bottin (1994) showed the presence of three exponential components of $t_{1/2}$ ≈ 500 ns, 20 μs and 100 μs during the transfer of electrons between the bound Fe-$S_{(A/B)}$ and ferredoxin in wild type PS-I from *Synechocystis* 6803. Figure 3c shows that in the PSI-D deleted mutant all of the fast first order phases are lost. However, upon reconstitution with previously isolated and purified PSI-D the fast first order phases return.

Discussion

In the work presented here we show the effect of deletion of PSI-D on the binding of PSI-C and PSI-E to the core complex and indicate the role of PSI-D in the first order reduction of soluble ferredoxin. Flash absorption studies of $P700^+$ measured at 820 nm show that the PSI-D deleted PS-I that there is some degradation of Fe-$S_{(A/B)}$. This is confirmed by low temperature EPR measurements which show the appearance of a 3Fe-4S type spectrum. The appearance of an 3Fe-4S signal indicates degradation of the 4Fe-4S centres but PSI-C is still stably bound as it remains associated to the core complex through out the lengthy solublisation of the complex with ß-DM. Illumination of Ascorbate/DCPIP samples at cryogenic temperatures results in EPR signals attributed to Fe-S_B alone. This is unusual as under these conditions one would normally expect a single electron transfer event from P700 to Fe-S_A. Illumination of these samples at room temperature and freezing under illumination also only gives the reduction of Fe-S_B. Illumination of samples at pH 10.0 in the presence of dithionite results in the reduction of Fe-S_X, Fe-S_A and Fe-S_B, although the relative amplitudes of the peaks is altered as compared to the wild type. It is unusual to require dithionite at pH 10.0 in order to reduce Fe-S_A, this suggests that the redox potential of Fe-S_A has altered in the PSI-D deleted mutant, possibly reflecting a change in its environment.

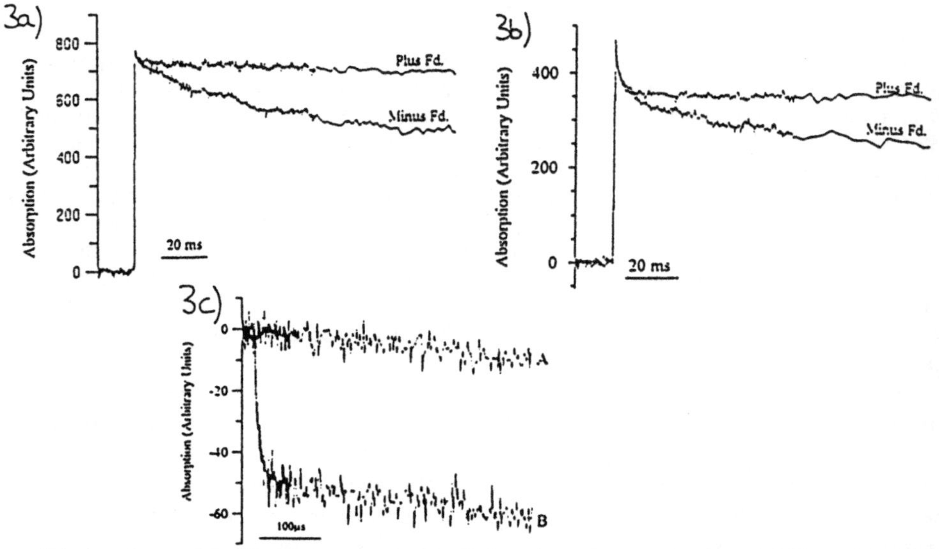

Figure 3a and b: Flash-induced absorption in willd type (a) and PSI-D deleted (b) PSI measured at 820 nm with or without soluble ferredoxin. The PSI reaction centres were suspended in 20 mM Tricine, pH 8, in the presence of 0.03 % b-DM, 30 mM NaCl, 5 mM $MgCl_2$, 6 mM DPIP + 1 mM sodium ascorbate.
Figure 3c: Differences between flash-induced absorption changes measured in the presence and in the absence of soluble ferredoxin at 580 nm. A) PSI-D deleted mutant, B) PSI-D deleted mutannt reconstituted with isolated PSI-D subunit. Experimental conditions as in fig 3a anb b.

Flash absorption studies of $P700^+$ measured at 820 nm in the presence and absence of ferredoxin indicate that following the deletion of PSI-D the PS-I complex retains the ability to reduce ferredoxin. However, a more reliable technique for monitoring the interaction between PS-I and soluble ferredoxin is the direct observation of electron transfer by monitoring change in absorbance at wavelengths between 480 nm and 600 nm . Under these conditions it becomes apparent that although the PSI-D deleted mutant can reduce ferredoxin it is a second order event. In the µs time scale, used for the measurement of first order reduction of ferredoxin by wild type PS-I (see Sétif and Bottin 1994), there is no reduction of ferredoxin. The first direct evidence for the requirement of PSI-D for the first order reduction of ferredoxin comes from PSI-D reconstitution experiments. Upon reconstitution, the reduction kinetics reverted from the second order kinetics to the fast first order kinetics observed in wild type PS-I. This clearly demonstrates that PSI-D is required for ferredoxin reduction in the nanosecand and microsecond time scale.

References

Bottin, H. and Lagoutte, B. (1992) Biochim. Biophys. Acta 1101, 48-56Fromme, P., Krauss, N., Hinrichs, W., Witt, I., Fromme, P., Pritzkow, W., Dauter, Z., Betzel, C., Wilson, K. S., Witt, H. T. & Saenger, W. (1993) Nature 361, 326-330
Rögner, M., Nixon, P. J., & Diner, B. A. (1990) J. Biol. Chem. 265, 6189-6196
Sétif, P. and Bottin, H. (1994) Biochemistry 33, 8495-8504

RECONSTITUTION OF THE 2Fe-2S CENTER INTO OVERPRODUCED RIESKE PROTEIN AND MUTATIONAL ANALYSIS OF INHIBITOR SPECIFICITY IN CYTOCHROME b6 IN CYANOBACTERIA.

Xiaonan Wu[1], Alexandre Tsapin[1], Yoon Shin Cho[1], Tou-Xang Lee[1], Sabine Metzger[3], Linda Genzlinger[1], Beatrice Holton[1], John Whitmarsh[2,3], and Toivo Kallas[1], [1]Dept. of Biology-Microbiology, Univ. Wisconsin, Oshkosh, WI, [2]USDA/ARS and [3]Biophysics, Univ. Illinois, Urbana, IL U.S.A.

1. Introduction

The cytochrome (cyt) b6-f complex in chloroplasts and cyanobacteria catalyzes the rate limiting quinol-oxidation step in photosynthetic electron transport and establishes a proton gradient across the thylakoid membrane [1, 2]. Four major proteins, cytochrome f (one c heme), cytochrome b6 (two b hemes), the Rieske Fe-S (one 2Fe-2S center), and subunit IV, encoded by the *petA, B, C,* and *D* genes, respectively, are required for activity as apparently are additional, smaller subunits [3, 4]. The related cyt b-c1 complex (complex III) occurs in mitochondria and many bacterial groups. To date, the only high-resolution structure for any component of cyt bf/bc1 complexes is that for the lumenal, heme-binding domain of cyt f [5]. Unresolved problems include the nature of binding sites for quinone substrates, pathways for electrons and protons, and the molecular basis for functional differences between the cyt b-c1 and b6-f complexes.

The *pet* genes for the cyt b6f complex have been cloned and sequenced from several cyanobacteria including *Nostoc* PCC 7906, *Synechococcus* PCC 7002, and *Synechocystis* PCC 6803 [ref. in 2]. However, genetic investigations in cyanobacteria are complicated by the apparent requirement of the cyt b6f complex for both photosynthetic and respiratory growth. All attempts at insertional inactivation have led to merodiploid, nonsegregant strains that carry both inactivated and wild-type *pet* gene copies [6, and D. Zarka, R. Alsaadi, and T. Kallas, unpublished]. We are pursuing two strategies to circumvent this limitation. The first entails overproduction of the Rieske subunit in *Escherichia coli* aimed at structural studies of this protein and *in vitro* reconstitution into Rieske-depleted cyt b6f complex. The latter has been accomplished in the case of the native spinach complex [7] and would provide a way to analyze mutant Rieske proteins. The second approach entails introduction of mutations into cyanobacterial *pet* genes that permit assembly of functional but altered cyt b6f complexes. Here we describe *in vitro* reconstitution of the characteristic 2Fe-2S center and g = 1.90 electron paramagnetic resonance (EPR) signal into overproduced Rieske protein and the construction and analysis of site-directed mutations in *petB* with altered sensitivity to inhibitors of quinol-oxidation.

2. Procedures

2.1 *Overproduction of the Nostoc Rieske protein*

Plasmid p3aR [8] contains the entire *Nostoc* PCC 7906 *petC* coding sequence [9] located downstream from the T7 RNA polymerase promoter and ribosome binding site of plasmid pET3a (Novagen, Madison, WI). Gene expression and protein synthesis were induced by addition of 0.4 mM isopropyl thio-β-galactoside (IPTG) to exponential cultures of *E. coli* BL21(DE3) bearing p3aR. Cell suspensions in 50 mM Tris-HCl pH 7.8, 10 mM EDTA, 1 mM PMSF were broken by French press. Pellet fractions (apparently inclusion bodies)

P. Mathis (ed.), Photosynthesis: from Light to Biosphere, Vol. II, 769–772.

from these lysates were washed several times in distilled water and contained predominantly the Rieske protein.

2.2 *Reconstitution of the 2Fe-2S center and EPR analysis*
Inclusion bodies were solubilized in 5.0 M guanidine-HCl, 25 mM MES, pH 6.8 and protein allowed to refold by dilution (1:10) into 50 mM MES pH 6.8, 0.03% β-dodecylmaltoside, 0.5% β-mercaptoethanol (β-ME), 1 mM PMSF, 1 mM amino caproic acid. After dialysis against the same buffer and then 50 mM Tris-HCl pH 8.3, 1% β-ME (or 1 mM DTT), 1 mM octylglucoside (OG), the protein solution (ca. 1-10 mg in 10 ml) was sparged with N_2 gas and anaerobic solutions of $FeCl_3$ and Na_2S (100 mM each) were added dropwise and the solution stirred under N_2 for 1-1.5 h. Protein was precipitated by addition of $(NH_4)_2SO_4$ to 50% of saturation, resuspended in 5-10 ml of 50 mM Tris-HCl pH 8.3, 1% β-ME (or 1 mM DTT), 20 mM OG, and dialyzed against the same buffer containing 1 mM OG. The procedure for reconstitution of Fe-S centers was essentially that of Malkin and Rabinowitz [10] and Parret et al. [11]. Samples were concentrated on Amicon, Centricon-10 (10,000 MWCO) units to 1-3 mg protein/ml and analyzed at liquid He temperature (15-20°K) in a Varian E-4 EPR spectrometer (National Biomedical ESR Center, Medical College of Milwaukee).

2.3 *Site-directed mutagenesis of the Synechococcus PCC 7002 petB gene*
Mutations in *petB* were made on plasmid pA6.2- (a derivative of pALTER-1) by oligonucleotide-directed mutagenesis according to the "Altered-Sites" procedure (Promega, Madison, WI). pA6.2- carries a 4.2 kb *petBD* fragment into which a 2 kb streptomycin/spectinomycin (Sm/Sp) resistance gene has been inserted ca. 1 kb 5' of *petB*. Mutations on pA6.2- (nonreplicating in *Synechococcus*) were introduced into *Synechococcus* by transformation with selection for Sm/Sp resistance. Integration of *petB* mutations and retention of wild-type (WT) alleles were tested by one of two polymerase chain reaction (PCR) methods. Each mutagenic oligonucleotide also created a new restriction site. PCR amplification of the *petB* region from *Synechococcus* transformants followed by restriction digestion represented one method. The second method was that of allele-specific primers [12]. Separate wild-type and mutant allele detection primers (differing only by their 3' base) were synthesized for each *petB* mutation. The WT primer sets would only amplify WT and the mutant primer sets only mutant DNA.

2.4 *Measurements of growth and plastoquinol to cyt f/c6 electron transfer rates*
Synechococcus growth rates were measured in A medium [13] bubbled with 1% CO_2, 39°C, 200 μE m^{-2} s^{-1} cool white fluorescent illumination. Plastoquinol to cyt f/c6 electron transfer rates were measured in intact cells at 39°C by time resolved, light-induced absorbance spectroscopy in a laboratory built instrument [14]. Cells (3-12 μM chl) were suspended in 10 mM NaCl, 5 mM HEPES pH 7.5, 10 mM $NaHCO_3$. Cyt f and c6 (= c553) were oxidized by a 100 ms flash (250 W Xenon lamp filtered by a red CS 2-58 blocking filter, Corning) and their rereduction monitored by following the decrease in O.D. at 557-540 nm. We did not resolve cyt f and b6 separately because transfer between them is rapid relative to the rate-limiting, quinol-oxidation step that we wanted to monitor.

3. Results and Discussion

3.1 *Reconstitution of the 2Fe-2S center and g=1.90 EPR signal into Nostoc Rieske protein*
Figure 1, panel A shows an overloaded gel of *Nostoc* Rieske protein that has been solubilized from inclusion bodies and reconstituted with Fe-S centers as described above. Such preparations contained predominantly Rieske protein (apparent MW ca. 22,000) and some of the additional bands represented aggregates or degradation products of the Rieske as shown in the immunoblot (lane 3) against a specific, Rieske antibody. EPR analysis of such reconstituted Rieske samples revealed the characteristic spectrum associated with the

high potential (ca +300 mV) 2Fe-2S Rieske center (Figure 1, panel B). In addition to the gy = 1.90 and gz = 2.03 peaks, a small depression at ca. g = 1.74 is also evident. This gx signal has been a useful diagnostic feature for cyt bc1 Rieskes but has seldom been detected from cyt b6f Rieskes. These data indicate that the native 2Fe-2S center has been restored *in vitro* to the overproduced Rieske protein and that such preparations should be useful for biochemical and biophysical analyses. Data not shown suggest that ca. 30% of the protein typically becomes reconstituted with Fe-S centers. The overproduced, reconstituted Rieske protein has now been purified by Whatman DE52 chromatography and migrates as a single band on overloaded, denaturing gels (not shown). To characterize the minimum protein domain required for stabilization of the 2Fe-2S center and retention of the unique Rieske redox and spectral properties, we have begun to make deletions from both the N- and C-termini and to overproduce these truncated proteins. Neither the addition to the N-terminus of 19 amino acids bearing a "6-His tag" for affinity purification nor deletion of 18 residues prevented reconstitution nor noticeably altered the Rieske EPR spectrum.

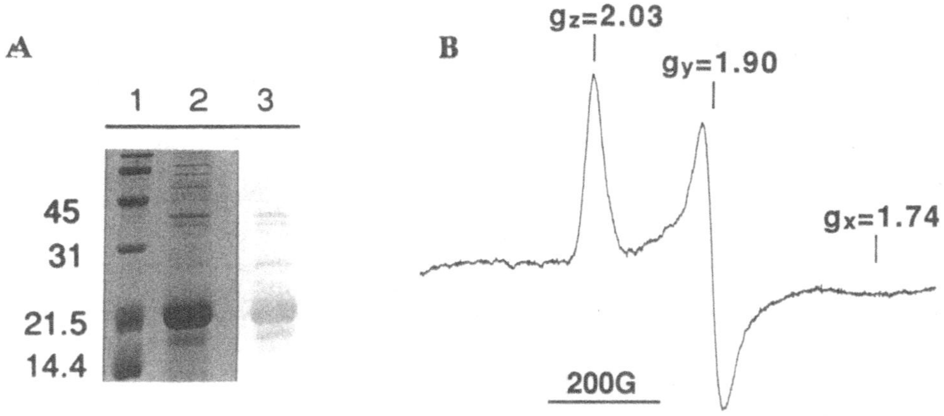

Figure 1. Panel A: Overproduced and reconstituted *Nostoc* Rieske protein. Lanes 1 and 2 show a stained SDS polyacrylamide gel and lane 3 an immunoblot of lane 2 reacted with a specific, *Nostoc* Rieske antibody. Lane 1: molecular weight standards (kDa). Lane 2: ca. 10 µl of protein from a 1.0 ml sample. *Panel B:* EPR spectrum of reconstituted Rieske protein (at 15°K) in 50 mM Tris-HCl pH 8.2, 0.01% Triton X-100, 1 mM DTT.

3.2 *Mutagenesis of petB in Synechococcus and characterization of inhibitor binding sites*
Because inactivating mutations are not allowed in cyanobacterial cyt b6f complexes, we first tested our mutagenesis strategy by constructing a "silent," BstUI mutation (this carries the upstream Sm/Sp tag and creates a new BstUI site in *petB* but does not alter the protein) and an expected lethal mutation, His202Gln, at one of the heme-liganding sites. As expected, the BstUI mutation could be introduced into the *Synechococcus* genome and resulted in segregational loss of all corresponding WT alleles. H202Q could be introduced into the genome but resulted in a merodiploid state with retention of both WT and mutant alleles. We next constructed Asp141Gly, Ser152Ala, Ser152Thr, and Ala147Gly mutations (positions Gly152, Thr163, and Gly158, respectively, in cyt b (bc1 complex) of *Rhodobacter capsulatus*) to test whether any of these could generate myxothiazol (myx) sensitivity in the naturally myx-resistant cyt b6f complex or affect the specificity of other inhibitors. G152D and G158A mutations in *R. capsulatus* resulted in myx and T163A in stigmatellin (stg) resistance [15]. Each of the D141G, S152A, S152T, and A147G

mutations integrated into the genome and replaced the corresponding WT alleles based on the analyses described above. The growth rates of these *Synechococcus* mutants and the BstUI strain in the absence of inhibitors were comparable to wild-type (ca. 3-4 h doubling at 39°C, 200 $\mu E\ m^{-2}\ s^{-1}$) as were the half-times for cyt f/c6 rereduction (ca. 9-12 ms at 39°C). Table I summarizes the results of cyt f/c6 rereduction measurements (I_{50} values) in the presence of myx, stg, and DBMIB. The D141G but not the A147G substitution increased sensitivity to myx as predicted by Daldal et al. [15]. S152A did not confer resistance to stg but, unexpectedly, the substitution of Thr for Ser (both hydroxylated) at position 152 slightly increased sensitivity. Possibly the larger Thr side group improves stg binding. Several of the mutations (S152A, S152T, A147G) produced small but consistent, increased resistance to DBMIB, the most potent and specific inhibitor of cyt b6f complexes. These data demonstrate the feasibility of site-specific mutagenesis of the cyt b6f complex in readily manipulated cyanobacteria and these particular mutations help to define quinol and inhibitor binding sites in this complex.

Table 1. Inhibitor Sensitivity in wild-type and mutant Synechococcus PCC 7002

	Inhibitor conc. (µM) for 50% inhibition of cyt f/c6 rereduction rate (I_{50})					
Inhibitor	wild-type	BstUI	D141G	S152A	S152T	A147G
DBMIB	0.15	0.15	0.15	0.33	0.32	0.50
Myxothiazol	insensitive	insensitive	20	insensitive	insensitive	insensitive
Stigmatellin	4	4	not tested	4	1.7	not tested

4. Acknowledgments

We are grateful to William Widger for a clone of the *Synechococcus petBD* operon and to the National Biomedical ESR Center for use of their facilities. This work was supported by NRI/USDA 94-37306-0802 and UW-O Faculty Development FDR298 awards to TK.

References

1. Cramer, W.A., Furbacher, P.N., Szczepaniak, A. and Tae, G-S. (1991) Curr Top Bioenerg 16, 179-222.
2. Kallas, T. (1994) in The Molecular Biology of Cyanobacteria (Bryant D.A.., ed.) pp.259-317, Kluwer Academic Publishers, Dordrecht, The Netherlands.
3. Pierre, Y. and Popot, J-L. (1993) C. R. Acad. Sci. Paris 316, 1404-1409.
4. Schmidt, C.L. and Malkin, R. (1993) Photosynth. Res. 38, 73-81.
5. Martinez, S.E., Huang, D., Szczepaniak, A., Cramer, W.A. and Smith, J.L. (1994) Structure 2, 95-105.
6. Osiewacz, H.D. (1992) Arch. Microbiol. 157, 336-342.
7. Adam, Z. and Malkin, R. (1987) FEBS Lett. 225, 67-71.
8. Holton, B., Middleton, K., Zarka, D, Alsaadi, R., and Kallas, T. (1992) in Research in Photosynthesis (Murata, N., ed.) Vol II, pp.567-570. Kluwer, Dordrecht.
9. Kallas,T., Spiller,S.,and Malkin,R.(1988) Proc. Natl. Acad. Sci. USA.85, 5794-5798
10. Malkin, R., and Rabinowitz, J.C. (1966) Biochem. Biophys. Res. Commun. 23, 822
11. Parret, K.G., Mehari, T., and Golbeck, J.H. (1990) Biochim. Biophys. Acta 1015, 341-352.
12. Fortina, P. Conant, R., Parrella, T., Rappaport, E., Scanlin, T., Schwartz, E., Robertson, J.M., and Surrey, S. (1992) Mol. Cell. Probes 6, 353-356.
13. Buzby, J.S., Porter, R.D., and Stevens, S.E.Jr. (1985) Science 230, 805-807.
14. Jones, R.W., and Whitmarsh, J. (1988) Biochim. Biophys. Acta 933, 258-268.
15. Daldal, F.,Tokito, M.K., Davidson,E. and Faham, M. (1989) EMBO J. 8, 3951-3961

^{1}H AND ^{15}N NMR SEQUENTIAL ASSIGNMENT, SECONDARY STRUCTURE AND TERTIARY FOLD OF [2Fe-2S] FERREDOXIN FROM *SYNECHOCYSTIS* sp. PCC 6803

Cécile Lelong, Pierre Sétif, Hervé Bottin, François André and Jean-Michel Neumann

CEA, DBCM, SBE and CNRS URA 1290, C.E. Saclay, 91191 Gif sur Yvette Cedex, France

The [2Fe-2S] chloroplast-type ferredoxins are distributed in various oxygenic photosynthetic organisms. They are small proteins, having a molecular mass of about 11 kDa and strongly acidic with highly homologous amino-acid sequences. They function as an electron donor in many metabolic pathways whereas most of them function as an electron acceptor to photosystem-I. Four different structures of cyanobacterial ferredoxins (Tsukihara et al., 1981, 1990; Rypniewski et al., 1991; Jacobson et al., 1993) as well as one from horsetail (Ikemizu et al., 1994) have been determined by X-ray crystallography at high resolution. All of them are very similar with the highest structural homology around the paramagnetic cluster and a high proportion of β sheet in the diamagnetic part of the protein.

Despite their small size, the structures of ferredoxins are difficult to determine by NMR due to the presence of the iron-sulfur cluster, which is paramagnetic at room temperature in any redox state (Fe^{2+}/ Fe^{3+} or Fe^{3+}/ Fe^{3+} for a [2Fe-2S] cluster). Many NMR studies have been performed on the chloroplast-type ferredoxins. Almost complete sequence-specific ^{13}C, ^{15}N and ^{1}H resonance assignments of the [2Fe-2S] vegetative and heterocyst ferredoxins from *Anabaena* 7120 were performed by Oh and coworkers (Oh & Markley, 1990a,b; Oh et al., 1990) and Chae et al. (1994) respectively. The identified secondary structure elements of the vegetative [2Fe-2S] Fd (Oh & Markley, 1990a,b) were in accordance with those found in the X ray structure of the homologous Fd from *Spirulina platensis* (Tsukihara et al., 1981). However, no solution structure has been modeled either for the vegetative nor for the heterocyst [2Fe-2S] Fd.

The ferredoxin of *Synechocystis* sp. PCC 6803 has been characterized and sequenced (96 amino acids including 6 cysteine residues) (Bottin & Lagoutte, 1992).

P. Mathis (ed.), Photosynthesis: from Light to Biosphere, Vol. II, 773–775.

Among the 6 cysteine residues, 4 are involved in the [2Fe-2S] cluster binding; the 2 others were not detected as free thiols, suggesting the presence of a disulfide bridge. This is consistent with the behavior of the electrophoretic mobility, which is largely modified in conditions appropriate for reducing a disulfide bridge (apparent M_r from 2.5 to 16.5 kDa) (Creighton, 1989). The preceding observations, concerning the putative disulfide bridge, stimulated the study of the solution structure of the *Synechocystis* 6803 ferredoxin by NMR.

Sequence-specific ^{1}H and ^{15}N assignment has been performed for the amino acid residues far from the paramagnetic cluster (distance higher than 8 Å). Inter-residue NOE constraints have allowed the identification of several secondary structure elements : one β sheet composed of four β strands, one α helix and two α helix turns.

The analysis of inter-residue NOEs suggests the existence of a disulfide bridge between the cysteine residues 18 and 85. Such a disulfide bridge has never been observed in plant-type ferredoxins. Structure modeling using the XPLOR program was performed with or without assuming the existence of a disulfide bridge. As a result, two structure families were obtained : both families show the same tertiry fold.

Due to the lack of NOE connectivities resulting from the paramagnetic effect from the [2Fe-2S] cluster, the structures were not well resolved in the region surrounding the [2Fe-2S] cluster, at both extremities of the α helix and the C and N terminus segments. In contrast, when taken separately, the β sheet and the α helix were well defined .

The tertiary fold of *Synechocystis* 6803 [2Fe-2S] ferredoxin shows the same structural arrangement as compared to that of the chloroplast-type ferredoxin extracted from *Anabaena variabilis*, *Aphanothece sacrum*. or *Anabaena* 7120 determined by X-ray cristallography (in particular, the main secondary structure elements). The major difference between *Synechocystis* 6803 ferredoxin and the other chloroplast-type ferredoxin structures is the probable presence of a disulfide bridge between Cys 18 and Cys 85. Such a disulfide bridge has never been described in chloroplast-type ferredoxin containing six cysteine residues. This disulfide bridge could be responsible for the distortions in one extremity of the β sheet. The β strands containing the cysteine residues are closer in the present model of *Synechocystis* 6803 ferredoxin than in the structure from *Anabaena variabilis* or *Anabaena* 7120 ferredoxins.

→ The present study proposes a tertiary fold of *Synechocystis* sp. PCC 6803 ferredoxin including a disulfide bridge between the cysteine residues 18 and 85. This work is the first report of a solution structure model of a plant-type [2Fe-2S] ferredoxin.

References

Bottin, H. & Lagoutte, B. (1992) Biochim. Biophys. Acta 1101, 48-56

Brünger, A.T., 1992, XPLOR 3.0 version, Yale University press, New Haven, CT

Chae, Y.K., Abildgaard, F., Moobery, E.S. & Markley, J.L. (1994) Biochemistry 3: 3287-3295

Chae, Y.K. & Markley, J.L. (1995) Biochemistry 34, 188-193

Creighton T.E. (1989) in Protein Structure, IRL Press, Oxford University Press, pp 15: 167

Ikemizu, S., Bando, M., Sato, T., Morimoto, Y. & Tsukihara, T. (1994) Acta cryst., D(: 167-174.

Jacobson, B.L., Chae, Y.K., Markley, J.L., Rayment, I. & Holden, H.M. (199: Biochemistry 32, 6788-6793

Oh, B.-H. & Markley J.L. (1990) Biochemistry 29, 3993-4004

Oh, B.-H. & Markley, J.L. (1990) Biochemistry 29, 4012-4017

Oh, B.-H., Moobery, E.S. & Markley, J.L. (1990) Biochemistry, 29, 4004-4011

Rypniewski, W.R., Breiter, M.M., Wesenberg, G., Oh, B.-H., Markley, J.L., Raymen I. & Holden, H.M. (1991) Biochemistry 30, 4126-4131

Tsukihara, T., Fukuyama, K., Nakamura, M., Katsube, Y., Tanaka, N., Kakudo, M Wada, K., Hase, T. & Matsubara, H. (1981) J. Biochem., 90, 1763-1773

Tsukihara, T., Fukuyama, K., Mizushima, M., Harioka, T., Kusunoki, M., Katsube, Y Hase, T. & Matsubara, H. (1990) J. Mol. Biol. 216, 399-410

INTERACTION OF THIOREDOXIN WITH CF_0CF_1

O. Schwarz, I. Ditmer-Knappmann and H. Strotmann, Institut für Biochemie der Pflanzen, Heinrich-Heine-Universität Düsseldorf, 40225 Düsseldorf, FRG

1. Introduction

Synthesis and hydrolysis of ATP in chloroplasts is catalyzed by a reversible proton-translocating ATPase (CF_0CF_1) which is a constituent of the thylakoid membrane. At high proton motive force induced by illumination with strong light, ATP formation is promoted, while in darkness or weak light ATP hydrolysis is thermodynamically favoured. However extensive loss of ATP as consequence of natural changes of light intensity is prevented by multiple regulation mechanisms.

Besides serving as the driving force of ATP synthesis, the proton gradient is necessary to convey CF_0CF_1 from an inactive to an active state [1]. This process which may be accompanied by the release of a tightly bound nucleotide from one of the three catalytic sites, seems to be related with a change of the conformation of the CF_1 subcomplex [2]. Furthermore, the activated enzyme may be modulated by reduction of a disulfide bond in the γ-subunit of CF_1 ("thiol modulation") [3]. The physiological reductand is a chloroplast thioredoxin [4]. In this paper we have investigated the specificity of the H^+-ATPase for the two chloroplast thioredoxins Td-m and Td-f.

With the thiol modulated CF_0CF_1 the rate of phosphorylation of thylakoids at non-saturating ΔpH is higher than with demodulated CF_0CF_1. This result was interpreted to indicate a lower energy requirement for enzyme activation of the thiol modulated compared with the non-modulated enzyme form [1]. On the other hand, the dissociation and Michaelis constants for the substrates ADP and phosphate are independent of the redox state of the H^+-ATPase [5]. Upon membrane deenergization the oxidized ("demodulated") enzyme is immediately deactivated whereas the reduced ("modulated") enzyme retains activity for several minutes [6].

In the present study we have analyzed the kinetics of thiol modulation as function of thioredoxin concentration and of the extent of the proton gradient. Moreover we have investigated the kinetics of ATPase deactivation at membrane deenergization following a modulation reaction carried out with saturating thioredoxin concentration, but different ΔpH. The results suggest the participation of more than the four known enzyme species ${}^{ox}E_{inactive}$, ${}^{ox}E_{active}$, ${}^{red}E_{inactive}$ and ${}^{red}E_{active}$.

2. Experimental

Thylakoids were isolated from freshly harvested spinach leaves as described in ref [7]. The experiments were conducted at controlled proton gradient by employing the ΔpH clamp technique [5]. The total reaction volume was 2.5 ml, the temperature was 20°C. The reaction medium contained 25 mM tricine buffer (pH 8.0), 50 mM KCl, 5 mM

P. Mathis (ed.), Photosynthesis: from Light to Biosphere, Vol. II, 777–780.

$MgCl_2$, 50 µM phenazine methosulfate, 50 nM valinomycin, 2 mM phosphoenol pyruvate, 20 U/ml salt-free pyruvate kinase and thylakoids corresponding to 25 µg chlorophyll/ml. The prechosen proton gradient was attained after less then 20 seconds of illumination with the activated controller device.
The kinetics of thiol modulation were analyzed by means of the generation of ATP hydrolyzing activity. To this end thioredoxin at indicated concentrations together with 1 mM DTT was added 20 seconds after the start of illumination. Then a mixture of 1 mM [γ-^{32}P]ATP, 5 mM phosphate and 0.5 µM nigericin was injected after different times and the light was switched off simultanously. Samples were taken 6,12,18 and 24 s later, deproteinized by 0.5 M $HClO_4$ and analyzed for [^{32}P]P_i [8].
To study the kinetics of deactivation, the thylakoids were illuminated to yield the indicated proton gradients in the presence of 3 µM thioredoxin and 1 mM DTT. After 120 s of illumination the ΔpH was discharged by switching off the light together with the addition of 0.5 µM nigericin. The ATP hydrolyzing activity was measured as before by injection of 1 mM [γ-^{32}P]ATP at different times. The velocities of ATP hydrolysis are relative measures of the activated reduced CF_0CF_1 [1].

3. Results and Discussion

3.1 Thioredoxin specificity of the thylakoid ATPase for thiol modulation

We investigated the specificity of the H^+-ATPase for different thioredoxins. To this end we measured the ATP hydrolyzing activity of thylakoids illuminated for two minutes in the presence of varying concentrations of Td-m, Td-f and E. coli thioredoxin (Td-b). Table 1 shows that the same concentrations of Td-f and Td-b cause equal efficiencies of modulation, but the efficiency of Td-m is much lower. About tenfold higher concentrations of Td-m are necessary to obtain the same ATPase activity. Galmiche et al. [9] measuring ATPase activities of ε-less CF_1, also found a higher efficiency of E.coli thioredoxin in comparison with Td-m. These results suggest that Td-f rather then Td-m is the physiological dithiol reductant and that Td-f can be fully replaced by Td-b.

Table 1:
Specificity of thiol modulation of CF_0CF_1 for thioredoxins

Td concentration [µM]	ATPase activity [µmol P_i/(mg chl h)]		
	Td-m	*Td-f*	*Td-b*
0.09	33	183	163
0.19	34	259	252
0.28	55	327	332

3.2 Kinetics of thiol modulation

The kinetics of thiol modulation were followed at light saturation as function of the concentration of reduced Td-b (Figure 1). Since the proton gradient was built up, i.e. the maximal number of enzyme molecules was already in the active oxidized state before thioredoxin was added, the evolution of ATPase activity is limited by the reduction process in the γ-subunit of CF_1 and the finally attained activity represents the maximal number of active enzyme molecules. Under the chosen conditions (presence of sufficient DTT), the reduction is practically irreversible. The initial velocities of the increase of the ATP hydrolyzing activity at the different Td concentrations then permit the calculation of the second order rate constant for reduction as 0.028 $\mu M^{-1}s^{-1}$. Appropriate measurements at limiting ΔpH show a lower initial increase of ATPase activity indicating a lower starting concentration of active oxidized enzyme. But the finally attained ATPase activity is also lower suggesting that the deactivation of reduced active to reduced inactive enzyme may play a progressive role with decreasing proton gradient (Table 2).

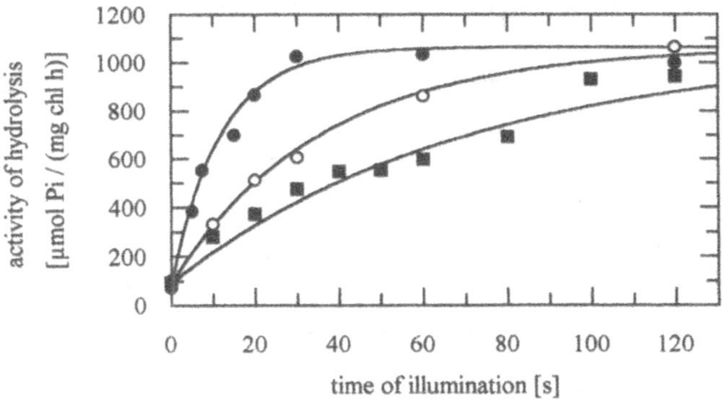

Figure 1: Time courses of thiol modulation at saturating ΔpH with 0.5µM Td-b (■), 1µM Td-b (o) and 3µM Td-b (●).

Table 2:
Parameters of thiol modulation at different extents of membrane energization.

membrane energization	final activity of ATP hydrolysis [µmol P_i/ (mg chl h)]	initial increase of ATPase activity [nmol P_i / (mg chl s^2 µMTd)]
saturating	1064	8.3
ΔpH = 2.7	478	2.1
ΔpH = 2.4	140	1.4

3.3 Kinetics of ATPase deactivation

Figure 2 shows the decay of the ATP hydrolyzing activity upon rapid relaxation of the proton gradient. The preceding thiol modulation was conducted with 3 µM reduced Td-b at either light saturation or at a lower proton gradient (ΔpH=2.7). When thiol modulation was carried out at ΔpH=2.7, deactivation of the ATPase is a monophasic process with a half time of about 270 s. When enzyme reduction was carried out at light saturation, however, the decay of ATPase activity is clearly biphasic. A rapid deactivation with a half time of 6 s is followed by a slow decay phase of 250 s half time. A biphasic decay of ATPase activity at light-dark transition was also observed after thiol modulation with the artificial reductand DTT [10]. The results suggest that at least two active reduced enzyme forms are involved, an unstable form $^{red}E^*_{active}$ which is formed and maintained at high proton gradients, and a metastable form $^{red}E_{active}$ formed at ΔpH ≤ 2.7 which decays only slowly at membrane deenergization. The finding that only the amplitude of the fast phase changes with ΔpH and that the decay time of the slow phase is independent of the previously employed proton gradient suggests that $^{red}E_{active}$ is a less active intermediate enzyme form which is passed through during the decay of the highly active form $^{red}E^*_{active}$. The complete sequence of reaction steps may be summarized as follows:

$$^{ox}E_{inactive} \underset{}{\overset{\Delta pH}{\rightleftharpoons}} {}^{ox}E_{active} \overset{Td}{\rightleftharpoons} {}^{red}E^*_{active} \underset{\Delta pH}{\rightleftharpoons} {}^{red}E_{active} \underset{\Delta pH}{\rightleftharpoons} {}^{red}E_{inactive}$$

At low proton gradient the first equilibrium may be on the side of $^{ox}E_{inactive}$ which could explain the slower conversion of $^{ox}E_{active}$ to $^{red}E^*_{active}$. Possibly another intermediate, but inactive form is involved in this transition. The form $^{red}E^*_{active}$ is converted to $^{red}E_{active}$ which therefore accumulates at low ΔpH.

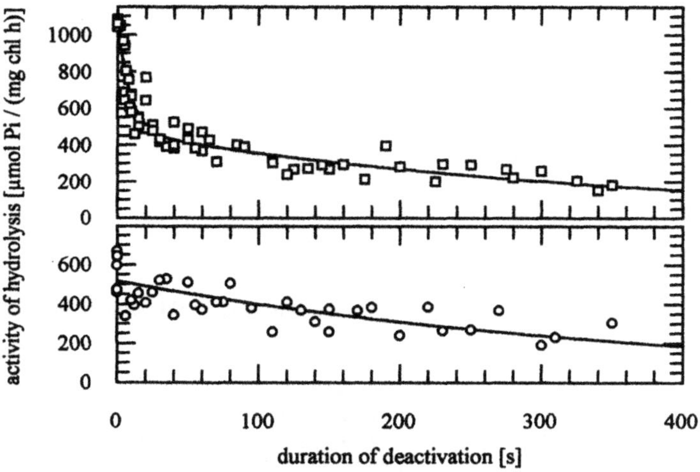

Figure 2: Kinetics of deactivation of the ATPase after thiol modulation at saturating energization (□) and ΔpH = 2.7 (○).

Acknowledgement
This work was supported by the Deutsche Forschungsgemeinschaft (Sonderforschungsbereich 189)

References

[1] Junesch, U. and Gräber, P. (1987) Biochim. Biophys. Acta 893, 275-288
[2] Lohse, D. and Strotmann, H. (1989) Biochim. Biophys. Acta 976, 94-101
[3] Schumann, J., Richter, M.L. and McCarty, R.E. (1985) J. Biol. Chem. 260, 11817-11823
[4] Mills, J.D., Mitchell, P. and Schürmann, P. (1980) FEBS Lett. 112/2, 173-177
[5] Kothen, G., Schwarz, O. and Strotmann, H. (1995) Biochim. Biophys. Acta 1229, 208-214
[6] McCarty, R.E. and Racker, E. (1968) J. Biol. Chem. 243, 129-137
[7] Strotmann, H. and Bickel-Sandkötter, S. (1977) Biochim. Biophys. Acta 460, 126-135
[8] Avron, M. (1960) Biochim. Biophys. Acta 40, 257-272
[9] Galmiche, J.M., Girault, G., Berger, G., Jacquot, J.P., Miginiac-Maslow, M., Wollman, E. (1990) Biochimie 72, 25-32
[10] Biaudet, P., Haraux, F., Strotmann, H. (1990) in: Current Research in Photosynthesis (Baltscheffsky, M, ed.), Vol. III, pp 141-144, Kluwer, Dordrecht.

ANALYSIS OF b-CYTOCHROME TITRATIONS IN TERMS OF INTERACTING REDOX COUPLES

Clemens Wagner and Dieter Walz
Biozentrum, University of Basel, Klingelbergstrasse 70
CH-4056 Basel, Switzerland

1. Introduction

The bf-complex in the membrane of thylakoids contains four redox centers, *viz.* the hemes b_L and b_H of cytochrome b, the Rieske FeS-center, and the heme of cytochrome f. In the putative structure of cytochrome b which includes eight transmembrane helices [1] the two hemes are surrounded by the four helices A to D. They are anchored via coordination to histidines of helices B and D and thus embedded in most likely very similar micro-environments. The heme rings are arranged in one plane perpendicular to the membrane surface, and separated by an edge to edge distance of 12 Å. In contrast, the heme of cytochrome f and the FeS-center are located in domains of their protein which protrude out of the membrane [1].

The redox centers in a protein such as the bf-complex define different states of the whole complex. In addition, and even more important is the possibility of interaction between the redox centers. Already in 1974 Malmström [2] has presented a formalism for redox complexes with interacting couples and used it for a reevaluation of data from the literature, thus demonstrating the relevance of interactions. However, with a few exceptions (see, e.g., refs. 3-5), this aspect is usually not taken into account, although molecular modeling of redox complexes with known structure can now provide data on interaction energies [6].

We present an algorithm for analyzing titration data of redox complexes with interaction. This algorithm avoids some oversimplifying assumptions made previously [4]. We apply it to two typical cases of cytochrome b titrations, and will show that the interaction energy found can be explained by electrostatic interaction.

2. Analysis of redox titrations

The parameters pertinent to the four redox states of cytochrome b are given in Table 1. An asterisk indicates that the parameter applies to a state with the other heme in the oxidized form. We assume that the ionic strength in the suspending medium is sufficiently large so that surface potentials can be neglected [3,4]. Due to the permeable redox mediators added the redox potential E_h at the site of the redox centers is then equal to that in the suspending medium. By means of the Nernst equation we obtain the following relations for concentration ratios, abbreviated by γ_i,

P. Mathis (ed.), Photosynthesis: from Light to Biosphere, Vol. II, 781–784.

$$\gamma_1 = c_2/c_4 = \exp\{F(E_h - E_L)/RT\} \qquad \gamma_2 = c_3/c_4 = \exp\{F(E_h - E_H)/RT\}$$

$$\gamma_3 = c_1/c_4 = \exp\{F[2E_h - (E_L + E_H + \Delta E_{int})]/RT\} \tag{1}$$

Here E_H and E_L denote the standard potentials of b_H and b_L with the other species in the reduced form, respectively. ΔE_{int} accounts for the change in standard potentials if the other species is in the oxidized form and thus indicates the change in interaction. R, T, and F are the gas constant, the absolute temperature and the Faraday constant, respectively.

Table 1. Parameters for the redox states of cytochrome b

redox state b_L	b_H	conc	ext.coeff. for wavelength pair λ_1, λ_2	average pK for H^+ binding
ox	ox	c_1	$\varepsilon^*_{L,o} + \varepsilon^*_{H,o}$	$pK_{o,o}$
ox	red	c_2	$\varepsilon_{L,o} + \varepsilon^*_{H,r}$	$pK_{o,r}$
red	ox	c_3	$\varepsilon^*_{L,r} + \varepsilon_{H,o}$	$pK_{r,o}$
red	red	c_4	$\varepsilon_{L,r} + \varepsilon_{H,r}$	$pK_{r,r}$

Because of different extinction coefficients (cf. Table 1) the change in redox state of the complex during a titration gives rise to a change in absorbance, which is usually measured as the absorbance difference $\Delta A_{\lambda1-\lambda2} = A(\lambda_1) - A(\lambda_2)$ between two wavelengths λ_1 and λ_2. With the extinction coefficients in Table 1 valid for the given wavelength pair the contribution ΔA_b of cytochrome b to $\Delta A_{\lambda1-\lambda2}$ amounts to, with d_{opt} denoting the optical path length,

$$\Delta A_b = [c_1(\varepsilon^*_{L,o} + \varepsilon^*_{H,o}) + c_2(\varepsilon_{L,o} + \varepsilon^*_{H,r}) + c_3(\varepsilon^*_{L,r} + \varepsilon_{H,o}) + c_4(\varepsilon_{L,r} + \varepsilon_{H,r})]\, d_{opt} =$$

$$[c_{tot}(\varepsilon^*_{L,o} + \varepsilon^*_{H,o}) + c_2\Delta\varepsilon^*_H + c_3\Delta\varepsilon^*_L + c_4(\Delta\varepsilon_L + \Delta\varepsilon_H)]\, d_{opt} \tag{2}$$

Here $\Delta\varepsilon_L = \varepsilon_{L,r} - \varepsilon_{L,o}$, $\Delta\varepsilon_H = \varepsilon_{H,r} - \varepsilon_{H,o}$, $\Delta\varepsilon^*_L = \varepsilon^*_{L,r} - \varepsilon^*_{L,o}$, and $\Delta\varepsilon^*_H = \varepsilon^*_{H,r} - \varepsilon^*_{H,o}$. Moreover $\varepsilon_{L,o} \approx \varepsilon^*_{L,o}$, $\varepsilon_{H,o} \approx \varepsilon^*_{H,o}$ was assumed, and the total concentration of cytochrome b is

$$c_{tot} = c_1 + c_2 + c_3 + c_4 \tag{3}$$

By means of Eqns. 1 and 3 Eqn. 2 can be transformed into

$$\Delta A_b = f_b \frac{1 + a(\gamma_1 + \mu\gamma_2)/(1 + \mu)}{1 + \gamma_1 + \gamma_2 + \gamma_3} + A_{o,b} \tag{4}$$

where

$$a = \Delta\varepsilon^*_L/\Delta\varepsilon_L = \Delta\varepsilon^*_H/\Delta\varepsilon_H \quad \text{and} \quad \mu = \Delta\varepsilon_L/\Delta\varepsilon_H = \Delta\varepsilon^*_L/\Delta\varepsilon^*_H \tag{5}$$

A value of $a \neq 1$ indicates an effect of interaction on light absorption, while μ indicates the ratio of the spectral contributions of the two hemes at the given wavelength pair. The quantity $f_b = c_{tot}(\Delta\varepsilon_L + \Delta\varepsilon_H)\, d_{opt}$ is a scaling factor, while $A_{o,b}$ includes $c_{tot}(\varepsilon^*_{L,o} + \varepsilon^*_{H,o})\, d_{opt}$ and a possible baseline shift. An additional redox couple present in the system which has a standard potential E_{add} in the titration range and absorbs at the given wavelength pair yields the contribution

$$\Delta A_{add} = f_{add} / \left[1 + \exp\{F(E_h - E_{add})/RT\} \right] + A_{o,add} \quad (6)$$

The effect of pH on redox titrations is assessed by H^+ binding to multiple sites on the protein. The treatment uses the concept of average protonation [7], which yields the average pK's listed in Table 1 [8] and pH-dependent correction terms for the γ_i. It is important to note that this concept does not require H^+ binding to occur at the redox centers.

The algorithm was applied to typical redox titration data taken from the literature [9,10] and the result is shown in Fig. 1. Since both hemes are chemically identical and located in similar micro-environments, it is reasonable to set $E_L = E_H = E_b$ and $\mu = 1$. Moreover, spectroscopic interaction is neglected, i.e. $a = 1$. The curves in Fig. 1 indicate the result of a parameter fit procedure, which yielded the following values for E_b and ΔE_{int}: -135 mV and 38 mV for the data in Fig. 1A, -67 mV and 21 mV for the data in Fig. 1B. Note that an additional redox couple representing cytochrome b_{559} was included in Fig. 1A but not in 1B, as was done by the corresponding authors. The values for the average pK's ranging between 6.7 and 7.2 show little dependence on the redox state, which indicates that the H^+ binding groups are most likely located on or near the protein surfaces facing the aqueous medium.

Our analysis clearly demonstrates that redox titrations of cytochrome b can equally well be interpreted in terms of *two interacting hemes with identical standard potential* and protonation at several sites on the protein not directly related to the redox centers. This is particularly interesting in the case of the data of Furbacher et al. [10] which were interpreted by the authors in terms of two non-interacting hemes with equal but pH-dependent midpoint potentials. The different standard potentials found for different thylakoid preparations may in part be due to different surface potentials, a parameter not yet included in our analysis. It may be

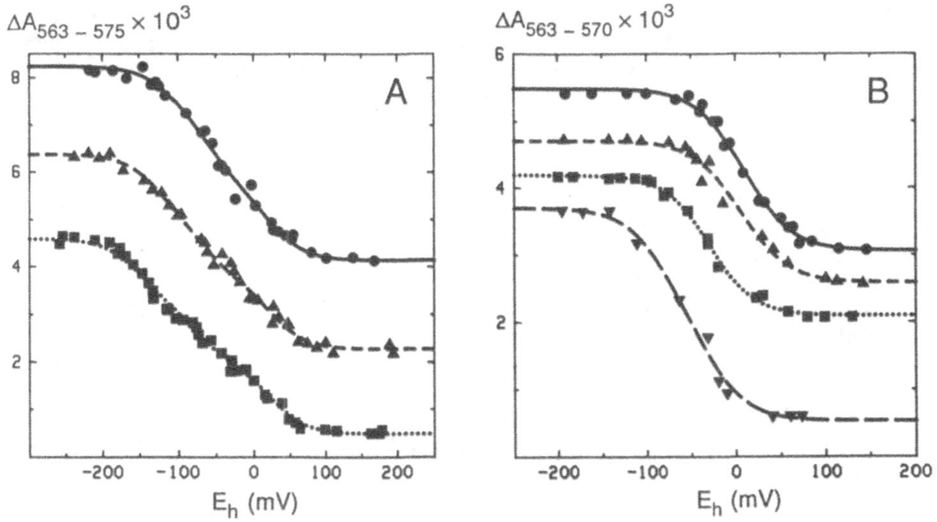

Fig. 1. Redox titration data taken from the literature and analyzed with the present algorithm. (A) Data from Kramer and Crofts [9]; pH-values are 6.3 (●), 7.1 (▲), and 8.75 (■). (B) Data from Furbacher et al. [10]; pH-values are 5.9 (●), 6.3 (▲), 7 (■), and 8 (▼).

added that our analysis works equally well with similar results when applied to other data to be found in the literature.

3. Electrostatic interaction

Since H^+ binding does not occur at the redox centers oxidized and reduced form of b_L or b_H differ by one elementary charge. Hence the reduction of one couple occurs at a different electric potential depending whether the other couple is oxidized or reduced. This gives rise to an electrostatic interaction of the redox couples and causes a difference ΔG_{int} in Gibbs free energy of the redox reaction of the couples. This also determines the difference ΔE_{int} in standard potentials due to interaction:

$$\Delta G_{int} = \Delta G(b_L)\big|_{bH,o} - \Delta G(b_L)\big|_{bH,r} = \Delta G(b_H)\big|_{bL,o} - \Delta G(b_H)\big|_{bL,r} = -F\,\Delta E_{int} \tag{7}$$

ΔG_{int} can be calculated from the work performed when moving an electron from infinity to the site of one heme in the presence of an elementary positive charge on the site of the other heme. From Coulomb's law it then follows that

$$\Delta E_{int} = \frac{e_o}{4\pi\varepsilon_o}\int_{\infty}^{d}\frac{dr}{r^2\varepsilon(r)} = 14.4\frac{V}{\text{Å}}\int_{\infty}^{d}\frac{dr}{r^2\varepsilon(r)} \tag{8}$$

Here d is the distance between the two hemes, while e_o and ε_o denote the elementary charge and the permittivity of vacuum, respectively. The relative permittivity $\varepsilon(r)$ is assumed to depend on the space coordinate r [11]. With $d \approx 18$ Å and $\varepsilon(r)$ proportional to r [11] one obtains $\Delta E_{int} \approx 40$ mV, which compares well with the values resulting from the analysis of the redox titrations. It is also in line with ΔE_{int} between 52 and 77 mV (i.e. ΔG_{int} between −52 and −77 meV) determined by Gunner and Honig [6] for the hemes in the reaction center of *Rhodopseudomonas viridis*. The distances between these hemes range from 14 to 16 Å.

References

1 Cramer, W.A. and Knaff, D.B. (1990) Energy Transduction in Biological Membranes, pp. 330-338, Springer-Verlag, New York
2 Malmström, B.G. (1974) Q. Rev. Biophys. 6, 389-431
3 Walz, D. (1979) Biochim. Biophys. Acta 505, 279-353
4 Walz, D. and Denis, M. (1984) in Charge and Field Effects in Biosystems (Allen M.J. et al., eds.) pp. 69-76, Abacus press, London
5 Santos, H., Moura, J.G., Moura, I.,LeGall, J. and Xavier, A.V. (1984) Eur. J. Biochem. 141, 283-296
6 Gunner, M. and Honig, B. (1991) Proc. Natl. Acad. Sci. USA 88, 9151-9155
7 Beroza, P., Fredkin, D.R., Okamura, M.Y. and Feher, G. (1995) Biophys. J. 66, 2233-2250
8 Wagner, C. (1995) PhD Thesis, University of Basel, Basel
9 Kramer, D.M. and Crofts, A.R. (1994) Biochim. Biophys. Acta 1184, 193-201
10 Furbacher, P.N., Girvin, M.E. and Cramer, W.A. (1989) Biochemistry 28, 8990-8998
11 Harvey, S.C. (1989) Proteins: Structure, Function, and Genetics 5, 78-92

8. Alternative electron pathways and regulation

PHOTOSYNTHETIC AND QUASI-PHOTOSYNTHETIC BACTERIA: A FOCUS ON SPECIES OF CURRENT RESEARCH INTEREST

HOWARD GEST
Photosynthetic Bacteria Group, Biology Department,
Indiana University, Bloomington, IN 47405, USA

1. Introduction

In the first section of this paper, I discuss how an understanding of non-oxygenic bacterial photosynthesis has influenced definition of the term photosynthesis and why it is now important to revise the definition. The second part brings attention to some noteworthy recent advances in research on anoxyphototrophs. In the following I refer to the latter simply as "photosynthetic bacteria" since the oxygenic "blue green bacteria " are distinguished by the term "cyanobacteria."

Before ca. 1930, many photosynthesis investigators apparently were not convinced that the photosynthetic bacteria were actually photosynthetic. This doubt stemmed from the prevailing definition of photosynthesis which included production of molecular oxygen as a major feature. Complicating the issue, there were also contradictory reports on whether or not the pigmented bacteria produced O_2. Eventually, it was firmly established that the bacteria do not produce oxygen and new questions arose, particularly: why don't photosynthetic bacteria produce O_2? This led to speculations on the comparative biochemistry of photosynthetic processes, which can now be seen as focusing on issues that were not central to the basic events in photosynthetic energy conversion. As we look back, we can see that much research and discussion on "comparative biochemistry" questions were propelled by contemporary definitions of photosynthesis (the history of concepts of the comparative biochemistry of oxygenic and anoxygenic photosyntheses is detailed in ref.1).

Definitions of scientific terms encapsulate much information resulting from many research efforts and it is instructive to review how the definition of photosynthesis has changed during the 20th century.

P. Mathis (ed.), Photosynthesis: from Light to Biosphere, Vol. II, 787–793.

2. Earlier Definitions of Photosynthesis

The term photosynthesis was first proposed by C.R. Barnes in 1893 (2) to identify the major plant process by which CO_2 and water are converted to sugars and molecular oxygen. It is remarkable that in the mid-1940's the definition remained essentially the same despite the well-established facts that bacterial photosynthesis does not necessarily involve reduction of CO_2 as a major process [many photosynthetic bacteria can grow photosynthetically on organic compounds as sole carbon sources] and O_2 is not produced. In 1963, Kamen (3) offered a revised general definition of photosynthesis which would include the bacterial process by avoiding specification of the carbon source used for photosynthetic growth and omitting formation of O_2 as a necessary product. His definition: "Photosynthesis is a series of processes in which electromagnetic energy is converted to chemical free energy which can be used for biosynthesis."

3. Quasi-photosynthetic Bacteria and a Revised Definition of Photosynthesis

Recent researches have raised the issue of an adequate definition of photosynthesis again. Thus, the term photosynthesis (or photosynthetic) has recently been misapplied in describing various aerobic bacterial species which produce bacteriochlorophyll a and carotenoids (even reaction centers!), but which appear in most instances to be incapable of using light as a sole, major, or even partial source of energy for growth. I have suggested that for the time being, at least, such organisms should be described as quasi-photosynthetic (4). A general definition of photosynthesis should obviously encompass both oxygenic and non-oxygenic photosyntheses, but exclude quasi-photosynthetic organisms. Accordingly, I have proposed the following definition (4): "Photosynthesis is a series of processes in which electromagnetic energy is converted to chemical energy used for biosynthesis of organic cell materials; a photosynthetic organism is one in which a major fraction of the energy required for cellular syntheses is supplied by light."

The foregoing definition focuses attention on unsolved physiological, biochemical, and evolutionary problems of interest. Why are quasi-photosynthetic bacteria unable to couple photochemical energy conversion to biosynthesis of cell materials? Have certain quasi-photosynthetic bacteria acquired photopigment biosynthesis genes by lateral gene transfer? Does the capacity of quasi-photosynthetic bacteria to synthesize photopigments merely reflect the evolutionary loss of other genes required for photosynthesis from photosynthetically-competent ancestors?

4. Photosynthetic Bacteria of Particular Interest

Many significant advances in understanding complex biochemical systems have emerged from detailed study of microbial species that appear to have "unusual" phenotypic properties. Frequently, the "unusual" property is merely an exaggeration in the level of a "normal" enzymatic activity or enzyme sequence, and this facilitates analysis. Sometimes, a species attracts attention because the organism grows rapidly and abundantly, and is hardy. Whatever the reason, particular species have become prominent for detailed study of particular aspects of bacterial photosynthesis.

4.1 *Rhodospirillum*

R. rubrum was the first photosynthetic bacterium isolated in pure culture. Following the end of World War II in 1945, it became the favorite organism for research on bacterial photosynthesis for some time, yielding a number of discoveries of fundamental importance. Among these were: (a) experiments showing that illumination of intact cells causes oxidation of endogenous cytochromes (5), (b) demonstration of photophosphorylation and light-stimulated redox reactions (6,7), and (c) first demonstration of N_2 fixation by a photosynthetic bacterium (8,9). [Later research showed that virtually all species of anoxyphototrophs also have the capacity to fix N_2]. The 1963 book *Bacterial Photosynthesis* (10) resulting from the first symposium on the topic has many more references in the subject index to *R. rubrum* than to any other photosynthetic bacterium.

In 1987, a remarkable new species of *Rhodospirillum* was isolated from Thermopolis Hot Springs (Wyoming) and given the species name *centenum* in recognition of the centennial of the first isolation of a photosynthetic bacterium in pure culture (i.e., *R. rubrum*). Although in many respects *R. centenum* appears to be a typical nonsulfur purple bacterium it has several properties rarely or never observed before in organisms of this category (11). For example: (i) under certain growth conditions, the cells are converted to thick-walled cysts which, in dry preparations, are quite resistant to desiccation and heat (11,12), and (ii) in this organism, atmospheric O_2 hardly represses synthesis of the photopigments during growth.

Attempts since 1883 to detect phototaxis of photosynthetic bacteria have given negative or ambiguous results. Thus, the occurrence of phototaxis in anoxyphototrophs has been consistently questioned or denied. However, genuine phototaxis of *R. centenum* colonies on agar was recently demonstrated (14,15,16). Since this bacterium is amenable to facile molecular biological/genetic manipulation it should prove useful as an experimental organism for identifying genes governing phototaxis and for analyzing details of mechanisms of cell movements. It should be noted that the linkage order for genes involved in Bchl and carotenoid biosyntheses is essentially the same in *R. centenum* as in *Rhodobacter capsulatus* and *Rhodobacter sphaeroides* (17). Conservation of the "Photosynthesis Gene Cluster" in these organisms implies either that there are structural or regulatory constraints that limit rearrangement of the gene cluster or that there may have been lateral transfer of the cluster among different species of anoxyphototrophs.

4.2 *Rhodobacter*

Rhodobacter capsulatus and its close relative *Rhodobacter sphaeroides* are expectedly very similar in physiological, biochemical and various other properties. Indeed, some investigators have confused them with one another (this can be avoided by consulting detailed "profiles" of the two organisms in refs. 18 and 19). These *Rhodobacter* species (formerly in the genus *Rhodopseudomonas*) have been particularly useful for exploring reaction center structure and function, regulation of photopigment and reaction center syntheses, and N_2 fixation. References to research on reaction centers and photopigment synthesis can be found in other papers in the proceedings of this Congress; for N_2 fixation and related aspects of nitrogen metabolism , see 20 and 21).

4.3 *Rhodopseudomonas*

There are indications that *Rhodopseudomonas* species are the most commonly found nonsulfur purple bacteria in soils and aqueous environments. *Rps. viridis* achieved great prominence due to the elegant studies of Deisenhofer et al. (22) who reported the detailed structure of its reaction center, the first structure at atomic resolution of a protein complex from a biological membrane. This achievement was recently followed up by McDermott et al. (23), who elucidated the crystal structure of the light-harvesting antenna complex (LH 2) from *Rps. acidophila.*

4.4 *Heliobacteria*

The first representative of the family *Heliobacteriaceae, Heliobacterium chlorum,* was isolated in 1983 (24), and four species are now known. These green anoxyphototrophs contain a novel form of Bchl, Bchl *g* (25,26), whose tetrapyrrole is essentially a hybrid of Bchl *b* and Chl *a*. It is of special interest that under certain conditions, *in vivo* Bchl *g* is converted to 8^1-hydroxy-Chl *a* (27). The reaction centers of contemporary heliobacteria resemble the reaction center of green plant Photosystem I (Fe-S type) and this lends support to the idea that the green plant photosystem complex evolved from an ancient fusion of components from two lines of anoxyphototrophs, namely, a "quinone type" reaction center from a purple bacterium with an "Fe-S type" reaction center from an early species of heliobacteria (28,29). The most recently identified species of *Heliobacterium, H. modesticaldum,* is a thermophilic organism isolated from hot springs and volcanic soils (30).

4.5 Thermophilic Photosynthetic Bacteria

Several moderately thermophilic anoxyphototrophs have been isolated in recent years from hot springs and other warm environments worldwide. These include the green sulfur bacterium *Chlorobium tepidum* (31), the purple sulfur bacterium *Chromatium tepidum* (32,33), and the aforementioned *Heliobacterium modesticaldum* (30). Hot spring anoxyphototrophs have yielded a number of new cell components of interest to the field of photosynthesis. Namely: the novel antenna complexes of *Chromatium tepidum* (34) whose absorption maxima lie further in the infrared than any antenna complex from organisms containing bacteriochlorophyll a ; thermally stable chlorosome components (35) in the rapidly growing species *Chlorobium tepidum;* thermally stable ribulose bisphosphate carboxylase (RuBisCO, Calvin cycle) from *Chromatium tepidum* (36,37), and thermally active nitrogenases for nitrogen fixation from *Chlorobium tepidum* (38) and *Heliobacterium modesticaldum* (30). In addition, a genetic system based on conjugation has been developed for *Chlorobium tepidum* (39) which should now permit genetic analysis of various unique aspects of photosynthesis found only in green sulfur bacteria. Thermophilic anoxyphototrophs have thus contributed to our understanding of photosynthetic diversity and have helped accelerate the pace of research on fundamental problems in photosynthesis.

Acknowledgement

Research of the author on photosynthetic bacteria is supported by U.S. National Science Foundation grant IBN-9303836.

References

1. Gest, H. (1993) Photosyn. Res. 35,87-96
2. Barnes, C.R. (1893) Bot. Gaz. 18, 403-411

3. Kamen, M.D. (1963) Primary Processes in Photosynthesis. Academic Press, New York and London
4. Gest, H. (1993) FEMS Microbiol. Lett. 112, 1-6
5. Duysens, L.N.M. (1954) Nature 173, 692-693
6. Frenkel, A.W. (1954) J. Am. Chem. Soc. 76, 5568-5569
7. Frenkel, A.W. (1958) J. Am. Chem. Soc. 80, 3479-3480
8. Gest, H. and Kamen, M.D. (1949) Science 109, 558-559
9. Kamen, M.D. and Gest, H, (1949) Science 109, 560
10. Gest, H., San Pietro, A. and Vernon, L.P. (eds) (1963) Bacterial Photosynthesis. Antioch Press, Yellow Springs, Ohio
11. Favinger, J., Stadtwald, R. and Gest, H. (1989) Ant. v. Leeuwenhoek J. Microbiol. 55, 291-296
12. Stadtwald-Demchick, R., Turner, F.R. and Gest, H. (1990) FEMS Microbiol. Lett. 67, 139-144
13. Yildiz, F.H., Gest, H. and Bauer,C.E. (1991) J. Bacteriol. 173, 5502-5506
14. Ragatz, L., Jiang, Z-Y., Bauer, C. and Gest, H. (1994) Nature 370, 104
15. Ragatz, L., Jiang, Z-Y., Bauer, C.E. and Gest, H. (1995) Arch. Microbiol. 163, 1-6
16. Gest, H. (1995) FEMS Microbiol. Revs., in press
17. Yildiz, F.H., Gest, H. and Bauer, C.E. (1992) Mol. Microbiol. 6, 2683-2691
18. Weaver, P.F., Wall, J.D. and Gest, H. (1975) Arch. Microbiol. 105, 207-216
19. Pellerin, N.B. and Gest, H. (1983) Current Microbiol. 9, 339-344
20. Hillmer, P. and Gest, H. (1977) J. Bacteriol. 129, 724-731; 732-739
21. Gest, H. (1994) Photosyn. Res. 40, 129-146
22. Deisenhofer, J., Epp, O., Miki, K., Huber, R. and Michel, H. (1985) Nature 318, 618-624
23. McDermott., G., Prince, S.M., Freer, A.A., Hawthornthwaite-Lawless, A.M., Papiz, M.Z., Cogdell, R.J. and Isaacs, N. W. (1995) Nature 374, 517-521
24. Gest, H. and Favinger, J.L. (1983) Arch. Microbiol. 136, 11-16
25. Brockmann, H., Jr. and Lipinski, A. (1983) Arch. Microbiol. 136, 17-19
26. Michalski, T.J., Hunt, J.E., Bowman, M.K., Smith, U., Bardeen, K., Gest,H., Norris, J. R. and Katz, J.J. (1987) Proc. Natl. Acad. Sci. (U.S.) 84, 2570-2574
27. Van de Meent, E.J., Kobayashi, M., Erkelens, C., van Peelen, P.A., Amesz, J. and Watanabe, T. (1991) Biochim. Biophys. Acta. 1058, 356-362
28. Blankenship, R.E. (1994) Ant. v. Leeuwenhoek J. Microbiol. 65, 311-329
29. Vermaas, W.F.J. (1994) Photosyn. Res. 41, 285-294

30. Kimble, L.K., Mandelco, L., Woese, C.R. and Madigan, M.T. (1995) Arch. Microbiol. 163, 259-267
31. Wahlund, T.M., Castenholz, R.W., Woese, C.R. and Madigan, M.T. (1991) Arch. Microbiol. 156, 81-91
32. Madigan, M.T. (1984) Science 225, 313-315
33. Madigan, M.T. (1986) Intl. J. Syst. Bacteriol. 36, 222-227
34. Garcia, D., Parot, P., Vermeglio, A. and Madigan, M.T. (1986) Biochim. Biophys. Acta 850, 390-395
35. Nozawa, T., Suzuki, M., Ohtomo, K., Morishita, H., Konami, H. and Madigan, M.T. (1991) Chem. Letts. 1991(9), 1641-1644
36. Heda, G.D. and Madigan, M.T. (1986) FEMS Microbiol. Lett. 51,45-50
37. Heda, G.D. and Madigan, M.T. (1989) Eur. J. Biochem. 184, 313-319
38. Wahlund, T.M. and Madigan, M.T. (1993) J. Bacteriol. 175, 474-478
39. Wahlund, T.M. and Madigan, M.T. (1995) J. Bacteriol. 177, 2583-2588

CONTROL AND ORGANIZATION OF ELECTRON TRANSPORT AND CARBON ASSIMILATION IN LEAVES

Agu Laisk, Tartu Ülikooli Molekulaar- ja Rakubioloogia Instituut, Riia tn. 181, Tartu, Estonia EE2400

1. Introduction

Photosynthetic electron transport chain is the following

1 2 3 4 5 6 7 8 9 10 11
P680*->Qa->PQ(H_2)->Cyt b_6f->PC->P700*->A0->A1->FeS->Fd->NADP(H)->BPGA ->... (1)
-> Triplet->O_2 -> O_2

The driving force for the flow is the product of the reducing potential of the excited chlorophylls *P*680* and *P*700* and the lifetime of the excitation. Under limited availability of photosynthetic acceptors, the too long-living excitation tends to be quenched by alternative electron acceptors, e.g. oxygen, resulting in singlet oxygen, superoxide radical or hydrogen peroxide [1]. Plants have developed processes that regulate the lifetime of excitation, to reach an optimum between the quantum yields of photosynthetic and alternative routes. At PS II a competitive nonphotochemical excitation quencher is induced in the antenna [2]. It has been suggested that a similar regulation occurs at PS I [3]. However, the mechanism of this regulation is probably different from that at PS II, due to the different pigment systems of the two photosystems. Possible mechanisms of the regulation of PS I efficiency are discussed below.

2. Material and Methods

Sunflower and tobacco plants were grown in growth chambers in Tartu and Canberra. Pumpkin was growing outdoors in Canberra. Similar fast-response gas exchange measurement systems were used in both Tartu and Canberra laboratories. In these systems gas mixtures of CO_2, O_2 and N_2 were preset by using capillaries and stabilized pressures to control flow rates. CO_2 concentration was measured by the infrared analyser LI 6262 (LiCor, Lincoln, Nebraska), O_2 concentration by the Zr-oxide analyser S-3A (Ametek, Pittsburgh, PA), water vapour by a micropsychrometer (Tartu) or by LI 6262 (Canberra).

P. Mathis (ed.), Photosynthesis: from Light to Biosphere, Vol. II, 795–800.

Half-response times for CO_2 and O_2 measurements were about 1 s. Leaf chambers were 4.3x4.3x0.3 cm^3, gas flow rate 20 cm^3 s^{-1}. The leaf transmission at 830 nm was measured using the PAM 101 fluorometer (H. Walz, Effeltrich, Germany). Specially designed fibre optic allowed to superimpose of 3 light sources (actinic, flashes, far-red) and to measure the optical signal. Data were recorded with a computer data-logger with 55 ms resolution.

3. Results

In theses experiments we investigated the 830 nm transmission, by recording its difference from the dark level, D830, in response to changes in CO_2 concentration and absorbed quantum flux density, Q. The signals were interpreted in terms of the photooxidation of *P*700. Under nonphotorespiratory conditions (1 or 0% O_2), the quantum yield of PS I was calculated in relation to all absorbed quanta as $Y_I = 4(P-R)/Q$,, where P is net CO_2 uptake or O_2 evolution rate and R is dark respiration in the light. R was assumed to be gradually suppressed toward light saturation of photosynthesis. In the experiment in Fig. 1 a sunflower leaf was transferred from less limiting to strictly limiting conditions for electron transport. During the following transient, high-intensity flashes of two different intensities where applied to obtain more oxidation of *P*700. Y_I is plotted vs. D830, a plot that theoretically is expected to be a straight line. Continuous lines were obtained under the initial less-limiting conditions (line 1) and 200 s (line 2) and 800 s (line 3) after the transition to the strictly limiting CO_2 . On each line the leftmost data point represents the quasi-steady-state and two other points were obtained under flashes. In order to calculate Y_I , it was assumed that the electron transport rate did not increase under the flashes, being CO_2 limited. Both interceptions of the lines with axes, Y_{Im} and $D830_m$ and the slope of the line changed during the experiment. The decrease of Y_{Im} was caused additively by variation of $D830_m$ and by variation of the slope of the Y_{Im} vs. D830 graph.

In another experiment, steady-state points were obtained at different light intensities (Fig. 2). From each steady-state, light intensity was decreased to 30 µEi and the O_2 evolution recorded after 5 s. The experiment was done in 10 ppm O_2 , so that respiration did not complicate the calculation of the quantum yield. Immediately after the transition to the low light, Y_{Im} was about 0.2 to 0.3, but after 4-5 min it increased to the steady-state of 0.48. Evidently, Y_{Im} was downregulated at higher light intensities and the control relaxed at the low light.

In the following experiments CO_2 uptake, O_2 evolution and 830 nm signal were measured during the dark-light induction of photosynthesis (Fig. 3). The initial peaks in the O_2 evolution and D830 were caused by fast electron transport, probably to the existing PGA. In the beginning of the induction photooxidised *P*700 increased proportionally with the CO_2 uptake rate, declining from the proportionality only a little before the rate approached the maximum. In the other experiment (Fig. 4), the O_2 evolution rate was recorded in a sunflower leaf during the dark-light induction. Periodically, light intensity was reduced to a value expected to become rate-limiting early during the induction. Unexpectedly, both O_2 evolution rates increased in parallel throughout the whole induction.

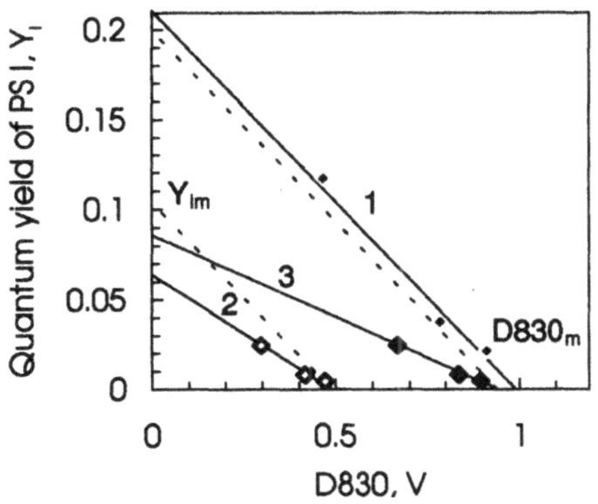

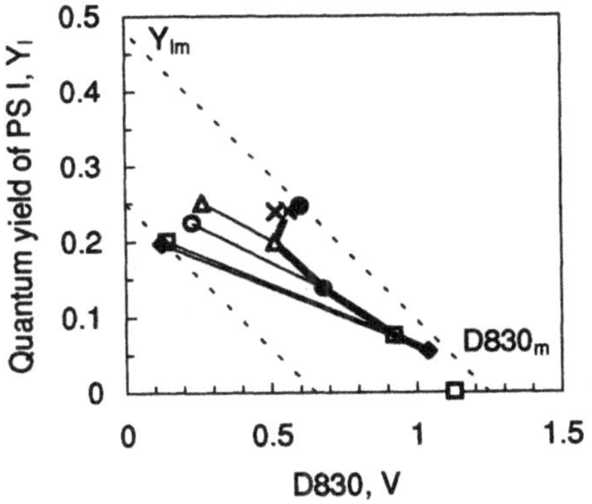

Fig. 1. The quantum yield of PS I, Y_I, plotted vs. D830, to show the variation of $D830_m$ and Y_{Im} during the adjustment to limited electron transport. Line 1 was measured from the initial state of 310 ppm CO_2 and 21% O_2. Line 2 was measured 200 s after the transition to 50 ppm CO_2 in 1.1% O_2 and line 3 was measured 800 s after the transition.

Fig. 2. Steady-state and momentary quantum yield Y_I plotted vs. D830 for a tobacco leaf. Steady-state points at different light intensities are joined with the bold line. At each steady-state, light was decreased to 30 µEi and O_2 evolution rate was recorded after 5 s. The so obtained transient data points are joined with the corresponding steady-state points with thin lines. Empty square was measured in far-red light. $D830_m$ was calculated considering that the far-red light contained 10% PS II light.

There was a basal O_2 evolution rate of 0.8 µmol m^{-2} s^{-1} that was present also in the complete absence of CO_2 and that became light-saturated at very low Q, probably related to nitrite reduction (dashed line). After this rate was subtracted from the both rates, the high- and low-light photosynthetic O_2 evolution rates increased almost proportionally.

4. Discussion

The extrapolations of the lines in Figs. 1 and 2 to the axes reveal $D830_m$, the signal corresponding to the maximum oxidisable *P*700, and Y_{Im}, the maximum quantum yield of PS I under the complete reduction of *P*700. In [3] we interpreted Y_{Im} as influenced by two factors. The parallel shift of the straight line (dotted lines in Fig. 1) shows the influence of the acceptor side limitation on Y_{Im}, while the change of the slope shows the influence of the varying excitation capture efficiency by PS I. If the change of the slope was not apparent, caused by rapid change of $D830_m$ together with Y_I, unresolved in our measurements, the possible mechanism of regulation of the excitation capture by PS I needs to be discussed. The complex of events that influence the quantum yield of PS I is shown in Fig. 5.

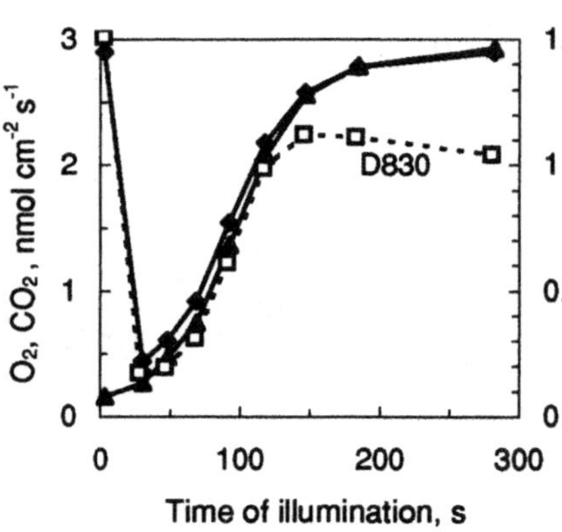

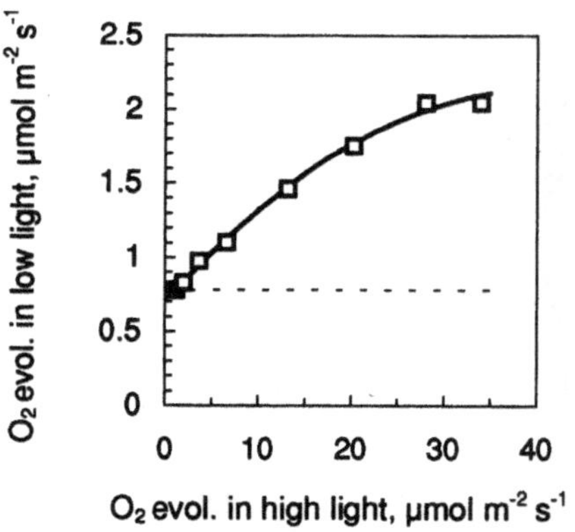

Fig. 3. Simultaneous measurements of O_2 evolution (diamonds), CO_2 uptake (triangles) and 830 nm transmission signal difference from the dark level (D830) during the dark-light induction of photosynthesis. Pumpkin leaf, CO_2 concentration 314 ppm, O_2 concentration 10 ppm, absorbed quantum flux density Q=139 nmol cm^{-2} s^{-1}, dark pre-exposure 18 min.

Fig.4. Parallel increase of O_2 evolution rates measured at Q= 1100 μmol m^{-2} s^{-1} (abscissae) and at Q = 50 μmol m^{-2} s^{-1} (ordinates) during the dark-light induction of photosynthesis. The induction was induced at the higher Q, at 38 μM CO_2 in N_2, but periodically Q was reduced for 5 s and the O_2 evolution rate was measured also at the low Q.

PS I emits very little fluorescence at room temperatures. It has been suggested that the long-wavelength chlorophylls P(>700) trap excitation from the bulk antenna P(<700) so rapidly that fluorescence of the bulk antenna is suppressed, but P(>700) itself is nonfluorescent (k_f is small) [4]. In this case the excitation transfer rate from P(>700) to P700 is important and the ratio of the rate-constants $k_c/(k_c+k_d)$ (Fig. 5) will determine the excitation capture efficiency by P700. The primary acceptors A_0 and A_1 are still very strong reducers. Besides photosynthetic acceptors, they would easily reduce O_2, provided that their lifetime in the reduced state is long enough. However, different from PS II, in PS I the electron is rapidly reversible from the primary acceptors A_0 and A_1, is better stabilised at FeS_a/FeS_b and becomes completely stable only on the following carriers [5]. The line denoting P700*/A_0 is drawn bold to emphasize that the primary charge separation is very fast. This is another possibility to explain why PS I is nonfluorescent, because recombination causes only a small delayed luminescence. The lifetime of the reduced A_0, A_1 and FeS is determined by the rate constants for photochemistry k_p and for recombination, k_r. The yield is determined by $k_p/(k_p+k_r)$. In PS I k_p is the rate constant for electron transfer from unstable to stable intermediate carriers and k_r is the rate constant for charge recombination. While the possibility that k_r may be regulated is not excluded, the variability of k_p is more evident.

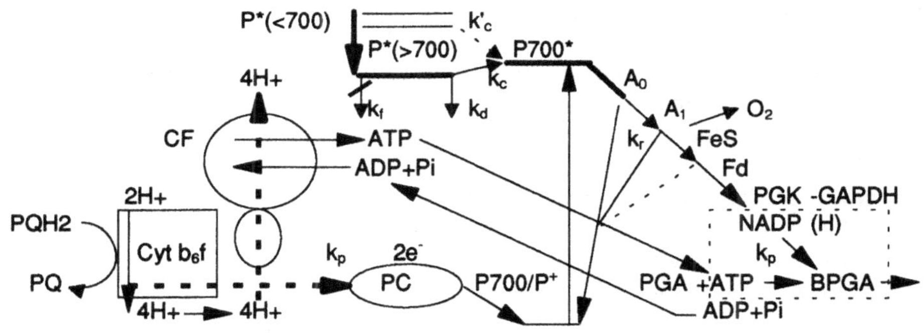

Fig. 5. Complex of photochemical and electron transport reactions determining the efficiency of PS I.

The rate-limiting step at the acceptor side of PS I is that of BPGA reduction at GAPDH, due to the limited availability of BPGA. Thus, k_p depends on the presence of BPGA, but the latter depends on the presence of ATP. On the other hand, it has been generally accepted that the photosynthetic control of PQH_2 oxidation is exercised by the counterpressure of protons, and the latter is built up when ATPase is rate-limiting [6]. Thus, both, the donor side resistance of PS I (bold dashed arrow through Cyt b_6f) and the acceptor side resistance of PS I (k_p of the PGK-GAPDH) simultaneously depend on the rate of the ATPase, but with a difference: while the acceptor side resistance depends on the ATP production rate, the donor side resistance depends on the proton consumption rate. Electrons can pass through Cyt b_6f but cannot be accepted at GAPDH when an insufficient number of protons was cotransported with them or when some protons leaked without coupled ATP synthesis. In this case the primary acceptors of PS I become reduced and the photosynthetic efficiency of PS I decreases. The acceptors can be oxidised only by oxidants that do not need coupled ATP, like O_2 (pseudocyclic)or PQ (cyclic pathway).

Thus, if the variation of the slope of the Y_I vs. D830 graph (Figs. 1 and 2) is real and not apparent, caused by simultaneous rapid change of $D830_m$, preference is given to the model where excitation is transferred to *P*700 via the nonfluorescent long-wavelength chlorophyll *P*(>700). If excitation is captured by *P*700 directly from the bulk antenna and fluorescence is quenched photochemically (dashed arrow in Fig. 5), then the excitation capture efficiency cannot be regulated. In the latter case it is difficult is to explain why the PS I acceptor side became temporarily more reduced after the transition to the low light (Fig. 2, the lower dashed line). It would require an additional assumption, e.g. that the Q-cycle operation was impaired while PQ was highly oxidised and electrons accumulated at the acceptor side of PS I because of insufficient proton cotransport.

The kinetic connection between the donor side (Cyt b_6f) and acceptor side (GAPDH) resistances of PS I is fast. This suggests that the actual pools participating in electron transport, including NADP(H) and Fd, are small. A structural model that accounts for this is that of electron channelling via enzyme supercomplexes or other small subcompartments with limited cross-diffusion. Experiments in Figs. 3 and 4 support this view. The gradual oxidation of electron carriers at the acceptor side of PS I (Fig. 3), but especially the parallel increase of the light-saturated and light-limited O_2 evolution rates during the induction is difficult to explain on an assumption of freely diffusible electron carriers. These kinetics suggest that a kind of "photosynthetosomes", capable for more or less complete photosynthetic electron transport and CO_2 uptake, are individually activated during the induction. This kinetic evidence is accompanied by findings that many carbon reduction enzymes are thylakoid-bound [7,8], has similarity to the situation at PS II [9] and extends the channels proposed in [10,11] further on to the acceptor side of PS I. Measurements of leaf absorbance at 830 nm with better than 1 ms time resolution during strong saturation pulses [12], accompanied by precise gas exchange measurements will lead to further progress in this problem.

Acknowledgment. The assistance of V. Oja at experiments in Figs. 1 and 4 is appreciated. Experiments in Figs. 2 and 3 were carried out while the author was a Visiting Fellow at Research School of Biological Sciences, Australian National University, Canberra, January-March 1995. This work was supported by Estonian Science Foundation grant 108 and International Science Foundation grants LCS000 and LLK100.

References

1 Badger M.R. (1985) Annu. Rev. Plant Physiol. 36, 27-53
2 Horton P., Ruban A.V. (1992) Photosynth. Res. 34, 375-385
3 Laisk A., Oja V. (1995) Photosynth. Res. (in press)
4 Govindjee (1973) in Practical Fluorescence. Theory, Methods, and Techniques (Guilbault G.G., ed) pp 543-575, Marcel Dekker, New York
5 Golbeck J.H. (1987) Biochim. Biophys. Acta 895, 167-204
6 Siggel U. (1974) in Proc. 3-rd Internat. Congr. on Photosynth., (Avron M.,ed) pp.645-654, Elsevier, Amsterdam
7 Süss K.-H., Arkona C., Manteuffel R. and Adler K. (1993) Proc. Natl. Acad. Sci. USA 90, 5514-5518
8 Adler K., Arkona C., Manteuffel R. and Süss K.-H. (1993) Cell Biol. Internat. 17, 213-220
9 Joliot P., Vermeglio A. and Joliot A. (1993) Biochim. Biophys Acta 1141, 151-174
10 Laisk A., Oja V., Walker D. and Heber U. (1992) Photosynthetica 27(4), 465-479
11 Laisk A. (1993) . Proc. R. Soc. Lond. B. 251, 243-251
12 Klughammer C. and Schreiber U. (1994) Planta 192, 261-268

ELECTRON TRANSPORT TO NITROGENASE IN THE PHOTOSYNTHETIC BACTERIUM *RHODOBACTER CAPSULATUS*.

Jouanneau Y., Hugo N., Armengaud J., Naud I., Meyer C., Willison J.C.
CEA/Grenoble - DBMS/Biochimie Microbienne (CNRS URA 1130 alliée à l'INSERM) - 17 rue des Martyrs - 38054 Grenoble cedex 9 - France

1. Introduction

Rhodobacter capsulatus is a photosynthetic bacterium, capable of utilizing dinitrogen (N_2) from the air as sole nitrogen nutrient. This ability, known as nitrogen fixation, is due to nitrogenase, a metalloenzyme which reduces N_2 to ammonia. The nitrogenase-catalyzed reaction is energetically costly, both in terms of ATP consumption and regeneration of low-potential reductants. In *R. capsulatus*, conversion of light energy through photosynthesis results in efficient supply of ATP and reductants to nitrogenase. Generation of reductants for nitrogenase (namely reduced ferredoxin) likely depends upon the proton motive force built up upon operation of the photosystem, but the mechanism by which reductants are actually produced is poorly understood. Hence, our studies have focused on the identification of the protein components involved in electron transport to nitrogenase. Multiple electron carriers, including six ferredoxins [1-7] and one flavodoxin [8] have been found in *R. capsulatus*. The possible function of these electron carriers is discussed on the basis of information derived from our biochemical as well as genetic studies. One ferredoxin, called FdI, has been shown to serve as physiological electron donor to nitrogenase [9]. The molecular interaction between FdI and nitrogenase is being studied by cross-linking experiments as well as site-directed mutagenesis. Besides, the isolation of other components participating in the electron transfer to nitrogenase has been undertaken. Based on genetic studies, a membrane-bound complex is likely to play a central role in this process. It is proposed that electrons produced by oxidation of organic carbon substrates are transferred to nitrogenase via the membrane-bound complex and FdI.

2. Properties and possible functions of six distinct ferredoxins from *R. capsulatus*

Six ferredoxins have been isolated from *R. capsulatus* so far. The general properties of these electron carriers are presented in Table 1. They can be divided into two groups based upon the nature of the iron-sulfur cluster(s) which is present at the active site. FdI, FdII and FdIII are dicluster ferredoxins containing two [4Fe-4S] clusters per polypeptide. FdII differs from the FdI and FdIII in having two unequivalent centers, with different mid-point redox potentials, one of which being of the [3Fe-4S] type [3,10]. The [4Fe-4S] center has one of the lowest mid-point potential known so far for a redox protein (Table 1). A FdII derivative containing two [4Fe-4S] clusters has been

P. Mathis (ed.), Photosynthesis: from Light to Biosphere, Vol. II, 801–805.

obtained by *in vitro* chemical insertion of iron and sulfur into the apoferredoxin [11]. This derivative was unstable and incompetent in electron transfer to nitrogenase.
FdIV, FdV and FdVI are representative of [2Fe-2S] ferredoxins. Little similarity was observed upon comparison of the six ferredoxin amino acid sequences, even when ferredoxins from the same group were compared. FdIV and FdV share some similarities with plant-type ferredoxins with respect to the amino acid sequence and the spectroscopic properties of the [2Fe-2S] center, but their mid-point redox potential is more than 130mV higher than that of their plant counterparts.

Properties	FdI	FdII	FdIII	FdIV	FdV	FdVI
Molecular mass	6,733	12,548	2 x 10,648	10,156	13,309	11,402
Type of Fe-S center	2 x [4Fe-4S]	[3Fe-4S] and [4Fe-4S]	4 x [4Fe-4S]	[2Fe-2S]	[2Fe-2S]	[2Fe-2S]
Mid-point redox potential	-510 mV	-430 mV and -630 mV	-440 mV	-275 mV	-220 mV	-255 mV
Structural gene	*fdxN*	*fdxA*	*fdxB*	*fdxC*	*fdxD*	*fdxE*
Nif -like regulation	Yes	No	Yes	Yes	Yes	No
Transfers electrons to nitrogenase	Yes	Yes	No	No	No	No

Table 1. Properties of the six ferredoxins from *Rhodobacter capsulatus*

All the ferredoxin-encoding genes have been cloned, and four of them were localized within regions of the chromosome where *nif* genes, responsible for nitrogen fixation, are clustered. FdI and FdIV are encoded by two contiguous genes *fdxN* and *fdxC*, respectively, that are cotranscribed [2,4,12]. They are associated with other genes involved in electron transport to nitrogenase [13]. The role of FdI is discussed in more detail below. FdIII is a dimeric ferredoxin with some unusual properties [5]. It has no known function, although the fact that its structural gene is co-transcribed with genes responsible for the nitrogenase FeMoCo biosynthesis suggests that it may be implicated in this process [14]. The gene coding for FdV was discovered upstream of the *nifHDK* operon encoding the nitrogenase polypeptides [15]. Although FdV is not required for N_2 fixation, it may provide nitrogenase with protection against oxidative damage as it is the case for a homologous ferredoxin from *Azotobacter vinelandii* [16].

The last two ferredoxins, FdII and FdVI, have functions unrelated to N_2 fixation. FdII is encoded by a gene that cannot be disrupted by mutagenesis, suggesting that this ferredoxin plays a role essential in the metabolism of *R. capsulatus* [12,17,18]. FdVI is structurally related to bacterial ferredoxins participating in the aerobic catabolism of aromatic or polycyclic compounds [7]. The *fdxE* gene encoding FdVI has recently been cloned. Sequence analysis of flanking DNA revealed that *fdxE* is probably monocistronic and that adjacent genes are not involved in the catabolism of aromatic compounds (Armengaud J., unpublished results). Experiments designed to isolate a *fdxE*-deleted mutant are in progress. Analysis of such a mutant should shed light on the function of this sixth ferredoxin.

3. Identification of the electron donor to nitrogenase in *R. capsulatus*

FdI is the more abundant ferredoxin in N_2 fixing bacteria [1,5]. The low value of its mid point redox potential (-510 mV, pH 7.6; Table 1) as well as *in vitro* experiments showing efficient electron transfer between FdI and nitrogenase, indicated that FdI functioned as a natural redox partner of nitrogenase. Genetic studies showed that disruption of the *fdxN* gene encoding FdI resulted in a dramatic reduction of N_2 fixation, although this ability was not completely lost [12,13]. Studies in our laboratory demonstrated that the defective phenotype of the *fdxN* mutant could be fully complemented by a plasmid carrying a copy of *fdxN* [9]. None of the other ferredoxins could functionally replace FdI, although FdII transfers electrons to nitrogenase *in vitro* [3]. The lack of FdI not only affected electron transport to nitrogenase but also the stability of the nitrogenase enzyme itself [9]. The residual nitrogenase activity observed in the *fdxN* mutant suggested that another electron carrier served as ancillary reductant of nitrogenase. A flavodoxin recently isolated from *R. capsulatus* [8] may be responsible for this activity. As the synthesis of the flavodoxin is stimulated upon iron deficiency, it is plausible that the flavodoxin may replace FdI when iron becomes limiting.

4. Molecular studies of the ferredoxin I - nitrogenase interaction

The interaction between nitrogenase and its natural electron donor FdI has been studied by cross-linking experiments. A covalently linked complex between FdI and dinitrogenase reductase has been obtained upon incubation of the two proteins with ethyl-(dimethylaminopropyl)-carbodiimide (EDC). Cross-linking was prevented by raising ionic strength, consistent with the notion that electrostatic forces are involved in the interaction. Peptide analysis is being performed to identify the amino acid residues participating in the cross-link.

The FdI-nitrogenase molecular interaction is also investigated using FdI mutant proteins obtained by site-directed mutagenesis. The aim is to identify residues of FdI that may participate in the interaction with dinitrogenase reductase. A three-dimensional model of FdI was elaborated after the crystallographic structure of a homologous dicluster ferredoxin. Then, variants of FdI were created by making substitutions of particular charged residues predicted to be exposed at the surface of the molecule. A mutant bearing a tyrosine instead of a methionine at position 2 (M2Y) was also constructed, because this residue is close to one [4Fe-4S] center and may influence its redox properties. Mutant proteins were produced, purified and found to be very similar to wild-type FdI with respect to their spectroscopic and redox properties. They also proved to be competent electron donors to nitrogenase *in vivo*, as judged from their ability to restore a wild-type phenotype when introduced in a *fdxN*-deleted mutant. However, two variants designated D36H and K27E/K28E, bearing a single and a double substitution respectively, showed altered kinetics of electron transfer to nitrogenase in *in vitro* assays. The M2Y variant was identical to wild-type FdI in this assay, although its mid-point redox potential was lowered by 20mV. The D36H ferredoxin exhibited a decreased affinity for dinitrogenase reductase by almost one order of magnitude, compared to natural FdI. In addition, it failed to form a cross-linked complex with the reductase upon incubation with EDC. It is concluded that the aspartate residue at position 36 in the FdI sequence plays an important role in the interaction between nitrogenase and its natural electron donor.

5. Involvment of a membrane-bound complex in the electron transport to nitrogenase

A new set of genes has recently been discovered in *R. capsulatus* that may encode components of an electron transport system to nitrogenase [13]. These genes organised in two adjacent *nif*-regulated operons have no known counterparts in other diazotrophs, and were designated *rnfA* to *rnfF* [13]. Four genes are predicted to code for transmembrane polypeptides, whereas two genes *rnfB* and *rnfC*, may specify iron-sulfur containing proteins. It is proposed that the *rnf* gene products assemble to form a membrane-bound complex, which would provide reductants to nitrogenase via FdI (Fig. 1).

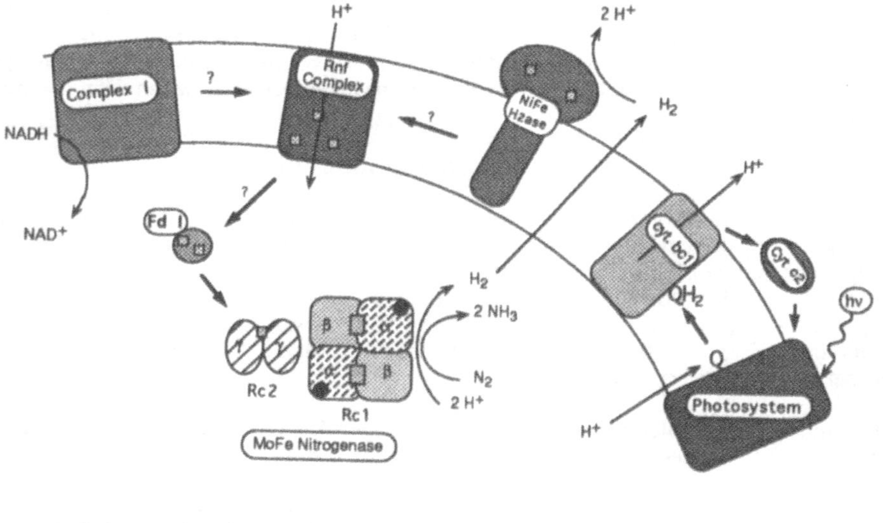

Figure 1. Scheme showing the hypothetical pathway of electrons towards nitrogenase in *Rhodobacter capsulatus*.

Preliminary analysis of membrane proteins by two dimensional electrophoresis indicated that the *rnf* genes are poorly expressed in *R. capsulatus*. It was therefore attempted to overexpress the *rnf* gene products in recombinant form in *E. coli*. This strategy proved useful for the overproduction and purification of a flavoprotein, encoded by a gene provisionally called ORF14 [13] that is part of an operon also composed of the *fdxC*, *fdxN* and *rnfF* genes. The flavoprotein contains one FMN group per 50 kDa subunit, and serves as electron acceptor to FdIV. Hence, FdIV and the flavoprotein appear to be functionally linked, consistent with the fact that their genes are contiguous.

Expression of putative membrane polypeptides, such as those encoded by *rnfA* and *rnfD*, yielded no appreciable synthesis of recombinant protein in *E. coli*. On the other hand, the *rnfB* and *rnfC* gene products were obtained in good yield and could be easily purified. RnfB was isolated in apoform and was subjected to reconstitution experiments by *in vitro* insertion of Fe-S clusters. A protein containing at least one [4Fe-4S] cluster was obtained that was unstable in air due to oxidative denaturation. RnfC expression gave rise to a truncated polypeptide of 25 kDa, which is half the size of the expected protein. Antibodies against purified *rnf* gene products will be used to localize them and assess their assembling into a functional complex in *R. capsulatus*.

References

1 Hallenbeck, P.C., Jouanneau, Y. and Vignais, P.M. (1982) Biochim. Biophys. Acta 681, 168-176
2 Schatt, E., Jouanneau, Y. and Vignais, P.M. (1989) J. Bacteriol. 171, 6218-6226
3 Jouanneau, Y., Meyer, C., Gaillard, J. and Vignais, P.M. (1990) Biochem. Biophys. Res. Commun. 171, 273-279
4 Grabau, C., Schatt, E., Jouanneau, Y. and Vignais, P.M. (1991) J. Biol. Chem. 266, 324-329
5 Jouanneau, Y., Meyer, C., Gaillard, J., Forest, E. and Gagnon, J. (1993) J. Biol. Chem. 268, 10636-10644
6 Armengaud, J., Meyer, C. and Jouanneau, Y. (1994) Biochem. J. 300, 413-418
7 Naud, I., Vinçon, M., Garin, J., Gaillard, J., Forest, E. and Jouanneau, Y. (1994) Eur. J. Biochem. 222, 933-939
8 Yakunin, A.F., Gennaro, G. and Hallenbeck, P.C. (1993) J. Bacteriol. 175, 6775-6780
9 Jouanneau, Y., Meyer, C., Naud, I. and Klipp, W. (1995) Biochim. Biophys. Acta (in press)
10 Armengaud, J. (1994) Ph.D. thesis. University of Grenoble, France
11 Armengaud, J., Gaillard, J., Forest, E. and Jouanneau, Y. (1995) Eur. J. Biochem. 231, 396-404
12 Saeki, K., Suetsugu, Y., Tokuda, K.I., Miyatake, Y., Young, D.A., Marrs, B.L. and Matsubara, H. (1991) J. Biol. Chem. 266, 12889-12895
13 Schmehl, M., Jahn, A., Meyer zu Vilsenforf, A., Hennecke, S., Masepohl, B., Schuppler, M., Marxer, M., Oezle, J. and Klipp, W. (1993) Mol. Gen. Genet. 241, 602-615
14 Moreno-Vivian, C., Hennecke, S., Pühler, A. and Klipp, W. (1989) J. Bacteriol. 171, 2591-2598
15 Willison, J.C., Pierrard, J. and Hübner, P. (1993) Gene 133, 39-46
16 Moshiri, F., Kim, J.W., Fu, C. and Maier, R.J. (1994) Mol. Microbiol. 14, 101-114
17 Duport, C. (1991) Ph.D. thesis. University of Grenoble, France
18 Duport, C., Jouanneau, Y. and Vignais, P.M. (1992) Mol. Gen. Genet. 231, 323-328

A PUTATIVE CO_2 TRANSPORTER IN *SYNECHOCYSTIS* PCC6803 DRIVEN BY NADPH DEHYDROGENASE-MEDIATED PHOTOSYSTEM-I CYCLIC ELECTRON FLOW

A. Katoh*, K.-S. Lee†, H. Fukuzawa†, K. Ohyama† and T. Ogawa*

*Biochemical Regulation, BioScience Center, Nagoya University, Nagoya 464-01, Japan and †Laboratory of Plant Molecular Biology, Department of Agricultural Chemistry, Kyoto University, Kyoto 606-01, Japan

1. Introduction

During the last several years there have been a number of important advances in the molecular analysis of the CO_2-concentrating mechanism in cyanobacteria [1]. These include the finding that complex-I type NADPH dehydrogenase (NDH) is involved in the transport of inorganic carbon (C_i; CO_2 and HCO_3^-) in *Synechocystis* PCC6803 [2-4]. It was inferred that this enzyme is an essential component of photosystem-1 cyclic electron flow which drives the C_i transport. Mi *et al.* [5-7] have verified this hypothesis by measuring the redox kinetics of P700 and chlorophyll fluorescence in *Synechocystis* PCC6803 and demonstrated that exogenous NADPH, photoreduced $NADP^+$ or ferredoxin donates electrons to P700 *via* plastoquinone in the spheroplasts or thylakoid membranes from the wild-type (WT) cells but not in those from the M55 mutant defective in *ndhB* [2]. They also showed that NADH was much less effective than NADPH as an electron donor [7]. The photosystem-1 cyclic electron flow mediated by NDH was postulated to supply ATP to drive the C_i transport.

The nature of C_i transporter is not known, but a number of evidence suggested the presence of separate transporters for CO_2 and HCO_3^-. For example, the HCO_3^- transport requires much higher concentration of sodium and has a lower affinity to the substrate than the CO_2 transport [8-13]. These differences in the properties between CO_2 and HCO_3^- transport enabled us to isolate mutants defective in CO_2 transport, which grew very slowly under the conditions unfavorable for HCO_3^- transport. We cloned and analyzed a DNA fragment of WT *Synechocystis* PCC6803 that, upon transformation, restores WT capabilities to both mutants. In this report we describe the properties of the mutants and a putative CO_2 transporter encoded by the gene (designated *cotA*) impaired in the mutants.

2. Procedure

Cells in a logarithmic phase of growth were mutagenized with N-methyl-N'-nitro-N-nitrosoguanidine (40 μg/ml) for 20 min in the light. After washing twice, the cells were grown under 3% CO_2 in air in normal BG11 medium [14] and then under air in the presence of ampicillin (100 μg/ml) in BG11 medium (pH7.0) where sodium concentration

P. Mathis (ed.), Photosynthesis: from Light to Biosphere, Vol. II, 807–812.

was reduced to 100 μM. Mutants were rescued as cells which grow normally on a BG11/plate under air but grew very slowly (SC1) or did not grow (SC2) on a BG11 (pH7.0) plate containing 100 μM sodium.

A clone that complements both mutants was isolated from a partial genomic library of WT *Synechocystis* constructed by ligating 3-4kbp DNA fragments of *Bam*HI digest into pKY184 vector [15]. A DNA fragment containing the contiguous region of the complementing clone was isolated from a λDASH library of WT *Synechocystis* by plaque hybridization. Complementation test was performed by the method of transformation reported by Dzelkalns and Bogorad [16]. After application of DNA (~50 ng) on BG11 plates (pH7) containing mutant cells in a top agar, the plates were incubated under 60 ppm CO_2 in the light. The transformants were visible in a week.

3. Result and Discussion

Sequencing of DNA in the clone that complemented SC1 and SC2 mutants and of DNA in the contiguous region revealed an open reading frame (ORF, designated *cotA*), which encodes a protein of 440 amino acids (Figs. 1, 2). An A to C substitution was found in *cotA* of SC1 that results in an Asn to His substitution in the deduced amino acid at residue 439 and a deletion of 7 bp nucleotides in *cotA* of SC2. The *cotA* product showed significant amino acid sequence homology to CemA found in an inner envelope membrane of pea chloroplasts [17, 18] and to the products of *cemA* homologues found in chloroplast genomes of other plants [19,20] (Fig. 2). The homology was much higher in the carboxyl-terminal region of 90 amino acid residues and showed at least 50% of sequence identity with CemA and its homologues.

Fig. 1. Restriction map of DNA that complements the mutants. The 1320-bp-nucleotide ORF (*cotA* gene) is indicated by the hatched bar, with an arrow showing the direction of *cotA*. B, *Bam*HI; Bg, *Bgl*II; E, *Eco*RI; P, *Pst*I; S; *Spe*I.

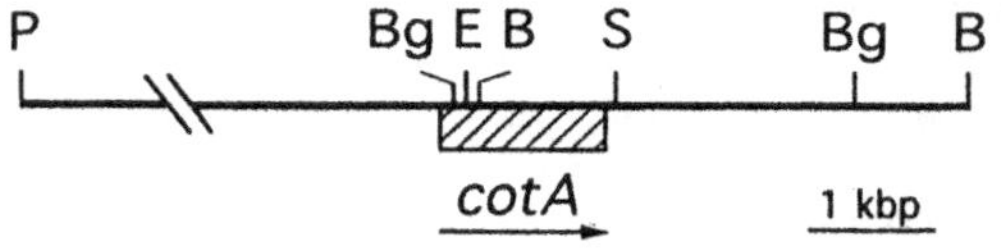

The SC1 and SC2 mutants grew as fast as the WT under air levels of CO_2 in BG11 medium. As expected, the growth of SC1 was much slower than the WT and SC2 was unable to grow under the conditions unfavorable for HCO_3^- transport; i.e. with 60 ppm CO_2 in normal BG11 medium (pH7.0) or with air in BG11 medium where sodium concentration was reduced to 100 μM (data not shown). Figure 3 shows the profiles of CO_2 exchange by the cells of WT, SC1 and SC2 grown in normal BG11 medium with air and indicates that the activity of CO_2 uptake in the mutants was about one fifth the activity of WT. The same result has been obtained with the SC1 mutant by measuring the uptake of $^{14}CO_2$ into the intracellular C_i pool; the activity of CO_2 uptake was much lower in the mutant than in the WT [4]. In contrast, there was no difference between the mutant

Fig. 2. Comparison of the deduced amino acid sequences for the products of *cotA* of *Synechocystis*, liverwort(MARPO) ORF434 [19], tobacco(TOBAC) ORF229 [20] and pea CemA [18]. :, Residues of plants identical to corresponding residues in *Synechocystis*. The European Molecular Biology laboratory data base was searched using the computer program developed by Person and Lipman [21]. ▾, Asn was substituted by His in the SC1 mutant. *, The histidine residue conserved among the products of the *Synechocystis* and plant genes.

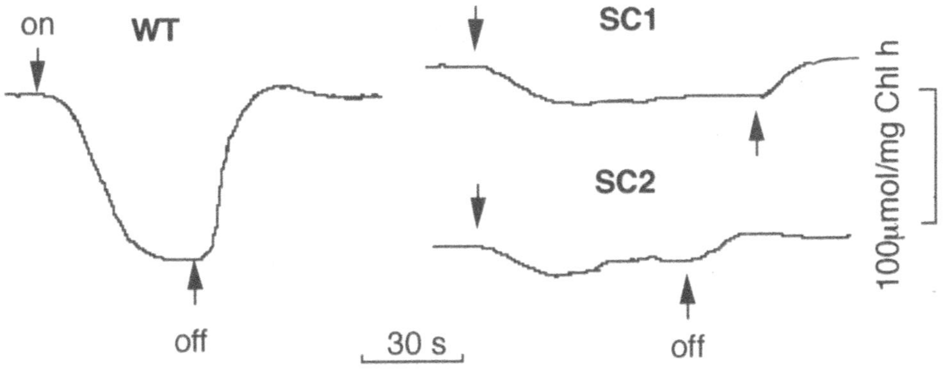

Fig. 3. The CO_2 exchange of air grown cells of WT, SC1 and SC2 upon switching the light on and off. Cells were suspended in Hepes-KOH buffer (pH7.0) containing 100 µM NaCl and air containing 340 ppm CO_2 was bubbled into the cell suspension in a reaction vessel and CO_2 concentration in the exchanged gas was measured by an infra-red CO_2 analyzer [22]. The observed CO_2 uptake under these conditions is due to CO_2 transport.

Fig. 4. The hydropathy profiles of the products of *cemA* and its homologues in pea, tobacco (TOBAC), liverwort (MARPO) and the *cotA* product of *Synechocystis*.

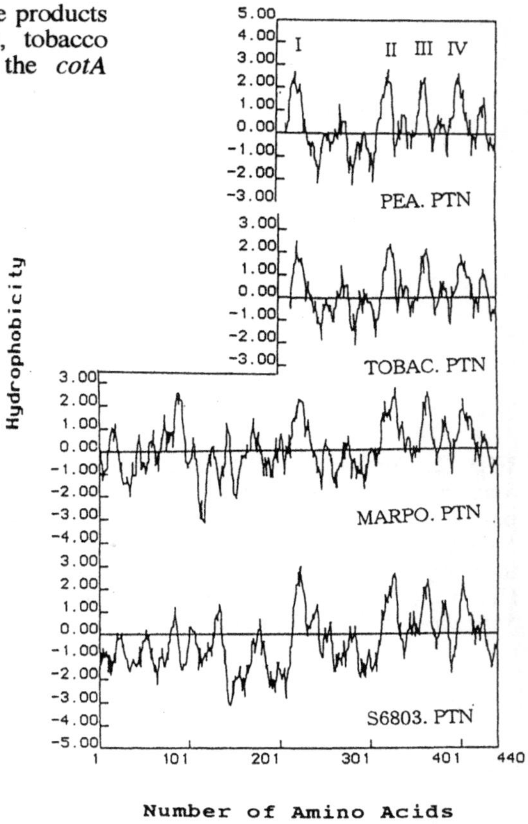

and the WT in the activity of $H^{14}CO_3^-$ uptake. Thus, the CO_2 transport system is specifically impaired in the mutant. These results clearly demonstrated that the inability of the mutants to transport CO_2 is a result of the impairment of the *cotA* gene.

The hydropathy profiles according to Kyte and Doolittle [23] showed 4 transmembrane domains from the middle to the carboxyl-terminal region of the protein (Fig. 4) and this structure was highly conserved among the products of *cotA*, *cemA* and the homologous ORFs. The *cotA* gene and the liverwort ORF434 have the additional 200 amino acids in the amino terminal regions, which might be lost in the tobacco and pea genomes in the course of evolution.

The *cotA* gene product could be a CO_2 transporter or a component of the CO_2 transporting system. NADPH dehydrogenase is required for energization of CO_2 transport [2-4]. Whether this enzyme is present in the cytoplasmic membrane must be carefully investigated but there is no doubt that the enzyme is present in the thylakoid membrane [3] and is an essential component of photosystem-I cyclic electron flow [4-7]. The NADPH-dehydrogenase-mediated cyclic electron transport is considered to produce ATP sufficient to drive the C_i transport. ATP binding motif was, however, not found in the *cotA* gene product. The CO_2 transporter may be composed of more than one subunits which include the *cotA* gene product and an ATP binding protein. One possible model is shown in Fig. 5, where CO_2 transporter consists of a CO_2 to HCO_3^- converter and a transporter functioning in cooperation with the converter. The *cotA* gene product may function as the converter and the ATP binding protein may transport the product. When CO_2 is transported into the cells, there is a stoichiometric extrusion of H^+ into the medium produced as a result of conversion of CO_2 to HCO_3^- [24]. ATP may also be consumed to extrude H^+ produced during the CO_2 transport.

CemA has been immunologically identified in the inner membrane of pea chloroplasts [17]. This suggests that the product of the *cemA* homologue in *Synechocystis*, *cotA*, may be located in the cytoplasmic membrane. An antibody raised against CemA did not crossreact with the cyanobacterial protein in Western blot analysis, which indicates that chloroplast CemA has some different characteristics from the cyanobacterial homologue, CotA.

Miller et al.[25] have proposed three models as the molecular mechanism for the CO_2 binding and the HCO_3^- formation steps in active CO_2 transport. These are the models of: 1) carbonic anhydrase-like moiety, 2) carbamate formation and 3) formation of transient metal complexes. The first and third models involve Zn and Ni, respectively, in the active site of the protein. The presence of the highly conserved histidine residue in the *cotA* gene product (asterisk in Fig.2) strongly suggests that the protein cherates a metal which plays a role in the conversion of CO_2 to HCO_3^-. The *cotA* gene has been expressed in *E. coli* and an antibody against the product is being raised. The use of the antibody will enable us to locate the product in the cell and also, by affinity chromatography, to isolate and purify the protein which will be subjected for biochemical and physiological analyses.

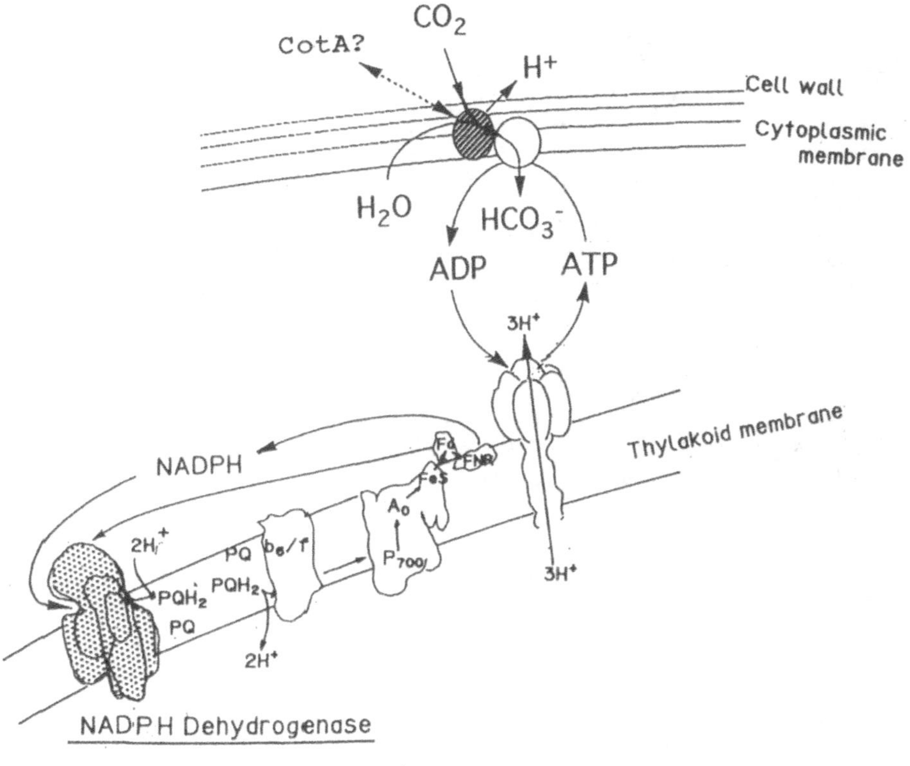

Fig. 5. A model of CO2 transporting system driven by NADPH-dehydrogenase-mediated photosystem-I cyclic electron flow.

REFERENCES

1. Kaplan, A. *et al.* (1995) in *The Molecular Biology of Cyanobacteria,* ed. Bryant, D.A. (Kluwer Academic Pub.), pp. 469-485.
2. Ogawa, T.(1991) *Proc. Natl. Acad. Sci. USA* **88**, 4275-4279.
3. Ogawa T (1992) *Plant Physiol*. **99**, 1604-1608
4. Ogawa, T. (1993) in *Photosynthetic Responses to the Environment*, eds.Yamamoto, H. & Smith, C. (American Society of Plant Physiologists, Rockville, MD), pp. 113-125.
5. Mi, H. *et al.* (1992) *Plant Cell Physiol.* **33,** 1233-1237.
6. Mi, H. *et al.* (1994) *ibid.* **35**, 163-173.
7. Mi, H. *et al.* (1995) *ibid.* **36,** 661-668.
8. Miller, A.G. & Canvin, D.T. (1984) *Plant Physiol.* **84**, 118-124.
9. Espie, G.S., Miller, A.G. & Canvin, D.T. (1988) *ibid.* **88**, 757-763.
10. Reinhold, L. *et al.* (1984) *ibid.* **76**, 190-192.
11. Espie, G.S., Miller, A.G. & Canvin, D.T. (1989) *ibid.* **91**, 387-394.
12. Espie, G.S., Miller, A.G. & Canvin, D.T. (1991) *ibid.* **97**, 943-953.
13. Yu, J-W., Price, G.D. & Badger, M.R. (1994) *Aust. J. Plant Physiol.* **21**, 185-195.
14. Stanier, R.Y. *et al..* **(1971)** *Bacteriol. Rev.* **35**, 171-205.
15. Ueguchi, C. & Itoh, K. (1992) *J. Bacteriol.* **174**, 1454-1461.
16. Dzelzkalns, V.A. & Bogorad, L. (1988) *EMBO J.* **7**, 333-338.
17. Sasaki, Y. *et al.* (1993) *FEBS Lett.* **316**, 93-98.
18. Willey, D.L. & Gray, J.C. (1990) *Plant Mol. Biol.* **15**, 347-356.
19. Ohyama, K. *et al.* (1986) Nature **322**, 572-574.
20. Shinozaki, K. *et al.* (1986) *EMBO J.* **5**, 2043-2049.
21. Pearson, W.R. & Lipman, D.J. (1988) *Proc. Natl. Acad. Sci.USA* **85**, 2444-2448.
22. Ogawa, T., Miyano, A. & Inoue, Y. (1985) *Biochim. Biophys. Acta* **808**, 77-84.
23. Kyte, J. & Doolittle, R.F. (1982) *J. Mol. Biol.*, **157**, 210-213.
24. Ogawa, T. & Kaplan, A. (1987) *Plant Physiol.* **83**, 888-891.
25. Miller, A.G., Espie, G.S. & Canvin, D.T. (1991) *Can. J. Bot.* **69**, 925-935.

Acknowledgements

We thank Dr. C. Ueguchi for providing us the pKY184 vector. This study was supported by Grants-in Aid for Scientific Research on Priority Areas (No. 0427103 to T.O. and No. 06259209 to H.F.) from the Ministry of Education, Science and Culture, Japan, and the New Energy and Industrial Technology Development Organization (NEDO), Japan.

O_2-DEPENDENT ELECTRON FLOW IN INTACT SPINACH CHLOROPLASTS: PROPERTIES AND POSSIBLE REGULATION OF THE MEHLER-ASCORBATE PEROXIDASE CYCLE

Ulrich Schreiber, Henning Hormann, Kozi Asada and Christian Neubauer
Lehrstuhl Botanik I, Universität Würzburg
Mittlerer Dallenbergweg 64; D-97082 Würzburg, Germany

1. Introduction

The transthylakoidal ΔpH plays a central role in the regulation of photosynthetic electron transport [1-3] and in the protection against photodamage by excess irradiation [4,5]. In recent years, numerous studies have been devoted to the elucidation of the molecular mechanisms of regulated radiationless energy dissipation correlated with ΔpH-formation and energy-dependent quenching of chlorophyll fluorescence [5-9]. Much less attention has been paid to the question of how the large ΔpH required for down-regulation of PSII is generated. Such down-regulation should be induced whenever the potential rate of charge separation in PSII exceeds the capacity of assimilatory flow. In this case, non-assimilatory electron transport coupled to proton translocation is required to increase the ΔpH to a "dissipative level" [2, 3, 7]. Major candidates for such energizing electron flow are nitrite- and oxaloacetate-reduction [10], cyclic PSI [11, 12] and the Mehler-Ascorbate Peroxidase Cycle [7,13-15]. Here we give a brief overview on the properties of the Mehler-Ascorbate Peroxidase Cycle, which has proven particularly effective in the formation of a regulatory ΔpH in intact spinach chloroplasts.

2. Material and Methods

Intact spinach chloroplasts were isolated from freshly harvested, greenhouse grown spinach [16]. The isolation procedure included a step of centrifugation through a layer of 40% (w/v) Percoll and 5 mM ascorbate was always present. The reaction medium was buffered at pH 8.0 and contained 15 mM ascorbate and 5 mM glycolaldehyde to inhibit carboxylation/oxygenation. 9-Aminoacridine (9-AA) was present at 1 µM.
Chlorophyll fluorescence was analyzed with a PAM Fluorometer (Walz, Effeltrich), as previously described [17]. 9-Aminoacridine fluorescence was measured with a pulse modulation fluorometer based on 1 µs xenon-flashes (Xe-PAM) [18]. Simultaneously with 9-AA fluorescence quenching, as a measure of ΔpH [19], electron transport was assessed polarographically (O_2-electrode) and by means of chlorophyll fluorescence quenching analysis [20,21].

P. Mathis (ed.), Photosynthesis: from Light to Biosphere, Vol. II, 813–818.

3. Results and Discussion

Definition of the Mehler-Ascorbate Peroxidase Cycle

While the principle role of O_2 as an electron acceptor in chloroplasts [22] and the function of ascorbate peroxidase in the detoxification of H_2O_2 [23, 24] have been known for some time, the concept of a specific regulatory cycle is relatively new and still awaits general acceptance. New understanding was initiated by mass spectrometric demonstration of tightly coupled photoreduction of $^{18}O_2$ and evolution of $^{16}O_2$ without $H_2{}^{18}O_2$ accumulation in the presence of an active ascorbate peroxidase [25]. This suggested that H_2O_2, photogenerated via univalent reduction of dioxygen in PSI and subsequent superoxide dismutation, is reduced to H_2O by ascorbate peroxidase using a photoreductant as electron donor. For the elucidation of the involved mechanisms chlorophyll fluorescence quenching analysis has been very useful [7, 21, 26] which allows the assessment of O_2-dependent charge separation rate at PSII, irrespective of whether this leads to a net O_2-change or not [20, 27]. Upon inhibition of CO_2-dependent O_2-evolution, fluorescence quenching analysis indicates the persistence of a substantial non-assimilatory flux, resulting in stimulation of non-photochemical quenching, which reflects increased ΔpH and non-radiative energy dissipation [7, 14, 17]. Arguments have been put forward [7, 14], which will be extended in the present contribution, that this energizing non-assimilatory flux does mainly involve electron transport in the Mehler-Ascorbate Peroxidase Cycle (MAP-cycle). The essential steps of the MAP-cycle are depicted schematically in Fig. 1. A central role is played by the ascorbate radical (monodehydroascorbate, MDA) which is formed upon H_2O_2-reduction by ascorbate (AsA) catalyzed by a membrane-bound ascorbate peroxidase (tAPX) [28]. This MDA is an excellent PSI acceptor [26, 28] and its Fd-dependent photoreduction [29] completes the cycle with full restoration of all involved components: There is as much O_2 consumed as O_2 evolved (by H_2O-splitting and superoxide-dismutation), the formed H_2O_2 is fully re-reduced to H_2O, and all ascorbate is restored by photoreduction of MDA.

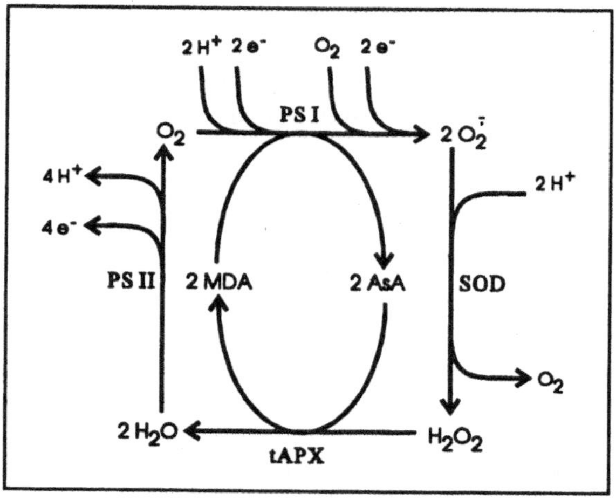

Figure 1: The Mehler - Ascorbate Peroxidase Cycle

Most importantly for the suggested physiological role, one turnover of this cycle (transient formation of one H_2O_2) consumes at least 8 quanta and leads to the transport of at least 16 H^+ (and possibly even 24 H^+) from the stroma into the thylakoid lumen. Hence, the MAP-cycle has a twofold potential for protection against photodamage: (1) by providing a valve-reaction in which O_2 and H_2O_2 (via MDA) serve as electron acceptors, (2) by creating a ΔpH which is not lowered by photophosphorylation, as the MAP-cycle does not consume ATP, in contrast to photorespiration [15,28].

Oxygen reduction and its activation by ΔpH

The MAP-cycle is initiated by O_2-reduction in the Mehler-reaction [22] which leads to superoxide formation and dismutation to H_2O_2 [31, 32]. Considering that it may withdraw reducing power from the Calvin cycle and that potentially harmful active oxygen species are formed as intermediates, strict regulation of its activity is essential. Hence, the starting reaction of O_2-reduction should be favored only if Calvin cycle activity is limiting. This is the case during the induction period of illumination after dark-adaptation [33] and upon excessive irradiation, which may be given even at moderate quantum flux densities whenever environmental stress factors lower the capacity of CO_2-fixation [21, 34]. The well-known Kautsky-effect of chlorophyll fluorescence induction is characterized by an initial peak of fluorescence yield, which can be suppressed by $NADP^+$ but not by O_2 [35]. It is followed by an O_2-dependent fluorescence decline which involves increases of photochemical and energy-dependent quenching [21]. This shows that O_2-reduction requires activation. Interestingly, this is not true when O_2-reduction is catalyzed by methyl viologen (MV) [35]. Information on the mechanism of activation has been obtained by comparative studies of the properties of the electron transport and ΔpH created by MV-catalyzed and uncatalyzed O_2-reduction [14, 17].

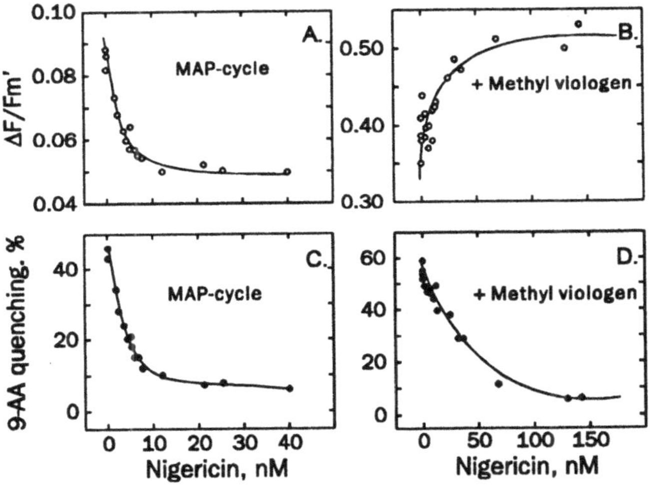

Figure 2: Effect of nigericin on the effective quantum yield of PSII assessed via ΔF/Fm' [27], and the ΔpH, as measured via 9-AA fluorescence quenching. Comparison of chloroplasts with active MAP-cycle and chloroplasts in the presence of methyl viologen. 650 μE/m^{2}s red irradiation.

The latter displays a remarkable pH 5 optimum, contrary to MV-dependent O_2-reduction with the usual pH 8 optimum [17, 36]. Furthermore, uncatalyzed O_2-dependent electron flow is stimulated by the ΔpH, whereas MV-dependent flow shows the usual suppression [14, 17]. These findings have led to the working hypothesis, that the naturally occurring uncatalyzed O_2-reduction, contrary to that mediated by MV, takes place within the thylakoid membrane, such that the formed superoxide can be either re-oxidized by oxidized PSI donors or first protonated (pKa = 4.8) and then dismutated to dioxygen and hydrogen peroxide [17]. Such a mechanism could explain the extreme sensitivity of the MAP-cycle to all factors affecting the integrity of the investigated chloroplasts and the potential extent of membrane acidification. In Fig.2 the effects of low concentrations of the uncoupler nigericin on O_2-dependent electron flow and on ΔpH formation are shown, comparing chloroplasts with MAP-cycle activity and chloroplasts in the presence of MV. It is apparent that the suppression of electron flow and of the ΔpH associated with MAP-cycle activity occurs at approximately 10 times lower nigericin concentration than the stimulation of MV-catalyzed O_2-reduction. Apparently, already very low concentrations of the uncoupler (I_{50} = 3 nM) can affect the activation of the MAP-cycle. These findings are important for two reasons: First, they call for great care in the treatment of chloroplasts and the use of certain chemicals when assessing MAP-cycle activity in isolated chloroplasts. For example, the fluorescing amine 9-AA and the inhibitor antimycin A also display some uncoupling action when used at too high concentrations, which may lead to misinterpretations. Second, they suggest the possibility of MAP-cycle regulation by the energy-status of the thylakoids. This aspect is further illustrated by the data in Fig. 3 which shows the effect of ADP-addition on O_2-dependent flow and on the associated ΔpH in class D chloroplasts in the absence of other acceptors. When upon ADP-addition the ΔpH is lowered by ATP-synthesis, there is not only a drop in membrane energization but also of MAP-cycle activity.

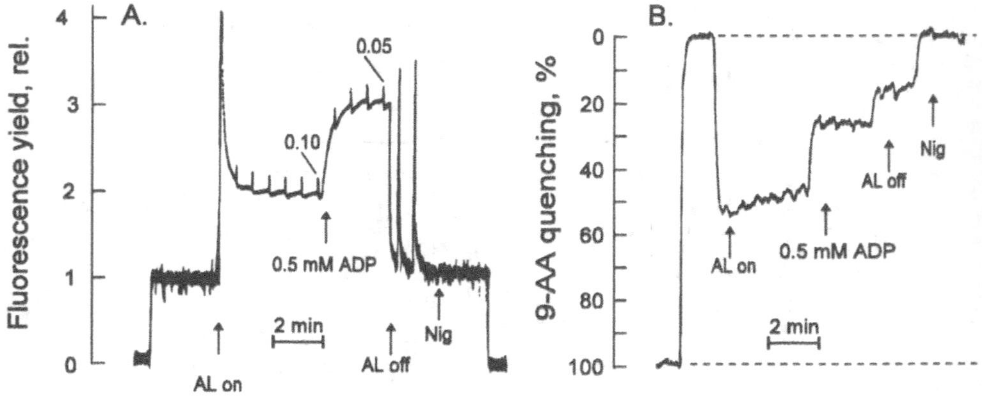

Figure 3: Effect of ADP-addition on MAP-cycle activity, as assessed via chlorophyll (A) and 9-AA fluorescence quenching (B). ADP lowers the effective PSII quantum yield ΔF/Fm' [27], from 0.1 to 0.05. Red actinic light (AL), 650 $\mu E/m^2s$

Requirements of O_2 and ascorbate for ΔpH-formation in intact chloroplasts

In the preceding section the role of the ΔpH, and particularly of internal acidification, for MAP-cycle activity was emphasized. On the other hand, besides its function as a valve reaction which consumes energy by electron cycling, the MAP-cycle can also contribute decisively to ΔpH-formation. Energy-dependent fluorescence quenching depends on the presence of dioxygen, with an apparent Km of 60 µM O_2 in intact spinach chloroplasts with high MAP-cycle activity [7]. This is confirmed by the data in Fig. 4 A which shows the dependence of ΔpH measured via 9-AA fluorescence quenching on the O_2-concentration in the medium, with half-maximal quenching around 60 µM O_2.

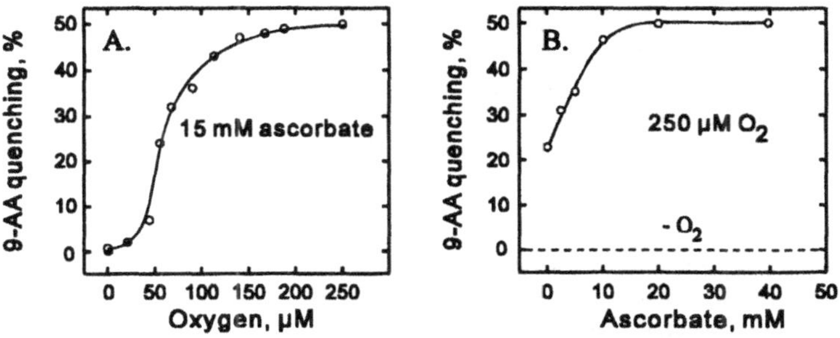

Figure 4: Oxygen and ascorbate requirement for light-induced ΔpH-formation in intact chloroplasts in the absence of acceptors expect those involved in the MAP-cycle. Light intensity, 1200 µE/m^{2}s, red light at 665 nm

While these data rule out that electron flow to other types of acceptors contributes significantly to ΔpH-formation in the given chloroplasts, there is the possibility of an O_2-requirement for poising cyclic electron flow [11, 12, 37, 38]. However, we tend to favor a major role of O_2 in ΔpH-formation by the MAP-cycle for the following reasons:

(1) Optimal redox-poising of cyclic flow requires much less O_2 than 60 µM [39].

(2) In intact chloroplasts with active MAP-cycle the relationship between 9-AA fluorescence quenching and linear electron flow (as assessed via ΔF/Fm' [27]) is almost identical to that in the presence of methyl viologen [14].

(3) In contrast to maize mesophyll thylakoids, spinach chloroplasts have a very low content of ferredoxin-reducible cyt b-559(Fd), which has been characterized as a carrier in cyclic PSI [40].

(4) As confirmed by the data in Fig.4 B., a high light-induced ΔpH in intact chloroplasts does not only require O_2 but ascorbate as well. Hence, the known ascorbate requirement for energy-dependent fluorescence quenching [41] results from its role in the ΔpH-formation as well as in the deepoxidation of violaxanthin to zeaxanthin [5, 9, 13, 34, 41]

In conclusion, substantial evidence has been accumulated suggesting that O_2-dependent electron flow in the MAP-cycle has the potential of generating a large ΔpH in intact spinach chloroplasts. Outstanding features of this cycle are its stimulation by ΔpH, with the consequence that no saturation at high light intensities is observed [14], and its sensitivity to the energy-status of the thylakoid membrane. On the basis of these features, the MAP-cycle may qualify for a important role in the down-regulation of PSII when irradiation is excessive. However, more experimentation will be necessary to confirm that these features are also relevant for regulation of photosynthesis under excessive illumination in intact leaves.

References

1 Heber, U., Egneus, H., Hanck, U., Jensen, M., and Köster, S. (1978) Planta 143, 41-49
2 Foyer, C., Furbank, R., Harbinson, J. and Horton, P. (1990) Photosynth. Res. 25, 83-100
3 Walker, D. (1992) New Phytol. 121, 325-345
4 Krause, G. and Behrend, U. (1986) FEBS Lett. 220, 298-302
5 Demmig-Adams, B. (1990) Biochim. Biophys. Acta 1020, 1-24
6 Weis, E. and Berry, J. (1987) Biochim. Biophys. Acta 894, 198-208
7 Schreiber, U. and Neubauer, C. (1990) Photosynth. Res. 25, 279-293
8 Krieger, G., Moya, I. and Weis, E. (1992) Biochim. Biophys. Acta 1102, 167-176
9 Horton, P., Ruban, A.V. Rees, D., Pascal, A.A., Noctor, G.D. and Young, A.J. (1992) FEBS Lett. 292, 1-4
10 Backhausen, J., Kitzmann, C. and Scheibe, R. (1994) Photosynth. Res. 42, 75-92
11 Heber, U. and Walker, D. (1992) Plant Physiol. 100, 1621-1626
12 Kobayashi, Y. and Heber, U. (1994) Photosynth. Res. 41, 419-428
13 Neubauer, C. and Yamamoto, H. (1992) Plant Physiol. 99, 1354-1361
14 Hormann, H., Neubauer, C. and Schreiber, U. (1994) Photosynth. Res. 41, 429-437
15 Osmond, C.B. and Grace, S.C. (1995) J. Exp. Bot. in press
16 Asada, K., Neubauer, C., Heber, U. and Schreiber, U. (1990) Plant Cell Physiol. 31, 557-564
17 Hormann, H., Neubauer, C., Asada, K. and Schreiber, U. (1993) Photosynth. Res. 37, 69-80
18 Schreiber, U. Neubauer C. and Schliwa U. (1993) Photosynth. Res. 36, 65-72
19 Schuldiner, S., Rottenberg, H. and Avron, M. (1971) Eur. J. Biochem. 25, 64-70
20 Schreiber, U., Schliwa, U. and Bilger, W. (1986) Photosynth. Res. 10, 51-62
21 Schreiber, U., Bilger, W. and Neubauer, C. (1994) In: Ecophysiology of Photosynthesis. (Schulze E.-D. and Caldwell, M.M., eds.) pp. 49-70, Springer-Verlag, Berlin
22 Mehler, A. (1951) Arch. Biochem. Biophys. 33, 65-77
23 Groden, D. and Beck, E. (1979) Biochim. Biophys. Acta 546, 1455-1462
24 Nakano, Y. and Asada K. (1981) Plant Cell Physiol. 22, 867-880
25 Asada, K. and Badger, M. (1984) Plant Cell Physiol. 25, 1169-1179
26 Neubauer, C. and Schreiber, U. (1989) Z. Naturforsch. 44c, 262-270
27 Genty, B., Briantais, J.M. and Baker, N.R. (1989) Biochim Biophys. Acta 990, 87-92
28 Miyake, C. and Asada, K. (1992) Plant Cell Physiol. 33, 541-553
29 Miyake, C. and Asada, K. (1994) Plant Cell Physiol. 35, 539-549
30 Osmond, C.B. (1981) Biochim. Biophys. Acta 693, 77-98
31 Elstner, E. and Kramer, R. (1973) Biochim. Biophys. Acta 314, 340-353
32 Asada, K., Kiso, K. and Yoshikawa, K. (1974) J. Biol. Chem. 249, 2175-2181
33 Radmer, R. and Kok, B. (1976) Plant Physiol. 58, 336-340
34 Björkman, O. and Demmig-Adams, B. (1994) In: Ecophysiolgy of Photosynthesis (Schulze, E.-D. and Caldwell, M.M., eds.) pp. 17-47, Springer-Verlag, Berlin
35 Neubauer, C. and Schreiber, U. (1989) Bot. Acta 102, 314-318
36 Schreiber, U., Reising, H. and Neubauer, C. (1991) Z. Naturforsch. 46c, 173-181
37 Arnon, D.J. and Chain, R.K. (1975) Proc. Natl. Acad. Sci. USA 72, 4961-4965
38 Ziem-Hanck, U. and Heber, U. (1980) Biochim. Biophys. Acta 591, 266-274
39 Heber, U. (1969) Biochim. Biophys. Acta 180, 302-319
40 Miyake, C., Schreiber, U. and Asada, K. (1995) Plant Cell Physiol. 36, 743-748
41 Neubauer, C., and Yamamoto, H.Y. (1992) Plant Physiol. 99, 1354-1361

BIOGENESIS OF CYTOCHROME COMPLEXES IN *Rhodobacter capsulatus*

Hans-Georg Koch, Olivia Hwang, Michael Grooms and Fevzi Daldal. Plant Science Institute, University of Pennsylvania, Department of Biology, Philadelphia, PA 19104-6018, USA

1. Introduction

In living cells there are many and diverse energy transduction mechanisms at work, utilized as metabolic pathways to provide energy for cellular functions. Despite the variety of these pathways, they all follow common themes in both structure and function. Specifically, they use a group of heme containing proteins, the cytochromes, ubiquitous in electron transfer pathways of nearly all organisms. Photosynthetic bacteria provide useful systems for studying the pathway of energy transduction and their regulation because of the diversity of their metabolism (1), and their well developed genetics (2). *Rhodobacter capsulatus* can use two different terminal oxidases for respiratory growth, the quinol oxidase (Q_{OX}) and the recently purified cyt *cbb*3 oxidase (3,4), which is a new type of cytochrome *c* oxidase. We used this new enzyme as a handle to study the biogenesis of cytochrome complexes in *Rb. capsulatus*. Considering, that cytochrome *c* oxidase is likely to be involved in the transition from aerobic to anaerobic growth, one would expect that cells have an exquisite mechanism for detecting environmental changes, and an ability to convey this information to a regulatory network for controlling the synthesis of various components involved in these pathways. To analyze this regulatory network, cytochrome oxidase-negative mutants of *Rb. capsulatus* were sought, and at least five different types of mutants were obtained based on genetic complementations and phenotypic characterizations.

2. Procedure

2.1 Molecular Genetic techniques

Molecular genetic techniques were used as described by Daldal *et al.* (5) and Sambrook *et al.* (6). The chromosomal libraries used were described previously (7). Mutagenesis was performed using ethyl methane sulfonate (EMS). Mutants were screened with the NADI reagent (8) for cytochrome *c* oxidase activity and for their ability to grow under photosynthetic conditions.

P. Mathis (ed.), Photosynthesis: from Light to Biosphere, Vol. II, 819–822.

2.3 Isolation of Chromatophores and SDS-PAGE Analysis

Chromatophores were prepared using 20 mM MOPS pH 7.0, 1 mM KCl buffer in the presence of 1 mM phenylmethylsulfonylfluoride as described previously (5). SDS-PAGE was performed according to Schaegger and von Jagow (9). Gels were stained either for proteins with Coomassie Brilliant Blue or with 3,3',5,5'-tetramethylbenzidine (TMBZ) and H_2O_2 to reveal the heme peroxidase activity of *c*-type cytochromes as described by Thomas et al. (10).

3. Results and Discussion

3.1 Isolation and Phenotype of Nadi⁻ Mutants

Using EMS- and Tn5-mutagenesis several Nadi⁻ mutants were isolated and their cytochrome *c* content was determined by the Schägger-type SDS-PAGE system (9). This analysis indicated that at least five different types of mutants were found (Table 1). Four distinct membrane-attached *c*-type cytochromes with M_r of 32, 31, 29 and 28 kDa are readily detected in chromatophore membranes of the wild type strain *Rb. capsulatus* MT1131 grown semiaerobically in MPYE-rich medium. Of these cytochromes the 31 kDa protein is the cyt c_1 of the cyt bc_1 complex, and the 29 kDa protein corresponds to the newly discovered membrane-associated cyt c_y (7). The two remaining 32- and 28 kDa cytochromes correspond to the di-heme cytochrome c_p and to the mono-heme cytochrome c_o, which are subunits of the cyt cbb_3 oxidase (4). In the *Rb. capsulatus* mutant M7 the 32 kDa protein is missing, while DB8 is devoid of both the 32 kDa and 28 kDa protein (Table 1). In comparison to the wild type strain MT1131, M4 and BK4 (Type II) are missing both the 32kDa and 28kDa protein. On the other hand, BK1 is characterized by the complete absence of all *c*-type cytochromes (Type IV) and is Ps⁻, while GK2, OH1, MR2 and IJ1 (Type V) are missing only the 32 kDa and 28 kDa protein and are Ps⁺. BK5 (Type III) contains all c-type cytochromes.

Table 1: Characterization of different NADI⁻ mutants of *Rb. capsulatus* [a].

Mutants	Phenotype		Complementation		Cytochromes				
	NADI	Ps[b]	p5T	p4A	c_p	c_1	c_y	c_o	c_2
Type I									
M7	-	+	+	-	-	+	+	+	+
DB8	-	+	+	-	-	+	+	-	+
Type II									
M4	-	+	-	+	-	+	+	-	+
BK4	-	+	-	+	-	+	+	-	+
Type III									
BK5	-	+	-	-	+	+	+	+	+
Type IV									
BK1	-	-	-	-	-	-	-	-	-
Type V									
OH1	-	+	-	-	-	+	+	-	+
GK2	-	+	-	-	-	+	+	-	+
IJ1	-	+	-	-	-	+	+	-	+
MR2	-	+	-	-	-	+	+	-	+

[a] for details see text ; [b] photosynthetic growth

3.2 Complementation of the Nadi⁻ Mutants

A chromosomal library was used to complement the NADI⁻ mutants M7 and M4, and two different plasmids, p5T and p4A, respectively, were isolated. p5T carries a 16.6 kb DNA fragment and also complements the mutant DB8 (Type I), while p4A contains a 6 kb BamHI-fragment in pRK404, and also complements the mutant BK4 (Type II). Restriction analysis of p4A and p5T showed no identical fragments, indicating that the two plasmids are different. The mutants BK5 (Type III), BK1 (Type IV), OH1, MR2, GK2 and IJ2 (Type V) can not be complemented by either p5T or p4A and are different in regards to their *c*-type cytochrome pattern, and therefore distinct (Table 1). These data anticipate that at least five different type of mutations affect the presence of an active cyt *cbb*3 oxidase. Several subclones of the plasmids p5T and p4A were constructed and tested for their ability to complement the appropriate mutants to define the location of the complementing gene(s). In the case of p5T, a 2.2 kb Hind III-BamHI-fragment (p5TΔH, Fig. 1) and for p4A, a region encompassing the 1.7 kb BglII fragment were identified as being sufficient to complement M7, and M4 , respectively. At present, the plasmids complementing the other types of mutants and the molecular nature of their defects remains unknown.

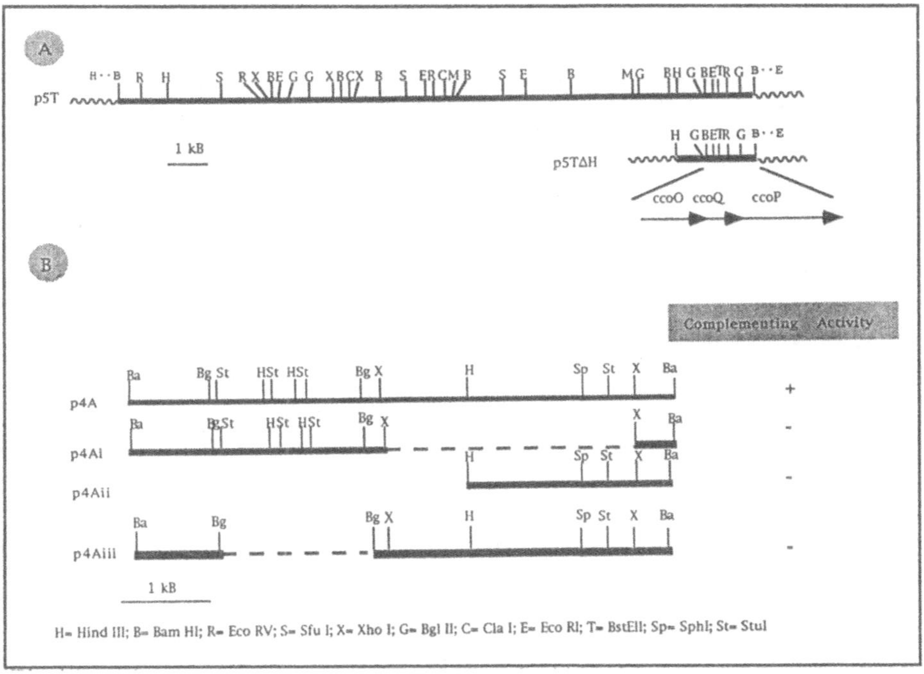

Fig.1 Restriction maps of the plasmids p5T and p4A. **(A)** Maps of p5T and p5TΔH; the orientation of the structural genes of cyt *cbb*3 within the 1.3 BamHI fragment is indicated by arrows. **(B)** Mapping of p4A; the fragments missing in various subclones are shown as dashed lines.

3.3 DNA Sequence Analysis of the Plasmids p5T and p4A

DNA sequence analysis was performed to determine the molecular basis of the complementing ability of the plasmids p5TΔH and p4A. First, the 1.25 kb BamHI-fragment of p5TΔH, was cloned into pBluescript (creating pMG1) and sequenced. The obtained sequence showed nearly 95% homology to part of the *ccoNOQP* gene cluster of *Rb. capsulatus* 37b4, encoding a *cb*-type cytochrome oxidase (11) and carries *ccoP*, *ccoQ* and the end of *ccoO*. The minimal fragment sufficient for the complementation (p5TΔH) is smaller than the proposed *ccoNOQP*-operon of 3.9 kb (11). Next, a MluI-BamHI-fragment 0.2 kb upstream of the HindIII-site (Fig.1) was cloned and partially sequenced. The obtained sequence showed no homology to *ccoN*, suggesting that apparently this gene is not located upstream of the *ccoOQP* cluster in *Rb. capsulatus* MT1131. Thus, possibly the genetic organization of the *cco* gene cluster is different between the strains 37b4 and MT1131.

3.4 Construction of strain MG1, lacking cyt *cbb*3 oxidase

The CcoQ gene product is a small polypeptide (predicted molecular weight: 6.5 kDa) of unknown function. The unique BstEII-site within the *ccoQ* gene located on the 1.25 kb BamHI fragment of pMG1, was used for insertion of a Kanamycin^R-cartridge. The obtained *ccoQ*::Km mutation was introduced into MT1131 using the gene transfer agent, and the Km^R mutant MG1 was tested for cyt *c* oxidase activity. Like M7G and DB8, MG1 was also NADI$^-$ and it complemented to Nadi$^+$ by the plasmids p5T and p5TΔH. Analysis of the cytochrome *c* profile of MG1 revealed that it is missing both the 32 kDa and 28 kDa protein, while its complementation with p5TΔH resulted in a cytochrome pattern identical to the wild type. This finding indicates that ccoQ is required for the steady-state presence of cyt c_o in chromatophore membranes (data not shown).

Acknowledgment: The work in this laboratory is supported by grants from NIH (GM 38 237) and DOE (91 ER 20052).

References

1 Ferguson, S.J., Jackson, J.B. and McEwan, A.G. (1987) FEMS Microbiol. Rev. **46**, 117-143
2 Donohue, T.J. and Kaplan, S. (1991) Methods Enzymol. **204**, 459-485
3 Marrs, B. and Gest, H. (1973) J. Bacteriol. **114**, 1045-1051
4 Gray, K., Grooms, M., Myllykallio, H., Moomaw, C., Slaughter, C. and Daldal, F. (1994) Biochemistry **33**, 3120-3127
5 Daldal, F., Cheng, S., Applebaum, J., Davidson, E. and Prince, R.C. (1986) Proc. Natl. Acad. Sci. USA, **83**, 2012-2016
6 Sambrook, J., Fritsch, E.F. and Maniatis, T. (1989) Molecular Cloning: A Laboratory Manual. 2nd edition. Cold Spring Harbor Laboratory Press, Cold Spring Harbor , N.Y.
7 Jenney, F.E. and Daldal, F. (1993) EMBO Journal **12**, 1283-1292
8 Keilin, D. (1966) The history of cell respiration and cytochrome. Cambridge University Press, Cambridge.
9 Schägger, H. and von Jagow, G. (1987) Anal. Biochem. **166**, 368-379
10 Thomas, P.E., Ryan, D. and Levin, W. (1976) Anal. Biochem. **75**, 168-176
11 Thöny-Meyer, L., Beck, C., Preisig, O. and Hennecke, H. (1994) Molec. Microbiol. **14**, 705-716

CHARACTERIZATION OF A DOUBLE DELETION MUTANT THAT LACKS CYTOCHROME C_6 AND CYTOCHROME M IN *SYNECHOCYSTIS* 6803

Sabine U. Metzger[1], Himadri B. Pakrasi[3], and John Whitmarsh[1,2]

[1]Center of Biophysics and Computational Biology, University of Illinois
[2]Photosynthesis Research Unit, USDA/ARS, Urbana, IL 61801
[3]Department of Biology, Washington University, St. Louis, MO 63130

1. Introduction

Recently, a novel c-type cytochrome (cyt M) has been reported by Malakhov et al. (1). Its predicted amino acid sequence reveals homology of its hydrophilic portion to water-soluble small c-type cytochromes involved in electron transfer processes in the respiratory chain in mitochondria, and in particular to cytochrome c_y, a membrane bound carrier in both photosynthetic and respiratory electron transport in *Rhodobacter sphaeroides* (2). However, deletion of the cyt M gene from *Synechocystis* did not affect photoautotrophic growth, respiration or oxygen evolution rates indicating that cyt M is not essential for normal cell growth (1).

In the cyanobacterium *Synechocystis* 6803 plastocyanin or cytochrome c_6 (cyt c_6), formerly known as cyt c_{553}, can transport electrons between the cytochrome bf complex and Photosystem I (PS I). Mutant strains of *Synechocystis* can be grown so that both plastocyanin and cyt c_6 are absent (3). The mutants grow photoautotrophically and sustain normal rates of oxygen evolution and dark respiration, indicating the presence of a third electron transport protein (4). We have measured the oxidation kinetics of cytochrome f and reduction kinetics of P700 in the mutant strains *in vivo* and found that electron transfer occurs 4-6 times slower than in WT cells. These results imply that the putative carrier is less efficient than plastocyanin or cyt c_6, possibly due to a different stoichiometry or to its intrinsic properties. However, we can not exclude the possibility of direct electron transfer from cytochrome f to P700.

We investigated the possibility that cyt M may be the third carrier, operating between the cyt bf complex and PSI by creating a mutant strain of *Synechocystis* that lacks cyt c_6, cyt M and plastocyanin when grown in the absence of copper.

2. Procedures

2.1 *Cell culture and DNA isolation.* Cells of *Synechocystis* 6803 were grown and total DNA extracted as described by Zhang et al. (1992, (3)). Southern blot hybridization was performed using a 470 bp cytM gene fragment as a probe.

2.2 *The Deletion/Insertion Plasmid Construct of the CytM Gene.* The strategy we followed was based on that of Malakhov *et al.* (1). Two oligonucleotides, which hybridize to a region upstream of the 5' end and to a region downstream of the 3' end of the open reading frame, were synthesized to generate a 1000 bp PCR clone of the cyt M gene from the genomic DNA of *Synechocystis* 6803. The PCR clone was digested with

P. Mathis (ed.), Photosynthesis: from Light to Biosphere, Vol. II, 823–826.

KpnI and *HindIII* and ligated into a pUC118 vector. To create the deletion/insertion construct, the plasmid was digested with *NcoI*, cutting out most of the coding region of the cytM gene (220 bp fragment). The chloramphenicol resistance marker cassette was ligated into the *NcoI* digested plasmid by Klenow fill-in reaction, thereby creating the cyt M deletion plasmid.

2.3 *Transformation of Synechocystis 6803.* *Synechocystis* 6803 was transformed by modification of the method of Williams (1988, 5). The mutant strain M109 that lacks the gene of cyt c_6 was transformed with the cyt M deletion plasmid to create a double deletion mutant M223 lacking both cytochrome c_6 and cytochrome M. The deleted cyt M gene was confirmed by PCR and by Southern blot analysis (not shown).

2.4 *Steady-state Electron Transport Rates.* The steady-state electron transport rates were measured with a Clark-type oxygen electrode as described by Zhang et al. (3)

2.5 *Flash-induced absorption spectroscopy.* For the measurement of P700 reduction, after dark-adaption, the sample was exposed to the measuring beam (laser-diode, 827 nm) for 10-20 ms prior to the laser flash (Nd-YAG laser, 532 nm, pulse width 8 ns, output energy 150 mJ). Upon the actinic laser flash, the oxidation and rereduction of P700 was monitored as an absorption increase and decrease, respectively, at 827 nm. Time-resolved light-induced absorbance changes in cytochromes were measured using a laboratory-built spectrophotometer as described elsewhere (3).

3. Results and Discussion

3.1 *Photoautotrophic growth.* The double deletion mutant, lacking both cyt c_6 and cyt M, grows photoautotrophically, both in the absence of copper which prevents the synthesis of plastocyanin, and in the presence of copper (Table 1).

Table 1 Growth rates of *Synechocystis* WT and mutant strains at 30°C

Strain	Doubling Time: light intensity 60 $\mu E \cdot m^{-2} \cdot s^{-1}$	Doubling Time: light intensity 270 $\mu E \cdot m^{-2} \cdot s^{-1}$
WT (-Cu)	16 ± 2 hrs	8 ± 1 hrs
M109 (-Cu)	18 ± 2 hrs	9 ± 1 hrs
M223 (-Cu)	24 ± 2 hrs	16 ± 2 hrs
M223 (+Cu)	20 ± 2 hrs	14 ± 2 hrs

However, in the double deletion strain M223 the growth curves reveal a significant lag phase as well as sensitivity to high growth light (Figure 1). This lag phase is not evident in either of the single deletion strains . The double mutant (M223) cultures were yellow or yellow/green, in contrast to the other strains which were green.

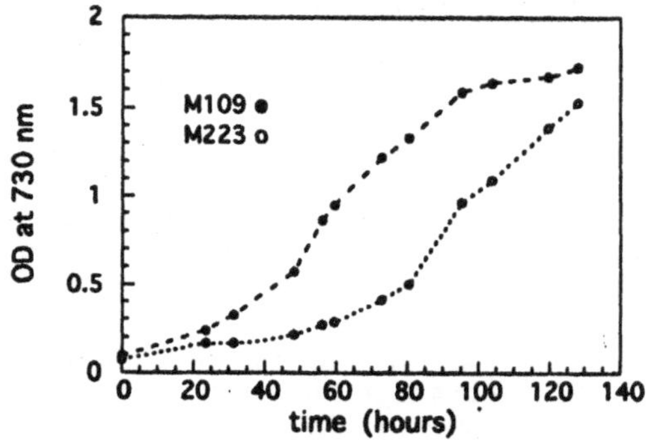

Fig.1 Growth curves of the single deletion mutant strain M109 and of the double deletion mutant strain M223 both grown in the absence of copper, at 30 °C and at a light intensity of 60 µE.

3.2 *Electron transfer kinetics.* The kinetics of P700 reduction *in vivo* in WT when grown either in the absence or presence of copper reveal that P700 is rapidly reduced with a half-time of 150-200 μs for 50% of P700 (Table 2). Similar kinetics are observed in the single mutants when either plastocyanin or cyt c_6 are present (Figure 2A). In contrast, the single cyt c_6 deletion mutant M109 grown in the absence of copper as well as the single plastocyanin deletion mutant M114 grown in the presence of copper that lack both plastocyanin and cyt c_6, display slower kinetics with only a minor somewhat variable fast phase of P700 reduction (Figure 2B). The kinetics of P700 reduction in M223 when grown in the presence of copper were identical to the WT, as expected since plastocyanin was present. In copper-depleted M223 cells however, P700 reduction was much slower than observed in WT cells (Figure 2C). In M223, reduction of P700 was monophasic exponential with a half-time of 8 ms.

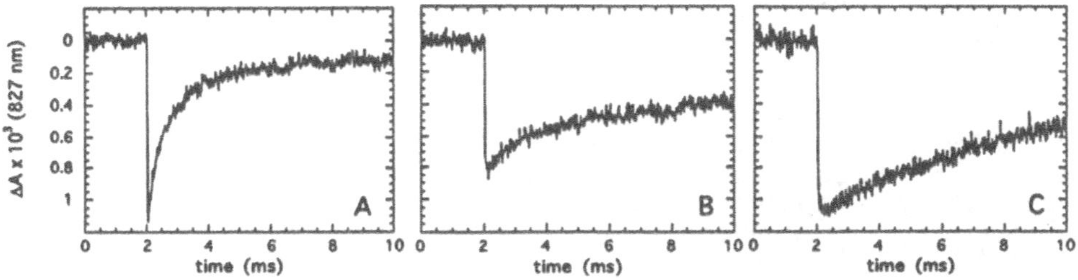

Fig.2 Kinetics of P700 reduction at 24 °C in whole cells from *Synechocystis* in a reaction medium containing 5 mM HEPES (pH 7.5), 10 mM NaCl, 10 mM NaCO3, and 50% ficoll. A: M114 cells grown without copper, Chl conc. 30 μM. B: M114 cells grown in 1 μM copper, Chl conc. 24 μM. C: M223 cells grown without copper, Chl conc. 36 μM.

3.3 *Light-induced cytochrome turn-over* In M223 cells grown in the presence or absence of copper, light-induced cytochrome turn-over due to cytochrome f could be observed when a long flash was given onto the sample to oxidize cytochrome f. Therefore, cytochrome f oxidation and reduction does occur in cells that lack cyt c_6, plastocyanin and cyt M (data not shown).

3.4 *Electron transport activities.* Steady-state electron transport rates were measured as light-induced oxygen evolution in the presence of bicarbonate. The electron transport rates in M223 grown without copper were similar to those in WT and M109 grown with or without copper (Table 2). The P700/Chl ratio of M223 cells grown without copper was similar to that of WT cells (P700/Chl:1/200). It is noteworthy that the P700 reduction half-time of 8 ms is the same as the rate-limiting step (plastoquinol oxidation) of WT cells which explains similar oxygen evolution rates. Therefore, it is unlikely that the slow rereduction rate of P700 could account for the slower growth rate.

Table 2 P700 reduction kinetics (at 24 °C) and steady state electron transport rates [H_2O to CO_2 (μmol O_2 · μmol chl^{-1} · hr^{-1}) at 28 °C)] in whole cells of WT and mutant strains of *Synechocystis* 6803

	Wt -Cu	Wt +Cu	M109 -Cu	M223 -Cu
P700 reduction kinetics	150 μs (50%) 2 ms (50%)	200 μs (60%) 2 ms (40 %)	700 μs (25%) 10 ms (75%)	8 ms (100%)
Photosynthetic Electron Transport Rate	260 ± 10	220 ± 20	220 ± 30	280 ± 30
Dark Respiration Rate	31 ± 10	25 ± 5	22 ± 6	32 ± 10

4. Conclusions

Our results show no evidence for a significant participation of cyt M to photosynthetic electron transfer. The fact that M223 grows photoautotrophically in the absence of copper implies that an additional alternate mechanism of electron transport between the cytochrome b/f complex and P700 must exist. Such a mechanism, either an as yet uncharacterized carrier protein or direct transfer from cytochrome f to P700, is necessary to account for the near-normal rates of oxygen evolution from this strain , and for the 9 ms phase of P700 reduction. The presence of plastocyanin, cyt c_6 and cyt M is therefore not essential for functional electron transfer. Nevertheless, in the absence of plastocyanin and cytochrome c_6, as well as in the absence of all three proteins, the photosynthetic electron transport between the cyt bf complex and Photosystem I is slowed down more than ten-fold. Furthermore, even in the absence of all three proteins, dark respiration rates are unaltered indicating that respiration can occur via pathways that exclude contribution of these electron carrier proteins.

Acknowledgments

We thank Dr. M. Malakhov and Prof. N. Murata for providing us with the cyt M deletion mutant strain ΔcytM.

References

1 Malakhov, M., Wada, H., Los, D., Semenenko, V., and Murata, N. (1994) J. Plant Physiol. 144, 259-264
2 Jenney, F. E. Jr. and Daldal, F. (1993) The EMBO Journal 12, 1283-1292
3 Zhang, L., Mc Spadden, B., Pakrasi, H. B. and Whitmarsh, J. (1992) J. Biol. Chem. 267, 19054-19059
4 Zhang, L., Pakrasi, H. B. and Whitmarsh, J. (1994) J. Biol. Chem. 269, 5036-5042
5 Williams, J.G.K. (1988) Methods in Enzymology 167, 766-778

THE PARTICIPATION OF CYCLIC ELECTRON TRANSPORT IN THYLAKOID MEMBRANE ENERGIZATION OF MAIZE MESOPHYLL CHLOROPLASTS

Boris N. Ivanov[1], and Gerald E. Edwards[2],
[1]Institute of Soil Sciences & Photosynthesis, Pushchino, Russia;
[2]Botany Department, Washington State University, Pullman, WA, USA.

1. Introduction

The input of cyclic electron flow (CEF) around PS1 in chloroplast energetics continues to be a topic under investigation as to its function and role in photosynthesis. There are no doubts that this cycle, which uses part of the electron transport chain from b_6f-complex to PS1 acceptors in the thylakoid membrane and ferredoxin outside the thylakoids, can provide membrane energization resulting in ATP synthesis [1,2,3]. However, the role of CEF as the source of membrane energization and ATP in intact systems needs to be proven in any special case. Evidence for CEF operation in the mesophyll chloroplasts of C_4 plants was obtained in several studies, but mostly under conditions where "only" the cyclic pathway operated in the absence of oxaloacetate (OAA), the main acceptor of noncyclic electron flow (NCEF) in these chloroplasts [4,5]. The most convincing evidence in favor of CEF operation *in vivo* was the decrease in rates of ATP dependent reactions in the presence of Antimycin A (Ant A), which only inhibits CEF at properly chosen concentrations [3]. Such an inhibition of ATP dependent CO_2 fixation by Ant A was observed with C_4 mesophyll chloroplast preparations containing phosphoenolpyruvate (PEP) carboxylase, with pyruvate + Pi as substrates in the absence of OAA [5]. This inhibition was associated with the ATP-dependent conversion of pyruvate and Pi to PEP, the latter being the substrate for PEP carboxylase. This communication presents new evidence for CEF participation in maize mesophyll chloroplast energetics based on the effects of Ant A.

2. Procedure

The experiments were conducted with intact maize mesophyll chloroplasts (intactness more 80%) isolated according to Jenkins and Russ [6]. The basic reaction medium (BRM) contained 0.3 M sorbitol, 50 mM HEPES or Tricine (pH 7.7 - 7.8), and 0.5 mM EDTA. The reactions were carried out at 30°C and at PPFD of 1200 µmol quanta m^{-2} s^{-1} provided from white light (except in 9-aminoacridine fluorescence measurements). Rates of O_2 exchange were measured in a Hansatech O_2 electrode system. Carbon fixation was measured according to Huber and Edwards[5] by determining rates of incorporation of $^{14}CO_2$ into acid-stable products. Phosphoenolpyruvate (PEP) and malate (MAL) were determined enzymatically in samples taken from the reaction mixture by measuring NADH oxidation in the presence of PEP carboxylase and malate dehydrogenase for PEP determination and by measuring NADP reduction in the presence of malic enzyme for

P. Mathis (ed.), Photosynthesis: from Light to Biosphere, Vol. II, 827–830.

MAL determination. Light induced changes of 9-aminoacridine (9AA) fluorescence were measured in the reaction mixture placed in a quartz cuvette; the filter combinations for actinic red light, exciting blue light and fluorescence provided an overlap of less than 15%. Ant A was dissolved in dimethyl sulfoxide. Chlorophyll was determined in 95% ethanol using the equations of Lichtenthaler [7].

3. Results and Discussion

The role of CEF in ATP production may be as important for the induction of photosynthesis in dark to light transitions, when the acceptor of NCEF is not present, as it is during steady state photosynthesis. In C_4 plants, ATP is needed for regeneration of PEP in order for CO_2 fixation to occur through PEP carboxylase, which generates oxaloacetate, an acceptor of NCEF. Induction of O_2 evolution is shown with isolated mesophyll chloroplasts of maize in the presence of pyruvate + Pi as substrates, and with addition of PEP carboxylase (Fig. 1). O_2 evolution requires the ATP-dependent conversion of pyruvate and Pi to PEP, the conversion of PEP + HCO_3^- to OAA and Pi by PEP carboxylase, and the reduction of OAA to Mal. It was found that the steady state rate of O_2 evolution induced under the above conditions was strongly affected by addition of Ant A to the reaction medium; a concentration of antibiotic as low as 1 μM severely decreased it (Fig. 1). This indicates that CEF may have a major role in induction of C_4 photosynthesis *in vivo*.

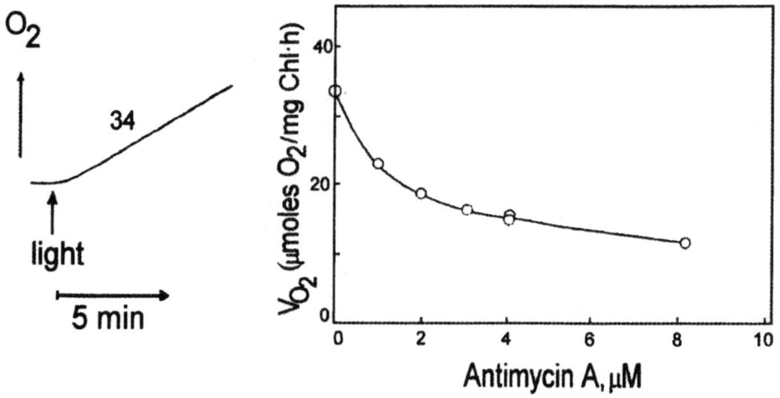

Figure 1. The effect of Antimycin A on induction of O_2 evolution during the illumination of intact maize mesophyll chloroplasts in the absence of oxaloacetate. Reaction mixture: BRM (see methods), 5 mM Pi, 6 mM $MgCl_2$, 525 units of catalase/ml, 10 mM pyruvate, 10 mM $NaHCO_3$, 1.2 units of PEP carboxylase (isolated from *Panicum maximum*), plus 5 mM glucose 6-P as activator, and with chloroplasts at 36 μg Chl/ml. Trace on the left shows control experiment without Ant A.

The above data demonstrates the participation of CEF in ATP production in the initial absence of an acceptor for NCEF. Further experiments were designed to test the possibility of the operation of CEF concurrently with NCEF. We investigated the effect of

Ant A on the generation of PEP and MAL, and specifically the PEP/MAL ratio, during the illumination of maize mesophyll chloroplasts in the presence of saturating concentrations of pyruvate and OAA (in the absence of PEP carboxylase). In these experiments the chloroplasts are catalyzing two partial reactions, the conversion of pyruvate (added as substrate) to PEP, which is dependent on ATP, and the conversion of OAA (added as substrate) to MAL, which is dependent on NADPH from NCEF. We found that Ant A decreased the ratio PEP/MAL by 20-30% at 1.5 μM concentration (Exp. 1, Table 1). It is also shown that the ratio of CO_2 fixed to O_2 evolved (Exp. 2 measured under the same experimental conditions except in the presence of PEP carboxylase and catalase) fell in the same proportion. These results can be explained by Ant A inhibition of CEF having a dual effect on these metabolic conversions. First, the decrease in the rate of PEP synthesis and the PEP/MAL ratio in the presence of Ant A provides evidence that CEF is contributing to the production of ATP which is utilized for PEP synthesis during NCEF. In Exp. 2, the rate of CO_2 fixation via PEP carboxylase is dependent on PEP production from pyruvate, while the rate of O_2 evolution is dependent on OAA reduction and NCEF. Thus, inhibition of the rate of PEP production by Ant A caused a decrease in the C/O_2 ratio. Interestingly, the rates of MAL formation and O_2 evolution were increased (10 to 15%) in the presence of Ant A, which suggests CEF contributes to membrane energization, the second effect, and causes some down-regulation of PSII activity [8].

TABLE 1. Effects of Antimycin A (Ant A) on light induced rates (μmol mg Chl^{-1} h^{-1}) of synthesis of phosphoenolpyruvate (PEP) and malate (MAL), on the rates of carbon fixation (C) and oxygen evolution (O_2), and on PEP/MAL and C/O_2 ratios in the reaction mixtures with intact maize mesophyll chloroplasts.

Experiment 1				Experiment 2			
Ant A [μM]	MAL	PEP	PEP/MAL	Ant A [μM]	C	O_2	C/O_2
0	430	318	0.74	0	196	131	1.5
1.5	495	277	0.56	1.5	183	152	1.2

The reaction mixture for Experiment 1 contained: BRM (see methods), 1 mM OAA, 10 mM pyruvate, 5 mM Pi, and chloroplasts with 15 μg Chl/ml. For Experiment 2, the reaction mixture also included: 6 mM $MgCl_2$, 5 mM glucose 6-P, 9 mM $NaH^{14}CO_3$, 525 units/ml of catalase, and 1.7 units of PEP carboxylase. The rates were determined from samples taken at 1.5 min and 4 min after onset of illumination.

To more clearly test whether Ant A can affect NCEF, experiments were conducted under basal conditions, i.e. in the absence of pyruvate and Pi, where the ΔpH generated under illumination is not used for ATP synthesis (Fig. 2). In this case, an almost twofold increase of OAA dependent O_2 evolution was observed at 1-2 μM Ant A. Under the same experimental conditions, the addition of Ant A caused a significant drop of light induced 9AA fluorescence quenching (results not shown); 9AA fluorescence quenching is commonly accepted as an indicator of membrane energization [9,10]. These results give the most convincing evidence that CEF energization of the thylakoid membrane of maize mesophyll chloroplasts occurs concurrently with NCEF.

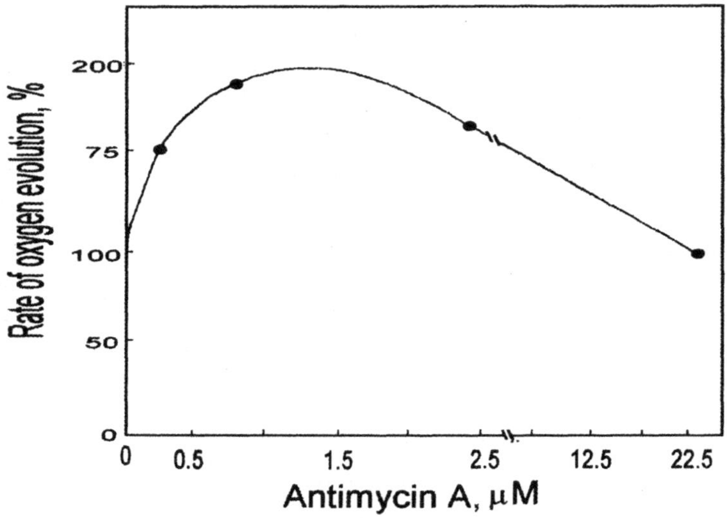

Figure 2. The concentration dependence of the Antimycin A affect on the rate of O_2 evolution during the oxaloacetate (OAA) reduction in intact maize mesophyll chloroplasts under basal conditions. Reaction mixture: BRM (see methods), 1.0 mM OAA, 525 units/ml of catalase, and with chloroplasts at 15 μg Chl/ml. 100% is 117 μmol O_2 mg Chl^{-1} h^{-1}. The rate of O_2 evolution at the highest Ant A concentration is the rate taken after 4 min of illumination (in this concentration region the rates progressively decreased after the second minute of illumination).

References

1. Arnon, D.I., Tsujimoto, H.Y., and McSwain, B.D. (1967). Nature, 214, 562-566.
2. Arnon, D.I. and Chain, R.K. (1979). FEBS Letters, 102, 133-138.
3. Moss, D.A. and Bendall, D.S. (1984). Biochem Biophys Acta, 767, 389-395.
4. Hatch, M.D. and Osmond, C.B. (1976) in Encyclopedia of Plant Physiology, v. 3 (Stocking C. R. and Heber U., eds.), pp. 144-184, Springer-Verlag, Berlin.
5. Huber, S.C. and Edwards, G.E. (1976). Biochem Biophys Acta, 449, 420-433.
6. Jenkins, C.L.D. and Russ, V.J. (1984). Plant Sci. Lett., 35, 19-24.
7. Lichtenthaler, H.K. (1987). Methods in Enzymology, 148, 350-382.
8. Heber, U., and Walker, D. (1992). Plant Physiol. 100, 1621-1626.
9. Fernyhough, P., Foyer, C., and Horton, P. (1983). Biochim. Biophys. Acta 725, 155-161.
10. Kobayashi, Y. and Heber, U. (1994). Photosynth. Res., 41, 419-428.

This research was supported by the CAST Program through the National Research Council and National Science Foundation Grant IBN 9317756.

STUDIES ON THE RELATIONSHIP BETWEEN PHOTOSYNTHETIC ELECTRON TRANSPORT AND CARBON METABOLISM

DZ Habash[1], MAJ Parry[1], MJ Paul[1], S Parmar[1], S Driscoll[1], J Knight[2], J Gray[2], A Keys[1] & DW Lawlor[1]. [1]Biochemistry and Physiology Dept., IACR-Rothamsted, Hertfordshire, AL5 2JQ, UK.[2]Dept. of Plant Sciences, University of Cambridge, Cambridge CB2 3EA, UK.

1. Introduction

The reduction in PSII activity, often termed 'downregulation', is a major process by which the photosynthetic apparatus responds to fluctuations in environmental conditions and changes in metabolism. The lowering of PSII activity is perceived as a dynamic and necessary process to ensure the protection of PSII from oxidative damage and to prevent overreduction of electron transport components under conditions of reduced RuBP carboxylation. The redox state of Q_A is in balance with the rate of utilisation of ATP and NADPH and changes in carbon metabolism feedback to PSII to enable the modulation or regulation of the rate of electron transport. Two studies are presented, using transgenic tobacco with a reduced complement of phosphoribulokinase (PRK) and wheat grown at elevated CO_2, which investigate the relationship between PSII activity, linear electron transport and RuBP carboxylation and oxygenation in intact leaves.

2. Procedure

Transgenic tobacco with either 5 or 100% complement of PRK were produced as described in (1) and grown at 25°C. Spring wheat was grown at 16°C under 350 (ambient CO_2) or 700 (elevated CO_2) μmol.mol^{-1} CO_2. Simultaneous measurements of PSII chlorophyll fluorescence and CO_2 assimilation were undertaken on intact attached leaves as in (2). Assimilation rates, intercellular CO_2 concentration,C_i, and transpiration rates were calculated as in (3). Linear electron transport rates, $J_1=(\Phi_C)(4)(PFD)$, and the partitioning of electron transport products to CO_2, $J_A=4(A+R_d)$, and O_2, $J_L=J_1-J_A$, were calculated from measurements of fluorescence and gas exchange in leaves at steady state photosynthesis under non-photorespiratory and photorespiratory conditions as detailed in (4,2). The partitioning of electron transport to O_2 assumes RuBP oxygenation as the only sink. The rate of RuBP carboxylation, $v_c=(A+R_d+0.5v_o)$, rate of RuBP oxygenation, $v_o=(J_L/6)$, mesophyll conductance to CO_2, $g_m=A/\{C_i-[(\Gamma_*(J_1+8(A+R_d))/(J-4(A+R_d))]\}$, the $[CO_2]$ at the carboxylation site, $C_c=C_i(A/g_m)$, the CO_2/O_2 specificity of Rubisco,$S_{c/o}$, from $v_c/v_o=S_{c/o}([C_c]/[O_2])$ (5) and J_1, A, C_i and leaf temperature were calculated for each PFD in a light-response curve with Γ_* and R_d determined as in (6). Fluorescence parameters and terminology were as in (2).

P. Mathis (ed.), Photosynthesis: from Light to Biosphere, Vol. II, 831–834.

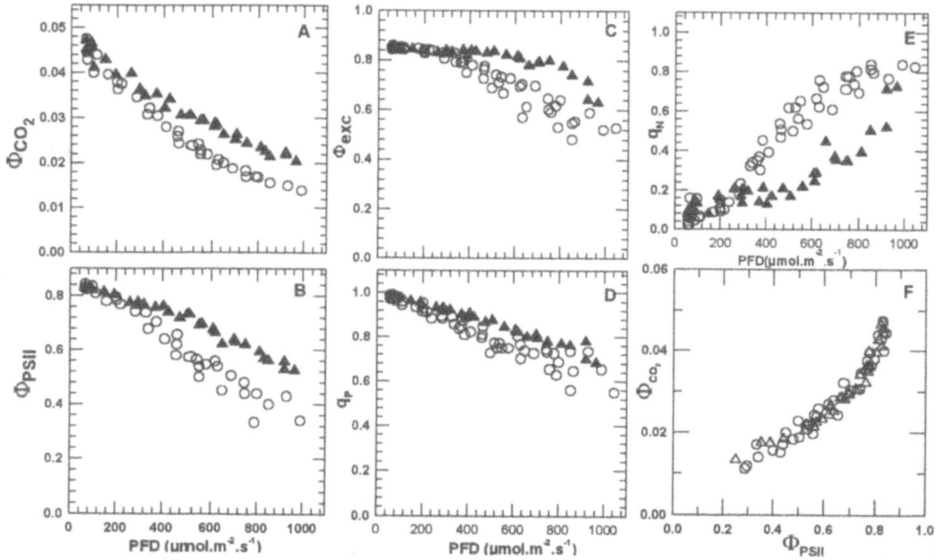

"Figure 1." The response of Φ_{CO2} **(A)**, Φ_{PSII} **(B)**, Φ_{exc} **(C)**, q_P **(D)** and q_N **(E)** to PFD in leaves at steady state photosynthesis and photorespiratory conditions in transformants with 100 %(▲) and 5% (○) PRK. **(F)** The relationship between Φ_{CO2} and Φ_{PSII} under physiological conditions in transformants with 100 %(△) and 5% (○) PRK.

3. Result and Discussion

Tobacco plants transformed with a 95% reduction in PRK showed a reduction in the relative quantum yield of carbon assimilation measured with PFD at or above those encountered during growth (Fig.1A). This is a result of the metabolic block in carbon assimilation due to the huge reduction in the concentration of PRK which becomes limiting with increasing light. This was also apparent as a lowering in the relative quantum yield of PSII electron transport (Fig.1B) which was caused by a reduction in the efficiency of energy transduction through open PSII centres,Φ_{exc},(Fig.1C). This relates to the onset of non-photochemical dissipation of energy as heat, expressed as an increase in the non-photochemical quenching of fluorescence,q_N,(Fig.1E). Photochemical quenching of fluorescence, q_p, which is an estimate of the pool of open PSII centres was also reduced (Fig.1D) in response to the lowered CO_2 assimilation rate. The relationship between Φ_{CO2} and Φ_{PSII} under photorespiratory conditions was similar in tobacco with 100 and 5% PRK (Fig.1F). However, this occurred at a different PFD, assimilation rate and electron transport rate (plots of these parameters as a function of Φ_{PSII}; not shown).In transformants with only 5% PRK there was a substantial reduction in total linear electron transport rate,J_1, electron transport dedicated to CO_2, J_A, and electron transport dedicated to O_2, J_L, (Fig.2A,B)compared to transformants with 100% PRK. The proportion of J_L/J_1 was also lower in the 5%PRK which was also apparent in the lower RuBP oxygenation/carboxylation ratio,v_o/v_c, (Fig.2C) pointing to a reduced rate of photorespiration.The specificity of Rubisco for CO_2/O_2, calculated *in vivo* from gas

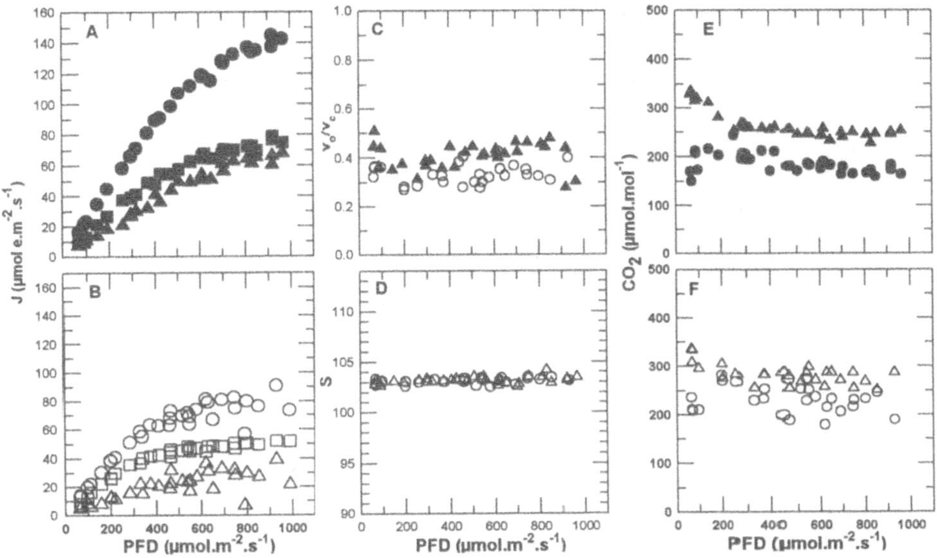

"Figure 2." The rates of linear electron transport, J_1, (circle), electron transport to CO_2, J_A, (square) and electron transport to O_2 ,J_L, (triangle) as a function of PFD for 100% **(A)** and 5% PRK **(B)** transformant tobacco. **(C)** The ratio of RuBP oxygenation/carboxylation as a function of PFD for 100% (▲) and 5% (○) PRK plants. **(D)** The CO_2/O_2 specificity of Rubisco as a function of PFD for 100% (△) and 5% (○) PRK plants. The concentration of CO_2 at the intercellular spaces (triangle) and at the site of carboxylation (circle) as a function of PFD for 100% **(E)** and 5% **(F)** PRK plants.

exchange and fluorescence data, (Fig.2D) and measured *in vitro* at 25°C (89 and 91 for 5% and 100% PRK) showed no change upon a reduction of PRK to 5%. The concentrations of CO_2 in the intercellular spaces,C_i, and at the site of carboxylation,C_c, were higher in transformants with 5% PRK compared to those with 100% PRK (Fig.2E,F). The lower v_o/v_c ratio in the 5% PRK plants may be due to the higher C_c and/or to the lower RuBP concentration measured in the 5% PRK plants (1).

The second study examined the component processes in photosynthesis in spring wheat grown for six weeks at elevated CO_2. Leaves developed at elevated CO_2 showed a higher net assimilation rate measured as a function of C_i (Fig.3A) which was more pronounced at the higher C_i indicating an increased capacity for RuBP regeneration. This enhancement in assimilation was also evident as a higher Φ_{PSII} indicating an increase in the relative quantum yield of PSII electron transport (Fig.3B). Partitioning linear electron transport to RuBP carboxylation and oxygenation showed a larger allocation to CO_2 (higher J_A) and a smaller allocation to O_2 (lower J_L) (Fig.3C) in leaves developed at elevated CO_2; as expected from the known kinetics of Rubisco in response to the higher C_c in leaves developed at elevated CO_2. Measurements of photosynthesis at ambient CO_2 for plants grown at either CO_2 treatment showed a consistently higher Φ_{CO2} in leaves developed at elevated CO_2 (Fig.3D) while little difference was obtained in Φ_{PSII} measured

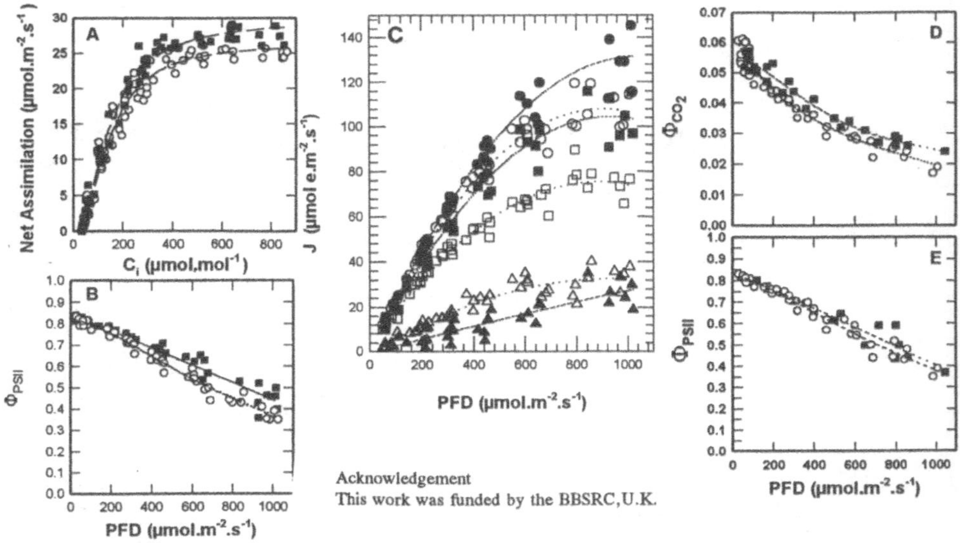

Acknowledgement
This work was funded by the BBSRC,U.K.

"Figure 3." The rate of net CO_2 assimilation as a function of C_i for wheat leaves developed at ambient (○) or elevated CO_2 (■) **(A)**. The response of Φ_{PSII} to PFD for wheat leaves developed at ambient (○) or elevated CO_2 (■) **(B)**. The rates of linear electron transport (circle), electron transport to CO_2 (square) and electron transport to O_2 (triangle) as a function of PFD for ambient (open symbols) and elevated (closed symbols) CO_2 grown leaves **(C)**. The response of Φ_{CO2} **(D)** and Φ_{PSII} **(E)** to PFD for wheat leaves developed at ambient (○) or elevated CO_2 (■) and measured at ambient CO_2.

at ambient CO_2 in plants grown at either [CO_2] (Fig.3E).

The two studies presented in this paper demonstrate the regulation of PSII as a primary process responding to changes in metabolism. PSII activity and the rate of electron transport are reduced in response to a lowered rate of carboxylation in tobacco transformants with low [PRK], while PSII activity and electron transport rates are increased in response to a greater demand for electron transport products in wheat leaves developed under elevated CO_2. The studies also demonstrate the potential in applying fluorescence and gas exchange measurements to study electron transport *in vivo.*

References

1. Paul, M.A.J., Knight, J.S., Habash, D., Parry, M.A.J., Lawlor, D.W., Barnes, S.A., Loynes, A., Gray, J.C. (1995) Plant J. 7(4),535-542
2. Habash, D.Z., Paul, M.J., Parry, M.A.J., Keys, A.J., Lawlor, D.W. (1995) Planta 197, in press
3. von Caemmerer, S., Farquhar, G.D. (1981) Planta 153,376-387
4. Ghashghaie, J., Cornic, G (1994) J. Plant Physiol. 143,643-650
5. Laing, W.A., Ogren, W.L., Hageman, R.H.(1974) Plant Physiol. 45,678-685
6. Brooks, A., Farquhar, G.D. (1985) Planta 165,397-406

PHOTOSYSTEM I-DEPENDENT CYCLIC ELECTRON FLOW CONTRIBUTES TO THE CONTROL OF PHOTOSYSTEM II IN LEAVES WHEN STOMATA CLOSE UNDER WATER STRESS

Ulvi Gerst, Ulrich Schreiber, Spidola Neimanis and Ulrich Heber,
Julius-von-Sachs-Institute, University of Würzburg,
Mittlerer Dallenbergweg 64, D-97082 Würzburg, Germany

1. Abstract

Although cyclic electron transport is not as necessary as a source of ATP in C3 photosynthesis as it is in C4 photosynthesis, it is important in providing the additional proton pressure in the intrathylakoid space which is needed for the control of the activity of photosystem II when stomata close under water stress. Together with assimilatory and photorespiratory CO_2 turnover, such control prevents the excessive reduction of electron carriers which is known to lead to the photoinactivation of electron transport in sunlight.

2. Introduction

During photosynthetic electron transport, protons are deposited in the thylakoid compartment of chloroplasts. The proton gradient formed across the thylakoid membrane is discharged as its energy is used in leaves for the synthesis of the ATP needed in carbon assimilation and photorespiration. A somewhat increased proton gradient is needed to initiate a protonation reaction inside the thylakoids which controls the activity of photosystem II (1,2). We were interested in the question which photosynthetic reactions contribute to the formation of a regulatory proton gradient in leaves.

3. Material and Methods

Leaves of *Prunus laurocerasus* and *Laurus nobilis* were obtained from plants growing at the Botanical Garden of the University of Würzburg. Gas exchange, chlorophyll fluorescence and light scattering were measured as described in refs. 2-5.

P. Mathis (ed.), Photosynthesis: from Light to Biosphere, Vol. II, 835–838.

4. Results and Discussion

Fig. 1 shows simultaneous recordings of photosynthetic carbon uptake (A), scattering of 535 nm light (B) and modulated chlorophyll fluorescence (C) of a leaf of *Prunus laurocerasus*. The leaf was illuminated first in the presence of 600 ppm CO_2 and then of 50 ppm CO_2. The photon flux density was 300 μmol m^{-2} s^{-1} (about 20 % of sunlight). As the light was turned on, both carbon assimilation and scattering of 535 nm light increased. Scattering indicates an alteration in membrane conformation which is caused by a large transthylakoid proton gradient (2,3). After its initial increase, scattering decreased as assimilation increased further. This indicates the progressive discharge of the transthylakoid proton gradient by ATP synthesis. It is important to note that effective carbon assimilation can proceed in rate-limiting light in the absence of increased light scattering. Thus, scattering indicates *excessive* energization of thylakoid membranes, not energization just sufficient to satisfy the energy requirements of assimilation and photorespiration. Indeed, scattering increased as net assimilation was lost during a decrease in the CO_2 concentration from 600 to 50 ppm. 1 s pulses of high intensity light given every 2 min on top of the background light now caused further spikes in the scattering response showing that the light scattering response was not yet saturated. CO_2 released by photorespiration actually escaped through opened stomata before it could be reassimilated (see dashed line in Fig. 1 which indicates the borderline between carbon gain and carbon loss in the experiment of Fig. 1). Darkening relaxed light scattering.

Assimilatory electron flow is only part of light-dependent electron flow. Total linear electron flow can be assessed by modulated chlorophyll fluorescence. The ratio of pulse-induced chlorophyll fluorescence to total fluorescence (F_v/F_m in Fig. 1C) is known to indicate the quantum efficiency of electron flow through PSII (6). Although net assimilation was completely lost, the remaining electron transport in 50 ppm CO_2 was still 60 % of the original electron transport in 600 ppm CO_2.

After the experiment shown on the left-hand side of Fig. 1 was completed, the leaf was exposed for 10 min to 6.8 % HCN in air. Subsequently transpiration (not shown), gas exchange, light scattering and chlorophyll fluorescence were measured as before. Poisoning with HCN, which was intended to inhibit carboxylase/oxygenase activity, caused partial closure of stomata, but this was insufficient to limit gas exchange. Surprisingly, HCN did not decrease respiratory CO_2 production in the dark. Illumination in the presence of 50 ppm CO_2 caused little (but detectable) CO_2 uptake. Light scattering increased much indicating maximum energization of the thylakoids. Apparently, reactions capable of pumping protons into the thylakoids such as linear electron transport to oxygen in the Mehler reaction and associated reactions (7) or cyclic electron transport around photosystem I (8,9) were still active after HCN poisoning. Addition of 600 ppm CO_2 decreased light scattering and increased carbon uptake by the poisoned leaf, but much less so than before poisoning. Some slow recovery of photosynthesis from HCN-poisoning was apparent (see slow increase in carbon uptake). Nevertheless, chlorophyll fluorescence indicated a dramatic decrease in linear electron transport compared to the unpoisoned state of the leaf. Electron transport in 600 ppm CO_2 was decreased after HCN-poisoning by 75 % and in 50 ppm CO_2 by 86 %. Inspite of this, non-photo-

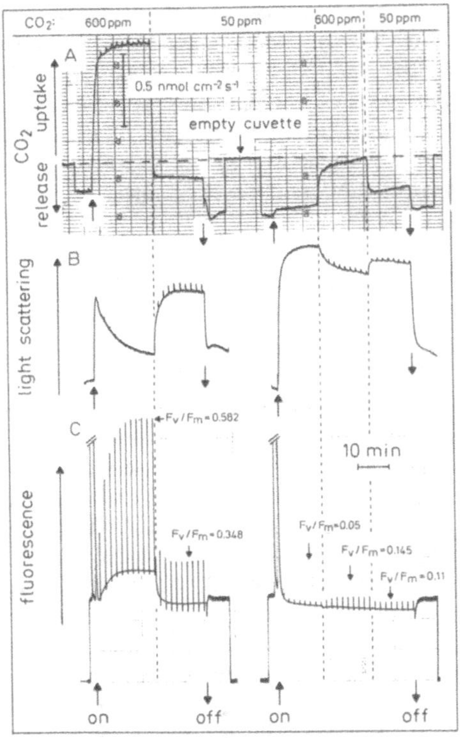

Figure 1. Simultaneous recordings of carbon dioxide exchange (A), scattering of 535 nm light (B) and modulated chlorophyll fluorescence (C) of a leaf of *Prunus laurocerasus*. 1 s pulses of red light (3100 µmol m^{-2} s^{-1}) were given before illumination and every 2 min during illumination on top of background red light of 300 µmol m^{-2} s^{-1}. F_v/F_m in (C) denotes the ratio of pulse-induced fluorescence spikes to total fluorescence. Large carbon uptake in 600 ppm CO_2 (left part of (A)) corresponds to 1.02 nmol cm^{-2} s^{-1}. For further explanation, see text.

chemical fluorescence quenching (i.e. dissipation of light energy as heat) was larger after than before HCN poisoning. The data suggest that most of the electron transport displayed by the unpoisoned leaf in 50 ppm CO_2 must be attributed to the interplay between photorespiration and assimilation (10).

The experiment shown in Fig. 2 was performed with a leaf of *Laurus nobilis* , because this leaf responded more rapidly to water stress than the *Prunus* leaf used in Fig. 1. When stomata close under water stress, intercellular CO_2 is likely to decrease to the CO_2 compensation point which in C3 plants is not far from 50 ppm CO_2, the concentration used in part of the *Prunus* experiment of Fig. 1. In Fig. 2, the quantum efficiency of electron flow through PSII is plotted against the quantum efficiency of electron flow through PSI. Linear electron flow uses both photosystems at a ratio of one to one. Indeed, as long as electron flow was slow during the opening of stomata in the light, the quantum efficiencies of electron flow through PSI and PSII were similar (see empty symbols, four measurements in the left part of Fig. 2). However, after stomata were forced to close by cutting the petiole in air (closed symbols), the quantum efficiency of electron flow through PSII decreased more than the quantum efficiency of electron flow through PSI indicating more charge separation in PSI than in PSII. Simultaneously, light scattering increased (see scattering change caused by a transition from 600 to 50 ppm

CO_2 in Fig. 1). Thus, the relative increase in charge separation in PSI observed when stomata close and linear electron transport slows down is accompanied by increased thylakoid energization. These results indicate that cyclic electron transport contributes substantially to ΔpH formation thereby controlling PSII and helping to prevent excessive reduction of electron carriers which causes photoinactivation. A different interpretation of the data would have to be considered, if PSI quantum capture efficiency were decreased by energization, a possibility which is presently under investigation.

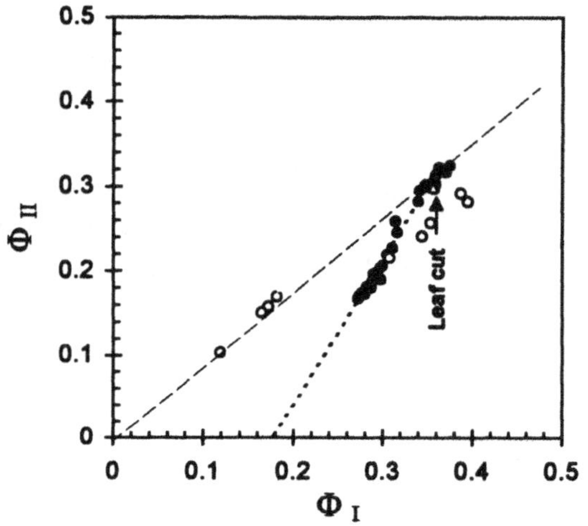

Figure 2. Relative quantum efficiencies of elecron flow through photosystem II (Φ_{II}) and photosystem I (Φ_I) of a laurel leaf. The leaf, still attached to the twig, was illuminated in air with 530 .$\mu mol \cdot m^{-2} \cdot s^{-1}$ red light until, after a prolonged period needed for stomatal openig and subsequent stomatal oscillations, photosynthesis had reached steady state (empty symbols). Its petiole was then cut in air to interrupt water supply and enforce progressive stomatal closure (closed symbols). Φ_I was measured using the method described in ref. 5 which avoids errors owing to acceptor-side limitation of electron flow through PSI.

References

1Horton, P. (1992) Photosynth. Res. 34, 106

2 Schönknecht, G., Neimanis, S., Katona, E., Gerst, U. and Heber, U. (1995) Proc. Natl. Acad. Sci. USA, in press

3 Heber, U. (1969) Biochim. Biophys. Acta 180, 302-319

4 Schreiber, U., Schliwa, U. and Bilger, W. (1986) Photosynth. Res. 10, 51-62

5 Klughammer, C. and Schreiber, U. (1994) Planta 192, 261-268

6 Genty, B., Briantais, G.M. and Baker, N.R. (1989) Biochim. Biophys. Acta 990, 87-92

7 Hormann, H., Neubauer, C. and Schreiber, U. (1994) Photosynth. Res. 41, 429-437

8 Heber, U. and Walker, D.A. (1992) Plant Physiol. 100, 1621-1626

9 Kobayashi, Y. and Heber, U. (1994) Photosynth. Res. 41, 419-428

10 Wu, J., Neimanis, S. and Heber, U. (1991) Bot. Acta 104, 283-291

OEC COMPLEX ENZYME AND PHOTOPHOSPHORYLATION INHIBITION BEHAVIOUR OF 1,1-DIPHENYLBOROXAZOLIDONE DERIVATIVES OF α-AMINOACIDS.

[1]Fernández -Vargas J., [2]Farfán N., [2]Castillo D., [2]Santillan R., [3]Cárabez A. and [1]Lotina-Hennsen B.[1]Facultad de Química y [3]Facultad de Medicina, Universidad Nacional Autónoma de México. 04510 México D.F., México.[2]Departamento de Química. Centro de Investigación y Estudios Avanzados del Instituto Politécnico Nacional. Apartado Postal 14-740,. 07000, México .D.F., México.

1. Introduction.

Boroxazolidones are neutral complexes obtained by reaction of α-aminoacids with diphenylborinic acid. Some of these compound have been shown to have biocidal activities [1]. It is known that half of the commercial herbicides inhibit photosynthesis. The common targets for herbicides are Q_B redox enzyme, b_6f complex or electron acceptors such as paraquat. As a part of our search for new herbicides we tested the effect of boroxazolidone derivatives from aspartic acid, glutamic acid, glycine, lysine, threonine and diphenyl-borinic acid on photosynthesis.

2. Materials and methods.

Diphenyl(glutamate -O, N) borane (DPHB$_1$), Diphenyl (lysinate -O, N) borane (DPHB$_2$), Diphenyl (glycinate -O, N) borane (DPHB$_3$), Diphenyl (threoninate -O, N) borane (DPHB$_7$), Diphenyl (aspartate -O, N) borane (DPHB$_8$) [2] were dissolved in dimethyl sulfoxide (DMSO). Chloroplasts were isolated from pea leaves (*Pisum sativum*) [3] and suspended, unless indicated otherwise, in an isolation medium containing: 100 mM sucrose, 5 mM $MgCl_2$, 10 mM KCl buffered with 30 mM Na^+-tricine at pH 8.0. methylviologen (MV) (0.05mM) was added as electron acceptor for Hill's reaction from water to MV, the reaction medium contained: 100 mM sucrose, 5 mM $MgCl_2$, 10 mM KCl and 30 mM tricine at pH 8.0. Chlorophyll, non-cyclic electron transport from water to MV, H^+-uptake, ATP-synthesis and Mg^{2+}-ATPase were determined as reported in the literature [3]. All reactions were initiated by illumination with a projector lamp. Electron transport from Photosystems I and II, water to silicomolibdate and diphenylcarbazide (DPC) to DCPIP or DCPIP/ferrycyanide (this last assay was performed on 0.8 M TRIS pH 8.0 treated thylakoids) were determined as reported [3].

3. Results and Discussion

We investigated the behaviour of boroxazolidones DPHB$_1$ DPHB$_2$ DPHB$_3$ DPHB$_7$ and DPHB$_8$ on different photosynthetic activities. Table 1 shows that DPHB$_2$ inhibed electron flow (basal, phosphorylating and uncoupled), ATP synthesis and proton uptake from water to MV in freshly lysed pea chloroplasts. The values of % inhibition of these activities were 17, 51, 65, 91 and 90 %, respectively at concentrations up to 1.2 mM.

P. Mathis (ed.), Photosynthesis: from Light to Biosphere, Vol. II, 839–842.

These results indicate that $DPHB_2$ acts as Hill's reaction inhibitor. Notice that when the thylakoid membranes are energized, the binding site for $DPHB_2$ is not available, in contrast to experiments where thylakoids were de-energized by uncoupling with NH_4Cl.

CONCENTRATION mM	B	P	U	S	H^+
NONE	486	669	1267	197	100
0.4	380	512	935	129	90
0.6	383	451	786	91	72
0.9	385	378	587	50	40
1.2	404	324	433	18	10

Table 1. Non-cyclic electron transport from water to MV (B= Basal, P= Phosphorylating, U= Uncoupled), S= ATP synthesis and H^+= Proton uptake as a function of $DPHB_2$ concentrations. Reaction mixture and other conditions as described in Material and methods. Rates of non-cyclic electron transport from water to MV (B, P, U) were expressed in $\mu eq\ e^-.h^{-1}.mg\ Chl^{-1}$ and for ATP synthesis and Proton uptake in $\mu eq\ ATP.h^{-1}.mg\ Chl^{-1}$ and $\mu eq\ H^+.h^{-1}.mg\ Chl^{-1}$ respectively.

In order to localize the site of $DPHB_2$ action, the electron flow from PS were determined. The electron transport of photosystem I from $DCPIP_{red}$ to MV, $TMQH_2$ to MV and PMS_{red} to MV were not affected by it.

	Addition mM	$H_2O \rightarrow DCPIP$ [%]	$H_2O \rightarrow SiMO$ [%]	$DPC \rightarrow DCPIP$ [%]
PS II	0	0	0	0
	0.4	18	25	0
	0.9	34	35	0
	1.2	74	76	0
	Addition mM	$DCPIP_{red} \rightarrow MV$ $\mu eq\ e^- \times h^{-1} \times mg\ chl^{-1}$	$PMS_{red} \rightarrow MV$ $\mu eq\ e^- \times h^{-1} \times mg\ chl^{-1}$	$TMQH_2 \rightarrow MV$ $\mu eq\ e^- \times h^{-1} \times mg\ chl^{-1}$
PS I	0	1933	2233	2400
	0.4	1933	1873	2214
	0.9	2040	1562	2133
	1.2	2400	1533	2000

Table 2. The photosystem II and I activities are given in percentage of electron flow inhibition compared to control as zero % inhibition. The concentrations used were: MV 50 μM, Ascorbate 300 μM, DBMIB 1 μM, [$Fe(CN_6)$] K_3 300 μM, Silicomolibdate 100 μM, DCPIP 100 μM, DPC 200 μM, DCMU 10 μM. Control values of electron flow in PS II from water to DCPIP, water to silicomolibdate and DPC to DCPIP were 622, 133, 120 respectively in $\mu eq\ e^-.h^{-1}.mg\ Chl^{-1}$. PS II electron flow were determined as described in materials and methods. Control values of electron flow from PS I from $DCPIP_{red}$ to MV; PMS_{red} to MV and $TMQH_2$ to MV were 1933, 2233, 2400 respectively in $\mu eq\ e^-.\ h^{-1}.\ mg\ Chl^{-1}$.

According to chemiosmotic theory a compound which inhibits any redox enzyme does not generate proton electrochemical gradient and therefore does not synthesize ATP, our results corroborated this statement.

The results show that, 1.2 mM $DPHB_2$ inhibited PSII electron transport from water to DCPIP, and water to silicomolibdate (Table 2) 86% and 89% respectively. To determine the site of electron transport inhibition from water to Q_A by $DPHB_2$, electron flow was measured from DPC to DCPIP in 0.8 M TRIS pH 8.0 poisoned chloroplasts. The results show that this span of electron transport was not affected by $DPHB_2$ (Table 2). Therefore $DPHB_2$ inhibited the water splitting enzyme, at least before DPC electron donation. The water-splitting enzyme system is a key event in photosynthesis. It is known that it is by far more labile than the PS II trapping center, and therefore various inhibitors and treatments are effective in selectively blocking or destroying the water oxidation step. The required concentration of each compound to inhibit water oxidation varies widely. The structure of these compounds also varies greatly.

These data may indicate that these compounds either inhibit at different redox steps of the oxidizing site of PS II or they have a different mechanism of action. The I_{50} of $DPHB_2$ (0.6 mM) is in the average range of various inhibitors at the same target Table 3.

Compound	Conc. (mM)	Oxygen evolution inhibition (%)	Ref.
Amonia	10	100	Izawa and Good 1972
Methylamine	100	100	Izawa and Good 1972
Hydroxylamine	1	100	Joliot 1966
Tris-washiing	800	60-80	Yamashita and Butler 1968
DCMU	0.2	80	Renger 1973
Cu^{2+}, Cd^{2+},Zn^{2+}	5×10^{-3}	80-100	Singh and Singh 1987
Cacacol	6×10^{-2}	100	Lotina-Hennsen et al. 1991
Diamines	1×10^{-2} -10	100	Lotina-Hennsen et al. 1987
Ivalin	2.8	50	Bernal-Morales et al. 1994
ANT2p	5×10^{-4}	80	Packham and Ford 1986
p-Diazonium benzenesulfonate	1.5	50	Giaquinta et al. 1974
Zinc	4.0	100	Tripathy and Mohanty 1980

Table 3. Effect of different compounds in the reaction of the water oxidation enzymes complex as reported in the literature

Table 4. shows the inhibitory effect of boroxazolidones ($DPHB_1$ $DPHB_7$ $DPHB_3$ $DPHB_8$) on ATP synthesis and phosphorylating electron flow associated with MV reduction/oxidation in a Mheler reaction in pea freshly lysed chloroplasts. $DPHB_1$ was the most effective inhibitor, the I_{50} was 1 mM. However, these compounds do not affect basal and uncoupled electron transport nor proton uptake. Moreover, inhibition of phosphorylating electron transport was totally reversed by 5 mM ammonium chloride, a known uncoupler.

These results indicate that these boroxazolidones act as a typical energy transfer inhibitors. As in the case of Dio-9, dicyclohexylcarbodiimide, phlorizin, antisera to CF_1 and piquerol A [4], boroxazolidones did not affect the light-induced proton uptake of chloroplasts even at concentrations that inhibited ATPsynthesis and phosphorylating electron flow (Table 4).

BASAL	NONE	0.4 mM	0.6 mM	0.9 mM	1.2 mM
DPHB1	143	125	118	109	104
DPHB7	514	554	571	571	600
DPHB3	520	520	520	520	520
DPHB8	132	114	111	115	127
PHOSPHORYLATING					
DPHB1	256	226	210	185	160
DPHB7	914	819	767	700	620
DPHB3	960	314	207	220	346
DPHB8	400	400	391	370	340
UNCOUPLED					
DPHB1	723	793	830	897	968
DPHB7	1657.5	1724	1737	1748	1750
DPHB3	1920	1993	2021	2038	2080
DPHB8	927	994	1021	1053	1083
ATP SYNTHESIS					
DPHB1	13676	13000	11052	7143	3118
DPHB7	205	189	169	120	79
DPHB3	9859	4050	2969	2898	3767
DPHB8	8456	5591	5001	5098	5114
PROTON UPTAKE					
DPHB1	100	75	76	78	96
DPHB7	160	140	120	100	67
DPHB3	100	78	83	109	160
DPHB8	97	250	262	241	198

Table 4. The photosynthetic activities were determined as described in table 1.

References

[1] Lang, K., et. al. (1962) Ger.Pat.,1,130,445, (1963) Chem Abstr. 58, 1488a.
[2] Farfán, N., Silva, D. and Santillan.R. (1993). Heteroatom. Chem. 4, 533-536.
[3] Calera, M. R., Soto, F., Sánchez, P.Bye, R., Hernández Bautista, B., Anaya, A. L., Lotina-Hennsen, B. and R. Mata, R. (1995) Phytochem.(in press)
[4] Mendoza, J. L., Jiménez, M. and B. Lotina-Hennsen, B. (1994) Pestic. Sci 40, 37-40(and references therein)
[5] Bernal-Morales et. al., (1994) Canadien J. Bot. 72, 177-.

THE pH-DEPENDENT REGULATION OF PHOTOSYNTHETIC ELECTRON TRANSPORT IN LEAVES

G. Schönknecht, S. Neimanis, U. Gerst & U. Heber
Lehrstuhl Botanik I der Universität Würzburg, Mittlerer Dallenbergweg 64
D-97082 Würzburg, Germany
E-mail: gerald@botanik.uni-wuerzburg.de

Abstract

Under conditions which allow high rates of photosynthesis, chlorophyll fluorescence and carbon assimilation were measured in the absence of photorespiration at different light intensities. Using 3 ATP/CO_2 and the known relationship between ATP synthesis rate and trans-thylakoid ΔpH, we calculated the light-dependent pH gradient across the thylakoid membrane (ΔpH) in intact leaves of spinach and barley. These ΔpH values were correlated with the photochemical (q_P) and non-photochemical (q_N) quenching of chlorophyll fluorescence, and with the quantum yields of PS II (Φ_{PSII}) and CO_2-fixation (Φ_{CO2}). At $\Delta pH > 2.1$, all parameters (q_P, q_N, Φ_{PSII}, and Φ_{CO2}) changed very steeply, following hexacooperative titration curves with slightly different pK_a values. In chlorophyll *b*-less mutants of barley (*chlorina-f2* 2800) the dependence of q_N on ΔpH was less steep compared to the wild-type.

Introduction

One important parameter of photosynthesis is the trans-thylakoid pH gradient (ΔpH). It powers ATP synthesis, facilitates energy dissipation and controls the electron transfer between photosystems II and I. Nothing is known about the magnitude of ΔpH in an intact leaf. In this study we calculate this trans-thylakoid ΔpH in leaves and present evidence that photosynthetic electron transport is regulated by a protonation reaction in the thylakoid lumen in a hexacooperative manner.

Materials and Methods

With spinach or barley we measured simultaneously CO_2 exchange and modulated chlorophyll fluorescence of the intact leaf at different light intensities. In order to obtain high rates of CO_2-assimilation and suppress photorespiration, these measurements were performed under 2000 µl/l CO_2 and 1 % O_2. From chlorophyll fluorescence data the photochemical (q_P) and non-photochemical (q_N) quenching of chlorophyll fluorescence (1) and the quantum yield of PS II (Φ_{PSII}) were calculated (2). For ATP hydrolysis measurments thylakoids were isolated. The ATP hydrolysis activity of the H^+-ATPase was measured in the presence of 30 % methanol and the liberated P_i was monitored via absorbance at 740 nm. For experimental details and a description of data analysis see (3).

P. Mathis (ed.), Photosynthesis: from Light to Biosphere, Vol. II, 843–846.

Results and Discussion

The trans-thylakoid ΔpH was calculated from the measured CO_2 uptake rates according to the following scheme:

$$\frac{CO_2}{Chl \cdot h} \Rightarrow \frac{ATP}{Chl \cdot h} \Rightarrow \frac{ATP}{ATPase \cdot s} \Rightarrow \Delta\,pH\,.$$

With a stoichiometry of 3 ATP/CO_2 the CO_2 uptake rate was recalculated into an ATP synthesis rate (per Chl). The Chl/H^+-ATPase stoichiometry was determined from measurements of the ATP hydrolysis rate of isolated thylakoids and used to recalculate ATP/(Chl · s) into ATP/(ATPase · s). Finally, the ΔpH was calculated from the ATP synthesis rate (per ATPase) using the data from Junesch and Gräber (4), which show a clear correlation between ATP synthesis rate and proton motive force (pmf). As the trans-thylakoid electrical potential at steady state is negligibly low we assumed ΔpH = pmf.

For spinach we measured an ATP hydrolysis rate of about 90 ATP/(1000 Chl · s). With an ATP hydrolysis rate of 90 ATP/(H^+-ATPase · s) for the membrane bound H^+-ATPase (4) this corresponds to a stoichiometry of 1000 Chl/H^+-ATPase. Fig. 1 shows the light-dependent increase of the trans-thylakoid ΔpH as calculated with this stoichiometry from measured CO_2 uptake. Because of the relatively flat dependence of ΔpH on the ATP synthesis rate an error of about 20 % for the Chl/H^+-ATPase stoichiometry results in only small errors for the calculated ΔpH (Fig. 1).

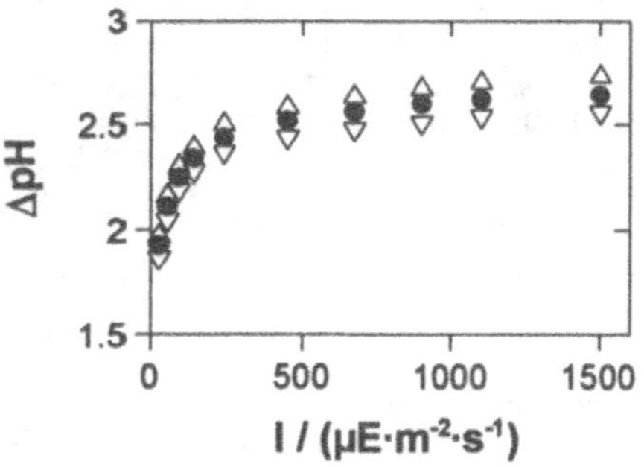

Figure 1: The calculated trans-thylakoid pH gradient (ΔpH) as a function of light intensity (I) for spinach. The different data symbols reflect three different chlorophyll to H^+-ATPase stoichiometries (△ 1200, ● 1000, ▽ 800) which were used to calculate the ΔpH.

With the data from Fig. 1, any other light-dependent parameter measured in parallel in the same experiment can be plotted as a function of the trans-thylakoid ΔpH. In Fig. 2 photochemical quenching (q_P) and non-photochemical quenching (q_N) as calculated from parallel chlorophyll fluorescence measurements are plotted as a function of trans-thylakoid ΔpH. Obviously both q_P and q_N displayed a very steep pH-dependence pointing to a cooperative titration process. Accordingly, the data (Fig's 2-4) were mathematically described (by non-linear regression analysis) by cooperative titration curves. It turned out that the pH-dependence of q_P as well as q_N could be described by a hexacooperative ($n \approx 6$) titration. The pK_a values for q_P were always between 0.11 to 0.16 pH units higher compared to q_N (Fig. 2). This difference was highly significant (Fig. 2b).

The steep pH-dependence of q_P and q_N allows a very effective down-regulation of the photosynthetic electron transport within a narrow pH range. This is meaningful and reflects physiological conditions. The slightly lower pK_a for q_N compared to q_P also makes good sense: the down-regulation of PS II by non-photochemical energy dissipation

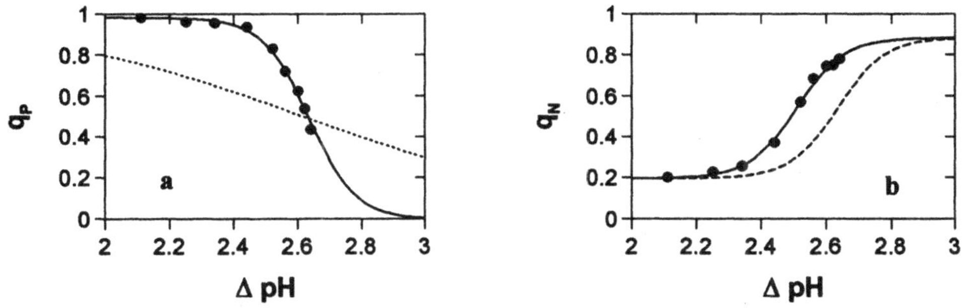

Figure 2: The **a**) photochemical (q_P) and the **b**) nonphotochemical (q_N) quenching of chlorophyll fluorescence as a function of the trans-thylakoid pH gradient (ΔpH) in an intact spinach leaf. Both data sets were mathematically described by cooperative titration curves (solid line), with a cooperativity of $n = 6$ and a midpoint of $pK_a = 2.63$ for q_P, and $n = 6$ and $pK_a = 2.51$ for q_N. In **a**) a titration curve with $n = 1$ (no cooperativity; ······), but otherwise the same parameters as for the solid line is given. For comparision of the pH-dependence of q_P and q_N in **b**) a hexacooperative titration curve with identical parameters but with $pK_a = 2.63$ as estimated for q_P is given (– – –).

increases before (at lower ΔpH) the photochemical energy conversion in PS II becomes limiting. In this way an over-reduction of the primary acceptor of PS II, Q_A, is prevented.

In Fig. 3 the pH-dependence of the quantum efficiencies of PS II (Φ_{PSII}) and CO_2-assimilation (Φ_{CO2}) are given. Both display a very steep, hexacooperative ($n \approx 6$) pH-dependence with nearly identical pK_a values. This pK_a ranged between the slightly different pK_a values for q_P and q_N. Obviously Φ_{PSII} and Φ_{CO2}, which reflect the gross regulation of photosynthetic electron transport, are determined by both photochemical and non-photochemical quenching (2).

When the same experiments as described above for spinach (Fig's 1-3) were carried out with barley, the same steep, hexacooperative pH-dependencies of q_P, q_N, Φ_{PSII}, and Φ_{CO2} were observed for the wild-type (Fig. 4, ○). With a chlorophyll *b*-less barley mutant however, the pH-dependence of q_N was significantly less steep ($n \approx 4$) (Fig. 4b), while q_P

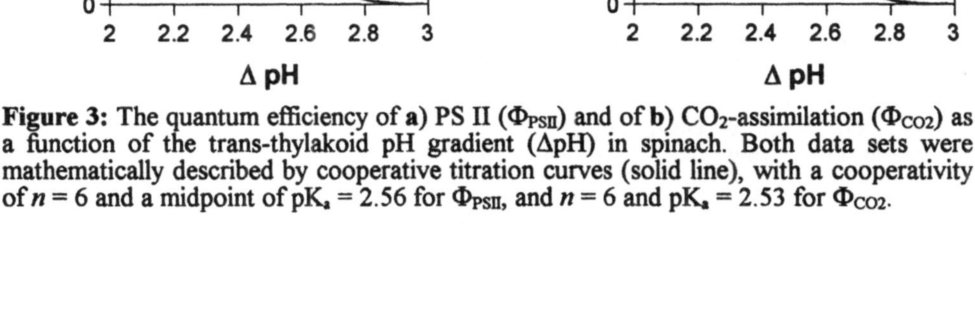

Figure 3: The quantum efficiency of **a**) PS II (Φ_{PSII}) and of **b**) CO_2-assimilation (Φ_{CO2}) as a function of the trans-thylakoid pH gradient (ΔpH) in spinach. Both data sets were mathematically described by cooperative titration curves (solid line), with a cooperativity of $n = 6$ and a midpoint of $pK_a = 2.56$ for Φ_{PSII}, and $n = 6$ and $pK_a = 2.53$ for Φ_{CO2}.

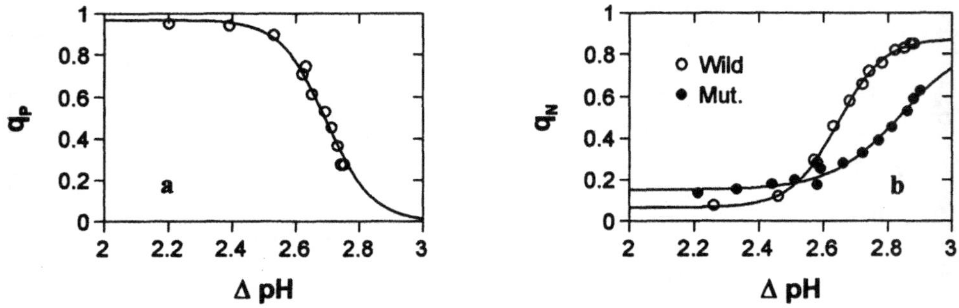

Figure 4: The **a**) photochemical (q_P) and the **b**) nonphotochemical (q_N) quenching of chlorophyll fluorescence as a function of the trans-thylakoid pH gradient (ΔpH) in intact barley leaves. All data sets were mathematically described by cooperative titration curves (solid lines). For the wild type (O) we estimated a cooperativity of $n = 6$ and a midpoint of $pK_a = 2.69$ for q_P, and $n = 6$ and $pK_a = 2.62$ for q_N, and for the chlorophyll *b*-less mutant (*chlorina-f2* 2800, •) $n = 4$ and $pK_a = 2.84$ for q_N.

displayed a hexacooperative pH-dependence in the mutant as well as in the wild-type. Obviously the changes in the light harvesting system of PS II caused by the mutant changed the pH-dependence of q_N (but not of q_P). It is tempting to speculate about a relation between the **trimeric** organization of the light harvesting complex of PS II and the **hexacooperative** pH-dependence of q_N.

Conclusions

Summarising our data may be interpreted in the following way: the photosynthetic electron transport (as reflected by q_P, q_N, Φ_{PSII}, and Φ_{CO2}) inside an intact leaf is regulated by the ΔpH (respectively the pH inside the thylakoid lumen) in a highly cooperative manner ($n \approx 6$) which allows a down-regulation within a very small pH range. The non-photochemical quenching (q_N) increases before (at lower ΔpH) the photochemical quenching (q_P) starts to decrease. The overall regulation (as reflected by Φ_{PSII} and Φ_{CO2}) has a midpoint (pK_a) at about pH 5.5, and the pH inside thylakoids hardly ever reaches pH 5.0 in an intact leaf. The light harvesting system of PS II is directly involved in the process of non-photochemical quenching (energy dissipation).

Acknowledgements

The financial support from the Deutsche Forschungsgemeinschaft (SFB 176 and SFB 251) is gratefully acknowledged.

References

1. van Kooten, O. & Snel, J.F.H. (1991) *Photosynth. Res.* **25:** 147-150
2. Genty, B., Briantais, J.-M. & Baker, N.R. (1989) *Biochim. Biophys. Acta* **990:** 87-92
3. Schönknecht, G., Neimanis, S., Katona, E., Gerst, U. & Heber, U. (1995) *Proc. Natl. Acad. Sci. USA*, in press
4. Junesch, U. & Gräber, P. (1991) *FEBS Lett.* **294:** 275-278

FERREDOXIN:NADP$^+$ REDUCTASE BOUND TO THE PSI-E SUBUNIT OF PHOTOSYSTEM I IS INEFFICIENT IN NADP$^+$-REDUCTION BUT CATALYZES THE REDUCTION OF QUINONES.

Hanne Linde Nielsen, Birgitte Andersen & Henrik Vibe Scheller
Plant Biochemistry Laboratory, Department of Plant Biology, Royal Veterinary and Agricultural University, 40 Thorvaldsensvej, DK-1871 Frederiksberg C, Copenhagen, Denmark

1. Introduction

Ferredoxin:NADP$^+$ oxidoreductase (FNR) in the chloroplasts of plants is associated with the thylakoid membranes. The identity of the binding site for FNR is controversial and several putative binding proteins have been reported [1-6]. We have devised a procedure for isolation of a PSI complex from barley with FNR specifically bound to the PSI-E subunit [7]. Inactivation of the *psaE* gene in cyanobacteria has little effect on photoautotrophic growth rates [8-10]. However, a *psaE$^-$* strain of *Synechococcus* sp. PCC 7002 was shown to be incapable of photoheterotrophic growth [10] and cyclic electron transport around PSI was impaired in the mutant [11]. A cyclic flow of electrons around PSI shares the components of linear electron transport from plastoquinone (PQ) to PSI. Between PSI and PQ, the cyclic electron flow is thought to involve ferredoxin and additional unidentified components. Experiments with inhibitors have suggested that FNR is involved in at least a fraction of the electron flow from ferredoxin to PQ [12-14]. Other experiments have shown that antibodies to FNR inhibit NADP$^+$-reduction but not cyclic electron transport [12,15, HV Scheller, unpublished]. Hosler and Yochum [16,17] have suggested that the seemingly contradictory results can be explained by two concurrent pathways for cyclic electron flow. They suggested that FNR is involved in an antimycin insensitive pathway. In a recent *in vivo* study it has been suggested that FNR is only indirectly involved in cyclic electron transport by providing NADPH as substrate for an NAD(P)H:PQ oxidoreductase [14].

Since *psaE$^-$* mutants are capable of ferredoxin reduction but not of cyclic electron transport, it appears that at least in cyanobacteria under laboratory conditions the major cyclic flow from ferredoxin involves PSI itself and not a distantly located ferredoxin:quinone reductase (FQR). In *Synechococcus* sp. there is evidence that FNR is bound to the phycobilisomes and not to PSI-E [18, B Nore and HV Scheller, unpublished]. In spite of this the three observations: 1) FNR is bound to PSI-E, 2) FNR is involved in cyclic electron flow, 3) *psaE$^-$* strains are incapable of cyclic electron flow, lead us to the hypothesis that the FNR/PSI-E complex constitutes a hitherto unidentified FQR. To investigate this possiblity we have characterized kinetic parameters of free isolated FNR and of FNR bound to PSI and studied the quinone reducing activity of FNR.

2. Materials and Methods

PSI, FNR, ferredoxin, and plastocyanin were isolated from seedlings of barley (*Hordeum vulgare* L.). PSI with bound FNR was isolated with the detergent *n*-decyl-ß-D-

P. Mathis (ed.), Photosynthesis: from Light to Biosphere, Vol. II, 847–850.

maltopyranoside [7]. Isolation of PSI devoid of FNR was carried out by a modification of the same procedure [19].

Photoreduction of $NADP^+$ was measured according to Naver et al. [19]. FNR was added either as bound to PSI or as free, intact FNR. No additional FNR was added when analyzing the bound FNR. When analyzing the free FNR, an equivalent amount of FNR-free PSI was included to drive the reaction. Under these conditions, no tight rebinding of FNR to the added PSI took place as evidenced by the absence of FNR in the pellet after ultracentrifugation. Diaphorase activity was measured in a reaction containing 20 mM Tricine pH 7.5, 10 mM $MgCl_2$, 120 μM DCPIP, 0-200 μM NADPH, and 10-25 nM FNR either as free, intact FNR or as PSI-bound FNR. Reduction of quinones was measured using NADPH as electron donor. The reaction mixture contained 20 mM Tris·HCl pH 7.8, 8 mM $MgCl_2$, 200 μM NADPH, 40 μM quinone, and 10-25 nM FNR. The oxidation of NADPH was monitored at 340-375 nm. Some of the quinones undergo absorption changes at 340-375 nm upon reduction. In these cases a resulting extinction coefficient was calculated from the oxidized minus reduced spectra.

3. Results and Discussion

The PSI complexes used contained about 0.5 mol FNR per mol P700. The Chl/P700 ratio was 180. PSI isolated with bound FNR according to the decylmaltoside procedure never contains stoichiometric amounts of FNR [7]. This could be due to a partial loss of FNR from PSI during isolation but could also reflect an *in vivo* situation where regulatory mechanisms ensure tight binding of FNR to a population of the PSI complexes. Thylakoid membranes contain one FNR per 250 Chl [4]. Thus, the 0.5 FNR/P700 in isolated PSI corresponds to about 25% of the FNR pool.

The activity of free and bound FNR with NADPH as substrate was analyzed by measurements of diaphorase activity (Fig. 1). For free FNR V_{max} =155 e^-/(FNR x s) and K_m=24 μM. For bound FNR the corresponding values were 14 e^-/(FNR x s) and 59 μM. In the assay for $NADP^+$-reduction, V_{max} was 104 and 11 e^-/(FNR x s) for free and bound FNR, respectively. Antimycin A at a concentration of 5 μM had no detectable effect in any of the assays.

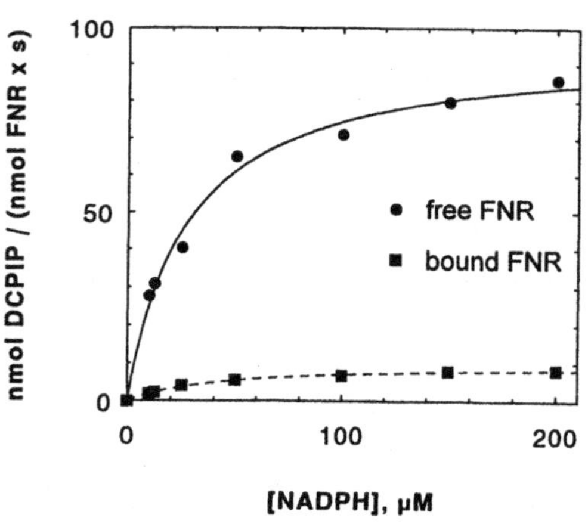

Fig. 1. Effect of NADPH concentration on diaphorase activity (NADPH→DCPIP) of free, intact FNR (●) and of bound FNR (■).

While some studies have shown the existence of antimycin insensitive cyclic electron transport, the significance of this path *in vivo* is not easily assessed in the absence of a specific inhibitor. However, in a recent study antimycin had little effect on the overall cyclic electron flow *in vivo* indicating that the antimycin insensitive path is important [14].

The ability of FNR to catalyze electron transport reactions is strongly dependent on whether FNR is bound to the PSI complex or present in an unbound state. Since $NADP^+$ is a much poorer substrate for the bound FNR it is likely that the binding site for $NADP^+$ is more or less buried in the membrane. Therefore, possible alternative substrates for bound FNR are quinones in the thylakoid membrane. The quinones which occur naturally in the thylakoid membrane, PQ and vitamin K_1, are too hydrophobic to be used in an aqueous *in vitro* assay with solubilized proteins. In order to analyze the specific requirements for activity, a number of different soluble quinones were tested (Table I). Addition of antimycin A had no effect on any of the activities measured.

Substrate	NADPH oxidation e^-/FNR s^{-1}	Relative activity (DCPIP=100)
phenyl-1,4-benzoquinone	1.4 ± 0.2	44
decyl-PQ	0	0
2,3-dimethoxy-5-methyl-benzoquinone (UQ-0)	1.3 ± 0.2	41
decyl-UQ	0	0
UQ-10	0	0
1,4-benzoquinone	0.3 ± 0.1	9
methyl-1,4-benzoquinone	0.16 ± 0.03	5
duroquinone	0.9 ± 0.1	28
vitamin K_3	1.5 ± 0.2	47
vitamin K_1	0	0
2-OH-1,4-naphtoquinone	0	0
1,4-naphtoquinone	2.3 ± 0.3	72
DCPIP	3.2 ± 0.3	100

Table I. *Quinone reduction by FNR bound to PSI. The values in the table (mean ± SE) are corrected for non-enzymatical oxidation of NADPH in the presence of the different quinones. The reactions contained 20 mM Tris (pH 7.8), 8 mM $MgCl_2$, 200 µM NADPH, 40 µM quinone, and 25 nM FNR.*

Since the solubility and diffusion characteristics of the quinones play a large role for the activity in the *in vitro* assay, care must be taken in conclusions as to the identitiy of the quinone acceptor which functions *in vivo*. Generally the naphtoquinones and the somewhat similar phenyl-benzoquinone were the best substrates. However, more PQ-like benzoquinones such as duroquinone and UQ-0 were also good acceptors. PQ is almost certainly a component in cyclic electron transport and therefore the most likely acceptor of electrons from FNR. The FNR bound to PSI-E is in rather close proximity to vitamin K_1. One of the two vitamin K_1 molecules in PSI constitutes the electron acceptor A_1 and has a very low midpoint potential. The other vitamin K_1 molecule of PSI could be a possible acceptor. However, numerous investigations of cyclic electron transport have implicated that plastoquinol oxidation is an obligate step in the pathway, since cytochrome b_6/f inhibitors such as DBMIB and stigmatellin abolish cyclic electron transport. No reduction of decyl-PQ could be detected. On the contrary, decyl-PQ added to the reaction mixture in the diaphorase assay (NADPH to DCPIP) with free FNR resulted in a strong inhibition ($I_{50} = 17 \pm 2$ µM) of the activity (Fig. 2). Vitamin K_1 also inhibited the activity, but not as much as decyl-PQ (Fig. 2). Contrary to this, neither decyl-UQ nor 2-OH-1,4-naphtoquinone inhibited the diaphorase activity.The lack of steady state reduction of

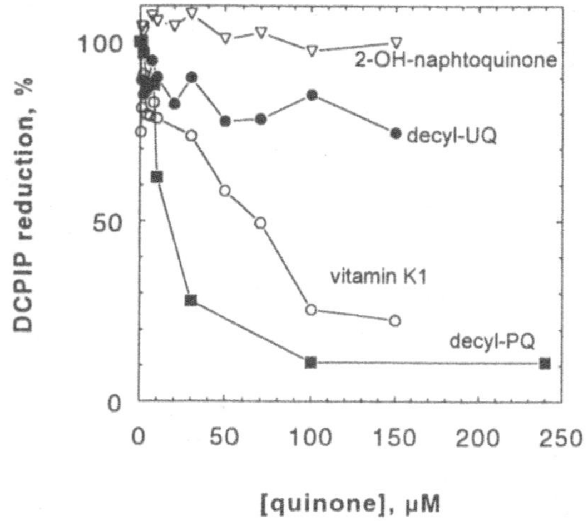

***Fig.** 2. Inhibition of diaphorase activity by quinones. The reaction contained 200 μM NADPH, 120 μM DCPIP and 10 nM free, intact FNR.*

decyl-PQ in spite of the high affinity of FNR for the quinone may be a result of the slow release of decyl-PQ from the complex into the aqueous medium. If this is the case, a more direct measurement of the activity of FNR as an FQR will require an assay employing reconstituted vesicles.

The highest rate obtained in the *in vitro* assay with the quinones is 2 e^-/FNR s^{-1} for the bound FNR. This value is comparable to the rate of rereduction of $P700^+$ by cyclic electron flow which has been measured to 2-5 s^{-1} [11,20, HV Scheller, unpublished results). Thus, the data presented here indicate that FNR could be directly involved in the cyclic electron transport and not simply by providing NADPH for an NADPH:PQ oxidoreductase as suggested in a study of *Chlamydomonas* [14].

4. References

1. Vallejos RH, Ceccarelli E, Chan R (1984) J Biol Chem **259**: 8048-8051
2. Ceccarelli EA, Chan RL, Vallejos RH (1985) FEBS Lett **190**: 165-168
3. Shin M, Ishida H, Nozaki Y (1985) Plant Cell Physiol **26**: 559-563
4. Matthijs HCP, Coughlan SJ, Hind G (1986) J Biol Chem **261**: 12154-12158
5. Berzborn RJ, Klein-Hitpass L, Otto J, Schünemann S, Oworah-Nkruma R, Meyer HE (1990) Z Naturforsch **45c**: 772-784
6. Nakatani S, Shin M (1991) Arch Biochem Biophys **291**: 390-394
7. Andersen B, Scheller HV, Møller BL (1992) FEBS Lett **311**: 169-173
8. Chitnis PR, Reilly PA, Miedel MC, Nelson N (1989) J Biol Chem **264**: 18374-18380
9. Rousseau F, Sétif P, Lagoutte B (1993) EMBO J **12**: 1755-1765
10. Zhao J, Snyder WB, Mühlenhoff U, Rhiel E, Warren PV, Golbeck JH, Bryant DA (1993) Mol Microbiol **9**: 183-194
11. Yu L, Zhao J, Mühlenhoff U, Bryant DA, Golbeck JH (1993) Plant Physiol **103**: 171-180
12. Shahak Y, Crowther D, Hind G (1981) Biochim Biophys Acta **636**: 234-243
13. Cleland RE, Bendall DS (1992) Photosynth Res **34**: 409-418
14. Ravenel J, Peltier G, Havaux M (1994) Planta **193**: 251-259
15. Böhme H (1977) Eur J Biochem **72**: 283-289
16. Hosler JP, Yochum CF (1985) Biochim Biophys Acta **808**: 21-31
17. Hosler JP, Yochum CF (1987) Plant Physiol **83**: 965-969
18. Schluchter W, Bryant DA (1992) product. Biochemistry **31**: 3092-3102
19. Naver H, Scott MP, Møller BL, and Scheller HV (1995) Physiol Plant, in press.
20. Maxwell PC, Biggins J (1976) Biochemistry **15**: 3975-3981

H_2O_2-INDUCED TRANSIENT OXIDATION OF A LIGHT-SENSITIVE CYT B-559 IN INTACT SPINACH CHLOROPLASTS IN THE PRESENCE OF ASCORBATE

Sebastian Heimann and Ulrich Schreiber
Lehrstuhl Botanik I, Universität Würzburg
Mittlerer Dallenbergweg 64, D-97082 Würzburg, Germany

1. Introduction

Despite of more than 25 years of research on cytochrome (cyt) b-559, its *in vivo* function is still not clear (1). Under physiological conditions cyt b-559, unlike the two cytochromes of the b/f-Rieske-complex, shows only slow and small light-induced absorption changes which are difficult to detect as they are overlapped by larger changes caused by cyt f and cyt b-563. In the past, measurements were therefore mostly done with thylakoids or PS II particles displaying low PS II donor side activity, under which conditions cyt b-559 does show light-induced oxidation, acting as an alternative donor to P_{680}^+. With the help of a recently developed, computer-controlled LED-Array-Spectrophotometer, light-induced and chemically induced absorbance changes of cytochromes in intact chloroplasts can be measured with high selectivity (2). In the present study, we have applied this system to study cyt b-559 redox changes in intact chloroplasts induced by the combination of H_2O_2 and light. Here we report on a cyt b-559 LP-fraction that can be oxidized by addition of H_2O_2, which is known to be transiently formed also *in vivo* under excessive irradiation.

2. Materials and Methods

Intact chloroplasts were isolated as described in (3), with additional 10 - 20 mM ascorbate in the isolation buffer. Intactness of the chloroplasts routinely was 90-98%, and chlorophyll concentration in the reaction medium was 40 µg/ml. With the applied LED-Array-Spectrophotometer absorbance changes at 16 wavelengths in the 540 - 600 nm region were measured quasi-simultaneously. Previously determined model spectra were used for the computer-assisted deconvolution of the absorbance changes (2).

3. Results and discussion

In the given preparation of intact spinach chloroplasts three different redox forms of cyt b-559 could be distinguished. One LP-fraction (ca. 15% of total cyt b-559), which was reduced in the presence of ascorbate as the sole redox buffer, could be partially oxidized by addition of small amounts of menadione [$E_{m,7}$ = -10 mV] (data not shown). The redox potential of this cyt b-559 LP-form therefore probably is lower than that of ascorbate [$E_{m,7}$ = +30 mV for $2e^-$-donation (4), see also below].

P. Mathis (ed.), Photosynthesis: from Light to Biosphere, Vol. II, 851–854.

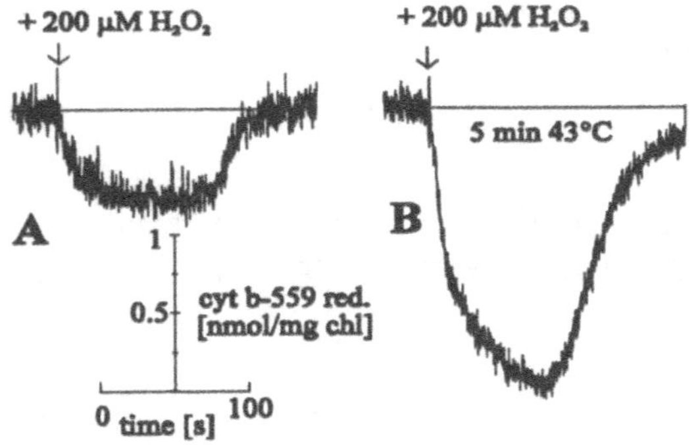

Figure 1: Transient oxidation of cyt b-559 in the dark by H_2O_2 in intact chloroplasts incubated with 5 mM ascorbate.
A: Control chloroplasts.
B: Chloroplasts after a heat-pretreatment (5 min at 43°C).
Both measurements carried out at 20°C.

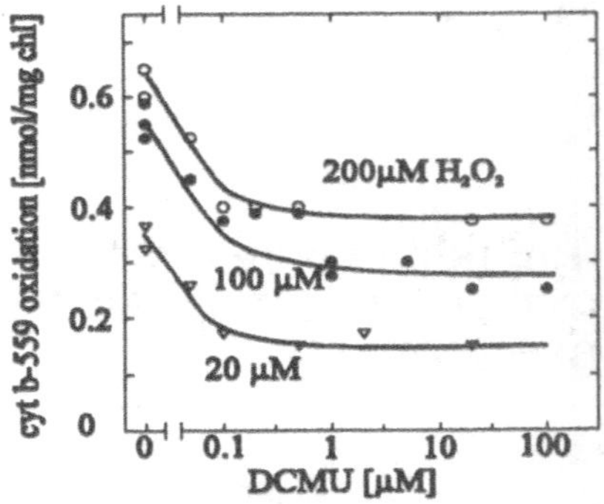

Figure 2: DCMU-concentration dependence for partial suppression of H_2O_2-induced cyt b-559 oxidation in the dark. Intact chloroplasts incubated for 5 min with 5 mM ascorbate and various DCMU-concentration before H_2O_2-addition.

In intact chloroplasts this cyt b-559 fraction could be transiently oxidized in the dark by 200 µM H_2O_2 in the presence of 5 mM ascorbate (Fig. 1A), with half-maximal oxidation occurring at 30 µM H_2O_2 (data not shown). The reductant of cyt b-559 after decomposition of the added H_2O_2 is probably identical to a product of ascorbate peroxidation, namely the monodehydroascorbate radical (MDA), which is a strong reductant as a $1e^-$-donor, whereas the redox potential of ascorbate itself as $1e^-$- donor is too high (4):

$$\text{MDA} \Leftrightarrow \text{dehydroascorbate} + e^- \qquad E_{m,7} = -210\text{ mV}$$
$$\text{ascorbate} \Leftrightarrow \text{MDA} + e^- + H^+ \qquad E_{m,7} = +330\text{ mV}$$

Short heat-pretreatment at temperatures above 40°C (but not higher than 65°C) increased the H_2O_2-sensitive fraction, reaching a maximum of ca. 45% of total cyt b-559 after 5 min at 49°C. In Fig. 1B an increased amplitude after short heat-treatment at 43°C is demonstrated. It may be concluded that H_2O_2-oxidizable cyt b-559 LP can form by modification of the major cyt b-559 HP which is a constituent of the PS II reaction center complex. The proximity to the PS II reaction center is also reflected by the sensitivity of H_2O_2-induced cyt b-559 LP oxidation to DCMU. As shown in Fig. 2, the H_2O_2-induced oxidation of cyt b-559 in the dark was partially inhibited by rather low DCMU-concentrations (maximal effect at 0.1 µM DCMU, e.g. at about one molecule per PS II). There is a DCMU-insensitive cyt b-559 oxidation which increases with H_2O_2-concentration, whereas the DCMU-sensitive fraction amounts to about 0.25 nmol/mg chl at all studied H_2O_2-concentrations. This argues for two different pathways of cyt b-559 oxidation by H_2O_2 in the dark, one of which involves the secondary PS II acceptor Q_B. H_2O_2-induced oxidation of cyt f in the dark, on the other side, was found to be enhanced by DCMU, especially in heat-pretreated chloroplasts (data not shown). From this we conclude that in the presence of H_2O_2 and ascorbate there is simultaneous oxidation of cyt f by H_2O_2 and re-reduction by ascorbate (or the MDA-radical) via the PQ-pool, with the latter involving Q_B, just like suggested for H_2O_2-induced oxidation of part of cyt b-559 LP. In view of the high redox potential of Q_B^- [$E_{m,7}$ = +290 mV for Q_B^-/Q_BH_2] (5), it appears possible that Q_B^- can accept an electron from reduced cyt b-559 LP. On the other hand, when cyt b-559 is oxidized, there should be rather a tendency for its reduction by Q_B^- [$E_{m,7}$ = -45 mV for Q_B^-/Q_B] (5).

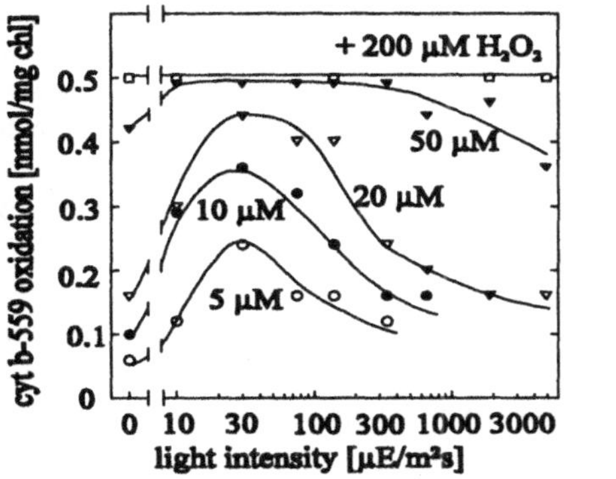

Figure 3: H_2O_2-induced oxidation of cyt b-559 in dependence of the intensity of steady-state background illumination. Intact chloroplasts in the presence of 5 mM ascorbate. Red actinic light peaking at 655 nm.

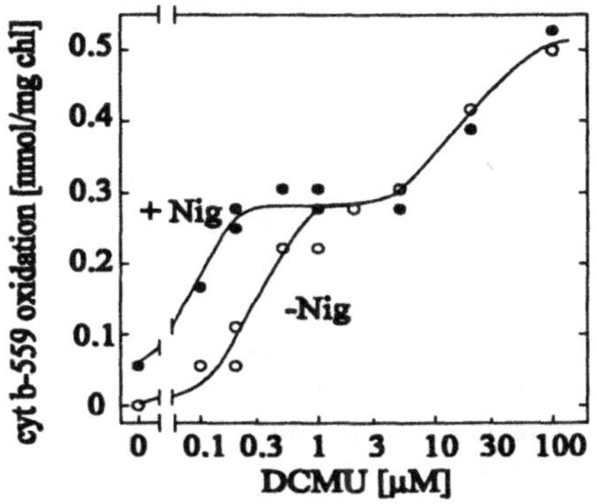

Figure 4: Stimulation of light-induced cyt b-559 oxidation by DCMU in the presence of 1 mM nitrite.
Intact chloroplasts +/- nigericin in the absence of external ascorbate. Cyt b-559 oxidation was assessed after 100 s illumination at 30 µE/m²s of 655 nm light.

In illuminated chloroplasts, the oxidation of cyt b-559 by small, non-saturating concentrations of H_2O_2 strongly depended on the applied light-intensity, with maximal oxidation observed around 30 µE/m²s (Fig. 3). 20 µM H_2O_2, for example, oxidized only 0.15 nmol cyt b-559/mg chl in the dark as well as at 5000 µE/m²s. But at 30 µE/m²s the same concentration of H_2O_2 led to oxidation of 0.45 nmol cyt b-559/mg chl.
When 1 mM nitrite was added as electron acceptor to intact chloroplasts, cyt b-559 oxidation could be also observed without addition of H_2O_2 at light intensities exceeding 50 µE/m²s, but only in the absence of externally added ascorbate (data not shown). Such light-induced cyt b-559 oxidation can be considerably **stimulated** by DCMU and then is also observed at lower irradiance, as shown in Fig. 4. Two well separated steps in the DCMU-concentration dependency are apparent, with the first step being enhanced by the uncoupler nigericin.
These findings are in agreement with the hypothesis that light-induced oxidation of cyt b-559 LP normally is counteracted by re-reduction via Q_BH_2, which is prevented by DCMU. Q_BH_2 can be formed either by 2e⁻ donation from ascorbate to Q_B or via photoreduction by PS II. Its accumulation is favored by the transthylakoidal proton gradient which is dissipated by nigericin. Furthermore, the ΔpH is known to drive reverse

electron flow in PS II (6), which may induce the coupled reduction of Q_A and of oxidized cyt b-559 LP by Q_BH_2.

There remains the question by which pathway cyt b-559 is oxidized during illumination in the experiment of Fig. 4. It cannot be ruled out that oxidation still involves H_2O_2 which may be formed, despite the presence of nitrite and DCMU, via univalent O_2-reduction and consequent superoxide dismutation at PS II centers (7). Another possibility is that light-induced cyt b-559 LP oxidation involves oxidants (D^+) at the PS II donor side, including P_{680}^+. Such oxidation in intact chloroplasts would normally be counteracted by simultaneous re-reduction by Q_BH_2 formed either via charge separation or Q_B-reduction by ascorbate. It should be emphasized that these considerations may apply only for a small fraction (15%) of cyt b-559 which can, however, be considerably increased by heat-treatment and other conditions weakening the PS II donor side and leading to cyt b-559 HP→LP transformation. All presented data are in agreement with the possibility of electron exchange between cyt b-559 LP and the various redox states of Q_B, as previously suggested for cyt b-559 HP (8). The following schemes summarize the possible pathways of cyt b-559 LP oxidation and reduction in the dark and in the light in intact chloroplasts (arrows indicating direction of e^--transfer), which appear likely in view of the presented results:

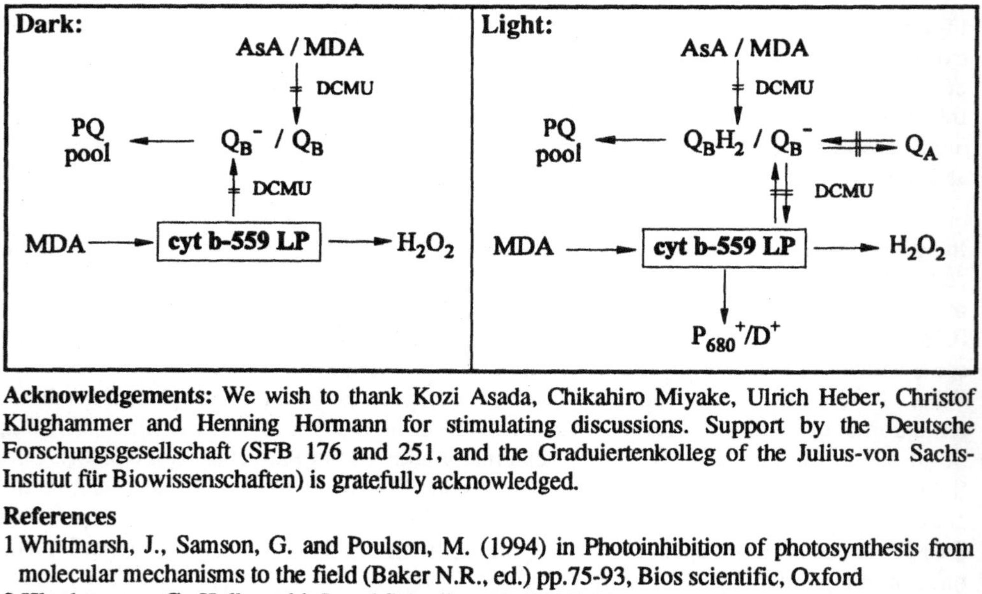

Acknowledgements: We wish to thank Kozi Asada, Chikahiro Miyake, Ulrich Heber, Christof Klughammer and Henning Hormann for stimulating discussions. Support by the Deutsche Forschungsgesellschaft (SFB 176 and 251, and the Graduiertenkolleg of the Julius-von Sachs-Institut für Biowissenschaften) is gratefully acknowledged.

References

1 Whitmarsh, J., Samson, G. and Poulson, M. (1994) in Photoinhibition of photosynthesis from molecular mechanisms to the field (Baker N.R., ed.) pp.75-93, Bios scientific, Oxford
2 Klughammer, C., Kolbowski, J. and Schreiber, U. (1990) Photosynth. Res. 25: 317-327
3 Hormann, H., Neubauer, C., Asada, K. and Schreiber U. (1993) Photosynth. Res. 37: 69-80
4 Njus, D. and Kelley, P.M. (1993) Biochim. Biophys. Acta 1144: 235-248
5 Diner, B.A., Petrouleas, V. and Wendoloski, J.J. (1991) Physiol. Plantarum 81: 423-436
6 Schreiber, U. and Avron, M. (1979) Biochim. Biophys. Acta 546: 436-447
7 Ananyev, G., Renger, G., Wecker, U. and Klimov, V. (1994) Photosynth. Res. 41: 327-338
8 Barabas, K., Kravcova, T. and Garab, G. (1993) Photosynth. Res. 6: 59-64

PS I FLUORESCENCE AT ROOM TEMPERATURE: POSSIBLE EFFECTS ON QUENCHING COEFFICIENTS

Erhard Pfündel, Institut für Pflanzengenetik und Kulturpflanzenforschung, Corrensstraße 3, D-06466 Gatersleben, Germany.
E-mail: erhard@magicalf.boerde.de

1. Introduction

Chlorophyll fluorescence measurements serve as a method to assess the state of photosynthesis *in vivo* without disturbing the leaf material [1]. The development of commercially available modulation fluorometers ensued the widespread use of fluorescence techniques in photosynthesis research. In combination with the saturating pulse method [1], modulation fluorometers provide a convenient way to calculate coefficients of photochemical and non-photochemical fluorescence quenching [2,3], and to estimate the quantum yield of non-cyclic electron transport of PS II [4].
These fluorescence parameters were derived under the assumption that PS I contribution to total fluorescence was negligibly small at physiological temperatures. In most modulation fluorometers, however, the spectral window of PS I and of PS II emission is used. Accordingly, it has been indicated that PS I fluorescence contributes significantly to the total fluorescence signal of modulation fluorometers [5,6,7].
Here, I present a simple model for the estimation of PS I contribution to total fluorescence. The model uses normalized variable fluorescence at room temperature and fluorescence emission spectra at 77K as experimental data. Plant species which differ in their PS I/PS II ratio were used. From the data, the maximum yield of photochemistry of PS II (Φ^{PSII}) is derived. In the case of a PAM 101 fluorometer, PS I fluorescence relative to total fluorescence is calculated. PSI-effects on quenching parameters are discussed.

2. Material and Methods

Zea mays and 12 species of the genus *Flaveria* were grown on garden mulch in the green house. The genus *Flaveria* includes species with C_3 photosynthesis and C_4 species (NADP-ME), the latter with elevated PS I/PS II ratios, and C_3-C_4 intermediate-type species. For all fluorescence measurements, fully developed leaves of plants that were dark-adapted for 8 to 12h were used. Room temperature fluorescence of intact leaves was measured with a PAM 101 fluorometer (Walz, Effeltrich, Germany). Initial fluorescence, F_0, was determined with 1.6 kHz modulation frequency and maximum fluorescence, F_M, was induced by a 0.7 s light pulse of 1400 $\mu mol \cdot m^{-2} \cdot s^{-1}$ intensity. Fluorescence emission spectra of diluted leaf powder [8] were recorded at 77K using a LS 50B fluorimeter (Perkin Elmer, Buckinghamshire, England). Bandwidths of excitation and emission were 15 nm and 4 nm, respectively. Stray light was reduced by a RG 1 red glass filter (Schott, Jena, Germany) placed between sample and photomultiplyer tube.

P. Mathis (ed.), Photosynthesis: from Light to Biosphere, Vol. II, 855–858.

2. Theoretical Considerations

2.1 *Room temperature fluorescence of leaves*

Normalized variable fluorescence is defined by $F_V/F_M = (F_M - F_0)/F_M$, where F_0 is obtained by excitation with weak modulated light and F_M is the modulated fluorescence in the presence of a saturating light pulse. If PS I and PS II contribute to fluorescence, then:

(eq 1) $$F_V/F_M = ((F_M^{PSI} + F_M^{PSII}) - (F_0^{PSI} + F_0^{PSII})) / (F_M^{PSI} + F_M^{PSII}) \quad .$$

In a dark-adapted leaf, PS II fluorescence is minimal (F_0^{PSII}) when PS II reaction center traps are open, and is maximal when they are closed (F_M^{PSII}). PS I fluorescence (F^{PSI}) is assumed to be constant for open and closed PS I reaction centers [9]. Thus:

(eq 2) $$F_V/F_M = (F_M^{PSII} - F_0^{PSII}) / (F_M^{PSII} + F^{PSI}) \quad .$$

Fluorescence of PS I, PS II equals the quanta absorbed by PS I, PS II ($A^{PSI,\,PSII}$), times a PS I, PS II-specific factor ($C^{PSI,\,PSII}$) that considers, e.g., fluorescence reabsorption, times the fluorescence yield of PS I, PS II ($\phi_F^{PSI,\,PSII}$):

(eq 3) $$F^{PSI,\,PSII} = A^{PSI,\,PSII} \cdot C^{PSI,\,PSII} \cdot \phi_F^{PSI,\,PSII} \quad .$$

Replacing in eq 2 the fluorescence terms F^{PSI}, F_0^{PSII}, and F_M^{PSII} by eq 3 formulated for PS I, and for PS II in the F_0 state and in the F_M state, yields after rearrangement:

(eq 4) $$F_V/F_M = (1/\Phi^{PSII}) + A \cdot C \cdot (1/\Phi^{PSII}) \cdot (\phi_F^{PSI} / \phi_{FM}^{PSII}) \quad ,$$

where $A = A^{PSI}/A^{PSII}$, $C = C^{PSI}/C^{PSII}$, and $\Phi^{PSII} = (\phi_{FM}^{PSII} - \phi_{F0}^{PSII})/\phi_{FM}^{PSII}$. ϕ_{F0}^{PSII} and ϕ_{FM}^{PSII} are the fluorescence yields of PS II with open and closed reaction centers, respectively.

2.2 *Low temperature fluorescence of diluted leaf powder*

Fluorescence spectra at 77K show maxima at 685 nm, *F*685, and 735 nm, *F*735. *F*685 is attributed to PS II. *F*735 is mainly caused by PS I and some PS II emission [10]. Thus:

(eq 5) $$F685 = \tilde{A}^{PSII} \cdot k_2 \cdot \phi_{F_M}^{PSII}$$

(eq 6) $$F735 = \tilde{A}^{PSI} \cdot k_1 \cdot \phi_F^{PSI} + \tilde{A}^{PSII} \cdot k_2' \cdot \phi_{F_M}^{PSII} \quad ,$$

where $k_2 = \tilde{\phi}_{F685}^{PSII} / \phi_{F_M}^{PSII}$, $k_1 = \tilde{\phi}_{F735}^{PSI} / \phi_F^{PSI}$, and $k_2' = \tilde{\phi}_{F735}^{PSII} / \phi_{F_M}^{PSII}$. Here, $\tilde{\phi}_{F685}^{PSII}$, $\tilde{\phi}_{F735}^{PSII}$ and $\tilde{\phi}_{F735}^{PSI}$ are fluorescence quantum yields at 77K of PS II at 685 nm, of PS II at 735 nm, and of PS I at 735 nm, respectively. $\tilde{A}^{PSI}$, $\tilde{A}^{PSII}$ are the quanta absorbed by PS I, PS II.

2.3 *Combination of equations for room temperature and low temperature fluorescence*

Replacing in eq. 4 the terms ϕ_{FM}^{PSII} and ϕ_F^{PSI} by eq 5 and eq 6, respectively, results in:

(eq 7) $$F_M/F_V = b + m \cdot F735/F685 \quad .$$

Thus, eq 7 relates F_M/F_V linearly to $F735/F685$, where the slope and Y-axis intercept are given by $m = A \cdot \tilde{A} \cdot C \cdot (1/\Phi^{PSII}) \cdot (k_2/k_1)$ and $b = 1/\Phi^{PSII} - m \cdot (k'_2/k_2)$, respectively. Here, $\tilde{A} = \tilde{A}^{PSII} / \tilde{A}^{PSI}$. Thus, the maximum yield of PS II, Φ^{PSII}, can be calculated:

(eq 8) $$\Phi^{PSII} = 1/(b + m \cdot (k'_2/k_2)) \quad .$$

Using eq 3, Φ^{PSII} can be expressed in fluorescence terms:

(eq 9) $$\Phi^{PSII} = (F_M^{PSII} - F_0^{PSII}) / F_M^{PSII} \quad .$$

The combination of eq 2 and eq 9 yields PS I fluorescence relative to F_M:

(eq 10) $$F^{PSI}/(F_M^{PSII} + F^{PSI}) = 1 - (F_V/F_M)/\Phi^{PSII} \quad .$$

3. Results

If optical properties and fluorescence yields are similar among various plant species, then eq 7 predicts that the fluorescence ratio F_M/F_V is linearly related to the fluorescence ratio $F735/F685$. For *Z. mays* and 12 species of *Flaveria*, Fig. 1 shows values of F_M/F_V from intact leaves at room temperature plotted against values of $F735/F685$ from diluted leaf powder at 77K. In agreement with eq 7, the two parameters were linearly related. The regression coefficients for the slope (*m*) and the Y-axis intercept (*b*) were 0.083 and 1.132, respectively. Using eq 8, the maximum yield of PS II (Φ^{PSII}) can be calculated from the parameters *m* and *b*, and the relative contribution of PS II to $F735$ (k'_2/k_2).

Among various PS II components separated by non-denaturating PAGE, CP47 exhibited the highest relative emission at 735 nm (k'_2/k_2 = 0.2; data not shown). For CP43 and for various components of LHC II, k'_2/k_2 ranged between 0.1 and 0.16. Calculated values of Φ^{PSII} were 0.877 and 0.871 for k'_2/k_2=0.1, and k'_2/k_2=0.2, respectively. PS I contribution to total room temperature fluorescence can be derived by using eq 10. For Φ^{PSII}=0.877, mean values of relative PS I emission were 5%, 7%, and 11% for C_3-plants (species A and B in Fig. 1), C_3-C_4-intermediate plants (C-H in Fig. 1) and C_4-plants, (I-M in Fig. 1), respectively.

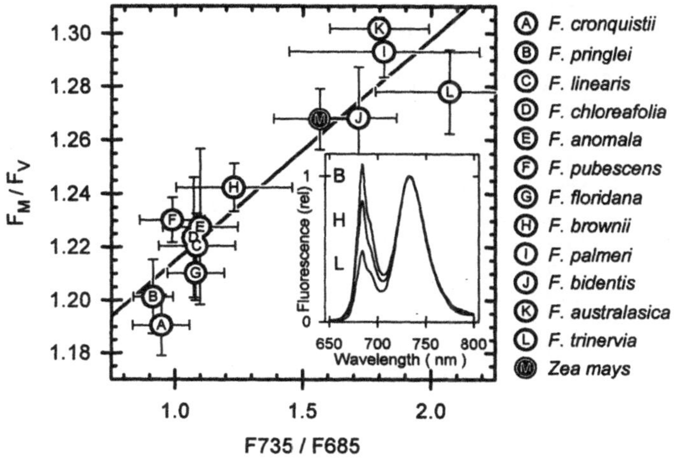

Fig. 1: Relation between F_M/F_V of leaves and $F735/F685$ of leaf powder. F_M/F_V was determined at room temperature and $F735/F685$ was derived from 77K emission spectra. 12 species of the genus *Flaveria* (A-L) and of *Z. mays* (M) were investigated. Data represent means of 3-6 individual plants (± SD). Slope and Y-axis intercept of linear regression analysis (solid line; correlation coefficient: r=0.925) are *m*=0.083 (± SE: 0.010) and *b*=1.132 (± SE: 0.014), respectively. Insert: normalized fluorescence emission spectra at 77K of *F. pringlei*, *F. brownii*, and *F. trinervia*.

4. Discussion

The linear relationship between F_M/F_V and $F735/F685$ (Fig. 1) suggests that the ratios of fluorescence yields (k_1, k_2, k'_2 in eq 7) and of optical properties ($\tilde{A}$, A, C in eq 7) are similar for the species investigated. It is reasonable to assume similar fluorescence quantum yields for fully developed leaves of the closely related species of the genus *Flaveria*. However, optical leaf properties might differ. C_4-plants of *Flaveria* are of the NADP-ME subtype which exhibits Kranz anatomy with enrichment of PS I in the bundle sheath compartment. Different leaf anatomy could result in different light gradients across the leaf and might also influence fluorescence reabsorption. However, the maximum photochemical yields of PS II derived from eq 7 (Φ^{PSII}: 0.87-0.88) agree well with the dark F_V/F_M values of 0.86-0.87 which were obtained when PS I fluorescence was largely excluded by optical filters [5,7]. In conclusion, the experimental data in Fig. 1 appear not to be seriously affected by variances in optical leaf properties.

PS I fluorescence relative to maximum fluorescence in dark-adapted leaves was calculated to be 5 and 11% for C_3 and C_4 plants, respectively. The effect of PS I upon the coefficient of non-photochemical quenching ([3]; denoted *NPQ* in Fig. 2) and upon the yield of non-cyclic electron transport ([4]; denoted *Yield* in Fig. 2), is estimated. For these considerations, maximal fluorescence in the light-adapted state (F'_M) is taken to range between 0.2 and 1 (Fig. 2A). In Fig. 2., all data are plotted against F'_M. F_M is set to 1. Constant values of 0.1 and 0.2 are assumed for F^{PSI} and for steady state fluorescence (F_S), respectively. For constant F^{PSI}, the coefficient of photochemical quenching [2] is not affected by F^{PSI}. For *NPQ* and *Yield*, the relative deviation of values calculated under the assumption F^{PSI}=0 from F^{PSI}-corrected data are identical (Fig. 2B). Fig. 2B illustrates that with decreasing F'_M the PS I-dependent relative deviation becomes increasingly amplified. The absolute deviation of the *Yield* remains limited because the corresponding absolute values decrease with increasing PS I effect (Fig. 2D). In the case of *NPQ*, however, PS I fluorescence introduces significant errors for high values, e.g., when *NPQ* for F^{PSI}=0 equals 2 than the PS I-corrected value was 2.9 (Fig. 2C). Thus, effects of PS I emission on coefficients of non-photochemical quenching should be taken into account when high F'_M quenching occurs.

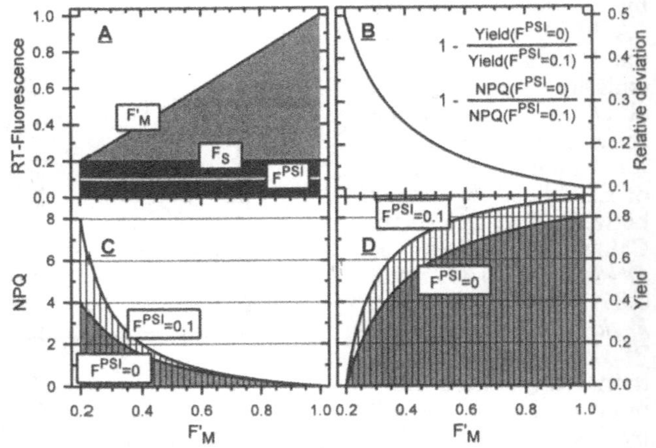

Fig. 2: Fluorescence coefficients in the presence and absence of PS I fluorescence. A: Fluorescence values used for model calculations. B: Relative deviation of *Yield* and *NPQ* calculated for F^{PSI}=0 from corresponding PS I-corrected values (F^{PSI}=0.1). C, D: *NPQ* and *Yield* calculated with F^{PSI}=0 and with F^{PSI}=0.1.

5. References

1 Schreiber, U., Bilger, W. and Neubauer, C. (1994) in Ecophysiology of Photosynthesis (Caldwell M.M., ed) pp. 49-70, Springer, Berlin, Germany

2 Schreiber, U., Schliwa, U. and Bilger, W. (1986) Photosynth. Res. 10, 51-62

3 Bilger, W. and Björkman, O. (1990) Photosynth. Res. 25, 173-186

4 Genty, B., Briantais, J.M. and Baker, N. (1989) Biochim. Biophys. Acta 990, 87-92

5 Adams III, W.W., Demmig-Adams, B., Winter, K. and Schreiber, U. (1990) Planta 180, 166-174

6 Falbel, T.G., Staehelin, L.A. and Adams III, W.W. (1994) Photosynth. Res. 42, 191-202

7 Genty, B., Wonders, J. and Baker, N.R. (1990) Photosynth. Res. 26, 133-139

8 Weis, E. (1986) Photosynth. Res. 6, 73-86

9 Butler W.L. (1978) Ann. Rev. Plant Physiol. 29, 345-378

10 Evans, E.H. and Brown, R.G. (1994) Photochem. Photobiol. B: Biol., 95-104

A NOVEL TEPIDOPTERIN OF A GREEN SULFUR BACTERIUM *CHLOROBIUM TEPIDUM*

Seung-Hyun Cho, Jong-Uk Na, Hwan Youn and Sa-Ouk Kang
Department of Microbiology, College of Natural Sciences, and Research Center for Molecular Microbiology Seoul National University, Seoul 151-742, Korea

1. INTRODUCTION

It was reported that several pterin compounds existed in plant and photosynthetic bacteria, and they were involved in photosynthesis: for example in plant, stimulation of photophosphorylation by the addition of reduced pterin compounds (1), stimulation of photooxidation of cytochrome C by pterin (2), the redox reaction of pterin compounds with ferredoxins (3), and photoreception and phototransduction of blue light near UV (4); in bacteria, prevention of the development of bacterial photosynthetic apparatus by inhibitor of pterin compounds in *Rhodospirillum* (5), regulation of carotenoid biosynthesis by pteridine inhibitors and biopterin in *Rhodospirillum* (6), the primary electron receptor like the photoreceptor in plant, and photoreduction by the crude fractions from *Rhodospirillum* or *Chromatium* (4,7). Especially, together with the above report that many cyanobacteria and purple bacteria produced various pterin glycoside, there was report that erythro-biopterin *N*-acylglucosamine was detected in green sulfur bacterim, *Chlorobium limicola* f. *thiosulfatophilum* NCIB 8327 (8). Therefore, pterin compounds may be involved in photosynthesis in green sulfur bacteria. In this paper, we report the purification and characterization of a novel pterin compound designated as tepidopterin in a green sulfur bacterium, *Chlorobium tepidum*, and also the changes of this compound during the growth.

2. MATERIALS AND METHODS

Microorganism and culture conditions--Chlorobium tepidum, kindly donated by R. Sirevag (University of Oslo, Norway) was grown photoautotrophically under the light intensity of 330 lux for 4 to 5 days at 45°C, according to the method by Pfennig (9) with some modifications in the concentration of sulfide and bicarbonate.

Chemicals--D-threo-biopterin was purchased from Dr. B. Schircks Lab., Switzerland, pterin, L-erythro-bioptein, sodium periodate, dimethylsulfoxide-D_6, Platinum oxide and b-*N*-acetylglucosamidase from Sigma, and sodium ascorbate from Merck. And all other chemicals used were of the highest quality generally available.

P. Mathis (ed.), Photosynthesis: from Light to Biosphere, Vol. II, 859–862.

Assay and purification--The cultured cells were harvested by centrifugation at 20,000 × g, and suspended in 10 volumes of 0.1 M phosphoric acid. The suspensions were sonicated at 4°C using a sonicator and the cell debris was removed by centrifugation. The supernatants were oxidized by the addition of 2N trichloroacetic acid (TCA) and iodine solution (1% I_2/2% KI). After recentrifugation, the oxidized iodine solutions were reduced by the addition of sodium ascorbate to the supernatant until the color disappeared. After the addition of sodium ascorbate, activated charcoal was added, and it was washed with distilled water and eluted with ethanol: 3% NH_4OH (1:1 v/v). The eluate was acidified by the addition of glacial acetic acid until the pH reached to 6.0, and concentrated by rotary evaporator. After Sep-Pak C18 and 0.2 mm nitrocellulose membrane clarification, the concentrated sample was injected onto a preparative DeltaPak C18 HPLC column (30 mm x 30 cm, Waters Associates, U.S.A.) and fractionated isocratically with 2% aqueous methanol at flow rate of 15 ml/min at 276 nm on the UV-visible absorption detector. The purified tepidopterin was concentrated by rotary evaporator and stored at -40°C for further study.

Structural determination--All ^{1}H- and ^{13}C-NMR spectra were obtained by the Bruker AMX-500 spectrometer, which was operated in the Fourier-transform mode at 301 K. Electrospray- and thermospray-mass spectrometry was performed using a VG Quattro quadrupole mass spectrometer. The circular dichroism spectra were obtained with a circular dichroism spectropolarimeter J720 (JASCO).

3. RESULTS AND DISCUSSION

Structural elucidation--NMR, Mass and CD spectra referred this tepidopterin to 1-*O*-(L-threo-biopterin-2'-yl)-b- *N*-acetylglucosamine, as shown in Fig. 1. In the history of discovery of pterin compounds in nature, although chemically synthesized L-threo-biopterin by Sugimoto (11,12), and ciliapterin which reported as L-threo-biopterin (13) but turned out to be D-monapterin (12) in *Tetrahynema*, were reported, but L-form was not. So, tepidopterin may be recorded as the first one having L-threo-biopterin moiety in nature.

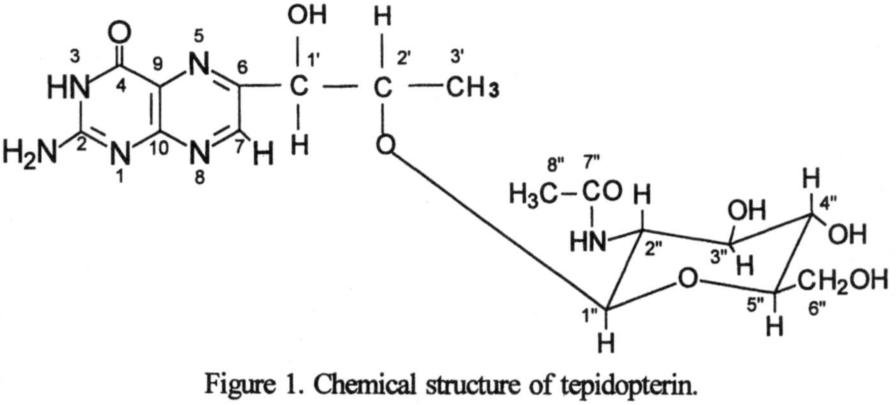

Figure 1. Chemical structure of tepidopterin.

The changes of tepidopterin during the growth of Chlorobium tepidum--Fig. 2 shows that maximum growth of *Chlorobium tepidum* was reached at 71.5 h accompanying the increace of tepidopterin, and after that time, stationary phase continued but intracellular tepidopterin decreased. The partially purified extracellular fluorophores were confirmed as pterin, threo-biopterin, and tepidopterin by the methods described above. Fig. 9B shows the changes of extracellular pterin compounds during growth. Pterin increased very much after maximum growth with some fluctuation at 100 h. Threo-biopterin increased slowly a little after 50 h but decreased a little after 200 h. Tepidopterin increased after 50 h with more variation compared with the two but decreased after about 100 h.

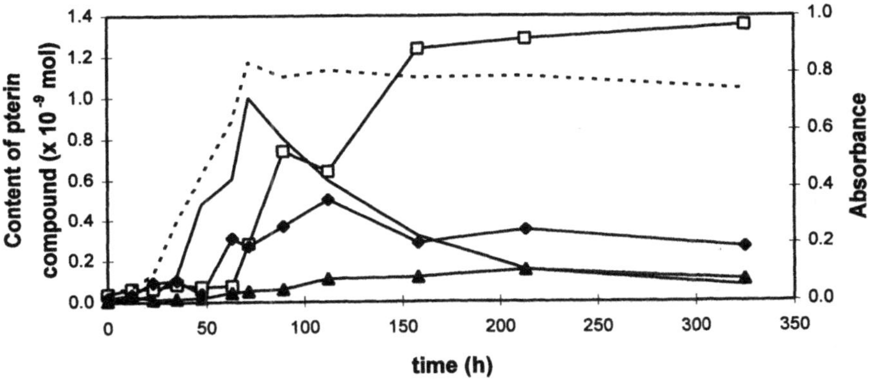

Fig. 2. Quantitation of intracellular tepidopterin and extracellular pterin compounds during the growth of *Chlorobium tepidum*. Changes of the contents of intracellular tepidopterin (–) was determined in the cell pellets. Growth rate (····) was calculated from the changes of bacteriochlorophyll contents of the cell pellets. Changes of the contents of extracellular pterin compounds were determined in culture broth. Pterin (—□—), tepidopterin (—♦ —), threobiopterin (—▲—).

The major redox state of tepidopterin inside the cell was tetrahydro-form deterimined above. So, this microorganism made tetrahydrotepidopterin increase during its growth, but expelled it after maximum growth. The correspondences of the growth and the increase of tetrahydrotepidopterin indicates that this compound may be important in growth although it is to be investigated what part of metabolism of growth it is involved in. However, the extrusion of this compound may be related to the environment which is want of substrates for the growth. The occurrence of threo-biopterin without b-*N*-acetylglucosamine moiety may result from the usage of sugar part although redox state is to be determined. The occurrence of pterin of which redox state was also not determined, may come from the oxidation of tetrahydrotepidopterin into one or two of the three redox states of pterin by physical factor such as light or chemical factor, or may come from the metabolism. Also there is the possibility that tetrahydrothreo-biopterin resulting from tetrahydrotepidopterin is oxidized into dihydropterin or pterin similar to the above report (14,15).

It is calculated that the concentration of tepidopterin inside the cell is about 1.79 mmol/g dry weight in the above purification step by using the estimation of 1 g wet weight to 0.2 g dry weight. And usually this microorganism is harvested at the rate of 1 g wet weight/ 1L culture, and by taking the above estimation of wet weight to dry weight and the rate of maximum 1 nmol tepidopterin/ 1ml culture in Fig. 2, so about 5 mmol/ g dry weight is calculated. The amount of tepidopterin inside the cell is estimated to 2-5 mmol/ g dry weight considering the loss during purification and the approximation of calculation. This amount is about 10 folds higher than biopterin content in pineal gland of rats (16), about from thousand to hundred-thousand times higher than folic acid, biotin, and thiamine in *Methanobacterium*, but comparable to the amount of methanopterin, methanofuran, CoM-SH which have a key role in electron transport in *Methanobacterium*; methane formation from carbon dioxide with H_2 (17). Judging from this fact, together with the increase of tepidopterin which corresponded with the growth, this compound is considered to play important role in growth of *Chlorobium tepidum*.

REFERENCES

1 Maclearn, F.I., Fujita, Y., Forrest, H.S. and Mayers, J. (1965) Science 179, 636-638
2 Maclearn, F.I., Fujita, Y., Forrest, H.S. and Mayers, J. (1966) Plant Physiol. 41, 774-779
3 Fuller, R.C. and Nugent, N.A. (1969) Proc. Natl. Acad. Sci., U.S.A. 63, 1311-1318
4 Galland, P. and Senger, H. (1988) Photochem. and Photobiol. 48, 811-820
5 Fuller, R.C., Kidder, G.W, Nugent, N.A., Dewey, V.C. and Rigopoulos, N. (1971) Photochem. Photobiol. 14, 359-371
6 Nugent, N.A. and Fuller, R.C. (1967) Science. 158, 922-924
7 Forrest, H.S. and van Baalen, C. (1970) Ann. Rev. Microbiol. 24, 91-108
8 Cha, K.W. (1994) Ph.D.-thesis, Seoul National University, Seoul, Korea
9 Pfennig, N. and Truper, H.G. (1984) in Bergey's Manual of Systematic Bacteriology. Vol.3. pp.1635-1709 (Staley, J.T., Bryant, M.P., Pfennig, N. and Holt, J.G., eds.). Williams and Wilkins, Baltimore
10 Wahlund, T.M., Woese, C.R., Castenholz, W.R. and Madigan, M.T. (1991) Arch.Microbiol. 156, 81-90
11 Klein, R., Thiery, R. and Tatischeff, I. (1990) Eur. J. Biochem. 187, 665-669
12 Klein, R. (1993) In Chemistry and Biology of Pteridines and Folates (Ayling, J.E., Nair, M.G. and Baugh, C.M., eds) pp. 43-46. Plenum Press, New York
13 Kidder, G.W. and Dewey, V.C. (1968) J. Biol. Chem. 243, 826-833
14 Armarego, W.L., Randles, F.D. and Taguchi, H. (1983) Eur.J.Biochem. 135, 393-403
15 Davis, M. D., Kaufman, S. and Milstien, S. (1988) Eur.J.Biochem. 173, 345-351
16 Fukushima, T. and Nixon, J.C. (1980) Anal. Biochem. 102, 176-188
17 White, R.H. (1991) J. Bacteriol. 173, 1987-1991

THE FUNCTION OF ASCORBIC ACID IN PHOTOSYNTHETIC PHOSPHORYLATION

Giorgio Forti and Gianluca Elli
Centro C.N.R. per la Biologia Cellulare e Molecolare delle Piante
Dipartimento di Biologia. Università degli Studi di Milano
Via Celoria, 26 20133 Milano, ITALY

1. Introduction

In isolated, stroma-free thylakoids ascorbate is oxidised by the membrane-bound ascorbate peroxidase (APX) in the presence of H_2O_2 to the free radical monodehydroascorbate (MDA) [1]. MDA is rapidly reduced by PS1 [1,2], and its reduction involves electron transport from H_2O to MDA through PS2, PS1 and the whole electron transport chain [2].
We demonstrate here that the electron transport from H2O to MDA is coupled to ATP synthesis and the ratio ATP/O_2 is 2, as in the case when NADP or methylviologen (MV) are the acceptors. It is also shown that APX fulfils the function of peroxidating ascorbate utilising H_2O_2 generated by the membrane-bound superoxide-dismutase (SOD) from $O_2^{\cdot-}$ produced in the Mehler reaction. We discuss a scheme of reaction sequence accounting for the catalytic role of ascorbate in scavenging the toxic species $O_2^{\cdot-}$ and H_2O_2 while providing the acceptor (MDA) for electron transport coupled to photophosphorylation.

2. Procedure

Thylakoids were isolated from freshly-harvested spinach leaves as previously described [2]. Electron transport was measured as the O_2 exchange in a Clark-type O_2 electrode, thermostated at 22°C. Actinic light of 630 to 710 nm was of saturating intensity. ATP was measured enzymatically, as the increase of absorbance at 340-390 nm upon addition of glucose, hexokinase, glucose-6-P-dehydrogenase and NADP to the stirred sample.

3. Results and Discussion

Table I shows that in the presence of H_2O_2, illuminated thylakoids evolve O_2 and synthesise ATP, with a ratio $ATP/2e^-$ of 1. The same ratio is observed when MV(table I),

P. Mathis (ed.), Photosynthesis: from Light to Biosphere, Vol. II, 863–866.

or NADP (not shown), is used as the acceptor. The rates of electron transport and of phosphorylation with MDA as the terminal acceptor were of ca. 60% the rates observed with MV or NADP [2].

Additions	O_2	ATP	ATP/O_2	ATP/$2e^-$
ASC 5 mM	72.4 ± 10.6	140.1 ± 33.0	1.93	0.96
MV 100 μM	-104.0 ± 38.0	222.0 ± 17.0	2.13	1.06
MV 100 μM +ASC 5 mM	N.D.	194.5 ± 2.5	N.D.	N.D.
MV 100 μM +ASC 40 mM	N.D.	231.2 ± 30.0	N.D.	N.D.

Table I. Coupling of ATP synthesis to MDA photoreduction. Lack of influence of ascorbate on phosphorylation coupled to MV reduction
Conditions: as under "Procedures", pH 8.0. [Chl]= 25 μg * ml^{-1}; [NaN_3]= 1 mM; [H_2O_2]= 0.5 mM (added at the onset of illumination, omitted in the presence of MV). The data are expressed in μmoles * mg^{-1} Chl * h^{-1}, ± standard deviation. In the case of MV dependent electron transport, SOD was added in the amount needed to lower to a constant value the rate of O_2 uptake. Light intensity was 60 mW * cm^{-2}

However, the rate of MDA photoreduction was found to be 8 to 10 times higher than the rate of the Mehler reaction. This is shown in table II: the rate of ATP synthesis was 23.3 μmoles * mg^{-1} Chl * h^{-1} with no additions (O_2 was the only acceptor), and 195 when ascorbate + H_2O_2 were added. When ascorbate was the only addition, the rate of ATP synthesis had the intermediate value of 49.3. The addition of CN^-, a known inhibitor of APX, suppressed almost completely the stimulation of phosphorylation by ascorbate and ascorbate + H_2O_2 (table II), while having no effect on the Mehler reaction-coupled phosphorylation. It is well known, on the other hand, that CN^- does not affect directly photosynthetic electron transport in thylakoids.

	Additions		
	None	ASC	ASC + H_2O_2
Control	23.3 ± 1.8	49.3 ± 4.6	195.0 ± 14.5
+ KCN 2mM	20.4 ± 1.2	26.0 ± 3.0	31.1 ± 3.8

Table II. Effects of cyanide, ascorbate and H_2O_2 on photophosphorylation
Conditions: as under "Procedures", pH 8.0. [Chl]= 10 μg * ml^{-1}; [ascorbate]= 5 mM; [H_2O_2]= 0.4 mM. The data are expressed in μmoles of ATP * mg^{-1} Chl * h^{-1}, ± standard deviation. Light intensity was 86 mW * cm^{-2}.

The following scheme describes the reaction sequence accounting for the role of ascorbate and of APX in photophosphorylation:

$$1)\ 2\ H_2O + 4\ O_2 \xrightarrow{PS2,\ PS1,\ LIGHT} 4\ O_2^{\cdot -} + O_2 + 4\ H^+$$

$$2)\ 4\ O_2^{\cdot -} + 4\ H^+ \xrightarrow{SOD} 2\ O_2 + 2\ H_2O_2$$

$$3)\ 4\ O_2^{\cdot -} + 4\ ASC \longrightarrow 4\ MDA + 4\ H_2O_2$$

$$4)\ 4\ H_2O_2 + 8\ ASC \xrightarrow{APX} 8\ MDA + 8\ H_2O$$

$$5)\ 6\ H_2O + 12\ MDA \xrightarrow{PS2,\ PS1,\ LIGHT} 12\ ASC + 3\ O_2$$

In the presence of ascorbate, its reaction with $O_2^{\cdot -}$ (reaction 3) competes with the disproportionation of $O_2^{\cdot -}$ by SOD (reaction 2), so inhibiting O_2 evolution, and more MDA and H_2O_2 are produced. The following reaction (4), catalysed by APX, produces more MDA. Finally, MDA is photoreduced by photosynthetic electron transport (reaction 5). The net result of the reaction sequence (1 + 3 + 4 + 5) is to increase by a factor of 4 the electron transport as compared with the Mehler reaction (reaction 1), and ATP synthesis will be increased correspondingly, because the rates of reactions (3), (4) and (5) are much higher than the rate of reaction (1). If the suppression of reaction (2) by ascorbate were incomplete, one should expect an intermediate stimulation of ATP synthesis, as indeed observed (tables II and III).

	Additions				
	None	Catalase	Catalase +SOD	Ascorbate	CAT +SOD + ASC
O_2	-10.70 ± 0.69	0.70 ± 0.70	1.98 ± 0.80	-2.34 ± 1.10	1.81 ± 0.40
ATP	10.40 ± 1.13	10.20 ± 1.49	10.90 ± 1.49	28.20 ± 4.30	14.60 ± 2.60
Effect of ASC	- -	- -	- -	+ 17.8	+ 4.4

Table III. Effects of ascorbate, SOD and catalase on O_2 uptake and ATP synthesis
Conditions: as under "Procedures", pH 8.0. [Chl]= 25 μg * ml^{-1}; [ascorbate]= 5 mM; SOD was added in excess of the amount required to disproportionate $O_2^{\cdot -}$ produced at a rate of 375 μmoles * mg^{-1} Chl * h^{-1}; added catalase activity was 4.39 μmoles O_2 * min^{-1}. All the data are expressed in μmoles * mg^{-1} Chl * h^{-1}, ± standard deviation. Light intensity was 60 mW * cm^{-2}.

In the presence of ascorbate the net O_2 balance should be zero because O_2 uptake due to reaction (1) is fully compensated by O_2 evolution linked to MDA reduction (reactions 1 to 5). This was indeed observed (table III): the addition of ascorbate drastically reduced light-dependent O_2 uptake, and stimulated ca. 3 times ATP synthesis. The addition of

catalase + SOD largely suppressed the effect of ascorbate on ATP synthesis (table III), as expected, because these enzymes destroy both $O_2^{\cdot-}$ and H_2O_2, so preventing the formation of MDA. A small amount of $O_2^{\cdot-}$ can escape destruction by both the membrane-bound and the added SOD, as previously reported [3]. This accounts for the residual O_2 uptake and stimulation of ATP synthesis by ascorbate (table III), and the incomplete suppression by CN^- of the ascorbate stimulation of phosphorylation (table II).
We conclude that the system APX-ascorbate has the important role of producing MDA, so providing the electron acceptor for phosphorylating electron transport, while scavenging the toxic species $O_2^{\cdot-}$ and H_2O_2. The high activity of thylakoid-bound APX [1], the high rate of MDA photoreduction (tables I and II, see also ref. [2]), and the fact that this system utilises both PS2 and PS1, make of it the best candidate to fulfil the role of compensating the stoichiometric shortage of ATP due to the unbalance between its production coupled to NADP reduction and the requirement for CO_2 assimilation.

References

1) Miyake C. and Asada K. (1992) Plant Cell Physiol. 33: 541-553.

2) Forti G. and Ehrenheim A. M. (1993) Biochim. Biophys. Acta 1183: 408-412

3) Takahashi M. and Asada K. (1988) Arch. Biochim. Biophys. 267: 714-722.

FERREDOXIN-DEPENDENT AND ANTIMYCIN A-SENSITIVE REDUCTION OF CYTOCHROME *B*-559 IN MAIZE THYLAKOIDS: PARTICIPATION OF A MENADIOL-REDUCIBLE CYTOCHROME *B*-559 IN CYCLIC ELECTRON FLOW

Chikahiro Miyake, Ulrich Schreiber[1], Hideto Miyoshi[2] and Kozi Asada
Res. Inst. Food Sci. Kyoto Univ., Uji., Kyoto, 611, JAPAN. [1]Lehrstuhl Botanik I, Univ. Würzburg, D-97082 Würzburg, GERMANY. [2]Faculty of Agriculture, Kyoto Univ., Kyoto, 606-01, JAPAN.

1. Introduction

Since the discovery of ferredoxin (Fd)-dependent cyclic electron flow around PSI in higher plants [1], two possible functions have been proposed, namely the adjustment of the photoproduction ratio of ATP/NADPH depending on the metabolic requirements of the dark reactions, and the down-regulation of PSII by the formation of a ΔpH leading to protection from photoinhibition [2].

Here we describe the identification of Cyt *b*-559, which is reducible by Fd under far-red light and by menadiol in the dark, as the mediator of Fd-cyclic electron flow around PS I in maize thylakoids [3]. With the help of various derivatives of antimycin A, the structural requirements of antimycin A for effective inhibition of Fd-cyclic electron flow are investigated.

2. Materials and Methods

Maize thylakoids from mesophyll chloroplasts were prepared as described in [3]. Fluorescence of 9-aminoacridine (9-AA) was measured with a Xe-PAM Fluorometer [4]. Absorbance changes that covered a range from 530 to 590 nm were measured with a laboratory-built LED-array spectrophotometer [5]. Derivatives of antimycin A were synthesized by Tokutake et al.[6].

3. Results and Discussion

3-1. *Participation of Cyt b-559 in Fd-cyclic electron flow* —— Maize thylakoids showed a high Fd-dependent quenching of 9-AA fluorescence by far-red light illumination, indicating the formation of ΔpH across thylakoid membranes by cyclic electron flow around PS I [3]. Further evidence for the operation of Fd-dependent cyclic electron flow was the inhibition of quenching of 9-AA fluorescence by antimycin A (I_{50} 0.2 μM).

We examined the effects of antimycin A and Fd on Cyts in maize mesophyll thylakoids. In the absence of Fd, an absorbance decrease at 554 nm was induced by far-red light, indicating the photooxidation of Cyt *f* (Fig. 1). On addition of Fd to the thylakoids an absorbance increase peaking at 560 nm was observed, indicating the reduction of Cyt *b*-559 by photoreduced Fd in PSI. The photoreduction of Cyt *b*-559 by Fd was reversed by antimycin A and the photooxidation of Cyt *f* was again observed. Furthermore, the reduced Cyt *b*-559 was oxidized under far-red light upon addition of antimycin A, indicating that antimycin A inhibits electron flow between Fd and Cyt *b*-559[3]. The concentration of antimycin A for half-maximal suppression of the absorbance increase at 560 nm was 0.2 μM, similar to that for the quenching of 9-AA fluoresence by far-red light. These results indicate that the Fd-reducible Cyt under far-red light is a mediator of Fd-cyclic electron flow, which hence is referred to as Cyt *b*-559(Fd). This was further supported by the oxidation of reduced Cyt *b*-559(Fd) by duroquinone ($E_{m,7}$ = +100 mV, close to the $E_{m,7}$ of plastoquinone), suggesting that the photoreduced Cyt *b*-559(Fd) can donate electrons to the plastoquinone in maize thylakoids [3].

P. Mathis (ed.), Photosynthesis: from Light to Biosphere, Vol. II, 867–870.

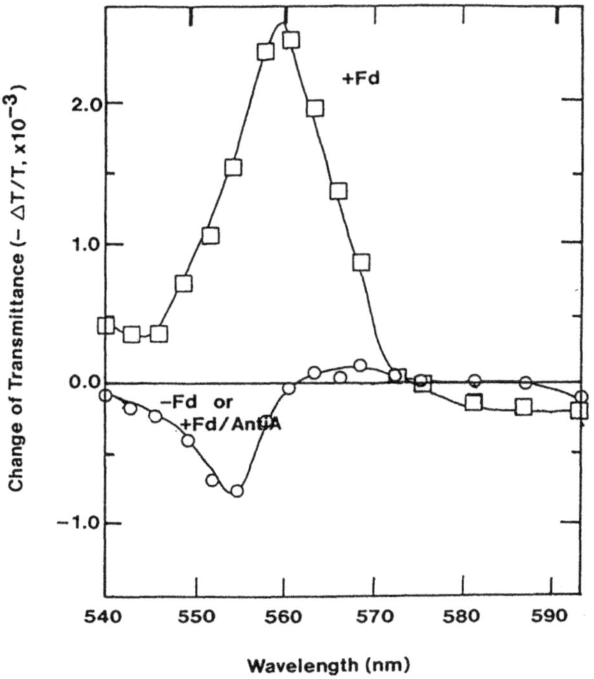

Fig. 1 *Fd-dependent reduction of Cyt b-559(Fd) of maize thylakoids by illumination with far-red light. The reaction mixture (2 ml) contained 50 mM HEPES-KOH (pH 7.6)/0.4 M sucrose/2 mM $MgCl_2$/10 mM KCl/thylakoid membranes (80 μg Chl). Where indicated 10 μM Fd and 1 μM antimycin A were added. Difference spectra between light [30 s after far-red light (>720 nm, 20 W m^{-2} s^{-1}) was turned on] and darkness are presented. Difference spectrum of -Fd was the same as that of +Fd/AntiA.*

On the basis of these results, we propose the following Fd-dependent pathway of cyclic electron flow: P700 ⇒ Fd ⇒ Cyt *b*-559(Fd) ⇒ plastoquinone ⇒ Cyt *b*/*f* ⇒ plastocyanin ⇒ P700.

3-2. *Molecular species and the amount of Cyt b-559(Fd) in maize mesophyll thylakoids* —— In PSII complexes Cyt *b*-559 exists in two forms, a hydroquinone ($E_{m,7}$ = 300 mV) reducible and an ascorbate ($E_{m,7}$ = 80 mV) reducible forms [7]. To the contrary, Cyt *b*-559(Fd) was reduced only by menadiol ($E_{m,7}$ = -10 mV)[3], suggesting that the molecular species of Cyt *b*-559(Fd) is different from Cyt *b*-559s in PSII. Cyt *b*-559(Fd) occupied 20% of the total Cyt *b*-559 in maize mesophyll thylakoids [3], as judged by the amount of menadiol-reducible Cyt *b*-559 and also of far-red light-reducible Cyt *b*-559(Fd) in the presence of Fd [3]. Assuming that the absorption coefficient at 559 nm of Cyt *b*-559(Fd) is the same as that of Cyt *b*-559 in PSII, the content of Cyt *b*-559(Fd) is estimated to be about 1 molecule per 1 molecule of P700 in maize mesophyll thylakoids.

3-3. *Structural requirement of antimycin A for the inhibition of the Fd-dependent reduction of Cyt b-559(Fd) by far-red light* —— Antimycin A is a powerful inhibitor of electron transport in the respiratory chain of mitochondria, where it suppresses electron transfer to ubisemiquinone from Cyt b_H in the Cyt *b*/*c1*-complex by binding to Cyt b_H [8]. The strucure of antimycin A consists of a 3-formylaminosalicylic acid (3-FA)-moiety linked via an amide bond to alkyl- and acyl-substituted dilactone-ring (D-ring)(Fig. 2). We examined the effects of antimycin A derivatives on the photoreduction of Cyt *b*-559(Fd) and ΔpH formation by far-red light in maize thylakoids, to elucidate the structural reuqirements for the inhibitory action of antimycin A on Fd-dependent cyclic electron flow.

First, the role of the D–ring of antimcyin A in the inhibition of the photoreduction of Cyt *b*–559(Fd) was examined. The derivatives where the D–ring of antimycin A is changed to 2,5–dibutylphenylbenzene (Compound A) and to di–n–octyl L–glutamate (Compound B) showed an inhibition of the photoreduction of Cyt *b*–559(Fd)(Fig. 2). On the other hand, derivatives in which the hydrophobicity of the D–ring of antimycin A is more decreased than in Compounds A and B, for example, with the D–ring being substituted by a benzene ring, showed very weak inhibitory effect on the photoreduction of Cyt *b*–559(Fd)(I_{50} = 60 μM). These results suggest that the D–ring of antimycin A interacts hydrophobically with the binding site of antimycin A in Cyt *b*–559(Fd) in maize thylakoids.

Compound	Structure	I_{50} (μM) Inhibition of 9-AA fluorescence quenching	I_{50} (μM) Inhibition of Cyt b559(Fd) photoreduction
Antimycin A	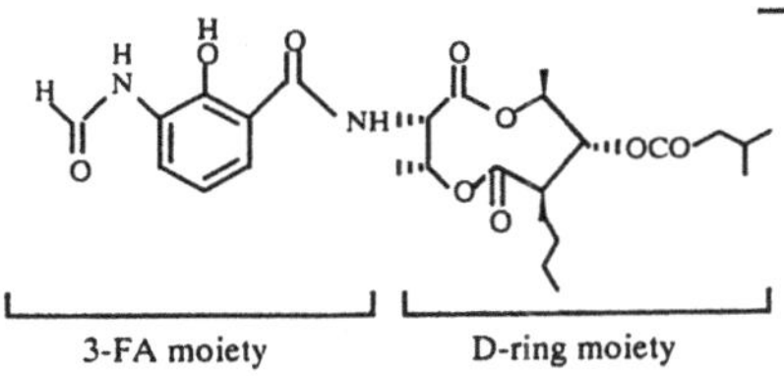	0.2	0.2
(A)	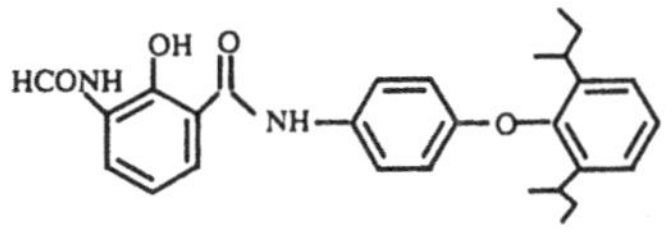	5.0	3.0
(B)	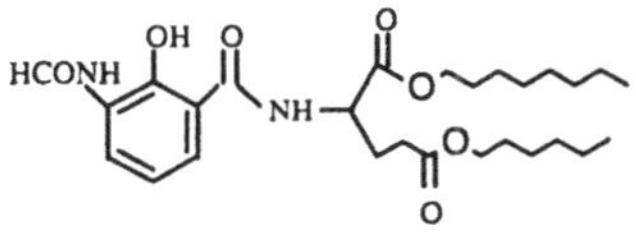	2.0	2.2
(C)	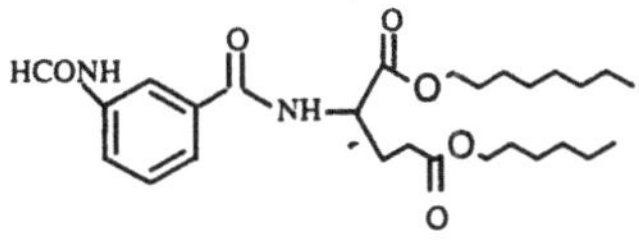	> 1000	> 1000
(D)	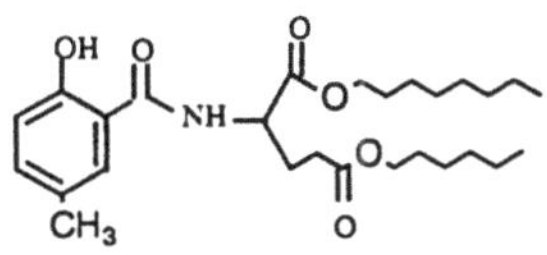	> 1000	> 1000

Fig. 2 *Structures of antimycin A and of various derivatives with the corresponding inhibitory effects on the photoreduction of Cyt b–559(Fd) and the quenching of 9–AA fluorescence by far–red light. I_{50} of each compound for the inhibitions of 9–AA fluorescence quenching and Cyt b–559(Fd) photoreduction by far–red light was determined as described in Materials and Methods.*

Next, the effect of the 3–FA moiety of antimycin A on the photoreduction of Cyt *b*–559(Fd) was examined. Compound C, where the hydroxyl group on the 3–FA moiety of Compound B was removed was ineffective suggesting that the hydroxyl group on the 3–FA moiety is indispensable for inhibition of Cyt *b*–559(Fd) photoreduction. Compound D, where formylamide group of the 3–FA moiety was removed, also showed low effectiveness suggesting that the 3–FA moiety is essential for inhibition of photoreduction of Cyt *b*–559(Fd)(Fig. 2). It was reported that the hydrogen atom of the hydroxyl group and the oxygen atom of the formylamide group on the 3–FA moiety of antimycin A interact with the binding site of antimycin A in the Cyt *b/c1*–complex of mitochondria via hydrogen bonds [6,9,10]. Thus, the hydrogen bonds formed by hydroxyl group and formylamide group of the 3–FA moiety of antimycin A would fit the molecule of antimycin A to its binding site in Cyt *b*–559(Fd) of maize mesophyll thylakoids, expressing its inhibitory action on the electron flow from Fd to Cyt *b*–559(Fd). All of the derivatives of antimycin A showed the same dose–dependency on both the Fd–dependent photoreduction of Cyt *b*–559(Fd) and the formation of ΔpH (Fig. 2). These results also strongly support that Cyt *b*–559(Fd) is a mediator of Fd–cyclic electron flow in maize mesophyll cells.

4. Concluding remarks —— In maize mesophyll thylakoids, menadiol–reducible, low redox potential Cyt *b*–559(Fd) is a mediator of electron transfer between reduced Fd and plastoquinones in the intersystem chain. Fd–dependent cyclic electron transport is characterized by its inhibition by antimycin A, which blocks electron flow from the photoreduced Fd to Cyt *b*–559(Fd), wheres it does not interact with the Cyt *b/f* complex [11,12,13].

The inhibitory action of antimycin A on the photoreduction of Cyt *b*–559(Fd) by far-red light, especially the roles of the 3–FA and D–ring moieties of antimycin A resemble very much those in the inhibition of the Q–cycle in the Cyt *b/c1*–complex of mitochondria, suggesting that the binding site of antimycin A in Cyt *b*–559(Fd) displays a high structural homology with the corresponding site in the Cyt *b/c1*–complex in mitochondria.

5. References

1 Tagawa, K. and Arnon, D.I. (1971) *Proc. Natl. Acad. Sci. USA* **49**, 567–572
2 Heber, U. and Walker, D.A. (1992) *Plant Physiol.* **100**, 1621–1626
3 Miyake, C., Schreiber, U. and Asada, K. (1995) *Plant Cell Physiol.* **36**, 743–748
4 Schreiber, U., Neubauer, C and Schliwa, U. (1993) *Photosynth. Res.* **36**, 65–72
5 Klughammer, C., Kolbowski, J. and Schreiber, U. (1990) *Photosynth. Res.* **25**, 317–327
6 Tokutake, N., Miyoshi, H., Nakazato, H. and Iwamura, H. (1993) *Biochim. Biophys. Acta* **1142**, 262–267
7 Wada, K. and Arnon, D.I. (1971) *Proc. Natl. Acad. Sci. USA* **68**, 3064–3068
8 Trumpower, B.L. (1990) *J. Biol. Chem.* **265**, 11409–11412
9 Tokutake, N., Miyoshi, H., Sataoh, T., Hatano, T. and Iwamura, H. (1994) *Biochim. Biophys. Acta* **1185**, 271–278
10 Miyoshi, H., Tokutake, N., Imaeda, Y. Akagi, T. and Iwamura, H. (1995) *Biochim. Biophys. Acta* (1995) **1229**, 149–154
11 Hauska, G., Hurt, E., Garbellini, N. Lockau, W. (1983) *Biochim. Biophys. Acta* **726**, 97–133
12 Moss, D.A. and Bendall, D.S. (1984) *Biochim. Biophys. Acta* **767**, 389–395
13 De Wolf, F.A., Krab, K., Visschers, R.W., De Waard, J.H. and Kraayenhof, R. (1988) *Biochim. Biophys. Acta* **936**, 487–503

CYCLIC ELECTRON TRANSPORT MEDIATED BY THYLAKOID-BOUND, NADPH-SPECIFIC PYRIDINE NUCLEOTIDE DEHYDROGENASE IN CYANOBACTERIA

Hualing Mi[1], Tsuyoshi Endo[1], Teruo Ogawa[2] and Kozi Asada[1]
[1]The Research Institute for Food Science, Kyoto University, Uji, Kyoto 611, Japan
[2]Bioscience Center, Nagoya University, Chikusa-ku , Nagoya 464-01, Japan

1. Introduction

In cyanobacteria, the respiratory electron transport chain is coupled with the photosynthetic electron transport chain, sharing several of its components. We have demonstrated that an NAD(P)H dehydrogenase (NDH), homologous to complex I in mitochondria, mediates not only respiratory electron flow but also cyclic electron flow around PS I in the cyanobacteria *Synechococcus* and *Synechocystis* (1,2), and presented further evidence for the participation of the NDH complex in the PS I cyclic electron transport chain in spheroplasts from wild-type (WT) and NDH-defective mutant cells of *Synechocystis* 6803 (3).

In the present work, we studied the cyclic electron transport around PS I using the thylakoid membranes from wild-type cells of *Synechocystis* sp. PCC 6803 and an NDH-defective mutant. We offer several lines of evidence in support of the hypothesis that the membrane-bound NDH functions as a mediator not only of the NADPH-dependent cyclic electron transport but also of ferredoxin-dependent cyclic electron transport around PS I (4).

2. Materials and Methods

2.1. Cell culture and preparation of thylakoid membranes - Wild-type (WT) cells of *Synechocystis* sp. PCC 6803 and cells of NDH-defective mutant M55, in which the *ndhB* gene for a subunit of NDH complex is inactive (5), were cultured photoautotrophically. The cells at the late logarithmic phase of growth were suspended in medium A that contained 25% (v/v) glycerol, 10 mM $MgCl_2$, 10 mM NaCl and 20 mM sodium phosphate buffer (pH 7.5). After incubation on ice for 1 h, the suspension was passed through a French pressure cell, and the resultant suspension was centrifuged at 5,000 x *g* for 10 min. The supernatant was centrifuged at 140,000 x *g* for 1 h, and the pelleted thylakoids were resuspended in the medium A.

2.2. Measurement of redox change of plastoquinone and P700 - The redox change of plastoquinone was monitored in terms of Chl fluorescence and that of P700 in terms of absorbance at 820 nm, with a PAM Chlorophyll Fluorometer (Walz, Effeltrich, Germany) equipped with an emitter-detector unit 101 ED for Chl fluorescence and ED 800T for P700.

2.3. Measurement of photooxidation of NAD(P)H - Photooxidation of NAD(P)H by thylakoid membranes under actinic light (> 620 nm, 250 W m^{-2}) was

P. Mathis (ed.), Photosynthesis: from Light to Biosphere, Vol. II, 871–874.

quantified by following the decrease in absorbance at 340 nm, with a spectrophotometer that could be cross-illuminated.

3. Results

3.1. Reduction of plastoquinone by NADPH - The yield of Chl fluorescence was increased by addition of NADPH to the thylakoid membranes from WT cells under weak actinic light (650 nm, 3 W m^{-2}), but the increase was not as great upon addition of NADH (Fig. 1). These results indicate that plastoquinone was reduced by electrons from reduced pyridine nucleotides. NADPH was preferred to NADH as an electron donor to the plastoquinone. The increase in Chl fluorescence caused by either NADPH or NADH was inhibited by rotenone, TTFA, capsaicin and $HgCl_2$, all of which are the inhibitors of NDH and succinate dehydrogenase in mammalian mitochondria. By contrast to the thylakoids from the WT cells, those from the NDH-defective mutant M55 showed little increase in Chl fluorescence upon the addition of either NADPH or NADH (Fig. 1). These results support our previous conclusion (3) that the donation of electrons from NADPH to the plastoquinone pool is mediated by NDH, and the NDH participating in the electron transfer from NADPH to plastoquinone is bound to the thylakoid membranes.

3.2. NADPH-mediated cyclic electron transport around PS I - Addition of $NADP^+$ to the WT thylakoid membranes did not cause any change in Chl fluorescence because little ferredoxin (Fd) was retained in the present thylakoid membranes. In the presence of a catalytic amount of Fd (3 μM), upon the addition of $NADP^+$ to the thylakoid membranes from WT cells, Chl fluorescence was transiently quenched, with a subsequent increase to a level above that recorded before the addition of $NADP^+$ (Fig. 2A). Upon illumination of the thylakoid membranes, the photoreduction of $NADP^+$ continued for 6 s, that is to say, just during the quenching of Chl fluorescence. Subsequently, NADPH was maintained at a plateau level of 13 μM, while the fluorescence remained at a high level. (Fig. 2B). The present kinetic data support the previous observations in spheroplasts (3) that the quenching by $NADP^+$ can be accounted for by its reduction, mediated by photoreduced Fd, and that the subsequent increase in Chl fluorescence is due to the reduction of plastoquinone by the NADPH that is photogenerated in PS I. The increase in the yield of Chl fluorescence was inhibited by rotenone (Fig. 2A),

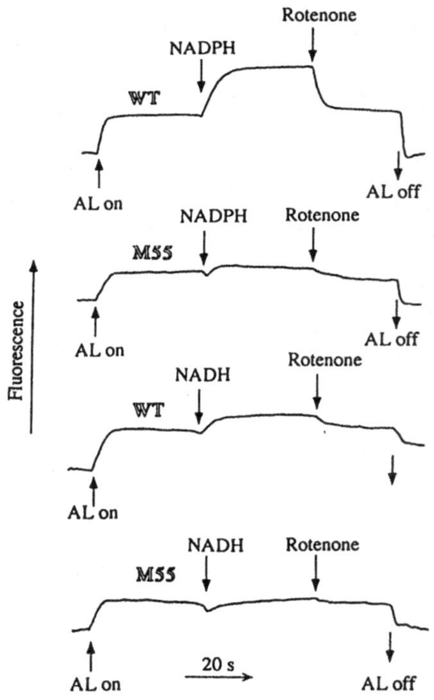

Fig. 1 Effects of NADPH, NADH and rotenone on the yield of chlorophyll fluorescence in thylakoids (20 μg Chl ml^{-1}) isolated from wild type (WT) and the NDH-deffective mutant M55. At the time indicated, 0.8 mM NADPH or NADH and 15 μM rotenone were added.

confirming the participation of the NDH complex in the electron donation from photogenerated NADPH to the intersystem chain and also the binding of the NDH complex to the thylakoids. However, the NADPH-induced increase in fluorescence yield was unaffected by either antimycin A at 2 - 20 μM or flavone at 1 - 50 mM (data not shown). By contrast to the thylakoids from WT cells, those from the NDH-defective mutant M55 showed only the quenching of Chl fluorescence by $NADP^+$ but not the subsequent increase (Fig. 2A), and NADPH was photoproduced continuously (Fig. 2B). These results do support our hypothesis that the NADPH-dependent cyclic flow of electrons around PS I is mediated by the thylakoid membrane-bound NDH complex.

3.3. Ferredoxin-mediated cyclic electron transport around PS I - The addition of a substrate amount of Fd (300 μM) to the thylakoid membranes also caused an increase in Chl fluorescence after a transient quenching, an indication that, in this system, photoreduced Fd can directly reduce plastoquinone. Again, no increase in Chl fluorescence was observed in the thylakoids from the NDH-defective mutant M55 cells (Fig. 3A). Under the same conditions, an increase in Chl fluorescence was observed after the actinic light had been turned off in the thylakoids from WT (Fig. 3B), indicating the reduction of the plastoquinone pool by the photoreductant accumulated during the illumination, as has also been observed in maize leaves (6).

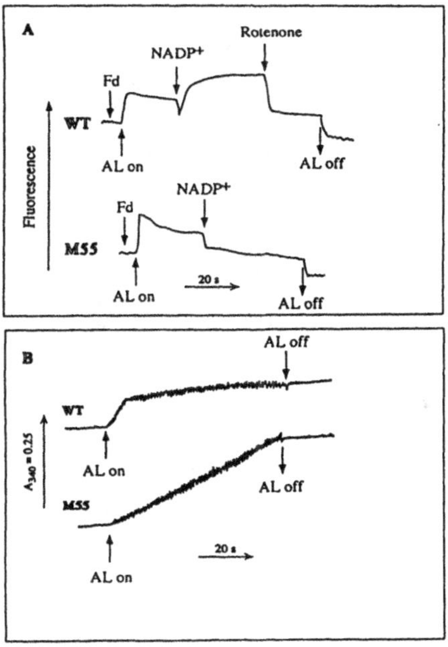

Fig. 2 Effects of $NADP^+$ (1 mM) on chlorophyll fluorescence in the presence of catalytic amount of ferredoxin (3 μM, Fd) in the thylakoids (10 μg Chl ml^{-1}) from WT and mutant M55 (Panel A). Photoproduction of NADPH from pre-added $NADP^+$ (0.5 mM) (Panel B)

The increase in Chl fluorescence after the actinic illumination had ceased was inhibited by rotenone and by DBMIB, but not by antimycin A, and it was not detected in the absence of Fd or in the NDH-defective mutant M55 (Fig. 3). The present results indicate that an antimycin A-insensitive, rotenone-sensitive, Fd-dependent cyclic electron flow is also mediated by NDH. The similar increase in fluorescence after extinction of the actinic light was observed also in the thylakoid membranes incubated with $NADP^+$ (1 mM) + ferredoxin (3 μM), in whole cells and in spheroplasts (data not shown), indicating that all these systems have active cyclic flow of electrons around PS I.

4. Discussion

In the present study, using the thylakoid membranes which did not contain soluble proteins such as Fd, transhydrogenase and respiratory electron mediators, we could show the direct donation of electrons from NADPH to NDH complex, which could not be clearly demonstrated in intact cells or spheroplasts. Effects of inhibitors of NDH could be evaluated for the first time without interference of the permeability barrier at cell membranes. Further, involvement of Fd in NDH-mediated cyclic flow of electrons and direct donation of electron from reduced Fd to NDH were also demonstrated taking advantage of the thylakoid preparation.

The NDH complex in the thylakoid membranes of *Synechocystis* PCC 6803 is similar to mitochondrial complex I in terms of its sensitivity to inhibitors. NADPH is a more effective donor to the membrane-bound NDH than is NADH. The various types of bacterial complex I have shown a low affinity for NADPH. Thus, the NADPH-specific complex I in cyanobacteria seems to be unique with respect to the specificity for the electron donor.

In addition to NADPH, the thylakoid-bound NDH could accept electrons from reduced Fd (Fig. 3). The ferredoxin-dependent cyclic electron flow in cyanobacteria is distinguished from that in plants that is sensitive to antimycin A and is mediated by cyt *b-559* (7).

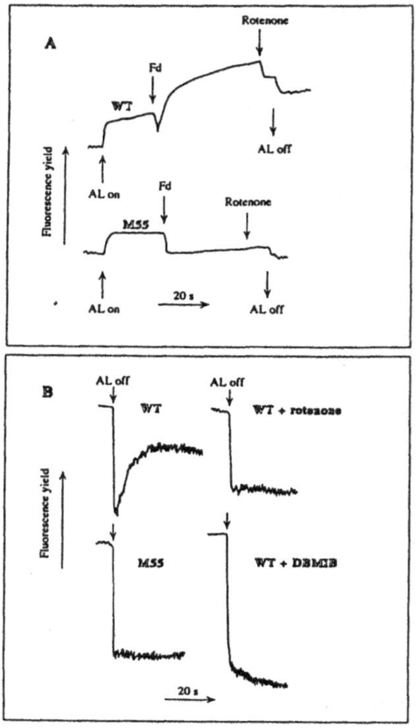

Fig. 3 Effects of substrate amount of Fd (300 μM), rotenone (5 μM) and DBMIB (10 uM) on chlorophyll fluorescence in thylakoids (20 μg Chl ml^{-1}) during illumination with red light (Panel A) and after the light was turned off (Panel B).

References

1. Mi, H., Endo, T., Schreiber, U. and Asada, K. (1992a) Plant Cell Physiol. 33: 1099-1105.
2. Mi, H., Endo, T., Schreiber, U., Ogawa, T. and Asada, K. (1992b) Plant Cell Physiol. 33: 1233-1237.
3. Mi, H., Endo, T., Schreiber, U., Ogawa, T. and Asada, K. (1994) Plant Cell Physiol. 35: 163-173.
4. Mi, H., Endo, T., Ogawa, T. and Asada, K. (1995) Plant Cell Physiol. 36: 661-668
5. Ogawa, T. (1991) Proc. Natl. Acad. Sci. USA 88: 4275-4279.
6. Asada, K., Heber, U. and Schreiber, U. (1993) Plant Cell Physiol. 34: 39-50.
7. Miyake, C., Schreiber, U. and Asada, K. (1995) Plant Cell Physiol. 36: 743-748

NON-PHOTOCHEMICAL QUENCHING OF CHLOROPHYLL FLUORESCENCE INDUCED IN THE COURSE OF TEMPERATURE SHIFT AND ACETATE ASSIMILATION IN *Chlamydomonas reinhardtii.*

Tsuyoshi. Endo[1], Ulrich. Schreiber[2] and Kozi. Asada[1]
[1]The Research Institute for Food Science, Kyoto University, Uji, Kyoto 611, Japan; [2]Lehlstuhl für Botanik I, Universität Würzburg, Mittlerer Dallenbergweg 64, 97082 Würzburg, Germany

1. Introduction

Down regulations of PS II by ΔpH dependent regulation and the transition from state 1 to state 2 (state 2 transition) can be visualized as non-photochemical quenching of chlorophyll fluorescence (qN). These regulatory mechanisms have been considered to function for protection against strong light. Recently, however, it has been shown that state 2 transition could be induced even in the dark by environmental stresses, such as inhibition of mitochondrial respiration (1) or hyperosmotic stress, in eukaryotic alga (2). Here, we show that in *Chlamydomonas reinhardtii*, temperature shift and addition of acetic acid in the dark induce both ΔpH-dependent and -independent qN. It is proposed that chlororespiratory electron input to the photosynthetic intersystem chain can form ΔpH across the thylakoid membranes and reduce the electron carriers in the intersystem chain, which respectively trigger ΔpH dependent regulation and state 2 transition.

2. Materials and Methods

2.1. Cell culture - *Chlamydomonas reinhardtii c-9* (IAM culture collection, Institute of Molecular and Cellular Biosciences, The University of Tokyo) was cultured photoautotrophically under fluorescent lamps at 5 W m^{-2} in the medium of Orth et al.(3) on a rotary shaker (100 rpm) at 27°C in 2% CO_2. Cells at an early stationary phase of growth that contained ca 30 μg Chl ml^{-1}, was used for the measurements of fluorescence and P700.

2.2. Measurements of chlorophyll fluorescence and P700 - Changes in absorbance of P700 at 820 nm were measured with a PAM Chlorophyll Fluorometer (Walz, Effeltrich, Germany), using the emitter-detector units ED 800T. Quenching analysis of Chl fluorescence was performed with the PAM Chlorophyll Fluorometer as previously (4,5) with a emitter-detector-cuvette assembly ED-101 US (6). Pulse-modulated excitation was obtained with a light emitting diode (650 nm), and modulated fluorescence was measured at l > 710 nm. Red actinic light was obtained from light emitting diodes with a peak emission at 660 nm. Saturating light pulses (1 s, 350 W m^{-2}) was applied with a xenon lamp (FL-103, Walz). These light sources were connected with the emitter-detector-cuvette assembly with multi-branched fiberoptics.

3. Results an Discussion

3.1. Temperature dependent quenching of Chl fluorescence - Figure 1 shows change of the maximum level of Chl fluorescence (*Fm'*) when temperature of the cell cuvette was increased from 28°C to 38°C. A drop of *Fm'* is observed, which is reversible to the original level by the temperature decrease, showing that the cells were not damaged irreversibly by the heat shock. The *Fm* quenching was sensitive to

P. Mathis (ed.), Photosynthesis: from Light to Biosphere, Vol. II, 875–878.

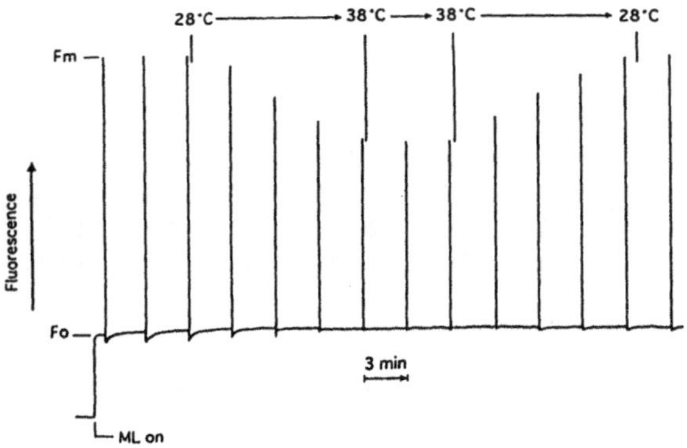

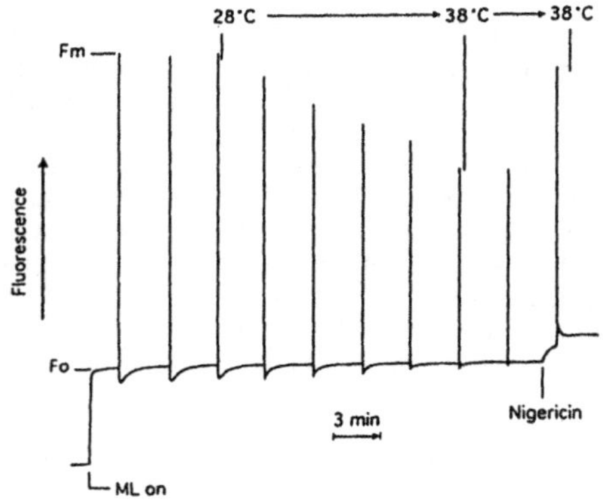

Fig.1 Induction of qN in *C. reinhardtii* at high temperature. Cell culture, dark adapted for at least 20 min at 28°C, were transferred to the PAM cuvette. The temperature of the cuvette was gradually increased from 28°C to 38°C, then the temperature was lowered again to 28,°C (upper panel) or 1 μM of nigericin was added (lower panel). Changes in *Fm'* were estimated by application of 1-s pulses of saturating light (350 W m^{-2}, Walz FL103) under a background measuring light (ML) which induced *Fo* level fluorescence.

uncoupling by the protonophore nigericin, indicating that the induced qN was energy quenching (qE) associated with ΔpH formation across the thylakoid membranes.

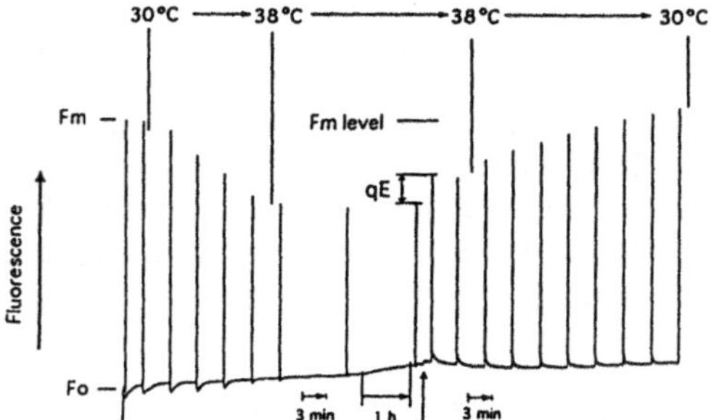

Fig. 2. Induction of quenchings induced by longer incubation at high temperature. The pretreatment and temperature shift to 38°C was the same as in Fig.1. After steady state level of *Fm* was atained, the cell suspension was kept at 38°C for 1 h. Nig: 1 μM nigericin.

Obviously, for an assessment of qN by the quenching coefficient qN=(*Fm*-*Fm*')/(*Fm*-*Fo*), *Fm* should be defined as the maximal value of *Fm*' which can be observed within the range of studied conditions, in this case, the maximum fluorescence yield at 28°C was difined as *Fm*. The drop of *Fm*' at high temperature was observed also in the light (upto 3 W m^{-2}), which by itself does not induce any qN, as well as in complete darkness without continuous application of measuring light (data not shown). Incubation at higher temperature than 40°C induced irreversible quenching of *Fm* which may reflects irreversible cell damage by the heat shock (data no shown).

3.2. qN induction upon longer incubation at high temperature- A longer incubation at 38°C did not induce further quenching (Fig. 2). In this case, however, the quenching was not totally sensitive to nigericin, indicating that qE and another type of quenching which was insensitive to uncoupling were overlapping. The lowered *Fm*' at 38°C increased again to the original level by the temperature decrease to 28°C, showing that the uncoupler-insensitive quenching of *Fm*' was not due to irreversible damage caused by the high temperature. Time course of induction of uncoupler sensitive and insensitive quenchings by the incubation at 38°C is shown in Figure 3. In order to avoid the actinic effect of repeated application of the saturating pulses, a single saturating pulse was used to determine *Fm*'. Overlapped induction of uncoupler sensitive and insensitive quenchings was observed. Induction of the uncoupler sensitive quenching attributable to qE occurred rapidly, but then the qE was gradually replaced by the uncoupler insensitive quenching. These observations are in agreement with the results in Figures 1 and 2. To date, three types of qN have been known; energy quenching (qE), quenching induced by state 2 transition (qT) and that induced by photoinhibition of PS II (qI). The reversibility of the quenching in the order of minutes and the low light intensity used in this study indicate that the uncouple -insensitive quenching observed is qT- type quenching attributable to the state 2 transition.

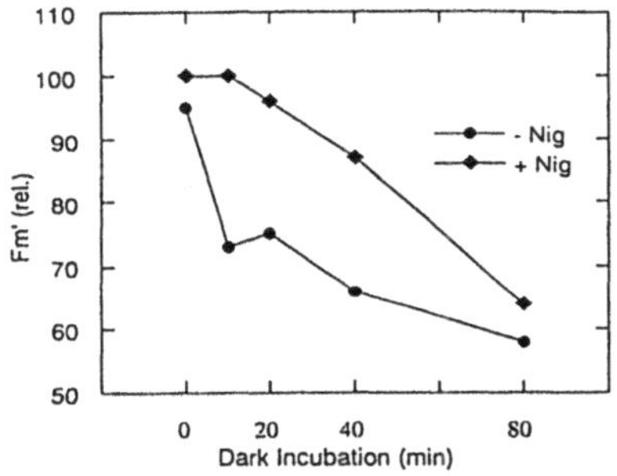

Fig. 3 Time course of qN induction in the presence or absence of the uncoupler nigericin (1 μM). *Fm*' estimated by application of a 1-s pulse of a saturating light are expressed relative to that in the dark adapted cells at 28°C in the presence of nigericin (1μM).

3.3. qN induction by acetate - Addition of acetic acid to cell culture induced a transient quenching of *Fm*' and increased the level of *F* (Fig. 4). The quenching was more evident at lower pH, which may be accounted for by the permeability barrier for anionic form to penetrate the plasma membranes. At pH 7 and 6, *Fm*' was recovered to the level before the addition of acetate in 30 min, but at pH 5, the recovery of *Fm*' was not observed upto 30 min (dat not shown). The quenching induced by acetic acid was biphasic which was most evident at pH 6 (F). The fast quenching and recovery was completed within a few min after addtition of acetic acid, and the slower phase within 20 min. The presence of nigericin slowed down quenching process in general and suppressed the fast phase of quenching -recovery, but the slow phase was not affected.

This indicates that the fast quenching is attributable to qE and the slow phase to qT. The pattern of qE and qT induction was similar to that found upon temperature shift.

The rereduction of p700+ after a pulse (5 ms) application, in the presence of DCMU was greatly stimulated by addition of acetate (data not shown). This stimulation of P700+ reduction can be considered as an indicator of reduced intersystem chain, and was also observed upon incubation at high temperature.

In conclusion, it has been shown that regardless of the light conditions, environmental changes, such as heat shock and addition of the respiratory substrate can regulate photosystem II through ΔpH dependent regulation and state 2 transition in *Chlamydomonas reinhardtii*. For triggering these regulation mechanisms the chlororespiratory electron input to the plastoquinone pool may play an important role.

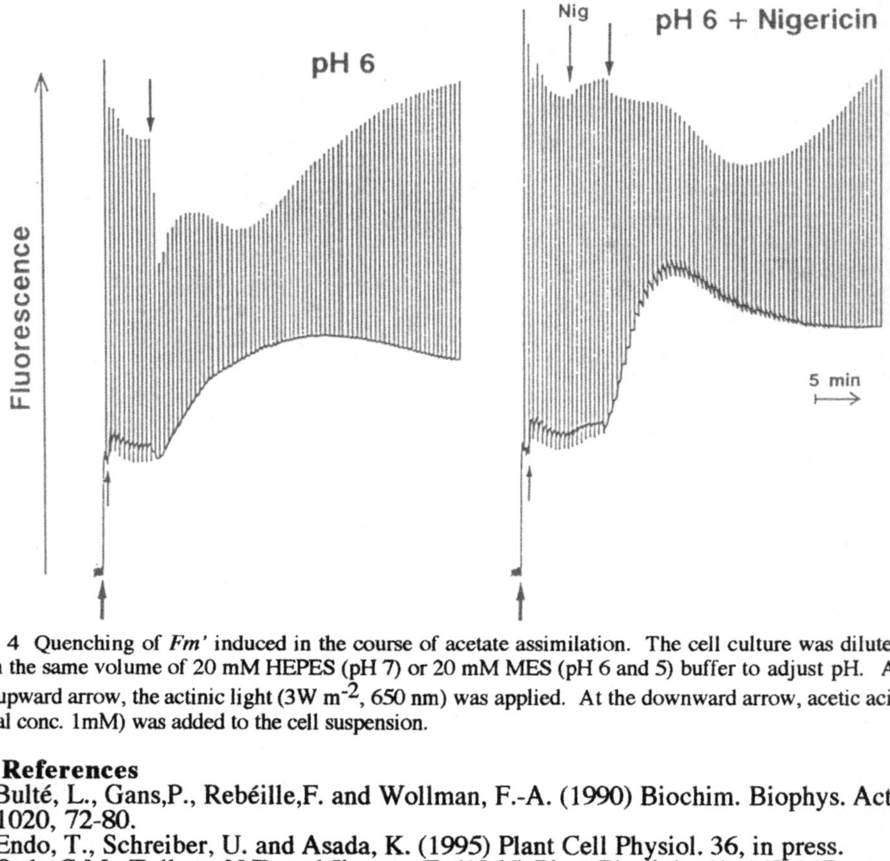

Fig. 4 Quenching of *Fm'* induced in the course of acetate assimilation. The cell culture was diluted with the same volume of 20 mM HEPES (pH 7) or 20 mM MES (pH 6 and 5) buffer to adjust pH. At the upward arrow, the actinic light (3W m^{-2}, 650 nm) was applied. At the downward arrow, acetic acid (final conc. 1mM) was added to the cell suspension.

4. References

1. Bulté, L., Gans,P., Rebéille,F. and Wollman, F.-A. (1990) Biochim. Biophys. Acta 1020, 72-80.
2. Endo, T., Schreiber, U. and Asada, K. (1995) Plant Cell Physiol. 36, in press.
3. Orth, G.M., Tolbert, N.E. and Jimenz, E. (1966) Plant Physiol., 41, 143-147.
4. Schreiber, U., Schliwa, U. and Bilger, W. (1986) Photosynth. Res. 10, 51-62.
5. Schreiber, U., Bilger, W. and Neubauer, C. (1994) in Ecological Studies(Schulze, E.-D. and Candwell, M.M. ed.) Vol. 100 pp.49-70, Springer, Berlin, Heiderberg.
6. Schreiber, U. Endo,T., Mi,H. and Asada, K. (1995) Plant Cell Physiol. 36, in press.

EVIDENCE FOR A MIGRATION OF *ndh* GENES FROM THE CHLOROPLAST TO THE NUCLEUS IN BLACK PINE

Sylvie Corneille, Stéphan Cuiné, Geneviève Guedeney, Michel Havaux and Gilles Peltier
CEA/Cadarache, Laboratoire d'Ecophysiologie de la Photosynthèse, DEVM, F-13108 Saint-Paul-lez-Durance, France.

1. Introduction

Complete sequencing of the plastid genome of different plant species (tobacco, liverwort, rice,...) has revealed the presence of open reading frames showing homologies with genes encoding for subunits of the mitochondrial NADH dehydrogenase or complex I (1). Chloroplastic *ndh* genes have been recently shown to encode for subunits of a thylakoid NAD(P)H dehydrogenase complex, likely involved in the reduction of plastoquinones (PQs) from a stromal NAD(P)H pool (2).

Recent sequencing of black pine (*Pinus thunbergii*) plastid genome showed that all *ndh* genes are non-functional in this plant species (3). Four *ndh* genes were completely lost, the other seven genes remaining as pseudogenes. This interesting finding raises the question of the physiological role of *ndh* genes (3). Are *ndh* genes and the encoded chloroplastic NAD(P)H dehydrogenase essential for photosynthesis ? In this case, one may conceive that *ndh* genes moved to the nucleus in some plant species during the evolution of plastids. On the other hand, if *ndh* gene products are absent from the chloroplast of some plant species, such as black pine, what are the consequences for the photosynthetic function ?

In the present work, we investigated whether the polypeptide products of *ndh* genes are present in black pine chloroplasts. We supply evidence for a migration of *ndh* genes from the chloroplast to the nucleus in this plant species.

2. Procedure

Black pine plantlets (*Pinus thunbergii*) were grown in a growth chamber under a light intensity of 350 $\mu E.m^{-2}.s^{-1}$. Chloroplasts were isolated from needles of 1 or 2 months old plants as described in (4). Extraction of the NAD(P)H dehydrogenase complex was carried out by partially solubilizing thylakoid proteins at 4°C, using 4% Triton X100. Non-denaturing gel electrophoresis and revelation of NAD(P)H dehydrogenase activity was performed as described in (2). Antibodies raised against synthetic peptides corresponding to well-conserved domains of chloroplastic genes *ndh* B and *ndh* I were prepared by Neosystem (France) and used for Western blot analysis (2).

P. Mathis (ed.), Photosynthesis: from Light to Biosphere, Vol. II, 879–882.

Plastid and genomic DNAs were prepared from chloroplasts and nuclei purified on a Percoll gradient according to the methods described in (4, 5, 6). DNA probes corresponding to the two exons of the chloroplastic tobacco *ndh* B gene were produced by PCR amplification of the PTB12 plasmid generously supplied by Dr. Sugiura. Hybridizations with labeled probes were carried out at 55°C and washes were performed at 55°C with 2X SSC, 0.1% SDS.

3. Results

Antibodies raised against synthetic peptides corresponding to well-conserved domains of the chloroplastic *ndh* B and *ndh* I genes respectively recognized a 28 kDa and a 18 kDa polypeptide in thylakoid membranes of black pine chloroplasts (Fig. 1), thus indicating that the corresponding gene products are likely present.

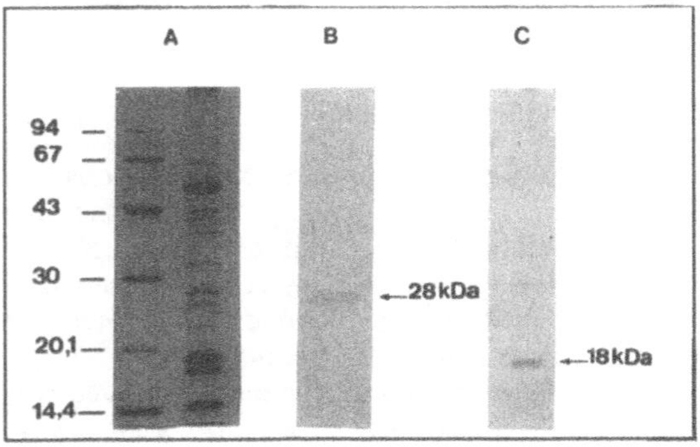

Figure 1. Black pine thylakoid proteins separated by SDS-PAGE and analyzed by Western blotting. (A) : thylakoid protein stained with Coomassie blue; Western blotting was carried out using an "anti-ndh B antibody" (B) and an "anti-ndh I antibody" (C).

To determine whether black pine *ndh* gene products are organized in a NAD(P)H dehydrogenase complex, as recently shown in potato chloroplasts (2), we carried out non-denaturing gel electrophoresis on Triton X100 solubilized thylakoid proteins and revealed the NAD(P)H dehydrogenase activity bands by following nitro-blue-tetrazolium (NBT) reduction. We observed two activity bands (A,B) when NADPH was used as an electron donor and one (C) when NADH was supplied.

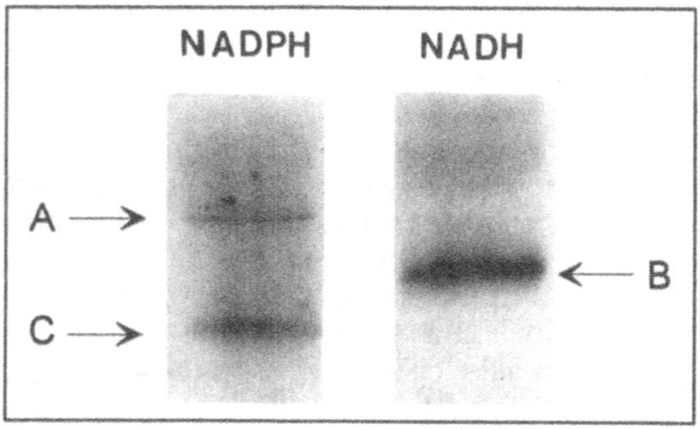

Figure 2. Revelation of NAD(P)H dehydrogenase activity using non-denaturing gel electophoresis of 4% Triton X100 solubilized thylakoid membranes of black pine. NAD(P)H dehydrogenase activities were revealed by following the formation of a blue precipatate occurring during NBT reduction in the presence of 1mM NADPH or NADH.

The different activity bands were subjected to a second dimension SDS/PAGE followed by a Western analysis using the "anti-ndhB" antibody. Band (A) and band (C) were recognized by the antibody, while band (B) was not. Band (C) was tentatively attributed to the diaphorase activity of FNR on the basis of the analogy with the case of potato (2).

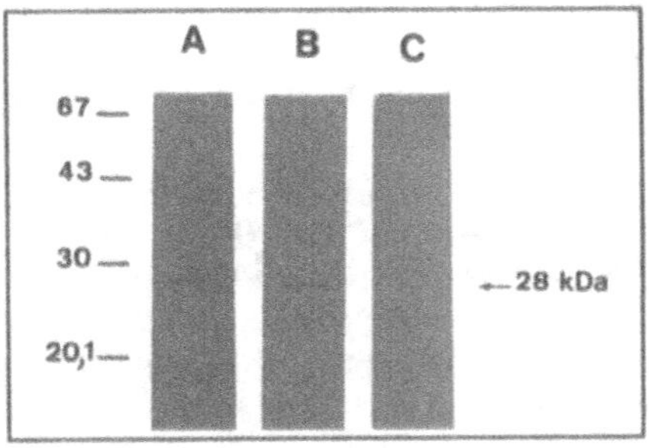

Figure 3 : Western blot analysis of the different NAD(P)H dehydrogenase activity bands shown on Fig. 2 following separation of the polypeptide components by a second dimension SDS-PAGE. (A) : activity band A; (B) : activity band B; (C) : activity band C. Western blotting was perfomed using an "anti-ndh B" antibody.

To determine whether the *ndh* B gene moved from the chloroplast to the nucleus in black pine, we performed Southern experiments on both black pine plastidial and nuclear DNAs (Fig. 4). We used two DNA probes, one corresponding to exon 1 (E1) of the tobacco *ndh* B gene, which is truncated in black pine, and the other corresponding to exon 2 (E2) which is absent from black pine chloroplast genome (3). Probe E1 hybridized with a single restriction fragment of the chloroplastic DNA corresponding to the truncated portion of the gene while probe E2 did not hybridize with any fragment. When used on nuclear DNA, both probes recognized a same-size restriction fragment thus showing that a copy of the *ndh* B gene is likely present in black pine nucleus.

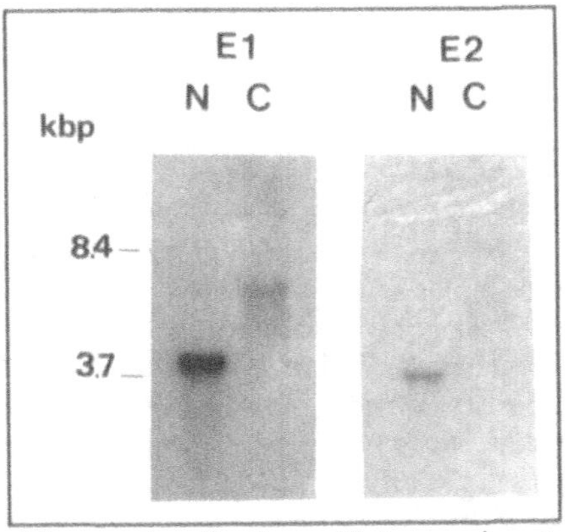

Figure 4 : Southern blot experiment performed on Bam H1 digests of chloroplast and nuclear black pine DNA using probes corresponding to exon 1 (E1) and exon 2 (E2) of the tobacco *ndh* B gene.

4. Discussion

We conclude from the present study that polypeptide products of both *ndh* B and *ndh* I genes are present in black pine chloroplasts. As shown for *ndh* B, other *ndh* gene products of black pine are likely components of a thylakoid NAD(P)H dehydrogenase complex, as recently reported in potato chloroplasts (2). Non-denaturing gel electrophoresis (Fig 2) and study of $P700^+$ re-reduction kinetics (not shown) indicate that the complex is fully functional an probably achieves PQ reduction from a stromal pool of NAD(P)H. Contrary to what was reported in potato chloroplasts, two different complexes, one more specific for NADPH oxidation and the other more specific for NADH oxidation, were observed. Interestingly, both bands were recognized by an "anti-ndh B" antibody, thus suggesting that both complexes contained *ndh* gene products as constitutive components. A mitochondrial origin of the activity band, due to a contamination by mitochondrial material was discarded since chloroplasts were purified on a Percoll gradient and since the "anti-ndh B" antibody used did not cross-react with mitochondrial proteins extracts (not shown). The nature and the role of each of these complexes will need further work to be determined.

Hybridizations of nuclear black pine DNA with a *ndh* B probe from tobacco chloroplast indicate that a copy of the *ndh* B gene is present in the nucleus. Since the corresponding chloroplastic gene is truncated, the nuclear gene probably encodes the 28 kDa polypeptide, component of the NAD(P)H dehydrogenase complex. We conclude that *ndh* genes probably migrated from the chloroplast to the nucleus in black pine. This event probably happened during the evolution of plastids, as it was the case for most of the genes of the photosynthetic cyanobacterial endosymbiont (7), presumably at the divergent points between gymnosperm and angiosperm (3).

The conservation of functional *ndh* genes in most photosynthetic plant species ranging from angiosperms to gymnosperms and bryophytes argues in favour of an essential role of these genes during photosynthesis. This role might be related to an involvement of the NAD(P)H dehydrogenase complex in chlororespiration, a respiratory activity discovered in chloroplasts. However, if chlororespiration is a well-established phenomenon in unicellular algae, its occurrence in higher plants chloroplasts is still a matter of debate (8). Alternatively, this role might be related to the supply by cyclic photophosphorylations of the extra-ATP which is needed for CO_2 assimilation. Indeed, *ndh* gene products have been recently shown to be associated to FNR as components of a NAD(P)H dehydrogenase complex which is likely involved in the cyclic electron pathway occurring around PS1 in physiological conditions (2).

References

1. Ohyama, K., Kohchi, T., Sano, T., Yamada, Y. (1988) TIBS 13, 19-22.
2. Guedeney, G., Corneille, S., Cuiné, S., Peltier, G. (1995) Xth Int. Photosynth. Cong. P. Mathis, ed. Kluwer Academic Publishers, The Netherlands (in press)
3. Wakasugi, T., Tsudzuki, J., Ito, S., Nakashima, K., Tsudzuki, T., Sugiura, M. (1994) Proc. Natl. Acad. Sci. USA 91, 9794-9798
4. Kondo, T., Ishibashi, T., Shibata, M. and Hirai, A. (1986) Plant Cell Physiol. 27, 741-744
5. Luthe, D.S. and Quatrano, R.S. (1980) Plant Physiol. 65, 305-308
6. Maniatis, T., Sambrook, J. and Fritsch, E.F. (1989) Molecular cloning; a laboratory manual, 2nd edition. Cold Spring Harbor, New York
7. Gray, M.W. (1989) Trends Genet. 5, 294-299
8. Peltier, G., Havaux, M., Ravenel, J. (1995) Xth Int. Photosynth. Cong. P. Mathis, ed. Kluwer Academic Publishers, The Netherlands (in press)

ndh B AND *ndh* I GENE PRODUCTS ARE ASSOCIATED TO FNR AS COMPONENTS OF A CHLOROPLASTIC NAD(P)H DEHYDROGENASE COMPLEX

Geneviève Guedeney, Sylvie Corneille, Stéphan Cuiné and Gilles Peltier
CEA/Cadarache, Laboratoire d'Ecophysiologie de la Photosynthèse, DEVM, F-13108 Saint-Paul-lez-Durance, France

1. Introduction

The plastid genome of higher plants has been reported to contain several open reading frames homologous to gene encoding for subunits of the mitochondrial NADH-ubiquinone oxidoreductase (1). These genes, called *ndh*, are transcribed (2) and the polypeptide products of *ndh* H *ndh* I and *ndh* K have been found in chloroplasts (3, 4, 5). It has been suggested that *ndh* genes may encode for subunits of a chloroplastic NAD(P)H dehydrogenase. However, although such a complex has been reported to occur in the chloroplast of microalgae (6, 7) the plastidic NAD(P)H dehydrogenase complex of higher plants has not been characterized and the participation of the *ndh* gene products to such a complex remains to be determined.

We report here the isolation of a thylakoid complex from a higher plant showing a NAD(P)H dehydrogenase activity and containing at least two of the *ndh* gene products (*ndh* B and *ndh* I) associated to ferredoxin NADP reductase (FNR). The physiological function of this complex is discussed in relation to its involvement in chlororespiration and/or cyclic electron flow around photosystem 1 (PS1).

2. Procedure

Potato plantlets (*Solanum tuberosum* L. cv Haig) were grown in a growth chamber under a light intensity of 350 $\mu E\ m^{-2}\ s^{-1}$. Intacts chloroplasts were isolated from leaves of 1 to 2 months old plants using a Percoll gradient (8). Stromal and thylakoid proteins were extracted, separated by SDS-PAGE and revealed by Coomassie blue or western blotting. For extraction of the NAD(P)H dehydrogenase, thylakoid membranes were washed using a buffer contaning 0.15 M NaCl to remove extrinsic proteins and further partially solubilized at 4°C in a 4% Triton X100 buffer. Non-denaturing gel electrophoresis was performed at 4°C in a 6% polyacrylamide gels containing 0.1% Triton X100. Staining of gels for NAD(P)H dehydrogenase activity was carried out using p-nitro-blue tetrazolium (NBT, 0.5 mg/ml) and 1 mM NADPH or NADH. Activity bands were excised, loaded on a 15% acrylamide gel and submitted to SDS-PAGE (7). Antibodies were raised against synthetic peptides derived from *ndh* B and *ndh* I gene sequences (Neosystem, France). Selected sequences were Y-T-K-K-D-V-R-S-N-E-A and D-F-Q-E-R-E-S-Y-D-M-L-G-I for the *ndh* B and for the *ndh* I gene, respectively. Antibodies against FNR were kindly provided by Dr. G.H. Schmidt (University of Bielefeld, Germany).

P. Mathis (ed.), Photosynthesis: from Light to Biosphere, Vol. II, 883–886.

3. Results

Antibodies raised against synthetic peptides corresponding to well-conserved domains of the two chloroplastic gene *ndh* B and *ndh* I respectively recognized a 32 and 18 kDa polypeptide in the thylakoid membranes of *Solanum tuberosum* (Fig. 1). Estimated sizes of *ndh* B and *ndh I* gene products deduced from the sequence of the corresponding chloroplastic tobacco genes are respectively of 39 and 18 kDa. A polypeptide of about 43 kDa, recognized by both antibodies, was attributed to a non-specific reaction. Pre-immune sera of both antibodies did not recognize any polypeptide (not shown).

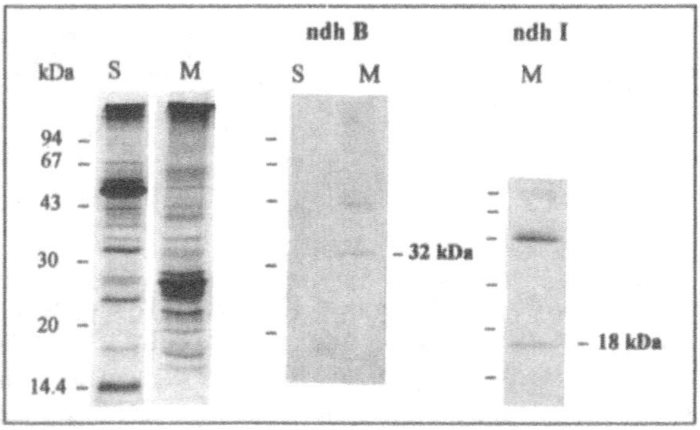

Figure 1. Western blot analysis of chloroplastic proteins from *Solanum tuberosum* separated by SDS-PAGE. (S) : soluble stromal proteins; (M) : thylakoid membrane proteins; (two left lanes) : staining with Coomassie blue; (ndh B) : Western analysis using an "anti-ndh B" antibody; (ndh I) : Western analysis using an "anti-ndh I" antibody.

In order to determine whether *ndh* gene products are organized in a thylakoid protein complex supporting NAD(P)H dehydrogenase activity, we carried out non-denaturing gel electrophoresis. NAD(P)H dehydrogenase activities was revealed by following NBT reduction. Two activity bands were observed in Triton X100 solubilized thylakoid membranes when NADPH was used as a substrate (Fig. 2). The same bands were observed, but at a lower intensity when NADH was supplied (not shown).

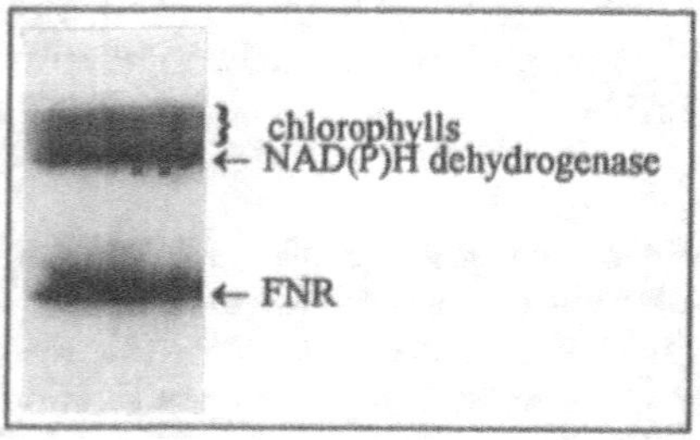

Figure 2. Non-denaturing gel electrophoresis of thylakoid membranes of *Solanum tuberosum* partially solubilized using 4% Triton X100. NADPH dehydrogenase activities were revealed by following NBT reduction in the presence of 1 mM NADPH as an electron donor. Activity bands (blue color) are shown by arrows. When 1 mM NADH was used instead of NADPH, the same activity bands were revealed, but at a much lower intensity.

After excision, the polypeptide composition of the different activity bands was analyzed by carrying out second dimension SDS-PAGE and Western blotting. The lower activity band, contained a 36 kDa polypeptide which was recognized by an "anti-FNR" antibody (not shown). This activity band, which was not recognized by any of the "anti-ndh" antibodies was attributed to the diaphorase activity of FNR. The upper activity band contained two polypeptides of 32 and 18 kDa, specifically recognized by the "anti-ndh B" and by the "anti-ndh I" antibody, respectively (Fig.3). Surprisingly, the anti-FNR antibody also recognized a 36 kDa polypeptide in the upper band (Fig. 3). Washing membrane with CHAPS according to the procedure described in ref. (9) removed part of the FNR from the complex, but the dehydrogenase activity declined at the same time (not shown). We conclude from these experiments that both *ndh* B and *ndh* I gene products are associated to FNR as components of a protein complex showing NAD(P)H activity. The MW of the complex was estimated by migration on a non-denaturing polyacrylamide gel gradient to about 670 kDa (data not shown).

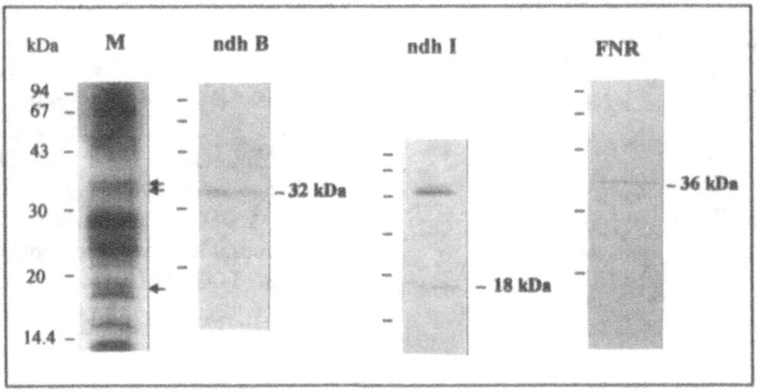

Figure 3. Second-dimension SDS/PAGE and Western analysis of the upper activity band shown on Fig. 2. (M) : revelation of proteins by silver staining; other lanes : Western analysis using antibodies directed against ndh B, ndh I gene products or against FNR.

When the upper activity band was subjected to SDS-PAGE and silver staining at least 10 polypeptide bands ranging from 14 to 67 kDa were revealed (Fig. 3). The 18 kDa, 32 kDa and 36 kDa bands likely correspond to ndh I, ndh B polypeptides and to and to FNR, respectively. Some of the unidentified bands likely correspond to other components of the NAD(P)H dehydrogenase complex, such as the other *ndh* gene products.

4. Discussion

In the present study, we report that both chloroplastic genes *ndh* B and *ndh* I encode for subunits of a membrane thylakoid complex showing NAD(P)H dehydrogenase activity. Previous studies have shown that chloroplastic *ndh* genes are transcribed (2). However, although the polypeptides products of three *ndh* genes have been identified (3, 4, 5), their involvement in a protein complex showing NAD(P)H dehydrogenase activity had not been demonstrated and the occurrence of such a complex in higher plant chloroplasts was serioulsy questioned (10). Initially, *ndh* genes have been identified on the basis of their homology with genes encoding for subunits of the NADH dehydrogenase complex I which uses NADH as a substrate within mitochondria. Within chloroplasts, NADPH is the main cofactor present and NADH concentration is probably very low. We suggest that the association of the *ndh* gene products to FNR allows this complex to use NADPH instead of NADH as a preferential substrate (Fig. 4). This would explain the better affinity of the complex for NADPH (Fig. 2). Interestingly, a similar hypothesis was recently proposed by Friedrich et al. (11) who noticed that the cyanobacterial complex was lacking the 3 subunits involved in the binding of NADH (12) and suggested that it might be the same for the chloroplastic complex. Association of FNR with an unidentified membrane component of 17.5 kDa was previously reported (9). Presumably, this unidentified polypeptide might be a *ndh* (for instance ndh I) gene product.

In a recent review, Bendall and Manasse (10) concluded that FNR associated to FQR (ferredoxin-PQ reductase) rather than a NAD(P)H dehydrogenase would be involved in the donation of electrons from the stroma to the plastoquinone (PQ) pool. In contrast to this view, it clearly appears from this work that the NAD(P)H dehydrogenase complex formed by the association of *ndh* gene products with FNR is most likely the functional complex involved in this reaction.

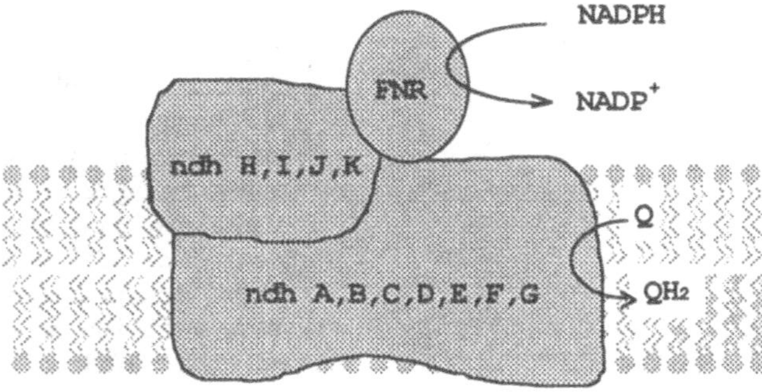

Figure 4. Hypothetical scheme of the subunit organization of the chloroplastic NAD(P)H dehydrogenase complex. This scheme is based on the published organizations of the cyanobacterial (12) and bacterial (13) NADH dehydrogenase complexes. We suggest that the association of a FNR subunit allows the complex to use NADPH instead of NADH as a preferential substrate.

Such an electron flow from NAD(P)H to the PQ pool may participate to chlororespiration, a respiratory activity discovered in the chloroplast of algae (7, 14). Under illumination, reduction of the PQ pool by NADPH would allow cyclic electron flow to occur around PS1. We suggest that the involvement of FNR in cyclic electron flow previously reported (10, 15) might be explained by the association of this enzyme to the NAD(P)H dehydrogenase complex.

By being able to carry out either NADP reduction from ferredoxin or PQ reduction from NADPH, FNR probably plays a central role during photosynthesis by regulating the partition between cyclic and non-cyclic electron pathways.

References

1. Ohyama, K., Kohchi, T., Sano, T. and Yamada, Y. (1988) TIBS. 13, 19-22
2. Matsubayashi, T. et al. (1987) Mol. Gen. Genet. 210, 385-393
3. Nixon, P.J., Gounaris, K., Coomber, S.A., Hunter, C.N., Dyer, T.A. and Barber, J. (1989) J. Biol. Chem. 264, 14129-14135
4. Wu, M. and Yang, J. (1989) Plant Cell 1, 551-557
5. Berger, S., Ellersiek, U., Westhoff, P. and Steinmüller, K. (1993) Planta 190, 25-31
6. Godde, D. (1982) Arch. Microbiol. 131, 197-202
7. Peltier, G. and Schmidt, G.W. (1991) Proc. Natl. Acad. Sci. USA 88, 4791-4795
8. Mills, W.R. and Joy, K.W. (1980) Planta 148, 75-83
9. Matthijs, H.C., Coughlan, S.J. and Hind, G. (1986) J. Biol. Chem. 261, 12 154-12158
10. Bendall, D.S. and Manasse, R.S. (1995) Biochim. Biophys. Acta 1229, 23-38
11. Friedrich, T., Steinmüller, K. and Weiss, H. (1995) FEBS Lett. 367, 107-111
12. Berger, S., Ellersiek, U., Kinzelt, D. and Steinmüller, K. (1993) FEBS Lett. 326, 246-250
13. Leif, H., Sled, V.D., Ohnishi, T., Weiss, H. and Friederich, T. (1995) Eur. J. Biochem. 230, 538-548
14. Bennoun, P. (1982) Proc. Natl. Acad. Sci. USA 79, 4352-4356
15. Ravenel, J., Peltier, G. and Havaux, M. (1993) Planta 193, 251-259

CHLORORESPIRATION IN UNICELLULAR GREEN ALGAE
Studies with mitochondrial respiration deficient mutants

Gilles Peltier, Michel Havaux and Jacques Ravenel
CEA/Cadarache, Laboratoire d'Ecophysiologie de la Photosynthèse, DEVM, F-13108 Saint-Paul-lez-Durance, France

1. Introduction

Chlorophyll fluorescence studies in unicellular algae have shown that the redox state of the plastoquinone (PQ) pool can be modified by O_2 and by respiratory inhibitors (1). This was interpreted by the existence of a chloroplastic respiratory chain (chlororespiration) interacting with the photosynthetic electron transport chain (1). However, the redox state of the PQ pool has been reported to be affected by changes in mitochondrial activity (2, 3), electrons entering chloroplasts via metabolic shuttles and reducing PQs most probably by a chloroplastic NAD(P)H dehydrogenase.

Evidence for a chlororespiratory activity has independently been obtained by monitoring O_2 exchange under flash excitation using a rapid O_2 electrode (4, 5). It was shown that PS1 activity was able to divert electrons away from a respiratory electron transport chain (4), while PS2 activity was able to supply electrons to the same respiratory process (5). This was interpreted as resulting from the interaction between chlororespiratory and photosynthetic electron transport chains, the two chains sharing some electron carriers (such as the PQ pool) in common (4, 5).

In order to clearly establish which phenomena can be attributed to interactions between chlororespiration and photosynthesis or to interactions between mitorespiration and photosynthesis, we studied *Chlamydomonas* mutants defective in mitochondrial cytochrome oxidase. We report that the flash-induced inhibition of respiration occurs in such mutants and is insensitive to inhibitors of the alternative mitochondrial respiration.

2. Material and Methods

Flash-induced O_2 exchange were measured at room temperature using a bare platinum electrode as described in ref. (4). Respiration rates were determined using an Clark-type O_2 electrode (Hansatech). Chlorophyll fluorescence measurements were performed using a PAM modulated fluorimeter (Walz). *Chlamydomonas* mutants defective in cyanide-sensitive cytochrome pathway (*dum*-3, *dum*-4, *dum*-11, *dum*-14 and *dum*-16) were generously supplied by Dr. R. Matagne (Université de Liège, Belgium) and were grown photoautotrophically as described in ref. (6).

3. Results

When algae are deposited in the dark on a bare platinum electrode, a steady state O_2 level is reached, corresponding to the equilibrium between respiratory processes, O_2 consumption by

P. Mathis (ed.), Photosynthesis: from Light to Biosphere, Vol. II, 887–890.

the electrode and O_2 diffusion through the gas liquid interphase. When dark-adapted cells of a *Chlamydomonas* mutant (*dum*-14) deficient in mitochondrial cytochrome oxidase were illuminated with a single turn-over flash, a transient increase in the O_2 level was observed (Fig. 1A). A similar transient was previously observed in wild-type (WT) cells and was attributed to the flash-induced inhibition of a chlororespiration (4). It was shown, by studying the effect of DCMU as well as *Chlamydomonas* mutants defective either in PS1 or in PS2, that inhibition of chlororespiration was due to the activity of PS1 (4). In the *dum*-14 mutant, the respiration rate was strongly inhibited (by about 80 %) by 0.4 mM salicyl hydroxamic acid (SHAM), an inhibitor of the alternative oxidase (Fig. 1B). This confirms that the cytochrome pathway was inoperative in the mutant and that the main respiratory process was the alternative pathway (6). However, although severely inhibiting mitochondrial respiration, SHAM had no effect on the flash-induced signal (Fig. 1A). Note that addition of SHAM to WT cells had no effect on the respiration rate (not shown).

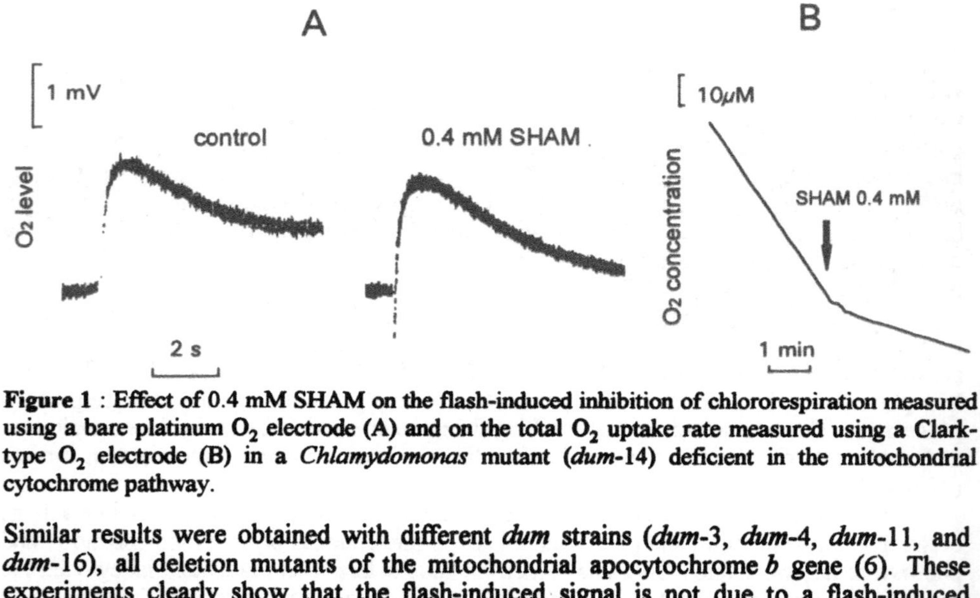

Figure 1 : Effect of 0.4 mM SHAM on the flash-induced inhibition of chlororespiration measured using a bare platinum O_2 electrode (A) and on the total O_2 uptake rate measured using a Clark-type O_2 electrode (B) in a *Chlamydomonas* mutant (*dum*-14) deficient in the mitochondrial cytochrome pathway.

Similar results were obtained with different *dum* strains (*dum*-3, *dum*-4, *dum*-11, and *dum*-16), all deletion mutants of the mitochondrial apocytochrome *b* gene (6). These experiments clearly show that the flash-induced signal is not due to a flash-induced inhibition of a mitochondrial process, thus confirming our previous interpretation that it is due to an inhibition of chlororespiration (4).

In order to estimate the electron flow rate involved in chlororespiration, we studied the effect of varying the flash frequency on the amplitude of the flash-induced O_2 signal (Fig. 2). For this purpose, 20 consecutive flashes (at a given frequency) were fired on DCMU-treated algae to reach a new steady state O_2 level corresponding to a partial inhibition of chlororespiration. Then the area induced by an additional single flash (applied at the same frequency) was measured. In control cells, a half-amplitude signal was observed when flashes were spaced by about 700 ms. This time was attributed to the half re-reduction time of the PQ pool in the dark. When algae were treated by SHAM, the half-time of re-reduction decreased to about 350 ms, thus showing that the inhibition of mitochondrial respiration and the associated increase in reducing power allowed electrons to re-enter the PQ pool more rapidly.

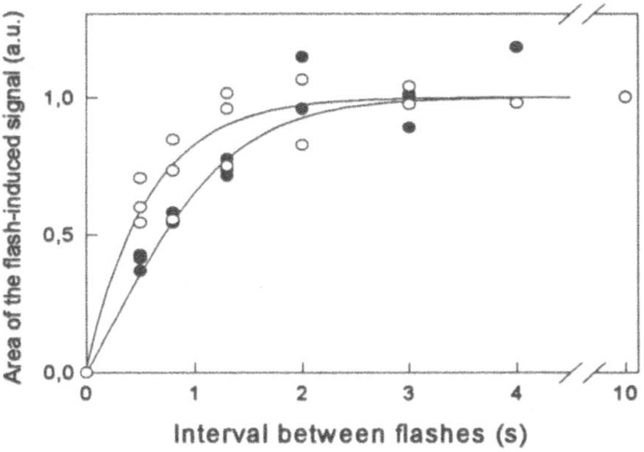

Figure 2 . Effect of the flash frequency on the flash-induced inhibition of chlororespiration in the *Chlamydomonas* mutant *dum*-14 treated with 10 μM DCMU. O_2 exchange were measured using a bare O_2 electrode. A first series of 20 consecutive flashes was used to reach a new stationary O_2 level before measuring the area of the flash-induced O_2 signals. (●) : control; (○) : treated with 0.4 mM SHAM.

We then investigated whether changes in the respiratory activity of *dum* mutants were able to alter the redox state of the PQ pool in the dark. For this purpose, we studied the effect of SHAM on the chlorophyll fluorescence levels F_0 and F_{max} measured using modulated fluorimetry. Figure 3 shows that addition of SHAM, which inhibited the respiration rate by about 70%, decreased both F_0 and F_{max} values by about 25 to 30%. This fluorescence quenching was not reversed following addition of the uncoupler nigericin (data not shown) and was attributed to a state transition from state 1 to state 2 as previously reported in WT cells treated with antimycin A and SHAM (2).

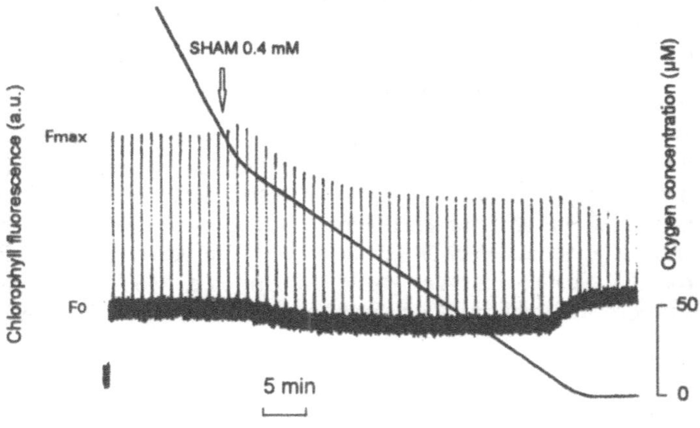

Figure 3. Effect of SHAM and anoxia on the chlorophyll fluorescence levels F_0 et F_{max} and on the respiration rate in the *Chlamydomonas* mutant *dum*-14. Algae were placed in a Clark-type O_2 electrode. Chlorophyll fluorescence level F_0 was measured with a dim modulated red light (non actinic) and F_{max} with a 800 ms flash of super-saturating white light.

Therefore, since the F_0 fluorescence level did not increase, this indicates that contrary to what was reported for WT cells (2), the inhibition of mitorespiration in the *dum* mutant had no significant effect on the redox state of the PQ pool. Upon anoxia, the apparent F_0 level rapidly increased. This effect was partially reversed by far-red illumination thus indicating an increase in the redox state of the PQ pool. Since mitochondrial respiration was likely completely inhibited by SHAM (we checked that higher SHAM concentrations had no additional inhibitory effect), this shows that inhibition of chlororespiratory O_2 uptake by anoxia induces a significant increase in the redox state of the PQ pool. Since the respiration rate of the *dum* mutant is about 50% of the WT (6) and since its respiration was inhibited by 70 to 80% by SHAM (Figs.1 & 3), this means that the remaining O_2 uptake represents about 10 to 15% of the respiration rate of WT cells, which is in the same range of our previous estimation of the chlororespiration rate (4). The increase in the F_0 level was followed by a progressive decrease in F_{max} (not reversed by nigericin), which was attributed to the completion of the state transition to state 2.

4. Discussion

We conclude from this study of *Chlamydomonas* mutants deficient in mitochondrial cytochrome oxidase that the flash-induced O_2 signal observed in dark-adapted algae is due to an inhibition of chlororespiration and not to an inhibition of mitorespiration. Therefore, since changes in chlorophyll fluorescence levels in response to respiratory inhibitors may reflect interactions between chloroplasts and mitochondria (2, 3) flash-induced O_2 signals can be regarded as the most conclusive evidence favoring the existence of a chlororespiration within algal chloroplasts.

The half-time of re-reduction of the partially oxidized PQ pool was significantly decreased when mitorespiration was inhibited, thus indicating that electrons can enter more readily the oxidized PQ pool. However, in contrast to what was reported for WT cells (2), the stationary redox level of the PQ pool in the dark remained unchanged in response to the inhibition of mitorespiration. This might be explained by the fact that *dum* mutants have a lower respiratory rate than WT cells (6). Alternatively, this could suggest a dynamic adjustment of the chlororespiratory electron flow to the redox state of the PQ pool. Transition from state 1 to state 2 is generally thought to be triggered by an increase in the redox state of the PQ pool. Low temperature fluorescence emission measurements are clearly needed to establish whether the F_{max} decrease observed in the *dum* mutant is due to a state transition which would have occurred in conditions where the PQ pool redox state remained unchanged.

Chlororespiration is now well-established in unicellular algae. However, its occurrence in higher plants chloroplasts is still a matter of debate. It was recently reported that flash-induced O_2 transients in chloroplasts were due to an interaction with chlororespiration similar to that occurring in algae (7). This appears however doubtful, since flash-induced O_2 signals reported in this study (a stimulation of O_2 uptake by PS1) were the opposite of that reported in algae (an inhibition of O_2 uptake by PS1) and can be more readily interpreted as resulting from PS 1-induced oxygen photoreduction (i.e. Mehler reaction). Clearly, additional work is needed to establish the occurrence of a chlororespiratory process in higher plant chloroplasts.

References

1 Bennoun, P. (1982) Proc. Natl. Acad. Sci. USA 79, 4352-4356
2 Gans, P., Rebeillé, F. (1990) Biochim. Biophys. Acta 1015, 150-155
3 Bennoun, P. (1994) Biochim. Biophys. Acta 1186, 59-66
4 Peltier, G., Ravenel, J., Verméglio, A. (1987) Biochim. Biophys. Acta 893, 83-90
5 Ravenel, J., Peltier, G. (1992) Biochim. Biophys. Acta 1101, 57-63
6 Dorthu, M.P., Remy, S., Michel, M.R., Colleaux, L., Breyer, D., Beckers, M.C., Englebert, S., Duyckaerts, C., Sluse, F.E., Matagne, R. (1992) Plant Mol. Biol. 18, 759-772
7 Gruszecki, W.I., Bader, K.P., Schmid, G.H. (1995) Biochim. Biophys. Acta 1188, 335-338

RESPIRATION AND THE SLOW SIGNAL OF APPARENT O_2 RELEASE UNDER FLASHING LIGHT IN *SYNECHOCYSTIS* SP. PCC 6803.

Pascal C. Meunier[1], Robert L. Burnap[2] & L. A. Sherman[1], [1]:Department of Biological Sciences, Purdue University, West Lafayette, IN 47906, USA; [2]:Department of Microbiology and Molecular Genetics, Oklahoma State University, Stillwater, OK 74078, USA

1. Introduction

In cyanobacteria, the respiratory and photosynthetic electron transport chains share common components. Thus, in the unicellular cyanobacterium *Synechocystis* sp. PCC 6803, brief flashes of light generate respiratory responses due to charge separations in Photosystems I and II (1). These responses have two components, an inhibition of respiration due to PSI activity, and a stimulation of respiration due to PSII activity (1). In a PSII-less mutant (triple *psbA* deletion) and a mutant unable to evolve O_2 (*psbC* insertion), the positive response is produced by the inhibition of respiration, is identical at each flash for low flash frequencies, and is much larger than that in the wild-type (1). The slow signal of O_2 exchange is inhibited by the combination of HQNO and DBMIB, inhibitors of the oxidation of plastoquinone (1). However, our results show that the wild-type responds differently to these inhibitors. In addition, it produces large respiratory responses with complex kinetics when subjected to 50 Hz flashes. The deconvoluted stimulation of respiration by PSII is proportional to the number of electrons transfered from PSII to the plastoquinone pool at each flash. The signals do not show a binary oscillation, as would be expected from the two electron gate Q_B, but a 4-step dependence that suggests increased misses in the S_2 to S_3 transition.

2. Materials and methods

2.1 Experimental

Synechocystis sp. PCC6803 wild-type was grown and maintained in enriched BG-11 media ("6714" media), and the experiments were performed with a centrifugal bare Pt electrode as described in (1).

2.2 Analysis of results

The peaks of the oxygen production responses were determined by averaging the signal over 8 ms. These values included a component due to the beginning of the slow signal, which peaked between 60 and 80 ms after the flash while using the 100 ms high-pass filter. This is expressed mathematically as:

$$Y_t^n = P_n + R_n/x \qquad (1)$$

P. Mathis (ed.), Photosynthesis: from Light to Biosphere, Vol. II, 891–894.

where "Y_t^n" denotes the total O_2 yield after the nth flash, evaluated 10-18 ms after the flash; "P_n" denotes the peak value of the photosynthetic O_2 signal after the nth flash; "R_n" is the peak value of the slow signal of apparent O_2 release (respiratory response) after the nth flash; "R_n/x" denotes the fraction 1/x of the amplitude of the slow signal of apparent O_2 release that was present 10-18 ms after the flash. We assumed that the shape of the slow signal remained the same throughout the experiment, with the superimposition of the fast signal of O_2 release. It follows that the fraction "1/x" (but not necessarily the value of "R_n/x") was constant throughout a single experiment.

The correction for the slow signal was performed as follows. Another average was calculated over 10 ms, from 65 to 75 ms after the flash (c.f. Fig.2). This second averaged signal consisted of two components, the tail of fast signal and the peak of the slow signal. This can be expressed as:

$$A_r^n = P_n/y + R_n \tag{2}$$

where "A_r^n" is the measured signal amplitude from 65 to 75 ms after the flash; "P_n" and "R_n" have the same significance as before; "P_n/y" denotes a fraction "1/y" of the amplitude of the photosynthetic O_2 signal (the tail of the signal). We assumed that the shape of the photosynthetic O_2 signal remained the same throughout the experiment, with the superimposition of the slow signal of apparent O_2 release. It follows that the fraction "1/y" was constant throughout a single experiment.

Since the fast signal of oxygen release was negligible on the first flash, we used the shape of this signal as a reference to determine the value of "x". The goal of the correction is to subtract the beginning of the slow signal at 10-18 ms after the flash from the fast signal. This was accomplished by the following calculation:

$$D_n = Y_t^n - A_r^n/x \tag{3}$$

where the "D_n" are to be analyzed as photosynthetic O_2 yields resulting from the S-state mechanism, and the other symbols have the same significance as before. The meaning of Eq. 3 is made clearer by replacing Eq. 1 and 2 into Eq.3, giving:

$$D_n = P_n(1 - 1/yx) \tag{4}$$

which shows that the deconvoluted amplitudes were directly proportional to the unknown "P_n". The "D_n" can be analyzed since the properties of the S-state mechanism are independent of how large a sample is used.

We determined the amplitudes of the respiratory transients by doing a linear regression of "Y_t^n" as a function of the "D_n". This yielded $Y_t^n \approx Y_p = a{*}D_n + b$; from that,

$$Y_t^n - Yp - b = P_n + R_n/x - a{*}P_n(1 - 1/yx) = R_n/x \tag{5}$$

assuming that $a=1/(1 - 1/yx)$, which should be true due to the regression.

3.Results and Discussion

The combined respiratory and photosynthetic responses in the wild-type *Synechocystis* sp. PCC 6803 are shown in Fig.1. The shape of the response to each flash changed from slow kinetics to fast kinetics when O_2 was produced during the splitting of water. Starting at 3 Hz, the slow respiratory responses developped additional kinetics as can be seen from the shift of the baseline. At 50 Hz, the photosynthetic O_2 yields were smaller owing to the smaller intensity of the flashes, and are seen on the graph as a thick black line. However, the respiratory responses display complex transients with two slow peaks whose amplitude, location and duration are variable depending on the physiological state of the cyanobacterium. Complex kinetics were also observed in O_2-evolution deficient mutants (1). This phenomenon may allow novel characterizations of electron transport pathways in cyanobacteria.

The use of the inhibitor HQNO (2-n-heptyl-4-hydroxyquinoline-N-oxide) stimulated the amplitudes of the respiratory transients several-fold and slowed the kinetics of the responses to single flashes (Fig.2). DBMIB (2,5-dibromo-3-methyl-6-isopropyl-p-benzoquinone, a.k.a. Dibromothymoquinone) also slowed the kinetics but did not significantly stimulate the amplitudes compared to HQNO. It also did not inhibit them, contrary to the inhibition expected from models of electron transport around PSI (2). By combining both inhibitors, the stimulation produced by HQNO was removed. By

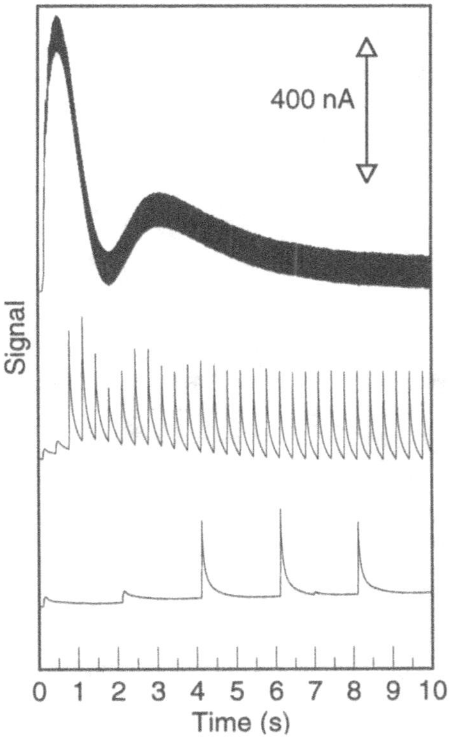

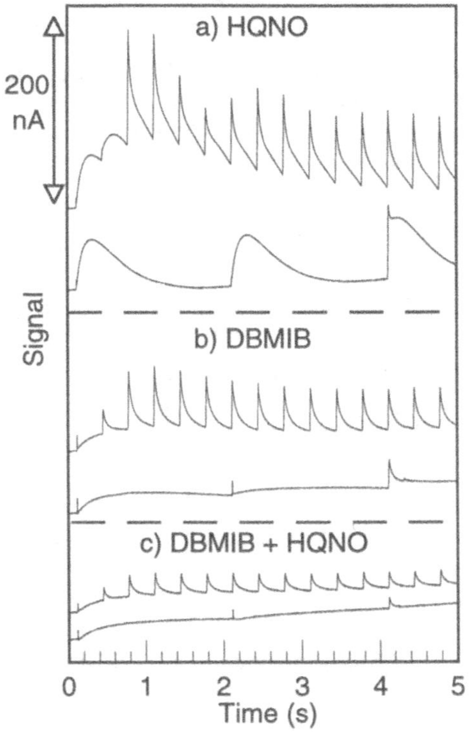

Fig.1 (left). O_2 yields in *Synechocystis* sp. PCC 6803 at flash frequencies of 0.5 Hz, 3 Hz and 50 Hz (bottom to top, respectively).

Fig. 2 (right). Effect of HQNO and DBMIB on the slow signal in *Synechocystis* sp. PCC 6803, at flash frequencies of 0.5 and 3 Hz.

contrast, in the *psbC* insertion mutant and in the triple *psbA* deletion mutant of *Synechocystis* sp. PCC 6803, the use of HQNO and DBMIB separately stimulated the amplitudes of the slow signal (1). The combination of both was also necessary to inhibit the slow signal (1).

The amplitudes of the respiratory transients were then deconvoluted as a function of the flash number, as described in Materials & methods (Fig.3). Provided that the deconvolution procedure is correct, these amplitudes are proportional to the number of electrons put out by PSII and thus to the quantum yield of PSII. The amplitudes show a clear 4-step dependency (in opposition to a binary one) indicating the contribution of the S-state mechanism. For reference, the flash number where most centers underwent the S-state transition S_2 to S_3 is also indicated. It can be seen that the S_2 to S_3 transition generates less electrons and probably has high misses, as was suggested previously (3).

In conclusion, the interplay of the photosynthetic and respiratory chains in cyanobacteria produces complex kinetics. Our results suggest that there is another pathway to PSI not using the cyt b_6/f complex, perhaps the one that is operational when the plastocyanin and the cytochrome c_{553} genes are deleted or not expressed (4). Alternatively there could be an additional plastoquinone binding site on the cyt b_6/f that is not affected by DBMIB. Our results also suggest that the pathway not affected by DBMIB is more active in the O_2-evolution deficient mutants, and that this pathway represents an adaptative advantage in the absence of PSII. Moreover, PSII properties can be studied from its output of electrons into the respiratory chain. Our results also support the proposal of a high miss probability on the S_2 to S_3 transition.

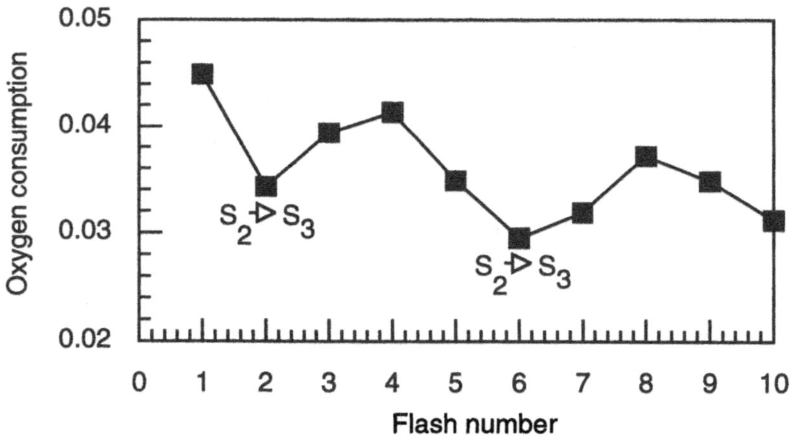

Fig. 3. O_2 consumption yields, deconvoluted from the 3 Hz data presented in (1).

References

1 Meunier, P.C., Burnap, R.L. and Sherman, L.A. (1995) Photosynth. Res. (in press)
2 Delrieu, M.J. (1974) Photochem. Photobiol. 20, 441-454
3 Yu, L., Zhao, J., Mühlendhoff, U., Bryant, D.A., Golbeck, J.H. (1993) Plant Physiol. 103, 171-180
4 Zhang, L., Pakrasi, H., Whitmarsh, J. (1994) J. Biol. Chem. 269, 5036-5042

OCCURRENCE OF AN ELECTRON CARRIER PROTEIN IN CHLOROPLAST ENVELOPE MEMBRANES

Y. Murata, Y. Saijoh*, Y. Arii*, A. Hatanaka* & M. Takahashi
Dept. Applied Biochem., Coll. Agric., Osaka Pref. Univ., Gakuen-cho, Sakai, 593, JAPAN; *Dept. Biol., Konan Univ., Okamoto, Higashinada, Kobe, 658, JAPAN

1. Introduction

Photosynthetic electron transport by thylakoid membranes generates pairs of oxidant and reductant resulting in the production of molecular oxygen and various reduced products. Most of the electrons are transferred to $NADP^+$ to reduce CO_2 in Calvin cycle and the rest is used to reduce O_2 generating O_2^- and ascorbate to detoxicate H_2O_2 which is a product from the disproportionation of O_2^- in chloroplasts by plastid superoxide dismutases (1). Besides above pathways, plants owe reductive assimilation of inorganic sulfur and nitrate to the photosynthetic reductants (2).

Most of these reactions occur inside the chloroplast except the first step in the nitrate assimilation, which is carried out in cytoplasm, nevertheless it depends on the photosynthetic generation of reductants (2). The reaction can be supported with reductants that are transported in the form of low molecular organic acids through envelope membranes (3).

Chloroplast envelopes border between the constituents of the plastid and cytoplasm and have channels for K^+ and H^+, incoming substrates such as phosphate and inorganic carbons, photosynthetic products, and precursor proteins of chloroplast (4). However, there have been no evidence for the presence of any electron transfer proteins. We found that photosynthetically generated electron leaks from thylakoid to cytosolic space through transmembrane electron transfer in envelope. This report first presents the occurrence of electron transfer in chloroplast envelope membranes that may support energy-requiring uptake of cytosolic substances into stromal space.

2. Materials and Methods

2.1 *Isolation of chloroplast envelope membranes*

Intact chloroplasts were isolated from spinach leaves using a Percoll density gradient centrifugation (5). Intactness of the chloroplast preparation was estimated from the percentage of the latent oxygen-evolving activity in the isotonic reaction medium containing 0.33 M sorbitol. Envelope membranes were isolated from frozen-thawed intact chloroplasts using a stepwise sucrose gradient centrifugation by the method of Cline et al. (5).

2.2 *Measurement of cytochrome (Cyt) c reduction*

P. Mathis (ed.), Photosynthesis: from Light to Biosphere, Vol. II, 895–898.

Photoreduction of Cyt *c* was measured at 550 nm with a reference at 540 nm with cross-continuous illumination of red light (> 650 nm; 12-14 mW cm^{-2}) as described previously (6).

Oxidation of Cyt *c* by envelope redox protein(s) was measured by a difference absorption spectrometry. One of the two compartments (0.45-cm light path length each) in the reference and sample cuvettes contained 0.95 ml of 50 mM HEPES-NaOH, pH 7.2, 1 mM EDTA, 1 mM KCN, and 20 μM reduced Cyt *c*. Remaining compartments were filled with 0.95 ml of 50 mM HEPES-NaOH, pH 7.2, that contained 1 mM EDTA and 1 mM KCN, unless otherwise specified and any absorbance differences between both cuvettes were nullified. The envelope membrane suspensions (50 μl) were put into the Cyt-*c* containing mixture in the sample cuvette but into the other compartment of the reference cuvette. Instead, 50 μl of the isolation buffer was added to the Cyt-*c* containing compartment of the reference cuvette and difference spectrum was measured for the wavelength from 600 to 350 nm.

3. Results and Discussion

3.1 *Reduction of Cyt* c *by illuminated chloroplasts*

Figure 1 *a* shows the reduction of Cyt *c* by illuminated chloroplasts in the isotonic medium. After the onset of illumination, a relatively rapid reduction was initiated and lasted for about 10 s followed by a slow steady-state reaction. Initial rate of the rapid phase of Cyt *c* reduction was 5 - 10 μmoles (mg of Chl)$^{-1}$ h^{-1} while that for the slow phase varied depending on the intactness of chloroplast preparation. In the hypotonic reaction medium, Cyt *c* was reduced without biphasic kinetics with a rate three-times as high as the steady-state rate observed in the isotonic medium (Fig. 1 *b*), which has been identified as a consequence of O_2^- production by thylakoid membranes (6).

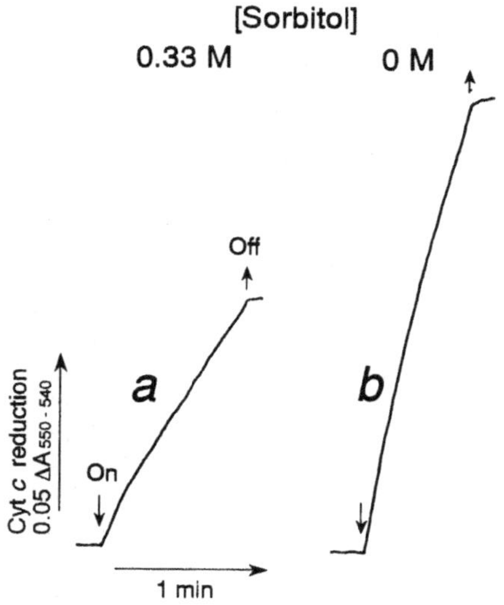

Fig. 1 *Cyt* c *reduction by illuminated spinach chloroplasts in isotonic medium* (a) *and in hypotonic one* (b). Downward and upward arrows indicate the times when the illumination was lighted on and off, respectively.

About 15% of the reducing equivalents that can be trapped by Cyt *c* leaks across the envelope membranes into extra plastid space. The Cyt *c* reduction was not inhibited by superoxide dismutase and the absence of Cyt *c* could not induce the production of H_2O_2, i. e., the spontaneous disproportionation of O_2^-, outside of chloroplast (data not shown). These findings suggests that Cyt c is reduced by envelope component(s) other than superoxide radicals.

3.2 *The presence of a redox protein in chloroplast envelope*

When reduced Cyt *c* was mixed with chloroplast envelopes, the absorption bands at the wavelengths of 550, 519, 417, and 314 nm, all of which are typical to the oxidation of Cyt c, were bleached (Fig. 2). Besides of above wavelengths, no absorbance change was observed in the wavelength range from 240 nm to 700 nm. The difference spectrum indicates the presence of redox component(s) reactive to Cyt *c* in chloroplast envelope membranes. The component may not show intense absorption changes characteristic to its reduction.

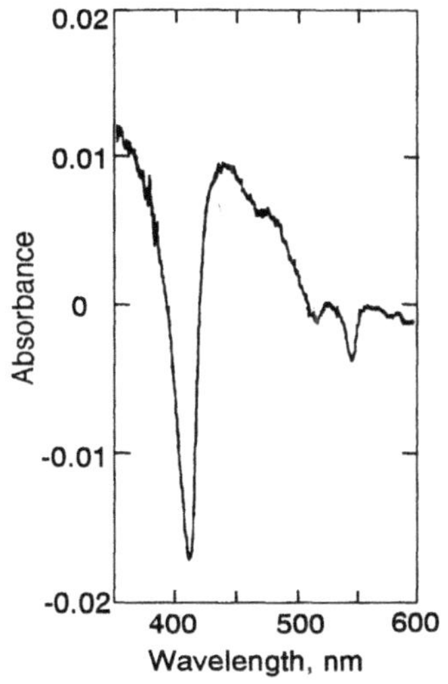

Fig. 2 *Oxidation of reduced Cyt* c *by envelope membrane protein(s).* Difference spectrum of 17.6 μM reduced Cyt c (92% reduced) with minus without envelope proteins (18.8 μg of protein ml^{-1}) was taken as described in the text.

The amount of oxidation of Cyt c was varied by the addition of different amounts of envelope membranes (Fig. 3, insets). E_h of the equilibrated reaction mixture was estimated from the redox ratio of Cyt c. The titration curve was fairly approximated to the Nernst equation with E_o' of 155 mV and the number of electron transferred (n) of 4 (Fig.3 a). The value of E_o' suggests that envelope redox protein has a quinone compound as the electron-carrying group. When the envelope proteins were solubilized with octyl-β-D-glucoside (OGP), E_o' slightly shifted negative by 12 mV and n became to 1 (Fig. 3 b).

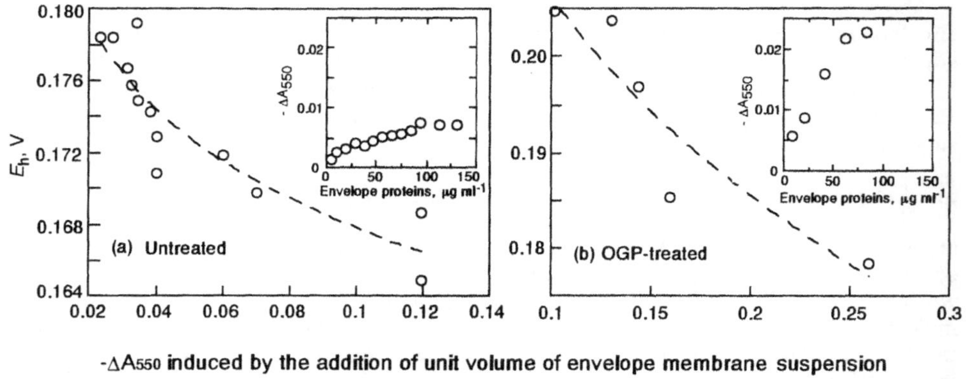

Fig. 3 Redox titration of envelope proteins. Oxidation of Cyt *c* was measured by the difference spectrum after the addition of 5 - 140 µl of untreated envelope membranes (a) and 10 - 100 µl of OGP-treated envelope membranes (b) to 22.1 and 16.3 µM Cyt *c* (97.6% reduced), respectively. E_h was calculated from the redox ratio of Cyt *c* in the reaction mixture assuming E_0' of 254 mV. Abscissa stands for the level of envelope proteins reduced by Cyt *c*, which are calculated from the redox titration data (insets).

The number of n higher than 2 indicates that an electron transferring complex consisted of more than two redox proteins reacts with reduced Cyt c. This complex may be dissociated when OGP is present. The function of envelope electron transfer is not clear yet, but we suspect that the electron transfer creates H^+ gradient to drive the transport processes across envelope membranes.

4. References

1 Asada, K. and Takahashi, M. (1987) in Photoinhibition (Kyle, D. J., Osmond, C. B., and Arntzen, C. J. eds.) pp.227-287, Elsevier, Amsterdam
2 Anderson, J. W. (1981) in The Biochemistry of Plants. Vol. 8 (Hatch, M. D. and Boardman, N. K. eds.) pp.473-500, Academic Press, New York
3 Flügge, U.-I. and Heldt, H. W. (1984) Trends Biochem. Sci. 9, 530-533
4 Douce, R. and Joyard, J. (1990) Annu. Rev. Cell. Biol. 6, 173-216
5 Cline, K., Andrews, J. Mersey, B., Newcomb, E. H. and Keegstra, K. (1981) Proc. Natl. Acad. Sci. U. S. A. 78, 3595-3599
6 Takahashi, M. and Asada, K. (1988) Arch. Biochem. Biophys. 267, 714-722

ESSENTIALITY OF BOTH CYTOCHROME AND ALTERNATIVE PATHWAYS OF RESPIRATION FOR PHOTOSYNTHESIS IN MESOPHYLL PROTOPLASTS OF PEA (*PISUM SATIVUM* L.)

K. Padmasree and A.S. Raghavendra
Department of Plant Sciences, School of Life Sciences,
University of Hyderabad, Hyderabad 500 046, INDIA.

1. Introduction

The mutually beneficial effects of photosynthesis and respiration in plant cells are believed to be due to marked interaction between chloroplasts and mitochondria, mediated through intracellular redox state and/or adenine/pyridine nucleotide levels (1,2). Mitochondrial oxidative metabolism is essential for photosynthetic carbon assimilation as revealed by the effects of oligomycin in leaves and protoplasts of barley (3-5), and studies with mutants of *Chlamydomonas* (6) and tobacco (7). Mitochondrial respiration not only benefits photosynthesis but also protects against photoinhibition, as demonstrated with mesophyll protoplasts of pea (8) and the cyanobacterial system of *Anacystis nidulans* (9).

Oxidative electron transport in plants occurs through both cytochrome pathway and an alternative cyanide resistant pathway (10). In the present work, the relative importance of these two pathways of oxidative electron transport (cytochrome and alternative pathways) in sustaining high rates of photosynthetic activity was examined by the use of antimycin A (inhibitor of cytochrome pathway), salicylhydroxamic acid (inhibitor of alternative pathway). The results were compared with the effects of oligomycin, which suppresses oxidative phosphorylation but has no direct effect on photosynthesis (3). Our results suggest that oxidative electron transport, through both cytochrome and alternative pathways, is important during the beneficial role of mitochondrial respiration in optimizing photosynthesis.

2. Materials and Methods

Pea (*Pisum sativum* L. cv. Arkel) plants were grown outdoor (natural photoperiod of approximately 12 h, average daily temperatures of 30 °C day/20 °C night). The

P. Mathis (ed.), Photosynthesis: from Light to Biosphere, Vol. II, 899–902.

mesophyll protoplasts were isolated from the first and second fully expanded leaves of 8-10 day old plants as already described (8). Photosynthetic and respiratory activities of protoplasts were determined by monitoring their oxygen evolution and uptake, respectively, using a Clark type O_2 electrode. Test compounds were included in the reaction medium so as to give the required final concentration, and protoplasts were pre-incubated in darkness at 25 °C for 5 min, before switching on the light. The levels of ATP and ADP were measured using enzymatic assays coupled to NADPH formation and NADH utilisation, respectively by a modified procedure of Stitt et al. (11).

3. Results and Discussion

Our observations confirm the concept that mitochondrial respiration is essential for optimal performance of photosynthesis in mesophyll protoplasts (1,2). When protoplasts were incubated with oligomycin (inhibitor of oxidative phosphorylation), antimycin A (inhibitor of cytochrome pathway) and SHAM (inhibitor of alternative pathway), the photosynthetic rate decreased markedly by >75%, while the decrease in respiratory rate was only marginal <30% (Fig. 1). There was a stimulation in protoplast respiration in presence of SHAM (Fig.1C). The inhibitory effect of SHAM on photosynthetic activities of leaf discs has earlier been noticed (12). Our results therefore point out that oxidative electron transport through both cytochrome and alternative pathways is essential for photosynthesis.

Oligomycin (inhibitor of oxidative phosphorylation) or antimycin A (inhibitor of cytochrome pathway) suppressed photosynthesis activity along with marked decrease in ATP and increase in ADP (Fig. 2A,B). The presence of SHAM (inhibitor of alternative

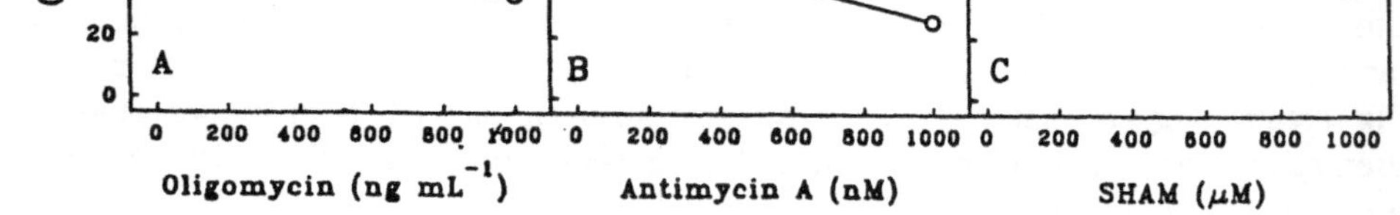

Figure 1. Effect of oligomycin, antimycin A or SHAM on photosynthetic oxygen evolution in light and respiratory oxygen uptake in darkness by mesophyll protoplasts. The rates of photosynthesis and respiration in the control sets were 165 $\pm$ 10.5 and 12 $\pm$ 1.1 μmol O_2 evolved/consumed mg^{-1} Chl h^{-1} respectively.

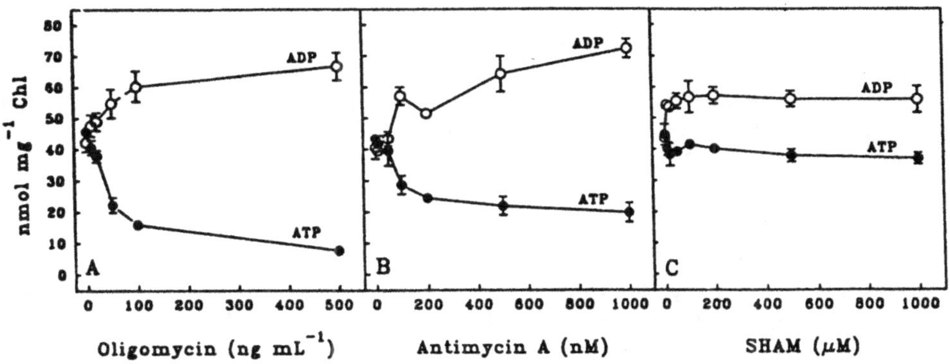

Figure 2. Effect of oligomycin, antimycin A or SHAM on the levels of ATP and ADP in pea mesophyll protoplasts.

pathway) caused only a marginal decrease in ATP (Fig. 2C), although the photosynthesis was suppressed by nearly 65% (Fig. 1C).

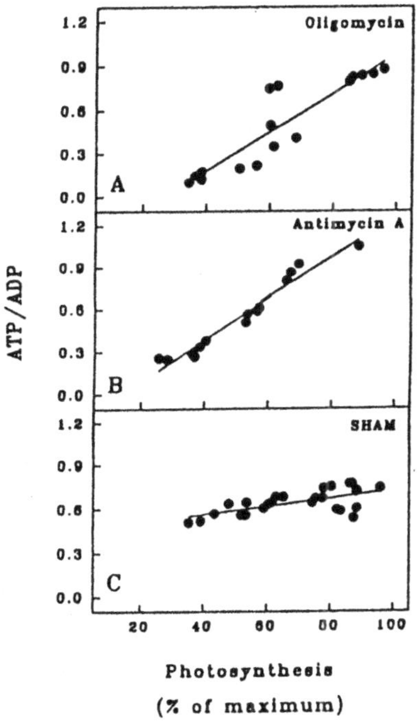

Figure 3. Correlation between photosynthesis and the ATP to ADP ratio in mesophyll protoplasts on incubation with oligomycin, antimycin or SHAM.

There was a sharp correlation between photosynthesis and the ratio of intracellular ATP to ADP in mesophyll protoplasts. The correlation between ATP to ADP ratio and photosynthetic rates of mesophyll protoplasts was very strong (r = 0.903 or 0.978) and statistically significant (P <0.001) in presence of oligomycin (Fig. 3A) or antimycin A (Fig. 3B). On the other hand, the correlation between photosynthesis and the ratio of ATP to ADP was very low (r = 0.508) in presence of SHAM (Fig. 3C). The alternative pathway can serve as a means of oxidising excess NADH, while its capability of ATP formation is much less than that of the cytochrome pathway (13). An indirect evidence in favour of ATP formation through alternative pathway is the stimulation of alternative respiration in roots by the uncouplers, FCCP or CCCP (14). This is the first report of a small but clear decrease in intracellular ATP due to SHAM and reveals the importance of electron transport through alternative pathway in optimizing photosynthetic performance in mesophyll protoplasts.

Recent studies suggest that at high photosynthetic fluxes the major function of mitochondrial oxidative metabolism is the oxidation of excess chloroplastic redox equivalents (5,8), while at low photosynthetic fluxes, the mitochondrial supply of ATP for cytosolic processes becomes important. We suggest that the mitochondrial oxidation of reduced equivalents is an important link leading to the optimization of photosynthesis in protoplasts.

4. Conclusions

Electron transport through both cytochrome and alternative pathways is important in exerting a beneficial effect on photosynthetic carbon assimilation. The alternative pathway of mitochondrial respiration is not only essential for optimal photosynthesis but also can modulate the intracellular ATP/ADP levels.

Acknowledgements

This work is supported by Career award grant (F.1/5-90 SR IV/SA II) from the University Grants Commission, New Delhi and KP was a recipient of UGC-SRF.

References

1. Raghavendra, A.S., Padmasree, K. and Saradadevi, K. (1994) Plant Sci. 97, 1-14
2. Krömer, S. (1995) Annu. Rev. Plant Physiol. Plant Mol. Biol. 46, 45-70
3. Krömer, S., Stitt, M. and Heldt, H.W. (1988) FEBS Lett. 226, 352-356
4. Krömer, S. and Heldt, H.W. (1991) Plant Physiol. 95: 1270-1275
5. Krömer, S., Malmberg, G. and Gardestöm, P. (1993) Plant Physiol. 102, 947-955
6. Lemaire, C., Wollman, F-A., and Bennoun, P. (1988) Proc. Natl. Acad. Sci. 85, 1344-1348
7. Hanson, K.R. (1992) Plant Physiol. 99, 276-283
8. Saradadevi, K. and Raghavendra, A.S. (1992) Plant Physiol. 99, 1232-1237
9. Shyam, R., Raghavendra, A.S. and Sane, P.V. (1993) Physiol. Plant. 88, 446-452
10. Siedow, J.N. and Berthold, D.A. (1986) Physiol. Plant. 66, 569-573
11. Stitt, M., Lilley, R.Mc.C., Gerhardt, R. and Heldt, H.W. (1989) in Methods in Enzymology (Fleischer S. and Fleischer B., eds.) Vol 174, pp. 518-552, Academic Press Inc, New York
12. Diethelm, R., Miller, M.G., Shibles, R. and Stewart, C.R. (1990) Plant Cell Physiol. 31, 179-185
13. Lambers, H. (1985) in Encylopaedia of Plant Physiology (Douce R. and Day D.A., eds.) New Series, Vol 18, pp. 418-473, Springer Verlag, Berlin
14. Lambers, H. (1990) in Plant Physiology, Biochemistry and Molecular Biology (Dennis D.T. and Turpin D.H., eds.) pp.124-143, Longman, Harlow, UK

ENHANCEMENT IN THE EFFICIENCY OF PHOTOSYSTEM II PHOTOCHEMISTRY IN *SYNECHOCOCCUS* PCC 7942 DUE TO PROLONGED PARTIAL INHIBITION OF PHOTOSYSTEM II

Madhulika Srivastava[1], Prasanna Mohanty[2] and Salil Bose[1]
[1]Centre for Biotechnology and [2]School of Life Sciences, Jawaharlal Nehru University, New Delhi 110067, India

1. Introduction

The light reactions of photosynthesis require coordinated functioning of various supramolecular complexes present in the thylakoids. However, several environmental and genetic conditions, like variations in light quality and intensity, nutritional stress, mutations affecting the light harvesting systems etc., lead to an imbalance in the rate of electron transport through the two photosystems. The cellular response to such a stress is mounted as structural and functional alterations in the photosynthetic apparatus. The alterations, presumably directed towards re-establishment of a balanced activity, are manifested both as short term reorganizational changes [1] and long term changes involving biosynthesis, assembly and degradation of various components of the thylakoids [2].

We have studied the nature of long term changes in the photosynthetic apparatus of cyanobacterium *Synechococcus* PCC 7942. A condition of prolonged imbalance in the rate of electron transport through photosystem I (PSI) and photosystem II (PSII) was imposed by growing the cells in presence of sub-lethal levels of a PSII inhibitor, 4-chloro-5-dimethylamino-2-phenyl-3-(2H)-pyridazinone, also known as SANDOZ 9785. SANDOZ 9785 is a DCMU like PSII inhibitor acting on the reducing side of PSII [3,4] with an I_{50} of 75 μM for cyanobacterial PSII [5]. It has been shown to effect changes in the excitation energy distribution in higher plants [6,7]. Considering major differences that exist between the higher plant and cyanobacterial photosynthetic apparatus, especially in terms of excitation energy distribution, it was of interest to study the long term alterations in the thylakoid membranes of *Synechococcus* PCC 7942 cells subjected to partial inhibiton of PSII.

2. Materials and Methods

Synechococcus PCC 7942 cells were grown in BG11 [8] under illumination with white light (45 $\mu mol.s^{-1}.m^{2}$) and continuous aeration. SANDOZ 9785 treatment was given by growing the cells in presence of 40 μM SANDOZ 9785 for 48 hours. Before measurement of any parameter, the cells were washed thoroughly to ensure complete removal of the inhibitor.

P. Mathis (ed.), Photosynthesis: from Light to Biosphere, Vol. II, 903–906.

PSII activity was measured in thylakoids as light dependent electron transport using Clark type O_2 electrode (Hansatech, Germany). The activity was assayed under varying light intensity using 0.1 mM DCBQ as the electron acceptor. Thylakoids were resuspended at a concentration of 5 μg chl/ml in the thylakoid isolation buffer (50 mM Hepes, pH 7.5, 50 mM $CaCl_2$, 25% glycerol and 1 mM PMSF). 1 μM DBMIB was also added in the reaction mixture.

Fluorescence induction was recorded using PAM chlorophyll fluorometer (Walz, Germany) in intact cells resuspended at a chlorophyll concentration of 5 μg/ml. Dark adapted cells were considered to be in State II. 5 minutes of illumination with red light (PAM 102, 650 nm) at intensity 20 $\mu mol.s^{-1}.m^2$ was used to induce State I transition.

3. Results and Discussion

Growth of *Synechococcus* PCC 7942 in presence of 40 μM SANDOZ 9785 led to a time dependent enhancement in the ratio of phycocyanin:chlorophyll. This was accompanied by an increase in the PSII rate measured in intact cells under illumination that preferrentially excited phycobilisomes (yellow) or chlorophyll antenna of PSII (blue) [9]. The nature of this stimulation was investigated further.

Variable fluorescence was measured in intact cells. F_o was measured using a modulated yellow light of extremely low intensity (PAM 101) and maximal fluorescence (F_p) was measured by giving a flash of saturating white light (1250 $\mu mol.s^{-1}.m^2$). Relative variable fluorescence, $(F_p-F_o)/F_o$ or F_v/F_o, exhibited approximately 60% increase due to SANDOZ 9785 treatment (Table 1) in both State I and State II adapted cells indicating an increased efficiency of light utilization by PSII.

Table 1. Effect of SANDOZ 9785 on relative variable fluorescence

	Relative variable fluorescence (F_v/F_o)		
	Control	Treated	increase
State II	0.14	0.23	64 %
State I	0.37	0.56	51 %

To determine whether the enhancement in efficiency of light utilization is due to changes occuring at the level of Chl *a* antenna or the phycobilisomes, PSII activity was measured in thylakoids. Figure 1 shows the dependence of PSII activity on light intensity as a double reciprocal plot. The y-intercept, representing the reciprocal of the rate of PSII activity under saturating light conditions, was similar for control and SANDOZ 9785 treated thylakoids. Since DCBQ oxidizes Q_A^- directly, presence of DCBQ ensured that these rates represent true rates of photochemistry uninfluenced by PQ pool. Hence, unaltered light saturated rate suggests that the number of PSII units did not change due to SANDOZ 9785 treatment.

The slope of the plot decreased for the SANDOZ 9785 treated thylakoids by 38% suggesting that either the functional antenna size or the quantum yield of photochemistry enhanced. The thylakoids were practically free of phycobilisomes. Hence, the enhancement in the efficiency of light utilization can be attributed to changes effected at the level of Chl *a* antenna.

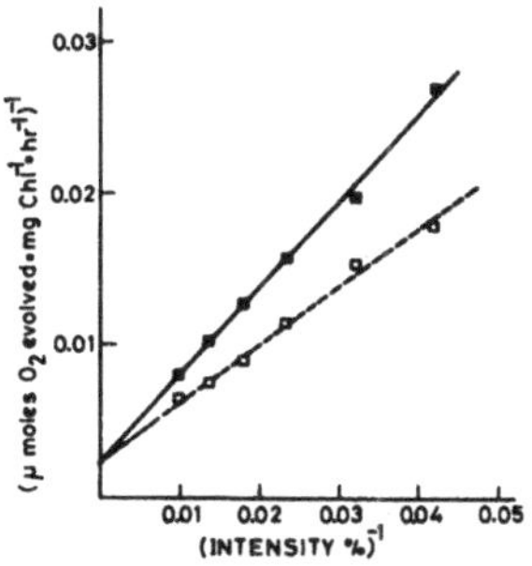

	1/slope	1/intercept
Control	1.83±0.26	356±82
Treated	2.96±0.21	357±38

Figure 1. Dependence of PSII activity ($H_2O \rightarrow$ DCBQ) of thylakoids on light intensity. (■) control, (□) treated. 100% intensity corresponds to 1300 μmol.s^{-1}.m^2 of white light. Accompanying table gives an average of parameters from five experiments.

Fluorescence induction under limiting light conditions was analysed to estimate the rate constant of PSII photochemistry. Fluorescence yield was measured by a modulated red light (PAM 101) of very low intensity. The induction was recorded (Figure 2A) in State I adapted cells, in presence of 15 μM DCMU, by providing red actinic light (650 nm) of limiting intensity (20 μmol.s^{-1}.m^2) from PAM 102. Using the analysis programme DA100, the kinetics of increase in the area (A) bounded by F_p line and F_t (fluorescence yield at time t) was derived. Subsequently, ln (A_{max} - A_t) was plotted as a function of time (Figure 2B). In presence of DCMU, reoxidation of Q_A^- is prevented and hence the slope of such a plot is indicative of the rate constant of PSII photochemistry [10]. The values of the relative slopes were -6.2±0.2 and -6.8±0.26 for control and SANDOZ 9785 treated cells, respectively. This small but significant change also indicates an increased efficiency of light energy utilization by PSII.

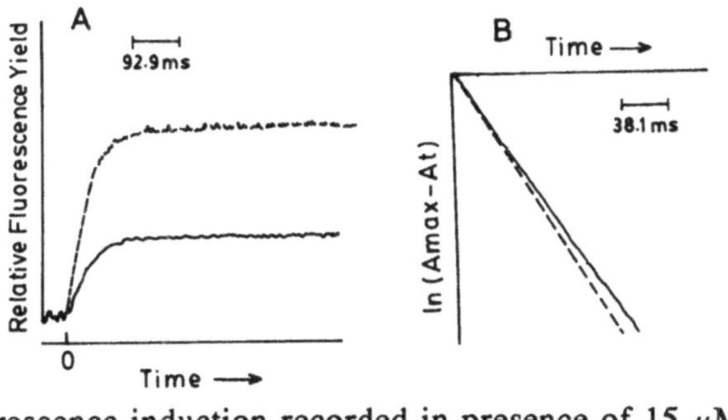

Figure 2. Fluorescence induction recorded in presence of 15 μM DCMU in State I adapted cells. A. The induction curve, B. plot of ln(A_{max}-A_t) during first 240 ms of induction. (——) control, (- - - -) treated. F_p was measured at 540 ms.

Photochemical efficiency is known to alter substantially in cyanobacteria during state transitions as a result of reorganizational changes occuring in the thylakoid membranes [1]. The change in variable fluorescence was, therefore, monitored as the control and SANDOZ 9785 treated cells underwent State I transition. A low intensity yellow (590 nm) modulated light (PAM 101) was used to monitor fluorescence yield during the course of State I transition induced by red light (650 nm, 20 μmol.s^{-1}.m^2) of PAM 102. Maximal fluorescence yield was periodically measured by superimposing saturating flashes of white light (1250 μmol.s^{-1}.m^2). Figure 3 demonstrates that neither the extent of energy redistribution that accompanied the State I transition nor the kinetics of state transition was significantly altered due to SANDOZ 9785 treatment. It appears, therefore, that the mechanism of SANDOZ 9785 induced enhancement in photochemical efficiency of PSII is distinct from that which operates during state transitions.

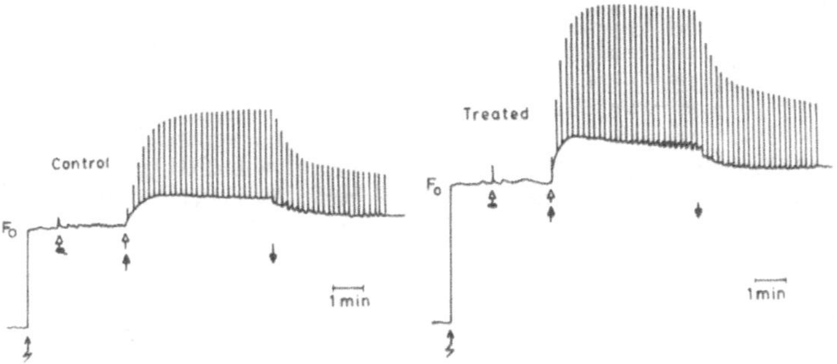

Figure 3. Slow fluorescence induction in *Synechococcus* 7942 cells during transition from dark adapted State II to State I by a red actinic light. (↕) modulated light on, (↑) single flash of saturating white light of 500 ms duration, (↑) series of saturating light flashes of duration 500 ms, (↑) actinic red light on, (↓) actinic red light off.

Prolonged partial inhibition of PSII in the cells of *Synechococcus* PCC 7942 due to the presence of SANDOZ 9785 led to partitioning of excitation energy in favour of PSII as evidenced by fluorescence emission spectra recorded at 77K [5]. Our results also suggest a decrease in the spillover of excitation energy from PSII to PSI [5]. It appears that the energy partitioned in favour of PSII is effectively being utilized for an enhancement in the photochemical efficiency.

Acknowledgements : We thank Council of Scientific and Industrial Research, India, for research grant # 09/0374/EMR to PM and fellowship to MS.

References

1. Biggins J and Bruce D (1989) *Photosynth. Res.* 20, 1-34
2. Melis A (1991) *Biochim. Biophys. Acta* 1058, 87-106
3. Joshi MK, Mohanty P, van Rensen JJS, and Bose S (1994) *J. Photochem. Photobiol. (B: Biology)* 23, 63-68
4. Droppa AM, Horvath G, Vass I and Demeter S (1981) *Biochim. Biophys. Acta* 638, 210-216
5. Srivastava M, Mohanty P and Bose S (1994) *Biochim. Biophys. Acta* 1186, 1-11
6. Mannan MR and Bose S (1985) *Photochem. Photobiol.* 41, 63-72
7. Bose S, Vijayan P, Santhanam R and Kandasamy MK (1992) *Biochim. Biophys. Acta* 1098, 352-358
8. Rippka R, Deruelles J, Waterbury JB, Herdman M and Stanier RY (1979) *J. Gen Microbiol.* 111, 1-10
9. Srivastava M, Bhaya D and Bose S (1994) *Photosynth. Res.* 41, 303-313
10. Melis A and Homman PH (1975) *Photochem. Photobiol.* 21, 431-435

CYCLIC ELECTRON TRANSFER, RESPIRATION AND SALT STRESS IN A NAD(P)H-DEHYDROGENASE DEFICIENT CLONE OF *SYNECHOCYSTIS* PCC6803.

R. JEANJEAN[1], M. HAVAUX[2], T. OGAWA[3], F. JOSET[1] and H.C.P. MATTHIJS[4]

[1] Unité de Métabolisme Energétique, LCB-CNRS, 13402 Marseille Cedex 20 (France); [2] Département de Physiologie Végétale et Ecosystèmes, CEA Cadarache, 13108 St-Paul-lez-Durance (France); [3] Bioscience Center, Nagoya University, Chikusa, Nagoya 464-01 (Japan); [4] BioCentrum Amsterdam, Laboratorium voor Microbiologie, Nieuwe Achtergracht 127, 1018-WS Amsterdam (The Netherlands).

1. Introduction

The freshwater cyanobacterium *Synechocystis* PCC6803 can grow in high NaCl concentrations. In such environmental conditions, it increases its dark respiration and photosystem I (PSI)-mediated cyclic electron transfer [1, 2]. A mutant of this strain, M55, defective in NAD(P)H-dehydrogenase, was obtained by disrupting the *ndhB* gene coding for a subunit of this complex [3]. This mutant can grow neither in low concentration of inorganic carbon nor in photoheterotrophy, and displays an impairment in respiration [3] and cyclic electron flow through PSI [3, 4, 5]. It is reported here that this mutant maintains its tolerance to salt thanks to the possibility to increase and modulate its PSI cyclic activity from an antimycin A sensitive (low salt) to a less antimycin A sensitive (high salt) pathway. The PSI cyclic electron flow at work in salt-adapted cells, Wild-Type and mutant, displays increased sensitivity to N-ethyl-maleimide (NEM). The possible routes for cyclic electron transfer around PSI present in high salt-adapted cells are discussed.

2. Procedure

2.1. Material. Wild-type (WT) *Synechocystis* PCC6803 and the M55 mutant [3] were grown as in [2].
2.2. PSI-mediated cyclic electron transfer was estimated by measurement of energy storage *in vivo* using the photoacoustic technique, as described elsewhere [6, 7, 8]. P700 redox state were monitored *via* cell absorbance changes around 830 nm [9].
2.3. Respiratory electron transfer. Dark oxygen uptake was carried out as in [2].
NADPH oxidation in thylakoids was monitored in the presence of decylubiquinone [2].

P. Mathis (ed.), Photosynthesis: from Light to Biosphere, Vol. II, 907–910.

3. Results and Discussion

3.1. Salt tolerance. The mutant M55 can tolerate the same maximal NaCl concentration as the WT, with generation times longer than the WT ones under equivalent conditions.
3.2. PSI-mediated cyclic electron transfer in standard and salt adapted cells. Photoacoustic spectroscopy enables real time measurement of the ongoing use of light energy [10]. It allows an estimation of the cyclic electron flow around PSI, since the energy storage (ES), unchanged when measured anaerobically or in the presence of DCMU, was practically abolished when methyl viologen was present. Table 1 shows the ES data, measured in low or high far-red light, for WT and M55 cells grown in standard or high salt media, and the effect of inhibitors on these values. As expected [4, 5], the mutant exhibited a very low ES, strongly inhibited by antimycin A (50 μM), whereas the WT ES was practically insensitive to this inhibitor. NEM had a weak effect on the ES of the WT when added alone. However, in combination with antimycin A, it produced a dramatic decrease of the ES of the WT. The decrease of ES by antimycin A in the M55 mutant is attribuable to an inhibition of cyclic electron flow around PSI. Respiration rates in the WT and the mutant were insensitive to antimycin A. A possible interpretation of these results could be the existence of two cyclic electron transfer pathways in the WT, one partly inhibited by NEM and the other by antimycin A, only the latter being available in the mutant. An antimycin A sensitive cyclic electron flow around PSI is present in chloroplasts of higher plants [11, 12, 13, 14], but its existence in cyanobacteria is still a matter of discussion [13, 14]. In salt-adapted WT [2] and mutant cells, the ES was significantly increased, and found highly sensitive to NEM. Its sensitivity to antimycin A was lower in salt-adapted than in standard M55 cells.
The half-time of P700 re-reduction after far-red light illumination was about 250-300 ms in the WT whereas it reached 900-1000 ms in the M55 mutant (Table 2). This parameter, regarded as an estimation of cyclic electron transfer around PSI [15], was insensitive to antimycin A in the WT. In contrast, incubation of the M55 mutant with this inhibitor produced a greater P700 re-reduction half-time. NEM displayed a slight effect on the half-time of P700 re-reduction in the WT, but when added concomitantly with antimycin A, it strongly increased this parameter. In salt-adapted cells, the half-time of P700 re-reduction, shorter than in standard conditions, was markedly affected by NEM in both the WT and the mutant. Half-time of P700 re-reduction in salt-adapted WT and mutant cells exhibited a weak sensitivity towards antimycin A.
Restoration of an efficient cyclic electron flow around PSI allows growth of salt-adapted M55 mutant cells in low inorganic carbon concentration (not shown), a result in good agreement with the role of this bioenergetic process in the uptake of inorganic carbon [14,16].

3.3. Dark respiration and NADPH oxidation by thylakoid preparations. In spite of its deficiency in respiration (20 % of the WT capacity) when grown under standard conditions, M55 increased its dark respiration two-fold when adapted to high salinity, though the final value remained lower than that measured in salt-adapted WT cells (200 %). NADPH oxidation has been monitored in the presence of decylubiquinone and of 2 mM $NADP^+$ (to partly inhibit NADPH oxidation by soluble ferredoxin-$NADP^+$ oxido-reductase (FNR) contaminating the membranes [17]), on thylakoids from standard or salt-adapted cells from the WT and the M55 mutant. Membranes from standard grown M55 cells displayed a NADPH oxidation capacity about 40% of that measured on WT membranes. In contrast, this activity reached the level of that measured in standard grown WT cells in preparations from salt-adapted mutant cells. The increase in NADPH oxidation capacity in salt-adapted cells of the M55 mutant cannot result from a «salt-induced» NAD(P)H-dehydrogenase activity, since M55 cells precultured in high salt medium are unable to grow photoheterotrophically. A membranous form of FNR displaying a higher molecular weight than the soluble form, is present in cyanobacteria and has a diaphorase activity [18]. This enzyme could act as electron carrier in cyclic electron transfer. It could contribute to the residual cyclic activity present under treatment of standard grown M55 mutant cells by antimycin A. Its involvement in the same bioenergetic process in salt-adapted cells could explain the observed changes in sensitivity towards antimycin A and NEM. The evidence of an increased amount of flavodoxin in salt-adapted *Synechocyctis* cells [19] supports the hypothesis of different cyclic electron pathways being induced in high versus standard salinity conditions.

TABLE 1: Photochemical energy storage (in % of absorbed light energy) measured in far-red light, of WT *Synechocystis* PCC6803 and mutant M55 grown in standard (0.05 M NaCl) or high salt (0.55M NaCl) media. Effects of different inhibitors.

	Standard conditions				High Salt Medium			
	WT	M55	WT	M55	WT	M55	WT	M55
Far-red light	$8\ W.m^{-2}$		$35\ W.m^{-2}$		$8\ W.m^{-2}$		$35\ W.m^{-2}$	
No addition	18	7	8	2.6	22	11	13	4.5
Antimycin A 50µM	17	3.5	7	0.5	20	9	12	3.8
NEM 1 mM	17	4	7	1.9	16	3	6.5	1
Antimycin A + NEM	10		5	<0.5	15	0.8	6	<0.5
Methyl Viologen	4							
Nitrogen atmosphere	17							

TABLE 2: Turnover of P700 in PSI-driven cyclic electron transfer in *Synechocystis* PCC6803 WT and M55 mutant grown in standard conditions (A) or in 0.55 M NaCl (B). Cells were illuminated with far-red light (>730nm; 6 $W.m^{-2}$). P700 turnover was determined by the half-time (in milliseconds) of dark reduction of $P700^{+}$ after switching off the far-red light. Effects of different inhibitors. Mean values for separate experiments (number in brackets) ± SD.

	WILD TYPE		MUTANT M55	
	A	B	A	B
No addition	280±50(8)	180±20(6)	980±60(5)	580±30(5)
Antimycin A 50 µM	310(2)	208±30(3)	1620±200(5)	680±80(3)
NEM 0.5 mM	360±60(8)	360±60(6)	1130±120(4)	1110±170(3)
Antimycin A+ NEM	620(2)	390±40(3)	1900(2)	1290±90(3)

References

1 Schubert H. and Hagemann M. (1990) FEMS Microbiol. Lett 71, 169-172.
2 Jeanjean R., Matthijs H.C.P., Onana B., Havaux M. and Joset F. (1993) Plant Cell Physiol. 34, 1073-1079.
3 Ogawa T. (1991) Proc. Natl. Acad. Sci. 88, 4275-4279.
4 Mi H., Endo T., Schreiber U., Ogawa T. and Asada K. (1992) Plant Cell Physiol. 33, 1233-1237.
5 Mi H., Endo T., Schreiber U., Ogawa T. and Asada K. (1994) Plant Cell Physiol. 35, 1233-1237.
6 Herbert S.K., Fork D.C. and Malkin S. (1990) Plant Physiol. 94, 926-934.
7 Havaux M., Lorrain L. and Leblanc R.M. (1989) FEBS Lett. 250, 395- 399.
8 Havaux M. (1992) Plant Cell Physiol. 33, 799-803.
9 Schreiber U., Klughammer C. and Neubauer C. (1988) Z. Naturforsch . 43, 686-698.
10 Malkin S. and Canaani O. (1994) Ann. Rev. Plant Physiol. Plant Mol. Biol. 45, 493-526.
11 O'Keefe D.P. (1988) Photosynth. Res. 17, 189-216.
12 Cleland R.E. and Bendall D.S. (1992) Photosynth. Res. 34, 409-418.
13 Manasse R.S. and Bendall D.S. (1993) Biochim. Biophys. Acta 1183, 361-368
14 Bendall D.S. and Manasse R.S. (1995) Biochim. Biophys. Acta 1229, 23-38
15 Maxwell P.C. and Biggins J. (1976) Biochemistry. 15, 3975-3981.
16 Ogawa, T., Miyano, A. and Inoue, Y. (1985) Biochim. Biophys. Acta 808, 77-84
17 Scherer S., Alpes I., Sadowski H. and Böger P. (1988) Arch. Biochim. Biophys. 267, 228-235.
18 Schluchter W.M. and Bryant D.A. (1992) Biochemistry 31, 3092-3102
19 Fulda S. and Hagemann M. (1995) J. Plant Physiol . in press.

POLYPHASIC CHLOROPHYLL *a* FLUORESCENCE TRANSIENT IN LEAVES EXPOSED TO ANAEROBIC CONDITIONS

Pierre Haldimann, Alaka Srivastava and Reto J. Strasser,
Laboratory of Bioenergetics, University of Geneva,
CH-1254, Jussy-Geneva, Switzerland

1. Introduction

The photosynthetic apparatus of plants can be affected by many stress factors such as chilling, heat and drought. The absence or low concentrations of CO_2 and O_2 represent another stress factor, since under these conditions fundamental physiological processes like carbon assimilation, photorespiration and mitochondrial respiration are inhibited. It is well known that under anaerobic conditions the plants become very prone to photoinhibition, even at low light intensities (1,2,3). Furthermore, experiments with anaerobiosis have suggested the presence in higher plants of a putative chlororespiratory pathway similar to that thought to exist in algae (4). The functional role of this pathway is supposed to be related to the need to reoxidize reduced pyridine nucleotide to sustain steady-state chloroplastic carbohydrate metabolism in the dark (4).

In this study we have used the fast chlorophyll (Chl) *a* fluorescence induction transients as a sensitive and non-invasive tool to characterize the effects of anaerobiosis in the dark and in the light on the photosynthetic activity of pea leaves.

2. Material and Methods

Pea plants (*Pisum sativum* L.) were grown in a greenhouse at 22°C/18°C (day/night) under natural sunlight. For the anaerobiosis treatments, leaves from several hours dark-adapted plants were placed in a gas tight plexiglas chamber on moistened filter paper and flushed with pure N_2 gas. The gas was bubbled through water before entering the chamber. The anaerobiosis treatment in the light was done at a photon flux density of 90 $\mu mol\ m^{-2}\ s^{-1}$ white actinic light. After the N_2 treatment, the leaves were kept in the air in the dark on moistened filter paper.Chl *a* fluorescence induction transients were measured at room temperature in the air with a Plant Efficiency Analyzer (PEA, Hansatech, Ltd, King's Lynn, Norfolk, England) as described previously (5). Chl *a* fluorescence was induced by illumination with a light intensity of 600 $W\ m^{-2}$ red actinic light (peak at 650 nm).

3. Results and Discussion

3.1 Anaerobiosis in the dark

It has been shown previously that when dark-adapted leaves are illuminated with actinic light of high intensity the fast Chl *a* fluorescence induction transient, when plotted on a logarithmic time scale, follows a typical polyphasic rise which has been called O-J-I-P (5). The O level

P. Mathis (ed.), Photosynthesis: from Light to Biosphere, Vol. II, 911–914.

corresponds to the initial fluorescence. The J (at about 2 ms) and the I (at about 30 ms) levels have been suggested to be related to the primary photochemical reactions and to the heterogeneity in the filling up of the plastoquinone pool, respectively. The P level corresponds to the state where all the plastoquinone molecules are in their reduced state.

Figure 1 shows that after exposing the leaves to N_2 in the dark the pattern of the fast Chl *a* fluorescence induction transient changes dramatically. Compared to the control leaves kept in air, which exhibited a typical O-J-I-P induction kinetic, the leaves exposed to N_2 showed a quick rise from O to J. The rise in the J level increased with increasing duration of the N_2 treatment. After a 120 min exposure of the leaves to N_2 the Chl *a* transient was restricted to an O-J (J=I=P) phase and resembled that observed for pea leaves poisoned with DCMU (5) which is known to block the electron transport beyond the primary quinone electron acceptor Q_A. The results suggest that anaerobiosis induces an inhibition of the electron transport after Q_A which resembles that imposed by DCMU.

However, contrary to a DCMU treatment the effect of anaerobiosis can be reverted simply by replacing the leaves to a normal atmosphere. Indeed, as shown in Fig. 1B, after two hours recovery in the air in the dark the Chl *a* fluorescence transients returned to the typical O-J-I-P pattern of untreated leaves.

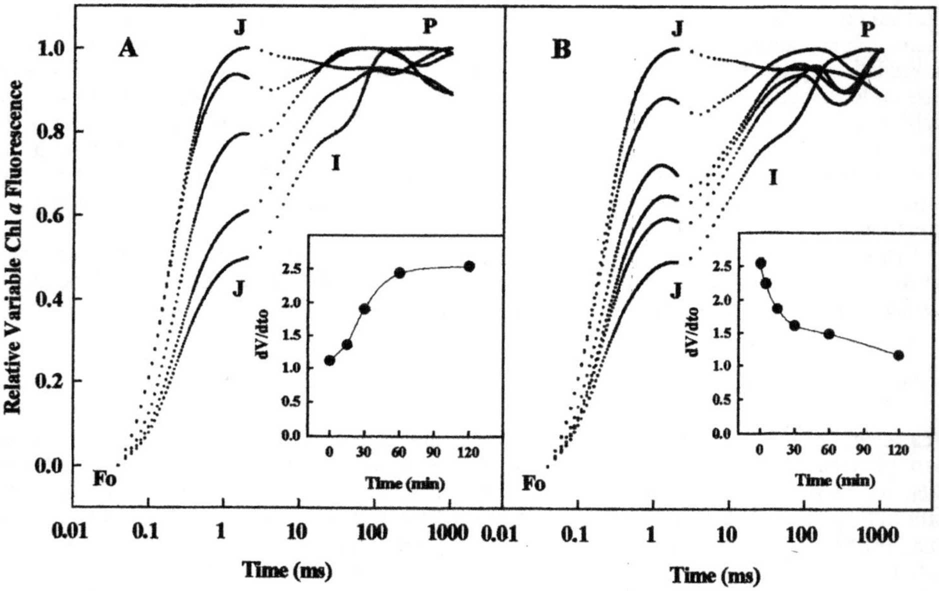

Fig. 1 Changes in the relative variable Chl *a* fluorescence $V_t = (F_t - F_0)/(F_m - F_0)$ of dark-adapted pea leaves (A) exposed to anaerobic conditions for 0, 15, 30, 60, and 120 min (from bottom to top) and (B) of pea leaves exposed to anaerobiosis for 120 min and then transferred to the air in the dark for 0, 5, 15, 30, 60, and 120 min (from top to bottom). Chl *a* fluorescence was induced by illuminating the leaves for 1 s at a light intensity of 600 W m^{-2}. The inserts show (A) the time course of the effect of anaerobiosis and (B) the time course of the recovery from anaerobiosis (120 min in N_2) in the air in the dark on the slope at the origin of the relative variable fluorescence. $dV/dto = (F_{300\mu s} - F_{100\mu s})/(F_m - F_{50\mu s})$.

The relative variable Chl *a* fluorescence is calculated as $V_t = (F_t - F_0)/(F_m - F_0)$ and can be taken as a measure of the fraction of Q_A being in its reduced state. The expression dV/dto = $(F_{300\mu s} - F_{100\mu s})/(F_m - F_{50\mu s})$ represents the variation of V at the beginning of the transient and is a measure of the relative rate of photochemistry (5). The insert in Fig. 1A shows that dV/dto increased with increasing duration of the N_2 treatment and that it almost leveled off after 60 min. After the transfer of the leaves to the air, dV/dto progressively returned back to the control value (Fig. 1B). These results further points to a reversible, anaerobiosis-induced inhibition of electron transport beyond Q_A.
It is possible that under anaerobic conditions there is an accumulation of a reductant which interacts with components of the electron transport chain leading to an anaerobic-dependent inhibition of the electron transport. One possibility would be that in the absence of O_2 there is an accumulation of reduced plastoquinone. An anaerobic-dependent increase in the Chl *a* fluorescence yield has been measured in spinach leaves and has been related to a dark reduction of the plastoquinone pool and its relaxation by reoxidation (4). It has been proposed that reduced pyridine nucleotide or an extrachloroplastic reductant could be involved in the dark reduction of the plastoquinone pool (4).

3.2 Anaerobiosis in the light

During anaerobiosis in the dark there was only a small decrease in the quantum yield of the primary photochemistry as measured by the ratio of variable to maximum fluorescence (F_v/F_m) (data not shown). In contrast, during anaerobiosis in the light the F_v/F_m ratio decreased progressively with increasing time. A 40% decrease in the F_v/F_m ratio was observed in the leaves kept in N_2 for 120 min as compared to the control leaves, indicating that the plants suffered from photoinhibition. The 120 min N_2 treated leaves recovered in their F_v/F_m ratio up to 90% of the control leaves after a recovery period of 120 min in the air in the dark.
Figure 2A shows the relative variable Chl *a* fluorescence transients (40 μs to 1 s) of 2 min dark-adapted pea leaves incubated in N_2 in the light for different periods of time. Anaerobiosis induced an increase in the J level and the I level was changed towards a longer time period and a lower intensity as compared to the dark-adapted control kept in air. Leaves illuminated in the air also showed an increase in the J level when the Chl *a* fluorescence transients were measured after a 2 min dark-adaptation, but in these leaves no significant changes appeared in the I level (data not shown). When the N_2 incubated leaves where transferred to the air in the dark, the J level reverted to the normal, while the I level stayed below that of untreated leaves, even after 60 min recovery (Fig. 2B). The inserts in Fig. 2 show that the primary rate of Q_A reduction (dV/dto) increased during the N_2 treatment (A), and decreased again after transferring the leaves to the air in the dark (B).
It has been suggested that anaerobiosis in the light lead to the accumulation of different forms of stabilized Q_A^- whith typical reoxidation kinetics in the dark ranging from seconds up to hours (6,7). Thus, it is possible that the reversible, anaerobiosis-dependent increase in the J level and in dV/dto could be due to the accumulation of reduced Q_A that cannot rapidly transfer the electrons to the second quinone electron acceptor Q_B. The slower filling up of the plastoquinone pool, as indicated by the change in the I level, may then be related to the accumulation of a population of reaction centers in which the rapid electron transfer from Q_A to Q_B is inhibited. Recovery in the air in the dark may be associated with a reestablishment of the electron transfer from Q_A to Q_B. The finding that the I level did not recover completely even after a prolonged period of recovery in the air in the dark might indicate that anaerobiosis in the light induces changes which are slowly reversible. Such changes might be related to protein conformational changes or to the formation and accumulation of modified photosystem II reaction centers (6,7).

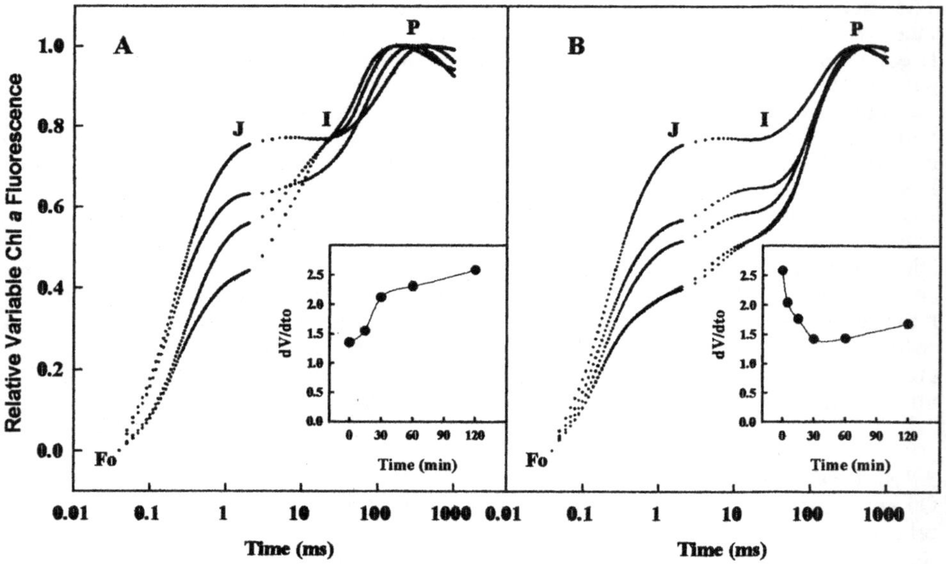

Fig. 2 Changes in the relative variable Chl *a* fluorescence $V_t = (F_t-F_o)/(F_m-F_o)$ of pea leaves (A) exposed to anaerobic conditions under a photon flux density of 90 µmol m^{-2} s^{-1} for 0, 15, 60, and 120 min (from bottom to top) and (B) of pea leaves exposed to anaerobiosis in the light for 120 min and then allowed to recover in the air in the dark for 0, 5, 15, 30, and 60 min (from top to bottom). Chl *a* fluorescence was induced by illuminating the leaves for 1 s at a light intensity of 600 W m^{-2} after a dark adaptation of 2 min. The inserts show (A) the time course of the effect of anaerobiosis and (B) the time course of the recovery from anaerobiosis (120 min in N_2) in the air in the dark on the slope at the origin of the relative variable fluorescence. $dV/dto = (F_{300\mu s} - F_{100\mu s})/(F_m - F_{50\mu s})$.

5. References

1 Krause, G.H. and Cornic, G. (1987) in Topics in Photosynthesis (Kyle, D.J., Osmond,C.B., and Arntzen, C.J., eds.) vol 9, pp.169-196, Elsevier, Amsterdam
2 Sundby, C. and Schiött, T. (1992) Photosynth. Res. 33: 195-202
3 Gong, H., Nilsen, S. and Allen, J.F. (1993) Biochim. Biophys. Acta 1142: 115-122
4 Harris, G.C. and Heber, U. (1993) Plant Physiol. 101: 1169-1173
5 Strasser, R.J., Srivastava, A. and Govindjee (1995) Photochem. Photobiol. 61: 32-42
6 Vass, I., Styring, S., Hundal, T., Koivuniemi, A., Aro, E.-M. and Andersson, B.(1992) Proc. Natl. Acad. Sci. USA 89: 1408-1412
7 Vass, I. and Styring, S. (1993) Biochemistry 32: 3334-3341

THE FAST FLUORESCENCE TRANSIENT OF *RHODOSPIRILLUM RUBRUM* IS POLYPHASIC OF THE TYPE O-K-J-I-P

Reto J. Strasser and Robin Ghosh
Laboratory of Bioenergetics, University of Geneva,
CH-1254, Jussy-Geneva, SWITZERLAND

1. Introduction

All phototrophic organisms emit chlorophyll *a* fluorescence while they are exposed to photosynthetically active light. It has been found that the fluorescence yield of dark adapted plants[1] or photosynthetic bacteria[2] is not constant upon illumination. This so-called variable fluorescence has been related to the primary photochemistry[3]. Its measurement *in vivo* at room or low temperature becomes a powerful tool to analyse the primary reactions of photosynthesis in higher plants and algae. Only a few reports exist about the fluorescence transients of photosynthetic bacteria. However, the main features of such transients have been reported[4] as early as for plants: Fluorescence intensity increases from a low initial value F_0 to a maximum F_P (fast fluorescence rise) and decreases thereafter to a stationary state F_S (slow decay), very similar in principal as in plants or oxygenic systems. Several features seem to be reflected in a full fluorescence transient which depends highly on the age and physiological state of the cells [5]. Some traces showed[6] that the fluorescence rise in bacteria can be polyphasic. The link of the primary photochemical redox reactions with the appearance of variable fluorescence has been established for bacteria[7] in a similar way as for plants. Many similarities between the observed fluorescence transients in photosynthetic bacteria, plants and cyanophyceae have been reported[8]. As in plants, the slow fluorescence decay from F_P to F_S is considered as reflecting conformational changes due to the adaptation of the sample to continuous light, while the fast fluorescence rise reflects the dynamics of several redox reactions leading to an intermediary stationary state at F_P. Recently[9] it has been observed with modulated fluorescence techniques an intermediate step between the initial F_0 and the maximal F_P fluorescence. Using fast non-modulated optoelectronic techniques we were able to detect several intermediate steps in the fast fluorescence rise between F_0 and F_P of *Rhodospirillum rubrum*[10]. Here we show for the first time traces of the fast fluorescence transients of photosynthetic bacteria with a high time resolution. The photochemical phase, the following redox phases and the adaptation phase (slow decay) to continuous light can well be separated. Due to the fact that new instruments with higher time resolution allow the detection of faster intermediate steps we introduced a terminology in alphabetical order for each step between F_P and F_0 as I, J, K etc.(from longer to shorter time) in oxygenic systems[11]. Due to the similarity of the fluorescence transient in photosynthetic bacteria with oxygenic systems we propose to use the same nomenclature. However, we note that the nomenclature e.g. O-K-J-I-P indicates only a succession of phenomenological steps in the fast fluorescence transient. Therefore, the molecular mechanistics behind these labelled steps have not to be the same for oxygenic systems (plants and cyanophyceae) and photosynthetic non-oxygenic bacteria.

P. Mathis (ed.), Photosynthesis: from Light to Biosphere, Vol. II, 915–918.

2. Material and Methods

Rhodospirillum rubrum S1 cells were grown anaerobically in M-medium photoheterotrophically with succinate as carbon source as described earlier[12].

A fiber optics device has been built into a growth bottle for bacteria under sterile conditions. Fast fluorescence measurements were done with 10 µs resolution from 40 µs to 10 s. The fluorescence was measured with 12 bits resolution and excited by 6 LEDs providing an intensity of 600 Wm^{-2} red (630 nm) light at the entrance of the fiber optics which guide lead the light from outside into the anaerobically kept bacteria suspension. The fluorescence was measured at 890 nm by a modified commercial fluorimeter (Plant Efficiency Analyzer, PEA, built by Hansatech instruments Ltd. King's Lynn Norfolk PE32 1JL, GB). A metallic plate spaced 2 mm from the fiber optics inside the cell culture guarantees that the light path was constant for all measurements. In this way real *fermenter-endoscopy* could be performed by daily measurements at any time without opening the anaerobic culture chamber.

3. Results and Discussion

On the following page the fluorescence transients of *R. rubrum* and pea leaves (insert) are shown on a logarithmic (top) and linear (bottom) time scale. Each sample was dark adapted for 1 hour and measured with the PEA instrument (crosses) during 1 (leaves) or 10 (bacteria) seconds. After a dark period of 10 seconds the same samples were measured again (open circles). The second illumination has an opposite effect on bacteria (lowering the fluorescence yield) than on plants (increased fluorescence yield). However, the dynamic characteristics of the full transient of plants and bacteria are very similar. A sequence of steps in the transient can be followed and labeled as O-K-J-I-P. The O to J transition has been assigned to the photochemical reactions. This is the only step which can be measured (data not shown) at 4°C or 77K. During the culture growth the fluorescence intensity at the K-level (up to 1 ms) is a useful measure to estimate the biomass of the culture. However, the shape of the transient - characterized by the fluorescence intensities at J (about 2 ms), I (about 200 ms) and P (about 2 s) - differ widely, depending on the physiological state and the growth conditions of the cells.

The fast phase O to J reflects the photochemical steps. It shows an exponential rise (under high light conditions) or a sigmoidal rise (under low light conditions). Consequently, the area growth $S = \int (F_{50ms} - F_t)\,dt$ versus $V_{(J)} = (F_t - F_{1ms}) / (F_{50ms} - F_{1ms})$ is a straight line for energetically separated antenna systems or a hyperbolic function for energetically grouped (connected) photosynthetic units.

Fig. 3 shows such a relationship experimentally for a bacteria culture grown under low light, which was pretreated 5 times (crosses) or 10 times (circles) with 10 seconds actinic light - 10 seconds dark intervals. A well pronounced sigmoidicity in the fluorescence transient V(t) and a hyperbolic-like behavior of the plot V_t versus S_t can be seen (Fig. 2).

4. Conclusion

We conclude: (1) The polyphasic chlorophyll *a* fluorescence rise O-K-J-I-P reflects the energetic dynamics of a chain of redox reactions independent of whether the system is oxygenic or not. (2) Fermenter-endoscopy is possible with photosynthetic bacteria. On line spectroscopy allows the biophysical description during growth. (3) Different types of behaviors of different mutants can be screened.

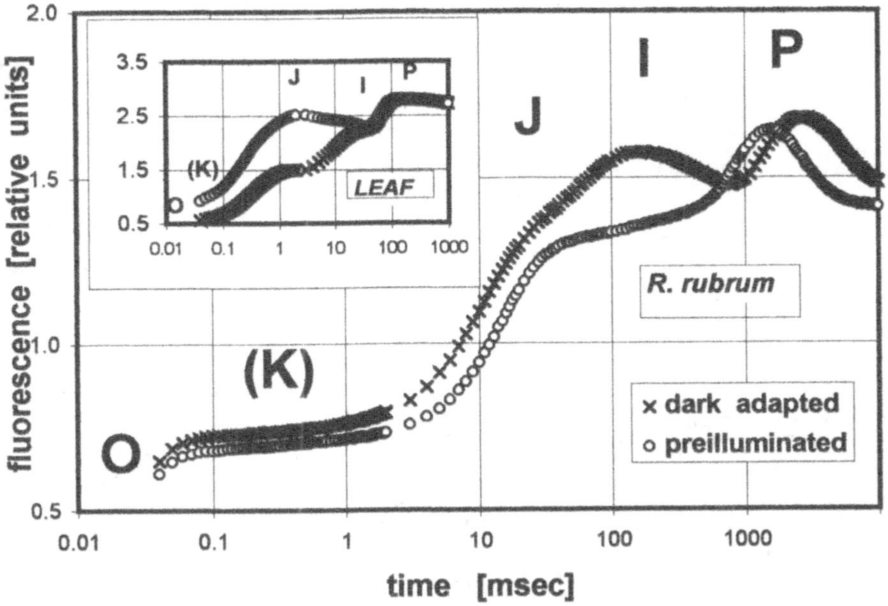
fluorescence [relative units]
time [msec]
O
(K)
J
I
P
R. rubrum
× dark adapted
○ preilluminated
LEAF
2.0
1.5
1.0
0.5
0.01
0.1
1
10
100
1000
3.5
2.5
1.5
0.5

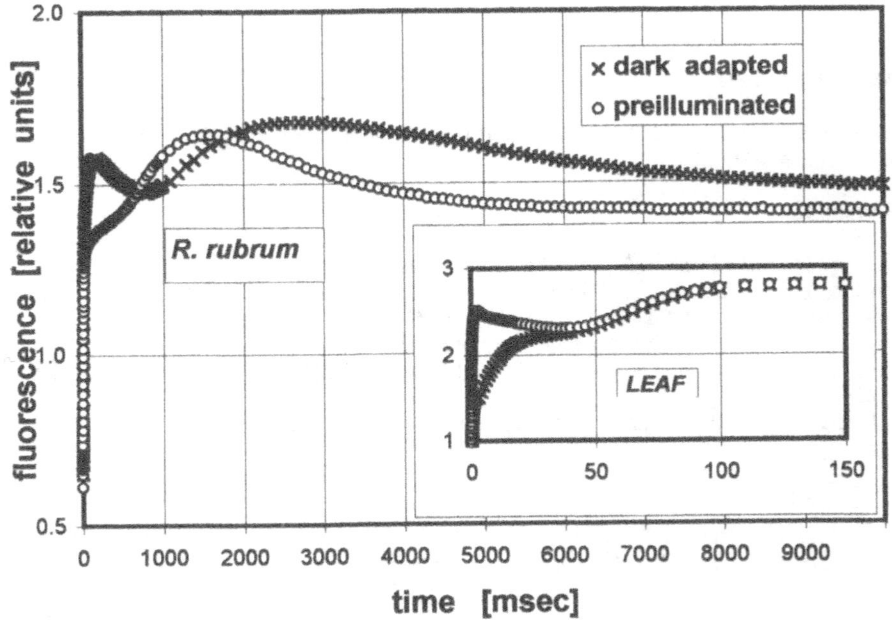
fluorescence [relative units]
time [msec]
× dark adapted
○ preilluminated
R. rubrum
LEAF
2.0
1.5
1.0
0.5
0
1000
2000
3000
4000
5000
6000
7000
8000
9000
3
2
1
50
100
150

Fig.3

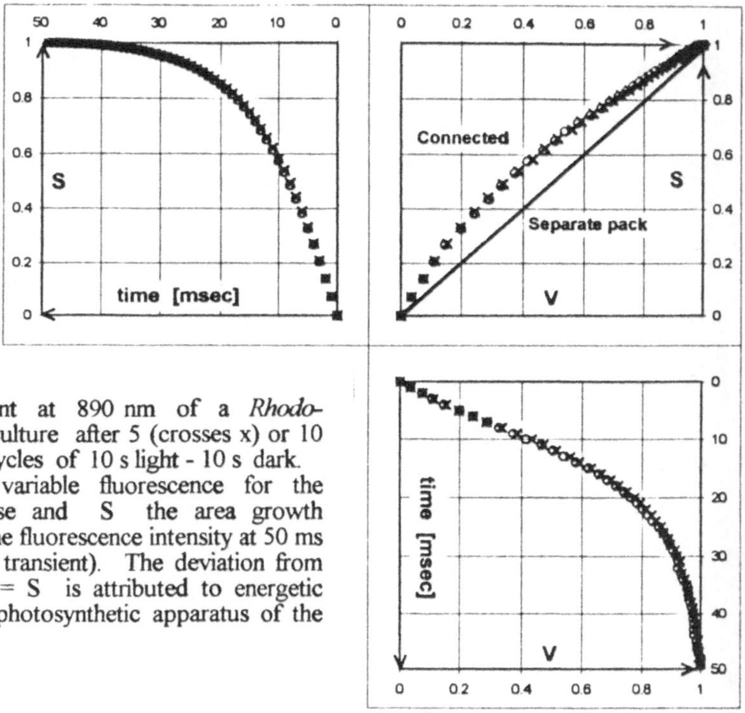

Fluorescence transient at 890 nm of a *Rhodospirillum rubrum* culture after 5 (crosses x) or 10 (circles) light-dark cycles of 10 s light - 10 s dark. V is the relative variable fluorescence for the photochemical phase and S the area growth (measured between the fluorescence intensity at 50 ms and the fluorescence transient). The deviation from the straight line V = S is attributed to energetic cooperativity in the photosynthetic apparatus of the bacteria.

4. Acknowledgement

This work was supported by Swiss National Foundation grants # 31.33678.92, # 5002-39816, # 5002-041801/1

5. References

1. Kautsky, H. and Hirsch, A. (1931) Naturwissenschaften 19: 964.
2. Wassink, F.C., Katz, E. and Dorrstein R. (1942) Enzymologia 10: 285-354.
3. Vredenberg, W.J. and Duysens, L.M.N. (1963) Nature 197: 355-357.
4. Clayton, R.K. (1966) Photochem. Photobiol 5, 679.
5. De Klerk, H., Govindjee, Kamen, M.D. and Lavorel, J. (1969) Proc. Natl. Acad. Sci. USA 62: 972-978.
6. Hoffman, W. and Metzner, H. (1971) 2d Int. Congress on Photosynthesis Stresa Vol I pp 369-371.
7. Duysens, L.M.N. (1986) In: Govindjee, J. Amesz and D.C. Fork (eds) Acad. press New York pp 13-16, pp 638.
8. Setlik, I., Waldburger-Schlapp, M. and Bachofen, R. (1989) In: Baltscheffsky, M. (ed) Current Research in Photosynthetis, Vol IV 17.341-17.344.
9. Joss, A., Mez, K., Känel, B., Hanselmann, K.W. and Bachhofen, R. (1994) J. Plant Physiol. 144: 333-338.
10. Strasser, R.J. and Ghosh, R. (1995) Experientia 51.
11. Strasser, R.J., Srivastava, A. and Govindjee (1995) Photochem. Photobiol. 61:32-42.
12. Ghosh, R., A. Hardmeyer, I. Thoener and R. Bachofer (1994) Applied and Environmental Microbiology, 60, 1698-1700.

NUMERICAL SIMULATION OF CHLOROPHYLL A FLUORESCENCE INDUCTION IN PLANTS

Alexandrina D.Stirbet[1,3], Govindjee[2], Bruno J.Strasser[1] and Reto J.Strasser[1]
[1]University of Geneva, 1254-Jussy, Geneva, SWITZERLAND;
[2]University of Illinois, Morril Hall, Urbana, IL 61801, USA;
[3]University of Bucharest, R-76900 Bucharest, ROMANIA

1. Introduction

Chlorophyll *a* fluorescence emitted by green plants reflects photosynthetic activity in a complex manner [1,2]. Illumination with continuous actinic light results in characteristic polyphasic time-course of the fluorescence yield, a phenomenon known as fluorescence induction or transient [2-4]. The fast phase of fluorescence induction, observed during the first second of illumination, has been the subject of theoretical treatment [5-9]. In this paper, we present simulation results obtained with a model centred on redox reactions of acceptor and donor sides of PSII (see Fig.1).

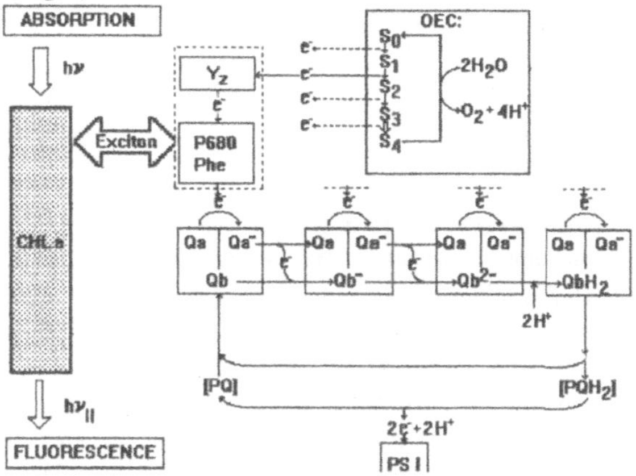

Fig. 1. Scheme of electron transport reactions considered in the model. CHL a: chlorophyll *a* antenna; OEC: oxygen evolving complex in S_n states; Y_Z: primary electron donor to $P680^+$; P680: reaction centre of PSII; Phe:pheophytin *a*;Q_A: primary bound plastoquinone; Q_B: secondary bound plastoquinone; PQ: plastoquinone from pool; PSI: photosystem I.

2. The model

The main assumptions of the model are: (1) Chl *a* variable fluorescence is related mainly to PSII kinetics [2]; (2) the net rate of Q_A reduction is higher in S_0 and S_1 than in S_2 and S_3 states of OEC [4]; (3) the relative variable fluorescence $V_t = (F_t-F_0)/(F_M-F_0)$ depends on the fraction of closed reaction centres $B(=[Q_A^-]/[Q_A]_{tot})$; the relation used is $V_t = B/(1+C(1-B))$, with $C = p_S F_V/F_0 = p_J/(1-p_J)$ (where p_S and p_J are the probabilities of connectivity between PSII units, based on the formula of Strasser (S) and Joliot (J) [10,11]); and (4) Chl *a* fluorescence is quenched by oxidised PQ [12]. For this type of quenching, the values of V_t were corrected by using Stern-Volmer relationship.

P. Mathis (ed.), Photosynthesis: from Light to Biosphere, Vol. II, 919–922.

The ordinary differential equations system, associated with the reactions considered in the model, was solved numerically with a specialised simulation software Gepasi [13]. The input parameters in the simulation program are the initial concentrations of reactants and the rate constants of reactions. The values of these parameters used in simulations, and the corresponding values from literature, are shown in Table 1.

Table 1. The values of input parameters used in simulations, and the corresponding values from literature (k is the rate constant of forward reaction; $t_{½}$ is the half time; and K_{eq} is the equilibrium constant).

Input parameters	*In simulations:* [1]*dark adapted*	[2]*light adapted*	*From literature*	
Number of molecules in PQ pool	5	5	5-10	[2]
$[S_1Q_AQ_B]_i:[S_0Q_AQ_B^-]_i$	0.85:0.15	0.8:0.2	0.87:0.13 (pH=7.5)	[19]
k (*or* $t_{½}$); K_{eq} *for* $Q_A^-Q_B \leftrightarrow Q_AQ_B^-$	3000 s^{-1}(230 μs);50	3000 s^{-1}(230 μs);50	100-200 μs; 20	[18]
k (*or* $t_{½}$); K_{eq} *for* $Q_A^-Q_B^- \leftrightarrow Q_AQ_B^{2-}$	1500 s^{-1}(460 μs);10	1500 s^{-1}(460 μs); 1	300-500 μs; 1-50	[18]
k (*or* $t_{½}$); K_{eq} *for* $Q_B^{2-} \leftrightarrow Q_BH_2$	1300 s^{-1}(530 μs);10	1300 s^{-1}(530 μs); 1	1300 s^{-1};10 or 1	[20]
k (*or* $t_{½}$); K_{eq} *for* $Q_BH_2 + PQ \leftrightarrow Q_B + PQH_2$	100 s^{-1}(1.3 ms); 1	100 s^{-1}(1.3 ms); 1	*1 ms; 1	[18]
k (*or* $t_{½}$) *for* $PQH_2 \rightarrow PQ$	70 s^{-1}(10 ms)	200 s^{-1} (3.5 ms)	2-5 ms/e^-	[18]
k (*or* $t_{½}$) *for* $S_3 \Rightarrow S_4 \rightarrow S_0$	700 s^{-1}(1 ms)	500 s^{-1}(1.3 ms)	1-1.3 ms 2 ms	[19] [18]

[1] leaves dark adapted 10 min before the measurement [14];
[2] leaves illuminated for 1h, and then dark adapted for 30 sec before the measurement [14];
*the half time of this second order reaction is: $t_{½} = \ln 2/(k^*[PQ]_i)$

The rate constants of Q_A reduction are not simple. They depend, among other parameters, on: the light intensity, cross section of the antenna of PS II units, the rate constants of the primary charge separation, and of the electron transfer from Phe to Q_A. Also, the reverse reactions represent the complex process of charge recombination. The rate constant value of Q_A reduction when OEC is in S_1 state (2000 s^{-1}) has been obtained empirically from DCMU simulated curves, knowing that in experiments V_{max} value is reached in 2 ms (for I = 600 W/m^2) [14].

3. Results and discussion

Simulated results are shown in Fig.2. The curves shown in Fig. 2A are obtained from simulations with input parameters presented in Table 1 for dark adapted plants. Curves 2-5 show the changes in time of the reaction centre concentrations of complexes with Q_A

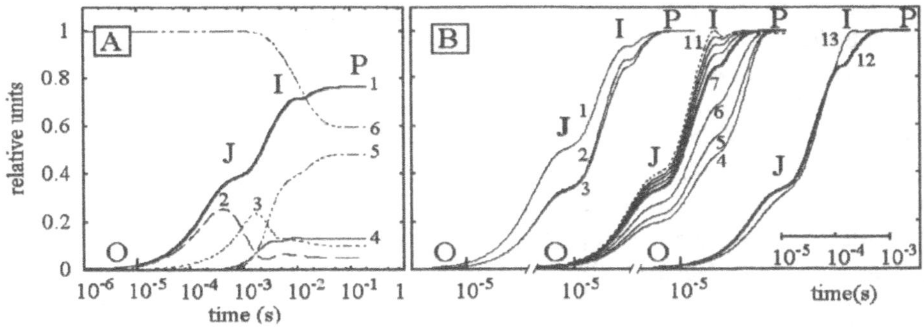

Fig. 2. **A** : typical curves obtained with input parameters for dark adapted condition presented in Table 1. Curves **1-6** ⇨ time variations of the concentrations of $[Q_A^-]_t$, $Q_A^-Q_B$, $Q_A^-Q_B^-$, $Q_A^-Q_B^{2-}$, $Q_A^-Q_BH_2$ and PQ/[PQ]$_{tot}$ respectively. **B** : Curves **1-3** ⇨ time variation of closed RCs, $[Q_A^-]_t$, the relative variable fluorescence V_t, and the relative variable fluorescence corrected for PQ quenching respectively. Curves **4-11** ⇨ time variation of the relative variable fluorescence, corrected for PQ quenching, for different values of the rate constant of oxidation of PQH_2 (0, 10, 30, 70, 100, 130, 170 and 300 s^{-1} respectively). Curves **12** and **13** ⇨ time variation of the variable fluorescence, corrected for PQ quenching, for dark adapted and light adapted values of input parameters respectively.

in its reduced form: $Q_A^-Q_B$, $Q_A^-Q_B^-$, $Q_A^-Q_B^{2-}$, and $Q_A^-Q_BH_2$ respectively. Curve 1 is the sum of these four redox forms of RCs ($[Q_A^-]_t$). We found that $Q_A^-Q_B$ is predominant at the J step. At the I and P steps all four forms of closed RCs are found in a mixture, $Q_A^-Q_BH_2$ being predominant at the P level. The PQ pool (curve 6) is reduced mostly in the I-P interval.

The effect of connectivity between PS II units, and nonphotochemical quenching by PQ, are presented in Fig.2B for dark adapted condition (curves 1, 2 and 3). The relative variable fluorescence was calculated with C = 1, corresponding to p_S = .25 and F_V/F_0 = 4 (or p_J = .5). It can be seen that PQ quenching lowers the I level.

The model presented here, in contrast to the earlier ones presented by Stirbet and Strasser [15,16], can simulate the dip after I (Fig.2B, curves 4-11), and after J (results not shown). The dip after I becomes deeper as the rate constant of plastoquinol oxidation increases (the PS I activity is enhanced). These results confirm the earlier concepts [17] that this dip is related to the dynamic equilibrium between the two photosystems. An increased number of PQ molecules in the pool also accentuates the dip (results not shown).

Curves 12 and 13 are the simulated transients obtained in dark adapted and light adapted conditions respectively. We have assumed that in light adapted condition the connectivity between PS II units is smaller (C = .25, corresponding to p_S = .125 and F_V/F_0 = 2 (or p_J = .2)). Moreover, we have used a lower value for the rate constant of Q_A reduction (1000 s^{-1} instead of 2000 s^{-1}), as the PSII absorption cross section is assumed to be reduced. In simulated curve for light adapted condition (curve 13 in Fig.2B) V_{max} level is obtained at I level, just as in the experimental data [14].

4. Concluding Remarks

In this model of the fast phase of Chl *a* fluorescence induction in plants, the rate constants of the reactions from the RC level are assumed to be dependent not only on the redox state of Q_A, but also on the S_n states of OEC. The shapes of the fluorescence transients have been simulated quite well, with the characteristic steps O, J, I, and P, as well as the dips. The dip after I is partially due to the S_n states, but is also correlated with the number of PQ molecules in the pool and the equilibrium dynamics between the two photosystems, being more accentuated as the number of PQ molecules is higher, and the activity of PSI is increased. The PQ nonphotochemical quenching influences the shape of the curves, decreasing I level. The simulated curves of fluorescence induction of the light adapted plants exhibit, just as in the experimental ones, only the O, J, and I levels.

We plan to apply this model to the analysis of experiments with different light intensities, preillumination, addition of inhibitors and other stress conditions. We hope to verify and predict dynamic parameters of the photosystems in photosynthesis through this non-invasive probe of chlorophyll *a* fluorescence.

Acknowledgement: Swiss National Fund (grant 31.33678.92) to R.J.Strasser.

1. Lavorel,J., and Etienne,A-L. (1977) Topics in Photosynthesis 2, 203-268
2. Govindjee (1995) Aust.J.Plant Physiol. 22, 131-160
3. Kautsky,H., and Hirsch,A (1931), Naturwissenschaften 48, 964
4. Dau,H. (1994) Photochem.Photobiol. 60, 1-23
5. Renger,G., and Schultze,A. (1985) Photobiochem.Photobiophys. 9, 79-87
6. Baake,E., and Strasser,R.J. (1990) in: Current Research in Photosynthesis (ed. Baltscheffsky, M.), Vol. 1, pp. 567-570, Kluwer Academic Publishers, Netherlands
7. Hsu,B.D. (1992) Biochim.Biophys.Acta 1140, 30-36
8. Hsu,B.D. (1993) Photosynthesis.Res. 36, 81-88
9. Trissl,H.-W., and Lavergne,J. (1995) Aust.J.Plant Physiol. 22, 183-193
10. Strasser,R.J. (1981) in: Structure and Molecular Organisation of the Photosynthetic Apparatus (Akoyunoglou,G., ed.), pp. 727-737, Philadelphia, PA
11. Joliot,A., and Joliot,P.(1964) C.R. Acad.Sci.Paris Ser. D 258, 4622-4625
12. Vernotte,C., Etienne,A.-L., and Briantais,J.-M. (1979) Biochim.Biophys.Acta 545, 519-527
13. Gepasi is a software package in the public domain (available from Internet). Author: Pedro Mendes, Dept.Biol.Sci, Univ.Wales, Aberystwyth, Dyfed, SY23 3DA, UK
14. Strasser,R.J., Srivastava,A., and Govindjee (1995) Photochem.Photobiol. 61, 32-42
15. Stirbet,A.D., and Strasser,R.J. (1995) Arch.Sci.Geneve 48, 41-60
16. Stirbet,A.D., and Strasser,R.J. (1995) Proceedings of the 1st International Symposium on mathematical modelling and simulation in agriculture & bio-industries, Brussels
17. Munday,J.C.M. Jr., and Govindjee (1969) Biophys.J. 9, 1-21
18. Crofts,A.R., and Wraight,C.A. (1983) Biochim.Biophys.Acta 726, 149-185
19. Haumann,M., and Junge,W. (1994) Biochemistry 33, 864-872
20. McPherson,P.H., Okamura,M.Y., and Feher,G. (1993) Biochim.Biophys.Acta 1144, 309-324

STUDIES ON CYCLIC ELECTRON FLOW INDUCED BY MIGRATION OF CYTOCHROME B_6/F COMPLEXES DURING STATE TRANSITIONS

D. Braun, V. Bolle, W. Rühle and A. Wild
Institute of General Botany, Johannes Gutenberg-University, 55099 Mainz, Germany

1.Introduction

Excess of PS II light results in processes called state 2 transition. This includes phosphorylation and migration of LHC II from grana- to stroma-membranes [1]. Some authors [2, 3] have observed that also cytochrome b_6f-complexes migrate into the stroma-thylakoids during state 2. This may be accompanied by an increase of cyclic electron flow [3]. We tried to prove this aspect by following the kinetics of cytochrome f oxidation and reduction with PS I light in stroma thylakoids with dual wavelength spectrometry. First we studied comparatively two methods of stroma-membrane purification. Isolated thylakoids were treated with Yeda-Press [4] or different Triton concentrations [5]. Secondly, we tried to measure the ferredoxin catalyzed cyclic electron flow in phosphorylated and non-phosphorylated thylakoid membranes by 9-Aminoacridine fluorescence quench.

2.Material and Methods

Thylakoids were isolated from spinach (*Spinacia oleracea*) according to a modified method of Jensen and Bassham [6] and plastocyanin as well as ferredoxin by a modified method of Ratajczak et al. [7].
Stroma-membranes were isolated from thylakoids 1.) by Yeda-Press treatment [4] or 2.) by Triton incubation [5] and then submitted to differential centrifugation. Vesicles pelleted at 40,000g for 30 min. are called Y40 or T40, respectively. The so called Y100 or T100, respectively, were sedimented at 100,000g for 60 min.. The Y40 and T40 include the grana-enriched fractions and the Y100 and T100 pure stroma-membranes.
Phosphorylation of thylakoids was induced in the presence of 0.2 mM ATP during 10 min. illumination with 6 W m^{-2} 660 nm light. Dephosphorylation was inhibited by addition of 10 mM NaF [2].
Kinetics of cytochrome f were measured by dual wavelength spectroscopy with a Sigma ZWS 11 spectrophotometer equipped with multibranched fiber optics according to Rühle et al. [8]. The membrane suspension was adjusted to a chlorophyll concentration of

P. Mathis (ed.), Photosynthesis: from Light to Biosphere, Vol. II, 923–926.

1 mg ml^{-1} and measured in a cuvette with an optical pathlength of 1.3 mm with addition of 0.1 mM methylviologen as electron acceptor and 10 mM hydroquinone to reduce cytochrome f completely in the dark. Total content of cytochrome f and P700 was determined from the ferricyanide-oxidized minus hydroquinone-reduced difference spectrum after solubilization with 1 % Triton X-100. The light driven proton gradient was followed by the quench of 9-Aminoacridine fluorescence with an Aminco DW2. The suspension adjusted to 20 µg chlorophyll ml^{-1}, was illuminated with light 1 > 695 nm. Cyclic electron flow was induced by addition of 1µM ferredoxin and inhibited by 8 µM antimycin A.

3.Results and Discussion

Tab 1 shows the Chl a/b and Cyt f/Chl ratios of thylakoids, phosphorylated and non-phosphorylated stroma-membranes. In purified stroma-lamellae the Chl a/b ratio is much higher than in thylakoids, the Cyt f/Chl ratio is lower. Compared with non-phosphorylated Y100, there is a decrease in the Chl a/b ratio and an increase in the Cyt f/Chl ratio in phosphorylated stroma-membranes. This decrease results in a migration of both LHC II and cytochrome b_6f-complexes into the stroma-thylakoids by phosphorylation during state 2 transitions. P700 is highly enriched in the stroma membranes, so that the amount of the PS I complexes exceeds that of the cytochrome b_6f complexes by a factor of 5 in non-phosphorylated and a factor of 4 in phosphorylated vesicles.
There is a similar decrease in Chl a/b for Y40 and T40 and the same increase for Y100 and T100. This indicates an equal grade of purification for both methods of preparing stroma membranes.

	thylakoids	Y100 non-phosph.	Y100 phosph.
Chl a/b	3.13	7.3	6.4 ***
Cytf/Chl [mmol/mol]	1.54	1.07	1.28 ***
P700/Chl [mmol/mol]	2.3	5.3	5.2 ***

Tab1: Pigment , cytochrome f and P700 contents in thylakoids and stroma membranes (Y100) prepared from non-phosphorylated and phosphorylated thylakoids. ***The differences between phosphorylated and non-phosphorylated membranes are highly significant ($p<0.001$).

The amplitudes of light 1 (700 nm) induced cytochrome f oxidation is depicted in Fig.1. In thylakoids all of the cytochrome f can be oxidized, but in the Yeda-Press treated vesicles Y40 only 54 % and in the Y100 only 49 % of the cytochrome f is oxidizable. This loss in interaction between the cytochrome b_6f complex and PS I can be due to a loss in plastocyanin. Addition of isolated plastocyanin (0.75 µM) restored some of the signal in the Y40 but not in the Y100. This can be interpreted by the formation of inside-out vesicles in the Y40 which are accessible to the externally added plastocyanin. The right-side-out vesicles of the Y100 are not affected by external plastocyanin addition. When 0.75 µM plastocyanin is present during the Yeda-Press treatment, it is able to prevent the loss in Y100 completely. The use of the detergent Triton X-100 to prepare stroma membranes leads also to a loss in interaction between cytochrome b_6f complex and PS I in T40 as well as in T100, when the detergent concentration is not optimal. In an optimal

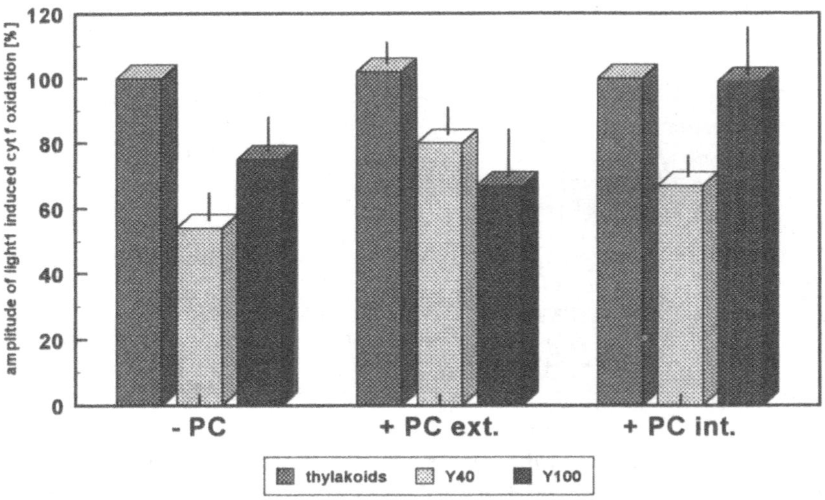

Fig.1: Amplitude of cytochrome f oxidation of vesicles produced by Yeda-Press treatment without, with external and internal plastocyanin addition; grana-enriched membranes are called Y40, stroma membranes Y100. The values are expressed as [%] of the cytochrome content of each membrane type determined by the ferricyanide-oxidized minus hydroquinone-reduced difference spectrum.

preparation all cytochrome f is oxidizable, but increasing the Triton concentration to 1.6 % results in a 50 % loss of the signal in T100 as well as in T40. The Triton method has the advantage that no plastocyanin is lost, but it may have a drawback in establishing a ferredoxin linked cyclic electron flow.

Fig.2 shows the 9-AA fluorescence quenching by thylakoids illuminated with light1 (> 695 nm). Without any addition a proton gradient is built up by the linear pseudocyclic type of electron flow resulting from a small part of PS II excitation by the short wavelength portion of the 695 nm light. This proton gradient is increased by the addition of ferredoxin, which enables an additional cyclic electron flow. Antimycin A is able to inhibit this enhancement almost completely. The post-illumination proton efflux occurred with a half-time of 31.5 ± 8.4 sec in non-phosphorylated and of 23.7 ± 7.5 sec in phosphorylated thylakoids. This uncoupling effect is highly significant ($p<0.001$). While the pseudocyclic electron transport is saturated with the highest light intensity used (28 W m^{-2}), the enhancement by ferredoxin is independent up to the lowest intensity (1.6 W m^{-2}). This shows that the cyclic electron flow is light saturated within the tested light interval. Phosphorylated thylakoids show a smaller part of the pseudocyclic proton gradient with each light intensity and a later saturation. This may be due to the smaller antenna size of PS II in the phosphorylated thylakoids. The enhancement of the proton gradient by ferredoxin is about the same in the phosphorylated as in the non-phosphorylated thylakoids, but the cyclic part of the proton flow calculated from the initial rates of post-illumination proton efflux is accellerated by about 30% in the phosphorylated compared to the non-phosphorylated thylakoids. In view of the uncoupling effekt the question is still open if an enhanced cyclic phosphorylation really occures in the state 2.

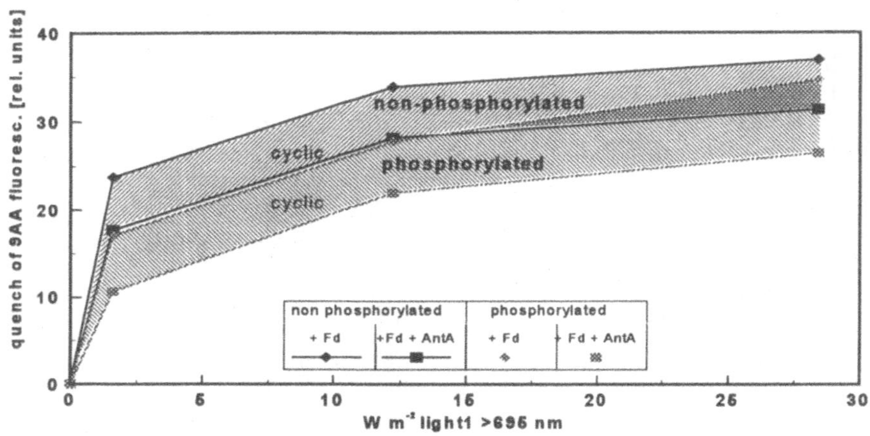

Fig 2: Amplitude of proton gradient with different light intensities in non-phosphorylated and phosphorylated thylakoids without, with 1 μM ferredoxin, and with 1μM ferredoxin + 8 μM antimycin A.

4.Summary

Stroma membranes can be prepared by Yeda-Press treatment or Triton incubation. The loss in interaction between PS I and the cytochrome b_6f complex after Yeda-Press treatment can be restored by plastocyanin addition. Nevertheless, it is not possible to establish a ferredoxin mediated electron flow in Y100 preparations. In thylakoids it is possible to demonstrate pseudocyclic and cyclic proton uptake. The post-illumination proton efflux is accellerated in phosphorylated compared non-phosphorylated thylakoids. Phosphorylated thylakoids produce a smaller proton gradient than non-phosphorylated thylakoids. Their pseudocyclic electron flow is saturated at higher light intensities. There is no difference of the cyclic part of the proton gradient, which is still saturated at low light intensities. In view of the uncoupling effect it is questionable if the cyclic phosphorylation is really enhanced in the state 2.

References

[1] Bennett J. (1991) Annu. Rev. Plant Physiol. Plant Mol. Biol. 42, 281-311
[2] Vallon O., Bulte L., Dainese P., Olive J., Bassi R. and Wollman FA (1991) Proc.Natl.Acad.Sci. USA 88, 8262-8266
[3] Anderson J.M. (1992) Photosynth.Res. 34, 341-357
[4] Anderson J.M. and Andersson B. (1982) Trends Biochem. Sci. 7, 288-292
[5] Van Leeuwen P.J., Nieveen N.C. and Van Garkom H.J. (1991) Photosynth.Res. 28, 149-153
[6] Jensen R.G. and Bassham J.A. (1966) Proc.Natl.Acad.Sci. 56, 1095-1099
[7] Ratajczak R., Mitchell R. and Haehnel W. (1988) Biochim Biophys Acta 933, 306-318
[8] Rühle W., Weinzettel A., Baur P., and Wild A. (1990) in Current Research in Photosynthesis (Baltscheffsky M., ed.) pp. II.8.827-830, Kluwer Academic Publishers, Dordrecht, The Netherlands

IN VIVO ESTIMATION OF CO_2/O_2 SPECIFICITY FACTOR OF RUBISCO AT VARYING TEMPERATURE

Jaleh Ghashghaie and Gabriel Cornic
Lab. d'Ecol. Vég., Bât. 362, Univ. Paris-Sud, 91405-Orsay, France.

1. Introduction

Measurements of chlorophyll fluorescence, together with net CO_2 uptake on intact leaves, makes it possible to estimate the photosynthetic electron transport rate *in vivo* (1, 2). It is also possible, using the same technique, to estimate the partitioning of electrons between CO_2 and O_2 reduction during photosynthesis of an intact C_3 leaf in normal air (3, 4, 5, 6, 7). This requires assumptions about the stoichiometry between the photosynthetic electron transport and the reduction of O_2 via the glycollate pathway (3, 4, 5).

The validity of the techniques used to estimate the photosynthetic electron transport on intact leaves and its partitioning between CO_2 and O_2 reduction can be tested by estimating the CO_2/O_2 specificity factor of Rubisco, S, (8) from the following relation:

$$v_C/v_O = S \cdot ([CO_2]/[O_2]) \qquad (1)$$

where, v_C and v_O are the rates of CO_2 and O_2 reduction and $[CO_2]$ and $[O_2]$ the concentrations, at a given temperature, of CO_2 and O_2 dissolved in the cells.

The aim of this work was to examine whether the specificity factor of Rubisco estimated *in vivo* from simultaneous measurements of gas exchange and chlorophyll fluorescence on intact leaves is compatible with the published values for the purified enzyme (9) at varying temperature. The experiments were conducted with intact leaves of french bean (*Phaseolus vulgaris* L.).

2. Procedure

Gas exchange was measured using an open system that has already been described (10) and leaf chlorophyll emission with a PAM 101 chlorophyll fluorometer (Walz). The assimilation chamber used allows simultaneous measurements of leaf net CO_2 uptake and chlorophyll fluorescence emission on the same part of the leaf. PPFD was 240 µmole photons . m^{-2} . s^{-1}.

P. Mathis (ed.), Photosynthesis: from Light to Biosphere, Vol. II, 927–930.

The method proposed by Genty et al. (2) was used to estimate the total electron transport rate. The linear relationship between the quantum yield of PSII electron transport (ΦPSII) and the quantum yield of leaf gross photosynthesis, Φ_C, (Φ_C is the net CO_2 assimilation plus dark respiration, R_D, or light respiration, R_L, divided by incident PPFD) was first established under non-photorespiratory conditions (Fig. 1). In these conditions, the photosynthetic electrons are mainly used to reduce CO_2. This " calibration curve " allows calculation of the total electron transport rate (J) of leaves under photorespiratory conditions i) by measuring the ΦPSII parameter, ii) incident PPFD and iii) by assuming that 4 electrons are used to reduce one molecule of CO_2:

$$J = \Phi_C \,.\, PPFD \,.\, 4 \qquad (2)$$

The total photosynthetic electron transport rate under photorespiratory conditions, can be expressed as follows:

$$J = 4(v_c + v_o) \qquad (3)$$

where, $v_c = A + R_d + 0.5v_o$ (4)

In this expression, R_d (day respiration) is the rate of CO_2 evolution in the light from processes other than photorespiration and $0.5v_o$ is the rate of CO_2 produced by the photorespiratory pathway (1 mole of CO_2 evolved for 2 moles of O_2 fixed). In equation 4, v_c represents a maximum estimation of carboxylation assuming that all the CO_2 produced is reassimilated. From equations 3 and 4, it follows that:

$$J = 4(A + R_d) + 6v_o \qquad (5)$$

where, $Jc = 4(A + R_d)$ (6)

and $Jo = 6v_o$ (7)

In the present work, light respiration ($R_L = R_d + 0.5v_o$) was estimated from the A/Ci curves at different temperatures and then v_c was calculated from eq. 4, Jo from 3 and 7 and Jc from 5 and 6. v_c and v_o calculated as described above were used to estimate the CO_2/O_2 specificity factor of Rubisco from eq. 1. Bunsen solubility coefficient at appropriate temperatures was used to calculate the concentrations of CO_2 and O_2 dissolved in the cell.

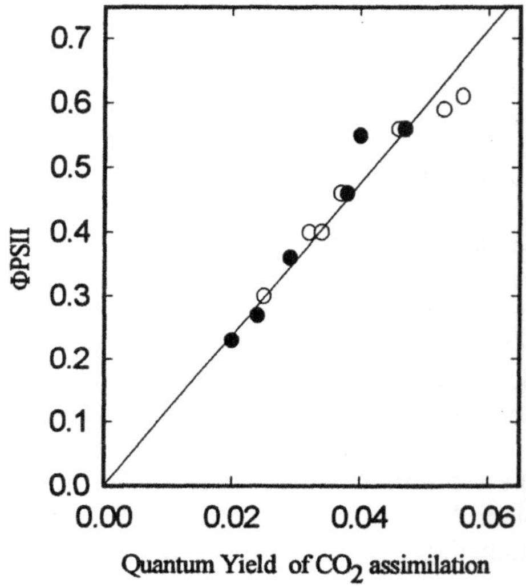

Fig. 1: The relationship between the quantum yield of PSII electron transport (ΦPSII) and the quantum yield of leaf gross photosynthesis (Φ_C) measured under non-photorespiratory conditions at different temperatures on intact leaves of *P. vulgaris* (21% O_2 and saturated CO_2 (closed symbols) and normal CO_2 with 2% O_2 (open symbols)).

3. Results and Discussion

The linear relationship between ΦPSII and Φ_C established under non-photorespiratory conditions is shown Fig. 1. These conditions were obtained using either normal CO_2 and 2% O_2 (Fig.1, open circles) or normal O_2 and saturating CO_2 (Fig. 1, closed circles). The relationship is similar for both cases.

Leaf net CO_2 uptake and quantum yield of PSII electron transport measured under both photorespiratory and non-photorespiratory conditions are presented in Fig. 2. As expected, leaf net photosynthesis is lower in normal air than in air with saturating CO_2 due to the oxygenation of RuBP in normal air. However, ΦPSII is similar in both photorespiratory and non-photorespiratory conditions for the whole range of temperatures used (Fig. 2B) indicating that the way the photosynthetic electrons are used in the chloroplast does not change the quantum yield of electron transport (11).

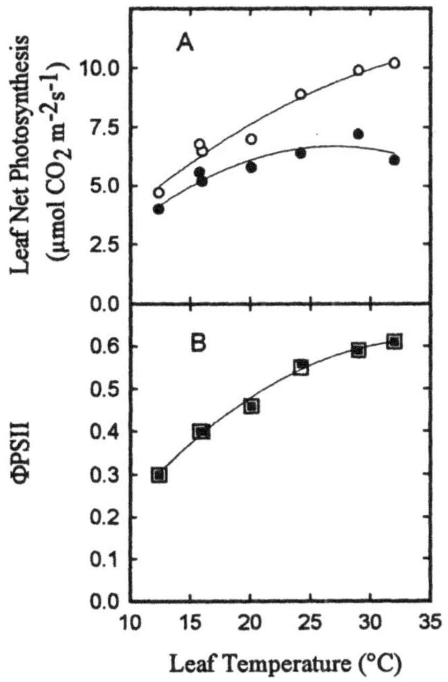

Fig. 2: A) Leaf net CO_2 assimilation and B) quantum yield of PSII electron transport (ΦPSII) measured on intact leaves of *P. vulgaris* as a function of leaf temperature. Closed symbols correspond to the measurements in normal air (350 ppm CO_2 and 21% O_2) and open symbols to thoses under non-photorespiratory conditions (21% O_2 and saturating CO_2).

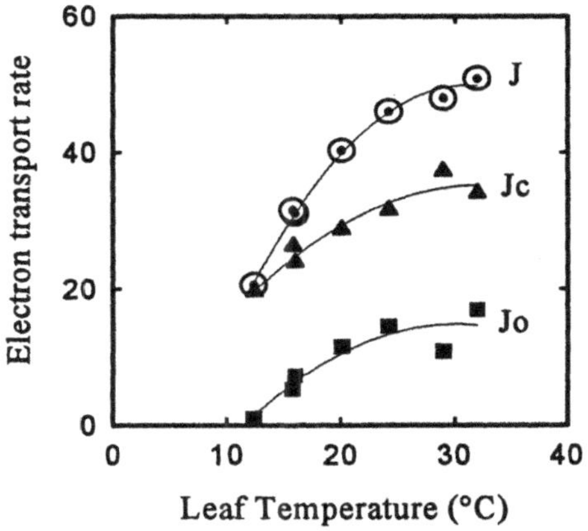

Fig. 3: The total photosynthetic electron transport rate (J) of intact leaves of *P. vulgaris* in normal air (●) and in air with saturating CO_2 (m) at varying leaf temperature. The rates of electron transport allocated to CO_2 (Jc, s) and to O_2 (Jo, ■) reduction are also presented (see Procedure).

Total electron transport rate (J) and its partitioning between CO_2 and O_2 reduction under photorespiratory conditions were calculated using the data of Fig. 2 and the " calibration curve " (Fig. 1) as explained in the Procedure section. As expected from Fig. 2, the total electron transport rate is similar in both airs (Fig. 3, closed and open circles) and J, Jc and Jo decrease with decreasing temperature.

The data of Fig. 3 were used to calculate v_C and v_O. The specificity factor of Rubisco (S) was then estimated from eq. 1. The results (Fig. 4) show that within a wide range of temperature (12 to 32 °C) the values of S estimated *in vivo* are in general agreement with the data of literature obtained *in vitro* using purified enzyme of spinach (9).

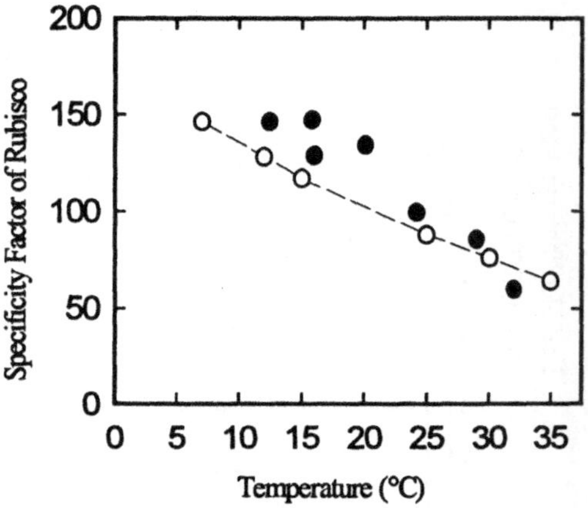

Fig. 4: CO_2/O_2 specificity factor of Rubisco calculated *in vivo* (●) from the data of Fig. 3. The open symbols (m) correspond to the *in vitro* values obtained using purified enzyme of spinach leaves (9).

Overall, it can be concluded that:

- The simultaneous measurements of gas exchange and chlorophyll fluorescence emission on intact leaves allow good estimations of i) the partitioning of photosynthetic electrons to CO_2 and O_2 reduction and of ii) the specificity factor of Rubisco.

- Within a wide range of temperatures, carboxylation and oxygenation of RuBP are the main processes consuming photosynthetic electrons (6, 7).

References

1 Weis, E. and Berry, J.A. (1987) Biochim. Biophys. Acta 894, 198-208
2 Genty, B.; Briantais, J.M. and Baker, N.R. (1989) Biochim. Biophys. Acra 990, 87-92
3 Peteson, R.B. (1989) Plant Physiol. 90, 1322-1328
4 Cornic, G. and Briantais, J.M. (1991) Planta 183, 178-184
5 Krall, J.P. and Edwards, G.E. (1992) Physiol. Plant. 86, 180-187
6 Laisk, A. and Sumberg, A. (1994) Plant Physiol. 106, 689-695
7 Ghashghaie, J. and Cornic, G. (1994) J. Plant Physiol. 143, 643-650
8 Laing, W.A.; Ogren, W.L. and Hageman, R.H. (1974) Plant Physiol. 54, 678-685
9 Jordan, D.B. and Ogren, W.L. (1984) Planta 161, 308-313
10 Cornic, G. and Ghashghaie, J. (1991) Planta 185, 255-260
11 Genty, B.; Goulas, B.; Dimon, G.; Peltier, G.; Briantais, J.M. and Moya, I. (1992) Res. in Photosynth. Vol. IV, 603-610

STATE TRANSITIONS IN THE UNICELLULAR RED ALGA *RHODELLA VIOLACEA*: PSII FLUORESCENCE QUENCHING.

E. Delphin, D. Kirilovsky, J.-C. Duval and A.-L. Etienne
URA 1810, CNRS, E.N.S., 46 rue d'Ulm, 75230 Paris Cedex O5, France.

1. INTRODUCTION

In many photosynthetic organisms, a rapid response to changes in illumination conditions is known as state transitions (1, 2). This regulatory process adjusts the distribution of excitation energy between Photosystem I and Photosystem II, upon preferential excitation of either one. In higher plants and green algae, it is allowed by the mobility of the membrane antenna LHC II. This mechanism involves the reversible phosphorylation of the LHC II-complex by a membrane kinase, controlled by the redox state of the plastoquinone pool (3).

In this study, state transitions were characterized in the unicellular marine red alga, *Rhodella violacea*, by low temperature fluorescence emission spectra. Kinetics of the process, spectra deconvolution, as well as effect of electron flow inhibitors and kinase/phosphatase inhibitors were studied to answer the open question about phosphorylation involvement in state transitions in red algae.

2. MATERIALS AND METHODS

R.violacea cells were grown in Erdschreiber medium, at 20°C under 40 μE m^{-2} s^{-1} white light, with a 16h light/ 8h dark photoperiod. Exponential phase cells were collected from 3 to 4 day old cultures.

State 1 was established with far-red (>720 nm) light, preferentially absorbed by Chl *a* PSI. Incubation in green light (550 nm) brought cells to state 2 (4). For all experiments, the light-induced transitions were accomplished by illumination for 5 min of a 4 ml sample. Then 150 μl were quickly filtered and the filter was immediatly plunged into liquid nitrogen. Cells were excited at 560 nm and fluorescence recorded on an Hitachi F-3010 Fluorescence Spectrophotometer.

When reagents were used, cells were adapted to state 1 or state 2, the reagent added, and cells transferred to the requested light. DCMU and DBMIB were added as a solution in ethanol at M and M respectively. Cells were incubated with staurosporine at 100 nM for 90 min and with NaF at 100 mM for 60 min.

The ratio PSII fluorescence emission to PSI fluorescence allowed to follow the kinetics of transition of *R.violacea* from the dark-adapted state to state 2. Cells were preadapted to darkness for 20 min, the fluorescence intensity at 695 nm and 720 nm was

P. Mathis (ed.), Photosynthesis: from Light to Biosphere, Vol. II, 931–934.

thus monitored at 77K after various periods of illumination with green light at room temperature.

The green alga *Chlamydomonas reinhardtii* was grown in TAP medium in the same culture conditions than *R.violacea*. Exponential phase cells were collected from 1 to 2 day old cultures. State 1 was obtained by 10 min illumination with far-red light. Cells were adapted to state 2 by incubation for 15 min in darkness under anaerobic conditions.

3. RESULTS AND DISCUSSION

Cells adapted to state 1 under far-red light showed a relative decrease of the fluorescence emission related to PSII (F685 and F695), while in state 2, under green light, this fluorescence was diminished. The ratio of Chl *a* PSII fluorescence to Chl *a* PSI fluorescence, i.e. F695/F720, is thus an indicator of light state. In our experiments, its characteristic values were 1,7-1,8 in state 1 and dark state and 1,2-1,3 in state 2.

F695/F720 was recorded in green light on cells preadapted to darkness. At the initial time the ratio was characteristic of state 1 (1,75). After 40 sec of green light illunination, the conversion to state 2 was accomplished, as indicated by the ratio F695/F720 decreased till the value 1,3 (Fig 1). This kinetic of transition is considerably faster than that observed in green plants and algae.

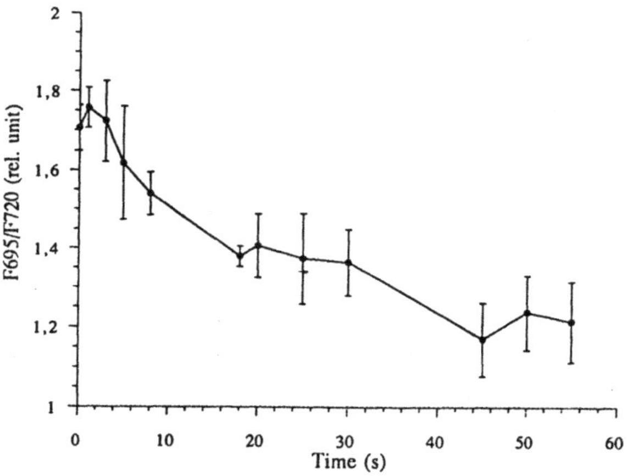

Figure 1: Transition to state 2 expressed as the evolution of the ratio F695/F720, versus the time of exposition to green light. Cells were preadapted to darkness for 20 min, the fluorescence intensity at 695 and 720 nm was thus monitored at 77K, after various periods of illumination with green light at room temperature.

The effects of DCMU and DBMIB, 2 reagents which perturb the electron transport chain (5), and therefore the redox level of plastoquinone, were observed. DCMU blocks the electron transport from PSII to plastoquinone, that becomes oxidized. DBMIB inhibits the transfer of electrons between the plastoquinone and the cytochrome b6/f complex. Plastoquinone is thus reduced. In cells adapted to state 2 with green light, addition of DCMU caused a significant shift towards state 1. In cells adapted to state 1 with far-red light, addition of DBMIB triggered transition to state 2. These results show that the change of the redox state of the plastoquinone pool is related to state transitions.

To determine if a kinase/phosphatase system is involved in state transitions in red algae, the action of a protein kinase inhibitor, staurosporine and a protein phosphatase inhibitor, NaF, have been tested on state1-state 2 transition and on state 2-state 1 transition respectively, *in vivo* in *R.violacea* (6). As a control test, we have tested the effect of these inhibitors on state transitions in the green alga *C.reinhardtii*, in which the involvement of protein phosphorylation has already been demonstrated (7).

R.violacea cells were adapted to state 1 by a 5 min irradiation with far-red light. After 90 min in presence of staurosporine they were transferred to green light for 5 min, leading to state 2, just as cells illuminated directly with green light with no treatment, indicating that staurosporine had no effect on state 1-state 2 transition (Fig 1a). Alternatively, state 2 transition was established by a 5 min exposition to green light. Ulterior transition towards state 1 occured under far-red light in absence or presence of NaF (data not shown). Figure 1b shows the effect of staurosporine on state 1-state 2 in *C.reinhardtii*. The cells of the green alga were adapted to state 1 under far-red light. After transfer to darkness and anaerobic conditions, cells were brought to state 2. However if cells were kept 90 min in presence of staurosporine under state 1 conditions, before transfer to conditions requested for state 2 conditions, they remained in state 1 (Fig 1b). Alternatively, cells adapted to state 2 and then illuminated with far-red light showed a transition to state 1. However, cells adapted to state 2 in presence of NaF (60 min incubation) remained in this state after subsequent transfer to state 1 conditions (data not shown).

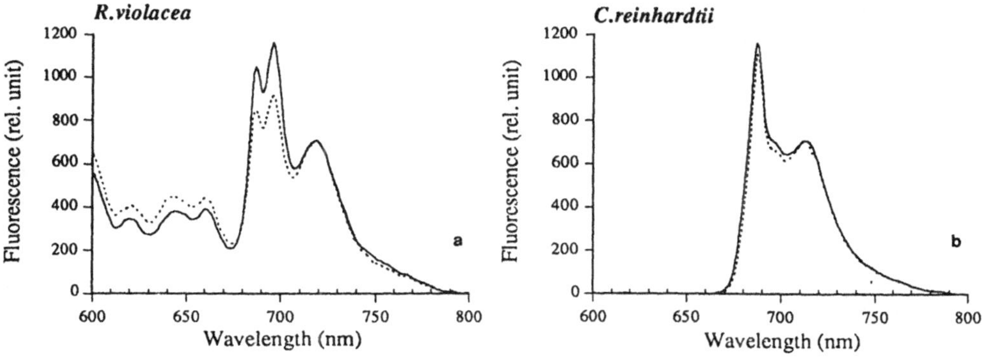

Figure 2: Effect of the inhibitor of protein kinase staurosporine, on the state 1-state 2 transition in *R.violacea* (a) and *C.reinhardtii* (b). Solid line, cells adapted to the requested state 1 conditions and incubated for 90 min with staurosporine; dashed line, above cells transferred to state 2 conditions, still in presence of staurosporine. Excitation wavelength: 560 nm for (a) and 440 nm for (b). Spectra were normalized to the PSI related peak.

The 77K emission fluorescence spectra of *R.violacea* in state 1 and state 2 were deconvoluted into 8 Gaussians. The areas of the Gaussians corresponding to F695 and

F720 were calculated in state 1 and state 2, and compared the Gaussian component characteristic of the added probe. The emission fluorescence of the probe (Fp) arise around 650 nm and correspond to *Spirulina* phycocyanin. The ratios "Fp"/"F695" and "Fp"/"F720" (Table 1) show that PSI fluorescence is similar in state 1 and state 2, while PSII fluorescence is higher in state 1 than in state 2 ("Fp"/"F695" lower).

Table 1: Comparison of F695 (Chl *a* PSII fluorescence) and F720 (Chl *a* PSI fluorescence) to Fp (probe fluorescence), in state 1 and state 2.

	STATE 1		STATE 2	
Ratio of areas	Fp/F695	Fp/F720	Fp/F695	Fp/F720
M	**1.196** ± 0.07	**0.482** ± 0.03	**1.794** ± 0.11	**0.502** ± 0.03

Bold figures denote the average value of each ratio, calculated from 5 experiments.

These results suggest that during state transitions, the PBS-antenna does not move from PSII to PSI, but that the excess of light energy reaching PSII might be diminished by non radiative quenching.

CONCLUSION:

Two main results were obtained in this study. First, in red algae, state transitions do not depend on a kinase/phosphatase system, suggesting that protein phosphorylations are not involved in this mechanism. Second, the decrease of light-energy reaching PSII does not correspond to its transfer to PSI.

We have concluded that the mechanisms leading to state transitions in red algae are different from those governing state transitions in green plants and algae, in which the reversible phosphorylation of LHC II (controlled by the redox level of plastoquinone) leads to its mobility between PSII and PSI. The kinetics of transitions we observed, faster than in green organims, are in agreement with this hypothesis. However, experiments with DCMU and DBMIB indicated that state transitions in *R.violacea* were sensitive to the redox state of plastoquinone. But, it does not necessarily imply that it is linked with a control on a kinase/phosphatase system.

REFERENCES:

1 Bonaventura, C. and Myers, J. (1969) Biochim. Biophys. Acta 189, 366-389
2 Murata, N. (1969) Biochim. Biophys. Acta 172, 242-251
3 Allen, J.F. (1992) Biochim. Biophys. Acta 1098, 275-335
4 Satoh, K. and Fork, D.C. (1983) Biochim. Biophys. Acta 722, 190-196
5 Mullineaux, C.W. and J.F. Allen (1990) Photosynth. Res. 23, 297-311
6 Delphin, E, Duval, J.-C. and Kirilovsky, D. (1995) submitted
7 Wollman, F.-A. and Delepaire, P. (1984) J. Cell Biol. 98, 1-7.

NADH:PLASTOQUINONE OXIDOREDUCTASE AS ADDITIONAL ELECTRON DONOR FOR PHOTOSYNTHESIS IN A CHROMOPHYTE ALGA, *PLEUROCHLORIS MEIRINGENSIS*

W. Seidel-Guyenot, J. Franke, C. Büchel
Institute of General Botany, University of Mainz, 55099 Mainz, FRG

1. Introduction

In the last decade numerous experimental evidences have been presented for the existence and operation of a respiratory redox chain in chloroplasts called chlororespiration (1, 2). During chlororespiration the plastoquinone (PQ) pool can be reduced in the dark *via* a dehydrogenase (NAD(P)H:PQ oxidoreductase, NPO) and oxidised at the expense of O_2 by an oxidase in the chloroplast. The electron transfer is coupled to a vectorial transport of protons across the thylakoid membrane.
Our previous results clearly demonstrated a pH-gradient across the thylakoid membrane of *Pleurochloris meiringensis* (Xanthophyceae) in the dark and a reduction of PQ independent of PSII (3). Here we give experimental evidence for a NPO in the chloroplasts of *P. meiringensis* and demonstrate the importance of this enzyme for the electron fluxes *in vivo*.

2. Material and Methods

Pleurochloris meiringensis Vischer (culture collection Göttingen, n° 830-3) was grown under 15 $\mu Em^{-2}s^{-1}$ white light as described previously (3). Isolation of chloroplasts was carried out as in (4). After osmotic breakage, thylakoids were suspended in 50 mM Tris/HCl-buffer (pH 7.5 for k_M determinations) and solubilized with 0.5% Triton-X-100 and 0.1% desoxycholate for 15 min on ice. After centrifugation (2000g, 5 min) the supernatant was used to assay NPO activity spectrophotometrically in the presence of 0.3 mM menadione according to (5). Chlorophyll concentration was determined in 90% acetone (6).
Light-dependent, methylviologen mediated O_2 consumption upon the addition of DCMU was measured with a clark-type electrode in isolated chloroplasts using different NAD(P)H concentrations according to (7).
P700-kinetics were recorded as changes in the absorbance at 820 nm with a PAM chlorophyll fluorometer (Walz, Effeltrich, Germany) using the emitter-detector unit ED 800T. Cells were harvested by sedimentation and used at a chlorophyll concentration of 0.2-0.4 mg/ml. CO_2-supply was ensured by 20 mM $KHCO_3$. Adaptation of the algae was carried out in the thermostated (20°C) cell by dark adaptation for 30 min or with white light of 160 $\mu Em^{-2}s^{-1}$, respectively. Light adapted cells were kept two minutes in the dark prior to measurement. Far-red (FR) light was selected through a long-pass filter (RG 695, Schott). Saturating multiple-turnover flashes

P. Mathis (ed.), Photosynthesis: from Light to Biosphere, Vol. II, 935–938.

(duration 50 ms) were provided from a xenon discharge lamp (XMT 103, Walz). Inhibitors were added after adaptation of the cells.

3. Results and Discussion

The *in vitro* assay of isolated thylakoid membranes of *P. meiringensis* revealed NAD(P)H:PQ oxidoreductase activity with a k_M of 7.3 μM at pH 7.5 for NADH (Fig. 1). In contrast to the enzyme of higher plants (8), the activity depended on the pH with a maximum at pH 9 (inset Fig. 1). High concentrations of rotenone (100 μM) were necessary to block the enzyme activity completely.

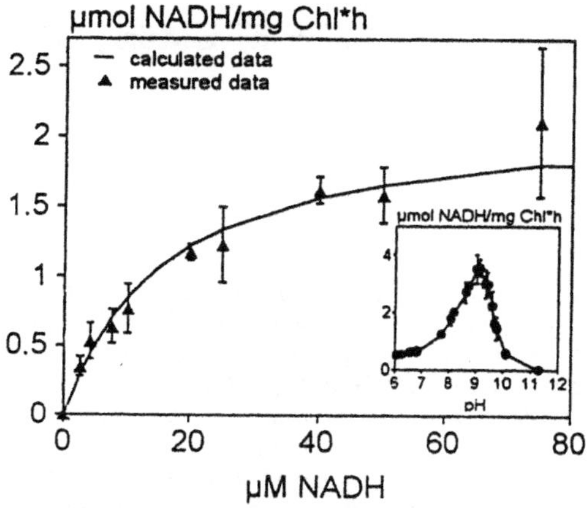

Figure 1. NPO activity as a function of NADH concentration. Mean values ± standard deviations at pH 7.5 are given. The calculated values were derived from the linear regression of the Lineweaver-Burk plot. Inset: activity of NPO versus pH.

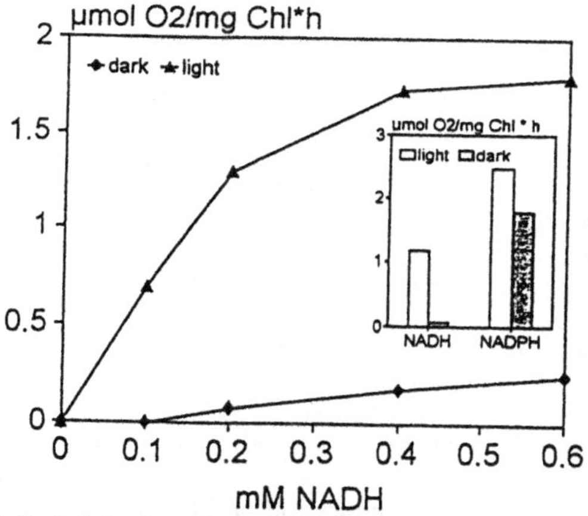

Figure 2. Methylviologen-dependent O_2 consumption in the chloroplasts of *P. meiringensis* upon addition of 20 μM DCMU in dependence on the NADH concentration. Illumination was carried out with white light of 400 $\mu Em^{-2}s^{-1}$. The inset compares O_2 consumption with 0.2 mM NADPH or NADH, respectively.

The crude extract showed an even higher affinity for NADPH as a substrate (k_M: 3.2 μM). This was mostly due to the diaphorase activity of ferredoxin:NADPH oxidoreductase (FNR) because of (i) its k_M, (ii) its pH optimum (pH 10) and (iii) because the NADPH driven, methylviologen mediated O_2 consumption upon the addition of DCMU was mostly independent of light (inset Fig. 2). In contrast, NADH driven O_2 consumption was mostly light dependent and, therefore, mediated by NPO. The NADPH and NADH consuming enzymes, FNR and NPO, differed significantly in their molecular weights as demonstrated by non-denaturing electrophoresis of the chloroplast extract and activity staining of gels (data not shown). Our data clearly demonstrate that the thylakoids of *P. meiringensis* exhibit NPO activity as reported from *Chlamydomonas* (5, 7, 9) and higher plants (8). Fig. 2 shows a maximal input of about 7 μmol e^-/mg Chl*h due to the action of NPO. Assuming a complete consumption of these electrons by an oxidase, the input of NPO would allow maximal 13% of total respiration to be due to chlororespiration. Direct measurements of the O_2 consumption in *P. meiringensis* yielded 6% (10).

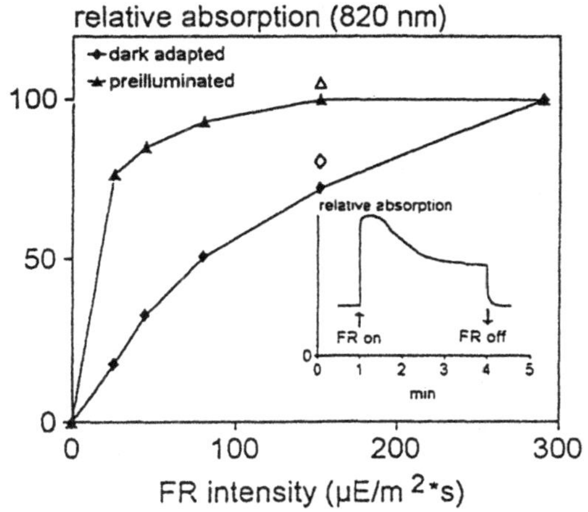

Figure 3. Relative oxidation of P700 versus FR intensity. The open symbols refer to measurements upon the addition of 20 μM DCMU. The inset demonstrates a time scan of dark adapted cells with 80 $\mu Em^{-2}s^{-1}$ FR intensity.

To study the electron fluxes *in vivo*, P700-oxidation kinetics were measured under FR light for separate excitation of PS I. In contrast to higher plants (11), PS I was re-reduced during continuous FR illumination of high intensities (inset Fig. 3) and extremely high quantum fluxes were necessary to saturate P700-oxidation especially in dark adapted cells (Fig. 3) Acceptor side limitation could be ruled out because the cells were supplied with sufficient HCO_3^- and because no difference in the oxidation was visible upon the addition of methylviologen prior to measurement. Therefore, the necessity of high FR intensities to saturate P700 oxidation must be due to an input of electrons into the intersystem chain during measurement as reported from cyanobacteria and explained by the interaction of photosynthesis and respiration in the same membrane (12).

However, the input of electrons could be due to i) linear electron transport from PS II, assuming the FR light to be actinic for PS II, ii) cyclic electron transport around PS I, and iii) chlororespiration. PSII is only slightly excitated as shown by the small enhancement of P700 oxidation upon the addition of DCMU (Fig. 3). For the same reason, different PSII antenna sizes of dark adapted and preilluminated cells (13) could not explain the need of higher quantum fluxes to saturate P700 oxidation in dark adapted cells. An electron input independent of PSII was further shown by the fast re-reduction of P700 after the offset of the FR illumination. Dark adapted cells had a $t_{1/2}$ of re-reduction of 370 ms, a value comparable to the halftimes in cyanobacteria (14), which remained constant upon the addition of DCMU. Only upon inhibition of NPO by adding 100 μM rotenone the re-reduction halftime was decelerated by 15% to 420 ms. The influence of NPO on PSI activity was further demonstrated by briefly disturbing the steady-state oxidation under high FR illumination by adding multiple turnover flashes of white light and calculating the reduction produced by the additional short illumination of both photosystems. Although DCMU completely inhibited PS II function, P700 reduction was only inhibited by 49.7% in comparison to control in dark adapted cells. Simultaneous addition of DCMU and rotenone increased the inhibition to 80.5%. Under these special illumination conditions, 50% of the electrons were supported by PSII, 19.5% by cyclic and 30.5% by NPO-mediated electron transport. In preilluminated cells, only 11.5% of electrons were supported by NPO. The decrease in NPO activity in preilluminated cells further demonstrates that NPO is acting in the chlororespiratory pathway mainly in the dark but also under low light conditions (3), therefore influencing not only induction of photosynthesis but also steady-state photosynthesis under low light. It is reasonable to assume that chlororespiration helps to maintain the redox and light $\Delta\mu_{H^+}$ values in a physiological range and, therefore, contributes to a larger flexibility of the plant to cope with environmental stresses.

References

1 Bennoun, P. (1982) Proc. Natl. Acad. Sci. USA 79, 4352-4356
2 Scherer, S.(1990) TIBS 15, 458-462
3 Büchel, C. and Wilhelm, C. (1990) Plant Physiol. Biochem. 28, 307-314
4 Seidel-Guyenot, W., Schwabe, C. and Büchel, C. (in preparation)
5 Peltier, G. and Schmidt, G.W. (1991) Proc. Natl. Acad. Sci. USA 88, 4791-4795
6 Jeffrey S.W. and Humphrey, G.F. (1975) Biochem. Biophysiol. Pflanzen 167, 191-194
7 Godde, D. (1982) Arch. Microbiol. 131, 197-202 (1982)
8 Cuello, J., Quiles, M.J., Albacete, M.E. and Sabater B (1995) Plant Cell Physiol. 36, 265-271
9 Godde, D. and Trebst, A. (1980) Arch. Microbiol. 127, 245-252
10 Büchel, C. and Garab, G. (1995) Planta, in press
11 Schreiber, U., Klughammer, C. and Neubauer, C. (1988) Z. Naturforsch. 4, 686-698
12 Mi, H., Endo, T., Schreiber, U. and Asada, K. (1992) Photosynth. Res. 33, 1099-1105
13 Büchel, C., Wilhelm, C. and Lenartz-Weiler, I. (1988) Botanica Acta 101, 306-310
14 Yu, L., Zhao, J., Mühlenhoff, U., Bryant, D.A. and Golbeck, J.H. (1993) Plant Physiol. 103, 171-180

CHARACTERIZATION OF ELECTRON CARRIERS, INCLUDING IRON-SULFUR PROTEINS, IN CHLOROPLAST ENVELOPE MEMBRANES. Montpellier, France, 20-25 August 1995

P. Jäger-Vottero*, A.J. Dorne*, J. Jordanov**, R. Douce* & J. Joyard*
* Laboratoire de Physiologie Cellulaire Végétale, DBMS
** Laboratoire de Spectroscopie des Complexes polymétalliques, DRFMC
URA CNRS 576, CEA, Centre d'Etudes Nucléaires de Grenoble, Université Joseph Fourier, F38054 Grenoble Cédex 9, France

1. Introduction

Envelope membranes play a central role in chloroplast membrane biogenesis[1] since they contain numerous enzymes involved in the biosynthesis of specific plastid membrane constituents. Important events in chloroplast biogenesis are the formation of polyunsaturated fatty acids and of colored carotenoids involving various desaturation steps which are poorly understood. *In vitro* experiments have demonstrated that quinones are involved in the sequential desaturation of phytoene[2] and that a ferredoxine:NADPH oxidoreductase could be involved in the functioning of a n-6 lipid-linked desaturase[3]. In both cases, molecular oxygen is the terminal electron acceptor. These results imply the presence in envelope membranes of an electron-transfer chain since intermediate redox compounds are necessary between NADPH, quinone and oxygen. To date, the only known envelope constituents that could play a role in electron transfer reactions are prenylquinones. Envelope membranes contain 4 to 12 μg prenylquinone (plastoquinone-9 and α-tocopherol) per mg protein[4]. Because semiquinone species are paramagnetic they are detectable by EPR spectroscopy. We have investigated the behaviour of envelope quinone by this technique, under reducing and oxidating conditions.

2. Procedures

Envelope membranes from intact spinach chloroplasts were prepared according to Douce and Joyard[5]. Envelope membranes (10 mg protein ml^{-1}) were stored in liquid nitrogen or lyophilized for pentane tratment. Pentane treatment of envelope membranes was adapted from Szarkowska[6]. EPR spectra were recorded on a Varian E109 spectrometer.

P. Mathis (ed.), Photosynthesis: from Light to Biosphere, Vol. II, 939–942.

Oxidation or reduction were performed on either native or pentane-treated envelope membranes. Chemical oxidation of envelope samples was achieved by ammonium and cerium nitrate whereas reduction was achieved with sodium dithionite (E'_0 = - 0.18 V), 5-deazaflavine/oxalate (E'_0 = - 0.65 V) and with physiological mediators, NADH and NADPH (E'_0 = - 0.32 V).

3. Results and Discussion

When analyzed by EPR spectroscopy, native envelope membranes show a signal in the g = 4.3 region and another complex one around g = 2 (fig. 1A). Interestingly, no signals were observed in the g = 3 (or 6) region. Spectra of pentane-treated envelope membranes were dominated by a large isotropic signal centered at g = 2.003 and showed a serie of signals with maxima at g values of 2.167, 2.077, 2.017, 1.961, 1.929 and 1.875 (fig. 1B). Some of them could be associated with Mn^{2+} as suggested by the typical 6 peaks component of the spectrum (fig. 1B). Manganese could be in part responsible for these signals since envelope membranes contain 0.030 ± 0.004 µg manganese per mg protein [7]. Nevertheless it is possible that some of these signals in the g = 2 region could be associated with iron-sulfur centers which can be revealed by reduction or oxidation.

The signal at g = 4.3 is characteristic of high spin Fe^{3+}, a cation that could be present either as "free" Fe^{3+} or associated with a protein. In our samples, this signal is probably not due to contaminating "free" Fe^{3+} since (a) its amplitude did not decrease when envelope membranes were washed with EDTA and (b) was still well visible when the spectra were recorded at 15 K instead of 4.2 K. Therefore, envelope membranes contain Fe^{3+} covalently linked to one (or several) protein.

The low spin hemes of cytochromes generally present an EPR signal in the g = 3 (and 6) region. Because of the sensitivity of the method, the lack of signal in this region strongly suggests that envelope membranes are devoid of cytochrome.

The signal at g = 2.003 has a band width of 12 ± 2 Gauss and can be associated with a semiquinone radical[8]. The presence of a stabilized semiquinone radical is rather surprising since in general they are transient species formed during oxidoreduction processes. The g = 2.003 signal arises from a rapidly relaxing radical as demonstrated by the rapid decrease of its amplitude when raising the temperature from 2.5 K to 46 K. The power saturation (from 0.01 to 50 mW) produced a biphasic curve. One component at g = 2.003 and a peak to peak width of 12 ± 2 Gauss saturates at 2 mW, whereas the other component centered at g = 2.017 has a larger peak to peak width of 21 ± 2 Gauss and does not saturate before 40 - 50 mW. The radical associated with the second signal is probably a flavosemiquinone.

EPR spectra of envelope membranes reduced by dithionite (E'_0 = - 0.18 V) revealed a signal centered at g = 2.057 (fig. 1C). The characteristics of this signal are those of an iron-sulfur protein but to date no iron-sulfur protein exhibiting such a high g value in the reduced state has been described in other biological systems (Cammack, 1992)[9]. More it

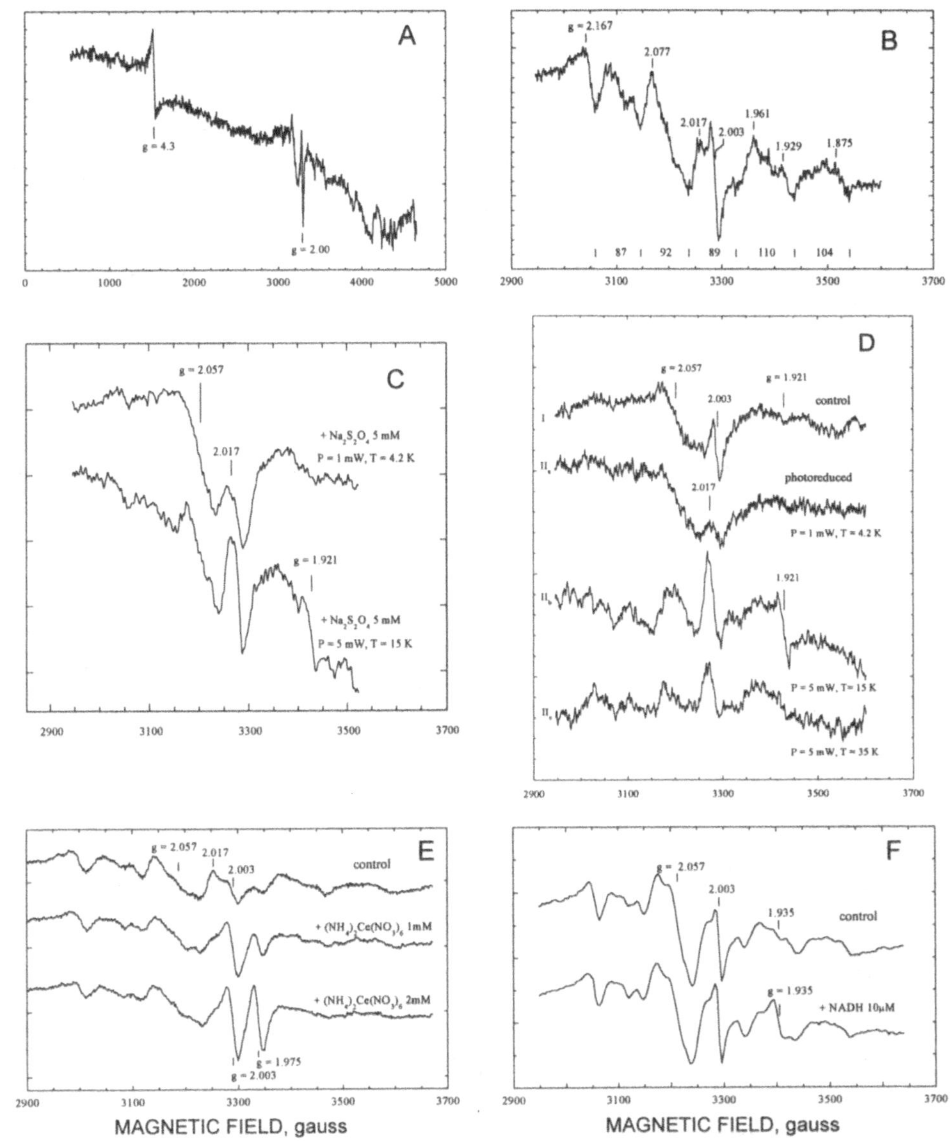

Figure 1. EPR spectra of native (**A** and **D**) and pentane-treated (**B**, **C**, **E** and **F**) envelope membranes from spinach chloroplasts. **A** and **D** : 10 mg protein/ml; **B-C-E-F** : 30 to 45 mg protein/ml. EPR conditions : 2500 ± 2500 Gauss (**A**) or 3300 ± 400 Gauss (**B-F**); Time constant : 0.128 s (**A**) or 0.25 s (**B-F**); Microwave power : 1 or 5 mW; Amplitude modulation 12.5 G; Frequency 9.22 Ghz. **C** : effect of dithionite; **D** : effect of photoreduction by 5-deazaflavine; **E** : Effect of oxidation with ammonium and cerium nitrate; **F** : effect of reduction with NADH.

was possible to reduce in native envelope preparation this iron-sulfur protein by addition of 10 mM NADPH (not shown).

Another iron-sulfur protein, the signal of which was centered at g = 1.921, was also revealed after dithionite reduction (fig. 1C) and under stronger reducing conditions (5-deazaflavine/oxalate, E'_0 = - 0.65 V) (fig. 1D). The relaxation properties of this center demonstrated that chloroplast envelope contain a [4Fe-4S] protein, paramagnetic when reduced with a global charge of +1, *i-e* a $[4Fe\text{-}4S]^{1+}$center.

When oxided by ammonium and cerium nitrate (E'_0 = +2.5 V), the EPR spectra (fig. 1E) of envelope membranes showed, in addition to the semiquinone signal at g = 2.003, an other signal at g = 1.975, attributable to a second semiquinone. One possible explanation is that these 2 semiquinones (named A and B) are in fact one split signal, the splitting being caused by a spin-spin interaction between them[10].

One additional Fe-S center, a $[2Fe\text{-}2S]^{1+}$center, associated with a signal centered at g = 1.935 was detected after reduction with NADH (fig. 1F). This center was not reduced by dithionite and 5-deazaflavine.

4. Conclusion

These results demonstrate that chloroplast envelope membranes contain several centers that could be elements of one or several electron transfer chain. The physiological significance of our results could be related to the presence of desaturase activities in envelope membranes. Obviously further work is needed to provide physiological and biochemical evidence for such an implication.

References

1 Joyard, J., Block, M.A. and Douce, R.(1991) *Eur J. Biochem.* 199, 489-509
2 Goodwin, T.W. (1993) *Meth. Enzymol.* 214, 330-340
3 Schmidt, H., Sperling, P. and Heinz, E. (1995) in *Plant Lipid Metabolism* (Kader, J.C., ed.) pp. 21-23, Kluwer Academic Publishers, Dordrecht, The Netherlands
4 Lichtenthaler, H.K., Prenzel, U., Douce, R. and Joyard, J. (1981) *Biochem. Biophys. Acta* 64, 99-105
5 Douce, R. and Joyard, J. (1982) in *Method in Chloroplast Molecular Biology* (Edelman, M, Hallick, R. and Chua, N.H. ed.) pp.239-256, Elsevier-North Holland, Amsterdam, The Netherlands
6 Szarkowska, L. (1966) *Arch. Biochem. Biophys.* 113, 519-525
7 Bowyer, J.R. and Ohnishi, T. (1985) in *Coenzyme Q* (Lenaz, G. ed.) pp. 409-432, John Wiley and Sons
8 Joyard, J. and Douce, R. (1976) *Physiol. Vég.* 14, 31-48
9 Cammack, R. (1992) *Adv. Inorg. Chem.* 38, 281-322
10 Salerno, J.C. and Ohnisni, T. (1980) *Biochem. J.* 192, 769-781

CHARACTERIZATION OF THE RC-BOUND CYTOCHROME AND STUDY OF THE ELECTRON TRANSFER IN *RHODOCYCLUS TENUIS*

Laure Menin, Pierre Parot, Barbara Schoepp, Pierre Richaud, and André Verméglio
CEA-DEVM/ LBC-CE Cadarache 13108 St Paul-lez-Durance Cedex, France

Abbreviations: cyt, cytochrome; HiPIP, High Potential Iron-Sulfur Protein; *Ru.*, *Rubrivivax*

1. INTRODUCTION

The photo-induced cyclic electron transfer including RC, bc_1 and c_2 participation was established *in vivo* for *Rhodopseudomonas. viridis* (1). More recently, it has been demonstrated in *Ru. gelatinosus* that the HiPIP is the electron donor which mediates electron transfer between the cyt bc_1 and the RC-bound tetrahemic cyt (2). In this paper, we have studied the cyclic photo-induced electron transfer in *Rhodocyclus tenuis*. This species contains, like most photosynthetic bacteria, a RC-bound tetrahemic cyt *c*, with two high midpoint potential (E_m) hemes (HP_1 and HP_2) and two low E_m hemes (LP_1 and LP_2) (3). Several periplasmic electron donors can be involved in the re-reduction of these hemes. Among these components can be found a HiPIP (300 mV) (4,5), a c_{551}, a $c_{549.5}$ and a cyt *c'* (6). In this work, we first performed the titration of the tetrahemic cytochrome in purified membranes in order to characterize the hemes in terms of E_m and to evaluate the possible electron transfers between each of these hemes and the different electron carriers. In a second part, the *in vivo* participation of the HiPIP and the cyt c_{551} has been investigated to specify their function by studying the photo-induced absorbance changes on whole cells of *Rc. tenuis* as a function of the redox conditions of the medium.

2. MATERIALS AND METHODS

Rc. tenuis cells were grown in Hutner medium at 30°C, in anaerobic conditions under continuous illumination (75 µmol of photons $m^{-2}s^{-1}$). A green carotenoidless culture has been produced by treatment with diphenylamine (10 µg/ml). Harvested cells were disrupted by French Press at 50 MPa. After elimination of intact cells by centrifugation (10000g, 10 min), membrane fragments were collected by ultracentrifugation (255000g, 90 min. Purification of the soluble carriers including HiPIP has been carried out according to the method previously described (2). Redox titration of the tetrahemic cytochrome was performed on purified membranes suspended in 25 mM Tris-HCl (pH 7). The redox mediators (diaminodurene, 1,2 naphthoquinone 4 sulfonic acid, 1,2 naphthoquinone, duroquinone, phenazine methosulfate, 2-OH 1,4 naphthoquinone) were added to a final concentration of 10 µM. Titration was performed by addition of small aliquots of sodium dithionite (10 mM) and measuring both the absorbance and photo-induced absorption changes in the α-band. Photo-induced absorption changes were recorded with a Joliot-type Flash-Kinetics-Spectrophotometer (7). Curve fitting was carried out with a non linear regression data analysis program (1987, BIOSOFT, Enzfitter 1.05).

P. Mathis (ed.), Photosynthesis: from Light to Biosphere, Vol. II, 943–946.

3. RESULTS

Characterization of the tetrahemic RC-bound cytochrome

a) Redox titration: fig.1 shows a typical redox titration curve of the RC-bound cytochrome hemes monitored at 554 nm on purified membranes, between 410 mV and -150 mV. The two main waves of the curve clearly show both the LP and HP hemes. n=1 Nernst curves (solid line) fit well the experimental points and allow the determination of the midpoint potentials of the two HP hemes, found to be +385 mV (HP_1) and 320 mV (HP_2) at pH 7. Redox potentials of the LP hemes are poorly resolved, but are roughly equal to +115 mV and 70 mV. Similar values have been obtained in the case of *Ru. gelatinosus* (130 mV and 70 mV for the LP hemes). The highest midpoint potential heme of *Rc. tenuis* (385 mV) is much higher than that of *Ru. gelatinosus* (320 mV).

b) Spectral kinetics: the significant difference between E_m of the HP and LP hemes allowed us to determine their spectral characteristics in the α-band (fig.1, B). When the potential is poised at 206 mV, only the HP hemes are reduced. The first laser flash oxidizes HP_2 with an α-band at 558 nm, whereas the second flash (xenon) oxidizes HP_1, whose α-band is centered at 557 nm. The α-bands of these two hemes are very similar with a shoulder at 554 nm. When membranes suspension is maintained at 60 mV, the two LP hemes are reduced. The first flash characterizes LP_2 (α-band centered at 552 nm), whereas the second flash induces LP_1 oxidation and the α-band is centered at 551 nm. These two LP hemes also have very similar spectra.

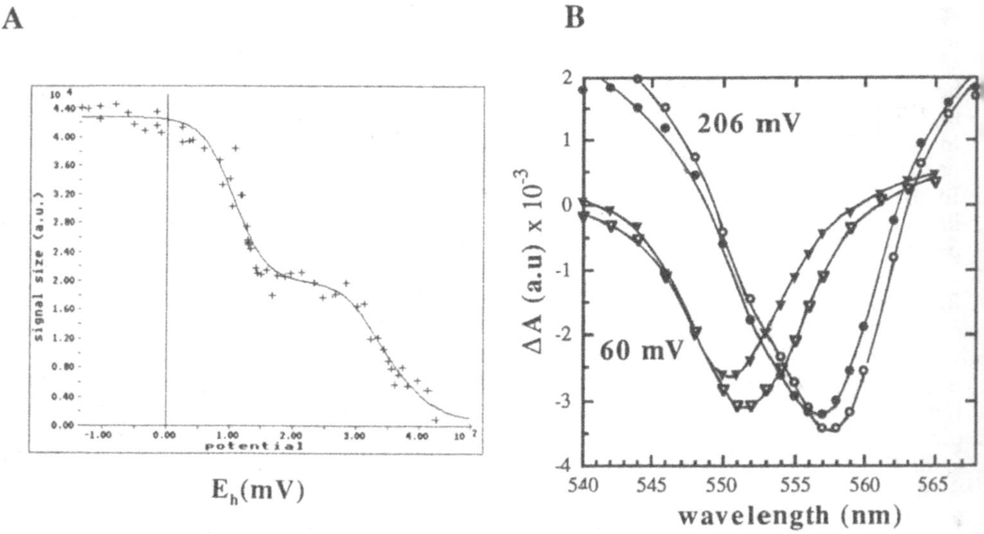

Fig.1 A) E_h dependence of the absorption amplitude measured at 554 nm in membranes of *Rc. tenuis* (Tris, HCl 25 mM pH 7). The continuous line corresponds to a fit with four n=1 Nernst equations where E_m (HP_1)= +385 mV, E_m (HP_2)= +320 mV, E_m (LP_1)= +115 mV, and E_m (LP_2)= +70 mV.
B) Difference absorption spectra obtained after a 1st laser flash (open symbol) and a 2nd xenon flash (filled symbol) at 206 mV (o) and 60 mV (∇). Redox potentials are relative to Normal Hydrogen Electrode. Flashes were spaced 50 ms apart.

Light-induced electron transfer in whole cells of Rc. tenuis

Reconstitution experiments on purified membranes have shown that the HiPIP and the cyt c_{551} can be physiological electron donors to the HP hemes (data not shown). Kinetics of re-reduction of the RC-bound cytochrome at 422 nm have been recorded for a suspension of whole cells with addition of benzoquinone (50 µM) under different conditions of redox potential. At very high potential (> 350 mV), the reduction is slow ($t_{1/2}$ = 20 ms) (fig.2 a), whereas decreasing the potential to 300 mV gives rise to a reduction kinetic which presents two phases, a fast one ($t_{1/2}$ = 350 µs) and a slow one ($t_{1/2}$ = 200 ms) (fig.2 b). The half-time of the fast re-reduction phase is of the same order of magnitude as those measured in *Ru. gelatinosus*. In addition, light-induced absorbance changes at 496 nm (fig.3) show a small absorbtion increase due to the oxidation of the HiPIP concomitant with the HP heme reduction at 422 nm. The fast phase amplitude is maximal for a redox potential equal to 300 mV (fig.2 insert), which corresponds to the midpoint potential of the HiPIP. In oxidizing conditions, the light-induced absorbance spectrum in the α-band shows a shift from 558 nm to 553 nm (fig.4 A (b)) which is complete in 5 ms. Moreover, there is little decrease in absorbance change amplitude during this shift. This electron transfer between two different cytochromes is also clearly visible in the γ-band (413 nm) and occurs with a $t_{1/2}$ equal to 275 µs (data not shown). This spectral shift is interpreted as an electron transfer from the soluble cyt c_{551} to the HP heme of the RC-bound tetrahemic cyt. At lower potentials (fig.4 A (a)), this shift is still present but the amplitude of the α-band after 5 ms is quite small and probably due to a remaining participation of the c_{551} in this redox condition. Because absorption of the HiPIP in this range of wavelength is zero, the amplitude falls between 30 µs and 5 ms. We conclude that upon these redox conditions (about 300 mV), the HiPIP functions as the electron donor to a HP heme. In order to know which of the HP hemes is reduced by the HiPIP or by the cyt c_{551}, spectra have

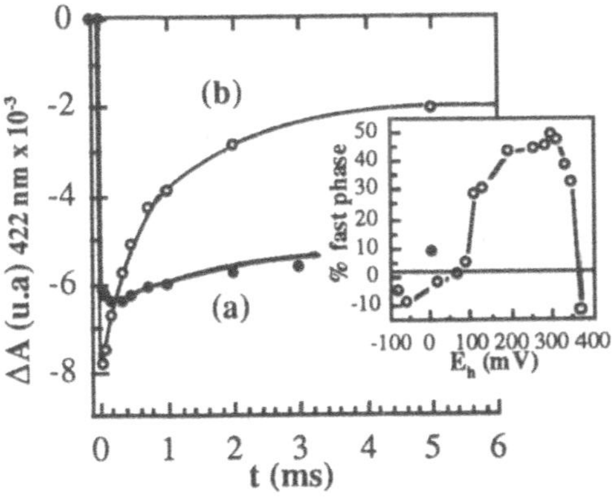

Fig.2 Kinetics of flash-induced ΔA on whole cells of *Rc. tenuis* with mediators and benzoquinone (50 µM) recorded at 422 nm for 2 redox potentials: 380 mV (a) and 290 mV (b) **INSERT:** % of fast phase between 50 µs and 1 ms as a function of E_h. Negative values are due to the onset of a slow oxidation phase before 1ms.

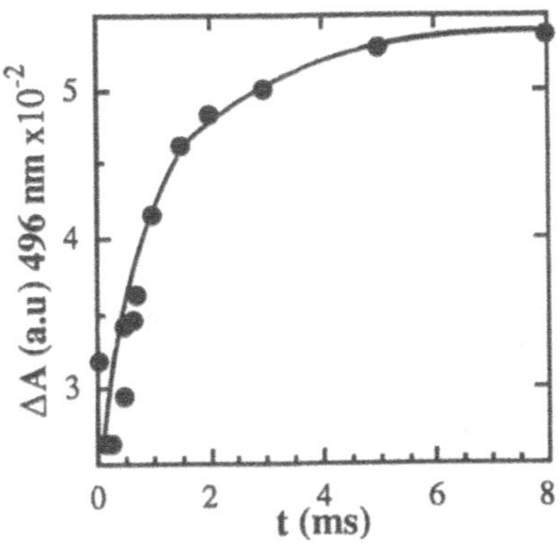

Fig.3 Kinetics of flash-induced absorbance changes recorded at 496 nm on whole green cells of *Rc. tenuis*.

been recorded between 554 and 563 nm in the two redox conditions (fig.4 B). A difference of 1 nm is visible and shows that the HiPIP reduces HP_2 (upper wavelength value in α-band), whereas the c_{551} reduces HP_1 (lower wavelength value in α-band).

4. CONCLUSION

We have found that, depending on redox conditions, the two HP hemes are able to reduce P^+ and work with two distinct soluble electron donors: the HiPIP is the electron donor of the HP_2 heme for redox potentials ranging from 120 mV to 350 mV. When the redox potential is increased, the c_{551} is the electron donor to the HP_1 heme and the kinetics of electron transfer is in the same order of magnitude than for the re-reduction of HP_2 heme by the HiPIP. These results clearly demonstrate the *in vivo* participation of these two electron donors in the photosynthetic electron transfer chain of *Rc. tenuis*. Moreover, the titration of the c_{551} (358 mV) confirms that this cyt is able to reduce the HP_1 heme (385 mV) in terms of redox potentials, whereas the HiPIP (300 mV) can give an electron to the HP_2 heme (320 mV).

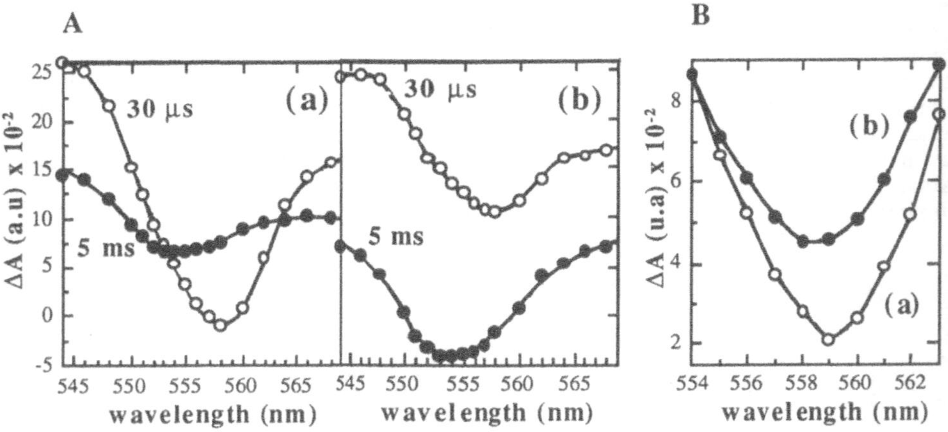

Fig.4 Light-induced absorption spectra (laser flash) on whole green cells of *Rc. tenuis* in the α-band **A**. after 30 µs and 5 ms in conditions of **(a)** potential below 350 mV **(b)** potential above 350 mV **B**. after 30 µs, between 554 nm and 563 nm **(a)** conditions as for A (a) **(b)** conditions as for A (b).

References

1 Garcia, D., Richaud, P., Vermeglio, A. (1993) *Biochim. Biophys. Acta* 1144, 295-301
2 Schoepp, B., Parot, P., Menin, L., Gaillard, J., Richaud, P., and Verméglio, A., *Biochemistry*, in press
3 Nitschke, W., Agalidis, I., and Rutherford, A.W. (1992) *Biochim. Biophys. Acta* 1100, 49-57
4 Holden, H.M., Meyer, T.E., Cusanovich, M.A., and Rayment, I. (1986) *J. Biol. Chem.* 261, 9, 4219-4220
5 Tedro, S.M., Meyer, T.E., and Kamen, M.D. (1985) *Arch. Biochem. Biophys.* 239 (1), 94-101
6 Bartsch, R.G. (1991) *Biochim. Biophys. Acta* 1058, 28-30
7 Joliot, P., Béal, D., and Frilley, B. (1980) *J. Chim. Phys.* 77, 209-216

GAS EXCHANGE AND MODULATED CHLOROPHYLL FLUORESCENCE TECHNIQUES FOR *IN VIVO* ASSESSMENTS OF ALTERNATIVE ELECTRON TRANSFER.

Fredrick F. and Lemeur R.
Department of Applied Ecology and Environmental Biology
Laboratory of Plant Ecology
Faculty of Agricultural and Applied Biological Sciences
Coupure links 653, 9000 Ghent, Belgium

1. Introduction

Quantitative relations between fluorescence emission and the quantum yield of linear electron transport have been proposed by Weis and Berry [1] and by Genty et al. [2]. The concerning relations can be used to predict rates of linear electron transport from fluorescence measurements. Gas exchange measurements, in addition, allow to assess the electron requirements associated with the photosynthetic carbon reductive (PCR) and oxydative (PCO) cycles [3].

In the present paper, we consider the possibility of using simultaneous chlorophyll and gas exchange measurements, performed under photorespiratory conditions, to evaluate the importance of alternative routes in dissipating of photosynthetic ATP and NADPH. The concept is illustrated with measurements carried out on *Schefflera arboricola* leaflets.

2. Procedure

The quantum yield of photochemistry in PSII was derived from measurements of chlorophyll fluorescence as $(F'_m-F_s)/F'_m$ [2]. The rate of linear electron transport (J_t) was calculated as $f_{PSII}(F'_m-F_s)/F'_m Q_a$, where f_{PSII} denotes the fraction of excitation energy that is absorbed by and transferred to antenna chlorophyll molecules of PSII and Q_a is the absorbed photosynthetic photon flux density.

The 'true CO_2 compensation point'(Γ_*), i.e. the chloroplast partial pressure of CO_2 (p_c) at which the rate of CO_2 uptake from carboxylation and the CO_2 loss from photorespiration are equal and the 'day respiration'(R_d), i.e. the rate of CO_2 loss from processes other than photorespiration, were derived from gas exchange measurements according to the method proposed by Brooks and Farquhar [4]. Rates of carboxylation (V_c) and of oxygenation (V_o) were derived from gas exchange measurements as $(A_n+R_d)p_c/(p_c-\Gamma_*)$ and $2\Gamma_*V_c/p_c$ [3], respectively, where A_n denotes the carbon dioxide uptake rate.

Hence, the following equation was applied to provide $f_{PSII}(1-f_{Alt})$, f_{Alt} denoting the proportion of the products of linear electron flow being utilized in alternative biochemical pathways :

P. Mathis (ed.), Photosynthesis: from Light to Biosphere, Vol. II, 947–950.

$$f_{PSII}\,(1 - f_{Alt}) = [\,\alpha_c V_c + \alpha_o V_o] \,/\, [\,((F_m' - F_s)/F_m')\,Q_a\,] \tag{1}$$

where α_c, α_o are stoichiometric coefficients indicating the number of electrons required to sustain the necessary ATP (or NADPH) for each carboxylation and oxygenation, respectively.

Simultaneous gas exchange and chlorophyll fluorescence measurements were performed on intact green *S. arboricola* leaflets (selected from eight, fully expanded but young leaves) under a wide range of cuvette partial pressures of CO_2, extending from 20 to 1000 μbar (1 μbar=0.1 Pa). A first series of measurements was carried out when the plants were well irrigated (matric potential of the peat substrate amounted to -1.5 kPa). Then, after ten days of withholding water (substrate matric potential had dropped to -15 kPa), a second series of measurements was carried out (i.e. on the same leaflets). Cuvette temperature was 27°C, the incident PPFD 380 $\mu mol.m^{-2}.s^{-1}$ and the relative humidity 60%.

The gas exchange measurements were conducted with an open flow, steady state gas-exchange system, i.e. a compact CO_2/H_2O porometer (type CQP-130, Walz, Germany) which was connected to a differential CO_2 infrared gas analyzer (Binos 100-4P, Rosemount, Germany) and to leaf cuvette (type PMK-10, Walz). Fluorescence was measured with a PAM fluorometer (type PAM-2000, Walz).

3. Results and discussion

In an initial approach, we assumed that f_{PSII} was constant (i.e. f_{PSII} = 0.5) and that cyclic photophosphorylation did not occur (i.e. α_c = 4.7 and α_o = 5.2 [5]). Hence, under conditions of low metabolic sink activity, i.e. 1) at low p_cs in leaflets of non-stressed plants and 2) at both low and high p_cs in leaflets of water stressed plants, calculated J_ts differed substantially from assessments of photosynthetic electron requirements ($\alpha_c V_c + \alpha_o V_o$) (Fig. 1). Values of $f_{PSII}(1-f_{Alt})$ showed considerable variation (Fig.2), revealing that either the assumption of a constant f_{PSII} was inappropriate or, alternatively, that f_{Alt} was adjusted so to match the rate of ATP generation and consumption.

Supposing, on the other hand, that the extra electron requirements for ATP formation, relative to the electrons required for NADPH formation, were satisfied by cyclic photophosphorylation, a value of four should be attributed to both α_c and α_o. So, recalculating $f_{PSII}(1-f_{Alt})$ with α_c=α_o=4, yielded even lower values for $f_{PSII}(1-f_{Alt})$ (Fig. 2).

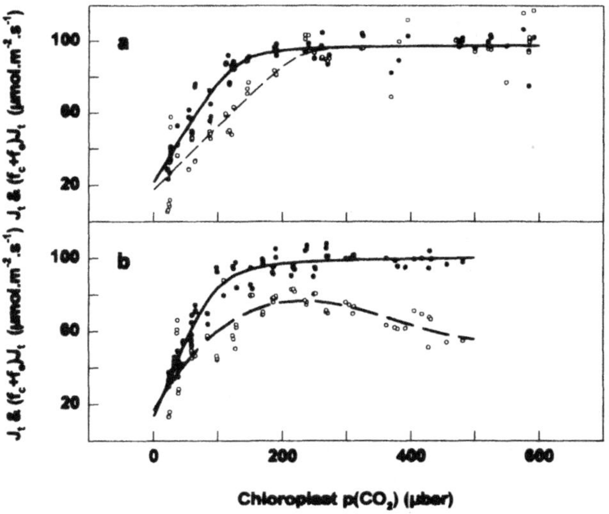

Figure 1. Rates of linear electron transport (J_t) derived from chlorophyll fluorescence measurements (i.e. f_{PSII} = 0.5)(filled circles) and rates of electron transport required to sustain necessary ATP for carboxylation and oxygenation ($4.7V_c+5.2V_o$) (hollow circles) as affected by the chloroplast partial pressure of CO_2, (chloroplast $p(CO_2)$) in non-stressed (a) and water stressed (b) S. arboricola leaflets. ($4.7V_c+5.2V_o$) is indicated on the Y-axes as $(f_o+f_c)J_r$

Of course, it can be questioned to which extent the variation in $f_{PSII}(1-f_{Alt})$ with changing p_c or with increasing water stress could be ascribed to changes in f_{Alt}. It can be argued that, inasmuch as f_{PSII} is adjusted so that an appropriate ratio of ATP/NADPH is provided (i.e. by maintaining a ratio of cyclic to linear electron flow of $(0.7V_c+1.2V_o)/(4V_c+4V_o)$), Δf_{PSII} can be approximated by $0.25\Delta[(0.7V_c+1.2V_o)/(4V_c+4V_o)]$, where Δ denotes a change in the respective variable (Fredrick, unpublished). Since $(0.7V_c+1.2V_o)/(4V_c+4V_o)$ varied from 0.14 to 0.25, absolute changes in f_{PSII} should have been smaller than 0.03. Hence, it was speculated that changes in f_{PSII} were insufficient to explain the observed variation in $f_{PSII}(1-f_{Alt})$. In other words, the observed changes in $f_{PSII}(1-f_{Alt})$ (Fig. 2) were partly attributable to changes in f_{Alt}, a finding which fits in with the general belief that under conditions of low metabolic activity, alternative biochemical pathways play an important role in dissipating excess of photosynthetic ATP and NADPH [6], [7] and [8].

The presented approach, which is based on simultaneous chlorophyll fluorescence and gas exchange measurements, offers perspectives for quantifying the importance of alternative routes in dissipating ATP and NADPH generated by linear electron flow. If indeed the excess of photosynthetic ATP and NADPH is used to some extent for the biosynthesis of secondary metabolites (such as isoprene), the presented approach could be further applied to identify the environmental conditions which favour biosynthesis of these metabolites.

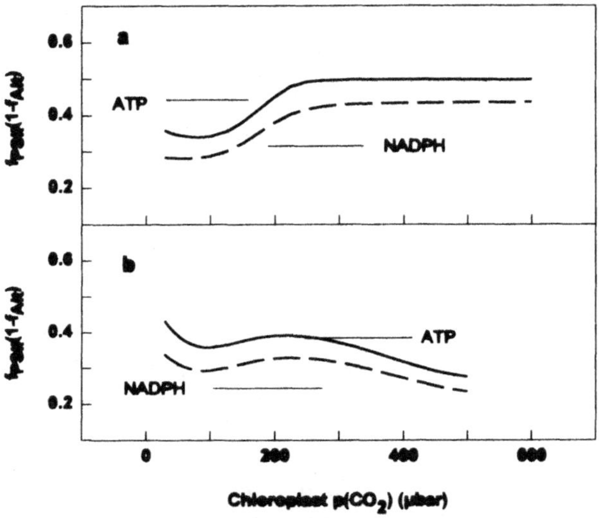

Figure 2 Assessments of $f_{PSII}(1-f_{AII})$ (Eq. 1) as a function of the chloroplast partial pressure of CO_2 (chloroplast $p(CO_2)$) in non-stressed (a) and water stressed (b) S. arboricola leaflets. ATP: α_c = 4.7, α_o = 5.2; NADPH: α_c = 4, α_o = 4.

References

1 Weis, E. and Berry, J.A. (1987) Biochim. Biophys. Acta 894, 198-208
2 Genty, B., Briantais, J.M. and Baker, N.R. (1989) Biochim. Biophys. Acta 990,87-92
3 von Caemmerer, S. and Farquhar, G.D. (1981) Planta 153, 376-387
4 Brooks, D. and Farquhar, G.D. (1985) Planta 165, 397-406
5 Sharkey, T.D. (1985) Bot. Rev. 51, 53-105
6 Stuhlfauth, T., Scheuermann R. and Fock, H.P. (1990) Plant Physiol. 92, 1053-1061
7 Fredrick, F., Lemeur, R. and Bogaert, G. (1995) Submitted to Plant Cell Environ.
8 Scheuermann, R., Biehler K., Stuhlfauth, T. and Fock, H.P. (1991) Photosynth. Res. 27, 189-197

PURIFICATION AND CHARACTERISATION OF AN NADH-DEHYDROGENASE (NDH-1) FROM *HELIOBACILLUS MOBILIS*

A. KOLPASKY, U. Mühlenhoff, A. Atteia* and W. Nitschke
Biochemistry of plants, Univ. Freiburg, FRG
*Dept. of Plant Biology, Göteborg University / Sweden

1. Introduction

During recent years the photosynthetic reaction centre (RC) in Heliobacteria was shown to be a typical representative of the RCI type photosystems (1). In whole cells, light induced electron transport was found to be cyclic around the RC and a cytochrome bc complex under a range of physiological conditions (2). In contrast to RCII-type photosystems, direct reduction of the quinone pool is considered not to occur in RCI-based electron transport chains. Therefore, the question arises via which pathway the reducing equivalents delivered into the cytoplasmatic phase are funnelled back into the membrane. A possible intermediate in this pathway would be an NADH:MK oxidoreductase. A respective enzyme activity was looked for in heliobacterial membranes and the responsible complex was partially purified and characterised.

2. Experimental Procedures

Membranes from *H.mobilis* at 10mg/ml total protein were solubilised using 17mM Hecameg, 500mM phosphate pH 7.4. Non-solubilised material was eliminated by centrifugation at 21000g for 15 minutes. The resulting supernant was fractionated on a 5-17% sucrose density gradient (20mM Hecameg, 500mM phosphate pH 7.4) and the gradient was harvested in 2ml aliquots. Activity was assayed spectroscopically following the decrease of the absorption band of NADH at 340nm using 0.2mM NADH as electron donor and 0.4 mM MK as electron acceptor (25mM Mops, pH 7.4). A 15% SDS-PAGE was performed according to Laemmli (3) and stained with CBB. Western Blots were decorated with antibodies raised against Complex I from *Neurospora crassa* (kindly provided by Dr. H. Weiss, Düsseldorf / FRG) and blots were developed using anti-rabbit IgG-antibodies linked with alkaline phosphatase. Decyl-MK was a kind gift by Dr. P. Rich (Bodmin / GB). All other chemicals are reagent grade.

3. Results

Fig.1a shows the distribution of NADH:MK oxidoreductase activity on the sucrose density gradient. A well defined band of high activity could be discerned roughly in the middle of the gradient. Fig.1b depicts the polypeptide composition of the aliquots obtained from the gradient as compared to total protein prior to fractionation (solubilized membranes).

P. Mathis (ed.), Photosynthesis: from Light to Biosphere, Vol. II, 951–954.

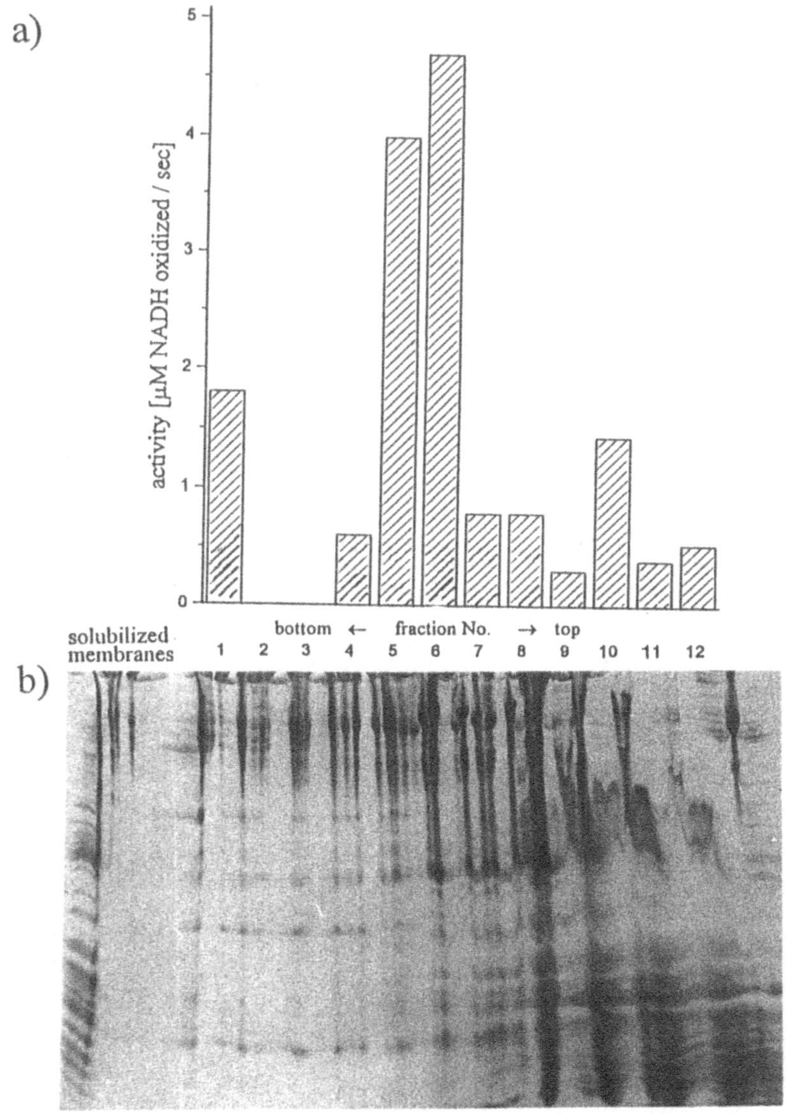

Figure 1a) Distribution of enzyme activity in the gradient
b) SDS-PAGE stained with CBB of respective aliquots from the gradient

Samples with high activity were pooled, concentrated in an Amicon cell (YM30) and a Western Blot using antibodies against Complex I from *Neurospora crassa* was performed (Fig.2). A significant crossreaction with this antibody was observed indicating that the partially purified activity results from a Complex I-related enzyme. In analogy to current

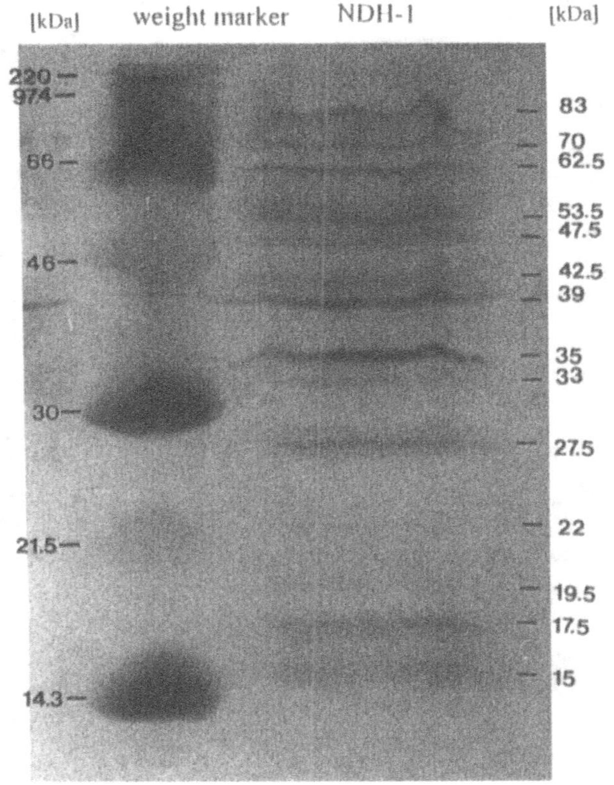

nomenclature for bacterial membrane-bound Complex I-type enzymes we denote this complex as NDH-1. Fig.3 depicts the characterisation of enzyme activity with respect to substrate specificity and inhibitor sensitivity. A clear-cut preference of NADH and naphtoquinones over NADPH and benzoquinones as donors and acceptors, respectively, was observed. The typical inhibitors of mammalian Complex I, Rotenone and KCN, were found to only partially suppress enzyme activity as also reported for other NDH-1's from micro-organisms (4,5). Highest inhibition is induced by NQNO, which is known as a potent inhibitor of the quinone-reducing Q_i-site of cytochrome bc complexes. The functional analogy of the Q_i-site and the quinone-reducing site on NDH-1 appears likely.

Figure 2) Immunostain of pooled samples with high activity as described in the text

4. Conclusions

The data shown above demonstrate that the heliobacterial NADH:MK-oxidoreductase activity arises from an enzyme belonging to the Complex I-family. Comparable to the other bacterial NDH-1 complexes, the heliobacterial enzyme appears to have a significantly simpler subunit composition than the mitochondrial Complex I. During the purification steps employed so far, the heliobacterial enzyme turned out to be rather stable therefore lending itself for further investigations. According to it's enzyme activity, NDH-1 is a possible candidate for mediation of cyclic photoinduced electron flow in Heliobacteria (Fig.4). Since mitochondrial Complex I is involved in several inherited diseases in humans, a better understanding of it's functional bases is desirable. The heliobacterial enzyme might be promising for such investigations.

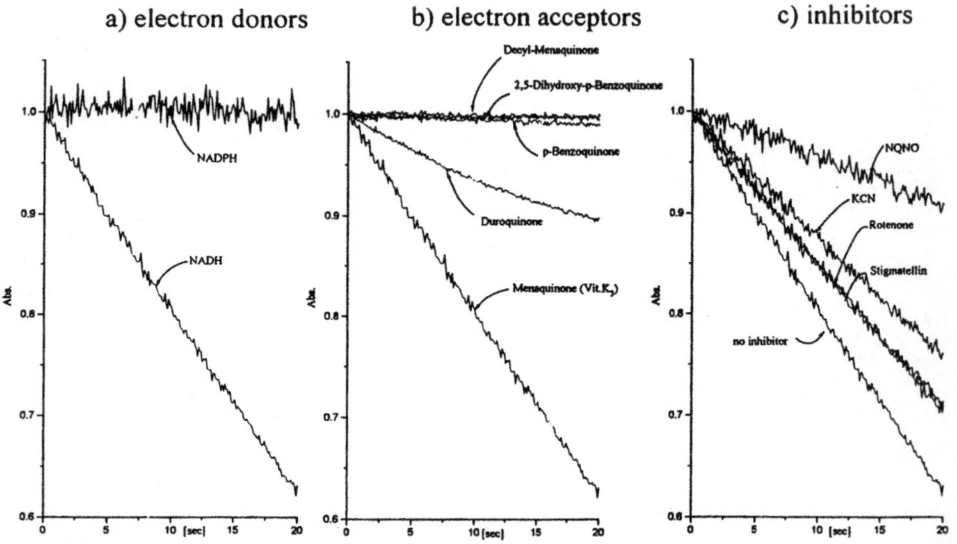

Figure 3) Effect of various substrates and inhibitors on donor:acceptor oxidoreductase activity of NDH-1

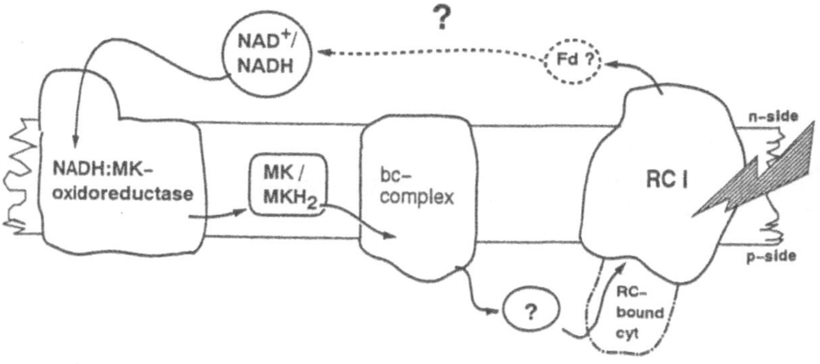

Figure 4) Hypothetical role of NDH-1 in photoinduced cyclic electron -flow in *H.mobilis*

References

(1)U. Liebl, M. Mockensturm-Wilson, J.T. Trost, D.C. Brune, R.E. Blankenship & W. Vermaas (1993), Proc.Natl.Acad.Sci. USA 90, 7124-7128
(2)D.M. Kramer, U. Liebl, B. Schoepp & W. Nitschke, unpublished
(3)U.K. Laemmli (1970), Nature 227, 680-685
(4)T. Yagi, K. Hon-nami & T. Ohnishi (1988), Biochemistry 27, 2008-2013
(5)U. Weidner, U. Nehls, R. Schneider, W. Fecke, H. Leif, A. Schmiede, T. Friedrich, R. Zensen, U. Schulte, T. Ohnishi & H. Weiss (1992), Biochim.Biophys.Acta 1101, 177-180

THE PERIPLASMIC SPACE IS THE SITE OF PHOTOSYNTHESIS IN THE PURPLE NONSULFUR BACTERIUM *RHODOBACTER SPHAEROIDES*

Izabella Prokhorenko[1,2] and Karl-Heinz Süss[1]
[1]Department of Molecular Cell Biology, Institute of Plant Genetics and Crop Plant Research, 06466 Gatersleben, Germany, [2]Permanent address: Institute of Photosynthesis and Soil Sciences, Russian Academy of Sciences, Pushtchino 142292, Moscow Region, Russia

1. INTRODUCTION

Purple nonsulfur bacteria (*Rhodospirillacea*) can grow photoheterotrophically on reduced carbon sources like butyrate and photolithoautotrophically on H_2/CO_2 (1). Under these conditions, growth depends on light energy which is used by the photosynthetic electron transport system and H^+-ATPase to form NADH and ATP. The Calvin cycle is the major pathway for CO_2 fixation where CO_2 is either taken up from the medium or supplied by the tricarboxylic acid cycle after oxidation of reduced carbon compounds (1). Although an immunocytochemical study has shown the key Calvin cycle enzyme ribulose-1,5-bisphosphate carboxylase/oxygenase (Rubisco) and chaparonin 60 to be localized around the cell envelope in the green photosynthetic bacterium *Chromatium vinosum* (2), only little is known on the cellular distribution and supramolecular organization of water-soluble and membrane-integral enzymes of the photosynthetic apparatus in purple nonsulfur bacteria.
In this study, immunoelectron microscopy has been employed to investigate the in situ localization of several proteins of the photosynthetic apparatus including Calvin cycle enzymes in *Rhodobacter sphaeroides*. Immunocytochemical evidence is provided indicating that Calvin cycle enzymes are confined to the periplasmic space and that H^+-ATPase and NAD^+ oxidoreductase are located primarily in the cell membrane of this bacterium.

2. MATERIALS AND METHODS

R. sphaeroides was grown for 6 d photoheterotrophically on butyrate/CO_2 or photolithoautotrophically on H_2/CO_2 (98.5% v/v/1.5% v/v) at a light intensity of ~500 $\mu E/cm^2$ (3). Cells were washed with 10% sorbitol/50 mM Tris-HCl, pH 8.0, and treated with lysozyme (0.5 mg lysozyme/ml cell suspension containing 0.5 g of wet cells) for 30 min to release periplasmic proteins. Lysozyme extracts were centrifuged for 15 min at 20.000 xg and protein precipitated with TCA from the supernatants (5% w/v). Protein was dissolved in 4% SDS/8 M urea/50 mM Tris-HCl, pH 8.0, and subjected to SDS/urea-polyacrylamide gel electrophoresis (SDS/urea PAGE) using 10-20% acrylamide gradient gels. SDS/urea-PAGE and immunoblotting were performed as described previously (4,5).
For immunoelectron microscopy, cells were fixed for 60 min with 3% glutaraldehyde in 20% glycerol/50 mM Tris-HCl, pH 8.0, prior to cryofixation of small cell pellets in liquid propane (-185°C). Cryosubstitution of cells with water-free methanol, embedding in LR White resin, and immunolabelling using protein A-15 nm gold particles were done as described previously (5). In control experiments, antibody was omitted or replaced with

P. Mathis (ed.), Photosynthesis: from Light to Biosphere, Vol. II, 955–958.

rabbit nonimmune IgG (Sigma). Sections were stained with uranyl acetate and lead citrate and examined in a Zeiss CEM 920A transmission electron microscope at 80 kV.

3. RESULTS AND DISCUSSION

Although Rubisco and chaparonin 60 were found to be associated with the cell envelope in a green photosynthetic bacterium (2), the cellular location of Calvin cycle enzymes in purple nonsulfur bacteria has not been determined. Photosynthetic electron transport and ATP synthesis in these bacteria is assumed to take place on intracytoplasmic chromatophores which evolve by infolding of the cell membrane into the cytoplasm (6). Therefore, photosynthesis is thought to be a cytoplasmic process. To test these assumptions, we have studied the in situ localization of several Calvin cycle enzymes, NAD^+ oxidoreductase, and H^+-ATPase in *R. sphaeroides* grown heterotrophically or photolithoautotrophically using the technique of immunoelectron microscopy. Specific antibodies raised against Calvin cycle components from *R. sphaeroides* (Rubisco forms I and II, and ribulosephosphate kinase, RPK) and equivalent enzymes from spinach chloroplasts cross-reacting with their bacterial counterparts (glyceraldehyde-3-phosphate dehydrogenase, GAPDH, fructosebisphosphate aldolase, transketolase, pentosephosphate epimerase) were used for immunoelectron microscopy. Antibodies raised against spinach chloroplast coupling factor, CF_1, and ferredoxin-$NADP^+$ reductase (FNR) cross-reacting with bacterial H^+-ATPase and NAD^+ oxidoreductase, respectively (K.-H. Süss, unpublished), have been also employed.

Figure 1 shows representative electron micrographs for the in situ location of Rubisco form I (L_8S_8) and RPK in photolithoauthotrophically grown cells of *R. sphaeroides*. These components appear to be confined to the periplasmic space rather than cytoplasm of cells. Similar results have been obtained for the minor Calvin cycle components GAPDH, aldolase, transketolase, and epimerase, but 70S-ribosomes were found to be localized exclusively to the cytoplasm of *R. sphaeroides* (I. Prokhorenko and K.-H. Süss, unpublished). Statistically, more than 70% of immunogold particles labelling the Calvin cycle enzymes Rubisco form I and II, RPK, aldolase, and GAPDH occurred in the periplasm. It should be noted, however, that in case of transketolase and epimerase a number of cells also showed significant enzyme labelling in the cytoplasm where up to 40% of the immunogold particles were counted. This may indicate the presence of isoenzyme forms of different cellular location and perhaps function. Although the in situ location of enzymes was virtually the same, *R. sphaeroides* grown photolithoautotrophically was found to possess smaller amounts of Calvin cycle enzymes than when grown under photoheterotrophical conditions. Immunoelectron microscopy also revealed that NAD^+ oxidoreductase (Fig. 1c) and H^+-ATPase (I. Prokhorenko and K.-H. Süss, unpublished) are located primarily in the cell membrane, whereas enzyme labelling on intracytoplasmic chromatophores was low irrespective of the phototrophic growth conditions employed. These observations led to the question as to whether Calvin cycle enzymes can be solubilized from bacteria by lysozyme/EDTA treatment. SDS-PAGE followed by immunoblotting of proteins from untreated bacteria, bacterial spheroplasts obtained by a combination of lysozyme/proteinase K treatment and lysozyme/EDTA extracts was used to investigate this problem. Lysozyme extracts that did not possess detectable amounts of chromatophores were found to contain several proteins including the Calvin cycle enzymes investigated. Preparations of spheroplasts still contained significant quantities of Calvin cycle enzymes, but these components were degraded either completely or cleaved into smaller peptides by proteinase K treatment. It has been con-

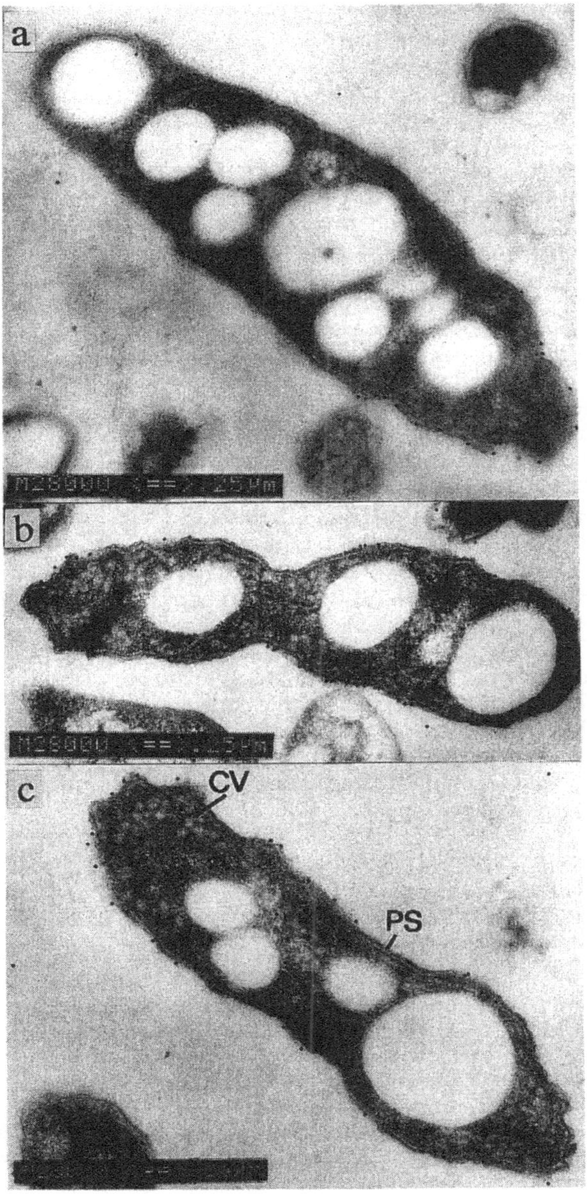

Figure 1: Immunogold localization (15 nm gold particles) of Rubisco form I (L_8S_8) (a), ribulose-5-phosphate kinase (b), and NAD^+ oxidoreductase (c) in photolithoautotrophically (H_2/CO_2) grown cells of *Rhodobacter sphaeroides*. The number of gold particles located in the periplasm/cytoplasm are as follows: a, 34/8; b, 41/3; c, 35/5. Note that immunogold particles labelling these enzymes are mainly localized to the periplasmic space. Control samples which were either labeled with or without nonimmune IgG (Sigma) prior to protein A-gold labelling did not show more than 2 gold particles per cell. PS, periplasmic space; CV, chromatophore vesicle.

cluded, therefore, that Calvin cycle enzymes are tightly associated with integral membrane proteins on the periplasmic surface of the cell membrane (K.-H. Süss, unpublished).
The periplasm, i. e. the space between the cell wall and cell membrane, is ~15 nm wide at its most constricted points which is roughly equivalent to the largest dimension (12 nm) of an hexadecameric (L_8S_8) Rubisco molecule (7). Besides Calvin cycle enzymes, the

periplasm is known to contain cytochrome c_2 which takes part in light-driven ATP synthesis coupled to reversed electron transport in *R. sphaeroides* (8). It appears, therefore, that periplasmic proteins form a thin, perhaps monomolecular protein layer on the outer surface of the cell membrane. Considering that NAD^+ oxidoreductases and H^+-ATPases are located primarily in the cell membrane (Fig. 1c), the enzymes catalyzing photosynthetic light and dark reactions appear to be closely neighboured in situ and might even form multienzyme complexes which are catalyzing sequential reactions in bacterial photosynthesis. Binding of Calvin cycle enzymes to cell membrane components is indicated both by their proteinase K sensitivity in spheroplasts and the finding that the lumen of chromatophores does not appear to contain these enzymes. The latter should be the case, however, if periplasmic Calvin cycle enzymes were able to diffuse freely and thus can be enclosed by cell membrane patches during the formation of chromatophores. Biochemical and immunocytochemical evidence has been obtained recently indicating that Calvin cycle enzymes and a form of FNR in chloroplasts of higher plants and the green alga *Chlamydomonas reinhardtii* assemble into multienzyme complexes which are associated with thylakoid membranes in situ (4,5). The spatial organization of sequential enzymes which is thought to be caused by complementary binding interactions may therefore provide the molecular basis for channeling of NADPH and ATP among membrane-integral and membrane-associated, water-soluble enzymes of the photosynthetic apparatus in chloroplasts of green plants, i. e. PSI-FNR-GAPDH and H^+-ATP synthase-RPK/PGK (4,5). The present results point to the possibility, that Calvin cycle enzymes in purple nonsulfur bacteria may be also associated with NADH-oxidoreductase and perhaps other electron transport proteins and H^+-ATPase on the periplasmic surface of the cell membrane to allow for nucleotide channeling among sequential enzymes. The occurrence of enzymes catalyzing the Calvin cycle in the periplasmic compartment is thought to be advantageous physiologically in that it enables photosynthetic bacteria to be highly competitive for uptake and fixation of CO_2 in natural environments. These results should stimulate further studies on the cellular location of other enzymes in photosynthetic bacteria including those of the Krebs cycle to shed light on unanswered questions regarding compartmentation, organization, and regulation of metabolic pathways contributing to phototrophic growth as well as the evolution of the photosynthetic apparatus.

ACKNOWLEDGEMENT. Antisera raised against bacterial Rubisco and RPK forms were a generous gift of Dr. T. R. Tabita, Columbus, OH. The technical assistance of Inge Glaser, Waltraut Panitz, and Bernhard Claus is gratefully acknowledged. This work was supported by a grant from Deutsche Forschungsgemeinschaft (Su 129/1-3).

REFERENCES

1 Tabita, F.R. (1988) Microbiol. Rev. 52, 155-189.
2 McFadden, B.A., Torres-Ruiz, J.A. and Franceschi, V.R. (1989) Planta 178, 297-302.
3 Tabita, F.R. and McFadden, B.A. (1974) J. Biol. Chem. 249,3459-3464.
4 Süss, K.-H., Arkona, C., Manteuffel, R. and Adler, K. (1993) Proc. Natl. Acad. Sci. USA 90, 5514-5518.
5 Süss, K.-H., Prokhorenko, I. and Adler, K. (1995) Plant Physiol. 107, 1387-1397.
6 Drews, G. and Oelze, J. (1981) Adv. Microb. Physiol. 22, 1-91.
7 Knight, S., Andersson, I. and Branden, C.-I. (1990) J. Mol. Biol. 69, 113-160.
8 Prince, R.C., Baccarini-Melandri, A., Hauska, G.A., Melandri, B.A. and Crofts, A.R. (1975) Biochim. Biophys. Acta 387, 212-227.

THE RELATIONSHIP BETWEEN ELECTRON TRANSPORT RATE THROUGH PS II AND CO_2 GAS EXCHANGE IN ANTARCTIC CRYPTOGAMS.

B. Schroeter[1], T.G.A. Green[2], L. Kappen[1], R.D. Seppelt[3] and K. Maseyk[2]
[1]Botanisches Institut, Universität Kiel, D-24098 Kiel, Germany; [2]Biological Sciences, University of Waikato, Hamilton, New Zealand; [3]Australian Antarctic Division, Kingston, Tasmania, Australia

1. Introduction

Recently the development of pulse-amplitude modulated chlorophyll fluorescence techniques provided an important tool for measurements of in vivo photosynthesis under field conditions (1, 2). Measurements of the photon use efficiency (ΦPSII) and the rate of apparent electron transport through PS II (ETR) can be used to monitor photosynthetic activity in situ under changing ambient light, temperature and water conditions. The close relationship between the quantum yield of CO_2 assimilation and the quantum yield of PS II photochemistry established for C3 and C4 plants (3) allowed the prediction of photosynthetic assimilation rates from measurements of chlorophyll fluorescence (4). The use of chlorophyll fluorescence techniques has been demonstrated also in lower plants in field and laboratory studies (5, 6, 7). The technique has been found particularly useful for studies of in situ photosynthetic activity in lichens and mosses under extreme environmental conditions (8, 9). However, a relationship between ΦCO_2 and ΦPSII is still to be confirmed in lower plants. In this paper we present data on the relationship between ΦPSII and ΦCO_2 in the moss *Bryum argenteum* and the lichen *Umbilicaria aprina* in continental Antarctica.

2. Procedure

CO_2 gas exchange and chlorophyll fluorescence measurements were carried out simultaneously using a minicuvette system (CMS400, Walz, FRG) with an infrared gas analyser in the open flow (10). Illumination was provided by a special halogen lamp with fibreoptics (FL-400, Walz, FRG) and neutral density filters were used to alter photosynthetic photon flux density (PPFD). The temperature controlled measuring cuvette was equipped with the fibreoptics probe of the chlorophyll fluorescence system (PAM-2000, Walz, FRG) permanently fixed about 10 mm distant from the sample. Standard routines were used to determine the effective PS II quantum yield (ΦPSII), relative electron transport rate (ETR = PPFD x ΦPSII) photochemical (qP) and non-photochemical (qN) quenching (7) at different temperature and PPFD combinations simultaneously with CO_2 exchange measurements (net photosynthesis, NP; dark respiration, DR). Gross photosynthesis (GP) was calculated as GP = NP + DR. All measurements were carried out with freshly collected material of the lichen *Umbilicaria aprina* Nyl. and the moss *Bryum*

P. Mathis (ed.), Photosynthesis: from Light to Biosphere, Vol. II, 959–962.

argenteum Hedw. in the field laboratory at Granite Harbour, continental Antarctica, (10) in November 1994. During the experiments the samples were kept at optimal moisture content, i.e. where maximal CO_2 exchange rates were achieved.

3. Results and Discussion

Figure 1 + 2 illustrate the PPFD dependency of CO_2 exchange and chlorophyll fluorescence parameters at +5°C in the moss *Bryum argenteum* and in the lichen *Umbilicaria aprina* respectively. In *B. argenteum* photosynthesis (GP, NP) was nearly saturated at PPFD of 250 μmol m^{-2} s^{-1} (Fig. 1). ETR almost exactly paralleled the changes in GP. Photochemical quenching (qP) declined rapidly with the increase in PPFD and qN showed a typical increase as saturation was exceeded. The PPFD dependency of photosynthesis reflects the habitat situation of *B. argenteum* at the beginning of summer (November) in continental Antarctica: most of the thalli were covered by snow of up to 1.5 m thickness which effectively reduced the PPFD inci-

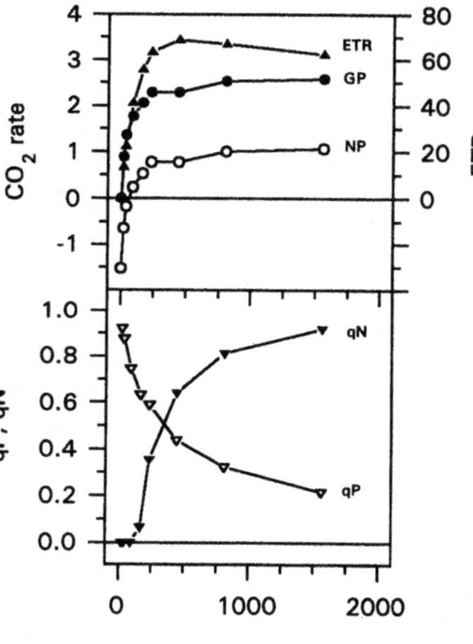

Fig. 1: PPFD dependency of CO_2 exchange (GP; NP) and chlorophyll fluorescence parameters (ETR; qP; qN) of *Bryum argenteum* measured simultaneously at 5°C. CO_2 rate in μmol CO_2 m^{-2} s^{-1}.

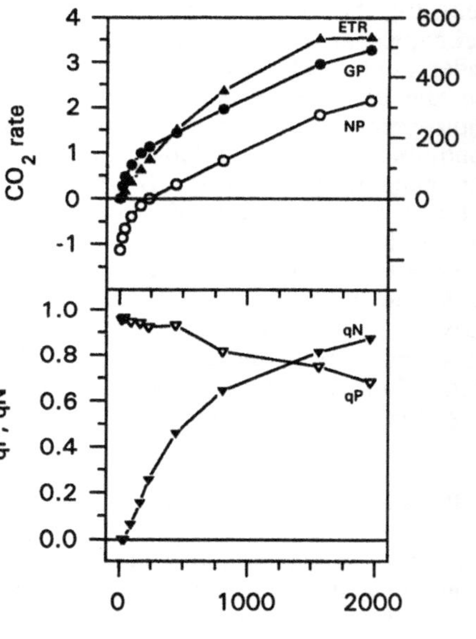

Fig. 2: PPFD dependency of CO_2 exchange (GP; NP) and chlorophyll fluorescence parameters (ETR; qP; qN) of *Umbilicaria aprina* measured simultaneously at 5°C. CO_2 rate in μmol CO_2 m^{-2} s^{-1}.

dent at the thallus level. During melt out *B. argenteum* is able to acclimatize to changing (increasing) PPFD conditons within a few days (T.G.A. Green, B. Schroeter, R.D. Seppelt, unpublished). In contrast, CO_2 exchange in *U. aprina* did not saturate with an increase in PPFD up to 2000 μmol m^{-2} s^{-1} (Fig. 2). Photochemical quenching (qP) declined slowly with increasing PPFD but remained at a high value compared with *B. argenteum*. Non-photochemical quenching was low at low PPFD and rose with increase of PPFD. At first sight it seems surprising that in *Umbilicaria aprina* no saturation of CO_2 exchange and ETR was found up to 2000 μmol m^{-2} s^{-1} PPFD at +5°C. However, PPFD is measured at the lichen surface and not at the level of the algal cells which are covered by the fungal cortex. PPFD is substantially lowered at the level of the algal chloroplasts depending on thickness and dark pigmentation of the cortex. *Umbilicaria aprina* is able to photosynthesize at temperatures down to -17°C and at subzero temperatures light saturation of CO_2 exchange was reached at PPFD of about 400 μmol m^{-2} s^{-1} (10). In continental Antarctica in *U. aprina* thallus temperatures of +5°C occur only during summer when high PPFD levels are frequent. Therefore, the PPFD dependency of CO_2 exchange and chlorophyll fluorescence as shown in Fig. 2 indicate that *U. aprina* is well adapted to summer conditions in continental Antarctica.

The relationship between chlorophyll fluorescence and CO_2 exchange parameters appears to be quite different in the moss and the lichen. If the photochemical efficiency of PSII (Φ_{PSII}) and apparent quantum yield of CO_2 assimilation (Φ_{CO_2}) are compared a strong linear relationship was found in *Bryum argenteum* in the temperature range from 0 to 25°C at various light levels (Fig. 3). This resembles the relationship at non-photorespiratory conditions in Higher Plants (3). By contrast, the lichen response is more complex; a negative correlation could be found if different

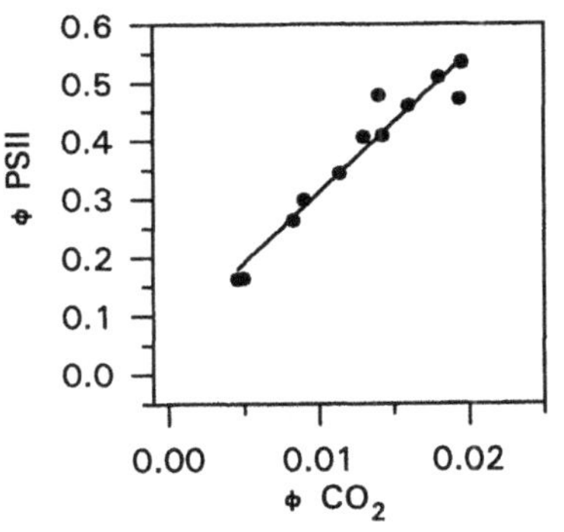

Fig. 3: Linear relationship between Φ_{PSII} and Φ_{CO_2} in *B. argenteum* (r^2=0.94) from 0°C to +25°C at various light levels.

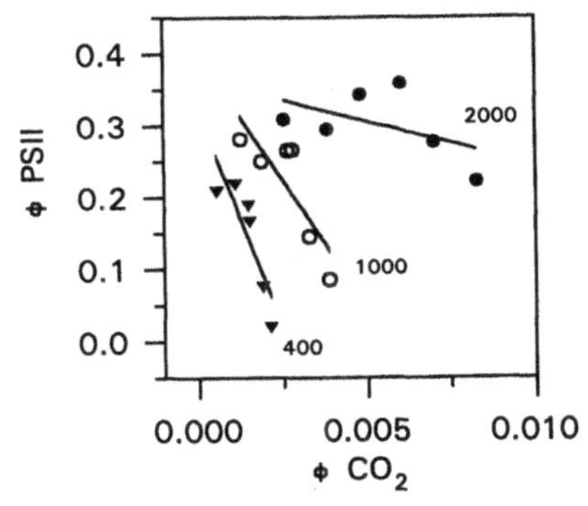

Fig. 4: Relationship between Φ_{PSII} and Φ_{CO_2} in *U. aprina* from 0°C to +25°C. Numbers indicate different levels of PPFD.

PPFD levels are separated (Fig. 4). It has to be pointed out that the lichen thallus as a poikilohydrous symbiosis of fungus and algae consists mainly of the non-photosynthesizing fungal symbiont. While both symbionts, fungus and alga, contribute to the CO_2 exchange regardless if calculated as NP or GP only the algal photosynthetic activity is reflected in the chlorophyll fluorescence measurements. Therefore, it seems reasonable that the temperature-dependent respiratory activity of the non-photosynthetic mycobiont causes the non-linear relationship between Φ_{CO_2} and Φ_{PSII} in *Umbilicaria aprina*.

4. References

1 Schreiber, U.; Schliwa, U.; Bilger, W. (1986) Photosyn. Res. 10, 51-62
2 Schreiber, U.; Bilger, W.; Neugebauer, C. (1994) in Ecophysiology of photosynthesis (Schulze E.D. & Caldwell M.M., eds.) pp.49-70, Springer, Berlin
3 Genty, B.; Briantais, J.-M,; Baker, N. (1989) Biochim. Biophys. Acta 990, 87-92
4 Edwards, G.E. and Baker, N.R. (1993) Photosyn. Res. 37, 89-102
5 Schroeter, B.; Kappen, L.; Moldaenke, C. (1991) Lichenologist 23, 253-265
6 Schroeter, B.; Green, T.G.A.; Seppelt, R.D.; Kappen, L. (1992) Oecologia 92, 457-462
7 Schroeter, B. (1994) Oecologia 98, 212-220
8 Hovenden, M.J. and Seppelt, R.D. (1995) Aust. J. Plant Physiol. 22, 321-330
9 Schroeter, B.; Schulz, F.; Kappen, L. (1995) in Antarctic Communities (Battaglia, B.; Valencia, J.; Walton, D.W.H., eds.), Cambridge University Press, in press
10 Schroeter, B.; Green, T.G.A.; Kappen, L.; Seppelt, R.D. (1994) Cryptogam. Bot. 4, 233-241

Grateful acknowledgement is made to the Deutsche Forschungsgemeinschaft (Schr 473/1-1) and the Waikato Antarctic Research Program for financial support and to NZAP and NSF VXE-6 for logistic support.

INDEX OF NAMES

GPSR Compliance

The European Union's (EU) General Product Safety Regulation (GPSR) is a set of rules that requires consumer products to be safe and our obligations to ensure this.

If you have any concerns about our products, you can contact us on ProductSafety@springernature.com

In case Publisher is established outside the EU, the EU authorized representative is:

Springer Nature Customer Service Center GmbH
Europaplatz 3
69115 Heidelberg, Germany

Batch number: 10761593

Printed by Printforce, the Netherlands